ISBN 978-0-265-83267-7
PIBN 10896150

FB &c Ltd, Dalton House, 60 Windsor Avenue, London, SW19 2RR.
Company number 08720141. Registered in England and Wales.

For support please visit www.forgottenbooks.com

OUR BUYERS' DIRECTORY WILL BE FOUND USEFUL IN BUYING.

CANADIAN MACHINERY AND MANUFACTURING NEWS

A monthly newspaper devoted to the manufacturing interests, covering in a practical manner the mechanical, power, foundry and allied fields. Published by The MacLean Publishing Company, Limited, Toronto, Montreal, Winnipeg, and London, Eng..

OFFICE OF PUBLICATION: 10 Front St. E., TORONTO

Vol. III. JANUARY, 1907 No. 1

SMALL TOOLS   MACHINE TOOLS

Taps, Dies

Reamers

Solid and Adjustable

Ratchet Drills

Milling Cutters

Solid and Inserted Blade

Boiler Punches

High Speed Drills

Gauges, etc.

PRATT & WHITNEY CO. HARTFORD Conn.

Agents for Canada

THE

Canadian Fairbanks Co.

LIMITED

Montreal Toronto Winnipeg Vancouver

No. 12—Two Spindle Profiling Machine

NO. 2—Hanging Milling Machine, overhanging arm.

Profiling Machines

Single and Double Spindle

Specially adapted for finishing parts of guns, sewing machines and any other accurate and interchangeable work, doing away with hand fitting.

ILLUSTRATED CATALOG ON REQUEST

Hand Milling Machines

adapted to milling the variety of duplicate parts incident to the manufacture of small machines.

VARIETY OF SIZES AND TYPES

New Type 18" Double Back Geared
ENGINE LATHE

If you are looking for the most complete and up-to-date **Lathe** on the market, investigate our new line of **Double Back Geared Engine Lathes.** For accuracy and handiness they are unexcelled. They are made extra heavy in all parts. Spindles are of extra large diameter and made of Crucible Steel. Cone is of large diameter and wide face. Quick Change Feed Mechanism gives four instantaneous changes of feed by simply throwing over lever in front of machine. These changes can be obtained without stopping machine.

We can make Immediate shipment of 18" Lathes with 8 and 10-ft. beds.

LONDON MACHINE TOOL CO., Limited

HAMITON, CANADA

Sales Offices: Traders' Bank Building, TORONTO, CANADA

HOLLOW HEXAGON TURRET LATHES—

No. 2 Machine—2¼ x 24-inch.

are used everywhere—— by the leading manufacturers throughout the world.

One customer in England now has THIRTY-SEVEN of them.

The interesting particulars regarding these rapid-production machines should interest you.

THE WARNER & SWASEY COMPANY

CLEVELAND, OHIO, U.S.A.

TURRET LATHES AND SCREW MACHINES.
BRASS-WORKING MACHINE TOOLS.

Agents in Canada:
A. R. WILLIAMS MACHINERY CO., Toronto
WILLIAMS & WILSON, Montreal

OUR 31-INCH SLIDING HEAD DRILL

is a very strong and powerful tool. It will drill 3-inch holes in the solid, or bore out cored holes up to 6 inches. Furnished in single machines, or in gangs. This Drill is a money-maker in any machine shop. Material and workmanship guaranteed. It will give you satisfactory and accurate service.

Details in our Catalogue N.

B. F. BARNES COMPANY

Carried in stock by our Ontario Agent
H. W. PETRIE, Toronto

ROCKFORD, ILL.

BENCH VISES

Made in Canada and a credit to any country

We put the same high-class workmanship on these Vises as we do on our other products.

Manufactured by

THE STEVENS COMPANY OF GALT, LIMITED

Galt, Ont., Canada.

Also Manufacturers of Drill Vises with Jig attachments Milling Machine Vises, Bench Vises and Pattern Makers' Universal Vises, etc.

The Time Saving Features

of MURCHEY NIPPLE and PIPE THREADING MACHINES count a lot when it comes to close figuring. One "Murchey" will cut more threads in a given time than any three machines of other makes, and produces absolutely uniform work.

The No. 3 machine, as shown, threads and reams at one operation, has capacity for pipe from 1½" to 4", inclusive, and will thread 300 four-inch short nipples complete in ten hours time. Dies open automatically, so no attention is required after the thread is once started. Gripping chuck jaws will not slip nor crush the pipe; all adjustments easily made. If you are interested in first-class Pipe Cutting Machines, you need our Catalogue, sent on request.

We also make the Murchey Improved Tapping Machine with Frictionless Driving Head

Murchey Machine and Tool Co. Corner 4th and Porter Streets DETROIT, MICH., U.S.A.

WHAT THE MERRELL MEANS TO YOU.

Simplicity of Construction.

Ease of Control and Operation.

Speed in Cutting either Iron or Steel—Rough or Smooth—Pipe.

Exactness—Not a machine leaves our factory until it has passed the MERRELL test, which means not until it has undergone every test for any kind of work it could be called upon to do—is the reason why we can give you this.

COMBINED HAND and POWER MERRELL

FOR FULL 30 DAYS FREE TRIAL.

We know it will do all your work in an easier, less costly and spredier manner.

This machine has great range of work. Our improved cutting-off tool cuts cleanly and quickly. Quick opening and closing die-head; encased gears; exceptionally fine steel chasers that may be released from threading while the machine is in motion.

The Head may be removed from the base and used as a portable hand machine. It will cut any size—any kind—of pipe up to 6-in.

Send a Postal for our Catalogue

THE MERRELL MFG. CO., Toledo, Ohio

THE CANADIAN FAIRBANKS CO.,
LIMITED
AGENTS:

Montreal Toronto Winnipeg Vancouver

Kerr Turbine Pumps for water works, condenser equipment and boiler feeding. These pumps are more economical than direct acting piston pumps, and require less attention, lubrication and space. No belts or gears. Speed adjustable to pressure desired. Regulation close. Start and stop quickly. For any capacity.

Turbine Blowers for gas works, foundries and forge shops. Air pressure 4 to 16 ounces. For any capacity.

Turbine Generators, 2 kw. to 300 kw. direct current and alternating current, 25 to 300 kw., 60 cycle.

The illustration shows an 18-inch condensing Kerr Steam Turbine coupled to a 4-inch 3-stage Worthington Pump. Capacity 200 g.p.m. against a head of 360 feet.

Main Office:

KERR TURBINE CO., Wellsville, N.Y.

Built in Canada by

WM. STEPHENSON

34 Bay Street N. HAMILTON, ONT.

Turret Lathes—Boring Mills

TWO catalogues are issued by us—one giving a full description of Gisholt Lathes; the other of Gisholt Boring Mills. Particular care has been used in editing these catalogs to point out the many labor and time saving devices embodied in the design of these machines. A large number of illustrations are given showing various kinds of work which may be finished on Gisholt Lathes. If you are interested in labor saving tools of this character let us send you our catalogs. We know you will find them interesting.

We also issue a Bulletin each month, telling of a new machine and appliances we are bringing out. If you are endeavoring to keep pace with the new things in labor saving tools, these Bulletins will be of value to you. We will be glad to send them on application.

Gisholt Machine Company

General Offices:	Works:
1309 Washington Ave.,	MADISON, WIS., U.S.A.
MADISON, WIS., U.S.A.	WARREN, PA. U.S.A.

Has Greater Productive Capacity

than any Upright ever built—our 30-in. RADIAL DRILL. Costs but a trifle more and is admirably adapted to light and medium manufacturing, being particularly effective in taking care of the ordinary work as found in the average shop.

Cone Drive and Swinging Table

Bickford 30"-36"-42" Radial Drill

A Few Features

12 speeds when driven through a speed box; 10 when a cone is used.

Tapping attachment incorporated in the design of the head.

Fitted with any one of six styles of tables.

Circular I-A for the asking

The Bickford Drill & Tool Company

Cincinnati, Ohio, U.S.A.

FOREIGN AGENTS—Schuchardt & Schutte, Berlin, Vienna, Stockholm, St. Petersburg, New York. Alfred H. Schutte, Cologne, Brussels, Liege, Paris, Milan, Bilbao, New York. Chas. Churchill & Co., Ltd., London, Birmingham, Manchester, Newcastle-on-Tyne and Glasgow.

CANADIAN AGENTS—H. W. Petrie, Toronto. Williams & Wilson, Montreal.

HEATING AND VENTILATING

Hot Blast Heating Systems for buildings of all sizes.
Patent Steam Heaters that give fully 25 per cent. more efficiency than others in the market.
Planing Mill Exhausters and ***Dust Separators*** of improved patterns.
Drying Systems for material of every description, blast system or otherwise.
Fans and Blowers for all purposes.

For Illustrated catalogue and particulars address

McEachren Heating and Ventilating Co.
GALT, ONT.

A SUITABLE HOLIDAY GIFT FOR YOUR FRIEND THE ENGINEER

He needs light to avoid danger.
Can be carried in the pocket.
No. 1 Flash Light

Price, $2.00. Express Prepaid

The Berlin Electrical Manufacturing Co.
472-474 King St. W., Toronto, Ont.

"BLISS"

INCLINABLE POWER PRESSES

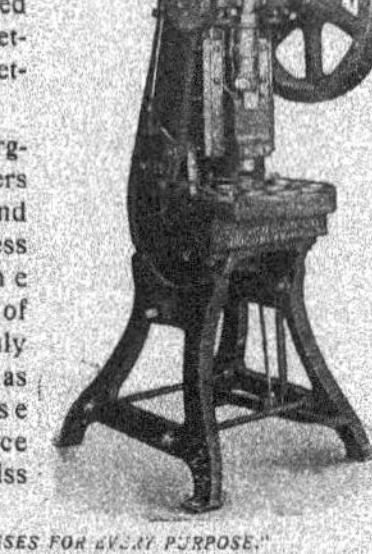

are made in 20 sizes, with fly wheel or gearing. They are adapted for tin, sheet-brass and sheet-steel work, etc.

We are the largest press builders in the world and owe our success largely to the "Bliss" Policy of putting out only such products as will increase public confidence in the E. W. Bliss Company.

"PRESSES FOR EVERY PURPOSE."

E. W. BLISS CO.,
20 ADAMS ST., BROOKLYN, N. Y.
Chicago Agts.: The Stiles-Morse Co., 138 Jackson Boulevard Chicago, Ill.

SUPERIOR

POWER PRESSES

Many new features embodied in this press. Write for prices and particulars on presses, dies, tools, and special machinery.

W. H. BANFIELD & SONS
120 Adelaide Street West, TORONTO.

Any recognized metal-working machinery firm can secure a sample

IMPERIAL CHUCK

We'll guarantee it to be the equal of any foreign-made chuck and better than any other Canadian-made chuck.

KER & GOODWIN
Manufacturers
BRANTFORD, CANADA

We make Hack Saws for men who want the best,—for men who have found that cheap Hack Saws overheat too quickly,—bind,—stick,—and break!

UNIVERSAL Hack Saws will not bind.

They will not stick.

They will not overheat or snap.

Can you afford to use inferior saws,— when you can buy **UNIVERSAL** Hack Saws for the same price?

Let us send you Samples and Prices.

West Haven Mfg. Co.
NEW HAVEN, CONN.

THE CANADIAN FAIRBANKS CO., Montreal, Toronto, Vancouver
RICE LEWIS & SON, Toronto

The Latest and Best Abrasive

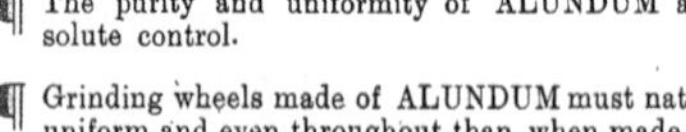

¶ The purity and uniformity of ALUNDUM are under absolute control.

¶ Grinding wheels made of ALUNDUM must naturally be more uniform and even throughout than when made of natural or artificial abrasives which vary in purity and uniformity.

¶ Norton Grinding Wheels are now made of ALUNDUM and are the fastest cutting grinding Wheels.

¶ We are ready to prove it and will send samples for test on most liberal terms.

NORTON COMPANY, Main Works: **WORCESTER, MASS.**

Niagara Falls New York Chicago

Canadian Agents { The Canadian Fairbanks Co., Montreal, Toronto, Winnipeg and Vancouver
F. H. Andrews & Son, Quebec, P. Q.

45

DOMINION HENDERSON BEARINGS LIMITED

NIAGARA FALLS, ONT.

ENGINEERS, TOOL-MAKERS, HIGH-CLASS MACHINISTS,

SOLE MANUFACTURERS OF THE

HENDERSON ROLLER BEARINGS

President - - - - W. Bowman
Managing Director - J. Dove-Smith
Secretary-Treasurer - J. Frank Adams

Roller bearings designed for any special purpose, automobile, axles, gears, etc.

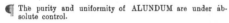

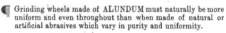

Old and New Tariffs Compared

Speech of Finance Minister Fielding giving the Government's reasons for the changes made—Comments of manufacturers and Jobbers.

Mr. Fielding, in his address in Parliment, referred to the fact that this was the second general revision under his regime. The first one was made in 1898. The Government wanted to encourage stability and not to be hasty in making changes. In the past it had resisted many applications for increases, and he did not believe any severe loss had resulted. So prosperous was the country no radical changes were required In changing the form of the schedules it was proposed to put articles under different classes according to their nature and duty. The free list schedule would be abolished, and each item without duty put in its class in the general tariff with the word "free" opposite, while other articles would have the rates of duties also placed opposite. Articles would be arranged in eleven groups, one of which would be "miscellaneous."

Under the old tariff there were practically four tariffs :

1. British Preference.
2. General Tariff.
3. Surtax.
4. French Treaty.

All these features would be retained, and in addition there would be an intermediary tariff. The general tariff would be substantially the tariff of the day, and not much above the present tariff, but there would be higher duties on a few items. The intermediate tariff, the new feature, would not be put into operation at once.

The British Preference.

The British preference would be adhered to. It had made Canada prominent, and the nation had derived an advantage from it, if it was derived from sentiment. British sales to Canada, which had decreased from 43 millions in 1890 to 29½ millions in 1897, when the preference was given, had now increased to 69 million dollars. The dutiable goods had increased from 20 to 52 millions.

The lower duties on British goods had put so much money into the pockets of the Canadian consumers, the saving to them since the preference was established, being estimated at $28,000,000. In addition, the low British prices had largely governed foreign sales in Canada.

The British preference schedule, while adhering to the general principle of one-third off, would have the duty applied to each specific article, and in some cases this would vary from 33 1-3 per cent. In some cases the preference would be less ; in some higher; but on the whole it would be more favorable to Britain.

Special account was being taken of British metals, and the preference on these would be increased, and would enable Britain to better meet severe competition.

In future articles qualifying for the preference would require to have had 25 per cent. of their value put into them in the shape of British labor. This would prevent foreign countries sending goods here through Britain without work of much cost being done on them here.

The Intermediate Tariff.

The intermediate tariff would grant rates of duties somewhat below those of the general tariff, and on articles averaging a 30 per cent. duty the reduction should amount to 10 per cent., but still leave a material advantage to countries enjoying the British preference. The object of this tariff was to afford Canada an instrument wherewith she could conduct negotiations with any country willing to give her more favorable conditions.

The Drawback Clause.

The object of the clause referring to drawbacks was to prevent the abuse of the free admission of articles granted for a particular purpose—chiefly manufacturing. As to a few of these about which there could be no difficulty, there would be no change, but other articles would have to pay the regular duty, and when it was proved that the particular purpose had been served, a rebate of 95 per cent. of the duty, in the shape of a drawback, would be made

Anti-Combine and Dumping Clauses.

The anti-combine clause adopted ten years ago would be retained. It had been a good thing in smashing the paper combine, which was followed by a reduction of 15 per cent. in duty, and though not often utilized, it had a wholesome effect. The clause provided or the abolition of duties where trusts charging extortionate prices were proved before a judicial committee to exist.

The dumping clause had also served a useful purpose. It would be so amended as to apply to articles on the free, as well as on the dutiable list. The special duty, which, except under certain conditions, has been equal to the difference between the fair market price of the article and the selling price, is, in the case of free goods, limited to 15 per cent.

German Surtax Stands.

The German surtax, would stand exactly as it was. Germany would be liable to the general tariff, plus a surtax of one-third. While there was no likelihood of early negotiations with Germany for a reduction from the general tariff, there was a lively hope of a better understanding as to the surtax. The imposition of the surtax was partly the result of a misunderstanding on the part of Germany.

Iron and Steel Bounties.

In discussing the new tariff and bounties on metals, Mr. Fielding said:

In 1897, we regarded it as important, that the cost of iron and steel to the consumer in Canada should not be too high, because iron and steel at moderate cost are the foundation of an immense variety of industrial enterprises. Therefore, we cut down the customs on iron and steel, and we arranged a system as follows:—On iron from native ore. $3 per ton; upon foreign ore, $2 per ton; upon steel, $3 per ton. These bounties were fixed on a sliding scale, and gradually diminished. Last year they were fifty-five per cent. of the original amount. This year, since the 1st of July, they stand at thirty-five per cent. of the original amount. They will expire on the 1st of July next, under existing legislation. As we have six months to run yet, that means an extension period of three years and a half, four years from the 1st of January, but one-half year is provided under the existing law.

We adopt to some extent a sliding scale. On pig iron manufactured from foreign ore for the calendar year 1907, $1.10 per ton; and for 1908, $1.10; for 1909, 70c.; for 1910, 40c. On steel ingots, we propose, in 1907, $1.65; in 1908, $1.65; in 1909, $1.65; in 1910, 60 cents. Iron bars are treated in the same way as steel ingots, but as they are not made in any considerable quantity, they are not of any practical importance.

To-day the bounty on iron from Canadian ore is 35 per cent. of $3, that is equal to $1.05 per ton. The bounty on iron from foreign ore is 35 per cent. of $3, which is 70 cents per ton. The difference to-day is 35 cents per ton. This difference is not sufficient to encourage native industry. Those interested have represented that if we could have kept that difference of $1 there would be some inducement, some larger prospect of the development of native ores. We are impressed with this argument, because while we wish to encourage the iron industry generally, we particularly want to encourage iron made from native ore in order to encourage the development of the ores of Canada. We propose, therefore, that the rates of bounty during these four years periods should be as follows:

For the year 1907 the bounty on iron from native ore will be $2.10, and on foreign ore, $1.10. For the next year we keep the figures the same, $2.10 for native ore, and $1.10 for foreign ore. For the third year the bounty on iron from native ore will be $1.70, whereas on iron from foreign ore it is reduced to 70 cents. Then for the fourth year we reduce the bounty on the native ore to 90 cents, and the bounty of the foreign ore to 40 cents, keeping up the difference of 50 cents between the two.

We propose that these bounties shall not apply to articles exported. There are rumors that the great steel trust of the United States may come into Canada to do business in the ordinary way, well and good, but if it comes here to make up iron and steel for export it might add largely to our burden. We think it is well to have it understood that we are giving these bounties for the encouragement of iron and steel for consumption in Canada, and if any parties undertake to export these articles they shall not be entitled to the bounty upon it.

Then there are bounties at present on angles, plates and iron rods. With respect to angles and plates there is a bounty of $3 and a duty of ten per cent. We have decided to abolish the bounty and allow these articles to fall as a rule into a third class at $7 per ton, subject to the preferential reduction. The bounty is at present $3 and the duty ten per cent.; we strike out the bounty and put them in with the usual tariff instead of the bounty in this case.

The result of the change is to give about the same degree of protection but to get rid of the bounties. Articles are now made in Canada to a considerable extent. We would be glad to do the same thing if we could find it convenient in relation to the bounty on iron rods. The difficulty there, however, is that iron rods are not iron articles. They are used by the manufacturers of other lines of goods, and if we impose a considerable duty on iron rods we will have to change the duties on the articles which are made from iron rods, and we do not wish at present to disturb that class of duties, so we continue the bounties we have at present. The other bounties are for fixed periods. The bounty on iron rods was not for any fixed period, and it is like a tariff item which may be changed from time to time.

Benefits of Bounty System.

This bounty question is one of widespread interest, and in some quarters strong objections have been taken to bounties, criticisms of a varied character. Some object to bounties because they are a protection, others because they think bounties are something which impose tremendous burdens upon the country for which the country receives no advantage. I think those who condemn bounties generally, do not stop to consider what would happen if we did not adopt this system. If we did not adopt the bounty system then unquestionably we should be obliged largely to increase the duties on iron, or to allow industries of a considerable character probably to close up, and we do not think that is desirable. Those who think that these bounties can be dispensed with have not perhaps given the matter as careful consideration as they might. These bounties have not been fruitless of good. They have accomplished a great deal in the development of the interests of the country. They have built up great establishments, and have given employment to thousands of men.

Then there is another view which is perhaps lost sight of. If you turn to the customs returns of ports, connected with the iron industry at Sault Ste. Marie, Midland, New Glasgow, Sydney and North Sydney, take all the various points where the Customs Department comes in touch with these iron industries, you will discover that during the period of these iron bounties there has been an enormous development of trade at these ports, the greater part of which is directly due to the development of these industries, and if you make a comparison between the customs revenues of those ports a few years ago, and the customs revenues as they have been under the influence of this bounty system, you will find that the increased collection of revenue at these ports is fully equal to every dollar that the Government ever paid by way of bounties, and so any impressions that may exist in the minds of any that the bounties are not in any way returning anything to the country is certainly a mistake.

British Preference on Metals.

The rates on lead manufactures, including lead pipe, lead shot and lead bullets, have been reduced about 5 per cent. and the preferential rate on lead in bars and sheets has also been reduced from 16 2-3 to 15 per cent.

In item 384 of the new tariff, an effort is made to divert trade to Great Britain, and at the same time to cheapen the cost to the consumer. This item reads:—Rolled iron or steel sheets, number fourteen gauge and thinner, n.o.p., Canada plates; Russia iron; flat galvanized iron or steel sheets; terne plate and rolled sheets of iron or steel, coated with zinc, spelter or other metal, of all widths or thicknesses, n.o.p., and rolled iron or steel hoop, hand scroll or strip, No. 14 gauge, and thinner, galvanized, or coated with other metal, or not, n.o.p. The articles are made free from Great Britain and dutiable from other countries. Tin plate is free under all tariffs, in the interests of manufacturers and consumers. The item has been enlarged by increasing the gauge of iron and steel sheets from seventeen to fourteen gauge, and the gauge of hoop, band, scroll and strip from eighteen to fourteen gauge.

The preferential rate on japanned ware, tinware and all manufacturers of tin, whether "hollowware" or not, also on manufactures of zinc and manufactures of aluminum is reduced from 16 2-3 per cent. to 15 per cent, while the general tariff rate of 25 per cent is continued.

Brass and copper wire, Britannia metal, nickel, silver, Nevada and German silver, etc., are all increased, but it will be observed that the ratio of preference to Great Britain is greater than it was before.

Boiler tubes were formerly dutiable at five per cent. general and 3 1-3 per cent. preference. They are now made free from Great Britain.

Galvanized iron or steel wire, 9, 12 and 13 gauge, formerly on free list is now made a dutiable at five per cent. under the general and intermediate tariffs and free from Great Britain. This is an item of very considerable importance, as over a million dollars' worth was imported last year, mostly from the United States.

Well-drilling machinery for boring and drilling for water is made free, whether made in Canada or not. It was formerly free, but the free admission was confined to such machinery as was not made in Canada. We think that boring for water is a matter of such importance and so we make the machinery free altogether.

A number of articles of iron and steel specified in old tariff are not named in new tariff. They fall under this general item, "manufactures of iron or steel, n.o.p." The principal exceptions are iron and steel castings, in the rough, formerly dutiable at 25 per cent, now 30 per cent.

OLD AND NEW TARIFFS COMPARED

The following table includes practically every article in which the machinery and kindred trades are interested. In the four columns are given the British preferential the intermediate, the new general and the old tariffs, the latter being given for the purpose of comparison. The German surtax of 33 1-3 per cent. is the only part of the tariff not included. In a study of the table it will be an easy matter to figure out the effect of the changes, remembering, of course, that the British preferential duty was formerly an even 33 1-3 per cent. whereas now each item stands by itself. For a more complete analysis, such as few will find necessary, it will be necessary to have a copy of the old tariff for reference as the classifications are entirely changed and the table is arranged on the basis of the present tariff.

Group 5—Pulp, Paper and Books.

	British Pref.	Intermediate	General	Former Tariff
Hemp paper, made on four cylinder machines and calendered to between .006 and .008-inch thickness, adapted for the manufacture of shot shells, primers, adapted for the manufacture of shot shells and cartridges, and felt board sized and hydraulic pressed and covered with paper or uncovered, adapted for the manufacture of gun wads	F	F	F	F
Strawboard, millboard and cardboard, tarred paper, felt board, sandpaper, glass or flint paper and emery paper or emery cloth, p.c.	15	22½	25	25

Group 6—Chemicals, Drugs, Oils and Paints.

	British Pref.	Intermediate	General	Former Tariff
Glue, liquid, powdered, or sheet, mucilage, gelatine, casein, adhesive paste and isinglass, p.c.	17½	25	27½	25
Lamp black, carbon black, ivory black and bone black	F	F	F	F
Ultramarine blue, dry or in pulp; whiting or whitening: Paris white and gilders' whiting; blanc fixel, satin white	F	F	F	F
Litharge	F	F	F	F
Dry red lead, orange mineral and zinc, white, p.c.	F	5	5	5
Dry white lead, p.c.	20	27½	30	30
White lead ground in oil, p.c.	25	32½	35	35
Ochres, ochrey earths, siennas and umbers, p.c.	10	12½	15	20
Oxides, fireproofs, rough stuff, fillers, laundry blueing and colors, dry, n.o.p., p.c.	15	20	22½	25
Liquid fillers, anti-corrosives and anti-fouling paints, and ground and liquid paints, n.o.p., p.c.	15½	25	27½	25
Paints and colors, ground in spirits, and all spirit varnishes and lacquers, per gallon	$1.00	$1.00	$1.00	$1.12½
Varnishes, lacquers, japans, japan driers, liquid driers and oil finish, n.o.p., per gallon	15c	15c	15c	20c
And per cent.	15	20	22½	20
Paris green, dry, p.c.	5	7½	10	10
Gold liquid paint, p.c.	15	22½	25	25
Putty, of all kinds, p.c.	12½	17½	20	20
Linseed or flaxseed oil, raw or boiled, per 100 lbs.	80c	$1.10	$1.20	25p.c.
Turpentine, raw or crude	F	F	F	F
Turpentine, spirits of	5	5	5	5
Illuminating oils, composed wholly or in part of the products of petroleum, coal, shale, or lignite, costing more than 30c. per gal., p.c.	15	17½	20	20
Lubricating oils, composed wholly or in part of petroleum, costing less than 25c per gallon, per gallon	1½c	2¼c	2½c	2½c
Crude petroleum, gas oils, other than naptha, benzine and gasoline, lighter than .8235, but not less than .775 specific gravity, at 60 degrees temperature, per gallon	1c	1¼c	1½c	1½c
Oils, coal and kerosene, distilled, purified, or refined, petroleum, and products of petroleum, n.o.p., per gallon	1¼c	2¼c	2½c	2½c
Gasoline or naphtha, under .775 specific gravity at 60 degrees temperature	F	F	F	F
Lubricating oils, n.o.p., and axle grease, per cent.	12½	17½	20	20

Group 7—Earths, Earthenware and Stoneware.

	British Pref.	Intermediate	General	Former Tariff
Fire brick, p.c.	5	7½	10	[illegible]
Building brick, paving brick, and manufactures of clay or cement, n.o.p., p.c.	12½	20	22½	20
Drain tiles, not glazed, p.c.	15	17½	20	20
Drain pipes, sewer pipes and earthenware fittings therefor, chimney linings or vents, chimney tops and inverted blocks, glazed or unglazed, earthenware tiles, n.o.p., p.c.	25	32½	35	35
Tiles or blocks of earthenware or of stone prepared for mosaic flooring, p.c.	20	27½	30	35
Closets, urinals, basins, lavatories, baths, bath tubs, sinks and laundry tubs of earthenware, stone, cement, or clay, or of other material, n.o.p., p.c.	20	30	35	30
Cement, Portland and hydraulic or water lime, in barrels, bags, or casks, the weight of the package to be included in the weight for duty, per 100 pounds	8c	11	12½	12½
Bags in which cement or lime mentioned in the next preceding item is imported, p.c.	15	20	20	20
Plaster of paris, or gypsum, calcined, and prepared wall plaster, the weight of the package to be included in the weight for duty, per 100 lbs.	8c	11c	12½c	12½c
Plaster of paris or gypsum, ground, not calcined, p.c.	10	12½	15	15
Clays, including china clay, fire clay and pipe clay not further manufactured than ground: gunister and sand; gravels: earths, crude only	F	F	F	F
Pumice, calcarous tufa, pumice stone and lava, not further manufactured than ground	F	F	F	F
Crucibles, clay or plumbago	F	F	F	F
Grindstones, not mounted, and not less than 36 inches in diameter, p.c.	10	12½	15	15
Grindstones, mounted or not, n.o.p. p.c.	17½	22½	25	25
Asbestos in any form other than crude, and all manufactures thereof, p.c.	15	22½	25	25
Plumbago, not ground or otherwise manufactured, p.c.	5	7½	10	10
Plumbago, ground, and manufactures of, n.o.p., and foundry facings of all kinds, p.c.	15	22½	25	25
Common and colorless window glass p.c.	7½	12½	15	15
Glass, in sheets and bent plate glass, n.o.p., p.c.	17½	22½	25	20
Plate glass, not beveled, in sheets or panes not exceeding seven square feet each, n.o.p., p.c.	7½	10	10	10
Plate glass, not beveled, in sheets or panes, exceeding seven square feet each and not exceeding twenty-five square feet each, n.o.p., p.c.	15	25	27½	25
Plate glass, n.o.p., p.c.	22½	30	35	35
Stained or ornamental glass windows, per cent.	20	27½	30	30
Glass demijohns or carboys, bottles, decanters, flasks, phials, glass jars and glass balls, lamp chimneys, glass shades or globes, cut, pressed or molded crystal or glass tableware, decorated or not, blown glass tableware, and manufactures of glass, n.o.p., p.c.	20	30	32½	30

Group 8—Metals and Manufactures Thereof.

	British Pref.	Intermediate	General	Former Tariff
Antimony, or regulus of, not ground, pulverized or otherwise manufactured...	F	F	F	F
Bismuth, metallic, in its natural state	F	F	F	F

	British Pref.	Inter-mediate	General	Former Tariff
Cinnabar, quicksilver, radium	F	F	F	
Tea lead	F	F	F	F
Lead, old, scrap, pig and block, p.c.	10	15	15	15
Lead, in bars and in sheets, p.c.	15	22½	25	25
Lead, manufactures of, n.o.p., p.c.	20	27½	30	35
Babbitt metal and type metal, in blocks, bars, plates, and sheets, p.c.	10	15	15	10
Phosphor tin and phosphor bronze in blocks, bars, plates, sheets and wire, per cent.	5	7½	10	10
Tin in blocks, pigs, bars, plates, or sheets, tin strip waste, and tin foil	F	F	F	F
Japanned ware, tinware, and all manufactures of tin, n.o.p., p.c.	15	22½	25	25
Zinc dust, zinc spelter and zinc in blocks, pigs, bars, rods, sheets and plates, and seamless drawn tubing of zinc	F	F	F	F
Zinc, manufactures of, n.o.p., p.c.	15	22½	25	25
Brass and copper, old and scrap, or in blocks, ingots, or pigs, copper and bronze in bars and rods, in coil, or otherwise, not less than six feet in length, unmanufactured, copper in strips, sheets or plates, not polished, planished, or coated, brass or copper tubing, in lengths of not less than six feet, and not polished, bent, or otherwise manufactured	F	F	F	F
Brass, in bars and rods, in coil or otherwise, not less than six feet in length, and brass in strips, sheets, or plates, not polished, planished, or coated, p.c.	5	7½	10	F
Brass wire, plain, p.c.	7½	10	12½	10
Copper wire, plain, tinned, or plated, per cent.	7½	10	12½	10
Brass and copper nails, tacks, rivets and burrs or washers; bells and gongs, n.o.p., and manufactures of brass or copper, n.o.p., p.c.	20	27½	30	30
Aluminum ingots, blocks, bars, rods, strips, sheets or plates; aluminum tubing in lengths not less than six feet, not polished, bent or otherwise manufactured	F	F	F	F
Manufactures of aluminum, n.o.p., p.c.	15	22½	25	25
Britannia metal, nickel silver, Nevada and German silver, manufactures of, not plated, n.o.p., p.c.	17½	27½	30	25
Gold, silver, and aluminum leaf; Dutch or Schlag metal leaf; brocade and bronze powder	15	25	27½	25
Articles consisting wholly or in part of sterling or other silverware, nickel-plated ware, gilt or electro-plated ware, n.o.p., manufactures of gold and silver, n.o.p., p.c.	22½	30	35	30
Platinum wire and platinum in bars, strips, sheets or plates	F	F	F	F
Clocks, watches, time recorders, watch glasses, clock and watch keys, clock cases and clock movements, p.c.	20	27½	30	25
Chronometers and compasses for ships	F	F	F	F
Blast furnace slag	F	F	F	F
Scrap iron and scrap steel, old and fit only to be remanufactured, being part of or recovered from any vessel wrecked in waters subject to the jurisdiction of Canada	F	F	F	F
Iron or steel scrap, wrought, being waste or refuse, including punchings, cuttings or clippings or iron or steel plates or sheets having been in actual use: crop ends of tie-plate bars, or of blooms, or of rails, the same not having been in actual use, per ton	70c	90c	$1.00	$1.00
But nothing shall be deemed scrap iron or scrap steel except waste or refuse iron or steel fit only to be remanufactured in rolling mills or furnaces, provided that articles of iron or steel, damaged in transit, if broken up under Customs supervision and rendered unsaleable except as scrap, may be entered for duty as scrap.				
Iron in pigs, iron kentledge, and cast scrap iron, ferro-silicon, ferro-manganese, and spiegeleisen, per ton	$1.50	$2.25	$2.50	$2.50
Iron or steel billets, weighing not less than sixty pounds per lineal yard: ingots, cogged ingots, blooms, slabs, puddled bars, and loops or other forms, n.o.p., less finished than iron or steel bars, but more advanced than pig iron, except castings, per ton	$1.75	$2.50	$2.75	$2.75
Rolled iron or steel angles, tees, beams, channels, girders and other rolled shapes or sections, not punched, drilled, or further manufactured than rolled, per ton	$4.25	$6.00	$7.00	$7.00
Bar iron or steel, rolled, whether in coils, rods, bars or bundles, comprising rounds, ovals, and squares and flats: steel billets, n.o.p., and rolled iron or steel hoop, band, scroll, or strip, twelve inches or less in width, number thirteen gauge and thicker, n.o.p., per ton	$4.25	$6.00	$7.00	$7.00
Rolled iron or steel beams, channels and angle-bars, weighing not less than forty pounds per lineal yard, and universal mill or rolled edge plates of steel, over twelve inches wide, for use in the manufacture of bridges, when imported by manufacturers of bridges, p.c.	5	10	10	10
Rolled iron or steel plates, not less than thirty inches in width, and not less than one-quarter of an inch in thickness, when imported by manufacturers of boilers for use in the manufacture of boilers, p.c.	5	10	10	10
Rolled iron or steel plates, not less than forty-eight inches in width and exceeding one-half inch in thickness, n.o.p., p.c.	5	10	10	10
Rolled iron or steel sheets or plates, sheared or unsheared, and skelp iron or steel, sheared or rolled in grooves, n.o.p., per ton	$4.25	$6.00	$7.00	$7.00
Skelp iron or steel, sheared or rolled in grooves, when imported by manufacturers, or wrought iron or steel pipe for use only in the manufacture of wrought iron or steel pipe in their own factories, p.c.	5	5	5	5
Rolled iron or steel sheets, number thirteen gauge and thinner, n.o.p., Canada plates, Russia iron, flat galvanized iron or steel sheets; terne plate and rolled sheets of iron or steel, coated with zinc, spelter or other metal, of all widths or thicknesses, n.o.p., and rolled iron or steel hoop, band, scroll, or strip, number fourteen gauge and thinner, galvanized or coated with other metal or not, n.o.p., p.c.	F	5	5	F
Chrome steel, p.c.	10	12½	15	15
Rolled iron or steel, bars, bands, hoops, scroll or strip, sheet or plate, of any size, thickness, or with galvanized or coated with any material or not, and cast steel when of greater value than three and a half cents per pound, n.o.p., p.c.	F	5	5	5
Swedish rolled iron and Swedish rolled steel nail rods, under half an inch in diameter, for the manufacture of horse-shoe nails, p.c.	10	12½	15	15
Iron and steel railway bars or rails of any form, punched or not, n.o.p., for railways, which term for the purposes of this item shall include all kinds of railways and tramways, even although they are used for private purposes only, and even although they are not used or intended to be used in connection with the business of common carrying of goods or passengers, per ton	$4.50	$6.00	$7.00	30p.c.

	British Pref.	Intermediate	General	Former Tariff
Iron or steel railway bars or rails, which have been in use in the tracks or railways in Canada and which have been exported from Canada and returned thereto after having been re-rolled, and weighing not less than fifty-six pounds per lineal yard when re-rolled, and which are to be used by the railway company importing them on their own tracks under regulations prescribed by the Minister of Customs, p.c.	25	25	25	25
Provided that the value for duty of such re-rolled rails shall be the cost of re-rolling the same. Provided also that whenever the Governor-in-Council is satisfied that a mill adapted and equipped for re-rolling such rails in substantial quantities has been established in Canada, the Governor-in-Council may by order-in-council to be published in the Canada Gazette abolish the duty specified in this item, and thereupon all such rails when imported shall be subject to such duty as otherwise provided in the Customs tariff.				
Railway fish and plates and tie plates, per ton	$5.00	$7.00	$8.00	$8.00
Switches, frogs, crossings, and intersections for railways, p.c.	20	30	32½	30
Iron or steel bridges or parts thereof, iron or steel structural work, columns, shapes or sections, drilled, punched or in any further state of manufacture than as rolled or cast, n.o.p., p.c.	22½	30	35	35
Springs, axles, axle bars, n.o.p., and axle blanks, and parts thereof, of iron or steel, for railway, tramway, or other vehicles, p.c.	22½	30	35	35
Forgings of iron or steel of whatever shape or size, or in whatever stage of manufacture, n.o.p., and steel shafting, turned, compressed, or polished, and hammered, drawn, or cold rolled iron or steel bars or shapes, n.o.p., p.c.	20	27½	30	30
Cast iron pipe of every description, per ton	$6.00	$7.00	$8.00	$8.00
Wrought or seamless iron or steel tubes for boilers, n.o.p., under regulations prescribed by the Minister of Customs, flues and corrugated tubes for marine boilers, p.c.	F	5	5	5
Seamless steel tubing valued at not less than 3½c per lb., rolled or drawn square tubing of iron or steel adapted for use in the manufacture of agricultural implements, p.c.	F	5	5	5
Tubes of rolled iron or steel, not joined or welded, not more than one half-inch in diameter, n.o.p., p.c.	5	7½	10	10
Wrought or seamless iron or steel tubing, plain or galvanized, threaded and coupled or not, over 4 inches in diameter, n.o.p., p.c.	10	12½	15	15
Wrought or seamless iron or steel tubing, plain or galvanized, threaded and coupled or not, 4 inches or less in diameter, n.o.p., p.c.	20	30	35	35
Iron or steel pipe or tubing, plain or galvanized, riveted, corrugated, or otherwise specially manufactured, including lock joint pipe, n.o.p., p.c.	20	27½	30	30
Wire-bound wooden pipe, p.c.	15	22½	25	25
Iron or steel fittings, for iron or steel pipe, of every description, p.c.	20	27½	30	30
Wire, crucible, cast steel, valued at not less than 6c per lb., p.c.	F	5	5	5
Galvanized iron or steel wire, Nos. 9, 12 and 13 gauge, p.c.	F	5	5	5
Buckthorn strip fencing, woven wire fencing, and wire fencing of iron or steel, n.o.p., not to include woven wire or netting made from wire smaller than No. 14 gauge, nor to include fencing of wire larger than No. 9 gauge, p.c.	10	12½	15	F
Wire of all metals and kinds, n.o.p., per cent.	15	17½	20	20
Wire, single or several, covered with cotton, linen, silk, rubber or other material, including cable so covered, n.o.p., p.c.	20	27½	30	30
Wire rope, stranded or twisted wire, clothes lines, picture, or other twisted wire, and wire cable, n.o.p., p.c.	17½	22½	25	25
Wire cloth or woven wire and wire netting of iron or steel, p.c.	20	27½	30	25
Coil chain, coil chain links, and chain shackles, of iron or steel, five-sixteenths of an inch in diameter and over, p.c.	5	5	5	5
Iron or steel nuts, washers, rivets, and bolts, with or without threads, nut bolt and hinge blanks, and T and strap hinges of all kinds, n.o.p., per 100 lbs.	75c	75c	75c	75c
And per cent.	10	20	25	25
Screws, commonly called "wood screws," or iron or steel, brass or other metal, including lag or coach screws, plated or not, and machine or other screws, n.o.p., p.c.	22½	30	35	35
Iron or steel cut nails and spikes (ordinary builder's), and railroad spikes, per 100 lbs.	30c	45c	50c	50c
Composition nails and spikes, and sheathing nails, p.c.	10	12½	15	15
Wire nails of all kinds, n.o.p., per 100 lbs.	40c	55c	60c	60c
Nails, brads, spikes and tacks of all kinds, n.o.p., p.c.	20	30	35	30
Wire cloth, or woven wire of brass or copper, p.c.	17½	22½	25	25
Iron or steel hollow ware, plain black, tinned or coated, n.o.p., and nickel and aluminum kitchen or household hollow-ware, n.o.p., p.c.	20	27½	30	30
Lamp springs, p.c.	7½	10	10	10
Lamps, side lights and head lights, lanterns, chandeliers, gas, coal oil, or other lighting fixtures, including electric light fixtures, or metal parts thereof, lava or other tips, burners, collars, galleries, shades, and shade holders, p.c.	20	27½	30	20
Gas meters and finished parts thereof, per cent.	22½	30	35	35
Fire engines and fire extinguishing machines, including sprinklers for fire protection, p.c.	22½	30	35	35
Threshing machine outfit, when consisting of traction or portable engine and separator, p.c.	15	17½	20	25
Hay loaders, potato diggers, horse powers, separators, n.o.p.; wind stackers, fodder or feed cutters, grain crushers, fanning mills, hay tedders, farm, road, or field rollers, post-hole diggers, snaths, and other agricultural implements, n.o.p., p.c.	15	22½	25	25
Shovels and spades, iron or steel, n.o.p.; shovel and spade blanks, and iron or steel cut to shape for the same, and lawn mowers, p.c.	20	30	32½	35
Stoves of all kinds, for coal, wood, oil, spirits, or gas, p.c.	15	22½	25	25
Belt pulleys of all kinds for power transmission, p.c.	15	25	27½	25
Telephone and telegraph instruments, electric and galvanic batteries, electric motors, dynamos, generators, sockets, insulators of all kinds, electric apparatus, n.o.p.; boilers, n.o.p.; and all machinery composed wholly or in part of iron or steel, n.o.p.; and integral parts of all machinery specified in this item, p.c.	15	25	27½	25
Manufactures, articles or wares of iron or steel, or of which iron and steel, or either, are the component materials				

	British Pref.	Inter-mediate	General	Former Tariff
of chief value, n.o.p., p.c.	20	27½	30	30
Anchors for vessels	F	F	F	F
Ingot molds, glass molds of metal, p.c.	5	7½	10	F
Iron sand or globules or iron shot and dry putty, adapted for polishing glass or granite, or for sawing stone	F	F	F	F
Steel bowls for cream separators and cream separators	F	F	F	F
Sundry articles of metal, as follows, when for use exclusively in mining or metallurgical operations, viz.: diamond drills, not including the motive power, coal-cutting machines, except percussion coal cutters, coal heading machines, coal augers, rotary coal drills, core drills, miners' safety lamps and parts thereof; also accessories for cleaning, filling and testing such lamps, electric or magnetic machines for separating or concentration iron ores, blast furnaces for the smelting of copper and nickels, converting apparatus for metallurgical processes in iron or copper; copper plates, plated or not, machinery for extraction of precious metals by the chlorination or cyanide processes; amalgam safes, automatic ore samplers, automatic feeders, retorts, mercury pumps, prometers, bullion furnaces, amalgam cleaners, blast furnace blowing engines, wrought iron tubing, butt or lap welded, threaded or coupled or not, over four inches in diameter, and integral parts of all machinery mentioned in this item	F	F	F	F
Machinery of a class or kind not made in Canada, for the manufacture of twine, cordage, or linen, or for the preparation of flax fibre	F	F	F	F
Well-drilling machinery and apparatus for boring and drilling for water	F	F	F	F
Iron or steel masts, or parts thereof, and iron or steel beams, angles, sheets, plates, knees, and cable chain, for wooden, iron, steel, composite ships and vessels, and iron, steel, or brass manufactures which at the time of their importation are of a class or kind not manufactured in Canada, when imported for use in the construction or equipment of ships or vessels under regulations prescribed by the Minister of Customs	F	F	F	F
Rolled round wire rods in the coil of iron or steel, not over three-eighths of an inch in diameter, when imported by wire manufacturers for use in making wire in the coil in their own factories	F	F	F	F
Materials under which enter into construction and form part of cream separators, when imported by manufacturers of cream separators to be used in their own factories for the manufacture of cream separators	F	F	F	F
Rolled steel for saws and for straw cutters, not tempered or ground, nor further manufactured than cut to shapes, without indented edges	F	F	F	F
Crucible sheet steel, eleven to sixteen gauge, 2½ to 18 inches wide, for the manufacture of mower and reaper knives, when imported by the manufacturers thereof for use exclusively in the manufacture of such articles in their own factories	F	F	F	F
Steel strip and flat steel wire, when imported into Canada by manufacturers of buckhorn and plain strip fencing for use exclusively in the manufacture of such articles in their own factories; and barbed fencing wire of iron or steel	F	F	F	F
Steel wire, Bessemer soft drawn spring of Nos. 10, 12, and 13 gauge, respectively, and home steel spring wire of Nos. 11 and 12 gauge, respectively, when imported by manufacturers of wire mattresses, to be used exclusively in the manufacture of such articles in their own factories	F	F	F	F
Cups, brass, being rough blanks for the manufacture of paper shells or cartridges, when imported by manufacturers of brass and paper shells and cartridges, for use exclusively in the manufacture of such articles in their own factories	F	F	F	F
Brass caps adapted for use in the manufacture of electric batteries	F	F	F	F
Iron tubing, lacquered or brass covered, not over two inches in diameter, and brass trimmings when imported by manufacturers of iron or brass bedsteads, for use exclusively in the manufacture of such articles in their own factories	F	F	F	F
Platinum crucibles	F	F	F	F
Platinum retorts, pans, condensers, tubing and pipe, and preparations of platinum, when imported by manufacturers of sulphuric acid, for use exclusively in the manufacture or concentration of sulphuric acid in their own factories	F	F	F	F
Steel balls, adapted for use on bearings of machinery and vehicles, p.c.	F	7½	10	10
Steel wood, p.c.	5	7½	10	10

Group 9—Wood and Manufacturers Thereof.

	British Pre	Inter-mediate	General	Former Tariff
House, office, cabinet or store furniture, of wood, iron or other material, in parts or finished; wire screens, wire doors and wire windows; cash registers; window cornices and cornice poles of all kinds: hair, spring and other mattresses; curtain stretchers; furniture springs and carpet sweepers, per cent.	20	27½	30	30

Group 10—Cotton, Flax, Hemp, Jute and Other Fibres.

	British Pre	Inter-mediate	General	Former Tariff
Hemp, dressed or undressed	F	F	F	F
Oakum or jute or hemp	F	F	F	F
Twine and cordage of all kinds, n.o.p., per cent.	20	22½	25	25
Sails for boats and ships, p.c.	15	22½	25	25
Lamp wicks, p.c.	17½	22½	25	25
Mats, door or carriage, other than metal, n.o.p., p.c.	25	30	35	35

Group 11—Miscellaneous.

	British Pre	Inter-mediate	General	Former Tariff
Asphalt or asphaltum solid; bone pitch, crude only; and resin or resin in packages of not less than 100 lbs.	F	F	F	F
Coal and pitch pine, Burgundy pitch; and coal and pine tar crude in packages of not less than fifteen gallons	F	F	F	F
Coal, anthracite; anthracite coal dust; coke	F	F	F	F
Bituminous slack coal; such as will pass through a ¾-inch screen, subject to regulations prescribed by the Minister of Customs, per ton	10c	12c	14c	15c
Bituminous round and run of mine, and coal, n.o.p.; per ton	35c	45c	53c	53c
Freight wagons, farm wagons, drays and sleighs, p.c.	17½	22½	25	25
Buggies, carriages, pleasure carts and vehicles, n.o.p., including automobiles and motor vehicles of all kinds; tires of rubber for vehicles of all kinds, fitted or not; cutters, children's carriages and sleds, and finished parts of all articles in this item, n.o.p., p.c.	22½	30	35	35
Provided that for duty purposes the minimum value of an open buggy shall be $40, and the minimum value of a covered buggy shall be $50.				
Railway cars or other cars, wheel-barrows, trucks, road or railway scrapers, and hand-carts, p.c.	20	27½	30	30

	British Pref.	Intermediate	General	Former Tariff
Bicycles and tricycles, n.o.p., p.c.	20	27½	30	30
Hides and skins, raw, whether dry, salted or pickled, and raw pelts	F	F	F	F
Fur tails, in the raw state	F	F	F	F
Fur skins of all kinds, not dressed in any manner	F	F	F	F
Belting, of leather, p.c.	12½	17	20	20
Belting, n.o.p., p.c.	20	25	27½	25
Harness and saddlery, including horse boots, p.c.	20	27½	30	30
Rubber cement, and all manufactures of India rubber and gutta percha, n.o.p., per cent.	15	25	27½	25
India rubber clothing, and clothing made waterproof with India rubber or gutta percha hose, and cotton or linen hose lined with rubber; rubber mats or matting and rubber packing, p.c.	22½	30	35	35
Gloves and mitts, of all kinds, p.c.	22½	30	35	35
Clothes wringers for domestic use, and parts thereof, p.c.	22½	30	35	35
Signs of any material other than paper, framed or not, letters and numerals of any material other than paper, p.c.	20	27½	30	30
Glycerine, when imported by manufacturers of explosives, for exclusively in the manufacture of such articles in their own factories	F	F	F	10
Nitro-glycerine, giant powder, nitro and other explosives, n.o.p., per lb.	1¾c	2¼c	2½c	3c
Blasting and mining powder, per lb.	1 1-3c	1¾c	2c	2c
Cannon, musket, rifle, gun and sporting powder and cannister powder, per lb.	2c	2¾c	3c	3c
Emery, in bulk, crushed or ground	F	F	F	F
Emery wheels, carborundum wheels, and manufactures of emery or of carborundum, p.c.	17½	22½	25	25
Junk, old; rags of cotton, linen, jute, hemp and wool; paper waste clippings, and waste of all kinds, n.o.p., except metallic; broken glass, or glass cullet	F	F	F	F
Fish hooks, for deep sea or lake fishing, not smaller in size than number 2.0; bank, cod, pollock and mackerel, fish lines; and mackerel, herring, salmon, seal, seine, mullet, net and trawl twine in hanks or coil, barbed or not, in variety of sizes and threads, including gilling thread in balls, and head ropes for fishing nets; manilla rope for holding traps in the lobster fishery; barked marline, and net morsels of cotton, hemp or flax; and deep-sea fishing nets or seines, when used exclusively for the fisheries, not to include hooks, lines or nets commonly used for sportsmen's purposes	F	F	F	F
Coverings, inside and outside, used in covering or holding goods imported therewith, shall be subject to the following provision, viz.—				
(a) Usual coverings containing free goods only; usual coverings, except receptacles capable of holding liquids, containing goods subject to a specific duty only, n.o.p.	F	F	F	F
(b) Usual covering containing goods subject to any ad valorem duty when not included in the invoice value of the goods they contain, p.c.	15	20	20	
(c) Provided that usual coverings containing goods subject to any ad valorem duty if included in the invoice value of the goods they contain, and not charged separately on the invoice, shall be subject to the same rate of duty ad valorem as the goods they contain, and may be combined with the goods for valuation and duty on the customs entry.				
(d) Provided further that receptacles capable of holding liquids, when containing goods, subject to a specific duty, shall be charged with the rate of duty to which the same would be subject if imported separately, except when the coverings and the goods contained therein are rated together in the tariff item;				
(e) Provided further that usual coverings designed for use other than in the bona fide transportation of the goods they contain, shall be charged with the rate of duty to which the same would be subject if imported separately;				
(f) Provided also that the term covering in this paragraph shall include packing boxes, crates, casks, cases, cartons, wrapping, sacks, bagging, rope, twine, straw or other articles used in covering or holding goods imported therewith, and the labor and charges for packing such goods, subject to regulations prescribed by the Minister of Customs;				
All goods not enumerated in this act as subject to any other rate of duty nor declared free of duty by this act, and not being goods the importation whereof is by this act or any other act prohibited, p.c.	15	17½	20	

Schedule B—Goods Subject to Drawback for Home Consumption.

	Drawback
Oil fuel and other articles, not machinery—when entering into the cost of binder twine manufactured in Canada	95 p.c.
Rolled iron, rolled steel and pig iron—when used in the manufacture of mowing machines, reapers, harvesters, binders, and attachments for binders	95 p.c.
Hemp bleaching compound—when used in the manufacture of rope	95 p.c.
Steel under one-half inch in diameter or under one-half inch square—when used in the manufacture of locks and knobs	95 p.c.
Steel cut to shape—when used in the manufacture of spoons	95 p.c.
Flat spring steel, steel billets and steel axle bars—when used in the manufacture of springs and axles for vehicles other than railway or tramway vehicles	95 p.c.
Spiral spring steel—when used in the manufacture of railway spiral springs	95 p.c.
Steel—when used in the manufacture of cutlery, files, augers, auger bits, hammers, axes, hatchets, scythes, reaping hooks, hoes, hay or straw knives, agricultural forks, hand rakes, skates, stove trimmings, bicycle chain and windmills	95 p.c.

PROHIBITED GOODS.

Goods manufactured or produced wholly or in part by prison labor, or which have been made within or in connection with any prison, jail or penitentiary, also goods similar in character to those produced in such institutions, when sold or offered for sale by any person, firm or corporation having a contract for the manufacture of such articles in such institutions, or by any agent of such person, firm or corporation, or when such goods were originally purchased from or transferred by any such contractor.

Metallic trading checks.

Any goods (a) which, if sold would be forfeited under the provisions of part of the Criminal Code; or (b) manufactured in any foreign state or country which bear any name or trade mark which is or purports to be the name or trade mark of any manufacturer, dealer or trader in the United Kingdom or in Canada, unless such name or trade mark is accompanied by a definite indication of the foreign state or country in which the goods were made or produced. Provided, that for the purposes of this item, if there is on any goods a name which is identical with or a colorable imitation of the name of a place in the United Kingdom or in Canada, such name, unless it is accompanied by the name of the state or country in which it is situate, shall, unless the Minister decides that the attaching of such name is not calculated to deceive (of which matter the Minister shall be the sole judge), be treated as if it was the name of a place in the United Kingdom or in Canada.

Iron and Steel Bounties.

Mr. Fielding's resolution on iron and steel bounties is as follows:

Resolved: (1) That it is expedient to repeal chapter 8 of the statutes of 1899 and chapter 60 of the statutes of 1903, from and after the 1st January, 1907.

(2) That it is expedient to provide that the Governor in Council may authorize the payment from the consolidated revenue fund of the following bounties on the under-mentioned articles manufactured in Canada for consumption therein, viz:

(a) In respect of pig iron manufactured from ore, on the proportion from Canadian ore produced during the calendar years — 1907, \$2.10 per ton; 1908, \$2.10 per ton; 1909, \$1.70 per ton; 1910, 90c per ton.

(b) In respect of pig iron manufactured from ore on the proportion from foreign ores produced during the calendar years — 1907, \$1.10 per ton; 1908, \$1.10 per ton; 1909, 70c per ton; 1910, 40c per ton.

(c) On puddled iron bars manufactured from pig iron made in Canada, during the calendar years—1907, \$1.65 per ton; 1908, \$1.65 per ton; 1909, \$1.05 per ton; 1910, 60c per ton.

(d) In respect of rolled round wire rods not over three-eighths of an inch in diameter, manufactured in Canada from steel produced in Canada from ingredients of which not less than fifty per cent. of the weight thereof consists of pig iron made in Canada, when sold to wire manufacturers for use in making wire in their own factories in Canada, on such wire rods made after the 31st December, 1906, \$6 per ton.

(e) In respect of steel ingots manufactured from ingredients of which not less than 50 per cent. of the weight thereof consists of pig iron made in Canada, on such ingots made during the calendar years—1907, \$1.65 per ton; 1908, \$1.65 per ton; 1909, \$1.05 per ton; 1910, 60c per ton. Provided that bounty shall not be paid on steel ingots from which steel blooms and billets for exportation from Canada are manufactured.

(3) That it is expedient to provide that the Governor in Council may make regulations to carry out the intentions of these resolutions.

(4) That it is expedient to provide that the Minister of Trade and Commerce shall be charged with the administration of the foregoing provisions.

OPINIONS ON THE NEW TARIFF

Canadian Fairbanks Co.

H. J. Fuller, whose views were incorrectly reported by a Montreal evening paper, has given his views on the recent changes in the tariff as follows:

(1) That I believe that the giving of the preferential tariff in favor of Great Britain is an entirely proper policy on the part of the Government; that Canada should buy, as far as possible, her surplus requirements from Great Britain.

(2) That however, the minimum tariff should at least be an equalizing tariff or, in other words, that the Canadian manufacturer should have at least an equal opportunity with the British manufacturer to supply the Canadian market.

(3) I do not believe that the tariff of 15 per cent. on machinery is fair protection to the Canadian manufacturer of these lines when the lower cost of labor, material, and the advantageous through freight rates are taken into consideration; to say nothing of the comparatively limited quantities in which he is able to manufacture.

(4) I believe that the increase in the duty to 27½ per cent. against American machinery is excessive, and that the many American manufacturers who have invested millions in their plants in Canada are entitled to at least some consideration when they are of necessity compelled to import from their affiliated plants in the United States some sizes and styles of machines which the present comparatively limited market in Canada does not enable them to make here.

It was in reference to this that I made mention of a number of Canadian-American concerns located here

Ship Building Interests.

Speaking of the new tariff, as affecting the shipbuilding interests of Canada, Frederic Nicholls, managing director of the Canadian General Electric Co., stated that the disabilities of the industry have been fully explained to the Finance Minister from time to time and all connected with it thought it would receive favorable consideration at his hands. The reason for this consideration being asked, is that there is no other such anomally existing throughout the entire Canadian tariff.

"Those of us who are manufacturing ships have to pay duty on nearly all the material that enters into the manufacture. For instance, if we manufacture engines, boilers or other accessories, we have to pay duty on raw materials and again, if we import finished engines or boilers we also have to pay duty.

"The same applies to everything included in fittings and furnishings, such as furniture, tableware, cutlery, glassware, and in fact, every detail of a ship. On the other hand, if a ship is imported from Great Britain, with furnishings complete it is allowed to enter Canada absolutely free of duty.

"We are hoping that this matter will receive separate and special consideration at the hands of the Finance Minister of the Government.

Forge Company's Opinion.

A manufacturing company in Canada whose principal business is the making of forgings giving its opinion states: "We are seriously affected by item 376 wherein the tariff on steel billets and blooms was raised 75c per ton. All our forgings or practically all are made from this material; therefore this change will not be conducive to lower priced forgings inasmuch as the other manufacturers in Canada, without excepton, are affected as seriously as we are, including the steel mills that make the manufacture of forgings a side issue."

Assists Gasoline Engine Makers.

J. R. Golden, general manager and secretary of the Canadian McVicker Engine Co., Galt Ont., states that now gasoline is placed on the free list it will be a great help to them in the manufacture of gasoline engines. Other changes do not affect them in any way.

Tariff Satisfactory.

The R. McDougall Co., of Galt, considers the changes in the new tariff on the whole satisfactory, as they would not expect the changes to be altogether in their favor.

Reacts Against Incandescent Lights.

R. B. Hamilton, managing director and the Packard Electric Co., St. Catharines, takes exception to the heavy increase on the duty on glass bulbs for incandescent lamps. He states that the changes which affect them principally are the placing of 10 per cent. duty on brass rods and sheets and the omission of any reference to specific items, such as iron castings in the rough, glass bulbs for incandescent lamps, etc. In the matter of the duty on brass, would explain that this was included in the tariff at the express request of a concern, which is at present not operating and which, at its best, could not begin to supply the requirements of Canadian users of these materials. The net result will, therefore, be a very heavy increase of expense to a large number of manufacturers. Were it possible to secure this material in satisfactory quality and quantities in Canada the manufacturers would probably have little, if any exception, to this item on the tariff. Under the circumstances, therefore it is causing a great deal of criticism.

Iron castings in the rough have heretofore been specifically rated at 25 per cent. In the new tariff they are not mentioned, and are, therefore, included in "Iron N.O.P. at 30 per cent.," which makes these raw materials dutiable at as high a rate as finished products.

The duty on glass bulbs for incandescent lamps was specifically placed at 10 per cent. under the old tariff, but no mention is now made of the item which, therefore, falls under the classification "Glass N.O.P." at 32½ per cent. This is a very serious matter to manufacturers of incandescent lamps, as the breakage and waste in this material is so great as to multiply the percentage of duty several times over. The change is therefore, placing Canadian manufacturers under a great burden, and with no corresponding benefit to any one, as the bulbs are not made in Canada, and could not be made here, since the market is not nearly sufficient to warrant the establishment of the extremely expensive plant required. Aside from these and a few other items of lesser importance to us, we have no criticism to offer on the new tariff.

Not Much Change.

H. W. Dorken of Dorken Bros. & Co., hardware importers, Montreal, when seen by a Hardware and Metal representative, remarked that, in his opinion, the Government had carried out a good policy in refusing to make any very radical changes in the tariff. So far as he had studied the tariff, although a few articles showed a higher duty, and some others a lower than before, conditions were practically the same. In most cases, the change did not amount to more than about 2½ per cent., which would not hurt anybody. "I was not pleased" said Mr. Dorken, "to see the German surtax retained, after its three years trial. Ours is almost entirely a German business, and our goods will still be subject to the additional duty, which, we had expected, would be removed. On the whole, however, the Government has done well, and has made an effort to adopt a tariff with as few clauses as possible, which will benefit the masses and not the few."

Paint Manufacturers' Object.

A. Ramsay, of the A. Ramsay & Son Co., did not care to express himself at any length, but said briefly: "The new provision for duties on linseed oil, seems to give a fairly good preference to England. The new tariff will not be satisfactory to paint manufacturers, as manufactured goods are not sufficiently protected to cover cost of raw material."

Brandram-Henderson thought that the Canadian linseed oil industry would be better protected under the proposed specific duty than when the old ad valorem duty was in force. Under the old conditions, the Canadian mills had to shut down, whenever prices in England went down very low. As Brandram-Henderson will have their own lead corroding plant in Montreal, by next Spring, they are not so hard hit by the failure to raise duties on paints, as are other firms.

Wise Action on Tin Plate.

W. E. Ramsay, manager of the Pedlar People's Montreal branch, said: "In framing this tariff, the Government have gone into the matter pretty thoroughly, and have done what they thought in the best interests of all, showing undue favors to none. Of course, there are errors and omissions in the original draft, which, if brought before the Minister in a clear comprehensive and unselfish manner, will be remedied. The Commission did well to listen to the appeals and petitions of the farmers, business men and manufacturers, against the imposition of a 33 1-3 per cent. duty on tinned, galvanized and black sheets. It would be most unwise to furnish employment for even 400 men, and tax seven millions to do so.

Fire Brick Duty a Mistake.

Mr. King, manager of F. Hyde & Co., dealers in cement and firebricks, was asked what effect the duty on fire brick would have. He answered that it would be a serious matter to the large iron and steel manufacturers, who yearly use several millions' of firebricks. "I think this duty is a mistake, and I do not understand why the Government have done it. It may have been because some dealers have been bringing in pressed brick from the United States, and entering it as fire brick, but, if so, these dealers will be punished at the expense of the manufacturers."

No Complaint to Make.

S. Brewer, secretary of the Thos. Robertson & Co., dealers in steamfitters' supplies and metals, said: "Taking it all in all, the tariff is very satisfactory. Of course, there are a few items which might be adjusted, but we realize that it would be impossible to frame a tariff which would suit everybody, and we ourselves have no serious complaints to make.

Mr. Howland Criticises Tariff.

"I am opposed to the new tariff for several reasons," said Peleg Howland, pres. of the Toronto Board of Trade, and head of the H. S. Howland & Sons Co., wholesale hardware firm. "I am not a believer in more than one tariff nor do I approve of the bonus system or the anti-dumping clause.:'

"The refusal of the Government to protect the tin-plate industry by a duty, should have been accompanied by a refusal to continue to bonus the iron and steel industries. The inconsistency of their position is shown by their bonusing the pig iron industry, and then granting a drawback of 95 per cent on pig iron imported by the farm implement manufacturers, thus cutting out the market for the furnaces. Why, too, make fish of one set of manufacturers and flesh of another? If any rebates are to be given, why not give to all? Why should wire manufacturers and foundrymen be treated differently to implement manufacturers? As a matter of fact the application of the new tariff practically puts wire manufacturers in the hands of the Canadian monoply and higher prices are certain to result.

"I cannot conceive that the makers of the tariff realize the working out of the anti-dumping clause, as a higher duty would bring more revenue and greatly lessen the temptation to beat the customs. The extra charge made by the anti-dumping clause is only paid once or twice, after which it will be retained by the foreign importer or given by him to his customer as commissions, etc. How can honest importers continue to pay the advance, when dishonest men arrange for commissions or other special inducements from manufacturers, with no possibility of discovery. The anti-dumping clause is merely a method of deceiving the public, and its more stringent application by giving the Finance Minister power to disontinue the 5 per cent leeway means a concealed increase of protection to Canadian manufacturers without increase of revenue.

"The new tariff will have the effect of increasing the prices on many such lines, as shaving brushes, lead pencils, etc., as well as the changes which are generally known.

Cheap Metals in Sight.

George T. Pepall, of M. & L. Samuel, Benjamin & Co., Toronto, considered that the general effect of the new tariff would be to benefit British industries, and that outside of iron and steel bars and tank plates metals would be very cheap in Canada in the future. He thought it was peculiar that the Government should single out brass rods, etc., as a dutiable article while copper was left free, especially as there is at present no working industry in Canada to be benefitted by the change.

Severe Blow to Merchants.

Ferg. McDonald, manager of the metal department of Rice Lewis & Son, Toronto, points out that the new tariff is decidedly unfair to metal merchants especially in the item of boiler plate which was formerly dutiable at 10 per cent. on 30 inch wide and over while now it is $7.00 per ton when bought by merchants, while boiler merchants can get it at the same old rate. The tendency of this change will be to take the business entirely out of the hands of the metal merchants as boiler manufacturers will buy their tubes and other supplies from the same source as they secure their boiler plates from, and as small boiler makers and repair men cannot import in large quantities and will be unable to buy from merchants to advantage under the new conditions they must go out of business and all the trade will fall into the hands of large boiler manufacturers. The same thing applies to the bridge making industry owing to the changes in duty on beams, angles, channels, etc. These articles are universally used building material and as they are not made in Canada it is peculiar that a duty should be imposed on them to the detriment of the building trade.

Little Change in Glass.

W. R. Hobbs of the Consolidated Plate Glass Co., Toronto, considers that the new tariff will have little effect in the glass trade so far as the demand from the hardware and builders is concerned. Common sheet glass and smaller sizes of plate are left unchanged and the only new addition is that British manufacturers secure a slight additional advantage on some lines of plates.

Tin Plate Agents Pleased.

Arthur J. Owen, Canadian manager of Franklin Saunders & Co., manufacturers of tin plate, etc., expressed himself as follows: "I have nothing to say against the new tariff. Had the proposed duty been placed on tin plate, it would have worked most adversely against tin can manufacturers and other users of tin plates, black sheets, etc. It is a matter of congratulation both for manufacturers and dealers interested in importing such materials from the Old Country that not only was no heavy duty imposed, but such duty as has existed was taken off so as to encourage British industries.

The Package Problem.

Alexander Gibb, metal and hardware importer, Montreal, said: "I am very well satisfied, so far as I understand the new tariff. Regarding packages, the tariff is not plain. Galvanized iron coming from Great Britain, is on the free list. Will there be a duty on the cases? Some people seem to think there will be, and this is about the only thing that is worrying the hardware trade. English hardware dealers are always in the habit of charging for cases, which are free from most United States manufacturers. Thus, if a duty is imposed on the cases, as has been done in certain cases in the past, there will be a discrimination against Great Britain, which I do not think Mr. Fielding intended.

Paint Men Dissatisfied.

C. C. Ballantyne of the Sherwin-Williams Company, Montreal, said: "Paint manufacturers are not satisfied with the new tariff as applied to their products. When it was proposed, a year and a half ago, to increase the duty on white lead, the paint manufacturers informed the Government that they would offer no objection to assisting a new industry to establish itself, so long as it was understood that the Government would make a corresponding increase in the duty on mixed paints so that we should have as good protection as we had when the duty on white lead was 5, and on the finished article 25 per cent. When we presented our suggestions to the Minister, we asked that the tariff should be raised from 25 to 35 per cent. on mixed paints which would have the same effect as when the duty was 5 per cent. on lead, and 25 per cent. on mixed paints.

The paint industry was perfectly satisfied with things as they were before the duty on white lead was imposed. The old duty on white lead was 5 per cent, and on lead ground in oil, 25 per cent. Now, the duty on the raw material is 20 per cent., and on ground white lead, 25 per cent., a difference of only 5 per cent.

I think a good many people will regret that the difference between the maximum and intermediate tariff is not greater. It is only 2½ per cent., which is but a flea bite. It is not enough to make U.S. industries fell that they will have to come here to establish branches, if they want to do business.

Increased Cost of Brass Goods.

Chas. Morrison of the James Morrison Brass Manufacturing Co., Toronto, says that his company have not had time yet to consider the changes made in the new tariff but that the addition of the 10 per cent. duty on brass rods, tubes and sheets would increase their cost of manufacture to the extent of the duty as all rods and tubes used by them are over six feet in length. The advance on sheets will probably force an increase in prices although this will be difficult on account of the recent advances caused by the extraordinary high price of copper now ruling. Mr. Morrison stated that if the change would result in inducing the Brass Rolling Mills at New Toronto to re-open no objection could be made to the new duty, as the new industry would certainly be entitled to a measure of protection. At present, however, the advance would apply to imported goods only as no brass rods or sheets are produced in Canada.

Brass Mills to Re-open.

The Canadian Brass Rolling Mills will re-open their works at New Toronto in January said a representative of the company to Hardware & Metal on Wednesday. "We assured the Government that if we were given protection we would re-open the mills and we have already prepared the foundation for a large addition to the plant and have considerable new machinery on the grounds. We will be producing rods and sheets early in the New Year in accordance with our promise to the Government.

CORRESPONDENCE COLUMN

Some Explanation Needed.

In view of the fact that Mr. P. B. Ball, Dominion commercial agent at Birmingham, had complained of the quality of Canadian calcium carbide being shipped into England, the matter was brought to the attention of the readers of Canadian Machinery in the last issue. Since then letters have been received from each of the three manufacturers of calcium carbide in this country giving the situation from their view point. The letters would indicate that either Mr. Ball has been misinformed to some extent or else guilty of hasty action in denouncing the general quality of Canadian calcium carbide that was being shipped to England. The letters are given as received and speak for themselves.

* * *

Dear Sir :—

In your issue for December 1903, page 451, we find an article in reference to poor carbide having been delivered in Great Britain. We agree with the statements made in this article, that it is a serious matter for Canadian manufacturers to put an inferior grade of their product upon the British market, and beg to state that the carbide referred to was not manufactured or sold by this company. Our works, as you are aware, are located at Merritton, Ontario.

Yours truly,
Willson Carbide Company, Limited.
D. D. McTavish, Secretary-Treasurer.

* * *

Dear Sir :—

In your issue of December an article is published entitled "Poor Calcium Carbide" and the writer quotes a report of Mr. F. D. Ball, the Dominion commercial agent at Birmingham, which alleges that the poor quality of calcium carbide is systematically and continually shipped from Canada for the English markets.

With all due deference to the Dominion agent I would state that his information is incorrect and is probably based on casual complaint against some one particular shipment Mr. Ball in recommending the Canadian Manufacturers to look into this condition which he alleges in making the statement that the quality shipped for the English market is not up to the standard required, evidently considers that Canadian manufacturers of calcium carbide give no attention to the quality of goods manufactured.

In refuting the charges against the quality of Canadian carbide I would state that the company which I represent has been shipping carbide for the British market for over two years and the fact that our sales during the present year were only limited to our capacity and the limitations of ocean transportation is somewhat o an indication that the carbide supplied is of the required quality.

This company has been selling large quantities of carbide to dealers in Manchester, Belfast, Dublin and Liverpool and our carbide is quite satisfactory. In many cases we have had tests made previous to shipments by analysts of Montreal and in every case the certificate received shows the quality considerably in advance of the standard set forth by the English Association.

The statement that in some instances carbide was found which produced gas which ran from 25 to 33 1-3 less than the standard amount has no reference to Shawinigan carbide. Other manufacturers in Canada may speak for themselves, but I am prepared to support my contention that such statements that are contained in your article have no foundation as applying to the output of this company.

I trust you will be good enough to give this statement as broad publicity as is warranted by your previous article.

Thanking you in anticipation, we are

Very truly yours,
Howard Murray,
Secretary-Treasurer.
The Shawinigan Carbide Co., Montreal.

* * *

Gentlemen :—

Our attention has been attracted by an article on page 451 of your issue for this month. We have read the item with interest and feel that you cannot express too strong disapprobation of the course of Canadian manufacturers who supply inferior articles for the export market, and more especially for the market in Great Britain. We wish to state that we have not sent any carbide to the Old Country for at least two years, and the material about which complaints were made could not have been supplied by us. We operate a large factory here at Ottawa, and mantain a uniformly high standard for all our goods whether they are for domestic or foreign markets. We would appreciate it if you would insert a note in your periodical to the effect that the carbide mentioned in your article of this month was not furnished by us.

Yours truly,
The Ottawa Carbide Co., Ltd.
L. Crannell, Sec-Treas.

Our Water Powers

Niagara Power in Toronto.

SINCE power was first turned on in Toronto from Niagara in November there has been some trouble in continuity of service of between 3,000 and 4,000 h.p. expected by the Toronto Electric Light Co. for their light and for motor power for the Street Railway only a small part is available each day after 3.30 p.m., the reason being that work on turbines, generators and the other necessary appliances is not nearly complete and the generators now ready cannot be operated, while the workmen are busy, to meet further the demand in Toronto the work on the turbines has been so arranged that when the Toronto Railways and the Electric Light Companies become urgent the day's work on the power house installations ceases and the current is turned on. At present the Toronto Railway Co. has turned over to the Electric Light Co. about 2,000 h.p. daily. This has made it possible for the Light Co. to use all its wires, pending the completion of the Electric Light Co.'s own transforming plant.

* * *

Vancouver's Water System.

The conservation of an adequate water supply for the city of Vancouver is a subject of towering importance to that city at the present time. The city officials visited the Government with a request that a reservoir be placed on the watershed of Seymour Creek that the supply might be protected from the depredations of lumbermen. As Vancouver grows larger it will require this water some day so that the city officials are wise in looking ahead to the future requirements of a great city as Vancouver is destined to become.

* * *

Power for Berlin.

The citizens of central western Ontario are greatly interested in the Power Commission's work in connection with supplying Niagara power to that district. At a meeting held on Dec. 5th, forty delegates from Galt, Guelph, Waterloo, Berlin, Acton and Mount Forest met in conference on the western power question in the Council Room of that city. Hon. Mr. Beck was present at the meeting, at which he was enthusiastically received, and there he clearly outlined the subject and gave a stirring address on the opportunity of the hour. A feature of the meeting was the endorsing of the work done by the Power Commission. The meeting pledged itself to do all in its power to secure all the necessary by-laws and the largest possible number of names in the forthcoming municipal elections. Mr. Cecil B. Smith gave an expert memorandum on the power situation in Berlin. The city is using 300 h.p. at present at a cost of $1,000.00 per month, which figured out $35.00 per h.p., considered on the maximum consumption, and $66.00 on the average consumption. The question was asked Mr. Beck if electrical power from Niagara would be cheaper than the power supplied by the present city plant, to which Mr. Beck replied that Niagara power would be much cheaper. A big saving will be effected as soon as the consumption grows and the city goes ahead.

* * *

Electricity for St. John.

The Grand Falls Power Company have offered to supply electrical power to the city of St. John, N.B., for 30 years at named maximum figures, which according to the company's estimates will effect the saving of $250,000 a year to users of power in St. John. It is proposed to develop sixty thousand horse power at the falls, with storage facility to increase the horse power by 25 per cent.

* * *

Calgary's Possibilities.

The water power proposition has been placed before the city of Calgary, that will not only involve an enormous outlay, but mean much to the future manufacturing interests of that City. It is one of the biggest propositions with which the city has yet been confronted. The proposition is to furnish power and light to the city by a company, and the city distribute it. W. M. Alexander of Brandon is interested in the scheme. He proposes to work for a street railway franchise which the city is at liberty to buy out in ten years. Mr. Alexander is a street railway man, and is conversant with the railway business.

A Forest Cascade on the L'Abime River, Cheticamp, Cape Breton.

CANADIAN MACHINERY
and Manufacturing News

A monthly newspaper devoted to machinery and manufacturing interests, mechanical and electrical trades, technical progress, construction and improvement, and to all users of power developed from steam, gas, electricity, compressed air and water in Canada.

The MacLean Publishing Company, Limited

JOHN BAYNE MACLEAN - - - *President*
W. L. EDMONDS - - - - *Vice-President*
F. S. KEITH, B.Sc. - - - - *Managing Editor*

Publishers of trade newspapers which circulate everywhere in Canada also in Great Britain, United States, West Indies, South Africa and Australia

OFFICES:

CANADA
MONTREAL - - - - 232 McGill Street, Telephone Main 1255
TORONTO - - - - 10 Front Street East, Telephone Main 2701
WINNIPEG - - - 511 Union Bank Building, Telephone 3726, F. R. Munro
BRITISH COLUMBIA - - - - Vancouver, Geo. S. B. Perry

GREAT BRITAIN
LONDON - - - - 88 Fleet Street, E.C., Telephone Central 12960, J. Meredith McKim
MANCHESTER - - - - 92 Market Street, H. S. Ashburner

UNITED STATES
CHICAGO - - - - 1001 Teutonic Bldg, J. Roland Kay

FRANCE
PARIS - - - Agence Havas, 8 Place de la Bourse

SWITZERLAND
ZURICH - - - - - Louis Wolt, Orell Fussli & Co.

SUBSCRIPTION RATE.

Canada, United States, $1.00 Great Britain, Australia and other colonies, 4s. 6d., per year; other countries, $1.50. Advertising rates on request.

Vol. III. JANUARY, 1907 No. 1

NEW ADVERTISERS IN THIS ISSUE:

Detroit Foundry Supply Co., Detroit.
Engineers' Equipment Co., Chicago.
The Fairbanks Co., Springfield, Ohio.
Gisholt Machine Co., Madison, Wis.
Murchey Machine Co., Detroit.
The Norton Company, Worcester, Mass.
Producer Gas Co., Toronto.

A Glad and Happy New Year

Apart from the exhilarating joy of mere existence that should be ever present there is in the Spirit of the New Year an added zest and interest in life. The past year with whatever of good or ill it may have brought is gone forever and another opens before us with all the freshness and gladness and glowing possibilities that the mind's eye may see fit to conjure.

May the coming New Year see for you the long delayed hope fulfilled and the cherished ambition realized. May this New Year be the happiest you have yet seen and full of that joy and gladness and content that go towards the Life Ideal.

THE CANADIAN TARIFF.

THE old rule that first impressions are always best does not hold good with the new Canadian tariff as the more time that is taken to digest its provisions, the more hidden changes come to the surface.

Many of the changes made are not only mystifying, but they are annoying without purpose, the gain for instance in the new regulation regarding packages—that packages on good paying ad valorem rates are dutiable at 20 per cent. where they are charged for separately on invoices—will be trivial to the Government in comparison to the trouble and extra expense it will cause importers particularly of light goods such as glass chimneys and globes.

Another change which will cause endless trouble is the change in the regulation providing that any article imported shall be subjected to the highest duty charged on any of its parts. That was plain and easily understood, but the new reading that duty will be charged on most important item contained in the article will lead to much controversy and varying decisions by different appraisers.

The contradictions in the new tariff are also many, the chief one being in first reducing the duty on mowers and binders from 20 to 17½ per cent., then giving in return a 95 per cent. rebate on all duties paid on iron and steel imported for goods manufactured for home consumption, after granting a continued bonus to the iron and steel industry. The reduction in duty pleases the farmers, while the rebate squares the implement manufacturers, but it also nullifies the advantage of the bonus to develop a home iron industry.

Without discussing the merits of the bonus system, other than to point out that its continuation for 3½ years will mean upwards of $10,000,000 to the existing iron and steel companies, this being a rather high price to pay to industries which are already fairly well established, we must look with disfavor on the introduction of the drawback system as applied to manufactures for home consumption as it grants favors to one industry not enjoyed by others. Why discriminate against wire manufacturers and foundrymen in favor of implement manufacturers?

The complete change in the classification of items makes the new tariff a brain-racker. An attempt has been made to group all items of a similar nature together but the effort has not been entirely successful as many items are entirely out of their groupings, earthenware items being in amongst metals and bronze powder, gold paint, etc., being nearly a hundred items away from other paints. Many suggestions for simplifying the tariff were made before the Tariff Commission, but few have evidently been adopted.

Taken as a whole the machinery and allied trades are not greatly affected by the tariff changes. A slight increase in the duty on machine tools, machinery parts and boilers has been off-set by a corresponding increase on the duty on raw materials. The iron and steel interests are asking that the implement makers' 95 per cent. rebate on pig iron be cut off. Rolling mill men are protesting against the new duty on steel billets and agitating for the old conditions on rolled and barred iron, sheet steel, etc. Metal merchants are protesting that the new tariff will force them out of the boiler plate and structural iron business.

TECHNICAL EDUCATION ENDORSED.

CONSIDERABLE significance is attached to the utterance of Premier Gouin of Quebec, at the recent banquet given by the Canadian Manufacturers' Association, to ex-President Ballantyne, especially as he outlined the policy of his Government with regard to technical schools.

For many years past, an agitation has been carried on by the manufacturers, trade unions and educationalists of the country, aiming at the establishment of institutions where mechanics and factory hands could learn the technicalities of their respective trades. Many schemes of education have been proposed tending to put students—machinists, boiler makers, plumbers, and a host of others—in touch with broad, general principles which they could readily adapt to the particular requirements of their work.

But the question of ways and means was the most difficult, and it was incessantly debated whether the responsibility for financing the schools should rest with the federal, provincial or civic Government. New hope was aroused, when, after the conference of Provincial Premiers, it was announced by Premier Gouin that the additional subsidy for Quebec would be largely devoted to education. Consequently, the supporters of technical schools, awaited with great interest, some official announcement in the matter.

And such is the construction which has been placed upon the latest speech of the Premier. Speaking to manufacturers, who are particularly interested in advancing the excellence of goods "made in Canada," by educating their workmen, he, first of all, warned them that they, themselves were neglecting their duty. In Germany, where technical education has played so prominent a part in the country's industrial supremacy, the crusade was begun by the manufacturers establishing and supporting schools. In this country, too, said Mr. Gouin, the manufacturers who have most at stake, should advance from resolutions to actions. Amid great applause, he promised liberal assistance from the Government, to such a movement. And now, the public await the action of the Quebec manufacturers. It's apparently "up to them."

THE STEAM TURBINE AND ITS FUTURE.

DURING the initial stages in the creation of any new form of engine, it is only natural that the discussion should be confined to a comparatively few leading engineers, and designers, for at the outset it must take men of imagination to foresee, and to mentally shepherd, the development. As the machine is used, and becomes less of a rarity, and as the practical men—the plodding, keep-it-going engineers, have more and more to meet emergencies, and to contrive expedients, the development takes on a new complexion, from the steady enforcement of a new point of view.

The theoretical engineer is, of course, much concerned with critical points in the principles of his engine, and it is almost inevitable that he should veil his ideas in the higher mathematics. The formula for a curve is in his estimation all important—and no doubt that bias of thought has a value in its place. But the standpoint of the practical men is somewhat different. The engineer making power, measures his engines with a coal shovel, and an oil can. While the live builder keeps busy figuring a compromise between time, cost, efficiency, appearance, and sales.

The drift of invention, engine room gossip, and shop experience, would all indicate that the steam-turbine is entering on a period of less mathematics, and more mechanical contrivance, and with this it is reasonable to expect that the steam turbine will experience a very rapid improvement in detail, and in methods of production.

At the present time the cost of turbine machinery is so considerable as, in a measure, to offset their economy, but this will improve from the consumers' point of view, from now on. The base patents of the leading types are expiring—and although the manufacturers are trying to maintain the monopoly by replacing them to some extent by improvements in auxiliaries, and details of construction, yet a glance over the advertising pages will show that makes, and makers, are even multiplying. The position of the Westinghouse-Parsons—and the Allis-Chalmers, Parsons, is an example on this continent. In Europe, the Ratean, the Riedler-Stumpf, and others, are extending the principle of De La Val's nozzle into powers, and sizes, hitherto monopolized by the Parson's. This all spells competition and education of the public.

But more important the manufacturer shows a tendency to cheaper, better wearing, or stronger forms. In this direction is the protecting wreath in the Allis-Chalmers, Parson's, or the sharp bevel edge point introduced in the building of the King's new yacht. Also simplification and improvement in the buffles at the main bearing to protect the condensers from atmosphere. Another example of the tendency, is the remodeling of the Curtis to a horizontal machine of moderate size, as now used by the Canadian General Electric.

IMPORTANCE OF A KNOWLEDGE OF COSTS.

THE element of cost is an important one in all branches of engineering, but in spite of this the average graduate of our technical colleges upon entering the field of his profession, has a very limited knowledge of the cost of engineering schemes, the cost of engineering materials and the cost of mechanical devices. His college years have been spent among theories and methods, and when upon graduating the business world demands results and asks in cold commercial terms "how much," he very often feels at a loss.

It is not desirable that our engineering colleges should have an elaborate course in estimating and cost finding, yet there should be something to direct the students' thoughts along this line. Unless this is done, he will probably neglect the commercial side of the profession to the detriment of his engineering training. Even if only a few suggestions were given as to the cost of engineering materials and their preparation it would be a step in the right direction. The student would then be led to recognize the importance of this phase of his work, and by the method of comparison could develop fairly comprehensive ideas as to the cost of different engineering enterprises and mechanical devices. And upon starting his career he would not be groping in the dark altogether in this respect.

A UNIQUE OFFER.

SOMETHING entirely new and original in the method of power plant installation has been adopted by the Producer Gas Co., of 11 Front St. East, Toronto. Their offer includes the giving away free of five suction gas plants, in which they merely divide with the users the net saving effected by their use. It implies great faith on the part of this company in the producer gas plant, and at the same time gives manufacturers time and opportunity for installing one of these plants without any class outlay.

Mechanical Reviews

OPINIONS AND STUDIES WORTH NOTING

DIFFERENTIAL GEAR IN CHAIN HOIST.

THE following sketches show the application of an epicylic train of gearing to one of the many chain hoists on the market.

Fig. 1 shows the usual arrangement of gears in the standard block. In this train B is held stationary on the shaft. D is loose on the shaft, but fast to the

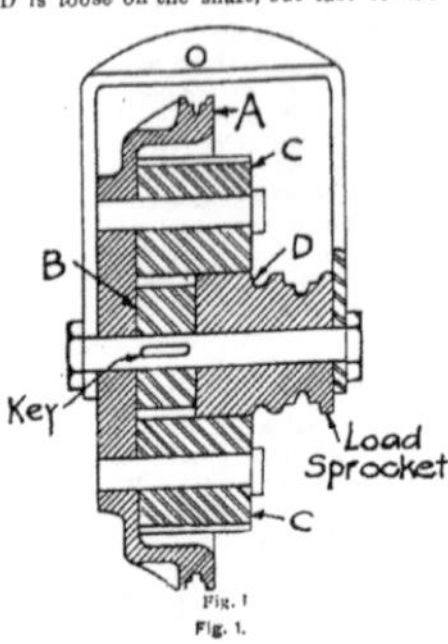

Fig. 1.

load sprocket. C, C are idlers meshing with both B and D, and turning loose on their supporting pin. A is the sproket over which the hand chain runs. B has one tooth more than D.

For example, say that D has 12 teeth and B 13 teeth. The number of teeth in C can be neglected as not affecting the speed. The result is that for every full turn of the load sprocket, sprocket A must make 12 revolutions. With 3 inches pitch diameter of the load sprocket, and 12 inches diameter of the sprocket A, then in order to raise a load up one foot 47 feet of hand chain will have to be hauled.

In connection with this chain block a little incident showing how essential is a knowledge of elementary mechanics in simple machine design will be noted. An old designer of various articles wished to make a high speed block from the one above described. The way he intended to get the high speed is shown in Fig. 2, and was by putting a pinion fast on the shaft with the hand chain sprocket, a, which meshes into a gear, f, loose on shaft. This gear was fast to a plate which held the idlers c, and c, and these were free to turn on their supporting pins.

The speed was stopped up even higher than I have indicated. I think it was to handle 500 pounds at about 200 feet per minute, a sort of dumbwaiter in disguise.

After a good deal of byplay, caused by a mechanic pronouncing the design N. G., he finally, by giving a list of about 16 patents he had received on various things, got the general manager to have a high speed block of that design made up. The gears used were cut

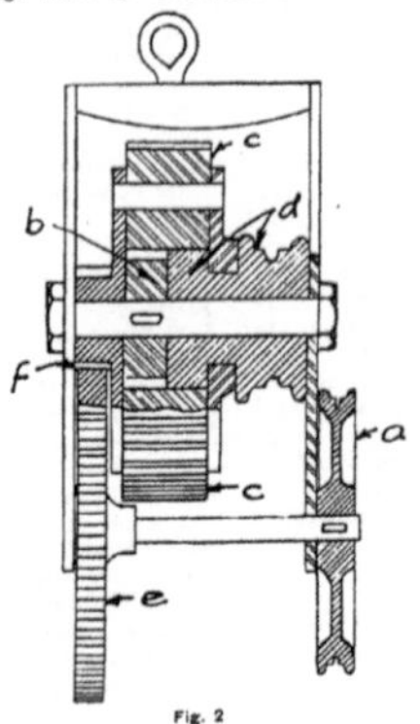

Fig. 2

gears of steel, and the whole block was well built.

Upon the day of the trial 50 pounds was attached to the load hook. Then a man attempted to haul it up. It took 500 pounds to raise the 50 pounds and even then it was not at 200 feet per minute.

When it was suggested that it would have been better to make it as in Fig. 3, as being less complicated, the designer took it as an insult.—Browning's Magazine.

VALVE SETTING.

THE opinions as to the correct settings of induction valves are many and in certain directions a very slight alteration will make a great difference to the running of the engine.

By consideration of the action of the atmospherically operated valve and its shortcomings, we should arrive at the best timing for the mechanically operated valve. The atmospheric valve requires considerable tension on its spring to close it, and, therefore, opens late; the tension, on its spring, increases as it opens, therefore it does not act regularly, but pulsates at slow speeds; it has considerable inertia, therefore if its spring is weak, to enable it to open correctly, it closes late (reducing the charge), and if sprung to close correctly reduces the charge by throttling.

The point of opening of the induction valve is not of very much consequence, as long as it is not open too early. The exhaust valve should be held open right at the top dead centre, or even a little over onto the suction stroke, so as to take full advantage of the scavenge due to the momentum of the gas through the valve. It is evidently useless to open the induction valve while the exhaust valve is open, or while the pressure in the cylinder is above atmospheric. The exhaust valve should close and the piston travel a short distance on the suction stroke before the valve opens. The valve can open quite late, provided the depression does not lift the exhaust valve, without causing very material

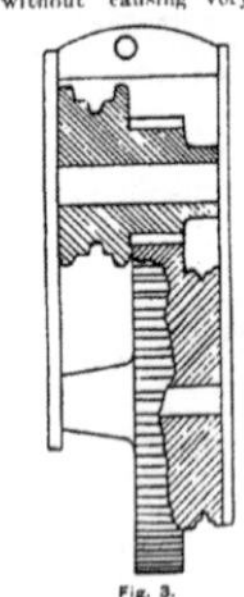

Fig. 3.

loss of power, as the difference of pressure increases the velocity of air, and with some carbureters improves the mixture at slow speeds.

The valve should be held open a short distance on the compression stroke, the distance depending on the speed the engine is to run. This does not cause a "back puff" such as is obtained with the atmospheric valve, but allows the cylinder to fill, while the piston velocity is low, due to the momentum of the gas in the mixture pipe. It is well in extreme cases to have the charge at a pressure slightly above atmospheric at the instant induction valve closes.

To insure that the closing valve follows the cam, the spring should be sufficiently strong to hold the valves closed against the full suction of the piston. This is, of course, a necessity in the case of the "variable lift."

The exact spring tension necessary varies with the compression ratio, but with ordinary engines it is about 11 pounds per square inch of valve area. The exhaust valve spring should be about the same strength to insure that it does not lift by suction, and to insure that it follows the cam.—The Autocar.

PROFILE OF BEVEL-GEAR TEETH.

By T. Mack.

THE number of teeth in any gear is based upon the working pitch diameter A as shown in illustration. If a gear is of 2½ inch circular pitch, and has a pitch diameter of 35.014 inches, it will have 44 teeth. However, the profile of the tooth is not to be developed as for a gear of this diameter or with a pitch radius equal to B, but as for a gear with a pitch radius equal to the back or conical radius C having a greater number of teeth. Thus a bevel gear as shown having 44 teeth, and the pitch radius B equal to 17.507 inches, would have a back radius C equal to 54.24 inches, and 136 teeth on its true outline. This number of teeth

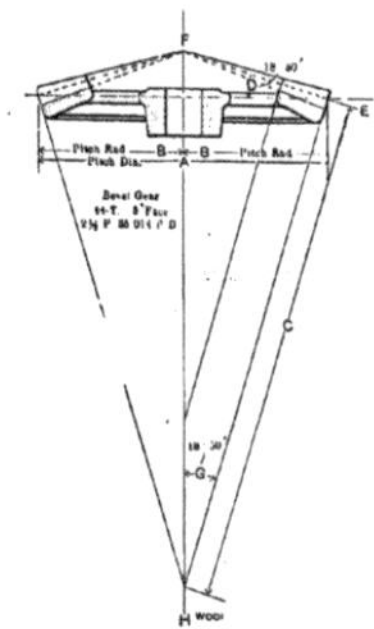

Profile of Bevel Gear Teeth.

of course determines the tooth profile.

Frequently gears are of such proportions that it is not convenient to determine the length of the back radius C by a layout. When such is the case it can be calculated in the following manner: If it is desired to calculate the back radius C, corresponding to a gear having a pitch radius equal to B, we first subtract from 90 degrees the angle D formed by the centre of tooth line E-F and the pitch diameter A. In this case that is 18 degrees 50 feet, and which is also the Angle G formed by the back radius of C and centre line of gear H F. From a table of natural sines take the sine of 18 degrees 50 feet, which is .32282, divide the working pitch radius B by this decimal and you get the length of the back radius C. Having obtained this radius, calculations are next made for the tooth profile in the ordinary manner.

PISTON RING GRINDER.

A MACHINE for finishing piston rings, particularly automobile and gas engine rings, is described. As will be gathered from the illustrations, it is intended only for grinding the periphery of the rings and not their sides. This it will do on rings up to 6 inches in diameter. To finish their circumference accurately, it is of special

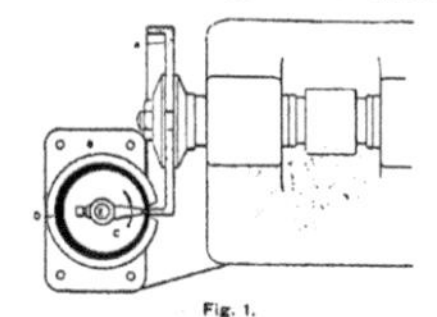

Fig. 1.

importance that the rings be ground under conditions similar to those that will exist in the engine; in other words, they must be ground while in tension. Hand-fitting that is required by the usual method of ring-grinding, is entirely avoided.

Figs. 1 and 2 give two elevations and a top view of the wheel and work, by referring to which the operation may be clearly understood.

The ring D is held in a shallow, cup-shaped piece B, which in turn is bolted to the movable slotted platen. The ring holder or cup-shaped piece does not turn around, but is bored out on the top side to about the diameter of the piston to take the ring. In the middle and driven from a worm-wheel below, there is located a vertical shaft E, on which is fastened a finger-like driver C, the end of which engages in the slot between the ends of the ring. It will be seen that when the vertical shaft is slowly revolved the ring will be moved inside the cup and its shape constantly changed as the high spots are ground off until the wheel touches all the way round, and then the ring is finished. To obtain a grinding contact the platen, ring and holder are moved bodily up to the face of a cupped emery wheel A, which gains access to the ring through a narrow notch cut out of the ring holder, as may be seen in the plan in Fig. 2.

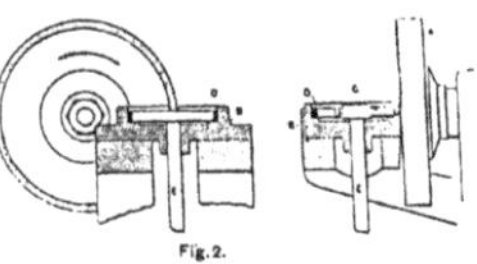

Fig. 2.

The ring-driving mechanism is driven by belt from the spindle to a shaft running through the column and connected at its opposite end to a worm and worm-wheel, and thence through universal joints and shafts to the ring driver. By changing the outer spindle pulley any desired ring speed can be obtained. The handle is used to engage and disengage the worm and worm-wheel, and in the latter condition the ring can easily be moved by the hand lent accord with the service required of wheel on the vertical shaft.

The machine is constructed in excelit. It is heavy and rigid throughout to take up vibration, and the head and the part on which the head rests are solid material; the column alone weighs 325 pounds. The bearings are dustproof, and are tapered so that they may be adjusted for wear. To the proper working of a machine of this character end play of the spindle is not permissible. To avoid it an adjustable tension arrangement has been placed at the end of the spindle opposite the emery wheel.

A ring which is turned larger than the bore of the cylinder, whether it be of uniform thickness, bored eccentrically or cast to a theoretical tension curve inside, will not have uniform tension or bearing on the surface of the cylinder until it is filed or scraped to a fit or let run until it wears itself to a bearing. The latter method of obtaining a fit is obviously undesirable, as it requires time, and in gas-engine manufacture, where the efficiency depends to a large extent on the compression, the rings must have perfect bearing. It is the contention of the manufacturer of this grinder that the only way to obtain such a perfect bearing is to grind the ring under the same conditions, or in tension, as it is subjected to under working conditions.

The grinding wheel is 10 in. in diameter, the spindle pulley runs at from 2,000 to 2,200 revs. per minute, the height of the spindle from the floor is 40 in., the base of the machine is 26 by 26 in., and the weight complete, without the countershaft 800lbs.—Iron Age.

Foundry Practice

MODERN IDEAS FOR THE TRADE.

SYSTEM IN FINDING FOUNDRY COSTS.

By D. C. Eggleston.

NOTWITHSTANDING the advance made in finding costs of production during the last few years it is generally true that the iron foundry has been neglected. Where the attention of cost system experts has been directed toward the iron foundry the tendency has been to try and adapt systems used in other departments. However, the work of the iron foundry is so different in nature from that of other departments that a radical departure must be made from the methods elsewhere employed.

In any iron foundry an analysis of the expense shows that it is incurred in connection with the cupola and several classes of work. The output of the former is in pounds of metal and of the latter in hours of labor. This suggests that the entire cupola expense should be prorated according to the number of pounds of castings produced and the expense against other classes of work on the basis of man-hours worked.

This method requires a division of expense between the cupola and other classes of work. In most cases no difficulty need be experienced in doing this; coke, fire-brick, ladles, lime stone, wood and some water are chargeable to cupola expense. The time of the foremen and foremen's clerks should be divided between cupola and other divisions of expense by estimate of the time spent on coke, pig iron, charging sheet and output reports. The administration expense not chargeable to the cupola should be distributed to other classes of work on the basis of the number of man hours in each. The rent expense includes all charges on account of grounds, buildings, heating, lighting and fire service. This division of expense is distributed on the basis of the number of square feet of floor space occupied by each class of work. Power expense should be distributed by estimate to the various classes of work.

The foundry clerk should keep a card file on which he records tools such as shovels and sieves received, given out (noting to which class of work), and the number on hand. Reference to the file facilitates the distribution of supplies. There are often some items concerning which there is doubt as to just which division of work they should be assessed. In such cases it is necessary to consult the foundry foreman that as accurate a division may be made as possible.

It is the best practice to assign letters indicative of the various rates of expense on the different classes of labor. Thus I. F. A. may be used to denote the rate on core work; I. F. B. on crane moulding, and so on. The foreman's clerk notes the proper letters on the workman's time ticket so that the expense can be added simultaneously with the productive labor to the cost of the hob. To the cost of metal the rate per pound for cupola expense can be added, thus giving the indirect expense incurred in connection with the cupola. The sum of productive labor, labor expense, raw material and cupola expense gives the total cost of producing the castings on an order.

In the statement of shop deliveries the productive labor, raw material and expense accounts should be credited and piece parts account is credited and the the casting is used on an order the the piece parts account debited. When proper account debited. Thus the cost system herewith described not only gives accurate costs of production, but also facilitates in keeping a check on the work of the iron foundry. If it is desired to cut down expense a study of the exhibit of iron foundry figures will suggest valuable economies. Comparative figures aid in showing wherein the increase or decrease lies. That the system herein described has been evolved to meet the requirements of the iron foundry in a large manufacturing works where it is in successful operation, recommends it to the attention of iron foundrymen.—Iron Age.

Berkshire Molding Machine.

BERKSHIRE MOLDING MACHINE.

IN MOST molding machines only one or two of the operations of making the mold are actually performed by the machine. In the Berkshire molding machine, manufactured by the Berkshire Manufacturing Company, Cleveland, Ohio, as the bulletin reads, "You push the lever, the machine does all the rest."

When operating the Berkshire machine the flask and bottom board are placed in position and the lever for starting the machine is thrown. The sand is riddled back of the machine and elevated by a bucket elevator to a suitable hopper above the machine.

The machine automatically carries the flask to the rear, where it is filled with sand, it then travels forward and in its course is met by the bottom board, supported upon the ram, which is forced down, ramming the sand. The vibrator is immediately thrown into action and at the same time lifting pins raise the flask off from the pattern. As the flask goes back to receive its charge of sand the bottom board is supported by suitable hooks, but as the ram comes down these hooks are drawn back so that the board remains upon the mold. All that the operator has to do is to lift the flask off and set it to one side, blow the sand from the table by means of the air hose, when all is ready for placing the other half of the flask and making the other half of the mold. The operator economizes time by placing the second half of the flask upon his machine and starting the machine before he starts to the floor with the half of the mold that has been completed. By the time he has placed the complete half of the mold on the floor, the second half has been finished by the machine, and is ready to be lifted off and carried to the floor. When using snap flasks and making light molds, the mold can be assembled on a bench by the side of the machine and only one trip to the floor made. In this case the molder can have two sets of flasks and arrange it so that the machine is ramming the drag for the second mold while he is carrying the first to the floor.

When making the cope the machine is so arranged that it cuts the sprue. In fact, the machine does all the work, so that all the operator has to do is to place the flask and board in position, throw the lever, and in eight seconds remove the finished half of the mold.

The pressure with which the ram presses the sand can be quickly adjusted, and as the flasks are filled automatically, all flasks are sure to be rammed alike. It is also possible to adjust the boards in such a way that the drag will be rammed harder than the cope.

One interesting point about the machine is that the pattern plates are heated by gas jets from beneath, so that the molds draw readily without the use of any parting material whatever. By the use of the proper grade of molding sand, very fine work can be made in this way without any facing, and the fact that no parting sand is introduced will greatly increase the life of the molding sand.

A PECULIAR CHILLED CASTING.*

By George C. Davis

THE casting to which I am about to call your attention presents some unusual features. For this reason I wish to describe it, even though I have no theory as to its cause, nor am I able to give all the data. The accompanying illustrations, Figs. 1 to 4, all show parts of the same casting. As shown in Fig. 1, it varies in thickness from ⅛ to ¾in., and in places has a wedge-shaped chill surrounded by grey iron on each side. A part of the casting, even in the light section, is grey without trace of chill, though it is hard. The analyses are given in the following table :—

	A. per cent.	B. per cent.
Silicon...	2.61	3.16
Sulphur......	0.106	0.102
Phosphorous...	0.60	0.74
Manganese	0.36	0.19
Graphitic carbon	3.40	0.19
Combined carbon		3.04

Owing to the difficulty in sampling, I am not able to present separate analyses of the grey and white portions. The figures given under A represent the casting illustrated in Fig. 1. The drillings were taken as near as possible to the chilled part. Sample B was taken at the foundry and represents the white portion from several castings made at a later date than A.

Before proceeding with the discussion I will give the method of charging, so far as I can, as follows :—

Diameter of cupola	43in.
Bed of cupola, broken coal	400lb.
Bed of cupola, coke	1400lb
Smelting charge, pig iron ...	4000lb.
Smelting charge, sprues......	2000lb.
Subsequent charges, coke......	350lb.
Subsequent charges, pig iron...	1600lb.
Subsequent charges, sprues...	1200lb.
Total melt, about	15 tons.

The first discussion I have been able to find on internally chilled castings is in vol. 17, p. 700, of the Transactions of the American Institute of Mechanical Engineers. An analysis is given which is as follows :—

	White inside Per cent.	Grey outside Per cent.
Silicon	2.74 ...	2.85
Combined carbon	2.55 ...	1.75
Graphitic carbon	1.31 ...	1.87
Manganese	0.50 ...	—

Phosphorous and sulphur are not given, but it is stated that no marked difference in these elements was found in the two samples. Internal chill is sometimes seen in pig iron, and this was at first thought to be due to segregation of sulphur or manganese. So far as I know, this has never been proved, and the analyses I can recall show little difference between the grey and white portions of the pig iron, except in the ratio of graphitic to combined carbon.

To return to the casting under discussion, the silicon in A is correct for light work, and should give good results; sample B is too high in this element, and I should expect a weak iron, but not a chill.

The sulphur in both cases is not excessive. In connection with this I would call your attention to the fact that while sulphur usually causes hard iron it does not always do so. I give in the table below two analyses of bad castings high in sulphur. These were water backs for ranges. They leaked under the hydraulic test, and when broken showed a porous fracture with reddish spongy spots; as indicated in the analyses, both were soft iron, yet they contain more sulphur and less silicon than A. Another water back casting, first class in every respect, showed 0.096 per cent. sulphur, or nearly as much as sample A. I do not think there is any reason to attribute the chill to sulphur. Following are the analyses of the defective castings :—

	Per cent.		Per cent.
Silicon	2.49	...	2.53
Sulphur	0.153	...	0.142
Phosphorous	0.68	...	0.73
Manganese	0.52	...	0.46
Graphitic carbon ...	2.94	...	2.63
Combined carbon....	0.24	...	0.59

Returning to the analyses given at the beginning of this paper, the phosphorus in both A and B is about as it should be; there is no reason to suppose it had anything to do with the chill.

The manganese evidently had nothing to do with the chill, as it is normal in A and low in B. The total carbon in A is correct for this class of work. In B it is a little low, as might be expected, owing to the high silicon, for in pig iron, as a rule, silicon increases and carbon decreases.

Another theory held by many is that the cause is in the management of the cupola and pouring. I quote the views of a well-known furnaceman, who writes writes me as follows :—

"I have noticed that this peculiar condition—grey on the exterior and white on the interior—oftentimes in very thin pieces of casting is due more to the melting of the iron than to the chemical composition of the iron. It cannot occur, however, with low silicon iron. So far as my experience goes it is characteristic of the use of high silicon iron which has been melted in a cupola with a cold bottom, enabling the iron to take up sulphur, or the iron has been well melted and run into cold ladles, under which condition certain chemical changes take place, which have been remedied by melting the iron hotter, thus reducing the percentage of sulphur, and also in tapping the iron into hot ladles instead of moderately warm ladles."

* Paper read before the Philadelphia Foundrymen's Association.

On the Art of Cutting Metals*

By F. W. Taylor

THIS paper, as presented to the Society by the president, Mr. F. W. Taylor, is of great length and in this article only the main features of his address are outlined. The experiments on the art of cutting metals which were carried on by Mr. Taylor and his colleagues, outlined in the paper, were continued through a period of twenty-six years, in various shops and at the expense of the owners of the shops. These shops include the Midvale Steel Co., Cramp Shipbuilding Co., Wm. Sellers & Co., the Link Belt Engineering Co. Dodge & Day, and the Bethlehem Steel Co., the latter more than all. In carrying out these experiments more than 10 machines have been fitted up at various times, all the machines used since 1894 being equipped with electric drive so as to obtain any desired cutting speed; and in the studying of these laws more than 800,000 pounds of steel and iron were cut into chips. It is estimated that between $150,000 and $200,000 have been spent upon this work, and it is significant that those who bore this expense had ample return in the increased output and economy in running their shops as a result of the experimental work. During the running of the experiments all the companies concerned and all who had anything to do with the experimental work, were bound by promise to the writer of the paper to secrecy concerning the results of the experiments, and it is a notable fact that through a period of 26 years none of the many men or companies connected with this work has broken a promise.

Some of the most important lessons learned by the author in the conducting of these experiments, particularly to young men, are:

1. Several men when heartily co-operating, even if of everyday calibre, can accomplish what would be next to impossible for any one man even of exceptional ability.

2. Expensive experiments can be successfully carried on by men without money, and the most difficult mathematical problems can be solved by very ordinary mathematicians, providing only that they are willing to pay the price in time, patience and hard work.

3. The old adage is again made good that all things come to him who waits, if only he works hard enough in the meantime.

The experiments described in this paper were undertaken by the author to obtain part of the information necessary to establish in a machine shop his system of management, the central idea of which is:

(a) To give each workman each day in advance a definite task, with detailed written instructions, and an exact time allowance for each element of the work.

(b) To pay extraordinary high wages to those who perform their tasks in the allotted time, and ordinary wages to those who take more than their time allowance.

There are three questions which must be answered each day in every machine shop by every machinist who is running a metal-cutting machine, such as a lathe, planer, drill press, milling machine, etc., namely:

(a) What tool shall I use?

(b) What cutting speed shall I use?

(c) What feed shall I use?

These investigations which were started with the definite purpose of finding the true answer to these questions under all the varying conditions of machine shop practice have been carried on up to the present time with this as the main object still in view.

The writer has confined himself almost exclusively to an attempted solution of this problem as it affects "roughing work"; i.e., the preparations of the forgings or castings for the final finishing cut, which is taken only in those cases where great accuracy or high finish is called for. Fine finishing cuts will not be dealt with. The principal object has been to describe the fundamental laws and principles which will enable them to do "roughing work" in the shortest time, whether the cuts are light or heavy, whether the work is rigid or elastic, and whether the machine tools are light and of small driving power or heavy and rigid with ample driving power.

In other words, the problem has been to take the work and machines as they are found in a machine shop, and by properly changing the countershaft speeds, equipping the shop with tools of the best quality and shapes, and then making a slide rule for each machine to enable an intelligent mechanic with the aid of these slide rules to tell each workman how to do each piece of work in the quickest time.

It is to be distinctly understood that this is not a vague, Utopian result, to be hoped for in the future, but that it is an accomplished fact, and has been the daily practice in our machine shops for several years; and that the three great questions, as to shape of tools, speed, and feed, above referred to, are

LET WIDTH OF TOOL = A AND RADIUS OF POINT = R, THEN

FOR BLUNT TOOL $R = \frac{1}{2}A - \frac{5}{32}''$

FOR SHARP TOOL $R = \frac{1}{2}A - \frac{3}{16}''$

For cutting hard steel and cast iron, these tools are ground to the following angles: Clearance angle 6°, back slope 8°, side slope 14°. (See paragraph 342)

For cutting medium steel and soft steel, these tools are ground to the following angles: Clearance angle 6°, back slope 8°, side slope 22°. (See paragraph 348)

EXACT OUTLINE OF CUTTING EDGE OF OUR STANDARD ROUND NOSE TOOLS FROM ½" x ½" to 2" x 3". (See paragraphs 325 to 331)

BLUNT TOOL FOR CUTTING HARD STEEL AND CAST IRON

SHARP TOOL FOR CUTTING MEDIUM AND SOFT STEEL

* A synopsis of a paper presented to the American Society of Mechanical Engineers, by F. W. Taylor, president, at the annual convention help in New York during the first week in December.

daily answered for all of the men in each shop far better by our one trained mechanic with the aid of his slide rule than they were formerly by the many machinists, each one of whom ran his own machine, etc., to suit his foreman or himself.

It may seem strange to say that a slide rule enables a good mechanic to double the output of a machine which has been run, for example, for ten years by a first-class machinist having exceptional knowledge of and experience with his machine, and who has been using his best judgment. Yet, our observation shows that, on the average, this understates the fact

To make the reason for this more clear it should be understood that the man with the aid of his slide rule is called upon to determine the effect which each of the twelve elements or variables given below has upon the choice of cutting speed and feed; and it will be evident that the mechanic, expert or mathematician does not live, who, without the aid of a slide rule or its equivalent, can hold in his head these twelve variables and measure their joint effect upon the problem.

These twelve elements or variables are as follows:

(a) The quality of the metal which is to be cut;

(b) The diameter of the work;

(c) The depth of the cut;

(d) The thickness of the shaving;

(e) The elasticity of the work and of the tool;

(f) The shape or contour of the cutting edge of the tool, together with its clearance and lip angles;

(g) The chemical composition of the steel from which the tool is made, and the heat treatment of the tool;

(h) Whether a copious stream of water, or other cooling medium, is used on the tool;

(j) The duration of the cut; i.e., the time which a tool must last under pressure of the shaving without being reground;

(k) The pressure of the chip or shaving upon the tool;

(l) The changes of speed and feed possible in the lathe;

(m) The pulling and feeding power of the lathe.

Broadly speaking, the problem of studying the effect of each of the above variables upon the cutting speed and of making this study practically useful, may be divided into four sections as follows:

(a) The determination by a series of experiments of the important facts or laws connected with the art of cutting metals.

(b) The finding of mathematical expressions for these laws which are so simple as to be suited to daily use.

(c) The investigation of the limitations and possibilities of metal cutting machines.

(d) The development of an instrument (a slide rule) which embodies, on the one hand, the laws of cutting metals and on the other, the possibilities and limitations of the particular lathe or planer, etc., to which it applies and which can be used by a machinist with-

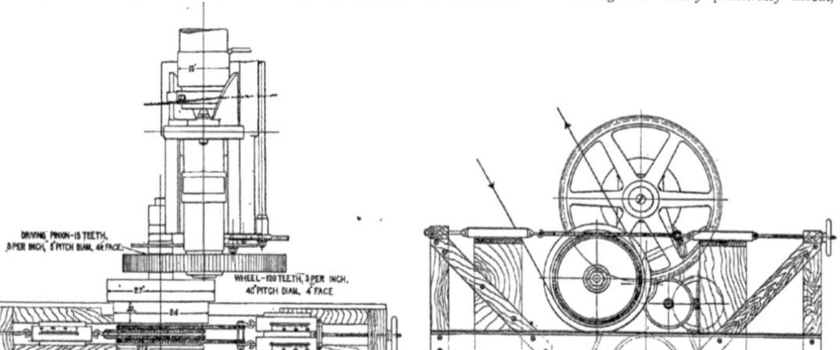

Fig. 1—Lathe Used in Determining the Pressure of the Chip on the Tool.

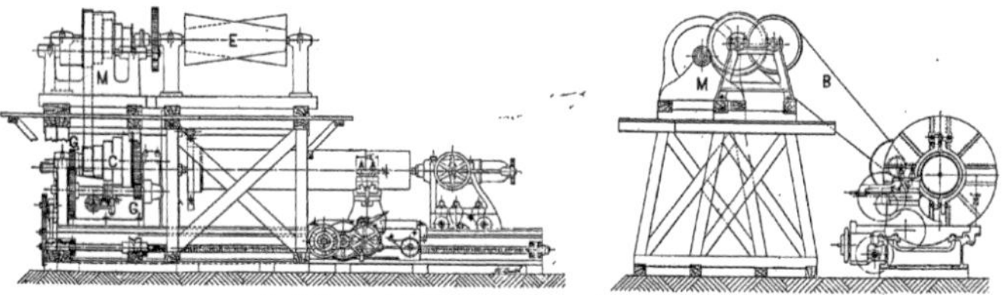

Fig. 2—Experimental Lathe at the Works of the Bethlehem Steel Co.

out mathematical training to quickly indicate in each case the speed and feed which will do the work quickest and best.

Important Steps in the Progress of the Experiments.

The writer has no doubt that many of the discoveries and conclusions which mark the progress of this work have been and are well known to other engineers, and we do not record them with any certainty that we were the first to discover or formulate them, but merely to indicate some of the landmarks in the development of our own experiments, which to us were new and of value. The following is a record of some of our more important steps:

(a) In 1881, the discovery that a round-nosed tool could be run under given conditions at a much higher cutting speed and therefore turn out much more work than the old-fashioned diamond-pointed tool.

(b) In 1881, the demonstration that, broadly speaking, the use of coarse feeds accompanied by their necessarily slow cutting speeds would do more work than the fine feeds with their accompanying high speeds.

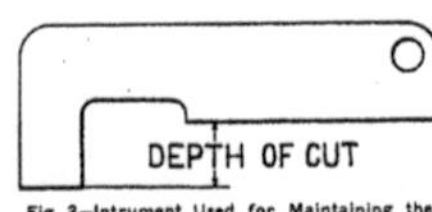

Fig. 3—Intrument Used for Maintaining the Depth of the Cut Uniform.

(c) In 1883, the discovery that a heavy stream of water poured directly upon the chip at the point where it is being removed from the steel forging by the tool, would permit an increase in cutting speed, and, therefore, in the amount of work done from 30 to 40 per cent.

(d) In 1883, the completion of a set of experiments with the round-nosed tools; first, with varying thickness of feed when the depth of the cut was maintained constant, to determine the effect of these two elements on the cutting speed.

(e) In 1883, the demonstration of the fact that the longer a tool is called upon the work continously under the pressure of a shaving, the slower must be the cutting speed, and the exact determination of the effect of the duration of the cut upon the cutting speed.

(f) In 1883, the development of the formulae which gave the mathematical expression to the two broad laws above referred to. Fortunately these formulae were of the type capable of logarithmic expression and therefore suited to the gradual mathematical development extending through a long period of years, which resulted in making our slide rules, and solved the whole problem in 1901.

(g) In 1883, the experimental determination of the pressure upon the tool required on steel tires to remove cuts of varying depths and thickness of shaving.

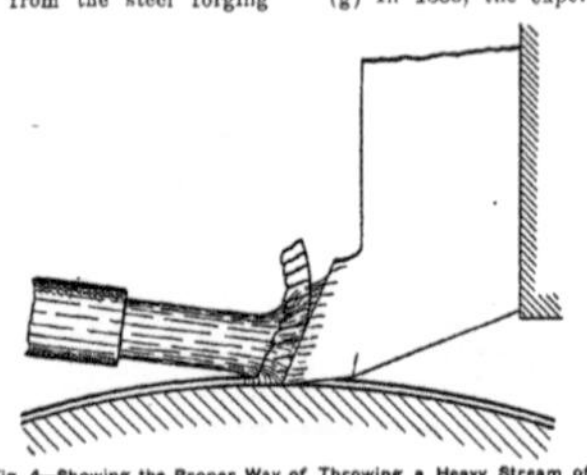

Fig. 4—Showing the Proper Way of Throwing a Heavy Stream of Water on Tool to Cool It.

(h) In 1883, the starting of a set of experiments on belting described in a paper published in Transactions, Vol. 15 (1894.)

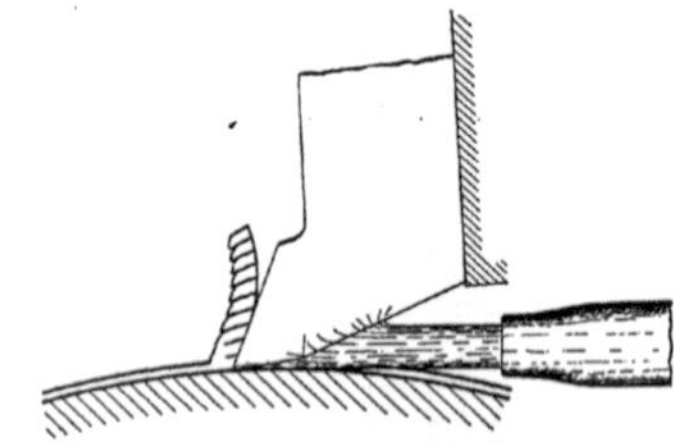

Fig. 5—Showing the Incorrect Way of Throwing Water on Tool.

(j) In 1883, the measurement of the power required to feed a round-nosed tool with varying depths of cut and thickness of shaving when cutting a steel tire.

(k) In 1884, the design of an automatic grinder for grinding tools in lots and the construction of a tool room for storing and isusng tools ready ground to the men.

(l) From 1885 to 1889, the making of a series of practical tables for a number of machines in the shops of the Midvale Steel Company, by the aid of which it was possible to give definite tasks each day to the machinists who were running machines, and which resulted in a great increase in their output.

(m) In 1886, the demonstration that the thickness of the chip or layer of metal removed by the tool has a much greater effect upon the cutting speed than any other element, and the practical use of this knowledge in making and cutting speed than any other element, and the practical use of this knowledge in making and putting into everyday use in our shops a series of broad-nosed cutting tools which enabled us to run with a coarse feed at as high a speed as had been before attained with round-nosed tools when using a fine feed, thus substituting, for a considerable portion of the work, coarse feeds and high speeds for our old maxim of coarse feeds and slow speeds.

(n) 1894 and 1895, the discovery that a greater proportional gain could be made in cutting soft metals through the use of tools made from self-hardening steels than in cuttng hard metals, the gain made by the use of self-harden-

speed, and therefore, in the amount of work done, of 16 per cent.

(t) In 1906, the discovery that by adding a small quantity of vanadium to tool steel to be used for making modern high speed chromium-tungsten tools heated to near the melting point, the hardness and endurance of tools, as well as their cutting speeds, are materially improved.

While many of the results of these experiments are both interesting and valuable, we regard as of by far the greatest value that portion of our experiments and of our mathematical work which has resulted in the development of the slide rules, i.e., the patient investigation and mathematical expression of the exact effects upon the cutting speed of such elements as the shape of the cutting edge of the tool, the thickness of the shaving, the depth of the cut, the quality of the metal being cut and the duration of the cut, etc. This work enables us to fix a daily task with a definite time allowance for each workman who is running a machine tool, and to pay the men a bonus for rapid work.

Belts and Individual Motor Drive.

A rather important discovery during the early experimental work was:

(1) That belting rules in common use furnished belts entirely too light for economy; and (2) that the proper way to take care of belting was to have each belt in a shop tightened at regular intervals with belt clamps especially fitted with spring balances, with which the tension of the belt was accurately weighed every time it was tightened, each belt being retightened each time to exactly the same tension.

In experiments subsequently tried extending over a period of nine years it was demonstrated that the ordinary rules gave belts only about one-half as heavy as should be used for economy.

Of late years there has been what may be almost termed a blind rush on the part of those who have wished to increase the efficiency of their machine shops towards driving each individual machine with an independent motor; and in this respect the writer makes a suggestion for which he thinks he may be accused of being old-fashioned or ultra-conservative.

The writer is firmly convinced through large personal observation in many shops and through having himself systematized two electrical works that in perhaps three cases out of four a properly designed belt drive is preferable to the individual motor drive for machine tools. There is no question that through a term of years the total cost, on the one hand of individual motor and electrical wiring, coupled with the maintenance and repairs, of this system will far exceed the first cost of properly designed shafting and belting plus maintenance and repairs (in most shops entirely too light belts and counter-shafts of inferior design are used, and the belts are not systematically cared for by one trained man and this involves a heavy cost for maintenance.) There is no question, therefore, that in many cases the motor drive means in the end additional complication and expense rather than simplicity and economy.

It is at last admitted that there is

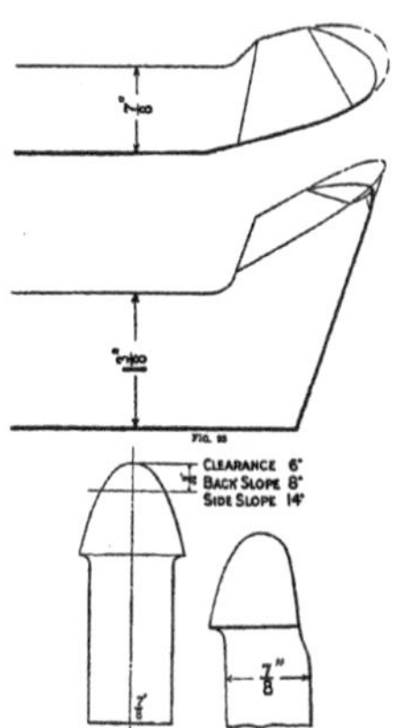

Fig. 7—The Standard ⅞-in. Tool With Which Most of Our Experiments Have Been Made.

little, if any, economy in power obtainable through promiscuous motor driving; and it will certainly be found to be a safe rule not to adopt an individual motor for driving any machine tool unless clearly evident and a large saving can be made by it.

(Concluded in Next Issue.)

The Ruby Silver Mining and Developing Company has been incorporated with a share capital of $500,000. The head office is to be at Hamilton. The company will do a mining, milling and reduction business.

Tricks of the Trade

Everybody is invited to contribute to this page. Send in your ideas, odd jobs you are doing, and anything of interest to fellow-workmen. Remuneration will be made for such contributions. [Ed.]

EXPANDING FINISHING CUTTER.

By B. Paney, Lachine Locks.

THE first screw is taper, through the cutter for adjusting the cutter when worn. The second is for tightening the cutter after being adjusted. A hole is drilled in cutter ¼ inch from end and a 1-16 inch cut is sawn through the cutter to the hole. A taper pin placed through the hole and bar. Square headed screws should be used on larger cutters. This method is particularly applicable for Capstain lathes where a set of brushes can be kept to fit the bars.

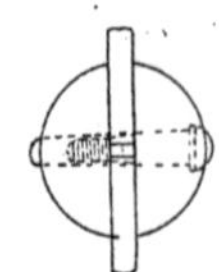

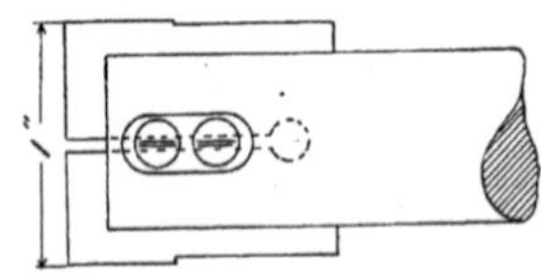

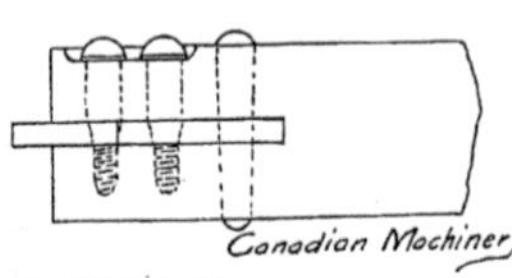

Expanding Finishing Cutter.

HOW TO MAKE A CHEAP, SERVICEABLE SNAP GAUGE.

F. E. Lauer, Niagara Falls, Ont.

THE snap gauge shown may be used on diameters ranging from 12 in. to 40 in. It is made from ¼x1 in. machine steel throughout. The first work on this gauge is done by the blacksmith when he bends the curved pieces. The first thing that is done in the tool room is to see that they are flat and then the spots are marked where the cross pieces are connected, the joint made here is half and half as drawn in section B and then drilled and rivetted. The tool steel gauge pieces are also put on as section B and are made like detail C, these gauge pieces are hardened and ground. A test gauge is made and then the snap gauge is ground to its proper measurement, the test gauge is always kept in the tool room and the snap gauge is always tested before going out into the shop. This gauge can be used to advantage where a standard must be maintained, as for pistons, rotors, etc. The approximate weight of the gauge shown in the drawing is 12 lbs.

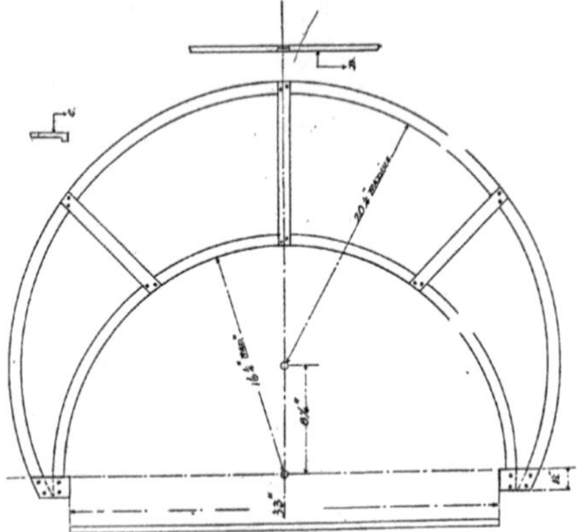

Making a Cheap Serviceable Snap Gauge.

LATHE ADJUSTMENT FOR BORING

By James C. Moore, Toronto.

HAVING occasion to bore out a number of connecting and eccentric rods, forked ends, I find my lathe face plate answered the purpose as a bolting surface. I made it adjustable by putting four studs in the saddle bolt slots as shown in the sketch. It will be found to take the place of horizontal boring mill work as the table can be elevated to any height. It will be found handy to bore out cylinders, segments, etc., and anything that one elevation is required for several pieces of the same work. By bolting an angle plate on another feature presents itself to the lathe man. In the same manner the jawed chuck can be put up and used as a vise.

ALZEN, A NEW METAL.

Alzen is the name given to a new metal, which is composed of two parts of aluminum and one part zinc, writes United States Consul-General Guenther, from Frankfort, Germany. It is said to equal cast-iron in strength, but is much more elastic. Alzen is superior because it does not rust as easily as does iron and it takes a high polish.

BOOK REVIEWS

PRACTICAL METAL TURNING.—A hand book for machinists, technical students and amateurs, by Joseph C. Horner, A.M.I.M.E., 404 pages, fully illustrated, price $3.50. New York: Norman W. Henley Publishing Company.

This book covers in a comprehensive manner the modern practice of machining metal parts in the lathe including the regular engine lathe, its essential designs, its uses, tools, attachments and the manner of holding the work and performing the operations. The modern engine lathe, its methods, tools and a great range of accurate work, the turret lathe, its tools and accessories and methods of performing its functions, as well as chapters on special work, grinding tool holders, speeds, modern tools, steels, etc. The author of this work is well known as a mechanical writer and is the author of several other standard books on mechanical practice. Principles and practice of the different branches of turning are considered and well illustrated. As designated by the author a feature of this book is the important section devoted to modern turret practice of the lathe itself, preference being given to the practice of turning rather than to the lathe design. The book is essentially designed from the engineers' point of view. Of its practicability there is no doubt.

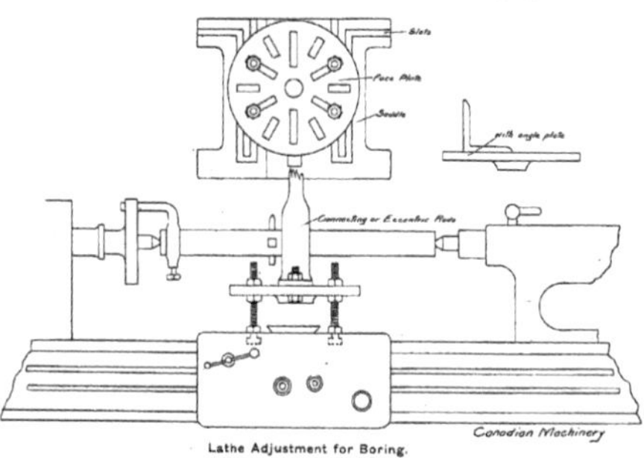

Lathe Adjustment for Boring.

PUNCHES, DIES AND TOOLS.—By Joseph V. Woodworth, M.E. A companion and reference volume to the author's elementary work, "Dies, Their Construction and Use for the Modern Working of Sheet Metals." A practical work of 500 pages, illustrated by nearly 700 engravings. Price, $4.00. New York: Norman W. Henley Publishing Company.

This book is a cyclopaedia of die making, punch making, die sinking, sheet metal working and making of special tools, devices and mechanical combinations for the punching, cutting, bending, forming and piercing, drawing, compressing, and assembling of sheet metal parts, and also articles of other materials in machine tools. This book is an entirely new work on the subject, nothing appearing in it that has previously appeared in the author's former book on the subject of dies. It is written by a practical man for the use of practical men interested in the working of sheet metals, the designing and constructing of punches and dies and the manufacturing of repetition parts and articles in presses. The designer, the machinist, the tool maker, the die maker and the manufacturer of sheet metal will find in this book one that fills a place heretofore vacant in the field of literature devoted to work of mechanics. The author himself is a designer of not a few of the tools illustrated in this book, which places him in a position of authority on the subject.

Heretofore, except through the pages of mechanical journals of the United States, there has been little literature on this subject, so that all interested will appreciate the fact that such literature is now available.

THE PRACTICAL ENGINEERS' POCKET BOOK, 1907.—Price, leather, gilt, with diary on ruled section paper, 1s. 6d., Published by the Technical Publishing Company, 287 Deansgate, Manchester, Eng.

The present edition of the "Practical Engineers' Pocket Book," marks a record for this publication. An exclusive revision has been made throughout and while deleting several pages of doubtful interest to the general user a considerable number of new notes have been added, among which are boiler setting, steam traps, engine covers, etc. A portion has been considered upon the steam turbine with suitable illustrations and revision and extension made on the engine room accessories, cotton belting, etc. The book as heretofore has had as its aim, the designer and engineer. It is a splendid little publication well worthy the consideration of the engineering profession.

THE DISTILLATION OF ALCOHOL.—By F. B. Wright, a practical hand book. The Distillation of Alcohol from Farm Products, including a process of melting, washing and macerating, fermenting and distilling alcohol from grain, beets, molasses, and the denaturing of alcohol for use in farm engine, automobiles, launch motors, and in heating and lighting, 194 pages, illustrated with diagrams. New York: Spon & Chamberlain.

As the author states, alcohol is a source of power, as a substitute for gasoline, petroleum and kindred hydrocarbines, was hardly known to the generality of Americans, until the passage of the Denaturing Act, by the last Congress. Alcohol then leaped into flame as a substitute for all the various forms of heat, hydrocarbine fuels. This is a timely volume on the subject, and is treated in a manner calculated to give the reader a grasp of the subject under consideration, and to show him the possibilities in this direction, in a manner probably before scarcely considered by the average reader. It shows its actual process of the distillation of alcohol, and its denaturing from a practical and manufacturing view point.

ELECTRICAL POCKET BOOK.—The Practical Engineers' Electrical Pocket Book for 1907. Price, 1s. 6d. net, leather, gilt, with diary on ruled section paper. Technical Publishing Company, 287 Deansgate, Manchester.

Further notes on other departments of engineering have been added to this year's pocket book. Such progress has been made in scientific and practical electricity, that this is not only possible but necessary. In some instances former contents have been amplified. The present edition is in keeping with the high standard of previous years, and is a valuable addition to concise literature and electrical engineering.

Power and Transmission

Steam *Gas* **Electricity** **Compressed Air** **Water**

AIR-COOLED ELECTRIC DRILLS.

ILLUSTRATIONS given herewith show views and sections of the single motor, two motor and three motor electric air - cooled drills manufactured by the Chicago Pneumatic Tool

Fig. 1—General View of Air Cooled Drill.

Company, Chicago. The single motor which is adapted to all classes of work for drilling iron and steel up to 1¼ inch in diameter as its name indicates is provided with one revolving electrical element or armature. The armature is of the slotted type, with wound coils of double silk covered magnet wire secured in place by wedges, no binding wire being used. The commutator is built up of hard drawn copper bars, insulated with amber mica, and the commutator sleeve is of machinery steel. Carbon

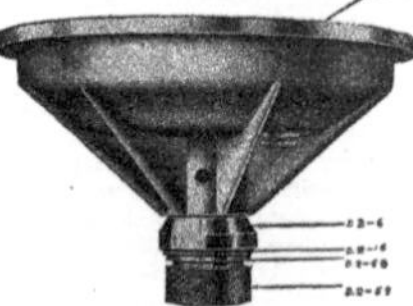

Fig 2—Drill Spindle with Ball Thrust Bearing.

brushes are used. The shaft is of high carbon steel, with the pinion cut as an integral part of it and is hardened and ground at the bearings. Ball bearings of the Hess-Bright type support the armature, provision being made for slight end play.

Machines capable of drilling holes larger than 1¼ in. in diameter are made in two and three motor types, for if one armature were used it would have to be quite large in diameter and the speed would have to be reduced to keep down the peripheral velocity. By the use of two or three small armatures properly geared to the main driving spindle a greater power can be secured for a given weight because the armatures can be run at higher speeds with perfect safety. Figs. 3 and 4 show sectional views of the two and three motor drills, respectively. Since the construction of these drills is somewhat analogous they can be described together.

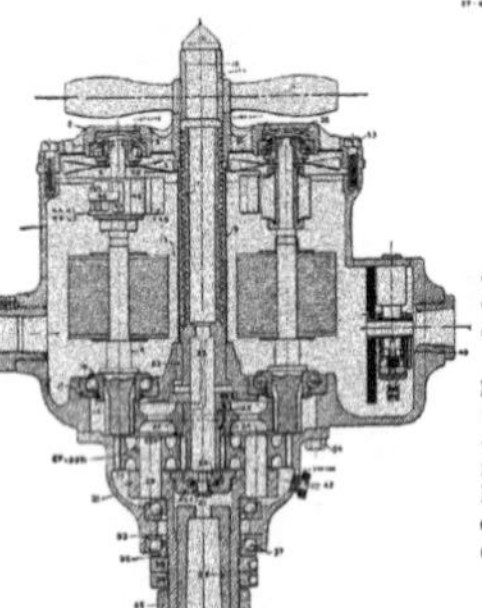

Fig 3—Sectional View of Two-Motor Drill.

Figs. 5 and 6 show the plan arrangement of the armatures and fields which are peculiar to the Duntley drills. The outer casing is of nonmagnetic material.

The fields are compound wound and a very low starting current is required, and at the same time a fairly close speed regulation is obtained. The fans are placed on the upper or commutator ends of the armatures, so that the air is drawn in and forced out at the lower end. The electrical construction of the parts, including the switch, is the same in general as for the single motor drills previously described. The planetary gear arrangement is similar, involving

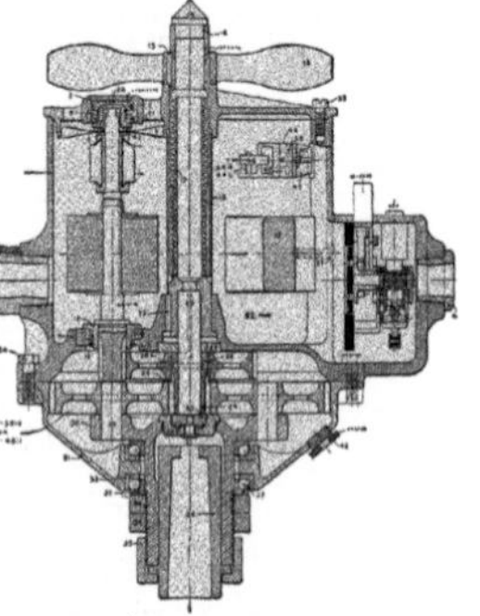

Fig. 4—Sectional View of Three Motor Drill.

only such modifications as are made necessary by the increase in the number of driving pinions.

The two and three motor drills are, as previously stated, adapted for heavy work and are used for drilling in iron or steel up to 2¼ in. and for boiler plate reaming and staybolt tapping. On large holes they have removed 2¼ cu. in. of cast iron per minute with comparative ease.

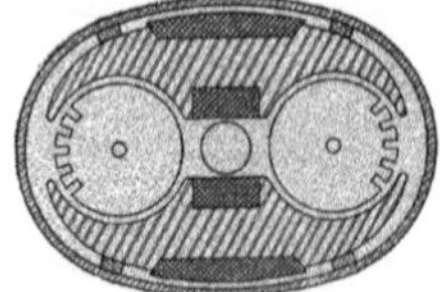

Fig. 5—Two-Motor Armature and Fields.

NEW GAS ENGINE BY-PRODUCT.

IT IS of interest to learn of an invention which has been perfected in Germany where, it is known, all matters connected with the economic utilization of gas power have received, and are still finding, the most careful attention of engineers. The subject was made public in an address recently made by Herr Hausser before a branch of the Society of German Engineers at Kaiserslautern, says F. E. Junge in Cassier's.

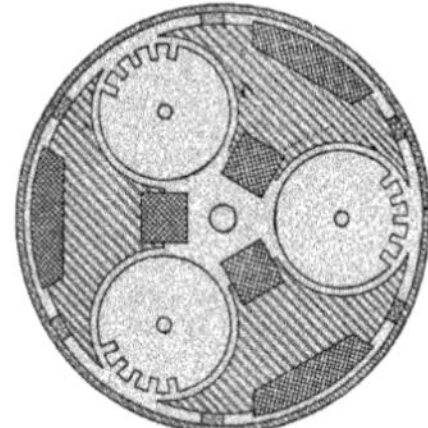

Fig. 6—Plan of Three-Motor Armature and Fields.

His aim is to produce nitric acid through explosive combustion, which, when properly directed, will create temperatures of several thousand degrees Centigrade.

An unusually attractive feature of the invention, and one that is likely to hasten its industrial exploitation, lies in the fact that the principle underlying it is extremely simple and that the process can be carried on by almost any owner of a gas engine and without increasing the initial cost, the operating expenses, the floor space and the complexity of the plant in any but a very slight degree.

It is to be hoped that the problem will commend itself to further examination and test, so that exact information on the commercial possibilities of the new process may soon be available. So far as one can judge at this early stage it does not seem to offer any practical difficulties in its execution. The formation of nitric acid in the engine cylinder can easily be avoided by keeping temperature in cylinder walls and piston above 120 degrees Centigrade, which is the dissociation limit of nitric acid. This, of course, is very easily accomplished.

Fig. 2—Coal Chute and Carrier.

In the evolution of gas power Hausser's invention presents a further important step ahead. The thermal superiority of the gas engine over all its competitors in the field of prime movers has by this time been acknowledged even by its most conservative antagonists, even such as keep only in touch with domestic achievements. Its mechanical efficiency has, by the proper employment of scientific methods of design, reached a level with that of the very best reciprocating steam engines.

In the iron and coal regions, and wherever waste gases of any kind are to be had, and where there is a market for available surplus power, it is the prime mover to be selected for driving blowing engines, rolling mills and pumps, if maximum industrial economy of a heat power plant is to be secured. Elsewhere its radius of action has been largely extended by the perfecting of gas producers, which will gasify the lowest grades of fuel with high efficiency and reliability. The new sphere of application opened up by the utilization of the chemical reaction of its working cycle discloses an outlook for further usefulness which is practically unlimited.

COAL-HANDLING EQUIPMENT.

WITH this issue are given illustrations and descriptions of a Jeffrey coal handling equipment for retail coal yards, which is now in operation at the coal yard of the Vandalia Coal Company at Chicago, Ill.

This system is known as the run-around V bucket combined conveying and elevating type, as will be seen from the general plan number 1, which shows a side, end and plan view.

The coal is carried about ninety feet, horizontally and about forty feet vertically, at the rate of from 50 to 75 tons per hour.

The equipment consists of 750 ton wooden storage pocket, a run-around combined elevator and conveyor, driving machinery, loading chutes, also a power shovel.

The method of operation is about as follows:—The coal is received in cars at one end of the building, it being unloaded from the cars by means of a Jeffrey power shovel and discharged directly into the elevating portion of the machine, which carries it into the upper storage pockets, distributes it the entire length by means of valves that are placed at intermediate points in the upper horizontal section of the coal handling apparatus.

There is an auxiliary storage below the upper pockets, where all of the surplus coal that the upper pockets cannot contain, is stored by means of the same conveying apparatus referred to.

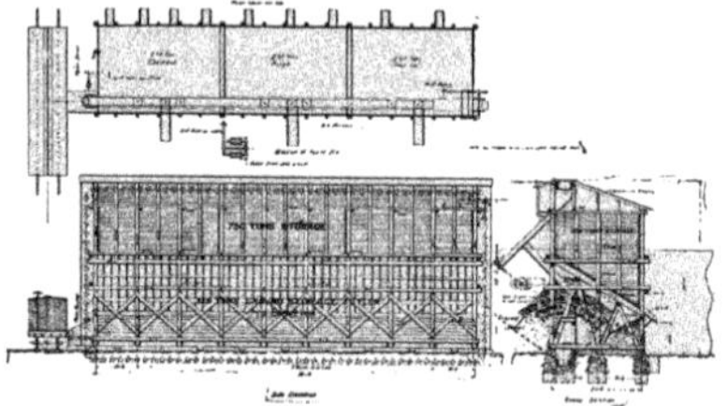

Fig.—1 Section View of Coal Handling Plant.

For reclaiming the coal from the lower storage, a tunnel is provided, through which the lower or return strand of the conveying apparatus travels. The tunnel is arranged with openings or gates, through which the coal is allowed to flow to the conveyor, and which latter carries it horizontally to one end of the building, elevates it and then carries it back into the upper storage bins. From these latter bins the coal is delivered to wagons by means of special combination coal chutes and screens. The slack from the coal, as it is being loaded into wagons, dropping through the screen and by means of a chute is carried under the storage bins.

Illustration number 2 shows the lower or return strand of the conveying apparatus with the gates and loading chutes ready to receive coal from the lower or auxiliary storage.

Illustration number 3 shows the loading chute and screens delivering coal to wagons.

The coal handling apparatus is constructed of two strands of 12 in. pitch steel thimble roller chain with special V buckets, the latter being designed to carry either horizontally or vertically.

The entire plant was installed complete by the Jeffrey Manufacturing Company, of Columbus, O., and is a model of its kind.

Fig 3 - Loading Coal Wagon.

This Company also makes a very complete line of coal handling apparatus for handling both large and small capacities and to suit almost any local conditions and requirements.

SYLVESTER AUTO THRESHER.

AN illustration is shown herewith of a monster thresher recently completed by the Sylvester Manufacturing Company, Lindsay, Ont., which is the product of three years hard work and constant planning on the part of this company. Mr. Richard Sylvester returned from Manitoba after getting this new machine in operation. In a letter written by N. Wolverton, Superintendent of Manitoba Experimental Farm, Brandon, Man., describing the machine the writer points out that one of these threshing outfits is now in operation on the experimental farm there. He states that a 25 horse power gasoline engine with two cylinders and a separator, are all upon one set of four wheels. The engine is under the separator. The driver sits in front directing the machine, which moves at the rate of about ¾ of a mile an hour through the field of grain, men pitching sheaves from the stooks upon a platform on either side of the separator.

Two men stand on this platform and pitch the sheaves, into the self-feeder, which is upon the top and in the centre of the machine. The grain is carried forward over and then under the cylinder, passing backward as in the usual way. The bagger stands on a platform, and when the bag is full ties and throws it on the ground, or the grain may be put in a wagon, which follows. The straw is dropped in a windrow behind the machine.

The machine has threshed barley and oats from the stook in the field, giving excellent satisfaction. It is now at the moment of writing threshing stacks of wheat. When threshing stacks, carriers are put on or a blower may be used. This is the first trial of the machine, which is new from the shops. The few stoppages have been occasioned entirely by adjustments of the machine, such as are incidental to almost any new machine during the first day or two of its trial. When moving, and not threshing, it has a speed of some three miles an hour. No fault can be found with principle involved in the new combination. I think the machine has been put together too hurriedly as they were anxious to get it here in the West before the stook threshing was done. The troubles can be remedied. I am inclined to believe that the machine will be used extensively in the Canadian West. The separator can be removed and the machine used as a traction for plowing or probably to run one of our 12-foot binders. There seems to be no danger from fire. Yesterday there was a very high wind blowing and the steam threshing outfits were compelled to stop work in this section of the country, but this machine continued to work without any difficulty.

The above cut shows the adaptability of this machine to actual conditions in the West. The Sylvester Manufacturing Company have a patent on this machine in Canada and the United States and other patents pending in this country, as well as foreign countries. The Sylvester Manufacturing Company intend to go into the manufacture of this on a large scale in the immediate future.

Sylvester Auto Thresher

TEST BLOWERS.

FOR testing blowers, to measure the volume, pressure and horse-power at pressures of from one to ten pounds per square inch, the following methods and formulae are adopted by the B. F. Sturtevant Co., Hyde Park, Mass

The volume of air discharged from an orifice or pipe is, theoretically, equal to the product of the velocity of the air flowing and the area of the orifice. Hence for the calculation of volume, the velocity is an important factor. To determine the velocity, the Pitot tube is commonly used as shown in the accompanying illustration. It should be inserted in the centre of a straight run of blast pipe within about ten feet of the blower. One part of the Pitot tube transmits the total pressure, which is the sum of the static pressure and the velocity pressure. The other part, in communication with the slats shown in illustration, transmits the static pressure. Evidently the difference is the velocity pressure. Each is connected to a water gauge which should show magnified readings, so that the difference may be accurately determined.

Great care should be exercised in measuring the velocity pressure, and the instruments should be carefully calibrated. In the ordinary blast pipe for conducting air from the blower to the cupola or furnace, the velocity should not exceed two or three thousand feet feet per minute. As this velocity corresponds to a pressure of only about 0.4 inch of water, the measurement requires care, but with good instruments the readings will be accurate enough for all practical purposes.

Volume.

The velocity pressure being known, the volume of free air passing through the pipe may be determined from the following formula:

$$V = av = \frac{60acP_1}{P}\sqrt{\frac{2gp}{d}},$$

in which

V=the volume of free air in cubic feet per minute,

c=coefficient of Pitot tube, which should be determined for each tube,

a=area of the pipe in square feet,

v=velocity in feet per minute,

2g=64.32,

p=velocity pressure in pounds per square foot; p is the difference between the two pressures observed on the Pitot tube,

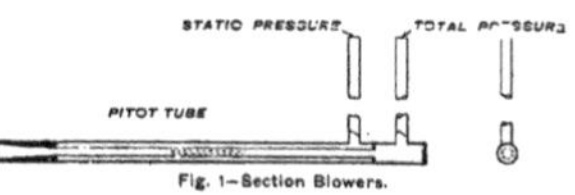

Fig. 1—Section Blowers.

d=density or weight per cubic foot of air at pressure, temperature, and humidity at point of observation,

P_1=absolute pressure of air in the pipe in pounds per square foot,

P =atmospheric pressure in pounds per square foot.

Horse-Power.

Assuming that the air is compressed without cooling, the horse power may be found from the following:

$$H.P. = \frac{VP\left[\left(\frac{P_1}{P}\right)^{\frac{1}{3}}-1\right]}{11,000.}$$

in which

V=volume of free air in cubic feet per minute, hs found above,

P=pressure of the atmosphere or suction pressure (absolute) in pounds per square foot,

P_1=pressure of compression (absolute) in pounds per square foot.

Formulae.

Including the preceding, there are four formulae sometimes used in computing the power required. Values obtained from these formulae have been placed in the form of curves and are shown in the accompanying engraving.

$$(1)\ H.P. = \frac{VPl_e\left(\frac{P}{P}\right)}{33000}$$

$$(2)\ H.P. = \frac{VP\left[\left(\frac{P_1}{P}\right)^{\frac{1}{3}}-1\right]}{11000}$$

$$(3)\ H.P. = \frac{V(P_1 - P)}{33000}$$

$$(4)\ H.P. = \frac{\text{lbs. per sq. in.} \times V}{200}$$

Formula No. 1 gives the horse power required when the air is cooled during compression as in the ordinary air compressor.

Formula No. 2 which has been explained, is used when it may be assumed that the air is compressed so quickly that it does not have time to cool the atmospheric temperature, as in nearly all blower work.

Formula No. 3, the ordinary "hydraulic" formula is ordinarily used for pressures up to five ounces.

Formula No. 4, is frequently used by makers of positive or rotary blowers for determining the horse power required for operating their machines.

TO REPAIR LEAKY VALVES.

When the brass seats of globe and angle valves become worn so that they leak badly the device here illustrated will be found useful. Remove the bonnet from the valve and clamp on the iron piece, A, which is made by cutting out a piece of sheet iron or steel as at B, and bending the three legs down and tapping to receive the three thumb screws.

A bushing, C, will be required, and should have a hole just large enough to admit the valve stem. By making a number of bushings of different sizes the device may be used on different sized valves.

To grind a valve, replace the bonnet with the jig as shown and put a little emery dust and oil on the valve seat. Then turn the stem, first in one direction and then in the opposite direction, at the time applying vertical pressure to make the emery take hold. Valves ground in this way are just as good as new and unless very badly worn or cut by the steam can be easily and quickly repaired.

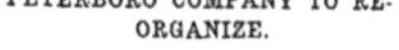

PETERBORO COMPANY TO REORGANIZE.

Owing to very heavy losses on some British Columbia contracts and the winding up of the affairs of the Ontario Bank, The Wm. Hamilton Manufacturing Co., of Peterboro, have gone into liquidation. The company has many contracts on hand at the present time, and it has been decided in order to get the necessary capital to carry on their

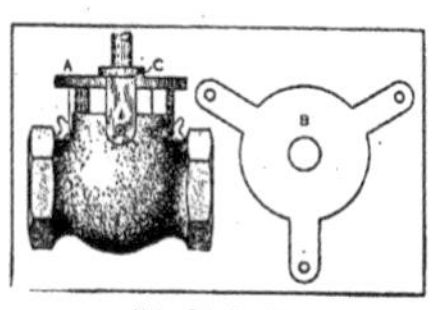

Valve Grinding Jig.

operations and to close up their account with the Ontario Bank, to reorganize. The capital is to be supplied to carry on present contracts and the company is to be reorganized on a permanent basis with larger capital. In the meantime everything is running smoothly in the factory. The 225 men at present employed will all continue as heretofore. The pay roll of this company is $1,200 a month and it is gratifying to the interests of Peterboro that it was not necessary to close down.

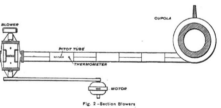

Fig. 2 - Section Blowers

Tenders are being asked for city supplies for Calgary, Alberta, and will be received by H. E. Tillis, city clerk, until the 31st of December, for the following materials: cement, lead pipe, corporation cocks, cast iron water pipe, sewer pipe.

Practical Questions and Answers

Tool Angles.

Ques.—Is there anything special in the hardening of a blanking die, not needed in the hardening of other tools?

Ans.—Great care should be taken in the hardening of a steel die, both in the heating and in the quenching. For the best results a gas furnace should be used, but when that is not obtainable charcoal will answer. To prevent the steel cracking around the screw and dowel holes in the die, they should be plugged with fire clay or asbestos. The dies is heated evenly to a cherry red, and then removed and dipped endwise into slightly warmed water, care being taken that it is dipped in straight and not moved around too much, as that has a tendency to cause the die to warp excessively. To temper the die, it is warmed immediately upon drawing from the water, and is placed upon a piece of cast iron heated red hot, and thus the hardness can be drawn to any desired temper, denoted by the coloring.

Water Annealing.

Ques.—There is a quick process of annealing called the "water annealing" process. Can you explain this process?

Ans.—There are several processes of "water annealing," as the quick processes are called, which give very satisfactory results when done by a practiced hand. Of these processes, the one is probably as good as the other, except that perhaps one person may have better success with one than with another. These quick methods of annealing may come in very handy at times, especially in repair shops, when a certain piece of work is required in a hurry.

One method is to heat the steel slowly to a dull cherry red, then remove it from the fire. As the steel cools down in the air the heat is tried with a piece of softwood, and when the steel has cooled so that the wood ceases to char, it is plunged quickly into an oil bath. This will leave the steel quite soft enough for easy machining.

Another method is to heat the steel slowly to a red heat, then allow it to lie in the ashes of the fire for a few minutes until it becomes almost black, and then drop it into soap suds and allow to cool.

Bevel Gear Cutting in Plain Milling Machine.

Ques.—Will you please explain the usual method of cutting bevel gears in a plain milling machine?

Ans.—The cutter is placed on the spindle of the milling machine in the ordinary way. A plain dividing head is used, splined to the face of the knee plate, which is bolted to the table of the machine. The spline in the knee plate must be cut to correspond with the angle of the gear, and thus there must be a separate spline provided for each angle of gear cut. Should there be a large variety of bevel gears to be cut on the same machine, necessitating a large number of different angles, it would be well topivot to the face of the knee plate a plain graduated plate carrying the dividing head. With this arrangement the head holding the gear blank can be set at any desired angle, and the necessity for cutting splines for each gear angle is obviated. This method of cutting bevel gears necessitates the constant attention of the operator, who is at the same time liable to make a number of errors, the dividing and angle setting being instances. To overcome these objections to using a plain milling machine for cutting spur and bevel gears full automatic gear cutting machines are manufactured, in which all the operations are automatic. Where there is considerab,e gear cutting to be done one of these machines is almost indispensible in modern practice.

Cutting Angle of Tools.

Ques.—Can you give me a short account of the development of the cutting angle for the different styles of cutting tools?

Ans.—In general the cutting angle of a tool depends upon the hardness of the substance to be cut, the softer the substance the smaller can be the cutting angle, but the cutting angle also depends upon the kind of substance. For instance cutting tools for brass are sometimes given negative rake in contrast to the considerable rake given tools designed for cutting steel. The knife is one of the keenest of cutting tools, with a cutting angle rarely exceeding 20 degrees, and in most cases much less than that. Chisels and similar tools for cutting such substances as wood, have a cutting or grinding angle varying from 20 to 25 degrees; and in this class of tools the sharpening facet is to be noticed. This is adopted as a matter of convenience in sharpening. In tools for cutting metals the question of tool angles assumes greater importance than with wood working tools. In these tools the edge is formed by the meeting of the upper and lower facets must be sufficiently wedge-like to penetrate and cut the material, but it must also be sufficiently strong to withstand the force of the cut and also to retain its keenness for a reasonable time. A consideration of this settles the cutting angle. Then so that there may not be undue friction between the tool and the work, and between the chip and that side of the tool against which it presses the end of the tool and the side are cut away, thus bringing other angle considerations into play. Thus the forming of a tool must always include the balancing of these various angles. The angle at the end of the tool is the angle of clearance, and varies in practice from 3 degrees to 15 degrees, and in some instances to 20 degrees. The smaller this angle can be with due regard to the elimination of friction, the better. Probably 10 per cent. is a general average. The angle at the upper face of the tool, known as top rake, governs the cutting action of the tool and also the movement of the severed chip. The greater this angle of rake the better for cutting fibrous metals such as steel and wrought iron, that is if it is not carried beyond the limit when it will dig into the metal and take a very uneven cut. For cast iron and similar crystalline metals this angle must be somewhat smaller than for steel. Then, too, this angle must not be increased too much or the resulting reduction of the angle between the top face and the end face will weaken the tool too much. The angle of rake will range from 35 degrees to 40 degrees for wrought iron or steel, 20 to 25 degrees for cast, and from 10 to 0 degrees for gun metal and similar metals and sometimes tools for brass cutting are given negative rake.

B. T. U. Values.

Would you kindly give me in "Question and Answers" a table of the heat values in B. T. U. of the various commercial grades of kerosene, gasoline, and denaturized alcohol. Also to which of these classes, if any, does the "petrol" used in Europe correspond?

Ans.—Authorities do not all agree as to the heat units contained in liquid hydro-carbons. Kerosene and gasoline may show by different tests as varying considerable in heat units per lb. There are different grades of kerosene and gasoline all of which show a value in heat units different from each other. The petrol used in Europe corresponds with our gasoline. It is known there as petroleum spirits or petrol.

Below is given a table from highly recognized authorities.

	B.T.U. per lb
Gasoline	11,000
Light petroleum	17,933
Heavy petroleum	19,210
Alcohol (pure)	12,100
Alcohol diluted 20 per cent. by weight with water	9,620

—Gas Power.

Machinery Development

Metal Working ***Special Apparatus*** ***Wood Working***

BAND RE-SAW WITH OUTSIDE BEARINGS.

ON examination these machines will be found in advance of any tool of their kind yet produced, in the way of simplicity of construction, elegance of design and adaptability to the work for which they are intended. They embody all the conveniences and attachments that are necessary or desirable for any kind of resawing within the capacity of the machine, and have ample power both on the blade and in the feed works for any reasonable demand, while all parts liable to breakage are reduced to a minimum, and a complete system of numbering and lettering applied and available when repairs are necessary. All the adjustments are easily and quickly made, and those necessary for a change from one class of work to another are all made from the operator's position at the working side of the machine.

The frame is box shape, very heavy, with broad base, which, when properly set, prevents all vibration of the machine when running. The bottom spindle is 2 11-16 in. diameter, has four bearings, two on frame, one outside of pulleys, and one outside of saw wheel. The top spindle is 2 5-16 in. diameter with double bearings. The wheels are of a form and dimensions which have been found correct from experience. They hang between heavy vertical columns which are rigidly bolted to the base. The top wheel is raised and lowered by screw, from front or back of machine, and may be adjusted independently when necessary to level the upper spindle. The lower wheel is very heavy, with a solid web in centre. The upper wheel is as light as possible, consistent with strength. The faces of both wheels are accurately finished on their own journals, ground in a special manner after the wheels are placed in position on the machine. Every wheel is tested for running accuracy before being shipped.

The feed works are driven by belt. The arrangement is the most simple possible, every adjustable part being within easy reach of the operator at his position. The rolls are driven by spur and bevel gears, all cased in, making the motion smooth and noiseless even at the fastest speed. Six feed rolls carry the stock to the saw all of which are driven by cut gearing. The right hand rolls are rigid in their boxes, but the left hand rolls are elastic so as to grip uneven stock and hold it firmly up against the rigid rolls, thus making a powerful feed even on very unequally sawn lumber.

The self centering attachment is so arranged that by one movement of a handle the rigid hand roll becomes elastic, so that you have a complete self-centering machine, or by one movement of the handle the right hand roll becomes rigid again. This special feature is our own invention and does away with the use of a wrench to make your machine self-centering. The rolls will tilt from 0 to one inch in six inches, allowing of the successful sawing of clap boards and other bevel work. They are held perfectly rigid without the using of a wrench. The blade is strained by special double acting knife balance levers with detachable weights, the strain being determined by figures on the weights. Blades up to 4 inches wide, 24 feet 6 inches long and from No. 19 to No. 22 gauge may be used. The guides have adjustable hardened steel back, flat roller, with front and side blocks, adjustable on planed ways. The lower guide is adjustable. The upper one is counter weighted and instantly adjustable vertically to the width of the stock. They also furnish a back guide to steady the saw on its upward travel.

Clark-Demill Band Re-Saw with Outside Bearings.

The saw runs within an inch of the centre of the back roll. When doing bevel sawing the feed works can be drawn back by one movement of a lever so that the saw cannot strike the rolls while they are tilted. The rolls cannot be tilted until the feed works are drawn back from the saw. Cleaners are provided on both wheels. A packing box is attached to the lower wheel guard, which should be kept filled with oily waste to soften the gum which often collects. It is then removed by scrapers fastened to the frame of the machine.

This machine is manufactured by Clark-Demill Co., Hespeler, Ont., from whom further information may be secured.

AUTOMATIC TWO WAY VISE.

DURING these days of wonderful engineering and mechanical accomplishments, the artisan is seldom awed by the birth of some new, economical, ingenious and almost human device. Gradually man is being superseded by his wonderful creations, the machine.

Though the machine shop itself has been revolutionized during the past few years, the vise and anvil have ever remained in their primitive state. But even the vise has reached that stage where it no longer remains the solid chunk of iron capable only of holding in one direction the clamped article between its jaws.

After years of experimental and practical tests, the Pittsburgh double swivel vise, manufactured by the Pittsburgh Automatic Vise and Tool Co. of Pittsburgh, Pa., has been perfected to the highest rung of mechanical ability and ingenuity. Due to its unique design, the vise is capable of turning in any direction, describing any angle of two complete circles. But it does not stop here, for by simply tightening the jaws upon the piece held, both swivels are automatically locked, not to be moved by any human means.

Possessing these marked advantages and original characteristics; it serves not only as a bench vise but as a universal rig for drill presses, planers, shapers and other machines where an article is desired to be held firmly and at any angle. For this purpose an extra base is provided, the vise being easily lifted from one base to the other without the undoing of a screw or bolt.

Fig. 1—Simplicity of the Several parts of the "PGH" Double Swivel Vise.

The vise itself is composed of eight sections or parts, the screw, front and rear jaws, two yokes, demale screw, friction plate and base. The base, which is adapted to be secured to any suitable support, is formed with an annular seat for the body of the vise. The body, consisting of two halves or yokes, provided each with a section or half of a hub seated in the base, are adapted to be slightly tilted in respect to each other and their seat, so that when they are drawn toward each other at their top they will be pressed outward below and thus locked frictionally in the base seat.

The rear jaw is adapted to be rotated about its axis within the body of the vise. It has two diameters, the larger one working in the front yoke and the smaller one in the rear. The friction plate is affixed to the outer extremity of the jaw member. This cap bears against the outer edge of the rear yoke and serves as a support and bearing for the sleeve or nut. This possesses a head which bears against the outside of the friction plate or cap and extends thence through same into the internal bore of the front jaw.

A special bearing alcove is placed on the screw between its head and fixed collar in front jaw, so as to serve as a reliable backing for the collar when the screw is rotated to open the jaws. This also gives the head of the screw a solid bearing against the full stock of the front jaw, therefore nothing to break or give way or become weakened under severe strains or pressure through the action of the screw.

Thus it will be understood that when the screw is tightened it tilts the body sections towards each other and locks them against rotation in the base. The inner jaw member has a splined connection with the inner jaw, thus preventing its rotation in respect thereto.

In accordance with its design the idea of strength is closely adhered to in its complete construction. The round slide bar, which adds phenomenal strength to the weakest portion of all other vises, is made of the finest cast or crucible steel, specially melted and mixed for this purpose. The jaws facings are all of tool steel, while the vise itself is composed of a semi steel. The screw instead of being a common, cast iron head cast upon a malleable or steel rod containing a thread, is a hammered steel forging made for this use alone.

Thus it possesses the distinction of being absolutely dust proof, non slipping and non breakable. Due to its ability of having its jaws revolved in any direction it serves as two vises, for if a jaw smaller than its face is desired, jaws are simply thrown to one side and the side of same used. If a stationary vise is desired the thumb screws on rear can be slightly turned and the two swivels tightly locked. By this means any desired friction can be obtained.

The single swivel vise, although possessing but one movement, is also a very marked advancement in vise construction. Unlike any other vise it is automatically locked by tightening on

Fig. 2—"PGH" Double Swivel Pipe Vise.

Fig. 3—"PGH" Double Swivel Vise.

the piece in the jaws. No levers, pins, screws or other "in the way" appliances. By lifting it from the base, the vise can readily be carried anywhere. The vise consists of but five parts, the rear and front jaws, screw, base and

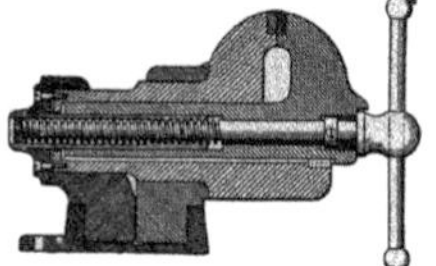

Fig. 4 Sectional View of "PGH" Double Swivel Vise.

locking dog. The latter is so constructed as to permit the screw to pass through its upper extremity. It is pivoted near its bottom upon an alloy steel bar supported by the base. Thus when the jaws are tightened upon an object the screw draws the dog towards them which naturally throws its bottom out against the walls of the base, thus locking it firmly.

In conformity with the originality and usefulness of the previously described tools, the pipe attachment for the single swivel vise, deserves mentioning. Unlike any other vise, the inventors have placed this important attachment on the rear end of the slide bar, out of the way at all times of the operator, the pipe jaws never interfering with the reception of work in the vise jaws proper. It consists of a tool steel attachment, milled perfectly to size, the two sides straddling the end of the slide bar. These are held in place by a bolt. A double linked chain, the pivot pins of which project beyond the sides of the links at each side, is attached at one end by a lug or pin to the body of the vise, below the slide bar. By placing the tube or pipe in position, the chain is brought around it and fastened on the lugs on the upper portion of the attachment. The screw is then given

Fig. 5—Sectional View of "PGH" Single Swivel Pipe Vise

a half turn which throws the draw bar back thus tightening the chain upon the pipe and at the same time locking the swivel movement. Its great advantages can readily be understood and appreciated by all mechanics. The necessity of opening the jaws sufficiently wide for the reception of the pipe is done away with as the length of the chain is sufficient to engage the pipe regardless of the position of the jaws. The range of sizes are also increased 100 per cent. over other combination pipe vises.

BENCH POWER PRESS.

HAVING had many enquiries for a bench power press that would do a large range of stamping work similar to what was formerly done in foot presses. W. H. Banfield & Sons, 120 Adelaide Street West, Toronto, have specially designed and built the power press shown in the accompanying illustration. The machine is particularly adapted for operating small cutting, perforating and forming dies, such as are extensively used in the manufacture of tinware, brass goods, burners, jewelry, pocket book trimmings, buttons, etc.

The total weight of this press is 425 pounds and in designing it special attention was given so as to get a strong, compact, and solid frame and not to interfere with a die space of considerable size. The following dimensions will give one an idea of the size and capacity of this press: Opening in the bed 3¾ in. by 4¾ in.; Width of opening in back 5½ in.; Distance back from centre of slide 4½ in.; And from bed to the slide 1 in.; Adjustment of slide 1½ in.; Distance between the gibs 5¼in.; Diameter and face of balance wheel 17 in. by 2¾ in.; Weight of balance wheel 76 pounds; Speed of balance wheel, per minute, 150 to 200 revolutions; Size of bolster plate is 8 in. by 9¼ in., by 1¼ in. thick and has a 2¼ inch round hole; Height to the centre of shaft 27½ in.; Height over all 36 in. The general design of this press is similar to the larger sizes of the best American make of presses, and is fitted with a right and left hand adjusting nut and a Stiles instantaneous clutch. The press can be easily and quickly adjusted from upright to incline position by simply loosening the clamp screws.

The combination of first-class workmanship and the best material obtainable make these presses of the greatest durability, accuracy and strength.

Fig. 6 Sectional View of "PGH" Double Swivel Pipe Vise.

SAW TOOTH NOTCHING PRESS.

FOR rapidly cutting teeth of circular disc saws, the machine shown in the accompanying illustration has recently been designed by the E. W. Bliss Company, Adams Street, Brooklyn, N.Y. A sample of the work done in the machine is shown on the floor. The machine is adapted for cutting the teeth of various sizes of saws of different diameters, and embodies several new and important features.

After the dies have been set and the press adjusted for the work, the operation is as follows: A plain blank is clamped between two discs which are fastened to the revolving spindle in

Fig. 7—"PGH" Single Swivel Vise with Patented Pipe Attachment.

Banfield Bench Power Press

front of the bed of the press. This spindle is rotated by a ratchet wheel located near the floor. The ratchet is actuated by a series of levers and cranks by a pair of bevel gears on the main shaft. The long vertical connecting rod has an adjusting screw for altering its throw, so that it may be accurately set for spacing the saw teeth to the required pitch. Approximate settings may be obtained by adjusting the end of the connecting rods in the slotted bell crank and pawl lever, the final adjustment being made by means of the screw already referred to. Depressing the hand levers sets the machine in operation, causing the saw disc to revolve, and the punch slide to descend, cutting a tooth at each stroke of the machine.

Bliss Saw Tooth Notching Press.

A particularly desirable feature of the machine is the ease with which its setting may be altered from one size disc to another. Heretofore, it has been necessary to remove a considerable part of the mechanism to change the indexing device. In this machine the indexing device is placed at the extreme lower end of the spindle where it can be easily removed by taking off one nut and a washer. To accommodate a different size disc, the spindle carrier is moved from or toward the punch by the hand wheel. The spindle carrier slides along a hinged table so that the teeth of the saw may be punched at the proper rake or angle. The table is secured in position by a clamping bolt in the fixture just below the lower die bed, engaged in the slot of a bracket which is bolted to the table. The slots ? the levers of the feeding mechanism are made use of to compensate for the swinging of the table.

The machine will cut teeth on discs of eight to twenty-four inches in diameter and will cut from the largest teeth to sixteen to the inch. The die and punch holders also have adjustments for setting them at any desired angle, making it possible to set the machine at practically any angle required in saw tooth punching of this class.

The machine is so arranged that saws of small diameters and of light gauge may be cut with the press driven as a flywheel press by placing the belt on the pulley rim which is fastened to the main gear. The gear in this case acts as a flywheel and the pinion and regular driving pulley and flywheel are thrown out of mesh. With the first arrangement the press runs at 80 strokes per minute and when geared at 40 strokes per minute.

FILE SHARPENING MACHINE.

THE American File Sharpening Co., 296 Broadway, New York, are placing on the market a machine for restoring worn out files to their original cutting edge. The illustration herewith shows the construction of the machine in operation. This machine can be operated by means of steam alone or a combination of steam and air can be used, the cut shows the air attachments. While it will operate successfully with 80 or 90 lbs. of steam the rapidity of doing its work is increased with 100 to 120 lbs. of steam. An abrasive known as carbo-flint is used in its operation and no sand blast process is used. An inexperienced workman may be taught the use of this machine in the course of a few hours. Its range includes everything from fine imported files to the coarse rasp, as well as hack saw blades. The capacity is from 200 to 300 files per day according to the size and condition in which they are received for re-sharpening. No acid or chemicals of any nature are used and the temper of the file is not drawn.

File Sharpening Machine.

Further information regarding these machines may be had from the American File Sharpener Co., 296 Broadway, New York.

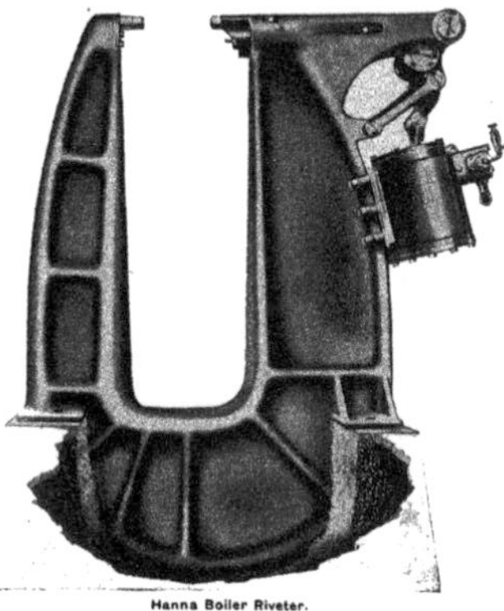

Hanna Boiler Riveter.

HANNA BOILER RIVETTER.

PNEUMATIC boiler and tank riveters have been developed by the Hanna Engineering Works, Chicago Ill., as shown in the accompanying illustration. This riveter has a stake of 21 inches diameter with 75¼ inches clearance. This style of machine was designed for driving rivets in small diameter tanks and sheets and is made in any reasonable reach and gap and for 50, 70, 80 and 100 tons pressure, capacity limited only by handling and healing arrangements. The size of stake of course, will vary with the length of reach and pressure to be exerted. When desired it is also furnished with flush top arrangements for doing close flanging work. This machine may be operated either portable or stationary in any position. The catalogue No. 3 of the Hanna Engineering Works will show by graphical illustrations the manner in which they are enabled to drive absolutely tight rivets with each and every stroke. It will be mailed upon request to all interested.

NEW SAFETY AUTOMATIC PRESS.

A NEW safety automatic inclinable press has been invented and is now being manufactured by Jas. B. Hall & Son, 112 Adelaide St., Toronto. This press is shown in the accompanying illustration. The two important features of the press are the safety automatic clutch and the fact that it is inclinable. This safety automatic clutch allows but one stroke of the press for each depression of the foot pedal. To obtain another stroke of the press the foot pedal must be released and again pressed down. The object of this feature is to reduce to a minimum the liability of accident to the operator.

The body of the press swings from the centre of the crank shaft bearings and does not affect the position of the fly-wheel or belt, and thus the press may be set while in motion. This feature is more conducive to perfect setting of dies in combination work.

Another feature of the press is the device for setting the ram to the desired height, it being a simple and secure device.

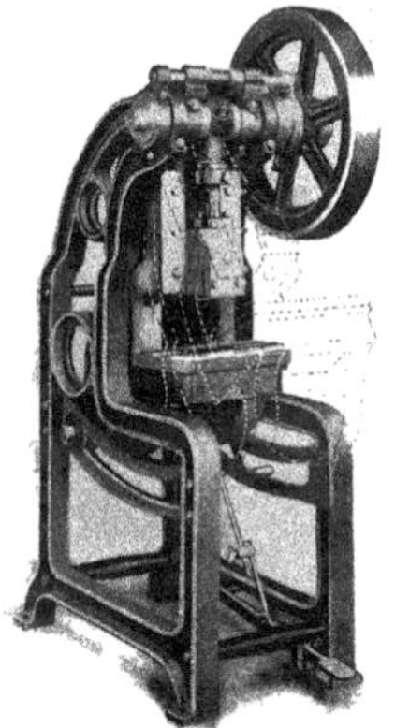

New Safety Automatic Press.

INTERESTING JAIL BREAKING TEST.

AN expert local steel and iron worker of Vicksburg, employed by the county to test the steel cages in the new county jail there has been able to find his way through the "unsawable" bars. This man, John Christian has been engaged on a contract with the Warren county to make an actual test as to the strength of the steel bars making the cages in the new jail. He has been able to saw through an inch of wrought steel with what is known as the ordinary No. 250 12 inch L. S. Starrett saw made in Athol, Mass. This test was made in the presence of the president of the Board of County Supervisors. It came as somewhat of a surprise to the county officials, as they had been formerly convinced by the many testimonials of the St. Louis firm that furnishes these bars that they were positively tool proof. As four cages of this kind are to go under a separate contract of something over $16,000.00 the company will have to come again and make a stronger bar for these cages.

ABOUT CATALOGUES

By mentioning Canadian Machinery to show that you are in the trade, a copy of any of these catalogues will be sent by the firm whose address is given.

WHITNEY CHAINS.—A number of bulletins have recently been issued by the Whitney Manufacturing Company, Hartford, Conn., describing various apparatus and paying special attention to the chains manufactured by this company. They include roller chains, on which special information is given in one of the bulletins. Other bulletins are devoted to keys and cutters, water tool grinder, milling machines, and tapping devices.

ROCKWELL FUEL OIL.—An 18-page catalogues issued by this company is devoted to fuel oil burning appliances as manufactured by the Rockwell Engineering Company, 26 Cortlandt St., New York, engineers and contractors. Some of their bulletins are devoted to their flue welding furnace, Rockwell heating machines, annealing and hardening furnace, metal melting furnace, brazing furnaces and pot furnaces.

THE HIGLEY METAL SAW.—A high-class catalogue of 47 pages, devoted to metal saws and grinders as manufactured by the Higley Machine Company, New York. The illustrations of the various machines are of a superior order and interesting information is given concerning the same.

HORIZONTAL BORING MACHINES.—These are described as the tools for accurate and quick boring in the catalogue of the Binsse Machine Company, Newark, N.J. Illustrations are given showing the various adaptations of the work of this machine. 40 pages, 5 by 7½.

FLEXIBLE TRANSMISSION. — The Coates Clipper Manufacturing Company, Worcester, Mass., are manufacturers of the Coates unite link flexible shaft. The many uses to which this shafting may be applied are mentioned and the illustrations of same given. Other Coates apparatus is described in this catalogue of 28 pages.

E. W. BLISS Co.—A 578 page cloth bound catalogue describes and illustrates very many designs of presses, dies and special machinery built by E. W. Bliss Company, Brooklyn, N.Y. Some fine half-tone engravings at the front illustrate the works of this company, both interior and exterior. The presses manufactured by E. W. Bliss, are known throughout the metal manufacturing world and have an excellent reputation wherever used. Many of these are to-day in operation in Canada with excellent results.

FROST & WOOD CALENDAR.—The annual calendar of the Frost & Wood Company, Smith's Falls, Ont., is to hand. As is usual with this firm a handsome calendar has been produced that will be an attraction to any office.

BALL BEARING JACKS.—A. O. Norton, 286 Congress Street, Boston, Canadian branch, Coaticook, Que., issue a catalogue devoted to special ball bearing and track jacks. These are handled in Canada by W. H. C. Musson & Co., Montreal, Que., from whom further information regarding these is obtainable.

HYDRO-ELECTRIC REPORT. — The third report of the Hydro-Electric Power Commission of the Province of Ontario is devoted to the Lake Huron and Georgian Bay Districts. It contains maps of these parts showing the location of water powers and adjacent industrial centres. As in the case with previous reports, data regarding the power required in the different localities and the cost of various lines has been calculated and tabulated. This is of particular interest to residents in this district as well as to all parts of Ontario.

FRANKLIN AIR COMPRESSORS.—The Chicago Pneumatic Tool Company's latest catalogue is devoted exclusively to a detailed description of the Franklin air compressors manufactured at their Franklin plant. Constructive principles are described in detail and information not hitherto published is given relative to several new designs of compressors, including the Corliss motor driven gas engine, and new pattern compressors of larger capacities. Considerable space is devoted to pumping by compressors describing the new Chicago water lathe. Numerous interesting tables and formulaes complete the contents of this valuable contribution to compressed air literature, which should be in the hands of all interested in the subject. The catalogue contains 116 pages, 6 x 9. Is mailed free on application to all interested by applying to this company.

SHARP FILES.—The catalogue on sharp files issued by the American File Company, of New York, which appeals to the people with the heading: Don't throw away your dull files, and goes on to illustrate and fully describe the new machine manufactured by them, also their Hercules folding ladder, which is set forth as useful for every purpose, such as painting, fire escapes, etc.

WIRE NAIL MACHINERY.—National Wire Nail Machinery is the subject of Bulletin No. 30, issued by the National Machinery Company, Tiffin, Ohio; a high grade of coated paper is used in this bulletin, on which the excellent engravings show to particular advantage. A line of samples of the nails made by these machines is also shown. Some of the illustrations give views of various parts of the mechanism. This bulletin should be in the hands of all interested in wire nail machinery.

SPLIT FRICTION CLUTCHES.—The Dodge Manufacturing Company, of Canada, have just issued a booklet describing Dodge Split Friction Clutches. This clutch is placed in position on the shaft without interference with the remainder of the equipment, and repairs and renewals are similarly facilitated. The price list is given. Cut off couplings are also described and illustrated.

HANNA RIVETER.—A thirty-four page catalogue, issued by the Hanna Engineering Works, Chicago. The various types are illustrated on half-tone paper, the pleasing effect being the interplacing of heavier stock containing the descriptive matter, the various riveters described include portable riveters, lever riveters, and riveting supplies. Illustrations of the various parts are also included.

HOWE SCALES.—A compact catalogue 4 x 6 inches, containing 175 pages, is devoted to the Howe scale manufactured by the Howe Scale Company, Rutland, Vermont. Besides platform scales and counter scales, heavier styles are described, including suspension scales for coal hoppers, bullock scales, car scales, transportation scales, railroad contract scales, etc.

MURCHEY MACHINE & TOOL CO.—Corner Fourth and Potter Streets, Detroit, Michigan, issue a catalogue devoted to the Murchey patent automatic collapsing taps, nipple and pipe thread machine, roller pipe cutters, tapping machines, etc. One of their important tapping machines is described in the machinery development department of this issue.

GAS DRIVEN ELECTRIC POWER.—A neat booklet containing the reprint of an article of the above title read before the Engineers' Society of Western Pennsylvania by J. R. Bibbins. This article deals with a Westinghouse gas engine installation at the power house of the Warren & Jamestown street railway. This article is an exceptionally interesting one; and the booklet can be obtained from the Westinghouse company's publishing department, Pittsburg, Pa.

PERSONAL MENTION

K. L. Aitken, consulting electrical engineer, has moved to larger quarters in 1003 Traders Bank Building, Toronto.

* * *

A Rising Contractor and Manufacturer.

Men about town, in Montreal, have, within the past few months, had frequent introduction to the firm name of "J. W. Harris Company, Limited," which has appeared in connection with so many large contracts for building and excavation. The president of this growing company, who are the patentees and manufacturers of the J. W. Harris Excavators and Trenching Machines, is J. W. Harris, whose likeness is presented on this page. Mr. Harris was born at Buffalo, N.Y., October, 20th, 1865, his mother being of French birth, and his father of Scotch descent. He was educated in Montreal at the Jacques Cartier Normal School, and learned his trade as plumber and steamfitter, with Brodeen & Lessard. In 1892 he became a partner in the firm of Lessard & Harris, general contractors, and undertook some large contracts, notably the C.P.R. "Angus 'shops," at Montreal. Mr. Harris has made remarkable progress in the commercial world within the past few years. In addition to being President and General Manager of the J. W. Harris Company, Limited, he is President of the Municipal Trenching Company, Vice-President of the Ross & Harris Company, railway contractors, Vice-President of the Montreal & Southern Counties Railway, President of the St. Lawrence Floating & Wrecking Company, and Director and Chairman of the Executive Committee of the Murray Bay Lumber & Pulp Company. Mr. Harris is a member of the Montreal Board of Trade and member of the Council, Chambre de Commerce. He is also a leading member of the Montreal Building Exchange.

J. W. HARRIS, Montreal.

* * *

Sales Manager Weds.

The marriage of Mr. N. S. Braden, sales manager of the Canadian Westinghouse Company, Hamilton, to Miss Mabel Greening, second daughter of Mr. and Mrs. S. O. Greening, of the same city, took place in the James Street Baptist Church there on Thursday, Nov. 29th. Miss Hattie Greening, sister of the bride, was maid of honor, and the bridesmaids were Miss Edna Greening, sister of the bride, Miss Gladys Zimmerman, of Hamilton, and Miss Agnes Ellis, of Hespeler. The groomsman was Mr. W. G. Cameron, of Cleveland, and Messrs. William Prince, Cleveland, H. D. Shute, Pittsburg, William Greening, Toronto, and H. M. Bostwick and Stuart MacDonald, of Hamilton, were the ushers. The gift of the groom to the bridesmaids were gold enameled pins, and to the groomsman and ushers gold enameled pearl stick pins. The ornaments of the bride included a magnificent twin setting diamond ring, also the gift of the groom. A reception was held at "Fonthill," the residence of the bride's parents, which was artistically decorated for the occasion. Gifts almost innumerable were received by the bride, including a sterling tea service from the Westinghouse staff at Montreal and a case of silver from the Pittsburg Westinghouse staff. Mr. and Mrs. Braden left in the evening foo the South on an extended trip and on their return will reside at 134 Duke Street, Hamilton. The newly married couple have hearty congratulations and the best wishes of Canadian Machinery.

COMPANIES INCORPORATED.

Douglas Bros. Company, Toronto, have been incorporated, as contractors for roofing and sheet metal work, with a share capital of $100,000, provisional directors of the company will be, T. Douglas, E. Douglas and G. R. Douglas, of Toronto.

The General Metal Foundry and Machinery Co., Montreal; capital, $199,000; purpose, to carry on the business of engineers, machinists boilermakers etc. The directors are: M. E. Lymburner, Sault Ste. Marie; L. L. Lymburner, J. E. Mathews, J. P. Mathews, all of Montreal.

A company is formed at Brockville for the purpose of manufacturing tools of all kinds. Also galvanizing and plating metals under the name of Canadian Billings & Spencer, with a share capital of $200,000. J. T. Gardner, J. H. Botsford and W. S. Buell of Brockville are directors of the company.

The General Metal Foundry and Machinery Company has been incorporated with a share capital of $199,000 with head office at Montreal. Directors of the company will be M. E. Lymburner, St. Agathe, Que., L. M. Lymburner, II. N. Lymburner, J. E. Mathews and J. B. Mathieu of Montreal and A. St. George of St. Paul, Que.

The W. J. Kemp Company have been incorporated with a share capital of $300,000, with head office at Stratford, for the purpose of manufacturing and dealing in agricultural and other implements and machinery. Provisional directors of the company are: W. J. Kemp, Stratford, N. J. Kemp and J. S. Lewis of Newark Valley, N.Y.

The Vulcan Company have been incorporated with a share capital of $100,000, and head office at London, Ontario, to manufacture furnaces, stoves, and other heating and cooking appliances. The provisional directors of the company will be: R. W. Shaw, E. W. M. Flock, and F. G. Mitchell, of London.

The Dominion Asbestos Company has been incorporated with a total capital stock of $500,000 for the purpose of developing mines, also to carry on business as electricians with head office at Montreal. Directors of the company will be H. H. Robertson, Boston, Mass., R. T. Hopper, F. H. Markey, W. W. Skinner and R. G. Grant, Montreal.

The Canadian Hardware and Manufacturing Company, has been incorporated with a share capital of $50,000, with heal office at Toronto, to carry on business as machinists, engineers, foundrymen, fitters, and smiths. The provisional directors will be: W. S. Harrison, W. H. Alderson, S. M. Linn, and W. J. McManus, of Toronto.

The Dominion Heating and Ventilating Company have been incorporated with a share capital of $40,000, to manufacture heating and ventilating systems, also iron, brass, aluminum, etc., machinery and casting, with head office in Hespeler. The provisional directors of the company will be, E. McNally, M. A. Secord and L. Reid of Galt.

C. Ormsby & Company, Toronto, has been incorporated, with a share capital of $50,000, for the purpose of manufacturing, metal roofing, sheeting, and metal building materials, also to engage in building operations, as contractors. The provisional directors of the company will be: C. E. Ormsby, W. J. Cook, and G. A. Graham, of Toronto.

The Percival Plow and Stove Company have been incorporated with a share capital of $200,000 for the purpose of manufacturing agricultural implements, stoves, and furnaces, with head office at Merrickville, Ont. Provisional directors of the company are: John B. Waddell, T. F. McKimm of Smith's Falls, Eli Stickney, R. C. Percival of Merrickville and W. R. Morris, Peterborough.

INDUSTRIAL PROGRESS

MACHINERY AND MANUFACTURING NEWS would be pleased to receive from any authoritative source industrial news of any sort, the formation or incorporation of companies, establishment or enlargement of mills, factories or foundries, railway or mining news, etc. All such correspondence will be treated as confidential when desired.

Another nitro-glycerine factory has been started at Petrolea.

Anton Bery has accepted a site for his proposed factory at Guelph.

The Torbrook iron mines, Halifax, N.S., have been sold for $250,000.

The Dominion Dash Company, Walkerville, Ont., will erect a new building there.

Ratepayers of Elmira, Ont., carried a by-law to expend $25,000 for waterworks.

Building permits issued in Winnipeg for the month of November total about $400,000.

The Gurney Tilden Co., of Hamilton, Ont., will erect a big warehouse at Edmonton.

The Sydney Cement Company, Sydney, N.S., will operate their plant all winter.

Wingham is after a $15,000 grant from Andrew Carnegie for a Public Library building.

The Marine Iron works, of Victoria, B.C., was damaged by fire to the extent of $5,000.

A new Presbyterian church will be erected in St. John, N. B., at the cost of $50,000.

Work has been started on the new warehouse for McClary Mfg. Co., in St. John, N.B.

Building permits issued in Toronto from December 8, to 15, amounted to over $116,000.

A new fire station is to be built in Winnipeg for the protection of C.P.R. shop's district.

The Schaake Machine Works, New Westminster, B.C., will spend $50,000 erecting new machine shops.

The Stevens Company of Galt, are to erect a new moulding plant in connection with other workshops.

Willian Gray & Company, Carriage manufacturers of Chatham, will build a $100,000 factory in Winnipeg.

The Atikokan Iron Company's blast furnace plant at Port Arthur will be ready in about six weeks time.

The village of Dublin is to be lighted by an acetylene gas plant now being installed by the Kurtz Company.

The contract for the new fire hall on Adelaide St., Toronto, was awarded last week for the sum of $59,535.

Fire did damage to the extent of $200 at the Ontario Tack Company's factory in Hamilton a few days ago.

Discoveries of tin deposits have been found on an island within 100 miles of Vancouver, by a German prospector.

The Canadian Ethinite Company of Niagara Falls, Ont., have commenced the construction of their factory building.

The F. B. Wood Company, Hamilton, will erect a large addition to their factory, making it twice its original size.

A brick and pulp mill will start at Prince Albert, Sask., it is expected that $100,000 will be invested by each concern.

The new nine storey building erected for the C. W. Lindsay & Company, of Montreal, was formally opened last week

The United States Mica & Milling Co. have made application for land in East Toronto for a manufacturing plant.

A new company with a capital stock of $50,000, will be formed in St. John, N.B., for the manufacture of hardwood flooring.

Every five days a new engine is turned out from the works of the Kingston Locomotive Co. Seven hundred men are employed.

John Glenn, Dungannon, is completing a machine which is an invention of his own, for the making of small cement tile for farm use.

A firm from Hartford, Conn., is negotiating with the town of Guelph with a view of establishing a forge works there.

The Manitoba Cartage and Warehousing Co., are applying for authority to increase their capital stock from $150,000 to $250,000.

Jones Bros. & Company, manufacturers of show cases, etc., Toronto, have increased their capital stock from $60,000 to $200,000.

A permit has been issued the Gerhard Heintzman Piano Company for the building of an addition to their Toronto factory, to cost $10,000.

The Phillips Manufacturing Company, Toronto, will erect a new two storey factory building containing 132,000 square feet of floor space.

The British Columbia Tie and Lumber Company's lumber mill at Prince Rupert, will have a capacity of 70,000 feet of lumber per day.

The Hamilton Bridge Company have secured the contract for all the iron work in the new C.P.R. bridge, across the Belly river, Lethbridge, Sask.

The Northern Navigation Company, which owned the steamer Monarch, recently lost on Isle Royale, will build a duplicate of their steamer Huronic.

The Macdonald Manufacturing Company, Toronto, have secured a permit for the building of a four storey addition to their factory, to cost $18,000.

The Dominion Carriage Company, Truro, N.S., has been organized with a $250,000 capital, and will erect buildings in the spring, employing 100 men.

Jenkins Bros., of New York, have ordered a 150 horse-power engine from the Robb Engineering Company for the new factory they are building in Montreal.

The large planing mill of Robertson & Hacketts, at Granville, B.C., was totally destroyed by fire, loss being estimatd at $40,000, partially covered by insurance.

The Robb Engineering Co. has received an order from the Allis-Chalmers-Bullock Co. for three 125 horse-power engines for the C.P.R. Hotel, Vancouver, B.C.

An English Cordage Company are making arrangements for the erection of a large warehouse in Brandon, Man., to be used as a distributing centre in the West.

A by-law has been prepared by the town council of Welland, Ont., to fix the assessment of the Robertson Machinery Co., of that place, at $1,000 for ten years.

A warehouse will be built at Saskatoon for the Manson, Campbell Co. and the Chatham Carriage Works, to be used as a distributing centre for their vehicles in the West.

The Lethbridge coal strike is at last over, the company granting the requests of the men. Work was started immediately as there is a great scarcity of coal in the West.

The Shaake Machine Works, of New Westminster, B.C., are making a large addition to their works to keep pace with the machine and iron work industry of that city.

Fire supposed to have originated from a workman's torch destroyed the C. N. R. boiler shop in Winnipeg and much valuable machinery. The loss is between $60,000 and $100,000.

Tenders are being asked for the various works required in the erection and completion of a grand stand also of a horticultural building on the Exhibition Grounds, Toronto.

The Grand Trunk Railway System and Geo. Okley & Sons, Toronto, have both received recently transfer and traveling trains from the Smart-Turner Machine Co., Hamilton.

The Vancouver Portland Cement Company are preparing to enlarge their premises, so as to have an output of 1,000 barrels per day. Their capacity is now 650 barrels per day.

A large iron foundry and machine shop is to be erected at Anaconda. B.C. Messrs. B. A. Isaacs and R. W. Hinton, of the Nelson Iron Works, are promoters of the enterprise.

The expenditure of a quarter of a million dollars at Stratford, by the G.

T.R., for the erection of a new station, and extending its locomotive repair workshops, has been decided upon.

O. P. Letchworth, expresses the opinion that the Brantford plant of the Pratt & Letchworth Company, will be increased rather than reduced, despite the fact that his company is about to have another plant in Montreal.

The Lufkin Rule Co., of Saginaw, Mich., has secured a site in Windsor, and will build a Canadian factory early next spring. Meanwhile the firm has secured temporary quarters and will begin building operations at once.

The Dominion Iron & Steel Company's rail mill at Sydney, which has been closed down for several weeks, has resumed operations. There is a sufficient stock of machinery on hand to keep the mill running double-shifted.

The National Cement Company of Durham, is in a very flourishing condition, having turned out 1,110 barrels per day all summer. They will construct a building for storage purposes, which will be 250 feet long and 50 feet wide.

The big brick addition to the Bertram shops at Dundas, is being rapidly pushed to completion. The big shop runs between the new molding shop and the main building. The firm has an ever-increasing demand for working space,

A disastrous fire occurred in Preston on Dec. 4, by which the car barns and rolling stock of the Galt, Preston & Hespeler and Preston and Berlin Electric Railways were totally destroyed, loss being estimated at about $75,000.

Barnhill & McLennan, of Amherst, N.S., have purchased a large lumber property at Etiomani, Sask., on which they intend building a 100 horse-power rotary sawmill. The machinery for the mill has been ordered from the Robb Engineering Co.

A company is to be established at Halifax, N.S., for the manufacture of railway box cars, flat cars, coal hoppers, and baggage cars. The plant will have a capacity of 1,500 cars a year, employing 250 men, with an annual wage outlay of $150,000.

Asbestos in payable quantity and quality has been discovered in the Transvaal. An expert who has examined the deposit states that it is of abnormal width and equal to the finest in Canada or Italy. An offer of £40 per ton has been received from Germany for the best quality.

Edmonton passed four money by-laws this week, totalling $275,000. Of this $121,000 was for paving, $119,000 for rails for the street railway; $45,000 for an incinerator, and $30,000 for the purchase of a third park site, two having previously been purchased. All the by-laws were passed by large majorities.

The Commercial Cement Company has been incorporated in Winnipeg with a capital of $150,000. The mills will be located on Boyne River where a good supply of suitable rock has been secured. Machinery has been ordered which will equip a mill with a capacity of 1,000 barrels per day.

One enterprising lumber firm in British Columbia, are building a small town on their limit, having erected houses for the married men, and plans are being prepared for the erection of a large hotel for the accommodation of the unmarried men, with a $2,000 bowling alley in the building.

A proposition to establish a large car building plant in St. John, N.B., is spoken of, both United States and Canadian capital being interested. If the proposition succeeds motor cars will be built. The promotors are figuring on developing power from Riverson Falls of St. John River.

A valuable discovery of iron ore was recently made in the Township of Snowdon, in the Haliburton District, about three-quarters of a mile from the line of the Irondale and Bancroft Railway. The deposit is 150 feet wide, and the ore a high grade magnetite. It is the largest body of ore yet discovered in the district.

An item in the last issue of Canadian Machinery made mention of a bearing company proposing to manufacture Wright's Roller Bearings in St. Catharines. This information was incorrect, as the only company having the right to manufacture Wright's Paper Rolling Bearings in Canada, is the Canadian Bearings Limited, of Hamilton.

The Dominion Carbonic Company has been incorporated with a share capital of $40,000, with head office Toronto, for the purpose of manufacturing gas, and dealing in all gas products, chemicals, and chemical products. Provisional Directors of the company will be R. J. Quale, and H. Woodbury, of Buffalo, N. Y.; Allen and J. J. Maclennan, of Toronto.

The Canada Foundry Company, of Toronto, have completed the erection of the structural steel for the west side of the Canada General Electric Company's new transformer building. Work will now begin on the east side wall, and in three week's time it is expected, the job of placing the roof on the building will be commenced. Operations will be continued all winter.

A company has been formed for the purpose of building and equipping cars for railways, also to manufacture all railway equipment and appliances. The corporate name of the company will be the National Car Company, with a share capital of $300,000, and head office at Toronto. The Provisional Directors of the company will be A. Stewart, T. H. Kilgore and L. Williams, of Toronto.

A steel industry capitalized at five millions, is the latest proposition to come before the Goderich Council, they state that the plant would employ about 1,500 hands, and occupy 100 acres. They ask certain bonuses, and the council have decided to take the matter up with the Canadian Pacific, with whom the steel interests have already communicated, and see what inducements can be offered.

New Sales Office.

To place them in a position to be able to better serve their customers in the Eastern States the Warner & Swasey Co., of Cleveland, Ohio, have opened a New York office in the Singer Building, 149 Broadway. Mr. H. L. Kinsley, formerly eastern representative, will be in charge, and make this his headquarters.

CONDENSED ADS.

RATES

One insertion—25c. for 20 words; 1c. a word for each additional word.

Yearly rate—$2.50 for twenty words or less, 10c a word for each additional word.

The above does not apply to notices under the head of "Machinery Wanted." These notices are inserted free for subscribers.

WESTERN REPRESENTATIVE WANTED.

AN agent in the west to handle Beardshaws' "Conqueror" high speed steel and drills, which are meeting with a favorable reception throughout Canada. Alex. Gibb, 13 St. John St., Montreal.

FOR SALE

$600.00 WILL buy Patterns for the following size gasoline engines: 3 and 6 horse power stationary; 5 horse power double opposed air cooled; 16 horse power double opposed cylinder; 12 horse power double upright, stationary; 10 horse power double upright auto or marine; 40 horse power, 4 cylinder marine All in good order. Have built several hundred high grade engines from these patterns. D. McKenzie & Co., London, Ont. (1)

SMOKE consumers are not necessary with producer gas power, because the smoke is turned into gas and used for power; no matter what fuel you use there is absolutely no smoke; write for details or call. W. Gillespie, 98 Front Street, Toronto.

CORBET Foundry and Machine Company, Owen Sound, Ont., make a specialty of marine and stationary engines and boiler repairs; also all kinds of sheet steel work, such as stacks, flumes, hoppers, etc.; write for prices.

TURBINE Water Wheels—The following is another proof that the Canadian turbine is the most economical wheel made:—"Almonte, Ontario, with a pair of 42" Canadian turbines, is practically the only town in the Ottawa valley having an extensive electric plant that was able to run to its full capacity throughout this dry summer, with water to spare." Charles Barber & Sons, Meaford, Ontario.

ASSIGNEE'S sale of steam planing mill and sash and door factory in the City of Woodstock, Ontario; first-class and old-established stand; only one other in city; will be sold by auction, Thursday, December 27th, at 2 p.m. H. B. Sproat, assignee, Woodstock, Ont.

"LITTLE GIANT" turbine water wheels, machine-cut wood and iron tooth gearing; power transmission machinery orders executed with despatch; get catalogue and prices. J. C. Wilson & Co., Glenora, Ont.

SITUATIONS VACANT

WANTED—Moulders and patternmakers—Accustomed to sawmill work. Apply, at once, to the Madison Williams Manufacturing Co., Lindsay, Ont.

FOREMAN Moulder and coremaker; thoroughly experienced in engine and machinery work; steady employment and good wages. Apply Clinton Thresher Co., Clinton, Ont.

WANTED—Mechanical draftsman (assistant). The Smart-Turner Machine Co., Limited, Hamilton, Ont.

Power Commission Letter.

Replying to the criticisms of the Hydro-Electro Power Commission's report in a letter in which they placed that the cost of power at $15 to $17.75 per h.p. at the city limits of Toronto, a letter has been written to the Board of Control of the city. They state that their investigations show that 25-cycle alternating current power at 550 or 220 volts could be furnished to the customers in the city of Toronto based on the maximum price furnished by the commission at from one cent per k.w. hour for the larger customers with overhead distribution, to 2 1-2 cents per k.w. hour for the smaller customers, receiving their power from the underground distribution.

The prices at present charged by the Toronto Electric Company vary from 2 2-3 cents per k.w. hour to 8 cents for smaller ones. It should, however, be stated that the present consumers of electric power would be under the necessity of exchanging their present direct current motors for 25-cycle alternating current ones, but it is evident that the estimated reduction in rates would afford them ample inducement to make the change.

It may also be said that the capital expenditure for an exclusive 25-cycle distribution would be very much less than that given in the first report of the Hydro-Electro Power Commission, probably not one-half.

As for arc lighting, the most modern type of enclosed arc lamps of the greatest power obtainable could be supplied at a cost of $47.25 per lamp per annum on the basis of a 1,400 light installation. In this connection I may say that recent improvements in street arc lighting would make it possible to reduce the cost to about $40 per lamp.

S.O.S. Annual Dinner.

The Engineering Society of the Faculty of Applied Science held their eighteenth annual dinner at the Rossin House on Dec. 7. The affair was a decided success, both financially and sociably. Among the guests from the engineering profession were:—C. H. Rust, city engineer, Toronto; M. J. Haney, contracting firm of Haney & Miller; E. Burk, Ontario Association Architects; A. N. Connor, Connor, Clark and Mons; F. S. Scott, consulting structural engineer, Toronto; and S. Gagne, secretary-treasurer, Toronto Branch of C.S.C.E.

S.P.S. Electrical Club.

For some time there has been a feeling among the senior mechanical and electrical students of the Faculty of Applied Science, University of Toronto, that the formation of a club for the discussing of subjects and the reading of papers of direct interest to them would be advantageous. This has culminated in the holding of a general meeting at which the S.P.S. Electrical Club was organized with the following officers: Hon. President, Dr. Galbraith; President, Mills Maclaughlan; Vice-Presiden,t F. R. Ewart; Secretary-Treasurer, J. C. Armer; Fouth Year Representative, N. P. F. Death; Third Year Representative, G. P. Coulter and C. N. Hutton; Critics, Prof. Rosebrugh, Prof. Angus, H. N. Price, B.A. Sc.; W. W. Gray, B.A., Sc.; H. G. Smith, B.A., Sc., and S. Dushman, B.A. The enthusiasm displayed at the organization meeting argues well for the success of the club.

Street Railway Convention.

At the King Edward Hotel, Toronto, on Dec. 6th and 7th, the semi-annual convention of the Canadian Street Railway Association was held. Nearly all the large companies in Canada were represented. A number of very instructive and interesting papers were contributed as follows: "Some of the Shops," by W. R. McRae, master mechanic of the Toronto Railway Company; "Track Construction, Maintenance and Repair," by A. M. Grantham, C.E., Toronto Railway Company; "Discipline," by J. E. Hutcheson, Ottawa Electrical Railway Co.; "Standardization of Equipment" by C. B. King, manager London Street Railway; "Freight and Mail," by E. F. Seixas; "Columbus Convention," by A. H. Royce, secretary of the association; "History of Electric Traction in the Province of Ontario," by Robert Clarke, Toronto Railway Company; "The Power House," by J. C. Rothery, of the East Liverpool, Ohio, Traction and Light Company.

New Car Company.

A million dollar car manufacturing plant has been proposed for Fort William, Ont., which is expected to give employment to between 400 and 1,000 men the year round. H. G. Brown, of Montreal, is interested in this and it is fully expected that construction will be begun in the spring.

Large Salary Increase.

The Yale & Towne Mfg. Co. have recently announced at a meeting of the president with the superintendent and foreman of the Stamfort plant an increase of wages and piece rates, which makes a total increase of $120,000 in one year. Although the average rate of wages and average rate of piece earnings in the Yale & Towne's works have been at no time greater the company recognized that under present industrial conditions a review of these rates is expedient. In the distribution of these voluntary additions to the earnings of the employes the company are giving special consideration to the length of time and quality of service and to individual efficiency of piece rates, as well as of day rates, which will be reviewed and where necessary will be readjusted.

Vice-President Satisfied.

Frank W. Morse, vice-president of the Grand Trunk Pacific announced at Winnipeg on Dec. 17th that they were well satisfied with the work of construction done during the past summer. Eighty per cent. of the construction proposed during the season was carried out. Next year is calculated to see large accomplishments in the way of construction and still more important work in the engineering department for the year 1907. Every precaution is being taken to provide against a scarcity of labor during the coming year, the company having representatives in all the important centres of population and using every means to provide the necessary number of men.

Clay Interests Convention.

The Clay Products Manufacturers' Association of Ontario, met in convention in the Rossin House in Toronto on Wednesday, Thursday, and Friday, December 12th, 13th and 14th. One of the most largely attended and most enthusiastic convention in the history of the association was held. The papers read were of a high order, and all calculated to educate the clay workers, and elevate their standing and grade of workmanship. The program of the convention included at the commencement a welcome to Toronto, by Mayor Coatsworth, followed by a lecture by Professor Coleman of the School of Practical Science of Toronto. Another important address given in the first afternoon of the convention was a paper by J. S. McCannell, manager Milton Pressed Brick Co., Milton, Ont., on "Mining and Preparation of Material." On the morning of the second day of the convention the delegates took a trip by special train to the plant of the Don Valley Brick Co., Todmorden, Ont. The remaining program included the following: Lecture on "Technical Education in its Relation to the Clay Industries," Prof. Edw. Orton, jr., Ohio State University, Columbus, Ohio; "Down Draft Kilns," Wm. McCrody, Lyons, Ont. At 8.15 p.m. a complimentary banquet was given to the members of C.C.P.M. by Toronto Clay Product Manufacturers. On Friday forenoon the following papers were given: "Uses of Exhaust Steam on a Brick Plant," A. M. Wickens, chief engineer Canadian Casualty & Boiler Insurance Co., Toronto; "Comparative Economy of Construction and Operation of Down Draft and Continuous Kilns," J. B. Miller, superintendent Don Valley Brick Co., Toronto; "Brick Spiel," H. de Joannis, editor "Brick," Chicago, Ill.; Lecture by Prof. Baker, Ontario School of Mining, Kingston, Ont; "Waste Heat Drying," Wm. Bailie, former superintendent La Prairie Brick Co., La Prairie, Que., followed by question drawer. A committee comprising of S. J. Fox, M.P.P., Wm. McCredie, J. B. Miller, J. S. McCarnel, C. H. Betchel A. W. Wright, J. B. Lochrie Wm. Handcock, J. Russell, and H. de Joannis was appointed to confer with Principal Galbraith of the School of Practical Science, in connection with the establishment of a course of Ceramics, at Toronto University. The following officers were elected:—President, J. B. Miller, Toronto; 1st Vice-President, J. S. McCannell, Milton; 2nd Vice-President, Charles Curtis, Peterborough; 3rd Vice-President, J. Cornhill, Chatham; Secretary-Treasuer, C. H. Bechtel, Waterloo; Executive, S. J. Fox, M.P.P., T. M. Mulligan, Harbard; Joseph Russell, Toronto; W. McCredie, Lindsay; David Martin, Thamesville. The association decided to hold its convention in Ottawa next year.

National Founders' Convention.

A record attendance contributed to the success of the convention held in New York on Nov. 14th and 15th by the National Founders' Association. The past season of the association had been marked by a determined struggle for the open shop. Some points brought out at this convention were that the molder must be the product of a four years' apprenticeship was exploded also that unskilled labor with the aid of molding machine could make foundry independent of union domination. Moreover it was found that the restriction of the machines used and the number of apprentices insisted upon by union interests must be swept away if the rapidly increasing need of foundry products was to be met. In an educational way the association has accomplished much good in the past year. Lessons taught by strikes during the past year were emphasized as follows:—

The absolute injustice of the minimum wage principle as applied to skilled labor.

The mistake of employes in opposing the introduction of labor saving machinery.

The rank injustice to the rising generation in limiting its opportunity of learning a trade.

No organization can long continue that stands unequivocally for reducing all mechanics to the same level; compelling the competent to divide earnings with the incompetent. This principle gradually but surely lowers the standard of the craft.

Neither can an organization long endure that persists in denying young men the opportunity this country affords them to develop along the line for which they are best fitted and receive their reward in proportion to the ability they possess and apply.

Also prominent among these lessons taught is the duty we, as manufacturers owe our loyal foreman. When such foremen have had this work in hand, the greatest progress has been made, and the foreman who cannot honestly perform the work assigned to him when defending his employers on such an issue as that presented this season is blind not only to his own interest, but to mankind in general.

It is manifestly our duty as proprietors and employers of labor, to look after both the welfare and attitude of our foremen, to understand them, to maintain their confidence, respect and strict loyalty. If such relations cannot be maintained then a change in foremen is imperative.

The greatest lesson taught is to our journeymen, who, if they expect to maintain an organization that will command the respect and good will of their employers must cause their unions to be dominated by the conservative, fair-minded element.

It has been demonstrated beyond question that foundries can be operated, castings produced and success obtained in the foundry business in spite of the effort of arrogant union officials to contrary.

When the workingmen of this continent learn the lesson of obedience to the law and constitution of the country in which they live, and to confine the functions of their organization within the limits of such laws and constitution 90 per cent. of the conflicts between capital and labor will disappear.

The greatest lesson to all, is that strikes such as we have faced this season are but a culmination of unwise, unreasonable and incompetent leadership of the masses of workmen, that they are merely a temporary conflict provided we will accept the means within our reach of dealing direct with and educating our own workmen instead of passing such dealings through the medium of the irresponsible agitator, who in ninety-nine cases out of one hundred distorts any statement made to him to suit his own opinions when presenting it to the members of his union.

The officers elected were:—President, O. P. Briggs Minneapolis; Vice-President Henry A. Carpenter, Providence; Secretary, F. W. Hutchings, Detroit; Treasurer, Western Trust & Savings Bank, Chicago; Honorary Member, W. H. Pfahler, Philadelphia. Besides these was a long list of district officers, those of the seventh division representing Canada being:—H. Cockshutt, Cock-Edgar McDougall, John McDougall, Caledonia Iron Works Co., Ltd. Montreal, Que.; R. J. Whyte, Frost & Wood Co., Ltd., Smith's Falls, Ont.; Frederic Nicholls Canada Foundry Co., Ltd., Toronto, and John M. Taylor, Taylor-Forbes Co., Ltd., Guelph, Ont.; C. Bermingham, Canada Locomotive Works, Kingston, was appointed a member of the Press Committee and Geo. W. Watts, Canada Foundry Co., Toronto, a member of Finance and Auditing. Among those from this side of the border at the convention were L. L. Anthes, Toronto Foundry Co., Toronto and John M. Taylor, Taylor-Forbes Co., Guelph.

Canada Metal Company Expands.

W. G. Harris of the Canada Metal Company, Toronto, has purchased the Toronto Ball Grounds occupying 3½ acres adjacent to the Grand Trunk and Canadian Pacific Railway in Parkdale and will next fall commence the erection of two buildings 40 feet wide by 500 feet in length, the railroad site being run between the two buildings which will be of concrete fire proof construction, one two storeys in height and the other one storey.

The Canada Metal Company have outgrown their present premises on William Street, the rapid development of their business having made necessary the present changes. They have recently constructed several furnaces for smelting copper, lead and zinc ores and intend to develop this branch of their trade so that it can be gone into on a larger scale when the new buildings are erected. The large floor space which will be available in the new premises will allow the company to still further develop their lead and tin rolling, galvanizing and tinning industry as well as taking up the smelting of ores.

The growth of the Canada Metal Company's business under Mr. Harris' management has been almost extraordinary, and they are to be congratulated upon securing such a splendid site for the enlargement of their business. The site is immediately adjoining the present plant of the Toronto Foundry Company, and of the land purchased a block of about 183 x 500 feet will be held for future developments.

Gasoline Company Insolvent.

The winding-up order was granted recently in respect to the McLachlan Engine Co., which was incorporated in the Ontario Company's Act in June, 1899. The McLachlan Company have been manufacturing at Swansea during the past year, but have been particularly unfortunate with their output.

Peterboro Works Closed.

Considerable surprise was evinced in manufacturing circles, when, on the motion of the Ontario Bank, Chief Justice Meredith granted a winding-up order of the Wm. Hamilton Manufacturing Co., of Peterboro. On Dec. 6th the company decided to admit its insolvency. The employes numbered 225 men, and the closing of this large works will mean a heavy set-back to the manufacturing interests in Peterboro. The company was organized in 1883, and have been carrying on an extensive manufacturing business, sawmills, mining machinery and water mills.

Big Cement Plant.

Southern Manitoba is looking forward to the establishment of an enormous cement industry which will envolve the investment of several hundred thousand dollars. The rights to a valuable depost in Southern Manitoba have been secured by a number of capitalists and the costly machinery for the production of a high-grade cement will be installed.

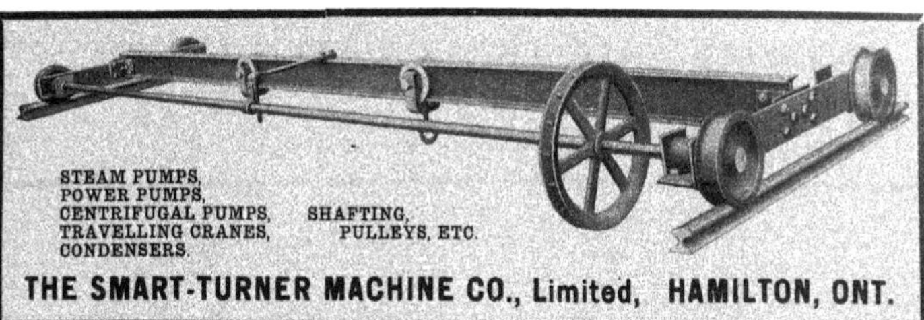

THE SMART-TURNER MACHINE CO., Limited, HAMILTON, ONT.

Potter & Johnston Manufacturing Automatics

will machine your duplicate parts from all kinds of castings, forgings and the bar with extreme accuracy and more economically than upon any other tools in the market.

Catalogue showing interesting specimens of parts finished on these tools free upon request.

ALSO UNIVERSAL SHAPERS

Potter & Johnston Machine Co., Pawtucket, R.I., U.S.A.

Canadian Representative: G. H. HOWARD, DUNDAS, ONT.

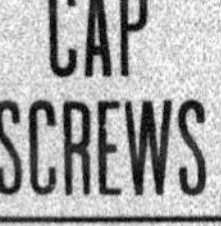

Square Head
Hexo Head
Filister Head
Finished
Case Hardened

COLD PRESSED NUTS
MADE IN
CANADA FOUNDRY COMPANY, LIMITED, TORONTO, ONT.
CANADA.
CANADA FOUNDRY COMPANY, LIMITED.

NUTS

Square
Hexagon
Blank
Semi-Finished
Finished

Head Office and Works: Toronto, Ont.

District Offices—Montreal, Halifax, Ottawa, Winnipeg, Vancouver, Rossland

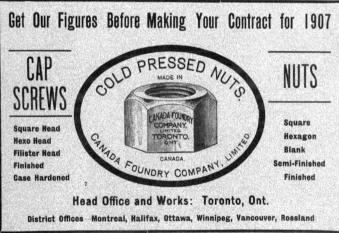

CANADIAN MACHINERY BUYERS' DIRECTORY

To Our Readers—Use this directory when seeking to buy any machinery or power equipment. You will often get information that will save you money.

To Our Advertisers—Send in your name for insertion under the heading of the lines you make or sell.

To Non-Advertisers—A nominal rate of $1 per line a year is charged non-advertisers.

Acids.
Canada Chemical Mfg. Co., London.

Abrasive Materials.
Baxter, Paterson & Co., Montreal.
The Canadian Fairbanks Co., Montreal.
Rice Lewis & Son, Toronto.
H. W. Petrie, Toronto.
Williams & Wilson, Montreal.

Air Brakes.
Canada Foundry Co., Toronto.
Canadian Westinghouse Co., Hamilton.

Air Receivers.
Allis-Chalmers-Bullock Montreal.
Canada Foundry Co., Toronto.
Canadian Rand Drill Co., Montreal.
Chicago Pneumatic Tool Co., Chicago.
Jenckes Machine Co., Sherbrooke, Que.
John McDougall Caledonian Iron Works Co., Montreal.

Alundum Scythes Stones
Norton Company, Worcester, Mass.

Arbor Presses.
Niles-Bement-Pond Co., New York.

Augers.
Chicago Pneumatic Tool Co., Chicago.
Rice Lewis & Son, Toronto.

Automobile Parts
Globe Machine & Stamping Co., Cleveland, Ohio.

Axle Cutters.
Butterfield & Co., Rock Island, Que.
A. B. Jardine & Co., Hespeler, Ont.

Axle Setters and Straighteners.
Canadian Buffalo Forge Co., Montreal.
Dominion Henderson Bearings, Niagara Falls, Ont.
A. B. Jardine & Co., Hespeler, Ont.

Babbit Metal.
Baxter, Paterson & Co., Montreal.
Canada Metal Co., Toronto.
Canada Machinery Agency, Montreal.
Rice Lewis & Son, Toronto.
Miller Bros. & Toms Montreal, Que.

Bakers' Machinery.
Pendrith Machinery Co., Toronto.

Barrels, Tumbling.
Buffalo Foundry Supply Co., Buffalo.
Detroit Foundry Supply Co., Detroit.
Dominion Foundry Supply Co., Montreal
Hamilton Facing Mill Co., Hamilton.
Globe Machine & Stamping Co., Cleveland, Ohio.
John McDougall Caledonian Iron Works Co., Montreal.
R. McDougall Co., Galt, Ont.
Miller Bros. & Toms, Montreal, Que.
J. W. Paxson Co., Philadelphia, Pa.
H. W. Petrie, Toronto.
The Smart-Turner Mach. Co., Hamilton.

Bars, Boring.
Dominion Henderson Bearings, Niagara Falls, Ont.
Hall Engineering Works, Montreal.
Miller Bros & Toms, Montreal.
Niles-Bement-Pond Co., New York.

Batteries, Dry.
Berlin Electrical Mfg. Co., Toronto.

Batteries, Flashlight.
Berlin Electrical Mfg. Co., Toronto.

Batteries, Storage.
Canadian General Electric Co., Toronto.
Chicago Pneumatic Tool Co., Chicago.
Rice Lewis & Son, Toronto.
John Millen & Son, Montreal.

Bearings, Roller.
Dominion Henderson Bearings, Niagara Falls, Ont

Bell Ringers.
Chicago Pneumatic Tool Co., Chicago.

Belting, Chain.
Baxter Paterson & Co. Montreal.
Canada Machinery Exchange, Montreal.
The Dodge Mfg. Co., Toronto.
Jeffrey Mfg. Co., Columbia, Ohio.
Link-Belt Eng. Co., Philadelphia.
Miller Bros. & Toms, Montreal.
Waterous Engine Works Co., Brantford.

Belting, Cotton.
Baxter, Paterson & Co., Montreal.
Canada Machinery Agency, Montreal.
Dominion Belting Co., Hamilton.
Rice Lewis & Son, Toronto.

Belting, Leather.
Baxter, Paterson & Co., Montreal.
Canada Machinery Agency, Montreal.
The Canadian Fairbanks Co., Montreal.
The Dodge Mfg. Co., Toronto.
McLaren, J. C., Montreal.
Rice Lewis & Son, Toronto.
H. W. Petrie, Toronto.
Williams & Wilson, Montreal.

Belting, Rubber.
Baxter, Paterson & Co., Montreal.
Canada Machinery Agency, Montreal.
Dodge Mfg. Co., Toronto.

Belting Supplies.
Baxter, Paterson & Co., Montreal.
The Dodge Mfg. Co., Toronto.
Rice Lewis & Son, Toronto.
H. W. Petrie, Toronto.

Bending Machinery.
John Bertram & Sons Co., Dundas, Ont.
Chicago Pneumatic Tool Co., Chicago.
London Mach. Tool Co., Hamilton, Ont.
National Machinery Co., Tiffin, Ohio.
Niles-Bement-Pond Co., New York.

Benders, Tire.
Canadian Buffalo Forge Co., Montreal.
A. B. Jardine & Co., Hespeler, Ont.
London Mach. Tool Co., Hamilton, Ont.

Blast Furnace Brick.
Detroit Foundry Supply Co., Detroit.
Hamilton Facing Mill Co., Hamilton.

Blowers.
Buffalo Foundry Supply Co., Montreal.
Canada Machinery Agency, Montreal.
Canadian Buffalo Forge Co., Montreal.
Detroit Foundry Supply Co., Detroit.
Hamilton Facing Mill Co., Hamilton.
Kerr Turbine Co., Wellsville, N.Y.
J. W. Paxson, Philadelphia, Pa.
Sheldon's Limited, Galt.

Blue Printing.
The Electric Blue Print Co., Montreal.

Boilers.
Canada Foundry Co., Limited, Toronto.
Canada Machinery Agency, Montreal.
Goldie & McCulloch Co., Galt.
Jenckes Machine Co., Sherbrooke, Que.
E. Leonard & Sons, London, Ont.
Manitoba Iron Works, Winnipeg.
John McDougall Calenonian Iron Works, Montreal.
H. W. Petrie, Toronto.
Robb Engineering Co., Amherst, N.S.
The Smart-Turner Mach. Co., Hamilton.
Waterous Engine Works Co., Brantford.
Williams & Wilson, Montreal.

Boiler Compounds.
Canada Chemical Mfg. Co., London, Ont.
Economical Mfg. & Supply Co., Toronto.
Hall Engineering Works, Montreal.
Hamilton Facing Mill Co., Hamilton.

Bolt Cutters.
John Bertram & Sons Co., Dundas, Ont.
London Mach. Tool Co., Hamilton.
National Machinery Co., Tiffin, Ohio.
Niles-Bement-Pond Co., New York.

Bolt and Nut Machinery.
John Bertram & Sons Co., Dundas, Ont.
Canada Machinery Agency, Montreal.
London Mach. Tool Co., Hamilton.
National Machinery Co., Tiffin, Ohio.
Niles-Bement-Pond Co. New York.

Bolts and Nuts, Rivets.
Baxter, Paterson & Co., Montreal.

Books, Mechanical.
Technical Pub. Co., Manchester.
The MacLean Pub. Co., Ltd., Toronto.
Spon & Chamberlain, New York

Boring and Drilling Machines.
American Tool Works Co., Cincinnati.
B. F. Barnes Co., Rockford, Ill.
John Bertram & Sons Co., Dundas, Ont.
Canada Machinery Agency, Montreal.
Canadian Buffalo Forge Co., Montreal.
A. B. Jardine & Co., Hespeler, Ont.
London Mach. Tool Co., Hamilton.
Niles-Bement-Pond Co., New York.

Boring Machine, Upright.
American Tool Works Co., Cincinnati.
John Bertram & Sons Co., Dundas, Ont.
London Mach. Tool Co., Hamilton.
Niles-Bement-Pond Co., New York.

Boring Machine, Wood.
Chicago Pneumatic Tool Co., Chicago.
London Mach. Tool Co., Hamilton.

Boring and Turning Mills.
American Tool Works Co., Cincinnati.
John Bertram & Sons Co., Dundas, Ont.
Canada Machinery Agency, Montreal.
London Mach. Tool Co., Hamilton.
Niles-Bement-Pond Co., New York.
H. W. Petrie, Toronto.

Box Puller.
A. B. Jardine & Co., Hespeler, Ont.

Brakes, Automobile,
Globe Machine & Stamping Co., Cleveland, Ohio.

Brass Foundry Equipment.
Detroit Foundry Supply Co., Detroit.

Brass Working Machinery.
Warner & Swasey Co., Cleveland, Ohio.

Brass Working Machine Tools.
Warner & Swasey Co., Cleveland, O.

Brushes, Foundry and Core.
Buffalo Foundry Supply Co., Buffalo.
Detroit Foundry Supply Co., Detroit.
Dominion Foundry Supply Co., Montreal
Hamilton Facing Mill Co., Hamilton

Brushes, Steel.
Buffalo Foundry Supply Co., Buffalo.

Bulldozers.
John Bertram & Sons Co., Dundas, Ont.
London Mach. Tool Co., Hamilton, Ont.
National Machinery Co., Tiffin, Ohio.
Niles-Bement-Pond Co., New York.

Calipers.
Baxter, Paterson & Co., Montreal.
Brown & Sharpe, Providence, R.I.
Rice Lewis & Son, Toronto.
John Millen & Son, Ltd., Montreal, Que.
L. S. Starrett & Co., Athol, Mass.
Williams & Wilson, Montreal.

Can Making Machinery
Canada Machinery Agency, Montreal.
Miller Bros & Toms, Montreal.

Cars, Foundry.
Buffalo Foundry Supply Co., Buffalo.
Detroit Foundry Supply Co., Detroit.
Dominion Foundry Supply Co., Montreal
Hamilton Facing Mill Co., Hamilton.

Castings, Grey Iron.
Allis-Chalmers-Bullock Montreal.
Dodge Mfg. Co., Toronto.
Hall Engineering Works, Montreal.
Jenckes Machine Co., Sherbrooke, Ont.
Laurie Engine & Machine Co., Montreal.
John McDougall Caledonian Iron Works Co., Montreal.
R. McDougall Co., Galt, Ont
Niagara Falls Machine & Foundry Co., Niagara Falls, Ont
Robb Engineering Co., Amherst, N.S.
Smart-Turner Machine Co., Hamilton.
Sylvester Mfg. Co., Lindsay, Ont.

Castings, Brass.
Chadwick Bros., Hamilton.
Hall Engineering Works, Montreal.
Niagara Falls Machine & Foundry Co. Niagara Falls, Ont.
Jenckes Machine Co., Sherbrooke.
R. McDougall Co., Galt
Robb Engineering Co., Amherst, N.S.

Castings, Semi-Steel.
Robb Engineering Co., Amherst, N.S.

Cement Machinery.
Allis-Chalmers-Bullock, Limited, Montreal
Jeffrey Mfg. Co., Columbus, Ohio.
John McDougall Caledonian Iron Works, Co., Montreal.
St. Lawrence Supply Co., Montreal.

Centreing Machines.
John Bertram & Sons Co., Dundas, Ont.
Dominion Henderson Bearings Niagara Falls, Ont.
Jeffrey Mfg. Co., Columbia, Ohio
London Mach. Tool Co., Hamilton, Ont.
Niles-Bement-Pond Co., New York.
Pratt & Whitney Co., Hartford, Conn.

Centres, Milling Machine.
The Stevens Co., Galt.

Centres, Planer.
American Tool Works Co., Cincinnati.
The Stevens Co., Galt.

Centres, Shaper.
The Stevens Co., Galt.

Centrifugal Pumps.
Canadian Buffalo Forge Co., Montreal.
John McDougall Caledonian Iron Works Co., Montreal.
R. McDougall Co., Galt.
D'Olier Engineering Co., New York.
Pratt & Whitney Co., Hartford, Conn.

Centrifugal Pumps—Turbine Driven.
Canadian Buffalo Forge Co., Montreal.
Kerr Turbine Co., Wellsville, N.Y.

Chaplets.
Buffalo Foundry Supply Co., Buffalo.
Detroit Foundry Supply Co., Detroit.
Hamilton Facing Mill Co., Hamilton.

Charcoal.
Buffalo Foundry Supply Co. Buffalo.
Detroit Foundry Supply Co., Detroit.
Dominion Foundry Supply Co., Montreal
Hamilton Facing Mill Co., Hamilton.

Chemicals.
Baxter, Paterson & Co., Montreal.
Canada Chemical Co., London.
Economical Mfg. & Supply Co., Toronto.

Chemists, Industrial.
Detroit Testing Laboratory, Detroit.

Chemists, Metallurgical.
Detroit Testing Laboratory, Detroit.

Chemists, Mining.
Detroit Testing Laboratory, Detroit.

Chucks, Ring Grinding.
A. B. Jardine & Co., Hespeler, Ont.
Chicago Pneumatic Tool Co., Chicago.

Chucks, Drill and Lathe.
American Tool Works Co., Cincinnati.
Baxter, Paterson & Co., Montreal.
John Bertram & Sons Co., Dundas, Ont.
Canada Machinery Agency, Montreal.
Ker & Goodwin, Brantford.
A. B. Jardine & Co., Hespeler, Ont.
London Mach. Tool Co., Hamilton.
John Millen & Son, Ltd. Montreal.
W. B. McLean & Co., Montreal.
Niles-Bement-Pond Co., New York.
H. W. Petrie, Toronto.
Rice Lewis & Son, Toronto.
Standard Tool Co., Cleveland.
Whitman & Barnes Co, St. Catharines.

Chucks, Planer.
American Tool Works Co., Cincinnati.
Canada Machinery Agency, Montreal.
W. B. McLean & Co., Montreal.
Niles-Bement-Pond Co., New York.

Chucking Machines.
American Tool Works Co., Cincinnati.
Niles-Bement-Pond Co., New York.
H. W. Petrie, Toronto.
Warner & Swasey Co., Cleveland, Ohio.

Circuit Breakers.
Allis-Chalmers-Bullock, Limited, Montreal
Canadian General Electric Co., Toronto.
Canadian Westinghouse Co., Hamilton.

Clippers, Bolt.

A. B. Jardine & Co., Hespeler, Ont.

Cloth and Wool Dryers.

Canadian Buffalo Forge Co., Montreal.
B. Greening Wire Co., Hamilton.
McEachren Heating and Ventilating Co., Galt, Ont.

Compressors, Air.

Allis-Chalmers-Bullock, Limited, Montreal
Canada Foundry Co., Limited, Toronto.
Canada Machinery Agency, Montreal.
Canadian Rand Drill Co., Montreal.
Canadian Westinghouse Co., Hamilton.
Chicago Pneumatic Tool Co., Chicago.
Detroit Foundry Supply Co., Detroit.
John McDougall, Caledonian Iron Works Co., Montreal
H. W. Petrie, Toronto.
The Smart-Turner Mach. Co., Hamilton.
Canada Machinery Agency, Montreal.
Canadian Westinghouse Co., Hamilton.
Hall Engineering Works, Montreal, Que.
London Mach. Tool Co., Hamilton.
R. McDougall Co., Galt, Ont.
Niles-Bement-Pond Co., New York.
H. W. Petrie, Toronto.
Pratt & Whitney Co., Hartford, Conn.
Williams & Wilson, Montreal.

Concentrating Plant.

Allis-Chalmers-Bullock, Montreal.

Concrete Mixers.

Jeffrey Mfg. Co., Columbus, Ohio.
Link-Belt Co., Philadelphia.

Condensers.

Canada Foundry Co., Limited, Toronto.
Canada Machinery Agency, Montreal.
Hall Engineering Works Montreal.
Smart-Turner-Machine Co., Hamilton.
Waterous Engine Co., Brantford.

Confectioners' Machinery.

Baxter, Paterson & Co., Montreal.
Pendrith Machinery Co., Toronto.

Consulting Engineers.

Charles Brandeis, Montreal.
Canadian White Co., Montreal.
Dominion Henderson Bearings, Niagara Falls, Ont.
Hall Engineering Works, Montreal.
Jules De Clercy, Montreal.
John S. Fielding, Toronto.
Miller Bros. & Toms, Montreal.
Roderick J. Parke, Toronto.
Plews & Trimingham, Montreal.
T. Pringle & Son, Montreal.
James C. Royce, Toronto.

Contractors.

Canadian White Co., Montreal.
D'Olier Engineering Co., New York.
Expanded Metal and Fireproofing Co., Toronto.
Hall Engineering Works Montreal.
Laurie Engine & Machine Co., Montreal.
John McDougall Caledonian Iron Works Co., Montreal.
Miller Bros. & Toms, Montreal.
Robb Engineering Co., Amherst N S.
The Smart-Turner Mach. Co., Hamilton.

Controllers and Starters Electric Motor.

Allis-Chalmers-Bullock, Montreal.
Canadian General Electric Co., Toronto.
Canadian Westinghouse Co., Hamilton.
T. & H. Electric Co., Hamilton.

Conveyor Machinery.

Baxter, Paterson & Co., Montreal.
The Dodge Mfg. Co., Toronto.
Jeffrey Mfg. Co., Columbus, Ohio.
Jenckes Machine Co., Sherbrooke, Que.
Link-Belt Co., Philadelphia.
John McDougall Caledonian Iron Works Co., Montreal.
Miller Bros. & Toms, Montreal.
Smart-Turner Machine Co., Hamilton.
Laurie Engine & Machine Co., Montreal.
Waterous Engine Works Co., Brantford.
Williams & Wilson, Montreal.

Coping Machines.

John Bertram & Sons Co., Dundas, Ont.
London Mach. Tool Co., Hamilton.
Niles-Bement-Pond Co., New York.

Core Compounds.

Buffalo Foundry Supply Co., Buffalo.
Detroit Foundry Supply Co., Detroit.
Dominion Foundry Supply Co., Montreal
Hamilton Facing Mill Co., Hamilton.

Core Ovens.

Detroit Foundry Supply Co., Detroit.
Hamilton Facing Mill Co., Hamilton.

Core Oven Bricks.

Buffalo Foundry Supply Co., Buffalo.
Detroit Foundry Supply Co., Detroit.
Hamilton Facing Mill Co., Hamilton.

Core Wash.

Buffalo Foundry Supply Co., Buffalo.
Detroit Foundry Supply Co., Detroit.
Hamilton Facing Mill Co., Hamilton.

Cotton Belting.

Baxter, Paterson & Co., Montreal.
Dominion Belting Co., Hamilton.

Cranes, Electric and Hand Power.

Canada Foundry Co., Limited, Toronto.
Dominion Foundry Supply Co., Montreal
Hamilton Facing Mill Co., Hamilton.
Link-Belt Co., Philadelphia.
John McDougall, Caledonian Iron Works Co., Montreal.
Miller Bros. & Toms, Montreal
Niles-Bement-Pond Co., New York.
Smart-Turner-Machine Co., Hamilton.

Crankshafts.

The Canada Forge Co., Welland.

Crushers, Rock or Ore.

Allis-Chalmers-Bullock, Montreal.

Contractors' Plant.

Allis-Chalmers-Bullock, Montreal.
John McDougall, Caledonian Iron Works Co., Montreal.
Niagara Falls Machine & Foundry Co., Niagara Falls, Ont.

Crank Pin Turning Machine

London Mach. Tool Co., Hamilton.
Niles-Bement-Pond Co., New York.

Crucibles.

Buffalo Foundry Supply Co., Buffalo
Detroit Foundry Supply Co., Detroit.
Dominion Foundry Supply Co., Montreal
Hamilton Facing Mill Co., Hamilton.
J. W. Paxson Co., Philadelphia.

Crucible Caps

Hamilton Facing Mill Co., Hamilton.

Crushers, Rock or Ore.

Allis-Chalmers-Bullock, Montreal.
Jeffrey Mfg. Co., Columbus, Ohio.

Cupolas.

Buffalo Foundry Supply Co., Buffalo.
Detroit Foundry Supply Co., Detroit.
Dominion Foundry Supply Co., Montreal
Hamilton Facing Mill Co., Hamilton.
J. W. Paxson Co. Philadelphia.
Sheldons Limited, Galt.

Cupola Blocks.

Detroit Foundry Supply Co., Detroit.

Cupola Blowers.

Buffalo Foundry Supply Co., Buffalo.
Canada Machinery Agency, Montreal.
Canadian Buffalo Forge Co., Montreal.
Detroit Foundry Supply Co., Detroit.
Hamilton Facing Mill Co., Hamilton.
McEachren Heating and Ventilating Co., Galt.
J. W. Paxson Co., Philadelphia.
Sheldon's Limited, Galt.
B. F. Sturtevant Co., Hyde Park, Mass.

Cutters, Coal.

Allis-Chalmers-Bullock, Montreal.

Cut Meters.

The Canadian Fairbanks Co., Montreal

Cutters, Flue

Chicago Pneumatic Tool Co., Chicago.
J. W. Paxson Co., Philadelphia.

Cutters, Milling.

Becker, Brainard Milling Machine Co. Hyde Park, Mass.
Brown & Sharpe, Providence, R.I.
Pratt & Whitney Co., Hartford, Conn.
Standard Tool Co., Cleveland.

Cutting-off Machines.

John Bertram & Sons Co., Dundas, Ont.
Canada Machinery Agency, Montreal.
London Mach. Tool Co., Hamilton.
H. W. Petrie, Toronto.
Pratt & Whitney Co., Hartford, Conn.

Cutting-off Tools.

Armstrong Bros. Tool Co., Chicago.
Baxter, Paterson & Co., Montreal.
London Mach. Tool Co., Hamilton.
H. W. Petrie, Toronto.
Pratt & Whitney, Hartford, Conn.
Rice Lewis & Son, Toronto.
L. S. Starrett Co., Athol, Mass.

Dies

Globe Machine & Stamping Co., Cleveland, Ohio.

Die Stocks

Hart Manufacturing Co., Cleveland, Ohio

Dies, Opening.

W. H. Banfield & Sons, Toronto.
Globe Machine & Stamping Co., Cleveland Ohio.
R. McDougall Co., Galt.
Pratt & Whitney Co., Hartford, Conn.

Dies, Sheet Metal.

W. H. Banfield & Sons, Toronto.
Globe Machine & Stamping Co., Cleveland, Ohio.

Die Sinkers

Globe Machine & Stamping Co., Cleveland, Ohio.

Dies, Threading.

John Millen & Son, Ltd., Montreal.

Dowel Pins.

Buffalo Foundry Supply Co., Buffalo.
Detroit Foundry Supply Co., Detroit.
Dominion Foundry Supply Co., Montreal
Hamilton Facing Mill Co., Hamilton.

Draft, Mechanical.

W. H. Banfield & Sons, Toronto.
Butterfield & Co., Rock Island, Que.
A. B. Jardine & Co., Hespeler.
Pratt & Whitney Co., Hartford, Conn.
Sheldons Limited, Galt.
Standard Tool Co., Cleveland.

Drawing Instruments.

W. B. McLean & Co., Montreal.
Rice Lewis & Son, Toronto.

Drawing Supplies.

The Electric Blue Print Co., Montreal

Drawn Steel, Cold.

Baxter, Paterson & Co., Montreal.
Miller Bros. & Toms, Montreal.
Union Drawn Steel Co., Hamilton.

Dredges, Gold, Dipper and Hydraulic.

Allis-Chalmers-Bullock Co., Montreal.

Drilling Machines, Arch Bar.

John Bertram & Sons Co., Dundas, Ont.
London Mach. Tool Co., Hamilton
Niles-Bement-Pond Co., New York.

Drilling Machines, Boiler.

American Tool Works Co., Cincinnati.
Canadian Buffalo Forge Co., Montreal.
John Bertram & Sons Co., Dundas, Ont.
Bickford Drill and Tool Co., Cincinnati.
The Canadian Fairbanks Co., Montreal.
A. B. Jardine & Co., Hespeler, Ont.
London Mach. Tool Co., Hamilton, Ont.
Niles-Bement-Pond Co., New York.
H. W. Petrie, Toronto.
Williams & Wilson, Montreal.

Drilling Machines Connecting Rod.

John Bertram & Sons Co., Dundas, Ont.
London Mach. Tool Co., Hamilton.
Niles-Bement-Pond Co. New York.

Drilling Machines, Locomotive Frame.

American Tool Works Co., Cincinnati.
B. F. Barnes Co., Rockford, Ill.
John Bertram & Sons Co., Dundas, Ont.
London Mach. Tool Co., Hamilton, Ont.
Niles-Bement-Pond Co., New York.

Drilling Machines, Multiple Spindle.

American Tool Works Co., Cincinnati.
B. F. Barnes Co., Rockford, Ill.
Baxter, Paterson & Co., Montreal.
John Bertram & Sons Co., Dundas, Ont.
Bickford Drill & Tool Co., Cincinnati.
Canada Machinery Agency Montreal.
London Mach. Tool Co., Hamilton, Ont.
Niles-Bement-Pond Co., New York.
H. W. Petrie, Toronto.
Williams & Wilson, Montreal.

Drilling Machines, Pneumatic

Canada Machinery Agency, Montreal.

Drilling Machines, Portable

Baxter, Paterson & Co., Montreal.
Canadian Buffalo Forge Co., Montreal.
A. B. Jardine & Co., Hespeler, Ont.
Niles-Bement-Pond Co., New York.

Drilling Machines, Radial.

American Tool Works Co., Cincinnati.
Baxter, Paterson & Co., Montreal
John Bertram & Sons Co., Dundas, Ont.
Bickford Tool & Drill Co., Cincinnati.
The Canadian Fairbanks Co., Montreal.
London Mach. Tool Co., Hamilton.
Niles-Bement-Pond Co., New York.
H. W. Petrie, Toronto.
Williams & Wilson, Montreal.

Drilling Machines, Suspension.

John Bertram & Sons Co., Dundas, Ont.
Canada Machinery Agency Montreal.
London Mach. Tool Co., Hamilton.
Niles-Bement-Pond Co. New York

Drilling Machines, Turret.

John Bertram & Sons Co., Dundas, Ont.
London Mach. Tool Co., Hamilton.
Niles-Bement-Pond Co., New York.

Drilling Machines, Upright.

American Tool Works Co., Cincinnati.
B. F. Barnes Co., Rockford, Ill.
Baxter, Paterson & Co., Montreal.
John Bertram & Sons Co., Dundas, Ont.
Canadian Buffalo Forge Co., Montreal.
The Canadian Fairbanks Co. Montreal.
A. B. Jardine & Co., Hespeler, Ont.
London Mach. Tool Co., Hamilton.
Niles-Bement-Pond Co., New York.
W. H. Petrie, Toronto.
Williams & Wilson Montreal.

Drills, Bench.

B. F. Barnes Co., Rockford, Ill.
Baxter, Paterson & Co., Montreal.
Canadian Buffalo Forge Co., Montreal.
London Mach. Tool Co., Hamilton.
Pratt & Whitney Co., Hartford, Conn.

Drills, Blacksmith.

Canada Machinery Agency, Montreal.
Canadian Buffalo Forge Co., Montreal.
A. B. Jardine & Co., Hespeler, Ont.
London Mach. Tool Co., Hamilton.
Standard Tool Co., Cleveland.
Whitman & Barnes Co., St. Catharines.

Drills, Centre.

Pratt & Whitney Co., Hartford, Conn.
Standard Tool Co., Cleveland, O.
L. S. Starrett Co., Athol, Mass.

Drills, Electric

B. F. Barnes Co., Rockford, Ill.
Baxter, Paterson & Co., Montreal.
Chicago Pneumatic Tool Co., Chicago.
Niles-Bement-Pond Co., New York.

Drills, Gang.

American Tool Works Co., Cincinnati.
B. F. Barnes Co., Rockford, Ill.
John Bertram & Sons Co., Dundas, Ont.
Pratt & Whitney Co., Hartford, Conn

Drills, High Speed.

Wm. Abbott, Montreal.
American Tool Works Co., Cincinnati.
B. F. Barnes Co., Rockford, Ill.
Pratt & Whitney Co., Hartford, Conn.
Standard Tool Co., Cleveland, O.
Whitman & Barnes Co., St. Catharines.

Drills, Hand.

Canadian Buffalo Forge Co., Montreal.
A. B. Jardine & Co., Hespeler, Ont.

Drills, Horizontal.

B. F. Barnes Co., Rockford, Ill.
John Bertram & Sons Co., Dundas, Ont.
Canada Machinery Agency, Montreal.
London Mach. Tool Co., Hamilton.
Niles-Bement-Pond Co., New York.

Drills, Jewelers.

Pendrith Machinery Co., Toronto.

Drills, Pneumatic.

Canada Machinery Agency, Montreal.
Chicago Pneumatic Tool Co., Chicago.
Niles-Bement-Pond Co., New York.

Drills, Radial.

American Tool Works Co., Cincinnati.
John Bertram & Sons Co., Dundas, Ont.
London Mach. Tool Co., Hamilton, Ont.
Niles-Bement-Pond Co., New York.

Drills, Ratchet.

Armstrong Bros. Tool Co., Chicago.
A. B. Jardine & Co., Hespeler.
Pratt & Whitney Co., Hartford, Conn.
Standard Tool Co., Cleveland.
Whitman & Barnes Co., St. Catharines.

Drills, Rock.

Allis-Chalmers-Bullock, Montreal.
Canadian Rand Drill Co., Montreal.
Chicago Pneumatic Tool Co., Chicago.
Jeffrey Mfg. Co., Columbus, Ohio.

Drills, Sensitive.

American Tool Works Co., Cincinnati.
B. F. Barnes Co., Rockford, Ill.
Canada Machinery Agency, Montreal.
Niles-Bement-Pond Co., New York.

Drills, Shop View.

John Bertram & Sons Co., Dundas, Ont.

Drills, Twist.

Baxter, Paterson & Co., Montreal.
Chicago Pneumatic Tool Co., Chicago
Alex. Gibb, Montreal.
A. B. Jardine & Co., Hespeler, Ont.
John Millen & Son, Ltd., Montreal.
Morse Twist Drill and Machine Co., New Bedford, Mass.
Pratt & Whitney Co., Hartford, Conn.
Standard Tool Co., Cleveland.
Whitman & Barnes Co., St. Catharines.

Drying Apparatus of all Kinds.

Canadian Buffalo Forge Co., Montreal.
McEachren Heating and Ventilating Co., Galt.
Sheldon's Limited, Galt
B. F. Sturtevant Co., Hyde Park, Mass

Dry Sand and Loam Facing.

Buffalo Foundry Supply Co., Buffalo.
Hamilton Facing Mill Co., Hamilton.

Dump Cars.

Canada Foundry Co., Limited, Toronto
Dominion Henderson Bearings, Niagara Falls, Ont.
Hamilton Facing Mill Co., Hamilton.

Jenckes Machine Co., Sherbrooke, Que
Link-Belt Eng. Co., Philadelphia.
John McDougall Caledonian Iron Works Co., Montreal.
Miller Bros. & Toms, Montreal.
Waterous Engine Co., Brantford.

Dust Separators.

Canadian Buffalo Forge Co., Montreal.
McEachren Heating and Ventilating Co., Galt.
Sheldon's Limited, Galt.

Dynamos.

Allis-Chalmers-Bullock, Montreal.
Canadian General Electric Co., Toronto.
Canadian Westinghouse Co., Hamilton.
D'Olier Engineering Co., New York.
Hall Engineering Works, Montreal, Que.
John Millen & Son, Ltd., Montreal.
Packard Electric Co., St. Catharines.
H. W. Petrie, Toronto.
B. F. Sturtevant Co., Hyde Park, Mass.
T. & H. Electric Co., Hamilton.
Volta Electric Repair Works, Toronto.

Dynamos—Turbine Driven.

Kerr-Turbine Co., Wellsville, N.Y.

Economizer, Fuel.

Canadian Buffalo Forge Co., Montreal.
Dominion Henderson Bearings, Niagara Falls, Ont.
McEachren Heating and Ventilating Co., Galt, Ont.
B. F. Sturtevant Co., Hyde Park, Mass.

Electrical Instruments.

Canadian Westinghouse Co., Hamilton.

Electrical Steel.

Baxter, Paterson & Co., Montreal.

Electrical Supplies.

Canadian General Electric Co., Toronto.
Canadian Westinghouse Co., Hamilton.
London Mach. Tool Co., Hamilton, Ont.
John Millen & Son, Ltd., Montreal.
Packard Electric Co., St. Catharines.
T. & H. Electric Co., Hamilton.
Volta Electric Repair Works, Toronto.

Electrically Driven Tools and Machinery.

AmericanTool Works Co., Cincinnati.
Baxter, Paterson & Co., Monteal.

Electrical Repairs

Canadian Westinghouse Co., Hamilton.
T. & H. Electric Co., Hamilton.
Volta Electric Repair Works, Toronto.

Elevator Buckets.

Jeffrey Mfg. Co., Columbus, Ohio.

Emery and Emery Wheels.

Baxter, Paterson & Co., Montreal.
Dominion Foundry Supply Co., Montreal
Hamilton Facing Mill Co., Hamilton.

Emery Wheel Dressers.

Baxter, Paterson & Co., Montreal.
Canada Machinery Agency, Montreal.
Diamond Saw & Stamping Works, Buffalo
Dominion Foundry Supply Co., Montreal
Hamilton Facing Mill Co., Hamilton.
John Millen & Son, Ltd., Montreal.
H. W. Petrie, Toronto.
Standard Tool Co., Cleveland.

Emery Wheel Dressers, Cutters.

Diamond Saw & Stamping Works, Buffalo.

Engineers and Contractors.

Canada Foundry Co., Limited, Toronto.
Canadian White Co. Montreal.
D'Olier Engineering Co., New York.
Hall Engineering Works, Montreal.
Jenckes Machine Co., Sherbrooke, Que.
Laurie Engine & Machine Co., Montreal.
Link-Belt Co., Philadelphia.
John McDougall, Caledonian Iron Works Co., Montreal.
Miller Bros. & Toms Montreal.
Robb Engineering Co., Amherst, N.S.
The Smart-Turner Mach. Co., Hamilton.

Engineers' Supplies.

Baxter, Paterson & Co., Montreal.
Hall Engineering Works, Montreal.
Rice Lewis & Son, Toronto.

Engines, Gas and Gasoline.

Baxter, Paterson & Co., Montreal.
Canada Machinery Agency, Montreal.
The Canadian Fairbanks Co., Montreal.
Canadian McVicker Engine Co., Galt.
Dominion Henderson Bearings, Niagara Falls, Ont.
Economic Heat, Light and Power Supply Co., Toronto.
W. Gillespie, Toronto.
The Goldie & McCulloch Co., Galt, Ont.
H. W. Petrie, Toronto.
The Smart-Turner Mach. Co., Hamilton.
Sylvester Mfg. Co., Lindsay, Ont.

Engines, Steam.

Allis-Chalmers-Bullock, Montreal
Canada Machinery Agency, Montreal.
Canadian Buffalo Forge Co., Montreal.
D'Olier Engineering Co., New York.
The Goldie & McCulloch Co., Galt, Ont.

E. Leonard & Sons, London, Ont.
Jenckes Machine Co., Sherbrooke, Que.
Laurie Engine & Machine Co., Montreal.
John McDougall Caledonian Iron Works, Montreal.
Miller Bros. & Toms, Montreal.
Robb Engineering Co., Amherst, N.S.
Sheldons Limited, Galt.
The Smart-Turner Mach. Co., Hamilton.
B. F. Sturtevant Co., Hyde Park, Mass.
Waterous Engine Works Co., Brantford.

Engravers.

Toronto Engraving Co., Toronto.

Exhaust Heads.

Canadian Buffalo Forge Co., Montreal.
McEachren Heating and Ventilating Co., Galt, Ont.
Sheldon's Limited, Galt, Ont.
B. F. Sturtevant Co., Hyde Park, Mass.

Expanded Metal.

Expanded Metal and Fireproofing Co. Toronto

Expanders.

A. B. Jardine & Co., Hespeler, Ont.

Pacing.

Detroit Foundry Supply Co., Detroit.

Fans, Electric.

Canadian Buffalo Forge Co., Montreal.
Canadian General Electric Co., Toronto
Canadian Westinghouse Co., Hamilton.
McEachren Heating and Ventilating Co., Galt, Ont.
Sheldon's Limited Galt, Ont.
The Smart-Turner Mach. Co., Hamilton.
B. F. Sturtevant Co., Hyde Park, Mass.

Pans, Exhaust.

Canadian Buffalo Forge Co., Montreal.
Detroit Foundry Supply Co., Detroit.
Dominion Foundry Supply Co., Montreal
Hamilton Facing Mill Co., Hamilton.
McEachren Heating and Ventilating Co., Galt.
Sheldon's Limited, Galt.
B. F. Sturtevant Co., Hyde Park, Mass.

Feed Water Heaters.

John McDougall, Caledonian Iron Works Co., Montreal.
The Smart-Turner Mach. Co., Hamilton

Files and Rasps.

Baxter, Paterson & Co., Montreal.
John Millen & Son, Ltd., Montreal.
Rice Lewis & Son, Toronto.
Nicholson File Co., Providence, R.I.
H. W. Petrie, Toronto.
Whitman & Barnes Co., St. Catharines.

Fillet, Pattern.

Baxter, Paterson & Co., Montreal.
Buffalo Foundry Supply Co., Buffalo.
Detroit Foundry Supply Co., Detroit.
Dominion Foundry Supply Co., Montreal
Hamilton Facing Mill Co., Hamilton.

Fire, Apparatus.

Waterous Engine Works Co., Brantford.

Fire Brick and Clay.

Baxter Paterson & Co., Montreal.
Buffalo Foundry Supply Co., Buffalo.
Detroit Foundry Supply Co., Detroit.
Dominion Foundry Supp'y Co., Montreal
Hamilton Facing Mill Co., Hamilton.
J. W. Paxson Co., Philadelphia.

Flash Lights.

Berlin Electrical Mfg. Co., Toronto.

Flour Mill Machinery.

Allis-Chalmers-Bullock, Montreal.
The Goldie & McCulloch Co., Galt, Ont.
John McDougall, Caledonian Iron Works Co., Montreal.
Miller Bros & Toms, Montreal.

Flue Rollers.

Chicago Pneumatic Tool Co., Chicago.

Forges.

Canada Foundry Co., Limited, Toronto
Canadian Buffalo Forge Co., Montreal.
Hamilton Facing Mill Co., Hamilton.
H. W. Petrie, Toronto.
Sheldon's Limited, Galt, Ont.
B. F. Sturtevant Co., Hyde Park, Mass.

Forgings, Drop.

John McDougall, Caledonian Iron Works. Co. Montreal.
H. W. Petrie, Toronto.
St. Lawrence Supply Co., Montreal.
Whitman & Barnes Co., St. Catharines

Forgings, Light & Heavy.

The Canada Forge Co., Welland.

Forging Machinery.

John Bertram & Sons Co., Dundas, Ont.
London Mach. Tool Co., Hamilton, Ont.
National Machinery Co., Tiffin, Ohio
Niles-Bement-Pond Co., New York.

Founders.

John McDougall, Caledonian Iron Works Co., Montreal
R. McDougall Co., Galt, Ont.

Niagara Falls Machine & Foundry Co., Niagara Falls, Ont.
The Smart-Turner Mach. Co., Hamilton.

Foundry Equipment.

Detroit Foundry Supply Co., Detroit.
Hamilton Facing Mill Co., Mill.
Hanna Engineering Works, Chicago.

Foundry Facings.

Buffalo Foundry Supply Co., Buffalo.
Detroit Foundry Supply Co., Detroit.
Hamilton Facing Mill Co., Hamilton.
J. W. Paxson Co., Philadelphia, Pa.

Friction Clutch Pulleys, etc.

The Dodge Mfg. Co., Toronto.
The Goldie & McCulloch Co., Galt.
Link-Belt Co., Philadelphia.
W. B. McLean & Co., Montreal.
Miller Bros. & Toms, Montreal.
Sylvester Mfg. Co., Lindsay, Ont.

Furnaces.

Detroit Foundry Supply Co., Detroit.
Dominion Foundry Supply Co., Montreal
Hamilton Facing Mill Co., Hamilton.

Galvanizing

Canada Metal Co., Toronto.

Gas Blowers and Exhausters.

Canadian Buffalo Forge Co., Montreal.
McEachren Heating and Ventilating Co., Galt, Ont.
D'Olier Engineering Co., New York.
Sheldons Limited, Galt.
B. F. Sturtevant Co., Hyde Park, Mass.

Gas Plants, Suction.

Baxter Paterson & Co., Montreal.
Dominion Henderson Bearings, Niagara Falls, Ont.
Economic Heat, Light and Power Supply Co., Toronto.
W. Gillespie, Toronto.

Gauges, Standard.

Brown & Sharpe, Providence, R.I.
W. B. McLean & Co., Montreal.
Pratt & Whitney Co., Hartford, Conn.

Gearing.

W. B. McLean & Co., Montreal.

Gear, Cutting Machinery.

Becker-Brainard Milling Mach. Co., Hyde Park, Mass.
Bickford Drill & Tool Co., Cincinnati.
Brown & Sharpe, Providence, R.I.
London Mach. Tool Co., Hamilton.
R. McDougall Co., Galt.
Niles-Bement-Pond Co., New York.
H. W. Petrie, Toronto.
Pratt & Whitney Co., Hartford, Conn.
Williams & Wilson, Montreal.

Gears, Angle.

Brown & Sharpe, Providence, R.I.
Chicago Pneumatic Tool Co., Chicago.
Laurie Engine & Machine Co., Montreal.
John McDougall, Caledonian Iron Works Co., Montreal.
Miller Bros. & Toms, Montreal.
Waterous Engine Co., Brantford.

Gears, Reducing.

Brown & Sharpe, Providence, R.I.
Chicago Pneumatic Tool Co., Chicago.
John McDougall, Caledonian Iron Works Co., Montreal.
Miller Bros. & Toms, Montreal.

Generators, Electric.

Allis-Chalmers-Bullock, Limited, Montreal
Canadian General Electric Co., Toronto
Canadian Westinghouse Co., Hamilton.
D'Olier Engineering Co., New York.
Hall Engineering Works, Montreal.
H. W. Petrie, Toronto.
B. F. Sturtevant Co., Hyde Park, Mass.
Toronto & Hamilton Electric Co., Hamilton.

Generators, Gas.

H. W. Petrie, Toronto.

Glass Bevelling Machines.

Pendrith Machinery Co., Toronto.

Graphite Paints.

P. D. Dods & Co., Montreal.

Graphite.

Detroit Foundry Supply Co., Detroit.
Dominion Foundry Supply Co., Montreal
Hamilton Facing Mill Co., Hamilton.

Grinders, Automatic Knife.

W. H. Banfield & Son, Toronto.

Grinders, Centre.

Niles-Bement-Pond Co., New York.
H. W. Petrie, Toronto.

Grinders, Cutter.

Becker-Brainard Milling Mach. Co., Hyde Park, Mass.
Brown & Sharpe, Providence, R.I.
John Millen & Son, Ltd., Montreal.
Pratt & Whitney Co., Hartford, Conn.
The Stevens Co., Galt, Ont.

Grinders, Tool.

Armstrong Bros. Tool Co., Chicago.
B. F. Barnes Co., Rockford, Ill.
Brown & Sharpe, Providence, R.I.
H. W. Petrie, Toronto.
The Stevens Co., Galt, Ont.
Williams & Wilson, Montreal.

Grinding Machines.

Brown & Sharpe, Providence, R.I.
The Canadian Fairbanks, Montreal.
Niles-Bement-Pond Co., New York.
Norton Company, Worcester, Mass.
H. W. Petrie, Toronto.

Grinding and Polishing Machines.

The Canadian Fairbanks Co., Montreal.
Dominion Henderson Bearings, Niagara Falls, Ont.
John Millen & Son, Ltd., Montreal.
Miller Bros. & Toms, Montreal.
Niles-Bement-Pond Co., New York.
H. W. Petrie, Toronto.
Pendrith Machinery Co., Toronto.

Grinding Wheels (Alundum)

Norton Company, Worcester, Mass.

Hack Saws.

Baxter, Paterson & Co., Montreal.
Canada Machinery Agency, Montreal.
The Canadian Fairbanks Co., Montreal.
John Millen & Son, Ltd, Montreal.
Niles-Bement-Pond Co., New York.
W. B. McLean & Co., Montreal.
H. W. Petrie, Toronto.
West Haven Mfg. Co., New Haven, Conn.
Williams & Wilson, Montreal.

Hack Saw Frames

Baxter, Paterson & Co., Montreal.
West Haven Mfg. Co., New Haven, Conn.

Hammers, Drop.

London Mach. Tool Co., Hamilton, Ont.
Miller Bros. & Toms Montreal.
Niles-Bement-Pond Co., New York.

Hammers, Rand—All Kinds.

Whitman & Barnes Co., St. Catharines.

Hammers, Steam.

John Bertram & Sons Co. Dundas, Ont.
London Mach. Tool Co., Hamilton, Ont.
Niles-Bement-Pond Co., New York.

Hangers.

Dominion Henderson Bearings, Niagara Falls, Ont.
The Goldie & McCulloch Co., Galt.
Miller Bros. & Toms, Montreal.
The Smart-Turner Mach. Co., Hamilton.
Waterous Engine Co., Brantford.
Pendrith Machinery Co., Toronto.

Hardware Specialties

Globe Machine & Stamping Co., Cleveland Ohio.

Heating Apparatus.

Canadian Buffalo Forge Co., Montreal
McEachren Heating and Ventilating Co., Galt, Ont.
Sheldon's Limited, Galt.
B. F. Sturtevant Co., High Park, Mass.

Hoisting and Conveying Machinery.

Allis-Chalmers-Bullock, Limited, Montrea
Jenckes Machine Co., Sherbrooke, Que.
Link-Belt Co., Philadelphia.
Miller Bros. & Toms, Montreal.
Niles-Bement-Pond Co., New York.
The Smart-Turner Mach. Co., Hamilton.
Waterous Engine Co., Brantford.

Hoists, Pneumatic.

Canadian Rand Drill Co., Montreal.
Chicago Pneumatic Tool Co. Chicago.
Dominion Foundry Supply Co., Montreal
Hamilton Facing Mill Co., Hamilton.

Hose.

Baxter, Paterson & Co., Montreal.

Hose, Air.

Canadian Rand Drill Co., Montreal.
Canadian Westinghouse Co., Hamilton.
Chicago Pneumatic Tool Co. Chicago.

Hose Couplings.

Canadian Rand Drill Co., Montreal.
Canadian Westinghouse Co., Hamilton.
Chicago Pneumatic Tool Co., Chicago.
Whitman & Barnes Co., St. Catharines.

Hose, Steam.

Allis-Chalmers-Bullock, Montreal.
Canadian Rand Drill Co., Montreal.
Canadian Westinghouse Co., Hamilton.

Hydraulic Accumulators.

Niles-Bement-Pond Co., New York.
The Smart-Turner Mach. Co., Hamilton.

Hydraulic Machinery.

Allis-Chalmers-Bullock, Montreal

John McDougall, Caledonian Iron Works Co., Montreal.
Niles-Bement-Pond Co., New York.

India Oil Stones.

Norton Company, Worcester, Mass.

Indicators, Speed.

L. S. Starrett Co., Athol, Mass.

Injectors.

Canada Foundry Co., Toronto.
The Canadian Fairbanks Co., Montreal.
Rice Lewis & Son, Toronto.
Penberthy Injector Co., Windsor, Ont.

Iron Cements.

Buffalo Foundry Supply Co., Buffalo.
Detroit Foundry Supply Co., Detroit.
Hamilton Facing Mill Co., Hamilton.

Iron Filler.

Buffalo Foundry Supply Co., Buffalo.
Detroit Foundry Supply Co., Detroit.
Dominion Foundry Supply Co., Montreal.
Hamilton Facing Mill Co., Hamilton.

Iron Tools.

John Bertram & Sons Co., Dundas, Ont.
A. B. Jardine & Co., Hespeler, Ont.
London Mach. Tool Co., Hamilton.
R. McDougall Co., Galt.
H. W. Petrie, Toronto.
The Stevens Co., Galt.

Key-Seating Machines.

B. F. Barnes Co., Rockford, Ill.
Niles-Bement-Pond Co., New York.

Lace Leather.

Baxter, Paterson & Co., Montreal.
Sadler & Haworth, Montreal.

Lamps, Arc and Incandescent.

Canadian General Electric Co., Toronto.
Canadian Westinghouse Co., Hamilton.
The Packard Electric Co., St. Catharines.

Lathe Dogs.

Armstrong Bros., Chicago.
Baxter, Paterson & Co., Montreal.
Pratt & Whitney Co., Hartford, Conn.

Lathes, Engine.

American Tool Works Co., Cincinnati.
B. F. Barnes Co., Rockford, Ill.
Baxter, Paterson & Co., Montreal.
John Bertram & Sons Co., Dundas, Ont.
Canada Machinery Agency, Montreal.
The Canadian Fairbanks Co., Montreal.
London Mach. Tool Co., Hamilton, Ont.
Niles-Bement-Pond Co., New York.
R. McDougall Co., Galt, Ont.
H. W. Petrie, Toronto.
Pratt & Whitney Co., Hartford, Conn.
The Stevens Co., Galt, Ont.

Lathes, Foot-Power.

American Tool Works Co., Cincinnati.
B. F. Barnes Co., Rockford, Ill.

Lathes, Screw Cutting.

B. F. Barnes Co., Rockford, Ill.
Baxter, Paterson & Co., Montreal.
Niles-Bement-Pond Co., New York.

Lathes, Automatic, Screw-Threading.

John Bertram & Sons Co., Dundas, Ont
London Mach. Tool Co., Hamilton, Ont
Pratt & Whitney Co., Hartford, Conn.

Lathes, Bench.

B. F. Barnes Co., Rockford, Ill.
London Mach. Tool Co., London, Ont.
Pratt & Whitney Co., Hartford, Conn.

Lathes, Turret.

American Tool Works Co., Cincinnati.
Baxter, Paterson & Co., Montreal.
John Bertram & Sons Co., Dundas, Ont.
Canada Machinery Agency, Montreal.
London Mach. Tool Co., Hamilton, Ont.
Niles-Bement-Pond Co., New York.
R. McDougall Co., Galt, Ont.
The Pratt & Whitney Co., Hartford, Conn.
The Stevens Co., Galt, Ont.
Warner & Swasey Co., Cleveland O.

Lathes, Buffing and Polishing.

Pendrith Machinery Co., Toronto.

Leather Belting.

Baxter, Paterson & Co., Montreal.
Canada Machinery Agency, Montreal.
The Canadian Fairbanks Co., Montreal
The Dodge Mfg. Co., Toronto.

Lime Stone Flux.

Hamilton Facing Mill Co., Hamilton.

Locomotives, Air.

Canadian Rand Drill Co., Montreal.

Locomotives, Electrical.

Canadian Westinghouse Co., Hamilton
Jeffrey Mfg. Co., Columbus, Ohio.

Locomotives, Steam.

Canada Foundry Co., Toronto.
Canadian Rand Drill Co., Montreal

Lubricating Plumbago.

Detroit Foundry Supply Co., Detroit.
Dominion Foundry Supply Co., Montreal
Hamilton Facing Mill Co., Hamilton.

Lumber Dry Kilns.

Canadian Buffalo Forge Co., Montreal
McEachren Heating and Ventilating Co., Galt, Ont.
H. W. Petrie, Toronto.
Sheldon's Limited, Galt, Ont.
B. F. Sturtevant Co., Hyde Park, Mass.

Machinery Dealers.

Baxter, Paterson & Co., Montreal.
Canada Machinery Agency, Montreal.
The Canadian Fairbanks Co., Montreal.
Dominion Henderson Bearings, Limited, Toronto.
Miller Bros. & Toms, Montreal.
H. W. Petrie, Toronto.
The Smart-Turner Mach. Co., Hamilton.
St. Lawrence Supply Co., Montreal.
Williams & Wilson, Montreal.

Machinery Designers.

Dominion Henderson Bearings, Niagara Falls, Ont.

Machinists.

W. H. Banfield & Sons, Toronto.
The Dodge Mfg. Co., Toronto.
Dominion Henderson Bearings, Niagara Falls, Ont.
Hall Engineering Works, Montreal.
Link-Belt Co., Philadelphia.
John McDougall, Caledonian Iron Works Co., Montreal.
Miller Bros. & Toms, Montreal.
Pendrith Machinery Co., Toronto.
Robb Engineering Co., Amherst, N.S.
The Smart-Turner Mach. Co., Hamilton.
Waterous Engine Co., Brantford.

Machinists' Small Tools.

Armstrong Bros., Chicago.
Brown & Sharpe, Providence, R.I.
Butterfield & Co., Rock Island, Que.
Diamond Saw and Stamping Works, Buffalo, N.Y.
Rice Lewis & Son, Montreal.
Pratt & Whitney Co., Hartford, Conn.
Standard Tool Co., Cleveland.
L. S. Starrett Co., Athol, Mass.
Williams & Wilson, Montreal.
Whitman & Barnes Co., St. Catharines.

Mailing Weights.

A. B. Jardine & Co., Hespeler, Ont.

Malleable Flask Clamps.

Buffalo Foundry Supply Co., Buffalo.

Mallet, Rawhide and Wood.

Buffalo Foundry Supply Co., Buffalo.
Detroit Foundry Supply Co., Detroit.

Mandrels.

Brown & Sharpe, Providence, R.I.
A. B. Jardine & Co., Hespeler, Ont.
Miller Bros. & Toms, Montreal.
The Pratt & Whitney Co., Hartford, Conn.
Standard Tool Co., Cleveland.
Whitman & Barnes Co., St. Catharines

Measuring Machines.

The Pratt & Whitney Co., Hartford, Conn

Mechanical Draft.

Canadian Buffalo Forge Co., Montreal.
H. W. Petrie, Toronto.
Sheldon's Limited, Galt.
B. F. Sturtevant Co., Hyde Park, Mass.

Metal Separators.

Toronto and Hamilton Electric Co., Hamilton.

Metallic Paints.

P. D. Dods & Co., Montreal.

Meters, Electrical.

Canadian Westinghouse Co., Hamilton.

Mill Machinery.

Baxter, Paterson & Co., Montreal.
The Dodge Mfg. Co., Toronto.
The Goldie & McCulloch Co., Galt, Ont.
Jenckes Machine Co., Sherbrooke, Que.
John McDougall, Caledonian Iron Works Co., Montreal.
H. W. Petrie, Toronto.
Robb Engineering Co., Amherst, N.S.
Waterous Engine Co., Brantford.
Williams & Wilson, Montreal.

Milling Attachments.

Becker-Brainard Milling Machine Co. Hyde Park, Mass.
John Bertram & Sons Co., Dundas, Ont.
Brown & Sharpe, Providence, R.I.
Niles-Bement-Pond Co., New York.
Pratt & Whitney, Hartford, Conn.
The Stevens Co., Galt, Ont.

Milling Machines, Horizontal.

Baxter, Paterson & Co., Montreal.
Becker-Brainard Milling Machine Co., Hyde Park, Mass.
John Bertram & Sons Co., Dundas, Ont.
Brown & Sharpe, Providence, R.I.
John Bertram & Sons Co., Dundas, Ont.
Canada Machinery Agency, Montreal.
London Mach. Tool Co., Hamilton, Ont.
Niles-Bement-Pond Co., New York.
Pratt & Whitney, Hartford, Conn.

Milling Machines, Plain.

American Tool Works Co., Cincinnati.
Becker-Brainard Milling Machine Co. Hyde Park, Mass.
Brown & Sharpe, Providence, R.I.
John Bertram & Sons Co., Dundas, Ont.
Canada Machinery Agency, Montreal.
The Canadian Fairbanks Co., Montreal.
London Mach. Tool Co., Hamilton, Ont.
Niles-Bement-Pond Co., New York.
H. W. Petrie, Toronto.
Pratt & Whitney Co., Hartford, Conn.
Williams & Wilson, Montreal.

Milling Machines, Universal.

American Tool Works Co., Cincinnati.
Becker-Brainard Milling Machine Co., Hyde Park, Mass.
Brown & Sharpe, Providence, R.I.
John Bertram & Sons Co., Dundas, Ont.
Canada Machinery Agency, Montreal.
The Canadian Fairbanks Co., Montreal.
London Mach. Tool Co., Hamilton, Ont.
Niles-Bement-Pond Co., New York.
H. W. Petrie, Toronto.
Williams & Wilson, Montreal.

Milling Machines, Vertical.

Becker-Brainard Milling Machine Co., Hyde Park, Mass.
Brown & Sharpe, Providence, R.I.
John Bertram & Sons Co., Dundas, Ont.
Canada Machinery Agency, Montreal.
London Mach. Tool Co., Hamilton, Ont.
Niles-Bement-Pond Co., New York.

Milling Tools.

Wm. Abbott, Montreal.
Becker-Brainard Milling Machine Co., Hyde Park, Mass.
Brown & Sharpe, Providence, R.I.
Dominion Henderson Bearings, Niagara Falls, Ont.
Geometric Tool Co., New Haven, Conn.
London Mach. Tool Co., Hamilton, Ont.
Pratt & Whitney Co., Hartford, Conn.
Standard Tool Co., Cleveland.
The Stevens Co., Galt, Ont.

Mining Machinery.

Allis-Chalmers-Bullock, Limited, Montreal
Canadian Rand Drill Co., Montreal.
Chicago Pneumatic Tool Co., Chicago.
The Dodge Mfg. Co., Toronto.
Jeffrey Mfg., Co., Columbus, Ohio.
Jenckes Machine Co., Sherbrooke, Que.
Laurie Engine & Machine Co., Montreal.
John McDougall, Caledonian Iron Works Co., Montreal.
Miller Bros. & Toms, Montreal.
T. & H. Electric Co. Hamilton.

Mixing Machines, Dough.

Pendrith Machinery Co., Toronto.

Mixing Machines, Special.

Pendrith Machinery Co., Toronto.

Model Tools.

Dominion Henderson Bearings, Niagara Falls Ont
Globe Machine & Stamping Co, Cleveland, Ohio.
The Stevens Co., Galt, Ont.
Wells Pattern and Model Works, Toronto

Motors, Electric.

Allis-Chalmers-Bullock Limited, Montreal
Canadian General Electric Co., Toronto.
Canadian Westinghouse Co., Hamilton.
Hall Engineering Works, Montreal.
The Packard Electric Co., St. Catharines.
B. F. Sturtevant Co., Hyde Park, Mass.
T. & H. Electric Co., Hamilton.

Motors, Air.

Canadian Rand Drill Co., Montreal.
Chicago Pneumatic Tool Co., Chicago.

Molders' Supplies.

Buffalo Foundry Supply Co., Buffalo.
Detroit Foundry Supply Co., Detroit.
Dominion Foundry Supply Co., Montreal
Hamilton Facing Mill Co., Hamilton

Molders' Tools.

Buffalo Foundry Supply Co., Buffalo.
Detroit Foundry Supply Co., Detroit.
Dominion Foundry Supply Co., Montreal
Hamilton Facing Mill Co., Hamilton.

Molding Machines.

Buffalo Foundry Supply Co., Buffalo.
Hamilton Facing Mill Co., Hamilton.
J. W. Paxson Co., Philadelphia, Pa.

Molding Sand.

T. W. Barnes, Hamilton.
Buffalo Foundry Supply Co., Buffalo.
Detroit Foundry Supply Co., Detroit.
Dominion Foundry Supply Co., Montreal
Hamilton Facing Mill Co., Hamilton.

Nippers, Stay Bolt.

Chicago Pneumatic Tool Co., Chicago.

Nut Tappers.

John Bertram & Sons Co., Dundas, Ont
A. B. Jardine & Co., Hespeler.
London Mach. Tool Co., Hamilton.
National Machinery Co., Tiffin, Ohio.

Oatmeal Mill Machinery.

The Goldie & McCulloch Co., Galt

Oils, Core.

Buffalo Foundry Supply Co. Buffalo.
Dominion Foundry Supply Co, Montreal
Hamilton Facing Mill Co., Hamilton.

Painting Machines, Pneumatic.

Chicago Pneumatic Tool Co., Chicago.

Patent Solicitors.

Hanbury A. Budden, Montreal.
Fetherstonhaugh & Blackmore, Montreal
Fetherstonhaugh & Co., Montreal.
Marion & Marion, Montreal.
Ridout & Maybee, Toronto.

Patterns.

John Carr, Hamilton.
The Dodge Mfg. Co., Toronto.
Hamilton Pattern Works, Hamilton.
John McDougall, Caledonian Iron Works Co., Montreal.
Miller Bros. & Toms Montreal.
The Stevens Co., Galt, Ont.
Wells' Pattern and Model Works, Toronto.

Pipe Cutting and Threading Machines.

Butterfield & Co., Rock Island, Que.
Canada Machinery Agency, Montreal.
A. B. Jardine & Co., Hespeler, Ont.
London Mach. Tool Co., Hamilton, Ont.
R. McDougall Co., Galt, Ont.
Niles-Bement-Pond Co., New York.

Pipe, Municipal.

Pacific Coast Pipe Co., Vancouver, B.C.

Pipe, Waterworks.

Pacific Coast Pipe Co., Vancouver, B.C

Pipe, Wooden.

Pacific Coast Pipe Co., Vancouver, B.C.

Planers, Standard.

American Tool Works, Cincinnati.
John Bertram & Sons Co., Dundas, Ont.
Canada Machinery Agency, Montreal.
The Canadian Fairbanks Co., Montreal.
London Mach. Tool Co., Hamilton Ont.
R. McDougall Co., Galt, Ont.
Niles-Bement-Pond Co., New York.
H. W. Petrie, Toronto.
Pratt & Whitney Co., Hartford, Conn.
Williams & Wilson, Montreal.

Planers, Rotary.

John Bertram & Sons Co., Dundas, Ont.
London Mach. Tool Co., Hamilton, Ont.
Niles-Bement-Pond Co., New York.

Power Hack Saw Machines.

Baxter, Paterson & Co., Montreal.
Diamond Saw and Stamping Works, Buffalo, N. Y.

Power Plants.

John McDougall Caledonian Iron Works Co., Montreal.
The Smart-Turner Mach. Co., Hamilton

Planing Mill Fans.

Canadian Buffalo Forge Co., Montreal.
McEachren Heating and Ventilating Co., Galt, Ont.
Sheldon's Limited, Galt, Ont.
B. F. Sturtevant Co., Hyde Park, Mass.

Plug Drillers, Pneumatic.

Canada Foundry Co., Limited, Toronto.
Canadian Rand Drill Co., Montreal.

Plumbago.

Buffalo Foundry Supply Co., Buffalo.
Detroit Foundry Supply Co., Detroit.
Dominion Foundry Supply Co., Montreal
Hamilton Facing Mill Co., Hamilton.
J. W. Paxson Co., Philadelphia, Pa.

Pneumatic Tools.

Allis-Chalmers-Bullock, Montreal.
Canadian Rand Drill Co., Montreal.
Chicago Pneumatic Tool Co., Chicago
Hamilton Facing Mill Co., Hamilton.
Hanna Engineering Works, Chicago.

Presses, Drill.

American Tool Works Co., Cincinnati.
Niles-Bement-Pond Co., New York.

Presses, Drop.

W. H. Banfield & Son, Toronto.
E. W. Bliss Co., Brooklyn, N.Y.
Canada Machinery Agency, Montreal.

Miller Bros. & Toms, Montreal.
Niles-Bement-Pond Co., New York

Presses, Hand.

E. W. Bliss Co., Brooklyn, N.Y.

Presses, Hydraulic.

Jenckes Machine Co., Sherbrooke, Que.
John Bertram & Sons Co., Dundas, Ont.
Laurie Engine & Machine Co., Montreal.
London Mach. Tool Co., Hamilton, Ont.
John McDougall Caledonian Iron Works Co., Montreal.
Miller Bros. & Toms, Montreal.
Niles-Bement-Pond Co., New York.

Presses, Power.

E. W. Bliss Co., Brooklyn, N.Y.
Canada Machinery Agency, Montreal.
Jenckes Machine Co., Sherbrooke, Que.
London Mach. Tool Co., Hamilton, Ont.
John McDougall Caledonian Iron Works Co., Montreal.
Niles-Bement-Pond Co., New York.

Pulp Mill Machinery.

Jeffrey Mfg. Co., Columbus, Ohio.
Jenckes Machine Co., Sherbrooke, Que.
Laurie Engine & Machine Co., Montreal.
John McDougall Caledonian Iron Works Co., Montreal.

Pulleys.

Baxter, Paterson & Co., Montreal.
Canada Machinery Agency, Montreal
The Canadian Fairbanks Co., Montreal.
The Dodge Mfg. Co., Toronto.
Dominion Henderson Bearings, Niagara Falls, Ont
The Goldie & McCulloch Co., Galt.
Jenckes Machine Co., Sherbrooke, Que.
Laurie Engine & Machine Co., Montreal.
Link-Belt Co., Philadelphia.
John McDougall Caledonian Iron Works Co., Montreal.
Miller Bros. & Toms, Montreal.
Pendrith Machinery Co., Toronto.
H. W. Petrie, Toronto.
The Smart-Turner Mach. Co., Hamilton.
Williams & Wilson, Montreal.
Waterous Engine Co., Brantford.

Pumps, Steam.

Allis-Chalmers-Bullock, Limited, Montreal
Canada Foundry Co., Toronto
Canada Machinery Agency, Montreal.
Canadian Buffalo Forge Co., Montreal.
D'Olier Engineering Co., New York.
Dominion Henderson Bearings, Niagara Falls, Ont.
The Goldie & McCulloch Co., Galt.
John McDougall Caledonian Iron Works, Montreal.
R. McDougall Co., Galt, Ont.
H. W. Petrie, Toronto.
St. Lawrence Supply Co., Montreal.
The Smart-Turner Mach. Co., Hamilton.
Sylvester Mfg. Co., Lindsay, Ont.
Waterous Engine Co., Brantford

Pumping Machinery.

Canada Foundry Co., Limited Toronto
Canada Machinery Agency, Montreal.
Canadian Buffalo Forge Co., Montreal.
Canadian Rand Drill Co., Montreal.
Chicago Pneumatic Tool Co., Chicago.
D'Olier Engineering Co., New York
Dominion Henderson Bearings, Niagara Falls, Ont
Hall Engineering Works, Montreal, Que.
Laurie Engine & Machine Co., Montreal.
London Mach. Tool Co., Hamilton, Ont.
John McDougall Caledonian Iron Works Co., Montreal
R. McDougall Co., Galt Ont.
St. Lawrence Supply Co., Montreal.
The Smart-Turner Mach. Co., Hamilton

Punches and Dies.

W. H. Banfield & Sons, Toronto.
Butterfield & Co., Rock Island.
Dominion Henderson Bearings, Niagara Falls, Ont
Globe Machine & Stamping Co.
A. B. Jardine & Co., Hespeler, Ont.
London Mach. Tool Co., Hamilton, Ont.
Miller Bros. & Toms, Montreal.
Pratt & Whitney Co., Hartford, Conn.
H. W. Petrie, Toronto.

Punches, Hand.

Canadian Buffalo Forge Co., Montreal

Punches, Power.

John Bertram & Sons Co., Dundas, Ont.
E. W. Bliss Co., Brooklyn, N.Y.
Canada Machinery Agency, Montreal.
Canadian Buffalo Forge Co., Montreal.
London Mach. Tool Co., Hamilton, Ont.
Niles-Bement-Pond Co., New York.

Punches, Turret.

London Mach. Tool Co., London, Ont
Taylor & McKenzie, Guelph.

Punching Machines, Horizontal.

John Bertram & Sons Co., Dundas, Ont.
London Mach. Tool Co., Hamilton, Ont.
Niles-Bement-Pond Co., New York.

Quartering Machines.

John Bertram & Sons Co., Dundas, Ont.
London Mach. Tool Co., Hamilton, Ont.

Rammers, Bench and Floor.

Buffalo Foundry Supply Co., Buffalo.
Detroit Foundry Supply Co., Detroit.
Hamilton Facing Mill Co., Hamilton.

Rapping Plates.

Detroit Foundry Supply Co., Detroit.
Hamilton Facing Mill Co., Hamilton.

Reamers.

Wm. Abbott, Montreal.
Baxter Paterson & Co., Montreal.
Brown & Sharpe, Providence, R.I.
Butterfield & Co., Rock Island.
Hanna Engineering Works, Chicago
A. B. Jardine & Co., Hespeler, Ont.
John Millen & Son, Ltd., Montreal.
Morse Twist Drill and Machine Co., New Bedford, Mass.
Pratt & Whitney Co., Hartford, Conn.
Standard Tool Co., Cleveland.
The Stevens Co., Galt, Ont.
Whitman & Barnes Co., St. Catharines.

Reamers, Steel Taper.

Butterfield & Co., Rock Island.
Chicago Pneumatic Tool Co., Chicago.
A. B. Jardine & Co., Hespeler, Ont.
John Millen & Son Ltd., Montreal.
Pratt & Whitney Co., Hartford, Conn.
Standard Tool Co., Cleveland.
Whitman & Barnes Co., St. Catharines

Rheostats.

Canadian General Electric Co., Toronto.
Canadian Westinghouse Co., Hamilton
Hall Engineering Works Montreal, Que.
T. & H. Electric Co., Hamilton,

Riddles.

Buffalo Foundry Supply Co., Buffalo
Detroit Foundry Supply Co., Detroit.
Hamilton Facing Mill Co., Hamilton.
J. W. Paxson Co., Philadelphia, Pa.

Riveters, Hydraulic.

London Mach. Tool Co., Hamilton, Ont.
Niles-Bement-Pond Co., New York.

Riveters, Pneumatic.

Hanna Engineering Works, Chicago.

Rolls, Bending.

John Bertram & Sons Co., Dundas, Ont.
London Mach. Tool Co., Hamilton, Ont.
Niles-Bement-Pond Co., New York.

Rotary Converters.

Allis-Chalmers-Bullock, Ltd., Montreal.
Canadian Westinghouse Co., Hamilton
Toronto and Hamilton Electric Co., Hamilton.

Safes.

Baxter, Paterson & Co., Montreal.
The Goldie & McCulloch Co., Galt, Ont

Sand, Bench.

Buffalo Foundry Supply Co., Buffalo.
Detroit Foundry Supply Co., Detroit
Hamilton Facing Mill Co., Hamilton.

Sand Blast Machinery.

Canadian Rand Drill Co., Montreal.
Chicago Pneumatic Tool Co., Chicago.
Detroit Foundry Supply Co., Detroit.
Hamilton Facing Mill Co., Hamilton.
J. W. Paxson Co., Philadelphia, Pa.

Sand, Brass.

Buffalo Foundry Supply Co. Buffalo.
Detroit Foundry Supply Co., Detroit.

Sand Fire.

Detroit Foundry Supply Co., Detroit.

Sand, Heavy, Grey Iron.

Buffalo Foundry Supply Co., Buffalo.
Detroit Foundry Supply Co., Detroit.
Hamilton Facing Mill Co., Hamilton.

Sand, Malleable.

Buffalo Foundry Supply Co., Buffalo,
Detroit Foundry Supply Co., Detroit.
Hamilton Facing Mill Co., Hamilton.

Sand, Medium Grey Iron.

Buffalo Foundry Supply Co., Buffalo.
Detroit Foundry Supply Co., Detroit.
Hamilton Facing Mill Co., Hamilton.

Sand, Plate.

Buffalo Foundry Supply Co. Buffalo.
Detroit Foundry Supply Co., Detroit.
Hamilton Facing Mill Co., Hamilton.

Sand Sifters.

Buffalo Foundry Supply Co., Buffalo.
Detroit Foundry Supply Co., Detroit.
Dominion Foundry Supply Co., Montreal
Hamilton Facing Mill Co., Hamilton.
Hanna Engineering Works, Chicago.

Sand, Stove Pipe.

Buffalo Foundry Supply Co. Buffalo.
Detroit Foundry Supply Co., Detroit.
Hamilton Facing Mill Co., Hamilton.

Saw Gummers.

Canada Machinery Agency, Montreal.
A. B. Jardine & Co., Hespeler, Ont.
Waterous Engine Co., Brantford.

Saw Machines, Power Hack.

Baxter Paterson & Co., Montreal.
Diamond Saw & Stamping Works, Buffalo
West Haven Mfg. Co., New Haven, Conn

Saw Mill Machinery.

Allis-Chalmers-Bullock, Limited, Montreal
Baxter, Paterson & Co., Montreal
Canada Machinery Agency Montreal.
Clark-Demill Co., Hespeler, Ont.
The Dodge Mfg. Co., Toronto.
Dominion Henderson Bearings, Niagara Falls, Ont
Goldie & McCulloch Co., Galt.
Miller Bros. & Toms, Montreal.
H. W. Petrie, Toronto.
Robb Engineering Co., Amherst, N.S.
Waterous Engine Works, Brantford.
Williams & Wilson, Montreal.

Sawing Machines, Metal.

Niles-Bement-Pond Co., New York.
West Haven Mfg. Co., New Haven, Conn.

Saws, Hack.

Baxter Paterson & Co., Montreal.
Canada Machinery Agency, Montreal.
Detroit Foundry Supply Co., Detroit.
Diamond Saw & Stamping Works, Buffalo.
London Mach. Tool Co., Hamilton
Rice Lewis & Son, Toronto.
John Millen & Son, Ltd., Montreal.
L. S. Starrett Co., Athol, Mass.
West Haven Mfg. Co., New Haven, Conn

Saws, Kitchen.

Diamond Saw & Stamping Works, Buffalo

Saws, Power Hack.

Diamond Saw & Stamping Works, Buffalo.
West Haven Mfg. Co., New Haven, Conn.

Screw Cutting Tools

Hart Manufacturing Co., Cleveland, Ohio

Screw Machines, Automatic.

Brown & Sharpe, Providence, R.I.
Canada Machinery Agency, Montreal.
London Mach. Tool Co., Hamilton, Ont.
Pratt & Whitney Co., Hartford, Conn.

Screw Machines, Hand.

Brown & Sharpe, Providence, R.I.
Canada Machinery Agency Montreal.
A. B. Jardine & Co., Hespeler.
London Mach. Tool Co., Hamilton, Ont.
Pratt & Whitney Co., Hartford, Conn.
Warner & Swasey Co., Cleveland, O.

Screw Plates.

Butterfield & Co., Rock Island, Que.
Hart Manufacturing Co. Cleveland, Ohio
A. B. Jardine & Co., Hespeler.

Second-Hand Machinery.

American Tool Works Co., Cincinnati.
Canada Machinery Agency, Montreal,
The Canadian Fairbanks Co., Montreal.
Goldie & McCulloch Co., Galt.
Machinery Exchange, Montreal.
Niles-Bement-Pond Co., New York.
H. W. Petrie, Toronto.
Robb Engineering Co., Amherst, N.S.
Waterous Engine Co., Brantford.
Williams & Wilson, Montreal.

Shafting.

Baxter, Paterson & Co., Montreal.
Canada Machinery Agency, Montreal.
The Canadian Fairbanks Co., Montreal.
The Dodge Mfg. Co., Toronto.
The Goldie & McCulloch Co., Galt, Ont
Jenckes Machine Co., Sherbrooke, Que.
Niles-Bement-Pond Co., New York.
Pendrith Machine Co., Toronto.
H. W. Petrie, Toronto.
Smart-Turner Machine Co., Hamilton.
Union Drawn Steel Co., Hamilton.
Waterous Engine Co., Brantford.

Shapers.

American Tool Works Co., Cincinnati.
John Bertram & Sons Co., Dundas, Ont
Canada Machinery Agency, Montreal.
The Canadian Fairbanks Co., Montreal.
London Mach. Tool Co., Hamilton, Ont.
Niles-Bement-Pond Co., New York.
H. W. Petrie, Toronto.
Pratt & Whitney Co., Hartford, Conn.
Williams & Wilson, Montreal.

Shearing Machine, Bar.

John Bertram & Sons Co., Dundas, Ont.
Canadian Buffalo Forge Co., Montreal.
A. B. Jardine & Co., Hespeler.
London Mach. Tool Co., Hamilton, Ont.
Niles-Bement-Pond Co., New York.

Shears, Bench.

Canadian Buffalo Forge Co., Montreal.

Shears, Continuous.

Canadian Buffalo Forge Co., Montreal.

Shears, Hand.

Canadian Buffalo Forge Co., Montreal.

Shears, Power.

John Bertram & Sons Co., Dundas, Ont
Canada Machinery Agency, Montreal.
Canadian Buffalo Forge Co., Montreal.
A. B. Jardine & Co., Hespeler, Ont.
Niles-Bement-Pond Co., New York.

Sheet Metal Goods

Globe Machine & Stamping Co., Cleveland, Ohio.

Shovels.

Baxter, Paterson & Co., Montreal.
Buffalo Foundry Supply Co., Buffalo.
Detroit Foundry Supply Co., Detroit.
Dominion Foundry Supply Co., Montreal
Hamilton Facing Mill Co., Hamilton.

Shovels, Steam.

Allis-Chalmers-Bullock, Montreal.

Sieves.

Buffalo Foundry Supply Co. Buffalo.
Detroit Foundry Supply Co., Detroit.
Dominion Foundry Supply Co., Montreal
Hamilton Facing Mill Co., Hamilton.

Silver Lead.

Buffalo Foundry Supply Co. Buffalo.
Detroit Foundry Supply Co., Detroit.
Dominion Foundry Supply Co., Montreal
Hamilton Facing Mill Co., Hamilton.

Sleeves, Reducing.

Chicago Pneumatic Tool Co., Chicago.

Slide Rests.

Niles-Bement-Pond Co., New York.

Smoke Connections.

Jenckes Machine Co., Sherbrooke, Que

Smoke Stacks.

Jenckes Machine Co., Sherbrooke, Que.

Snap Flasks

Buffalo Foundry Supply Co., Buffalo.
Detroit Foundry Supply Co., Detroit
Dominion Foundry Supply Co., Montreal
Hamilton Facing Mill Co., Hamilton.

Soapstone.

Buffalo Foundry Supply Co., Buffalo.
Detroit Foundry Supply Co., Detroit.
Dominion Foundry Supply Co., Montreal
Hamilton Facing Mill Co., Hamilton.

Special Machinery.

W. H. Banfield & Sons, Toronto.
Baxter, Paterson & Co., Montreal
John Bertram & Sons Co., Dundas, Ont.
Dominion Henderson Bearings, Niagara Falls Ont
Globe Machine & Stamping Co., Cleveland Ohio.
Hanna Engineering Works, Chicago.
Laurie Engine & Machine Co., Montreal.
London Mach. Tool Co., Hamilton, Ont.
R. McDougall Co., Galt, Ont.
Pendrith Machinery Co., Toronto.
H. W. Petrie, Toronto.
The Smart-Turner Mach. Co., Hamilton
The Stevens Co., Galt.
Waterous Engine Co., Brantford.

Special Machines and Tools.

Dominion Henderson Bearings, Niagara Falls Ont
Pratt & Whitney, Hartford, Conn.
The Stevens Co., Galt, Ont.

Special Manufacturing.

Dominion Henderson Bearings, Niagara Falls, Ont.
Globe Machine & Stamping Co., Cleveland.
Miller Bros. & Toms, Montreal.
Pendrith Machine Co., Toronto.
The Stevens Co., Galt, Ont.

Special Repairs.

Dominion Henderson Bearings, Niagara Falls, Ont.

Speed Changing Countershafts.

The Canadian Fairbanks Co., Montreal
The Dodge Mfg. Co., Toronto.

Spike Machines.

National Machinery Co., Tiffin, O.
The Smart-Turner Mach. Co., Hamilton.

Spray Cans.

Detroit Foundry Supply Co., Detroit.
Dominion Foundry Supply Co., Montreal
Hamilton Facing Mill Co., Hamilton.

Sprue Cutters.

Dominion Foundry Supply Co., Montreal
Detroit Foundry Supply Co., Detroit.
Hamilton Facing Mill Co., Hamilton.

Stamp Mills.

Allis-Chalmers-Bullock, Limited, Montreal

Stampings, Sheet Metal.

Baxter, Paterson & Co., Montreal.

Steam Hot Blast Apparatus

Canadian Buffalo Forge Co., Montreal.
McEachren Heating and Ventilating Co., Galt.
Sheldon's Limited, Galt.
B. F. Sturtevant Co., Hyde Park, Mass.

Steam Plants.

Jenckes Machine Co., Sherbrooke, Que.

Persons addressing advertisers will kindly mention having seen their advertisement in this paper.

Directory of Consulting Engineers, Patent Attorneys, Architects and Contractors.

JOHN S. FIELDING
Mem. Soc. C.E., West Penn., '87
Consulting Engineer
DAMS, MILLS, BRIDGES, MACHINERY
Room 2, 15 Toronto Street, Toronto, Ont.

HANBURY A. BUDDEN
Advocate · Patent Agent.
New York Life Building MONTREAL.
Cable Address, BREVET, MONTREAL.

CONSULTING ENGINEERS
should have their card in this page. It will be read by the manufacturers of Canada :: :: :: ::
CANADIAN MACHINERY
Montreal. Toronto. Winnipeg.

T. Pringle & Son
HYDRAULIC, MILL & ELECTRICAL ENGINEERS
FACTORY & MILL CONSTRUCTION A SPECIALTY.
Coristine Bldg., St. Nicholas St., Montreal.

RODERICK J. PARKE
A.M. Can. Soc. C.E. A.M. Amer. Inst. E.E.
CONSULTING ELECTRICAL ENGINEER
INDUSTRIAL, STEAM AND ELECTRIC POWER PLANTS DESIGNED. TESTS. REPORTS.
51-53 JANES BLDG., TORONTO, CAN.
Long Distance Telephones—Office and Residence.

JAMES C. ROYCE, S.B.
(Mass. Inst. of Tech.)
MECHANICAL AND ELECTRICAL ENGINEER
GAS POWER PLANTS, STEAM POWER PLANTS, OPEN HEARTH STEEL PLANTS, ELECTRIC RAILROADS
Inventions Perfected, Special Examinations and Tests, Reports
51-2 Janes Bldg., Cor. King and Yonge Sts. TORONTO, CAN.
Telephones { Office, Main 4652. Residence, Junc. 81.

OPAL GLASS TILING
FOR WALLS OF
MACHINERY AND POWER HOUSES
Most approved material.
TORONTO PLATE GLASS IMPORTING CO'Y
PLATE AND WINDOW GLASS
135 to 143 Victoria St., · Toronto

CHARLES BRANDEIS,
A. M. CAN. SOC. C.E.
MEM. AMER. ELECTRO-CHEMICAL SOC., ETC.
CONSULTING ENG.NEER
Estimates, Plans and Supervision of Hydraulic and Steam-Electric Light, Power and Railroad Plants, Specifications, Reports, Switchboard Designs, Complete Factory Installations, Electric Equipment of Mines and Electro-Chemical Plants.
Long Distance Telephone, Main 3256.
Cable Address, Brandeis, Montreal.
W. U. Code Univ-Edition.
62-63 Guardian Building MONTREAL

PATENTS *TRADE MARKS AND DESIGNS*
PROCURED IN ALL COUNTRIES
Special Attention given to Patent Litigation
Pamphlet sent free on application.
RIDOUT & MAYBEE *103 BAY STREET TORONTO*

FETHERSTONHAUGH & BLACKMORE
PATENTS
TRADE MARKS.
Liverpool & London & Globe Building, Montreal.
Fetherstonhaugh & Dennison, 9 Toronto St. Toronto. Our reference book for clients "The Prospective Patentee." Also Ottawa, Washington and Winnipeg.

PATENTS PROMPTLY SECURED
We solicit the business of Manufacturers, Engineers and others who realise the advisability of having their Patent business transacted by Experts. Preliminary advice free. Charges moderate. Our Inventor's Adviser sent upon request. Marion & Marion, New York Life Bldg, Montreal; and Washington, D.C., U.S.A.

PATENTS THAT PROTECT
FETHERSTONHAUGH & CO.
Patent Solicitors & Experts
Fred. B. Fetherstonhaugh, M.E., Barrister-at-law and Counsel and expert in Patent Causes. Charles W. Taylor, B.Sc., formerly Examiner in Can. Patent Office.
MONTREAL CAN. LIFE BLDG.
TORONTO HEAD OFFICE, CAN. BANK COMMERCE BLDG.

PRESS CLIPPINGS
About any subject or business. We read nearly every paper in Canada, and can supply you with what the papers have to say about anything you are interested in.
——WRITE FOR TERMS——
CANADIAN PRESS CLIPPING BUREAU
10 Front Street East, · · · TORONTO.

THE ELECTRIC BLUE PRINT CO.
Blue Prints · Positive Blue Prints
Black Prints · Multi-color Prints
Largest and finest plant.
Best finished work in Canada at lowest prices.
40 Hospital St., · MONTREAL

THE DETROIT TESTING LABORATORY
1111 Union Trust Building
DETROIT, MICH.
Experienced Foundry Chemists and Metallurgists. Iron Mixtures for all classes of Castings our Specialty. Reasonable charges for Analyses.

J. A. DeCew
Chemical Engineer
Sun Life Building, MONTREAL
Industrial Plants and Processes sold and installed. Apparatus and Materials for Water Purification. Free Analyses to prospective purchasers.

JULES DE CLERCY, M. E. Gas Engineer
Expert knowledge of all classes of Engines and Producer Plants. If you are thinking of installing a Gas Engine or Producer Plant write me. If your plant is not working satisfactorily I can help you.
413 Dorchester St. · MONTREAL, QUE.

60 YEARS' EXPERIENCE

TRADE MARKS
DESIGNS
COPYRIGHTS &C.
Anyone sending a sketch and description may quickly ascertain our opinion free whether an invention is probably patentable. Communications strictly confidential. HANDBOOK on Patents sent free. Oldest agency for securing patents.
Patents taken through Munn & Co. receive *special notice*, without charge, in the
Scientific American.
A handsomely illustrated weekly. Largest circulation of any scientific journal. Terms, $3 a year; four months, $1. Sold by all newsdealers.
MUNN & Co. 361 Broadway, **New York**
Branch Office, 625 F St., Washington, D. C.

There is only one Typewriter of which this has ever been said: the
REMINGTON
Great and enduring is the merit on which such reputation is built.
THE REMINGTON TYPEWRITER CO., 33 Adelaide St. E., Toronto
AND EVERYWHERE.

DUNTLEY ELECTRIC DRILLS
Franklin Air Compressor. Class O.S.C.
We manufacture our Franklin Compressors in more than 100 styles and sizes.
"BOYER" and "KELLER" Chipping Hammers
CANADIAN PNEUMATIC TOOL CO., LIMITED
TORONTO
MONTREAL
BRANCH OF
CHICAGO
CHICAGO PNEUMATIC TOOL COMPANY
NEW YORK
SALES AGENTS FOR CANADA: H. J. HOLDEN & CO., Limited, MONTREAL
"Little Giant" Drills made in 13 sizes
Our Corliss Valve Drills are the most powerful Air Drills ever made, weight considered.
Chicago Stay Bolt Chucks avoid the squaring of ends. They are very rapid and efficient.
"BOYER" and "KELLER" Riveting Hammers

BROWN & SHARPE MFG. CO.
PROVIDENCE, R.I., U.S.A.
B. & S. Cutters
Their wide reputation for
ACCURACY and RELIABILITY
has been established by the extreme carefulness exercised in their manufacture
The requisite of every B. & S. Cutter is
QUALITY
36 varieties, 3700 sizes kept in stock
APPLY FOR A CUTTER LIST AT
Aikenhead Hardware, Limited - Toronto, Canada
H. W. Petrie - Toronto, Canada
The Canadian Fairbanks Co. - Montreal, Canada
Williams & Wilson - Montreal, Canada
Wood, Vallance & Co. - Hamilton, Canada
Thompson & Sutherland - New Glasgow, Nova Scotia

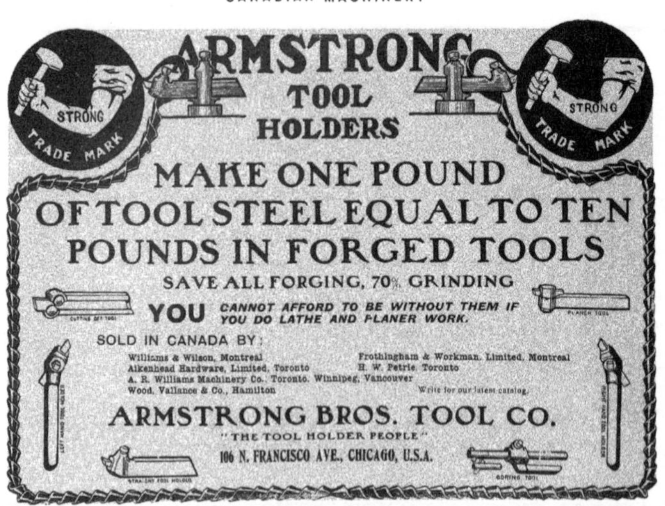

CLASS "N" ADJUSTABLE COLLAPSING TAPS

You can save Taps, Threads, Time and Money by using our Class "N" Adjustable Collapsing Taps. When the required depth of hole is tapped, the chasers automatically recede and the tap can instantly be withdrawn.

This saves the time of backing out, prevents injury to the threads and to the cutting edges of the tap, which gives longer and better service.

This tool is made for use on brass finishing lathes, screw machines and on any turret head

ASK FOR CATALOGUE.

The Geometric Tool Co., New Haven, Conn.

Canadian Agents: WILLIAMS & WILSON, Montreal, Que. **(Westville Station) U.S.A.**

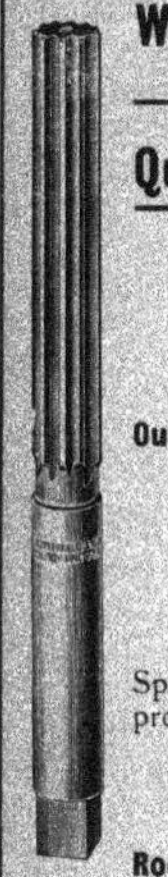

We're Not Much on Talk —but we're Strong on Quality

We don't supply everything—but everything we do supply is 1st class

Our **Specialty** embraces

Reamers
Taps
Dies and
Screw Plates

Special construction and finest steel produce the BUTTERFIELD QUALITY.

BUTTERFIELD & CO.

Rock Island - Canada

Twist Drills

Standard Drills are used by the largest manufacturers, with the best of success. We claim Highest Quality and our goods will verify our claim. If you have never used them send us a trial order. We are prepared to meet any condition of service.

DRILLS WITH LUBRICATING HOLES.

If you want to make a hole at a maximum of speed, and at the same time with the greatest accuracy, use one of our drills with holes from shank to point for conveying oil or other lubricant to the point of the drill. The general advantages of this form of lubricated drill over the ordinary twist drill are the greatly increased speed possible in drilling, the assistance given by the lubricant in the discharge of the chips, its durability and high cutting efficiency. The sizes commonly used are from ⅜ inch to 2½ inches diameter. Our drills have the lubricating hole through the solid metal, this latest improvement in lubricating drills being made solely by us under patents which we own. The special advantage of this drill over the other makes of drills with lubricating holes is that it is composed of one solid piece of steel; there is no part to work loose or wear out, the whole of the drill having the same life. In manufacturing large quantities of a single articles having a drilled hole, one of our drills is invaluable. This is because of its long life, the speed at which it can be run without getting too hot for use, and the time saved by not having to withdraw the drill from the hole to cool it or to remove the chips.

Morse Twist Drill and Machine Co.,

New Bedford, Mass., U.S.A.

Drills,
Reamers,
Cutters,
Chucks,
Taps,
Dies,
Machinery and Machinists' Tools.

SAVE TIME ***SAVE MONEY***

Beardshaw's HIGH SPEED STEEL and DRILLS.

"CONQUEROR" Brand

Once tried always used.

Beardshaw's High Speed Steel in PROFILE SECTIONS require **no** forging. Compare the **saving** in weight with ordinary sections.

All kinds of High-Grade Tool Steel supplied

An 8 ft. bar of
1-in. Vee weighs 8 lbs.
1-in. Fluted weighs 11½ lbs.
1-in. Star weighs 12 lbs.
1-in. Round weighs 21 lbs.
1-in. Octagon weighs 22½ lbs.
1-in. Square weighs 27 lbs.

Write for Catalogue

ALEXANDER GIBB, 13 St. John St., - - MONTREAL

ALLIS-CHALMERS-BULLOCK LIMITED MONTREAL
ELECTRIC PLANTS
ALLIS-CHALMERS-BULLOCK LIMITED MONTREAL
Two of our 75 K.W. direct current generators and auxiliary apparatus forming the lighting plant at Lewis Bros., Limited, wholesale hardware, Montreal. For detailed description of these generators see Bulletin 1049. For a list of our various products see the classified index of this journal.
ALLIS-CHALMERS-BULLOCK, Limited
Head Office and Works, Montreal. District Offices: Toronto, Traders Bank Bldg; Montreal, Sovereign Bank Bldg; Winnipeg, 251 Notre Dame Ave.; Nelson, Josephine St.; Vancouver, 416 Seymour St.; New Glasgow, N.S., Telephone Bldg.

CANADIAN MACHINERY AND MANUFACTURING NEWS

A monthly newspaper devoted to the manufacturing interests, covering in a practical manner the mechanical, power, foundry and allied fields. Published by The MacLean Publishing Company, Limited, Toronto, Montreal, Winnipeg, and London, Eng.

OFFICE OF PUBLICATION: 10 Front St. E., TORONTO

Vol. III. FEBRUARY, 1907 No. 2

CLEVELAND AUTOMATIC MACHINE CO.

CLEVELAND, OHIO, U.S.A.

EASTERN REPRESENTATIVE—J. B. Anderson, 2450 North Thirtieth Street, Philadelphia, Pa.

WESTERN REPRESENTATIVE—H. E. Nunn, 22 Fifth Avenue, Chicago, Ill.

FOREIGN REPRESENTATIVES—Chas. Churchill & Co., London, Manchester, Birmingham, Newcastle-on-Tyne and Glasgow. Messrs. Schuchardt & Schutte, Berlin, St. Petersburg and Stockholm. Alfred H. Schutte, Cologne, Brussels, Liege, Paris, Milan and Bilbao.

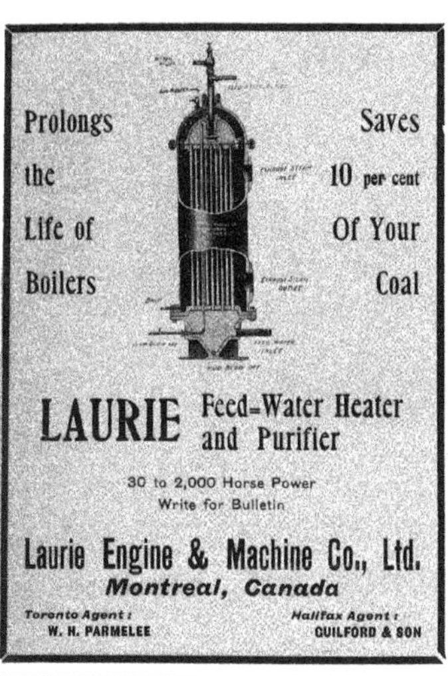

Norton Grinding Wheels Made of Alundum

meet the requirements in **all** kinds of grinding. The purity and uniformity of **ALUNDUM** are under absolute control. Grinding wheels made of it can be accurately duplicated at any time and maintain their standard of work. We are ready to supply all sizes, shapes and grits that modern practice calls for. Let us know something about your work, and we will tell you the grade of wheel suitable for it.

Send for Booklet "Alurdum; Its Invention and Use"

Norton Company

Main Works and General Offices

WORCESTER, MASS.

Alundum Plant, Niagara Falls

CANADIAN AGENTS

The Canadian Fairbanks Co.
Montreal, Toronto, Winnipeg and Vancouver

F. H. Andrews & Son
Quebec, P.Q.

55

Mechanical Draft

An application of fans for securing the highest efficiency of fuel and the greatest steaming capacity of boilers—Fiske.

THERE is nothing new or irrational about Mechanical Draft—it dates back from the first use of bellows, and is best appreciated in the latest and largest power plants.

Buffalo Electric "B" Volume Exhauster

MECHANICAL DRAFT means a positive control over your boilers. No dependence on weather conditions. No fixed limit to boiler capacity. Quick response to demands for steam.

A typical Buffalo Induced Draft Plant.

Left Hand Bottom Horizontal Discharge Fan with Cylinder below shaft engine.

MECHANICAL DRAFT means no tall, expensive chimneys. Perfect combustion of low-grade fuels. Realizes, in connection with fuel economizers, the highest possible boiler efficiency.

MECHANICAL DRAFT has long ceased to be an experiment, but is well defined and accepted engineering practice. The best results are only possible when installations are designed and executed by specialists in this service. Bring your draft problems to us.

CANADIAN BUFFALO FORGE CO., Limited

MONTREAL, QUE.

STURTEVANT BLOWERS
FOR
FOUNDRY CUPOLAS

These blowers have been in use for nearly fifty years and enjoy a world-wide reputation for efficiency and endurance.

B. F. Sturtevant Co.

Boston, Mass.

General Office and Works, Hyde Park, Mass.

NEW YORK PHILADELPHIA CHICAGO
CINCINNATI LONDON

Designers and Builders of Heating, Ventilating, Drying and Mechanical Draft Apparatus; Fan Blowers and Exhausters; Rotary Blowers and Exhausters; Steam Engines, Electric Motors and Generating Sets; Pneumatic Separators, Fuel Economizers; Forges, Exhaust Heads, Steam Traps, Etc.

401

The Canada Chemical Manufacturing Company, Limited

MANUFACTURERS OF

Commercial Quality **Acids and Chemicals** Chemically Pure Quality

ACIDS—Sulphuric, Muriatic, Nitric, Mixed, Acetic, Phosphoric, Hydrofluoric.

CHEMICALS—Salt Cake, Glauber's Salts, Soda Hypo, Silicate, Sulphide, Epsom Salts, Blue Vitriol, Alumina Sulphate, Lime Bisulphite, Nitrite of Iron, C.T.S. and Calcium Acid Phosphate.

Chemical Works and Head Office	Sales Office	Warehouses
LONDON	TORONTO	TORONTO nd MONTREAL

Return Tubular and Locomotive Fire-Box Boilers Tanks, Stacks, Rivetted Pipe.

THE MANITOBA IRON WORKS, Limited, WINNIPEG, MAN.

WIRE ROPE

All Kinds and Sizes, and

Standard and Lang's Patent Lay.

Prices Right   Prompt Shipments

Rope Fittings
Rope Grease

The B. GREENING WIRE CO.
LIMITED
Hamilton, Ont.   **Montreal, Que.**

SEND FOR booklet "X" and bulletins 50, 52, 57, 58

containing facts which prove

RENOLD SILENT CHAIN

to be the correct solution of the maximum power-transmission at minimum cost problem.

LINK-BELT COMPANY

PHILADELPHIA, CHICAGO, INDIANAPOLIS

NEW YORK	PITTSBURG	ST. LOUIS	BOSTON
299 Broadway.	1501 Park Bldg.	Missouri Trust Bldg.	154 Federal St

DENVER	NEW ORLEANS
Lindrooth, Shubart & Co.	Wilmot Machinery Co.

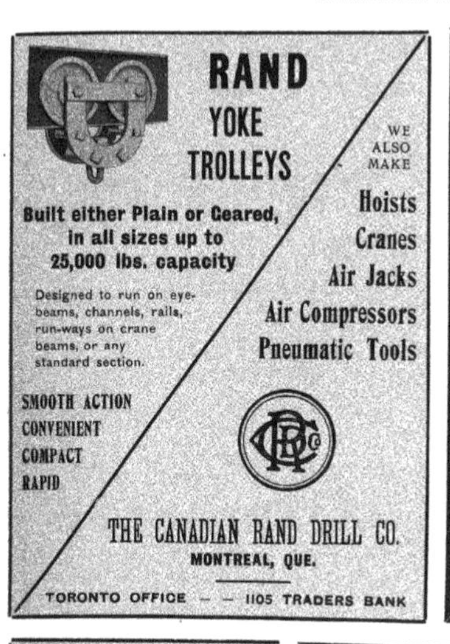
RAND
YOKE
TROLLEYS
Built either Plain or Geared, in all sizes up to 25,000 lbs. capacity
Designed to run on eye-beams, channels, rails, run-ways on crane beams, or any standard section.
SMOOTH ACTION
CONVENIENT
COMPACT
RAPID
WE ALSO MAKE
Hoists
Cranes
Air Jacks
Air Compressors
Pneumatic Tools
THE CANADIAN RAND DRILL CO.
MONTREAL, QUE.
TORONTO OFFICE — — 1105 TRADERS BANK

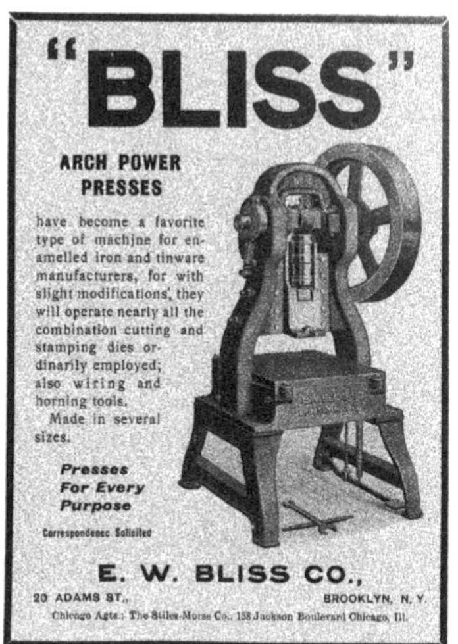
"BLISS"
ARCH POWER PRESSES
have become a favorite type of machine for enamelled iron and tinware manufacturers, for with slight modifications, they will operate nearly all the combination cutting and stamping dies ordinarily employed; also wiring and horning tools.
Made in several sizes.
Presses For Every Purpose
Correspondence Solicited
E. W. BLISS CO.,
20 ADAMS ST., BROOKLYN, N. Y.

SUPERIOR
POWER PRESSES
Many new features embodied in this press. Write for prices and particulars on presses, dies, tools, and special machinery.
W. H. BANFIELD & SONS
120 Adelaide Street West, TORONTO.

The Quality
of
IMPERIAL
CHUCKS
is apparent.
If you saw one you would see the quality.
Sample sent Free
KER & GOODWIN
Manufacturers
BRANTFORD, CANADA

A New Industry in the West

BY GEORGE B. PERRY

WHAT may be classed as a new industry in the West, and one which has been gradually finding field for development as the towns of Western Canada have grown in the past four years, is the manufacture of water pipe from the famous Douglas fir of British Columbia. This material lends itself peculiarly to the construction of stave pipe, because of its wonderful quality, being so clear and straight grained and free from all defects. It is possible to secure a class of timber in B.C. which is absolutely free from knots, pitch pockets, shakes, or other defects common in almost every other class of timber. This perfect timber is used in making machine-banded wood stave and water pipe.

The product, which is being turned out by the factory of the Pacific Coast Pipe Company and the Canada Pipe Company in Vancouver, is finding favor in the entire West, and has even been making its way in Eastern Canada. As a material at once cheap and reliable and satisfactory for water-works construction in towns and cities, the wood stave pipe is recommending itself to those numerous towns in the new West which find themselves with perhaps very moderate means, compelled to instal an entire new system of water supply. The conditions in the West are such that very young communities have to undertake public works of this kind which much older towns in the East have not yet thought of.

Naturally, when such heavy responsibilities are to be undertaken, with their burden of taxation, economy as well as efficiency is a factor of very great importance. In this way, many towns in the new provinces of Western Canada are being enabled, through using wood stave pipe, to instal very complete and high-class water-works systems at very low cost. The number of such towns where modern improvements are demanded increases year by year with the wonderful growth that characterizes the West.

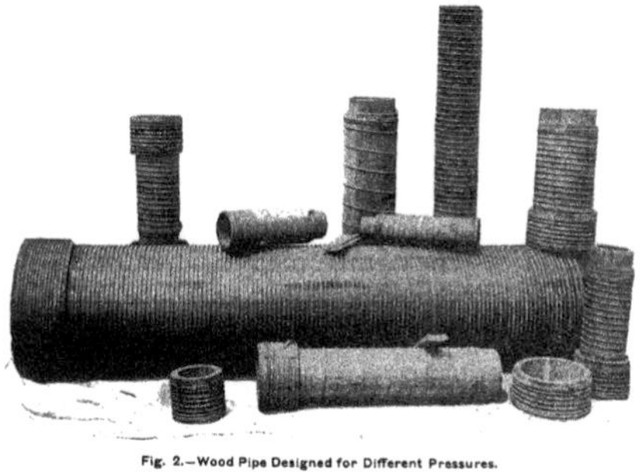

Fig. 1.—Making Wood Stave Pipe in plant of Pacific Coast Pipe Co., Vancouver, B.C.

Machine-banded wood stave pipe has been constructed in the United States for many years. The first factory was in Eastern New York State, and many towns have systems installed with this product many years ago. In the West, the wood stave pipe of machine-banded construction has for many years been made in the State of Washington, where the class of timber is practically the same as that in B.C. It is this splendid material, which makes it possible to turn out such a high-class pipe at a price so much lower than any other pipe made. It is in fact a specialty, and no other timbered region in Canada can ever turn out the class of product which is being made in Vancouver from the Douglas fir.

Fig. 2.—Wood Pipe Designed for Different Pressures.

The machine-banded pipe is made in all sizes from 2 to 24 inches, internal diameter, and is also constructed for all pressures, from mere conduit pipe, suitable for irrigation, up to several hundred feet head. In making the pipe of these different pressures, there is no difference in the quality of the material, the banding being the only change, the construction being exactly similar. In all cases the pipe is positively guaranteed to stand the pressure for which it is supplied.

Many eminent engineers, including some of the most prominent consulting engineers in water-works construction in Eastern Canada, have expressed themselves as heartily in favor of the wood stave pipe. Its use is so extensive now that there is no question of experiment in it, and the fact that it has been used so many years with extreme satisfaction is proof that the claims of the manufacturer are well substantiated.

The Pacific Coast Pipe Company, Limited, have well equipped and very conveniently arranged factories in Vancouver, their headquarters, and many Canadian engineers, and members of the Canadian Manufacturers' Association inspected the works the past summer, on the occasion of the visits of the engineers and the manufacturers to the Coast. The process of manufacture is simple, yet part of it is a specialty, and skilled men are employed, while not a section of pipe is turned out which has not had the most watchful supervision in the various steps of manufacture. This care and excellent workmanship is the pride of the manufacturers and they unhesitatingly refer all inquirers to the many cities and towns which have used their product in the three seasons it has been made, since the factory was established.

Fig. 3.—High Pressure Wood Stave Pipe with Flange Coupling.

The Business Situation

BY B. E. WALKER, President Canadian Bank Commerce

WHILE we are enjoying an extraordinary prosperity, there are signs of a strain which must bring trouble if disregarded. We are a borrowing country, and we cannot be reminded of this too often. As we fix capital in new structures, public or private, railways, buildings, etc., some one must find the capital in excess of what we can ourselves provide out of the saleable products of our labor. The number of countries willing to buy our securities has been steadily increasing, but we must not be blind, as we sometimes seem to be, to the fact that our power to build depends largely on whether these countries have surplus capital to invest. By means of the cable the trading nations of the world have been brought very near together, and while nations of the world have been averted, and the adjustment of capital to the world's needs has been greatly improved, still for the same reason world-wide trouble in the money markets sometimes arises with a suddenness which is alarming to those at least who are not watching for the signs. We are passing through such a period just now, happily without a general breakdown, but unless we mend our ways we are not likely to escape a similar or still worse condition next autumn which may wreck our fair prosperity. Europe is bearing the enormous cost of two great wars, both in the loss of capital actually destroyed and also in the loss to individuals from the decline in the values of the national securities of the countries interested in the wars. And since these wars, losses on an unexampled scale have occurred by earthquakes and fire. The volume of trade and the unusual amount of building in many countries have at the same time vastly increased the amount of capital required. This has been accompanied by a steady rise in prices throughout the world, and by a most pronounced and widespread advance in the scale of personal expenditure. It is true that it has also been accompanied by the greatest production of gold and of other commodities, but the effect of the various influences has naturally been to put upon the money markets a strain which has only just failed to cause a general breakdown of credit. To make the outlook still more serious, the United States, and other less important countries, including Canada, contemplate expenditures on a very large scale for railway and other building. This, then, is a time for every prudent man to survey carefully his financial position. If he has debts he should consider how he will pay them if he should have to face world-wide stringency in money. Has he assets which the world needs for daily use, or assets which will sell only when the sun is shining? If he is happily in easy conditions as to debt, he will, if he is wise, consider every circumstance arising in his business which tends towards debt instead of towards liquidation. As for those who are plunging in real estate at inflated prices and in mining stocks, nothing, we presume, but the inevitable collapse which follows these seasons of mania will do any good.

Fig. 4.—Low Pressure Wood Stave Pipe, Self Coupling.

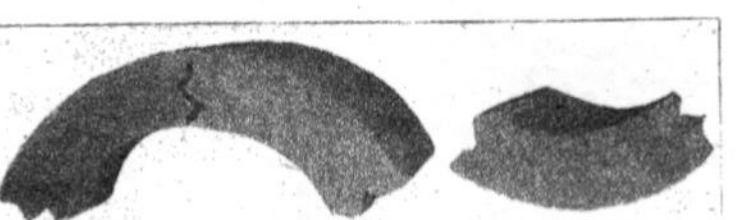

Fig. 5.—Wood Stave, Showing Joint.

BIG FIRMS AMALGAMATE.

Particulars of the amalgamation of the Canada Screw Company, established 1866, and the Ontario Tack Company, established in 1887, two of Hamilton's largest industries, have been announced. Nearly $2,000,000 is involved in the deal. The plants of both concerns are to be considerably enlarged and new lines added to their products. Application has been made for a new charter in which the capital stock will be greatly increased in order to allow for the extension of the works.

The new company will be known as the Canada Screw Co., and the officers will be: Cyrus A. Birge, president; Chas. Alexander, vice-president; F. H. Whitton, general manager; W. F. Coote, secretary-treasurer; James O. Callaghan, director of works. The other directors will be Hon. Senator Gibson and Charles S. Wilcox.

Industrial Waters: An Important Feature in Manufacturing

By J. A. De Cew, B.A. Sc, Chemical Engineer

WHEN a marked industrial growth is taking place in a new country and many factors are exerting their influence either for or against the development of different localities or districts, it is very interesting to note the rivalry that often exists between various towns and cities, any of which may be selected as the site for a coming industry.

When a manufacturer happens to be looking for a location for his plant, he naturally desires to select a locality which will offer him the cheapest production, and he is more often influenced by such a consideration than by any artificial stimulus that may be offered.

Therefore, after such questions as raw materials, market and transportation have been considered, the effect of local conditions on the cost of manufacture is of vital consequence.

As compared with the other natural advantages of a manufacturing region, the availability of a pure and cheap water supply is of the greatest economic importance. The amount of water that is used in some industries is hard to conceive when thinking of it in gallons and it is only when water has to be purchased that this factor is thought of at all, as there are many industries that would hardly exist at all if they were obliged to consider the water they use as an item of expense. However, it is these very manufacturers that are most seriously concerned with the quality of the water that they use, for a very small content of an injurious ingredient may seriously affect the quality of the material treated when such large quantities of water are used.

We have many localities in Canada where the water is relatively soft owing to the insolubility of the rocks and soil over which the streams and rivers pass. Yet these same waters often carry in solution ferruginous and organic matters which are equally deleterious to many industries. On the other hand are many streams whose course lies mainly through a limestone region and these always contain what are commonly called hard waters.

In Eastern Canada the quality of our average river water is not considered bad from an industrial standpoint and this is one of the reasons that the manufacturer so often ignores the fact that even moderately bad water may do a lot of injury when used in sufficient quantity.

Boiler Feed Waters.

There is one use for water which nearly all manufacturers have in common and that is in the production of steam. In this operation a more or less impure water is continually coming into the boiler while nothing but pure water is passing out in the form of steam and it is common experience that what we consider to be pretty good water will soon leave behind considerable residue. The question to determine from the standpoint of boiler economy is, how bad will the water have to be before it will pay to treat it for the removal of the scale-forming solids prior to its use in steam production. This of course depends upon the relation between the loss sustained from using the natural water and the cost of purification, and whenever the matter is treated as being worth consideration the probabilities are always in favor of the methods of purifying.

It is not only in the production of steam that water softeners are found necessary, but also in the washing treatments of many materials.

In dying, printing, bleaching, wool-washing, paper-making, soap-making, laundrying and tanning, calcarious waters are found to be very injurious and troublesome. Therefore, these industries are often confronted with various losses and difficulties in addition to the lost efficiency of their boilers, owing to the chemical action of these hard waters upon their manufactured product. When an industry of this kind happens to be established in a locality where soft water is not obtainable, it is generally found profitable to purify a large portion of the water supply.

Soap Works.

In the manufacture of soap a water containing bicarbonates of calcium and magnesium should never be used without purification, otherwise the insoluble soaps that these metals form will become incorporated and injure the quality of the product. On the other hand the water may contain chlorides or soluble sulphates and in this case it is very difficult to effect a complete saponification. A water containing organic or earthy matters in suspension is not fit for soap-makers use until it has undergone a careful filtration.

In fact the importance of using a pure soft water in making soap is evinced by the fact that one of the daily uses and functions of soap is to make water soft, for this has to be accomplished before a lather can be formed.

Laundries.

The production of soft water by means of soap is anything but an economical method however, and when we find that a hard water is used in the laundry we also find that the amount of soap consumed is many times that actually doing useful work on the clothes. Now if an attempt is made to soften the water while in use by adding sodium carbonate, a precipitate of calcium carbonate is deposited on the fabric and this is the cause of many blotches and spots that are difficult to remove.

Owing to this it is absolutely necessary in laundry work that any purification required must be prior to the operation of washing.

Tanneries.

In the process of tanning large quantities of water are used both in the unhairing treatment and in the tan-pit.

In the unhairing process the greasy matter secreted by the small hair glands is saponified by means of quick-lime and if the water used should contain the bi-carbonate of calcium in solution, it is immediately acted on by the free lime and the insoluble carbonate of calcium precipitated. This then becomes deposited in the dermic tissue and interferes with the absorption of the tannin into the cells of the hide thus retarding the tanning process as well as causing brown stains in the leather.

If a hard water should be used in tan-pits, the bi-carbonates of lime and magnesia become converted into tannates of lime and magnesia which oxidize rapidly in the air to form secondary products of tan-oxylic and tan-omelanic acids and these produce a brown coloration in the leather. When a water is used in tanning contains salts of iron in solution, these will, in contact with air, become quickly transformed from the ferrous to the ferric state and then combine with the tannin in the leather giving the black coloration of the tannate of iron.

These iron salts, however, can be removed by a special treatment in the same apparatus that removes the lime and magnesium salts. This treatment of ferruginous waters is one that might be of interest to the tanners of Canada for the writer has been informed that they often experience difficulties from the use of waters of this kind.

Paper Mills.

In the manufacture of paper the quality of water used has an important bearing upon the color and sizing properties.

In the case of a ferruginous water a brown tinge is given to the paper by the precipitated oxides. With a hard water carrying calcium and magnesium salts, a well sized paper is difficult to obtain for a quantity of the resinates of these metals are formed which have no sizing value whatever.

The importance of a good water supply is so vital to this industry that very few paper mills are located where the water is of an unfavorable character. Some mills, however, are unfortunately situated in this regard and to them a proper purification of a portion of the water is preferable to the losses in manufacture that would otherwise result.

Sugar Refineries.

The character of the water used in the production of sugar has a very marked effect upon the processes employed.

For instance, a calcareous water may interfere with the process of diffusion by depositing the insoluble carbonate, while it also damages the residuary products and diminishes their value as a cattle food.

Moreover, the use of a hard water will cause a deposit on the heating surface of the boiling or evaporating apparatus and as a result the efficiency of these machines is greatly reduced.

This case is then analogous to the others mentioned and the same remedy should be applied.

Methods of Water Purification.

Many chemical and mechanical methods have been suggested and employed for the treatment of unsuitable waters.

For the removal of solids in suspension, gravity or mechanical filtration is generally employed and the present tendency is towards the increased use of those mechanical filters that combine the maximum of filtering area with the minimum of floor space required. These conditions are found ideally represented in the filter press, which, when used for water has a very large capacity.

Waters containing salts in solution can often be partially purified by boiling, but this is much more costly than the precipitation of the salts with chemical reagents. If, however, chemical materials alone are employed, they must either be used simultaneously with other processes or sufficient time allowed for the subsidence of the flocculent matter precipitated by the treatment.

Either of these methods has its disadvantages and in consequence an apparatus is generally used which will both assist the chemical reaction required and also remove the impurities thrown out of solution.

These apparatus which are of various designs are called water softeners, and are the only true solution for the difficulties arising from the use of bad water. Some of them, however, are of much better design than others and give much better results.

In order to assist the reader in choosing a plant of this kind in case it should be required, let us consider some of the essential features of an apparatus of this kind. In the first place, the plant should be so designed that the maximum amount of impurities can be removed with the minimum of effort and expense. The completeness of the purification effected will depend however, upon the accuracy with which the proper amount of softening chemicals is supplied to the water treated, these required proportions being determined by previous analysis. The economy of operation and the capacity of the plant are influenced very largely by the design of the apparatus and the methods employed.

Some softeners are so constructed that a complete chemical re-action and precipitation of the salts cannot be obtained without the use of heat, while other machines are able to obtain the same result in cold water. The latter is, without doubt, a distinct advantage for although it may be said that hot water is required for boiler use yet the fact remains that considerable heat is always lost in radiation before the water can be consumed.

Moreover, an apparatus of this kind should be as nearly as automatic as possible and preferably continuous as well, for it is only in this way that labor and attention can be nearly eliminated.

Some plants of this kind have been known to reduce the hardness of a water from 30 degrees to one half of a degree and at the same time remove all of the suspended impurities in the water.

When results like this can be obtained the advantages of and the necessity for, a system of this kind becomes very evident to any manufacturer who is not blessed with an ideal water supply and who gives the question his consideration.

At The Noon Hour

THE WAY OF IT.

"You see, he was whirled around a shaft at the mill and pretty seriously hurt. Now he's suing the boss for $500 damages."

"Well, that's the way of the whirled."

A Long Wait.

A LONG WAIT.

Tired of waiting for back pay which was several weeks overdue, one of the draftsmen in a railway construction office in Canada made the drawing on a large scale as represented herewith, the staff sending it with their compliments to the superintendent of the party. It had the desired effect and before many days everything was adjusted satisfactorily and the men were happy.

KEEP A GOIN'.

If you strike a thorn or rose
Keep a Goin' !
If it hails or if it snows
Keep a Goin' !
Tain't no use to sit and whine
When the fish ain't on your line ;
Bait your hook and keep on tryin'
Keep a Goin' !

If the weather kills your crop
Keep a Goin' !
When you tumble from the top
Keep a Goin' !
S'pose you're out of every dime,
Gettin' broke ain't any crime ;
Tell the world you're feeling prime
Keep a Goin' !

When it looks like all is up
Keep a Goin' !
Drain the sweetness from the cup
Keep a Goin !
See the wild birds on the wing,
Hear the bells that sweetly ring ;
When you feel like sighin'—sing,
Keep a Goin' !

A new collegiate institute will be erected at Brandon, Man., at the cost of $65,000.

Just About Ourselves

Just two years old.

Yes, that is all in years, but youth and age in the newspaper realm are of different import to length of days in human existence.

Twenty-one years must come and go before the human being can claim full maturity. As many days may suffice to bring a newspaper to the state where its definite policy, field and destiny are apparent, but its further growth and development are limited only by the possibilities of its further usefulness and influence.

* * *

When Canadian Machinery started two years ago there was no other mechanical paper in the field, so that it met with a quick and ready response from those interested. From the first issue its success seemed assured, and the publishers are grateful to those who not only recognized that there was a field for such a paper as Canadian Machinery, but who have been loyal from the start and are still numbered as its best friends. Those who have been in closest touch with it have been most ready to express their appreciation of its growth from month to month.

Results speak for themselves and we do not hesitate to let Canadian Machinery as it appears from issue to issue speak for itself. It has now a recognized place in trade journalism that is secure. This has been attained not only by the most diligent effort on the part of the publishers but by suggestions from time to time from friends both far and near. We are always on the lookout for suggestions whereby the paper may be of greater interest and value and appreciate criticisms from any of our readers.

* * *

Its advertising value was strikingly illustrated very recently in an actual test with other industrial and technical newspapers in the Canadian field. A manufacturer of a certain type of apparatus wished to make a special offer of a somewhat unusual kind and to do this he advertised in the same space in the different industrial nespapers in this country. When the results of direct enquiries were counted up it was found that those from Canadian Machinery were more than twice the number received from any other and equalled almost the entire amount of all the rest put together. No more striking proof of its value as an advertising medium could be had than this, as probably at no other time has such an actual test been made whereby a direct judgment on merits could be assured. It is a direct testimony to the value of Canadian Machinery's influence and we are satisfied that the experience of the advertiser in question has been the experience of many others.

* * *

A few of the kind expressions of opinion received at our office during the past few weeks are of interest as illustrating an appreciation entirely unsolicited. "I find it a very valuable paper," E. Auger, Bowmanville, Ont. "I am pleased to tell you that I take great pleasure and interest in reading Canadian Machinery," L. Bailey, Hamilton, Ont. "I have been a constant reader of your paper, Canadian Machinery, for some time and have gained considerable information through the same, especially from the articles on spur and bevel gears," S. W. Jennings, Montreal, Que. "Your paper is all right," the Gardner Governor Co., Quincy, Ill.

Another tribute to Canadian Machinery received recently from the Detroit Foundry Supply Co., Detroit, Mich., explains the feeling of that firm towards us when placing advertising last month: "In connection with this ad., it may be of interest for you to know that this is the first advertisement the Detroit Foundry Supply Co. has resorted to outside of our own paper, the 'Blast,' and is complimentary to Canadian Machinery. However we do not wish you to feel that we are 'handing' you anything, as we can appreciate your progressive and hustling capabilities."

Just one more reference. In a letter from Mr. J. Meredith McKim, manager of the London office of the MacLean Publishing Co., in which he encloses a letter from the London County Council, he states: "The L. C. C. is a very large institution and must employ several thousand hands in its various departments and it is therefore a compliment that the paper should receive an inquiry of this sort. Of course I agree with you it is a well deserved compliment." He stated further having just concluded arrangements with the secretary of the Royal Exchange, Glasgow, Scotland, to supply them with a copy of Canadian Machinery until further notice. This institution is very strict regarding what papers they will accept, so that this is a further tribute.

* * *

As Canadian Machinery stands to-day it comes far short of our aims and ideals. In fact we are planning and preparing to increase its usefulness by the addition of another feature not yet incorporated but which as soon as arrangements can be made will be exploited. We are constantly looking for articles on practical and industrial subjects, systems, cost keeping, methods of reducing labor and any of the many features that are new in our modern industrial progress to improve and further increase the scope of the features already adopted.

THE REAL SOCIETY.

Our chief want in life is, somebody who shall make us do what we can. This is the service of a friend. With him we are easily great. There is a sublime attraction in him to whatever virtue is in us. How he flings wide the doors of existence! What questions we ask of him! What an understanding we have! How few words are needed! It is the only real society. | *Emerson.*

BOOK REVIEWS.

THE SLIDE RULE.—A practical manual by Charles N. Pickworth, Whitworth scholar, author of The Indicator, etc., being the tenth edition of this book. 105 pages, illustrated; price, $1.00. New York: D. Van Nostrand Co., 23 Murray St.

The call for another edition of this the first English work on the modern slide rule, making the tenth to appear, testifies somewhat to the popularity of the author's treatise. Although the working knowledge of the slide rule may be obtained by rote it is indisputable that a comprehensive knowledge of the rudiments of logarithmic computation is necessary to enable the operator to take an intelligent grasp of the subject. A good deal of new matter has been included in this edition, besides a number of new illustrations. New forms of slide rules and a series of special instruments of recent introduction possessing novel features have been included. The practical problems as applied to the slide rule are not only of importance but of extreme interest.

REINFORCED CONCRETE, in three parts. Part (1), Methods of Calculation, by Albert W. Buel, C.E. Part (2), Representative Structures. Part (3). Methods of Construction, by Chas. S. Hill. Second edition revised and enlarged. New York: The Engineering News Publishing Co.

Since the publication of the first edition of this book the edition was practically exhausted before the work of revision was commenced, showing the popularity to which it attained, being an evidence of the need of such a publication as well as an acknowledgment by those interested in concrete work of the manner in which this volume covered the field. The book is not a result of collaboration on the part of the authors, but is really two books in one. Part one, dealing with the methods of calculation, and parts two and three, dealing with the details of and the methods employed in constructing various classes of reinforced concrete structure. In part one the economic uses and property of reinforced concrete are dealt with, followed by chapters on beams and theories of flexure, columns, retaining walls, etc., and tests and designs of arches. Part two takes up and illustrates representative examples of reinforced concrete constructions, dealing with bridges and culverts, tanks and reservoirs and walls, dams, chimneys, etc. The first chapter in part three describes the material employed in the fabrication of reinforced concrete and throughout the chapter are dealt with the methods of construction of various classes of work.

CANADIAN MACHINERY
and Manufacturing News

A monthly newspaper devoted to machinery and manufacturing interests, mechanical and electrical trades, technical progress, construction and improvement, and to all users of power developed from steam, gas, electricity, compressed air and water in Canada.

The MacLean Publishing Company, Limited

JOHN BAYNE MACLEAN - - - *President*
W. L. EDMONDS - - - - *Vice-President*
F. S. KEITH, B.Sc. - - - *Managing Editor*

Publishers of trade newspapers which circlate everywhere in Canada also in Great Britain, United States, West Indies, South Africa and Australia

OFFICES:

CANADA:
MONTREAL - - - - 232 McGill Street, Telephone Main 1255
TORONTO - - - - 10 Front Street East, Telephone Main 2701
WINNIPEG - - - 511 Union Bank Building, Telephone 3726, F. R. Munro
BRITISH COLUMBIA - - - - Vancouver, Geo. S. B. Perry

GREAT BRITAIN
LONDON - - - - 88 Fleet Street, E.C., Telephone Central 12960, J. Meredith McKim
MANCHESTER - - - - 92 Market Street, H. S. Ashburner

UNITED STATES
CHICAGO - - - - 1001 Teutonic Bldg, J. Roland Kay

FRANCE
PARIS - - - Agence Havas, 8 Place de la Bourse

SWITZERLAND
ZURICH - - - - - Louis Wolt, Orell Fussli & Co.

SUBSCRIPTION RATE.

Canada, United States, $1.00, Great Britain, Australia and other colonies, 4s. 6d., per year; other countries, $1.50. Advertising rates on request.

Subscribers who are not receiving their paper regularly will confer a favor on us by letting us know. We should be notified at once of any change in address, giving both old and new.

Vol. III. FEBRUARY, 1907 No. 2

ALLEGED HOSTILITY TO U.S. MAGAZINES.

FOR some time past enquiries have been received from United States manufacturers and others regarding our position on the question of an alleged movement in Canada to keep out United States magazines. Regarding this we wish to state very emphatically that there is no such organized movement and any action that has been taken has not originated with publishers in Canada. Regarding our own position in the matter we are very much opposed to any movement reacting against the best publications in the United States, believing that Canadians should have the most liberal access to the best reading possible. A portion of a letter written by Col. J. B. MacLean, president of the MacLean Publishing Co., publishers of Canadian Machinery, who is himself a member of the executive of the Canadian Manufacturers' Association, to Mr. J. P. Murray, chairman of the Commercial Intelligence Committee of the C.M.A., will show the stand he takes in the matter.

"There is a matter which I think might be brought before the Commercial Intelligence Committee. The Canadian Government propose shutting out U.S. publications, trade newspapers, magazines and that sort of thing. I would probably benefit by such a policy more than anyone else in the country, owning as I do so many newspapers, but I think it is a very short-sighted policy on the part of the Government. By all means let them shut out the cheap and useless magazines, but I think it is to the advantage of our people, particularly of our business men, manufacturers, and their employes, to read the best that is published in the United States and Great Britain that they may be kept right up to the times in every particular in every department of manufacturing and selling of goods. We are doing a lot in this way in our own papers, but we cannot cover all the fields and we never hesitate to recommend or advise any of our readers anywhere in Canada to subscribe for and read the best publications in the United States.

"I think this is a matter the association should deal with and point out to the Government the inadvisability of such a course."

Considerable activity has been noticeable of late in the Post Office Department of both the United States and Canada, which has led up to the present condition of affairs. The most important move has been the abrogation by Canada of the post convention between the United States and Canada in so far as that convention relates to second-class matter. This will take effect next Spring. The original instrument was ratified in 1888, but it was amended two years ago so as to provide "for the right of each administration to decline to transmit through its mails, except when duly prepaid by stamps affixed in the country of origin at the rate applicable to miscellaneous printed matter, such newspapers and periodicals as it would decline to transmit through its mails under the statutory newspaper and periodical privileges accorded to publishers and newsdealers, if such newspapers and periodicals were published in its own country."

In abrogating the convention, Canada notifies the United States that, if by legislation new regulations are framed controlling the second-class matter in the United States, it will be prepared to enter upon negotiations for another convention relating to this class of matter.

Thus it will be seen that Canada is still ready to make another treaty as soon as the United States Government, by legislation, frames new regulations controlling second-class matter. It is over this point that the whole difficulty has apparently arisen, giving effect that Canada is antagonistic to United States publications. She is merely waiting until the United States Government frames new laws in this matter, at which time there is not the shadow of a doubt but that a suitable treaty will be made permitting of the entrance into Canada of reputable United States magazines on the same terms as they are received through the mails of the United States.

More than this, we have reason to believe on the authority of an official of the Canadian Post Office Department that it was on the suggestion of a high official in the U.S. Post Office Department at Washington that the Canadian Government has taken the action referred to above. This was evidently done to assist the U.S. authorities in some scheme which has not yet been made public.

READY FOR CANADIAN TRADE.

MR. L. Gerald Freedman of Melbourne, Australia, is at present in Canada studying the possibilities of Canadian trade with Australia. Mr. Freedman declares that an excellent market is open in a sister colony for Canadian manufactured articles, and expressed surprise that more effort was not made on the part of manufacturers in this country to secure some of this valuable trade. He emphasized the fact that the annual purchases of Australia amounted to $425,000,000 and that at present United States was aggressively supplying the demands in which Canadians should, at least, be taking an active part. The commercial conditions are such that an increased exchange of goods between the two countries should be materialized. At the present time motors of all kinds are very much in demand, agricultural implements, machine tools, power apparatus, leather products, furniture and all kinds of woodwork. He is satisfied that should Canadian manufacturers make a move in this direction they can without any great effort secure a large share of the Australasian trade.

CONTINUED INDUSTRIAL BOOM.

GREAT as was the industrial expansion in Canada during 1906 there is already every evidence that this will be greatly overshadowed by the progress of 1907. Already there is promise of the greatest year in railroad construction in the history of the country, almost every province in the Dominion receiving a share, but being particularly true of the west. Many large works are under way, particularly in the metal industries, giving promise of a development in this line hitherto unparalleled. Many United States firms have already intimated their intention to establish branch manufacturing plants in this country. Enough of interest has already been expressed and enough of confidence already assured to bring about what, under ordinary circumstances, might be considered an industrial boom, but what in Canada is the steady, progressive and solid development of a country whose resources are just being recognized.

One feature in connection with this is quite noticeable, being the stimulated interest on the part of municipalities to secure industries in their midst. An awakening along this line has been experienced that is reacting greatly to the benefit of towns that were hitherto indifferent regarding new industries. There are possibilities of development in the country at present that would warrant every town becoming a city, and with proper attention to their interests the citizens therein have every chance of either creating new industries or attracting them within their limits.

SENSIBLE TALK.

IT IS seldom that a manufacturer comes out in opposition to the protectionist policy, but R. T. Crane, of Chicago, takes thus unusual position in a recent article, discussing U.S. Senator J. J. Hill's recent plea for reciprocity in Canada. Mr. Crane is probably the largest manufacturer of valves and heating supplies in America, one of his company's many warehouses being located in Winnipeg.

Mr. Crane contends that "a large amount of trade with the Canadians has already been lost, and that much more is likely to be lost, by the stupid and short-sighted policy of our Congress in making the Dingley tariff law a bar to reciprocal trade relations, not only with Canada, but with every other foreign country. We deliberately took a course opposed to reciprocity at a time when Canada appeared to be entirely satisfied with its trade relations with this country, and thus lost the enormous advantage we then possessed, and it is now too late for us to try to re-establish our former position, much less to take advantage of the greatly improved conditions in the Dominion. In this case we have antagonized a neighbor that, by its geographical situation, ought to have been one country with the United States. There is no more sense in putting up trade barriers between Canada and this country than there would be in putting them up between Illinois and Indiana. Once we had committed ourselves to the Dingley tariff, we antagonized the Canadians, and they then swung sharply toward a fiscal policy of protection. This policy is now thoroughly and generally approved, and there is no prospect that the Canadians will change it."

Continuing, Mr. Crane points out that to change Canada's fiscal policy would be unfair to both Canadian manufacturers and to Americans, who have established factories in Canada, with the expectation that the policy would be permanent. He further argues that the high protective policy in the States is forcing manufacturers there to pay high prices for raw materials, this increase in manufacturing costs placing American goods at a disadvantage in foreign markets.

CONTENTS

Milling Machine Attachments.

THE UNIVERSAL SPIRAL HEAD.—By John Edgar

(Continued from last issue.)

II.

IN THE former article the system of plain indexing was described. In this article we will take up the three remaining systems: The Direct, Compound, and Differential.

Direct Indexing.

The direct indexing system consists of the process of turning the spindle direct, without the interposition of any intermediates, as with plain indexing, the division being obtained by revolving the spindle and dropping a pin into a plate fast to it, the whole process being in direct dealing with the spindle. In ordinary cases the worm is dropped out of mesh with the worm gear, leaving the spindle free to be revolved at will depending entirely upon the hold that the under pin has on the plate. The plate is fastened to the front end of the spindle and is shown at (h) and the pivot (i). The number of holes is usually small and is in all cases such that it is divisible by the common divisors; four, six and eight. This system is used for indexing very low numbers, and then only in cases where accuracy is not important. The necessity of dropping the worm out of mesh makes this system undesirable, although most heads are so arranged.

Compound Indexing.

The compound system of indexing is unimportant, as it is little used and has the very undesirable quality of being complicated and that exact results cannot be obtained in many cases, it being in a measure an approximate method.

In machines fitted for the compound method the index head has to be provided with two index pins, one on either side of the plate, and which must be used in conjunction in such a manner as to obtain a compound movement of the crank and plate.

One pin, that on the back of the plate, is so arranged that it may be adjusted so that any circle of holes may be used and at the same time hold the plate stationary when it is in the hole. The other pin is that in the index crank and is also adjustable. The plates must in this case be drilled through so that the pin may be used from behind.

It will be found that in plain indexing prime numbers and others not given in the tables can be obtained approximately by that (the plain) system. Thus in our example of 73 divisions in the former article we find from the table that 72 is the nearest number and is obtained by 15 holes in the 27 circle, each division is so much greater than it ought to be by 1-73 of a division. If the plate could be given a movement against the crank 1-73 of 15-27 turn for each division or at each indexing we would obtain the desired 73 divisions. This is the principle of compound indexing. The process as described above is very simple but other items enter into the problem such as limits in the number of holes in the plate, and the dividing of the periphery of a piece of work into a number of equal spaces whose number is prime is by no means a simple matter even by this compound method, it being necessary in some cases to go around the work a number of times in order to get the desired result, that is, the process is to set the index pin and take readings which give a lower number than that desired and by compounding the two movements get the number wanted by repeating the indexing until the work has revolved a stated number of times, instead of once as in plain indexing.

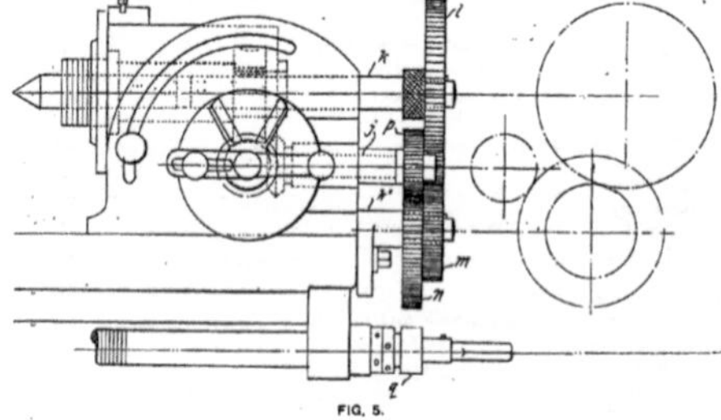

FIG. 5.

Thus, referring to our example again, we would set the spacer and pin to index 6 turns and 28 holes in the 47-circle and the back pin to take 1 hole in the 49-circle, working the plate in the opposite direction to that in which we turn the crank and repeat the operation until we had revolved the spindle twelve times. This method would give us a division within one ten-thousandth, or each division would be one seventy-third of one ten-thousandth too long, which is well within the limits of shop requirements. An exposition of how these prime numbers are figured would require so much space and would be of so little use to the busy shop man that no attempt will be made here to show any, and as the differential method, to be described in full, is much simpler and answers all the requirements which the compound method was devised to fill, the need of the latter method is overcome.

In the description of the mechanism of the head we stopped when the point had been reached where the head was complete so far as ordinary conditions of indexing were taken into consideration. We will now proceed with the description.

Mechanism of Head.

If on the inner side of the index plate we fix a bevel gear so that the plate and gear act as one and at the same time are loose on the worm shaft, and connect them by another bevel gear to the shaft, (Fig. 5), and by connecting this shaft and the lead screw by gearing, we have means of revolving the spindle of the head by power. The motion of this shaft and of the bevel gears and plate, is transmitted to the spindle by locking the plate and index crank together by means of the detent pin. This is the manner in which the spindle is driven in spiral milling. If on the other hand the index pin is withdrawn, the plate will still be driven by the gearing, but the spindle will remain stationary.

In plain indexing the plate is held in a stationary position by the back pin. If this pin is withdrawn and the index pin is also withdrawn the plate is free to revolve, carrying with it the shaft (j), which is connected with it by the bevel gears. By reversing the operation the plate may be revolved by means of the shaft (j). This at once gives us a method by which we can make up the desired advancement of the plate in order to obtain the 73 divisions in our example. This advancement must be made by steps at each indexing and the plate held stationary at the interval between the indexings. This can be accomplished by gearing the spindle and the worm shaft (j), together with gears of the proper ratio to give the desired movement of the plate at each indexing of the work. This is accomplished by placing a plug (k) into the rear end of the spindle, which is provided with a taper hole, into which the taper end of the plug fits. The plug is made to receive the regular change gears on the outer end and is connected to the worm shaft (j), as shown in Fig. 5, by the gears, (l, m, n and p).

Differential Indexing.

The action of the device is as follows: The index crank is turned, causing the worm, worm gear and spindle to revolve. This in turn causes the plug (k) to revolve, turning the worm shaft the desired amount by means of the change gears, which, through the bevel gears, advances or retards the plate the fractional turn necessary to make up for the loss or gain in the indexing of the crank. The use of this device automatically makes up the error in the division which is being obtained by means of the ordinary plain indexing methods. The process is exactly the same as plain indexing so far as the operator is concerned, for after the gears are in place the simple movement of the crank, as in plain indexing, is made to take the place of the compound movements necessary in compound indexing. And by this method with the proper gearing we can obtain any division desired. This is differential indexing.

In machines which are equipped with this method of indexing the index table is so arranged that the readings given show the arrangement of gears for all numbers up to 360 degrees where they are necessary, otherwise the reading is given as in plain indexing.

Looking at one of these tables we find for our example of 73 divisions that we must take a 49-circle and move 28 holes at each indexing, and also that on the spindle plug (k) must be placed a 48-tooth gear; on the shaft (j) a 28-tooth gear, and that two idlers must be used in order to reach and have the plate turn opposite to the motion of the crank. The number of idlers needed depends on the head being used, the number given above being for example only.

If we move the crank 28 holes in a 49-circle, we move the spindle through $\frac{28}{49} \times \frac{1}{40}$ of a revolution, which is $\frac{1}{70}$, or if the gears were not in place we would, by carrying out the process, index 70 divisions. But we desire 73, so the plate must be moved in the opposite direction to that of the crank, enough so that the spindle loses 3-70 of a turn, or in other words, the plate must be given a backward movement equal to 3 divisions or 3x28 holes in the 49 circle.

The ratio of the gearing to connect the stud (k) and the shaft (j) must be proportioned so as to accomplish this. One division is obtained by 28 holes on the 49 circle, so that 3 divisions equal three times that amount, which is

$$3 \times \frac{28}{49} = \frac{84}{49} \text{ or } \frac{12}{7}$$

The gear on the spindle stud (k) must have 12 teeth to every 7 in the gear on the worm stud (j), which is met by the gears given in the table, namely, 48 and 28, respectively. After the gears are in position all that is necessary is to manipulate the crank as in plain indexing, the process being simple in comparison with the compound movements necessary in the compound method. By this differential method there is less danger of making blunders, which are easy enough made without increasing the ratio of such chances.

The gearing used in the above example is simple, while Fig. 5 shows compound gearing in place. The compounding of gears is done to increase the ratio of the driver to the driven when the distance between the centres is short, so short that simple gearing would not be practical. In compound gearing we have two simple trains of gears which when used in combination give the desired ratio, thus if we had two simple trains whose ratios were 1 to 3 and 1 to 5, respectively, we would have by using as a compound a whole ratio of 1 to 15, or by inversing one train we would obtain the ratio of 3 to 5. The compound gears are keyed to a sleeve so that they act in unison. The gear (n) is termed the "first gear on stud," and the gear (m) the "second gear on stud." The gear (p) is called the "gear on worm" because it drives the worm and the gear (l), the "gear on spindle." Sometimes it is necessary to have a compound gear and still have the driver and driven gear turn in opposite directions. This is accomplished by placing an idle gear between the gear on worm and the second gear on stud, or between the gear on spindle and the first gear on stud.

Care should be exercised in having the gears properly in mesh so that no backlash occurs between the teeth, as it is liable to lead to error in the division.

It will be noticed that the spindle cannot be elevated at an angle when this arrangement is used, making it impossible to cut bevel gears by this method, nor can it be used in cutting spiral gears or other work of the character, as the worm shaft is connected with the spindle, making a locked train, even could connection be made with the screw which under the conditions shown is impossible. However, the method presents enough merits to allow these few unimportant points being scored against it.

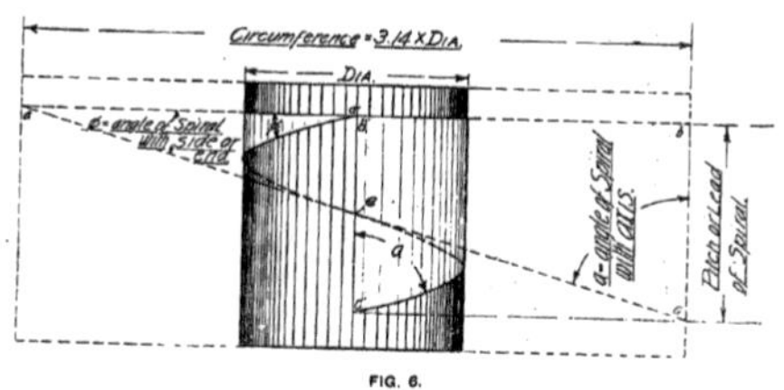

FIG. 6.

Spiral Cutting.

Having described the principles of and the means by which we may calculate the necessary data for obtaining any number of divisions, either by the method of plain indexing or the differential method, we will next turn our attention to the problem of spiral cutting and explain the principles of and give a few simple rules and methods by which the necessary gears may be found and the other data that is necessary in setting the machine for this class of work.

The principle on which the spiral head depends is the same as that applied to the lathe in chasing threads, but the action is just the reverse; the spindle drives the screw on the lathe, but the spiral head spindle is driven from the lead screw of the milling machine.

(To be Continued.)

On the Art of Cutting Metals

By F. W. Taylor—(Continued from last issue.)

The author of the paper thinks that the time is fast going by for the great personal or individual achievement of any one man standing alone and without the help of those around him. And the time is coming when all great things will be done by the co-operation of many men in which each man performs that function for which he is best suited, each man preserves his own individuality and is supreme in his particular function, and each man at the same time loses none of its originality and proper personal initiative, and yet is controlled by and must work harmoniously with many other men.

He has great confidence in the good judgment and common sense of mechanics in the long run, but as a class they are extremely conservative, and if left to themselves their progress from older towards better methods will be exceedingly slow; and his experience is that rapid improvement can only be brought about through constant and heavy pressure from those who are over them. It must be said therefore that to get any great benefit from the laws derived from these experiments, the slide rules prepared by the author and his colleagues must be used, and these slide rules will be of but little, if any, value under the old style of management, in which the machinist is left with the final decision as to what shape of tool, depth of cut, speed, and feed, he will use. The slide rules cannot be left at the lathe to be banged about by the machinist. They must be used by a man with reasonably clean hands, and at a table or desk, and this man must write his instructions as to speed, feed, depth of cut, etc., and send them to the machinist well in advance of the time that the work is to be done.

A long time will be required in any shop to bring about this radically new order of things; but in the end the gain is so great that the author says, without hesitation, that there is hardly a machine shop in the country whose output cannot be doubled through the use of these methods. And this applies not only to large shops, but also to comparatively small establishments.

Results of Experiments.

The conclusions arrived at as a result of the experiments along different lines are given in brief form, as stated by the author, under the explanatory headings:

Standard Shop Tools.

In our practical experience in managing shops we have found it no easy matter to maintain at all times an ample supply of cutting tools ready for immediate use by each machinist, treated and ground so as to be uniform in quality and shape; and the greater the variety in the shape and size of the tools, the greater becomes the difficulty of keeping always ready a sufficient supply of uniform tools. Our whole experience, therefore, points to the necessity of adopting as small a number of standard shapes and sizes of tools as practicable. It is far better for a machine shop to err upon the side of having too little variety in the shape of its tools rather than on that of having too many shapes.

Tip, Clearance and Angle of Tools.

Contrary to the opinion of almost all novices in the art of cutting metals, the clearance angle and the back slope and side slope angles of a tool are by no means the most important elements in the design of cutting tools, their effect for good or evil upon the cutting speed and even upon the pressure required to remove the chip being much less than is ordinarily attributed to them.

The following are our conclusions regarding the clearance angle of the tool.

(a) For standard shop tools to be ground by a trained grinder or on an automatic grinding machine, a clearance angle of 6 degrees should be used for all classes of roughing work.

(b) In shops in which each machinist grinds his own tools a clearance angle of from 9 to 12 degrees should be used.

The following are the conclusions arrived at regarding the angles at which tools should be ground:

(a) For standard tools to be used in a machine shop for cutting metals of average quality: Tools for cutting cast iron and the harder steels, beginning with a low limit of hardness, of about carbon 0.45 per cent., say, with 100,000 pounds tensile strength and 18 per cent. stretch, should be ground with a clearance angle of 6 degrees, back slope 14 degrees, giving a lip angle of 68 degrees.

(b) For cutting steels softer than, say, carbon 0.45 per cent. having about 100,000 pounds tensile strength and 18 per cent. stretch, tools should be ground with a clearance angle of 6 degrees, back slope of 8 degrees, side slope of 22 degrees, giving a lip angle of 61 degrees.

(c) For shops in which chilled iron is cut a lip angle of from 86 to 90 degrees should be used.

(d) In shops where work is mainly upon steel as hard or harder than tire steel, tools should be ground with a 5 degrees, side slope 9 degrees, giving a clearance angle of 6 degrees, back slope lip angle of 74 degrees.

(e) In shops working mainly upon soft steels, say, carbon 0.10 per cent. to 0.15 per cent., it is probably economical to use tools with lip angles keener than 61 degrees.

(f) The most important consideration in choosing the lip angle is to make it sufficiently blunt to avoid the danger of crumbling or spalling at the cutting edge.

(g) Tools ground with a lip angle of about 54 degrees cut softer qualities of steel, and also cast iron, with the least pressure of the chip upon the tool. The pressure upon the tool, however, is not the most important consideration in selecting the lip angle.

Forging and Grinding Tools.

The following are the important conclusions arrived at upon this subject:

(a) The shapes into which the tools are dressed and the ordinary methods of dressing them are highly uneconomical, mainly because they can be ground only a few times before requiring redressing.

(b) The tool steel from which the tool is to be forged should be one and one-half times as deep as it is wide.

(c) To avoid the tendency of the tool to upset in the tool post or under, under pressure of the cut, the cutting edge and the nose of the tool should be set well over to one side of the tool.

(d) Tool builders should design lathes boring mills, etc., with their tool posts set down lower than is customary below the centre of the work.

(e) In choosing the shape for dressing a tool, that shape should be given the preference in which the largest amount of work can be done for the smallest combined cost of forging and grinding.

(f) Forging is much more expensive than grinding, therefore the tool should be designed so that it can be ground: (a) the greatest number of times with a single dressing; and (b) with the smallest cost each time it is ground.

(g) The best method of dressing a tool is to turn its end up high above the body of the tool. Tools can be entirely dressed by this means in two heats.

The following are the important conclusions arrived at with reference to grinding tools :

(a) More tools are ruined in every machine shop through over heating in grinding than from any other cause.

(b) The most important consideration is how to grind tools rapidly without overheating them.

(c) To avoid overheating, a stream of water amounting to five gallons per minute should be thrown, preferably at a slow velocity, directly on the nose of the tool where it is in contact with the emery wheel.

(d) To avoid overheating where tools are ground by hand or with an automatic tool grinder, the surface of the tool should never be allowed to fit closely against the surface of the grindstone. To prevent this, tools should be constantly moved or wobbled about during the operation of grinding.

(e) To lessen the danger of overheating on the emery wheel and to promote rapid grinding, tools should be dressed so as to leave the smith shop with a clearance angle of about 20 degrees, while 6 degrees only is needed for cutting.

(f) Flat surfaces upon tools tend far more than curved surfaces to heat tools in grinding.

(g) Tools with keen lip angles (i. e., steel side slopes) are much more expensive to grind that blunt lip angles.

(h) It is economical to use an automatic tool grinding machine even in a small shop.

(j) There is little economy in an automatic grinder for any shop unless standard shapes have been adopted for tools, and a large supply of tools is kept always on hand in a first-class tool room so that tools of exactly the same shape can be ground in quite large batches or lots.

(k) Corundum wheels made of a mixture of grit size No. 24 and size No. 30 are the most satisfactory for grinding ordinary shop tools.

(l) In grinding flat surfaces skilful hand grinders invariably keep the tool wobbling about on the face of the grindstone in order to avoid heating.

Cooling the Tool With Heavy Stream of Water.

Cooling the nose of a tool by throwing a heavy stream of water or other fluid directly upon the chip at the point where it is being removed from the steel forgings enables the operator to increase his cutting speed about 40 per cent. The economy realized through this simple expedient is so large that it is a matter of the greatest surprise that the experimenters on the art of cutting metals have entirely overlooked this source of gain.

The following are the important conclusions arrived at as to the effect on the cutting speed of cooling the tool with a heavy stream of water :

(a) With high speed tools a gain of 40 per cent can be made in cutting steel or wrought iron by throwing in the most advantageous way a heavy stream of water upon the tool.

(b) A heavy stream of water (3 gallons per minute) for a 1-inch by 2½-inch tool and a smaller quantity as the tool grows smaller, should be thrown directly upon the chip at the point where it is being removed from the forging by the tool. Water thrown upon any other part of the tool or the forging is much less efficient.

(c) With modern high speed tools a gain of 16 per cent. can be made by throwing a heavy stream of water on the chip in cutting cast iron.

The gain for the different types of tools in cutting steel is :

(a) Modern high speed tools, 40 per cent.;

(b) Old style self-hardening tools, 33 per cent.;

(c) Carbon tempered tools, 25 per cent.

Chatter of the Tool.

The following are the general conclusions arrived at in regard to this question :

(a) Too small lathe-dogs or clamps or an imperfect bearing at the points at which the clamps are driven by face plate produce vibration.

(b) To avoid chatter, tools should have cutting edges with curved outlines and the radius of curvature of the cutting edges should be small in proportion as the work to be operated on is small.

(c) Chatter can be avoided, even in tools with straight cutting edges by using two or more tools at the same time in the same machine.

(d) The bottom of the tool should have a true, solid bearing on the tool support which should extend forward almost directly beneath the cutting edge.

(e) The body of the tool should be greater in depth than its width.

(f) It is sometimes caused by badly made or fitted gears.

(g) Shafts may be too small in diameter or too great in length.

(h) Chatter of the tool necessitates cutting speeds from 10 to 15 per cent slower than those taken without chatter whether tools are run with or without water.

The Effect of Feed and Depth of Cut on Cutting Speed.

The following are the principal conclusions arrived at on this subject :

(a) With any given depth of cut metal can be removed faster, i.e., more work can be done, by using the combination of a coarse feed with its accompanying slower speed than by using a fine feed with its accompanying higher speed.

(b) The cutting speed is affected more by the thickness of the shaving than by the depth of the cut. A change in the thickness of the shaving has about three times as much effect on the cutting speed as a similar or proportional change in the depth of the cut has upon the cutting speed. Dividing the thickness of the shaving by 3 increases the cutting speed 1.8 times while dividing the length that the shaving bears on the cutting edge by 3 increases the cutting speed 1.27 times.

(c) Expressed in mathematical terms the cutting speed varies with our standard round-nosed tool approximately in inverse proportion to the square root of the thickness of the shaving or of the feed, i.e., S varies with $\sqrt{F}$ approximately.

(d) With the best modern high speed tools, varying the feed and the depth of the cut causes the cutting speed to vary in practically the same ratio whether soft or hard metals are being cut.

(e) The same general formula expresses the laws for the effect of depth of cut and feed upon the speed, the constants only requiring to be changed. This is a matter of very great importance, and it enables us to use a single slide rule as a means of finding the proper combination of speed and depth of cut and feed for all qualities of metal which may be cut.

(f) The same general type of formula expresses the laws governing the effect of the feed and depth of cut upon the cutting speed when using our different sized standard tools. This is also fortunate as it simplifies matematical work in the final solution of the speed problem.

COPPER FAMINE.

A copper famine is impending owing to the enormous demand for the metal at home and abroad. The demand is due to the great growth of electrical work all over the world. Manufacturers of electrical machinery say that the demand for their products is unprecedented. Authorities in the copper trade say it is doubtful if the output of copper this year will show a material increase over that of 1905. The scarcity of skilled miners has handicapped the larger producers and consequently has reduced production. It is also likely that the Chilian production will be materially reduced by the recent earthquake.

ABOUT CATALOGUES

By mentioning Canadian Machinery to show that you are in the trade, a copy of any of these catalogues will be sent by the firm whose address is given.

WIRE NAIL MACHINERY. Bulletin No. 30 of the National Machinery Co., illustrating and describing the National wire nail machinery.

INTEGRATING WATTMETERS. Booklet of the Canadian Westinghouse Co., Hamilton, illustrating and describing type C. integrating wattmeters.

STARTING AND FIELD RHEOSTATS. Circular No. 1139 of the Canadian Westinghouse Co., Hamilton, illustrating and describing their starting and field rheostats.

AKRON FRICTION CLUTCH. Catalogue of the Akron Clutch Co., Akron, O., illustrating and describing in detail the Akron friction clutch. The catalogue is standard size.

CALYX CORE DRILLS. Catalogue No. 91, issued by Ingersoll, Rand Co., New York, describing the Davis calyx diamondless core drill. Views are shown of this drill in actual operation.

CONE GAS GENERATOR. Catalogue of the Cone Gas Machine Co., Detroit, Mich., illustrating and describing the Cone gas generator. It is well illustrated and contains a detail description of the plant.

ENGINE LATHES. A well illustrated catalogue issued by the Fairbanks Machine Tool Co., Springfield, Ohio: describing their engine lathes. The illustrations are of detail parts as well as of general views.

GAS ENGINES. Handsome and well-illustrated catalogue of the Bruce-Meriam-Abbott Co., Cleveland, O., description of their gas and gasoline engines for electric lighting, pumping and general power purposes.

GAS ENGINES. A high-class 28-page catalogue is being issued by the Bruce-Meriam-Abbott Co., of Cleveland, Ohio. This firm makes a wide variety of high-class gas and gasoline engines, several types of which are illustrated.

UNIVERSAL CUTTER AND TOOL GRINDER. Standard sizes catalogue of the Dayton Machine & Tool Works, Dayton, O., illustrating and describing their universal cutter and tool grinder. The catalogue is well illustrated.

FRICTION CLUTCH. Folder describing the Clarke improved friction clutch now being manufactured in Canada by William and J. G. Greey, 2 Church St., Toronto. This is a line which has lately been taken up by this firm.

FOUR SPECIALTIES. A folder, issued by F. L. Patterson & Co., 86 Victoria St., Toronto, Ont., describing four steam specialties manufactured by them. These include feed water heater and purifier, a steam separator, exhaust heads and Patterson belt pump.

STILL MORE SUGGESTIONS. As in previous years the Canadian Westinghouse Co., have issued an attractive hanger containing pithy paragraphs for manufacturers and business men. It makes an attractive wall hanger and a world of wisdom is contained in the paragraphs.

CANADIAN FAIRBANKS CO. Catalogue F. of the Canadian Fairbanks Co., Ltd., descriptive of power transmission appliances. This is their general power transmission catalogue, and is very complete in every respect, containing 111 pages, 6x9 inches, and a complete index.

PNEUMATIC TOOLS. The Globe Pneumatic Engineering Co., 150 Queen Victoria St., London, E.C., have sent catalogue describing some of their many lines of pneumatic tools. The illustrations are good and many tools are shown in actual operation. Forty-three pages, 8¼x10¼.

SMART & BROWN, of Erith, Kent, Eng., are sending out a folder illustrating their milling and wheel cutting attachments, also folder describing the colonial three-cylinder motor manufactured by this firm, being recommended as an ideal motor to drive a centrifugal pump or dynamo.

HIGH GRADE MACHINE TOOLS. An interesting little 24-page catalogue issued by the London Machine Tool Co., Ltd., Hamilton, Can., illustrating and describing various lines of their high-grade machine tools. Any one interested in machine tools will receive a copy on application to the above firm.

UNIVERSAL TOOL GRINDER. A very handsome and well illustrated catalogue illustrating and describing the Universal tool grinder of the Gisholt Machine Co., Madison, Wis. The catalogue is printed in two colors, and contains valuable information as well as fine illustrations of this grinder.

HANDY DESK CALENDAR. The Owen Sound Iron Works Co., engineers, founders, machinists and boiler makers, of Owen Sound, Ont., have issued a very attractive desk calendar, containing an adjustable card for the various months. It is of convenient size and makes an attractive addition to any desk.

FANCY CALENDAR. Allis-Chalmers-Bullock, of Montreal, have issued their 1907 calendar in the form of their well-known trademark shield. Besides a bird's-eye view of their works it contains various coats of arms of the different provinces in colors. The whole presents a handsome appearance, the general effect being pleasing in the neatness and general artistic arrangement.

BOILER FEED WATER HEATERS. The Loew Mfg. Co., Cleveland, Ohio, have issued a 31-page catalogue descriptive of their feed water heaters. Several lines are illustrated, views being given showing the large-size work done by this firm. These heaters possess some very distinctive features, as outlined in the catalogue, which may be had on application to the above firm.

WIRES AND CABLES. Supply catalogue, section 7, of the Canadian General Electric Co., Toronto, Ont. This catalogue is printed the same size as their regular engineering bulletin and may be filed in the same holder. Various styles of insulated wires and cables are illustrated and described. At the back some very interesting information is given, including tables of wiring, resistances and the rules of the National Board of Fire Underwriters.

THE WESTINGHOUSE DIARY for 1907. The Canadian Westinghouse Co. have the thanks of many of their friends, who were remembered at the beginning of the year with a copy of their handsome, high-class little diary for the year 1907. Besides a regular diary, with pages for addresses and memoranda, it contains a great deal of valuable information in the electrical field, the same being in concise form and handy for ready reference. A copy of this will no doubt be highly prized by those fortunate enough to receive one.

SHEFFIELD CALENDAR. A very neat calendar of a most sensible sort has been issued by Sybry, Searls & Co., Ltd., Cannon Steel Works, Sheffield. The upper part of the card shows three views of the interior and exterior of the works, and the calendar portion has the business man's week from Monday to Saturday on separate sheets with small blanks for memoranda each day. The reverse side contains a great deal of information of interest to metal workers. No doubt the publishers would be glad to send copies to any readers who make application.

Mechanical Reviews

OPINIONS AND STUDIES WORTH NOTING

A FOUNDRY RIG.

By Thomas Mack.

THE accompanying illustrations show a convenient sweeping rig, to which may be attached various strike boards, the striking edges of which contain radial sections of the pieces to be swept up. As the stroke boards are rigidly and permanently secured together with cleats, they can be attached to or removed from the rig very quickly as the case may require, no special adjusting being necessary. The advantages of the rig can readily be seen, for if a casting is

Fig. 1

to be duplicated all that is necessary is to go to the pattern storage, get the strike boards, and attach them to the rig.

The blocks B-B, upon the cleats which secure the boards together, guide the parts into their correct position upon the rig.

The rig is shown as applied to the sweeping up of a green sand mold for producing a dome casting, as shown in half section and half elevation in Fig. 1. The assembled arrangement of the rig with the strike boards attached is shown in Fig. 2. At the right of the spindle the operation of sweeping the exterior form of the dome is shown in progress. Upon this sweep surface the cope is subsequently rammed up. At the left of the spindle the rig is shown in the reverse position or upside down, as when set for sweeping off the metal thickness or striking up the interior form of the dome. Of course, the striking up of the interior form takes place after the ramming up and lifting off of the cope.

The separate parts of the arrangement are shown in Fig. 3. The rig itself is shown at the left, and consists of a rigid frame made from 3½ x 1¼-inch material, the frame being strengthened by cross braces, as shown. The spindle sleeves are of the hinge design, as this form of sleeve allows the rig to be attached to or detached from the spindle without disturbing the latter ; that is, the spindle can be set up and adjusted and the rig attached afterward. The sleeve is shown in greater detail above the frame. At the right of Fig. 3 are shown the strike boards as they would appear when removed from the rig. Care should be taken in attaching boards together to see that they are rigidly secured to the cleats with screws and a spot of glue at the extreme ends of the cleats. There are two advantages in this manner of gluing. First, any shrinkin or swelling that may take place in the boards will be toward or from the glue spots at the ends of the cleats and will prevent the striking edge of the boards being distorted. Second, if glue was applied to the portion of the cleats extending across the boards, warping of the boards would be the result. In fitting the boards to the frame the latter is laid down and the boards with the striking edge dressed to form, placed in about their correct position and the cleats attached. Next the centering pin shown directly above the striking boards in Fig. 3 is used in locating the blocks B-B, Figs. 2 and 3. These blocks should be securely fastened in place with glue and screws. The centering pin is usually turned up from well-seasoned hardwood, its diameter being the same as that of the spindle. A portion of the centre is cut out to half the diameter and the center line of the pin scribed thereon. By showing the pin through the sleeves as if it were the spindle and turning the flat side of the pin into the plane of the striking boards, the center line upon the pin becomes immediately available for setting the boards.—The Foundry.

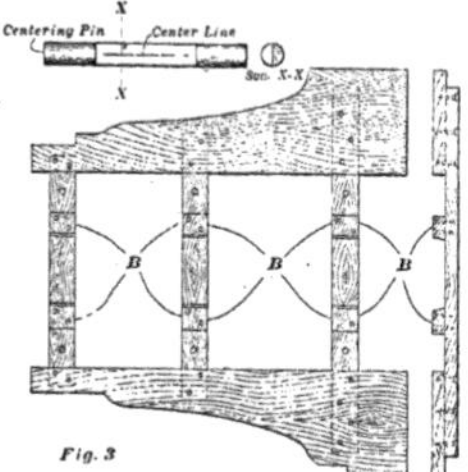

Fig. 3

Cleats
Sleeve
Spindle
Collar
B
B
B
Spindle Socket

Fig. 2

POWER REQUIRED BY VARIOUS MACHINE TOOLS.

APPROXIMATE power ratings of electric motors which have been applied and actually operated on the machine tools mentioned are given in the following table. By referring to this table anyone can arrive approximately at the power required for operating any particular tool.

Niles 96 in. driving-wheel lathe, chain connected, 10 hp. Group, overhead belted. Niles 42 in. engine lathe, emery wheel; Craven quartering machine, 5 hp. Niles double axle lathe, chain connected, 7½ hp. Niles 48-in. car wheel lathe, chain connected, 10 hp. Sellers planer 54-in x 54 in., belted overhead, 10 hp.

Niles 8-ft. boring mill, chain connected, 10 hp.

Niles 8-ft. wheel lathe with quartering attachment, variable-speed motor, chain connected, 10 hp.

For driving quartering attachment to this lathe, two motors, chain connected, 3 hp.

Fairbain, Kennedy & Naylor 72-in. driving wheel lathe, belted overhead, 7½ hp.

Putnam 79-in. driving-wheel lathe, belted overhead, 7½ hp.

Niles 96-in. driving-wheel lathe, chain connected, 10 hp.

Group, belted overhead. Becker Brainerd No. 6 vertical miller; Becker Brainerd No. 7 vertical miller; emery wheel; Niles 30-in. shaper; Bement Niles No. 2 Cotter drill; Niles vertical boring mill; Niles 24-in. drill press; Niles 4 spindle drill, 15 hp.

Niles double axle lathe, chain connected, 7½ hp.

Niles 48-in. car-wheel borer, for driving tool, chain connected, 7½ hp.

For driving hoist—geared, 3 hp.

Niles 48-in. hydraulic wheel press, belted overhead, 5 hp.

Group, belted overhead. Niles 6-spindle drill; Niles 24-in. shaper; Bement Niles 15 in. slotter; Niles gap lathe 24 in. x 48 in.; Niles lathe 28 in. x 14-ft. Niles lathe 18 in. x 8 ft., 15 hp.

Bement Niles 18-in. vertical slotter, chain connected, 10 hp.

Tool room group, belted overhead. Brown & Sharpe milling machine; S. C. Wright 16 in. lathe; Bement & Son 9-in. shaper Niles-Bement-Pond No. 2 milling machine; Niles-Bement-Pond cutter grinder; Sellers twist drill grinder, 7½ hp.

Niles 7-ft. boring mill, belted overhead, 7½ hp.

Group, belted overhead. Bement 32 in drill press; Lodge & Davis 22-in. lathe; Gisholt 21-in. turret lathe; Niles-Bement-Pond Co. 20-in. lathe, 10 hp.

Group, belted overhead. Pond engine lathe, 36 in. x 17-ft.; Pond engine lathe, 36 in. x 20 ft.; Niles engine lathe, 30 in. x 16-ft.; Gisholt universal tool grinder, 10 hp.

Group, belted overhead. Niles spindle drill; Newton cold saw; Niles-Bement-Pond lathe, 20 in. x 10 ft.; 3 Jones & Lawson turret lathes, 2 in. x 24 in.; Gisholt 24-in. turret lathe; No. 2 water emery wheel, 15 hp.

Group. Gisholt 28 in. turret lathe; Niles-Bement-Pond Co. 24-in. drill press, 10 hp.

Pond planer, 32-in x 10-ft. belted overhead, 5 hp.

Niles 30-in. x 30-in. planer, geared 5 hp.

Group, belted overhead. Gisholt 34 in. turret lathe; Niles No. 2½ horizontal boring mill; Niles 22-in. lathe; Lodge & Davis 18-in. lathe; Le Blond No. 3 milling machine, 15 hp.

Group, belted overhead. Niles 20-in. shaper; Niles-Bement-Pond Co., 12-in. pipe machine; Niles 16 ft. radial drill press, 5 hp.

Niles 60 in. x 60 in. planer, overhead belted, 15 hp.

Pond 48 in. x 48 in. planer, chain connected, 10 hp.

Niles 60-in. horizontal boring mill, variable speed motor, chain connected, 3 hp.

Group, overhead belted. Brass room, Lodge & Davis lathe; 2 Niles 18-in. lathes; 2 Fox 18-in. lathes; 2 Niles-Bement-Pond Co. 18-in. lathes; 2 Fox 16 in. lathes, and emery wheel, 10 hp.

Group, overhead belted. Pond lathe, 28-in. x 14 ft; Wm. Bement 16-in. slotter; Wm. Bement 12 in. slotter; 2 Niles lathes, 26 in. x 12 ft.; Niles 24-in. shaper; Bement Niles 12-in. shaper; Springfield Mfg. Co.'s No. 5 oscillating surface grinder; Springfield Mfg. Co.'s face and variety grinder, 15 hp.

Group, overhead belted. Pnd lathe, 28-in x 20-ft.; Niles-Bement-Pond 21-in. drill press; Niles screw machine; Niles engine lathe, 36 in. x 15 ft.; Bement 24 in. drill press; No. 2 water emery wheel; Niles extension lathe; Fitchburg 32-in. lathe; Bement 50-in. drill press; centreing machine, 15 hp. Niles 30-in. doubled headed shaper, belted overhead, 5 hp. Tin Shop. Bliss press and rolls, belted overhead, 10 hp. Group, belted. Niles 4 spindle drill with sliding table for drilling mud rings; Bement Niles 10 in. vertical drilling machine; Putnam 18 in. drilling machine, 75 hp. No. 2 Long & Allsatter double punch and shear, chain connected, 7½ hp. Group. Niles 7 ft. plate-bending roll; Ajax 12-ft. plate bending roll, motor on floor, chain connected to countershaft and belted to tools, 20 hp. Kneeland No. 1 R lever shear for 1½-in. square steel, belted overhead, 5 hp. Lloyd Booth No. 6 lever shear for scrap geared, 7½ hp. Williams, White & Co.'s No. 2 bulldozer, geared, 5 hp. Ajax forging roll, motor on floor, belted, 10 hp. No. 10 Sturtevant steel pressure blower motor on floor, belted to overhead countershaft, belted to blower, 40 hp. Morgan double punch and shear, chain connected, 10 hp. Newton No. 3 bar cold saw, chain connected, 5 hp. Group. Blakeslee 1-in. forging machine; 2 Blakeslee 2 in. forging machines; 1½ in. forging machine, on floor, belted to overhead countershaft, 40 hp. Group. 2 Acme 1½ in. 6 spindle nut tappers; Oliver 8 spindle nut tapper; Acme 2 in. double bolt cutter; Lewis & Oliver 1-in. double bolt cutter; National 1½-in. double bolt cutter; 2 Acme 1½-in. double bolt cutters; Acme 5 in. single bolt cutter, motor on floor, belted to overhead shaft, Acme 1½ in. single bolt cutter 30 hp. Flue shop. 2 Hartz flue welders; combined flue cutter and tester, overhead belted, 7½ hp. Outdoor flue rattler, 34 in. diameter by 16 ft., motor on floor in separate inclosure, belted, 7½ hp. Tank shop. Small Gramo Klausman band saw, overhead belted, 5 hp. Physical laboratory. Testing machine, motor on floor, belted, 5 hp. Brass foundry. 2 car brass boring machines; Root positive blower; rattler; overhead belted, 10 hp. Stores dept. Incline movable platform, variable speed motor chain connected, 3 hp.—El. World & Eng.

VALUE OF BABBITT METAL.

Babbitt metal is a composition that is used extensively as a substitute for brass in the lining of journal bearings, says the Valve World. To make such a bearing, the shaft for which it is required is placed in position in its recess or cavtiy, and then the melted babbitt metal is poured in around it and allowed to cool, thus forming the bearing for the shaft. The value of babbitt metal for this purpose lies in the fact that it has all the merits of brass as a wearing material, and at the same time melts at a much lower temperature; and also being of a very fluid nature when melted, it may be poured into the cavities around the shaft and the expensive operation of fitting brass to these places thus be dispensed with. What is called the "genuine," or superior babbitt metal is described under the heading of Tin Alloys. But there are a variety of cheaper grades of this material, in which lead is very extensively used alloyed with antimony, or with antimony and tin. While these compositions are not so good as the "genuine," they are of sufficient merit for a large variety of cheap machinery, and are extensively used in the repairing of agricultural implements, and other grades of cheap machinery. An immense quantity of the combination of lead and antimony is produced by the silver smelters, a market for which is found in the production of babbitt metal.

Foundry Practice

MODERN IDEAS FOR THE TRADE.

FERRO-ALLOYS.

By A. E. Outerbridge, Jr.

THE use of ferro-alloys in the foundry is likewise extending. The simple method first described in the address given before the metallurgical section of the institute a year ago for softening iron for castings to any desired degree has been continued, with exceedingly beneficial results.

The speed of turning pulleys has been largely increased and the time required to complete the machine work upon them has been reduced, and in other light castings as well. In addition to the material improvement in the strength of the metal by this treatment, there is a decided decrease in the shrinkage, so that some castings of irregular shape which are difficult to make without cracking in cooling, owing to unequal strains, are now made without this tendency by this process. Other advantages accruing from the addition of a very small amount of high-grade ferro-silicon (containing about 50 per cent. silicon) have been observed. Among these may be mentioned the cleaning action upon the molten metal which the alloy effects by its deoxiding influence. Singularly, also, it is found that commercially pure silicon does not produce these results, neither does the ordinary grade of ferro-silicon, containing about 20 per cent. of silicon.

The samples cast in sand are all perfectly gray, but there is an appreciable difference in the fracture, the treated metal showing softer iron than the untreated sample.

The metal selected for this particular test was comparatively low in silicon, consequently having comparatively high chilling property, but this fact would never be suspected from an examination of the fracture of the test sample cast in sand. Very few ptactical founders are aware of the large variation in chilling property of their iron, sometimes of presumably similar grades, melted on different days, or even of the wide variations which may occur in different portions of one heat. The chill cup test here shown is an invaluable instructor in this matter, and without its aid the beneficial effects of adding high-grade ferro-silicon to molten cast iron could not have been so readily detected and exhibited. These chill magnifiers have been used with great benefit in daily practice more than twenty years. They are far more effective than the ordinary chill tests made in car wheel foundries. Such tests are only applicable for very high chilling iron, while the cup tests here described are only intended for use with comparatively low chilling irons.

In the case of pure silicon the specific gravity of the material is too low and its melting point too high to permit it to become incorporated with the molten

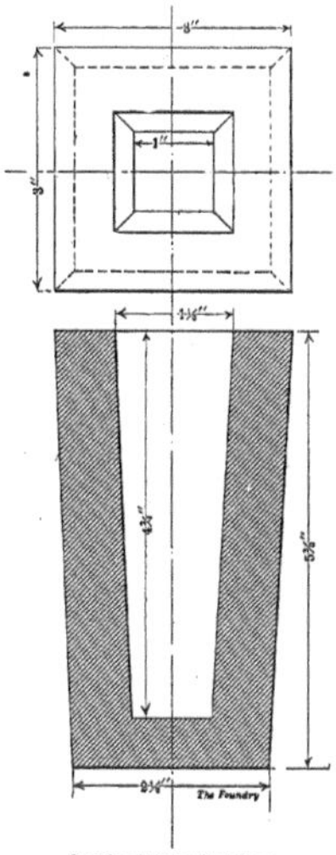

Cup for Chilling Test Bars.

iron in the ladle, in the other case the proportion of silicon is too low to permit a sufficient amount to be dissolved in the iron to produce a radical change in its quality without causing dull iron in the ladle.

There is no doubt but that when the merits, simplicity, certainty of action and other advantages of this process of treating molten iron in the ladle are better known and appreciated it will come into very extensive use, for it enables the founder to modify the character of his iron to suit individual castings, a matter of considerable importance and value. This is accomplished without expense, for the cost of adding silicon to the iron in this manner is actually less than by the usual method of adding an equal amount of silicon to the iron in the form of pig iron, comparatively high in silicon, added in the cupola. For example, pig iron containing six per cent. of silicon costs at the present time, let us say, $22.50 per ton, and we may estimate for comparative purposes $3.75 per unit of silicon; ferro-silicon containing 50 per cent. of silicon costs at the present time in powdered form $100 per ton, or $2 per unit of silicon on the same basis of calculation.

Furthermore, there is no loss of silicon when added in the ladle, as there is when melted in the cupola, and this partly accounts for the fact that a given quantity of silicon, when added in the ladle, is much more effective as a softener than the same quantity charged into the cupola. Silicon added in the cupola always weakens the iron, while silicon added in the ladle always strengthens it.

CRANDALL CUPOLA.

IN designing this cupola Mr. L. A. Crandall had in view economy of fuel, the most perfect combustion, and fluid iron in the shortest space of time. In placing this cupola on the market the Detroit Foundry Supply Co. of Detroit, Mich., and Windsor, Ontario, are offering the foundry public something with many points of excellence and superiority.

The air chamber being on the inside rather than outside of the shell enables the forming a "bosh" along the same lines as a blast furnace. With the best point of combustion in the centre, it naturally follows that in this cupola with the air chamber in the inside of the shell much nearer centre is attained than with straight line cupolas. In the bed of the cupola the greatest amount of fuel is used. The area of a straight lined cupola of 53 inches in diameter, inside 6 inches lining, is 1385.4 square inches. The standard size cupola requires a bed of coke to 20 inches above the upper tuyeres, which will be about 52 inches for this size. This height multiplied by 1385.4 gives cubic feet of area, equals 7,204 cubic inches and this divided by 65 gives the weight of coke necessary for the bed (65 cubic inches is the space a pound of coke will occupy in a cupola). A Crandall cupola 55 inches in diameter has an average diameter inside of 6 inches brick lining of 37 inches or 1075.2 square inches in area. Distance bed to 20 inches above tuyeres is 40 inches. This height multiplied by area give 4,300 cubic inches, which, divided by 65, gives 662 lbs. of coke, or a saving of 447 lbs. to each heat.

This cupola at the melting point is the same diameter inside of brick lining as in the straight lined cupola, and will melt equally as much iron per hour with the eceonomy in coke to the users, advantage. Having a slight taper from bottom of charging door to melting point any possibility of charges hanging up is prevented. The charging door made of heavy screening will not allow of heat coming through the charging door owing to radiation of wire. Nor will it burn out as will a solid door, making it very convenient for the charger or melter as he is able at all times to see into the cupola without the annoyance of excessive heat.

Mr. Crandall, the inventor of this cupola, is not a theorist but has manufactured, operated and sold cupolas for 20 years, which places him an authority in the cupola business. Further information regarding this will be given by the Detroit Foundry Supply Company, Detroit, Mich.

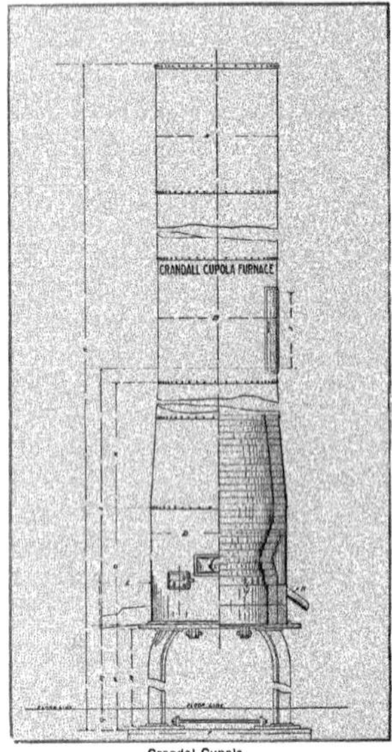

Crandal Cupola.

NEW WAY TO TREAT SCRAP BRASS.

A method of treating scrap brass has been patented by Henry J. Krebs of Wilmington, Del. (U.S. Patent 831,010, Sept. 11, 1906), which may be of interest to scrap manufacturers who desire to investigate the recovering of the zinc which always volatilizes when brass is melted. While this process is intended to treat the brass primarily for the recovery of the oxide of zinc, the same principle may be used for the recovery of the oxide of zinc in brass melting.

The principle of the appliance and process will be seen in the illustration. A cupola (1) is used for melting the brass. The blast of air that is used in the cupola oxidizes the zinc in the brass and the oxide of zinc passes to the chamber (2). This chamber is intended to divert the oxide of zinc in the cooling chamber (5). Auxiliary cooling chambers (6) are also provided through which the fumes pass. These chambers are directly connected to an exhaust fan (8) and then the fumes pass to a tower (15) which is filled with brickwork similar to an acid tower. An acid solution, say sulphuric acid, is sprayed through pipes 13 and 14 and converts the oxide of zinc into sulphate of zinc. From the tower the fumes pass to bags (10) which catch any excess of oxide of zinc which passes through the tower. A valve (12) may control the passage of the fumes so that only a portion is

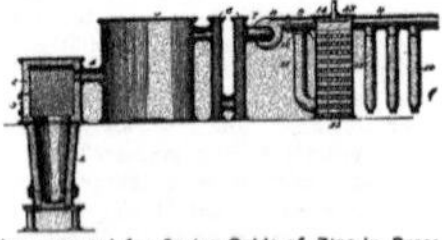

Arrangement for Saving Oxide of Zinc in Brass Melting.

obtained as sulphate of zinc while the rest are obtained in the bags.

In order to use an appliance of this nature in the recovery of the oxide of zinc which escapes in brass melting, it would be necessary to use an exhaust fan in the flue of the furnace as natural draught would otherwise be choked so that it would be difficult to obtain the necessary heat in the melting furnaces.

AN ENQUIRY.

W. J. Hepburne, 223 McCaul St., Toronto, wishes to know where he can purchase finished gyb keys. There are several manufacturers of these in Canada, who will no doubt be anxious to attend to Mr. Hepburne's requirements.

The election of officers of the Canadian Railway Club held recently resulted as follows: Mr. W. Kennedy, master mechanic, Grand Trunk Railway, President; Mr. W. R. McRae, master mechanic, Toronto Railway Company, First Vice-President, and Mr. R. Preston, master mechanic, Canadian Pacific Railway, Second Vice-President.

The workman who takes a real interest in his work doesn't have to spend much time looking for a job.

Tricks of the Trade

Everybody is invited to contribute to this page. Send in your ideas, odd jobs you are doing, and anything of interest to fellow-workmen. Remuneration will be made for such contributions. [Ed.]

LIFTING BY THE HEAD.

By R. Manley Orr, Brantford.

A FEW days ago I saw an engineer trying to lift the cap off his main bearing with a rope and a couple of pulleys which he had rigged up as shown in Fig. 1. He pulled on the rope at A till he lifted himself clear off the floor but could not raise the cap. By jerking down with his whole weight he could just jerk up the cap about one-quarter of an inch; and there he was stuck. After he had tugged away in vain for a while I suggested that he make a change in his rig and attach the pulley C to the beam and the pulley B to the cap and then get up on the beam himself and try pulling up; just the reverse of the arrangement which he had. He soon had the change made as shown in Fig. 2, and found that the cap came up easily. After a little study

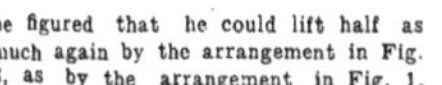

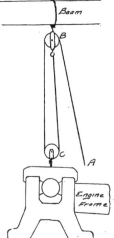

Fig. 1.

he figured that he could lift half as much again by the arrangement in Fig. 2, as by the arrangement in Fig. 1. That engineer was myself and I am learning that in working about machinery one can sometimes lift more with their head than with their hands.

OILING LOOSE PULLEYS.

By L. Failey, Hamilton.

UNDER the head of Tricks of the Trade I would like to say for the benefit of those who may have had some trouble with loose pulleys that I have had a lot of

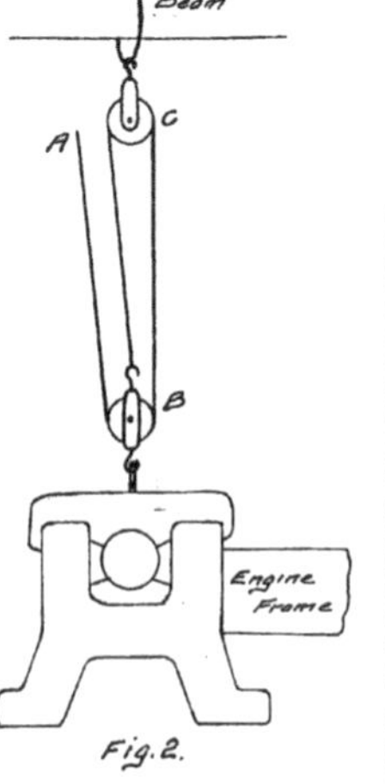

Fig. 2.

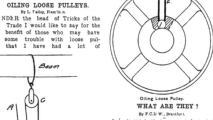

difficulty with loose pulleys that run at a high speed in getting the oil to stay in and oil the shaft properly, but at last I struck a scheme that has proved to be just the thing, and a glance at the accompanying sketch will explain the trick.

It will be seen at Fig. 1 is a hole drilled right through the hub of the pulley and Fig. 2 is a piece of ¼-inch brass tube with a grub screw in the outer end to keep the oil from coming out. The hole that is drilled in the hub just breaks through into the pulley bore and a piece of candle wick pulled through the hole absorbs the oil that is put into the brass tube. And so the wick coming in contact with the shaft keeps a perfect lubrication on the frame.

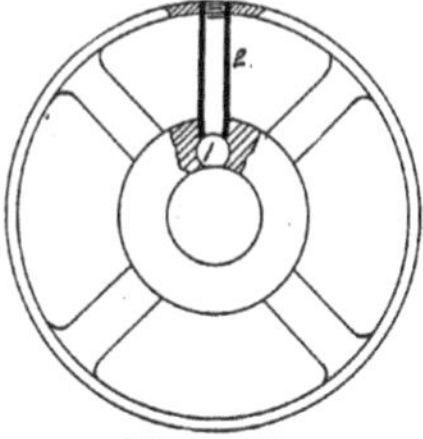

Oiling Loose Pulley.

WHAT ARE THEY ?

By F.C.D W., Brantford.

I don't think that many at first sight (even if they could see the pieces themselves) would be able to spot just what they were, or where they came from. These two bits shown in the figure are two pieces of common babbit metal taken from a broken bearing. Some time ago I had occasion to oversee the removal of a 7-inch shaft from the power house of the Western Counties Power Co.'s power house. On removing the cap from one of the large bearings we found the babbit very badly broken. A cul-de-sac had been formed by a piece falling down until it rested against the babbit in the body of the bearing and in this space we found these two pieces. One is almost a perfect sphere and the other is perfectly cylindrical, having one end pointed and the other rounding off. At first I couldn't see just how these were formed, but the point on the one piece gave me a clue. The piece that had fallen down had been over two anchor holes, and these two pieces had been sheared off when it was forced down by the shaft, and had fallen out

What Are They?

of the holes to be turned around by the shaft in every direction until the sphere was formed. The cylindrical piece was just too long to be turned in more than one direction, hence it could only assume the cylindical form.

Practical Questions and Answers

Production of Pisten Rings.

Ques.—What are the different operations involved in the making of piston and piston valve rings?

Ans.—The making of piston and piston valve rings involves several processes of an interesting nature. The machining operation commences on a cylindrical casting, which casting always has lugs cast on one end by which it can be chucked to the face-plate of a lathe to be bored and turned. In ordinary practice the boring of the outside and inside diameters of the cylinder is done as two separate operations. But both of these operations can be proceeded with simultaneously by employing two cutter bars and tools. The cylinder is held to the face-plate, and the two cutter-bars are strapped to a flanged plate, which is tongued and fitted into a T slot on the tool rest. The cutter-bars are placed parallel with the length of the cylinder, one being inside and one outside the cylinder. The tools are fitted into the ends of the cutter-bars, and are adjusted to the cut by means of adjusting screws operating on the cutter bars.

The inside and outside surfaces of the casting having been turned up, the face of the cylinder is trued up. Then the casting is ready for the cutting off of the rings. This may be done with an ordinary parting tool, or with a double-tongued parting tool, one tongue to regulate the width of the ring cut-off. The ring, having been cut off, the new face exposed by the operation is trued up, and another ring parted from the cylinder. Some of the double-tongued parting tools are made adjustable.

The unfinished side of the rings are trued by chucking them in the jaws of an ordinary lathe chuck and in a special expanding chuck made for the purpose.

The cylinder from which packing rings are parted is always turned larger than the bore of the cylinder in which the rings are to work. The amount of this enlargement is equal to one-third of whatever metal is cut away in splitting the ring. The old practice of splitting rings was to cut a piece out of the ring in a straight line, but at an angle to the surface of the ring. The ring was then sprung into the cylinder, and the high places filed down until there was a fit. But that practice is now obsolete. The modern practice is to divide the amount to be cut out between the two sides of the ring. By this method the ring is equal to a solid ring, if the joint is well made. The rings are now chucked, and turned to fit the cylinder. The most efficient way to hold the rings together while they are being chucked is by means of a clamp, instead, as is done often, riveting together the tongues formed in the rings by the cutting. The rings are usually turned in pairs.

Sometimes packing rings are made thick on one side and thin on the other side of their diameter, and then afterwards split on the thinnest side, on the supposition that when the ring is sprung into place in the cylinder it will expand equally in all directions, thereby maintaining a true circle at all parts of the diameter. This form of packing ring has, however, fallen into disuse, they not having given the satisfaction expected of them.

Size of Ropes and Pulleys for Rope.

Ques.—Can you give any hints regarding the relative size of pulleys and ropes for the most efficient rope drive?

Ans.—The size and number of ropes as well as the least diameter of the pulley for any given power are points of importance, and should be considered for each case. It is better to use small ropes, consistent with not having too many in the space available. There is less internal friction with small ropes than large ones, and hence greater efficiency and endurance. For a main drive where the pulleys are seldom less than 5 feet in diameter, 1¾-inch ropes are most usually employed. The least size of the pulley should not be less than thirty times the diameter of the rope.

Allowance for the Machining of Forgings

Ques.—In good practice what allowance is made for the machining of forgings?

Ans.—There is considerable difference of opinion regarding the allowance necessary for the machining of forgings, but a good rule to be adopted is to allow ¼-inch for forgings up to 5 inches in diameter, ⅜-inch for pieces ranging from 6 inches to 8 inches, ½-inch for 9 inches to 10 inches, and 1 inch for 1 foot.

High Compression in Gas Engines.

Ques.—Can you furnish me with data regarding the high compression of the explosive mixture in gas engines?

Ans.—In these columns of the September, 1906, issue, some information is given on this subject, and some experiments conducted by Prof. Burstall, of Birmingham University, are quoted. Additional information along this line is quoted from R. E. Mathot's book on Gas Engines and Producer Gas Plants." He says:

In actual practice the problem of high pressures is apparently very difficult of solution, and many of the best firms still cling to old ideas. The reason for this course is perhaps to be found in the fact that certain experiments which they made in raising the pressures resulted in discouraging accidents. The explosion-chambers became over heated; valves were distorted; and premature ignition occurred. Because the principle underlying high pressures was improperly applied, the results obtained were poor.

High pressures cannot be used with impunity in cylinders not specially designed for their employment; and this is the case with most engines of the older type, among which may be included most English, French, and particularly of American construction. In American engines notably, the explosion-chamber, the cylinder and its jacket are generally cast in one piece, so that it is very difficult to allow for the free expansion of certain members with the high and unequal temperature to which they are subjected.

Some builders have attempted to use high pressures without concerning themselves in the least with a modification of the explosive mixture. The result has been that, owing to the richness of the mixture, the explosive pressure was increased to a point far beyond that for which the parts were designed. Sudden starts and stops in operation, overheating of the parts and even breaking of crank-shafts, were the results. The engines had gained somewhat in power, but no progress had been made in economy of consumption, although this was the very purpose of increasing the compression.

High pressures render it possible to employ poor mixtures and still ensure ignition. A quality of street gas, for example, which yields one horsepower per hour with 17.5 cubic feet, and a mixture of 1 part gas and 8 of air compressed to 78 pounds per square inch, will give the same power as 14 cubic feet of the same gas mixed with 12 parts of air and compressed to 171 pounds per square inch.

"Scavenging" of the cylinder, a practice which engineers of modern ideas seem to consider of much importance, is better effected with high pressures, for the simple reason that the explosion-chamber, at the end of the return stroke, contains considerably less burnt gases when its volume is smaller in proportion to that of the cylinder.

In impoverishing the mixture to meet the needs of high pressures, the explosive power is not increased and in practice hardly exceeds 365 to 427 pounds per square inch. With the higher pressures thus obtained there is no reason for subjecting the moving parts to greater forces.

The increase in temperature of the cylinder-head and of the valves, due wholly to high compression, is perfectly counteracted by an arrangement which most designers seem to prefer, and which consists in placing the mixture and exhaust valves in a passage forming a kind of anti-chamber completely surrounded by water. The immediate vicinity of this water assures the perfect and equal cooling of the valve seats. .... It is advisable to run the producer gas engine "colder" than the older street-gas type, in which the more economic speed is that at which the water emerges from the jacket at about a temperature of 104 degrees F. It would seem advisable to meet the requirements of piston lubrication by reducing to a minimum the quantity of heat withdrawn by the circulating water.

For street-gas engines, however, the cylinders should be worked at the highest possible temperature consistent with the requirements of lubrication. It should not be forgotten that, in large engines fed with producer gas, economy of consumption is a secondary consideration, because of the low quality of fuel required.

ACTIVITY AT NIAGARA FALLS.

The Canadian Nut & Bolt Co. have decided to build a factory at Niagara Falls to employ 50 men. Already they have contracts for a large supply of nuts and bolts for the Canadian Ramapo Iron Works, which is also established in that city. A factory for the manufacture of automatic fire shutters is also to be started. Silver plating has been undertaken by the Wm. Rogers Co. The Niagara Falls Machine & Foundry Co. have secured a large contract to supply cast and wrought iron work for the new Detroit River tunnel, and another to supply fire hydrants to the city of Toronto. Their plant will be enlarged at once.

ELECTRICAL ASSOCIATION QUESTION BOX

A. A. Dion, of Ottawa, has sent to the members of Canadian Electrical Association question blanks to be filled in with any questions members of the association desire to have answered. This feature of the meetings of the Canadian Electrical Association has in the past been of the greatest help and benefit to its members, and the editor is looking forward to an even more liberal interest than heretofore in this department.

PERSONAL MENTION

Mr. Peter Eyermann, one of the foremost gas engine authorities, designers and engineers from Germany, has been appointed chief of the Engineering Dept. of the DuBois Iron Works.

Professor Louis Hurdt of McGill University, Montreal, together with Wm. Kennedy, hydraulic engineer of Montreal, and H. N. Ruttan, city engineer of Winnipeg, have been appointed a Power Committee by the city of Winnipeg, to act as a consulting board in connection with the power projects of that city.

The appointment of Mr. A. Gaboury to the position of superintendent of the Montreal Street Railway is an evidence of the possibilities before a young man of intelligence and energy. In twelve years Mr. Gaboury has risen step by step from a most humble position in the company to be superintendent of the entire system.

Mr. Charles Brandeis, consulting engineer of Montreal, has recently been appointed controlling engineer to the Minister of Public Works and Labor at Quebec. This appointment means a tribute to the high standing Mr. Brandeis is achieving in professional work. He is at present in Cuba on a two weeks' trip.

Mr. S. W. Smith, formerly with the Canadian Westinghouse Co. in their Montreal office, has accepted the position of Ontario representative for the Packard Electric Co., of St. Catharines, with headquarters at Toronto. Mr. Smith has many friends in Toronto and throughout Ontario, who will be glad to notice his appointment to this position and it is safe to predict that the interests of the firm will be well looked after under the management of Mr. Smith.

Mr. W. H. Banfield, of W. H. Banfield & Sons of 120 Adelaide St. West, Toronto, has just returned from a trip to the United States, and whilst there he purchased a complete outfit of canning machinery for the Western Canning Co., Ltd., of Winnipeg. The bulk of the order was placed with Torris Wold & Co., of Chicago, Ill., the machinery purchased being similar to that used in the Swift Co. Banfield & Sons are making a specialty of this class of goods, and have already done considerable business in the various canning factories in the Dominion. They are also making arrangements to represent Torris Wold, for Canada.

T. Gerald Freedman, a director of the Buyers and Sellers of Australasia Co., Ltd., with head office at Melborne, Australia, spent some days in Toronto recently and was a welcome visitor at the office of Canadian Machinery. Mr. Freedman has just completed a tour of the United States studying manufacturing conditions, and came to Canada with a view to getting in touch with the manufacturing interests of this country, being particularly interested in the development of Canadian export trade to Australia. The American office of this company is at 1269 Broadway, New York, where they will be glad to receive communications from any manufacturers in Canada desirous of increasing or opening up trade with Australia.

Successful Draftsmanship.

Many men wish over and over again that they had a working knowledge of drafting and a more intimate insight into the reading of mechanical drawings. This is particularly true of mechanical men who have not had an earlier opportunity of making good along this line.

MR. FRED. DOBE.

It has created a longing on the part of some to better their position, and oftentimes a feeling of embarrassment at not being able to make a simple sketch or a correct drawing.

Mr. Fred Dobe, M.E. and chief draftsman of the Engineers' Equipment Company, 97 Washington Street, Chicago, has for a long time been making a practice of giving personal and individual instructions in complete architectural as well as mechanical drawing and designing, and is prepared to accept a few more students old or young. His instructions are given by mail, but must not be compared with the ordinary correspondence school instructions, as all the work is laid out personally by himself and prepared especially for individual requirements. With this method he is able to satisfy and educate any experienced or absolutely inexperienced man, who is willing to better himself. He sells no diplomas, but insists on your work being a more practical diploma and to do the talking, and guarantees by contract to qualify you in a few months' instruction to be able to hold a first-class draftsman's position and instruct you until you are competent. He furnishes free this month only, without extra cost, a complete and highest grade full drawing outfit, including finest set of German silver tools worth $13.85. His book, "Successful Draftsmanship," size 6x9, is sent free with full particulars for four cents in stamps to cover cost of mailing.

New District Manager.

Recognition of his ability and excellent work done in the service of the company has come to Mr. W. Archie Duff, assistant manager of the Montreal Office of the Canadian Westinghouse Co., in his promotion to the western managership with headquarters at Winnipeg. Mr. Duff has spent many years able service in the Westinghouse interests and this well

MR. W. A. DUFF,
New District Manager of Canadian Westinghouse Company.

merited promotion is appreciated by his many friends throughout the country. Mr. Duff is a graduate of McGill University in electrical engineering, and his past experience together with natural ability in engineering work eminently qualify him for the position he now holds. Mr. Duff is well known in the electrical engineering field and carries to Winnipeg the best wishes of all his friends in the east. He is a brother of Mr. Melville Duff of the C. P. R. Steamship Service.

Mr. Thos. R. Loudon, B. Sc., of Toronto, has been appointed general manager and secretary of the Canadian McVicker Engine Co., of Galt, Ont. Mr. Loudon's previous experience should admirably fit him for this position.

Prominent Manufacturer Dead.

Manufacturers throughout Canada will regret to learn of the rather sudden death a week ago of Leonard McGlashan, of Niagara Falls. With a party of relatives and friends he had gone to California a month ago to enjoy the milder winter climate of the Pacific coast, but on his arrival there was taken ill with typhoid fever, the disease which caused his death. Mr. McGlashan was 61 years of age and besides a widow, leaves a daughter and one son, L. Lee McGlashan, of the McGlashan-Clarke company, Niagara Falls, manufacturers of cutlery and silverware.

Mr. McGlashan was probably the wealthiest citizen of Niagara Falls and was one of the first users of power from Niagara River, besides being interested in many enterprises, including mines and railways in the southwestern states and Mexico, and was a large real estate owner. He was until recently the chief owner of the Ontario Silver Company, which had a large factory at Thorold, then at Humberstone and finally located in Niagara Falls, where it is an important industry. Mr. McGlashan sold out his interest to the International Silver Company some time ago, and since then the McGlashan-Clarke Company was formed and built the large works at Niagara Falls, which commenced operations only a month or so ago.

A Manufacturing Mayor.

At the election for the office of mayor for Owen Sound on Jan. 7th, Mr. Mathew Kennedy of Wm. Kennedy & Sons of that city was elected to the position by a large majority. With Mr. Kennedy's broad business experience and large interests there and public spiritness the the citizens of Owen Sound are assured of their interests being well looked after for the coming year.

VOTES ON PUBLIC IMPROVEMENTS

The councils of a large number of Ontario cities and towns submitted by-laws at the municipal elections on Jan. 7, to raise money by way of debentures for various improvements. In the majority of cases these by-laws were carried.

Southampton decided to raise $5,000 to extend the waterworks system.

Napanee passed by-laws to improve the town's sewerage.

Waterloo voted 390 for and 74 against the sewer commission proposal, and also carried a by-law for a $23,000 loan for sewer farm improvements.

St. Catharines voted favorably for extension of the waterworks.

Listowel decided by 17 votes to raise $14,500 to construct new bridges.

Brantford carried sewer extension by-law.

Almonte decided to appoint an electric light commission.

Kincardine carried the Grand Trunk station, Coleman, and Park by-law.

Woodstock endorsed by-laws to establish a parks commission and to raise $20,000 for sewer extensions.

In a quiet vote Guelph took a jump forward in its municipal government by deciding in favor of commissions to manage its waterworks and gas and electric light plants. The vote for the waterworks commission was 1,132 for, 3'6 against; for gas and electric light commission, 1,104 for 371 against, and to loan $25,000 to enable Morlock Bros. to extend their factory, 966 for, 159 against.

Renfrew, by a vote of 136 to 103, decided to raise $6,000 for sewer extensions, and the county system of good roads was carried, 171 to 65.

THE LATE LEONARD McGLASHAN.

Bowmanville decided to have its waterworks managed by a commission.

ENGINEERS' CLUB OFFICERS.

At the election of officers at the Engineers' Club of Toronto, which took place Thursday, Jan. 10th, the following were elected: President, C. B. Smith; First Vice-President, J. G. Cing; Second Vice-President, A. B. Barry; Secretary, Willis Chipman; Treasurer, John S. Fielding; Chairman of Rooms Committee, C. M. Caniff; Chairman of Library Committee, A. F. Macallum; Chairman of Papers Committee, R. G. Black; Auditors, W. E. Douglas and W. H. Patton.

Power and Transmission

Steam *Gas* *Electricity* *Compressed Air* *Water*

DODGE FRICTION CLUTCHES.

DURING the working hours, when the engine or other prime mover is in continuous motion, some of the driven machinery may be required to run all the time; all of it is presumably supposed to run some of the time; power users from the remainder of the system, then the shipping belt is not the thing.

These conditions, as well as any others wherein convenience of manipulation for any item or section of a system is a factor, are satisfied most readily, and ling or cut-off clutch is properly used.

Thus, by intelligent planning of any new system or alteration of an old one, this desirable feature of providing means for throwing any unit or section into or out of service at will may be secured.

FIG. 1.

FIG. 3.

but not all of it can commonly be expected to run all the time.

The ordinary shipping belt is an entirely satisfactory arrangement for starting and stopping individual machines of comparatively light power requirements, and even heavier ones whose sometimes only, by the use of the friction clutch in some form. For moderate powers, in cases where the periods of idleness are comparatively small, the pulley, sheave, sprocket or gear may be fitted to the sleeve of a clutch on the continuously moving shaft. Larger power units are properly cared for only by the quill, running in its own bearings, the shaft passing freely through its hollow centre. Where division of a line of shafting into successive sections will satisfy the conditions, the clutch coup-

Waste of power, undue wear, loose pulley nuisances, squealing of belts in shipping and other familiar troubles almost inseparable from the "good old way" of doing things, are largely eliminated when the Dodge split friction clutch is properly used.

Fig. 1 represents the Dodge slip friction clutch. In this great frictional power is secured from the very nature of its construction, consisting of a disc into which are driven hard maple blocks, presenting end grain to inside and outside driving plates and powerful levers for forcing the plates into contact with the wooden block. Power is thus transmitted to the clutch sleed or to the

FIG. 2.

FIG. 4.

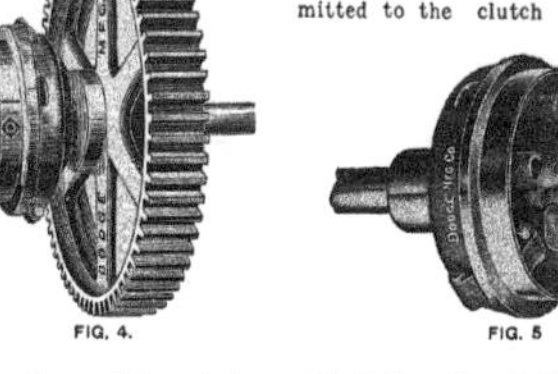

FIG. 5

continuity of operation is only occasionally disturbed. But when larger power units — whether individual machines, groups of machines, or whole departments of a plant—frequently or even at longer intervals must be isolated as

next shaft section, if the clutch be used as a coupling.

Fig. 2 shows a phantom view of a 60-inch heavy Dodge slip friction clutch capable of transmitting 1,200 h.p. at 250 revolutions per minute.

Fig. 3 shows the clutch used in combination with an iron pulley, and Fig. 4 with a spur gear on the extended sleeve. Fig. 5 shows the Dodge split friction clutch used as a cut-off coupling, and Fig. 6 Dodge solid friction clutch and cut-off coupling. Further information regarding these may be had from the Dodge Mfg. Co., Toronto.

FIG. 6.

SIGHT FEED OIL PUMP.

A SIGHT feed oil pump manufactured by the Sight Feed Oil Pump Co., of Milwaukee, Wis., is shown in the accompanying illustration. This pump was built for use on a 22-inch by 48-inch double tandem horizontal gas engine of 1,000 h.p. The pump as shown lubricates one side of the engine, that is to say, a pair of tandem cylinders, requiring therefore two such outfits to lubricate the entire engine of 1,000 h.p. The apparatus consists essentially of two 4-foot pumps, mounted upon one large oil tank. Each one of these four feed pumps has the driving mechanism so arranged that two feeds are driven by each lever, that is to say, each pump is really divided in two. The small lay shaft just below the pump has four driving pins set 90 degrees apart and is driven by a chain from the main engine valve lay shaft in synchorism with the same. The sprocket wheel on the valve lay shaft is held by a clamp and may be advanced or retarded to produce the required timing of the pump.

Sight Feed Oil Pump.

By this combination it is enabled to introduce the oil upon the moving piston at the instant it will do most good. Two feeds are led to each end of each cylinder and the oil is introduced on the piston through two check valves of special construction, which are placed at an angle of 30 degrees each side of the vertical. The pumps are so timed that delivery of oil through these valves is commenced just as the piston comes underneath the oil port, and continues during the time that the piston travels toward the end of the cylinder and until this port is again uncovered, thus thoroughly lubricating the piston rings. The reciprocating motion of the piston then distributes this oil over the cylinder walls from end to end.

FEED WATER HEATERS.

FEED water heaters in two types as manufactured by the Loew Mfg. Co., Cleveland, Ohio, are illustrated herewith, these being the Loew vertical open boiler feed water heater and

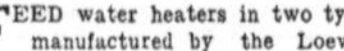

FIG. 1—Loew Vertical Open Boiler, Water Heater and Purifier.

purifier, in sizes about 800 h.p., and the Loew closed boiler feed water heater, vertical type, showing the groups of tubes.

The standard open type is used in connection with engines exhausting free to the atmosphere. The other style, known as the receiving type, is used in connection with engines exhausting into

FIG. 2—Loew Vertical Type Closed Feed Water Heater.

heating systems of various kinds, being equipped with a steam trap. The tendency heretofore has been to filter the feed water downwards which causes the substances held in suspension to be deposited on the filter bed, forming a white scale over the surface of and clogging the filter bed. In the Loew the filter bed is located directly above the second storage chamber, supported on a cast iron frame. The pump draws the water upward through the filter bed, thus overcoming the disadvantages that exist in other heaters.

The closed heater consists of a series of groups of tubes incased in an outer cylindrical shell, which serves as an enclosure for the exhaust steam that surrounds the tube. The tubes are arranged in groups so that the water passes six times the length of the tube through the exhaust steam in making its journey from water inlet to water outlet. The materials entering into the construction of the closed heater are principally heavy cast iron and brass, the shell, legs and caps being of cast iron and the tubes of straight, seamless drawn brass. These brass tubes have been found to effect a ready transmission of heat from the exhaust steam to the water. These feed water heaters and purifiers are made in sizes ranging from 50 h.p. to 4,000 h.p. An interesting catalogue describing these has just been issued by the Loew Mfg. Co., Cleveland, Ohio, who will supply anyone with a copy on application.

VARIABLE SPEED GEAR.

A VARIABLE speed gear of the expanding pulley type is shown in the accompanying illustration as manufactured by the S. & S. Engineering Co., of Brooklyn, N.Y., who have opened up a Canadian branch at 255 Jackson street west, Hamilton, Ont.

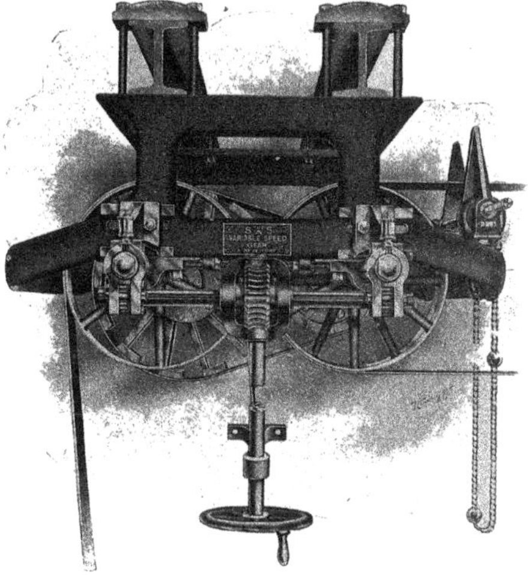

Fig. 1—End View of Variable Speed Countershaft.

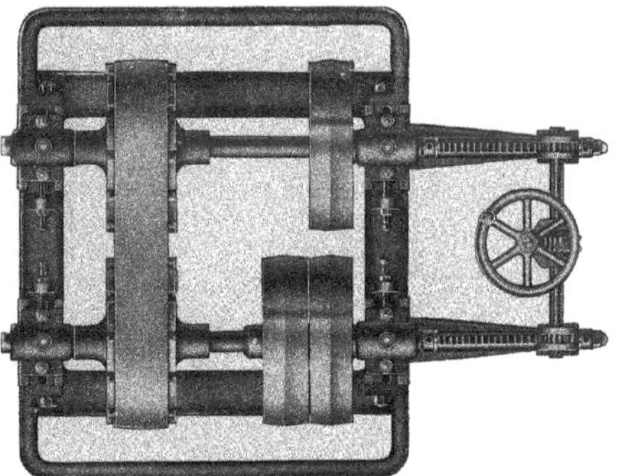

Fig. 2—View from Underneath.

Fig. 1 shows an end view of the variable speed countershaft and Fig. 2 a view from underneath. This variable speed gear lately perfected will transmit any amount of power, and vary the speed at the ratio of approximately 4 to 1. Each size takes up a minimum amount of space consistent with the amount of power handled. The handle wheel shown may be placed in a position most suitable for operators' requirements.

The variation of speed is effected by means of two solidly constructed expansion pulleys, belted together, each of which is capable of nearly doubling itself in diameter. The diameters of these pulleys altering simultaneously give the required change of speed, as the diameter of one increases, the other decreases. The expansion or contraction of each pulley is brought about in the following manner: Referring to detailed view of the shaft on opposite page, it will be seen that the hub of this expanding pulley is composed of cast iron disks or plates and that when all are assembled and bolted together the hub is of large diameter. The pairs of spokes riveted to each rim section are staggered and slide in corresponding slots, machined in these said plates. The spokes have teeth cut on their edges for part of their length which mesh with a broad pinion which is revolved by the insertion or withdrawal of an inner shaft with helical key ways cut in it. This inner shaft is prevented from turning within the tubular shaft by two oppositely placed keys, which not only serve to key fast pulley to tubular shaft, but project through same and fit into the straight grooves of the inner shaft.

Machinery Development

Metal Working **Special Apparatus** **Wood Working**

NEW FORM OF TAPPING MACHINE.

The tapping machine shown in the accompanying illustration has been designed by the Murchey Machine & Tool

Murchey Tapping Machine.

Co., Detroit, for the rapid production of steam and gas pipe fittings, as well as for special work.

Its characteristics are simplicity, strength, convenience in handling and great power.

Some of the more important points possessed by this machine may be enumerated as follows: A patented, frictionless head, which is of very great importance to the successful work of a tapping machine, easy travel of the spindle and large saving of taps and power. The machine shown in the cut has a table with a lateral and transverse movement which quickly and easily centers the work to the cored hole, another point of great importance in a tapping machine. The machine is arranged to use automatic collapsing taps with an adjustable stop, arranged to come in contact with a sliding collar which expands the chasers on the upward travel of the spindle. This is counterbalanced, and when the tap automatically collapses, it is quickly withdrawn, and is again ready for use. This avoids the necessity of reversing or stopping the machine, and effects a saving in labor of about 50 per cent. The machine being equipped with a pair of 4 x 16 clutch pulleys, permits of the machine being used either for solid taps or right and left hand threads. The machine is furnished, if desired, with a lead screw, which is arranged to feed any pitch of thread to suit requirements, or it can be fitted with a lever feed, hand worm feed or power feed and automatic stop, according to requirements.

Any kind of table may be furnished with the machine to meet the requirements of the work.

The principal dimensions of the machine with a range of ½ to 5 inches, inclusive, are shown in the following: Spindle, diameter, 2 ½ inches; Hole in spindle, 27-16 diameter by 4 inches deep; Centre of spindle to column, 14 inches; Size of table, 14 by 22 inches; Height of table from floor, 26 inches; Total height from floor, 7 ft. 8 inches; Floor space required, 2 ft. by 5 ft.

This machine can be furnished with special chucks for gripping flanges, or fittings, and can also be built in sizes to take up to 12 inch pipe.

WIRE NAIL MACHINES.

ILLUSTRATIONS are given herewith showing features of the National wire nail machines manufactured by the National Machinery Co., Tiffin, Ohio. One of the special features of these machines is the high speed at which they operate. The movements are such that even at their highest speed there is a minimum of jarring or racking. The cutting-off dies are carried in heavy cast steel arms that rock or oscillate on a shaft, as seen in figure 2, and as the motion is transmitted from cranks on the main shaft by pitmans the result is a very smooth action, the effect of the pitmans being to reverse the direction of movement of the cut-off dies without any jar.

The movement resembles that of a shear, resulting in better points on the nails and longer life to the cutting dies than is possible in machines having sliding cutters moving on parallel lines. The power is applied to the shear arms in line with the cutting strains, distributing the pressure on the bearings perfectly and minimizing wear. The

FIG. 1—National Wire-Nail Machine.

heading tool is carried in a substantial cross head that is also driven by a pitman from crank in the main shaft. Provision is made everywhere for easily compensating for wear; all bearings may be kept snug, thus permitting the setting of dies to just the right posi-

FIG. 2—Die Mechanism of Wire Nail Machine.

tion, i.e., so that their wear will result principally from contact with the wire and not from striking against each other.

The cut-off dies are held in substantial holders that may be removed bodily from the machine, thus permitting the dies being readily replaced. Adjustments in two directions are provided for the lower die and in three directions for the upper one. The lower die holder has a screw device by means of which it may be sensitively adjusted in either direction horizontally at right angles to the wire; the upper holder has this also, and in addition a rack-and-pinion-actuated wedge device that provides a sensitive adjustment laterally with the wire. The grip dies are held in a box provided with screw controlled taper gibs. This box may be quickly removed for changing dies. It is adjustable laterally and horizontally by screw devices; it is backed by a taper wedge which is moved in either direction by a screw, and permits gauging the amount of stock for the nail's head to a nicety. All adjusting screws are hardened steel.

"AMERICAN" VARIABLE SPEED PLANERS.

THE necessity for "reduction in shop cost" having advanced to such a degree of late and the requirements of machine tools to meet these conditions, has led The American Tool Works Company, Cincinnati, Ohio, U.S.A., to develop a line of variable speed planers. In reducing planing costs to a minimum, the up-to-date works manager often requires variable cutting speeds to cover different classes of work in hand. The "American" variable speed planers are built to suit all these requirements. They build two types of variable speed planers; one with four (4) cutting cutting speeds, and the other with two (2) cutting speeds, each type having a constant speed return of the platen which is greatly in excess of its respective highest cutting speed. A description of these two types is given below.

These speeds are obtained through a speed box, in the designing of which the aim has been, efficiency, simplicity, and durability. The four (4) speeds are taken from a carefully chosen range, calculated to cover all requirements of modern planer work. These, with the constant high-speed return of the platen, insure the greatest working economy. The speed box is securely bolted to the top of housings, these being of special design and very substantial. It is symetrical in appearance and being completely closed insures the greatest economy of oil consumption and absolute freedom from flying oil. All shafts in the speed box run in long bronze bushed bearings perfectly lubricated by the "ring" or "dynamo" system of oiling. Frictions have been omitted, thereby eliminating trappy construction and slippage, thus increasing the pulling power of the planer. The gears run in oil, thus giving the longest life and least noise to these parts. They are of ample proportions, wide face and course pitch. The pinions are all steel and cut integral with their shafts. Gears and pinions are cut with special cutters, incurring the greatest accuracy, longest life and minimum noise. Any of the four (4) speeds may be obtained by the simple manipulation of two convenient levers. A safety locking device prevents the engaging of two conflicting speeds at the same time. Index plate shows clearly how to obtain the desired speed.

FIG. 3-Cut Lever Construction of Wire Nail Machine.

The driving pulleys have fly-wheel rims, the momentum of which reduces to a minimum all shocks to the driving mechanism due to intermittent cutting and at reversing also insuring a steady

Variable Speed Planer.

even pull at the cutting tool. They are perfectly balanced, running without the least vibration even on the highest speeds. This, coupled with a scientifically designed and accurately built planer in all its details, results in a finished job of planing so smooth and perfect that it requires the least, if any, attention from the vise hand in fitting. Belt or electric motor drive may be furnished, the latter being shown above. When motor driven, should the motor become disabled, the driving pinion on outside of speed box may be replaced by a pulley and the planer driven by belt from a countershaft, or another motor conveniently placed. The flexibility of this construction insures the constant use of the machine at all times. When belt driven, the planer may be readily converted into motor-driven at any time after installation.

POWER PRESS WITH SPECIAL FEED.

HARDLY any industry having employed a hand fed power press for producing all or part of its output that has not considered the attaching of some automatic device which would not only increase the production but also remove the possibility of danger to the operator. This fact is perhaps nowhere more apparent than in the machine illustrated, which has only recently been designed for automatically cutting hack saw blades from 8 inches to 20 inches long, made out of regular saw blade steel approximately 1-16-inch thick.

Formerly two distinct methods have been employed for this work. The first was to cut the strips in a gang slitter. This was found to be unsatisfactory because the slitting operation would so twist and distort the strips that it was with difficulty they were straightened. Then a plain power press with a shearing die and an adjustable back gauge was used. The sheet was fed by hand against this gauge, and at each stroke of the press a piece was cut off of proper width. This method answered fairly well, but the output was not sufficiently large. Then, again, there existed a certain amount of danger to the operator as he came toward the end of the sheet he was cutting.

To overcome all of these difficulties the E. W. Bliss Co., 20 Adams street, Brooklyn, N.Y. has built the machine already mentioned. It is automatic in its action and operates as follows: The

Power Press with Special Feed.

sheet which is to be cut into strips is clamped in the carrier on the feed table. Depressing the treadle causes the slide to descend, operating a set of pawls, which, in connection with a rack device, advances the sheet uniformly at stated intervals until it has been cut entirely into strips. The feed can be adjusted to advance the sheet at ¼-inch to 1-inch at each stroke, varying by sixteenths. After the sheet has been cut up the hand lever at the side is raised to release the pawls and the carrier is pulled back into position by hand. The sheet is held down by a very heavy spring clamp, which securely holds it during the cutting. The feed table is 20 inches square.

THE LITTLE GIANT CORNER DRILL.

THE Chicago Pneumatic Tool Co. have just placed on the market a new "Little Giant" drill for corner work. From past experience in air tool practice, the firm hold that this new drill surpasses any other drill yet devised for drilling in close quarters and in corners particularly; the machine having been designed especially for work in the latter class. Parts for No. 4 Little Giant drill interchange with the new drill mentioned, thus insuring quick repairs.

This new tool weighs but 35 lbs.; capacity 1½-inch (but in emergency cases will drive 2-inch twist drills with very satisfactory results); spindle speed when running light, 150 R.P.M.; under load with 80 lbs. air pressure, 100 R.P.M.; distance from end of socket to end of feed screw when run down, 5¾-inches; length of feed, 2 inches; distance from centre of spindle to outside of housing, 1 5-16 inches.

In addition to the fact that it is the most powerful drill ever devised, weight considered, it possesses advantages over other designs of corner drills, owing to the spindle being driven by gears instead of ratchet principle, which insures steady and constant spindle movement.

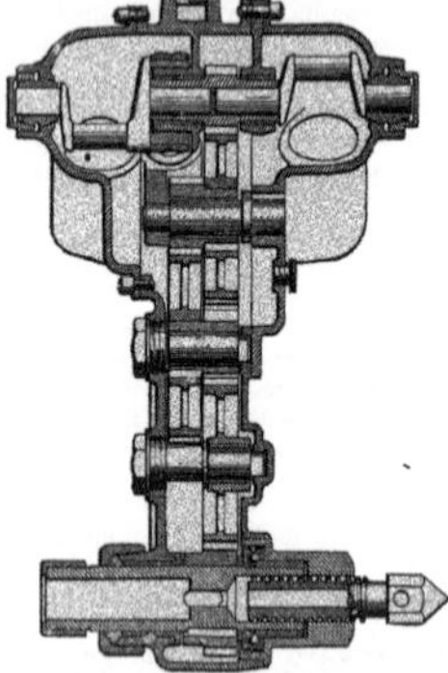

Little Giant Corner Drill.

The accompanying illustration gives a very clear idea as to the general design which it will be noted is neat and compact.

Industrial Progress

CANADIAN MCHINERY AND MANUFACTURING NEWS would be pleased to receive from any authoritative source industrial news of any sort, the formation or incorporation of companies, establishment or enlargement of mills, factories or foundries, railway or mining news, etc. All such correspondence will be treated as confidential when desired

Civil Engineers' Convention.

The twenty-first annual meeting of the Canadian Society of Civil Engineers is being held in Montreal on Jan. 29th, 30th and 31st. The Tuesday meeting will include the nomination of scrutineers, election of officers and the address of the retiring president, H. M. Lumsden.

On Wednesday the members are invited to visit the works of the Canada Car Company at Turcot, the works of the Simplex Railway Appliance Company at Rockfield, and the Canadian Rubber Company's works at Notre Dame Street east.

By the courtesy of the Montreal Street Railway, cars will be in waiting at the Windsor Hotel, and will leave at 9.30 a.m. sharp for Rockfield. The members will be the guests of the Canada Car Company at luncheon. In the evening there will be the annual dinner in the Windsor Hotel.

On Thursday there will be a meeting for the reception of reports of scrutineers and concluding the business of the annual meeting, and also a meeting of the newly-elected council.

Another Canadian Industry.

The latest manufacturing firm in the United States to establish a plant in Canada is the Gilson Mfg. Co., of Port Washington, Wis., who are now building a plant at Guelph, Ont., for the manufacture of gasoline engines. Although there are many manufacturers of gasoline engines in this country there is plenty of room for the establishment of a plant as is now under way for the Gilson Mfg. Co. This firm make a specialty of a 1 h.p. air cooled engine that has come in for considerable attention, having been eminently successful wherever being installed and operated whether of this size or other sizes manufactured by the company.

Power Equipment Wanted.

The Alvinston Power Co., of Alvinston, Ont., are in the market for new engines and dynamos for their plant in that town. The manager is H.R. Carruthers, who will be glad to receive quotations from various manufacturers.

Brandon, Man., is giving a thirty year contract to a United States firm, the National Lighting & Heating Co., to supply gas to the city. They will erect a gas plant at a cost of $100,000.

Enlarging Their Business.

Hutchison Bros., expert electricians and machinists, of Victoria, B.C., have recently enlarged their business and propose erecting fully equipped plant on most modern lines. They will manufacture special machinery and conduct business of general engineering work, manufacturing gas and oil engines of their own design. Automobile shops will also be run in connection with the works. The officers of the newly reorganized company are President, W. G. Winterburn; Vice-President, R. H. Hutchison; and Managing Director, D. C. Hutchison.

Machine Shops for Vancouver.

The British Columbia Marine Railway Co. is being enlarged during the present year by the addition of a large machine shop similar to those now in existence in the company's yards at Esquimalt, where 300 men are employed. A very large amount of new machinery will be required for this installation.

Important New Industry.

Since the incorporation of the Canadian Billings & Spencer Co. as a Canadian branch of the Billings & Spencer Co., of Hartford, Conn., with a capital of $200,000, an investigation of the suitability of the various towns has been made by the manager, Mr. J. Gill Gardner, of Brockville. Up to the present no decision has been reached, but as soon as satisfactory arrangements have been made work will commence at once on a modern and first-class plant to manufacture drop forgings and drop forging machinery, making a specialty of auto and bicycle parts. The Billings & Spencer Co. are one of the oldest and largest firms making drop forgings in the United States. They manufacture a large per cent. of the forgings for the United States Ordnance Department.

Change Due to Fire.

The Dodge Manufacturing Co's. Montreal branch premises at 419 St. James St., were recently damaged by fire. The Dodge Co., however, promptly secured temporary premises at 364 to 368 St. James St. Full and complete stocks of the company's lines in power transmission machinery were at once shipped from the works at Toronto, thus enabling business to proceed as usual. The Dodge Manufacturing Co. are Canada's largest pulley manufacturers, and their facilities and resources were evidenced in the prompt manner in which the stock was replenished.

Railway Progress for 1907.

During the coming year railway building in Canada will be carried forward on a still greater scale. It is proposed to build 3,314 miles of railway, the amount expended by the various companies as follows: Canadian Pacific, $23,000,000; Canadian Northern, $15,000,000; Grand Trunk, $5,000,000; Grand Trunk Pacific, $7,000,000: Great Northern, $10,000,000; and Temiskaming and Northern Ontario, $2,000,000. The Grand Trunk Pacific is not the only big work under way. The C. P. R. is staking out a line along the shores of Lake Ontario, passing through Bowmanville, Oshawa and Whitby. The Canadian Northern has surveyed a line that runs somewhat north of the G.T.R. and projected C.P.R. lines, and application has just been made at Ottawa for a Canadian Northern line from Toronto to Windsor. It is reported that next year the G.T.R. will jump into the important job of double-tracking its line from Toronto to North Bay, while strong arguments will be advanced to the Ontario Government to extend the Temiskaming line to Toronto. All these proposals are for tracks to cover long distances, and anyone of these undertakings would have caused a sensation ten years ago. To-day we treat them as matters of course. The Temiskaming line will be extended to form a junction with the Grand Trunk Pacific.

Large Power Available.

At the Empire Club luncheon held in Toronto on Dec. 27th, Mr. Cecil B. Smith gave a talk on Niagara and hydro-electric power. He instanced that the investigations of the Power Commission showed that from five to six million h.p. may be available to the people of the province of Ontario even during the dry weather season.

Reviewing some of the work done in developing water powers in other countries, he instanced Italy, where the Government was spending forty million dollars in electrical development enterprises. In France and Switzerland similar Government enterprises were being undertaken. In California electrical power was now being transmitted 250 miles. Canada, he said, was particularly fortunate in regard to immense water powers suitable for permanent power development.

Interesting Legal Case.

Mr. Justice Britton decided rather an interesting case on January 11th in the Assize Court, in that of Scott v. Barchard. W. H. Barchard & Son, Toronto, were the defendants in an action for $2,500 damages for injuries sustained to the thumb of the right hand of Richard Scott, a machinist. The main point urged by plaintiff's counsel was that the circular saw at which he was working was unprotected by a guard, as is the case in most mills, and several machinists from various box factories gave evidence to the effect that they had never been in one where guards were not used on the saws. Mr. W. H. Barchard in the witness box stated that he had never heard of guards being used, and had never seen one similar to the model shown in court. In answer to a question of counsel for the plaintiff, he said that he took no trade journals whatever. In reply to the judge's questions, the jury brought in a finding as follows: "Did the accident happen by reason of the absence of a guard on defendant's rip saw, at which the plaintiff was working?" "No." "If so, were the defendants negligent in not having that saw securely guarded?"

"No." "Could the plaintiff, by the exercise of reasonable care, have avoided the accident?" "Yes." "What damages?" "None." "If the defendant could have avoided the accident what could he have done?" "Kept the table-top clear of lumber on the left side of the saw."

New Car Construction Plant.

The unprecedented railway development now going on in Canada is attracting the attention of outside capitalists, and the exceptional advantages of the suburbs of Montreal prove most alluring to them.

It is said that what will be the largest car construction plant in Canada is shortly to be erected near Montreal West. The new works are to have a capacity of turning out from 50 to 60 cars per day.

The estimated cost is five million dollars. In addition to the car shops proper, there will be a forge shop, axle, truck and bolster shop, and wheel and casting foundry.

The formalities of organizing the company have not yet been attended to, and even the name is yet to be decided upon, but operations will be started at once, and it is promised that the works will be turning out cars early in the coming autumn.

New Machine Shop.

The Polson Iron Works, in order to keep pace with the industrial growth of Canada, are again planning large additions to their works. One of these additions will be a new machine shop 80 x 350 feet, south of the office buildings, the foundations for walls and machine tools being already laid. The new shop will have a 20-ft. gallery on either side, for the lighter class of work, with two 40-ft. electric travelling cranes, and will be equipped with the best tools on the market. The present machine shop will be added to the boiler shop.

Tenders for Grand Trunk Pacific.

The second call for tenders to undertake the transcontinental railway construction was gazetted in Ottawa on January 5th. Contractors will be given until Feb. 14th to tender on the five new sections of the line. These are from Grand Falls, N.B. to the Quebec boundary, about 62 miles. From the Quebec boundary to Quebec Bridge, about 150 miles. Fom La Tuque, Que. west to Weymontachene, about 45 miles, and from a point about 8 miles west of the Abitibi River east for 150 miles.

Power Pointers.

In his speech in Toronto regarding the power by-law not then passed, Hon. Adam Beck brought out the following points:

The commission's estimate of $17.50 per h.p. as the maximum cost of power delivered in Toronto is absolutely reliable. Probably that figure is too high.

On Toronto's present consumption of power a saving of at least $500,000 per year can be made. With the increased amount of power used, if the price of power were lowered, this saving will be immensely greater.

Every citizen will benefit, and not the manufacturers alone.

An immense impetus will be given to industrial development in Toronto.

The enterprise will be from the first on a self-supporting basis. The users of power or light will bear all the expense.

After the signing of the contract power can be supplied to consumers within eighteen months.

The contract can be for a term varying from ten to forty years, at the option of the city.

There will be no danger of an interrupted service, and an auxiliary steam plant to meet emergencies will not be necessary.

The enthusiastic manner in which the citizens of Toronto voted in favor of the by-law will give Mr. Beck and his colleagues a chance to demonstrate what can actually be done along this line.

Engineers' Club Dinner.

The annual dinner of the Engineers' Club of Toronto, was held on Jan. 3rd, at their club rooms, 96 King St. West, being attended by some 75 members. A toast to "Our Country" was proposed by Mr. F. L. Somerville, president of the club, being responded to by Dr. Galbraith of the School of Practical Science. T. H. White, chief engineer of the Canadian Northern Railway, who also responded, declared that he had seen every portion of Canada from ocean to ocean and his pride in the country increased with his knowledge of it. A toast to the "City of Toronto" was responded to by C. A. Goade, who stated that if the plans were carried out Toronto would undoubtedly become the most beautiful in Canada. A toast to the "Sister Societies" was responded to by Cecil B. Smith, representing the Canadian Society of Civil Engineers, and Mr. Dillon-Mills, representing the Canadian Institute, and Mr. McKenzie representing the Student Society of S.P.S.

Owen Sound Activity.

The metal working interests of Owen Sound are unusually busy at the present time, although the present season is considered the slackest of the year. The Owen Sound Iron Works, who are making a specialty of pumps and cement working machinery, have a large number of contracts on hand to supply various cement plants throughout the country with new machinery. These include two wet grinding tube mills, two large duplex slurry pumps, besides shafting pulleys, etc., for the Owen Sound Portland Cement Co. Three duplex steel frame pumps, one single acting slurry pump, three feed pumps, rotary coal dryer vertical cooler, coal stock hoppers, raw stock hoppers, two steel smoke stacks, 84' high and thousand h.p. feed water heater for the Colonial Cement Co., of Wiarton. Other apparatus on hand include orders from the Sun Portland Cement Co., the Imperial Cement Co., Raven Lakes Portland Cement Co.

At the works of William Kennedy & Sons the plant is working in full time. One of the large undertakings includes ten horizontal water wheels, 49" diameter for a 9' head for the Corporation of Morrisburg. Eleven water wheels have just been turned out for the St. Raymond Paper Co., being installed at their plant at St. Raymond, Que. The work is at present under way for three water wheels for the Penman Mfg. Co. of Paris, Ont. The steel casting department is working to its utmost limit and plans are under way for large increase in casting.

The Canadian Scientific Apparatus Co. have established a plant at 929 Yonge St., Toronto, for the manufacture of dies and tools, together with instruments of precision, surgical instruments, electrical and special machinery.

Large Outlay for Machinery.

The Dey Concentrating Co., of New York, one of the largest concerns in the world, making a specialty of the treatment of graphite, have purchased twenty properties in the Laurentian district between Labelle and Buckingham. The company has spent $500,000 in the equipment of machinery to be used in the development of these properties, which extend over 100 miles of territory. It is understood that the graphite flakes obtained from various points on the Labelle line are the largest and best in the market for the refining of precious metals, being absolutely free from any of the impurities what cause defects in the crucibles.

Frost and Wood Fully Equipped.

Since suffering the loss of almost their entire plant a year ago the Frost & Wood Co., of Smith Falls, have built one of the most up-to-date manufacturing establishments in the country. It has been modelled along the most modern plans and will enable this company to turn out their work with greater economy and increased perfection.

In the new machinery electricity plays a prominent part, it having been decided to run the entire works by electrical energy, the machine tools being run from individual and group driving purchased from the shops of the Canadian Westinghouse Co., including one 230 K. W. 440 volts, revolving field engine, type A. C., generator complete; one 140 K. W. 140 volts revolving field engine, type A. C., generator complete; one 6 K. W. belt-driven 125 volts exciter generator; one 10 K. W. belt-driven 125 volts exciter generator; one switchboard, blue Vermont marble, 5 panels; two 75 horsepower induction motors; two 40-horsepower induction motors; three 30-horsepower induction motors; four 20-horsepower induction motors; three 20-horsepower induction motors; three 10-horsepower induction motors, and four 7½-horsepower induction motors, and in addition to this the very latest methods have been adopted throughout the plant, and it is expected that this firm will go ahead at a faster rate than ever.

New Industry for New Glasgow.

Another new industry is to be started in New Glasgow for the manufacture of steel castings. A company with a capital of $200,000 has secured a site near the steel works and will at once commence the erection of suitable buildings.

Will Enlarge Works.

E. Leonard & Sons, London, have taken out a building permit for an addition to the machine shop, on York St., between Colborne and Waterloo. The addition will be 54 by 60 feet.

London defeated the waterworks by-law.

Bolton will raise $2,500 for cement walks.

The new Normal School at Hamilton will cost $50,000.

Thorold carried a waterworks by-law by a majority of 12.

The Welland Electric Street Railway Company is applying for a charter at Welland, Ont.

A project is on foot to establish a large railway car building plant in St. John, N.B.

The Mickle Dyment Lumber Company will erect a new factory in Brantford at the cost of $10,000.

The building returns of Calgary and suburbs, for the year 1906, will be in the vicinity of $3,000,000.

A new cement roundhouse has just been completed at Brantford for the G.T.R. at the cost of $8,000.

A lime industry is spoken of for Port Arthur, and a location is being looked for on which to start the plant.

The DuBois Iron Works, of DuBois, Pa., have taken over the entire business of the Lazier Engine Co., of Buffalo, N.Y.

The Woodstock Electric Light Company have been given the contract for electric lighting at Woodstock, N.B.

An application has been made to the Council of Brandon for a franchise to install a gaslight and power plant in that city.

The building permits issued in Vancouver for 1906, total $4,084,840, which is more than double those of two years ago.

J. I. Case Co., the large American threshing machinery manufacturers, will establish Canadian headquarters at Regina.

The Shipway Iron, Bell and Wire Company, of Toronto, are considering the moving of their plant to Niagara Falls.

The Waterous Engine Works, of Brantford, have taken out a building permit for a $4,000 addition to their main building.

The Imperial Coal Company, of Beersville, N.B., are planning to establish a brick-making plant with a capacity of 20,000 per day.

The blast furnace plant of the Atikokan Iron Company in Port Arthur, under construction all summer, is now almost complete.

The Board of Trade of Raymond, Alta., have appointed a committee to look into the question of installing an electric light plant.

The Canadian General Electric Company are installing an electric plant at Dorchester, N.B., for street and commercial lighting.

The Winton Automobile Company will establish a branch factory in Toronto. It is also intended to establish factories in England.

The Canada Screw Company will begin extensive building operations in the spring, doubling the capacity of their plant in Hamilton.

The Phillips Manufacturing Company of Toronto have plans out for a large factory, which will be erected for them in the spring.

F. C. Filer, of the Northern Electric & Manufacturing Company, of Montreal, is establishing a branch of his company in Winnipeg.

A large saw mill will be erected by the Graham Island Lumber Company, at Massett Harbor, B.C., with a cutting capacity of 250,000 feet daily.

The B. F. Graham Lumber Company has purchased the Taylor, Pattison Mill on Victoria Arm. The price paid was in the neighborhood of $50,000.

Over one hundred thousand dollars has been subscribed by Sydney, N.S., citizens for stock in the new railway construction plant, in that city.

The Stanley Smelting Company, of Maine, have received a license to manufacture iron, steel, maganese, coke, copper, lumber, etc., in Ontario.

The Miller Reversible Gasoline Engine Co., of Toronto, have removed from their quarters at 39 Sherbourne street to 569 Yonge street, of the same city.

Owing to the scarcity of brick the new telephone building at Edmonton is to be of concrete up to the street level, and of cement blocks above that.

Great interest is being manifested in the iron resources of the Port Arthur district, a great number of prospectors being out for American capitalists.

The Massey-Harris Company, Toronto, have taken out a permit for a small addition to their building. The Toronto Carpet Company will also enlarge.

The city of Winnipeg is making additions to its power plant which will necessitate buying more boilers and electric generators, together with auxiliary apparatus.

The Hamilton Steel & Iron Co. have just installed two duplex outside pack plunger pumps with pot valves, the output of the plant of the Smart-Turner Machine Co.

The sum of $600,000 has been subscribed in Quebec for the starting of a cement industry. The land has been purchased, and works will be started immediately.

The plants of the Tilson Manufacturing Company and the Standard Fitting and Valve Company, of Guelph, will be rapidly pushed to completion in the early Spring.

The contract for erecting the Meisel Manufacturing Company's new plant at Port Arthur has been awarded and work will be commenced on the buildings at once.

The Canadian Northern Railway has purchased about 110 acres of land on the Don flats, Toronto, to be used for roundhouses, car sheds, repair shops, cold storage plant and other requirements of the railway.

A large metal manufacturing company is negotiating to locate in Calgary. They offer to build a plant costing $100,000 and employ from sixty to seventy hands.

The Temiskaming & Northern Ontario Railway Commission is asking for tenders for the construction of general offices at North Bay, at the cost of $25,000 or $30,000.

The Majestic Wire Fencing Co., Detroit, Mich., will erect a factory at Victoria, B.C., for the manufacture of concrete reinforcements at a cost of about $40,000.

The Dominion Sewer Pipe Company have acquired property in Hamilton and will erect a factory, for the manufacture of sewer pipes, flue linings, gutter pipe and wall coping.

The Economic Power, Light and Heat Supply Co., York street, Toronto, have installed a 25 h.p. National Gas Engine in the Rossin House, Toronto, for electric lighting purposes.

Peterboro carried a by-law to provide $21,000 additional to the $40,000 already in the hands of the Board of Education to erect a new collegiate institute, costing $61,000.

Another industry has been located at Welland, by John A Reeb, of Port Colborne, and W. J. Sommervile of Welland, who will manufacture wood fibre wall plaster, and cement tile.

The Chapman Double Ball Bearing Co. recently ordered from the London Machine Tool Co., a 600 ton belt driven, drawing and stamping press—one of the most powerful machines used for this purpose.

The new works of the Plymouth Cordage Company, Welland, will be operated by electricity, the Cataract Power Company, of Hamilton, having secured a contract to supply 1,200 h.p.

The ratepayers of Wolfville, N.S., are considering the acquisition of the electric plant for both street and commercial lighting. If acquired, an all-night service will probably be adopted.

A group of Montreal merchant princes with a total capital of $20,000,000, propose to supply gas and electricity to the city of Montreal in competition with the existing Montreal Light, Heat & Power Co.

The pipe works of New Westminster, B.C., are rapidly nearing completion. The main building is 138 feet by 68 feet. It is expected that the company will be turning out pipe by the middle of next month.

The Commercial Cement Company, of Carman, Man., have started to instal their new machinery, and work will be pushed with all haste so as to have their plant in operation in the early spring.

The plant of the Goderich Engine and Bicycle Co. has been purchased by the Rogers Manufacturing Co., of Toronto. General foundry and railway specialty work will be done in the shops.

The Canadian General Electric Company, Peterboro, will shortly let contracts for a two-storey addition, 250 feet long, to their plant. The new addition will be for the lamp and wire departments.

The Manitoba Iron Works, have under construction ten double drum, double cylinder, steam hoists. These are for use on the Transcontinental Railway contract, and are the first ever manufactured in Winnipeg.

The Grand Falls Power Co., St. John, N.B., have offered, if given the exclusive right of supplying the city with electric power for 30 years, to supply power at a much lower rate than at present exists in St. John.

The Dyment Foundry Company, of Barrie, have a few men working to complete some orders which were on hand at time of fire. When these are finished the men will be let go and the buildings closed up.

The citizens of Sydney have voted to grant a bonus of $50,000 to the new Rolling Mills Company. The company will begin building operations as soon as possible.

There were erected in Montreal during 1906, 1,484 new buildings, valued at $7,745,023, as compared with 1,145 valued at $4,779,380 for the previous year. Ten structures exceeded in estimated cost $100,000 each.

The Great Northern Railway has announced a new freight tariff, effective December 27, making voluntary reduction in rates from the south to Winnipeg, Portage la Prairie, Brandon, Prince Albert and Edmonton.

Large consignments of sisal have been landed at Halifax this season by the Mexican Liner Sokoto. It is estimated that one hundred carloads per month will be imported via Halifax for Ontario manufacturers of rope.

A scheme proposed by the London, England, County Council, in which it is prepared to serve an area of 451 miles with electrical energy at an enormous industrial and commercial gain, will involve the expenditure of $47,000,000.

New Liskeard, Ont., will invite tenders for the erection of a new bridge over the Wabbi River. It will be a steel span for a bridge seventy-five feet long, driveway twenty-two feet, and a walk on each side five feet wide.

Promoters from Harvey, Illinois, have been at Windsor looking over a site on which they may establish a big manufacturing plant if the United States Steel Company carries out its proposal to build near Sandwich.

The Dominion Pipeworks and the Patterson Boiler Works of New Westminster, B.C., are in course of construction, and work will be started, on two other important industries, the Crystal Glass Works, and the soap factory.

The Vulcan Machine & Tool Co. have recently opened a machine shop at 116 Adelaide St. W., Toronto, and although they have been in operation only a short time, they have secured many contracts and have been kept busy since starting.

The Standard Fuel Manufacturing Company of Fairview, Halifax, have decided to move their works to Amherst, N.S. The company manufactures a fire lighter composed of sawdust and a chemical mixture formed into briquettes.

New York capitalists are in communication with the municipal council of Goderich, with a view to building a $5,000,000 steel plant. The concern would employ fifteen hundred hands, and would occupy one hundred acres of land.

Power will be generated by next fall on the Aroostook Falls in New Brunswick. Already the tunnel is well under way and a lot of heavy machinery being set up. Besides selling power to New Brunswick it is proposed also to transmit some of it to Maine.

The contract for the locomotive shops at Moncton, N.B., for the Intercolonial Railway, to replace those recently burned, has been awarded to a Montreal firm, at a price of about half a million dollars, and will be built entirely of concrete and steel.

The International Acheson Graphite Company of Niagara Falls, N.Y., are now manufacturing a soft artificial graphite which may be used for electrotyping, lubrication and stove polish. A new discovery by E. G. Acheson has made this possible.

P. W. Sotham, engineer of the Hydro-Electric Power Commission, reported to the Board of Control in Toronto that the cost of commercial light will be 16 cts. per k.w. hr., and residential lighting 5 cts., while the Toronto Electric Light Co. charge 25 cts. per k. w. hr.

The Canadian White Company made a proposition to take over the street railway franchise at Edmonton, Alta., but it did not receive favorable consideration, as the council have in view the operation of an electric street railway as a municipal enterprise.

A tunnel crossing the English Channel is assured. The Channel Tunnel Co. has been formed, half of the work to be done by British capitalists and the other half by French capitalists at a cost of 16,000,000 pounds sterling. The total length of the tunnel will be about 22 miles.

The contract has been let for the building of the Berg. Sand, Lime and Brick Company, in Brantford. The buildings will consist of a two-storey main building 50 feet square, and another building 100 feet long. The new buildings, when completed, will cost $5,000.

The Taylor, Forbes Company's foundry at Guelph suffered $5,000 damage by fire on January 6th. The fire only temporarily stopped the work in the radiator foundry, about 100 men being out of employment for about two days. The loss was fully covered by insurance.

The largest and heaviest single block of granite ever sent from the United States into Canada has just been shipped to Montreal and will be fashioned into a memorial to be erected in honor of the late Hon. Raymond Prefontaine, Minister of Marine and Fisheries.

Hon. H. R. Emmerson, at a Liberal banquet in St. John, N.B., on December 28th, announced that Canada will spend over $62,000,000 on railways during the coming year. It means the incoming of thousands of workers, and the opening up of thousands of square miles of new territory.

The Economic Power, Light and Heat Supply Co., York street, Toronto, are installing a 90 h.p. Pintsch Suction Gas Producer plant and National Gas Engine at Christie-Brown's, Toronto. They are also putting in a 110 h.p. Pintsch Suction Producer at the new works of the Brantford Screw Co.

The National Car Co. will establish their works at Whitby. These works are to supply the needs of the proposed electric line of the Huron and Ontario Railway. The town will bonus the company to the extent of $2,000 yearly for 12 years, and the works will be exempt from municipal taxes.

The Barchard box factory on Duke Street, Toronto, was damaged by fire on January 11th, to the extent of about $1,500. The cause is said to have been an over-heated bearing, which ignited some saw-dust. All the belts and main driving shaft were destroyed, and the saw and printing machine damaged.

The Canada General Electric Company and the Canada Foundry Company who are just completing additions to their buildings at Peterboro and Toronto, are again required to enlarge to meet with the ever increasing demand for their products, which are largely used in construction work.

A new fire-clay lining for stoves has been invented by Mr. Werther, formerly superintendent of the open hearth furnaces of the Dominion Iron and Steel Company. The basis of the preparation is Sydney cement, and the cost of this improved lining will be no greater than that of the ordinary iron material.

A big deposit of high grade iron ore has been discovered near Desbarats, Ont., a few miles below the Soo. It is said to be as good as any in the Lake Superior country. It lies close to the water, and can be shipped to southern lake ports easily, making a much shorter haul than from Lake Superior.

The Bell Telephone Company have contracted for 83,000 poles for delivery next year for the extension of telephone lines in Canada. This represents 2,500 miles of construction. The rural telephone service of the province of Manitoba will be doubled in extent. There are some 2,000 farmers connected at present.

The Ontario Iron and Steel Company's plant, which has been under construction for some months at Welland, will be a much larger industry than was expected. The plans now are for an expenditure of $300,000, and the works will employ about five hundred men. It is expected they will be completed by June 1st.

The Munro Wire Works, in Winnipeg, have extended their plant to meet with the increased business. This new building cost about $30,000, and is set apart for the production of all kinds of wire mattresses and the handling of iron beds. The Northwestern Brass Co. are erecting a steel building, 80x100 feet, at the cost of $40,000.

The Safety Explosives Company has been incorporated with a share capital stock of $300,000 for the purpose of manufacturing explosives and chemicals of every nature. The head office will be at Montreal, Que., the directors being W. H. Evans, R. W. Withycomb, W. A. Wier, W. J. Wright and A. W. C. Macalister of Montreal.

Duplex boiler feed pumps have been supplied within the past month by Smart-Turner Machine Co., of Hamilton, to T. W. Sims, Little Current, Ont., the Nova Scotia Fertilizer Co., Halifax, N.S., the Canada Steel Goods Co., Hamilton, the McLachlan Carriage Co., of Oshawa, E. E. Stevens, Halifax, N.S., the Metropolitan Oil & Soap Co., Toronto.

The Producers Torpedo Company has been incorporated with a share capital of $15,000, for the purpose of manufacturing nitro-glycerine, dynamite and other high explosives. The head office will be at Leamington, the provisional directors being G. W. Benson, E. Wigle and E. Winter, of Leamington, and Wm. Fleming, of Kingsville.

Nine new industries have been located in Toronto during 1906. These are the Kindall Bed and Mattress Co., St. Louis; Blanchite Process Paint Co., New York; Canada Bolt and Nut Co., Boston; De Sauga Silk Co., St. Etienne, France; Chemical Laboratories; Dominion Carriage Works, Berlin Electrical Manufacturing Company.

A pair of 42" Canadian turbines have just been installed in the plant of the Collins Inlet Lumber Co. to drive a gang of saw mills and run a box factory. These turbines working under a head of 36' give 318 h.p. were installed by C. Barber & Sons, Meaford, who are also installing another pair of 36" horizontal turbines for the Delhi Light & Power Co.

The city of Montreal is interesting itself in the water power of the Beauharnois canal, and recently a deputation from the city interviewed Sir Wilfrid Laurier on the subject. Sir Wilfrid expressed the opinion that if the power were leased to any private company the latter would first have to guarantee to furnish Montreal, Beauharnois and Valleyfield with whatever power they require at market prices.

The Sydney Cement Company have decided to enlarge their plant, which is situated on ground leased by them from the Dominion Iron & Steel Company, between two railroads, a position of considerable advantage for deliveries of raw material and for shipment of its products. The cement is made from slag which is procured from the blast furnaces of the Dominion Iron & Steel Company.

A million and a half will be spent by the Montreal Light, Heat & Power Co. on the Soulanges Canal for development. It is expected to have this completed by the early fall, whereby 1,500 h.p. will be added to the electric capacity of the company. The contract for the dam and canal work is to be executed by the Canadian White Co. The Dominion Bridge Co. are building the steel gates, Allis-Chalmers-Bullock the hydraulic machinery and water wheels, and the Canadian Westinghouse the electrical machinery.

The Hamilton and Fort William Navigation Company of Hamilton, have placed an order with the Canadian Shipbuilding Company for an 8,500-ton freight steamer, to be built for the iron ore, coal, and wheat trade. This steamer, which will be one of the largest on the lakes, will be built at the shipbuilding company's Port Colborne shipyard. The machinery will be built in Toronto.

A charter has been granted by the Manitoba Government to the Manitoba Rolling Mills Company, with a capital of $100,000, to take over and operate the Kirkwood Iron & Steel Rolling Mills, of Winnipeg. These mills were built by T. M. Kirkwood, Toronto, who has lately associated with him the president and officers of the United States Horseshoe Company, of Erie, Pa., Mr. Kirkwood will be the vice-president of the company.

Welland, Ont., Town Council, have given the contract for street lighting to the Stark Electric System, Limited, of Toronto. The contract is for 40 arc lights of 2,000 c.p., and the price is $40 per light per year, all public buildings belonging to the corporation to be lighted free. The rate for private lighting is to be 8 cents per kilowatt, with 25 per cent. discount if paid by the 15th of each month.

R. L. Worthington, who has severed his connections with the Vulcan Iron Works, of Winnipeg, entertained the foreman and office staff at a tea on Jan. 7, when he was presented with a beautiful combination desk and book case. Mr. Worthington is resigning his position as draughtsman after six years with the Vulcan Works, to accept a similar position with Kelly Bros. & Mitchell, Winnipeg.

Winter doesn't stop building in Canada as is intanced by the large number of permits for new buildings which have been issued during December and the present month in all parts of Canada. For instance, in Toronto alone permits issued totalled to $961,000 in December and from Jan. 1 to 8 the amount was $367,000. The figures given by us in this issue in several cases cover the totals for the past year and are gratifying from every standpoint. Even more pleasing, however, is the certainty that 1907 will show even larger totals than the big records made in 1905 and 1906.

Samples of rock containing copper, found around the Moon River, in the district of Parry Sound, have been sent to New York for examination, and capitalists there lost no time in providing all necessary capital for thoroughly testing the find. Some of the recent assays from the present workings show well the composite character of the vein matter. They gave silver 20 ounces, copper 22 per cent., gold from $10 to $500 per ton, with values in platinum, cobalt and zinc. One of the largest crystals of mica that perhaps has ever been mined has been found in one of these mines, it being about the size of an ordinary door, and required no less than four drill holes to dislodge it. The mica found, though dark in color, contains no iron, and improved in quality with depth.

At Fort William gigantic works are being constructed for the Canadian Iron & Foundry Co., where employment will be given to 500 men. Seven large buildings are now under way, these being of steel and concrete. The company own 25 acres of land and have 1,000 ft. of river frontage. They will handle their own ore, etc., by water, necessitating the building of a large dock equipped with a modern coal and ore handling plant, which is now under way. The size of the new buildings under construction are:

Pipe foundry	221 ft. x 55 ft.
Dressing shed	202 ft. x 63 ft.
Pipe coremaking room	145 ft. x 63 ft.
Cupola room	116 ft. x 55 ft.
Grey iron foundry	141 ft. x 55 ft.
Car wheel foundry	141 ft. x 55 ft.
Machine shop	140 ft. x 55 ft.

New Foundry and Machine Shop.

New Liskeard is to have a new foundry and machine shop in the very near future. H. M. McEwen of Carleton Place, for many years with Gillies Bros., has purchased a site and will commence work on the foundry as soon as the snow is off the ground. It is proposed to put in an up-to-date machine shop and foundry equipment and to put on a staff of first-class machinists, molders, boilermakers and blacksmiths.

Magnetic Arc Lamps.

The town of Westmount, Que., awarded a contract to the Canadian General Electric Co. of Toronto, for the entire equipment necessary for the installation of 150 magnetic-arc lamps. This system was completed and put in operation last month, and the results far exceeded the expectations, the streets of Westmount now being the best lighted in Canada, and also the first by magnetic arc lamps.

These lamps give a very powerful, steady, white light, the arcs being much wider than the older types and therefore cast less shadow. They also consume about one-third less current, and have a much longer life.

Galt's Growth.

Galt has given substantial evidence in the way of growth by the erection of 112 dwelling houses. The manufacturing part of the town, has also shown a decided increase, some new firms starting, and a number of old ones adding considerable floor space, 314,710 square feet being added to 15 enterprises, divided viz.: Galt Knitting Company, 87,500 square feet; MacGregor, Gourlay Company, 84,845 square feet; Malleable Iron Company, 36,991 square feet; P. W. Gardiner & Son, 22,400 square feet; Sheldons, Limited, 21,000 square feet; R. McDougall Company, 13,200 square feet; Galt Robe Company, 10,360 square feet; Cowan & Company, 10,230 square feet; McVicker Engine Company, 9,548 square feet; Box Factory, 4,050 square feet; J. J. Stevens Company, 4,100 square feet; Victoria Wheel Works, 3,600 square feet; Peter Hay Knife, 1,600 square feet. The total amount of money expended for building purposes in Galt in 1906, was $560,000.

Canadian Shipbuilding.

A well organized and energetic movement is on foot in the Maritime Provinces to build up the steel shipbuilding industry. A statement summing up the argument in favor of a tonnage bonus on steel ships built in Canada was lately submitted to the Tariff Commission, and the Dominion Government has been memorialized to grant such a bonus on a basis of $6 per gross ton. The Boards of Trades and municipal councils of Halifax, Dartmouth and other places have taken the matter up, and much interest is felt regarding the manner in which the request for a bonus will be treated at Ottawa. Years ago in Eastern Canada, shipbuilding assumed very large proportions, but with the advent of steel construction it rapidly declined. Steel vessels can be built cheaper in British yards on account of the cheaper iron, coal, and labor there Those who agitate for a bonus on steel ships point out that all maritime nations give protection to their shipping. There is an increasing demand in Canada for steel vessels for both coastwise and lake trade, but it is claimed that Canadian yards cannot compete with British and foreign builders in this line of construction. It is pointed out that Canadian built tonnage has decreased in less than 30 years from 183,000 to 33,000, while the entries in and out from sea have increased from six to sixteen millions, the coastwise shipping from ten to forty-five millions, and the Great Lakes tonnage has more than doubled. There are now in Canada, besides several smaller firms, four large steel shipbuilding companies—the Algoma Steel Company, Sault Ste. Marie; the Canadian Iron Furnace Company, Montreal; the Collingwood Shipbuilding Company, Collingwood; and the Canadian Shipbuilding Company, Toronto. These yards are not all fully and steadily employed, and it is suggested that while many millions of dollars are being expended in Canada in nearly every industry, including iron and steel, it is time we should build our own ships and do our own carrying trade.

Companies Incorporated

Otonabee Power Co., Ltd., has been granted power to increase the capital stock of the company from $200,000 to $600,000.

The Burlington Masonic Hall Company, has been incorporated with a share cpaital of $10,000 for the erection of a hall for the Burlington Masonic Lodge.

Butler Bros.-Hoff Company, incorporated under the laws of the State of New York, have received a license to carry on work, as builders, contractors and decorators in Ontario.

Gibson Mfg. Co., Ltd., Guelph, capital $50,000, purpose to manufacture and deal in gasoline engines. Directors are: John Gibson, H. H. Bolens, Port Washington, Wis.; C. L. Dunbar and J. W. Lyon, Guelph.

The W. I. Kemp Co., Ltd., Stratford, capital $300,000, purpose to manufacture agricultural machinery. The directors are: W. I. Kemp, Stratford; N. J. Kemp, J. S. Lewis and J. S. Kemp, Newark Valley, N. Y.

Dominion Car & Foundry Co., Ltd., capital stock $5,000,000, to manufacture cars, rolling stock, for railways. Directors, W. B. Kelley, R. P. Lamont, W. W. Butler, G. McAvity and A. H. Chave. head office, Montreal.

The Lufkin Rule Company, a corporation incorporated under the laws of the State of Michigan, have been granted a license to manufacture all kinds of measuring articles, tools, machinery, etc., in Ontario.

The Craig Harness Company has been incorporated with a share capital of $100,000, and head office to be at Ottawa. The provisional directors will be R. Craig, J. H. Cameron, and C. H. Clendenning, of Ottawa.

St. Catharines Drilling Co., Ltd., share capital $40,000, incorporated to drill for natural gas, oil, petroleum, salt and similar products. Directors, N. L. Nelson, P. I. Price, H. H. Collier, with head office at St. Catharines, Ont.

The Meisel Mfg. Co., Ltd., Port Arthur, capital $350,000, purpose to manufacture and repair all kinds of machinery. The directors are: F. W. Woods, G. C. Meisel, R. Mc. M. Meisel, all of Port Huron, and James Conmee, Port Arthur.

Central Foundry, Limited, has been incorporated with a share capital of $200,000.00, to carry on a general foundry business, with head office at the city of Toronto. Directors to be H. L. Bowers, H. T. Bush and R. C. Donald.

The Shedrick Rigby Co., Montreal, have been incorporated with a capital of $20,000, to manufacture machinery, electric appliances, etc. The charter members include C. E. Shedrick, J. S. Rigby and P. C. Ryan, Montreal.

The Collingwood Shipping Co., capital share $90,000, incorporated to operate steamships, elevators, etc. Directors to be W. T. Allan, M. Brophey, J. R. Arthur, W. A. Hogg and W. Carmichael, with head office at Collingwood, Ont.

The Gundy-Clapperton Co., Toronto, have been incoporated with a capital of $40,000, to manufacture cut glass, silver, jewelry, etc. The provisional directors include N. F. Gundy, H. G. Clapperton and W. H. Wise, Toronto.

Mussens Ltd., Montreal, capital $500,000, purpose to carry on the business of W. H. C. Mussen & Co., dealers in railway supplies, etc. The directors are: W. H. C. Mussen, Geo. Boulter, G. G. Foster, C. G. McKinnon and W. R. Staveley, all of Montreal.

Cobalt and New Ontario Prospectors, Developers and Investors, Ltd., Toronto, capital $500,000, purpose to carry on the business of a mining, milling, reduction and development company. The directors are: J. Q. Ross, A. W. Holmested, F. H. Potts and A. R. Bickerstaff, all of Toronto.

The Oil Well Supply Company, a corporation incorporated under the laws of the State of Pennsylvania, have received a license to deal in iron and steel, or any other metals, or any articles of commerce from metal or wood in Ontario.

The Clifton Sand, Gravel & Construction Co., St. Catharines, Ont., have been incorporated with a capital of $250,000, to manufacture stone, gravel, gas, oil, etc. The provisional directors include P. I. Price, H. Yale, and G. F. Peterson, St. Catharines, Ont.

The Western Rubber & Apparatus Co., Winnipeg, Man., have been incorporated with a capital of $25,000 to manufacture rubber goods, fire apparatus, etc. The provisional directors include R. M. McLeod, A. A. Andrews, and J. H. Anderson, Winnipeg, Man.

The Well Machine and Wind Mill Company, has been incorporated with $20,000 capital; provisional directors: C. S. Tyrrell, merchant; J. H. Inkster, agent; H. W. Hutchinson, manager; W. S. Evans, financial agent; J. A. Machray, barrister-at-law, all of Winnipeg.

The Standard Concrete Construction Company has been incorporated with a share capital of $100,000 for general building construction, with head office to be in Toronto. Directors will be F. Rielly, G. Sutherland, E. Denton, J. B. Bartram and E. A. Scott, all of Toronto.

The Sherbrooke Novelty Company has been incorporated with a share capital of $20,000 for the purpose of manufacturing electrical and mechanical appliances, household utensils and smallwares. The head office will be at Sherbrooke, Que., and the directors will be E. R. Ebbitt and A. C. Snowden, Montreal; R. A. Wright, A. T. Boydell and J. A. Swan, Sherbrooke.

The Canadian General Industrial and Development Company, has been incorporated with a share capital of $150,000, with head office at Chatham, for the purpose of sinking wells, boring for oil, natural gas, etc. The provisional directors of the company will be O. B. Sargent, J. M. McCoig, and W. A. Hadley, of Chatham.

Bechtels, Limited, has been incorporated with a share capital of $75,000, for the purpose of manufacturing brick, tile blocks, and cement products. The head office will be at Waterloo, and provisional directors will be: B. E. Bechtel, W. B. Bechtel, W. J. Watson, C. E. Whyard of Waterloo; C. H. Bechtel, of Berlin, and P. A. Watson, of Galt.

The Western Canada Development Company has been incorporated with a $1,000,000 capital, to carry on business as miners, coal miners, oil producers, refiners and gas makers, in any part of the Dominion of Canada with head office to be at Winnipeg, incorporating J. S. Hough, A. C. Ferguson, C. Williams, E. B. Lindsay and W. M. Graham, all of Winnipeg.

The Schultz Manufacturing Company has been incorporated with a share capital of $50,000 to purchase the business of Ernest Schultz, known as The Schultz Manufacturing Company, makers of stamped, pressed and spun metal goods, castings, lamps, lanterns, etc., with head office at Hamilton and directors to be E. P. Schultz, A. H. Brittain and E. A. Schultz, of Hamilton.

Province of Quebec, District of Bedford } SUPERIOR COURT.

IN THE MATTER OF

The Philipsburg Milling Co.

IN LIQUIDATION.

NOTICE IS HEREBY GIVEN that in pursuance of an order granted by the said Superior Court, all the assets of the above Company in Liquidation, comprising the building in which the Mill is situate, and all the plant and accessories, will be sold at Public Auction, to the highest bidder, either *en bloc* or in lots to suit purchasers, on the ***ninth day of February*** next, at ***two o'clock in the afternoon,*** at the premises of the Company in the ***Village of Philipsburg, Que.***

The Mill is completely equipped for a general milling business as well as for the manufacture of prepared foods, and both the building itself and the entire plant are practically new, having been in use only a few months.

The plant includes among other things: 18 H.P. Gasoline Engine, Clark Grinder, Corn and Cob Crusher, Stones, Smutter, Bolts, etc.

For further particulars apply to the undersigned.

W. FREDERIC KAY,
Liquidator, Philipsburg, Que.

CONDENSED ADS.

RATES

One insertion—25c. for 20 words; 1c. a word for each additional word.

Yearly rate—$2.50 for twenty words or less, 10c a word for each additional word.

The above does not apply to notices under the head of "Machinery Wanted." These notices are inserted free for subscribers.

WESTERN REPRESENTATIVE WANTED.

AN agent in the west to handle Beardshaw's "Conqueror" high speed steel and drills, which are meeting with a favorable reception throughout Canada. Alex. Gibb, 13 St. John St., Montreal.

FOR SALE

$600.00 WILL buy Patterns for the following size gasoline engines: 3 and 6 horse power stationary; 5 horse power double opposed air cooled; 16 horse power double opposed cylinder; 12 horse power double upright, stationary; 10 horse power double upright auto or marine; 40 horse power, 4 cylinder marine. All in good order. Have built several hundred high grade engines from these patterns. D. McKenzie & Co., London, Ont. (1)

A COMPLETE LINE OF ROADMAKING machinery, consisting of rock crushers, road graders, scrapers, dumping wagons and Robinson's cement tile moulds; send for a booklet. Climax Road Machine Co., Hamilton.

FOR SALE—A fully equipped Wire Mill and Nail Factory as a going concern, with about 500 tons of wire rods; factory situated on 150 ft. of waterfront with trackage optional to buyer. B.C. Wire and Nail Co., Limited, Vancouver, B.C. (2)

BELTING, RUBBER, CANVAS AND LEATHER, Hose, Packing, Blacksmith's and Mill Supplies at lowest prices. N. Smith, 138 York Street, Toronto. (2tf)

SITUATIONS VACANT

WANTED—AT ONCE—MACHINIST and planer hands; also general blacksmith; state experience and wages. The Waterloo Manufacturing Co., Waterloo, Ont.

WANTED — FIRST-CLASS MACHINIST — with some experience on iron patterns. Gurney Foundry Co., Limited, Toronto Junction.

WANTED — AT ONCE — FIRST-CLASS TOOL grinder, for Landis and Hendy-Norton machines; also several first-class machinists for lathe work. Apply Canada Cycle & Motor Co., Limited, Toronto Junction.

SITUATIONS WANTED

ENGINEER AND DRAUGHTSMAN (25), salt and salt making machinery; small marine and mill; technical training; requires position. A. M. Mere Heath, Davenham, Cheshire, England.

THE
WESTINGHOUSE
COMPANIES
ENCOMPASS
THE GLOBE

CANADIAN MACHINERY BUYERS' DIRECTORY

To Our Readers—Use this directory when seeking to buy any machinery or power equipment. You will often get information that will save you money.

To Our Advertisers—Send in your name for insertion under the heading of the lines you make or sell.

To Non-Advertisers—A nominal rate of $1 per line a year is charged non-advertisers.

Acids.
Canada Chemical Mfg. Co., London.

Abrasive Materials.
Baxter Paterson & Co., Montreal.
The Canadian Fairbanks Co., Montreal.
Rice Lewis & Son, Toronto.
Norton Co., Worcester, Mass.
H. W. Petrie, Toronto.
Williams & Wilson, Montreal.

Air Brakes.
Canada Foundry Co., Toronto.
Canadian Westinghouse Co., Hamilton.

Air Receivers.
Allis-Chalmers-Bullock Montreal.
Canada Foundry Co., Toronto.
Canadian Rand Drill Co., Montreal.
Chicago Pneumatic Tool Co., Chicago.
Jenckes Machine Co., Sherbrooke, Que.
John McDougall Caledonian Iron Works Co., Montreal.

Alundum Scythes Stones
Norton Company, Worcester, Mass.

Arbor Presses.
Niles-Bement-Pond Co., New York.

Augers.
Chicago Pneumatic Tool Co., Chicago.
Rice Lewis & Son, Toronto.

Automatic Machinery.
Cleveland Automatic Machine Co., Cleveland.
Potter & Johnston Machine Co., Pawtucket R. I.

Automobile Parts
Globe Machine & Stamping Co., Cleveland, Ohio.

Axle Cutters.
Butterfield & Co., Rock Island, Que.
A. B. Jardine & Co., Hespeler, Ont.

Axle Setters and Straighteners.
Canadian Buffalo Forge Co., Montreal.
Dominion Henderson Bearings, Niagara Falls, Ont.
A. B. Jardine & Co., Hespeler, Ont.

Babbit Metal.
Baxter Paterson & Co., Montreal.
Canada Metal Co., Toronto.
Canada Machinery Agency Montreal.
Greey Wm. & J. G., Toronto.
Rice Lewis & Son, Toronto.
Miller Bros. & Toms Montreal, Que.

Bakers' Machinery.
Greey Wm. & J. G., Toronto.
Pendrith Machinery Co., Toronto.

Barrels, Tumbling.
Buffalo Foundry Supply Co., Buffalo
Detroit Foundry Supply Co., Windsor
Dominion Foundry Supply Co., Montreal
Hamilton Facing Mill Co., Hamilton.
Globe Machine & Stamping Co, Cleveland Ohio.
John McDougall Caledonian Iron Works Co., Montreal.
R. McDougall Co., Galt, Ont.
Miller Bros & Toms, Montreal, Que.
J. W. Paxson Co., Philadelphia, Pa.
H. W. Petrie, Toronto.
The Smart-Turner Mach. Co., Hamilton.

Bars, Boring.
Dominion Henderson Bearings, Niagara Falls, Ont.
Hall Engineering Works, Montreal.
Miller Bros & Toms Montreal.
Niles-Bement-Pond Co., New York.

Batteries, Dry.
Berlin Electrical Mfg Co., Toronto.

Batteries, Flashlight.
Berlin Electrical Mfg Co., Toronto.

Batteries, Storage.
Canadian General Electric Co, Toronto.
Chicago Pneumatic Tool Co., Chicago.
Rice Lewis & Son, Toronto.
John Millen & Son, Montreal.

Bearings, Roller.
Dominion Henderson Bearings, Niagara Falls Ont.

Bell Ringers.
Chicago Pneumatic Tool Co., Chicago.

Belting, Chain.
Baxter Paterson & Co. Montreal.
Canadian Fairbanks-Morse Co. Montreal.
The Dodge Mfg. Co., Toronto.
Greey, Wm. & J. G., Toronto.
Jeffrey Mfg. Co., Columbus, Ohio.
Link-Belt Eng. Co., Philadelphia.
Miller Bros. & Toms Montreal.
Waterous Engine Works Co., Brantford.

Belting, Cotton.
Baxter, Paterson & Co., Montreal.
Canada Machinery Agency, Montreal.
Dominion Belting Co., Hamilton.
Greey, Wm. & J. G., Toronto.
Rice Lewis & Son, Toronto.

Belting, Leather.
Baxter Paterson & Co., Montreal.
Canada Machinery Agency, Montreal.
The Canadian Fairbanks Co., Montreal.
The Dodge Mfg. Co., Toronto.
Greey, Wm. & J. G., Toronto.
McLaren, J. C., Montreal.
Rice Lewis & Son, Toronto.
H. W. Petrie, Toronto.
Williams & Wilson, Montreal.

Belting, Rubber.
Baxter, Paterson & Co., Montreal.
Canada Machinery Agency, Montreal.
Dodge Mfg. Co., Toronto
Greey, Wm. & J. G., Toronto.

Belting Supplies.
Baxter Paterson & Co. Montreal.
The Dodge Mfg. Co. Toronto.
Greey Wm. & J. G., Toronto.
Rice Lewis & Son, Toronto.
H. W. Petrie, Toronto.

Bending Machinery.
John Bertram & Sons Co., Dundas, Ont.
Chicago Pneumatic Tool Co., Chicago.
London Mach. Tool Co., Hamilton, Ont.
National Machinery Co., Tiffin, Ohio.
Niles-Bement-Pond Co., New York.

Benders, Tire.
Canadian Buffalo Forge Co., Montreal.
A. B. Jardine & Co., Hespeler, Ont.
London Mach. Tool Co., Hamilton, Ont.

Blast Furnace Brick.
Detroit Foundry Supply Co., Detroit
Hamilton Facing Mill Co., Hamilton.

Blowers.
Buffalo Foundry Supply Co., Montreal.
Canada Machinery Agency, Montreal.
Canadian Buffalo Forge Co., Montreal.
Detroit Foundry Supply Co., Windsor
Hamilton Facing Mill Co., Hamilton.
Kerr Turbine Co., Wellsville, N.Y.
J. W. Paxson, Philadelphia, Pa.
Sheldon's Limited, Galt
Sturtevant, B. F., Co., Hyde Park, Mass.

Blue Printing.
The Electric Blue Print Co., Montreal.

Boilers.
Canada Foundry Co., Limited, Toronto.
Canada Machinery Agency, Montreal.
Goldie & McCulloch Co., Galt.
Jenckes Machine Co., Sherbrooke, Que.
E. Leonard & Sons, London, Ont.
Manitoba Iron Works, Winnipeg.
John McDougall Caledonian Iron Works, Montreal.
Owen Sound Iron Works Co., Owen Sound.
H. W. Petrie, Toronto.
Robb Engineering Co. Amherst, N.S.
The Smart-Turner Mach. Co., Hamilton.
Waterous Engine Works Co., Brantford.
Williams & Wilson, Montreal.

Boiler Compounds.
Canada Chemical Mfg. Co., London, Ont.
Hall Engineering Works, Montreal.
Hamilton Facing Mill Co., Hamilton.

Bolt Cutters.
John Bertram & Sons Co., Dundas, Ont.
London Mach. Tool Co., Hamilton.
National Machinery Co., Tiffin, Ohio.
Niles-Bement-Pond Co., New York.

Bolt and Nut Machinery.
John Bertram & Sons Co., Dundas, Ont.
Canada Machinery Agency Montreal.
London Mach. Tool Co., Hamilton.
National Machinery Co., Tiffin, Ohio.
Niles-Bement-Pond Co., New York.

Bolts and Nuts, Rivets.
Baxter, Paterson & Co., Montreal.

Books, Mechanical.
Technical Pub. Co., Manchester.
The MacLean Pub. Co., Ltd., Toronto.

Boring and Drilling Machines.
American Tool Works Co., Cincinnati.
B. F. Barnes Co., Rockford, Ill.
John Bertram & Sons Co., Dundas, Ont.
Canada Machinery Agency, Montreal.
Canadian Buffalo Forge Co., Montreal.
A. B. Jardine & Co., Hespeler, Ont.
London Mach. Tool Co., Hamilton.
Niles-Bement-Pond Co., New York.

Boring Machine, Upright.
American Tool Works Co., Cincinnati.
John Bertram & Sons Co., Dundas, Ont.
London Mach. Tool Co., Hamilton.
Niles-Bement-Pond Co., New York.

Boring Machine, Wood.
Chicago Pneumatic Tool Co., Chicago.
London Mach. Tool Co., Hamilton.

Boring and Turning Mills.
American Tool Works Co., Cincinnati.
John Bertram & Sons Co., Dundas, Ont.
Gisholt Machine Co., Madison, Wis.
Canada Machinery Agency, Montreal.
London Mach. Tool Co., Hamilton.
Niles-Bement-Pond Co., New York.
H. W. Petrie, Toronto.

Box Puller.
A. B. Jardine & Co., Hespeler, Ont.

Brakes, Automobile,
Globe Machine & Stamping Co., Cleveland, Ohio.

Brass Foundry Equipment.
Detroit Foundry Supply Co., Detroit.

Brass Working Machinery.
Warner & Swasey Co., Cleveland, Ohio.

Brass Working Machine Tools.
Warner & Swasey Co., Cleveland O.

Brushes, Foundry and Core.
Buffalo Foundry Supply Co., Buffalo.
Detroit Foundry Supply Co., Windsor.
Dominion Foundry Supply Co., Montreal
Hamilton Facing Mill Co., Hamilton.

Brushes, Steel.
Buffalo Foundry Supply Co., Buffalo.

Bulldozers.
John Bertram & Sons Co., Dundas, Ont.
London Mach. Tool Co., Hamilton, Ont.
National Machinery Co., Tiffin, Ohio.
Niles-Bement-Pond Co., New York.

Calipers.
Baxter, Paterson & Co., Montreal.
Brown & Sharpe, Providence, R.I.
Rice Lewis & Son, Toronto.
John Millen & Son, Ltd., Montreal, Que.
L. S. Starrett & Co., Athol, Mass.
Williams & Wilson, Montreal.

Can Making Machinery
Canada Machinery Agency, Montreal.
Miller Bros & Toms, Montreal.

Cars, Foundry.
Buffalo Foundry Supply Co., Buffalo.
Detroit Foundry Supply Co., Windsor
Dominion Foundry Supply Co., Montreal
Hamilton Facing Mill Co., Hamilton.

Castings, Brass.
Chadwick Bros., Hamilton.
Greey Wm. & J. G., Toronto.
Hall Engineering Works, Montreal.
Jenckes Machine Co., Sherbrooke.
Kennedy Wm., & Sons, Owen Sound.
McDougall R. Co., Galt
Niagara Falls Machine & Foundry Co. Niagara Falls, Ont.
Owen Sound Iron Works Co., Owen Sound.
Robb Engineering Co., Amherst, N.S.

Castings, Grey Iron.
Allis-Chalmers-Bullock Montreal.
Dodge Mfg Co., Toronto.
Greey, Wm. & J. G., Toronto.
Hall Engineering Works, Montreal.
Jenckes Machine Co., Sherbrooke, Ont.
Kennedy Wm. & Sons, Owen Sound
Laurie Engine & Machine Co., Montreal.
John McDougall Caledonian Iron Works Co., Montreal.
R. McDougall Co., Galt, Ont.
Niagara Falls Machine & Foundry Co., Niagara Falls, Ont.
Owen Sound Iron Works Co., Owen Sound.
Robb Engineering Co., Amherst, N.S.
Smart-Turner Machine Co., Hamilton.

Castings, Steel.
Kennedy, Wm., & Sons, Owen Sound.

Castings, Semi-Steel.
Robb Engineering Co., Amherst, N.S.

Cement Machinery.
Allis-Chalmers-Bullock Limited, Montreal
Greey Wm. & J. G., Toronto.
Jeffrey Mfg. Co., Columbus, Ohio.
John McDougall Caledonian Iron Works, Co., Montreal
Owen Sound Iron Works Co., Owen Sound
St. Lawrence Supply Co., Montreal.

Centreing Machines.
John Bertram & Sons Co., Dundas, Ont.
Dominion Henderson Bearings Niagara Falls, Ont.
Jeffrey Mfg. Co., Columbus Ohio
London Mach. Tool Co., Hamilton, Ont.
Niles-Bement-Pond Co., New York.
Pratt & Whitney Co., Hartford, Conn.

Centres, Planer.
American Tool Works Co., Cincinnati.

Centrifugal Pumps.
Canadian Buffalo Forge Co., Montreal.
John McDougall Caledonian Iron Works Co., Montreal
R. McDougall Co., Galt.
D'Olier Engineering Co., New York.
Pratt & Whitney Co., Hartford, Conn.

Centrifugal Pumps—Turbine Driven.
Canadian Buffalo Forge Co., Montreal.
Kerr Turbine Co., Wellsville, N.Y.

Chaplets.
Buffalo Foundry Supply Co., Buffalo.
Detroit Foundry Supply Co, Windsor.
Hamilton Facing Mill Co., Hamilton.

Charcoal.
Buffalo Foundry Supply Co. Buffalo.
Detroit Foundry Supply Co. Windsor.
Dominion Foundry Supply Co., Montreal
Hamilton Facing Mill Co., Hamilton.

Chemicals.
Baxter, Paterson & Co., Montreal.
Canada Chemical Co., London.

Chemists' Machinery.
Greey Wm & J. G., Toronto.

Chemists, Industrial.
Detroit Testing Laboratory, Detroit.

Chemists, Metallurgical.
Detroit Testing Laboratory, Detroit.

Chemists, Mining.
Detroit Testing Laboratory, Detroit.

Chucks, Ring Grinding.
A. B. Jardine & Co., Hespeler, Ont.
Chicago Pneumatic Tool Co., Chicago.

Chucks, Drill and Lathe.
American Tool Works Co., Cincinnati.
Baxter, Paterson & Co., Montreal.
John Bertram & Sons Co., Dundas, Ont.
Canada Machinery Agency, Montreal.
Ker & Goodwin, Brantford.
A. B. Jardine & Co., Hespeler, Ont.
London Mach. Tool Co., Hamilton.
John Millen & Son, Ltd. Montreal.
Niles-Bement-Pond Co., New York.
H. W. Petrie, Toronto.
Rice Lewis & Son, Toronto.
Standard Tool Co., Cleveland.
Whitman & Barnes Co., St. Catharines.

Chucks, Planer.
American Tool Works Co., Cincinnati.
Canada Machinery Agency, Mon real.
Niles-Bement-Pond Co., New York.

Chucking Machines.
American Tool Works Co., Cincinnati.
Niles-Bement-Pond Co., New York.
H. W. Petrie, Toronto.
Warner & Swasey Co., Cleveland, Ohio.

Circuit Breakers.
Allis-Chalmers-Bullock, Limited, Montreal
Canadian General Electric Co., Toronto.
Canadian Westinghouse Co., Hamilton.

Clippers, Bolt.
A. B. Jardine & Co., Hespeler, Ont.

Cloth and Wool Dryers.
Canadian Buffalo Forge Co., Montreal.
B. Greening Wire Co., Hamilton.
McEachren Heating and Ventilating Co., Galt, Ont.

Collectors, Pneumatic.
Sturtevant, B. F., Co., Hyde Park, Mass.

Compressors, Air.
Allis-Chalmers-Bullock, Limited, Montreal
Canada Foundry Co., Limited, Toronto.
Canada Machinery Agency, Montreal.
Canadian Rand Drill Co., Montreal.
Canadian Westinghouse Co., Hamilton.
Chicago Pneumatic Tool Co., Chicago.
Detroit Foundry Supply Co., Windsor.
John McDougall, Caledonian Iron Works Co., Montreal
H. W. Petrie, Toronto.
The Smart-Turner Mach. Co., Hamilton.
Canada Machinery Agency, Montreal.
Canadian Westinghouse Co., Hamilton.
Hall Engineering Works, Montreal, Que.
London Mach. Tool Co., Hamilton.
R. McDougall Co., Galt, Ont.
Niles-Bement-Pond Co., New York.
H. W. Petrie, Toronto.
Pratt & Whitney Co., Hartford, Conn.
Williams & Wilson, Montreal.

Concentrating Plant.
Allis-Chalmers-Bullock, Montreal.
Greey, Wm. & J. G., Toronto.

Concrete Mixers.
Jeffrey M'f'g. Co., Columbus, Ohio.
Link-Belt Co., Philadelphia.

Condensers.
Canada Foundry Co., Limited, Toronto.
Canada Machinery Agency, Montreal.
Hall Engineering Works Montreal.
Smart-Turner-Machine Co., Hamilton.
Waterous Engine Co., Brantford.

Confectioners' Machinery.
Baxter, Paterson & Co., Montreal.
Greey, Wm & J. G., Toronto.
Pendrith Machinery Co., Toronto.

Consulting Engineers.
Charles Brandeis, Montreal.
Canadian White Co., Montreal.
Dominion Henderson Bearings, Niagara Falls, Ont.
Hall Engineering Works, Montreal.
Jules De Clercy, Montreal.
John S. Fielding, Toronto.
Miller Bros. & Toms, Montreal.
Roderick J. Parke, Toronto.
Plews & Trimingham, Montreal.
T. Pringle & Son, Montreal.
James C. Royce, Toronto.

Contractors.
Canadian White Co., Montreal.
D'Olier Engineering Co., New York.
Expanded Metal and Fireproofing Co., Toronto.
Hall Engineering Works Montreal.
Laurie Engine & Machine Co., Montreal.
John McDougall Caledonian Iron Works Co., Montreal.
Miller Bros. & Toms, Montreal.
Robb Engineering Co., Amherst, N S.
The Smart-Turner Mach. Co., Hamilton.

Controllers and Starters Electric Motor.
Allis-Chalmers-Bullock, Montreal.
Canadian General Electric Co., Toronto.
Canadian Westinghouse Co., Hamilton.
T. & H. Electric Co., Hamilton.

Conveying Systems.
Sturtevant, B. F., Co., Hyde Park, Mass.

Conveyor Machinery.
Baxter, Paterson & Co., Montreal.
The Dodge Mfg. Co., Toronto.
Greey, Wm. & J. G., Toronto
Jeffrey Mfg. Co., Columbus, Ohio.
Jenckes Machine Co., Sherbrooke, Que.
Link-Belt Co., Philadelphia.
John McDougall Caledonian Iron Works Co., Montreal.
Miller Bros. & Toms, Montreal.
Smart-Turner Machine Co., Hamilton.
Laurie Engine & Machine Co., Montreal.
Waterous Engine Works Co., Brantford.
Williams & Wilson, Montreal.

Coping Machines.
John Bertram & Sons Co., Dundas, Ont.
London Mach. Tool Co., Hamilton.
Niles-Bement-Pond Co., New York.

Core Compounds.
Buffalo Foundry Supply Co., Buffalo.
Detroit Foundry Supply Co., Windsor.
Dominion Foundry Supply Co., Montreal
Hamilton Facing Mill Co., Hamilton.

Core Ovens.
Detroit Foundry Supply Co., Windsor.
Hamilton Facing Mill Co., Hamilton.

Core Oven Bricks.
Buffalo Foundry Supply Co., Buffalo.
Detroit Foundry Supply Co., Windsor.
Hamilton Facing Mill Co., Hamilton.

Core Wash.
Buffalo Foundry Supply Co., Buffalo.
Detroit Foundry Supply Co., Windsor.
Hamilton Facing Mill Co., Hamilton.

Cotton Belting.
Baxter, Paterson & Co., Montreal.
Dominion Belting Co., Hamilton.
Greey, Wm. & J. G., Toronto.

Couplings.
Owen Sound Iron Works Co., Owen Sound

Cranes, Electric and Hand Power.
Canada Foundry Co., Limited, Toronto
Dominion Foundry Supply Co., Montreal
Hamilton Facing Mill Co., Hamilton.
Link-Belt Co., Philadelphia.
John McDougall, Caledonian Iron Works Co., Montreal.
Miller Bros. & Toms, Montreal.
Niles-Bement-Pond Co., New York.
Owen Sound Iron Works Co., Owen Sound
Smart-Turner-Machine Co., Hamilton.

Crankshafts.
The Canada Forge Co., Welland.

Crushers, Rock or Ore.
Allis-Chalmers-Bullock, Montreal.

Contractors' Plant.
Allis-Chalmers-Bullock, Montreal.
John McDougall, Caledonian Iron Works Co., Montreal.
Niagara Falls Machine & Foundry Co., Niagara Falls, Ont.

Crank Pin Turning Machine
London Mach. Tool Co., Hamilton.
Niles-Bement-Pond Co., New York.

Crucibles.
Buffalo Foundry Supply Co., Buffalo.
Detroit Foundry Supply Co., Windsor
Dominion Foundry Supply Co., Montreal
Hamilton Facing Mill Co., Hamilton.
J. W. Paxson Co., Philadelphia.

Crucible Caps
Hamilton Facing Mill Co., Hamilton.

Crushers, Rock or Ore.
Allis-Chalmers-Bullock, Montreal.
Jeffrey Mfg. Co., Columbus, Ohio.

Cupolas.
Buffalo Foundry Supply Co., Buffalo.
Detroit Foundry Supply Co., Windsor
Dominion Foundry Supply Co., Montreal
Hamilton Facing Mill Co., Hamilton.
J. W. Paxson Co., Philadelphia.
Sheldons Limited, Galt.

Cupola Blocks.
Detroit Foundry Supply Co., Detroit.

Cupola Blowers.
Buffalo Foundry Supply Co., Buffalo.
Canada Machinery Agency, Montreal.
Canadian Buffalo Forge Co., Montreal.
Detroit Foundry Supply Co., Windsor
Hamilton Facing Mill Co., Hamilton.
McEachren Heating and Ventilating Co., Galt.
J. W. Paxson Co., Philadelphia.
Sheldon's Limited, Galt.
B. F. Sturtevant Co., Hyde Park, Mass.

Cutters, Coal.
Allis-Chalmers-Bullock, Montreal.

Cut Meters.
The Canadian Fairbanks Co., Montreal

Cutters, Flue
Chicago Pneumatic Tool Co., Chicago.
J. W. Paxson Co., Philadelphia.

Cutters, Milling.
Becker, Brainard Milling Machine Co. Hyde Park, Mass.
Pratt & Whitney Co., Hartford, Conn.
Standard Tool Co., Cleveland.

Cutting-off Machines.
John Bertram & Sons Co., Dundas, Ont.
Canada Machinery Agency, Montreal.
London Mach. Tool Co., Hamilton.
J. W. Petrie, Toronto.
Pratt & Whitney Co., Hartford, Conn.

Cutting-off Tools.
Armstrong Bros. Tool Co., Chicago.
Baxter, Paterson & Co., Montreal.
London Mach. Tool Co., Hamilton.
H. W. Petrie, Toronto.
Pratt & Whitney, Hartford, Conn.
Rice Lewis & Son, Toronto.
L. S. Starrett Co., Athol, Mass.

Dies
Globe Machine & Stamping Co., Cleveland, Ohio.

Die Stocks
Hart Manufacturing Co., Cleveland, Ohio

Dies, Opening
W. H. Banfield & Sons, Toronto
Globe Machine & Stamping Co., Cleveland Ohio.
R. McDougall Co., Galt.
Pratt & Whitney Co., Hartford, Conn.

Dies, Sheet Metal.
W. H. Banfield & Sons, Toronto.
Globe Machine & Stamping Co., Cleveland, Ohio

Die Sinkers
Globe Machine & Stamping Co., Cleveland, Ohio

Dies, Threading,
John Millen & Son, Ltd, Montreal.

Dowel Pins.
Buffalo Foundry Supply Co., Buffalo.
Detroit Foundry Supply Co., Windsor.
Dominion Foundry Supply Co., Montreal
Hamilton Facing Mill Co., Hamilton.

Draft, Mechanical.
W. H. Banfield & Sons, Toronto.
Butterfield & Co., Rock Island, Que.
A. B. Jardine & Co., Hespeler.
Pratt & Whitney Co., Hartford, Conn.
Sheldons Limited, Galt.
Sturtevant, B. F., Co., Hyde Park, Mass.

Drawing Instruments.
Rice Lewis & Son, Toronto.

Drawing Supplies.
The Electric Blue Print Co., Montreal

Drawn Steel, Cold.
Baxter, Paterson & Co., Montreal.
Greey Wm. & J. G., Toronto
Miller Bros. & Toms, Montreal.
Union Drawn Steel Co., Hamilton.

Dredges, Gold, Dipper and Hydraulic.
Allis-Chalmers-Bullock Co., Montreal.

Drilling Machines, Arch Bar.
John Bertram & Sons Co., Dundas, Ont.
London Mach. Tool Co., Hamilton
Niles-Bement-Pond Co., New York.

Drilling Machines, Boiler.
American Tool Works Co., Cincinnati.
Canadian Buffalo Forge Co., Montreal.
John Bertram & Sons Co., Dundas, Ont.
Bickford Drill and Tool Co., Cincinnati.
The Canadian Fairbanks Co., Montreal.
A. B. Jardine & Co., Hespeler, Ont.
London Mach. Tool Co., Hamilton, Ont.
Niles-Bement-Pond Co., New York.
H. W. Petrie, Toronto.
Williams & Wilson, Montreal.

Drilling Machines Connecting Rod.
John Bertram & Sons Co., Dundas, Ont.
London Mach. Tool Co., Hamilton.
Niles-Bement-Pond Co., New York.

Drilling Machines, Locomotive Frame.
American Tool Works Co., Cincinnati.
B. F. Barnes Co., Rockford, Ill.
John Bertram & Sons Co., Dundas, Ont.
London Mach. Tool Co., Hamilton, Ont.
Niles-Bement-Pond Co., New York.

Drilling Machines, Multiple Spindle.
American Tool Works Co., Cincinnati.
B. F. Barnes Co., Rockford, Ill.
Baxter, Paterson & Co., Montreal.
John Bertram & Sons Co., Dundas, Ont.
Bickford Drill & Tool Co., Cincinnati.
Canada Machinery Agency Montreal.
London Mach. Tool Co., Hamilton, Ont.
Niles-Bement-Pond Co., New York.
H. W. Petrie, Toronto.
Williams & Wilson, Montreal.

Drilling Machines, Pneumatic
Canada Machinery Agency, Montreal.

Drilling Machines, Portable
Baxter, Paterson & Co., Montreal.
Canadian Buffalo Forge Co., Montreal.
A. B. Jardine & Co., Hespeler, Ont.
Niles-Bement-Pond Co., New York.

Drilling Machines, Radial.
American Tool Works Co., Cincinnati.
Baxter, Paterson & Co., Montreal.
John Bertram & Sons Co., Dundas, Ont.
Bickford Tool & Drill Co., Cincinnati.
The Canadian Fairbanks Co., Montreal.
London Mach. Tool Co., Hamilton.
Niles-Bement-Pond Co., New York.
H. W. Petrie, Toronto.
Williams & Wilson, Montreal.

Drilling Machines, Suspension.
John Bertram & Sons Co., Dundas, Ont.
Canada Machinery Agency Montreal.
London Mach. Tool Co., Hamilton.
Niles-Bement-Pond Co., New York

Drilling Machines, Turret.
John Bertram & Sons Co., Dundas, Ont.
London Mach. Tool Co., Hamilton.
Niles-Bement-Pond Co., New York.

Drilling Machines, Upright.
American Tool Works Co., Cincinnati.
B. F. Barnes Co., Rockford, Ill.
Baxter, Paterson & Co., Montreal.
John Bertram & Sons Co., Dundas, Ont
A. B. Jardine & Co., Hespeler, Ont.
London Mach. Tool Co., Hamilton.
Niles-Bement-Pond Co., New York.
W. H. Petrie, Toronto.
Williams & Wilson, Montreal.

Drills, Bench.
B. F. Barnes Co., Rockford, Ill.
Baxter, Paterson & Co., Montreal.
Canadian Buffalo Forge Co., Montreal
London Mach. Tool Co., Hamilton.
Pratt & Whitney Co., Hartford, Conn.

Drills, Blacksmith.
Canada Machinery Agency, Montreal.
Canadian Buffalo Forge Co., Montreal.
A. B. Jardine & Co., Hespeler, Ont.
London Mach. Tool Co., Hamilton.
Standard Tool Co., Cleveland.
Whitman & Barnes Co., St. Catharines

Drills, Centre.
Pratt & Whitney Co., Hartford, Conn.
Standard Tool Co., Cleveland, O.
L. S. Starrett Co., Athol, Mass.

Drills, Electric
B. F. Barnes Co., Rockford, Ill.
Baxter, Paterson & Co., Montreal.
Chicago Pneumatic Tool Co., Chicago.
Niles-Bement-Pond Co., New York.

Drills, Gang.
American Tool Works Co., Cincinnati.
B. F. Barnes Co., Rockford, Ill.
John Bertram & Sons Co., Dundas, Ont.
Pratt & Whitney Co., Hartford, Conn.

Drills, High Speed.
Wm. Abbott, Montreal.
American Tool Works Co., Cincinnati.
B. F. Barnes Co., Rockford, Ill.
Pratt & Whitney Co., Hartford, Conn.
Standard Tool Co., Cleveland, O.
Whitman & Barnes Co., St. Catharines.

Drills, Hand.
Canadian Buffalo Forge Co., Montreal.
A. B. Jardine & Co., Hespeler, Ont.

Drills, Horizontal.
B. F. Barnes Co., Rockford, Ill.
John Bertram & Sons Co., Dundas, Ont.
Canada Machinery Agency, Montreal.
London Mach. Tool Co., Hamilton.
Niles-Bement-Pond Co., New York.

Drills, Jewelers.
Pendrith Machinery Co., Toronto.

Drills, Pneumatic.
Canada Machinery Agency, Montreal.
Chicago Pneumatic Tool Co., Chicago.
Niles-Bement-Pond Co., New York.

Drills, Radial.
American Tool Works Co., Cincinnati.
John Bertram & Sons Co., Dundas, Ont.
Bickford Drill & Tool Co., Cincinnati
London Mach. Tool Co., Hamilton, Ont.
Niles-Bement-Pond Co., New York.

Drills, Ratchet.

Armstrong Bros. Tool Co., Chicago.
A. B. Jardine & Co., Hespeler.
Pratt & Whitney Co., Hartford, Conn.
Standard Tool Co., Cleveland.
Whitman & Barnes Co., St. Catharines.

Drills, Rock.

Allis-Chalmers-Bullock, Montreal.
Canadian Rand Drill Co., Montreal.
Chicago Pneumatic Tool Co., Chicago.
Jaffrey Mfg. Co., Columbus, Ohio.

Drills, Sensitive.

American Tool Works Co., Cincinnati.
B. F. Barnes Co., Rockford, Ill.
Canada Machinery Agency, Montreal.
Niles-Bement-Pond Co., New York.

Drills, Shop View.

John Bertram & Sons Co., Dundas, Ont.

Drills, Twist.

Baxter, Paterson & Co., Montreal.
Chicago Pneumatic Tool Co., Chicag
Alex. Gibb, Montreal.
A. B. Jardine & Co., Hespeler, Ont.
John Millen & Son, Ltd., Montreal.
Morse Twist Drill and Machine Co., New Bedford, Mass.
Pratt & Whitney Co., Hartford, Conn.
Standard Tool Co., Cleveland.
Whitman & Barnes Co., St. Catharines.

Drying Apparatus of all Kinds.

Canadian Buffalo Forge Co., Montreal.
Greey, Wm. & J. G., Toronto
McEachren Heating and Ventilating Co., Galt.
Sheldon's Limited, Galt
B. F. Sturtevant Co., Hyde Park, Mass

Dry Kiln Equipment.

Sturtevant, B. F., Co., Hyde Park, Mass.

Dry Sand and Loam Facing.

Buffalo Foundry Supply Co., Buffalo.
Hamilton Facing Mill Co., Hamilton.

Dump Cars.

Canada Foundry Co., Limited, Toronto
Dominion Henderson Bearings, Niagara Falls, Ont.
Greey, Wm. & J. G., Toronto
Hamilton Facing Mill Co., Hamilton.
John McDougall, Caledonian Iron Works Co., Montreal.
Niles-Bement-Pond Co., New York.

Jenckes Machine Co., Sherbrooke, Que
Link-Belt Eng. Co., Philadelphia.
John McDougall Caledonian Iron Works Co., Montreal
Miller Bros. & Toms, Montreal.
Owen Sound Iron Works Co., Owen Sound
Waterous Engine Co., Brantford.

Dust Separators.

Canadian Buffalo Forge Co., Montreal.
Greey, Wm. & J. G., Toronto
McEachren Heating and Ventilating Co., Galt.
Sheldon's Limited, Galt.

Dynamos.

Allis-Chalmers-Bullock, Montreal.
Canadian General Electric Co., Toronto.
Canadian Westinghouse Co., Hamilton.
Consolidated Electric Co., Toronto
D'Olier Engineering Co., New York.
Hall Engineering Works, Montreal, Que.
John Millen & Son, Ltd., Montreal.
Packard Electric Co., St. Catharines.
H. W. Petrie, Toronto.
B. F. Sturtevant Co., Hyde Park, Man.
T. & H. Electric Co., Hamilton.

Dynamos—Turbine Driven.

Kerr-Turbine Co., Wellsville, N.Y.

Economizer, Fuel.

Canadian Buffalo Forge Co., Montreal.
Dominion Henderson Bearings, Niagara Falls, Ont.
McEachren Heating and Ventilating Co., Galt, Ont.
B. F. Sturtevant Co., Hyde Park, Man.

Electrical Instruments.

Canadian Westinghouse Co., Hamilton.

Electrical Steel.

Baxter, Paterson & Co., Montreal.

Electrical Supplies.

Canadian General Electric Co., Toronto.
Canadian Westinghouse Co., Hamilton.
London Mach. Tool Co., Hamilton, Ont.
John Millen & Son, Ltd., Montreal.
Packard Electric Co., St. Catharines.
T. & H. Electric Co., Hamilton.

Electrically Driven Tools and Machinery.

American Tool Works Co., Cincinnati.
Baxter Paterson & Co., Monteal.

Electrical Repairs

Canadian Westinghouse Co., Hamilton.
T. & H. Electric Co., Hamilton.

Elevator Buckets.

Greey, Wm. & J. G., Toronto
Jeffrey Mfg. Co., Columbus, Ohio.

Emery and Emery Wheels.

Baxter, Paterson & Co., Montreal.
Dominion Foundry Supply Co., Montreal
Hamilton Facing Mill Co., Hamilton.

Emery Wheel Dressers.

Baxter, Paterson & Co., Montreal.
Canada Machinery Agency, Montreal.
Dominion Foundry Supply Co., Montreal
Hamilton Facing Mill Co., Hamilton.
John Millen & Son, Ltd., Montreal.
H. W. Petrie, Toronto.
Standard Tool Co., Cleveland.

Engineers and Contractors.

Canada Foundry Co., Limited, Toronto.
D'Olier Engineering Co., New York.
Greey, Wm. & J. G., Toronto
Hall Engineering Works, Montreal.
Jenckes Machine Co., Sherbrooke, Que.
Laurie Engine & Machine Co., Montreal.
Link-Belt Co., Philadelphia.
John McDougall, Caledonian Iron Works Co., Montreal.
Miller Bros. & Toms Montreal.
Robb Engineering Co., Amherst, N.S.
The Smart-Turner Mach. Co., Hamilton.

Engineers' Supplies.

Baxter, Paterson & Co., Montreal.
Greey, Wm. & J. G., Toronto
Hall Engineering Works, Montreal.
Rice Lewis & Son, Toronto.

Engines, Gas and Gasoline.

Baxter, Paterson & Co., Montreal.
Canada Machinery Agency, Montreal.
The Canadian Fairbanks Co., Montreal.
Canadian McVicker Engine Co., Galt.
Dominion Henderson Bearings, Niagara Falls, Ont.
Economic Heat, Light and Power Supply Co., Toronto.
Gilson Mfg. Co., Guelph
The Goldie & McCulloch Co., Galt, Ont.
H. W. Petrie, Toronto.
Producer Gas Co., Toronto
The Smart-Turner Mach. Co., Hamilton.

Engines, Steam.

Allis-Chalmers-Bullock, Montreal
Canada Machinery Agency, Montreal.
Canadian Buffalo Forge Co., Montreal.
D'Olier Engineering Co., New York.
The Goldie & McCulloch Co., Galt, Ont.
E. Leonard & Sons, London, Ont.
Jenckes Machine Co., Sherbrooke, Que.
Laurie Engine & Machine Co., Montreal.
John McDougall Caledonian Iron Works, Montreal.
Miller Bros. & Toms, Montreal.
Robb Engineering Co., Amherst, N.S.
Sheldons Limited, Galt.
The Smart-Turner Mach. Co., Hamilton.
B. F. Sturtevant Co., Hyde Park, Mass.
Waterous Engine Works Co., Brantford.

Exhaust Heads.

Canadian Buffalo Forge Co., Montreal.
McEachren Heating and Ventilating Co., Galt, Ont.
Sheldon's Limited, Galt, Ont.
B. F. Sturtevant Co., Hyde Park, Mass.

Expanded Metal.

Expanded Metal and Fireproofing Co. Toronto

Expanders.

A. B. Jardine & Co., Hespeler, Ont.

Facing.

Detroit Foundry Supply Co., Detroit.

Fans, Electric.

Canadian Buffalo Forge Co., Montreal.
Canadian General Electric Co., Toronto
Canadian Westinghouse Co., Hamilton.
McEachren Heating and Ventilating Co., Galt, Ont.
Sheldon's Limited Galt, Ont.
The Smart-Turner Mach. Co., Hamilton.
B. F. Sturtevant Co., Hyde Park, Mass.

Fans, Exhaust.

Canadian Buffalo Forge Co., Montreal.
Detroit Foundry Supply Co., Windsor.
Dominion Foundry Supply Co., Montreal
Greey, Wm. & J. G., Toronto
Hamilton Facing Mill Co., Hamilton.
McEachren Heating and Ventilating Co., Galt.
Sheldon's Limited, Galt.
B. F. Sturtevant Co., Hyde Park, Mass.

Feed Water Heaters.

John McDougall, Caledonian Iron Works Co., Montreal.
The Smart-Turner Mach. Co., Hamilton

Files and Rasps.

Baxter, Paterson & Co., Montreal.
John Millen & Son Ltd., Montreal.
Rice Lewis & Son, Toronto.
Nicholson File Co., Providence, R.I.
H. W. Petrie, Toronto.
Whitman & Barnes Co., St. Catharines.

Fillet, Pattern.

Baxter, Paterson & Co., Montreal.
Buffalo Foundry Supply Co., Buffalo.
Detroit Foundry Supply Co., Windsor.
Dominion Foundry Supply Co., Montreal
Hamilton Facing Mill Co., Hamilton.

Fire, Apparatus.

Waterous Engine Works Co., Brantford.

Fire Brick and Clay.

Baxter Paterson & Co., Montreal.
Buffalo Foundry Supply Co., Buffalo.
Detroit Foundry Supply Co., Windsor.
Dominion Foundry Supply Co., Montreal
Hamilton Facing Mill Co., Hamilton.
J. W. Paxson Co., Philadelphia.

Flash Lights.

Berlin Electrical Mfg. Co., Toronto.

Flour Mill Machinery.

Allis-Chalmers-Bullock, Montreal.
Greey, Wm. & J. G., Toronto
The Goldie & McCulloch Co., Galt, Ont.
John McDougall, Caledonian Iron Works Co., Montreal.
Miller Bros. & Toms, Montrea.

Flue Rollers.

Chicago Pneumatic Tool Co., Chicago.

Forges.

Canada Foundry Co., Limited, Toronto.
Canadian Buffalo Forge Co., Montreal.
Hamilton Facing Mill Co., Hamilton.
H. W. Petrie, Toronto.
Sheldon's Limited, Galt, Ont.
B. F. Sturtevant Co., Hyde Park, Mass.

Forgings, Drop.

John McDougall, Caledonian Iron Works. Co., Montreal.
H. W. Petrie, Toronto.
Whitman & Barnes Co., St. Catharnes

Forgings, Light & Heavy.

The Canada Forge Co., Welland.

Forging Machinery.

John Bertram & Sons Co., Dundas, Ont.
London Mach. Tool Co., Hamilton, Ont.
National Machinery Co., Tiffin, Ohio
Niles-Bement-Pond Co., New York.

Founders.

Greey Wm. & J. G., Toronto
John McDougall, Caledonian Iron Works Co., Montreal
R. McDougall Co., Galt, Ont.
Niagara Falls Machine & Foundry Co. Niagara Falls, Ont.
The Smart-Turner Mach. Co., Hamilton.

Foundry Equipment.

Detroit Foundry Supply Co., Windsor.
Hamilton Facing Mill Co., Mill.
Hanna Engineering Works, Chicago.

Foundry Facings.

Buffalo Foundry Supply Co., Buffalo.
Detroit Foundry Supply Co., Windsor.
Hamilton Facing Mill Co., Hamilton.
J. W. Paxson Co., Philadelphia, Pa.

Friction Clutch Pulleys, etc.

The Dodge Mfg. Co., Toronto.
The Goldie & McCulloch Co., Galt.
Greey, Wm. & J. G., Toronto
Link-Belt Co., Philadelphia.
Miller Bros. & Toms, Montreal.

Friction Hoists.

Owen Sound Iron Works Co., Owen Sound

Furnaces.

Detroit Foundry Supply Co., Windsor.
Dominion Foundry Supply Co., Montreal
Hamilton Facing Mill Co., Hamilton.

Galvanizing

Canada Metal Co., Toronto.

Gas Blowers and Exhausters.

Canadian Buffalo Forge Co., Montreal.
McEachren Heating and Ventilating Co., Galt, Ont.
D'Olier Engineering Co., New York.
Sheldon's Limited, Galt.
B. F. Sturtevant Co., Hyde Park, Man.

Gas Plants, Suction.

Baxter Paterson & Co., Montreal.
Dominion Henderson Bearings, Niagara Falls, Ont.
Economic Heat, Light and Power Supply Co., Toronto.
Producer Gas Co., Toronto
Williams & Wilson, Montreal

Gauges, Standard.

Brown & Sharpe, Providence, R.I.
Pratt & Whitney Co., Hartford, Conn.

Gearing.

Greey, Wm. & J. G. Toronto

Gear, Cutting Machinery.

Becker-Brainard Milling Mach. Co., Hyde Park, Mass.
Bickford Drill & Tool Co., Cincinnati.
Brown & Sharpe, Providence, R. I.
Kennedy, Wm., & Sons, Owen Sound
London Mach. Tool Co., Hamilton.
R. McDougall Co., Galt
Niles-Bement-Pond Co., New York.
H. W. Petrie, Toronto.
Pratt & Whitney Co., Hartford, Conn.
Williams & Wilson, Montreal.

Gears, Angle.

Brown & Sharpe, Providence, R.I.
Chicago Pneumatic Tool Co., Chicago.
Greey, Wm. & J. G., Toronto
Laurie Engine & Machine Co., Montreal.
John McDougall, Caledonian Iron Works Co., Montreal.
Miller Bros. & Toms, Montreal.
Waterous Engine Co., Brantford.

Gears, Cut.

Kennedy, Wm., & Sons, Owen Sound

Gears, Iron.

Kennedy, Wm., & Sons, Owen Sound

Gears, Mortise.

Kennedy, Wm., & Sons, Owen Sound

Gears, Reducing.

Brown & Sharpe, Providence, R.I.
Chicago Pneumatic Tool Co., Chicago.
Greey, Wm. & J. G., Toronto
John McDougall, Caledonian Iron Works Co., Montreal.
Miller Bros. & Toms, Montreal.

Generating Sets.

Sturtevant, B. F., Co., Hyde Park, Mass.

Generators, Electric.

Allis-Chalmers-Bullock, Limited, Montreal
Canadian General Electric Co., Toronto
Canadian Westinghouse Co., Hamilton.
D'Olier Engineering Co., New York.
Hall Engineering Works, Montreal.
H. W. Petrie, Toronto.
B. F. Sturtevant Co., Hyde Park, Mass.
Toronto & Hamilton Electric Co., Hamilton.

Generators, Gas.

H. W. Petrie, Toronto.

Glass Bevelling Machines.

Pendrith Machinery Co., Toronto.

Graphite Paints.

P. D. Dods & Co., Montreal.

Graphite.

Detroit Foundry Supply Co., Windsor.
Dominion Foundry Supply Co., Montreal
Hamilton Facing Mill Co., Hamilton.

Grinders, Automatic Knife.

W. H. Banfield & Son, Toronto.

Grinders, Centre.

Niles-Bement-Pond Co., New York.
H. W. Petrie, Toronto.

Grinders, Cutter.

Becker-Brainard Milling Mach. Co., Hyde Park, Mass.
Brown & Sharpe, Providence, R.I.
John Millen & Son, Ltd., Montreal.
Pratt & Whitney Co., Hartford, Conn.

Grinders, Tool.

Armstrong Bros. Tool Co., Chicago.
B. F. Barnes Co., Rockford, Ill.
Brown & Sharpe, Providence, R.I.
H. W. Petrie, Toronto.
Williams & Wilson, Montreal.

Grinding Machines.

Brown & Sharpe, Providence, R.I.
The Canadian Fairbanks, Montreal.
Niles-Bement-Pond Co., New York.
Norton Company, Worcester, Mass.
H. W. Petrie, Toronto.

Grinding and Polishing Machines.

The Canadian Fairbanks Co., Montreal.
Dominion Henderson Bearings, Niagara Falls, Ont.
Greey, Wm. & J. G., Toronto
John Millen & Son, Ltd., Montreal.
Miller Bros. & Toms, Montreal.
Niles-Bement-Pond Co., New York.
H. W. Petrie, Toronto.
Pendrith Machinery Co., Toronto.

Grinding Wheels (Alundum)

Norton Company, Worcester, Mass.

Hack Saws.

Baxter, Paterson & Co., Montreal.
Canada Machinery Agency, Montreal.
The Canadian Fairbanks Co., Montreal.
John Millen & Son, Ltd., Montreal.
Niles-Bement-Pond Co., New York.
H. W. Petrie, Toronto.
Williams & Wilson, Montreal.

Hammers, Drop.

London Mach. Tool Co., Hamilton, Ont.
Miller Bros. & Toms Montreal.
Niles-Bement-Pond Co., New York.

Hammers, Rand—All Kinds.

Whitman & Barnes Co., St. Catharines.

Hammers, Steam.

John Bertram & Sons Co. Dundas, Ont.
London Mach. Tool Co., Hamilton, Ont.
Niles-Bement-Pond Co., New York.

Hangers.

Dominion Henderson Bearings, Niagara Falls, Ont.
The Goldie & McCulloch Co., Galt.
Greey, Wm. & J. G., Toronto
Kennedy, Wm., & Sons, Owen Sound
Miller Bros. & Toms, Montreal.
Owen Sound Iron Works Co., Owen Sound
The Smart-Turner Mach. Co., Hamilton.
Waterous Engine Co., Brantford.
Pendrith Machinery Co., Toronto.

Hardware Specialties

Globe Machine & Stamping Co., Cleveland Ohio.

Heating Apparatus.

Canadian Buffalo Forge Co., Montreal
McEachren Heating and Ventilating Co., Galt, Ont.
Sheldon's Limited, Galt.
B. F. Sturtevant Co., High Park, Mass.

Hoisting and Conveying Machinery.

Allis-Chalmers-Bullock Limited, Montreal
Jenckes Machine Co., Sherbrooke, Que.
Link-Belt Co., Philadelphia.
Miller Bros. & Toms, Montreal.
Niles-Bement-Pond Co., New York.
The Smart-Turner Mach. Co., Hamilton.
Waterous Engine Co., Brantford.

Hoists, Pneumatic.

Canadian Rand Drill Co., Montreal.
Chicago Pneumatic Tool Co. Chicago.
Dominion Foundry Supply Co., Montreal
Hamilton Facing Mill Co., Hamilton.

Hose.

Baxter, Paterson & Co., Montreal.

Hose, Air.

Canadian Rand Drill Co., Montreal.
Canadian Westinghouse Co., Hamilton
Chicago Pneumatic Tool Co. Chicago.

Hose Couplings.

Canadian Rand Drill Co., Montreal.
Canadian Westinghouse Co., Hamilton.
Chicago Pneumatic Tool Co., Chicago.
Whitman & Barnes Co., St. Catharines.

Hose, Steam.

Allis-Chalmers-Bullock, Montreal.
Canadian Rand Drill Co., Montreal.
Canadian Westinghouse Co., Hamilton.

Hydraulic Accumulators.

Niles-Bement-Pond Co., New York.
The Smart-Turner Mach. Co., Hamilton.

Hydraulic Machinery.

Allis-Chalmers-Bullock, Montreal
Barber, Chas., & Sons, Meaford

India Oil Stones.

Norton Company, Worcester, Mass.

Indicators, Speed.

L. S. Starrett Co., Athol, Mass.

Injectors.

Canada Foundry Co., Toronto.
The Canadian Fairbanks Co., Montreal.
Rice Lewis & Son, Toronto.
Penberthy Injector Co., Windsor, Ont.

Iron Cements.

Buffalo Foundry Supply Co., Buffalo.
Detroit Foundry Supply Co., Windsor.
Hamilton Facing Mill Co., Hamilton.

Iron Filler.

Buffalo Foundry Supply Co., Buffalo.
Detroit Foundry Supply Co., Windsor.
Dominion Foundry Supply Co. Montreal
Hamilton Facing Mill Co., Hamilton.

Iron Tools.

John Bertram & Sons Co., Dundas, Ont.
A. B. Jardine & Co., Hespeler, Ont.
London Mach. Tool Co., Hamilton.
R. McDougall Co., Galt.
H. W. Petrie, Toronto.

Key-Seating Machines.

B. F. Barnes Co., Rockford, Ill.
Niles-Bement-Pond Co., New York.

Lace Leather.

Baxter, Paterson & Co., Montreal.

Lamps, Arc and Incandescent.

Canadian General Electric Co., Toronto.
Canadian Westinghouse Co., Hamilton.
The Packard Electric Co., St. Catharines.

Lathe Dogs.

Armstrong Bros., Chicago
Baxter, Paterson & Co. Montreal.
Pratt & Whitney Co., Hartford, Conn.

Lathes, Engine.

American Tool Work Co., Cincinnati.
B. F. Barnes Co., Rockford, Ill.
Baxter, Paterson & Co. Montreal
John Bertram & Sons Co., Dundas, Ont.
Canada Machinery Agency, Montreal.
The Canadian Fairbanks Co., Montreal.
London Mach. Tool Co., Hamilton, Ont.
Niles-Bement-Pond Co., New York.
R. McDougall Co., Galt, Ont.
H. W. Petrie, Toronto.
Pratt & Whitney Co., Hartford, Conn.

Lathes, Foot-Power.

American Tool Works Co., Cincinnati.
B. F. Barnes Co., Rockford, Ill.

Lathes, Screw Cutting.

B. F. Barnes Co., Rockford, Ill.
Baxter, Paterson & Co., Montreal.
Niles-Bement-Pond Co., New York.

Lathes, Automatic, Screw-Threading.

John Bertram & Sons Co., Dundas, Ont
London Mach. Tool Co., Hamilton, Ont
Pratt & Whitney Co., Hartford, Conn.

Lathes, Bench.

B. F. Barnes Co., Rockford, Ill.
London Mach. Tool Co., London, Ont.
Pratt & Whitney Co., Hartford, Conn.

Lathes, Turret.

American Tool Works Co., Cincinnati.
Baxter, Paterson & Co., Montreal
John Bertram & Sons Co., Dundas, Ont.
Canada Machinery Agency, Montreal.
Gisholt Machine Co., Madison, Wis.
London Mach. Tool Co., Hamilton, Ont.
Niles-Bement-Pond Co., New York.
R. McDougall Co., Galt, Ont.
The Pratt & Whitney Co., Hartford, Conn.
Warner & Swasey Co., Cleveland O.

Lathes, Buffing and Polishing.

Pendrith Machinery Co., Toronto.

Lath Mill Machinery.

Owen Sound Iron Works Co., Owen Sound

Leather Belting.

Baxter, Paterson & Co., Montreal
Canada Machinery Agency, Montreal.
The Canadian Fairbanks Co., Montreal
The Dodge Mfg. Co., Toronto.

Lime Stone Flux.

Hamilton Facing Mill Co., Hamilton.

Locomotives, Air.

Canadian Rand Drill Co., Montreal.

Locomotives, Electrical.

Canadian Westinghouse Co., Hamilton
Jeffrey Mfg. Co., Columbus, Ohio.

Locomotives, Steam.

Canada Foundry Co., Toronto.
Canadian Rand Drill Co., Montreal.

Lubricating Plumbago.

Detroit Foundry Supply Co., Detroit.
Dominion Foundry Supply Co., Montreal
Hamilton Facing Mill Co., Hamilton.

Lumber Dry Kilns.

Canadian Buffalo Forge Co., Montreal
McEachren Heating and Ventilating Co., Galt, Ont.
H. W. Petrie, Toronto.
Sheldon's Limited, Galt, Ont.
B. F. Sturtevant Co., Hyde Park, Mass.

Machinery Dealers.

Baxter, Paterson & Co. Montreal.
Canada Machinery Agency, Montreal.
The Canadian Fairbanks Co., Montreal.
Dominion Henderson Bearings, Limited, Toronto.
Miller Bros. & Toms, Montreal.
H. W. Petrie, Toronto.
The Smart-Turner Mach. Co., Hamilton.
Williams & Wilson Montreal.

Machinery Designers.

Dominion Henderson Bearings, Niagara Falls, Ont.
Greey, Wm. & J. G., Toronto

Machinists.

W. H. Banfield & Sons, Toronto.
The Dodge Mfg. Co., Toronto.
Dominion Henderson Bearings, Niagara Falls, Ont.
Greey, Wm. & J. G. Toronto
Hall Engineering Works Montreal.
Link-Belt Co., Philadelphia.
John McDougall, Caledonian Iron Works Co., Montreal.
Miller Bros. & Toms Montreal.
Pendrith Machinery Co., Toronto.
Robb Engineering Co., Amherst, N.S.
The Smart-Turner Mach. Co., Hamilton.
Waterous Engine Co., Brantford.

Machinists' Small Tools.

Armstrong Bros., Chicago.
Brown & Sharpe, Providence, R.I.
Butterfield & Co., Rock Island, Que.
Rice Lewis & Son, Montreal.
Pratt & Whitney Co., Hartford, Conn.
Standard Tool Co., Cleveland.
L. S. Starrett Co., Athol, Mass.
Williams & Wilson, Montreal.
Whiteman & Barnes Co., St. Catharines.

Mailing Weights.

A. B. Jardine & Co., Hespeler, Ont.

Malleable Flask Clamps.

Buffalo Foundry Supply Co., Buffalo.

Mallet, Rawhide and Wood.

Buffalo Foundry Supply Co., Buffalo.
Detroit Foundry Supply Co., Detroit.

Mandrels.

Brown & Sharpe, Providence, R.I.
A. B. Jardine & Co., Hespeler, Ont.
Miller Bros. & Toms, Montreal
The Pratt & Whitney Co., Hartford, Conn.
Standard Tool Co., Cleveland.
Whitman & Barnes Co., St. Catharines

Measuring Machines.

The Pratt & Whitney Co., Hartford, Conn.

Marine Work.

Kennedy, Wm., & Sons, Owen Sound

Mechanical Draft.

Canadian Buffalo Forge Co., Montreal.
H. W. Petrie, Toronto.
Sheldon's Limited, Galt.
B. F. Sturtevant Co., Hyde Park, Mass.

Metal Separators.

Toronto and Hamilton Electric Co., Hamilton.

Metallic Paints.

P. D. Dods & Co., Montreal.

Meters, Electrical.

Canadian Westinghouse Co., Hamilton.

Mill Machinery.

Baxter, Paterson & Co., Montreal.
The Dodge Mfg. Co., Toronto.
Greey Wm. & J. G. Toronto
The Goldie & McCulloch Co., Galt, Ont.
Jenckes Machine Co., Sherbrooke, Que.
John McDougall Caledonian Iron Works Co., Montreal.
H. W. Petrie, Toronto.
Robb Engineering Co., Amherst, N.S.
Waterous Engine Co., Brantford.
Williams & Wilson, Montreal.

Milling Attachments.

Becker-Brainard Milling Machine Co. Hyde Park, Mass.
John Bertram & Sons Co., Dundas, Ont.
Brown & Sharpe, Providence, R.I.
Niles-Bement-Pond Co., New York.
Pratt & Whitney, Hartford, Conn.

Milling Machines, Horizontal.

Baxter, Paterson & Co., Montreal.
Becker-Brainard Milling Machine Co. Hyde Park, Mass.
John Bertram & Sons Co., Dundas, Ont.
Brown & Sharpe, Providence, R.I.
John Bertram & Sons Co., Dundas, Ont.
Canada Machinery Agency, Montreal.
London Mach. Tool Co., Hamilton, Ont.
Niles-Bement-Pond Co., New York.
Pratt & Whitney, Hartford, Conn.

Milling Machines, Plain.

American Tool Works Co, Cincinnati.
Becker-Brainard Milling Machine Co. Hyde Park, Mass.
Brown & Sharpe, Providence, R.I.
John Bertram & Sons Co., Dundas, Ont.
Canada Machinery Agency, Montreal.
The Canadian Fairbanks Co, Montreal.
London Mach. Tool Co., Hamilton, Ont.
Niles-Bement-Pond Co., New York.
H. W. Petrie, Toronto.
Pratt & Whitney Co., Hartford, Conn.
Williams & Wilson, Montreal

Milling Machines, Universal.

American Tool Works Co., Cincinnati.
Becker-Brainard Milling Machine Co., Hyde Park, Mass.
Brown & Sharpe, Providence, R.I.
John Bertram & Sons Co., Dundas, Ont.
Canada Machinery Agency, Montreal.
The Canadian Fairbanks Co., Montreal.
London Mach. Tool Co., Hamilton, Ont.
Niles-Bement-Pond Co., New York.
H. W. Petrie, Toronto.
Williams & Wilson, Montreal.

Milling Machines, Vertical.

Becker-Brainard Milling Machine Co., Hyde Park, Mass.
Brown & Sharpe, Providence, R.I.
John Bertram & Sons Co., Dundas, Ont.
Canada Machinery Agency, Montreal.
London Mach. Tool Co., Hamilton, Ont.
Niles-Bement-Pond Co., New York.

Milling Tools.

Wm. Abbott, Montreal.
Becker-Brainard Milling Machine Co., Hyde Park, Mass.
Brown & Sharpe, Providence, R.I.
Dominion Henderson Bearings, Niagara Falls, Ont.
Geometric Tool Co., New Haven, Conn.
London Mach. Tool Co., Hamilton, Ont.
Pratt & Whitney Co., Hartford, Conn.
Standard Tool Co., Cleveland.

Mining Machinery.

Allis-Chalmers-Bullock Limited, Montreal
Canadian Rand Drill Co., Montreal.
Chicago Pneumatic Tool Co., Chicago.
The Dodge Mfg. Co., Toronto.
Jeffrey Mfg. Co., Columbus, Ohio.
Jenckes Machine Co., Sherbrooke, Que.
Laurie Engine & Machine Co., Montreal.
John McDougall, Caledonian Iron Works Co., Montreal.
Miller Bros. & Toms, Montreal.
T. & H. Electric Co. Hamilton.

Mixing Machines, Dough.

Greey, Wm. & J. G., Toronto
Pendrith Machinery Co., Toronto.

Mixing Machines, Special.

Greey, Wm. & J. G., Toronto
Pendrith Machinery Co., Toronto.

Model Tools.

Dominion Henderson Bearings, Niagara Falls Ont.
Globe Machine & Stamping Co., Cleveland, Ohio.
Wells Pattern and Model Works, Toronto

Motors, Electric.

Allis-Chalmers-Bullock Limited, Montreal
Canadian General Electric Co., Toronto.
Canadian Westinghouse Co., Hamilton.
Consolidated Electric Co., Toronto
Hall Engineering Works, Montreal.
The Packard Electric Co., St. Catharines.
B. F. Sturtevant Co., Hyde Park, Mass.
T. & H. Electric Co., Hamilton.

Motors, Air.

Canadian Rand Drill Co., Montreal.
Chicago Pneumatic Tool Co., Chicago.

Molders' Supplies.

Buffalo Foundry Supply Co., Buffalo.
Detroit Foundry Supply Co., Windsor.
Dominion Foundry Supply Co., Montreal
Hamilton Facing Mill Co., Hamilton

Molders' Tools.

Buffalo Foundry Supply Co. Buffalo.
Detroit Foundry Supply Co. Windsor
Dominion Foundry Supply Co. Montreal
Hamilton Facing Mill Co., Hamilton.

Molding Machines.

Buffalo Foundry Supply Co. Buffalo.
Hamilton Facing Mill Co., Hamilton.
J. W. Paxson Co., Philadelphia, Pa.

Molding Sand.

T. W. Barnes, Hamilton.
Buffalo Foundry Supply Co., Buffalo.
Detroit Foundry Supply Co., Windsor
Dominion Foundry Supply Co., Montreal
Hamilton Facing Mill Co., Hamilton.

Nippers, Stay Bolt.

Chicago Pneumatic Tool Co., Chicago.

Nut Tappers.

John Bertram & Sons Co., Dundas, Ont
A. B. Jardine & Co., Hespeler.
London Mach. Tool Co., Hamilton.
National Machinery Co., Tiffin, Ohio.

Oatmeal Mill Machinery.

Greey, Wm. & J. G., Toronto
The Goldie & McCulloch Co., Galt

Oils, Core.

Buffalo Foundry Supply Co., Buffalo.
Dominion Foundry Supply Co., Montreal
Hamilton Facing Mill Co., Hamilton.

Paint Mill Machinery.

Greey, Wm. & J. G., Toronto

Painting Machines, Pneumatic.
Chicago Pneumatic Tool Co., Chicago.

Patent Solicitors.
Hanbury A. Budden, Montreal.
Fetherstonhaugh & Blackmore, Montreal
Fetherstonhaugh & Co., Montreal.
Marion & Marion, Montreal.
Ridout & Maybee, Toronto.

Patterns.
John Carr, Hamilton.
The Dodge Mfg. Co., Toronto.
Hamilton Pattern Works, Hamilton.
John McDougall Caledonian Iron Works Co., Montreal
Miller Bros. & Toms, Montreal.
Wells' Pattern and Model Works, Toronto.

Rolls, Bending.
John Bertram & Sons Co., Dundas, Ont.
London Mach. Tool Co., Hamilton, Ont.
Niles-Bement-Pond Co., New York.

Rolls, Chilled Iron.
Greey, Wm. & J. G., Toronto

Rolls, Sand Cast.
Greey, Wm. & J. G., Toronto

Rotary Blowers.
Sturtevant, B. F., Co., Hyde Park, Mass.

Rotary Converters.
Allis-Chalmers-Bullock, Ltd., Montreal.
Canadian Westinghouse Co., Hamilton
Toronto and Hamilton Electric Co., Hamilton.

Safes.
Baxter Paterson & Co., Montreal
The Goldie & McCulloch Co., Galt, On

Sand, Bench.
Buffalo Foundry Supply Co., Buffalo.
Detroit Foundry Supply Co., Windsor
Hamilton Facing Mill Co., Hamilton.

Sand Blast Machinery.
Canadian Rand Drill Co., Montreal.
Chicago Pneumatic Tool Co., Chicago.
Detroit Foundry Supply Co., Windsor
Hamilton Facing Mill Co., Hamilton.
J. W. Paxson Co., Philadelphia, Pa.

Sand, Brass.
Buffalo Foundry Supply Co., Buffalo
Detroit Foundry Supply Co., Windsor

Sand Fire.
Detroit Foundry Supply Co., Detroit.

Sand, Heavy, Grey Iron.
Buffalo Foundry Supply Co., Buffalo.
Detroit Foundry Supply Co., Windsor
Hamilton Facing Mill Co., Hamilton.

Sand, Malleable.
Buffalo Foundry Supply Co., Buffalo.
Detroit Foundry Supply Co., Windsor
Hamilton Facing Mill Co., Hamilton.

Sand, Medium Grey Iron.
Buffalo Foundry Supply Co., Buffalo.
Detroit Foundry Supply Co., Windsor
Hamilton Facing Mill Co., Hamilton.

Sand, Plate.
Buffalo Foundry Supply Co., Buffalo.
Detroit Foundry Supply Co., Windsor
Hamilton Facing Mill Co., Hamilton.

Sand Sifters.
Buffalo Foundry Supply Co., Buffalo.
Detroit Foundry Supply Co., Windsor
Dominion Foundry Supply Co., Montreal
Hamilton Facing Mill Co., Hamilton.
Hanna Engineering Works, Chicago.

Sand, Stove Pipe.
Buffalo Foundry Supply Co., Buffalo.
Detroit Foundry Supply Co., Windsor
Hamilton Facing Mill Co., Hamilton.

Saw Gummers.
Canada Machinery Agency, Montreal.

Saw Machines, Power Hack.
Baxter Paterson & Co., Montreal.

Saw Mill Machinery.
Allis-Chalmers-Bullock, Limited, Montreal
Baxter, Paterson & Co., Montreal
Canada Machinery Agency, Montreal.
The Dodge Mfg. Co., Toronto.
Dominion Henderson Bearings, Niagara Falls, Ont
Goldie & McCulloch Co., Galt.
Greey, Wm. & J. G., Toronto
Miller Bros. & Toms, Montreal
Owen Sound Iron Works Co., Owen Sound
H. W. Petrie, Toronto.
Robb Engineering Co. Amherst,
Waterous Engine Works, Brantfo
Williams & Wilson, Montreal.

Sawing Machines, Metal.
Niles-Bement-Pond Co. New York.

Saws, Hack.
Baxter Paterson & Co., Montreal.
Canada Machinery Agency, Montreal.
Detroit Foundry Supply Co., Windsor
London Mach. Tool Co., Hamilton
Rice Lewis & Son, Toronto.
John Millen & Son, Ltd., Montreal.
L. S. Starrett Co., Athol, Mass.

Screw Cutting Tools
Hart Manufacturing Co., Cleveland, Ohio

Screw Machines, Automatic.
Brown & Sharpe, Providence, R.I.
Canada Machinery Agency, Montreal.
Cleveland Automatic Machine Co., Cleveland, Ohio
London Mach. Tool Co., Hamilton, Ont.
Pratt & Whitney Co., Hartford, Conn.

Screw Machines, Hand.
Brown & Sharpe, Providence, R.I.
Canada Machinery Agency Montreal.
A. B. Jardine & Co., Hespeler.
London Mach. Tool Co., Hamilton, Ont.
Pratt & Whitney Co., Hartford, Conn.
Warner& Swasey Co., Cleveland, O.

Screw Plates.
Butterfield & Co., Rock Island, Que.
Hart Manufacturing Co. Cleveland, Ohio
A. B. Jardine & Co., Hespeler.

Second-Hand Machinery.
American Tool Works Co., Cincinnati.
Canada Machinery Agency, Montreal.
The Canadian Fairbanks Co., Montreal.
Goldie & McCulloch Co., Galt.
Machinery Exchange, Montreal.
Niles-Bement-Pond Co., New York.
H. W. Petrie, Toronto.
Robb Engineering Co., Amherst, N S.
Waterous Engine Co., Brantford.
Williams & Wilson, Montreal.

Shafting.
Baxter, Paterson & Co., Montreal.
Canada Machinery Agency, Montreal
The Canadian Fairbanks Co., Montreal.
The Dodge Mfg. Co., Toronto.
The Goldie & McCulloch Co., Galt, Ont
Greey Wm. & J. G. Toronto
Jenckes Machine Co., Sherbrooke, Que.
Kennedy Wm. & Sons, Owen Sound
Niles-Bement-Pond Co., New York.
Owen Sound Iron Works Co., Owen Sound
Pendrith Machine Co., Toronto.
H. W. Petrie, Toronto.
Smart-Turner Machine Co., Hamilton.
Union Drawn Steel Co., Hamilton.
Waterous Engine Co., Brantford.

Shapers.
American Tool Works Co., Cincinnati.
John Bertram & Sons Co., Dundas, Ont
Canada Machinery Agency, Montreal.
The Canadian Fairbanks Co., Montreal.
London Mach. Tool Co., Hamilton, Ont.
Niles-Bement-Pond Co., New York.
H. W. Petrie, Toronto.
Potter & Johnston Machine Co., Pawtucket, R.I.
Pratt & Whitney Co., Hartford, Conn.
Williams & Wilson, Montreal.

Shearing Machine, Bar.
John Bertram & Sons Co., Dundas, Ont.
Canadian Buffalo Forge Co., Montreal.
A. B. Jardine & Co., Hespeler.
London Mach. Tool Co., Hamilton, Ont.
Niles-Bement-Pond Co., New York.

Shears, Bench.
Canadian Buffalo Forge Co., Montreal.

Shears, Continuous.
Canadian Buffalo Forge Co., Montreal.

Shears, Hand.
Canadian Buffalo Forge Co., Montreal.

Shears, Power.
John Bertram & Sons Co., Dundas, Ont.
Canada Machinery Agency, Montreal.
Canadian Buffalo Forge Co., Montreal.
A. B. Jardine & Co., Hespeler, Ont.
Niles-Bement-Pond Co., New York.

Sheet Metal Goods
Globe Machine & Stamping Co., Cleveland, Ohio.

Sheet Steel Work.
Owen Sound Iron Works Co., Owen Sound

Shingle Mill Machinery.
Owen Sound Iron Works Co., Owen Sound

Shop Trucks.
Owen Sound Iron Works Co., Owen Sound

Shovels.
Baxter, Paterson & Co., Montreal.
Buffalo Foundry Supply Co., Buffalo.
Detroit Foundry Supply Co., Detroit.
Dominion Foundry Supply Co., Montreal
Hamilton Facing Mill Co., Hamilton.

Shovels, Steam.
Allis-Chalmers-Bullock, Montreal.

Sieves.
Buffalo Foundry Supply Co. Buffalo.
Detroit Foundry Supply Co., Windsor
Dominion Foundry Supply Co., Montreal
Hamilton Facing Mill Co., Hamilton.

Silver Lead.
Buffalo Foundry Supply Co., Buffalo.
Detroit Foundry Supply Co., Windsor
Dominion Foundry Supply Co., Montreal
Hamilton Facing Mill Co., Hamilton.

Sleeves, Reducing.
Chicago Pneumatic Tool Co., Chicago.

Slide Rests.
Niles-Bement-Pond Co., New York.

Smoke Connections.
Jenckes Machine Co., Sherbrooke, Que

Smoke Stacks.
Jenckes Machine Co., Sherbrooke, Que.

Snap Flasks
Buffalo Foundry Supply Co., Buffalo.
Detroit Foundry Supply Co., Windsor
Dominion Foundry Supply Co., Montreal
Hamilton Facing Mill Co., Hamilton.

Soapstone.
Buffalo Foundry Supply Co., Buffalo.
Detroit Foundry Supply Co., Windsor
Dominion Foundry Supply Co., Montreal
Hamilton Facing Mill Co., Hamilton.

Special Machinery.
W. H. Banfield & Sons, Toronto.
Baxter Paterson & Co., Montreal.
John Bertram & Sons Co., Dundas, Ont.
Dominion Henderson Bearings, Niagara Falls Ont
Globe Machine & Stamping Co., Cleveland Ohio.
Greey, Wm. & J. G., Toronto
Hanna Engineering Works, Chicago.
Laurie Engine & Machine Co., Montreal.
London Mach. Tool Co., Hamilton, Ont.
R. McDougall Co., Galt, Ont.
Pendrith Machinery Co., Toronto.
H. W. Petrie, Toronto.
The Smart-Turner Mach. Co., Hamilton
Waterous Engine Co., Brantford.

Pipe Cutting and Threading Machines.
Butterfield & Co., Rock Island, Que.
Canada Machinery Agency, Montreal.
A. B. Jardine & Co., Hespeler, Ont.
London Mach. Tool Co., Hamilton, Ont.
R. McDougal Co., Galt, Ont.
Niles-Bement-Pond Co., New York.

Pipe, Municipal.
Canadian Pipe Co., Vancouver, B.C.
Pacific Coast Pipe Co., Vancouver, B.C.

Pipe, Waterworks.
Canadian Pipe Co., Vancouver, B.C.
Pacific Coast Pipe Co., Vancouver, B.C

Pipe, Wooden.
Canadian Pipe Co., Vancouver, B.C.
Pacific Coast Pipe Co., Vancouver, B.C.

Planers, Standard.
American Tool Works, Cincinnati.
John Bertram & Sons Co., Dundas, Ont.
Canada Machinery Agency, Montreal.
The Canadian Fairbanks Co., Montreal.
London Mach. Tool Co., Hamilton, Ont.
R. McDougall Co., Galt, Ont.
Niles-Bement-Pond Co., New York.
H. W. Petrie, Toronto.
Pratt & Whitney Co., Hartford, Conn.
Williams & Wilson, Montreal.

Planers, Rotary.
John Bertram & Sons Co., Dundas, Ont.
London Mach. Tool Co., Hamilton, Ont.
Niles-Bement-Pond Co., New York.

Power Hack Saw Machines.
Baxter, Paterson & Co., Montreal.

Power Plants.
John McDougall Caledonian Iron Works Co., Montreal
The Smart-Turner Mach. Co., Hamilton

Planing Mill Fans.
Canadian Buffalo Forge Co., Montreal.
McEachren Heating and Ventilating Co., Galt, Ont.
Sheldon's Limited, Galt, Ont.
B. F. Sturtevant Co., Hyde Park, Mass.

Plug Drillers, Pneumatic.
Canada Foundry Co., Limited, Toronto.
Canadian Rand Drill Co., Mont

Plumbago.
Buffalo Foundry Supply Co., Buffalo.
Detroit Foundry Supply Co., Windsor
Dominion Foundry Supply Co., Montreal
Hamilton Facing Mill Co., Hamilton,
J. W. Paxson Co., Philadelphia, Pa.

Pneumatic Separators.
Sturtevant, B. F., Co., Hyde Park, Mass.

Pneumatic Tools.
Allis-Chalmers-Bullock, Montreal.
Canadian Rand Drill Co., Montreal.
Chicago Pneumatic Tool Co., Chicago
Hamilton Facing Mill Co., Hamilton.
Hanna Engineering Works, Chicago.

Presses, Drill.
American Tool Works Co., Cincinnati.
Niles-Bement-Pond Co., New York.

Presses, Drop.
W. H. Banfield & Son, Toronto.
E. W. Bliss Co., Brooklyn, N.Y.
Canada Machinery Agency, Montreal.
Laurie Engine & Machine Co., Montreal
Miller Bros. & Toms, Montreal.
Niles-Bement-Pond Co., New York.

Presses, Hand.
E. W. Bliss Co., Brooklyn, N.Y.

Presses, Hydraulic.
Jenckes Machine Co., Sherbrooke, Que.
John Bertram & Sons Co., Dundas, Ont.
Laurie Engine & Machine Co., Montreal.
London Mach. Tool Co., Hamilton, Ont.
John McDougall Caledonian Iron Works Co. Montreal.
Miller Bros. & Toms, Montreal.
Niles-Bement-Pond Co., New York.
Perrin, Wm. R., & Co., Toronto

Presses, Power.
E. W. Bliss Co., Brooklyn, N.Y.
Canada Machinery Agency, Montreal.
Jenckes Machine Co., Sherbrooke, Que.
London Mach Tool Co., Hamilton, Ont.
John McDougall Caledonian Iron Works Co., Montreal.
Niles-Bement-Pond Co., New York.

Presses Power Screw.
Perrin, Wm. R., & Co., Toronto

Pulp Mill Machinery.
Greey, Wm. & J. G., Toronto
Jeffrey Mfg Co., Columbus, Ohio.
Jenckes Machine Co., Sherbrooke, Que.
Laurie Engine & Machine Co., Montreal.
John McDougall Caledonian Iron Works Co., Montreal.
Waterous Engine Works Co., Brantford

Pulleys.
Baxter, Paterson & Co., Montreal.
Canada Machinery Agency, Montreal.
The Canadian Fairbanks Co., Montreal.
The Dodge Mfg. Co., Toronto.
Dominion Henderson Bearings, Niagara Falls, Ont.
The Goldie & McCulloch Co., Galt.
Greey, Wm. & J. G., Toronto
Jenckes Machine Co., Sherbrooke, Que.
Kennedy, Wm. & Sons, Owen Sound
Laurie Engine & Machine Co., Montreal.
Link-Belt Co., Philadelphia.
John McDougall Caledonian Iron Works Co., Montreal.
Miller Bros. & Toms, Montreal.
Owen Sound Iron Works Co., Owen Sound
Pendrith Machinery Co., Toronto.
H. W. Petrie, Toronto.
The Smart-Turner Mach. Co., Hamilton.
Williams & Wilson, Montreal.
Waterous Engine Co., Brantford.

Pumps, Steam.
Allis-Chalmers-Bullock, Limited, Montreal
Canada Foundry Co., Toronto.
Canada Machinery Agency, Montreal.
Canadian Buffalo Forge Co., Montreal.
D'Olier Engineering Co., New York.
Dominion Henderson Bearings, Niagara Falls, Ont.
The Goldie & McCulloch Co., Galt.
John McDougall Caledonian Iron Works, Montreal.
R. McDougall Co., Galt, Ont.
H. W. Petrie, Toronto.
St. Lawrence Supply Co., Montreal.
The Smart-Turner Mach. Co., Hamilton.
Waterous Engine Co., Brantford.

Pumping Machinery.
Canada Foundry Co., Limited, Toronto
Canada Machinery Agency, Montreal.
Canadian Buffalo Forge Co., Montreal.
Canadian Rand Drill Co., Montreal.
Chicago Pneumatic Tool Co., Chicago.
D'Olier Engineering Co., New York.
Dominion Henderson Bearings, Niagara Falls, Ont
Hall Engineering Works, Montreal, Que.
Laurie Engine & Machine Co., Montreal.
London Mach. Tool Co., Hamilton, Ont.
John McDougall Caledonian Iron Works Co., Montreal.
R. McDougall Co., Galt Ont.
The Smart-Turner Mach. Co., Hamilton.

Punches and Dies.
W. H. Banfield & Sons, Toronto.
Butterfield & Co., Rock Island.
Dominion Henderson Bearings, Niagara Falls, Ont.
Globe Machine & Stamping Co.
A. B. Jardine & Co., Hespeler, Ont.
London Mach. Tool Co., Hamilton, Ont.
Miller Bros. & Toms, Montreal.
Pratt & Whitney Co., Hartford, Conn.
H. W. Petrie, Toronto.

Punches, Hand.
Canadian Buffalo Forge Co., Montreal

Punches, Power.
John Bertram & Sons Co., Dundas, Ont.
E. W. Bliss Co., Brooklyn, N.Y.
Canada Machinery Agency, Montreal.
Canadian Buffalo Forge Co., Montreal.
London Mach. Tool Co., Hamilton, Ont.
Niles-Bement-Pond Co., New York.

Punches, Turret.
London Mach. Tool Co., London, Ont.

Punching Machines, Horizontal.
John Bertram & Sons Co., Dundas, Ont.
London Mach. Tool Co., Hamilton, Ont.
Niles-Bement-Pond Co., New York.

Quartering Machines.
John Bertram & Sons Co., Dundas, Ont.
London Mach. Tool Co., Hamilton, Ont.

Rammers, Bench and Floor.
Buffalo Foundry Supply Co., Buffalo.
Detroit Foundry Supply Co., Windsor
Hamilton Facing Mill Co., Hamilton.

Rapping Plates.
Detroit Foundry Supply Co., Windsor
Hamilton Facing Mill Co., Hamilton.

Reamers.
Wm. Abbott, Montreal.
Baxter, Paterson & Co., Montreal.
Brown & Sharpe, Providence, R.I.
Butterfield & Co., Rock Island.
Hanna Engineering Works, Chicago.
A. B. Jardine & Co., Hespeler, Ont.
John Millen & Son, Ltd., Montreal.
Morse Twist Drill and Machine Co., New Bedford, Mass.
Pratt & Whitney Co., Hartford, Conn.
Standard Tool Co., Cleveland.
Whitman & Barnes Co., St. Catharines.

Reamers, Steel Taper.
Butterfield & Co., Rock Island.
Chicago Pneumatic Tool Co., Chicago.
A. B. Jardine & Co., Hespeler, Ont.
John Millen & Son Ltd., Montreal.
Pratt & Whitney Co., Hartford, Conn.
Standard Tool Co., Cleveland.
Whitman & Barnes Co., St. Catharines.

Rheostats.
Canadian General Electric Co., Toronto.
Canadian Westinghouse Co., Hamilton.
Hall Engineering Works, Montreal, Que.
T. & H. Electric Co., Hamilton,

Riddles.
Buffalo Foundry Supply Co., Buffalo.
Detroit Foundry Supply Co., Windsor
Hamilton Facing Mill Co., Hamilton.
J. W. Paxson Co., Philadelphia, Pa.

Riveters, Hydraulic.
London Mach. Tool Co., Hamilton, Ont.
Niles-Bement-Pond Co., New York.

Riveters, Pneumatic.
Hanna Engineering Works, Chicago.

Special Machines and Tools.
Dominion Henderson Bearings, Niagara Falls Ont.
Pratt & Whitney, Hartford, Conn.
The Stevens Co., Galt, Ont.

Special Manufacturing.
Dominion Henderson Bearings, Niagara Falls, Ont.
Globe Machine & Stamping Co., Cleveland.
Miller Bros. & Toms, Montreal.
Pendrith Machine Co., Toronto.
The Stevens Co., Galt, Ont.

Special Repairs.
Dominion Henderson Bearings, Niagara Falls, Ont.

Speed Changing Countershafts.
The Canadian Fairbanks Co., Montreal
The Dodge Mfg. Co., Toronto.

Spike Machines.
National Machinery Co., Tiffin, O.
The Smart-Turner Mach. Co., Hamilton

Spray Cans.
Detroit Foundry Supply Co., Windsor
Dominion Foundry Supply Co., Montreal
Hamilton Facing Mill Co., Hamilton.

Sprue Cutters.
Dominion Foundry Supply Co., Montreal
Detroit Foundry Supply Co., Windsor
Hamilton Facing Mill Co., Hamilton.

Stamp Mills.
Allis-Chalmers-Bullock, Limited, Montreal

Stampings, Sheet Metal.
Baxter, Paterson & Co., Montreal.

Steam Hot Blast Apparatus
Canadian Buffalo Forge Co., Montreal.
McEachren Heating and Ventilating Co., Galt.
Sheldon's Limited, Galt.
B. F. Sturtevant Co., Hyde Park, Mass.

Steam Plants.
Jenckes Machine Co., Sherbrooke, Que.

Steam Separators.
McEachren Heating and Ventilating Co., Galt.
Robb Engineering Co., Montreal.
Sheldon's Limited, Galt.
Smart-Turner Mach. Co., Hamilton
Waterous Engine Co., Brantford.

Steam Specialties.
Engineering Specialties Co., Belfast, Ireland.
McEachren Heating and Ventilating Co., Galt.
Sheldon's Limited, Galt.

Steam Traps.
Canada Machinery Agency, Montreal.
Engineering Specialties Co., Belfast, Ireland.
McEachren Heating and Ventilating Co., Galt.
Pendrith Machinery Co., Toronto.
Sheldon's Limited, Galt.
B. F. Sturtevant Co., Hyde Park, Mass.

Steel Pressure Blowers
Buffalo Foundry Supply Co., Buffalo.
Canada Machinery Agency, Montreal.
Canadian Buffalo Forge Co., Montreal.
Buffalo Foundry Supply Co., Buffalo.
Dominion Foundry Supply Co., Montreal
Hamilton Facing Mill Co., Hamilton.
McEachren Heating and Ventilating Co., Galt.
J. W. Paxon Co., Philadalph'a, Pa.
Sheldon's Limited, Galt.
B. F. SturtevantCo., Hyde Park, Mass.

Steel Tubes.
Baxter, Paterson & Co., Montreal.
John Millen & Son, Montreal.

Steel, High Speed.
Wm. Abbott, Montreal.
Canadian Fairbanks Co., Montreal.
Alex. Gibb, Montreal.
Jessop, Wm., & Sons, Sheffield, Eng.
B. K. Morton Co., Sheffield, Eng.
Williams & Wilson, Montreal.

Stocks and Dies
Hart Manufacturing Co., Cleveland, Ohio
Pratt & Whitney Co., Hartford, Conn.

Stone Cutting Tools, Pneumatic
Allis-Chalmers-Bullock, Ltd., Montreal
Canadian Rand Drill Co., Montreal.

Stone Surfacers
Chicago Pneumatic Tool Co., Chicago.

Stove Plate Facings.
Buffalo Foundry Supply Co., Buffalo
Detroit Foundry Supply Co., Windsor.
Dominion Foundry Supply Co., Montreal
Hamilton Facing Mill Co., Hamilton.

Swage, Block.
A. B. Jardine & Co., Hespeler, Ont.

Switchboards.
Allis-Chalmers-Bullock, Limited, Montreal
Canadian General Electric Co., Toronto.
Canadian Westinghouse Co., Hamilton
Hall Engineering Works, Montreal, Que
Toronto and Hamilton Electric Co., Hamiton.

Talc.
Detroit Foundry Supply Co., Windsor.
Hamiton Facing Mill Co., Hamilton.
Buffalo Foundry Supply Co., Buffalo.

Tapping Machines and Attachments.
American Tool Works Co., Cincinnati.
John Bertram & Sons Co., Dundas, Ont.
Bickford Drill & Tool Co., Cincinnati.
Canada Machinery Agency, Montreal.
The Geometric Tool Co., New Haven.
A. B. Jardine & Co., Hespeler.
London Mach. Tool Co., Hamilton, Ont
Murchey Machine & Tool Co., Detroit.
Niles-Bement-Pond Co., New York.
H. W. Petrie, Toronto.
Pratt & Whitney, Cincinnati, O.
L. S. Starrett Co., Athol, Mass.
Williams & Wilson, Montreal.

Tapes, Steel.
Rice Lewis & Son, Toronto.
John Millen & Son, Ltd., Montreal.
L. S. Starrett Co., Athol, Mass.

Taps.
Baxter, Paterson & Co., Montreal.
Whitman & Barnes Co., St. Catharines.

Taps, Collapsing.
The Geometric Tool Co., New Haven.

Taps and Dies.
Wm. Abbott, Montreal.
Baxter, Paterson & Co., Montreal.
Butterfield & Co., Rock Island, Que.
The Geometric Tool Co., New Haven.
A. B. Jardine & Co., Hespeler, Ont.
Rice Lewis & Son, Toronto.
John Millen & Son, Ltd., Montreal.
Pratt & Whitney Co., Hartford, Conn.
Standard Tool Co., Cleveland.
L. S. Starrett Co., Athol, Mass.

Testing Laboratory.
Detroit Testing Laboratory, Detroit.

Testing Machines.
Detroit Foundry Supply Co., Windsor.
Dominion Foundry Supply Co., Montreal
Hamilton Facing Mill Co., Hamilton.

Thread Cutting Tools.
Hart Manufacturing Co., Cleveland, Ohio

Tiling, Opal Glass.
Toronto Plate Glass Importing Co., Toronto.

Time Switches, Automatic, Electric.
Berlin Electric Mfg. Co., Toronto.

Tinware Machinery
Canada Machinery Agency, Montreal.

Tire Upsetters or Shrinkers
Canadian Buffalo Forge Co., Montreal.
A. B. Jardine & Co., Hespeler, Ont.

Tool Cutting Machinery
Canadian Rand Drill Co., Montreal.

Tool Holders.
Armstrong Bros. Tool Co., Chicago.
Baxter, Paterson & Co., Montreal.
Fairbanks Co., Springfield, Ohio.
John Millen & Son, Ltd., Montreal.
H. W. Petrie, Toronto.
Pratt & Whitney Co., Hartford, Conn.

Tool Steel.
Wm. Abbott, Montreal.
Wm. Jessop, Sons & Co., Toronto.
Canadian Fairbanks Co., Montreal
B. K. Morton & Co., Sheffield, Eng.
Williams & Wilson, Montral.

Torches, Steel.
Baxter, Paterson & Co., Montreal
Detroit Foundry Supply Co., Windsor.
Dominion Foundry Supply Co., Montreal
Hamilton Facing Mill Co., Hamilton.

Transformers and Convertors
Allis-Chalmers-Bullock, Montreal.
Canadian General Electric Co., Toronto.
Canadian Westinghouse Co., Hamilton.
Hall Engineering Works, Montreal, Que.
T. & H. Electric Co., Hamilton.

Transmission Machinery
Allis-Chalmers-Bullock, Montreal.
The Canadian Fairbanks Co., Montreal.
The Dodge Mfg. Co., Toronto.
Greey, Wm. & J. G., Toronto.
Laurie Engine & Machine Co., Montreal.
Link-Belt Co., Philadelphia.
Miller Bros. & Toms Montreal.
H. W. Petrie, Toronto.
The Smart-Turner Mach. Co., Hamilton.
Waterous Engine Co., Brantford.

Transmission Supplies
Baxter, Paterson & Co., Montreal.
The Canadian Fairbanks Co., Montreal.
The Dodge Mfg. Co., Toronto.
Wm. & J. G. Greey, Toronto.
The Goldie & McCulloch Co., Galt.
Miller Bros. & Toms, Montreal
Pendrith Machinery Co., Toronto.
H. W. Petrie, Toronto.

Trolleys
Canadian Rand Drill Co., Montreal.
John Millen & Son, Ltd., Montreal.
Miller Bros. & Toms, Montreal.

Trucks, Dryer and Factory
Canadian Buffalo Forge Co., Montreal.
Greey, Wm. & J. G., Toronto.
McEachren Heating and Ventilating Co., Galt, Ont.
Sheldon's Limited, Galt, Ont.

Tube Expanders (Rollers)
Chicago Pneumatic Tool Co., Chicago.
A. B. Jardine & Co., Hespeler.

Turbines, Steam
Allis-Chalmers-Bullock, Limited, Montreal
Canadian General Electric Co., Toronto.
Canadian Westinghouse Co., Hamilton.
D'Olier Engineering Co., New York.
Jenckes Machine Co., Sherbrooke, Que.
Kerr Turbine Co., Wellsville, N.Y.

Turntables
Detroit Foundry Supply Co., Windsor.
Dominion Foundry Supply Co., Montreal
Hamilton Facing Mill Co., Hamilton.

Turret Machines.
American Tool Works Co., Cincinnati.
John Bertram & Sons Co., Dundas, Ont.
The Canadian Fairbanks Co., Montreal.
London Mach. Tool Co., Hamilton, Ont.
Niles-Bement-Pond Co., New York.
H. W. Petrie, Toronto.
Pratt & Whitney Co., Hartford, Conn.
Warner & Swasey Co., Cleveland, Ohio.
Williams & Wilson, Montreal.

Upsetting and Bending Machinery.
John Bertram & Sons Co., Dundas, Ont
A. B. Jardine & Co., Hespeler.
London Mach. Tool Co., Hamilton, Ont.
National Machinery Co., Tiffin, O.
Niles-Bement-Pond Co., New York.

Valves, Blow-off.
Chicago Pneumatic Tool Co., Chicago.

Valves, Back Pressure.
McEachren Heating and Ventilating Co., Galt, Ont.
Sheldon's Limited, Galt.

Valves, Reducing.
McEachren Heating and Ventilating Co., Galt, Ont.

Vaults.
The Goldie & McCulloch Co., Galt.

Ventilating Apparatus.
Canada Machinery Agency, Montreal
Canadian Buffalo Forge Co., Montreal.
McEachren Heating and Ventilating Co., Galt, Ont
Sheldon & Sheldon, Galt, Ont.
B. F. Sturtevant Co., Hyde Park, Mass.

Vises.
Baxter, Paterson & Co., Montreal.
Butterfield & Co., Rock Island, Que.
Rice Lewis & Son, Toronto
John Millen & Son, Ltd., Montreal.
Miller Bros. & Toms, Montreal.

Vises, Planer and Shaper.
American Tool Works Co., Cincinnati, O.
A. B. Jardine & Co., Hespeler, Ont.
John Millen & Son, Ltd., Montreal.
Niles-Bement-Pond Co., New York.

Washer Machines.
National Machinery Co., Tiffin, Ohio.

Water Power Plants.
Greey, Wm. & J. G., Toronto.
Jenckes Machine Co., Sherbrooke, Que.

Water Wheels.
Allis-Chalmers-Bullock Co., Montreal.
Barber, Chas., & Sons, Meaford.
The Goldie & McCulloch Co., Galt, Ont.
Greey, Wm. & J. G., Toronto.
Canada Machinery Agency, Montreal.
Jenckes Machine Co., Sherbrooke, Que.
Wm. Kennedy & Sons, Owen Sound.

Water Wheels, Turbine.
Kennedy, Wm., & Sons, Owen Sound

Wheelbarrows.
Baxter, Paterson & Co., Montreal.
Buffalo Foundry Supply Co., Buffalo
Detroit Foundry Supply Co., Windsor.
Dominion Foundry Supply Co., Montreal
Hamilton Facing Mill Co., Hamilton.

Winches, Hand.
Pendrith Machinery Co., Toronto.

Window Wire Guards.
Expanded Metal and Fireproofing Co. Toronto.
B. Greening Wire Co., Hamilton, Ont.

Wire Chains.
The B. Greening Wire Co., Hamilton.

Wire Cloth and Perforated Metals.
Expanded Metal and Fireproofing Co. Toronto.
B. Greening Wire Co., Hamilton, Ont.

Wire Guards and Railings.
Expanded Metal and Fireproofing Co. Toronto.
B. Greening Wire Co. Hamilton, Ont.

Wire Nail Machinery.
National Machinery Co., Tiffin, Ohio.

Wire Rope.

B. Greening Wire Co., Hamilton, Ont.

Wood-working Machinery.

Baxter, Paterson & Co., Montreal
Canada Machinery Agency, Montreal.
The Canadian Fairbanks Co., Montreal.
Goldie & McCulloch Co., Galt.
H. W. Petrie, Toronto.
Waterous Engine Works Co., Brantford.
Williams & Wilson, Montreal.

Wrenches, Adjustable Tap.

Baxter, Paterson & Co., Montreal.
Butterfield & Co., Rock Island, Que.
A. B. Jardine & Co., Hespeler Ont.
Standard Tool Co., Cleveland.

Wrenches, Machinists.'

Baxter, Paterson & Co., Montreal.
Whitman & Barnes Co., St. Catharines.

Wrenches, Ratchet.

Whitman & Barnes Co., St. Catharines

Wrenches, Universal.

Whitman & Barnes Co., St. Catharines

Wrought Iron and Steel Washers.

Baxter, Paterson & Co., Montreal.

ALPHABETICAL INDEX.

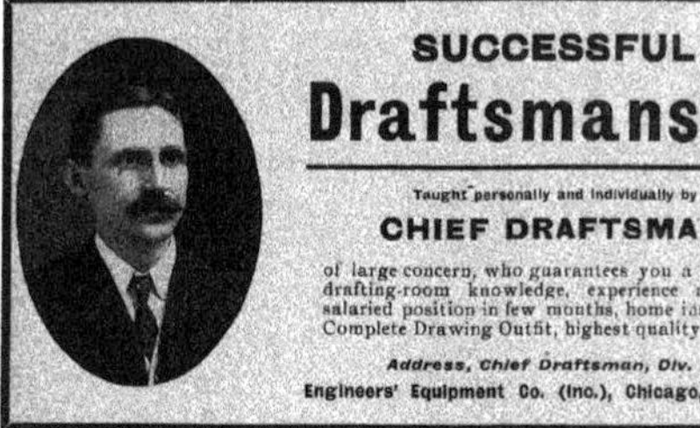

SUCCESSFUL
Draftsmanship
Taught personally and individually by
CHIEF DRAFTSMAN
of large concern, who guarantees you a first-class drafting-room knowledge, experience and high salaried position in few months, home instruction. Complete Drawing Outfit, highest quality, free.
Address, Chief Draftsman, Div. 22
Engineers' Equipment Co. (Inc.), Chicago, Ill., U.S.A.

VALUE, $13 85
FREE TO MY STUDENTS
Delivered at once

An intelligent Engineer uses nothing but the **GENUINE PENBERTHY INJECTOR.** He has had experience with just-as-good kind before.

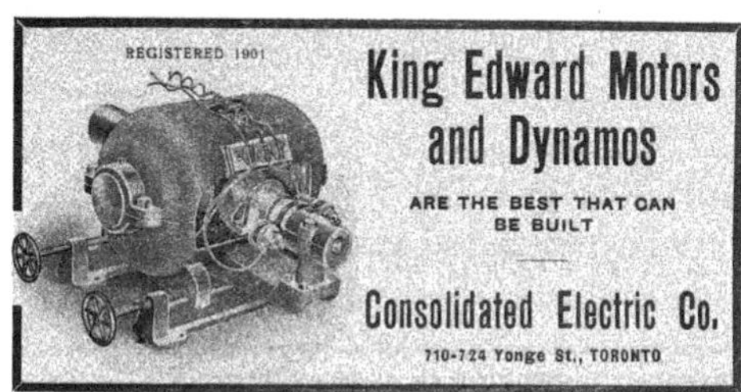

Persons addressing advertisers will kindly mention having seen their advertisement in Canadian Machinery

OWEN SOUND IRON WORKS, LIMITED
OWEN SOUND, ONT.

Cement Mill Machinery, Boiler and Steel Tank Work of all kinds, Grey Iron and Brass Castings

---

PATTERNS
WELLS' PATTERN AND MODEL WORKS

Tel. Main 3581. (HARRY WELLS, Proprietor.)

Patterns and Models made in wood and metal for Engines, Pumps, Furnaces, Agricultural, Electrical and Architectural Works and Machines of every description.

35 Richmond St. E., Toronto

Don't let your moulders swear, get your patterns made by

JOHN CARR
14 Mary St., HAMILTON
Phone 2755

We should have your business.

If prompt delivery, reasonable prices and best of workmanship appeal to you, we should have your pattern orders.

SEND FOR PATTERN CLASSIFICATION CARD

THE HAMILTON PATTERN WORKS
134 Bay St. N., HAMILTON, ONT.
Phone, 2664.

FOR SALE

1	10-H.P. Buffalo 4-cylinder 4-cycle Marine Motor, good as new	$250.00
1	34-H.P. Trebert, complete with reversing wheel, used only one season. Price	450.00
1	6-H.P. Gillies Stationary Gasolene Engine, will sell cheap.	
1	220-Volt 2 H.P. Electric Motor, London make, new	50.00
1	¼-H.P. 110 Volt Motor, "American" make, with speed controllers, new..	25.00
1	Peck, Stowe & Wilcox flexible shaft, 8 ft. long, with drill press and breast drill attachment, little used	25.00
1	Vertical Power Drill, cheap.	
1	13-in. Turning Lathe, cheap.	
1	Tinsmith's Metal Roll, cheap.	
1	18-ft. x 4-ft. 9-in. Beam Launch, equipped with a 2½ H.P. Norwalk Engine, clinker built, launched Sept. '06. A bargain	225.00
1	26-ft. Open Launch, with 10 H.P. Buffalo Motor, fully equipped, speed 9 miles per hour, cheap.	

PRICE AND DESCRIPTION ON APPLICATION.

HALLIDAY ELECTRIC CO.
KINGSTON, ONT.

WM. KENNEDY & SONS
HIGH-CLASS

Steel Gears of all kinds
Iron Gears
Grey Iron, Steel and
Brass Castings

MARINE WORK receives every attention

WM. KENNEDY & SONS, Limited
Owen Sound, Ont.

The Canadian Turbine Water Wheel

80 per cent. under all working conditions.

Exceptionally Durable

Chas. Barber & Sons
Meaford, Ont.

ISLAND CITY

YOUR MACHINE PAINT

is possibly not up to the standard. Perhaps it does not give you the satisfaction you would like.

WE manufacture a splendid line of paints for the special use of machinery manufacturers who require a good, easily applied, hard wearing, and beautifying paint.

IT is made in paste form and can be used as a filler, applying with a knife. For brushing purposes it may be thinned with turpentine.

OUR PAINT has great covering capacity, spreads freely and leaves a good surface.

Write for Sample Cards

P. D. DODS & CO., MONTREAL, TORONTO, WINNIPEG AND VANCOUVER

Directory of Consulting Engineers, Patent Attorneys, Architects and Contractors.

JOHN S. FIELDING

Mem. Soc. C.E., West Penn., '87

Consulting Engineer

DAMS, MILLS, BRIDGES, MACHINERY

Room 2, 15 Toronto Street, Toronto, Ont.

HANBURY A. BUDDEN

Advocate — Patent Agent.

New York Life Building — MONTREAL.

Cable Address, BREVET, MONTREAL.

CONSULTING ENGINEERS

should have their card in this page. It will be read by the manufacturers of Canada :: :: :: ::

CANADIAN MACHINERY

Montreal. Toronto. Winnipeg.

T. Pringle & Son

HYDRAULIC, MILL & ELECTRICAL ENGINEERS

FACTORY & MILL CONSTRUCTION A SPECIALTY.

Coristine Bldg., St. Nicholas St., Montreal.

RODERICK J. PARKE

A.M. Can. Soc. C.E. A.M. Amer. Inst. E.E.

CONSULTING ELECTRICAL ENGINEER

INDUSTRIAL STEAM AND ELECTRIC POWER PLANTS DESIGNED. TESTS. REPORTS.

51-53 JANES BLDG., TORONTO, CAN

Long Distance Telephones—Office and Residence.

Buffing and Polishing Lathes

PENDRITH MACHINERY CO.

74 W. Adelaide Street, Toronto

OPAL GLASS TILING

FOR WALLS OF

MACHINERY AND POWER HOUSES

Most approved material.

TORONTO PLATE GLASS IMPORTING CO'Y

PLATE AND WINDOW GLASS

135 to 143 Victoria St., - Toronto

CHARLES BRANDEIS,

A. M. Can. Soc. C.E.

Mem. Amer. Electro-Chemical Soc., etc.

CONSULTING ENGINEER

Estimates, Plans and Supervision of Hydraulic and Steam-Electric Light, Power and Railroad Plants, Specifications, Reports, Switchboard Designs, Complete Factory Installations, Electric Equipment of Mines and Electro-Chemical Plants.

Long Distance Telephone, Main 3256.
Cable Address, Brandeis, Montreal.
W. U. Code Univ-Edition.

62-63 Guardian Building — MONTREAL

PATENTS *TRADE MARKS AND DESIGNS*

PROCURED IN ALL COUNTRIES

Special Attention given to Patent Litigation

Pamphlet sent free on application.

RIDOUT & MAYBEE *103 BAY STREET TORONTO*

FETHERSTONHAUGH & BLACKMORE

PATENTS

TRADE MARKS.

Liverpool & London & Globe Building. Montreal.

Fetherstonhaugh & Dennison, 9 Toronto St. Toronto. Our reference book for clients "The Prospective Patentee." Also Ottawa, Washington and Winnipeg.

PATENTS PROMPTLY SECURED

We solicit the business of Manufacturers, Engineers and others who realize the advisability of having their Patent business transacted by Experts. Preliminary advice free. Charges moderate. Our Inventor's Adviser sent upon request. Marion & Marion, New York Life Bldg, Montreal; and Washington, D.C., U.S.A.

PATENTS THAT PROTECT

FETHERSTONHAUGH & CO.

Patent Solicitors & Experts

Fred. B. Fetherstonhaugh, M.E., Barrister-at-law and Counsel and expert in Patent Causes. Charles W. Taylor, B.Sc., formerly Examiner in Can. Patent Office.

MONTREAL CAN. LIFE BLDG.

TORONTO HEAD OFFICE, CAN. BANK COMMERCE BLDG.

PRESS CLIPPINGS

About any subject or business. We read nearly every paper in Canada, and can supply you with what the papers have to say about anything you are interested in.

——WRITE FOR TERMS——

CANADIAN PRESS CLIPPING BUREAU

10 Front Street East, - - - TORONTO.

THE ELECTRIC BLUE PRINT CO.

Blue Prints — Positive Blue Prints
Black Prints — Multi-color Prints

Largest and finest plant.

Best finished work in Canada at lowest prices.

40 Hospital St., - MONTREAL

THE DETROIT TESTING LABORATORY

1111 Union Trust Building

DETROIT, MICH.

Experienced Foundry Chemists and Metallurgists. Iron Mixtures for all classes of Castings our Specialty. Reasonable charges for Analyses.

J. A. DeCew

Chemical Engineer

Sun Life Building, MONTREAL

Industrial Plants and Processes sold and installed. Apparatus and Materials for Water Purification. Free Analyses to prospective purchasers.

JULES DE CLERCY, M. E. Gas Engineer

Expert knowledge of all classes of Engines and Producer Plants. If you are thinking of installing a Gas Engine or Producer Plant write me. If your plant is not working satisfactorily I can help you.

413 Dorchester St. - MONTREAL, QUE.

PLEWS & TRIMINGHAM.

Consulting Engineers.

ELECTRICAL INSPECTION BUREAU AND TESTING LABORATORY.

40 HOSPITAL & 22 ST. JOHN STS. MONTREAL.

---

60 YEARS' EXPERIENCE

PATENTS

TRADE MARKS
DESIGNS
COPYRIGHTS &C.

Anyone sending a sketch and description may quickly ascertain our opinion free whether an invention is probably patentable. Communications strictly confidential. HANDBOOK on Patents sent free. Oldest agency for securing patents.

Patents taken through Munn & Co. receive *special notice*, without charge, in the

Scientific American.

A handsomely illustrated weekly. Largest circulation of any scientific journal. Terms, $3 a year; four months, $1. Sold by all newsdealers.

MUNN & Co. 361 Broadway, New York

Branch Office, 625 F St., Washington, D. C.

"IT SELLS ON ITS REPUTATION"

There is only one Typewriter of which this has ever been said: the

REMINGTON

Great and enduring is the merit on which such reputation is built.

THE REMINGTON TYPEWRITER CO., 33 Adelaide St. E., Toronto

AND EVERYWHERE.

Persons addressing advertisers will kindly mention having seen their advertisement in this paper.

NEW AND UP-TO-DATE

BOLT AND NUT MACHINERY

INCLUDING

Bolt Cutters, Nut Tappers, Bolt Headers, Upsetting and Forging Machines, Wire Nail and Spike Machines and Bulldozers.

Send for Catalogue M.

NATIONAL MACHINERY CO., Tiffin, Ohio, U.S.A.

Canadian Agents: H. W. PETRIE, Toronto, Ont. WILLIAMS & WILSON, Montreal, Que.

THE

ELECTRICAL TIMES,

LIGHTNING

8 Breams Buildings, Chancery Lane
LONDON, ENGLAND

The Organ of the

ELECTRICAL

and Allied Industries

Circulates All Over the World

It is a Sixpenny Paper for Twopence; and is more widely read than any other British Electrical Journal. Means business all the time.

Annual Subscription:
Post Free, 13/6, Foreign; 10/6, Home.
Published weekly at 2d.

WE HAVE THE BEST APPLIANCES.
WE MAKE THEM RIGHT AND AT
A REASONABLE PRICE.

A. B. JARDINE & CO., Hespeler, Ont.

BEST
LEATHER
BELT
MADE

THE J. C. McLAREN BELTING Co.
OAK
EXTRA
TANNED
EST
1856

TORONTO
ST. JOHN, N.B.

MONTREAL

WINNIPEG
VANCOUVER

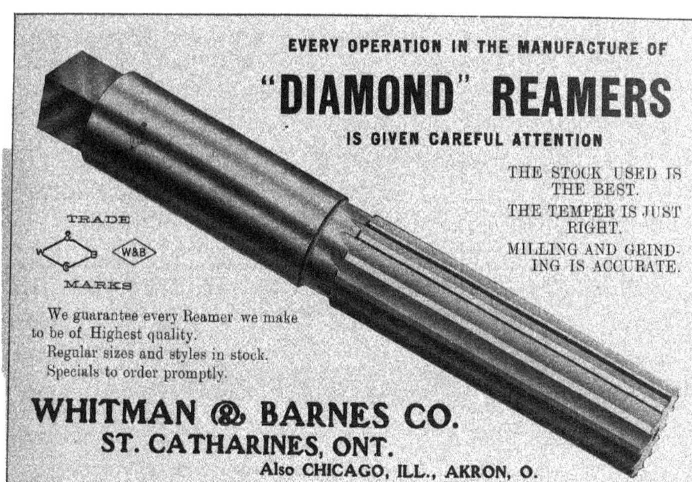
EVERY OPERATION IN THE MANUFACTURE OF
"DIAMOND" REAMERS
IS GIVEN CAREFUL ATTENTION
THE STOCK USED IS THE BEST.
THE TEMPER IS JUST RIGHT.
MILLING AND GRINDING IS ACCURATE.
TRADE MARKS
W&B
We guarantee every Reamer we make to be of Highest quality.
Regular sizes and styles in stock.
Specials to order promptly.
WHITMAN & BARNES CO.
ST. CATHARINES, ONT.
Also CHICAGO, ILL., AKRON, O.

Money-Tending Information

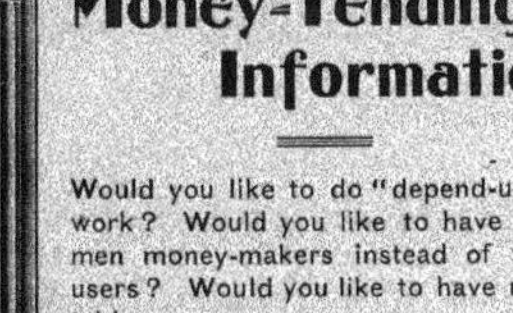

Would you like to do "depend-upon" work? Would you like to have your men money-makers instead of time-users? Would you like to have reputable, never-go-wrong

Reamers, Taps, Stocks, Dies, Screw Plates, General Tools—

Would you? If so, place your name and address on a post card and mail it to

BUTTERFIELD & CO.
Rock Island - Quebec

STANDARD TWIST DRILLS

Results Count

In long and continued service, which is the final test, Standard Tools prove their worth. Why not use tools that can substantiate their claim of

HIGHEST QUALITY?

THE STANDARD TOOL CO.
CLEVELAND, OHIO. NEW YORK, N.Y.

ON YOUR SATISFACTION DEPENDS OUR PROSPERITY.

"MORSE" TOOLS

ARE MADE TO SATISFY.

Arbors, Chucks, Cutters, Counterbores, Countersinks, Dies, Drills, Gauges, Machines, Mandrels, Mills, Reamers, Screw-plates, Sleeves, Sockets, Taps, Taper Pins, Wrenches.

MORSE TWIST DRILL & MACHINE CO.
NEW BEDFORD, MASS., U. S. A.

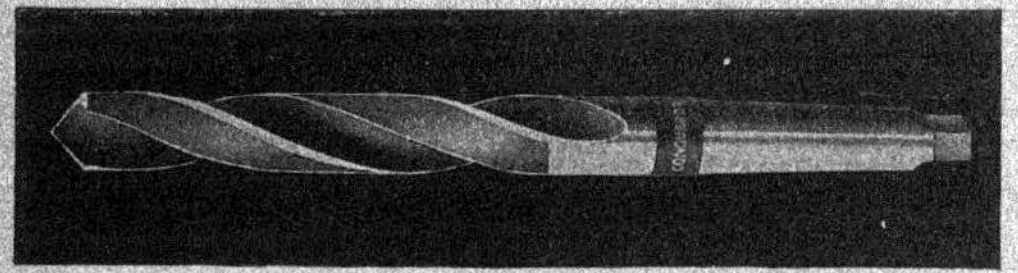

Beardshaw's "Conqueror" HIGH SPEED STEEL AND HIGH SPEED DRILLS

GIVE BEST RESULTS of any on the market.

ALEXANDER GIBB, 13 St. John St. - MONTREAL

Responsible Agents Wanted in the West

GASOLINE TORCH

TORCHES

of All Kinds

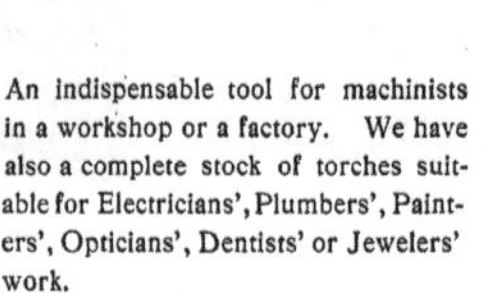

An indispensable tool for machinists in a workshop or a factory. We have also a complete stock of torches suitable for Electricians', Plumbers', Painters', Opticians', Dentists' or Jewelers' work.

GASOLINE FORGES

Write for prices

RICE LEWIS & SON

TORONTO LIMITED

BOOKS FOR ENGINEERS

DRAUGHTSMEN, SCIENCE STUDENTS, ETC.

Sent Post Free to any Address, at home or abroad, at Published Price

Just Published Second Edition, with numerous specimen Workshop Cost Forms, price 12/6 net, post free.

WORKSHOP COSTS (for Engineers and Manufacturers), by Sinclair Pearn and Frank Pearn (Directors of Messrs. Frank Pearn and Co., Limited, West Gorton Ironworks, Manchester), with a preface by G. Croydon Marks, Assoc. Memb. Inst. C. E., M.I.M.E., etc.

	NET PRICE s.	d.
The Fan; including the Theory and Practice of Centrifugal and Axial Fans, by Ch. H. Innes, M.A.	4	0
The Proportions and Movement of Slide Valves, by W. D. Wansbrough	4	6
Machines and Tools Employed in the Working of Sheet Metals, by R. B. Hodgson	4	6
Governors and Governing Mechanism, by Hall	2	6
Specification of a Modern Lancashire Boiler and its Seating (200 lbs. pressure), by "Inspector"	5	
Continuous Current Dynamos and Motors and their Control, by W. R. Kelsey	5	
The Resistance and Power of Steamships, Atherton and Mellanby	5	0
Notes on Construction and Working of Pumps, Marks	3	6
Modern Ironfoundry Practice:		
Part I., Hand Moulding, Bale	5	0
Part II., Machine Moulding, Bale	3	6
Modern Gas and Oil Engines, by F. Grover. 3rd Edition	5	0
The Indicator and its Diagrams, by Chas. Day. 3rd Edition	4	6
The Chemistry of Materials of Engineering, by A. H. Sexton. 2nd Edition	5	0

	NET PRICE s.	d.
The Management of Small Engineering Workshops, Barker	7	6
Problems in Machine Design, by Chas. Innes. 2nd Edition	4	6
Heat and Heat Engines; a Treatise on Thermo-Dynamics. Popplewell	6	0
Centrifugal Pumps, Turbines and Water Motors. 3rd Edition	4	6
Application of Graphic Methods to the Design of Structures. 2nd Edition	5	0
Engineering Estimates and Cost Accounts, Burton. 2nd Edition	3	0
Graphic Methods of Engine Design, Barker. 2nd Ed.	3	6
Injectors: Theory, Construction and Working, Pullen. 3rd Edition	3	6
Construction of Cranes and Lifting Machinery, Marks. 2nd Edition	3	
Marine Engineers: Their Qualifications and Duties	5	
A.B.C. of the Differential Calculus, Wansbrough	3	6
The Theta-Phi Diagram Applied to Steam, Gas, Oil, and Air Engines	3	0
Mechanical Engineering Materials, by Marks	1	6
The Naval Engineer and Command of the Sea, Burton	2	6

THE TECHNICAL PUBLISHING CO., LIMITED, 287 Deansgate, Manchester, Eng.; **D. VAN NOSTRAND CO.,** 23 Murray St. and 27 Warren St., New York, **and all Booksellers.**

INDUCTION MOTORS
ALLIS-CHALMERS Bullock LIMITED MONTREAL
ALLIS-CHALMERS Bullock LIMITED MONTREAL
One of our 300 h.p. Induction Motors driving an air compressor in the power house of the Canadian Copper Co., Copper Cliff, Ont. These Motors are adapted to the operation of all kinds of machinery, but especially cement works, brick yards, powder mills, paint factories, flour mills, paper mills, mines, etc.
For a list of our various products, see the classified index of this journal.
ALLIS-CHALMERS-BULLOCK, Limited
Head Office and Works, Montreal. District Offices: Toronto, Traders Bank Bldg; Montreal, Sovereign Bank Bldg; Winnipeg, 251 Notre Dame Ave.; Nelson, Josephine St.; Vancouver, 416 Seymour St.; New Glasgow, N.S., Telephone Bldg.

CANADIAN MACHINERY AND MANUFACTURING NEWS

A monthly newspaper devoted to the manufacturing interests, covering in a practical manner the mechanical, power, foundry and allied fields. Published by The MacLean Publishing Company, Limited, Toronto, Montreal, Winnipeg, and London, Eng.

OFFICE OF PUBLICATION: 10 Front St. E, TORONTO

Vol. III. MARCH, 1907 No. 3

GOLDIE CORLISS ENGINES

VERTICAL AND HORIZONTAL, SIMPLE, TANDEM COMPOUND AND CROSS COMPOUND

A Few Special Features

Rigidity of Frame

Ring Oiling Outer Bearing

Hardened Steel Pins throughout

Good General Design

21 and 42 x 30 Vertical Compound Goldie Corliss Engine

Send for Catalogue

Bulletin No. 8 describes Girder Frame Engines

Bulletin No. 3 describes Heavy Duty Engines

BUILT BY THE GOLDIE & McCULLOCH CO., LIMITED

GALT —— ONTARIO —— CANADA

Western Branch—248 McDERMOTT AVE., WINNIPEG, Man. Quebec Agents—ROSS & GREIG, MONTREAL, Que.

WE MAKE Wheelock Engines, Corliss Engines, Ideal High Speed Engines, Boiler Pumps, Flour Mill Machinery, Oatmeal Mill Machinery, Wolf Gyrators, Emery Choppers, Woodworking Machinery, Shingle Machinery, Heading and Stave Machinery, Wood Rim Split Pulleys, Iron Pulleys, Shafting Hangers, Friction Clutch Couplings, Friction Clutch Pulleys, Safes, Vaults, Vault Doors.

Westinghouse

Westinghouse Type S Motor Driving Hilles & Jones Punch and Shears.

MACHINE TOOL MOTORS

No one part of a motor-driven machine tool has a more important bearing upon its satisfactory performance than the driving motor; with motors of

Westinghouse Quality

correctly applied, you have every assurance that the tool will develop its highest efficiency.

We maintain a corps of motor-drive experts; we offer their advice and assistance, they are worth money to every machine tool user.

Canadian Westinghouse Co., Limited

General Office and Works, HAMILTON, ONTARIO

Traders Bank Building **TORONTO**	For Particulars Address Nearest Office **HAMILTON**	Sovereign Bank of Canada Bldg. **MONTREAL**
152 Hastings Street **VANCOUVER**	922-923 Union Bank Bldg. **WINNIPEG**	134 Granville Street **HALIFAX**

Has Greater Productive Capacity

than any Upright ever built—our 30-in. RADIAL DRILL. Costs but a trifle more and is admirably adapted to light and medium manufacturing, being particularly effective in taking care of the ordinary work as found in the average shop.

Cone Drive and Swinging Table

Bickford 30"-36"-42"Radial Drill

A Few Features

12 speeds when driven through a speed box; 10 when a cone is used.

Tapping attachment incorporated in the design of the head.

Fitted with any one of six styles of tables.

Circular I-A for the asking

The Bickford Drill & Tool Company

Cincinnati, Ohio, U.S.A.

FOREIGN AGENTS—Schuchardt & Schutte, Berlin, Vienna, Stockholm, St. Petersburg, New York. Alfred H. Schutte, Cologne, Brussels, Liege, Paris, Milan, Bilbao, New York. Chas. Churchill & Co., Ltd., London, Birmingham, Manchester, Newcastle-on-Tyne and Glasgow.

CANADIAN AGENTS—H. W. Petrie, Toronto. Williams & Wilson, Montreal.

NORTON JACKS

ARE MADE IN CANADA AND SAVE YOU DELAY AND DUTY OVER AMERICAN JACKS

They are Ratchet Screw Jacks with Ball Bearings to reduce the friction. Working parts are protected from grit and rust. Superior to Hydraulic Jacks for all classes of service. Simple, durable, and guaranteed in every respect. Absolutely safe. Will work in any position. Standard on the leading railroads.

Made in 50 styles; 8 to 70 tons capacity

Send for complete illustrated catalogue and discount.

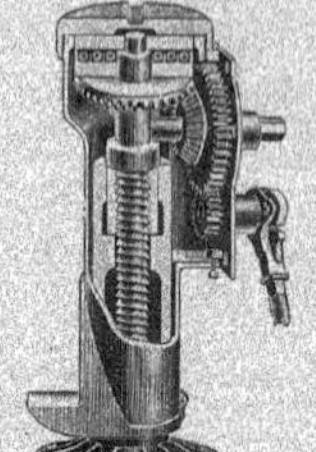

Sectional View of 70-Ton Bridge Jack

25-Ton Foot Lift Jack

35-Ton Plain Jack

Traversing Jack 10 to 60 Tons

Track Jack 10 and 15 Tons

Journal Box Jack 15 Tons

MADE ONLY BY

A. O. NORTON, Inc., Coaticook, Prov. Quebec

Stock carried by MUSSENS, Limited, Montreal

DOES SHOP COST COUNT WITH YOU?

IF SO, READ THIS

14-in. to 62-in. Swing. Belt or Motor Driven

The "AMERICAN" LATHE with "PATENTED" ALL-GEAR HEAD and SINGLE PULLEY DRIVE is endowed with

Enormous Power for Heavy Feeds

EXAMPLES The 20-inch size will remove 7 1-2 lbs. cast iron in one (1) minute.
A cone pulley requiring 2 hours and 40 minutes on a regular lathe, was finished in 50 minutes on this powerful *"Geared Head Lathe."*

Catalog "C" now ready

THE AMERICAN TOOL WORKS CO.

700 to 750 CULVERT STREET · CINCINNATI, OHIO

LATHES PLANERS SHAPERS RADIAL DRILLS

Write direct or to Agents H. W. PETRIE, Montreal; A. R. WILLIAMS MACHINERY CO., Toronto, Winnipeg, Vancouver

THE

Becker-Brainard

AUTOMATIC GEAR-CUTTING MACHINE

is automatic in all movements. You set the machine and pay no further attention to it until the gear is finished.

Gear accuaracy is assured by an unusually large dividing wheel. We have a few of the 36" machines in stock for immediate delivery.

Becker-Brainard Milling Machine Co.

HYDE PARK, MASS., U.S.A.

CANADIAN AGENTS:

A. R. WILLIAMS MACHINERY CO., LTD., TORONTO, WINNIPEG and VANCOUVER.

H. W. PETRIE'S

PARTIAL LIST OF

Snaps in Machinery on Hand

Bought before prices advanced. 2,000 Machines in stock to choose from

Machine Tools

Lathes

New 36" x 16' Engine Lathe
" 32" x 18' " "
Refitted 30" x 14' " "
New 24" x 16' " "
" 21" x 10' " "
Refitted 21" x 12' " "
New 20" x 14' " "
" 20" x 16' " "
Refitted 20" x 10' " "
" 20" x 8' " "
New 18" x 8' " "
" 16" x 10' " "
" 15" x 6' " "
" 13" x 8' " "
" 20" x 42" x 14' Gap Lathe
" 24" x 40" x 12' " "
Refitted 18" x 30" x 12' " "
" Large face Lathe.

Planers

Refitted 60" x 60" x 11'
" 42" x 42" x 26
" 30" x 30" x 8
New 26" x 26" x 5
Refitted 26" x 25" x 7
New 24" x 24" x 6½
Refitted 24" x 24" x 6
" 24" x 24" x 3
22" x 18" x 5
New 20" x 20" x 5'

Drills

New 72" Full Universal Radial
Refitted 42" Niles Radial
" 80" B G Radial
" 36" B.G. Drill
New 24" Cincinnati with tapping attach
Large stock of 23" and smaller drills

No. 2 Mitts & Merrill Key Seater
36" Gould & Eberhardt Gear Cutter
Nos. 3½ and 4 Grunard Arbor Presses
New 1" and 1½" Bolt Cutter
60-lb. Palmer Hammer
2—No. 25 Cushioned Hilse Hammers

Punches and Shears

4—New 15" Punch and Shear
" 12" " " "
Large Alligator Shear

Engines

2- 18" x 48" Harris-Corliss
1—17" x 42" Brown
2—16" x 30" Corliss
9½" x 14½" x 12" McIntosh & Seymore Engine Tandem
8" x 14" x 14" Ri e Tandem Compound Engine
20" x 24", plain slide valve
16" x 24" " " "
15½" x 24" " " "
12" x 16" " " "
11" x 24" " " "

Boilers

Horizontal Stationary Tubular Boilers

	Approx. H.P.
72" x 14', 114 3" tubes, plug hat dome	125
60" x 17' 6", 54 4" " " "	85
52" x 14', 64 3½ " " " "	80
6 " x 13' 9" 84 3" " " " "	75
60" x 15' 4", 80 3" " " " "	75
60" x 12', 74 3" " " " "	60
54" x 14' 4", 64" 3" " " " "	65
56" x 12', 60 3" " " " "	55
48" x 13' 6", 44 3" " " " "	50
48" x 16' 1", 52 3" " " " "	46
46" x 13', 53 3" " " " "	45
48" x 13' 10" 52 3" " " " "	45
48" x 13' 6", 42 3" x 12 tubes, man hole dome	40
44" x 13' 8", 48 3" x 12' tubes, plug hat dome	43
44" x 14½', 51 3" tubes, plug hat dome	40
44" x 14' 6", 40 3" " " " "	36
41" x 12', 46 3" " " " "	35
44" x 11' 9", 42 3" tubes, manhole dome	35
44" x 11' 6", 43 3" " " "	35
44" x 11' 4", 46 3" " " "	33
44" x 11' 3", 31 3" " " "	30
42" x 12', 2 13" tubes, no dome	30
38" x 15', 34 3" tubes, plug hat dome	28
38" x 13', 33 2" tubes	25
38" x 12', 28 3" tubes, manhole dome	25
36" x 12', 26 3" tubes, plug hat dome	25
38" x 10' 28 3" tubes, no dome	20
32" x 11', 22 3" tubes plug hat dome	18
38" x 8" 26 3" tubes	15
36" x 11', 24 3" tubes, plug hat dome	16
30" x 8', 18 2" tubes	12
30" x 8', 24 3" tubes, plug hat dome	
26" x 10¼", 11 3" tubes, no dome	11
26" x 7½", 21 2½" tubes, plug hat dome	8
22" x 5' 10", 14 2" and 1 3" tubes, no dome	6
18" x 36", 9 2" and 4 3" tubes	2
15¼" x 27", 21 2" and 1 3½" tubes	1½

Vertical Boilers

All sizes from 1 to 60-H.P.

H. W. PETRIE, TORONTO and MONTREAL

Write for stock list containing full list of machinery

Canadian Agents:

The Canadian Fairbanks Co., Limited, Montreal, Toronto, Winnipeg and Vancouver.

F. H. Andrews & Son, Quebec, P.Q.

BLOWERS

Heating
Ventilating
Drying Forges
Mechanical Draft
Blowers.
Exhausters
Steam Specialties

HEATING AND VENTILATION
by Fan System

Natural Draft and Blower Dry Kilns
For Lum'er, Brick, etc.
Cars, Trucks, Rail, etc.

ENGINES
both Vertical and Horizontal

Shavings Exhaust System

SHELDONS, Limited, Engineers and Manufacturers **Galt, Ont., Canada**

Montreal Office, 325 St. James St. Phone Main 2919

For Isolated Lighting Plants

the Sturtevant Generating Sets are especially adapted because of their high efficiency and compact construction. These distinctive features of the type here shown were developed under rigid specifications.

A 100 K. W. vertical generating set can be located in a space measuring less than 6 feet wide, 9½ feet long and 8 feet high; will operate at a combined efficiency of over 86 per cent and will consume not exceeding 31 pounds of water per K. W. hour.

We also build 31 other sizes and styles of generating sets under 100 K.W. All but the largest can be delivered completely set up and ready for operation. Send for Bulletin 63.

B. F. Sturtevant Co., Boston, Mass.

General Office and Works, Hyde Park, Mass.

NEW YORK PHILADELPHIA CHICAGO CINCINNATI LONDON

Designers and Builders of Heating, Ventilating, Drying and Mechanical Draft Apparatus; Fan Blowers and Exhausters; Rotary Blowers and Exhausters Steam Engines, Electric Motors and Generating Sets, Pneumatic Separators, Fuel Economizers; Forges; Exhaust Heads, Steam Traps, Etc. 479

RENOLD SILENT CHAIN

is used almost exclusively on the standard motors because of its supremacy in the power-transmitting field. Adapted to short centres; damp, hot or oily places; and will run in either direction. Action always smooth and positive.

Write for Booklet "J" and Special Bulletins 50, 52, 57, 58.

LINK-BELT COMPANY

PHILADELPHIA, CHICAGO, INDIANAPOLIS

NEW YORK 299 Broadway. PITTSBURGH 1501 Park Bldg. BOSTON 84 State St. BUFFALO 321 Ellicott Square Bldg.

Kerr Turbine Pumps for water works, condenser equipment and boiler feeding. These pumps are more economical than direct acting piston pumps, and require less attention, lubrication and space. No belts or gears. Speed adjustable to pressure desired. Regulation close. Start and stop quickly. For any capacity.

Turbine Blowers for gas works, foundries and large shops. Air pressure 4 to 16 ounces. For any capacity.

Turbine Generators, 2 kw. to 300 kw. direct current and alternating current, 25 to 300 kw., 60 cycle.

The illustration shows an 18-inch condensing Kerr Steam Turbine coupled to a 4-inch 3-stage Worthington Pump. Capacity 200 g.p.m. against a head of 360 feet.

Main Office:

KERR TURBINE CO., Wellsville, N.Y.

Built in Canada by

WM. STEPHENSON

34 Bay Steet N. *HAMILTON, ONT.*

WATEROUS ENGINE WORKS CO., Limited

BRANTFORD, CANADA

MANUFACTURERS OF:

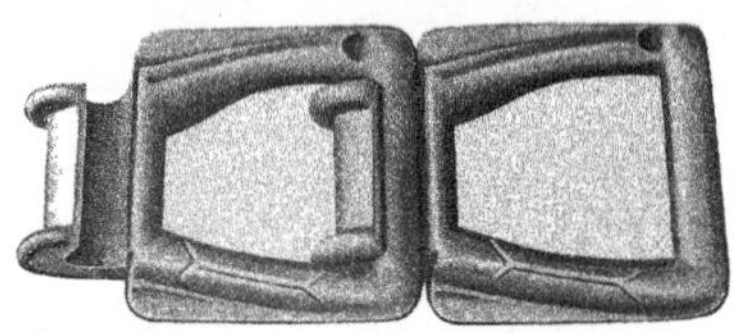

We carry the largest stock of the "Original" Ewart Chain-Belting in Canada. Also Forged, Wrought and Malleable Chain of all descriptions.

Saw Mill and Pulp Mill Machinery,
Engines, Boilers,
Fire Apparatus,
Brick Machinery, Elevator and
Conveyor Machinery,
Chain Belting, etc.

WRITE US for FULL Particulars, Prices and Catalogues

VERTICAL ENCLOSED ENGINES

We are prepared to supply high speed vertical enclosed engines of the English type, up to 750 horse power.

The oiling system of these engines is under a pressure of 15 lbs. to the square inch, ensuring a copious supply of oil to all bearings.

They have Robb-Armstrong automatic governors and Sweet balanced valves, giving the best regulation and highest economy.

Robb Engineering Co., Limited, Amherst, N.S.

DISTRICT OFFICES
320 Ossington Avenue, TORONTO; Wm. McKay, Manager.
Bell Telephone Building, MONTREAL; Watson Jack, Manager.
355 Carlton Street, WINNIPEG ; J. F. Porter, Manager

BOILERS

Return Tubular
"McDougall" Water Tube
Lancashire
Marine

TANKS

Water Tanks
Penstocks
Coal and Cement Bins and Hoppers

MACHINERY

Complete Power Plants designed and installed

The John McDougall Caledonian Iron Works Co., Limited

MONTREAL, QUE.

McDOUGALL

LATHES AND PIPE SCREWING MACHINES

ARE

MACHINE TOOLS

of latest construction and design, and users of same have good words for us always. We like to show these Tools working, for that is what they are made for and generally this is enough to satisfy.

CIRCULARS YOURS FOR THE ASKING

THE R. McDOUGALL CO., Limited
GALT, CANADA

THE DEMAND OF THE AGE

That's the only thing the shrewd business man will **not** ignore.

Rapidity, simplicity, accuracy and strength are the four essentials of all labor-reducing machinery.

THE NEW MURCHEY TAPPING MACHINE

is representative of the genius and progress of our age, in-as-much as it is a time-saver, a labor-saver and, hence, **a sure and certain money-saver.**

The New Murchey has many patented devices which go to make it the premier tapping machine on the market.

A POST CARD BRINGS PARTICULARS

MURCHEY MACHINE & TOOL CO.
DETROIT, MICH., U.S.A.

Quick Work

When and where quick work is needed.
That means economy.

THIS COMBINATION HAND AND POWER MERRELL PIPE THREADING AND CUTTING MACHINE,

can be operated by a child.

That means economy.

The head may be removed from the base and used as a portable hand machine.

That means economy.

This **MERRELL** will cut any kind of pipe—any style or pitch of thread—up to 12 inches—under all conditions. We will send it to you on 30 days free trial if you wish. Why not write us for further particulars and prices.

The Merrell Mfg., Co., 12 Curtis St.,
TOLEDO, O.

THE CANADIAN FAIRBANKS CO., LIMITED
AGENTS
Montreal Toronto Winnipeg Vancouver

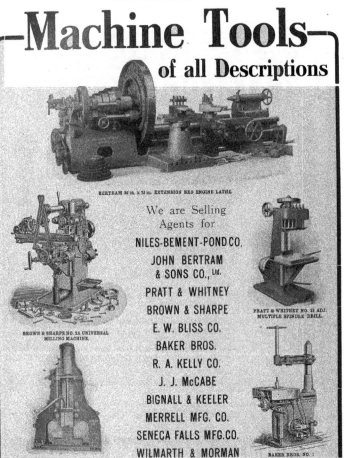

Modern Canadian Manufacturing Plants.

ARTICLE XXI.—Collingwood Shipbuilding Co., Collingwood, Ont.

Canadian Shipbuilding.

WHEN steel supplanted wood in the building of ships, Canada dropped her position in an industry in which she then held an important place. Through the neglect to .ollow up the opportunity offered in this line the Canadian Government failed to foster an industry which, had it been developed, would have meant millions of dollars being kept in the country which have gone to other nations to supply the ships necessary for our large and growing commerce.

The example of Great Britain along this line was one well worthy of being followed. As a pioneer and leading manufacturing country she early developed shipbuilding and for many years practically controlled this industry. The Government of Great Britain, professedly for the purpose of her navy, yet designedly to encourage, develop and protect her trade and commerce on the seas, directly assisted and aided the British shipbuilders by the expenditure of millions of pounds sterling in contracts to the British shipbuilders, and further by the grant allowed on all merchant ships so constructed that in case of necessity they could be commissioned for naval service by the British Government. Other nations slowly awoke to the fact that they were helping to build up one of the greatest industries existing, in Great Britain, and since then have taken steps to establish shipbuilding within their own limits. The United States has spent, and is spending, enormous sums to encourage this feature of her industrial growth. Germany, France, Japan, Russia and Austria have all likewise taken steps in this direction. Canada's position along this line, as far as the Government is concerned, has been one of passive inactivity. The result of this is that our transportation by water has practically been usurped by other nations and in particular by the United States.

Collingwood Shipbuilding Co., Collingwood—Interior View of Machine Shop.

Collingwood Shipbuilding Co., Collingwood—View from the Harbor.

Collingwood Shipbuilding Co., Collingwood—One End of Punch Shed.

It goes without saying that Great Britain can manufacture ships much more cheaply than Canada. This for several reasons. In the first place, raw material. such as steel plates, shapes, bars, rivets and the like are furnished at their very doors, while Canadian shipbuilders have to pay freight thereon. Skilled labor is more plentiful and the cost of labor less. Coal supply is procured at less cost. The British Government and other Governments give British shipbuilders large contracts, enabling them to operate on a large scale and hence economically. The strongest point of all, however, is that British ships come into Canada free of duty and many are sold here, and at the same time are allowed to engage in Canadian transportation without restraint.

Coupled with this is the fact that the Canadian shipbuilder is penalized or restricted with a duty ranging from 25 per cent. to 30 per cent. on a large portion of the raw materials and parts necessary to shipbuilding, which have to be imported.

In 1902 the shipbuilders of Canada sent a memorial to the Government asking for a bonus or a subsidy to encourage this trade. On Nov. 22nd, 1906 the Tariff Commission was supplied with further information from prominent citizens of Toronto, Halifax and Dartmouth, showing the argument in the case of bonusing shipbuilding in Canada.

It is pointed out that no attempt has been made to meet the new conditions created by the use of steel. The industries operating in Canada are doing so heavily handicapped, and cannot hope to compete with British builders.

In Spite of Circumstances.

Regardless of the fact that shipbuilding in Canada is heavily handicapped, there are to be found several shipbuilding yards making steady progress, and one in particular, the Collingwood Shipbuilding Company, of Collingwood, Ont., that has of late years made rapid strides and employs to-day nearly 1,000 men. From various views shown some idea of the scope of this plant and the work done therein may be ascertained. The largest boat ever turned out in

Collingwood Shipbuilding Co., Collingwood—Hoisting a 400-H.P. Engine.

Canada, the "Midland Prince," with a length over all of 486 feet, was launched only a few months ago from its ways in this yard and is now being fitted up and will be completed before the opening of navigation this season.

The Works.

Besides the large dry dock and large space for outdoor construction, the permanent buildings of this company include a mould loft, punch shed, furnace shop, blacksmith shop, pattern shop,

Collingwood Shipbuilding Co., Collingwood—Hoisting-50 ton Marine Boiler on Board New 6,000-ton Steel Lake Steamer.

large modern boiler shop, foundry and machine shop of the same design, and permanent store rooms and office building.

Personnel of Company.

The company was reorganized in 1905 under the present management, which includes the following officers: President, Alex. McDougall, Duluth; vice-president, Thos. Long, Toronto; secretary-treasurer, Sandford Lindsay, Toronto; manager, Jas. M. Smith; assistant manager, D. B. Brown; chief engineer, F. E. Courtice; naval architect, P. W. Kelty.

Engineering Features.

In these works is manufactured everything used in connection with a ship, except, of course, the rolling of the various plates and angles, and some of the auxiliary machinery. Even to the engines and boilers, propeller wheels and shafts, the work is done within the limits of the company's yards.

The power for the plant is developed by a 300 h.p. tandem-Corliss engine, and immediately transformed into electrical energy by means of two 250 volt, 150 k.w. continuous current generators. From these power is transmitted to all parts of the plant where individual motors are used.

Compressed air plays an important part in a shipbuilding yard and without its use the modern methods could not be attempted. The continuous rap, rap, rap, of the shipbuilding yards is from dozens of air riveters. By means of the same agency chipping, caulking and drilling are accomplished. Not only in the actual shipbuilding but in the boiler, machine shop and foundry is it of service.

Machine and Boiler Shop and Foundry.

The boiler shop is equipped with a 60 ton hydraulic hoist of the company's own design, so arranged that all the movements of a 60 ton boiler can be easily controlled by one man, with three operating levers. Here will also be found a set of vertical rolls with an ultimate capacity of a 2 in. plate 12 ft. wide, operated by a separate double cylinder steam engine.

The hydraulic rivetter is the largest in Canada, being 11 feet vertical gap and is of the five stage type, having a range of pressure on the rivet as follows, 35 tons, 70 tons, 105 tons, 140 tons, 175 tons.

This building, as well as the foundry and machine shop, is of steel construction throughout, each is 120 ft. wide by

Collingwood Shipbuilding Co., Collingwood—1,000-H.P. Engines and Boilers.—Tug "Emerson," M.T.Co.

80 ft. long, with provision for doubling the length at any time. Each is equipped with a 35 ton electric traveling crane with 60 ft. span. Both the foundry and machine shop are likewise equipped with up-to-date machinery. A spe-

cial feature of the foundry is a 60 in. cupola with a capacity of 11 tons per hour.

Enormous Yard Crane.

Plans have already been made for the installation in the very near future of a 10 ton crane with a total travel of 560 ft. It will have a bridge span of 72 ft. with extensions at each end reaching over the dry dock at one end and over the yard at the other, giving a total bridge span of 122 feet. It is 60 ft. to the underside of the bridge girder to enable the crane to be moved over the entire length of ship under construction.

Some Work Done.

In speaking of the work done in the Canadian shipbuilding yards and comparisan made with work of other countries, it must be borne in mind that conditions here are altogether different and that shipbuilding, referring to the supply of labor only, is at a handicap. It is necessary to turn out the work with men unaccustomed to shipbuilding. Many of them have followed different trades or arrived with no particular trade and must be broken in. Under such conditions the turning out of a Midland Prince in eight months time is a creditable feat. She is the largest Canadian ship afloat on fresh or salt water service.

Some of the dimensions of this boat, together with a mention of other work turned out, will be found of interest. The Midland Prince, with a length of 486 ft. over all and 466 ft. between perpendiculars. The moulded breadth is 55 ft. and depth 31 ft. The engines are triple expansion, vertical with 42 in. stroke, cylinders being 23 in., 38½ in., and 63 in., respectively. Installed in this boat are the two largest boilers ever built in Canada. They are 15 ft. 6 in. diam., by 12 ft., each with three 51 in. furnaces and working under a pressure of 183 pounds per square inch. Weight of each boiler complete is 65 tons.

On the ways at the present time are three steel scows for the Dominion Government, each 144 ft. long, with a capacity of 500 cubic yards. Nearing completion is a tug for the Dominion Government 115 ft. over all, breadth 23 ft., depth, 11ft., engine triple expansion, 15 in., 25 in., and 41 in., and 25 in. stroke. One boiler 13 ft. 6 in., by 11 ft., 180 pounds per square inch.

Another steel boat is in course of manufacture for the Farrar Transportation Company, being 406 ft. over all, 50 ft. breadth and 28 ft. deep. The engines are triple expansion, 21 in., 35 in. and 57 in. by 42 in. stroke, with two boilers 14 ft. 6 in. by 12 ft., at a pressure of 180 pounds. Work is also under way changing the steel schooner Agawa to a steamer for the Algoma Central Steamship Co. This ship was built by this company in 1902, dimensions 389 ft. long, 48 ft. beam and 26½ ft. deep. She will be completely equipped as a modern lake type steamer throughout in machinery, cabins, etc. She will be driven by two Scotch marine boilers 14 ft. dia. by 12 ft. long, 180 lbs. working pressure. Engines are vertical triple expansion, 20 in. by 33½ in. by 55 in., by 40 in. stroke.

Collingwood Shipbuilding Co., Collingwood—Rivetting up the Largest Boiler Ever Built in Canada. 15 feet 6 inches diameter x 12 feet long; 185 lbs. working pressure.

A REMARKABLE WRECK.

An extraordinary and expensive cement advertisement was furnished by a recent shipwreck. The ship Socoa, bound for San Francisco with a cargo of cement for use in the rebuilding of the city, was wrecked off the Lizard on the Cornwall coast. The ship struck a rock, which tore a large hole in her side, and remained fastened as upon a pivot. When the salvage crew arrived to see about taking the Socoa from her perilous position, says The Cement Age, the men found a remarkable condition of affairs. The water had entered the hold and its action upon the cargo had caused the cement to set. It had accommodatingly set hard around the rock that pierced the side of the ship, which now remains fastened there permanently in its unique position. The entire cargo has become as hard as stone, and nothing can be done with the ship except to dismantle as much of the wood as can be removed. The hull will remain there for many years as a conspicuous advertisement for the cement manufacturers.

That Electric Merger

NOT since the general public interest in hydro-electric power became a live issue in the minds of manufacturers and others in Ontario has any question arisen arousing more public interest than what has followed in the train of the incorporation of the Dominion Power & Transmission Co. by Dominion charter, with an authorized capital of $25,000,000. An air of mystery surrounded the incorporation and the only names mentioned in connection therewith have been those of Hon. J. M. Gibson and W. C. Hawkins, general manager of the Cataract Power Co. of Hamilton. Following closely the granting of the charter came the announcement that this immense company had absorbed the Cataract Power Co. of Hamilton, with its many subsidiary companies. The situation immediately assumed the aspect of a huge stock watering machine in which the rights of the Ontario Legislature were jeopardized and the possibility of the Hydro-Electric Commission of Ontario carrying out its programme nullified. The Ontario Government is up in arms and the Toronto Board of Control has taken action to have the Union of Ontario Municipalities (formed with the purpose of supplying cheap power to the various districts) discuss the question.

By gaining a Dominion charter and absorbing companies under the control of the province it would appear that this company was seeking to overcome provincial jurisdiction in its operations. Besides controlling electric power, the company also aims at operating smelting works, lumber mills, telephone and telegraph wires.

The Hamilton Cataract, Power, Light & Traction Co., with its stock at par value of $1,700,000, paying an annual dividend of $51,000, has by its absorption in the Dominion Power and Transmission Co., been increased to $5,100,000.

The Cataract Company owns and operates the following companies: (1) The Hamilton Electric Light & Cataract Power Company; (2) The Cataract Power Company of Hamilton; (3) The Hamilton Electric Light & Power Company; (4) The Electrical Power & Manufacturing Company; (5) The Dundas Electric Company; (6) The Hamilton Street Railway; (7) The Hamilton Radial Electric Railway; (8) The Hamilton & Dundas Street Railway; (9) The Lincoln Electric Light & Power Company.

Such companies are under the direct control of the Provincial Government, and both parties in the Provincial Legislature are prepared to stand by the rights of the province in such matters and oppose the Dominion Government as to their right to issue the charter given.

Discussing the question on the floor of the House recently, Premier Whitney reported as follows:

For several years there had been a steadily increasing list of applications to the Dominion Parliament for charters for companies declaring them to be to the general advantage of Canada, and which, necessarily, removed these companies beyond the jurisdiction of the province. He continued:

"That takes the enterprises as all honorable gentlemen are aware, out of our jurisdiction. It has been alleged—and, it may be truly alleged, it is a question that will have to be decided in the courts, I suppose—that all possible subjects, all possible enterprises of a business nature with reference to which a company may be chartered, can be taken out of the jurisdiction of this province by the insertion of those magical words, 'for the general advantage of Canada.'

"While it is so asserted, as a matter of law there has never been a decision. If that assertion is correct our jurisdiction here will simply be a nominal jurisdiction. As I understand it, Mr. Speaker, unless we disregard comity entirely, we would perhaps have no standing more than that we could object to the granting of original charters by the Dominion. But when we have instances like that quoted by my hon. friend, where a great blanket corporation is created, taking power to hide within its folds a number of short local electrical roads, which have been chartered by the Province of Ontario, and remove them from our jurisdiction; when that is being done I think the time has arrived for the Province of Ontario to interfere. Companies of that description come here. They get powers given to them which they cannot get from any power on earth except the Ontario Government. They exercise those powers, they make use of them, and they get by means of the exercise of those Provincial powers into a position where they have a financial standing.

"Then they go to the Dominion Parliament and ask to be put in this class 'for the general advantage of Canada,' which takes them out of the jurisdiction of the Province of Ontario, which created them, out of the jurisdiction of the Ontario Railway and Municipal Board, and out of the jurisdiction of this Legislature.

"I say, and I weigh my words well, that this Government will not submit unless it is compelled to. It is possible that it may be compelled. It is possible in other words, and I make no charge against the Dominion Government, but from what I have heard and a little of what I have seen I am afraid it will not be of much use to ask for reasonable terms from the Railway Committee of the House of Commons. I am not prepared to make charges against the Dominion Government, because I am not well informed of the facts which would justify any charges. But in reference to that class of companies where people come and get the great opportunities and privileges which they desire in order to use our highways, where they come to secure our endorsement of their agreements with municipalities, and then go to the Dominion Parliament to escape from the jurisdiction under which they came into existence, I think I have a right to say that any course which the Legislature of this Province may find it necessary to pursue will be a justifiable course.

"Hon. gentlemen will understand that neither myself nor any other member of the Government has had time since this charter was first brought to our attention to consider carefully and decide what our rights in the premises are. Therefore I will not attempt to suggest any remedy at this time. If all else fails us probably we have the power of taxation."

Hon. Geo. P. Graham, Leader of the Opposition, stated that the Liberal party would always be found where the interests of the province were jeopardized by Federal aggression, on the side of provincial rights, which evoked hearty enthusiasm from both sides of the House.

An interesting sequel is looked forward to as in this question financial, manufacturing, industrial and commercial interests are involved.

It is said that the cement works at Longue Point, near Montreal, operated by James Morgan, have been sold to the Fordwick Co., of Virginia, which has decided to establish a second plant in the vicinity of Montreal, costing in the neighborhood of $2,000,000, with a capacity of 6,000 barrels per day. The Fordwick Co. will also build large docks to store coal from Sydney. The engineering firm of W. S. Barstow & Co., of New York, are preparing the plans and specifications, and have given the building contract to the Canadian White Co., of Montreal. The new plant will employ from 1,000 to 2,000 men.

Civil Engineers Convention.

ON Jan. 29, 30 and 31, at their quarters in Montreal, the Canadian Society of Civil Engineers held its annual convention, which was largely attended by men from every province in the Dominion. A departure from the custom took place at this convention to the effect of devoting the time to business of the society rather than, as was usual, having papers read and discussed. It had been found in the past that too little time was left for the regular business of the society when a large portion of the convention was taken up with the discussion of papers.

The first days' meeting was devoted to the report of the Council and counting the ballots for the election of officers. The result of the elections was as follows: President, W. W. McLea, Walbank, vice-president and chief engineer of the Montreal Light, Heat & power Co.

Vice-presidents: Messrs. M. J. Butler, chief engineer railways and canals, Ottawa; Phelps Johnson, manager of the Dominion Bridge Company, and J. S. Dennis, chief engineer of the C. P. R. irrigation works, Calgary.

Council: Messrs. G. A. Keefer, Vancouver; D. McPherson, assistant chief engineer Transcontinental Railway, Ottawa; G. H. Duggan, second vice-president Dominion Coal Co., Sydney; C. E. W. Dodwell, engineer Public Works, Halifax; C. H. Rust, city engineer, Toronto; W. McNab, assistant engineer, G.T.R.; W. F. Tyre, Montreal; E. V. Johnson, Ottawa; W. H. Breethaupt, Berlin; J. A. Jamieson, R. S. Lea, R. A. Ross, G. J. Desbarats, John Kennedy, F. P. Gutelius, assistant chief engineer C. P. Railway; W. H. Laurie, Prof. R. J. Durley, L. G. Papineau and H. Hardman, all of Montreal, and A. A. Dion, of Ottawa.

This concluded the business of the convention, which adjourned shortly after noon. Later in the day the newly elected council met and appointed standing committee for the year.

On Wednesday, through the courtesy of the Montreal Street Railway, special cars were provided for the party, leaving Windsor Hotel at 9 a.m., whence they were conveyed to the works of the Simplex Railway Appliance Co., at Blue Bonnett. An interesting hour was spent going through this new plant, where bolsters and special apparatus for the manufacture of railway trucks are made. The economy of production in this plant was at once apparent. The latest devices and methods of transportation have been adopted to reduce the cost of manufacturing to a minimum. The oil blast furnaces were a feature of these works, one of their strong points being that they are completely under the control of the operator, regarding the supply of heat. From there the party journeyed to the enormous plant of the Canada Car Co., at Cote St. Paul, where they were entertained in a most hospitable manner to a bountiful luncheon.

Several hours were spent inspecting this enormous establishment, where from 12 to 20 freight cars are turned out complete every day. The president

MR. W. McLEA WALBANK
Newly-Elected President of the Canadian Society of Civil Engineers.

of the company announced that the building of the works had been done in record time, being accomplished during one of the severest winters Montreal has had, with deep snow to interfere with the operations, and hard frost to retard the laying of the concrete work, with which the walls are entirely composed. The works were built by Canadians, by Canadian capital, and, as far as possible, Canadian material, which is also true of the equipment. When the works started it was not expected to run the entire plant for some time, but already the company have contracts in sight to keep them running at full capacity for over a year.

Continuing their trip, the civil engineers journeyed back to the city and on through the works of the Canadian Rubber Co., where they had a chance of seeing the various processes in connection with the manufacture of rubber goods, including rubber belting, matting, hose, horse pads, water bottles, foot balls, clothing, boots, shoes and rubbers.

In the evening the annual banquet of the society was held at the Windsor Hotel. About two hundred members were present to celebrate the twenty-first birthday of the society. M. J. Butler, of Ottawa, Deputy Minister of Railways and Canals, occupied the chair in the absence of the retiring president, H. S. Lumsden, who called upon Mr. G. W. Stevens, president of the Harbor Commission, to propose the toast. Mr. Stevens congratulated the society and proceeded to point out what a debt the country owed the engineering profession for its advancement. They had done much that the country should be proud of. It was only necessary to mention the names of such men as Page, Shanley Gzowski and Keefer, to show the kind of engineers Canada had produced. Major Stewart Howard briefly replied to the toast, after which Mr. Butler proposed our guests, replied to by Mr. C. H. Catalli, president of the Chambre de Commerce, and Mr. King, of the Canada Car Co. The retiring president and council were proposed by Mr. C. E. Dodwell. Other toasts proposed were the sister professions, the visiting members, the press.

On the last day of the convention various matters in connection with the work of the society were discussed. The resolution was brought in appointing a committee on the standardization of cement, to collect all possible data and report to the society. J. A. Jamieson, of Montreal, introduced a resolution regarding sending a memorial to the Government regarding the selling of cement. According to the present custom cement is sold in barrels of 325, 350 and 375 pounds weight. It was suggested fixing the standard as a sack of 100 pounds, and that cement be sold by weight rather than by the barrel. This was heartily endorsed by the society.

The report of the year showed the affairs of the society in a most flourishing condition, both financially and as regards increased membership.

The Gzowski for 1906 was awarded to Walter J. Frances for his paper on "Mechanical Locks in Canada."

CANADIAN MACHINERY
and Manufacturing News

A monthly newspaper devoted to machinery and manufacturing interests, mechanical and electrical trades, the foundry, technical progress, construction and improvement, and to all users of power developed from steam, gas, electricity, compressed air and water in Canada.

The MacLean Publishing Company, Limited

JOHN BAYNE MACLEAN - - President
W. L. EDMONDS - - Vice-President
H. V. TYRRELL - - Manager

OFFICES:

CANADA
MONTREAL - 232 McGill Street, Phone Main 1255
TORONTO - 10 Front Street East, Phone Main 2701
WINNIPEG, 511 Union Bank Building, Phone 3726, F. R. Munro
BRITISH COLUMBIA - Vancouver, Geo. S. B. Perry

UNITED STATES
CHICAGO - 1001 Teutonic Bldg., J. Roland Kay

GREAT BRITAIN
LONDON - 88 Fleet Street, E.C., Phone Central 12960, J. Meredith McKim
MANCHESTER - 92 Market Street, H. S. Ashburner

FRANCE
PARIS - Agence Havas, 8 Placede la Bourse

SWITZERLAND
ZURICH - Louis Wolt, Orell Fussli & Co.

SUBSCRIPTION RATE.

Canada, United States, $1.00, Great Britain, Australia and other colonies, 4s. 6d., per year; other countries, $1.50. Advertising rates on request.

Subscribers who are not receiving their paper regularly will confer a favor on us by letting us know. We should be notified at once of any change in address, giving both old and new.

V. l. III. MARCH, 1907 No. 3

CONTENTS

UNIFORM LEGISLATION FOR STEAM BOILERS.

ALL the large manufacturers of boilers in Canada are unanimous in their opinion that legislation in reference to land and stationary boilers is all too backward. As the situation now exists we have a Dominion Marine Act, covering marine boilers but in no way affecting the use of other boilers in the various provinces. The fact that British Columbia has a law similar to the Dominion Marine Act shows the possibility of the same being applied to every province. As stated in the letter from Mr. C. H. Waterous, of Brantford, published in this issue, Alberta, Saskatchewan and Manitoba each have boiler laws different from the other, and less strict than the British Columbia law. Ontario has no special legislation regarding land or stationary boilers, and Quebec has a law different from the others, while the other provinces are similarly situated.

A valuable suggestion comes from Mr. James M. Smith, manager of the Collingwood Shipbuilding Co., regarding the appointing of a committee consisting of Messrs. F. E. Courtice, of the Collingwood Shipbuilding Co.; Arthur E. Spotton, of Goldie & McCullough, Galt; John Gerrall, Consulting Engineer, Toronto, and John J. Main, of the Polson Iron Works, Toronto, with power to seek assistance if necessary, with a view to framing a standard set of rules for stationary boiler building. This is a suggestion along the right line and, no doubt, could be acted upon with little difficulty. We would suggest to the gentlemen named that the question be taken up at once. It is an entirely different proposition going to the Dominion Government and asking for legislation along this line with nothing definite to propose, to being in a position to present a set of rules drawn out by men in an authoritative position. In the first case the Government would have nothing definite to act upon, and in the latter case they would have no reasonable excuse to set the matter aside.

Why let this matter drag on any longer? There is nothing like prompt and definite action in a situation of this kind, and we know that all manufacturers would like to see the question settled immediately. Canadian Machinery will heartiy co-operate with any effort made in this connection, and would appreciate any suggestion whereby this could be attained immediately. If the gentlemen named, together with others that might be, would be willing to spend a little time we are sure that success would inevitably follow, and a cause of annoyance removed for all time to come.

UPHOLDING PROVINCIAL RIGHTS.

THE formation of a twenty-five million dollar company with a Dominion charter and the powers granted therein, with what has followed in its train, has caused an upheaval in parliamentary circles in Canada. This is the result of a company's being able by Dominion charter to acquire rights vested under the jurisdiction of the Provincial Legislature. Following the incorporation of the Dominion Power & Transmission Co., with the above-mentioned enormous capital, came the absorption of the Cataract Power Co. of Hamilton with its subsidiary companies. These latter include four or five electric railways and several power companies, all enjoying their privileges under provincial control. The Ontario Government contend, and rightly so, that the Dominion parliament cannot grant a company such sweeping rights as to absorb companies not existing under charter and take them out of the control of the province. The fight between the two governments is now on, involving a question of vital importance to every province in the Dominion. The Ontario Government is determined not to give in in the matter, both sides of the House being unanimous on the question. The Ontario Government has already commenced legislation covering the point in question that will be of a rather sweeping nature. It is hoped, however, the matter will be adjusted between the two governments satisfactorily to the best interests of the whole community.

COMMERCE FOLLOWS ADVERTISING.

AT the Canadian Club of Toronto, on Monday, Feb. 18th, Mr. John A. Cooper, stated that the old proverb, "Commerce follows the flag" had given way to a newer one, "Commerce follows advertising." This, he stated, was shown most emphatically in connection with the difference between British and United States trade with Canada. It is well known that the imports from the United States are many times greater than the imports from Canada, apart from geographical reasons. This is explained by the fact that Canadians have been educated to using United States manufactured goods from the medium of wholesale advertising in publications read by Canadians. An incident was stated of his going into a drug store and asking for a certain line of British made tooth powder. The druggist had never heard of it and produced four different kinds all manufactured in the United States. He said he must handle these because the people asked for them.

Why do people ask for goods of United States manufacture? Because they learn of them, see them advertised, become acquainted to some extent with their merit from publicity, and then become regular customers.

This is a pointed instance of the negligence of British manufacturers to a proper appreciation of Canadian trade. If appreciative, they do not employ the methods to any universal extent that are being found not only satisfactory but essential on the part of manufacturers of the United States.

If the British manufacturer wishes to secure Canadian trade there is nothing surer. "He must advertise his goods in Canada."

SELLING CEMENT BY WEIGHT.

A GOOD move was made at the recent convention of the Canadian Society of Civil Engineers when the members unanimously endorsed the selling of cement by weight. A committe was appointed to present a memorial to the Canadian Government petitioning them to legislate in the matter. At present the bulk of cement placed on the market is in barrels supposed to contain a standard weight when in reality they range all the way from three hundred and twenty-five to three hundred and seventy-five pounds per barrel. Ordering cement by the barrel is a misnomer unless the weight is specificially stated which in most cases it is not. Then on a contract of several thousand barrels the difference in weight may assume a serious amount. It was stated that the Dominion Government was one of the worst offenders contributing to such a condition of affairs. Being an enormous consumer of cement very few official contracts specify the amount a barrel shall contain.

In view of the fact that cement is much more easily handled in sacks and that one hundred pounds weight constitutes a reasonable package, it is proposed to suggest that this become universal practice in Canada. There seems no reasonable excuse that the Government should not fall in with the idea. It will not only remedy an existing evil in regard to weight but will standardize the hundred pound package which is a convenient one and which should meet with general approval not only on the part of contractors and cement users, but from the manufacturers themselves.

INADEQUATE PROTECTION AGAINST EXHAUST STEAM.

THOSE who have occasion to frequent the business sections of large cities cannot fail to have noticed that through careless discharging of exhaust steam from buildings there frequently descends upon passers-by a regular shower of condensed steam, with most annoying consequences. The nuisance referred to is common to all large cities, but in New York the newspapers are taking the matter up and pressing for the adoption of proper ordinances by which the municipal authorities shall have the power to restrict it.

Readers of this paper will be quick to recognize that the difficulty lies in the fact that the engineers in many of the large office buildings are not well acquainted with the fact that the use of exhaust heads would put an end to the discomfort they are causing the public. In all ower plants which are run, non-condensing exhaust heads should be considered an essential part of the equipment.

IMPORTANT ELECTRICAL ADVANCEMENT.

RECENTLY two announcements have been made of very great importance to applied electricity as concerning everyday life. On Feb. 11th, his sixtieth birthday, Thomas A. Edison announced that he had finally perfected his storage battery for electric motor vehicles. He proposes letting the public see them during the coming summer, believing that he has attained the highest point of efficiency. According to his own statement he has tried this battery under the severest circumstances, and the result warranting his prophecying that the new storage battery will solve the problem of vehicle transportation in great cities. Should Mr. Edison's confidence in his new invention not be over rated he will thus have conferred another boon to humanity, particularly to city dwelling mankind.

A few weeks previous to the above event Professor Parker, of Columbia University, and Walter G. Clark, electrical engineer, announced the substitution of helion for carbon in the incandescent filliment, with the result of reducing electrical lighting to one-third its present cost. This has been the result of several years' labor with the definite object in view of finding a more economical substance for incandescent filliment. The result of their researches is the substance which they have named helion, which is composed largely of silicon. At present it is expected that this lamp will cost more than the other, but besides consuming only a fraction of the current will last twice as long.

Cylinder Relief Valves*

By R. E. Johnson

WHEN I undertook the task of writing a paper on relief valves, I certainly had no idea of the extent of the subject, and to my surprise, as soon as I started I found out I was going to have a pretty hard task, to get up a paper that would bring out both the technical and practical points in regard to the benefits and use of a relief valve to the steam-working parts of a locomotive. I wrote to mechanical engineers of a number of different railways, both in the United States and England, for their views on, and experiences with, relief valves. Some of them gave me a few simple experiences, but none of them cared to talk about their benefits and functions, only saying they were used for relieving any undue pressure from various causes in the cylinders and steam chests. One gentleman, a superintendent of motive power of a foreign road, said, until I wrote him he had never taken the time to go into the matter; they were using them on the engines on his road, and they were doing the work they were supposed to do, of course, he said, "We have different sizes for different sized cylinders." What I consider one of the most important things in regard to relief valves, is the size of valve in comparison size of cylinder, yet, in answer to all my letters, not one gave anything definite about the sizes.

In looking up the meaning of relief valves in the dictionary, it explained that, "Relief valves are valves fitted to a boiler, steam-pipe, or cylinder, to relieve any excess pressure that might endanger the delicate and most important parts of a steam engine." By this definition, safety valves could be dealt with under the same heading as cylinder relief valves, but I am only going to take up the latter.

When a locomotive or, in fact, any steam engine is designed, the cylinders are made to stand a given pressure, which is, of course, above the boiler pressure—pressure causen by compression of steam in the cylinders. Now, on the cylinder walls is not where the pressure is greatest, but on the cylinder heads, as we find in the study of "Strength of Materials." The tendency of a cylinder subjected to internal pressure is to fail, or rupture, in the direction of its length.

Not so very long ago, when relief valves for cylinders were not used, a great deal of trouble was experienced by broken cylinder heads; the bore of cylinders was covered with grit and dirt, and the valves and seats were cut and grooved, causing delay to and expense against engines everywhere. Of course something had to be done to remedy this, and relief valves came into use. They were practically the same as our loaded spring valves of to-day, but there was no standard governing their size, and they were tried in different places about cylinders and steam chests. It was what was needed, and they began to improve in design until we have to-day a very nearly perfect relief valve.

About broken cylinder heads: Take an engine without relief valves running about 30 miles per hour, under a good head of steam. Something occurs that necessitates the train coming to a stop within a short distance. The engineer immediately reverses his engine and gives her steam in back gear. With this sudden strain, and if there is any condensation in cylinder in nine cases out of ten something has to go; if it is not the cylinder heads it is the steam chest; whereas, if the engine was equipped with relief valves, all that undue pressure would have been relieved. Another thing, in places where feed water is bad, and boilers dirty, the boiler primes, and water is carried into cylinders. There you will find, without relief valves, another case of broken cylinder heads; of course, if the engineer opened his cylinder cocks as soon as the boiler began to prime he might be able to save himself, but cylinder cocks are not to be relied upon in regard to size of outlet. A case happened to a switching engine in England: A heavy freight train was to be moved a short distance by a switcher. The engineer could not move the train at first. With wide open throttle he pulled the reverse lever from forward corner to back corner, and then, after taking up the slack, he threw her again into extreme forward gear. Before he had time to place his hand on the throttle smash went his cylinder head, and on getting down to examine it, he also found the valve rod bent just outside the steam chest. The mechanical engineer of the road told me that he had all his switching engines equipped with relief valves after that. I have noticed on our roads in Canada and the United States that it is a habit practised too extensively by the engineers on taking up the slack of a heavy train they leave the throttle open, while they reverse suddenly, instead of waiting a couple of minutes to allow the pressure in cylinders to subside. No matter if an engine is equipped with relief valves they won't stand much of that sort of usage.

Another little incident occurred on our own road not so very long ago: A dead engine, not uncoupled, was run around the yard and switched finally on the turntable. During the movements it had developed enough air pressure in the boiler to be able to work it into the shop. In this case the reverse lever was used to do this. Now, this engine was not equipped with relief valves, and during the movements it had sucked into cylinders, steam chest, and steam ways, sand, cinders, and dirt out of the smoke-box, thereby causing needless expense in taking out pistons and valves to have them cleaned. It would probably have been a surprise to them had they seen the inside of those cylinders, valves and steam ways. All this foreign matter tends quickly to destroy the usefulness of valves and piston rings by jamming and cementing rings and heads solid. Speaking about drawing foreign substances into cylinders: I was in conversation with a gentleman on the train going from Avonmouth to Bristol, and in talking about relief valves (I might say he told me he was a master mechanic on a railway in South Africa), he said if running engines without relief valves treated him the way one engine had, he hoped they would never come into general use. He had some trouble with an engine's valves, and in taking out the valve he found a groove running lengthwise on seat and valve face, and what was more a pebble, as he thought, about the size of a large pea, embedded in the seat. He took the pebble to have it analyzed, and it was found to be a very fine diamond. As to how that happened to get there it is hard to say, but he puts it down to the engine not having relief valves.

Now, in summing up, we have a few benefits from relief valves if properly applied:

1. They prevent foreign matters from smoke-box being drawn into steam ways.
2. They prevent broken and leaky steam joints or cylinder heads by compressed condensation.
3. They make free-running engines when drifting.
4. They reduce the cost of maintenance on pistons and valves.

Of course, with the good braking appliances now on our engines, the old practice of "plugging" is about eliminated, and the danger of failures from this source is minimized. As there is still though a great deal of it to be

* Read at Canadian Pacific Railway Club, Montreal.

done, it would be advisable to put a stop to it once and for all.

The usefulness of relief valves is greater on a rolling division than on a level section of road, as on the latter steam is in constant use, while on the former "shut-off" and drifting down grades bring out the value of the relief valve.

The automatic relief valve, as applied to many of our engines, gives trouble by stem of valve breaking, and allowing the valve to unseat itself, and also by valve seat becoming defaced. The combined relief and pop valve gives abnormal trouble by steam leaks, especially on high pressure side of compound engines. This is a very serious matter from the fact that in severe cold weather a small leak makes a cloud of steam, and the enginemen are (when slow) running in a fog. This could be avoided in some cases, where space permits, by applying the improved relief valve (described later) which allows the escape steam to go to the atmosphere by way of smoke stack.

The size of relief valves is governed by proportion of cylinder, but in some cases I have seen an 18 in. cylinder with the same size relief valve as a 35 in. cylinder. For my part, I do not think this is right. The latter should be increased two to one. The cost of maintenance of a good relief valve, that can take care of its own escape steam, would be exceedingly small, while the cost of maintenance of the combined pop and relief valve in cold weather is very high, for the reason that in the summer the leaks are not visible. A question came up one day, "Does a cylinder relief valve allow all the water in the cylinder to escape?" The answer was found to be "No, the valve will clear the cylinder of all the water except the quantity due to the clearance between the top and bottom of the cylinder and the cylinder cover, or end, as the case may be." Before going any further I should like to give an account of a small experience I made with relief valves. For a couple of voyages on the S.S. "Montcalm" I had charge of the refrigerating plant, and the engines were equipped with loaded spring relief valves, one on each end of both low and high pressure cylinders. The steam pipe from boiler to engine was about 90 feet long, which caused a fair amount of condensation. Underneath the relief valves I placed small tins to catch the water blown out. In one hour and thirty-seven minutes I had a good quart of water. I then blocked up the relief valves and waited a few minutes until water began to pound in cylinders, and I noticed that the engine was slowing down. I immediately set valves free, and in a very few minutes everything was running ship-shape. The next time we stopped the machine I, on the quiet, had to put in two new studs in the high pressure cylinder head, because I broke two in trying to tighten head up after my little experiment.

So far as I can ascertain, there is no definite rule for giving the proper sizes of relief valves for different sized cylinders, and I think it would be well worth while to carry out some experiments with a view to establishing such a rule —using different sized valves with the same pipe connections to the cylinder. An experiment I have thought of that could be tried with spring loaded valves is: Have a connection or electric contact between the valve and cab, so that when the valve lifts it will ring a bell in the cab. Let someone take very exact notes on every position of lever in cab—more especially of reduction of train line, and air reservoir pressures in bringing the train to a stop—then, if possible, under same conditions (by cooperation of engineer) block the valve so that it will not lift—and take the same notes—at the same time noting the difference in momentum of train while drifting into a station; also take cards off the engine with the relief valves closed, and running free with valves in normal condition.

ABOUT CATALOGUES

By mentioning Canadian Machinery to show that you are in the trade, a copy of any of these catalogues will be sent by the firm whose address is given.

FOUNDRY SUPPLIES.—The J. D. Smith Foundry Supply Co., Cleveland, have issued a most complete catalogue of over 250 pages describing practically everything necessary to the modern foundry. It is No. 37 of their publications.

ELECTRIC DRILLS, ETC.—The Chicago Pneumatic Tool Co.'s electrical tool catalogue No. 21 covers electric drills, grinders, drilling stands, etc. It consists of 60 pages, and is said to be one of the most complete ever shown by a single manufacturer. A copy will be furnished on request.

AIR AND GAS COMPRESSORS.—Catalogue H-36 of the Ingersoll-Rand Co., New York, contains 64 pages of well-illustrated information on Ingersoll-Sergeant Class "H" air and gas compressors. Leaflet 6-A concerns Haeseler and Imperial pneumatic tools.

GEAR CUTTERS.—The catalogue issued by the Brooke Tool Manufacturing Co., Birmingham, contains prices of Involute gear cutters in both best cast steel and the company's special high speed steel, arranged side by side for easy comparison. This firm is now making the shallow or stub form of tooth for motor car gears, electric crane gears, and other quick running machinery, which, it is claimed, runs much more smoothly than the usual length of tooth.

STEAM HAMMERS.—The David Bell Engineering Works, Buffalo, in Catalogue No. 806, give a lot of information concerning Bell's Steam Hammers, including general directions for the foundation and setting of the different styles of hammer.

THE J. W. HARRIS COMPANY., Montreal are getting out their new catalogue for 1907, with complete illustrations of the "Harris Excavator," which will be of considerable interest to contractors.

MACHINE TOOLS.—A very complete catalogue of machine tools, including upright drills, gang drills, lathes, tool grinders, etc., has been issued by the B. F. Barnes Co., Rockford, Ill.

BELTING.—A revised price list of Maple Leaf belting has been issued by the Dominion Belting Co., Hamilton. With it is enclosed a list of firms throughout Canada who use this make.

GRINDING WHEELS.—An interesting booklet on the manufacture of "Alundum," used in the making of Norton grinding wheels comes from the Norton Company, Worcester, Mass., and Niagara Falls, N.Y.

THE METRIC SYSTEM.—The Decimal Association, of London, Eng., have issued in pamphlet form Lord Kelvin's views on the advantages of the metric system, together with the opinion of other eminent men, and explanatory and comparative tables.

HYDRANTS AND STEAM SPECIALTIES.—Two pamphlets, No.s 440 and 425, issued by the Canada Foundry Co., deal respectively with Walker hydrants and steam specialties, the latter including reducing valves, steam traps, injectors, and feed water controllers.

Uniform Boiler Legislation

WHY there is not uniform legislation in Canada in regard to stationary and land boilers is somewhat of a mystery and forms a vexatious problem not only to manufacturers of steam boilers but to users as well. The fact that a boiler can be made in Quebec and used there that will not be accepted in Ontario, and furthermore one acceptable in Ontario might not be allowed in British Columbia shows that the situation needs remedying. The Canadian Manufacturers' Association is taking up the question, but so far nothing has been done. To ascertain the opinion of manufacturers on this subject a letter was written to some of the manufacturers of boilers in Canada, and the opinions given from several representative firms show the unanimity of opinion in this respect. British Columbia is the only province that claims a particular steam boilers' inspection act, which while very stringent is impartial, and subjects all boilers to Government inspection. Some opinions of manufacturers are herewith quoted, followed by some details of the situation in British Columbia.

Some Opinions.

F. S. Keith, Esq., Managing Editor, "Canadian Machinery," 10 Front St. E., Toronto, Ont.

At the present time the Dominion Government has a Marine Boiler Law which governs the manufacture and use of all boilers used in marine work. This means that all marine boilers used in Canada are made to conform to one general law, so that a marine boiler to be used in British Columbia must be the same in practically every particular as the same boiler would be if used in any of the other provinces.

For stationary or land boilers there is no general law, but each province has a law of its own. British Columbia has a law that is similar to the Dominion Marine Act—every boiler used in British Columbia, with the exception of marine boilers, must conform to this law. Alberta, Saskatchewan and Manitoba each have boiler laws different from the other, and less strict than the British Columbia law. Ontario has no special legislation regarding land or stationary boilers. Quebec has a law different from the others, and the other provinces are similarly situated

It would, I think, be very much in the interests of both the users and manufacturers of boilers in Canada if a uniform act were to be enacted that would govern the manufacture of boilers for all the provinces If the Marine Act were adopted both for stationary as well as marine boilers used in Canada, then every boiler used in Canada would be a first-class boiler. The consumer would be assured of getting an article that would be serviceable and safe in use, and manufacturers would be all placed on a plane, so far as quality of boilers is concerned, and would so be able to carry boilers in stock which would be suitable for any part of the country.

As it is now, a boiler that would be suitable for Ontario would not pass British Columbia inspection, and a boiler that would pass Quebec inspection would probably not pass inspection in most of the other provinces.

I am of the opinion that if the Dominion Government would take up with the provinces this question, and arrange, if possible, for a uniform act, it would be a very desirable thing for the country.

Yours truly,
The Waterous Engine Works Co., Ltd.,
C. H. WATEROUS,
Manager.

* * *

In reply to your letter of the 16th inst. re inspection of boilers, would say that in our opinion it would be desirable, from the standpoint of boiler makers, and also that of the user, if some universal inspection legislation could be put through.

We are afraid, however, that it would be very difficult to reconcile various conflicting interests.

Yours truly,
Canada Foundry Co., Limited,
J. J. ASHWORTH,
Sales Manager.

* * *

Canadian Machinery & Manufacturing News, 10 Front St. East, City.

Gentlemen :—

We have before us your favor of the 6th inst. re legislation affecting the inspection of boilers, and in reply beg to say that we are heartily in favor of the universal inspection of boilers for the whole of the Dominion rather than legislation by the separate provinces. British Columbia is the only province to adopt laws regarding the inspection of stationary boilers, and were each province to pass legislation to this end it is not likely that it would be identical, and to manufacturers like ourselves who are shipping boilers to all parts of the Dominion, it would create great confusion trying to keep track of the laws of the various provinces. The remedy to our mind would be a universal inspection by the Dominion Government.

Yours truly,
Polson Iron Works, Limited,
A. H. JEFFRY,
Secretary.

* * *

Dear Sir :—

In answer to your favor of the 6th inst., we most decidedly favor a uniform inspection of boilers throughout the Dominion. We believe that representatives of the different provinces should get together and unite on one standard inspection act and have the same act put in force in every province of the Dominion. The Engine & Boiler Section of the Canadian Manufacturers' Association has taken this matter up and is moving in this direction. It is not necessary for us to point out the difficulties and disadvantages under the present system. We believe most decidedly that one uniform standard should be adopted for all the provinces of the Dominion and anything that you can do to further this will be appreciated by all the manufacturers and users of boilers.

Yours truly,
The Goldie & McCullough Co., Limited.

* * *

A Definite Suggestion.

Re the question of uniform law covering the building and inspection of stationary or land boilers, I think there should be a uniform law over the whole Dominion, and that it should be under Government supervision, and that the law should be made as simple as possible, with as few conditions or allowances to each rule as could possibly be made to have it all very simple and bussinesslike, so that there would be no question of leaving anything to the intelligence or desire of the boiler builders or inspectors, other than the judgment of workmanship. I would suggest that it would be a very good idea to appoint Mr. F. E. Courtice, of our works, Mr. Arthur K. Spotton of Goldie-McCullough, Galt, Mr. John Gerell, Consulting Engineer, Toronto, and Mr. Main of the Polson Iron Works, Toronto, as a committee, with power to seek assistance if necessary, with a view to framing a standard set of rules for stationary boiler building.

Yours very truly,
Collingwood Shipbuilding Co., Limited.
JAMES M. SMITH,
Manager.

Inspection Requirements.

Every boiler built in British Columbia may be inspected during construction.

The maker must furnish the Chief Inspector of Boilers with a copy of the drawings from which such boiler is being built, and such information as will enable him to calculate the safe working pressure. The same drawings and information must also be supplied by the maker of any boiler to be used in British Columbia, no matter where it may be built. Owners of boilers must every year make provision for inspection, and the engineers in charge of such boilers must assist the inspector in his examination, and point out to him any defects they know or believe to exist, under penalty of having their certificate revoked for neglect to comply with this provision of the Act.

Notice by Agents, Importers and Purchasers.

Agents of manufacturers, or importers of boilers, new or second-hand, must send notice to the Chief Inspector of all boilers imported by them, together with the drawings and information mentioned above. The penalty for default is from $100 to $500. Any person purchasing a steam boiler, either new or second-hand, must notify the Chief Inspector of such purchase, giving name of seller, and stating when and where it can be inspected. Any person installing and putting such a boiler into operation without giving such notice is liable to a penalty of from $100 to $500.

Inspection Fee.

For the first inspection of each boiler the minimum inspection fee is from $5 to $40, according to its horse-power. For subsequent yearly inspection the fee is $5 for each boiler up to 25 horsepower. Where the horse-power exceeds that, an additional charge is made of twenty cents for each horse-power up to 50 horse-power; fifteen cents for each of the next twenty-five horse-power; ten cents for each of the next twenty-five; five cents for each of the next fifty, and two and one-half cents for each additional horse-power. Where two boilers belonging to the same plant are made ready for inspection at the same time, a reduction of ten per cent. in the fees for yearly inspections is allowed; where three boilers are made ready the reduction is twenty per cent., and where four or more boilers are made ready the reduction is thirty per cent.

Note.—The legislative enactments and regulations regarding the construction, use and inspection of steam boilers in British Columbia can be obtained from J. Beck, Inspector of Steam Boilers and Machinery, Victoria, B.C.

Ten Things.

There are ten things for which no one has ever yet been sorry.

For doing good to all.
For speaking evil of none.
For hearing before judging.
For thinking before speaking.
For holding an angry tongue.
For being kind to the distressed.
For asking pardon for all errors.
For being patient toward everybody.
For stopping the ears to a tale-bearer.
For disbelieving most of the evil reports.

Crowded Out.

"Johnny, where is your mechanical engine?"

"Pop's got it."

"How about your automatic trip-hammer?"

"Uncle Bill won't let me have it."

"Your Japanese top?"

"Uncle Jim's playing with that."

"Well, you seem to be in the way in the nursery. Guess you'd better go into the library for awhile."—Louisville Courier-Journal.

At The Noon Hour

Wanted—Position as Fireman.

I am a man industrious;
The reason that I'm idle thus
Is just because I cannot find
A job that's suited to my mind
Quite gladly back to work I'd go
If I could find a job, you know,
That really, truly suited me
And with my system would agree.
I'd like to fire some power machine
That operates with gasoline.

No Waste.

"Not an ounce of waste in a ton of coal, in smoke or otherwise," was the way the advertisement read: "Household rights for fifty cents by mail." Fifty cents duly enclosed brought the information, "Burn the coal, can the smoke and eat the ashes," neatly printed on a card.

Growth of Western Cities.

Apropos of the mushroom growth of new towns on the western frontier, a locomotive engineer relates the following:

"One day I was driving my engine across the prairie when suddenly a considerable town loomed up ahead where nothing had snowed up the day before.

"'What town's this?' says I to my fireman.

"'Blamed if I know,' says Bill. 'It wasn't here when we went over the road yesterday.'

"Well, I slowed down, and directly we pulled into the station, where over five hundred people were waiting on the platform to see the first train come in.

"The conductor came along up front and says to me:

"'Jim, first we know we'll be running by some important place. Get this town on your list and I'll put a brakeman on the rear platform to watch out for towns that spring up after the train gets by!'"—Minneapolis Journal.

Mike.—"Pat, kin yez tell me what kapes them bricks together?"

Pat.—"Sure, Mike; it's the mortar."

Mike.—"Not by a dom sight; that kapes them apart."—Harper's Weekly.

Bunkoed.

An agent came around to me—
(He said he was a "gent");
To sell to me machinery
He said was his intent.

I bought a forge, asbestos lined,
He said he'd guarantee;
A patent kind he said I'd find
A model forge to be.

Now o'er that forge my spirit grieves.
And weeps my faithful pard—
For he believes it has the heaves.
Because it breathes so hard.

I'd like to meet that cuss, by George,
Who sold that forge to me;
Make him disgorge. It was no forge—
It was a forgery.

—Modern Machinery.

New Intercolonial Shops.

As soon as the snow begins to disappear work on the Intercolonial Railway Co.'s shops at Moncton, N.B., will be pushed with all possible speed. The new buildings will be constructed of cement and steel. Between 600 and 700 men will be engaged at the work during the coming season.

Proposed Changes in the New Tariff.

HON. Mr. Fielding has given notice of about 120 changes in the new tariff schedule brought down in Parliament on November 29th last. No official explanation of the changes will be given until the resolutions are moved in the House. For the most part they are of a technical or relatively unimportant character, designed to meet the objections of manufacturers and others on minor points.

Changes in Iron Duties.

Some important changes are made in the duties in iron and steel. A new item is introduced comprising flat-eye bar blanks, not punched or drilled or rolled; edge plates of steel over twelve inches wide for use in steel structural work or car construction, the preferential tariff being fixed at $2 per ton, the intermediate at $2.75, and the general at $3.

On scrap iron and similar material the British preference has been reduced to 50 cents per ton. The duty on rolled iron or steel beams and other rolled shapes other than railway bars or rails has been fixed at $2 per ton under preferential, $2.75 intermediate, and $3 general. Boiler plate not more than thirty inches wide, wrought seamless tubes for boilers, wire riggings, and steel wire to be used in the manufacture of rope have been placed on the free list.

The changes, as they affect the hardware trade, are as follows:

Item 261 is amended to place spirits of turpentine on the free list instead of dutiable at 5 per cent.

Item 270 provides for a reduction in the general tariff on crude petroleum, gas oils other than naphtha, benzine and gasoline, the duty to be one and a half cents a gallon instead of two and a half cents.

Item 281 is changed to read:—"Fire brick of a class and kind not made in Canada, preferential free, others, 5 per cent. each." The present tariff rates are 5, 7½, and 10 per cent.

Item 316A is a new one. It reads:—"Incandescent lamp globes for use in the manufacture of incandescent lamps, and mantle stocking for gas-lights, 5, 7½ and 10 per cent."

Duties on Glass.

Item 326 adds to the dutiable articles of glass of all kinds, bottles, lamp chimneys, globes, etc., the words, "blown glass, table and other cut glass ware."

Item 344 change to read:—"Tinware, japanned or not, and all manufactures of tin, n.o.p., 15, 22½ and 25 per cent."

Item 355, which provides for the free importation of Britannia metal and German silver, strikes out the words "or bars" when it applies to the free importation of German silver in bars.

Item 274, on iron or steel scrap, wrought, being waste or refuse, including punchings, cuttings, or clippings of iron or steel plates or sheets having been in actual use; crop ends of tin plate, bars, or of blooms or of rails, the same not having been in actual use, the British preferential has been reduced from seventy to fifty cents.

Item 379 has been changed to read "rolled iron or steel beams, channels, angles and other rolled shapes of iron, steel not punched, drilled or further manufactured than rolled, weighing not less than thirty-five pounds per lineal yard, not being square, flat, oval or round shapes, and not being railway bars or rails, per ton preferential, $2; intermediate, $2.75; general, $3."

Item 379 is a new item, as follows: "Flat eye bar blanks, not punched or drilled, and universal mill or rolled plates of steel over twelve inches wide, for use exclusively in the manufacture of bridges or of steel structural work or of car construction, per ton, preferential, $2; intermediate, $2,75; general, $3."

Boiler Plate Free to Manufacturers.

Item 380 is changed to read as follows:—"Boiler plate of iron or steel, not less than thirty inches in width and not less than a quarter of an inch in thickness, for use exclusively in the manufacture of boilers, under regulations by the Minister of Customs, is made free in all the tariffs."

Item 381 has been changed to read:—"Rolled iron or steel plates, not less than thirty inches in width and not less than ¼ of an inch in thickness, n.o.p., per ton, preferential, $2; intermediate, $2.75; general, $3."

The item formerly was "Plates of 48 inches preferential, 5 per cent.; intermediate, 10 per cent., and general, 10 per cent."

On item 384 "strips, polished or not," have been added after "steel sheets." Preferential, free; intermediate, 5 per cent.; general, 7½ per cent. "Flat galvanized iron or steel sheets" have been struck out.

Item 386 has been made to read:—"Rolled iron or steel and cast steel, in bars, bands, hoops, scroll, strip, sheet or plate, of any size, thickness or width, galvanized or coated with any material or not, and steel blanks for the manufacture of milling cutters, when of greater value than 3½ cents per pound: preferential, free; intermediate and general, 5 per cent. each."

Item 387 is a new one, and reads:—"Steel in bars or sheets, to be used exclusively in the manufacture of shovels, when imported by manufacturers of shovels: preferential, 10 per cent.; intermediate, 12½ per cent.; general, 15 per cent."

Item 395—Wrought or seamless iron or steel tubes for boilers, n.o.p., under regulations prescribed by Minister of Customs, flues and corrugated tubes for marine boilers, are made free in all three tariffs. It was: Preferential, free; intermediate and general, 5 per cent. each.

Item 397—Tubes of rolled iron or steel not joined or welded, not more than 1½ inches in diameter, n.o.p., have been made free in all tariffs.

Increase on Coil Chain.

Item 410—Coil chain, coil chain links and chain shackles of iron or steel five-sixteenths of an inch in diameter and over; Preferential, 5 per cent. as at present; intermediate, increased from 5 per cent. to 7½ per cent., and general from 5 per cent. to 10 per cent.

Item 411—Malleable sprocket or link belting chain is made free in all the tariffs. It was 15 per cent., intermediate 17½, and general 20 per cent.

Item 451A is new. "Stoves, urns of metal and dove tails, chaplets and hinge tubes of tin for use in the manufacture of stoves, preferential 5 per cent., intermediate 7½ per cent., general 10 per cent."

Item 460—after the words "processes in" the words "iron or copper" are struck out and the word "metals" inserted instead.

Item 461A is new. "Iron or steel pipes, not butt or lap welded, and wire-bound wooden pipe not less than 30 inches internal diameter, when for use exclusively in alluvial gold-mining, preferential 5 per cent., intermediate 7½ per cent., general 10 per cent.

Item 462, which formerly included only "blast furnace slag trucks," is amended to include blowers of steel of a class or kind not made in Canada, for use in the smelting of ores or in the reduction, separation or refining of metals, rotary kilns, revolving roasters and furnaces, of metals of a class or kind not made in Canada, designed for roasting ore, mineral, rock or clay, blast furnace, slag trucks and slag pots of a kind not made in Canada, made free in all tariffs.

Item 524A is new. Seamless cotton or linen duck, in circular form of a class or kind not made in Canada, for use in the manufacture of hose pipe, made free in all tariffs.

Item 682—After the words Manila rope insert "not exceeding 1½ inches in circumference"; before the words "fishing nets" strike out the words "deep sea."

Drawback Increased to 99 Per Cent.

Schedule "B" covering goods subject to drawback for home consumption is amended so as to make the portion of duty (not including special or dumping duty) payable as drawback, 99 per cent. instead of 95 per cent.

Item 1,002 in this schedule is changed to read: "Malleable iron castings and pig iron," instead of "rolled iron, rolled steel and pig iron."

In item 1,009 all the words after augur bits are struck out, and bit braces is added to the othem items in which drawback is payable.

The following items are also added to schedule "B" as subject to payment of drawbacks:

1,014—Nickel, nickel silver and German silver in bars, rods, strips, sheets and plates, when used in the manufacture of spoons and cutlery, a drawback of 65 per cent.

Item 1,015—Rolled angles of iron or steel, nine and ten gauge, not over one and a half inches wide, and used in the manufacture of bedsteads, are subject to a drawback of 99 per cent.

Item 1,017—Lap-welded tubing of iron or steel, not less than two and a half inches diameter, threaded and coupled or not, testing 1,000 pounds pressure to the square inch, when used in oil or natural gas wells, and for transmission of natural gas under high pressure from gas wells to points of distribution, drawback 99 per cent.

Item 1,018—Machinery imported prior to July 1st, 1908, and other articles not machinery, when entering into the cost of tin plate manufactured in Canada, drawback 99 per cent.

Item 1,019—Bituminous coal, when imported by proprietors of smelting works and converted at the works into coke for the smelting of metals from ores, drawback 99 per cent.

Milling Machine Attachments.

THE UNIVERSAL SPIRAL HEAD.—By John Edgar

THE calculation of gears for this kind of work, whether on the lathe or milling machine, has always been a source of trouble for the average machinist. It is the intention here to give just such simple rules as can be easily remembered and referred to in case they are forgotten.

It has been explained in the former article that the worm gear on the spindle has forty teeth and that the worm is single threaded so that the ratio of worm shaft to spindle is 40 to 1, and it will require forty turns of the worm shaft to cause the spindle to make one complete revolution. The lead screw of the universal milling machine has four threads per inch, or it is one-quarter of an inch pitch, and will move the table one-quarter of an inch for every turn. If the worm shaft of the head and the screw of the machine are so geared together that they revolve at the same speed it will require forty turns of the screw to revolve the spindle one revolution. Turning the screw forty turns will move the table along lengthwise ten inches, and when geared so, the spiral that would be traced has a lead of 10 inches. This is known as the "lead of the machine." This lead of the machine is a constant quantity with all standard heads and milling machines and forms a basis on which to make our calculations for the gear ratio necessary to cut a given spiral. Thus, if we have a lead of eight inches to cut, we proceed as follows. The machine, when set with even gears, cuts a 10-inch lead, therefore the ratio of the driver to the driven must be as 10 is to 8 in order that the 8-inch lead may be obtained, or we may reason thus: In moving the table lengthwise eight inches, the screw must turn 8 × 4 = 32 turns, and in order to obtain one complete revolution of the head spindle the worm shaft must turn 40 revolutions, therefore the ratio of turns of worm shaft to those of the screw is as 40 is to 32. And a 40-tooth gear may be used on the screw and a 32-tooth gear on the worm shaft. If we have no such gears or only one of them in the set, we must multiply each term by the same number until we have a pair of numbers which correspond to numbers of teeth in two of the set of change gears. When two gears can be used as figured above they may be brought into connection by either one or two idle gears according to the "hand" of the spiral to be cut. This arrangement is known as simple gearing and can be used in low pitches which do not differ from 10 inches by a very large amount. The above may be placed in the form of a rule which will be much easier to apply and also to refer to.

Rule for finding gears for spiral cutting, (simple gearing).

$$\frac{\text{Lead of machine}}{\text{Lead to be cut}} = \frac{\text{Driving gear}}{\text{Driven gear.}}$$

Where the lead of the machine is 10 we have,

$$\frac{10}{\text{Lead to be cut}} = \frac{\text{Driving gear}}{\text{Driven gear.}}$$

Our last example expressed as above is,

$$\frac{10}{8} = \frac{40}{32}$$

Here we see that the first number has been multiplied by 4 to obtain the last, which is the process in finding any ratio required. Thus let it be required to cut a lead of 16 inches,

$$\frac{10}{16} = \frac{20}{32} = \frac{40}{64} = \frac{60}{96} \text{ or } \frac{30}{48}$$ here we have used as common multipliers 2, 4, 6 and 3, respectively, showing the different combinations that can be used. This could be carried on indefinitely.

When the difference between the required lead and that of the machine is great, the case becomes more complicated and compounding of the gears will be necessary to obtain the desired ratio. The rule for this case is,

$$\frac{\text{Lead of machine}}{\text{Lead to be cut}} = \frac{\text{Driving gears}}{\text{Driven gears.}}$$

Let it be required to cut a lead of 45 inches. Applying the latter rule we have,

$$\frac{10}{45} = \frac{\text{driving gears}}{\text{driven gears}} = \frac{2}{5} \times \frac{5}{9} = \frac{24}{60} \times \frac{30}{54}$$

Here we find that we cannot use simple gearing, because in order to have the small gear of a reasonable diameter the large gear would be of such large proportions that it would be impossible to have them mesh on the short centre distance of the two shafts. So we compound the gearing. The best method of going about the problem of choosing the necessary gears is to break up the fraction in the first number, in this case — 450 so as to obtain two fractions whose product will equal the original. Thus in the example, we have factored the numerator into its two even factors 2 and 5, and the denominator into its factors 5 and 9. By multiplying the terms of each fraction by a common multiplier we arrive at fractions which, while they are not altered in value, give numbers which equal the number of teeth in the gears that can be used. In the above example we have multiplied the terms of the first fraction $\frac{2}{5}$ by the common multipliers 12 and obtained 24 and 60 as the possible gears, and the second fraction by 6 and obtained for gears 30 and 54 teeth. The driver gears are always the "second gear on stud" and the "gear on the worm," and the driving gears the "gear on screw" and the "second gear on stud."

Where accuracy is not important and where the lead has no real significance, as in milling cutters, twist drills, etc., where the diameter is small and the pitch or lead long, an error of one or two per cent. is allowable.

In setting the milling machine table when about to mill spiral work it is necessary to know what the angle of the spiral is with the axis. This angle can be found by laying out a triangle, having as its short legs the circumference of the work and the lead of the spiral, and the angle the long leg makes with the leg representing the lead is the angle of the spiral with the axis. Such a triangle is shown in Fig. C, (a-c-b). In the figure is shown a cylinder upon which is wrapped this triangle showing the spiral path (a-e-c) followed by the cutter when cutting the lead which the machine is geared to cut. The angle of the spiral can also be found by trigonometry by the following rule:

Rule of finding angle of spiral with axis.

$$\frac{\text{circumference}}{\text{lead of spiral}} = \text{tangent of angle.}$$

The angle may be found from any table of trigonometrical functions.

Sometimes it is desired to find the lead of spiral corresponding to a given angle by changing the rule to accommodate this case we have,

$$\text{lead of spiral} = \frac{\text{circumference}}{\text{tangent of angle.}}$$

Or we may by constructing a triangle beginning at the point (a) in Fig. 6, laying out the two sides which meet at this point and diverging by the angle which is obtained by subtracting the angle of spiral with the axis from 90, making the line representing the circumference equal to that quantity; and at the point (b), and at right angles to the circumference line (ab), draw the line (bc) and you have the lead of spiral in this line whose length is (bc).

Tricks of the Trade

Everybody is invited to contribute to this page. Send in your ideas, odd jobs you are doing, and anything of interest to fellow-workmen. Remuneration will be made for such contributions. [Ed.]

HANDY INDICATOR RIG.

By F. C. D. W., Brantford.

HEREWITH is a drawing of a very handy indicator rig which I thought might be of interest to your readers; possibly solve the question for some one who, perhaps, has been looking for some such rig. One of its strongest claims for recognition is the fact that when taken apart it fills very little space, thus being convenient for the itinerant engineer. Another claim is the facility with which it can be "lined up" and adjusted to suit the stroke, etc., of the engine to be indicated.

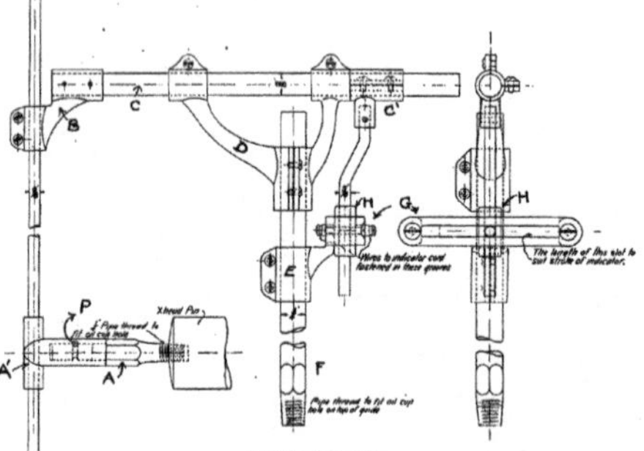

Handy Indicator Rig.

Mr. G. Pickles, foreman of the engine department of the Waterous Engine Works Co., Brantford, is responsible for the design, and to him I am indebted for the sketches from which I have made my drawing. The castings are all of brass and the rods are of cold rolled steel. The standard is threaded to fit the hole tapped for the grease cup, which is generally in the centre of the guide and the motion is taken from the crossheal pin in the manner shown.

A HANDY TOOL.

By James C. Moore, Toronto.

A VERY handy tool I have had in almost constant use while at small lathe work I made from a piece of chisel steel ⅜ octagon by ten inches long. I drilled a 7-16 hole in 6 inches on one side to hold very small boring and threading tools. On same side I made a square hole to receive ¼ tool stand ⅜ from end and held fast by same ¼ set screw as holds round steel, so it can be used as a larger boring tool. On the other end I have it angled off to receive ¼ tool pieces to go up to a corner, as can be seen by sketch. This octagon steel is preferred to round steel as it can be readily used in same position for the square bars and can be packed up to any desired height in tool post much easier.

A Handy Tool.

SHOP ETHICS.

Where the Dignity of Labor Comes in.

By C. J. Stuart, Montreal.

IN these modern days when we are accustomed to hear so much about "the dignity of labor," it should be frankly stated that there is no dignity whatever in being driven or rushed about the work in question. To be alert and a quick worker is all right, but no one can truthfully say that either the justifiable speeding up of a "shirker," or the equally undignified sight of a good man being hustled by an ignorant or selfish slave driver, is in any way edifying to behold.

Perhaps the fact is that this phrase has something of the Eastern figurative exaggeration, like "the camel that never tires." The laborer originally intended—a workman who glories in his work—is far removed from our commonplace idea of labor. Be it also understood that office men and "intellectuals" are within the terms of our text as much as the field reapers, the hammer-and-tongs men, and others raising a dust by the sweat of their brow.

The dignity of labor is the dignity of excellence—the dignity of a man who knows he is doing his honest best; or the dignity of a good man justly proud of good work. It is the good work that dignifies the laborer—not the pose; nor his little ticket of membership; nor the job.

UNUSUAL BORING OPERATION.

By G. R. Lang.

WE are sending you under separate cover a photograph of, we believe, an unusual operation for the boring mill. The out board bearing or pedestal shown was too large for any of our horizontal boring bars, which we had fitted with special heads, for boring out spherical bearings, so we rigged up the vertical mill as shown. The left hand head and ram were locked stationary, the right hand cross head was left free to move at will on the cross rail. The ram of this head, which did the cutting, was connected to the left hand stationary ram with a link as shown. The distance between the pivot holes in this link are the same as the radius of the bore. The rams were adjusted to height, so as to bring this link in a horizontal position when the cutting tool was half way through the bore. When the cut is started, the link, ram and tool are in the position as shown, the vertical feed is applied to the ram, and as it feeds downward it describes a circular motion. The tendency during the first half is to bring all the lost motion to the one side; after the tool passes the centre, however, the link is pulling, instead of pushing, the ram, which causes the slack in the parts to shift in the other direction. This causes a little irregularity in the outline of the bore.

To avoid this, on the finishing cuts the tool is run to the centre of the cut, the machine is stopped, the pin which holds the connecting link to the ram is removed, the tool run to the bottom of the cut, by hand, the pin again inserted and the cut run from the bottom up to

Unusual Boring Operation

the centre. With a little care almost a perfect sphere is produced. An 18-inch pin gauge started at the two diagonal corners was turned through with only a variation of .003 inches at any point. The bushing or bearing for this pedestal was also cored and turned on the mill, using the same connecting link, the tool, of course, being on the outside of the work instead of inside. A great many of these pedestals have been done in this way, the average time being about five hours each. This, however, was due to the good facilities of this shop for quickly and securely chucking their large work. As will be noticed in the illustration, turn-buckles are used for drivers, and T-bolt heads and studs are used instead of belts for clamping the work to the table.

We would mention in connection with the job just described that it is necessary to set the tool so that the cutting point is central with the boring mill table, measured in a horizontal line from the face of the cross rail; if it is not, the tool will cut a radius larger than that described by the moving pin.

GRINDING CRANK PIN.

By N. C. Johnson, Owen Sound.

When grinding a crank pin in an engine disc, by hand, throw the crank disc to position "A" and grind firmly for a few minutes until the emery has found a true bearing with the pin.

Throw the crank disc to position "B" and repeat as in "A."

Throw to "C" and "D" and do same.

This is an improvement on the old way of grinding a pin with the crank in one position, and it keeps the pin from wearing the lower side of the hole by excessive grinding with emery.

My experience in both cases, gives a better result by throwing the crank than by leaving it in one position, and also saves time, making a truer job.—N. C. JOHNSON.

BANQUET AND PRESENTATION.

On the occasion of his leaving to become manager of the western interests of the Canadian Westinghouse Co., as mentioned in last issue, Mr. Archibald W. Duff, of Montreal, was the recipient of an appreciation on Saturday, Feb. 2nd, from business associates and personal friends to the number of over sixty, in the form of a banquet held at the Canada Club. The chair was occupied by Mr. Edgar McDougall, president of the Allis-Chalmers-Bullock Co. Mr. Claud Hickson presented Mr. Duff with a gold watch, suitably inscribed, as a token from his Montreal friends. The present was brought in a large box, which, on being opened, was found to contain a very small-sized colored boy, who held the watch in his hand, and duly handed it to Mr. Duff, much to the amusement of those present.

WINS THE GZOWSKI MEDAL.

At the annual meeting of the Canadian Society of Civil Engineers recently held in Montreal, the Gzowski gold medal for 1906 was awarded to Mr. Walter J. Francis, C.E. (Tor. Univ.) M. Can. Soc. C.E., M. Am. Soc. E., of the Dominion Engineering and Construction Company, Limited, Montreal.

The Gzowski gold medal is awarded annually by the Canadian Society of Civil Engineers for the best paper read before the society during the preceding twelve months, and the medal for 1906 was won by Mr. Francis for his masterly paper on "Mechanical Lift Locks in Canada." The paper deals with the construction of the famous hydraulic lift locks on the Trent Canal at Peterborough, Ont., in the building of which Mr. Francis played a most prominent part, not only in the actual construction work, but also in the designing and planning of the scheme. The building of the hydraulic lift locks at Kirkfield, Ont., is also dealt with, and the paper is not only valuable from an engineering and scientific standpoint, but is also of great interest to the lay reader, giving as it does many interesting details concerning one of the most notable engineering feats accomplished in Canada in recent years, and showing the possi-

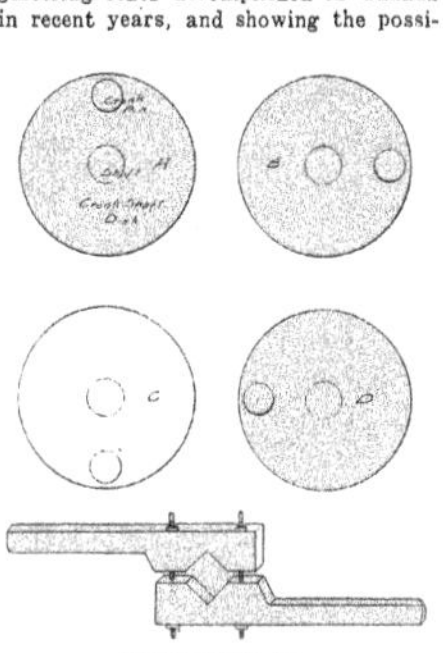

Grinding Crank Pin.

bilities of concrete for hydraulic and general construction work.

The Gzowski Medal Fund was established in 1892 by Colonel Sir Casimir S. Gzowski, when he retired from the presidency of the Canadian Society of Civil Engineers, and is controlled by the governing body of that society, the medal only being awarded in the event of papers of sufficient merit being submitted during the year, and the medal now being made for presentation to Mr. Francis is the fourteenth which has been awarded since the institution of the fund.

Mr. Francis, although a Torontonian by birth, is a Montrealer by adoption, and is now assistant manager and chief engineer for the Dominion Engineering and Construction Company of Montreal, and as such has sole charge of all their engineering and construction work.

Mechanical Reviews

OPINIONS AND STUDIES WORTH NOTING

SHOP TOOLS AND METHODS.

FIG. 1 shows a box tool, used in this case in the bench-lathe tail-stock. The bushing a, of which different sizes can be used in the same box, is made to fit any size wire required. Fig. 2 shows an end view of the cutting tool b in its relation to the bushing; it is held by a set screw on the flattened top.

Provision for Adjusting the Box Tool.

The tool is adjusted to the cut by the screw c, Fig. 1, which screw has a flange coming between the head of screw d and the body e, thus allowing it to act as a regular feed screw.

The stop or end-adjustment screw f can be held with screw g; or by slitting and expanding as shown at h. This latter scheme gives a fine tension.

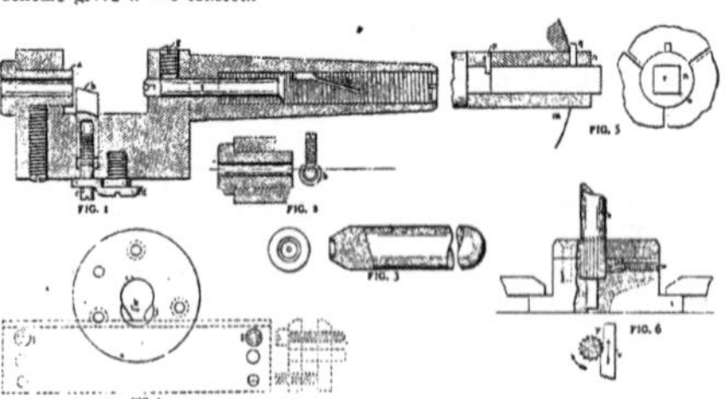

The body e is of mild steel about ¾x1¼ inches, centered and turned to shape and milled across for the opening.

Punch for Face-Plate Work.

I carry in stock a number of circular punches, like Fig. 3, mostly for transfer work on the drill press. Incidentally I use them for locating work on a face plate where it is required to remove stock at the side of a hole already made. A job of this kind is shown in Fig. 4, which represents a round die. I bore out and finish the round part i of the die, then to save the filing the irregular part clear through, I turn out the part of circle j at the back of the die. I locate the centre k when laying the die out, and, finding this with the pump centre of the bench lathe, bore out the back clearance j the first thing.

How the Punch is Used.

Using the parellel bars, shown by dotted lines, to transfer the irregular form of the die from the face to the back, I hold the punch, Fig. 2, against the work and mark around it, moving it to suit me, and then strike it with the hammer for an impression. Then I catch the punch in a chuck which fits the tail-stock spindle of the lathe, and use it to locate the work by the stamped mark until the clamps are secured on the face plate.

This is only an off-hand way to get an approximation to the true shape without plugging the hole i and finding a centre k for circle j.

Transferring Lines.

I sometimes get an opposite parallel line by holding the work in a small Vanderbeek or other drill vise, resting it on the surface plate and transferring from front to back with the surface gauge.

Again I put small pieces in a Billings & Spencer vise which has parallel jaws, line the work in front with the front jaw, and use the back jaw as a straight edge to get a line opposite the first on the back of the piece.

A Pair of Filing Bars.

The parallel bars shown by the dotted lines in Fig. 4 have been used for years on work, when I wished to file through a die quickly to line. They are of mild steel with beveled edges, which are tinted with blue stone and replaned or milled, as needed. The screws l clamp the work and the other screws place the bars parallel with it. while the dowel pins keep the edges of the bars parallel.

Adapter for Square Stock.

Fig. 5 illustrates an adapter for square stock to be held in a spring chuck m. A piece of round stock n is bored centrally and filed out square to right size and split at o for closing on the work. The pin p gauges the work in the adapter and the pin q gauges the latter in the chuck. In this way the work r can be readily duplicated as to length.

Milling Small Stock.

Fig. 6 shows a slabbing device for sizing square stock such as that held in the adapter in Fig. 5.

The milling cutter s may be held in a drill press and the guide block t fastened to the table. The stock is then pushed through the opening, as seen at u, and against the cutter as at v.

The screws w hold the gib x which gauges the thickness to which the stock is cut, and the face of the guide block in the rear of the cutter should correspond to the milled face of the work at y.

CAMS vs. ECCENTRICS FOR GAS-ENGINE VALVE GEAR — THE TWO-STROKE CYCLE CHAMPIONED.

FOR the sake of helping to attain perfection in gas-engine design, I am taking the liberty to give my opinion on a small portion of the said article. In Fig. 105 (reproduced herewith as Fig. 1), a drawing is given showing the Klein arrangement of valve-operating mechanism for the inlet valve. I will admit this is extremely ingenious in design, but I cannot agree with the writer that it forms a cheaper construction than the old cam arrangement. Unfortunately, an eccentric is not adapted to causing a quick motion at the end of its stroke without resorting to some sort of cam arrangement between it and the valve stem; then why not use the cam arrangement directly instead of with an eccentric? The half speed of the eccentric would seem to present difficulties in the way of properly adjusting and timing at each end of the eccentric stroke, and the cheaper construction is certainly that of the cam arrangement.

The accompanying sketch, Fig. 2, will serve to show the simplicity and cheap construction of a pure cam motion, and while there is one pivoted joint more on the cam design than on the eccentric shown, it is a cheap form of joint, and if a slide was substituted for the joints 3 and 4 the two joints would be done away with.

In addition to the joints mentioned there is the double-cam lever which the roller rides on, and which is more than equal to the cost of making a cam, as the latter is an extremely cheap part to manufacture, with tools arranged for the purpose. Then there is the addition of the eccentric to make, which represents about the excess of cost of the Klein arrangement over the old-style cam device. This is not intended as a criticism of design, but to show that users of the direct cam arrangement can offer a cheaper construction and one with as long life as the ecentric style for two-stroke-cycle engines.

In reference to the undecided advantages of the two- and four-stroke cycles over each other, there appears to be a growing appreciation of the large two-stroke-cycle engine, which type should have the support of engineers and designers more than it has had. The future will, no doubt, witness a great many two-cycle installations. It would seem also that an ideal design would be a double cylinder vertical with the lower end of the cylinders used for compressing the charge, instead of exploding the charge at both ends for the sake of accessibility and cheapness of construction, the only difficulty being in the successful cooling of the piston.—L. J. MONAHAN, in Power.

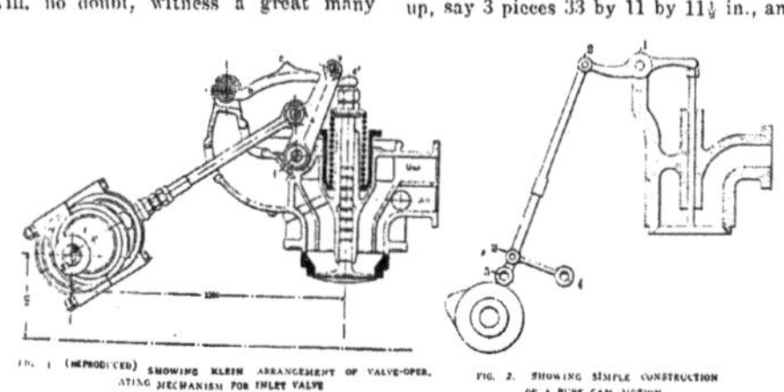

FIG. 1 (REPRODUCED) SHOWING KLEIN ARRANGEMENT OF VALVE-OPERATING MECHANISM FOR INLET VALVE

FIG. 2. SHOWING SIMPLE CONSTRUCTION OF A PURE CAM MOTION

MAKING A PATTERN FOR A SPUR-WHEEL BLANK.

THE following is a quick way of making a pattern for a spur-wheel blank or fly-wheel. We will take for example a wheel 44½ in. outside diameter, 40 in. inside diameter, 4½ in. face, six arms ⅞ in. thick, feathers and boss as shown at Fig. 1.

First get out three pieces for the arms, 45 in. long, planed to 7 in. in width and ⅞ in. in thickness, and lock them together as shown by dotted lines in Fig. 2. Set out on them the outside diameter width of arms and radii as Fig. 2. Take them apart to the bandsaw and pare away the surplus stuff of the arms at B, Fig. 2. Roughly bandsaw out the segments, leaving ¼ in. for turning off on the wood-turning lathe. Lay the arms on a true table; then trim, glue, and screw to the arms the first layer of the segments Fig. 1. and A Fig. 2.

Turn the pattern over, and trim, glue, and screw the intermediate segments. Strike on them the line D, and finish off by paring away the surplus stuff. It is now straightforward work to complete the building up of the rim, taking care to sink the outside screws to avoid turning off the heads in the lathe.

While the glue is drying, turn up the bosses and prepare the feathers or ribs, fitting them up to the bosses.

If there is not a faceplate (a wooden one) in the shop, we must knock up one—anything between 30 and 38 in. in diameter being a suitable size for this particular job, though of course it will come in for a hundred other jobs. Glue up, say 3 pieces 33 by 11 by 11½ in., and bandsaw to the largest diameter they will hold up to.

Glue and screw well an 11 by 1½ in. batten right across the middle of the back, and screw or bolt on the iron faceplate. Rather than turn up the face of plate right across the true, it is better practice to screw on a ring of segments 4 or 5 in. wide, as in Fig. 3. At any future time this narrow face can be re-trued up in a few minutes, and when worn out can be replaced by a new one, thus saving the body of plate from being constantly turned down and becoming useless.

Having faced up the ring on the faceplate, mark on it a circle of say, 30 in. diameter, and also mark on both sides of the arms of the pattern a corres-

PATTERN FOR A SPUR-WHEEL BLANK.—FIG. 1.

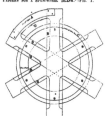

PATTERN FOR A SPUR-WHEEL BLANK.—FIG. 2.

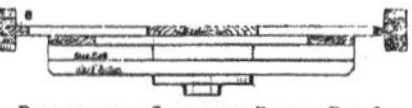

PATTERN FOR A SPUR-WHEEL BLANK.—FIG. 3.

ponding circle C, Fig. 2, and scribe up marks with a square on the edges of the arms.

Now screw the pattern by six stout screws with washers under the heads, through the arms, on to the faceplate, taking great care that the scriber marks on the arms exactly cover the 30in. circle on the faceplate. Having turned up the outside face and one side, turn the inside of the rim down to a sharp corner on the arm, allowing a full 3-16 in. for taper (B, Fig. 3). Take out the screws, reverse the pattern, and complete the turning of the other side.

Glue in a ½ in. leather fillet at B, Fig. 3, and at B, Fig. 1. This fillet is very handy to take out if at any time the inside diameter of the wheel is required to be made smaller, by laying in pieces all round and re-gluing in the leather; or, to be made larger, by paring out.

All that now remains to be done is to screw on the bosses, fit in the feathers and pare them out to A, Fig. 1. Slightly round the edges and brad in place.

It is as well to make an extra feather for the moulder to mend up with if needed, to save pulling one off the pattern. Screw on the print rather than peg it on, so that the boss can be turned down and used for another job. The Mechanical World.

Foundry Practice

MODERN IDEAS FOR THE TRADE.

HEAVY STAMPINGS TO REPLACE CASTINGS.

STEEL STAMPINGS have recently begun to displace iron, brass, malleable and steel castings for a number of uses. The accompanying illustrations, Figs, 1 and 2, show two large stampings of heavy material, and Fig. 3 shows the punch press designed for producing these stampings. The tool is built by the Toledo Machine & Tool Co., Toledo.

Fig. 1 is a stamping from ¼-inch steel plate. The blank required was 20 inches diameter. The stamping is formed and the centre cut and flanged in two operations. The outer flange is 2½ inches high and the inner flange 1 inch high.

FIG. 2.—STAMPING MADE ON TOLEDO PRESS.

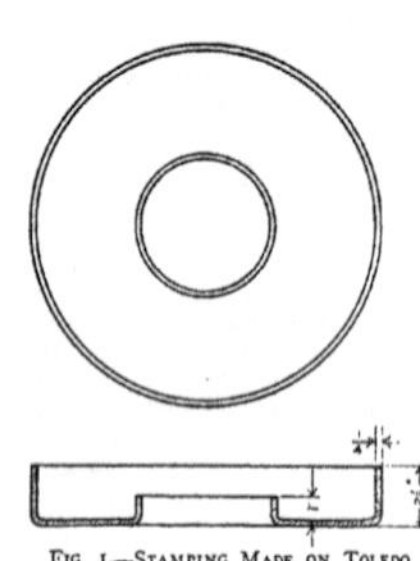

FIG. 1.—STAMPING MADE ON TOLEDO PRESS.

Fig. 2 was stamped from ½--inch plate, and required a blank 12 inches diameter. The centre opening was punched and flanged in three operations. the flange being two inches high, as shown.

The press which did this work was built for the Crosby Co., Buffalo, which makes a specialty of producing stampings of this character for a wide range of work. The press has a capacity of 12,000 tons pressure. It is of massive proportions, as may be judged from the size of the man standing alongside. The ram is operated by a single pitman driven through a train of spur gears and a jaw clutch. The frame is of cast iron, is made in one piece and weighs 42,800 pounds. The distance from the bed to the slide with adjustment up is 31 inches. The stroke is 14 inches, while the adjustment of the slide is six inches. The crank shaft at the bearing is 13½ inches in diameter. The gearing is massive and powerful, the main driving gear being 92 inches diameter by 14 inches face. Its weight is 9,000 pounds. The proportion of the gearing is 40.1. The press is provided with a fly wheel 60 inches in diameter and weighing 2,400 pounds. The distance from the floor to the top of the large gear is 14 feet 8 inches, and the bed is 37 inches square. The total weight of the machine is 100,000 pounds.

USES OF ALUMINUM.

THE United States produces more aluminum than any other country in the world, according to the metallurgical experts at Washington, and it is likely to be the first nation to employ the metal in its coinage. Next fall, it is reported, the Director of the Mint will experiment with aluminum coins as substitutes for the bronze one cent piece at present in use.

Only a few years ago considerations of expense would have forbidden any such action. In 1885 a pound of the metal cost $200. So late as 1889 it sold for $4.50 a pound. Now it is quite within the probabilities that in another year it will be profitably produced for 35 cents a pound.

Nature has been lavish enough in this instance, but man is slow to learn her secret. In its various compounds—and it enters into composition of a very large number of minerals—aluminum forms about one-twelfth of the crust of the earth.

Ordinarily it is extracted from the oxide alumina, which is the chief constituent of common clay. Until a few years ago, however, only expensive processes have served to separate the metal from the combinations of which it is found.

During the last decade American scientists and inventors have overcome many of the old obstacles. The first article ever made of the metal was in compliment to Louis Napoleon who had helped St. Clair Neville, the first manufacturer—a table for the baby prince imperial. At present it seems impossible to name a use to which the metal may not be put.

Aluminum is lighter than glass, and only one-fourth as heavy as silver, which it somewhat resembles. Though it is as hard as zinc, scarcely any metal is equally malleable and ductile. It can be drawn into the finest wire and hammered into the thinnest sheets.

If sold at a low price aluminum, which does not tarnish, will largely supplant silver for decorating purposes. It is an excellent conductor and might take the place of copper in electric plants. Since no metal save steel of the highest grade is, weight for weight, its equal in strength, it would compete with the former in almost every manufacturing industry.

France and Germany and the United States have built torpedo boats of aluminum, finding that its lightness insures speed, and that it resists corrosion and galvanic action better than any other metal. Yachts, rowboats, bicycles, motorcars, surgical instruments, skates and cooking utensils have already been made from it.

PERSONAL MENTION

Mr. H. A. McEwen of Carleton Place, has bought the machinery in the Chatsworth Foundry, and is negotiating for the purchase of the property.

Mr. J. W. Harris, head of the J. W. Harris Co., Montreal, was in New York recently. He leaves shortly for Winnipeg and other western points.

Mr. Ford Watson of the Sheldon Co., Galt, Ont., has been spending the past eight weeks in Sussex, installing a shaving exhaust system for the Sussex Mfg. Co.

Mr. K. J. C. Zinck, late of the Algoma Central Railway, has accepted a position in the railway supply department of the Canadian Fairbanks Co., at Montreal.

Mr. A. P. Horsman, manager of the Gurney Standard Metal Company, of Calgary, Alta., was in Toronto during the past month conferring with the head office officials.

Mr. A. Merner, formerly manager of the Waterloo Manufacturing Co., has purchased the Alliston foundry, which has been closed for some years. A new company is being formed.

Mr. Edgar G. Powell, of J. W. Harris Co., Montreal, was in Winnipeg at the beginning of the month on business for the firm. He called at the Winnipeg office of Canadian Machinery on Feb. 4.

Mr. F. Hammar, of the firm of Schuchardt & Schutte, New York, visited Montreal at the beginning of February and was introduced to the trade by Mr. J. J. Sophus, Montreal representative of the firm.

On leaving the employ of the Vulcan Iron Works, Winnipeg, to accept a position with Kelly Bros. & Mitchell, Mr. George Duross was presented by the Vulcan employes with an address and handsome Morris chair.

Mr. Wm. Hall, of the Hall Engineering Works, Montreal, has returned from a trip through the large manufacturing centres of Great Britain, and reports several new agencies, as well as great prosperity everywhere.

Mr. W. A. Kenning has resigned his position as sales manager of the Miller-Morse Hardware Co., Winnipeg, and has been succeeded by Mr. Louis J. Blackwood, who was formerly with Merrick, Anderson & Co.

Mr. Harry A. Norton, vice-president of the Norton Jack Co., of Boston, was a recent caller at the Montreal office of Canadian Machinery. Mr. Norton has just returned from a trip through India, Egypt, and the Holy Land.

Mr. Gordon McKeag, of Winnipeg, formerly of Fergus, has been appointed manager of the Anchor Wire Fence Company of that city. Mr. McKeag has represented this firm on the road for the past seven years, and is deserving of promotion.

Mr. A. W. McConnachie, one of the directors of the Chicago Pneumatic Tool Co., was a visitor at the office of Canadian Machinery recently. He reports enormously increasing trade both in Great Britain and Canada.

The Angola Furnace Co., of Angola, Ind., is moving to Decatur. The company has been reorganized and the new firm will be under the general management of Mr. James Doherty, , formerly secretary and treasurer of the Doherty Mfg. Co., Sarnia.

Messrs. H. G. Eadie and H. P. Douglas, of Montreal, have formed a partnership under the name of Eadie & Douglas, to carry on an extensive building and contracting business. Mr. Douglas was formerly with the Canadian White Co., and he and his partner will make a strong business team.

Mr. W. A. Hastings, of Hamilton, was in Montreal attending the meeting of heads of departments in the various branches of the Canadian Iron and Foundry Co. The 25 representatives present were afterwards banquetted by the president, Mr. Edgar McDougall, and vice-president, Mr. T. J. Drummond.

Mr. J. F. Macgregor, for eight years treasurer of the Galt Machine Works Benefit Society, recently resigned the position. He has been succeeded by Mr. George Grover. On February 8th Mr. Macgregor was presented with a bronze statute and an address by the members of the society in recognition of his services.

Sir Benjamin Baker, the eminent engineer of the Forth bridge; Sir Wm. White, past president of the Institution of Civil Engineers of Great Britain, and Mr. Octave Chanute, one of the foremost and one of the oldest civil engineers in the United States, have been elected honorary members of the Canadian Society of Civil Engineers.

Mr. W. H. Wiggs, of the Mechanics Supply Co., Quebec, was in Toronto last month attending the 35th annual convention of the Ontario and Quebec Young Men's Christian Associations. Mr. Wiggs is one of the most ardent workers in Y.M.C.A. work in Canada, and was honored by being elected president of the Ontario and Quebec Associations at the convention.

The annual meeting of the Williams Mfg. Co. was held at the office of the company, St. Henri (Montreal) on the 13th inst. Bartlett McLennan, the president, occupied the chair. The old board of directors was re-elected, and at a subsequent meeting, Bartlett McLennan was chosen as president and Chas. W. Davis, vice-president and managing director. Alex. Dube was appointed secretary.

* * *

Canadian Honored.

Dr. Howard T. Barnes, a graduate of McGill, has been chosen to succeed Dr. Rutherford as professor of physics in that university. It was announced at a special meeting of the board of governors at McGill recently that Dr. Howard T. Barnes, D. Sc. F.R.S.C., had received this appointment. He will succeed to a position held in recent years by two men whose services to science have been recognized the world over, Professor Callendar and Dr. Rutherford. Dr. Barnes has become one of the chief authorities on certain branches of physics. His researches extending over several years, on the specific heat of water, have become a classic, and after occupying the attention of the Royal Society of London at a special session, were made the basis of the report on this subject, to the conference of physicists at the last Paris Exhibition. His work on "Ice Formation and Frazil," on which he has recently published the first authoritative book, has important bearings on the utilization of water powers in northern countries, such as Canada and Russia; and his recording thermometers are coming into use in the regulation of furnaces and other branches of manufacture.

* * *

Montreal Agent Marries.

A wedding of considerable interest to the Montreal trade took place Thursday afternoon when J. Sophus, Canadian representative of Schuchardt & Schutte of New York and Berlin, was married to Miss Jennie Colquhoun of Manchester, England. The ceremony took place at St. Thomas' Church, and was performed by the Rev. Canon Renaud. While distinctly private, a large number of immediate friends attended, and many handsome gifts were received. The young couple left on a short tour to the Southern States and on their return will take up their residence in Montreal.

J. Sophus has been in Montreal a little more than a year and a half during which time he has made steady progress. Formerly he was superintendent of large works in Manchester, England, and before that again had been engaged as a traveler for machinery and metals in Europe, visiting Germany and Russia as well as other countries. Gifted as a linguist he has exceptional advantages, and as Canadian representative of Schuchardt & Schutte he deserves congratulations. This firm have branches in China, Japan, Russia, France, India, and other places, having also immense headquarters in New York City. They have confidence in the Canadian market and are spending large sums this year to promote their interests. Canadian Machinery supplements the many good wishes extended to Mr. Sophus on his new step.

Machinery Development

Metal Working **Special Apparatus** **Wood Working**

PONDEROUS STEAM SHOVEL.

STEAM shovels have become an important adjunct to the development of a country, particularly in connection with the building of railroads and other heavy work. The Canada Foundry Co. are now manufacturing in a special department the Bucyrus steam shovel. The illustration given herewith shows more graphically than words the speed at which these shovels excavate the earth. The picture was taken as a string of flat cars were backing down a track besides one of the monster shovels, which kept on working as the cars passed it, and on each car was placed a load from the dipper. The facilities of the Canada Foundry Company are such that they are able to construct this class of heavy machinery in large quantities and they have removed the reproach that this class of work all had to be imported because it was not manufactured in Canada.

These shovels are of a ponderous size, weighing 90 tons apiece. Their capacity is shown by the fact that three loads of the dipper will more than fill a flat car. The building of these shovels is a new industry as far as this country is concerned, an appreciation of which is shown by the fact that the Canada Foundry Co. have received contracts for these shovels from the Canadian Pacific Railway, the Canadian Northern Railway and several firms of railway contractors.

View Showing Bucyrus Steam Shovel at Work.

SPECIAL STEAM HAMMERS.

STEAM hammers of a special design are being placed on the market by the David Bell Engineering Works, an illustration of one being shown herewith. One of the main features of these hammers is that they are provided with reinforced guides which help to make up the main frame. This improvement consists in making the main casting at the point to which the lower end of the slides are bolted, fully fifty per cent. stronger and heavier than heretofore, and increasing greatly the strength by the addition of a vertical flange on the corss web tying the lower end of the slide faces together in the main casting. This greatly strengthens the point where the most strain comes. The stiffening of the column adds to the resistance to shearing at any point in the column.

The main working valve of the balanced piston type, works without friction in a removable bushing, which has steam port edges finished, thus giving uniform operation. Throttle valve is of the circular form, ground with emery in construction, to its seat and held tight by steam pressure. The 1,600 pound and larger sizes have quadrant connection to throttle valve, similar to that on main lever. Stuffing box or expansion joint connections are in cylinder casting, for steam and exhaust pipes. The valve motion is extremely simple, with few working parts, and giving the most accurate and sensitive control to the blow, not connected except by sliding contact with the hammer head. It is free from all shock or jar of blow, is therefore of the most durable and efficient construction, and requires no attention except proper lubrication. The hammer head is of hammered steel, with piston rod. The hammer head itself is set at the proper angle in the main frame casting, so as to permit drawing and finishing work either way of the dies, without the work interfering with the column. Head finished from the solid, having gang milled "V" grooves, which work in adjustable "V" slides, which are also gang milled, with special formed cutters. This gives perfect bearing to the working surfaces in contact. The slides are of heavy construction, through bolted to the main frame, with spring washers on bolts, and are fitted iron and iron against taper gib extending entire length, and arranged for independent adjustment. Piston rod fitted to taper hole in head, the jam of which taper constitutes the real hold, but with a safety pin to give warning in case of rod getting loose.

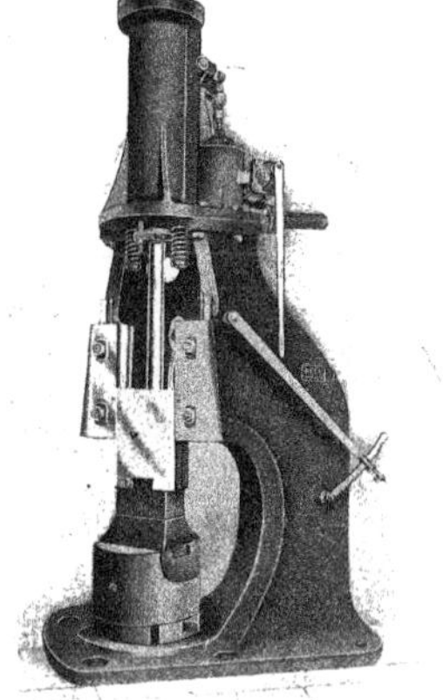
Bell Special Steam Hammer.

BLISS AUTOMATIC PRESS.

IN the accompanying half-tone is illustrated a small automatic press for making a six pointed, star shaped tobacco tag. The automatic mechanism embodies two distinct types of feeds, viz., a double roll feed and a lateral feed. The double roll feed is operated by means of a rod connection with a shaft at one end and with the feed crank at the other. Motion is given to the lateral feed by the movement of the slide in connection with the series of levers.

Two operations, which are necessary for making the tags, are performed at each stroke of the press after the first. The work is done as follows: Strips of stock of proper width and gauge are placed between the guide plates, the forward hand wheel is then turned so as to carry the strip well under the rolls. The press is now ready to do this work. A pressure on the foot treadle starts the press and causes the stock to be automatically advanced to its proper position by means of the first pair of rolls. The plunger carrying the blanking die and forming die will then descend, blanking out the tag, and the lateral feed will carry the cut blank under the forming die, where it is operated upon by the second stroke of the press as is customary in all regular double dies. The back roll feed is for carrying away the scrap and also for keeping proper tension on the stock. The capacity of this press is about sixty complete tags per minute. It weighs complete, as shown, about 700 lbs., and has been built by the E. W. Bliss Co., 20 Adams Street, Brooklyn, N.Y.

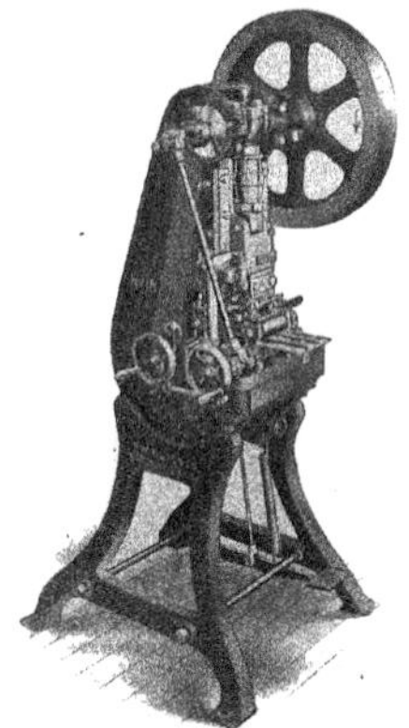
Bliss Automatic Press.

FRICTION PIPE WRENCH.

A SIMPLE friction wrench for manipulating polished brass and nickel-plated pipes, writes John Weldon, in Popular Mechanics, and shown in the accompanying sketch, is constructed of the handle, A, a piece of oak wood one inch thick, three inches wide, 14 inches long, with end of same rounded as shown at B; a piece of canvas, C, so folded as to make three thicknesses and the same width as handle and about 18 inches long, so that it will encircle all ordinary work. It is fastened about 3 inches from end of handle by means of two brass plates 3 inches square and five countersunk bolts as shown.

To use the wrench pass the canvas around the pipe or other cylindrical object and tuck the folded end between the canvas and wood, as shown in the sketch. Then, when the handle is pulled down, the canvas will tighten around the pipe and under all ordinary conditions produce sufficient friction to turn it.

The use of this wrench will not mar

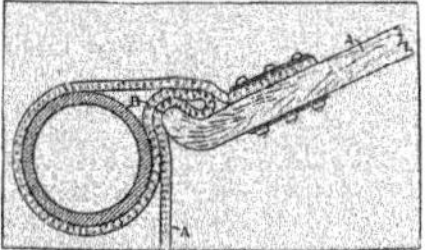
Home-Made Wrench.

the surface of the work, nor bend the pipe, because the compression is distributed evenly over nearly the entire circumference.

NIPPLE AND PIPE THREADING MACHINE.

THE machine shown in the accompanying half-tone has recently been placed on the market by the Murchey Machine & Tool Co., and the makers claim that it is the only double head machine of its kind that has capacity to cut 2 4-in. threads at one operation. It has been designed to meet the requirements of manufacturers engaged in the manufacturing of pipe nipples or long lengths of pipe, and will cut 4 times as many threads as the old style machine.

The die heads on this machine are steel and of new design. There are 6 chasers held rigid in slots by face ring, the head is in two parts, and opens automatically from the centre by action of the reamer coming in contact with the end of the pipe when the thread has reached its proper length, combining the process of reaming internal end of pipe with the automatic action of releasing the die, thus insuring perfectly reamed pipes and uniform lengths of threads, regardless of the position of the pipe in the vise.

The vise jaws have exceptionally long bearing surfaces which makes this machine a very successful machine for threading close nipples.

There is a separate reamer furnished for every size pipe within the range of the machine. The lead screw attachment is one of the most important improvements recently put on this machine. In operating, this dispenses with the usual method of starting threads by hand pressure; with the lead screw the operator simply puts the pipe in the vice grips and throws in the lead screw, no further attention is required, the operation is repeated in the other vise and when the first thread has reached its proper length the dies open automatically and the same action releases the lead screw. The combination is extremely simple as well as positive and effective. No special care is required in adjustments and every detail is so constructed that it is practically fool proof with safeties provided in case of unforseen accidents.

It has two three-step cone pulleys, diameters 12 in., 14 in. and 16 in., with of speed, 3 changes with the cones and 2 changes with the gears. The gear ra- 3½ in. belt surface. It has 6 changes tio is 25 to 1 for the heavy work. The motor shown on the machine is a 3½ horsepower motor made by the Triumph Electric Co. of Cincinnati. The makers of this improved nipple machine claim that it will easily produce 700 4 in. threads in 10 hours.

Nipple and Pipe Threading Machine.

HANNA SPECIAL LATTICE RIVETER.

THIS machine was designed for doing very close lattice work, and will drive ⅞-inch rivets in 10 inch channels 4 inches between the flanges. It has a reach of 15 inches and a gap of 12 inches, and the frame through the gap is 15 inches wide and 3¾ inches at its thickest point, tapering both ways. This will permit the riveter to swing from side to side in latticing in the very small space of 4 inches.

Special Lattice Riveter.

The working mechanism of this machine has the regular Hanna motion, which means that it will exert a predetermined uniform known pressure for each and every stroke throughout the last half of piston travel, thereby driving absolutely tight rivets. This feature is distinct in the Hanna riveter alone.

The total output of coal in Nova Scotia for 1906 amounted in round numbers to 5,170,000 tons, an increase of 488,000 over 1905. The Cape Breton mines gave an increase of 352,000 tons, Pictou 88,500. Cumberland decrease, 34,506, Inverness 80,000,

Power and Transmission

Steam Gas Electricity Compressed Air Water

THE CLARKE FRICTION CLUTCH.

WM. & J. G. GREEY, 2 Church St., Toronto, have recently undertaken the manufacture of the Clarke friction clutch, which embodies distinctive features. The clutch is simple in design, and having few parts is not liable to get out of order. It is exceedingly strong and rigid and is claimed to have a larger frictional surface than any other clutch of its size. In it the oiling facilities have been well attended to, and the design is such that the few wearing parts may be quickly adjusted to take up all slackness.

The entire pressure is exerted uniformly over the whole surface, the pressure being not greater on the outside edge than on the inside. Being compact it occupies a very small space on the shaft. It can be regulated to shift when a given load is exceeded. This point is invaluable to some users when an overload endangers breaking valuable machinery. When the clutch is engaged it is self contained and requires no pins or strings to hold it at work. It is manufactured in sizes from 3 to 125 h.p.

THE GASOLINE ENGINE.

How to Treat Worn Valves, Cams and Cam Rollers and Gears.

A BAD running engine often indicates a worn condition of the valves, cams or the gears that actuate them. In very many cases of reported trouble we have found such a condition. Either the valve, cam, or the roller which travels over it, or both, are found badly worn on their surface, or the roller may become worn in its bearing where it operates on its spindle, or the spindle itself becomes worn. Another not infrequent thing is to find that the roller, on account of dirt, gum or being drawn up too snugly, gets stuck on its spindle and ceases to revolve as it should, and the result is that a flat place is worn on the roller at the one point only in its circumference which comes in contact with the cam. A cam roller is intended to roll as it passes over the cam, thereby reducing the friction on either to a minimum, and unless it does roll, a flat place will be the result. It is quite important, therefore, to take frequent notice of the cam and cam rollers while the engine is in operation as well as while it is at rest. Sometimes the cam roller may appear all right at rest, but soon after the engine is put to work it may get stuck, and the resulting friction very quickly damages it to such an extent that it cannot fill its office effectively. You see that when a roller gets stuck on its spindle it has to slide over the high part of the cam, which does the lifting of the valve, instead of rolling over as is intended. The sliding movement is also a grinding one, which quickly results in damage. The effect of a worn cam or cam roller can be seen when we carefully consider what they are naturally required to do. It is of course known that the cam is intended to open the valve as well as to allow it to close. The valve must open and close at the proper time, or in harmony with the piston movement. It must also in opening lift high enough from its seat to give the valve port a free clearance. We will assume for instance that a valve-cam becomes worn right at the point where it comes in contact with the roller to begin opening the valve. The least bit of wear at this point causes the valve to open a little later and the more the wear the greater the delay in opening the valve, consequently the valve is prevented from operating harmoniously and at the proper time with the piston movements, and its functions and usefulness are impaired just to the extent of the wear on the cam. If the cam becomes worn on its entire surface it not only opens the valve too late, but allows it to close too early and does not lift it high enough from its seat. A flat place on a cam roller, even if the cam itself is not worn, will have the same effect, exactly.

Let us consider what effect this delayed and reduced valve movement will have on the running of the engine. The exhaust valve, for instance, has for its function the relief of the cylinder from its internal pressure by letting out into the open air the burnt gases after they have pretty well spent their force against the piston on its outward and working stroke. And inasmuch as there yet remains quite a pressure at the end of the working stroke the valve should open before the end of the piston's working stroke, at about four-fifths of the stroke, in order to let out and reduce, as near as possible, the pressure within the cylinder to atmospheric pressure by the end of the working stroke. If the valve should not open until the piston started on its exhaust movement, the pressure within could not be relieved fast enough for the piston travel, and it would be compelled to

The Clarke Friction Clutch Coupling.

crowd against a pressure on its entire exhaust stroke. It would take quite a liberal percentage of the power that the previous impulse from the gas explosion had stored in the rims of the flywheel to overcome this pressure against the piston during its exhaust stroke, and when it should have a free movement without practically any opposition. Therefore, in order to save instead of wasting this power, the exhaust valve should be opened when the piston has completed about four-fifths of its working stroke. This will allow a good share of the remaining pressure to escape through the exhaust valve while the piston is completing its working stroke, and there will be no perceptible resistance to its exhaust movement, because the exhaust valve remainsure and burnt gases, so that by the time the piston has reached the end of the exhaust stroke the space within the cylinder behind the piston has reached the point of atmospheric pressure. The exhaust valve then should open before the end of the working stroke of the piston and remain open during the entire exhaust stroke, about 1 1-5 piston movement in all. Any wear on the cam or cam roller that gives the valve less advantage that this is a direct hindrance to power output from the engine. The valve must open in time so that the pressure within the cylinder has dropped nearly if not quite to atmospheric pressure by the time the piston begins its exhaust stroke, so that it may have a free and unobstructed movement on this idle stroke.—American Blacksmith.

(To Be Continued.)

VIBRATING STUFFING BOX.

THIS stuffing box automatically adjusts itself to any out of line movement of the piston rod or stem. It will be observed that the stuffing box is arranged within a casing and is held against the ground ball joint ring by means of springs, assisted by the steam pressure from the cylinder, keeping the joints tight, and preventing leakage. A clearance is provided between the stuffing box and the interior of the casing, and between the rear surface of the stuffing box and the cylinder head. This permits the stuffing box to move laterally relative to the casing, to compensate for out of line movement of the rod or stem, and to rock on the curved face, or ball joint, to adapt itself to any angular movement of the rod.

The spacing rings at each end of the packing work in connection with the packing, holding the stuffing box out of contact with the moving parts, thus preventing wear.

On Dec. 1st, 1906, one of these stuffing boxes was placed on the main piston rod of a 100 h.p. engine in the Butterfield Power Bldg., Detroit. When the stuffing box was applied to this engine the piston rod was placed out of line so as to have an angular, as well as a lateral movement of more than 1-16 in., and it has been found that the stuffing box "floats with the rod" without any resistance, performing perfect work.

In this instance the stuffing box is packed with plastic metallic packing, although it will be seen that any suitable packing can be used.

Since the stuffing box was placed in operation it has been inspected by many mechanical experts who have expressed much interest in its operation, and all have pronounced it a most practical and economic device.

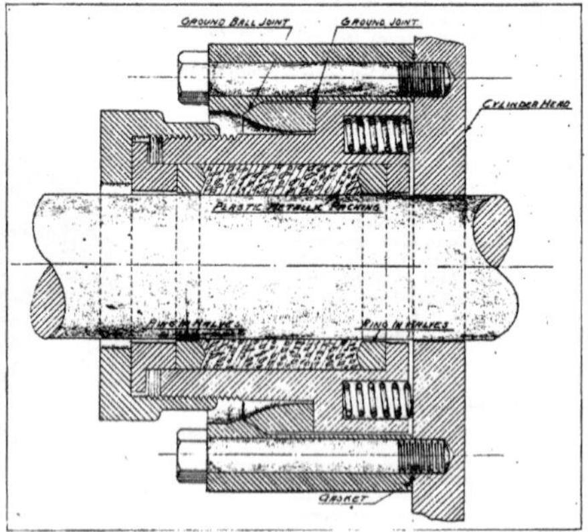

Vibrating Stuffing Box.

The Steel Mill Packing Company of Detroit, Mich., are equipping their factory with machinery for manufacturing these vibrating stuffing boxes and will soon be prepared to place them on the market.

STEAM ECONOMY IN ROLLING MILLS.

A RATEAU exhaust steam accumulator and turbine has been installed at the Hallside Works of the Steel Company of Scotland to use the exhaust steam from one high pressure cogging machine with two cylinders, each 40 x 60 in., one finishing train engine with two cylinders, each 42 x 60 in., two small mill engines and four steam hammers, delivering a total of about 41,000 lb. of steam per hour. The power is used for lighting and to drive rolls, saws, sand blast apparatus and machine tools. The accumulator, which has two compartments, separates from the water any oil carried over, and maintains the water in energetic circulation. It has a diameter of 11 ft. 6 in. and a length of 34 in. The turbine is rated at 700 h.p. and has 11 sets of blades, with wheels 40 in. in diameter. It develops its full power at a speed of 1,500 revolutions per minute, and when working at an overload of 10 per cent. the inlet pressure is never over 12 lb. absolute. The vacuum is 28 in. The generator was designed to deliver 2,000 amperes, and has carbon brushes and a commutator provided with special ventilating ducts through the centre. The ventilation is set up by means of a fan blade attached to each bar. Tests of the unit have shown a consumption of 66.4 lb. of steam per kilowatt hour at one-tenth load, and with an absolute inlet pressure under 3 lb. per square inch. At 200 k.w. load and an absolute admission pressure of 5.35 lb., the consumption was 42 lb. per unit. A half load and an admission pressure of 8.25 lb. absolute, the steam used was 37 lb. per unit. With a load of 450 k.w. and an admission pressure of 11.4 lb. absolute, the figure was 36.6 lb. The vacuum fell during these tests from 28.7 in. to 27.9 in., as the load increased.

Companies Incorporated

A company has been formed in Kingston for the manufacture of brick. It will be called the Perfect Brick and Tile Company, having a capital of $50,000.

The Dominion Power & Transmission Co., capitalized at $25,000,000, has been incorporated to develop water and other powers for the generation of electric, steam, pneumatic, and hydraulic power, with head office at Toronto.

The British Columbia Car Co., capitalized at $1,000,000, to manufacture freight cars, street cars, etc. Directors, A. C. Flumerfelt, J. G. Woods, F. Buscombe, Alfred Kelly and M. M. Campbell. Head office, Vancouver, B.C.

Messrs. Ney, Camp and Company have been incorporated with a capital of $40,000 for the purpose of manufacturing furniture, etc. The head office will be at Stratford, and provisional directors being, W. J. Ney, N. W. Camp and J. H. Bamber, Stratford.

The Gamer Manufacturing Company has been incorporated with a share capital of $50,000 for the purpose of manufacturing bedsteads, etc. The head office will be at Chesley, the directors being: Arthur Gamer, Ada Gamer, A. Harrod, T. E. Devitt and F. Gillon of Weston, Ont.

The Collingwood Shipping Co. has been incorporated with a capital of $90,000, for the purpose of constructing elevators, wharves, docks, warehouses, etc. The head office will be at Collingwood, and provisional directors being, W. T. Allan, M. Brophy and W. A. Hogg, of Collingwood.

The Standard Concrete Construction Company, Toronto, has been incorporated with a capital of $100,000, for the purpose of general building and construction. The head office will be at Toronto, and provisional directors being, F. Rielly, J. B. Bartram and E. A. Scott, Toronto.

M. McKenzie & Company have been incorporated with a capital of $75,000 for the purpose of manufacturing railway, marine and contractors' supplies. The head office will be at Montreal, the directors being, W. D. Hamilton, A. A. Lunan, A. Dunn, L. Lahaye and A. Lunan of Montreal.

The Wingold Stove Company has been incorporated with a capital stock of $40,000, for the purpose of manufacturing stoves, ranges, furnaces, etc. The head office will be at Winnipeg, the directors being: F. B. Blanchard, O. Gensmer, R. Kellow, L. Blanchard and I. A. Mackay, of Winnipeg.

Love Brothers, has been incoruorated with a share capital of $150,000 for the purpose of manufacturing building materials and supplies. The head office will be at Toronto, the provisional directors being: P. Love, H. W. Love, E. G. Long, F. Orford, W. J. Coulter, and E. J. Barton of Toronto.

The A. Workman and Company has been incorporated with a share capital of $60,000 for the purpose of manufacturing and dealing in hardware. The head office will be at Ottawa, the directors being T. Workman, A. Workman, and A. A. Whillians, of Ottawa.

The Brandon and Robertson Manufacturing Company has been incorporated with a capital of $75,000, for the purpose of manufacturing implements, machinery, etc., the head office to be at Brandon, Man., and provisional directors being, W. Brandon, J. A. Robertson and R. J. Brandon, of Brandon, Man.

The Eastern Construction Company has been incorporated with a share capital of $1,000,000, for the purpose of manufacturing machinery for construction purposes. The head office will be at Ottawa, the directors being: J. Gillespie, H. H. Short, F. H. Honeywell, S. B. Johnston and J. B. Prendergast of Ottawa.

Trussed Concrete Steel Company of Canada, has been incorporated with a share capital of $200,000, to manufacture materials used in connection with concrete reinforcement. The head office will be at Walkerville, the directors being G. Kahn, D. C. Raymond and L. Wiman, of Toronto.

The M. McKenzie Company has been incorporated with a capital stock of $75,000 for the purpose of manufacturing mill, marine, railway and contractors' supplies. The head office will be in Montreal, the directors being: U. D. Hamilton, A. A. Lunan, A. Dunn and L. Lahaye of Montreal, and Alex. Lunan of Huntingdon, Que.

The Montreal Wood Mosaic Flooring Company has been incorporated, with a capital stock of $5,000, for the purpose of manufacturing wood flooring, window and door screens, steel mats, etc. The head office will be at Montreal, the directors being, A. McLean, Buffalo, and D. H. McLennan, C. Stewart and R. W. Barclay, of Montreal.

The Crescent Machine Company has been incorporated with a share capital of $20,000 for the purpose of manufacturing machinery, implements, rolling stock and hardware. The head office will be at Montreal, the directors being: C. M. Gardiner, C. D. Drabble, W. A. Patterson. H. S. Williams and W. Bovey, of Montreal.

The Dominion Smelters, Limited, has been incorporated with a share capital of $1,000,000 for the purpose of carrying on business as miners and refiners. The head office will be in Sault Ste. Marie, Ont., the provisional directors being H. H. Muggley, W. S. Peters, and T. J. MacCune, of Oshkosh, Wisconsin.

The Georgian Bay Oil Company has been incorporated with a share capital of $1,000,000 for the purpose of carrying on business as gas producers and petroleum oil refiners. The head office will be in Fort Erie, the provisional directors being S. Johnston, F. B. Johnston, F. R. MacKelcan, A. J. Thomson, and R. H. Parmenter, of Toronto.

The North Shore Transportation and Wreckage Company, Quebec, has been incorporated, with a capital stock of $250,000, for the purpose of manufacturing vessels, steamboats, machinery, etc., and to construct wharves, piers, warehouses, etc. The head office will be in Quebec, the directors being, J. A. Fafard, O. C. Bernier and A. Tegnon, of Quebec.

The Theodore Lefebvre and Company has been incorporated with a capital stock of $80,000, for the purpose of manufacturing all kinds of paints, oils, chemicals and drugs. The head office will be at Montreal, the directors being C. A. M. Lefebvre, L. M. T. Lefebvre, R. B. Lefebvre, and L. S. Lefebvre, of Montreal.

The Consolidated Bicycle and Motor Company, Winnipeg, Man., has been incorporated, with a capital of $60,000, for the purpose of manufacturing bicycles, motor cars, boats, pumps, etc. The head office will be at Winnipeg, provisional directors being C. T. Cruikshank, J. A. Hudson and J. S. St. Mars, all of Winnipeg.

The Foreign Rail Joint Company has been incorporated with a share capital of $50,000 for the purpose of manufacturing rails, rail joints, angle bars, and all kinds of railway supplies and railway contractors' supplies. The head office will be at Toronto, the provisional directors being J. S. Lovell, W. Bain, R. Gowans, and E. W. McNeill.

The Canada Arms and Rifle Sights Company has been incorporated with a share capital of $200,000 for the purpose of manufacturing guns, rifles, and other small arms; also to acquire the business of the Mitchell Rifle Sight Company. The head office will be at Toronto, the provisional directors being H. Dixon, T. Cocking, and R. Staite, of Toronto.

Cameron & Company has been incorporated with a share capital of $100,000 for the purpose of manufacturing furniture, doors, sash, pulp, turpentine, wood alcohol, etc. The head office will be at Ottawa, the provisional directors being: U. A. Cameron, R. G. Cameron, G. C. Edwards, G. Cameron of Ottawa, J. H. Gates of Burlington, and Hugh McLean of Buffalo, N.Y.

"Warden King, Limited," has been incorporated with a capital stock of $1,000,000 for the purpose of manufacturing all kinds of heating apparatus, soil pipes, steam fittings, builders', plumbers', and steamfitters' supplies. The head office will be in Montreal, the directors being J. C. King, L. A. Payette, R. C. McMichael, of Montreal, and W. Grieg and F. G. Bush, of Westmount, Que.

The Canadian Rand Company has been incorporated with a capital stock of $500,000, for the purpose of taking over the Canadian Rand Drill Company and continue in the manufacture of air compressors, rock drills, pumps, pneumatic tools, etc. The head office will be at Sherbrooke, Que., the directors being, G. Doubleday, New York; E. Webber, of Montreal; S. W. Jenckes, W. Fanwell and H. D. Lawrence, of Sherbrooke, Que.

Industrial Progress

CANADIAN MCHINERY AND MANUFACTURING NEWS would be pleased to receive from any authoritative source industrial news of any sort, the formation or incorporation of companies, establishment or enlargement of mills, factories or foundries, railway or mining news, etc All such correspondence will be treated as confidential when desired

An opera house will be erected at New Liskeard, Ont.

Universal Systems, Toronto, have decided to erect a plant in Montreal.

A new hotel will be erected at Kenora, Ont., at the cost of $100,000.

A new Y.M.C.A. building will be erected in Montreal at the cost of $300,000

A market building will be erected at Berlin, Ont., at a cost of about $20,000.

A new Baptist church will be built on Pape Avenue, Toronto, at a cost of $4,500.

Plans are being prepared for the erection of a new hotel at Kingston to cost $120,000.

An up-to-date hotel building is going to be erected at Woodstock at the cost of $60,000.

The Vancouver Portland Cement Company, operating at Tod Inlet, B.C., are enlarging their plant.

The Mickle-Dyment planing factory at Brantford will have a $10,000 extension made in the Spring.

The Peterboro Steel Rolling Mills Company, Peterboro, are considering the erection of a new plant.

The Metallic Roofing Company of Toronto, will enlarge their factory on King and Dufferin streets.

A new hotel will be built on the corner of York and Front streets, Toronto, by G. Percival, Toronto.

The C.P.R. will erect car-shops at London, Ont., and will probably employ about five hundred hands.

An English syndicate will absorb many large steel plants in B.C., and also take over great coal and iron areas.

Nathaniel Dyment, proprietor of the Dyment Foundry Company, Barrie, died at his home in Barrie this week.

The Dominion Radiator Company has taken out a permit for the erection of a plant to cost $160,000, in Toronto.

A grant of $50,000 has been made for the erection of a hygienic institute at London by the Provincial Government.

The Ottawa Steel Casting Company have applied for an increase of $100,000 to their capital stock, making it $350,-000.

The Corundum Wheel Company, of Hamilton, have taken out a permit for an addition to their factory to cost $1,000.

The Northwest Brass Company, Winnipeg, Man., are erecting a new steel building, 100x80 feet, at a cost of about $10,000.

Tenders are asked for the erection of a large factory building on Carlaw Ave., Toronto, for the Philips Manufacturing Company.

James Woods, of the Woods Company, manufacturers of lumberman's supplies, Ottawa, will erect a $200,000 factory at Winnipeg.

The Hamilton Incubator Company have secured a permit for the erection of a brick factory in Hamilton, at the cost of $6,000.

Plans have been completed for a handsome new three-storey building to be erected by the Oddfellows of Strathcona, Alberta.

The Dominion Carriage Company, Truro, N.S., recently organized, with a capital of $250,000, will erect new building in the Spring.

Kurze and McLean Company have procured the lease of a factory in Stratford, and will go into the manufacture of acetylene gas plants.

Rhodes, Curry & Co., of Amherst, N.S., have been awarded the contract for the new round-house at Halifax The price is about $80,000.

The North Amercian Refining and Smelting Company will likely erect their smelter in Thorold, instead of St. Catharines, as at first intended.

A new theatre will be erected on Richmond street, Toronto, at a cost of about $75,000. The promoters include Rust and Weber of New York city.

The Meisel shops at Port Arthur will be erected almost entirely of glass. Preparations are being made for rushing the construction work in the spring.

John E. Wilson, of St. John, N.B., will erect a large addition of 40x100 feet to his foundry, and will go extensively into the manufacture of stoves.

Negotiations are being carried on for the establishment of a match factory in Lindsay. J. D. Mantion, formerly of Hull, Que., is promoter of the industry.

The Grand Trunk Pacific Company has called for tenders at Victoria, B.C., for the erection of a new hotel at Prince Rupert, costing in the neighborhood of $50,000.

The Lake Superior Corporation steel rail mill made a new high record in the month of January, turning out 19,285 tons, the previous high record being 16,800 tons.

The statement for the foreign trade of Canada for the six months ending with December 31, shows an increase in exports of $12,890,000, and in imports of $30,124,000.

The National Car Company, Halifax, N.S., recently incorporated with a capital of $1,000,000 intend erecting a plant for the construction of from 15 to 25 cars per day.

The David Spencer Company will erect an eight-storey building in Vancouver, B.C., to be used for manufacturing purposes. The cost estimated at about $200,000.

The National Car Company, Toronto, have accepted the site of 5 acres offered by the town of Whitby, for their car works. Work will be begun on the new shops at once.

A new Y.M.C.A. building will be erected in Ottawa, containing 97 dormitories. It will be of steel construction and fireproof throughout, costing in the neighborhood of $200,000.

Plans are now ready for the new dam and bridge at Buckhorn, on the Trent Canal. Tenders will be called in a short time, and it is estimated that the work will cost about $35,000.

The Massey-Harris Company, of Brantford, are making extensive additions to their plant by building a 100-foot addition to their foundry, and also enlarging their blacksmith shop.

The county of Wentworth has decided to build a house of refuge. The building will cost between $35,000 and $40,000, and it is proposed to have a farm of 125 acres in connection.

The Ottawa Car Company will extend their plant and branch out into the manufacture of steam railway rolling stock construction. The new shops will cover several acres of land.

The contract for the erection of the Peterborough armouries has been let. The successful tenderer is Geo. H. Proctor, Sarnia, and the figure of the contract signed is $125,190.

An employee of the Canada Foundry, Toronto, was working on a crane doing some riveting on Wednesday, when he fell twenty feet, fracturing his skull. He is expected to recover.

The Toronto Electric Light Company will apply to the Ontario Government for an increase in its capital stock of $1,000,000, which will bring the total capitalization to $4,000,000.

The Iron and Brass Manufacturing Company, of Weston, have now located in Chesley. A new building 300 feet long has been erected, and the company will employ 75 hands.

A sawmill with a capacity of 40,000 feet a day will be erected by the B.C. Mills, Timber & Trading Co. in Burnaby Lake, B.C. The contract is let to Mitchell & Ferris, of Vancouver.

The railway construction works and rolling mills, which have been proposed for Sydney, C. B., will be erected in the Spring, as sufficient capital for the financing of the scheme is available.

The production of copper in the United States has increased from 27,000 tons in 1880 to 436,000 in 1906, and the United States now furnishes over 57 per cent. of the world's supply.

Wm. Patton, contractor, London, has assigned.

E. Roy & Co., contractors, Montreal, have dissolved.

Grimard & Bontin have registered in Montreal as contractors.

Musson's, Ltd., of Montreal, have opened a new branch office in Quebec.

A $20,000 foundry will shortly be established in Calgary by Minnesota capital.

The assets of La Forderie Generale, Varennes, Que., are to be sold on March 7th.

A. Monaghan's machine shop at Kinkora, P.E.I., was destroyed by fire on Feb. 13.

Extensions are to be made to the works of the Niagara Falls Machine & Foundry Co.

The W. P. McNeil Co.'s foundry and bridge works at New Glasgow were burned on Feb. 5th.

The machine shop at the Halifax dry dock, with all its contents, was destroyed by fire on Jan. 29th.

The Toronto Board of Control is applying to the Legislature for power to expropriate the electric light plant.

The Ottawa Terminal Company will seek incorporation with the obj?ct of building a central railway depot.

The Olds gasoline engine works at Sherbrooke, was damaged by fire. The loss was covered by insurance.

The Powassan Lumber Company are installing a Canadian turbine in one of their mills at Powassan, Ontario.

The Schaake Machine Works, New Westminster, have commenced work on their new pattern shops and office.

The foundry of the Maritime Engineering Co., Moncton, N.B., was almost totally destroyed by fire on Feb. 12.

The Ontario Power Co., will enlarge its plant and at once install two more generators at a cost of half a million dollars.

The number of directors of the Northern Electric and Manufacturing Co., Montreal, has been increased from seven to nine.

$25,000,000 is the amount involved in those portions of the Grand Trunk Pacific for which the contracts are now being let.

The foundry of the B.C. Marine Railway, at Esquimalt, was destroyed by fire on Feb. 2. The loss was between $4,000 and $5,000.

Fire caused $5,000 damage to the foundry and moulding shop of C. Wilson & Sons, Limited, scale manufacturers, Toronto, on Feb. 2nd.

The Simonds Canada Saw Company, Montreal, have removed their office and factory in Toronto from 265 King St. west to 105 Adelaide St. east.

The Monerch Brass Works, Port Colborne, is erecting a new foundry building. Augustine & Son, of the same town, are enlarging their planing mill.

A bridge building plant is to be established at Sault Ste. Marie, Ont., under the name of the Algoma Bridge Co. A. W. Hillier will be manager, and 100 hands will be employed at the start.

The Dominion Government has been asked to aid a projected railway from Sudbury to Cobalt, to open up the splendid mining section of that district.

St. John is to have a new factory for the manufacture of a security fastener. The machinery is already being installed and a number of hands will be employed.

Mr. E. D. Cleghorn intends to establish a factory in Orillia for the manufacture of screws, bolts, etc., to be used principally in the construction of automobiles.

The factory of the Boston & Lockport Block Co., manufacturers of marine pumps and blocks, 100 Condor St. E., Boston, was partially destroyed by fire on Feb. 1st.

A sand lime brick plant is to be established in Brantford by John Mann & Sons and the American Clay Machinery Co. The capacity will be 30,000 brick per day.

Wm. Baird, Woodstock, has moved his machine shop from Victoria street to Dundas street. A new building has been erected, and additional machinery will shortly be installed.

Ottawa proposes purchasing the works and rights of the Metropolitan Electric Co., at Britannia, where it is claimed 10,000 h.p. can be developed for an additional expenditure of $250,000.

Permits have been issued in Toronto for the erection of a grand stand and horticultural buildings at the Exhibition grounds. The former will cost $216,465 and the horticultural building $90,000.

The production of lead in Canada for the year 1906 amounts to 26,000 tons. The production for 1905 was 27,000 tons. The record year was 1900, when a total of 30,000 tons was reached.

The Canada Cycle and Motor Company, of Toronto Junction, has purchased property in Ottawa, and will erect showrooms, and repair shop, expecting to be open for business in the Spring.

The rail mill of the Lake Superior Corporation has been turning out 700 to 750 tons a day. The rail output this month is expected to run 18,000 tons against 16,700 tons the previous high record.

A large bank building will be erected at Victoria, B.C., by the Royal Bank of Canada. The Bank of Nova Scotia have also purchased property on King street, Toronto, and will erect a bank building.

The Link-Belt Co., Philadelphia, has acquired a new office location at 84 State street, Boston, from which the future business of its drive chain department in New England will be looked after.

The Dominion Steel Company's Sydney rail mill in January broke the record for the greatest monthly production of steel rods, Their mill turned out 7,960 tons, the former record being 7,000 tons.

A $110,000 hotel will be erected by E. J. Fader, of the Torpedo Touring and Tug Company, at New Westminster. The new building will be known as the Hotel Russel and is to be completed by October, 1907.

Another rich discovery of iron ore has been made near Port Arthur, which has proven to be the most extensive and highest grade ore found in that section. Samples of the red-ore will assay 60 per cent. iron.

The Canadian-Mexican Pacific Steamship service will carry samples of general merchandise free of charge to the various ports touched by them in order to show what Canada and Mexico have to offer each other.

A new grain elevator with a capacity of about 10,000,000 bushels will be erected at Port Arthur for the Grand Trunk Pacific Railway. The contract has been awarded to Barnett and McQueen of Port Arthur.

A new company to manufacture cement in Montreal proposes to turn out 2,500 barrels a day. This will be one of the largest cement works on the continent. The contract for the buildings has already been let.

The Petrolea Association of Stationary Engineers was formed at a meeting held on Feb. 13, and the following officers were elected: Hon. Pres., J. E. Armstrong; Pres., H. A. Day; Vice-Pres., F. Stonehouse.

The Jackson Co., of Goderich, commenced operations in their new quarters in Goderich about the middle of February, beginning with a staff of 20 hands. The company's specialty will be the manufacture of boys' clothing.

"Electro Metals, Limited," is the name of a New York concern which purposes to build and operate an electric ore-reducing plant in the Niagara Peninsula. English capital from New Castle, Eng., is largely interested.

The Canadian Pacific Railway Co. will build a new depot at Calgary, Alta. It is estimated that the structure will cost $200,000, and outside of Winnipeg will be the finest on their western lines, and one of the finest on the whole system.

When the Galt Malleable Iron Works opened in October last 25 men were employed. Last pay-day at the works 132 regularly employed workmen received their pay envelopes, an increase of over 400 per cent. in the number of employes.

The plant of the Chapman Double Ball Bearing Co., Toronto, has been moved from Pearl street to Sorauren avenue. The new premises and equipment are up-to-date and complete in every particular. The office address of the company will still be Scott street.

The Canadian Pacific Railway will make large and important additions to their shops at Winnipeg, involving the expenditure of half a million dollars, as soon as the weather will permit of commencing operations. Nine buildings are included in the plans.

The plant of the Canadian Glass Works, located on the outskirts of Montreal, and employing about 150 hands, was burned to the ground on Wednesday. The loss is in the neighborhood of $20,000 on the plant and equipment and is covered by insurance.

The Imperial Steel and Barb Wire Co., of Collingwood, is looking for a suitable point in the west to establish a plant for the manufacture of barb wire and nails. The company is looking for inducements, guaranteeing to employ 100 to 200 men all the year round.

At the annual meeting of the International Portland Cement Co., held in Ottawa this week, the shareholders approved the recommendation of the directors to double the capacity of the plant at Hull. At present it is turning out 2,000 barrels of cement per day.

A number of final year students in electrical engineering at McGill University visited the works of the Canadian Westinghouse Co., in Hamilton, on Feb. 18. A tour of the works was made and luncheon was served by the company. The party also visited Decew Falls.

The Canadian branch of the American Locomotive Works at Montreal, is largely extending its structural steel plant, and has so many orders booked for this kind of steel that even the enlarged plant will be kept running to its fullest capacity throughout the year.

A large, four-storey building in Hamilton will be erected in the Spring, and will be sub-let to small manufacturers. This proposition should meet with great success, as small manufacturers are frequently handicapped by being unable to get suitable buildings in starting out.

The main building of the new plant which Jenkins Bros., New York, valve manufacturers, are erecting in Montreal, will be 200x50 feet. There will also be a foundry 150x64 feet, and an engine and boiler room 60x45 feet. Later, the company expects to build a foundry 100x60 feet.

The Stratton Babbit & Car Bearing Co., of Moncton., N.B., decided at their annual meeting to accept a proposition from the city of Montreal to remove its factory there. The following directors were elected: Messrs. Simpson, E. M. Jones, James Flannagan, J. R. Stratton and R. A. Borden.

The Dominion Bearing Company, of St. Catharines, are manufacturing automobile equipment, engines, running gear and bearings for supply to a selling concern. The company will shortly place on the market an improved roller bearing. About eighteen men are now employed in the works.

W. H. Banfield & Sons, of Toronto, are supplying the Massey-Harris Co. with four 1,000-pound drop hammers. This order was secured in competition and is a compliment to the ability of this firm to undertake such work on a large scale. Each of these hammers will weight 20,000 pounds.

The total amount paid to date in iron and steel bounties in Canada is $8,814,833, of which $998,000 went to the Soo Company; $1,416,469 to the Hamilton companies; $151,095 to the Deseronto company; $1,369,158 to the Nova Scotia Steel Co., at Sydney, and $3,466,519 to the Dominion Steel Co.

The King Edward Hotel, Toronto, is to be enlarged this spring, when a two-storey addition will be added. This will provide 200 more rooms, 100 on each flat. A lot of new machinery will be installed for lighting and other purposes, and for power to run a number of up-to-date labor-saving machines.

The Nova Scotia Steel and Coal Company will carry out their original intention of erecting another furnace. Already orders have been received which far exceed the capacity of the plant, and in order to keep up with the increasing demand the company will have to enlarge their works by another furnace.

One of the largest sawmills on the Pacific coast is to erected shortly at Vancouver. The contract for the machinery of the new plant has been let to the Prescott Co., of Menominee, and the contract is let to E. H. Heaps & Co. The new plant will have a capacity to handle logs ten feet in diameter and 110 feet long.

A $10,000,000 corporation is in process of formation with a view to entering into competition with the Pittsburg Reduction Co., the heaviest producer of aluminum in the United States. It is reported that the enterprise will erect a plant along the Cumberland River, in Kentucky, at a cost approximating $3,000,000.

The Ham & Nott Company, of Brantford, are about to build a large factory in Ottawa. The new factory will manufacture bee-keeping supplies, spring beds, and screen doors, and will employ over one hundred hands while in full operation. Arrangements are now being made with regard to a suitable site and other features.

A shipbuilding company is to be formed at Kingston, with a site on the west side of the dry dock. The company will likely have a capital of $100,000. Already the companies existing here are full of business, and there is easily a field for increased and larger operations. The new company would be handy for repairing vessels that enter the dock.

An agreement has been signed with the town council of Welland by E. Billings, of the Spencer Company, of Hartford, Conn., the largest manufacturers of drop forgings in the United States, to locate there. The branch will be known as the Canadian Billings and Spencer Company, and plans for a plant to cost about $100,000 will be prepared at once.

The Sanderson-Harold Company, of Paris, manufacturers of refrigerators, screen doors, etc., contemplate building a large addition to their factory, making it double the capacity of the present one. The move is compulsory in order to keep pace with the growth of business. 70 hands are now employed and the factory has been running day and night for months.

The following directors were elected at the annual meeting of the Shawinigan Water & Power Co., held in Montreal on Jan. 28: President, Senator Mackay; vice-president, Mr. J. E. Aldred; directors, Messrs. John Joyce, W. R. Warren, H. H. Melville, Thos. McDougall, Denis Murphy, William Mackenzie, J. N. Greenshields, and Howard Murray, secretary.

The Montreal Steel Works made a record showing in the fiscal year ending Dec. 31. The financial statement submitted at the annual meeting shows that the management has placed the company in a very strong position. Net earnings for the year amounted to $196,097, and after the payment of dividends on preferred stook there was earned over 20 per cent. on the common stock.

The Huber Mfg. Co., of Marion, Ohio, is establishing a branch for the manufacture of agricultural implements at Portage la Prairie, Manitoba. To accomplish this an amalgamation has taken place between the company and the Portage Iron & Machine Co. When final arrangements have been made this company will have one of the most complete machine establishments in the west.

The Dominion Graphite Company, of New York, will erect a large mill, representing an outlay of nearly a million dollars for the treatment of graphite, to be mined in the Laurentian ranges near Buckingham, Que. Not long ago the Dry Concentrating Company, of New York, announced their intention of erecting a smelter with the same object in view, and both plants are now in course of construction.

A pair of 42 in. horizontal Canadian turbines have just been installed in the plant of the Collins Inlet Lumber Company to drive their saw mills and box factory. These turbines will work under a head of 29 feet, developing 450 h.p. They were furnished by Chas. Barber & Sons, Meaford, Ontario, patentees and sole makers of these turbines. A pair of 36-inch horizontal Canadian turbines are also being installed for the Delhi Light & Power Co.

Montreal is to have a big new cotton mill. W. T. Whitehead, formerly of the Dominion Textile Company, and for many years identified with the cotton interests, has been successful in forming a company with a capital of $1,500,000, under the title of the Mount Royal Spinning Company. All the capital has been subscribed. The new mill, which will be situated either at Maisonneuve or on the Lachine Canal, will, it is expected, be in operation in eighteen months.

The Hamilton Radial Railway will make application to Parliament during the present session for an amendment to its charter, permitting of an extension from Mimico to Toronto, the purchase of the Brantford & Hamilton Electric Railway, Hamilton, Grimsby & Beamsville Railway, or others entering Hamilton, and the construction of a line from Brantford to the Detroit river, an international bridge over the Niagara, and a ferry system on the Detroit river are also included.

Ham and Nott, manufacturers of refrigerators and screen doors, Brantford, will likely build another plant, double the size of present one. Owing to being unable to make satisfactory arrangements with Ottawa it is probable they will not go there. The company at present employ 125 hands and are working day and night. In the event of the management maintaining their head plant at Brantford, the number of employees will be doubled.

A delegation composed chiefly of members of the legal profession waited on the Hon. Frank Cochrane, Minister of Lands, Forests and Mines, on behalf of their clients, to disapprove of a plan to impose a tax per acre on mining lands. He intimated that at least part of the revenue derived from taxation of the mines of the province would be applied to encourage the smelting and re-

RED STAR BOILER COMPOUND
BOILER COMPOUND
It's the RED STAR Quality you want.
It's the best.
It is sold under guarantee.
You can try it.
It is 25 Years Old, not an experiment.
BOILER COMPOUND FEEDERS
Have you read our booklets about feeding Compound?
Write for a booklet.
They are full of information.
They tell how and why we loan them.
They are free, but worth the asking.
Lake Erie Boiler Compound Co.
BUFFALO, N.Y.
Canadian Branch: TORONTO
TRADERS' BANK BUILDING
BRANCHES
New York
171 Broadway
Philadelphia
4253 Leidy Ave.
New Orleans, La.
518 Camp St.
Rochester, N.Y.
St. Louis, Mo.
Boston
153 Milk Street
Pittsburg
331 Fourth Ave.
Newark, N.J.
15 Fillmore St.
Louisville, Ky.
Kansas City, Mo.
Chicago
65 Wabash Ave.
Cincinnati
70-71 Mitchell Bldg.
Kalamazoo, Mich.
210 S. Pitcher St.
Marshalltown, Ia.
Roanoke, Va.

fining of ores in Ontario, as it would increase in the value of mining properties, besides allowing their owners to work deposits of ore of lower grade than was profitable at present.

The McClary Manufacturing Co., after investigation and comparison of cost, have decided to install a producer gas plant for manufacturing purposes in their factory at London, instead of making use of Niagara power.

Enormous Industry.

Commissioner of Industries Thompson, of Toronto, is in touch with the principals of the Paris Industrial Co., the Compagnie Des Metaux Unitas, of Paris, France. This firm is being represented in Canada at the present by Mr. Alfred Yeates, who hopes to establish a large branch of their company in this country. It will mean the employment of between 1,000 and 3,000 hands, for the manufacture of mining tools of all kinds. This firm holds the rights of a patent process of steel for which great things are claimed. The Paris works of this company employ 7,000 hands, and a branch in Alsace 3,000. Every effort is being made to procure this industry for Toronto, and Commissioner Thompson hopes it will definitely locate. Brantford is also trying to prove its desirability as a site for the proposed works.

Increasing Rapidly.

Since commencing operations the Galt malleable iron works has almost daily increased its pay roll, which now amounts to 140 men. The directors have already decided to enlarge the works and plans have been drawn for an extension. It is contemplated thus to double the capacity of the plant and by August it is expected there will be 300 men at work in the shops.

Disastrous Fire.

A serious conflagration took place at the works of Good, Shapley & Muir, of Brantford, entailing a loss of about $35,000. The three-storey western section of the plant, including the repair shop, the wood department on the ground floor, the paint shop and wood shop, and the tin and bee supplies department, are practically destroyed. The machine shop was slightly scorched, but did not receive serious injury. The origin of the fire was a heated journal on a small engine in the repair department on the ground floor. The portion of the works not destroyed by fire will continue running, although about 35 men will be thrown out of employment for some time.

To Manufacture Cars.

The British Columbia Car Co. has been incorporated in Vancouver, with a capital of $1,000,000, to engage in the manufacture of freight and street cars and other kindred work. It is recognized that there is a splendid opening in Vancouver for the establishment of a car building industry. This has been shown particularly during the past season, where a shortage of cars has existed on all lines. As a large part of the lumber entering into the construction of these cars comes from British Columbia the company should be in a position to manufacture with greater economy.

New Agencies Secured.

The London Machine Tool Co., 1206 Traders' Bank building, Toronto, have recently secured the sole agency for Canada for Potter & Johnston, Pawtucket, R.I., manufacturers of automatic chucking lathes and universal shapers; and also for the Murphy Machine & Tool Co., of Detroit, manufacturers of tapping machines, automatic nipple machines, and automatic collapsing taps and dies. These new articles are in keeping with the numerous other lines for which the London Machine Tool Co. are agents, and their acquisition speaks well for the progressiveness of the firm.

LEATHER BELTING IN CANADA.

THE J. C. McLaren Belting Co., of Montreal, are the pioneer manufacturers of leather belting in Canada, having been established by the late J. C. McLaren in the year 1856, 51 years ago. Later on the manufacture of card clothing and reeds was taken up, and having always had a regard for the quality and workmanship of their products, the building up of a very large business was the natural connecting result. Other important features which have helped to place the firm in the high position it now occupies are: The courteous treatment of customers, and the absolute guarantee of satisfaction that goes with goods sent out of the factory.

F. A. JOHNSON,
Director and General Manager, J. C. McLaren Belting Co.

After the death of the founder of the firm, the business was successfully carried on by his son, D. W. McLaren, until his death a year ago last August, when the management came under the hands of F. A. Johnson, who for some years previously had been connected with the company. Mr. Johnson is an energetic business man as well as a competent manager, and dating from the day that he took over the "big chair" the business has forged steadily ahead.

The firm now employs a large number of travelling salesmen, who cover the country from coast to coast, besides having agencies or branches, where complete stocks are carried, in twelve Canadian cities, not counting the head office, at 292 St. James St., Montreal.

Agencies and Branches.

St. John, N.B., Canadian Oil Co.; Hamilton, Alexander Hardware Co.; Ottawa, W. F. Colston Co.; Toronto (Branch), 50 Colborne St.; London, Ont., A. Westman; Winnipeg (Branch), Telfer Building; Chatham, Ont., Park Bros.; Calgary, Great West Saddlery Co.; Edmonton, Great West Saddlery Co.; Regina, H. W. Laird & Co.; Prince Albert, A. W. Woodman; Vancouver, Canadian Fairbanks Co.

General Manager.

The active head of this firm is Mr. F. A. Johnson, who, as general manager, is developing the business at a rapidly increasing rate.

In the early eighties he entered the lumber business, and for seven years followed this line, becoming thoroughly acquainted with the business in all its branches, from the time the logs were cut in the woods until they were sawn into lumber. The next few years he engaged in the mining and milling of ores in California, when he returned to Canada, having accepted the position of assistant manager of the English Portland Cement Co., at Marlbank, Ont. Serving a number of the same interests his field was transferred to Montreal in connection with the mining and milling of asbestos for the Standard Asbestos Co., of which company he became a director and secretary-treasurer

Some years ago, while remaining a director of the Standard Asbestos Co., he became actively associated with The J. C. McLaren Belting Co., of Montreal, the oldest firm in the leather belting business in Canada, succeeding to the position of director and general manager on the death of Mr. D. W. McLaren. The policy of producing the highest class of goods, and fair dealing, which has characterized this business from its inception, and to which it owes its continued success, is being faithfully carried out. This, coupled with Mr. Johnson's long experience in the using, as well as the manufacture, of belting, should bespeak for the firm a very large share of the general prosperity Canada is enjoying to-day.

J. R. MILFORD
Superintendent Card Clothing Department

WM. LINTON
Traveller Central Ontario

W. S. BROCK
Manager Winnipeg Branch

JOHN EDWARDS
Factory Superintendent

A. DOUGLAS McARTHUR
Manager Toronto Branch

WM. S. CAMPBELL
Accountant

STANLEY J. GONIN
Traveller Montreal and vicinity

J. C. FAIR
Traveller Maritime Provinces.

A. G. HEWTON
Traveller Eastern and Western Ontario

Sales Staff of J. C. McLaren Belting Company, Montreal

Brilliant Banquet.

The employes' banquet of the Waterous Engine Works Co., Brantford, held on the evening of Feb. 8th, was a brilliant success. This has come to be an annual institution in the lives of those connected with this company, and from what has been accomplished in this direction has evidently come to stay. The committee in charge were: J. Whiting, A. Styles, C. Hay, George Baird, W. Strie, J. Fyle, D. McLean, M. Blacker, T. R. Logan and G. Pickles. The hall was handsomely decorated for the occasion, a feature of the decorations being a steel model of a fire ladder truck intertwined by ferns and carnations. At the head tables were seated W. G. Raymond, chairman of the evening; Mrs. C. H. Waterous, Judge Hardy, Mrs. Logan Waterous, Mr. C. A. Waterous, Mrs. D. J. Waterous, Mr. Julius Waterous, Mrs. Judge Hardy, Mr. C. H. Waterous, president of the company, Mrs. Julius Waterous, Mr. D. J. Waterous, vice-president, Mrs. C. A. Waterous, Mr. Logan Waterous and Mrs. A. T. Waterous.

As toast-master and chairman of the evening Mr. W. G. Raymond was the man in the place. After the toast to the King had been heartily received, a toast to our guests was responded to by Mr. Chas. Waterous and Mr. D. J. Waterous. Canada was ably upheld by His Honor Judge Hardy. The ladies were not forgotten, a toast being ably responded to by Mr. W. Giddens. In the absence of Mr. Cockshutt, "Constitutional Liberty" was responded to by G. Pickles; after which a hearty vote of thanks was tendered to all responsible for the success of the evening. A pleasant incident of the affair was the presentation by Miss Farrell, on behalf of the employes, of a beautiful vase full of scarlet roses to Mrs. C. H. Waterous.

Montreal Busy.

Montreal will have a busy 1907. During the year the Simplex Car Works, along the Lachine Canal, are to be greatly extended, so as to give double the ordinary capacity. The Canada Car Company have already more orders than they can attend to, and may also have to extend. In the same neighborhood are the Allis-Bullock-Chalmers machinery works, the Wire Works, and the Radiator factory, while the new cotton company, which is to build a factory at Lachine in the Spring or Summer, is materializing. In the east end, the Locomotive Works at Longue Pointe are to be extended at a cost of $1,000,000. In the neighborhood of the Angus shops a new town has grown up, and this will be added to during the Spring and Summer, for the C.P.R. cannot overtake its own orders at these shops, which are the largest and best equipped on the continent.

The Tallest Building.

Contracts have been awarded for the erection in New York City of what will be the tallest building in the world. This is the proposed new Singer building on Broadway, north of Liberty Street. It will require for the work 6,000 tons of structural steel. The building will have thirty-six regular floors, on top of which will be a dome including four floors, the fortieth floor being 550 feet above the curb. This is a part of the transformation of a city into an abode of cliff-dwellers. The building will be 625 feet high, the tallest skyscraper in the city, and will have wind anchors so that it may be firmly braced against every gale. The wind pressure, on account of the structure's great altitude, will be tremendous, and for that reason the building is to be literally tied to its foundations by an ingenious arrangement of steel rods. They will be three and a half inches in diameter, and descend for nearly fifty feet into the concrete which forms the caissons resting on solid rock eighty-five feet below the curb. The lowest rod has on the end of it a great anchor plate to which it is secured.

Dundas Weddings.

A happy event occurred on Feb. 6th at the home of Mr. Henry Bertram, of Dundas, senior member of the firm of John Bertram & Sons. The occasion was the marriage of his eldest daughter, Miss Jean Bertram to Dr. Alex. F. Pirie of Costa Rica. The ceremony was performed by Rev. S. H. Gray on the evening of an ideal Winter day, in the large drawing-room of the bride's home, being witnessed by one hundred guests. After a short stay in Dundas the newly married couple will set out for their future home in Costa Rica.

On Wednesday evening, Feb. 20th, an equally pleasing event took place, when Mr. James Bennett Bertram, youngest son of the late Mr. John Bertram, was married to Mary, elder daughter of the Hon. Thomas Bain. The Rev. S. H. Gray, of Knox Church, again officiated. The bridesmaid was Miss Helen Bain, the bride's sister, and Dr. W. T. Wilson acted as groomsman. Mr. and Mrs. Bertram left late in the evening for New York and Philadelphia, and on their return will occupy their house on Sydenham Street, Dundas.

Canadian Machinery joins in the good wishes of both the newly-married couples' many friends.

MACHINERY FOR EXPORT

On Admirality and War Office Lists.

IN STOCK. Ready for immediate delivery.
For Sale, on Hire, or on Purchase Hire.

MACHINERY

Reliable, Second-hand and New, including Portable and Semi-Portable Engines, also Locomotive Engines, Engines and Boilers combined, also Horizontal, Vertical, Traction, Winding, Gas, and Oil Engines. Road Rollers, Plenty of Cornish, Lancashire, Loco-Type, Vertical and other Boilers, great variety. Saw Benches. Stone Breakers. Pulsometers and other Pumps. Steam Winches. Cranes and other Lifting Machinery.

Inquiries and Orders receive careful and prompt attention.

Send for copy of "Phillips' Monthly Machinery Register," the best medium in the world for the purchase and sale of New and Second-hand Machinery of every description.

Note the Only Addresses:

CHARLES D. PHILLIPS,

Emlyn Engineering Works, NEWPORT, Mon., and Emlyn Works, GLOUCESTER.

Established 39 Years

PATENT RIGHTS FOR SALE

A U. S. firm manufacturing Graphoil Lubricators and Steam Cylinder Graphite Cups would like to correspond with reliable party with a view to introducing their goods on a commission basis in Canada, or would negotiate for the sale, in whole or in part, of Canadian patent rights. Address in first instance Box 51L, CANADIAN MACHINERY, Toronto.

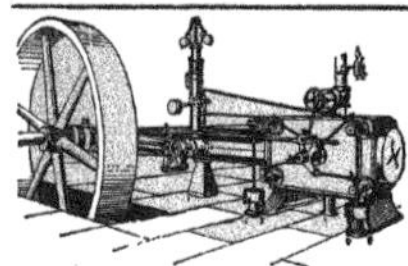

Leonard Corliss Engines

Peerless Self-oiling and Leonard Automatics for Electrical and Manufacturing purposes. Standard Return and Water Tube Boilers, all sizes. Open and Closed Exhaust Steam Heaters, Pumps, Etc., for Complete Steam Plants.

E. Leonard & Sons, London, Ont.

Figures talk strong and convincingly in favor of the Webster Heaters. There they stand in indisputable array. We hold the proofs ready to mail to you. Satisfy yourself regarding the accuracy of all the claims we make for

THE WEBSTER

FEED-WATER HEATER
PURIFIER AND FILTER

The type shown in the cut gives the greatest heating power and highest temperature without waste. It utilizes every drop of partially heated water, thereby adding greatly to its economy. It cannot waste steam. It does save fuel. And—the *general effectiveness* of this type is admitted by our competitors.

Let us mail you particulars.

DARLING BROTHERS, LIMITED

Toronto Montreal Winnipeg

MR. BUSINESS MAN:

Have you read

THE BUSY MAN'S MAGAZINE?

It is to meet the requirements of busy men like yourself that this MAGAZINE is being published. It ransacks each month the hundreds of excellent magazines and periodicals from all parts of the world, and reproduces the timely, interesting, entertaining and helpful articles every business man should read.

Each issue contains the cream of the world's publications—160 pages of select, up-to-date reading matter by the greatest writers of the age.

This is styled by its readers as "The Necessary Magazine." We want you to get a sample copy.

The MacLean Publishing Company
LIMITED

TORONTO - CANADA

CANADIAN MACHINERY BUYERS' DIRECTORY

To Our Readers—Use this directory when seeking to buy any machinery or power equipment. You will often get information that will save you money.

To Our Advertisers—Send in your name for insertion under the heading of the lines you make or sell.

To Non-Advertisers—A nominal rate of $1 per line a year is charged non-advertisers.

Acids.
Canada Chemical Mfg. Co., London.

Abrasive Materials.
Baxter, Paterson & Co., Montreal.
The Canadian Fairbanks Co., Montreal.
Rice Lewis & Son, Toronto.
Norton Co., Worcester, Mass.
H. W. Petrie, Toronto.
Williams & Wilson, Montreal.

Air Brakes.
Canada Foundry Co., Toronto.
Canadian Westinghouse Co., Hamilton.

Air Receivers.
Allis-Chalmers-Bullock Montreal.
Canada Foundry Co., Toronto.
Canadian Rand Drill Co., Montreal.
Chicago Pneumatic Tool Co., Chicago.
Jenckes Machine Co., Sherbrooke, Que.
John McDougall Caledonian Iron Works Co., Montreal.

Alloy, Ferro-Silicon.
Paxon, J. W., Co. Philadelphia.

Alundum Scythes Stones
Norton Company, Worcester, Mass.

Arbor Presses.
Niles-Bement-Pond Co., New York.

Augers.
Chicago Pneumatic Tool Co., Chicago.
Rice Lewis & Son, Toronto.

Automatic Machinery.
Cleveland Automatic Machine Co. Cleveland.
Potter & Johnston Machine Co., Pawtucket, R. I.

Automobile Parts
Globe Machine & Stamping Co., Cleveland, Ohio.

Axle Cutters.
Butterfield & Co., Rock Island, Que.
A. B. Jardine & Co., Hespeler, Ont.

Axle Setters and Straighteners.
A. B. Jardine & Co., Hespeler, Ont.
Standard Bearings, Ltd., Niagara Falls, Ont.

Babbit Metal.
Baxter, Paterson & Co., Montreal.
Canada Metal Co., Toronto.
Canada Machinery Agency, Montreal.
Greey, Wm. & J. G., Toronto.
Rice Lewis & Son, Toronto.
Lumen Bearing Co., Toronto.
Miller Bros. & Toms, Montreal, Que.

Bakers' Machinery.
Greey, Wm. & J. G., Toronto.
Pendrith Machinery Co., Toronto.

Baling Presses.
Perrin, Wm. R., & Co., Toronto.

Barrels, Steel Shop.
Cleveland Wire Spring Co., Cleveland.

Barrels, Tumbling.
Buffalo Foundry Supply Co., Buffalo.
Detroit Foundry Supply Co., Windsor
Dominion Foundry Supply Co., Montreal
Hamilton Facing Mill Co., Hamilton.
Globe Machine & Stamping Co., Cleveland, Ohio.
John McDougall Caledonian Iron Works Co., Montreal.
R. McDougall Co., Galt, Ont.
Miller Bros. & Toms, Montreal, Que.
Northern Engineering Works, Detroit.
J. W. Paxson Co., Philadelphia, Pa.
H. W. Petrie, Toronto.
The Smart-Turner Mach. Co., Hamilton.

Bars, Boring.
Hall Engineering Works, Montreal.
Miller Bros. & Toms, Montreal.
Niles-Bement-Pond Co., New York.
Standard Bearings Ltd., Niagara Falls, Ont.

Batteries, Dry.
Berlin Electrical Mfg. Co., Toronto.

Batteries, Flashlight.
Berlin Electrical Mfg. Co., Toronto.

Batteries, Storage.
Canadian General Electric Co. Toronto.
Chicago Pneumatic Tool Co., Chicago.
Rice Lewis & Son, Toronto.
John Millen & Son, Montreal.

Bearing Metals.
Lumen Bearing Co., Toronto

Bearings, Roller.
Standard Bearings, Ltd. Niagara Falls, Ont.

Bell Ringers.
Chicago Pneumatic Tool Co., Chicago.

Belting, Chain.
Baxter, Paterson & Co. Montreal.
Canada Machinery Exchange, Montreal.
Greey, Wm. & J. G., Toronto.
Jeffrey Mfg. Co., Columbia, Ohio.
Link-Belt Eng. Co., Philadelphia.
Miller Bros. & Toms, Montreal.
Waterous Engine Works Co., Brantford.

Belting, Cotton.
Baxter, Paterson & Co., Montreal.
Canada Machinery Agency, Montreal
Dominion Belting Co., Hamilton.
Greey, Wm. & J. G., Toronto.
Rice Lewis & Son, Toronto.

Belting, Leather.
Baxter, Paterson & Co., Montreal
Canada Machinery Agency, Montreal.
The Canadian Fairbanks Co., Montreal.
Greey, Wm. & J. G., Toronto.
McLaren, J. C., Montreal.
Rice Lewis & Son, Toronto.
H. W. Petrie, Toronto.
Williams & Wilson, Montreal.

Belting, Rubber.
Baxter, Paterson & Co., Montreal.
Canada Machinery Agency, Montreal.
Greey, Wm. & J. G., Toronto.

Belting Supplies.
Baxter, Paterson & Co., Montreal.
Greey, Wm. & J. G., Toronto.
Rice Lewis & Son, Toronto.
H. W. Petrie, Toronto.

Bending Machinery.
John Bertram & Sons Co., Dundas, Ont.
Chicago Pneumatic Tool Co., Chicago.
London Mach. Tool Co., Hamilton, Ont.
National Machinery Co., Tiffin, Ohio.
Niles-Bement-Pond Co., New York.

Benders, Tire.
A. B. Jardine & Co., Hespeler, Ont.
London Mach. Tool Co., Hamilton, Ont.

Blast Furnace Brick.
Detroit Foundry Supply Co., Detroit
Hamilton Facing Mill Co., Hamilton.

Blowers.
Buffalo Foundry Supply Co., Montreal.
Canada Machinery Agency, Montreal.
Detroit Foundry Supply Co., Windsor
Hamilton Facing Mill Co., Hamilton.
Kerr Turbine Co., Wellsville, N.Y.
J. W Paxson, Philadelphia, Pa.
Sheldon's Limited, Galt
Sturtevant, B. F., Co., Hyde Park, Mass.

Blast Gauges—Cupola.
Paxson, J. W., Co. Philadelphia.

Blue Printing.
The Electric Blue Print Co., Montreal.

Boilers.
Canada Foundry Co., Limited, Toronto.
Canada Machinery Agency, Montreal.
Goldie & McCulloch Co., Galt.
Jenckes Machine Co., Sherbrooke, Que.
E. Leonard & Sons, London, Ont.
Manitoba Iron Works, Winnipeg.
John McDougall Calenonian Iron Works, Montreal.
Owen Sound Iron Works Co., Owen Sound.
H. W. Petrie, Toronto.
Robb Engineering Co., Amherst, N.S.
The Smart-Turner Mach. Co., Hamilton.
Waterous Engine Works Co., Brantford.
Williams & Wilson, Montreal.

Boiler Compounds.
Canada Chemical Mfg. Co., London, Ont.
Hall Engineering Works, Montreal.
Lake Erie Boiler Compound Co., Toronto

Bolt Cutters.
John Bertram & Sons Co., Dundas, Ont.
London Mach. Tool Co., Hamilton.
National Machinery Co., Tiffin, Ohio.
Niles-Bement-Pond Co., New York.

Bolt and Nut Machinery.
John Bertram & Sons Co., Dundas, Ont.
Canada Machinery Agency, Montreal.
London Mach. Tool Co., Hamilton.
National Machinery Co., Tiffin, Ohio.
Niles-Bement-Pond Co. New York.

Bolts and Nuts, Rivets.
Baxter, Paterson & Co., Montreal.

Books, Mechanical.
Technical Pub. Co., Manchester.
The MacLean Pub. Co., Ltd., Toronto.

Boring and Drilling Machines.
American Tool Works Co., Cincinnati.
B. F. Barnes Co., Rockford, Ill.
John Bertram & Sons Co., Dundas, Ont.
Canada Machinery Agency, Montreal.
A. B. Jardine & Co., Hespeler, Ont.
London Mach. Tool Co., Hamilton.
Niles-Bement-Pond Co., New York.

Boring Machine, Upright.
American Tool Works Co., Cincinnati.
John Bertram & Sons Co., Dundas, Ont.
London Mach. Tool Co., Hamilton.
Niles-Bement-Pond Co., New York.

Boring Machine, Wood.
Chicago Pneumatic Tool Co., Chicago.
London Mach. Tool Co., Hamilton.

Boring and Turning Mills.
American Tool Works Co., Cincinnati.
John Bertram & Sons Co., Dundas, Ont.
Gisholt Machine Co., Madison, Wis.
Canada Machinery Agency, Montreal.
London Mach. Tool Co., Hamilton.
Niles-Bement-Pond Co., New York.
H. W. Petrie, Toronto.

Box Puller.
A. B. Jardine & Co., Hespeler, Ont.

Boxes, Steel Shop.
Cleveland Wire Spring Co., Cleveland.

Boxes, Tote.
Cleveland Wire Spring Co., Cleveland

Brakes, Automobile,
Globe Machine & Stamping Co., Cleveland, Ohio.

Brass Foundry Equipment.
Detroit Foundry Supply Co., Detroit.
Paxson, J. W., Co., Philadelphia

Brass Working Machinery.
Warner & Swasey Co., Cleveland, Ohio.

Brass Working Machine Tools.
Warner & Swasey Co., Cleveland, O.

Brushes, Foundry and Core.
Buffalo Foundry Supply Co., Buffalo.
Detroit Foundry Supply Co., Windsor.
Dominion Foundry Supply Co., Montreal
Hamilton Facing Mill Co., Hamilton
Paxson, J. W., Co., Philadelphia

Brushes, Steel.
Buffalo Foundry Supply Co., Buffalo.
Paxson, J. W., Co., Philadelphia

Buckets, Hoisting and Foundry.
Paxson, J. W., Co, Philadelphia

Bulldozers.
John Bertram & Sons Co., Dundas, Ont.
London Mach. Tool Co., Hamilton, Ont.
National Machinery Co., Tiffin, Ohio.
Niles-Bement-Pond Co., New York.

Calipers.
Baxter, Patterson & Co., Montreal.
Brown & Sharpe, Providence, R.I.
Rice Lewis & Son, Toronto.
John Millen & Son, Ltd., Montreal, Que.
L. S. Starrett & Co., Athol, Mass.
Williams & Wilson Montreal.

Can Making Machinery
Canada Machinery Agency, Montreal.
Miller Bros. & Toms, Montreal.

Cars, Foundry.
Buffalo Foundry Supply Co., Buffalo.
Detroit Foundry Supply Co., Windsor
Dominion Foundry Supply Co., Montreal
Hamilton Facing Mill Co., Hamilton.
Paxson, J. W., Co., Philadelphia

Case Hardening Bone.
Paxson, J. W., Co, Philadelphia

Castings, Aluminum.
Lumen Bearing Co., Toronto

Castings, Brass.
Chadwick Bros., Hamilton.
Greey, Wm. & J. G. Toronto.
Hall Engineering Works, Montreal.
Jenckes Machine Co., Sherbrooke.
Kennedy, Wm., & Sons, Owen Sound.
Lumen Bearing Co., Toronto
McDougall R. Co, Galt.
Niagara Falls Machine & Foundry Co. Niagara Falls, Ont.
Owen Sound Iron Works Co., Owen Sound.
Robb Engineering Co, Amherst, N.S.

Castings, Grey Iron.
Allis-Chalmers-Bullock Montreal.
Greey, Wm. & J. G., Toronto.
Hall Engineering Works, Montreal.
Jenckes Machine Co., Sherbrooke, Que.
Kennedy, Wm., & Sons, Owen Sound.
Laurie Engine & Machine Co., Montreal.
Manson Mfg. Co., Thorold, Ont
John McDougall Caledonian Iron Works Co., Montreal.
R. McDougall Co., Galt, Ont.
Niagara Falls Machine & Foundry Co., Niagara Falls, Ont
Owen Sound Iron Works Co., Owen Sound.
Robb Engineering Co., Amherst, N.S.
Smart-Turner Machine Co., Hamilton.

Castings, Phosphor Bronze.
Lumen Bearing Co., Toronto

Castings, Steel.
Kennedy, Wm., & Sons, Owen Sound.

Castings, Semi-Steel.
Robb Engineering Co., Amherst, N.S.

Cement Machinery.
Allis-Chalmers-Bullock, Limited, Montreal
Greey, Wm. & J. G., Toronto
Jeffrey Mfg. Co., Columbus, Ohio.
John McDougall Caledonian Iron Works, Co., Montreal
Owen Sound Iron Works Co., Owen Sound
St. Lawrence Supply Co., Montreal.

Centreing Machines.
John Bertram & Sons Co., Dundas, Ont.
Jeffrey Mfg Co., Columbia, Ohio
London Mach. Tool Co., Hamilton, Ont.
Niles-Bement-Pond Co., New York.
Pratt & Whitney Co., Hartford, Conn.
Standard Bearings, Ltd., Niagara Falls, Ont

Centres, Planer.
American Tool Works Co., Cincinnati.

Centrifugal Pumps.
Manson Mfg. Co., Thorold, Ont
John McDougall Caledonian Iron Works Co, Montreal.
R. McDougall Co., Galt.
D'Olier Engineering Co., New York.
Pratt & Whitney Co., Hartford, Conn.

Centrifugal Pumps—Turbine Driven.
Kerr Turbine Co., Wellsville, N.Y.

Chaplets.
Buffalo Foundry Supply Co., Buffalo.
Detroit Foundry Supply Co., Windsor.
Hamilton Facing Mill Co., Hamilton.
Paxson, J. W., Co., Philadelphia

Charcoal.
Buffalo Foundry Supply Co., Buffalo.
Detroit Foundry Supply Co., Windsor.
Dominion Foundry Supply Co., Montreal
Hamilton Facing Mill Co., Hamilton.
Paxson, J. W., Co., Philadelphia

Chemicals.
Baxter, Paterson & Co., Montreal.
Canada Chemical Co., London.

Chemists' Machinery.
Greey, Wm. & J. G., Toronto.

Chemists, Industrial.
Detroit Testing Laboratory, Detroit.

Chemists, Metallurgical.
Detroit Testing Laboratory, Detroit.

Chemists, Mining.
Detroit Testing Laboratory, Detroit.

Chucks, Ring Grinding.
A. B. Jardine & Co., Hespeler, Ont.
Chicago Pneumatic Tool Co., Chicago.

Chucks, Drill and Lathe.
American Tool Works Co., Cincinnati.
Baxter, Paterson & Co., Montreal.
John Bertram & Sons Co., Dundas, Ont.
Canada Machinery Agency, Montreal.
Ker & Goodwin, Brantford.
A. B. Jardine & Co., Hespeler, Ont.
London Mach. Tool Co., Hamilton.
John Millen & Son, Ltd., Montreal.
Niles-Bement-Pond Co., New York.
H. W. Petrie, Toronto.
Rice Lewis & Son, Toronto.
Standard Tool Co., Cleveland.
Whitman & Barnes Co., St. Catharines.

Chucks, Planer.
American Tool Works Co., Cincinnati.
Canada Machinery Agency, Montreal.
Niles-Bement-Pond Co., New York.

Chucking Machines.
American Tool Works Co., Cincinnati.
Niles-Bement-Pond Co., New York.
H. W. Petrie, Toronto.
Warner & Swasey Co., Cleveland, Ohio.

Circuit Breakers.
Allis-Chalmers-Bullock, Limited, Montreal
Canadian General Electric Co., Toronto.
Canadian Westinghouse Co., Hamilton.

Clippers, Bolt.
A. B. Jardine & Co., Hespeler, Ont.

Cloth and Wool Dryers.
B. Greening Wire Co., Hamilton.
Dominion Heating and Ventilating Co., Hespeler.

Collectors, Pneumatic.
Sturtevant, B. F., Co., Hyde Park, Mass.

Compressors, Air.
Allis-Chalmers-Bullock, Limited, Montreal
Canada Foundry Co., Limited, Toronto.
Canada Machinery Agency, Montreal.
Canadian Rand Drill Co., Montreal.
Canadian Westinghouse Co., Hamilton.
Chicago Pneumatic Tool Co., Chicago.
Detroit Foundry Supply Co., Windsor.
John McDougall, Caledonian Iron Works Co., Montreal
H. W. Petrie, Toronto.
The Smart-Turner Mach. Co., Hamilton.
Canada Machinery Agency, Montreal.
Canadian Westinghouse Co., Hamilton.
Hall Engineering Works, Montreal, Que.
London Mach. Tool Co., Hamilton.
R. McDougall Co., Galt, Ont.
Niles-Bement-Pond Co., New York.
H. W. Petrie, Toronto.
Pratt & Whitney Co., Hartford, Conn.
Williams & Wilson, Montreal.

Concentrating Plant.
Allis-Chalmers-Bullock, Montreal.
Greey, Wm. & J. G., Toronto.

Concrete Mixers.
Jeffrey M'f'g. Co., Columbus, Ohio.
Link-Belt Co., Philadelphia.

Condensers.
Canada Foundry Co., Limited, Toronto.
Canada Machinery Agency, Montreal.
Hall Engineering Works, Montreal.
Smart-Turner-Machine Co., Hamilton.
Waterous Engine Co., Brantford.

Confectioners' Machinery.
Baxter, Paterson & Co., Montreal.
Greey, Wm. & J. G., Toronto.
Pendrith Machinery Co., Toronto.

Consulting Engineers.
Charles Brandeis, Montreal.
Canadian White Co., Montreal.
Hall Engineering Works, Montreal.
Jules De Clercy, Montreal.
John S. Fielding, Toronto.
Miller Bros. & Toms, Montreal.
Roderick J. Parke, Toronto.
Plews & Trimingham, Montreal.
T. Pringle & Son, Montreal.
Standard Bearings, Ltd., Niagara Falls, Ont.

Contractors.
Canadian White Co., Montreal.
D'Olier Engineering Co., New York.
Expanded Metal and Fireproofing Co., Toronto.
Hall Engineering Works, Montreal.
Laurie Engine & Machine Co., Montreal.
John McDougall Caledonian Iron Works Co., Montreal.
Miller Bros. & Toms, Montreal.
Robb Engineering Co., Amherst, N.S.
The Smart-Turner Mach. Co., Hamilton.

Contractors' Supplies.
Manson Mfg. Co., Thorold, Ont.

Controllers and Starters Electric Motor.
Allis-Chalmers-Bullock, Montreal.
Canadian General Electric Co., Toronto.
Canadian Westinghouse Co., Hamilton.
T. & H. Electric Co., Hamilton.

Conveying Systems.
Sturtevant, B. F., Co., Hyde Park, Mass.

Converters, Steel.
Northern Engineering Works, Detroit.
Paxson, J. W., Co., Philadelphia

Conveyor Machinery.
Baxter, Paterson & Co., Montreal.
Greey, Wm. & J. G., Toronto
Jeffrey Mfg. Co., Columbus, Ohio.
Jenckes Machine Co., Sherbrooke, Que.
Link-Belt Co., Philadelphia.
John McDougall Caledonian Iron Works Co., Montreal.
Miller Bros. & Toms, Montreal.
Smart-Turner Machine Co., Hamilton.
Laurie Engine & Machine Co., Montreal.
Waterous Engine Works Co., Brantford.
Williams & Wilson, Montreal.

Coping Machines.
John Bertram & Sons Co., Dundas, Ont.
London Mach. Tool Co., Hamilton.
Niles-Bement-Pond Co., New York.

Core Compounds.
Buffalo Foundry Supply Co., Buffalo.
Detroit Foundry Supply Co., Windsor.
Dominion Foundry Supply Co., Montreal
Hamilton Facing Mill Co., Hamilton.
Paxson, J. W., Co., Philadelphia

Core-Making Machines.
Paxson, J. W., Co., Philadelphia

Core Ovens.
Detroit Foundry Supply Co., Windsor.
Hamilton Facing Mill Co., Hamilton.
Paxson, J. W., Co., Philadelphia

Core Oven Bricks.
Buffalo Foundry Supply Co., Buffalo.
Detroit Foundry Supply Co., Windsor.
Hamilton Facing Mill Co., Hamilton.
Paxson, J. W., Co., Philadelphia

Core Taper and Wax Wire.
Paxson, J. W., Co., Philadelphia

Core Wash.
Buffalo Foundry Supply Co., Buffalo.
Detroit Foundry Supply Co., Windsor.
Hamilton Facing Mill Co., Hamilton.
Paxson, Co., J. W., Philadelphia

Cotton Belting.
Baxter, Paterson & Co., Montreal.
Dominion Belting Co., Hamilton.
Greey, Wm. & J. G., Toronto.

Couplings.
Owen Sound Iron Works Co., Owen Sound

Cranes, Electric and Hand Power.
Canada Foundry Co., Limited, Toronto
Dominion Foundry Supply Co., Montreal
Hamilton Facing Mill Co., Hamilton.
Link-Belt Co., Philadelphia.
John McDougall, Caledonian Iron Works Co., Montreal.
Miller Bros. & Toms, Montreal.
Niles-Bement-Pond Co., New York.
Northern Engineering Works, Detroit
Owen Sound Iron Works Co., Owen Sound
Paxson, J. W., Co., Philadelphia
Smart-Turner-Machine Co., Hamilton.

Crankshafts.
The Canada Forge Co., Welland.
St. Clair Bros., Galt

Crushers, Rock or Ore.
Allis-Chalmers-Bullock, Montreal.

Contractors' Plant.
Allis-Chalmers-Bullock, Montreal.
John McDougall, Caledonian Iron Works Co., Montreal.
Niagara Falls Machine & Foundry Co., Niagara Falls, Ont.

Crank Pin Turning Machine
London Mach. Tool Co., Hamilton.
Niles-Bement-Pond Co., New York.

Crucibles.
Buffalo Foundry Supply Co., Buffalo.
Detroit Foundry Supply Co., Windsor
Dominion Foundry Supply Co., Montreal
Hamilton Facing Mill Co., Hamilton.
J. W. Paxson Co., Philadelphia.

Crucible Caps
Hamilton Facing Mill Co., Hamilton.
Paxson, J. W., Co., Philadelphia

Crushers, Rock or Ore.
Allis-Chalmers-Bullock, Montreal.
Jeffrey Mfg. Co., Columbus, Ohio.

Cupolas.
Buffalo Foundry Supply Co., Buffalo.
Detroit Foundry Supply Co., Windsor
Dominion Foundry Supply Co., Montreal
Hamilton Facing Mill Co., Hamilton.
Northern Engineering Works, Detroit
J. W. Paxson Co., Philadelphia.
Sheldons Limited, Galt.

Cupola Blast Gauges.
Paxson, J. W., Co., Philadelphia

Cupola Blocks.
Detroit Foundry Supply Co., Detroit.
Hamilton Facing Mill Co., Hamilton
Northern Engineering Works, Detroit
Paxson, J. W., Co., Philadelphia

Cupola Blowers.
Buffalo Foundry Supply Co., Buffalo.
Canada Machinery Agency, Montreal.
Detroit Foundry Supply Co., Windsor
Hamilton Facing Mill Co., Hamilton.
McEachren Heating and Ventilating Co., Galt.
Northern Engineering Works, Detroit
Paxson, J. W., Co., Philadelphia
Sheldon's Limited, Galt.
B. F. Sturtevant Co., Hyde Park, Mass

Cutters, Coal.
Allis-Chalmers-Bullock, Montreal.

Cut Meters.
The Canadian Fairbanks Co., Montreal

Cutters, Flue
Chicago Pneumatic Tool Co., Chicago.
J. W. Paxson Co., Philadelphia.

Cutters, Milling.
Becker, Brainard Milling Machine Co. Hyde Park, Mass.
Pratt & Whitney Co., Hartford, Conn.
Standard Tool Co., Cleveland.

Cutting-off Machines.
John Bertram & Sons Co., Dundas, Ont.
Canada Machinery Agency, Montreal.
London Mach. Tool Co., Hamilton.
H. W. Petrie, Toronto.
Pratt & Whitney Co., Hartford, Conn.

Cutting-off Tools.
Armstrong Bros. Tool Co., Chicago.
Baxter Paterson & Co., Montreal.
London Mach. Tool Co., Hamilton.
H. W. Petrie, Toronto.
Pratt & Whitney, Hartford, Conn.
Rice Lewis & Son, Toronto.
L. S. Starrett Co., Athol, Mass.

Dies
Globe Machine & Stamping Co., Cleveland, Ohio.

Die Stocks
Hart Manufacturing Co., Cleveland, Ohio

Dies, Opening
W. H. Banfield & Sons, Toronto
Globe Machine & Stamping Co., Cleveland, Ohio.
R. McDougall Co., Galt.
Pratt & Whitney Co., Hartford, Conn.

Dies, Sheet Metal.
W. H. Banfield & Sons, Toronto.
Globe Machine & Stamping Co., Cleveland, Ohio

Die Sinkers
Globe Machine & Stamping Co., Cleveland, Ohio.

Dies, Threading.
Hart Mfg. Co., Cleveland
John Millen & Son, Ltd., Montreal.

Dowel Pins.
Buffalo Foundry Supply Co., Buffalo.
Detroit Foundry Supply Co., Windsor.
Dominion Foundry Supply Co., Montreal
Hamilton Facing Mill Co., Hamilton.

Draft, Mechanical.
W. H. Banfield & Sons, Toronto.
Butterfield & Co., Rock Island, Que.
A. B. Jardine & Co., Hespeler.
Pratt & Whitney Co., Hartford, Conn.
Sheldon's Limited, Galt.
Sturtevant, B. F., Co., Hyde Park, Mass.

Drawing Instruments.
Rice Lewis & Son, Toronto.

Drawing Supplies.
The Electric Blue Print Co., Montreal

Drawn Steel, Cold.
Baxter, Paterson & Co., Montreal.
Greey, Wm. & J. G., Toronto
Miller Bros. & Toms, Montreal.
Union Drawn Steel Co., Hamilton.

Dredges, Gold, Dipper and Hydraulic.
Allis-Chalmers-Bullock Co., Montreal.

Drilling Machines, Arch Bar.
John Bertram & Sons Co., Dundas, Ont.
London Mach. Tool Co., Hamilton
Niles-Bement-Pond Co., New York.

Drilling Machines, Boiler.
American Tool Works Co., Cincinnati.
John Bertram & Sons Co., Dundas, Ont.
Bickford Drill and Tool Co., Cincinnati.
The Canadian Fairbanks Co., Montreal.
A. B. Jardine & Co., Hespeler, Ont.
London Mach. Tool Co., Hamilton, Ont.
Niles-Bement-Pond Co., New York.
H. W. Petrie, Toronto.
Williams & Wilson, Montreal.

Drilling Machines Connecting Rod.
John Bertram & Sons Co., Dundas, Ont.
London Mach. Tool Co., Hamilton.
Niles-Bement-Pond Co. New York.

Drilling Machines, Locomotive Frame.
American Tool Works Co., Cincinnati.
B. F. Barnes Co., Rockford, Ill.
John Bertram & Sons Co., Dundas, Ont.
London Mach. Tool Co., Hamilton, Ont.
Niles-Bement-Pond Co., New York.

Drilling Machines, Multiple Spindle.
American Tool Works Co., Cincinnati.
B. F. Barnes Co., Rockford, Ill.
Baxter, Paterson & Co., Montreal.
John Bertram & Sons Co., Dundas, Ont.
Bickford Drill & Tool Co., Cincinnati.
Canada Machinery Agency, Montreal.
London Mach. Tool Co., Hamilton, Ont.
Niles-Bement-Pond Co., New York.
H. W. Petrie, Toronto.
Williams & Wilson, Montreal.

Drilling Machines, Pneumatic
Canada Machinery Agency, Montreal.

Drilling Machines, Portable
Baxter, Paterson & Co., Montreal.
A. B. Jardine & Co., Hespeler, Ont.
Niles-Bement-Pond Co., New York.

Drilling Machines, Radial.
American Tool Works Co., Cincinnati.
Baxter, Paterson & Co., Montreal.
John Bertram & Sons Co., Dundas, Ont.
Bickford Tool & Drill Co., Cincinnati.
The Canadian Fairbanks Co., Montreal.
London Mach. Tool Co., Hamilton.
Niles-Bement-Pond Co., New York.
H. W. Petrie, Toronto.
Williams & Wilson, Montreal.

Drilling Machines, Suspension.
John Bertram & Sons Co., Dundas, Ont.
Canada Machinery Agency, Montreal.
London Mach. Tool Co., Hamilton.
Niles-Bement-Pond Co., New York

Drilling Machines, Turret.
John Bertram & Sons Co., Dundas, Ont.
London Mach. Tool Co., Hamilton.
Niles-Bement-Pond Co., New York.

Drilling Machines, Upright.
American Tool Works Co., Cincinnati.
B. F. Barnes Co., Rockford, Ill.
Baxter, Paterson & Co., Montreal.
John Bertram & Sons Co., Dundas, Ont
A. B. Jardine & Co., Hespeler, Ont.
London Mach. Tool Co., Hamilton.
Niles-Bement-Pond Co., New York.
W. H. Petrie, Toronto.
Williams & Wilson Montreal.

Drills, Bench.
B. F. Barnes Co., Rockford, Ill.
Baxter, Paterson & Co., Montreal.
London Mach. Tool Co., Hamilton.
Pratt & Whitney Co., Hartford, Conn.

Drills, Blacksmith.
Canada Machinery Agency, Montreal.
A. B. Jardine & Co., Hespeler, Ont.
London Mach. Tool Co., Hamilton.
Standard Tool Co., Cleveland.
Whitman & Barnes Co., St. Catharines.

Drills, Centre.
Pratt & Whitney Co., Hartford, Conn.
Standard Tool Co., Cleveland, O.
L. S. Starrett Co., Athol, Mass.

Drills, Electric
B. F. Barnes Co., Rockford, Ill.
Baxter Paterson & Co., Montreal.
Chicago Pneumatic Tool Co., Chicago.
Niles-Bement-Pond Co., New York.

Drills, Gang.
American Tool Works Co., Cincinnati.
B. F. Barnes Co., Rockford, Ill.
John Bertram & Sons Co., Dundas, Ont.
Pratt & Whitney Co., Hartford, Conn.

Drills, High Speed.
Wm. Abbott, Montreal.
American Tool Works Co., Cincinnati.
B. F. Barnes Co., Rockford, Ill.
Pratt & Whitney Co., Hartford, Conn.
Standard Tool Co., Cleveland, O.
Whitman & Barnes Co., St. Catharines.

Drills, Hand.
A. B. Jardine & Co., Hespeler, Ont.

Drills, Horizontal.
B. F. Barnes Co., Rockford, Ill.
John Bertram & Sons Co., Dundas, Ont.
Canada Machinery Agency, Montreal.
London Mach. Tool Co., Hamilton.
Niles-Bement-Pond Co., New York.

Drills, Jewelers.
Pendrith Machinery Co., Toronto.

Drills, Pneumatic.
Canada Machinery Agency, Montreal.
Chicago Pneumatic Tool Co., Chicago.
Niles-Bement-Pond Co., New York.

Drills, Radial.
American Tool Works Co., Cincinnati.
John Bertram & Sons Co., Dundas, Ont.
Bickford Drill & Tool Co., Cincinnati.
London Mach. Tool Co., Hamilton, Ont.
Niles-Bement-Pond Co., New York.

Drills, Ratchet.
Armstrong Bros. Tool Co., Chicago.
A. B. Jardine & Co., Hespeler.
Pratt & Whitney Co., Hartford, Conn.
Standard Tool Co., Cleveland.
Whitman & Barnes Co., St. Catharines.

Drills, Rock.
Allis-Chalmers-Bullock, Montreal.
Canadian Rand Drill Co., Montreal.
Chicago Pneumatic Tool Co., Chicago.
Jaffrey Mfg. Co., Columbus, Ohio.

Drills, Sensitive.
American Tool Works Co., Cincinnati.
B. F. Barnes Co., Rockford, Ill.
Canada Machinery Agency, Montreal.
Niles-Bement-Pond Co., New York.

Drills, Shop View.
John Bertram & Sons Co., Dundas, Ont.

Drills, Twist.
Baxter, Paterson & Co., Montreal.
Chicago Pneumatic Tool Co., Chicago.
Alex. Gibb, Montreal.
A. B. Jardine & Co., Hespeler, Ont.
John Millen & Son, Ltd., Montreal.
Morse Twist Drill and Machine Co., New Bedford, Mass.
Pratt & Whitney Co., Hartford, Conn.
Standard Tool Co., Cleveland.
Whitman & Barnes Co., St. Catharines.

Drying Apparatus of all Kinds.
Dominion Heating & Ventilating Co., Hespeler
Greey, Wm. & J. G., Toronto
Sheldon's Limited, Galt
B. F. Sturtevant Co., Hyde Park, Mass

Dry Kiln Equipment.
Sturtevant, B. F., Co., Hyde Park, Mass.

Dry Sand and Loam Facing.
Buffalo Foundry Supply Co., Buffalo.
Hamilton Facing Mill Co., Hamilton.
Paxson, J. W., Co., Philadelphia

Dump Cars.
Canada Foundry Co., Limited, Toronto
Greey, Wm. & J. G., Toronto
Hamilton Facing Mill Co., Hamilton.
John McDougall, Caledonian Iron Works Co., Montreal.
Niles-Bement-Pond Co., New York.
Standard Bearings, Ltd., Niagara Falls, Ont.
Jenckes Machine Co., Sherbrooke, Que
Link-Belt Eng. Co., Philadelphia.
John McDougall Caledonian Iron Works Co., Montreal.
Miller Bros. & Toms, Montreal.
Owen Sound Iron Works Co., Owen Sound
Paxson, J. W., Co., Philadelphia
Waterous Engine Co., Brantford.

Dust Separators.
Greey, Wm. & J. G., Toronto
Dominion Heating and Ventilating Co., Hespler.

Paxson, J. W., Co., Philadelphia
Sheldon's Limited, Galt.

Dynamos.
Allis-Chalmers-Bullock, Montreal.
Canadian General Electric Co., Toronto.
Canadian Westinghouse Co., Hamilton.
Consolidated Electric Co., Toronto.
D'Olier Engineering Co., New York.
Hall Engineering Works, Montreal, Que.
John Millen & Son, Ltd., Montreal.
Packard Electric Co., St. Catharines.
H. W. Petrie, Toronto.
B. F. Sturtevant Co., Hyde Park, Mass.
T. & H. Electric Co., Hamilton.

Dynamos—Turbine Driven.
Kerr-Turbine Co., Wellsville, N.Y.

Economizer, Fuel.
Dominion Heating & Ventilating Co. Hespeler
Standard Bearings, Ltd., Niagara Falls.
B. F. Sturtevant Co., Hyde Park, Mass.

Electrical Instruments.
Canadian Westinghouse Co., Hamilton.

Electrical Steel.
Baxter, Patterson & Co., Montreal.

Electrical Supplies.
Canadian General Electric Co., Toronto.
Canadian Westinghouse Co., Hamilton.
London Mach. Tool Co., Hamilton, Ont.
John Millen & Son, Ltd., Montreal.
Packard Electric Co., St. Catharines.
T. & H. Electric Co., Hamilton.

Electrically Driven Tools and Machinery.
American Tool Works Co., Cincinnati.
Baxter Paterson & Co., Montreal.

Electrical Repairs
Canadian Westinghouse Co., Hamilton.
T. & H. Electric Co., Hamilton.

Elevator Buckets.
Greey, Wm. & J. G., Toronto
Jeffrey Mfg. Co., Columbus, Ohio.

Elevators, Foundry.
Northern Engineering Works, Detroit

Emery and Emery Wheels.
Baxter, Paterson & Co., Montreal.
Dominion Foundry Supply Co., Montreal
Hamilton Facing Mill Co., Hamilton.
Paxson, J. W., Co., Philadelphia

Emery Wheel Dressers.
Baxter Paterson & Co., Montreal.
Canada Machinery Agency, Montreal.
Dominion Foundry Supply Co., Montreal
Hamilton Facing Mill Co., Hamilton.
John Millen & Son, Ltd., Montreal.
H. W. Petrie, Toronto.
Paxson, J. W., Co., Philadelphia
Standard Tool Co., Cleveland.

Engineers and Contractors.
Canada Foundry Co., Limited, Toronto
D'Olier Engineering Co., New York.
Greey, Wm. & J. G., Toronto
Hall Engineering Works, Montreal.
Jenckes Machine Co., Sherbrooke, Que.
Laurie Engine & Machine Co., Montreal.
Link-Belt Co., Philadelphia.
John McDougall Caledonian Iron Works Co., Montreal.
Miller Bros. & Toms Montreal.
Robb Engineering Co., Amherst, N.S.
The Smart-Turner Mach. Co., Hamilton.

Engineers' Supplies.
Baxter, Paterson & Co., Montreal.
Greey, Wm & J. G., Toronto
Hall Engineering Works, Montreal.
Rice Lewis & Son, Toronto.

Engines, Gas and Gasoline.
Baxter, Paterson & Co., Montreal.
Canada Machinery Agency, Montreal.
The Canadian Fairbanks Co., Montreal.
Canadian McVicker Engine Co., Galt.
Economic Heat, Light and Power Supply Co., Toronto.
Gilson Mfg. Co., Guelph
The Goldie & McCulloch Co., Galt, Ont.
H. W. Petrie, Toronto.
Producer Gas Co., Toronto
The Smart-Turner Mach. Co., Hamilton.
Standard Bearings, Ltd., Niagara Falls.

Engines, Steam.
Allis-Chalmers-Bullock, Montreal
Canada Machinery Agency, Montreal.
D'Olier Engineering Co., New York.
The Goldie & McCulloch Co., Galt, Ont.
E. Leonard & Sons, London, Ont.
Jenckes Machine Co., Sherbrooke, Que.
Laurie Engine & Machine Co., Montreal
Manson Mfg. Co., Thorold Ont
John McDougall Caledonian Iron Works, Montreal.
Miller Bros. & Toms, Montreal.
Robb Engineering Co., Amherst, N.S.
Sheldons Limited, Galt.
The Smart-Turner Mach. Co., Hamilton.
B. F. Sturtevant Co., Hyde Park, Mass.
Waterous Engine Works Co., Brantford.

Exhaust Heads.
Dominion Heating & Ventilating Co., Hespeler
Sheldon's Limited, Galt, Ont.
B. F. Sturtevant Co., Hyde Park, Mass.

Expanded Metal.
Expanded Metal and Fireproofing Co. Toronto

Expanders.
A. B. Jardine & Co., Hespeler, Ont.

Fans, Electric.
Canadian General Electric Co., Toronto
Canadian Westinghouse Co., Hamilton.
Dominion Heating & Ventilating Co., Hespeler
Paxson, J. W., Co., Philadelphia
Sheldon's Limited Galt, Ont.
The Smart-Turner Mach. Co., Hamilton.
B. F. Sturtevant Co., Hyde Park, Mass.

Fans, Exhaust.
Canadian Buffalo Forge Co., Montreal.
Detroit Foundry Supply Co., Windsor.
Dominion Foundry Supply Co., Montreal
Dominion Heating & Ventilating Co., Hespeler.
Greey, Wm. & J. G., Toronto
Hamilton Facing Mill Co., Hamilton.
Paxson, J. W., Co., Philadelphia
Sheldon's Limited, Galt.
B. F. Sturtevant Co., Hyde Park, Mass.

Feed Water Heaters.
Darling Bros., Montreal
Laurie Engine & Machine Co., Montreal
John McDougall, Caledonian Iron Works Co., Montreal.
The Smart-Turner Mach. Co., Hamilton

Files and Rasps.
Baxter, Paterson & Co., Montreal.
John Millen & Son Ltd., Montreal.
Rice Lewis & Son, Toronto.
Nicholson File Co., Providence, R.I.
H. W. Petrie, Toronto.
Whitman & Barnes Co., St. Catharines.

Fillet, Pattern.
Baxter, Paterson & Co., Montreal.
Buffalo Foundry Supply Co., Buffalo.
Detroit Foundry Supply Co., Windsor.
Dominion Foundry Supply Co., Montreal
Hamilton Facing Mill Co., Hamilton.

Fire, Apparatus.
Waterous Engine Works Co., Brantford.

Fire Brick and Clay.
Baxter Paterson & Co., Montreal.
Buffalo Foundry Supply Co., Buffalo.
Detroit Foundry Supply Co., Windsor.
Dominion Foundry Supply Co., Montreal
Hamilton Facing Mill Co., Hamilton.
J. W. Paxson Co., Philadelphia.
Toronto Pottery Co., Toronto

Flash Lights.
Berlin Electrical Mfg. Co., Toronto.

Flour Mill Machinery.
Allis-Chalmers-Bullock, Montreal.
Greey, Wm. & J. G., Toronto
The Goldie & McCulloch Co., Galt, Ont.
John McDougal, Caledonian Iron Works Co., Montreal.
Miller Bros. & Toms, Montreal.

Flue Rollers.
Chicago Pneumatic Tool Co., Chicago.

Forges.
Canada Foundry Co., Limited, Toronto.
Hamilton Facing Mill Co., Hamilton.
H. W. Petrie, Toronto.
Sheldon's Limited, Galt, Ont.
B. F. Sturtevant Co., Hyde Park, Mass.

Forgings, Drop.
John McDougall, Caledonian Iron Works Co., Montreal.
H. W. Petrie, Toronto.
St. Clair Bros., Galt
Whitman & Barnes Co., St. Catharnes

Forgings, Light & Heavy.
The Canada Forge Co., Welland.

Forging Machinery.
John Bertram & Sons Co., Dundas, Ont.
London Mach. Tool Co., Hamilton, Ont.
National Machinery Co., Tiffin, Ohio
Niles-Bement-Pond Co., New York.

Founders.
Greey Wm. & J. G., Toronto
John McDougall, Caledonian Iron Works Co., Montreal
R. McDougall Co., Galt, Ont.
Niagara Falls Machine & Foundry Co., Niagara Falls, Ont.
The Smart-Turner Mach. Co., Hamilton.

Foundry Equipment.
Detroit Foundry Supply Co., Windsor.
Hamilton Facing Mill Co., Mill.
Hanna Engineering Works, Chicago.
Northern Engineering Works Detroit
Paxson, J. W., Co., Philadelphia

Foundry Facings.
Buffalo Foundry Supply Co., Buffalo.
Detroit Foundry Supply Co., Windsor.
Hamilton Facing Mill Co., Hamilton.
J. W. Paxson Co., Philadelphia, Pa.

Friction Clutch Pulleys, etc.
The Goldie & McCulloch Co., Galt.
Greey, Wm. & J. G., Toronto
Link-Belt Co., Philadelphia.
Miller Bros. & Toms, Montreal.

Friction Hoists.
Owen Sound Iron Works Co., Owen Sound

Furnaces.
Detroit Foundry Supply Co., Windsor.
Dominion Foundry Supply Co., Montreal
Hamilton Facing Mill Co., Hamilton.
Paxson, J. W., Co. Philadelphia

Furnace Linings.
Paxson, J. W., Co., Philadelphia

Galvanizing
Canada Metal Co., Toronto.

Gas Blowers and Exhausters.
D'Olier Engineering Co., New York.
Dominion Heating & Ventilating Co. Hespeler
Sheldons Limited, Galt.
B. F. Sturtevant Co., Hyde Park, Mass.

Gas Plants, Suction.
Baxter Paterson & Co., Montreal
Economic Heat, Light and Power Supply Co., Toronto.
Producer Gas Co., Toronto
Standard Bearings, Ltd., Niagara Falls.
Williams & Wilson Montreal

Gauges, Standard.
Brown & Sharpe, Providence, R.I.
Pratt & Whitney Co., Hartford, Conn.

Gearing.
Greey, Wm. & J. G. Toronto

Gear, Cutting Machinery.
Becker-Brainard Milling Mach. Co., Hyde Park, Mass.
Bickford Drill & Tool Co., Cincinnati.
Brown & Sharpe, Providence, R. I.
Greey, Wm. & J. G., Toronto
Kennedy, Wm., & Sons, Owen Sound
London Mach. Tool Co., Hamilton.
R. McDougall Co., Galt
Niles-Bement-Pond Co., New York.
H. W. Petrie, Toronto.
Pratt & Whitney Co., Hartford, Conn.
Williams & Wilson, Montreal.

Gears, Angle.
Brown & Sharpe, Providence, R.I.
Chicago Pneumatic Tool Co., Chicago.
Greey, Wm. & J. G., Toronto
Laurie Engine & Machine Co., Montreal.
John McDougall, Caledonian Iron Works Co., Montreal.
Miller Bros. & Toms, Montreal.
Waterous Engine Co., Brantford.

Gears, Cut.
Kennedy, Wm., & Sons, Owen Sound

Gears, Iron.
Greey, Wm & J. G., Toronto
Kennedy, Wm., & Sons, Owen Sound

Gears, Mortise.
Greey, Wm. & J. G., Toronto
Kennedy, Wm., & Sons, Owen Sound

Gears, Reducing.
Brown, David & Sons, Huddersfield, Eng
Brown & Sharpe, Providence, R.I.
Chicago Pneumatic Tool Co., Chicago.
Greey, Wm. & J. G., Toronto
John McDougall, Caledonian Iron Works Co., Montreal.
Miller Bros. & Toms, Montreal.

Generating Sets.
Sturtevant, B. F., Co., Hyde Park, Mass.

Generators, Electric.
Allis-Chalmers-Bullock, Limited, Montreal
Canadian General Electric Co., Toronto
Canadian Westinghouse Co., Hamilton.
D'Olier Engineering Co., New York.
Hall Engineering Works, Montreal.
H. W. Petrie, Toronto.
B. F. Sturtevant Co., Hyde Park, Mass.
Toronto & Hamilton Electric Co., Hamilton.

Generators, Gas.
H. W. Petrie, Toronto.

Glass Bevelling Machines.
Pendrith Machinery Co., Toronto.

Graphite Paints.
P. D. Dods & Co., Montreal.

Graphite.
Detroit Foundry Supply Co., Windsor.
Dominion Foundry Supply Co., Montreal
Hamilton Facing Mill Co., Hamilton.
Paxson, J. W., Co., Philadelphia

Grinders, Automatic Knife.
W. H. Banfield & Son, Toronto.

Grinders, Centre.
Niles-Bement-Pond Co., New York.
H. W. Petrie, Toronto.

Grinders, Cutter.
Becker-Brainard Milling Mach. Co., Hyde Park, Mass.
Brown & Sharpe, Providence, R.I.
John Millen & Son, Ltd., Montreal.
Pratt & Whitney Co., Hartford, Conn.

Grinders, Tool.
Armstrong Bros. Tool Co., Chicago.
B. F. Barnes Co., Rockford, Ill.
Brown & Sharpe, Providence, R.I.
H. W. Petrie, Toronto.
Williams & Wilson, Montreal.

Grinding Machines.
Brown & Sharpe, Providence, R.I.
The Canadian Fairbanks, Montreal.
Niles-Bement-Pond Co., New York.
Norton Company, Worcester, Mass.
Paxson, J. W., Co., Philadelphia.
H. W. Petrie, Toronto.

Grinding and Polishing Machines.
The Canadian Fairbanks Co., Montreal.
Greey, Wm. & J. G., Toronto
John Millen & Son, Ltd., Montreal.
Miller Bros. & Toms, Montreal.
Niles-Bement-Pond Co., New York.
H. W. Petrie, Toronto.
Pendrith Machinery Co., Toronto.
Standard Bearings, Ltd., Niagara Falls.

Grinding Wheels (Alundum)
Norton Company, Worcester, Mass.

Hack Saws.
Baxter, Paterson & Co., Montreal.
Canada Machinery Agency, Montreal.
The Canadian Fairbanks Co., Montreal.
John Millen & Son, Ltd., Montreal.
Niles-Bement-Pond Co., New York.
H. W. Petrie, Toronto.
Williams & Wilson, Montreal.

Hammers, Drop.
London Mach. Tool Co., Hamilton, Ont.
Miller Bros. & Toms, Montreal.
Niles-Bement-Pond Co., New York.

Hammers, Rand—All Kinds.
Whitman & Barnes Co., St. Catharines.

Hammers, Steam.
John Bertram & Sons Co., Dundas, Ont.
London Mach. Tool Co., Hamilton, Ont.
Niles-Bement-Pond Co., New York.

Hangers.
The Goldie & McCulloch Co., Galt.
Greey, Wm. & J. G., Toronto
Kennedy, Wm., & Sons, Owen Sound
Miller Bros. & Toms, Montreal.
Owen Sound Iron Works Co., Owen Sound
The Smart-Turner Mach. Co., Hamilton.
Waterous Engine Co., Brantford.
Pendrith Machinery Co., Toronto.
Standard Bearings, Ltd., Niagara Falls, Ont.

Heating Apparatus.
Dominion Heating & Ventilating Co., Hespeler
Sheldon's Limited, Galt.
B. F. Sturtevant Co., High Park, Mass.

Hoisting and Conveying Machinery.
Allis-Chalmers-Bullock, Limited, Montreal
Greey, Wm. & J. G., Toronto
Jenckes Machine Co., Sherbrooke, Que.
Link-Belt Co., Philadelphia.
Miller Bros. & Toms, Montreal.
Niles-Bement-Pond Co., New York.
Northern Engineering Works, Detroit
The Smart-Turner Mach. Co., Hamilton.
Waterous Engine Co., Brantford.

Hoists, Electric.
Northern Engineering Works, Detroit

Hoists, Pneumatic.
Canadian Rand Drill Co., Montreal.
Chicago Pneumatic Tool Co., Chicago.
Dominion Foundry Supply Co., Montreal
Hamilton Facing Mill Co., Hamilton.
Northern Engineering Works, Detroit

Hose.
Baxter, Paterson & Co., Montreal.

Hose, Air.
Canadian Rand Drill Co., Montreal.
Canadian Westinghouse Co., Hamilton.
Chicago Pneumatic Tool Co., Chicago.
Paxson, J. W., Co., Philadelphia

Hose Couplings.
Canadian Rand Drill Co., Montreal.
Canadian Westinghouse Co., Hamilton.
Chicago Pneumatic Tool Co., Chicago.
Paxson, J. W., Co., Philadelphia
Whitman & Barnes Co., St. Catharines.

Hose, Steam.
Allis-Chalmers-Bullock, Montreal.
Canadian Rand Drill Co., Montreal.
Canadian Westinghouse Co., Hamilton.
Paxson, J. W. Co., Philadelphio

Hydraulic Accumulators.
Niles-Bement-Pond Co., New York.
Perrin, Wm. R., Co., Toronto.
The Smart-Turner Mach. Co., Hamilton.

Hydraulic Machinery.
Allis-Chalmers-Bullock, Montreal.
Barber, Chas., & Sons, Meaford

India Oil Stones.
Norton Company, Worcester, Mass.

Indicators, Speed.
L. S. Starrett Co., Athol, Mass.

Injectors.
Canada Foundry Co., Toronto.
The Canadian Fairbanks Co., Montreal.
Rice Lewis & Son, Toronto.
Penberthy Injector Co., Windsor, Ont.

Iron Cements.
Buffalo Foundry Supply Co., Buffalo.
Detroit Foundry Supply Co., Windsor.
Hamilton Facing Mill Co., Hamilton.
Paxson, J. W., Co., Philadelphia

Iron Filler.
Buffalo Foundry Supply Co., Buffalo.
Detroit Foundry Supply Co., Windsor.
Dominion Foundry Supply Co., Montreal
Hamilton Facing Mill Co., Hamilton.
Paxson, J. W., Co., Philadelphia.

Iron Tools.
John Bertram & Sons Co., Dundas, Ont.
A. B. Jardine & Co., Hespeler, Ont.
London Mach. Tool Co., Hamilton.
R. McDougall Co., Galt.
H. W. Petrie, Toronto.

Jacks.
Norton, A. O., Coaticook, Que.

Kegs, Steel Shop.
Cleveland Wire Spring Co., Cleveland

Key-Seating Machines.
B. F. Barnes Co., Rockford, Ill.
Niles-Bement-Pond Co., New York.

Lace Leather.
Baxter, Paterson & Co., Montreal.

Ladles, Foundry.
Northern Engineering Works, Detroit

Lamps, Arc and Incandescent.
Canadian General Electric Co., Toronto.
Canadian Westinghouse Co., Hamilton.
The Packard Electric Co., St. Catharines.

Lathe Dogs.
Armstrong Bros., Chicago
Baxter, Paterson & Co., Montreal.
Pratt & Whitney Co., Hartford, Conn.

Lathes, Engine.
American Tool Work Co., Cincinnati.
B. F. Barnes Co., Rockford, Ill.
Baxter, Paterson & Co., Montreal
John Bertram & Sons Co., Dundas, Ont.
Canada Machinery Agency, Montreal.
The Canadian Fairbanks Co., Montreal.
London Mach. Tool Co., Hamilton, Ont.
Niles-Bement-Pond Co., New York.
R. McDougall Co., Galt, Ont.
H. W. Petrie, Toronto.
Pratt & Whitney Co., Hartford, Conn.

Lathes, Foot-Power.
American Tool Works Co., Cincinnati.
B. F. Barnes Co., Rockford, Ill.

Lathes, Screw Cutting.
B. F. Barnes Co., Rockford, Ill.
Baxter, Paterson & Co., Montreal.
Niles-Bement-Pond Co., New York.

Lathes, Automatic, Screw-Threading.
John Bertram & Sons Co., Dundas, Ont.
London Mach. Tool Co., Hamilton, Ont.
Pratt & Whitney Co., Hartford, Conn.

Lathes, Bench.
B. F. Barnes Co., Rockford, Ill.
London Mach. Tool Co., London, Ont.
Pratt & Whitney Co., Hartford, Conn.

Lathes, Turret.
American Tool Works Co., Cincinnati.
Baxter, Paterson & Co., Montreal.
John Bertram & Sons Co., Dundas, Ont.
Canada Machinery Agency, Montreal.
Gisholt Machine Co., Madison, Wis.

London Mach. Tool Co., Hamilton, Ont.
Niles-Bement-Pond Co., New York.
R. McDougall Co., Galt, Ont.
The Pratt & Whitney Co., Hartford, Conn.
Warner & Swasey Co., Cleveland, O.

Lathes, Buffing and Polishing.
Peudrith Machinery Co., Toronto.

Lath Mill Machinery.
Owen Sound Iron Works Co., Owen Sound

Leather Belting.
Baxter Paterson & Co., Montreal
Canada Machinery Agency, Montreal.
The Canadian Fairbanks Co., Montreal
Greey, Wm. & J. G., Toronto

Lime Stone Flux.
Hamilton Facing Mill Co., Hamilton.
Paxson, J. W., Co., Philadelphia

Locomotives, Air.
Canadian Rand Drill Co., Montreal.

Locomotives, Electrical.
Canadian Westinghouse Co., Hamilton
Jeffrey Mfg. Co., Columbus, Ohio.

Locomotives, Steam.
Canada Foundry Co., Toronto.
Canadian Rand Drill Co., Montreal

Lubricating Plumbago.
Detroit Foundry Supply Co., Detroit.
Dominion Foundry Supply Co., Montreal
Hamilton Facing Mill Co., Hamilton.
Paxson, J. W., Co., Philadelphia

Lumber Dry Kilns.
Dominion Heating & Ventilating Co., Hespeler.
H. W. Petrie, Toronto.
Sheldon's Limited, Galt, Ont.
B. F. Sturtevant Co., Hyde Park, Mass.

Machinery Dealers.
Baxter, Paterson & Co., Montreal.
Canada Machinery Agency, Montreal.
The Canadian Fairbanks Co., Montreal.
Miller Bros. & Toms, Montreal.
H. W. Petrie, Toronto.
The Smart-Turner Mach. Co., Hamilton.
Standard Bearings, Ltd., Niagara Falls.
Williams & Wilson, Montreal.

Machinery Designers.
Greey, Wm. & J. G., Toronto
Standard Bearings, Ltd., Niagara Falls.

Machinists.
W. H. Banfield & Sons, Toronto.
Greey, Wm. & J. G. Toronto
Hall Engineering Works, Montreal.
Link-Belt Co., Philadelphia.
John McDougall, Caledonian Iron Works Co., Montreal.
Miller Bros. & Toms, Montreal.
Paxson, J. W., Co., Philadelphia
Pendrith Machinery Co., Toronto.
Robb Engineering Co., Amherst, N.S.
The Smart-Turner Mach. Co., Hamilton.
Standard Bearings, Limited, Niagara Falls, Ont.
Waterous Engine Co., Brantford.

Machinists' Small Tools.
Armstrong Bros., Chicago.
Brown & Sharpe, Providence, R.I.
Butterfield & Co., Rock Island, Que.
Rice Lewis & Son, Montreal.
Pratt & Whitney Co., Hartford, Conn.
Standard Tool Co., Cleveland.
L. S. Starrett Co., Athol, Mass.
Williams & Wilson, Montreal.
Whiteman & Barnes Co., St. Catharines.

Malleable Flask Clamps.
Buffalo Foundry Supply Co., Buffalo.
Paxon, J. W., Co., Philadelphia

Mallet, Rawhide and Wood.
Buffalo Foundry Supply Co., Buffalo.
Detroit Foundry Supply Co., Detroit.
Paxson, J. W., Co., Philadelphia

Mandrels.
Brown & Sharpe, Providence, R.I.
A. B. Jardine & Co., Hespeler, Ont.
Miller Bros. & Toms, Montreal
The Pratt & Whitney Co., Hartford, Conn.
Standard Tool Co., Cleveland.
Whitman & Barnes Co., St. Catharines

Measuring Machines.
The Pratt & Whitney Co., Hartford, Conn.

Marine Work.
Kennedy, Wm., & Sons, Owen Sound.

Mechanical Draft.
Paxson, J. W., Co., Philadelphia
H. W. Petrie, Toronto.
Sheldon's Limited, Galt.
B. F. Sturtevant Co., Hyde Park, Mass.

Metallic Paints.
P. D. Dods & Co., Montreal.

Meters, Electrical.
Canadian Westinghouse Co., Hamilton.

Mill Machinery.
Baxter, Paterson & Co., Montreal.
Greey, Wm. & J. G., Toronto
The Goldie & McCulloch Co., Galt, Ont.
Jenckes Machine Co., Sherbrooke, Que.
John McDougall, Caledonian Iron Works Co., Montreal.
H. W. Petrie, Toronto.
Robb Engineering Co., Amherst, N.S.
Waterous Engine Co., Brantford.
Williams & Wilson, Montreal.

Milling Attachments.
Becker-Brainard Milling Machine Co. Hyde Park, Mass.
John Bertram & Sons Co., Dundas, Ont.
Brown & Sharpe, Providence, R.I.
Niles-Bement-Pond Co., New York.
Pratt & Whitney, Hartford, Conn.

Milling Machines, Horizontal.
Baxter, Paterson & Co., Montreal.
Becker-Brainard Milling Machine Co. Hyde Park, Mass.
John Bertram & Sons Co., Dundas, Ont.
Brown & Sharpe, Providence, R.I.
John Bertram & Sons Co., Dundas, Ont.
Canada Machinery Agency, Montreal.
London Mach. Tool Co., Hamilton, Ont.
Niles-Bement-Pond Co., New York.
Pratt & Whitney, Hartford, Conn.

Milling Machines, Plain.
American Tool Works Co., Cincinnati.
Becker-Brainard Milling Machine Co. Hyde Park, Mass.
Brown & Sharpe, Providence, R.I.
John Bertram & Sons Co., Dundas, Ont.
Canada Machinery Agency, Montreal.
The Canadian Fairbanks Co., Montreal.
London Mach. Tool Co., Hamilton, Ont.
Niles-Bement-Pond Co., New York.
H. W. Petrie, Toronto.
Pratt & Whitney Co., Hartford, Conn.
Williams & Wilson, Montreal.

Milling Machines, Universal.
American Tool Works Co., Cincinnati.
Becker-Brainard Milling Machine Co., Hyde Park, Mass.
Brown & Sharpe, Providence, R.I.
John Bertram & Sons Co., Dundas, Ont.
Canada Machinery Agency, Montreal.
The Canadian Fairbanks Co., Montreal.
London Mach. Tool Co., Hamilton, Ont.
Niles-Bement-Pond Co., New York.
H. W. Petrie, Toronto.
Williams & Wilson, Montreal.

Milling Machines, Vertical.
Becker-Brainard Milling Machine Co., Hyde Park, Mass.
Brown & Sharpe, Providence, R.I.
John Bertram & Sons Co., Dundas, Ont.
Canada Machinery Agency, Montreal.
London Mach. Tool Co., Hamilton, Ont.
Niles-Bement-Pond Co., New York.

Milling Tools.
Wm. Abbott, Montreal.
Becker-Brainard Milling Machine Co., Hyde Park, Mass.
Brown & Sharpe, Providence, R.I.
Geometric Tool Co., New Haven, Conn.
London Mach. Tool Co., Hamilton, Ont.
Pratt & Whitney Co., Hartford, Conn.
Standard Tool Co., Cleveland.
Standard Bearings, Limited, Niagara Falls, Ont.

Mining Machinery.
Allis-Chalmers-Bullock, Limited, Montreal
Canadian Rand Drill Co., Montreal.
Chicago Pneumatic Tool Co., Chicago.
Jeffrey Mfg. Co., Columbus, Ohio.
Jenckes Machine Co., Sherbrooke, Que.
Laurie Engine & Machine Co., Montreal.
John McDougall, Caledonian Iron Works Co., Montreal.
Miller Bros. & Toms, Montreal.
T. & H. Electric Co. Hamilton.

Mixing Machines, Dough.
Greey, Wm. & J. G., Toronto
Pendrith Machinery Co., Toronto.
Paxson, J. W., Co., Philadelphia

Mixing Machines, Special.
Greey, Wm. & J. G., Toronto
Pendrith Machinery Co., Toronto.
Paxson, J. W., Co., Philadelphia

Model Tools.
Globe Machine & Stamping Co, Cleveland, Ohio.
Standard Bearings, Ltd., Niagara Falls.
Wells Pattern and Model Works, Toronto

Motors, Electric.
Allis-Chalmers-Bullock Limited, Montreal
Canadian General Electric Co., Toronto
Canadian Westinghouse Co., Hamilton.
Consolidated Electric Co., Toronto
Hall Engineering Works, Montreal.
The Packard Electric Co., St. Catharines.
B. F. Sturtevant Co., Hyde Park, Mass.
T. & H. Electric Co., Hamilton.

Motors, Air.
Canadian Rand Drill Co., Montreal.
Chicago Pneumatic Tool Co., Chicago.

Molders' Supplies.
Buffalo Foundry Supply Co., Buffalo.
Detroit Foundry Supply Co., Windsor.
Dominion Foundry Supply Co., Montreal
Hamilton Facing Mill Co., Hamilton.
Paxson, J. W., Co., Philadelphia

Molders' Tools.
Buffalo Foundry Supply Co., Buffalo.
Detroit Foundry Supply Co., Windsor
Dominion Foundry Supply Co., Montreal
Hamilton Facing Mill Co., Hamilton.
Paxson, J. W., Co., Philadelphia

Molding Machines.
Buffalo Foundry Supply Co., Buffalo.
Hamilton Facing Mill Co., Hamilton.
J. W. Paxson Co., Philadelphia, Pa.

Molding Sand.
T. W. Barnes, Hamilton.
Buffalo Foundry Supply Co., Buffalo.
Detroit Foundry Supply Co., Windsor
Dominion Foundry Supply Co., Montreal
Hamilton Facing Mill Co., Hamilton.
Paxson, J. W., Co., Philadelphia

Nippers, Stay Bolt.
Chicago Pneumatic Tool Co., Chicago.

Nut Tappers.
John Bertram & Sons Co., Dundas, Ont
A. B. Jardine & Co., Hespeler.
London Mach. Tool Co., Hamilton.
National Machinery Co., Tiffin, Ohio.

Oatmeal Mill Machinery.
Greey, Wm. & J. G., Toronto
The Goldie & McCulloch Co., Galt

Oils, Core.
Buffalo Foundry Supply Co., Buffalo.
Dominion Foundry Supply Co., Montreal
Hamilton Facing Mill Co., Hamilton.
Paxson, J. W., Co., Philadelphia

Paint Mill Machinery.
Greey, Wm. & J. G., Toronto

Pans, Lathe.
Cleveland Wire Spring Co., Cleveland
Paxson, J. W., Co., Philadelphia

Pans, Steel Shop.
Cleveland Wire Spring Co. Cleveland

Patent Solicitors.
Hanbury A. Budden, Montreal.
Fetherstonhaugh & Blackmore, Montreal
Fetherstonhaugh & Co., Montreal.
Marion & Marion, Montreal.
Ridout & Maybee, Toronto.

Patterns.
John Carr, Hamilton.
Hamilton Pattern Works, Hamilton.
John McDougall, Caledonian Iron Works Co., Montreal
Miller Bros. & Toms Montreal.
Wells' Pattern and Model Works, Toronto.

Pipe Cutting and Threading Machines.
Butterfield & Co., Rock Island, Que.
Canada Machinery Agency, Montreal.
Hart Mfg. Co., Cleveland.
A. B. Jardine & Co., Hespeler, Ont.
London Mach. Tool Co., Hamilton, Ont.
R. McDougall Co., Galt, Ont.
Niles-Bement-Pond Co., New York.

Pipe, Municipal.
Canadian Pipe Co., Vancouver, B.C.
Pacific Coast Pipe Co., Vancouver, B.C.

Pipe, Waterworks.
Canadian Pipe Co., Vancouver, B.C.
Pacific Coast Pipe Co., Vancouver, B.C

Pipe, Wooden.
Canadian Pipe Co., Vancouver, B.C.
Pacific Coast Pipe Co., Vancouver, B.C.

Planers, Standard.
American Tool Works, Cincinnati.
John Bertram & Sons Co., Dundas, Ont.
Canada Machinery Agency, Montreal.
The Canadian Fairbanks Co., Montreal.
London Mach. Tool Co., Hamilton, Ont.
R. McDougall Co., Galt, Ont.
Niles-Bement-Pond Co., New York.
H. W. Petrie, Toronto.
Pratt & Whitney Co., Hartford, Conn.
Williams & Wilson, Montreal.

Planers, Rotary.
John Bertram & Sons Co., Dundas, Ont.
London Mach. Tool Co., Hamilton, Ont.
Niles-Bement-Pond Co., New York.

Power Hack Saw Machines.
Baxter, Paterson & Co., Montreal.

Power Plants.
John McDougall Caledonian Iron Works Co., Montreal.
The Smart-Turner Mach. Co., Hamilton

Planing Mill Fans.
Dominion Heating & Ventilating Co., Hespeler
Sheldon's Limited, Galt, Ont.
B. F. Sturtevant Co., Hyde Park, Mass.

Plug Drillers, Pneumatic.
Canada Foundry Co., Limited, Toronto.
Canadian Rand Drill Co., Montreal

Plumbago.
Buffalo Foundry Supply Co., Buffalo.
Detroit Foundry Supply Co., Windsor
Dominion Foundry Supply Co., Montreal
Hamilton Facing Mill Co., Hamilton.
J. W. Paxson Co., Philadelphia, Pa.

Pneumatic Separators.
Paxson, J. W., Co., Philadelphia
Sturtevant, B. F., Co., Hyde Park, Mass.

Pneumatic Tools.
Allis-Chalmers-Bullock, Montreal.
Canadian Rand Drill Co., Montreal.
Chicago Pneumatic Tool Co., Chicago
Hamilton Facing Mill Co., Hamilton.
Hanna Engineering Works, Chicago.
Paxson, J. W., Co., Philadelphia

Presses, Drill.
American Tool Works Co., Cincinnati.
Niles-Bement-Pond Co., New York.

Presses, Drop.
W. H. Banfield & Son, Toronto.
E. W. Bliss Co., Brooklyn, N.Y.
Canada Machinery Agency, Montreal.
Laurie Engine & Machine Co., Montreal
Miller Bros. & Toms, Montreal.
Niles-Bement-Pond Co., New York.

Presses, Hand.
E. W. Bliss Co., Brooklyn, N.Y.

Presses, Hydraulic.
John Bertram & Sons Co., Dundas, Ont.
Laurie Engine & Machine Co., Montreal.
London Mach. Tool Co., Hamilton, Ont.
John McDougall Caledonian Iron Works Co., Montreal.
Miller Bros. & Toms, Montreal.
Niles-Bement-Pond Co., New York.
Perrin, Wm. R., & Co., Toronto

Presses, Power.
E. W. Bliss Co., Brooklyn, N.Y.
Canada Machinery Agency, Montreal.
London Mach. Tool Co., Hamilton, Ont.
John McDougall Caledonian Iron Works Co., Montreal.
Niles-Bement-Pond Co., New York.

Presses Power Screw.
Perrin, Wm. R., & Co., Toronto

Pulp Mill Machinery.
Greey, Wm. & J. G., Toronto
Jeffrey Mfg. Co., Columbus, Ohio.
Laurie Engine & Machine Co., Montreal.
John McDougall Caledonian Iron Works Co., Montreal.
Waterous Engine Works Co., Brantford

Pulleys.
Baxter, Paterson & Co., Montreal.
Canada Machinery Agency, Montreal
The Canadian Fairbanks Co., Montreal.
The Goldie & McCulloch Co., Galt.
Greey, Wm. & J. G., Toronto
Laurie Engine & Machine Co., Montreal.
Link-Belt Co., Philadelphia.
John McDougall Caledonian Iron Works Co., Montreal.
Miller Bros. & Toms, Montreal.
Owen Sound Iron Works Co., Owen Sound
Pendrith Machinery Co., Toronto.
H. W. Petrie, Toronto.
The Smart-Turner Mach. Co., Hamilton.
Standard Bearings, Ltd., Niagara Falls,
Williams & Wilson, Montreal.
Waterous Engine Co., Brantford.

Pumps, Hydraulic.
Perrin, Wm. R. & Co., Toronto

Pumps, Steam.
Allis-Chalmers-Bullock, Limited, Montreal
Canada Foundry Co., Toronto.
Canada Machinery Agency, Montreal.
D'Olier Engineering Co., New York.
The Goldie & McCulloch Co., Galt.
John McDougall Caledonian Iron Works, Montreal.
R. McDougall Co., Galt, Ont.
H. W. Petrie, Toronto.
St. Lawrence Supply Co., Montreal.
The Smart-Turner Mach. Co., Hamilton.
Standard Bearings, Ltd., Niagara Falls,
Waterous Engine Co., Brantford.

Pumping Machinery.
Canada Foundry Co., Limited, Toronto
Canada Machinery Agency, Montreal.
Canadian Rand Drill Co., Montreal.
Chicago Pneumatic Tool Co., Chicago.
D'Olier Engineering Co., New York.
Hall Engineering Works, Montreal, Que.
Laurie Engine & Machine Co., Montreal.
London Mach. Tool Co., Hamilton, Ont.
John McDougall Caledonian Iron Works Co., Montreal.
R. McDougall Co., Galt, Ont.
The Smart-Turner Mach. Co., Hamilton
Standard Bearings, Ltd., Niagara Falls,

Punches and Dies.
W. H. Banfield & Sons, Toronto.
Butterfield & Co., Rock Island.
Globe Machine & Stamping Co.
A. B. Jardine & Co., Hespeler, Ont.
London Mach. Tool Co., Hamilton, Ont.
Miller Bros & Toms, Montreal.
Pratt & Whitney Co., Hartford, Conn.
H. W. Petrie, Toronto.
Standard Bearings, Ltd., Niagara Falls.

Punches, Power.
John Bertram & Sons Co., Dundas, Ont.
E. W. Bliss Co., Brooklyn, N.Y.
Canada Machinery Agency, Montreal.
Canadian Buffalo Forge Co., Montreal.
London Mach. Tool Co., Hamilton, Ont.
Niles-Bement-Pond Co., New York.

Punches, Turret.
London Mach. Tool Co., London. Ont.

Punching Machines, Horizontal.
John Bertram & Sons Co., Dundas, Ont.
London Mach. Tool Co., Hamilton, Ont.
Niles-Bement-Pond Co., New York.

Quartering Machines.
John Bertram & Sons Co., Dundas, Ont.
London Mach. Tool Co., Hamilton, Ont.

Rammers, Bench and Floor.
Buffalo Foundry Supply Co., Buffalo.
Detroit Foundry Supply Co., Windsor
Hamilton Facing Mill Co., Hamilton.
Paxson, J. W., Co., Philadelphia

Rapping Plates.
Detroit Foundry Supply Co., Windsor
Hamilton Facing Mill Co., Hamilton.
Paxson, J. W., Co., Philadelphia

Raw Hide Pinions.
Brown, David & Sons, Huddersfie'd, Eng.

Reamers.
Wm. Abbott, Montreal.
Baxter, Paterson & Co., Montreal.
Brown & Sharpe, Providence, R.I.
Butterfield & Co., Rock Island.
Hanna Engineering Works Chicago.
A. B. Jardine & Co., Hespeler, Ont.
John Millen & Son, Ltd., Montreal.
Morse Twist Drill and Machine Co., New Bedford, Mass
Pratt & Whitney Co., Hartford, Conn.
Standard Tool Co., Cleveland.
Whitman & Barnes Co., St. Catharines.

Reamers, Steel Taper.
Butterfield & Co., Rock Island.
Chicago Pneumatic Tool Co., Chicago.
A. B. Jardine & Co., Hespeler, Ont.
John Millen & Son Ltd., Montreal.
Pratt & Whitney Co., Hartford, Conn.
Standard Tool Co., Cleveland.
Whitman & Barnes Co., St. Catharines.

Rheostats.
Canadian General Electric Co., Toronto.
Canadian Westinghouse Co., Hamilton.
Hall Engineering Works, Montreal, Que.
T. & H. Electric Co., Hamilton.

Riddles.
Buffalo Foundry Supply Co., Buffalo
Detroit Foundry Supply Co., Windsor
Hamilton Facing Mill Co., Hamilton.
J. W. Paxson Co., Philadelphia, Pa.

Riveters, Hydraulic.
London Mach. Tool Co., Hamilton, Ont.
Niles-Bement-Pond Co., New York.

Riveters, Pneumatic.
Hanna Engineering Works, Chicago.

Rolls, Bending.
John Bertram & Sons Co., Dundas, Ont.
London Mach. Tool Co., Hamilton, Ont.
Niles-Bement-Pond Co., New York.

Rolls, Chilled Iron.
Greey, Wm. & J. G., Toronto

Rolls, Sand Cast.
Greey, Wm. & J. G., Toronto

Rotary Blowers.
Paxson, J. W., Co., Philadelphia
Sturtevant, B. F., Co., Hyde Park, Mass.

Rotary Converters.
Allis-Chalmers-Bullock, Ltd., Montreal.
Canadian Westinghouse Co., Hamilton
Paxson, J. W., Co., Philadelphia
Toronto and Hamilton Electric Co., Hamilton.

Safes.
Baxter, Paterson & Co., Montreal.
The Goldie & McCulloch Co., Galt.

Sand, Bench.
Buffalo Foundry Supply Co., Buffalo.
Detroit Foundry Supply Co., Windsor
Hamilton Facing Mill Co., Hamilton.
Paxson, J. W., Co., Philadelphia

Sand Blast Machinery.
Canadian Rand Drill Co., Montreal.
Chicago Pneumatic Tool Co., Chicago.
Detroit Foundry Supply Co., Windsor
Hamilton Facing Mill Co., Hamilton.
J. W. Paxson Co., Philadelphia, Pa.

Sand, Brass.
Buffalo Foundry Supp'y Co., Buffalo.
Detroit Foundry Supply Co., Windsor
Paxson, J. W., Co., Philadelphia

Sand Fire.
Detroit Foundry Supply Co., Detroit.
Paxson, J. W., Co., Philadelphia

Sand, Heavy, Grey Iron.
Buffalo Foundry Supply Co., Buffalo.
Detroit Foundry Supply Co., Windsor
Hamilton Facing Mill Co., Hamilton.
Paxson, J. W., Co., Philadelphia

Sand, Malleable.
Buffalo Foundry Supply Co., Buffalo.
Detroit Foundry Supply Co., Windsor
Hamilton Facing Mill Co., Hamilton.
Paxson, J. W., Co., Philadelphia

Sand, Medium Grey Iron.
Buffalo Foundry Supply Co., Buffalo.
Detroit Foundry Supply Co., Windsor
Hamilton Facing Mill Co., Hamilton.
Paxson, J. W., Co., Philadelphia

Sand, Plate.
Buffalo Foundry Supply Co., Buffalo.
Detroit Foundry Supply Co., Windsor
Hamilton Facing Mill Co., Hamilton.
Paxson, J. W., Co., Philadelphia

Sand Sifters.
Buffalo Foundry Supply Co., Buffalo.
Detroit Foundry Supply Co., Windsor
Dominion Foundry Supply Co., Montreal
Hamilton Facing Mill Co., Hamilton.
Hanna Engineering Works, Chicago.
Paxson, J. W., Co., Philadelphia

Sand, Stove Plate.
Buffalo Foundry Supply Co., Buffalo.
Detroit Foundry Supply Co., Windsor
Hamilton Facing Mill Co., Hamilton.
Paxson J. W., Co., Philadelphia

Saw Mill Machinery.
Allis-Chalmers-Bullock, Limited, Montreal
Baxter, Paterson & Co., Montreal.
Canada Machinery Agency, Montreal.
Goldie & McCulloch Co., Galt.
Greey, Wm. & J. G., Toronto
Miller Bros. & Toms, Montreal
Owen Sound Iron Works Co., Owen Sound
H. W. Petrie, Toronto.
Robb Engineering Co., Amherst.
Standard Bearings, Ltd., Niagara Falls, Ont.
Waterous Engine Works, Brantford
Williams & Wilson, Montreal.

Sawing Machines, Metal.
Niles-Bement-Pond Co. New York.
Paxson, J. W., Co., Philadelphia

Saws, Hack.
Baxter Paterson & Co., Montreal.
Canada Machinery Agency, Montreal.
Detroit Foundry Supply Co., Windsor
London Mach. Tool Co., Hamilton
Rice Lewis & Son, Toronto.
John Millen & Son, Ltd., Montreal.
L. S. Starrett Co., Athol, Mass.

Screw Cutting Tools
Hart Manufacturing Co., Cleveland, Ohio

Screw Machines, Automatic.
Brown & Sharpe, Providence, R.I.
Canada Machinery Agency, Montreal.
Cleveland Automatic Machine Co., Cleveland, Ohio
London Mach. Tool Co., Hamilton, Ont.
Pratt & Whitney Co., Hartford, Conn.

Screw Machines, Hand.
Brown & Sharpe, Providence, R.I.
Canada Machinery Agency Montreal.
A. B. Jardine & Co., Hespeler.
London Mach. Tool Co., Hamilton, Ont.
Pratt & Whitney Co., Hartford, Conn.
Warner & Swasey Co., Cleveland, O.

Screw Plates.
Butterfield & Co., Rock Island, Que.
Hart Manufacturing Co., Cleveland, Ohio
A. B. Jardine & Co., Hespeler.

Second-Hand Machinery.
American Tool Works Co., Cincinnati.
Canada Machinery Agency, Montreal.
The Canadian Fairbanks Co., Montreal.
Goldie & McCulloch Co., Galt.
Machinery Exchange, Montreal.
Niles-Bement-Pond Co., New York.
H. W. Petrie, Toronto.
Robb Engineering Co., Amherst, N S.
Waterous Engine Co., Brantford.
Williams & Wilson, Montreal.

Shafting.
Baxter, Paterson & Co., Montreal.
Canada Machinery Agency, Montreal
The Canadian Fairbanks Co., Montreal.
The Goldie & McCulloch Co., Galt, Ont
Greey, Wm. & J. G. Toronto
Jenckes Machine Co., Sherbrooke, Que.
Kennedy, Wm., & Sons, Owen Sound
Niles-Bement-Pond Co., New York.
Owen Sound Iron Works Co., Owen Sound
Pendrith Machine Co., Toronto.
H. W. Petrie, Toronto.
Smart-Turner Machine Co., Hamilton.
Union Drawn Steel Co., Hamilton.
Waterous Engine Co., Brantford.

Shapers.
American Tool Works Co., Cincinnati.
John Bertram & Sons Co., Dundas, Ont
Canada Machinery Agency, Montreal.
The Canadian Fairbanks Co., Montreal.
London Mach. Tool Co., Hamilton, Ont.
Niles-Bement-Pond Co., New York.
H. W. Petrie, Toronto.
Potter & Johnston Machine Co., Pawtucket, R.I.
Pratt & Whitney Co., Hartford, Conn.
Williams & Wilson, Montreal.

Shearing Machine, Bar.
John Bertram & Sons Co., Dundas, Ont.
A. B. Jardine & Co., Hespeler.
London Mach. Tool Co., Hamilton, Ont.
Niles-Bement-Pond Co., New York.

Shears, Power.
John Bertram & Sons Co., Dundas, Ont.
Canada Machinery Agency, Montreal.
A. B. Jardine & Co., Hespeler, Ont.
Niles-Bement-Pond Co., New York.
Paxson, J. W., Co., Philadelphia

Sheet Metal Goods
Globe Machine & Stamping Co., Cleveland, Ohio.

Sheet Steel Work.
Owen Sound Iron Works Co., Owen Sound

Shingle Mill Machinery.
Owen Sound Iron Works Co., Owen Sound

Shop Trucks.
Greey, Wm. & J. G., Toronto
Owen Sound Iron Works Co., Owen Sound
Paxson, J. W., Co., Philadelphia

Shovels.
Baxter, Paterson & Co., Montreal.
Buffalo Foundry Supply Co., Buffalo.
Detroit Foundry Supply Co., Detroit.
Dominion Foundry Supply Co., Montreal
Hamilton Facing Mill Co., Hamilton.
Paxson, J. W., Co., Philadelphia

Shovels, Steam.
Allis-Chalmers-Bullock, Montreal.

Sieves.
Buffalo Foundry Supply Co. Buffalo.
Detroit Foundry Supply Co., Windsor
Dominion Foundry Supply Co., Montreal
Hamilton Facing Mill Co., Hamilton.
Paxson, J. W. Co., Philadelphia

Silver Lead.
Buffalo Foundry Supply Co. Buffalo.
Detroit Foundry Supply Co., Windsor
Dominion Foundry Supply Co., Montreal
Hamilton Facing Mill Co., Hamilton.
Paxson, J. W., Co., Philadelphia

Sleeves, Reducing.
Chicago Pneumatic Tool Co., Chicago.

Snap Flasks
Buffalo Foundry Supply Co., Buffalo.
Detroit Foundry Supply Co., Windsor
Dominion Foundry Supply Co., Montreal
Hamilton Facing Mill Co., Hamilton.
Paxson, J. W., Co., Philadelphia

Soapstone.
Buffalo Foundry Supply Co., Buffalo.
Detroit Foundry Supply Co., Windsor
Dominion Foundry Supply Co., Montreal
Hamilton Facing Mill Co., Hamilton.
Paxson, J. W., Co., Philadelphia

Solders.
Lumen Bearing Co., Toronto

Special Machinery.
W. H. Banfield & Sons, Toronto.
Baxter, Paterson & Co., Montreal
John Bertram & Sons Co., Dundas, Ont.
Globe Machine & Stamping Co., Cleveland Ohio.
Greey, Wm. & J. G., Toronto
Hanna Engineering Works, Chicago.
Laurie Engine & Machine Co., Montreal.
London Mach. Tool Co., Hamilton, Ont.
R. McDougall Co., Galt, Ont.
Pendrith Machinery Co., Toronto.
H. W. Petrie, Toronto.
The Smart-Turner Mach. Co., Hamilton
Standard Bearings, Ltd., Niagara Falls,
Waterous Engine Co., Brantford.

Special Machines and Tools.
Paxson, J. W., Philadelphia
Pratt & Whitney, Hartford, Conn.
Standard Bearings, Ltd., Niagra Falls,

Special Manufacturing.
Globe Machine & Stamping Co., Cleveland.
Miller Bros. & Toms, Montreal.
Pendrith Machine Co., Toronto.
Standard Bearings, Ltd., Niagara Falls.

Speed Changing Countershafts.
The Canadian Fairbanks Co., Montreal

Spike Machines.
National Machinery Co., Tiffin, O.
The Smart-Turner Mach. Co., Hamilton

Spray Cans.
Detroit Foundry Supply Co., Windsor
Dominion Foundry Supply Co., Montreal
Hamilton Facing Mill Co., Hamilton.
Paxson, J. W., Co., Philadelphia

Springs, Automobile.
Cleveland Wire Spring Co., Cleveland

Springs, Coiled Wire.
Cleveland Wire Spring Co., Cleveland

Springs, Machinery.
Cleveland Wire Spring Co., Cleveland

Springs, Upholstery.
Cleveland Wire Spring Co., Cleveland

Sprue Cutters.
Detroit Foundry Supply Co., Windsor
Dominion Foundry Supply Co., Montreal
Hamilton Facing Mill Co., Hamilton.
Paxson J. W., Co., Philadelphia

Stamp Mills.
Allis-Chalmers-Bullock, Limited, Montreal
Paxson, J. W., Co., Philadelphia

Steam Hot Blast Apparatus
Dominion Heating & Ventilating Co. Hespeler, Ont.
Sheldon's Limited, Galt.
B. F. Sturtevant Co., Hyde Park, Mass.

Steam Separators.
Dominion Heating & Ventilating Co., Hespeler
Robb Engineering Co., Montreal.
Sheldon's Limited, Galt.
Smart-Turner Mach. Co., Hamilton
Waterous Engine Co., Brantford.

Steam Specialties.
Dominion Heating & Ventilating Co., Hespeler
Sheldon's Limited, Galt.

Steam Traps.
Canada Machinery Agency, Montreal.
Dominion Heating & Ventilating Co., Hespeler
Pendrith Machinery Co., Toronto.
Sheldon's Limited, Galt.
B. F. Sturtevant Co., Hyde Park, Mass.

Steel Pressure Blowers
Buffalo Foundry Supply Co., Buffalo.
Dominion Foundry Supply Co., Montreal
Dominion Heating & Ventilating Co., Hespeler
Hamilton Facing Mill Co., Hamilton.
J. W. Paxson Co., Philadelphia, Pa.
Sheldon's Limited, Galt.
B. F. Sturtevant Co., Hyde Park, Mass.

Steel Tubes.
Baxter, Paterson & Co., Montreal.
John Millen & Son, Montreal.

Steel, High Speed.
Wm. Abbott, Montreal.
Canadian Fairbanks Co., Montreal.
Alex. Gibb, Montreal.
Jessop, Wm., & Sons, Sheffield, Eng.
B. K. Morton Co., Sheffield, Eng.
Williams & Wilson, Montreal.

Stocks and Dies
Hart Manufacturing Co., Cleveland, Ohio
Pratt & Whitney Co., Hartford, Conn.

Stone Cutting Tools, Pneumatic
Allis-Chalmers-Bullock, Ltd., Montreal
Canadian Rand Drill Co., Montreal.

Stone Surfacers
Chicago Pneumatic Tool Co., Chicago.

Stove Plate Facings.
Buffalo Foundry Supply Co., Buffalo
Detroit Foundry Supply Co., Windsor.
Dominion Foundry Supply Co., Montreal
Hamilton Facing Mill Co., Hamilton.
Paxson, J. W., Co., Philadelphia

Swage, Block.
A. B. Jardine & Co., Hespeler, Ont.

Switchboards.
Allis-Chalmers-Bullock, Limited, Montreal
Canadian General Electric Co., Toronto.
Canadian Westinghouse Co., Hamilton
Hall Engineering Works, Montreal, Que.
Toronto and Hamilton Electric Co., Hamilton.

Talc.
Buffalo Foundry Supply Co., Buffalo.
Detroit Foundry Supply Co., Windsor.
Hamilton Facing Mill Co., Hamilton.
Paxson, J. W., Co., Philadelphia

Tapping Machines and Attachments.
American Tool Works Co., Cincinnati.
John Bertram & Sons Co., Dundas, Ont.
Bickford Drill & Tool Co., Cincinnati.
The Geometric Tool Co., New Haven.
A. B. Jardine & Co., Hespeler.
London Mach. Tool Co., Hamilton, Ont
Murchey Machine & Tool Co., Detroit.
Niles-Bement-Pond Co., New York.
H. W. Petrie, Toronto.
Pratt & Whitney, Cincinnati, O.
Williams & Wilson, Montreal.

Tapes, Steel.
Rice Lewis & Son, Toronto.
John Millen & Son, Ltd., Montreal.
L. S. Starrett Co., Athol, Mass.

Taps.
Baxter, Paterson & Co., Montreal.
Whitman & Barnes Co., St. Catharines.

Taps, Collapsing.
The Geometric Tool Co., New Haven.

Taps and Dies.
Wm. Abbott, Montreal.
Baxter, Paterson & Co., Montreal.
Butterfield & Co., Rock Island, Que.
The Geometric Tool Co., New Haven.
A. B. Jardine & Co., Hespeler, Ont.
Rice Lewis & Son, Toronto.
John Millen & Son, Ltd., Montreal.
Pratt & Whitney Co., Hartford, Conn.
Standard Tool Co., Cleveland.
L. S. Starrett Co., Athol, Mass.

Testing Laboratory.
Detroit Testing Laboratory, Detroit

Testing Machines.
Detroit Foundry Supply Co., Windsor.
Dominion Foundry Supply Co., Montreal
Hamilton Facing Mill Co., Hamilton.
Paxson, J. W., Co., Philadelphia

Thread Cutting Tools.
Hart Manufacturing Co., Cleveland, Ohio

Tiling, Opal Glass.
Toronto Plate Glass Importing Co., Toronto.

Time Switches, Automatic, Electric.
Berlin Electric Mfg. Co., Toronto.

Tinware Machinery
Canada Machinery Agency, Montreal.

Tire Upsetters or Shrinkers
A. B. Jardine & Co., Hespeler, Ont.

Tool Cutting Machinery
Canadian Rand Drill Co., Montreal.

Tool Holders.
Armstrong Bros. Tool Co., Chicago.
Baxter, Paterson & Co., Montreal.
Fairbanks Co., Springfield, Ohio.
John Millen & Son, Ltd., Montreal.
H. W. Petrie, Toronto.
Pratt & Whitney Co., Hartford, Conn.

Tool Steel.
Wm. Abbott, Montreal.
Wm. Jessop, Sons & Co., Toronto.
Canadian Fairbanks Co., Montreal
B. K. Morton & Co., Sheffield, Eng.
Williams & Wilson, Montreal.

Torches, Steel.
Baxter, Paterson & Co., Montreal.
Detroit Foundry Supply Co., Windsor.
Dominion Foundry Supply Co., Montreal
Hamilton Facing Mill Co., Hamilton.
Paxson, J. W., Co., Philadelphia

Transformers and Convertors
Allis-Chalmers-Bullock, Montreal.
Canadian General Electric Co., Toronto.
Canadian Westinghouse Co., Hamilton.
Hall Engineering Works, Montreal, Que.
T. & H. Electric Co., Hamilton.

Transmission Machinery
Allis-Chalmers-Bullock, Montreal.
The Canadian Fairbanks Co., Montreal.
Greey, Wm. & J. G., Toronto.
Laurie Engine & Machine Co., Montreal.
Link-Belt Co., Philadelphia.
Miller Bros. & Toms, Montreal.
Paxson, J. W., Co., Philadelphia
H. W. Petrie, Toronto.
The Smart-Turner Mach. Co., Hamilton.
Waterous Engine Co., Brantford.

Transmission Supplies
Baxter, Paterson & Co., Montreal.
The Canadian Fairbanks Co., Montreal.
Wm. & J. G. Greey, Toronto.
The Goldie & McCulloch Co., Galt.
Miller Bros. & Toms, Montreal
Pendrith Machinery Co., Toronto.
H. W. Petrie, Toronto.

Trolleys
Canadian Rand Drill Co., Montreal.
John Millen & Son, Ltd., Montreal.
Miller Bros. & Toms, Montreal
Northern Engineering Works, Detroit

Trolley Wheels.
Lumen Bearing Co., Toronto

Trucks, Dryer and Factory
Dominion Heating & Ventilating Co., Hespeler
Greey, Wm. & J. G., Toronto.
Northern Engineering Works, Detroit
Sheldon's Limited, Galt, Ont.

Tube Expanders (Rollers)
Chicago Pneumatic Tool Co., Chicago.
A. B. Jardine & Co., Hespeler.

Turbines, Steam
Allis-Chalmers-Bullock Limited, Montreal
Canadian General Electric Co., Toronto.
Canadian Westinghouse Co., Hamilton.
D'Olier Engineering Co., New York.
Kerr Turbine Co., Wellsville, N.Y.

Turntables
Detroit Foundry Supply Co., Windsor.
Dominion Foundry Supply Co., Montreal
Hamilton Facing Mill Co., Hamilton.
Northern Engineering Works, Detroit
Paxson, J. W., Co., Philadelphia

Turret Machines.
American Tool Works Co., Cincinnati.
John Bertram & Sons Co., Dundas, Ont.
The Canadian Fairbanks Co., Montreal.
London Mach. Tool Co., Hamilton, Ont.
Niles-Bement-Pond Co., New York.
H. W. Petrie, Toronto.
Pratt & Whitney Co., Hartford, Conn.
Warner & Swasey Co., Cleveland, Ohio.
Williams & Wilson, Montreal.

Upsetting and Bending Machinery.
John Bertram & Sons Co., Dundas, Ont.
A. B. Jardine & Co., Hespeler.
London Mach. Tool Co., Hamilton, Ont.
National Machinery Co., Tiffin, O.
Niles-Bement-Pond Co., New York.

Valves, Blow-off.
Chicago Pneumatic Tool Co., Chicago.

Valves, Back Pressure.
Dominion Heating & Ventilating Co. Hespeler
Sheldon's Limited, Galt.

Vaults.
The Goldie & McCulloch Co., Galt.

Ventilating Apparatus.
Canada Machinery Agency, Montreal
Dominion Heating & Ventilating Co., Hespeler, Ont.
Sheldon & Sheldon, Galt, Ont.
B. F. Sturtevant Co., Hyde Park, Mass.

Vises.
Baxter, Paterson & Co., Montreal.
Butterfield & Co., Rock Island, Que.
Rice Lewis & Son, Toronto
John Millen & Son, Ltd., Montreal.
Miller Bros & Toms, Montreal.

Vises, Planer and Shaper.
American Tool Works Co., Cincinnati, O.
A. B. Jardine & Co., Hespeler, Ont.
John Millen & Son, Ltd., Montreal.
Niles-Bement-Pond Co., New York.

Washer Machines.
National Machinery Co., Tiffin, Ohio.

Water Wheels.
Allis-Chalmers-Bullock Co., Montreal.
Barber, Chas., & Sons, Meaford.
Canada Machinery Agency, Montreal.
The Goldie & McCulloch Co., Galt, Ont.
Greey, Wm. & J. G., Toronto.
Wm. Kennedy & Sons, Owen Sound.

Water Wheels, Turbine.
Kennedy, Wm., & Sons, Owen Sound

Wheelbarrows.
Baxter, Paterson & Co., Montreal.
Buffalo Foundry Supply Co., Buffalo
Detroit Foundry Supply Co., Windsor.
Dominion Foundry Supply Co., Montreal
Hamilton Facing Mill Co., Hamilton.
Paxson, J. W., Co., Philadelphia

Winches, Hand.
Pendrith Machinery Co., Toronto.

Window Wire Guards.
Expanded Metal and Fireproofing Co. Toronto.
B. Greening Wire Co., Hamilton, Ont.
Paxson, J. W., Co. Philadelphia

Wire Chains.
The B. Greening Wire Co., Hamilton.

Wire Cloth and Perforated Metals.
Expanded Metal and Fireproofit. Co. Toronto.
B. Greening Wire Co., Hamilton. Ont.
Paxson, J. W., Co, Philadelphia.

Wire Guards and Railings.
Expanded Metal and Fireproofing Co. Toronto.
B. Greening Wire Co. Hamilton, Ont.

Wire Nail Machinery.
National Machinery Co., Tiffin, Ohio.

Wire Rope.
B. Greening Wire Co., Hamilton, Ont.

Wood-working Machinery.
Baxter, Paterson & Co., Montreal
Canada Machinery Agency, Montreal.
The Canadian Fairbanks Co., Montreal.
Goldie & McCulloch Co., Galt.
H. W. Petrie, Toronto.
Waterous Engine Works Co., Brantford.
Williams & Wilson, Montreal.

ALPHABETICAL INDEX.

OWEN SOUND IRON WORKS, LIMITED
OWEN SOUND, ONT.

Cement Mill Machinery, Boiler and Steel Tank Work of all kinds, Grey Iron and Brass Castings

---

PATTERNS
WELLS' PATTERN AND MODEL WORKS
Tel. Main 3581. (HARRY WELLS, Proprietor.)
Patterns and Models made in wood and metal for Engines, Pumps, Furnaces, Agricultural, Electrical and Architectural Works and Machines of every description.
35 Richmond St. E., Toronto

Don't let your moulders swear, get your patterns made by
JOHN CARR
14 Mary St., HAMILTON
Phone 2755

We should have your business.

If prompt delivery, reasonable prices and best of workmanship appeal to you, we should have your pattern orders.

SEND FOR PATTERN CLASSIFICATION CARD
THE HAMILTON PATTERN WORKS
134 Bay St. N., HAMILTON, ONT.
Phone, 2664.

FOR SALE

1	10-H.P. Buffalo 4-cylinder 4-cycle Marine Motor, good as new	$250.00
1	34-H.P. Trebert, complete with reversing wheel, used only one season. Price	450.00
1	6-H.P. Gillies Stationary Gasolene Engine, will sell cheap.	
1	220-Volt 2 H.P. Electric Motor, London make, new	50.00
1	½-H.P. 110 Volt Motor, "American" make, with speed controllers, new..	25.00
1	Peck, Stowe & Wilcox flexible shaft, 8 ft. long, with drill press and breast drill attachment, little used	25.00
1	Vertical Power Drill, cheap.	
1	13-in. Turning Lathe, cheap.	
1	Tinsmith's Metal Roll, cheap.	
1	18-ft. x 4-ft. 9-in. Beam Launch, equipped with a 2½ H.P. Norwalk Engine, clinker built, launched Sept. '06. A bargain	225.00
1	26-ft. Open Launch, with 10 H.P. Buffalo Motor, fully equipped, speed 9 miles per hour, cheap.	

PRICE AND DESCRIPTION ON APPLICATION.

HALLIDAY ELECTRIC CO.
KINGSTON, ONT.

WM. KENNEDY & SONS
HIGH-CLASS
Steel Gears of all kinds
Iron Gears
Grey Iron, Steel and
Brass Castings
MARINE WORK receives every attention

WM. KENNEDY & SONS, Limited
Owen Sound, Ont.

The Canadian Turbine Water Wheel

The only purely Canadian Water Wheel of exceptional value, and ahead of all foreign makes or copies of foreign wheels made in Canada.

Our proof is practical tests

Chas. Barber & Sons
Meaford, Ont.

Directory of Consulting Engineers, Patent Attorneys, Architects and Contractors.

JOHN S. FIELDING
Mem. Soc. C.E., West Penn., '87
Consulting Engineer
DAMS, MILLS, BRIDGES, MACHINERY
Room 2, 15 Toronto Street, Toronto, Ont.

HANBURY A. BUDDEN
Advocate — Patent Agent.
New York Life Building — MONTREAL.
Cable Address, 'BREVET, MONTREAL.

CONSULTING ENGINEERS
should have their card in this page. It will be read by the manufacturers of Canada :: :: :: ::
CANADIAN MACHINERY
Montreal. Toronto. Winnipeg.

T. Pringle & Son
HYDRAULIC, MILL & ELECTRICAL ENGINEERS
FACTORY & MILL CONSTRUCTION A SPECIALTY.
Coristine Bldg., St. Nicholas St., Montreal.

RODERICK J. PARKE
A.M. Can. Soc. C.E. A.M. Amer. Inst. E.E.
CONSULTING ELECTRICAL ENGINEER
INDUSTRIAL STEAM AND ELECTRIC POWER PLANTS DESIGNED. TESTS. REPORTS.
51-53 JANES BLDG., TORONTO, CAN
Long Distance Telephones—Office and Residence.

Buffing and Polishing Lathes
PENDRITH MACHINERY CO.
74 W. Adelaide Street, Toronto

OPAL GLASS TILING
FOR WALLS OF
MACHINERY AND POWER HOUSES
Most approved material.
TORONTO PLATE GLASS IMPORTING CO'Y
PLATE AND WINDOW GLASS
135 to 143 Victoria St., - Toronto

CHARLES BRANDEIS,
A. M. CAN. SOC. C.E.
MEM. AMER. ELECTRO-CHEMICAL SOC., ETC.
CONSULTING ENGINEER
Estimates, Plans and Supervision of Hydraulic and Steam-Electric Light, Power and Railroad Plants, Specifications, Reports, Switchboard Designs, Complete Factory Installations, Electric Equipment of Mines and Electro-Chemical Plants.
Long Distance Telephone, Main 3256.
Cable Address, Brandeis, Montreal.
W. U. Code Univ-Edition.
62-63 Guardian Building — MONTREAL

PATENTS *TRADE MARKS AND DESIGNS*
PROCURED IN ALL COUNTRIES
Special Attention given to Patent Litigation
Pamphlet sent free on application.
RIDOUT & MAYBEE 103 BAY STREET TORONTO

FETHERSTONHAUGH & BLACKMORE

Liverpool & London & Globe Building. Montreal.

Fetherstonhaugh & Dennison, 9 Toronto St. Toronto. Our reference book for clients "The Prospective Patentee." Also Ottawa, Washington and Winnipeg.

PATENTS PROMPTLY SECURED
We solicit the business of Manufacturers, Engineers and others who realize the advisability of having their Patent business transacted by Experts. Preliminary advice free. Charges moderate. Our Inventor's Adviser sent upon request. Marion & Marion, New York Life Bldg, Montreal; and Washington, D.C., U.S.A.

PATENTS THAT PROTECT
FETHERSTONHAUGH & CO.
Patent Solicitors & Experts
Fred. B. Fetherstonhaugh, M.E., Barrister-at-law and Counsel and expert in Patent Causes. Charles W. Taylor, B.Sc., formerly Examiner in Can. Patent Office.
MONTREAL CAN. LIFE BLDG.
TORONTO HEAD OFFICE, CAN. BANK COMMERCE BLDG.

PRESS CLIPPINGS
About any subject or business. We read nearly every paper in Canada, and can supply you with what the papers have to say about anything you are interested in.
——WRITE FOR TERMS——
CANADIAN PRESS CLIPPING BUREAU
10 Front Street East, - - - TORONTO.

THE ELECTRIC BLUE PRINT CO.
Blue Prints — Positive Blue Prints
Black Prints — Multi-color Prints
Largest and finest plant.
Best finished work in Canada at lowest prices.
40 Hospital St., - MONTREAL

THE DETROIT TESTING LABORATORY
1111 Union Trust Building
DETROIT, MICH.
Experienced Foundry Chemists and Metallurgists. Iron Mixtures for all classes of Castings our Specialty.
Reasonable charges for Analyses.

J. A. DeCew
Chemical Engineer
Sun Life Building, MONTREAL
Industrial Plants and Processes sold and installed. Apparatus and Materials for Water Purification. Free Analyses to prospective purchasers.

JULES DE CLERCY, M. E. Gas Engineer
Expert knowledge of all classes of Engines and Producer Plants. If you are thinking of installing a Gas Engine or Producer Plant write me. If your plant is not working satisfactorily I can help you.
413 Dorchester St. - MONTREAL, QUE.

PLEWS & TRIMINGHAM.
Consulting Engineers.
ELECTRICAL INSPECTION BUREAU · AND TESTING LABORATORY.
40 HOSPITAL & 22 ST JOHN STS. MONTREAL.

TRADE MARKS
DESIGNS
COPYRIGHTS &C.
Anyone sending a sketch and description may quickly ascertain our opinion free whether an invention is probably patentable. Communications strictly confidential. HANDBOOK on Patents sent free. Oldest agency for securing patents.
Patents taken through Munn & Co. receive *special notice*, without charge, in the
Scientific American.
A handsomely illustrated weekly. Largest circulation of any scientific journal. Terms, $3 a year; four months, $1. Sold by all newsdealers.
MUNN & Co. 361 Broadway, New York
Branch Office, 625 F St., Washington, D. C.

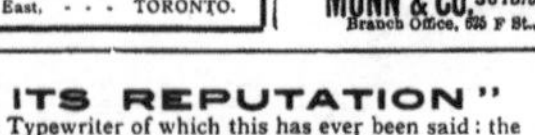

"IT SELLS ON ITS REPUTATION"
There is only one Typewriter of which this has ever been said: the
REMINGTON
Great and enduring is the merit on which such reputation is built.

THE REMINGTON TYPEWRITER CO., 33 Adelaide St. E., Toronto
AND EVERYWHERE.

with your

Cas or Gasoline Engines

and have never used

"BEST" Dry Batteries

it's your own fault.

Meet us at the Automobile and Sportsmen's Exhibition, space 93, April 6th to 13th, 1907, and we will show you how to avoid such troubles.

For sale by all leading dealers

The Berlin Electrical Mfg. Co.
LIMITED

Dept. M. **474 King St. W., Toronto, Ont.**

Cheapest Power Known

PINTSCH SUCTION GAS PLANTS
NATIONAL GAS ENGINES

No Smoke	Simplest
No Pressure	Easy Starting
No Danger	Low Cost
No Leakage	Loose Liners
No Gas Holder	Heavy Fly Wheel
No Noise	Special Electric Light Types
No Boiler	
No Chimney	

Installations in successful operation total over 200,000 b.h.p in units from 10 to 2,000 h.p.

SEND FOR CATALOGUE AND PRICES

The Economic Power, Light and Heat Supply Co., Limited

40 York Street **TORONTO, ONT.**

N. J. HOLDEN CO., 86 McGill St., MONTREAL
Agents for the Province of Quebec.

McVICKER ENGINES

Efficiency with Simplicity

is what we claim.

Only one-third the usual number of parts used in construction.

Perfect balance while running.

No experience needed to operate our engine.

WRITE FOR CATALOGUE

The Canadian McVicker Engine Company
LIMITED
GALT - - - ONTARIO

MOTORS AND DYNAMOS

OF THE HIGHEST GRADE

BUILT FOR HARD SERVICE
EFFICIENCY AND DURABILITY

We make them and guarantee them to be as good as we say they are.

The accompanying cut shows our

OPEN-TYPE FOUR-POLE DYNAMO OR MOTOR

for belt driving. It is of latest multipolar design, has coil-wound armatures, is substantial in construction, sparkless running and of the highest efficiency.

Built in sizes from
15 to 60 horse power

We should like to quote you prices

THE KAY ELECTRIC DYNAMO & MOTOR CO., Limited

Head Office, 1170 DUNDAS ST.
Factory, TORONTO JUNCTION

TORONTO, CANADA

JANDUS ARC LAMPS

The JANDUS Interchangeable Arc Lamp offers itself on grounds of

Simplicity

Economy of Installation
Economy of Operation

It embodies all of the features required by the best modern practice.

SOLE CANADIAN REPRESENTATIVES

THE PACKARD ELECTRIC CO.
LIMITED

ST. CATHARINES

MONTREAL WINNIPEG

NEW AND UP-TO-DATE

BOLT AND NUT MACHINERY

INCLUDING

Bolt Cutters, Nut Tappers, Bolt Headers, Upsetting and Forging Machines, Wire Nail and Spike Machines and Bulldozers.

Send for Catalogue M.

NATIONAL MACHINERY CO., Tiffin, Ohio, U.S.A.

Canadian Agents: H. W. PETRIE, Toronto, Ont. WILLIAMS & WILSON, Montreal, Que.

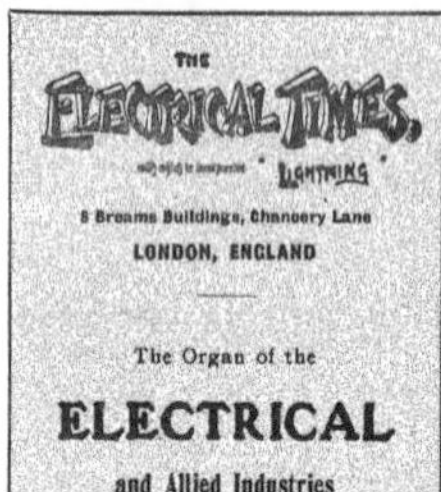

5 Breams Buildings, Chancery Lane
LONDON, ENGLAND

The Organ of the

ELECTRICAL

and Allied Industries

Circulates All Over the World

It is a Sixpenny Paper for Twopence; and is more widely read than any other British Electrical Journal. Means business all the time.

Annual Subscription:
Post Free, 13/6, Foreign; 10/6, Home.
Published weekly at 2d.

STAY BOLT TAPS

WE WANT YOUR ORDERS. HAVING FIRST-CLASS APPLIANCES, WE MAKE THESE TAPS RIGHT, AT A REASONABLE PRICE.

A. B. JARDINE & CO., Hespeler, Ont.

WORK AND PRICES RIGHT

GALVANIZING

ONT WIND ENGINE & PUMP CO.
TORONTO, ONT. LIMITED

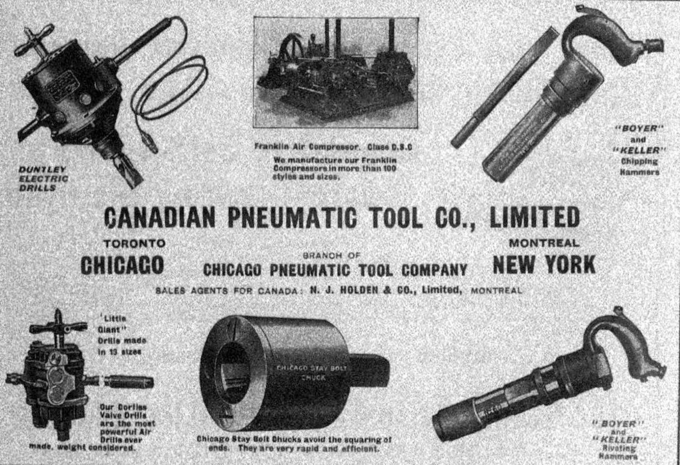

DUNTLEY ELECTRIC DRILLS
Franklin Air Compressor. Class C.S.C
We manufacture our Franklin Compressors in more than 100 styles and sizes.
"BOYER" and "KELLER" Chipping Hammers
CANADIAN PNEUMATIC TOOL CO., LIMITED
TORONTO
MONTREAL
CHICAGO
BRANCH OF
CHICAGO PNEUMATIC TOOL COMPANY
NEW YORK
SALES AGENTS FOR CANADA: N. J. HOLDEN & CO., Limited, MONTREAL
"Little Giant" Drills made in 13 sizes
Our Corliss Valve Drills are the most powerful Air Drills ever made, weight considered.
CHICAGO STAY BOLT CHUCK
Chicago Stay Bolt Chucks avoid the squaring of ends. They are very rapid and efficient.
"BOYER" and "KELLER" Riveting Hammers

First Cost is the Last Thing to Consider, if You Want Tool Holders That Earn Money.

The United States Planer Tools will save their cost in less than a month.

The wide range of work these tools are capable of being used for can only be learned by actual use.

The shank is of steel, drop forged, and case-hardened after being finished.

As the shank is very rigid, larger sized cutters are used than in other tools, thus permitting cuts as heavy as the machine will stand.

A support for the cutter extends all around the head of the shank. This does not interfere with under cutting, or getting close into a corner.

By means of the large number of serrations in washer and shank, a very close adjustment can be made and retained when cutter is removed for grinding, thus effecting a saving in time necessary to re-adjust other makes of similar tools.

The Fairbanks Co.

Springfield, Ohio.

SEND TO-DAY FOR CATALOGUE 6.

ENGINEERING:

AN ILLUSTRATED WEEKLY JOURNAL

Edited by WILLIAM H. MAW and JAMES DREDGE

PRICE SIXPENCE.

"ENGINEERING," besides a great variety of Illustrated Articles relating to Civil, Mechanical, Electrical and Military Engineering and Notes of General Professional Interest, devotes a considerable space in each issue to the illustration and description of all matters connected with the PRACTICAL APPLICATION OF PHYSICAL SCIENCE, and with its Wide and Influential Circulation in all parts of the world, this Journal is an

UNRIVALLED MEDIUM FOR ADVERTISEMENTS

SUBSCRIPTIONS (Home, Foreign and Colonial)

"ENGINEERING" can be supplied, direct from the Publisher, post free for Twelve Months, at the following rates, **payable in advance**:—

For the United Kingdom	£1 9 2
For all places abroad:—	
Thin Paper Copies	£1 16 0
Thick "	£2 0 6

Subscribers residing abroad are strongly recommended to order copies on thick paper, as the illustrations are necessarily very much less effective in the thin paper edition, and to remit, when possible, by Post Office Order made payable at Bedford Street, Strand, London, W.C. When Foreign Subscriptions are so forwarded, advice should be sent to the Publisher.

Foreign and Colonial Subscribers receiving Incomplete Copies through Newsagents are requested to communicate the fact to the Publisher, together with the Agent's Name and Address.

Cheques should be crossed "Union of London and Smiths Bank, Limited, Charing Cross Branch."

All accounts are payable to "ENGINEERING, LIMITED."

ADVERTISEMENTS.

Prepaid Advertisements on Wrapper, **classified** under headings: Tenders—Appointments Open—Situations Wanted—Partnerships—Wanted, &c.—For Sale—Auction Sales—Publications—Miscellaneous—Patent Agents—&c.

By the Line—Four Lines or under, 3s., each additional Line, 8d. The Line averages Seven Words.

By the Inch—8s. per Inch Single Column, One Insertion.

5% *allowed on* 6, 10% *on* 13, 15% *on* 26, *and* 20% *on* 52 *Insertions.*

Advertisements intended for insertion in the current week's issue must be delivered not later than 5 p.m. on Thursday.

For Displayed Adveritsements, Special Terms are given.

Offices of "ENGINEERING," 35 & 36, Bedford Street, STRAND, LONDON, W.C.

OTIS ELEVATORS

FOR ALL DUTIES

Electric, Hydraulic, Belt, Steam and Hand Power

MANUFACTURED BY

OTIS-FENSOM ELEVATOR COMPANY

LIMITED

Head Office:
TORONTO, ONT.

Works:
HAMILTON, ONT.

ALLIS-CHALMERS-BULLOCK LIMITED MONTREAL
INDUCTION MOTORS
ALLIS-CHALMERS-BULLOCK LIMITED MONTREAL
One of our 300 h.p. Induction Motors driving an air compressor in the power house of the Canadian Copper Co., Copper Cliff, Ont. These Motors are adapted to the operation of all kinds of machinery, but especially cement works, brick yards, powder mills, paint factories, flour mills, paper mills, mines, etc.
For a list of our various products, see the classified index of this journal.
ALLIS-CHALMERS-BULLOCK, Limited
Head Office and Works, Montreal. District Offices: Toronto, Traders Bank Bldg; Montre
Sovereign Bank Bldg; Winnipeg, 251 Notre Dame Ave.; Nelson, Josephine St.;
Vancouver, 416 Seymour St.; New Glasgow, N.S., Telephone Bldg.

CANADIAN MACHINERY AND MANUFACTURING NEWS

A monthly newspaper devoted to the manufacturing interests, covering in a practical manner the mechanical, power, foundry and allied fields. Published by The MacLean Publishing Company, Limited, Toronto, Montreal, Winnipeg, and London, Eng.

OFFICE OF PUBLICATION: 10 Front St. E., TORONTO.

Vol. III. APRIL, 1907 No. 4

6-inch Single Punch and Shearing Machine

20-inch Combined Punch and Shearing Machine

Can make immediate shipment on the following Machine Tools:

30x30x8-ft. Planer
22x14-ft. All Geared Engine Lathe
2-inch Single Bolt Cutter
1-inch Single Bolt Cutter
6-ft. Plain Radial Drill, Speed Box
16-inch Crank Shaper
36x60x16-ft. Gap Lathe
10-inch Sensitive Drills

We solicit your inquiries for anything in Machine Tool line

2-inch Double Bolt Cutter

No. 3 Double Crank Cutting Press

London Machine Tool Co., Limited, Hamilton, Canada

Sales Office, Room 1206 Traders Bank Building, Toronto, Canada

IF YOU USE GEARS

INSIST ON HAVING THEM CUT ON THE

Automatic Gear Hobbing Machine

50%

OF POWER YOU WASTE NOW SAVED

HOW?

ALL gears cut on the generating principle have perfect teeth.

IF YOU CUT GEARS

YOU CANNOT AFFORD TO USE ANY OTHER MACHINE THAN THE

Automatic Gear Hobbing Machine

50%

OF TIME SAVED

HOW ?

The cutting is a continuous performance; no indexing, stopping or starting.

It is the only machine on the generating principle on the market.

BAXTER, PATERSON & CO.

TEL. M 847 | 102 St. Antoine Street, - MONTREAL

Has Greater Productive Capacity

than any Upright ever built—our 30-in. RADIAL DRILL. Costs but a trifle more and is admirably adapted to light and medium manufacturing, being particularly effective in taking care of the ordinary work as found in the average shop.

. Cone Drive and Swinging Table

Bickford 30"-36"-42" Radial Drill

A Few Features

12 speeds when driven through a speed box; 10 when a cone is used.

Tapping attachment incorporated in the design of the head.

Fitted with any one of six styles of tables.

Circular 1-A for the asking

The Bickford Drill & Tool Company

Cincinnati, Ohio, U.S.A.

FOREIGN AGENTS—Schuchardt & Schutte, Berlin, Vienna, Stockholm, St. Petersburg, New York. Alfred H. Schutte, Cologne, Brussels, Liege, Paris, Milan, Bilbao, New York. Chas. Churchill & Co., Ltd., London, Birmingham, Manchester, Newcastle-on-Tyne and Glasgow.

CANADIAN AGENTS—H. W. Petrie, Toronto. Williams & Wilson, Montreal.

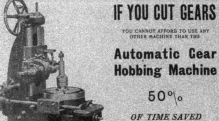

"AMERICAN"

VARIABLE SPEED PLANERS

BELT OR MOTOR DRIVEN
IN

STANDARD or WIDENED PATTERNS

FOUR (4) SPEEDS
OBTAINED THRU SPEED BOX

TWO (2) SPEEDS
OBTAINED THRU COUNTERSHAFT

EACH TYPE WITH CONSTANT SPEED RETURN

27 in. to 72 in. between Housings.

PARTICULARLY ADAPTED TO A WIDE RANGE OF WORK

Catalog "C" now ready

THE AMERICAN TOOL WORKS CO.

700 to 750 CULVERT STREET - CINCINNATI, OHIO, U.S.A.

LATHES PLANERS SHAPERS RADIAL DRILLS

Write direct or to H. W. PETRIE, Montreal; A. R. WILLIAMS MACHINERY CO., Toronto, Winnipeg, Vancouver

Becker-Brainard

Milling Machines

and

Milling Cutters

Guarantee ACCURACY and ECONOMY in Production

Becker-Brainard Milling Machine Co.

Hyde Park, Mass, U.S.A. Branch Offices: THE BOURSE, Phila, Pa.; WILLIAMSON BLDG., Cleveland, O.

Canadian Agents: A. R. WILLIAMS MACHINERY CO., LTD., TORONTO and MONTREAL, CAN. Agents, London, Eng.: SELIG SONNENTHAL & CO.

326

FROM
200 to 500 %
That is the actual saving effected by using our machines.
¶ All we ask is that you send us samples or drawings of your work.
¶ When you receive our estimate you will KNOW that your duplicate parts have been costing you too much, and we will have your order for our machines.
"The Pre-eminence of the
CLEVELAND
is indisputable"
Here's the Proof:
247
Piece No. 247
Material, Tool Steel, made on a 2" Cleveland Automatic Plain Machine, drawing full size. Output per hour, 8; this means all operations for completing this piece. Actual cost of labor, from 8 to 9 mills for each piece. Send along samples or drawings. We guarantee our output. One man can operate from 4 to 6 machines.
Piece No. 132
132
Material, Machine Steel Bar, made on a 3¼" Cleveland Automatic Turret Machine, drawing full size. Time of all operations for completing this piece, 15 minutes. Actual cost of labor, from 1½ to 2 cents.
AS WE HAVE NO CANADIAN REPRESENTATIVE, WE SOLICIT CORRESPONDENCE AND ORDERS DIRECT
CLEVELAND AUTOMATIC MACHINE CO.
CLEVELAND, OHIO, U.S.A.
EASTERN REPRESENTATIVE—J. B. ANDERSON, 2450 North Thirtieth Street, Philadelphia, Pa.
WESTERN REPRESENTATIVE—H. E. NUNN, 22 Fifth Avenue, Chicago, Ill.
FOREIGN REPRESENTATIVES—CHAS. CHURCHILL & CO., London, Manchester, Birmingham, Newcastle-on-Tyne and Glasgow. MESSRS. SCHUCHARDT & SCHUTTE, Berlin, Vienna, St. Petersburg and Stockholm. ALFRED H. SCHUTTE, Cologne, Brussels, Liege, Paris, Milan and Bilbao.

RENOLD SILENT CHAIN

is used almost exclusively on the standard motors because of its supremacy in the power-transmitting field. Adapted to short centres; damp, hot or oily places; and will run in either direction. Action always smooth and positive.

Write for Booklet "J" and Special Bulletins 50, 52, 57, 58.

LINK-BELT COMPANY

PHILADELPHIA, CHICAGO, INDIANAPOLIS

NEW YORK 299 Broadway. PITTSBURGH 1501 Park Bldg. BOSTON 84 State St. BUFFALO 321 Ellicott Square Bldg.

Kerr Turbine Pumps for water works, condenser equipment and boiler feeding. These pumps are more economical than direct acting piston pumps, and require less attention, lubrication and space. No belts or gears. Speed adjustable to pressure desired. Regulation close. Start and stop quickly. For any capacity.

Turbine Blowers for gas works, foundries and forge shops. Air pressure 4 to 16 ounces. For any capacity.

Turbine Generators, 2 kw. to 300 kw. direct current and alternating current, 25 to 300 kw., 60 cycle.

The illustration shows an 18-inch condensing Kerr Steam Turbine coupled to a 4-inch 3-stage Worthington Pump. Capacity 200 g.p.m. against a head of 360 feet.

Main Office:

KERR TURBINE CO., Wellsville, N.Y.

Built in Canada by

WM. STEPHENSON

34 Bay Steet N. *HAMILTON, ONT.*

High Speed is Essential

in modern lifting, but low repair bills are also essential. In choosing a high speed hoist get the easiest to operate, the fastest and most durable. Greatest ease and speed may be proven in a 30-day trial if you choose. Durability takes years to prove, but examine the construction, note the cut steel gears, substantial parts thoroughly protected from dirt, and then order

Round Spur Geared Blocks

Made in 15 sizes, from 500 pound to 40,000 pound capacity. We can refer you to many satisfied users in the leading machine and railroad shops in Canada.

Frothingham & Workman, Limited
Montreal

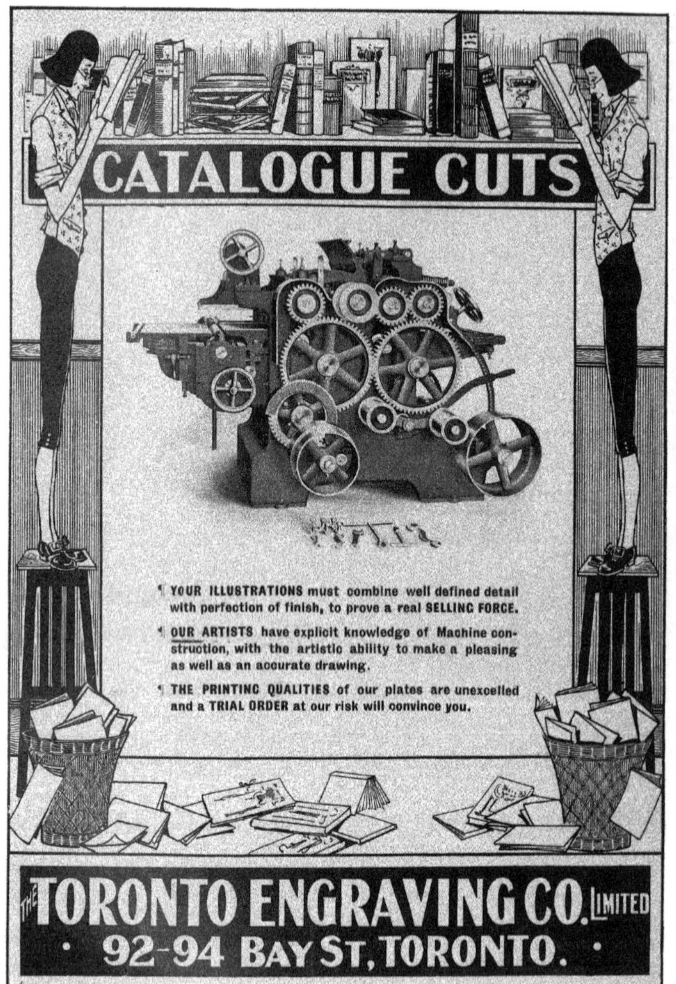

CATALOGUE CUTS
¶ YOUR ILLUSTRATIONS must combine well defined detail with perfection of finish, to prove a real SELLING FORCE.
¶ OUR ARTISTS have explicit knowledge of Machine construction, with the artistic ability to make a pleasing as well as an accurate drawing.
¶ THE PRINTING QUALITIES of our plates are unexcelled and a TRIAL ORDER at our risk will convince you.
THE TORONTO ENGRAVING CO. LIMITED
· 92-94 BAY ST, TORONTO. ·

Drawing $125 to $250 monthly require more than Schoolroom Knowledge.

As Chief Draftsman of Eng'g firm I know exactly the Quality and Quantity of experience and practical training you must have, and will prepare you in few months home instructions for **above paying positions,** which I **furnish free anywhere.**

Complete Highest Grade **Drawing Outfit,** everything included, with **German Silver Set of Tools, value $13.85, FREE** next ten days.

Address, CHIEF DRAFTSMAN, Div. 22 ENG'S EQUIP'T CO., Inc., Chicago.

Steam Turbine Series Pump for High Pressures

STEAM TURBINE MACHINERY

MOTORS
DYNAMOS
PUMPS

D'Olier Engineering Co.

74 CORTLANDT STREET, - - NEW YORK, U.S.A.

BLOWERS

HEATING AND VENTILATION
by Fan System

Natural Draft and Blower Dry Kilns
For Lumber, Brick, etc.
Cars, Trucks, Rail, etc.

Heating
Ventilating
Drying Forges
Mechanical Draft
Blowers.
Exhausters
Steam Specialties

ENGINES

both Vertical and Horizontal

Shavings
Exhaust
System

SHELDONS, Limited, Engineers and Manufacturers Galt, Ont., Canada

Montreal Office, 325 St. James St. Phone Main 2919

DuBois Gas Engines

"THE IMPROVED TYPE"

We guarantee minimum fuel consumption. Every working part is readily accessible. Valves can be removed and replaced in three minutes time. Our improved Fly Ball Governor is instantly sensitive to variable loads. Valve trouble impossible. All highly heated gases relieved from the cylinder through an auxiliary exhaust.

The Speed, Gas, Air and Electric Spark adjustable while engine is running.

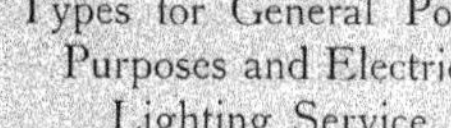

Sizes 5 to 300 H. P. designed to operate on

Gas	Gasoline
Kerosene	Crude Oil
Alcohol	Distillate

1907 Catalogue just issued explains fully

DuBois Iron Works

807 Brady St., DU BOIS, PENNA.

To Gas Engine Builders

We carry the largest stock in the Dominion of Gas Engine

Accessories

Lubricators, Spark Plugs, Ignition Cable, Dry Cells, Ignition Dynamos.

The

Astor Magneto

dispenses with batteries or dynamos.

You should decidedly see our catalogue and get our prices before placing orders elsewhere.

John Millen & Son, Limited

321 ST. JAMES ST. MONTREAL — 132 BAY ST. TORONTO

Cheapest Power Known

PINTSCH SUCTION GAS PLANTS
NATIONAL GAS ENGINES

No Smoke	Simplest
No Pressure	Easy Starting
No Danger	Low Cost
No Leakage	Loose Liners
No Gas Holder	Heavy Fly Wheel
No Noise	Special Electric Light Types
No Boiler	
No Chimney	

Installations in successful operation total over 200,000 b.h.p. in units from 10 to 2,000 h.p.

SEND FOR CATALOGUE AND PRICES

The Economic Power, Light and Heat Supply Co., Limited

40 York Street TORONTO, ONT.

N. J. HOLDEN CO., 66 McGill St., MONTREAL
Agents for the Province of Quebec.

Modern Canadian Manufacturing Plants.

ARTICLE XXII.—Plant of Sheldons Limited.

SOME four years ago the partnership company of Sheldon & Sheldon was organized to buy and take over the plant and machinery and equipment of the McEachren Heating and Ventilating Company of Galt. The old plant was added to in the way of machine tools and other devices, and the business was increased in every possible direction. For three years under this name the company was operated, when it was deemed advisable to organize a stock company, under the title of "Sheldons, Limited." A Dominion charter was obtained, and officers were appointed as follows:

President, W. D. Sheldon.
Vice-president, S. R. Sheldon.
Sec-treasurer, J. P. Stewart.

Immediately on the organization of this new company the extensions and additions contemplated were entered upon, and in the course of from two to three months the buildings were erected complete.

This company was organized with a capital stock of $200,000, a good part of this being new capital put into the company for the erection of additions and the installation of new machinery. Since that time the company's business has increased very materially, and they have at present on their books orders for some of the largest fans which have yet been used in this country, among them being orders for wheels running from 14 to 18 feet in diameter.

General Lay-out.

The works are situated on the corner of West Main and Cedar streets. The main factory is built in the form of an L, the short leg of the L running north and south on West Main street, and the new extension, which was erected in 1906, extending in a westerly direction on Cedar street, on the street line.

The engine and boiler room and electric light equipment are situated in a brick building about midway up the length of the new extension on the north side. On the west side of this boiler and engine room extends the foundry.

The machine shop, or main building, is built in two floors, the ceilings having a height of 15 feet. Both floors are served by travelling cranes. The size of the main building is, on the short leg of the L, about 120 x 65 feet, two floors, and the long leg about 140 x 65 feet. The foundry is about 45 feet in width by about 120 feet long, and is served with a jib crane for handling the iron.

The offices are situated in the corner of the main building. The main offices are on the ground floor, with the draughting room above. These are divided into private and general offices, and are fitted with modern heating and sanitary arrangements, and are equipped with vaults for correspondence, drawings, etc.

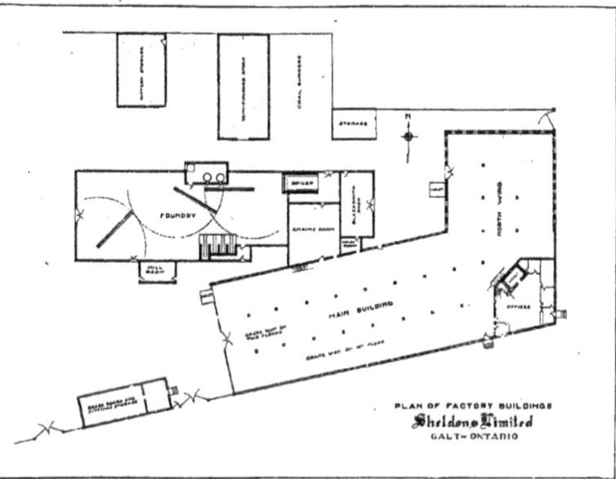

Plan of Plant—Sheldons Ltd., Galt.

Manufacture of Heaters.

In the north wing of the main building are manufactured the heaters used in connection with the fan system of heating and ventilating. This department employs about ten men, and their entire time is spent in constructing and testing these coils. The pipe is cut with a high speed rotary cutter, with automatic measuring attachment, and is threaded by a semi-automatic, double-head pipe-threading machine. The elbows on the standing pipes of these heaters are put on by a special machine made by Sheldons, Limited, themselves, which does the work much better and far more quickly than it could be done by the old method of hand connection. The bases are all constructed on slatted floors built over concrete basins, and as each section is completed it is tested to a hydrostatic pressure of 150 pounds per square inch, and made tight under this pressure.

Immediately adjoining this department is the main room in which is done all the machine work on the castings as they come from the foundry. In this department are seen lathes of different sizes, as manufactured by the Mac-Gregor-Gourlay Co., Galt; R. McDougall Co., Galt; Bertram & Son, Dundas, Ont. These lathes are all operated at high speed, and utilize high speed steel cutters. The new additions to the machine shop and equipment this year comprise: A large boring mill, made by Bertram & Son, Dundas, for boring out engine frames and cylinders; one large vertical boring mill for turning up engine work and fan parts; one key-seating machine for cutting necessary key ways in the fan spiders; two radial drills of the latest design; one large riveting and punching machine, and two large punching and shearing machines, made by the London Machine Tool Company, Hamilton. They are also installing a large Landis grinder, and have recently ordered a new shaper of one of the largest sizes made by MacGregor-Gourlay Company, and a large radial drill, of special make.

They also have a large MacGregor-Gourlay planer, universal milling machine, special boring and tapping machine for bases, a large pipe cutting and threading machine for working on pipe up to six or eight inch, and about six small and medium-sized drill presses.

The heavy machines, such as the horizontal boring mill, vertical boring mill, planer, large lathes and Landis grinder will be situated in the centre bay of the main buiding, which is served by a travelling crane.

Tool Room.

To the machine shop has also been added a tool-making equipment for supplying all the necessary jigs and special tools for the operation of the above-mentioned machines.

The tool making is done on the north side of the main building, and they are at present busy in putting up and partitioning off this department, and are about to install a tool-checking system for keeping track of all the jigs and tools which are used.

Erecting Floors.

The main working floor is situated on side of the main building, and they are entire length. On this floor are fitted all the engines, forges, fans and fan wheels, bearings and steam specialties. This bay is also served by a travelling crane for the convenience of handling and working on the large engine frames, used for operating some of the heavy fans. In the centre of this floor is situated a large testing plate built upon solid concrete, which is constructed right down to bed rock. This is served with steam from the boiler, and is used for testing and experimenting with all engines, every one of which is operated in the shop for at least two days, under load; and while under operation the speed, power and other particulars are noted. The cards taken from these engines are kept on record.

View of Erecting Floor; Sheldons Ltd., Galt.

In the machine and fitting department, which occupies the entire ground floor of the main building and the new wing, are employed in the neighborhood of fifty-five men, consisting of machine operators, fitters, and their helpers.

Sheet Metal and Pattern Department.

The upstairs of the main building and the north wing is occupied by a pattern department, the heavy sheet metal working department, and the galvanized iron department.

The pattern room is fitted with the necessary wood working machinery required for pattern making. In the pattern department are employed six men.

The sheet metal department is fitted with punches, rotary slitting shears, brakes, large slitting shear for cutting a ten-foot sheet in one operation, and other special tools and jigs for the manufacture and construction of the steel shells of the fans. This department is equipped to work all sheet metals up to and including plate of ½ inch in thickness.

The large fan wheels are also constructed in this department, and are made on special jigs constructed for the purpose. The wheels, when constructed, are conveyed to the testing shaft on the lower floor, where they are given a running balance, which is the only practical method of balancing, so as to ensure continuous operation without creating undue noise and resistance. This department employs in the neighborhood of thirty men.

In this department is made and constructed all the galvanized iron work used in the erection and equipment of the heating systems. Nearly every installation will include from 1,000 lbs. upwards of this light galvanized iron work. Employed in this manner are from fifteen to twenty men, with their helpers.

To the north of the main building is

A Corner in the Machine Shop; Sheldons Ltd., Galt.

the boiler and engine room. This part is entirely new, and consists of a 75 h.p. Wheelock engine and 115 h.p. boiler, both made by the Goldie & McCulloch Co., Galt. The main driving shaft is equipped with Chapman double ball-bearings. In the engine room is also situated the heating system, consisting of a 5,000-ft. heater, 140-inch fan and direct connected 7x10 engine. This heating system was made by Sheldons themselves, and installed complete. The electric lighting system for the factory is also situated in the engine room, and consists of a 35 kilowatt Canadian Westinghouse generator. This generator will shortly be operated by one of Sheldons' own make of automatic horizontal side crank engines. The switch board is also of the Canadian Westinghouse make, and is complete with switches for all departments.

View of Sheet Metal Department; Sheldons Ltd., Galt.

The Foundry.

The foundry building is a one-floor building about 45-ft. wide and 120-ft. long. In this department are employed twenty or twenty-five moulders with the necessary laborers to help them. In the foundry a special device is used for chilling the tread of the wheels used in the construction of brick and lumber cars, of which this firm makes a large quantity. These chills consist of collapsible cast iron frames which fit around the wheel mould between the upper and lower half of the sand impression. The foundry equipment includes three large core ovens, fitted with steel doors and down draft combustion chambers. These ovens are of the firm's own design and construction.

The cupola used in the foundry is made by the firm, is 30 inches in diameter inside the lining, and has a maximum melting capacity of five tons per hour. The usual run in the foundry is about two and a half tons of iron per day. The cupola is situated in a specially constructed room, which is entirely fire-proof, the charging floor being made entirely of steel. The walls are solid 12-inch brick, and the roof is made of steel and galvanized iron.

View of Foundry; Sheldons Ltd., Galt.

The coke and iron is conveyed from the yard to the charging floor by means of an elevator, made by the Otis-Fensom Elevator Co., of Hamilton. In the yard are several store houses, erected for the storage of valves and pipe fittings, for steel plates, channels, I-beams and angle iron. There is a special house erected for the storage of finished and semi-finished machines, and also a large metal clad building for the storage of patterns. There is a blacksmith shop, about 25 x 40 feet, in which are employed two blacksmiths.

The shipping is at present all done on heavy trucks, but there is every indication of their obtaining shortly switch connections and inter-switch facilities with the C.P.R. and the Grand Trunk.

The buildings are constructed of brick, with wooden posts and 3-inch flooring. There is a large area of glass surface, and the roof is of the Monitor type, which gives an extra supply of light on the upper floor.

They have installed the Dey time check system, that is, they use the Dey time clock, and the men ring in their time as they arrive at and leave the works. They have also a cost system installed in the factory, by which they keep track of the time each man spends on certain work, every order passing through the shop with an order number, which the workman puts on the card issued to him for that purpose, and when the machine is completed these order numbers and cards are gathered from the shop and checked over by the superintendent, then passed into the office, where the cost is made up. These orders are checked every day.

PERSONAL MENTION

E. F. Valiquet, superintendent of J. W. Harris Co., excavators, Montreal, was born in 1865 at St. Martin, Que. In 1877 he came to Montreal and served his time as machinist in Miller Bros. and Toms' factory. He then went to the United States, where he stayed about five years, coming back to Montreal in 1892, where he accepted the position of

E. F. VALIQUET, MONTREAL.

chief engineer for W. Rutherford & Sons. He was foreman of Lymburner & Mathews, also for Darling Bros.' machine shops, Montreal, and the Canadian Portland Cement Co., of Marlbank, Ont. The abattoir at the Montreal stock yards was erected under his supervision. Mr. Valiquet has also held the position of chief engineer of the Locomotive and Machine Co., Longue Point, (Montreal), Que., and head millwright for the Weber Elevator, and his present position with the J. W. Harris Co., Ltd., Montreal, where he has been for the past two years. He is at present instructor for the Association of Stationary Engineers Montreal; boiler inspector for the province of Quebec, and provincial examiner for engineers.

Mr. J. Conlin, first vice-president of the International Machinists' Union, was in Toronto on March 17th.

Mr. Chas. McDonald, manager of the St. John Iron Works, St. John, N.B., left for England on March 15th.

Mr. Lindsay Malcolm, M.A., formerly of Guelph, and a gold medalist of Queen's University, has been appointed city engineer, of Stratford, at a salary of $1,-300.

Mr. Wm. Banks, who is leaving for Toronto, where he will enter business for himself, was presented with a gold watch and address by the employes of the Cockshutt plough works, Brantford, with which he has been connected for many years.

The engagement of Mr. Roland W. Robb, of the Robb Engineering Co., Amherst, N.S., to Miss Mabel Pugsley, daughter of Mr. Robert Pugsley, is announced.

Mr. W. Risdon, manager of the Erie Iron Works, St. Thomas, has sold out his interests in the company to Mr. George Spackman, western representative of the Massey-Harris Co.

It is reported in Sydney that Mr. F. P. Jones, general manager of the Dominion Iron and Steel Co., will shortly resign, and be succeeded by Mr. C. S. Cameron, at present comptroller of the company.

Mr. Harry A. Norton, of A. O. Norton, Boston, Mass., who has just returned from a trip round the world in the interests of Norton jacks, has again sailed for Europe, and will be away three months.

Mr. W. C. Groening, superintendent of the Pere Marquette shops in St. Thomas, has been transferred to Grand Rapids, Mich. His successor is Mr. F. C. Pickers, formerly shop fareman at Grand Rapids.

Mr. G. J. Bury, Winnipeg, general manager of the western lines of the C. P.R., has been offered the position of general manager of the Chicago, Rock Island and Pacific Railroad Company, at a salary of $30,000.

Mr. Geo. T. Moss has severed his connection with the Canada Screw Factory, Hamilton, going to the Canadian Westinghouse Co. Before leaving he was presented with a gold locket by the employes of the screw factory.

Mr. Chas. Spellman, a well-known young traveler for the Canada Forge Works, Welland, Ont., was caught by some machinery while in the works on March 22, and almost instantly killed. His home was in Titusville, Pa.,

The employes of the R. H. Smith saw works, St. Catharines, on March 15th presented Mr. H. A. Blair with a fob and locket on the occasion of his leaving the assistant managership of the firm to take a similar position with the R. Hoe Saw Works, New York.

Mr. R. F. Harrison, of London, Eng., is in Montreal to take over the sales management of the F. W. Reddaway Belting Co. Mr. Harrison comes here with a well-established record in the Old Country, and he will have complete control of the Canadian business.

The office staff of the Goold-Shapley-Muir Co., Brantford, on March 4th presented Mr. W. J. Craig with an umbrella, and Mrs. Craig with a cake dish, on the occasion of Mr. Craig's retiring from the firm with the removal of the bee-keepers' supply branch to Ottawa.

Harry C. Hoefinghoff, president of the Bickford Drill & Tool Co., Cincinnati, Ohio, died on March 2nd, after an operation for appendicitis. Mr. Hoefinghoff was born in Cincinnati in 1871, and was one of the most prominent young men engaged in the manufacture of machine tools in the country. His affability won for him a wide circle of friends, among both his business associates and fellow-townsmen, and he was considered one of the most promising young business men of Cincinnati. A wife and three children survive.

Mr. J. K. Macdonald, for many years in business as a machinist at 318 Craig street west, Montreal, died on March 11th, at the age of 66. He was a native of Campbelltown, Scotland, and came to Montreal forty-five years ago. He leaves a widow, three sons, and three daughters.

At the annual convention of the American Railway Engineering and Maintenance of Way Association at Chicago, lately, William McNab, assistant engineer of the Grand Trunk Railway, was elected vice-president. This means that next year he will become the president, and that probably the next annual convention will be held in Montreal.

Mr. John G. Sullivan, who has been assistant chief engineer on the Panama canal for over a year, has been appointed superintendent of construction by the Canadian Pacific Railway for all lines east of Fort William, with his office in Toronto. For a number of years Mr. Sullivan had charge of construction for the C.P.R. in the west, and then left that position to go to Panama. He will enter upon his new work immediately.

Mr. John D. Maclennan, C.E., a Canadian engineer, who has distinguished himself in the United States, died suddenly in New York on Feb. 26th. He

W. J. FRANCIS
of the Dominion Bridge and Construction Co.
Winner of the Gzowski Medal.

was the eldest son of Mr. Roderick Maclennan, C.E., Toronto. He began his engineering career with the C.P.R. during its construction, and on the completion of the work went to the United States, where he built the Florida Coast Railway. Later he had charge of the construction of the filtration plant for the waterworks of Washington, and other public works of importance, including the investigation of the Philadelphia filtration system. Mr. Maclennan was in his 51st year.

Producer Gas vs. Niagara Power

By Kendall Greenwood, B.A.Sc.

Electric power at the prices quoted by the Ontario Power Commission is conceded to be probably the cheapest form of energy for manufacturing obtainable at the present time.

This is only true, however, within the district within a radius of 50 miles of the city of Hamilton. Outside this radius there is a source of energy considerably cheaper than Niagara power and that is producer gas power.

Gas power is becoming more and more a factor in the generation of electric energy. There are at the present time a great many producer gas plants in operation. Gas engine makers are working double shifts in order to keep up with their orders for gas plants.

With producer gas it is possible to generate electric power for driving motors in factories, lighting and other purposes, and to utilize the waste heat from the gas engine exhaust to raise steam for heating and other uses.

Suppose a manufacturer has a factory and that he requires a fairly steady load of 100 h.p. for 10 hours for 300 days per year; and that he has decided to use motors and electric lighting. what will be his best source of energy, and what will it cost?

The manufacturer goes to the city of Toronto and finds that Niagara power will cost as follows, for the above mentioned load:

.35 cent per horse-power per hour as a meter rate.

.40 cent per horse-power per hour flat rate whether he uses the power or not.

Total, .75 cent per horse-power hour.

His power would therefore cost for 300,000 horse-power hours at .75 cent, $2,250.

In addition to this there are the fixed charges and maintenance on his motor equipment. This will be the same in either case and has not therefore been taken into consideration.

The power user then goes to the producer gas manufacturer and finds that a 100 horse-power suction gas engine and producer plant will cost installed

....	$5,700
and an electric generator, etc...	2,300
Total	$8,000

The fixed charges for this plant will be as follows:

Interest, at 6 per cent.	$480
Depreciation, engine, 7 per cent.	294
" producer, 4 per cent.	54
" electric generator 10 per cent.	230
Labor (one man, $50 month)	600
Repairs, etc.	100
Total	$1,758

The fuel and oil charges will be as follows:

A suction gas plant will burn a pound of good pea anthracite per horse-power per hour when running at or near full load; and at half load the consumption will be about 1.5 lbs. Pea anthracite costs in Toronto about $4.20 per ton.

Fuel used may be calculated as follows:

H.P. Hrs. Days Lbs.

$$\frac{100 \times 10 \times 300 \times 1}{2,000} \text{ eq. 150 tons per year.}$$

Allowing 3 lbs. for stand-by losses per hour we have,

$$\frac{3 \times 14 \times 300}{2,000} \text{ eq. 6 tons per annum.}$$

Total fuel will then be 156 tons at $4.20	$655
Oil and waste, etc., will cost ...	120
	$775

Total cost for power per annum:

Fixed charges on plant	$1,758
Fuel, oil and waste, etc.	775
Total	$2,533

Total cost per horse-power per annum	$25.33

Reducing to same basis as rates for Niagara power we have, for 300,000 horse-power hours:

Fixed charged per h.p. hour...	.586 cent
Fuel, charges per h.p. hour...	.258 cent
Total	.844 cent

Hence, we find that producer gas power only costs less than one-tenth of one cent more per horse-power hour than Niagara power for the same plant capacity.

While the cost of power in a one hundred horse-power gas plant is very slightly higher than Niagara power in Toronto there are advantages to be gained from the use of producer gas which would more than offset the extra cost.

Niagara power is as yet very much too expensive to use as a source of heat for buildings; consequently a user of Niagara power has to use coal directly for heating purposes for six months in the year.

In a 100 h.p. gas plant there are about 3,500 b.t.u. available per pound of coal burned, or 350,000 b.t.u. per hour in the exhaust.

By exhausting this heat through the tubes of an ordinary steam boiler it is possible, assuming an efficiency of 60 per cent., to raise 216 pounds of water to steam at 70 pounds boiler pressure per hour. That is, there are available in the exhaust gases from the 100 h.p. gas engine, 7 boiler horse-power during the time that the engine is running. 10 hours per day.

By arranging this boiler with a grate, steam may be kept up during the night by burning coal in the usual manner.

Thus, it is possible to recover 7 per cent. of the capacity of the plant and to save $175 per year, and so reduce the fuel cost of the power to the comparatively small amount of $600. The total cost per horse-power per annum is thus reduced to $23.58.

The writer has charged the whole of the engineer's time to the cost of power. It is quite fair that a patron of this time should be charged to other work, as it has been found that about three hours per day is sufficient to attend to the whole plant in medium-sized installations.

It can now be easily seen that producer gas can easily compete with hydro-electric power in all portions of the country, with the exception of the Niagara peninsula and vicinity.

APPOINT AGENTS IN BRITISH COLUMBIA.

It will be of interest to the trade in British Columbia to know that John Millen & Son, Ltd., Montreal, have appointed Messrs. A. G. Urquhart & Co., DeBeck block, 336 Hastings St., Vancouver, B.C., as their agents for British Columbia. Urquhart & Co. will carry a stock of supplies and accessories most in demand by the automobile and motor boat trade, as well as a stock of bicycles and bicycle sundries.

OLD AGE DEFIED.

THE Hungarian chemist Brunn, claims to have discovered a liquid chemical compound which renders certain kinds of matter proof against the effects of time. He asserts that it doubles the density of nearly every kind of stone and renders it water proof. It imparts to all metals qualities which defy oxygen and rust. The professor says that while traveling in Greece some twenty-five years ago, he noticed that the mortar in stones of ruins which were known to be over 2,000 years old, was as hard, fresh, and tenacious as if it had been made only a year. He secured a piece of mortar and has been working on it ever since until now, when he says, he has discovered the secret. His discovery, he claims, will, at least double the life of metal exposed to the air, such as bridges, railroads, vessels, and tanks

CANADIAN MACHINERY
and Manufacturing News

A monthly newspaper devoted to machinery and manufacturing interests, mechanical and electrical trades, the foundry, technical progress, construction and improvement, and to all users of power developed from steam, gas, electricity, compressed air and water in Canada.

The MacLean Publishing Company, Limited

JOHN BAYNE MACLEAN - - President
W. L. EDMONDS - - - Vice-President
H. V. TYRRELL - - - - Manager

OFFICES:

CANADA
MONTREAL - 232 McGill Street, Phone Main 1255
TORONTO - 10 Front Street East, Phone Main 2701
WINNIPEG, 511 Union Bank Building, Phone 3726, F. R. Munro
BRITISH COLUMBIA - Vancouver, Geo. S. B. Perry

UNITED STATES
CHICAGO - 1001 Teutonic Bldg., J. Roland Kay

GREAT BRITAIN
LONDON - 88 Fleet Street, E.C., Phone Central 12960, J. Meredith McKim
MANCHESTER - 92 Market Street, H. S. Ashburner

FRANCE
PARIS - Agence Havas, 8 Place de la Bourse

SWITZERLAND
ZURICH - - - Louis Wolf, Orell Fussli & Co.

SUBSCRIPTION RATE.

Canada, United States, $1.00, Great Britain, Australia and other colonies, 4s. 6d., per year; other countries, $1.50. Advertising rates on request.

Subscribers who are not receiving their paper regularly will confer a favor on us by letting us know. We should be notified at once of any change in address, giving both old and new.

Vol. III. APRIL, 1907 No. 4

CONTENTS

A FITTING EDUCATION FOR THE MECHANIC.

ONE of the big problems in the manufacturing industry of the present day is that of securing well trained mechanics. Direct evidence of the seriousness of this problem is the interest which so many large firms and also associations and societies are taking in the question of apprenticeship courses and technical training. Skilled mechanics cannot be produced within a day's notice. They must get a thorough training in some way or another; and the ever increasing demand for these men in Canada, as the country expands, makes the problem, as to how best to train young men and boys in mechanical lines to fill positions as necessity demands, a most vital one to the manufacturer.

It is impossible to draw a distinct line between the engineer and the mechanic. Looking at the question broadly, it might be said that the engineer is a mechanic, who, because of greater ability or because of closer application to work, has reached the higher rung of the ladder of success in mechanics; but it is true that frequently the best of engineers are not trained mechanics, and frequently the best and brightest of mechanics would make very poor engineers. The good engineer must be something more than a mechanic. The relative merits of college men and practically trained men as engineers are often a matter for discussion; but this cannot be discussed to advantage except in relation to some one particular case.

The question of training for mechanics, however, is apart from this. Because of the great advances in methods of manufacture during recent years, the good mechanic of to-day requires a better general education than in years gone by, and also a more complete technical training. In metal working industries the technical training required is along the line of mathematics used in simple designing, in elementary mechanics, and in mechanical drawing, together with information of a general nature concerning the line of work to be followed, which should deal particularly with things which cannot be picked up in the shop.

Quite a large number of firms in Canada and the United States have apprenticeship courses, which include a technical as well as practical training; and this certainly appears to be one of the best methods of solving this problem. At the same time the Government, civic authorities, boards of trade, and associations, such as the Canadian Manufacturers' Association, should do all in their power o promote the cause of technical education in Canada; and if this matter is gone into with enthusiasm we will, in time, have as complete a system in Canada as exists in Germany, and other countries. This will not only elevate the plane of the mechanic, but will assist in making the product of Canadian manufacturers equal to that of manufacturers in any country. It is gratifying to see that this question is receiving very considerable attention from different organizations in Canada, and also from the Government and civic authorities.

At the last meeting of the Central Railway and Engineering Club of Canada, held in Toronto in the latter part of March, the subject of technical education was discussed; only a short time ago the technical committee of the Canadian Manufacturers' Association made a personal visit to the one technical school in Toronto for information regarding the conduct of the work there; and within the last few months this subject has been discussed and rediscussed in numerous mechanical societies in the United States. This shows that extraordinary interest is being taken in this subject at the present time; and it is to be hoped that the results will be worthy of the efforts which are being put forth.

A plan to further the cause of technical education is being tried by Mr. J. Dove-Smith, of Standard Bearings, Limited, Niagara Falls, Ont. He has arranged with the Collegiate Institute there, to include a course in the plant of the Standard Bearings as part of the regular curriculum of the Collegiate. The enthusiasm with which this has been taken up by the students at the Collegiate is ample evidence of the success of this idea in this instance at any rate; and there seems no reason that similar courses should not be arranged for in other towns and cities in Canada. This idea should certainly have good results, as it will serve to acquaint young men and boys with the many attractive features in mechanics; and the minds of those boys who would make a success at such work are set in the right direction. Surely the originator of this idea is to be congratulated.

THE ENGINEER AND THE STEAM TURBINE.

THERE is an idea prevalent among engineers that anyone with a certificate can turn on steam and run a turbine outfit. The notion probably arises from the known good behaviour of many turbines in use ; and turbines in running order certainly give little trouble in themselves. But this very desirable feature is assured only by unfailing vigilance. Little matters that would pass without the least thought or apprehension in a pump, send the engineers on watch in a turbine ship all on edge.

Tradition says that the night before the "Ark" put to sea Noah, who was anxious about the refrigerator and condensed water for the animals, called up his chief engineer for sailing orders, and amongst other general instructions gave expression to the classic motto " and don't get caught." With turbine machinery more than ever the marine engineer recognizes this as his golden rule. Coal bills, side commissions, bad temper, talking back to the shore captain, anything, almost, will be forgiven him, but at sea the engines must not stop. Now the things that stop the turbines accidentally are measured in thousandths of an inch. Old line engineers do not at once recognize this limited clearance, because they do not see the necessity. The necessity is there, however. The great attention paid to shades of wear in the thrust block, to sounding the "dummies" and "the finger pieces" are all things that cannot be done with the turbine running, but on which good running depends, for the expansion when heating up takes effect from the "adjustment block" or thrust block and is cumulative at the far end of the turbine rotor, where the inter blade clearance will be correspondingly exaggerated. So the first row of blades is adjusted by feeling to .002 inch—this further down the expansion runs up to nearly .10 inch—and puts a limit to the length of a drum or rotor, apart from what bearing will carry.

Then a hot bearing on a crank shaft will call for attention, but it cannot in the ordinary course of events work the destruction that hot bearings are capable of in a turbine, because in the long and narrow clearance between rotor and casing the slightest distortion arising from a hot journal would let thousands of blades into trouble. To safeguard this contingency, they have cooled and forced lubrication, but the very ease of such a device is a trap for the ordinary man. It fails when least expected ; and because it fails only once in a great while, we get tired expecting it—every ten minutes—but to get caught means—other troubles.

In two years time the Mediterranean service between Marseilles and Alexandria will see the commission of four large turbiners now in the yards, and already the superintendent is looking about for careful men with turbine experience. The chief engineer of a 12-knot freighter, when he heard it mentioned, laughed and said that in two years time the service would be full of turbine men, and would-be turbine men, that any good company could have an unlimited choice of. But marine superintendents in a position to know take a different view. They seem to think that as turbine steamships multiply the best men will be at a premium. The necessity of careful and resourceful men becoming more and more apparent, managers will feel that not every chief with an "extra first" certificate is an available man for turbines. From the owners' point of view a chief engineer's value is gauged not by the troubles he has gone through, but by all those he has avoided.

In the next place good "chiefs" and "seconds" knowing others to have gone from reciprocating ships to turbiners will not be willing to take a subordinate position for the sake of a little experience that they may underrate. To them it will look like winning their way up again. There is also a somewhat short limit to the promotion of the men now serving in turbine ships, for it is not every engineer, no matter what his experience may be, who is capable of taking charge of men and machinery in a modern passenger ship. It may be quite a generation before supply and demand gets on even terms again, as far as the engineer and the marine turbine are concerned. Meanwhile plums will go to those far sighted individuals who most adapt themselves to the new conditions and who are prepared to sacrifice present positions to qualify for new responsibilities.

These remarks apply in a much less degree to the land turbine—in central station work the turbine is not subjected to the severe trying-out process of machinery at sea—nor could stationary engineers, if transplanted, be expected to successfully meet the many demands on a marine engineer entailed by conditions in the sea service, and the ways of seafaring men.

ON THE GROUND FLOOR.

CANADA has reached a stage in her development, when opportunities are most favorable for the establishment of industries, which will in years to come be regarded as of foremost importance. In metaphorical language, the national structure has risen above the foundation and basement stages to the ground floor. The door of opportunity stands open, affording an easy entrance to all who seek admission.

If we may consider the development of Canada to be in any degree analogous to that of the United States, it is apparent that the firms, which make a good start in the years of early development, will be the strong ones when the country comes to maturity. In the case of Canada, her growth will be even more rapid than that of the United States, for the reason that we, in this country, are able to benefit from the experience of the Republic. In knowledge, if not in experience, we stand to-day her equal.

The importance of getting in on the ground floor must not be minimized. In coming years certain work must be done. The firms which will do this work will undoubtedly be those that have shown by past accomplishments their ability to perform it. New industrial concerns may be called into being as the years go by and as the exigencies of the situation demand, but none the less is it true that the old-established firms, when properly managed, will enjoy a favor not accorded to less experienced competitors.

Industrial concerns outside Canada can share in Canadian business as well. For many years to come it will be impossible for this country to supply its own needs, particularly in the matter of those classes of machinery for which the demand is comparatively small. To these concerns we would reiterate our advice that it is important to get in on the ground floor. By watching this market carefully, by constant advertising in the best mediums and by personal canvass on occasion, this purpose can well be accomplished.

Power Required to Drive Twist Drills†

A Diagram for Finding the Power Consumed in Drilling Cast Iron

By C. S. Frary and E. A. Adams

As a result of the recent experiments with high-speed drills it was found that the thrust necessary to push a drill through a piece of metal was very much greater than had been supposed. Hence it would be necessary to build very much heavier and stronger machines in order to use the drills up to their full capacity.

Two years ago the Worcester Polytechnic Institute, in accordance with this fact, designed and built a new 10-inch drill press. The machine as it is now standing in the shops of the institute, consists of a box-shaped upright post somewhat similar to the parabolic posts of the latest planers. The spindle is driven by a steel mitre gear meshing with a rawhide gear. There are four belt speeds ranging from 328 to 525 r.p.m. The feeding mechanism is the weakest point in the whole construction of the drill press. It consists of a chain drive from the jack shaft, to a large sprocket, which is connected to a cone of gears by means of a universal joint actuating a small gear, which can be shifted from the lowest step of the cone to the highest, thus giving a range of feeds from 0.0045 to 0.0225 inch per revolution of the spindle. The drill is so designed that at the end of the hole which is being drilled a reversing clutch is thrown in and the drill will be drawn out of the work. As the drill was finally built the quick return ratio was about 6:1 at the highest feed and 29:1 at the lowest. It was designed for a return speed of 0.225 inch per revolution of the spindle. In other figures the return speed was about 112 inches per minute or about 2 inches per second. The machine was actually built and run with that return speed. The only trouble experienced was that the spindle after reversing, instead of stopping, threw in the feed clutch and started to feed into the hole which it had already drilled. Outside of the few things mentioned the drill press was a very strong machine and behaved extremely well under the severe test to which it was subjected, during the few twist-drill tests which could be made in the limited time at the disposal of the experimenters.

The Dynomometer.

The dynamometer, for registering the thrust and the twisting moment of the twist drill, was very simple and efficient. A hollow piston with a round top to form a table, was scraped to fit a cast-iron cylinder. The cylinder was filled with heavy cylinder oil, and had an ordinary pressure gauge tapped into the lower end. The gauge read pressure per square inch and in order to get the thrust of the drill it was necessary to multiply by the area of the piston, which was about 20 square inches. To measure the twisting moment a steel band, fastened to the enlarged top of the piston, was connected to an indicator spring by means of a steel rod screwed into the bottom of the indicator piston. Since the area of the indicator piston was only ½ square inch, and the force was applied direct in this case, the indicator spring had to be rated at ⅛ its usual value when used in a steam indicator. The movement of the drum of the indicator was obtained by passing a cord over a pulley, which was attached to the carriage of the spindle, and then fastening the end to a projecting arm on the dynamometer. Taking the average force as registered by the indicator diagram and multiplying by the radius of the round table gave the twisting moment. In none of the tests did the moment exceed 350 inch pounds, which was obtained with a ⅞-inch drill running at a speed of 328 r.p.m. and feeding 0.0225 inch per revolution.

In order to be sure that the thrust given by the dynamometer was correct, a calibration curve was found by applying loads on the piston of the dynamometer and reading the pressure recorded by the gauge. Every reading of the thrust during the drill tests was corrected for static friction in the dynamometer by referring to this calibration curve.

No. of Test.	Diameter of Drill	Thickness of Web.	Angle of Lip.	Hardness of Test Piece.	Speed.		Thrust—lbs. for Feeds in Inches per Revolution.								
					R.p.m.	Ft. p. m.	0.0225	0.0203	0.0176	0.0156	0.0135	0.0113	0.009	0.0068	0.0045
16	½"	0.078"	59°	15	525	66	1142	995	832	738	640	576	454	405	320
17	9/16"	0.086"	59°	15	525	74	1540	1363	1192	1020	860	732	618	476	379
18	⅝"	0.093"	59°	15	450	72	1180	1079	916	817	738	668	582	497	405
19	11/16"	0.098"	59°	15	386	68	1296	1130	935	778	700	643	515	466	373
20	¾"	0.104"	59°	15	386	71	1218	1075	914	820	736	860	585	508	416
21	⅞	0 1.5'	59°	15	328	73	1543	1406	1260	1110	963	849	742	635	529

VARIATION OF THRUST WITH DIAMETER OF DRILL.

Fig. 1.

Thickness of Drill Webb.

During tests made by Prof. W. W. Bird and Mr. Fairfield in 1904, it was found that there was a great variation of the thrust obtained from drills of the same diameter working under the same conditions. On measuring the thickness of the web of the drill they were found to vary quite widely for the same diameter of drill even among drills of the same manufacture. After considerable study Mr. Fairfield deduced from common practice the following law: Beginning with a drill of zero diameter and assuming a thickness of web equal to 1-64 inch, the thickness of web should increase 1-64 inch for every increase of ⅛ inch in the diameter of the drill.

In all of the tests made this law was followed as far as possible. Drills made of Novo steel were used varying from ½ to ⅞ inch by sixteenths, and as far as they could be obtained conformed to the law proposed by Mr. Fairfield.

Among the many possible tests which could be made with twist drills only a few were taken up. All tests were made on cast iron, keeping the surface speed of the different-sized drills as nearly constant as the four speeds of the drill press would allow, and changing the feed.

The Test Pieces.

The test pieces were 10 x 2 x 1½ inches poured horizontally, being gated at three places on one side. This was to insure a fairly even hardness throughout the entire length of the specimen. Before making a test each specimen was tested for hardness by drilling a ⅝-inch hole and stopping the drill just below the surface of the specimen. The pressure registered by the gauge was taken as an indication of the hardness of the specimen.

† From the "Journal" of the Worcester Polytechnic Institute.

A test consisted of drilling three holes for each of the nine feeds at a constant surface speed of the drill of about 70 feet per minute. Three specimens were chosen of the same relative hardness and one hole drilled in each for the different feeds, and readings of thrust and twist being taken for each. This was done for all of the drills and formed one part of the experiment.

All of the above tests were made with the tests where the speed was kept constant the thrust was very nearly proportional to the feed.

With a constant surface speed the thrust is very nearly proportional to the diameter of the drill.

With the different angles of the lip the thrust decreases from 75 down to 45 degrees, and then increases for any further decrease in the angle. The moment is quite different. In fact it cannot be said that the moment is in any way proportional to the angle of the lip

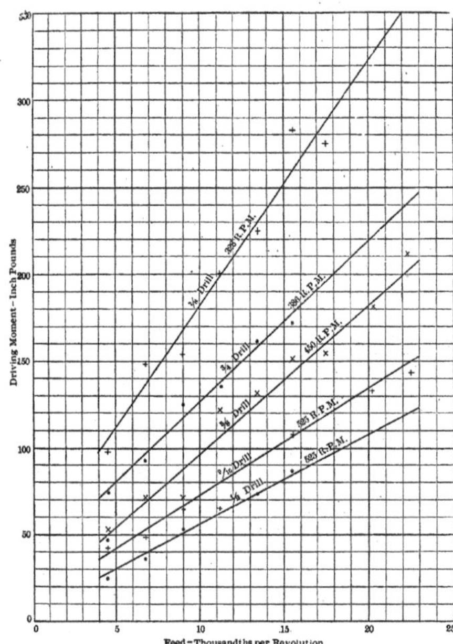

Fig. 2—Relation of Driving Moment to Amount of Feed for Different Size Drills.

a constant angle of the lip of the drill of 59 degrees. Several tests were made with a ⅝-inch drill by keeping the speed constant and varying the angle of the lip from 37½ to 70 degrees, and measuring the thrust and moment for the nine different speeds.

General Conclusions.

As a result of these tests there are several interesting conclusions. In all although it is nearly proportional to the feed.

From a study of the variation of the thrust with the angle of the lip it would appear that the 45-degree angle ought to give the best results in practical machine shop work. There is only one instance, known by the experimenters, in which the 45-degree angle is given the preference over the older angle of 59 degrees. That is in the case of William Sellers & Company, Philadelphia.

Results of Tests.

(The results of these tests are shown in the accompanying table and diagrams.

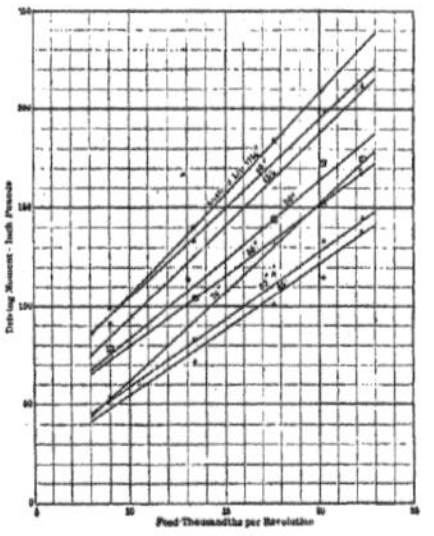

Fig. 3—Relation of Moment to Angle of Lip and Feed; Diam. of Drill, 5-8 in.: Speed. 450 R.P.M.

The diagrams were prepared by the American Machinist from tables prepared by the writers of this paper. These diagrams speak for themselves.—Editor.)

TOOLS BY MAIL.

The Mechanics' Supply Co., Quebec, make a specialty of securing mail orders for small tools; and they have a special department for building up this side of their business. To facilitate this buying of small tools by mail, the company issue special sheets and catalogues, which may be had upon application to the company.

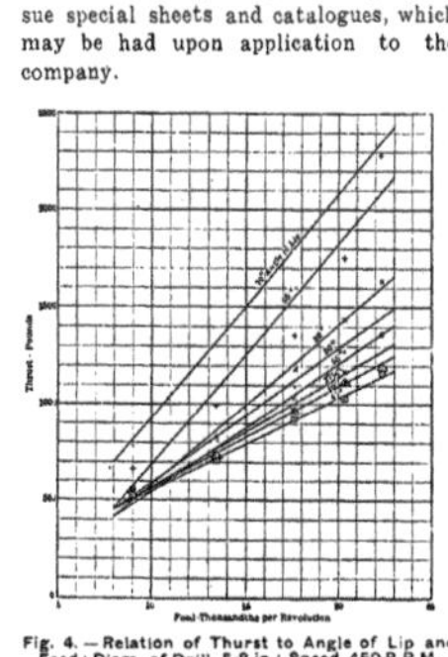

Fig. 4. — Relation of Thurst to Angle of Lip and Feed; Diam. of Drill, 5-8 in.; Speed. 450 R.P.M.

Kelly Bros. & Co., Kenora, have received the contract to erect a $100,000 round-house for the C.P.R. at Kenora.

Consideration of Canadian Patent Act

Compulsory Manufacturing Requirements of the Canadian Patent Act, Viewed in Reference to Dependent Patents.

By Egerton R. Case, Toronto.

IT OCCURS to me that it will be of considerable interest to analyze the Canadian Patent Act in reference to compulsory manufacture of patented inventions. It is quite clear that if a man manufactures under his patent within two years from date of same or within an authorized extension of that period he will maintain the validity of his patent. But this manufacture under certain patents is not so simple a matter as it may at first seem. In order to fully understand what I mean by the last sentence, I would invite the reader's attention to the following:

Under Sub-section (a) of Section 4 of the Act to amend the Patent Act dated August 13th, 1903, the owner of a patent must work the invention covered thereby within two years of the date of the patent, or within an authorized extension of that period. In place of the foregoing section, if the invention belongs to the following class, it may be placed under Section 7, provided the commissioner of patents' assent is obtained, which relieves the owner of the patent from the necessity of manufacturing his invention in order to keep valid his patent covering same.

The following is the class of patents which may be placed under section 7: Certain patents for an art or process; certain patents for improvements on a patented invention when both patents are not held by the same person; patents for certain appliances or apparatus used in connection with railways, telegraph, telephone, and lighting systems, and other works usually under the control of public or large private corporations, and which appliances or apparatus cannot be installed or constructed without the consent of such corporations, and certain patents for inventions which are manufactured or constructed only to order, and are not, according to custom, carried in stock.

When a patent has been placed under Section 7, any person may apply to the commissioner for a license to construct, use, and sell the invention covered thereby, and if the commissioner is satisfied that the reasonable requirements of the public in reference to the invention covered by the patent in question have not been satisfied by reason of the neglect or refusal of the owner of the patent to make, use, construct, or sell the invention covered thereby, he may make an order requiring the owner of the patent to grant a license to the person applying for same upon suitable terms as to royalties and security for payment as the commissioner deems just. If the owner of the patent refuses or neglects to comply with the commissioner's order within three months after date of same, his patent becomes null and void.

When we come to view the requirements of our Patent Act as to compulsory manufacture in reference to a certain class of patent covering an improvement of an invention covered by a prior patent, a difficulty arises, and the object of this paper is to find out how the owner of a patent covering an invention, (an improvement on an invention covered by a prior patent) can comply with the requirements of the Patent Act in order to keep his patent valid. For the purpose of this paper, Class A patents are primary or containing patents; that is, patents that are broad enough to include any improvements that may be made in any particular invention covered by this class of patents during its lifetime; Class B patents are those that are subject to Class A patents, but yet which may be marketed by being merely added to the invention covered by Class A patents without necessitating the reconstruction or the reorganization of this invention, and Class B-C patents are those which are contained in or are subject to Class A patents and which cannot be used therewith without the reconstruction or the reorganization of the invention covered by Class A patents. In investigating the question which called this article into being, I shall first consider Class B patents in their relationship to Class A patents, and then afterwards consider Class B-C patents in their relationship to Class A patents.

In the class of inventions which may be placed under Section 7, before enumerated, it will have been noticed that when an invention is an improvement on a previously patented invention, and both patents are not held by the same party, the commissioner may order (presumably, no matter what the nature of the invention may be) that the patent covering said invention be made subject to Section 7 instead of to Sub-section (a) of Section 4.

For argument's sake, we will suppose that a Class B patent has been placed under Section 7; now a demand for the invention arises; how is the owner of the Class B patent going to supply the demand? As he may lawfully purchase the invention covered by Class A patent and add his invention thereto, and sell the resulting article as his own, he can supply the demand for his invention and maintain the validity of his patent. Even if Class A patent be placed under Section 7 and the owner thereof be not manufacturing under same, the owner of Class B patent, (whether his patent be subject to Section 7 or no) may obtain a license under Section 7 and manufacture the invention covered by Class A patent and add his own invention thereto and sell the resulting article as his own.

In considering Class B patents in connection with their relationship to Class A patents, no great difficulty arises, but when we come to consider the relationship of Class B-C patents relative to Class A patents, a difficulty certainly arises. In order to clearly illustrate this relationship, we will first consider the position of a Class B-C patent relative to a Class A patent placed under Section 7, when a demand arises for the invention covered by the first class of patent, and there be no manufacture under Class A patent. If the owner of Class B-C patent applies for a compulsory license to manufacture the invention covered by Class A patent, he may obtain same. But it is very questionable if the commissioner is empowered to grant a license of such a nature that the invention covered by Class A patent may be reorganized or reconstructed before the invention covered by Class B-C patent can be used therewith. If it be that the commissioner have the power of granting such a license, then in the supposed case the owner of Class B-C patent can maintain the validity thereof.

Again, if Class A patent be subject to Section 7, and the owner thereof be supplying the demand therefor, can the commissioner compel him to grant a license of the necessary character to the owner of Class B-C patent? If he have the power, then Class B-C patent remains valid; but if he have not that power, Class B-C patent expires. But it is clear that the commissioner has not that power, when the demand under Class A patent is being supplied.

Again, if Class A patent be not subject to Section 7, the commissioner has no power to compel the owner thereof to grant licenses. Now, in this case, if the owner of Class A patent will not grant the owner of Class B-C patent a license to manufacture to supply the demand, this latter class of patent must expire at the end of the term allowed for manufacture. Therefore it will be seen that the validity of Class B-C patent depends upon the owner of Class A patent.

Let us now consider the position of Class B-C patent relative to Class A patent not subject to Section 7, when the first class of patent has been made subject to Section 7 and the owner of Class A patent wishes to manufacture under Class B-C patent. Upon demand, the owner of Class A patent can obtain the necessary license under Section 7 relative to Class B-C patent; under these circumstances, Class B-C patent remains valid.

But, if the owner of Class A patent conspires to create a demand for the invention covered by Class B-C patent, knowing that manufacture thereunder cannot legally be carried on without a license from him, what is the poor owner of Class B-C patent going to do? If he manufacture without a proper license, he is at once liable to damages for infringement. Yet the demand for the invention covered by Class B-C patent may still be kept up, and unless this is satisfied the patent must surely lapse.

It is quite right and proper that Class A patents should convey to their owners broader rights and privileges relative to the invention covered thereby than is conveyed to the owners of Class B-C patents. But under our Patent Act as it now exists, it is submitted that under certain well-defined conditions the owners of Class B-C patents suffer a great hardship, and the capital they may have invested therein is liable to be rendered valueless. As a result of the premature lapsing of a Class B-C patent under our present Patent Act, I would call the reader's attention to the benefits that accrue to the owner of a Class A patent. Suppose a Class A patent is thirteen years and eleven months old. The longest period of exclusive right that can be enjoyed by the owner of a Class B-C patent is four years. Now as Class A patent expires one month after the expiration of the Class B-C patent, the owner of Class A patent is in a much stronger and better position than the owner of Class B-C patent, because having had possession of the Canadian market for eighteen long years, he has old and wide connections and business channels that give him an enormous advantage over his competitors, although his patent has lapsed. If the manufacture of his old invention requires an expensive plant, it is not very likely that competitors will try and wrest his market away from him when they consider this fact, and the other facts before stated. Now, having a properly-equipped plant, he at once commences the manufacture of the invention covered by Class B-C patent and by modernizing his goods, he is still able to defy competition, and so have a further and unfair monopoly of the market. He has had his monopoly of the market for a certain line of goods, and it is unfair that it should be continued at the expense of others. As the Patent Act now stands, the owner of a Class A patent is favored at the expense of the owner of a Class B-C patent. It is quite right and proper that the owner of Class A patent should control his invention during the lifetime of his patent, but it certainly is a great hardship that the Patent Act now works an injustice to holders of Class B-C patents.

Everything points strongly to the antiquity of Sub-section (a) of Section 4 of the Patent Act as amended August 13, 1903. Canadian industries will be amply protected and the holders of Class B-C patents will receive fair tratment if the requirement as to importation be retained, and all patents granted issue subject to compulsory license, under proper regulations. Our Patent Act should place all inventions on an equal footing, and if for no other reason, the requirement as to manufacture within a limited time should be stricken therefrom, so as to cover the conditions controlling Class B-C patents. Of course if Class B-C patents be placed under Section 7 and no demand be made for the invention covered thereby, it will remain valid. No man will know, as a general rule, whether or no his patent is subject to another patent unless he has had this question thoroughly looked into. It is natural to suppose that if he knew beforehand what status his Class B-C patent occupied under the Patent Act, he certainly would not spend time and money in developing same. As this question becomes better understood, it is clearly apparent that as our Patent Act now stands, it is discouraging the production of inventions of a certain class, whereas it should offer the fullest encouragement to all classes of inventions, so that all inventors and manufacturers may have an equal chance of reaping benefits from their efforts.

From the foregoing, it will be understood that when a Class A patent is a recent, or a comparatively recent one, the owner thereof may come into possession of a very valuable asset, by reason of the expiration of a Class B-C patent, as, through the deficiency of the Patent Act as it now stands, the public (although the Class B-C patent has expired) will not have the right to exploit same, because of its dependency upon Class A patent.

From the above facts stated relative to the relationship of Class B-C patents to Class A patents, a number of questions arise. What consideration has the Government given the owner of Class B-C patent that compensates him for paying the Government fees, and for disclosing a valuable invention? Would not the interests of Canada be as well protected if the requirement as to manufacture of our Patent Act as it now stands, were entirely done away with? This said section certainly works an injustice to the owners of Class B-C patents. It is clear to one who has followed this question carefully, that the placing of Class B-C patents under Section 7 will not solve the difficulty. As the industrial activity in a country increases, inventors and manufacturers watch each others' inventions very closely, as they know that their business will survive the competition of rivals if they can manufacture and market their productions so as to net them a greater profit than is made by their rivals. As a result of this keen competition, inventors and manufacturers study very carefully different inventions and do their best to improve thereon. So when an improvement is made on any invention and covered by a Class B-C patent, the Patent Act should give this improvement the full protection accorded to inventions covered by Class A patents.

Suppose a foreigner controls a Class A patent in the case above supposed, does not the Patent Act now actually operate in his favor as against a Canadian, the owner of a Class B-C- patent? Is this condition fair? The vast majority of Canadian patents are taken out by foreigners, and, as the Patent Act now stands, it is conceivable that they might largely control the industrial affairs in Canada, which state of things is not one to be appreciated by Canadians. I submit, that if we retain the importation and compulsory license clauses, and remove the antiquated clause relative to manufacture as set forth in Sub-section (a) of Section 4 of the Act to amend the Patent Act dated August 13, 1903.

In several European countries, patents of addition (Class B and Class B-C patents), expire with the expiration of the parent patent (Class A patent). But such a law is not a just one, because, a manufacturer who has had the market is placed on a much better footing than any rival, for the reasons before set forth.

(Continued next issue.)

Gas, Gas Engines, and Gas Producers[†]

By J. Seton Gray, Can. Soc. C.E.

THE thermal efficiency of a modern steam plant is about 10 per cent. and that of a gas engine plant about 20 per cent. This is the principal reason for the large amount of work that has been done in recent years on the gas engine.

Again, while gas can be manufactured

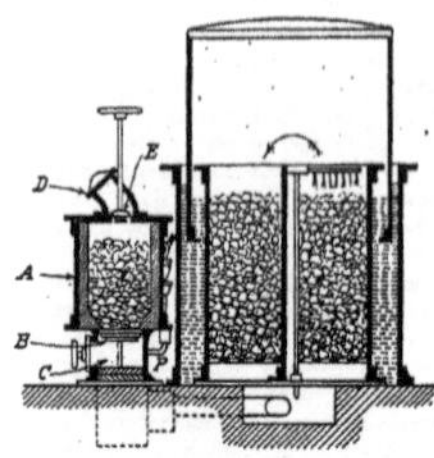

Fig. 1—Dowson Plant.

in a gas producer for about three cents per 1,000 cubic feet, city lighting gas costs between 50 and 100 cents per 1,000 cubic feet. This explains why so much attention is being paid to the development of the gas producer for the making of power gas.

The many methods adopted for the production of power gas may be divided into three general classes.

1. When a carbonaceous such as coal is heated in a closed retort, gases are given off which may be collected and used for power. This method is that at present in use for the manufacture of illuminating gas.

2. If steam be blown through a mass of incandescent fuel, a combustible gas is produced. In this process the fuel is kept incandescent by a blast of air, the steaming and blowing periods being intermittent.

3. If steam and air together be uninterruptedly blown into incandescent fuel, a gas, containing hydrogen and carbon monoxide, is produced continuously. The amounts of steam and air are regulated so as to keep the fuel at a fairly constant temperature.

Gas produced by the first method, while generally of high calorofic value, is costly, as already mentioned. Water gas, produced by the second method, is much cheaper; but as it contains much hydrogen, is a very inflammable gas, and on this account cannot be used with the high compression pressures now employed in gas engine practice. These high pressures are necessary if a high efficiency is required of an engine.

The third method, however, gives a cheap gas, well suited for gas engine work, and will be treated more fully in this thesis.

Nearly all fuels containing carbon can be used for the production of producer gas. It must be noted, however, that if the gas is going to be used in a gas engine cylinder, it must be of uniform quality. It must also contain no tar or other impurities in order to avoid trouble with the valves.

To Mr. Dowson we owe the first successful producer gas plant. His plant was designed to work with anthracite coal. This coal is non-caking, and, being nearly pure carbon, contains very few densible hydrocarbons or tar. It is therefore an ideal fuel for the gas producer.

As this producer embodies in the best way the fundamental principles necessary for the production of gas from solid fuel, it will be described here.

Fig. 1 is a diagrammatic section of a Dowson gas producer, taken from "The Gas and Oil Engine," by D. Clerk. As may be seen, the producer consists of a cylindrical casing (A), lined with fire-brick, and having at the bottom fire bars (B), above a closed ash pit ((C). The upper part of the generator is closed by a metal plate on which is mounted a fuel hopper (D), having an internal bell valve (E) operated from the outside.

To begin operations, the upper cover is removed from the hopper (D), the bell valve is opened; a fire is built upon the bars (B), and air is forced through it by the steam jet (N) and the pipe (P). Fuel is slowly added from above till the whole mass is incandescent and fills the producer to a depth of about 18 inches at least. During this heating process, gases are given off by way of the open hopper and are ignited there. Care must be taken not to inhale this gas. (It contains much CO and is very poisonous, but when burned it is harmless.) When the fuel is incandescent the inner and outer valves of the hopper are closed, and the gas flows by a pipe through cooling and scrubbing devices, finally finding its way to the gas holder through the coke scrubber formed within it. From the gas holder the gas flows through another scrubber, as shown by the arrow, and thence to the engine.

In 1895 Renier, a Frenchman, after much experimental work with a Dowson gas producer, succeeded in making the engine draw its own supply of gas direct from the gas generator, without the use of a gas holder. While this plant was not a commercial success, it was the forerunner of the modern suction gas producer.

The Dowson gas producer has already been described. The following is a description of a suction gas producer; the difference between the two systems may be seen from the descriptions: The adjoining sketch (Fig. 2), shows the general construction of this piece of apparatus. To set it in operation a fire is built on the grate. The door at the front of the producer allows this to be

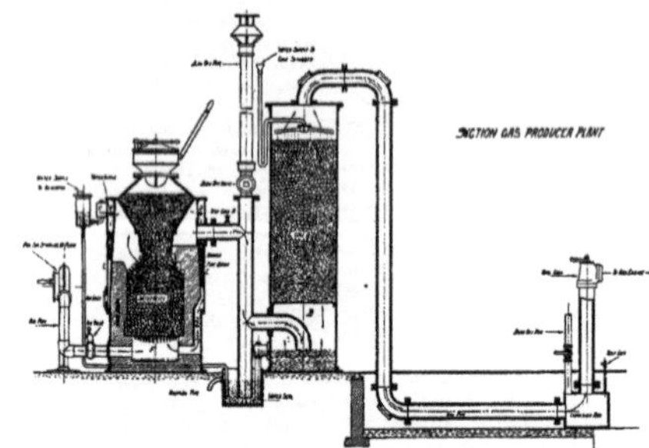

Fig. 2—Suction Gas Plant.

† Read before the Mechanical Section of the Canadian Society of Civil Engineers, Feb. 27, 1907.

easily done. A quantity of paraffine waste, or other combustible material, is placed inside the generator, and lighted. Broken wood is then added through the hopper at the top, and this is followed by coal until a bright fire is burning. The air for combustion is supplied by a small hand fan (B). As soon as the fire is burning brightly the water necessary for the production of the gas is turned on.

This gas is tested by opening a small test cock (A), and applying a light. If the gas burns with a blue flame it is ready for use. The gas is then allowed to enter the scrubber (B), and to pass through the expansion box to the engine. Before being allowed to enter the engine the gas is again tested, and if it still burns with a blue flame the engine may be started. The hand fan is stopped as soon as the engine starts work, and thereafter the whole plant becomes automatic.

During the suction stroke of the engine a vacuum is set up in the engine cylinder. To fill this vacuum air and steam are drawn into the generator, are there turned into combustible gases, which pass through the scrubber to the engine.

For the production of the necessary steam a water supply (C), is fixed the generator. This supplies water to the vaporisor (D). This vaporisor is just a pipe perforated with a number of holes. Every time the engine draws gas, which it does in proportion to the load, a small quantity of water is drawn out of the vaporisor, falls on a quantity of hot refractory material (E) in the casing, and is there converted into steam. This steam is caught up by the current of air drawn in at the air valve. The mixture of air and gas passes round the casing to a space (F) under the fire bars, from whence it is drawn up through the body of incandescent fuel, where it is turned into gas. The gas passes from the generator through a water seal at the bottom of the coke scrubber, passes through the closely-packed coke, where its tar and other impurities are extracted, and then goes to the engine cylinder. A continuous stream of water is allowed to fall in a spray over the coke, whereby the gas is cooled to the normal temperature.

The theory of the chemical action that goes on in the gas producer is as follows: When air comes in contact with glowing carbon we get, first of all, the burning of the carbon to form carbon dioxide, according to the equation $C+O_2=CO_2$. Now carbon at a high temperature is a very strong reducing agent, so when the CO_2, already formed, comes in contact with the mass of glowing carbon through which it must pass, the CO_2 carbon dioxide, is reduced to CO, carbon monoxide, according to the equation $C+CO_2=2CO$. This carbon monoxide is a combustible gas, very poisonous, and can be used in the gas engine.

When steam comes in contact with glowing carbon, it is reduced by the action of the carbon to CO and 2H, according to the equation $C+H_2O=H_2+CO$.

The producer plant last described is known as the suction plant, because it is kept in continuous operation by the suction stroke of the engine. Another plant also in common use is known as the pressure plant. A well-known example of such a plant is the Dowson plant, already described. The chemical reactions in this case are exactly similar to those already described when dealing with the suction producer. The principal differences between the two systems are that the pressure system requires a steam boiler in which to generate the steam required for operation, also, that with this system a gas holder, in which to store the gas, is required.

There is one serious objection to the pressure system which is not found in the other system. The whole producer is under pressure. If, then, there are any leaks between the producer and the engine, gas will escape. As has already been mentioned, this gas is very poisonous, and may cause the death of a careless operator. The suction producer is full proof in this respect, because the pressure in the system is less than atmospheric. If, then, there are any leaks in the system, all that can happen is that air will pas into the system through these leaks. While this will reduce the efficiency of the plant, it can do no other damage. The fact that no boiler or gas holder is required in the suction producer is a great point in its favor, because simplicity is the essence of good engineering, especially when we are dealing with a machine which is going to be put under the care of men who have no engineering knowledge or skill.

It takes only a few minutes to start up a small gas producer plant. The Highland and Agricultural Society of Scotland made some tests in Glasgow in 1905, to ascertain the time required to get an engine and producer plant to full

DIAGRAM SHOWING HOW THE HEAT IN 1 LB. OF COAL IS UTILIZED IN THREE CHARACTERISTIC PLANTS.

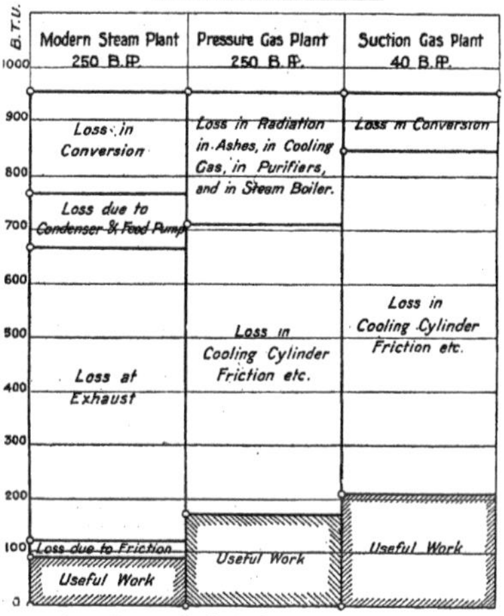

Fig. 3—Heat Diagram.

working load, starting with the producer empty and cold. Only two men were allowed to start each plant on trial. The results were as follows:

Builder of Plant.	Capacity of Plant B. H.P.	Time to start.	Remarks.
Campbell Gas Engine Co	18	13 min.	
" " "	8	17½ min.	
Crossley Bros.	24	15½ min.	
Industrial Engin'ring Co.	10		Started but was compelled to stop because water was shut off at main.
National Gas Engine Co.	20	48 min.	Excessive time caused by defective igniter.
" " "	10	15¾ min.	
Messrs. Tangyes, Ltd ...	21	16 min.	
" " "	12	12½ min.	

These tests were all made starting with the producer cold. The average value for a small plant is about 15 minutes. But in actual practice the fire is banked up when the producer is not in operation as, for instance, when it stands over night. When this is done the engine can be running in about seven minutes from the start. According to Mr. J. Emerson Dowson, the stand-by losses due to this banking up of the fire amount, in a moderate-sized plant, to about three lbs. of coal per hour.

Comparison of Steam and Producer Plants.

Dealing with efficiency, the adjoining diagram shows very clearly the values obtained from three characteristic plants.

Column 1 shows how the heat is used up in a modern steam plant of 250 h.p. The total heat contained in the coal was 952 b.t.u. Of this there is a loss in conversion of 20 per cent., much of the heat in the fuel passing up the chimney. There is a further loss of 10 per cent. in the feed pump, in condensation, and in radiation. Of the remaining 70 per cent., 57½ per cent. is lost in the engine exhaust; and after making an allowance for friction losses, we find that out of a total of 100 per cent., only 10 per cent. is converted into actual work.

While this value of 10 per cent. may be obtained in test, in actual everyday practice the efficiency will be still less.

Column 2 shows how the heat is used up in a pressure producer plant of 250 h.p. There is, first of all, a loss of 25½ per cent. in radiation, in ashes, in gas coolers, and in steam boiler. Of the 74½ per cent. which goes forward to the engine, 33.2 per cent. is lost in cooling the engine cylinder, 20 per cent. is lost in exhaust, and after allowing for friction we find that 18 per cent. of the available heat is given as actual work in this system.

Column 3 shows the distribution of the losses in a 40 h.p. suction producer system. It may be taken for granted that the losses in a 250 h.p. system will be smaller. Of the total heat in the coal 89 per cent. is transferred in the gas to the engine cylinder. After deducting engine losses, as in the last case, we find that 23 per cent. of the available heat is transferred into mechanical work. The efficiencies then are as follows:

Steam, 10 per cent.; pressure producer 18 per cent.; suction producer plant, 23 per cent.

There are one or two other points worthy of mention while dealing with the efficiency question. In a small steam plant the loss due to bad stoking is often quite considerable; in a producer plant there is very little such loss.

With regard to stand-by losses, as before mentioned, these are about three lbs. of coal per hour in a moderate-sized producer plant, whereas, according to results obtained by Mr. Dowson, this loss is about 71.5 lbs. of coal per hour in a steam plant of the same size. When we consider that most are idle for about 199 hours every week, we see how great will be the difference in coal due to stand-by loss.

This loss is small in producer plants because, since very little air is passing through the fire, when the fire is banked up in the gas generator that piece of apparatus is turned into a slow combustion furnace.

With regard to the efficiencies mentioned on the last page, there is a point of great practical interest which is too often overlooked. When we say that the efficiency of a steam plant is 10 per cent., while that of a gas producer plant is 23 per cent., we mean, among other things, that, for the work equivalent of 10 tons of coal, we must not only buy 100 tons, but we must also pay for the labor of handling this, and also for storage space. With the producer plant the calculations are made only on 23 tons of coal. A similar relation holds in the disposal of the ashes.

Dealing with the problem of fluctuating loads, the following test was made on a suction producer plant by Messrs. Crompton & Co., Ltd., of London, Eng. A gas engine was run for four hours with a load of 10 h.p., then a load of 80 h.p. was thrown on suddenly. The plant immediately responded, and hardly a flicker was noticed in the lights supplied from the engine. It is to be borne in mind also that this was done without the use of a gas holder. This can be done by any well-designed producer plant, and is a performance that an engine working from a steam boiler would find very hard to beat.

Coming now to the problem of attendance. It is found that a complete producer and gas engine of 100 b. h.p. capacity requires the labor of one man for two hours each day, to keep it in first-class running condition. Everyone who has run a steam boiler and engine plant knows the troubles that are constantly turning up. The sanitary authorities complain that so much smoke is being thrown into the air, or the injector fails to operate, and the water begins to creep dawn in the water gauge. The boiler has to be cleaned out twice a year; there is a large amount to be paid for insurance, for inspection, for wear and tear of fire bars and other fittings, for the repair of leaks in joints caused by the high pressure of the whole system, and, finally, there is the knowledge that some day the whole plant may take it into its head to go out by the roof.

Place.	K.W. hrs.	K.W. per lb. coal.	Lubric. oil per K.W. hr.	Cotton waste per K.W. hr.	H.P. capacity
Clausthal	219,150	.36	5.6	.558	100
Cransee	51,847	.36	5.62	.558	100
Neumarkt	125,076	.37	7.22	.442	80
Neurode	119,800	.24	3.03	1.42	160
Reichenbach	49,437	.3	6.20	.50	60
Soberheim	95,326	.36	7.36	3.50	100
Schoenberg	58,514	.35	12.8	1.30	80
Schwetz	82,312	.398	4.81	1.06	160
Walserorde	60,352	.32	8.24	2.03	70
Weinenden	71,828	.24	6.97	1.98	160
Karlsruhe	70,766	.45	5.14	1.08	100
Werden	102,716	.36	6.07	2.99	90

Contrast with this the gas producer. It is clean and efficient, there is very little water to handle (¾ gallon per b. h.p., a steam plant requires 4 gallons per b. h.p.), and very little attention required. The whole system, except the engine, is subject to pressures of only a few lbs. per square inch. There is no chimney to build and maintain, nor is there any smoke nuisance. When the facts are considered that the system is cheap to instal, and also cheap to operate, one has to wonder at the slow growth of the gas producer industry compared with what it might be.

It has been argued against the producer plant that gas engines are not very good for the operation of electric generators. As an example to contradict this statement, in Granada, Spain, three single cylinder gas engines were installed rated at 80 b. h.p. each. These engines drive alternators in parallel, and have supplied the whole city with light for the last two years. Now, there are gas engines on the market in which the problem of balancing has been very carefully dealt with. If the result mentioned above could be obtained with single cylinder engines, there should be no difficulty in obtaining satisfactory operation from the modern three cylinder engines.

The following is taken from Dr. Oskar Nagel:

Several years of experience have shown that the gas power plants are fully as reliable as the best steam plants, and have the advantage of much greater economy. The following table of results of plants built in Austria and Germany, on the Koerting system, bears out his statement. It is to be noted with regard to these results that the plants are small, and were working only a short time each day, and that stand-by losses are taken into account.

A few facts and figures at this point should prove of interest.

A small joiner shop was driven by a 30 h.p. motor. This motor was replaced by a 60 h.p. suction producer plant, built by the National Gas Engine Co. The electric drive cost $13.20 per week. The same work was done by the producer at $2.88 per week for fuel, $5.32 for labor, oil, etc., also interest on capital and depreciation. This shows net saving of $6 per week. It also shows that the gas engine is going to be a keen competitor of the central station (electric).

The National Gas Engine Co. installed two complete suction plants in the south of Scotland in a factory which had been buying power from a central station. Before the installation was made the engineers in charge of the work made the following calculation of probable saving: The 160,000 units of electricity ewquired per annum, at 3½ cents per unit, gives a power bill of $5,[illegible] per annum. For the same power, with a suction producer, assuming one lb. coal, gives one b. h.p. hr., 108 tons of coal per annum would be required. At $3 per ton the fuel for the producer costs $324 per annum. Allowing 10 per cent. for interest on capital and depreciation, $171 for labor, $120 for oil and sundries, the total expenditure is $1,881 per annum, and the saving due to the adoption of the producer plant $3,119 per annum. The cost of engine and producer plant, complete, was $6,240, the electrical equipment cost $3,360 for two dynamos, switchboard, and wiring, or the total of $9,660. A saving of $3,119 per annum would therefore pay for the whole plant in three years.

It is interesting to mark the effect of the perfecting of the gas producer on the gas engine industry. Messrs. Thorneycroft & Co., shipbuilders, London, England, have just fitted a number of canal barges with suction producer plants and gas engines. The results of these have been so satisfactory that they are going to try them on coasting and merchant vessels. The British Admiralty is making experiments to find what are the limitations in using it for naval work.

The greatest advantage the producer has for marine work is that the amount of coal to be carried is greatly reduced.

In a country like Canada, where farming is done on a large scale, and where the power users are so scattered as to prohibit the building of central stations for power and light, the gas producer ought to have a large and increasing use. Messrs. Tangye, Ltd., of Birmingham, England, have put on the market a portable gas engine and producer plant to meet the demand of the farmers.

About ten years ago attention was drawn to the fact that a large amount of power is available in gas which is usually thrown away from blast furnaces. It had been stated on good authority that 468 h.p. may be developed per ton of iron produced per hour. In the United States of America alone there were produced in 1905 23,000,000 tons of pig iron. This is equivalent to an available power of 1,225,000 h.p.

The chief difficulty to be overcome in the use of this blast furnace gas in a gas engine is that connected with the removal of the large quantities of dust which it contains. Another difficulty sometimes met in a small plant, namely, that the gas is very variable in quality, is overcome in large plants by mixing the gas obtained from several furnaces together. However, the trouble due to this is not very great in a well-designed plant. The gas, which is very hot when it leaves the furnace, is usually cooled in the process of extracting the dust.

The quantity of dust in the gas varies greatly with the kind of ore and coal used. For instance, the Cockerill Co., Belgium, had a 200 h.p. gas engine running at their works in Seraing for three years without any special provision being made for the elimination of the dust contained in the gas. During all that time the engine never had to be cleaned on account of dust, although it was running night and day.

On the other hand, this same company, at their works in Differdingen, experienced trouble right from the start with some 600 h.p. engines which they installed. Investigation showed that the Differdingen gas contained four or five grammes of dust per cubic metre of gas, while the Seraing gas contained only from .25 gm. to .5 gm.

Experience shows that the furnaces using hematite ores give a gas containing very little dust, and what dust there is settles very easily, even in short lengths of pipe. Oolitic ores, on the other hand, give a gas containing much dust, which passes quite readily with the gas through long lengths of pipe.

There are two ways at present in use for the purification of blast furnace gases.

1. Passing the gas through scrubbers containing sawdust or coke, in exactly the same way as is done with producer gas.

2. The gas is caused to pass through a contrifugal fan. A jet of water enters the axis of the fan, and is driven outwards in the form of a fine spray. This spray of water gathers up all the dust in the gas. This latter method was tried in the Differdingen plant already mentioned, with the result that while the gas contained 4 gms. dust per cubic metre when it left the furnace, it held only .25 gm. after passing through the fan, and could then be used with success in the engines.

The calorific value of this gas, as might be expected, is very variable. It may be taken that rich gas means poor operation in the blast furnace, while poor gas represents good operation therein. The average calorific value is about 110 b.t.u. per cubic feet, and an average analysis shows CO, 28 per cent.; H. 2.5 per cent.; CO_2, 7.25 per cent.; N. 61.3 per cent.

This paper has been an endeavor to point out a few of the merits of the gas producer. The subject can only be taken up in a general way, because there are so many conditions to be met in the problem of power generation, that each case must be taken up separately, nevertheless the success of the producer, during the few years in which it has been developed, makes a thorough knowledge of this piece of apparatus necessary to anyone who pretends to be up to date in power plant work. The small producer plant has a field in sparsely-settled districts which cannot be as well met by any other existing piece of apparatus. The blast furnace engines are also growing in popularity as their operation becomes better known. As has already been pointed out, there is a great deal of power going to waste at present in existing blast furnaces. If the whole of the power from Niagara were utilized, it would only give three times the h.p. that is thrown away by the blast furnaces of the United States of America alone. From these facts one is encouraged to believe that the gas engine industry has a bright future before it, now that the gas producer has proved to be a commercial success.

BOOK REVIEWS

MODERN STEAM ENGINEERING. In theory and practice, by Gardener D. Hiscox, M.E., 487 pages and 405 specially made engravings and diagrams; price $3; New York: the Norman W. Henly Publishing Co.

The author of this book is already well known from former publications written by him, such as "Gas, Gasoline and Oil Engines," and "Compressed Air in All its Applications." The book is designed for engineers, firemen, electricians and steam users, and for a practical theatise approaches nearer the ideal than any book issued heretofore on the subject. It fully describes and illustrates the property and use of steam for the generation of power in the various types of engines in use. The slide valve, high speed, Corliss, compound multi-expansion engines, and their valve gear, the DeLaval, Parsons, Curtiss and other turbines are fully described and illustrated. Forty-two tables of the properties and application of steam and its various uses are included. An important and very valuable feature of this book is the questions and answers submitted after each chapter. There are nearly two hundred of these, such as might be asked by an examining board in connection with steam practice. Thus the book is not only a guide but a teacher. Many books of a much more exhaustive character have been written on this subject, but for the size of the volume the author seems to have hit the mark fairly well.

BOILER WATERS, Scale, Corrosion, Foaming. By William Wallace Christie, M.A.S.M.E., consulting engineer, author of Chimney Formulae and Tables, etc. 235 pages, 77 illustrations, price $3.00 net. New York: D. Van Nostrand Co.

After the title page the author has a panel containing four boiler proverbs, as follows: "A steam boiler is a steam generator, not a kettle for chemical reaction." "Get, if possible, a supply of clean, soft, natural water." "The only compound to put into a boiler is pure water." "Oxygen, the most useful element, is, when free in boilers, a most destructive corrosive element." The author outlines as an axiom in the preface that the relative value of one boiler to another may in many cases be measured by its scale-forming propensity with a given water. Purify this water and all boilers come much nearer a uniform value per unit of heating surface. The book contains ten chapters dealing somewhat exhaustively with the subject at hand. Great importance is attached to the question of the softening of the water to be used, it being intimated that such water cannot produce scale or corrosion.

THE COPPER HAND BOOK.—A manual of the Copper Industry of the World, filed and published by Horace J. Stevens, Houghton, Michigan, 1,116 pages, $5.00 in buckram binding with gilt top, $7.50 in full morocco gilt.

This is the only publication devoted exclusively to the copper industry. Although encyclopaedic in scope it is written throughout in plain language. The major portion of the book is devoted to a chapter describing important mines, and listing every copper mining company of importance. 4626 mines and companies are listed in this book, there being 777 more titles than in the preceding annual edition. It is the only publication in the mining field that gives ratings to the mines and companies. The good companies are recommended, and the fraudulent companies, of which there are many, are denounced in plain language, the facts upon which, first ratings are based, are given in detail. The publisher makes an unusual offer of sending this book on a week's approval prepaid to any address in the world, without any advance payment, and the publisher states that of many thousands of the books sent out, less than 5 per cent. have been returned as unsatisfactory, and less than 3 per cent. of the books retained, remain unpaid for.

ABOUT CATALOGUES.

BORING MACHINES.—Horizontal boring machines are described and illustrated in catalogue and treatise No. 4, 1905, of the Binsse Machine Co., Newark, N.J.

SAND SIFTERS.—Battle Creek Sand Sifters, worked by steam, air, belt and hand power, are described in a pamphlet issued by the Hamilton Facing Mill Co.

LIGHT AND HEAT.—Vol. I., No. 5, of Lux, "a miniature magazine of light literature," issued by the Nernst Lamp Co., Pittsburg, is to hand, and consists of 15 pages of hints in the electric lighting and heating line.

PNEUMATIC TOOLS AND FORGE.—Thor pneumatic tools are shown working in various capacities by a number of very clear photographs in catalogue No. 7, just received from the Independent Pneumatic Tool Co., Chicago and Aaron, Ill., A pamphlet enclosed treats of the Gunnell pneumatic forge.

ELECTRIC TRAVELLERS AND OTHER CRANES.—A bulletin illustrating a number of installations of cranes and other equipment furnished various railways covering a wide range of service, with numerous illustrations from photographs, has been issued by the Whiting Foundry Equipment Co., of Harvey, Ill.

BULLETIN No. 139, which has just been issued by the B. F. Sturtevant Co., of Boston, Mass., in its engineering series, presents a full line of generating sets driven by direct-connected vertical enclosed engines with forced lubrication. The published list contains 14 sizes, ranging from 3 to 50 K. W. in output; the former driven by a 3½ x 3 engine and the latter by a 12 x 10. All of the engines were especially designed for generator driving.

SOMETHING NEAT. One of the most attractive calendar souvenirs issued for the coming year is that of H. J. Logan, bookbinders' and printers' machinery, 136 Bay St., Toronto. The calendar is mounted on an attractive aluminum background containing the names of some of the lines manufactured by this company. Alongside is a handsome thermometer, which may be read with ease at some distance. It makes one of the most attractive desk calendars that have come to this office.

Mechanical Reviews

OPINIONS AND STUDIES WORTH NOTING

AUTOMATIC REVERSING MECHANISM FOR GRINDING MACHINE.

THE problem which led to the design of the reversing mechanism shown in the accompanying cut was to get a grinding machine for grinding the bores of cast steel car wheels, the bore being about 1½ inch diameter and larger. Owing to the hardness of the steel used in the wheels, no roughing cut could be taken before the grinding operation. This necessitated that sometimes up to ¾-inch on the diameter had to be ground out, and in order to make the operating cost of the machine as low as possible it was equipped with automatic reverse for the back and forth motion of the emery wheel, and with automatic feed of the cut.

One new feature of the machine in question is that the emery wheel is driven by an independent motor, which is mounted on the emery wheel rest so that it travels with the emery wheel. The reversing of the traveling motion is obtained by a driving bevel gear, A, and an engaging double pinion, B, which can slide back and forth on the shaft, but keyed to it in order to drive. When one end of the pinion engages the driving gear the carriage moves forward, and when the other end of the pinion engages the driving gear the carriage moves backward. This back and forth sliding motion of the pinion is caused by a system of levers, a plunger acted upon by a spring, and a shaft which is fastened to the bed of the machine and is equipped with two collars. By changing the distance between these two collars the length of the traveling motion is changed.

The illustration shows the pinion engaged so that the carriage will move backward. As the carriage moves, the lever C comes in contact with the rear collar. The lever moves and pushes the plunger upward, compressing the spring. During the first half of the period the link E does not move, owing to the oblong slot H, and gear B remains fully engaged to the driving gear. When the levers C and D have passed the central position, the pressure of the spring comes into action and pushes the plunger downward. This moves the lever D, and by link E the motion is transferred to the lever F, which causes the pinion to slide, disengaging one end and engaging the other. The carriage reverses and starts to move forward. In order to obtain the feed of the emery wheel the lever G in connected to the reversing mechanism. At the end of the lever two strings are fastened, which are led up to the ceiling, and over sheaves lead down one to each of two ratchets at the end of the feed screw. These ratchets are arranged so that when the carriage reverses at the rear end one of the ratchets feeds, and when the carriage reverses at the front end the other ratchet feeds the cut.—S. K., in "Machinery," N.Y.

Do not waste time, for that is the stuff life is made of.—Franklin

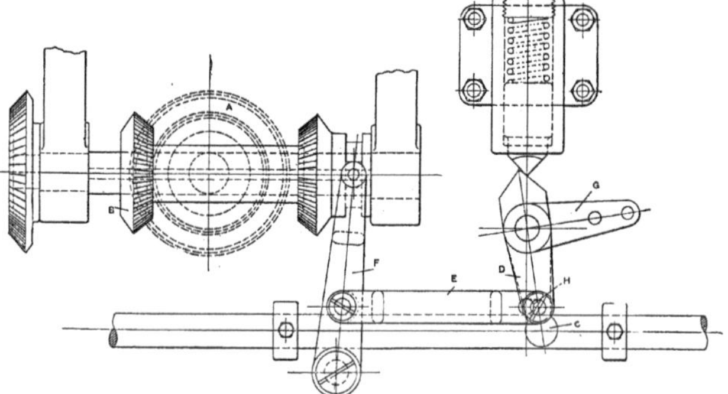

Automatic Reversing Mechanism for Grinding Machine.

A COMPACT TOOL CHEST.

I SEND you a design of a tool chest for machinists and tool-makers, which I find to be the best chest for use in a fine machine shop, as it takes little or no bench room. The body is made of ½-inch boards. The lid marked B is cut on an angle and hinged in the centre of the top, like the design. The cover, marked A, is hung on hinges inside of the lid B, giving a space between A and the inside of the lid B, and on the inside of A for calipers, dividers, or any such tools, to be held under brass clips, and a space in front of A, as shown in the design, for scales, straight edges, centre gauges, and drill gauges.

A piece of heavy felt or leather fastened on the face of A, as shown in the design, stops the chafing of the tools on the drawers. These can be held in position under small, round-headed, brass screws. The front and the bottom of the drawers, also the backs, are to be made 3-16-inch thick. The ends should not be less than ¼-inch thick, and a groove 3-16x¼-inch should be cut for runways, leaving the space E at the bottom for heavy tools such as hammers, monkey wrenches, etc., thus leaving the drawers for other tools.

This chest can be carried by a shawl strap, the design for the handle of which is shown above. This handle can be made of brass or any other metal. This chest, if tools are properly arranged, will hold all that any machinist or toolmaker will require, as there is practically no unused space. I have had one of these in use for two years, and it meets all requirements.—H. F. Benitez, in American Machinist.

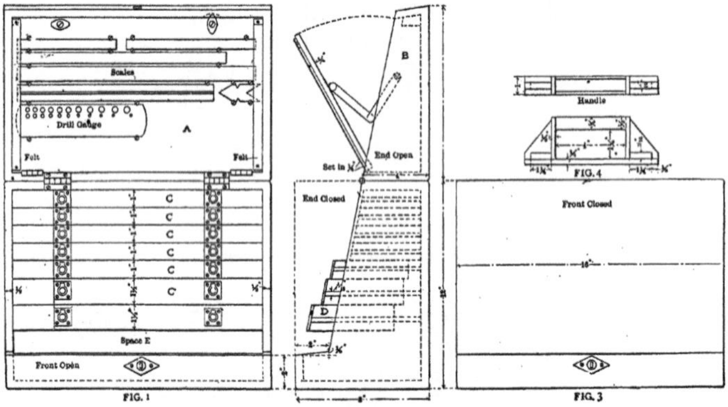

FIG. 1 FIG. 2 FIG. 3

A Compact Tool Chest.

WHEEL FLANGE OILER.

THE curves of street railways are usually so sharp that it is necessary to keep them well lubricated in order to prevent undue wear and screeching of the wheels as they grind against the tracks. Aside from the expense of keeping the curves lubricated the practice of greasing them is extremely objectionable to pedestrians who are liable to soil and ruin their clothing by contact with the oily substance. The accompanying engraving shows a method of obviating this objection which consists in oiling the flanges of the wheels instead of the tracks. The oiling device is arranged to be brought into operation at the will of the motorman so that the lubricant is applied only when needed and

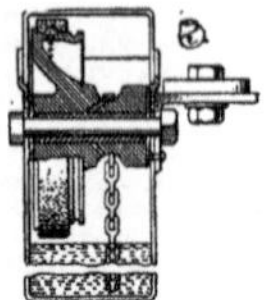

Wheel Flange Oiler.

where needed, because it is on the flanges that most of the friction occurs. This oiler is not limited to street railways, but is also applicable to the curves of steam railways where much power has been uselessly spent because heretofore it has not been the practice to oil these curves. The device comprises an oil chamber in which a wheel is mounted to rotate. The wheel is provided with a wick which is seated like a tire on the outer rim. Oil holes lead through the rim of the wheel to the wick. The wheel does not touch the oil in the chamber, but a loose chain which hangs in the oil is turned by the wheel and serves to feed the lubricant to the rim whence it passes through the holes to the wick. In use the device is mounted on a bracket in such position that it can be swung against the wheel flange by the operation of a lever. A portion of the oil chamber is cut away permitting contact of the wick with the flange, and thus causing the wheel in the oil chamber to rotate and feed the oil to the wick. The inventors of this improved oiler are Messrs. F. S. Baird and E. W. Carroll, of Congress, Arizona.—Scientific American.

KYNOCH, LIMITED, PRINTERS AS WELL AS MANUFACTURERS.

Those who know Kynoch, Limited, simply as manufacturers of gas plants, gas engines, etc., and who receive their printed matter from time to time, may not have noticed that every catalogue, circular, etc., bears the imprint of the Kynoch press, and is done at their works in Witton, Birmingham, England. The literature printed by them is certainly designed with considerable dignity and printed very tastefully.

A booklet issued by them, which may be of considerable interest to readers of Canadian Machinery, is one on "Cheapest Power on Earth," giving much information on the economy to be effected by the use of gas plants. Copies of this booklet are available by application to Kynochs, Limited.

A man may hide himself from you in every other way, but he cannot in his work; there you have him to the utmost.—Ruskin.

Tricks of the Trade

Everybody is invited to contribute to this page. Send in your ideas, odd jobs you are doing, and anything of interest to fellow-workmen. Remuneration will be made for such contributions. [Ed.]

FEATURES IN GAS ENGINE CONSTRUCTION.

By S. L. Fear, Toronto.

DURING the past few years the growth of the gas engine has been so rapid that the great tendency has been toward new things and new arrangements, rather than toward a perfecting of apparatus in operation. But now that the gas engine has come to stay, and we are not looking for any radical changes, it is up to the machine shop to raise the standard, and increase the efficiency of this new motor. There are now many shops turning out machines, and each shop holds to certain ways of casting, forging, machining, or assembling; and these are held to be best. As to which is the best—that is a matter of discussion.

Casting of Cylinders.

Let us look into a few of these differences, and take firstly the casting of cylinders. Of course we must first decide whether we will make the casting with the water jacket integral with or separate from the cylinder. This is a point of discussion. Casting the jacket separate provides an easy and cheap renewal in case it is cracked by the water freezing; on the other hand, the jacket cast integral saves the packing of an extra joint, which is troublesome. Casting the jacket integral seems to be in favor, so then let us take the casting of a cylinder with water jacket integral.

Our next point of difference or question is: Shall we cast the head separate from, or integral with, the cylinder? In the case of the head being cast integral, the machining or boring of the cylinder is the difficulty; while in the cylinder with head cast separate, ease of inspection of piston is provided, and the boring bar is used with less difficulty. In assembling, however, we have a joint to pack, which joint must first have been machined and then provided with the necessary stud bolts.

Before passing further we might note that some shops place a steel lining inside the cast cylinder; for which they claim the advantage that a new liner may be more easily put in place of the old one, than the cylinder could be rebored; that upon replacing the old liner and piston the engine is as good as new.

Then, to return to our casting of a water-jacketed cylinder, shall we cast the head integral or separate from the cylinder? Since we are to cast the jacket integral with the cylinder, we must look after a way of cleaning the jacket from the sand core, and if we make the head separate, the core can easily be removed. Then we must consider the valve seats. Shall we place these in pockets at the sides, or shall we place them in the cylinder head? Providing pockets, either both on one side or one on each side, complicates the patterns, and increases the core work; while if the head is made spherical and the valves placed at an angle of 90 degrees, or if the head is made flat and valves placed with stems vertical, the core work and casting are both simplified.

The material used in cylinder casting is, of course, the best grey cast iron. One or two firms anneal their cylinders

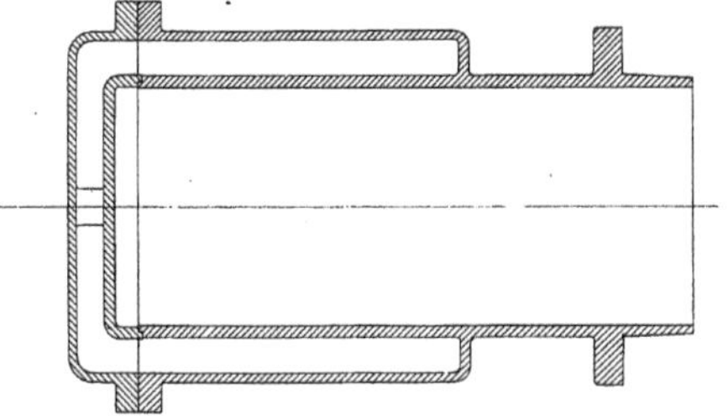

Fig. 1—Section of Cylinder.

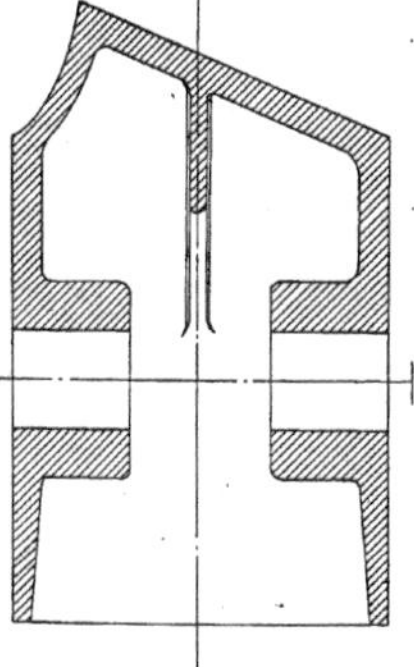

Fig. 2—Two Cycle Piston Head.

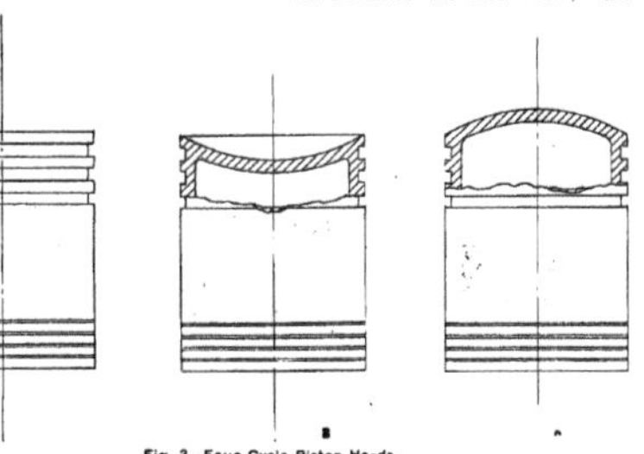

Fig. 3—Four Cycle Piston Heads.

in order that all harmful stains, that may be caused in the casting by cooling, are eliminated.

Gas Engine Pistons.

The piston used in gas engines is of the trunk type, single ended; in some cases a double ended piston is used, but then it must be water cooled. The trunk type of piston, of a good quality of grey cast iron, is provided with snap rings, also of good grey cast.

The shape of the head varies. Fig. 2 shows a two-cycle piston head, Fig. 3 shows three shapes of a four-cycle piston, (a) a flat head, (b) the sunk head, and (c) the crowned head. Pistons shown in (b) and (c) are not often used, since they cause internal strains in the piston not in its direction of motion.

The bearings of the pin are usually equally divided between the piston and connecting rod. At the bottom of the piston oil grooves are provided to lubricate the cylinder.

The manufacture of the piston rings is usually as follows: A casting is made in the shape of a cylinder, about 1-4 of an inch larger than the rings on the outside diameter, and 1-4 of an inch smaller on the inside diameter. The casting is then placed in the lathe and roughed out to an outside diameter, a little larger than the finished diameter. It is then cut in the miller, so that one side will slip past the other. The ring is then clamped on the face-plate of the finishing lathe, contracted as it will be in the cylinder; and then finished to .003 ins.—.005 ins. larger than the cylinder diameter.

The rings are kept in position on the piston by dowel pins, in such a manner that their joints are staggered around the circumference, and thus tend to minimize the leakage of gas past the joints.

The piston pin is usually solid, but sometimes hollow. Bushings for wear are in some cases provided in the piston, and then the connecting rod clamps the pin; or in others, bushings are provided for the connecting rod, and the pin is held by set screws in the piston bearings. The pin is ground accurately to size.

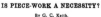

Fig. 4—Detail of Piston and Connecting Rod.

The Connecting Rod.

The connecting rod is almost always a forging. In some cases the piston end is milled square and a wedge provided in connection with the bearings, to take up the wear, as shown in Fig. 5. The crank end in this illustration is of the usual marine type, which is commonly used with large vertical gas engines. The bolts holding the bearing to the rod may be quite light, since there is no tensile strain on the return stroke.

Fig. 5—Double Cylinder Crank Shaft.

Gas Engine Cranks.

Figures 6 and 7 illustrate the two and three-throw cranks, as used in gas engine work. The cranks are forgings with the balancing weights bolted onto the arms. The two-throw crank has cranks at 360 degrees; and the moving parts are therefore not balanced in themselves, but the explosions are balanced. The three-throw crank has the best balance of the cranks commonly used, both dynamically and kinetically, in that the mechanical motions of the pistons and the cycles of operation overlap.

Fig. 6—Triple Cylinder Crank Shaft.

IS PIECE-WORK A NECESSITY?

By G. C. Keith.

WHY is a piece-work or premium system necessary? This question has come up many times. For a solution of this problem it is only necessary to compare two shops, one with and one without some system of rewarding their men according to their capacities, and note the difference in cost. Why is it that some men in our offices receive ten or fifteen thousand dollars a year and others only a few hundred or probably a thousand? Is not the answer that the former have a greater capacity for work than others? Would we not sooner have lost five thousand soldiers than Lord Roberts in South Africa? I may safely say, yes. Now here we have a man worth five thousand other men. On a smaller scale, we find this condition equally true in the shops. We have men ingenious enough to see an easy way to do certain jobs and thus accomplish twice or three times as much as certain others. Then there are others who never appear to accomplish anything, so in all fairness should not the man who uses his brain be rewarded? Should not the man who accomplishes the most work receive the most pay? Leaving the employer out of the question altogether and in fairness to the men, such a system is necessary, and we believe that the premium and piece-work systems are the best solutions to the [illegible]tion that we have to-day. When such a system is adopted we find less power going to waste, the output is largely increased, men avoid accidents and delays and the men's intelligence is quickened.

Have a Shop System.

A general shop system has been hinted at, a general plan that may be followed, but the managers must make the system they adopt their own. Success in obtaining the true costs will be aided by the system, a system applied in your own way to suit your own conditions. System alone, however, will not keep the work on time. It supplies the means by which it may be done and it remains with the superintendent to know the value and get the results. It is like an indicator diagram on an engine, and just as the indicator diagram shows whether or not the work is being properly and efficiently done, so with a proper cost system. A certain amount of experimenting may be necessary to get the facts wanted into proper shape so that the superintendent can see at a glance the efficiency of the plant and defects in operation, things he could not get either by inspection or operation. A system cannot be jumped at in a day, but the time and money will be well spent perfecting such a system, and when brought into working shape it will be only then that the manufacturer will realize what a valuable asset a cost system is to the firm.

Dealing With Men.

I have mentioned that increasing the pay of a mechanic in the shape of higher wages as piece-work or premium, tends to quicken his intelligence. Men who have made a scientific study of economic problems, capital and labor, tell us that a first-class man with favorable conditions will accomplish from three to four times as much work as an average man. Then the premium and piece-work systems tend to make first-class men. It develops them. And when a man is fitted for a higher grade of work give it to him. Do not put a first-class mechanic on work that a lower-priced man could do. No one would ever think of hitching a trotter such as "Dan Patch" to a plow. It would unfit the horse for faster work. So with a mechanic. He would be unfitted for finer work. Good men work at high speed if paid well. Of course there must necessarily be a limit to the wages. An example of this was seen in Kingston, Ontario, a few years ago when probably the largest stove foundry in Ontario was compelled to close its doors. The workmen struck for higher wages and obtained their requests. The second time the men walked out, and probably against the employer's better judgment, he again granted them their requests, but warned them that another strike meant closing the works. The workmen thought only of the "enormous" profits and went out again. They had forgotten about the enormous expenses, and the manufacturer was compelled to close down. Lack of knowledge, mutual suspicion and misunderstanding between men and their employers is probably the greatest drawback to the piece-work system. Probably it is because of this misunderstanding and suspicion on the part of the men that we have what has been termed "soldiering." That is, taking an hour to do a certain job that should be done without difficulty in much less time, probably half an hour or twenty minutes. Some men are naturally slow and easy. They have been used to putting in time and it is only after many examples and probably some external pressure that they develop a more rapid gait.

BORING CAR WHEELS.

How to bore forty car wheels per day, in a lathe, by a very inexpensive and simple method.

Make a casting like figure (A), in the accompanying illustration, and bolt to the face place of the lathe; then place the car wheel in the casting, and tap all around until the wheel is firmly in place. Place the cutter-bar through the cored hole of the car wheel, and then into the gig behind. Place the other end of cutter-bar in the tail-stock centre and ram through. Pull the bar out; and loosen the wheel from the casting, which is finished with the one cut.

For the cutter-bar, take an ordinary piece of 2-in. steel and turn 8 inches down to 1 15-16 inches, making the end fit the hole in the gig. Then slot the bar for the cutter, making one side taper to suit the key.

For the cutter, take a piece of 3-8 x 1 1-2-in. steel and cut off 2 1-4-in. long. Then place in the bar and turn to 2 3-16 inches. Then grind clearance and temper.

A person wishing to bore different size holes in car wheels must have a complete set of cutters.

N. C. JOHNSON.

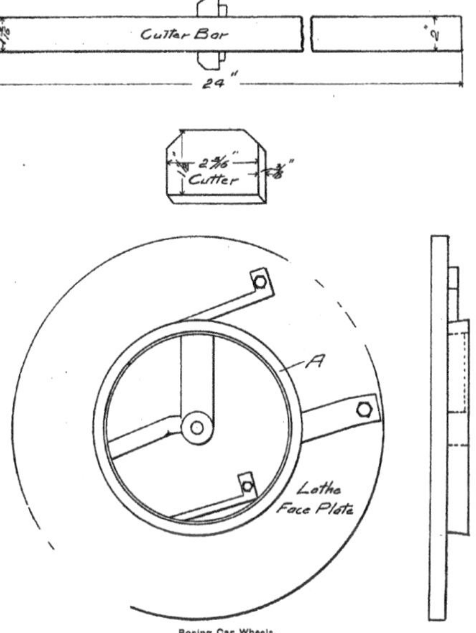

Boring Car Wheels.

Practical Questions and Answers

Question— Will you please give information and formulas used in the design of the main parts of small gas engines?

Answer—According to the authority these formulas differ considerably. For a simple stationary gas engine, trunk piston and single acting, some simple and handy formulas have been compiled at Cornell University, the constants and co-efficients having been obtained for the most part as a result of an investigation of current practice in gas engine design.

The following symbols will be used:

d=diameter of cylinder in inches.
l=length of stroke in inches.
p=is the maximum pressure during normal operation, this varies from 250 to 350 pounds, but an average value is 300 pounds per square inch.

First there is the thickness of the cylinder walls. This is represented by t, and it depends upon the safe working stress for cast iron, s.

$$t=\frac{pd}{2s}+\frac{1}{4}$$

For this class of work a safe value of s is 2,450, and substituting:

$$t=0.000204\ p\ d+\frac{1}{4}$$

Substituting the average value of p, we have:

$$t=\frac{d}{16}+\frac{1}{4}$$

The thickness of the jacket wall, T, and of the water jacket space, j, are determined to a great extent by consideration of moulding and casting, but also depend on the thickness of the cylinder walls, t.

$$T=0.6\ t \text{ and } j=1.25\ t.$$

An emperical formula is given for the number of studs in the cylinder head, q.

$$q=\frac{2}{3}\ d+2$$

the nearest whole number, of course, being used.

The outside diameter of the studs, o, may be obtained from this formula, s being maximum working stress allowable at the root of the thread:

$$o=\frac{1}{\sqrt{0.7\ s}}d\sqrt{\frac{p}{q}}$$

Considering the value of s to be 7,800,

$$o=0.0135\ d\sqrt{\frac{p}{q}}$$

If p has the value, 300, and q is 8, this formula reduces to:

$$o=d/12$$

The length of the piston, L, is found as follows:

$$L=\left(\frac{\pi}{4}.\frac{0.22}{b}\right)\frac{p\ d}{u}, \text{ where}$$

b=bearing pressure on the projected area of piston, L d sq. inches.
and u=ratio of the length of the connecting rod to the radius of the crank,

The usual value for u is 5 and for b, 7 pounds per sq. inch. Then,

$$L=0.025\ p\frac{d}{5}$$

giving p its usual value, this reduces to:

$$L=1\frac{1}{2}\ d$$

The bearing pressure p on the projected area of the piston is due to thrust of the power stroke and the weight of reciprocating parts.

The thickness of the rear wall of the cylinder, z, is given by:

$$z=\left(\frac{0.41}{\sqrt{s}}\right)d\sqrt{p}$$

If this wall is ribbed, as it usually is, a high value of s may be allowed, 5,320.

Then,

$$z=0.00562\ d\ \sqrt{p},$$

or giving p its value, 300,

$$z=d/10$$

The length and diameter of the piston pin, l″ and d″, are given by:

$$d''=d^{4}\sqrt{\frac{\pi}{4\ s\ b}}\sqrt{p}$$

and $l''=d''\sqrt{\frac{\pi\ s}{4\ b}}$, when the piston pin is considered as a beam, and

s=stress in pin, and
b=bearlng pressnre on projected area of pin.

Allowing s=10500 and b=2800, $d''=0.0218\ d\sqrt{p}$ and $l''=1.75\ d''$, and this again reduces to:

$$d''=0.21d \text{ and } l''=1.75d''$$

The area of the mid-section of the connecting rod, a, is given by:

$$a=\frac{k}{44,560}\ p\ d^2\left(1+\frac{0,00012c^2}{r^2}\right)$$

where k=factor of safety for the connecting rod
c=distance from centre to centre of rod
and r=radius of gyration of the middle section

Giving k the value 3.9,

$$a=0.0000875\ p\ d^2\left(1+\frac{0.00012c^2}{r^2}\right)$$

If the cross-section is circular, having a diameter D, $r^2=\frac{D^2}{16}$. If it is rectangular, having a height, h, $r^2=\frac{h^2}{12}$. This part of the formula, $\left(\frac{1+0.00012c^2}{r^2}\right)$, usually equals 1.6, and in this case the diameter of the mid-section is:

$$D=0.23d$$

This equation gives the diameter of the crank pin, $d'''=\sqrt[3]{(4/s)\ Mpd^3}$, where s=working stress and M=is the arm of the effective bending moment on the crank-pin for the stress caused by the reaction on the main bearing due to the explosion, and=m−($\frac{3}{8}$ l‴+$\frac{1}{4}$l′), where m = $\frac{1}{2}$ distance from centre to centre of main bearings; l‴=length of crank-pin journal and l′=length of main bearing journal. Allowing s=10,600

$$d'''=\sqrt[3]{0.000379Mpd^3},$$

and making p=300 and m=0.6d, as it usually is:

$$d'''=0.41d$$

The length of the crank-pin journal, l‴,

$$l'''=\left(\frac{0.14^{5}\pi}{4b}\right)\frac{pd^2}{d'''} \text{ where}$$

d‴=diam of crank-pin.
b=average bearing pressure on the projected area of the crank-pin (d‴ l‴), due to the average value of the load during a complete cycle. b is about 14.5 per cent. of the maximum load due to the explosion.

Assuming the value of b to be 213, the formula becomes:

$$l'''=0.000\,35p\frac{d^2}{d'''}$$

If p is 300 and if d‴ is 0.41d,

$$l'''=0.95d'''$$

If x be the thickness of the throws of a centre crank engine and y the breath, then

$$x=\frac{5}{8}d''' \text{ and } y=2\frac{1}{8}x$$

The diameter of the crank shaft on main bearings, d, is given by:

$$d'=\sqrt[3]{\left(\frac{4}{s}\right)\ pd^2M^1}, \text{ where}$$

M′=0.325 l′+0.090 l, being the arm of the effective bending moment on the crank-shaft at the inner edge of the main bearing for the stress caused by the reaction on the main bearing due to explosion,
and l′=length of main bearing journal.

Giving s the value 9,500,

$$d'=\sqrt[3]{0.000422pd^2M^1}$$

and if p=300 and M^1=0.4d as is usual,

$$d'=\frac{5}{8}d$$

The length of the main bearing journal l′ is:

$$l'=\left(\frac{\pi}{24b}\right)\frac{pd^2}{d'}$$

Here b is usually 123, and then,

$$l'=0.001068p\frac{d^2}{d'}$$

If p=300, and d′=$\frac{5}{8}$d, then

$$l'=2\frac{1}{4}d'$$

The outside diameter of the flywheel, D, is:

$$D=\left(\frac{12v}{\pi}\right)\frac{1}{N}, \text{ where}$$

v=velocity of rim in feet per minute, and N=r.p.m.

A safe value of v is 3,220 feet per minute, and then,

$$D=\frac{12,300}{N}$$

The weight of the flywheel is:

$$W=\frac{272,300,000000}{F}\frac{A}{D^2N^3}, \text{ where}$$

F=speed fluctuation co-efficient, about 0.054 usually.
B=rated brake horse-power.
D=diameter of the wheel, and
N=r.p.m.

Substituting for F=.054, we have

$$W=5,000,000,000,000\frac{H}{D^2N^3}$$

Substituting for $D=\frac{12,300}{N}$,

$$W=33,000\frac{A}{N}$$

It will be noticed that these formulae, by assuming certain average values for certain quantities, are simplified to a remarkable degree. These formulæ have been elaborated upon and are more fully explained in "Elements of Gas Engine Design," by S. A. Moss, formerly at Cornell University.

Foundry Practice

MODERN IDEAS FOR THE TRADE.

Science in the Foundry*

By William H. Hearne

THE selection of pig iron, its mixing and melting in the foundry, are becoming a matter of much more interest than formerly. Twenty-five years ago there was not a foundry that had a chemist in its employ, and scarcely a blast furnace running on foundry iron had a laboratory. Foundry pig iron was bought, sold, and mixed, entirely by its appearance and the experience the foundryman had had with the same brand previously. At that time this was a comparatively safe and satisfactory method to pursue, for the reason that at that time the ores and fuel from which each furnace made its product were largely drawn from local sources, and the name of the district in which the furnace was located told from what the iron was made.

The great growth of the iron industries since that time, the development of the railroad systems and the increased facilities the railroads offer for the assembling of materials, the discovery and development of coking-coal fields, the development of the Lake Superior ore fields, and the cheapness with which this ore can be transported on the Great Lakes to widely distant points have changed this situation entirely.

This makes it necessary for the foundryman to know either the analysis of the pig iron, or the ore and fuel that are used.

The chemistry of foundry practice is one of elimination, and this has been the cause of a,most limitless specifications. Really there is practically no positive knowledge, but all knowledge is acquired by experiments, and most of these experiments have been made by men having preconceived ideas of the results to be attained. This, of course, is natural, and is probably the only way to go about it. The result is that an experiment is made with iron running silicon, 2.50, phosphorus 0.80, sulphur 0.035, manganese, 0.40 and the experimenter discovers that he has made a satisfactory casting. Consequently, when he wants an iron for this kind of work he demands iron of exactly this analysis, not knowing whether or not, if he had iron of silicon 2.00, phosphorus 1.10, sulphur 0.05, manganese 1.25, he would get the same result and possibly a much better and sounder casting; for the phosphorus is if anything a greater fluidizer than the silicon, and the manganese is a powerful oxidizer. When the second iron is melted he would lose at least one-third of the manganese, and with it would go the sulphur from his iron and possibly some picked up from his fuel. This will be especially true if he will tap his iron from the cupola into a big mixing ladle so as to give the manganese time to act.

The fact of the matter is that the foundry cupola is a very poor thing in which to make a mixture of iron, and very little reaction goes on there beyond the loss of a little silicon and the taking on of some sulphur. Besides, the slags of most cupolas are too acid, and much manganese is lost which ought to be carried forward into the mixing ladle, when it could be used to eliminate the sulphur. There is not nearly enough lime or fluorspar used by foundrymen. If the ordinary foundryman would introduce a little fluorspar into his molten iron, he would be surprised at the amount of sulphur he would smell.

There are four things necessary for a good foundry mixture: fluidity, soundness, softness and strength. How is fluidity to be attained? First, by hot melting; second, by the presence in the pig iron of certain alloys—carbon, silicon, phosphorus and manganese—and by freedom from sulphur. Practically pure iron is white and will not run at all, but will melt into a sponge and burn before it will run.

In ordinary foundry pig iron there is about 93 per cent. of metallic iron and about 7 per cent. of alloys, and it is the proportion of these alloys to each other that determines the character and grade of the iron.

The real softness of a casting depends on the total amount of carbon and on the proportion of graphitic to combined carbon. This every foundryman knows, and the real problem is how to get the proper proportion. Every foundryman knows that one sure way is to anneal the casting or to cool it slowly. This is expensive and only to be done when he has failed to handle his mixture or his pouring properly.

In pig iron the proportion of free, or graphitic, to combined carbon depends entirely on the rate of cooling, and the slightest accident at the time of casting will affect this proportion and will change iron which should be 2X foundry to 2 plain or 3 foundry. In the old days many blast-furnace men took advantage of this fact and made their pig molds short and fat so as to open the grain. As a result iron, that in silicon and sulphur would have been 2 plain, by cooling slowly became 2X foundry. Really, however, there are more chances for the furnaceman to lose than to gain, as anything which impedes or interferes with the running of the iron will tend to combine the carbon, and it is a common thing to get 1 X, 2 X, and 2 plain or 3 foundry iron in the same cast, in which case all three grades will have the same silicon, sulphur, phosphorus and total carbon, and will vary only in the proportion of combined to graphitic. All this relates to the mechanical causes of the proportion of the carbon.

Chief among the foreign elements in pig or cast iron that affect the carbons chemically is silicon. This element has the property of throwing out of solution the carbon in the graphitic form, and many melters seem to think it the only cure for all hard iron. Everything else being equal, it is safe to say that an iron carrying 2.50 per cent. silicon is a softer iron than one carrying 2 per cent. silicon. Silicon is not in itself a softener; it acts so only by affecting the carbons. An iron with 10 per cent. silicon can scarcely be drilled. Pig iron with 4 per cent. silicon will not contain as much total carbon as one of 2.50 per cent., as the silicon unites with the iron and robs it of some of its carbon; so that a high silicon pig iron never has the dark, open fracture of a 2.50 to 3 per cent. silicon iron.

Sulphur is the most dangerous enemy of soft, sound castings, as it not only tends to combine the carbon, but by doing this also causes the grain to close up quickly, and to retain in little holes or sacks gases which might and should

* Abstracted from a paper presented before the American Association for the Advancement of Science at New York.

escape. In this way it causes the casting to pit and be porous.

If iron can be kept fluid long enough the sulphur will all rise through the iron and pass off, as can easily be proved by drilling holes in any piece of iron of reasonable size, poured in open sand. If three holes are drilled, one above the other, the proportion of sulphur in each set of drillings will vary as the height. This was demonstrated ten years ago by Guy R. Johnson in a series of more than a hundred experiments. The problem, then, is to give the sulphur a chance to get away before it is locked in the casting. There are many ways to do this, but the simplest and easiest is by using an iron carrying a fair percentage of manganese, with a large mixing ladle. In special cases it can also be done by putting into the mixing ladle a small amount of pulverized ferro-manganese, or a small amount of fluorspar or of aluminum. Any of these additions will cause the iron in the ladle to boil, and during this boiling the sulphur will be thrown off, and, as your molder will say, the iron is cleaned. Being free from sulphur, your casting will be soft, as it will have the proper proportion of graphitic carbon. Most of the so-called patent fluxes for making semi-steel castings are nothing but pulverized ferro-manganese, fluorspar, and pulverized high ferro-silicons. Every foundryman can obtain these alloys himself and make his own fluxes.

There is a brand of iron made in Virginia, from ores mined on the property smelted with coke, made by the company that demonstrated the futility of grading pig iron by silicon and sulphur alone. A certain foundryman took an iron made in Western Alabama, analyzing: Silicon, 3.25 per cent.; phosphorus, 0.90 per cent.; sulphur, 0.05 per cent.; an iron from Birmingham, analyzing about the same, except that the phosphorus was about 0.75; and the Virginia iron, analyzing: Silicon, 1.50 per cent.; phosphorus, 1.15 per cent.; manganese, 1.00 per cent.; sulphur, 0.-005 per cent.

The three irons were melted under identical circumstances and were poured into molds 1 inch wide, ½ inch thick and 14 inches long, for a fluidity test. The first iron filled the mold for 9¾ inches, the second for 8½ inches, the third for 14 inches, and the foundryman said he thought if he could have got the end of the mold out quickly enough, it would have run to the end of the shop. This of course was an enthusiastic exaggeration, but the test settled in that foundryman's mind that phosphorus and manganese were elements that affected the fluidity of iron fully as much as silicon; also that an iron that had this fluidity must also be low in sulphur. The product of this furnace is used almost entirely by foundries making steam, gas, and air cylinders, ammonia valves, and other castings that need to be dense and free from pinholes and yet soft enough to be reamed and cut.

The casting of pig iron in chill molds has very little, if any, effect on the character of the iron, especially if the foundryman uses a mixing ladle. It does not affect the total carbon at all; if anything it rather increases the amount of this element, for the reason that the surface of the pig iron is cooled quickly, in this way preventing the throwing off of "kish," which is so evident in a cast house when running "hot" iron. It somewhat increases the total combined carbon, owing to the sudden cooling or chilling of the surface; but if the iron contains the proper amount of alloys and is melted hot and tapped into a mixing ladle the casting will be as satisfactory in every respect as though the pig iron had been sand cast. In addition the chilled casting eliminates one source of trouble, i.e., the sand or other dirt in which the sand-cast pig iron is run.

A test was made several years ago by taking a pig of sand-cast iron, weighing it, pickling it, and then re-weighing. By calculation it was found that the particular brand of metal carried 70 pounds to the ton of sand. This adds just so much to the impurities to be contended with. In addition, it is a fact that iron free from sand melts more quickly than iron coated with it.

The Foundry Core Room [†]

By Geo. H. Wadsworth, Supt. Falls Rivet and Machine Company, Cuyahoga Falls, Ohio.

FROM personal observations of the writer, the majority of core rooms have but very little attention whatever given them from the principal heads of the foundry departments. A great many core rooms in times gone by were located in some out of the way place, or in some dark corner where it was a hard job to find them at all. This applies principally to small jobbing foundries, and we must give credit to modern foundries which have been built within recent years, and in which we can say that the core rooms are as well taken care of as regards room, light and equipment, as any other part of the foundry. But there is still room for improvement in a great many modern core rooms by the installation of core room equipment and appliances, and the giving of the necessary care and instructions, as well as in any other part of the foundry.

Customs are now changing. Since the installation of core room machinery our core makers are getting more familiar with it, and how to take care of it. How did this come about? We all know that in a machine shop custom educates boys to use the oil can and take care of the machines they are operating, and this is just as necessary in a core room. Was it done when core machines were first being introduced in the core room? I say no. I have personally gone into a core room after a machine had been installed and found it had never been oiled since it was received, and when an oil can was asked for it would take from fifteen to twenty minutes to find one, and when found the probabilities are that the oil would not be fit for use. But times are changing. When the men and boys take as much care of their tools and equipment in the core room as is the case in the machine shop, core machines and equipment will last considerably longer and will give a great deal more satisfaction, and it will certainly educate the core makers to have a more observing mind.

The oil can has been merely used as an example to show that small things have not been taken into consideration in the core room. Now, if so little attention is given to small matters, how much is given to the larger ones which are more costly. Core making machinery and equipment is becoming more generally used every day, and from the building of one type and operating of same, which machine made cores from ⅜ in. to 1½ in., we have to-day five different sizes and types of machines, which all originated from the ⅜ in. to 1½ in., which is generally known as the sausage-stuffing type. This machine was built and put on the market about August, 1901. The same machine was afterwards increased in size and became a standard at 2½ in. for two or three years, until the writer designed and built a machine to make up to 5 in., for which he afterwards made larger dies and increased it to 7 in. At that time it was supposed that this would have to be a power machine, but we will be pleased to demonstrate to you gentlemen to-day the making by hand of a 7 in. core 24 in. long in from 15 to 20 seconds. Another type is our speed changing core machine which has a slip gear and can be changed in speed for small and large cores.

† A paper read before the New England Foundrymen's Association, Boston, February 13th.

Machinery Development

Metal Working ***Special Apparatus*** ***Wood Working***

NEW LONDON MACHINE TOOL LATHE.

THE accompanying illustration is of a new lathe, which has just been placed on the market by the London Machine Tool Co., Hamilton, Ont. As may be seen from the illustration, this lathe is built in the most up-to-date style; and has been designed for use with high speed steel. The illustration shows that power is transmitted through the spindle from the countershaft through a single pulley and gearing enclosed in the head. For each change of belt speed there are four changes of spindle speed. A three-speed countershaft is provided, thus allowing twelve changes of spindle speed, ranging from 12 r.p.m. to 300 r.p.m. On the 22-in. lathe the driving belt is 5 ins. wide, and the driving pulley 14 ins. in diameter.

The spindles are of crucible steel, accurately ground; and the spindle bearings are of self-oiling lumen bronze. Spindle thrust is taken up by ball bearings working against hardened steel rings.

A feature of the feed mechanism is the automatic stop, which throws out both rod and screw feed. This feature is particularly valuable when working up to a shoulder, or in cutting threads up to a shoulder. Forty changes of rod, screw and cross feeds can be obtained. There is a safety device for preventing the rod and screw feeds from becoming locked together. The compound rest and cross sides are provided with micrometer valves. The carriage is very heavy, and has very long bearings on the V's, as may be seen.

Prize Drafting Outfit.

New London Machine Tool Lathe.

The tail-stock is heavy; and the spindle in the tail-stock is looked from beneath, thus preventing its being thrown out of line when clamped.

Well-arranged time is the surest mark of a well-arranged mind.—Pitman.

COMPLETE DRAUGHTING OUTFIT AS PRIZE.

The accompanying illustration shows a complete draughting outfit, including a full set of German silver instruments, which Fred. W. Dobe, M.E., chief draughtsman, Engineers' Equipment Co., Chicago, is offering to his students as a premium for the best set of drawings made by them during their course.

This system of home instruction in draughting, as inaugurated by Mr. Dobe, is modelled after the leading practical schools of this kind in Germany.

IMPROVED GAP ENGINE LATHE.

AN improved gap engine lathe has just been placed on the market by the Rahn-Carpenter Co., Cincinnati, Ohio, which is shown in the accompanying illustration. This lathe is designed especially to meet the requirements of repair shops or other shops where there is a large range and variety of work to be done.

The gap bed in this lathe is rigidly constructed so as to be strong enough for any work within the range of the lathe. The gap allows the working of pieces of large diameter within the range of the gap, as satisfactorily as in a lathe of much larger proportions. An extra piece is furnished to fit into the gap when the full swing of the lathe is not desired; this gives the carriage a full bearing on the V's when the tool is being used close to the head stock.

The lathe has both longitudinal and cross feeds ranging from twelve to one hundred; and will cut threads from 2 to 40 per inch. The lathe is back geared, with a ratio of 10 to 1.

These lathes are built in sizes 16-23-in. swing, 18-25-in. swing, and 20-26-in. swing.

*Improvements in Industrial Hygiene**

THE past decade has witnessed a most remarkable advance in the improvement of working conditions for the operators in our manufacturing establishments. While the manufacturer has, to a great extent, voluntarily introduced such improvements, and has witnessed in their introduction an increase in the efficiency and longevity of his workers, the pressure of the law has in many cases been necessary to force him to meet the requirements.

Just now exhibitions are being held in various cities where appliances for safeguarding the person and the health of employees are displayed. Important among the devices for improving atmospheric conditions in crowded, hot, dusty, ill-smelling or badly ventilated work-rooms is the fan blower.

Although the blower is very generally installed in connection with a heater as part of a mechanical system of ventilation and heating, local application as is usually necessary where the vitiation is excessive, in the metal working trades, for instance, the fan finds a broad field of application.

In the brass foundry hoods may be introduced over the brass furnaces, and ventilation secured with certainty by the introduction of an exhaust fan. This should properly be of the cased or enclosed centrifugal type, the disc or propeller type being only suitable for operating against slight resistances.

Rattling and tumbling barrels may be rendered innocuous by exhausting through hollow trunions or from housings built over the tumblers. An exhaust system in connection with the sand blasting room will greatly improve the conditions under which the workman operates.

Probably the greatest need of exhaust ventilation in the metal industry exists in connection with buffing, grinding and polishing processes. The high speed of rotation of the wheel tends to throw the fine particles of dust or lint directly into the atmosphere which soon becomes surcharged therewith. Each wheel is usually enclosed in an individual hood rigidly supported and connecting through an exhaust pipe to a horizontal main laid just behind the grinders and thence to the fan. Each connection is provided with a blast gate.

The fan speed required for an exhausting system depends largely upon the material to be removed. When properly applied a vacuum of from one to one and a half ounces is sufficient with a proper system of hooding to handle all dust in installations such as are here presented. If, however, the wheels are not properly hooded or the system not properly laid out, even a four ounce vacuum may hardly prove sufficient to give satisfactory results. Hence it behooves one to accept the advice of those experienced in this class of work. When it is known that the power of a fan increases as the cube of the speed it becomes evident that a slight saving in the cost of the initial installation may be quickly absorbed in the added expense to operate the fan at unnecessarily high pressure. A fan at four ounce speed will require eight times the power necessary to operate it at one ounce speed. In numerous instances existing installations have been remodelled, larger fans installed, new hoods introduced and such a marked saving made in the running expense as to pay for the new installation inside of a year.

Improved Gap Engine Lathe with Compound Rest.

* Supplied by the B. F. Sturtevant Company.

RECENT PROGRESS IN ELECTRIC FURNACES.

At the fifth annual meeting of the Society of Practical Science recently, a paper was read by Mr. S. Dushham on the recent progress in electric furnaces. He was of the opinion that where water power could be obtained, the iron ores of Ontario, which are refractory, and high in sulphur, could be smelted in the electric furnace at a cost that would compete with the blast furnace.

Power and Transmission

Steam Gas Electricity Compressed Air Water

NEW TYPE OF GAS ENGINE FOR MARINE WORK.

THE following description is given by Commander A. B. Willets, in the Journal of American Society of Naval Engineers of a new type of gas engine for marine work:

There will be no future for the single-acting motor for large powers in marine work. The manifest advantages in the matter of weight and arrangement for variations in cruising powers, as well as in greater effectiveness in distribution in the double-acting type, will make it unquestionably the only one to be considered in the development of this class of machinery for sea-going craft. The difficulties at first experienced in arranging for the proper cooling of pistons and rods, and in devising efficient packing for the rod stuffing boxes, have been overcome, and, as before noted, examples of double-acting gasoline motors of 500 horse-power are already in satisfactory operation afloat.

It will be seen in the accompanying illustration that the valves are positively operated, water-cooled, and balanced. The pistons are cooled by water circulating through the cross-head, up a tube in the piston rod, providing a return circuit around this tube. A regular reversing controlling cylinder is used with compressed air instead of steam, bringing the reversing under perfect control. Means can also be provided by which, with one motion of the lever, all the lower exhaust valves can be lock opened and all the lower inlet valves can be lock closed instantly, thus changing the engine at once to a single-acting six-cylinder motor, or, in other words, cutting down to half power with one movement and without any substantial decrease in economy. Furthermore, the two units of three cylinders each can be unclutched, so that the after three cylinders can be operated alone single acting, which actually reduces the power to one-quarter of the original. Taking these parts into consideration, together with the fact of being able (with a liquid primary fuel) to get under way at a moment's notice under full power and maintain absolute regularity of speed as long as the fuel lasts, there seems to be little more to be desired toward securing an ideal marine motor, and, even should producer gas be used, the same control of the variations of the power is maintained, although the ability to instantly start from cold port conditions is not secured, there being, of course, some delay in getting the fire properly under way in the retort.

The carbureter of this engine is controlled by the throttle-valve lever at the reversing cylinder. The inlet valves for compressed air are on the back, and not shown, but are merely applied in this type to the lower ends of the three after cylinders, so that the three upper ends of these cylinders and the entire three

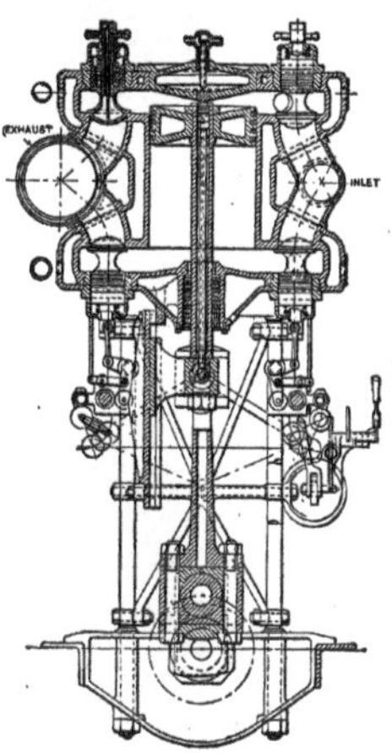

New Type of Gas Engine for Marine Work.

forward cylinders are constantly "on" gasoline, either in the go-ahead or backing position of the cam shafts; and, as the sparking continues in all cylinders during the action of reversal, the very moment the compressed air starts the engine in reverse the other cylinders immediately take up action with gasoline, and the engine is running on explosive fuel immediately, so that the action of the compressed air is not practically more than the shifting into reverse motion and starting the engine in the desired direction. The moment this results the reverse lever is shifted to the full-throw notch, which restores the lower ends of the after cylinders to gasoline. This reversal is practically instantaneous. All the cam shafts are positively driven by gearing from the main shaft, and, consequently, the adjustments of the cams are exact, and the tripping of the valves in this arrangement is accomplished almost noiselessly.

The lubrication of this motor is also a very interesting feature. Forced feed with sight-feed adjustment is accomplished through plunger boxes shown on the after end of the motor, the upper box being for delivery to the centre of each cylinder, and the lower box for delivery to the middle of the piston-rod stuffing boxes, the plunger pumps for each of the small pipes being worked by a lever with a small connecting rod to a pin on the end of the exhaust cam shaft, which gives sufficient rotary motion for the desired lift, the lift being also further regulated by a slot in the plunger lever, to which the connecting rod is attached. The lubrication for the crossheads and crank pins is accomplished by a small plunger pump worked by a gear at the forward end of the exhaust-cam shaft, similar to the gasoline pump shown at the forward end of the inlet.valve cam shaft in the photograph, the oil being kept in a constant flow from the crank pit into which it falls, up through a strainer box between the two sets of three cylinders (and just seen in the photograph), back of the middle branching of the inlet pipe, and fed by gravity from this tank to the journals; a surplus of oil being constantly circulated by this method and kept constantly strained without waste. The journals are of ample dimensions and give no trouble whatever from heating with this system of lubrication.

PETROL ENGINE OMNIBUSES.

Three single deck omnibuses equipped with 40 h.p., 4-cylinder petrol engines, have been built by Halleys' Industrial Motors, Ltd., Glasgow, Scotland, for public service in Canada. Two gearing, three forward speeds of 4, 8 and 12 miles per hour, and one reverse speed of 3½ miles per hour, can be obtained.

NEW BULLOCK GENERATOR.

A NEW generator is being placed on the market by the Allis-Chalmers Co. and will soon be available on the Canadian market through Allis-Chalmers-Bullock of Montreal, who have the rights of all lines manufactured by the parent companies. This arrangement gives the Canadian firm the advantage of the extensive engineering staff and its broad experience for the working out of any problem involved in mechanical or electrical engineering. Any machine brought out by the United States company is thus immediately at the disposal of the Canadian firm, the same being true of any improvement on a machine or any mechanical or electrical contrivance of any kind. The latest from the works of the parent company is a small type generator, which is shown in the accompanying illustrations. The views showing the pole pieces, a section of the stationary field, the brush oscillator, the armature core and commutator with temporary shaft and the generator complete, except shaft, will give a better idea of this new machine than many pages of descriptive matter.

A Section of the Stationary Field.

MASSIVE ROLLING MILL ENGINES.

Two of the four large Allis-Chalmers rolling mill engines which have been purchased by the Carnegie Steel Co. for the Edgar Thompson Works, Pittsburg, Pa., have now been completely installed. These engines are of immense size, being the largest engines for rolling mill service yet built. They have cylinders 50 and 78 inches in diameter, with 60-inch stroke. The frame and slide of one of the engines was cast in a single piece and weighs 104 tons.

Armature Core and Commutator with Temporary Shaft.

The valve gear used on these engines is the Reynolds Corliss type, driven by two eccentrics and designed for long range cut-off. The high pressure cylinder is placed in tandem behind the low pressure. The engines operate non-condensing under a steam pressure of 150 pounds, and at a speed of about 75 r. p. m. Under these conditions a maximum indicated h.p. of 8,000 is developed by each engine.

Small Type "I" Generator (Without Shaft).

Industrial Progress

CANADIAN MCHINERY AND MANUFACTURING NEWS would be pleased to receive from any authoritative source industrial news of any sort, the formation or incorporation of companies, establishment or enlargement of mills, factories or foundries, railway or mining news, etc. All such correspondence will be treated as confidential when desired.

A million dollar drydock is talked of for Port Arthur.

The Northern Plumbing and Heating Co. has dissolved.

Young Bros., Almonte, have sold out their foundry business.

The Down Draft Furnace Co., of Galt, will build a large addition.

Lamontagne & Bayard have registered in Montreal as contractors.

Jas. H. Edmonds & Co., metal spinners, Toronto, have assigned.

Joseph Guilbault & Co., have registered in Montreal as contractors.

A million dollar refinery to treat Cobalt ores is talked of for Hull.

A large addition has been begun by the Ottawa Steel Castings Co.

A new furniture factory employing 40 or 50 men will be erected at Elmira.

The Reeves Engine Co. is planning to build a large warehouse in Regina.

Beaucage & Beaucage have registered in Deschambault, Que., as contractors.

A grist mill and elevator will be erected in Canora at a cost of $30,000.

Fort Erie is asking tenders for the installation of an electric lighting system.

The Berlin (Ont.) Hotel Co. has been incorporated, with a capital of $250,000.

A shipbuilding company is to be formed at Kingston, with a capital of $100,000.

The Alberta Pacific Elevator Co., of Calgary, will erect an elevator in Bawlf, Alta.

A new factory is being built for G. A. Eastman, trunk manufacturer, Winnipeg.

A bicycle and automobile factory will be erected by McKinnon Bros., in Port Arthur.

A large brewery is to be built in Medicine Hat by E. F. Sterner, of Milwaukee.

Moose Jaw may spend $160,000 on improvements to the electric light plant this year.

The Washburn Crosby Co., of Minneapolis, will erect an 8,000-barrel mill at Keewatin.

The Petrolea Gas Co. and the Electric Light, Heat and Power Co., have amalgamated.

A new boiler shop has been erected on Bridge St., Ottawa, by the Capital Boiler Works.

The Dominion Smelters Co. will build a customs smelter at Parry Sound, to cost $1,500,000.

Lendon Bros. Leamington, are extending their premises by the erection of a new warehouse.

The Niagara Falls Machine and Foundry Co. has been awarded the contract for hydrants for Hamilton, at $40.50 per hydrant.

A new boiler manufactory has been started in temporary premises on Humber Bay, Toronto.

The core-room of the malleable iron works in Brantford, was damaged by fire on March 4.

The Muirhead Milling Co. will erect a 200-barrel flour mill at Port Arthur to cost $100,000.

Dresden, Ont., defeated a by-law to aid a machinery manufacturing industry on March 18.

The Crescent Machine Co. has registered in Montreal, with Charles M. Gardiner as president.

McCamus, McKelvie & Dowser, New Liskeard, will open a sash and door factory at Englehart.

The Wortman & Ward Company, London, are adding largely to their premises and plant.

G. R. Charleson, Minnedosa, has closed his machine shop, to build a new and more capacious one.

Tenders for the erection of a soap factory in Regina for J. M. Young & Co., have been called for.

The roundhouse of the Canadian Pacific Railway Co., Havelock, Ont., will be doubled in capacity.

An apartment block to contain one hundrex suites is to be erected at a cost of $300,000 in Winnipeg.

The Cline Furniture Co., Stratford, has under consideration extensive additions to its factory.

Christie, Brown & Co., Toronto, will extend their premises by a six-storey and basement factory.

Wetaskiwin ratepayers have voted to spend $140,000 for waterworks and $30,000 for electric lighting.

The Ellison Milling & Elevator Co. is calling for tenders for a $75,000 mill and elevator for Calgary.

W. H. Stirling's brass foundry at 28 Water Street, St. John, N.B., was damaged by fire on March 20.

A $15,000 or $20,000 apartment house is to be erected on Queen street east by McCarthy & Co., Toronto.

Two sawmills, with a total capacity of 160,000 feet of lumber, are being built on False Creek, Vancouver.

The Ottawa Car Co. are looking for a suitable site at Hintonburg for the location of a new factory.

The Canadian Westinghouse Co., Hamilton, will build a large extension to its present plant this year.

The Dominion Car & Foundry Co., Limited, has registered in Montreal, with W. V. Kelly president.

The General Brass Works, Limited, Toronto, are building a factory and foundry on Sterling Avenue.

D. Gavin, Vancouver, has been granted a permit for the erection of a $16,000 brick warehouse in that city.

Welland has carried a by-law granting a site and certain concessions to the Billings & Spencer Forge Co.

The Grand Trunk has placed an order for 3,000 tons of steel for new shops with the American Bridge Co.

The Canadian Flax Cordage Co. have asked St. Marys for a loan of $5,000 to extend their plant in that town.

Preparations for an enlargement to the works of the Goold, Shapley, Muir Co., Brantford, are under way.

Alliston has carried a by-law granting a bonus of $5,000 to the Merner Manufacturing Co., of Waterloo.

The contractors have been notified to begin work on the new buildings for the Gibson Manufacturing Co., Guelph.

The Moose Jaw Electric Co. has lately commenced operations in the supply and contract business in that town.

Chas. Rogers & Sons, Toronto, furniture and mantel manufacturers, are building an addition to their factory.

A permit has been granted to the Ottawa Wine Vault Co. for the erection of a stone building at a cost of $28,000.

D. Whitney will have a large brick and stone business block erected at Lethbridge, costing from $10,000 to $15,000.

The Canadian Niagara Power Co. has been granted permission to increase its bonds from $8,000,000 to $10,000,000.

Nerlich & Co., Toronto, have been granted a permit for a two-storey addition to their warehouse, to cost $35,000.

A permit has been granted to the A. J. Burton Saw Co., Vancouver, for the erection of a $3,300 addition to its plant.

The addition which is being built by the McGregor-Gourlay Co., Galt, is expected to be ready for occupation in June.

A large extension will be made to the car works planing mill of the Rhodes, Curry & Co., Amherst, N.S., in the spring.

The Montreal & Lake Superior SS. Co. has decided to order another boat of the type of the Wahcondah and Neepawah.

The International Portland Cement Co. have decided to duplicate their plant at Hull. The new capacity will be 4,500 barrels.

Gordon, Mackay & Co., are erecting a five-storey clothing factory at the corner of Queen and Crawford Sts., Toronto.

The Silliker Car Company has bought three acres of land in Halifax, and the building of shops is expected to begin shortly.

The Polson Iron Works Co., Toronto, have been given the contract for a new hydraulic dredge for the city to cost $72,000.

James Coristine & Co., Limited, will erect a seven-storey office building on Lemoine St., Montreal, at a cost of $82,000.

The Robertson Machinery Co., St. Catharines, will erect a large warehouse and go into the wholesale hardware business.

A soil pipe manufactory is about to locate in Galt. The McDougall Co.'s moulding shop, King St., is the probable site.

Robert Hamilton & Co., Vancouver, have ordered two hand-power traveling cranes from the Smart-Turner Machine Co., Hamilton.

The Canadian Pacific Railway will build a new depot at Calgary during the coming season. The cost is estimated at $200,000.

The Dunbar Machinery Manufacturing Co., of Woodstock, N.B., is considering the advisability of moving its plant to Gibson.

Smith Bros. & Wilson have been awarded the contract for the erection of Ald. T. Wilkinson's new warehouse in Regina to cost $12,000.

The Polson Iron Works have been awarded the contract for the construction of a $12,000 hydraulic pump for the city of Toronto.

A $30,000 bridge shop, 400 x 70 feet, will be erected on Sorauren Avenue, Toronto, by the Dominion Bridge Company, of Montreal.

Application has been made for a winding up order in connection with the York Mfg. Co., Toronto, makers of laundry machinery.

The Dominion Power and Transmission Co. will erect car shops in Hamilton to build all the cars needed for its many different lines.

The Pope Manufacturing Co., of Wisconsin, may establish a plant in West Selkirk, Man., for the manufacture of gasoline launches, etc.

Moose Jaw is considering an expenditure of $163,000 for additions and improvements to the municipal electric light and power plant.

The Grand Trunk Ry. are making big extensions to the Point St. Charles works, with a view to manufacturing their own rolling stock.

The Canadian Inspection Co., Limited, inspecting and testing engineers, have found it necessary to open a branch office in Toronto.

The Canada Foundry Co. is asking tenders for the plumbing work connected with the new machine shop now being completed by the firm.

The T. Eaton Co., Toronto, have been granted a permit for the erection of an eight-storey building, at the corner of Yonge and Albert streets.

The Northern Turpentine Co. has been incorporated with a capital of $250,000 for the establishment of a large turpentine plant at Killaloe, Ont.

The John Inglis Co., Toronto, have been awarded the contract for a million gallon pumping engine for the pumping station on Toronto Island.

The Public Works Committee of Calgary have reported in favor of the installation of a municipal street railway system at a cost of $250,000.

Cayuga (Ont.) is in communication with a company, capitalized at $500,000, which is looking for a favorable site for the establishment of a smelter.

A cordage company will be established at Valleyfield, Que., arrangements having been made with the Montreal Cotton Co. for supply of electric power.

The A. Darling Co., Toronto, has secured a permit to erect a nine-storey warehouse at the corner of Spadina Ave. and Adelaide St., to cost $100,000.

The smelting capacity of the Granby Consolidated at Phoenix will be increased one-third, thus making provision for a daily capacity of 4,200 tons.

The King Radiator Co., Toronto, who secured a site for their radiator works in Ashbridge Bay, Toronto, will commence building operations at once.

W. E. Bonner, of the Non-resident Owners' Realty Co., Winnipeg, will erect a new hotel at Strathcona this summer, costing between $10,000 and $15,000.

The town of St. Andrews, N. B., has asked permission of the legislature to borrow $40,000 for an electric street lighting system, sewers, wharves, etc.

W. H. Banfield & Sons, Toronto, have added a 28-inch Smith & Mills shaper to their machine shop. They state that they are rushed to death with orders.

In future the factory of the Peterborough Pre-Payment Meter Co., will be operated practically as a branch of the Allis-Chalmers-Bullock Co., of Montreal.

The Consolidated Electric Co., Toronto, has installed seven "King Edward" motors in the new building of the Booth Copper Co., on Sumach Street, Toronto.

The Smart-Turner Machine Co. has purchased the Zealand's wharf property in Hamilton, and will erect a factory there for the manufacture of motor boats.

Work was commenced on March 8 on the addition to the Galt Malleable Iron Works. When this is completed the capacity of the works will be about doubled.

The largest private carshops in Canada are expected to be located in Medicine Hat. It is said that building operations will be commenced early this spring.

The T., H. & B. Ry. will replace the workshops, recently destroyed by fire, at Hamilton, with a brick building to cost $10,000. George E. Mills has the contract.

The Scientific Brick Co., Toronto, is applying for a winding-up order in the affairs of the Modern Brick and Stone Co. The latter Co. was capitalized at $200,000.

Shaw Bros., Dauphin, Man., have recently ordered from the Robb Engineering Co. one 100 h.p. Robb-Armstrong automatic engine for their plant at Dauphin.

An order has been placed with the Carnegie Steel Company, of Pittsburg, for more than 800 tons of steel rails to enter into the construction of an electric street railway system in Edmonton.

The machinists employed by the Grand Trunk Railway, have asked the company for an all round increase, and in the event of a refusal a strike is anticipated.

The Rat Portage Lumber Co. will install a $100,000 plant in the $250,000 mill at Harrison, B.C., purchased three years ago, and cutting will commence in the spring.

The Canada Foundry Co., Toronto, have since January 1, delivered two ten wheel locomotives and one switching locomotive to the Canadian Northern Railway Co.

The Vancouver Board of Trade has passed a resolution calling on the Dominion Government to build a 250,000-bushel elevator in that city to store Alberta wheat.

The ratepayers of Wetaskiwin unanimously passed debenture by-laws to raise $150,000 for the installation of waterworks and $30,000 for electric light extension.

The Daisy Boiler and Radiator Mfg. Co., of Montreal, has made application to the Board of Control of Toronto, for a suitable site on which to build a branch factory.

Chas. Rogers & Sons have been granted a permit to erect a three-storey factory adjacent to their present premises at the corner of Defoe and Tecumseh streets, Toronto.

The Alberta Brick Co. has been formed at Medicine Hat, with a capital of $30,000. The president of the company is W. C. Harris, with W. A. Wyatt as secretary-treasurer.

The London Brass Works Co., having found their business to have outgrown their premises, on Carling St., have purchased a site in the East End for a new building.

The Iron Moulders' and Core-Makers' Union of Toronto at a meeting on March 10 decided to ask for an increase of 25 cents per day, making their minimum wage $3 a day.

The Northwestern Brass Co., Winnipeg, representing an investment of $10,000, will be turning out its products in less than two months, and will employ from 60 to 75 hands.

The Goldie & McCulloch Co. has definitely decided to erect a new safe works in Galt. The company also purchased some acres of land on which to build dwellings for employes.

A $100,000 company, to be known as the Amherst Red Stone Quarry Co., will develop the red stone quarry near Amherst, N.S. A sawing plant and steam shovel will be installed.

A factory for the manufacture of automobiles and bicycles will be erected at Port Arthur. McKinnon Bros. have secured a site and building operations will commence at once.

The Galt, Preston & Hespeler and the Preston & Berlin Electric Railway Companies will spend $110,000 in double tracking the line from Preston to Galt, building new cars, and installing more power producing machinery.

The Hamilton Steel and Iron Co., Hamilton, Ont., have taken out a permit to build a $300,000 addition to its blast furnace and a $50,000 addition to its open hearth steel plant.

The Sydney Foundry & Machine Works may add a steel plant for the manufacture of all kinds of castings, and the company will probably be formed into a joint stock concern.

The new Bessemer converter being installed at the Dominion Iron and steel works at Sydney, is nearing completion. This converter will increase the output by 50,000 or 60,000 tons a month.

The National Spring and Wire Co., of the State of Michigan, has secured the right to do business in Ontario. A. W. Marquis, of St. Catharines, has been appointed as their attorney for Ontario.

Representatives of the Crandall Cutlery Co. of Pennsylvania, have been making investigations in connection with the establishment of a Canadian branch Dundas, Ont., is under consideration.

The Montreal Street Railway Co. will make large extensions to its lines, comprising the Montreal Park and Island Railway and the Montreal Terminal Railway, during the coming summer.

The Whitman & Barnes Co. have decided to retain the St. Catharines branch of the firm, and erect new buildings. The city council will see that better railway facilities are provided.

North Vancouver is likely to have a large engineering works. A company is negotiating with the North Vancouver Ferry and Power Co. for land suitable for the establishment of such a plant.

Hespeler has granted a loan of $15,000 to the Hespeler Hoisting Machinery Co. It is thought that the Parkin Elevator Co., of Hamilton, may join the Hespeler company and move to that town.

The Ontario Interurban Railway Co. will apply for incorporation shortly, with the object of building an electric line between Belleville and Picton, with branches at Weller's Bay and Sand Banks.

Definite arrangements having been made by the Canada Stove Co., Ottawa, for their amalgamation with the Canada Radiator Co., Lachine, a large new factory will be built in Ottawa this spring.

The New Brunswick legislature is in communication with the Mackenzie & Mann syndicate regarding a proposition by the latter to construct a railway down the valley of the St. John River to Westfield.

The J. W. Woods Mfg. Co.'s plant, Ottawa, will be moved to Hull, Que. Mr. Woods has granted a $40,000 contract with Holbrook and Sutherland, who will undertake the erection of the new factory.

The T. & No. O. Railway Commission let the contract to the Montreal Locomotive Works for six new locomotives. The aggregate cost will be about $125,000. Delivery must be made by the end of October.

Fire broke out in the machine shop of the McGregor-Banwell Fence Co., Walkerville, on March 16. It spread to the main building, and caused a loss of between seventy and eighty thousand dollars.

The Standard Sand & Machine Co., Cleveland, O., is building a complete plant for handling sand for continuous moulding. This is to be installed in the plant of the Standard Ideal Co., Port Hope, Canada.

The Somerville Co., Limited, will erect a $40,000 building on the property leased by them in Ashbridge's Marsh, Toronto, if arrangements can be made with the harbor commissioners for the necessary filling-in.

The Ontario Hydro-Electric Commission has received two requests for estimates of the cost of electrical energy. One is from Collingwood for prices on 3,000 h.p. and the other from Uxbridge for prices on 600 h.p.

The Canadian White Co., Montreal, have received a contract from the Fordwick Co., New York, for the construction at Kingston of a million-dollar cement plant, with an initial capacity of 2,500 barrels per day.

The city of Victoria is about to call for tenders for the installation of two large pumps for the city waterworks system at Elk Lake. They will be purchased on the recommendation of Water Commissioner Raymur.

A group of capitalists have made an offer to the council of Fort William to establish a million dollar shipbuilding plant, which will employ between 350 and 500 men, if the city gives them proper encouragement.

Building will be begun in connection with the enlargement of the Grand Trunk shops in Stratford as soon as the weather will permit. The general plan will be along the lines of the shops at Battle Creek, Mich.

The Canadian Boomer & Boschert Press Co., Limited, Montreal, are furnishing a large hydraulic press to the Peterboro works of the Canadian General Electric Co., for use in the manufacture of transformers.

It is understood that a Canadian branch of the Berlin Manufacturing Co. of Beloit, Wis., manufacturers of woodworking machinery, has been organized, and that Hamilton has been chosen as the site for a large plant.

The Pittsburgh Reduction Company, which enjoys a monopoly of the aluminum industry in America, is contemplating the erection of a large plant near Mille Roches on the St. Lawrence, at a cost of several million dollars.

The "Allied Industries," believed to consist of the Canadian Fire Engine Co., of London, Ont., and two other large firms, will establish a plant in St. Louis (Montreal), if arrangements now in progress turn out as expected.

The town of Prince Albert, Sask., has accepted tenders amounting to $26,237 for additions to the electric light plant, the successful tenderers being Goldie & McCulloch, the Allis, Chalmers, Bullock Co., and the Canada Foundry Co.

The property of the Jackson Wagon Works has been acquired for the Galt factory of the Canadian Brass Manufacturing Co. Mr. Le Favor, the representative of the company, says the work will be rushed to completion.

The Scott Machine Co., one of the latest additions to London's manufacturing concerns, is rapidly building up a large business, and there is every indication that before long they will have to look for more extensive quarters.

F. Oliver, manager of the Perrin Plow Co.; B. Knapp, Morrisburg, Ont., and R. S. Harder, Aurora, Ont., have purchased the plant of the Rideau Electric Co., Smith's Falls, Ont., and will go in for stove manufacturing.

It is stated that, the Great Lake Engineering Works, Detroit, Mich., and the Toledo Shipbuilding Co., Toledo, Ohio, will consolidate and that they will erect a large shipyard plant in Canada at some point opposite Detroit.

Plans have been filed for the extension of the Niagara, St. Catharines and Toronto Railway from Fort Erie to Lundy's Lane. Plans have also been filed for an electric line from Niagara-on-the-Lake to St. David's and St. Catharines.

It is expected in Thorold, Ont., that the North America Refining & Smelting Co. will locate their smelter in Thorold Township, a short distance from the town. The township will vote on a by-law to exempt the concern from taxes.

The amalgamated business of the Canada Screw Co. and the Ontario Tack Co. may be removed from Hamilton to Welland. Negotiations are on for the change but the company state that no definite decision has yet been arrived at.

The Dominion Government will enlarge its accommodation at Ottawa by the erection of two buildings on the eastern side of the canal. One building will have a floor area of 300,000 square feet, and cost $2,000,000; the other an area of 100,000 square feet, costing $750,000.

A $1,000,000 cement plant to produce 5,000 barrels per day, and to be running in less than a year is being promoted by J. L. Howard, president of the Western Fuel Co., Nanaimo and San Francisco. The site for the proposed works is at Kendall.

The Tidman Silver and Aluminum Co., Toronto, has been incorporated, with a capital of $40,000, to manufacture and deal in aluminum ware, silverware and metal novelties. Provisional directors of the company are: F. A. Lewis, E. Gillis and M. W. Mayhr.

Rhodes, Curry & Co., of Amherst, have received an order to build forty three-ton store cars for the Dominion Iron and Steel Company. The Wallace Stone Company have also placed an order with the same firm for eleven cars of the same type.

The Eastern Townships Bank have filed plans for a 10-storey fireproof building to be erected at the corner of St. James and McGill Sts., Montreal, at a cost of $300,000. The Peter Lyall Co. have the contract, Cox and Amos being the architects.

The capital of the International Gas Appliance Co. has been increased from $49,000 to $200,000, and the name of the company has been changed to "International Lighting Appliances, Limited." The number of directors has also been increased to five.

Twelve prominent men of North Bay, Haileybury, and other New Ontario towns are applying for a charter, under the name of the Silver Belt Electric Railway Co., to build 35 miles of lines in the districts mentioned. The proposed capital is $1,500,000.

The Kitchener Lumber Co., of Kitchener, B.C., have recently ordered a large mill from the Robb Engineering Co., consisting of one 100 h.p. Robb-Armstrong slide valve engine, one 100 h.p. Robb-Mumford-Brady portable boiler, rotary mill, gang edger, etc.

A bill to give the Dominion Iron and Steel Company the right to form subsidiary companies to carry on any trade or business within the limits of the company, and to guarantee the stock of these companies, has been introduced in the Nova Scotia Legislature.

R. Bigley, manufacturer of stoves and furnaces, and Toronto agent for the Buck Stove Co., Brantford, has purchased land on the Weston Road, near Toronto Junction, and purposes building a factory to cost $7,000, employing at first 50 men, afterwards 120 men.

The Niagara Falls Machine & Foundry Co. has abandoned its plan to enlarge its present works, and instead will build an entirely new plant on a new site. The new works will be about twice as large as the present one and will be modern in every respect.

Mr. Maury, an expert engineer of Peoria, Ill., who was employed by the water commissioners of London, Ont., to investigate the matter of an adequate water supply, has recommended the combined river and spring water plan. This plan would cost $429,000.

The Colonial Engineering Co. has been incorporated to carry on the business of electrical, mechanical and civil engineers and contractors, and to manufacture and deal in machinery connected with such work. The capital of the company is $125,000, and the head office will be in Montreal.

Morlock Bros., furniture manufacturers, Guelph, Ont., will double the capacity of their factory this year. One new four-storey building, 100 x 65 feet, will be erected, and two stories, 100 x 45 feet, will be added to another. The total cost of the buildings will be about $18,000.

The Wolverine Brass Works, of Grand Rapids, have decided to locate a branch in Toronto. The manager, who was in Toronto last month, says the duty levied on United States goods makes it impossible to export at a profit into Canada, and hence the decision to open a branch here.

The Hamilton and Fort William Nav. Co. have placed an order with the Canadian Shipbuilding Co. for an 8,000-tons steamer, for the upper lake coal and ore trade. The vessel will be 460 feet long, 53 feet beam, and a 29 feet draught. Quadruple expansion engines and Scotch boilers will be installed.

The blast furnace plant of the Atikokan Iron Co., Port Arthur, Ont., will shortly be put in operation. The structural work which has been going on for the past two years has been completed, and the furnace is now ready to erect. The plant will turn out about 150 tons of pig iron in 24 hours.

A 70 x 40 foot addition has just been completed by the Canadian Corundum Wheel Co., of Hamilton, containing a new $16,000 list kiln of 12 ft. inside capacity. This increase in equipment is necessary in order to meet continued growth of business. It is the third enlargement of the plant in three years.

The enlargement to the lighting plant of the town of Barrie, commenced last May, has been completed, and the new equipment is now running in service. Mr. K. L. Aitken, the consulting engineer, made an extensive series of tests during the week of March 11th, and all the apparatus has since been accepted.

H. Krug & Co. are building a large addition to their chair factory in Berlin, Ont. When completed, the new building will be 150 x 85 feet, four stories of brick mill construction. About 100 feet of the addition is now built, the other 50 feet will require the tearing down of a part of the old factory.

The Imperial Steel & Wire Company offer to construct at Hamilton a wire drawing mill with a capacity of one hundred tons of wire daily, at an approximate cost of $200,000, if the city will guarantee the principal and interest of the first $100,000 and grant a site of ten acres of land with a river frontage of 150 feet.

The Borden-Canadian Co. have recently opened an office at 166 Adelaide Street West, Toronto, and will soon put on the Canadian market their new Rapid Worker hand stock for threading pipe. Several new features are embraced in this stock, and it is thought that a ready sale will be found for it in both this country and Europe.

The contract for grading the new railway between St. Marys and Embro has been let to John E. Webb, Toronto. The Dominion Steel Co. will provide the rails. One large bridge will be necessary in order to cross the Thames at St. Marys. The line, which is 15 miles in length, will be operated by the Canadian Pacific Railway.

C. F. May, of Toronto, representing a $1,500,000 company, is seeking terms from the city of Kingston for the establishment of a plant for the smelting and refining of Cobalt ores. W. W. Hadley and John Grey, of Guelph, are also mentioned in connection with a similar undertaking, having secured options on zinc ore lands in Frontenac.

The International Veneer and Lumber Co., of Philadelphia, will establish a factory at Arnprior if given by the town a free site of two acres, a subvention of $2,000, exemption from taxation for a period of ten years, free water during a period of ten years for a 250 horse-power boiler, and assistance from the council in placing $8,000 stock.

The Canadian Fairbanks Co., Montreal, have ordered from the Robb Engineering Co., two 45 h.p. and one 65 h.p. portable boilers. These boilers are for use in British Columbia, and are the second lot ordered by this company within the last two months. They are known as the Robb-Mumford-Brady portable boilers and are becoming very popular in British Columbia for sawmill work.

The annual report of the Canadian Westinghouse Co., Ltd., shows for the year ended Dec. 31, net earnings of $346,961, increase $126,416; dividends, $179,550, increase $30,450; balance $167,411, increase $95,966; reserve for depreciation, $120,000, leaving a surplus of $47,411; previous surplus, $232,041; balance carried forward, Jan. 1, 1907, $279,452.

The Ontario Government is installing a Duplex brass-fitted pump at Massey, Ont., which is being supplied by the Smart-Turner Machine Co., of Hamilton. The Robb Rngineering Works, Amherst, N.S.; the Doty Engine Works, Goderich, and the Miramichi Machine and Foundry Co. are among other purchasers of Duplex pumps during the past few weeks.

The Canada Mfg. Co. are spending between $7,000 and $8,000 on enlargements to their factory at Berlin, Ont. An old, unused wing of the present factory is being reconstructed and enlarged, and will be equipped for the manufacture of high class mahogany furniture. A new dry kiln, tin covered, on concrete foundation, with a capacity of about 20,000 feet, will be added this spring.

An important transfer of property took place last week, when Frank Oliver manager of the Perrin Plow Co.; B. Knapp, Morrisburg, and R. S. Harder, Aurora, purchased the plant of the Rideau foundry, Smith's Falls, and all stock on hand. The new company will commence operations at once in the manufacture of stoves. The stock on hand amounts to $10,000, including about 500 completed stoves.

The Intercolonial Railway Company have issued a new edition of their map of Ontario, Quebec and the Maritime Provinces. The map is tastefully colored, clearly printed, and contains along the margin views of interesting scenes on the route of the famous Maritime Express. They will be pleased to send one free of charge to any reader of this paper upon request to Mr. J. J. McConiff, City Passenger Agent, Montreal.

One of the new factories lately contributed to Canada by Uncle Sam is the Anthony Fence Company, of Walkerville, Ont. This new concern is an outgrowth of a similar business established at Tecumseh, Michigan, and although one of the youngest of American fence companies, having been under way only about three years, it now ranks as one of the most important of its kind. The Walkerville business is financed largely by Canadians.

To meet the growth of the city the Vancouver Gas Co. is planning large increase of its plant. The present location of the producing plant is so restricted that a new site has to be secured for the proposed larger plant. The company now has an expert on the ground, L. L. Merrifield, M. Inst.M.E., C.G.S., of Toronto, and this gentleman is to go over the whole situation and make recommendations for the enlargement of the present capacity of the gas plant.

The new Gantry crane building for the Collingwood Shipbuilding Co. is now well under way. The crane has a travel north and south of 560 feet and east and west 122 feet. Its hoist is 60 feet. The main hoist motor is of 40 h.p. designed to lift 10 tons from 50 to 60 feet per minute. The bridge travel motor is of 75 h.p. and is capable of driving the entire bridge at 500 feet per minute, while the cross trolley motor is of 10 h.p. for a speed of 250 feet per minute. There is also an auxiliary hoist with a 5 h.p. motor, which will be used for quick lifts below 2 tons.

After experimenting with steel flat cars and steel passenger, postal, baggage and Pullman cars, railway officials are devoting their attention to steel box cars. It is claimed that these would be at least 12 per cent. lighter than the standard box car, and this and other advantages will probably induce some of the leading railways to place large orders in the near future. The new steel box car built by the Union Pacific, which is the first of the kind, is being watched with interest by representatives of various roads.

The firm name of the machine business so successfully carried on during the past two years by Mervin Armstrong at 166 Adelaide Street West, Toronto, has been changed to Armstrong & Hatch, Arthur E. Hatch having entered into partnership with Mr. Armstrong. Mr. Hatch, who has been on the road for the past six years, representing the Smart Bag Co., Ltd., resigned his position some days ago to take charge of the office in connection with his new work, thus allowing Mr. Armstrong to devote his whole time and attention to the practical end of the business. The capacity of the plant has recently been largely increased by the installation of new and larger machinery, and still other equipment is coming forward. For some time past the shop has been running night and day.

Convention of Foundrymen's Association.

The annual convention of the American Foundrymen's Association and the exhibits of the Foundry Supply Association, which are run in conjunction with the convention, will be held in Philadelphia on May 20th-24th. The exhibits of the Supply Association will include all classes of foundry equipment and supplies; and one of the most interesting exhibits will be that of oil melting furnaces, and also molding machines.

In the past Canadian foundrymen have taken considerable interest in these conventions; and probably this year there will be the usual representation at the convention from Canada.

There is some thought of holding the next annual convention in Toronto, if possible; but owing to the tariff regulations some of the American supply houses are of the opinion that it would be practically impossible to have the exhibits in Toronto. In consequence it has been considered that the exhibit of the Supply Association could be made in Niagara Falls, N.Y., while the convention proper of the Foundrymen's Association could be held in Toronto. Then it would be advisable to hold some of the sessions in Niagara Falls and some in Toronto; and as the distance between the two places is short, this would be feasible.

Companies Incorporated

The Massive Corundum Co., of Niagara Falls, has received a charter.

The National Clothing Co. is incorporated, with a capital of $20,000. The head office will be in Montreal.

The German Canadian Smelting and Refining Co. has been incorporated under provincial charter with a capital of $1,000,000.

The Dominion Quarry Company has been incorporated, with a capital of $20,000, the head office to be in Montreal. The company will carry on a general quarry and contracting business.

The Kaine & Bird Transportation Co. has been incorporated, with a capital of $100,000, and head office in Quebec, to build and operate vessels and carry on a general transportation business.

The King Electrical Works have been incorporated, with a capital of $30,000, and head office in Montreal, to deal in brass, copper and other metals, and to manufacture electrical machinery and fixtures.

The McFarlane and Douglas Co. have been incorporated under provincial charter, with a capital of $100,000, to carry on the business of founders and machinists and the manufacture of builders' supplies.

The Toronto Automobile Co., Toronto, has been incorporated, with a capital of $40,000, to manufacture automobiles, motor cars, bicycles, cycles, etc. The provisional directors include J. S. Tomenson, C. H. McArthur, and B. Browne, Toronto.

The Tidman Silver & Aluminum Works, Toronto, have been incorporated, with a capital of $40,000, to manufacture aluminum ware, silverware, metals, etc. The provisional directors include H. D. McCormick, F. A. Lewis and D. A. Rose, Toronto.

The Deseronto Furniture Co., Deseronto, Ont., has been incorporated, with a capital of $65,000, to manufacture lumber, timber, furniture, etc. The provisional directors include J. Dalton, Deseronto, Ont.; W. S. Love, and I. F. Love, Napanee, Ont.

Canada West Manufacturers, Limited, Winnipeg, Man., have been incorporated, with a capital of $50,000, to manufacture machinery, engines, implements, vehicles, etc. The provisional directors include G. Bingham, J. A. Cowan, and W. J. Cummings, Winnipeg, Man.

The Petrolea Bridge Co., Petrolea, Ont., has been incorporated, with a capital of $40,000, to manufacture structural steel, bridge working machinery, concrete, cement, etc. The provisional directors include J. Fraser, T. Johnstone and I. Greenizen, Petrolea, Ont.

W. C. McIntyre, H. M. Marler, A. Racine, and W. T. Whitehead, of Montreal, with C. W. Trenholme, Westmount, have been incorporated under the name of the Mount Royal Spinning Company. The capital is placed at $3,000,000, and the head office will be in Montreal.

A. C. McMaster, G. R. Geary, F. D. Byers and others, have been incorporated under the name of Electrical Specialties, Limited, to carry on the business of electricians, mechanical engineers and manufacturers. The capital is $300,000, and the head office will be in Toronto.

The Carter-Hall-Aldinger Co., Winnipeg, Man., has been incorporated, with a capital of $50,000, to manufacture machinery, tools, contractors' and builders' supplies, etc. The provisional directors include W. H. Carter, P. C. Locke, Winnipeg, Man., and A. H. Aldinger, Chicago, Ill.

The business of T. Pringle & Son, hydraulic engineers, is to be taken over by a company newly incorporated under the name of T. Pringle & Son, Limited, and capitalized at $250,000. The head office will be in Montreal. A general engineering and construction business will be carried on.

Philip Lahee, Arthur Morin and others have been incorporated under the name of the Dominion Electric Manufacturing Co., to carry on the business of electrical and mechanical engineers and manufacturers. The capital is fixed at $20,000, and the head office of the company will be in Montreal.

The International Steel Company of Canada has been given a charter to carry on a general business as engineers, machinists, and manufacturers of machinery of all kinds. The new company will take over the business at present carried on by the St. Lawrence Supply Co., Montreal, and the head office will be in that city. The capital is $500,000.

The Canadian Fire Extinguisher Co., Toronto, has been incorporated, with a capital of $40,000, to manufacture fire preventative appliances, electric light and gas fittings, machinery, motors, dynamos, switch boards, engines, pumps, brass, lead, valves, pipes, faucets, etc. The provisional directors include F. W. C. Dickson, J. Murphy and W. Lauder, Toronto.

A charter has been granted to W. D. Long, G. H. Bisby and G. F. Glassco, merchants; Edward Gurry, mechanic; F. W. Gates, insurance agent, and P. D. Crerar, barrister, all of Hamilton, who will purchase the patent issued to Edward Gurry for an improvement in machines for moulding pig iron. The firm name will be Gurry Patents, Limited, and the capital is $40,000.

A charter has been granted to H. B. Muir, P. L. Shaw, G. McL. Baynes and others, who will engage in the printing and publishing business, under the name of Contractors' Reports, Limited, with the main object of conducting a system of reports for the use of contractors, builders, architects and others. The head office will be in Montreal, and the capital of the company is $20,000.

The Smart Bag Co. has been incorporated, with $2,500,000 capital, to manufacture jute, cotton, paper, and other bags, to deal in burlaps, ropes and twines, and articles of a similar nature, as well as to construct and maintain the various power plants necessary for such a business. The Smart Bag Co., Limited, and the Canadian Bag Co., Limited, are to be absorbed in the new organization. The head office will be in Montreal.

PIG IRON FOR ENGLISH TRADE.

It is stated by J. B. Jackson, in the Trade & Commerce Reports, that the grades of Canadian pig iron which are arriving at Liverpool at present are of an exceptionally high order; and that English manufacturers always give Canadian pig iron the preference over the common warrant iron produced at Middlesborough. He also states that the firms in Liverpool are almost unanimous in the opinion that if Canadian exporters would make a closer study of the requirements of the British market, and produce a more uniform graded metal, their trade would increase ten-fold, and that they would get 50 per cent. more for it than they do at present.

MAKING SOLDERING FLUX.

Chloride of zinc is very extensively used as a flux in soft soldering. It is an excellent one, too, and nothing has yet been found to take its place. It also has the advantage of cheapness. The action of chloride of zinc in soldering is based upon the fact that it dissolves the oxides of tin and lead upon the solder, and produces a clean surface for uniting with the metal to be soldered.

The mistake that is frequently made in making soldering flux is to add water. The stronger the flux can be made, and still remain in the liquid condition, the better it will be. As it is the chloride of zinc that does the work, the presence of water is a detriment. The water also has the advantage of producing spattering. The action of a weak flux in cleaning the surface of the molten solder is far inferior to that of a strong solution. In order to produce the best results, the flux should be made by dissolving zinc in strong muriatic acid until it will take up no more.

The C.P.R. will erect a $20,000 station at Souris.

Condensed or "Want" Advertisements

RATES

One insertion—25c. for 20 words; 1c. a word for each additional word.

Yearly rate—$2.50 for twenty words or less, 10c a word for each additional word.

The above does not apply to notices under the head of "Machinery Wanted." These notices are inserted free for subscribers.

SITUATIONS WANTED

ENGINEER AND DRAUGHTSMAN (25), salt and salt making machinery; small marine and mill; technical training; requires position. A. M. Mere Heath, Davenham, Cheshire, England.

ASSISTANT GENERAL MANAGER, McGill graduate, desires change of position. Twelve years' practical experience along engineering, machine, foundry and boiler shop lines. Best of references. Address Box 25, CANADIAN MACHINERY, Toronto. [3]

MACHINERY FOR SALE.

BELTING, RUBBER, CANVAS AND LEATHER, Hose, Packing, Blacksmith's and Mill Supplies at lowest prices. N. Smith, 138 York Street, Toronto. (2tf)

HAVING replaced our steam plant with electrical equipment at all of our mines and smelters, we now have for sale, air compressors, hoists, boilers equipped with underfeed stokers, pumps, Hereschoff furnaces, blowing engines, etc.; all of which can be inspected at Copper Cliff and shipped promptly; prices f.o.b. here; terms cash. The Canadian Copper Co., Copper Cliff, Ont.

EMERY STANDS—We have a quantity of these ready for delivery at exceptionally low figures.

GASOLINE ENGINES—Marine and stationary; 1½ to 20 h.p.

STEAM ENGINES—Second-hand Westinghouse; 20 h.p.

WASHERS—One thousand pounds 1 3-16 inches. Apply G. P. Wallington & Co., 11 Front street east, Toronto.

MACHINISTS WANTED

MACHINISTS—We want two or three first-class fitters. Apply Mr. Shand, Dodge Manufacturing Co., Toronto Junction.

MACHINISTS WANTED—We have increase our plant and desire floor and vise hands for day work, and lathe and planer hands for day and night; good wages and steady employment for A1 hands. Address or apply The John Bertram & Sons Co., Limited, Dundas, Ont.

WANTED—Two machinists, one moulder and boiler-maker. John McCrae, Lindsay, Ont.

WANTED—Machinist—Must understand engine repairing. Bruce Agricultural Works, Teeswater.

BUSINESS CHANCE.

WANTED—First-class foundry and machine company to buy interest in high-class patented machine, weighing seven tons and to manufacture, sell and instal same; this is a good proposition and will stand investigation. Apply Box 26, CANADIAN MACHINERY, Toronto.

FOREMAN WANTED

FOREMAN—Machine shop making hoisting engines and dredges; must understand modern tools and high-speed steel; state experience, age and salary. M. Beatty & Sons, Limited, Welland, Ont.

MOULDERS WANTED

WANTED—First-class foundry man; to take entire charge on stove plate; must understand cupola, core making and willing to do everything in connection with the foundry; to the right man a steady position guaranteed. Niagara Falls Heating and Supply Co., Limited, Niagara Falls, Ont.

FLOOR MOULDERS WANTED—Wages 25c. to 30c. per hour. Apply to G. R. Craig, master mechanic, the Canadian Copper Co., Copper Cliff, Ont.

PATENT RIGHTS FOR SALE

A U. S. firm manufacturing Graphoil Lubricators and Steam Cylinder Graphite Cups would like to correspond with reliable party with a view to introducing their goods on a commission basis in Canada, or would negotiate for the sale, in whole or in part, of Canadian patent rights. Address in first instance Box 51L, CANADIAN MACHINERY, Toronto.

CANADIAN MACHINERY BUYERS' DIRECTORY

To Our Readers—Use this directory when seeking to buy any machinery or power equipment.
You will often get information that will save you money.
To Our Advertisers—Send in your name for insertion under the heading of the lines you make or sell.
To Non-Advertisers—A nominal rate of $1 per line a year is charged non-advertisers.

Acids.
Canada Chemical Mfg. Co., London.

Abrasive Materials.
Baxter, Paterson & Co., Montreal.
The Canadian Fairbanks Co., Montreal.
Rice Lewis & Son, Toronto.
Norton Co., Worcester, Mass.
H. W. Petrie, Toronto.
Williams & Wilson, Montreal.

Air Brakes.
Canada Foundry Co., Toronto.
Canadian Westinghouse Co., Hamilton.

Air Receivers.
Allis-Chalmers-Bullock Montreal.
Canada Foundry Co., Toronto.
Canadian Rand Drill Co., Montreal.
Chicago Pneumatic Tool Co., Chicago.
John McDougall Caledonian Iron Works Co., Montreal.

Alloy, Ferro-Silicon.
Paxon, J. W, Co. Philadelphia.

Alundum Scythe Stones
Norton Company, Worcester, Mass.

Arbor Presses.
Niles-Bement-Pond Co., New York.

Augers.
Chicago Pneumatic Tool Co., Chicago.
Rice Lewis & Son, Toronto.

Automatic Machinery.
Cleveland Automatic Machine Co. Cleveland
National-Acme Mfg. Co., Cleveland
Potter & Johnston Machine Co., Pawtucket, R. I.

Automobile Parts
Globe Machine & Stamping Co., Cleveland, Ohio.

Axle Cutters.
Butterfield & Co., Rock Island, Que.
A. B. Jardine & Co., Hespeler, Ont.

Axle Setters and Straighteners.
A. B. Jardine & Co., Hespeler, Ont.
Standard Bearings, Ltd, Niagara Falls

Babbit Metal.
Baxter, Paterson & Co., Montreal.
Canada Metal Co., Toronto.
Canada Machinery Agency, Montreal.
Greey, Wm. & J. G, Toronto.
Rice Lewis & Son, Toronto.
Lumen Bearing Co., Toronto.
Mechanics Supply Co, Quebec, Que.
Miller Bros. & Toms, Montreal, Que.

Bakers' Machinery.
Greey, Wm. & J. G., Toronto.
Pendrith Machinery Co., Toronto.

Baling Presses.
Perrin, Wm. R., & Co, Toronto.

Barrels, Steel Shop.
Cleveland Wire Spring Co., Cleveland.

Barrels, Tumbling.
Buffalo Foundry Supply Co., Buffalo.
Detroit Foundry Supply Co., Windsor
Dominion Foundry Supply Co., Montreal
Hamilton Facing Mill Co., Hamilton.
Globe Machine & Stamping Co, Cleveland, Ohio.
John McDougall Caledonian Iron Works Co., Montreal.
R. McDougall Co., Galt, Ont.
Miller Bros. & Toms, Montreal, Que.
Northern Engineering Works, Detroit.
J. W. Paxson Co., Philadelphia, Pa.
H. W. Petrie, Toronto.
The Smart-Turner Mach. Co., Hamilton.

Bars, Boring.
Hall Engineering Works, Montreal.
Miller Bros. & Toms Montreal.
Niles-Bement-Pond Co., New York.
Standard Bearings Ltd., Niagara Falls

Batteries, Dry.
Berlin Electrical Mfg. Co., Toronto.
Mechanics Supply Co., Quebec, Que.

Batteries, Flashlight.
Berlin Electrical Mfg. Co., Toronto.
Mechanics Supply Co., Quebec Que.

Batteries, Storage.
Canadian General Electric Co. Toronto
Chicago Pneumatic Tool Co., Chicago.
Rice Lewis & Son, Toronto.
John Millen & Son, Montreal.

Bearing Metals.
Lumen Bearing Co., Toronto.

Bearings, Roller.
Standard Bearings, Ltd. Niagara Falls

Bell Ringers.
Chicago Pneumatic Tool Co., Chicago.
Mechanics Supply Co., Quebec, Que.

Belting, Chain.
Baxter, Paterson & Co. Montreal.
Canada Machinery Exchange, Montreal.
Greey, Wm. & J. G., Toronto.
Jeffrey Mfg. Co., Columbia, Ohio.
Link-Belt Eng. Co., Philadelphia.
Miller Bros. & Toms, Montreal.
Waterous Engine Works Co., Brantford.

Belting, Cotton.
Baxter, Paterson & Co., Montreal.
Canada Machinery Agency, Montreal.
Dominion Belting Co., Hamilton.
Rice Lewis & Son, Toronto.

Belting, Leather.
Baxter, Paterson & Co., Montreal.
Canada Machinery Agency, Montreal.
The Canadian Fairbanks Co., Montreal.
Greey, Wm. & J. G., Toronto.
McLaren, J. C., Montreal.
Rice Lewis & Son, Toronto.
H. W. Petrie, Toronto.
Williams & Wilson, Montreal.

Belting, Rubber.
Baxter, Paterson & Co., Montreal.
Canada Machinery Agency, Montreal.
Greey, Wm. & J. G., Toronto.

Belting Supplies.
Baxter, Paterson & Co., Montreal.
Greey, Wm. & J. G., Toronto.
Rice Lewis & Son, Toronto.
H. W. Petrie, Toronto.

Bending Machinery.
John Bertram & Sons Co., Dundas, Ont.
Chicago Pneumatic Tool Co., Chicago.
London Mach. Tool Co., Hamilton, Ont.
National Machinery Co., Tiffin, Ohio.
Niles-Bement-Pond Co., New York.

Benders, Tire.
A. B. Jardine & Co., Hespeler, Ont.
London Mach. Tool Co., Hamilton, Ont.

Blowers.
Buffalo Foundry Supply Co., Montreal.
Canada Machinery Agency, Montreal.
Detroit Foundry Supply Co, Windsor
Hamilton Facing Mill Co., Hamilton.
Kerr Turbine Co., Wellsville, N.Y.
Mechanics Supply Co., Quebec, Que.
J. W. Paxson, Philadelphia, Pa.
Sheldon's Limited, Galt.
Sturtevant, B. F., Co., Hyde Park, Mass.

Blast Gauges—Cupola.
Paxson, J. W., Co. Philadelphia.
Sheldons, Limited, Galt

Blue Printing.
The Electric Blue Print Co., Montreal.

Blow-Off Tanks.
Darling Bros., Ltd., Montreal.

Boilers.
Canada Foundry Co., Limited, Toronto.
Canada Machinery Agency, Montreal.
Goldie & McCulloch Co., Galt.
E. Leonard & Sons, London, Ont.
John McDougall Calenonian Iron Works, Montreal.
Manitoba Iron Works, Winnipeg.
Mechanics Supply Co., Quebec, Que.
Owen Sound Iron Works Co., Owen Sound
H. W. Petrie, Toronto.
Robb Engineering Co. Amherst, N.S.
The Smart-Turner Mach. Co., Hamilton.
Waterous Engine Works Co., Brantford.
Williams & Wilson, Montreal.

Boiler Compounds.
Canada Chemical Mfg. Co., London, Ont.
Hall Engineering Works, Montreal.
Lake Erie Boiler Compound Co., Toronto

Bolt Cutters.
John Bertram & Sons Co., Dundas, Ont.
London Mach. Tool Co., Hamilton.
Mechanics Supply Co, Quebec, Que.
National Machinery Co., Tiffin, Ohio.
Niles-Bement-Pond Co., New York.

Bolt and Nut Machinery.
John Bertram & Sons Co., Dundas, Ont.
Canada Machinery Agency, Montreal.
London Mach. Tool Co., Hamilton.
National Machinery Co., Tiffin, Ohio.
Niles-Bement-Pond Co. New York.

Bolts and Nuts, Rivets.
Baxter, Paterson & Co., Montreal.
Mechanics Supply Co., Quebec, Que.

Boring and Drilling Machines.
American Tool Works Co., Cincinnati.
B. F. Barnes Co., Rockford, Ill.
John Bertram & Sons Co., Dundas, Ont.
Canada Machinery Agency, Montreal.
A. B. Jardine & Co., Hespeler, Ont.
London Mach. Tool Co., Hamilton.
Niles-Bement-Pond Co., New York.

Boring Machine, Upright.
American Tool Works Co., Cincinnati.
John Bertram & Sons Co., Dundas, Ont.
London Mach Tool Co., Hamilton.
Niles-Bement-Pond Co., New York.

Boring Machine, Wood.
Chicago Pneumatic Tool Co., Chicago.
London Mach. Tool Co., Hamilton.

Boring and Turning Mills.
American Tool Works Co., Cincinnati.
John Bertram & Sons Co., Dundas, Ont.
Gisholt Machine Co., Madison, Wis.
Canada Machinery Agency, Montreal.
London Mach. Tool Co., Hamilton.
Niles-Bement-Pond Co., New York.
H. W. Petrie, Toronto.

Box Puller.
A. B. Jardine & Co., Hespeler, Ont.

Boxes, Steel Shop.
Cleveland Wire Spring Co, Cleveland.

Boxes, Tote.
Cleveland Wire Spring Co., Cleveland.

Brass Foundry Equipment.
Detroit Foundry Supply Co., Detroit.
Paxson, J. W., Co., Philadelphia

Brass Working Machinery.
Warner & Swasey Co., Cleveland, Ohio.

Brass Working Machine Tools.
Warner & Swasey Co., Cleveland, O.

Brushes, Foundry and Core.
Buffalo Foundry Supply Co., Buffalo.
Detroit Foundry Supply Co, Windsor.
Dominion Foundry Supply Co., Montreal
Hamilton Facing Mill Co., Hamilton.
Mechanics Supply Co., Quebec, Que.
Paxson, J. W., Co., Philadelphia

Brushes, Steel.
Buffalo Foundry Supply Co., Buffalo.
Paxson, J. W., Co., Philadelphia

Bulldozers.
John Bertram & Sons Co., Dundas, Ont.
London Mach. Tool Co., Hamilton, Ont.
National Machinery Co., Tiffin, Ohio.
Niles-Bement-Pond Co., New York.

Calipers.
Baxter, Patterson & Co., Montreal.
Rice Lewis & Son, Toronto.
Mechanics Supply Co., Quebec, Que.
John Millen & Son, Ltd., Montreal, Que.
L. S. Starrett & Co., Athol, Mass.
Williams & Wilson Montreal.

Cars, Foundry.
Buffalo Foundry Supply Co., Buffalo.
Detroit Foundry Supply Co., Windsor
Dominion Foundry Supply Co., Montreal
Hamilton Facing Mill Co., Hamilton.
Paxson, J. W., Co., Philadelphia

Case Hardening Bone.
Paxson, J. W., Co, Philadelphia

Castings, Aluminum.
Lumen Bearing Co., Toronto

Castings, Brass.
Chadwick Bros., Hamilton.
Greey, Wm. & J. G. Toronto.
Hall Engineering Works, Montreal.
Kennedy, Wm., & Sons, Owen Sound.
Lumen Bearing Co., Toronto
McDougall R. Co, Galt.
Niagara Falls Machine & Foundry Co. Niagara Falls, Ont.
Owen Sound Iron Works Co., Owen Sound.
Robb Engineering Co, Amherst, N.S.

Castings, Grey Iron.
Allis-Chalmers-Bullock Montreal.
Greey, Wm. & J. G, Toronto.
Hall Engineering Works, Montreal.
Kennedy, Wm., & Sons, Owen Sound.
Laurie Engine & Machine Co., Montreal.
Manson Mfg. Co., Thorold, Ont
John McDougall Caledonian Iron Works Co., Montreal.
R. McDougall Co., Galt, Ont.
Niagara Falls Machine & Foundry Co., Niagara Falls, Ont
Owen Sound Iron Works Co., Owen Sound.
Robb Engineering Co., Amherst, N.S.
Smart-Turner Machine Co., Hamilton.

Castings, Phosphor Bronze.
Lumen Bearing Co., Toronto

Castings, Steel.
Kennedy, Wm., & Sons, Owen Sound.

Castings, Semi-Steel.
Robb Engineering Co., Amherst, N.S.

Cement Machinery.
Allis-Chalmers-Bullock, Limited, Montreal
Greey, Wm. & J. G., Toronto
Jeffrey Mfg. Co., Columbus, Ohio.
John McDougall Caledonian Iron Works, Co., Montreal
Owen Sound Iron Works Co., Owen Sound

Centreing Machines.
John Bertram & Sons Co., Dundas, Ont.
Jeffrey Mfg Co., Columbia, Ohio
London Mach. Tool Co., Hamilton, Ont.
Niles-Bement-Pond Co., New York.
Pratt & Whitney Co., Hartford, Conn.
Standard Bearings, Ltd., Niagara Falls

Centres, Planer.
American Tool Works Co., Cincinnati.

Centrifugal Pumps.
Manson Mfg. Co., Thorold, Ont
John McDougall Caledonian Iron Works Co, Montreal.
R. McDougall Co., Galt.
D'Olier Engineering Co., New York.
Pratt & Whitney Co., Hartford, Conn.

Centrifugal Pumps—Turbine Driven.
Kerr Turbine Co., Wellsville, N.Y.

Chaplets.
Buffalo Foundry Supply Co., Buffalo.
Detroit Foundry Supply Co., Windsor.
Hamilton Facing Mill Co., Hamilton.
Paxson, J. W., Co., Philadelphia

Charcoal.
Buffalo Foundry Supply Co., Buffalo.
Detroit Foundry Supply Co., Windsor.
Dominion Foundry Supply Co., Montreal
Hamilton Facing Mill Co., Hamilton.
Paxson, J. W., Co., Philadelphia

Chemicals.
Baxter, Paterson & Co., Montreal.
Canada Chemical Co., London.

Chemists' Machinery.
Greey Wm. & J. G., Toronto.

Chemists, Industrial.
Detroit Testing Laboratory Detroit.

Chemists, Metallurgical.
Detroit Testing Laboratory, Detroit

Chemists, Mining.
Detroit Testing Laboratory, Detroit.

Chucks, Ring Grinding.
A. B. Jardine & Co., Hespeler, Ont.
Chicago Pneumatic Tool Co., Chicago.

Chucks, Drill and Lathe.
American Tool Works Co., Cincinnati.
Baxter, Paterson & Co., Montreal.
John Bertram & Sons Co., Dundas, Ont.
Canada Machinery Agency, Montreal.
Ker & Goodwin, Brantford.
A. B. Jardine & Co., Hespeler, Ont.
London Mach. Tool Co., Hamilton.
Mechanics Supply Co., Quebec, Que.
John Millen & Son, Ltd., Montreal.
Niles-Bement-Pond Co., New York.
H. W. Petrie, Toronto.
Rice Lewis & Son, Toronto.
Standard Tool Co., Cleveland.

Chucks, Planer.
American Tool Works Co., Cincinnati.
Canada Machinery Agency, Montreal.
Niles-Bement-Pond Co., New York.

Chucking Machines.
American Tool Works Co., Cincinnati.
Niles-Bement-Pond Co., New York.
H. W. Petrie, Toronto.
Warner & Swasey Co., Cleveland, Ohio

Circuit Breakers.
Allis-Chalmers-Bullock, Limited, Montreal
Canadian General Electric Co., Toronto.
Canadian Westinghouse Co., Hamilton.

Clippers, Bolt.
A. B. Jardine & Co., Hespeler, Ont.

Cloth and Wool Dryers.
Dominion Heating and Ventilating Co., Hespeler.
B. Greening Wire Co., Hamilton.
Sheldons Limited, Galt

Collectors, Pneumatic.
Sheldons Limited, Galt
Sturtevant, B. F. Co., Hyde Park, Mass.

Compressors, Air.
Allis-Chalmers-Bullock, Limited, Montreal
Canada Foundry Co., Limited, Toronto.
Canada Machinery Agency, Montreal.
Canadian Rand Drill Co., Montreal.
Canadian Westinghouse Co., Hamilton.
Chicago Pneumatic Tool Co., Chicago.
Darling Bros., Ltd., Montreal
Detroit Foundry Supply Co., Windsor.
John McDougall, Caledonian Iron Works Co. Montreal
H. W. Petrie, Toronto.
The Smart-Turner Mach. Co., Hamilton.
Hall Engineering Works, Montreal, Que.
London Mach. Tool Co., Hamilton.
R. McDougall Co., Galt, Ont.
Niles-Bement-Pond Co., New York.
H. W. Petrie, Toronto.
Pratt & Whitney Co., Hartford, Conn.
Williams & Wilson, Montreal.

Concentrating Plant.
Allis-Chalmers-Bullock, Montreal.
Greey, Wm. & J. G., Toronto.

Concrete Mixers.
Jeffrey M'g. Co., Columbus, Ohio.
Link-Belt Co., Philadelphia.

Condensers.
Canada Foundry Co., Limited, Toronto.
Canada Machinery Agency, Montreal.
Hall Engineering Works Montreal.,
Smart-Turner-Machine Co., Hamilton.
Waterous Engine Co., Brantford.

Confectioners' Machinery.
Baxter, Paterson & Co., Montreal.
Greey, Wm. & J. G., Toronto.
Pendrith Machinery Co., Toronto.

Consulting Engineers.
Charles Brandeis, Montreal.
Canadian White Co., Montreal.
Hall Engineering Works, Montreal.
Jules De Clercy, Montreal.
John S. Fielding, Toronto.
Miller Bros. & Toms, Montreal.
Roderick J. Parke, Toronto.
Plews & Trimingham, Montreal.
T. Pringle & Son, Montreal.
Standard Bearings, Ltd., Niagara Falls, Ont

Contractors.
Canadian White Co., Montreal.
D'Olier Engineering Co, New York.
Expanded Metal and Fireproofing Co., Toronto.
Hall Engineering Works Montreal.
Laurie Engine & Machine Co., Montreal.
John McDougall Caledonian Iron Works Co., Montreal
Miller Bros. & Toms, Montreal.
Robb Engineering Co Amherst N S.
The Smart-Turner Mach. Co., Hamilton.

Contractors' Supplies.
Manson Mfg. Co., Thorold, Ont.

Controllers and Starters Electric Motor.
Allis-Chalmers-Bullock, Montreal.
Canadian General Electric Co., Toronto.
Canadian Westinghouse Co., Hamilton.
T. & H. Electric Co., Hamilton.

Conveying Systems.
Sheldons Limited, Galt
Sturtevant, B. F., Co., Hyde Park, Mass.

Converters, Steel.
Northern Engineering Works, Detroit.
Paxson, J. W., Co., Philadelphia

Conveyor Machinery.
Baxter, Paterson & Co, Montreal.
Greey Wm. & J. G., Toronto
Jeffrey Mfg. Co., Columbus, Ohio.
Link-Belt Co., Philadelphia.
John McDougall Caledonian Iron Works Co., Montreal.
Miller Bros. & Toms, Montreal.
Smart-Turner Machine Co., Hamilton.
Laurie Engine & Machine Co., Montreal.
Waterous Engine Works Co., Brantford.
Williams & Wilson, Montreal.

Coping Machines.
John Bertram & Sons Co., Dundas, Ont.
London Mach. Tool Co., Hamilton.
Niles-Bement-Pond Co., New York.

Core Compounds.
Buffalo Foundry Supply Co., Buffalo.
Detroit Foundry Supply Co., Windsor.
Dominion Foundry Supply Co, Montreal
Hamilton Facing Mill Co., Hamilton.
Paxson, J. W., Co., Philadelphia

Core-Making Machines.
Paxson, J. W., Co., Philadelphia

Core Ovens.
Detroit Foundry Supply Co., Windsor.
Hamilton Facing Mill Co., Hamilton.
Paxson, J. W., Co., Philadelphia
Sheldons Limited, Galt

Core Oven Bricks.
Buffalo Foundry Supply Co., Buffalo.
Detroit Foundry Supply Co., Windsor.
Hamilton Facing Mill Co., Hamilton.
Paxson, J. W., Co., Philadelphia

Core Taper and Wax Wire.
Paxson, J. W, Co., Philadelphia

Core Wash.
Buffalo Foundry Supply Co., Buffalo.
Detroit Foundry Supply Co., Windsor.
Hamilton Facing Mill Co., Hamilton.
Paxson, Co., J. W., Philadelphia

Couplings.
Owen Sound Iron Works Co, Owen Sound

Cranes, Electric and Hand Power.
Canada Foundry Co., Limited, Toronto
Dominion Foundry Supply Co., Montreal
Hamilton Facing Mill Co., Hamilton.
Link-Belt Co., Philadelphia.
John McDougall, Caledonian Iron Works Co., Montreal.
Miller Bros. & Toms Montreal.
Niles-Bement-Pond Co., New York.
Northern Engineering Works, Detroit
Owen Sound Iron Works Co, Owen Sound
Paxson, J. W., Co, Philadelphia
Smart-Turner-Machine Co., Hamilton.

Crankshafts.
The Canada Forge Co., Welland.
St. Clair Bros., Galt

Crushers, Rock or Ore.
Allis-Chalmers-Bullock. Montreal.

Contractors' Plant.
Allis-Chalmers-Bullock, Montreal.
John McDougall, Caledonian Iron Works Co., Montreal.
Niagara Falls Machine & Foundry Co., Niagara Falls, Ont.

Crank Pin Turning Machine
London Mach. Tool Co., Hamilton.
Niles-Bement-Pond Co., New York.

Crucibles.
Buffalo Foundry Supply Co, Buffalo.
Detroit Foundry Supply Co, Windsor.
Dominion Foundry Supply Co., Montreal
Hamilton Facing Mill Co., Hamilton.
J. W. Paxson Co., Philadelphia.

Crucible Caps
Hamilton Facing Mill Co., Hamilton.
Paxson, J. W., Co., Philadelphia

Crushers, Rock or Ore.
Allis-Chalmers-Bullock, Montreal.
Jeffrey Mfg. Co., Columbus, Ohio.

Cup Grease.
Mechanics Supply Co., Quebec

Cupolas.
Buffalo Foundry Supply Co., Buffalo.
Detroit Foundry Supply Co., Windsor
Dominion Foundry Supply Co., Montreal
Hamilton Facing Mill Co., Hamilton.
Northern Engineering Works, Detroit
J. W. Paxson Co. Philadelphia.
Sheldons Limited, Galt.

Cupola Blast Gauges.
Paxson, J. W., Co., Philadelphia
Sheldons Limited, Galt

Cupola Blocks.
Detroit Foundry Supply Co., Detroit.
Hamilton Facing Mill Co., Hamilton
Northern Engineering Works Detroit
Ontario Lime Association Toronto
Paxson, J. W., Co, Philadelphia
Toronto Pottery Co., Toronto

Cupola Blowers.
Buffalo Foundry Supply Co., Buffalo.
Canada Machinery Agency, Montreal.
Detroit Foundry Supply Co., Windsor
Dominion Heating and Ventilating Co., Hespeler
Hamilton Facing Mill Co., Hamilton.
Northern Engineering Works, Detroit
Paxson J. W, Co., Philadelphia
Sheldon's Limited, Galt.
B. F. Sturtevant Co., Hyde Park, Mass.

Cutters, Flue
Chicago Pneumatic Tool Co., Chicago.
J. W. Paxson Co., Philadelphia.

Cutters, Milling.
Becker, Brainard Milling Machine Co. Hyde Park, Mass.
Pratt & Whitney Co., Hartford, Conn.
Standard Tool Co., Cleveland.

Cutting-off Machines.
John Bertram & Sons Co., Dundas, Ont.
Canada Machinery Agency, Montreal.
London Mach. Tool Co., Hamilton.
H. W. Petrie, Toronto.
Pratt & Whitney Co., Hartford, Conn.

Cutting-off Tools.
Armstrong Bros. Tool Co., Chicago.
Baxter Paterson & Co, Montreal.
London Mach. Tool Co., Hamilton.
Mechanics Supply Co., Quebec, Que.
H. W. Petrie, Toronto,
Pratt & Whitney, Hartford, Conn.
Rice Lewis & Son, Toronto.
L. S. Starrett Co., Athol, Mass.

Damper Regulators.
Darling Bros., Ltd., Montreal

Dies
Globe Machine & Stamping Co., Cleveland, Ohio.

Die Stocks
Hart Manufacturing Co., Cleveland, Ohio
Mechanics Supply Co., Quebec

Dies, Opening
W. H. Banfield & Sons, Toronto
Globe Machine & Stamping Co., Cleveland Ohio.
R. McDougall Co., Galt.
Pratt & Whitney Co., Hartford, Conn.

Dies, Sheet Metal.
W. H. Banfield & Sons, Toronto.
Globe Machine & Stamping Co., Cleveland, Ohio

Die Sinkers
Globe Machine & Stamping Co., Cleveland Ohio.

Dies, Threading,
Hart Mfg. Co., Cleveland
Mechanics Supply Co., Quebec, Que.
John Millen & Son, Ltd, Montreal.

Draft, Mechanical.
W. H. Banfield & Sons, Toronto.
Butterfield & Co., Rock Island, Que.
A. B. Jardine & Co., Hespeler.
Mechanics Supply Co., Quebec, Que.
Pratt & Whitney Co., Hartford, Conn.
Sheldons Limited, Galt.
Sturtevant, B. F., Co., Hyde Park, Mass.

Drawing Instruments.
Rice Lewis & Son, Toronto.
Mechanics Supply Co, Quebec, Que.

Drawing Supplies.
The Electric Blue Print Co., Montreal.

Drawn Steel, Cold.
Baxter, Paterson & Co., Montreal.
Greey Wm. & J. G., Toronto
Miller Bros. & Toms, Montreal.
Union Drawn Steel Co., Hamilton.

Drilling Machines, Arch Bar.
John Bertram & Sons Co., Dundas, Ont.
London Mach. Tool Co., Hamilton
Niles-Bement-Pond Co., New York.

Drilling Machines, Boiler.
American Tool Works Co., Cincinnati.
John Bertram & Sons Co., Dundas, Ont.
Bickford Drill and Tool Co., Cincinnati.
The Canadian Fairbanks Co., Montreal.
A. B. Jardine & Co., Hespeler, Ont.
London Mach. Tool Co., Hamilton, Ont.
Niles-Bement-Pond Co., New York.
H. W. Petrie, Toronto.
Williams & Wilson, Montreal

Drilling Machines Connecting Rod.
John Bertram & Sons Co., Dundas, Ont.
London Mach. Tool Co., Hamilton
Niles-Bement-Pond Co. New York.

Drilling Machines, Locomotive Frame.
American Tool Works Co., Cincinnati.
B. F. Barnes Co., Rockford, Ill.
John Bertram & Sons Co., Dundas, Ont.
London Mach. Tool Co., Hamilton, Ont.
Niles-Bement-Pond Co., New York.

Drilling Machines, Multiple Spindle.
American Tool Works Co., Cincinnati.
B. F. Barnes Co., Rockford, Ill.
Baxter, Paterson & Co., Montreal.
John Bertram & Sons Co., Dundas, Ont.
Bickford Drill & Tool Co., Cincinnati.
Canada Machinery Agency Montreal.
London Mach. Tool Co., Hamilton, Ont.
Niles-Bement-Pond Co., New York.
H. W. Petrie, Toronto.
Williams & Wilson, Montreal.

Drilling Machines, Pneumatic
Canada Machinery Agency, Montreal.

Drilling Machines, Portable
Baxter, Paterson & Co., Montreal.
A. B. Jardine & Co., Hespeler, Ont.
Mechanics Supply Co., Quebec, Que.
Niles-Bement-Pond Co., New York.

Drilling Machines, Radial.
American Tool Works Co, Cincinnati.
Baxter, Paterson & Co., Montreal.
John Bertram & Sons Co., Dundas, Ont.
Bickford Tool & Drill Co., Cincinnati.
The Canadian Fairbanks Co., Montreal.
London Mach. Tool Co., Hamilton.
Mechanics Supply Co., Quebec, Que.
Niles-Bement-Pond Co., New York.
H. W. Petrie, Toronto.
Williams & Wilson, Montreal.

Drilling Machines, Suspension.
John Bertram & Sons Co., Dundas, Ont.
Canada Machinery Agency Montreal.
London Mach. Tool Co., Hamilton.
Niles-Bement-Pond Co New York

Drilling Machines, Turret.
John Bertram & Sons Co., Dundas, Ont.
London Mach. Tool Co., Hamilton.
Niles-Bement-Pond Co., New York.

Drilling Machines, Upright.
American Tool Works Co., Cincinnati.
B. F. Barnes Co., Rockford, Ill.
Baxter, Paterson & Co, Montreal.
John Bertram & Sons Co., Dundas, Ont
A. B. Jardine & Co., Hespeler, Ont.
London Mach. Tool Co., Hamilton.
Mechanics Supply Co., Quebec, Que.
Niles-Bement-Pond Co., New York.
W. H. Petrie, Toronto.
Williams & Wilson, Montreal.

Drills, Bench.
B. F. Barnes Co., Rockford, Ill.
Baxter, Paterson & Co. Montreal.
London Mach. Tool Co., Hamilton.
Mechanics Supply Co. Quebec, Que.
Pratt & Whitney Co., Hartford, Conn.

Drills, Blacksmith.
Canada Machinery Agency, Montreal.
A. B. Jardine & Co., Hespeler, Ont.
London Mach. Tool Co., Hamilton.
Mechanics Supply Co., Quebec, Que.
Standard Tool Co., Cleveland.

Drills, Centre.
Mechanics Supply Co., Quebec, Que.
Pratt & Whitney Co., Hartford, Conn
Standard Tool Co, Cleveland, O.
L. S. Starrett Co., Athol, Mass.

Drills, Electric
B. F. Barnes Co., Rockford, Ill.
Baxter Paterson & Co., Montreal.
Chicago Pneumatic Tool Co., Chicago.
Niles-Bement-Pond Co., New York.

Drills, Gang.

American Tool Works Co., Cincinnati.
B. F. Barnes Co., Rockford, Ill.
John Bertram & Sons Co., Dundas, Ont.
Pratt & Whitney Co.. Hartford, Conn.

Drills, High Speed.

Wm. Abbott, Montreal.
American Tool Works Co., Cincinnati.
B. F. Barnes Co., Rockford, Ill.
Pratt & Whitney Co., Hartford, Conn.
Standard Tool Co., Cleveland, O.

Drills, Hand.

A. B. Jardine & Co., Hespeler, Ont.

Drills, Horizontal.

B. F. Barnes Co., Rockford, Ill.
John Bertram & Sons Co., Dundas, Ont.
Canada Machinery Agency, Montreal.
London Mach. Tool Co., Hamilton.
Niles-Bement-Pond Co., New York.

Drills, Jewelers.

Pendrith Machinery Co., Toronto.

Drills, Pneumatic.

Canada Machinery Agency, Montreal.
Chicago Pneumatic Tool Co., Chicago.
Independent Pneumatic Tool Co., Chicago Ill.
Niles-Bement-Pond Co., New York.

Drills, Radial.

American Tool Works Co., Cincinnati.
John Bertram & Sons Co., Dundas, Ont.
Bickford Drill & Tool Co., Cincinnati
London Mach. Tool Co., Hamilton, Ont.
Niles-Bement-Pond Co., New York.

Drills, Ratchet.

Armstrong Bros. Tool Co., Chicago.
A. B. Jardine & Co., Hespeler,
Mechanics Supply Co., Quebec, Que.
Pratt & Whitney Co., Hartford, Conn.
Standard Tool Co., Cleveland.

Drills, Rock.

Allis-Chalmers-Bullock, Montreal.
Canadian Rand Drill Co., Montreal.
Chicago Pneumatic Tool Co., Chicago.
Jaffrey Mfg. Co., Columbus, Ohio.

Drills, Sensitive.

American Tool Works Co., Cincinnati.
B. F. Barnes Co., Rockford, Ill.
Canada Machinery Agency, Montreal.
Niles-Bement-Pond Co., New York.

Drills, Twist.

Baxter, Paterson & Co., Montreal.
Chicago Pneumatic Tool Co., Chicag
Alex. Gibb, Montreal.
A. B. Jardine & Co., Hespeler, Ont.
Mechanics Supply Co., Quebec, Que.
John Millen & Son, Ltd., Montreal.
Morse Twist Drill and Machine Co., New Bedford, Mass.
Pratt & Whitney Co., Hartford, Conn.
Standard Tool Co., Cleveland.

Drying Apparatus of all Kinds.

Dominion Heating & Ventilating Co., Hespeler
Greey, Wm. & J. G., Toronto
Sheldon's Limited, Galt
B. F. Sturtevant Co., Hyde ss

Dry Kiln Equipment.

Sheldons Limited, Galt
Sturtevant, B. F., Co., Hyde Park, Mass.

Dry Sand and Loam Facing.

Buffalo Foundry Supply Co., Buffalo.
Hamilton Facing Mill Co., Hamilton.
Paxson, J. W., Co., Philadelphia

Dump Cars.

Canada Foundry Co., Limited, Toronto
Greey, Wm. & J. G., Toronto
Hamilton Facing Mill Co., Hamilton.
John McDougall, Caledonian Iron Works Co., Montreal.
Niles-Bement-Pond Co., New York.
Standard Bearings, Ltd., Niagara Falls.
Jenckes Machine Co., Sherbrooke, Que
Link-Belt Eng. Co., Philadelphia.
John McDougall Caledonian Iron Works Co., Montreal.
Miller Bros. & Toms, Montreal.
Owen Sound Iron Works Co., Owen Sound
Paxson, J. W., Co., Philadelphia
Waterous Engine Co., Brantford.

Dust Separators.

Greey, Wm. & J. G., Toronto
Dominion Heating and Ventilating Co., Hespler.
Paxson, J. W., Co., Philadelphia
Sheldon's Limited, Galt.

Dynamos.

Allis-Chalmers-Bullock, Montreal.
Canadian General Electric Co., Toronto.
Canadian Westinghouse Co., Hamilton.
Consolidated Electric Co., Toronto
D'Olier Engineering Co., New York.
Hall Engineering Works, Montreal, Que.
Mechanics Supply Co., Quebec, Que.
John Millen & Son, Ltd., Montreal.
Packard Electric Co., St. Catharines.
H. W. Petrie, Toronto.
B. F. Sturtevant Co., Hyde Park, Man.
T. & H. Electric Co., Hamilton.

Dynamos—Turbine Driven.

Kerr-Turbine Co., Wellsville, N.Y.

Economizer, Fuel.

Dominion Heating & Ventilating Co., Hespeler
Standard Bearings, Ltd., Niagara Falls.
B. F. Sturtevant Co., Hyde Park, Man.

Electrical Instruments.

Canadian Westinghouse Co.. Hamilton.
Mechanics Supply Co., Quebec, Que.

Electrical Steel.

Baxter, Patterson & Co., Montreal.

Electrical Supplies.

Canadian General Electric Co., Toronto.
Canadian Westinghouse Co., Hamilton.
London Mach. Tool Co., Hamilton, Ont.
Mechanics Supply Co., Quebec, Que
John Millen & Son. Ltd., Montreal.
Packard Electric Co., St. Catharines.
T. & H. Electric Co., Hamilton.

Electrical Repairs

Canadian Westinghouse Co., Hamilton.
T. & H. Electric Co., Hamilton.

Elevator Buckets.

Greey, Wm. & J. G., Toronto
Jeffrey Mfg. Co., Columbus, Ohio.

Elevators, Foundry.

Northern Egineering Works, Detroit

Elevators—For any Service.

Darling Bros., Ltd., Montreal

Emery and Emery Wheels.

Baxter, Paterson & Co., Montreal.
Dominion Foundry Supply Co., Montreal
Hamilton Facing Mill Co., Hamilton.
Paxson, J. W., Co., Philadelphia

Emery Wheel Dressers.

Baxter, Paterson & Co., Montreal.
Canada Machinery Agency, Montreal.
Desmond-Stephan Mfg. Co., Urbana, Ohio
Dominion Foundry Supply Co., Montreal
Hamilton Facing Mill Co., Hamilton.
Mechanics Supply Co., Quebec, Que.
John Millen & Son, Ltd., Montreal.
H. W. Petrie, Toronto.
Paxson, J. W., Co., Philadelphia
Standard Tool Co., Cleveland.

Engineers and Contractors.

Canada Foundry Co., Limited, Toronto
Darling Bros., Ltd., Montreal
D'Olier Engineering Co., New York.
Greey, Wm. & J. G., Toronto
Hall Engineering Works, Montreal.
Laurie Engine & Machine Co., Montreal.
Link-Belt Co., Philadelphia.
John McDougall, Caledonian Iron Works Co., Montreal.
Miller Bros. & Toms Montreal.
Robb Engineering Co., Amherst, N.S.
The Smart-Turner Mach. Co., Hamilton.

Engineers' Supplies.

Baxter, Paterson & Co., Montreal.
Greey, Wm. & J. G., Toronto
Hall Engineering Works, Montreal.
Mechanics Supply Co., Quebec, Que.
Rice Lewis & Son, Toronto.

Engines, Gas and Gasoline.

Baxter, Paterson & Co., Montreal.
Canada Machinery Agency Montreal.
The Canadian Fairbanks Co., Montreal.
Canadian McVicker Engine Co., Galt.
Du Bois Engine Works, Du Bois, Pa.
Economic Heat, Light and Power Supply Co., Toronto.
Gilson Mfg. Co., Guelph
The Goldie & McCulloch Co., Galt, Ont.
Ontario Wind Engine & Pump Co. Toronto
H. W. Petrie, Toronto.
Producer Gas Co., Toronto
The Smart-Turner Mach. Co., Hamilton.
Standard Bearings, Ltd., Niagara Falls.

Engines, Steam.

Allis-Chalmers-Bullock, Montreal
Canada Machinery Agency, Montreal.
D'Olier Engineering Co., New York.
The Goldie & McCulloch Co., Galt, Ont.
E. Leonard & Sons, London, Ont.
Laurie Engine & Machine Co., Montreal.
Manson Mfg. Co., Thorold Ont
John McDougall Caledonian Iron Works, Montreal.
Miller Bros. & Toms, Montreal.
Robb Engineering Co., Amherst, N.S.
Sheldons Limited, Galt.
The Smart-Turner Mach. Co., Hamilton.
B. F. Sturtevant Co., Hyde Park, Mass.
Waterous Engine Works Co., Brantford.

Exhaust Heads.

Darling Bros., Ltd., Montreal.
Dominion Heating & Ventilating Co., Hespeler
Sheldon's Limited, Galt, Ont.
B. F. Sturtevant Co., Hyde Park, Mass.

Expanded Metal.

Expanded Metal and Fireproofing Co. Toronto

Expanders.

A. B. Jardine & Co., Hespeler, Ont.

Fans, Electric.

Canadian General Electric Co., Toronto
Canadian Westinghouse Co., Hamilton.
Dominion Heating & Ventilating Co., Hespeler
Mechanics Supply Co., Quebec, Que.
Paxson, J. W., Co., Philadelphia
Sheldon's Limited Galt, Ont.
The Smart-Turner Mach. Co., Hamilton.
B. F. Sturtevant Co., Hyde Park, Mass.

Fans, Exhaust.

Canadian Buffalo Forge Co., Montreal.
Detroit Foundry Supply Co., Windsor.
Dominion Foundry Supply Co., Montreal
Dominion Heating & Ventilating Co., Hespeler.
Greey, Wm. & J. G., Toronto
Hamilton Facing Mill Co., Hamilton.
Paxson, J. W., Co., Philadelphia
Sheldon's Limited, Galt.
B. F. Sturtevant Co., Hyde Park, Mass.

Feed Water Heaters.

Darling Bros., Montreal
Laurie Engine & Machine Co., Montreal
John McDougall, Caledonian Iron Works Co., Montreal.
The Smart-Turner Mach. Co., Hamilton

Files and Rasps.

Baxter, Paterson & Co., Montreal.
Mechanics Supply Co., Quebec, Que.
John Millen & Son. Ltd., Montreal.
Rice Lewis & Son, Toronto.
Nicholson File Co., Providence, R.I.
H. W. Petrie, Toronto.

Fillet, Pattern.

Baxter, Paterson & Co., Montreal.
Buffalo Foundry Supply Co., Buffalo.
Detroit Foundry Supply Co., Windsor.
Dominion Foundry Supply Co., Montreal
Hamilton Facing Mill Co., Hamilton.

Filters, Oil.

Darling Bros., Ltd., Montreal

Fire Apparatus.

Waterous Engine Works Co., Brantford.

Fire Brick and Clay.

Baxter Paterson & Co., Montreal.
Buffalo Foundry Supply Co., Buffalo.
Detroit Foundry Supply Co., Windsor.
Dominion Foundry Supply Co., Montreal
Hamilton Facing Mill Co., Hamilton.
Ontario Lime Association, Toronto
J. W. Paxson Co., Philadelphia.
Toronto Pottery Co., Toronto

Flash Lights.

Berlin Electrical Mfg. Co., Toronto.
Mechanics Supply Co., Quebec, Que.

Flour Mill Machinery.

Allis-Chalmers-Bullock, Montreal.
Greey, Wm. & J. G., Toronto
The Goldie & McCulloch Co., Galt, Ont.
John McDougall, Caledonian Iron Works Co., Montreal.
Miller Bros. & Toms, Montreal.

Forges.

Canada Foundry Co., Limited, Toronto.
Hamilton Facing Mill Co., Hamilton.
Mechanics Supply Co., Quebec, Que.
H. W. Petrie, Toronto.
Sheldon's Limited, Galt, Ont.
B. F. Sturtevant Co., Hyde Park, Mass.

Forgings, Drop.

John McDougall, Caledonian Iron Works Co., Montreal.
H. W. Petrie, Toronto.
St. Clair Bros., Galt

Forgings, Light & Heavy.

The Canada Forge Co., Welland.
Hamilton Steel & Iron Co., Hamilton

Forging Machinery.

John Bertram & Sons Co., Dundas, Ont.
London Mach. Tool Co., Hamilton, Ont
National Machinery Co., Tiffin, Ohio
Niles-Bement-Pond Co., New York.

Founders.

Greey, Wm. & J. G., Toronto
John McDougall, Caledonian Iron Works Co., Montreal
R. McDougall Co., Galt, Ont.
Niagara Falls Machine & Foundry Co., Niagara Falls, Ont.
Manson Mfg. Co., Thorold, Ont.
The Smart-Turner Mach. Co., Hamilton.

Foundry Equipment.

Detroit Foundry Supply Co., Windsor.
Hamilton Facing Mill Co., Mill.
Hanna Engineering Works, Chicago.
Northern Engineering Works, Detroit
Paxson, J. W., Co., Philadelphia

Foundry Facings.

Buffalo Foundry Supply Co., Buffalo.
Detroit Foundry Supply Co., Windsor.
Hamilton Facing Mill Co., Hamilton.
J. W. Paxson Co., Philadelphia, Pa.

Friction Clutch Pulleys, etc.

The Goldie & McCulloch Co., Galt.
Greey, Wm. & J. G., Toronto
Link-Belt Co., Philadelphia.
Miller Bros. & Toms, Montreal.

Furnaces.

Detroit Foundry Supply Co., Windsor.
Dominion Foundry Supply Co., Montreal
Hamilton Facing Mill Co., Hamilton.
Paxson, J. W., Co., Philadelphia

Galvanizing

Canada Metal Co., Toronto.
Ontario Wind Engine & Pump Co., Toronto

Gas Blowers and Exhausters.

D'Olier Engineering Co., New York.
Dominion Heating & Ventilating Co. Hespeler
Sheldon's Limited, Galt.
B. F. Sturtevant Co., Hyde Park, Man.

Gas Plants, Suction.

Baxter Paterson & Co., Montreal
Economic Heat, Light and Power Supply Co., Toronto.
Producer Gas Co., Toronto
Standard Bearings, Ltd., Niagara Falls.
Williams & Wilson, Montreal

Gauges, Standard.

Mechanics Supply Co., Quebec, Que.
Pratt & Whitney Co., Hartford, Conn.

Gearing.

Greey, Wm. & J. G., Toronto

Gear Cutting Machinery.

Baxter, Patterson & Co., Montreal
Becker-Brainard Milling Mach. Co., Hyde Park, Mass.
Bickford Drill & Tool Co., Cincinnati.
Greey, Wm. & J. G., Toronto
Kennedy, Wm., & Sons, Owen Sound
London Mach. Tool Co., Hamilton.
R. McDougall Co., Galt.
Niles-Bement-Pond Co., New York.
H. W. Petrie, Toronto.
Pratt & Whitney Co., Hartford. Conn.
Williams & Wilson, Montreal.

Gears, Angle.

Chicago Pneumatic Tool Co., Chicago.
Greey, Wm. & J. G., Toronto
Laurie Engine & Machine Co., Montreal.
John McDougall, Caledonian Iron Works Co., Montreal.
Miller Bros. & Toms, Montreal.
Waterous Engine Co., Brantford.

Gears, Cut.

Kennedy, Wm., & Sons, Owen Sound

Gears, Iron.

Greey, Wm & J. G., Toronto
Kennedy, Wm., & Sons, Owen Sound

Gears, Mortise.

Greey, Wm. & J. G., Toronto
Kennedy, Wm., & Sons, Owen Sound

Gears, Reducing.

Brown, David & Sons, Huddersfield, Eng
Chicago Pneumatic Tool Co., Chicago.
Greey, Wm. & J. G., Toronto
John McDougall, Caledonian Iron Works Co., Montreal.
Miller Bros. & Toms, Montreal.

Generating Sets.

Sturtevant, B. F., Co., Hyde Park, Mass.

Generators, Electric.

Allis-Chalmers-Bullock, Limited, Montreal
Canadian General Electric Co., Toronto
Canadian Westinghouse Co., Hamilton
D'Olier Engineering Co., New York.
Hall Engineering Works, Montreal.
H. W. Petrie, Toronto.
B. F. Sturtevant Co., Hyde Park, Mass.
Toronto & Hamilton Electric Co., Hamilton.

Generators, Gas.

H. W. Petrie, Toronto.

Glass Bevelling Machines.

Pendrith Machinery Co., Toronto.

Graphite Paints.

P. D. Dods & Co., Montreal.

Graphite.

Detroit Foundry Supply Co., Windsor.
Dominion Foundry Supply Co., Montreal
Hamilton Facing Mill Co., Hamilton.
Mechanics Supply Co., Quebec, Que.
Paxson, J. W., Co., Philadelphia

Grinders, Automatic Knife.

W. H. Banfield & Son, Toronto.

Grinders, Centre.

Niles-Bement-Pond Co., New York.
H. W. Petrie, Toronto.

Grinders, Cutter.

Becker-Brainard Milling Mach. Co., Hyde Park, Mass.
John Millen & Son, Ltd., Montreal.
Pratt & Whitney Co., Hartford, Conn.

Grinders, Tool.

Armstrong Bros. Tool Co., Chicago.
B. F. Barnes Co., Rockford, Ill.
H. W. Petrie, Toronto.
Williams & Wilson, Montreal.

Grinding Machines.

Brown & Sharpe, Providence, R.I.
The Canadian Fairbanks, Montreal.
Niles-Bement-Pond Co., New York.
Norton Company, Worcester, Mass.
Paxson, J. W., Co., Philadelphia
H. W. Petrie, Toronto.

Grinding and Polishing Machines.

The Canadian Fairbanks Co., Montreal.
Greey, Wm. & J. G., Toronto
Independent Pneumatic Tool Co., Chicago, Ill.
John Millen & Son, L'td., Montreal.
Miller Bros. & Toms, Montreal.
Niles-Bement-Pond Co., New York.
H. W. Petrie, Toronto.
Pendrith Machinery Co., Toronto.
Standard Bearings, Ltd., Niagara Falls

Grinding Wheels (Alundum)

Norton Company, Worcester, Mass.

Hack Saws.

Baxter, Paterson & Co., Montreal.
Canada Machinery Agency, Montreal.
The Canadian Fairbanks Co., Montreal.
Mechanics Supply Co., Quebec, Que.
John Millen & Son, Ltd., Montreal.
Niles-Bement-Pond Co., New York.
H. W. Petrie, Toronto.
Williams & Wilson, Montreal.

Hack Saw Frames.

Mechanics Supply Co., Quebec, Que.

Hammers, Drop.

London Mach. Tool Co., Hamilton, Ont.
Miller Bros. & Toms Montreal.
Niles-Bement-Pond Co., New York.

Hammers, Steam.

John Bertram & Sons Co. Dundas, Ont.
London Mach. Tool Co., Hamilton, Ont.
Niles-Bement-Pond Co., New York.

Hangers.

The Goldie & McCulloch Co., Galt.
Greey, Wm. & J. G., Toronto
Kennedy, Wm., & Sons, Owen Sound
Miller Bros & Toms, Montreal.
Owen Sound Iron Works Co., Owen Sound
The Smart-Turner Mach. Co., Hamilton.
Waterous Engine Co., Brantford.
Pendrith Machinery Co., Toronto.
Standard Bearings Ltd., Niagara Falls

Hardware Specialties.

Mechanics Supply Co., Quebec, Que.

Heating Apparatus.

Darling Bros., Ltd., Montreal
Dominion Heating & Ventilating Co., Hespeler
Mechanic Supply Co., Quebec, Que.
Sheldons Limited, Galt.
B. F. Sturtevant Co., High Park, Mass.

Hoisting and Conveying Machinery.

Allis-Chalmers-Bullock, Limited, Montrea
Greey, Wm & J. G., Toronto
Link-Belt Co., Philadelphia.
Miller Bros. & Toms, Montreal.
Niles-Bement-Pond Co., New York.
Northern Engineering Works, Detroit
The Smart-Turner Mach. Co., Hamilton.
Waterous Engine Co., Brantford,

Hoists, Electric.

Northern Engineering Works, Detroit

Hoists, Pneumatic.

Canadian Rand Drill Co., Montreal.
Chicago Pneumatic Tool Co. Chicago.
Dominion Foundry Supply Co., Montreal
Hamilton Facing Mill Co., Hamilton.
Northern Engineering Works, Detroit

Hose.

Baxter, Paterson & Co., Montreal.
Mechanics Supply Co., Quebec, Que.

Hose, Air.

Canadian Rand Drill Co., Montreal.
Canadian Westinghouse Co., Hamilton.
Chicago Pneumatic Tool Co. Chicago.
Paxson, J. W., Co., Philadelphia

Hose Couplings.

Canadian Rand Drill Co., Montreal.
Canadian Westinghouse Co., Hamilton.
Chicago Pneumatic Tool Co., Chicago.
Mechanics Supply Co., Quebec, Que.
Paxson, J. W., Co., Philadelphia

Hose, Steam.

Allis-Chalmers-Bullock, Montreal.
Canadian Rand Drill Co., Montreal.
Canadian Westinghouse Co., Hamilton.
Mechanics Supply Co., Quebec, Que.
Paxson, J. W. Co., Philadelphia

Hydraulic Accumulators.

Niles-Bement-Pond Co., New York.
Perrin, Wm. R., Co., Toronto.
The Smart-Turner Mach. Co., Hamilton.

Hydraulic Machinery.

Allis-Chalmers-Bullock, Montreal.
Barber, Chas., & Sons, Meaford

India Oil Stones.

Norton Company, Worcester, Mass.

Indicators, Speed.

Mechanics Supply Co., Quebec, Que.
L. S. Starrett Co., Athol, Mass.

Injectors.

Canada Foundry Co., Toronto.
The Canadian Fairbanks Co., Montreal.
Desmond-Stephan Mfg. Co., Urbana, Ohio
Mechanics Supply Co., Quebec, Que.
Rice Lewis & Son, Toronto.
Penberthy Injector Co., Windsor, Ont.

Iron and Steel Bars and Bands.

Hamilton Steel & Iron Co., Hamilton

Iron Cements.

Buffalo Foundry Supply Co., Buffalo.
Detroit Foundry Supply Co., Windsor.
Hamilton Facing Mill Co., Hamilton.
Paxson, J. W., Co., Philadelphia

Iron Filler.

Buffalo Foundry Supply Co., Buffalo.
Detroit Foundry Supply Co., Windsor.
Dominion Foundry Supply Co. Montreal
Hamilton Facing Mill Co., Hamilton.
Paxson, J. W., Co., Philadelphia.

Jacks.

Norton, A. O., Coaticook, Que.

Kegs, Steel Shop.

Cleveland Wire Spring Co., Cleveland

Key-Seating Machines.

B. F. Barnes Co., Rockford, Ill.
Niles-Bement-Pond Co., New York.

Lace Leather.

Baxter, Paterson & Co., Montreal.

Ladles, Foundry.

Northern Engineering Works, Detroit

Lamps, Arc and Incandescent.

Canadian General Electric Co., Toronto.
Canadian Westinghouse Co., Hamilton.
The Packard Electric Co., St. Catharines.

Lathe Dogs.

Armstrong Bros., Chicago
Baxter, Paterson & Co., Montreal.
Pratt & Whitney Co., Hartford, Conn.

Lathes, Engine.

American Tool Work Co., Cincinnati.
B. F. Barnes Co., Rockford, Ill.
Baxter, Paterson & Co., Montreal
John Bertram & Sons Co., Dundas, Ont.
Canada Machinery Agency, Montreal.
The Canadian Fairbanks Co., Montreal.
London Mach. Tool Co., Hamilton, Ont.
Niles-Bement-Pond Co., New York.
R. McDougall Co, Galt, Ont.
H. W. Petrie, Toronto.
Pratt & Whitney Co., Hartford, Conn.

Lathes, Foot-Power.

American Tool Works Co., Cincinnati.
B. F. Barnes Co., Rockford, Ill.

Lathes, Screw Cutting.

B. F. Barnes Co., Rockford, Ill.
Baxter, Paterson & Co., Montreal.
Niles-Bement-Pond Co., New York.

Lathes, Automatic, Screw-Threading.

John Bertram & Sons Co., Dundas, Ont
London Mach. Tool Co., Hamilton, Ont
Pratt & Whitney Co., Hartford, Conn.

Lathes, Bench.

B. F. Barnes Co., Rockford, Ill.
London Mach. Tool Co., London, Ont.
Pratt & Whitney Co., Hartford, Conn.

Lathes, Turret.

American Tool Works Co., Cincinnati.
Baxter, Paterson & Co., Montreal
John Bertram & Sons Co., Dundas, Ont.
Gisholt Machine Co., Madison, Wis.
London Mach. Tool Co., Hamilton, Ont.
Niles-Bement-Pond Co., New York.
The Pratt & Whitney Co., Hartford, Conn.
Warner & Swasey Co., Cleveland, O.

Lathes, Buffing and Polishing.

Pendrith Machinery Co., Toronto.

Lath Mill Machinery.

Owen Sound Iron Works Co., Owen Sound

Leather Belting.

Baxter, Paterson & Co., Montreal.
Canada Machinery Agency, Montreal.
The Canadian Fairbanks Co., Montreal
Greey, Wm. & J. G., Toronto

Lime Stone Flux.

Hamilton Facing Mill Co., Hamilton.
Paxson, J. W., Co, Philadelphia

Locomotives, Air.

Canadian Rand Drill Co., Montreal.

Locomotives, Electrical.

Canadian Westinghouse Co., Hamilton
Jeffrey Mfg. Co., Columbus, Ohio.

Locomotives, Steam.

Canada Foundry Co., Toronto.
Canadian Rand Drill Co., Montreal

Lubricating Plumbago.

Detroit Foundry Supply Co., Detroit.
Dominion Foundry Supply Co., Montreal
Hamilton Facing Mill Co., Hamilton.
Paxson, J. W., Co., Philadelphia

Lumber Dry Kilns.

Dominion Heating & Ventilating Co., Hespeler.
H. W. Petrie, Toronto.
Sheldon's Limited, Galt, Ont.
B. F. Sturtevant Co., Hyde Park, Mass.

Machinery Dealers.

Baxter, Paterson & Co., Montreal.
Canada Machinery Agency, Montreal.
The Canadian Fairbanks Co., Montreal
Miller Bros. & Toms, Montreal.
H. W. Petrie, Toronto.
The Smart-Turner Mach. Co., Hamilton.
Williams & Wilson Montreal.

Machinery Designers.

Greey, Wm. & J. G., Toronto
Standard Bearings, Ltd., Niagara Falls.

Machinists.

W. H. Banfield & Sons, Toronto.
Greey, Wm. & J. G. Toronto
Hall Engineering Works, Montreal.
Link-Belt Co., Philadelphia.
John McDougall, Caledonian Iron Works Co., Montreal.
Miller Bros. & Toms Montreal.
Paxson, J. W., Co, Philadelphia
Pendrith Machinery Co., Toronto.
Robb Engineering Co., Amherst, N.S.
The Smart-Turner Mach. Co., Hamilton.
Standard Bearings, Ltd, Niagara Falls
Waterous Engine Co., Brantford.

Machinists' Small Tools.

Armstrong Bros., Chicago.
Butterfield & Co., Rock Island, Que.
Mechanics Supply Co., Quebec, Que.
Rice Lewis & Son, Montreal.
Pratt & Whitney Co., Hartford, Conn.
Standard Tool Co., Cleveland.
L. S. Starrett Co., Athol, Mass.
Williams & Wilson, Montreal.

Malleable Flask Clamps.

Buffalo Foundry Supply Co, Buffalo.
Paxon, J. W., Co., Philadelphia

Malleable Iron Castings.

Galt Malleable Iron Co., Galt

Mallet, Rawhide and Wood.

Buffalo Foundry Supply Co., Buffalo.
Detroit Foundry Supply Co., Detroit.
Paxson, J. W., Co., Philadelphia

Mandrels.

A. B. Jardine & Co., Hespeler, Ont.
Miller Bros & Toms, Montreal.
The Pratt & Whitney Co., Hartford, Conn.
Standard Tool Co., Cleveland.

Marine Work.

Kennedy, Wm., & Sons, Owen Sound

Metallic Paints.

P. D. Dods & Co., Montreal.

Meters, Electrical.

Canadian Westinghouse Co., Hamilton.
Mechanics Supply Co, Quebec, Que.

Mill Machinery.

Baxter, Paterson & Co., Montreal.
Greey, Wm. & J. G., Toronto
The Goldie & McCulloch Co., Galt, Ont.
John McDougall, Caledonian Iron Works Co., Montreal.
H. W. Petrie, Toronto.
Robb Engineering Co., Amherst, N.S.
Waterous Engine Co., Brantford.
Williams & Wilson, Montreal.

Milling Attachments.

Becker-Brainard Milling Machine Co. Hyde Park, Mass.
John Bertram & Sons Co., Dundas, Ont.
Niles-Bement-Pond Co., New York,
Pratt & Whitney, Hartford, Conn.

Milling Machines, Horizontal.

Baxter, Paterson & Co., Montreal.
Becker-Brainard Milling Machinery Co. Hyde Park, Mass.
John Bertram & Sons Co., Dundas, Ont.
London Mach. Tool Co., Hamilton, Ont.
Niles-Bement-Pond Co., New York.
Pratt & Whitney, Hartford, Conn.

Milling Machines, Plain.

American Tool Works Co., Cincinnati.
Becker-Brainard Milling Machine Co. Hyde Park, Mass.
John Bertram & Sons Co., Dundas, Ont.
Canada Machinery Agency, Montreal.
The Canadian Fairbanks Co., Montreal.
London Mach. Tool Co., Hamilton, Ont.
Niles-Bement-Pond Co., New York.
H. W. Petrie, Toronto.
Pratt & Whitney Co., Hartford, Conn.
Williams & Wilson, Montreal.

Milling Machines, Universal.

American Tool Works Co., Cincinnati.
Becker-Brainard Milling Machine Co., Hyde Park, Mass.
John Bertram & Sons Co., Dundas, Ont.
Canada Machinery Agency, Montreal.
The Canadian Fairbanks Co., Montreal.
London Mach. Tool Co., Hamilton, Ont.
Niles-Bement-Pond Co., New York.
H. W. Petrie, Toronto.
Williams & Wilson, Montreal.

Milling Machines, Vertical.

Becker-Brainard Milling Machine Co., Hyde Park, Mass.
Brown & Sharpe, Providence, R.I.
John Bertram & Sons Co., Dundas, Ont.
Canada Machinery Agency, Montreal.
London Mach. Tool Co., Hamilton, Ont.
Niles-Bement-Pond Co., New York.

Milling Tools.

Wm. Abbott, Montreal.
Becker-Brainard Milling Machine Co., Hyde Park, Mass.
Geometric Tool Co., New Haven, Conn.
London Mach. Tool Co., Hamilton, Ont.
Pratt & Whitney Co., Hartford, Conn.
Standard Tool Co., Cleveland.
Standard Bearings, Ltd., Niagara Falls

Mining Machinery.

Allis-Chalmers-Bullock, Limited, Montreal
Canadian Rand Drill Co., Montreal.
Chicago Pneumatic Tool Co., Chicago.
Jeffrey Mfg. Co., Columbus, Ohio.
Laurie Engine & Machine Co., Montreal.
John McDougall, Caledonian Iron Works Co., Montreal.
Miller Bros. & Toms, Montreal.
T. & H. Electric Co. Hamilton.

Mixing Machines, Dough.

Greey, Wm. & J. G., Toronto
Pendrith Machinery Co., Toronto.
Paxson, J. W., Co., Philadelphia

Mixing Machines, Special.

Greey, Wm. & J. G., Toronto
Pendrith Machinery Co., Toronto.
Paxson, J. W., Co., Philadelphia

Model Tools.

Globe Machine & Stamping Co., Cleveland, Ohio.
Standard Bearings, Ltd., Niagara Falls.
Wells Pattern and Model Works, Toronto

Motors, Electric.

Allis-Chalmers-Bullock Limited, Montreal
Canadian General Electric Co., Torono
Canadian Westinghouse Co., Hamilton.
Consolidated Electric Co, Toronto
Hall Engineering Works, Montreal.
Mechanics Supply Co., Quebec, Que.
The Packard Electric Co., St. Catharines.
B. F. Sturtevant Co., Hyde Park, Mass.
T. & H. Electric Co., Hamilton.

Motors, Air.

Canadian Rand Drill Co., Montreal.
Chicago Pneumatic Tool Co., Chicago.

Molders' Supplies.

Buffalo Foundry Supply Co., Buffalo.
Detroit Foundry Supply Co., Windsor.
Dominion Foundry Supply Co., Montreal
Hamilton Facing Mill Co., Hamilton.
Mechanics Supply Co., Quebec, Que.
Paxson, J. W., Co., Philadelphia

Molders' Tools.

Buffalo Foundry Supply Co., Buffalo.
Detroit Foundry Supply Co., Windsor
Dominion Foundry Supply Co., Montreal
Hamilton Facing Mill Co., Hamilton.
Mechanics Supply Co., Quebec, Que.
Paxson, J. W., Co., Philadelphia

Molding Machines.

Buffalo Foundry Supply Co., Buffalo.
Hamiton Facing Mill Co., Hamilton.
J. W. Paxson Co., Philadelphia, Pa.

Molding Sand.

T. W. Barnes, Hamilton.
Buffalo Foundry Supply Co., Buffalo.
Detroit Foundry Supply Co., Windsor
Dominion Foundry Supply Co., Montreal
Hamilton Facing Mill Co., Hamilton.
Paxson, J. W., Co., Philadelphia

Nut Tappers.

John Bertram & Sons Co., Dundas, Ont
A. B. Jardine & Co., Hespeler.
London Mach. Tool Co., Hamilton.
National Machinery Co., Tiffin, Ohio.

Oatmeal Mill Machinery.

Greey, Wm. & J. G., Toronto
The Goldie & McCulloch Co., Galt

Oils, Core.

Buffalo Foundry Supply Co., Buffalo.
Dominion Foundry Supply Co., Montreal
Hamilton Facing Mill Co., Hamilton.
Paxson, J. W., Co., Philadelphia

Oil Extractors.

Darling Bros., Ltd., Montreal

Paint Mill Machinery.

Greey, Wm. & J. G., Toronto

Pans, Lathe.

Cleveland Wire Spring Co., Cleveland
Paxson, J. W., Co., Philadelphia

Pans, Steel Shop.

Cleveland Wire Spring Co., Cleveldnd

Patent Solicitors.

Hanbury A. Budden, Montreal.
Fetherstonhough & Blackmore, Montreal
Fetherstonhaugh & Co., Montreal.
Marion & Marion, Montreal.
Ridout & Maybee, Toronto.

Patterns.

John Carr, Hamilton.
Galt Malleable Iron Co., Galt
Hamilton Pattern Works, Hamilton.
John McDougall, Caledonian Iron Works Co., Montreal
Miller Bros. & Toms Montreal.
Wells' Pattern and Model Works, Toronto.

Pig Iron.

Hamilton Steel & Iron Co., Hamilton

Pipe Cutting and Threading Machines.

Butterfield & Co., Rock Island, Que.
Canada Machinery Agency, Montreal.
Hart Mfg. Co., Cleveland
A. B. Jardine & Co., Hespeler, Ont.
London Mach. Tool Co., Hamilton, Ont.
R. McDougall Co., Galt, Ont.
Niles-Bement-Pond Co., New York.

Pipe and Fittings.

Mechanics Supply Co., Quebec, Que.

Pipe, Municipal.

Canadian Pipe Co., Vancouver, B.C.
Pacific Coast Pipe Co., Vancouver, B.C.

Pipe, Waterworks.

Canadian Pipe Co., Vancouver, B.C.
Pacific Coast Pipe Co., Vancouver, B.C

Planers, Standard.

American Tool Works, Cincinnati.
John Bertram & Sons Co., Dundas, Ont.
Canada Machinery Agency, Montreal.
The Canadian Fairbanks Co., Montreal.
London Mach. Tool Co., Hamilton, Ont.
R. McDougall Co., Galt, Ont.
Niles-Bement-Pond Co., New York.
H. W. Petrie, Toronto.
Pratt & Whitney Co., Hartford, Conn.
Williams & Wilson, Montreal.

Planers, Rotary.

John Bertram & Sons Co., Dundas, Ont.
London Mach. Tool Co., Hamilton, Ont.
Niles-Bement-Pond Co., New York.

Planing Mill Fans.

Dominion Heating & Ventilating Co., Hespeler
Sheldon's Limited, Galt, Ont.
B. F. Sturtevant Co., Hyde Park, Man.

Plumbago.

Buffalo Foundry Supply Co., Buffalo.
Detroit Foundry Supply Co., Windsor
Dominion Foundry Supply Co., Montreal
Hamilton Facing Mill Co., Hamilton,
Mechanics Supply Co., Quebec, Que.
J. W. Paxson Co., Philadelphia, Pa.

Pneumatic Separators.

Paxson, J. W., Co., Philadelphia
Sheldons Limited, Galt
Sturtevant, B. F., Co., Hyde Park, Mass.

Pneumatic Tools.

Allis-Chalmers-Bullock, Montreal.
Canadian Rand Drill Co., Montreal.
Chicago Pneumatic Tool Co., Chicago
Hamilton Facing Mill Co., Hamilton.
Hanna Engineering Works, Chicago.
Independent Pneumatic Tool Co. Chicago, Ill.
Paxson, J. W., Co., Philadelphia

Power Hack Saw Machines.

Baxter, Paterson & Co., Montreal.

Power Plants.

John McDougall Caledonian Iron Works Co., Montreal.
The Smart-Turner Mach. Co., Hamilton

Power Plant Equipments.

Darling Bros., Ltd., Montreal

Presses, Drop.

W. H. Banfield & Son, Toronto.
E. W. Bliss Co., Brooklyn, N.Y.
Canada Machinery Agency, Montreal.
Laurie Engine & Machine Co., Montreal
Miller Bros. & Toms, Montreal.
Niles-Bement-Pond Co., New York.

Presses, Hand.

E. W. Bliss Co., Brooklyn, N.Y.

Presses, Hydraulic.

John Bertram & Sons Co., Dundas, Ont.
Laurie Engine & Machine Co., Montreal.
London Mach. Tool Co., Hamilton, Ont.
John McDougall Caledonian Iron Works Co., Montreal.
Miller Bros. & Toms, Montreal.
Niles-Bement-Pond Co., New York.
Perrin, Wm. R., & Co., Toronto

Presses, Power.

E. W. Bliss Co., Brooklyn, N.Y.
Canada Machinery Agency, Montreal.
London Mach. Tool Co., Hamilton, Ont.
John McDougall Caledonian Iron Works Co., Montreal.
Niles-Bement-Pond Co., New York.

Presses Power Screw.

Perrin, Wm. R., & Co., Toronto

Pressure Regulators.

Darling Bros., Ltd., Montreal

Pulp Mill Machinery.

Greey, Wm. & J. G., Toronto
Jeffrey Mfg. Co., Columbus, Ohio.
Laurie Engine & Machine Co., Montreal.
John McDougall Caledonian Iron Works Co., Montreal.
Waterous Engine Works Co., Brantford

Pulleys.

Baxter, Paterson & Co., Montreal.
Canada Machinery Agency, Montreal.
The Canadian Fairbanks Co., Montreal.
The Goldie & McCulloch Co., Galt.
Greey, Wm. & J. G., Toronto
Laurie Engine & Machine Co., Montreal.
Link-Belt Co., Philadelphia.
John McDougall Caledonian Iron Works Co., Montreal.
Miller Bros. & Toms, Montreal.
Owen Sound Iron Works Co., Owen Sound
Pendrith Machinery Co., Toronto.
H. W. Petrie, Toronto.
The Smart-Turner Mach. Co., Hamilton.
Standaad Bearings, Ltd., Niagara Falls.
Williams & Wilson, Montreal.
Waterous Engine Co., Brantford.

Pumps.

Ontario Wind Engine & Pump Co., Toronto

Pump Governors.

Darling Bros., Ltd., Montreal

Pumps, Hydraulic.

Perrin, Wm. R. & Co., Toronto

Pumps, Steam.

Allis-Chalmers-Bullock, Limited, Montreal
Canada Foundry Co., Toronto.
Canada Machinery Agency, Montreal.
Darling Bros., Ltd., Montreal
D'Olier Engineering Co., New York.
The Goldie & McCulloch Co., Galt.
John McDougall Caledonian Iron Works, Montreal.
R. McDougall Co., Galt, Ont.
H. W. Petrie, Toronto.
The Smart-Turner Mach. Co., Hamilton.
Standard Bearings, Ltd., Niagara Falls,
Waterous Engine Co., Brantford.

Technical Books

Dies

Their Construction and Use for the Modern Working of Sheet Metals.

By Joseph V. Woodsworth

Large 8vo.; 384 pages; 505 illustrations; bound in cloth; price, *$3.00*

The New Metal Worker Pattern Book

A Treatise on Pattern-Cutting as Applied to all Branches of Sheet Metal Work.

By Geo. W. Kittredge

Cloth bound; price postpaid, *$5.00*

Modern Machine Shop Tools

Their Construction, Operation and Manipulation, including both Hand and Machine Tools.

By W. H. Vandervoort, M.E.

Large 8vo.; 555 pages; 673 illustrations; bound in cloth; price, *$4.00*

Modern Steam Engineering

By Gardner D. Hiscox

Price, *$3.00*

This is the most complete practical book issued for the Stationery Engineer. Just issued.

Compressed Air

Its Production, Uses and Appliances.

By Gardner D. Hiscox

Large 8vo.; 820 pages; 545 illustrations; bound in cloth; price, *$5.00*

Locomotive Catechism

By Robert Grimshaw

Illustrated by 222 engravings and 12 folding plates; price, *$2.00*

Address Technical Book Dept.:

The MacLean Publishing Co., Limited

Montreal Toronto Winnipeg

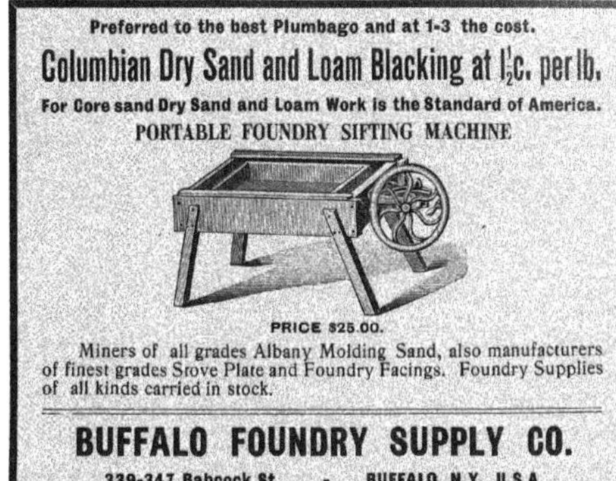

FINEST QUALITY

GREY-IRON CASTINGS

LIGHT OR HEAVY

OUR PLANT IS EQUIPPED WITH THE BEST APPARATUS FOR HIGH-GRADE WORK.

We make a specialty of burning necks on rolls.

High Speed Engines, Centrifugal Pumps and Contractors' Supplies

SEND FOR AN ESTIMATE.

The Manson Mfg. Co., Limited

THOROLD, ONT.

Pumping Machinery.

Canada Foundry Co., Limited, Toronto
Canada Machinery Agency, Montreal.
Canadian Rand Drill Co., Montreal.
Chicago Pneumatic Tool Co., Chicago.
Darling Bros., Ltd., Montreal
D'Olier Engineering Co., New York.
Hall Engineering Works, Montreal, Que.
Laurie Engine & Machine Co., Montreal.
London Mach. Tool Co., Hamilton, Ont.
John McDougall Caledonian Iron Works Co., Montreal.
R. McDougall Co., Galt, Ont.
The Smart-Turner Mach. Co., Hamilton.
Standard Bearings, Ltd., Niagara Falls,

Punches and Dies.

W. H. Banfield & Sons, Toronto.
Butterfield & Co., Rock Island.
Globe Machine & Stamping Co.
A. B. Jardine & Co., Hespeler, Ont.
London Mach Tool Co., Hamilton, Ont.
Miller Bros & Toms, Montreal.
Pratt & Whitney Co., Hartford, Conn.
H. W. Petrie, Toronto.
Standard Bearings, Ltd., Niagara Falls,

Punches, Hand.

Mechanics Supply Co., Quebec, Que.

Punches, Power.

John Bertram & Sons Co., Dundas, Ont.
E. W. Bliss Co., Brooklyn, N.Y.
Canada Machinery Agency, Montreal.
Canadian Buffalo Forge Co., Montreal.
London Mach. Tool Co., Hamilton, Ont.
Niles-Bement-Pond Co., New York.

Punches, Turret.

London Mach. Tool Co., London, Ont.

Punching Machines, Horizontal.

John Bertram & Sons Co., Dundas, Ont.
London Mach. Tool Co., Hamilton, Ont.
Niles-Bement-Pond Co., New York.

Quartering Machines.

John Bertram & Sons Co., Dundas, Ont.
London Mach. Tool Co., Hamilton, Ont.

Railway Spikes and Washers.

Hamilton Steel & Iron Co., Hamilton

Rammers, Bench and Floor.

Buffalo Foundry Supply Co., Buffalo.
Detroit Foundry Supply Co., Windsor
Hamilton Facing Mill Co., Hamilton.
Paxson, J. W., Co., Philadelphia

Rapping Plates.

Detroit Foundry Supply Co., Windsor
Hamilton Facing Mill Co., Hamilton.
Paxson, J. W., Co., Philadelphia

Raw Hide Pinions.

Brown, David & Sons, Huddersfield, Eng.

Reamers.

Wm. Abbott, Montreal.
Baxter, Paterson & Co., Montreal.
Butterfield & Co., Rock Island.
Hanna Engineering Works, Chicago.
A. B. Jardine & Co., Hespeler, Ont.
Mechanics Supply Co., Quebec, Que.
John Millen & Son, Ltd., Montreal.
Morse Twist Drill and Machine Co., New Bedford, Mass.
Pratt & Whitney Co., Hartford, Conn.
Standard Tool Co., Cleveland.

Reamers, Steel Taper.

Butterfield & Co., Rock Island.
Chicago Pneumatic Tool Co., Chicago.
A. B. Jardine & Co., Hespeler, Ont.
John Millen & Son, Ltd., Montreal.
Pratt & Whitney Co., Hartford, Conn.
Standard Tool Co., Cleveland.
Whitman & Barnes Co., St. Catharines.

Rheostats.

Canadian General Electric Co., Toronto.
Canadian Westinghouse Co., Hamilton.
Hall Engineering Works, Montreal, Que.
T. & H. Electric Co., Hamilton,

Riddles.

Buffalo Foundry Supply Co., Buffalo.
Detroit Foundry Supply Co., Windsor
Hamilton Facing Mill Co., Hamilton.
J. W. Paxson Co., Philadelphia, Pa.

Riveters, Pneumatic.

Hanna Engineering Works, Chicago.
Independent Pneumatic Tool Co., Chicago, Ill.

Rolls, Bending.

John Bertram & Sons Co., Dundas, Ont.
London Mach. Tool Co., Hamilton, Ont.
Niles-Bement-Pond Co., New York.

Rolls, Chilled Iron.

Greey, Wm. & J. G., Toronto

Rolls, Sand Cast.

Greey, Wm. & J. G., Toronto

Rotary Blowers.

Paxson, J. W., Co., Philadelphia
Sturtevant, B. F., Co., Hyde Park, Mass.

Rotary Converters.

Allis-Chalmers-Bullock, Ltd., Montreal.
Canadian Westinghouse Co., Hamilton.
Paxson, J. W., Co., Philadelphia
Toronto and Hamilton Electric Co., Hamilton.

Safes.

Baxter, Paterson & Co., Montreal.
The Goldie & McCulloch Co., Galt,

Sand, Bench.

Buffalo Foundry Supply Co., Buffalo.
Detroit Foundry Supply Co., Windsor
Hamilton Facing Mill Co., Hamilton.
Paxson, J. W., Co., Philadelphia

Sand Blast Machinery.

Canadian Rand Drill Co., Montreal.
Chicago Pneumatic Tool Co., Chicago.
Detroit Foundry Supply Co., Windsor
Hamilton Facing Mill Co., Hamilton.
J. W. Paxson Co., Philadelphia, Pa.

Sand, Brass.

Buffalo Foundry Supply Co., Buffalo.
Detroit Foundry Supply Co., Windsor
Paxson, J. W., Co., Philadelphia

Sand Fire.

Detroit Foundry Supply Co., Detroit.
Paxson, J. W., Co., Philadelphia

Sand, Heavy, Grey Iron.

Buffalo Foundry Supply Co., Buffalo.
Detroit Foundry Supply Co., Windsor
Hamilton Facing Mill Co., Hamilton.
Paxson, J. W., Co., Philadelphia

Sand, Malleable.

Buffalo Foundry Supply Co., Buffalo,
Detroit Foundry Supply Co., Windsor
Hamilton Facing Mill Co., Hamilton.
Paxson, J. W., Co., Philadelphia

Sand, Medium Grey Iron.

Buffalo Foundry Supply Co., Buffalo.
Detroit Foundry Supply Co., Windsor
Hamilton Facing Mill Co., Hamilton.
Paxson, J. W., Co., Philadelphia

Sand, Plate.

Buffalo Foundry Supply Co., Buffalo.
Detroit Foundry Supply Co., Windsor
Hamilton Facing Mill Co., Hamilton.
Paxson, J. W., Co., Philadelphia

Sand Sifters.

Buffalo Foundry Supply Co., Buffalo.
Detroit Foundry Supply Co., Windsor
Dominion Foundry Supply Co., Montreal
Hamilton Facing Mill Co., Hamilton.
Hanna Engineering Works, Chicago.
Paxson, J. W., Co., Philadelphia

Saw Mill Machinery.

Allis-Chalmers-Bullock, Limited, Montreal
Baxter, Paterson & Co., Montreal
Canada Machinery Agency, Montreal.
Goldie & McCulloch Co., Galt.
Greey, Wm. & J. G., Toronto
Miller Bros. & Toms, Montreal
Owen Sound Iron Works Co., Owen Sound
H. W. Petrie, Toronto.
Robb Engineering Co., Amherst.
Standard Bearings, Ltd., Niagara Falls,
Waterous Engine Works, Brantford
Williams & Wilson, Montreal.

Sawing Machines, Metal.

Niles-Bement-Pond Co., New York.
Paxson, J. W., Co., Philadelphia

Saws, Hack.

Baxter Paterson & Co., Montreal.
Canada Machinery Agency, Montreal
Detroit Foundry Supply Co., Windsor
London Mach. Tool Co., Hamilton
Mechanics Supply Co., Quebec, Que.
Rice Lewis & Son, Toronto.
John Millen & Son, Ltd., Montreal.
L. S. Starrett Co., Athol, Mass.

Screw Cutting Tools

Hart Manufacturing Co., Cleveland, Ohio
Mechanics Supply Co., Quebec, Que.

Screw Machines, Automatic.

Canada Machinery Agency, Montreal.
Cleveland Automatic Machine Co., Cleveland, Ohio
London Mach. Tool Co., Hamilton, Ont.
National-Acme Mfg. Co., Cleveland.
Pratt & Whitney Co., Hartford, Conn.

Screw Machines, Hand.

Canada Machinery Agency, Montreal.
A. B. Jardine & Co., Hespeler.
London Mach. Tool Co., Hamilton, Ont.
Mechanics Supply Co., Quebec, Que.
Pratt & Whitney Co., Hartford, Conn.
Warner & Swasey Co., Cleveland, O.

Screw Machines, Multiple Spindle.

National-Acme Mfg. Co., Cleveland

Screw Plates.

Butterfield & Co., Rock Island, Que.
Hart Manufacturing Co. Cleveland, Ohio
A. B. Jardine & Co., Hespeler
Mechanics Supply Co., Quebec, Que.

Screws, Cap and Set.

National-Acme Mfg. Co., Cleveland

Screw Slotting Machinery, Semi-Automatic.

National-Acme Mfg. Co., Cleveland

Second-Hand Machinery.

American Tool Works Co., Cincinnati.
Canada Machinery Agency, Montreal.
The Canadian Fairbanks Co., Montreal.
Goldie & McCulloch Co., Galt.
Machinery Exchange, Montreal.
Niles-Bement-Pond Co., New York.
H. W. Petrie, Toronto.
Robb Engineering Co., Amherst, N S.
Waterous Engine Co., Brantford.
Williams & Wilson, Montreal.

Shafting.

Baxter, Paterson & Co., Montreal.
Canada Machinery Agency, Montreal.
The Canadian Fairbanks Co., Montreal.
The Goldie & McCulloch Co., Galt, Ont
Greey, Wm., & J. G. Toronto
Kennedy Wm., & Sons, Owen Sound
Niles-Bement-Pond Co., New York.
Owen Sound Iron Works Co., Owen Sound
Pendrith Machine Co., Toronto.
H. W. Petrie, Toronto.
Smart-Turner Machine Co., Hamilton.
Union Drawn Steel Co., Hamilton.
Waterous Engine Co., Brantford.

Shapers.

American Tool Works Co., Cincinnati.
John Bertram & Sons Co., Dundas, Ont
Canada Machinery Agency, Montreal.
The Canadian Fairbanks Co., Montreal.
London Mach. Tool Co., Hamilton, Ont.
Niles-Bement-Pond Co., New York.
H. W. Petrie, Toronto.
Potter & Johnston Machine Co., Pawtucket, R.I.
Pratt & Whitney Co., Hartford, Conn.
Williams & Wilson, Montreal.

Shearing Machine, Bar.

John Bertram & Sons Co., Dundas, Ont.
A. B. Jardine & Co., Hespeler.
London Mach. Tool Co., Hamilton, Ont.
Niles-Bement-Pond Co., New York.

Shears, Power.

John Bertram & Sons Co., Dundas, Ont.
Canada Machinery Agency, Montreal.
A. B. Jardine & Co., Hespeler, Ont.
Niles-Bement-Pond Co., New York.
Paxson, J. W., Co., Philadelphia

Sheet Metal Goods

Globe Machine & Stamping Co., Cleveland, Ohio.

Sheet Steel Work.

Owen Sound Iron Works Co., Owen Sound

Shingle Mill Machinery.

Owen Sound Iron Works Co., Owen Sound

Shop Trucks.

Greey, Wm. & J. G., Toronto
Owen Sound Iron Works Co., Owen Sound
Paxson, J. W., Co., Philadelphia

Shovels.

Baxter, Paterson & Co., Montreal.
Buffalo Foundry Supply Co., Buffalo.
Detroit Foundry Supply Co., Detroit.
Dominion Foundry Supply Co., Montreal
Hamilton Facing Mill Co., Hamilton.
Mechanics Supply Co., Quebec, Que.
Paxson, J. W., Co., Philadelphia

Shovels, Steam.

Allis-Chalmers-Bullock, Montreal.

Sieves.

Buffalo Foundry Supply Co., Buffalo.
Detroit Foundry Supply Co., Windsor
Dominion Foundry Supply Co., Montreal
Hamilton Facing Mill Co., Hamilton.
Paxson, J. W., Co., Philadelphia

Silver Lead.

Buffalo Foundry Supply Co., Buffalo.
Detroit Foundry Supply Co., Windsor
Dominion Foundry Supply Co., Montreal
Hamilton Facing Mill Co., Hamilton.
Paxson, J. W., Co., Philadelphia

Sleeves, Reducing.

Chicago Pneumatic Tool Co., Chicago.

Snap Flasks

Buffalo Foundry Supply Co., Buffalo.
Detroit Foundry Supply Co., Windsor
Dominion Foundry Supply Co., Montre
Hamilton Facing Mill Co., Hamilton.
Paxson, J. W., Co., Philadelphia

Soapstone.

Buffalo Foundry Supply Co., Buffalo.
Detroit Foundry Supply Co., Windsor
Dominion Foundry Supply Co., Montreal
Hamilton Facing Mill Co., Hamilton.
Paxson, J. W., Co., Philadelphia

Solders.

Lumen Bearing Co., Toronto

Special Machinery.

W. H. Banfield & Sons, Toronto.
Baxter, Paterson & Co., Montreal
John Bertram & Sons Co. Dundas, Ont
Globe Machine & Stamping Co., Cleveland Ohio.
Greey, Wm. & J. G., Toronto
Hanna Engineering Works, Chicago.
Laurie Engine & Machine Co., Montreal.
London Mach. Tool Co., Hamilton. Ont.
R. McDougall Co., Galt, Ont.
Pendrith Machinery Co., Toronto.
H. W. Petrie, Toronto.
The Smart-Turner Mach. Co., Hamilton
Standard Bearings, Ltd., Niagara Falls,
Waterous Engine Co., Brantford.

Special Machines and Tools.

Paxson, J. W., Philadelphia
Pratt & Whitney, Hartford, Conn.
Standard Bearings, Ltd., Niagra Falls,

Special Milled Work.

National-Acme Mfg. Co., Cleveland

Speed Changing Countershafts.

The Canadian Fairbanks Co., Montreal

Spike Machines.

National Machinery Co., Tiffin, O.
The Smart-Turner Mach. Co., Hamilton

Spray Cans.

Detroit Foundry Supply Co., Windsor
Dominion Foundry Supply Co., Montrea
Hamilton Facing Mill Co., Hamilton.
Paxson, J. W., Co., Philadelphia

Springs, Automobile.

Cleveland Wire Spring Co., Cleveland

Springs, Coiled Wire.

Cleveland Wire Spring Co., Cleveland

Springs, Machinery.

Cleveland Wire Spring Co., Cleveland

Springs, Upholstery.

Cleveland Wire Spring Co., Cleveland

Sprue Cutters.

Detroit Foundry Supply Co., Windsor
Dominion Foundry Supply Co., Montreal
Hamilton Facing Mill Co., Hamilton.
Paxson, J. W., Co., Philadelphia

Stamp Mills.

Allis-Chalmers-Bullock, Limited, Montrea
Paxson, J. W., Co., Philadelphia

Steam Hot Blast Apparatus

Dominion Heating & Ventilating Co. Hespeler, Ont.
Sheldon's Limited, Galt.
B. F. Sturtevant Co., Hyde Park, Mass.

Steam Separators.

Darling Bros., Ltd., Montreal.
Dominion Heating & Ventilating Co., Hespeler
Robb Engineering Co., Montreal.
Sheldon's Limited, Galt.
Smart-Turner Mach. Co., Hamilton
Waterous Engine Co., Brantford.

Steam Specialties.

Darling Bros., Ltd., Montreal
Dominion Heating & Ventilating Co., Hespeler
Sheldon's Limited, Galt.

Steam Traps.

Canada Machinery Agency, Montreal.
Darling Bros., Ltd., Montreal
Dominion Heating & Ventilating Co., Hespeler.
Mechanics Supply Co., Quebec, Que.
Pendrith Machinery Co., Toronto.
Sheldon's Limited, Galt.
B. F. Sturtevant Co., Hyde Park, Mass.

Steam Valves.

Darling Bros., Ltd., Montreal

Steel Pressure Blowers

Buffalo Foundry Supply Co., Buffalo.
Dominion Foundry Supply Co., Montreal
Domininion Heating & Ventilating Co., Hespeler
Hamilton Facing Mill Co., Hamilton.
J. W. Paxon Co., Philadelphia, Pa.
Sheldon's Limited, Galt.
B. F. SturtevantCo., Hyde Park, Mass.

Steel Tubes.

Baxter, Paterson & Co., Montreal.
Mechanics Supply Co., Quebec, Que.
John Millen & Son, Montreal.

Our Triple Hem
Steel Shop Barrels and Boxes
Pay for themselves

Write us for catalogue and prices.

High Grade Coiled Wire Springs for all purposes

The Cleveland Wire Spring Co. Cleveland, Ohio.

The Globe Tilting Tumbling Barrel

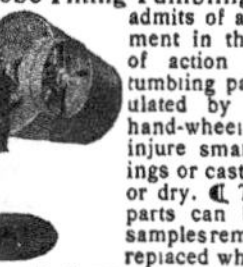

admits of an adjustment in the degree of action on the tumbling parts—regulated by the side hand-wheel. Doesn't injure small stampings or castings, wet or dry. ℂ Tumbling parts can be seen, samples removed and replaced while running; and easily dumped into a small receptacle. ℂ Send for Booklet E.

The Globe Machine & Stamping Co.
Hamilton Avenue, Cleveland, O.

W. JACKMAN & Co., Caxton House, Westminster, London, Eng.

CHADWICK BROS.
BRASS FOUNDERS

Manufacturers of
Cast, Spun and Sheet Metal Goods; Brass, Copper and Aluminum Castings; Electro Plating;
Brass and Nickel Railings; Brass Fittings of all kinds
Engineers' and Plumbers' Brass Goods;
All kinds of Brass Turnings; Bolts, Nuts, Screws, etc.
Brass and Metal Goods of every description made to order;
Gas and Gasoline Engine Supplies.

CHADWICK BROS
Factory: HAMILTON, ONT. (Oak Ave.)

Wax Wire for Core Vent

The cheapest and most perfect Core Vent ever devised. Leaves the vent hole clear, clean and unobstructed. Perfectly flexible under all temperatures. Sizes from one and one-half to nine millimeter in diameter. Put up in 1-lb. packages, continuous lengths. Cost per foot, one-fifth of a cent up. Samples sent on application.

For highest quality of Foundry Facings and Supplies, write us

THE HAMILTON FACING MILL CO., Limited
FOUNDRY OUTFITTERS
HAMILTON, Ontario. MONTREAL, Quebec.

Economy Steel Wire Wheel Brush

Operated on an ordinary Emery Grinder

The use of an "Economy" means saving of time and money and a better finished product in almost every plant where metal articles are made.

One man with an Economy Wheel Brush will do more and better work than five men by other methods.

Patented June 5th, 1906

Sectional View, showing construction

The Hub is made of malleable iron and consists of a pair of face plates with connecting sleeve nut in the centre. The wire part, or brush proper, is made in sections, each section being held in position by a malleable dividing ring.
When brush part is worn out, new sections can be substituted for the worn-out ones by simply unscrewing the sleeve nut.

THE OSBORN MANUFACTURING CO.
CLEVELAND, O.

Delays are Dangerous

They often read serious financial loss. You cannot afford to place your orders for

PIPE, CORE, STOVE PLATE and MACHINERY MOLDING SAND and LAKE ROOFING GRAVEL

with any old concern. Sometimes you want your sand orders filled mighty quick. Send for prices and samples. Quick deliveries a specialty.

T. W. BARNES, HAMILTON, ONT.

Steel, High Speed.
Wm. Abbott, Montreal.
Canadian Fairbanks Co., Montreal.
Alex. Gibb, Montreal.
Jessop, Wm., & Sons, Sheffield, Eng.
B. K. Morton Co., Sheffield, Eng.
Williams & Wilson, Montreal.

Stocks and Dies
Hart Manufacturing Co., Cleveland, Ohio
Mechanics Supply Co., Quebec, Que.
Pratt & Whitney Co., Hartford, Conn.

Stone Cutting Tools, Pneumatic
Allis-Chalmers-Bullock, Ltd., Montreal
Canadian Rand Drill Co., Montreal.

Stone Surfacers
Chicago Pneumatic Tool Co., Chicago.

Stove Plate Facings.
Buffalo Foundry Supply Co., Buffalo
Detroit Foundry Supply Co., Windsor.
Dominion Foundry Supply Co., Montreal
Hamilton Facing Mill Co., Hamilton.
Paxson, J. W., Co., Philadelphia.

Swage, Block.
A. B. Jardine & Co., Hespeler, Ont.

Switchboards.
Allis-Chalmers-Bullock, Limited, Montreal
Canadian General Electric Co., Toronto.
Canadian Westinghouse Co., Hamilton
Hall Engineering Works, Montreal, Que.
Mechanics Supply Co., Quebec, Que.
Toronto and Hamilton Electric Co., Hamilton.

Talc.
Buffalo Foundry Supply Co., Buffalo.
Detroit Foundry Supply Co., Windsor.
Hamilton Facing Mill Co., Hamilton.
Paxson, J. W., Co., Philadelphia.

Tapping Machines and Attachments.
American Tool Works Co., Cincinnati.
John Bertram & Sons Co., Dundas, Ont.
Bickford Drill & Tool Co., Cincinnati.
The Geometric Tool Co., New Haven.
A. B. Jardine & Co., Hespeler.
London Mach. Tool Co., Hamilton, Ont.
Murchey Machine & Tool Co., Detroit.
Niles-Bement-Pond Co., New York.
H. W. Petrie, Toronto.
Pratt & Whitney, Cincinnati, O.
L. S. Starrett Co., Athol, Mass.
Williams & Wilson, Montreal.

Tapes, Steel.
Rice Lewis & Son, Toronto.
Mechanics Supply Co., Quebec, Que.
John Millen & Son, Ltd., Montreal.
L. S. Starrett Co., Athol, Mass.

Taps.
Baxter, Paterson & Co., Montreal.
Mechanics Supply Co., Quebec, Que.

Taps, Collapsing.
The Geometric Tool Co., New Haven.

Taps and Dies.
Wm. Abbott, Montreal.
Baxter, Paterson & Co., Montreal.
Butterfield & Co., Rock Island, Que.
The Geometric Tool Co., New Haven.
A. B. Jardine & Co., Hespeler, Ont.
Rice Lewis & Son, Toronto.
Mechanics Supply Co., Quebec, Que.
John Millen & Son, Ltd., Montreal
Pratt & Whitney Co., Hartford, Conn.
Standard Tool Co., Cleveland.
L. S. Starrett Co., Athol, Mass.

Testing Laboratory.
Detroit Testing Laboratory, Detroit

Testing Machines.
Detroit Foundry Supply Co., Windsor.
Dominion Foundry Supply Co., Montreal
Hamilton Facing Mill Co., Hamilton.
Paxson, J. W., Co., Philadelphia

Thread Cutting Tools.
Hart Manufacturing Co., Cleveland, Ohio
Mechanics Supply Co., Quebec, Que.

Tiling, Opal Glass.
Toronto Plate Glass Importing Co., Toronto.

Time Switches, Automatic, Electric.
Berlin Electric Mfg. Co., Toronto.

Tinware Machinery
Canada Machinery Agency, Montreal.

Tire Upsetters or Shrinkers
A. B. Jardine & Co., Hespeler, Ont.

Tool Cutting Machinery
Canadian Rand Drill Co., Montreal.

Tool Holders.
Armstrong Bros. Tool Co., Chicago.
Baxter, Paterson & Co., Montreal.
Fairbanks Co., Springfield, Ohio.
John Millen & Son, Ltd., Montreal.
H. W. Petrie, Toronto.
Pratt & Whitney Co., Hartford, Conn.

Tool Steel.
Wm. Abbott, Montreal.
Wm. Jessop, Sons & Co., Toronto.
Canadian Fairbanks Co., Montreal
B. K. Morton & Co., Sheffield, Eng.
Williams & Wilson, Montreal.

Torches, Steel.
Baxter, Paterson & Co., Montreal.
Detroit Foundry Supply Co., Windsor.
Dominion Foundry Supply Co., Montreal
Hamilton Facing Mill Co., Hamilton.
Paxson, J. W., Co., Philadelphia

Transformers and Convertors
Allis-Chalmers-Bullock, Montreal.
Canadian General Electric Co., Toronto.
Canadian Westinghouse Co., Hamilton.
Hall Engineering Works, Montreal, Que.
T. & H. Electric Co., Hamilton.

Transmission Machinery
Allis-Chalmers-Bullock, Montreal.
The Canadian Fairbanks Co., Montreal.
Greey, Wm. & J. G., Toronto.
Laurie Engine & Machine Co., Montreal.
Link-Belt Co., Philadelphia.
Miller Bros. & Toms Montreal.
Paxson, J. W., Co., Philadelphia
H. W. Petrie, Toronto.
The Smart-Turner Mach. Co., Hamilton.
Waterous Engine Co., Brantford.

Transmission Supplies
Baxter, Paterson & Co., Montreal.
The Canadian Fairbanks Co., Montreal.
Wm. & J. G. Greey, Toronto.
The Goldie & McCulloch Co., Galt.
Miller Bros. & Toms, Montreal
Pendrith Machinery Co., Toronto.
H. W. Petrie, Toronto.

Trolleys
Canadian Rand Drill Co., Montreal.
John Millen & Son, Ltd., Montreal.
Miller Bros. & Toms, Montreal.
Northern Engineering Works Detroit

Trolley Wheels.
Lumen Bearing Co., Toronto

Trucks, Dryer and Factory
Dominion Heating & Ventilating Co., Hespeler
Greey, Wm. & J. G., Toronto.
Northern Engineering Works, Detroit
Sheldon's Limited, Galt, Ont.

Tube Expanders (Rollers)
Chicago Pneumatic Tool Co., Chicago.
A. B. Jardine & Co., Hespeler.
Mechanics Supply Co., Quebec, Que.

Turbines, Steam
Allis-Chalmers-Bullock, Limited, Montreal
Canadian General Electric Co., Toronto.
Canadian Westinghouse Co., Hamilton.
D'Olier Engineering Co., New York.
Kerr Turbine Co., Wellsville, N.Y.

Turntables
Detroit Foundry Supply Co., Windsor.
Dominion Foundry Supply Co., Montreal
Hamilton Facing Mill Co., Hamilton.
Northern Engineering Works, Detroit
Paxson, J. W., Co., Philadelphia

Turret Machines.
American Tool Works Co., Cincinnati.
John Bertram & Sons Co., Dundas, Ont.
The Canadian Fairbanks Co., Montreal.
London Mach. Tool Co., Hamilton, Ont.
Niles-Bement-Pond Co., New York.
H. W. Petrie, Toronto.
Pratt & Whitney Co., Hartford, Conn.
Warner & Swasey Co., Cleveland, Ohio.
Williams & Wilson, Montreal.

Upsetting and Bending Machinery.
John Bertram & Sons Co., Dundas, Ont.
A. B. Jardine & Co., Hespeler.
London Mach. Tool Co., Hamilton, Ont.

National Machinery Co., Tiffin, O.
Niles-Bement-Pond Co., New York.

Valves, Blow-off.
Chicago Pneumatic Tool Co., Chicago.
Mechanics Supply Co., Quebec, Que.

Valves, Back Pressure.
Darling Bros., Ltd., Montreal
Dominion Heating & Ventilating Co. Hespeler
Mechanics Supply Co., Quebec, Que.
Sheldon's Limited, Galt.

Valve Reseating Machines.
Darling Bros., Ltd., Montreal

Ventilating Apparatus.
Canada Machinery Agency, Montreal
Darling Bros., Ltd., Montreal
Dominion Heating & Ventilating Co., Hespeler, Ont.
Sheldon & Sheldon, Galt., Ont.
B. F. Sturtevant Co., Hyde Park, Mass.

Vises, Planer and Shaper.
American Tool Works Co., Cincinnati, O.
A. B. Jardine & Co., Hespeler, Ont.
John Millen & Son, Ltd., Montreal.
Niles-Bement-Pond Co., New York.

Washer Machines.
National Machinery Co., Tiffin, Ohio.

Water Tanks.
Ontario Wind Engine & Pump Co., Toronto

Water Wheels.
Allis-Chalmers-Bullock Co., Montreal.
Barber, Chas., & Sons, Meaford.
Canada Machinery Agency, Montreal.
The Goldie & McCulloch Co., Galt, Ont.
Greey, Wm. & J. G., Toronto.
Wm. Kennedy & Sons, Owen Sound.

Water Wheels, Turbine.
Barber, Chas., & Sons, Meaford, Ont.
Kennedy, Wm., & Sons, Owen Sound

Wheelbarrows.
Baxter, Paterson & Co., Montreal.
Buffalo Foundry Supply Co., Buffalo
Detroit Foundry Supply Co., Windsor.
Dominion Foundry Supply Co., Montreal
Hamilton Facing Mill Co., Hamilton.
Paxson, J. W., Co., Philadelphia

Winches, Hand.
Pendrith Machinery Co., Toronto.

Wind Mills.
Ontario Wind Engine & Pump Co., Toronto

Window Wire Guards.
Expanded Metal and Fireproofing Co Toronto.
B. Greening Wire Co., Hamilton, Ont.
Paxson, J. W., Co. Philadelphia

ALPHABETICAL INDEX.

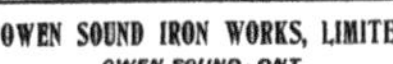

OWEN SOUND IRON WORKS, LIMITED
OWEN SOUND, ONT.

Cement Mill Machinery, Boiler and Steel Tank Work of all kinds, Grey Iron and Brass Castings

WM. KENNEDY & SONS

HIGH-CLASS
Steel Gears of all kinds
Iron Gears
Grey Iron, Steel and
Brass Castings

MARINE WORK receives every attention

WM. KENNEDY & SONS, Limited
Owen Sound, Ont.

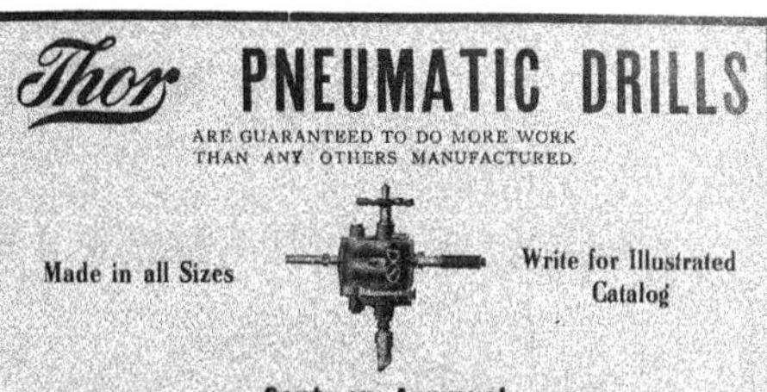

Sent on Approval

Transportation charges paid both ways if they do not prove satisfactory

Independent Pneumatic Tool Co.

First National Bank Building
CHICAGO

170 Broadway
NEW YORK CITY

MALLEABLE IRON CASTINGS

WE ALSO MAKE METAL PATTERNS

Galt Malleable Iron Co., Limited - Galt, Ontario

Directory of Consulting Engineers, Patent Attorneys, Architects and Contractors.

JOHN S. FIELDING
Mem. Soc. C.E., West Penn., '87
Consulting Engineer
DAMS, MILLS, BRIDGES, MACHINERY
Room 2, 15 Toronto Street, Toronto, Ont.

HANBURY A. BUDDEN
Advocate Patent Agent.
New York Life Building MONTREAL.
Cable Address, BREVET MONTREAL.

CONSULTING ENGINEERS
should have their card in this page. It will be read by the manufacturers of Canada :: :: :: ::
CANADIAN MACHINERY
Montreal. Toronto. Winnipeg.

T. Pringle & Son
HYDRAULIC, MILL & ELECTRICAL ENGINEERS
FACTORY & MILL CONSTRUCTION A SPECIALTY.
Coristine Bldg., St. Nicholas St., Montreal.

RODERICK J. PARKE
A.M. Can. Soc. C.E. A.M. Amer. Inst. E.E.
CONSULTING ELECTRICAL ENGINEER
INDUSTRIAL STEAM AND ELECTRIC POWER PLANTS DESIGNED. TESTS. REPORTS.
51-53 JANES BLDG., TORONTO, CAN
Long Distance Telephones—Office and Residence.

J. A. DeCew
Chemical Engineer
Sun Life Building, MONTREAL
Industrial Plants and Processes sold and installed.
Apparatus and Materials for Water Purification.
Free Analyses to prospective purchasers.

PLEWS & TRIMINGHAM.
Consulting Engineers.
ELECTRICAL INSPECTION BUREAU · AND TESTING LABORATORY.
40 HOSPITAL & 22 ST. JOHN STS. MONTREAL.

JULES DE CLERCY, M. E. Gas Engineer
Expert knowledge of all classes of Engines and Producer Plants. If you are thinking of installing a Gas Engine or Producer Plant write me. If your plant is not working satisfactorily I can help you.
413 Dorchester St. - MONTREAL, QUE.

CHARLES BRANDEIS,
A. M. CAN. SOC. C.E.
MEM. AMER. ELECTRO-CHEMICAL SOC., ETC.
CONSULTING ENGINEER
Estimates, Plans and Supervision of Hydraulic and Steam-Electric Light, Power and Railroad Plants, Specifications, Reports, Switchboard Designs, Complete Factory Installations, Electric Equipment of Mines and Electro-Chemical Plants.
Long Distance Telephone, Main 3256.
Cable Address, Brandeis, Montreal.
W. U. Code Univ-Edition.
62-63 Guardian Building MONTREAL

PATENTS *TRADE MARKS AND DESIGNS*
PROCURED IN ALL COUNTRIES
Special Attention given to Patent Litigation
Pamphlet sent free on application.
RIDOUT & MAYBEE *103 BAY STREET TORONTO*

FETHERSTONHAUGH & BLACKMORE
PATENTS
TRADE MARKS.
Liverpool & London & Globe Building. Montreal.
Fetherstonhaugh & Dennison, 9 Toronto St. Toronto. Our reference book for clients "The Prospective Patentee." Also Ottawa, Washington and Winnipeg.

PATENTS PROMPTLY SECURED
We solicit the business of Manufacturers, Engineers and others who realize the advisability of having their Patent business transacted by Experts. Preliminary advice free. Charges moderate. Our Inventor's Adviser sent upon request. Marion & Marion, New York Life Bldg, Montreal; and Washington, D.C., U.S.A.

PATENTS THAT PROTECT
FETHERSTONHAUGH & CO.
Patent Solicitors & Experts
Fred. B. Fetherstonhaugh, M.E., Barrister-at-law and Counsel and expert in Patent Causes. Charles W. Taylor, B.Sc., formerly Examiner in Can. Patent Office.
MONTREAL CAN. LIFE BLDG.
TORONTO HEAD OFFICE, CAN. BANK COMMERCE BLDG.

OPAL GLASS TILING
FOR WALLS OF
MACHINERY AND POWER HOUSES
Most approved material.
TORONTO PLATE GLASS IMPORTING CO'Y
PLATE AND WINDOW GLASS
135 to 143 Victoria St., - Toronto

Buffing and Polishing Lathes
PENDRITH MACHINERY CO.
74 W. Adelaide Street, Toronto

THE ELECTRIC BLUE PRINT CO.
Blue Prints Positive Blue Prints
Black Prints Multi-color Prints
Largest and finest plant.
Best finished work in Canada at lowest prices.
40 Hospital St., - MONTREAL

THE DETROIT TESTING LABORATORY
1111 Union Trust Building
DETROIT, MICH.
Experienced Foundry Chemists and Metallurgists.
Iron Mixtures for all classes of Castings our Specialty.
Reasonable charges for Analyses.

1906
The Thirty-Third Year in the History of the
Remington Typewriter
Shows an Increase in Output of
31%
Over the Previous Year

Typewriters come and typewriters go
But the Remington runs on forever

Remington Typewriter Company

Are you interested in any of the lines that are advertised?
A Post Card will bring you price list and full information.
Don't forget to mention Canadian Machinery.

An intelligent Engineer uses nothing but the **GENUINE PENBERTHY INJECTOR.** He has had experience with just-as-good kind before.

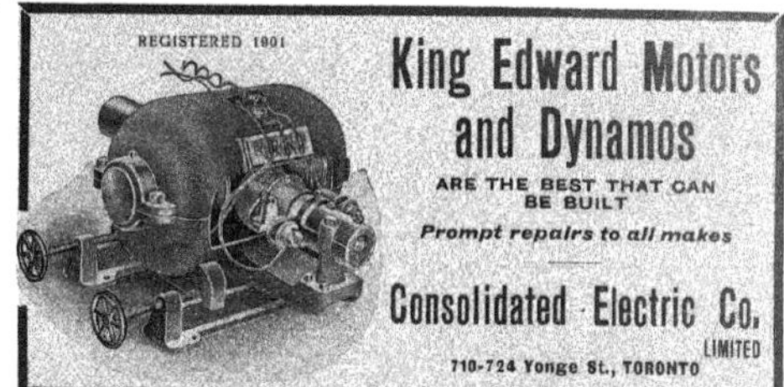

REGISTERED 1901
King Edward Motors and Dynamos
ARE THE BEST THAT CAN BE BUILT
Prompt repairs to all makes
Consolidated Electric Co.
LIMITED
710-724 Yonge St., TORONTO

The Engineering Times

A Weekly Newspaper of Commercial Engineering

32 PAGES

Price Twopence
Annual Subscription, 10s. 10d.

CHIEF OFFICE:
6, Bouverie St., London, E.C.

SCOTLAND:
Mr. Alex. Forbes
118a, Renfield St., Glasgow

"BEST" DRY BATTERIES

have no equal in their line

FOR IGNITION WORK

A single set will prove it.
Have you tried them?

SPACE 93 AT SPORTSMAN'S SHOW
APRIL 6th to 13th

For sale by all leading dealers.

The Berlin Electrical Mfg. Co., Limited
474 King Street West,
TORONTO

ALTERNATE CURRENT DYNAMOS and MOTORS

FOR ALL CIRCUITS

TORONTO & HAMILTON ELECTRIC CO.
99-101 McNab Street, HAMILTON, ONT.

The Eugene F. Phillips Electrical Works, Limited

Montreal, Canada

TORONTO STORE - - - 67 ADELAIDE STREET EAST

Magnet and Trolley Wire

Bare Copper Wire Weatherproof Wire

NEW AND UP-TO-DATE

BOLT AND NUT MACHINERY

INCLUDING

Bolt Cutters, Nut Tappers, Bolt Headers, Upsetting and Forging Machines, Wire Nail and Spike Machines and Bulldozers.

Send for Catalogue M.

NATIONAL MACHINERY CO., Tiffin, Ohio, U.S.A.

Canadian Agents: H. W. PETRIE, Toronto, Ont. WILLIAMS & WILSON, Montreal, Que.

THE

ELECTRICAL TIMES,

"LIGHTNING"

5 Breams Buildings, Chancery Lane

LONDON, ENGLAND

The Organ of the

ELECTRICAL

and Allied Industries

Circulates All Over the World

It is a Sixpenny Paper for Twopence; and is more widely read than any other British Electrical Journal. Means business all the time.

Annual Subscription:

Post Free, 18/6, Foreign; 10/6, Home. Published weekly at 2d.

TAPS AND DIES

The best material, the greatest care exercised in every process of manufacture, guided by long experience, result in the best possible Taps and Dies.

It Pays to Use the Best Tools

A. B. JARDINE & CO., Hespeler, Ont.

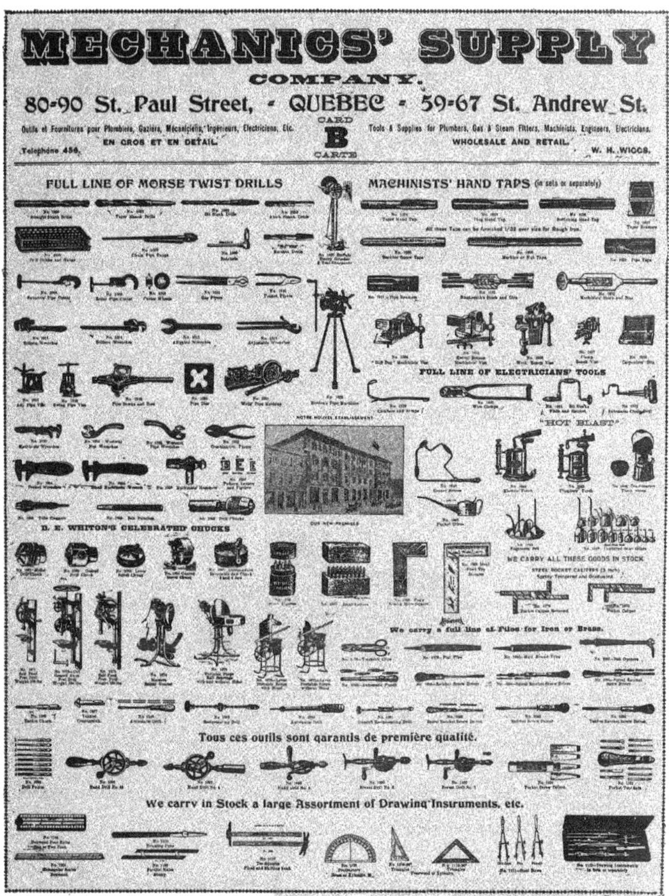
MECHANICS' SUPPLY
COMPANY.
80-90 St. Paul Street, - QUEBEC - 59-67 St. Andrew St.
Outils et Fournitures pour Plombiers, Gaziers, Mécaniciens, Ingénieurs, Electriciens, Etc.
EN GROS ET EN DETAIL.
CARD
B
CARTE
Tools & Supplies for Plumbers, Gas & Steam Fitters, Machinists, Engineers, Electricians.
WHOLESALE AND RETAIL.
Telephone 456.
W. H. WIGGS.
FULL LINE OF MORSE TWIST DRILLS
MACHINISTS' HAND TAPS (in sets or separately)
FULL LINE OF ELECTRICIANS' TOOLS
"HOT BLAST"
D. E. WHITON'S CELEBRATED CHUCKS
WE CARRY ALL THESE GOODS IN STOCK
We carry a full line of Files for Iron or Brass.
Tous ces outils sont garantis de première qualité.
We carry in Stock a large Assortment of Drawing Instruments, etc.
Be sure to give Figure and Card Number when ordering.
CARD "B"

Reamers, Taps and Dies
Screw Plates
We are now HEADQUARTERS in Canada for REAMERS
We have installed the necessary machinery for making all kinds of Reamers. The standard of quality will be kept as high as we have set for our TAPS, DIES and SCREW PLATES.
Best of Methods and Materials
BUTTERFIELD & CO., Rock Island, P.Q.

THE STANDARD TOOL CO'S
STANDARD TOOL CO.
Milling Cutters
Made of either
Carbon or High Speed Steel
Formed cutters, also special cutters with inserted Blades. Tell us your needs.
OFFICE and FACTORY
Cleveland, Ohio
EASTERN SALESROOM
94 Reade St., New York

Armstrong Tool Holders
Have saved years of time and thousands of tons of Tool Steel, here's TWO NEW THINGS will prove an equally good investment.
DROP FORGED OF STEEL
THE ARMSTRONG IMPROVED TOOL POST
Combines the strength and holding power of the strap and stud tool clamp with the convenience of the "open side" and ordinary set screw tool post. A slight turn of one nut releases tool or sets it solidly.
No trouble with "stripping" or "upset" screws.
ARMSTRONG IMPROVED TOOL POST
The ARMSTRONG THREE-BAR BORING TOOL
INCREASES PRODUCTION AND REDUCES COST OF TOOL MAINTENANCE
CUTTERS CANNOT JAR LOOSE
HIGH SPEEDS AND BIG FEEDS
ONLY SET THEM TIGHTER
SOLD IN CANADA BY:—Williams & Wilson, Montreal; Aikenhead Hardware, Limited, Toronto; A. R. Williams Machinery Co., Toronto, Winnipeg, Vancouver; Wood, Vallance & Co., Hamilton; Frothingham & Workman, Limited, Montreal; H. W. Petrie, Toronto.
WRITE FOR OUR LATEST CATALOG.
ARMSTRONG BROS. TOOL CO., 106 N. FRANCISCO AVE. CHICAGO, U.S.A.
"THE TOOL HOLDER PEOPLE"

ALLIS-CHALMERS-Bullock-LIMITED MONTREAL
MINING PLANT
ALLIS-CHALMERS-Bullock-LIMITED MONTREAL
This is a View of Three out of Ten of our Ingersoll Steam Drills Engaged in Rock Excavation. For full particulars of these Drills see Catalogue 81
AIR Compressors, Coal Cutters, Concentrating and Cyanide Plants, Engines, Gold and Silver Mills, Gold Dredges, Gyratory and Jaw Crushers, Hoisting Engines, Huntington Mills, Prospecting Mills, Pumps, Rock and Ore Breakers, Rock Drills, Roasting and Smelting Furnaces, Stamp Mills, Tube Mills, Steam Turbines, Turbine Water Wheels, Electric Generators and Motors.
ALLIS-CHALMERS-BULLOCK, Limited
Head Office and Works, Montreal; District Offices: Toronto, Traders Bank Bldg; Montreal, Sovereign Bank Bldg; Winnipeg, 251 Notre Dame Ave.; Nelson, Josephine St.; Vancouver, 416 Seymour St.; New Glasgow, N.S., Telephone Bldg.

YOU WILL SAVE MONEY IN BUYING IF YOU REFER TO OUR BUYERS' DIRECTORY

CANADIAN MACHINERY AND MANUFACTURING NEWS

A monthly newspaper devoted to the manufacturing interests, covering in a practical manner the mechanical, power, foundry and allied fields. Published by The MacLean Publishing Company, Limited, Toronto, Montreal, Winnipeg, and London, Eng.

OFFICE OF PUBLICATION: 10 FRONT STREET EAST, TORONTO

Vol. III. MAY, 1907 No. 5

Lincoln Milling Machines,

Thread Milling Machines

Double Column Milling Machines.

Automatic Screw Machines,

Hand Screw Machines

Die-Sinking Machines,

Gang Drilling Machines

Adjustable Multi-Spindle Drills.

Cutting-off Machines,

Turret Lathes,

Engine Lathes,

Bench Lathes,

Gear Cutters,

Profiling Machines

Rifling Machines

Gun-Barrel and Tube Drilling Machines,

Centering Machines.

Pratt & Whitney Four-Spindle Sensitive Drill

We build Sensitive Drills with one, two, three or four spindles for drilling holes up to 5/16-inch in Diameter. Vertical movement of spindles by hand lever. Three speeds with clamp stops, which are instantly adjustable. Thrust of spindles received on ball-bearings. Spindles and posts counterbalanced by coiled springs, which are in cases.

Quick Delivery on Some Sizes.

Pratt & Whitney Co., HARTFORD, Conn., U.S.A.

AGENTS FOR CANADA

The Canadian Fairbanks Co., Limited

MONTREAL TORONTO WINNIPEG VANCOUVER

26x42x12-ft. Gap Lathe

27-in. Double Back Geared Engine Lathe

We furnish complete equipments for Railway, Locomotive and General Machine Shops. We solicit your valued inquiries for anything in the following lines:

GAP LATHES.	PUNCH AND SHEARING MACHINES.	AXLE LATHES.
ENGINE LATHES.	UNIVERSAL RADIAL DRILLS.	PRESSES.
PLANERS.	PLAIN RADIAL DRILLS.	BULL-DOZERS.
SHAPERS.	PLAIN MILLING MACHINES	STEAM HAMMERS.
HOR. BORING MILLS.	UNIVERSAL MILLING MACHINES.	DROP HAMMERS.
VERT. BORING MILLS.	SLAB MILLING MACHINES.	CAR WHEEL BORERS.

UPRIGHT DRILLS. FORGING MACHINERY.
DRIVING WHEEL LATHES. HYDROSTATIC WHEEL PRESSES.

24x24x6-ft. Iron Planer

20-in. Travelling Head Shaper

London Machine Tool Co., Limited, Hamilton, Canada

Sales Office, Room 1206 Traders Bank Building, Toronto, Canada

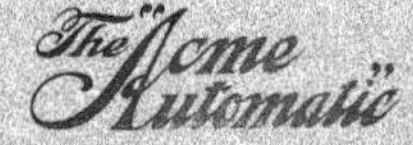

Multiple Spindle Screw Machine

Operates on FOUR bars at a time. Performs EIGHT OR MORE operations simultaneously.

Produces a finished piece in the time required for the longest single operation or less.

Uses but ONE set of tools.

Occupies ONLY the same floor space as ANY screw machine.

Requires ONLY the same attention as a single spindle machine.

Write for new booklet

Built in sizes from [illegible]-inch to 2[illegible]-inch capacity

Will increase your output—Will save you money

The National-Acme Mfg. Co.

Cleveland, - Ohio

Branch Offices—
NEW YORK
CHICAGO

General Foreign Representatives—
SCHUCHARDT & SCHUTTE
ALFRED H. SCHUTTE

IF YOU USE GEARS

INSIST ON HAVING THEM CUT ON THE

Automatic Gear Hobbing Machine

50 %

OF POWER YOU WASTE NOW SAVED

HOW?

ALL gears cut on the generating principle have perfect teeth.

IF YOU CUT GEARS

YOU CANNOT AFFORD TO USE ANY OTHER MACHINE THAN THE

Automatic Gear Hobbing Machine

50 %

OF TIME SAVED

HOW ?

The cutting is a continuous performance; no indexing, stopping or starting.

It is the only machine on the generating principle on the market.

BAXTER, PATERSON & CO.

TEL. M 847 *102 St. Antoine Street,* - *MONTREAL*

18,500 Automatic Screw Machines

Made and sold in fourteen years—and doing the work for which they were designed—turning out duplicate parts for the purchasers at a saving of from 200 to 500 per cent.

¶ Why don't you get in line—increase your output and effect a saving on every piece?

¶ It won't cost you much to mail us your blueprints or parts and let us tell you what our machines will save you.

Whether from Bar or Castings the "CLEVELAND" will do it for you.

Piece No. 227. Material, Machine Steel Bar, made on a 3¼-in. Cleveland Automatic Turret Machine, drawing full size. Output per hour, 4; this means all operations for completing this piece. Actual cost of labor, from 1½ to 2 cents for each piece. Send along samples or drawings. We guarantee our outputs. One man can operate from 4 to 6 machines.

Piece No. 260. Material, Machine Steel Bar, made on a ⅜-in Cleveland Automatic Plain Machine, drawing full size. Output per hour, 120; this means all operations for completing this piece. Actual cost of labor, from 1¼ to ⅔ mills for each piece. Send along sample or drawings. We guarantee our outputs. One man can operate from 4 to 6 inches.

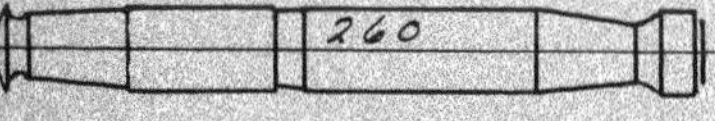

AS WE HAVE NO CANADIAN REPRESENTATIVE, WE SOLICIT CORRESPONDENCE AND ORDERS DIRECT

CLEVELAND AUTOMATIC MACHINE CO.

CLEVELAND, OHIO, U.S.A.

EASTERN REPRESENTATIVE—J. B. ANDERSON, 2450 North Thirtieth Street, Philadelphia, Pa.
WESTERN REPRESENTATIVE—H. E. NUNN, 22 Fifth Avenue, Chicago, Ill.
FOREIGN REPRESENTATIVES—CHAS. CHURCHILL & CO., London, Manchester, Birmingham, Newcastle-on-Tyne and Glasgow. MESSRS. SCHUCHARDT & SCHUTTE, Berlin, Vienna, St. Petersburg and Stockholm. ALFRED H. SCHUTTE, Cologne, Brussels, Liege, Paris, Milan and Bilbao.

MURCHEY NIPPLE AND PIPE THREADING MACHINES

Our No. 3 Machine

are four times as rapid as the ordinary old style pipe machines and the most efficient and convenient tools of their class. Thread and ream at one operation. Automatic dies insure perfect threads without any attention after the work is once started. Specially designed gripping chuck jaws. Improved nipple holders. With lead screw and automatic trip. Capacity, 700-4 threads complete in 10 hours.

WRITE FOR LATEST CATALOGUE

Murchey Machine and Tool Co., Fourth and Porter Sts. DETROIT, MICH.

Sole Canadian Agents—LONDON MACHINE TOOL CO., 1206 Traders Bank Building, TORONTO

SUPERIOR

POWER PRESSES

Many new features embodied in this press. Write for prices and particulars on presses, dies, tools, and special machinery.

W. H. BANFIELD & SONS

120 Adelaide Street West, TORONTO.

NORTON GRINDING WHEELS

MADE OF

ALUNDUM

Meet the Requirements for ALL Kinds of Grinding.

¶ There is no investment that will pay a larger percentage of profit than a modern, up-to-date Grinding Department equipped with Norton Grinding Wheels and Grinding Machinery. Years of special study have given us the experience that will help us to help you.

Send to nearest office for Booklet on Alundum.

NORTON COMPANY,

Main Works, Worcester, Mass., U.S.A.

ALUNDUM Plant, Niagara Falls, N.Y.

NEW YORK: 26 Cortlandt St., Havemeyer Bldg. CHICAGO: 48 So. Canal St.

CANADIAN AGENTS:

The Canadian Fairbanks Co., Limited, Montreal, Toronto, Winnipeg and Vancouver.

F. H. Andrews & Son, Quebec, P.Q.

63

52-inch Vertical Mill—Motor Drive.

GISHOLT TOOLS

THE profitable use of Turret-Lathes depends largely upon the design and construction of the tool equipment. Without the knowledge of how to tool them properly, Turret Lathes would be of little more value than engine lathes. Long experience in making Turret Lathes and an equally long experience in making the tools for finishing work requiring boring, turning and facing, has given us a fund of information that will be invaluable to the manufacturer contemplating the finishing of this class of work. We have made tools for over 10,000 different pieces, and it is seldom indeed that a job now comes to us with which we are not already familiar.

This knowledge of tool outfits will enable us to tell you exactly what you may expect of our machine. If you have work of this class, it will be to your interest to take it up with us. Our Turret Lathe Catalog will give you a very good idea of the great range of Gisholt Lathes.

Gisholt 21″ Lathe.

Gisholt Machine Co.

WORKS

MADISON, Wis. WARREN, Pa.

GENERAL OFFICES

1315 Washington Street, Madison, Wis., U.S.A.

Foreign Agents:

Alfred H. Schutte, Cologne Brussels, Liege, Paris, Milan, Bilbao, Barcelona, Schuchardt & Schutte, Berlin, Vienna, St. Petersburg, Stockholm, Copenhagen, C. W. Burton, Griffiths & Co., London, England.

Morse Twist Drills

are now made of

High-Speed Steel

of their own special brand, and are guaranteed as to quality of material and workmanship.

They have all the necessary hardness to give great wearing quality, and at the same time retain the requisite toughness which reduces breakage to a minimum.

They are made in taper shank, straight shank, taper length and jobber's length, and wire gauge drills.

Let us quote you on Morse High-Speed Drills.

Hardware Montreal

F.&W.

Frothingham & Workman, Limited Montreal

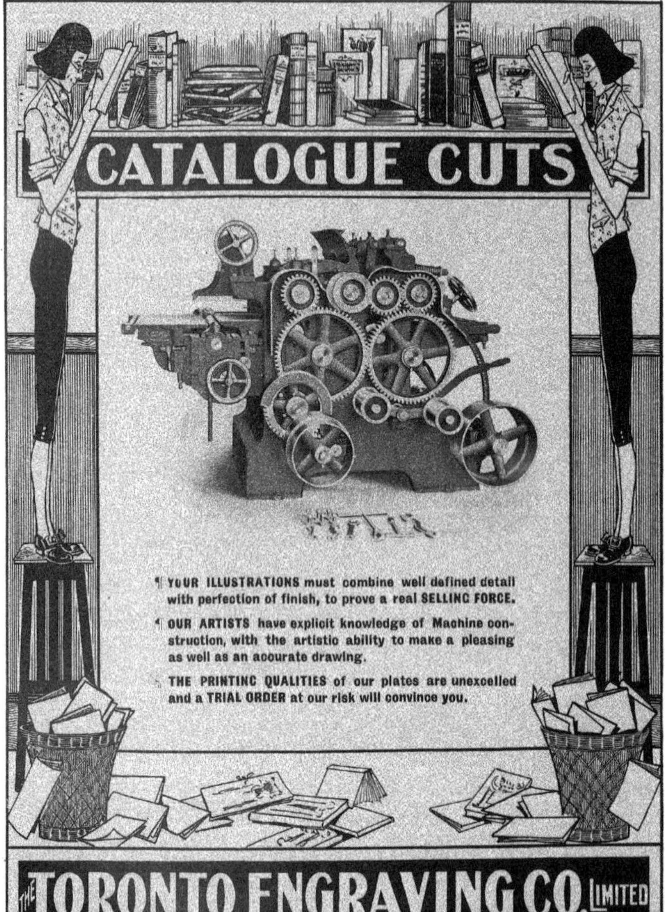

CATALOGUE CUTS
¶ YOUR ILLUSTRATIONS must combine well defined detail with perfection of finish, to prove a real SELLING FORCE.
¶ OUR ARTISTS have explicit knowledge of Machine construction, with the artistic ability to make a pleasing as well as an accurate drawing.
¶ THE PRINTING QUALITIES of our plates are unexcelled and a TRIAL ORDER at our risk will convince you.
THE TORONTO ENGRAVING CO. LIMITED
· 92-94 BAY ST, TORONTO. ·

H. W. PETRIE

BOILERS OF ALL KINDS

TUBULAR
10 to 200 H.P.

FIRE BOX
8 to 150 H.P.

WATER TUBE
100 to 400 H.P.

VERTICAL
1 to 80 H.P.

Inquire at Petrie's before buying

MACHINERY

of any kind, new or second-hand

Write for Stock List and Supply Catalogue

H. W. PETRIE, *TORONTO and MONTREAL*

The Canada Chemical Manufacturing Company, Limited

MANUFACTURERS OF

Commercial Quality **Acids and Chemicals** Chemically Pure Quality

ACIDS—Sulphuric, Muriatic, Nitric, Mixed, Acetic, Phosphoric, Hydrofluoric.

CHEMICALS—Salt Cake, Glauber's Salts, Soda Hypo, Silicate, Sulphide, Epsom Salts, Blue Vitriol; Alumina Sulphate Lime Bisulphite, Nitrite of Iron, C.T.S. and Calcium Acid Phosphate.

Chemical Works and Head Office	Sales Office	Warehouses
LONDON	TORONTO	TORONTO and MONTREAL

Return Tubular and Locomotive Fire-Box Boilers Tanks, Stacks, Rivetted Pipe.

THE MANITOBA IRON WORKS, Limited, WINNIPEG, MAN.

JEFFREY ELECTRIC LOCOMOTIVES

FOR MINE HAULAGE

Fully Described in Bulletins Nos. 10, 11 & 12, FREE.

If interested in Elevating, Conveying, Drilling, Screening, Crushing or Coal Handling Machinery will be glad to send you catalogues.

FOR FREIGHT HAULAGE

THE JEFFREY MFG. COMPANY

COLUMBUS, OHIO, U.S.A.

Canadian Branch: Cor. Cote and Lagauchetiere Sts., MONTREAL, Can.

RENOLD SILENT CHAIN

is used almost exclusively on the standard motors because of its supremacy in the power-transmitting field. Adapted to short centres; damp, hot or oily places; and will run in either direction. Action always smooth and positive.

Write for Booklet "J" and Special Bulletins 50, 52, 57, 58.

LINK-BELT COMPANY

PHILADELPHIA, CHICAGO, INDIANAPOLIS

NEW YORK	PITTSBURGH	BOSTON	BUFFALO
299 Broadway.	1501 Park Bldg.	84 State St.	321 Ellicott Square Bldg

KYNOCH
SUCTION GAS ENGINES
AND
PRODUCER PLANTS

We have been operating one of these Engines and Plants for a year. We propose installing a larger Engine and Plant in our new warerooms now being built. The saving in expense will **Pay for the Engine in Two Years.**

WRITE FOR CATALOGUES AND PRICES

WILLIAMS & WILSON
320-326 St. James St., Montreal

Equip your factory with the Hill Patent Collar oiling Bearings.

Miller Bros. & Toms
Montreal,
Quebec

Leonard Corliss Engines

Peerless Self-oiling and Leonard Automatics for Electrical and Manufacturing purposes. Standard Return and Water Tube Boilers, all sizes. Open and Closed Exhaust Steam Heaters, Pumps, Etc., for Complete Steam Plants.

E. Leonard & Sons, London, Ont.

Telegraphic Address: "HALLTYNE"

WORKS: Phone Main 256

THE HALL ENGINEERING WORKS

14 COTE ST., MONTREAL

ENGINEERS, BOILERMAKERS, BRASS & IRON FOUNDERS, COPPER SMITHS & BLACKSMITHS, SHIP REPAIRS A SPECIALTY

Agents for J. & E. Hall's Refrigerating Machinery, Atlas Preservative Co. & United Asbestos Co. of LONDON, ENG.

THOS. HALL, M. N. E. C. I. E. S. Phone West 1737
Late Supt. Engineer Messrs. Elder Dempster & Co. & Can. Pac. Rly. Co., London

T. O. SINCLAIR, M. I. M. E. Phone West 917

WATER SUPPLY

for Sprinkler System, Water Works, Fire Pressure, etc., etc.

Steel and Wood Tanks, Steel Substructures, Steel Flagstaffs, Bell Towers, Etc.

WINDMILLS, GASOLINE ENGINES, HYDRAULIC RAMS

ONTARIO WIND ENGINE and PUMP CO.
Winnepeg TORONTO Montreal

McVICKER ENGINES

Efficiency with Simplicity

is what we claim.

Only one-third the usual number of parts used in construction.

Perfect balance while running.

No experience needed to operate our engine.

WRITE FOR CATALOGUE

The Canadian McVicker Engine Company LIMITED
GALT - - - ONTARIO

Some Features of Large Gas Engines

By Robert W. Angus, B.A.Sc. †

ALTHOUGH gas engines had been developed and constructed to some extent for many years, yet it was not until 1862 that the conditions governing their economical working were discovered by M. Beau de Rochas, a French engineer.

After a careful scientific study of the subject he found that high efficiency depended, amongst other things, on producing a high pressure at the beginning of expansion, a result which could be obtained by compressing the charge of fuel before ignition. This process had not been previously employed, and it is largely to it that we owe the success of the gas engine, although at the time of its discovery, the importance of compression was not appreciated, the development of the gas engine being along different lines, as is shown by the appearance, among others, of the Otto and Langen engine in 1866.

Beau de Rochas also described the cycle to be carried on in the cylinder according to the following scheme: (1) Outward stroke of piston drawing in charge of fuel; (2), inward stroke of piston compressing the charge; (3), firing of the charge with piston at the inner end of stroke; (4), expansion of charge after ignition during second outward stroke of piston; (5), discharge of products of combustion during second inward stroke of piston. This cycle will at once be recognized as the one carried out in all "four cycle" engines, so called because four strokes of the piston are required to complete the cycle.

The cycle thus described was first put to practical use in an engine built by Otto in 1876, and on this account has been frequently named the "Otto cycle," although to Otto belongs the honor only of constructing an engine working on Beau de Rochas' cycle.

It quickly displaced all other cycles and put the gas engine at once in a firm position as a successful heat engine.

Since the time of Otto, the mechanical details of gas engines have been improved, and it is not often that a machine reaches the same state of perfection in so short a time, yet the original cycle is now very commonly used, and probably the great majority of engines of the present day employ it.

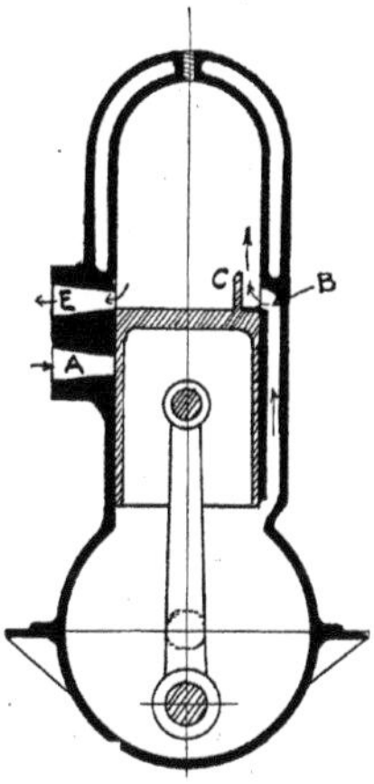

Fig. 2—Two Cycle Engine.

The engine of this type is so common in form and in such general use that it is unnecessary to describe it in detail, but it may be pointed out that the principal improvements have been in the way of substituting simpler inlet valves for the slide valves in the original engine, and in the substitution of more convenient and positive methods of ignition, and in suitable arrangements of parts for the development of large powers.

Small engines on the Beau de Rochas cycle are made single-acting, and this is also true of engines built by several continental firms in sizes up to 600 h.p. and even higher, and this, in view of the fact that there is only one power stroke per two revolutions for each cylinder used.

So long as the space occupied by the engine, or the steadiness of speed are not serious considerations, there can be little doubt that the single-acting, four-cycle engine possesses many advantages in the way of accessibility of the parts, notably the piston; and the greater simplicity of construction. But these engines have very large cylinders for the power developed, and must be fitted with enormously heavy fly-wheels to assure the steadiness of running required for the driving of electric machines. The size of the fly-wheel may, of course, be much reduced by using two cylinders in the engine, but the space occupied is rather excessive.

For large powers, therefore, some other method must be adopted, and at present there are two plans: (1), Either to make the "four-cycle" engines double-acting, or (2) to use some other cycle in which the number of power strokes bears a larger proportion to the total number of strokes than one to four. A cycle of the latter type was invented in 1880 by Mr. Dugald Clerk, in which a power stroke is secured in the working cylinder for two piston strokes, engines working in this way being called for obvious reasons "two-cycle" engines.

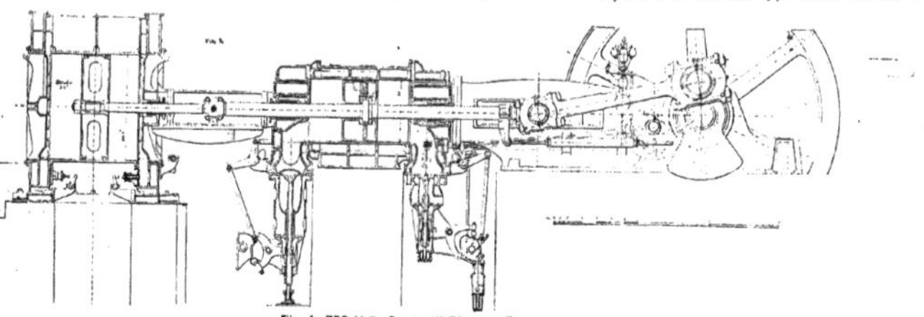

Fig. 1—750 H.P. Cockerill Blowing Engine.

† Professor of Mechanical Engineering, University of Toronto.

Very large engines are frequently of the "two-cycle" type, and when made also double-acting two power strokes per revolution, are secured as in the double-acting steam engine.

An illustration of a large double-acting "four-cycle" engine is shown at Fig. 1, which represents a Cockerill engine direct connected to a blowing cylinder for blast furnace work. It will be at once noticed that this type of engine is more or less complicated, because a crosshead must be used and special water-cooling arrangements devised for cooling the valves, piston-rod, piston and cylinder, and also that a very carefully made stuffing box, which is extremely long in the engine illustrated, must be used to prevent leakage under pressures, which may be as high as or higher than 300 pounds per square inch. The arrangements for cooling water, as well as the valves, and the mechanism for operating them, are so clearly shown that no explanation is necessary.

pumps, which slightly compress the gas and air separately or together and deliver them to the cylinder at the end of the expansion stroke in such a way as to drive out the burnt products of combustion and replace these by a fresh charge of mixture ready to be compressed at the next stroke. In many engines of this type, especially larger ones, a charge of air is sent in ahead of the mixture, in order to prevent contact of the latter with the hot gases, as this is apt to cause explosion of the entering charge. This cycle is now extensively applied, as it may be used in small engines as well as in large ones and in all cases requires a smaller cylinder for a given power. Certain of the valves are also dispensed with and smaller fly-wheels may be used for the same steadiness in running, but the economy of the smaller engines at least, is not so great as in the "four-cycle" engines.

Fig. 2 shows an outline of a small en-

piston closes the port (A), and when the piston has moved down enough to open port (B), the charge of fuel passes into the cylinder, drives out the previously burnt gases, and is ready for compression on the next upward stroke. Evidently since (B) and (E) are open at the same time, some of the charge may pass out into the exhaust port and be lost. This is partly prevented by a baffle plate (C), cast on the piston. The simplicity and compactness of this engine commend it, so that it is quite often preferred to more economical machines.

In large engines economy is, however, usually a prime consideration, and hence they are not nearly so simple in construction as those described above.

Fig. 3 illustrates the Korting two-cycle double-acting engine, which is built in America by the De La Vergne Machine Co., of New York. The engine is built with one cylinder, in sizes from 500 to 1,500 h.p., all double-acting. Re-

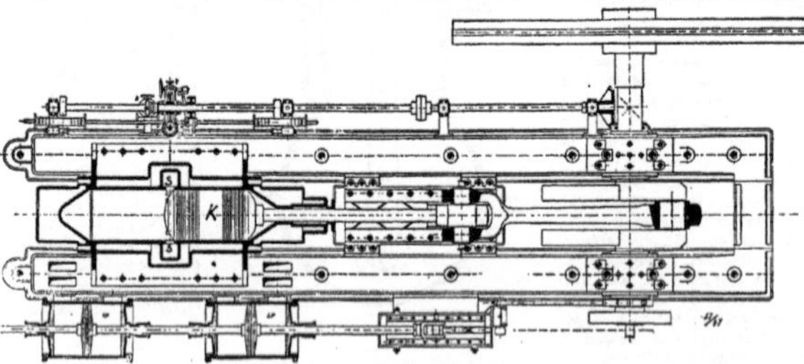

Fig. 3A—Korting Double-Acting Gas Engine.

This company builds engines up to 5,000 h.p., and has constructed single-cylinder double-acting engines up to 1,600 h.p., although for this power frequently two tandem double-acting cylinders are used, and for large powers two or more cylinders are also used.

Many other illustrations might be given of similar engines, as they are very common, but space will not permit. Small engines are very rarely made double-acting, as the arrangements required for cooling the pistons and rods, etc., would make the engine too complicated.

Coming now to the two-cycle engine it may be stated briefly that the ignition takes place at the inner dead point, and the compression and expansion strokes are the same as in the other cycle, except that the exhaust port is opened before the expansion is completed. The suction and exhaust strokes are not used, their place being taken by one or more separate compressing

gine of this type, which is quite commonly used in launches and for other purposes, the fuel being usually gasoline. In this engine there are no valves, and for the compressing pump a closed crank case is substituted in which the mixture is compressed. Beginning with the piston at the bottom of the stroke and on the point of ascending, the port (B) is open and the cylinder is filled with a charge of fuel which was compressed in the crank case on the previous down stroke. As soon as the lower edge of the piston rises above the port (A) the mixture is drawn into the crank case and compression takes place in the cylinder as soon as the top of the piston rises above the exhaust port (E). When the piston reaches its highest position the charge is fired and the downward or power stroke takes place, the burnt products discharging as soon as the piston uncovers the port (E). During this downward stroke the gas in the crank case is slightly compressed after the

ferring now to the figures, the piston (K) is seen to be very long, having rings at each end. There are two pumps, one (GP) for gas, and the other (LP) for the air, the gas and air being delivered separately at a pressure of about 9 lbs., and mixed as they enter the cylinder through the valves on top. The pump valves are so adjusted that the air pump first discharges a quantity of air into the cylinder, to sweep out the burnt products and prevent pre-ignition of the explosive gases by contact with the hot burnt gases. The gas and air pumps then discharge together into the cylinder, so that the explosive mixture follows a fresh charge of air, and not a charge of hot gas. A detail of the piping connections is given in the lower figure.

Starting with the piston at the crank end of the cylinder as shown, the exhaust ports (S) are uncovered, and the burnt gases from the previous stroke pass out of the head end. This is further facilitated by opening the back inlet valve (E), which admits air followed by the explosive mixture from the pumps. At the beginning of the return stroke the inlet valve is closed and the charge compressed and ignited near the dead point.. The gas expands and is discharged as soon as the piston uncovers the exhaust ports (S). The action in the other end of the cylinder is exactly the same, and the engine thus receives two impulses per revolution. Evidently the length of the piston must be about the same as its stroke in order that the exhaust may be properly regulated, as no valves are used for this purpose. The piston, piston rods, etc., are water-cooled.

The Oechelhauser engine is shown in Fig. 4, since there are some features in its design worthy of note. It is also of the two-cycle type, and is probably more commonly used for driving dynamos than for any other purpose.

In the figure it will be noticed that there are three cranks, the outer pair being at 180 degrees to the inner one. The latter is connected directly to the piston in the power cylinder. The outer cranks are connected to a single crosshead behind the cylinder, and from this crosshead a second piston is driven in the working cylinder, the latter being open at both ends except for the pistons. It will also be noticed that these pistons always travel in opposite directions. To the crosshead is also connected a double-acting pump, which is used for compressing air in the end nearest the working cylinder, and the explosive mixture in the other end. The power cylinder contains the exhaust port (E), which is controlled by the left-hand piston, and ports (A) and (M) for the air and mixture, respectively, these

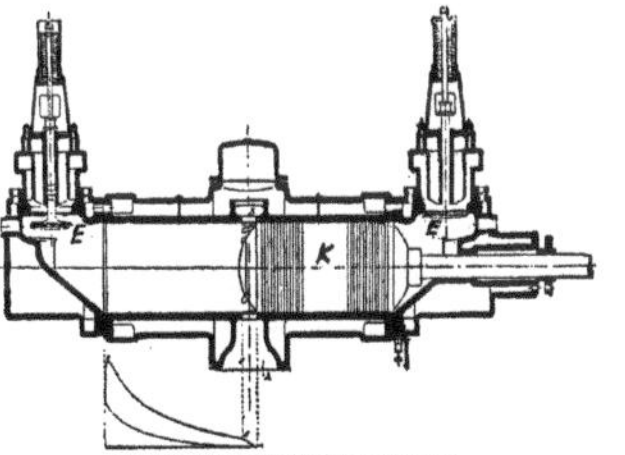

Fig. 3B—View of Cylinder and Valves.

ports being controlled by the right-hand piston. Pipes marked (A) contain air; (G) gas; (M) explosive mixture.

The method of operation is as follows: Assuming the two pistons in the positions shown, the port (E) is open, and the burnt gases are escaping through it, the ports (A) and (M) are also open, the former, which evidently opens in advance of (M), admitting first a charge of air, which is followed by a charge of mixture on the opening of (M). As the revolution continues the pistons approach each other, closing the ports (E) (A), (M), and compressing the charge, which is fired when the pistons are clos-

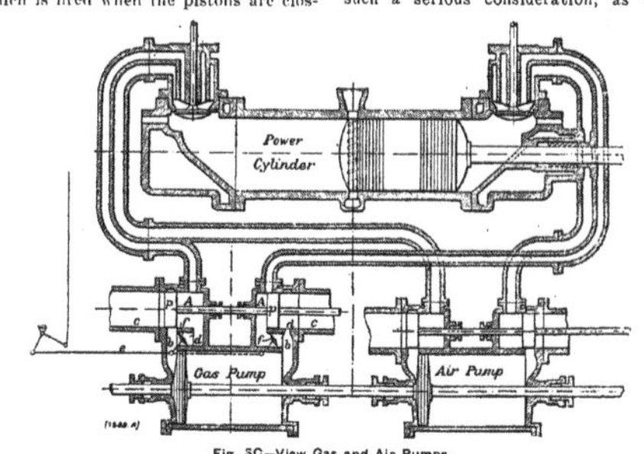

Fig. 3C—View Gas and Air Pumps.

est together. The pistons are then driven apart and the work done, the pistons uncovering the ports at the end of the stroke, and the cycle is repeated. Such an engine is, of course, very long, and there are some objections to the shape of the crank, shaft, etc., but on the whole the machine is simple in construction, does not require valves, and an explosion is obtained per revolution, no stuffing boxes being required in the working cylinder. It is possible that some of the working fluid may pass from (M) through (E) when both are open, but by careful design this loss may be made very small.

In the latest types of the Oechelhauser engine the tendency is to compress the gas and air separately and mix them as they enter the cylinder.

Fair comparison between the two classes of engines is rather difficult because it is not easy to get a good basis from which to judge. For launches and similar purposes the desirable condition is frequently that an engine shall be as simple as possible, positive in its action, and that it shall take up a minimum space. Economy of fuel is not such a serious consideration, as there is usually plenty of storage capacity for the supply of fuel necessary to make the short trips for which these launches are so commonly used. For such purposes, therefore, the two-cycle engines are quite popular, chiefly on account of their compactness and smoothness of running, and because there are no adjustments of valves to be made and little to go wrong, making them thus easily managed by inexperienced persons. Many persons prefer the larger four-cycle machines for this purpose, however, and apparently the decision as to the type depends on the reputation of the particular engine.

For large machines there are always experienced engineers in charge, and hence the choice of machine is governed by the first cost, the cost of operation and of maintenance, reliability of action and space occupied.

The two-cycle engine requires one or two extra pumps, which make it more complicated, but in connection with them it must be remembered that the pressures are small, and therefore these parts may be made very light, so that probably the friction loss is less than that resulting from the suction and exhaust strokes of the four-cycle engine. This difference again is to be set against the power required in compressing the gases.

The construction of the working cylinder is also simplified in the two-cycle type by the absence of exhaust valves and valve gears, which are a cause of trouble on account of the effect of heat on them, and valves are dispensed with altogether in the Oechelhauser engine.

At the same time two-cycle engines may be decreased in efficiency on account of some of the mixture passing out through the exhaust ports before the latter are closed, a difficulty which may be minimized in design. The reduced size of flywheels in the two-cycle engine has been previously mentioned, this being due to the fact that the turning moment of the shaft is subject to much smaller variations than in the other cases.

Some information may also be obtained from the following tables of dimensions of engines, some of which are from the builders' catalogues, and some from a paper by Mr. H. A. Humphrey, read before the Institute of Mechanical Engineers, January, 1901.

Dimensions of Some Large Gas Engines.

Type of engine or name of builder	I.H.P.	Two or four cycle.	Number of cylinders.	Diameter of cylinders .... inches.	stroke ... .. inches.	Revs. per min.	Kind of gas.	Fl. or space occupied .... .. feet.	Remarks.
Crossley........	600	4	2	30	36	135	Blast furnace	....	....
Crossley	530 B.H.P.	4	2	28	36	130	Coal	27x16	....
Premier..........	650	4	2	28	30	150	Mond	....	....
Nurnberg	545 B.H.P.	4	2	24	36	115	....	46x15	Double acting tandem cylinders.
Cockerill.........	600	4	1	51.2	55.1	94	Blast furnace	36x20	Mechanical efficiency on test 73.1%
Westinghouse ...	621	4	3	25	30	149	Natural	....	Mechanical efficiency on test 89%.
Korting	600 B.H.P.	2	1	27.5	48	90	....	43x21	....
Oechelhauser	600	2	2	18.9	31.5 each piston	135	Blast furnace	....	....

This table has been made up for sizes as near 600 h.p. as possible.

It may be of interest to state that the mechanical efficiency of a Cockerill engine by test was 81 per cent., the power developed being about 870 i.h.p., the engine having but one cylinder.

A test on a Korting engine with a double-acting cylinder 29.7 in. diam., by 55.1 ins. stroke, reported in "Power" for January, 1907, gives 607 b.h.p., and 789 i.h.p. in the power cylinder, with an i.h.p. of about 88 accounted for in the pumps. The total efficiency of the engine was 77 per cent.

With regard to the conditions affecting efficiency, the methods of ignition, the methods of starting large engines, and the special devices used in scavenging and governing, much might be said, but space does not permit. It is well to keep in mind, however, that gas engines are in the field to stay, that they are steadily being improved in construction and reliability of action, and that they possess greater possibilities in the way of efficiency than steam engines.

Builders are now able to guarantee that the consumption of coal per hour per brake horse-power will not exceed one pound, even on engines considerably under 100 h.p., where a suction producer is used. These suction plants also take up very little room.

It seems, however, that there is still room for improvement, possibly by a radical departure from the original cycle, and that greater developments along this line may yet be expected.

CORRESPONDENCE.

The following letter received at the editorial office of Canadian Machinery from the Northern Electrical Manufacturing Co., Madison, Wis., quoted in part, calls attention to a situation the existence of which we had not hitherto known. Any further information on the subject from readers of this paper will be appreciated.

Gentlemen:—

"We have found a very antagonistic sentiment against United States machinery on the part of the Canadian consumers.

In fact some stationers seem to carry

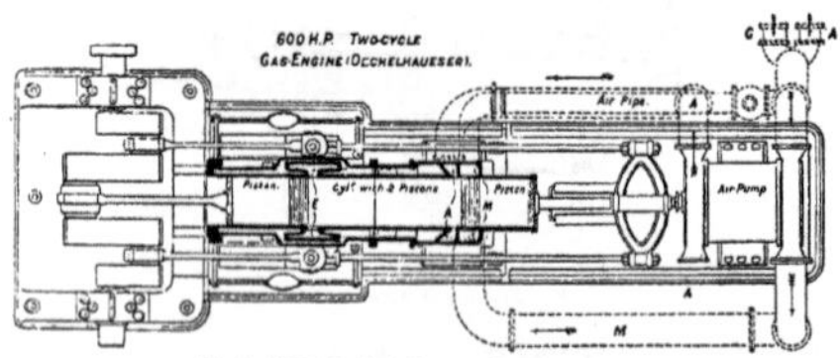

Fig. 4.—600 H.P. Oechelhauser Two Cycle Engine.

rubber stamps furnished to Canadian manufacurers reading, 'We buy in Canada.' Several instances of this kind have come to the writer's attention recently and as the type of the stamp is uniform it thus appears to be an instance of the very general sentiment.

Very truly yours,
Northern Electrical Mfg. Co.

In illustrating the extent of the influence of Canadian Machinery, a letter received from the Scottish Highlands enquiring for this paper should be of interest.

Kinlochlevenonach,
January 5th, 1907.

Gentlemen,—Please let me know on what conditions you supply Canadian Machinery and Manufacturing News. Address,

WILLIAM McLARNON,
Kinlochlevenonach,
Inverness-shire, Scotland.

Regarding Boiler Inspection.

Canadian Machinery,
Toronto, Ont.

In a general way we do favor the universal inspection of boilers for the whole Dominion provided the inspection law is a good one and properly carried out. We find the inspection laws of the various provinces where they are enforced as a benefit as tending to bring about a higher class of boiler work, but, at present, there is considerable variation in the regulations of different sections and in the manner they are carried out, which causes considerable trouble to the boiler manufacturer who is supplying boilers to all the various provinces. It would be much better if there was a Federal law applying to all the provinces, or if the various provinces would adopt the Dominion Act, so as to bring about uniformity which would enable the boiler manufacturer to build a standard line of boilers.

Hoping these expressions of opinion may be of use to you, we remain,

Yours truly,
ROBB ENGINEERING CO. LTD.
D. W. Robb, president.

*Economy of the Electric Motor**

By Kenneth Drinkwater

THE economy of electric power is now generally recognized but it may be interesting to you to know wherein this economy lies; how this motive power may be best adapted and to what a variety of uses it may be applied.

With regard to economy, perhaps the best illustration of this is in the industrial establishment where it is necessary to transmit power to a large number and great variety of machines. This can be done by employing a large motor to transmit power by means of shafting, belts and pulleys, or by using separate motors for each machine or group of machines. The greater economy is, of course, obtained by the use of separate motors.

When power is transmited by belts and shafts, the percentage of the total horse-power output of the engine or motor which is actually useful at the productive machine, is at the best only about 77 per cent. That is to say, at least 23 per cent. of your available power is lost in transmission. These figures are based on horse-power required to drive the shafting and belting when running light. There is good reason to believe that these losses increase with the increase of load.

The disadvantages of long lines of shafting were realized long before electricity offered a way out of the difficulty. Unless the greatest care be used in keeping the shafting in alignment and the journals well oiled, the power wasted in friction of bearings and belts becomes a very large percentage of the total power received from the prime mover.

The friction loss is continuous even when, as is often the case in large shops, a single particular tool, like a large boring mill, has to run when all the rest of the shop is shut down.

The system of electric drive is the solution of this problem. It offers numerous advantages; the system is easily installed; and it is readily extended and maintained.

In the case of individual drives, the highest possible economy is obtained because the motor furnishes power only as required by the work of that particular machine.

There are, undoubtedly, numerous special requirements which can best be fulfilled by use of the individual drive, but in the vast majority of cases the application of an individual motor to each tool carries the matter too far.

Small motors like small engines are less efficient than large ones, besides costing very much more in proportion to the power delivered. The general plan which meets with the approval of the best engineers is to employ individual drives for tools which require variable speed drive. In addition to this it is often desirable to drive individually on account of the comparatively large amount of power involved in a single machine, or as is often the case, when convenience of location makes it desirable to isolate a machine, then the individual drive is indispensable.

Consider the case of a large works consisting of a number of shops. Evidently one steam engine or turbine cannot be used to drive the shafting in all the shops. It is necessary then to either install one large central plant and distribute the steam to engines located in each of the shops or in the case of a very large works to install a number of boiler plants.

The first cost of such installations is enormous; the cost of maintenance and attendance is very high and then there is always the possibility, in the case of a central plant, of some accident occurring which will close down the whole shop. Of course this is partially got over by duplicating the power plants, but then think of the cost.

The electrically operated works of today offers a striking contrast to this lay-out. The motive power is obtained either by installing electric generators or by getting it from outside sources. It is then transmitted by means of wires to all parts of the shop or shops, doing away with a large percentage of the loss formerly occurring in pipes, shafts, etc., since the loss in electric transmission is very small indeed in comparison.

This is got over in large up-to-date plants by the fact that alternating current is used for the majority of the works, and in the case of a breakdown, connections may be made in a very short time to outside sources. In small plants, the current is nearly always taken from the supply companies, and the risk of a shut-down, for any length of time, is very remote.

We now come to the varied uses to which the motor may be applied.

The decided advantages of the application of motors to machine tools in industrial work has been thoroughly exemplified in machine shop practice today. The conditions under which machine tools operate are so varied that it is impossible to make any general statement covering all of the possible operating conditions, but some of the individual conditions are always important, as, for instance, the character of the work machined, kind of material cut, shape of the cutting tool, quality of the tool steel, method of treating tool steel. All of these should be taken into account, to intelligently fit a motor to any machine tool.

The factors which have had more to do with the recent impetus given to the study of rapid production than any others are the high speed steels and the variable speed electric motors.

In the case of a machine requiring variable speed, the variable speed motor offers decided advantages in the way of rapid and economical production. With the old method of speed variation, by means of cone pulleys or nests of gears, only large increments of speed are obtainable. This invariably means that tools cannot be worked up to their limit of productive capacity. The variable speed motor may, in some cases, actually decrease the cost of the machine tool by eliminating extremely bulky and expensive mechanical speed changing devices.

Up to now I have not distinguished between the alternating current and the direct current motor. It might be of interest to know something about the desirability and economy of employing the one or the other.

The alternating current motor is singularly adaptable to planers, slotters, shapers or tools of a similar nature, in which reciprocating motion is employed, provided variable cutting speed is not an object. It is obvious that on the quick reverse with machines of this character, unless the motor is abnormally large, or a fly wheel be employed, there is imposed a considerable momentary overload upon the motor. These overloads will be more readily taken care of by the alternating current motor, than by the direct current motor, for the reason that overloads on direct current motors, if severe, will, if the overload capacity of the motor is not adequate, be accompanied by flashing at the brushes.

The alternating current motor is also singularly adaptable to grinding operations, to certain classes of wood-working machinery and to tools situated in places that are exposed to moisture, acid fumes or inflammable materials.

For group driving it is specially desirable, and for the individual drive where the conditions are of a definite character the outfit is ideal in simplicity of construction and general reliability.

* Abstracted from an address by Kenneth Drinkwater before the Y.M.C.A., Montreal, Feb. 1907.

CANADIAN MACHINERY
and Manufacturing News

A monthly newspaper devoted to machinery and manufacturing interests, mechanical and electrical trades, the foundry, technical progress, construction and improvement, and to all users of power developed from steam, gas, electricity, compressed air and water in Canada.

The MacLean Publishing Company, Limited

JOHN BAYNE MACLEAN - - President
W. L. EDMONDS - - - Vice-President
H. V. TYRRELL - - - - Manager

OFFICES:

CANADA
MONTREAL - - 232 McGill Street, Phone Main 1255
TORONTO - 10 Front Street East, Phone Main 2701
WINNIPEG, 511 Union Bank Building, Phone 3726, F. R. Munro
BRITISH COLUMBIA - Vancouver, Geo. S. B. Perry

UNITED STATES
CHICAGO - 1001 Teutonic Bldg. J. Roland Kay

GREAT BRITAIN
LONDON - 88 Fleet Street, E.C. Phone Central 12960, J. Meredith McKim
MANCHESTER - 92 Market Street, H. S. Ashburner

FRANCE
PARIS - Agence Havas, 8 Place de la Bourse

SWITZERLAND
ZURICH - - - Louis Wolf, Orell Fussli & Co.

SUBSCRIPTION RATE.

Canada, United States, $1.00 Great Britain, Australia and other colonies, 4s. 6d., per year; other countries, $1.50. Advertising rates on request.

Subscribers who are not receiving their paper regularly will confer a favor on us by letting us know. We should be notified at once of any change in address, giving both old and new.

Vol. III. MAY, 1907 No. 5

CONTENTS

CONVENTION OF AMERICAN FOUNDRYMEN.

THE annual convention of the American Foundrymen's Association and the Foundry Supply Association will be held in Philadelphia from May 20 to May 24. Something was said last issue about the exhibits of the supply association; but the topic of chief interest to Canadian foundrymen at the present time is the probability of the convention being held in Toronto in 1908. At the New York convention in 1906, the invitation was extended by Toronto to the association to hold their convention in 1908 in that city. This was discussed at the Cleveland convention, and now arrangements are being made whereby all the particulars may be laid before the association at Philadelphia. Because of the difficulty and expense of moving the exhibits across the border, it is proposed to hold the convention jointly at Toronto and Niagara Falls, and the sittings of the convention being divided between the two cities; but it is possible that the convention will be held in Toronto altogether, and in regard to the bringing of the exhibits across the border, the Department of Customs, Ottawa, have made the following statement: "I beg to say that such machinery and apparatus as is imported for exhibition purposes only, will be allowed entry, subject to its immediate exportation at the end of the exhibition. Any articles, however, which are brought into Canada for sale will require to have duty paid thereon,"

In order that the convention might be held in Canada, it would be necessary for the Canadian foundrymen to form themselves into an association in order that there might be unity of action. The Canadian Manufacturers' Association are investigating this question, and it is probable that this will result in some sort of an understanding between the Foundry Foremen's Association in Hamilton and Montreal, so that a definite invitation may be extended to the American association by the Canadian Manufacturers' Association or by a united body of foundrymen.

The importance to the Canadian foundrymen of having this convention in Canada is great. The educative value of such a convention would be immense. The exhibits of the Supply Association are up-to-date and representative of all modern foundry appliances. The papers read before the convention are prepared by men of wide experience, and are, therefore, interesting and instructive. Then the personal element, that is, the gathering together of foundrymen from all over the country in social and business intercourse, resulting in the free interchange of ideas and experiences, which is of extreme value to the advancement of the science of founding and to the individual foundrymen, is of great importance.

It is only within the last few years that founding has been carried on in a way befitting its industrial importance; and the big factor in this advancement has been education. The convening of foundrymen annually, the presentation of papers, and displaying of up-to-date appliances, as is done at conventions is the most interesting and effective channel of education. Eleven years ago the American Foundrymen's Association was organized at Philadelphia, and undoubtedly the advancement of founding has been due in no small measure to the influence of this association.

The industrial expansion of Canada is but in its infancy; the prospects, for future expansion are dazzling. Founding will play an important part in this expansion. Therefore, Canadian foundrymen and Canadian manufacturers should be in a position now to develop the founding industry along up-to-date lines. At present labor-saving appliances are very scarce in Canadian foundries, and without doubt, one of the chief reasons for this is that those in charge have not come face to face with the numerous appliances which are now on the market. The holding of the convention in Toronto would be a means of bringing manufacturers and foundrymen into closer touch with modern ideas and efficient appliances in foundry practice; and besides being of great educative value would probably be the means of enthusing those interested in the formation of a Canadian Foundrymen's Association, in which many prominent foundrymen in Canada are much interested.

It is to be hoped that Canada will be well represented at the coming convention at Philadelphia, and that it will be possible to arrange for the holding of the next convention in Toronto, or in Toronto and Niagara Falls.

A WARNING.

THERE are several men who appear to be systematically going about the country soliciting subscriptions for the MacLean Trade Newspapers and Magazines and are pocketing the proceeds where they are able to land a subscriber. Several people have been victimized in this way. Some of these men are not confining their efforts to the MacLean papers, but, from enquiries we have received from one or two of our contemporaries in the United States, are practising the same methods in regard to their publications.

Business men and others are cautioned to look out for these fraudulent canvassers, and we would take it as a great favor if they would notify us by wire, at our expense, should they meet any of these men, as it is our intention to prosecute them if we once get the hands of the law upon them.

No one is authorized to solicit subscriptions for us unless he has our regular subscription forms and a letter signed by us giving him the right to receive moneys on our behalf. Our friends, the travelers, frequently send us subscriptions, and, of course, where a merchant knows a traveler to be a representative of a reputable house he may be sure that the subscription he may take will reach its proper destination.

We trust that our friends throughout the country will lend us their assistance in endeavoring to land these fraudulent solicitors. The names of the men and the localities in which they were last heard of are: C. H. Raymond, Berlin, Ont.; J. C. Murray, Farnham, Que.; McDonald, St. Hyacinthe, Que. There is another who operated in Galt, whose name we have not yet ascertained.

DEPARTMENT OF RAILWAYS AT McGILL.

THE McGill University have sent out a circular describing the courses in transportation which have been talked of for some time, and towards which the large railways are contributing. On the Advisory Board of this department are the following prominent men: Sir Thomas Shaughnessy, Chas. M. Hays, R. B. Angus, C. J. Fleet, E. B. Greenshields, Henry T. Bovey, dean of engineering, and Prof. Clarence Morgan. The department will be ready for the enrolment of students in the fall of the year.

There are three courses which may be followed in this department, namely, the operating department or executive offices, the motive power department, the engineering department, and that of maintenance of way. The entrance examinations and the first two years of these three courses are the same as for any other course in engineering at McGill. During the other two years of the first named course, there will be specialization in subjects such as, economics, elementary law, English, freight service, railway engineering, railway operation, stenography and structural engineering. The other two courses will follow closely the third and fourth years in the regular courses of mechanical engineering and civil engineering, respectively.

During many years to come, there will be immense expansion of railways in Canada. There will be numerous positions of great and minor importance to be filled; and there are probably more chances of important and useful careers in railwaying in Canada then there are in any other industrial field. The courses will therefore be popular; and good work will no doubt be done if the courses are properly handled. From the names upon the Advisory Board this seems assured.

This is an indication that the railways recognize the value of a thorough training in fundamentals; and it is another step in the advancement of technical training.

ARTICLES FOR FOUNDRYMEN.

COMMENCING in this issue is a series of articles on Modern Foundry Practice, by Samuel Groves. The writer of these articles has been in close touch with foundry practice throughout his entire career in the United States and England; and the numerous ideas concerning modern foundry practice, which he has gathered, and which he is bringing out in these articles, will be of great value to foundrymen in this country. The first article deals largely with the historic side of foundry practice, but in the succeeding articles, the writer will plunge right into his subject of modern practice.

A PRESIDENT AT LAST.

THE University of Toronto has at last a president. The Board of Governors, after carefully surveying the field of available men, both here and in the Old Country, have selected Robert A. Falconer, M.A., B.D., LL.D., Litt. D., Halifax, as president. Dr. Falconer has had a splendid academic career, and is spoken of in high esteem by those who know him. He, therefore, comes to his most important position well recommended.

It is stated that the salary to be received by the new president will far exceed that previously paid, and $10,000 is mentioned as the probable sum. The responsibility of the office is great, and there is scope for the most important work, not the least of which is the influence a man in such a position can exercise over the men who are to be the future strength or weakness of Canada. A man fitted for this most important office is certainly worthy of a much higher salary than has heretofore been paid.

McGILL'S LOSS.

IN THE destruction by fire of the engineering building and medical building, McGill University has received a severe blow. The loss cannot be altogether estimated in actual dollars and cents, since some of the equipment cannot be replaced; and the inconvenience to which the authorities will be placed this fall will be great. But a monetary estimate of the loss in the engineering building has been placed at three-quarters of a million, with insurance of $420,000. This means an immense handicap to the university at a time when it can little afford it, especially in the engineering line. It is to be hoped that public spirited men with means at their disposal will come to the aid of McGill, as they have done in the past.

The engineering building was erected, equipped and endowed by Sir Wm. MacDonald, and was opened in 1893. This was the most perfectly equipped science building on the continent, with one exception, the recently completed science building at the University of Pennsylvania.

The Workman annex to the engineering building, although saved from destruction by the fire doors, was considerably damaged by smoke and water.

The Milling Machine vs. The Planer

By John Edgar

THE milling machine is slowly but surely encroaching on the field of the old-time favorite, the planer. It is not doing so with the ease that it takes to say the words, but with much opposition from those whom it will most benefit; hence the following.

It is not so much that the milling

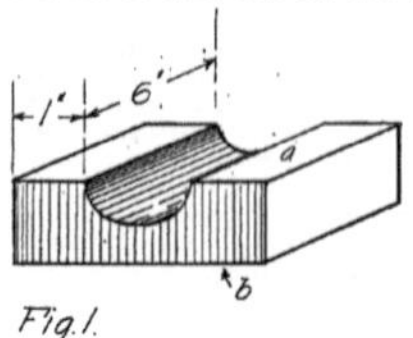

Fig. 1.

machine needs a champion that I take up the banner in its favor, nor is it on account of any real antagonism towards its opponents, but because it is struggling along, and making fair progress, against that foe of all new ideas, conservatism.

Many take as an issue the equipment necessary to the successful operation of the milling machine. Is not this theme sometimes set forth in a much exaggerated light? Equipment we must have, but is it not necessarily so with any machine tool for its successful operation? Even the planer—handy as it is—cannot claim much as a record breaker without some special tools or fixtures. Then why should there be all this hue and cry about such matters when comparing the results obtained with the milling machine.

Then again, we are reminded that cutters need sharpening occasionally and require some care in handling. Is there not a far better chance for the proper care and handling of the cutters used on the milling machine with less need of the personal element than is the case with the cutting tools of the planer? This sort of argument may sound rash to some, but on a little consideration on their part it will be recognized as the truth.

Why is it necessary to associate an elaborate tool-room with the handling of the milling machine? Surely in present day practice as found in the ordinary plant, where modern methods are appreciated, we have on hand all the necessary facilities for the proper sharpening of the cutters used on the milling machine; and where it is used to a larger extent than that which would be termed common practice the extra amount of trouble, if such it may be termed, would be warranted by the increased output and the reduction in the cost of products that the miller is sure to show where properly used.

There can be no reasonable argument against the milling machine on account of the inability to obtain the cutters necessary or on the score that it is impossible to produce satisfactory results from home-made equipment, as there are in the market to-day several firms who are ready to give the service of their specialized plants to produce the best possible results in shape of cutters at no risk to the buyer. It is possible to get from such a source just such a cutter as the requirements of the special case call for.

When fitting up for the milling of a job the first cost is large and may seem exorbitant to many, but when the cost is taken per piece the results are more favorable to the milling machine than would be looked for under the circumstances by those unfamiliar with this kind of work.

Fig. 2.

But it would surprise many, who are at present not familiar with the milling machine and its characteristics, to see what a favorable showing can be made with but a very meagre equipment. Many a job that does not warrant special cutters and fixtures may be handled on a milling machine to advantage and even under such circumstances give the planer a good run for the money. It is because many know the milling machine only by its reputation as a tool-room machine, fit for but the lightest work and such jobs that cannot be done on any other style of machine that the foregoing opinions are held. When you come down to the bare facts, there are but few jobs that a man can think of that cannot be done in some way on the milling machine of the universal type. Many, we must admit, would be rather out of place and impracticable, but others that are at present considered typical of the lathe and planer or shaper would, if it were only given a chance, be turned over to the milling machine as rightfully belonging within its field.

Is there in the present state of the arts a more economic method of removing superfluous metal than the milling machine? It would be foolish to contest the fact that, on the milling machine, we can clear a larger area of the superfluous metal for a dollar than can be done on the planer; so that we, in comparing the two methods, have sifted the matter down to the quality of the surface obtained. What better surface could be desired than that left by a milling cutter when run at a high surface speed, removing a light chip at a reasonably fine feed? In fact, the grade of surface obtained is easily made to conform with the requirements by adjusting the feed and speed at which it is run. The cutter may be of the axial type or an ordinary face mill, but the results in either case are all that could be desired.

It has been my good fortune for a number of years to be situated where I could watch the growth of the milling machine and the planer side by side in their range of usefulness. The growth of the former in favor is beyond a doubt, and this growth has not been due to any sentimental reasons, but from that purely business reason—increased production for a minimum outlay. It might be mentioned here that had any especial pains been taken to push the miller the results would have been much more striking. The fact cannot be too loudly proclaimed that for equal outlay the miller can show the planer its heels on work that at present is almost exclusively done on the latter machine.

What the milling machine has accomplished in the small arms shop, the typewriter factory and the electrical works it can parallel on the heavier work of the engine works and other manufactories, where money-saving methods are sought and where rapid production of accurate work is required.

It is noticeable that the opposition the milling machine meets grows weaker with the installation of every machine.

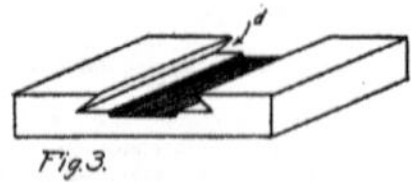

Fig. 3.

It needs no other advertisement than its own work. But we must not look forward to the day when the milling machine will completely supersede the planer, for the latter has its own field which requires a machine of its peculiarities. But we may look forward to the day, and that day is not far off, when the milling machine will come into its own with all the glory that should crown such an event.

Facts to Substantiate Claims.

Having up to this point dealt in generalities, let us get down to facts that tend to show the truth of the foregoing claims. There seems to be a strong tendency to restrict the milling machine to just that class of work on which it has made its reputation and for which it was originally invented. It is on that class of work that competition has completely died out. Few, who have had no special experience with the milling machine, will concede the point that it, as a producer, is fit for any other class of work than that irregular work found in gun shops, typewriter factories and electrical instrument works. It is just on this point that I wish to take issue.

Let us take a few figures to show as a comparison the time required to machine a certain job, that shown in Fig. 1, on both the planer and milling machine. We will first deal with the top, (A). If we place this work on the planer—one having two cross heads—in two strings of 12 each, allowing one inch between each piece for clamping, we would have a stroke of about 7 feet. Suppose that the planer runs on the cut at 30 feet per minute, which is above the average, and has a return speed of 50 feet. The time consumed for each stroke would be 7 ÷ 30 feet for cut, and 7 ÷ 50 for the return, or, roughly, a total of one-third of a minute. Allowing a feed of tool of 1-32 inch we would have a total of 64 strokes of actual cutting. Allowing nothing for the time taken in running the tool across the gap, either by hand or power, we have as the time necessary for machining the piece, 21 1-3 minutes.

The feed of table of a milling machine necessary to accomplish the same amount of work under similar conditions would be 8[illegible] ÷ [illegible]7.3 = 4 inches per minute, which is very reasonable on such a cut.

On a milling machine the number of strings of caps operated on at the same time is only restricted by the power and size of the machine. Two strings could be milled at the same time on a medium size column type machine, say a No. 3, while with a large size available, three or possibly four could be easily handled. Now in milling we have the alternative of short or long strings of caps. With shorter strings more frequent settings are required, but this is done without perceptibly increasing the time per piece, as would obviously be the case in planing. In fact, the shorter strings are to be advised on the miller as it allows of time between setting for the cutters to cool, increasing the run between grindings. Where the operator is running two or more machines and where the pieces are short, a longer string is necessary.

When we come to compare the time consumed in machining the base we have figures that speak well for the miller on work that one hears much claimed for as typical planer work. The width of this piece is, say, six inches, and when set as for the former operation the time consumed would be 64 minutes on a planer, whereas the miller could walk along at a feed of but 1.3 inches in order to equal that time; but as a matter of fact a feed of 3 inches would not tax the miller very heavily. This 3-inch feed would be obtained by the use of ordinary plain slab milling cutters. At that feed the surface ought to be sufficiently smooth and accurate for most cases. On wide surfaces such as this it is advantageous to sometimes resort to face milling, using a large inserted tooth face-mill either in a vertical spindle machine which is peculiarly adapted for that class of work, or in a horizontal spindle machine. The feed in such a case could easily be raised to 7 inches per minute and yet leave a surface which should satisfy the requirements of almost every case. The time required to machine the base with a face-mill would be,—allowing a 6-inch over-run of cutter,—5 × 6 ÷ 7 = 4.2 ÷ 7 minutes, or 1¼ minutes per piece, as against 64 ÷ 1[illegible] = 5 [illegible]-3 minutes, the time per piece for doing the job on the planer, a trifle less than one-fifth as long for the miller as for the planer with a finish of the same quality.

The above figures have been for ordinary carbon steel; and had the high speed steels been considered the time would be reduced considerably, both for planer and miller, to the advantage of the latter.

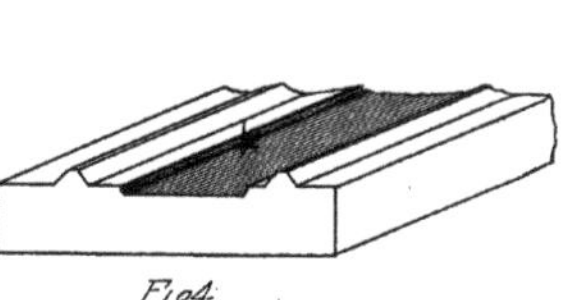

Fig. 4.

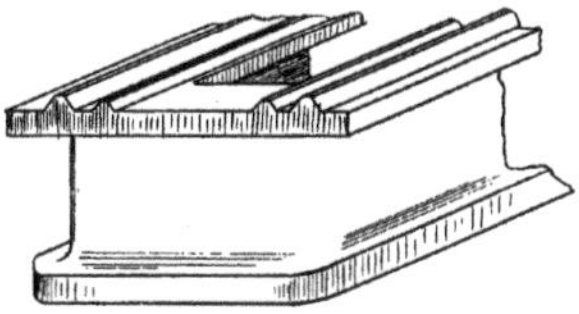

Fig. 5.

The example given above is for plain work; but where the miller shows the planer its heels is on more irregular work, as shown in Fig. 2, which is similar to that in Fig. 1, except that the cap has a side fit (C). The job of fitting on a planer is a tedious one at best and one which is never of a duplicate character, while on the milling machine this job would be tackled with gang cutters and each piece would be a duplicate of every other as to the fit, and at a very great saving in time. Fig. 3 is another piece of work for which the miller is especially fitted, in that it is possible to produce work that is accurately formed. The angle (D) has long been an irksome job for the planer-hand, but it is handled as easily on the miller as is a plain surface.

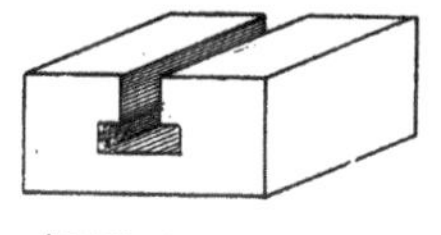

Fig. 6.

Who but those who have "hooked" out a tee-slot on a planer can appreciate the gain and the ease with which the operation is performed on a miller. Such a job is illustrated in Fig. 6. So apt are we to look with prejudice upon an innovation in shop practice that it is hard to convince without actual trial, that, on the work shown in Figs. 4 and 5—a planer table and a lathe bed, respectively—the milling machine is a keen competitor of the planer. But what better job could be handed to a milling machine! Here we recognize the desirability of duplicate, accurate work. What easier way of producing such work is there than with the miller! A job of this kind would of necessity be handled with gang cutters. The removal of the majority of the stock would be done first and the casting allowed to season, as is the practice when planing. Then, when required, the finishing chip may be taken, removing with sharp cutters only that metal which is necessary to take the wind out of the surface. You say that it would take too much care on the part of the operator; no more so than should be taken at the planer. There is this much to be said: that once the work is set and the machine started, the operator is at liberty to turn the most of his attention to some other duties; not so with the planer, as the operator must hang over the machine and give his undivided attention to the work on that machine, in order to obtain satisfactory results.

Equipment of Miller.

As mentioned above, the equipment is a point on which we hear a lot of argument. Why go to extremes in comparing the points of one method with another? Some cases warrant a very extensive equipment while others do not. "What should be called a good equipment, and what would it cost?" would be a matter-of-fact question. Take the horizontal spindle column type machine, and with a good assortment of cutters amounting to a matter of 75 or 100 dollars; with this you would be prepared to tackle any ordinary work that would be liable to present itself. The selection is, of course, dependent upon the size of the machine and general class of work it is to be used on. The vertical machine is, by the nature of its design, capable of handling a very large class of work with a much more limited equipment than is the horizontal machine. In its case a few well assorted end mills and a large face mill or two would be quite ample for most cases and is in fact all that is necessary for the successful operation of the machine. The amount and quality of work depends under such circumstances much upon the operator, as is the case with the planer. It seems to be the common thought that the milling machine may be placed in the hands of incompetent workmen without affecting the results. It is far from the truth; the results depend much on the operator in ordinary work. Unless the work is of a special nature and is handled by the aid of special equipment, the personal factor enters very much into the question. Thus unless we have these special features we must make up the deficiency by employing more intelligent help. We do so in the case of the planer, why not in the case of the miller also?

It is to be hoped that in the near future we will see an impartial division of work between the planer and its competitor, the milling machine, to the glory of each. This will surely take place as the milling machine begins to fill in the gaps and shows what a cost reducer it really is.

Said the Turbine to the Coal Pile

By C. J. S., Montreal.

WITH the coming of the steam turbine we were told that it was the engine of the future; that is but a few years ago. Again, only the other day, Dr. Riedler, speaking to the Society of German Engineers, said, "The turbine is no longer the motor of the future:—it is the steam engine of the present." There is certainly much to support such a view. At the same time, the present steam turbine has a few weak spots; and the reciprocating engine a few strong points. Yet, in its way, it is curious that those stoutest and most uncompromising in defence of the older form of power plant, are perhaps least competent, by training or experience, to give a sound opinion. We can understand how men used to only one type of machine would be prejudiced; but the engineers running turbines—the marine engineers in particular—are men who have served their time with triple or quadruple expansion engines. To them turbines mean a change, with a whole engine pathology to relearn; yet not one of these men would willingly go back to pistons, on an even choice. From the engineer's point of view, it is being more and more recognized that the turbines take more observant attention, but less work.

Then in the all-important matter of coal consumption, engine-room gossip has it that the Allan Line's "Virginian" was doing the trip from side to side last season on a third less coal than the crack Empress boats, and at a shade better sea speed. It is generally credited that she has never had the steam the turbines could use—not even on her trial run, when she logged within a fraction of 20 knots.

It is said that on her best trips last year the fires were burning 28 lbs. of coal per sq. ft. of grate area, for days on end, probably a record in itself, for the New York fliers think they are working hard when they do 25 lbs.; but even at this limit of a stoker's endurance the boilers could not hold the turbines going without throttling; and, of course, in a Parsons turbine, throttling means lost work.

The tendency to give their turbines too much initial steam area seems to be a failing of Parsons' marine designers, and has gained for their machines the reputation of "great steam eaters."

In the case of the Cunard "Carmania" the steam consumption was so excessive that it was found necessary to bend in the blades of the first few rows, that the boilers might hold the requisite working pressure. This expedient, while it improved the working condition, probably affected the efficiency adversely. Rumor had it that "too many cooks" and "the broth" was part of the trouble. The Cunards did things in a large way, and organized a "board of advisors," of very distinguished attainments. While this may have assisted Parsons in a well-merited advertisement, it is claimed it gave no material assistance to the shop's designers. It is to be sincerely hoped that both lessons have found application in the new 25-knotters, now being engined. But, apart from these much-noised-abroad tendencies to overwork the boilers, both land and marine turbines have shown remarkable overall economy. That is to say, while the steam consumption is large for their size, it is used most efficiently.

Although figures are somewhat jealously guarded from competing lines, yet it is known that in some of the passages of the "Carmania" and the "Virginian," the coal economy has surpassed even the best trial performances of triple expansion engines of any size, and not excluding special naval trials—not, as a rule, to be compared with the conditions of a working voyage. When we have the great 14 ft. dia. rotors in the turbines of the "Lusitania" and her sister ship, we may expect a performance that will silence once for all carping criticism of the turbine, for marine propulsion at least. For there is every reason to believe that the economical advantage of the turbine increases with its size and horsepower.

For central station work, there is only one engine-room growl, to keep the generator cool. It is claimed the proximity of the electric generator to radiating steam surfaces, and heat transmission through the shaft is not good for the output

Consideration of Canadian Patent Act

Compulsory Manufacturing Requirements of the Canadian Patent Act, Viewed in Reference to Dependent Patents.

By Egerton R. Case, Toronto.

(Continued from last issue).

From the foregoing paper, we obtain the following deductions : That the owner of a Class B-C patent is in the following predicament : (a) The Patent Act punishes him by destroying the validity of his patent if manufacture thereunder is not commenced within the authorized period, and (b) the same Act offers him no positive relief for compulsory non-manufacture. He is left, as best he may, to make suitable arrangements with the owner of the Class A patent, in order to keep valid his patent. Under certain conditions, the owner of Class B-C patent is at the mercy of the owner of Class A patent, and for the reasons before set forth, this should not be the case. To the most heedless it is quite evident that his interests should have proper recognition under our Patent Act.

Because of the foregoing disabilities to which Class B-C patents are liable, we reach the following conclusions : firstly, that as our Patent Act now stands, it cannot, under certain conditions, give protection to Class B-C patents beyond two years from their date, or an authorized extension of that time, (supposing the commissioner's extension of the two-year period is valid as applied to a Class B-C patent) ; secondly, that the owner of a Class A patent is treated more favorably than the owner of a Class B-C patent ; thirdly, that the owner of a Class A patent has an unfair advantage over the public generally in that he can manufacture under a Class B-C patent as soon as same lapses, thus deriving greater benefits than he ought to derive, because (a) as he owns Class A patent, and although Class B-C patent has lapsed, he has, as a consequence, the monopoly of the invention covered by Class B-C patent, and (b) as a consequence he can market the two inventions merged into one article, or he can manufacture the invention covered by Class A patent possessed of its own identity, and so make the public pay double tribute to him. The full usefulness of Class B-C patent does not belong to the public until the Class A patent expires, which is quite right and proper ; but it is clear that the owner of Class A patent should not be in a position to enjoy the fruits of the invention covered by Class B-C patent after the manner that our Patent Act now permits.

This paper is not written with the view to altogether point out how our Patent Act may be amended to give ample protection to all classes of inventions, as it is primarily for the purpose of pointing out wherein our Patent Act is very faulty. It is suggested that our Patent Act be amended by striking out Sub-section (a) of Section 4, and providing a section whereby when it is impossible to manufacture under a Class B-C patent without necessarily infringing a Class A patent, that this Class B-C patent be protected from nullification. To provide for Class B-C patents, we can adopt a portion of the French Patent Act, as the latter part of paragraph 2, article 32, provides for an excuse for non-working. The above or similar provisions are necessary in our Patent Act as it now stands, for the reasons amply set forth hereinbefore.

I have looked into this question relative to Class B-C patents in foreign countries very carefully, and here state results of my inquiry.

In the United States, the owner of a patent is under no obligation to manufacture, nor is he compelled to grant licenses; therefore it will be seen that the holder of a Class B-C patent, while he cannot manufacture under his patent, is not wrested of same because of his enforced inactivity. Suppose a Class A United States patent is fourteen years old. Now the owner of Class B-C patent will only have to wait three years until he is perfectly free to start manufacturing. He may manufacture the invention covered by Class A patent, and market it, or he may incorporate his invention covered by Class B-C patent in the invention covered by Class A patent by re-organizing or re-constructing this latter and market the product. What a grand privilege the owners of Class B-C patents in the United States enjoy as against the severity inflicted upon the owners of the same class of patents in Canada by our Patent Act as it now stands !

In Great Britain, a patent cannot be upset through non-user or non-working.

In Belgium, Class B-C patents will expire with the expiration of Class A patents.

In France, the law courts have considered the question relative to Class A and Class B-C patents, but same has not been decided in such a manner that definitive information can be had in relation thereto. As a general rule, the French law cannot oblige the owner of Class B-C patent to work same when the same law would punish him for so doing. In order to provide for Class B-C patents, Paragraph (a), Article 32, of the French law provides for an excuse for the non-working of this class of patents. But if it could be proved that the owner of Class B-C patent could have obtained a license from the owner of Class A patent, and could consequently have worked his patent, it is more than likely that the French law courts would upset Class B-C patent.

Neither in Norway nor Sweden does the Patent Act of these countries make special provisions for Class B-C patents. But in Denmark, the Danish Patent Act empowers the commissioner of patents to grant Class B-C patents. As far as I have been able to ascertain, no court in the above-mentioned Scandinavian countries has been called upon to decide as to how the working provisions of their respective Patent Acts are to be complied with in case the owner of Class A patent will not grant licenses on reasonable terms.

As regards Austria, Article 21 of the Austrian Patent Law states as follows :

"The patentee of an invention which cannot be worked without using a previous patented invention, has the right to ask from the owner of the prior patent the grant of a permission to use the same, provided that three years have elapsed since the date of publication in the patent journal of the previously-granted patents, and provided that the latter invention is of considerable industrial importance.

"The license granted entitles the owner of the prior patent to ask on his part for a license from the owner of the later patent, which gives him power to use the later invention, provided, however, that the later invention is in real connection with the previous invention."

From the above, one will readily understand that the owner of Class B-C patent may compel the owner of Class A patent to grant him a license.

There is no provision in the Italian, nor in the Hungarian patent laws, similar to that of Article 21 of the Austrian patent law. But in Italy and Hungary the owner of Class B-C patent would comply with the working requirement of their respective patent laws if within the term fixed by the patent laws of the said countries, he have advertisements published or direct offers made to Italian or Hungarian manufacturers (as the case may be) by which the holder of Class B-C patent would prove his readiness to have his invention industrially worked in those countries, and to permit the use of same by the grant of licenses to manufacturers, Italian and Hungarian. It would seem that it would also be wise for the owner of Class B-C patent to make direct offers of a license to the owner of Class A patent.

In Spain and Portugal there appear to be no special provisions in the respective Patent Acts for those countries covering the relationship of a Class B-C patent to a Class A patent.

As regards the German Patent Law, I have not been able to find out that it covers fully the question I have brought up relative to the relationship of Class A and Class B-C patents. It would seem that the general opinion is held in Germany that if the improvement covered by Class B-C patent is a material one surpassing the invention covered by Class A patent in its technical results, and if the owner of Class B-C patent be using only a small part of the invention covered by Class A patent, then in such a case the owner of Class A patent would not be entitled to prevent the owner of Class B-C patent from using the invention covered by Class A patent. On the other hand, if the invention covered by the Class B-C patent is of a very small character, then, of course, the owner of Class A patent would be fully justified in not granting a license to the owner of Class B-C patent, because the invention covered by Class B-C patent would almost wholly derive its benefits from the invention covered by Class A patent.

Montmorency Falls, Quebec, One of Our Water Powers.

As regards Switzerland, at the time the Federal Parliament discussed the Patent Act, it was clearly shown that the interpretation to be given to Section 3, Article 9 thereof, is that the working of a Swiss patent may take place in any country. Therefore the manufacture of an invention patented in Switzerland, in the United States or Canada, for instance, will comply with the conditions of the Swiss law, which requires that an invention be worked within three years from the date of application of the patent covering same.

The Japanese law requires that an invention be worked within three years from date of issue of the patent.

My agents in France, closely associated with the International Congress of Industrial Property, recently write me, and state that: "In the near future, it is expected that the continental nations of Europe will temper, if not entirely abolish, the working clause contained in their patent laws. At the last meeting of the international association, held this year (1906) at Milan, Italy, a resolution was passed advocating the abolition of that clause and its being replaced by the so-called compulsory license system, thus suppressing forfeiture for non-working. It now remains for this resolution to be taken into account at the next meeting of the delegates of the states comprised in the National Union, and if the delegates adopt the suggestion, the resolution will become law by way of a further additional Act to the convention of 1883."

For a long time prior to the above-quoted letter, I had in mind the preparation of a paper calling attention to the disadvantages of our Patent Act as it now stands.

What I have herein stated applies to arts, machines, manufactures, and compositions of matter.

The present chiefs of our Canadian patent office are to be highly commended for having adopted the compulsory license clause, and it is to be hoped that they will, in the near future, so amend our Patent Act as to relieve the holders of Class B-C patents of the disabilities under which they now labor.

Foundry Practice

MODERN IDEAS FOR THE TRADE.

Modern Foundry Practice

By Samuel Groves

THE art of founding in metals is of very ancient origin. History reveals it as being in existence at the earliest dawn of civilization. In Gen. iv., 2, Tubal Cain is described as the instructor of every artificer in brass and iron.

An historical retrospect showing the lines along which founding has progressed from primitive times up to now, would make an interesting story; but our purpose is entirely practical, hence, while a study of the rule-of-thumb methods practiced in ancient foundries might serve to pass a pleasant hour, it certainly would not be as profitable as a careful exposition of the science of founding in well equipped, modern workshops. Before going right to the heart of our subject, however, I may say that there is something almost pathetic in the passing of the old-fashioned, illiterate, but skilled resourceful moulder of the ancient regime; who knew nothing about metallography: cementite, pearlite, etc.; chemistry: carbon, silicon, etc.; static pressure on moulds; or the scientific values of silica, alumina, etc. in sands; but who, the moment a complex pattern of magnitude was placed before him, knew how to design a flask for it, grade and manipulate the necessary sands and facings; fix cores; select metals, and finally produce a sound casting with a skin fair to see. James Hemphill, founder of the firm which has built nearly every large rolling mill plant in the United States, was once asked, what, in his opinion, constituted a good moulder? He replied: "The best moulder is the man who has the greatest number of tricks up his sleeve." This sagacious, picturesque type of moulder of the old school, is fast disappearing, and a new order of technically trained mechanic is taking his place.

Fig. 1—An Ancient Foundry.

In connection with the Carnegie Technical School, Pittsburgh, is a model foundry, equipped with cupolas, moulding machines and labor-saving appliances, under the supervision of scientifically trained teachers, specially selected for their practical knowledge of founding. Similar departments are in existence at Purdue and Cornell Universities; showing that American educationalists are fully alive to the economic importance of this branch of industry. In a little while there will be no difficulty in finding men thoroughly equipped theoretically and practically to fill postions of trust in great foundries of the United States. Alas, Canada is doing nothing whatever to train her young men along these lines. But something should be done and done quickly, too, if the Dominion is to hold its own in the industrial competition of the world! We have already commenced to develope the great mineral resourees of the country, and in manufacturing should begin where the United States and European countries leave off.

The aim of the following series of articles will be, to set forth graphically, the latest methods and developments in modern foundry practice; in the hope that they may be of practical value to many a manufacturer, foundryman and engineer of works. And if they do something towards attracting the attention of the directors of our technical institutions to the rank folly of utterly neglecting the subject of founding, in our halls of learning, another object will have been achieved.

The Renaissance of Ironfounding.

In days gone by, the mechanical genius of the world was largely concentrated on the invention of special tools and labor-saving appliances for handling and finishing rough castings in the machine shop; while the foundry where the castings were made, was left to struggle fitfully along with its antiquated cupola, erratic fan, dangerous drop ladle, stationary jib crane, wasteful core oven, primitive hand riddle, flexible wooden flasks and costly patterns—all operated and applied by rule-of-thumb. During the last ten years, however, a startling change has come over the scene. The foundry, like all things suitable and necessary to the ascent of man, has been passing through the evolutionary process. Dark, gloomy, comfortless foundries, stifling hot in Summer, intensely cold in Winter, and devoid of adequate ventilation, are no longer universal; for in the great industrial centres may now be seen well-lighted and admirably ventilated shops, with walls of brick, or concrete-steel, and equipped with electric traveling cranes, trundle-geared ladles, sand sifting machines, drop bottom cupolas, tumbling barrels, pneumatic riddles, rammers, chippers, hoists and mould cleaners, portable mould driers, cycloidal blowers, pyrometers, rolling doors for ovens, core-making machine, testing apparatus, and wonderfully conceived moulding ma-

Fig 2—Nasmyth's Ladle.

chines—which are fast transforming the iron founders' craft from an art into a science.

In the larger of the 5,000 or more foundries in the United States, most of the machines and appliances enumerated are to be seen; but it is a deplorable fact, that in both Great Britain and Canada, these labor-saving devices are—with the exception of an isolated shop or two—conspicuous by their absence. The remarkable progress made in the United States has taken place since the establishment of the American Foundrymen's Association, in 1896. At the annual conventions manufacturers or their representatives have taken every advantage of the new ideas advanced, and economic problems solved in the papers read and discussed before this powerful organization, by West, Keep, Moldencke, Mumford and other well-known foundry experts; hence, the transformation scenes beheld in many

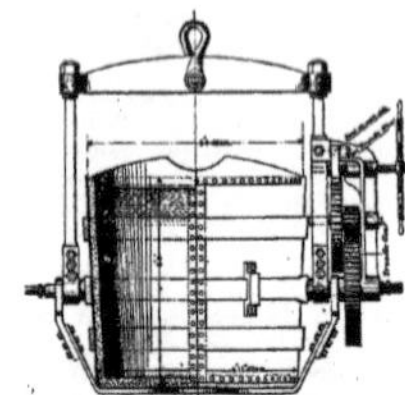

Fig. 3—An 18-ton Trundle Geared Ladle.

an old and fine equipment in many a new American foundry to-day. Instead of the crude methods of handling material and chaotic condition of things which characterized the average foundry twenty years ago; one can now perceive a rapidity of action, systematic method of handling and general orderliness which is manifestly conducive to the more economic production of castings. The first cost of these labor-saving installations is, no doubt, considerable, but those who have taken the risks are now reaping a rich reward in the shape of continually increasing profits. Many an engineer in charge of a manufacturing plant is regretting that he so long neglected the foundry, since he is now getting better castings, increasing the output, decreasing the percentage of loss and producing at a much lower cost.

Nasmyth, The Pioneer.

Prior to 1838, the hand power jib crane was practically the only labor-saving device to be seen in the foundry. About the date mentioned, however, James Nasmyth—immortalized in Smiles' "Lives of the Engineers"—invented the application of the worm and gear for mechanically operating foundry ladles (Fig. 2). He never made anything out of the patent, for he had drawings of the device prepared and distributed to the principal existing foundries free; since he looked upon his invention not only as a labor-saving, but as a life-saving device. In the pouring of heavy castings, serious accidents had occurred by the overturning of large ladles full of molten metal. Well do I remember as a boy in the "seventies," seeing five-ton ladles operated by hand, with six men on the tilting and steering handles. When the metal splashed out of the runner gate over the laborers, it was no uncommon thing to see them scamper like rats, leaving the skilled moulder controlling the pouring operation, with the whole resistance thrown on his muscles. Sometimes the strain was too much, and over went the ladle with its molten contents.

A still further advance in ladle mechanism was the introduction of the well-known trundle gear, in which the first motion shaft pinion has only two teeth. The first I ever saw was in the Walker Manufacturing Company's, Cleveland, Ohio. In 1895 the writer designed the 18-ton ladle with trundle gear, shown in (Fig. 3).

By means of this ladle, together with an old 9-ton ladle—which he remodelled and fitted with the trundle gear operating mechanism as illustrated in detail (Fig. 4)—castings were poured weighing over 20 tons.

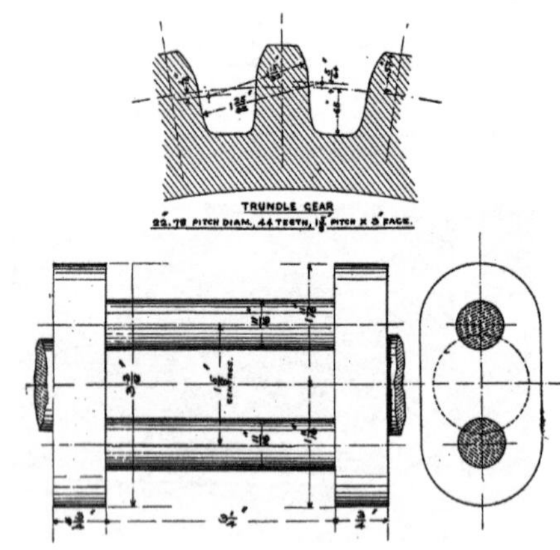

DETAIL OF STEEL TRUNDLE PINION.

Fig. 4.

The mechanical advantages of the trundle gear over the worm gear as applied to ladles, are manifest. In the first place, the efficiency of the worm and worm wheel is the lowest of all forms of gearing, since they consume by friction one-quarter to two-thirds of the power received. True, it is safe, but the trundle gear is equally safe when rightly designed, and much quicker in action, while the objectionable "drop" common to all crane ladies, may by this device, be reduced to a minimum. In fact, we have in the trundle pinion an ideal prime mover and mode of motion for metal pouring.

The only instance of trouble that I ever heard of, was on a ladle where the pins were defectively secured to the housings, and the trundle pinion itself placed on the second motion shaft instead of the first, subjecting it to the abnormal stresses transmitted by the compounding of the gears, thus abrogating largely the mechanical advantages claimed. I well remember the pleasure pictured on the face of a well-known Pennsylvania consulting engineer, as he witnessed the rapidity with which one man on the "pilot" wheel, reversed the 18-ton ladle (Fig. 3) when about 12 1-2 tons of metal had been poured into a mould, and the foreman cried "Up!" Indeed, there is as much difference between a properly designed trundle-geared ladle, and one fitted with the old Nasmythian worm gear as there is between the antiquated, slow-running horse cars and the swift-moving electric cars of our modern streets.

Before passing from the subject of ladles, one may opportunely refer to a most reprehensible custom practiced in some foundries to-day, which deserves a passing word, and that is, the daubing and lining of comparatively large crane ladles with common clay wash and fire clay only. All ladles—five ton capacity and upwards—should be lined with split fire bricks 8x4x2. By this plan, not only is the danger of burning the shell avoided, but the metal may be kept hot much longer, since the latent heat is not dissipated by radiation. It is true, that a brick lining is a laborious job; but it may be kept clean and intact a long while if a few handfuls of Dr. Kirk's flux is sprinkled on the bottom and around the sides of the ladle before the metal is tapped out of the cupola.

(To be continued.)

Manganese in Low Silicon Iron

By J. F. Gaffney †

IN cast iron having from 0.60 per cent. to 0.70 per cent. silicon, the addition of manganese above 1.38 per cent. gradually hardens the metal, the combined carbon and the chill increasing with the addition of manganese. When the manganese is high and the casting large enough to be grey, the fracture is open and coarse, and the graphite scales very large and crystalline. High manganese to a certain extent prevents the absorption of sulphur from coke.

When manganese is below 1.38 per cent. (this percentage is only approximate and depends largely upon the percentage of silicon present; the lower the silicon, the sooner the manganese will commence hardening the iron) its action is different: it softens the iron, lowers the combined carbon and decreases the chill; this effect is more marked where the sulphur is high.

Manganese may be used either for decreasing or increasing the chill, depending upon the nature of the mixture and the amount of manganese used. Manganese can be used to advantage in low silicon and chilling iron in the following cases:

First: In mixtures where the per cent. of scrap is large and the sulphur necessarily high, (this will occur in a car wheel mixture, where usually a large portion of old metal is used) the results of this increase in manganese would be a lower sulphur, lower combined carbon, less chill and greater strength.

† The Dominion Foundry Supply Co.

Second: Very often chilled plates are required, having hard chilled faces and soft backs suitable for planing. Manganese added in the right proportion will reduce the tendency to mottle and make a comparatively soft graphitic back.

In all cases where chilling irons are melted and the sulphur is over 0.07 per cent., the iron can be strengthened by the use of ferro-manganese or pig iron having a high percentage of manganese.

In the manufacture of large hydraulic cylinders the manganese should be kept low, as it is necessary to have a close mottled iron to withstand the pressure and prevent leakage.

Phosphorus has practically no effect upon the grade of iron as determined by fracture. As phosphorus increases the total carbon decreases, and the normal amount of carbon varies from about 3.75 per cent. in normal hematites down to about 3.30 per cent. in normal Cleveland irons containing 4.5 per cent. phosphorus; that is, with an increase of 1.5 per cent. phosphorus there is an increase of 0.45 per cent. of total carbon. Phosphorus has no effect upon combined carbon, but as the phosphorus increases the iron shows a lighter color, due to the phosphite of iron present.

Usual Composition of Castings.

Castings for machinery and general engineering work vary greatly in composition in different districts, but as a general rule the phosphorus averages about 1 per cent. Manganese usually varies from 0.3 per cent. to 1.5 per cent. but the average is about 0.7 per cent. Silicon may vary from 1.0 per cent. to 3.0 per cent.; but the average is 1.8 per cent. Total carbon varies to a very great extent. In some cases the total carbon is entirely in the form of graphite; in others the greater part may be combined. It may vary largely in the same casting. Sulphur usually varies from 0.03 per cent. to 0.15 per cent. In special engineering castings where great strength is required, total carbon, manganese, silicon and phosphorus are all low usually.

For castings required to be soft and tough, hematite iron alone is used. Iron for special castings is in some cases remelted once or twice before casting. This is only necessary when a higher grade of iron than is required is being used. Where reasonable care is taken in the weighing of all materials charged into the furnace, the results will be quite as regular without this remelting. The effect of remelting in the cupola is equivalent to lowering the grade. Usually silicon is decreased from 0.2 per cent. to 0.3 per cent. Sulphur is usually increased to the extent of from 0.02 per cent. to 0.04 per cent. About 25 per cent. of the manganese is lost. Total carbon increases or decreases according to the amount present, whether below or above the normal.

Phosphorus.

A very slight amount of phosphorus is absorbed from fuel by iron in melting. It is practically a constant quantity, the small increase being proportionate to the amount of phosphorus in the fuel.

Manganese.

Manganese loses in passing through a cupola. Its action is that of a protector from oxidation, and the amount lost in passing through depends upon the amount of blast and also upon the percentage of sulphur in the fuel. The greater the amount of air forced into the molten metal the more manganese is lost as an oxide. The greater the percentage of sulphur in the fuel the greater will be the percentage of manganese passing into the slag as a sulphide. Manganese is not united with iron as a compound; it is rather alloyed with it, having practically no affinity for the iron.

Carbon.

Iron may lose or gain carbon in melting. This is governed by the composition of the iron and the conditions or method of melting. An iron while passing down a cupola has a portion of its carbon burnt off, the quantity depending upon the amount of blast and the time which the iron takes to pass through the zone of the cupola where the oxidizing influences are at work. After having a portion of its carbon removed, the molten iron comes into contact with the incandescent coke in the bed, and from this it absorbs carbon. The hotter the iron and the longer it remains in contact with this fuel the more carbon it will absorb. If the amount of fuel is sufficient, the iron may absorb more carbon than it loses in passing the tuyeres. If, however, the blast be heavy, and the ratio of fuel to iron small, the iron will lose more carbon in passing through the oxidizing zone than it can possibly pick up from the bed. Melting with high blast and low percentage of coke will cause a loss of carbon, while low blast and high percentage of coke will cause a gain.

The original composition of the pig iron has much to do with this loss or gain. An iron high in carbon, if melted in a cupola with a small amount of fuel, will lose in carbon; while a pig iron low in carbon, melted in the cupola with a relatively large amount of coke, will gain in carbon. Thus a low carbon tends to gain, a high carbon tends to lose, when melted in a cupola.

It is more correct to say that a low carbon, low silicon iron, will gain in carbon. The action of carbon and of silicon on cast iron are almost identical, although the results are different. Pure iron will absorb carbon up to 6.67 per cent.; it will absorb silicon up to 23 per cent. One part of carbon is then about three and one-half times as effective in cast iron as is one part of silicon. Iron in the melted state may be considered a solution in which carbon and silicon each reduces the power of the iron to dissolve the other. For every rise of 0.1 per cent. in pig iron made under the same conditions, there will be a corresponding decrease of 0.35 per cent. in silicon, and vice versa.

Summing up, a small percentage of fuel, high blast, high carbon and high silicon tend to cause loss of carbon in remelting. A low blast, large percentage of fuel, low carbon and low silicon tend to cause iron to gain carbon in melting.

Prof Turner's Experiments.

In cast iron, graphite for crushing strength should be under 2.0 per cent., for general strength it should be about 2.8 per cent.; for strength and softness it should be about 3.00 per cent.; for softness over 3.10 per cent., when the total carbon is 3.4 per cent.

Combined carbon gives maximum general strength at 0.4 per cent.

Do not dare to live without some clear intention toward which your living shall be bent. Mean to be something with all your might.—Phillips Brooks.

*Iron Melting With Induction Furnace**

By Dr. Richard Moldenke

FOR some time past foundrymen have been interested in the developments of electric smelting so far as it might be applied to their industry. As the necessary information as to whether iron may be melted in a commercial way can only be given by our electrical friends, perhaps a few remarks on the requirements involved may be of use to those who have apparatus and processes adapted for foundry work.

The average foundryman of to-day has constant calls for steel castings along with his regular routine work in grey iron. This, on account of the rapid introduction of the steel casting into machine construction. These steel castings he must sub-let to the steel foundries. The latter are looking for tonnage, and do not like to fill up with small quantities of comparatively light work. Hence high prices, which cut the foundryman's profit. In a similar way the foundryman has to deal with malleable castings. The malleable castings founder has both steel and grey iron put up to him, and it may not lie very far in the future when the steel founder may be asked to produce both grey iron and steel castings along with his steel. It is to be understood, of course, that this is all brought about in making contracts with concerns who place their work by the year, or the thousand tons, be the castings what they may during the period of contract.

Now, very few foundrymen are equipped for this. They strive to place these outside lines with other people at the smallest loss to themselves. Yet if there were a convenient and easily operated process, many of them would install it immediately to take care of just such conditions.

The smelting, or rather plain melting, of iron electrically has always seemed to me ideal in its way. We do not want to produce chemical changes in our mixtures if we can help it. We want only to melt quick, produce very hot iron, and punish the metal as little as we can. Every time we melt the metal under present conditions we hurt it somewhat, the degree of the damage done depending upon a number of conditions, both chemical and physical. We counteract this by additions of steel to reduce the total carbon, or selecting the silicon content in such a way that with the reduction in this element incident to the process, a strong iron results. If, however, we could have a process which in no way changes the composition, we could put into the melt just what we want out of it, and one of the serious difficulties of foundry metallurgy would be solved. Again, if we could regulate the temperature in such a way that the iron is not overheated while melting, but can be heated up very high afterwards, we could obviate the oxidation of the metal during the melt, and in addition remove any existing evil of that kind by the use of ferro-manganese in the melt when it has been brought to practically a steel temperature, at which point the ferro-manganese will do its work.

The induction furnace, it seems to me, fills these requirements, and I would like to see more work done along the line for the foundry. I may be in error, but it would seem to me that scrap of all kinds, properly selected, is all that need be melted, cleaned by some ferro-alloy, and then cast in the usual way. The enormous steel production will always yield scrap enough to supply the demand for small steel castings, once a process of this kind can be made to work commercially, and I would be very glad to be of assistance in bringing a process of this kind to the attention of the foundry world.

Taking the non-ferrous metals. A brass foundry would have abundant use for an electric smelting process, if run on the lines laid out above. The melting loss in the brass foundry in zinc and tin is a great one, and one that runs up into money quickly. A clean, wasteless process, as the electrical one should be, would be a boon to the industry. There are many foundries which could melt within short periods of time during the day or night when their plant is not used for its regular purposes. Hence only the electrical apparatus proper might be required, the necessary power being available.

The foundry has practically every melting process under the sun in use for making the various classes of castings. We see the open-hearth, the Bessemer and the cupola, the air furnace and the crucible in operation everywhere. Yet every metallurgist knows that the crucible process gives the best metal, if you can stand the cost. Now, the electrical furnace, if the melting proper is conducted so that the advantages of the crucible process are retained; that is, the temperature not allowed to exceed safe limits, and the metal kept from oxidizing influences, is bound to give the highest class of product, and it seems to me with the greatest ease of manipulation.

I would therefore urge the study of the line of work in contradistinction to the smelting of steel from ore; in other words, to simply melt, instead of com-

* This article, written by Dr. Richard Moldenke, Watchung, N.J., in Electrochemical and Metallurgical Industry, should be of considerable interest to Canadian foundrymen as a suggestion from a man who has had long experience in foundry affairs.—Editor.

pleting the cycle of operations in a metallurgic process. The great foundry industry is ready to take this up once the commercial feasibility has been demonstrated.

A PORTABLE CORE OVEN.

MANY attempts have been made to design an economical and at the same time efficient portable core oven. Most cores are baked in heavy brick structures which take up an unduly large amount of space, and which cannot be moved or shifted.

Several types of ovens have been put on the market, which are equipped with swinging doors. The objection to all ovens of this type is that there is a lare amount of waste space, due to the difference between the area of the quadrant and the area of a square having sides equal to the radius of the quadrant. One objection to the quadrant or semicircle type of oven is that it is not suitable for drying long cores such as are made on core machines, for the clearance which must be left along the curved edge of the shelf still further reduces the area that can be covered with core plates. The objection to all ovens of the stationary shelf type is that the core plates have to be handled into and out of the oven through openings, which interfere with the draft or heat conditions and thus retard the baking of the cores in the oven.

The accompanying illustration is of a new drawer oven, which is self-contained and economical. The one shown is equipped with four drawers, supported by rollers at the back and by pipe guides at the front. The drawers are removed from the oven by slipping a pair of handles into the pipes shown in the front, lifting them slightly and running them forward by means of the wheels on the inside at the back of the shelves, the action being much like that of a wheelbarrow. The shelves are self-sealing, in that when they are drawn forward a plate closes the opening, so that the baking of the cores on the other shelves goes on without interruption. In one of the illustrations the shelves are shown open and drawn to different distances, so as to illustrate the variety of cores which can be baked in an oven at one time. The large cores on the lower shelf are 7 inches in diameter.

This oven was run for an entire day on one bucketful of fine coke or breeze, swept up in the coke bin. The consumption of fuel in an oven of this kind would naturally be less than the other types, as the space is all utilized to the greatest possible extent, there being no ducts or passages in which radiation takes place.

The fact that the fronts of the doors drop down, as shown, enables the core plates to be taken out easily.

These ovens can be distributed about the core department, thus minimizing the distance the core makers have to walk to put their cores in the oven, and a rearrangement of the department can be effected very easily, as the portable oven is no more difficult to move than the core bench.

One point in this oven to which especial attention should be called, is the provision for supplementary shelves dividing the main drawers into divisions.

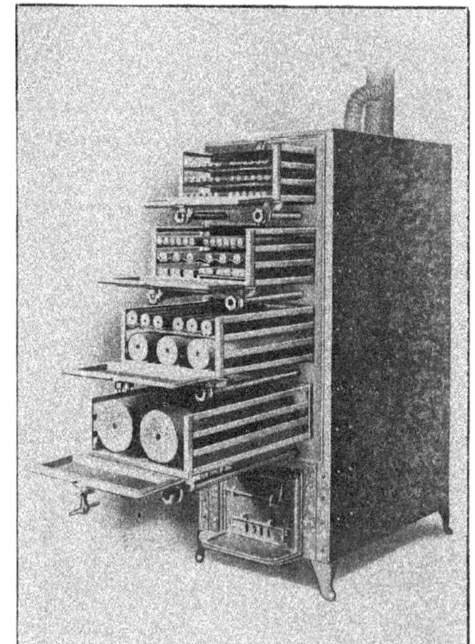

A New Portable Core Oven.

As shown in the illustration, the two highest drawers are divided into three shelves each, the next into two and the bottom one is arranged without any supplementary shelves. The facility with which these different divisions can be put in will enable the operator to dry any size of cores readily.

This device was designed by George H. Wadsworth, of Cuyahoga Falls, Ohio, who has applied for patent on the same, and it is manufactured by The Falls Rivet & Machine Co., Cuyahoga Falls, Ohio.

NEW FOUNDRY GETTING INTO SHAPE.

When the alterations now being made to the Port Hope radiator factory are complete, the Central Foundry Co., which is to occupy the premises, will have one of the most up-to-date establishments of the kind in Canada. A considerable amount of machinery has already been received. The Corliss engine has been manufactured by Messrs. Goldie & McCulloch, and the cupola blowers by the Sheldon Manufacturing Co., of Galt. The track and steel cars are manufactured by the Atlas Steel Car Co., of Cleveland, O., and the Hammant Steel Car and Engineering Co., of Hamilton. The sand conveyors and mixers are manufactured by the Standard Sand and Machine Co., of Cleveland, O. The electric generators are manufactured by the Canadian General Electric Co., and the pipe mills are being manufactured by the W. W. Sly Co., of Cleveland, O. All patterns and flasks are being manufactured by the firm itself.

Mechanical Reviews

OPINIONS AND STUDIES WORTH NOTING

THE EVOLUTION OF A TAPPING OPERATION.

THE three sketches shown represent the jigmakers' efforts to find a quick and satisfactory way to tap the zinc nut which is shown full size. The nuts were to be made in large quantities.

The first attempt was by means of a spring chuck in the tapping fixture, Fig.

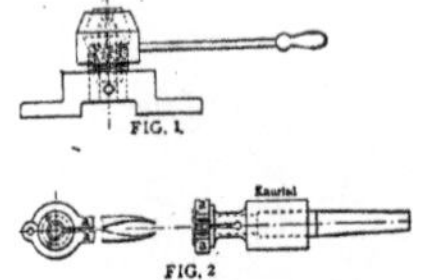

1, which was used on a vertical-spindle tapping machine. A spring knock-out, not shown, was used for ejecting the nut. This method was very slow, and the chuck gave much trouble by clogging with chips.

As some more productive way became necessary, it was decided to utilize an old Fox lathe for the purpose, and the device in Fig. 2 was made. This is a spring chuck of special shape, provided with a taper shank to fit tail spindle and closed upon the work by means of the hinged handles (a.a.) The tap was made with the shank as long as the lathe would accommodate, the end of the shank milled with three V-shaped to remove, and the necessity of having a good left hand that would not tire with the constant gripping of the handles.

The tail spindle of the lathe was of square section, so that the last fixture was made to clamp upon the end of the spindle. The body of this fixture was of machine steel, and to avoid the labor of sinking a square hole in the end of it, a hole was bored somewhat larger than the diagonal of the square. Next, two pieces of steel were roughly shaped to semi-circular section, and a 90-degree V planed in the flat side, and then clamped onto the spindle and the outside turned to fit the bore of the fixture, which was split and clamped by the screw (a).

The front part of the body was planed away, as shown in the sketch, and fitted with the hinged piece, (b) side plates (c.c.), lever (d), and link (e), so that raising the lever would swing up the piece (b) on its hinge pin (f) and open up in front.

After being fitted properly, the whole was placed in its future working position and centred by the tool in the headstock and afterward chucked in the lathe, trued up by this centre, and bored grooves for about an inch and held securely by a turn of the hand in a No. 2 Almond drill chuck in the live spindle, the chuck jaws fitting into the grooves in the tap shank.

After the tail spindle had been relieved of its screw feed so as to slide freely the operation was as follows: With the left hand in position to close chuck, the right hand picks up and inserts a nut and then grips the nurled end of the chuck and pushes ahead until the tap passes through the nut, when the chuck is released and returned to the first position by both hands, the nut remaining on the shank of the tap, after which the right hand is free to begin again. The tapping is continued until the tap shank is full, when the lathe is stopped and the nuts slipped off the shank into the box.

It was this necessity of using both hands for tapping and reversing that caused this method to be abandoned and the fixture, Fig. 3, designed. There were other defects also, such as the chuck filling up with chips, which were not easy by the split bushing (g), also bored out in the rear of the bushing for clearance. The bushing (g) is hardened, and any form of small nut may be tapped in the fixture by making the bushing to suit, the two halves being held in by screws. The side plates are cut away at (h) and the oblong hole (j) is also worked out to allow the chips an easy egress. The stop (k) prevents the link from passing over the centre.

In operation the right hand remains holding the upturned end of lever (d), the lever is raised, a nut inserted by the left hand, the lever pressed down, clamping the nut, and the fixture is pushed ahead by the right hand, which, as soon as the nut is tapped, raises the lever and draws back the fixture for another nut.

The same tap was used with No. 2 and No. 3 fixtures, so that each hand performs a separate function instead of both being required, as with the methods 1 and 2. Another gain is that a handful of nuts may be picked up by the left hand and one selected by thumb and finger ready to insert, while the tapping and reversing is being done by the right. By the previous methods each nut must be picked up singly, dur-

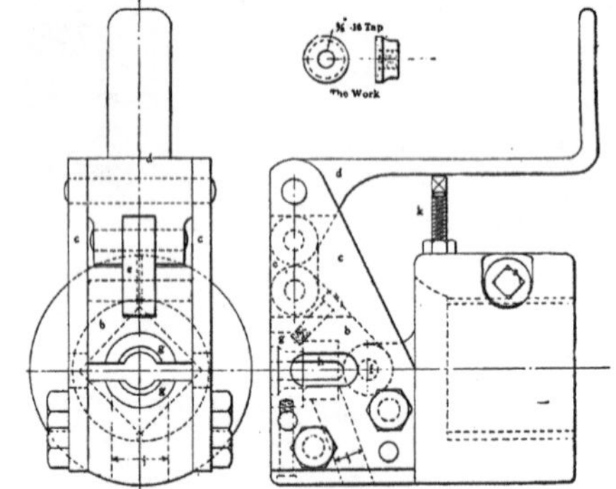

Fig. 3.—The Successful Fixture—The Evolution of a Tapping Operation.

ing which time the machine and other hand were idle.

As the same small boy operated each of these fixtures in turn, his output may be taken to indicate the merit of each method. With the first fixture the best day's work was about 12 gross in 10 hours; with the second, about 20 gross; and with the last one about 43 gross.

The same price obtained with first and second fixtures, while with No. 3 it was reduced 40 per cent. per gross. —Raymond Grant, in American Machinist.

PILOT WHEEL WITH TELESCOPIC ARMS.

I RECENTLY had a quantity of heavy drilling to do on a hand turret lathe without power feed, and found that if I could have longer arms on the pilot wheel, I could get a better leverage, use high-speed steel drills, and take a pretty heavy cut. On examination I found that, owing to the small space between the end of the arms and the tray (shown at X, Fig. 1), I could not fit longer arms without some alteration.

Of course, I could have overcome this difficulty by extending the pilot-wheel out till the arms cleared the outside of the end tray, but the idea of telescopic arms occurred to me, and I thought they would look better than the long, solid arms, also they would not be so much in the way.

Fig. 1 is a section of the bed, showing the telescopic arms in their normal position.

How They Were Made.

Fig. 2 shows the construction of one arm, all four being made in the same manner. (A) is the hub of the wheel, (b) is a cast-iron sleeve, turned and drilled

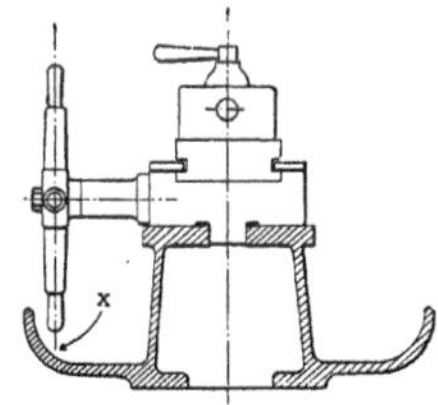

SECTION THROUGH BED OF LATHE

Fig. 1—Pilot Wheel with Telescopic Arms.

to carry the steel arm (c). The spring (d) keeps the arm pulled towards the centre, and is only just strong enough to hold it in against its own weight when hanging vertically downward. The plug (e) is driven in merely to keep the arm in the sleeve when assembling.

It requires very little effort with one hand to lift or pull the arm out, and the pressure put on for drilling keeps it out until released, when, of course, it immediately springs back to original position.—"Conrad," in American Machinist.

NEW METHOD OF MILLING THE FLUTES OF REAMERS.

IN MILLING the flutes of reamers it is customary to mill them so that the cutting edges will not come an equal distance from one another. This prevents chattering and permits the use of an even number of flutes. The difficulties encountered in milling the flutes on unequal distances, or breaking up the flutes as it is commonly termed in the shop, are that if all the grooves are milled to the same depth the remaining land evidently will be wider in the case where the distance from cutting edge to

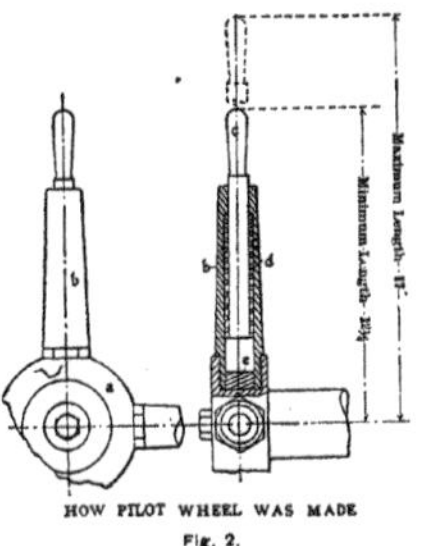

HOW PILOT WHEEL WAS MADE

Fig. 2.

cutting edge is larger than it will be in the case where this distance is smaller. To overcome this it would, of course, be possible to mill the flutes deeper between the cutting edges, which are further apart to insure that the width of the land would be equal in all cases. That this is impracticable when fluting reamers in any large quantities is easily apprehended, as it would necessitate raising or lowering the milling machine table for each flute being cut. In the Zeitschrift fur Werkzeugmaschinen und Werkzeuge, a method is shown employed by the large machine tool firm of Ludwig Loewe & Co., Berlin, Germany. The principle of this method is clearly shown in the accompanying cut. A formed cutter, eccentrically relieved, is employed which, instead of forming only the flutes, forms the actual land of the reamer, thus insuring that every land becomes equally wide with the others. The depth of the flute is determined by the depth of the portion of the cutter in front of the cutting edge of the reamer and it is easily seen that all the flutes will be equally deep.

That this method will be more expensive than the one commonly employed, in which the lands are permitted to become wide or narrow according to the

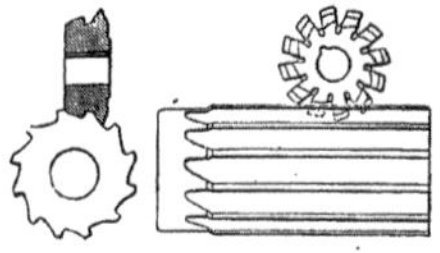

amount the flutes are broken up, is evident, but it cannot be disputed that the general appearance of the reamer will be greatly improved. The greater expense in making reamers in this manner will depend on two factors. In the first place, the eccentrically relieved cutter will cost more to produce than the ordinary fluting cutter. In the second place, the cutting speed cannot be as high with a cutter of this description as it could be with an ordinary milling cutter. On the other hand, it is possible not only to gain the advantages mentioned above in regard to width of land and depth of flute, but incidentally there is also gained the possibility of giving to the flute a more correct form to answer the requirements of strength as well as ship room, which are often by necessity overlooked on account of the straight sides forming the flutes which are necessary to adopt when using the ordinary straight-sided fluting cutter, with milling cutter teeth of the common shape. While it cannot be expected that this method will be used to any great extent on account of its drawbacks from a commercial point of view, it is ingenious and well worth attention.—Machinery, New York.

DEFECTS OF METRIC SYSTEM.

At a recent banquet of the Empire Club, Toronto, Prof. N. F. Dupuis, of Queen's University, discussed the metric system, and criticized the movement on foot to make the metric system of measurement universal by criticizing the system itself.

It was shown that from the scientific standpoint the metric system was far from perfect. After pointing out that a wide interval existed between the metric and decimal system, the professor showed that the decimal system was adopted accidentally and that it was not the most perfect one which could be devised. In all the figures under 20, only one, 14, would give a worse base than ten. It was immaterial what length should be taken for a unit.

Centrifugal Pumps

By Henry F. Schmidt

ALTHOUGH for years the centrifugal pump has been looked upon as a very inefficient method of pumping large volumes of water at low heads, generally not exceeding twenty-five feet, during the past eight or ten years considerable progress has been

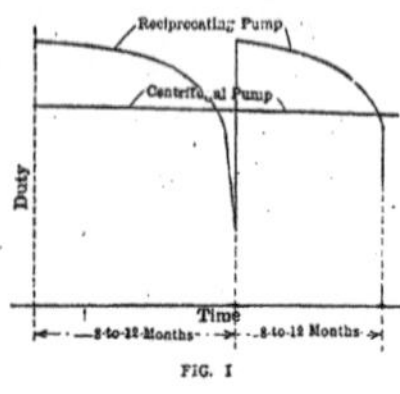

FIG. 1

made in Europe toward increasing the efficiency of this type of pump, at the same time improving the design so as to deal with very much higher heads. The result is that it is now possible under favorable conditions to obtain efficiencies of from 75 to 80 per cent., even at heads of from five to six hundred feet. This result is accomplished by two expedients, the high pressures being obtained by the use of multiple stages, that is, by delivering the discharge from one pump to the suction of the next, and so on, thus adding together the pressures created in the separate pumps; the high efficiency being in a great measure due to the addition of diffusion vanes placed around the impeller in such a manner that part of the kinetic energy imparted to the water by the impeller is converted into the potential or pressure form.

While this work has been going on in Europe for a long period, America has been backward in realizing the possibilities in the development of these pumps, and it is only within the past two or three years that it has received

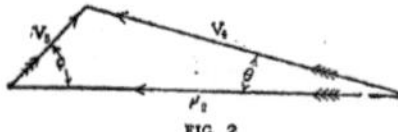

FIG. 2

serious attention for other than low-lift work. Now, however, a number of American manufacturers have placed pumps of this description on the market, albeit with one exception they have followed the designs of the European builders very closely.

In the light of past developments, it is hardly optimistic to predict that within the next few years the turbine centrifugal pump, driven either by a steam turbine or high-speed gas engine, will replace the present type of plunger pump for municipal water works, hydraulic-elevator service, fire pumps, and, in fact, in nearly every service where the volume of water to be handled is large enough so that a high efficiency can be obtained in the centrifugal pump.

Having made this statement, it is proper to submit the reasons why, as follows: First, small cost; second, small space required; third, the efficiency has been found from experience to remain practically constant over years of continuous use; fourth, absence of valves and plungers which need to be packed; fifth, only one moving part and nothing that can get out of order, consequently practically no attendance needed; sixth, almost "fool proof" construction, it being possible to shut the delivery or suction valves completely when pump is running, without injury to the pump or piping, since the pressure can only raise about 10 per cent. under these conditions; seventh, it is practically automatic, because if the pressure on the line decreases slightly,

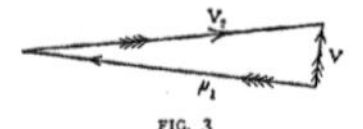

FIG. 3

the volume of water is greatly increased, while if the pressure is slightly increased—showing that the demand is not so great—the volume delivered is reduced, and this entirely independent of pressure-controlling devices or attention from the man in charge; eighth, no more oil is required than for an electric motor of the power required to drive the pump; ninth, practically no depreciation need be charged against a centrifugal pump, where the water is fairly clean, other than that required to rebabbitt the bearings when worn; tenth, steady pressure and flow of water.

The result of these conditions is such that in the case of a centrifugal pump to deliver 1,600 gallons per minute against a 350-foot head for a hydraulic elevator plant, notwithstanding that the duty guaranteed for a three-cylinder compound pump was twice that for a centrifugal pump, when the interest on the investment, oil, valve renewals, etc., were taken into account, the centrifugal pump showed itself superior by a little over two hundred dollars a year. This brings out a point to which I wish to call the attention of anyone who may have to compare the yearly costs of a plunger pump and a centrifugal pump. It is the common custom to figure the annual cost for coal on the assumption that the efficiency of the pump remains constant for the entire year, but this assumption is far from true. In the case of the plunger pump, as the valves begin to wear they begin to leak and do not work as well as at first, the leakage increasing quite rapidly toward the end of the life of the valves, until finally one or more valves give way entirely, and there is a sudden drop in the duty of the pump. This is shown graphically in Fig. 1, in which the curved lines represent the duty of a plunger pump, while the horizontal line nearly parallel

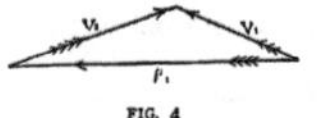

FIG. 4

to the base represents the duty of a centrifugal pump, the efficiency of which remains practically constant. The value of a pump is proportional to the area under this curve, so it is seen from the diagram that the all-round yearly value of the plunger pump is not nearly what it appears to be.

Some will no doubt argue that in a well-cared-for pump this state of affairs would never exist, but this is extremely doubtful, partly because this wear of valves goes on unnoticed until a valve breaks and causes a pound, and partly because the valves are usually rather difficult to get at, and therefore the average engineer will not take the trouble to replace a valve until it becomes absolutely necessary. Furthermore, it must be remembered that in general the man in charge of a plant is only an "average engineer." For this reason I think that in making the yearly estimate of the running costs the plunger pump should be credited with at least 5 to 10 per cent. lower efficiency than is shown on the face of the guarantee.

Owing to the fact that it is very difficult to obtain a cheap and economical steam turbine to run at the speeds re-

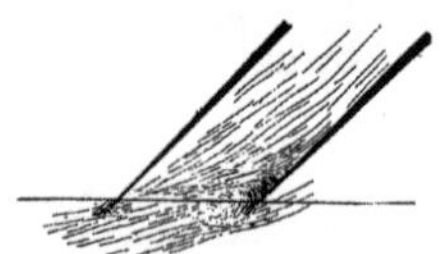

FIG. 5

quired, that is, under 2,000 revolutions per minute, little has been done so far in the driving of centrifugal pumps by this type of motor, except those in which gearing is employed. The reason for this is that the makers of pumps have found that the efficiency of the pump is seriously impaired when a high speed is employed, and it is only within the last year or so that speeds of even 1,800 revolutions per minute have been reached. This drop in the efficiency is found to be caused by an effect similar to what is termed "cavitation" in screw propellers, which refers to the vacuum formed ahead of the screw, due to the water not being able to follow the blades. In pumps, nearly the same thing happens in that the water cannot flow into the impeller fast enough, although in this case it is probably not entirely due to the water not being able to follow the blades, but also due to the blades throwing the water away from them. This is because the water enters the impeller at right angles to the direction of motion, causing shock which has a tendency to pile the water up ahead of the blade and thus throw part of the water away from the blade, instead of causing it to enter.

The velocity diagrams for a centrifugal pump with the blades turned backward are shown in Figs. 2 and 3, in

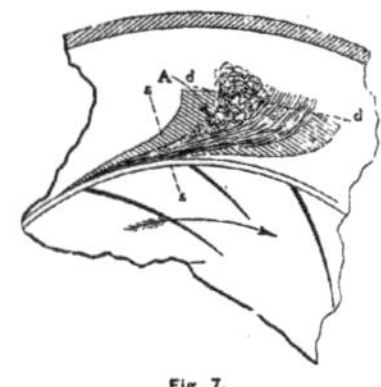

Fig. 7.

which μ_1 is the velocity of the inside of the impeller, V is the intake velocity which is at right angles to μ_1 and V_2 is the velocity of the water relative to the blade. Fig. 2 is the diagram for the exit, μ_2 being the velocity of the circumference of the impeller, V_3 the velocity of the water relative to the frame of the machine. In Fig. 5 is shown the effect which was spoken of above, namely, the piling action due to the water not entering tangentially to the blade.

From this, the writer wishes to make a rather novel suggestion: that of placing guide blades inside of the impeller, as shown diagrammatically in Fig. 6. The object of this will be quite clear from what has already been said. The action of these internal guides is shown in Fig. 4. Here, instead of the water entering the impeller at right angles to the direction of motion, it enters with a greater velocity and at an angle, as shown by V_1. Consequently, the velocity V_2 is made much smaller and the angle between it and μ_1 is made greater, so that it is possible by proper design to have the water enter the impeller nearly if not quite tangentially.

The final result of this is that the speed of the inside of the impeller may be increased considerably, thus permitting much higher speed of rotation to be adopted, and still obtain the same efficiency, if not a higher efficiency, for reasons which will presently be given. The velocity which it is possible to give V_1 will probably be from thirty-five to forty feet per second, which means that the blade velocity may be increased a similar amount, although of course it is needless to say that the lifting power of the pump would be reduced and it would therefore be necessary to place the pump lower, so that the water would nearly run into it by gravity, but this would be objectionable only in comparatively rare cases.

If the water can be made to enter the pump, higher speeds of rotation are advantageous to economy, for, with a given head, the diameter of the impellers may be made smaller, which means smaller rotation losses due to the friction of the water on the disks of the impellers. The power thus lost varies as the cube of the number of revolutions per minute, and the fifth power of the diameter of the impeller, or, very nearly by the formula,

$$HP = \left(\frac{N}{100}\right)^3 \frac{D^5}{1000}$$

in which N is the number of revolutions per minute, and D is the diameter in feet. Consequently, if the number of revolutions is increased and the diameter is decreased, the power lost in rotating the impellers is reduced, and furthermore, which amounts to the same as increased efficiency, the cost is reduced, since the higher the number of revolutions, the smaller the pump, and less interest chargeable against the installation.

Lower cost of construction brings up wishes to make. Instead of casting the impellers and blades in one piece, and finishing the surface by hand, the writer proposes making the blades out of sheet steel or bronze and bending them to propore shape on a form, then finishing them on an emery wheel, after which they are set in a mold and cast into the impeller disks and hub when the molten metal is run in the mold. In this way, the cost of finishing the im- which in the end is equal to a gain in efficiency. The diffusion vanes can be made in a similar manner and cast with the casing, and as two sides of the channel formed are polished, the sides would not need to be finished, which of course, also applies to the impeller.

Fig. 7 shows the usual form of diffusion vane employed in turbine pumps, and the writer wishes to criticize this form of diffusion vane for not following the natural stream lines of the water as it leaves the impeller. It is evident that the water leaving the impeller would naturally tend to flow in nearly straight lines, and hence in the form of vane shown in Fig. 7, there would be a decided tendency to form eddies at A, which would materially reduce the effective area of the passage at (b b). Now, the area at (a a) should be to the area at (b b) inversely as the velocities at those points, in order that the velocity of the water may be reduced from the high rate at which it leaves the impeller to that allowable in the passages, and thereby the kinetic energy converted into potential energy. If, however, part of the area at (b) is ineffective, the reduction in velocity will not be so great, or, rather, though the velocity is reduced, the kinetic energy which it represents is only partially converted into potential energy, the remainder being lost in eddies. Of course, there will always be a certain loss from ed-

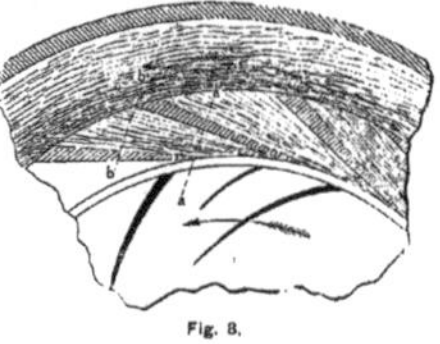

Fig. 8.

dies, but the sharp turn at (A) will tend to increase it greatly.

To avoid this, the writer proposed the form shown in Fig. 8, in which the water can more nearly follow its natural course, and which presents no sharp turns which are bound to form eddies. Another point in this connection is the number of diffusion vanes provided. In most commercial pumps now in the market the number of vanes is from six to twelve, which if applied to an impeller of large diameter makes the passage too wide, and this is also conducive to losses by eddies. If the construction of Fig. 8 were employed, and the blades formed and cast in the mold, it would be a very simple matter to increase greatly the number of vanes, although of course the opposite extreme must not be approached, since then there would be too great a friction loss.

The reason for the reduced chances for eddies when the passages are narrow is easily seen by approaching the limit when the space between the walls would be so small that a molecule could just pass through them without a chance for side movement.—Power.

Power and Transmission

Steam Gas Electricity Compressed Air Water

THE GRAPHOIL LUBRICATING ATTACHMENT.

FOR very many purposes there is no better lublicating agept than graphite in combination with oil to act as a carrier. Although this is well known the difficulties attending the satisfactory use of the combination have limited its practical adoption. These are due to its specific gravity being greater than that of the most viscous oil. As a result no mixture of graphite and oil can be made to be used as a lubricating compound, since the graphite persists in precipitating, defeating the

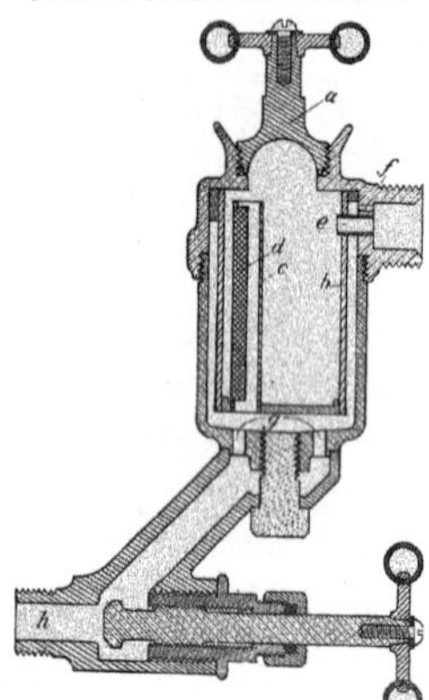

The Graphoil Lubricating Attachment.

maintaining of a uniform mixture. Either the materials must be continually agitated, or the oil must be fed in such a way as to take up a percentage of graphite, the two being brought together as near as possible to the surface that is to be lubricated, so that the graphite will be continuously carried to these surfaces and the mixture allowed at no time to come to rest long enough for the graphite to separate from the oil. The latter is the principle employed in the Graphoil lubricating specialties, manufactured by the Comstock Engine Company, Brooklyn, N.Y., one of which is herewith illustrated.

This consists of a graphite cup, a sectional view being shown. This cup is attached to the cylinder lubricator in such a way that oil and graphite are introduced into the steam pipe, and from thence to the steam chest in the same way as the oil alone in ordinary sight lubricators.

The construction of the cup, as can be seen from the illustration, consists of an inner and an outer cup separated by a steam space, which serves to equalize the pressure in and around the inner cup (b), and also serves to prevent condensation in the latter. Inside the inner cup is an upright brass tube (c) open at the top and bottom, and cut away on one side, where it is covered with wire cloth (d). The graphite is placed in the inner cup (b) after removing the filler plug (a), and rests in a compact mass around the tube (c), with its upper surface exposed.

When the Graphoil cup is connected with an oil lubricator the oil flows into the inner cup, passes over the surface of the graphite, taking up some of it, then goes through the meshes of the wire screen (d) and down the tube (c) to the bottom of the outer cup. From there it drains through the holes (g) into the passage leading to the outlet (h). Each drop of oil fed contains the right proportion of graphite, generally 1½ per cent., Dixon's No. 2 flake graphite being the grade preferred.

The attachment is constructed entirely of brass, the exterior parts being castings, the brass cup (b) a piece of seamless tubing closed at one end with a header, and the small tube (c), also of brass, cut away, as before mentioned, and covered with brass gauze of 24 mesh, No. 30 wire. At (f) are three holes surrounding the oil inlet (e), the purpose of which is to connect the space surrounding the inner cup with the cavity where the oil enters, to balance the pressure so that the oil will flow by gravity.

It is claimed for the Graphoil cup that it will save from 25 to 60 per cent. in cylinder oil. A direct consequence is a less amount of oil in the steam, which is of advantage when the steam is condensed and returned to the boilers, as it reduces the oil in the boiler feed. Further claims are that it will protect the packing and prolong its life, decrease the friction, reduce wear on the engine and increase the effective horse-power. As has been already shown, it is simple in construction and has no mechanism to get out of order. It cannot feed too much graphite. When desired the cup may be used with no graphite in it, so that oil alone is delivered through the cup. The cups are particularly adapted for use on air compressors, cylinders of locomotives, multiple oiler systems, eccentric troughs, wipers, etc. Their use on high or low speed

Fig. 1—Lazier Gas Engine.

steam engines, where the movement is irregular or not continuous, as on blower engines, elevator or steam pumps, etc., insures thoroughly efficient lubrication at a greatly reduced cost, and effectively prevents the groaning and laboring peculiar to engines doing service under such conditions.

Someone asked Thomas A. Edison, "Don't you believe that genius is inspiration?" "No," he replied; "genius is perspiration."

NEW TYPE OF GAS ENGINE.

THE accompanying illustration is of a new type of gas engine, designed by Arthur H. Lazier, and built by the Lazier Gas Engine Co., Buffalo.

Quite a number of new ideas have been incorporated in this engine, which makes it of considerable interest. The valves are operated mechanically. A double system of ignition, which can be run either separately or in combination, has been provided, and it is so arranged that the time of firing can be changed while the engine is in operation. Another feature of the engine is the running of cams and the cam shaft in oil. The system of operating the cam shaft is also new, there being but four gears in the entire machine. These gears are spiral and run in oil, which makes their operation almost noiseless. The method of taking up lost motion in the connecting rod is shown in the cross section view of the engine. The "splash" system of lubrication is used in the engine.

Fig. 2—Two Cylinder Lazier Gas Engine.

This engine, it is claimed by the makers, can be arranged to operate on almost any kind of fuel, including alcohol, kerosene, distillate, illuminating gas or producer gas.

This engine is made in sizes from 200 to 300 h.p., and owing to its simplicity of parts the makers claim to be able to build it at a comparatively low cost.

PNEUMATIC REVERSIBLE DRILL.

THE accompanying illustration is of a reversible pneumatic drill made by the Independent Pneumatic Tool Co., Chicago. The operation of these drills is clearly shown in the illustration. The motive power is of the 4-cylinder reciprocating piston type, the pistons being connected to an opposed crankshaft. The drills are equipped with the Corliss valve motion, which allows the live air to be magazined and controlled up to within ⅛ of an inch from the cylinder; this air when released quickly, acts on the piston instantaneously. This construction allows no air to pass through except what is absolutely needed; and thus economy of operation is effected.

All joints in the case have been dispensed with, excepting one between the gear case and the cylinder, thereby simplifying construction, assisting in keeping the working parts in line, and preventing leakage. The telescopic feed with which this drill is equipped is one of its particular features, since it gives a very large range.

In drills requiring great power, compound gearing is employed, thus insuring the required power with slow speed.

Pneumatic Reversible Drill.

SHRINKAGE AND CONTRACTION.

By Ernest C. Morehead.

THERE are but few founders who recognize the distinction between shrinkage and contraction, but consider it one and the same thing. However, some distinction is necessary.

The actions of shrinkage and contraction are distinct in their nature and are divided by a slight expansion which takes places at the moment of solidification.

Shrinkage is the decrease in volume of the liquid metal until it becomes solid, at which time comes the action of expansion, but this expansion is not as great as the previous shrinkage and is rarely noticed.

After this expansion the casting again decreases in size and continues to do so until it reaches the temperature of the air. This is known as contraction. Thus, they are two separate actions and should be recognized by distinctive terms.

Feeding of the mold is still another important operation which is not given due consideration. For instance, step inside the foundry and you will see a molder using a feed-head that will solidify almost as soon as the casting is poured, although the casting itself will remain in a fluid state some minutes after the head is permanently set.

You may pass to still another molder whom you will see insert a cold ¾-inch feeding-rod in a 2-inch head and very often the rod, which has a chilling effect, will draw iron from the casting rather than aid in feeding. Many cases of this kind could be cited that are every-day occurrences in most every foundry, but let us pass and get down to the proper feeding of the mold.

In feeders 2 inches in diameter a rod should not be used more than ⅜-inch in diameter.

For heads 4 inches to 6 inches, use ⅝ to ¾-inch rods and for feeders above 6 inches, rods ¾ to 1-inch can be used.

All feeding-rods should be well heated in a ladle of hot metal before being introduced in the feeding-head. The iron in the feeding-head is usually dull, an account of its having flowed from the bottom of the mold to fill the feeders, so the feeding-rod should pass through the casting to a depth that would insure hot iron being fed to the part that is going to shrink. Where feeding-rods are used plenty of hot iron should be on hand to keep the feeder open. Molders disregard this important point and frequently you see them chilling the head with water rather than feeding it with hot iron.—Obermayer Bulletin.

Machinery Development

Metal Working **Special Apparatus** **Wood Working**

OWEN PLAIN MILLING MACHINE.

THE Owen Machine Tool Co., Springfield, Ohio, have recently remodelled their plain milling machine. Two sizes of this new line, known as No. 2B and No. 3B, have been placed on the market, and the accompanying illustration shows the No. 2B.

As may be seen from the illustration, special attention has been paid to stiffness and rigidity in the improved design. The machine has a geared key, no chain being used between the spindle and the feed screw, as in the former designs.

The feed changes, of which there are thirty-two, may be obtained while the machine is in motion without the slightest injury to the working parts, the handles for controlling these changes are always in easy reach of the operator, and the feeds are automatic in all directions. The usual telescopic drive is eliminated, being replaced by vertical and horizontal shafts and sliding bevel gears.

The table has a double bearing, being fitted both above and below the dovetailed slide. This tends to keep it in good alignment even when working at the extreme of its motion, at the same time preventing it from cramping, and thus allowing it to work freely. All the gearing throughout the machine is of steel. The spindle is of crucible steel, running in phosphor bronze boxes provided with means of compensation for wear. The back gears are single in the machine shown, and double in the No. 3B size, giving respectively 12 and 18 changes of speed with three-step cone and two-speed countershaft. The overhanging arm is of solid steel, carrying an arbor support lined with a bronze bushing.

For the No. 2B machine shown in the cut, the longitudinal movement is 28 inches; cross feed, 7½ inches; vertical feed, 19½ inches. The largest diameter of the cone is 11⅞ inches and it has four steps for a 3-inch belt. The spindle is bored for a No. 10 B. & S. taper. The net weight of the machine is 2,850 pounds. The dimensions for the No. 3B machine are as follows: Longitudinal movement, 38 inches; transverse movement, 11 inches; vertical movement, 20¾ inches. The largest diameter of the cone is 12 7-16 inches for a 3½-inch belt. The spindle is bored for a No. 11 B. & S. taper. The net weight of the machine is 4,300 pounds.

Owen Plain Milling Machine.

A THREE BAR BORING TOOL.

A THREE-BAR boring tool, as shown in Fig. 1, has just been placed on the market by Armstrong Bros. Tool Co., Chicago. As shown in the illustration it is a combination of post and holder, which is made of bar steel. The holder has a T-head fitted in the tool post slot, by which it is clamped by the nut at top,

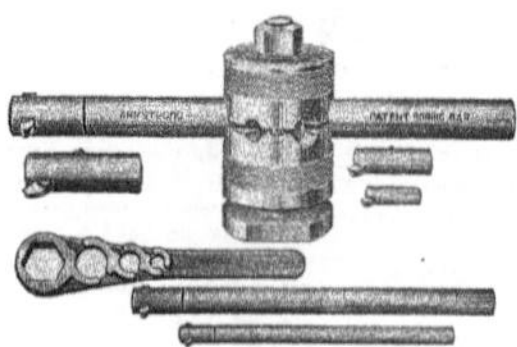

Fig. 1.—Three Bar Boring Tool.

Fig. 2—Three Bar Boring Tool.

which also serves to clamp the bars in place. As indicated by the name, there are three bars of different diameters. The fact that but a single turn of the wrench is necessary to release both the bar and the holder facilitates the quick changing of bars, thus allowing the operator to use the bar most suitable for his work with very little time lost in changing. The wrench shown has one opening for the nut and one each for tightening the cutters in the three sizes of bars furnished with the tool. The tightening of these cutters is affected in such a way that the pressure of the cut tends to hold them firmly in position. Fig 2 illustrates this tool in position on a lathe bed.

COMBINED BRAKE AND FOLDER.

A COMBINED brake and folder has recently been placed on the market by the Niagara Machine & Tool Works, Buffalo, which is shown in the accompanying illustration. Several new features are incorporated in this machine, the chief of which is its suitability for making curved forms, the bars being self-adjusting for such work. Another feature of the folder is that it enables the operator to turn close locks. The working length of this machine is 52 inches, and has a capacity for No. 18 iron and lighter.

The self-adjusting feature may be enlarged upon. By means of this feature curves, ogees, etc., can be formed with ease and neatness without adjusting the upper and lower planting bars independently.

These bars adjust themselves to the proper position in relation to the pivots of the bending bar for the bend to be made at the same time that they are moved apart the proper distance, regardless of whether a sharp lock is to be turned around the edge of the upper clamping bar or the sheet is to be bent around inserted semi-circular forms. The benefits are a saving of time and trouble and a certainty of obtaining good and uniform work. If required, the upper bar can be disconnected and raised independently. The upper and lower clamping bars will open up to 4 inches.

The following attachments are furnished with the machine: A steel blade with a sharp edge at an angle of 40 degrees to be used for cornice work, another with an edge having an angle of 20 degrees, for folding hooks for lock seams on iron up to 20 gauge and a wooden semi-circular form of ¾-inch radius. Back gauges are provided for bends up to 26 inches wide and two adjustable pointer gauges, used to indicate the proper position of the sheet according to prick marks in forming curved members. These gauges can be removed when not wanted. There is an angle stop on the right side to limit the motion of the bending bar in case a number of bends are to be made at the same angle. A blade is fastened in front of the bending bar to straighten it for ordinary work. This blade can be removed in making narrow bends. The back gauge is fastened to the lower clamping bar, and if it should be in the way can readily be removed. If desired a front gauge can be attached to the bending bar.

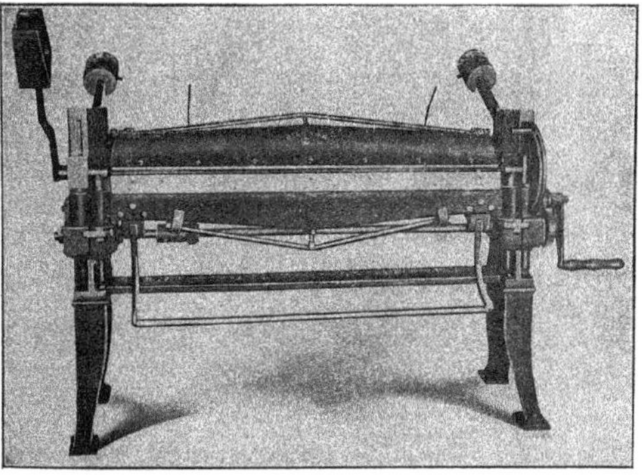

Combined Brake and Folder.

AN IMPROVED TOOL POST.

AN improved style of tool post, shown in the accompanying illustration, has just been placed on the market by Armstrong Bros. Tool Co., Chicago. This tool post combines in itself the strength and holding power of the strap and stud tool clamp, with the convenience of the open slide and ordinary set screw tool posts.

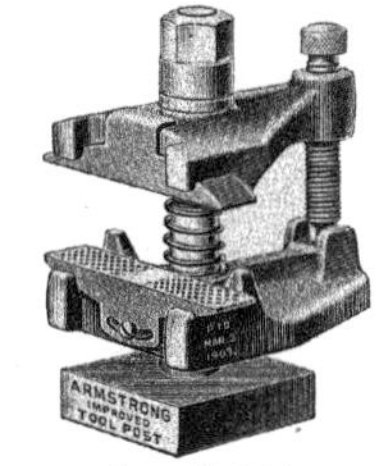

Improved Tool Post.

This tool post is made of drop forge steel throughout, and, as shown in the illustration, consists of a pair of jaws carrying tilting, clamping faces; these jaws are pressed apart by a spring and may be clamped together by the T-head bolt which passes through them and into the slot of the tool block. An adjustable screw furnishes the back support for the clamping action.

It is claimed for this tool post that it is stronger and stiffer than the ordinary type, will not slip or allow the tool to chatter, and will consequently do more work. It will work up close to the chuck, and has a great range of adjustment. The open side permits a rapid change and adjustment for tools. It will not cut or tear the tool shank and is thus especially adapted for use with tool holders. By using V-blocks fitted to this tool post, boring bars of various diameters can be conveniently held.

THE ROAD TO SUCCESS.

A young man who really and earnestly desires to succeed should never waste any time in dissipation, not even in so-called harmless dissipation.

He should, of course, allow himself the necessary amount of recreation and rest, and he should try to live a healthy, regular life.

He should try to acquire regular habits; that is, sleep and eat at the same hours every day and night, so as to keep in perfect physical health.

Then he should make it a rule every week to put by a certain amount of his earnings and acquire the habit of saving.

A NEW FRICTION CLUTCH.

THE accompanying illustration is of a new friction clutch which has been patented and is being manufactured by John Rieppel, Cowanesque, Pa.

(A) is the shaft, (B) is the pulley, to the arms of which the friction ring (C) is bolted. (D) is a sleeve secured to (A), upon it the hub of the pulley (B) is journaled. A collar (E) at one end prevents the pulley moving end-ways in that direction, while a flange (F) on the other end of (D) prevents it moving in the opposite direction. The lugs of (F) have a pair of clutch shoes (G) pivotally bolted to each of them by the bolts (N). The clutch shoes (G) are lagged with suitable frictional material, such as wood.

The ends of the clutch shoes (G) furthest from the pivots (N) are provided with angular projections (H), through these projections are holes for the bolts (I). Two of these angular ends (H) are threaded to receive the threaded sleeves (J) which are provided with a flange and pilot holes for adjusting them in or out of (H). These two members (H) are also split and the two bolts (K) passing through them at right angles to these splits are used to clamp the sleeves (J) when their adjustment is correct. The other two members (H) are provided with cam faces (L) which fit similar faces on the two levers (M). The other ends of the levers (M) are pivoted to the ends of the link (O), the other ends of which are pivoted to the sliding collar (P) which is movable longitudinally on the shaft (A). The collar (P) is provided with a groove in which works the split ring (Q) having trunnions by which it is connected to a shifter lever (not shown).

The collar (P) has a lug (R) on one side of it, through which is a clearance hole to receive the bolt (S) screwed into (F). The object of (S) is to limit the movement of the collar (P) on the shaft (A). In practice the shoes are normally out of frictional engagement with the ring (C) on the pulley (B).

When it is desired to throw the clutch into engagement, the collar (P) is moved away from (F), this forces the links (O) up at right angles with the shaft, which operation turns the levers (M) on their bearings on the pins (I). The cam faces (L) and (M), swing the clutch shoes (G), on the bolts (N), and force them into frictional contact with the inner surface of the friction ring (C). Throwing the collar (P) in the opposite direction releases the clutch shoes (G), the springs (T) then draw the clutch shoes back out of contact with the friction ring (C).

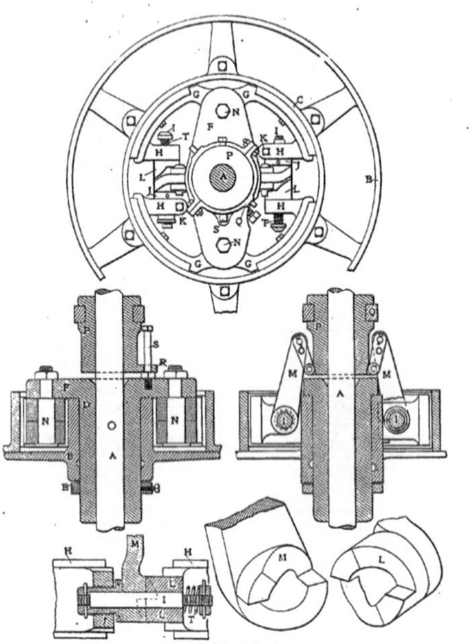

A New Style of Clutch.

STANDARD ENGINE DESIGN.

THE design of reciprocating steam engines can now be safely said to have reached a stage of more or less standardization. The proportions of the various details to give the best results under given working conditions can be found in any standard work on the subject. This standardization can be accounted for, not only by the number of years which the reciprocating engine has been a standard article of manufacture, but perhaps more from the fact that the reciprocating steam engine, as an engine, is not patented, and consequently designers from all parts of the world have been devoting their energies to arrive at the solution fotse b'' ergies to arrive at the solution of "best practice." Had the development of this class of machinery, however, been in the hands of one man or one firm only, no matter how clever they might have been, it is too much to expect that they could have attained the same degree of perfection that now xists. Designers differ largely in their views, and it is only by the failures and successes of the schemes of these various engineers that we have been able to deduce standard formulae for the proportions of the respective parts of steam engines.

W. J. A. LONDON.

LUBRICATING GAS ENGINE CYLINDERS.

A method of lubricating the cylinders of gas engines is given in "Graphite" and depends on the finishing cut on the piston. This is made with a diamond-pointed tool with slow feed enough to produce very fine lines on the surface of the piston or what might be called a finely corrugated surface. This is filled with graphite, with the result that it faces up smooth with the two materials in contact with the surface, a small amount is required to properly lubricate the cylinder and make up the small amount that is constantly being carried out of it.

BOOK REVIEWS

BUILDING MECHANICS' READY REFERENCE.—A carpenters' and woodworkers' edition, by H. D. Ritchey, Superintendent of Construction U. S. Public Buildings, 119 illustrations, 226 pages, price $1.50 net, New York, John Wiley & Sons.

This book is intended as a ready reference manual for the use of the carpenter and wood worker. It is not filled with complicated formula or diagrams, but the entire matter made plain and simply stated, appealing directly to the man for whom it is intended.

THINGS THAT ARE USUALLY WRONG. — By John E. Sweep: 52 pages, illustrated, with drawings and diagrams. New York: Hill Publishing Co. In the preface, the author when stating that the book would be useful to mechanical and engineering students, machine designers and inventors, states that should those who buy the books be disappointed they can console themselves with the fact that it did not cost much. A few, however, will be disappointed with this volume as it contains some first-class kinks, and any mechanical man desirous of improving himeslf and his methods can peruse this little volume with considerable interest and profit.

MODERN MILLING MACHINES. Their design, construction and working. A handbook for practical men and engineering students, by Joseph G. Horner. 269 illustrations, 300 pages; price $4; New York: Norman W. Henley Publishing Co., 132 Nassau street.

So important has the milling machine become in metal work manufacture of late years that a book on the subject is very opportune. The author is well known in connection with various mechanical subjects, which he treats in a thoroughly practical manner from the view point of years of experience. Milling machines have become highly specialized, and the work of milling is now subdivided into different groups, just as that of turning is, ranging from very plain to very difficult work. The opening chapter deals with the leading elements of milling machine design and construction, followed by a chapter on plain and universal machines. This leads us to special attachments and special machines, after which a chapter is devoted to cutters, an important feature of the milling machine practice. The remaining chapters deal with various milling operations, showing the scope of the milling machine in modern practice. The book is well written, is clear and practical, and can be recommended to mechanics and superintendents in search of fuller information regarding milling machine practice.

MODERN PLUMBING ILLUSTRATED, by R. M. Starbuck. 384 large quarto pages, with 55 full-page and detailed engravings. A standard work for plumbers, architects, builders, property owners, and boards of health and plumbing examiners. Price $4; New York: The Norman W. Henley Publishing Co.

This book is an outgrowth of, and an improvement upon, the author's series of fifty blue print charts known as "Modern Plumbing Illustrated." The present volume deals with high-grade modern plumbing, showing detail work and complete plumbing systems. Each one of the 55 plates is accompanied by several pages of descriptive matter, in some instances as many as eight or ten, the text forming a critical and concise treatment to each phase of the work under discussion. While the book is somewhat ambitious in its scope, it covers well what it sets out to accomplish, and should prove invaluable to all interested in the most approved methods of plumbing construction, drainage and ventilation of dwelling apartments and public buildiigs.

ABOUT GAS ENGINES.—We have before us a volume entitled, "Questions and Answers from the Gas Engine," published by the Gas Engine Publishing Co., of Cincinnati, Ohio. This book has been compiled from the "Answers to Inquiries" column of "the Gas Engine," and consists of the more interesting and valuable inquiries which have appeared in that journal for the past eight years. The inquiries relate to the design, construction, operation and repair of gas and gasoline engines, for stationary, marine and automobile use, and the answers were made by some of the best recognized authorities on the various subjects, in America and Europe.

This volume measures about 5 x 7 inches, and contains about 275 pages and should prove of value to any who are interested in gas engine construction or repairing.

PERSONAL MENTION.

Louis Boissinot, of Dorval & Boissinot, machinists, Montreal, is dead.

Mr. A. D. Bayne, manager of the Canadian Westinghouse Co., Hamilton, has left for Europe.

Mr. Clarence Bullis, of Minneapolis, has arrived in Regina to accept the management of the Saskatchewan Automobile and Gasoline Engine Company.

Mr. J. H. Dohner, for a number of years with the National Cash Register Co., has accepted a position as assistant general superintendent with the National-Acme Manufacturing Co.

Mr. W. G. Rogers, secretary-treasurer of the Erie Iron Works, St. Thomas, has disposed of his interest in the company to Mr. G. F. Spackman, who recently purchased Mr. J. Risdon's stock in the company.

Mr. J. D. Sullivan, superintendent of Mackenzie & Mann's Halifax & Southwestern Railway, dropped dead of heart trouble on April 11th. He was about 35 years old, and was formerly in the C.P.R. service in Ontario.

Mr. K. L. Aitken. consulting engineer, of Toronto, presented for discussion a paper entitled "Synchronous Converters vs. Motor Generator Sets," by Mr. Paul M. Lincoln, at the April meeting of the Toronto branch of the American Institute of Electrical Engineers.

Mr. R. J. Sutton, late of the Labatt Mfg. Co., London, Ont., has returned to the United States, and is now located at 73 Eppirt street, East Orange, N.J. He intends to make a specialty of introducing the premium plan of paying for labor in machine and other shops.

Thos. Moore, sales manager of the Belleville Rolling Mills, was in Toronto a few weeks ago. He reports the demand for their horseshoes so great that the company has been compelled to take two of its travelers off the road and to double the milling capacity. The company is also installing bolt and nut machinery.

Mr. Wilfred H. Parmelee, who for the past year and a half has conducted a manufacturers' agency in Toronto, has lately associated with him Mr. R. E. Nicholson, and the business will hereafter be conducted under the name of Parmelee & Nicholson, with offices at 1314 Traders' Bank building, Toronto. Mr. Nicholson was with Wood, Vallance & Co. for about three years, having previously been connected with Rice Lewis & Co., while Mr. Parmelee was for five years manager for W. R. Perrin & Co., Ltd., and the present joining of forces should make a strong combination. Among the manufacturers already represented by the firm are the Laurie Engine & Machine Co., Ltd., Montreal, makers of engines, heaters and power presses; the Canadian Boomer & Boschert Press Co., Ltd., Montreal, hydraulic knuckle-joint presses; P. E. Shantz, Preston, trucks; the Bourne, Fuller Co., Cleveland, bar iron, steel, etc.; Glaholm & Robson, Sunderland, England, wire rope; J. D. McArthur & Co., Brockville, leather belting; Burleigh & Weeks, Whitehall, N.Y., moulding sand; McKeefrey & Co., Leetonia, Ohio, coke. In addition to acting as agents for the above, the firm is a direct importer of English fullers' earth.

ABOUT CATALOGUES

By mentioning Canadian Machinery to show that you are in the trade, a copy of any of these catalogues will be sent by the firm whose address is given.

THE CHATHAM MOTOR CAR CO. issues a neat booklet descriptive of the Chatham Motor Car, manufactured by the company.

MARINE ENGINES. — Fairbanks-Morse marine engines are dealt with in a catalogue recently issued by the Canadian Fairbanks Co., Limited.

"SOMETHING ELECTRICAL FOR EVERYBODY" is the title of catalogue twenty-two of the Manhattan Electrical Supply Co., New York.

SILICON ALLOY.—"Outerbridge" silicon alloy, for softening, strengthening and cleaning cast iron, is the subject dealt with in bulletin No. 16 of the J. W. Paxson Co., Philadelphia.

A CATALOGUE AND PRICE LIST of Armstrong tool holders and other machine shop specialties, in compact pocket form, is No. 14, issued by the Armstrong Bros. Tool Co., Chicago.

STORAGE BATTERIES FOR STATIONARY SERVICE, manufactured by the Westinghouse Machine Co., East Pittsburg, Pa., are dealt with in Catalogue S., recently issued by the company.

THE TORONTO POTTERY CO. issues a neat and comprehensive catalogue of high grade fire brick. Several interesting and valuable tables are included in the booklet, which is nicely printed and of handy size.

"AXLES AND AN ECONOMY EFFESTED, is the title of a booklet issued by Canadian Bearings, Limited, Hamilton. It describes the Wright taper roller bearing axle, and shows how and why superiority is claimed for it over other designs.

THE STANDARD CHEESE BOX BENDER is described and illustrated in a four-page folder issued by the manufacturer, J. T. Schell, Alexandria, Ont. This is one of three pamphlets covering cheese box machinery.

CONTRACTORS' HOISTING ENGINES.—Various styles of Lidgerwood hoisting engines for general contracting purposes are illustrated and described in Bulletin 200, recently sent us by Allis-Chalmers-Bullock, Limited, Montreal.

THE DUBOIS IRON WORKS, Dubois, Pa., in an attractively-bound booklet of 35 pages, describes and illustrates DuBois horizontal gas engines designed to operate on natural, illuminating and producer gas, gasoline, alcohol, kerosene, and crude oil.

GASOLINE, GAS, AND ALCOHOL ENGINES.—The Gilson Mfg. Co. is issuing a new catalogue illustrating and describing Gilson engines. These engines will be shipped from the new Canadian factory now being built by the firm in Guelph, within the next few months.

PAINT MACHINERY.—A new sectional catalogue of paint machinery, color, ink, and printers' roller-making appliances, issued by Wm. & J. G. Greey, Toronto, is to hand. It consists of 80 pages, and gives most complete information regarding everything in these lines.

A MOST COMPLETE CATALOGUE is that of the American Clay Machinery Co., Bucyrus, Ohio. It consists of 230 well illustrated pages, handsomely bound in cloth. Brick, tile and sewer pipe machinery of all kinds, dryers, cars, conveyors, etc., are fully described, and the book should certainly prove a trade-winner.

THE AMERICAN WIRE ROPE NEWS, for January, issued by the American Wire Rope Co., Chicago, shows the company's products used in the rigging of fishing vessels, examples of Gloucester craft being scattered through the pages. Elevator work is also treated of, illustrations being given of the system in use in "The Rookery," Chicago.

"EVERYTHING YOU NEED IN YOUR FOUNDRY," is treated of in general catalogue No. 40, recently issued by the S. Obermayer Co., Cincinnati. It is a book of 370 pages, and could hardly be more complete, both illustrations and text being ample and explicit. The Obermayer Co.'s goods are handled in Canada by the Dominion Foundry Supply Co., Montreal.

THE BAKER WATER SOFTENER, in India rubber and gutta percha machinery are two lines dealt with in catalogues recently issued by David Bridge & Co., engineers and machinists, Castleton Iron Works, Manchester. The last mentioned booklet includes also belt-sewing machines, hoists, Heywood & Bridge's patent friction clutch, and other specialties made by the firm.

A NOVEL ADVANCE AGENT, in the form of an attractively-printed folder, comes from the T. R. Almond Mfg. Co., Brooklyn. Through the cover, by means of an aperture cut in the shape of an Almond drill chuck, is seen a half-tone photo of the company's representative. On opening the folder, his request for an appointment meets the eye of the prospective buyer. The folder should certainly prove a good introduction for the drummer.

OPENING SALES OFFICES.

In order to keep pace with the requirements of their increase in business, the John McDougall Caledonian Iron Works Co., Limited, of Montreal, have opened sales offices at the following places: Montreal, 82 Sovereign Bank Bldg.; Toronto, 810 Traders Bank Bldg.; Winnipeg, 251 Notre Dame St.; Vancouver, 416 Seymour St.; Nelson, Josephine St.; New Glasgow, N.S., Telephone Bldg.. Their principal products are waterworks equipment and all kinds of hydraulic and mill machinery.

THE AUTO SHOW.

The second annual Automobile and Sportsmen's Show was held in the Montreal Arena, April 6th to 13th, and was a success in every way. All available space in the large building was occupied to the best advantage, by displays of motor cars, motor boats, bicycles, automobile accessories, sporting goods, etc., and the buzz of machinery and tooting of horns sent an unaccountable thrill down the visitor's spine.

Among the exhibits worthy of note was a magnificent display of automobile and motor boat accessories, by John Millen & Son, Ltd., of Montreal, and Toronto, among which were to be found gas lamps and searchlights, carburetors, coils, horns, cable, accumulators, voltmeters, batteries, etc.

Adjoining the Millen booth was that of the Chapman Double Ball Bearing Company, with a magnificent display of shafting, wheels, etc., in motion, besides various other devices to demonstrate their double ball bearings. This exhibition was one of the features of the show.

The Berlin Electrical Mfg. Co., Ltd., of Toronto, had an attractive display of batteries and electrical novelties, the feature of which was a light, run by one of their storage batteries.

The Canadian Rubber Co., of Montreal, were very much to the fore and the display of the new "Canadian Clincher" automobile tire was a credit to Canadian industries.

John Forman, Montreal, had on display among his large stock of gasoline engine ingition specialties, the K. W. igniter, of which he has recently secured the sole Canadian agency.

Industrial Progress

CANADIAN MCHINERY AND MANUFACTURING NEWS would be pleased to receive from any authoritative source industrial news of any sort, the formation or incorporation of companies, establishment or enlargement of mills, factories or foundries, railway or mining news, etc. All such correspondence will be treated as confidential when desired.

The G.T.R. will build a subway at Port Hope.

W. G. Paton, machinist, Wingham, has gone out of business.

A waterworks plant will be erected at Welland, to cost $70,000.

The new Vancouver courthouse to be erected will cost $300,000.

The Jenking-Leslie Brass Mfg. Co., of Montreal, has dissolved.

The Imperial Glass Co., Toronto, will erect a factory on Mutual St.

A charter has been granted to the Richelieu Foundry Co., Sorel.

The Anchor Wire Fence Co. will erect a new factory in Stratford.

The G.T.R. will increase their staff and enlarge their shops at London.

The Manhattan Asbestos Co., Black Lake, Quebec, has been dissolved.

The Geo. Foster & Sons Co., Brantford, will erect a $25,000 warehouse.

The Sanitary Packing Co. will build a $20,000 canning factory at Weston.

The Toronto Street Railway Co. is installing a Sturtevant heating apparatus.

Joseph A. Nadeau, proprietor of a sawmill in Carleton, Que., has assigned.

Johnston & McGregor, Smiths Falls, Ont., are advertising their foundry for sale.

The R. Watt Machine Works, Ridgetown, had their foundry damaged by fire.

A $10,000 apartment house will be erected on Broadview Ave., Toronto, this Spring.

The Adams Wagon Works, Brantford, will erect a new brick warehouse to cost $8,000.

The Red Cedar Lumber Co. will build a new mill on the waterfront at Vancouver.

The Sawyer-Massey Company, Hamilton, will erect a warehouse at Saskatoon.

The T. Eaton Co. will build a store on Portage Avenue, Winnipeg, to cost $110,000.

The C.P.R. will rebuild elevator D, at Fort William, which was destroyed last Fall.

The Georgian Bay plant of the Canada Iron Furnace Co., at Midland, was burned out.

The McKinnon Dash and Metal Co., St. Catharines, will erect a chain factory there.

Disston, Son & Co., saw mfrs., Toronto, are making a $15,000 addition to their mills.

The Western Fuel Co., Nanaimo, will erect a new sawmill with a capacity of 15,000 feet.

The Slingsby Mfg. Co., Galt, will make a $10,000 extension to their plant this spring.

A new threshing machine manufactory will be erected at Alliston, Ont., by A. Merner.

The Canadian Westinghouse Co., Hamilton, will erect a $40,000 addition to its plant.

The Hotel Dieu, Kingston, is to be remodelled and made into a mica works for Kent Bros.

The Welland Vale Manufacturing Co. will build a central establishment at St. Catharines.

The Waderlow Iron Works, Leamington, is considering the advisability of moving to Dunnville.

The Stevens Co., Galt, has now its own moulding shop. The first cast was taken on March 28th.

Fire caused $25 damage to the office of the American Abell Engine Co., Toronto, on April 15th.

Calgary capitalists are promoting a $600,000 Rocky Mountain Cement Co., with works at Frank.

The Hall Elevator Company and the Alberta Pacific Grain Co. will build mills at Killam, Alta.

The Welland Vale Mfg. Co., Welland Ont., will build a central establishment in St. Catharines, Ont.

The Toronto Theatre Co. have secured a permit to build the Gayety Theatre on Richmond St. west.

The foundry of the R. Watt Machine Works, Ltd., Ridgetown, Ont., was damaged by fire last month.

The Canada Permanent Loan Co. will erect a three-storey office building at Regina, to cost $60,000.

The C.P.R. will build a branch line between Peterboro and Victoria Harbor, to cost $3,000,000.

The Elk Lumber Co. will build a new lumber mill near Nelson, B.C., to handle annually 12,000,000 feet.

The entire plant of the Georgian Bay Engineering Works, Midland, was wiped out by fire on April 6th.

The McLaughlin Carriage Co., Oshawa, Ont., will erect a new factory for the manufacture of automobiles.

The new physics building of Toronto University will be equipped with a Sturtevant ventilating fan.

A large bolt and fishplate concern is asking concessions with a view to erecting a plant in Three Rivers.

Henry Pickleman has registered in Montreal under the name of the Dominion Mill Stock and Metal Co.

Berlin ratepayers have voted to purchase the Berlin & Waterloo Street Railway, at a cost of $76,200.

The Canada Screw Co. are to enlarge their works at Hamilton, having decided not to locate at Welland.

The Hillman Copper Co. has obtained a charter, the head office of the company to be in Sault Ste. Marie.

The Gurney Foundry Co., Toronto, have secured a permit to erect a one-storey addition to their Toronto Junction plant, to cost $4,000.

Hamilton has decided to ask for estimates on the cost of establishing a municipal electric lighting plant.

The Lake of the Woods Milling Co., Keewatin, will soon have their new mill, costing $1,250,000, in operation.

The Listowel Foundry & Machine Shop has been opened up by George Melrose, with a skilled staff of mechanics.

Rapid progress is being made in the construction of Canadian Iron & Foundry Co.'s shops at Fort William.

The Meisel Mfg. Co., Port Arthur, are building harvester works at the east end of the town to cost $250,000.

The village of Alliston has passed a by-law to grant a bonus of $5,000 to the Merner Mfg. Co., of Waterloo.

The Coupe Mfg. Co., Toronto, which has been engaged in the manufacture of laundry supplies, has been wound up.

A. M. Fraser will erect a twelve-storey bank and office building in Winnipeg. J. D. Atcheson is the architect.

The by-law for the purpose of developing electrical power for Shelburne, Ont., to cost $60,000 was defeated.

Improvements and additions costing about $500,000 will be made to the King Edward Hotel, Toronto, this Summer.

Damage to the extent of $600 was done to the brass works of W. Coulter & Sons, Toronto, on Tuesday, March 12.

The Canadian Underskirt Co. have secured a permit to erect a three-storey factory addition at Queen and Noble Sts.

The Economical Power, Light & Heat Co., Toronto, has purchased a new rotary type of Sturtevant high pressure blower.

It is reported that the works of the Canadian Ramapo Iron Works, Niagara Falls, will be ready to begin operations in June.

The American Steel Wire Co., Cleveland, Ohio, will erect a branch factory at Lethbridge, Alta., at a cost of about $2,000,000.

The Radcliffe Lumber Co., Duluth, have bought blocks of land on Vancouver Island and will erect a large mill there.

Hyslop Bros., Toronto, have secured a permit for the erection of a $30,000 garage at corner of Shuter and Victoria street.

The water commissioners of Niagara Falls have asked for $40,000 for extensions and improvements to the waterworks system.

The contract for the steel pipe for the north Rosedale sewer has been awarded to John Inglis Co., their tender being $4,480.

A wood-working plant is being built at Fort William by the Seamans-Kent

Co., Meaford, to cost $100,000, and to employ 100 men.

It is expected that tenders will shortly be asked for the construction of the C.P.R.'s new line between New Westminster and Eburne.

The large apartment house being erected at the corner of Yonge and Bloor Sts., Toronto, is nearing completition. It will cost $125,000.

The Manufacturers' Life Insurance Co., Toronto, will build a 12-storey block facing on King, Bay and Melinda Sts., to cost $600,000.

A two-storey concrete factory, 100 feet in length and 50 feet wide, will be erected in Stratford for the manufacture of Anchor wire fences.

Mechanical draft equipment is being furnished by the B. F. Sturtevant Co., of Boston, for the Toronto Bolt & Forging Co., Swansea, Ont.

The Freyseng Cork Co., Toronto., will build a three-storey addition to their factory at the corner of Queen and Sumach streets, to cost $7,000.

A plant for the manufacture of car wheels and water pipes will be built at Fort William by the Canada Iron Foundry Co. costing $250,000.

The Packard Electric Co., St. Catharines, Ont., has ordered an 8½-ton traveling crane from the Smart-Turner Machine Co., Limited, Hamilton.

Thessalon, Ont., has voted a bonus of $10,000 to the Saginaw Lumber & Salt Co. to assist the company to build and operate a saw mill in that town.

The Canada Foundry Co. has been awarded the contract for Toronto's supply of cast iron pipe during the twelve months terminating April 1st, 1908.

The rail mill of the Dominion Iron and Steel Co., Sydney, is at present occupied on a contract for 37,000 tons of 100-pound rails for the Grand Trunk.

The premises of the Standard Tin Co. and J. W. Delainey, machinery and tool makers; 160-162 Duke street, Toronto, were badly damaged by fire on April 16.

A permit has been issued for the erection in Montreal of a five-storey building by Williams & Wilson, dealers in machinery supplies, to cost $47,000.

A wire mill will be erected at Lethbridge to cost about $2,000,000, being a branch of the American Steel Wire Co., Cleveland, O. It will employ 1,000 men.

The Canadian Northern Railway will spend considerable during the coming year on repair shops and other buildings required by the company in Winnipeg.

The Mississippi Iron Works at Almonte, conducted for 32 years by Young Bros., has been sold to H. C. Bowland, D. Williams, and W. Glover, all of Almonte.

A tram line from New Westminster to Chilliwack is practically assured. Surveys have been completed, and it is expected that construction will begin this summer.

Stenhouse Brothers will erect a large machine shop in Fort William, capable of doing all kinds of heavy work, vessel repairing, etc. The contract for the building has been let to Cusson & Matthieu.

At the recent International Exhibition held in Liege, Belgium, the Armstrong Bros. Tool Co. was awarded a medal for its exhibit of lathe and planer tool holders.

Messrs. Franklin & Ball, promoters of a bolt and screw industry, are endeavoring to secure favorable terms from the town of Waterford, with the idea of locating there.

The Atikokan Iron Co., Port Arthur, has built blast furnaces and plant for roasting ore, and a large dock. The entire cost is $1,300,000, and will employ 500 men.

The Confederation Life Association, Toronto, has taken out a permit to erect an eight-storey office building on the corner of Queen and Victoria streets to cost $200,000.

The Western Iron Works, of Winnipeg, commenced operations of March 26th, when the first casting was done in the fine, modern plant erected by the company in Elmwood.

Battleford has carried a by-law to raise $90,000 for municipal waterworks, sewerage and electric lighting. Tenders closed on April 15, and the work is to be commenced at once.

A 400 foot addition, six storeys in height, is being made by the National-Acme Mfg. Co., Cleveland, Ohio, to its present factory, which itself is one-eighth of a mile long.

The Dominion Bridge Company has taken over the bridge and structural steel department of the Locomotive and Machine Company, whose works are located at Longue Pointe.

The Smart-Turner Machine Co., Limited, Hamilton, are building a single outside packed plunger pump for the corporation of Calgary, to be used in the waterworks system.

The premises of Fred. Thompson & Co., manufacturing and contracting electrical engineers, 326 Craig street west, Montreal, were damaged by fire to the extent of $3,000, on April 6th.

It is proposed to build an electric railway between Fort Frances and Duluth, when the power dam at Fort Frances and the Great Northern power plant at Duluth are completed.

The Grand Trunk will erect two new elevators, one at Fort William, the other at Tiffin, each having a capacity of two and a half million bushels. They will cost over a million dollars.

Work has begun on the new five-storey fireproof head office building which the Canadian General Electric Co. is erecting, at a cost of $400,000, on the corner of King and Simcoe streets, Toronto.

Bechtels, Limited, Waterloo, and the Anglo-American Leather Co., Huntsville, Ont., have ordered automatic feed pumps and receivers from the Smart-Turner Machine Co., Limited, Hamilton.

The King Electrical Works, Montreal, has been incorporated, with a capital of $30,000, to carry on the business of manufacturers of and dealers in electrical machinery, brass, copper and other metals.

The Montreal city council has approved of a scheme which will assure a supply of 50,000,000 gallons of water to the city per day. The work will cost $2,132,000, and will take four years to complete.

The Saskatchewan Electric Co., Regina, is succeeded by General Electric, Limited. The new company has a capital of $25,000, and branches will probably be established in Saskatoon and Prince Albert.

It is probable that car works for the building of electric and steam cars will be built in St. Thomas. A number of enterprising citizens have the matter in hand, and the project is now almost an assured success.

The famous steam engine of the Ogilvie Flour Mills, Winnipeg, one of the largest in Canada, and once one of the sights of the city, has been discarded, and electricity is now the motive power employed in the mills.

The National Rolling Mill Co. has been successful in making financial arrangements with the town of Sydney, N.S., for location there, and as soon as a suitable location is selected construction work will begin.

The Sutherland Rifle Sight Co. has asked tenders for the erection of a factory building at New Glasgow, N.S. The main building will be 152x42 feet, in addition to which there will be the power building and office.

A public meeting of St. Thomas citizens has endorsed a scheme to give $5,000 bonus toward the establishment in the city of a Canadian branch factory of a large go-cart concern, of Elkhart, Ind. A by-law will be submitted.

The Asbestos Mfg. Co., of London, Ont., has been incorporated with a capital of $25,000 to manufacture and deal in asbestos. The promoters are H. V. Everham, R. Van Z. Mattison, jr., and R. Van Z. Mattison, sr., all of London.

A large United States concern connected with the marble and onyx industry, is in communication with Mr. Joseph Thompson, commissioner of industries for the city of Toronto, regarding the establishment of a Canadian branch.

W. V. Hunt, of the British Columbia Electric Railway Co., Vancouver, B.C., has been appointed engineer in charge of the work of installing new water power units at Lake Buntington. Three units of 100,000 h.p. each will be installed, at a cost of about $250,000.

The contract for the construction of twenty bridges, from Trout River to Fort William, on the Lake Superior division of the C.P.R., has been awarded to Frank Munro, contractor, of Montreal. The work includes excavation, cement work and pile driving.

Hamilton city council has passed a by-law to raise $62,000 to be spent on extensions to the waterworks system. Of this, $5,000 will be spent on meters, $27,000 on mains, $18,000 on services, and $12,000 on extensions to the filtering basins.

Gajnon & Garon and A. Henuset & Co., Montreal, have registered as contractors. Harold G. Eadie, chief agent for the White Fireproof Construction

Co., has registered in the same city, as also Larose & Berthiaume, sash and door manufacturer.

A. Berg & Sons, Toronto, are installing a sand-lime brick plant of 20,000 capacity for the Schultz Bros. Co., Limited, Brantford. It will be in operation at an early date. A similar plant is also being installed by the firm for the Silica Brick & Lime Co., Victoria.

Morgan Harris, Brantford, has purchased the machine shop of J. B. Rouse, at 22 Dalhousie street, and will be joined by Chas. Herod, of the Goold, Shapley & Muir Co., in a general repairing and tool manufacturing business. Considerable new machinery will be installed.

The Dominion Heating & Ventilating Co., Hespeler, has just shipped a fan, heater, a number of cars, and other brickyard equipment to Samuel Watson, brick manufacturer, Orillia. The company is also supplying the Amalgamated Oil Co., of Petrolea, with a complete brickyard outfit.

The Crocker-Wheeler Co., of Ampere, N.J., has opened an office in Birmingham, Ala., owing to the inability of the New Orleans and Baltimore offices to cope with increasing trade. Mr. B. A. Schroder will be in charge of the branch, which will be located in the Woodward building.

A patent on a new blast furnace has been secured by H. W. Hixon, Victoria Mines, Ont. The main features of the furnace comprise a lining of refractory material, and an air-jacket, constituting a sunstitute for the water-jacket formerly used, and through which the air passes on its way to the twyers.

The International Harvester Co. has placed a contract, through Laurie & Lamb, Montreal, the Canadian representatives, for Foster superheaters to be installed in twenty 350-h.p. Sterling boilers. The Copper Queen Consolidated Mining Co. has also placed an order for superheaters in eight Sterling boilers of 400 and 450 h.p.

The F. C. Hunt Plumbing Co., of London, have gone into the supply manufacturing business, having established a foundry at Hensall, about forty miles north of the city. At present they are confining their operations to the manufacture of pumps, soil pipes and fittings. They report trade very good, with some large contracts in prospect.

Construction work has been begun on the Robertson Machinery Co.'s new factory in Welland. The main building, of brick, will be 150x60 feet, and one storey in height. It will be divided into a machine shop, 75x60 ; foundry, 50x60, and a clearing room and storehouse, 25x60. A storehouse, 40x150, of corrugated iron, will be parallel to the main building.

The firm of W. H. Laurie & Co., Montreal, consulting and contracting engineers, has issued a circular to the trade to the effect that the firm name has been changed from W. H. Laurie & Co., to Laurie & Lamb, although there will be no change in the personnel of the firm. The company will, as heretofore, make a specialty of power generation, transmission and utilization.

Hon. S. N. Parent, president of the Transcontinental Railway Commission, was waited upon by a deputation of Quebec citizens on March 29, and interviewed respecting the advisability of forming a company to erect large car shops in that city. Mr. Parent encouraged the proposal, and believed that a large order might be looked for from the Transcontinental Railway.

Duplex pumps are being supplied by the Smart-Turner Machine Co., of Hamilton, to the Victoria Harbor Lumber Co., Whitefish, Ont.; the Midland Engine Works Co., Midland; the E. Long Mfg. Co., Orillia; Jas. Ready, Fairville, N.B.; the Delhi Mfg. Co., Delhi, Ont.; the Berlin Lion Brewery; the Royal Hamilton Yacht Club and the Montreal Reduction & Smelting Co., Trout Mills, Ont.

Work has commenced on the new buildings for the Standard Fitting and Valve Co., in Guelph. The machinery building, 70 x 200 feet, is to be completed within the next month, and the foundry, 80 x 164, within one month thereafter. The core room, ovens and japan room will be 25 feet by 50, and the engine and boiler rooms 38 by 40. The cost of buildings and equipment will total $100,000.

The new foundry building for the Niagara Falls Hardware and Foundry Co., for which tenders have been taken, is to be 80 feet by 170 feet, of iron frame, with concrete or brick between the steel uprights near the ground, and windowz above. There will be a 25 or 30-ton crane. Bids have not yet been asked foe construction of the building for offices, pattern rooms and machine shops. It will be 50 or 60 feet by 175 to 180 feet.

A Vancouver firm, Evans, Coleman & Evans, secured from the Winnipeg city council recently a large order for cast iron pipe, amounting to some 6,000 tons. This large contract, the largest expected to be open for competition in Canada this year, was secured against competitors representing nearly all the makers of cast iron pipe in Britain as well as the three Canadian pipe foundries. Cast iron has advanced seriously in price and competition has therefore been keen and close.

The Norton Company, Worcester, Mass., and Niagara Falls, N.Y., manufacturers of grinding wheels made of alundum, and other abressive specialties, is to erect a large addition to its Worcester works. The building designated as plant 2, will be extended about 300 feet in length by 111 feet in width, which will more than double the present capacity. This will be fully equipped with kilns, mixing machines, shaving machines, etc., so as to permit of a large increase in output.

The new building which the Goold, Shapley & Muir Co. is to erect in Brantford, will be on an extensive scale, doubling the present capacity of the company. They consist of a machine shop, warerooms and wood shop, and will cost about $60,000. The machine shop will be erected to the west of the present main plant, and will be 132x54 feet. It will be a one-storey brick structure and will be equipped with a travelling crane 38 feet long, running the entire length of the building.

Baines & Peckover, hardware dealers and manufacturers' agents, have moved from 126 Bay street, Toronto, to 98 Esplanade street east. The new premises are more suitable in every way for carrying the heavy line of goods dealt in by the firm, and have been thoroughly remodelled into the most comfortable and capacious office and warehouse quarters. The fact that the new building is so much closer to shipping facilities than were the old premisrs should prove of great advantage to the firm.

Work on the Hamilton Steel & Iron Co.'s new 300-ton furnace has been making good progress with the favorable weather of the past few weeks, and it is expected that everything will be in running order by July. The total furnace capacity of the plant will then be 500 tons per day. The firm has recently added railway spikes to its list of products. The spikes are manufactured from open hearth basic steel bar, and as this department is an entirely new feature of the company's work, and therefore not yet too hard pressed with orders, it should help considerably those looking for rapid delivery in this line during the present busy times.

The Starr Manufacturing Co., The Dartmouth Rolling Mills, and the Dartmouth Machine and Forge Co., have been amalgamated. In addition to the lines hitherto manufactured by the Starr Manufacturing Co., the amalgamated companies will manufacture merchant bar iron in rounds, squares and flats, wrought iron and steel forgings, carriage axles, saw mill machinery, and structural iron work. The officers of the new company are: J. C. Mackintosh, president; Jas. Simmonds, vice-president; T. Ritchie, G. E. Faulkner, F. H. Oxley, E. D. Adams, H. E. Hill, directors.

After May 1st the Chicago branch of the Niles-Bement-Pond Co. will occupy its new offices in the Commercial National Bank building. The Pratt & Whitney Company will abandon its show room and offices at 46-48 South Canal street, and will combine its machinery sales department with that of the Niles-Bement-Pond Company. The show room and stock of Pratt & Whitney small tools, and the small tools sales department, will be located on the ground floor of the new Plamondon building, Clinton and Monroe streets. Mr. Geo. F. Mills, who has for several years looked after the interests of these companies in the Chicago territory, will continue as manager of the Chicago offices.

The Scott Machine Co. is one of the busiest of London's new concerns. They have just installed in the Tecumseh House, which has recently been overhauled and improved throughout, a laundry and fan outfit for the kitchen, to be operated by one of the Scott Co.'s own gas engines. The company report all iron lines very busy, the only difficulty being annoying delay in the delivery of material. They have secured the contract for overhauling the presses of the Southam Printing Company, a work that will keep a number of men busy a couple of months. Elevator business is quiet with the firm, but from the number of inquiries being made the outlook cannot but be regarded as cheering.

CONDENSED ADS.

RATES

One insertion—25c. for 20 words; 1c. a word for each additional word.

Yearly rate—$2.50 for twenty words or less, 10c a word for each additional word.

The above does not apply to notices under the head of "Machinery Wanted." These notices are inserted free for subscribers.

SITUATION VACANT.

WANTED—Young man well experienced in machinery supply business able to sell and knowing thoroughly all requirements of the trade. Salary, commission and share of profits. Apply Box 32, CANADIAN MACHINERY, Toronto.

SITUATIONS WANTED

ASSISTANT GENERAL MANAGER, McGill graduate, desires change of position. Twelve years' practical experience along engineering, machine, foundry and boiler shop lines. Best of references. Address Box 25, CANADIAN MACHINERY, Toronto. [3]

MACHINERY FOR SALE.

BELTING, RUBBER, CANVAS AND LEATHER, Hose, Packing, Blacksmith's and Mill Supplies at lowest prices. N. Smith, 138 York Street, Toronto. (2tf)

BUSINESS CHANCE.

WANTED—First-class foundry and machine company to buy interest in high-class patented machine, weighing seven tons and to manufacture, sell and instal same; this is a good proposition and will stand investigation. Apply Box 26, CANADIAN MACHINERY, Toronto.

A GOOD cash paying electric wiring business; a small capital only required; selling out owing to ill health. For particulars apply to Box 31, CANADIAN MACHINERY, Toronto. [8]

Companies Incorporated

The Brockville Malleable Iron Co., Brockville, Ont., has obtained a charter.

The Ford Automatic Fire Shutter Co. has been incorporated, with head office in Niagara Falls.

"The Niagara Iron and Steel Co." has been incorporated, with head office at Toronto and capital stock of $750,000.

"Industrial Development Company, of Canada" has been incorporated, with head office at Hull and capital stock of $1,000,000.

The Lacoste Ship-Brake Co. has been incorporated, with $45,000 capital, and head office in Montreal, to manufacture a ship-brake, invented by Louis Lacoste.

The Bawden Machine & Tool Co., Limited, Toronto, has been incorporated with a capital of $100,000, the provisional directors being W. B. Mudie, Geo. Tillie, and C. W. Mitchell.

The Iroquois Mfg. Co., Iroquois, Ont., has been incorporated, with a capital of $40,000, to manufacture machines, tools, utensils, etc. The provisional directors include R. S. Smart, J. Moffatt and T. R. E. MacInnes, Ottawa.

A limited liability company has been organized under provincial charter to take over the business of the Bawden Machine & Tool Co., Toronto, under the same name, with a capital of $100,000, and head office in Toronto.

The Imperial Supply Co. has been incorporated, with head office in Montreal, and a capital of $100,000, to manufacture and sell railway, marine and contractors' supplies, and to act as agents for other makers of the same lines.

The Niagara Iron & Steel Co., of Toronto, has been incorporated with a capital of $1,000,000 to carry on the business of rolling mills and the manufacture of iron and steel rails and a general smelting and refining business.

W. J. Bishop, contractor; W. C. Strachan, manufacturer, and others have been incorporated as The Rexford Bishop, Limited, to carry on a general contracting business, with head office in Montreal. The company is capitalized at $100,000.

The St. Thomas Automobile & Garage Co., St. Thomas, Ont., has been incorporated, with a capital of $25,000, to manufacture automobiles, bicycles, etc. The provisional directors include G. M. Baldwin, A, E, Thomas and T. Hall, St. Thomas, Ont.

The Niagara Iron & Steel Co. has been incorporated, with a capital of $1,000,000 for the manufacture of steel rails and other iron and steel commodities. The incorporators include Hon. J. K. Kerr, J. A. Paterson, Wm. Davidson and G. F. McFarland.

The Labatt Manufacturing Co., London, manufacturing plumbers' supplies, has been incorporated with a capital of $200,000. The provisional directors are: John Labatt and H. F. Labatt, London; H. J. Wood and D. W. Wilson, Buffalo; S. C. Newburn, Hamilton.

The Gurry Patents Co., Hamilton, has been incorporated with a capital of $40,000, to purchase and use the patent of Edward Gurry for a new pig iron moulding machine. The provisional directors are W. D. Long, P. D. Crerar, and E. Gurry, all of Hamilton.

The Doon Twine & Cordage Co., of Doon, Ontario, has been incorparated with a capital of $250,000 for the manufacturing of twines, cordage, and upholstering stock. The provisional directors are: A. Forster, E. G. Perine, J. Stauffer, H. Krug and G. A. Clare.

The Brockville Malleable Iron Co. has been incorporated, the provisional directors being, H. A. Stewart, J. A. Mackenzie, J. I. Mallery, J. Connolly, M. H. Harrison and Dr. Clark. Buildings will be erected immediately and the manufacturing will commence in the fall.

V. J. Hughes, broker; Harold Rolph, engineer, with others, have been incorporated, under the firm name of Metcalf Engineering, Limited, with a capital of $25,000, to carry on the business of civil, mechanical and electrical engineers, contractors and inspectors. The head office will be in Montreal.

J. W. Dowker, A. Dowker, Minnie G. Dowker, J. E. Arnold and W. F. Pratt, have been incorporated, under the name of the Dowker Brick Co., to take over the brickmaking business now carried on by Alem Dowker, the head office of the company to be in Fort Frances, and the capital to be $40,000.

J. A. McRae, Niagara Falls; W. H. Chandler and J. H. McNeil, Toronto, all contractors, with F. H. Markey and R. C. Grant, Montreal, have been incorporated, under the name of McRae, Chandler & McNeil, with a capital of $100,000, to carry on a general contracting business. The head office will be in Montreal.

William Stone, Alfred Jephcott, James P. Murray, Thomas Roden and L. V. Dusseau, all of Toronto, have been incorporated, under the name of Automatic Sprinklers, Limited, to manufacture and install automatic sprinklers and other devices for protection against fire. The capital is placed at $100,000, and Toronto will be the chief place of business.

The incorporation of Westinghouse, Church, Kerr & Company, of Canada, is gazetted, the head office of the company to be in Montreal, and the capital to be $2,500,000, the objects of the concern being to carry on the business of machinists, builders, electrical, civil and hydraulic engineers; to manufacture and deal in engines, machinery, boilers and rolling stock, and to do a general engineering and construction business.

In the course of a short time Sydney C.B., will have another important industry in the shape of a brass foundry, which will give employment to at least 75 men. The company, which will be known as Shaw & Mason, has been organized provisionally and is now seeking incorporation. The company will be capitalized at $100,000. The following are the provisional directors: E. E. Shaw, T. P. Mason, H. C. Burchell, G. A. R. Rowlings, Alex. Johnston, Norman McDonald and J. A. Young. The business of the company will be the manufacture of brass goods of all descriptions and certain lines of cast iron.

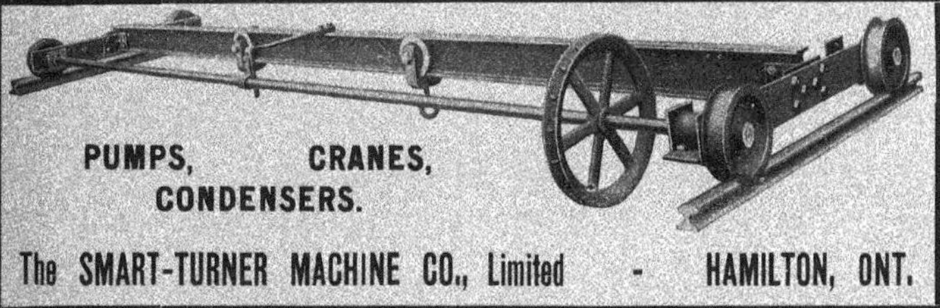
PUMPS, CRANES,
CONDENSERS.
The SMART-TURNER MACHINE CO., Limited - HAMILTON, ONT.

NO ASHES TO BURN
IF YOU USE A
"PREMIER"
GAS ENGINE
and when installed with a Canada Gas Producer it will save more coal than burning ashes.
LET US TELL YOU ABOUT IT.
CANADA FOUNDRY COMPANY, Limited
Head Office and Works: TORONTO, ONT.
District Offices: Montreal Halifax Ottawa Winnipeg Vancouver Rossland

PATTERNS
WELLS' PATTERN AND MODEL WORKS

Tel. Main 3581. (HARRY WELLS, Proprietor.)
Patterns and Models made in wood and metal for Engines, Pumps, Furnaces, Agricultural, Electrical and Architectural Works and Machines of every description.

35 Richmond St. E., Toronto

We should have your business.

If prompt delivery, reasonable prices and best of workmanship appeal to you, we should have your pattern orders.

SEND FOR PATTERN CLASSIFICATION CARD

THE HAMILTON PATTERN WORKS

134 Bay St. N., HAMILTON, ONT.
Phone, 2384.

MACHINERY FOR EXPORT

On Admiralty and War Office Lists

IN STOCK. Ready for immediate delivery.
For Sale, on Hire, or on Purchase Hire.

MACHINERY

Reliable, Second-hand and New, including Portable and Semi-Portable Engines, also Locomotive Engines, Engines and Boilers combined, also Horizontal, Vertical, Traction, Winding, Gas, and Oil Engines. Road Rollers. Plenty of Cornish, Lancashire, Loco-Type, Vertical and other Boilers, great variety. Saw Benches. Stone Breakers. Pulsometers and other Pumps. Steam winches. Cranes and other Lifting Machinery.

Inquiries and Orders receive careful and prompt attention.

Send for copy of "Phillips' Monthly Machinery Register," the best medium in the world for the purchase and sale of New and Second-hand Machinery of every description.

Note the Only Addresses:

CHARLES D. PHILLIPS,

Emlyn Engineering Works, NEWPORT, Mon., and Emlyn Works, GLOUCESTER.

Established 39 Years

Hydraulic Presses
Power Screw Presses
Filter Presses

William R. Perrin AND Company, Limited,
TORONTO, Canada.

The Canadian Turbine Water Wheel

Ample power, economically generated is the assurance that goes with every Canadian Turbine. It is for you to find out that the Canadian Turbine has no equal. Our customers will help you to decide this question and it will pay you to investigate the matter.

Chas. Barber & Sons, Meaford, Ont.
Turbine Water Wheels, Governors and Gearing Only.
Established 1867

CHADWICK BROS.
BRASS FOUNDERS

Manufacturers of
Cast, Spun and Sheet Metal Goods; Brass, Copper and Aluminum Castings; Electro Plating; Brass and Nickel Railings; Brass Fittings of all kinds Engineers' and Plumbers' Brass Goods; All kinds of Brass Turnings; Bolts, Nuts, Screws, etc. Brass and Metal Goods of every description made to order; Gas and Gasoline Engine Supplies.

CHADWICK BROS.
Factory: HAMILTON, ONT. (Oak Ave.)

LUMEN BEARING COMPANY

Brass Founders

Buffalo, U.S.A.
Toronto, Can.

APPRENTICES

can Bank Money in a few weeks.

Work will take but a few hours a week.

You Need no Money because we provide the capital.

We will start you in business by sending you your first stock without charge. The money from the sales of these will be used in buying new stock.

If you will try it, we will send supplies at once. Your profits are large.

IN ADDITION

there are bonus prizes: a watch is one of these bonuses.

Some boys and youths are earning a large sum each week as extra money.

WRITE AT ONCE TO

The MacLean Pub. Company
10 Front St. E., TORONTO

Electrical Machinery Co. of Toronto

Motors and Dynamos
Our Specialty

Manufacturers
Designers
Engineers
for all
ectrical Apparatus

Office and Works, 113-115 Simcoe Street. Phone M. 6639

CANADIAN MACHINERY BUYERS' DIRECTORY

To Our Readers—Use this directory when seeking to buy any machinery or power equipment. You will often get information that will save you money.
To Our Advertisers—Send in your name for insertion under the heading of the lines you make or sell.
To Non-Advertisers—A nominal rate of $1 per line a year is charged non-advertisers.

Acids.
Canada Chemical Mfg. Co., London

Abrasive Materials.
Baxter Paterson & Co., Montreal.
The Canadian Fairbanks Co., Montreal.
Rice Lewis & Son, Toronto.
Norton Co., Worcester, Mass.
H. W. Petrie, Toronto.
Williams & Wilson, Montreal.

Air Brakes.
Canada Foundry Co., Toronto.
Canadian Westinghouse Co., Hamilton.

Air Receivers.
Allis-Chalmers-Bullock Montreal.
Canada Foundry Co., Toronto.
Canadian Rand Drill Co., Montreal.
Chicago Pneumatic Tool Co., Chicago.
John McDougall Caledonian Iron Works Co., Montreal.

Alloy, Ferro-Silicon.
Paxon, J. W., Co., Philadelphia.

Alundum Scythe Stones
Norton Company, Worcester, Mass.

Arbor Presses.
Niles-Bement-Pond Co., New York.

Augers.
Chicago Pneumatic Tool Co., Chicago.
Frothingham & Workman Ltd., Montreal
Rice Lewis & Son, Toronto.

Automatic Machinery.
Cleveland Automatic Machine Co. Cleveland.
National-Acme Mfg. Co., Cleveland
Potter & Johnston Machine Co., Pawtucket, R. I.

Automobile Parts
Globe Machine & Stamping Co., Cleveland, Ohio.

Axle Cutters.
Butterfield & Co., Rock Island, Que.
A. B. Jardine & Co., Hespeler, Ont.

Axle Setters and Straighteners.
A. B. Jardine & Co., Hespeler, Ont.
Standard Bearings, Ltd , Niagara Falls

Babbit Metal.
Baxter, Paterson & Co., Montreal.
Canada Metal Co., Toronto.
Canada Machinery Agency, Montreal.
Frothingham & Workman Ltd., Montreal
Greey Wm & J. G , Toronto.
Rice Lewis & Son, Toronto.
Lumen Bearing Co., Toronto.
Mechanics Supply Co , Quebec, Que.
Miller Bros. & Toms. Montreal, Que.

Bakers' Machinery.
Greey, Wm. & J. G., Toronto.

Baling Presses.
Perrin, Wm. R., & Co., Toronto.

Barrels, Steel Shop.
Cleveland Wire Spring Co., Cleveland.

Barrels, Tumbling.
Buffalo Foundry Supply Co , Buffalo.
Detroit Foundry Supply Co , Windsor
Dominion Foundry Supply Co., Montreal
Hamilton Facing Mill Co., Hamilton.
Globe Machine & Stamping Co , Cleveland Ohio.
John McDougall Caledonian Iron Works Co , Montreal.
R. McDougall Co., Galt, Ont.
Miller Bros & Toms, Montreal, Que.
Northern Engineering Works Detroit.
J. W Paxson Co., Philadelphia, Pa.
H. W. Petrie, Toronto.
The Smart-Turner Mach. Co., Hamilton.

Bars, Boring.
Hall Engineering Works, Montreal.
Miller Bros & Toms Montreal.
Niles-Bement-Pond Co., New York.
Standard Bearings Ltd., Niagara Falls

Batteries, Dry.
Berlin Electrical Mfg Co., Toronto.
Mechanics Supply Co., Quebec, Que.

Batteries, Flashlight.
Berlin Electrical Mfg Co., Toronto.
Mechanics Supply Co., Quebec Que

Batteries, Storage.
Canadian General Electric Co. Toronto
Chicago Pneumatic Tool Co , Chicago.
Rice Lewis & Son, Toronto.
John Millen & Son, Montreal.

Bearing Metals.
Lumen Bearing Co., Toronto.

Bearings, Roller.
Standard Bearings, Ltd. Niagara Falls

Bell Ringers.
Chicago Pneumatic Tool Co., Chicago.
Mechanics Supply Co., Quebec, Que.

Belting, Chain.
Baxter, Paterson & Co. Montreal.
Canada Machinery Exchange, Montreal.
Greey, Wm. & J. G., Toronto.
Jeffrey Mfg. Co., Columbus, Ohio.
Link-Belt Eng. Co., Philadelphia.
Miller Bros. & Toms Montreal.
Waterous Engine Works Co., Brantford.

Belting, Cotton.
Baxter, Paterson & Co., Montreal.
Canada Machinery Agency, Montreal.
Dominion Belting Co., Hamilton.
Rice Lewis & Son, Toronto.

Belting, Leather.
Baxter, Paterson & Co., Montreal.
Canada Machinery Agency, Montreal.
The Canadian Fairbanks Co., Montreal.
Frothingham & Workman Ltd., Montreal
Greey, Wm. & J. G., Toronto.
McLaren, J. C., Montreal.
Rice Lewis & Son, Toronto.
H. W. Petrie, Toronto.
Williams & Wilson, Montreal.

Belting, Rubber.
Baxter, Paterson & Co., Montreal.
Canada Machinery Agency Montreal.
Frothingham & Workman, Ltd., Montreal
Greey, Wm. & J. G., Toronto.

Belting Supplies.
Baxter Paterson & Co , Montreal.
Greey Wm. & J. G , Toronto.
Rice Lewis & Son, Toronto.
H. W. Petrie, Toronto.

Bending Machinery.
John Bertram & Sons Co., Dundas, Ont.
Chicago Pneumatic Tool Co., Chicago.
London Mach. Tool Co., Hamilton, Ont.
National Machinery Co., Tiffin, Ohio.
Niles-Bement-Pond Co., New York.

Benders, Tire.
A. B. Jardine & Co., Hespeler, Ont.
London Mach. Tool Co., Hamilton, Ont.

Blowers.
Buffalo Foundry Supply Co., Montreal.
Canada Machinery Agency, Montreal.
Detroit Foundry Supply Co , Windsor
Hamilton Facing Mill Co., Hamilton.
Kerr Turbine Co., Wellsville, N.Y.
Mechanics Supply Co , Quebec Que.
J. W Paxson, Philadelphia, Pa.
Sheldon's Limited, Galt
Sturtevant, B. F., Co., Hyde Park, Mass.

Blast Gauges—Cupola.
Paxson, J. W., Co. Philadelphia
Sheldons, Limited, Galt

Blocks, Tackle.
Frothingham & Workman Ltd , Montreal

Blocks, Wire Rope.
Frothingham & Workman, Ltd., Montreal

Blue Printing.
The Electric Blue Print Co., Montreal.

Blow-Off Tanks.
Darling Bros., Ltd., Montreal.

Boilers.
Canada Foundry Co., Limited, Toronto.
Canada Machinery Agency, Montreal.
Goldie & McCulloch Co., Galt.
E. Leonard & Sons, London, Ont.
John McDougall Caledonian Iron Works, Montreal.
Manitoba Iron Works, Winnipeg.
Mechanics Supply Co , Quebec, Que.
Owen Sound Iron Works Co., Owen Sound
H. W. Petrie, Toronto.
Robb Engineering Co Amherst, N S.
The Smart-Turner Mach. Co., Hamilton.
Waterous Engine Works Co., Brantford.
Williams & Wilson, Montreal.

Boiler Compounds.
Canada Chemical Mfg. Co., London, Ont.
Hall Engineering Works, Montreal.
Lake Erie Boiler Compound Co., Toronto

Bolt Cutters.
John Bertram & Sons Co., Dundas, Ont.
London Mach. Tool Co., Hamilton.
Mechanics Supply Co , Quebec , Que.
National Machinery Co., Tiffin, Ohio.
Niles-Bement-Pond Co., New York.

Bolt and Nut Machinery.
John Bertram & Sons Co., Dundas, Ont.
Canada Machinery Agency Montreal.
London Mach. Tool Co., Hamilton.
National Machinery Co., Tiffin, Ohio.
Niles-Bement-Pond Co. New York.

Bolts and Nuts, Rivets.
Baxter, Paterson & Co., Montreal.
Mechanics Supply Co , Quebec, Que.

Boring and Drilling Machines.
American Tool Works Co., Cincinnati.
B. F. Barnes Co., Rockford, Ill.
John Bertram & Sons Co., Dundas, Ont.
Canada Machinery Agency Montreal.
A. B. Jardine & Co., Hespeler, Ont.
London Mach. Tool Co., Hamilton.
Niles-Bement-Pond Co., New York.

Boring Machine, Upright.
American Tool Works Co., Cincinnati.
John Bertram & Sons Co., Dundas, Ont.
London Mach Tool Co., Hamilton.
Niles-Bement-Pond Co., New York.

Boring Machine, Wood.
Chicago Pneumatic Tool Co., Chicago.
London Mach. Tool Co., Hamilton.

Boring and Turning Mills.
American Tool Works Co., Cincinnati.
John Bertram & Sons Co., Dundas, Ont.
Gisholt Machine Co , Madison, Wis.
Canada Machinery Agency, Montreal.
London Mach. Tool Co., Hamilton.
Niles-Bement-Pond Co., New York.
H. W. Petrie, Toronto.

Box Puller.
A. B. Jardine & Co., Hespeler, Ont.

Boxes, Steel Shop.
Cleveland Wire Spring Co , Cleveland.

Boxes, Tote.
Cleveland Wire Spring Co., Cleveland

Brass Foundry Equipment.
Detroit Foundry Supply Co., Detroit.
Paxson, J. W., Co., Philadelphia

Brass Working Machinery.
Warner & Swasey Co., Cleveland, Ohio.

Brushes, Foundry and Core.
Buffalo Foundry Supply Co., Buffalo.
Detroit Foundry Supply Co , Windsor.
Dominion Foundry Supply Co., Montreal
Hamilton Facing Mill Co., Hamilton.
Mechanics Supply Co , Quebec, Que.
Paxson, J W., Co., Philadelphia

Brushes, Steel.
Buffalo Foundry Supply Co., Buffalo.
Paxson, J. W., Co., Philadelphia

Bulldozers.
John Bertram & Sons Co., Dundas, Ont.
London Mach. Tool Co., Hamilton, Ont.
National Machinery Co., Tiffin, Ohio.
Niles-Bement-Pond Co., New York.

Calipers.
Baxter, Paterson & Co., Montreal.
Frothingham & Workman Ltd., Montreal
Rice Lewis & Son, Toronto.
Mechanics Supply Co., Quebec, Que.
John Millen & Son, Ltd , Montreal, Que.
L. S. Starrett & Co., Athol, Mass.
Williams & Wilson Montreal.

Cars, Foundry.
Buffalo Foundry Supply Co., Buffalo.
Detroit Foundry Supply Co., Windsor
Dominion Foundry Supply Co., Montreal
Hamilton Facing Mill Co., Hamilton.
Paxson, J. W., Co., Philadelphia

Castings, Aluminum.
Lumen Bearing Co., Toronto

Castings, Brass.
Chadwick Bros., Hamilton.
Greey Wm. & J. G. Toronto.
Hall Engineering Works, Montreal.
Kennedy Wm., & Sons, Owen Sound.
Lumen Bearing Co., Toronto
McDougall R. Co , Galt
Niagara Falls Machine & Foundry Co. Niagara Falls, Ont.
Owen Sound Iron Works Co., Owen Sound.
Robb Engineering Co , Amherst, N.S.

Castings, Grey Iron.
Allis-Chalmers-Bullock Montreal.
Greey Wm & J. G , Toronto
Hall Engineering Works, Montreal.
Kennedy, Wm , & Sons, Owen Sound.
Laurie Engine & Machine Co., Montreal.
Manson Mfg. Co., Thorold. Ont
John McDougall Caledonian Iron Works Co., Montreal.
R. McDougall Co., Galt, Ont.
Niagara Falls Machine & Foundry Co., Niagara Falls Ont
Owen Sound Iron Works Co., Owen Sound.
Robb Engineering Co., Amherst, N.S.
Smart-Turner Machine Co., Hamilton.

Castings, Phosphor Bronze.
Lumen Bearing Co., Toronto

Castings, Steel.
Kennedy, Wm., & Sons, Owen Sound.

Castings, Semi-Steel.
Robb Engineering Co., Amherst, N.S.

Cement Machinery.
Allis-Chalmers-Bullock Limited, Montreal
Greey, Wm. & J. G., Toronto
Jeffrey Mfg. Co., Columbus, Ohio.
John McDougall Caledonian Iron Works, Co., Montreal
Owen Sound Iron Works Co., Owen Sound

Centreing Machines.
John Bertram & Sons Co., Dundas, Ont.
Jeffrey Mfg Co., Columbus, Ohio
London Mach. Tool Co., Hamilton, Ont.
Niles-Bement-Pond Co., New York.
Pratt & Whitney Co., Hartford, Conn.
Standard Bearings, Ltd., Niagara Falls

Centres, Planer.
American Tool Works Co., Cincinnati.

Centrifugal Pumps.
Manson Mfg Co., Thorold, Ont
John McDougall Caledonian Iron Works Co , Montreal
R. McDougall Co., Galt.
D'Olier Engineering Co., New York.
Pratt & Whitney Co., Hartford, Conn.

Centrifugal Pumps—Turbine Driven
Kerr Turbine Co., Wellsville, N.Y.

Chain, Crane and Dredge.
Frothingham & Workman, Ltd., Montreal

Chaplets.

Buffalo Foundry Supply Co., Buffalo.
Detroit Foundry Supply Co., Windsor.
Hamilton Facing Mill Co., Hamilton.
Paxson, J. W., Co., Philadelphia

Charcoal.

Buffalo Foundry Supply Co., Buffalo.
Detroit Foundry Supply Co., Windsor.
Dominion Foundry Supply Co., Montreal.
Hamilton Facing Mill Co., Hamilton.
Paxson, J. W., Co., Philadelphia

Chemicals.

Baxter, Paterson & Co. Montreal.
Canada Chemical Co., London.

Chemists' Machinery.

Greey, Wm. & J. G., Toronto.

Chemists, Industrial.

Detroit Testing Laboratory, Detroit.

Chemists, Metallurgical.

Detroit Testing Laboratory, Detroit

Chemists, Mining.

Detroit Testing Laboratory, Detroit.

Chucks, Ring Grinding.

A. B. Jardine & Co., Hespeler, Ont.
Chicago Pneumatic Tool Co., Chicago.

Chucks, Drill and Lathe.

American Tool Works Co., Cincinnati.
Baxter, Paterson & Co., Montreal.
John Bertram & Sons Co., Dundas, Ont.
Canada Machinery Agency, Montreal.
Frothingham & Workman, Ltd., Montreal
Ker & Goodwin, Brantford.
A. B. Jardine & Co., Hespeler, Ont.
London Mach. Tool Co., Hamilton.
John Millen & Son, Ltd., Montreal.
Niles-Bement-Pond Co., New York.
H. W. Petrie, Toronto.
Rice Lewis & Son, Toronto.
Standard Tool Co., Cleveland.

Chucks, Planer.

American Tool Works Co., Cincinnati.
Canada Machinery Agency, Montreal.
Niles-Bement-Pond Co., New York.

Chucking Machines.

American Tool Works Co., Cincinnati.
Niles-Bement-Pond Co., New York.
H. W. Petrie, Toronto.
Warner & Swasey Co., Cleveland, Ohio

Circuit Breakers.

Allis-Chalmers-Bullock, Limited, Montreal
Canadian General Electric Co., Toronto.
Canadian Westinghouse Co., Hamilton.

Clippers, Bolt.

Frothingham & Workman Ltd., Montreal
A. B. Jardine & Co., Hespeler, Ont.

Cloth and Wool Dryers.

Dominion Heating and Ventilating Co., Hespeler.
B. Greening Wire Co., Hamilton.
Sheldons Limited, Galt

Collectors, Pneumatic.

Sheldons Limited, Galt
Sturtevant, B. F., Co., Hyde Park, Mass.

Compressors, Air.

Allis-Chalmers-Bullock, Limited, Montreal
Canada Foundry Co., Limited, Toronto.
Canada Machinery Agency, Montreal.
Canadian Rand Drill Co., Montreal.
Canadian Westinghouse Co., Hamilton.
Chicago Pneumatic Tool Co., Chicago.
Darling Bros., Ltd., Montreal
Detroit Foundry Supply Co., Windsor.
John McDougall, Caledonian Iron Works Co., Montreal
H. W. Petrie, Toronto.
The Smart-Turner Mach. Co., Hamilton.
Hall Engineering Works, Montreal, Que.
London Mach. Tool Co., Hamilton.
Niles-Bement-Pond Co., New York.
H. W. Petrie, Toronto.
Pratt & Whitney Co., Hartford, Conn.
Williams & Wilson, Montreal.

Concentrating Plant.

Allis-Chalmers-Bullock, Montreal.
Greey, Wm. & J. G., Toronto.

Concrete Mixers.

Jeffrey M'f'g. Co., Columbus, Ohio.
Link-Belt Co., Philadelphia.

Condensers.

Canada Foundry Co., Limited, Toronto.
Canada Machinery Agency, Montreal.
Hall Engineering Works Montreal.
Smart-Turner-Machine Co., Hamilton.
Waterous Engine Co., Brantford.

Confectioners' Machinery.

Baxter, Paterson & Co., Montreal.
Greey, Wm & J. G., Toronto.
Pendrith Machinery Co., Toronto.

Consulting Engineers.

Charles Brandeis, Montreal.
Canadian White Co., Montreal.
Hall Engineering Works, Montreal,
Jules De Clercy, Montreal.
John S. Fielding, Toronto.
Miller Bros. & Toms, Montreal.
Roderick J. Parke, Toronto.
Plews & Trimingham, Montreal.
T. Pringle & Son, Montreal.
Standard Bearings, Ltd., Niagara Falls

Contractors.

Canadian White Co., Montreal.
D'Olier Engineering Co., New York.
Expanded Metal and Fireproofing Co. Toronto.
Hall Engineering Works Montreal.
Laurie Engine & Machine Co., Montreal.
John McDougall Caledonian Iron Works Co., Montreal
Miller Bros. & Toms, Montreal.
Robb Engineering Co., Amherst N S.
The Smart-Turner Mach. Co., Hamilton.

Contractors' Plant.

Allis-Chalmers-Bullock, Montreal.
John McDougall, Caledonian Iron Works Co., Montreal.
Niagara Falls Machine & Foundry Co., Niagara Falls, Ont.

Contractors' Supplies.

Manson Mfg. Co., Thorold, Ont.

Controllers and Starters Electric Motor.

Allis-Chalmers-Bullock, Montreal.
Canadian General Electric Co., Toronto.
Canadian Westinghouse Co., Hamilton.
T. & H. Electric Co., Hamilton.

Conveying Systems.

Sheldons Limited, Galt
Sturtevant, B. F., Co., Hyde Park, Mass.

Converters, Steel.

Northern Engineering Works, Detroit.
Paxson, J. W., Co., Philadelphia

Conveyor Machinery.

Baxter, Paterson & Co., Montreal.
Greey Wm & J. G., Toronto
Jeffrey Mfg. Co., Columbus, Ohio.
Link-Belt Co., Philadelphia.
John McDougall Caledonian Iron Works Co., Montreal.
Miller Bros. & Toms, Montreal.
Smart-Turner Machine Co., Hamilton.
Laurie Engine & Machine Co., Montreal.
Waterous Engine Works Co., Brantford.
Williams & Wilson, Montreal.

Coping Machines.

John Bertram & Sons Co., Dundas, Ont.
London Mach. Tool Co., Hamilton.
Niles-Bement-Pond Co., New York.

Core Compounds.

Buffalo Foundry Supply Co., Buffalo.
Detroit Foundry Supply Co., Windsor.
Dominion Foundry Supply Co., Montreal
Hamilton Facing Mill Co., Hamilton.
Paxson, J. W., Co., Philadelphia

Core-Making Machines.

Paxson, J. W., Co., Philadelphia

Core Ovens.

Detroit Foundry Supply Co., Windsor.
Hamilton Facing Mill Co., Hamilton.
Paxson, J. W., Co., Philadelphia
Sheldons Limited, Galt

Core Oven Bricks.

Buffalo Foundry Supply Co., Buffalo.
Detroit Foundry Supply Co., Windsor.
Hamilton Facing Mill Co., Hamilton.
Paxson, J. W., Co., Philadelphia

Core Wash.

Buffalo Foundry Supply Co., Buffalo.
Detroit Foundry Supply Co., Windsor.
Hamilton Facing Mill Co., Hamilton.
Paxson, Co., J. W., Philadelphia

Couplings.

Owen Sound Iron Works Co, Owen Sound

Cranes, Electric and Hand Power.

Canada Foundry Co., Limited, Toronto
Dominion Foundry Supply Co., Montreal
Hamilton Facing Mill Co., Hamilton.
Link-Belt Co., Philadelphia.
John McDougall, Caledonian Iron Works Co., Montreal.
Miller Bros. & Toms, Montreal,
Niles-Bement-Pond Co., New York.
Northern Engineering Works Detroit
Owen Sound Iron Works Co, Owen Sound
Paxson, J. W., Co, Philadelphia
Smart-Turner-Machine Co., Hamilton.

Crank Pin.

Sight Feed Oil Pump Co., Milwaukee, Wis.

Crankshafts.

The Canada Forge Co., Welland.
St. Clair Bros., Galt

Crabs.

Frothingham & Workman, Ltd., Montreal

Crank Pin Turning Machine

London Mach. Tool Co., Hamilton.
Niles-Bement-Pond Co., New York.

Cross Head Pin.

Sight Feed Oil Pump Co., Milwaukee, Wis.

Crucibles.

Buffalo Foundry Supply Co., Buffalo.
Detroit Foundry Supply Co., Windsor
Dominion Foundry Supply Co., Montreal
Hamilton Facing Mill Co., Hamilton.
J. W. Paxson Co., Philadelphia.

Crucible Caps

Hamilton Facing Mill Co., Hamilton.
Paxson, J. W., Co., Philadelphia

Crushers, Rock or Ore.

Allis-Chalmers-Bullock, Montreal.
Jeffrey Mfg. Co., Columbus, Ohio.

Cup Grease.

Mechanics Supply Co., Quebec

Cupolas.

Buffalo Foundry Supply Co., Buffalo.
Detroit Foundry Supply Co., Windsor
Dominion Foundry Supply Co., Montreal
Hamilton Facing Mill Co., Hamilton.
Northern Engineering Works, Detroit
J. W. Paxson Co. Philadelphia.
Sheldons Limited, Galt.

Cupola Blast Gauges.

Paxson, J. W., Co., Philadelphia
Sheldons Limited, Galt

Cupola Blocks.

Detroit Foundry Supply Co., Detroit.
Hamilton Facing Mill Co., Hamilton
Northern Engineering Works. Detroit
Ontario Lime Association Toronto
Paxson, J. W., Co, Philadelphia
Toronto Pottery Co., Toronto

Cupola Blowers.

Buffalo Foundry Supply Co., Buffalo.
Canada Machinery Agency, Montreal.
Detroit Foundry Supply Co., Windsor
Dominion Heating and Ventilating Co., Hespeler
Hamilton Facing Mill Co., Hamilton.
Northern Engineering Works, Detroit
Paxson J. W., Co., Philadelphia
Sheldon's Limited, Galt.
B. F. Sturtevant Co., Hyde Park, Mass.

Cutters, Flue

Chicago Pneumatic Tool Co., Chicago.
J. W. Paxson Co., Philadelphia.

Cutters, Milling.

Becker, Brainard Milling Machine Co. Hyde Park. Mass.
Frothingham & Workman, Ltd., Montreal
Pratt & Whitney Co., Hartford, Conn.
Standard Tool Co., Cleveland.

Cutting-off Machines.

John Bertram & Sons Co., Dundas, Ont.
Canada Machinery Agency, Montreal.
London Mach. Tool Co., Hamilton.
H. W. Petrie, Toronto.
Pratt & Whitney Co., Hartford, Conn.

Cutting-off Tools.

Armstrong Bros. Tool Co., Chicago.
Baxter Paterson & Co, Montreal.
London Mach. Tool Co., Hamilton.
Mechanics Supply Co., Quebec, Que.
H. W. Petrie, Toronto.
Pratt & Whitney, Hartford, Conn.
Rice Lewis & Son, Toronto.
L. S. Starrett Co., Athol, Mass.

Damper Regulators.

Darling Bros., Ltd., Montreal

Dies

Globe Machine & Stamping Co., Cleveland, Ohio.

Die Stocks

Hart Manufacturing Co., Cleveland, Ohio
Mechanics Supply Co., Quebec

Dies, Opening

W. H. Banfield & Sons, Toronto
Globe Machine & Stamping Co., Cleveland Ohio.
R. McDougall Co., Galt.
Pratt & Whitney Co., Hartford Conn.

Dies, Sheet Metal.

W. H. Banfield & Sons, Toronto.
Globe Machine & Stamping Co., Cleveland. Ohio

Dies, Threading,

Frothingham & Workman, Ltd., Montreal
Hart Mfg. Co., Cleveland
Mechanics Supply Co., Quebec, Que.
John Millen & Son, Ltd, Montreal.

Draft, Mechanical.

W. H. Banfield & Sons, Toronto.
Butterfield & Co., Rock Island, Que.
A. B. Jardine & Co., Hespeler
Mechanics Supply Co., Quebec, Que.
Pratt & Whitney Co., Hartford, Conn.
Sheldon's Limited, Galt.
Sturtevant, B. F., Co., Hyde Park, Mass.

Drawing Instruments.

Rice Lewis & Son. Toronto.
Mechanics Supply Co, Quebec, Que.

Drawing Supplies.

The Electric Blue Print Co., Montreal

Drawn Steel, Cold.

Baxter Paterson & Co., Montreal.
Greey, Wm. & J. G., Toronto
Miller Bros. & Toms, Montreal.
Union Drawn Steel Co., Hamilton.

Drilling Machines, Arch Bar.

John Bertram & Sons Co., Dundas, Ont.
London Mach. Tool Co., Hamilton
Niles-Bement-Pond Co., New York.

Drilling Machines, Boiler.

American Tool Works Co., Cincinnati.
John Bertram & Sons Co., Dundas, Ont.
Bickford Drill and Tool Co., Cincinnati.
The Canadian Fairbanks Co., Montreal.
A. B. Jardine & Co., Hespeler, Ont.
London Mach. Tool Co., Hamilton, Ont.
Niles-Bement-Pond Co., New York.
H. W. Petrie, Toronto.
Williams & Wilson, Montreal

Drilling Machines Connecting Rod.

John Bertram & Sons Co., Dundas, Ont.
London Mach. Tool Co., Hamilton.
Niles-Bement-Pond Co. New York.

Drilling Machines, Locomotive Frame.

American Tool Works Co., Cincinnati.
B. F. Barnes Co., Rockford, Ill.
John Bertram & Sons Co., Dundas, Ont.
London Mach. Tool Co., Hamilton, Ont.
Niles-Bement-Pond Co., New York.

Drilling Machines, Multiple Spindle.

American Tool Works Co., Cincinnati.
B. F. Barnes Co., Rockford, Ill.
Baxter, Paterson & Co., Montreal.
John Bertram & Sons Co., Dundas, Ont.
Bickford Drill & Tool Co., Cincinnati.
Canada Machinery Agency, Montreal.
London Mach. Tool Co., Hamilton, Ont.
Niles-Bement-Pond Co., New York.
H. W. Petrie, Toronto.
Williams & Wilson, Montreal.

Drilling Machines, Pneumatic

Canada Machinery Agency, Montreal.

Drilling Machines, Portable

Baxter, Paterson & Co., Montreal.
A. B. Jardine & Co., Hespeler, Ont.
Mechanics Supply Co., Quebec, Que.
Niles-Bement-Pond Co., New York.

Drilling Machines, Radial.

American Tool Works Co, Cincinnati.
Baxter, Paterson & Co. Montreal
John Bertram & Sons Co., Dundas, Ont.
Bickford Tool & Drill Co., Cincinnati.
The Canadian Fairbanks Co., Montreal.
London Mach. Tool Co., Hamilton.
Mechanics Supply Co., Quebec, Que.
Niles-Bement-Pond Co., New York.
H. W. Petrie, Toronto.
Williams & Wilson, Montreal.

Drilling Machines Suspension.

John Bertram & Sons Co., Dundas, Ont.
Canada Machinery Agency Montreal.
London Mach. Tool Co., Hamilton.
Niles-Bement-Pond Co New York

Drilling Machines, Turret.

John Bertram & Sons Co., Dundas, Ont.
London Mach. Tool Co., Hamilton.
Niles-Bement-Pond Co., New York.

Drilling Machines, Upright.

American Tool Works Co., Cincinnati.
B. F. Barnes Co., Rockford, Ill.
Baxter, Paterson & Co, Montreal.
John Bertram & Sons Co., Dundas, Ont
A. B. Jardine & Co., Hespeler, Ont.
London Mach. Tool Co., Hamilton.
Mechanics Supply Co., Quebec, Que.
Niles-Bement-Pond Co., New York.
W. H. Petrie, Toronto.
Williams & Wilson Montreal.

Drills, Bench.

B. F. Barnes Co., Rockford, Ill.
Baxter, Paterson & Co. Montreal.
London Mach. Tool Co., Hamilton.
Mechanics Supply Co., Quebec, Que.
Pratt & Whitney Co., Hartford, Conn.

Drills, Blacksmith.

Canada Machinery Agency, Montreal.
Frothingham & Workman, Ltd., Montreal
A. B. Jardine & Co., Hespeler, Ont.
London Mach. Tool Co., Hamilton.
Mechanics Supply Co., Quebec, Que.
Standard Tool Co., Cleveland.

Drills, Centre.

Mechanics Supply Co., Quebec, Que.
Pratt & Whitney Co., Hartford, Conn.
Standard Tool Co., Cleveland, O.
L. S. Starrett Co., Athol, Mass.

Drills, Electric

B. F. Barnes Co., Rockford, Ill.
Baxter, Paterson & Co., Montreal.
Chicago Pneumatic Tool Co., Chicago.
Niles-Bement-Pond Co., New York.

Drills, Gang.

American Tool Works Co., Cincinnati.
B. F. Barnes Co., Rockford, Ill.
John Bertram & Sons Co., Dundas, Ont.
Pratt & Whitney Co., Hartford, Conn.

Drills, High Speed.

Wm. Abbott, Montreal.
American Tool Works Co., Cincinnati.
B. F. Barnes Co., Rockford, Ill.
Frothingham & Workman Ltd., Montreal
Pratt & Whitney Co., Hartford, Conn.
Standard Tool Co., Cleveland, O.

Drills, Hand.

A. B. Jardine & Co., Hespeler, Ont.

Drills, Horizontal.

B. F. Barnes Co., Rockford, Ill.
John Bertram & Sons Co., Dundas, Ont.
Canada Machinery Agency, Montreal.
London Mach. Tool Co., Hamilton.
Niles-Bement-Pond Co., New York.

Drills, Pneumatic.

Canada Machinery Agency, Montreal.
Chicago Pneumatic Tool Co., Chicago.
Independent Pneumatic Tool Co., Chicago, Ill.
Niles-Bement-Pond Co., New York.

Drills, Radial.

American Tool Works Co., Cincinnati.
John Bertram & Sons Co., Dundas, Ont.
Bickford Drill & Tool Co., Cincinnati
London Mach. Tool Co., Hamilton, Ont.
Niles-Bement-Pond Co., New York.

Drills, Ratchet.

Armstrong Bros. Tool Co., Chicago.
Frothingham & Workman, Ltd., Montreal
A. B. Jardine & Co., Hespeler.
Mechanics Supply Co., Quebec, Que.
Pratt & Whitney Co., Hartford, Conn.
Standard Tool Co., Cleveland.

Drills, Rock.

Allis-Chalmers-Bullock, Montreal.
Canadian Rand Drill Co., Montreal.
Chicago Pneumatic Tool Co., Chicago.
Jeffrey Mfg. Co., Columbus, Ohio.

Drills, Sensitive.

American Tool Works Co., Cincinnati.
B. F. Barnes Co., Rockford, Ill.
Canada Machinery Agency, Montreal.
Dwight Slate Machine Co., Hartford, Conn.
Niles-Bement-Pond Co., New York

Drills, Twist.

Baxter, Paterson & Co., Montreal.
Chicago Pneumatic Tool Co., Chicago
Frothingham & Workman Ltd., Montreal
Alex. Gibb, Montreal.
A. B. Jardine & Co., Hespeler, Ont.
Mechanics Supply Co., Quebec, Que.
John Millen & Son, Ltd., Montreal.
Morse Twist Drill and Machine Co., New Bedford, Mass.
Pratt & Whitney Co., Hartford, Conn.
Standard Tool Co., Cleveland.

Drying Apparatus of all Kinds.

Dominion Heating & Ventilating Co., Hespeler
Greey, Wm. & J. G., Toronto
Sheldon's Limited, Galt
B. F. Sturtevant Co., Hyde Park, Mass.

Dry Kiln Equipment.

Sheldons Limited, Galt
Sturtevant, B. F., Co., Hyde Park, Mass.

Dry Sand and Loam Facing.

Buffalo Foundry Supply Co., Buffalo.
Hamilton Facing Mill Co., Hamilton.
Paxson, J. W., Co., Philadelphia

Dump Cars.

Canada Foundry Co., Limited, Toronto
Greey, Wm. & J. G., Toronto
Hamilton Facing Mill Co., Hamilton.
John McDougall, Caledonian Iron Works Co., Montreal.
Niles-Bement-Pond Co., New York.
Standard Bearings, Ltd., Niagara Falls.
Jenckes Machine Co., Sherbrooke, Que
Link-Belt Eng. Co., Philadelphia.
John McDougall Caledonian Iron Works Co., Montreal
Miller Bros. & Toms, Montreal.
Owen Sound Iron Works Co., Owen Sound
Paxson, J. W., Co., Philadelphia
Waterous Engine Co., Brantford.

Dust Separators.

Greey, Wm. & J. G., Toronto
Dominion Heating and Ventilating Co., Hespeler.
Paxson, J. W., Co., Philadelphia
Sheldon's Limited, Galt.

Dynamos.

Allis-Chalmers-Bullock, Montreal.
Canadian General Electric Co., Toronto.
Canadian Westinghouse Co., Hamilton.
Consolidated Electric Co., Toronto.
Hall Engineering Works, Montreal, Que
Mechanics Supply Co., Quebec, Que.
John Millen & Son, Ltd., Montreal.
Packard Electric Co., St. Catharines.
H. W. Petrie, Toronto.
B. F. Sturtevant Co., Hyde Park, Mass.
T. & H. Electric Co., Hamilton.

Dynamos—Turbine Driven.

Kerr-Turbine Co., Wellsville, N.Y.

Economizer, Fuel.

Dominion Heating & Ventilating Co., Hespeler
Standard Bearings, Ltd., Niagara Falls.
B. F. Sturtevant Co., Hyde Park, Mass.

Electrical Instruments.

Canadian Westinghouse Co., Hamilton.
Mechanics Supply Co., Quebec, Que.

Electrical Steel.

Baxter, Patterson & Co., Montreal.

Electrical Supplies.

Canadian General Electric Co., Toronto.
Canadian Westinghouse Co., Hamilton.
London Mach. Tool Co., Hamilton, Ont.
Mechanics Supply Co., Quebec, Que
John Millen & Son, Ltd., Montreal.
Packard Electric Co., St. Catharines.
T. & H. Electric Co., Hamilton.

Electrical Repairs

Canadian Westinghouse Co., Hamilton.
T. & H. Electric Co., Hamilton.

Elevator Buckets.

Greey, Wm. & J. G., Toronto
Jeffrey Mfg. Co., Columbus, Ohio.

Elevators, Foundry.

Northern Engineering Works, Detroit

Elevators—For any Service.

Darling Bros. Ltd., Montreal

Emery and Emery Wheels.

Baxter, Paterson & Co., Montreal.
Dominion Foundry Supply Co., Montreal
Frothingham & Workman Ltd., Montreal
Hamilton Facing Mill Co., Hamilton.
Paxson, J. W., Co., Philadelphia

Emery Wheel Dressers.

Baxter, Paterson & Co., Montreal.
Canada Machinery Agency, Montreal.
Desmond-Stephan Mfg. Co., Urbana, Ohio
Dominion Foundry Supply Co., Montreal
Frothingham & Workman Ltd., Montreal
Hamilton Facing Mill Co., Hamilton.
Mechanics Supply Co., Quebec, Que.
John Millen & Son, Ltd., Montreal.
H. W. Petrie, Toronto.
Paxson, J. W., Co., Philadelphia
Standard Tool Co., Cleveland.

Engineers and Contractors.

Canada Foundry Co., Limited, Toronto.
Darling Bros., Ltd., Montreal
Greey, Wm. & J. G., Toronto
Hall Engineering Works, Montreal.
Laurie Engine & Machine Co., Montreal.
Link-Belt Co., Philadelphia.
John McDougall, Caledonian Iron Works Co., Montreal.
Miller Bros. & Toms, Montreal.
Robb Engineering Co., Amherst, N.S.
The Smart-Turner Mach. Co., Hamilton.

Engineers' Supplies.

Baxter, Paterson & Co., Montreal
Frothingham & Workman Ltd., Montreal
Greey, Wm. & J. G., Toronto
Hall Engineering Works, Montreal.
Mechanics Supply Co., Quebec, Que.
Rice Lewis & Son, Toronto.

Engines, Alcohol.

Du Bois Iron Works, Du Bois, Pa.

Engines, Gas and Gasoline.

Baxter, Paterson & Co., Montreal.
Canada Foundry Co., Toronto.
Canada Machinery Agency, Montreal.
The Canadian Fairbanks Co., Montreal.
Canadian McVicker Engine Co., Galt.
Du Bois Iron Works, Du Bois, Pa.
Economic Heat, Light and Power Supply Co., Toronto.
Gilson Mfg. Co., Guelph
The Goldie & McCulloch Co., Galt, Ont.
Ontario Wind Engine & Pump Co., Toronto
H. W. Petrie, Toronto.
Producer Gas Co., Toronto
The Smart-Turner Mach. Co., Hamilton
Standard Bearings, Ltd., Niagara Falls.

Engines, Hoisting.

Du Bois Iron Works, Du Bois, Pa.

Engines, Oil.

Du Bois Iron Works, Du Bois, Pa.

Engines, Pumping.

Du Bois Iron Works, Du Bois, Pa.

Engines, Steam.

Allis-Chalmers-Bullock, Montreal
Canada Machinery Agency, Montreal.
The Goldie & McCulloch Co., Galt, Ont.
E. Leonard & Sons, London, Ont.
Laurie Engine & Machine Co., Montreal.
Manson Mfg. Co., Thorold, Ont
John McDougall Caledonian Iron Works, Montreal.
Miller Bros. & Toms, Montreal.
Robb Engineering Co., Amherst, N.S.
Sheldons Limited, Galt.
The Smart-Turner Mach. Co., Hamilton.
B. F. Sturtevant Co., Hyde Park, Mass.
Waterous Engine Works Co., Brantford.

Exhaust Heads.

Darling Bros., Ltd., Montreal.
Dominion Heating & Ventilating Co., Hespeler
Sheldon's Limited, Galt, Ont.
B. F. Sturtevant Co., Hyde Park, Mass.

Expanded Metal.

Expanded Metal and Fireproofing Co. Toronto

Expanders.

A. B. Jardine & Co., Hespeler, Ont.

Fans, Electric.

Canadian General Electric Co., Toronto.
Canadian Westinghouse Co., Hamilton.
Dominion Heating & Ventilating Co. Hespeler
Mechanics Supply Co., Quebec, Que.
Sheldon's Limited Galt, Ont.
The Smart-Turner Mach. Co., Hamilton.
B. F. Sturtevant Co., Hyde Park, Mass.

Fans, Exhaust.

Canadian Buffalo Forge Co., Montreal.
Detroit Foundry Supply Co., Windsor.
Dominion Foundry Supply Co., Montreal
Dominion Heating & Ventilating Co., Hespeler.
Greey, Wm. & J. G., Toronto
Hamilton Facing Mill Co., Hamilton.
Paxson, J. W., Co., Philadelphia
Sheldon's Limited, Galt.
B. F. Sturtevant Co., Hyde Park, Mass.

Feed Water Heaters.

Darling Bros., Montreal
Du Bois Iron Works, Du Bois, Pa.
Laurie Engine & Machine Co., Montreal
John McDougall, Caledonian Iron Works Co., Montreal.
The Smart-Turner Mach. Co., Hamilton

Files and Rasps.

Baxter, Paterson & Co., Montreal.
Frothingham & Workman, Ltd., Montreal
Mechanics Supply Co., Quebec, Que.
John Millen & Son, Ltd., Montreal.
Rice Lewis & Son, Toronto.
Nicholson File Co., Providence, R.I.
H. W. Petrie, Toronto.

Fillet, Pattern.

Baxter, Paterson & Co., Montreal.
Buffalo Foundry Supply Co., Buffalo.
Detroit Foundry Supply Co., Windsor.
Dominion Foundry Supply Co., Montreal
Hamilton Facing Mill Co., Hamilton.

Fire Apparatus.

Waterous Engine Works Co., Brantford.

Fire Brick and Clay.

Baxter Paterson & Co., Montreal.
Buffalo Foundry Supply Co., Buffalo.
Detroit Foundry Supply Co., Windsor.
Dominion Foundry Supply Co., Montreal
Hamilton Facing Mill Co., Hamilton.
Ontario Lime Association, Toronto
J. W. Paxson Co., Philadelphia.
Toronto Pottery Co., Toronto

Flash Lights.

Berlin Electrical Mfg. Co., Toronto.
Mechanics Supply Co., Quebec, Que.

Flour Mill Machinery.

Allis-Chalmers-Bullock, Montreal.
Greey, Wm. & J. G., Toronto
The Goldie & McCulloch Co., Galt, Ont.
John McDougall, Caledonian Iron Works Co., Montreal.
Miller Bros. & Toms, Montreal.

Forges.

Canada Foundry Co., Limited, Toronto.
Frothingham & Workman Ltd., Montreal
Hamilton Facing Mill Co., Hamilton.
Mechanics Supply Co., Quebec, Que.
H. W. Petrie, Toronto.
Sheldon's Limited, Galt, Ont.
B. F. Sturtevant Co., Hyde Park, Mass.

Forgings, Drop.

John McDougall, Caledonian Iron Works Co., Montreal.
H. W. Petrie, Toronto.
St. Clair Bros., Galt

Forgings, Light & Heavy.

The Canada Forge Co., Welland.
Hamilton Steel & Iron Co., Hamilton

Forging Machinery.

John Bertram & Sons Co., Dundas, Ont.
London Mach. Tool Co., Hamilton, Ont
National Machinery Co., Tiffin, Ohio
Niles-Bement-Pond Co., New York.

Founders.

Greey, Wm. & J. G., Toronto
John McDougall, Caledonian Iron Works Co., Montreal
R. McDougall Co., Galt, Ont.
Niagara Falls Machine & Foundry Co., Niagara Falls, Ont.
Manson Mfg. Co., Thorold, Ont
The Smart-Turner Mach. Co., Hamilton.

Foundry Equipment.

Detroit Foundry Supply Co., Windsor.
Hamilton Facing Mill Co., Hamilton
Hanna Engineering Works, Chicago.
Northern Engineering Works, Detroit
Paxson, J. W., Co., Philadelphia

Foundry Facings.

Buffalo Foundry Supply Co., Buffalo.
Detroit Foundry Supply Co., Windsor.
Hamilton Facing Mill Co., Hamilton.
J. W. Paxson Co., Philadelphia, Pa.

Friction Clutch Pulleys, etc.

The Goldie & McCulloch Co., Galt.
Greey, Wm. & J. G., Toronto
Link-Belt Co., Philadelphia.
Miller Bros. & Toms, Montreal.

Furnaces.

Detroit Foundry Supply Co., Windsor.
Dominion Foundry Supply Co., Montreal
Hamilton Facing Mill Co., Hamilton.
Paxson, J. W., & Philadelphia Co.,

Galvanizing

Canada Metal Co., Toronto.
Ontario Wind Engine & Pump Co., Toronto

Gas Blowers and Exhausters.

Dominion Heating & Ventilating Co. Hespeler
Sheldon's Limited, Galt.
B. F. Sturtevant Co., Hyde Park, Mass.

Gas Plants, Suction.

Baxter Paterson & Co., Montreal
Economic Heat, Light and Power Supply Co., Toronto.
Producer Gas Co., Toronto
Standard Bearings, Ltd., Niagara Falls.
Williams & Wilson, Montreal

Gauges, Standard.

Mechanics Supply Co., Quebec, Que.
Pratt & Whitney Co., Hartford, Conn.

Gearing.

Greey, Wm. & J. G. Toronto

Gear Cutting Machinery.

Baxter, Patterson & Co., Montreal
Becker-Brainard Milling Mach. Co., Hyde Park, Mass.
Bickford Drill & Tool Co., Cincinnati.
Dwight Slate Machine Co., Hartford, Conn.
Greey, Wm. & J. G., Toronto
Kennedy, Wm., & Sons, Owen Sound
London Mach. Tool Co., Hamilton.
R. McDougall Co., Galt
Niles-Bement-Pond Co., New York.
H. W. Petrie, Toronto.
Pratt & Whitney Co., Hartford, Conn.
Williams & Wilson, Montreal.

Gears, Angle.

Chicago Pneumatic Tool Co., Chicago.
Greey, Wm. & J. G., Toronto
Laurie Engine & Machine Co., Montreal.
John McDougall, Caledonian Iron Works Co., Montreal.
Miller Bros. & Toms, Montreal.
Waterous Engine Co., Brantford.

Gears, Cut.

Kennedy, Wm., & Sons, Owen Sound

Gears, Iron.

Greey, Wm. & J. G., Toronto
Kennedy, Wm., & Sons, Owen Sound

Gears, Mortise.

Greey, Wm. & J. G., Toronto
Kennedy, Wm., & Sons, Owen Sound

Gears, Reducing.

Brown, David & Sons, Huddersfield, Eng
Chicago Pneumatic Tool Co., Chicago.
Greey, Wm. & J. G., Toronto
John McDougall, Caledonian Iron Works Co., Montreal.
Miller Bros. & Toms, Montreal.

Generators, Electric.

Allis-Chalmers-Bullock, Limited, Montreal
Canadian General Electric Co., Toronto
Canadian Westinghouse Co., Hamilton.
D'Olier Engineering Co., New York.
Ball Engineering Works, Montreal.
H. W. Petrie, Toronto.
B. F. Sturtevant Co., Hyde Park, Mass.
Toronto & Hamilton Electric Co. Hamilton.

Generators, Gas.

H. W. Petrie, Toronto.

Graphite Paints.

P. D. Dods & Co., Montreal.

Graphite.

Detroit Foundry Supply Co., Windsor.
Dominion Foundry Supply Co., Montreal
Hamilton Facing Mill Co., Hamilton.
Mechanics Supply Co., Quebec, Que.
Paxson, J. W., Co., Philadelphia

Grinders, Automatic Knife.

W. H. Banfield & Son, Toronto.

Grinders, Centre.

Niles-Bement-Pond Co., New York.
H. W. Petrie, Toronto.

Grinders, Cutter.

Becker-Brainard Milling Mach. Co., Hyde Park, Mass.
John Millen & Son, Ltd., Montreal.
Pratt & Whitney Co., Hartford, Conn.

Grinders, Tool.

Armstrong Bros. Tool Co., Chicago.
B. F. Barnes Co., Rockford, Ill.
H. W. Petrie, Toronto.
Williams & Wilson, Montreal.

Grinding Machines.

Brown & Sharpe, Providence, R.I.
The Canadian Fairbanks, Montreal.
Niles-Bement-Pond Co., New York.
Norton Company, Worcester, Mass.
Paxson, J. W., Co., Philadelphia
H. W. Petrie, Toronto.

Grinding and Polishing Machines.

The Canadian Fairbanks Co., Montreal.
Greey, Wm. & J. G., Toronto
Independent Pneumatic Tool Co., Chicago, Ill.
John Millen & Son, Ltd., Montreal.
Miller Bros. & Toms, Montreal.
Niles-Bement-Pond Co., New York.
H. W. Petrie, Toronto.
Pendrith Machinery Co., Toronto.
Standard Bearings, Ltd., Niagara Falls

Grinding Wheels (Alundum)

Norton Company, Worcester, Mass.

Hack Saws.

Baxter, Paterson & Co., Montreal.
Canada Machinery Agency, Montreal.
The Canadian Fairbanks Co., Montreal.
Frothingham & Workman, Ltd., Montreal
Mechanics Supply Co., Quebec, Que.
John Millen & Son, Ltd., Montreal
Niles-Bement-Pond Co., New York.
H. W. Petrie, Toronto.
Williams & Wilson, Montreal.

Hack Saw Frames.

Mechanics Supply Co., Quebec, Que.

Hammers, Drop.

London Mach. Tool Co., Hamilton, Ont.
Miller Bros. & Toms, Montreal.
Niles-Bement-Pond Co., New York.

Hammers, Steam.

John Bertram & Sons Co., Dundas, Ont.
London Mach. Tool Co., Hamilton, Ont.
Niles-Bement-Pond Co., New York.

Hangers.

The Goldie & McCulloch Co., Galt
Greey, Wm. & J. G., Toronto
Kennedy, Wm., & Sons, Owen Sound
Miller Bros. & Toms, Montreal.
Owen Sound Iron Works Co., Owen Sound
The Smart-Turner Mach. Co., Hamilton.
Waterous Engine Co., Brantford.
Pendrith Machinery Co., Toronto.
Standard Bearings Ltd., Niagara Falls

Hardware Specialties.

Mechanics Supply Co., Quebec, Que.

Heating Apparatus.

Darling Bros., Ltd., Montreal
Dominion Heating & Ventilating Co., Hespeler
Mechanics Supply Co., Quebec, Que.
Sheldons Limited, Galt.
B. F. Sturtevant Co., High Park, Mass.

Hoisting and Conveying Machinery.

Allis-Chalmers-Bullock Limited, Montreal
Greey, Wm. & J. G., Toronto
Link-Belt Co., Philadelphia.
Miller Bros. & Toms, Montreal.
Niles-Bement-Pond Co., New York.
Northern Engineering Works, Detroit

The Smart-Turner Mach. Co., Hamilton.
Waterous Engine Co., Brantford.

Hoists, Chain and Rope.

Frothingham & Workman Ltd., Montreal

Hoists, Electric.

Northern Engineering Works, Detroit

Hoists, Pneumatic.

Canadian Rand Drill Co., Montreal.
Chicago Pneumatic Tool Co., Chicago.
Dominion Foundry Supply Co., Montreal
Hamilton Facing Mill Co., Hamilton.
Northern Engineering Works, Detroit

Hose.

Baxter, Paterson & Co., Montreal
Mechanics Supply Co., Quebec, Que.

Hose, Air.

Canadian Rand Drill Co., Montreal.
Canadian Westinghouse Co., Hamilton.
Chicago Pneumatic Tool Co., Chicago.
Paxson, J. W., Co., Philadelphia

Hose Couplings.

Canadian Rand Drill Co., Montreal.
Canadian Westinghouse Co., Hamilton.
Chicago Pneumatic Tool Co., Chicago.
Mechanics Supply Co., Quebec, Que.
Paxson, J. W., Co., Philadelphia

Hose, Steam.

Allis-Chalmers-Bullock, Montreal.
Canadian Rand Drill Co., Montreal.
Canadian Westinghouse Co., Hamilton.
Mechanics Supply Co., Quebec, Que.
Paxson, J. W., Co., Philadelphia

Hydraulic Accumulators.

Niles-Bement-Pond Co., New York.
Perrin, Wm. R., Co., Toronto.
The Smart-Turner Mach. Co., Hamilton.

Hydraulic Machinery.

Allis-Chalmers-Bullock, Montreal.
Barber, Chas., & Sons, Meaford

India Oil Stones.

Norton Company, Worcester, Mass.

Indicators, Speed.

Mechanics Supply Co., Quebec, Que.
L. S. Starrett Co., Athol, Mass.

Injectors.

Canada Foundry Co., Toronto.
The Canadian Fairbanks Co., Montreal.
Desmond-Stephan Mfg. Co., Urbana, Ohio
Frothingham & Workman Ltd., Montreal
Mechanics Supply Co., Quebec, Que.
Rice Lewis & Son, Toronto.
Penberthy Injector Co., Windsor, Ont.

Iron and Steel.

Frothingham & Workman, Ltd., Montreal

Iron and Steel Bars and Bands.

Hamilton Steel & Iron Co., Hamilton

Iron Cements.

Buffalo Foundry Supply Co., Buffalo.
Detroit Foundry Supply Co., Windsor.
Hamilton Facing Mill Co., Hamilton.
Paxson, J. W., Co., Philadelphia

Iron Filler.

Buffalo Foundry Supply Co., Buffalo.
Detroit Foundry Supply Co., Windsor.
Dominion Foundry Supply Co., Montreal
Hamilton Facing Mill Co., Hamilton.
Paxson, J. W., Co., Philadelphia.

Jacks.

Frothingham & Workman, Ltd., Montreal
Norton, A. O., Coaticook, Que.

Kegs, Steel Shop.

Cleveland Wire Spring Co., Cleveland

Key-Seating Machines.

B. F. Barnes Co., Rockford, Ill.
Niles-Bement-Pond Co., New York.

Lace Leather.

Baxter, Paterson & Co., Montreal.

Ladles, Foundry.

Frothingham & Workman Ltd., Montreal
Northern Engineering Works, Detroit

Lamps, Arc and Incandescent.

Canadian General Electric Co., Toronto.
Canadian Westinghouse Co., Hamilton.
The Packard Electric Co., St. Catharines.

Lathe Dogs.

Armstrong Bros., Chicago
Baxter, Paterson & Co., Montreal.
Pratt & Whitney Co., Hartford, Conn.

Lathes, Engine.

American Tool Works Co., Cincinnati.
B. F. Barnes Co., Rockford, Ill.
Baxter, Paterson & Co., Montreal
John Bertram & Sons Co., Dundas, Ont.
Canada Machinery Agency, Montreal.

The Canadian Fairbanks Co., Montreal.
London Mach. Tool Co., Hamilton, Ont.
Niles-Bement-Pond Co., New York.
H. W. Petrie, Toronto.
Pratt & Whitney Co., Hartford, Conn.

Lathes, Foot-Power.

American Tool Works Co., Cincinnati.
B. F. Barnes Co., Rockford, Ill.

Lathes, Screw Cutting.

B. F. Barnes Co., Rockford, Ill.
Baxter, Paterson & Co., Montreal.
Niles-Bement-Pond Co., New York.

Lathes, Automatic, Screw-Threading.

John Bertram & Sons Co., Dundas, Ont
London Mach. Tool Co., Hamilton, Ont
Pratt & Whitney Co., Hartford, Conn.

Lathes, Bench.

B. F. Barnes Co., Rockford, Ill.
London Mach. Tool Co., London, Ont.
Pratt & Whitney Co., Hartford, Conn.

Lathes, Turret.

American Tool Works Co., Cincinnati.
Baxter, Paterson & Co., Montreal
John Bertram & Sons Co., Dundas, Ont.
Gisholt Machine Co., Madison, Wis.
London Mach. Tool Co., Hamilton, Ont.
Niles-Bement-Pond Co., New York.
The Pratt & Whitney Co., Hartford, Conn.
Warner & Swasey Co., Cleveland O.

Leather Belting.

Baxter, Paterson & Co., Montreal.
Canada Machinery Agency, Montreal.
The Canadian Fairbanks Co., Montreal
Greey, Wm. & J. G., Toronto

Lime Stone Flux.

Hamilton Facing Mill Co., Hamilton.
Paxson, J. W., Co., Philadelphia

Locomotives, Air.

Canadian Rand Drill Co., Montreal.

Locomotives, Electrical.

Canadian Westinghouse Co., Hamilton
Jeffrey Mfg. Co., Columbus, Ohio.

Locomotives, Steam.

Canada Foundry Co., Toronto.
Canadian Rand Drill Co., Montreal.

Lubricating Plumbago.

Detroit Foundry Supply Co., Detroit.
Dominion Foundry Supply Co., Montreal
Hamilton Facing Mill Co., Hamilton.
Paxson, J. W., Co., Philadelphia

Lubricators, Force Feed.

Sight Feed Oil Pump Co., Milwaukee, Wis.

Lumber Dry Kilns.

Dominion Heating & Ventilating Co., Hespeler.
H. W. Petrie, Toronto.
Sheldon's Limited, Galt, Ont.
B. F. Sturtevant Co., Hyde Park, Mass.

Machinery Dealers.

Baxter, Paterson & Co., Montreal.
Canada Machinery Agency, Montreal.
The Canadian Fairbanks Co., Montreal.
Miller Bros. & Toms, Montreal.
H. W. Petrie, Toronto.
The Smart-Turner Mach. Co., Hamilton.
Williams & Wilson, Montreal.

Machinery Designers.

Greey, Wm. & J. G., Toronto
Standard Bearings, Ltd., Niagara Falls.

Machinists.

W. H. Banfield & Sons, Toronto.
Greey, Wm. & J. G., Toronto
Hall Engineering Works, Montreal.
Link-Belt Co., Philadelphia.
John McDougall, Caledonian Iron Works Co., Montreal.
Miller Bros. & Toms, Montreal.
Paxson, J. W., Co., Philadelphia
Pendrith Machinery Co., Toronto.
Robb Engineering Co., Amherst, N.S.
The Smart-Turner Mach. Co., Hamilton.
Standard Bearings, Ltd., Niagara Falls
Waterous Engine Co., Brantford.

Machinists' Small Tools.

Armstrong Bros., Chicago.
Butterfield & Co., Rock Island, Que.
Frothingham & Workman Ltd., Montreal
Mechanics Supply Co., Quebec, Que.
Rice Lewis & Son, Montreal.
Pratt & Whitney Co., Hartford, Conn.
Standard Tool Co., Cleveland.
L. S. Starrett Co., Athol, Mass.
Williams & Wilson, Montreal.

Malleable Flask Clamps.

Buffalo Foundry Supply Co., Buffalo.
Paxon, J. W., Co., Philadelphia

Malleable Iron Castings.

Galt Malleable Iron Co., Galt

Mallet, Rawhide and Wood.

Buffalo Foundry Supply Co., Buffalo.
Detroit Foundry Supply Co., Detroit.
Paxson, J. W., Co., Philadelphia

Mandrels.

A. B. Jardine & Co., Hespeler, Ont.
Miller Bros. & Toms, Montreal
The Pratt & Whitney Co., Hartford, Conn.
Standard Tool Co., Cleveland.

Marking Machines.

Dwight Slate Machine Co., Hartford, Conn.

Marine Work.

Kennedy, Wm., & Sons, Owen Sound

Metallic Paints.

P. D. Dods & Co., Montreal.

Meters, Electrical.

Canadian Westinghouse Co., Hamilton.
Mechanics Supply Co., Quebec, Que.

Mill Machinery.

Baxter, Paterson & Co., Montreal.
Greey, Wm. & J. G., Toronto
The Goldie & McCulloch Co., Galt, Ont.
John McDougall, Caledonian Iron Works Co., Montreal.
H. W. Petrie, Toronto.
Robb Engineering Co., Amherst, N.S.
Waterous Engine Co., Brantford.
Williams & Wilson, Montreal.

Milling Attachments.

Becker-Brainard Milling Machine Co. Hyde Park, Mass.
John Bertram & Sons Co., Dundas, Ont.
Niles-Bement-Pond Co., New York.
Pratt & Whitney, Hartford, Conn.

Milling Machines, Horizontal.

Baxter, Paterson & Co., Montreal.
Becker-Brainard Milling Machinery Co. Hyde Park, Mass.
John Bertram & Sons Co., Dundas, Ont.
London Mach. Tool Co., Hamilton, Ont.
Niles-Bement-Pond Co., New York.
Pratt & Whitney, Hartford, Conn.

Milling Machines, Plain.

American Tool Works Co., Cincinnati.
Becker-Brainard Milling Machine Co. Hyde Park, Mass.
John Bertram & Sons Co., Dundas, Ont.
Canada Machinery Agency, Montreal.
The Canadian Fairbanks Co., Montreal.
London Mach. Tool Co., Hamilton, Ont.
Niles-Bement-Pond Co., New York.
H. W. Petrie, Toronto.
Pratt & Whitney Co., Hartford, Conn.
Williams & Wilson, Montreal.

Milling Machines, Universal.

American Tool Works Co., Cincinnati.
Becker-Brainard Milling Machine Co., Hyde Park, Mass.
John Bertram & Sons Co., Dundas, Ont.
Canada Machinery Agency, Montreal.
The Canadian Fairbanks Co., Montreal.
London Mach. Tool Co., Hamilton, Ont.
Niles-Bement-Pond Co., New York.
H. W. Petrie, Toronto.
Williams & Wilson, Montreal.

Milling Machines, Vertical.

Becker-Brainard Milling Machine Co., Hyde Park, Mass.
Brown & Sharpe, Providence, R.I.
John Bertram & Sons Co., Dundas, Ont.
Canada Machinery Agency, Montreal.
London Mach. Tool Co., Hamilton, Ont.
Niles-Bement-Pond Co., New York.

Milling Tools.

Wm. Abbott, Montreal.
Becker-Brainard Milling Machine Co., Hyde Park, Mass.
Geometric Tool Co., New Haven, Conn.
London Mach. Tool Co., Hamilton, Ont.
Pratt & Whitney Co., Hartford, Conn.
Standard Tool Co., Cleveland.
Standard Bearings, Ltd., Niagara Falls

Mining Machinery.

Allis-Chalmers-Bullock, Limited, Montreal
Canadian Rand Drill Co., Montreal.
Chicago Pneumatic Tool Co., Chicago.
Jeffrey Mfg. Co., Columbus, Ohio
Laurie Engine & Machine Co., Montreal.
John McDougall, Caledonian Iron Works Co., Montreal.
Miller Bros. & Toms, Montreal.
T. & H. Electric Co., Hamilton.

Mixing Machines, Dough.

Greey, Wm. & J. G., Toronto
Pendrith Machinery Co., Toronto.
Paxson, J. W., Co., Philadelphia

Mixing Machines, Special.

Greey, Wm. & J. G., Toronto
Pendrith Machinery Co., Toronto.
Paxson, J. W., Co., Philadelphia

Model Tools.

Globe Machine & Stamping Co., Cleveland, Ohio.
Standard Bearings, Ltd., Niagara Falls.
Wells Pattern and Model Works, Toronto

Motors, Electric.

Allis-Chalmers-Bullock Limited, Montreal
Canadian General Electric Co., Torono
Canadian Westinghouse Co., Hamilton.

Consolidated Electric Co. Toronto
Hall Engineering Works, Montreal.
Mechanics Supply Co., Quebec, Que.
The Packard Electric Co., St. Catharines.
B. F. Sturtevant Co., Hyde Park, Mass.
T. & H. Electric Co., Hamilton.

Motors, Air.

Canadian Rand Drill Co., Montreal.
Chicago Pneumatic Tool Co., Chicago.

Molders' Supplies.

Buffalo Foundry Supply Co, Buffalo.
Detroit Foundry Supply Co., Windsor.
Dominion Foundry Supply Co., Montreal
Hamilton Facing Mill Co., Hamilton
Mechanics Supply Co., Quebec, Que.
Paxson J. W., Co., Philadelphia

Molders' Tools.

Buffalo Foundry Supply Co., Buffalo.
Detroit Foundry Supply Co., Windsor
Dominion Foundry Supply Co., Montreal
Hamilton Facing Mill Co., Hamilton.
Mechanics Supply Co., Quebec, Que.
Paxson, J. W., Co, Philadelphia

Molding Machines.

Buffalo Foundry Supply Co., Buffalo.
Hamiton Facing Mill Co., Hamilton.
J. W. Paxson Co., Philadelphia, Pa.

Molding Sand.

T. W. Barnes, Hamilton.
Buffalo Foundry Supply Co., Buffalo.
Detroit Foundry Supply Co., Windsor
Dominion Foundry Supply Co., Montreal
Hamilton Facing Mill Co., Hamilton.
Paxson, J. W., Co., Philadelphia

Nut Tappers.

John Bertram & Sons Co., Dundas, Ont
A. B. Jardine & Co., Hespeler.
London Mach. Tool Co., Hamilton.
National Machinery Co., Tiffin, Ohio.

Oatmeal Mill Machinery.

Greey, Wm. & J. G., Toronto
The Goldie & McCulloch Co., Galt

Oilers, Gang.

Sight Feed Oil Pump Co., Milwaukee, Wis.

Oils, Core.

Buffalo Foundry Supply Co. Buffalo.
Dominion Foundry Supply Co, Montreal
Hamilton Facing Mill Co., Hamilton.
Paxson, J. W., Co., Philadelphia

Oil Extractors.

Darling Bros., Ltd., Montreal

Paint Mill Machinery.

Greey, Wm. & J. G., Toronto

Pans, Lathe.

Cleveland Wire Spring Co., Cleveland
Paxson, J. W., Co., Philadelphia

Pans, Steel Shop.

Cleveland Wire Spring Co., Cleveldnd

Patent Solicitors.

Hanbury A. Budden, Montreal.
Fetherstonhough & Blackmore, Montreal
Fetherstonhaugh & Co., Montreal.
Marion & Marion, Montreal.
Ridout & Maybee, Toronto.

Patterns.

John Carr, Hamilton.
Galt Malleable Iron Co., Galt
Hamilton Pattern Works, Hamilton.
John McDougall, Caledonian Iron Works Co., Montreal.
Miller Bros. & Toms Montreal.
Wells' Pattern and Model Works, Toronto.

Pig Iron.

Hamilton Steel & Iron Co., Hamilton

Pipe Cutting and Threading Machines.

Butterfield & Co., Rock Island, Que.
Canada Machinery Agency, Montreal.
Frothingham & Workman, Ltd., Montreal
Hart Mfg. Co., Cleveland
A. B. Jardine & Co., Hespeler, Ont.
London Mach. Tool Co., Hamilton, Ont.
R. McDougall Co., Galt, Ont.
Niles-Bement-Pond Co., New York.

Pipe, Municipal.

Canadian Pipe Co., Vancouver, B.C.
Pacific Coast Pipe Co., Vancouver, B.C.

Pipe, Waterworks.

Canadian Pipe Co., Vancouver, B.C.
Pacific Coast Pipe Co., Vancouver, B.C

Planers, Standard.

American Tool Works, Cincinnati.
John Bertram & Sons Co., Dundas, Ont.
Canada Machinery Agency, Montreal.
The Canadian Fairbanks Co., Montreal.
London Mach. Tool Co., Hamilton, Ont.
Niles-Bement-Pond Co., New York.
H. W. Petrie, Toronto.
Pratt & Whitney Co., Hartford, Conn.
Williams & Wilson, Montreal.

Planers, Rotary.

John Bertram & Sons Co., Dundas, Ont.
London Mach. Tool Co., Hamilton, Ont.
Niles-Bement-Pond Co., New York.

Planing Mill Fans.

Dominion Heating & Ventilating Co., Hespeler
Sheldon's Limited, Galt, Ont.
B. F. Sturtevant Co., Hyde Park Man.

Plumbago.

Buffalo Foundry Supply Co., Buffalo.
Detroit Foundry Supply Co., Windsor
Dominion Foundry Supply Co., Montreal
Hamilton Facing Mill Co., Hamilton,
Mechanics Supply Co., Quebec, Que.
J. W. Paxson Co., Philadelphia, Pa.

Pneumatic Tools.

Allis-Chalmers-Bullock, Montreal.
Canadian Rand Drill Co., Montreal.
Chicago Pneumatic Tool Co., Chicago
Hamilton Facing Mill Co., Hamilton.
Hanna Engineering Works, Chicago.
Independent Pneumatic Tool Co. Chicago, Ill.
Paxson, J. W., Co., Philadelphia

Power Hack Saw Machines.

Baxter, Paterson & Co., Montreal.
Frothingham & Workman, Ltd., Montreal

Power Plants.

John McDougall Caledonian Iron Works Co., Montreal.
The Smart-Turner Mach. Co., Hamilton

Power Plant Equipments.

Darling Bros., Ltd., Montreal

Presses, Drop.

W. H. Banfield & Son, Toronto.
E. W. Bliss Co., Brooklyn, N.Y.
Canada Machinery Agency, Montreal.
Laurie Engine & Machine Co., Montreal
Miller Bros. & Toms, Montreal
Niles-Bement-Pond Co., New York

Presses, Hand.

E. W. Bliss Co., Brooklyn, N.Y.

Presses, Hydraulic.

John Bertram & Sons Co., Dundas, Ont.
Laurie Engine & Machine Co., Montreal.
London Mach. Tool Co., Hamilton, Ont.
John McDougall Caledonian Iron Works Co., Montreal.
Miller Bros. & Toms, Montreal.
Niles-Bement-Pond Co., New York.
Perrin, Wm. R., & Co., Toronto

Presses, Power.

E. W. Bliss Co., Brooklyn, N.Y.
Canada Machinery Agency, Montreal.
Laurie Engine & Machine Co., Montreal
London Mach. Tool Co., Hamilton, Ont.
John McDougall Caledonian Iron Works Co., Montreal.
Niles-Bement-Pond Co., New York.

Presses Power Screw.

Perrin, Wm. R., & Co., Toronto

Pressure Regulators.

Darling Bros., Ltd., Montreal

Producer Plants.

Canada Foundry Co., Toronto

Pulp Mill Machinery.

Greey, Wm. & J. G., Toronto
Jeffrey Mfg Co., Columbus, Ohio.
Laurie Engine & Machine Co., Montreal.
John McDougall Caledonian Iron Works Co., Montreal.
Waterous Engine Works Co., Brantford

Pulleys.

Baxter, Paterson & Co., Montreal.
Canada Machinery Agency, Montreal
The Canadian Fairbanks Co., Montreal.
The Goldie & McCulloch Co., Galt.
Greey, Wm. & J. G., Toronto
Laurie Engine & Machine Co., Montreal.
Link-Belt Co., Philadelphia.
John McDougall Caledonian Iron Works Co., Montreal.
Miller Bros. & Toms, Montreal.
Owen Sound Iron Works Co., Owen Sound
Pendrith Machinery Co., Toronto.
H. W. Petrie, Toronto.
The Smart-Turner Mach. Co., Hamilton.
Standaad Bearings Ltd., Niagara Falls,
Williams & Wilson, Montreal.
Waterous Engine Co., Brantford.

Pumps.

Laurie Engine & Machine Co., Montreal
Ontario Wind Engine & Pump Co., Toronto

Pumps, Boiler Feed.

Du Bois Iron Works, Du Bois, Pa.

Pumps, Boiler Test.

Du Bois Iron Works Du Bois, Pa.

Pumps, Circulating.

Du Bois Iron Works, Du Bois, Pa.

Pumps, Contractors.
Du Bois Iron Works Du Bois, Pa

Pumps, Electric.
Du Bois Iron Works, Du Bois, Pa.

Pumps, Elevator.
Du Bois Iron Works, Du Bois, Pa.

Pumps, Engine Gas and Oil.
Du Bois Iron Works, Du Bois, Pa.

Pump Governors.
Darling Bros., Ltd., Montreal

Pumps, Hydraulic.
Laurie Engine & Machine Co., Montreal.
Perrin, Wm. R. & Co., Toronto

Pumps, Irrigation.
Du Bois Iron Works, Du Bois, Pa.

Pumps, Marine.
Du Bois Iron Works, Du Bois, Pa.

Pumps, Mine (Cylinder & Heads).
Du Bois Iron Works, Du Bois Pa.

Pumps, Oil.
Sight Feed Oil Pump Co., Milwaukee, Wis.

Pumps, Power.
Du Bois Iron Works, Du Bois, Pa.

Pumps, Steam.
Allis-Chalmers-Bullock, Limited, Montreal
Canada Foundry Co., Toronto.
Canada Machinery Agency, Montreal.
Darling Bros., Ltd., Montreal
D'Olier Engineering Co., New York.
Du Bois Iron Works Du Bois, Pa.
The Goldie & McCulloch Co., Galt.
John McDougall Caledonian Iron Works, Montreal.
H. W. Petrie, Toronto.
The Smart-Turner Mach. Co., Hamilton.
Standard Bearings, Ltd., Niagara Falls.
Waterous Engine Co. Brantford.

Pumps, Tank.
Du Bois Iron Works, Du Bois, Pa.

Pumping Machinery.
Canada Foundry Co., Limited, Toronto
Canada Machinery Agency, Montreal.
Canadian Rand Drill Co., Montreal.
Chicago Pneumatic Tool Co., Chicago.
Darling Bros., Ltd., Montreal
D'Olier Engineering Co., New York.
Hall Engineering Works, Montreal, Que.
Laurie Engine & Machine Co., Montreal.
London Mach. Tool Co., Hamilton, Ont.
John McDougall Caledonian Iron Works Co., Montreal.
R. McDougall Co., Galt, Ont.
The Smart-Turner Mach. Co., Hamilton
Standard Bearings, Ltd., Niagara Falls.

Punches and Dies.
W. H. Banfield & Sons, Toronto.
Butterfield & Co., Rock Island.
Globe Machine & Stamping Co.
A. B. Jardine & Co., Hespeler, Ont.
London Mach. Tool Co., Hamilton, Ont.
Miller Bros. & Toms, Montreal.
Pratt & Whitney Co., Hartford, Conn.
H. W. Petrie, Toronto.
Standard Bearings, Ltd., Niagara Falls,

Punches, Hand.
Mechanics Supply Co., Quebec, Que.

Punches, Power.
John Bertram & Sons Co., Dundas, Ont.
E. W. Bliss Co., Brooklyn, N.Y.
Canada Machinery Agency, Montreal.
Canadian Buffalo Forge Co., Montreal.
London Mach. Tool Co., Hamilton, Ont.
Niles-Bement-Pond Co., New York.

Punches, Turret.
London Mach. Tool Co., London, Ont

Punching Machines, Horizontal.
John Bertram & Sons Co., Dundas, Ont.
London Mach. Tool Co., Hamilton Ont.
Niles-Bement-Pond Co., New York.

Quartering Machines.
John Bertram & Sons Co., Dundas, Ont.
London Mach. Tool Co., Hamilton, Ont.

Railway Spikes and Washers.
Hamilton Steel & Iron Co., Hamilton

Rammers, Bench and Floor.
Buffalo Foundry Supply Co., Buffalo.
Detroit Foundry Supply Co., Windsor
Hamilton Facing Mill Co., Hamilton.
Paxson, J. W., Co., Philadelphia

Rapping Plates.
Detroit Foundry Supply Co., Windsor
Hamilton Facing Mill Co., Hamilton.
Paxson, J. W., Co., Philadelphia

Raw Hide Pinions.
Brown, David & Sons, Huddersfield, Eng.

Reamers.
Wm. Abbott, Montreal.
Baxter, Paterson & Co., Montreal.
Butterfield & Co., Rock Island.
Frothingham & Workman, Ltd., Montreal
Hanna Engineering Works Chicago.
A. B. Jardine & Co., Hespeler, Ont.
Mechanics Supply Co., Quebec, Que.
John Millen & Son, Ltd., Montreal.
Morse Twist Drill and Machine Co., New Bedford, Mass
Pratt & Whitney Co., Hartford, Conn.
Standard Tool Co., Cleveland.

Reamers, Steel Taper.
Butterfield & Co., Rock Island.
Chicago Pneumatic Tool Co., Chicago.
A. B. Jardine & Co., Hespeler, Ont.
John Millen & Son, Ltd., Montreal.
Pratt & Whitney Co., Hartford, Conn.
Standard Tool Co., Cleveland.
Whitman & Barnes Co., St. Catharines.

Rheostats.
Canadian General Electric Co., Toronto.
Canadian Westinghouse Co., Hamilton.
Hall Engineering Works, Montreal, Que.
T. & H. Electric Co., Hamilton,

Riddles.
Buffalo Foundry Supply Co., Buffalo.
Detroit Foundry Supply Co., Windsor
Hamilton Facing Mill Co., Hamilton.
J. W. Paxson Co., Philadelphia, Pa.

Riveters, Pneumatic.
Hanna Engineering Works, Chicago.
Independent Pneumatic Tool Co., Chicago, Ill.

Rolls, Bending.
John Bertram & Sons Co., Dundas, Ont.
London Mach. Tool Co., Hamilton, Ont.
Niles-Bement-Pond Co., New York.

Rolls, Chilled Iron.
Greey, Wm. & J. G., Toronto

Rolls, Sand Cast.
Greey, Wm. & J. G., Toronto

Rotary Blowers.
Paxson, J. W., Co., Philadelphia
Sturtevant, B. F., Co., Hyde Park, Mass.

Rotary Converters.
Allis-Chalmers-Bullock, Ltd., Montreal.
Canadian Westinghouse Co., Hamilton.
Paxson, J. W., Co. Philadelphia
Toronto and Hamilton Electric Co., Hamilton.

Safes.
Baxter Paterson & Co., Montreal.
The Goldie & McCulloch Co., Galt.

Sand, Bench.
Buffalo Foundry Supply Co., Buffalo.
Detroit Foundry Supply Co., Windsor
Hamilton Facing Mill Co., Hamilton.
Paxson J. W., Co., Philadelphia

Sand Blast Machinery.
Canadian Rand Drill Co., Montreal.
Chicago Pneumatic Tool Co., Chicago.
Detroit Foundry Supply Co., Windsor
Hamilton Facing Mill Co., Hamilton.
J. W. Paxson Co., Philadelphia, Pa.

Sand, Heavy, Grey Iron.
Buffalo Foundry Supply Co., Buffalo.
Detroit Foundry Supply Co., Windsor
Hamilton Facing Mill Co., Hamilton.
Paxson, J. W., Co., Philadelphia

Sand, Malleable.
Buffalo Foundry Supply Co., Buffalo.
Detroit Foundry Supply Co., Windsor
Hamilton Facing Mill Co., Hamilton.
Paxson, J. W., Co., Philadelphia

Sand, Medium Grey Iron.
Buffalo Foundry Supply Co., Buffalo.
Detroit Foundry Supply Co., Windsor
Hamilton Facing Mill Co., Hamilton.
Paxson, J. W., Co., Philadelphia

Sand Sifters.
Buffalo Foundry Supply Co., Buffalo.
Detroit Foundry Supply Co., Windsor
Dominion Foundry Supply Co., Montreal
Hamilton Facing Mill Co., Hamilton.
Hanna Engineering Works, Chicago
Paxson, J. W., Co. Philadelphia

Saw Mill Machinery.
Allis-Chalmers-Bullock, Limited, Montreal
Baxter, Paterson & Co., Montreal.
Canada Machinery Agency, Montreal.
Goldie & McCulloch Co., Galt.
Greey, Wm. & J. G., Toronto
Miller Bros. & Toms, Montreal
Owen Sound Iron Works Co., Owen Sound
H. W. Petrie, Toronto.
Robb Engineering Co. Amherst.
Standard Bearings, Ltd., Niagara Falls
Waterous Engine Works, Brantford
Williams & Wilson, Montreal.

Sawing Machines, Metal.
Niles-Bement-Pond Co. New York.
Paxson, J. W., Co., Philadelphia

Saws, Hack.
Baxter Paterson & Co., Montreal.
Canada Machinery Agency, Montreal.
Detroit Foundry Supply Co., Windsor
Frothingham & Workman, Ltd., Montreal
London Mach. Tool Co., Hamilton
Mechanics Supply Co., Quebec, Que.
Rice Lewis & Son, Toronto.
John Millen & Son, Ltd., Montreal.
L. S. Starrett Co., Athol, Mass.

Screw Cutting Tools
Hart Manufacturing Co., Cleveland, Ohio
Mechanics Supply Co., Quebec, Que.

Screw Machines, Automatic.
Canada Machinery Agency, Montreal.
Cleveland Automatic Machine Co., Cleveland, Ohio
London Mach. Tool Co., Hamilton, Ont.
National-Acme Mfg. Co., Cleveland.
Pratt & Whitney Co., Hartford, Conn.

Screw Machines, Hand.
Canada Machinery Agency Montreal.
A. B. Jardine & Co., Hespeler.
London Mach. Tool Co., Hamilton, Ont.
Mechanics Supply Co. Quebec, Que.
Pratt & Whitney Co., Hartford, Conn.
Warner & Swasey Co., Cleveland, O.

Screw Machines, Multiple Spindle.
National-Acme Mfg. Co., Cleveland

Screw Plates.
Butterfield & Co., Rock Island, Que.
Frothingham & Workman, Ltd., Montreal
Hart Manufacturing Co. Cleveland, Ohio
A. B. Jardine & Co., Hespeler
Mechanics Supply Co., Quebec, Que.

Screws, Cap and Set.
National-Acme Mfg. Co., Cleveland

Screw Slotting Machinery, Semi-Automatic.
National-Acme Mfg. Co., Cleveland

Second-Hand Machinery.
American Tool Works Co., Cincinnati.
Canada Machinery Agency, Montreal.
The Canadian Fairbanks Co., Montreal.
Goldie & McCulloch Co., Galt.
Machinery Exchange, Montreal.
Niles-Bement-Pond Co., New York.
H. W. Petrie, Toronto.
Robb Engineering Co., Amherst, N.S.
Waterous Engine Co., Brantford
Williams & Wilson, Montreal.

Shafting.
Baxter, Paterson & Co., Montreal
Canada Machinery Agency, Montreal
The Canadian Fairbanks Co., Montreal.
Frothingham & Workman, Ltd., Montreal
The Goldie & McCulloch Co., Galt, Ont
Greey, Wm. & J. G. Toronto
Kennedy, Wm., & Sons, Owen Sound
Niles-Bement-Pond Co., New York.
Owen Sound Iron Works Co., Owen Sound
Pendrith Machine Co., Toronto.
H. W. Petrie, Toronto.
Smart-Turner Machine Co., Hamilton.
Union Drawn Steel Co., Hamilton.
Waterous Engine Co., Brantford.

Shapers.
American Tool Works Co., Cincinnati.
John Bertram & Sons Co., Dundas, Ont
Canada Machinery Agency, Montreal.
The Canadian Fairbanks Co., Montreal.
London Mach. Tool Co., Hamilton, Ont.
Niles-Bement-Pond Co., New York.
H. W. Petrie, Toronto.
Potter & Johnston Machine Co., Pawtucket, R.I.
Pratt & Whitney Co., Hartford, Conn.
Williams & Wilson, Montreal.

Shearing Machine, Bar.
John Bertram & Sons Co., Dundas, Ont.
A. B. Jardine & Co., Hespeler.
London Mach. Tool Co., Hamilton, Ont.
Niles-Bement-Pond Co., New York.

Shears, Power.
John Bertram & Sons Co., Dundas, Ont
Canada Machinery Agency, Montreal.
A. B. Jardine & Co., Hespeler, Ont.
Niles-Bement-Pond Co., New York.
Paxson, J. W., Co., Philadelphia

Sheet Metal Goods
Globe Machine & Stamping Co., Cleveland, Ohio.

Sheet Steel Work.
Owen Sound Iron Works Co., Owen Sound

Shingle Mill Machinery.
Owen Sound Iron Works Co., Owen Sound

Shop Trucks.
Greey, Wm. & J. G., Toronto
Owen Sound Iron Works Co., Owen Sound
Paxson, J. W., Co., Philadelphia

Shovels.
Baxter, Paterson & Co., Montreal.
Buffalo Foundry Supply Co., Buffalo.
Detroit Foundry Supply Co., Detroit.
Frothingham & Workman, Ltd., Montreal
Dominion Foundry Supply Co., Montreal
Hamilton Facing Mill Co., Hamilton.
Mechanics Supply Co., Quebec, Que.
Paxson, J. W., Co., Philadelphia

Shovels, Steam.
Allis-Chalmers-Bullock, Montreal.

Sieves.
Buffalo Foundry Supply Co. Buffalo.
Detroit Foundry Supply Co., Windsor
Dominion Foundry Supply Co., Montreal
Hamilton Facing Mill Co., Hamilton.
Paxson, J. W., Co. Philadelphia

Silver Lead.
Buffalo Foundry Supply Co., Buffalo.
Detroit Foundry Supply Co., Windsor
Dominion Foundry Supply Co., Montreal
Hamilton Facing Mill Co., Hamilton.
Paxson, J. W., Co., Philadelphia

Sleeves, Reducing.
Chicago Pneumatic Tool Co., Chicago.

Snap Flasks
Buffalo Foundry Supply Co., Buffalo.
Detroit Foundry Supply Co., Windsor
Dominion Foundry Supply Co., Montreal
Hamilton Facing Mill Co., Hamilton.
Paxson, J. W., Co., Philadelphia

Soapstone.
Buffalo Foundry Supply Co., Buffalo.
Detroit Foundry Supply Co., Windsor
Dominion Foundry Supply Co., Montreal
Hamilton Facing Mill Co., Hamilton.
Paxson, J. W. Co., Philadelphia

Solders.
Lumen Bearing Co., Toronto

Special Machinery.
W. H. Banfield & Sons, Toronto.
Baxter, Paterson & Co., Montreal
John Bertram & Sons Co. Dundas, Ont
Globe Machine & Stamping Co., Cleveland Ohio.
Greey, Wm. & J. G., Toronto
Hanna Engineering Works, Chicago.
Laurie Engine & Machine Co., Montreal.
London Mach. Tool Co., Hamilton, Ont.
R. McDougall Co., Galt, Ont.
Pendrith Machinery Co., Toronto.
H. W. Petrie, Toronto.
The Smart-Turner Mach. Co., Hamilton
Standard Bearings, Ltd., Niagara Falls,
Waterous Engine Co., Brantford.

Special Machines and Tools.
Paxson, J. W., Philadelphia
Pratt & Whitney, Hartford, Conn.
Standard Bearings, Ltd., Niagra Falls,

Special Milled Work.
National Acme Mfg. Co., Cleveland

Speed Changing Countershafts.
The Canadian Fairbanks Co., Montreal

Spike Machines.
National Machinery Co., Tiffin, O.
The Smart-Turner Mach. Co., Hamilton

Spray Cans.
Detroit Foundry Supply Co., Windsor
Dominion Foundry Supply Co., Montreal
Hamilton Facing Mill Co., Hamilton
Paxson, J. W., Co., Philadelphia

Springs, Automobile.
Cleveland Wire Spring Co., Cleveland

Springs, Coiled Wire.
Cleveland Wire Spring Co., Cleveland

Springs, Machinery.
Cleveland Wire Spring Co., Cleveland

Springs, Upholstery.
Cleveland Wire Spring Co., Cleveland

Sprue Cutters.
Detroit Foundry Supply Co., Windsor
Dominion Foundry Supply Co., Montreal
Hamilton Facing Mill Co., Hamilton.
Paxson J. W., Co., Philadelphia

Stamp Mills.
Allis-Chalmers-Bullock, Limited, Montreal
Paxson, J. W., Co., Philadelphia

Steam Hot Blast Apparatus
Dominion Heating & Ventilating Co. Hespeler, Ont.
Sheldon's Limited, Galt.
B. F. Sturtevant Co., Hyde Park, Mass.

Steam Separators.

Darling Bros., Ltd., Montreal.
Dominion Heating & Ventilating Co., Hespeler
Robb Engineering Co., Montreal.
Sheldon's Limited, Galt.
Smart-Turner Mach. Co., Hamilton
Waterous Engine Co., Brantford.

Steam Specialties.

Darling Bros., Ltd., Montreal
Dominion Heating & Ventilating Co., Hespeler
Sheldon's Limited, Galt.

Steam Traps.

Canada Machinery Agency, Montreal.
Darling Bros., Ltd., Montreal
Dominion Heating & Ventilating Co., Hespeler
Mechanics Supply Co., Quebec, Que.
Pendrith Machinery Co., Toronto.
Sheldon's Limited, Galt.
B. F. Sturtevant Co., Hyde Park, Mass.

Steam Valves.

Darling Bros., Ltd., Montreal

Steel Pressure Blowers

Buffalo Foundry Supply Co., Buffalo.
Dominion Foundry Supply Co., Montreal
Dominion Heating & Ventilating Co., Hespeler
Hamilton Facing Mill Co., Hamilton.
J. W. Paxson Co., Philadelphia, Pa.
Sheldon's Limited, Galt.
B. F. Sturtevant Co., Hyde Park, Mass.

Steel Tubes.

Baxter, Paterson & Co., Montreal.
Mechanics Supply Co., Quebec, Que.
John Millen & Son, Montreal.

Steel, High Speed.

Wm. Abbott, Montreal.
Canadian Fairbanks Co., Montreal.
Frothingham & Workman Ltd., Montreal
Alex. Gibb, Montreal.
Jessop Wm., & Sons, Sheffield, Eng.
B. K. Morton Co., Sheffield, Eng.
Williams & Wilson, Montreal.

Stocks and Dies

Hart Manufacturing Co., Cleveland, Ohio
Mechanics Supply Co., Quebec, Que.
Pratt & Whitney Co., Hartford, Conn.

Stone Cutting Tools, Pneumatic

Allis-Chalmers-Bullock, Ltd., Montreal
Canadian Rand Drill Co., Montreal.

Stone Surfacers

Chicago Pneumatic Tool Co., Chicago.

Stove Plate Facings.

Buffalo Foundry Supply Co., Buffalo
Detroit Foundry Supply Co., Windsor.
Dominion Foundry Supply Co., Montreal
Hamilton Facing Mill Co., Hamilton.
Paxson, J. W., Co., Philadelphia

Swage, Block.

A. B. Jardine & Co., Hespeler, Ont.

Switchboards.

Allis-Chalmers-Bullock, Limited, Montreal
Canadian General Electric Co., Toronto.
Canadian Westinghouse Co., Hamilton
Hall Engineering Works, Montreal, Que.
Mechanics Supply Co., Quebec, Que.
Toronto and Hamilton Electric Co., Hamilton.

Talc.

Buffalo Foundry Supply Co., Buffalo.
Detroit Foundry Supply Co., Windsor.
Hamilton Facing Mill Co., Hamilton.
Paxson, J. W., Co., Philadelphia

Tanks, Oil.

Sight Feed Oil Pump Co., Milwaukee, Wis.

Tapping Machines and Attachments.

American Tool Works Co., Cincinnati.
John Bertram & Sons Co., Dundas, Ont.
Bickford Drill & Tool Co., Cincinnati.
The Geometric Tool Co., New Haven.
A. B. Jardine & Co., Hespeler.
London Mach. Tool Co., Hamilton, Ont
Murchey Machine & Tool Co., Detroit
Niles-Bement-Pond Co., New York.
H. W. Petrie, Toronto.
Pratt & Whitney, Cincinnati, O.
L. S. Starrett Co., Athol, Mass.
Williams & Wilson, Montreal.

Tapes, Steel.

Frothingham & Workman, Ltd., Montreal
Rice Lewis & Son, Toronto.
Mechanics Supply Co., Quebec, Que.
John Millen & Son, Ltd., Montreal.
L. S. Starrett Co., Athol, Mass.

Taps.

Baxter, Paterson & Co., Montreal.
Mechanics Supply Co., Quebec, Que.

Taps, Collapsing.

The Geometric Tool Co., New Haven.

Taps and Dies.

Wm. Abbott, Montreal.
Baxter, Paterson & Co., Montreal.
Butterfield & Co., Rock Island, Que.
Frothingham & Workman Ltd., Montreal
The Geometric Tool Co., New Haven.
A. B. Jardine & Co., Hespeler, Ont.
Rice Lewis & Son, Toronto.
Mechanics Supply Co., Quebec, Que.
John Millen & Son, Ltd., Montreal.

Pratt & Whitney Co., Hartford, Conn.
Standard Tool Co., Cleveland.
L. S. Starrett Co., Athol, Mass.

Testing Laboratory.

Detroit Testing Laboratory, Detroit

Testing Machines.

Detroit Foundry Supply Co., Windsor.
Dominion Foundry Supply Co., Montreal
Hamilton Facing Mill Co., Hamilton.
Paxson, J. W., Co., Philadelphia

Thread Cutting Tools.

Hart Manufacturing Co., Cleveland, Ohio
Mechanics Supply Co., Quebec Que.

Tiling, Opal Glass.

Toronto Plate Glass Importing Co., Toronto.

Time Switches, Automatic, Electric.

Berlin Electric Mfg. Co., Toronto.

Tinware Machinery

Canada Machinery Agency, Montreal.

Tire Upsetters or Shrinkers

A. B. Jardine & Co., Hespeler, Ont.

Tool Cutting Machinery

Canadian Rand Drill Co., Montreal.

Tool Holders.

Armstrong Bros. Tool Co., Chicago.
Baxter Paterson & Co., Montreal.
Fairbanks Co., Springfield, Ohio.
John Millen & Son, Ltd., Montreal.
H. W. Petrie, Toronto.
Pratt & Whitney Co., Hartford, Conn.

Tool Steel.

Wm. Abbott, Montreal.
Frothingham & Workman Ltd., Montreal
Wm. Jessop, Sons & Co., Toronto.
Canadian Fairbanks Co., Montreal
B. K. Morton & Co., Sheffield, Eng.
Williams & Wilson, Montral.

Torches, Steel.

Baxter, Paterson & Co., Montreal.
Detroit Foundry Supply Co., Windsor.
Dominion Foundry Supply Co., Montreal
Hamilton Facing Mill Co., Hamilton.
Paxson, J. W., Co., Philadelphia

Transformers and Convertors

Allis-Chalmers-Bullock, Montreal.
Canadian General Electric Co., Toronto.
Canadian Westinghouse Co., Hamilton.
Hall Engineering Works, Montreal, Que.
T. & H. Electric Co., Hamilton.

Transmission Machinery

Allis-Chalmers-Bullock, Montreal.
The Canadian Fairbanks Co., Montreal.
Greey, Wm. & J. G., Toronto.
Laurie Engine & Machine Co., Montreal.
Link-Belt Co., Philadelphia.
Miller Bros. & Toms, Montreal.
Paxson, J. W., Co., Philadelphia
H. W. Petrie, Toronto.
The Smart-Turner Mach. Co., Hamilton.
Waterous Engine Co., Brantford.

Transmission Supplies

Baxter, Paterson & Co., Montreal.
The Canadian Fairbanks Co., Montreal.
Wm. & J. G. Greey, Toronto.
The Goldie & McCulloch Co., Galt.
Miller Bros. & Toms, Montreal
Pendrith Machinery Co., Toronto.
H. W. Petrie, Toronto.

Trolleys

Canadian Rand Drill Co., Montreal.
John Millen & Son, Ltd., Montreal.
Miller Bros. & Toms, Montreal.
Northern Engineering Works, Detroit

Trolley Wheels.

Lumen Bearing Co., Toronto

Trucks, Dryer and Factory

Dominion Heating & Ventilating Co., Hespeler
Greey, Wm. & J. G., Toronto.
Northern Engineering Works, Detroit
Sheldon's Limited, Galt, Ont.

Tube Expanders (Rollers)

Chicago Pneumatic Tool Co., Chicago.
A. B. Jardine & Co., Hespeler.
Mechanics Supply Co., Quebec, Que.

Turbines, Steam

Allis-Chalmers-Bullock, Limited, Montreal
Canadian General Electric Co., Toronto.
Canadian Westinghouse Co., Hamilton.
D'Olier Engineering Co., New York.
Kerr Turbine Co., Wellsville, N.Y.

Turntables

Detroit Foundry Supply Co., Windsor.
Dominion Foundry Supply Co., Montreal
Hamilton Facing Mill Co., Hamilton.
Northern Engineering Works, Detroit
Paxson, J. W., Co., Philadelphia

Turret Machines.

American Tool Works Co., Cincinnati.
John Bertram & Sons Co., Dundas, Ont.
The Canadian Fairbanks Co., Montreal.
London Mach. Tool Co., Hamilton, Ont.
Niles-Bement-Pond Co., New York.
H. W. Petrie, Toronto.
Pratt & Whitney Co., Hartford, Conn.
Warner & Swasey Co., Cleveland, Ohio.
Williams & Wilson, Montreal.

ALPHABETICAL INDEX.

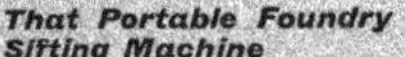

Will Save You Its Cost Several Times Annually.

Write for Circular and Prices.

Don't Throw Money Away!

Most foundry owners do that very thing. Why should you—or any man in business for dollars and cents—use Plumbago when

Columbian Dry Sand and Loam Blacking

will do the same work, and do that work better, at **one-third the cost?**

Foundry Supplies of all kinds ready for delivery.

Buffalo Foundry Supply Co., 339-347 Babcock Street, BUFFALO, N.Y.

Delays are Dangerous

They often read serious financial loss. You cannot afford to place your orders for

PIPE, CORE, STOVE PLATE and MACHINERY MOLDING SAND and LAKE ROOFING GRAVEL

with any old concern. Sometimes you want your sand orders filled mighty quick. Send for prices and samples. Quick deliveries a specialty.

T. W. BARNES, HAMILTON, ONT.

WHITING CUPOLAS and LADLES

PLUMBAGO

FOR ALL PURPOSES

Foundry Facings

AND

Foundry Supplies

THE DOMINION FOUNDRY SUPPLY CO.

TORONTO, ONT. MONTREAL, QUE. Limited

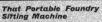

OWEN SOUND IRON WORKS, LIMITED
OWEN SOUND, ONT.

Cement Mill Machinery, Boiler and Steel Tank Work of all kinds, Grey Iron and Brass Castings

WM. KENNEDY & SONS
HIGH-CLASS
Steel Gears of all kinds
Iron Gears
Grey Iron, Steel and
Brass Castings
MARINE WORK receives every attention

WM. KENNEDY & SONS, Limited
Owen Sound, Ont.

Thor PNEUMATIC DRILLS

ARE GUARANTEED TO DO MORE WORK THAN ANY OTHERS MANUFACTURED.

Made in all Sizes

Write for Illustrated Catalog

Sent on Approval

Transportation charges paid both ways if they do not prove satisfactory

Independent Pneumatic Tool Co.

First National Bank Building
CHICAGO

170 Broadway
NEW YORK CITY

MALLEABLE IRON CASTINGS

WE ALSO MAKE METAL PATTERNS

Galt Malleable Iron Co., Limited - Galt, Ontario

Directory of Consulting Engineers, Patent Attorneys, Architects and Contractors.

Are you interested in any of the lines that are advertised ?
A Post Card will bring you price list and full information.
Don't forget to mention Canadian Machinery.

UNION DRAWN STEEL COMPANY, Limited

Manufacturers of

BRIGHT FINISHED STEEL

Largest in Canada. Best Equipped on the Continent. Capacity twice present requirements of the trade.

Rounds, Flats, Squares and Hexagons

LARGE STOCK

Send for Price List

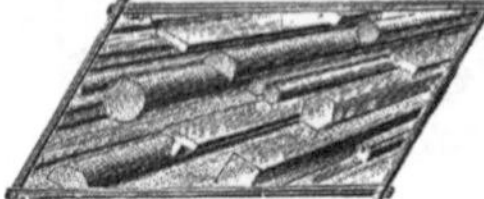

Office and Works:

HAMILTON, CANADA

PERPETUAL SATISFACTION

That's what Webster Heaters mean to thousands; because they represent the genius and experience of the entire heating industry. Built on scientific principles they stay built.

THE WEBSTER

FEED-WATER HEATER

PURIFIER AND FILTER

The type shown in the cut is known as the Webster Class "E. C." Heater, made in sizes ranging from 500 to 1,500 h.p. This type assures economy while affording the greatest heating power and highest temperature. It cannot help saving fuel because it is impossible for it to waste steam.

We have interesting particulars to mail interested parties.

DARLING BROTHERS, LIMITED

Toronto Montreal Winnipeg

ISLAND CITY

Graphite Paints

IT is well known that Graphite makes one of the best paints as a preservative of exposed metal work. It is the best paint to use for iron castings or girders on account of its filling properties thereby giving it a good covering.

Machine Paints

BESIDES the ordinary, our machine paints are made in paste form and can be used as a filler, applied with a knife or thinned with turpentine and brushed on, spreads freely and leaves a good surface.

Ask Us for One of Our Catalogues

P. D. Dods & Co., Montreal and Toronto

An intelligent Engineer uses nothing but the GENUINE PENBERTHY INJECTOR. He has had experience with just-as-good kind before.

REGISTERED 1901

King Edward Motors and Dynamos

ARE THE BEST THAT CAN BE BUILT

Prompt repairs to all makes

Consolidated Electric Co. LIMITED

710-724 Yonge St., TORONTO

The Engineering Times

A Weekly Newspaper of Commercial Engineering

32 PAGES

Price Twopence

Annual Subscription, 10s. 10d.

CHIEF OFFICE:

6, Bouverie St., London, E.C.

SCOTLAND:

Mr. Alex. Forbes

118a, Renfield St., Glasgow

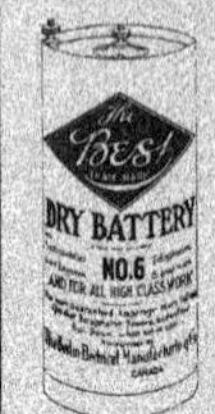

"BEST" DRY BATTERIES

have no equal in their line

FOR IGNITION WORK

A single set will prove it.
Have you tried them?

For sale by all leading dealers.

The Berlin Electrical Mfg. Co., Limited
474 King Street West,
TORONTO

DO YOU DO IT?

When you read an advertisement that interests you, why not write the advertiser and get more information? It will help you to keep posted. Try it. The advertisers in **Canadian Machinery** will appreciate it.

WRITE THEM. TRY IT.

LIGHTING and POWER PLANTS

MOTORS FOR DIRECT CONNECTION TO MACHINES

TORONTO & HAMILTON ELECTRIC CO.

99-101 McNab Street, HAMILTON, ONT.

Eugene F. Phillips Electrical Works, Limited

MONTREAL - TORONTO

CANADA

Bare and Insulated Electric Wire	Railway Feeder and Trolley Wire
Electric Light Line Wire, Incandescent and Flexible Cords	Americanite, Magnet, Office and Annunciator Wires Cables for Aerial and Underground Use

MOTORS AND DYNAMOS

OF THE HIGHEST GRADE

BUILT FOR HARD SERVICE
EFFICIENCY AND DURABILITY

We make them and guarantee them to be as good as we say they are.

The accompanying cut shows our

OPEN-TYPE FOUR-POLE DYNAMO OR MOTOR

for belt driving. It is of latest multi-polar design, has coil-wound armatures, is substantial in construction, sparkless running and of the highest efficiency.

Built in sizes from
15 to 60 horse power

We should like to quote you prices

THE KAY ELECTRIC DYNAMO & MOTOR CO., Limited

Head Office, 1170 DUNDAS ST.
Factory, TORONTO JUNCTION

TORONTO, CANADA

JANDUS ARC LAMPS

The JANDUS Interchangeable Arc Lamp offers itself on grounds of

Simplicity

Economy of Installation

Economy of Operation

It embodies all of the features required by the best modern practice.

SOLE CANADIAN REPRESENTATIVES

THE PACKARD ELECTRIC CO. LIMITED

ST. CATHARINES

MONTREAL WINNIPEG

TOOLS AND SUPPLIES
for
Plumbers, Gas and Steam Fitters
Machinists and Electricians
APPAREILS ETC
pour
Plombiers, Gaziers, Mecaniciens
Ferblantiers et Electriciens
WHOLESALE AND RETAIL
MECHANICS SUPPLY CO.
EN GROS ET EN DETAIL
Showrooms: 80-90 St. Paul St.
QUEBEC
Entrepot: 59-67 Rue St. Andre

No Frost Breaks, no Corrosion No Electrolysis
CANADIAN PIPE COMPANY LTD.
HEAD OFFICE VANCOUVER B.C. P.O.BOX 915.
It is easily and cheaply laid. Its carrying capacity is never decreased by rust.
GALVANIZED WIRE WOUND WOODEN PIPE
CANADIAN PIPE COMPANY, Limited
PHONE 1642
VANCOUVER, B.C.
P.O. BOX 915

WOOD STAVE WATER PIPE
Machine-banded made from choice British Columbia or Douglas Fir, absolutely clear and free from knots or defects, wound with heavy gauge galvanized steel wire. For
Water Works, Power Plants, Irrigation, Mining and all Hydraulic Propositions, where conduits or pressure pipe are required.
Reliable, durable, economical, efficient, easy to instal. Costs very much less than any other sort of water pipe, and with greater efficiency in use.
Write for Catalogue and Circulars.
Estimates also given on continuous stave pipe of large diameters.
Wood Pipe Designed for Different Pressures.
PACIFIC COAST PIPE COMPANY
Limited
P.O. Box 563
1551 Granville Street (Next Bridge) - VANCOUVER, B.C.

In Buying Taps and Reamers
Consider
First, Quality
Second, Accuracy
Third, Durability
Last, Price
The Name
Butterfield
Stamped
on
A
TAP,
REAMER
or any
other TOOL
is sufficient
guarantee of the
three first considerations.
Prices on Application
BUTTERFIELD & CO., Rock Island, P.Q.

THE STANDARD
TOOL CO'S
Twist Drills
Made of either
Carbon or High Speed Steel
Every operation in their manufacture is the result of years of study and practical experience. Buy them and you can be assured of tools that will give you satisfactory service.
OFFICE and FACTORY
Cleveland, Ohio
EASTERN SALESROOM
94 Reade St., New York

Armstrong Tool Holders

ALLIS-CHALMERS-Bullock-LIMITED MONTREAL
TRANSFORMERS
ALLIS-CHALMERS-Bullock-LIMITED MONTREAL
FOR the standard line of Lighting Transformers, Allis-Chalmers-Bullock, Limited, employ the "core" type, now almost universally recognized as the best all-round type for lighting service. These Transformers are oil-filled, self-cooled. Cast iron tanks are used up to 10-k.w. The cases above 10-k.w. are made of corrugated terne sheet.
FOR FURTHER PARTICULARS SEE BULLETIN 300.
ALLIS-CHALMERS-BULLOCK, Limited
Head Office and Works, Montreal; District Offices: Toronto, Traders Bank Bldg; Montreal, Sovereign Bank Bldg; Winnipeg, 251 Notre Dame Ave.; Nelson, Josephine St.; Vancouver, 416 Seymour St.; New Glasgow, N.S., Telephone Bldg.

YOU WILL SAVE MONEY IN BUYING IF YOU REFER TO OUR BUYERS' DIRECTORY

CANADIAN MACHINERY AND MANUFACTURING NEWS

A monthly newspaper devoted to the manufacturing interests, covering in a practical manner the mechanical, power, foundry and allied fields. Published by The MacLean Publishing Company, Limited, Toronto, Montreal, Winnipeg, and London, Eng.

OFFICE OF PUBLICATION: 10 FRONT STREET EAST, TORONTO

Vol. III. JUNE, 1907 No. 6

SMALL TOOLS

Taps, Dies

Reamers
Solid and Adjustable

Ratchet Drills

Milling Cutters
Solid and Inserted Blade

Boiler Punches

High Speed Drills

Gauges, etc.

PRATT & WHITNEY CO. HARTFORD Conn.

Agents for Canada

THE

Canadian Fairbanks Co.
LIMITED

Montreal Toronto
Winnipeg Vancouver

No. 12—Two Spindle Profiling Machine.

No. 2—Hand Milling Machine, overhanging arm.

MACHINE TOOLS

Profiling Machines

Single and Double Spindle

Specially adapted for finishing parts of guns, sewing machines and any other accurate and interchangeable work, doing away with hand fitting.

ILLUSTRATED CATALOG ON REQUEST

Hand Milling Machines

adapted to milling the variety of duplicate parts incident to the manufacture of small machines.

VARIETY OF SIZES AND TYPES

Becker-Brainard

Milling Machines

and

Milling Cutters

Guarantee ACCURACY and ECONOMY in Production

Becker-Brainard Milling Machine Co.

Hyde Park, Mass, U.S.A. Branch Offices: THE BOURSE, Phila, Pa.; WILLIAMSON BLDG., Cleveland, O.

Canadian Agents: A. R. WILLIAMS MACHINERY CO., LTD., TORONTO and MONTREAL, CAN. Agents, London, Eng.: SELIG SONNENTHAL & CO.

326

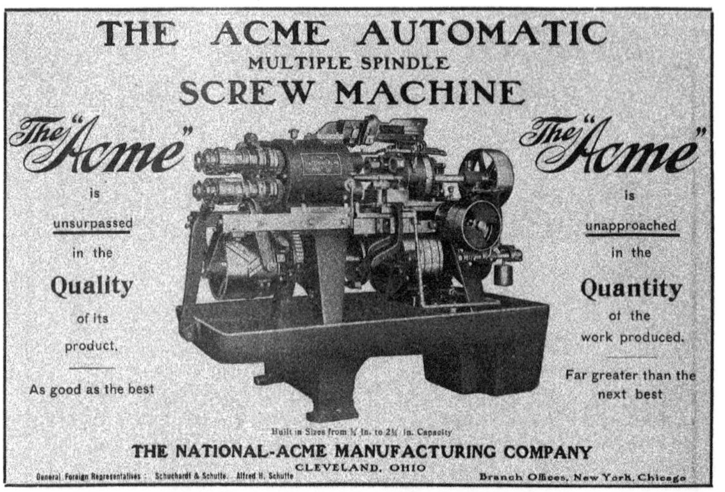
THE ACME AUTOMATIC
MULTIPLE SPINDLE
SCREW MACHINE
The "Acme"
is
unsurpassed
in the
Quality
of its
product.
As good as the best
The "Acme"
is
unapproached
in the
Quantity
of the
work produced.
Far greater than the
next best
Built in Sizes from ½ In. to 2¼ In. Capacity
THE NATIONAL-ACME MANUFACTURING COMPANY
CLEVELAND, OHIO
General Foreign Representatives: Schuchardt & Schutte. Alfred H. Schutte
Branch Offices, New York, Chicago

Will Cut
Plate, Bar
and
Profile Iron
Powerful,
Durable,
Handy
Portable
Universal Shears
FOR
HAND POWER
CUTS WITHOUT
DEFORMING
THE PROFILE
WRITE FOR PARTICULARS
BAXTER, PATERSON & CO.
102 St. Antoine Street
Montreal

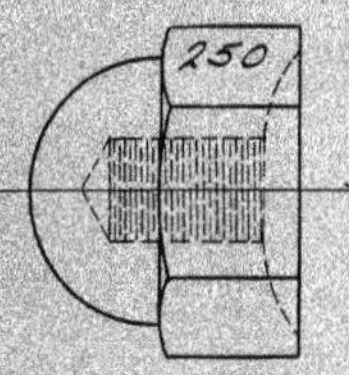

Piece No. 250. Material, Brass Rod, made on a 2-inch Cleveland Automatic Turret Machine; drawing full size. Output per hour, 45; this means all operations for completing this piece. Actual cost of labor, from 1 to 1⅓ mills for each piece. Send along samples or drawings. We guarantee our outputs. One man can operate from 4 to 6 machines.

Save from 200% to 500%

You can do that by installing our automatic screw machines.

Cleveland Automatics are built for all kinds of work, and are guaranteed to produce the finished part from bar or castings and effect a saving of time in the process.

You positively cannot lose anything by mailing us blue prints or samples of your work, and allow us to demonstrate the Cleveland's ability as a money-saver.

"The Pre-eminence of the **CLEVELAND** is Indisputable"

Piece No. 195. Material, Cast Iron, using our tilting magazine attachment on a 2-inch Cleveland Automatic Turret Machine; drawing full size. Output per hour, 10; this means all operations for completing this piece. Actual cost of labor, from 8 to 9 mills for each piece. Send along samples or drawings. We guarantee our outputs. One man can operate from 4 to 6 machines.

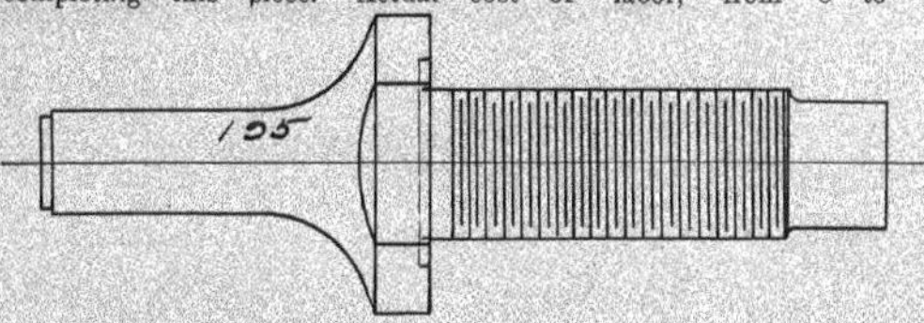

AS WE HAVE NO CANADIAN REPRESENTATIVE, WE SOLICIT CORRESPONDENCE AND ORDERS DIRECT

CLEVELAND AUTOMATIC MACHINE CO.

CLEVELAND, OHIO, U.S.A.

EASTERN REPRESENTATIVE—J. B. ANDERSON, 2450 North Thirtieth Street, Philadelphia, Pa.
WESTERN REPRESENTATIVE—H. E. NUNN, 22 Fifth Avenue, Chicago, Ill.
FOREIGN REPRESENTATIVES—CHAS. CHURCHILL & CO., London, Manchester, Birmingham, Newcastle-on-Tyne and Glasgow. MESSRS. SCHUCHARDT & SCHUTTE, Berlin, Vienna, St. Petersburg and Stockholm. ALFRED H. SCHUTTE, Cologne, Brussels, Liege, Paris, Milan and Bilbao.

"BLISS"

STRAIGHT SIDED PRESSES

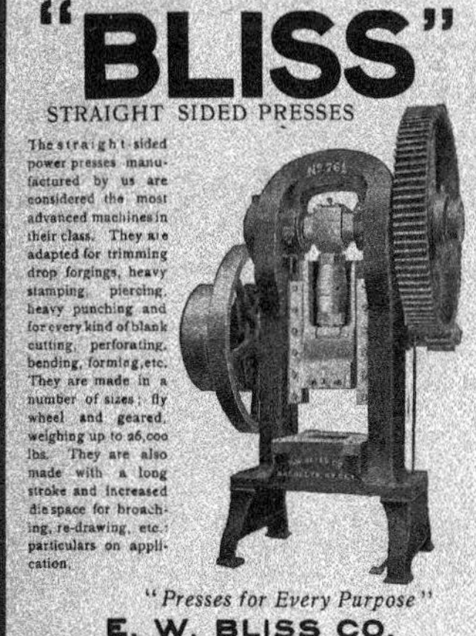

The straight-sided power presses manufactured by us are considered the most advanced machines in their class. They are adapted for trimming drop forgings, heavy stamping, piercing, heavy punching and for every kind of blank cutting, perforating, bending, forming, etc. They are made in a number of sizes; fly wheel and geared, weighing up to 26,000 lbs. They are also made with a long stroke and increased die space for broaching, re-drawing, etc.; particulars on application.

"Presses for Every Purpose"

E. W. BLISS CO.

20 ADAMS ST., BROOKLYN, N. Y.

Chicago Agts.: The Stiles-Morse Co., 138 Jackson Boulevard, Chicago, Ill.

POTTER & JOHNSTON

MANUFACTURING AUTOMATICS

will machine your duplicate parts from all kinds of castings, forgings and the bar with extreme accuracy and more economically than upon any other tools in the market.

Catalogue showing interesting specimens of parts finished on these tools free upon request.

ALSO UNIVERSAL SHAPERS

POTTER & JOHNSTON MACHINE CO.

PAWTUCKET, R.I. - - U.S.A.

Gisholt 30-inch Mill

Quick Deliveries

On a limited number of 30-inch Vertical Boring Mills we are in a position to make early deliveries.

Headstock driven by 4-step cone pulley; sixteen table speeds; eight feeds; micrometer index dials reading to .001" on all feeds; automatic feed tripping device for tripping any feed at a predetermined point; all gears encased. Mill belt feed. Write for catalog describing Gisholt Mills from 30" to 72".

Gisholt Machine Co.

Works: Madison, Wis.; Warren, Pa.

General Offices:
1315 Washington Street,
Madison, Wis., U.S.A.

Foreign Agents:

Alfred H. Schutte, Cologne, Brussels, Liege, Paris, Milan, Bilbao, Barcelona, Schuchardt & Schutte, Berlin, Vienna, St. Petersburg, Stockholm, Copenhagen, C. W. Burton, Griffiths & Co., London, England.

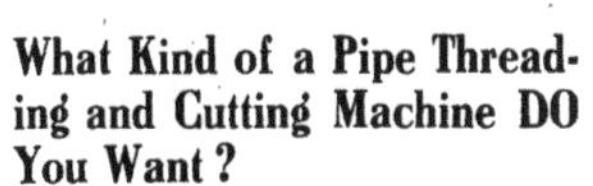

What Kind of a Pipe Threading and Cutting Machine DO You Want?

One that will thread any size or kind of pipe easily, quickly and cheaply?

One that can be operated by a mere boy?

One that will allow of very quick changes from size to size of pipe?

One that will forget to break down?

All this—and more?

Then our answer to you is investigate Merrell Pipe Threading and Cutting Machines—and do it right away.

We're so sure of our machines that we send them on thirty days free trial. The risk is all ours—will you write?

The Merrell Mfg., Co., 12 Curtis St.,
TOLEDO, O.

THE CANADIAN FAIRBANKS CO., LIMITED
AGENTS
Montreal Toronto Winnipeg . Vancouver

MURCHEY DOUBLE HEAD RAPID NIPPLE and PIPE THREADING MACHINE

With variable speed motor directly attached

Glad to mail you our Catalogue

MURCHEY MACHINE & TOOL CO.
DETROIT, MICH., U.S.A.
1306 Sole Canadian Agents: The London Machine Tool Co.
Traders Bank Building, Toronto.

Morse Reamers
as well as
Morse Twist Drills

possess all the desirable features that perfect Reamers and Twist Drills should have.

Try them and prove the truth of our assertion.

We have a large assortment of these two styles always in stock. We can send you an illustrated catalog showing many other kinds and sizes.

No. 115
SOLID
REAMER

No. 117
SHELL REAMER

Catalog and prices sent on request.

F. & W. Hardware Montreal

F&W

Frothingham & Workman, Limited
Montreal

"LAURIE" POWER PRESSES

Laurie Engine & Machine Co., Limited
MONTREAL

Toronto Agents: Parmelee & Nicholson
Halifax Agents: Guilford & Son

Steel Converters

Flockton, Tompkin & Co. LTD.

Newhall Steel Works

SHEFFIELD

Makers of

Specialty AIR HARDENING TOOL STEEL for High Speed Work

ENGINEERS' TOOL STEEL, for Machine Tools, Chisels, etc.

DRILL AND JUMPER STEEL, for Miners and Contractors

SOLID STEEL HAMMERS, of any shape or size for Miners and Engineers

SPRING STEEL, SECTION BARS

Purchasing Agents wanted in principal Canadian Cities.

NORTON JACKS

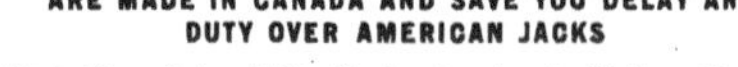

ARE MADE IN CANADA AND SAVE YOU DELAY AND DUTY OVER AMERICAN JACKS

They are Ratchet Screw Jacks with Ball Bearings to reduce the friction. Working parts are protected from grit and rust. Superior to Hydraulic Jacks for all classes of service. Simple, durable, and guaranteed in every respect. Absolutely safe. Will work in any position. Standard on the leading railroads.

Made in 50 styles; 8 to 70 tons capacity

Send for complete illustrated catalogue and discount.

Sectional View of 70-Ton Bridge Jack

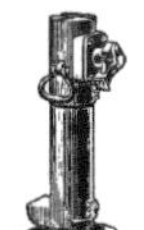

25-Ton Foot Lift Jack

35-Ton Plain Jack

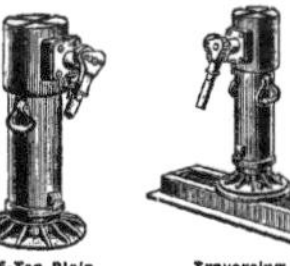

Traversing Jack 10 to 60 Tons

Track Jack 10 and 15 Tons

Journal Box Jack 15 Tons

MADE ONLY BY

A. O. NORTON, Inc., Coaticook, Prov. Quebec

Stock carried by MUSSENS, Limited, Montreal

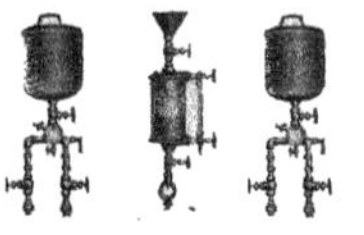

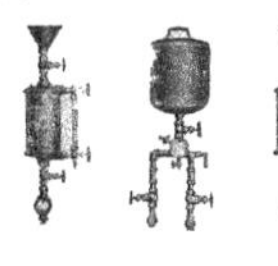

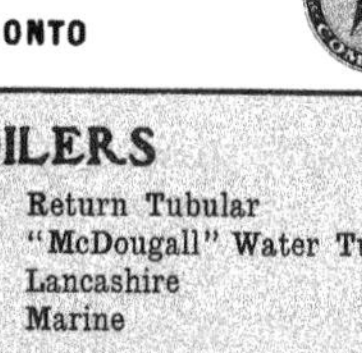

WE MANUFACTURE A FULL LINE

of air moving machinery for creating pressures up to 10 pounds per square inch.

We are in a position to advise without prejudice as to type.

Our interest centres in the success of the installation.

B. F. STURTEVANT COMPANY,

BOSTON, MASS.

General Office and Works, Hyde Park, Mass.

New York Philadelphia Chicago London

Designers and Builders of Heating, Ventilating, Drying and Mechanical Draft Apparatus; Fans, Blowers and Exhausters; Steam Engines, Electric Motors and Generating Sets; Fuel Economizers; Forges, Exhaust Heads, Steam Traps, Etc.

568

BLOWERS

HEATING AND VENTILATION
by Fan System

Natural Draft and Blower Dry Kilns
For Lumber, Brick, etc. Cars, Trucks, Rail, etc.

Heating
Ventilating
Drying Forges
Mechanical Draft
Blowers.
Exhausters
Steam Specialties

ENGINES
both Vertical and Horizontal

Shavings Exhaust System

SHELDONS, Limited, Engineers and Manufacturers **Galt, Ont., Canada**

Montreal Office, 325 St. James St. Phone Main 2919

MACHINE TOOLS

Bertram 18-in. Double Back Geared Engine Lathe.

of all kinds for all purposes.

Prompt delivery if ordered immediately

WE ARE THE SELLING AGENTS FOR THE FOLLOWING MANUFACTURERS:

Niles-Bement-Pond Co.
The John Bertram & Sons Co., Ltd.
Pratt & Whitney
Brown & Sharpe
E. W. Bliss Co.
The R. McDougall Co., Ltd.
R. A. Kelly Co.
J. J. McCabe
Bignall & Keeler
Baker Bros.
Merrell Mfg. Co.
Seneca Falls Mfg. Co.
Wilmarth & Morman
Niagara Machine and Tool Co.
S. A. Woods Machine Co.
American Wood-working Machinery Co.

Bertram 12-18-in. Boring and Turning Mill

The Canadian Fairbanks Co., Ltd.

MONTREAL TORONTO WINNIPEG VANCOUVER

The Production of Pig Iron

Illustrated in the new blast furnaces of the Hamilton Steel and Iron Co.

In Canada the production of pig iron is assuming large proportions. Few realize the amount of iron which is now being turned out from Canadian furnaces; but the building of the new 300-ton blast furnace by the Hamilton Steel and Iron Co. is ample evidence of the importance of this industry in Canada. The development of the iron industry in Canada, and particularly in Ontario, can be well shown by a short history of the development of the Hamilton Steel and Iron Co., Hamilton.

In 1896 the Hamilton Blast Furnace Co. erected a blast furnace on the picturesque shore of Burlington Bay, and in February of that year the first pig iron made in the Province of Ontario was produced. The furnace was designed to give a production of 150 tons per day, and during the first few months of its operation the many difficulties incidental to any new enterprise were encountered. Men accustomed to blast furnace work could not be found in Ontario, and green men had to be trained for the work. The use of Canadian ore alone was found not to be satisfactory, and a mixture of imported ore had to be adopted. In addition to this the price of pig iron during the two years following the establishment of the furnace was probably the lowest ever recorded in the history of the iron trade.

The company, however, did not lose heart. In June, 1899, the Hamilton Blast Furnace Co. amalgamated with the Ontario Rolling Mills Co., Hamilton, under the name of the Hamilton Steel and Iron Co., Ltd., with a capital of $2,000,000. At the time of the amalgamation the plant consisted of one blast furnace, two rolling mills and a forge and washer plant. Immediately after the consolidation the blast furnace was altered so as to increase the output from 150 tons to 200 tons per day. An open hearth steel plant and a steel rolling mill were added. Since then additional furnaces have been added to the open hearth plant, one of which has but very recently been completed.

The company have now under construction a second blast furnace, which, it is expected, will be in operation by August. The furnace is up-to-date in every respect, but, before describing this furnace plant in detail, something of general interest might be said concerning the operation of the blast furnace in the smelting of iron ores.

The Blast Furnace in Theory.

Everyone knows in a general way the construction and operation of a blast furnace. Iron ore is the oxides of iron. In the blast furnace the oxygen is removed, leaving the iron. The furnace consists of an immense stack, as shown in Fig. 1 of the Hamilton Steel and Iron Co. The parts of the furnace are the hearth, the bosh and the stack. The

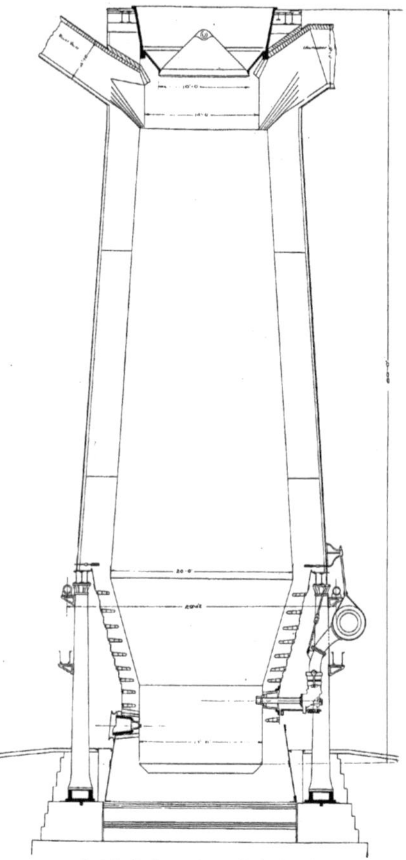

Fig. 1—The New Furnace—Hamilton Steel and Iron Co.

stack is the upper part of the furnace; the bosh is the bulging part, and the hearth is the lower part. The ore, coke, limestone and other necessary commodities are charged at the top of the furnace. Air is admitted at the bottom; the burning of the fuel creates a great heat. The mixture at the top is heated, gradually sinks down, and is melted in the bosh. The melted iron runs down to the bottom of the hearth, where the slag is removed and the iron poured. In brief that is the operation of the blast furnace.

Styles of Ores.

There are several styles of ores, including: Magnetite, composition, Fe O and $Fe_2 O_3$, 72.4 per cent. iron; hematite composition, $Fe_2 O$, 70 per cent. iron, soft ore and hard ore; Limonite, composition $Fe_2 O_3$, 60 per cent. iron; siderite, composition Fe O and CO_2, 48.3 per cent. iron. There are two other ores, both siderite, one mixed with clay and the other with carbonaceous matter. In commerce, ores are known chiefly by names, significant of the section in which the ore is obtained. The most of Canadion ore is hard hematite.

At Hamilton a mixture of ores is used, the best mixture for the different irons required being determined by practice.

Kinds of Pig Iron.

There are two distinct classes of pig iron—white iron and grey iron. In white iron, the carbon and the iron are chemically combined, while in the grey foundry iron the carbon is mixed mechanically with the iron. Pig iron must contain carbon and silicium to give it the desired properties, and it always contains manganese, phosphorus and sulphur.

Uses of Constituents in Charge.

Each constituent in the charge to the furnace has its use. The ore is there to be smelted; the fuel (coke) is to supply heat and to supply the reducing agent, carbon; the limestone (CaO, CO_2) combines with the excess of silica (from the clay in the ore), and is also supposed to unite with the sulphur. The presence of sulphur weakens the iron, and to get rid of it is one of the problems in the blast furnace practice.

Kinds of Fuel Used.

In Canada and the United States, 86 per cent. of the fuel used in smelting is coke, 12 per cent. is anthracite and 2 per cent. charcoal. The advantages of coke as fuel are its rigidity, strength and porosity. Coke makes the stock porous and prevents it from crushing together and forming arches, etc.

Reactions in the Furnace.

According to the reactions which are supposed to take place, the furnace is divided into five zones. The first, and upper, zone is the warming zone, where the CO_2 is driven off the limestone and the moisture from the ore and fuel is evaporated. This makes the ore more porous. The next zone is the reducing zone, where some such reaction as this takes place:

$$CO + Fe_2 O_3 = CO_2 + 2FeO$$

$$\text{and } FeO + CO = CO_2 + Fe.$$

Thus the iron (Fe) is free. The carbon monoxide (CO) comes from the union of the oxygen in the air with the carbon in the fuel.

The third zone is the carbonizing zone. Here the iron is in a spongy mass, and absorbs carbon from the coke. The melting point of the iron is thus lowered. Next is the melting zone. Here the iron melts, and there is a reduction of silica (SiO_2) and manganese dioxide (MnO_2) There is also a further absorption of carbon by the iron. The fifth and bottom zone is the burning zone, where the excess of carbon is burned out of the iron. From this zone the sla~ is removed and the metal ruu off.

Slags and Their Composition.

The slags are chiefly composed of silicates of three substances, namely aluminum oxide ($Al_2 O_3$) calcium oxide (CaO) and magnesium oxide (MgO). These are thus:

$$\left.\begin{array}{l} Ae_2 O_3 \\ CaO \\ MgO \end{array}\right\} Si\, O_2$$

The composition of the slag predetermines what the proportions of the different ores, coke and limestone should be. Thus, having decided what style of slag is to be produced, the composition of the mixture for the charge can be worked out theoretically. This, of course, will be moderated by practice considerably.

The Air Supply, the Ore Mixture and the Gas.

There are three chief things in the blast furnaces which go through regular cycles of operation, namely, the ore mixture, or stock, the air blast, and the resulting gas.

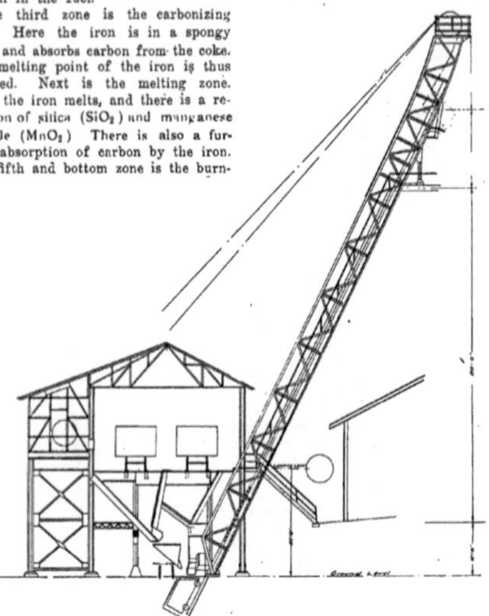

Fig. 2—The Skep and Ore Bins.

Cycle of Stock.

The different constituents of the stock are taken from bins, weighed and mixed. The mixture is conveyed by means of the skip to the top of the furnace. Here it is dumped on to a bell. This bell is dropped, allowing the mixture to fall into a chamber on top of a second bell. The first bell is then closed and the second one dropped, which allows the mixture to fall into the furnace. This two-bell arrangement prevents the escape of gas from the furnace. The metal is tapped from the hearth of the furnace into a runner. From this runner the metal is poured into sand moulds, or is transferred to a pig casting machine.

In the new furnace at Hamilton the weighing and mixing of the different ores and constituents is done automatically. The mixture is conveyed up the skip and is dumped automatically. The new furnace is provided with a modern double bell arrangement. A sectional view of the skip and ore bin is shown in Fig. 2.

The furnace is 85 ft. high, having a 20 ft. bosh and 13 ft. hearth. The top of the furnace is provided with explosion doors, as shown in Fig. 1, and the bell is operated by hydraulic cylinders. The stock is distributed by an apparatus driven with an electric motor.

Cycle of Air Blast.

From the blowing engines the air blast is conveyed to the hot stoves through the cold blast main. In the stoves the air is heated and passes to the bustle pipe, through the hot blast main. The bustle pipe is a large pipe which completely surrounds the hearth of the furnace. From the bustle pipe the air is led through the tuyeres into the furnace, where it produces combustion.

These stoves are immense circular fire brickwork constructions. The hot gas

Fig. 4—Foundation of Furnace, Showing Walls of Cast House.

Fig. 5 Foundation of Three Stoves and Furnace, Showing Old Furnace and Boiler House.

Fig. 6—Erection of Columns Upon Which the Furnace Stands.

Fig. 7—Three Stoves Almost Completed, and the Furnace Partially Erected.

Fig. 8—Steel Construction of Ore Bins, Showing Stoves and Furnace in Background.

Fig. 9—The Furnace and Stoves as They are Now.

coming from the furnace is burned in these stoves, and the checker brick work absorbs the heat. When one stove becomes sufficiently hot, the gas is led into a second stove and a cold blast connection made to the hot stove. This provides for the heating of the air.

The Hamilton furnace is provided with three stoves, with provision for a fourth. These stoves are of the two passage type, and are each 90 by 20 ft. The valves and valve seats of these furnaces are of the latest design.

The blowing engines are operated by steam. The boiler house contains ten sterling boilers, the total capacity of 3,500 h.p. The blowing engines are compound condensing, built by the Tod Engine Co., of Youngstown, Ohio.

Cycle of Gas.

The gas, which consists chiefly of carbon monoxide and carbon dioxide, is brought from the top of the furnace to the gas main through a large pipe called the down-comer. In this pipe are devices for catching the dust which the gas contains. The gas is then mixed with air and burned in the cold stoves as before described. The burnt gases travel up a flue to the open air.

The Use of Gas for Blowing Engines.

Very often the gas which comes from the furnace is used for other purposes besides that of heating the cold stoves. For instance, it is sometimes used in gas engines for the operation of the blowers. Such a use of gas results in the saving of fuel which would otherwise be used in the production of steam for the steam blowing engines. Practical consideration however sometimes makes it more advantageous to use steam blowing engines.

Construction of Furnace and Stoves.

The outside shell of the furnace consists of steel sheets. The furnace is lined with fire brick, and rests, as is shown in some of the illustrations, on cast iron columns. The hearth is built up inside these columns. The hearth is the hottest part of the furnace, and in order to keep down the temperature as much as possible, and thus lengthen the life of the hearth, it is water cooled. The tuyeres and tuyere connections are also water jacketed. In the Hamilton furnace the capacity of the pumping plant which supplies the water for these cooling jackets will be twelve million gallons per day. The ordinary requirements for the whole plant is about six million gallons. This will be pumped by electrically driven pumps. Auxiliary steam pumps are provided in case of accident to the electrically driven pumps. The electrical power will be furnished by two Canadian Westinghouse engine generators.

The stoves are also built up of steel sheets. They are lined, and the sections are built up of fire brick.

Everything in connection with the furnace is being built in the most substantial manner. The foundations are of concrete resting on over one thousand piles. This work was done by Rowan & Elliott, St. Catharines, Ont. The contractors for the steel of the stack and stoves, dust catcher, cold and hot blasts, were the John Inglis Co., Toronto. The Hamilton Bridge Works, Hamilton, furnished and erected the cast house, skip hoist, and ore bins. Tallman & Sons, of Hamilton, are responsible for the brass and copper work, including the tuyeres, and Purdy, Mansell & Co., Toronto, have put in all the steam and water lines.

The plans for the new blast furnace were furnished by Frank C. Roberts & Co., Philadelphia, who have designed some of the largest and best furnaces in America.

The ores, coke and limestone will all be handled through steel bins, and will be delivered at the bottom of the double skip hoist by electric cars.

The pigs in this plant will be cast in sand. A Brown hydraulic pig breaker will be installed, and each bed of iron will be carried by an electric crane traveling the whole length of the cast house, and will be delivered to the breaker, where each thrust of the ram will break two pigs and the connecting sow into six pieces; these pieces will drop into a chute, which will deliver them into a railway car for shipment.

New Furnace to Produce Foundry Iron.

The new furnace will be engaged solely on the production of foundry pig iron. It will be the largest furnace in Canada engaged in this work. As soon as the new furnace is blown in, the present furnace will be used for producing basic iron for the open hearth furnaces. It will be taken to them in a molten condition. A fourth furnace has just been added to the open hearth furnace plant, which brings the capacity up to one hundred thousand tons of open hearth steel per year.

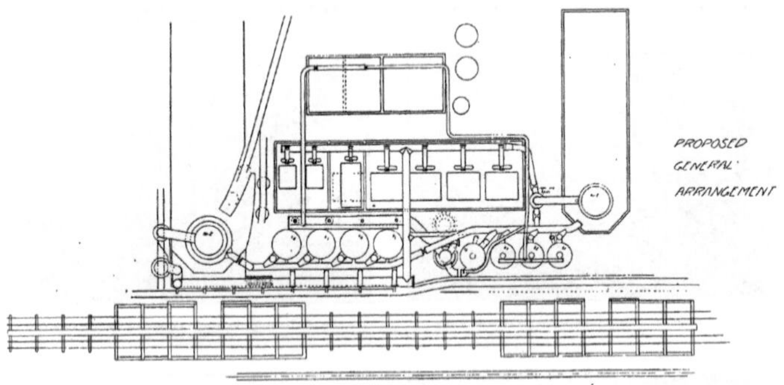

Fig. 3—Ground Plan of Works, Showing Both Old and New Furnaces.

PERSONAL MENTION.

Mr. D. Lorne McGibbon has succeeded Mr. H. C. Miner as president of the Consolidated Rubber Co., Montreal.

Fred A. Geier, president of the Cincinnati Milling Machine Co., has been elected president of the Bickford Drill & Tool Co., Cincinnati, Ohio. H. C. Hoefinghoff, recently deceased, Frank Hushart becomes general manager.

In the death of John M. Dixon, organizer of the Canadian branch of the Amalgamated Society of Engineers, the engineers of Canada and of Ontario particularly will suffer a great loss. Mr. Dixon's life has been one which should inspire and help the rising generation of engineers.

Mr. Charles F. Wolfe, formerly general superintendent of the Waterous Engine Works, Brantford, has accepted a position as inspector with the Canadian Boiler and Casualty Insurance Co., of Toronto. He was presented with a complimentary address and a cabinet of silver, the gift of the employes of the company, on the evening of April 26th. Mr. Wolfe was with the Waterous Company for 35 years.

Mr. J. W. Campbell, manager of the contracts and sales department of the Canadian General Electric and Canada Foundry Companies, was tendered a farewell banquet at the King Edward Hotel, Toronto, on April 23rd, on the occasion of his leaving the position which he has held for 15 years to assume the office of managing director of the Alberta Portland Cement Co., Calgary.

H. Somerville and D. D. Van Every have opened an office at Hamilton, Ont., as consulting and supervising engineers and factory architects. These men have had considerable experience in modern branches of structural, mechanical and electrical engineering, having been engaged upon the design and construction of the following shops: C.P.R. Angus Shops, Montreal, Canada Car Co., Montreal, Lackawanna Steel Co., Buffalo, N. Y., the John Bertram & Sons Co., Dundas, Ont., and M. Beatty & Sons Co., Welland. There is surely a broad field for work such as Somerville & Van Every have undertaken; and Canadian Machinery wishes them every success.

ALUMINUM ALLOYS.

When alloyed with copper, aluminum acts similarly to zinc, but much more strongly, so that an addition of 1 per cent. aluminum produces as much effect as 3.5 per cent. of zinc. Aluminum bronzes are much stronger than ordinary brasses, but those containing 10 per cent. or more of aluminum are so hard that they cannot be worked.

BOOK REVIEWS

PRODUCER GAS AND GAS PRODUCERS. A treatise on producer gas and producer gas plants by Samuel S. Wyer, M.E.; published by the Engineering & Mining Journal, New York: cloth-bound, 296 pages, 9 x 6.

This book is written in a style to be of value to those who may not necessarily be familiar with the laws of thermo-chemistry; the first few chapters of the book are devoted to the explanation of a few of the fundamental laws of the subject, which are used throughout the book. Following these chapters are several on the chemical reactions which take place in the producer, together with a number of computations of the generation and utilization of heat in the producer. The action of steam in the producer is dealt with quite completely. In these chapters the efficiency and fuel requirements of the plants are dealt with.

The author gives a concise historical account of the development of the producer gas plant from the early date of 1669; and this is followed by a complete description of modern pressure and suction gas plants, including descriptions of the detailed construction and action of parts of producers which is not commonly given in the current literature sent out by the different firms. He also devotes some space to the description of gas cleaning operations, and the devices used by different firms for this purpose. Several chapters are devoted to the description of special uses of producer gas, such as firing of steam boilers, firing kilns, and for firing different types of melting furnaces. The operation and testing of the plants is given considerable attention, and a complete data sheet for a test is furnished. The book is completed by what the author calls a "Bibliography of Gas Producers," which consists of a chronological record of what has been written about producer gas or producer gas plants since 1841.

In view of the very considerable interest which is being taken in this subject at the present time, and that the book is written in a way which can be easily understood by anyone who would be interested in the subject, this book will be of considerable general value.

THE COMMERCIAL ORGANIZATION OF ENGINEERING FACTORIES. An exposition of modern practice with forms and precedents for the use of directors, secretaries, managers, accountants, cashiers, and all students of industrial economy; by Henry Spencer; 221 pages, illustrated with tables and blank forms; price 10s. 6d.; published by Spon & Chamberlain, Liberty St., New York.

This book is to one who is interested in the commercial side of engineering as the mechanical handbook is to the mechanical engineer. It gives useful information in concise form which will be of very great value in the formation of companies to carry on engineering work, in the arranging of methods of manufacturing, and in the systematizing of the manufacturing plant generally. It contains chapters on the filing and looking after correspondence, methods of contracting and looking after contracts, conducting of the receiving department and store room, the estimating department, advertising department, manner of looking after orders received, the drawing office, the operation of the costs department, methods of keeping track of the cost of the plant and buildings and the depreciation of the plant and estate, the forwarding department, the accountant's department, the cashier's department, the necessity of complete organization in regard to selling agents and travelers, the secretaries' department.

To a student looking for general information on this subject, this book will be of very great value; and to those having charge of work such as outlined in the contents of the book, it should prove of very considerable interest and value, the work being full of hints for such men in the conduct of their work.

AMERICAN STATIONARY ENGINEERING.—Facts, rules, and general information gathered from thirty years' practical experience as running, erecting and designing engineer. By W. E. Crane. Published by The Derry-Collard Co., New York. 285 pages, 8x5½.

This book is essentially for the practical man. As the author says in his preface, the work is a plain talk on everyday work about engines, boilers and their accessories. It is largely a record of personal experiences of the author, illustrated and explained so as to be of actual assistance to any engineer in the operation of his power plant. Scientific and mathematical discussions are avoided; and all formulas which have been used are simplified to a remarkable extent. The book is well written, and is easily read. Very few books on the market cover this subject in such a practical manner. It will undoubtedly be a popular book with the engineer; and it will also be of very great value to the young and inexperienced engineer.

CANADIAN MACHINERY
and Manufacturing News

A monthly newspaper devoted to machinery and manufacturing interests, mechanical and electrical trades, the foundry, technical progress, construction and improvement, and to all users of power developed from steam, gas, electricity, compressed air and water in Canada.

The MacLean Publishing Company, Limited

JOHN BAYNE MACLEAN - - President
W. L. EDMONDS - - - Vice-President
H. V. TYRRELL - - - Business Manager
J. C. ARMER, B.A.Sc., - Managing Editor

OFFICES:

CANADA
MONTREAL - 232 McGill Street, Phone Main 1255
TORONTO - 10 Front Street East, Phone Main 2701
WINNIPEG, 511 Union Bank Building, Phone 3773, F. R. Munro
BRITISH COLUMBIA - Vancouver, Geo. S. B. Perry

UNITED STATES
CHICAGO - 1001 Teutonic Bldg., J. Roland Kay

GREAT BRITAIN
LONDON - 88 Fleet Street, E.C., Phone Central 12460, J. Meredith McKim
MANCHESTER - 92 Market Street, H. S. Ashburner

FRANCE
PARIS - Agence Havas, 8 Place de la Bourse

SWITZERLAND
ZURICH - - - Louis Wol, Orell Fussli & Co.

SUBSCRIPTION RATE.

Canada, United States, $1.00, Great Britain, Australia and other colonies 4s. 6d., per year; other countries, $1.50. Advertising rates on request.

Subscribers who are not receiving their paper regularly will confer a favor on us by letting us know. We should be notified at once of any change in address, giving both old and new.

VOL. III. JUNE, 1907 No. 6

CONTENTS

TO OUR READERS.

We desire the co-operation of all our readers in making Canadian Machinery of the greatest educative value and interest. Each one can assist towards this end by letting us know his views concerning topics of particular or general interest, by sending us sketches and descriptions of out-of-the-ordinary jobs which he has performed, and, in fact, by letting us hear of anything that would be of interest to others like himself. For such contributions we gladly give suitable remuneration. In this way can readers make the paper of more personal interest to themselves and others.

Canadians naturally take a deep interest in seeing Canadian enterprises forging ahead. To make Canadian Machinery a paper worthy of the approbation of the Canadian mechanical world, we need the personal assistance of Canadians. We can get writers, but we wish Canadian writers and Canadian correspondents. Will you not assist us?

SPECIALIZED APPRENTICESHIP—A NOTE OF WARNING.

Specialized apprenticeship is proposed by the apprenticeship committee of the National Machine Tool Builders. At the recent convention of that association a system of specialized apprenticeship was submitted. This system is simply the training of an apprentice in a special line of work, such, for instance, as the operating of a boring mill in certain class of work, instead of along broad lines in machine shop practice. In this way a man can become productively efficient in a very short time. He can earn good wages in a shorter time than in the general apprenticeship. His value to the employer increases rapidly. The wages during the course are better than in the general course. As manufacturing economics are now, this is good for the employer. He wishes a large number of specialists. From the apprentice's standpoint the system is good. He can start his course at a reasonable wage—a wage upon which he can live; and in a short time he can earn good wages.

Then there is the general aprpenticeship. The time is longer. The wages during that time are smaller. The apprentice has his choice. If he is ambitious and can afford it, he will take the general course: If good pay in a short time is the object, the specialized course suits him. A combination of these has been tested in the Bullard Machine Tool Works, and found satisfactory. So Mr. Bullard says.

So far all seems well, but there is another side. There is a danger in this—a danger that the race of trained mechanics may degenerate into a race of machine operators or specialists. The progress of the manufacturing art is due to the work of thousands of thinking mechanics. The more machine specialists, the fewer all-round mechanics. This will have its effect. The machine operator, who is a specialist only, is lost outside his own line. His view is narrow. He is handicapped. His ambition dies for lack of means of expression. For large manufacturing plants perhaps this is good. System regards men as machines. The more like machines men are, the better for system—but the worse for the man and the worse for the mechanical trade.

But there is still another side to this, as shown in the Bullard works. There men who had been laborers became machine operators through this specialized course. Lack of opportunity, lack of ambition or lack of ability prevented them from becoming mechanics. They became machine operators—specialists. They bettered themselves. That is well.

The combination of the specialized and general apprenticeship as suggested by the committee is undoubtedly good. In very many cases it would be to the advantages of employer and employes. But let us not forget that our modern methods of specialized manufacture may be detrimental to the advancement of manufacturing methods.

In the future we may see manufactories operated like clockwork; each man performing his work as a wheel or pinion in the large machine; but we may also find but few men to think, to originate, to plan, to do their part in lifting manufacturing to a higher plane.

MORE ATTENTION SHOULD BE GIVEN PAPERS.

At the recent convention of the American Foundrymen's Association it was quite evident that the close proximity of the convention hall to the Supply Association exhibit was not advantageous. The one interfered with the other, and the convention suffered. One of the chief objects in the presentation of papers is to bring out discussion on the topic of the paper. Owing to frequent interruptions and to delay in commencing the proceedings of each session, there was not the discussion on the numerous subjects that there should have been. Close application cannot be given to the reading of papers unless the convention room is free from interruptions. For intelligent discussion there must be close application.

At next year's convention the meeting room should be arranged at some distance from the exhibits. The noise of the exhibits would not penetrate to the convention rooms, and there would be fewer interruptions than at the Philadelphia convention. The importance of the convention proceedings should be impressed upon the members. This would be influential in securing a better attendance. These objects should be worked for at next convention: Opening of sessions on time; a better attendance; no interruptions, and much more discussion.

AMERICAN FOUNDRYMEN'S ASSOCIATION HAS A WORTHY OBJECT.

When we see fourteen hundred men gather together in convention from far and near, the natural query is: Why? What do they gain? That number of foundrymen and foundry foremen gathered at Philadelphia at the recent convention. Why did they come? What did they gain? These questions are soon answered. The convention provided a means for social and business intercourse between a large number of men drawn together by the ties of a common calling. It provided a means for the interchange of ideas and knowledge of foundry practice.

Papers on subjects of great moment were read by men who have made a special study of such subjects. Opportunity for discussion was afforded. The Supply Association made an excellent exhibit of foundry machinery and supplies. This afforded a means for studying the latest appliances and ideas and comparing one with another. From the papers, from the discussion, or from the exhibits much useful knowledge and many useful ideas were stored up by the visiting foundrymen for future use. That is reason enough.

The objects of the Foundrymen's Association are purely educative. The one aim is to elevate founding to a higher plane. Since the inception of the association a great deal has been done towards fulfilling this aim.

The association is doing good work. It deserves hearty support. That Canadian foundrymen realize this is evident. Over fifty Canadians were at Philadelphia. Next year the convention will be held in Canada. Canadian foundrymen will unite to make this convention a more successful one than has been.

LESSENING THE JAR OF MACHINERY.

J. Derôme, in a recent issue of the Revue Scientifique (Paris, April 20), discusses the problem of the suppression of vibration in industrial installations. As he says, although it is not one of the most pressing questions, it is, nevertheless, one that has not yet received satisfactory solution.

Municipal authorities, as a rule, make it one of the conditions of installation that the machinery be built on a solid bed of masonry. In a number of cases this masonry has been laid in "made" ground, which, of course, does not fully meet the requirements of the authorities.

The most satisfactory solution reached is probably the recent invention of Mr. Jolivet, which depends on the following principle:

"In place of relying on the elasticity of rubber or cloth, which decreases as time goes on, the machine is placed on a series of numerous metal springs of steel, grouped in parallel between sheets of iron, the whole constituting an elastic plate or bed. The plate thus made up, whose total thickness does not exceed two inches, may be placed under the masonry foundation, or between two parts thereof, or between the machine and the foundation.

"The springs, generally of coachspring form, are, in the interior of the plate, surrounded with some insulating material such as felt, whose purpose is to damp the vibrations of the metal. The number of the springs and their individual elasticity are determined by the nature of the installation."

SOMETHING TO BOAST OF.

"A workman's proudest boast, is the pay well earned, for work well done." Take note of the implied grounds of satisfaction. "Good work—at an honest price"—a job a craftsman can look at with self-complacency, a cost at which the buyer gets a bargain. Surely the best of business foundations for a reputation, in shop, office, or out-door gang—and therefore something to be proud of; also a good profitable basis for both producer and customer. "Good work cheap at the price," has built both reputation and many strong concerns. It is a little motto Canadians might take for their industrial battle cry and if they live up to it it's a talisman that is sure to win.

Concerning Tool Rooms

A Practical Tool Room System for an average size shop, such as the writer has seen in successful operation.

By Geo. D. MacKinnon

It is the desire of the writer of this article, not to describe an ideal system which might be put into operation in some ideal shop, having ideal men in all its departments, from the manager down to the water boy—the only kind of an establishment where such a system could be put into useful service—but one combining a number of the best features of those with which he has come in contact—and the purpose of the writer will probably be better served by a description of a system suitable for a machine shop of, say, one hundred to one hundred and fifty men, of which class we have not a few in Canada.

Each system, however, must vary in detail with the shop and the class of work done; for when one is doing a very special class of work, special tools are required, and the system is not as complicated as in a shop doing miscellaneous machine work.

Best Location for Tool Room.

In the general economy of the works, the location of the tool room is an important factor; and it should be situated where it will be most convenient for the majority of the men in the shop. If the shop consists of a central bay with two side floors, the best location for the tool room will be about the centre of one of the sides, either in an annex to the main building, or else occupying a portion of one of the side floors. The former location is the more desirable.

Tool Room Should Be Enclosed.

But wherever it is located it should be entirely enclosed, so as to make it separate and distinct from the rest of the shop, for in this as much as in anything else, the efficiency of the tool room depends. There should be no access to it to the shop employe, except for good and sufficient reasons; and in some cases the writer has found it decidedly better to make the rule absolute, and prevent all but tool room employes entering. This is especially the case where the check system is in use, and the workman is held strictly accountable for his tools. This rule is for the purpose of protecting him as much as anything else.

Construction of the Tool Room.

A tool room, inside the shop, constructed with the lower portion of the walls made of matched lumber about five feet high, and with heavy wire netting above, serves the purpose as well as anything else, and has the advantage of thorough ventilation and lighting—two necessary features which all shop managers admit, but for which very many neglect to provide.

Necessary Equipment of Tool Rooms.

It is generally admitted that it is necessary to have in connection with the tool room an equipment of machines capable of taking care of all new tool work and the ordinary run of repairs; but of what this equipment should consist will depend upon the scope and organization of the tool department. In a shop such as the above the writer has found it advisable to combine several departments, which, in larger works, are usually separate.

THE STEWART CO., LIMITED
MACHINE SHOP
TOOL LIST

Name................................ ....

Reg. No............ Date...................

All blue prints, sketches and gauges must be returned to the Tool Room before 1.00 p.m. Saturday. Any man failing to return these will be fined at the rate of 5 cents for each article not returned. All damage or loss of tools will be charged to the owner of the check held as receipt unless such damage or loss is proved to have been unavoidable.

Each workman will be held responsible for the following list, at the prices annexed :

NO.	NAME	PRICE
	Bolts	15
	Brush	15
	Carriers	30
	Centre punches	.05
	....	....
	....	....
	....	....
	....	....
	....	....
	....	....
	Tools, Lathe	25
	Wrenches	35

Other tools needed can be obtained from the Tool Room on leaving a check as receipt.

All Repair and Tool Room Work Under One Foreman.

These are, besides the small tool department, the machine and motor repair department, belt fixing gang, and shafting oilers; and it is well to have all this work done under the superintendence of a competent foreman, who will also have charge of tool shop. In this way one man is responsible for the running of the machines, and whenever it may be necessary to take down, say, a countershaft to rebush pulleys, rebabbitt boxes, etc., on account of lack of oil, he knows just where to place the responsibility. This is also true in the case of belt repairs; and it is not then necessary to look all around the shop for the one who is supposed to attend to that work, with the consequent loss of time of both the machine and the operator. He is located at once from the tool room as he will naturally work under the orders of the tool room foreman, who will know of his whereabouts.

With such a scope and organization, machines suitable for the accommodation of the different classes of tools and repairs peculiar to the shop, are readily decided upon.

These should be located in one end of the tool room, and partitioned off by wire netting from the tool serving department.

Run Tool Room With Separate Motor.

If electric power is in use throughout the shops it is desirable that the tool shop be run with a separate motor, so that repairs may be made here after hours without running, possibly, a long line of shafting to accomplish the result.

Should be Toolsmith in Tool Room.

In connection with the tool shop the writer has found that it not only is a great convenience, but is economical besides, to have a tool-smith in this department; then all new forged tools can be made here, and all tempering, hardening, etc., be done. In a shop of this size one smith and a helper are able to take care of all such work, and, being under the charge of a responsible foreman, there is a minimum of delay in filling rush tool orders—a very important item in shop economy.

Tool Grinding Done in Tool Room.

Whether the machine equipment will comprise universal tool grinding machines will depend upon whether the work being done in the shop is of a special class permitting a large number of parts to be put through at one time on a piece-work rate, or whether it is of a varied class done by day labor. If the former, a tool grinding machine is usually a necessity, for men do not care to stop—and in fact should not be required to shop—their machines to grind tools, with the consequent delay in production. But in the latter case they usually prefer to grind their tools to suit themselves; and I have often seen a machine hand going direct to the emery wheel, with a most carefully ground tool, which he had just procured from the tool room, simply because it did not measure up exact to his own idea of rake and clearance.

Detail Equipment of Tool-Room.

Having then decided upon the tool shop and the scope of the tool department, it is an easy matter to decide upon the equipment of the tool serving section of this department. This equipment is one which should be as simple as it can possibly be made. It's true that in some shops, doing a very high grade of work, very carefully fitted compartments are frequently necessary for the proper storing of special and expensive tools, jigs, gauges, etc.; but when it comes to having a special slot in a special case for a certain ordinary machine bolt—such, as I have seen arranged—with other things in proportion—then the working of the system is going to clip a little more off the net profits of the concern. The simpler we can arrange our tools, so that they can be selected and served to the workman with the least possible delay, the better the system will be.

Combined Check and Charge System Best.

It has been the writer's experience that a combined check and charge system for the issuing of tools is desirable, and this is carried out in this way: Whenever a man engages to work in the shop, to whom it is necessary to serve tools, a tool list is made out for him. This is numbered either with his time register number or a special tool room number. The former is to be preferred, if it can be so arranged with the time department. This list is as per Fig. 1, and is filed away in its numerical order. A bunch of, say, twelve brass checks, bearing the same number as the slip, is given to him, together with a duplicate tool list; and these checks are charged to him on these lists.

Then, should the man desire a number of standard tools which he is to retain for an extended period, these, too, are charged to him. His bunch of checks is then available for special tools. It is desirable that the number of checks in the tool room belonging to each man be kept as small as possible, as there is then much less liability of their being misplaced and lost.

With this system it is necessary that when a man is quiting work, and before he is paid off, he must secure from the tool room foreman a clearance note, which is addressed to the time-keeper and which states whether he has or has not returned all his tools. If he has settled satisfactorily with the tool room the note is marked O.K.; but if not, the missing tools are noted, and charges made for them from standard charge list—which amounts are deducted from any moneys due him.

Of course it is not to be supposed that every man can return all the tools he has worked with for months past, and a good deal of discretion must be used in such cases so that no injustice will be done the men.

(To be continued).

Specialization in Apprenticeship Courses

Two Systems of Apprenticeship Proposed by the Committee on Apprenticeship at the Machine Tool Builders' Convention—one General and one Special.

Two systems of apprenticeship, one system, general, covering the entire machinists' trade, as apprenticeship has always done, and the other a special system, covering a single branch of the trade, were proposed by the committee on apprenticeship at the convention of the National Association of Machine Tool Builders, at Fortress Munroe, May 14 and 15. At the present time there is the demand for specialists in the machine shop, and this has led to the testing of this special apprenticeship course. The report of this committee was brought in by E. P. Bullard, Jr. This special system of apprenticeship is the result of experiments tried in the Bullard Machine Tool Works.

The idea at the base of the special apprenticeship system is that in the present condition of the machine-manufacturing industry the chief demand is for specialists, and also that, by reason of the shortened time in which a man may become proficient in a single branch of the trade, it is possible to pay materially higher wages to special than to general apprentices, and thus attract many young men who must have a living wage, but who cannot live on the wages which can be paid to general apprentices.

With both systems it is proposed to supply the apprentice with a set of tools at the beginning of his apprenticeship, which are to remain the property of the employer until the completion of the term, at which time they are to be given to the apprentice.

The apprentices at the completion of their term should be presented with diplomas, naming the shop in which they have worked and the special branch of the trade to which they have been apprenticed.

Wage Rate to Apprentices.

The length of the general apprenticeship course was recomemnded to be three years. The wages paid during those three years was to be 8, 10 and 13 cents per hour respectively. In the special apprenticeship course the time would be shorter, and the wage rates higher, these varying with the class of work.

Thus, for turning and planing the course is to cover two years, the rate of wages advancing each six months, and being 12, 14, 16 and 18 cents. For the vertical and horizontal boring mill the term is to be 1½ years, the rate advancing at the end of each six months, and being 15, 17 and 20 cents. In the case of drilling and milling, the term is to comprise a single year only.

In the Bullard Works, Mr. E. P. Bullard says it has been found feasible to train very efficient men by this specialized system of apprenticeship; and it is possible to greatly increase the earning capacity of such apprentices in a very short time.

Some of those taken in at the Bullard Works in this special course were beyond the usual age of apprenticeship, the applicants being men of natural ability, but working as laborers only, and a good proportion of them have increased their earning capacity threefold. In other cases it has been possible to obtain farmers' sons who were unable to take the general course, because of the lower wages, but who, by reason of the self-sustaining wages offered at the beginning of the special courses, have been able to take those courses and have done so.

The systems are so arranged that an apprenticeship may be transferred from the special to the general course, with suitable credit for the time served. In the Bullard shop in some cases young men had begun with a special course, saved up money from their wages, and then taking the general course, living for a time partially upon their savings.

Mr. Bullard's idea was that eventually the general apprenticeship course would become largely a means of training men and traveling representatives.

Producer Gas, Its Manufacture and Use

Relative Cost of Steam and Gas Compared

By J. R. Armer, Grand Trunk Railway, Toronto.

At the last meeting of the Central Railway and Engineering Club of Canada. Mr. J. R. Armer gave a most instructive paper on "Producer Gas and its Manufacture." This paper is long, and a few of the outstanding features have been abstracted.—(Editor.)

After telling of the manufacture of different kinds of gas, including coal gas, used for illuminating and other purposes, and blast furnace gas, the writer of the paper went right into the subject of the manufacture of producer gas.

The methods of manufacturing producer gas may be classified under two headings, according to the way in which the draft is created, namely, induced or forced.

Fuels Used in Gas Making.

Almost any carbonaceous fuel may be used in the manufacture of producer gas, including charcoal, coke, hard coal, lignite, peat, wood, sawdust and even city refuse. If coal is used, it should be of good quality. It should have a low percentage of ash, and should not clinker. In case the gas is to be used in gas engines or in the manufacture of iron, it should contain a low percentage of sulphur. The coal should be as nearly uniform in size as possible, and not too large. The fuels used in suction producer gas plants, which are the most common in small units, are usually coke or anthracite coal. These fuels do not clinker, and they produce a gas containing a low percentage of tar. Lignite and peat have been used in Europe in pressure producers, and the gas has been used for gas engines in many instances.

Some of the devices used for cleaning are as follows: liquid scrubbers, rotary scrubbers, deflectors and absorbers. The abstraction of tar is one of the greatest difficulties met with in cleaning producer gas, and as its elimination is absolutely necessary for the best results in gas engine practice, the means adopted for its removal are very important considerations.

Where there is an excessive amount of tar some builders use rotary scrubbers, or tar extractors, in addition to the wet scrubbers. They claim very satisfactory results. Other makers, instead, advocate having the fires so arranged that the tendency is for the tar to be converted into a permanent gas in the producer. To accomplish this the products of distillation are passed through a hot bed of fuel.

One of the important points about a producer gas plant is the short time required in starting, and the little attention required when it is standing idle; even after several hours' idleness the plant may be started and be working at its full capacity in a few minutes.

The labor required to operate with a pressure producer plant is about the same as that required for a steam plant of similar size, while with a suction producer plant the labor required is much less.

The amount of water used in a producer plant is not more than half that required in a steam plant of similar size. If the cooling water, used in the scrubbers, be used over and over again this will be greatly reduced.

No more skill is required to handle a producer gas power plant than is needed for a steam plant of similar size; but it may require some time for men trained to handle steam plants to become accustomed to gas engines and gas producers.

Operation of Producer Gas Plants.

For best results, plants should be kept clean; pipes should be kept gas tight, and arranged so that they may be cleansed of dust and tar; scrubbers should be inspected periodically, to see if the coke or other material is clean enough to assure efficient working. They should be cleaned once in six or eight months, according to conditions.

The fire should be drawn from the producer about once a week, and the clinkers removed; if the fires are rekindled immediately so that the fire brick lining is not chilled, the life of the lining will be prolonged considerably.

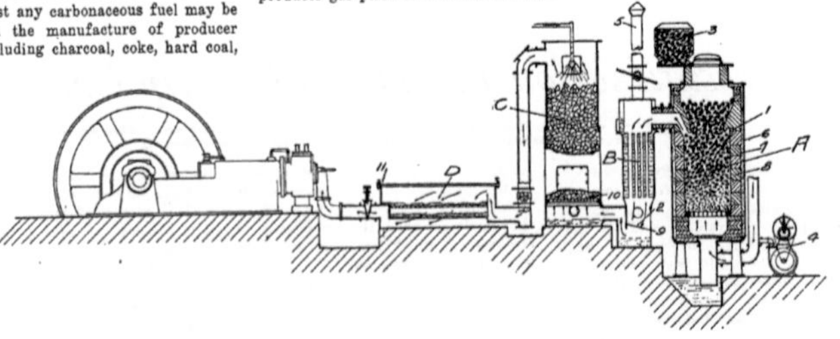

TYPICAL SUCTION GAS PLANT

The regulation of the steam supply to the producer is of considerable importance. In pressure producers the regulation is not very difficult, but in suction producers, where the air supply in the producer depends upon the load on the engine, it requires special attention. The amount of steam going into the suction producer must always be proportional to the amount of gas used by the engine, so that the fires will neither become too hot nor be dampened. Different firms have different methods of regulating this.

Gas Power Plants vs. Steam Power Plants.

In the case of the gas producer, the gas is conveyed to the engine cylinder, where almost perfect combustion is obtained. The losses in this case are occasioned mainly by the heat radiated from the producer and connecting pipes, in the heat lost in cooling the gas, in cooling the engine cylinder, and in that lost in the exhaust from the engine. The amount of energy lost in this way compared to the amount originally in the coal is very large; the proportional loss, however, is much smaller than in a steam plant.

In a steam plant all the combustion takes place in the fire box, under poorly regulated conditions, an excessive flow of air always being required which carries a large proportion of the heat up the stack, together with unburnt gases, and particles of carbon in the form of smoke. The remaining heat is absorbed by the boiler in generating steam, or is radiated from the fire box; further losses occur through the radiation from the steam pipes and engines, and finally there is the large loss from the exhaust on the engine.

The following heat losses in a modern steam plant, when compared with the losses in the gas plant, will give an idea of where the greater efficiency of a gas power plant, over a steam plant, comes in: Out of 13,500 B. T. U., the value per pound of coal used in the fire box, 135 B. T. U. were lost in ashes, 675 in radiation from the boiler, 2,970 in chimney loss, 190 in auxiliary exhaust, 520 in radiation from steam pipes and engine, 7,737 in the exhaust from the condenser, 1,273 were converted into power. Considering the mechanical efficiency of the engine as 93 per cent., this would give 1,176 B. T. U. as delivered from the engine in B. H. P. By comparing 1,176 B. T. U. with the original 13,500 B. T. U., the efficiency of the whole plant is 8.7 per cent.

As a comparison with this, the following figures have been taken from a report on tests made with the Crossley suction gas plants and gas engines by consulting engineers and others, and read before the Manchester Association of Engineers in connection with a paper on this subject.

In the tests reported, the thermal efficiency of the plants ran as high as 24.95 per cent. One of these tests was made with a plant at Milford-on-Sea Electric Works, which at the time of making the test had been in operation for some months. The efficiency of this plant was low in comparison with some of the results obtained, as will be seen, so that the comparison is conservative. The fuel used was anthracite coal having a heating value of 14,895 B.T.U. per lb. The indicated horse power of the engine was 46.25 I.H.P. The test lasted 3 hrs. 49 mins. The total amount of coal used was 133.5 lbs.; the coal used per I.H.P. hour was .785. Assuming that the mechanical efficiency of the engine was 85 per cent. the amount of coal per B.H.P. would be .89 lbs. The value in B.T.U. of .89 lbs. of coal would be 13,256 B.T.U., which when compared with 2,545, the B.T.U. value equivalent to 1-H.P., gives the efficiency of the whole plant including the mechanical efficiency of the engine as 19.35 per cent. If the losses in this case be divided up according to the usual proportion of loss in the different parts of the suction producer power plant they will be as follows: Out of 13,256, the B.T.U. value of the fuel used, 2,651 are lost in ashes, radiation, cooling of gas, etc., 4,445 in exhaust and 3,166 in cooling the cylinder; the remaining 2,994 are converted into power; the efficiency of the engine being 85 per cent., 2,545 of these are converted into B.H.P.

In comparing these two plants, the efficiency of the steam plant being 8.7 per cent., and that of the gas plant being 19.35 per cent., we find a saving of 55 per cent. in favor of the suction producer power plant. In neither of these cases has allowance been made for standby losses, which favor the steam plant.

Regarding fuel required per H.P., manufacturers make numerous claims, many of which have been substantiated. The claims average about 1 lb. of anthracite per H.P. per hour, and vary from 0.6 to 1 1-3 lbs. per B.H.P. per hour. The 0.6 figure is pretty low, but numerous tests have shown a consumption as low as .8 lbs. anthracite per B.H.P. per hour.

Considering the high thermal efficiency and the many other points of advantage about a producer gas power plant, its outlook for the future is very bright. It has passed the experimental stage and has become a formidable competitor of all classes of power. Its field of usefulness will doubtless be widened greatly and already it is being considered as a means of propelling vessels, driving traction engines and for developing power under many other conditions.

Discussion on Paper.

In the discussion of this paper several points of interest were brought out. Mr. Samuel Groves said, that the pioneers in this country in this work were the French Canadians. To his knowledge over ten suction gas plants had been installed in Quebec. In England they are ahead of both Canada and the United States in the utilization of the gas engine. The United States, however, have beaten the world in the use of the gas producer. The best practical results in Canada have been achieved at Wellington, Prince Edward County, Ont. There, with a 65 h.p. suction gas plant, electrical energy is being supplied at $20 per h.p. year, 10 hr. day; 18 arc lamps for street lighting, at $530 per annum. The stores and residences pay one-half cent per night for continuous incandescent light. Mr. Groves spoke of the electric furnace as having a bright future; and brought out the fact that the waste gases produced in the electric furnace are comparatively pure, and are, therefore, very suitable for use in the gas engine. He predicted that the suction gas producer system would play an important part in opening out and developing industries in the great Northwest, were lignites abound.

Mr. R. J. Goudy pointed out that the economy of the gas engine operated by producer gas lies in the almost direct conversion of the energy in the coal into work. He said that the United States navy were adapting the gas producer and gas engine in naval work.

Mr. J. J. Fletcher said that he was opposed to gas producer power plants simply because it was against his profession; he is a steam man. If the gas producer power plants are to be our source of power, he would have to learn the profession over again in the direction of gas producer power plants.

RICE LEWIS & SONS' MACHINERY DEPARTMENT.

Rice Lewis & Son's Co., Toronto, have started a new department in their business, a machine tool and general machinery department. In branching out into this line, the company were of the opinion that there was a large field for such an undertaking. They were of the opinion that there should be closer relationship between the buyer and the seller. They have arranged so that intending purchasers may obtain reliable information as to the best machine for their purpose. Their salesmen will be men well acquainted with machinery which they sell, so that they may advise as to the best machine for the work required of it, should the purchaser desire such information.

The company have secured a number of good agencies, to which they will be continually adding. At present their chief lines will be machine tools and power machinery. Later they will handle woodworking machinery. They will stock a sample line of machinery. It will be their object to make deliveries as quickly as market conditions will allow. New machinery will receive their chief attention.

This new department is in charge of Mr. Rossiter Kellogg.

In this new undertaking the company have behind them the reputation which they have gained in lines more or less closely related to machinery; and their new department will be conducted in a way befitting this good reputation.

Foundry Practice

MODERN IDEAS FOR THE TRADE.

Modern Foundry Practice

By Samuel Groves.—Article II.

After Nasmyth's ladle, in 1838, came Jackson's invention of his gear-moulding machine in 1855. Just as the worm-geared ladle was the initial step towards introducing applied mechanics into the foundry, with a view of the labor-saving in the handling of material; so, the advent of the spur wheel moulding machine marks the transition from crude, rule-of-thumb practice to more scientific founding, and greater economy in the production of castings.

Jackson's Gear-Moulding Machine

for the specific purpose of moulding toothed wheels was a stroke of genius; since it represented not only an immense advance in foundry practice, as we shall see, but supplied a long-felt want. In those days, as now, toothed wheels were in universal demand in workshop, factory, mill, and mine. The first cast-iron gear wheel was made at Coalbrookdale, Staffordshire, England, in 1782. From that time until 1855, the conventional method of manufacturing was to mould them from full patterns;

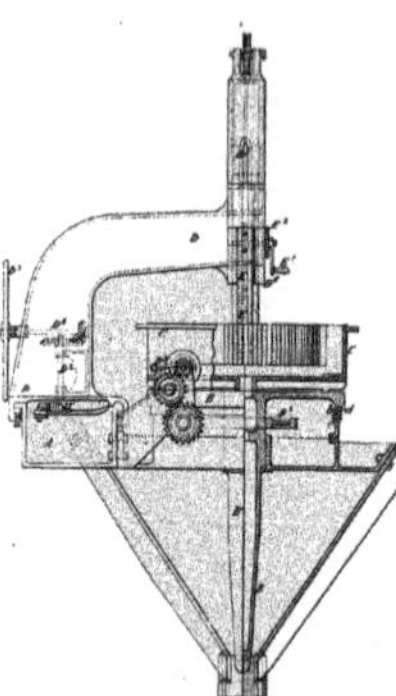

Fig. 5—Jackson's Gear Moulding Machine.

but it had been found that the constant rapping and soaking with water caused the patterns to warp and get out of shape. Moreover, with a view of aiding the withdrawal of the pattern from the sand, it was found necessary to allow draft on the face—leaving in the casting a permanent defect; for a gear with tapered face is wrong mechanically; and many accidents were due to this cause. Another serious defect of this hand-moulding system, was the irregularity in the pitch of the teeth—caused by patching; for experience proved that in spite of rapping, water soaking, and draft, it was almost impossible to lift a many toothed pattern from the sand without bringing away lumps of facing from the tooth moulds.

A gear wheel not true in diameter and face, and with teeth unequally spaced, is practically useless as a mechanical appliance. Under the old regime, it

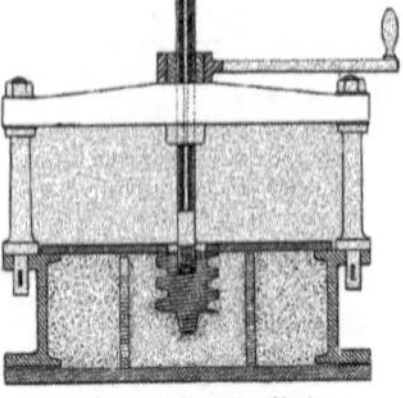

Fig. 6—Worm Moulding Machine

was well nigh a hopeless task to turn out a gear wheel without one or other of the defects indicated; hence, there was general dissatisfaction with the foundries. But the coming of Jackson's moulding machine brought about a complete revolution; since it provided an effective remedy for the defects inherently associated with the prevailing system of moulding (1) by rendering obsolete, costly full patterns; (2) by doing away with draft; and (3) by guaranteeing perfect concentricity, and equal spacing of teeth. And yet, notwithstanding these manifest advantages, so conservative were the owners of foundries generally, that in 1892 the engineering concerns in Great Britain and the United States having gear-moulding machines, could be counted on one hand; while in Canada there was not one. About the year 1896, however, there seemed bright prospects of a widespread adoption of gear-moulding machines;

Fig. 7—Stripping Plate and Drawback Machine.

when all at once came a startling counter action, similar to another phenomenal change in engineering practice which took place a year or two earlier.

At the beginning of the last decade, 1890, it seemed as though cable railways and rope drive traveling cranes were firmly entrenched, when almost without warning electricity came in, and swept away in two years, both systems. With like suddenness came in high-grade tool steels, together with milling and gear-cutting machines, ensuring perfect gears having faces parallel to the axis, and teeth spaced with contoured with mathematical regularity. This innovation enabled the factory, rolling mill and steelworks engineer to design machines and appliances on a scale of strength and magnitude undreamed of previously; for they could now use large steel gears having the precision and exactitude of clockwork, instead of the coarsely-moulded specimens produced from full pattern, or cores in the foundry. As a consequence, the moulding machine as applied to the making of gear wheels, received a setback; it almost seemed as though its days were numbered. Time, however, has proved—even in the case of gears to be cut—the utility of moulding the forms by machinery—especially those of magnitude. Steel has not displaced cast-iron in the manufacture of gears—except in the case of high powers; and from the signs of the times is not likely to do so. In both cases, particularly the latter, there is a strong reaction in favor of the gear-moulding machine. The reasons

Fig. 8—Universal Gear Moulding Machine.

are obvious. Even where the teeth are to be cut, it is better not to cast a solid rim, since in so doing initial stresses are set up in the neck of the arms next the rim, due to unequal shrinkage, caused by lack of uniformity in section. Besides, in cutting the teeth out of the solid rim, the hard, close-grained, combined carbon next the skin is taken away, leaving the coarse grained softer metal as a wearing surface. In 1896 the writer made a plea in favor of this view, before the American Foundrymen's Convention at Philadelphia.

Now, even firms making a specialty of gear cutting, are using moulding machines, and instead of casting solid rims in their gears, are leaving a film of metal around the teeth contour, thus avoiding unequal shrinkage, and at the same time getting the advantage of the close-grained metal as a wearing surface—in this way increasing the life of the gears. In the United States and Great Britain, there is hardly a steel foundry in existence without a series of gear-moulding machines. Another advantage of this system is, that the simple pattern rig for hundreds of gears can be stored in a comparatively

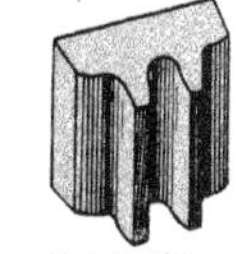

Fig. 9—Tooth Block.

small building; for all that is required is a tooth block, and two core boxes for arms and centre core, whereas the area needed for an equivalent number of full patterns would cover an acre of ground, be an expensive first cost and would depreciate very rapidly.

Inasmuch as Canada is destined for great industrial development, and gear-moulding in both cast-iron and steel, is likely to play an important part in the "good time coming," it may be deemed an opportune moment to set forth with as much brevity as is consistent with clearness, the principles and latest practice in the moulding of gear wheels by machinery.

Moulding a Worm.

The common method of moulding a worm is, to form the mould by means of a pattern made in halves. Now, these pattern halves can not be withdrawn from the sand direct, owing to the overhanging of the thread; they must be twisted out. Anyone who has tried to do it knows how difficult it is —patching and mending invariably follow. This trouble, together with the enlargement of the mould due to rapping,

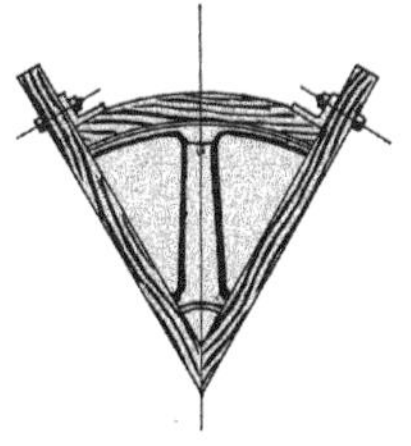

ARM CORE BOX.

Fig. 10.

added to the unsightly fin which appears on the casting at the parting line of the pattern, renders this old plan altogether unsatisfactory. But the machinery moulder has completely overcome these difficulties by bringing to his aid a simple mechanical appliance (Fig. 6).

A brass pattern is chased in the lathe, leaving the necessary allowance for contraction. This pattern, with tapered core print at lower end, and tapped hole in top for receiving lifting rod, is set vertically in the flask, hand packed with suitable facing, rammed around and scraped off. This done, a plate having a hole in the centre of the same shape and rake as thread of worm, and also a cross bow with guiding gland in middle, is secured to top of flask. The lifting rod, with left-handed screw, is then inserted into the pattern, and the pattern screwed out of the mould up through the slot in cover plate; the mould of each thread being self-slicked in the process of withdrawal. The cope is then put on, rammed, gated, and the mould poured. Thus, by a simple method of machine moulding, duplicate worms, perfectly smooth and true, can

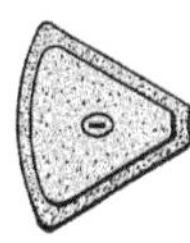

Fig. 11—Green Sand Arm Core.

be cast in greatly increased quantities and at comparatively small cost.

Moulding Smaller Gears and Pinions.

It would not pay to mould pinions and smaller gears required in quantities on a revolving table machine like that shown in Fig. 7; the process would be altogether too slow. Neither is it economical to mould the blanks in the foundry and cut the teeth in the machine shop. A more satisfactory plan is to prepare a cut toothed iron brass, or aluminum pattern, and mould from same on a simple, draw-back mould forming machine, similar to that shown in Fig. 7.

With scientifically selected sands and facings, and care in the operation of the machines, a laborer can make these gears almost as fast as a skilled moulder can turn out the blanks. The castings will be superior to cut gears; for while the flanks of the teeth are smooth and clean, and the pitch true, they will have the additional advantage of the hard skin of combined carbon as a wearing surface.

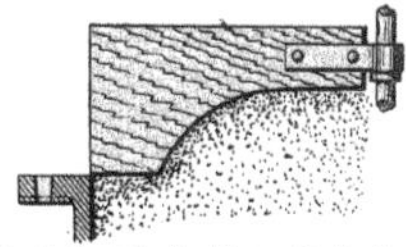

Fig. 12—Sweep for Bevel Reverse Mould: Cope.

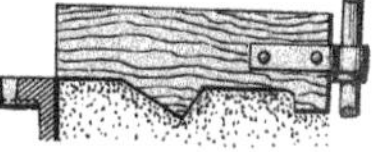

Fig. 13 Sweep for Bevel Bed Drag.

Universal Gear Moulding.

Gears over 12 inches and up to 72 inches in diameter, are generally made on a revolving table machine like Fig. 8;

with either straight or bevelled faces; and with either external or internal teeth, the shape of which may be either spur, staggered, helical or worm. Supposing a wheel 40 inches in diameter, having 84 teeth 1½-inch pitch and 4½-inch face is required, a full pattern would take at least five days to make, and cost from $20 to $25; whereas, all that would be required for machine moulding would be a single tooth block pattern (Fig 9); a sweep for tooth backing; and a radial box for green sand arm cores. (Fig 10). These latter would take from 1½ to 2 days in the making, and cost less than one half the price of a full pattern. The centre core in both cases would be made in a standard core box. From this the reader will readily perceive, that by the machine moulding process, the gear could be made and installed before the full pattern was ready for the sand—a manifest advantage in the matter of breakdowns in factories.

The "modus operandi" of moulding a 40-inch wheel on the "universal" machine is as follows: The cast-iron flask with planed points is placed on the table and carefully levelled, so that it is perfectly horizontal. The bottom plate is then covered with a layer of coke, for venting, and covered with floor sand which is well rammed and vented. The inside of the perforated flask body is then daubed with clay wash and packed with floor sand; supplemented with a refractory facing of a grade suited to the general metal section of the gear. By means of a sweep with a radial face piece and judicious ramming, a substantial wall of sand is left, forming a backing upon which the series of tooth space moulds are to be built. This done, the tooth block pattern, having been adjusted properly, is lowered into the mould, so that it touches both the bottom and backing. Thereupon, the moulder with a bucket of selected facing near him, sifts the sand into the pattern cavity with one hand, and rams, by means of a short rod of iron, with the other; taking care to firmly secure the tooth space mould to the backing by means of nails. After pushing a vent wire down through the tooth into the coke below, soaking the top with water and sleeking with a trowel, the leather on top of pattern is gently tapped with a wooden mallet; then the moulder takes hold of the hand wheel behind the pattern, gives a half turn and the tooth block is withdrawn inwards towards the centre of the flask, clear of the point of the tooth space mould. Without delay, the operator lifts his hand to the star handle connected to the rack and pinion on the balanced vertical spindle above, gives a few turns and the tooth block pattern is hoisted out of the flask. The pattern is then cleaned. Whereupon, the moulder steps to the left, gives a certain number of turns to the handle on the index plate, and the table revolves just enough to give the requisite pitch of the teeth. Straightway, the tooth block is lowered again into the mould, and by means of the wheel on the eccentric drawback device, is pushed forward to the mould backing —only a distance of one pitch away from the inter-tooth space form already made. The process of ramming, etc., is repeated and repeated, until the inner circle of the mould is completely filled with projecting tooth space forms. The flask is then carried into the core oven for drying. So far, provision has only been made for moulding the teeth; we now come to the rim, arms and hub.

In the best modern practice, the arms of gears are made in green sand (see Fig. 11); designed for arms of I shaped section. Seeing these cores have to be hung in the cope, it is necessary to build up the core on a grid, in which is cast a lifting hook. The central part is made of coarse gravel sand, and cinders, from which small diagonal vents are carried to a main vent in the top. These sectional cores are skin dried in the core oven, or by means of the portable drying apparatus, and coated with a refractory solution; the aim being to prevent the burning of the silica in the green sand, and at the same time to ensure a smooth and clean surface on the casting.

The cope for a spur gear has a plain joint surface; the only exceptions being the apertures for centre core, arm core lifting rods, overflow and runner gates, hence, is generally made on a striking plate adjoining the machine. The runner gate is invariably placed around the centre core, with two or three inlet holes, the object being, to deliver the molten metal into the hub, from whence it flows out through all the arms simultaneously to the rim and teeth, thus ensuring a more homogeneous casting.

Gears with staggered, helical, or worm teeth are all made in like manner; but bevels are made somewhat differently. The drag is filled with hard rammed sand and a form coincident with back of curved arms is swept as per Fig. 12, and coated with parting sand. The cope is then put on, rammed with sand, vented and removed. The sand in drag which formed the reverse mould, is then dug out, and by means of another sweep (Fig. 13), the lower face and backing upon which the inter-tooth space moulds are to be built by machine is formed. The bevelled tooth block is then adjusted, and the operation of machine moulding the teeth begins; the process being similar to that of spur tooth moulding, except that instead of the pattern being withdrawn horizontally, it is hoisted vertically: otherwise the processes are identical.

To the uninitiated, the operations described may seem somewhat complicated; in actual practice, however, the work is comparatively simple. Having made one of each of the various kinds of toothed wheels on the machine, subsequent work is only duplicate as regards method; hence, any intelligent, conscientious moulder, gifted with patience and a capacity for exact work, can soon become expert.

Standard Methods for Analysis of Iron

Report of Committee of American Foundrymen's Association

At the annual convention of the association in 1905, this committee reported a method for the determination of silicon in iron, and last year added methods for determining total carbon and sulphur. The committee now adds methods for determining graphitic carbon, manganese and phosphorus, thus including all the determinations usually made on iron in which occasion for difference between the buyer and seller is apt to arise. The report will therefore include all the methods decided upon.

We would also like to call attention to the following quotations from the report of 1905, which indicated the intentions of these methods.

"In recommending the above method, it was recognized that it is almost an impossibility to get chemists to use a standard method in their daily work. Hence the above method, as recommended, is intended primarily as a check method in case of dispute between laboratories, or as between buyer and seller.

"Hence a method, accurate in every point was sought, shortness being sacrificed to some extent to insure accuracy or the chance of error by a careless operator. Little in the above is left to the judgment of the chemist.

"It will be further recognized that in the purchase and sale of pig iron or castings under specification, that standard methods are essential in order to allow the parties of both parts to make their determinations with the assurance that, on the score of method, they are on the same footing."

We wish also to emphasize the ideas involved in the selection of members of this committee—that is to have representatives from commercial laboratories as well as works chemists, and to have members from different sections of the country. The success in carrying out these ideas will be seen by the following list of members :

Andrew A. Blair, Booth, Garrett & Blair, Philadelphia.
H. E. Diller, chemist, Western Electric Co., Chicago.
H. E. Field, metallurgical engineer, Macintosh, Hemphill & Co., Pittsburg.
R. F. Flinterman, chief chemist, McCormick division, International Harvester Co., Chicago.
Allan P. Ford, metallurgist, Eaton, Cole & Burnham Co., Bridgeport, Conn.
J. O. Handy, chief chemist, Pittsburg Testing Laboratory, Pittsburg.
J. R. Harris, chief chemist, Tennessee Coal Iron & Railroad Co., Birmingham, Ala.
H. C. Loudenbeck, chief chemist, Westinghouse Air Brake Co., Pittsburg.
R. S. MacPherran, chief chemist, Allis-Chalmers Co., Milwaukee.
W. G. Scott, chief chemist, J. I. Case Threshing Machine Co., Racine, Wis.
Henry Souther, Henry Souther Engineering Co., Hartford, Conn.
Prof. Thomas B. Stillman, Stevens Institute of Technology, Hoboken, N.J.

Determination of Silicon.

"Weigh one gramme of sample, add 30 c. c. nitric acid (1.13 sp. gr.); then 5 c. c. sulphuric acid (conc). Evaporate on hot plate until all fumes are driven off. Take up in water and boil until all ferrous sulphate is dissolved. Filter on an ashless filter, with or without suction pump, using a cone. Wash once with hot water, once with hydrochloric acid, and three or four times with hot water. Ignite, weigh, and evaporate with a few drops of sulphuric acid and 4 or 5 c. c. of hydrofluoric acid. Ignite slowly and weigh. Multiply the difference in weight by 4702, which equals the per cent. of silicon."

Determination of Sulphur.

Dissolve slowly a three gram sample of drillings in concentrated nitric acid in a platinum dish covered with an inverted watch glass. After the iron is completely dissolved, add two grams of potassium nitrate, evaporate to dryness and ignite over an alcohol lamp at red heat. Add 50 c. c. of a one per cent. solution of sodium carbonate, boil for a few minutes, filter, using a little paper pulp in the filter if desired, and wash with a hot one per cent. sodium carbonate solution. Acidify the filtrate with hydrochloric acid, evaporate to dryness, take up with fifty c. c. of water and two c. c. of concentrated hydrochloric acid, filter, wash and after diluting the filtrate to about 100 c. c. boil and precipitate with barium chloride. Filter, wash well with hot water, ignite and weigh as barium sulphate, which contains 13,733 per cent. of sulphur.

Determination of Phosphorus.

Dissolve two grams sample in fifty c. c. nitric acid (sp. gr. 1.13), add 10 c. c. hydrochloric acid and evaporate to dryness. In case the sample contains a fairly high percentage of phosphorus it is better to use half the above quantities. Bake until free from acid, re-dissolving in twenty-five to thirty c. c. of concentrated hydrochloric acid, dilute to about sixty c. c., filter and wash. Evaporate to about twenty-five c. c., add twenty c. c. concentrated nitric acid, evaporate until a film begins to form, add thirty c. c. of nitric acid (sp. gr. 1.20) and again evaporate until a film begins to form. Dilute to about 150 c. c. with hot water and allow it to cool. When the solution is between 70 degrees and 80 degrees C. and fifty c. c. of molybdate solution. Agitate the solution a few minutes, then filter on a tarred Gulch crucible having a paper disc at the bottom. Wash three times with a three per cent. nitrate acid solution and twice with alcohol. Dry at 100 degrees to 105 degrees C. to constant weight. The weight multiplied by 0.0163 equals the per cent. of phosphorus in a one gram sample.

To make the molybdate solution add one hundred grams molybdate acid to 250 c. c. water, and to this add 150 c. c. ammonia, then stir until all is dissolved and add 65 c.c. nitric acid (1.42 sp. gr.). Make another solution by adding 400 c. c. concentrated nitric acid to 1100 c. c. water, and when the solutions are cool, pour the first slowly into the second with constant stirring and add a couple of drops of ammonium phosphate.

Determination of Manganese.

Dissolve one and one-tenth grams of drillings in twenty-five c. c. nitric acid (1.13 sp. gr.), filter into an Erlenmeyer flask and wash with thirty c. c. of the same acid. Then cool and add about one-half gram of bismuthate until a permanent pink color forms. Heat until the color has disappeared, with or without the precipitation of manganese dioxide, and then add either sulphurous acid or a solution of ferrous sulphate until the solution is clear. Heat until all nitrous oxide fumes have been driven off, cool to about 15 degrees C.; add an excess of sodium bismuthate—about one gram—and agitate for two or three minutes. Add fifty c. c. water containing thirty c. c. nitric acid to the liter, filter on an asbestos filter into an Erlenmeyer flask, and wash with fifty to one hundred c. c. of the nitric acid solution. Run in an excess of ferrous sulphate and titrate back with potassium permanganate solution of equal strength. Each c. c. of N-10 ferrous sulphate used is equal to 0.10 per cent. of manganese.

This determination requires considerable apparatus ; so in view of putting as many obstacles out of the way of its general adoption in cases of dispute your committee has left optional several points which were felt to bring no chance of error into the method.

The train shall consist of a pre-heating furnace, containing copper oxide (Option No. 1) followed by caustic potash (1.20 sp. gr.), then calcium chloride, following which shall be the combustion furnace in which either a porcelain or platinum tube may be used (Option No. 2). The tube shall contain four or five inches of copper oxide between plugs of platinum gauze, the plug to the rear of the tube to be at about the point where the tube extends from the furnace. A roll of silver foil about two inches long shall be placed in the tube after the last plug of platinum gauze. The train after the combustion tube shall be anhydrous cupric sulphate, anhydrous cuprous chloride, calcium chloride, and the absorption bulb of potassium hydrate (sp. gr. 1.27) with prolong filled with calcium chloride. A calcium chloride tube attached to the aspirator bottle shall be connected to the prolong.

In this method a single potash bulb shall be used. A second bulb as sometimes used for a counterpoise being more liable to introduce error than correct error in weight of the bulb in use, due to change of temperature or moisture in the atmosphere.

The operation shall be as follows : To one gram of well mixed drillings add 100 c. c. of potassium copper chloride solution and 7.5 c. c. of hydrochloric acid (conc.). As soon as dissolved as shown by the disappearance of all copper, filter on previously washed and ignited asbestos. Wash thoroughly the beaker in which the solution was made with 20 c. c. of dilute hydrochloric acid (1.1), pour this on the filter and wash the carbon out of the beaker by means of a wash bottle containing dilute hydrochloric acid (1.1) and then wash with warm water until all the acid is washed out of the filter. Dry the carbon at a temperature between 95 and 100 degrees C.

Before using the apparatus a blank shall be run and if the bulb does not gain in weight more than 0.5 milligram, put the dried filler into the ignition tube and heat the pre-heating furnace and the part of the combustion furnace containing the copper oxide. After this is heated start the aspiration of oxygen or air at the rate of three bubbles per second, to show in the potash bulb. Continue slowly heating the combustion tube by turning on two burners at a time, and continue the combustion for thirty minutes if air is used ; twenty minutes if oxygen is used. (The Shimer crucible is to be heated with a blast lamp for the same length of time).

When the ignition is finished turn off the gas supply gradually so as to allow the combustion tube to cool off slowly and then shut off the oxygen supply and aspirate with air for ten minutes. Detach the potash bulb and prolong, close the ends with rubber caps and allow it to stand for five minutes, then weigh. The increase in weight multiplied by 0.27273 equals the percentage of carbon.

The potassium copper chloride shall be made by dissolving one pound of the salt in one liter of water and filtering through an asbestos filter.

Option No. 1.—While a pre-heater is greatly to be desired, as only a small percentage of laboratories at present use them, it was decided not to make the use of one essention to this method; subtraction of the weight of the blank to a great extent eliminating any error which might arise from not using a preheater.

Option No. 2.—The Shimer and similar crucibles are largely used as combustion furnaces and for this reason it was decided to make optional the use of either the tube furnace or one of the standard crucibles. In case the crucible is used it shall be followed by a copper tube 3-16 inch inside diameter and ten inches long, with its ends cooled by water jackets. In the centre of the tube shall be placed a disk of platinum gauze, and for three or four inches in the side towards the crucible shall be silver foil and for the same distance on the other side shall be copper oxide. The ends shall be plugged with glass wool, and the tube heated with a fish tail burner before the aspiration of air is started.

Graphite.

Dissolve one gram sample in thirty-five c. c. nitric acid (1.13 sp. gr.) filter on asbestos, wash with hot water, then with potassium hydrate (1.1 sp. gr.) and finally with hot water. The graphite is then ignited as specified in the determination of total carbon.

Convention of American Foundrymen

Convention to be Held in Toronto Next Year.

At the closing sessions of the American Foundrymen's convention, which was held in Philadelphia from May 20 to 24, Toronto was recommended as the convention city for next year. The decision was left in the hands of the executive. Dr. Moldenke, secretary of the association, says Toronto will be the next place of meeting. The decision rests chiefly with him. At the closing session of the Supply Association, the decision as to the next place for holding their exhibits was left to the executive; but a strong recommendation to hold the exhibits in the same city and in conjunction with the convention of the American Foundrymen's Association was passed. Thus it is practically decided that the foundrymen's convention and the exhibit of the Foundry Supply Association will be in Toronto in 1908.

It was at first planned to hold the convention in Toronto, and to have the exhibits in Niagara Falls, N.Y. This, however, was finally decided inadvisable.

To Canadian foundrymen this decision regarding the place of next year's convention is of prime importance; and to the large number of Canadians present at the convention it is gratifying that their efforts in that direction have been influential. At the closing sessions of both the Foundrymen's Association and the Supply Association, L. L. Anthes, Toronto Foundry Co., Toronto, pointed out in a convincing manner the many reasons for holding the entire convention in Toronto.

Large Number of Canadians Present.

The increasing interest which Canadian manufacturers and Canadian foundrymen are taking in the advancement of founding in Canada was evidenced by the large number of Canadians who attended the convention in Philadelphia. Forty-four Canadians registered, and there were many there who did not register. Those registered are, in order of registration :—

T. J. Best, the Best Steel Casting Co., Quebec.
David Reid, Canadian Westinghouse Co., Hamilton.
J. K. H. Pope, London, Ont.
J. B. Walton, McClary Mfg. Co., London, Ont.
Samuel Terrill, Guelph, Ont.
H. D. Reed, Doty Engine Works Co., Goderich, Ont.
L. L. Anthes, Toronto Foundry Co., Toronto.
J. C. Armer, editor Canadian Machinery, Toronto.
A. E. Pipher, Sanitary Ideal Mfg. Co., Port Hope, Ont.
J. K. Moffat, Moffat Stove Worvs, Weston, Ont.
Arthur W. White, Geo. White Sons' Co., London, Ont.
Geo. A. Burman, King Radiator Co., Toronto.
F. H. Stoneman, Wortman & Ward Mfg. Co., London, Ont.
R. J. Cluff, King Radiator Co., Toronto.
E. B. Fleury, Hamilton Facing Mills Co., Hamilton.
W. J. Thompson, Hamilton Facing Mills Co., Hamilton.
H. V. Tyrrell, manager Canadian Machinery, Toronto.
Wm. Yellowley, Canadian Locomotive Co., Kingston, Ont.
Robert Agnew, Canadian Locomotive Co., Kingston, Ont.
J. S. Taylor, Paris, Ont.
Wm. Surdam, Smith's Falls, Ont.
J. R. Nichol, Smith's Falls, Ont.
J. P. Hockin, Taylor-Forbes Co., Guelph, Ont.
Geo. C. Wilson, Taylor-Forbes Co., Guelph, Ont.
A. J. Calw. Pease Foundry Co., New Toronto, Ont.
A. G. Storie, Ontario Malleable Iron Co., Oshawa, Ont.
J. S. Storie, Oshawa Steam & Gas Fitting Co., Oshawa, Ont.

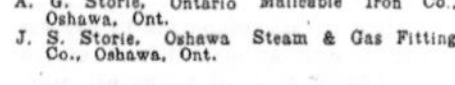

Group of Delegates at Convention of American Foundrymen, Philadelphia, May 20-24.

F. G. Truil, Ontario Malleable Iron Co., Oshawa, Ont.
G. S. Trubell, Empire Mfg. Co., London, Ont.
A. J. Palmer, Empire Mfg. Co., London, Ont.
J. H. McGregor, Galt, Ont.
H. Hertfelder, Dodge Mfg. Co., Toronto.
W. G. Reid, Toronto.
A. H. Tallman, Tallman & Sons., Hamilton.
Robert-Hunter, Goldie & McCulloch Co., Galt.
N. K. B. Patch, Toronto.
R. J. Hopper, Brantford, Ont.
G. W. Duecharme, Montreal.
J. A. Lagnon, Montreal.
Geo. H. Weaver, Dominion Foundry Supply Co., Montreal.
Kenneth Falconer, Gunn, Richards & Co., Montreal.
T. Ben. Bennett, D. Maxwell & Sons., St. Marys, Ont.

Among the Canadians present who did not register were:—E. C. Gurney, Gurney Foundry Co., Toronto; J. Wright and P. McMichael, Dominion Radiator Co., Toronto.

Aims and Objects of the Association.

This shows what an interest Canadians take in the American Foundrymen's Association. The holding of the convention in Toronto next year will result in increased interest. For the benefit of those who are not acquainted with the association, it may be well to discuss its aims and objects.

The American Foundrymen's Association was organized at Philadelphia ten years ago with the purpose in view of placing founding on a higher industrial plane. Its aims are purely educative. To Dr. Moldenke is due the chief credit for the success which the association has had. Of late years there have been affiliated associations formed—the Foundry Supply Association, the Associated Foundry Foremen and the newly organized Brass Founders' Association. Each year a convention is held at which papers pertaining to foundry practice are read and discussed, and at which the Supply Association make exhibits. This convention provides an opportunity for the foundrymen to meet in social intercourse, to exchange ideas and to become familiar with all the latest machinery and appliances which are placed on the market.

The advancement in the founding industry in the least ten years is due in no small measure to the work of the American Foundrymen's Association.

This Year's Convention Most Successful.

The convention at Philadelphia this year has been a most successful one. The papers presented to the association were good. The movement to organize an affiliated brass association was successful. The exhibit of foundry machinery and foundry supplies was the best that has ever been made in America.

Officers of Foundrymen's Association.

The officers of the American Foundrymen's Association for the ensuing year are as follows:

President, S. G. Flagg, Jr.

Vice-president, 1st district, C. J. Caley.

Vice-president, 2nd district, John S. Barr.

Vice-president, 3rd district, H. E. Field.

Vice-president, 4th district, J. H. Whiting.

Vice-president, 5th district, A. K. Beckwith.

Vice-president, 6th district, T. W. Sheriff.

Vice-president, 7th district, J. P. Golden.

Vice-president, 8th district, L. L. Anthes.

Secretary-treasurer, R. Moldenke.

The officers of the Associated Foundry Foremen for the coming year are as follows:

President, James F. Webb, Lake Shore & Michigan Southern Railroad shops, Elkhart, Ind.

First Vice-president, W. S. McQuinnin, Allegheny Foundry Company, Warren, Pa.

Second Vice-president, W. O. Steele, Bateman Mfg. Company, Grenlock, N.J.

Secretary and Treasurer, F. C. Everitt, J. L. Mott Company, Trenton, N.J.

Under the constitution of the Associated Foundry Foremen the presidents of the local associations become vice-presidents of the national association. These are the following: W. F. Brunan, Erie, Pa.; Thos. Glasscock, Milwaukee, Wis.; E. W. Smith, Chicago, Ill.; Hugh McPhee, New York, and other eastern territory; W. A. Keller, Indianapolis, Ind.; Hugh McKenzie, Cleveland, Ohio; W. A. Perrine, Philadelphia, Pa.; David Reid, Hamilton, Ont.; H. J. Holmes, Cincinnati.

Hugh McPhee, the retiring president, was elected an honorary member of the association. A committee was appointed to revise some portions of the constitution and by-laws, and the changes will be submitted to letter ballot.

Group of Canadians at Convention of American Foundrymen.

Officers of Brass Association.

The officers of the Brass Founders' Association, formed at this convention, are as follows:

President, Charles J. Caley.

Secretary, Andrew M. Fairlie, Philadelphia.

Treasurer, J. H. Sheeler.

Officers of Foundry Supply Association.

The newly elected officers of the Foundry Supply Association are as follows:

President, E. H. Mumford, E. H. Mumford Co., Philadelphia.

1st Vice-president, W. P. Shepard, Rogers, Brown & Co.

2nd Vice-president, W. S. Quigley, Rockwell Engineering Co.

3rd Vice-president, H. E. Atwater, Osborne Mfg. Co.

4th Vice-president, Wm. Chambers, Garden City Sand Co.

Treasurer, J. S. McCormick, J. S. McCormick Co.

Secretary, H. M. Lane, Cleveland.

Trustees—W. E. Kanavel, Interstate Sand Co.; E. J. Woodison, Detroit Foundry Supply Co.; W. J. Thompson, Hamilton Facing Mills Co.; F. W. Perkins, Arcade Mfg. Co.; and Geo. Rayner, the Carborundum Co.

Notes of the Convention.

The convention was a famous success. The exhibits of the Supply Association were the best ever made.

* * *

Dr. Moldenke and H. M. Lane were hard-worked men. They worked day and night and night and day.

* * *

Joseph A. Walker, vice-president of the Joseph Dixon Crucible Co., died during the convention.

* * *

The shad dinner at Washington Park will long be remembered.

* * *

A feature of the after-smoker musical performance was the singing of the Canadian national anthem by a number of enthusiastic Canucks.

* * *

Canadians were in evidence everywhere, and at all times Canadians make themselves felt wherever they go.

EXHIBITS OF SUPPLY ASSOCIATION.

A splendid exhibit was made by the Supply Association, each member doing his share to contribute to the splendid effect. Practically all the exhibits were educative. Nearly all were in actual operation. There were a large number of new machines shown and new ideas brought out. Moulding machines and core-making machines were the features of the exhibition. Quite a study could be made of the different principles involved in the machines; and there was ample opportunity afforded to judge of the relative merits of each. There were also many other new features of decided interest.

W. W. SLY MFG. CO.—A complete exhibit of tumbling barrels, cinder mills, dust arresters, core sand separators, resin grinders and other foundry equipment. One of the features of the exhibit was the burnt sand cleaner. This machine is illustrated and described in another part of this issue. The exhaust tumbling barrels have some important features, among them being unlimited guarantee as to strength and wearing qualities of the pinions and bearings, the automatic opening and closing of the damper in the exhaust pipe as the barrels are thrown into and out of commission. The dust arrester was also an attractive part of the exhibit. This arrester exhausted dust from the tumbling barrels, and expelled the clean air.

DETROIT FOUNDRY SUPPLY.—Represented by E. J. Woodison, L. A. Crandall and W. R. Beers. Owing to the going astray of shipment they were unable to show their complete exhibit, which was to have consisted of foundry supplies, the Crandall cupola and furnaces. Mr. Woodison was influential in having the exhibit in Toronto next year. During the convention the company sold all the apparatus they had on exhibit, as well as several Crandall cupolas. This firm distributed a neat souvenir, consisting of a pocket-book.

GUNN, RICHARDS & CO.—Represented by Kenneth Falconer in the interests of costs systems.

C. DRUCKLEIB.—Exhibit of sand blast apparatus for cleaning castings. This apparatus differs from other sand blast apparatus in its application of sand at three different points by means of special jets working in unison. These machines are made in Canada by the Canadian Rand Co.

CHICAGO PNEUMATIC TOOL CO.—Represented by G. A. Barden, B. H. Tripp, F. C. Severin, J. M. Towle, W. H. S. Bateman, H. B. Griner, P. R. Severin and Howard Small. They had a comprehensive exhibit of pneumatic machinery.

HANNAH ENGINEERING WORKS.—Represented by W. L. Laib. The exhibit consisted of shakers and pneumatic riveters. The shaker has several distinctive features.

CLEVELAND WIRE SPRING CO.—Represented by J. W. Campbell and C. H. Errieson. Exhibit consisted of steel barrels and boxes for handling small castings in the foundry, and for similar purposes in a manufacturing plant. Souvenir, a neat note book.

OSBORNE MFG. CO.—Represented by H. R. Atwater, vice-president; F. D. Jacobs, sales manager; C. V. Jacobs, J. C. Boyton. They had a splendid display of their line of goods, brushes and other foundry supplies, in the form of pyramids. Their souvenir, hat and clothes brushes.

LINK BELT ENGINEERING CO.—Represented by Arthur D. Shaw and T. B. O'Neil. The exhibit consisted of a conveying machine for handling sand, coke, etc., in the foundry. This is a demonstration of the ease of handling materials in the foundry with suitable appliances. The Reynold silent chain was included in the exhibit.

J. W. PAXON.—Represented by I. F. Kreiner, designer of exhibit; E. M. Taggart, J. Goehring, J. H. Joulin, Canadian and New England representative; M. Carr, A. G. Warren, M.E.; H. M. Rougher, treasurer and advertising manager; Howard Evans, vice-president, and Geo. Moore, moulding machine man. Largest exhibit in the building, consisting of all kinds of foundry supplies, including the New Glenwood Roevover moulding machine. An exhibition of the manufacture of brushes and sieves was given. Souvenir, a pair of dividers.

ARCADE MFG. CO., Freeport, Ill.—Three moulding machines, equipped with different styles of pattern plates, showing the different styles of work which is handled by these machines. Neat pin tray as souvenir.

BERKSHIRE MFG. CO., Cleveland, O.—Standard Berkshire moulding machines in operation, moulding radiators, hot water pipes, etc. Puzzle match-box as souvenir.

PH. BONVILLAIN & E. RONCERAY, Paris.—Exhibited a full line of their universal moulding machines in operation, with several styles of pattern plates. There was also shown several styles of intricate castings made from their moulds.

CYRUS BORGNER, Philadelphia.—Fire brick and other foundry supplies.

A. BUCH'S SONS CO., Elizabethtown, Pa.—Exhibit of moulding machines.

CAMDEN COKE CO., Philadelphia.

CORTLAND CORUNDUM WHEEL CO., Cortland, N.Y.—Exhibit of grinding machines. Souvenir, grinding stone in leather case.

CROCKER-WHEELER CO., Ampere, N.J.

DIAMOND CLAMP AND FLASH CO., Richmond, Ind.—Exhibit of three styles of core machines, with which a very varied style of core can be made. Representatives are: W. N. Gartside and Jos. H. Gartside, traveling representative.

JOSEPH DIXON CRUCIBLE CO., Jersey City, N.J.—Represented by A. L. Haasis, W. J. Coane, Neville Roane, W. A. Housten and Frank Krug. A fine display of crucibles. Souvenir, a box of pencils.

STANLEY DOGGETT, New York.—Exhibited "Perfection" parting compound.

EDWARD J. ETTING, Philadelphia.—An exhibit of cupolas, tumbling barrels, ladles, etc., and a magnetic separator.

FALLS RIVET & MACHINE CO., Cuyahoga Falls, Ohio.—Exhibit of the Wadsworth core making machinery including the improved core machine; and also the new Wadsworth core ovens, which have but recently been placed on the market. There was also shown a gagger mould and a friction drive tumbling barrel. Souvenir, a cigar case.

FOX MACHINE CO., Grand Rapids, Mich.—Exhibit of Fox trimmers.

FOUNDRY SPECIALTY CO., Cincinnati, Ohio.—Practical demonstrations were made of the use of a new line of compounds for parting and facings for brass and iron work.

THE GARDEN CITY SAND CO., Chicago.—An exhibit of samples of a complete line of mounding sands.

THE GOLDSCHMIDT THERMIT CO., New York.—There was demonstrated the use of thermit for various purposes to which it is put, including welding, repair of large castings, purification of iron in the ladle, etc.

EDWIN HARRINGTON, SON & CO., Philadelphia.—An exhibit of Peerless hoists, for use in foundries.

HERMAN PNEUMATIC MACHINE CO., Zelienople, Pa.—An exhibit of three different styles of moulding machines, together with samples of work which had been done with them.

A. E. HOERMANN, New York.—An exhibit of sand blast apparatus and a complete line of Green pneumatic hammers, sand rammers, etc.

THE INTERSTATE SAND CO., Cleveland, O.—An exhibit of foundry sands. Souvenir, handsome watch fob.

KILLING MOULDING MACHINE CO., Davenport, Ia.—Represented by E. J. Potter and M. K. Weigel. An exhibit of moulding machines and automatic steel clasps, floor plate, steel flasks and new power-driven sand riddle. Souvenir, a handsome paper weight and dice box.

ARTHUR KOPPEL CO., New York.—An exhibit of the industrial railway system. Souvenir, note book.

W. W. LINDSAY & CO., Philadelphia.—A complete line of foundry chaplets, etc.

MACPHAIL FLASK & MACHINE CO., Chicago.—An exhibit of steel flasks and a new type of galvanized sheet steel bottom board.

C. E. MILLS OIL CO., Syracuse, N.Y.—An exhibit of core oil and Mills Day core compound.

MITCHELL-PARKS MFG. CO., St. Louis.—Exhibit of their gravity moulding machine in operation.

MONARCH EMERY & CORUNDUM WHEEL CO., Camden, N.J.—Exhibit of emery and corundum wheels.

MONARCH ENGINEERING & MANUFACTURING CO., Baltimore.—Exhibit of a tilting furnace, a Monarch blower, a ladle heater, and other foundry supplies.

E. H. MUMFORD CO., Philadelphia.—An exhibit of moulding machines and their new Mumford sand supply and screening apparatus for moulding machines.

J. S. McCORMICK CO., Pittsburg.—An exhibit of moulding machines, the Dings magnetic separator, the McCormick continuous sand mixer and other foundry supplies.

S. OBERMAYER CO., Cincinnati.—Represented by J. T. Thorner, vice-president; S. T. Johnston, J. E. Evans, W. M. Fitzpatrick, C. M. Goldman, E. D. Frohman. Showed a new line of foundry ladles.

OLIVER MACHINERY CO., Grand Rapids, Mich.—A Universal saw bench, a band saw, and a trimmer.

OTTO GAS ENGINE WORKS, Philadelphia.—Their horizontal gas engine supplied power to the various exhibitors.

THOS. W. PANGBORN CO., New York.—An exhibit of foundry supplies, including sand blast apparatus, core machines, and sand sifters.

PETTINOS BROS., Bethlehem, Pa.—An exhibit of lubricating and crucible flake graphites, foundry dust, blackings, etc., and various grades and styles of plumbago.

HENRY E. PRIDMORE, Chicago.—Exhibit of four moulding machines, two of the rock-over drop type, and two standard stripping plate machines.

RANDALL TRAM RAIL CO., Philadelphia.—An exhibit of the Moyer tram rail system.

ARTHUR E. RENDLE, New York.—Exhibit of skylights and fireproof windows.

ROCKWELL ENGINEERING CO., New York.—An exhibit of the Rockwell melting furnace.

J. D. SMITH FOUNDRY SUPPLY CO., Cleveland.—An exhibit of foundry supplies, including the new sprue cutter and Cleveland moulding machines. Souvenir, handsome memorandum case.

P. H. & F. M. ROOTS CO., Connersville, Ind.—An exhibit of foundry blowers.

R. B. SEIDEL, INC., Philadelphia.—Exhibit of crucibles. Souvenir, watch fob.

SMYTH SWOBODA & CO., New York.—An exhibit of the "Lycopart" parting.

STANDARD SAND & MACHINE CO., Cleveland.—An exhibit of their continuous automatic sand and mixing machine.

THE TABOR MANUFACTURING CO., Philadelphia.—An exhibit of the Tabor moulding machine and the Taylor Newbold saws.

B. F. STURTEVANT CO., Hyde Park, Mass.—An exhibit of their high pressure rotary type blower, the interior of which was shown by the blower being partly dismantled. There were also shown various other styles of blowers.

U. S. GRAPHITE CO., Philadelphia.—An exhibit of graphite products. Souvenir, a set of playing cards.

THE WHITING FOUNDRY EQUIPMENT CO., Harvey, Ill.—Furnished space for the reception and entertainment of friends and guests.

THE WILBRAHAM, GREEN BLOWER CO., Philadelphia.—An exhibit of foundry blowers, motor driven.

YALE & TOWNE MFG. CO., New York.—An exhibit of their portable electric hoists and their triplex, duplex and differential train blocks in operation.

Points about Compressors and Pneumatic Tools

The Field of Compressed Air; Interesting points concerning Compressors; Care of Pneumatic Tools.

By W. P. Bissinger, Chicago Pneumatic Tool Co.

During the past five years much progress has been made toward standardizing operations with air tools, hitherto accomplished by different means in different shops; also we are witnessing the realization of prophecies regarding the extension and widespread use of air power that seemed wildly extravagant when they were made. From the air brake pump to the Corliss compressor is a wide span, yet we have seen the air power plant develop from one to the other during the past ten years.

I shall merely mention in passing the familiar applications of the air hammer for chipping, riveting, scaling, or any service demanding a rapid, hard, percussive blow; the air motor for drilling, reaming, flue rolling, turning car journals, or any kind of work requiring rotative power—the air hoist, either in the form of a motor or straight lift cylinder and piston. Likewise the pneumatic jack, the sand blast, the stay bolt nipper, the paint burner and the paint sprayer, the air cleaning nozzle and numerous other special uses devised for this readiest of powers.

In the early days of air utilization much was heard about the rivalry between pneumatic power and electricity for transmitting energy from its source

Air Compressor with Mechanically Operated Intake Valves.

of inception to its point of consumption and each had its votaries and advocates more or less extreme in their partisanship.

To-day each has found its recognized and legitimate sphere of usefulness, and frequently both forms of power are made to co-operate in the achievement of the best results.

Compressors Now More Efficient.

It is natural to expect that larger compression units would produce compressed air with less relative expenditure of power, since the opportunity exists for refinement of design and construction not possible in the smaller compressors. But so great has been the demand for a reduction in the cost of compressed air production that contemporary compressors in all sizes are vastly superior to the earlier patterns in use when air tools were first introduced. This has been accomplished through an intelligent appreciation by the builders of air compressing machinery of the more arduous service which compressors undergo, to which they have responded by designing machines adequate in weight, strength, bearing surface, valve areas, and automatic regulation, to the severer conditions. Users of compressed air in the selection of their compressors have contributed an influence equally beneficial by exercising a discriminating knowledge and judgment (expensively obtained at the school of experience) which they did not possess when compressors first supplanted air brake pumps.

Evil of Overrating Compressors.

An evil to which not over-scrupulous compressor builders have contributed is the over-rating of compressor capacities. As is well known, compressors are rated according to their piston displacement, being the cylinder area multiplied by the piston speed. From this result deductions due to clearance losses and heat expansion must be made to arrive at the actual volume of air delivered. These losses necessarily vary according to the quality of the machine, emphasizing the inevitable conclusion that greater initial investment pays a handsome dividend.

Beyond an arbitrary limit, however, compressor ratings at high speeds mean nothing but deception, since greater displacement than the air valve area permits is not possible even if structural strength and bearing surface are adequate, which they rarely are.

Experienced compressor users realize thoroughly the desirability of providing machines of size sufficient to deliver the requisite yield at moderate working speed, reducing cost of maintenance and greatly prolonging the life of the machine.

But while these recent prosperous times have enabled a great betterment of compressor equipment, there is much yet to be accomplished. There still remain air brake pumps in use supplying shop requirements and plants with compressors inadequate or obsolete, or both, are still altogether too numerous. An investigation applied to individual conditions would develop results so surprising as to overcome all ordinary objections to the increased investment, and instances where compressors have been shown to earn their cost within one to two years after installation are by no means rare.

Compressors Driven by Steam Engine, Electric Motor and Gas Engine.

For railroad shops and industrial establishments the steam driven compressor is of course most generally employed, though many more motor driven compressors are used than formerly, conditions being favorable.

The demand is steadily growing for compressors of moderate capacity for use where steam to drive them is not available. Motor driven compressors meet this requirement if electrical current is obtainable and gasoline engine driven compressors where neither steam nor electricity may be had. Such machines are highly useful for maintenance of way and bridge construction and repair at junction points. The necessity of testing the air brake equipment of cars received from other lines has also created a comfortable demand for self-contained gasoline driven compressors with engine and compressor mounted upon one bed, the engine driving the compressor by gear or silent chain.

Compressors up to 200 cubic feet per minute capacity for a terminal air pressure of 100 lbs. per square inch are usually of the single cylinder type, double acting in that they compress air at each of the strokes, displacing the cylinder contents twice per revolution. Above 200 cubic feet per minute two stage compressors show an economy that should not be disregarded. Formerly duplex compressors having two simple compressing cylinders were common, but except for low air pressures, that do not warrant compounding, these have been superseded by the two stage type.

The steam cylinders have balanced slide valves, with independent adjustable cut-off usually provided for cylin-

End View of Compressor with Mechanically Operated Valves.

ders of twelve inches diameter or larger. Steam cylinders are compounded generally when the steam pressure at throttle is sufficient to warrant. The limit of capacity at which the Corliss valve should supplant the slide valve varies according to the appropriation available and the cost of fuel. However, Corliss compressors of 1,500 and 2,000 cubic feet per minute capacity are much more frequently installed than formerly.

Wasted Air Means Wasted Money.

Effectual measures should be adopted to prevent the useless waste of compressed air, so frequently encountered in shops. Since the atmosphere we breathe is free, its value when compressed, represented by the cost of maintenance of plant and power expended in compressing it, is too lightly regarded, with the result that it becomes a favorite means of blowing dust from machines, work benches and garments, and a plaything in the hands of the michchievous, frequently with dangerous and sometimes with fatal results. For cleaning armatures and delicate machinery, compressed air is often an indispensable agent, but strict orders and severe penalties should prevent its indiscriminate and unauthorized waste.

Keep Air Free From Dirt.

Dirt and grit in compressed air clogs the necessarily delicate working parts of the pneumatic tools and the inevitable excessive wear on valves and other parts causes rapid depreciation in value and power. To avoid this it is desirable to screen or filter the air at the compressor intake, but care must be observed not to have the screen of too fine a mesh or the air supply will become partially throttled. Occasional instances have occurred where compressor capacity has been markedly curtailed from this cause. As most of the air tools now sold are provided with individual strainers, trouble from dirt in the tools is not frequent. These strainers also must be kept clean or the tool will appear to have lost its power.

Lubricating Compressors and Tools.

The lubrication of compressors and pneumatic tools is a feature deserving careful attention. A too frequent mistake is made by using in air cylinders of compressors, oil intended for steam cylinders. Such oil is of low flash point, whereas the power lubrication of air cylinders demands a light oil of high flash point and of very best quality. Oil of poor grade and low flash point becomes vaporized in air cylinders and is discharged with the air without effecting lubrication.

Oil should be fed to air cylinders slowly and sparingly, as too much oil will clog the air valves, causing them to stick and give trouble. Air valves should be examined and cleaned at intervals by washing in kerosene or naphtha. When this is done the valves should be removed from the compressor. Engineers have been known to introduce kerosene through the air inlet pipe, an effective method of cleansing dirty valves, but sometimes equally effective in producing an explosion, since the oil forms a fine spray or mist which when compressed with the air produces a condition similar to that in the cylinder of an oil engine.

The plan of feeding soap suds into the air cylinder through the lubricator is excellent for keeping valves clean, but when this is done, oil should be fed through afterward to prevent rust.

The lubrication of pneumatic tools is of equal importance. One cannot do better than obtain and use one of the several brands of oil furnished by pneumatic tool makers who have made a special study of the requirements. Such oil is necessarily light, and under no circumstances should a heavy oil be used, as the cooling effect of the expanding air would cause it to clog the tool parts and prevent the free movement of the parts.

Pneumatic hammers should be carefully cleaned after using and kept submerged in a tank of oil when not in service. An excellent device for effectively lubricating pneumatic tools is an automatic oiler inserted in the supply hose about twenty inches from the tool with oil proof hose between oiler and tool, which, operating on the principle of an atomizer, enables the flow of the lubricant to be regulated to a nicety.

Equipment of Manufacturing Plants

Economy can be Effected by Correct Placing of Machinery.

By G. C. Keith, B.A.Sc.

In erecting a building for the manufacturing of a certain machine or article, too much care cannot be taken to construct the building to suit the work. An architect should be engaged who is thoroughly conversant with the character of the work. Heads of departments should be brought together and schemes decided upon. This should also be done in the placing of the machinery. The equipment is of more importance than the building, and the proper placing of machinery should not be neglected.

Just as in locating a plant, transportation is one of the most important factors in settling on a manufacturing site, so in the shop. The machinery should be so arranged that the castings, etc., will have to be carried as short distances and handled as few times as possible. In a certain large manufacturing plant in Ontario thousands of dollars have been spent shifting and re-arranging the machinery. This not only wastes considerable time, but it is accomplished only at a large cost.

Before placing the machines, positions should be carefully studied. Every time a change is made the profits are lessened. The proper placing of machinery is a great factor in keeping up the efficiency of the plant.

Work Should be Standardized.

As soon as possible the work should be standardized. Gauges should be made as well as jigs to assist in quick manufacturing. A toolroom is necessary, well equipped with good machinery and first-class men. Here the plans are made for the best way to do any special job, and jigs are made for the work.

In some machine shops it has always been the rule: "One machine—one tool—one man," but shop superintendents and foremen should recognize that a great saving is accomplished by doubling up. For this reason we now have one-headed planers replaced by those with four heads; we have the automatic nut-tapping and screw-cutting machines with three and four heads, and we have the turret-lathe.

Let me cite the saving there is in roller bearing tie-pins alone by the use of the automatic turret-lathe. In one machine shop it takes a boy, working at the rate of about five cents a hundred, milling the ends of the tie-pins. He puts them in a jig, one at a time, tightens it up and forces it into a revolving mill by means of a lever, doing one end and then the other. He must then loosen the jig to take the pin out. Before the milling can be done a man must cut off the pins in lengths in the shears. Now let us look at a different shop. Here twelve automatic turret-lathes stand in a row. The wire for the pins is fed into the machine from a coil, the ends are milled for the rings and the pins are cut to their proper length. These twelve machines are looked after by one man.

Where there are thousands of small articles to be made, as in some sewing machine, bicycle and agricultural works, such machinery is essential. Such an installation costs considerable, but the saving effected much more than pays the interest on the investment.

Tricks of the Trade

Send in your ideas, sketches or jobs you are doing, and anything of interest to your fellow-workmen. Remuneration will be made for such contributions.

WAYS AND MEANS OF DOING THINGS

STUD HOLDER FOR BOLT CUTTING MACHINE.

The accompanying cut shows a stud holder intended to hold short studs in a bolt-cutting machine when one end of the stud has already been threaded. Referring to the cut, Figs. 1 and 3 show end views and Fig. 2 a sectional view of the device. The body (H) is made of machine steel, and is first drilled crosswise to receive the jaws (J), shown in Fig. 6. These jaws are made in one piece and fastened in place by the screws (FF). Then the body with this piece in place, is chucked and bored out for the plunger (P), a detail of which is shown in Fig. 4, and is, at the same time bored and threaded for the machine steel bushings (N). After this is done, the jaws are taken out, cut in two, and the slots for the screws (FF) made a little longer to allow the jaws to freely adjust themselves to the plunger point. The jaws and plunger are made of tool steel and hardened. The square bushings shown can be made of square cold rolled stock and the body of the same kind of round stock, no outside finish being necessary.

The assembled sectional view shows a bushing for 1-inch studs. Fig. 5 shows a bushing for ¾-inch studs. The inside end of the bushing is counterbored for the button (E), which is a hardened tool steel piece held in place by the two screws (SS), which prevent it from falling out, but allow it to slide in about 1-16-inch, the same as the plunger slides in the 1-inch bushing. The threads on the insides of the bushings on all sizes are made the length of the diameter of the stud. The stud to be cut is screwed

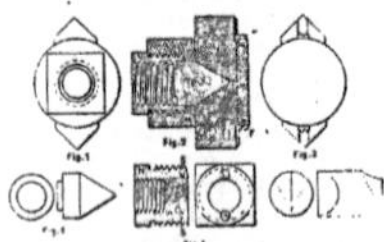

Stud Holder for Bolt Cutting Machine.

loosely down against the plunger (P) or the button (E), as the case may be, and the jaws of the vise tightened up against the jaws of the holder, which, in turn, force the plunger and, in sizes smaller than 1-inch, the button against the end of the stud, locking it, and also securing the holder in the vise. Opening the vise allows the button, plunger and jaws to go back to their former places, and the stud can be easily removed.

M. H. BALL, in Machinery.

EMERY WHEEL DRESSER.

The cut herewith shows a simple emery wheel dresser made from an ordinary bent piece of band iron with four or five tool steel washers between the ends. A small bolt passes through the washers and the band iron holding it together. If the wheels of an ordinary dresser are worn out ordinary tool steel washers may be inserted as these will last just as long and are a great deal cheaper

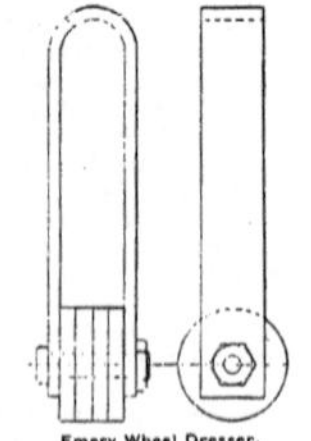

Emery Wheel Dresser.

than the wheels bought especially for the purpose.—Roy B. Demming, in Machinery.

TURRET TOOLS FOR ROUGHING AND FINISHING PINIONS.

The cut herewith shows a tool which I designed recently for turning and facing pinions on a Warner & Swasey screw machine. It consists of a cast iron overhanging arm carrying a roughing and finishing tool attached in the ordinary way to the turret with two bolts. A cast iron head fitted to the spindle holds the arbor on which the pinion is turned. This arbor is made of tool steel, hardened and ground, and fitted with Woodruff keys, as shown, to prevent the arbor and pinion from turning on the seat while the cuts are taken. The bushing shown in the arm steadies the arbor during the operation. The hole for the bushing was bored in the machine, bringing it perfectly in line. The tools used are 1 x ½-inch high speed steel cutters. The nut shown is for backing off the pinion after the job is

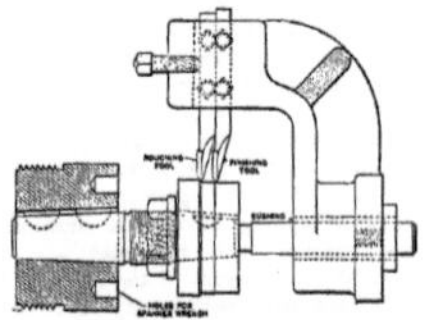

Turret Tools for Roughing and Finishing Pinions.

completed. The pinions are first chucked in the same machine, bored, and reamed with a taper reamer. They are then faced on the end with a tool in the cross slide, after which they are put on the fixture shown and turned. Afterward they are faced on the other end to the proper length with a tool in the cross slide.

C. W. PUTNAM, in Machinery.

EDUCATION AND THE TRADES.

In his last annual message, President Roosevelt said:

"It should be one of our prime objects as a nation, so far as feasible, constantly to work toward putting the mechanic, the wage-worker who works with his hands, on a higher plane of efficiency and reward * * * Unfortunately, at the present, the effect of some of the work in the public schools is in the exactly opposite direction. If boys and girls are trained merely in literary accomplishments, to the total exclusion of industrial, manual, and technical training, the tendency is to unfit them for industrial work and to make them reluctant to go into it, or unfitted to do well if they do go into it."

Words coming from the lips of so great a publicist as President Roosevelt and delivered on such a significant occasion, have a pregnancy well worth our consideration. In the above quoted words the President is calling the nation's attention to an ominous fact, that of the scarcity of skilled and experienced mechanics and the inadequacy of machinery for training them. As citizens of America we have been very slow to see this fact. We are just now beginning to realize the gravity of the situation.

In a large majority of cases where applicants for any vacant position are asked for, the demand is made that the applicant state his experience. Hitherto, many bright, ambitious and capable boys, who, though having no experience in the trade in which they wish to enter, are willing to start "at the bottom of the ladder," if even they may secure a start in that particular trade, are discouraged by being "turned down" because of the company's rigid insistence upon the experience principle. President Roosevelt, readily realizing the distressing vicissitudes of fortune in every vocation, whether profession or trade, went to the root of the question by putting his finger on the public school system. He justly deplored the undue attention given in the schools to "training in merely literary accomplishments, to the total exclusion of industrial, manual, and technical training."

In such an age as ours, when the mechanical spirit is abroad, it is a gross absurdity to preclude the possibility of children, who have not sufficient mental capacity to make a success of any profession of earning the necessities of life through some trade, by training them in merely literary accomplishment, entirely losing sight of the fact that those entering trades need as much training and preparation as the others. It is to be sincerely hoped that such an unhappy illusion as the one just named will be speedily dispelled. Such illusions prove disastrous to the prosperity and welfare of a nation.

Men before President Roosevelt, however, saw the necessity of establishing schools wherein to give boys experience in their chosen trades before actually entering the manufactories, thereby better equipping them for their trades and increasing the possibility of rapid promotion. Such schools are in successful operation in some of the larger centres of population in the United States. New York, Boston, Pittsburg, Philadelphia, Indianapolis, and Springfield, Mass., have their free evening trade schools. As will be noticed, the sessions of these schools are held in the evening, and are meant primarily to prepare boys of sixteen years and up, who have left school, to enter the trades. The teachers are men actually engaged during the day in mechanical work and thus well fitted to train the boys. So far, these schools have proved to be invaluable factors in enhancing the commercial conditions of the republic. They are largely attended, the one opened in Philadelphia this year having an attendance of over five hundred.

The central aim of all successful educators has been, and is, to make all education concrete and practical. The predominant materialism of our age demands it. In the city of Toronto we have two large schools which emphasize pre-eminently the practical side of education. The Technical School and the School of Practical Science* are model institutions. They do a splendid work in their own fashion, but they do not go far enough. Technical instruction has reached a very high standard of perfection in these schools. Their work would be more universally satisfactory if mechanical instruction was introduced and imparted. It is a notable fact that college—and school—trained mechanics often make deplorable failures when it comes to applying the training and knowledge they have acquired to actual and real conditions. In a conversation with a mechanic of five years' experience the other day, the present writer was told that the experience of a man in the shops was eminently preferable to the training gotten in a technical or science school. The mechanic told an incident of a young man "just out of college with all his diplomas," being assigned a certain work in the shop where he worked. After a great deal of work with discouraging results, the college-trained mechanic came to the manager and acknowledged his inability to satisfactorily complete the work.

*Our correspondents' plea for the advancement of technical and mechanical knowledge is a worthy one. This subject is now receiving great attention from all quarters. But distinction should be made between trade schools and schools or colleges of Applied Science. The School of Practical Science is a misleading name. It is now, however, the Faculty of Applied Science of the University of Toronto. Its purpose is to give a thorough training in mathematics and applied science to fit one to take up one of the branches of engineering. There is no pretense of giving a practical training.—(Editor.)

Would it not have been eminently better for that young man if he had had the privilege of attending a trade school such as are in successful operation in the United States and gotten his training from men actually engaged in the trade instead of from college professors so well versed in theory but in a number of cases so deficient in practical training. Our educational system is a practicable system; let us strive to make it more practicable by collaborating in the establishment of trade schools, and give encouragment and aid to those boys who have been "turned down" because of inexperience.

READER.

AN IMPROVED TOOL POST.

The illustration shows a lathe tool post that possesses decided superiority over any now in common use.

Figs. 1 and 2 show a square-nose tool

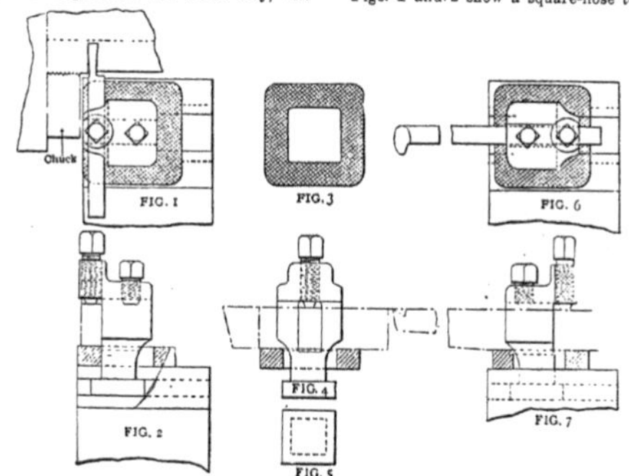

An Improved Tool Post.

firmly held within 1-16 inch of the chuck jaws. Fig. 3 is the tool-post ring.

Figs. 4 and 5 show the square end of the tool post a good sliding fit in the tool slide, permitting any of the four sides to face the work.

Figs. 6 and 7 show a boring tool that cannot slip.

The advantage shown in Fig. 1 where a parting tool could work within 1-16 inch of the chuck jaw is quite sufficient even alone to recommend the tool post.

I have seen many parting tools, square-nose and thread tools, that had been broken simply because the tilting gib caused the tool to lose its side clearance; these tools were all of the off-set type.

Offset tools are a poor make-shift and should have no place in the up-to-date shop; nor should the tilting gib.—F. Rattek, in American Machinist.

Developments in Machinery

Metal Working Wood Working Power and Transmission

NEW TURRET LATHE.

The new Pratt & Whitney 2¼ x 26 in. open turret lathe is a universal machine suitable for doing a great variety of work from the bar and on forgings and castings, without continually requiring special appliances and expensive cutting tools. To accomplish this purpose many new features, including a cross sliding turret, have been introduced.

This machine possesses practically all the flexibility and adaptability of the engine lathe. The extreme rigidity, powerful spindle drive, quick changes of speeds and feeds, heavy cross feeding turret and numerous adjustable stops, admit of narrower limits of error, as well as a marked reduction in the cost over work produced on the ordinary turret or engine lathe.

This machine has a stiff head, with constant speed arranged for either direct-connected motor or countershaft drive by means of a single pulley. The turret is mounted on a slide, having both positive power and hand longitudinal and traverse directions. The machine is recommended for bar work up to 2¼ in. diameter by 26 in. long, for castings up to 14 in. diameter and for cylindrical operations on work within these dimensions.

An unusually heavy spindle of special steel, with cylindrical bearings, runs in bronze split sleeves. The thrust of spindle is against an independent upright, cast solid with the head, and insures against any springing tendency under heavy end cutting strains. Provision is made for taking up wear of spindle and end thrust.

The direction and variation of the spindle to the work holding spindle are obtained by levers operating friction clutches. It is impossible to connect more than one set of gears with the spindle and main driving shaft at the same time.

New Pratt & Whitney Turret Lathe.

The gears are of extra heavy pitch and of ample width to safely withstand the hardest usage. The head, which is stationary, is of box construction, the gears running continually in oil.

Eight variations of speed are provided and by using the two-speed countershaft this may be doubled. All of the controlling levers and connections are within easy reach of the operator. The spindle can be instantly stopped by movement of any lever on head stock.

The rod chuck may be operated while machine is running and has extraordinary gripping power.

The collet jaws are supported up to their outer end, which is particularly desirable in forming work from the cross slide.

The complete chuck can be readily removed from the spindle when combination lathe chucks or special face plates for castings are to be substituted.

A positive screw feeding device automatically feeds the rod forward to its stop.

The bar may be round, square, hexagon, or any irregular cross section, and need not necessarily be free from scale, as there are no delicate parts or complicated gearing to become clogged.

A follower bar is furnished which enables short pieces of stock to be as conveniently handled as long bars, and at the same time serves to keep such pieces concentric with the spindle.

An efficient stock stop for gauging the length of stock is provided, which, when not in use, is moved forward and swung upward, so as not to interfere with the turret tools.

The turret revolves about a large conical stud held firmly in the cross slide.

The various tools may be accurately located and with rigid backing, so that the heaviest cuts can be taken without the slightest spring or backward movement. The tools are held in place by straps and are backed up by uprights cast solid with the turret. Severe tests have proved this to be a superior method of unyieldingly holding the tools against all tortional and backward strains.

The locking bolt is directly under the cutting tool and is horizontal, thereby overcoming the tendency of a vertical bolt to lift the turret from its seat.

Indexing can be accomplished at all positions of the cross slide, and is automatic, although the turret may be rotated to any position by hand.

One of the most important features in this open turret lathe is the compound turret with power and hand feeds and adjustable stops which are conveniently located.

The longitudinal turret slide travels on large raised V's, is provided with gibs its full length, and a binder which permits the slide to be firmly clamped to the bed at any point within its travel.

The power longitudinal feed is positive in both directions, and has six changes, any of which can be instantly set by movement of lever.

There are six automatic longitudinal stops and six supplementary stops, which give two positions to each turret tool. If necessary, all twelve stops may be used for one or all tools in the turret, making it possible to effectively cover all requirements. The stops are held in a heavy steel bracket, which may be moved along the front of the bed and clamped where desired. In case it is desired to run through a few special pieces of work, the automatic stops may be dispensed with and the supplementary stops used in their place without the necessity of disturbing adjustments.

The distance from the axis of the spindle to the turret tool is altered by traversing the turret slide. This arrangement permits ample support for long bars and if machine is belt driven, gives an unvarying belt tension. Motor can be mounted on head without difficulty.

The cross slide has both hand and power feed. There are six variations of the power feed in either direction. Eight distinct adjustable cross stops are provided, which may be used in any combination desired.

Bed and pan are made in one single casting and have U-shaped cross webbing, insuring rigidity. Generous provision for oil and chips is provided.

A variety of turret tools adapted to meet practically all the various requirements are furnished with this machine to order.

THE BURKE TWO-SPINDLE BENCH MILLER.

Accompanying is an illustration of a two-spindle bench milling machine which was originally built by the Burke Machinery Co., Cleveland, Ohio, as a specialty for one of the large manufacturers of typewriters, for the special purpose of milling both ends of the type bars at one time. The usefulness of such a machine for general bench work has led the manufacturers to place it on the market regularly.

The spindles rotate in different directions, so that the work can be clamped to the table and raised between the cuttters, the cutting on each end being done with a downward cut. The adjustment of the heads permits a great variety of work to be handled, not only in the millling of the two ends of a piece simultaneously, but in ordinary surface milling, where substantially the machine will have the capacity of two single millers.

The table of the miller has a longitudinal movement of 6 in., a transverse movement of 2½ in. and a vertical movement of 5½ in. The working surface of the table is 4 x 12 in. The maximum distance between the bed and the spindles is 6½ in., and the minimum distance 4 in. The largest diameter of the cone driving pulley is 6 in. and the smallest 3½ in. The spindles have holes to take No. 9 B. & S. tapers. The machine stands 25 in. high over all and weighs, complete with countershaft, about 400 lbs.

The driving pulley is on the back of the column and may be partly seen in the illustration. On the same shaft with this pulley is a large flanged pulley, which is connected by a continuous belt to the two spindle driving pulleys. The belt passes successively around one of the spindle driving pulleys, around the main driving pulley, around the second spindle driving pulley and around an idler pulley below the driving pulley back to the first spindle driving pulley. The idler is adjustable, to take up the slack of the belt caused by moving the heads. The shafts of the spindle driving pulleys carry gears which drive the spindles, on the left side directly and on the right side through an intermediate gear to give the reverse rotation to that spindle. The gears are of cast iron, the spindles of crucible steel and the bearings of phosphor bronze.

The longitudinal feed, which is used more for positioning the work than anything else, is operated, as shown in the engraving, by a ratchet lever and pinion meshing a rack on the lower side of the table. The feeding movement is usualy obtained by lifting the table vertically between the cutters, these cutters being adjustable sideways, as before stated. The table, saddle and knee are raised vertically by the hand wheel shown, which operates through gearing to drive a gear engaging in a rack on the face of the column.

The Burke Two-Spindle Bench Miller.

All the spindle bearings are equipped with wick oilers and suitable oil boxes cast beneath the bearings. In case it should be desirable to attach weights to the knee to counterbalance it the small grooved pulleys on the side of the column are provided, over which the cords supporting the weights are hung.

FRICTION BENCH DRILL.

The accompanying illustration is of a neat and simple friction bench drill, which has been placed on the market by Krug & Crosby, Hamilton, Ont. The friction drive allows the speed to be changed instantly while the drill is in motion, which is done by releasing the screw on the stop rod and moving the friction wheel to or away from the centre of the driving disc as greater or less speed is required.

This machine will drill to the centre of 14-inch radius.

Friction Bench Drill.

HOPPES FEED WATER HEATER.

Accompanying are illustrations which show sectional views of the improved open feed water heater and purifier, made by the Hoppes Manufacturing Co., Springfield, Ohio. Fig. 1 is a side view, showing the arrangement of pans and the oil catcher. Fig. 2 is an end view, showing the delivery of the water to the feed troughs and pans. The shell of the heater is cylindrical, and the heads are "bunked," a design well calculated to resist the pressure. The feed troughs, pans and bottom of heater are made of cast iron, while the shell is made of pig iron. In vast "T" type, however, the entire shell, as well as before mentioned parts, is made of cast iron.

The normal storage of water in the bottom of the heater is shown in the end sectional view, but where more storage is desired, several layers of the bottom pans are left out and the maximum water line is fixed at the centre of the shell by raising the overflow dam. This dam is shown at the rear end and controls the overflow and acts as a skimmer for same. The overflow pipe is attached near the bottom of the back

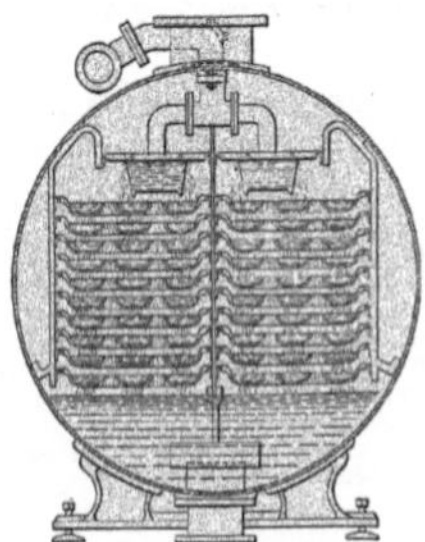

Fig. 1—Hoppes Feed Water Heater.

head. The water is admitted to the heater by a balance regulating valve and regulated by a float in a separate float chamber.

In operation the water is delivered into the feed-troughs through the feed-pipe shown at the top and is evenly distributed by means of discs having uniform openings placed between the flanges of the tees in the feed-pipe and the L-shaped connections extending down into the feed-troughs. The water overflows from the troughs into the pans and overflowing the sides it follows along the under side of the pans in a thin film to the centre of each trough before dropping down into the next pans. While following the under sides of the pans, the exhaust steam will come in contact with the water and heat it to the full temperature of the exhaust. All the solids which may be held in solution in the water will, when liberated by the heat, form mostly on the under side of the pans, while matter in suspension will be deposited inside of the pans. As the water continues to follow the under side of the pans, or the lime already formed, the full efficiency of the machine is realized until ready to clean. The oil catcher provided is very large and efficient.

Fig. 1—Hoppes Feed Water Heater.

In open heaters of large size the regulation and distribution of the water is a very important matter, as to be effective the steam should come in direct contact with the water, and to do this the passage of the water should be gentle and as free from agitation as possible. The principle of flowing the water evenly over the under side of the pans in these heaters brings this about in a most practical manner.

By the use of the artificial regulation provided in this heater, the water is not only evenly distributed to the pans, but two or more heaters may be used in multiple by providing a single regulating valve in the main feed-pipe controlled by a float attached to an equalizing pipe at the bottom.

COAL HANDLING MACHINERY.

The problem of supplying fuel to large boiler houses has received a great deal of attention during the past few years, and several types of machines have been built for the rapid unloading of fuel barges. The type which is obtaining considerable success in the eastern part of the country is the two-man steeple with horizontal boom. These towers have been made both of steel and timber construction, some a combination frame with timber struts and fitted with steel booms.

One of the examples of recent installation of this character is the steel tower erected at the plant of a large sugar refinery in Philadelphia. This tower is shown in the accompanying view, which brings out the rather heavy style of construction. This equipment is de-

Jeffrey Coal Handling Mechanism.

signed to unload barges which come in a rather narrow slip, and which lie head on to the unloading tower. The coal is elevated, crushed, screened, weighed and delivered to the belt conveyer, which carries it to the boiler plant, located some 300 feet away. This work requires the services of two men on the tower; and one, and part of the time two, men in the barge to help clean up and guide the bucket during the last part of the unloading.

The tower proper is an all steel structure made up of lattice channel columns, with stiff bracing. The structure contains five distinct floors or levels in it, made up of heavy planks laid on steel floor beams. All floors have, in addition to the heavy planks, a wearing surface made of 1-inch hardwood flooring laid diagonally. The first or lower floor contains the hoisting engine, and the engine for driving the crusher; the second floor contains automatic scales; the third floor a four-roll crusher; the fourth floor, receiving hopper; the fifth floor is at the level of the boom which supports the trolley, and there is, in addition, a crow's-nest around the head sheave which allows easy access for the inspection of the bearings located at this point. All the floors which are outside of the house proper are well provided with hand railings. The cabs protecting the operators are two in number, located at opposite sides of the tower.

In one cab is placed the bucket man who controls the hoisting, lowering, opening and closing of bucket; in the other the trolley man who controls the motion of the trolley along the boom. These cabs are very well provided with windows; in fact, three sides are all glass. The covering used is heavy galvanized corrugated sheet, made of re-rolled iron.

Extending from the lower engine floor to the fourth, or hopper, floor is a housing made of heavy corrugated galvanized sheet iron. The hoisting machinery used to handle the bucket and trolley consists of one Lidgerwood Steeple Standard hoisting engine fitted with two cylinders 15 x 24 inches, direct connected to 30-inch drums equipped with 54-inch double cone friction automatic brakes. The trolley engine consists of a pair of 10 x 12-inch engines, direct connected to a 16-in drum fitted with cone friction and band brakes.

The arrangement of levers for controlling the hoisting engines is very simple, and each operator has but three to handle, one for each hand and one for his foot; the foot lever in each case controlling the throttle on the engine. The bucket with which this tower is equipped is the second of its type which has been placed in service along the eastern coast, but still is only one of a number of the large group which are working successfully in handling sand, gravel, broken stones, coal and other materials at different points throughout the interior of the country and along the Pacific slope. It differs in many points from the ordinary clam shells with which our readers are, no doubt, familiar, and will handle successfully materials which the clam shells cannot penetrate. In this particular instance the equipment will have to handle large lump coal more than half the time, and in order to obtain the desired capacity the bucket must be able to handle the large coal as rapidly as the small. The equipment is designed for continuous capacity of 120 tons per hour from barge to receiving hopper. The hoist from barge at mean tide water to the hopper is about 60 feet, and the length of travel is 57 feet.

In the hands of expert runners, with this height of hoist and travel, coal can be unloaded at the rate of 200 tons per hour. The equipment was designed and built by the Jeffrey Mfg. Co., Columbus, Ohio, with Canadian office on Cote street, Montreal.

The view was taken while the tower was in process of construction, and does not show the covering which was afterwards placed upon the steel frames.

New Pipe Threading Machine.

NEW PIPE THREADING AND CUTTING MACHINE.

A new size pipe threading and cutting off machine, known as the P.D.Q.C. No. 6, has recently been brought out by the Bignall & Keeler Mfg. Co., Edwardsville, Ill. The machine is particularly adapted for shops having large quantities of pipe of one size to thread at one time. It is equipped with a quick operating chuck, controlled by a hand wheel and pinion which engages in a segment gear on the end of the cone shifting arm. The cone slides freely on the arbor; as it is moved forward rollers on the ends of the chuck arms roll up on the surface of the cone, and the arms being thereby spread apart tighten the jaws on the pipe. When the cone is retracted, springs draw the jaws away from the pipe. As this gripping chuck can be operated while the machine is running, the jaws being set for a given size of the pipe, an entire lot can be threaded without stopping the machine. The steel jaws in the chuck are graduated, which facilitates the setting for a given size of pipe.

The die head is of the peerless type, as used on the machines of similar type, manufactured by this company, in which the dies can be instantly released from the pipe after the pipe is threaded. The cutting-off tool is held in the slide in the front of the die stand and a reaming tool for removing the burr from the pipe is also provided.

The rear chuck is provided with three independent jaws with which fittings can be made up, and also a bushing for holding the pipe central without gripping the pipe.

The drive is from a four-step cone pulley at the back which, in connection with compound shifting gears, affords eight changes of speed. The machine can be arranged to be driven by belt or motor. An automatic oil pump in the bed of the machine supplies oil to both the dies and the cutting off tool.

The machine illustrated occupies a floor space of 50 x 120 in., and weighs in the neighborhood of 7,500 lbs. Ten other sizes of the machines are made, ranging in pipe capacities from 1¼ to 6-in. diameter, inclusive.

THE EMMERT UNIVERSAL VISE.

A new type of vise which is rather out of the ordinary is being placed on the market by the Canadian Fairbanks Co., known as the Emmert Universal vise.

As the name suggests, this vise is universal, and has every possible adjustment which the workman could find convenient. Thus quick work in shaping,

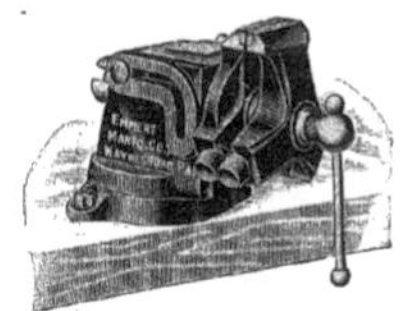

Emmert Universal Vice.

finishing and fitting is facilitated greatly. While in service the vise can be swung into any desired position without necessitating the releasing of its hold on the article gripped within the jaws. The vertical position gives the advantage of a full circle swing, as does also the horizontal position, and any intermediate position.

Pattern makers' vises of the Emmert type have seven and six pairs of jaws, while the metal worker's vise has five pairs.

This vise is used very extensively throughout Great Britain and the United States, but this is its first introduction in Canada.

THE INJECTOR SAND BLAST.

Accompanying is an illustration of the injector sand blast apparatus as made by C. Druchlieb, New York, and in Canada by Canadian Rand Co., Montreal. This injector sand blast apparatus is an application of the injector principle, as commonly applied to steam injectors,

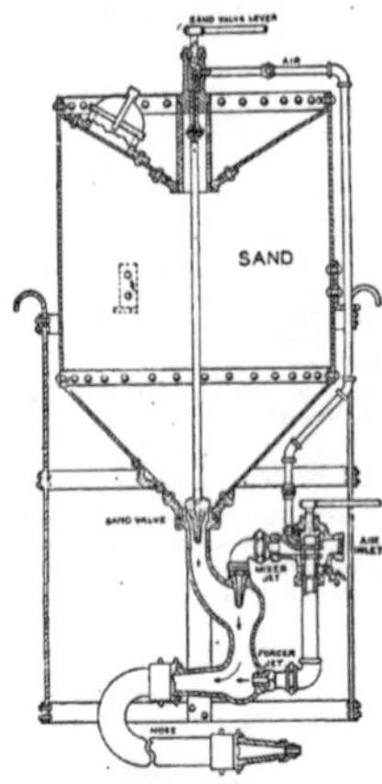

Injector Sand Blast Apparatus.

to the sand blast, the highest possible velocity necessary being imparted to the mixture of compressed air and sand by the several air jets, by means of which the quantity of the air can be controlled and directed, an essential feature to be found only in this apparatus.

With it castings can be cleaned, scale and rust removed from steel structural work, dirt and weather stains from the stone and brick work of buildings, and other and similar work performed, with the minimum amount of air, sand and pressure.

The injector sand blast apparatus comprises a steel or other sand tank suitably supported and provided with lugs for convenience in slinging. The top is depressed, allowing the tank to be easily filled, and the sand filling hole is closed by a special locking device. On the top of the sand tank is a small bonnet through which is led a vertical hollow stem terminating in the sand valve at the bottom of the tank. To balance the pressure on the sand, the air pressure is admitted to the sand tank through the vertical pipe shown outside thereof connected to the bonnet, the vertical stem being perforated.

Extending vertically through the sand valve is an air jet, admitting the compressed air into the injector casting, and producing a vacuum, which causes the sand to flow with evenness and regularity.

Below this jet is the mixing chamber, provided with a second vertical air jet, connected to the direct air supply, acting upon the mixture of air and sand and putting it into action, its velocity being increased by contracting the walls of the casting.

Below the mixing chamber is a third and horizontal air jet from the main air supply, forming with the enclosing casting, a forcer which delivers the mixed air and sand, now in very active motion, sharply and with full vigor to the work in hand, through a line of rubber hose provided with a special nozzle.

At one side of the sand tank and convenient to the hand is a special cock admitting and regulating the air supply consecutively to the sand tank, the central pipe and the several nozzles in the sand valve, the mixing chamber and forcer, or to all simultaneously, the supply of air being received from an air tank or compressor. A drain cock is provided to remove any condensation from the air.

The hose is connected to the mixing chamber by a patented, specially designed heavy hose coupling.

AN AIR-COOLED GASOLINE ENGINE.

The accompanying is an illustration of the air cooled small size gasoline engine manufactured by the Gilson Manufacturing Co., Ltd., Guelph, Ont. These engines are made in sizes from 1 to 5½ h.p.

The flanges around the cylinder provide a cooling surface of approximately 1,000 square inches in the 1½ h.p. engine and 1,400 square inches in the 2½ h.p. engine. Larger sizes are, in addition, provided with a fan with baffle plates which throws a current of air directly against the cylinder head. This fan can also be supplied to smaller engines, if desired.

The sparking plug, the intake valve and the exhaust valve, are all located in the cylinder head, and are, therefore, well cooled by the current of air.

Special attention is paid to the detail construction of this engine, compactness, neatness and accessibility of parts, being important features. The engines are so balanced that they can be operated without fastening the skibs, upon which the engine stands, to the floor.

Air Cooled Gas Engine.

STATIONARY ENGINEERS' CERTIFICATES.

The Legislature of Ontario at its recent session passed an act respecting stationary engineers in which engineers and employers are alike interested. Briefly stated, its provisions are that, after the 1st day of July, 1908, no engineer will be allowed to operate or have charge of a stationary steam plant of 50 horse power or upwards who does not hold a Government certificate. There are three classes of engineers to whom certificates will be granted without the applicant having to undergo an examination; first, those who on the 20th of April, 1907 (the date on which the act was passed), held certificates from an association of stationary engineers in Ontario, or a marine or locomotive engineers' certificate; second, engineers who on the above date were in charge of a plant of 25 horse power or over in Ontario; third, engineers who had at any time previous to the passing of this act, not less than two years' experience in the operation of such a plant in the province. Those who cannot qualify as above will have to pass the examinations which will hereafter be prescribed by the board of examiners.

Those interested may obtain a copy of the act and application forms for certificates by addressing the secretary, Department of Agriculture, Toronto.

Enough study has been spent on the subject to determine that men especially suited to any particular kind of labor, if supplied with proper implements and intelligently directed, can do on an average, at least three times as much as the average workman does, if the limiting factor is physical exertion and, if assured proper compensation, will do so day after day.

ABOUT CATALOGUES

By mentioning Canadian Machinery to show that you are in the trade, a copy of any of these catalogues will be sent by the firm whose address is given.

STEAM SPECIALTIES.—A little illustrated booklet issued by Darling Bros., Ltd., manufacturers of steam specialties, Mantreal. This booklet describes the different specialties made by this firm.

ELECTRIC PROPELLER FANS.—Bulletin 146 of the Sturtevant engineering series issued by the B. F. Sturtevant Co., Hyde Park, Mass., illustrating and describing their electric propeller fans.

J. T. SCHELL.—Three bulletins on the standard heading matcher, the standard heading turner, and the improved saw mill outfits, manufactured by J. T. Schell, Alexandria, Ont. These bulletins are illustrated and are of standard size.

GISHOLT LATHES.—An attractive catalogue of the Gisholt Machine Co., Madison, Wis., dealing with their turret lathes, boring mills, and tool grinders. This catalogue is well gotten up and well illustrated, and is 8 by 10 inches.

GARDNER'S GRINDERS.—A standard size, illustrated catalogue, descriptive of "Gardner's Improved" disc grinders, band polishing steel, disc wheel circles, and sectional wheel chucks, made by the Gardner Machine Co., Beloit, Wis.

STEAM BOILERS.—A catalogue issued by the Robb Engineering Co., Limited, Amherst, N.S., illustrating and describing the Robb-Mumford boiler. The catalogue is well illustrated and contains some interesting and useful data on steam boilers.

THE ROLLINS ENGINE.—Bulletin No. 15 of the Rollins Engine Co., 29 Mason St., Nashua, N.H. This bulletin gives detail description and illustration of the Rollins engine. Each important part is very clearly illustrated with half-tone engravings.

FOUNDRY FURNISHINGS.—A general catalogue of foundry supplies published by the Detroit Foundry Supply Co., Detroit, Mich. This catalogue is a handsome and most complete one; 431 pages, 9 x 6 in. It is well illustrated and contains a complete index.

THE BOWSER OIL TANKS.—Catalogue of S. F. Bowser & Co., Fort Wayne, Ind., illustrating and describing the Bowser self-measuring oil tanks and pumps for various purposes. The catalogue is well illustrated. The Toronto branch of this company is at 530 Front St. West.

THE MARION FEED-WATER HEATER AND PURIFIER.—Catalogue of the Marion Incline Filter & Heater Co., Marion, Ohio. This catalogue illustrates and describes in an effective way the operation of their feed-water heating and purifying apparatus, and points out its advantages.

BRICK MACHINES.—The McIntosh Brick Machine Co., of Goderich, is sending out some good literature to prove the merit of sand-cement pressed brick, and the machine which makes it. A pamphlet of convincing testimonials is enclosed with a neat and attractive catalogue.

BELL ENGINES AND THRESHERS.—A complete catalogue of the Robert Bell Engine & Thresher Co., Ltd., Seaforth, Ont., illustrating and describing their traction, portable and stationary engines, locomotive and stationary boilers, separators, saw mill machinery, belting, threshermen's and mill supplies; 88 pages, 5½ by 9½ inches.

ROBB-ARSTRONG ENGINES. — A complete and well illustrated catalogue of the Robb-Armstrong engines, giving a detailed description, with illustrations, of their manufacture. These engines include Corliss and slide valve engines, horizontal and vertical, simple and compound. A very complete description of parts is given.

PNEUMATIC APPLIANCES. — The general catalogue of pneumatic appliances manufactured by the Curtis Manufacturing Co., St. Louis. This catalogue includes a description of their air compressors, air hoists, bridge cranes, traveling cranes, trolley systems, jib cranes, and pneumatic elevators. The catalogue is well illustrated and is of standard size.

FIRE BRICK.—A very handsome catalogue published by Harbison-Walker Refractories Co., Pittsburg, Pa., containing useful information in connection with the use of silica, manganese, chrome and fire clay brick. The book is full of illustrations and contains a large number of tables, as well as space at the back for notes, memoranda, etc.

GENERAL ALTERNATING CURRENT DATA are given in Bulletin 74 just published by the Crocker-Wheeler Co., Ampere, N.J., entitled "Engine Type A. C. Generators." Views are shown of plants where Crocker Wheeler alternators have been installed, and details are given describing the design of generators developed by the company.

MACHINE MOULDING.—A most complete catalogue issued by MM. Ph. Bonvillain and E. Ronceray, 9 and 11 Rue des Envierges, Paris, with descriptive treatment of moulding sand, the universal system of pattern plate making, the universal hydraulic moulding plant, etc. This firm are exhibiting their machine at the Philadelphia convention of the American Foundrymen's Association.

ALMOND PRODUCT.—A very neat little catalogue issued by the T. R. Almond Mfg. Co., 83-85 Washington St., Brooklyn, N.Y., illustrating and describing the Almond product, including "Almond" drill chucks, right-angle transmission devices, turret heads, flexible arms for electric lights, electrical steel tubing. Illustrations are very suggestive, in that the important parts are brought out very prominently.

FROM NOVA SCOTIA.—Rhodes, Curry & Co., Amherst, N.S., have sent a very useful little souvenir, in the form of a card case and memo book. This is made of handsome maroon leather and lined with tan calfskin. The cover is free from advertising with the exception of the firm name and address, which appear in gold lettering. The above firm will be pleased, as long as the supply lasts, to send one to any in the trade who will mention Canadian Machinery when writing.

NOTES ON HIGH SPEED STEEL.—This is the title of an interesting little booklet on "Conqueror Tool Steel," issued by J. Beardshaw & Son, Baltic Steel Works, Sheffield, Eng.

This book contains much valuable information for users of tool steel, such as: Instructions for working "Conqueror" H. V. tool steel; forging; hardening of turning and planing tools; hardening of drills, milling cutters and tools of intricate shape; annealing; grinding, etc., besides interesting and instructive notes on ordinary tool steel.

Alexander Gibb, manufacturers' agent, 13 St. John St., Montreal, is sole Canadian agent for "Conqueror" tool steel, and he will be pleased to supply copies of this booklet to any in the trade who will mention Canadian Machinery when writing.

Industrial Progress

CANADIAN MACHINERY AND MANUFACTURING NEWS would be pleased to receive from any authoritative source industrial news of any sort, the formation or incorporation of companies, establishment or enlargement of mills, factories or foundries, railway or mining news, etc. All such correspondence will be treated as confidential when desired.

A large sawmill will be erected at New Westminster.

A new two-storey box factory will be erected in Toronto.

A large addition is being built to the Guelph Carpet Mills.

The Whitman-Barnes Co.'s works will remain in St. Catharines.

The London Fence Machine Co., London, will build this year.

The Toronto Engraving Co. will erect a new five-storey building.

Head & Co., bakers, will erect a $10,000 plant at Fort William.

Work on the new I.C.R. shops at Moncton is going on rapidly.

A $20,000 addition will be made to the Kuntz brewery in Waterloo.

The W. H. Malkin Co., Vancouver, will erect a new warehouse.

D. Fraser & Sons, Fredericton, N.B., will rebuild the Aberdeen mill.

Large additions are to be made to the Fraser River Tannery Company.

A new brickyard has been opened up in Wetaskiwin, Alt., by A. Geuz.

A new technical college will be erected at Halifax, to cost $100,000.

Mills & Anticknap are arrnaging to build a large tannery in Welland.

Efforts are being made to get the Ottawa Car Co. to move to Toronto.

The Phillips Mfg. Co., Toronto, will erect a factory at a cost of $83,000.

W. E. Shantz and H. Quelsdorf will build a furniture factory in Berlin.

The West Kootenay Power Co. have taken over the Cascade Power Co.

The electric light plant, Indian Head, will be extended at a cost of $25,000.

The Alberta Pacific Elevator Co. are building a large elevator at Calgary.

The Nelson Iron Works have started their new foundry at Greenwood, B.C.

Tenders for waterworks in Quebec, Regina and Thorold are being called for.

The Dominion Smelter Co. have decided to erect a smelter at Parry Sound.

The Bell Telephone Co. will erect an entirely new plant in Chatham, Ont.

Lymburner & Matthews, brass founders, Montreal, have dissolved partnership.

Mitchell & Phelan, have been registered in Montreal as iron and brass moulders.

The Breadner Mfg. Co., Ottawa, will erect a three-storey factory this summer.

A new carriage factory and warehouse costing $30,000, will be built at Winnipeg.

Frankel Bros., Toronto, will erect a two-storey warehouse, at a cost of $85,000.

The Gorman, Eckert Co., London, are spending upwards of $60,000 on a new plant.

Dyment, Baker & Co., London, are preparing plans for an extension this year.

The Krug Furniture Co., Berlin, will make a $10,000 extension to their factory.

The Fowler Canning Co., Hamilton, is doubling its capacity at a cost of $76,000.

The Ault & Wibourg Co. will build an ink factory in Toronto, at a cost of $35,000.

It is proposed to build a geological museum building for Toronto University.

A new winter fair building for Brandon, costing $30,000, will be built this summer.

The works of the London Fence Co., Portage la Prairie, Man., have been burned.

The Canadian Pacific Railway Co. may erect a station and a hotel at Fort William.

A power plant will be installed in the village of Streetsville, Ont., at a cost of $20,000.

The Stratford Gas Co., Stratford, will erect a large addition to their present plant.

The G.T.R. have let contracts for an enormous addition to their present shop at Stratford.

The E. W. Gillett Co., Toronto, will make a $30,000 extension to their present plant.

The Frontenac Gas Co., Quebec, will install a new plant at a cost of about $1,000,000.

E. Leonard & Sons, London, Ont., will erect a new plant in that city in the near future.

The Napanee canning factory is erecting a large addition to its already extensive plant.

The E. W. Gillett Co., Toronto, will build a four-storey brick warehouse, to cost $12,000.

The General Brass Co., Toronto, will build a one-storey brick factory at a cost of $9,000.

The Standard Fitting and Valve Company will build a $100,000 factory in Guelph this year.

The Sanitary Packing Co., Weston, Ont., will erect a new plant at a cost of about $20,000.

The Christie, Brown Co., Toronto, will erect a six-storey brick factory, costing $37,500.

Heintzman & Company, Toronto Junction, are building a large addition to their factory.

Wm. Berry, Brantford, will erect a factory for the manufacture of towels in Tilsonburg, Ont.

Extensive additions are to be made to the Vulcan Boiler Works at New Westminster, B.C.

The Gananoque Belt Company, Gananoque, is spending $20,000 in additions and new machinery.

New Grand Trunk freight offices and additions to the freight sheds will be built at Brantford.

The Sydney Foundry Co., Sydney, C. B., are going to erect several new steel structure buildings.

The Montreal Rolling Mills Co., Montreal, have increased their capital from $816,000 to $1,200,000.

Several requests have been made for the right to develop power at the Big Chute, Severen River.

G. B. Reid, Toronto, will erect a two-storey brick factory on Sherbourne St., at a cost of $12,000.

The Grundy, Clapperton Co., Ltd., are to erect a factory for the manufacture of cut glass in Toronto.

Calgary ratepayers have authorized a loan of $250,000 to establish a municipal electric street railway.

The Virden Manufacturing Company, Virden, Man., will erect workshops in that place this season.

The Laprairie Brick Company, Laprairie, Que., are building extensive additions to their plant.

An addition to the roundhouse of the Canadian Pacific Railway Co. at Lethbridge, will cost $18,000.

The Malone-Manning wood fibre factory, which was burned out at Owen Sound, will be rebuilt.

The Gurney Foundry Company will expend $4,000 on an addition to their Toronto Junction plant.

The Imperial Varnish Co., Toronto, suffered heavily by fire caused by the explosion of a paint mill.

The Revelstoke Sawmills Co., Revelstoke, B.C., are to erect a sash and door factory at Moosejaw.

The Collingwood Shipbuilding Co., Collingwood, will double the capacity of their extensive plant there.

An electrical smelter for the treatment of iron ores will be erected on the shores of Burrard Inlet, B.C.

The Gundy, Clapperton Company, Toronto, will erect a cut glass manufacturing plant in that city.

Damage to the extent of $6,000 was done by fire to the Longue Pointe Cement works, near Montreal.

Steps are being taken by the Government to utilize the peat bags in Ontario and Quebec for fuel purposes.

The Canada Pride Wrought Iron Range Co., Toronto, will erect a $5,500 frame and metal factory in that city.

The Lantz Marble Company of Buffalo, are to erect a factory to cost $40,000 in the east end of Toronto.

A Bradshaw & Co., Toronto, will erect a factory at Toronto Junction for the manufacture of garments.

The Canadian Iron and Foundry Co., St. Thomas, Ont., are building a two-story brick pattern storage house.

The Canadian Pacific Railway have

let the contract for the construction of a large machine shop at Brandon.

C. K. Milne, Hamilton, has patented a new art or process of treating and preserving iron patterns from rust.

The Gananoque Spring and Axle Co. have been authorized to increase their capital stock from $150,000 to $300,000.

The Intercolonial Railway Co. are installing three 100 h.p. Robb-Mumford boilers in their shops at Moncton, N.B.

The growth of the business of the Copp Foundry, Fort William, for the first three months of this year was 69 per cent.

The Frontenac Gas Company, of Quebec city, will install a new plant this summer at a cost of one million dollars.

The Granby Co.'s smelter in British Columbia has been shut down for some time, due to the scarcity of coal and coke.

The Guelph Axle Works will extend its factory during the present year, the plans calling for a building 40 by 76 feet.

The town of Chatham are considering the installing of a gas engine plant in the civic lighting and water works station.

The Jobin, Marrin Co., Winnipeg, have found it necessary to double the capacity of their present large warehouse.

The Queen City Oil Company, Toronto, have filed plans for a $15,000 extension to their oil works and warehouses.

The Construction and Paving Co., Toronto, will build a two-storey iron asphalt paving factory at a cost of $12,000.

The veneer factory of Mulhall & Co., of Sundridge, Ont., was totally destroyed by fire on April 10th. The loss is $16,000.

The Bell Telephone Co. will build a branch exchange in Winnipeg, to cost $40,000, and a stores building, to cost $36,000.

H. Disston & Son, Toronto, have been granted a permit to erect a saw factory on Adelaide Street, to cost $10,000.

A meat packing plant will be built in Edmonton this year, at a cost of $500,000, and a brewery plant at a cost of $250,000.

The Battle Creek Health Food Co., London, have machinery on the way that will double the capacity of their new plant.

The saw mill owned by B. Grier, and the building occupied by the Bonner Leather Co., Montreal, have been destroyed by fire.

A large power house will be built in connection with the Canadian Pacific Railway hotel at Victoria, B.C., at a cost of $40,000.

The city engineer, Windsor, is advertising for tenders for the construction of 115,000 feet of cement walks to be laid this season.

The Gurney Foundry Co. will spend $4,000 in increasing the capacity of the moulding department in its works at Toronto Junction.

The London Street Railway Co., London, will erect an addition to their building and improve their equipment at a cost of $50,000.

The Amherst Foundry Co., Amherst, N.S., suffered loss by fire to the extent of $75,000. The enamelling department was completely destroyed.

A sub-committee of the board of works, Hamilton, have been appointed to look into the cost of establishing a municipal lighting plant.

W. H. Acton and T. W. Suddaby, of Gananoque, will establish a harness factory at Kingston, to be known as the Kingston Harness Works.

The new plant at North Battleford, Sask., will soon be in operation. Sand, lime, brick and artificial Indiana sandstone will be manufactured.

Messrs. Clarke & Clarke, Toronto, will erect their new factory at the head of Manning avenue. They sold their old site to the Canadian Northern.

The West Park Brick Company will Strathcona, Alta., have commenced the manufacture of ordinary brick, and later will establish a pressed brick plant.

Harry Cates, proprietor of the Brandon, Man., Pump and Windmill Works, has decided to build a large new brick factory and warehouse in that city.

The new machinery for the American-Canadian Oil Co., to be used in the development of the company's property at Morinville, B.C., is being installed.

The Brockville Light and Power Co., Brockville, Ont., are installing a new 72 inches by 18 feet return tubular boiler, made by the Robb Engineering Co.

The International Portland Cement Co., of Hull, Que., has increased its capital from $1,000,000 to $1,250,000, and its capacity to 3,000 barrels per day.

The Stanley Smelting Works will erect a new smelter at Kingston, where they will be granted a free site of five acres and ten years' exemption from taxation.

The American Institute of Mining Engineers will hold its annual convention in Toronto, opening July 23. The members will visit Temiskaming, Cobalt and Sudbury.

A syndicate at Virden, Man., will erect a large building for the manufacture of the Whiteford weighing machine, patents for which have been secured only recently.

Work on the plant of the National Rolling Mills, Sydney, N.S., will begin at once. This company has been organized by C. V. Wetmore and F. A. Crowell.

The Dow Cereal & Milling Co., Pilot Mound, Man., intend building a large oatmeal and cereal mill at Edmonton this summer, and are at present figuring on the cost.

The Thames River, Ont. at Big Bend, has been examined by H. G. Acres, assistant engineer of the Hydro-electric Power Commission, with a view to power development.

T. R. Booth, Ottawa, has started work on a new sulphide and paper board mill at the Chaudiere. The construction will be of concrete and the total cost $75,000.

Work on the plant of the National Rolling Mills, Limited, at Sydney, N.S., will begin at once. This company has been organized by C. V. Wetmore and F. A. Crowell.

The Perfection Power Pressed Cement Machine Co. have erected a plant in the Spring Ridge Sand Pits, B.C. They manufacture power pressed cement building blocks.

The ratepayers of Tillsonburg almost unanimously carried a by-law to loan William Berry, of Brantford, $15,000 with which to erect a large towelling and textile factory.

At the big industrial exhibition to be held in Winnipeg from the 13th to the 20th of July, 10,800 square feet will be set aside for a special display of "Made in Winnipeg" articles.

The third unit of the Electrical Development Company's power station at Niagara Falls has been installed and tested. It develops 13,000 h.p. A fourth unit is being completed.

Fire in the premises of J. A. Dawson, dealer in electrical and street railway supplies, 745 Craig St., Montreal, caused damage to the extent of over $15,000 on the evening of April 26.

The Anchor Wire Fence Co., of Stratford, are planning to build a new factory this summer. The proposed building is two storeys high and 100 by 50 feet. It will be built of concrete.

The long hoped for industrial awakening of Halifax seems to be approaching. The I.C.R. are erecting an extensive roundhouse, and the Silliker Car Works, are installing an up-to-date plant.

The City of London has asked the Ontario Railway and Municipal Board to approve of a by-law authorizing the issue of debentures to the amount of $25,250 for the extension of its waterworks system.

The Anchor Wire Fence Company, Stratford, are planning to build a new factory this sum mer. The proposed building will be two storeys high and 100 by 50 feet. It will be built of concrete.

The Alberta Portland Cement Co., Calgary, are installing a 500 k.w. direct connected generator to a Robb-Armstrong cross compound Corliss engine, 150 r.p.m., for lighting and power purposes.

The Robb Engineering Co., Amherst, N.S., have recently supplied the town of Truro, N.S., with a 150 h.p. Robb-Mumford water tube boiler, a duplex feed pump and a Robb feed water heater.

The Pacific Coast Mills & Timber Co. are planning to erect a large mill on this side of the international boundary. The site has not been chosen. Mr. B. H. Silver is opening up an office in Vancouver.

The old Frost & Wood factory in Oshawa, latterly occupied by John Stacey as a planing mill and sash and door factory, was destroyed by fire on April 26. The loss amounted to $20,000, with insurance of $7,000. The factory will be rebuilt.

On account of finds of rich clay in Morris, Man., a company has been formed to operate a modern brickyard. Building operations will commence immediately under supervision of Mr. Windsor.

The International Heating and Lighting Co., Cleveland, Ohio, have secured an option on a site at Portage la Prairie and will begin the erection of a gas plant as soon as the deal can be put through.

Phillips and Buttorff Manufacturing Co., of Nashville, Ten., have recently added to their foundry equipment a 16 ton per hour Newton cupola, manufactured by the Northern Engineering Works, Detroit.

A company for the manufacture of freight cars is being organized at St. Thomas, Ont., with a capital of $400,000. It is said that the company can obtain a contract which will keep it busy for ten years.

Plans for a two-storey brick factory 50 x 150 feet, have been approved by the Canadian Brass Manufacturing Company, Galt. It is expected that building operations will be proceeded with at an early date.

The Dominion Iron & Steel Co., Sydney, are erecting 350 coke ovens. These, with the 500 now in use, makes 850. A new record was made on May 23. The three smelters on double shifts turned out 900 tons of metal.

The Alberta Government have set aside $200,000 for the construction of telephone lines into the province. Already a large number of long distance lines are under construction and several others are being planned.

The Page-Hersey Iron, Lead and Tube Co., Guelph, will immediately extend their present plant by the erection of a large new building, which, when completed and equipped, will double that firm's already large output.

Nova Scotia Steel & Coal Co., have purchased a large deposit of iron ore in Brazil. This is the first of Canadian industries to acquire foreign ore deposits. This ore deposit in Brazil is said to be one of the richest known of.

It is quite probable that an extensive car plant will be erected in Moncton, N.B. The Board of Trade are quite enthusiastic about it, recommending the city council to grant free water, light and a tax exemption for twenty years.

Scottstown, Ont., is bidding for industries. The municipal council has, by resolution, authorized a bonus of $10,000 as an inducement to any manufacturing industry that will invest $15,000 in a plant and pay out at least $12,500 annually in wages.

North Battleford, Sask., is to have a $75,000 sand lime brick works. Work on the factory will commence immediately and in two months the plant will be turning out 20,000 bricks a day. The Schwartz Brick Co., of New York City, are the promoters.

The Longue Pointe Cement Works, Longue Pointe, Que., suffered severely from fire. The whole interior of the factory was gutted. The works were of recent construction and equipped with modern machinery. A large addition was in process of construction.

A pressed steel car wheel plant is about to be established at Montreal, according to recent report. There are distinct advantages in this class of wheels over the usual cast wheels. Graham Fraser is the capitalist who is said to be behind the enterprise.

The Restigouche Woodworking Company's factory, of Dalhousie, N.B., was completely destroyed by fire last month. The boilers and engines were saved, but a large amount of manufactured lumber was consumed. The loss is estimated at $80,000, with insurance of $40,000.

A Canadian shipbuilding plant will be established at Fort William. The company which has it in charge is made up of prominent Chicago and Canadian capitalists. It is understood that an investment of $2,000,000 will be made for construction and working expenses.

The Diamond Coal Co. Calgary, are installing a 10-ton hand-power traveling crane supplied by the Smart-Turner Machine Co., Hamilton. The Standard Soap Co., Calgary, are putting in a single effect evaporator, built by the Smart-Turner Machine Co., Hamilton, Ont.

Mills & Anticknap, hide dealers and tanners, Welland, have been making experiments during the past few months, and have succeeded in perfecting a new method for tanning sole leather. They will build a large tanning establishment in that town, to carry out their new ideas.

The Canadian Pacific Sulphite Co., Swanson Bay, B.C., will shortly invite tenders for five large boilers and a large amount of heavy machinery for paper-making. The boilers will be of the horizontal tubular type, each seventy-two inches in diameter by twenty-two feet in length.

The Sydney & Glace Bay Railway Co., Sydney, C.B., are installing a new engine and boiler plant consisting of a compound Robb-Armstrong vertical engine, direct connected to 250 k.w. generator, a 250 h.p. return tubular boiler and a 500 h.p. Robb feed water heater and steam separator.

The new plant of the Michigan Copper and Brass Rolling Mills of Detroit will have two 5-ton, 3-motor electric traveling northern cranes. These cranes are of the low type, direct current design and have already been installed by the manufacturers, Northern Engineering Works, Detroit, Mich.

The town of Campbellford are voting on a by-law to authorize the issue of debentures to the amount of $15,000 for granting a bonus by way of a loan to Dickson Bros. to aid them in the erection of suitable buildings for the manufacture of steel bridges and other structural work within the said town.

One of the features of the exhibition at Atlantic City during the covention of Master Car Builders & Master Mechanics from June 12th to the 19th, will be the demonstration of the results to be achieved from the use of the file sharpener of the American File Sharpener Co., 296 Broadway, New York.

Dr. R. R. Stoner, Minneapolis, president of the Stoner Land Co., states that the company will establish a big brick plant near Medicine Hat. The capacity to start with will be 50,000 bricks a day, rapidly increasing to 100,000, and the class of brick manufactured will be pressed or repressed, and probably paving brick.

Owing to the difficulty of getting prompt shipments of mining machinery in the Cobalt district, there is a movement on foot to establish manufacturing plants in that district for the production of such machinery. It is estimated that $200,000 worth of mining machinery was shipped into the Cobalt during the past year.

Great extensions are being made to the Canadian Locomotive Works, Kingston, which will call for an expenditure of $300,000. The new power house will be equipped with the most modern equipment, and will cost $100,000. Some 600 men are working night and day now, and before two years 1,000 men will be busy in the works.

A plant for the extraction of by-products from wood will be established in Victoria, B.C. R. N. Calkins, a mechanical engineer, has been carrying on investigations for some time, and he has met with sufficient encouragement to cause him to organize a company for the extraction of tar, wood spirits, turpentine, and other products from fir wood.

The Huber Manufacturing Company, Marion, Ohio, is planning the establishment of a plant for the manufacture of agricultural implements at Portage la Prairie, Man., and to accomplish this have amalgamated with the Portage Iron and Machine Company. The new company expects to have one of the most complete machine establishments in the west.

Because they cannot get pig iron fast enough to supply the plant without buying in the American market at exorbitant prices, the Algoma Steel Company will at once commence the erection of a $1,000,000 blast furnace. Superintendent Lewis says it will be the largest furnace in Canada, and its erection will be followed at once with a big coke plant to supply the steel works.

McKenzie, Mann & Co., Toronto, Ont., have recently purchased from the Robb Engineering Co., Amherst, N.S., some boiler and engine room equipment, consisting of a Robb-Armstrong engine, direct connected to a 200 k.w. generator, a 10-inch by 10-inch vertical engine, two 67 inches by 18 feet return tubular boilers, a 250 h.p. Robb feed water heater and two duplex feed pumps.

Among the new buildings to be erected at Fort William during 1907 are the Grand Trunk Pacific elevator, 7,000,000 bushels, cost, $2,000,000 ; shipyards and dry dock, $1,000,000 ; Imperial Steel & Iron Works, $500,000 ; Consolidated Elevator Co., $200,000 ; Canada Iron & Foundry Co., $200,000 ; Ogilvie Flour Mills Co., $175,000 ; Canadian Pacific Railway coal docks and sheds, $77,500 ; Muirshead & Block elevator, $20,000.

What will undoubtedly by the largest installation of steam superheaters in Canada up to date, is that just contracted for by the Dominion Iron and Steel Co., Sydney, viz., thirty "Foster" superheaters to be placed under eighteen Babcock & Wilcox type, and twelve Cahall type boilers of 250 h.p. each, a total of 7,500 h.p. This order was placed in the hands of Messrs. Laurie & Lamb, of Montreal.

The contract for the new Central Power Station at Washington, D.C., which has just been let by the United States Government, will mean the establishment of the largest plant of its kind in the U.S. The contract calls for sixteen 600 h.p. high pressure Atlas water tube boilers, and four 2,000 k.w. Westinghouse-Parsons turbines. The consulting engineers of the work are J. G. White & Co., of New York.

By-laws granting exemption and land to the Stanley Lead Company, Toronto, and the Grey & Hedley Zinc Company, on smelters to be established at Kingston were carried by the ratepayers of the Limestone City on Monday. The two companies will use part of the city's water lots below Cataraqui bridge. Buildings to cost $140,000 will be erected, and work will start at once. Lead and zinc will be received from the mines in North Frontenac, within easy distance of Kingston.

Among the orders received by the Smart-Turner Machine Co., Hamilton, for their standard duplex pumps are: the Midland Towing and Wrecking Co., Midland; Revillon Bros., Nipigon, Ont.; the Polson Iron Works, Toronto; the Port Credit Brick Co., Port Credit; R. J. Morrow, Collingwood; the Huntsville Lumber Co., Huntsville; the Highlander Lumber Co.; the Mineral Range Iron Mining Co., Bessemer, Ont.; the John Inglis Co., Limited, Toronto; the Atchison Co., Hamilton, Ont.; the Standard Soap Co., Calgary.

Iron ore assaying 61 per cent. hematite, has been discovered in Boggy Creek Valley, near Roblin, Manitoba, the fortunate prospectors being M. J. Galvin, Toronto, and T. Wagner, Buffalo, nephew of Senator Wagner, the Michigan iron magnate. Credit for the discovery belongs to Mr. Wagner, who made it while assisting in the construction of the Canadian Northern Railway Company's trestle across Boggy Creek several years ago. Since that time he and his associates have been quietly prospecting, and are confident that exceptionally rich deposits have now been found.

Medicine Hat has been getting many new industries of late. The latest is a powder company known as "The J. C. Mitchell Smokeless Powoer Company, of Canada." The company starts out with a capital of $100,000 and will manufacture powder said to be far superior to anything in the market in the explosive line. The new powder, called "Mitchellite," was invented by J. C. Mitchell after twenty years of experimenting. As soon as machinery can be procured, a plant with a capacity of 20,000 pounds per day will be installed. In the meantime 1,500 pounds per day will be manufactured by hand.

Chain Factory for Canada.

Application has been made to the Canadian Government by the Standard Chain Co., for a Canadian charter. As soon as this is granted the company propose to erect a plant at Walkerville, Ont., for the manufacture of chain. In this plant all sizes and grades of coil chain, log chain and harness chain will be manufactured. This plant, it is thought, will be in operation early in the fall.

Alex. Gibb, one of Montreal's leading manufacturers' agents, and who has represented the Standard Chain Co. in Canada for some time, was the promoter of this enterprise, and deserves considerable credit, in thus promoting industrial expansion of Canada. He will be on the Canadian board of directors, and will probably manage the sales of the company.

Equipping Rolling Mills With Electric Motors.

The Tennessee Coal & Iron Co. has recently placed an order with the Crocker-Wheeler Co., Ampere, N.J., for the complete electric motor equipment for its new steel rail mill at Birmingham, Ala. This order comprises fifteen of the special rolling mill motors which the Crocker-Wheeler Co. build.

Addition to the Canadian Corundum Wheel Co.'s Plant.

The Canadian Corundum Wheel Co., Hamilton, Ont., have just started fires in another large new kiln. This will enable the company to give better deliveries than they have been able to do in the past. There is a very large demand for vitrified wheels at the present time.

Large Order From the United States.

The Robb Engineering Co., Amherst, N.S., have recently received a large order from Vanderbeek & Sons., Jersey City, for a power-house equipment, and also other machinery. This equipment includes 14-inch. by 14-inch. Robb-Armstrong engine, a 72-inch. by 16-foot return tubular boiler, a 50-foot stack, a duplex steam pump, a complete, a complete rotary mill, and a complete lath mill.

Rolling Mills at Sydney.

The site has finally been decided upon for the National Rolling Mills Co., at Sydney, N.S.; and from a commercial point of view, it is almost ideal. Construction will begin immediately. The buildings at first will not cover all the property, leaving plenty of room for expansion. At present materials for railway construction and car building will be manufactured, but other departments will be added as the demand increases.

Wants Reciprocity in Coal.

Chas. S. Hamlin, in an address before the Pittsburg Traffic Club, said:

"The people of Massachusetts earnestly desire, among other things, that coal be made reciprocally free of duty between Canada and the United States.

"Canada buys more than three tons of bituminous coal from us for every ton we buy from her. On every ton we import from Canada we have to pay duty to our Government, and similarly every ton of bituminous coal imported into Canada must pay duties to the Canadian Government. It is not an over-estimate to state that last year more than $1,000,000 of duties were paid to the Canadian Government by Canadian railroads alone on bituminous coal imported from the United States."

McKenzie & Mann to Build a Blast Furnace.

McKenzie & Mann have made application to the City of Toronto for a site in the vicinity of Ashbridge's Bay, supposedly for the erection of a blast furnace. It is understood, however, that plans are on foot for a more extensive plant than a blast furnace, probably locomotive and repair works for their railway. Many Canadian shops are at work on the Canadian Northern Railway Co.'s large equipment orders, calling for a total outlay of more than five million dollars. With the exception of the dining and sleeping cars all of this business goes to the Canadian works.

The claims of the Dominion Iron & Steel Co. for bounties on their product during the months of January, February, March and April call for a total payment of almost $400,000.

Foundry and Machine Shop for Sydney.

The latest addition to Sydney's industries is the Sydney Foundry & Machine Works, Limited, which was recently incorporated, with Jas. Clarke, as president; W. E. Clarke, secretary-treasurer, and H. C. Burchell, J. A. Young, J. T. Burchell, A. A. McIntyre and Henry McDonald, directors. The new company has taken over the business carried on during the past nineteen years by Messrs. J. and W. E. Clarke, known as the Sydney Foundry & Machine Works. Arrangements have already been completed for the installation of modern forging machines and a complete plant for the manufacture of steel castings, bolts, rivets, spikes, etc. New buildings will also be erected. The capital stock of the company is $100,000.

Neat Envelope Opener.

Flockton, Tompkin & Co., Limited, Newhall Steel Works, Sheffield, makers of the "Cat" brand of steel for engineers, miners, etc., are sending out a very neat envelope opener; an article always useful to the man at the desk. It is of bright polished steel, somewhat in the form of an open pocket-knife, with the name of the makers and the "Cat" brand neatly brought out on the solid handle.

To Enlarge Brass Foundry.

W. R. Cuthbert & Co., Montreal, brass founders, have obtained a vacant lot adjoining their property and purpose, during the present year, to enlarge their premises and give employment to about five hundred additional men. The improvements will cost the firm $100,000.

Companies Incorporated

The Provincial Construction Co., Ltd., Montreal; capital, $20,000; purpose, to carry on the business of contracting; directors, J. P. Pauze, H. Baurgard, and others, all of Montreal.

Canada Office Furniture Co., Ltd., of Montreal; capital, $35,000; purpose, to manufacture and deal in furniture, etc.; directors, D. Wishart, D. E. Turner, F. A. Wishart, all of Montreal.

Northern Oil and Gas Co., Montreal; to carry on a general refinery business; capital, $90,000; incorporators, J. M. Fortier, M. Marchand, L. M. Fortier, J. A. Mann and S. W. Ridenour.

Shaw & Mason, Limited, Sydney, N. S., has been organized with a capital of $100,000 to manufacture brass and cast-iron fittings. The company will be the only one of its kind in Nova Scotia.

The Wm. Strachan Co., Ltd., Montreal; capital, $100,000; purpose, to manufacture soaps, perfumes, etc.; directors, G. W. MacDougall, L. MacFarlane and C. A. Pope, all of Montreal.

Frame and Hay Fence Co., Stratford; to manufacture and deal in wire fences, gates, and other fencing materials; capital, $100,000; incorporators: R. S. Frame, D. D. Hay, R. B. Murray.

Structural Steel Co., Ltd., Montreal; capital, $500,000; purpose, to carry on bridge building, and to manufacture structural iron; directors, T. Johnson, W. C. McIntyre and others, Montreal.

The Folding Box Co., Ltd., Owen Sound; capital, $49,000; purpose, to manufacture folding wooden boxes; directors, Matthew Kennedy, Norman Ross and A. G. McKay, all of Owen Sound.

The Mergenthaler Co., Ltd., Toronto; capital, $40,000; purpose, to manufacture and deal in linotype, type casting and composing machines; directors, J. D. Montgomery, R. A. Montgomery and

Crown Canister Co., Dundas; to manufacture and deal in canisters, dies, wire and paper goods and metal ware; capital, $40,000; incorporators: R. R. Gamey, S. Metcalfe, W. B. Bentley, Jas. Watt, Jr.

The S. Cote Motor Co., Ltd., Montreal; capital, $20,000; purpose, to manufacture and deal in gas and gasoline engines, steam engines, etc.; directors, Simeon Cote, A. Meunier and R. LucPleur, all of Montreal.

The Perrin Shocker Mfg. Co., New Liskeard, Ont.; capital, $100,000; purpose, to manufacture and sell agricultural implements.; directors, Thos. McCamus, Wm. J. Emmerson, and J. L. Brown, all of New Liskeard.

J. W. Harris Manufacturing Co., Ltd., Montreal; capital, $1,000,000; purpose, to carry on business as engineers, shipbuilders, founders and contractors; directors, Thos. Craig, W. B. Powell, H. C. Mussen and J. W. Harris, contractor, all of Montreal.

The Standard Sanitary Mfg. Co., of Canada, Ltd., Toronto; capital, $20,000; purpose, to manufacture cast iron and enamel ware, brass goods, soil pipe, and other plumbers' supplies; directors, G. S. Hodgson, H. W. Maw, C. T. Gillespie, all of Toronto.

The Twin City Oil Company, Berlin, Ontario; capital, $40,000; to manufacture and deal in all kinds of oils, greases and compounds. Provisional directors: V. O. Phillips, J. A. Phillips, R. Richmond, C. N. Huether and Elizabeth Ann Phillips.

The Dictograph Company of Canada, with head office in Toronto, has been formed, with a capital of $250,000. The company have been authorized to manufacture dictographs, electrical, acoustic, and office specialties, as well as labor saving devices of all descriptions.

The St. Simeon Lumber Co., Ltd., St. Hyacinthe, Que.; capital, $200,000; purpose, to carry on the lumber trade in all its branches, including the building and operating of saw mills, etc.; directors, Ovide Brouillard, Michael Archambault, J. P. P. Robert, J. A. Tellier, all of St. Hyacinthe.

Standard Fitting & Valve Co., Ltd., Guelph; capital, $100,000; purpose, to manufacture cast and malleable iron fittings, valves, etc., and to carry on business as foundrymen and machinists; directors, Henry Aird, G. W. Aird, E. S. Platt, Troy, U.S.A., and J. M. Taylor, G. D. Forbes, Guelph.

National Refining Company, Toronto; capital, $20,000; to carry on the business of buying, selling, smelting and refining ores and metals, and to deal in dental and other trade supplies. Provisional directors: W. M. McTavish, R. A. McTavish and R. J. Dunlop.

The Wilson Automobile Co., Ltd., Ottawa; capital, $145,000; purpose, to carry on the business of mechanical engineers and manufacturers of automobiles, motor boats, engines and other machinery; directors, B. S. Wilson, H. R. Wilson and G. H. Wilson, all of Montreal, and S. H. McKay, of Ottawa.

Sterling Gas Company, Port Colborne, Ontario; capital, $40,000; to obtain oil or gas by drilling for the same or by any other means and to pipe the same and to deal in and with oil or gas. Provisional directors: M. A. Reeb, C. E. Steel and Mary Elizabeth Reeb.

The Ontario Steel Tubular Axle Co., Belleville, has been incorporated with a capital of $20,000 to manufacture and deal in steel tubular axles, and to carry on a foundry machine shop. The provisional directors are: H. P. Thomas, R. E. Colling, J. S. McKeown, W. J. Thomson, J. W. Boyd and J. C. Panter.

Imperial Supply Co., Montreal; to manufacture railway, marine and contractors' supplies, either in metal or wood; capital, $100,000; incorporators: Herbert H. Bradfield, Harry H. Bradfield, W. R. Duckworth, H. G. Myers, C. A. Myers.

Harriston Stove Co., Harriston, Ont., to manufacture stoves, ranges, heaters, furnaces, tools, machinery, etc., etc., and to take over as a going concern the Canada Stove Works of the same town; incorporators: J. E. Cave, F. Blacker, F. Burger, Jas. F. Hinde, Anson Spotton.

Directory of Consulting Engineers, Patent Attorneys, Architects and Contractors.

CANADIAN MACHINERY BUYERS' DIRECTORY

To Our Readers—Use this directory when seeking to buy any machinery or power equipment. You will often get information that will save you money.

To Our Advertisers—Send in your name for insertion under the heading of the lines you make or sell.

To Non-Advertisers—A nominal rate of $1 per line a year is charged non-advertisers.

Acids.
Canada Chemical Mfg. Co., London.

Abrasive Materials.
Baxter, Paterson & Co., Montreal.
The Canadian Fairbanks Co., Montreal.
Rice Lewis & Son, Toronto.
Norton Co., Worcester, Mass.
H. W. Petrie, Toronto.
Williams & Wilson, Montreal.

Air Brakes.
Canada Foundry Co., Toronto.
Canadian Westinghouse Co., Hamilton.

Air Receivers.
Allis-Chalmers-Bullock Montreal.
Canada Foundry Co., Toronto.
Canadian Rand Drill Co., Montreal.
Chicago Pneumatic Tool Co., Chicago.
John McDougall Caledonian Iron Works Co., Montreal.

Alloy, Ferro-Silicon.
Paxon, J. W, Co. Philadelphia.

Alundum Scythe Stones
Norton Company, Worcester, Mass.

Arbor Presses.
Niles-Bement-Pond Co., New York.

Augers.
Chicago Pneumatic Tool Co., Chicago.
Frothingham & Workman, Ltd., Montreal
Rice Lewis & Son, Toronto.

Automatic Machinery.
Cleveland Automatic Machine Co. Cleveland.
National-Acme Mfg. Co., Cleveland
Potter & Johnston Machine Co., Pawtucket, R. I.

Automobile Parts
Globe Machine & Stamping Co., Cleveland, Ohio.

Axle Cutters.
Butterfield & Co., Rock Island, Que.
A. B. Jardine & Co., Hespeler, Ont.

Axle Setters and Straighteners.
A. B. Jardine & Co., Hespeler, Ont.
Standard Bearings, Ltd, Niagara Falls

Babbit Metal.
Baxter Paterson & Co., Montreal.
Canada Metal Co., Toronto.
Canada Machinery Agency, Montreal.
Frothingham & Workman Ltd., Montreal
Greey Wm. & J. G, Toronto.
Rice Lewis & Son, Toronto.
Lumen Bearing Co., Toronto.
Mechanics Supply Co, Quebec, Que.
Miller Bros. & Toms, Montreal, Que.

Bakers' Machinery.
Greey, Wm. & J. G., Toronto.

Baling Presses.
Perrin, Wm. R., & Co., Toronto.

Barrels, Steel Shop.
Cleveland Wire Spring Co., Cleveland.

Barrels, Tumbling.
Buffalo Foundry Supply Co., Buffalo.
Detroit Foundry Supply Co, Windsor
Dominion Foundry Supply Co., Montreal
Hamilton Facing Mill Co., Hamilton.
Globe Machine & Stamping Co, Cleveland, Ohio.
John McDougall Caledonian Iron Works Co., Montreal.
R. McDougall Co., Galt, Ont.
Miller Bros. & Toms, Montreal, Que.
Northern Engineering Works, Detroit.
J. W. Paxson Co., Philadelphia, Pa.
H. W. Petrie, Toronto.
The Smart-Turner Mach. Co., Hamilton.

Bars, Boring.
Hall Engineering Works, Montreal.
Miller Bros. & Toms, Montreal.
Niles-Bement-Pond Co., New York.
Standard Bearings, Ltd., Niagara Falls

Batteries, Dry.
Berlin Electrical Mfg. Co., Toronto.
Mechanics Supply Co., Quebec, Que.

Batteries, Flashlight.
Berlin Electrical Mfg. Co., Toronto.
Mechanics Supply Co., Quebec Que

Batteries, Storage.
Canadian General Electric Co. Toronto
Chicago Pneumatic Tool Co., Chicago.
Rice Lewis & Son, Toronto.
John Millen & Son, Montreal.

Bearing Metals.
Lumen Bearing Co., Toronto.

Bearings, Roller.
Standard Bearings, Ltd. Niagara Falls

Bell Ringers.
Chicago Pneumatic Tool Co., Chicago.
Mechanics Supply Co., Quebec, Que.

Belting, Chain.
Baxter, Paterson & Co. Montreal.
Canada Machinery Exchange, Montreal.
Greey, Wm. & J. G., Toronto.
Jeffrey Mfg. Co., Columbia, Ohio.
Link-Belt Eng. Co., Philadelphia.
Miller Bros. & Toms, Montreal.
Waterous Engine Works Co., Brantford.

Belting, Cotton.
Baxter, Paterson & Co., Montreal.
Canada Machinery Agency, Montreal.
Dominion Belting Co., Hamilton.
Rice Lewis & Son, Toronto.

Belting, Leather.
Baxter, Paterson & Co., Montreal.
Canada Machinery Agency, Montreal.
The Canadian Fairbanks Co., Montreal.
Frothingham & Workman Ltd., Montreal
Greey, Wm. & J. G., Toronto.
McLaren, J. C., Montreal.
Rice Lewis & Son, Toronto.
H. W. Petrie, Toronto.
Williams & Wilson, Montreal.

Belting, Rubber.
Baxter, Paterson & Co., Montreal.
Canada Machinery Agency, Montreal.
Frothingham & Workman, Ltd., Montreal
Greey, Wm. & J. G., Toronto.

Belting Supplies.
Baxter, Paterson & Co, Montreal.
Greey, Wm. & J. G., Toronto.
Rice Lewis & Son, Toronto.
H. W. Petrie, Toronto.

Bending Machinery.
John Bertram & Sons Co., Dundas, Ont.
Chicago Pneumatic Tool Co., Chicago.
Rice Lewis & Son, Toronto.
London Mach. Tool Co., Hamilton, Ont.
National Machinery Co., Tiffin, Ohio.
Niles-Bement-Pond Co., New York.

Benders, Tire.
A. B. Jardine & Co., Hespeler, Ont.
London Mach. Tool Co., Hamilton, Ont.

Blowers.
Buffalo Foundry Supply Co., Montreal.
Canada Machinery Agency, Montreal.
Detroit Foundry Supply Co, Windsor
Hamilton Facing Mill Co., Hamilton.
Kerr Turbine Co., Wellsville, N.Y.
Mechanics Supply Co., Quebec, Que.
J. W. Paxson, Philadelphia, Pa.
Sheldon's Limited, Galt.
Sturtevant, B. F., Co., Hyde Park, Mass.

Blast Gauges—Cupola.
Paxson, J. W., Co. Philadelphia.
Sheldons, Limited, Galt

Blocks, Tackle.
Frothingham & Workman, Ltd, Montreal

Blocks, Wire Rope.
Frothingham & Workman, Ltd., Montreal

Blue Printing.
The Electric Blue Print Co., Montreal.

Blow-Off Tanks.
Darling Bros., Ltd., Montreal.

Boilers.
Canada Foundry Co., Limited, Toronto.
Canada Machinery Agency, Montreal.
Goldie & McCulloch Co., Galt.
E. Leonard & Sons, London, Ont.
John McDougall Calenonian Iron Works, Montreal.
Manitoba Iron Works, Winnipeg.
Mechanics Supply Co., Quebec, Que
Owen Sound Iron Works Co., Owen Sound.
H. W. Petrie, Toronto.
Robb Engineering Co., Amherst, N.S.
The Smart-Turner Mach. Co., Hamilton.
Waterous Engine Works Co., Brantford.
Williams & Wilson, Montreal.

Boiler Compounds.
Canada Chemical Mfg. Co., London, Ont.
Hall Engineering Works, Montreal.
Lake Erie Boiler Compound Co., Toronto

Bolt Cutters.
John Bertram & Sons Co., Dundas, Ont.
London Mach. Tool Co., Hamilton.
Mechanics Supply Co., Quebec, Que.
National Machinery Co., Tiffin, Ohio.
Niles-Bement-Pond Co., New York.

Bolt and Nut Machinery.
John Bertram & Sons Co., Dundas, Ont.
Canada Machinery Agency, Montreal.
Rice Lewis & Son, Toronto.
London Mach. Tool Co., Hamilton.
National Machinery Co., Tiffin, Ohio.
Niles-Bement-Pond Co. New York.

Bolts and Nuts, Rivets.
Baxter, Paterson & Co., Montreal.
Mechanics Supply Co., Quebec, Que.

Boring and Drilling Machines.
American Tool Works Co., Cincinnati.
B. F. Barnes Co., Rockford, Ill.
John Bertram & Sons Co., Dundas, Ont.
Canada Machinery Agency, Montreal.
A. B. Jardine & Co., Hespeler, Ont.
London Mach. Tool Co., Hamilton.
Niles-Bement-Pond Co., New York.

Boring Machine, Upright.
American Tool Works Co., Cincinnati.
John Bertram & Sons Co., Dundas, Ont.
London Mach. Tool Co., Hamilton.
Niles-Bement-Pond Co., New York.

Boring Machine, Wood.
Chicago Pneumatic Tool Co., Chicago.
London Mach. Tool Co., Hamilton.

Boring and Turning Mills.
American Tool Works Co., Cincinnati.
John Bertram & Sons Co., Dundas, Ont.
Gisholt Machine Co., Madison, Wis.
Canada Machinery Agency, Montreal.
Rice Lewis & Son, Toronto.
London Mach. Tool Co., Hamilton.
Niles-Bement-Pond Co., New York.
H. W. Petrie, Toronto.

Box Puller.
A. B. Jardine & Co., Hespeler, Ont.

Boxes, Steel Shop.
Cleveland Wire Spring Co., Cleveland.

Boxes, Tote.
Cleveland Wire Spring Co., Cleveland.

Brass Foundry Equipment.
Detroit Foundry Supply Co., Detroit.
Paxson, J. W., Co., Philadelphia

Brass Working Machinery.
Warner & Swasey Co., Cleveland, Ohio.

Brushes, Foundry and Core.
Buffalo Foundry Supply Co., Buffalo.
Detroit Foundry Supply Co, Windsor.
Dominion Foundry Supply Co., Montreal
Hamilton Facing Mill Co., Hamilton.
Mechanics Supply Co., Quebec, Que.
Paxson, J. W., Co., Philadelphia

Brushes, Steel.
Buffalo Foundry Supply Co., Buffalo.
Paxson, J. W., Co., Philadelphia

Bulldozers.
John Bertram & Sons Co., Dundas, Ont.
London Mach. Tool Co., Hamilton, Ont.
National Machinery Co., Tiffin, Ohio.
Niles-Bement-Pond Co., New York.

Calipers.
Baxter, Patterson & Co., Montreal.
Frothingham & Workman, Ltd., Montreal
Rice Lewis & Son, Toronto.
Mechanics Supply Co., Quebec, Que.
John Millen & Son, Ltd., Montreal, Que.
L. S. Starrett & Co., Athol, Mass.
Williams & Wilson Montreal.

Cars, Foundry.
Buffalo Foundry Supply Co., Buffalo.
Detroit Foundry Supply Co., Windsor
Dominion Foundry Supply Co., Montreal
Hamilton Facing Mill Co., Hamilton.
Paxson, J. W., Co., Philadelphia

Castings, Aluminum.
Lumen Bearing Co., Toronto

Castings, Brass.
Chadwick Bros., Hamilton.
Greey, Wm. & J. G., Toronto.
Hall Engineering Works, Montreal.
Kennedy, Wm., & Sons, Owen Sound.
Lumen Bearing Co., Toronto
McDougall R. Co, Galt.
Niagara Falls Machine & Foundry Co. Niagara Falls, Ont.
Owen Sound Iron Works Co., Owen Sound.
Robb Engineering Co, Amherst, N.S.

Castings, Grey Iron.
Allis-Chalmers-Bullock Montreal.
Greey, Wm. & J. G., Toronto.
Hall Engineering Works, Montreal.
Kennedy, Wm., & Sons, Owen Sound.
Laurie Engine & Machine Co., Montreal.
Manson Mfg. Co., Thorold, Ont
Maxwell, David, & Sons, St. Marys.
John McDougall Caledonian Iron Works Co., Montreal.
R. McDougall Co., Galt, Ont.
Niagara Falls Machine & Foundry Co., Niagara Falls, Ont
Owen Sound Iron Works Co., Owen Sound.
Robb Engineering Co., Amherst, N.S.
Smart-Turner Machine Co., Hamilton.

Castings, Phosphor Bronze.
Lumen Bearing Co., Toronto

Castings, Steel.
Kennedy, Wm., & Sons, Owen Sound.

Castings, Semi-Steel.
Robb Engineering Co., Amherst, N.S.

Cement Machinery.
Allis-Chalmers-Bullock, Limited, Montreal
Greey, Wm. & J. G., Toronto
Jeffrey Mfg. Co., Columbus, Ohio.
John McDougall Caledonian Iron Works, Co., Montreal
Owen Sound Iron Works Co., Owen Sound

Centreing Machines.
John Bertram & Sons Co., Dundas, Ont.
Jeffrey Mfg Co., Columbus, Ohio
London Mach. Tool Co., Hamilton, Ont.
Niles-Bement-Pond Co., New York.
Pratt & Whitney Co., Hartford, Conn.
Standard Bearings, Ltd., Niagara Falls

Centres, Planer.
American Tool Works Co., Cincinnati.

Centrifugal Pumps.
Manson Mfg. Co., Thorold, Ont
John McDougall Caledonian Iron Works Co, Montreal.
R. McDougall Co., Galt.
D'Olier Engineering Co., New York.
Pratt & Whitney Co., Hartford, Conn.

Centrifugal Pumps—Turbine Driven
Kerr Turbine Co., Wellsville, N.Y.

Chain, Crane and Dredge.
Frothingham & Workman, Ltd., Montreal

Chaplets.

Buffalo Foundry Supply Co., Buffalo.
Detroit Foundry Supply Co., Windsor.
Hamilton Facing Mill Co., Hamilton.
Paxson, J. W., Co., Philadelphia

Charcoal.

Buffalo Foundry Supply Co. Buffalo.
Detroit Foundry Supply Co., Windsor.
Dominion Foundry Supply Co., Montreal
Hamilton Facing Mill Co., Hamilton.
Paxson, J. W., Co., Philadelphia

Chemicals.

Baxter, Paterson & Co., Montreal.
Canada Chemical Co., London.

Chemists' Machinery.

Greey, Wm & J. G., Toronto.

Chemists, Industrial.

Detroit Testing Laboratory, Detroit.

Chemists, Metallurgical.

Detroit Testing Laboratory, Detroit

Chemists, Mining.

Detroit Testing Laboratory, Detroit.

Chucks, Ring Grinding.

A. B. Jardine & Co., Hespeler, Ont.
Chicago Pneumatic Tool Co., Chicago.

Chucks, Drill and Lathe.

American Tool Works Co., Cincinnati.
Baxter, Paterson & Co., Montreal.
John Bertram & Sons Co., Dundas, Ont.
Canada Machinery Agency, Montreal.
Frothingham & Workman, Ltd., Montreal
Ker & Goodwin, Brantford.
A. B. Jardine & Co., Hespeler, Ont.
London Mach. Tool Co., Hamilton.
John Millen & Son, Ltd. Montreal.
Niles-Bement-Pond Co., New York.
H. W. Petrie, Toronto.
Rice Lewis & Son, Toronto.
Standard Tool Co., Cleveland.

Chucks, Planer.

American Tool Works Co., Cincinnati.
Canada Machinery Agency, Montreal.
Niles-Bement-Pond Co., New York.

Chucking Machines.

American Tool Works Co., Cincinnati.
Niles-Bement-Pond Co., New York.
H. W. Petrie, Toronto.
Warner & Swasey Co., Cleveland, Ohio

Circuit Breakers.

Allis-Chalmers-Bullock, Limited, Montreal
Canadian General Electric Co., Toronto.
Canadian Westinghouse Co., Hamilton.

Clippers, Bolt.

Frothingham & Workman, Ltd. Montreal
A. B. Jardine & Co., Hespeler, Ont.

Cloth and Wool Dryers.

Dominion Heating and Ventilating Co., Hespeler.
B. Greening Wire Co., Hamilton.
Sheldons Limited, Galt

Collectors, Pneumatic.

Sheldons Limited, Galt
Sturtevant, B. F., Co., Hyde Park, Mass.

Compressors, Air.

Allis-Chalmers-Bullock, Limited, Montreal
Canada Foundry Co., Limited, Toronto.
Canada Machinery Agency, Montreal.
Canadian Rand Drill Co., Montreal.
Canadian Westinghouse Co., Hamilton.
Chicago Pneumatic Tool Co., Chicago.
Darling Bros., Ltd., Montreal
Detroit Foundry Supply Co., Windsor.
John McDougall, Caledonian Iron Works Co., Montreal
H. W. Petrie, Toronto.
The Smart-Turner Mach. Co., Hamilton.
Hall Engineering Works, Montreal, Que.
London Mach. Tool Co., Hamilton.
Niles-Bement-Pond Co., New York.
H. W. Petrie, Toronto.
Pratt & Whitney Co., Hartford, Conn.
Williams & Wilson, Montreal.

Concentrating Plant.

Allis-Chalmers-Bullock, Montreal.
Greey, Wm. & J. G., Toronto.

Concrete Mixers.

Jeffrey M'f'g. Co., Columbus, Ohio.
Link-Belt Co., Philadelphia.

Condensers.

Canada Foundry Co., Limited, Toronto.
Canada Machinery Agency, Montreal.
Hall Engineering Works Montreal.
Smart-Turner-Machine Co., Hamilton.
Waterous Engine Co., Brantford.

Confectioners' Machinery.

Baxter, Paterson & Co., Montreal.
Greey, Wm & J. G., Toronto.
Pendrith Machinery Co., Toronto.

Consulting Engineers.

Charles Brandeis, Montreal.
Canadian White Co., Montreal.
Hall Engineering Works, Montreal.
Jules De Clercy, Montreal.
John S. Fielding, Toronto.
Miller Bros. & Toms, Montreal.
Roderick J. Parke, Toronto.
Plews & Trimingham, Montreal.
T. Pringle & Son, Montreal.
Standard Bearings, Ltd., Niagara Falls

Contractors.

Canadian White Co., Montreal.
D'Olier Engineering Co., New York.
Expanded Metal and Fireproofing Co. Toronto.
Hall Engineering Works Montreal.
Laurie Engine & Machine Co., Montreal.
John McDougall Caledonian Iron Works Co., Montreal.
Miller Bros. & Toms, Montreal.
Robb Engineering Co., Amherst N S.
The Smart-Turner Mach. Co., Hamilton.

Contractors' Plant.

Allis-Chalmers-Bullock, Montreal.
John McDougall, Caledonian Iron Works Co., Montreal.
Niagara Falls Machine & Foundry Co., Niagara Falls, Ont.

Contractors' Supplies.

Manson Mfg. Co., Thorold, Ont.

Controllers and Starters Electric Motor.

Allis-Chalmers-Bullock, Montreal.
Canadian General Electric Co., Toronto.
Canadian Westinghouse Co., Hamilton.
T. & H. Electric Co., Hamilton.

Conveying Systems.

Sheldons Limited, Galt
Sturtevant, B. F., Co., Hyde Park, Mass.

Converters, Steel.

Northern Engineering Works, Detroit.
Paxson, J. W., Co., Philadelphia

Conveyor Machinery.

Baxter, Paterson & Co., Montreal.
Greey, Wm. & J. G., Toronto
Jeffrey Mfg. Co., Columbus, Ohio.
Rice Lewis & Son, Toronto.
Link-Belt Co., Philadelphia.
John McDougall Caledonian Iron Works Co., Montreal.
Miller Bros. & Toms, Montreal.
Smart-Turner Machine Co., Hamilton.
Laurie Engine & Machine Co., Montreal.
Waterous Engine Works Co., Brantford.
Williams & Wilson, Montreal.

Coping Machines.

John Bertram & Sons Co., Dundas, Ont.
London Mach. Tool Co., Hamilton.
Niles-Bement-Pond Co., New York.

Core Compounds.

Buffalo Foundry Supply Co., Buffalo.
Detroit Foundry Supply Co., Windsor.
Dominion Foundry Supply Co., Montreal
Hamilton Facing Mill Co., Hamilton.
Paxson, J. W., Co., Philadelphia

Core-Making Machines.

Paxson, J. W., Co., Philadelphia

Core Ovens.

Detroit Foundry Supply Co., Windsor.
Hamilton Facing Mill Co., Hamilton.
Paxson, J. W., Co., Philadelphia
Sheldons Limited, Galt

Core Oven Bricks.

Buffalo Foundry Supply Co., Buffalo.
Detroit Foundry Supply Co., Windsor.
Hamilton Facing Mill Co., Hamilton.
Paxson, J. W., Co., Philadelphia

Core Wash.

Buffalo Foundry Supply Co., Buffalo.
Detroit Foundry Supply Co., Windsor.
Hamilton Facing Mill Co., Hamilton.
Paxson, Co., J. W., Philadelphia

Couplings.

Owen Sound Iron Works Co., Owen Sound

Cranes, Electric and Hand Power.

Canada Foundry Co., Limited, Toronto
Dominion Foundry Supply Co., Montreal
Hamilton Facing Mill Co., Hamilton.
Link-Belt Co., Philadelphia.
John McDougall, Caledonian Iron Works Co., Montreal.
Miller Bros. & Toms, Montreal.
Niles-Bement-Pond Co., New York.
Northern Engineering Works, Detroit
Owen Sound Iron Works Co., Owen Sound
Paxson, J. W., Co., Philadelphia
Smart-Turner-Machine Co., Hamilton.

Crank Pin.

Sight Feed Oil Pump Co., Milwaukee, Wis.

Crankshafts.

The Canada Forge Co., Welland.
St. Clair Bros., Galt

Crabs.

Frothingham & Workman, Ltd., Montreal

Crank Pin Turning Machine

London Mach. Tool Co., Hamilton.
Niles-Bement-Pond Co., New York.

Cross Head Pin.

Sight Feed Oil Pump Co., Milwaukee, Wis.

Crucibles.

Buffalo Foundry Supply Co., Buffalo.
Detroit Foundry Supply Co., Windsor
Dominion Foundry Supply Co., Montreal
Hamilton Facing Mill Co., Hamilton.
J. W. Paxson Co., Philadelphia.

Crucible Caps

Hamilton Facing Mill Co., Hamilton.
Paxson, J. W., Co., Philadelphia

Crushers, Rock or Ore.

Allis-Chalmers-Bullock, Montreal.
Jeffrey Mfg. Co., Columbus, Ohio.

Cup Grease.

Mechanics Supply Co., Quebec

Cupolas.

Buffalo Foundry Supply Co., Buffalo.
Detroit Foundry Supply Co., Windsor
Dominion Foundry Supply Co., Montreal
Hamilton Facing Mill Co., Hamilton.
Northern Engineering Works, Detroit
J. W. Paxson Co., Philadelphia.
Sheldons Limited, Galt.

Cupola Blast Gauges.

Paxson, J. W., Co., Philadelphia
Sheldons Limited, Galt

Cupola Blocks.

Detroit Foundry Supply Co., Detroit.
Hamilton Facing Mill Co., Hamilton
Northern Engineering Works Detroit
Ontario Lime Association Toronto
Paxson, J. W., Co., Philadelphia
Toronto Pottery Co., Toronto

Cupola Blowers.

Buffalo Foundry Supply Co., Buffalo.
Canada Machinery Agency, Montreal.
Detroit Foundry Supply Co., Windsor
Dominion Heating and Ventilating Co., Hespeler
Hamilton Facing Mill Co., Hamilton.
Northern Engineering Works, Detroit
Paxson J. W., Co., Philadelphia
Sheldon's Limited, Galt.
B. F. Sturtevant Co., Hyde Park, Mass.

Cutters, Flue

Chicago Pneumatic Tool Co., Chicago.
J. W. Paxson Co., Philadelphia.

Cutters, Milling.

Becker, Brainard Milling Machine Co. Hyde Park, Mass.
Frothingham & Workman, Ltd., Montreal
Pratt & Whitney Co., Hartford, Conn.
Standard Tool Co., Cleveland.

Cutting-off Machines.

John Bertram & Sons Co., Dundas, Ont.
Canada Machinery Agency, Montreal.
London Mach. Tool Co., Hamilton.
H. W. Petrie, Toronto.
Pratt & Whitney Co., Hartford, Conn.

Cutting-off Tools.

Armstrong Bros. Tool Co., Chicago.
Baxter Paterson & Co., Montreal.
London Mach. Tool Co., Hamilton.
Mechanics Supply Co., Quebec, Que.
H. W. Petrie, Toronto.
Pratt & Whitney, Hartford, Conn.
Rice Lewis & Son, Toronto.
L. S. Starrett Co., Athol, Mass.

Damper Regulators.

Darling Bros., Ltd., Montreal

Dies

Globe Machine & Stamping Co., Cleveland, Ohio.

Die Stocks

Curtis & Curtis Co., Bridgeport, Conn.
Hart Manufacturing Co., Cleveland, Ohio
Mechanics Supply Co., Quebec

Dies, Opening

W. H. Banfield & Sons, Toronto
Globe Machine & Stamping Co., Cleveland Ohio.
R. McDougall Co., Galt.
Pratt & Whitney Co., Hartford Conn.

Dies, Sheet Metal.

W. H. Banfield & Sons Toronto.
Globe Machine & Stamping Co., Cleveland, Ohio

Dies, Threading.

Frothingham & Workman, Ltd., Montreal
Hart Mfg. Co., Cleveland
Mechanics Supply Co., Quebec, Que.
John Millen & Son, Ltd., Montreal.

Draft, Mechanical.

W. H. Banfield & Sons, Toronto.
Butterfield & Co., Rock Island, Que.
A. B. Jardine & Co., Hespeler
Mechanics Supply Co., Quebec, Que.
Pratt & Whitney Co., Hartford, Conn.
Sheldon's Limited, Galt.
Sturtevant, B. F., Co., Hyde Park, Mass.

Drawing Instruments.

Rice Lewis & Son, Toronto.
Mechanics Supply Co., Quebec, Que.

Drawing Supplies.

The Electric Blue Print Co., Montreal.

Drawn Steel, Cold.

Baxter, Paterson & Co., Montreal.
Greey, Wm. & J. G., Toronto
Miller Bros. & Toms, Montreal.
Union Drawn Steel Co., Hamilton.

Drilling Machines, Arch Bar.

John Bertram & Sons Co., Dundas, Ont.
London Mach. Tool Co., Hamilton
Niles-Bement-Pond Co., New York.

Drilling Machines, Boiler.

American Tool Works Co., Cincinnati.
John Bertram & Sons Co., Dundas, Ont.
Bickford Drill and Tool Co., Cincinnati.
The Canadian Fairbanks Co., Montreal.
A. B. Jardine & Co., Hespeler, Ont.
London Mach. Tool Co., Hamilton, Ont.
Niles-Bement-Pond Co., New York.
H. W. Petrie, Toronto.
Williams & Wilson, Montreal

Drilling Machines Connecting Rod.

John Bertram & Sons Co., Dundas, Ont.
London Mach. Tool Co., Hamilton.
Niles-Bement-Pond Co. New York.

Drilling Machines, Locomotive Frame.

American Tool Works Co., Cincinnati.
B. F. Barnes Co., Rockford, Ill.
John Bertram & Sons Co., Dundas, Ont.
London Mach. Tool Co., Hamilton, Ont.
Niles-Bement-Pond Co., New York.

Drilling Machines, Multiple Spindle.

American Tool Works Co., Cincinnati.
B. F. Barnes Co., Rockford, Ill.
Baxter, Paterson & Co., Montreal.
John Bertram & Sons Co., Dundas, Ont.
Bickford Drill & Tool Co., Cincinnati.
Canada Machinery Agency Montreal.
London Mach. Tool Co., Hamilton, Ont.
Niles-Bement-Pond Co., New York.
H. W. Petrie, Toronto.
Williams & Wilson, Montreal.

Drilling Machines, Pneumatic

Canada Machinery Agency, Montreal.

Drilling Machines, Portable

Baxter, Paterson & Co., Montreal.
A. B. Jardine & Co., Hespeler, Ont.
Mechanics Supply Co., Quebec, Que.
Niles-Bement-Pond Co., New York.

Drilling Machines, Radial.

American Tool Works Co., Cincinnati.
Baxter Paterson & Co., Montreal
John Bertram & Sons Co., Dundas, Ont.
Bickford Tool & Drill Co., Cincinnati.
The Canadian Fairbanks Co., Montreal.
London Mach. Tool Co., Hamilton.
Mechanics Supply Co., Quebec, Que.
Niles-Bement-Pond Co., New York.
H. W. Petrie Toronto.
Williams & Wilson, Montreal.

Drillingg Machines. Suspension.

John Bertram & Sons Co., Dundas, Ont.
Canada Machinery Agency Montreal.
London Mach. Tool Co., Hamilton.
Niles-Bement-Pond Co., New York

Drilling Machines, Turret.

John Bertram & Sons Co., Dundas, Ont.
London Mach. Tool Co., Hamilton.
Niles-Bement-Pond Co., New York.

Drilling Machines, Upright.

American Tool Works Co., Cincinnati.
B. F. Barnes Co., Rockford, Ill.
Baxter, Paterson & Co., Montreal.
John Bertram & Sons Co., Dundas, Ont
Dwight Slate Machine Co., Hartford, Conn.
A. B. Jardine & Co., Hespeler, Ont.
Rice Lewis & Son, Toronto.
London Mach. Tool Co., Hamilton.
Mechanics Supply Co., Quebec, Que
Niles-Bement-Pond Co., New York.
W. H. Petrie, Toronto.
Williams & Wilson Montreal.

Drills, Bench.

B. F. Barnes Co., Rockford, Ill.
Baxter, Paterson & Co., Montreal.
London Mach. Tool Co., Hamilton.
Mechanics Supply Co., Quebec Que
Pratt & Whitney Co., Hartford, Conn.

Drills, Blacksmith.

Canada Machinery Agency, Montreal.
Frothingham & Workman, Ltd., Montreal
A. B. Jardine & Co., Hespeler, Ont.
London Mach. Tool Co., Hamilton.
Mechanics Supply Co., Quebec, Que.
Standard Tool Co., Cleveland.

Drills, Centre.
Mechanics Supply Co. [Quebec], Que.
Pratt & Whitney Co., Hartford, Conn.
Standard Tool Co., Cleveland, O.
L. S. Starrett Co., Athol, Mass.

Drills, Electric
B. F. Barnes Co., Rockford, Ill.
Baxter Paterson & Co., Montreal.
Chicago Pneumatic Tool Co., Chicago.
Niles-Bement-Pond Co., New York.

Drills, Gang.
American Tool Works Co., Cincinnati.
B. F. Barnes Co., Rockford, Ill.
John Bertram & Sons Co., Dundas, Ont.
Pratt & Whitney Co., Hartford, Conn.

Drills, High Speed.
Wm. Abbott, Montreal.
American Tool Works Co., Cincinnati.
B. F. Barnes Co., Rockford, Ill.
Frothingham & Workman, Ltd., Montreal
Pratt & Whitney Co., Hartford, Conn.
Standard Tool Co., Cleveland, O.

Drills, Hand.
A. B. Jardine & Co., Hespeler, Ont.

Drills, Horizontal.
B. F. Barnes Co., Rockford, Ill.
John Bertram & Sons Co., Dundas, Ont.
Canada Machinery Agency, Montreal.
London Mach. Tool Co., Hamilton.
Niles-Bement-Pond Co., New York.

Drills, Pneumatic.
Canada Machinery Agency, Montreal.
Chicago Pneumatic Tool Co., Chicago.
Independent Pneumatic Tool Co., Chicago Ill.
Niles-Bement-Pond Co., New York.

Drills, Radial.
American Tool Works Co., Cincinnati.
John Bertram & Sons Co., Dundas, Ont.
Bickford Drill & Tool Co., Cincinnati
London Mach. Tool Co., Hamilton, Ont.
Niles-Bement-Pond Co., New York.

Drills, Ratchet.
Armstrong Bros. Tool Co., Chicago.
Frothingham & Workman, Ltd. Montreal
A. B. Jardine & Co., Hespeler.
Mechanics Supp'y Co., Quebec, Que.
Pratt & Whitney Co., Hartford, Conn.
Standard Tool Co., Cleveland.

Drills, Rock.
Allis-Chalmers-Bullock, Montreal.
Canadian Rand Drill Co., Montreal.
Chicago Pneumatic Tool Co., Chicago.
Jaffrey Mfg. Co., Columbus, Ohio.

Drills, Sensitive.
American Tool Works Co., Cincinnati.
B. F. Barnes Co., Rockford, Ill.
Canada Machinery Agency, Montreal.
Dwight Slate Machine Co., Hartford, Conn.
Niles-Bement-Pond Co., New York

Drills, Twist.
Baxter, Paterson & Co., Montreal.
Chicago Pneumatic Tool Co., Chicago
Frothingham & Workman, Ltd., Montreal
Alex. Gibb, Montreal.
A. B. Jardine & Co., Hespeler, Ont.
Mechanics Supply Co., Quebec Que.
John Millen & Son, Ltd., Montreal.
Morse Twist Drill and Machine Co., New Bedford, Mass.
Pratt & Whitney Co., Hartford, Conn.
Standard Tool Co., Cleveland.

Drying Apparatus of all Kinds.
Dominion Heating & Ventilating Co., Hespeler
Greey, Wm. & J. G., Toronto
Sheldon's Limited, Galt
B. F. Sturtevant Co., Hyde Park, Mass.

Dry Kiln Equipment.
Sheldons Limited, Galt
Sturtevant, B. F., Co., Hyde Park, Mass.

Dry Sand and Loam Facing.
Buffalo Foundry Supply Co., Buffalo.
Hamilton Facing Mill Co., Hamilton.
Paxson, J. W., Co., Philadelphia

Dump Cars.
Canada Foundry Co., Limited, Toronto
Greey, Wm. & J. G., Toronto
Hamilton Facing Mill Co., Hamilton.
John McDougall, Caledonian Iron Works Co., Montreal.
Niles-Bement-Pond Co., New York.
Standard Bearings, Ltd., Niagara Falls.
Jenckes Machine Co., Sherbrooke, Que
Link-Belt Eng. Co., Philadelphia.
John McDougall Caledonian Iron Works Co., Montreal
Miller Bros. & Toms, Montreal.
Owen Sound Iron Works Co., Owen Sound
Paxson, J. W., Co., Philadelphia
Waterous Engine Co., Brantford.

Dust Separators.
Greey, Wm. & J. G., Toronto
Dominion Heating and Ventilating Co., Hespeler
Paxson, J. W., Co., Philadelphia
Sheldon's Limited, Galt.

Dynamos.
Allis-Chalmers-Bullock, Montreal.
Canadian General Electric Co., Toronto.
Canadian Westinghouse Co., Hamilton.
Consolidated Electric Co., Toronto
Electrical Machinery Co., Toronto.
Hall Engineering Works, Montreal, Que.
Mechanics Supply Co., Quebec, Que.
John Millen & Son, Ltd., Montreal.
Packard Electric Co., St. Catharines.
H. W. Petrie, Toronto.
B. F. Sturtevant Co., Hyde Park, Mass.
T. & H. Electric Co., Hamilton.

Dynamos—Turbine Driven.
Kerr-Turbine Co., Wellsville, N.Y.

Economizer, Fuel.
Dominion Heating & Ventilating Co., Hespeler
Standard Bearings, Ltd., Niagara Falls.
B. F. Sturtevant Co., Hyde Park, Mass.

Electrical Instruments.
Canadian Westinghouse Co., Hamilton.
Mechanics Supply Co., Quebec, Que.

Electrical Steel.
Baxter, Patterson & Co., Montreal.

Electrical Supplies.
Canadian General Electric Co., Toronto.
Canadian Westinghouse Co., Hamilton.
London Mach. Tool Co., Hamilton, Ont.
Mechanics Supp'y Co., Quebec, Que.
John Millen & Son, Ltd., Montreal.
Packard Electric Co., St. Catharines.
T. & H. Electric Co., Hamilton.

Electrical Repairs
Canadian Westinghouse Co., Hamilton.
T. & H. Electric Co., Hamilton.

Elevator Buckets.
Greey, Wm. & J. G., Toronto
Jeffrey Mfg. Co., Columbus, Ohio.

Elevators, Foundry.
Northern Engineering Works, Detroit

Elevators—For any Service.
Darling Bros. Ltd., Montreal

Emery and Emery Wheels.
Baxter, Paterson & Co., Montreal.
Dominion Foundry Supply Co., Montreal
Frothingham & Workman Ltd., Montreal
Hamilton Facing Mill Co., Hamilton.
Paxson, J. W., Co., Philadelphia

Emery Wheel Dressers.
Baxter Paterson & Co., Montreal.
Canada Machinery Agency, Montreal.
Desmond-Stephan Mfg. Co., Urbana, Ohio
Dominion Foundry Supply Co., Montreal
Frothingham & Workman Ltd., Montreal
Hamilton Facing Mill Co., Hamilton.
International Specialty Co., Detroit.
Mechanics Supply Co., Quebec, Que.
John Millen & Son, Ltd., Montreal.
H. W. Petrie, Toronto.
Paxson, J. W., Co., Philadelphia
Standard Tool Co., Cleveland.

Engineers and Contractors.
Canada Foundry Co., Limited, Toronto.
Darling Bros., Ltd., Montreal
Greey, Wm. & J. G., Toronto
Hall Engineering Works, Montreal
Laurie Engine & Machine Co., Montreal.
Link-Belt Co., Philadelphia.
John McDougall, Caledonian Iron Works Co., Montreal.
Miller Bros. & Toms, Montreal.
Robb Engineering Co., Amherst, N.S.
The Smart-Turner Mach. Co., Hamilton.

Engineers' Supplies.
Baxter, Paterson & Co., Montreal
Frothingham & Workman Ltd., Montreal
Greey, Wm. & J. G., Toronto
Hall Engineering Works, Montreal.
Mechanics Supply Co., Quebec, Que.
Rice Lewis & Son, Toronto.

Engines, Alcohol.
Du Bois Iron Works, Du Bois, Pa

Engines, Gas and Gasoline.
Baxter, Paterson & Co., Montreal.
Canada Foundry Co., Toronto.
Canada Machinery Agency, Montreal.
The Canadian Fairbanks Co., Montreal.
Canadian McVicker Engine Co., Galt.
Du Bois Iron Works, Du Bois, Pa.
Economic Heat, Light and Power Supply Co., Toronto.
Gilson Mfg. Co., Guelph
The Goldie & McCulloch Co., Galt, Ont.
Rice Lewis & Son, Toronto.
Ontario Wind Engine & Pump Co., Toronto
H. W. Petrie, Toronto.
Producer Gas Co., Toronto
The Smart-Turner Mach. Co., Hamilton
Standard Bearings, Ltd., Niagara Falls.

Engines, Hoisting.
Du Bois Iron Works, Du Bois, Pa.

Engines, Oil.
Du Bois Iron Works, Du Bois, Pa.

Engines, Pumping.
Du Bois Iron Works, Du Bois, Pa.

Engines, Steam.
Allis-Chalmers-Bullock, Montreal
Belliss & Morcom, Birmingham, Eng.
Canada Machinery Agency, Montreal
The Goldie & McCulloch Co., Galt, Ont.
E. Leonard & Sons, London, Ont.
Rice Lewis & Son, Toronto
Laurie Engine & Machine Co., Montreal.
Manson Mfg. Co., Thorold, Ont.
John McDougall Caledonian Iron Works, Montreal.
Miller Bros. & Toms, Montreal.
Robb Engineering Co., Amherst, N.S.
Sheldons Limited, Galt.
The Smart-Turner Mach. Co., Hamilton.
B. F. Sturtevant Co., Hyde Park, Mass.
Waterous Engine Works Co., Brantford.

Exhaust Heads.
Darling Bros., Ltd., Montreal
Dominion Heating & Ventilating Co., Hespeler
Sheldon's Limited, Galt, Ont.
B. F. Sturtevant Co., Hyde Park, Mass.

Expanded Metal.
Expanded Metal and Fireproofing Co., Toronto

Expanders.
A. B. Jardine & Co., Hespeler, Ont.

Fans, Electric.
Canadian General Electric Co., Toronto.
Canadian Westinghouse Co., Hamilton.
Dominion Heating & Ventilating Co., Hespeler
Mechanics Supp'y Co., Quebec, Que.
Sheldon's Limited, Galt, Ont.
The Smart-Turner Mach. Co., Hamilton.
B. F. Sturtevant Co., Hyde Park, Mass.

Fans, Exhaust.
Canadian Buffalo Forge Co., Montreal.
Detroit Foundry Supply Co., Windsor.
Dominion Foundry Supply Co., Montreal
Dominion Heating & Ventilating Co., Hespeler.
Greey, Wm. & J. G., Toronto
Hamilton Facing Mill Co., Hamilton.
Paxson, J. W., Co., Philadelphia
Sheldon's Limited, Galt.
B. F. Sturtevant Co., Hyde Park, Mass.

Feed Water Heaters.
Darling Bros., Montreal
Du Bois Iron Works, Du Bois, Pa.
Laurie Engine & Machine Co., Montreal
John McDougall, Caledonian Iron Works Co., Montreal.
The Smart-Turner Mach. Co., Hamilton.

Files and Rasps.
Baxter, Paterson & Co., Montreal.
Frothingham & Workman, Ltd., Montreal
Mechanics Supply Co., Quebec, Que.
John Millen & Son, Ltd., Montreal.
Rice Lewis & Son, Toronto.
Nicholson File Co., Providence, R.I.
H. W. Petrie, Toronto.

Fillet, Pattern.
Baxter, Paterson & Co., Montreal.
Buffalo Foundry Supply Co., Buffalo.
Detroit Foundry Supply Co., Windsor.
Dominion Foundry Supply Co., Montreal
Hamilton Facing Mill Co., Hamilton.

Fire Apparatus.
Waterous Engine Works Co., Brantford.

Fire Brick and Clay.
Baxter Paterson & Co., Montreal
Buffalo Foundry Supply Co., Buffalo.
Detroit Foundry Supply Co., Windsor.
Dominion Foundry Supply Co., Montreal
Hamilton Facing Mill Co., Hamilton.
Ontario Lime Association, Toronto
J. W. Paxson Co., Philadelphia.
Toronto Pottery Co., Toronto

Flash Lights.
Berlin Electrical Mfg. Co., Toronto.
Mechanics Supply Co., Quebec, Que.

Flour Mill Machinery.
Allis-Chalmers-Bullock, Montreal.
Greey, Wm. & J. G., Toronto
The Goldie & McCulloch Co., Galt, Ont.
John McDougall, Caledonian Iron Works Co., Montreal.
Miller Bros. & Toms, Montreal.

Forges.
Canada Foundry Co., Limited, Toronto.
Frothingham & Workman Ltd., Montreal
Hamilton Facing Mill Co., Hamilton.
Mechanics Supply Co., Quebec, Que.
H. W. Petrie, Toronto.
Sheldon's Limited, Galt, Ont.
B. F. Sturtevant Co., Hyde Park, Mass.

Forgings, Drop.
John McDougall, Caledonian Iron Works Co., Montreal.
H. W. Petrie, Toronto.
St. Clair Bros., Galt

Forgings, Light & Heavy.
The Canada Forge Co., Welland.
Hamilton Steel & Iron Co., Hamilton

Forging Machinery.
John Bertram & Sons Co., Dundas, Ont.
London Mach. Tool Co., Hamilton, Ont.
National Machinery Co., Tiffin, Ohio
Niles-Bement-Pond Co., New York.

Founders.
Greey, Wm. & J. G., Toronto
John McDougall, Caledonian Iron Works Co., Montreal
R. McDougall Co., Galt, Ont.
Niagara Falls Machine & Foundry Co., Niagara Falls Ont.
Manson Mfg. Co., Thorold, Ont.
The Smart-Turner Mach. Co., Hamilton.

Foundry Equipment.
Detroit Foundry Supply Co., Windsor.
Hamilton Facing Mill Co., Hamilton
Hanna Engineering Works, Chicago.
Northern Engineering Works, Detroit
Paxson, J. W., Co., Philadelphia

Foundry Facings.
Buffalo Foundry Supply Co., Buffalo
Detroit Foundry Supply Co., Windsor.
Hamilton Facing Mill Co., Hamilton.
J. W. Paxson Co., Philadelphia Pa.

Friction Clutch Pulleys, etc.
The Goldie & McCulloch Co., Galt.
Greey, Wm. & J. G., Toronto
Link-Belt Co., Philadelphia.
Miller Bros. & Toms, Montreal.

Furnaces.
Detroit Foundry Supply Co., Windsor.
Dominion Foundry Supply Co., Montreal
Hamilton Facing Mill Co., Hamilton.
Paxson, J. W., Co., Philadelphia Co.

Galvanizing
Canada Metal Co., Toronto.
Ontario Wind Engine & Pump Co., Toronto

Gas Blowers and Exhausters.
Dominion Heating & Ventilating Co. Hespeler
Sheldons Limited, Galt.
B. F. Sturtevant Co., Hyde Park, Mass.

Gas Plants, Suction.
Baxter Paterson & Co., Montreal
Economic Heat, Light and Power Supply Co., Toronto
Producer Gas Co., Toronto
Standard Bearings, Ltd., Niagara Falls
Williams & Wilson, Montreal

Gauges, Standard.
Mechanics Supply Co., Quebec, Que.
Pratt & Whitney Co., Hartford, Conn.

Gearing.
Greey, Wm. & J. G. Toronto

Gear Cutting Machinery.
Baxter, Patterson & Co., Montreal
Becker-Brainard Milling Mach. Co., Hyde Park, Mass.
Bickford Drill & Tool Co., Cincinnati.
Dwight Slate Machine Co., Hartford, Conn.
Greey, Wm. & J. G., Toronto
Kennedy Wm. & Sons, Owen Sound
London Mach. Tool Co., Hamilton.
R. McDougall Co., Galt
Niles-Bement-Pond Co., New York.
H. W. Petrie, Toronto.
Pratt & Whitney Co., Hartford, Conn.
Williams & Wilson, Montreal.

Gears, Angle.
Chicago Pneumatic Tool Co., Chicago.
Greey, Wm. & J. G., Toronto
Laurie Engine & Machine Co., Montreal.
John McDougall, Caledonian Iron Works Co., Montreal.
Miller Bros. & Toms, Montreal.
Waterous Engine Co., Brantford.

Gears, Cut.
Kennedy, Wm., & Sons, Owen Sound

Gears, Iron.
Greey, Wm. & J. G., Toronto
Kennedy, Wm., & Sons, Owen Sound

Gears, Mortise.
Greey, Wm. & J. G., Toronto
Kennedy, Wm., & Sons, Owen Sound

Gears, Reducing.
Brown, David & Sons, Huddersfield Eng.
Chicago Pneumatic Tool Co., Chicago.
Greey, Wm. & J. G., Toronto
John McDougall, Caledonian Iron Works Co., Montreal.
Miller Bros. & Toms, Montreal.

Generators, Electric.

Allis-Chalmers-Bullock, Limited, Montreal
Canadian General Electric Co., Toronto
Canadian Westinghouse Co., Hamilton.
D'Olier Engineering Co., New York.
Hall Engineering Works, Montreal.
H. W. Petrie, Toronto.
B. F. Sturtevant Co., Hyde Park, Mass.
Toronto & Hamilton Electric Co. Hamilton.

Generators, Gas.

H. W. Petrie, Toronto.

Graphite Paints.

P. D. Dods & Co., Montreal.

Graphite.

Detroit Foundry Supply Co., Windsor.
Dominion Foundry Supply Co., Montreal
Hamilton Facing Mill Co., Hamilton.
Mechanics Supply Co., Quebec, Que.
Paxson, J. W., Co., Philadelphia

Grinders, Automatic Knife.

W. H. Banfield & Son, Toronto.

Grinders, Centre.

Niles-Bement-Pond Co., New York.
H. W. Petrie, Toronto.

Grinders, Cutter.

Becker-Brainard Milling Mach. Co., Hyde Park, Mass.
John Millen & Son, Ltd., Montreal.
Pratt & Whitney Co., Hartford, Conn.

Grinders, Tool.

Armstrong Bros. Tool Co., Chicago.
B. F. Barnes Co., Rockford, Ill.
H. W. Petrie, Toronto.
Williams & Wilson, Montreal.

Grinding Machines.

Brown & Sharpe, Providence, R.I.
The Canadian Fairbanks, Montreal.
Rice Lewis & Son, Toronto.
Niles-Bement-Pond Co., New York.
Norton Company, Worcester, Mass.
Paxson J. W., Co., Philadelphia
H. W. Petrie, Toronto.

Grinding and Polishing Machines.

The Canadian Fairbanks Co., Montreal.
Greey, Wm. & J. G., Toronto
Independent Pneumatic Tool Co., Chicago, Ill.
John Millen & Son, Ltd., Montreal.
Miller Bros. & Toms, Montreal.
Niles-Bement-Pond Co., New York.
H. W. Petrie, Toronto.
Pendrith Machinery Co., Toronto.
Standard Bearings, Ltd., Niagara Falls

Grinding Wheels (Alundum)

Norton Company, Worcester, Mass.

Hack Saws.

Baxter, Paterson & Co. Montreal.
Canada Machinery Agency, Montreal.
The Canadian Fairbanks Co., Montreal.
Frothingham & Workman, Ltd., Montreal
Mechanics Supply Co., Quebec, Que.
John Millen & Son, Ltd., Montreal.
Niles-Bement-Pond Co., New York.
H. W. Petrie, Toronto.
Williams & Wilson, Montreal.

Hack Saw Frames.

Mechanics Supply Co., Quebec, Que.

Hammers, Drop.

London Mach. Tool Co., Hamilton, Ont.
Miller Bros. & Toms Montreal.
Niles-Bement-Pond Co., New York.

Hammers, Power.

Rice Lewis & Son, Toronto.

Hammers, Steam.

John Bertram & Sons Co. Dundas, Ont.
London Mach. Tool Co., Hamilton, Ont.
Niles-Bement-Pond Co., New York.

Hangers.

The Goldie & McCulloch Co., Galt.
Greey, Wm. & J. G., Toronto
Kennedy, Wm., & Sons Owen Sound
Miller Bros. & Toms, Montreal
Owen Sound Iron Works Co., Owen Sound
The Smart-Turner Mach. Co., Hamilton.
Waterous Engine Co., Brantford.
Pendrith Machinery Co., Toronto.
Standard Bearings Ltd., Niagara Falls

Hardware Specialties.

Mechanics Supply Co. Quebec Que.

Heating Apparatus.

Darling Bros., Ltd., Montreal
Dominion Heating & Ventilating Co., Hespeler
Mechanics Supply Co., Quebec, Que.
Sheldons Limited, Galt.
B. F. Sturtevant Co., High Park, Mass.

Hoisting and Conveying Machinery.

Allis-Chalmers-Bullock, Limited, Montreal
Greey, Wm. & J. G., Toronto
Link-Belt Co., Philadelphia.
Miller Bros. & Toms, Montreal.
Niles-Bement-Pond Co., New York.
Northern Engineering Works, Detroit
The Smart-Turner Mach. Co., Hamilton.
Waterous Engine Co., Brantford.

Hoists, Chain and Rope.

Frothingham & Workman, Ltd., Montreal

Hoists, Electric.

Northern Engineering Works, Detroit

Hoists, Pneumatic.

Canadian Rand Drill Co., Montreal.
Chicago Pneumatic Tool Co. Chicago.
Dominion Foundry Supply Co., Montreal
Hamilton Facing Mill Co., Hamilton.
Northern Engineering Works, Detroit

Hose.

Baxter, Paterson & Co., Montreal
Mechanics Supply Co., Quebec, Que.

Hose, Air.

Canadian Rand Drill Co., Montreal.
Canadian Westinghouse Co., Hamilton.
Chicago Pneumatic Tool Co. Chicago.
Paxson, J. W., Co., Philadelphia

Hose Couplings.

Canadian Rand Drill Co., Montreal.
Canadian Westinghouse Co., Hamilton.
Chicago Pneumatic Tool Co., Chicago.
Mechanics Supply Co., Quebec, Que.
Paxson, J. W., Co., Philadelphia

Hose, Steam.

Allis-Chalmers-Bullock, Montreal.
Canadian Rand Drill Co., Montreal.
Canadian Westinghouse Co., Hamilton.
Mechanics Supply Co., Quebec Que.
Paxson, J. W. Co., Philadelphia

Hydraulic Accumulators.

Niles-Bement-Pond Co., New York.
Perrin, Wm. R., Co. Toronto.
The Smart-Turner Mach. Co., Hamilton.

Hydraulic Machinery.

Allis-Chalmers-Bullock, Montreal.
Barber, Chas. & Sons, Meaford

India Oil Stones.

Norton Company, Worcester, Mass.

Indicators, Speed.

Mechanics Supply Co. Quebec, Que.
L. S. Starrett Co., Athol, Mass.

Injectors.

Canada Foundry Co., Toronto.
The Canadian Fairbanks Co., Montreal.
Desmond-Stephan Mfg. Co., Urbana, Ohio
Frothingham & Workman Ltd., Montreal
Mechanics Supply Co. Quebec, Que.
Rice Lewis & Son, Toronto.
Penberthy Injector Co., Windsor, Ont.

Iron and Steel.

Frothingham & Workman, Ltd., Montreal

Iron and Steel Bars and Bands.

Hamilton Steel & Iron Co., Hamilton

Iron Cements.

Buffalo Foundry Supply Co., Buffalo.
Detroit Foundry Supply Co., Windsor.
Hamilton Facing Mill Co., Hamilton.
Paxson, J. W., Co., Philadelphia

Iron Filler.

Buffalo Foundry Supply Co., Buffalo.
Detroit Foundry Supply Co., Windsor.
Dominion Foundry Supply Co. Montreal
Hamilton Facing Mill Co., Hamilton.
Paxson, J. W., Co., Philadelphia.

Jacks.

Frothingham & Workman, Ltd., Montreal
Norton, A. O., Coaticook, Que.

Kegs, Steel Shop.

Cleveland Wire Spring Co., Cleveland

Key-Seating Machines.

B. F. Barnes Co., Rockford, Ill.
Niles-Bement-Pond Co., New York.

Lace Leather.

Baxter, Paterson & Co., Montreal.

Ladles, Foundry.

Frothingham & Workman Ltd., Montreal
Northern Engineering Works, Detroit

Lamps, Arc and Incandescent.

Canadian General Electric Co., Toronto.
Canadian Westinghouse Co., Hamilton.
The Packard Electric Co., St. Catharines.

Lathe Dogs.

Armstrong Bros., Chicago
Baxter, Paterson & Co., Montreal.
Pratt & Whitney Co., Hartford, Conn.

Lathes, Engine.

American Tool Work Co., Cincinnati.
B. F. Barnes Co., Rockford, Ill.
Baxter, Paterson & Co., Montreal
John Bertram & Sons Co., Dundas, Ont.
Canada Machinery Agency, Montreal.
The Canadian Fairbanks Co., Montreal.
London Mach. Tool Co., Hamilton, Ont.
Niles-Bement-Pond Co., New York.
H. W. Petrie, Toronto.
Pratt & Whitney Co., Hartford, Conn.

Lathes, Foot-Power.

American Tool Works Co., Cincinnati.
B. F. Barnes Co., Rockford, Ill.

Lathes, Screw Cutting.

B. F. Barnes Co., Rockford, Ill.
Baxter, Paterson & Co., Montreal.
Niles-Bement-Pond Co., New York.

Lathes, Automatic, Screw-Threading.

John Bertram & Sons Co., Dundas, Ont
London Mach. Tool Co., Hamilton, Ont
Pratt & Whitney Co., Hartford, Conn.

Lathes, Bench.

B. F. Barnes Co., Rockford, Ill.
London Mach. Tool Co., London, Ont.
Pratt & Whitney Co., Hartford, Conn.

Lathes, Turret.

American Tool Works Co., Cincinnati.
Baxter, Paterson & Co., Montreal
John Bertram & Sons Co., Dundas, Ont.
Gisholt Machine Co., Madison, Wis.
London Mach. Tool Co., Hamilton, Ont.
Niles-Bement-Pond Co., New York.
The Pratt & Whitney Co., Hartford, Conn.
Warner & Swasey Co., Cleveland O.

Leather Belting.

Baxter, Paterson & Co., Montreal
Canada Machinery Agency, Montreal.
The Canadian Fairbanks Co., Montreal
Greey, Wm. & J. G., Toronto

Lime Stone Flux.

Hamilton Facing Mill Co., Hamilton.
Paxson, J. W., Co., Philadelphia

Locomotives, Air.

Canadian Rand Drill Co., Montreal.

Locomotives, Electrical.

Canadian Westinghouse Co., Hamilton
Jeffrey Mfg. Co., Columbus, Ohio.

Locomotives, Steam.

Canada Foundry Co., Toronto.
Canadian Rand Drill Co., Montreal

Lubricating Plumbago.

Detroit Foundry Supply Co., Detroit.
Dominion Foundry Supply Co., Montreal
Hamilton Facing Mill Co., Hamilton.
Paxson, J. W., Co., Philadelphia

Lubricators, Force Feed.

Sight Feed Oil Pump Co., Milwaukee, Wis.

Lumber Dry Kilns.

Dominion Heating & Ventilating Co., Hespeler.
H. W. Petrie, Toronto.
Sheldon's Limited, Galt, Ont.
B. F. Sturtevant Co., Hyde Park, Mass.

Machinery Dealers.

Baxter, Paterson & Co., Montreal.
Canada Machinery Agency, Montreal.
The Canadian Fairbanks Co., Montreal.
Miller Bros. & Toms, Montreal.
H. W. Petrie, Toronto.
The Smart-Turner Mach. Co., Hamilton.
Williams & Wilson Montreal.

Machinery Designers.

Greey, Wm. & J. G., Toronto
Standard Bearings, Ltd., Niagara Fal's.

Machinists.

W. H. Banfield & Sons, Toronto.
Greey, Wm. & J. G. Toronto
Hall Engineering Works Montreal.
Link-Belt Co., Philadelphia.
John McDougall, Caledonian Iron Works Co., Montreal.
Miller Bros. & Toms Montreal.
Paxson, J. W., Co., Philadelphia
Pendrith Machinery Co., Toronto.
Robb Engineering Co., Amherst, N.S.
The Smart-Turner Mach. Co., Hamilton.
Standard Bearings, Ltd. Niagara Falls
Waterous Engine Co., Brantford.

Machinists' Small Tools.

Armstrong Bros., Chicago.
Butterfield & Co., Rock Island, Que.
Frothingham & Workman Ltd., Montreal
Mechanics Supply Co., Quebec, Que.
Rice Lewis & Son, Montreal.
Pratt & Whitney Co., Hartford, Conn.
Standard Tool Co., Cleveland.
L. S. Starrett Co., Athol, Mass.
Williams & Wilson, Montreal.

Malleable Flask Clamps.

Buffalo Foundry Supply Co., Buffalo.
Paxon, J. W., Co., Philadelphia

Malleable Iron Castings.

Galt Malleable Iron Co., Galt

Mallet, Rawhide and Wood.

Buffalo Foundry Supply Co., Buffalo.
Detroit Foundry Supply Co., Detroit.
Paxson, J. W., Co., Philadelphia

Mandrels.

A. B. Jardine & Co., Hespeler, Ont.
Miller Bros. & Toms, Montreal
The Pratt & Whitney Co., Hartford, Conn.
Standard Tool Co., Cleveland.

Marking Machines.

Dwight Slate Machine Co., Hartford, Conn.

Marine Work.

Kennedy, Wm., & Sons, Owen Sound

Metallic Paints.

P. D. Dods & Co., Montreal.

Meters, Electrical.

Canadian Westinghouse Co., Hamilton.
Mechanics Supply Co., Quebec, Que.

Mill Machinery.

Baxter, Paterson & Co., Montreal.
Greey, Wm. & J. G., Toronto
The Goldie & McCulloch Co., Galt, Ont.
John McDougall, Caledonian Iron Works Co., Montreal.
H. W. Petrie, Toronto.
Robb Engineering Co., Amherst, N.S.
Waterous Engine Co., Brantford.
Williams & Wilson, Montreal.

Milling Attachments.

Becker-Brainard Milling Machine Co. Hyde Park, Mass.
John Bertram & Sons Co., Dundas, Ont.
Niles-Bement-Pond Co., New York.
Pratt & Whitney, Hartford, Conn.

Milling Machines, Horizontal.

Baxter, Paterson & Co., Montreal.
Becker-Brainard Milling Machinery Co. Hyde Park, Mass.
John Bertram & Sons Co., Dundas, Ont.
London Mach. Tool Co., Hamilton, Ont.
Niles-Bement-Pond Co., New York.
Pratt & Whitney, Hartford, Conn.

Milling Machines, Plain.

American Tool Works Co., Cincinnati.
Becker-Brainard Milling Machine Co. Hyde Park, Mass.
John Bertram & Sons Co., Dundas, Ont.
Canada Machinery Agency, Montreal.
The Canadian Fairbanks Co., Montreal.
London Mach. Tool Co., Hamilton, Ont.
Niles-Bement-Pond Co., New York.
H. W. Petrie, Toronto.
Pratt & Whitney Co., Hartford, Conn.
Williams & Wilson, Montreal.

Milling Machines, Universal.

American Tool Works Co., Cincinnati.
Becker-Brainard Milling Machine Co., Hyde Park, Mass.
John Bertram & Sons Co., Dundas, Ont.
Canada Machinery Agency, Montreal.
The Canadian Fairbanks Co., Montreal.
London Mach. Tool Co., Hamilton, Ont.
Niles-Bement-Pond Co., New York.
H. W. Petrie, Toronto.
Williams & Wilson, Montreal.

Milling Machines, Vertical.

Becker-Brainard Milling Machine Co., Hyde Park, Mass.
Brown & Sharpe, Providence, R.I.
John Bertram & Sons Co., Dundas, Ont.
Canada Machinery Agency, Montreal.
London Mach. Tool Co., Hamilton, Ont.
Niles-Bement-Pond Co., New York.

Milling Tools.

Wm. Abbott, Montreal.
Becker-Brainard Milling Machine Co., Hyde Park, Mass.
Geometric Tool Co., New Haven, Conn.
London Mach. Tool Co., Hamilton, Ont.
Pratt & Whitney Co., Hartford, Conn.
Standard Tool Co., Cleveland.
Standard Bearings, Ltd., Niagara Falls

Mining Machinery.

Allis-Chalmers-Bullock, Limited, Montreal
Canadian Rand Drill Co., Montreal.
Chicago Pneumatic Tool Co., Chicago.
Jeffrey Mfg. Co., Columbus, Ohio.
Laurie Engine & Machine Co., Montreal.
Rice Lewis & Son, Toronto.
John McDougall, Caledonian Iron Works Co., Montreal.
Miller Bros. & Toms, Montreal.
T. & H. Electric Co. Hamilton.

Mixing Machines, Dough.

Greey, Wm. & J. G., Toronto
Pendrith Machinery Co., Toronto.
Paxson, J. W., Co., Philadelphia

Mixing Machines, Special.

Greey, Wm. & J. G., Toronto
Pendrith Machinery Co. Toronto.
Paxson, J. W., Co., Philadelphia

Model Tools.

Globe Machine & Stamping Co., Cleveland, Ohio.
Standard Bearings, Ltd., Niagara Falls.
Wells Pattern and Model Works, Toronto

Motors, Electric.

Allis-Chalmers-Bullock Limited, Montreal
Canadian General Electric Co., Toronto
Canadian Westinghouse Co., Hamilton.
Electrical Machinery Co., Toronto.

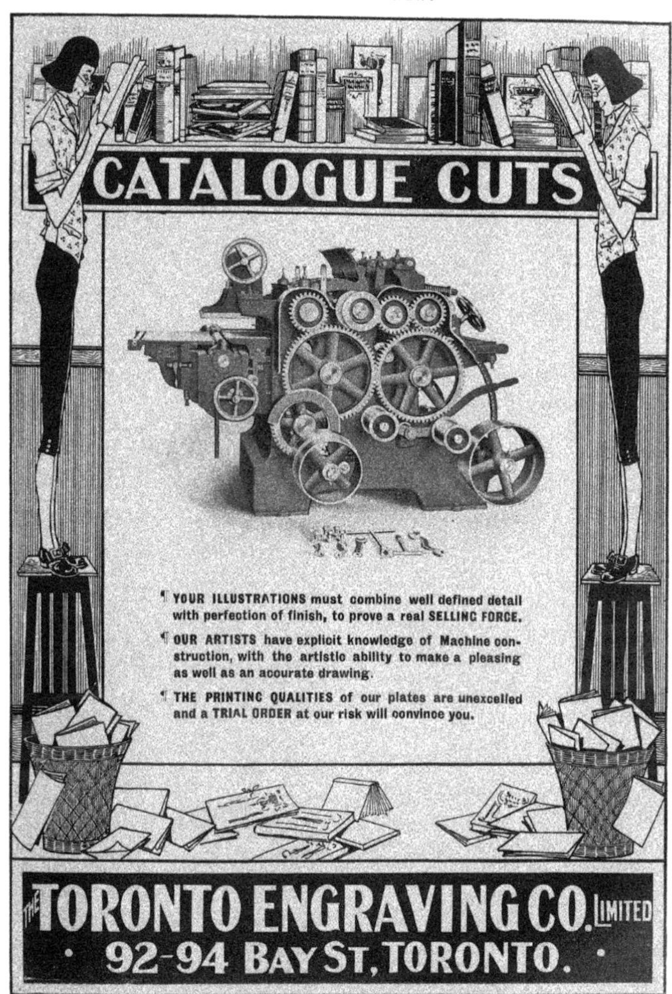
CATALOGUE CUTS
¶ YOUR ILLUSTRATIONS must combine well defined detail with perfection of finish, to prove a real SELLING FORCE.
¶ OUR ARTISTS have explicit knowledge of Machine construction, with the artistic ability to make a pleasing as well as an accurate drawing.
¶ THE PRINTING QUALITIES of our plates are unexcelled and a TRIAL ORDER at our risk will convince you.
THE TORONTO ENGRAVING CO. LIMITED
· 92-94 BAY ST, TORONTO. ·

Consolidated Electric Co., Toronto
Hall Engineering Works, Montreal.
Mechanics Supply Co., Quebec, Que.
The Packard Electric Co., St. Catharines.
B. F. Sturtevant Co., Hyde Park, Mass.
T. & H. Electric Co., Hamilton.

Motors, Air.
Canadian Rand Drill Co., Montreal.
Chicago Pneumatic Tool Co., Chicago.

Molders' Supplies.
Buffalo Foundry Supply Co., Buffalo.
Detroit Foundry Supply Co., Windsor.
Dominion Foundry Supply Co., Montreal
Hamilton Facing Mill Co., Hamilton
Mechanics Supply Co., Quebec, Que.
Paxson J. W., Co., Philadelphia

Molders' Tools.
Buffalo Foundry Supply Co. Buffalo.
Detroit Foundry Supply Co., Windsor
Dominion Foundry Supply Co., Montreal.
Hamilton Facing Mill Co., Hamilton.
Mechanics Supply Co., Quebec, Que.
Paxson, J. W., Co., Philadelphia

Molding Machines.
Buffalo Foundry Supply Co. Buffalo.
Hamilton Facing Mill Co., Hamilton.
J. W. Paxson Co., Philadelphia, Pa.

Molding Sand.
T. W. Barnes, Hamilton.
Buffalo Foundry Supply Co., Buffalo.
Detroit Foundry Supply Co., Windsor
Dominion Foundry Supply Co., Montreal
Hamilton Facing Mill Co., Hamilton.
Paxson, J. W., Co., Philadelphia

Nut Tappers.
John Bertram & Sons Co., Dundas, Ont
A. B. Jardine & Co., Hespeler.
London Mach. Tool Co., Hamilton.
National Machinery Co., Tiffin, Ohio.

Nuts.
Canada Nut Co., Toronto

Oatmeal Mill Machinery.
Greey Wm. & J. G., Toronto
The Goldie & McCulloch Co., Galt

Oilers, Gang.
Sight Feed Oil Pump Co., Milwaukee, Wis.

Oils, Core.
Buffalo Foundry Supply Co. Buffalo
Dominion Foundry Supply Co., Montreal
Hamilton Facing Mill Co., Hamilton.
Paxson, J. W., Co., Philadelphia

Oil Extractors.
Darling Bros., Ltd., Montreal

Paint Mill Machinery.
Greey, Wm. & J. G., Toronto

Pans, Lathe.
Cleveland Wire Spring Co., Cleveland
Paxson, J. W., Co., Philadelphia

Pans, Steel Shop.
Cleveland Wire Spring Co., Cleveland.

Patent Solicitors.
Hanbury A. Budden, Montreal.
Fetherstonhaugh & Blackmore, Montreal
Marion & Marion, Montreal.
Ridout & Maybee, Toronto.

Patterns.
John Carr, Hamilton.
Galt Malleable Iron Co., Galt
Hamilton Pattern Works, Hamilton.
John McDougall Caledonian Iron Works Co., Montreal
Miller Bros. & Toms Montreal.
Wells' Pattern and Model Works, Toronto.

Pig Iron.
Hamilton Steel & Iron Co., Hamilton

Pipe Cutting and Threading Machines.
Butterfield & Co., Rock Island, Que.
Canada Machinery Agency, Montreal.
Curtis & Curtis Co., Bridgeport, Conn.
Frothingham & Workman, Ltd., Montreal
Hart Mfg. Co., Cleveland
A. B. Jardine & Co., Hespeler, Ont.
London Mach. Tool Co., Hamilton, Ont.
R. McDougall Co., Galt, Ont.
Niles-Bement-Pond Co., New York.

Pipe, Municipal.
Canadian Pipe Co., Vancouver B.C.
Pacific Coast Pipe Co., Vancouver, B.C.

Pipe, Waterworks.
Canadian Pipe Co., Vancouver, B.C.
Pacific Coast Pipe Co., Vancouver, B.C.

Planers, Standard.
American Tool Works, Cincinnati.
John Bertram & Sons Co., Dundas, Ont.
Canada Machinery Agency, Montreal.
The Canadian Fairbanks Co., Montreal.
Rice Lewis & Son, Toronto.
London Mach. Tool Co., Hamilton, Ont.
Niles-Bement-Pond Co., New York.
H. W. Petrie, Toronto.
Pratt & Whitney Co., Hartford, Conn.
Williams & Wilson, Montreal.

Planers, Rotary.
John Bertram & Sons Co., Dundas, Ont.
London Mach. Tool Co., Hamilton, Ont.
Niles-Bement-Pond Co., New York.

Planing Mill Fans.
Dominion Heating & Ventilating Co., Hespeler
Sheldon's Limited, Galt, Ont.
B. F. Sturtevant Co., Hyde Park Man.

Plumbago.
Buffalo Foundry Supply Co., Buffalo.
Detroit Foundry Supply Co., Windsor
Dominion Foundry Supply Co., Montreal
Hamilton Facing Mill Co., Hamilton,
Mechanics Supply Co., Quebec, Que.
J. W. Paxson Co., Philadelphia, Pa.

Pneumatic Tools.
Allis-Chalmers-Bullock, Montreal.
Canadian Rand Drill Co., Montreal.
Chicago Pneumatic Tool Co., Chicago
Hamilton Facing Mill Co., Hamilton.
Hanna Engineering Works, Chicago.
Independent Pneumatic Tool Co. Chicago, Ill.
Paxson, J. W., Co., Philadelphia

Power Hack Saw Machines.
Baxter, Paterson & Co., Montreal.
Frothingham & Workman, Ltd., Montreal

Power Plants.
John McDougall Caledonian Iron Works Co., Montreal
The Smart-Turner Mach. Co., Hamilton

Power Plant Equipments.
Darling Bros., Ltd., Montreal

Presses, Drop.
W. H. Banfield & Son, Toronto.
E. W. Bliss Co., Brooklyn, N.Y.
Canada Machinery Agency, Montreal.
Laurie Engine & Machine Co., Montreal
Miller Bros. & Toms, Montreal
Niles-Bement-Pond Co., New York

Presses, Hand.
E. W. Bliss Co., Brooklyn, N.Y.

Presses, Hydraulic.
John Bertram & Sons Co., Dundas, Ont.
Laurie Engine & Machine Co., Montreal.
London Mach. Tool Co., Hamilton, Ont.
John McDougall Caledonian Iron Works Co., Montreal
Miller Bros. & Toms Montreal.
Niles-Bement-Pond Co., New York.
Perrin, Wm. R. & Co., Toronto

Presses, Power.
E. W. Bliss Co., Brooklyn, N.Y.
Canada Machinery Agency, Montreal.
Laurie Engine & Machine Co., Montreal
London Mach. Tool Co., Hamilton, Ont.
John McDougall Caledonian Iron Works Co., Montreal.
Niles-Bement-Pond Co., New York.

Presses Power Screw.
Perrin, Wm. R., & Co., Toronto

Pressure Regulators.
Darling Bros., Ltd., Montreal

Producer Plants.
Canada Foundry Co., Toronto

Pulp Mill Machinery.
Greey, Wm. & J. G. Toronto
Jeffrey Mfg. Co., Columbus, Ohio.
Laurie Engine & Machine Co., Montreal.
John McDougall Caledonian Iron Works Co., Montreal.
Waterous Engine Works Co., Brantford

Pulleys.
Baxter Paterson & Co. Montreal.
Canada Machinery Agency, Montreal.
The Canadian Fairbanks Co., Montreal.
The Goldie & McCulloch Co., Galt.
Greey, Wm. & J. G., Toronto
Laurie Engine & Machine Co., Montreal.
Link-Belt Co., Philadelphia
John McDougall Caledonian Iron Works Co., Montreal
Miller Bros. & Toms, Montreal.
Owen Sound Iron Works Co., Owen Sound
Pendrith Machinery Co., Toronto
H. W. Petrie, Toronto.
The Smart-Turner Mach. Co., Hamilton.
Standard Bearings Ltd., Niagara Falls.
Williams & Wilson, Montreal.
Waterous Engine Co., Brantford.

Pumps.
Laurie Engine & Machine Co., Montreal
Ontario Wind Engine & Pump Co., Toronto

Pumps, Boiler Feed.
Du Bois Iron Works, Du Bois, Pa.

Pumps, Boiler Test.
Du Bois Iron Works Du Bois, Pa.

Pumps, Circulating.
Du Bois Iron Works, Du Bois, Pa.

Pumps, Contractors.
Du Bois Iron Works Du Bois Pa

Pumps, Electric.
Du Bois Iron Works, Du Bois, Pa

Pumps, Elevator.
Du Bois Iron Works, Du Bois, Pa

Pumps, Engine Gas and Oil.
Du Bois Iron Works, Du Bois, Pa.

Pump Governors.
Darling Bros. Ltd., Montreal

Pumps, Hydraulic.
Laurie Engine & Machine Co., Montreal
Perrin, Wm. R. & Co., Toronto

Pumps, Irrigation.
Du Bois Iron Works, Du Bois, Pa.

Pumps, Marine.
Du Bois Iron Works, Du Bois, Pa.

Pumps, Mine (Cylinder & Heads).
Du Bois Iron Works, Du Bois Pa.

Pumps, Oil.
Sight Feed Oil Pump Co. Milwaukee, Wis.

Pumps, Power.
Du Bois Iron Works, Du Bois, Pa.

Pumps, Steam.
Allis-Chalmers-Bullock, Limited, Montreal
Canada Foundry Co., Toronto.
Canada Machinery Agency, Montreal.
Darling Bros., Ltd., Montreal
D'Olier Engineering Co., New York.
Du Bois Iron Works Du Bois, Pa.
The Goldie & McCulloch Co., Galt.
John McDougall Caledonian Iron Works, Montreal.
H. W. Petrie, Toronto.
The Smart-Turner Mach. Co., Hamilton.
Standard Bearings Ltd., Niagara Falls.
Waterous Engine Co. Brantford

Pumps, Tank.
Du Bois Iron Works, Du Bois, Pa.

Pumping Machinery.
Canada Foundry Co., Limited Toronto
Canada Machinery Agency, Montreal.
Canadian Rand Drill Co., Montreal.
Chicago Pneumatic Tool Co., Chicago.
Darling Bros., Ltd., Montreal
D'Olier Engineering Co., New York.
Hall Engineering Works, Montreal, Que.
Laurie Engine & Machine Co., Montreal.
London Mach. Tool Co., Hamilton, Ont.
John McDougall Caledonian Iron Works Co., Montreal.
R. McDougall Co., Galt Ont.
The Smart-Turner Mach. Co., Hamilton
Standard Bearings, Ltd., Niagara Falls.

Punches and Dies.
W. H. Banfield & Sons, Toronto.
Butterfield & Co., Rock Island.
Globe Machine & Stamping Co.
A. B. Jardine & Co., Hespeler, Ont.
London Mach. Tool Co., Hamilton, Ont.
Miller Bros. & Toms, Montreal.
Pratt & Whitney Co., Hartford, Conn.
H. W. Petrie, Toronto.
Standard Bearings, Ltd., Niagara Falls.

Punches, Hand.
Mechanics Supply Co., Quebec, Que.

Punches, Power.
John Bertram & Sons Co., Dundas, Ont.
E. W. Bliss Co., Brooklyn, N.Y.
Canada Machinery Agency, Montreal.
Canadian Buffalo Forge Co., Montreal.
London Mach. Tool Co., Hamilton, Ont.
Niles-Bement-Pond Co., New York.

Punches, Turret.
London Mach. Tool Co., London, Ont.

Punching Machines, Horizontal.
John Bertram & Sons Co., Dundas, Ont.
London Mach. Tool Co., Hamilton Ont.
Niles-Bement-Pond Co., New York

Quartering Machines.
John Bertram & Sons Co., Dundas, Ont.
London Mach. Tool Co., Hamilton, Ont.

Railway Spikes and Washers.
Hamilton Steel & Iron Co., Hamilton

Rammers, Bench and Floor.
Buffalo Foundry Supply Co., Buffalo.
Detroit Foundry Supply Co., Windsor
Hamilton Facing Mill Co., Hamilton.
Paxson, J. W., Co., Philadelphia

Rapping Plates.
Detroit Foundry Supply Co., Windsor
Hamilton Facing Mill Co., Hamilton.
Paxson, J. W., Co., Philadelphia

Raw Hide Pinions.
Brown, David & Sons Huddersfield, Eng

Reamers.
Wm. Abbott, Montreal.
Baxter Paterson & Co., Montreal.
Butterfield & Co., Rock Island.
Frothingham & Workman Ltd., Montreal
Hanna Engineering Works, Chicago
A. B. Jardine & Co., Hespeler, Ont.
Mechanics Supply Co., Quebec, Que.
John Millen & Son, Ltd., Montreal.
Morse Twist Drill and Machine Co., New Bedford, Mass.
Pratt & Whitney Co., Hartford, Conn.
Standard Tool Co., Cleveland.

Reamers, Steel Taper.
Butterfield & Co., Rock Island.
Chicago Pneumatic Tool Co., Chicago.
A. B. Jardine & Co., Hespeler, Ont.
John Millen & Son Ltd., Montreal.
Pratt & Whitney Co., Hartford, Conn.
Standard Tool Co., Cleveland.
Whitman & Barnes Co., St. Catharines.

Rheostats.
Canadian General Electric Co., Toronto.
Canadian Westinghouse Co., Hamilton.
Hall Engineering Works, Montreal, Que.
T. & H. Electric Co., Hamilton.

Riddles.
Buffalo Foundry Supply Co., Buffalo.
Detroit Foundry Supply Co., Windsor
Hamilton Facing Mill Co., Hamilton.
J. W. Paxson Co., Philadelphia, Pa.

Riveters, Pneumatic.
Hanna Engineering Works, Chicago.
Independent Pneumatic Tool Co., Chicago, Ill.

Rolls, Bending.
John Bertram & Sons Co., Dundas, Ont.
London Mach. Tool Co., Hamilton, Ont.
Niles-Bement-Pond Co., New York.

Rolls, Chilled Iron.
Greey, Wm. & J. G., Toronto

Rolls, Sand Cast.
Greey, Wm. & J. G., Toronto

Rotary Blowers.
Paxson, J. W., Co., Philadelphia
Sturtevant, B. F. Co., Hyde Park Mass.

Rotary Converters.
Allis-Chalmers-Bullock, Ltd., Montreal.
Canadian Westinghouse Co., Hamilton
Paxson, J. W., Co., Philadelphia
Toronto and Hamilton Electric Co., Hamilton.

Safes.
Baxter Paterson & Co., Montreal
The Goldie & McCulloch Co., Galt.

Sand, Bench.
Buffalo Foundry Supply Co., Buffalo.
Detroit Foundry Supply Co., Windsor
Hamilton Facing Mill Co., Hamilton.
Paxson J. W., Co., Philadelphia

Sand Blast Machinery.
Canadian Rand Drill Co., Montreal.
Chicago Pneumatic Tool Co., Chicago.
Detroit Foundry Supply Co., Windsor
Hamilton Facing Mill Co., Hamilton.
J. W. Paxson Co., Philadelphia, Pa.

Sand, Heavy, Grey Iron.
Buffalo Foundry Supply Co., Buffalo.
Detroit Foundry Supply Co., Windsor
Hamilton Facing Mill Co., Hamilton.
Paxson, J. W., Co., Philadelphia

Sand, Malleable.
Buffalo Foundry Supply Co., Buffalo.
Detroit Foundry Supply Co., Windsor
Hamilton Facing Mill Co., Hamilton.
Paxson, J. W., Co., Philadelphia

Sand, Medium Grey Iron.
Buffalo Foundry Supply Co., Buffalo.
Detroit Foundry Supply Co., Windsor
Hamilton Facing Mill Co., Hamilton.
Paxson, J. W., Co., Philadelphia

Sand Sifters.
Buffalo Foundry Supply Co., Buffalo.
Detroit Foundry Supply Co., Windsor
Dominion Foundry Supply Co., Montreal
Hamilton Facing Mill Co., Hamilton.
Hanna Engineering Works, Chicago.
Paxson, J. W., Co., Philadelphia

Saw Mill Machinery.
Allis-Chalmers-Bullock, Limited, Montreal
Baxter, Paterson & Co., Montreal
Canada Machinery Agency, Montreal.
Goldie & McCulloch Co., Galt.
Greey, Wm. & J. G. Toronto
Miller Bros. & Toms, Montreal
Owen Sound Iron Works Co., Owen Sound
H. W. Petrie, Toronto.
Robb Engineering Co., Amherst.
Standard Bearings, Ltd., Niagara Falls
Waterous Engine Works, Brantford
Williams & Wilson, Montreal.

Sawing Machines, Metal.
Niles-Bement-Pond Co. New York.
Paxson, J. W., Co., Philadelphia

Saws, Hack.
Baxter Paterson & Co., Montreal.
Canada Machinery Agency, Montreal.
Detroit Foundry Supply Co., Windsor
Frothingham & Workman, Ltd., Montreal
London Mach. Tool Co., Hamilton
Mechanics Supply Co., Quebec, Que.
Rice Lewis & Son, Toronto.
John Millen & Son, Ltd., Montreal.
L. S. Starrett Co., Athol, Mass.

Screw Cutting Tools
Hart Manufacturing Co., Cleveland, Ohio
Mechanics Supply Co., Quebec, Que.

Screw Machines, Automatic.
Canada Machinery Agency, Montreal.
Cleveland Automatic Machine Co., Cleveland, Ohio
London Mach. Tool Co., Hamilton, Ont.
National-Acme Mfg. Co., Cleveland.
Pratt & Whitney Co., Hartford, Conn.

Screw Machines, Hand.
Canada Machinery Agency Montreal.
A. B. Jardine & Co., Hespeler.
London Mach. Tool Co., Hamilton, Ont.
Mechanics Supply Co. Quebec, Que.
Pratt & Whitney Co., Hartford, Conn.
Warner& Swasey Co., Cleveland, O.

Screw Machines, Multiple Spindle.
National-Acme Mfg. Co., Cleveland

Screw Plates.
Butterfield & Co., Rock Island, Que.
Frothingham & Workman, Ltd., Montreal
Hart Manufacturing Co. Cleveland, Ohio
A. B. Jardine & Co., Hespeler
Mechanics Supply Co., Quebec, Que.

Screws, Cap and Set.
National-Acme Mfg. Co., Cleveland

Screw Slotting Machinery, Semi-Automatic.
National-Acme Mfg. Co., Cleveland

Second-Hand Machinery.
American Tool Works Co., Cincinnati.
Canada Machinery Agency, Montreal.
The Canadian Fairbanks Co., Montreal.
Goldie & McCulloch Co., Galt.
Machinery Exchange, Montreal.
Niles-Bement-Pond Co., New York.
H. W. Petrie, Toronto.
Robb Engineering Co., Amherst, N.S.
Waterous Engine Co., Brantford.
Williams & Wilson, Montreal.

Shafting.
Baxter, Paterson & Co., Montreal.
Canada Machinery Agency, Montreal
The Canadian Fairbanks Co., Montreal.
Frothingham & Workman L'd., Montreal
The Goldie & McCulloch Co., Galt, Ont
Greey, Wm., & J. G., Toronto
Kennedy Wm., & Sons, Owen Sound
Niles-Bement-Pond Co., New York.
Owen Sound Iron Works Co., Owen Sound
Pendrith Machine Co., Toronto.
H. W. Petrie, Toronto.
Smart-Turner Machine Co., Hamilton.
Union Drawn Steel Co., Hamilton.
Waterous Engine Co., Brantford.

Shapers.
American Tool Works Co., Cincinnati.
John Bertram & Sons Co., Dundas, Ont
Canada Machinery Agency, Montreal.
The Canadian Fairbanks Co., Montreal.
Rice Lewis & Son, Toronto.
London Mach. Tool Co., Hamilton, Ont.
Niles-Bement-Pond Co., New York.
H. W. Petrie, Toronto.
Potter & Johnston Machine Co., Pawtucket, R.I.
Pratt & Whitney Co., Hartford, Conn.
Williams & Wilson, Montreal.

Shearing Machine, Bar.
John Bertram & Sons Co, Dundas, Ont.
A. B. Jardine & Co., Hespeler.
London Mach. Tool Co., Hamilton, Ont.
Niles-Bement-Pond Co., New York.

Shears, Power.
John Bertram & Sons Co., Dundas, Ont.
Canada Machinery Agency, Montreal.
A. B. Jardine & Co., Hespeler, Ont
Niles-Bement-Pond Co., New York
Paxson, J. W., Co., Philadelphia

Sheet Metal Goods
Globe Machine & Stamping Co, Cleveland, Ohio.

Sheet Steel Work.
Owen Sound Iron Works Co., Owen Sound

Shingle Mill Machinery.
Owen Sound Iron Works Co., Owen Sound

Shop Trucks.
Greey, Wm. & J. G., Toronto
Owen Sound Iron Works Co., Owen Sound
Paxson, J. W., Co., Philadelphia

Shovels.
Baxter, Paterson & Co., Montreal.
Buffalo Foundry Supply Co., Buffalo.
Detroit Foundry Supply Co., Detroit.
Frothingham & Workman, Ltd., Montreal
Dominion Foundry Supply Co., Montreal
Hamilton Facing Mill Co., Hamilton.
Mechanics Supply Co., Quebec, Que.
Paxson, J. W., Co., Philadelphia

Shovels, Steam.
Allis-Chalmers-Bullock, Montreal.

Sieves.
Buffalo Foundry Supply Co., Buffalo.
Detroit Foundry Supply Co., Windsor
Dominion Foundry Supply Co., Montreal
Hamilton Facing Mill Co., Hamilton.
Paxson, J. W., Co., Philadelphia

Silver Lead.
Buffalo Foundry Supply Co., Buffalo.
Detroit Foundry Supply Co., Windsor
Dominion Foundry Supply Co., Montreal
Hamilton Facing Mill Co., Hamilton.
Paxson, J. W., Co., Philadelphia

Sleeves, Reducing.
Chicago Pneumatic Tool Co., Chicago.

Snap Flasks
Buffalo Foundry Supply Co., Buffalo.
Detroit Foundry Supply Co., Windsor
Dominion Foundry Supply Co., Montreal
Hamilton Facing Mill Co., Hamilton.
Paxson, J. W., Co., Philadelphia

Soapstone.
Buffalo Foundry Supply Co., Buffalo.
Detroit Foundry Supply Co., Windsor
Dominion Foundry Supply Co., Montreal
Hamilton Facing Mill Co., Hamilton.
Paxson, J. W., Co., Philadelphia

Solders.
Lumen Bearing Co., Toronto

Special Machinery.
W. H. Banfield & Sons, Toronto.
Baxter, Paterson & Co., Montreal
John Bertram & Sons Co. Dundas, Ont
Globe Machine & Stamping Co., Cleveland, Ohio.
Greey, Wm. & J. G., Toronto
Hanna Engineering Works, Chicago.
Laurie Engine & Machine Co., Montreal.
London Mach. Tool Co., Hamilton, Ont.
R. McDougall Co., Galt, Ont.
Pendrith Machinery Co., Toronto.
H. W. Petrie, Toronto.
The Smart-Turner Mach. Co., Hamilton
Standard Bearings, Ltd., Niagara Falls,
Waterous Engine Co., Brantford.

Special Machines and Tools.
Paxson, J. W., Philadelphia
Pratt & Whitney, Hartford, Conn.
Standard Bearings, Ltd., Niagara Falls,

Special Milled Work.
National Acme Mfg. Co., Cleveland

Speed Changing Countershafts.
The Canadian Fairbanks Co., Montreal

Spike Machines.
National Machinery Co., Tiffin, O.
The Smart-Turner Mach. Co., Hamilton

Spray Cans.
Detroit Foundry Supply Co., Windsor
Dominion Foundry Supply Co., Montreal
Hamilton Facing Mill Co., Hamilton.
Paxson, J. W., Co., Philadelphia

Springs, Automobile.
Cleveland Wire Spring Co., Cleveland

Springs, Coiled Wire.
Cleveland Wire Spring Co., Cleveland

Springs, Machinery.
Cleveland Wire Spring Co., Cleveland

Springs, Upholstery.
Cleveland Wire Spring Co., Cleveland

Sprue Cutters.
Detroit Foundry Supply Co., Windsor
Dominion Foundry Supply Co., Montreal
Hamilton Facing Mill Co., Hamilton.
Paxson, J. W., Co., Philadelphia

Stamp Mills.
Allis-Chalmers-Bullock, Limited, Montreal
Paxson, J. W., Co., Philadelphia

Steam Hot Blast Apparatus
Dominion Heating & Ventilating Co. Hespeler, Ont.
Sheldon's Limited, Galt.
B. F. Sturtevant Co., Hyde Park, Mass.

Steam Separators.
Darling Bros., Ltd., Montreal.
Dominion Heating & Ventilating Co., Hespeler
Robb Engineering Co., Montreal.
Sheldon's Limited, Galt.
Smart-Turner Mach. Co., Hamilton
Waterous Engine Co., Brantford.

Steam Specialties.
Darling Bros., Ltd., Montreal
Dominion Heating & Ventilating Co., Hespeler
Sheldon's Limited, Galt.

Steam Traps.
Canada Machinery Agency, Montreal.
Darling Bros., Ltd., Montreal
Dominion Heating & Ventilating Co., Hespeler
Mechanics Supply Co., Quebec, Que.
Pendrith Machinery Co., Toronto.
Sheldon's Limited, Galt.
B. F. Sturtevant Co., Hyde Park, Mass.

Steam Valves.
Darling Bros., Ltd., Montreal

Steel Pressure Blowers
Buffalo Foundry Supply Co., Buffalo.
Dominion Foundry Supply Co., Montreal
Dominion Heating & Ventilating Co., Hespeler
Hamilton Facing Mill Co., Hamilton.
J. W. Paxon Co., Philadelphia, Pa.
Sheldon's Limited, Galt.
B. F. Sturtevant Co., Hyde Park, Mass.

Steel Tubes.
Baxter, Paterson & Co., Montreal.
Mechanics Supply Co., Quebec, Que.
John Millen & Son, Montreal.

Steel, High Speed.
Wm. Abbott, Montreal.
Canadian Fairbanks Co., Montreal.
Frothingham & Workman, Ltd., Montreal
Alex. Gibb, Montreal.
Jessop, Wm., & Sons, Sheffield, Eng.
B. K. Morton Co., Sheffield, Eng.
Williams & Wilson, Montreal.

Stocks and Dies
Hart Manufacturing Co., Cleveland, Ohio
Mechanics Supply Co., Quebec, Que.
Pratt & Whitney Co., Hartford, Conn.

Stone Cutting Tools, Pneumatic
Allis-Chalmers-Bullock, Ltd., Montreal
Canadian Rand Drill Co., Montreal.

Stone Surfacers
Chicago Pneumatic Tool Co., Chicago.

Stove Plate Facings.
Buffalo Foundry Supply Co., Buffalo
Detroit Foundry Supply Co., Windsor.
Dominion Foundry Supply Co., Montreal
Hamilton Facing Mill Co., Hamilton.
Paxson, J. W., Co., Philadelphia

Swage, Block.
A. B. Jardine & Co., Hespeler, Ont.

Switchboards.
Allis-Chalmers-Bullock, Limited, Montreal
Canadian General Electric Co., Toronto.
Canadian Westinghouse Co., Hamilton
Hall Engineering Works, Montreal, Que.
Mechanics Supply Co., Quebec, Que.
Toronto and Hamilton Electric Co., Hamilton.

Talc.
Buffalo Foundry Supply Co., Buffalo.
Detroit Foundry Supply Co., Windsor.
Hamilton Facing Mill Co., Hamilton.
Paxson, J. W., Co., Philadelphia

Tanks, Oil.
Sight Feed Oil Pump Co., Milwaukee, Wis.

Tapping Machines and Attachments.
American Tool Works Co., Cincinnati.
John Bertram & Sons Co., Dundas, Ont.
Bickford Drill & Tool Co., Cincinnati.
The Geometric Tool Co., New Haven,
A. B. Jardine & Co., Hespeler.
London Mach. Tool Co., Hamilton, Ont
Murchey Machine & Tool Co., Detroit.
Niles-Bement-Pond Co., New York.
H. W. Petrie, Toronto.
Pratt & Whitney, Cincinnati, O.
L. S. Starrett Co., Athol, Mass.
Williams & Wilson, Montreal.

Tapes, Steel.
Frothingham & Workman, Ltd., Montreal
Rice Lewis & Son, Toronto.
Mechanics Supply Co., Quebec, Que.
John Millen & Son, Ltd., Montreal.
L. S. Starrett Co., Athol, Mass.

Taps.
Baxter, Paterson & Co., Montreal.
Mechanics Supply Co., Quebec, Que.

Taps, Collapsing.
The Geometric Tool Co., New Haven.

Taps and Dies.
Wm. Abbott, Montreal.
Baxter, Paterson & Co., Montreal.
Butterfield & Co., Rock Island, Que.
Frothingham & Workman, Ltd., Montreal
The Geometric Tool Co., New Haven.
A. B. Jardine & Co., Hespeler, Ont.
Rice Lewis & Son, Toronto.
Mechanics Supply Co., Quebec, Que.
John Millen & Son, Ltd., Montreal.
Pratt & Whitney Co., Hartford, Conn.
Standard Tool Co., Cleveland.
L. S. Starrett Co., Athol, Mass.

Testing Laboratory.
Detroit Testing Laboratory, Detroit

Testing Machines.
Detroit Foundry Supply Co., Windsor.
Dominion Foundry Supply Co., Montreal
Hamilton Facing Mill Co., Hamilton.
Paxson, J. W., Co., Philadelphia

Thread Cutting Tools.
Hart Manufacturing Co., Cleveland, Ohio
Mechanics Supply Co., Quebec Que.

Tiling, Opal Glass.
Toronto Plate Glass Importing Co., Toronto.

Time Switches, Automatic, Electric.
Berlin Electric Mfg. Co., Toronto.

Tinware Machinery
Canada Machinery Agency, Montreal.

Tire Upsetters or Shrinkers
A. B. Jardine & Co., Hespeler, Ont.

Tool Cutting Machinery
Canadian Rand Drill Co., Montreal.

Tool Holders.
Armstrong Bros. Tool Co., Chicago.
Baxter, Paterson & Co., Montreal.
Fairbanks Co., Springfield, Ohio.
John Millen & Son, Ltd., Montreal.
H. W. Petrie, Toronto.
Pratt & Whitney Co., Hartford, Conn.

Tool Steel.
Wm. Abbott, Montreal.
Flockton, Tompkin & Co., Sheffield, Eng.
Frothingham & Workman, Ltd., Montreal
Wm. Jessop, Sons & Co., Toronto.
Canadian Fairbanks Co., Montreal
B. K. Morton & Co., Sheffield, Eng.
Williams & Wilson, Montral.

Torches, Steel.
Baxter, Paterson & Co., Montreal.
Detroit Foundry Supply Co., Windsor.
Dominion Foundry Supply Co., Montreal
Hamilton Facing Mill Co., Hamilton.
Paxson, J. W., Co., Philadelphia

Transformers and Convertors
Allis-Chalmers-Bullock, Montreal.
Canadian General Electric Co., Toronto.
Canadian Westinghouse Co., Hamilton.
Hall Engineering Works, Montreal, Que.
T. & H. Electric Co., Hamilton.

Transmission Machinery
Allis-Chalmers-Bullock, Montreal.
The Canadian Fairbanks Co., Montreal.
Greey, Wm. & J. G., Toronto.
Laurie Engine & Machine Co., Montreal.
Link-Belt Co., Philadelphia.
Miller Bros. & Toms Montreal.
Paxson, J. W., Co., Philadelphia
H. W. Petrie, Toronto.
The Smart-Turner Mach. Co., Hamilton.
Waterous Engine Co., Brantford.

Transmission Supplies
Baxter, Paterson & Co., Montreal.
The Canadian Fairbanks Co., Montreal.
Wm. & J. G. Greey, Toronto.
The Goldie & McCulloch Co., Galt.
Miller Bros. & Toms Montreal
Pendrith Machinery Co., Toronto.
H. W. Petrie, Toronto.

Trolleys
Canadian Rand Drill Co., Montreal.
John Millen & Son, Ltd., Montreal.
Miller Bros. & Toms, Montreal.
Northern Engineering Works Detroit

Trolley Wheels.
Lumen Bearing Co., Toronto

Trucks, Dryer and Factory
Dominion Heating & Ventilating Co., Hespeler
Greey, Wm. & J. G., Toronto.
Northern Engineering Works, Detroit
Sheldon's Limited, Galt, Ont.

Tube Expanders (Rollers)
Chicago Pneumatic Tool Co., Chicago.
A. B. Jardine & Co., Hespeler.
Mechanics Supply Co., Quebec, Que.

Turbines, Steam

Allis-Chalmers-Bullock Limited, Montreal
Canadian General Electric Co., Toronto.
Canadian Westinghouse Co., Hamilton.
D'Olier Engineering Co., New York.
Kerr Turbine Co., Wellsville, N.Y.

Turntables

Detroit Foundry Supply Co., Windsor.
Dominion Foundry Supply Co., Montreal
Hamilton Facing Mill Co., Hamilton.
Northern Engineering Works, Detroit
Paxson, J. W., Co., Philadelphia

Turret Machines.

American Tool Works Co., Cincinnati.
John Bertram & Sons Co., Dundas, Ont.
The Canadian Fairbanks Co., Montreal.
London Mach. Tool Co., Hamilton, Ont.
Niles-Bement-Pond Co., New York.
H. W. Petrie, Toronto.
Pratt & Whitney Co., Hartford, Conn.
Warner & Swasey Co., Cleveland, Ohio.
Williams & Wilson, Montreal.

Upsetting and Bending Machinery.

John Bertram & Sons Co., Dundas, Ont.
A. B. Jardine & Co., Hespeler.
London Mach. Tool Co., Hamilton, Ont.
National Machinery Co., Tiffin, O.
Niles-Bement-Pond Co., New York.

Valves, Blow-off.

Chicago Pneumatic Tool Co., Chicago.
Mechanics Supp y Co., Quebec, Que.

Valves, Back Pressure.

Darling Bros., Ltd., Montreal
Dominion Heating & Ventilating Co. Hespeler
Mechanics Supply Co., Quebec, Que.
Sheldon's Limited, Galt.

Valve Reseating Machines.

Darling Bros. Ltd., Montreal

Ventilating Apparatus.

Canada Machinery Agency, Montreal
Darling Bros., Ltd., Montreal
Dominion Heating & Ventilating Co., Hespeler, Ont.
Sheldon's Ltd., Galt.
B. F. Sturtevant Co., Hyde Park, Mass.

Vises, Planer and Shaper.

American Tool Works Co., Cincinnati, O.
Frothingham & Workman, Ltd., Montreal
A. B. Jardine & Co., Hespeler, Ont.
John Millen & Son, Ltd., Montreal.
Niles-Bement-Pond Co., New York.

Washer Machines.

National Machinery Co., Tiffin, Ohio.

Water Tanks.

Ontario Wind Engine & Pump Co., Toronto

Water Wheels.

Allis-Chalmers-Bullock Co., Montreal.
Barber, Chas., & Sons, Meaford.
Canada Machinery Agency, Montreal.
The Goldie & McCulloch Co., Galt, Ont.
Greey, Wm. & J. G., Toronto.
Wm. Kennedy & Sons, Owen Sound.

Water Wheels, Turbine.

Barber, Chas., & Sons, Meaford, Ont.
Kennedy, Wm., & Sons, Owen Sound

Wheelbarrows.

Baxter, Paterson & Co., Montreal.
Buffalo Foundry Supply Co., Buffalo
Frothingham & Workman Ltd., Montreal
Detroit Foundry Supply Co., Windsor.
Dominion Foundry Supply Co., Montreal
Hamilton Facing Mill Co., Hamilton.
Paxson, J. W., Co., Philadelphia

Winches.

Frothingham & Workman, Ltd., Montrea

Wind Mills.

Ontario Wind Engine & Pump Co., Toronto

Window Wire Guards.

Expanded Metal and Fireproofing Co Toronto.
B. Greening Wire Co., Hamilton, Ont.
Paxson, J. W., Co. Philadelphia

Wire Chains.

The B. Greening Wire Co., Hamilton.

Wire Cloth and Perforated Metals.

Expanded Metal and Fireproofing Co. Toronto.
B. Greening Wire Co., Hamilton. Ont.
Paxson, J. W., Co., Philadelphia.

Wire Guards and Railings.

Expanded Metal and Fireproofing Co. Toronto.
B. Greening Wire Co. Hamilton, Ont.

Wire Nail Machinery.

National Machinery Co., Tiffin, Ohio.

Wire Rope.

Frothingham & Workman, Ltd. Montreal
B. Greening Wire Co., Hamilton, Ont.

Wood Boring Machines. Pneumatic.

Independent Pneumatic Tool Co., Chicago, Ill.

Wood-working Machinery.

Baxter, Paterson & Co., Montreal
Canada Machinery Agency, Montreal.
The Canadian Fairbanks Co., Montreal.
Goldie & McCulloch Co., Galt.
H. W. Petrie, Toronto.
Waterous Engine Works Co., Brantford.
Williams & Wilson, Montreal.

ALPHABETICAL INDEX.

CASTINGS GREY IRON
IN ROUGH OR MACHINED.
PATTERNS WOOD OR IRON.
ESTIMATES FURNISHED.
DAVID MAXWELL & SONS
ST. MARY'S, ONT.

FIRE BRICK
"Every Quality." "For Every Purpose."
The TORONTO POTTERY Co., Limited
FACTORIES IN OHIO
TORONTO, CANADA

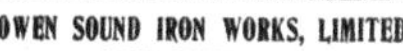

OWEN SOUND, ONT.

Cement Mill Machinery, Boiler and Steel Tank Work of all kinds, Grey Iron and Brass Castings

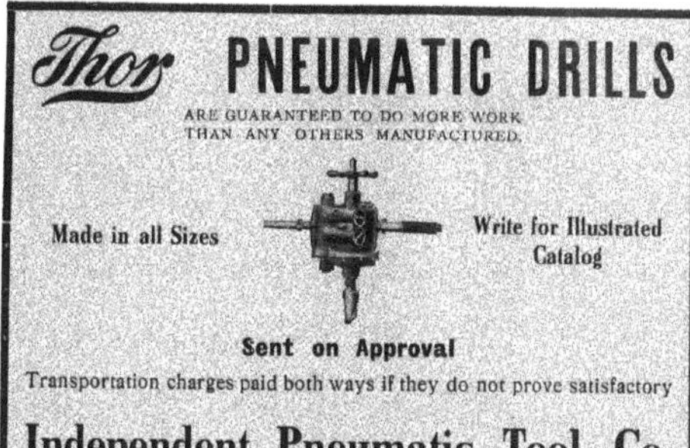

WM. KENNEDY & SONS

HIGH-CLASS

Steel Gears of all kinds
Iron Gears
Grey Iron, Steel and
Brass Castings

MARINE WORK receives every attention

WM. KENNEDY & SONS, Limited
Owen Sound, Ont.

American Pattern Shop Practice

By H. J. McCASLIN.

JUST ISSUED. An up-to-date, practical work on pattern-making, based upon the author's own experience, treating of molding and core making as well as patternmaking. Special attention is given to core box work. The following subjects are fully dealt with in a clear, concise way:

Engine Patterns — **Cearing**
Molding and Cores — **Representative Patterns**
Sweep Work — **Hints, Suggestions and Rigs**

BOUND IN CLOTH. FULLY ILLUSTRATED.

PRICE $3.00, POSTPAID.

Technical Book Dept., MACLEAN PUBLISHING CO, 10 Front St. E. - To onto

"Economo" Emery Wheel Dresser

A Substitute for the Black Diamond

"ECONOMO" ECONOMO COMBINATION

PRICE: Economo, $4.00 each; Economo Combination, $4.50 each.

Will true or shape any wheel except the very hardest and carborundum. The most efficient Roughing Dresser on the market—never gets dull from use.

Made of an abrasive nearly as hard as the Black Diamond. The Combination dresser is the roughing dresser placed in the "handle end" of the "Regular." It is made with corrugated tool steel discs and excels in wear and efficiency any dresser heretofore made. These tools are guaranteed satisfactory or money refunded. Sent on 10-day trial.

INTERNATIONAL SPECIALTY CO.
363 Holden Ave. *DETROIT, MICH.*

Equip your factory with the Hill Patent Collar oiling Bearings.

Miller Bros. & Toms

Montreal,
Quebec

MALLEABLE IRON CASTINGS

WE ALSO MAKE METAL PATTERNS

Galt Malleable Iron Co., Limited - Galt, Ontario

WIRE DOOR MATS

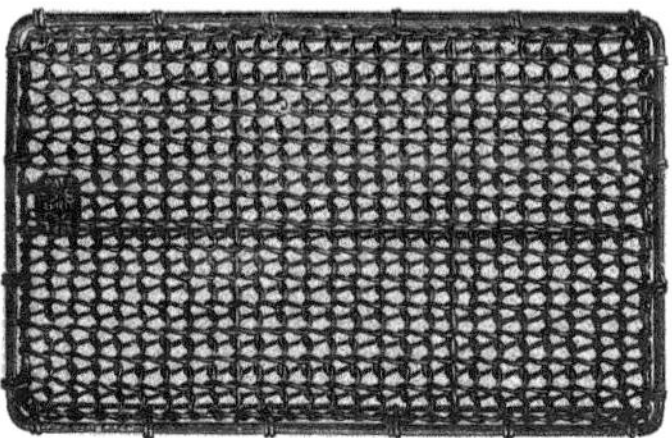

The cleanest, most durable and perfect sanitary mat made. Has been on the market for over 20 years. All sizes and shapes made to order. For stock sizes, see catalogue.

The B. Greening Wire Co., Limited
HAMILTON, ONT. MONTREAL, QUE.

A MONEY SAVER

Money saved is money made. A wasteful heating system makes serious inroads into the net profits. We do not say that you should instal a Webster Heater, but we do say that you should know more about

THE WEBSTER
FEED-WATER HEATER PURIFIER AND FILTER

and investigate the solidity of the claims we make for it before investing a dollar in any heating system manufactured anywhere at any price.

The type shown in the :ut does save money. It cannot waste steam. It does not waste fuel. It is economic even although it affords the highest temperature. *You* should know more about it.

DARLING BROTHERS, LIMITED
Toronto Montreal Winnipeg

Kerr Turbine Pumps for water works, condenser equipment and boiler feeding. These pumps are more economical than direct acting piston pumps, and require less attention, lubrication and space. No belts or gears. Speed adjustable to pressure desired. Regulation close. Start and stop quickly. For any capacity.

Turbine Blowers for gas works, foundries and forge shops Air pressure 4 to 16 ounces. For any capacity.

Turbine Generators, 2 kw. to 300 kw. direct current and alternating current, 25 to 300 kw., 60 cycle.

The illustration shows an 18-inch condensing Kerr Steam Turbine coupled to a 4-inch 3-stage Worthington Pump. Capacity 200 g.p.m. against a head of 360 feet.

Main Office:

KERR TURBINE CO., Wellsville, N.Y.

Built in Canada by

WM. STEPHENSON

34 Bay Steet N. *HAMILTON, ONT.*

Link-Belt Pivoted Overlapping-Bucket Conveyor

carries coal to bunkers or floor of boiler-room, and removes ashes to storage or cars. Overlapping joints of the buckets prevent spill.

View shows conveyor in Power House, Chicago and West Indiana Ry. Co., 47th St., Chicago. **Write for particulars.**

LINK-BELT COMPANY

PHILADELPHIA, CHICAGO, INDIANAPOLIS

NEW YORK 299 Broadway. PITTSBURGH 1501 Park Bldg. ST. LOUIS Missouri Trust Bldg. SEATTLE 440 New York Block

JANDUS ARC LAMPS

The JANDUS Interchangeable Arc Lamp offers itself on grounds of

Simplicity

Economy of Installation

Economy of Operation

It embodies all of the features required by the best modern practice.

SOLE CANADIAN REPRESENTATIVES

THE PACKARD ELECTRIC CO. LIMITED

ST. CATHARINES

MONTREAL WINNIPEG

PINTSCH SUCTION GAS PLANTS
NATIONAL GAS ENGINES

No Smoke	Simplest
No Pressure	Easy Starting
No Danger	Low Cost
No Leakage	Loose Liners
No Gas Holder	Heavy Fly Wheel
No Noise	Special Electric
No Boiler	Light Types
No Chimney	

Installations in successful operation total over 200,000 b.h.p. in units from 10 to 2,000 h.p.

SEND FOR CATALOGUE AND PRICES

The Economic Power, Light and Heat Supply Co., Limited

40 York Street TORONTO, ONT.

N. J. HOLDEN CO., 66 McGill St., MONTREAL
Agents for the Province of Quebec.

WE MAKE IT

BABBITT METAL

Imperial Babbitt

IS BEST FOR HIGH-SPEED AND HIGH-PRESSURE WORK.

IT CAN'T BE BEAT.

Every Pound Backed by a Guarantee.

WE ARE SPECIALISTS

and can give you a bearing metal to suit any class of work.

WRITE US. TRY US.

THE CANADA METAL CO., LIMITED
Toronto, Ont.

McVICKER Efficiency with Simplicity ENGINES

is what we claim.

Only one-third the usual number of parts used in construction.

Perfect balance while running.

No experience needed to operate our engine.

WRITE FOR CATALOGUE

The Canadian McVicker Engine Company LIMITED

GALT - - - ONTARIO

TELEGRAPHIC ADDRESS: "HALLTYNE"

WORKS: PHONE MAIN 258

THE HALL ENGINEERING WORKS

14 COTE ST., MONTREAL

ENGINEERS, BOILERMAKERS, BRASS & IRON FOUNDERS, COPPER SMITHS & BLACKSMITHS, SHIP REPAIRS A SPECIALTY

Agents for J. & E. Hall's Refrigerating Machinery.
Atlas Preservative Co. & United Asbestos Co.
of LONDON, ENG.

THOS. HALL, M. N. E. C. I. E. S. — Phone West 1737

Late Supt. Engineer Messrs. Elder Dempster & Co. & Can. Pac. Rly. Co., London

T. O. SINCLAIR, M. I. M. E. — Phone West 917

SIGHT FEED OIL PUMP COMPANY

New York, N.Y.
Mr. William Martin, 141 Broadway

Boston, Mass.
Mr. H. E. Rundlett, 141 Milk St.

Pittsburgh, Pa.
Laughlin, Green Co. 326 Fourth Ave.

Scranton, Pa.
Scranton Supply & Machinery Co., 131 Wyoming Ave.

Hazleton, Pa.
Hazleton Machinery & Supply Co.

Allentown, Pa.
The William H. Taylor Company

Chicago, Ill.
Mr. Alex. Hodgart, 1207 Fisher Bldg.

Birmingham, Ala.
Mr. James J. O'Rourke, 511 Brown-Marx Bldg.

San Francisco, Cal.
Phoenix Tool & Valve Co. 131 Beale St.

Montreal, Canada
Mr. Homer Taylor, Temple Bldg.

Model "G," eighteen-feed, automatic oil pump, three kinds of oil.

MILWAUKEE, WIS., U.S.A.

UNION DRAWN STEEL COMPANY, Limited

Manufacturers of

BRIGHT FINISHED STEEL

Largest in Canada. Best Equipped on the Continent. Capacity twice present requirements of the trade.

Rounds, Flats, Squares and Hexagons

LARGE STOCK

Send for Price List

Office and Works:

HAMILTON, CANADA

FOR FREIGHT HAULAGE

THE JEFFREY MFG. COMPANY

COLUMBUS, OHIO, U.S.A.

Canadian Branch: Cor. Cote and Lagauchetiere Sts., MONTREAL, Can.

It is easily and cheaply laid. Its carrying capacity is never decreased by rust.

GALVANIZED WIRE WOUND WOODEN PIPE

CANADIAN PIPE COMPANY, Limited

PHONE 1642 **VANCOUVER, B.C.** **P.O. BOX 915**

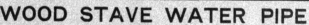

WOOD STAVE WATER PIPE

Wood Pipe Designed for Different Pressures.

Machine-banded made from choice British Columbia or Douglas Fir, absolutely clear and free from knots or defects, wound with heavy gauge galvanized steel wire. For

Water Works, Power Plants, Irrigation, Mining and all Hydraulic Propositions, where conduits or pressure pipe are required.

Reliable, durable, economical, efficient, easy to instal. Costs very much less than any other sort of water pipe, and with greater efficiency in use.

Write for Catalogue and Circulars.

Estimates also given on continuous stave pipe of large diameters.

PACIFIC COAST PIPE COMPANY

P.O. Box 563 **Limited**

1551 Granville Street (Next Bridge) - VANCOUVER, B.C.

STAR PROFILE

SAVES WEIGHT

FLUTED PROFILE

Beardshaw's Conqueror Steel

supplied in

Water Hardening and Air Hardening Quality

THE MOST ECONOMICAL HIGH SPEED STEEL

for Chisels, Rimers, Milling Cutters, etc.

ALEXANDER GIBB

13 St. John Street - - MONTREAL

for Lathe Tools of all descriptions and Special Drills.

WELL KNOWN BRANDS MADE IN CANADA BY

AMERICAN

ARCADE

KEARNEY and FOOT

McCLELLAN

Nicholson Co.

GLOBE

EAGLE

GREAT WESTERN

J. B. SMITH

Dominion Works, Port Hope, Ont.

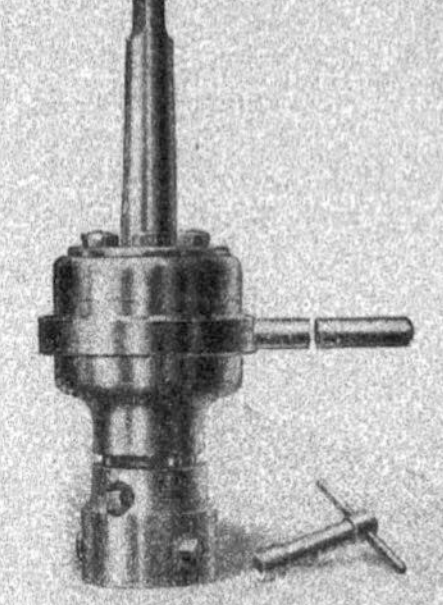

THE GEOMETRIC IMPROVED REVERSING TAP HOLDER

Can Be Applied to Any Drill Press or Lathe

Works equally well in either a horizontal or vertical position.

This Tapper will turn your Drill Press or Speed Lathe into a fine Tapping Machine at little cost, as no reversing belt is required. Let us send you one on trial. If you do not like it, it may be returned at our expense.

ASK FOR BOOKLET

The Geometric Tool Co., New Haven, Conn.

Canadian Agents: WILLIAMS & WILSON, Montreal, Que. (Westville Station) U.S.A.

REAMERS
MANUFACTURED BY
Butterfield & Co.
will Stand the Hardest Kind of Work and
their Accuracy will Satisfy the Most Critical Mechanic.
These two qualities combined make First Quality.
Butterfield & Co.
Rock Island, P.Q.

The Standard Tool Co's
Drill Chucks
STANDARD TOOL CO.
THE STANDARD IMPROVED
Made in five sizes. Capacity up to ¼ in., ⅜ in., ½ in., ¾ in., 1 in.
We furnish arbors for these chucks with shanks 1-2 in., 5-8 in., 3-4 in. and 1 in. diam., and with taper shanks; or we will thread the hole in the hub to fit any special templet at a small cost.
VERY STRONG AND SUBSTANTIAL
OFFICE and FACTORY
Cleveland, Ohio
EASTERN SALESROOM
94 Reade St., New York

Armstrong Tool Holders
Have saved years of time and thousands of tons of Tool Steel, here's TWO NEW THINGS will prove an equally good investment.
DROP FORGED OF STEEL
THE ARMSTRONG IMPROVED TOOL POST
Combines the strength and holding power of the strap and stud tool clamp with the convenience of the "open side" and ordinary set screw tool post. A slight turn of one nut releases tool or sets it solidly.
No trouble with "stripping" or "upset" screws.
ARMSTRONG IMPROVED TOOL POST
The ARMSTRONG THREE-BAR BORING TOOL
INCREASES PRODUCTION AND REDUCES COST OF TOOL MAINTENANCE
CUTTERS CANNOT JAR LOOSE
HIGH SPEEDS AND BIG FEEDS
ONLY SET THEM TIGHTER
SOLD IN CANADA BY:—Williams & Wilson, Montreal; Aikenhead Hardware, Limited, Toronto; A. R. Williams Machinery Co., Toronto Winnipeg, Vancouver; Wood, Vallance & Co., Hamilton; Frothingham & Workman, Limited, Montreal; H. W. Petrie, Toronto.
WRITE FOR OUR LATEST CATALOG.
ARMSTRONG BROS. TOOL CO., 106 N. FRANCISCO AVE. CHICAGO, U.S.A.
"THE TOOL HOLDER PEOPLE"

NEW AND UP-TO-DATE

BOLT AND NUT MACHINERY

INCLUDING

Bolt Cutters, Nut Tappers, Bolt Headers, Upsetting and Forging Machines, Wire Nail and Spike Machines and Bulldozers.

Send for Catalogue M.

NATIONAL MACHINERY CO., Tiffin, Ohio, U.S.A.

Canadian Agents: H. W. PETRIE, Toronto, Ont. WILLIAMS & WILSON, Montreal, Que.

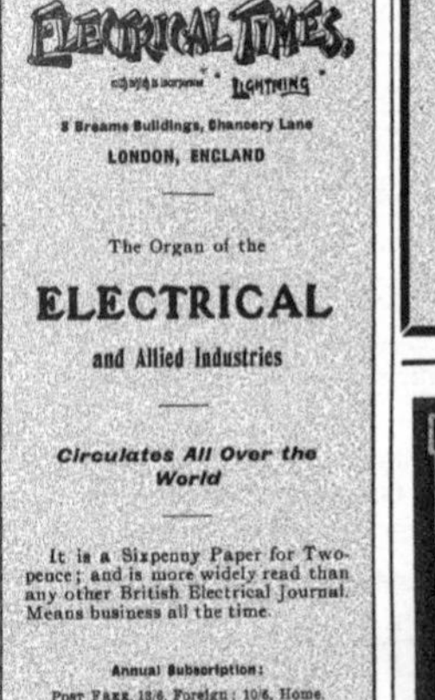

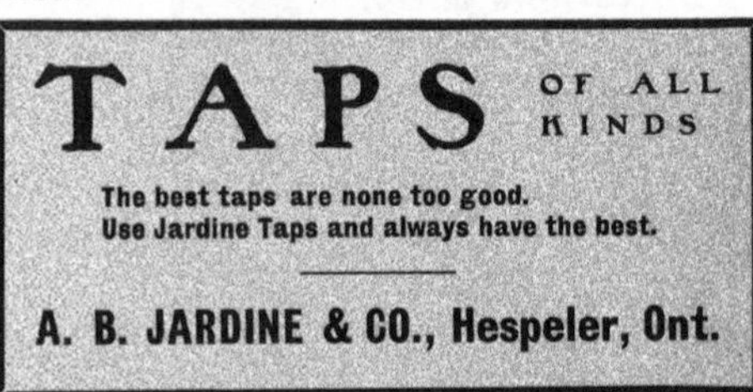

TAPS OF ALL KINDS
The best taps are none too good.
Use Jardine Taps and always have the best.
A. B. JARDINE & CO., Hespeler, Ont.

IS HANDIEST CHEAPEST STRONGEST FOR

Floors, Roofs, Partitions, and all Varieties of Concrete Slab Work

CATALOGUES—SAMPLES FREE

Expanded Metal & Fireproofing Co. Limited
100 King Street West, TORONTO

DO YOU KNOW

RAYO

is the new name of

ALPHA HIGH SPEED STEEL?

In stock—ANNEALED and NOT ANNEALED

with BAINES & PECKOVER, - - Toronto, Ont.
E. G. PRIOR & CO., - - - Victoria, B.C.

Canadian Representative: D. W. CLARK, P.O. Box 521, TORONTO

As good as "NOVO"

A claim made by other makers which has yet to be proved.

Have you tried the new Sectional "NOVO" Steel for Lathe, Planer and Cut-off Tool ?

Requires NO FORGING

A Saving of 35 to 45 per cent.

WILLIAM ABBOTT
334 St. James St., MONTREAL

JESSOP'S BEST TOOL STEEL "ARK" High-Speed Steel

THE FAVORITE BRANDS WITH USERS OF GOOD STEEL.
A LARGE ASSORTMENT OF SIZES IN STOCK.
JESSOP'S HIGH-GRADE FILES AND RASPS.

80 Bay, St., Toronto, Ontario, Chas. L. Bailey, Agent.
Reid-Newfoundland Company, St. John's, Newfoundland.
Jas. Robertson Co., Ltd. Montreal, Quebec
Jas. Robertson Co., Ltd., St. John, New Brunswick.

WM. JESSOP & SONS, Limited, Manufactory, SHEFFIELD, ENGLAND.

STARRETT TOOLS

ARE THE STANDARD FOR ACCURACY, WORKMANSHIP, DESIGN AND FINISH

Starrett Micrometers

Features of Superiority—Patented

Our micrometers have a more exact and easier way of adjustment than by the old method of a movable anvil. This is obtained by placing over the barrel a thin, graduated sleeve, which carries the base or zero line, instead of having this line marked on the barrel itself. This sleeve may be turned by means of a small spanner wrench to bring the zero line correct to compensate for wear. The thin sleeve also keeps dirt from the screw. A knurled locking nut contracting a split bushing around the spindle tightens and keeps the spindle central and true, or by a slight turn locks it firm, making a solid gauge when desired.

Our 176 page Catalogue shows many styles of Micrometers, from ½ in. to 12 in., and in Metric from 30 mm. to 300 mm.

Send for Free Catalogue, No. 173, of Fine Mechanical Tools

THE L. S. STARRETT CO., ATHOL, MASS., U. S. A.

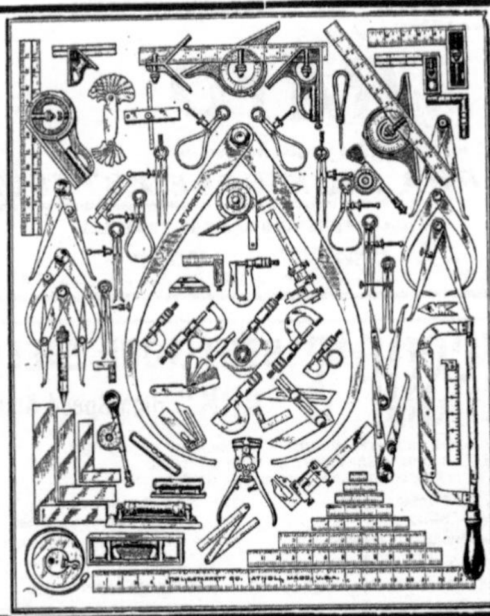

Machinist Tools

Tools of the finest quality; best material; unequalled workmanmanship. Warranted accurate and satisfactory.

Any Tool the Machinist can Require

Our large and varied stock is complete in every respect. The quality of these goods is so superior, it needs no mention.

Rice Lewis & Son
LIMITED
TORONTO

Books for Engineers, Draughtsmen, Science Students, Etc.

Sent Post Free to any Address, at home or abroad, at Published Price

Just Published, Second Edition, with numerous specimen Workshop Cost Forms, price 12/6 net, post free.

WORKSHOP COSTS (for Engineers and Manufacturers), by Sinclair Pearn and Frank Pearn (Directors of Messrs. Frank Pearn and Co., Limited, West Gorton Ironworks, Manchester), with a preface by G. Croydon Marks, Assoc. Memb. Inst. C. E., M.I.M.E., etc.

	NET PRICE s.	d.
Air Compressors and Blowing Engines, by Chas. H. Innes, M.A.	4	6
The Fan; including the Theory and Practice of Centrifugal and Axial Fans, by Ch. H. Innes, M.A.	4	0
The Proportions and Movement of Slide Valves, by W. D. Wansbrough	4	6
Machines and Tools Employed in the Working of Sheet Metals, by R. B. Hodgson	4	6
Governors and Governing Mechanism, by Hall	2	6
Specification of a Modern Lancashire Boiler and its Seating (200 lbs. pressure), by "Inspector"	5	0
Continuous Current Dynamos and Motors and their Control, by W. R. Kelsey	5	0
The Resistance and Power of Steamships, Atherton and Mellanby	5	0
Notes on Construction and Working of Pumps, Marks	3	6
Modern Ironfoundry Practice:		
Part I., Hand Moulding, Bale	5	0
Part II., Machine Moulding, Bale	3	6
Modern Gas and Oil Engines, by F. Grover. 4th Edition	5	0
The Indicator and its Diagrams, by Chas. Day. 3rd Edition	4	6
The Chemistry of Materials of Engineering, by A. H. Sexton. 2nd Edition	5	0

	NET PRICE s.	d.
The Management of Small Engineering Workshops, Barker	7	6
Problems in Machine Design, by Chas. Innes. 2nd Edition	4	6
Heat and Heat Engines; a Treatise on Thermo-Dynamics. Popplewell	6	0
Centrifugal Pumps, Turbines and Water Motors. 4th Edition	4	6
Application of Graphic Methods to the Design of Structures. 2nd Edition	5	0
Engineering Estimates and Cost Accounts, Burton. 2nd Edition	3	0
Graphic Methods of Engine Design, Barker. 2nd Ed.	3	6
Injectors: Theory, Construction and Working, Pullen. 3rd Edition	5	6
Construction of Cranes and Lifting Machinery, Marks. 2nd Edition	3	6
Marine Engineers: Their Qualifications and Duties	5	0
A.B.C. of the Differential Calculus, Wansbrough	3	0
The Theta-Phi Diagram Applied to Steam, Gas, Oil, and Air Engines	3	0
Mechanical Engineering Materials, by Marks. 2nd Edition	2	6
The Naval Engineer and Command of the Sea, Burton	2	6

THE TECHNICAL PUBLISHING CO., LIMITED, 287 Deansgate, Manchester, Eng. ; **D. VAN NOSTRAND CO.,** 23 Murray St. and 27 Warren St., New York, **and all Booksellers.**

CANADIAN MACHINERY AND MANUFACTURING NEWS

A monthly newspaper devoted to the manufacturing interests, covering in a practical manner the mechanical, power, foundry and allied fields. Published by The MacLean Publishing Company, Limited, Toronto, Montreal, Winnipeg, and London, Eng.

OFFICE OF PUBLICATION: 10 FRONT STREET EAST, TORONTO

Vol. III. JULY, 1907 No. 7

SMALL TOOLS

Made in Canada

HIGHEST GRADE OBTAINABLE

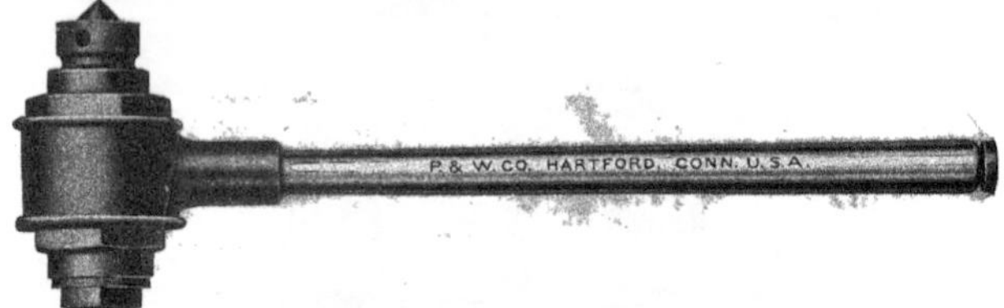

Write to Nearest Office for New Small Tool Catalogue.

Pratt & Whitney Company

of Canada, Limited, Dundas, Ontario.

SALES AGENTS—THE CANADIAN FAIRBANKS COMPANY, LIMITED
Montreal Toronto Winnipeg Vancouver

SIDE CRANK IDEAL SHOWING EXTENDED SUB BASE AND SHAFT FOR GENERATOR

SIDE CRANK IDEAL ENGINES

A HIGH SPEED ENGINE SPECIALLY DESIGNED FOR DIRECT CONNECTION

ASK FOR CATALOGUE AND PRICES

BUILT BY **THE GOLDIE & McCULLOCH CO.,** LIMITED

GALT — ONTARIO — CANADA

Western Branch—248 McDERMOTT AVE., WINNIPEG, Man., Quebec Agents—ROSS & GREIG, MONTREAL, Que.

In Case of Repairs

Westinghouse Motor Driving Horizontal Boring Mill.

With Individual Motor Drive only the individual tool need be out of service.

With Belt Drive all the tools on the line shafting must be shut down, and often the entire plant.

Westinghouse Motors are successfully driving every class of machine tool.

Canadian Westinghouse Co., Limited

General Office and Works, HAMILTON, ONTARIO

For Particulars Address Nearest Office

Traders Bank Building, TORONTO
152 Hastings Street, VANCOUVER
922-923 Union Bank Bldg., WINNIPEG
Sovereign Bank of Canada Bldg., MONTREAL
134 Granville Street, HALIFAX

A GOOD INVESTMENT

If the ablest practical men in the machinery manufacturing industry were asked to name the best investment in their line the answer would be

The 22-inch ALL-GEARED LATHE

manufactured by **The London Machine Tool Company, Limited,** and we're ready to prove to you that they would be right.

Do you believe in carrying on business in the least expensive way? Then this Lathe will satisfy you.

It is a money **saver** because of its **time** saving qualities. This Lathe has many features which make for Lathe perfection.

Head consists of six gears of wide face and coarse pitch, working into positive clutches, giving four changes of speed in geometrical progression. With three-speed countershaft, twelve speeds are obtained.

The spindles are manufactured from high carbon crucible steel and first bored from the solid stock roughed out on the lathe, final finishing cuts taken, then put on grinder and ground true to standard sizes.

But—Our literature gives full particulars of this **Money-Saving Machine** and we'd like your opinion, as a practical man, of our last achievement.

***IMPORTANT*—We can make quick delivery on above size.**

The London Machine Tool Company LIMITED

HAMILTON, ONTARIO

Sales Office, Room 1206 Traders' Bank Building, Toronto, Can.

Why Continue to Use Antiquated Methods for Producing Your Duplicate Parts?

Do you know that with the aid of our special attachments you can Form, Drill, Tap, Ream, Cross Drill, Slot, Slab, Knurl, Recess, Letter, and make Eccentric parts on our machines?

Whether from the bar, castings or forgings we can save you money. Here are two samples of what we can do for you.

"The Pre-eminence of the

CLEVELAND

is indisputable"

Piece No. 256.—Material, Machine Steel Bar, made on a 1¼-in. Cleveland Automatic Plain Machine, drawing full size. Output per hour, 20; this means all operations for completing this piece. Actual cost of labor, from 3 to 3½ mills for each piece. Send along samples or drawings. We guarantee our outputs. One man can operate from 4 to 6 machines.

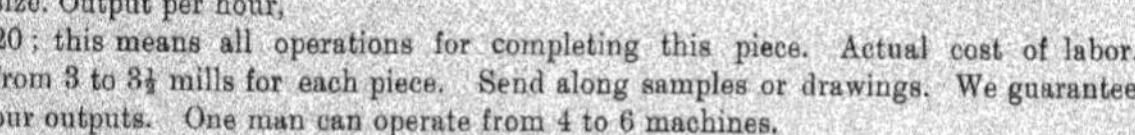

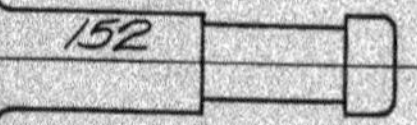

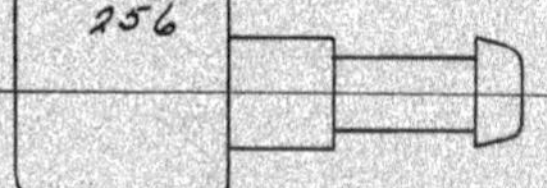

Piece No. 152.—Material, Brass Castings, made on a 2-in. Cleveland Automatic Turret Machine, using our tilting magazine, drawing full size. Time of all operations for completing this piece, 4 minutes. Actual cost of labor, from 4 to 5 mills. One man can operate 4 to 6 machines.

AS WE HAVE NO CANADIAN REPRESENTATIVE, WE SOLICIT CORRESPONDENCE AND ORDERS DIRECT

CLEVELAND AUTOMATIC MACHINE CO.

CLEVELAND, OHIO, U.S.A.

EASTERN REPRESENTATIVE—J. B. ANDERSON, 2450 North Thirtieth Street, Philadelphia, Pa.

WESTERN REPRESENTATIVE—H. E. NUNN, 22 Fifth Avenue, Chicago, Ill.

FOREIGN REPRESENTATIVES—CHAS. CHURCHILL & CO., London, Manchester, Birmingham, Newcastle-on-Tyne and Glasgow. MESSRS. SCHUCHARDT & SCHUTTE, Berlin, Vienna, St. Petersburg and Stockholm. ALFRED H. SCHUTTE, Cologne, Brussels, Liege, Paris, Milan and Bilbao.

H. W. PETRIE'S

Partial List of Machinery on Hand for Immediate Delivery:

Can quote interesting prices. Write for complete stock list.

H. W. PETRIE, *TORONTO and MONTREAL*

MURCHEY DOUBLE HEAD RAPID NIPPLE and PIPE THREADING MACHINE

With variable speed motor directly attached

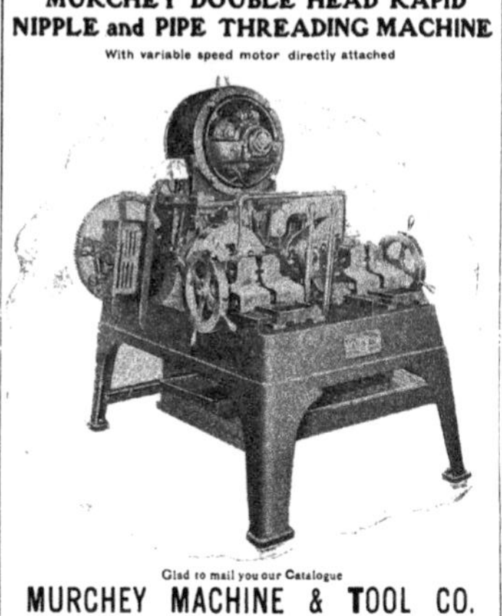

Glad to mail you our Catalogue

MURCHEY MACHINE & TOOL CO.

DETROIT, MICH., U.S.A.

12'6 Sole Canadian Agents: The London Machine Tool Co. Traders Bank Building, Toronto.

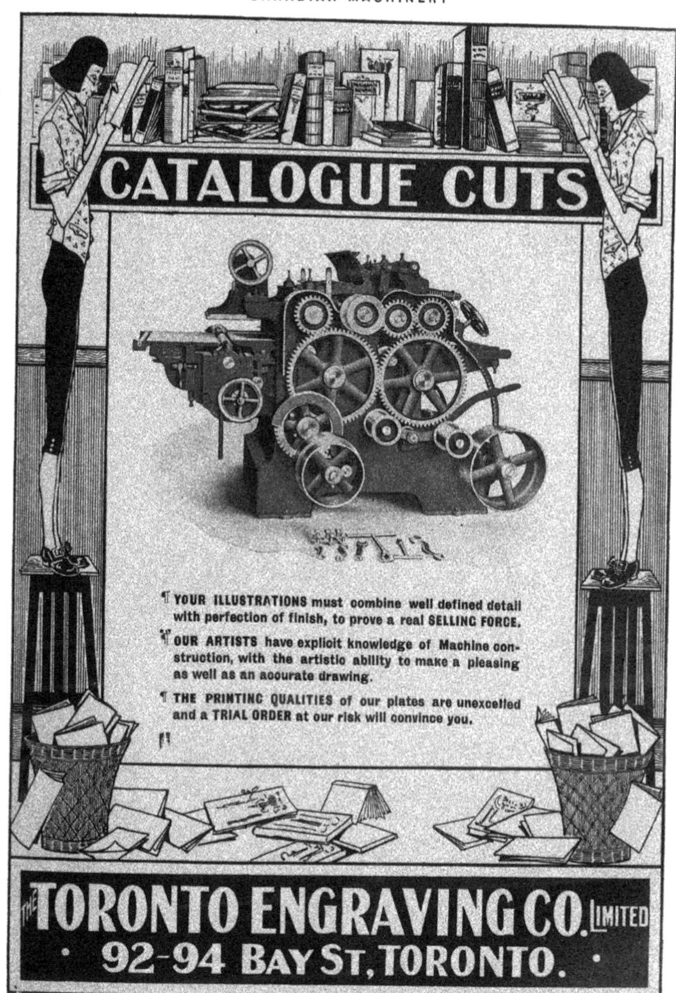
CATALOGUE CUTS
¶ YOUR ILLUSTRATIONS must combine well defined detail with perfection of finish, to prove a real SELLING FORCE.
¶ OUR ARTISTS have explicit knowledge of Machine construction, with the artistic ability to make a pleasing as well as an accurate drawing.
¶ THE PRINTING QUALITIES of our plates are unexcelled and a TRIAL ORDER at our risk will convince you.
THE TORONTO ENGRAVING CO. LIMITED
· 92-94 BAY ST, TORONTO. ·

WATER SUPPLY

for Sprinkler System, Water Works, Fire Pressure, etc., etc.

Steel and Wood Tanks, Steel Substructures, Steel Flagstaffs, Bell Towers, Etc.

WINDMILLS, GASOLINE ENGINES, HYDRAULIC RAMS

ONTARIO WIND ENGINE and PUMP CO.
LIMITED
Winnipeg TORONTO Montreal

TELEGRAPHIC ADDRESS: "HALLTYNE" — WORKS PHONE MAIN 256

THE HALL ENGINEERING WORKS

14 COTE ST., MONTREAL

ENGINEERS, BOILERMAKERS, BRASS & IRON FOUNDERS, COPPER SMITHS & BLACKSMITHS, SHIP REPAIRS A SPECIALTY

Agents for J. & E. Hall's Refrigerating Machinery,
Atlas Preservative Co. & United Asbestos Co.
of LONDON, ENG.

THOS. HALL, M. N. E. C. I. E. S. — Phone West 1737
Late Supt. Engineer Messrs. Elder Dempster & Co. & Can. Pac. Rly. Co., London

T. O. SINCLAIR, M. I. M. E. — Phone West 917

SIGHT FEED OIL PUMP COMPANY

New York, N.Y.
Mr. William Martin, 141 Broadway

Boston, Mass.
Mr. H. E. Rundlett, 141 Milk St.

Pittsburgh, Pa.
Laughlin, Green Co. 326 Fourth Ave.

Scranton, Pa.
Scranton Supply & Machinery Co., 131 Wyoming Ave.

Hazleton, Pa.
Hazleton Machinery & Supply Co.

Allentown, Pa.
The William H. Taylor Company

Chicago, Ill.
Mr. Alex. Hodgart, 1207 Fisher Bldg.

Birmingham, Ala.
Mr. James J. O'Rourke, 511 Brown-Marx Bldg.

San Francisco, Cal.
Phoenix Tool & Valve Co. 131 Beale St.

Montreal, Canada
Mr. Homer Taylor, Temple Bldg.

Model "G," eighteen-feed, automatic oil pump, three kinds of oil.

MILWAUKEE, WIS., U.S.A.

Established over 50 years. — Telegrams: "Belliss" Birmingham. A.B.C. 5th edition.

BELLISS & MORCOM

ENGINEERS LIMITED

Birmingham, - - - England

ENGINES

DIRECT, ROPE OR BELT DRIVE for

Electric Power, Lighting, Traction and Mill Driving

ALSO

Condensing Plants

5,000 H.P. Installed in Canada

Representing our Two-Crank Compound Type Engine

ADVANTAGES — Self-Lubrication; Simplicity; Strength; Experience; Low Cost; Reliability; Minimum Maintenance; Compactness; Good Governing; Durability; Maintained Efficiency; Quiet Running.

3 300 Engines supplied or building, representing upwards of 600,000 H.P., in sizes ranging from 5 to 2,500 B.H.P. of Compound and Triple-Expansion Types.

SOLE AGENTS FOR CANADA:

LAURIE & LAMB, 212 Board of Trade Buildings
MONTREAL, CANADA

Telegraphic Address. "Larco" Montreal. Codes: A.B.C. 5th Edition and Western Union.

OWEN SOUND IRON WORKS, LIMITED
OWEN SOUND, ONT.

Cement Mill Machinery, Boiler and Steel Tank Work of all kinds, Grey Iron and Brass Castings

---

The Engineering Times

The Weekly Newspaper of Commercial Engineering

32 PAGES

Price Twopence.

Annual Subscription, 10s. 10d.
Foreign - - 13s. 0d.

ADVERTISEMENTS.

Prepaid Advertisements on Wrapper, classified under headings: Tenders—Appointments Open—Situations Wanted—Partnerships—Wanted—For Sale, &c.

By the Line—Four Lines or under, 2/-; each additional Line, 6d.; the Line averages Six Words.

By the Inch, 6/- per Inch, Single Column, One Insertion.

SUBJECT TO DISCOUNT

20% off 13 Insertions; 25% off 26 insertions; 33⅓% off 52 insertions

CHIEF OFFICE: 6, Bouverie St., LONDON, E.C.

Leonard Corliss Engines

Peerless Self-oiling and Leonard Automatics for Electrical and Manufacturing purposes. Standard Return and Water Tube Boilers, all sizes. Open and Closed Exhaust Steam Heaters, Pumps, Etc., for Complete Steam Plants.

E. Leonard & Sons, London, Ont.

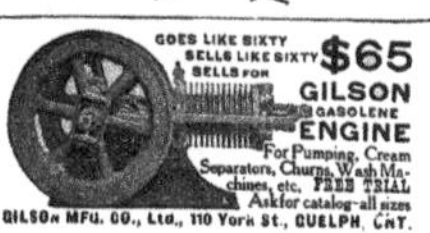

An intelligent Engineer uses nothing but the GENUINE PENBERTHY INJECTOR. He has had experience with just-as-good kind before.

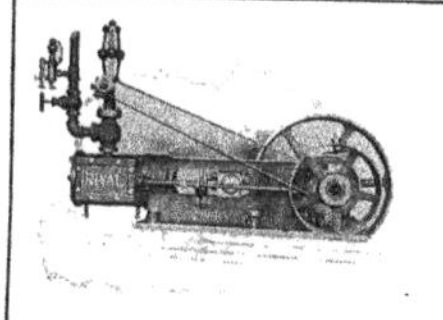

"RIVAL" ENGINES

MEDIUM PRICE
MEDIUM SPEED
MEDIUM SIZE

Write for Stock List.

Laurie Engine & Machine Co., Limited
MONTREAL

Toronto Agents: Parmelee & Nicholson
Halifax Agents: Guilford & Son

BLOWERS

HEATING AND VENTILATION
by Fan System

Natural Draft and Blower Dry Kilns
For Lumber, Brick, etc. Cars, Trucks, Rail, etc.

Heating
Ventilating
Drying Forges
Mechanical Draft Blowers.
Exhausters
Steam Specialties

ENGINES
Both Vertical and Horizontal

Shavings Exhaust System

SHELDONS, Limited, Engineers and Manufacturers **Galt, Ont, Canada**

Montreal Office, 325 St. James St. Phone Main 2919

Steel Converters

Flockton, Tompkin & Co.

Newhall Steel Works LTD.

SHEFFIELD

Makers of

Specialty AIR HARDENING TOOL STEEL for High Speed Work

ENGINEERS' TOOL STEEL, for Machine Tools, Chisels, etc.

DRILL AND JUMPER STEEL, for Miners and Contractors

SOLID STEEL HAMMERS, of any shape or size for Miners and Engineers

SPRING STEEL, SECTION BARS

Purchasing Agents wanted in principal Canadian Cities.

"Pioneer" Pressed Steel Shaft Hangers

Two hundred per cent. more rigid than Cast Iron Hangers—Unbreakable —It's a Fact

15-16 in. to 3 15-16 in. diameter; 8 in. to 30-in. drop

Fifty per cent. lighter—much handier than Cast Iron Hangers and cheaper to erect. All sizes in stock

Selling Agents

Canada Machinery Agency

W. H. NOLAN, Prop. 298 St. James Street, MONTREAL

Shantz No. 4 Pipe Cutting and Threading Machine

THIS machine is fitted with the **"Borden Patent Solid Adjustable Die Head"** which is the easiest operated and most perfect mechanism invented. Any degree of accuracy may be obtained simply by the adjustment of an adjusting screw upon which the operating lever rests. Any variation may be obtained to suit any make of fitting, whether standard or not.

No backing off; just raise the lever and the pipe is released.

Change of dies easily and quickly made.

Universal chucks for gripping the pipe are placed on the front and rear ends of the arbor

In buying a Pipe Threading Machine the shrewd buyer demands **Rapidity, Accuracy, Strength** and **Simplicity.** These qualities, with the many exclusive features, make the Shantz No. 4 a great labor and time saver.

It will pay you to get full particulars. Drop us a card.

I. E. SHANTZ & CO. BERLIN CANADA

MACHINE TOOLS

BERTRAM VERTICAL MILLING MACHINE

We carry a large stock of High-Grade

MACHINE TOOLS

at all of our Branches, and are therefore in a position to make very prompt delivery on your requirements.

Are You in the Market for Anything in this Line? If so, don't forget that

WE ARE THE SELLING AGENTS FOR THE FOLLOWING MANUFACTURERS:

Niles-Bement-Pond Co.
The John Bertram & Sons Co., Ltd.
Pratt & Whitney
Brown & Sharpe
E. W. Bliss Co.
The R. McDougall Co., Ltd.
R. A. Kelly Co.
J. J. McCabe
Bignall & Keeler
Baker Bros.
Merrell Mfg. Co.
Seneca Falls Mfg. Co.
Wilmarth & Morman
Niagara Machine & Tool Co.
S. A. Woods Machine Co.
American Woodworking Machinery Co.

BERTRAM 30-inch HIGH SPEED LATHE—MOTOR DRIVEN

The Canadian Fairbanks Co., Ltd.

MONTREAL TORONTO WINNIPEG VANCOUVER

Convention of Waterworks Association

Large Number of Prominent Engineers Present—Many Interesting Papers Read and Discussed—Social Enjoyment in Trolley Ride and Moonlight Excursion—Display of Waterworks Supplies.

During the week of June 16 to June 22, there was gathered in Toronto a large number of prominent engineers from Canada and the United States at the twenty-seventh annual convention of the American Waterworks Association. In all there were five hundred active members, associate members and guests registered at the convention headquarters, in the King Edward hotel. Of these, two hundred were active members. This was a busy convention, evening, as well as morning and afternoon sessions, being held.

Group of Delegates Attending Convention of American Waterworks Association, Toronto, June 17-22.

The trolley ride around the city, on Wednesday, and the moonlight excursion on Thursday evening were practically the only digressions the members took from work.

The ladies were suitable entertained.

There was a goodly representation of manufacturers and supply men. The exhibits consisted chiefly of brass goods.

Conducting a Municipal Waterworks System.

The opening day, the second day, of the convention, was taken up in speeches of welcome, and in the discussion of the best methods of conducting a municipal waterworks system.

A strong argument in favor of the installation of meters on all water services was presented by Mr. W. Volhardt, of Staten Island, N.Y. He urged the necessity of meters, both on account of the desirability of curtailing waste and of giving a square deal between all classes of consumers, and advocated the installation of a universal system at the beginning.

Mr. George W. Rafter followed with a paper on "The Cost of Meters in Rochester, N.Y.," showing that the use of meters did not reduce the use of water. He made the deduction that water takers by meter paid 100 per cent. profit to the city at 14 cents per 1,000 gallons, while those taking water at fixture rates received it at far less than cost. These deductions were questioned by Superintendent Little, of the Rochester waterworks, and also by Mr. Fisher, Rochester.

James L. Tighe, of Holyoke, Mass., read a paper on "Water Consumption, Waste and Meter Rates." In Holyoke, the waterworks, although owned by the city, pay taxes just the same as a private corporation. The water consumption there increased from 85 gallons per capita in 1895, to 142 gallons in 1903, and 150 or 160 in 1904. A more vigilant inspection succeeded in reducing this to 114 gallons in 1906. This showed the ill-effect of allowing extravagances in the use of water.

Mr. Saunders, of London, raised the question of the economy of restricting the consumption of water when it cost more to do it than to pump all the people wanted. Mr. Voithardt, in reply, gave an experience in which after putting in meters, he had got back in fourteen months the entire cost of installing the meter system.

Dangers of Filter Plants.

J. M. Diven read a paper on "The Care of a Mechanical Filter Plant." He considered a filtered water supply as not an unmixed blessing, as much depended on how it was operated. There was a chance that the people would be worse off than before, as filter beds made excellent breeding grounds for bacteria, and water users were liable to forego all ordinary precautions and depend on the filtration plant alone. A properly trained man was required for a filter, one who understood its workings and had some knowledge of bacteriology.

Detecting Waterworks Losses.

At the evening session Mr. Edward S. Cole, of New York, gave a paper on the pitometer and waterworks losses, which was illustrated by lantern slides. The question of waste he considered the most important on account of the great expense entailed by it, but with the portable Pilot tube, he considered that there was now no excuse for permitting underground losses.

Producer Gas Plants.

The progress in economics in steam consumption was discussed by Capt. H. G. H. Tarr, of Buffalo. The only hope of relief from the heavy expense he hoped to find in the gas engine. This he illustrated by the record of a gas plant established at Poughkeepsie, N.Y., in 1905, which had given great satisfaction. The percentage of waste in the extraction of power from coal was estimated to be 90 per cent., and the saving with a gas plant was shown to be very great.

Mr. F. A. Barbour, of Boston, told of pumping water by a producer gas plant at St. Stephen, N.B.

Cost of Electrically Operated and Steam Pumps.

The morning session of the third day of the convention was taken up pretty fully in the discussion of electrically driven turbine pumps in comparison, as to cost, with steam pumps. The election of officers and the choosing of the next convention city occupying the rest of the session.

The paper by Henry L. Lyon, deputy water commissioner, of Buffalo, contended that the cost per million gallons of water pumped by electrically-driven engines was $4.84, while the steam-driven pumps required $5.22 per million gallons.

Messrs. Wm. J. Sando, consulting engineer, Milwaukee; Chas. A. Hague, consulting engineer, New York, and D. W. French, superintendent, Weehawken, N.J., claimed that the experience of Buffalo could not be accepted as a general preposition. It was pointed out that the proximity of Buffalo to the Falls, and the consequent cheapness of electrical power there, weakened the argument in favor of the electrical pumps. In Boston, where coal costs $3.87 a ton, as against $2.50 in Buffalo, equalizing the head pumped against, the price of pumping was only $4, as against $5.22 in Buffalo. Seven municipal and 15 individual pumping stations in the U.S. pump water cheaper than Buffalo, notwithstanding the favorable conditions existing at Buffalo.

Anticipating a lively discussion at the American Waterworks Association on the relative merits of steam and electric pumping, Mr. John McDougall, Caledonian Iron Works Co., Ltd., distributed among the members a bulletin, giving the official test of the three-stage 14-inch Worthington turbine, driven by a 400 h.p. Allis-Chalmers-Bullock motor installed at the McGairsh street pumping station of the Montreal waterworks.

Election of Officers.

Officers were elected as follows:—

Geo. H. Felix, Reading, Pa., president; D. W. French, Hoboken, N.J., 1st vice-president; Dr. William P. Mason, Troy, N.Y., 2nd vice-president; Jerry O'Shaughnessy, Columbus, Ohio, 3rd vice-president; Alexander Milne, St. Catharines, Ont., 4th vice-president; Charles Henderson, Waterloo, Iowa, 5th vice-president; John H. Diven, Charleston, S.C., secretary-treasurer. The finance committee was re-elected.

Next Convention at Washington.

After an exciting time Washington was chosen as the place for the convention in 1908. The cities bidding for this were: Washington, Chicago, Milwaukee, Philadelphia and Atlantic City.

Other papers read and discussed were: High Duty vs. Low Duty Pumping Engines, by Irving H. Reynolds, Youngstown, O.; and, A Peculiar Instance of Contamination of Ground Water, by Wm. P. Mason, Troy, N.Y.

The afternoon was taken up in a trolley ride around the city and to several of the parks. The evening session consisted of interesting illustrated papers on waterworks stations in other countries, given by Allen Hazen, New York, and Geo. A. Johnson, New York.

Flow of Water in Pipes.

The morning and afternoon sessions of the fourth day were taken up in the presentation and discussion of papers, and in the reports of committees. A moonlight excursion, given by the city of Toronto, was much enjoyed by the delegates.

The committee on waterworks standards submitted specifications for cast iron water pipe and special castings. The report, after some discussion, was returned to the committee, who are to report further next year.

Nicholas S. Hill, C.E., of New York, gave a paper on tuberculation and the flow of water in pipes, a subject which he thought had been very much neglected. He arranged the deposits which lessen the carrying capacity of water pipes into three classes—incrustations, growth on the inner surface of iron pipes and accumulations in inverts, hollows and dead ends. He was satisfied that tuberculation would not occur where the iron was protected from contact with the water. Slime, pipe moss and sponge growths were treated from a scientific standpoint, all being dependent on organisms carried in the pipes for their food. Efficient filtration would remove these deposits by starvation. The application of coal tar pitch was considered the best coating ever devised for pipes.

Dr. Geo. F. Whitney, New York, gave a number of experiences in cleaning water mains, showing how great increases in the water pressure had been obtained in various cities.

The question of cleaning water mains was further discussed by C. H. Campbell, of Atlanta, who read the paper of Col. Park Woodward. It stated that water works systems were too often put in with too little provision for the future growth of the town or city. Thus the mains were often too small within a few years but, in many cases, it would be found cheaper to clean them than to lay new ones. A test in Atlanta had shown that the cleansing of the mains had saved the laying of an additional main, the increase in supply being 65 per cent., and the pumping engine showing an increase in delivering capacity of over 45 per cent.

National Water Boards.

The evil effect of allowing political considerations to influence the management of a water works system was presented by Secretary Diven, in a paper on "How can Politics be Eliminated from Municipal Water Works?" He said that long study and practice were required to make a good water works manager, and if the changes in civic administration meant a new man after every election, there was no opportunity of the manager learning his business. He denounced the system whereby ward workers were rewarded by positions for which they were unfit, and favored the appointment of water commissions which would have absolute control of the water works plant and also of the sewage system. These commissioners should elect state boards, and from the latter a national board should be selected to control the inter-state water supply.

Mr. Henry L. Lyon, of Buffalo, gave a paper on a plan for interchange between low and high service systems. The Buffalo water works, he said, were started in 1848 and purchased by the city 20 years later. Now there was a reservoir containing over 116 million gallons, and the daily consumption was 130 million gallons. There were ten pumps, and the necessity of changing these from one service to the other created a problem in taking care of the distribution pipes, the solution of which was described.

Mr. H. S. Baker, of Chicago, told of the remodeling of the discharge pipes at the Springfield avenue pumping station, where the system was entirely remodeled without interfering with the work of the station. It was done by laying the discharge in consecutive steps, so that at no time were more than two pumps out of service. The present pumping capacity of the Chicago water works was given at 625 million gallons daily.

Mr. Harry A. Lord, of Ogdensburg, described a break in the water main in the bed of the Oswegatchie River, and the means used in repairing it. His paper was read, in Mr. Lord's absence, by the secretary, the writer having been attacked by rheumatism.

Proper Water Rates.

"The Proper Rates for Domestic and Public Service," was the subject of the paper by Mr. John Ericson, of Chicago. He thought that on account of local conditions varying, the rates applicable to several cities could not be averaged to provide a safe basis for charges by any corporation.

Dr. W. P. Mason, of Troy, gave some notes as to the action of water upon lead and zinc, saying that a water user should know what possible damage might result to his supply by reason of its being conveyed through metallic piping. It was shown that water containing traces of zinc had been used for long periods without injurious effects. Mr. R. J. Thomas, of New Lowell, cited a case where 40 cases of poisoning resulted from lead in a driven well.

"The St. Louis Method of Purification," was the subject of an interesting paper by Mr. Wilson F. Montfort, St. Louis, who stated that the water was purified by sedimentation basins. After it passed the first basin the water was fairly well clarified.

Dr. John Galbraith, LL.D., Toronto University, was made honorary member of the association.

Large Number of Papers.

On the closing day of the convention a large number of short papers were presented.

In the paper on "Dowsing," by Mr. Charles Anthony, jr., Bahia Blanca, Argentine Republic, who was not present, it was stated that the action of the "divining rod" was due to radial activity.

"Repairing a Broken Force Main," by Mr. C. W. Wiles, Delaware, Ohio, related the personal experience of the writer, whose chief difficulty in repairing a three or four foot main was that the water was running through the course at the time.

Mr. T. W. Davey, Middletown, N.Y., spoke of "Stripping Reservoir Land, or Some Methods Used for the Prevention of the Growth of Algae."

Other subjects disposed of were:—"Some Personal Waterworks Experiences," Mr. L. N. Case, Duluth; "Some Experiences Met With in the Management of a Small Waterworks Plant," Mr. Howard L. Williams, Luddington, Mich.; "A Wheel Pump," Mr. H. F. Dunham, New York; "The Operation of a Pump While Submerged by Flood," Mr. E. Forest Williams, Du Quesne, Pa.; "Rates and Regulations for Private Fire Protection in Atlanta, Ga.," Mr. Park Woodward, Atlanta; "Special Fire Protection Rates and Regulations, Elmira, N.Y.," Mr. J. W. Dinen, Charleston, S.C.; "Connection From Public Water Supply for Private Fire Protection Service," Mr. A. W. Hardy, Chicago.

Exhibits of Brass Goods.

The only Canadian manufacturers to exhibit were the Jas. Morrison Brass Mfg. Co., and the Canadian Fairbanks Co., both of Toronto. The latter had a small display of valves and other brass goods, while the Morrison Co. had one of the largest exhibits at the convention, showing a full line of their corporation waterworks service brass works, as well as a number of engineers' clocks, recording pressure gauges, etc. Chas. World and A. Betton, Ontario sales representatives, were in charge, and all Canadian delegates were reminded that "Made in Canada" waterworks brass goods could be secured.

The Mueller Mfg. Co., Decatur, Ill., had probably the largest representation of any firm, they being represented by Messrs. Fred. Mueller, Henry Mueller and wife, and H. F. Clark, of Decatur; Oscar Mueller and wife, New York, and R. M. Hastings, Buffalo, N.Y., their Ontario representative. They had a very large exhibit, comprising all classes of water, gas and plumbing brass goods.

The Hays Mfg. Co., of Erie, Pa., were represented by C. J. Nagle, the Canadian sales agent of their company, and well-known to the trade throughout Canada. Amongst the leading lines included in their display were the Hays extension service boxes, corporation cocks, lead goose necks, iron body clean outs, and general waterworks and gas

Full Lines on Display.

The Glauber Brass Mfg. Co., of Cleveland, had a nice exhibit in charge of I. Herzbrun, special traveling representative. Their display consisted mostly of waterworks brass goods, a complete line of inverted curb stops and stop and wastes being shown, from the smallest to the largest sizes. Other lines on display were different styles and sizes of corporation cocks, goose-necks for water connection, stop and stop waste cocks for plumbers' use, service and meter cocks for gas companies, and an assortment of compression and fuller work, both brass finished and nickel-plated. "They never leak and they always work easy" is the Glauber motto.

United Brass Mfg. Co., Cleveland, O., was represented by W. J. and S. P. Schoenberger, of Cleveland. They had a similar line of brass goods on the display table, including a number of United lead goose-necks, the joints of which were wiped by the Corcoran wipe joint machine, sold by this company. A demonstration of the splendid work of this machine was also made at one of the departments of the Toronto waterworks. The Messrs. Schoenberger were kept busy explaining the work and system of operation of the wipe joint machine, it creating considerable interest amongst the delegates.

Other exhibitors included the Waterworks Equipment Co., 180 Broadway, New York, lead wool for caulking joints under water; Worthington Pump Co., New York, meters; National Water Main Cleaning Co., 27 William street, New York, iron pipe cleaners; Thompson Meter Co., Brooklyn, N.Y., meters; the Pito Meter Co., New York, testing machines; Neptune Meter Co., New York, meters; Cancos Mfg. Co., Philadelphia, black squadron packing; the Union Water Meter Co., Worcester, Mass., and the Wykoff Supply Co., Elmira, N.Y. The last named company exhibited a section of wooden water pipe bound by metal straps, which had been in service under 30 lbs. of pressure regularly for 42 years.

The Pumping Engine Record.

The world's record for economy and efficiency of waterworks pumping engines is held by the plant at Bissell's Point, St. Louis, Mo., built by Allis-Chalmers Co., of Milwaukee. The duty reached at the official test was 181,-068,605 foot pounds. In order that engineers may know exactly how these figures were reached the company printed in bulletin form complete details of the test. These and other bulletins containing information not usually made public, but of great value to those interested in waterworks, were distributed among the members of the American Waterworks Association by Allis-Chalmers-Bullock, Limited, in a handsome souvenir cover.

This firm and the John McDougall Caledonia Iron Works were represented by John S. Maclean.

Learn to hide your aches and pains under a pleasant smile. No one wants to know if you have the earache, headache, or rheumatism.

The Milling Machine Vise

Uses of the Flat and Swivel Vises--Methods of setting Jaws of Vises where very accurate work is necessary.

By John Edgar

The character of the work handled on the milling machine makes it necessary to have an effective means of holding the work securely under the heavy strains incidental to the milling operation. That common tool, the vise, is found in several different variations in this service. We find it in several forms, viz.: The flat vise, the swivel vise, the table vise and special vises of several kinds.

Flat Milling Machine Vise.

The flat vise is the one most used and is shown in Fig. 1. The design shown is similar to those placed on the market by several manufacturers, and represents good practice. This vise is of very solid construction, well able to stand the heavy strains of the cut as well as the initial strain necessary to grip the work firmly in the jaws. These vises are very limited in range, which is necessarily the result of compactness and strength. The following table gives the ranges of the commercial sizes now available:

Dimensions of Flat Vises.

Width of jaws.	Depth of jaws.	Jaws open.
5 ins.	1½ ins.	3 ins.
6 "	1⅝ "	3¾ "
7 "	1¾ "	4½ "
8 "	2 "	6 "

These flat vises, as seen, consist of a base, or backjaw (a) and a sliding jaw (b), which is moved by the screw (c), which has a square shank for the use of a crank wrench. The sliding and back jaws are fitted with false or removable jaws of steel which may be substituted by jaws of special shape for special work. The use of these special jaws will be fully explained later.

The uses to which this vise is put are numerous; but before going into its use let us become acquainted with the other forms of vise that are used in milling machine work.

The swivel vise is nothing more than a flat vise mounted on a rotary or swivel base, which may be a component part of the vise or only an attachment making a combination vise that can be used as a flat vise and, when needed, as a rotary or swivel vise. This attachment is merely a circular plate with a circular tee-slot cut in its top for bending the vise in any position by means of tee-bolts. This base is commonly graduated in degrees so that by means of the index mark the vise may be set at any desired angle. This combination makes for most purposes a solid construction, which is not, as a rule, obtained with the ordinary swivel vise, as the former sets lower and offers better opportunity for bolting down the vise in the angular position.

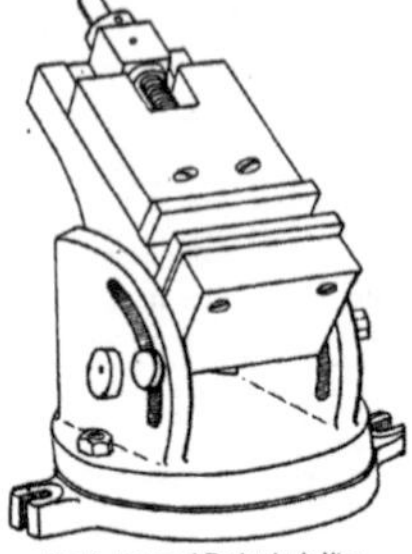

Fig. 2 Universal Toolmaker's Vice.

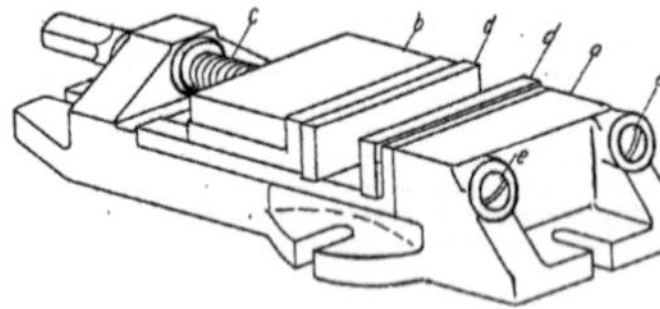

Fig. 1 – Plain Flat Milling Machine Vice.

Fig. 3 – Table Vice for use on Milling Machine.

The flat vise is provided with two slots at right angles to each other, by means of which the vise may be set with the jaws either parallel or at right angles to the arbor. This is accomplished by means of tongues placed in these slots and in the tee-slots of the table. The rotary base is also provided with slots so that it may be always placed in a position relative to the table slots that the graduation marks will coincide with the index mark on the vise itself.

In Fig. 2 we have a special form of swivel vise which has the peculiarity of swivelling in a vertical plane as well as in the horizontal. This vise is much used by toolmakers and comes in handy in making form tools and such work as the toolmaker is constantly running up against. The swivelling movements of this vise are also measured by means of graduations, giving the degree of angular displacement.

As with the flat vise, this style is also provided with the false jaws, and for the same purpose. Some makers go still farther in the design of the toolmakers vise and provide means by which it is possible to swivel the vise jaws at an angle in the inclined plane. This vise has a greater range of usefulness, but is more expensive and less compact and rigid.

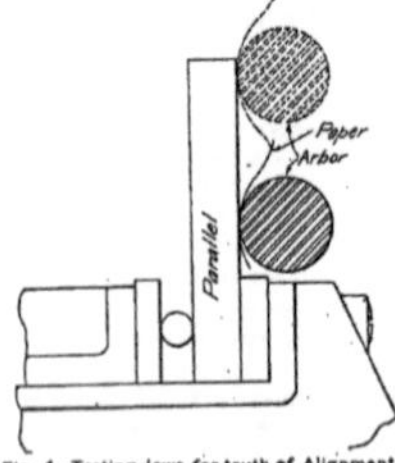

Fig. 4 – Testing Jaws for truth of Alignment.

Flat Vise of Great Range.

In order to overcome the limited range encountered in the commercial sizes of the flat vise, the vise shown in Fig. 3 was devised. The range of this vise is only limited by the length of the table of the machine on which it is used. This table vise consists of two jaws, (a) and (b), and two stops (c) and (c1). The jaw (a) is bolted down to the table securely by the three bolts in the table slots. This jaw is provided with a tongue that engages with the centre slot in the table in order to ensure the face of the jaw being at right angles with the feed of the table. The jaw (b) is made so that it may be fastened to the table in the same manner as the jaw (a) with the exception that but one bolt is used. This jaw is also made so that it may be adjusted by means of the screws (d1) and (d2), which act upon the stop blocks (c) and (c1). It is obvious that this style of vise is only available for holding work lengthwise of the table. It has been most used in connection with the vertical spindle machines.

For Special Work Vises Should be Tested.

It is the aim of the manufacturers of vises for milling machines to produce a tool that is built on "right lines;" that is, with the jaws at right angles with the tongues in the base and at right

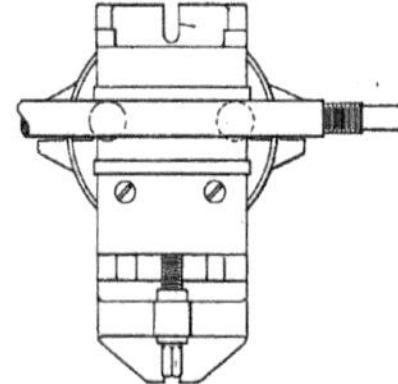

Fig. 5—Setting Vise with Jaws Parallel with the Milling Machine Arbor.

angles in a vertical plane with the base or top of the table on which it sets. Therefore, we may be reasonably safe in accepting a vise that is in fairly good condition as accurate enough for most purposes. When, however, we have a job on hand that calls for a little more care than the ordinary run of such work, it would be best to test the vise in all ways to make sure that the work will not be spoilt through inaccuracies of the vise.

The vertical truth of the jaws may be tested by the method shown in Fig. 4, where (a) is a parallel pinched between the jaws and the round piece of drill rod or other round stock. Now bring the arbor against the parallel in the lowest position and then with the arbor in the same relative position lengthwise of the table drop the knee until the extreme end of the parallel is reached, and if the arbor is still in contact with the parallel and does not press any harder against it in one position than in the other the jaw may be accepted as at right angles with the top of the table. By the use of a piece of thin paper inserted between the arbor and the parallel the pressure between the two may be better tested. This should be tried first on one jaw, then on the other, to make sure that they are parallel. It is more important to have the back jaw right than it is that the sliding or adjustable jaw be so.

In case an inaccuracy is discovered, the jaws may be shimmed up sufficiently to bring them square.

To prove the truth of the jaws in a direction parallel with the spindle we may raise the knee so that the arbor will come into the opening between the jaws with its centre about half-way in the gap, and then by pinching two pieces of thin paper or two other pieces of material of equal thickness between the arbor and the jaw at each end of it we may test the parallelism of one with the other by the "bite" of the two pieces of paper. In this position the index mark of the swivel base, if the vise is a swivel vise, should be at the zero mark on the graduated circle.

Testing Vise for Vertical Miller.

Should the vise be used on a vertical spindle machine the parallelism with the motion of the table or the carriage may be tested by means of a plug placed in the spindle hole, and by the paper test being applied first at one end of the jaw and then at the other, as shown in Fig. 5 by the dotted lines.

To Have Jaws at Right Angles to Axis of Cutter.

Should it be desired to set the vise so that the jaws would be at right angles with the axis of the cutter, we would proceed as in Fig. 6, and by means of the square and the paper test the squareness of the jaws with the centre of the arbor. The dotted circles show how the vertical spindle machine may be used to test the setting of the vise when the square is used.

It is well to remark here that in making these tests with the arbor it must run reasonably true, as much depends on the truth of the arbor for the accuracy of the setting obtained.

These Tests Necessary for Special Tool Making.

As mentioned above, these refinements are very rarely gone into in the use of the milling machine vise, but there are some jobs that the toolmaker has to do that occasionally call for extreme care and skilful handling. In such a case it would be best to remove the tongues from the flat vise, if it is used, and set it by the methods just explained.

To Increase Holding Power of Jaws.

When holding work in the vise jaws, especially where heavy cuts are taken parallel with the jaws, the placing of a strip of paper between the work and the jaws will increase the holding power of the latter considerably, and for most work this will not alter the practical accuracy of the results enough to make any particular difference. The thinner the paper is, the better the results.

In holding round bars or shafting in the vise, the use of a V-block jaw will be found advantageous. When we have a lot of round work of the same diameter to slot or mill in the vise, it will be found advantageous to use a holder made of a block of iron with a hole bored lengthwise and split along one side, so that when passed over the work and held in a vise the work is held securely without in any way marring the surface. Tubing may be held in this way without the danger of crushing it out of shape.

When we have a flat piece of work which requires a milling operation to

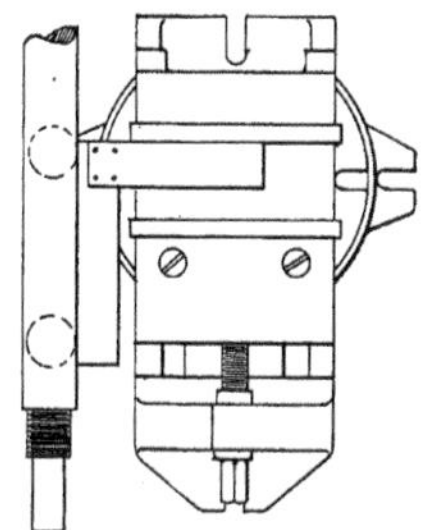

Fig. 6—Setting Vise with Jaws at Right Angles to Milling Machine Arbor.

form the edge it is necessary to hold the work well up to the cutter in order to avoid any chattering, which is inevitable unless it is so supported. This is done by making a set of jaws to follow closely the shape of the cut and substitute them in place of the regular jaws by means of the screws, as mentioned in the foregoing part of the article.

Vise Best Means for Holding Work.

There are other means of holding work on the milling machine, but for work that is of such a size and shape that it is available for holding in the vise, there is no more efficient or effective way of doing so. For quick short cuts the vise offers the best advantages as a holding medium. Some special uses for the vise will be taken up later.

CANADIAN MACHINERY
and Manufacturing News

A monthly newspaper devoted to machinery and manufacturing interests, mechanical and electrical trades, the foundry, technical progress, construction and improvement, and to all users of power developed from steam, gas, electricity, compressed air and water in Canada.

The MacLean Publishing Company, Limited

JOHN BAYNE MACLEAN - - President
W. L. EDMONDS - - - Vice-President
H. V. TYRRELL - - - Business Manager
J. C. ARMER, B.A.Sc., - Managing Editor

OFFICES:

CANADA
MONTREAL - 232 McGill Street, Phone Main 1255
TORONTO - 10 Front Street East, Phone Main 2701
WINNIPEG, 511 Union Bank Building, Phone 3726, F. R. Munro
BRITISH COLUMBIA - Vancouver, Geo. S. B. Perry

UNITED STATES
CHICAGO - 1001 Teutonic Bldg., J. Roland Kay

GREAT BRITAIN
LONDON - 88 Fleet Street, E.C., Phone Central 12960, J. Meredith McKim
MANCHESTER - 92 Market Street, H. S. Ashburner

FRANCE
PARIS - Agence Havas, 8 Place de la Bourse

SWITZERLAND
ZURICH - - - Louis Wolf, Orell Fussli & Co.

SUBSCRIPTION RATE.

Canada, United States, $1.00. Great Britain, Australia and other colonies, 4s. 6d., per year; other countries, $1.50. Advertising rates on request.

Subscribers who are not receiving their paper regularly will confer a favor on us by letting us know. We should be notified at once of any change in address, giving both old and new.

Vol. III. JULY, 1907 No. 7

NEW ADVERTISERS IN THIS ISSUE.

Canadian Pilling Co., Montreal.
Carborundum Co., Niagara Falls, N.Y.
Chapman Double Ball Bearing Co., Toronto.
Colonial Engineering Co., Montreal.
Doggett, Stanley, New York.
Eberhardt Bros. Machine Co., Newark, N.J.
Foundry Specialty Co., Cincinnati, Ohio.
Gas & Electric Power Co., Toronto.
Lew Mfg. Co., Cleveland, Ohio.
Partamol Co., New York.
Shantz, I. E., & Co., Berlin, Ont.
Ely, W. W., Mfg. Co., Cleveland, Ohio.
Somerville & Van Every, Hamilton.
United States Graphite Co., Saginaw, Mich.

CONTENTS

CANADA HANDICAPPED BY LACK OF EXPERIENCE.

An article on the slow growth of the Dominion by a young Canadian now resident in the States, is attracting a great deal of attention, and is being criticized by many of the newspapers.

So far as we have seen the Canadian papers have contented themselves with attacking the New York writer and his theories as to the cause, but none have themselves satisfactorily explained why Canadian progress has been so very slow, why American growth has been so very rapid, and why so large a percentage of our business young men and women have gone to the States, and having gone there, why they have made much greater successes than their equally clever and brainy companions who remained in Canada.

Some of the writer's statements and deductions are absurd, but in the main his article is absolutely true. Until recently Canada has not progressed. The men in Canada who have done things; the Canadians who have gone abroad and done better things and whose opinions are worth while, say that Canada's failure to progress is not due to British connection, which is a great advantage, or to any lack of ability on the part of the Canadians at home, but to a lack of experience, and to the lack of encouragement and inspiration which experience gives.

Many of these writers say that Canada has suffered through lack of capital. Excepting in stringencies such as we have at present, Canada has had all the capital needed for any development on which her citizens have ever found it necessary to embark. Where Canadians have had the experience, encouragement and inspiration to do things, they have been done, and generally done better than in the United States, and the progress has been greater. Take our dairying for example and the industries, such as pork packing, allied with this; we have outclassed the Americans in both quantity and quality. We have done so because we were shown how; we have surpassed those who taught us. We will do the same in many other lines of manufacture.

To succeed Canadians must have the experience, encouragement and inspiration of those who are successful themselves; this can only be got by personal contact or by reading the publications from which these can be gleaned. Yet the Canadian Post Office, by abrogating the postal arrangement with the United States, has done its best to shut out the best trade and technical newspapers, magazines, etc., from which thousands of Canadians have been getting inspiration in the last few years to do things. They have been doing, and the country is growing as it never did before in population and wealth. The best interests of the Dominion require that we have these publications. Our development will suffer from lack of them.

The MacLean newspapers have more to gain than any other concern in Canada by shutting out American publications; yet they have all along taken the broad ground that what was best for the nation was, in the long run, best for them, and if they cannot hold their own field against United States competitors, they do not deserve it.

ENGINEERING COURSES AT WINNIPEG.

Recognizing that the Canadian Northwest will require a large number of men with engineering training to carry on the immense amount of engineering work which will be entailed by the quick growth of the country, the University Council of Manitoba University will organize courses in civil, mechanical, and electrical engineering.

The rapid growth of the Applied Science Faculty of the University of Toronto is ample evidence of the field for such an undertaking by the Manitoba University. At present there are quite a number of students from Western Canada attending science in Toronto. No doubt there will be plenty of material for the new department in the Manitoba University.

In time the University will have the full equipment

required for complete courses. At present it is without even the beginnings of such an equipment. Inasmuch, however, as the first two years in these courses consists mainly of theoretical instruction and drafting work, it is proposed to go ahead at once next October. The whole matter of the equipment of the University for these courses, with shop facilities not merely for forging, but for foundry work and the other necessary practical work, will have to be taken up by the University Commission when it is appointed.

It is thought that arrangements can be made with the machine shops and foundries in the city for the accommodation of the students for practical work.

In embarking upon this new enterprise, the University should secure strong men for the conduct of these new departments. Upon the personnel of the teaching staff will depend to no small extent the success of the venture and the success of the graduating students.

PAPER SHOULD STAND UPON ITS INTRINSIC MERITS.

An expression of opinion as published by "Paper Trade Journal," strikes at the core of the weakness of a good many mechanical and engineering papers as well as trade papers. This paper says:

> "No stronger evidence of the valuelessness of a trade paper as an advertising medium can be presented than that of soliciting advertising upon the strength of sales of machinery or other goods which may be brought about through the influence (?) of the paper's representative. It is at once a frank confession of the lack of the essential qualities upon which the advertising value of any publication must of necessity depend—a bona fide subscription list and that influence which comes through prestige."

Many an advertiser has been practically held up in this way by advertising representatives. In some rare instances the advertiser may secure sufficient graft orders to make the investment worth while; but that is only the case when the graft has been handed over before the manufacturer signs the advertising order. Once the advertising man has secured his advertising contract his interest in that special case evaporates. A new interest is then taken up, and all the graft is directed into a new channel.

It is absurd to think that implied or expressed promises should result in anything to the manufacturer after his signature to the advertising order has been secured. All inquiries which may come to papers adopting such a policy, will of a surety be used as bait for new advertisers. In the big majority of cases the manufacturer who allows himself to be thus beguiled is badly sold.

There is another thing which must be considered by the manufacturer. "The publication which cannot secure advertising patronage upon its merits is certainly not worth considering." No publisher of reliable papers will permit this graft. It is only the paper which cannot secure and hold its advertisers on the strength of its intrinsic value, that resort to such methods of obtaining advertising orders.

A policy of that kind spells ruin for the paper sooner or later. A sound paper cannot be built upon such an insecure foundation.

Such a paper is useless to the manufacturer as an advertising medium.

MALLEABLE IRON INDUSTRY.

There is not a more busy industry in Canada at the present time than the malleable iron. All the manufacturers of malleable castings have more work on hand than they can handle. Capacities of present plants are being enlarged, and projects for new ones are being formed. The making of malleable castings is a tricky business. Not all heats turn out well, even in experienced plants. To get the carbon in the iron combined chemically to form white iron without getting it hard requires careful mixing and careful manipulation.

In view of these facts, the opinions expressed by Dr. Richard Moldenke in the article, "The Malleable Foundry as a Business Venture," published in the foundry department of this issue, will be of particular interest to Canadian foundrymen and Canadian manufacturers.

Dr. Moldenke is a recognized authority on malleable iron, and his words should carry some conviction. He advises the manufacturer not to establish malleable castings plants unless there is at least five tons of castings to be made daily. He reviews all the considerations which should be taken up before embarking upon such an enterprise, either to supply a manufacturer's own needs or to do a jobbing trade. These considerations are based upon conditions existing in times of prosperity, like the present, and also upon conditions existing when times are bad.

In succeeding issues we will publish similar articles on "The Grey Iron Foundry as a Business Venture," and "The Steel Foundry as a Business Venture," by Thomas D. West and by W. M. Carr, respectively, both men specialists in their lines.

CANADIAN INVESTMENTS IN BRITAIN.

Canada during the next few years will need considerable capital to carry on her enterprises. She will have to look for this capital outside her own Dominion. Great Britain and the United States are the countries chiefly interested. The large percent of the money must come from these countries.

In view of this fact it is interesting to note what Frederic Nichols says concerning the stand British capitalists take in this matter. He has just returned from a visit to the Old Country, and says:

> "The general feeling among British investors is very favorable to Canadian enterprises in view of the fact that their own home railway and South African securities, which have been their favorite investment heretofore, are receiving but scant attention at the present time."

This is well. Already Canadian enterprises are feeling the tightness of money somewhat. Unless the situation is relieved before it becomes more stringent, there may be a reaction, with accompanying bad times. It is gratifying to know that Britain's capitalistic sentiment is in favor of Canadian investment.

Suction Gas: Its Manufacture and Use

ECONOMY OBTAINED IN SMALL PLANT INSTALLED IN FLOUR MILL UNDER OPERATING CONDITIONS.

Article III. By Emil Stern.

The two articles on suction gas in the October and December, 1906, issues of this paper dealt largely in a descriptive way with the subject. The first article showed general features of these plants; while the latter described a number of individual plants as made by modern manufacturers.

The purpose of this article is to draw conclusions from the two preceding ones.

It can safely be said that in the course of the next few years Canada will see a fight between ancient and modern prime moving systems, as all the European countries have seen, Germany first, France and England following. On this continent the riches of the natural sources of power have delayed the development of gas producers for some time. The United States first started to develop them, and recently some Canadian manufacturers have started to build gas producers, some of them trying their own design, others adopting the safer way of manufacturing under a license and from drawings of old established European firms, thus securing for the Canadian buyer the advantages of home-made goods, as well as the advantage of well tried systems, the reliability of which has been proven by years of experience.

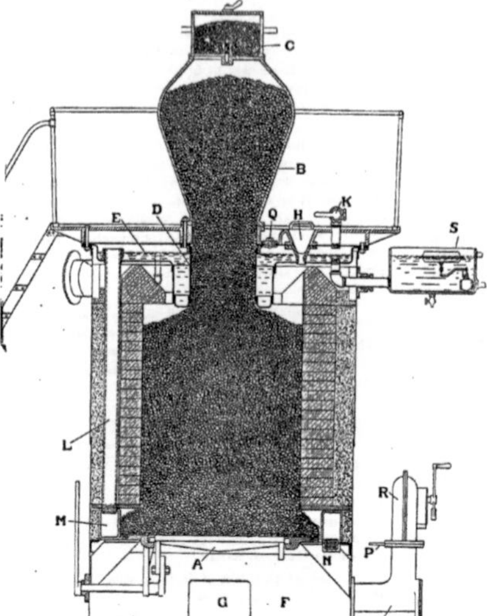

A Style of Suction Plant with Internal Vaporizer.

Economy of Suction Plants.

If we try to find out what has made the suction gas producer such a complete success we find that many points have to be considered, chiefly economy.

The manufacturer has several ways of obtaining power. The easiest way, the most expensive, too, is to buy the power from a central station, installing an electric motor and paying 2½c. or more per b.h.p. hour, which totals up to $75 per year, (3,000 hours for one b.h.p.). The gasoline engine in the country does not stand much better.

Recently the writer saw a gas power installation that, better than all scientific explanations, shows the superiority of the gas producer in all cases where power is required for any length of time. I do not include in this consideration small power installations required only one or two hours every few days. In these cases, and similar ones, city gas, gasoline or central station electric power are much more suitable, as the fuel economy would be greatly decreased by the stand-by losses. The case I am referring to is that of a chopping mill equipped with an 11-inch blade feed chopper in a small town in Ontario where no power was available from the town. A miller decided to put in a 35 h.p. gasoline engine, which worked quite satisfactorily as far as just the working was concerned ; but when the miller compared his expenses with the incoming money he found that nothing was left for him. This average daily result was :

Incoming money—	
250 bags at 5c.	$ 12.50
Expenses—	
Gasoline, 30 gals. at 25c.	$ 7.50
Help	1.25
Lunch for two	1.00
Oil and waste	.25
	$10.00

This left $2.50 for himself, not figuring on rent, insurance, interest and depreciation. This result was so discouraging that he at last stopped the mill altogether. However, a Toronto firm installed for him a suction gas producer to run his old gasoline engine; the result is as follows : The capacity of his engine was decreased from 250 bags a day to 200 bags. This is accounted for by the fact that producer gas, being a less rich gas than that produced from gasoline, requires a larger engine for the same power.

Incoming money—	
200 bags at 5c.	$10.00
Expenses—	
Hard coal, 400 lbs. at $5 per ton	$ 1,00
Help	1.25
Lunch for two	1,00
Oil and waste	.25
	$3.50

This leaves a profit of $6.50 a day, a satisfactory figure for the size of the mill. Comparing the figures for gasoline and suction gas it is evident that gasoline as a fuel costs six times as much, or 500 per cent. more, than suction gas.

According to the above statement relative costs are as follows :

To chop 1 bag of feed with gasoline 3c

To chop 1 bag of feed with suction gas ..... ½c

These figures are for a small plant, and are taken from the results of actual running for several months, and not

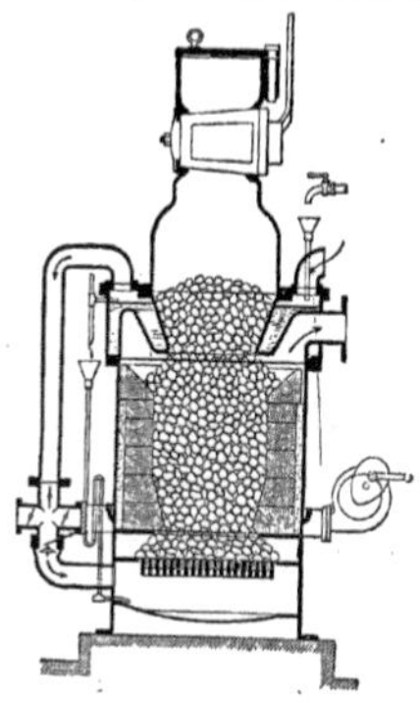

Producer with Internal Vaporizer, and with Cock Valve in Hopper.

from a short test, which is always liable to be somewhat out.

Suction Plant Pays for Itself in Two Years.

It is a fact that a producer gas power installation will pay for itself within two years, even at an initial cost of

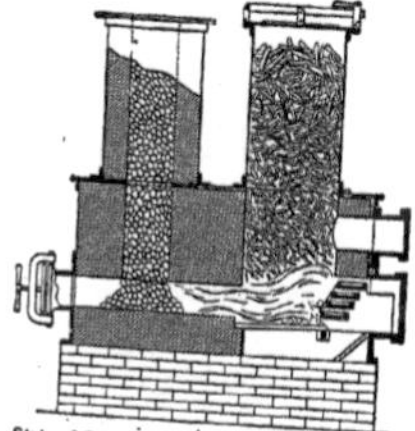

Style of Producer for burning wood fuel—the Riche Combustion Producer.

$50 per b.h.p. for large plants, and as high as $100 per b.h.p. for small plants.

Central Stations Will Install Producer Gas Plants.

It is said that after a while the only way in which the large central electric stations can be made to pay will be to install producer gas plants and do away with their present steam plants. Then the selling price per kilo-watt could be reduced, and all power users could derive profit from this more economical way of generating power.

Better Utilization of Heat in Gas Plant Than in Steam Plant.

The economy of the gas producer in connection with the gas engine depends altogether upon the better utilization of the heat contained in the coal. A steam boiler and engine plant does not utilize more than 6 to 18 per cent. of the heat energy contained in the coal. A producer gas and gas engine plant, on the other hand, turns 20 to 33 per cent. of the heat contained in the coal into mechanical energy. We can easily imagine the reason for this poor economy with steam as compared with producer gas, when we consider that with the gas engine, one medium is done away with altogether. When we put coal in our steam boiler and ignite it, gas is generated and burned above the coal, heating and vaporizing the water for use in the engines. To make suction gas, coal is used in the same way, but instead of the gas being burned under the boiler to produce steam, it is taken to the gas engine and burned there directly. The combustion in the engine is practically a complete one, and thus smoke is eliminated. This, in my idea, is the greatest consideration of all, meaning the saving of national wealth represented in unburnt coal, and the doing away with the ugliest feature of all our cities, the black cloud of smoke hanging over them, carrying dirt and sickness.

PERSONAL MENTION.

Mr. P. A. Kerr, formerly of the Peterboro Shovel and Tool Company, has been appointed superintendent of the Otonabee Power Co., of the same place.

J. Hutton and C. Harris, of the C.P.R. engineering staff on the construction of the Guelph and Goderich Railway, have left Goderich, Ont., for China, where they will be engaged on railway survey work.

Dr. R. K. McClung, who has been senior demonstrator in physics in McGill University, Montreal, for the past three years, has been appointed professor of physics in Mount Allison University, Sackville, N.B.

Chas. H. Danser, who has for the past two years been chief engineer of the Provincial Public Works Department of Manitoba, has been appointed deputy minister of that department. He will still retain the position of chief engineer in connection with his position as deputy minister.

Mr. Thomas MacFarlane, Dominion analyst, who was shortly to retire from public service, died very suddenly this month in Ottawa. The deceased was one of the most eminent analytical chemists in Canada, and had been in the public service for over 25 years. Prior to coming to Canada he was a mining engineer in Norway.

S. H. Jones, president of the St. Thomas Brass Co., St. Thomas, Ont.; the Canadian Bronze Co., Montreal, and the Northwestern Brass Co., Winnipeg, has removed his headquarters from Buffalo, N.Y., to 2 Rector street, New York City. The Magnus Metal Co., which has its headquarters in Buffalo, and of which Mr. Jones is general manager, has sold out to the National Lead Co., a Guggenheimer concern, and its headquarters are now 111 Broadway, New York.

Mr. W. E. Richardson has resigned his position as general superintendent of erection for the Westinghouse Machine Co., to take charge of the Sight Feed Oil Pump Co.'s business at Milwaukee. Mr. Richardson, who is president of this company and inventor of the oil pumps they manufacture, was for sixteen years in the employ of the Allis-Chalmers Co., and has for the past two years been with the Westinghouse Co., in charge of the erection of all the engines, turbines, etc., made by them. A very wide circle of friends in engineering fields will be interested to learn of the Oil Pump Co.'s good fortune.

WISHES WATER POWER PLANT EQUIPMENT.

As an indication of how Canadian Machinery is consulted when manufacturers are installing new machinery and equipment, the following letter from H. Corby Distillery Co., Ltd., Belleville, Ont., is presented :

Belleville, Ont., June 20th, 1907.
The Canadian Machinery,
10 Front St. E.,
Toronto.

Dear Sirs,—Will you please let me have a copy of your paper.

Owing to our recent fire, we are compelled to install a new water power plant, and wish to get the names of the firms who supply the necessary equipment. Yours truly,

H. CORBY DISTILLERY CO., LTD.

It is probable that before long the wooden telegraph pole will be replaced by the concrete. The average life of a wooden pole is seven years; whereas the concrete pole lasts indefinitely.

According to the Ironmonger, Mr. E. L. Rinmann, of the University of Upsala, claims to have discovered a new process for the electrical extraction of aluminium from blue clay, by which the cost of production is reduced to about one-quarter of the present rate.

How an Apprentice Became Superintendent

Boy Started Mechanical Career at Fourteen and is Superintendent of Large Plant at Thirty-two - Diligent Reading and Steady Work along a Special Line Brought Success.

Man is the architect of his own career. Few are not themselves responsible for their success or failure. Upon the surface some appear handicapped in the fight; but when conditions are analyzed completely the supposed handicap is often found a help.

Determination to succeed, perseverence in carrying out plans, and good hard work along specialized lines are significant of success.

There is inspiration for the young fellow starting at the bottom in the fact that very many men occupying important positions and drawing large salaries have been there before him.

The bottom is a good place to start. One is then personally acquainted with all the rungs of the success ladder, from the bottom up. A great thing, this personal acquaintance!

"Ever since a youngster, I have been determined to succeed. I never lost sight of this object. I read everything I could lay my eyes on. I never went back on a job. My motto was and is: 'Anything that can be done, I can do!' This refers to my special line. Outside of that I am not an authority, I must take for granted what other men say."

This is what Mr. A. Grass, superintendent for Whitman & Barnes Mfg. Co., St. Catharines, Ont., says concerning the means of his advancement from an apprentice to the position he now holds. Grass' experiences are ample evidence of how perseverence can make fortune smile He has done what any other young mechanic can do, if he sets his mind on it and works with everlasting perseverence.

Grass started his career as a mechanic when a lad of fourteen years in wiring baled hay. Later he ran the engine used in this work. That early he showed the spirit of progress. His liking for mechanical work led him to start as an apprentice in a marine machine shop in St. Catharines. It was while there that he took to reading mechanical papers. He was not satisfied with one or two, but read many. That early in life he realized the importance of specializing. He was much interested in the making and using of dies. He studied their construction; he studied their use. He read everything he came across concerning them. Result! He is now works superintendent of a large plant where the forging of tools is the chief work.

From the place of his apprenticeship, Grass went to Thorold to undertake blanking die work for bicycle works. That class of die work did not appeal to him, and he did not remain long. He went to Whitman & Barnes, getting the position because of his knowledge of die work. There he proved himself. His undertakings were successful. He knew about die work. He had ideas, and carried them out. There was a place for such a man. He was made superintendent.

Nor has it taken a lifetime to do this. Grass is now only thirty-two.

The boy as an apprentice saw in himself the man as a superintendent. He had faith in himself and in his possibilities. He succeeded.

This man started out with a very slight education. His start in life was the one he gave himself, His opportunities were those he made for himself. One would say that he was handicapped. But in this case, as in many others, the handicap did not hold the man down. He himself attributes his success to specializing along one line, to keen observation of everything he saw, to close reading of matter pertaining to his special work in all mechanical papers he could obtain, and to his perseverence in carrying work to a successful issue. He says, not in a boasting, but in a matter-of-fact way, that he has never been floored by a job; and he has had some difficult propositions thrown at him.

Grass has the power of concentrating his mind on a problem; and now he is often to be seen in his office lying back in his chair with his feet on the desk, staring into vacancy. He is thinking.

Grass has trained himself not to worry. If something goes wrong in the plant, causing considerable loss, he does not worry. But he, at the same time, sees that it cannot occur again.

REPORT OF NORTHERN WATER POWERS.

The fifth report of the Hydro-electric Power Commission, dealing with the Algoma, Thunder Bay and Rainy River districts, was made public the other day. The areas north of the height of land are not dealt with in the report owing to the sparseness of the population and the lack of demand for power there. These may be dealt with later when the construction of the Transcontinental and the extension of the Temiskaming & Northern Ontario Railways are accomplished facts. In reference to the Algoma and Thunder Bay districts the report says the Spanish, Vermilion and Mississauga rivers, flowing into Lake Huron along the north shore, "are the most important as regards size and hydro-electric possibilities, numerous water powers admitting of more or less economical development existing on these rivers." Along the north shore of Lake Superior there are many fine water powers with, at present, little economic value, the Nepigon, Kaministiquia and Current Rivers being the exception. This is due to their being within "transmission distance of the rapidly increasing power markets of Fort William and Port Arthur, and, especially in the case of the Nepigon, being of sufficient extent to use for grinding the pulp output of the Nepigon watershed."

Some details are given of the power developments now in existence in the two districts named. In this connection it is stated that at the Canada Copper Company's power station at Turbine, on the Spanish River, which transmits to the company's plant at Copper Cliff, has a capacity of 5,400 horse power, and additional units are to be installed this year to bring the capacity up to 10,800 horse power in all. The Kaministiquia Power Company's plant at Kakabeka Falls, on the Kaministiquia River, has, the report says, a partial installation of 10,000 horse power, which is shortly to be doubled.

Three principal transmission schemes are assumed as best meeting the present and probable future power requirements for a number of points, namely, a line from Dog Lake to Port Arthur, Fort William and vicinity; a line from Cameron Rapids on the Nepigon, to Port Arthur, Fort William and vicinity; a line from the Slate Falls, on the Mississauga, to Thessalon and Bruce Mines. Some of the estimated figures as to cost for power delivered to the sub-stations, ready for distribution, are: At Port Arthur for 12,382 horse power, $9.10; for half the quantity named, $12.60 per horse power per annum for twenty-four hour power. At Bruce Mines from Slate Falls for 3,300 horse power, $14.72, and for half the quantity $22 per horse power. At Thessalon, from Slate Falls, 150 horse power $27.10, and for half the quantity $30.92 per horse power.

Of Rainy River the report says: "This district is probably the most copiously watered of any in the province, water powers being abundant, and in many cases of considerable magnitude." The whole of it is embraced by one drainage area, "being part of an immense system, the run-off of which reaches tide-water in Hudson's Bay by way of Winnipeg River, Lake Winnipeg and the Nelson River." The town of Kenora is the only power market in the district at the present time. Fort Frances, owing to its geographical situation as a border town, and to the hydraulic possibilities of the Couchiching Falls, "will doubtless play an important part in the future industrial development of the district."

Development of a Crude Oil Engine

A concise history of how the writer solved the many problems in developing his idea of an internal combustion engine to operate on crude oil — Radically new ideas worked out, absolute failures at first, but finally developed in a commercially successful engine.

By H. Addison Johnston

Among the more important reasons for the low "commercial efficiency" or "power per dollar ratio" attained by the present well-known types of liquid fuel internal combustion motors, a few may be briefly stated as follows:

1. A refined product is used for fuel.
2. A large percentage of the heat developed by the combustion of the fuel is absorbed by the water jacket and thus lost.
3. The limit of allowable compression pressures is very low, due to the danger of pre-ignition of the mixed air and gas during compression.
4. The lowering of compression pressures in throttle governing engines, when running under full load, increases the fuel consumption.

Other serious defects in the ordinary systems are due, first, to the electric ignition devices being complicated and liable to disorder, and second, to the carburetors or equivalent mechanisms being uncertain and failing to give correct mixtures under variable loads, or when starting the engines.

It was evident to me that if all grades of heavy unrefined oils were to be used, a vaporizing system would be of no use, as a large percentage of the constituents of heavy oils will not vaporize, but merely form a thick deposit of tar and carbon in the vaporizing chamber. Hence some other system had to be devised.

Having had some experience with oil furnaces for heating steel in which heavy crude oil was broken up into a fine spray with compressed air, and in that state completely burned, with no sign of residue, I was firmly convinced that just as perfect combustion would take place in an engine cylinder under proper conditions.

Reducing Water Jacket Losses.

The only method of reducing the water jacket loss, seemed to lie in the use of a combustion chamber and cylinder that needed no water jacket, but it was plainly apparent that if the cooling system were eliminated, the charge would reach a very high temperature during compression, and hence the fuel and the air should not be allowed to come into contact until the moment for ignition had arrived.

Mathematically, the requirements would call for the walls of the combustion chamber to be of a non-conducting material, with no capacity to retain heat; but as far as the writer's present knowledge extends, no heat insulating or non-conducting material has been found, which will serve for the interior lining of a cylinder, though many experiments have been made for the purpose of finding one. Hence it has been necessary to fall back to the engine's old stand-by, cast iron, and to use an asbestos packing to prevent the escape of heat.

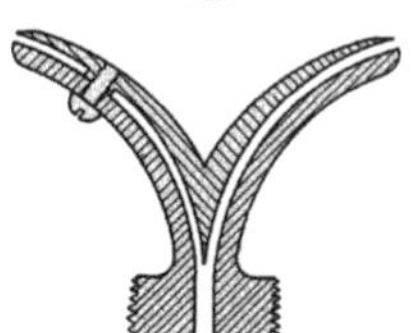

Fig. 2—Johnston Crude Oil Engine.

The Johnston Crude Oil Engine.

The problem, then, as it presented itself to me, was, to discover what new features could be devised, and what old devices eliminated, so that the resulting product would be an engine which would, as far as possible, be subject to none of the undesirable features mentioned above.

This descent from the theoretical, ideal chamber, with non-conducting walls, to the practical cast iron asbestos-packed cylinder, is one of the many approximations that have to be adopted in working out our mathematical equations with commercial materials.

The iron absorbs a certain amount of heat from the burning charge, which it gives out to the succeeding fresh charges and thus in a small degree detracts from our theoretical efficiency; but, as will be shown later, this works out to the general benefit of the proposition.

Consideration of Limitations in Compression.

The limits noted as to the degree of compression allowable in the ordinary gas engine cycle would cease to apply in the new cycle if the fuel and air were not to be in contact during compression.

If then, the fuel and air were not to be in contact during compression and the combustion must take place before expansion, the fuel must be injected at the end of the compression stroke and be ignited immediately.

Ignition and Vaporizing Solved.

The question of ignition has now solved itself to a certain extent. After the engine was warm and running there would be sufficient heat retained from the previous charge to ignite the fresh fuel immediately when injected.

As the fuel was now to be injected in-

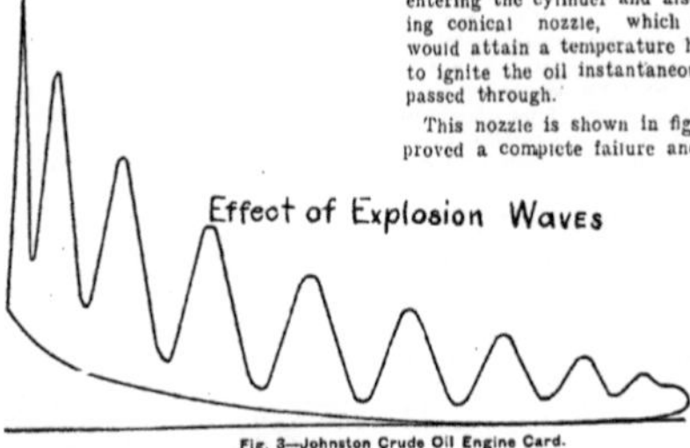

Fig. 3—Johnston Crude Oil Engine Card.

to the already compressed air, and burned, instead of being vaporized, mixed and exploded; the matter of vaporizers, or carburetors, needed no attention and the only requisite was a pump which would automatically measure and deliver to the engine at the proper time each working stroke, the quantity of oil required by the load.

As there is no explosive mixture question to consider, the air supply does not need to be regulated with the oil, and hence the compression pressure will remain practically constant.

An Experimental Engine Built.

After working out these ideas, to demonstrate their value, it was necessary to construct an engine embodying them. I, therefore, made complete working drawings for an 8-in. x 8-in. engine, and set out to find capital for building the machine. After a very great deal of trouble and the lapse of several months, I secured what I estimated would be a sufficient amount of capital to construct the engine and try out the principles involved. But, as is usual in experimental work, the original amount raised proved quite insufficient and a further considerable sum of money was expended before the success of the invention was assured.

Figure 1 is a sectional elevation of the engine as originally designed, while Fig. 5 shows the construction now used.

Though it was originally intended to operate the engine on a two-stroke cycle, compressing the air charge alone to its full compression pressure in the crank end of the cylinder, the mechanical difficulties in working out this cycle were so serious that it was decided to omit the feature of preliminary compression and to arrange the engine to work on an ordinary four-stroke cycle.

Experimental Work in Spraying Devices.

I had designed a compressed air spraying device to break up the fuel before entering the cylinder and also a diverging conical nozzle, which I expected would attain a temperature high enough to ignite the oil instantaneously as it passed through.

This nozzle is shown in figure 2. It proved a complete failure and was the first of a long series of devices intended to be attached to the cylinder head in front of the injected fuel, to cause ignition and distribution of the fuel.

The method of trying out these devices was as follows: A jump spark plug was inserted in the cylinder head, and the necessary electric connections made, gasoline was fed into the air supply pipe and the engine started up by hand on gasoline.

After a few minutes' running the ignition became automatic and the spark was shut off. The point of ignition gradually advanced as the internal ignition devices became hotter, until it was much too early and caused pounding. The peculiar fact was discovered that if more gasoline was now turned on, the pounding ceased entirely, and by feeding a heavy excess of gasoline, the engine could be run until the lighter parts of the interior of the cylinder showed red hot.

During the latter part of these runs on gasoline, the exhaust was as black as the smoke from a fresh-fired locomotive. When the temperature became so high that we could no longer run on gasoline, the oil was turned on, through the sprayer.

For two months the engine would not continue running five minutes after the gasoline was shut off.

Much time was spent at this period in experimenting with different substances for the interior lining of the cylinder; but no non-metallic substance could be found suitable for the work.

Sprayers were tried by the dozen, all the ignition devices that ingenuity could contrive. At last by a process of elimination, certain forms were found to be more favorable than others, certain natural laws made themselves apparent, and were gradually formulated, thus establishing guiding lines for future experiments.

Difficulties Met with in Operating Engine and Change of Design.

The first indicator cards showed a mean effective pressure of only 10 lbs. This was gradually worked up until 44 lbs. was reached. At this stage the engine would continue to run, but only if the load were varied to suit it. If the load were too light the engine stopped; if it were too heavy it stopped; a variation of the load, ten or fifteen per cent. either way, resulted in a stoppage.

Normal Card
M.E.P. 95 lbs.

Fig. 4 Johnston Crude Oil Engine.

At this time it was decided to make considerable changes in the design of the cylinder and combustion chamber. A new air pump was also built. The m.e.p. was then worked up to 78 lbs. My theoretical card had shown 101 lbs., so further experiments were necessary. Another long period of trying-out followed. The m.e.p. rose by small steps until 95 lbs. was reached. This is as high as is now reached in average work, though individual cards are sometimes obtained up to 105 lbs. The highest card ever obtained from the engine showed 120 lbs. m.e.p.

Difficulty in Making Engine Run on Light Load.

During all this period of development, the engine would never run on a very light load. If the load were thrown off while the engine was running, it would very soon begin to pound, the pounding in a very few minutes becoming so severe, as to make it advisable to shut off the oil. If it were not shut off, the engine stopped after a few minutes in any case.

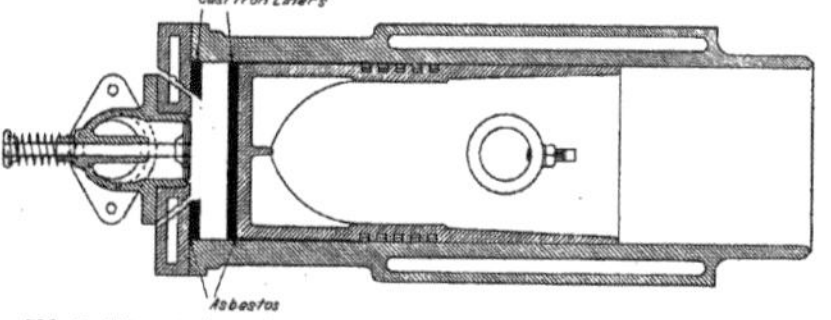

FIG. 5.—This cut shows the construction now used on cylinders up to 12-inches in diameter. It will be noted that while no water-jacketed surface is presented to the burning charge, the part of cylinder in which the rings run is kept cool. On full load the temperatures of the cast iron liners approach a low red heat, but as they do not touch any part except those to which they are fastened, and do not have to stand any strain, no harm results.

Over eighteen months elapsed between the time when the engine first ran on oil and the time when it could be started and run on no load. The engine may now be started and run for any period on no load and yet will take care of full load if thrown on at any time.

Trouble With Explosion Waves in Cylinder.

While the no-load difficulty was bad enough, the worst trouble was due to explosion waves forming in the cylinder. This trouble did not show itself until after the m.e.p. had been worked up to about 50 lbs. The external evidence of these explosion waves was a pound in the cylinder, varying from a mild thump, which could easily be mistaken for a loose bearing on the shaft or rod, to a terrible metallic bang, which sounded as if the whole engine was being pounded to pieces with a steam hammer. The violence of these waves was so great, that in many cases the pencil lead in the indicator would be snapped off on each side of the holder, immediately when the cock was opened.

Figure 3 is a reproduction of a card, showing the effect of the waves on an indicator card, while figure 4 shows the shape of normal cards.

It has taken two years of the most tedious and discouraging work to find out why these waves are formed and how to prevent them forming. It is now possible to design engines so that explosion waves will not form; but the chance of a designer striking the right combination accidentally is less than one in a thousand.

It has generally been the opinion among gas engine experts that explosion waves are largely due to pockets in the combustion chamber, but in all engines experimented with, there have been no pockets of any kind. Neither, as may seem strange to those accustomed to gas engines, does the time of injection of the fuel make very much difference, the waves form apparantly almost as easily when the fuel is injected after the crank is past the centre as when it is before it.

No Accumulation of Tar from Use of Crude Oil.

The principal cause of failure of the previous attempts to use crude oil, or crude residues, in an internal combustion engine, lies in the tendency of tar and other carbonaceous material to accumulate in the vaporizers, cylinders, valve chambers, and other parts of the mechanism. It was freely predicted by those in authority that the Johnston oil engine would be subject to the same difficulties; and among all the experts to whom the proposition was submitted in the beginning, the number who at all appreciated the advantages of the new system could be counted on the fingers of one hand.

In over three years of constant experimental work, there has never been the slightest trouble from this cause, though every kind of cheap heavy oil that could be obtained has been tested in the engine.

The whole of the fuel is absolutely burned and the exhaust is clear and colorless if the engine is properly adjusted and not overloaded. It was found that a considerable overload may be carried, if an extra large supply of oil is fed; the exhaust under these circumstances is quite brown, but there is never the least tendency for deposits to form anywhere in the system.

Grades of Oil Used in Engine.

About 90 per cent. of the oil used in our tests is a fuel oil, supplied by a Canadian oil company. It is a residue from crude oil, after the gasoline, benzine, distillates and kerosene have been extracted and thus is the cheapest of any obtainable fuel.

It was a dark, greenish black in color. The specific gravity is about .875 flash point, and open test 170 degrees F.

The engine has been run on crude oil fron Texas, crude oil from Petrolea, cheap fuel oil from anywhere bought in the open market, kerosene, benzine and gasoline. No adjustments in the engine are necessary in changing from one fuel to another, nor is there any appreciable difference in the number of gallons of the different fuels required to produce a given amount of power.

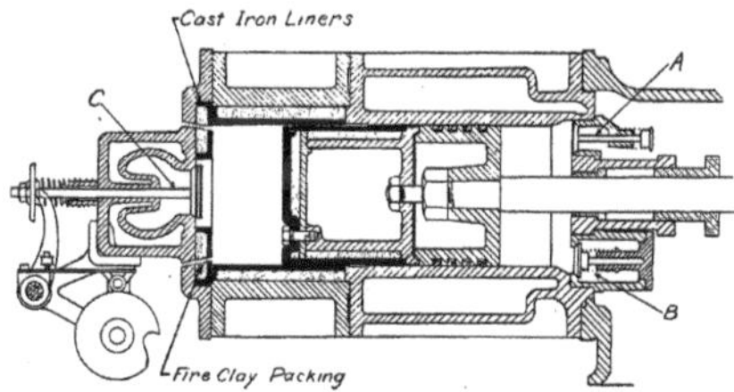

FIG. 1.—This cut shows the engine as originally designed. The air charge was to be taken into the crank end of the cylinder through the valve (A), compressed to 150 lbs. on the return stroke of the piston and discharged through the valve (B) into a reservoir not shown. From the reservoir a pipe extended to the valve (C), and just before the piston started on the working stroke the valve (C) opened and allowed the compressed air to enter the working cylinder. After the cylinder had received its full charge of air the oil was to be injected in a spray and ignited immediately. Owing to mechanical difficulties this system was not used and the engine was run on an ordinary four-stroke cycle.

The first tests of fuel economy showed a consumption of 1¼ gallons per 10 h.p. hours, about six months later, 1 gallon.

The consumption now guaranteed by

the Johnston Oil Engine Co., Ltd., of Toronto, Canada, on their 9-in. x 12-in. engine is ¾-gallon of any petroleum product per 10 h.p. hours. The best test or record to date is a consumption of 3-3 gallons per 10 h.p. hours.

Governing of Engine.

The speed regulation is very simply worked out. As no particular proportions of oil and air are required, no explosive mixture being formed, but the oil being simply burned, it is necessary to control only the quantity of oil injected per working stroke, while the air supply may be left constant. The oil pump is controlled by the governor and measures out each charge of oil to suit the load.

Any type of governor may be used, and as the movement of a small cam, ¼ of an inch, is the total motion necessary for complete control, the speed variation can be reduced to a very low percentage.

Starting the Engine.

As noted above, the original method of starting the engine was to use a gasoline explosive mixture, and an electric spark, continuing to run on gasoline until the combustion chamber was hot. There were several serious objections to this method, the principal one being the extremely violent pre-ignition explosions, which occurred just after the gasoline was shut off. As explained before, when running to heat up the engine, it was necessary to feed a great excess of gasoline, to prevent pre-ignition. When the gasoline was shut off, the proportion of gasoline to air fell rapidly with each succeeding stroke, and when the mixture became approximately correct, the resulting explosions occurring at the latter part of the compression stroke, were simply terrific.

The original intention had been to heat the interior of the combustion chamber, in starting, with a plumber's torch, and after the experimental work had progressed sufficiently, so that there was a reasonable possibility of the engine starting when turned over, the torch was used to raise the parts to the temperature necessary for the ignition of the first charge.

This method was fairly satisfactory, but it had the disadvantage that it was necessary to remove the torch from the cylinder, and plug the opening before the engine could be started. If the least time were lost in putting in the plug, or in getting the engine turned over, the parts would be cooled off, and the whole routine had to be gone through again.

After a considerable number of devices had been tried out, the "thimble" starting ignitor, such as is now used, was evolved. As the name indicates, this device is in the form of a thimble and is about one inch in diameter. It is secured to the engine in such a position, that some of the oil, as it is injected into the cylinder, strikes the interior walls of the thimble.

In starting the engine, the flame of a torch is played upon the exterior of the thimble, until it is red hot, the time required averaging three minutes. The engine is then started and the torch is removed afterwards. This method is perfectly reliable and is better than any other for inexperienced operators.

As it was necessary to equip every engine with a compressor, to pump air for spraying the oil, the use of compressed air for starting naturally suggested itself. This starting mechanism was very easily worked out for the second engine, and all engines now built are supplied with an extra air tank for starting. The starting tank may be pumped up to any desired point while the engine is running, and will hold its pressure indefinitely.

To start, it is only necessary to open a valve and move a small lever. The absolute certainty with which an engine of this type will start, is a feature which appeals most strongly to gas engineers.

There are only three conditions to be fulfilled, to make the starting of one of these engines certain, and these conditions are such that it is possible to determine them beforehand. It is not necessary to try to start to see whether the conditions are correct.

The three conditions are: There must be at least 200 lbs. air pressure in the tanks; there must be oil in the sprayer, and the thimble must be red hot.

There is a pressure gauge on the tank; there is a try-cock on the oil line; when the thimble looks red it is hot enough.

The latest achievement in connection with the engine is the successful application of the jump spark for starting. This is the only heavy oil engine in the world, which can be started cold, with a sparker. The ordinary jump spark plug will not fire the oil, but by combining a little theory with a whole lot of practice, one has been designed which will do the work.

The current is usually left on for two or three minutes when starting, after which the engine is hot enough to run without it.

The Engine as it is Now.

To sum up the matter in a few words, the engine is now commercially perfect.

It will operate on any liquid fuel.

It is not subject to either ignition or carburetor troubles.

It is self-starting and reliable.

No part of the engine is subject to deterioration, with the possible exception of the ignitor, which may have to be renewed, in from six to twelve months, at a cost of a few cents.

For marine work, a compressed air reverse is fitted, which is absolutely certain and positive in action.

These engines are now being built commercially by the Johnston Oil Engine Co., Ltd., in Toronto, Canada, and it is expected that arrangements will soon be made for their manufacture in the United States.

Well-Known Men at Corner Stone Laying of King Radiator Company's Factory.

GROUP OF WELL-KNOWN MEN.

The corner-stone laying of the King Radiator Co.'s factory a fortnight ago brought together a number of well known supply men, as will be seen in the accompanying illustration. The "twins," "Bob" Cluff and "Fred" Somerville, will be recognized in the front of the group on the left side. The second to the left is Architect F. H. Herbert, and the third, M. J. Quinn, of Cluff Bros. W. J. Cluff is wielding the trowel on the corner of the building. D. McArthur, of the Standard Ideal, is in a characteristic pose, with a cigar against his white vest, and over his left shoulder, Lorne Somerville, of Somervilles, Limited, has tried to disguise himself behind a broad smile. The third from the right side is Dr. McMahon, financially interested in the King Radiator Company, and next him is George Burman, who is to be superintendent of the radiator works.

Tricks of the Trade

WAYS AND MEANS OF DOING THINGS

Send in your ideas, sketches or jobs you are doing, and anything of interest to your fellow-workmen. Remuneration will be made for such contributions.

AN INSERTED TOOTH FACE MILLING CUTTER.

By Robert P. Keith.

After having placed an order with a prominent milling machine manufacturer in the east, the question next arising was: "What kind of a face mill will we use on this machine?"

The machine in question is of the duplex type; independently driven spindles; independently adjustable, vertically and horizontally, and it is intended that this machine shall drive eight-inch cutters, removing from 1-32-inch to ⅛-inch from the surfaces of gray iron castings; the width of these castings presented to the cutters being approximately 7⅞-inch maximum, to 3½-inch minimum.

Inserted Tooth Milling Cutter.

After considering the matter it was decided that it would pay us to design a special miller for our purposes. Thus we went ahead, backed by the courage of our own convictions in the matter, and designed a pair of mills as shown in the accompanying illustrations.

The soft steel forgings or blanks were secured for each of the bodies and turned up, bored and threaded to fit the right and left spindles on the machine. Chunks of cast iron were then gripped in the lathe chuck, bored and threaded to fit snugly on the ends of the spindle of the lathe, on which they were then mounted. The rough ends of these chunks were then turned and threaded to fit the threads already in the cutter blanks, which were then finished all over, running on this cast iron adapter. This procedure was followed out with both mills, the making of the cast iron adapter—as it might be called—being necessary in order to mount the blank on its own centre for the outside turning.

Pieces of steel were then turned up, threaded on one end to fit the blanks and turned down to the proper size, and tapered on opposite end to fit the hollow taper arbor of our milling machine indexing centre. Holes were drilled and tapped out in the small taper ends of these pieces of steel, so that we might introduce a cap screw from the back end of the milling centre and draw these studs or arbors in the indexing device snugly. This being done, the milling cutter blank was secured on to the threaded nose by a dowel pin after having been screwed on tightly, and the work of milling begun.

The rest of the operations need no especial description, unless it might be, perhaps, the grinding of the cutters, which is accomplished in an ordinary cutting grinder, the cutters being ground tangentially instead or radially as usual. Particular attention is called to the steel used for the cutters, namely, ⅝-inch by ¾-inch patent "Z" bar section Novo, which, it will be seen, requires but little grinding in order to clear the cut.

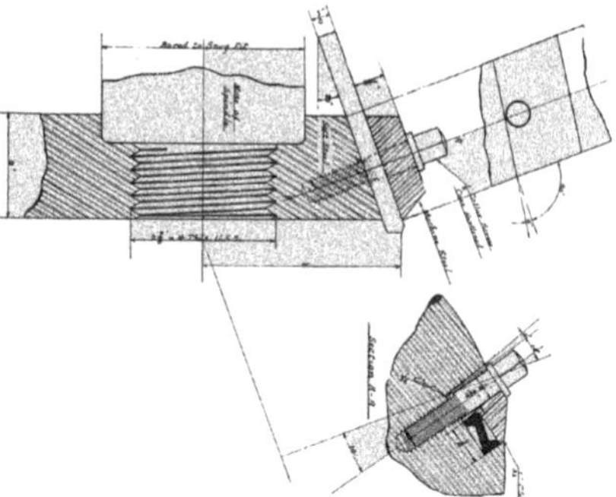

Working Drawing of Inserted Tooth Face Milling Cutter

Attention is also called to the feature of giving the cutter both top rake (8 degrees) and side rake (14 degrees); most designers being satisfied with the former. After the completion of the work and the mounting of a piece on the mill for the trial cut, the time had evidently arrived for someone to either disappear or else loosen up his clothes to make room for expansion. The mills during the trial cut were running at 25 R.P.M., at a feed of .112 per revolution, which would mean a table feed of 2.57 inches per minute, and a cutting speed of 52.4 feet per minute, the depth of cut being approximately 3-16 inches.

This cut was taken without trembling, chattering or noise, and after having taken several such cuts it was evident that a somewhat slower speed would be better, and, consequently, this was reduced to 23 R.P.M., at which the mills are now running very satisfactorily indeed.

Not being satisfied that our expenditure in the way of time and labor on these mills had been altogether wise, a standard design of inserted tooth mill with taper pin binder and ¼-inch x 1-inch inserted teeth was tried out on this same machine. This mill would not take over 1-16-inch cut at the same rate of feed and speed without chattering, and, therefore, quickly dulling the teeth, which were 18 in number, against 12 in the mill we made. This cutter would also easily stall the machine, whereas the other mills do not seem to tax the belts seriously with our cutters. The standard cutters shove or scrape off a very fine cut per tooth, whereas our mills turned off a curly chip, as a properly ground lathe tool should do.

Finally, the points worthy of consideration are the manner of wedging in the cutters, as shown in figure 2, the unusual size of the teeth, the unusually small number of teeth, and the extra thickness of blank in which these teeth are held, together with the feature of top and side rake, as well as clearance, which was ground to 5 degrees.

Courtesy is due the liberal policy of the management in executing this work, as well as the interest and co-operation of foreman and toolmakers.

It is now time for the woodmen to appear with their axes.

FIXTURE FOR CUTTING KEYWAYS ON THE SHAPER.

By Gorman Dizer.

The accompanying drawings represent a splining bar and fixture that were designed for holding work, especially for to be splined; but the advent of the bar and fixture herein described quite changed the aspect of the universe to the man on the job, inasmuch as splining cutters became almost a pleasure, and incidentally a few dollars were put in the bosses' pocket.

Fig. 1 shows a side view of the bar complete, fig. 2 a top view, fig. 4 an end view, and fig. 3 a side and end view of the cutter.

The bar was made of tool steel packhardened and ground at (d), fig. 1, to a working fit in the holes in the cutter blanks, which were reamed 0.006 inch under size to allow for grinding after hardening.

At (e), fig. 1, the bar was turned about 1-32-inch smaller than at (d), to not forced out of line with the shaper ram.

The cutter (f) was made of tool steel and was left rather hard, as it was well supported against any bending strains by being fitted closely to the slot milled completely through the bar. The hole in the cutter was made a running fit on the pin (i), the pin being a drive fit in the bar. The flat spring (g, fig. 1) was held in place by two screws (h) and pressed against the side of a notch (j) let in the cutter (fig. 3), the pressure of the spring tending to throw the cutter upward, away from the work, and the turning of the knurled screw (e) tending to force it down, the point of the screw bearing against the rounded back of the cutter.

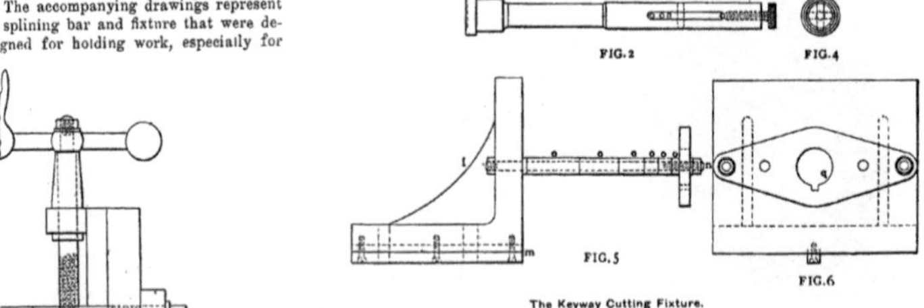

The Keyway Cutting Fixture.

Fig. 7—Fixture in Position on the Shaper.

splining milling cutters, but which could be used for other work.

In the shop where this fixture was made and used it had been the practice to mount the cutter blanks, before milling the teeth, in a chuck or upon a face plate in the lathe and cut the spline with a tool mounted upon the carriage, the carriage being run back and forth by hand. This proved an arm-breaking and cuss-word evoking job, as there were a lot of them constantly coming along do away with unnecessary friction and trouble in fitting. At (b) the bar was ground to fit the hole in the clapper block of the shaper in which the bar was used, the thickness of the shoulder at (a) being fitted very carefully to the depth of the recess in the clapper block, so that when the bar was put in place and the clapper block bound down by two screws (u), fig. 7, the bar was held securely against any turning or end movement, and at the same time was

The cutter had clearance from the cutting edge upward, and from the cutting edge backward the clearance was very slight, so that the cutter could be sharpened repeatedly by grinding off in front, without changing the size of the spline materially.

The fixture for holding the work, shown in side view fig. 5 and front view fig. 6, consisted of the cast iron angle plate, k, (fig. 5), the long shoulder bolts (n), the collars (o) and the clamp (p). The angle plate had a hole bored through it ⅛-inch larger than the largest splining bar used. The clamp (p) had a hole of similar size.

In operation, the screw, e, (fig. 1) was turned back until the cutter came up inside the bar. The clapper of the shaper was then raised, the regular tool post removed and the splining bar put in its place. The clapper was then fastened down securely by the two screws, as shown in fig. 7, binding the bar firmly in place.

The knee of the shaper was then brought to such a height and lateral position that, when the angle plate was put on, the hole in the latter registered approximately concentric with the bar. The angle plate was then put on, the bar passing through the hole in it, and fastened down by three hexagonal-head screws, the key in the bottom of the plate fitting in one of the slots in the shaper bed, locating the plate laterally and at right angles to the bar.

The shaper ram was then brought forward until the entire working length of the bar was in front of the angle plate, the clamp and collars removed, the cutter blanks slipped on the bar and pushed back against the angle plate, the clamp replaced, and the nuts screwed home, binding the cutter blanks firmly against the plate.

The stroke of the shaper was then adjusted so that the cutter came past the end of the blanks on both the forward and back strokes, and the shaper started.

The tool cut on the backward stroke of the shaper ram, the belt being crossed. As the bar came through the front end of the blanks, the screw was given a slight turn to the right, forcing the cutter down to the work; on the return stroke the cutter was pulled through the whole length of all the blanks, the chips flying out back of the angle plate. At each succeeding stroke of the shaper ram, the cutter was set down a little farther, there being ample opportunity for the fingers to grasp the screw, turn it and let go from the time it first protruded through the blanks until it was drawn back by the return stroke of the ram. Oil was occasionally squirted in at the top of the slot in which the tool worked, and, running down around the bar and the edge of the tool, lubricated both the cut and the bar in the blanks.

Fig. 7 shows the bar and the fixture in place on the shaper end and a set of miscellaneous blanks clamped on in readiness to be splined.

The collars shown on the bolts in the face plate were not used when the fixture was worked up to its capacity, but were provided to be put on between the clamp and the nuts when blanks aggregating a shorter length than those shown in the drawing were to be splined.

One lot of blanks could be removed and another lot put in place in a few minutes, as there was no adjusting to do aside from simply stringing the blanks on the bar and fastening with the clamp.

As everything was rigidly braced and there was no possibility of the cutter springing away from the work, a cut as heavy as the shaper belt would pull could be taken at each stroke.

The braces on the angle plate were so spaced that the shaper head could come between them right up to the angle plate, thus doing away with any extra length of bar.

This device worked to perfection. Three sizes of bars were made, for 7/8-inch, 1-inch and 1 1/4-inch holes, and 1/8-inch, 5-32-inch and 3-16-inch keyways.—American Machinist.

DRILL PRESS SLIP COLLETS.

By F. E. Lauer.

With the use of the following tools on drill presses where different sizes of drills are being constantly used on repetition work, they will prove themselves a time saver, and at the same time dispense with the abuse drill collets and drills usually get. The holder (a) fits into the spindle of the drill press, and all sizes of drills smaller than the bore of the drill press may be operated in it. We may use three different sizes of standard taper shank drills with this holder by using three collets like cut (b), which are reamed to suit Nos. 1, 2 and 3 standard taper shank drills. A man will have a drill in each all the time, if necessary, and sometimes it is necessary to have two or more collets, with the same number of taper shanks, to do the work he has got. Its usefulness can be readily seen, however, when one is able to just lift the collar which is on the outside of the holder while it is running and the collet and drill will drop out. It will be seen that when the collar 1, fig. (a), is raised it has a greater space for the head of the pin, and when the weight of the collet (b) rests on the pins they slip back, just allowing it to be withdrawn; when another is to be inserted, the rounded corner at the tang end of (b) will push back the pins when the collar is raised and the pins will locate themselves immediately when the collar is dropped. Another feature is the short tang, with its rounded corner, which corresponds with the pins in the holder. Going back to the holder, it will be seen that there is no slot drilled for the tang of the collet, this being dispensed with by hardened steel pins, so it can be readily seen that it is not an expensive tool to make.

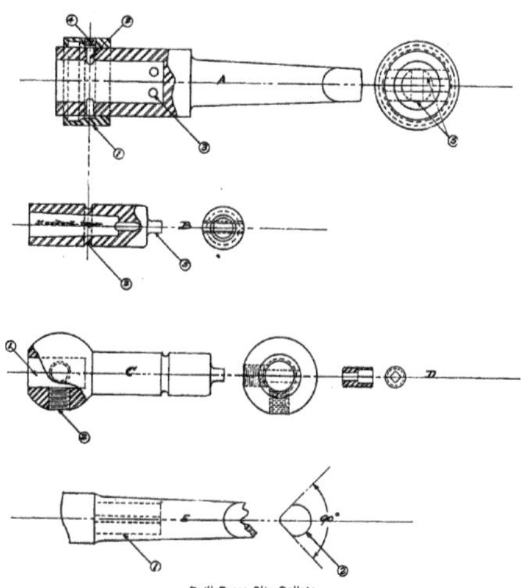

Drill Press Slip Collets.

4, fig. (a), is a tapped hole for inserting the two small hardened pins. 5, fig (a), shows the two pins placed, showing the box for the tang to fit in and the dotted lines; in the background is the small end of the shank and tang. The collet (b) is from 005-inch to 010-inch smaller than the bore in the holder, so as to insure freeness.

Fig (c) is the collet for holding reamers and straight shank drills, and the feature of this holder is the binding screws being at right angles to each other and the use of a large square hole set screw, as fig. (b). These screws are usually about 18 pitch and, being hollow, bind the shank over a much larger area than the ordinary screw and reduces the torsional strain a great deal. Another great difficulty to overcome is the constant breaking of tangs of drills and then never getting attended to properly, and if these drills were turned and milled as the dotted lines in fig. (e) and fig. 2 (e), and used in the collet (c), it will be found that they will outlive the drill with the tang.

A FEW TEST DEFECTS OF INTERNAL COMBUSTION ENGINES.

By H. J. Church.

I propose to outline here a few of the faults which are met somewhat frequently in the testing of gas and gasoline engines. The defects may be roughly divided into two classes: First, those which are due to defective water jackets; second, those connected with the actual running of the engine.

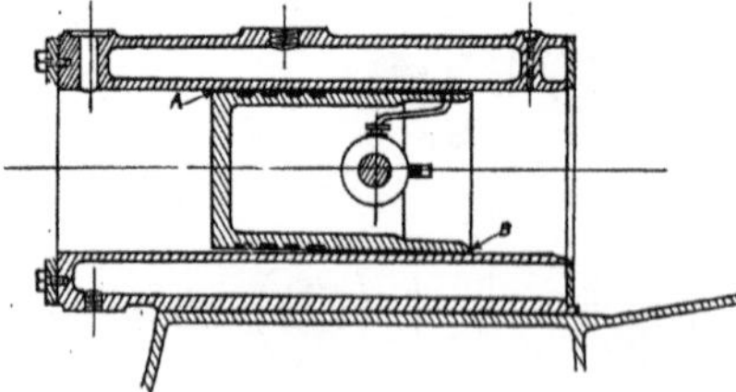

Fig. 1.—Showing Piston Bearing at A and B.

Leaks in Water Jackets.

Under the first heading will come all leaks in cylinder jacket, head and exhaust shell. Those in the cylinder water jacket are usually the result of the presence of porous metal, and if not of an extremely bad type, may be overcome by the careful use of the peaning hammer. Often what at first appears a really bad leak can be entirely stopped. Usually after peaning we give the cylinder a treatment of sal-ammoniac. The sal-ammoniac is dissolved in hot water, in the proportion of one pound sal-ammoniac to five gallons of water. After standing for twenty-four hours or longer (a week is none too much to obtain the full benefit of the treatment) the cylinder is emptied, allowed to dry off, and then is retested. The action of sal-ammoniac on cast iron is too well known to go into detail.

The troubles in the exhaust shells and cylinder heads generally necessitated the scraping of same. One plan was tried with good results with the heads. The leaks here usually occurred in the suction valve port. The heads were stripped and sent away to be galvanized. This completely overcame the trouble. The great disadvantage was the remachining necessary after the treatment.

Troubles in Running Engines.

The piston will usually require more or less attention. Knocking, thumping and blowing are the three troubles that gave us most care. They can be usually overcome by the judicious removal of the high spots, which are observable after the engine has been running some time. Frequently most of the knocking is caused by the back end of the piston expanding more than the front portion, owing to its proximity to the explosive charge. In general, it is safe to allow no bearing at all on the portion of the piston in the rear of the first ring.

One instance of blowing met with was rather interesting. It occurred along the lower portion of the piston. On removing the piston we found the rear end bearing hard on the top of the cylinder, at point (a) in fig. 1, and the front end at the point (b). In reality, the piston was slightly out of parallel with the cylinder. This is shown, greatly exaggerated, in the figure. The removal of the two high spots overcame the trouble.

The pistons should also be examined to find if the dowel pins which hold the piston rings in position are a tight fit. This may be done by filling the piston with gasoline. If there should be the slightest leak the gasoline will usually work its way through.

The crank pin brasses in our vertical engines frequently give us a lot of trouble. Often the presence of grit in the crank case oil is the cause. In a batch of 4 horse-power engines, which we tested, all the crank pin brasses ran hot. We tried all ordinary means to overcome this, but were unsuccessful. At last we removed the cylinder and placed a straight-edge across the top of the crank case, as shown in fig. II. We then measured the distance (a) between the straight-edge and the crank, and also

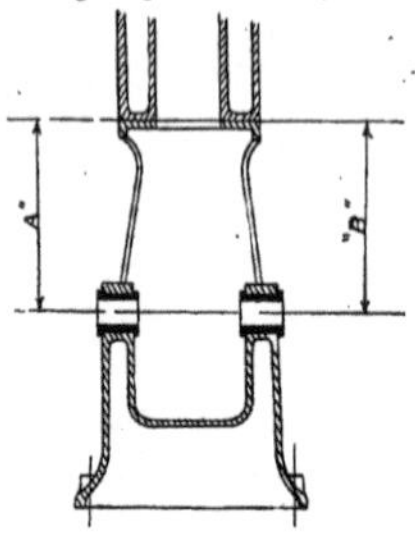

Fig. 2—Showing Bed Out of Alignment.

the dimension (b) on the opposite side. There was a difference of 1-16-inch between the two dimensions. The result of this was that the cylinder was tilted slightly out of line. This was sufficient to cause the brasses to run hot, as on squaring up the crank case with the cylinder the brasses ran perfectly cool.

A SAFE TOOLMAKERS' VISE.

By C. W. Putnam.

Everyone who has to clamp finished work in a vise knows the disadvantages of the ordinary jaw caps of copper or lead. They usually fall out at the wrong time and are seldom true enough to clamp a piece of work exactly even.

The strips of sheet steel are screwed to the side of the vise just under the jaws themselves. These carry the lower

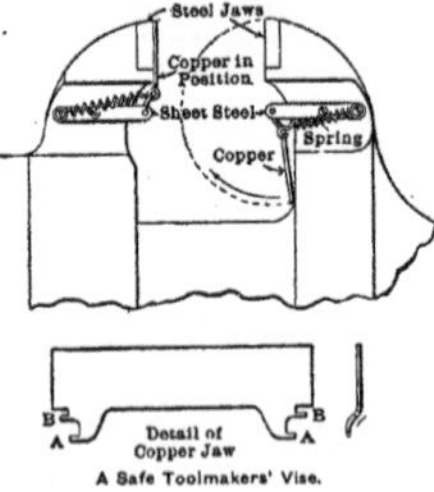

A Safe Toolmakers' Vise.

pins or trunnions of the copper jaws (aa), while the springs hook on to (bb).

The sketch shows the left-hand copper face in position and the right face swung down out of the way. It's a good scheme. Try it.—American Machinist.

Developments in Machinery

Metal Working Wood Working Power and Transmission

NEW TYPE OF MILLING MACHINE.

Herewith is illustrated a new Becker-Brainard milling machine, made in two styles, back-geared and not back-geared respectively, known as No. 25 and No. 26. In their design special attention has been paid to the requirements of the manufacturer of small machine parts which are produced in large quantities, such as found in small arms, type-writers, sewing machines and electrical works. In bringing out the new by interchanging the feed driving pulleys on the back of the machine, giving in combination eight changes, from .007 to .100.

The table is operated by worm and hobbed rack, the worm being driven by means of a worm gear of large size and worm of coarse pitch and of correspondingly high efficiency. The arrangement for disengaging the feed is by a new and novel worm mechanism by which the worm is thrown out of mesh with the of 4 to 1 ratio, allowing it to be returned to the original position in the least possible time.

The knee of the new model has been lengthened sufficiently so that a harness brace may be used for the arbor and still have a cross range for the table equal to that of the old style machines. This harness is especially worthy of notice and makes for convenience as well as rapidity. It consists of a brace which is gibbed to the knee slide; a

Becker-Brainard Milling Machine, No. 25.

Becker-Brainard Milling Machine, No. 26.

model especial attention has been given to the feed works, that they may be able to withstand the full power of the driving belt, and at the same time give good service in the rough usage to which these machines are subjected. This new feed is driven by belts which get their motion from the spindle of the machine by means of a train of gears, so arranged that the velocity of the belt is sufficient to drive all feeds that the main belt will stand. The changes of the feed are obtained by four-step cones and

gear, and leaves in a path at right angles with the axis of same, overcoming the objection of the old style gravity drop worm of clinging to the gear by friction alone. It also equalizes the wear on the worm gear teeth. The worm is also engaged and thrown out of mesh by the same lever, making in all a neat, convenient and positive method of automatically disengaging the feed and stopping the travel of the table at a predetermined point. The table is also supplied with a hand quick return

clamp that is fastened to the arbor support yoke in a manner that allows it to be swivelled around its centres, allowing the brace to be removed without removing any bolts. This clamp is made fast to the brace by friction, which gives a more rigid hold than the old style bolt washer and slot arrangement, at the same time allowing of a much stiffer brace. The convenience of this device will be at once appreciated by the operator. The arm, which is a solid steel bar, is adjustable lengthwise.

These machines are equipped with a rigid box knee and with a telescopic elevating screw, allowing the machine to be set in any position without regard to beams or floor construction, as the screw does not project below the floor line.

The base of these new model machines has been designed on the same lines as the other Becker-Brainard machines, which are extra heavy, absorbing all vibration.

The spindle cone and back gears are of the standard Becker-Brainard design, the spindle bearing being cylindrical in form, the wear being taken up by concentric compensating bronze boxes.

The appearance of the machine has in no way been neglected, since new patterns were made throughout. Great care was given to the symmetrical appearance of the machines as a whole, all corners being well rounded and the graceful outlines speak for themselves.

The ranges of these machines are as follows: Longitudinal feed, 34 inches; cross feed, 8 inches, and vertical adjustment, 18 inches.

Large Size Renold Silent Chain.

RENOLD SILENT CHAIN DRIVES.

As showing the rapidly increasing appreciation of the many advantages of Renold silent chain drives for high-speed power transmission, we mention two recent applications to different conditions in widely separated parts of the country.

To the Seattle Brewing and Malting Company, Seattle, Wash., belong the distinction of installing the largest Renold silent chain drive in this country—2 325 h.p. parallel transmission to refrigerating machinery; and under the supervision of Wm. M. Piatt, assistant city engineer, Columbia, S.C., a single strand Renold silent chain drive has been put in to operate a Worthington centrifugal pump from a horizontal turbine at a speed of 1,755 feet per minute.

The two chains for the Brewing Company are run side by side on single wheels 14 feet from centre to centre, the load being uniformly distributed by Dodge spring-centres on the driven wheel. The speed is 1,100 feet a minute. The pump drive measures 11 feet 7 inches centre to centre, and has a capacity of 225 h.p.

A Renold silent chain gear, as it is termed by the American manufacturers, the Link-Belt Company, consists of the silent chain and especially cut sprocket wheels. The links, stamped from high-carbon steel, are held together by hardened steel pins and bushings, which make a practically indestructible joint and one which provides the greatest possible bearing surface for a given width of chain.

It is due to the peculiar formation of the links and their method of contact with the wheels that the defects and racket common to the ordinary drive-chain are overcome; the stretch that may result from wear of the teeth is compensated for by the chain automatically assuming a larger pitch diameter while providing the surest possible contact with the driving and driven wheel, one half the circumference of each wheel being always in mesh.

As this continues to be the case, even when the chain becomes worn, it is evident that its maximum efficiency will be maintained although the demand may be severe enough to completely incapacitate ordinary drives.

The chain will run in either direction with uniform, quiet and positive action and is not adversely affected by hot or cold temperature, damp or oily situations.

AUTOMATIC ENGINE STOP.

There is a general impression among engine owners that the governor on the engine is a safety device, and is capable of providing for all emergencies; this is a fallacy, as statistics show that about 200 reciprocating engines and several turbines and synchronous convertors have run away in the past twenty months and all had governors.

The governor is designed to keep the speed of an engine constant under variations of load, but as most governors are arranged to operate by centrifugal force they cannot act instantly, as when the load goes off suddenly; it is for this reason that there is real need for a safety device entirely apart from the governor that will act quickly in case of emergency.

Fig. 1—Automatic Engine Stop.

Some of the causes of recent flywheel accidents are: breaking of main belt; loose governor pulley; reversal of dynamo; sticking governor; but chiefly the breaking or slipping of a governor belt and the sudden relief of the load.

While insurance pays the property damages and personal injuries, it does not make good the losses from non-production pending repairs to the plant, nor the wage loss of the employes. Obviously, then, the remedy lies in the providing of such means as will prevent accidents.

There is still another use for the same device; namely to shut down the engine from a distant part of the plant by pressing a switch. One stop is designed to be attached to the governor column, and upon being electrically tripped opens a steam valve allowing steam to enter a small cylinder against a piston, which raises the governor balls to their maximum position, thereby cutting off the supply of steam to the cylinder. In connection with a speed limit device,

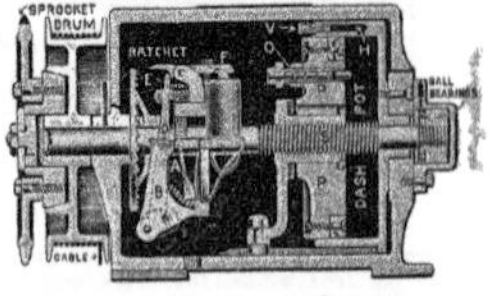

Fig. 2—Automatic Engine Stop.

this device is known as the Corliss type of Monarch Engine Stop System.

The speed limit consists of a centrifugal device connected to the engine shaft by a chain-drive; it is in circuit with the engine-stop and can be set at any predetermined speed; when this speed is reached, electric contact is made by tripping the engine stop.

Another type of engine-stop operates directly upon the throttle-valve, and can be easily applied to any engine without interfering with its regular work. This is called the Monarch type of the Monarch Engine Stop and Speed Limit System, Fig. 1. It is bolted to the engine frame at any convenient place and is attached to the valve-stem by means of a sprocket wheel and chain. As the valve is opened, a cable, to one end of which is attached a weight, is wound on a drum on the stop and is held by a pawl which engages in a ratchet wheel. When the stop is tripped electrically, the weight is released, revolving the sprocket wheel and thus closing the valve, a dash-pot in the stop forming a cushion to prevent jamming the valve.

In Fig. 2 will be seen a wire cable passing around this drum which is attached to a weight serving as power to operate the stop.

Inside the case and fastened to the shaft is a ratchet wheel engaged by a pawl (B), which prevents the weight from turning the shaft in the direction that closes the valve. This pawl is held in engagement with the ratchet by vertical lever (D), which is held in position by the left end of the armature lever (F). The magnets are placed in circuit with an electric current, and when the circuit is closed, the armature end of the lever (F) is pulled down, releasing the upper end of the vertical lever (D) which also serves as a hammer, striking a lug on the pawl (B) throwing it out of engagement with the ratchet, thus allowing the shaft of the stop to revolve and close the valve by means of sprocket chain attached to the sprocket wheel of the stop, engaging a similar sprocket wheel attached to the throttle valve stem, the weight, as before mentioned, furnishing the power.

At the opposite or right end of the stop is a dash-pot, as illustrated in Fig. 4, which consists of a cylinder into which the piston (P) fits closely. On this end of the shaft is cut a square threaded screw (S), passing through a nut fastened in the centre of the piston (P), so that as the shaft revolves by means of this screw the piston is carried into the cylinder and the air behind its inner face is compressed, forming a perfect cushion. The speed at which the stop acts may be accurately adjusted by turning the by-pass valve (V) that governs the amount of air that is forced through the air passage (H) as the piston moves in. Below this by-pass valve (V) and in piston (P) is located the releasing valve (O); this releasing valve (O), which is opened by contact against the bottom of the dash-pot, is adjustable. After the piston has cushioned and the throttle valve of the engine is near its seat, the compressed air in chamber (H) escapes through this releasing valve (O), allowing the throttle valve to take its seat softly, but yet with sufficient force to close it tightly.

Fig. 3—Application of Monarch Automatic Speed Limit.

THE "IMPERIAL" MOTOR HOIST.

Accompanying is an illustration of the "Imperial" motor hoist specially designed by the Canadian Rapid Drill Co., Montreal, for use in foundries and machine shops, where the handling of flasks and the erection of machinery requires smoothness of action and accuracy of control. It is simple in design and has small outside dimentions.

Imperial Motor Hoist.

The company's hoist consists of one high pressure, three-cylinder motor, geared to a hoisting drum provided with lifting rope and hook. The controlling valve is positive action and accurately controlled. An exclusive feature is the automatic lock which holds the load in arny position without the use of a brake.

Special attention has been paid to producing a machine which will stand the hardest usage at the hands of unskilled operators, with the least amount of care and attention. The motor is fitted with ground steel working parts throughout, and the transmission gear runs in an oil filled case. These hoists are built in capacities from 1,000 to 10,000 pounds.

NEW EBERHARDT GEAR CUTTER.

Accompanying are photos showing No. 2B automatic spur and bevel gear cutting machine, designed and manufactured by Eberhardt Brothers Machine Company, Newark, N.J.

The machine is suitable for such work as lathe and milling machine change gears, feed and adjusting spur and bevel gears, milling of face clutches, cutters and saws, and, in fact, all sylindrical or conical work requiring automatic milling, where either accuracy or rapid production, or both, are essential.

The machine has a capacity of 24-inch diameter, 6-inch face and 8 diametral pitch in steel at a good feed. The construction follows the general design of the line of machines of this company except where the size of the machine allows of different application of the same principles. This is shown in the bevel gear drive to the cutter-spindle, providing a very efficient drive, to give additional stiffness to the slide when raised.

A screw is provided for adjusting the lower slide toward or from the column, to allow for different lengths of hubs. A dial, graduated to thousandths, facilitates this setting. Graduated dials are also provided on the indexing worm, and on the cutter spindle bearing, for rolling and shifting when cutting bevel and mitre gears, and on the depth adjusting screw, for setting the proper depth to be cut.

The indexing mechanism is positive, and operates the master wheel, which is large for this size machine, as may be seen.

The outside support to the work arbor is adjustable for different lengths of arbors, but is always centred opposite the work spindle. This is especially con-

New Eberhardt Gear Cutter

especially adaptable to the high spindle speeds required by high steel cutters. The changes of speed are obtained by means of gears, immediately driving the bevel pinion.

A segment, graduated in degrees, provides for cutting bevel and mitre gears. The segment is operated by means of a worm, meshing in the teeth in its periphery, thus allowing convenient adjustment. The slide can be raised to 90 degrees, and the machine is very suitable for such work as milling face clutches, etc.

A strap is shown in the front view of the machine, which is used as a brace

venient in a machine of this class, and allows rapid setting.

The photo shows the work spindle with a 60 deg. centre and dog driver. This is furnished with the machine, and is of use for such work as milling flutes in taps and reamers; for cutting gears on ordinary lathe mandrel; for cutting pinions solid with the shaft, etc. The centre has a taper shank fitting the work spindle hole, and is drawn in and forced out positively, by means of a bolt operated by a handle at the back of the work head. The usual taper and shoulder nut arbors can be used for cutting gears, the hole in the work spindle being No. 10 B. & S. taper. The spindle is of machine steel.

The cutter arbor is solid with the cutter spindle on this size, and takes cutters with 1-inch hole. The chips are caught in the box on the side of the machine, and the oil runs through, being caught in the ample reservoir formed around the frame of the machine. The oil pump affords a constant stream of cutting lubricant, and can be adjusted to regulate the supply.

TRO-THERMIC PROCESS.

The final report of the commission appointed by the Canadian Government in 1903 to investigate the different electro-thermic processes of smelting iron ores and the making of steel in Europe, has been published. As is known, this commission went to Europe, and in 1904 a report of their investigations was issued. The commission was then authorized to conduct experiments on Canadian ores. This was done, and this report is a complete one regarding the experimental work done at Sault Ste. Marie under the supervision of Eugene Haanel, superintendent of mines.

The report contains a detailed statement of the work done and all the measurements made, of the analysis of the pig and slags produced and of the iron ores employed. Plans are given of two commercial electric furnaces which have recently been patented.

Foundry Practice

MODERN IDEAS FOR FOUNDRYMEN

*Consideration of Foundry Design**

By F. A. Coleman, Cleveland.

A foundry is a machine composed of a number of parts, each part having a certain duty which should be performed in harmony with the other parts. This machine should be housed in buildings erected to suit the machine.

Raw materials are received or fed into the machine at certain places and the output is finished castings. We think this statement covers the foundry proposition; if it does, we are led to believe there are some peculiar methods employed by many foundrymen.

A foundryman desiring to buy an engine does not go to the engine builder and ask for an engine large enough to fill a building 30 x 40 feet, but he tells him he wants a certain power and type of machine. Nevertheless many foundries start with the size of the building and work backwards to the machine they desire to house. It is this illogical method of building foundries that has tempted the writer to prepare this paper on "Foundry Design," not with the idea of trying to write a treatise on foundry buildings, but of trying to help in the design of the foundry machine.

Many statements made herein will undoubtedly be questioned by foundrymen because they will not line up in all cases with certain exact and known conditions, but general conditions are being discussed and necessarily general statements must be made; therefore the attempt will be made to treat the subject in a very broad manner.

One expecting to enter the foundry business may have a certain fixed sum of money to invest and desires to produce the best results from this investment; or a certain limited piece of property is to be used and it is necessary to produce the greatest output from the area; or it is desired to produce a certain tonnage of castings. Ordinarily, any one of the above three conditions will explain the beginning of a new foundry, and a careful study will show that in no one of them should the building be allowed to become the determining factor in the design of the machine.

The writer claims that the buildings are drawbacks to the foundry; they keep out light, complicate the ventilation, and are in the way generally.

Problems of Foundry Design.

It seems as if there are but three problems to be solved in designing a foundry, which may be stated as follows:

I. Given a certain sum of money, to produce the largest and most efficient foundry, or

II. Given a certain area of land, to produce the greatest output efficiently at the least expense.

III. Given a certain desired output, to produce the highest efficiency at the least expense.

The general problem is to produce the greatest result with the smallest expense. This of course is what all foundrymen are seeking to do.

There are many kinds of foundries, and while it is not really necessary in discussing this subject to name them, yet it may be of interest to see how many different lines of foundry work are carried on.

A classification according to metals follows:

I. Iron.
II. Steel.
III. Brass.
IV. Aluminum.

I. Iron foundries produce grey and malleable iron castings.

II. Steel foundries produce acid, basic and crucible steel.

III. Brass foundries produce all kinds of alloy castings, of which we may assume copper as the base.

IV. Aluminum foundries are almost universally a part of brass foundries and produce alloys containing high percentages of aluminum.

Classifying according to methods of melting and avoiding the unusual or experimental methods we have:

I.—IRON—
- Cupola furnaces.
- Air furnaces.

II.—STEEL—
- Open-hearth.
- Converter.
- Crucible.

III.—BRASS—
- Crucible.
- Direct melting.

IV.—ALUMINUM—
- Crucible.
- Direct melting.

Classifying as to fuels we have:

I.—Iron—In the cupola furnace—
- Coke.
- Coal.
- Charcoal.

In the air furnace—
- Coal.
- Gas.

II.—STEEL—In the open-hearth—
- Gas.
- Oil.

Iron melted in the cupola for the converter, with cold blast.

In the crucible furnace—
- Soft coal.
- Oil.
- Gas.

III.—BRASS—In the crucible furnace.—
- Coke.
- Hard coal.
- Soft coal.
- Oil.
- Gas.

In direct melting furnace—
- Oil.
- Gas.

IV.—ALUMINUM—In the crucible furnace—
- Oil.
- Gas.
- Coke.

Direct melting—
- Coal.
- Oil.
- Gas.

Classifying according to character of work:

I.—IRON—
- Jobbing—light work.
- Jobbing—heavy work.
- Architectural.
- Hardware.
- Stove plate.
- Soil pipe.
- Water and gas pipe.
- Car wheel.

*Paper presented before Philadelphia convention of American Foundrymen's Association.

Pipe fittings.
Machinery.
Electrical.

II.—STEEL—
Jobbing.
Electrical.
Machinery.
Marine.
Railroad.

III.—BRASS—
Jobbing.
Ornamental and art.
Lighting fixtures.
Architectural.
Hardware.
Plumbers' and sanitary supplies.
Steam and hydraulic.
Marine.
Bell.
Automobile.

IV.—ALUMINUM—
Jobbing.
Automobile.

These classifications are not supposed to be complete or exact, but on examining them it will be readily seen that many of these points enter into the foundry design. The foundryman should therefore settle a number of them before considering going ahead with a foundry proposition.

Classification of Work.

The foundryman should determine as closely as possible the class of work he expects to make and the process he expects to use in making it; if the class of work to be made is definitely known, the problem is simplified; too much importance cannot be laid upon this fact because the foundry can then be designed to meet known conditions.

It makes no difference what a foundry may make, all foundries have three general operations which are :

A—Melting
B—Molding
C—Cleaning.

A.—All materials, processes, equipment and labor, necessary to produce molten metal in the molds are included under this head.

B.—All materials, processes, equipment and labor necessary to produce a form of mold for receiving the melted metal come under this head ; as cores are a part of the mold, core making is included.

C.—All materials, processes, equipment and labor necessary to remove adhering molding metal and excess metal not a part of the actual pattern come under this head.

According to this arrangement the melting cost would be the cost of the metal in the mold; the molding cost would be the cost of the molds and cores ready for the metal; the cleaning cost would include the removal of the rough castings from the mold, making it like the actual pattern and delivering to the shipping room; a number of overhead expenses would, of course, be chargeable to these three operations.

Automobile and Marine Engine Cylinders.

We may assume the case of a foundryman who desires to make automobile and marine engine cylinders; he either knows from his experience or finds on investigation that he can produce the best castings with the air furnaces; this of course determines his method of melting.

He next finds that he can depend, or will have to depend, upon the business of certain customers, and that it will be to his advantage to make the work upon some style of molding machine.

If he builds his foundry upon this last assumption, and finds the business comes as he expected, then he will get a certain output; if he has been wrong in his assumption, and he has to resort to bench or floor molding, then his output per molder will be reduced.

The first assumption would require a certain melting capacity per molder, while the last would require a smaller melting capacity per molder.

There are of course many points involved in the ouput of a foundry, but it may be roughly stated that the output divided by the area will be constant for the same line of work.

There are exceptions to this statement, of course. In a brass shop making sanitary work in iron flasks, the molds if poured flat may be stacked three and four high; if poured on end they cannot be stacked and the output of a molder will require a much larger area.

The endeavor is being made to show that many things are peculiar to each metal and to each method, and therefore all these points should be taken up and settled as a part of the preliminary work. It is admitted, that at times no one can determine what should be done, but it is to the foundrymen's interest to reduce the elements of the unknown and the uncertain as much as they can.

Foundry Layout.

It can now be assumed that the foundryman has definitely decided to make a certain line of work in a certain way, and is ready to design a foundry machine which will give him the greatest output of good castings for the least cost.

He should have a map of the property showing streets, alleys, railroads, etc., if the property is or may be affected by buildings adjoining, and these should be noted.

If the area is large and the surface is irregular it may be necessary to make a topographical survey of the tract, but it can be set down as a fact that a careful study of the property itself will show that certain advantages may be gained by designing the foundry with reference to these property conditions.

A building 70 x 100 feet may be designed and do very well on a ten-acre lot, but will be a failure on a lot 70 x 100 feet, surrounded by high buildings in the city.

The amount of metal to be melted to produce the tonnage of finished castings, and the number of molders necessary to handle the metal are next determined.

From this we find the area required for the furnaces, molders, core makers, core ovens and cleaning; the areas and capacities of the storages for metals and fuels are found ; the storage of the flasks, molding and core sands, etc., may be determined; furnaces, molders', core makers', and cleaners' supplies should have a place and require areas which should be determined; pattern and machine shops; pattern storage, locker and wash room; closets and urinals; sand preparation rooms, shipping room, room for holding castings to see if orders are completed ; foundry foreman's office ; main office; power plant and laboratory; it may be necessary to consider all of these.

Certain operations requiring floor space are necessary to every foundry; some of the things noted are needed by some foundries, and all such areas should be determined as closely as possible.

The determination of the various areas required, and such other information as is peculiar to each foundry, places the foundryman in position to start the actual layout of the foundry.

Arrangement of Areas.

The problem now is to make an arrangement of these various areas which will permit of the movement of the materials through the various processes of melting, molding and cleaning, and deliver the finished castings to the shipping room with the smallest amount of labor, the minimum amount of equipment, and the lowest cost.

The storages for all materials received should be close to a central unloading point where they are to be used. The first statement is particularly true of foundries served by sidings.

Fuel and metal should be very accessible to the furnaces. Sands should be convenient to the core room and the molding floors. They may be passed through the sand preparation rooms and then be distributed to the core room and the various molding floors. Likewise the core ovens should be close to the fuel, and the core room located so as to serve the foundry without unnecessary handling of cores.

Core Departments.

In foundries served by cranes, the ovens should be arranged so that the cores may be handled from the cars by the main bay cranes into place.

This does not mean that the main cranes should be used to load cars, which may be done by core bay cranes. It is not profitable to bake large and small cores in the same oven; the best results are obtained, both in baking and in fuel saving, by baking bench cores in small ovens and floor cores in large ovens. If possible the small core racks should be placed on the division line between the core room and the foundry and accessible to both.

Charging Floors.

In the foundries where the charges are hoisted, the charging floor should be large enough to carry at least one day's supply to avoid shut downs. In heavy shops running two or more cupolas side by side, they should be spaced so that two cranes may serve at the same time; pits should not be located in front of the cupolas.

Provision should be made for the handling of the ladles at the cupolas by small cranes so as not to tie up the main bay cranes.

It is best to pull the heavy and mold cars by haulages independent of the cranes; the heavy cranes should not be called upon to do all kinds of work, which can be better done by lighter and cheaper equipment.

All areas should be regular in form and not cut up, posts on columns should be avoided and should be kept off the molding floors or placed at division lines between bays or floors.

Taking up the form of construction best adapted to foundry purposes, it would seem that of two types, the one which will give the greatest amount of light and the best ventilation should be the one to adopt.

Crane loads and roof loads should be carried directly through columns to the foundations; there is no reason for requiring the actual walls to carry loads.

Light and Ventilation.

Every foundry requires all the light it can get and ideal results are obtained when there is nothing but glass from the sills to the eaves and a continuous line of glass from end to end.

The lanterns or monitors should be wide and high; a narrow monitor is about as expensive as a wide one and not as effective.

The general form of construction as above recommended is so simple and good that it does not seem necessary to mention it, and yet the majority of foundries are much more costly and do not produce as good results.

In addition, whitewash or cold water white paint are worth many times the cost.

If the foundries will wash the windows but once a year they should do it in October and not in the spring; many a foundry can declare a dividend with the money wasted on lighting, when nature stands ready to do it for nothing if only given the chance.

Do not, under any consideration, build any kind of a foundry to do any kind of work with less than 14 feet outside walls. A single bay foundry 50 feet wide should be not less than 16 feet high on the sides.

When floors are to be paved in the foundries, good concrete will give first-class results. The writer is fully aware that the above two statements will be very severely criticized.

Power.

Foundry power is a question which interests many. Shops requiring a small amount of power in the cities can often buy power from the electric companies to good advantage; this is particularly true when there is large fluctuation in the quantity used.

The number of producer gas engines will increase in the foundries as there are many points in their favor; probably the majority of foundries use less than 100 horsepower. The building of successful bituminous coal gas producers in units as small as 125 horsepower, together with the present reliability of gas engines, are strong factors in making this increase.

Lockers and Wash Rooms.

The modern foundries consider locker and wash rooms as necessary; they should be used for washing and dressing rooms only, and should be in charge of some one at the time when the men are going to and leaving work.

The closets and urinals should be located so that it is not necessary for the men to leave their floors; the saving in lost or loafing time is worth the additional cost.

In the larger plants a room for the holding of patterns on their way to the foundry is a convenience. Patterns are thoroughly overhauled to know that they are in condition, with all parts, core boxes, etc., accounted for.

Cleaning Department.

In foundries running a large number of pieces of the same pattern, a casting cleaning room has many points in its favor; castings are turned into this room from the foundry; they are counted and inspected and any shortage must be supplied by the foundry; this avoids congesting the shipping room with incomplete orders. In shops having factories, this room really becomes a store house where requisitions from the factory are filled.

The foundry foreman's office should be located convenient to the foundry to give a good view of the plant; in larger shops requiring a number of sub-foremen, this is not so important.

Distribution of Power.

The distribution of power throughout the plant by motors does away with the necessity of making the location of the power plant a factor in the design of the foundry. The pattern shops, flask makers' shop and machine shop are not an actual part of the foundry, and while it may be necessary to consider them, nevertheless, the foundry layout should not take these into serious consideration.

The actual foundry provides for melting, molding and cleaning; everything else which may be necessary is subsidiary and should give way.

The writer desires, however, to qualify the above statements with reference to foundry plants which are designed to occupy the entire area of the property, and where the greatest output must be secured from this area. Under such conditions everything must be considered as a part of the whole and it becomes somewhat a case of give and take between the various departments.

The east with its older foundries and closely settled localities has been, in many lines of work, forced or led into the construction of foundry plants of two or more stories.

The west is beginning to do the same thing under the same stress of conditions; the melting, molding and generally the core making are carried on on the top floor; cleaning, sorting and shipping and other operations being done on the lower floors; such foundries are generally engaged in light work only.

Building Materials.

Of the materials of construction it may be said that brick is mostly used with wood or steel framing. Fire proof construction is increasing; the foundry, with its furnaces, ovens, hot metal and careless class of employes is peculiarly subject to fires; that this fact is recognized by nearly all foundries is shown by the construction of the fire proof pattern storages.

Oftentimes the destruction of the plant itself during the rush season is fully as serious as the loss of the patterns; this is true of jobbing foundries where customers generally insist upon the delivery of their castings, fire or no fire.

In small or light shops, floors should be planned to give each molder a certain area which can be served very readily with metal either by trolley system or some other method which will take the metal quickly and without confusion to each area.

The great need for designing plants not only for the present, but for the future is shown by the large number of foundries, which have a main foundry and a vast litter of dog houses; many foundries are not thoroughly christened until they add such a dog house. But they are expensive; they break up the routine, require extra foremen and generally rob the main foundry of its share of light.

The writer believes that the foundryman should thoroughly study his plans on paper; each problem has an answer which can be more cheaply worked out on paper than on the ground.

It should be remembered that profit is the ultimate end; profit produced by making good castings at a low cost.

The fixed charges should be kept down. Bad design resulting in the expensive handling of materials means useless fixed charges, limited output and of course too high a cost.

Many an expensive investment in equipment or makeshifts could have been avoided in many foundries by the proper designing of the original plant. Many foundrymen now keenly realize that what appeared like a saving in construction has resulted in a heavy cost for operation.

*The Malleable Foundry as a Business Venture**

By Dr. Richard Moldenke, Watchung, N.J.

Hardly a week passes but that I receive inquiries regarding the cost of establishing malleable castings plants. The answer I have to write invariably is: "Unless you have at least five tons daily to make—don't."

When one has personally gone through the building of a great establishment, and has the making of hunreds of thousands of tons of malleable castings behind him, caution is the order of the day, and unless the quantity is sufficient and the deliveries from outside unsatisfactory, it is wisest to buy from existing plants rather than to venture upon the little known fields of the heat treatment of white cast iron, for that is what the malleable casting industry comes to in the final analysis.

As a rule the requests for information come from those iron foundries which get orders for plenty of grey iron castings with a sprinkling of malleable, and possibly some steel castings also. The high price they have to pay outside for small lots naturally interferes with their legitimate profit, and hence the desire to produce for themselves. I believe that when the electric furnace is perfected, or made commercially possible, that a different aspect will be given to the case, but at present we have to take things as they are.

The question of embarking into the malleable casting business resolves itself into a market for the work to be produced, and the location to make that work at the lowest cost. It is very rare that a location is picked out unless the territory tributary to it can support a comfortably live malleable plant.

*Paper presented before Philadelphia convention of American Foundrymen's Association.

Those concerns which use a large quantity of these castings are the natural pioneers for a given region. The fact that they can make money where they are would be a pretty good guarantee that they can either save money or get their desired deliveries by going into the malleable casting business. Moreover, their profit as manufacturers will carry them over the two or more years of loss in starting the malleable plant.

Advantages and Disadvantages.

To sum up briefly the advantages and disadvantages, I give the following:

(1)—The expanding market for malleable castings, and the great room for improvement in most of the producing works, gives a living chance to the man to go into the business, provided he understands it thoroughly.

(2)—In times of prosperity, plants in existence, where great caution is exercised, must turn away large quantities of offered work, being unable to produce it.

(3)—The moment the malleable castings works are full, deliveries are so desperately bad and slow that attention is turned directly to new enterprises coming in.

(4)—The adaptability of the malleable casting to specialties in railroad, agricultural and hardware lines, is such that the beginner has a good chance to succeed, provided his specialty is good.

(5)—Where absolutely good material is required, it pays the consumer of malleable castings to go into making these himself, provided he has enough tonnage. Only he who knows how easily one can slip up on the quality of the malleable casting, also knows how often this is done, and the customer is not the wiser.

(6)—The malleable product from its adaptability to specialties, and comparatively light castings at that, means a production in quantity of each separate piece. Hence it is not a question of half a dozen castings, but oftentimes several hundred thousand, or in other words, continuous production, with comparatively well regulated and systematized shop cost.

Now the disadvantages follow from the above:

(1)—The expanding market, while making room, also leaves very empty shops in dull times, for the existing plants usually expand with each high wave of prosperity and manage to hold the desirable customers.

(2)—Where the work turned away by the full plants is taken by the beginner, it is at either lower prices, or if at high prices, is lost again when these fall. The ability of the founder who commences, to hold customers who may have been exasperated elsewhere, or his ability to produce cheap enough to cut under, or most likely, his having a specialty which pays the freight, will make or break him.

(3)—The high capital required is a great detriment, being about four times as much as for the same tonnage in grey iron for the works alone. This means more than appears on the surface. The malleable casting process is intricate, the chances of loss great, and it takes fully two years to get a desirable class of men in the shop, after all the bums of the country have passed through it. Hence the chances of no profit, if not worse, for a long time, which must be met and carried until success appears. To give an instance, a rough calculation of the cost of a shop to produce 50 tons of castings daily would come up to say $250,000. By the time that shop would be actually producing the 50 tons daily at a normal shop cost, over double the amount of money will have been expended. Then, however, with a good market, the shop will pay handsomely.

(4)—The high grade of skill required. While the process as we now know it is comparatively simple, we have to reckon with so many factors that an expert in malleable castings is a scarce article. With the range of composition for the iron for various thicknesses of the castings very close, a white iron to start with, with its double contraction, and danger of shrinkage in the interior, with the necessity of chills, the limit to the size and weight, the expansion in the anneal, etc., it is a pretty delicate task to turn out large quantities of various classes of castings, and have them all high grade. Hence the high salaries the important men must be paid, and the eternal vigilance required in the works.It must be remembered that your day's work goes into the anneal, and you do not see it for a week. The next day's work the same, and hence a mistake in the temperature of the ovens and you have ruin for possibly a week before you find it out. I remember the slipping off of the night annealer to a nearby speak-easy, with the increase in the natural gas pressure just at that time, which meant thousands of dollars' of damage for two or three hours' inattention.

(5)—The fierce competition that is met with. This is queer but nevertheless the case. In spite of all the troubles the malleable casting man has to contend with, he makes himself more by cutting his competitor's prices, until all have very little left. This is primarily the case, because to run a shop successfully and to hold an expensive organization, everybody must be busy. The first aim therefore is tonnage, to keep the shop producing at maximum capacity all the time, practically regardless of prices sometimes. Hard times drop prices in the malleable shop far quicker than in the grey iron establishment. I remember the time when castings were sold to the railroads at a shade less than two cents a pound, delivered. When it is remembered that malleable castings ought to bring a cent a pound more than the corresponding grey iron work, it will be seen what a new man is up against in starting when prosperity is on the wane.

The further reason for this fierce competition is that nearly everybody in this line has a specialty, or several of them, which sell under patents and for as high a price as the community will stand. Hence this velvet can cover a deficit to keep the rest of the works in operation. This, of course, is hard on the man who has no specialty, and hence again, the necessity of having abundant capital until such a specialty is found.

To sum up then, with a good specialty or a number of castings used in the manufacture of well paid for articles, and a sufficient daily tonnage, it is good policy to go into the malleable casting industry as a business venture; but with less than say five tons daily sale, and with insufficient working capital, and men not thoroughly trained in this line, it is better to let it alone.

THE TWENTIETH CENTURY MOLDING MACHINE.

That the pneumatic molding machine is ever forging to the front as one of the indispensable adjuncts of successful foundry practice, and is successfully used to reduce the cost of production in the foundry all will acknowledge, but the cost of installing machines of this kind has deterred the small foundry from doing so.

A very successful machine in use which can be installed at moderate cost is known as the Twentieth Century Molding Machine, patented by J. F. Webb, of Elkhart, Indiana.

The variety of work this machine is capable of handling when placed in the hands of a molder, or handy man, is little understood.

It has often been said: "Some of the best goods come in small parcels."

The Twentieth Century Molding Machine occupies a small amount of space, is moderate in cost, and might be well called the little giant, or giant wonder, as it is a powerful machine in small compass.

The inventor's ideas shown in the machine are to have a portable pneumatic molding machine in order to obtain the best results.

The machine is portable in order to save the operator the hard labor of carrying the molds the distance he is required to in using a stationary machine, and to have a positive pressure in order to ram every mold the same, and not have one too hard, and the other not hard enough, and another important feature is to have a machine that will not require repairs, as repairs are costly in time and money.

The time the machine stands idle means a curtailment of production, and this in turn means profits lost, and no foreman likes to see the cost of castings mounting upward, and have his employer considering the loss of profits, and customers are apt to become impatient at delays in delivery of castings, when rushed with orders.

The inventor's aim has been to produce a machine simple in construction, moderate in cost, powerful yet compact, one easily understood, and manipulated by molder, or handy man, and easily adjusted to a great variety of patterns, requiring no elaborate equipment, and embodying in every respect the ideas of a practical molder.

When in use the molder pulls the car toward him, places his flask having a plate on which are the pattern between cope and drag on the table, riddles his sand on the pattern, and fills the flask, places the bottom board in position and rolls the flask over, and proceeds with the cope in the same way, then places presser board on top of cope, and by giving the table a slight push, moves it under the cylinder. With his right hand he throws a lever and the air is introduced into the cylinder, and drag and cope are pressed at the same time and piston returns up into the cylinder ready for the next mold. The operator then pulls the flask toward him and starts the vibrator in motion, lifts off his cope, setting it to one side, and lifts the plate having the patterns on it, and the mold is ready for closing.

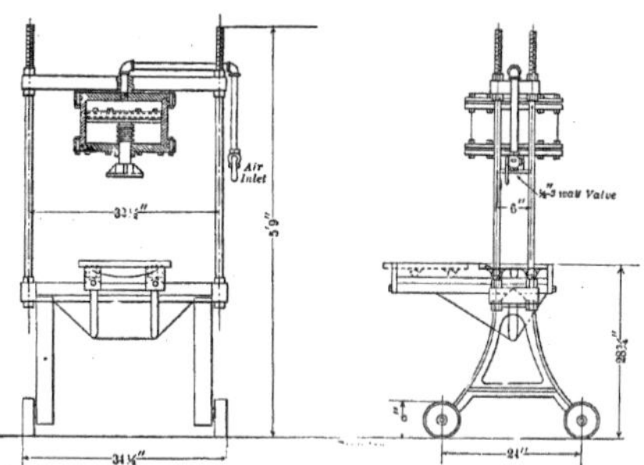

Twentieth Century Molding Machine.

In order to use a deeper flask the nuts above the upper crosspiece are screwed up to the desired height and crosspiece raised up to them, and nuts underneath screwed up to it, thus raising the pressure head, and allowing a deeper flask to be pressed.

This machine, ready for shipment, weighs only six hundred pounds. Yet a pressure of ten thousand pounds is exerted with an air pressure of 100 lbs. per square inch, and with an ordinary 12 in. x 16 in. flask 40 lbs. is enough.

Selling agents: J. S. McCormick Co., Pittsburgh, Pa.

AMONG THE SOCIETIES

SOCIETY OFFICERS.

Canadian Mining Institute.

President, George R. Smith, Thetford Mines, Quebec ; secretary, H. Mortimer Lamb, Victoria, B.C.; treasurer; J. Stevenson Brown, Montreal.

Toronto Branch A. I. E. E.

Chairman, R. G. Black ; vice-chairman, K. L. Aitken, secretary, R. T. McKeen.

Marine Engineers.

Grand president, E. J. Henning, Toronto ; grand secretary, Neil J. Morrison, St. John, N.B.

Canadian Electrical Association.

President, R. G. Black, Toronto ; vice-presidents, R. S. Kelch, Montreal, W. R. Ryerson, Niagara Falls, Ont.; secretary-treasurer, T. S. Young, Confederation Life Building, Toronto.

Canadian Association of Stationary Engineers.

President, Joseph Ironsides, Hamilton ; vice-president, Ed. Grandbois, Chatham ; secretary, W. L. Outhwaite, Toronto.

Ontario Association of Stationary Engineers.

President, A. M. Wickens, Toronto ; vice-president, W. A. Sweet ; board of directors, W. Blackgrove, W. A. Sweet, A. M. Wickens and A. E. Edkins ; treasurer and registrar to be appointed by board of directors.

Canadian Railway Club.

President, W. D. Robb ; secretary, James Powell, chief draughtsman motive power department, G. T. R.; treasurer, S. S. Underwood, chief draughtsman car department, G. T. R.

Engineers' Club of Toronto.

President, C. B. Smith ; secretary, Willis Chipman, Rooms 96 King St. W.

Canadian Society of Civil Engineers.

President, W. McLean Walbank ; secretary-treasurer, C. H. McLeod, Rooms, 877 Dorchester street, Montreal.

Society of Chemical Industry

Chairman, Prof. W. H. Ellis ; vice-chairmen, H. H. Van Der Linde, Milton ; H. Hersey, A. McGill ; hon.-secy., Alfred Burton, 44 York Street, Toronto.

Ontario Land Surveyors.

President, Otto J. Klotz, Ottawa ; vice-pres., Thos. Fawcett, Niagara Falls ; secy.-treas., Capt. Killaly Gamble, Toronto.

CONVENTION OF AMERICAN SOCIETY OF MECHANICAL ENGINEERS.

The fifty-ninth annual convention of the American Society of Mechanical Engineers was held at Indianapolis on May 28-31. There was a registered attendence of about three hundred.

While somewhat small in point of numbers, the meeting was eminently successful, the papers being of a uniformly high quality and the discussions thereon of unusual interest. A noticeable fact in regard to these discussions was that those who took part in them were men eminently qualified to discuss the subjects under review, and with the result that much information was brought to light.

Large Number of Interesting Papers.

The committee on papers had arranged for a large number of papers on special subjects, chief among which were automobiles and superheated steam. Some of these were prepared by non-members of the society upon invitation of the committee on papers, the authors being experts in their particular lines. Among the papers were: "Collapsing Pressure of Bessemer Steel Lap-welded Steel Tubes," Prof. R. T. Stewart; "The Balancing of Pumping Engines," A. F. Nagles; "The Economy of the Long Kiln," E. C. Soper; "Ball Bearings," Henry Hees; "Air Cooling of Automobile Engines," John Wilkinson; "Materials for Automobiles," Ellwood Hayes; "Special Auto Steel," Thos. J. Fay; "European Motor Cars," B. D. Gray; "Cost of Heating Storehouses," H. O. Lacount; several papers on "Superheated Steam," a synopsis of which appears in this issue of Canadian Machinery.

Report on Machine Screws.

The Committee on Machine Screws made their report.

This committee, which has been in existence for about five years, has been engaged in the work of preparing standard specifications for machine screws. It has presented several reports to the society, but the discussion provoked by each report caused it to be referred back to the committee for further consideration. The revised report recommends a number of changes in the screw gauge outside diameters, enabling the United States standard form of thread to be used without departing from the present pitch diameters, which are large on the small screws and small on the large ones. It was moved and carried that the society accept the report and the committee be discharged.

Visiting Plants.

An afternoon and evening was devoted by the convention to inspection visits to a number of manufacturing plants in Indianapolis, including Attis Engine Works, the Nordyke & Mormon Co., the National Motor Vehicle Co., and the Parry Mfg. Co. Suitable entertainment was provided for the ladies during the convention.

CONVENTION OF RAILWAY MEN.

About 7,000 association members, special guests and exhibitors registered at the forty-first annual convention of Master Car Builders, Master Mechanics, and Railway Manufacturers' Associations at Atlantic City in June. At the same time there was a meeting of the secretaries of the different railway clubs. James R. Powell represented the Canadian Railway Club, Montreal. Subjects of interest were presented and discussed and officers for the ensuing year were elected as follows: Chairman, James R. Powell, Montreal; vice-chairman, F. O. Robinson, Richmond, Va.; secretary-treasurer, Harry D. Vought, New York.

Best Display Ever Made.

The supply men made an exhibit of railway appliances, which for variety and completeness has not been equalled.

Officers of Master Car Builders' Association.

Many things of interest were brought up and discussed by the car builders. Officers were elected as follows:

President, G. N. Dow, mechanical inspector, Lake Shore, Cleveland; first vice-president, R. F. McKenna, M.C.B., Lackawanna, Scranton; treasurer, John Kirby, Adrian, Mich..

Railway Master Mechanics.

The 40th annual convention of the American Railway Mechanics' Association was held at the same time, and was the most successful yet held. Officers were elected as follows:

President, William McIntosh, Central Railroad of New Jersey, Jersey City; first vice-president, H. H. Vaughan, Canadian Pacific, Montreal; treasurer, Angus Sinclair, New York.

CANADIAN RAILWAY CLUB MEETS

At the annual meeting of the Canadian Railway Club the following officers were elected:

President, W. D. Robb, superintendent motive power, G.T.R.; first vice-president, L. R. Johnson, assistant superintendent of motive power, C.P.R.; second vice-president, H. H. Vaughan, assistant to vice-president, C.P.R.; Executive Committee:—W. E. Fowler, T. McHattie, A. A. Maver, W. M. Dietrich, W. H. Evans, Professor Morgan and G. T. Bell. Audit Committee:—A. G. Walker, W. H. Gault, J. S. N. Dougall, J. Powell, chief draughtsman motive power department, G.T.R., was re-elected secretary, and S. S. Underwood, of the G.T.R., treasurer.

SOME THINGS TO LEARN.

Learn to laugh. A good laugh is better than medicine.

Learn to attend strictly to your own business—a very important point.

Learn how to tell a story. A well-told story is as welcome as a sunbeam in a sickroom.

Learn to stop grumbling. If you cannot see any good in this world keep the bad to yourself.

Learn to keep your own troubles to yourself. The world is too busy to care for your ills and sorrows.

ABOUT CATALOGUES

By mentioning Canadian Machinery to show that you are in the trade, a copy of any of these catalogues will be sent by the firm whose address is given.

WATERWORKS SUPPLIES—An attractive catalogue, a souvenir of the waterworks convention, held in Toronto, enumerating the different articles manufactured by the Jas. Morrison Brass Mfg. Co., Toronto. The catalogue is illustrated, and gives, with prices, a complete list of such articles as are used by waterworks in the laying of services in cities and towns.

HYDRAULIC MACHINE TOOLS.—A large catalogue of hydraulic machine tools, electrical and other special machinery, issued by Henry Berry & Co., Ltd., Croydon Works, Leeds, Eng. This catalogue is an exceptional one in every way. It is very large and very complete, containing 300 pages, 9 by 12 inches; bound cloth. Heavily coated paper is used, so that the fine half-tone illustrations reproduce exceptionally well. The machinery illustrated and described in this catalogue includes hydraulic pumps of different kinds, hydraulic pressure accumulators, hydraulic riveters of very many styles, hydraulic wall, jib and traveling cranes, electric traveling cranes, hydraulic forging presses, hydraulic flanging presses, stamping presses, wheel presses, and many other kinds, hydraulic bending machines, hydraulic packing presses, pipe bending rolls, boiler flue flanging machines, hot-iron saws, hydraulic shearing machines, hydraulic punching machines, hydraulic working valves, and many other specialties.

ROCK DRILLS.—An artistically produced catalogue of "Chicago Giant" rock drills and kindred appliances, is being sent to the trade by the Chicago Pneumatic Tool Company. The book is printed in colors on high grade paper and contains ninety-six pages of matter referring to rock drills. The text is well written and fully explanatory, and is embellished with half-tone engravings illustrating the "Chicago Giant" rock drills and views of parts, followed by several pages devoted to rock drill steels, and an interesting description of the method of lubrication used in the "Chicag o Giant," one of the distinguishing features of the drill. Several pages are devoted to Franklin Air Compressors, another of the company's products, followed by illustrations and descriptions of the "Baby Giant" or one man rock drills, and scenes at work. Catalogue No. 22 is the title of the book.

Copies will be forwarded upon request by addressing the Chicago Pneumatic Tool Company, Fisher Building, Chicago, or 95 Liberty Street, New York, N.Y.

BINDING MACHINE.—We have received from Wm. O. Greenway, manufacturers' agent, 13 St. John street, Montreal, a ten-page circular, printed on fine coated paper, describing and illustrating the "Kennedy" patent binding machine and the work it does. This machine is used by general engineers, electrical engineers, gas fitters, plumbers, etc., for binding all kinds of pipe and iron.

THE 16TH EDITION of Thomas Noakes & Sons' catalogue and price-list of London made steam-fittings for marine, locomotive and stationary engines has just been published, and a copy is received. We have no previous edition to compare this with, but we compliment the publishers on this attractive list, which is conveniently compiled and neatly printed on excellent stock. From cover to cover it runs upward of 123 pages. On application to 4 and 5 Osborn Place, Brick Lane, London, E., England, a copy may be obtained by any subscriber to this paper.

SCREW CLAMPS—A little illustrated booklet, issued by the Jas. L. Taylor Mfg. Co., Bloomfield, N.J. This booklet fully describes, with price lists, the different clamps made by this firm.

PUMPING ENGINES — A handsome catalogue, issued by the Allis-Chalmers-Bullock, Ltd., Montreal, describing with clear illustrations the different types of pumping engines manufactured by this firm. It is gotten up as a souvenir of the American Waterworks convention, held this year in Toronto, and contains Bulletin No. 1,609, which gives, with illustrations, an official report of the tests recently made on three 20,000,000-gallon pumping engines, manufactured by this company, and installed at Bissell's Point pumping station, St. Louis. A complete description of this plant is given, and the book is well bound, and very neat throughout.

STEEL SPECIALTIES—A neat little catalogue, issued by the Cleveland Wire Springs Co., Cleveland, Ohio, illustrating and describing their patented steel specialties.

BOOK REVIEWS.

MODERN AMERICAN LATHE PRACTICE. A practical work on the American lathe, by Oscar E. Perrigo, M.E.; completely illustrated, 424 pages; price $2.50; published by the Norman W. Henley Publishing Co., New York, N.Y.

The author in this volume has treated in a very comprehensive way the history and development of the lathe from its earliest form up to present-day practice; he has briefly discussed its effects upon manufacturing, and has described its practical use for the various classes of work for which it is suitable; and he has also compared from a practical standpoint the different lathes as made by the several American firms at the present day. The contents of the book may be summed up as follows: The American lathe, its origin and development, its design, its various types, including engine lathes, heavy lathes, high speed lathes, turret lathes, and special lathes; the application of the electric motor to the lathe; lathe attachments; lathe work; lathe tools; rapid change gear mechanism; speeds and feeds; power for cutting tools; and lathe testing.

This work will be of exceptional interest to anyone who is deeply interested in lathe practice, as one very seldom sees such a complete treatise on a subject as this is on the lathe. It is a book which is easily read, as it is altogether descriptive, and the numerous illustrations make this descriptive work easy to follow.

AMERICAN PATTERN SHOP PRACTICE.—by H. J. McCaslin. Published by The Frontier Co., Cleveland, O. Sold in Canada by The MacLean Publishing Co. Price, $3.00.

The writer of this book has treated his subject in a somewhat different manner from that adopted by other writers on the subject, the book being for the greater part a record of work with which the author has come into personal contact. The book is divided into six sections, each section dealing with a particular branch of the subject. These sections are: engine patterns; molding and cores; sweep work; gearing; representative patterns; hints; suggestions and rigs.

The chief point of difference in this book compared with others on the same subject is that the molding and core-making as well as the pattern-making has been considered in each case. This makes the discussion very complete.

Special attention has been devoted to the more intricate and important parts of pattern-making, such as core-box work.

LOCOMOTIVES, SIMPLE, COMPOUND AND ELECTRIC.—By H. C. Regan, locomotive engineer; fifth edition, revised and enlarged. John Wiley & Sons, New York. Sold in Canada by The MacLean Publishing Co., Toronto. 932 pages, 8x5½. $3.50.

This is the fifth edition of this well-known work, revised to include the latest developments in steam and electric locomotives. The additional matter in regard to steam locomotives includes a full description of the principle of compounding, including the balanced four-cylinder compound, and the steam superheater. There is a complete discussion of the construction and operation of electric locomotives, including the single phase system, using single phase motors, the polyphase system; using the induction motor; the three phase system, using rotary converters, and direct current motors, and the three-wire direct current system.

The discussion and descriptions throughout the book are practical in character; and practically no theoretical discussions are entered upon. This book is a most complete one on its subject.

IRON AND STEEL MANUFACTURE. The principles and practice of iron and steel manufacture, by Walter Macfarlane; illustrated; 266 pages, 8 x 5 inches; published by Longmans, Green & Co., New York: price, $1.00.

The book is designed to be of value to students in metallurgy, for metallurgists and for engineers. The subject has been treated by considering first the finished material and then the ores, the author's experience showing this to be the best system. Each chapter is self contained. Several chapters have been revised by experts in the subject treated.

The contents of the book include the treatment of the puddling process, crucible steel, tool steel, Bessemer and Siemens processes of steel making, foundry practice, iron and steel castings, malleable castings, case hardening, blast furnace practice from ores and their composition to finished pig iron.

INDUSTRIAL AND TECHNICAL EDUCATION. Report of the Commission on Industrial and Technical Education submitted to the Senate and House of Representatives of Massachusetts; published by Columbia University, Teachers' College, New York.

The volume contains a detailed report of its work in connection with the investigation of the condition of technical and industrial education holding in the State of Massachusetts. Public hearings were held in all the important industrial centres, and much valuable information concerning of all classes of industries was gathered. The question in the report was considered from two standpoints, that of the industries and the adult man on the one hand, and the child on the other.

Importance of Advertising

Engineering advertisements should be written by men well acquainted with Engineering and Mechanics.

Editor, Canadian Machinery and Manufacturing News,
Toronto, Canada.

Dear Sir,—May I, through the valuable columns of your paper, call the attention of those manufacturers, who are interested in engineering industries, to a very important feature of their business, namely, the sales department, which has incorporated with it their publicity department ?

When referring to this department, it is perhaps necessary to show from what sources, and in what percentage, the total volume of business originates ; and I think it is generally believed that 50 per cent. of the total business originates from personal solicitation ; 25 per cent. from advertising, directly and indirectly, and the balance cannot be accounted for, but which is closely allied with the former.

Advertising space being in such constant demand, numbers of publications have sprung up, and the advertisements are simply put there for some reason or another, best known to the manufacturer, without very much attention being paid to the design and the information which the advertisement ought to contain.

As a reader depends a great deal upon the advertising columns for information, it ought to be that this information should be more specific, instead of simply referring to a trade mark or a heavy cut of a particular machine or appliance, and as an engineer's knowledge of a machine depends upon the information he can get concerning it, in many cases, he has to depend upon the arguments of one manufacturer, because he has neither time nor the inclination to delve through large quantities of literature which relate collectively to the subject in question.

Regarding catalogues and literature, there is much room for improvement in this respect, as on account of the great competition that exists, a manufacturer must embody in his literature something more than the superlative term, and as his time is so taken with other features of his business, it remains for him to employ the best talent to explain and illustrate his particular products.

In view of the large expenditure a manufacturer is obliged to make in this direction, he ought to make the most of his space and literature, and these sources of publicity should be part of a complete treatise or work explaining his products.

My experience in many parts of the world leads me to believe, that in many cases, advertising is looked upon as so much of a speculation, which may or may not pay, but no one can tell whether it does or not ; so why bother ; and catalogues and literature looked upon as the secret of their business.

If, instead of trying to outbid each other for the heaviest type in the advertising columns, a manufacturer would try and show his prospective customers by a lucid explanation, how they could benefit by using his machine or appliance, it would have a tendency to increase his sales.

The time has come when engineering advertising ought to become a special business itself ; and it is in no way to be allied with general advertising, except from an artistic point of view, and those people who handle it should be engineers of experience, whose ideas do not run on caricatures, or those things which are not consistent with good business, as it is not possible for a man to explain and illustrate the merits of a farinaceous food one day, and the next, a semi-automatic capstan lathe.

Many large concerns have their own publicity department, which is an excellent thing, but the idea should always be borne in mind, that it is necessary to be a specialist in the business you are thinking of becoming advertising expert for, before you take up advertising; although in many cases, the applicant is an advertising man first, and learns the business afterwards.

As advertising, direct and indirect, is responsible for such a large amount of business, why cannot it be more valuable, and be the means of reducing the expense of personal solicitation ?

Respectfully yours,

ALGERNON LEWIN CURTIS,
3706 Spruce St., Philadelphia,
Pa., and Chatteris, Eng.

The "ad." is mightier than the cut price.

Industrial Progress

CANADIAN MACHINERY AND MANUFACTURING NEWS would be pleased to receive from any authoritative source industrial news of any sort, the formation or incorporation of companies, establishment or enlargement of mills, factories or foundries, railway or mining news, etc. All such correspondence will be treated as confidential when desired.

Building Operations.

A new armory will be erected at Belleville, to cost $90,000.

A new school will be erected at Port Arthur, Ont., to cost $30,000.

A new school will be erected at Nelson, B.C., at a cost of $60,000.

The Smart Bag Co. will erect a factory on Logan avenue, Toronto.

The General Brass Co., Toronto, will build a factory to cost about $9,000.

The Alberta-Pacific Elevator Co. will build a large elevator at Calgary.

A new reference library, to be erected soon in Toronto, will cost $260,000.

The Massey-Harris Co. will erect a warehouse at Saskatoon, to cost $50,000.

The Canadian Pacific Railway Co. will erect a new depot at Treherne, Man.

The British-American Oil Refineries, Ltd., will establish a refinery in Vancouver.

New car construction works are likely to be erected shortly in Victoria, B.C.

A new steel bridge is to be built over the Pembina river at Snowflake, Man.

Rhodes, Curry & Co. have added a rolling mill to their works at Amherst, N.S.

The G. W. Green Co., Peterboro, have added a large molding shop to their plant.

The lumber mill of W. E. Laking, Cloverdale, B.C., has been destroyed by fire.

The Corby distillery was destroyed by fire during the month, at a loss of $250,000.

A new Roman Catholic church will be erected at Peterboro at once, to cost $40,000.

The Sylvester Bros.' Mfg. Co., Lindsay, Ont., will erect a warehouse at Brandon, Man.

A new Wycliffe College will be erected on Hoskin avenue, Toronto, to cost $280,000.

The saw mill of J. O. Gilbert & Son, Bishop's Crossing, Que., has been destroyed by fire.

A new Roman Catholic Cathedral, to cost $200,000, will be built at Hamilton, Ont.

A one-hundred-barrel flour mill will be erected at Rosthern, Sask., at a cost of [illegible].

Mayor Smith, Regina, Sask., will erect a business block there, at a cost of $40,000.

A new threatre will be erected on St. Catherine street, Montreal, to cost about $68,000.

A bonus has been granted John Hayne & Co. to locate a stair factory at Marmora, Sask.

W. P. Housen, Lethbridge, Alta., will erect a saw-mill on the Belly river, to cost $50,000.

A post office and customs house is being erected at Owen Sound, Ont., at a cost of $75,000.

John Leckie, Toronto, will erect a four-storey warehouse, on Wellington street, in that city.

The Massey-Harris Co. will erect a warehouse in Saskatoon this summer, at a cost of $45,000.

A branch of the Bank of Nova Scotia will be erected at Windsor, N.S., at a cost of $15,000.

The Grand Trunk Railway will erect new freight offices and freight sheds at Brantford, Ont.

The Manitoba Linseed Oil Mills, Ltd., are likely to erect an $80,000 plant at St. Boniface, Man.

H. E. Bond & Co., Toronto, are to erect a five-storey warehouse at Toronto, at a cost of $80,000.

Brown & Co., contractors and builders, St. Marys, intend to erect a box factory in that town.

The ratepayers of Carman, Man., will vote on a by-law to raise $16,000 for a waterworks system.

Messrs. McPherson & Wilband, formerly of Vancouver, are erecting a saw-mill at Abbotsford, B.C.

The wood-working factory of G. S. Eddy, Bathurst, N.S., has been destroyed by fire. Loss about $15,000.

The saw mill of Davis & Kennedy, at Lindsay, has been totally destroyed by fire, at a loss of $15,000.

The Dominion Coal Co.'s output for May totalled 328,917 tons, and the total shipments 279,374 tons.

Substantial enlargements are shortly to be made to the Pere Marquette Railway shops at St. Thomas.

MacGregor & MacIntyre will erect a one-storey iron works on Pearl street, Toronto, to cost $6,000.

The saw-mill of H. C. McCall, Wooler, Ont., has been completely destroyed by fire, with great loss.

J. R. Booth will add a $75,000 sulphide and paper board mill to his big mills at the Chaudiere, Ottawa.

The Construction & Paving Co., Toronto, will erect a two-storey asphalt paving factory, to cost $12,000.

The grist and lumber mills of Jules Patry, near Papineauville, Que., have been completely destroyed by fire.

The warehouse of the Imperial Varnish Co., Toronto, was recently damaged by fire to the extent of $1,000.

Fire lately did damage to the extent of $250,000 to the plant of the Canadian Shipbuilding Co., Fort Erie.

The Merchants Bank of Canada will erect a large building on the present site of the Palace hotel, Vancouver, B.C.

The Bell Telephone Co. have decided to install an entire new plant in Chatham, and place their wires underground through the city.

The Canadian Locomotive Co., Kingston, will spend $300,000 on extensions to their plant during the coming year.

Tenders have been asked for the erection of five steel bridges in district "B" of the Transcontinental Railway.

Fire recently destroyed the grist and saw mill owned by Fred. C. Barrett, near Strathroy. Loss, $8,000.

John Inglis & Co. will build a one-storey galvanized iron blacksmith shop on Strachan avenue, Toronto.

Fire broke out in the elevator shaft of the Smith Table Co. plant, at Mohawk, N.Y., causing $100,000 damage.

Damage by fire to the extent of $1,750 was done to the furniture factory of A. M. Souter & Co., Hamilton.

The wood-working factory of Wm. Gardhouse, Carman, Man., was recently destroyed by fire. Loss about $4,000.

A new paper mill has been added to the plant of the pulp mills which have been operating at Milton Falls, N.S.

The Dominion Transport Co., Toronto, will erect a double-decker steel and concrete stable, to cost about $40,000.

The American Steel Mine Co., Cleveland, Ohio, will erect a factory near Lethbridge, Alta., at a cost of $2,000,000.

A large factory and molding shop, for the manufacture of furnaces, will be erected soon by R. Bigley, at Toronto.

Plans have been prepared for new city hall and market buildings to be erected at Brantford, to cost about $70,000.

Fort William will vote on by-laws to raise $94,000 for the extension of telephone, electric light and water systems.

The woodworking factory at Bathurst, N.B., owned by Geo. S. Eddy, suffered by fire recently to the extent of $15,000.

The Carriage Mountings Co., Ltd., Toronto, has commenced operations on new factory buildings at Niagara Falls, Ont.

A new shingle and lumber mill is to be erected soon by J. W. Davis & Son, on the Fraser river, near New Westminster.

The congregation of the Mill Street Presbyterian church, Port Hope, will erect a new edifice, to cost about $6,500.

The Board of Education, Sydney, C.B., are about to erect a new high school building at that place, to cost $40,000.

A new business block, three storeys high, will be erected by the Toronto Savings and Loan Co., in the city of Peterboro.

The factory of the Malone-Manning Wood Fibre Co., Owen Sound, which was recently destroyed by fire, will be rebuilt.

The sash and door factory of the Tobin Mfg. Co., of Bromptonville, Que., suffered loss by fire recently to the extent of $65,000.

Tenders will shortly be called for the erection of a five-storey addition, 140x83 feet, to the Agricultural College, Winnipeg, Man.

The Maritime Mfg. Co., clothing manufacturers, Pugwash, N.S., will erect a three-storey brick factory, 75x40 feet, at that place.

The Belleville Hardware Company's box manufacturing plant, at Belleville, was this month damaged by fire to the extent of $12,000.

A plant, for the manufacture of cement brick has been established in New Westminster, by W. R. Gordon, and is now in operation.

Grey & Hedley, Guelph, and the Stanley Smelting Works are to erect smelting plants shortly on Cataraqui Bay, near Kingston.

The Toronto Engraving Co. are to erect a five-storey office building, at the corner of Adelaide and Duncan streets, to cost $30,000.

A branch factory of the American Machinery and Export Co., saw-mill machinery manufacturers, will likely be erected at Windsor, Ont.

The Minerva Mfg. Co., Ltd., Toronto, are erecting a five-storey mill construction building, 150x60 feet, on Bathurst street, in that city.

The Golden West Soap Co. have purchased a site in Burrard Inlet, Vancouver, B.C., upon which to erect a factory, at a cost of $150,000.

The Colonial Investment and Loan Co., having purchased a site on King street, east, Toronto, will erect a 12-storey office building there.

The J. I. Case Mfg. Co., of Racine, Wis., are reported to have issued bonds to erect factories in Fort William, Ont., and St. Petersburg, Russia.

It is the intention of the Provincial Government to proceed immediately with the construction of a telephone system in the city of Brandon.

Two new fire halls will be erected at Edmonton, Alta., at a cost of $41,000. Each hall is to have a chemical engine and hose wagon and ladders.

The town of Parry Sound is submitting a by-law to the people to grant a bonus of $100,000 for the establishing of a smelter at Parry Sound.

The Sarnia Board of Trade are considering a proposition from United States capitalists to establish a large pipe and tube industry at that place.

The St. Charles Condensing Co., Ingersoll, propose establishing a branch of their business at Tillsonburg, if satisfactory conditions can be secured.

A syndicate of Canadian and United States capitalists are reported to have decided on establishing a steel plant at Walkerville, to employ 500 hands.

D. G. Loomis & Sons, contractors, Montreal, will establish a large brick-making plant at Cote St. Paul, Montreal, on property just purchased for $76,168.

The Brunette Saw Mills Company's offices at Sapperton, New Westminster, together with the store and post office, were recently destroyed by fire, at loss of $500,000.

The Niagara Falls Machine and Foundry Co., Niagara Falls, Ont., are contemplating building new premises, and making additions to their plant and equipment.

Somerville, Ltd., manufacturers of plumbers' and steamfitters' supplies, are about to erect a new brass foundry at Toronto, to cost in the neighborhood of $200,000.

The International Heating & Lighting Co., Cleveland, Ohio, have secured a site at Portage la Prairie, Man., and will commence the erection of a gas plant there.

The Walkerville Land and Building Co., Walkerville, have let the contract to V. Williamson for the erection of three brick blocks, composed of four houses each.

The city council of Sarnia, Ont., has ratified, provisionally, a by-law to advance $10,000 and a free site to the Standard Chain Works, of Pittsburg, to locate there.

The entire plant, together with the finished stock, of the Canada Woodenware Company, Ltd., Hampton, N.B., has been recently destroyed by fire. Loss, $40,000.

The Temiskaming Railway Commission have awarded to O'Boyle Bros. Construction Company the contract for the railway stores building to be erected at North Bay.

The Clark Foundry Company will build new

foundry works at Sydney, C.B. The building will be of steel and concrete, and will be equipped with the most modern machinery.

The Victoria Gas Co., Victoria, B.C., intend making considerable extensions to their plant this season. The company now has 20 miles of mains, and 5 more are to be added.

The contract for the construction of the entire plant of the National Rolling Mills Co., Sydney, C.B., has been awarded E. A. Walberg. The expenditure will exceed $500,000.

Williams & Wilson, Montreal, are building a large addition to the premises on St. James street, Montreal. The new building will be 53x100 feet, four storeys and basement.

The Nova Scotia Steel and Coal Co. are erecting a new forge building, 240x75 feet, which will be fitted with the best machinery necessary for the production of 200 axles per day.

The Clark Foundry Co. will build a new foundry works at Sydney, C.B. The building, which will be of steel and concrete, will be equipped with the most modern machinery.

The International Heating and Lighting Co., Cleveland, O., will start, at once, on the installation of their plant in Portage la Prairie. They will spend some $10,000 on this plant.

Application is now being made by A. M. Fraser for approval of plans for a 14-storey structure, to be erected on Main street, Winnipeg. The building will be 170 feet in height.

Wilson & Co., of Petrolea, oil experts, have leased about 1,000 acres of land in Caradoc township, about eight miles from Strathroy, and already have an outfit at work near Longwood.

The Cranbrook Electric Light Co. has been organized, and will build a dam and power house on the St. Mary's river, for the purpose of supplying the town with electric light and power.

Work will be commenced immediately on the erection of the brass and iron foundry works of Shaw & Mason, Ltd., at Sydney, C.B. The contract price is said to be in the vicinity of $10,000.

The Cutknife Trading & Realty Co., North Battleford, Sask., will erect a sand-lime brick plant at a cost of $35,000. Besides sand-lime brick, the plant will turn out artificial Indiana sandstone.

The excavation for the new Frontenac gas works at St. Malo, Que., has commenced and the work is being rapidly pushed to completion. Mr. Emmerson McMillan is president of the new company.

Lacombe, on the Calgary and Edmonton line of the C.P.R., has a building boom on its hands. Over $200,000 worth will be undertaken this year. A number of fine business blocks are now going up.

A large furniture factory, for the manufacture of case goods, is being built in Peterboro, and expect to begin operations at once. The directors are, T. W. Oke and E. J. Holcomb. The capital is $10,000.

The Peter Hamilton Mfg. Co., of Peterboro, have suffered a loss by fire to the extent of $10,000. The paint shop and upper storeroom were completely destroyed. It is the intention to rebuild immediately.

Chatham citizens have subscribed the necessary amount for the building of the lake extension of the Chatham, Wallaceburg and Lake Erie Electric Railway. The road is controlled by American capitalists.

The Colonial Weaving Co., Peterboro, propose making substantial enlargements to their plant. The city council has agreed to give them a site adjacent to railway siding, subject to a vote of the people.

On the recommendation of the Minister of Education, the Government has decided to grant $5,000 towards the establishment of a technical school at Sault Ste. Marie, Ont. The new building will cost about $20,000.

J. H. G. Russell, architect, Winnipeg, is asking tenders for an addition to the warehouse of the G. F. Stephens Co., in that city. The addition will be of brick, with stone foundations, and four storeys high.

The new plant of the Canada Nut Co., Toronto, which is being built on the Toronto, Niagara and St. Catharines Railway, just out of Niagara Falls, Ont., is well under way. It is being built entirely of concrete.

The new plant of the Rapid Tool Co., Ltd., Peterboro, is completed, and has been in operation a month. The building is equipped with the latest machinery for the manufacture of augers and all wood-working tools.

The C.P.R. will shortly make large additions to their shops at Winnipeg. The new buildings will cost 100,000, and will include a power house, car shops, planing mill, new tender shops, pattern shops and a foundry.

The plant of the Canadian Ramapo Iron Works, Niagara Falls, Ont., is now nearing completion. The machinery is now being installed, and they are delayed by the non-appearance of the electrical equipment.

Montreal is to have another theatre, to be opened next fall. The new theatre will be situated at the corner of St. Urbain and St. Catherine streets, and will be a modern fireproof structure, capable of seating 1,800 people.

Permits have been issued by the city architect for the erection of a three-storey building by the Horological Institute, Toronto, to cost $8,500, and for a two-storey brick warehouse, to be built for the Imperial Glass Works, to cost $14,000.

A by-law to grant a loan of $20,000 to the Wolverine Brass Goods Mfg. Co., Grand Rapids, has been carried by the ratepayers of Chatham. This company will establish a large plant there, to be in running order by January, 1908, and to employ 100 men.

The Canadian Distilleries Co. is being formed by the Goodernams, the Walkers, Weiser, and other large Canadian distilleries, to erect a distillery at Winnipeg. It will be one of the largest in Canada, and will call for the expenditure of $200,000.

The Peterboro Boiler and Radiator Co.'s factory at Hastings, Ont., is fast nearing completion, and by the middle of July operations will be commenced in the manufacturing line. The machinery is now being installed. Head office is at Peterboro.

Werlick Bros. & Co., manufacturers of pianolas, have purchased the old Preston electric power house. Besides remodelling this building, they will put up quite an addition. They are at present in a part of the building occupied by the Crown Furniture Co.

The Red River Valley Brick Corporation may establish a brick works at Vancouver. It is proposed to erect a plant similar to their present one at Grand Forks, N.D., which has an output of $200,000 pressed bricks daily. Suitable clay has been found.

A big plan is on foot for the establishment of pulp and paper mills in Canada. Five million dollars is the capital, backed by Harmsworth and other Englishmen. Property consisting of fifteen hundred square miles in Exploits valley, Newfoundland, has been secured.

The large reinforced concrete building, which the Canadian General Electric Co. are adding to their plant at Peterboro, is about completed. The machinery and traveling cranes are now being installed. Some three or four hundred additional hands will be required when the work is completed.

The International Harvester Co. have decided to make a large extension to their Hamilton works. An entirely new plant will be constructed, hardly as extensive as the present one, but nevertheless complete and on a large scale. Gas and gasoline engines will be manufactured for the Canadian trade.

There is a proposition before the town of Goderich for the development of electrical energy from the Black Hole on River Maitland, and the building of an electrical road from Goderich north to Lucknow and another branch to Amberley. Some time ago the town passed a by-law granting $15,000 for this purpose.

Shaw & Mason, brass founders, have prepared plans for a frame factory to be erected at Sydney N.S. The floor space will occupy fifteen thousand square feet. Besides the brass foundry there will be an iron foundry and nickel-plating and sheet metal work will be carried on. The machinery will be operated by electricity.

The Stratford Mill Building Co., Stratford, has been awarded the contract for the Medicine Hat Milling Co., for machinery to double their capacity, at a cost of $80,000. The Milling Company is installing a 100-h.p. steam plant and many large machines of the very latest improved type. The first car of machinery is to be shipped not later than the 15th of July. The ratepayers of Medicine Hat voted a loan of $20,000.

Mining Operations.

The Dominion Antimony Co. is about to erect at West Gore a $50,000 mill for the treatment of ore.

The Plum Coulee Milling Co. will erect a large flour mill in Arcola, Sask., to replace the one destroyed by fire last year.

A big discovery of copper has been made on Skincuttle Inlet, in Queen Charlotte Islands, and it is said the discoverers have turned down an offer of $600,000 for the property.

It is thought that the city of Sydney, C.B., will shortly move for the erection of a new building for high school purposes, which will include facilities for a proposed technical school at that place.

The directors of the Superior mine, near Sault Ste. Marie, announce that they will build a short line from their mine, on the Ottawa Central, west of the Soo., to the Algoma Central. The cost will be $75,000.

The iron mines of the west coast of Texada Island, B.C., are to be immediately opened, and within sixty days steady shipments of ore to Irondale smelter will be made. Jas. A. Moore, of Seattle, is furnishing the capital, and work will be started immediately on the construction of a wharf at Gillies' Bay.

Oil of the highest quality has been discovered in the southwestern section of Alberta province. Huge gushers are now flowing from the earth, and J. S. Williams, an oil expert from Summerlands, Cal., who has inspected the district, has stated he believes a huge underground oil lake exists along the eastern slope of the Rocky Mountains.

It is learned on good authority that the big coal concern organized to operate the areas in and around Port Morien, the North Atlantic Collieries, Ltd., has been successfully floated. The leading promoters and directors of this company, which was organized in 1903, are, B. F. Pearson, Halifax; H. M. Whitney, Boston, and G. E. Drummond, Montreal.

Railway Construction.

Edmonton is installing a street railway system.

Large car works will be established at Bridgewater, N.S., by Canadian and New York capitalists.

The Canadian Northern Railway Co. are considering the erection of machine shops at Longue Point, Que.

The G.T.P. Railway are building a steel concrete bridge to span the Saskatchewan river at Saskatoon.

The Michigan Central Railway Co. are building a new subway in St. Thomas, Ont., at a cost of $66,000.

The C.N.R. and G.T.R. officials are about to decide upon final plans and layout for a joint terminal at Winnipeg.

Work on the new tramway from Lac du Bonnet to Point du Bois has commenced. Mr. Wm. Newman is the contractor.

The Cape Breton Electric Co., Sydney C.B., are laying an electric transmission line between Sydney and North Sydney.

The work of building another section of the double track between Fort William and Winnipeg has been commenced by the C.P.R.

The Winnipeg Electric Street Railway will go ahead at once and double-track Portage and William avenues, and also the Belt line.

The city council, Toronto, have decided to erect a crib-work along the lake front opposite the exhibition grounds, at a cost of $45,700.

The Board of Control, Toronto, invite tenders up to July 16th, for the erection of steel railway bridges for the Lansdowne avenue subway.

The Atlantic, Quebec and Western Railway has asked permission to extend its lines from Metapedia to Edmonton, on the St. John river.

It is said the C.P.R. will replace their wooden passenger cars with cars of steel. The Pennsylvania Railway Co. are placing 2,000 such cars on their line.

The Transcontinental Railway Commission is calling for tenders for the construction of eight miles of the line in New Brunswick, running easterly from Chipman.

Considerable additions are being just now made to the rolling stock of the Grand Trunk, in the delivery of some five thousand freight cars of various kinds.

Contracts for the construction of 6,000 steel cars, costing $6,500,000, have been placed by the United States Steel Corporation, with the Pressed Steel Car Co., the Standard Steel Car Co., and the American Car and Foundry Co.

Prompted by a demand for better steel rails that will make railroading safer, officials of the Carnegie Steel Co. will operate open-hearth furnaces exclusively at the Braddock, Pa., plant. Two new open-hearth furnaces are now being erected.

The Grand Trunk Pacific Railway Company will this month begin the construction of the first of its lines in British Columbia. The line on which the work will be started will be a branch that will tap the main line half way between Hazelton and Prince Rupert, and will run southward to Kitamaat, on the coast.

Announcement is made that the deal, with a syndicate of Pittsburg capitalists, for the control of the Brantford Street Railway, the Grand Valley Railway, and the Woodstock & Ingersoll Railway, has gone through. It is expected to reconstruct the present line to Paris and Galt, and also to put on a better service. The total mileage eventually will be 125 miles, and the sum involved is some $3,000,000.

Miscellaneous.

Jean Gauthier, contractor, Montreal, has assigned.

A waterworks system will be installed at Reliance, Alta.

An electric light system will be installed at Summerlea, Que.

A post office will be erected at Knowlton, Ont., at a cost of $5,000.

Welland, Ont., has decided to spend $70,000 on a new waterworks system.

Calgary endorsed a by-law recently to raise $125,000, to install municipal 'phones.

Vancouver, B.C., will spend about $750,000 on the extension of its waterworks system.

Messrs. McAra and Mowat, Regina, Sask., will erect an office block, at a cost of $30,000.

The saw-mill of E. Jull, Norwich, Ont., was recently destroyed by fire. Loss about $2,000.

The Amherst Foundry Co., Ltd., suffered a loss by fire last month to the extent of $75,000.

The Western Quebec Milling Co. has been registered in Laprairie, to do a general milling business.

The Canadian Pacific Railway Co. will build a line from Hamilton, Ont., to Guelph Junction, Ont.

The premises of the Union Abattoir Co., Montreal, were destroyed by fire recently, at a loss of $100,000.

The Brewing and Malting Co., Lethbridge, Alta., will erect an addition to their plant, to cost $35,000.

Tenders have been asked for the installation of a complete new fire alarm system in the town of Welland.

The Robert Bell Engine & Thresher Co., Seaforth, Ont., will erect a large warehouse at Moose Jaw, Sask.

The Canadian Fairbanks Co. have opened a branch office in Quebec. R. Miquelon has been appointed manager.

A copper mine has been discovered near Woodstock, N.B., about one hundred yards from the St. John river.

Fire did damage to the extent of $2,000 to the Victoria Foundry at Ottawa, and the plant has been closed down.

The new factory of the London Fence Co., Portage la Prairie, Man., has been damaged by fire to the extent of $75,000.

The limber mill of Coultier & Irish, at Carthage, Ont., together with all the machinery, has been destroyed by fire. Loss, $15,000.

The lumber mill of B. Grier & Co., and the tannery of the Bonner Leather Co., Montreal, were destroyed by fire. Loss, $35,000.

A large power house will be erected in connection with the Canadian Pacific Railway hotel at Victoria, B.C., to cost about $40,000.

The city engineer, Windsor, Ont., is calling for tenders for the construction of 115,000 feet of cement walks, to be laid in that city.

The Renfrew Power Co. have increased the capacity of their power plant from 400 h.p. to 1,000 h.p., at an additional cost of $30,000.

The Niagara Falls Machine and Foundry Co., Niagara Falls, Ont., have a large contract on hand from Toronto and Hamilton for hydrants.

The ratepayers of Sherbrooke, Que., are to vote on a by-law to raise $200,000, for the purpose of developing water power at Westbury, Ont.

D. Conroy, Peterboro, Ont., has been awarded the contract for the construction od the new bridge and dam at Buckhorn, Ont., to cost about $56,000.

Tenders for lighting by electricity, the streets of the city of Hamilton, for five or ten years, will be received by the city clerk, up to noon, July 2nd, 1907.

The city of Peterboro is negotiating with the Light and Power Co., of that place, to purchase its extensive business, which also includes the gas works.

The S. W. & A. Railway Co., Windsor, Ont., have decided to build a power house next to the plant of the Canadian Salt Co., and use exhaust steam from the salt works.

The Niagara Engine Works, Ltd., Niagara-on-the-Lake, having made an assignment, the plant and property will be offered for sale, subject to a reserve bid, on July 18th, 1907.

Negotiations are being carried on between Windsor, Ont., and the American Machinery & Export Co., regarding the establishment of a Canadian plant in Windsor by that company.

The ratepayers of Campbellford, Ont., have voted favorably on a by-law to loan Messrs. Jas. and Geo. T. Dickson, $15,000, to aid them in establishing a bridge works in that town.

The Hamilton Cataract Light and Power Co. have filed plans with the Ontario Railway Board for a transmission line from Hamilton, passing through Barton and Salt Lake townships.

A syndicate of Amherst and Antigonish men acquired control of timber limits in Newfoundland and Labrador, comprising 450,000 acres. The purchase price was a quarter of a million dollars.

The Supreme Heating Co. have started active operations on their plant at Welland. The main building, 202x70 feet, will be commenced immediately, and the other building will be added later.

The Manitoba Rolling Mills have commenced operations in their Winnipeg plant. The president of the corporation is Mr. L. A. McElroy, also president of the United States Horseshoe Co., Erie, Pa.

The Radcliffe Lumber Company, Duluth, have recently obtained extensive timber limits from the E. & N. Railway in the Alberni district, B.C. The company will build one of the largest saw-mills on the coast on the Alberni canal.

Coal has recently been discovered in the vicinity of Saskatoon, Sask., and L. G. Colder, of that town, has just purchased the necessary machinery at Minneapolis for the purpose of properly investigating as to the value of the new find.

The Northern Engineering Works, of Detroit, Mich., have recently furnished the General Electric Co., Schenectady, N.Y., foundry department, with a No. 84 Newten cupola, with stack 74 feet high. This cupola will have a capacity of 16 tons per hour.

The iron and steel bounties paid by the Dominion Government for the nine months' period to the 1st of April totalled $1,299,801, made up as follows: On pig iron $385,231; on steel ingots, $575,259; on puddled iron bars, $313, and on manufactures of steel, $338,998.

The city of Vancouver has just purchased from the Seagrave Co., of Walkerville, Ont., the first automobile fire apparatus to be used in Canada. The outfit consists of two hose wagons and a double 60-gallon chemical engine, all of which will be in automobile form.

The furnaces of the Dominion Iron and Steel Co., Sydney, C.B., poured last month some 28,128 tons, an increase of 5,125 tons over the best previous record. To do this 907 tons had to be poured daily, which indicates the large and steady demand now made for pig iron.

A great shortage of labor exists at Fernie, B.C., where it has become impossible to obtain men for outside work. Development work in the coal properties is being seriously retarded; the mills cannot get men for the busn and the railroads are exceedingly short of laborers.

The Dominion Power and Transmission Company have filed plans with the Ontario Railway and Municipal Board, for a new transmission line from Hamilton to Decew Falls, Ont. It is said the new line will be completed next winter. The line will be used only as an auxilliary, and when anything is wrong with the other two lines, it will be called into requisition.

Among the orders received by the Robb Engineering Co., Amherst, for their celebrated engines are: N. & M. Smith, Halifax; Columbia River Lumber Co., Golden, B.C.; Temiskaming and Northern Ontario Railway; Canadian General Electric Co.; Canadian Westinghouse Co., Hamilton; Somerville, Ltd., Toronto; W. P. McNeil & Co., New Glasgow; Canadian Rand Drill Co., Sherbrooke.

Among the orders received by Darling Bros., Ltd., Montreal, for their Webster Feed Water Heater, Purifier and Filter, are: 2,500 horse-power machine for J. R. Booth, Ottawa; 2,500 horse-power machine for Dominion Coal Co., Glace Bay; 2,000 horse-power machine for T. Eaton Co., Toronto. This is the fourth order this firm has received from the Dominion Coal Co., all the machines being over 800 horse-power.

In the new pipe foundry of the United States Cast Iron Pipe and Foundry Co., Burlington, N.J., four 5-ton special electric Northern jib cranes have been installed, two of them of the column type and two of the transfer jib type. These cranes will assist the overhead traveling cranes, and they are of high speed design, especially suited to pipe foundry service. This company also installed a 20-ton electric Northern traveling crane.

The Scottdale plant of the U.S. Cast Iron Pipe and Foundry Co. have recently added four No. 84 Newten patent cupolas to the melting equipment of their pipe foundry. This gives them in all six cupolas of this make. Each cupola has a stack of approximately 57 feet high and an average capacity of about 16 tons per hour, giving approximately 96 tons hourly capacity. These cupolas were furnished by the Northern Engineering Works, Detroit, Mich.

Among the orders received by the Smart-Turner Machine Co., Hamilton, for their standard duplex pumps, are: R. H. Buchanan & Co., Montreal; Canadian Canners, Belle River; Dominion Heating & Ventilating Co., Hespeler; Robb Engineering Works, Amherst; Stoney Lake Navigation Co., Young's Point; Messrs. Donaldson & Wark, Hamilton; Brompton Pulp and Paper Co., East Angus, P.Q.; Canada Tin Plate Co., Morrisburg; Doty Engine Works, Goderich; the Corporation of Hamilton, Ont.; Messrs. West & Peachy, Simcoe.

* * *

English Gauge Glasses.

Baxter, Paterson & Co., Montreal, have secured the agency of the "invicta" and "Adamant" gauge glasses, manufactured by the S. C. Bishop Co., St. Helens, England. These glasses are tested to 5,000 pounds hydraulic pressure. The special feature of these glasses is the fused ends, which give them a greater power of endurance under high pressure.

* * *

Krugg & Crosby to Make Motor Boats.

Krugg & Crosby, Hamilton, are now making gasoline motor boats, under the name of the Guarantee Motor Co. They are making a 2½ and 5 horse-power single cylinder, and a 5 and 10 horse-power double cylinder engine. These engines are reversible, and have a special guarantee.

They are also making a specialty of engine cylinder castings.

* * *

Fire at Greey's Foundry.

The plant of Wm. and J. G. Greey, Toronto, was recently damaged by fire to the extent of $10,000, the east section of the machine shop being destroyed. Besides the damage done to the finished stock and to the working machinery, most of which was badly wrecked, considerable loss was entailed by water, which soaked the belting all through the plant. The intention is to rebuild immediately, the loss being covered by insurance.

* * *

Copper Concentrator at the Soo.

Ex-Mayor Frank Parry, of the Michigan Soo.; Geo. Kemp, F. C. Smith, and other capitalists, under the name of the Superior Copper Company, with a capital of $2,000,000, will erect a big concentrating plant, on the Ontario side of the river, at Sault Ste. Marie. The plant, which will have a capacity of 400 tons a day, will be the largest of its kind in Ontario, and is to be followed in the near future by other big industries. The company will also build a railway, five miles long, from Algoma Central to their Superior mine.

* * *

Canada's Trade Report for 1906-7.

The returns just published of the trade of Canada for the twelve months ended March 31, 1906, and the twelve months ended March 31, 1907, show the exports to Great Britain during the twelve months ended March, 1907, amounted to $135,469,430, an increase of $8,930,931 over the exports of the previous twelve months. Exports to the United States totalled $109,772,944, an increase of $20,145,508. Canada's total exports to all countries amounted to $272,206,606, an increase of $32,082,960. Imports from Great Britain for the twelve months ended March, 1907, were, $83,229,256, an increase of $16,089,287, over the previous twelve months. From the United States our imports were $208,721,601, an increase of $39,791,452, for the same twelve months. Canada's total imports from all countries were $340,374,745, an increase of $60,046,036.

* * *

Addition to Safe Works.

J. & J. Taylor, safe manufacturers, 145 Front St. E., Toronto, have just completed and are at present moving into an important addition to their already large plant. The new building is a handsome brick and steel structure 242x60 feet, on cement foundations reinforced by extra heavy concrete piers. The new factory addition, which will cost about $50,000, was necessitated by the growing demand for safes, fire-proof vaults, etc., in keeping with the growth and development of the country. It is of first-class factory construction, one storey with gallery, steel trussed roof with sheet metal outside covering. A forty-ton electric traveling crane is being installed so that the heaviest work may be handled with facility. Power will be supplied by three direct current electric motors running three separate lines of shafting, the latter having been installed. For two years past this company have been running their factory night and day to keep up with orders. About 300 men are employed and 50 more will be required.

Companies Incorporated.

The Provincial Construction Co., Montreal: capital, $20,000; purpose, to carry on contracting business. Incorporators: L. Beauchamp, Montreal, and others.

The Louison Lumber Co., Ltd., Sayabec, Que.; capital, $200,000; purpose, to carry on a lumber business. Directors: A. C. Dutton, Springfield, Mass., and others.

Pacific Coal Mines, Ltd., capital, $6,000,000; purpose, to carry on a mining, milling and reduction business. Incorporatoes, J. S. Lovell, accountant, Toronto, and others.

J. E. Sauve, Ltd., Montreal; capital, $40,000; purpose, to carry on lumber business. Directors: J. E. Sauve, manufacturer; R. Sauve, machinist, and others, Montreal.

Norton Telephone Mfg. Co., Toronto; capital, $40,000; purpose, to manufacture electrical supplies, telephones, etc. Provisional directors: J. E. Day, J. M. Ferguson, A. W. Bixel, Toronto.

Parker Car Heating Co., London, Ont.; capital, $16,000; purpose, to manufacture car heating devices, etc. Provisional directors: T. Parker, J. M. McEvoy, E. J. Dawson, London, Ont.

Snap Co., Ltd., Montreal; capital, $90,000; purpose, to manufacture soaps, chemicals, etc. Provisional directors: T. J. Clark, A. L. Malone, barrister-at-law, and others, all of Toronto.

A charter has been granted a corporation of American capitalists, with $5,000,000 capital, to manufacture and sell the devices invented by John H. Parsons, for the consumption of smoke.

Ontario Copper Co., Toronto; capital, $3,000,000; purpose, to carry on a mining, milling and reduction business. Provisional directors are: J. M. Ewing, A. G. Ross and W. S. Edwards, Toronto.

Farley Phillips Co., Montreal; capital, $45,000; purpose, to manufacture cut glass and similar ware. Incorporators are, G. Phillips, W. Smith, Estelle Coriveau and E. Gill, all of Montreal.

Anderson Logging Co., Ltd., Aberdeen, B.C.; capital, $50,000; purpose, to do general lumbering business, and to take over as a going concern the firm of Strand & Jennierson, of the same town.

The Vulcan Portland Cement Co., Ltd., Montreal; capital, $2,500,000; purpose, to manufacture Portland cement. Provisional directors: G. W. Macdougall, K.C.; L. Macfarlane, and others, Montreal.

Kingston Harness Works, Kingston; capital, $40,000; purpose, to manufacture and sell harness and other leather goods. Provisional directors: T. W. Suddaby, W. H. Acton and Wm. Bennett, Kingston.

Louison Lumber Co., Sayabec, Que.; capital, $200,000; purpose, to manufacture lumber, timber, pulp, etc. Provisional directors: H. H. Brodie, A. W. Cameron, Montreal; A. C. Dutton, Springfield, Mass.

Ideal Oak Leather Co., Toronto; capital, $50,000; purpose, to carry on the business of tanners and manufacturers of leather. Provisional directors are: Arthur Gate, W. H. Smith and Geo. Kerwin, Toronto.

Newbury Canning & Preserving Co., Newbury, Ont.; capital, $30,000; purpose, to carry on a canning and preserving business. Provisional directors: J. L. Heatherington, P. B. Robertson, and C. Rush, Newbury.

Chesley Furniture Co., Chesley, Ont.; capital, $40,000; purpose, to manufacture furniture, railway fixtures, rugs, carpets, machinery, etc. Provisional directors; W. G. Durst, J. Hauser and W. Krug, Chesley, Ont.

The Kingston Harness Works, Kingston; capital, $40,000; purpose, to manufacture harness, leather goods, etc. Provisional directors are: T. W. Suddaby, C. F. Suddaby, Kingston, Ont., and J. Bedrad, Gananoque, Ont.

MacLean Cream Separator Co., Sarnia; capital, $10,000; purpose, to manufacture cream separators. Directors, R. V. LeSueur, L. A. MacLean, D. W. H. Lucas, E. A. McDonald and J. W. Hamilton, all of Sarnia.

Faircloth Art Glass and Decorating Co., Toronto; capital, $40,000; purpose, to manufacture art and stained glass, wall paper, etc. Provisional directors; R. T. Faircloth, J. M. Faircloth and G. A. Pringle, Toronto.

The North American Mineral and Lumber Co., Montreal; capital, $49,000; purpose, to prospect for minerals, and to erect saw mills, etc., in carrying on lumber trade. Directors: H. P. Adams, H. J. Cassard, Montreal.

Tillbury Town Gas Co., Chatham; capital, $40,000; purpose, to prospect for and develop oil and gas wells. Provisional directors: R. E. Kiser, G. W. Holmes, Chatham; J. A. Tremblay and J. S. Richardson, Tilbury.

Dominion Exploration and Development Co., Toronto; capital, $1,000,000; purpose, to carry on a mining, milling and reduction business. Provisional directors are: R. J. Tough, R. D. Moorhead and R. H. Patterson, Toronto.

Laidlaw Lithographing Co., Hamilton; capital, $100,000; purpose, to carry on the business of printers, lithographers, bookbinders, etc. Provisional directors: R. Laidlaw, C. R. McCullough and J. I. McLaren, Hamilton, Ont.

The Glen Hayes Coal Mining and Development Co., Winnipeg, Man.; capital, $1,000,000; purpose, to deal in coal, oil, coke, and other metals. Incorporators: G. A. Bull, R. S. Armstrong, Winnipeg; S. L. Head, Rapid City, Man.

The River Plate Shipping Co., Ltd., Montreal; capital, $150,000; purpose, to carry on the business of a shipping company. Incorporators: C. A. McCollough, New York; W. A. Tait, Arlington, Mass., and G. I. Dewar, Ottawa.

Canadian Logging Tool Co., Ltd., Sault Ste. Marie; capital, $15,000; purpose, to manufacture lumberman's tools, etc. Incorporators are: David Wolf, Grand Rapids; Wm. Latta, Wm. Davis, E. S. Gough, and others, all of Chicago.

The Lake Superior Iron & Steel Co., Ltd., Sault Ste. Marie; capital, $1,000,000; purpose, to manufacture and deal in iron, steel, nickel and other products. The provisional directors are: C. W. Thompson, H. B. Johnson, E. Perryson, and H. Todhunter.

Glen Hayes Coal Mining and Development Co., Winnipeg; capital, $[illegible]; purpose, to mine, sell and deal in coal, coke and other metals. Incorporators are, George A. Bull, R. S. Armstrong, H. C. Hamelin, of Winnipeg, and Sampson Leslie Head, of Rapid City.

Cody Manufacturing Co., Sarnia; capital, $50,000; purpose, to deal in and manufacture all articles of iron, brass, copper, and other metals, and to do a machine shop and foundry business. Provisional directors: A. S. Cody, R. D. Cody and Thos. French, Sarnia.

Portage la Prairie Construction Co., Ltd., Portage la Prairie; capital, $[illegible]; purpose, to manufacture builders' and contractors' supplies and carry on the business of builders and contractors. Provisional directors, Henry Stephens, Wm. Armstrong, Wm. Richardson and Hugh Armstrong, all of Portage la Prairie.

Manufacturers and Manufacturers' Agents

It will pay you to watch our condensed column each month. There are many money-making propositions brought to your attention here. You may find just what you are looking for.

RATES

One insertion—25c. for 20 words; 1c. a word for each additional word.

Yearly rate—$2.50 for twenty words or less, 10c. a word for each additional word.

The above does not apply to notices under the head of "Machinery Wanted." These notices are inserted free for subscribers.

BUSINESS CHANCES.

A GOOD cash paying electric wiring business; a small capital only required; selling out owing to ill health. For particulars apply to Box 31, CANADIAN MACHINERY, Toronto. [8]

MACHINISTS—$5,000 buys at par stock in established manufacturing company, western city; salesmen, machinists, blacksmiths, accountants; here's a chance; assets exceed two hundred thousand and company solid financially; write us if interested. Locators, Winnipeg.

FOR SALE.

BELTING, RUBBER, CANVAS AND LEATHER, Hose, Packing, Blacksmith's and Mill Supplies at lowest prices. N. Smith, 138 York Street, Toronto. (2tf)

CORBETT FOUNDRY & MACHINE CO., Owen Sound, make stacks, tanks, hoppers, etc.; write for prices.

GREY IRON CASTINGS.

WE make all kinds of grey iron castings; give us a trial. Rogers Manufacturing Co., Goderich, Ont.

CASTINGS, machine work of all kinds, the pattern-making undertaken under the supervision of skilled engineers; works outside city; country prices; special terms to the trade. G. P. Wellington & Co., 11 Front Street east, Toronto.

MANUFACTURERS' AGENTS WANTED.

SHEFFIELD manufacturers of tool steel, spring steel, section bars, etc., as advertised on page 22, wish to correspond with reliable Canadian firms with view to sole agencies in certain territories. Address "Newhall," care CANADIAN MACHINERY, 88 Fleet Street, London, Eng.

SITUATION VACANT.

WANTED—Young man well experienced in machinery supply business able to sell and knowing thoroughly all requirements of the trade. Salary, commission and share of profits. Apply Box 32, CANADIAN MACHINERY, Toronto.

SITUATIONS WANTED.

ADVERTISER, shortly emigrating, seeks position as correspondence clerk and accountant. Well up in all details of engineering and allied business; ledger, balancing, shorthand and typewriting; excellent references. Lister, Ipswich Road, Ardleigh, Colchester, Eng.

ASSISTANT GENERAL MANAGER, McGill graduate, desires change of position. Twelve years' practical experience along engineering, machine, foundry and boiler shop lines. Best of references. Address Box 25, CANADIAN MACHINERY, Toronto. [3]

ENGINEER, Scotchman (33), single, all round millwright and draughtsman; been superintending constructional work and 9 years works manager of iron foundries; excellent references; open for anything in line; any part of Canada. Box 330, CANADIAN MACHINERY, 88 Fleet St. E.C., London, Eng. [7]

YOUNG man representing five large American manufacturers desires to make connection with a Canadian machine supply house or hardware firm wishing to open a mill supply department. Box 24M, CANADIAN MACHINERY, Montreal. [7]

Motors and Launches

I have the following new complete launches about ready for delivery:

Power Skiff, 20' by 4'; fine finish, equipped with 2 h.p. Erd Motor, $190.00.

Launch, 22', 4' 6" beam, cedar planking, turtle back deck, torpedo stern, equipped with 7½ h.p. 2 cylinder Erd motor, speed 10 miles an hour, $375.00.

Launch, 25' by 6' beam, cedar planking, quartered oak deck and coaming, equipped with 10 h.p. 2 cylinder Erd motor, speed 11 miles per hour, $550.00.

Erd motors from 2 h.p. to 30 h.p., from $65.00 to $600.00, also one 10 h.p. Buffalo 4 cylinder 4 cycle marine motor, as good as new, $250 00.

One 34 h.p. Trebert, complete with reversing wheel, used only one season, $400.00.

One 6 h.p. Gillies stationary gasoline engine, cheap.

One 18' by 4' 9" clinker built launch, equipped with 2½ h.p. Norwalk engine, launched Sept. 1906, speed 8 miles per hour, a bargain at $225.00.

For further information address

K. A. CAMERON

Marine Architect

345 King St. W. - KINGSTON, ONT.

CANADIAN MACHINERY BUYERS' DIRECTORY

To Our Readers—Use this directory when seeking to buy any machinery or power equipment. You will often get information that will save you money.

To Our Advertisers—Send in your name for insertion under the heading of the lines you make or sell.

To Non-Advertisers—A nominal rate of $1 per line a year is charged non-advertisers.

Acids.

Canada Chemical Mfg. Co., London

Abrasive Materials.

Baxter, Paterson & Co., Montreal.
The Canadian Fairbanks Co., Montreal.
Rice Lewis & Son, Toronto.
Norton Co., Worcester, Mass.
H. W. Petrie, Toronto.
Williams & Wilson, Montreal.

Air Brakes.

Canada Foundry Co., Toronto.
Canadian Westinghouse Co., Hamilton.

Air Receivers.

Allis-Chalmers-Bullock Montreal.
Canada Foundry Co., Toronto.
Canadian Rand Drill Co., Montreal.
Chicago Pneumatic Tool Co., Chicago.
John McDougall Caledonian Iron Works Co., Montreal.

Alloy, Ferro-Silicon.

Paxon, J. W., Co. Philadelphia.

Alundum Scythe Stones

Nort n Company, Worc ster, Mass.

Arbor Presses.

Niles-Bement-Pond Co., New York.

Augers.

Chicago Pneumatic Tool Co., Chicago.
Frothingham & Workman Ltd., Montreal
Rice Lewis & Son, Toronto.

Automatic Machinery.

Cleveland Automatic Machine Co. Cleveland.
National-Acme Mfg. Co., Cleveland
Potter & Johnston Machine Co., Pawtucket, R. I.

Automobile Parts

Globe Machine & Stamping Co., Cleveland, Ohio.

Axle Cutters.

Butterfield & Co., Rock Island, Que.
A. B. Jardine & Co., Hespeler, Ont.

Axle Setters and Straighteners.

A. B. Jardine & Co., Hespeler, Ont.
Standard Bearings, Ltd., Niagara Falls

Babbit Metal.

Baxter, Paterson & Co., Montreal.
Canada Metal Co., Toronto.
Canada Machinery Agency, Montreal.
Frothingham & Workman Ltd., Montreal
Greey, Wm. & J. G., Toronto.
Rice Lewis & Son, Toronto.
Lumen Bearing Co., Toronto.
Mechanics Supply Co., Quebec, Que.
Miller Bros. & Toms, Montreal, Que.

Bakers' Machinery.

Greey, Wm. & J. G., Toronto.

Baling Presses.

Perrin, Wm. R., & Co., Toronto.

Barrels, Steel Shop.

Cleveland Wire Spring Co., Cleveland.

Barrels, Tumbling.

Buffalo Foundry Supply Co., Buffalo.
Detroit Foundry Supply Co., Windsor
Dominion Foundry Supply Co., Montreal
Hamilton Facing Mill Co., Hamilton.
Globe Machine & Stamping Co., Cleveland, Ohio.
John McDougall Caledonian Iron Works Co., Montreal.
Miller Bros. & Toms, Montreal, Que.
Northern Engineering Works, Detroit.
J. W. Paxson Co., Philadelphia, Pa.
H. W. Petrie, Toronto.
Sly, W. W., Mfg. Co., Cleveland
The Smart-Turner Mach. Co., Hamilton.

Bars, Boring.

Hall Engineering Works, Montreal.
Miller Bros. & Toms, Montreal.
Niles-Bement-Pond Co., New York.
Standard Bearings Ltd., Niagara Falls

Batteries, Dry.

Berlin Electrical Mfg. Co., Toronto.
Mechanics Supply Co., Quebec, Que.

Batteries, Flashlight.

Berlin Electrical Mfg. Co., Toronto.
Mechanics Supply Co., Quebec Que.

Batteries, Storage.

Canadian General Electric Co. Toronto
Chicago Pneumatic Tool Co., Chicago.
Rice Lewis & Son, Toronto.
John Millen & Son, Montreal.

Bearing Metals.

Lumen Bearing Co., Toronto.

Bearings, Roller.

Standard Bearings, Ltd. Niagara Falls

Bell Ringers.

Chicago Pneumatic Tool Co., Chicago.
Mechanics Supply Co., Quebec, Que.

Belting, Chain.

Baxter, Paterson & Co. Montreal.
Canada Machinery Exchange, Montreal.
Greey, Wm. & J. G., Toronto.
Jeffrey Mfg. Co., Columbia, Ohio.
Link-Belt Eng. Co., Philadelphia.
Miller Bros. & Toms, Montreal.
Waterous Engine Works Co., Brantford.

Belting, Cotton.

Baxter, Paterson & Co., Montreal.
Canada Machinery Agency, Montreal.
Dominion Belting Co., Hamilton.
Rice Lewis & Son, Toronto.

Belting, Leather.

Baxter, Paterson & Co., Montreal.
Canada Machinery Agency, Montreal.
The Canadian Fairbanks Co., Montreal.
Frothingham & Workman Ltd., Montreal
Greey, Wm. & J. G., Toronto.
McLaren, J. C., Montreal.
Rice Lewis & Son, Toronto.
H. W. Petrie, Toronto.
Williams & Wilson, Montreal.

Belting, Rubber.

Baxter, Paterson & Co., Montreal.
Canada Machinery Agency, Montreal.
Frothingham & Workman, Ltd., Montreal
Greey, Wm. & J. G., Toronto.

Belting Supplies.

Baxter Paterson & Co., Montreal.
Greey, Wm. & J. G., Toronto.
Rice Lewis & Son, Toronto.
H. W. Petrie, Toronto.

Bending Machinery.

John Bertram & Sons Co., Dundas, Ont.
Chicago Pneumatic Tool Co., Chicago.
Rice Lewis & Son, Toronto
London Mach. Tool Co., Hamilton, Ont.
National Machinery Co., Tiffin, Ohio.
Niles-Bement-Pond Co., New York.

Benders, Tire.

A. B. Jardine & Co., Hespeler, Ont.
London Mach. Tool Co., Hamilton, Ont.

Blowers.

Buffalo Foundry Supply Co., Montreal.
Canada Machinery Agency, Montreal.
Detroit Foundry Supply Co., Windsor
Hamilton Facing Mill Co., Hamilton.
Kerr Turbine Co., Wellsville, N.Y.
Mechanics Supply Co., Quebec, Que.
J. W. Paxson, Philadelphia, Pa.
Sheldon's Limited, Galt.

Blast Gauges—Cupola.

Paxson, J. W., Co. Philadelphia.
Sheldons, Limited, Galt

Blocks, Tackle.

Frothingham & Workman, Ltd., Montreal

Blocks, Wire Rope.

Frothingham & Workman, Ltd., Montreal

Blue Printing.

The Electric Blue Print Co., Montreal.

Blow-Off Tanks.

Darling Bros., Ltd., Montreal.

Boilers.

Canada Foundry Co., Limited, Toronto.
Canada Machinery Agency, Montreal.
Goldie & McCulloch Co., Galt.
E. Leonard & Sons, London, Ont.
John McDougall Caledonian Iron Works, Montreal.
Manitoba Iron Works, Winnipeg.
Mechanics Supply Co., Quebec, Que.
Owen Sound Iron Works Co., Owen Sound
H. W. Petrie, Toronto.
Robb Engineering Co., Amherst, N.S.
The Smart-Turner Mach. Co., Hamilton.
Waterous Engine Works Co., Brantford.
Williams & Wilson, Montreal.

Boiler Compounds.

Canada Chemical Mfg. Co., London, Ont.
Hall Engineering Works, Montreal.
Lake Erie Boiler Compound Co., Toronto

Bolt Cutters.

John Bertram & Sons Co., Dundas, Ont.
London Mach. Tool Co., Hamilton.
Mechanics Supply Co., Quebec, Que.
National Machinery Co., Tiffin, Ohio.
Niles-Bement-Pond Co., New York.

Bolt and Nut Machinery.

John Bertram & Sons Co., Dundas, Ont.
Canada Machinery Agency, Montreal.
Rice Lewis & Son, Toronto.
London Mach. Tool Co., Hamilton.
National Machinery Co., Tiffin, Ohio.
Niles-Bement-Pond Co. New York.

Bolts and Nuts, Rivets.

Baxter, Paterson & Co., Montreal.
Mechanics Supply Co., Quebec, Que.

Boring and Drilling Machines.

American Tool Works Co., Cincinnati.
B. F. Barnes Co., Rockford, Ill.
John Bertram & Sons Co., Dundas, Ont.
Canada Machinery Agency, Montreal.
A. B. Jardine & Co., Hespeler, Ont.
London Mach. Tool Co., Hamilton.
Niles-Bement-Pond Co., New York.

Boring Machine, Upright.

American Tool Works Co., Cincinnati.
John Bertram & Sons Co., Dundas, Ont.
London Mach. Tool Co., Hamilton.
Niles-Bement-Pond Co., New York.

Boring Machine, Wood.

Chicago Pneumatic Tool Co., Chicago.
London Mach. Tool Co., Hamilton.

Boring and Turning Mills.

American Tool Works Co., Cincinnati.
John Bertram & Sons Co., Dundas, Ont.
Gisholt Machine Co., Madison, Wis.
Canada Machinery Agency, Montreal.
Rice Lewis & Son, Toronto.
London Mach. Tool Co., Hamilton.
Niles-Bement-Pond Co., New York.
H. W. Petrie, Toronto.

Box Puller.

A. B. Jardine & Co., Hespeler, Ont.

Boxes, Steel Shop.

Cleveland Wire Spring Co., Cleveland.

Boxes, Tote.

Cleveland Wire Spring Co., Cleveland

Brass Foundry Equipment.

Detroit Foundry Supply Co., Detroit.
Paxson, J. W., Co., Philadelphia

Brass Working Machinery.

Warner & Swasey Co., Cleveland, Ohio.

Brushes, Foundry and Core.

Buffalo Foundry Supply Co., Buffalo.
Detroit Foundry Supply Co., Windsor.
Dominion Foundry Supply Co., Montreal
Hamilton Facing Mill Co., Hamilton.
Mechanics Supply Co., Quebec, Que.
Paxson, J. W., Co., Philadelphia

Brushes, Steel.

Buffalo Foundry Supply Co., Buffalo.
Paxson, J. W., Co., Philadelphia

Bulldozers.

John Bertram & Sons Co., Dundas, Ont.
London Mach. Tool Co., Hamilton, Ont.

National Machinery Co., Tiffin, Ohio.
Niles-Bement-Pond Co., New York.

Cable Grease.

United States Graphite Co., Saginaw, Mich.

Calipers.

Baxter, Patterson & Co., Montreal.
Frothingham & Workman Ltd., Montreal
Rice Lewis & Son, Toronto.
Mechanics Supply Co., Quebec, Que.
John Millen & Son, Ltd., Montreal, Que.
L. S. Starrett & Co., Athol, Mass.
Williams & Wilson Montreal.

Carbon.

United States Graphite Co., Saginaw, Mich.

Cars, Foundry.

Buffalo Foundry Supply Co., Buffalo.
Detroit Foundry Supply Co., Windsor
Dominion Foundry Supply Co., Montreal
Hamilton Facing Mill Co., Hamilton.
Paxson, J. W., Co., Philadelphia

Castings, Aluminum.

Lumen Bearing Co., Toronto

Castings, Brass.

Chadwick Bros., Hamilton.
Greey, Wm. & J. G., Toronto.
Hall Engineering Works, Montreal.
Kennedy, Wm., & Sons, Owen Sound.
Lumen Bearing Co., Toronto
Niagara Falls Machine & Foundry Co., Niagara Falls, Ont.
Owen Sound Iron Works Co., Owen Sound.
Robb Engineering Co., Amherst, N.S.

Castings, Grey Iron.

Allis-Chalmers-Bullock Montreal.
Greey, Wm. & J. G., Toronto.
Hall Engineering Works, Montreal.
Kennedy, Wm., & Sons, Owen Sound.
Laurie Engine & Machine Co., Montreal.
Manson Mfg. Co., Thorold, Ont
Maxwell, David, & Sons, St. Marys.
John McDougall Caledonian Iron Works Co., Montreal.
R. McDougall Co., Galt, Ont.
Niagara Falls Machine & Foundry Co., Niagara Falls, Ont
Owen Sound Iron Works Co., Owen Sound.
Robb Engineering Co., Amherst, N.S.
Smart-Turner Machine Co., Hamilton.

Castings, Phosphor Bronze.

Lumen Bearing Co., Toronto

Castings, Steel.

Kennedy, Wm., & Sons, Owen Sound.

Castings, Semi-Steel.

Robb Engineering Co., Amherst, N.S.

Cement Machinery.

Allis-Chalmers-Bullock, Limited, Montreal
Greey, Wm. & J. G., Toronto.
Jeffrey Mfg. Co., Columbus, Ohio
John McDougall Caledonian Iron Works, Co., Montreal
Owen Sound Iron Works Co., Owen Sound

Centreing Machines.

John Bertram & Sons Co., Dundas, Ont.
Jeffrey Mfg. Co., Columbia, Ohio
London Mach. Tool Co., Hamilton, Ont.
Niles-Bement-Pond Co., New York.
Pratt & Whitney Co., Hartford, Conn.
Standard Bearings, Ltd., Niagara Falls

Centres, Planer.

American Tool Works Co., Cincinnati.

Centrifugal Pumps.

Manson Mfg. Co., Thorold, Ont
John McDougall Caledonian Iron Works Co., Montreal.
Pratt & Whitney Co., Hartford, Conn.

Centrifugal Pumps—Turbine Driven

Kerr Turbine Co., Wellsville, N.Y.

Chain, Crane and Dredge.

Frothingham & Workman, Ltd., Montreal

Chain Lubricants.

United States Graphite Co., Saginaw, Mich.

Chaplets.

Buffalo Foundry Supply Co., Buffalo.
Detroit Foundry Supply Co., Windsor.
Hamilton Facing Mill Co., Hamilton.
Paxson, J. W., Co., Philadelphia

Charcoal.

Buffalo Foundry Supply Co. Buffalo.
Detroit Foundry Supply Co., Windsor.
Dominion Foundry Supply Co., Montreal
Hamilton Facing Mill Co., Hamilton.
Paxson, J. W., Co., Philadelphia

Chemicals.

Baxter, Paterson & Co. Montreal.
Canada Chemical Co., London.

Chemists' Machinery.

Greey Wm & J. G., Toronto.

Chemists, Industrial.

Detroit Testing Laboratory, Detroit.

Chemists, Metallurgical.

Detroit Testing Laboratory, Detroit

Chemists, Mining.

Detroit Testing Laboratory, Detroit.

Chucks, Ring Grinding.

A. B. Jardine & Co., Hespeler, Ont.
Chicago Pneumatic Tool Co., Chicago.

Chucks, Drill and Lathe.

American Tool Works Co., Cincinnati.
Baxter, Paterson & Co., Montreal.
John Bertram & Sons Co., Dundas, Ont.
Canada Machinery Agency, Montreal.
Frothingham & Workman Ltd., Montreal
Ker & Goodwin, Brantford.
A. B. Jardine & Co., Hespeler, Ont.
London Mach. Tool Co., Hamilton.
John Millen & Son, Ltd. Montreal.
Niles-Bement-Pond Co., New York.
H. W. Petrie, Toronto.
Rice Lewis & Son, Toronto.
Standard Tool Co., Cleveland.

Chucks, Planer.

American Tool Works Co., Cincinnati.
Canada Machinery Agency, Montreal.
Niles-Bement-Pond Co., New York.

Chucking Machines.

American Tool Works Co., Cincinnati.
Niles-Bement-Pond Co., New York.
H. W. Petrie, Toronto.
Warner & Swasey Co., Cleveland, Ohio

Circuit Breakers.

Allis-Chalmers-Bullock, Limited, Montreal
Canadian General Electric Co., Toronto.
Canadian Westinghouse Co., Hamilton.

Clippers, Bolt.

Frothingham & Workman, Ltd. Montreal
A. B. Jardine & Co., Hespeler, Ont.

Cloth and Wool Dryers.

Dominion Heating and Ventilating Co., Hespeler.
B. Greening Wire Co., Hamilton.
Sheldons Limited, Galt

Collectors, Pneumatic.

Sheldons Limited, Galt

Compressors, Air.

Allis-Chalmers-Bullock, Limited, Montreal
Canada Foundry Co., Limited, Toronto.
Canada Machinery Agency, Montreal.
Canadian Rand Drill Co., Montreal.
Canadian Westinghouse Co., Hamilton.
Chicago Pneumatic Tool Co., Chicago.
Darling Bros., Ltd., Montreal
Detroit Foundry Supply Co., Windsor.
John McDougall, Caledonian Iron Works Co., Montreal
H. W. Petrie, Toronto.
The Smart-Turner Mach. Co., Hamilton.
Hall Engineering Works, Montreal, Que.
London Mach. Tool Co., Hamilton.
Niles-Bement-Pond Co., New York.
H. W. Petrie, Toronto.
Pratt & Whitney Co., Hartford, Conn.
Williams & Wilson, Montreal.

Concentrating Plant.

Allis-Chalmers-Bullock, Montreal.
Greey, Wm. & J. G., Toronto.

Concrete Mixers.

Jeffrey M'f'g. Co., Columbus, Ohio.
Link-Belt Co., Philadelphia.

Condensers.

Canada Foundry Co., Limited, Toronto.
Canada Machinery Agency, Montreal.
Hall Engineering Works Montreal.
Smart-Turner-Machine Co., Hamilton.
Waterous Engine Co., Brantford.

Confectioners' Machinery.

Baxter, Paterson & Co., Montreal.
Greey, Wm. & J. G., Toronto.

Consulting Engineers.

Hall Engineering Works, Montreal.
Jules De Clercy, Montreal.
Miller Bros. & Toms, Montreal.
Roderick J. Parke, Toronto.
Plews & Trimingham, Montreal.

T. Pringle & Son, Montreal.
Somerville & Van Every, Hamilton
Standard Bearings, Ltd., Niagara Falls

Contractors.

Expanded Metal and Fireproofing Co. Toronto.
Hall Engineering Works Montreal.
Laurie Engine & Machine Co., Montreal.
John McDougall Caledonian Iron Works Co., Montreal.
Miller Bros. & Toms, Montreal.
Robb Engineering Co., Amherst N S.
The Smart-Turner Mach. Co., Hamilton.

Contractors' Plant.

Allis-Chalmers-Bullock, Montreal.
John McDougall, Caledonian Iron Works Co., Montreal.
Niagara Falls Machine & Foundry Co., Niagara Falls, Ont.

Contractors' Supplies.

Manson Mfg. Co., Thorold, Ont.

Controllers and Starters Electric Motor.

Allis-Chalmers-Bullock, Montreal.
Canadian General Electric Co., Toronto.
Canadian Westinghouse Co., Hamilton.
T. & H. Electric Co., Hamilton.

Conveying Systems.

Sheldons Limited, Galt

Converters, Steel.

Northern Engineering Works, Detroit.
Paxson, J. W., Co., Philadelphia

Conveyor Machinery.

Baxter, Paterson & Co., Montreal.
Greey Wm. & J. G., Toronto
Jeffrey Mfg. Co., Columbus, Ohio.
Rice Lewis & Son, Toronto
Link-Belt Co., Philadelphia.
John McDougall Caledonian Iron Works Co., Montreal.
Miller Bros. & Toms, Montreal.
Smart-Turner Machine Co., Hamilton.
Laurie Engine & Machine Co., Montreal.
Waterous Engine Works Co., Brantford.
Williams & Wilson, Montreal.

Coping Machines.

John Bertram & Sons Co., Dundas, Ont.
London Mach. Tool Co., Hamilton.
Niles-Bement-Pond Co., New York.

Core Compounds.

Buffalo Foundry Supply Co., Buffalo.
Detroit Foundry Supply Co., Windsor.
Dominion Foundry Supply Co., Montreal
Hamilton Facing Mill Co., Hamilton.
Paxson, J. W., Co., Philadelphia
United States Graphite Co., Saginaw, Mich.

Core-Making Machines.

Paxson, J. W., Co., Philadelphia

Core Ovens.

Detroit Foundry Supply Co., Windsor.
Hamilton Facing Mill Co., Hamilton.
Paxson, J. W., Co., Philadelphia
Sheldons Limited, Galt

Core Oven Bricks.

Buffalo Foundry Supply Co., Buffalo.
Detroit Foundry Supply Co., Windsor.
Hamilton Facing Mill Co., Hamilton.
Paxson, J. W., Co., Philadelphia

Core Sand Cleaners.

Sly, W. W., Mfg. Co., Cleveland

Core Wash.

Buffalo Foundry Supply Co., Buffalo.
Detroit Foundry Supply Co., Windsor.
Hamilton Facing Mill Co., Hamilton.
Paxson, Co., J. W., Philadelphia

Couplings.

Owen Sound Iron Works Co., Owen Sound

Cranes, Electric and Hand Power.

Canada Foundry Co., Limited, Toronto
Canadian Pilling Co., Montreal
Dominion Foundry Supply Co., Montreal
Hamilton Facing Mill Co., Hamilton.
Link-Belt Co., Philadelphia.
John McDougall, Caledonian Iron Works Co., Montreal.
Miller Bros. & Toms Montreal.
Niles-Bement-Pond Co., New York.
Northern Engineering Works Detroit
Owen Sound Iron Works Co., Owen Sound
Paxson, J. W., Co., Philadelphia
Smart-Turner-Machine Co., Hamilton.

Crank Pin.

Sight Feed Oil Pump Co., Milwaukee, Wis.

Crankshafts.

The Canada Forge Co., Welland.
St. Clair Bros., Galt

Crabs.

Frothingham & Workman, Ltd., Montreal

Crank Pin Turning Machine

London Mach. Tool Co., Hamilton.
Niles-Bement-Pond Co., New York.

Cross Head Pin.

Sight Feed Oil Pump Co., Milwaukee, Wis.

Crucibles.

Buffalo Foundry Supply Co., Buffalo.
Detroit Foundry Supply Co., Windsor
Dominion Foundry Supply Co., Montreal
Hamilton Facing Mill Co., Hamilton.
J. W. Paxson Co., Philadelphia.

Crucible Caps

Hamilton Facing Mill Co., Hamilton.
Paxson, J. W., Co., Philadelphia

Crushers, Rock or Ore.

Allis-Chalmers-Bullock, Montreal.
Jeffrey Mfg. Co., Columbus, Ohio.

Cup Grease.

Mechanics Supply Co., Quebec

Cupolas.

Buffalo Foundry Supply Co., Buffalo.
Detroit Foundry Supply Co., Windsor
Dominion Foundry Supply Co., Montreal
Hamilton Facing Mill Co., Hamilton.
Northern Engineering Works, Detroit
J. W. Paxson Co. Philadelphia.
Sheldons Limited, Galt.

Cupola Blast Gauges.

Paxson, J. W., Co., Philadelphia
Sheldons Limited, Galt

Cupola Blocks.

Detroit Foundry Supply Co., Detroit.
Hamilton Facing Mill Co., Hamilton
Northern Engineering Works Detroit
Ontario Lime Association Toronto
Paxson, J. W., Co., Philadelphia
Toronto Pottery Co., Toronto

Cupola Blowers.

Buffalo Foundry Supply Co., Buffalo,
Canada Machinery Agency, Montreal.
Detroit Foundry Supply Co., Windsor
Dominion Heating and Ventilating Co., Hespeler
Hamilton Facing Mill Co., Hamilton.
Northern Engineering Works, Detroit
Paxson J. W., Co., Philadelphia
Sheldon's Limited, Galt.

Cutters, Flue

Chicago Pneumatic Tool Co., Chicago.
J. W. Paxson Co., Philadelphia.

Cutters, Milling.

Becker, Brainard Milling Machine Co. Hyde Park, Mass.
Frothingham & Workman Ltd., Montreal
Pratt & Whitney Co., Hartford, Conn.
Standard Tool Co., Cleveland.

Cutting-off Machines.

John Bertram & Sons Co., Dundas, Ont.
Canada Machinery Agency, Montreal.
London Mach. Tool Co., Hamilton.
H. W. Petrie, Toronto.
Pratt & Whitney Co., Hartford, Conn.

Cutting-off Tools.

Armstrong Bros. Tool Co., Chicago.
Baxter Paterson & Co., Montreal.
London Mach. Tool Co., Hamilton.
Mechanics Supply Co., Quebec Que.
H. W. Petrie, Toronto.
Pratt & Whitney, Hartford, Conn.
Rice Lewis & Son, Toronto.
L. S. Starrett Co., Athol, Mass.

Damper Regulators.

Darling Bros., Ltd., Montreal

Dies

Globe Machine & Stamping Co., Cleveland, Ohio.

Die Stocks

Curtis & Curtis Co., Bridgeport, Conn.
Hart Manufacturing Co., Cleveland, Ohio
Loew Mfg. Co., Cleveland, Ohio
Mechanics Supply Co., Quebec

Dies, Opening

W. H. Banfield & Sons, Toronto
Globe Machine & Stamping Co., Cleveland Ohio.
Pratt & Whitney Co., Hartford, Conn.

Dies, Sheet Metal.

W. H. Banfield & Sons Toronto.
Globe Machine & Stamping Co., Cleveland, Ohio

Dies, Threading,

Frothingham & Workman, Ltd., Montreal
Hart Mfg. Co., Cleveland
Mechanics Supply Co., Quebec, Que.
John Millen & Son, Ltd., Montreal.

Draft, Mechanical.

W. H. Banfield & Sons, Toronto.
Butterfield & Co., Rock Island, Que.
A. B. Jardine & Co., Hespeler
Mechanics Supply Co., Quebec, Que.
Pratt & Whitney Co., Hartford, Conn.

Sheldon's Limited, Galt.

Drawing Instruments.

Rice Lewis & Son, Toronto.
Mechanics Supply Co., Quebec, Que.

Drawing Supplies.

The Electric Blue Print Co., Montreal

Drawn Steel, Cold.

Baxter, Paterson & Co., Montreal.
Greey, Wm. & J. G., Toronto
Miller Bros. & Toms, Montreal.
Union Drawn Steel Co., Hamilton.

Drilling Machines, Arch Bar.

John Bertram & Sons Co., Dundas, Ont.
London Mach. Tool Co., Hamilton
Niles-Bement-Pond Co., New York.

Drilling Machines, Boiler.

American Tool Works Co., Cincinnati.
John Bertram & Sons Co., Dundas, Ont.
Bickford Drill and Tool Co., Cincinnati.
The Canadian Fairbanks Co., Montreal.
A. B. Jardine & Co., Hespeler, Ont.
London Mach. Tool Co., Hamilton, Ont.
Niles-Bement-Pond Co., New York.
H. W. Petrie, Toronto.
Williams & Wilson, Montreal

Drilling Machines Connecting Rod.

John Bertram & Sons Co., Dundas, Ont.
London Mach. Tool Co., Hamilton.
Niles-Bement-Pond Co. New York.

Drilling Machines, Locomotive Frame.

American Tool Works Co., Cincinnati.
B. F. Barnes Co., Rockford, Ill.
John Bertram & Sons Co., Dundas, Ont.
London Mach. Tool Co., Hamilton, Ont.
Niles-Bement-Pond Co., New York.

Drilling Machines, Multiple Spindle.

American Tool Works Co., Cincinnati.
B. F. Barnes Co., Rockford. Ill.
Baxter, Paterson & Co., Montreal.
John Bertram & Sons Co., Dundas, Ont.
Bickford Drill & Tool Co., Cincinnati.
Canada Machinery Agency Montreal.
London Mach. Tool Co., Hamilton, Ont.
Niles-Bement-Pond Co., New York.
H. W. Petrie, Toronto.
Williams & Wilson, Montreal.

Drilling Machines, Pneumatic

Canada Machinery Agency, Montreal.

Drilling Machines, Portable

Baxter, Paterson & Co., Montreal.
A. B. Jardine & Co., Hespeler, Ont.
Mechanics Supply Co., Quebec, Que.
Niles-Bement-Pond Co., New York.

Drilling Machines, Radial.

American Tool Works Co., Cincinnati.
Baxter, Paterson & Co., Montreal.
John Bertram & Sons Co., Dundas, Ont.
Bickford Tool & Drill Co., Cincinnati.
The Canadian Fairbanks Co., Montreal.
London Mach. Tool Co., Hamilton.
Mechanics Supply Co., Quebec, Que.
Niles-Bement-Pond Co., New York.
H. W. Petrie, Toronto.
Williams & Wilson, Montreal.

Drilling Machines. Suspension.

John Bertram & Sons Co., Dundas, Ont.
Canada Machinery Agency Montreal.
London Mach. Tool Co., Hamilton.
Niles-Bement-Pond Co. New York

Drilling Machines, Turret.

John Bertram & Sons Co., Dundas, Ont.
London Mach. Tool Co., Hamilton.
Niles-Bement-Pond Co., New York.

Drilling Machines, Upright.

American Tool Works Co., Cincinnati.
B. F. Barnes Co., Rockford, Ill.
Baxter, Paterson & Co., Montreal.
John Bertram & Sons Co., Dundas, Ont
Dwight Slate Machine Co., Hartford, Conn.
A. B. Jardine & Co., Hespeler, Ont.
Rice Lewis & Son, Toronto.
London Mach. Tool Co., Hamilton.
Mechanics Supply Co., Quebec, Que.
Niles-Bement-Pond Co., New York.
W. H. Petrie, Toronto.
Williams & Wilson Montreal.

Drills, Bench.

B. F. Barnes Co., Rockford, Ill.
Baxter, Paterson & Co., Montreal.
London Mach. Tool Co., Hamilton.
Mechanics Supply Co. Quebec Que.
Pratt & Whitney Co., Hartford, Conn.

Drills, Blacksmith.

Canada Machinery Agency Montreal.
Frothingham & Workman, Ltd., Montreal
A. B. Jardine & Co., Hespeler, Ont.
London Mach. Tool Co., Hamilton.
Mechanics Supply Co., Quebec, Que.
Standard Tool Co., Cleveland.

Drills, Centre.

Mechanics Supply Co. e, Que.
Pratt & Whitney Co., Hartford, Conn.
Standard Tool Co , Cleveland, O.
L. S. Starrett Co., Athol, Mass.

Drills, Electric

B. F. Barnes Co., Rockford, Ill.
Baxter Paterson & Co., Montreal.
Canadian Pilling Co., Montreal
Chicago Pneumatic Tool Co., Chicago.
Niles-Bement-Pond Co., New York.

Drills, Gang.

American Tool Works Co., Cincinnati.
B. F. Barnes Co., Rockford, Ill.
John Bertram & Sons Co., Dundas, Ont.
Pratt & Whitney Co., Hartford, Conn.

Drills, High Speed.

Wm. Abbott, Montreal.
American Tool Works Co., Cincinnati.
B. F. Barnes Co., Rockford, Ill.
Frothingham & Workman, Ltd., Montreal
Pratt & Whitney Co., Hartford, Conn.
Standard Tool Co., Cleveland, O.

Drills, Hand.

A. B. Jardine & Co., Hespeler, Ont.

Drills, Horizontal.

B. F. Barnes Co., Rockford, Ill.
John Bertram & Sons Co., Dundas, Ont.
Canada Machinery Agency, Montreal.
London Mach. Tool Co., Hamilton.
Niles-Bement-Pond Co., New York.

Drills, Pneumatic.

Canada Machinery Agency, Montreal.
Chicago Pneumatic Tool Co., Chicago.
Independent Pneumatic Too. Co., Chicago, New York
Niles-Bement-Pond Co., New York.

Drills, Radial.

American Tool Works Co., Cincinnati.
John Bertram & Sons Co., Dundas, Ont.
Bickford Drill & Tool Co., Cincinnati
London Mach. Tool Co., Hamilton, Ont.
Niles-Bement-Pond Co., New York.

Drills, Ratchet.

Armstrong Bros. Tool Co., Chicago.
Frothingham & Workman, Ltd., Montreal
A. B. Jardine & Co., Hespeler.
Mechanics Supply Co., Quebec, Que.
Pratt & Whitney Co., Hartford, Conn.
Standard Tool Co., Cleveland.

Drills, Rock.

Allis-Chalmers-Bullock, Montreal.
Canadian Rand Drill Co., Montreal.
Chicago Pneumatic Tool Co., Chicago.
Jaffrey Mfg Co., Columbus, Ohio.

Drills, Sensitive.

American Tool Works Co., Cincinnati.
B. F. Barnes Co., Rockford, Ill.
Canada Machinery Agency, Montreal.
Dwight Slate Machine Co., Hartford, Conn.
Niles-Bement-Pond Co., New York

Drills, Twist.

Baxter, Paterson & Co., Montreal.
Chicago Pneumatic Tool Co., Chicago
Frothingham & Workman, Ltd., Montreal
Alex. Gibb, Montreal.
A. B. Jardine & Co., Hespeler, Ont.
Mechanics Supply Co., Quebec Que.
John Millen & Son, Ltd., Montreal.
Morse Twist Drill and Machine Co., New Bedford, Mass.
Pratt & Whitney Co., Hartford, Conn.
Standard Tool Co., Cleveland.

Drying Apparatus of all Kinds.

Dominion Heating & Ventilating Co., Hespeler
Greey, Wm. & J. G., Toronto
Sheldons Limited, Galt

Dry Kiln Equipment.

Sheldons Limited, Galt

Dry Sand and Loam Facing.

Buffalo Foundry Supply Co. Buffalo.
Hamilton Facing Mill Co., Hamilton.
Paxson, J. W., Co., Philadelphia

Dump Cars.

Canada Foundry Co., Limited, Toronto
Greey, Wm. & J. G., Toronto
Hamilton Facing Mill Co., Hamilton.
John McDougall, Caledonian Iron Works Co., Montreal.
Niles-Bement-Pond Co., New York.
Standard Bearings, Ltd., Niagara Falls.
Link-Belt Eng. Co., Philadelphia.
John McDougall, Caledonian Iron Works Co., Montreal.
Miller Bros. & Toms, Montreal.
Owen Sound Iron Works Co., Owen Sound
Paxson, J. W., Co., Philadelphia
Waterous Engine Co., Brantford.

Dust Arresters.

Sly, W. W., Mfg. Co., Cleveland

Dust Separators.

Greey, Wm. & J. G., Toronto
Dominion Heating and Ventilating Co., Hespeler.
Paxson, J. W., Co., Philadelphia
Sheldon's Limited, Galt.

Dynamos.

Allis-Chalmers-Bullock, Montreal.
Canadian General Electric Co., Toronto.
Canadian Westinghouse Co., Hamilton.
Consolidated Electric Co., Toronto
Electrical Machinery Co., Toronto.
Gas & Electric Power Co., Toronto
Hall Engineering Works, Montreal, Que.
Mechanics Supply Co., Quebec, Que.
John Millen & Son, Ltd., Montreal.
Packard Electric Co., St. Catharines.
H. W. Petrie, Toronto.
T. & H. Electric Co., Hamilton.

Dynamos—Turbine Driven.

Kerr-Turbine Co., Wellsville, N.Y.

Economizer, Fuel.

Dominion Heating & Ventilating Co., Hespeler
Standard Bearings, Ltd., Niagara Falls

Electrical Instruments.

Canadian Westinghouse Co., Hamilton.
Mechanics Supply Co., Quebec, Que.

Electrical Steel.

Baxter, Patterson & Co., Montreal.

Electrical Supplies.

Canadian General Electric Co., Toronto.
Canadian Westinghouse Co., Hamilton.
London Mach. Tool Co., Hamilton, Ont.
Mechanics Supply Co., Quebec, Que.
John Millen & Son, Ltd., Montreal.
Packard Electric Co., St. Catharines.
T. & H. Electric Co., Hamilton.

Electrical Repairs

Canadian Westinghouse Co., Hamilton.
T. & H. Electric Co., Hamilton.

Elevator Buckets.

Greey, Wm. & J. G., Toronto
Jeffrey Mfg. Co., Columbus, Ohio.

Elevators, Foundry.

Northern Egineering Works, Detroit

Elevators—For any Service.

Darling Bros. Ltd., Montreal

Emery and Emery Wheels.

Baxter, Paterson & Co., Montreal.
Dominion Foundry Supply Co., Montreal
Frothingham & Workman Ltd., Montreal
Hamilton Facing Mill Co., Hamilton.
Paxson, J. W., Co., Philadelphia

Emery Wheel Dressers.

Baxter Paterson & Co., Montreal.
Canada Machinery Agency, Montreal.
Desmond-Stephan Mfg. Co., Urbana, Ohio
Dominion Foundry Supply Co., Montreal
Frothingham & Workman Ltd., Montreal
Hamilton Facing Mill Co., Hamilton.
International Specialty Co., Detroit.
Mechanics Supply Co. Quebec, Que.
John Millen & Son, Ltd., Montreal.
H. W. Petrie, Toronto.
Paxson, J. W., Co., Philadelphia
Standard Tool Co., Cleveland.

Engineers and Contractors.

Canada Foundry Co., Limited, Toronto.
Darling Bros., Ltd., Montreal
Greey, Wm. & J. G., Toronto
Hall Engineering Works, Montreal.
Laurie Engine & Machine Co., Montreal.
Link-Belt Co., Philadelphia.
John McDougall Caledonian Iron Works Co., Montreal.
Miller Bros. & Toms Montreal.
Robb Engineering Co., Amherst, N S.
The Smart-Turner Mach. Co., Hamilton.

Engineers' Supplies.

Baxter, Paterson & Co., Montreal.
Frothingham & Workman, Ltd., Montreal
Greey, Wm. & J. G., Toronto
Hall Engineering Works Montreal.
Mechanics Supply Co., Quebec, Que.
Rice Lewis & Son, Toronto.

Engines, Gas and Gasoline.

Baxter, Paterson & Co., Montreal.
Canada Foundry Co., Toronto.
Canada Machinery Agency Montreal.
The Canadian Fairbanks Co., Montreal.
Canadian McVicker Engine Co., Galt.
Gilson Mfg. Co., Guelph
The Goldie & McCulloch Co., Galt, Ont.
Rice Lewis & Son, Toronto
Ontario Wind Engine & Pump Co. Toronto
H. W. Petrie, Toronto.
The Smart-Turner Mach. Co., Hamilton
Standard Bearings, Ltd., Niagara Falls

Engines, Steam.

Allis-Chalmers-Bullock, Montreal
Belliss & Morcom, Birmingham, Eng.
Canada Machinery Agency, Montreal.
The Goldie & McCulloch Co., Galt, Ont.
E. Leonard & Sons, London, Ont.
Rice Lewis & Son, Toronto.
Laurie Engine & Machine Co., Montreal.
Manson Mfg. Co., Thorold Ont
John McDougall Caledonian Iron Wo Montreal.
Miller Bros. & Toms, Montreal.
Robb Engineering Co., Amherst, N.S.
Sheldons Limited, Galt.
The Smart-Turner Mach. Co., Hamilton.
Waterous Engine Works Co., Brantford.

Exhaust Heads.

Darling Bros., Ltd., Montreal.
Dominion Heating & Ventilating Co., Hespeler
Sheldons Limited, Galt, Ont.

Expanded Metal.

Expanded Metal and Fireproofing Co. Toronto

Expanders.

A. B. Jardine & Co., Hespeler, Ont.

Fans, Electric.

Canadian General Electric Co., Toronto,
Canadian Westinghouse Co., Hamilton.
Dominion Heating & Ventilating Co. Hespeler
Mechanics Supply Co., Quebec, Que.
Sheldons Limited Galt, Ont.
The Smart-Turner Mach. Co., Hamilton.

Fans, Exhaust.

Canadian Buffalo Forge Co., Montreal.
Detroit Foundry Supply Co., Windsor.
Dominion Foundry Supply Co., Montreal
Dominion Heating & Ventilating Co., Hespeler.
Greey, Wm. & J. G., Toronto
Hamilton Facing Mill Co., Hamilton.
Paxson, J. W., Co., Philadelphia
Sheldons Limited, Galt.
B. F. Sturtevant Co., Hyde Park, Mass.

Feed Water Heaters.

Darling Bros, Montreal
Laurie Engine & Machine Co. Montreal
John McDougall, Caledonian Iron Works Co., Montreal.
The Smart-Turner Mach. Co., Hamilton

Files and Rasps.

Baxter, Paterson & Co., Montreal.
Frothingham & Workman, Ltd., Montreal
Mechanics Supply Co., Quebec Que.
John Millen & Son, Ltd., Montreal.
Rice Lewis & Son, Toronto.
Nicholson File Co., Port Hope
H. W. Petrie, Toronto.

Fillet, Pattern.

Baxter, Paterson & Co., Montreal.
Buffalo Foundry Supply Co., Buffalo.
Detroit Foundry Supply Co., Windsor.
Dominion Foundry Supply Co., Montreal
Hamilton Facing Mill Co., Hamilton.

Fire Apparatus.

Waterous Engine Works Co., Brantford.

Fire Brick and Clay.

Baxter Paterson & Co., Montreal.
Buffalo Foundry Supply Co., Buffalo.
Detroit Foundry Supply Co., Windsor.
Dominion Foundry Supply Co., Montreal
Hamilton Facing Mill Co., Hamilton.
Ontario Lime Association Toronto
J. W. Paxson Co. Philadelphia.
Toronto Pottery Co., Toronto

Flash Lights.

Berlin Electrical Mfg. Co., Toronto.
Mechanics Supply Co., Quebec, Que.

Flour Mill Machinery.

Allis-Chalmers-Bullock, Montreal.
Greey, Wm. & J. G., Toronto
The Goldie & McCulloch Co., Galt, Ont.
John McDougall, Caledonian Iron Works Co., Montreal.
Miller Bros. & Toms, Montreal.

Forges.

Canada Foundry Co., Limited, Toronto.
Frothingham & Workman Ltd., Montreal
Hamilton Facing Mill Co., Hamilton.
Mechanics Supply Co., Quebec, Que.
H. W. Petrie, Toronto.
Sheldons Limited, Galt, Ont.

Forgings, Drop.

John McDougall, Caledonian Iron Work Co., Montreal.
H. W. Petrie, Toronto.
St. Clair Bros., Galt

Forgings, Light & Heavy.

The Canada Forge Co., Welland.
Hamilton Steel & Iron Co., Hamilton

Forging Machinery.

John Bertram & Sons Co., Dundas, Ont.
London Mach. Tool Co., Hamilton, Ont
National Machinery Co., Tiffin, Ohio
Niles-Bement-Pond Co., New York.

Founders.

Greey, Wm. & J. G., Toronto.
John McDougall, Caledonian Iron Works Co., Montreal
Niagara Falls Machine & Foundry Co., Niagara Falls, Ont.
Manson Mfg. Co., Thorold, Ont.
Maxwell, David, & Sons, St. Marys
The Smart-Turner Mach. Co., Hamilton.

Foundry Equipment.

Detroit Foundry Supply Co., Windsor.
Hamilton Facing Mill Co., Hamilton
Hanna Engineering Works Chicago.
Northern Engineering Works Detroit
Paxson, J. W., Co., Philadelphia

Foundry Parting.

Partomol Co., New York
Foundry Specialty Co., Cincinnati
Stanley Doggett, New York

Foundry Facings.

Buffalo Foundry Supply Co., Buffalo.
Detroit Foundry Supply Co., Windsor.
Hamilton Facing Mill Co., Hamilton.
J. W. Paxson Co., Philadelphia, Pa.
United States Graphite Co., Saginaw, Mich.

Friction Clutch Pulleys, etc.

The Goldie & McCulloch Co., Galt.
Greey, Wm. & J. G., Toronto
Link-Belt Co., Philadelphia.
Miller Bros. & Toms, Montreal.

Furnaces.

Detroit Foundry Supply Co., Windsor.
Dominion Foundry Supply Co., Montreal
Hamilton Facing Mill Co., Hamilton.
Paxson, J. W., g Philadelphia Co.,

Galvanizing

Canada Metal Co., Toronto.
Ontario Wind Engine & Pump Co., Toronto

Gas Blowers and Exhausters.

Dominion Heating & Ventilating Co. Hespeler
Sheldons Limited, Galt.

Gas Plants, Suction.

Baxter Paterson & Co., Montreal
Colonial Engineering Co., Montreal
Standard Bearings, Ltd., Niagara Falls
Williams & Wilson, Montreal

Gauges, Standard.

Mechanics Supply Co., Quebec, Que.
Pratt & Whitney Co., Hartford, Conn.

Gearing.

Eberhardt Bros. Machine Co., Newark, N.J.
Greey, Wm. & J. G., Toronto

Gear Cutting Machinery.

Baxter, Patterson & Co., Montreal
Becker-Brainard Milling Mach. Co., Hyde Park, Mass.
Bickford Drill & Tool Co., Cincinnati.
Dwight Slate Machine Co., Hartford, Conn.
Eberhardt Bros. Machine Co., Newark, N.J.
Greey, Wm. & J. G., Toronto
Kennedy, Wm., & Sons, Owen Sound
London Mach. Tool Co., Hamilton.
Niles-Bement-Pond Co., New York.
H. W. Petrie, Toronto.
Pratt & Whitney Co., Hartford, Conn.
Williams & Wilson, Montreal.

Gears, Angle.

Chicago Pneumatic Tool Co., Chicago.
Greey, Wm. & J. G., Toronto
Laurie Engine & Machine Co., Montreal.
John McDougall, Caledonian Iron Works Co., Montreal.
Miller Bros. & Toms, Montreal.
Waterous Engine Co., Brantford.

Gears, Cut.

Kennedy, Wm., & Sons, Owen Sound

Gear Grease.

United States Graphite Co., Saginaw, Mich.

Gears, Iron.

Greey, Wm. & J. G., Toronto
Kennedy, Wm., & Sons, Owen Sound

Gears, Mortise.

Greey, Wm. & J. G., Toronto
Kennedy, Wm., & Sons, Owen Sound

Gears, Reducing.

Brown, David & Sons, Huddersfield, Eng
Chicago Pneumatic Tool Co., Chicago.
Greey, Wm. & J. G., Toronto
John McDougall, Caledonian Iron Works Co., Montreal.
Miller Bros. & Toms, Montreal.

Generators, Electric.

Allis-Chalmers-Bullock, Limited, Montreal
Canadian General Electric Co., Toronto
Canadian Westinghouse Co., Hamilton.
Hall Engineering Works, Montreal.
H. W. Petrie, Toronto.
Toronto & Hamilton Electric Co. Hamilton.

Generators, Gas.

H. W. Petrie, Toronto.

Graphite Paints.

P. D. Dods & Co., Montreal.
United States Graphite Co., Saginaw, Mich.

Graphite.

Detroit Foundry Supply Co., Windsor.
Dominion Foundry Supply Co., Montreal
Hamilton Facing Mill Co., Hamilton.
Mechanics Supply Co., Quebec, Que.
Paxson, J. W., Co., Philadelphia
United States Graphite Co., Saginaw, Mich.

Grinders, Automatic Knife.

W. H. Banfield & Son, Toronto.

Grinders, Centre.

Niles-Bement-Pond Co., New York.
H. W. Petrie, Toronto.

Grinders, Cutter.

Becker-Brainard Milling Mach. Co., Hyde Park, Mass.
John Millen & Son, Ltd., Montreal.
Pratt & Whitney Co., Hartford, Conn.

Grinders, Tool.

Armstrong Bros. Tool Co., Chicago.
B. F. Barnes Co., Rockford, Ill.
H. W. Petrie, Toronto.
Williams & Wilson, Montreal.

Grinding Machines.

The Canadian Fairbanks, Montreal.
Rice Lewis & Son, Toronto.
Niles-Bement-Pond Co., New York.
Paxson, J. W., Co., Philadelphia
H. W. Petrie, Toronto.

Grinding and Polishing Machines.

The Canadian Fairbanks Co., Montreal.
Greey, Wm. & J. G., Toronto
Independent Pneumatic Tool Co., Chicago, Ill.
John Millen & Son, Ltd., Montreal.
Miller Bros. & Toms, Montreal
Niles-Bement-Pond Co., New York.
H. W. Petrie, Toronto.
Fendrith Machinery Co., Toronto.
Standard Bearings, Ltd., Niagara Falls

Grinding Wheels

Carborundum Co., Niagara Falls
Norton Company, Worcester, Mass.

Hack Saws.

Baxter, Paterson & Co., Montreal.
Canada Machinery Agency, Montreal.
The Canadian Fairbanks Co., Montreal.
Frothingham & Workman, Ltd., Montreal
Mechanics Supply Co., Quebec, Que.
John Millen & Son, Ltd., Montreal.
Niles-Bement-Pond Co., New York.
H. W. Petrie, Toronto.
Williams & Wilson, Montreal.

Hack Saw Frames.

Mechanics Supply Co., Quebec, Que.

Hammers, Drop.

London Mach. Tool Co., Hamilton, Ont.
Miller Bros. & Toms, Montreal.
Niles-Bement-Pond Co., New York.

Hammers, Power.

Rice Lewis & Son, Toronto.

Hammers, Steam.

John Bertram & Sons Co., Dundas, Ont.
London Mach. Tool Co., Hamilton, Ont.
Niles-Bement-Pond Co., New York.

Hangers.

The Goldie & McCulloch Co., Galt.
Greey, Wm. & J. G., Toronto
Kennedy, Wm., & Sons, Owen Sound
Miller Bros. & Toms, Montreal.
Owen Sound Iron Works Co., Owen Sound
The Smart-Turner Mach. Co., Hamilton.
Waterous Engine Co., Brantford.
Standard Bearings Ltd., Niagara Falls

Hardware Specialties.

Mechanics Supply Co., Quebec, Que.

Heating Apparatus.

Darling Bros., Ltd., Montreal
Dominion Heating & Ventilating Co., Hespeler
Mechanics Supply Co., Quebec, Que.
Sheldons Limited, Galt.

Hoisting and Conveying Machinery.

Allis-Chalmers-Bullock, Limited, Montreal
Greey, Wm. & J. G., Toronto
Link-Belt Co., Philadelphia.
Miller Bros. & Toms, Montreal.
Niles-Bement-Pond Co., New York.
Northern Engineering Works, Detroit
The Smart-Turner Mach. Co., Hamilton.
Waterous Engine Co., Brantford.

Hoists, Chain and Rope.

Frothingham & Workman, Ltd., Montreal

Hoists, Electric.

Northern Engineering Works, Detroit

Hoists, Pneumatic.

Canadian Rand Drill Co., Montreal.
Chicago Pneumatic Tool Co., Chicago.
Dominion Foundry Supply Co., Montreal
Hamilton Facing Mill Co., Hamilton.
Northern Engineering Works, Detroit

Hoists, Portable and Stationery.

Canadian Pilling Co., Montreal

Hose.

Baxter, Paterson & Co., Montreal
Mechanics Supply Co., Quebec, Que.

Hose, Air.

Canadian Rand Drill Co., Montreal.
Canadian Westinghouse Co., Hamilton.
Chicago Pneumatic Tool Co., Chicago.
Paxson, J. W., Co., Philadelphia

Hose Couplings.

Canadian Rand Drill Co., Montreal.
Canadian Westinghouse Co., Hamilton.
Chicago Pneumatic Tool Co., Chicago.
Mechanics Supply Co., Quebec, Que.
Paxson, J. W., Co., Philadelphia

Hose, Steam.

Allis-Chalmers-Bullock, Montreal.
Canadian Rand Drill Co., Montreal.
Canadian Westinghouse Co., Hamilton.
Mechanics Supply Co., Quebec, Que.
Paxson, J. W. Co., Philadelphia

Hydraulic Accumulators.

Niles-Bement-Pond Co., New York.
Perrin, Wm. R., Co., Toronto.
The Smart-Turner Mach. Co., Hamilton.

Hydraulic Machinery.

Allis-Chalmers-Bullock, Montreal.
Barber, Chas., & Sons, Meaford

India Oil Stones.

Norton Company, Worcester, Mass.

Indicators, Speed.

Mechanics Supply Co., Quebec, Que.
L. S. Starrett Co., Athol, Mass.

Injectors.

Canada Foundry Co., Toronto.
The Canadian Fairbanks Co., Montreal.
Desmond-Stephan Mfg. Co., Urbana, Ohio
Frothingham & Workman Ltd., Montreal
Mechanics Supply Co., Quebec, Que.
Rice Lewis & Son, Toronto.
Penberthy Injector Co., Windsor, Ont.

Iron and Steel.

Frothingham & Workman, Ltd., Montreal

Iron and Steel Bars and Bands.

Hamilton Steel & Iron Co., Hamilton

Iron Cements.

Buffalo Foundry Supply Co., Buffalo.
Detroit Foundry Supply Co., Windsor.
Hamilton Facing Mill Co., Hamilton.
Paxson, J. W., Co., Philadelphia

Iron Filler.

Buffalo Foundry Supply Co., Buffalo.
Detroit Foundry Supply Co., Windsor.
Dominion Foundry Supply Co., Montreal
Hamilton Facing Mill Co., Hamilton.
Paxson, J. W., Co., Philadelphia.

Jacks.

Frothingham & Workman, Ltd., Montreal
Norton, A. O., Coaticook, Que.

Kegs, Steel Shop.

Cleveland Wire Spring Co., Cleveland

Key-Seating Machines.

B. F. Barnes Co., Rockford, Ill.
Niles-Bement-Pond Co., New York.

Lace Leather.

Baxter, Paterson & Co., Montreal.

Ladles, Foundry.

Frothingham & Workman Ltd., Montreal
Northern Engineering Works, Detroit

Lamps, Arc and Incandescent.

Canadian General Electric Co., Toronto.
Canadian Westinghouse Co., Hamilton.
The Packard Electric Co., St. Catharines.

Lathe Dogs.

Armstrong Bros., Chicago
Baxter, Paterson & Co., Montreal.
Pratt & Whitney Co., Hartford, Conn.

Lathes, Engine.

American Tool Works Co., Cincinnati.
B. F. Barnes Co., Rockford, Ill.
Baxter, Paterson & Co., Montreal.
John Bertram & Sons Co., Dundas, Ont.
Canada Machinery Agency, Montreal.
The Canadian Fairbanks Co., Montreal.
London Mach. Tool Co., Hamilton, Ont.
Niles-Bement-Pond Co., New York.
H. W. Petrie, Toronto.
Pratt & Whitney Co., Hartford, Conn.

Lathes, Foot-Power.

American Tool Works Co., Cincinnati.
B. F. Barnes Co., Rockford, Ill.

Lathes, Screw Cutting.

B. F. Barnes Co., Rockford, Ill.
Baxter, Paterson & Co., Montreal.
Niles-Bement-Pond Co., New York.

Lathes, Automatic, Screw-Threading.

John Bertram & Sons Co., Dundas, Ont
London Mach. Tool Co., Hamilton, Ont
Pratt & Whitney Co., Hartford, Conn.

Lathes, Bench.

B. F. Barnes Co., Rockford, Ill.
London Mach. Tool Co., London, Ont.
Pratt & Whitney Co., Hartford, Conn.

Lathes, Turret.

American Tool Works Co., Cincinnati.
Baxter, Paterson & Co., Montreal.
John Bertram & Sons Co., Dundas, Ont.
Gisholt Machine Co., Madison, Wis.
London Mach. Tool Co., Hamilton, Ont.
Niles-Bement-Pond Co., New York.
The Pratt & Whitney Co., Hartford, Conn.
Warner & Swasey Co., Cleveland O.

Leather Belting.

Baxter, Paterson & Co., Montreal.
Canada Machinery Agency, Montreal.
The Canadian Fairbanks Co., Montreal
Greey, Wm. & J. G., Toronto

Lime Stone Flux.

Hamilton Facing Mill Co., Hamilton.
Paxson, J. W., Co., Philadelphia

Locomotives, Air.

Canadian Rand Drill Co., Montreal.

Locomotives, Electrical.

Canadian Westinghouse Co., Hamilton
Jeffrey Mfg. Co., Columbus, Ohio.

Locomotives, Steam.

Canada Foundry Co., Toronto.
Canadian Rand Drill Co., Montreal

Locomotive Turntable Tractors.

Canadian Pilling Co., Montreal

Lubricating Graphite.

United States Graphite Co., Saginaw, Mich.

Lubricants.

United States Graphite Co., Saginaw, Mich.

Lubricating Plumbago.

Detroit Foundry Supply Co., Detroit.
Dominion Foundry Supply Co., Montreal
Hamilton Facing Mill Co., Hamilton.
Paxson, J. W., Co., Philadelphia

Lubricators, Force Feed.

Sight Feed Oil Pump Co., Milwaukee, Wis.

Lumber Dry Kilns.

Dominion Heating & Ventilating Co., Hespeler.
H. W. Petrie, Toronto.
Sheldons Limited, Galt, Ont.

Machinery Dealers.

Baxter, Paterson & Co., Montreal.
Canada Machinery Agency, Montreal.
The Canadian Fairbanks Co., Montreal.
Miller Bros. & Toms, Montreal.
H. W. Petrie, Toronto.
The Smart-Turner Mach. Co., Hamilton.
Williams & Wilson Montreal.

Machinery Designers.

Greey, Wm. & J. G., Toronto
Standard Bearings, Ltd., Niagara Falls.

Machinery Lubricators.

United States Graphite Co., Saginaw, Mich.

Machinists.

W. H. Banfield & Sons, Toronto.
Greey, Wm. & J. G. Toronto
Hall Engineering Works, Montreal.
Link-Belt Co., Philadelphia.
John McDougall, Caledonian Iron Works Co., Montreal.
Miller Bros. & Toms Montreal.
Paxson, J. W., Co., Philadelphia
Robb Engineering Co., Amherst, N.S.
The Smart-Turner Mach. Co., Hamilton.
Standard Bearings, Ltd, Niagara Falls
Waterous Engine Co., Brantford.

Machinists' Small Tools.

Armstrong Bros., Chicago.
Butterfield & Co., Rock Island, Que.
Frothingham & Workman Ltd., Montreal
Mechanics Supply Co., Quebec, Que.
Rice Lewis & Son, Montreal.
Pratt & Whitney Co., Hartford, Conn.
Standard Tool Co., Cleveland.
L. S. Starrett Co., Athol, Mass.
Williams & Wilson, Montreal.

Malleable Flask Clamps.

Buffalo Foundry Supply Co., Buffalo.
Paxson, J. W., Co., Philadelphia

Malleable Iron Castings.

Galt Malleable Iron Co., Galt

Mallet, Rawhide and Wood.

Buffalo Foundry Supply Co., [illegible]
Detroit Foundry Supply Co., [illegible]
Paxson, J. W., Co., Philadelphia

Mandrels.

A. B. Jardine & Co., Hespeler Ont.
Miller Bros & Toms, Montreal
The Pratt & Whitney Co., Hartford, Conn.
Standard Tool Co., Cleveland

Marking Machines.

Dwight Slate Machine Co., Hartford, Conn.

Marine Work.

Kennedy, Wm., & Sons, Owen Sound

Metallic Paints.

P. D. Dods & Co., Montreal.

Meters, Electrical.

Canadian Westinghouse Co., Hamilton.
Mechanics Supply Co., Quebec, Que.

Mill Machinery.

Baxter, Paterson & Co., Montreal.
Greey Wm. & J. G., Toronto
The Goldie & McCulloch Co., Galt, Ont.
John McDougall, Caledonian Iron Works Co., Montreal.
H. W. Petrie, Toronto.
Robb Engineering Co., Amherst, N.S.
Waterous Engine Co., Brantford.
Williams & Wilson, Montreal.

Milling Attachments.

Becker-Brainard Milling Machine Co. Hyde Park, Mass.
John Bertram & Sons Co., Dundas, Ont.
Niles-Bement-Pond Co., New York,
Pratt & Whitney, Hartford, Conn.

Milling Machines, Horizontal.

Baxter, Paterson & Co., Montreal.
Becker-Brainard Milling Machinery Co. Hyde Park, Mass.
John Bertram & Sons Co., Dundas, Ont.
London Mach. Tool Co., Hamilton, Ont.
Niles-Bement-Pond Co., New York.
Pratt & Whitney, Hartford, Conn.

Milling Machines, Plain.

American Tool Works Co., Cincinnati.
Becker-Brainard Milling Machine Co. Hyde Park, Mass.
John Bertram & Sons Co., Dundas, Ont.
Canada Machinery Agency, Montreal.
The Canadian Fairbanks Co., Montreal.
London Mach. Tool Co., Hamilton, Ont.
Niles-Bement-Pond Co., New York.
H. W. Petrie, Toronto.
Pratt & Whitney Co., Hartford, Conn.
Williams & Wilson, Montreal.

Milling Machines, Universal.

American Tool Works Co., Cincinnati.
Becker-Brainard Milling Machine Co., Hyde Park, Mass.
John Bertram & Sons Co., Dundas, Ont.
Canada Machinery Agency, Montreal.
The Canadian Fairbanks Co., Montreal.
London Mach. Tool Co., Hamilton, Ont
Niles-Bement-Pond Co., New York.
H. W. Petrie, Toronto.
Williams & Wilson, Montreal.

Milling Machines, Vertical.

Becker-Brainard Milling Machine Co., Hyde Park, Mass.
Brown & Sharpe, Providence, R.I.
John Bertram & Sons Co., Dundas, Ont.
Canada Machinery Agency, Montreal.
London Mach. Tool Co., Hamilton, Ont.
Niles-Bement-Pond Co., New York.

Milling Tools.

Wm. Abbott, Montreal.
Becker-Brainard Milling Machine Co., Hyde Park, Mass.
Geometric Tool Co., New Haven, Conn.
London Mach. Tool Co., Hamilton, Ont.
Pratt & Whitney Co., Hartford, Conn.
Standard Tool Co., Cleveland
Standard Bearings, Ltd., Niagara Falls

Mining Machinery.

Allis-Chalmers-Bullock, Limited, Montreal
Canadian Rand Drill Co., Montreal.
Chicago Pneumatic Tool Co., Chicago.
Jeffrey Mfg. Co., Columbus, Ohio.
Laurie Engine & Machine Co., Montreal.
Rice Lewis & Son, Toronto.
John McDougall, Caledonian Iron Works Co., Montreal.
Miller Bros. & Toms, Montreal.
T. & H. Electric Co. Hamilton.

Mixing Machines, Dough.
Greey, Wm. & J. G., Toronto
Paxson, J. W., Co., Philadelphia

Mixing Machines, Special.
Greey, Wm. & J. G., Toronto
Paxson, J. W., Co., Philadelphia

Model Tools.
Globe Machine & Stamping Co., Cleveland, Ohio.
Standard Bearings, Ltd., Niagara Falls.
Wells Pattern and Model Works, Toronto

Motors, Electric.
Allis-Chalmers-Bullock Limited, Montreal
Canadian General Electric Co., Toronto
Canadian Westinghouse Co., Hamilton.
Electrical Machinery Co., Toronto.
Consolidated Electric Co., Toronto
Gas & Electric Power Co., Toronto
Hall Engineering Works, Montreal.
Mechanics Supply Co., Quebec, Que.
The Packard Electric Co., St. Catharines.
T. & H. Electric Co., Hamilton.

Motors, Air.
Canadian Rand Drill Co., Montreal.
Chicago Pneumatic Tool Co., Chicago.

Molders' Supplies.
Buffalo Foundry Supply Co., Buffalo.
Detroit Foundry Supply Co., Windsor.
Dominion Foundry Supply Co., Montrea
Hamilton Facing Mill Co., Hamilton.
Mechanics Supply Co., Quebec, Que.
Paxson J. W., Co., Philadelphia

Molders' Tools.
Buffalo Foundry Supply Co., Buffalo.
Detroit Foundry Supply Co., Windsor
Dominion Foundry Supply Co., Montreal
Hamilton Facing Mill Co., Hamilton.
Mechanics Supply Co., Quebec, Que.
Paxson, J. W., Co., Philadelphia

Molding Machines.
Buffalo Foundry Supply Co., Buffalo.
Hamilton Facing Mill Co., Hamilton.
J. W. Paxson Co., Philadelphia, Pa.

Molding Sand.
T. W. Barnes, Hamilton.
Buffalo Foundry Supply Co., Buffalo.
Detroit Foundry Supply Co., Windsor
Dominion Foundry Supply Co., Montreal
Hamilton Facing Mill Co., Hamilton.
Paxson, J. W., Co., Philadelphia

Nut Tappers.
John Bertram & Sons Co., Dundas, Ont
A. B. Jardine & Co., Hespeler.
London Mach. Tool Co., Hamilton.
National Machinery Co., Tiffin, Ohio.

Nuts.
Canada Nut Co., Toronto

Oatmeal Mill Machinery.
Greey Wm. & J. G., Toronto
The Goldie & McCulloch Co., Galt

Oilers, Gang.
Sight Feed Oil Pump Co., Milwaukee, Wis.

Oils, Core.
Buffalo Foundry Supply Co., Buffalo.
Dominion Foundry Supply Co., Montreal
Hamilton Facing Mill Co., Hamilton.
Paxson, J. W., Co., Philadelphia

Oil Extractors.
Darling Bros., Ltd., Montreal

Paint Mill Machinery.
Greey, Wm. & J. G., Toronto

Pans, Lathe.
Cleveland Wire Spring Co., Cleveland
Paxson, J. W., Co., Philadelphia

Pans, Steel Shop.
Cleveland Wire Spring Co., Cleveland

Patent Solicitors.
Hanbury A. Budden, Montreal.
Fetherstonhaugh & Blackmore, Montreal
Marion & Marion, Montreal.
Ridout & Maybee, Toronto.

Patterns.
John Carr, Hamilton.
Galt Malleable Iron Co., Galt
Hamilton Pattern Works, Hamilton
John McDougall, Caledonian Iron Works Co., Montreal.
Miller Bros. & Toms, Montreal.
Wells' Pattern and Model Works, Toronto.

Pig Iron.
Hamilton Steel & Iron Co., Hamilton

Pipe Cutting and Threading Machines.
Butterfield & Co., Rock Island, Que.
Canada Machinery Agency, Montreal.
Curtis & Curtis Co., Bridgeport, Conn.
Frothingham & Workman, Ltd., Montreal
Hart Mfg. Co., Cleveland
A. B. Jardine & Co., Hespeler, Ont.
Loew Mfg. Co., Cleveland, Ohio
London Mach. Tool Co., Hamilton, Ont.
Niles-Bement-Pond Co., New York.
Shantz, I. E., & Co., Berlin, Ont.

Pipe, Municipal.
Canadian Pipe Co., Vancouver, B.C.
Pacific Coast Pipe Co., Vancouver, B.C.

Pipe, Waterworks.
Canadian Pipe Co., Vancouver, B.C.
Pacific Coast Pipe Co., Vancouver, B.C.

Planers, Standard.
American Tool Works, Cincinnati.
John Bertram & Sons Co., Dundas, Ont.
Canada Machinery Agency, Montreal.
The Canadian Fairbanks Co., Montreal.
Rice Lewis & Son, Toronto.
London Mach. Tool Co., Hamilton, Ont.
Niles-Bement-Pond Co., New York.
H. W. Petrie, Toronto.
Pratt & Whitney Co., Hartford, Conn.
Williams & Wilson, Montreal.

Planers, Rotary.
John Bertram & Sons Co., Dundas, Ont.
London Mach. Tool Co., Hamilton, Ont.
Niles-Bement-Pond Co., New York.

Planing Mill Fans.
Dominion Heating & Ventilating Co., Hespeler.
Sheldons Limited, Galt, Ont.

Plumbago.
Buffalo Foundry Supply Co., Buffalo.
Detroit Foundry Supply Co., Windsor
Dominion Foundry Supply Co., Montreal
Hamilton Facing Mill Co., Hamilton.
Mechanics Supply Co., Quebec, Que.
J. W. Paxson Co., Philadelphia, Pa.
United States Graphite Co., Saginaw, Mich.

Pneumatic Tools.
Allis-Chalmers-Bullock, Montreal.
Canadian Rand Drill Co., Montreal.
Chicago Pneumatic Tool Co., Chicago
Hamilton Facing Mill Co., Hamilton.
Hanna Engineering Works, Chicago.
Independent Pneumatic Tool Co. Chicago, New York.
Paxson, J. W., Co., Philadelphia

Power Hack Saw Machines.
Baxter, Paterson & Co., Montreal.
Frothingham & Workman, Ltd., Montreal

Power Plants.
John McDougall Caledonian Iron Works Co., Montreal.
The Smart-Turner Mach. Co., Hamilton

Power Plant Equipments.
Darling Bros., Ltd., Montreal

Presses, Drop.
W. H. Banfield & Son, Toronto.
E. W. Bliss Co., Brooklyn, N.Y.
Canada Machinery Agency, Montreal.
Laurie Engine & Machine Co., Montreal
Miller Bros. & Toms, Montreal.
Niles-Bement-Pond Co., New York

Presses, Hand.
E. W. Bliss Co., Brooklyn, N.Y.

Presses, Hydraulic.
John Bertram & Sons Co., Dundas, Ont.
Laurie Engine & Machine Co., Montreal.
London Mach. Tool Co., Hamilton, Ont.
John McDougall Caledonian Iron Works Co., Montreal.
Miller Bros. & Toms, Montreal.
Niles-Bement-Pond Co., New York.
Perrin, Wm. R., & Co., Toronto

Presses, Power.
E. W. Bliss Co., Brooklyn, N.Y.
Canada Machinery Agency, Montreal.
Laurie Engine & Machine Co., Montreal
London Mach. Tool Co., Hamilton, Ont.
John McDougall Caledonian Iron Works Co., Montreal.
Niles-Bement-Pond Co., New York.

Presses Power Screw.
Perrin, Wm. R., & Co., Toronto

Pressure Regulators.
Darling Bros., Ltd., Montreal

Producer Plants.
Canada Foundry Co., Toronto

Pulp Mill Machinery.
Greey, Wm. & J. G., Toronto
Jeffrey Mfg Co., Columbus, Ohio.
Laurie Engine & Machine Co., Montreal.
John McDougall Caledonian Iron Works Co., Montreal.
Waterous Engine Works Co., Brantford

Pulleys.
Baxter, Paterson & Co., Montreal.
Canada Machinery Agency, Montreal.
The Canadian Fairbanks Co., Montreal.
The Goldie & McCulloch Co., Galt.
Greey, Wm. & J. G., Toronto
Laurie Engine & Machine Co., Montreal.
Link-Belt Co., Philadelphia
John McDougall Caledonian Iron Works Co., Montreal.
Miller Bros. & Toms, Montreal.
Owen Sound Iron Works Co., Owen Sound
H. W. Petrie, Toronto.
The Smart-Turner Mach. Co., Hamilton.
Standard Bearings Ltd., Niagara Falls,
Williams & Wilson, Montreal.
Waterous Engine Co., Brantford.

Pumps.
Laurie Engine & Machine Co., Montreal
Ontario Wind Engine & Pump Co., Toronto

Pump Governors.
Darling Bros., Ltd., Montreal

Pumps, Hydraulic.
Laurie Engine & Machine Co., Montrea
Perrin, Wm. R. & Co., Toronto

Pumps, Oil.
Sight Feed Oil Pump Co., Milwaukee, Wis.

Pumps, Steam.
Allis-Chalmers-Bullock, Limited, Montreal
Canada Foundry Co., Toronto.
Canada Machinery Agency, Montreal.
Darling Bros., Ltd., Montreal
The Goldie & McCulloch Co., Galt.
John McDougall Caledonian Iron Works, Montreal.
H. W. Petrie, Toronto.
The Smart-Turner Mach. Co., Hamilton.
Standard Bearings, Ltd., Niagara Falls,
Waterous Engine Co. Brantford.

Pumping Machinery.
Canada Foundry Co., Limited, Toronto
Canada Machinery Agency, Montreal.
Canadian Rand Drill Co., Montreal.
Chicago Pneumatic Tool Co., Chicago.
Darling Bros., Ltd., Montreal
Hall Engineering Works, Montreal, Que.
Laurie Engine & Machine Co., Montreal.
London Mach. Tool Co., Hamilton, Ont.
John McDougall Caledonian Iron Works Co., Montreal.
The Smart-Turner Mach. Co., Hamilton
Standard Bearings, Ltd., Niagara Falls,

Punches and Dies.
W. H. Banfield & Sons, Toronto.
Butterfield & Co., Rock Island.
Globe Machine & Stamping Co.
A. B. Jardine & Co., Hespeler, Ont.
London Mach Tool Co., Hamilton, Ont.
Miller Bros. & Toms, Montreal.
Pratt & Whitney Co., Hartford, Conn.
H. W. Petrie, Toronto.
Standard Bearings, Ltd., Niagara Falls,

Punches, Hand.
Mechanics Supply Co., Quebec, Que.

Punches, Power.
John Bertram & Sons Co., Dundas, Ont.
E. W. Bliss Co., Brooklyn, N.Y.
Canada Machinery Agency, Montreal.
Canadian Buffalo Forge Co., Montreal.
London Mach. Tool Co., Hamilton, Ont.
Niles-Bement-Pond Co., New York.

Punches, Turret.
London Mach. Tool Co., London, Ont.

Punching Machines, Horizontal.
John Bertram & Sons Co., Dundas, Ont.
London Mach. Tool Co., Hamilton, Ont.
Niles-Bement-Pond Co., New York.

Quartering Machines.
John Bertram & Sons Co., Dundas, Ont.
London Mach. Tool Co., Hamilton, Ont.

Railway Spikes and Washers.
Hamilton Steel & Iron Co., Hamilton

Rammers, Bench and Floor.
Buffalo Foundry Supply Co., Buffalo.
Detroit Foundry Supply Co., Windsor
Hamilton Facing Mill Co., Hamilton.
Paxson, J. W., Co., Philadelphia

Rapping Plates.
Detroit Foundry Supply Co., Windsor
Hamilton Facing Mill Co., Hamilton.
Paxson, J. W., Co., Philadelphia.

Raw Hide Pinions.
Brown, David & Sons, Huddersfield, Eng

Reamers.
Wm. Abbott, Montreal.
Baxter Paterson & Co., Montreal.
Butterfield & Co., Rock Island.
Frothingham & Workman, Ltd., Montreal
Hanna Engineering Works, Chicago.
A. B. Jardine & Co., Hespeler, Ont.
Mechanics Supply Co., Quebec, Que.
John Millen & Son, Ltd., Montreal.
Morse Twist Drill and Machine Co., New Bedford, Mass.
Pratt & Whitney Co., Hartford, Conn.
Standard Tool Co., Cleveland.

Reamers, Steel Taper.
Butterfield & Co., Rock Island.
Chicago Pneumatic Tool Co., Chicago.
A. B. Jardine & Co., Hespeler, Ont.
John Millen & Son Ltd., Montreal.
Pratt & Whitney Co., Hartford, Conn.
Standard Tool Co., Cleveland.
Whitman & Barnes Co., St. Catharines.

Rheostats.
Canadian General Electric Co., Toronto.
Canadian Westinghouse Co., Hamilton.
Hall Engineering Works, Montreal, Que.
T. & H. Electric Co., Hamilton,

Riddles.
Buffalo Foundry Supply Co., Buffalo.
Detroit Foundry Supply Co., Windsor
Hamilton Facing Mill Co., Hamilton.
J. W. Paxson Co., Philadelphia, Pa.

Riveters, Pneumatic.
Hanna Engineering Works, Chicago
Independent Pneumatic Tool Co., Chicago Ill.

Rolls, Bending.
John Bertram & Sons Co., Dundas, Ont.
London Mach. Tool Co., Hamilton, Ont.
Niles-Bement-Pond Co., New York.

Rolls, Chilled Iron.
Greey, Wm. & J. G., Toronto

Rolls, Sand Cast.
Greey, Wm. & J. G., Toronto

Rotary Blowers.
Paxson, J. W., Co., Philadelphia

Rotary Converters.
Allis-Chalmers-Bullock, Ltd., Montreal.
Canadian Westinghouse Co., Hamilton
Paxson, J. W., Co. Philadelphia
Toronto and Hamilton Electric Co., Hamilton.

Rubbing and Sharpening Stones.
Norton Co., Worcester, Mass.

Safes.
Baxter Paterson & Co., Montreal.
The Goldie & McCulloch Co., Galt,

Sand, Bench.
Buffalo Foundry Supply Co., Buffalo.
Detroit Foundry Supply Co., Windsor
Hamilton Facing Mill Co., Hamilton.
Paxson J. W., Co., Philadelphia

Sand Blast Machinery.
Canadian Rand Drill Co., Montreal.
Chicago Pneumatic Tool Co., Chicago.
Detroit Foundry Supply Co., Windsor
Hamilton Facing Mill Co., Hamilton.
J. W. Paxson Co., Philadelphia, Pa.

Sand, Heavy, Grey Iron.
Buffalo Foundry Supply Co., Buffalo.
Detroit Foundry Supply Co., Windsor
Hamilton Facing Mill Co., Hamilton.
Paxson, J. W., Co., Philadelphia

Sand, Malleable.
Buffalo Foundry Supply Co., Buffalo.
Detroit Foundry Supply Co., Windsor
Hamilton Facing Mill Co., Hamilton.
Paxson, J. W., Co., Philadelphia

Sand, Medium Grey Iron.
Buffalo Foundry Supply Co., Buffalo.
Detroit Foundry Supply Co., Windsor
Hamilton Facing Mill Co., Hamilton.
Paxson, J. W., Co., Philadelphia

Sand Sifters.
Buffalo Foundry Supply Co., Buffalo.
Detroit Foundry Supply Co., Windsor
Dominion Foundry Supply Co., Montreal
Hamilton Facing Mill Co., Hamilton.
Hanna Engineering Works, Chicago.
Paxson, J. W., Co., Philadelphia

Saw Mill Machinery.
Allis-Chalmers-Bullock, Limited, Montreal
Baxter, Paterson & Co., Montreal.
Canada Machinery Agency, Montreal.
Goldie & McCulloch Co., Galt.
Greey, Wm. & J. G., Toronto
Miller Bros. & Toms, Montreal
Owen Sound Iron Works Co., Owen Sound
H. W. Petrie, Toronto.
Robb Engineering Co., Amherst,
Standard Bearings Ltd., Niagara Falls
Waterous Engine Works, Brantford
Williams & Wilson, Montreal.

Sawing Machines, Metal.
Niles-Bement-Pond Co. New York.
Paxson, J. W., Co., Philadelphia

Saws, Hack.
Baxter Paterson & Co., Montreal.
Canada Machinery Agency, Montreal.
Detroit Foundry Supply Co., Windsor
Frothingham & Workman, Ltd., Montreal
London Mach. Tool Co., Hamilton
Mechanics Supply Co., Quebec, Que.
Rice Lewis & Son, Toronto.
John Millen & Son, Ltd., Montreal.
L. S. Starrett Co., Athol, Mass.

Screw Cutting Tools
Hart Manufacturing Co., Cleveland, Ohio
Mechanics Supply Co., Quebec, Que.

Screw Machines, Automatic.
Canada Machinery Agency, Montreal.
Cleveland Automatic Machine Co., Cleveland, Ohio
London Mach. Tool Co., Hamilton, Ont.
National-Acme Mfg. Co., Cleveland.
Pratt & Whitney Co., Hartford, Conn.

Screw Machines, Hand.
Canada Machinery Agency Montreal.
A. B. Jardine & Co., Hespeler.
London Mach. Tool Co., Hamilton, Ont.
Mechanics Supply Co., Quebec, Que.
Pratt & Whitney Co., Hartford. Conn.
Warner& Swasey Co., Cleveland, O.

Screw Machines, Multiple Spindle.
National-Acme Mfg. Co., Cleveland

Screw Plates.
Butterfield & Co., Rock Island, Que.
Frothingham & Workman, Ltd., Montreal
Hart Manufacturing Co. Cleveland, Ohio
A. B. Jardine & Co., Hespeler
Mechanics Supply Co., Quebec, Que.

Screws, Cap and Set.
National-Acme Mfg. Co., Cleveland

Screw Slotting Machinery, Semi-Automatic.
National-Acme Mfg. Co., Cleveland

Second-Hand Machinery.
American Tool Works Co., Cincinnati
Canada Machinery Agency, Montreal.
The Canadian Fairbanks Co., Montreal.
Goldie & McCulloch Co., Galt.
Machinery Exchange, Montreal.
Niles-Bement-Pond Co., New York.
H. W. Petrie, Toronto.
Robb Engineering Co., Amherst, N.S.
Waterous Engine Co., Brantford.
Williams & Wilson, Montreal.

Shafting.
Baxter, Paterson & Co., Montreal.
Canada Machinery Agency, Montreal
The Canadian Fairbanks Co., Montreal.
Frothingham & Workman, Ltd., Montreal
The Goldie & McCulloch Co., Galt, Ont
Greey, Wm., & J. G., Toronto
Kennedy, Wm., & Sons, Owen Sound
Niles-Bement-Pond Co., New York.
Owen Sound Iron Works Co., Owen Sound
H. W. Petrie, Toronto.
Smart-Turner Machine Co., Hamilton.
Union Drawn Steel Co., Hamilton.
Waterous Engine Co., Brantford.

Shapers.
American Tool Works Co., Cincinnati.
John Bertram & Sons Co., Dundas, Ont
Canada Machinery Agency, Montreal.
The Canadian Fairbanks Co., Montreal.
Rice Lewis & Son, Toronto.
London Mach. Tool Co., Hamilton, Ont.
Niles-Bement-Pond Co., New York.
H. W. Petrie, Toronto.
Potter & Johnston Machine Co., Pawtucket, R.I.
Pratt & Whitney Co., Hartford, Conn.
Williams & Wilson, Montreal.

Shearing Machine, Bar.
John Bertram & Sons Co., Dundas, Ont.
A. B. Jardine & Co., Hespeler.
London Mach. Tool Co., Hamilton, Ont.
Niles-Bement-Pond Co., New York.

Shears, Power.
John Bertram & Sons Co., Dundas, Ont
Canada Machinery Agency, Montreal.
A. B. Jardine & Co., Hespeler, Ont.
Niles-Bement-Pond Co., New York.
Paxson, J. W., Co., Philadelphia

Sheet Metal Goods
Globe Machine & Stamping Co., Cleveland, Ohio.

Sheet Steel Work.
Owen Sound Iron Works Co., Owen Sound

Shingle Mill Machinery.
Owen Sound Iron Works Co., Owen Sound

Shop Trucks.
Greey, Wm. & J. G., Toronto
Owen Sound Iron Works Co., Owen Sound
Paxson, J. W., Co., Philadelphia

Shovels.
Baxter, Paterson & Co., Montreal.
Buffalo Foundry Supply Co., Buffalo.
Detroit Foundry Supply Co., Detroit.
Frothingham & Workman, Ltd., Montreal
Dominion Foundry Supply Co., Montreal
Hamilton Facing Mill Co., Hamilton.
Mechanics Supply Co., Quebec, Que.
Paxson, J. W., Co., Philadelphia

Shovels, Steam.
Allis-Chalmers-Bullock, Montreal.

Sieves.
Buffalo Foundry Supply Co., Buffalo.
Detroit Foundry Supply Co., Windsor
Dominion Foundry Supply Co., Montrea
Hamilton Facing Mill Co., Hamilton.
Paxson, J. W., Co., Philadelphia

Silver Lead.
Buffalo Foundry Supply Co., Buffalo.
Detroit Foundry Supply Co., Windsor
Dominion Foundry Supply Co., Montreal
Hamilton Facing Mill Co., Hamilton.
Paxson, J. W., Co., Philadelphia

Sleeves, Reducing.
Chicago Pneumatic Tool Co., Chicago.

Snap Flasks
Buffalo Foundry Supply Co., Buffalo.
Detroit Foundry Supply Co., Windsor
Dominion Foundry Supply Co., Montre
Hamilton Facing Mill Co., Hamilton.
Paxson, J. W., Co., Philadelphia

Soapstone.
Buffalo Foundry Supply Co., Buffalo.
Detroit Foundry Supply Co., Windsor
Dominion Foundry Supply Co., Montreal
Hamilton Facing Mill Co., Hamilton.
Paxson, J. W., Co., Philadelphia

Solders.
Lumen Bearing Co., Toronto

Special Machinery.
W. H. Banfield & Sons, Toronto.
Baxter, Paterson & Co., Montreal.
John Bertram & Sons Co. Dundas, Ont
Globe Machine & Stamping Co., Cleveland, Ohio.
Greey, Wm. & J. G., Toronto
Hanna Engineering Works, Chicago.
Laurie Engine & Machine Co., Montreal.
London Mach. Tool Co., Hamilton. Ont.
Pendrith Machinery Co., Toronto.
H. W. Petrie, Toronto.
The Smart-Turner Mach. Co., Hamilton
Standard Bearings, Ltd., Niagara Falls,
Waterous Engine Co., Brantford.

Special Machines and Tools.
Paxson, J. W., Philadelphia
Pratt & Whitney, Hartford, Conn.
Standard Bearings, Ltd., Niagara Falls,

Special Milled Work.
National-Acme Mfg. Co., Cleveland

Speed Changing Countershafts.
The Canadian Fairbanks Co., Montreal

Spike Machines.
National Machinery Co., Tiffin, O.
The Smart-Turner Mach. Co., Hamilton

Spray Cans.
Detroit Foundry Supply Co., Windsor
Dominion Foundry Supply Co., Montreal
Hamilton Facing Mill Co., Hamilton.
Paxson, J. W., Co., Philadelphia

Springs, Automobile.
Cleveland Wire Spring Co., Cleveland

Springs, Coiled Wire.
Cleveland Wire Spring Co., Cleveland

Springs, Machinery.
Cleveland Wire Spring Co., Cleveland

Springs, Upholstery.
Cleveland Wire Spring Co., Cleveland

Sprue Cutters.
Detroit Foundry Supply Co., Windsor
Dominion Foundry Supply Co., Montreal
Hamilton Facing Mill Co., Hamilton.
Paxson, J. W., Co., Philadelphia

Stamp Mills.
Allis-Chalmers-Bullock, Limited, Montreal
Paxson, J. W., Co., Philadelphia

Steam Hot Blast Apparatus
Dominion Heating & Ventilating Co. Hespeler, Ont.

Steam Separators.

Darling Bros., Ltd., Montreal.
Dominion Heating & Ventilating Co., Hespeler
Robb Engineering Co., Montreal.
Sheldon's Limited, Galt.
Smart-Turner Mach. Co., Hamilton
Waterous Engine Co., Brantford.

Steam Specialties.

Darling Bros., Ltd., Montreal
Dominion Heating & Ventilating Co., Hespeler
Sheldon's Limited, Galt.

Steam Traps.

Canada Machinery Agency, Montreal.
Darling Bros., Ltd., Moutreal
Dominion Heating & Ventilating Co., Hespeler.
Mechanics Supply Co., Quebec, Que.
Sheldons Limited, Galt.

Steam Valves.

Darling Bros., Ltd., Montreal

Steel Pressure Blowers

Buffalo Foundry Supply Co., Buffalo.
Dominion Foundry Supply Co., Montreal
Domininion Heating & Ventilating Co., Hespeler
Hamilton Facing Mill Co., Hamilton.
J. W. Paxon Co., Philadelphia, Pa.
Sheldon's Limited, Galt.

Steel Tubes.

Baxter, Paterson & Co., Montreal.
Mechanics Supply Co., Quebec, Que.
John Millen & Son, Montreal.

Steel, High Speed.

Wm. Abbott, Montreal.
Canadian Fairbanks Co., Montreal.
Frothingham & Workman, Ltd., Montreal
Alex. Gibb, Montreal.
Jessop, Wm., & Sons, Sheffield, Eng.
B. K. Morton Co., Sheffield, Eng.
Williams & Wilson, Montreal.

Stocks and Dies

Hart Manufacturing Co., Cleveland, Ohio
Mechanics Supply Co., Quebec, Que.
Pratt & Whitney Co., Hartford, Conn.

Stone Cutting Tools, Pneumatic

Allis-Chalmers-Bullock, Ltd., Montreal
Canadian Rand Drill Co., Montreal.

Stone Surfacers

Chicago Pneumatic Tool Co., Chicago.

Stove Plate Facings.

Buffalo Foundry Supply Co., Buffalo
Detroit Foundry Supply Co., Windsor.
Dominion Foundry Supply Co., Montreal
Hamilton Facing Mill Co., Hamilton.
Paxson, J. W., Co., Philadelphia

Swage, Block.

A. B. Jardine & Co., Hespeler, Ont.

Switchboards.

Allis-Chalmers-Bullock, Limited, Montreal
Canadian General Electric Co., Toronto.
Canadian Westinghouse Co., Hamilton
Hall Engineering Works, Montreal, Que.
Mechanics Supply Co., Quebec, Que.
Toronto and Hamilton Electric Co., Hamiton.

Talc.

Buffalo Foundry Supply Co., Buffalo.
Detroit Foundry Supply Co., Windsor.
Hamiton Facing Mill Co., Hamilton.
Paxson, J. W., Co., Philadelphia

Tanks, Oil.

Sight Feed Oil Pump Co., Milwaukee, Wis.

Tapping Machines and Attachments.

American Tool Works Co., Cincinnati.
John Bertram & Sons Co., Dundas, Ont.
Bickford Drill & Tool Co., Cincinnati.
The Geometric Tool Co., New Haven,
A. B. Jardine & Co., Hespeler.
London Mach. Tool Co., Hamilton, Ont
Murchey Machine & Tool Co., Detroit.
Niles-Bement-Pond Co., New York.
H. W. Petrie, Toronto.
Pratt & Whitney, Cincinnati, O.
L. S. Starrett Co., Athol, Mass.
Williams & Wilson, Montreal.

Tapes, Steel.

Frothingham & Workman, Ltd., Montreal
Rice Lewis & Son, Toronto.
Mechanics Supply Co, Quebec, Que.
John Millen & Son, Ltd, Montreal.
L. S. Starrett Co., Athol, Mass.

Taps.

Baxter, Paterson & Co, Montreal.
Mechanics Supply Co., Quebec, Que.

Taps, Collapsing.

The Geometric Tool Co., New Haven.

Taps and Dies.

Wm. Abbott, Montreal.
Baxter, Paterson & Co., Montreal.
Butterfield & Co., Rock Island, Que.
Frothingham & Workman, Ltd., Montreal
The Geometric Tool Co., New Haven.
A. B. Jardine & Co., Hespeler, Ont.
Rice Lewis & Son, Toronto.
Mechanics Supply Co., Quebec, Que.
John Millen & Son, Ltd, Montreal.
Pratt & Whitney Co., Hartford, Conn.
Standard Tool Co., Cleveland.
L. S. Starrett Co., Athol, Mass.

Testing Laboratory.

Detroit Testing Laboratory, Det roit

Testing Machines.

Detroit Found.y Supply Co., Windsor.

Dominion Foundry Supply Co., Montreal
Hamilton Facing Mill Co., Hamilton.
Paxson, J. W., Co., Philadelphia

Thread Cutting Tools.

Hart Manufacturing Co, Cleveland, Ohio
Mechanics Supply Co., Quebec, Que.

Tiling, Opal Glass.

Toronto Plate Glass Importing Co., Toronto.

Time Switches, Automatic, Electric.

Berlin Electric Mfg. Co., Toronto.

Tinware Machinery

Canada Machinery Agency, Montreal.

Tire Upsetters or Shrinkers

A. B. Jardine & Co., Hespeler, Ont.

Tool Cutting Machinery

Canadian Rand Drill Co., Montreal.

Tool Holders.

Armstrong Bros. Tool Co., Chicago.
Baxter Paterson & Co., Montreal.
John Millen & Son, Ltd, Montreal.
H. W. Petrie, Toronto.
Pratt & Whitney Co., Hartford, Conn.

Tool Steel.

Wm. Abbott, Montreal.
Flockton, Tompkin & Co., Sheffield, Eng.
Frothingham & Workman, Ltd., Montreal
Wm. Jessop, Sons & Co., Toronto.
Canadian Fairbanks Co., Montreal
B. K. Morton & Co., Sheffield, Eng.
Williams & Wilson, Montral.

Torches, Steel.

Baxter, Paterson & Co., Montreal.
Detroit Foundry Supply Co., Windsor.
Dominion Foundry Supply Co., Montreal
Hamilton Facing Mill Co., Hamilton.
Paxson, J. W., Co., Philadelphia

Transformers and Convertors

Allis-Chalmers-Bullock, Montreal.
Canadian General Electric Co., Toronto.
Canadian Westinghouse Co., Hamilton.
Hall Engineering Works, Montreal, Que.
T. & H. Electric Co., Hamilton.

Transmission Machinery

Allis-Chalmers-Bullock, Montreal.
The Canadian Fairbanks Co., Montreal.
Greey, Wm. & J. G., Toronto.
Laurie Engine & Machine Co., Montreal.
Link-Belt Co., Philadelphia.
Miller Bros. & Toms, Montreal.
Paxson, J. W., Co., Philadelphia
H. W. Petrie, Toronto.
The Smart-Turner Mach. Co., Hamilton.
Waterous Engine Co., Brantford.

Transmission Supplies

Baxter, Paterson & Co., Montreal.
The Canadian Fairbanks Co, Montreal.
Wm. & J. G. Greey, Toronto.
The Goldie & McCulloch Co., Galt.
Miller Bros. & Toms, Montreal.
H. W. Petrie, Toronto.

Trolleys

Canadian Rand Drill Co., Montreal.
John Millen & Son, Ltd., Montreal.
Miller Bros. & Toms, Montreal.
Northern Engineering Works Detroit

Trolley Wheels.

Lumen Bearing Co., Toronto

Trucks, Dryer and Factory

Dominion Heating & Ventilating Co., Hespeler
Greey, Wm. & J. G., Toronto.
Northern Engineering Works, Detroit
Sheldon's Limited, Galt, Ont.

Tube Expanders (Rollers)

Chicago Pneumatic Tool Co., Chicago.
A. B. Jardine & Co, Hespeler.
Mechanics Supply Co., Quebec, Que.

Turbines, Steam.

Allis-Chalmers-Bullock, Limited, Montreal
Canadian General Electric Co., Toronto.
Canadian Westinghouse Co., Hamilton.
Kerr Turbine Co., Wellsville, N.Y.

Turntables

Detroit Foundry Supply Co., Windsor.
Dominion Foundry Supply Co., Montreal
Hamilton Facing Mill Co., Hamilton.
Northern Engineering Works, Detroit
Paxson, J. W., Co., Philadelphia

Turret Machines.

American Tool Works Co., Cincinnati.
John Bertram & Sons Co., Dundas, Ont.
The Canadian Fairbanks Co., Montreal.
London Mach. Tool Co., Hamilton, Ont.
Niles-Bement-Pond Co., New York.
H. W. Petrie, Toronto.
Pratt & Whitney Co., Hartford, Conn.
Warner & Swasey Co., Cleveland, Ohio.
Williams & Wilson, Montreal.

Upsetting and Bending Machinery.

John Bertram & Sons Co., Dundas, Ont.
A. B. Jardine & Co., Hespeler.
London Mach. Tool Co., Hamilton, Ont.
National Machinery Co., Tiffin, O.
Niles-Bement-Pond Co., New York.

Valves, Blow-off.

Chicago Pneumatic Tool Co., Chicago.
Mechanics Supply Co., Quebec, Que.

ALPHABETICAL INDEX.

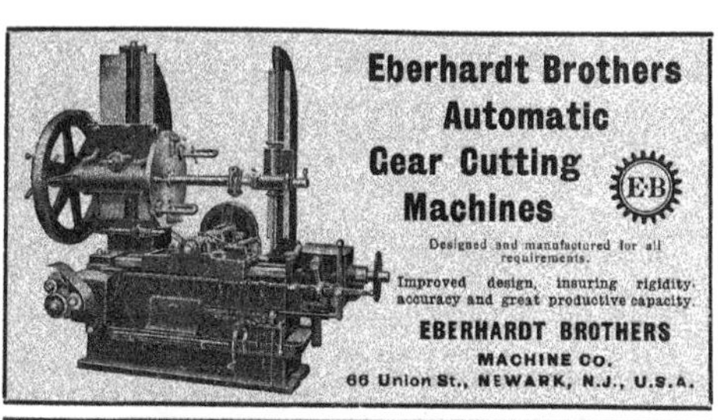

Eberhardt Brothers Automatic Gear Cutting Machines

Designed and manufactured for all requirements.

Improved design, insuring rigidity, accuracy and great productive capacity.

EBERHARDT BROTHERS MACHINE CO.
66 Union St., NEWARK, N.J., U.S.A.

WM. KENNEDY & SONS

HIGH-CLASS

Steel Gears of all kinds
Iron Gears
Grey Iron, Steel and
Brass Castings

MARINE WORK receives every attention

WM. KENNEDY & SONS, Limited
Owen Sound, Ont.

PIPE CUTTING AND THREADING MACHINES

FORBES Patent Die Stock

For Bench or Trench, Hand or Power. Range from ⅛ to 15 inches.
PORTABLE, SIMPLE, STRONG, DURABLE.
WRITE FOR CATALOG.

THE CURTIS & CURTIS COMPANY,
36 Garden Street, Bridgeport, Connecticut

DO YOU DO IT?

When you read an advertisement that interests you, why not write the advertiser and get more information? It will help you to keep posted. Try it. The advertisers in **Canadian Machinery** will appreciate it.

WRITE THEM. TRY IT.

If you were at Philadelphia, you doubtless saw our Cleary Air Cooled Gagger Board and our Roller Bearing Sand Sifter

If you have air in your shop, you will find that a bountiful supply of gaggers will always be on hand and many a casting saved that is otherwise lost through lack of gaggers to hold up the sand.

For mixing facing or core sand by hand or power the Detroit Revolving Sand Sifter will do the work of four men. Did you ever think of making money this way?

Price—Large size, $30.00; small size, $23.00; f.o.b. Windsor.

Price, $27.00; f.o.b. Windsor

Exclusive Canadian Sales Agents Perfection Parting—Splendid substitute for Lycapodium, at less than half the Price

THE DETROIT FOUNDRY SUPPLY CO.

Facing Mills—DETROIT, MICH. *Warehouses—DETROIT and WINDSOR*

T. Pringle & Son

HYDRAULIC, MILL & ELECTRICAL ENGINEERS

FACTORY & MILL CONSTRUCTION A SPECIALTY.

Coristine Bldg., St. Nicholas St., Montreal.

RODERICK J. PARKE

A.M. Can. Soc. C.E. A.M. Amer. Inst. E.E.

CONSULTING ELECTRICAL ENGINEER

INDUSTRIAL STEAM AND ELECTRIC POWER PLANTS DESIGNED. TESTS. REPORTS.

51-53 JANES BLDG., TORONTO, CAN
Long Distance Telephones—Office and Residence.

J. A. DeCew

Chemical Engineer

Sun Life Building, MONTREAL

Industrial Plants and Processes sold and Installed.
Apparatus and Materials for Water Purification.
Free Analyses to prospective purchasers.

PLEWS & TRIMINGHAM.

Consulting Engineers.

ELECTRICAL INSPECTION BUREAU AND TESTING LABORATORY.

40 HOSPITAL & 22 ST. JOHN STS. MONTREAL.

JULES DE CLERCY, M. E. Gas Engineer

Expert knowledge of all classes of Engines and Producer Plants. If you are thinking of installing a Gas Engine or Producer Plant write me. If your plant is not working satisfactorily I can help you.

413 Dorchester St. - MONTREAL, QUE.

HANBURY A. BUDDEN

Advocate Patent Agent.
New York Life Building MONTREAL.
Cable Address, BREVET MONTREAL.

THE ELECTRIC BLUE PRINT CO.

Blue Prints Positive Blue Prints
Black Prints Multi-color Prints

Largest and finest plant.
Best finished work in Canada at lowest prices.

40 Hospital St., - MONTREAL

PATENTS *TRADE MARKS AND DESIGNS*

PROCURED IN ALL COUNTRIES

Special Attention given to Patent Litigation

Pamphlet sent free on application.

RIDOUT & MAYBEE *103 BAY STREET TORONTO*

PATENTS PROMPTLY SECURED

We solicit the business of Manufacturers, Engineers and others who realize the advisability of having their Patent business transacted by Experts. Preliminary advice free. Charges moderate. Our Inventor's Adviser sent upon request. Marion & Marion, New York Life Bldg, Montreal; and Washington, D.C., U.S.A.

PATENTS THAT PROTECT

FETHERSTONHAUGH & CO.

Patent Solicitors & Experts

Fred, B. Fetherstonhaugh, M.E., Barrister-at-law and Counsel and expert in Patent Causes. Charles W. Taylor, B.Sc., formerly Examiner in Can. Patent Office.

MONTREAL CAN. LIFE BLDG.

TORONTO HEAD OFFICE, CAN. BANK COMMERCE BLDG.

SOMERVILLE & VAN EVERY

CONSULTING AND SUPERVISING ENGINEERS

Reports and Valuations, Power Developments
Manufacturing Plants Designed and Erected
Equipments Installed

Federal Life Bldg., HAMILTON, ONT.

THE DETROIT TESTING LABORATORY

1111 Union Trust Building

DETROIT, MICH.

Experienced Foundry Chemists and Metallurgists.
Iron Mixtures for all classes of Castings our Specialty.
Reasonable charges for Analyses.

LANE & CARR

Foundry and Metallurgical

ENGINEERS

Steel Foundry, Gray Iron and Malleable Iron Practice,

Including furnace designs, plant designs and Metallurgical consultation.

W. M. CARR, 120 Liberty St., New York City

H. M. LANE, 1137 Schofield Bldg., Cleveland, O.

Canadian Machinery and Manufacturing News

is a medium through which you can talk to 95 per cent. of the citizens of Canada who must be consulted when Machinery, Tools, Power Equipment and Industrial Supplies are to be bought. If your announcement does not appear regularly you are missing opportunities for increased business.

Let us Send you Rates and Further Particulars

Persons addressing advertisers will kindly mention having seen their advertisement in this paper.

BOYER AND KELLER RIVETING AND CHIPPING HAMMERS

AND

"LITTLE GIANT" DRILLS

HAVE DEMONSTRATED THEIR SUPERIORITY OVER ALL OTHERS

Examine the vital parts—note the quality of material used in their construction and simplicity of mechanical parts, then give them a trial on your own work. Tools shipped free for demonstrative purposes to responsible parties.

DUNTLEY

AIR COOLED PORTABLE

ELECTRIC TOOLS

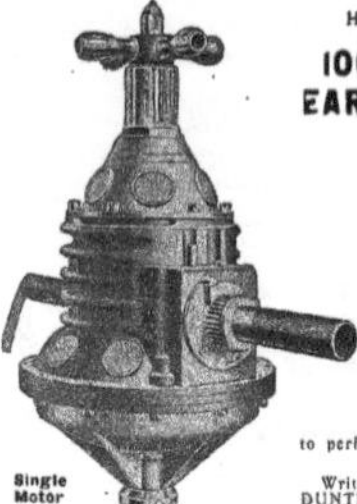

Single Motor Type Drill

HAVE A RECORD OF

100 PER CENT. EARNING POWER

HAVE SAVED THEIR FIRST COST

in a few months for many purchasers.

EVERY TOOL GUARANTEED

to perform the service for which it is designed.

Write for catalogue describing DUNTLEY ELECTRIC DRILLS, HOISTS, GRINDERS, BUFFERS AND BLOWERS.

CANADIAN PNEUMATIC TOOL COMPANY

TORONTO MONTREAL

BRANCH OF

CHICAGO PNEUMATIC TOOL COMPANY

CHICAGO NEW YORK

SALES AGENTS FOR CANADA:

N. J. HOLDEN & CO., Limited, MONTREAL

Rand Sheet Pile Driver

Operated either by compressed air or steam.

Does the work of from 15 to 20 men and requires only two men for its operation.

Once having been placed on the top of a pile, it drives down rapidly and evenly with little attention from the operator and with little danger of splitting the end of the pile.

Canadian Rand Company,
LIMITED

58 Sovereign Bank Building, - Montreal, Que.

The Place to Secure THE BEST

CASTINGS

GREY IRON

BRASS BRONZE

ALUMINUM

ROUGH AND MACHINED

Hoisting Engines
Power Hoists

ELECTRIC, STEAM, BELT AND HAND POWER

Derricks and Contractors' Machinery

FOR FURTHER PARTICULARS ADDRESS

NIAGARA FALLS MACHINE AND FOUNDRY COMPANY, LIMITED

NIAGARA FALLS, Ont.

ISLAND CITY

A NEW TRANSFER CAR

For Street Car purposes, operated by only one man, with a single lever. No motors required. New design.

ST. CLAIR BROS., - - - - GALT, ONT.

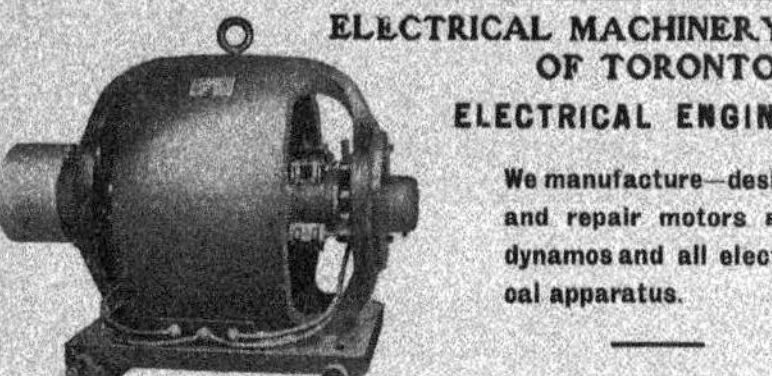
Type B—Direct Current 5 to 75 Horse Power

ELECTRICAL MACHINERY CO. OF TORONTO

ELECTRICAL ENGINEERS

We manufacture—design and repair motors and dynamos and all electrical apparatus.

Office and Works
113-115 Simcoe St., Toronto

Hydraulic Presses
Power Screw Presses
Filter Presses

William R. Perrin AND Company, Limited, TORONTO, Canada.

REGISTERED 1901

King Edward Motors and Dynamos

ARE THE BEST THAT CAN BE BUILT

Prompt repairs to all makes

Consolidated Electric Co. LIMITED
710-724 Yonge St., TORONTO

ALTERNATE CURRENT DYNAMOS and MOTORS

FOR ALL CIRCUITS

TORONTO & HAMILTON ELECTRIC CO.
99-101 McNab Street, HAMILTON, ONT.

The Eugene F. Phillips Electrical Works, Limited

Montreal, Canada

TORONTO STORE - - - 67 ADELAIDE ST. EAST

Magnet and Trolley Wire
Bare Copper Wire Weatherproof Wire

Satisfaction

and

Accuracy Guaranteed

REAMERS

bearing our name are preferred by mechanics who wish to excel in their work. Our name is a guarantee of quality.

Butterfield & Co.

Rock Island, P. Q.

THE STANDARD TOOL CO'S REAMERS

All Kinds for All Purposes

Accuracy is essential in Reamers. Standard Reamers are subjected to careful micrometer inspection all through the various operations, with the result that we can guarantee them to be accurate.

OFFICE and FACTORY
Cleveland, Ohio

EASTERN SALESROOM
94 Reade St., New York

Armstrong Tool Holders

Have saved years of time and thousands of tons of Tool Steel, here's TWO NEW THINGS will prove an equally good investment.

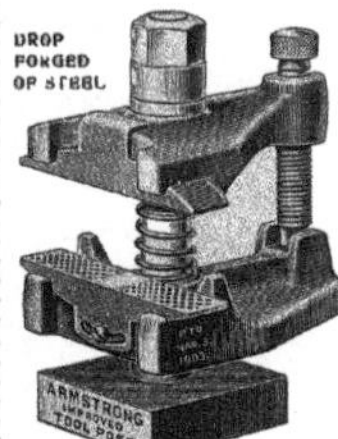

THE ARMSTRONG IMPROVED TOOL POST

Combines the strength and holding power of the strap and stud tool clamp with the convenience of the "open side" and ordinary set screw tool post. A slight turn of one nut releases tool or sets it solidly.

No trouble with "stripping" or "upset" screws.

The ARMSTRONG THREE-BAR BORING TOOL

INCREASES PRODUCTION AND REDUCES COST OF TOOL MAINTENANCE

CUTTERS CANNOT JAR LOOSE—

HIGH SPEEDS AND BIG FEEDS —ONLY SET THEM TIGHTER

SOLD IN CANADA BY:—Williams & Wilson, Montreal; Aikenhead Hardware, Limited, Toronto; A. R. Williams Machinery Co., Toronto, Winnipeg, Vancouver; Wood, Vallance & Co., Hamilton; Frothingham & Workman, Limited, Montreal; H. W. Petrie, Toronto.

WRITE FOR OUR LATEST CATALOG.

ARMSTRONG BROS. TOOL CO., 106 N. FRANCISCO AVE. CHICAGO, U.S.A.

"THE TOOL HOLDER PEOPLE"

NEW AND UP-TO-DATE

BOLT AND NUT MACHINERY

INCLUDING

Bolt Cutters, Nut Tappers, Bolt Headers, Upsetting and Forging Machines, Wire Nail and Spike Machines and Bulldozers.

Send for Catalogue M.

NATIONAL MACHINERY CO., Tiffin, Ohio, U.S.A.

Canadian Agents: H. W. PETRIE, Toronto, Ont. WILLIAMS & WILSON, Montreal, Que.

SPECIAL TAPS

We have first-class appliances for making Special Taps. You can depend on work and material being right. Send proper specifications and you will get just what you want at a reasonable price.

We solicit your orders.

A full stock of regular taps kept on hand.

A. B. JARDINE & CO., Hespeler, Ont.

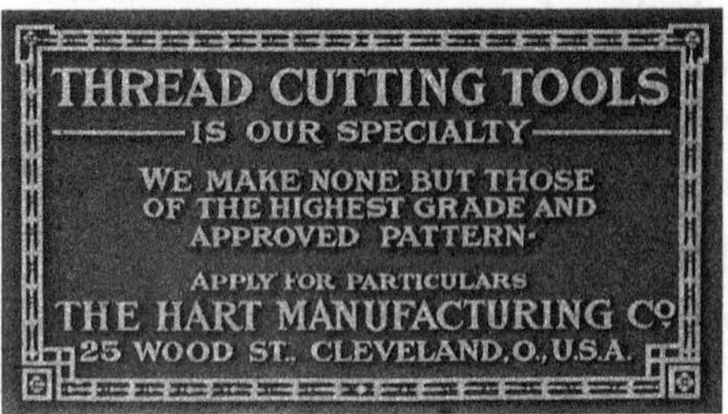

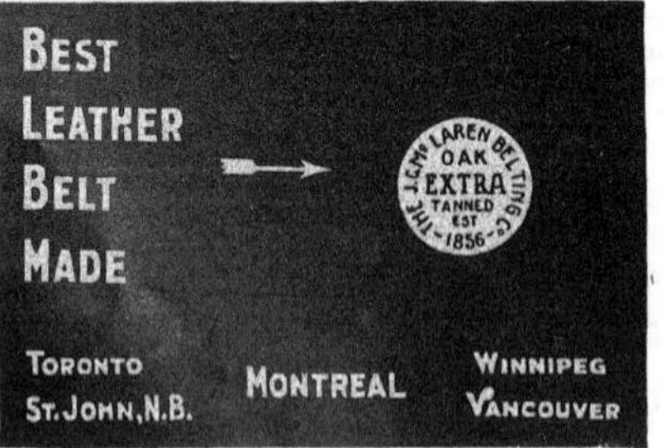

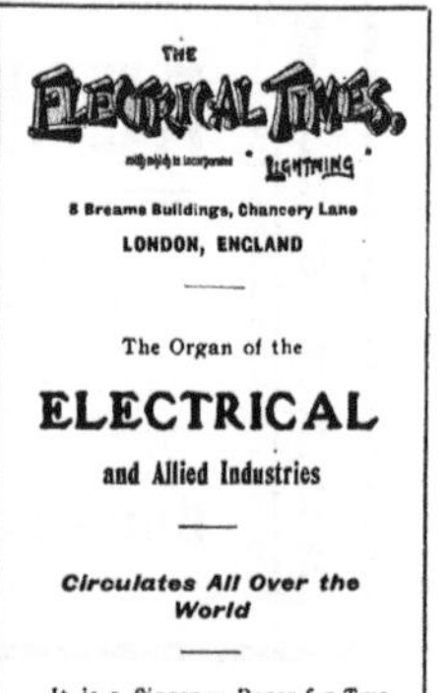

Circulates All Over the World

It is a Sixpenny Paper for Twopence; and is more widely read than any other British Electrical Journal. Means business all the time.

Annual Subscription:

Post Free, 13/6, Foreign; 10/6, Home. Published weekly at 2d.

YOU WILL SAVE MONEY IN BUYING IF YOU REFER TO OUR BUYERS' DIRECTORY

CANADIAN MACHINERY AND MANUFACTURING NEWS

A monthly newspaper devoted to the manufacturing interests, covering in a practical manner the mechanical, power, foundry and allied fields. Published by The MacLean Publishing Company, Limited, Toronto, Montreal, Winnipeg, and London, Eng.

OFFICE OF PUBLICATION: 10 FRONT STREET EAST, TORONTO

Vol. III. AUGUST, 1907 No. 8

The New Automatic Sizing Grinder

A perfectly developed machine for cylindrical work up to 5 inches in diameter and 48 inches in length.

New 5x48 Pratt & Whitney Automatic Sizing Grinder

Accurate Grinding with Unskilled Labor

The design is such as to produce accurate work without the attention of a skilled mechanic. An arm which rests on the work being ground is set with a micrometer screw to make an electric contact when the work is ground to desired size. The wear of the emery wheel has no effect whatever, as the machine will continue to feed until the electric contact is obtained. When work is nearly to size, machine changes to a very fine feed which is thrown out entirely when work is ground to exact size.

Pratt & Whitney Company

Hartford, Conn., U.S.A

SALES AGENTS—THE CANADIAN FAIRBANKS COMPANY, LIMITED

Montreal Toronto Winnipeg Vancouver

THE "AMERICAN" RADIAL DRILL

2½' to 7' Arm—Belt or Motor Driven.

CATALOG "C" NOW READY

HAS BROKEN ALL RECORDS FOR RAPID DRILLING

A 1¼ in. Hole thru 7½ in. Cast Iron in One Minute is what the 4-foot "American" Radial does.

THERE ARE GOOD REASONS FOR THIS ENORMOUS POWER.

Built with Speed Box or Cone Pulley Drive.

☞ **NOTICE** To receive new catalogue when issued, machine tool users please send postal with company's name and mechanical officer in charge thereof.

THE AMERICAN TOOL WORKS CO.

700 to 750 CULVERT STREET - CINCINNATI, OHIO, U.S.A.

LATHES — PLANERS — SHAPERS — RADIAL DRILLS

Write Direct or to Agents: H. W. PETRIE, Montreal; A. R. WILLIAMS MACHINERY CO., Toronto, Winnipeg, Vancouver. 9

Becker - Brainard

Milling Machines

and

Milling Cutters

Guarantee ACCURACY and ECONOMY in Production

Becker - Brainard Milling Machine Co.

Hyde Park, Mass, U.S.A. Branch Offices: THE BOURSE, Phila, Pa.; WILLIAMSON BLDG., Cleveland, O.

Canadian Agents: A. R. WILLIAMS MACHINERY CO., LTD., TORONTO and MONTREAL, CAN. Agents, London, Eng.: SELIG SONNENTHAL & CO.

326

The Acme is a high-grade special tool, designed and adapted to the elimination of all unnecessary time in the manufacture of screw machine product. It has been proven the most economical machine tool in the market for the manufacture of duplicate parts from the bar.

The product of the Acme is as good as the best in quality, and far surpasses all others in quantity.

The Acme way is to produce a finished piece in the length of time required to perform the longest single operation.

Your enquiries are solicited. Estimates made on your samples will convince you of the economy of the Acme in your factory on your work.

OUR NEW DESCRIPTIVE BOOKLET ON REQUEST.

THE

ACME AUTOMATIC

MULTIPLE SPINDLE

SCREW MACHINE

Built by

THE National-Acme Mfg. Co.

CLEVELAND, O.

Branch Offices—
NEW YORK
BOSTON
CHICAGO

General Foreign Representatives—
ALFRED H. SCHUTTE
SCHUCHARDT & SCHUTTE

POTTER & JOHNSTON

MANUFACTURING AUTOMATICS

will machine your duplicate parts from all kinds of castings, forgings and the bar with extreme accuracy and more economically than upon any other tools in the market.

Catalogue showing interesting specimens of parts finished on these tools free upon request.

ALSO UNIVERSAL SHAPERS

POTTER & JOHNSTON MACHINE CO.

PAWTUCKET, R.I. - - U.S.A.

AUTOMATIC GANG DRILLS

BUILT FOR BUSINESS

These machines have independent self-operating spindles. Built in two, three, four and six spindle styles, 14-inch, 20-inch and 23-inch swing.

They are the latest machines in shop practice for most rapid and economical output.

Complete catalogue ready for those interested.

B. F. BARNES COMPANY

ROCKFORD, ILL.

European Branch:

149 QUEEN VICTORIA STREET, LONDON, E.C.

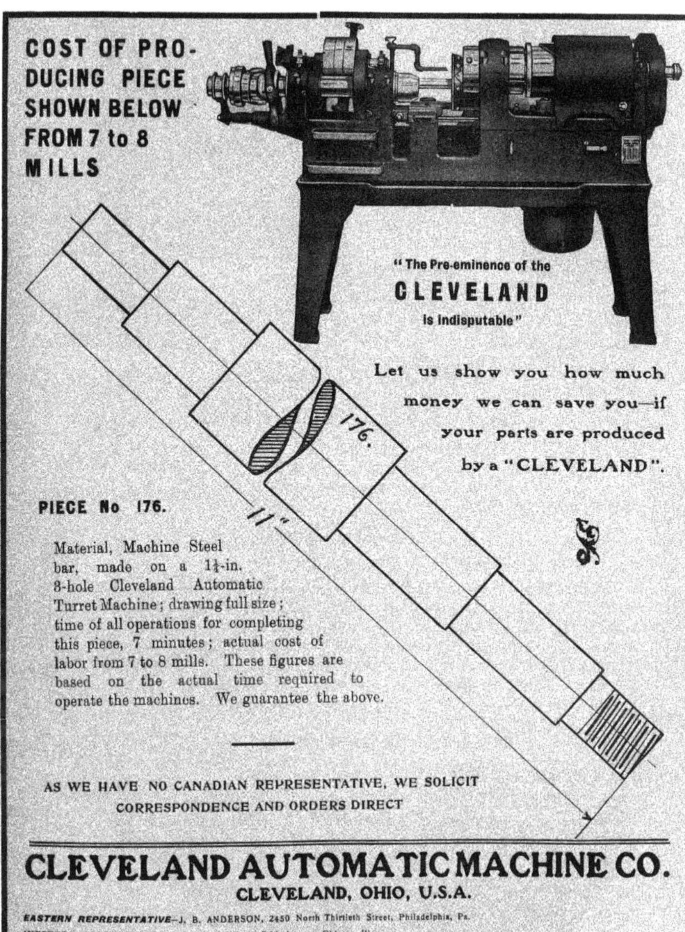
COST OF PRO-
DUCING PIECE
SHOWN BELOW
FROM 7 to 8
MILLS
" The Pre-eminence of the
CLEVELAND
is indisputable"
Let us show you how much
money we can save you—if
your parts are produced
by a "CLEVELAND".
176.
11"
PIECE No 176.
Material, Machine Steel bar, made on a 1¼-in. 8-hole Cleveland Automatic Turret Machine; drawing full size; time of all operations for completing this piece, 7 minutes; actual cost of labor from 7 to 8 mills. These figures are based on the actual time required to operate the machines. We guarantee the above.
AS WE HAVE NO CANADIAN REPRESENTATIVE, WE SOLICIT
CORRESPONDENCE AND ORDERS DIRECT
CLEVELAND AUTOMATIC MACHINE CO.
CLEVELAND, OHIO, U.S.A.
EASTERN REPRESENTATIVE—J. B. ANDERSON, 2450 North Thirtieth Street, Philadelphia, Pa.
WESTERN REPRESENTATIVE—H. E. NUNN, 22 Fifth Avenue, Chicago, Ill.
FOREIGN REPRESENTATIVES—CHAS. CHURCHILL & CO., London, Manchester, Birmingham, Newcastle-on-Tyne and Glasgow. MESSRS. SCHUCHARDT & SCHUTTE, Berlin, Vienna, St. Petersburg and Stockholm. ALFRED H. SCHUTTE, Cologne, Brussels, Liege, Paris, Milan and Bilbao.

One Gisholt Equals Four Engine Lathes

ON MEDIUM sized castings and forgings requiring boring, turning and facing, we have demonstrated time and again the saving of from 50% to 100% over engine lathe methods.

We will Exhibit at Canadian National Exhibition

Machinery Hall—near West Entrance. Will have a Gisholt Lathe, Vertical Boring Mill and Tool Grinder. These tools represent the most approved designs in modern machine tools of their classes. They are recognized as The Standard Types. If you are a user of machine tools or are interested in them, it will be well worth your time to spend a few moments at our stand.

GISHOLT MACHINE COMPANY

GENERAL OFFICES:
1309 Washington Ave., Madison, Wis., U. S. A.

WORKS:
Madison, Wis.
Warren, Pa.

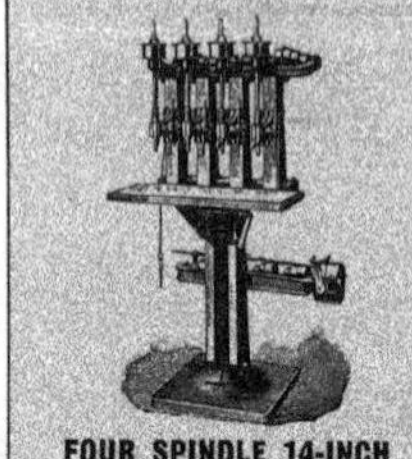

RAYO

is the new name of ALPHA **HIGH SPEED STEEL**

Agents: BAINES & PECKOVER, - - Toronto, Ont.
E. G. PRIOR & CO., - - - Vancouver, B.C.

Canadian Representative: D. W. CLARK, P.O. Box 521, TORONTO, Ont.

REINFORCED CONCRETE

This season's prices now arranged
With enquiry state quantity

EXPANDED METAL

IS HANDIEST / CHEAPEST / STRONGEST FOR

Floors, Roofs, Partitions, and all Varieties of Concrete Slab Work

CATALOGUES—SAMPLES FREE

Expanded Metal & Fireproofing Co. Limited
100 King Street West, TORONTO

JESSOP'S

BEST TOOL STEEL
"ARK" High-Speed Steel

THE FAVORITE BRANDS WITH USERS OF GOOD STEEL.
A LARGE ASSORTMENT OF SIZES IN STOCK.
JESSOP'S HIGH-GRADE FILES AND RASPS.

80 Bay, St., Toronto, Ontario,
Chas. L. Bailey, Agent.

Jas. Robertson Co., Ltd.
Montreal, Quebec

Reid-Newfoundland Company
St John's, Newfoundland.

Jas. Robertson Co., Ltd.,
St. John, New Brunswick.

WM. JESSOP & SONS, Limited, Manufactory, SHEFFIELD, ENGLAND.

"INTRA" STEEL

Made by
Messrs. Jonas & Colver, Ltd., SHEFFIELD, ENG.

Specially recommended for

Taps, Dies, Punches, Chisels, Screw-Cutting Tools, Etc.

Combines toughness and durability with a cutting power superior to highest grades of carbon steel, at less price.

IN STOCK

WILLIAM ABBOTT
334 St. James St., MONTREAL

Steel Converters

Flockton, Tompkin & Co. LTD.

Newhall Steel Works
SHEFFIELD

Makers of

Specialty AIR HARDENING TOOL STEEL for High Speed Work

ENGINEERS' TOOL STEEL, for Machine Tools, Chisels, etc.

DRILL AND JUMPER STEEL, for Miners and Contractors

SOLID STEEL HAMMERS, of any shape or size for Miners and Engineers

SPRING STEEL, SECTION BARS

Purchasing Agents wanted in principal Canadian Cities.

"Pioneer"
Pressed Steel Shaft Hangers

Two hundred per cent. more rigid than Cast Iron Hangers—Unbreakable —It's a Fact

15-16 in. to 3 15-16 in. diameter; 8 in. to 30-in. drop

Fifty per cent. lighter—much handier than Cast Iron Hangers and cheaper to erect. All sizes in stock

Selling Agents

Canada Machinery Agency

W. H. NOLAN, Prop. 298 St. James Street, MONTREAL

TELEGRAPHIC ADDRESS: "HALLTYNE" — WORKS PHONE MAIN 256

THE HALL ENGINEERING WORKS

14 COTE ST., MONTREAL

ENGINEERS, BOILERMAKERS, BRASS & IRON FOUNDERS, COPPER SMITHS & BLACKSMITHS, SHIP REPAIRS A SPECIALTY

Agents for J. & E. Hall's Refrigerating Machinery.
Atlas Preservative Co. & United Asbestos Co. of LONDON, ENG.

THOS. HALL, M. N. E. C. I. E. S. — Phone West 1737
Late Supt. Engineer Messrs. Elder Dempster & Co. & Can. Pac. Rly. Co., London

T. O. SINCLAIR, M. I. M. E. — Phone West 917

SIGHT FEED OIL PUMP COMPANY

New York, N.Y. Mr. William Martin, 141 Broadway
Boston, Mass. Mr. H. E. Rundlett, 141 Milk St.
Pittsburgh, Pa. Laughlin, Green Co. 326 Fourth Ave.
Scranton, Pa. Scranton Supply & Machinery Co., 131 Wyoming Ave.
Hazleton, Pa. Hazleton Machinery & Supply Co.

Allentown, Pa. The William H. Taylor Company
Chicago, Ill. Mr. Alex. Hodgart, 1207 Fisher Bldg.
Birmingham, Ala. Mr. James J. O'Rourke, 521 Brown-Marx Bldg.
San Francisco, Cal. Phoenix Tool & Valve Co. 131 Beale St.
Montreal, Canada Mr. Homer Taylor, Temple Bldg.

Model "G," eighteen-feed, automatic oil pump, three kinds of oil.

MILWAUKEE, WIS., U.S.A.

Established over 50 years. — Telegrams: "Belliss" Birmingham. — A.B.C. 5th edition

BELLISS & MORCOM

ENGINEERS LIMITED

Birmingham, - - - England

Representing our Three-Crank Expansion Type Engine

ENGINES

DIRECT, ROPE OR BELT DRIVE for

Electric Power Lighting, Traction and Mill Driving

ALSO

Condensing Plants

5,000 H.P. Installed in Canada

ADVANTAGES — Self-Lubrication; Simplicity; Strength; Experience; Low Cost; Reliability; Minimum Maintenance; Compactness; Good Governing; Durability; Maintained Efficiency; Quiet Running.

3,300 Engines supplied or building, representing upwards of 600,000 H.P., in sizes ranging from 5 to 2,500 B.H.P. of Compound and Triple-Expansion Types.

SOLE AGENTS FOR CANADA:

LAURIE & LAMB, 212 Board of Trade Buildings
MONTREAL, CANADA

Telegraphic Address. "Larco" Montreal. Codes: A.B.C. 5th Edition and Western Union.

OWEN SOUND IRON WORKS, LIMITED

OWEN SOUND, ONT.

Cement Mill Machinery, Boiler and Steel Tank Work of all kinds, Grey Iron and Brass Castings

The Engineering Times

The Weekly Newspaper of Commercial Engineering

32 PAGES

Price Twopence.

Annual Subscription, 10s. 10d.
Foreign - - 13s. 0d.

ADVERTISEMENTS.

Prepaid Advertisements on Wrapper, classified under headings: Tenders—Appointments Open—Situations Wanted—Partnerships—Wanted—For Sale, &c.

By the Line—Four Lines or under, 2/-; each additional Line, 6d.; the Line averages Six Words.

By the Inch, 8/- per Inch, Single Column, One Insertion.

SUBJECT TO DISCOUNT

20% off 13 Insertions; 25% off 26 insertions; 33⅓% off 52 insertions

CHIEF OFFICE: 6, Bouverie St., LONDON, E.C.

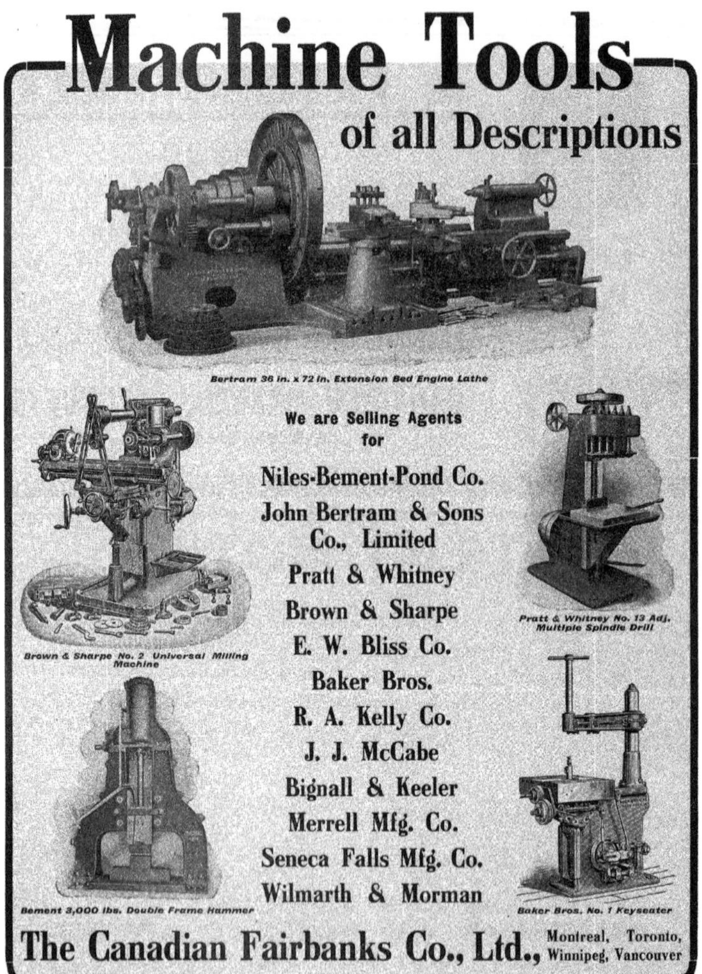
Machine Tools
of all Descriptions
Bertram 36 in. x 72 in. Extension Bed Engine Lathe
We are Selling Agents
for
Niles-Bement-Pond Co.
John Bertram & Sons
Co., Limited
Pratt & Whitney
Brown & Sharpe
E. W. Bliss Co.
Baker Bros.
R. A. Kelly Co.
J. J. McCabe
Bignall & Keeler
Merrell Mfg. Co.
Seneca Falls Mfg. Co.
Wilmarth & Morman
Brown & Sharpe No. 2 Universal Milling Machine
Pratt & Whitney No. 13 Adj. Multiple Spindle Drill
Bement 3,000 lbs. Double Frame Hammer
Baker Bros. No. 1 Keyseater
The Canadian Fairbanks Co., Ltd., Montreal, Toronto, Winnipeg, Vancouver

How to Heat and Drive Steel Rivets

Awarded First Prize in Champion Contest at the Joint Convention of the International Railway Master Boiler Makers' Association and the Master Steam Boiler Makers' Association, held in Cleveland, O., May 21-23, 1907.

By James Crombie†

In riveting, no matter whether by hand or machine, there are several matters that must first receive attention before any rivets can be driven or a satisfactory job made.

We have first the assembling of the work. The plates must be bolted together and all the rivet holes reamed out to correct size, i.e., 1-16 inch larger than the size of rivet to be used. Reamers are made with about 3 inches of tapered end and 4 inches of straight section for each different size of hole. This insures that the hole will be parallel, or the same size all the way through. After reaming, the work should be taken apart and all burrs and pieces of cuttings left between the plates by the reamer scraped off and the two surfaces made perfectly clean. The outer sides of holes should also be gone over with a broad countersink drill to remove the burrs from the edge of the hole. The work should then be bolted up close and any joints warmed and closed up with hammers or squeezed in the hydraulic machine.

Rivets.

Rivets should be of best quality of open-hearth steel, extra soft grade, ultimate strength 45,000 to 55,000 pounds, elastic limit not less than one-half of the ultimate strength.

$$\text{elongation} = \frac{1{,}400{,}000}{\text{ultimate strength}}$$

maximum phosphorus .06 per cent., maximum sulphur .04 per cent.., cold and quench bends 180 degrees flat on itself without fracture on outside of bent portion. On being nicked and broken off the steel should be of a light gray color, showing a close-grained, silky fracture. A course-grained, dark-colored rivet steel should be looked on with suspicion, as the chances are that after being riveted up the rivet head will drop off. Excess of sulphur or excess of phosphorus will make the steel "short" or brittle, causing the rivets to break. We had a lot of trouble with one consignment of rivets. After being driven the heads or points of a large percentage of these rivets would drop off, not immediately after being driven, nor when the plate edge was being calked, but 4, 5 or 10 hours after they were finished. Some of these broken rivets, 4 inches long, were backed out of the hole, they were bent 180 degrees on themselves cold without any sign of fracture. The trouble was apparently excess of phosphorus, causing segregations in spots in the steel. Where these spots were the rivet would break off easily and the material was very brittle. These rivets went into the scrap heap in double-quick time.

Heating the rivets does not depend so much upon the kind of fire or fuel as upon the rivet heater. Given a good, willing heater, and no matter how adverse the conditions are, he is always there with a good hot rivet and the work goes swinging along. Good work can be obtained from a coal or coke fire, either with a side or bottom blast. There is always the danger in these fires of overheating the rivets, or they may be left in the fire too long while the squad is changing or during the meal hour. The rivets will then have a heavy scale whether it will be best to leave them in. I believe in having them out if at all possible, either by cutting off the heads and backing them out with a punch and on them and be wasted, becoming too small to fill the hole properly; this should be avoided.

Rivets that are sparking hot or show signs of overheating should be discarded. If such rivets should be driven through any mistake or any heads show signs of being burnt on examination after the job is finished, the question arises, whether more injury will be caused to the plates by cutting those rivets out or hammer or drilling them out. There should be no danger of burnt rivets, although if too many rivets are put into the furnace at one time and the work is lagging behind, the rivets will quickly get scaled and waste.

I have got good results from an oil furnace with any baffle brick, using pitch oil and allowing the flame to blow right on to the rivets. This was principally on large sized rivets. I believe that the best results are obtained from a good

Illustration of Producer Gas Plant, the economy of which was spoken of in the article on Producer Gas in the July issue, by Emil Stern. This is a Pintsch Plant installed by the Economic Power, Light & Heat Supply Co.

† Foreman boilermaker of Sawyer & Massey Co., Hamilton.

oil furnace with a baffle-brick or fire wall using compressed air. We use oil rivet forges burning fuel oil, or gas oil with compressed air, and get nice, soft rivets, just a bright cherry red or white-hot if desired. This style of forge leaves nothing to be desired. It has a fire-wall, heats rivets quickly, and does not cost much to run. The oil and air is mixed before entering the furnace, and then passes through a small opening at the side of the furnace. Combustion then takes place, the flame striking the baffle-brick or fire-wall, then passing over the top of the fire-wall and heating the chamber where the rivets are. The heat then radiates from the top down upon the rivets.

Rivets that to be driven by hand or by an air hammer require more heat than rivets for the hydraulic riveter. A cherry red will be best for a hydraulic machine, but for an air hammer or hand riveting the rivets should be a light yellow, turning to white, but not sparking. That would be about 1,500 degrees F. for hydraulic and 2,000 degrees for hand-driven rivets.

Machine Riveting.

There are several kinds of machines for riveting. I have seen good riveting done on an old steam riveter, each rivet receiving from two to three blows. This style of machine has gone out of date and is seldom seen now.

We have our good friend the hydraulic riveter or "Bull." This machine may be considered ideal both as to operation and effect. The water is supplied in pipes at a pressure of 1,500 pounds or more per square inch. In the best class of machines the pressure is delivered through different valves, so that the operator has control of the pressure at all times, reducing or increasing the pressure according to the size of rivet. The best pressure is 100 tons per square inch of rivet section. The water, after the rivet is squeezed, passes back to the tank again, and is used over again, so that there is very little waste of water. The full pressure is available throughout the full length of the stroke, so that the rivet is properly upset, thoroughly filling the hole from end to end, and also squeezing the plates close, holding them for several seconds while the rivet is contracting or cooling off.

For heavy work, the machine should have a plate-closing device, so that the plates can be held close together before any pressure is applied to the rivet, the rivet dies being controlled by separate levers and working in the centre of the plate-closing device yet separate from it.

A good hydraulic crane is best suited for handling the work at the riveter. It is quick in action and very sensitive, and the hole is quickly centred with the rivet dies. All levers should be handy for the operator, the same operator controlling the levers for both crane and riveter. The work should be hung fair. A little time spent on this will be gained later on and temper saved, too.

After the work has been cleaned and bolted up it is squeezed up and all the bolts tightened. The boiler should then be tacked at each quarter, then at each eighth, and afterwards at every third or fourth hole. The remaining rivets are then driven, going right along around the boiler, allowing the full pressure of the machine to remain for a few seconds on each rivet as it is driven. At the longitudinal seam or butt-joint the end should be tacked first, to prevent springing, then at the centre, after which the rivets may be driven in succession. The operator should be given to understand that it is not the number of holes he can fill up, but the quality of the work on the testing floor that counts. The danger to be looked for is the crushing or scalloping of the plate in front of the rivet, through excessive pressure or through too much being planned off the plate edge. Also the point may fail by not being properly proportioned, by the plate breaking along the line of rivets by the shearing out of the rivets or by the shearing of the plate in front of the rivets, and if the work is not hanging fair the rivets may be driven offside, thus necessitating cutting out.

All rivets should be put into the holes from the outside, as it is much quicker and requires a man less on the job. A man inside the boiler is apt to get hurt, too.

There is also the pneumatic lever type of machine, the hydro-pneumatic type; the pneumatic toggle-joint machine giving a full pressure only at end of stroke, and the portable hydraulic type similar to the "Bull." The latter is handy for riveting around the ends of boilers or furnace mouths, or around the flanges of flues in Lancashire or Cornish boilers. The water, after it is discharged from this machine, does not return to the tank as in the fixed riveter, but is conducted through a rubber hose to the nearest sewer. This type is wasteful as regards water.

I pin my faith on the hydraulic riveters, as first-class work can be obtained from them. The rivets do not require such an intense heat, therefore there is no risk of burnt rivets, and there need be no guess work about the pressure applied to each rivet.

I use high-speed air hardening steel for rivet dies on the "Bull." The initial cost is high, but they are all right. I have driven 180,000 rivets with four pairs of dies made from this steel, and they have not been to the tool room since we started to use them. They are still sharp on the edge and are only beginning to show slight signs of wear. The most that can be got from a pair of ordinary cast steel rivet dies is about 4,000 rivets, while these, I believe, will average 50,000 each.

Another good riveter is the air hammer or "Gun," working with compressed air at a pressure of 80 to 100 pounds per square inch. With this machine a careless riveter, or one wanting to make a show, may drive a short rivet, too short to properly fill the hole, merely blinding the work by allowing his hammer handle to describe a large circle, thereby spreading the rivet instead of allowing it to upset into the hole. If there is enough stock in the rivet and the rivet is driven fair into the hole, these hammers will give a good tight rivet that will not require calking. With the air hammer it is necessary to hammer up the plate in front of the riveter and to have the work all thoroughly closed up, as there is no squeeze as in the hydraulic machine.

Hand Riveting.

In work to be riveted by hand it is better to have all holes countersunk The countersink must be narrow and deep, from 1-16 to 1-8 inch of the bottom of the hole in the first plate, then if corrosion or wasting occurs the rivet and plate will both wear equally and the strength of the rivet will not be impaired. Where a shallow, broad countersink is used there is more chance of the rivet breaking off than where a narrow deep one is used. The rivets may be snapped by hand, but this practice has gone out, as the rivets are not so reliable as when countersunk.

The best practice is to finish on the third rivet, that is, we have our squad all ready to start with hammers, the hole is hammered close and the first rivet is put in and hammered down but not finished, the second hole is plied close, the rivet hammered down as before, the same with the third and fourth. The fifth rivet is then put in, the sixth hole plied close and our squad goes back to the third rivet, giving it a few heavy blows, then finishing it off with their riveting hammers. Thus they work right along, always finishing on the third rivet back.

The Holder-on.

It is hard to give any hard and fast rule regarding holding up the rivets, as there are so many different classes of boiler work and so many different tools to be used around a boiler in holding up. We have our dolly bars and water-space hammers, and also long-handled hammers, supported on a hook. Where possible it is best to use the compressed air holder-on. In all these things we must be guided by circumstances, always endeavoring to get the head well up or the rivet will not be tight.

Opinions Concerning "Ashes Burning"

"Cannot burn ashes," says C. S. Miner—Coal in ashes only that burns—Better to screen ash pile than to use brine and acid solution—Formula probably useless.

In these days the motto is to waste nothing. It is proverbial that in the pork packing industry nothing is lost but the squeel. There is the same tendency in all other industries. A striking example of this principle is evidenced in the by-product producer gas plant. No wonder then, that, when the cobbler in Altona, Pa., published his ash burning formula, the whole country was stirred! Thousands of tests were made by householders and others with results which led many to pronounce the scheme a success. From the first chemists and those acquainted with the laws of cumbustion scouted the idea that there could be anything in the scheme, but these opinions carried not much weight in the face of the many experiments which showed that the mixture as prescribed would burn. But few records of tests made upon a common sense, commercial basis, were published; but the few that were showed that not only was there nothing to be gained, but that the salt and acid in the mixture might play havoc with the iron grates.

For the benefit of the readers of Canadian Machinery, some data on this subject has been gathered which shows the trend of opinion among those who would be chiefly affected by this scheme.

It was not until shares in a company formed to manufacture a compound which, when mixed with one part coal and two parts ashes, would produce from 50 to 150 per cent. more heat than could be secured from the best coal, that the United States Government took any notice of the affair. But when this absurd propaganda was brought to light, there was need for investigation. Following is a report made by Carl Shelby Miner, Director of the Miner Laboratories, Chicago, upon the question:

Silver Mines, Port Arthur—Cut Loaned by W. G. MacFarlane.

Statement by Director of Miner Laboratories.

"Let me begin the discussion by stating explicitly that you cannot burn ashes. Ashes are a mineral matter which has absolutely no fuel value. Ordinary ashes, however, contain a small amount of unburned coal, and this coal can be burned under proper conditions. Ashes and coal are of variable composition, but, for the purpose of this discussion, we will assume that ordinary hard coal contains 10 per cent. of ash, by which we mean absolutely incombustible material; and the ashes from a well-regulated furnace contain 25 per cent. of unburned coal, which is a liberal estimate. There would then remain 13 1-3 per cent. of "ashes," of which 3 1-3 would be coal and 10 per cent. genuine ash. Therefore, the loss if these ashes are thrown away is 3 1-3 per cent. of the coal used. In the form of a table this statement appears on the following page.

"So, if we admit that this method will enable us to burn all the coal in the ashes, and that is the limit of its possibilities, we may hope to save 1-33 per cent. of our coal bills. The advocates of this formula cannot claim that it contains any heat-producing material except the oxalic acid, and that is present in so small a quantity as to be negligible in making this calculation. The water and salt have no fuel value. There is, therefore, absolutely no basis for the claim that ashes and coal mixed with this compound are a better fuel than coal alone.

"This is not by any means the first attempt to produce a compound that will increase the combustion of coal in stoves or furnaces. Some years ago a combustion powder for this purpose was exploited in Germany, which appeared to be a great success, for when it was thrown into a fire the flame brightened up at once and appeared to burn more briskly. The unsympathetic government chemists, however, analyzed the powder and, finding it to be only common table salt, branded the whole thing as a fake. The salt is retained in the present formula. There have, however, been formulae devised more satisfactory than that one. The basis of all of them is some material which will release oxygen, the gas necessary to promote combustion, in close proximity to the coal in the ashes. This result is obtained by mxing the ground ashes with the compound before putting them into the

Coal contains—		
Combustible material	90 per cent.	
Incombustile material or ash	10 per cent.	
Ashes contain—		
Combustible material	25 per cent.,	equivalent to 3 1-3 p.c. of coal used
Incombustible material	75 per cent.,	equivalent to 10 p.c. of coal used

furnace. In the cobbler's formula the oxalic acid is probably intended to furnish oxygen. It contains oxygen, but it also contains carbon and hydrogen in amounts more than sufficient to combine with its oxygen, and it therefore uses up oxygen instead of furnishing it. The water is the only constituent of any possible value in this formula. It is composed of hydrogen and oxygen and at high temperatures breaks up into these two gases and so might furnish oxygen for the combustion of the coal in the ashes with which it has been mixed.

"The scheme of mixing water with ashes to burn coal out of them has been known for years and has been used by householders all over the country. It is covered by a United States patent for "a method of producing complete combustion of the mixture." Comparative tests made with ashes mixed with water and ashes mixed according to the cobbler formula show practically identical results. In both cases there was left about 7 per cent. of unconsumed coal in the ashes. It therefore seems clear from both theory and practice that there is no new and useful improvement covered by this formula, but it is, as rehashing of an old and time-tried method.

"The probability is that neither method has any real value, but that the results obtained are due entirely to the repeated burning to which the ashes are subjected. It is reasonable to assume that if the ashes are fed back into the furnace often enough, with plenty of fresh coal, the coal in them will eventually be consumed."

There is a statement by a man who should know. Ashes cannot burn. All that can be done is to consume the coal that is left in the ashes, and C. S. Miner expresses the doubt whether the salt and oxalic acid assist the combustion to any great extent.

Opinions of Canadian Manufacturers.

Opinions on this subject from some large coal consumers in Canada were sought by Canadian Machinery. These speak for themselves:

Walkerville, July 10, 1907.
Canadian Machinery,
10 Front St. east, Toronto, Ont.:

Dear Sirs,—We acknowledge the receipt of your favor of the 5th instant, and in reply thereto we would say that we have not experimented in any way with the ashes burning scheme referred to. We have, however, noticed several articles in connection with the matter in the daily press, and we are inclined to think that there is very little advantage to be gained by adopting the system.

Yours very truly,
Hiram Walker & Sons, Limited.

Montreal, July 10, 1907.
Canadian Machinery,
Toronto, Ont.:

Gentlemen,—We duly received yours of the 5th inst., and in reply to your inquiry would say that we have given very little consideration to the question of burning ashes, as the problem with us is to so consume our coal that all the ashes left therein would not be worth thinking about, and that is without doubt the case with us at the present time, as the percentage of coal, or carbon, in our ashes is infinitesimal; further, would say that we have no faith whatever that any mixture of salt, etc., is of the slightest use—in fact in our opinion it is a detriment.

From this you will understand that, in our opinion, if any concern has a large percentage of coal in their ashes the cheapest and most economical way for them to do would be to sift the coal out and thereby save a lot of trouble, labor and waste of good coal in heating up ashes and clinkers.

Yours very truly,
The Dominion Oil Cloth Co.,
Limited.

Toronto, July 12, 1907.
The Canadian Machinery, City:

Dear Sirs,—Your favor of the 5th was duly received, in reply beg to say, we have not made any tests of the schemes you refer to. If you know of anything that is of any material value we should be glad to hear from you.

Respectfully, yours,
The Otto Higel Co., Limited.

Hamilton, July 17, 1907.
Canadian Machinery, Toronto, Ont.:

Gentlemen,—In the burning of bituminous slack there is always a certain amount of loss of good coal due to its falling through the grates with the ashes. These ashes mixed with one part of good coal, then moistened with saft water, into which has been added a little oxalic acid, will give an intense heat when thrown on top of a good fire. The result is the formation of a layer of molten slag over the top of the fire, which completely cuts off all draft and deadens the fire, necessitating cleaning after each firing. This, of course, is a great drawback and prevents the practice being carried on under a steam boiler.

It was our intention to make tests on sample pieces of steel plate, placed in the smoke flue and among the tubes, to determine whether the products of combustion given off by the ash mixtures had any corrosive effect, but owing to the difficulty of keeping up the fires due to the above mentioned cakening of the ash mixture, we were unable to carry the experiment to a completion.

Yours truly,
Canadian Westinghouse Co.
Limited.

Toronto, July 6, 1907.
Messrs. Canadian Machinery and Manufacturing News, 10 Front St. east, city.

Dear Sirs,—Yours of the 5th inst. received in reference to burning of ashes. We think the idea a good one so far as a steady warm fire is desired—not one that is to be pushed.

We find that, in the use of the ash-burning material, it is a considerable advantage as, while the ash itself is dead, there is a large quantity of cinders which, with the process, one is able to obtain a most satisfactory fire for steady purposes.

Yours truly,
The Canada Metal Co., Limited.

Hamilton, July 6, 1907.
Canadian Machinery, Toronto, Ont.:
Dear Sir,—In reply to your letter of the 5th inst. we would say that we have made no experiments with burning coal ashes which amount to anything.
Yours truly,
The Hamilton Steel & Iron Co., Limited.

Toronto, July 11, 1907.
Canadian Machinery, 10 Front St. east, Toronto:
Dear Sirs,—Your letter of July 5th to the Canadian General Electric Co. regarding question of test of ash burning has been referred to me. In answer would say that at the works of the Canada Foundry Co. we have made no experiments to test the efficiency of this scheme.
Yours very truly,
Canada Foundry Co., Limited.

Experiment Showed Loss of 15.9 Per Cent.

The test made by R. C. Harris, City Property Commissioner, Toronto, was very complete, and was conducted upon a sound basis. This was an evaporative test under one of the boilers in the City Hall. The following is a record of test as given by Mr. Harris:

"Three thousand one hundred and fifty-six pounds of slack were mixed with 2,441 of ashes, the whole being saturated with 51 gallons of water, in which were dissolved 25 pounds of common salt and two pounds of oxalic acid. The fuel was fired automatically by an underfeed stoker with forced draft. During this trial the average gauge pressure was 73, but fluctuated rapidly between 65 and 80, and it was only by continuously driving the fan supplying the draft that we were enabled to hold the steam at this average. It was apparent that good combustion was not taking place, inasmuch as the top of the fire was at all times far short of incandescence. The ash in the mixture packed, with the result that the air supply requisite to properly support combustion was deficient. At the conclusion of the test there was no apparent detrimental affect on boilers and grates.

"In the second trial, firing straight bituminous slack by means of underfeed stoker with forced draft the average gauge pressure maintained was 83, with scarcely any fluctuation, the fan being driven but a portion of the time.

"The cost of evaporating 1,000 pounds of water from and at 212 degrees F. with the ash admixture and chemical solution was 20.62 cents., while in the second test, with coal only, it was but 17.78 cents. This shows that to evaporate 1,000 pounds of water from and at 212 degrees F. with the mixture of coal, ash and chemical solution it costs 2.84 cents more than with the coal untreated, demonstrating a loss of 15.9 per cent. in the use of the treated fuel. If further allowance were made for the additional amount of steam needed to continuously drive the fan in the first test, the loss would be considerably greater.

"We found that it was absolutely impossible to hold the steam with a preparation of one part of coal to three of ashes. It was also necessary during the run to at times replenish the fire with coal unmixed with ashes. The indications throughout the whole of the test where the fuel was mixed with ashes were altogether unsatisfactory."

Prof. W. H. Ellis, professor of applied chemistry in the School of Practical Science, University of Toronto, laughed to scorn the idea that ashes burn. He said, "it is an absolutely fixed equation that ashes, or the residium left after complete combustion, have no heating power whatever, and no possible treatment can make them produce heat, whether in conjunction with fresh coal or otherwise."

Our Water Powers; Falls in Current River Park, Ont.—Cut Loaned by W. G. MacFarlane.

CANADIAN MACHINERY

and Manufacturing News

A monthly newspaper devoted to machinery and manufacturing interests, mechanical and electrical trades, the foundry, technical progress, construction and improvement, and to all users of power developed from steam, gas, electricity, compressed air and water in Canada.

The MacLean Publishing Company, Limited

JOHN BAYNE MACLEAN - - President
W. L. EDMONDS - - - Vice-President
H. V. TYRRELL - - - Business Manager
J. C. ARMER, B.A.Sc., - Managing Editor

OFFICES:

CANADA
MONTREAL - 232 McGill Street, Phone Main 1255
TORONTO - 10 Front Street East, Phone Main 2701
WINNIPEG, 511 Union Bank Building, Phone 3726, F. R. Munro
BRITISH COLUMBIA - Vancouver, Geo. S. B. Perry

UNITED STATES
CHICAGO - 1001 Teutonic Bldg. J. Roland Kay

GREAT BRITAIN
LONDON - 88 Fleet Street, E.C. Phone Central 12960, J. Meredith McKim
MANCHESTER - 92 Market Street, H. S. Ashburner

FRANCE
PARIS - Agence Havas, 8 Place de la Bourse

SWITZERLAND
ZURICH - - - Louis Wolf, Orell Fussli & Co.

SUBSCRIPTION RATE.

Canada, United States, $1.00, Great Britain, Australia and other colonies, 4s. 6d., per year; other countries, $1.50. Advertising rates on request.

Subscribers who are not receiving their paper regularly will confer a favor on us by letting us know. We should be notified at once of any change in address, giving both old and new.

Vol. III. AUGUST, 1907 No. 8

CONTENTS

CANADIAN SECURED THE PRIZE.

In this issue of Canadian Machinery there appears an article entitled: "How to Heat and Drive Steel Rivets," by James Crombie, foreman boilermaker of the Sawyer & Massie Co., Hamilton. The article was awarded first prize at the recent joint convention of the International Railway Boiler Makers' Association and the Master Steam Boiler Makers' Association at Cleveland, for which he received $50 in gold. Mr. Crombie is to be congratulated, and his friends and fellow Canadian mechanics feel proud of him.

Very often the men who are in the best position to impart information along mechanical lines, so far as the information itself is concerned, are handicapped by lack of ability to express themselves clearly. When a man has information which is valuable and knows how to impart it, he is fortunate.

A great deal more writing upon mechanical subjects has been done in the United States than in Canada. The demand, made by publishers of mechanical papers and mechanical books in the States led to the development of many writers. In Canada this has not been so to the same extent. Therefore, all the greater credit is due a Canadian when he wins in such a game against American competition.

To the inexperienced, describing some article, telling about some kink, or writing an article on some mechanical subject, comes rather difficult at first, but upon some little practice comes more naturally. It is surprising how one can think he knows a lot about some particular subject or thing, but upon attempting to write concerning it, he often finds himself quite uncertain of many details. Writing is a splendid education in that it teaches one to make thorough investigation to get at all the details before forming opinions in his own mind concerning it. It's a splendid thing to read, but a better thing to write.

In practically every little or big shop in Canada, there must be devices which have been originated in that shop to do some special job or to act as a substitute for something not contained in the equipment of that shop. Why should not somebody in these shops assist himself and others by making sketches of such devices, writing a short description of them and their uses and having them published? Canadian Machinery is always looking for just such material. The remuneration received for such contributions would pay amply for the time spent, and there would be the additional satisfaction of knowing that one was educating oneself, and at the same time helping some fellow mechanic out of a tight place by the suggestion.

JULY METAL MARKET.

Throughout the past month has ruled the usual dullness characteristic of the midsummer season, although there have been one or two exceptions. During the earlier part of July the metal market was quiet, but about the 15th a revival came in the demand for iron and steel. The foundries did not remain idle as long as was expected, since those companies who were awaiting shipments from Europe, which, in a number of cases, were fulfilments of orders given two months previous, were anxious to take advantage of the prospective enlivening of the market. During the week commencing July 15, arrivals of cargoes aggregating between 7,000 and 8,000 tons of iron were reported. It was not long before this heavy tonnage was distributed.

The English iron market has been subjected through the month to many fluctuations, caused in many cases mostly by speculation. In the first week of July prices were weak and little call was experienced. The next fortnight registered a strong recovery, and again, now, there are indications of a relapse. The American market has been dull throughout the month.

Ingot tin has been steadily weakening. The corner in the London market which held up transactions for some time was broken in the middle week of July, and since then prices in the English market have steadily declined from £18 to a minimum of £8. Premiums have been greatly reduced and local market conditions are in a much healthier state. The ruling price for Lamb, Flag, and Straits tin in the local market during the month has been $44.

With the arrival of H. H. Rogers, the copper king, at New York City, about the 1st of July, the tension existing in the American copper market was broken. Much uncertainty had existed previous to Mr. Rogers' arrival, and little buying was done. Prices were immediately lowered and the call rapidly increased.

Supplies in the local market throughout July have been adequate to the call, which was small, owing to the uncertainty which ruled despite the amelioration in American conditions.

Much confidence is expressed by dealers in the probability of immediate enlivenment of business. Hopefulness is the general attitude at present.

PROGRESS OF PRODUCER GAS.

Undoubtedly there is economy to be secured with producer gas plants and engines. It is not now all hearsay and manufacturers' claims in Canada, as it was three or four years ago. There are many plants now in operation in Canada, and manufacturers or users of power who are thinking of installing plants can see for themselves exactly what producer gas plants are doing.

Not all the plants, which have been installed in Canada, are operating as satisfactorily as claimed in the catalogue of the manufacturer, but there are now a sufficient number installed which are giving excellent satisfaction to prove that the producer gas plant in general is not at fault, but that, in a case where a plant does not operate satisfactorily, the fault lies with the operator or with the design of that particular patent.

When one sees the volt needle swinging to the tune of 25 or 30 volts every few seconds in the power line from a dynamo operated by producer gas engine and plant, one naturally thinks that the engine governor must be off on a "hunting" trip; but when one peeps into the gas generator and notes that it is about half full of fuel, and sees the gas burning above the fuel bed, one no longer blames the engine governor, but the engineer in charge of the plant; when, however, the engineer tells you that if he fills the generator with fuel, as it should be, the engine stops from lack of gas, one would lay the blame to lack of gas space in the generator, but hesitate, lest you again make a mistake. Then one arrives at the safe conclusion that something is wrong with the design of the gas space or other part of the generator, or that the engineer in charge does not know his business. Then one wonders that the company who made the installation does not have a thorough investigation before its reputation in the producer gas plant line is damaged.

But, as was said before, such an occurrence does not reflect discredit upon producer gas plants in general.

For the benefit of manufacturers who may be interested in producer gas plants we hope to publish next month illustrations of several producer gas plant installations in Canada together with necessary description of the plant and engine, and as complete a record as possible of what a manufacturer is chiefly interested in, namely, fuel consumption under ordinary running conditions, attendance required, amount of cleaning necessary and other minor points about the operating of the plant which may be of interest.

OPPORTUNITIES IN BUILDING CONTRACTORS' MACHINERY.

While at the present time there are big opportunities for building up a manufacturing business of almost any kind in Canada, yet there are special opportunities in some lines. One of these appears to be the manufacture of contractors' machinery. Building is the first industry affected by the rapid growth of a country. Now the builder and contractor uses a great deal more labor saving machinery than he did a few years ago. There will be an immense amount of this machinery needed for many years to come. There is a large field for a firm in its manufacture.

The working up of such a business can well be started in a small way—in fact in a repair machine shop and foundry. A start is the repairing and job work for local contractors and builders. Good work is done—more and more work is dug up—great perseverance is shown—the horizon broadens—a bigger and better equipped shop is necessary—the firm is on its way to prosperity.

A business we have in mind was built up just in that way. Now the firm have one of the largest and most modern plants in Canada. Another we know has made a start. Two progressive young fellows are at the head of it. They are looking for the gradual development into a large business.

Concerning Tool Rooms *

A Practical Tool Room System for an Average Size Shop, such as the Writer has seen in Successful Operation.

BY Geo. D. MacKinnon, B.A. Sc.

ARTICLE II.

To facilitate the settling up with men quiting work as well as to reduce the time of the one who may have to look up their checks, it is very desirable that the checks be located on properly arranged check boards, instead of being scattered on pins all over the room; and it is a good plan, and a convenient one, too, for the tool window tender to have all these check boards located within his reach near the window. It is surprising the number of tools that can be checked for on a small board. Such an arrangement may consist of twelve boards, and these can be arranged systematically so that a minimum of time is lost in looking for or placing a check; and this also reduces the time lost by a man returning a tool since his check can be handed to him immediately instead of him having to wait until the tender walks down the room to the tool case and back again, such as would occur if the check were placed on the case in which the tool is kept.

Type of Tool .... Shop Number of Tool ....
Name of Maker .... Tool Located in ....
Bought from .... Kind of Drive ....
Makers Number

Requisition Number	Purchase Number	Price.	Date Ordered	Date Delivered	Date Erected	Rate per hour	Rate in force From To	Tool Scrapped or sold	Price Obtained

Remarks:

Fig. 1.

Damage to Tools Should Be Reported.

It frequently happens that tools are destroyed in service, sometimes through accident and sometimes through carelessness. The former is for the most part unavoidable, although the number of so-called accidents would be considerably reduced by strict attention to business; but one must expect a certain amount of carelessness since one cannot expect that every man in the works will take a deep interest in the work he is doing; and lack of interest is akin to carelessness.

It is one of the objects of the shop superintendent to keep his operating expenses as low as possible, consistent with efficiency; and in the matter of damage to tools, machines, etc., he can be greatly assisted by the tool room. Various methods are suggested for the cutting down of this portion of the shop expense, and one very satisfactory way is to have the attention of the superintendent or manager drawn to the facts of these damages. A convenient way of doing this is to have the man who does the damage report the same on a special form, such as Fig. 2. This form is obtained from the tool room, is filled out by the workman, signed by his foreman, and returned with the broken tool to the tool room. The tool room foreman notes the extent of the damage, the probable cost of repair, etc., and each morning all reports of the previous day are sent to the manager's office for his disposal.

THE STEWART CO., LTD.

REPORT OF IMPERFECT WORK OR DAMAGE DONE.

Shop .... Date 190

Order No. .... Machine No. ....
Damage done to Machine or Work

Man who did damage ....
Foreman in charge ....
Reported by ....
Company's loss

Time received ....
Signed .... Signed ....
Remarks:

Fig. 2.

When and When Not to Dock for Broken Tools.

The writer does not advocate the principle of docking a workman's wages for damage done if said man is working on an hourly rate, except when the damage is done through gross carelessness; but such a policy is worthy of consideration when a man is on a piece rate and is pushing his machine for all it will do. But in such a case an understanding should be arrived at with him when he is starting on his piece rate. But dock or no dock, the mere fact of the damage being brought to the attention of the manager and entered against the man's record, is generally sufficient to reduce carelessness to a minimum.

Record of Damage Should Be Kept.

Records should be kept in the tool room of all damage done, and in a shop of this size it requires a very small amount of the foreman's time to

* Article I. appeared in the June issue of Canadian Machinery.

record such damages. In the matter of small tools it is sufficient for all purposes to enter the damage in day-book form, totalling up at the end of each month. But in the case of machine tools a card system is desirable. This may have the form of Fig. 3 (front and back), one card being used for each machine. When it becomes necessary to purchase a new machine it does not usually take a manager long to decide after looking over these card records, of the most economical make; and such a record is a valuable one.

REPAIRS

Date of Break Down	Repair Completed	Cost of Repair	Extent and Nature of Damage

Fig. 3.

Window Tender Should Be Capable Man

We often find that the man who has been given the position of window tender is one who is content to draw the pay usually allotted to an ordinary laborer. I have no hesitation in saying that the employment of such a man in such a position is a dead loss to the company; and the one who holds this important position should be a capable mechanic. He should know when a tool is or is not suitable for the work intended; and he should be able to "fit up" tools for those requiring them without it being necessary for the man wanting the tools to explain to him just how they should be fitted up. The writer has frequently seen a crowd of highly paid mechanics standing at a tool room window waiting to be served, while the attention of the man behind the window —who by the way was drawing 12½ cents per hour—was taken up by one of the number endeavoring to secure a tool suitable for his job, which tool might have been given to him at once by one who himself had had experience along those lines. It is an easy matter to figure the loss to the company under such circumstances.

Check System for Tools Used in Erecting

Shops which do any amount of outside erection work, know what a loss there generally is through the losing and destruction of tools, tackle, etc. When one figures it out the extent of this loss is often surprising, and so long as the men are allowed to pick up tools, etc., indiscriminately, pack them up and take them out of the shops without any check on them, so long will the loss be large. It requires very little time to fill out a form such as Fig. 4, which is left with the tool room foreman as a receipt for the tools, and checked off by him when they are returned. Since the erecting engineer knows that the people at home are interested in the tools which he has out he is more apt to take care of them and to see that they are all gathered up before he starts for home.

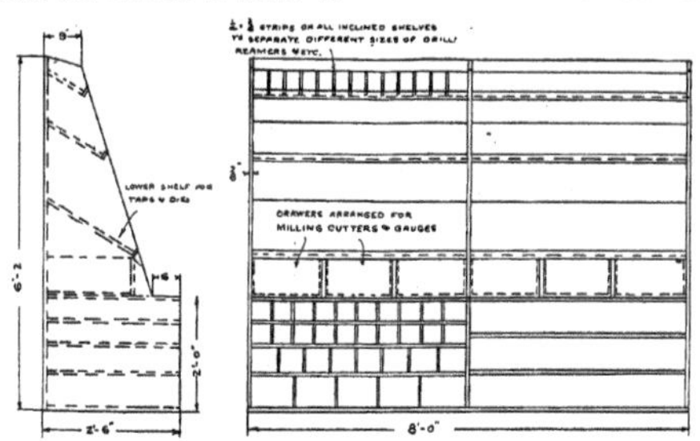

Fig. 5—Tool Case for Tool Room.

Stocking and Arrangement of Tools in Tool Room.

The stocking of the different tools in the tool room is a matter concerning which a good deal of consideration

THE STEWART CO., LIMITED.

Tool List.

Date ....,190

Firm name ....

Address ....

Job ....

Tools taken or sent out for use of ......

No. taken out.	NAME	No. Ret'd by workman.	No. to be ret'd by firm.
	Air Chippers		
	" Drills		
	" Hose, Diameter and length		
	" Rivetters		
	Bear Punches		
	" " Handles		
	Beading Tools		
	Backing Out Punches		
	Bolting Up Wrenches		
	Bolts and Nuts		
	Caulking Tools		
	Chisels—Hand		
	Chains		
	Chain Blocks		
	Clamps		
	Cold Sets		
	Crow Bars		
	Dolly Bars		
	Drill Collets		
	Drift Pins		
	Files		
	Flat Drills		
	Flogging Hammer		
	Hand Hammers		
	Holding-on Hammers		
	Jack Screws		
	Machine Taps		
	" Dies		
	Monkey Wrenches		
	Old Men		
	Patch Bolts		
	Pinch Bars		
	Pipe Dies		
	" Stocks		
	" Cutters		
	" Wrenches		
	" Tongs		
	Ratchets		
	Reamers		
	Rivet Blocks		
	Rivet Forges		
	Rivetting Hammers		
	Rivet Snaps—Air		
	" " Hand		
	Rope		
	Rope Blocks		
	Side Sets		
	Sledges		
	Stay Bolt Taps		
	" " Reamers		
	Test Pump		
	Tongs		
	Tools for Air Chippers		
	Tube Expanders		
	Twist Drills		
	Vises		
	Washers		

Fig. 4.

should be given, for it is not at all uncommon to see all sorts of tools piled up in a case without semblance of order. Needless to say, it is impossible to keep track of tools when they are kept in this way. It should not be necessary for the attendant to take more than a glance at a case to assure himself that a certain tool is in or out; and this is very easily accomplished if a little thought is given to arrangement before fitting up the tool room, though I regret to note that the rule seems to be to get the case built first and arrange them as best you may after they are in place.

Fig. 5 shows a case arranged for twist drills, reamers and milling cutters. The different sizes are arranged in order and in the case of reamers, a slight beading separates one reamer from the other. This case may also be used very conveniently for hand taps and dies, the latter being placed on wooden pins or on a recessed board, similar to those for milling cutters. When tools are arranged in this way there is more inducement for the attendants to keep them clean than where "any old place" is good enough for them.

Ordinary forged tools are conveniently kept in a case with square pigeon holes, the upper portion of the case being arranged for small tools, the central portion for large tools, and the lower portion for mandrels, plug centres and other heavy and bulky tools. Ordinary flat drills are also conveniently kept in such a case. Plug and ring gauges may be neatly arranged in sets either in boxes with hinged covers, or in double slide drawers somewhat after the fashion of those used in vertical letter filing cabinets.

Snap gauges may be placed on one end of a tool case and held in place —the smaller inside the larger—by means of wooden or iron buttons; while drill sockets may be arranged in a similar position in rows—two wooden strips for each row being placed across the end of the case—the upper one being perforated to accommodate the different sized sockets, and the lower one solid to act as a stop.

Other cases will be required, besides those mentioned above, to accommodate tools of a miscellaneous character, which will depend upon the style of work done in the shop; but these can usually be readily decided upon. Special cases will be required for jigs, fixtures, etc., and there should always be arranged where they are easily accessible and not liable to damage.

In many shops oil is served to the men from the store rooms, or they are allowed to help themselves from tank or barrel. In a later article the writer hopes to take up this matter of oil and show how it is possible to affect a large saving by having it served out from the tool room.

In conclusion, I desire to say to those in control of manufacturing establishments: look carefully into the running of your tool room, for on them the efficiency of your shop depends more largely than on any other portion of the works.

System of Machine Shop Costs

A description of the system of carrying on production and estimating costs in the plant of the Chapman Double Ball Bearing Co.

By J. C. ARMER.

It is very essential in designing a system for carrying out work in a machine shop, that simplicity be kept prominently in view. The simpler any system is, which at the same time will do efficient work, the better; but it is allowable to have office systems more elaborate than shop systems. The man in the shop who has to carry out the big part of a shop system, is not in a position to do anything very elaborate in this line, and yet the record from the mechanic of work done should be of such form as not to necessitate too much recording work in the superintendent's office.

In present day practice it is not suffcient that only the cost of a machine as a whole be known; but the detail cost of each part should be recorded. With such a record, it is possible to cut down the cost to a minimum. The material used in, and the cost of, each individual part can be investigated fully at any time. Should the cost of production run above the average, the cause can be traced and investigated, by comparing with previous costs.

There is still another feature which

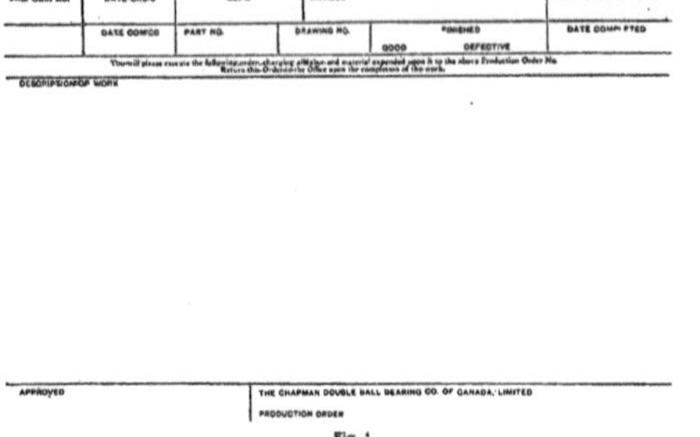

PRO. ORD. NO.	DATE ORD'D	DEPT.	ARTICLE		DATE TO BE COMPLETED
	DATE COM'CD	PART NO.	DRAWING NO.	FINISHED GOOD / DEFECTIVE	DATE COMPLETED

You will please execute the following order, charging all labor and material expended upon it to the above Production Order No. Return this Order to the Office upon the completion of the work.

DESCRIPTION OF WORK

APPROVED | THE CHAPMAN DOUBLE BALL BEARING CO. OF CANADA, LIMITED
PRODUCTION ORDER

Fig. 1

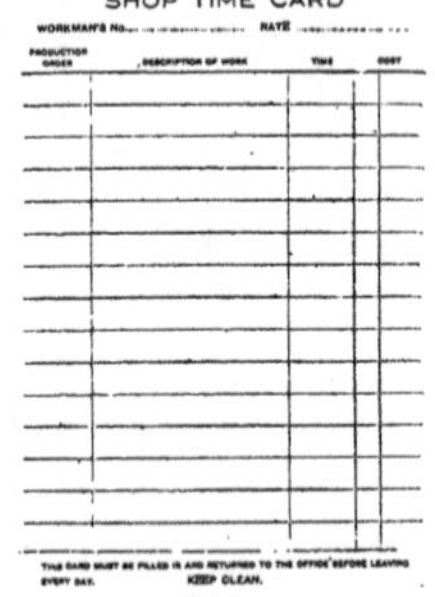

SHOP TIME CARD

WORKMAN'S NO. .... RATE ....

PRODUCTION ORDER	DESCRIPTION OF WORK	TIME	COST

THIS CARD MUST BE FILLED IN AND RETURNED TO THE OFFICE BEFORE LEAVING EVERY DAY. KEEP CLEAN.

Fig. 2

should be incorporated in a machine shop system. It should be possible to compare the cost of each individual part of two similiar machines when the two are made in different lots. Thus one might be made in a batch of five while another in a batch of twenty-five. The records should show the advantage to each part in cost of putting through twenty-five instead of five machines, as well as the difference of cost in the two completed machines.

A Simple yet Complete System.

A shop system is used in the plant of the Chapman Double Ball Bearing Co., Toronto, which is both simple and effective. This system was designed and applied by Mr. W. E. Cane, superintendent; and the writer is indebted to Mr. W. J. Murray and to Mr. Cane for the information contained in this article.

The Production Order Form.

When an order comes into the busi-

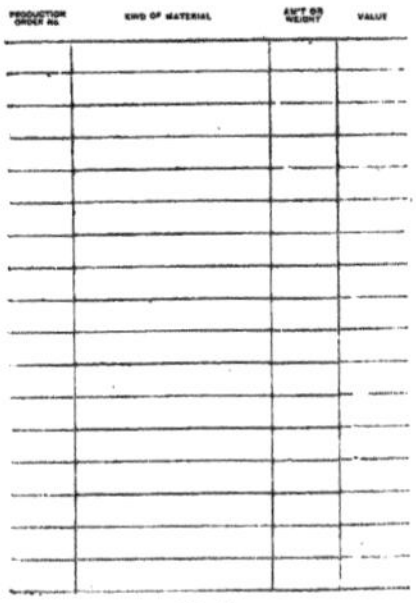

PRODUCTION ORDER No.	KIND OF MATERIAL	AM'T OR WEIGHT	VALUE

Fig. 3

ness office, it is recorded, and transferred to the superintendent's office. The superintendent sizes up the order, and makes out production orders for the different parts, with the number of parts required. The production order blanks are on a pad in triplicate, two being thin paper, and the third a heavy card. All three are of different colors. Carbon paper serves for duplicating. The style of this production order is shown in Fig. 1. One of these orders goes to the foreman, another is attached to the requisite blue prints to be handed to the mechanic, and the third, the stiff card, is kept on file in the superintendent's office.

The details of this production order should be noted. The completeness of the record which can be kept on this card is shown by the heading required to be filled out.

The Shop Time Cards.

Each mechanic in the shop is supplied with shop time cards, as illustrated in Figs. 2 and 3, Fig. 3 being the back of the card. The mechanic receives a production order to carry out, together with the necessary blue prints. This cost of each, computed from the time and his rate of pay.

Cost Data Kept on Back of Production Order Card.

This card, completed, is deposited in the superintendent's office, where data

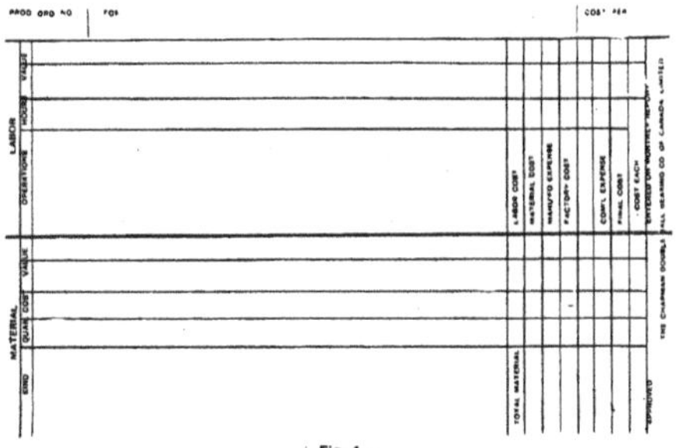

Fig. 4

production order will contain orders for one or more pieces of work to be done. The mechanic first secures materials from the stock room, data concerning which is noted on the blue print. He enters the number of the production order on the back of his time card, as in Fig. 3, the kind of the material and the amount. The value is filled in either by himself

from it can be filled in on the production order card, which was filed in the office (Fig. 1), such as date when completed, good or defective work and signature of approval by the superintendent. The production card is then turned over, which appears as is shown in Fig. 4, and data for this particular production order is filled in from the

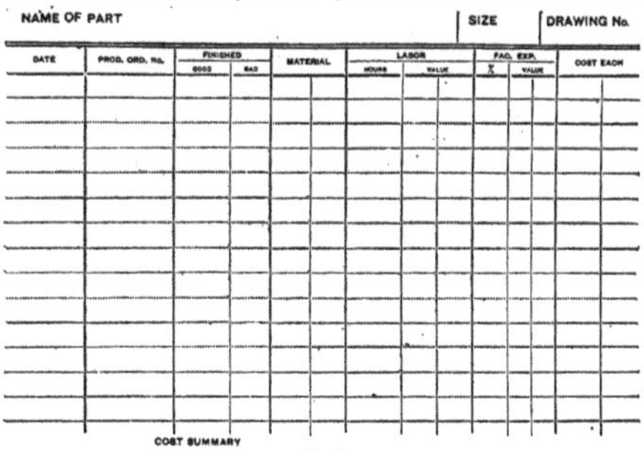

NAME OF PART | SIZE | DRAWING No.

DATE	PROD. ORD. No.	FINISHED		MATERIAL	LABOR		FAC. EXP.		COST EACH
		GOOD	BAD		HOURS	VALUE	%	VALUE	

COST SUMMARY

Fig. 5

or someone appointed to do it. He starts work and fills in the blank spaces on the front of his card, such as his number, his rate of wages, the number of the production order, a description of the different operations occasioned by that order, the time taken for each, and the

shop time card, (Figs. 2 and 3.) Thus a complete record of material and labor used in fulfilling any production order is kept on the back side of the production order card. This can be totalled as shown at bottom of card; and the total cost of each article in the order can be determined by filling in the proper proportion of manufacturing expense, factory cost, commercial expense, etc., which is arrived at through the office system.

One production order may be executed by any number of men, according to the number of operations and the kinds of operations required. They will all work from the same production order, but each will use his own time card. Then entries are made from each time card on the back of the production order card in the office.

Summary of Costs of Individual Parts For Different Orders.

In order to have a concise record of the cost of different parts of a mechanism, there is what is known as a cost summary card, as in Fig. 5. This card is filled out from the back of the production order card (Fig. 4), and also from the front, with name of part, etc., material, labor and cost. A card is kept for each piece. From this summary card the relative cost of any one piece for all different production orders can at once be seen.

This system has been in operation at the plant of the Chapman Double Ball Bearing Co. since they started operations in their new plant. It has been tested thoroughly and found very satisfactory.

ABOUT CATALOGUES.

CRANES—Circular No. 76 of the Whiting Foundry Co., Harvey, Ill., fully describing, with illustrations, their different makes of cranes and foundry equipment.

TURBINE PUMPS—Bulletin No. 101, issued by the John McDougall Iron Works Co., Montreal, illustrating and describing their electrically driven turbine pumps.

MOLDING MACHINES—Catalogue of the Henry E. Pridmore Co., Chicago, Ill. This catalogue illustrates and describes fully the drop molding machines manufactured by this company.

HOISTING ENGINES—A catalogue issued by Allis-Chalmers-Bullock, Ltd., Montreal, illustrating and describing fully the Lidgerwood hoisting engines, for mining purposes, for which they are sole Canadian agents.

—BALLS—											
RECEIVED				DELIVERED				BALANCE			
[illegible]	[illegible]	[illegible]	[illegible]	[illegible]	[illegible]	[illegible]	[illegible]	[illegible]	[illegible]	[illegible]	[illegible]

STOCK REPORT FOR WEEK ENDING.............. 19...

Fig. 6

Record of Stock Kept.

There is also a neat system of keeping track of stock in the factory. With this system a complete record of stock is kept on cards, such as in Fig. 6, which are revised weekly. Thus by simply looking over the stock cards, the condition of the stock can be found at a moment's notice. These cards are kept in the stock room, and one for each kind of stock is delivered to the superintendent's office every week. Fig. 6 shows the stock card for balls for the bearings. As these balls are delivered from the stock room, they are entered on the card. This means very little work, yet forms a complete record. Similiar cards are filled out for other items of stock.

All Cards Kept on File.

In the office of the superintendent, all these cards are filed in filing cabinets or on files. The summary cards are kept in a cabinet.

ELECTRICAL EQUIPMENT—Pamphlet issued by the Westinghouse Electric & Manufacturing Co., describing the electrical equipment at the Hornell shops of the Erie Railroad. It consists of 24 pages, 6x9 inches, fully illustrated.

STORAGE BATTERIES—An attractive catalogue of the Westinghouse Machine Co., East Pittsburg, Pa., dealing with their different types of storage batteries. Each type is clearly illustrated, and the catalogue is very neat throughout.

MOLDING MACHINES—A descriptive catalogue, issued by the Arcade Manufacturing Co., Freeport, Ill., illustrating their modern molding machines. The catalogue is well gotten up and includes a number of testimonial letters received by this firm.

GOLDSCHMIDT THERMIT CO. — A handsome catalogue, issued by the Goldschmidt Thermit Co., New York, illustrating and describing the Thermit welding process, together with pamphlet giving full instructions for welding wrought iron and steel rods by the Thermit method.

BLOWERS AND GAS EXHAUSTERS—A catalogue of blowers and gas exhausters, manufactured by the Wilbraham-Green Blower Co., Philadelphia. This catalogue gives the points of merit, and fully describes their blowers and exhausters. It is well illustrated, and is of standard size.

SHIP BUILDING—A neatly bound catalogue of Shaw, Hunter & Wigham Richardson, Ltd., Wallsend shipyard, Wallsend-on-Tyne. This book describes, with fine illustrations, what this firm has done in the shipbuilding line, and also includes a short history of the company, with a detailed list of shipowners for whom vessels have been built. It consists of 40 pages, 6x10 inches, with a strong cover. This is a credit to the firm.

GENERATORS AND MOTORS—A well gotten up catalogue, of the General Electric Co., Westeras, Sweden, entitled, "What We Can Do." This book describes, with half-tone illustrations, the different types of generators and motors manufactured by this reliable firm, and also gives views of a number of power plants, where their motors and generators have been installed. There is shown an illustration of a machine built and installed for the Hamilton Steel & Iron Co.

PERSONAL NOTES.

R. J. Cottrel has been appointed acting locomotive foreman on the G.T.R.

J. H. Mills, formerly road foreman of locomotives on the C.P.R., has been appointed district master mechanic with headquarters at Montreal.

W. H. Towner has been appointed acting general foreman on the Grand Trunk Railway at Toronto shops, vice Mr. J. C. Gerden, resigned.

The headquarters of Mr. S. Phipps, master mechanic of the Pacific Division of the Canadian Pacific Railway, have been changed from Revelstoke to Vancouver, B.C.

A. R. Mackay, formerly an engineer with the Electric Development Co., Niagara Falls, Ont., has this week assumed the position of advertising manager for the Canadian Fairbanks Co., Montreal.

E. E. Austin, formerly road foreman of locomotives on the Canada Pacific Railway, has been appointed district master mechanic of the Pacific Division on the same road with office at Nelson, B.C.

The Control of Superheated Steam *

The Best Materials for the Control of the High Temperatures of Superheated Steam—Iron Castings for Fittings or Valves Unsuitable—Cast Steel and Nickel very Suitable Metals—Best Designs of Valves, Joints, Etc.

By M. W. Kellogg

Since the introduction of superheated steam as a large factor of economy in stationary power plant use, the question of what type of material is best for the proper controlling of the resulting high temperatures has caused a great deal of investigation and interest.

In the following discussion of materials, some reasons will be given which are the results of experience and test, and other facts which we have accumulated from reliable sources will be shown. This article treats particularly of what might be called in a general way, piping systems, which systems are made up of pipe, fittings, valves, and the necessary details connected therewith, such as joints, gaskets, etc, and are taken up separately.

Pipe Considerations for Superheated Steam.

There can be little question as to the matter of pipe except quality. Of course, welded wrought iron or steel pipe is successful, but the difference in the quality of pipe under different conditions is very material. As in nearly all instances in a superheated steam station, the old-fashioned screwed joint is not satisfactory, it is necessary to "work" the pipe—that is, weld, van stone, etc.—to make either a welded, van stone, or other joint of the same general description.

The accompanying illustration (fig. 1) is what is known as a "van stone joint."

For this work, the pipe made from open hearth steel is a great deal the best for manufacturing reasons, because it can be properly "worked," there being less carbon, and the quality is much more uniform. Bessemer steel pipe will very often act in a satisfactory manner, but one is never sure that Bessemer will run evenly, and, therefore, troubles may result.

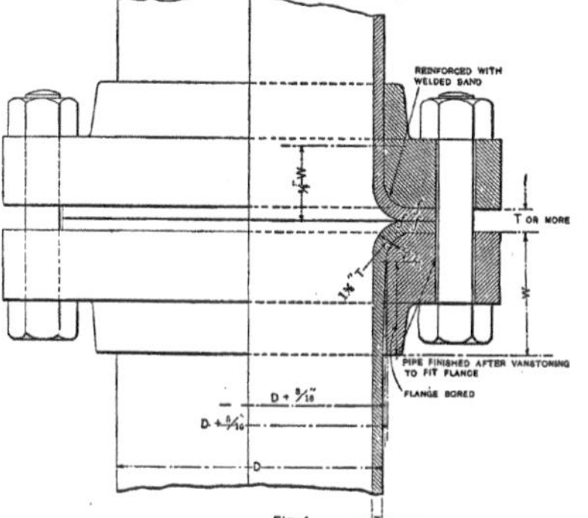

Fig. 1.

It is practically impossible to "work" wrought iron pipe. In making this "van stone joint," the pipe will almost surely split very badly, not only at the weld, but also around its outer circumference.

On the other hand a good quality of wrought iron will cut and thread more easily with standard pipe machines and dies than a steel pipe, and a Bessemer steel pipe will thread much more easily than open hearth steel pipe.

A great many manufacturers have great difficulty in threading open hearth steel pipe, because dies are not set as they should be for open hearth steel. The die in a pipe machine should be set at a greater angle, with the radius of the pipe passing through the point of contact of the die, for soft steel than for other kinds.

Full weight pipe is perfectly suitable for any temperature and working pressure up to 250 pounds, as long as the pipe is not made thin at any point by cutting or threading.

Suitable Fittings.

The general designs of fittings are very satisfactory for use with superheated steam, with the one exception, that very few manufacturers include what is known as the "long fillet" between the body of the fitting and the flange. This is a very desirable point, due to the fact that at this place there is the greatest strain from shrinkage in the molds, which also tends to develop porous spots. Most large users of this type of material have learned this thoroughly and design their fittings specially; the chief difference in their design from that of the general manufacturer merely covering this point. The quality of the material in fittings, however, is a very important thing in connection with superheated steam.

The latest practice is to do away with fittings entirely on high pressure steam lines and put what are known as "nozzles" on the piping itself. This is accomplished by welding wrought steel pipe on the side of another section, so as to accomplish the same result as a fitting. In this way rolled or cast steel flanges and a van stone or welded joint can be used. This method has three distinct advantages, to wit:

(a) The quality of the metal used, for reasons explained hereafter when the subject of the effect of heat on metals is taken up.

(b) The lightening of the entire work.

(c) The doing away with a great many joints.

As a general average, at least 50 per cent. of the joints can be left out, and sometimes this portion runs up as high as 60 or 70 per cent., according to the layout of the system.

If this method is employed, substantial welds must be made, not only to stand the pressures required but also the strains; this is accomplished successfully in Germany, England, and the United States.

* Abstract from a paper presented before the American Society of Mechanical Engineers at the Indianapolis meeting, May, 1907.

Design of Valves.

It is important to have a good design of valve. I believe that nearly any of the designs made by the good manufacturers are entirely suitable; such as a broken or solid wedge valve of the ordinary type, under the condition that all machine work is done thoroughly and the quality of metal used is satisfactory for the purpose intended.

It may be interesting to note here the effect of a large range of temperatures on a short piece of steel. By calculation, a piece of steel six inches long, heated 500 degrees, will expand 0.019 inch. This figure is put down to show how variations in the coefficient of expansion of metals by heat have a large effect on the permanency of a valve staying tight, and it can readily be seen that a small proportion of the distance given is sufficient to cause trouble.

Metals and Temperatures.

I find that different authorities vary slightly in their statement as to what temperatures different metals will stand with good results. German authorities state that cast iron should not be used above 480 degree F. Other authorities allow us to go as high as 575 degrees F. Above these temperatures in cast iron the limit of elasticity is reached with a pressure varying from 140 to 175 pounds. Under such conditions the material is strained and does not resume its former shape, and eventually shows surface cracks, which continue to grow until it lets go. These temperatures and pressures also lead in time to a shrinkage of all parts, and to a structural alteration of the metal which results in leakages in valves at the seatings. Therefore, it would seem that iron castings are unsuitable for both fittings and valves to be used in any superheated steam work. While they may last for some time, after a few years' use the metal becomes very weak and some cast iron has reached the point in weakness where if it were merely tapped lightly with a hammer it would break into pieces.

The only adaptable metal I believe to be cast steel. The results of tests by Bach on this metal for the effect of temperature are such that at 572 degrees F., the reduction in breaking strength only amounts to about 1.1 per cent. and at 752 degrees F. to about 7.8 per cent. Therefore, it seems that this metal is practically capable of withstanding all pressures and temperatures up to at least 800 degrees F., without showing any appreciable weakness.

The influence of high temperatures on bronze, etc., is very material. At ordinary temperatures this metal has a breaking strength of about 34,100 pounds per square inch and an elongation of 36 per cent. At 572 degrees F. the breaking strength falls to about 19,500 pounds per square inch and the elongation to 11.5 per cent. At 662 degrees F., which is quite a common temperature, as it leaves the superheaters, the breaking strength of bronze only amounts to 12,200 pounds per square inch and the elongation at the breaking point is only approximately 1¾ per cent. This seems to eliminate entirely brass or bronze of ordinary composition for use with highly superheated steam.

The effect of temperature on nickel is very similar to that of cast steel and in consequence this material is very suitable for use in connection with highly superheated steam. Bach recommends that bronze alloys be done away with for use on steam lines above temperature of about 390 degrees F. Even neglecting the special quality of nickel seatings, on account of the great toughness of this metal and the methods which can be used for securing rings of this substance to the valves and conical surfaces, it has the special advantage of having the coefficient of contraction and expansion with temperature almost exactly the same as that of cast steel, so that no slackness of the rings occurs and the valves remain absolutely steam tight. There are instances in which valves constructed with nickel seatings have been satisfactorily used with steam temperatures up as high as 932 degrees F.

Seats, discs, and bushings made of brass or plain bronze do not retain their shape.

For spindles on superheated steam work I strongly recommend nickel steel, which holds its shape and does not deteriorate with high temperatures.

Seatings in valves should not only be screwed in but also pinned in addition, using a fine thread which is very long, to give a tight joint. Seats should also have a flange on the top that makes a joint with the body when screwed down, which prevents the tendency to leak through.

I think it is generally acknowledged that the old fashioned screwed joint, no matter how well made, would not be suitable for superheated steam work. This leaves for discussion two general types, viz.: welded joints, and what are generally known as van stone or climax joints; that is, any joint where the pipe is turned over the face of the flanges.

In welding a flange on a piece of pipe, great care must be taken to see that the weld is perfect because of the unequal thicknesses of the metals to be so welded. If the weld is thoroughly made, this type of joint is very good, although for erection purposes, due to the fact that the flanges cannot swivel, it does not equal the turned-over joint as mentioned above. The manufacturing expenses in making a welded joint is very good, although for erection purposes, due to the fact that the flanges cannot swivel, it does not equal the turned-over joint as mentioned above. The manufacturing expenses in making a welded joint are also much more for the same type of work accomplished, on account of the necessity of doing all finishing work after all rough work, such as welding and bending, has been completed. Therefore, the cost of welded joints is greater, not only for the work done but because of the increased expense in finishing on account of the necessity of employing methods different from those where the flanges, etc., were all finished before the joints were made, as is possible on the turned-over joint mentioned above.

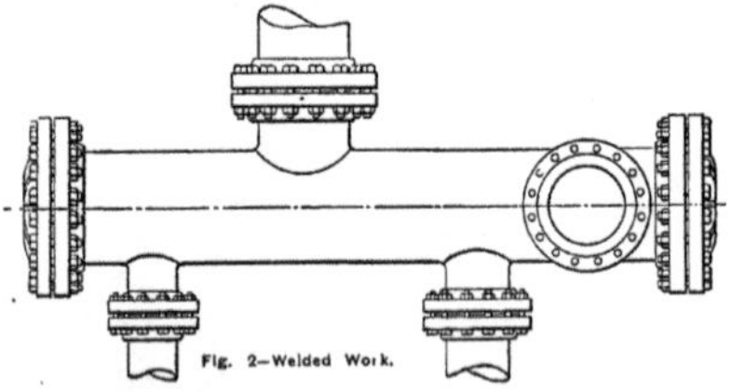
Fig. 2—Welded Work.

In regard to the turned-over or van stone joint, the quality of its manufacture seems to us the most important feature. This joint can be made in a careless way where the pipe is in no way thickened up and only faced on the front. A joint of this kind does not give good results, principally for two reasons:

(a) The thinness of the metal on the turned over portion; and

(b) On account of the recesses left between the back of the pipe on the turned over portion and the flange, due to the pipe not being finished at this point.

The writer believes, however, if this

joint is properly made, it is equal to the welded joint as a manufactured article and superior to the welded joint as an article for erection.

To have this type well made, the pipe on the end should be thickened up in an amount sufficient so that after the joint is turned over there will be enough metal left to face the turned over portion on the front, on the outer edge, and on the back. We of course take for granted that the flanges are finished on the front. After the work above mentioned is done, the pipe should be as thick on the turned over portion as the original thickness, or very close to it. Increasing the thickness of the pipe on the end before going through the operation is done in several ways. In a general way, I consider any of the methods satisfactory. The point made of facing the turned over portion of the pipe on the back is an exceedingly important one, much more so than most people seem to realize. I have known instances where it has been found impossible to make a ground joint, for no other reason.

In reference to making up a joint, I believe that the face of all flanges or pipe where a joint should be made ought to be given a fine tool finish and have the face level, and then use a gasket of some description. A perfectly made ground joint is a good thing but it is very expensive, and it is hard to get the average contractor to furnish it in a perfectly workmanlike manner. Also, after it is so done, it is liable not to stay tight, on account of the tremendous expansion and contraction causing such strains that the joints are liable to open up, particularly when the pressure is taken off the plant. The simple expansion and contraction on the bolts that make up a joint would cause this.

Gaskets for High Temperatures.

There are large numbers of gaskets manufactured of all types and descriptions. It is very hard to take up this subject and be fair to each of the manufacturers, for the reason that practically no one can and has had experience with every type made to judge for himself, and hearsay would lead us to suppose that all of them are at one time perfect and at other times useless.

I have used a great many different types of gaskets, however, and have obtained the best results with a corrugated soft Swedish steel gasket with "Smooth-on" applied, and with the McKim gasket, which is of copper or bronze surrounding asbestos. The ordinary corrugated copper gasket is a very popular make and has been used a great deal. On superheated steam, usually sad results follow. There seems to be some peculiar action that causes this, as on superheated steam lines a corrugated copper gasket will in time pit out in some part of the flange nearly through the entire gasket. I have heard a great many reasons given for the cause, such as electrical action, disintegration, etc.

The wear of a gasket depends largely on the method of pulling up bolts on flanges. In fact I believe that a great many troubles have occurred because of imperfect erection. If joints are pulled up entirely on one side and left loose on the other, and then taken up on that side, trouble with the gasket is almost certain. The bolts should be taken up gradually all around the flange. The experience of the erecting crews on high class superheated steam lines is an exceedingly important thing. The average steamfitter is not suited to this type of work. He has had experience with lower pressures and less important tasks and after a piece of work is erected by him it is customary to find a great many leaks which are usually only eradicated after the whole joint has been broken and properly repaired. All these troubles can be eliminated by using only steamfitters experienced in the type of work under consideration.

Discussion of Paper at Convention.

There was some discussion on this paper at the convention, and Mr. Foster did not agree with the author in regard to the difference between pipe lines carrying superheated and saturated steam.

A pipe line carrying a moderate superheat, say up to 500 degrees Fahr. would not differ materially from one carrying 150 pounds saturated steam. Mr. Foster did not agree with Mr. Kellogg's conclusions in respect to gaskets, nor to the use of cast iron, rather favoring cast iron.

J. Roland Brown commended the use of van stone joints, but advocated at least one threaded and flanged joint in the line for convenience of make-up in the field. More pipe lines are injured by springing in the field because they do not agree with the drawings than by any other cause. The method of installing gaskets has more to do with their success than the style of gasket. Mr. Brown spoke of three successful types of gaskets, a long fire asbestos gasket held with a small quantity of binder and coated with graphite; a joint made by grinding together a male and female flange and inserting a soft steel corrugated gasket covered with graphite and oil; and two flanges with a groove into which a cast iron ring is inserted. This last type is especially good on heavy fittings.

A paper was presented by R. P. Bolton on the above subject. His paper criticizes the existing practice of applying superheaters to boilers and states that by merely placing a superheater coil in a certain part of a gas passage of a boiler and connecting a steam supply to it is by no means to be regarded as a complete solution of the problem. The operation of an ordinary boiler furnace is subjected to so many variable elements in the nature of the fuel, its combustion, draft regulations, door openings, ash and clinker accumulations and the personal equation that an apparatus as sensitive as a superheater should be protected from these varying influences. The disappointments experienced in superheated steam operations of high temperature are due to these influences and fluctuations in the steam output of the boiler, which will also affect the degree of superheat. Furnace conditions should be such as to eliminate all fluctuations in the flow and temperature of gases, or the heater must be placed so that the flow of the gas volume can be controlled.

In regard to this paper, J. R. Brown stated that where the superheater is located where it is not exposed to more than 1,000 degrees Fahr. it is unnecessary to flood the superheater when raising steam if fair judgment is used. When a superheater of the boiler setting type goes out of commission, it will cause the loss of only one unit and the ratio of superheating to water heating surface in the rest of the plant remains the same. When an independently-fired superheater goes out, it reduces all or a large part of the plant to saturated steam conditions. In large units, the boiler setting type of superheater, as now designed, does not average over 175 degrees superheat, although some are sold to develop 200 degrees. In order to get the higher degree of superheat, it is necessary to increase the height of the boiler setting.

BOILER MANUFACTURER DEAD.

Arthur G. Booth, secretary-treasurer of the Booth Copper Co., Toronto, died recently.

Mr. Booth was but thirty-eight years of age, and had been ailing for about a year.

Besides his wife and children, his father, Geo. Booth, president of the Booth Copper Co., and two brothers, Walter C. and Clarence H. Booth, survive him; the former is vice-president of the above company and the latter head of the Pressed Steel Bath Co., Detroit, Mich. The late Mr. Booth has had a long and successful business career and was at one time secretary of the Steel Clad Bath Co., Toronto.

Success depends as much upon the use of our off working time as it does upon our hours of labor.

Tricks of the Trade

WAYS AND MEANS OF DOING THINGS

Send in your ideas, sketches or jobs you are doing, and anything of interest to your fellow-workmen. Remuneration will be made for such contributions.

FOR BORING ARMATURE BEARINGS.

This is a chuck used to great advantage by the Toronto Street Railway Company for boring out babbited arma-

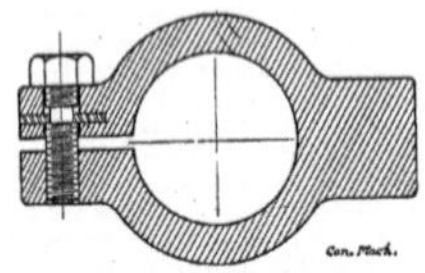

Boring Armature Bearings.

ture bearings. The chuck is of ordinary design, but has one unique feature—that instead of being bored out 1-32 in. larger to admit of insertion of bearing and then being clamped down, as is customary, it is bored to exact diameter of bearing, and by means of a slot cut as shown at A, and a piece of bent plate inserted, against which the shoulder on the clamping screw acts, when screw is backed out, the chuck will open sufficiently and quickly to allow of bearing to be extracted. It is especially useful where a great number of bearings of the same diameter have to be bored and it can be seen that the bearings will always be bored on the dead centre. The lug on the opposite side is put on simply to balance the chuck in the lathe, and can also be used when the thread in the other lug becomes stripped, or otherwise useless.

JIG FOR DRILLING LATHE HANDLE

By A. Allen.

Herewith is presented a jig for drilling a three-ball handle for a lathe, and the jig can be modified to be of value in drilling similar pieces.

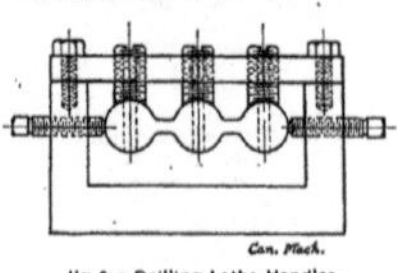

Jig for Drilling Lathe Handles.

It consists of a cast iron base to which is fastened a cast iron plate by means of cap screws. In this plate are tapped three steel bushes whose inner diameter is the same as the diameter of the hole to be drilled. The drill is guided by these bushes and, as can be seen, the holes will be strictly parallel and at correct centres, as centres are always fixed beforehand. The handle, as seen, is held in position by set screws. It has been used to great advantage in the shop of John Bertram & Sons, Dundas, Ont., and also in that of the Toronto Gas and Gasoline Engine Co.

BRUSH HOLDER ADJUSTING RIG.

By H. Hutchinson.

The accompanying illustration is of a jig used in the machine shop of the Toronto Railway Company as a brush holder adjusting rig.

The jig consists of a metal bedplate to which is bolted a cylinder the same size in diameter as the average commutator. In the centre of this cylinder is a post about 4 inches in diameter. The cylinder is slotted to receive a steel insert the same size as a carbon brush, and the post is also slotted to allow the steel insert to enter. An adjustable bracket is also bolted to the bedplate to permit of regulating the distance between the centres of the different types of motors. The complete brush holder is then bolted to this bracket and, if correct, the steel insert will easily pass through carbon-holder, cylinder and pocket in the slot in the centre part of the jig. By these means proper distances and angles are secured.

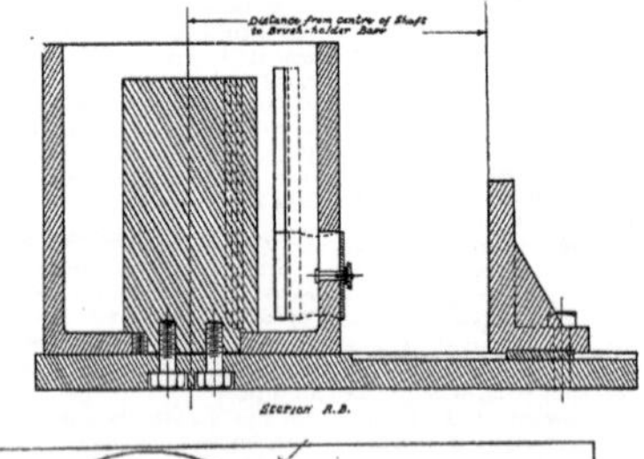

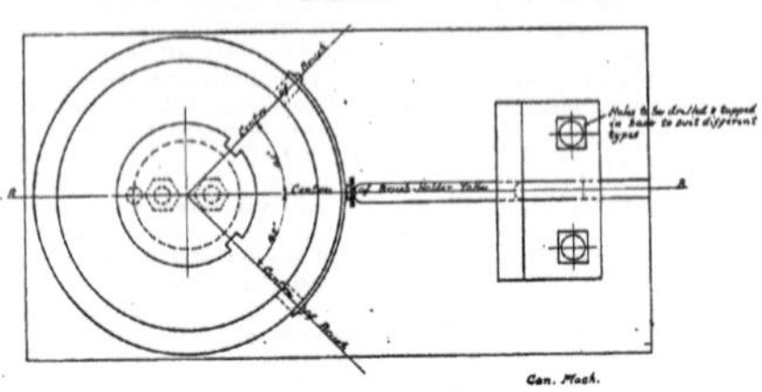

Jig for Boring Armature Bearings.

LEARN TO ESTIMATE.

By L. Marteaux.

To make a correct estimate is not easy, and it takes a good deal of experience. There is a great difference between an estimate and a guess.

However, there is to my knowledge not much difficulty in fixing the price of tools which are made for shop use only. The cost of such tools is a secondary matter, as the profit is to come from their product. An estimate is of more importance where the tools are sold, and where the competition of others in the same line of business has to be met.

In answer to an advertisement of ours a diemaker applied for work; in asking him the usual questions, he told me that he had been in business for himself, but had given it up. He was put to work and proved to be a good mechanic.

He had finished a die and, when I was inspecting it, he asked : "Do you mind telling me how much this die is worth?" To which I answered : "How much do you think?" He pondered a little and then he said : "Is it ten dollars?" I could not help laughing. His time on the job amounted to $13.55; material and blacksmith, $4.80; total $18.35, not including shop expenses or profit. I could not help telling him that he had probably done better for his customers than for himself.

This shows that a man may be a good mechanic and still not be able to know the value of his work, which is the first requirement for estimating.

Quite a number of men don't think it worth while to figure up how much time they put on a job. They don't seem to think that some day they might be asked to estimate. I believe that every man should have a note-book in which to keep the time spent on each job, and in doing so he will be able to judge intelligently the value of the work he has done.

When making my first estimate, I figured how many hours it would take me to do the job myself. Being only a few days in the shop, I had had no chance to find out what my help could do. But I was determined to find this out, so I decided to make my quotations in duplicate for some time to come.

To compare the result obtained with my quotation, I entered every order on another memorandum, noting the price quoted, the time for doing the work, and the name of the man or men doing it, also their rate of wages, as in accompanying table:

This may seem a laborious job, but it is easy. It takes only a few minutes every morning and it is well worth doing. After the job is finished I can tell in a minute the result. I can tell by it if my judgment in the ability of my men is correct. The results have been in most cases satisfactory, excepting where things happened which were beyond control. I have continued the use of the memorandum books, and they have accumulated to 14 in number. They contain a lot of information as to prices, methods of doing certain work, and mistakes. Yes, I have recorded my mistakes, too.

In the course of years I have gained a lot of experience in estimating and otherwise, but (I will not say I am entirely fool-proof) of this fact I am reminded when I look at my memorandums, from every page I can see that time and patience are required to get experience, and that this experience cannot be bought at so much per pound.—American Machinist.

Order No. 80643.	Article.	Price.	To Whom it Was Sold.
Name of Workmen.	**Hours.**		
C. B.	10-10-10-10	= 40 at 30 cents......	$12.00
P. D.	3- 6-10- 6	= 25 at 27 cents......	4.75
T. G.	6-10-10- 3	= 28 at 20 cents......	5.60
M. L.	6-10- 4- 3	= 23 at 20 cents......	4.60
			$26.95
Blacksmith........			4.00
Material........			8.25
Total........			$39.10

JIG FOR FINISHING ECCENTRICS AT ONE SETTING.

By W. C. Murphy.

Anybody who has had anything to do with finishing eccentrics has undoubtedly run up against the difficulties in such jobs. As I have some time been doing this work on an improved turret lathe, I have myself realized some of the difficulties attached to this job. Recently our general foreman devised a simple jig for boring and turning locomotive eccentrics at one setting.

It consists of two plates, the main plate as shown in Fig. 1 and the sliding plate as shown in Fig. 2.

The main plate is secured to the machine by means of the boss on the back fitting into the hole in the chuck (the jaws having been removed) with three cap screws C C C through the plate into the chuck. The centre slot is to fit a like tongue on the sliding plate. The short slots on either side are to receive the four bolts (x x) which hold the sliding plate to the main plate and are made long enough to allow a shifting of the sliding plate equal to one-half the throw of the eccentrics to be machined. The oblong hole in the centre of the sliding plate is made wide enough to clear the boring bar and long enough to permit the sliding of the plate equal to one-half the throw of the eccentric. Holes D D shown in Fig. 2 are stop-pin holes which match up with holes D D in the plate, Fig. 1. These holes are located at distances from each other equal to one-half the throw of the eccentrics. The holes E E, etc., are to receive centres E and are located in such a manner as to strike the solid part of the eccentric. Holes H are for studs

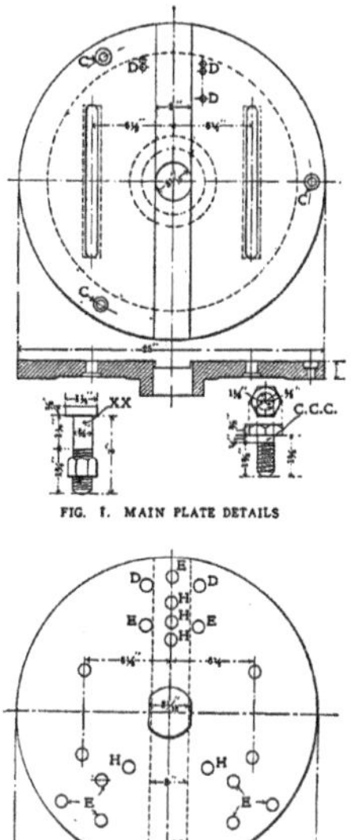

FIG. 1. MAIN PLATE DETAILS

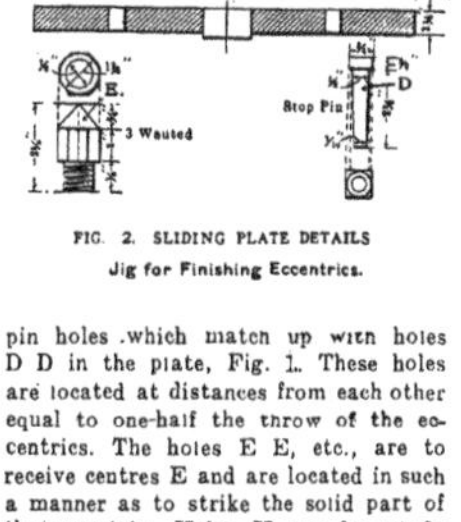

FIG. 2. SLIDING PLATE DETAILS

Jig for Finishing Eccentrics.

used to clamp the work to the face plate in the usual manner.—American Machinist.

OILING BEARINGS.

By E. S. Newton.

I am prompted to give a bit of my experience along this line. Some 20 to 22 years ago we made an engine to run an electric light plant, with cylinder 24x48 inches, which was a pretty large one for our small shop to tackle. Soon after it was running, it developed an inordinate desire to heat the main bearing. This bearing was 12 inches in diameter by 21 inches long; the quarter boxes were babbited, and the bottom shoe was of phosphor bronze 8 inches wide, giving 168 square inches to carry about 18,000 pounds, which would not seem excessive, only a little over 100 pounds to the square inch. The box was equipped with two sight feed oil cups, and had an opening on top 6x16 inches down to the shaft. The first regular run it made was only a few hours long—a "moonlight run"—and the engineer managed to keep it going. Soon the runs lengthened, and then the trouble grew serious. One morning after an all-night

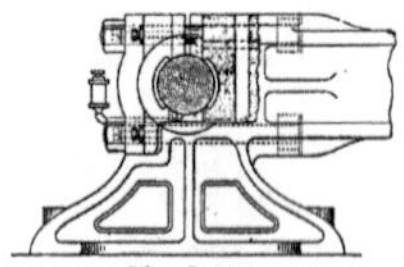

Oiling Bearings.

run I went over to the plant and it seemed as though there was a barrel of tallow on the floor—tallow mixed with black lead, oil and water (only the water didn't mix), and the box was then too hot to handle. Something had to be done and as I was practically responsible for the outfit, I felt a trifle worried. I don't know as it does any good to look at a thing that "acts up" but I did look at it a few minutes and then asked the engineer to take off the cap and take out the bronze shoe and send it to the shop.

I cut a groove 20½ inches long as near the edge of the shoe as was practicable and put in the pipe and sight feed oil cup, shown in the cut. We got it back in time for that evening's run, and from that time till the engine was replaced by a larger one, some four or five years later, that bearing was never any warmer than the air in the room. As first made, this bronze shoe was beveled to carry oil under the shaft, and grooved diagonally according to the time-honored custom. We noticed one curious thing about it; if the oil cups on top were set to feeding, oil would run out over the top of the added oil cup at the left, just as much as was delivered by the two on top; stopping the oil from the top resulted in the new oil cup instantly becoming operative again. The engine ran "over;" had it run "under," I would have cut the groove at the other edge, and carried oil to it by drilling all way through the shoe.—Machinery, N.Y.

A STORY OF EDUCATION.

By W. D. Forbes.

A young man came into a machine shop and got a job as lathe hand. He was good at his work and ran his lathe at proper speeds on various diameters when the material was considered, and he took big cuts and fine ones in their proper places, earning his sixteen dollars a week.

On Saturday nights he with many of his shopmates took in a vaudeville show and contentment, to a very large degree, was his lot. Every now and then he had to cut threads and he would look at a little brass plate on the headstock of his lathe and it would tell him what gears he had to put on the stud and on the screw, and there was no trouble at all in getting the right thread.

He often heard about the "angles of threads," their "pitch," "the area" of a piston and the "tensile strength" of steel and iron and other metals, of "plans" and "elevation," but just what they all meant was not clear to him.

One day the foreman gave him a job which required 11½ threads to the inch and his little plate failed him. He told the foreman that his lathe would not cut 11½ threads. The foreman asked why it would not? He did not know. He thought he remembered that a lathe he had seen did cut 11½ threads.

The foreman looked at the little plate, figured a half minute and told him to use certain gears and to his great surprise the job came out all right, and to him the foreman became a great mathematician.

Some time later again the little plate failed and the foreman was out looking after a break-down job. Here was real trouble. The work was wanted at once and he could not do it. An apprentice boy who had helped him now and then in taking off his chuck and putting long pieces between centres, asked what was the trouble and was told. In a few minutes the proper gears had been figured out by the apprentice and with fear and trembling used, and "mortification" set in when the job came out all right.

A few nights later the lathe hand was at an evening school with the apprentice awkwardly working over a drawing board. In six months he had mastered the mysteries of "plans" and "elevation" and "assembled drawings" and what had been so hard before to understand in blue prints was now clear, and the words "angles" and "areas" conveyed real meaning to him. He could figure out just how many feet the surface of a shaft was making in a minute and what size piston rod should be used for a given area of piston with a given steam pressure, and he could cut any thread wanted without ever looking at the little brass plate.

He was so carried away by the desire for knowledge that he began to feel very sorry for the rest af his fellow workmen and he induced many to go to night school.

One day he asked the foreman for a raise in wages and got it the next week. This acted like a spur to a horse and the value of education and knowledge was practically proved. He talked education at noon to his shopmates and before he knew it he was shouting its value like a campaign orator. He besought all to study, learn, and thereby rise to higher planes so that they could all become foremen and proprietors.

Soon he was made foreman of a room, he never went to vaudeville shows but took in all educational lectures and he continued to extol the value of education. He did not, however, find being a foreman as easy a job as he thought it was and he began to understand why a foreman was so anxious to get the work out quickly and swore so when a job was spoiled or delayed.

One thing troubled him, however, a good deal and that was the fact of not being any happier than he was before he gained knowledge—the ability to figure out threads and areas, and to know what π stood for. To be able to figure threads and almost everything else was nice but he worried because he could not tell the difference between an involute and an epicycloidal form of tooth and more because so many jobs were spoiled by the men and that was something he never did before he was a foreman.

He still talked education to his men. One day while doing so at noon he pointed out of the window to two men digging a ditch in the street and said "Knowledge is what you want. Look at those two men at work in the street at a dollar and a half a day. With knowledge they could lift themselves above such work and earn twice as much!" Mike, the oiler, who sat by his side smoking his clay pipe, looked up and said, "If we all got education who would dig the ditches?"—American Machinist.

Developments in Machinery

Metal Working Wood Working ***Power and Transmission***

NEW QUICK CHANGE MILLING MACHINE.

The Adams Company, Dubuque, Iowa, are placing upon the market the Farewell quick change milling machine that embodies several features not found in machines in this class.

While it is a hand or lever fed machine, it is intended not only to do the light work usualy done on hand-fed machines, but to do accurately and more quickly much of the work usually done on a shaper, on larger and more expensive power-fed milling machines, on the profiler and on the cam cutting machines.

The table has an inclined bearing surface of 1½x23½ inches, running the entire length of the table and two vertical surfaces 1½x23½ inches, also the entire length of the table. This feature is claimed to be conducive to the least lateral lost motion and liability to chatter that it is possible to secure in a freely moving slide. The weight of the table and all upon it tend to hold the sliding surfaces in constant contact. The table has a finished or working surface of 8x18 inches, unusually wide to accommodate bulky work and to give wide transverse range of adjustment to chucks and vises.

buckle holds the swinging cone pulley shaft in position to keep the spindle belt taut in all positions of the head.

Six spindle speeds are obtainable. The spindle is of forged crucible steel and is hollow its entire length. The spindle bearings are bronze and are tapering. Provision is made for a quick and sensitive adjustment of the spindle bearings in the quill.

The pendant is provided with a centre controlled by thumb screw to steady long arbors. Can cutting guide pin may be secured in the other end of pendant.

The vise furnished with the machine has jaws 6 inches wide, 2 inches deep

Two Views of New Quick Change Milling Machine.

With carbon steel cutters, when the tooth contact was comparatively slow, power feeds were required to get the slow, steady feeds necessary to produce true, smooth surfaces; but with high speed steel milling cutters, with the tooth contact nearly quadrupled, the surface produced by hand-fed mechanism is satisfactory and the rapidity of operation, particularly on short cuts, greatly increased.

Instead of following the common practice of providing the vertical and transverse adjustments in this milling machine, by interposing double or treble sliding elements between the work table and the column of the milling machine, which makes rigidity of the table impossible, the table of the Farwell quick change milling machine hangs and slides upon a vertical rail. There is but one slide between the table and the column.

The spindle head is counterbalanced by weights in the column. Part of these weights, like scale weights, may be removed so the weight of the head will feed the cutter down into the work by gravity, or additional weights may be added so the cutter will feed up, the weights being heavier than the head. The vertical movement of 12 inches is provided.

As both the vertical and transverse feed of the cutter is independent of the table, or the weight of the work upon the table, small cutters may be sensitively fed both in and down, without danger of breaking cutters.

The drive of the spindle is by a 4-inch belt from the swinging cone pulley shaft and by a 2½-inch belt from the cone on swinging shaft to the countershaft.

A distance rod with adjusting turn

and opens 3 inches without reversing. By reversing the sliding jaw the vise will hold work 6 inches wide. The vise is exceedingly heavy and is provided with engaging flanges so it may be secured in the quick change clamps in two vertical, as well as two horizontal positions.

The three-jawed chuck furnished with the machine is 6 inches diameter, has two sets of jaws and will take 1 9-16 inches through the centre. The chuck is mounted on a heavy angle plate which may be secured by the quick change clamps in two vertical and in a horizontal position. The chuck has 24 stops or graduations which are very convenient in spacing cutters, laying out key ways on quarters or opposite sides of a shaft, or for squaring or milling hexagon heads. This chuck is also very convenient for circular miling.

The quick change features, whereby

the positioning of the cutter in relation to the work, the quick locking or releasing of the various slides, the quick change of the position of the feed levers and the quick change clamps for the vises, chucks and fixtures are peculiarities of this machine.

Quick Change Features.

The feed levers not only move the table, spindle or head slides, but also locks these slides by a right angle movement of the lever. This is of great convenience in profiling or when feeding in two directions. By this arrangement the table may be locked, the cutter fed into the work to proper depth, then locked and then the table released and the longitudinal feed begun, all without the use of wrenches or without the operator removing his hand from either of the levers.

The lever, when moved to the locking position is disengaged from the pinion it controls, so the lever may be swung to a new position. The act of unlocking the slide re-engages the lever and the pinion in the new position.

Profiling and Cam Cutting.

The spindle head is provided with an arm extending back and in line with the spindle in which a guide pin may be secured either 4, 6, 8 or 10 inches from the spindle centre. A guide pin may also be secured in the overhanging arm pendant either 2½ inches or 3½ inches from spindle centre.

A roller the size of the cutter on either of these guide pins will follow a pattern secured to the table, chuck or angle plate, while the cutter will duplicate the form on the work. The power of the spindle drive and the rigidity of this machine permits of heavier cuts with larger cutters than are usually used on profiling machines.

IMPROVED BATH GRINDER.

Improvements which have been made in the bath grinder have greatly increased the efficiency of the machine. The accompanying illustrations show the new machine. It will be noticed that a marked improvement has been made in the spindle head over the old pattern. The head is much heavier and is bored tapering to receive split boxes. The large ends of the taper boxes are at the ends of the head, giving rigidity to the spindle at both ends, upon which the wheels are mounted. Surrounding the spindle box on the right-hand end of the spindle head is a projecting square flange, to which the wheel hood is clamped. This flange is independent of the box bearing. It will be noticed in clamping the hood to the head that the clamp screw for binding same is in a straight line under the spindle, making a rigid safety clamp in case the wheel breaks.

Fig. 1—Improved Bath Grinder.

Fig. 2—Improved Bath Grinder.

Fig. 1 shows our extension arbor support, for surface grinding. On the top of the front face of the spindle head is ¾-inch slot for locating the support. The arm has a tongue on the back side which fits into the groove on the spindle head and is clamped to same by two screws. On the projecting end of the arm will be noticed a phosphur bronze bearing, the lower end of which is adjustable to the spindle. This assures absolute rigidity and allows no spring in the spindle when doing surface work. Mounted on the end of the arm is a wheel hood arranged for holding the water spout. In this case, the water spout is used on either wheel hood as shown.

The spindle head is elevated by the vertical handwheel on the top of the machine. A graduated dial reads the vertical movement to .001 inch, reading the clearance when grinding cutters. The head spindle is driven by a horizontal belt and any tension of belt desired is obtained by the adjustment of nuts, shown on the rod on each side of the cone frame.

Another improvement is the lengthening of the bearings of the table slide, this adds greatly to the rigidity of the machine. On the left hand side of the apron, opposite the two handwheels, is shown one of the combined levers for reversing the machine in either direction or locking the carriage at any desired position. By this construction, in all kinds of grinding, it requires only one hand of the operator to reverse or stop the machine, leaving the other hand free to adjust the work, as in cutter grinding. This is a most desirable feature on grinding machines.

Capacity of machine as follows: from

Forbes Patent Die Stock.

centre of spindle to top of swivel plate, takes 36 inches between centres and will swing 9½ inches or 14 inches. Will surface grind 36 inches by 8 inches. Machine complete with all attachments weighs 3,500 pounds. These machines are made by the Bath Grinder Co., Fitchburg, Mass.

FORBES PATENT DIE STOCK.

All who have had experience in the cutting and threading of wrought iron pipe, and especially the larger sizes, know well the difficulty with which it is attended. Either a hand die stock with its long handles, or a cumbersome, heavy and expensive machine must be used; in the former case it is rarely that anything larger than 4-inch pipe is attempted, and even then requires four men to do the work, and very hard work it is; in the latter case, the machine is not portable and all the pipe has to be carted to the machine, and as they take up so much room and are so very heavy, it is unpracticable to use in confined places or in isolated mills.

Forbes Patent Die Stock.

Herewith is illustrated the Forbes Patent Die Stock, manufactured by The Curtis & Curtis Co., of Bridgeport, Conn., which it is claimed meets the requirements of the case without the above disadvantages. The machine consists of a die carrying gear supported and surrounded by a shell and actuated by a small pinion embedded in the side of the shell and working on the large gear with the pipe vise attached to the back of the machine. To operate it, the pipe is placed through the pipe vise with the end to be threaded against the back of the dies. The die carrying gear is then revolved by means of a crank on the end of the pinion; as the dies revolve, the gear is drawn back into the shell and the dies are thus brought on to the pipe. These dies are opening and adjustable to any variations of fittings and when the thread is cut, they can be opened and the pipe taken out without running back or stopping the machine. In cutting off pipe the gear is shoved back in the shell and held by a stop so as to give it a rotary without a traveling motion. A blade cutter is then inserted in the gear which is automatically feed forward as the gear revolves.

These machines are made in a great variety of sizes, to meet almost any range desired, from ¼ I.D. to 15 O.D. for hand or power. All hand machines can be made into power machines by the addition of a cast iron base, with the necessary gearing and countershafting, therefore, they can be run as power machine in a shop, or taken from the base and used on outside work as a hand machine.

When desired, a direct connected engine or motor is connected with them, so that they can be run directly from a steam or electric current, without the necessity of having power available.

ROBERTSON UNIVERSAL DRILL PRESS.

One of the important features of the drill press manufactured by the Robertson Manufacturing Co., Buffalo, N.Y., is the universal adjustment given to the drill table. As may be seen in the cut, the table can be rotated about a horizontal axis, thus making it possible to drill a hole at any angle with the surface by which the work is fastened down. The value of this feature has been demonstrated continuously in the plant of the builders in the past few years. It was originally designed to perform certain operations in their product, and proved to be such a success that they have decided to build a complete line of drill presses with this feature. Their long experience with it enables them to present it in practical form.

The knee is raised by a crank fitted to a steel shaft, with a pinion milled from the solid, meshing with a rack on the column. The rack is also of steel, with cut teeth. The universal joint for the work table is so designed as to give a rigid support to the table, with provision for drawing all bearings tightly together with the clump screws shown. A lock bolt is provided for setting the table accurately in 45 and 90 degree positions. The spindle is of special high carbon steel, carefully fitted, with the thrust taken by fibre collars. The hole is a No. 3 Morse taper. The column is heavy and secured to the base by a clamping bolt. The base is provided with T slots for clamping heavy work.

The machine may be provided in any of the following forms: With universal table or solid knee; with lever feed; with wheel and lever feed and quick return; wheel and lever feed, quick re-

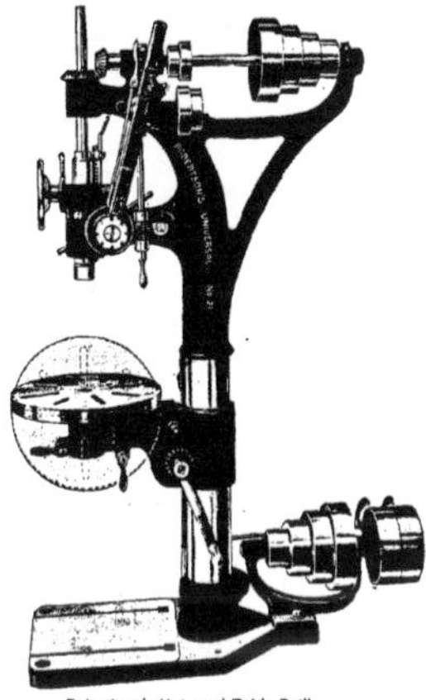

Robertson's Universal Table Drill.

turn, power feed, automatic stop; it may also be furnished with or without back gears as desired. The bevel gears are planed from the solid metal and are provided with guards. The machine shown will drill to the centre of a 42-inch circle.

DRILL-MAKER AND SHARPENER.

The illustration herewith represents the Word Brothers drill-making and sharpening machine. It consists of a cast-iron bed plate with two power hammers, a vertical and horizontal, mounted thereon, and a dead block, carried by guide rods, which are attached to the bed plate.

The hammers are actuated by compressed air and are equipped with automatic controlling valves which permit the operator to handle both hammers by a single foot treadle, which is pivoted under the hollow of his foot, thereby enabling him to operate the machine without moving his foot from one place. This dead block is supported by heavy guide rods securely bolted to the bed plate. It is moved by a screw; which, in turn, is driven by a friction gear running six hundred revolutions per minute. The dead block is adjusted to gauge the drill bits and to properly shape the wings of a new drill.

Drill Making and Sharpening Machine.

By a pressure of the foot on the toe of the treadle the horizontal hammer is brought into action, the intensity and speed of the blows are governed by the pressure of the foot on the treadle; by releasing the pressure of the foot on the toe of the treadle the hammer is stopped by the controlling valve, which acts instantly, shutting off the supply of air to the hammer, bringing the piston to a retracted position and the foot treadle to a level, or central, position. By a pressure of the foot on the heel of the treadle the vertical hammer is started and stopped in the same manner. The foot treadle is always brought to a central or neutral position by the controlling valves when the pressure of the operator's foot is removed or is resting equally on each end of the treadle, and the pistons of the hammers, carrying the dies or hammer heads, are always brought to the upward or backward stroke and held there. These controlling valves prevent one hammer from coming into action while the other hammer is in use. Their importance, in a machine of this kind, is illustrated by the fact that the time required by them to stop one hammer, hang it up safely, and start the other hammer under a full head of air is one-tenth of a second.

In line with and moving to and from the horizontal hammer is the dead block, a massive casting faced with tool steel, which the drills or steels rest upon and which receives the impact of the horizontal hammer when it is in action. the different lengths of steels, accurately and quickly, by a slight pressure on a lever, bringing it forward or backward, as desired.

Between the horizontal hammer and the dead block are adjustable split dollies which work on the cutting edges of a drill, and which forge and sharpen all sized drills without any change. The horizontal hammer strikes these dollies rapidly, which are driven against the drill bit or steel, and which rebound slightly at every blow, permitting the scale from the steel to fall clear of the work, while the dollies are thus being struck they are opened and closed by a cam lever attached to the guides carrying them. This opening and closing of the dollies, while they are working on a drill or a blank piece of steel, prevents undue buckling or bending of the steel, as the dollies work only on two wings of a drill bit at a time, the other two wings of the bit bracing and supporting the wings thus being worked upon. This causes the metal in the wings of the drill to be equally distributed and hammered alike, so that the whole bit is a homogeneous mass of equal density and compactness, permitting the drill to take a high temper and stand up to hard work without broken corners or battered edges.

The vertical hammer carries dies suitable for the work to be performed and, like the dollies described above, they do not have to be changed in handling the different sized drills. The functions of the vertical hammer are to swage and gauge the drill bits and to properly shape the wings of a new drill.

The weight of the machine is four thousand pounds, they being compactly and rigidly built to stand the hard service required of them. Their capacity is from six hundred to eight hundred drills per day. The machine forges new drills almost as rapidly as it sharpens old ones.

The machines are manufactured in Rossland, British Columbia; Houghton, Michigan, and San Francisco, California.

MOTOR-DRIVEN HACK SAW.

Application of the electric motor to a hack saw is illustrated herewith in that of the Robertson Manufacturing Co., Buffalo.

Because of the many uses and applications of these machines compact and close connection was desirable, and was obtained by mounting the motor under the bed at the rear end, where it occupies no extra space. The starting box is mounted on the outside of the legs, also at the rear end. The pinion on the motor and the intermediate pinion are of fibre, metal bushed for set screws. The other gears are cut from solid metal. The motor is of 1/4 h.p. and is furnished for either direct or alternating current.

The saw is of the company's regular type, having an automatic stop gravity feed and quick starting clutch. The one illustrated has an adjustable frame capable at its greatest extension of using a 17-in. blade for cutting stock up to 8x8 inch. For smaller work a shorter

Motor Driven Hack Saw.

blade can be used by adjusting the outer arm on the top bar. The equipment is one especially useful for cutting long stock in all shapes.

CIGAR HOLDER SOUVENIR.

A handsome cigar holder is being sent out as a souvenir by the Ludlow Valve Manufacturing Co., Troy, N.Y. No doubt the company will be pleased to send this souvenir to readers of Canadian Machinery if, when writing, they mention this paper.

Foundry Practice

MODERN IDEAS FOR FOUNDRYMEN

An Example of Loam Molding

A detailed description of the making and pouring of a loam mold for a six-foot pipe with a seven-foot rectangular branch.

By Jabez Nall.

ARTICLE I.

While the method of making molds for castings, by the use of a brick structure faced with loam, has in the present-day foundry practice been discarded to same extent in favor of the "skin-dried" mold, because of the less amount of labor and time required to produce the casting, with almost equal results, there are still many times when it is the only practical way to make castings of certain design. While the foundry expense by this method may be high, there is often a corresponding saving in the pattern shop, and the finished casting is usually of a high grade.

There is another feature of the loam mold, that when the shape of the casting permits, the same mold can be used more than once; while we may not term it a permanent mold, we have known as many as 8 to 10 castings to be made from the same outside mold, with nothing more done to it, than the burnt blacking brushed off while hot, and a fresh coat applied, with possibly a little patching with fire clay. A new core is of course being provided each time. In such a case this style of molding cheapens production. Other points in its favor will be brought out later on.

It is therefore essential that the young pattern-maker, as well as the young molder, should make himself thoroughly familiar, not only with the principles, but with the details of this method of the production of castings.

Fig. 1.

Pattern Work Required for 6-foot Pipe.

The example given in this article of a 6-ft. pipe with a 7-ft. rectangular branch, 1 in. metal thickness, has one or two special features that will be of interest. The work of the pattern-maker, on such a casting is not difficult or intricate and consists of the making of four sweeps or loam boards, as shown in Fig. 1: (1) Seat sweep; (2) Sweep for outer casing; (3) Sweep for centre core; (4) Sweep for face of branch core.

There is also the making of a skeleton pattern for the branch as shown in Fig. 2, which is a plan view on centre line. The upper half shows it used as a pattern for the mold, with ring for core print attached. The lower half of view shows the same used as core box. This branch is made in halves dowelled together, and when used as a pattern a loose ring turned true on the outside is made for core print. This is made 12 in. wide to suit the brickwork, and is likewise cut in halves. Another ring of equal width turned true on the inside is made for a core box; this is left whole and the two halves of pattern secured to it, and used as indicated in Fig. 2.

The Finished Casting.

The bottom flange is made in 12 segments, as shown. The cut mentioned, as at A, is not necessary in this case; but this flange may be used again on some other job, molded in reverse order, when this cut enables the molder to pull in the flange.

Better for Skeleton Pattern and Loam Board Not to Fit Closely.

Some patternmakers spend much unnecessary time and labor in carefully fitting even a skeleton pattern to a loam board; when it is better that they do not fit too close. Take this branch for instance, it is necessary that the three staves on the cardinal centres of each half be correct, and these are easily got from the layout, but the rest should be well clear of the board and only approximately correct in the cut. The application of the principles of projection will readily give the lengths of each of these without any further trying. We have known one man to spend ten days on a job, and work harder than another man, who made a similar but larger one in three days. This may seem exaggeration, but is fact, and represents the difference between the cut and try and cut and try again system, and working to a correct line. But this, as Kipling says, is another story.

Making the Mold.

In commencing the mold it is necessary that the molder have his plans fully matured; he must know all that he is going to do from laying the first point. A generous allowance of "mud" is placed between each brick.

In building any loam mold it is important that the brick be not too close, as to prevent proper venting. This done, it is daubed up with a coarse loam, well rubbed in by the hand, afterwards left for a few hours to stiffen before being given a thin coat of fine loam and placed in the oven to dry.

Other Rigging for Mold.

The molder then gives his attention to the other "rigging" required. This includes a lifting ring for lower half of mold, having three lugs for lifting purposes, one side being carried out to allow the building of the brickwork for the branch as shown in Fig. 5, and two rings for the joint of mold at centre line of branch. These two rings may be cast in the same open sand mold, by the flange on branch, and three supporting rings for the centre core, these being about ¾ in. thick and 7 in. wide, allowing 2 in. of space between the outer edge and face of core; these supporting rings are also split to allow for placing in position without removing braces from spindle, and to allow for contraction of casting. This rigging can be readily traced by referring to Figures 3 and 4.

In addition to the foregoing, an arbor or lifting plate for the branch core must be made, as shown in Figures 3, 4 and 5. This is provided with two loops for lifting hooks, another hole cast in the outside plate, allowing for a third hitch. A turnbuckle can then be used with this greatly facilitating the setting of the branch core.

This completes the rigging required for the job, the 5-ft. pulley ring, and

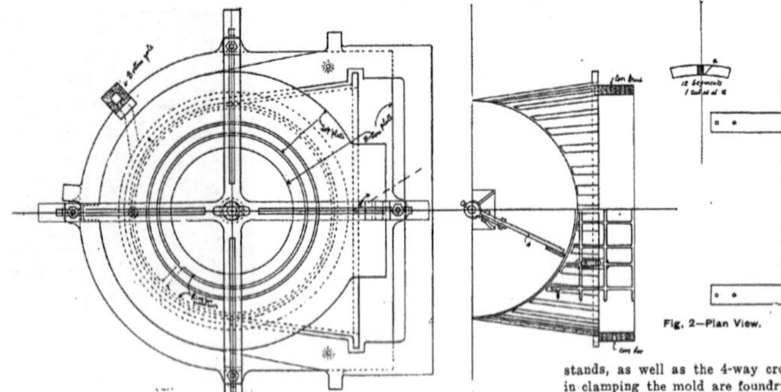

Fig. 5—Plan View.

Fig. 2—Plan View.

stands, as well as the 4-way cross used in clamping the mold are foundry stock.

Handling and Erection of Mold.

The foundation for the mold, being dry, is placed in a pit of sufficient depth, to allow for the easy pouring and handling of casting. This would be governed by the conditions of the foundry as regards condition of floor, head-room, crane capacity, etc.

This plate, being set and levelled, and the spindle being readjusted, board No. 2 is attached, the lower flange set, and the joint covered with paper to ensure a true parting. The lower lifting ring is set in a bed of coarse loam, and on this the outer wall is built to the height of the lower edge of the branch flange. The half of the skeleton pattern, with core print attached, is set up and braced in place, the building of brickwork continued to the centre line, the loaming up of the branch keeping pace with the building of brickwork. At the top the parting ring is set to hold

brick to taking out the casting. Usually he makes his own "rigging," and the first thing needed is a bottom, or foundation plate, which should be of ample size to support the brickwork for the centre core and the outside walls, including the branch; this should be about 3 in. thick and should be provided with lugs have cored holes for the binding bolts as shown in Figures 3, 4 and 5. This, as well as all the other plates, is cast in open sand. The spindle having been placed in the socket, plumbed up and secured at the top by the wall braces, sweep No. 1 is adjusted and three rows of brick laid on to form seat of mold. It is best that this be made deep, as shown, to ensure the mold returning exactly to the same place and to prevent any accident by straining at this making it of sufficient depth for the two, first pouring the lifting ring for the top half of mold, which must be strong enough to carry the superstructure without springing, and provided with three lifting lugs similar to bottom lifting ring. After the iron is set, parting sand is thickly strewn on, and the other ring poured on top about 1 in. thick. Three holes are cast in these plates in which socket and pins for dowels are afterwards inserted, thus bringing the two halves of mold correctly together after being separated.

A top or covering plate is made having bolting lugs to match foundation plate, and also having eight holes for the pop gates at top of mold, two holes for risers, one to come in top flange and the other at the highest point of

the structure together and make a level joint. The face of the mold is then given a coat of coarse loam ½ in. to ¾ in. thick, well rubbed in by the hand, and its face regulated by the sweep board. After this is set a finish coat is applied as usual by the sweep.

The joint being sufficiently set, the upper lifting ring is bedded in place and the top half of mold made in like manner. Care must be given to the arching of brickwork over the branch. This and the outer wall should be not less than 9 in. or a full brick in thickness. We have known a lighter wall to result in the loss of the casting from the fact that the wall was not strong enough to resist the ramming outside, notwithstanding that when the mold was closed it showed up correct in thickness when tried by gauges. The wall crowded in at the top until, instead of 1 in., only 7-16 in. was left, and it was impossible to know this until it was found by the casting cracking as a result of strain owing to uneven thickness of metal.

Care should be taken to support the small amount of brick behind the branch flange by ½-in. iron rods, and, for easing the vent at these points, a few straws cut square at the ends, laid in among the loam between the bricks, will make excellent pipes to carry off the gases. The outer casing complete, with the top flange swept in, the whole is given a good coat of blacking. The top half lifted off and placed in the oven to dry, the branch pattern withdrawn, the lower half is also lifted away and dried. Board No. 2 having been taken off, spindle board No. 3 is adjusted and the centre core built up in position; after finishing it is covered with sheet iron and dried in place.

In order that the two cores may be built at the same time another spindle is set up and the branch pattern with core ring attached is braced plumb and level. Board No. 4 is attached to this spindle as shown in Fig. 2. The face of this board is set at ¼ in. greater radius than radius of centre core; this allows of the two cores abutting in the centre without crushing the weak points of core when being set in mold.

In making the branch core it is necessary that some false work be built to form a bed as far as the core arbor reaches; or that the arbor be placed reverse side up and filled in with wedged brick and loamed up to a segment, then turned over and placed in position. The brick in the open spaces at the ends should be secured to the arbor by wiring them together. The arch is then carried over as indicated in Fig. 4. Two holes must be left at the top for insertion of lifting hooks. The face of core is built up a half brick wide and loamed up to the board, the thin points supported by rods running well back between the bricks. The other part of the core is filled with dry brick, and the holes for lifting hooks are filled in, after the core is placed in the mold.

The top or covering plate having been placed reverse side up, it is given a coat of loam, holes being left for the pop gates and risers aforementioned. It is then dried. The mold can now be assembled as shown in Figures 3 and 4.

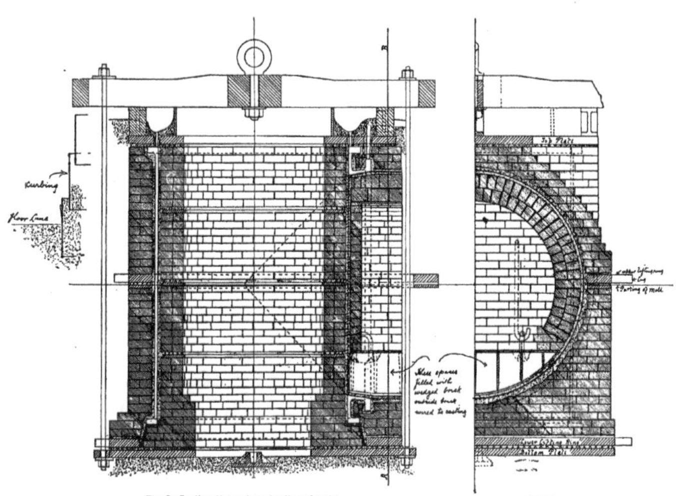
Fig. 3—Section through centre line of mold

Fig. 4

Molding the casting this way gives the molder a chance to watch his metal thickness and see that no crush or other accident happens as he goes along.

The whole being secured to the satisfaction of the molder, the outside ramming is now made, gate cores being set, as indicated in Fig. 5, connecting with the lower flange, vents being cared for by stakes or rods placed while ramming and drawn up and afterwards removed. A straw rope connected to any special part will carry the vent in any direction desired. This outside ramming having been carried up to about a foot below the top of foundry floor, curbing is set, as indicated to the left of Fig. 3, and the ramming carried up to the top of mold, the bottom gate being connected to the top by dry sand cores. Plugs are placed in the risers and top gates aforementioned and runner and flow-off built, the latter in any convenient direction.

Pouring and After Handling.

A double pouring-basin is made having the level for bottom gate about 4 in. lower than the one for the top gates. In pouring the metal the bottom gate is poured first until about 2 ft. high in the mold, when the ladle is tipped or basin flowed over and the bulk of metal poured in from the top. The bottom gate set as shown gives a whirling motion to the metal, and in coming from the top the metal does not have a chance to cut the sand and breaks up any slag or impurity that has entered the mold, and prevents any cold shot.

About an hour, or less, after pouring, the brickwork of the centre core should be split to allow for the contraction and prevent any shrinkage strain in the thin metal. If the metal were thicker this might not be necessary. This operation must be begun at the bottom. To do this it is necessary that a man enter the mold from the top. No man should be left in for more than five minutes at a time, and as the whole operation will take fully fifteen minutes, three or four changes will have to be made. It is an undesirable and dangerous job an account of the gas, and could easily be done away with by use of a cheap mechanical device; but tools cost money and in the foundry labor is cheap.

This accomplished, we can only wait to see the casting as shown in photo, Fig. 6.

The Grey Iron Foundry as a Business Venture †

By Thomas D. West, Sharpsville, Pa.

This article describes what a man embarking in the grey iron foundry business, should know before he begins what methods to pursue after he has decided to start, and what obstacles he is likely to encounter. Each sentence forms the text for a paper in itself, but taking the first, the main point which should be considered in starting a foundry, we have the following:

First—The great lack of good skilled labor.

Second—Influences retarding the training of a plentiful supply of skilled mechanics.

Third—Independence and insubordination of many of the few skilled and respectable men available that may be obtained after extra inducements are made for their services.

Fourth—Difficulties and losses arising from trying to do work with the riffraff that a starter is compelled to employ in beginning operations.

Fifth—The slow attainment of discipline and organization in a shop's working force.

Sixth—Losses through defects developed in machinery and tools caused by trying them out and bringing them into good working order.

Seventh—The advantage that sellers of appliances and purchasers of castings take of ignorant beginners.

Eighth—Extra risks with consequent losses by accidents incurred through the want of experience in operating machinery and shop appliances.

† Paper presented before American Foundrymen's Association.

Ninth—Difficulties in obtaining practical and good executive managers and foremen.

Tenth—Losses and embarrassment caused by the narrow range of vision of hired managers, as compared with that of the man who may have his all at stake striving for a foothold.

Eleventh—Inability of proprietors, inexperienced in actual molding, mixing and melting, to avoid mistakes in getting the best managers and foremen, the windy blow-hard being often better liked than the one of true merit because of inability to rightly judge.

Twelfth—The underestimation of the obstacles and uncertainties that every beginner must encounter.

The twelve points just given are as a whole encountered much more in brisk than in dull times, and in times of very great prosperity can be the means of driving the most experienced workers and financial managers to the wall.

A beginner's fight to exist can be such, even with sufficient working capital, that he is brought to the point of despair of success by reason of the great odds and seemingly insurmountable difficulties that will confront him.

The difficulty of obtaining good overseers as well as skilled labor, which is most pronounced in good times, is alone one of the twelve factors that demands more consideration than a reasonable working capital, and often knocks out the best of experience, good management and hard work.

We could take up every one of the twelve points and fill a number of pages with a discussion of them, but will let it go with the simple statement. Aside from the great need of skill, the general public's great lack of comprehension as to the demands on skill and experience needed by our molders in connection with the making of castings is about the next greatest evil to injure many starters and the trade in general. This poor conception that the general public has of the foundry business has caused many novices to undertake the work and there is one instance that recently came to the writer's notice in which a bookkeeper who never lifted a rammer put all his hard earned money into a small foundry and hired a farm hand to run it for him while he maintained his salaried position. This new foundry owner was a bright and modern financier and when his working capital was exhausted he did not close his shop until he had involved others in his downfall.

There is one feature difficult to comprehend and that is the great difficulty of making men inexperienced in the business realize its hazard as well as the fact that skill is required to make good castings. Of course almost any fairly intelligent man can make a good casting, of almost any character, allowing him time to experiment, and where there is sufficient capital and cupola capacity to keep re-melting his bad castings. Few beginners, however, have sufficient assets to support this experimental work.

The novice's inability to understand the requirements for making good castings at the first pouring is such that even men that can successfully make small and medium castings day in and day out believe that they are competent to be trusted with the making of most any heavy castings. When this belief can be found so prevalent, as it is with molders throughout the world, is it any wonder that many who have walked through a foundry a few times get the idea that all is easy, and that nothing else is required than a few patterns and sufficient muscle to pound sand, under the control of an intelligent business head to make a foundry pay, ever forgetting that intelligence is one requisite and experience another.

The question of the methods a man should pursue, who has decided to start a foundry is answered as follows:

First—He should make a thorough investigation to discover if there is a demand for the class of castings he intends to make, and what competition exists.

Second—The selection of a site that will be central to his market, and in building to start on as small a scale as practical, but planned for extensions that can be carried out without tearing down too much of the existing plant.

Third—In the search for machinery and appliances he should endeavor, as far as possible, to get the experience of others. The A. F. A. convention should be visited for its information and its exhibits inspected to note and investigate the latest improvements.

Fourth—In buying machinery and appliances he should avoid, as far as possible, the purchase of second-hand tools. The new has all its life, whereas the old is more or less worn out, and will not as a rule give good service for the money expended.

Fifth—One seeking orders, and not competent to quote prices, should not be ashamed to seek his manager's or foreman's advice.

Sixth—In starting to make castings don't be in a hurry to fill your shop with work. Go slow, feel your way, and your chances for error or failure will be greatly decreased.

Seventh—Be sure to have sufficient working capital and good chances of obtaining a surplus over and above what is considered necessary.

As to what a man should know before he begins is rather a broad question to ask an experienced founder, with the expectation of getting any answer other than that he learn to be a thorough molder as well as a business manager. No man who intends to be at the head of a foundry can be too thorough in the actual work of molding, mixing and melting, and he should have executive and business qualifications.

There is no trade demanding so much skill and experience wherein so many enter who are wholly deficient in practical knowledge of its operations as a foundry business. It is a case of the many trying to make two wrongs make one right, and because some learn of an inexperienced man having made a success of his entry into the field they think they can do likewise.

Thirty years ago, few, if any, ever thought of going into the foundry business if they were not so practical that they could discharge any molder and take hold of his work in a more masterly manner. To-day we find bankers, lawyers, and even clerical men, who have assumed the management of foundries. We must ever remember that brains and intelligence do not stand for experience. If the day ever comes when practical education is given greater recognition, we will have the best that is possible in the foundry president or manager, as well as in the employe, but not before.

NEW MOLDING MACHINE.

Accompanying is an illustration of the molding machine recently placed on the market by J. N. Paxon Co., Philadelphia, the Glenwood Rockover machine.

The chief advantage of this machine, as outlined by the makers, is that the regular hand patterns can be used by fastening them to a board, if they are flat backs, or if irregular shape, by making a cope and drag hard match, the pattern being used for making all the drags; then being placed in the match for making copes, at the same time not interfering with their usefulness as a hand pattern. The parts of the machine are all interchangeable. It is made as low as possible, so that it is convenient for a man to "tread off a mold," also by dropping the mold away from the pattern, as is done in this case, it leaves the mold resting on the frame at a height that is very convenient to be taken away.

Glenwood Rock-Over Moulding Machine.

The machine is arranged so when the mold is rolled over in position to have the pattern drawn, the mold carrier is rigidly locked in a position exactly at right angles with the ways. It is also furnished with means of adjustment for wear, so that the match carrier can always be kept in this position, and as it is locked it allows the operator to give his undivided attention to tne next operation, and after the pattern is drawn, the locking is released by a foot treadle, placed convenient for the operator. The mold carrier being counterbalanced by springs with a takeup, so that more or less tension can be readily applied to the mold carrier, to counterbalance the different weight of match.

The machine is equipped with a very simple and effective evening device (as shown in illustration). By a hand wheel convenient for the operator, this evening device is released, and at the same time the table is brought up under the bottom board, and without any resistance, is allowed to take a seat firmly under the board, and by revolving the hand wheel, it is then clamped firmly into position, so that if a mold is of uneven weight, on the different sides, it will fall away from the pattern in a perfectly straight line.

The lever for dropping the mold is arranged on a toggle joint, so considerable movement of the lever is necessary at the starting of the molding, in order to move the mold very little, insuring thereby an even drop, and on account of the leverage, it is very easily manipulated by one hand of the operator, allowing him with the other hand to give attention to the clamp on the evening device, and if a vibrator is used, also to operate this device at the same time.

CONVENTION NUMBER OF BLAST.

The June issue of "Blast," a periodical issued by the Detroit Foundry Supply Co., Detroit, is devoted to a resume of happenings at the Philadelphia convention of the American Foundrymen's Association. The editor of this number has shown that while attending very much to business at the convention he was far from being oblivious to the humorous happenings. Those not having already received a copy of the issue should do so by applying to the Detroit Foundry Supply Co., mentioning this paper.

ASSOCIATION OFFICERS

AMERICAN FOUNDRYMEN'S ASSOCIATION.

President—S. G. Flagg, Jr.

Vice-Presidents—C. J. Caley, John S. Burr, H. E. Field, J. H. Whiting, A. K. Beckwith, T. N. Sheriff, J. P. Golden, L. L. Anthes.

Secy.-Treas.—Richard Moldenke, Watchung, N.J.

BRASS FOUNDERS' ASSOCIATION.

President—Chas. J. Caley.

Secretary—Andrew M. Fairlie, Philadelphia.

Treasurer—J. H. Sheeler.

FOUNDRY SUPPLY ASSOCIATION.

President—E. H. Mumford, E. H. Mumford Co., Philadelphia.

1st Vice-President—W. P. Shepard, Rogers, Brown & Co.

2nd Vice-President—W. S. Quigley, Rockwell Engineering Co.

3rd Vice-President—H. E. Atwater, Osborne Mfg. Co.

4th Vice-President—Wm. Chambers, Garden City Sand Co.

Treasurer—J. S. McCormick, J. S. McCormick Company.

Secretary—H. M. Lane, Cleveland.

Trustees—W. E. Kanavel, Interstate Sand Co.; E. J. Woodison, Detroit Foundry Supply Co.; W. J. Thompson, Hamilton Facing Mills Co.; F. W. Perkins, Arcade Mfg. Co.; and Geo. Rayner, the Carborundum Co.

AMONG THE SOCIETIES

SOCIETY OFFICERS.

Canadian Mining Institute.

President, George R. Smith, Thetford Mines, Quebec; secretary, H. Mortimer Lamb, Victoria, B.C.; treasurer, J. Stevenson Brown, Montreal.

Toronto Branch A. I. E. E.

Chairman, R. G. Black; vice-chairman, K. L. Aitken, secretary, R. T. McKeen.

Marine Engineers.

Grand president, B. J. Henning, Toronto; grand secretary, Neil J. Morrison, St. John, N.B.

Canadian Electrical Association.

President, R. G. Black, Toronto; vice-presidents, R. S. Kelch, Montreal, W. R. Ryerson, Niagara Falls, Ont.; secretary-treasurer, T. S. Young, Confederation Life Building, Toronto.

Canadian Association of Stationary Engineers.

President, W. A. Sweet, Hamilton; vice-president, Jos. Ironsides, Hamilton; secretary, W. L. Outhwaite, Toronto.

Canadian Railway Club.

President, W. D. Robb; secretary, James Powell, chief draughtsman motive power department, G. T. R.; treasurer, S. S. Underwood, chief draughtsman car department, G. T. R.

Engineers' Club of Toronto.

President, C. B. Smith; secretary, C. M. Canniff, 100 King Street W.; treasurer, John S. Fielding, 15 Toronto Street.

Canadian Society of Civil Engineers.

President, W. McLean Walbank; secretary-treasurer, C. H. McLeod. Rooms, 877 Dorchester street, Montreal.

Society of Chemical Industry

Chairman, Prof. W. H. Ellis; vice-chairmen, H. H. Van Der Linde, Milton; H. Hersey, A. McGill; hon.-secy., Alfred Burton, 44 York Street, Toronto,

Ontario Land Surveyors.

President, Otto J. Klötz, Ottawa; vice-pres., Thos. Fawcett, Niagara Falls; secy.-treas., Capt. Killaly Gamble, Toronto.

CONVENTION OF STATIONARY ENGINEERS.

The eighteenth annual convention of the Canadian Association of Stationary Engineers will be held in Guelph from the 13th to the 15th of August, inclusive. The city will assist the local engineers in making this convention a success.

AMERICAN CHEMICAL SOCIETY.

The semi-annual meeting of the American Chemical Society was held in the Chemistry and Mining Building of the University of Toronto on the last days of June. Many important papers were presented and discussed, most of which were of a very scientific character. A visit was made to Guelph to inspect the Ontario Agricultural College, and a party took a trip to Cobalt, where there is much to interest such a body of men.

The paper which was of the greatest interest to the commercial world at present was that presented by E. G. Acheson, the Acheson Graphite Co., Niagara Falls, N.Y. and Ont., on "Deflocculated Graphite." Mr. Acheson has developed the manufactuer of artificial graphite in the electric furnace, and in the last year or two he has developed a method of suspending very finely powdered graphite in oil for purposes of lubrication. He is arranging for manufacturing based on this new method.

The American Chemical Society has a membership of over 3,000, and ranks in standing with kindred societies in Great Britain and Germany. It differs from these, however, in that industrial chemists as well as students of pure science are included in its membership. In fact, industrial chemists form a large percentage of its members.

There were quite a number of Canadian chemists present at the convention, and no small compliment was paid Canadian chemists by the president when he said in his opening address that the society's meeting in Toronto was an acknowledgement of work accomplished in chemistry by Canadians.

American Institute of Mining Engineers

Summer Meeting of A.I.M.E.—Reception by City and Provincial Government—Visit to Cobalt District.

The summer meeting of the American Institute of Mining Engineers was held in Toronto, July 22-29, with headquarters at King Edward. The members were received by the city and Provincial Government. They held business sessions the first two days, then left on special government train for Cobalt and districts nearby, where they inspected different mines.

What the Institute is For.

The American Institute of Mining Engineers is one of the most important technical societies in the world. It was organized in 1871, and has a membership of over 4,000. While most of the members are residents of North America, its membership embraces leading mining engineers, metallurgists and geologists of nearly all countries in which the mineral industry has been developed.

The president of the institute usually holds office for two years. John Hays Hammond the eminent mining engineer, was elected to the office during the present year. The secretary of the institute during the past 36 years has been Dr. R. W. Raymond, one of the world's most renowned mining men.

The headquarters of the institute are in New York. The United Engineering Building, opened about a year ago, which cost $1,500,000, a gift of Andrew Carnegie, now houses the institute, together with the sister societies of Mechanical and Electrical Engineers.

In addition to the publication of technical papers, the institute yearly organizes excursions to various mining centres. In 1901, for example, the excursion consisted of two special trains from New York City to Mexico and return. In 1905 the excursion of the institute was to British Columbia and the Canadian Yukon. On the invitation of the Iron and Steel Institute the American Institute held its summer meeting in England in 1906.

Reception and Business Meetings.

It was through the efforts of Hon. Frank Cochrane and Professor W. G. Miller, that the meeting of the institute was held in Canada. The Government and civic authorities welcomed and gave receptions to the members, and entertained them while in Toronto. At the business sessions several important papers were presented and some valuable addresses given. Among the papers were: Destruction of Salt Industry at Salton, Cal., by Prof. W. P. Blake; Some Reflections on Secrecy in the Arts, by Dr. James Douglas, New York; in which the author advocated that there be less secrecy maintained regarding industrial processes; Coal Briquetting in the United States, by Edward U. Parker, Washington.

Should Be No Railway Subsidies.

Dr. James Douglas, a Canadian now living in New York, entered a strong protest against the subsidizing of railroads in Canada. That was one feature that had always struck him as objectionable. In the United States they had never experienced difficulty in getting money with which to build a railroad if from an economic and commercial point of view a railroad was required. When they bought the votes of a district by subsidizing a railway they introduced into their politics the very worst possible element of corruption. As a Canadian he thought a protest should be made against the practice. On the other side of the line they were suffering from congestion on their railroads, but that state of matters would never come in Canada so long as they continued the nefarious method of depending on the Government for assistance instead of depending on themselves.

An address on the Tar Sands of Athabasca was given by Dr. Robt. Bell, Ottawa. He knew of no other place in the world in which there was so great an accumulation of tar sand, formed by the outpouring of petroleum, as in the valley of the Athabasca. The sandbeds, he estimated, covered 1,350 square miles and had an average depth of 150 feet. They formed a valuable asset in that they could be used in the manufacture of oil and fuel, paving and roofing.

The electric air drill, an invention of great importance to mines, and one upon which Mr. Edison and Professor Elihu Thompson spent hundreds of thousands of dollars and worked for years, was explained by Mr. W. L. Saunders, New York. Mr. Edison was baffled, and declared that the electric drill was a failure. Professor Thompson continued his efforts, and it was to him that the credit of the invention of the electric air drill was due. It had been a long and costly battle, but the drilling of rock by electricity was no longer a theory, but a fact.

Cobalt Mines Unique.

Prof. W. G. Miller, the provincial geologist, read a paper on the subject of the Cobalt mineral area. The district, he said, was unique. There was no other place on the North American continent where the association of minerals was of the same character, including high values of silver, nickel and arsenic. Only in Saxony, Bohemia and France were similar formations to be found. Prof. Miller referred to the large production in Ontario of corundum and mica. The greatest nickel mine in the world—the Creighton Mine—was situated also in the province.

The Cobalt region had gone ahead because of the silver alone for the Cobalt, nickel and arsenic had been practically lost. By perfecting the methods of refining there was no doubt that the ore would bring higher prices. There was no exact information but Prof. Miller estimated that the production of silver would be 7,000 tons this year.

Trip to Cobalt.

The members of the Institute went to Cobalt as the guests of the Provincial Government, by special train, and spent several days inspecting mines around Cobalt and other districts.

TORONTO ENGINEERS' OUTING

On July 26 the annual outing of the Engineers' Club, of Toronto, took place. It consisted of a trip by private electric car to Jackson's Point on the Metropolitan Railway. Stops were made at different points of interest along the newly constructed line. At the Point members interested themselves by boating, fishing, quoits, etc. An enjoyable day was spent.

Industrial Progress

CANADIAN MACHINERY will be pleased to receive as confidential industrial news of any description the incorporation of companies, establishment or enlargement of plants, railway, mining or municipal news.

New Plants and Additions.

A steel keg company may locate in Welland, Ont.

A gas company has been formed at Waskada, Man.

The new Rideau rink at Ottawa will cost $20,000.

A malleable iron works will be established at London, Ont.

The Labatt Co., London, will soon erect some new buildings.

Westman & Baker are to erect an $8,000 factory in Toronto.

The Maritime Engineering Co., Moncton, N. B., have sold out.

Chadwick Bros. are just installing a four-spindle cock grinder.

The Hull Electric Co., Hull, Que., is making additions to its plant.

An up-to-date pumping station is to be installed in Cobalt, Ont.

The Albion Stove Works, Victoria, B.C., are extending their foundry.

The Clark Foundry Co. will erect a large foundry in Sydney, N.S.

The Fairbanks Scale Co. will erect a large plant at Sherbrooke, Que.

Swift, Copeland & Co., Montreal, are building a factory in that city.

McIlwraith & Austin, founders, Listowel, Ont., have dissolved partnership.

A steel bridge will be built across the Humber river at Lambton, Ont.

The Winnipeg Casket Co. will build a factory at Winnipeg to cost $40,000.

The lift lock on Trent Valley canal at Kirkfield, Ont., has been opened.

A large car shops will be built in New Westminster, B.C., costing $200,000.

The plant of the Standard Chain Co., at Sarnia, Ont., will cost $40,000.

There is a splendid opening for a foundry and machine shop in Prince Albert.

The Freyseng Cork Co. are making a $7,000 addition to their Toronto plant.

The Berlin Machine Co., Hamilton, will erect a brick factory costing $150,000.

J. J. Heffron & Co., mattress manufacturers, Toronto, have been burned out.

Adam Beck will build a large extension to his box factory at London, Ont.

The Silica brick making plant at Parson's Bridge, B.C., is in running order.

The Vancouver Structural Steel Works will erect a large plant at Vancouver.

Joseph Gray, Peterboro, Ont., will move and considerably enlarge his lath mill.

The sawmill plant of A. Cooper, at Ouimet, Ont., has been destroyed by fire.

The Alberta Brick Co., Medicine Hat, turned out their first lot of bricks lately.

A Toronto brick company will develop the sand deposits near Vancouver, B.C.

Picard & Lalonde, machinists, Montreal, are spending $13,000 on new machinery.

The Dominion Cartridge Co., Montreal, will erect a factory at Brownsville, Que.

Michigan interests are contemplating the erection of a pulp mill at Sarnia, Ont.

The Electric Repair & Contracting Co., Montreal, will move to larger premises.

The warehouses of the Berlin Omnibus Company were totally destroyed by fire.

The Colonial Weaving Co., Peterboro, will build a large addition to their plant.

The Waterloo Threshing Company may build a $200,000 plant at Portage la Prairie.

T. Dexter & Son, London, are remodelling their flour mill to 200 barrels capacity.

The Canadian Iron and Foundry Co., St. Thomas, Ont., will extend their works.

The International Snow Plow Co., Stratford, will manufacture steel box and flat cars.

The sewage disposal plant at the Woodbine, Toronto, is now run by Niagara power.

Sydney Smart, contractor, has installed a compressed brick plant at Malfort, Sask.

Carley & Wellard have commenced a furniture and cabinet making business in Winnipeg.

The Ham & Nott Co., Brantford, will double the size of their plant, at a cost of $40,000.

The Welland Vale Manufacturing Co. will greatly increase its plant in St. Catharines.

Safes made of concrete, reinforced with twisted steel, have been found very satisfactory.

J. B. Rouse & Co., Brantford, have sold their machine shops to Harris, Herod & Co.

The sawmill and electric light business of Geo. Leighton, Harriston, Ont., is for sale.

The International Steel Co., Montreal, have opened a branch at 50 Adelaide St., Toronto.

The warehouse and sheds of the Imperial Oil Co., Brandon, were destroyed by fire recently.

A large brick making plant will be erected at Cote St. Paul, Que., by D. G. Loomis & Sons.

The power house at Lake Ontario park, Kingston, Ont., was destroyed by fire recently.

The Lethbridge Electric Light Co., are installing a Babcock & Wilcox water-tube boiler.

The Morden, Man., Electric Light Co. will move their plant to the Okanagan Valley, B. C.

By a recently discovered process, alcohol can be made from peat at a cost of 6 cents per gallon.

The Polson Iron Works, Toronto, are building a dredge for the Pacific coast to cost $150,000.

Operations have commenced in the new glass factory of the Crystal Glass Co., at Sapperton, B.C.

Plans are being prepared for a $75,000 structure for the Canada Steel Goods Co., in Hamilton.

The Canada Spool Cotton Co., Maisonneuve, Que., are considering the erection of a new plant.

The International Heating and Light Co. will erect a gas plant in Edmonton costing over $1,000,000.

The shingle mill of G. Le Clair at Hastings, B.C., was damaged by fire to the extent of $25,000.

A large factory is to be erected at Lindsay, Ont., by Mr. Kennedy, late of Kennedy, Davis & Son.

The Canadian Cutlery Co. wants to build a plant in Grimsby, Ont. A loan of $25,000 is wanted.

The Canada Screw Co., Hamilton, are contemplating an addition to their plant to cost $150,000.

The Belleville Rolling Mills are running night and day, not being able to keep up with the demand.

The Otto Higel Co., Toronto, are installing Chapman double ball bearings in their new addition.

A plant for the automatic making of fruit cans has been installed by the Acme Can Works, Montreal.

The Canadian Government will establish a wireless station on the Government steamer "Inadra."

The Copper Smelting Co., Pictou, N.S., recently turned out a carload of copper matte worth $600.

The Colonial Weaving Co., Peterboro, will erect a new factory and install machinery costing $30,000.

The Montreal Pipe Company, Londonderry, N. S., will install a plant for the manufacture of car wheels.

The Goold, Shapley & Muir Co., of Brantford, will erect an addition to their plant, costing $27,000.

Keenan Bros., Owen Sound, are installing Chapman Double Ball Bearings in their new box factory.

The new mill of the Lake of the Woods Milling Co. at Keewatin, which cost $1,250,000, is is operation.

The Waterton Land and Power Co. will erect a saw and planing mill at Lethbridge, Alta., to cost $40,000.

Fire in the foundry of William Coulter & Sons, Toronto, damaged the premises to the extent of $400.

The Standard Shirt Co., Montreal, will erect a new factory on Delorimer avenue, Montreal, to cost $100,000.

The acid plant of the Dominion Pulp Mill at Chatham was completely destroyed by fire; damage, $11,000.

Messrs. Babcock & Wilcox, Montreal, are supplying 2,000 h.p. additional B. & W. boilers, with superheaters and chain grate automatic stokers to the Canadian Pacific Railway Co.'s Angus shop.

The Nyo Canning Co. and Cotton Shingle Mill Co., Vancouver, were destroyed by fire at a loss of $10,000.

The machine shop of John H. Hall, Brantford, will in future be under the name of John H. Hall & Sons.

The Berlin Machinery & Tool Co., Beloit, Wis., have commenced the erection of their plant at Hamilton.

The Canadian Billings & Spencer Co. are completing their plant in Welland. They will make drop forgings.

The Northern Engineering and Supply Co. have commenced work on their new Fort William, Ont., building.

The Montreal Terra Cotta Lumber Co., Montreal, are equipping their plant at Maisonneuve with electric motors.

The Canadian Machine Telephone Co., Toronto, is starting work on the installation of its plant at Lindsay.

A fixed assessment has been granted by Midland to the Canada Iron Foundry Co. and Midland Engine Works.

All the material for the smelter at Trout Mills, Ont., is on the ground and erection will commence immediately.

Bemiss Bros. Bag Co., Boston, will erect an enormous plant at Welland, Ont., the plant alone costing $1,500,000.

The Canada Woodenware Company, whose plant was recently burned at Hampton, may build at Fairville, N.B.

The Maritime Mfg. Co., clothing manufacturers, Pugwash, N.S., will erect a three-storey brick factory, 75x40 feet.

The Canada Woodenware Co., whose plant at Hampton, N.B., was destroyed by fire, will rebuild at Fredericton, N.B.

The Rainy River Development Co., Fort Francis, Ont., are erecting a power plant to generate 65,000 horsepower.

The Montreal Rolling Mills will erect a large galvanizing plant. They are also increasing the capacity of their wire mill.

Work is progressing rapidly on the construction of the power plant of the Stave Lake Power Co., Stave Lake, B.C.

O. H. Bell and G. Harold, Owen Sound, will erect a $18,000 furniture factory in that town. The town has granted a loan.

The Wilcox Hardware Manufacturing Company, of London, is contemplating the installation of a large molding plant.

A company, capitalized at $150,000, is being formed in Pembroke, Ont., for the manufacture of alumina and its by-products.

The Hayne Milling Co., Bridgen, Ont., have let the contract for the remodelling of their mill to R. Whitelaw, Brantford.

An air compressor of 5,000 horsepower capacity is being installed on the Montreal river. The air will be piped to Cobalt.

The Renfrew Power Co., Renfrew, Ont., have increased the output of their plant to 1,000 horsepower, at a cost of $30,000.

The furniture factory of William Cyr, at Ottawa East, was destroyed by fire. The loss is placed at $8,000, insured for $7,000.

Work has been commenced on the new wagon works of T. Brayshaw, Victoria, to cost [illegible]. Electricity will be used throughout.

A plant for the extraction of gas from straw to be used for fuel and power will soon be erected at Portage la Prairie, Man.

The Canadian Westinghouse Co. are equipping extensive additions to their plant in Hamilton with Chapman double ball bearings.

The Standard Chemical Co. will build a plant in Toronto, to cost $50,000. This company will also erect a plant at Delorimeter, Que.

J. Ballantine & Co., Preston, Ont., recently added to their plant a 20-ton planer, manufactured by the MacGregor-Gourlay Co., Galt.

The Colonial Engineering Co., Montreal, are installing a producer gas engine in the plant of the Frame & Hay Fence Co., Stratford.

T. C. Brewer, Woodstock, N.B., has the contract for the two new masonry piers for the Fredericton highway bridge, costing $20,000.

A mill for the manufacture of all kinds of iron pipe will be erected at Welland, Ont., in connection with the Ontario Iron & Steel Co.

The O. C. King Lumber Co., Humbolt, have sold their entire hardware, lumber and machinery business to the Saskatchewan Elevator Co.

C. Doutre is arranging for the installation of a chain of Government-owned and operated wireless stations on the British Columbia coast.

The Dominion Steel Co., N.S., suffered severely by fire recently. Their loading pier at Wabana, Nfld., was destroyed by fire. Much machinery and coal was destroyed; loss, $50,000.

The National Spring and Wire Co. are installing machinery for the manufacture of springs in their new premises at St. Catharines, Ont.

A committee is at work at Prince Albert, Sask., trying to find a method for developing power from the Saskatchewan and its tributaries.

The Sydney Foundry & Machine Co. are contemplating the removal of their works to Halifax. North Sydney is anxious to secure the industry.

The new sand lime brick plant of Schultz Bros., Brantford, which was installed by A. Berg & Son, Toronto, turns out 20,000 bricks per day.

A Buck Eye trenching machine is at work on the Moose Jaw gravity water system. It does the work of 150 men and is giving complete satisfaction.

A factory for the manufacture of steel office supplies and a grain cleaner will be erected at Portage la Prairie by Beamans & Co., Minneapolis, Minn.

The Ruud Mfg. Co., Pittsburgh, manufacturers of automatic hot water heaters, have established a Canadian branch at 155 King St. W., Toronto.

Somerville, Limited, Toronto, have placed an order with the Robb Engineering Co., Amherst, N.S., for a 200 horsepower Robb-Armstrong Corliss engine.

The Dominion Radiator Co., Toronto, are offering for sale their yard and buildings, including foundry and machine shop on Dufferin street, Toronto.

The Canadian Brass Co., Galt, have their factory almost completed. Most of the machinery, which is of the most improved type, has been installed.

A contract has been signed by the city of Ottawa and the Ottawa and Hull Electric Power Co. by which power is supplied to the city at $15 per horsepower.

The Chignecto & Amherst Electric Power Co. will soon have their power plant at Chignecto coal mine, Maccan, N.S., in operation and transmit power to Amherst.

A firm manufacturing gas engines is desirous of obtaining a site in Barrie, Ont. They want a loan of $25,000. A sash and door factory also want to establish there.

Wm. Oliver will erect a [illegible] factory at Lethbridge. He will also form a company and build a $14,000 plant for the manufacture of cement bricks.

MacKenzie & Mann are contemplating the erection of a large smelter and other industries at Ashbridges Bay. The initial outlay will be in the neighborhood of $30,000,000.

The Chapman Double Ball Bearing Co., Toronto, received the contract for the complete equipment of the new plant of the Standard Valve & Fitting Co., Guelph, Ont.

The Hartley Foundry Company is asking the city of Brantford for a fixed assessment. They will erect a plant to make tools and machinery fittings. They will employ 25 men.

The Cape Breton Electric So., will erect a new power plant at North Sydney, N.S. Good progress is being made on the construction of their transmission lines to Sydney.

The T. H. Taylor Company, Chatham, Ont., have purchased the old binder twine plant from the M. J. Wilson Cordage Company, and will convert the factory into a woolen mill.

The Dennis Wire & Iron Works, London, have found it necessary to increase their facilities. They will immediately proceed with the erection of an addition to their present factory.

The Canadian Locomotive Co., Kingston, Ont., have placed an order with the Chapman Double Ball Bearing Co. for the complete equipment of their plant at Kingston with double ball bearings.

A company has been formed for building and operating a large sawmill on the north arm of the Fraser River, in British Columbia. The president of the company is R. H. McKee, Vancouver.

The permits issued at Hamilton for new industries for the first six months of 1907 aggregated $700,000. The new blast furnace and buildings of the Hamilton Steel and Iron Co., cost $300,000.

The Cascapedia Trading Co., of the States, will locate in Dalhousie, N.B., and build a saw pulp and shingle mill there if they can have a free site. The plant will cost $1,000,000, and employ 3,000 men.

Chadwick Bros. have secured the contract for brass corporation work for the city of Quebec, have just completed a contract for the city of Hamilton, and are now working on brass work for city of Toronto.

A great combination of iron and steel manufacturers is being formed in Great Britain to combat American and German competition, control the British trade and dominate the steel industry of the world.

The new molding shop of the Stevens Co., Galt, is now in full swing. The large jib crane is in its place and the firm are doing jobbing work on castings, besides making all necessary for their own use.

The new machine shop of the Northern Engineering & Supply Co., at Fort William, Ont., was opened recently. The plant is equipped with modern machinery and can handle any kind of repair or machine work.

There are rumors that a third hydro-electric power company will be formed in British Columbia, for developing power on the Cheakamus River, at the head of Howe Sound: 100,000 h.p. could be transmitted to Vancouver.

At the present time the Dunbar iron industry of Woodstock, N.S., is negotiating with the city of Fredericton, and is likely to remain there and amalgamate with the foundry firm of McFarlane, Thompson & Anderson.

Messrs. W. Knack, H. Anderson, H. H. Jackson, F. Bye and C. Howard have purchased the Hill sash and door factory at Markdale, Ont., and after installing some modern machinery will manufacture furniture of all kinds.

Swift & Co. will build an enormous packing establishment at Edmonton, to cost $1,000,000. Other factories will be build in connection to utilize the by-products. These are soap and glue factories, tanneries and button works.

The Dominion Engineering and Construction Co., Montreal, have completed the machine shop for the Canada Foundry Co., Davenport, Ont., and the new transformer building for the Canadian General Electric Co., Peterboro, Ont.

The McLaughlin Carriage Co., Oshawa, are equipping their new automobile factory with Chapman double ball bearings. This company was one of the first in Canada to install these bearings, and has had them in use over three years.

E. A. Wallberg, Montreal, is the contractor for the following: Intercolonial Railway shops, Moncton, N.B., and Halifax, N.S.; Prince Edward Island Railway shops, Charlottetown, P.E.I.; Intercolonial Railway engine house and freight yard, Pictou, N.S.

A large furniture factory is being built in Deseronto by John Dalton & Co. It is a four-storey concrete building, one hundred and fifty feet long by fifty feet wide. When completed it will employ about one hundred hands. The machinery is now being installed.

The Robertson Machinery Company have recently closed a deal by which Welland gains another substantial industry. Kischell & Co., of Toronto, are a going concern, and their amalgamation with the Robertson Machinery Co. brings to Welland a lot of new business.

Water power rights at Healey Falls, Ont., have been leased by the Ontario Government to the Northumberland-Durham Power Co. Three thousand horsepower is to be developed within two years, and the company are to begin work of development within four months after June first.

The Brantford Screw Co., Brantford, Ont., have installed a 250 horsepower gas engine run by natural gas. It was manufactured by Struthers & Wells, Warren, Pa. The Adams Wagon Co., Brantford, and the Woods Milling Co., St. George, Ont., have installed similar engines.

The J. I. Case Mfg. Co., Racine, Wis., have issued bonds, the proceeds of which will be used to build a factory in Fort William, Ont. The company have owned 192 acres of land outside the town for several years but it is only recently that their intention has been made public.

Chadwick Bros. have just installed one of the latest and most up-to-date patterns of automatic lathes from the National Acme Mfg. Co., of Cleveland, with four spindles, cross drilling and threading rolling attachment for making all kinds of brass nuts, screws, studs and bolts.

The Sherlock Manning Organ Co., of London, Ont., are erecting a new factory, which, when completed, will more than double their present output. The new building will have a floor space of 300 square feet more than the present one, making in all a total floor area of 40,000 square feet.

The Regina Machine & Iron Works Co., Regina, capitalized at $100,000, are shortly to erect a large foundry and machine works. The building, which will cost $20,000, will be fitted with a large amount of up-to-date machinery, thus enabling the company to handle practically all classes of iron, steel or machine work. The officers of the company are: P. Cooper, E. Osborne, R. Reid, J. Reid, J. F. Embury, and J. A. Kerr, all of Regina.

The Toronto Street Railway Co. have installed in their plant, a 200-ton hydraulic wheel press, for pressing wheels on and off axles, which was made by John Bertram & Sons, Dundas, Ont. Owing to the additional work now to be done, since taking in the York Radial Railway, a heavier type of press was needed.

The Brockville Malleable Iron Company is being organized in Brockville, with local capital. The directors purpose to establish a plant capable of reducing either five tons or ten tons of castings daily. The five-ton plant complete, including $104,000 for working capital, is estimated to require $35,000, while a ten-ton plant is estimated at $50,000.

The Macdonald Co., Grand Forks, have two large contracts on hand at present for the Granby Co. They are now working at a large tank with a capacity of 100,000 gallons. The other contract is for a mammoth flue dust chamber. This flue will be 13x15 feet inside measurement and elevated 22 feet above the feed floor. 300 tons of steel will be used in the construction, which is the heaviest piece of steel work ever attempted in this country.

Municipal Undertakings.

Souris, Man., will build a hospital costing $12,000.

Welland, Ont., will spend $20,000 on improving its roads.

Montreal will spend $2,000,000 in bettering its water supply.

Chatham will improve both its water and lighting service.

The town of Truro, N.S., is installing a new fire alarm system.

Summerlea, Que., is to have a municipal electric light system.

Five Government telephone lines are being erected in Alberta.

Cardston has voted $20,000 for water works and electric lighting.

The town of Oakland, Man., is inviting tenders for a steel bridge.

Work on the $60,000 sewerage system for Selkirk, Man., is under way.

Moose Jaw is to extend its municipal electric lighting plant and system.

Tenders are called for a bridge over the Saugeen river at Teeswater, Ont.

A $10,000 extension will be made to the waterworks system at Vancouver.

The city engineer of Ottawa is calling tenders for a municipal asphalt plant.

The town council of Morden, Man., is to install an electric lighting system.

Port Arthur will raise $360,000 for bridges, waterworks and street car service.

Wetaskiwin, Alta., has decided upon the construction of water and sewer works.

The Ingersoll Board of Trade is trying to secure a large steel plant for that town.

Bridgeburg, Ont., has in contemplation a sewage system that will cost about $125,000.

Goderich, Ont., is asking for $28,000 to improve the waterworks and lighting systems.

The water committee of Montreal is asking for $17,000 to extend the system to Rosemount.

The waterworks system of Chilliwack, B.C., has been taken over by a company for $34,000.

The city council of St. John, N.B., will call tenders for waterworks extension to cost $30,000.

A by-law will be voted upon in Sherbrooke, Que., for the development of power at Westbury, Ont.

Raymond, Alta., will improve its water system and install a municipal electric light system.

Brandon, Man., has voted $50,000 for a waterworks system, and $150,000 for fire equipment.

The Ottawa Electric Commission have ordered a 750 horsepower generater for the municipal plant.

Campbellford, Ont., has voted a loan of $15,000 to Jas. and Geo. Dickson for a bridge works.

Niagara Falls is to have an improved waterworks system on which $40,000 is to be expended.

Berlin has acquired interests in the Berlin & Waterloo Street Railway Co. to the extent of $76,000.

Kingston recently passed a by-law exempting two smelting companies from taxation and granting them a free site. They will smelt lead and zinc.

Sherbrooke, Que., will vote on a by-law to raise $200,000 for the purpose of developing water power.

Dunnville, Ont., has decided to grant a franchise to M. L. Perry to supply electricity to that town.

A by-law will be submitted to the town of Ingersoll for $35,000 to take over the waterworks plant.

A by-law will be submitted to the ratepayers of Niagara Falls, Ont., to raise $3,000 for a public library.

Lachine, Que., will borrow $50,000 for the purpose of improving the waterworks and sewerage services.

The town of Campbellford, Ont., is calling tenders for a water turbine, generators and transmission line.

Alymer, Ont., will vote on by-laws to purchase the waterworks, $50,000, and $60,000 for a sewage system.

Fort William will vote on bylaws to raise $84,000 for the extension of telephone, electric light and water systems.

The Ottawa Water Works Committee will install a duplicate set of pumps to give a daily capacity of 30,000 gallons.

The contract for the construction of the waterworks system at Simcoe has been let to T. M. Cullon, of Huntsville.

A by-law has been presented by the town of Newmarket, Ont. for an extension of its municipal electric lighting system.

Fort William, Ont., has voted $94,000 for the extension of the municipal waterworks, telephone and electric light systems.

The ratepayers of Estevan, Sask., will vote on a by-law to raise $82,000 for municipal electric light and waterworks systems.

The Municipal Construction Co., Regina, has secured the contract for installing a $50,000 waterworks system at Battleford, Sask.

The towns of Collingwood, Midland and Penetanguishene have applied for municipal rights to develop electricity at the Severn river.

The corporation of Ottawa may take over the Ottawa Electric Company's plant and lines and amalgamate them with the municipal lighting enterprise.

The Ottawa electric commission is making arrangements for taking over the street lighting system of the Ottawa Electric Co. $24,000 is the price offered.

The following by-laws were carried at Medicine Hat: $75,000 for waterworks extension, $20,000 for improvements to the natural gas scheme and $60,000 for fire alarms.

The Peterboro, Ont. council have decided to apply to the Hydro-Electric Power Commission to acquire water power privileges and construct the necessary works to supply cheap power.

The contract for laying sewer pipes at Fernie, B.C., has been let to Hugh Macdonald, Victoria, the contract price being $30,000. Tenders for a disposal plant will be asked for shortly.

The city of Medicine Hat is increasing the capacity of its pumping plant by 1,200,000 gallons per day. Two new pumps will be installed operated by gas engines, natural gas being used.

By-laws were carried by large majorities to assist the location in Sarnia of the Standard Chain Company, of Pittsburg, and the Jenks Dresser Company, manufacturers of iron bridges.

The Canadian General Electric Company have been awarded the contract for transformers, etc., for the municipal plant at Winnipeg, at $2,810. The Packard Electric Company, St. Catharines, obtained the contract for 100 arc lamps and cut-outs at $2,700.

Halifax intends to spend $768,500 on municipal work, as follows: Pavements, $75,000; sidewalks, $150,000; improving fire department, $70,000; water extension, $135,000; sewerage, $50,000; school houses, $95,588; street extension and various purposes, $122,500.

It is proposed that the municipalities of Orangeville and Shelburne combine to develop the available water power at Hornings Mill. The amount of power which could be developed would be between 200 and 400 horse power. The cost of development has been estimated at $85,000.

Medicine Hat is installing a modern waterworks system. The Roberts Co., of Philadelphia, have the contract for a filtration plant, consisting of two 1,000,000 gallon units. The contract was also let for a huge stand-pipe with capacity of 500,000 gallons. This has already been erected.

The following by-laws were passed by the ratepayers of Fort William: To raise $68,000 for the purchase of land to be used for industrial sites; $94,000 to extend the water, light and telephone systems, distributed as follows: $36,000 for telephone, $32,000 for water, $26,000 for light; $18,000 for a fire hall and police station at Westfort; $3,750 to enlarge the central fire hall; $20,000 to purchase a site for the plant of the Imperial Steel and Wire Company.

The Winnipeg city council is in favor of adopting the Red river as a source of water supply. It is estimated that 15,000,000 gallons per day is required. The total estimated cost of the necessary work is $1,545,000. The cost of pumping and softening the water is estimated at $40, per million gallons.

Thirteen by-laws were carried at Port Arthur, Ont. They included the construction of two concrete and steel bridges, the purchase of street cars, improvements to the Current river, extension of the waterworks, erection of a police station, double tracking the street railway for seven miles, establishing of an incinerating plant, erection of a new car barns and isolation hospital and park improvements. A plebiscite on development of Dog Lake Falls carried.

Mining Notes.

Extensive additions are planned for the Granby mines, B.C.

A coal mine is being opened up at Mosher, B.C. by the C.P.R.

A valuable vein of copper has been struck near Larder Lake, Ont.

There are now twelve blast furnaces at work at the Boundary Mines, B.C.

The Nova Scotia Steel and Coal Company, Sydney, N.S., are opening a new mine.

The Boundary Smelters, B.C., are turning out copper at the rate of 4,500,000 pounds daily.

Gold, silver, iron and copper have been found on the G.T.P. at the north of Lake Superior.

P. T. Thornton, Brandon, has purchased the Wilcox coal mine at Taber, Alta., for $250,000.

The property of the Ontario Nickel Copper Company, near Sudbury, has been bought by a syndicate.

The Maritime Coal & Power Company have started operations in their new power plant at Amherst, N.S.

A mining company will be formed at Sturgeon Lake, Ont., to operate the Wyndego mine, near Kenora, Ont.

A huge reduction and refining works will shortly be erected on the property of the La Rose mine, Cobalt, Ont.

A deposit of copper claimed to be the second largest in Canada has been found north of Sault Ste. Marie.

The British Columbia Copper Company, Greenwood, B.C., have recently placed their [illegible] smelter in operation.

The Dominion Iron and Steel Company, has started a Bessemer plant which will increase its monthly output to 30,000 tons.

Coal has been found near the Yellowhead Pass, B.C., where the G.T.P. and Canadian Northern propose crossing the rockies.

Tin bearing ore with a high percentage of tin has been discovered in Lunenburg County, N.S. It will be developed by H. Peers, Halifax.

An enormous concentrator is being erected at the Blue Bell mine, near Ainsworth, B.C. Other reducing machinery is also being installed.

R. [illegible]. Anderson, of Minneapolis, has bought the lead property of J. W. Andrew, near Port Arthur. Development will commence immediately.

A very good report has been handed in about the deposits of iron ore which lie 30 miles north of Sudbury. There are enormous deposits of very rich ore. The Canadian Northern will open up the mine.

The Consolidated Mining & Smelting Company, B.C., have acquired the properties of the Phoenix Amalgamated. Active operations will at once begin and a twenty-drill air compressor will be installed.

The Nova Scotia Steel & Coal Company have awarded the contract to D. Sutherland, North Sydney, for the construction of 2½ miles of railway, costing $50,000.

The plant for obtaing iron ore in the Nepigon district, Ont., from sand will be opened again by the North Shore Reduction Co. S. N. Smith, Minneapolis, is the promoter.

The Dominion Coal Company's shipments of June 30th broke all records, 20,000 tons being shipped to the St. Lawrence alone. The total shipments for the week amounted to 64,024 tons.

Of the fourteen blast furnaces at the Boundary smelters, twelve are now in operation, handling 5,000 or 6,000 tons of ore per day. The Dominion Copper Company have two smelters running and will soon have eight stacks in full blast.

A bituminous shale called stellarite has been found in Pictou County, N.S. It is very rich in oil, yielding as high as 100 gallons of clarified oil to the ton. A company of New York capitalists will develop the mineral.

Owing to the shortage of ore cars the Dominion Copper Company, B.C., will build an aerial tramway from their mine to the Boundary smelter. The operations of the Granby mine have also been curtailed by the car shortage.

The mining in British Columbia has increased at a tremendous rate. The copper mines have the greatest revenue, but until recently gold had the lead. The following table shows the relation between the mining resources of British Columbia and the remaining provinces:

	British Columbia	All other Provinces
Gold	$5,579,039	$ 306,032
Silver	1,897,320	3,539,497
Copper	8,288,565	2,318,095
Lead	2,667,578	111,980
Nickel	None	8,948,934
Coal and Coke	5,458,044	14,245,032

This gives a total of $23,980,526 for British Columbia, against $31,193,876 of the other provinces. The figures are the total for 1906.

Railroad Construction.

The C.P.R. will build a round-house in Saskatoon.

The new station at Charlottetown, P.E.I., was opened recently.

The new Intercolonial station at Moncton, N.B., will cost $45,000.

The C.P.R. will build a line from Victoria Harbor to Peterboro.

The C.P.R. will build subways over Herny and Elgin Sts., Brantford.

The Grand Trunk subway at Lansdowne Ave., Toronto, will cost $53,443.

The Grand Trunk Pacific is graded for seventy miles west of Saskatoon.

Grading has been commenced on the G.T.P. east of Portage la Prairie.

Work is progressing rapidly on the G.T.P. from Saskatoon to Edmonton.

The Guelph and Goderich branch of the C.P.R. has been opened as far as Blyth.

Steps are under way for the use of electricity on the C.P.R. grades in the West.

The British Columbia Electric Railway will extend its system at Fort Langley.

Work has commenced at Fort Simpson on the Pacific and Hudson's Bay Railway.

The C.P.R. will spend $257,000 in Saskatoon, Man. The new station will cost $50,000.

The C.P.R. will call for tenders for a steel bridge at Letbbridge, to cost $1,000,000.

The Canadian Pacific are double-tracking their line between Fort William and Winnipeg.

Work has commenced on the construction of the Kingston, Ottawa Railway, at Rideau.

The Scott Machine Co., London, are constructing two test boring machines for the C.P.R.

The Strathcona-Alberta Radial Tramway Co. are anxious to operate a street railway system.

The Halifax & South Western Railway have let the contract for a steel bridge near Halifax.

The Hamilton Steel and Iron Co. have commenced making railroad spikes from basic steel bars.

The C.P.R. will erect a power house in connection with their new hotel at Victoria to cost $40,000.

The Quebec, Montreal and Southern Railway has been purchased by the Delaware & Hudson R.R.R.

Vancouver is to be connected by trolley with Tacoma. Slone & Webster have bought the franchise.

The C.P.R. are reducing the grades in the Crow's Nest Pass and otherwise improving the roadbed.

The C.P.R. will replace the wooden trestle over the Rideau River at Merrickvelle with a steel bridge.

Plans are out for a new $40,000 station and a twenty-one stall round-house for Estevan, Sask., for the C.P.R.

The Canadian Northern are looking forward to the completion of the Brandon-Regina line in the near future.

The Canadian Northern will run a branch from Fernie up the Elk River. This will open up some good coal areas.

Operations have commenced on the erection of the new car and erecting shops of the Prince Edward Island Railway.

The contract for the construction of a three hundred thousand dollar pier at Vancouver has been awarded by the C.P.R.

Thomas Merrill, Duluth, will build a logging railway 16 miles long, through his lumber claims in British Columbia.

It is understood in railway circles that a universal signal system will be adopted on all American and Canadian roads.

The new owners of the Woodstock, Thames Valley and Ingersoll Railway will construct the line between Ingersoll and Woodstock.

The work of electrifying the Sarnia tunnel is progressing rapidly. The machinery is on the ground and will soon be installed.

Plans are under way for the erection of a union station at Portage la Prairie by the G.T.P. and Canadian Northern Railways.

The contract for the construction of 50 miles of the C.P.R. on the Shedo extension has been awarded to J. G. Hargrave & Co., Winnipeg.

The C.P.R. will construct fifty giant locomotives for the use on the mountain division. They will have a draw bar pull of 40,000 pounds.

It is stated that the Canadian Northern will use electricity developed by the Slave Lake Power Co., to operate its Vancouver-Seattle line.

The contract for 65,000 tons of steel rails for the National Transcontinental Railway has been awarded to the Dominion Iron & Steel Co., Sydney, N.S.

Operations are soon to be commenced on the electric line from Bellingham, Wash., to British Columbia. E. W. Purdy, Bellingham, is one of the directors.

A large roundhouse is being constructed at Smith's Falls, Ont. by the C.P.R. It will have capacity for 22 of the largest engines that the company have.

The Canadian Northern Railway is to erect its new Winnipeg shops in Fort Rouge. The total cost of the buildings will be $76,000. The architect is Samuel Brown.

The contracts for the five steel bridges on eastern Quebec section of the Natianal Transcontinental Railway have been awarded to the Dominion Bridge Co., Montreal.

Work is going on for the relaying of the Grand Trunk double tracks between Montreal and Toronto with hundred pound rails, the old eighty pound steel rails being taken up. This will make the heaviest rails over such a section of track in Canada.

The two chief projects in railway construction this season in British Columbia will be the new electric line from New Westminster to Eburne. The other will be started as soon as contracts are let. Both lines will be operated by the B.C. Electric Railway Co.

The Northern Empire Railroad Co. is the latest aggregation of Americans organized for the purpose of building a railroad into Canada. It will cross the international boundary near Cardston, Alberta, and from thence will run north to Fort McMurray on the Athabasca River.

The contract for the Eburne-Westminster branch of the C.P.R. has been awarded to J. B. Bright, Vancouver. The British Columbia Electric Railway Co. will electrify and operate the line. This company will also take over the branch from Westminster Junction to New Westminster.

The Government has approved of contracts for Intercolonial locomotives and rolling stock to the amount of about $1,000,000. The locomotives are chiefly heavy freighters. A large order for cars was divided between the Crossen Car Co., of Cobourg, Ont., and Rhodes, Curry & Co., of Amhearst.

In the near future the C.P.R. will operate its trains from Fort William west, for about 30 miles, and also over the heavy grades in the mountains, by electric moguls. For the Fort William grade the power will be drawn from Kakabeka Falls, while the Bow, the Kicking Horse, the Beaver, the Illecillewalt, the Eagle and the Fraser will furnish ample power for the mountains.

Companies Incorporated.

The Hub Spoke and Bent Goods Co., of Sarnia, have changed their name to Loughead, Limited.

The Argenteuil Lumber Co., Morin Flats, Que.: capital, $20,000; to deal in lumber and construct and maintain wharves, piers, etc.

Buffalo Larder Gold Mines, Limited, Toronto: capital, $2,000,000. Provisional directors: C. S. MacInnis, C. C. Robinson, all of Toronto.

The Northern Reduction Co., Toronto, capital, [illegible]; to mine, mill and reduce ores. Incorporators: W. C. Mackay, T. D. Byers, A. N. Morine, T. Pottage, all of Toronto.

Morden, Limited, Brandon, Manitoba; capital, $50,000; to manufacture gas and electric fixtures. Provisional directors: M. W. Morden, J. Watson and J. B. Nobel, all of Brandon.

The Jones Safety Device Co., Hamilton; capital, $50,000; to do a machinery business in Hamilton. Incorporators: C. Readman, Grace Hewson, Florence Philips, all of Toronto.

The Elgin Cobalt Mining Co., St. Thomas, Ont.; capital, $200,000; to carry on mining, milling and reduction of ores. Directors: J. H. Courtenay, W. H. King and J. T. Utter, all of St. Thomas.

The Ridgetown Canning Co., Ridgetown, Ont.; capital, $100,000; to carry on a fruit and vegetable canning business. Provisional directors: C. B. Kenleyside, B. V. Hole, G. E. Coleman, London, Ont.

Canadian Lithographic Stone Co., Montreal; capital, $500,000; to carry on a mining, milling and reduction business. Incorporators: R. B. Hutcheson, F. W. Hibbard, W. J. Ross, all of Westmount, Que.

Cobalt Confederation Mines, Limited; capital, $3,000,000, Cobalt; to carry on the business of mining, milling, reduction and development. Provisional directors: J. S. Bosquet, M. McLeod, R. Herron, F. C. Powell and H. Hortman, all of Cobalt.

Missisquoi Marble Co., Missisquoi, Que., capital $500,000; to deal in marble, tile brick and carry on the business of house decorators. Directors: J. T. Shearer, R. J. Dale, S. H. Ewing, S. Carsley, W. Mann and H. Timmis, all of Montreal.

Dillons, Limited, Montreal, capital $49,000; to manufacture and deal in drugs, chemicals and dye-stuffs. Directors: G. A. Dillon, R. P. Dillon, R. Taschereau, S. Dore and R. Genest, all of Montreal.

The Hydro-Electric Construction Co., Toronto; capital, $50,000; contractors, hydraulic and electrical engineers. Provisional directors: A. Keith, A. G. Lawrence and W. R. Wadsworth, all of Toronto.

The Red Jack Mining Co., Midland; capital, $500,000; to carry on the operations of a mining, milling reduction and development company. Provisional directors: J. S. McDonald, A. B. Thompson and T. J. Wilson.

Montreal Gold and Silver Mining Co., Cobalt; capital, $10,000; to carry on all branches of mining, milling reduction and development. Provisional directors: O. Baker, W. J. Sutherland and J. C. Armstrong, all of Cobalt.

The Electric Smelters, Ottawa; capital, $18,000; to deal in electric smelters and supplies. Directors: J. H. Reid, Cornwall; G. P. Borphy, E. D. Lafleur, J. C. Scott, all of Ottawa, and S. L. Tingley, Providence, U.S.

The Galetta Electric Power and Milling Co.; capital, $100,000; head office, Arnprior, Ont.; to develop electrical energy. Provisional directors: T. Moran, M. Sullivan, both of Arnprior, and J. Brennan, of Sand Point, Ont.

The Ontario Metal Novelty Manufacturing Co.; capital, $100,000; head office, Toronto; to manufacture metal novelties, machine tools, dies, etc. Provisional directors: E. Currie, M. Campbell, J. S. Woodhouse, all of Toronto.

An Ontario license has been granted to John S. Metcalfe Co., Illinois, to construct elevators and manufacture all kinds of machinery. R. S. McFarlane, Midland, Ont., will be the company's attorney, with a capital of $50,000.

A company to be known as Pressed Bricks, Limited, has been formed in Edmonton for the manufacture of pressed bricks by the Berg process. The officers are: President, D. R. Fraser; vice-president, C. May; directors, A. Brown, C. Gallagher, J. H. Gariepy, A. J. Manson and J. Macdonald; secretary-treasurer, H. J. Helliwell, all of Edmonton.

Building Notes.

Swift, Copland & Co., Montreal, are building a factory in Montreal.

The Manitoba Government will erect a telephone exchange at Brandon.

A high school will be erected at Smith's Falls, Ont., to cost $45,000.

The Diamond Glass Co., Toronto, will build an elevator costing $20,000.

Tenders are invited for the construction of an armory at Walkerton, Ont.

The C.P.R. is asking for tenders for the erection of the station at Calgary.

The Royal Bank of Canada will erect a building in Toronto, costing $200,000.

The Bank of Montreal will erect a branch at Portage la Prairie to cost $55,000.

The Adams Wagon Co. will erect a brick warehouse at Brantford costing $8,000.

An addition is to be made to the King Edward Hotel, Toronto, costing $500,000.

The congregation of Knox Church, Galt, are to erect a schoolhouse costing $25,000.

St. Anne's Church congregation, Toronto, is to erect a brick edifice to cost $30,000.

A company is being formed for the purpose of erecting a $75,000 hotel at Lethbridge.

Queen's University, Kingston, Ont., will erect a biological building, to cost $35,756.

It is stated that a Toronto syndicate will erect a large hotel on the Caer Howell site, Toronto, to cost $300,000.

The building permits issued in Regina for June include R. H. Williams' block, $40,000, and A. W. Smith's block, $35,000.

A modern apartment house will be erected at Vancouver costing $350,000. Champion & Pound are the local representatives of the company.

Among the buildings being erected at Regina are: F. N. Darke, business block to cost $110,000; Masonic Temple, $50,000; J. W. Smith, [illegible] & Brown, block, $35,000.

A syndicate has been organized, including Messrs. Campion and Pound to erect a ten-storey apartment house in Vancouver, to cost $250,000.

Work has been startdd by the Mortimer Company, Ottawa, on a $25,000 addition to its premises. The addition is to be used for the printing rooms.

A new one hundred barrel flour mill will be erected in Rosthern, Sask. this summer by the Rosthern Flour Mills Company, costing in the neighborhood of $20,000.

The Dominion Radiator Works will build a large chimney in connection with their works in Toronto. The stack will be 125 feet high, contain 25,000 bricks and cost $20,000.

An enormous rink will be erected at Halifax, N.S. to cost $35,000. The Commercial Cable Co. will build a $9,000 cable tank on their premises. The Silliker Co. will build an $5,000 paint shop.

The Great Northern Railway will erect a new station at Brandon to cost $30,000. The Brandon Construction Company are the contractors. A new winter fair and stock pavilion will also be erected costing $24,500.

The Lands Produce & Cold Storage Company are building a large cold storage plant at St. John, N.B., to cost $60,000. Apple factories will be erected in the Annapolis Valley and a canning factory at Kentville.

The following buildings are being erected at Saskatoon: The National Trust Co., $3[illegible],000; Canadian Bank of Commerce, $65,000; the Great West Furniture Co., $50,000; the Sutherland Block, $10,000; the Alexandra School, $35,000.

The following buildings are being put up in Hamilton: New armories, costing $250,000; Herkimer Baptist church, to cost $23,000; Home for [illegible], [illegible]; Isolation Hospital, $75,000; a new building in connection with the Home for Consumptives to cost $10,000.

Among the more important buildings being erected at Prince Albert, Sask., are: McKay & Adams' brick block, $30,000; P. D. Tyerman, hotel, $25,000; J. Sanderson, brick block, $30,000, and Presbyterian Church, $45,000.

The Colonial Investment and Loan Company will erect a skyscraper on Yonge St., Toronto. The plans call for a building of steel, brick and terra cotta, ten stories high, to cost about $150,000, exclusive of the cost of the land.

Statistics gathered from 37 towns throughout the west show that buildings actually under way therein amount to $22,500,000, exclusive of Winnipeg. In Edmonton alone $6,000,000 will be expended in new structures. This wonderful development, great as it is, would be one-third more were it not for the prohibitive rates charged for lumber. Wages are uniformly high.

Among the buildings being erected at Prince Albert are the following: Sanderson & Knox block, $55,000; Pollock block, $10,000; Moore block, $40,000; Prince Albert Times block, $5,000; G. W. Baker, residence, $12,000; hi[illegible] school, $100,000; fire hall, $10,000; Baker & Eaton block, $40,000; R. C. Church, $[illegible]0,000; S. J. Donaldson, residence, $5,000 and A. MacDonald, residence, $8,000.

To Build Ornamental Iron Works Plant.

The Canadian Ornamental Iron Works, 35 Yonge St. Arcade, Toronto, are preparing plans for a new plant for the manufacture of ornamental iron work, including iron fence, stairs, etc.

Starts Operations.

The new factory of the Gilson Manufacturing Co., Limited, at Guelph, Ont., Canada, is now ready to start operations. The plant is modern and is equipped in an up-to-date manner.

Greeys Again in Running Order.

Fire did damage to the plant of Wm. & J. G. Greey, Toronto, recently, but repairs have now been made and everything is in running order again.

Goldie-McCulloch to Make Producer Gas Plants.

The Goldie-McCullough Co., Galt, Ont., are installing machinery for the manufacture of gas engines. At first engines of 160 horsepower will be built, these will be given a thorough test before other classes are constructed. Producer gas plants will be manufactured in connection with the engines.

Fairbanks' Offices Renovated

The head offices of the Canadian Fairbanks Co., Montreal, have been renovated to very good effect. The wood finishings are done in stained chestnut, which give it a very homelike appearance. The flat throughout is illuminated by the most improved system of enclosed arc lamps.

Hamilton Tool and Optical Reorganize.

The name of The Hamilton Tool & Optical Company, Limited, has been changed to "Hamilton Tool Company, Limited," and the office and factory will shortly be removed from 80 Murray St. to much larger premises on the corner of Catharine and Barton Sts., where in future the business of the company will be carried on.

Extension to Moose Jaw's Electrical Plant.

The following extensions are proposed by the city council of Moosejaw to the municipal electrical plant: two boilers and a smoke stack; a boiler house with a capacity of 2,000 horsepower in boilers. One year from now it is proposed to add two more boilers, a 500 kilowatt steam turbine, together with the necessary equipment.

The city council are considering the introduction of a by-law for the raising of debentures of $90,000.

Pintsch Producer Just Installed.

The installation of a 100 h.p. producer gas plant and engine, has just been completed by the Economic Power, Light & Heat Supply Co., Toronto, for the Christie, Brown & Co., Toronto. The installation consists of a Pintsch producer and a National gas engine. The fuel used in the producer is gas work's coke, which has proved very satisfactory fuel. The plant has been operated, and works very well indeed. Some figures regarding the cost of operating the plant will be given later.

Installation of Producer Gas Plants.

The Colonial Engineering Co., Montreal, are under contract to install two 100 h.p. Hornsby-Stockport suction gas plants and engines for Ames Holden, Limited, Montreal; one 100 h.p. equipment for Lamontague, Limited, Montreal; one 100 h.p. equipment for the Empire Manufacturing Co., London; a 200 h.p. municipal lighting plant equipment for Chatham; and also other smaller plants.

This company are doing well considering that they have been in business such a short time.

A Large Pumping Engine.

Mather & Platt, Limited, Manchester, Eng., of whom Drummond, McCall & Co., Montreal, are Canadian representatives, have been awarded an order from the Montreal Water & Power Co. for a large turbine pumping engine, to be installed at the St. Gabriel pumping station next January. This new engine is claimed to be the largest pumping engine in Canada, if not in America. Its capacity is 15,000,000 gallons per day with a pressure of 175 pounds, or a head of 406 feet. The pump is driven by one of Mather & Platt's own motors, the entire pumping apparatus will be arranged on one bed of plate.

French Company Open Offices in Philadelphia.

Ph. Bouvillain & E. Ronuray, which firm exhibited their universal system of machine moulding at the Philadelphia convention of the American Foundrymen's Association, have opened offices at 1315 Race St., Philadelphia, from where they will handle their United States and Canadian business. They have put a number of machines in operation at this address on a considerable variety of pattern plates. This they think is the best illustration of what can be made on molding machines.

CANADIAN MACHINERY BUYERS' DIRECTORY

To Our Readers—Use this directory when seeking to buy any machinery or power equipment. You will often get information that will save you money.
To Our Advertisers—Send in your name for insertion under the heading of the lines you make or sell.
To Non-Advertisers—A nominal rate of $1 per line a year is charged non-advertisers.

Acids.
Canada Chemical Mfg. Co., London

Abrasive Grains & Powders
Carborundum Co., Niagara Falls, N.Y.

Abrasive Materials.
Baxter, Paterson & Co., Montreal.
The Canadian Fairbanks Co., Montreal.
Rice Lewis & Son, Toronto.
Norton Co., Worcester, Mass.
H. W. Petrie, Toronto.
Carborundum Co., Niagara Falls, N.Y.
Williams & Wilson, Montreal.

Air Brakes.
Canada Foundry Co., Toronto.
Canadian Westinghouse Co., Hamilton.

Air Receivers.
Allis-Chalmers-Bullock Montreal.
Canada Foundry Co., Toronto.
Canadian Rand Drill Co., Montreal.
Chicago Pneumatic Tool Co., Chicago.
John McDougall Caledonian Iron Works Co., Montreal.

Alloy, Ferro-Silicon.
Paxon, J. W., Co. Philadelphia.

Alundum Scythe Stones
Norton Company, Worcester, Mass.

Arbor Presses.
Niles-Bement-Pond Co., New York.

Augers.
Chicago Pneumatic Tool Co., Chicago.
Frothingham & Workman, Ltd., Montreal
Rice Lewis & Son, Toronto.

Automatic Machinery.
Cleveland Automatic Machine Co. Cleveland.
National-Acme Mfg. Co., Cleveland
Potter & Johnston Machine Co., Pawtucket, R. I.

Automobile Parts
Globe Machine & Stamping Co., Cleveland, Ohio.

Axle Cutters.
Butterfield & Co., Rock Island, Que.
A. B. Jardine & Co., Hespeler, Ont.

Axle Setters and Straighteners.
A. B. Jardine & Co., Hespeler, Ont.
Standard Bearings, Ltd, Niagara Falls

Babbit Metal.
Baxter, Paterson & Co., Montreal.
Canada Metal Co., Toronto.
Canada Machinery Agency, Montreal.
Frothingham & Workman Ltd., Montreal
Greey Wm. & J. G., Toronto.
Rice Lewis & Son, Toronto.
Lumen Bearing Co., Toronto.
Mechanics Supply Co., Quebec, Que.
Miller Bros. & Toms, Montreal, Que.

Bakers' Machinery.
Greey, Wm. & J. G., Toronto.

Baling Presses.
Perrin, Wm. R., & Co., Toronto.

Barrels, Steel Shop.
Cleveland Wire Spring Co., Cleveland.

Barrels, Tumbling.
Buffalo Foundry Supply Co., Buffalo.
Detroit Foundry Supply Co., Windsor
Dominion Foundry Supply Co., Montreal
Hamilton Facing Mill Co., Hamilton.
Globe Machine & Stamping Co, Cleveland, Ohio.
John McDougall Caledonian Iron Works Co., Montreal.
Miller Bros. & Toms, Montreal, Que.
Northern Engineering Works, Detroit.
J. W. Paxson Co., Philadelphia, Pa.
H. W. Petrie, Toronto.
Sly, W. W., Mfg. Co., Cleveland
The Smart-Turner Mach. Co., Hamilton

Bars, Boring.
Hall Engineering Works, Montreal.
Miller Bros. & Toms, Montreal.
Niles-Bement-Pond Co., New York.
Standard Bearings Ltd., Niagara Falls

Batteries, Dry.
Berlin Electrical Mfg. Co., Toronto.
Mechanics Supply Co., Quebec, Que

Batteries, Flashlight.
Berlin Electrical Mfg. Co., Toronto.
Mechanics Supply Co., Quebec Que

Batteries, Storage.
Canadian General Electric Co. Toronto
Chicago Pneumatic Tool Co., Chicago.
Rice Lewis & Son, Toronto.
John Millen & Son, Montreal.

Bearing Metals.
Lumen Bearing Co., Toronto.

Bearings, Roller.
Standard Bearings, Ltd. Niagara Falls

Bell Ringers.
Chicago Pneumatic Tool Co., Chicago.
Mechanics Supply Co., Quebec, Que.

Belting, Chain.
Baxter, Paterson & Co. Montreal.
Canada Machinery Exchange, Montreal.
Greey, Wm. & J. G., Toronto.
Jeffrey Mfg. Co., Columbia, Ohio.
Link-Belt Eng. Co., Philadelphia.
Miller Bros. & Toms, Montreal.
Waterous Engine Works Co., Brantford.

Belting, Cotton.
Baxter, Paterson & Co., Montreal.
Canada Machinery Agency, Montreal.
Dominion Belting Co., Hamilton.
Rice Lewis & Son, Toronto.

Belting, Leather.
Baxter Paterson & Co., Montreal.
Canada Machinery Agency, Montreal.
The Canadian Fairbanks Co., Montreal.
Frothingham & Workman Ltd, Montreal
Greey, Wm. & J. G., Toronto.
McLaren, J. C., Montreal.
Rice Lewis & Son, Toronto.
H. W. Petrie, Toronto.
Williams & Wilson, Montreal.

Belting, Rubber.
Baxter, Paterson & Co., Montreal.
Canada Machinery Agency, Montreal.
Frothingham & Workman, Ltd., Montreal
Greey, Wm. & J. G, Toronto.

Belting Supplies.
Baxter Paterson & Co, Montreal.
Greey, Wm. & J. G., Toronto.
Rice Lewis & Son, Toronto.
H. W. Petrie, Toronto.

Bending Machinery.
John Bertram & Sons Co., Dundas, Ont.
Chicago Pneumatic Tool Co., Chicago.
Rice Lewis & Son, Toronto
London Mach. Tool Co., Hamilton, Ont.
National Machinery Co., Tiffin, Ohio.
Niles-Bement-Pond Co., New York.

Benders, Tire.
A. B. Jardine & Co., Hespeler, Ont.
London Mach. Tool Co., Hamilton, Ont

Blowers.
Buffalo Foundry Supply Co., Montreal.
Canada Machinery Agency, Montreal.
Detroit Foundry Supply Co, Windsor
Dominion Foundry Supply Co., Montreal
Hamilton Facing Mill Co., Hamilton.
Kerr Turbine Co., Wellsville, N.Y.
Mechanics Supply Co., Quebec, Que.
J. W. Paxson, Philadelphia, Pa.
Sheldon's Limited, Galt
Wilbraham-Green Blower Co, Philadelphia

Blast Gauges—Cupola.
Paxson, J. W., Co. Philadelphia.
Sheldons, Limited, Galt

Blocks, Tackle.
Frothingham & Workman, Ltd, Montreal

Blocks, Wire Rope.
Frothingham & Workman, Ltd., Montreal

Blue Printing.
The Electric Blue Print Co., Montreal.

Blow-Off Tanks.
Darling Bros., Ltd., Montreal.

Boilers.
Canada Foundry Co., Limited, Toronto.
Canada Machinery Agency, Montreal.
Goldie & McCulloch Co., Galt.
E. Leonard & Sons, London, Ont.
John McDougall Caledonian Iron Works, Montreal.
Manitoba Iron Works, Winnipeg.
Mechanics Supply Co., Quebec, Que.
Owen Sound Iron Works Co., Owen Sound
H. W. Petrie, Toronto.
Robb Engineering Co. Amherst, N S.
The Smart-Turner Mach. Co., Hamilton.
Waterous Engine Works Co., Brantford.
Williams & Wilson, Montreal.

Boiler Compounds.
Canada Chemical Mfg. Co., London, Ont.
Hall Engineering Works, Montreal.
Lake Erie Boiler Compound Co., Toronto

Bolt Cutters.
John Bertram & Sons Co., Dundas, Ont.
London Mach. Tool Co., Hamilton.
Mechanics Supply Co., Quebec, Que
National Machinery Co., Tiffin, Ohio.
Niles-Bement-Pond Co., New York.

Bolt and Nut Machinery.
John Bertram & Sons Co., Dundas, Ont.
Canada Machinery Agency, Montreal.
Rice Lewis & Son, Toronto.
London Mach. Tool Co., Hamilton.
National Machinery Co., Tiffin, Ohio.
Niles-Bement-Pond Co. New York.

Bolts and Nuts, Rivets.
Baxter, Paterson & Co., Montreal.
Mechanics Supply Co., Quebec, Que.

Boring & Drilling Machines
American Tool Works Co., Cincinnati.
B. F. Barnes Co., Rockford, Ill.
John Bertram & Sons Co., Dundas, Ont.
Canada Machinery Agency Montreal.
A. B. Jardine & Co., Hespeler, Ont.
London Mach. Tool Co., Hamilton.
Niles-Bement-Pond Co., New York.

Boring Machine, Upright.
American Tool Works Co., Cincinnati.
John Bertram & Sons Co., Dundas, Ont.
London Mach Tool Co., Hamilton.
Niles-Bement-Pond Co., New York.

Boring Machine, Wood.
Chicago Pneumatic Tool Co., Chicago.
London Mach. Tool Co., Hamilton.

Boring and Turning Mills.
American Tool Works Co., Cincinnati.
John Bertram & Sons Co., Dundas, Ont.
Gisholt Machine Co., Madison, Wis.
Canada Machinery Agency, Montreal.
Rice Lewis & Son, Toronto.
London Mach. Tool Co., Hamilton.
Niles-Bement-Pond Co., New York.
H. W. Petrie, Toronto.

Box Puller.
A. B. Jardine & Co., Hespeler, Ont.

Boxes, Steel Shop.
Cleveland Wire Spring Co, Cleveland.

Boxes, Tote.
Cleveland Wire Spring Co., Cleveland

Brass Foundry Equipment.
Ph. Bonvillain & E. Ronceray, Philadelphia
Detroit Foundry Supply Co., Detroit.
Dominion Foundry Supply Co., Montreal
Paxson, J. W., Co., Philadelphia

Brass Working Machinery.
Warner & Swasey Co., Cleveland, Ohio.

Brushes, Foundry and Core.
Buffalo Foundry Supply Co., Buffalo.
Detroit Foundry Supply Co, Windsor.
Dominion Foundry Supply Co., Montreal
Hamilton Facing Mill Co., Hamilton.
Mechanics Supply Co., Quebec, Que.
Paxson, J. W., Co., Philadelphia

Brushes, Steel.
Buffalo Foundry Supply Co., Buffalo.
Dominion Foundry Supply Co, Montreal
Paxson, J. W., Co., Philadelphia

Bulldozers.
John Bertram & Sons Co., Dundas, Ont.
London Mach. Tool Co., Hamilton, Ont

National Machinery Co., Tiffin, Ohio.
Niles-Bement-Pond Co., New York.

Cable Grease.
United States Graphite Co., Saginaw Mich.

Calipers.
Baxter, Patterson & Co., Montreal.
Frothingham & Workman, Ltd., Montreal
Rice Lewis & Son, Toronto.
Mechanics Supply Co., Quebec, Que.
John Millen & Son, Ltd, Montreal, Que.
L. S. Starrett & Co., Athol, Mass.
Williams & Wilson Montreal.

Carbon.
Dominion Foundry Supply Co., Montreal
United States Graphite Co., Saginaw, Mich.

Cars, Foundry.
Buffalo Foundry Supply Co., Buffalo.
Detroit Foundry Supply Co., Windsor
Dominion Foundry Supply Co., Montreal
Hamilton Facing Mill Co., Hamilton.
Paxson, J. W., Co., Philadelphia

Castings, Aluminum.
Lumen Bearing Co., Toronto

Castings, Brass.
Chadwick Bros., Hamilton.
Greey, Wm. & J. G. Toronto.
Hall Engineering Works, Montreal.
Kennedy, Wm., & Sons, Owen Sound.
Lumen Bearing Co., Toronto
Niagara Falls Machine & Foundry Co. Niagara Falls, Ont.
Owen Sound Iron Works Co., Owen Sound.
Robb Engineering Co, Amherst, N.S.

Castings, Grey Iron.
Allis-Chalmers-Bullock Montreal.
Greey, Wm. & J. G, Toronto
Hall Engineering Works, Montreal.
Kennedy, Wm., & Sons, Owen Sound
Laurie Engine & Machine Co., Montreal.
Manson Mfg. Co., Thorold, Ont
Maxwell, David, & Sons, St. Marys.
John McDougall Caledonian Iron Works Co., Montreal.
Niagara Falls Machine & Foundry Co., Niagara Falls, Ont
Owen Sound Iron Works Co., Owen Sound.
Robb Engineering Co., Amherst, N.S.
Smart-Turner Machine Co., Hamilton.

Castings, Phosphor Bronze.
Lumen Bearing Co., Toronto

Castings, Steel.
Kennedy, Wm., & Sons, Owen Sound.

Castings, Semi-Steel.
Robb Engineering Co., Amherst, N.S.

Cement Machinery.
Allis-Chalmers-Bullock, Limited, Montreal
Greey, Wm. & J. G., Toronto
Jeffrey Mfg. Co., Columbus, Ohio.
John McDougall Caledonian Iron Works, Co., Montreal
Owen Sound Iron Works Co., Owen Sound

Centreing Machines.
John Bertram & Sons Co., Dundas, Ont.
Jeffrey Mfg Co., Columbus, Ohio
London Mach. Tool Co., Hamilton, Ont.
Niles-Bement-Pond Co., New York.
Pratt & Whitney Co., Hartford, Conn.
Standard Bearings, Ltd., Niagara Falls

Centres, Planer.
American Tool Works Co., Cincinnati.

Centrifugal Pumps.
Manson Mfg. Co., Thorold, Ont
John McDougall Caledonian Iron Works Co, Montreal.
Pratt & Whitney Co., Hartford, Conn.

Centrifugal Pumps—Turbine Driven
Kerr Turbine Co., Wellsville, N.Y.

Chain, Crane and Dredge.
Frothingham & Workman, Ltd., Montreal

Chain Lubricants.
United States Graphite Co., Saginaw, Mich.

"MORSE" TOOLS

Drills, Reamers, Cutters, Chucks, Taps, Dies, Machines, Machinists' Tools, Arbors, Counter-bores, Countersinks, Gauges, Mandrels, Mills, Screw Plates, Sleeves, Sockets, Taper Pins, Wrenches.

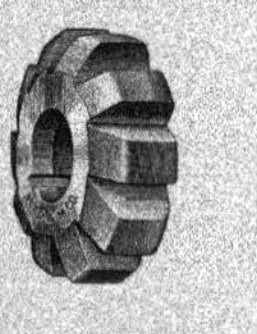

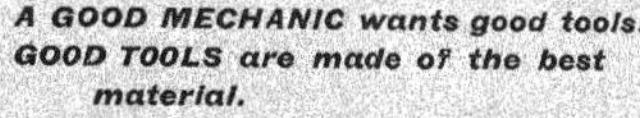

A GOOD MECHANIC wants good tools. GOOD TOOLS are made of the best material.

The BEST MATERIAL is given to capable workmen.

CAPABLE WORKMEN make honest tools.

Therefore MORSE TOOLS GIVE SATISFACTION.

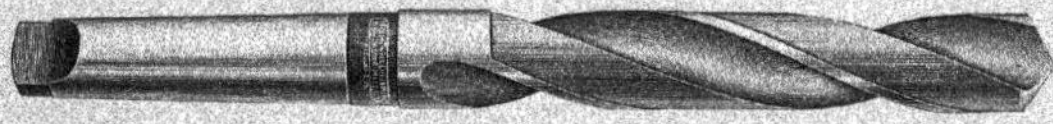

Morse Twist Drill and Machine Co.

NEW BEDFORD, MASS., U.S.A.

Chaplets.
Buffalo Foundry Supply Co., Buffalo.
Detroit Foundry Supply Co., Windsor.
Dominion Foundry Supply Co., Montreal
Hamilton Facing Mill Co., Hamilton.
Paxson, J. W., Co., Philadelphia

Charcoal.
Buffalo Foundry Supply Co. Buffalo.
Detroit Foundry Supply Co., Windsor.
Doggett, Stanley, New York
Dominion Foundry Supply Co., Montreal
Hamilton Facing Mill Co., Hamilton.
Paxson, J. W., Co., Philadelphia

Charcoal Facings.
Doggett, Stanley New York

Chemicals.
Baxter, Paterson & Co., Montreal.
Canada Chemical Co., London.

Chemists' Machinery.
Greey Wm & J. G., Toronto

Chemists, Industrial.
Detroit Testing Laboratory Detroit.

Chemists, Metallurgical.
Detroit Testing Laboratory, Detroit

Chemists, Mining.
Detroit Testing Laboratory, Detroit.

Chucks, Ring Grinding.
A. B. Jardine & Co., Hespeler, Ont.
Chicago Pneumatic Tool Co., Chicago.

Chucks, Drill and Lathe.
American Tool Works Co., Cincinnati.
Baxter, Paterson & Co., Montreal.
John Bertram & Sons Co., Dundas, Ont.
Canada Machinery Agency, Montreal.
Frothingham & Workman, Ltd., Montreal
Hamilton Tool Co., Hamilton, Ont.
Ker & Goodwin, Brantford.
A. B. Jardine & Co., Hespeler, Ont.
London Mach. Tool Co., Hamilton.
John Millen & Son, Ltd., Montreal.
Niles-Bement-Pond Co., New York.
H. W. Petrie, Toronto.
Rice Lewis & Son, Toronto.
Standard Tool Co., Cleveland.

Chucks, Planer.
American Tool Works Co., Cincinnati.
Canada Machinery Agency, Montreal.
Niles-Bement-Pond Co., New York.

Chucking Machines.
American Tool Works Co., Cincinnati.
Niles-Bement-Pond Co., New York.
H. W. Petrie, Toronto.
Warner & Swasey Co., Cleveland, Ohio

Circuit Breakers.
Allis-Chalmers-Bullock, Limited, Montreal
Canadian General Electric Co., Toronto.
Canadian Westinghouse Co., Hamilton.

Clippers, Bolt.
Frothingham & Workman, Ltd. Montreal
A. B. Jardine & Co., Hespeler, Ont.

Cloth and Wool Dryers.
Dominion Heating and Ventilating Co., Hespeler.
B. Greening Wire Co., Hamilton.
Sheldons Limited, Galt

Collectors, Pneumatic.
Sheldons Limited, Galt

Compressors, Air.
Allis-Chalmers-Bullock, Limited, Montrea
Canada Foundry Co., Limited, Toronto.
Canada Machinery Agency, Montreal.
Canadian Rand Drill Co., Montreal.
Canadian Westinghouse Co., Hamilton.
Chicago Pneumatic Tool Co., Chicago.
Darling Bros., Ltd., Montreal
Detroit Foundry Supply Co., Windsor.
John McDougall, Caledonian Iron Works Co., Montreal
H. W. Petrie, Toronto.
The Smart-Turner Mach. Co., Hamilton.
Hall Engineering Works, Montreal, Que.
London Mach. Tool Co., Hamilton.
Niles-Bement-Pond Co., New York.
H. W. Petrie, Toronto.
Pratt & Whitney Co., Hartford, Conn.
Williams & Wilson, Montreal.

Concentrating Plant.
Allis-Chalmers-Bullock, Montreal.
Greey, Wm. & J. G., Toronto.

Concrete Mixers.
Jeffrey M'f'g. Co., Columbus, Ohio.
Link-Belt Co., Philadelphia.

Condensers.
Canada Foundry Co., Limited, Toronto.
Canada Machinery Agency, Montreal.
Hall Engineering Works Montreal.
Smart-Turner-Machine Co., Hamilton.
Waterous Engine Co., Brantford.

Confectioners' Machinery.
Baxter, Paterson & Co., Montreal.
Greey, Wm & J. G., Toronto.

Consulting Engineers.
Hall Engineering Works Montreal.
Jules De Clercy, Montreal.

Miller Bros. & Toms, Montreal.
Roderick J. Parke, Toronto.
Plews & Trimingham, Montreal.
T. Pringle & Son, Montreal.
Somerville & Van Every, Hamilton
Standard Bearings, Ltd., Niagara Falls

Contractors.
Expanded Metal and Fireproofing Co. Toronto.
Hall Engineering Works Montreal.
Laurie Engine & Machine Co., Montreal
John McDougall Caledonian Iron Works Co., Montreal.
Miller Bros. & Toms, Montreal.
Robb Engineering Co. Amherst N.S.
The Smart-Turner Mach. Co., Hamilton.

Contractors' Plant.
Allis-Chalmers-Bullock, Montreal.
John McDougall, Caledonian Iron Works Co., Montreal.
Niagara Falls Machine & Foundry Co., Niagara Falls, Ont.

Contractors' Supplies.
Manson Mfg. Co., Thorold, Ont.

Controllers and Starters Electric Motor.
Allis-Chalmers-Bullock, Montreal.
Canadian General Electric Co., Toronto.
Canadian Westinghouse Co., Hamilton.
T. & H. Electric Co., Hamilton.

Conveying Systems.
Sheldons Limited, Galt

Converters, Steel.
Northern Engineering Works, Detroit.
Paxson, J. W., Co., Philadelphia

Conveyor Machinery.
Baxter, Paterson & Co., Montreal.
Greey Wm & J. G., Toronto
Jeffrey Mfg. Co., Columbus, Ohio.
Rice Lewis & Son, Toronto.
Link-Belt Co., Philadelphia.
John McDougall Caledonian Iron Works Co., Montreal.
Miller Bros. & Toms, Montreal.
Smart-Turner Machine Co., Hamilton.
Laurie Engine & Machine Co., Montreal.
Waterous Engine Works Co., Brantford.
Williams & Wilson, Montreal.

Coping Machines.
John Bertram & Sons Co., Dundas, Ont.
London Mach. Tool Co., Hamilton.
Niles-Bement-Pond Co., New York.

Core Compounds.
Buffalo Foundry Supply Co., Buffalo.
Detroit Foundry Supply Co., Windsor.
Dominion Foundry Supply Co., Montreal
Hamilton Facing Mill Co., Hamilton.
Paxson, J. W., Co., Philadelphia
United States Graphite Co., Saginaw, Mich.

Core-Making Machines.
Paxson, J. W., Co., Philadelphia

Core Ovens.
Detroit Foundry Supply Co., Windsor.
Dominion Foundry Supply Co., Montreal
Hamilton Facing Mill Co., Hamilton.
Paxson, J. W., Co., Philadelphia
Sheldons Limited, Galt

Core Oven Bricks.
Buffalo Foundry Supply Co., Buffalo.
Detroit Foundry Supply Co., Windsor.
Dominion Foundry Supply Co., Montreal
Hamilton Facing Mill Co., Hamilton.
Paxson, J. W., Co., Philadelphia

Core Sand Cleaners.
Sly, W. W., Mfg. Co., Cleveland

Core Wash.
Buffalo Foundry Supply Co., Buffalo.
Detroit Foundry Supply Co., Windsor.
Dominion Foundry Supply Co., Montreal
Hamilton Facing Mill Co., Hamilton.
Paxson, Co., J. W. Philadelphia

Couplings.
Owen Sound Iron Works Co., Owen Sound

Cranes, Electric and Hand Power.
Canada Foundry Co., Limited Toronto
Canadian Pilling Co., Montreal
Dominion Foundry Supply Co., Montreal
Hamilton Facing Mill Co., Hamilton.
Link-Belt Co., Philadelphia.
John McDougall, Caledonian Iron Works Co., Montreal.
Miller Bros. & Toms, Montreal.
Niles-Bement-Pond Co., New York.
Northern Engineering Works, Detroit
Owen Sound Iron Works Co., Owen Sound
Paxson, J. W., Co., Philadelphia
Smart-Turner-Machine Co., Hamilton.

Crank Pin.
Sight Feed Oil Pump Co., Milwaukee, Wis.

Crankshafts.
The Canada Forge Co., Welland.
St. Clair Bros., Galt

Crabs.
Frothingham & Workman, Ltd., Montreal

Crank Pin Turning Machine
London Mach. Tool Co., Hamilton.
Niles-Bement-Pond Co., New York.

Cross Head Pin.
Sight Feed Oil Pump Co., Milwaukee, Wis.

Crucibles.
Buffalo Foundry Supply Co., Buffalo
Detroit Foundry Supply Co., Windsor
Dominion Foundry Supply Co., Montreal
Hamilton Facing Mill Co., Hamilton.
J. W. Paxson Co., Philadelphia.

Crucible Caps
Dominion Foundry Supply Co Montreal
Hamilton Facing Mill Co., Hamilton.
Paxson, J. W., Co., Philadelphia

Crushers, Rock or Ore.
Allis-Chalmers-Bullock, Montreal.
Jeffrey Mfg. Co., Columbus, Ohio.

Cup Grease.
Mechanics Supply Co., Quebec

Cupolas.
Buffalo Foundry Supply Co., Buffalo.
Detroit Foundry Supply Co., Windsor
Dominion Foundry Supply Co., Montreal
Hamilton Facing Mill Co., Hamilton.
Northern Engineering Works, Detroit
J. W. Paxson Co., Philadelphia.
Sheldons Limited, Galt.

Cupola Blast Gauges.
Dominion Foundry Supply Co., Montreal
Paxson, J. W., Co., Philadelphia
Sheldons Limited, Galt

Cupola Blocks.
Detroit Foundry Supply Co., Detroit.
Hamilton Facing Mill Co., Hamilton
Northern Engineering Works, Detroit
Ontario Lime Association Toronto
Paxson, J. W., Co., Philadelphia
Toronto Pottery Co., Toronto

Cupola Blowers.
Buffalo Foundry Supply Co., Buffalo.
Canada Machinery Agency, Montreal.
Detroit Foundry Supply Co., Windsor
Dominion Foundry Supply Co Montreal
Dominion Heating and Ventilating Co., Hespeler
Hamilton Facing Mill Co., Hamilton.
Northern Engineering Works, Detroit
Paxson J. W., Co., Philadelphia
Sheldon's Limited, Galt.

Cutters, Flue
Chicago Pneumatic Tool Co., Chicago.
J. W. Paxson Co., Philadelphia.

Cutters, Milling.
Becker, Brainard Milling Machine Co. Hyde Park, Mass.
Frothingham & Workman Ltd., Montreal
Hamilton Tool Co., Hamilton, Ont.
Pratt & Whitney Co., Hartford, Conn.
Standard Tool Co., Cleveland.

Cutting-off Machines.
John Bertram & Sons Co., Dundas, Ont.
Canada Machinery Agency, Montreal.
London Mach. Tool Co., Hamilton.
H. W. Petrie, Toronto.
Pratt & Whitney Co., Hartford, Conn.

Cutting-off Tools.
Armstrong Bros. Tool Co., Chicago.
Baxter Paterson & Co., Montreal.
London Mach. Tool Co., Hamilton.
Mechanics Supply Co. Quebec Que.
H. W. Petrie, Toronto.
Pratt & Whitney, Hartford, Conn.
Rice Lewis & Son, Toronto.
L. S. Starrett Co., Athol, Mass.

Damper Regulators.
Darling Bros., Ltd., Montreal

Dies
Globe Machine & Stamping Co., Cleveland, Ohio.
Woelfle Bros., Berlin, Ont.

Die Stocks
Curtis & Curtis Co., Bridgeport, Conn.
Hart Manufacturing Co., Cleveland, Ohio
Loew Mfg. Co., Cleveland, Ohio
Mechanics Supply Co., Quebec

Dies, Opening
W. H. Banfield & Sons, Toronto
Globe Machine & Stamping Co., Cleveland Ohio.
Pratt & Whitney Co., Hartford, Conn.
Woelfle Bros Berlin, Ont.

Dies, Sheet Metal.
W. H. Banfield & Sons Toronto.
Globe Machine & Stamping Co., Cleveland, Ohio

Dies, Threading.
Frothingham & Workman, Ltd., Montreal
Hart Mfg. Co., Cleveland
Mechanics Supply Co., Quebec, Que.
John Millen & Son Ltd., Montreal.

Draft, Mechanical.
W. H. Banfield & Sons, Toronto.
Butterfield & Co., Rock Island, Que.

A. B. Jardine & Co., Hespeler
Mechanics Supply Co., Quebec, Que.
Pratt & Whitney Co., Hartford, Conn.
Sheldon's Limited, Galt.

Drawing Instruments.
Rice Lewis & Son, Toronto.
Mechanics Supply Co., Quebec, Que.

Drawing Supplies.
The Electric Blue Print Co., Montreal

Drawn Steel, Cold.
Baxter, Paterson & Co., Montreal.
Greey Wm. & J. G., Toronto
Miller Bros. & Toms, Montreal.
Union Drawn Steel Co., Hamilton.

Drilling Machines, Arch Bar.
John Bertram & Sons Co., Dundas, Ont.
London Mach. Tool Co., Hamilton
Niles-Bement-Pond Co., New York.

Drilling Machines, Boiler.
American Tool Works Co., Cincinnati.
John Bertram & Sons Co., Dundas, Ont.
Bickford Drill and Tool Co., Cincinnati.
The Canadian Fairbanks Co., Montreal.
A. B. Jardine & Co., Hespeler, Ont.
London Mach. Tool Co., Hamilton, Ont.
Niles-Bement-Pond Co., New York.
H. W. Petrie, Toronto.
Williams & Wilson, Montreal

Drilling Machines Connecting Rod.
John Bertram & Sons Co., Dundas, Ont.
London Mach. Tool Co., Hamilton.
Niles-Bement-Pond Co. New York.

Drilling Machines, Locomotive Frame.
American Tool Works Co., Cincinnati.
B. F. Barnes Co., Rockford, Ill.
John Bertram & Sons Co., Dundas, Ont.
London Mach. Tool Co., Hamilton, Ont.
Niles-Bement-Pond Co., New York.

Drilling Machines, Multiple Spindle.
American Tool Works Co., Cincinnati.
B. F. Barnes Co., Rockford, Ill.
Baxter, Paterson & Co., Montreal.
John Bertram & Sons Co., Dundas, Ont.
Bickford Drill & Tool Co., Cincinnati.
Canada Machinery Agency Montreal.
London Mach. Tool Co., Hamilton, Ont.
Niles-Bement-Pond Co., New York.
H. W. Petrie, Toronto.
Williams & Wilson, Montreal.

Drilling Machines, Pneumatic
Canada Machinery Agency, Montreal.

Drilling Machines, Portable
Baxter, Paterson & Co., Montreal.
A. B. Jardine & Co., Hespeler, Ont.
Mechanics Supply Co., Quebec, Que.
Niles-Bement-Pond Co., New York.

Drilling Machines, Radial.
American Tool Works Co., Cincinnati.
Baxter, Paterson & Co., Montreal
John Bertram & Sons Co., Dundas, Ont.
Bickford Tool & Drill Co., Cincinnati.
The Canadian Fairbanks Co., Montreal.
London Mach. Tool Co., Hamilton.
Mechanics Supply Co., Quebec, Que.
Niles-Bement-Pond Co., New York.
H. W. Petrie, Toronto.
Williams & Wilson, Montreal.

Drilling Machines, Suspension.
John Bertram & Sons Co., Dundas, Ont.
Canada Machinery Agency Montreal.
London Mach. Tool Co., Hamilton.
Niles-Bement-Pond Co. New York

Drilling Machines, Turret.
John Bertram & Sons Co., Dundas, Ont.
London Mach. Tool Co., Hamilton.
Niles-Bement-Pond Co., New York.

Drilling Machines, Upright.
American Tool Works Co., Cincinnati.
B. F. Barnes Co., Rockford, Ill.
Baxter, Paterson & Co., Montreal.
John Bertram & Sons Co., Dundas, Ont
Dwight Slate Machine Co., Hartford, Conn.
Hamilton Tool Co., Hamilton, Ont.
A. B. Jardine & Co., Hespeler, Ont.
Rice Lewis & Son, Toronto.
London Mach. Tool Co., Hamilton.
Mechanics Supply Co., Quebec, Que.
Niles-Bement-Pond Co., New York.
W. H. Petrie, Toronto.
Williams & Wilson, Montreal.

Drills, Bench.
B. F. Barnes Co., Rockford, Ill.
Baxter, Paterson & Co., Montreal.
Hamilton Tool Co., Hamilton, Ont.
London Mach. Tool Co., Hamilton.
Mechanics Supply Co. Quebec Que.
Pratt & Whitney Co., Hartford, Conn.

Drills, Blacksmith.
Canada Machinery Agency, Montreal.
Frothingham & Workman, Ltd., Montreal
A. B. Jardine & Co., Hespeler, Ont.
London Mach. Tool Co., Hamilton.
Mechanics Supply Co., Quebec, Que.
Standard Tool Co., Cleveland.

Drills, Centre.

Mechanics Supply Co. [illegible], Que.
Pratt & Whitney Co., Hartford, Conn.
Standard Tool Co., Cleveland, O.
L. S. Starrett Co., Athol, Mass.

Drills, Electric

B. F. Barnes Co., Rockford, Ill.
Baxter, Paterson & Co., Montreal.
Canadian Pilling Co., Montreal
Chicago Pneumatic Tool Co., Chicago.
Niles-Bement-Pond Co., New York.

Drills, Gang.

American Tool Works Co., Cincinnati.
B. F. Barnes Co., Rockford, Ill.
John Bertram & Sons Co., Dundas, Ont.
Pratt & Whitney Co., Hartford, Conn.

Drills, High Speed.

Wm. Abbott, Montreal.
American Tool Works Co., Cincinnati.
B. F. Barnes Co., Rockford, Ill.
Frothingham & Workman, Ltd., Montreal
Pratt & Whitney Co., Hartford, Conn.
Standard Tool Co., Cleveland, O.

Drills, Hand.

A. B. Jardine & Co., Hespeler, Ont.

Drills, Horizontal.

B. F. Barnes Co., Rockford, Ill.
John Bertram & Sons Co., Dundas, Ont.
Canada Machinery Agency, Montreal.
London Mach. Tool Co., Hamilton.
Niles-Bement-Pond Co., New York.

Drills, Pneumatic.

Canada Machinery Agency, Montreal.
Chicago Pneumatic Tool Co., Chicago.
Independent Pneumatic Tool Co., Chicago, New York
Niles-Bement-Pond Co., New York.

Drills, Radial.

American Tool Works Co., Cincinnati.
John Bertram & Sons Co., Dundas, Ont.
Bickford Drill & Tool Co., Cincinnati
London Mach. Tool Co., Hamilton, Ont.
Niles-Bement-Pond Co., New York.

Drills, Ratchet.

Armstrong Bros. Tool Co., Chicago.
Frothingham & Workman, Ltd., Montreal
A. B. Jardine & Co., Hespeler.
Mechanics Supply Co., Quebec, Que.
Pratt & Whitney Co., Hartford, Conn.
Standard Tool Co., Cleveland.

Drills, Rock.

Allis-Chalmers-Bullock, Montreal.
Canadian Rand Drill Co., Montreal.
Chicago Pneumatic Tool Co., Chicago.
Jaffrey Mfg. Co., Columbus, Ohio.

Drills, Sensitive.

American Tool Works Co., Cincinnati.
B. F. Barnes Co., Rockford, Ill.
Canada Machinery Agency, Montreal.
Dwight Slate Machine Co., Hartford, Conn.
Niles-Bement-Pond Co., New York.

Drills, Twist.

Baxter, Paterson & Co., Montreal.
Chicago Pneumatic Tool Co., Chicago
Frothingham & Workman, Ltd., Montreal
Alex. Gibb, Montreal.
A. B. Jardine & Co., Hespeler, Ont.
Mechanics Supply Co., Quebec Que.
John Millen & Son, Ltd., Montreal.
Morse Twist Drill and Machine Co., New Bedford, Mass.
Pratt & Whitney Co., Hartford, Conn.
Standard Tool Co., Cleveland.

Drying Apparatus of all Kinds.

Dominion Heating & Ventilating Co., Hespeler
Greey, Wm. & J. G., Toronto
Sheldons Limited, Galt

Dry Kiln Equipment.

Sheldons Limited, Galt

Dry Sand and Loam Facing.

Buffalo Foundry Supply Co. Buffalo.
Dominion Foundry Supply Co., Montreal
Hamilton Facing Mill Co., Hamilton.
Paxson, J. W., Co., Philadelphia

Dump Cars.

Canada Foundry Co., Limited, Toronto
Dominion Foundry Supply Co., Montreal
Greey, Wm. & J. G., Toronto
Hamilton Facing Mill Co., Hamilton.
John McDougall, Caledonian Iron Works Co., Montreal.
Niles-Bement-Pond Co., New York.
Standard Bearings, Ltd., Niagara Falls.
Link-Belt Eng. Co., Philadelphia.
John McDougall Caledonian Iron Works Co., Montreal.
Miller Bros. & Toms, Montreal.
Owen Sound Iron Works Co., Owen Sound
Paxson, J. W., Co., Philadelphia
Waterous Engine Co., Brantford.

Dust Arresters.

Sly, W. W., Mfg. Co., Cleveland

Dust Separators.

Greey, Wm. & J. G., Toronto
Dominion Heating and Ventilating Co., Hespeler.
Paxson, J. W., Co., Philadelphia
Sheldon's Limited, Galt.

Dynamos.

Allis-Chalmers-Bullock, Montreal.
Canadian General Electric Co., Toronto.
Canadian Westinghouse Co., Hamilton.
Consolidated Electric Co., Toronto
Electrical Machinery Co., Toronto.
Gas & Electric Power Co., Toronto
Hall Engineering Works, Montreal, Que.
Mechanics Supply Co., Quebec, Que.
John Millen & Son, Ltd., Montreal.
Packard Electric Co., St. Catharines.
H. W. Petrie, Toronto.
T. & H. Electric Co., Hamilton.

Dynamos—Turbine Driven.

Kerr-Turbine Co., Wellsville, N.Y.

Economizer, Fuel.

Dominion Heating & Ventilating Co. Hespeler
Standard Bearings, Ltd., Niagara Falls.

Electrical Instruments.

Canadian Westinghouse Co., Hamilton.
Mechanics Supply Co., Quebec, Que.

Electrical Steel.

Baxter, Patterson & Co., Montreal.

Electrical Supplies.

Canadian General Electric Co., Toronto.
Canadian Westinghouse Co., Hamilton.
London Mach. Tool Co., Hamilton, Ont.
Mechanics Supply Co., Quebec, Que
John Millen & Son, Ltd., Montreal.
Packard Electric Co., St. Catharines.
T. & H. Electric Co., Hamilton.

Electrical Repairs

Canadian Westinghouse Co., Hamilton.
T. & H. Electric Co., Hamilton.

Elevator Buckets.

Greey, Wm. & J. G., Toronto
Jeffrey Mfg. Co., Columbus, Ohio.

Elevators, Foundry.

Dominion Foundry Supply Co., Montreal
Northern Engineering Works, Detroit

Elevators—For any Service.

Darling Bros. Ltd., Montreal

Emery and Emery Wheels.

Baxter, Paterson & Co., Montreal.
Dominion Foundry Supply Co., Montreal
Frothingham & Workman Ltd., Montreal
Hamilton Facing Mill Co., Hamilton.
Paxson J. W. Co. Philadelphia

Emery Wheel Dressers.

Baxter Paterson & Co., Montreal.
Canada Machinery Agency, Montreal.
Desmond-Stephan Mfg. Co., Urbana, Ohio
Dominion Foundry Supply Co., Montreal
Frothingham & Workman Ltd., Montreal
Hamilton Facing Mill Co., Hamilton.
International Specialty Co., Detroit.
Mechanics Supply Co., Quebec, Que.
John Millen & Son, Ltd., Montreal.
H. W. Petrie, Toronto.
Paxson, J. W., Co., Philadelphia
Standard Tool Co., Cleveland.

Engineers and Contractors.

Canada Foundry Co., Limited, Toronto.
Darling Bros., Ltd., Montreal
Greey, Wm. & J. G., Toronto
Hall Engineering Works, Montreal.
Laurie Engine & Machine Co., Montreal.
Link-Belt Co., Philadelphia.
John McDougall, Caledonian Iron Works Co., Montreal.
Miller Bros. & Toms Montreal.
Robb Engineering Co., Amherst, N S.
The Smart-Turner Mach. Co., Hamilton.

Engineers' Supplies.

Baxter, Paterson & Co., Montreal.
Frothingham & Workman, Ltd., Montreal
Greey, Wm. & J. G., Toronto
Hall Engineering Works, Montreal.
Mechanics Supply Co., Quebec, Que.
Rice Lewis & Son, Toronto.

Engines, Gas and Gasoline.

Baxter, Paterson & Co., Montreal.
Canada Foundry Co., Toronto.
Canada Machinery Agency, Montreal.
The Canadian Fairbanks Co., Montreal.
Canadian McVicker Engine Co., Galt.
Gilson Mfg. Co., Guelph
The Goldie & McCulloch Co., Galt, Ont.
Rice Lewis & Son, Toronto.
Ontario Wind Engine & Pump Co. Toronto
H. W. Petrie, Toronto.
The Smart-Turner Mach. Co., Hamilton
Standard Bearings, Ltd., Niagara Falls

Engines, Steam.

Allis-Chalmers-Bullock, Montreal
Belliss & Marcom, Birmingham, Eng.
Canada Machinery Agency, Montreal.
The Goldie & McCulloch Co., Galt, Ont.
E. Leonard & Sons, London, Ont.
Rice Lewis & Son, Toronto.
Laurie Engine & Machine Co., Montreal.
Manson Mfg. Co., Thorold Ont
John McDougall Caledonian Iron Works Montreal.
Miller Bros. & Toms, Montreal.
Robb Engineering Co., Amherst, N.S.
Sheldons Limited, Galt.
The Smart-Turner Mach. Co., Hamilton.
Waterous Engine Works Co., Brantford.

Exhaust Heads.

Darling Bros., Ltd., Montreal.
Dominion Heating & Ventilating Co., Hespeler
Sheldons Limited, Galt, Ont.

Expanded Metal.

Expanded Metal and Fireproofing Co. Toronto

Expanders.

A. B. Jardine & Co., Hespeler, Ont.

Fans, Electric.

Canadian General Electric Co., Toronto.
Canadian Westinghouse Co., Hamilton.
Dominion Heating & Ventilating Co. Hespeler
Mechanics Supply Co., Quebec, Que.
Sheldons Limited Galt, Ont.
The Smart-Turner Mach. Co., Hamilton.

Fans, Exhaust.

Canadian Buffalo Forge Co., Montreal.
Detroit Foundry Supply Co., Windsor.
Dominion Foundry Supply Co., Montreal
Dominion Heating & Ventilating Co., Hespeler.
Greey, Wm. & J. G., Toronto
Hamilton Facing Mill Co., Hamilton.
Paxson, J. W., Co., Philadelphia
Sheldons Limited, Galt.
B. F. Sturtevant Co., Hyde Park, Mass.

Feed Water Heaters.

Darling Bros., Montreal
Laurie Engine & Machine Co., Montreal
John McDougall, Caledonian Iron Works Co., Montreal.
The Smart-Turner Mach. Co., Hamilton

Files and Rasps.

Baxter, Paterson & Co., Montreal.
Frothingham & Workman, Ltd., Montreal
Mechanics Supply Co., Quebec, Que.
John Millen & Son, Ltd., Montreal.
Rice Lewis & Son, Toronto.
Nicholson File Co., Port Hope
H. W. Petrie, Toronto

Fillet, Pattern.

Baxter, Paterson & Co., Montreal.
Buffalo Foundry Supply Co., Buffalo.
Detroit Foundry Supply Co., Windsor.
Dominion Foundry Supply Co., Montreal
Hamilton Facing Mill Co., Hamilton.

Fire Apparatus.

Waterous Engine Works Co., Brantford.

Fire Brick and Clay.

Baxter Paterson & Co., Montreal.
Buffalo Foundry Supply Co., Buffalo.
Detroit Foundry Supply Co., Windsor.
Dominion Foundry Supply Co., Montreal
Hamilton Facing Mill Co., Hamilton.
Ontario Lime Association Toronto
J. W. Paxson Co., Philadelphia.
Toronto Pottery Co., Toronto

Flash Lights.

Berlin Electrical Mfg. Co., Toronto.
Mechanics Supply Co., Quebec, Que.

Flour Mill Machinery.

Allis-Chalmers-Bullock, Montreal.
Greey, Wm. & J. G., Toronto
The Goldie & McCulloch Co., Galt, Ont.
John McDougall, Caledonian Iron Works Co., Montreal.
Miller Bros. & Toms, Montreal.

Forges.

Canada Foundry Co., Limited, Toronto.
Frothingham & Workman, Ltd., Montreal
Hamilton Facing Mill Co., Hamilton.
Mechanics Supply Co., Quebec, Que.
H. W. Petrie, Toronto.
Sheldons Limited, Galt, Ont.

Forgings, Drop.

John McDougall, Caledonian Iron Work Co., Montreal.
H. W. Petrie, Toronto.
St. Clair Bros., Galt

Forgings, Light & Heavy.

The Canada Forge Co., Welland.
Hamilton Steel & Iron Co., Hamilton

Forging Machinery.

John Bertram & Sons Co., Dundas, Ont.
London Mach. Tool Co., Hamilton, Ont
National Machinery Co., Tiffin, Ohio
Niles-Bement-Pond Co., New York.

Founders.

Greey, Wm. & J. G., Toronto
John McDougall, Caledonian Iron Works Co., Montreal.
Niagara Falls Machine & Foundry Co., Niagara Falls, Ont.
Manson Mfg. Co., Thorold, Ont.
Maxwell, David & Sons, St. Marys
The Smart-Turner Mach. Co., Hamilton.

Foundry Coke.

Baird & West, Detroit

Foundry Equipment.

Ph. Bonvillain & E. Ronceray, Philadelphia
Detroit Foundry Supply Co., Windsor.
Dominion Foundry Supply Co., Montreal
Hamilton Facing Mill Co., Hamilton
Hanna Engineering Works, Chicago.
Northern Engineering Works, Detroit
Paxson, J. W., Co., Philadelphia

Foundry Parting.

Doggett, Stanley, New York
Dominion Foundry Supply Co., Montreal
Partomol Co., New York
Foundry Specialty Co., Cincinnati
Stanley Doggett, New York

Foundry Facings.

Buffalo Foundry Supply Co., Buffalo.
Detroit Foundry Supply Co., Windsor.
Doggett Stanley, New York
Dominion Foundry Supply Co., Montreal
Hamilton Facing Mill Co., Hamilton.
J. W. Paxson Co., Philadelphia, Pa.
United States Graphite Co., Saginaw, Mich.

Friction Clutch Pulleys, etc.

The Goldie & McCulloch Co., Galt.
Greey, Wm. & J. G., Toronto
Link-Belt Co., Philadelphia.
Miller Bros. & Toms, Montreal.

Furnaces.

Detroit Foundry Supply Co., Windsor.
Dominion Foundry Supply Co., Montreal
Hamilton Facing Mill Co., Hamilton.
Paxson, J. W., Philadelphia Co.

Galvanizing

Canada Metal Co., Toronto.
Ontario Wind Engine & Pump Co., Toronto

Gas Blowers and Exhausters.

Dominion Heating & Ventilating Co. Hespeler
Sheldons Limited, Galt.

Gas Plants, Suction.

Baxter Paterson & Co., Montreal
Colonial Engineering Co., Montreal
Standard Bearings, Ltd., Niagara Falls
Williams & Wilson, Montreal

Gauges, Standard.

Mechanics Supply Co., Quebec, Que.
Pratt & Whitney Co., Hartford, Conn.

Gearing.

Eberhardt Bros. Machine Co., Newark, N.J.
Greey, Wm. & J. G. Toronto

Gear Cutting Machinery.

Baxter, Patterson & Co., Montreal
Becker-Brainard Milling Mach. Co., Hyde Park, Mass.
Bickford Drill & Tool Co., Cincinnati.
Dwight Slate Machine Co., Hartford, Conn.
Eberhardt Bros. Machine Co., Newark, N.J.
Greey, Wm. & J. G., Toronto
Kennedy, Wm., & Sons, Owen Sound
London Mach. Tool Co., Hamilton.
Niles-Bement-Pond Co., New York.
H. W. Petrie, Toronto.
Pratt & Whitney Co., Hartford, Conn.
Williams & Wilson, Montreal.

Gears, Angle.

Chicago Pneumatic Tool Co., Chicago.
Greey, Wm. & J. G., Toronto
Laurie Engine & Machine Co., Montreal.
John McDougall, Caledonian Iron Works Co., Montreal.
Miller Bros. & Toms, Montreal.
Waterous Engine Co., Brantford.

Gears, Cut.

Kennedy, Wm., & Sons, Owen Sound

Gear Grease.

United States Graphite Co., Saginaw Mich.

Gears, Iron.

Greey, Wm. & J. G., Toronto
Kennedy, Wm., & Sons, Owen Sound

Gears, Mortise.

Greey, Wm. & J. G., Toronto
Kennedy, Wm., & Sons, Owen Sound

Gears, Reducing.

Brown, David & Sons, Huddersfield, Eng
Chicago Pneumatic Tool Co., Chicago.
Greey, Wm. & J. G., Toronto
John McDougall, Caledonian Iron Works Co., Montreal.
Miller Bros. & Toms, Montreal.

Generators, Electric.
Allis-Chalmers-Bullock, Limited, Montreal
Canadian General Electric Co., Toronto
Canadian Westinghouse Co., Hamilton.
Hall Engineering Works, Montreal.
H. W. Petrie, Toronto.
Toronto & Hamilton Electric Co. Hamilton.

Generators, Gas.
H. W. Petrie, Toronto.

Graphite Paints.
P. D. Dods & Co., Montreal.
United States Graphite Co., Saginaw, Mich.

Graphite.
Detroit Foundry Supply Co., Windsor.
Doggett, Stanley, New York
Dominion Foundry Supply Co., Montreal
Hamilton Facing Mill Co., Hamilton.
Mechanics Supply Co. Quebec, Que.
Paxson, J. W., Co., Philadelphia
United States Graphite Co., Saginaw, Mich.

Grinders, Automatic Knife.
W. H. Banfield & Son, Toronto.

Grinders, Centre.
Niles-Bement-Pond Co., New York.
H. W. Petrie, Toronto.

Grinders, Cutter.
Becker-Brainard Milling Mach. Co., Hyde Park, Mass.
John Millen & Son, Ltd., Montreal.
Pratt & Whitney Co., Hartford, Conn.

Grinders, Tool.
Armstrong Bros. Tool Co., Chicago.
B. F. Barnes Co., Rockford, Ill.
H. W. Petrie, Toronto.
Williams & Wilson, Montreal.

Grinding Machines.
The Canadian Fairbanks, Montreal.
Rice Lewis & Son, Toronto.
Niles-Bement-Pond Co., New York.
Paxson, J. W., Co., Philadelphia
H. W. Petrie, Toronto.

Grinding & Polishing Machines
The Canadian Fairbanks Co., Montreal.
Greey, Wm. & J. G., Toronto
Independent Pneumatic Tool Co., Chicago, Ill.
John Millen & Son Ltd., Montreal.
Miller Bros. & Toms, Montreal.
Niles-Bement-Pond Co., New York.
H. W. Petrie, Toronto.
Pendrith Machinery Co., Toronto.
Standard Bearings, Ltd., Niagara Falls

Grinding Wheels
Carborundum Co., Niagara Falls
Norton Company, Worcester, Mass.

Hack Saws.
Baxter, Paterson & Co. Montreal.
Canada Machinery Agency, Montreal.
The Canadian Fairbanks Co., Montreal.
Frothingham & Workman, Ltd Montreal
Mechanics Supply Co., Quebec, Que.
John Millen & Son, Ltd., Montreal.
Niles-Bement-Pond Co., New York.
H. W. Petrie, Toronto.
Williams & Wilson, Montreal.

Hack Saw Frames.
Mechanics Supply Co., Quebec, Que.

Hammers, Drop.
London Mach. Tool Co., Hamilton, Ont.
Miller Bros. & Toms Montreal.
Niles-Bement-Pond Co., New York.

Hammers, Power.
Rice Lewis & Son, Toronto.

Hammers, Steam.
John Bertram & Sons Co. Dundas, Ont.
London Mach. Tool Co., Hamilton, Ont.
Niles-Bement-Pond Co., New York.

Hangers.
The Goldie & McCulloch Co., Galt.
Greey, Wm. & J. G., Toronto
Kennedy, Wm., & Sons, Owen Sound
Miller Bros. & Toms, Montreal.
Owen Sound Iron Works Co., Owen Sound
The Smart-Turner Mach. Co., Hamilton.
Waterous Engine Co., Brantford.
Standard Bearings, Ltd., Niagara Falls

Hardware Specialties.
Mechanics Supply Co., Quebec, Que.

Heating Apparatus.
Darling Bros., Ltd., Montreal
Dominion Heating & Ventilating Co., Hespeler
Mechanics Supply Co., Quebec, Que.
Sheldons Limited, Galt.

Hoisting and Conveying Machinery.
Allis-Chalmers-Bullock, Limited, Montreal
Greey, Wm & J. G., Toronto
Link-Belt Co., Philadelphia.
Miller Bros. & Toms, Montreal.
Niles-Bement-Pond Co., New York.
Northern Engineering Works, Detroit
The Smart-Turner Mach. Co., Hamilton.
Waterous Engine Co., Brantford.

Hoists, Chain and Rope.
Frothingham & Workman Ltd., Montreal

Hoists, Electric.
Northern Engineering Works, Detroit

Hoists, Pneumatic.
Canadian Rand Drill Co., Montreal.
Chicago Pneumatic Tool Co., Chicago.
Dominion Foundry Supply Co., Montreal
Hamilton Facing Mill Co., Hamilton.
Northern Engineering Works, Detroit

Hoists, Portable & Stationery.
Canadian Pilling Co., Montreal

Hose.
Baxter, Paterson & Co., Montreal
Mechanics Supply Co., Quebec, Que.

Hose, Air.
Canadian Rand Drill Co., Montreal.
Canadian Westinghouse Co., Hamilton.
Chicago Pneumatic Tool Co., Chicago.
Paxson, J. W., Co., Philadelphia

Hose Couplings.
Canadian Rand Drill Co., Montreal.
Canadian Westinghouse Co., Hamilton.
Chicago Pneumatic Tool Co., Chicago.
Mechanics Supply Co., Quebec, Que.
Paxson, J. W., Co., Philadelphia

Hose, Steam.
Allis-Chalmers-Bullock, Montreal.
Canadian Rand Drill Co., Montreal.
Canadian Westinghouse Co., Hamilton.
Mechanics Supply Co., Quebec, Que.
Paxson, J. W. Co., Philadelphia

Hydraulic Accumulators.
Ph. Bonvillain & E. Ronceray, Philadelphia
Niles-Bement-Pond Co., New York.
Perrin, Wm. R., Co. Toronto.
The Smart-Turner Mach. Co., Hamilton.

Hydraulic Machinery.
Allis-Chalmers-Bullock Montreal
Barber, Chas., & Sons, Meaford

India Oil Stones.
Norton Company, Worcester, Mass.

Indicators, Speed.
Mechanics Supply Co. Quebec, Que.
L. S. Starrett Co., Athol, Mass.

Injectors.
Canada Foundry Co., Toronto.
The Canadian Fairbanks Co., Montreal.
Desmond-Stephan Mfg. Co., Urbana, Ohio
Frothingham & Workman Ltd., Montreal
Mechanics Supply Co. Quebec, Que.
Rice Lewis & Son, Toronto.
Penberthy Injector Co., Windsor, Ont.

Iron and Steel.
Frothingham & Workman, Ltd., Montreal

Iron and Steel Bars and Bands.
Hamilton Steel & Iron Co., Hamilton

Iron Cements.
Buffalo Foundry Supply Co., Buffalo.
Detroit Foundry Supply Co., Windsor.
Dominion Foundry Supply Co. Montreal
Hamilton Facing Mill Co., Hamilton.
Paxson, J. W., Co., Philadelphia

Iron Filler.
Buffalo Foundry Supply Co., Buffalo.
Detroit Foundry Supply Co., Windsor.
Dominion Foundry Supply Co. Montreal
Hamilton Facing Mill Co., Hamilton.
Paxson, J. W., Co., Philadelphia

Jacks.
Frothingham & Workman, Ltd., Montreal
Norton, A. O., Coaticook, Que.

Kegs, Steel Shop.
Cleveland Wire Spring Co., Cleveland

Key-Seating Machines.
B. F. Barnes Co., Rockford, Ill.
Niles-Bement-Pond Co., New York.

Lace Leather.
Baxter, Paterson & Co., Montreal.

Ladles, Foundry.
Dominion Foundry Supply Co., Montreal
Frothingham & Workman Ltd., Montreal
Northern Engineering Works, Detroit

Lamps, Arc and Incandescent.
Canadian General Electric Co., Toronto.
Canadian Westinghouse Co., Hamilton.
The Packard Electric Co., St. Catharines.

Lathe Dogs.
Armstrong Bros., Chicago
Baxter, Paterson & Co., Montreal.
Pratt & Whitney Co., Hartford, Conn.

Lathes, Engine.
American Tool Work Co., Cincinnati.
B. F. Barnes Co., Rockford, Ill.
Baxter, Paterson & Co., Montreal.
John Bertram & Sons Co., Dundas, Ont.
Canada Machinery Agency, Montreal.
The Canadian Fairbanks Co., Montreal.
London Mach. Tool Co., Hamilton, Ont.
Niles-Bement-Pond Co., New York.
H. W. Petrie, Toronto.
Pratt & Whitney Co., Hartford, Conn.

Lathes, Foot-Power.
American Tool Works Co., Cincinnati.
B. F. Barnes Co., Rockford, Ill.

Lathes, Screw Cutting.
B. F. Barnes Co., Rockford, Ill.
Baxter, Paterson & Co., Montreal.
Niles-Bement-Pond Co., New York.

Lathes, Automatic, Screw-Threading.
John Bertram & Sons Co., Dundas, Ont
London Mach. Tool Co., Hamilton, Ont
Pratt & Whitney Co., Hartford, Conn.

Lathes, Bench.
B. F. Barnes Co., Rockford, Ill.
London Mach. Tool Co., London, Ont.
Pratt & Whitney Co., Hartford, Conn.

Lathes, Turret.
American Tool Works Co., Cincinnati.
Baxter, Paterson & Co., Montreal.
John Bertram & Sons Co., Dundas, Ont.
Gisholt Machine Co., Madison, Wis.
London Mach. Tool Co., Hamilton, Ont.
Niles-Bement-Pond Co., New York.
The Pratt & Whitney Co., Hartford, Conn.
Warner & Swasey Co., Cleveland O.

Leather Belting.
Baxter, Paterson & Co., Montreal
Canada Machinery Agency, Montreal.
The Canadian Fairbanks Co., Montreal
Greey, Wm. & J. G., Toronto

Lime Stone Flux.
Hamilton Facing Mill Co., Hamilton.
Paxson, J. W., Co., Philadelphia

Locomotives, Air.
Canadian Rand Drill Co., Montreal.

Locomotives, Electrical.
Canadian Westinghouse Co., Hamilton
Jeffrey Mfg. Co., Columbus, Ohio.

Locomotives, Steam.
Canada Foundry Co., Toronto.
Canadian Rand Drill Co., Montreal

Locomotive Turntable Tractors.
Canadian Pilling Co., Montreal

Lubricating Graphite.
Dominion Foundry Supply Co. Montreal
United States Graphite Co., Saginaw, Mich.

Lubricants.
United States Graphite Co., Saginaw, Mich.

Lubricating Plumbago.
Detroit Foundry Supply Co., Detroit.
Dominion Foundry Supply Co., Montreal
Hamilton Facing Mill Co., Hamilton
Paxson, J. W., Co., Philadelphia

Lubricators, Force Feed.
Sight Feed Oil Pump Co., Milwaukee, Wis

Lumber Dry Kilns.
Dominion Heating & Ventilating Co., Hespeler.
H. W. Petrie, Toronto.
Sheldons Limited, Galt, Ont.

Machinery Dealers.
Baxter, Paterson & Co. Montreal.
Canada Machinery Agency, Montreal.
The Canadian Fairbanks Co., Montreal.
Miller Bros. & Toms, Montreal.
H. W. Petrie, Toronto.
The Smart-Turner Mach. Co., Hamilton.
Williams & Wilson Montreal.

Machinery Designers.
Greey, Wm. & J. G., Toronto
Standard Bearings, Ltd., Niagara Falls.

Machinery Lubricators.
United States Graphite Co., Saginaw, Mich.

Machinists.
W. H. Banfield & Sons, Toronto.
Greey, Wm. & J. G. Toronto
Hall Engineering Works, Montreal.
Link-Belt Co., Philadelphia.
John McDougall, Caledonian Iron Works Co., Montreal.
Miller Bros. & Toms Montreal.
Paxson, J. W., Co., Philadelphia
Robb Engineering Co., Amherst, N.S.
The Smart-Turner Mach. Co., Hamilton.
Standard Bearings, Ltd. Niagara Falls
Waterous Engine Co., Brantford.

Machinists' Small Tools.
Armstrong Bros., Chicago.
Butterfield & Co., Rock Island, Que.
Frothingham & Workman Ltd., Montreal
Mechanics Supply Co. Quebec, Que.
Rice Lewis & Son, Montreal.
Pratt & Whitney Co., Hartford, Conn.
Standard Tool Co., Cleveland.
L. S. Starrett Co., Athol, Mass.
Williams & Wilson, Montreal.

Malleable Flask Clamps.
Buffalo Foundry Supply Co., Buffalo.
Dominion Foundry Supply Co., Montreal
Paxson, J. W., Co., Philadelphia

Malleable Iron Castings.
Galt Malleable Iron Co., Galt

Mallet, Rawhide and Wood.
Buffalo Foundry Supply Co., Buffalo.
Detroit Foundry Supply Co., Detroit.
Dominion Foundry Supply Co., Montreal
Paxson, J. W., Co., Philadelphia

Mandrels.
A. B. Jardine & Co., Hespeler, Ont.
Miller Bros. & Toms, Montreal.
The Pratt & Whitney Co., Hartford, Conn.
Standard Tool Co., Cleveland.

Marking Machines.
Dwight Slate Machine Co., Hartford, Conn.

Marine Work.
Kennedy, Wm., & Sons, Owen Sound

Metallic Paints.
P. D. Dods & Co., Montreal.

Meters, Electrical.
Canadian Westinghouse Co., Hamilton.
Mechanics Supply Co., Quebec, Que.

Mill Machinery.
Baxter, Paterson & Co., Montreal.
Greey, Wm. & J. G., Toronto
The Goldie & McCulloch Co., Galt, Ont.
John McDougall Caledonian Iron Works Co., Montreal.
H. W. Petrie, Toronto.
Robb Engineering Co., Amherst, N.S.
Waterous Engine Co., Brantford.
Williams & Wilson, Montreal.

Milling Attachments.
Becker-Brainard Milling Machine Co. Hyde Park, Mass.
John Bertram & Sons Co., Dundas, Ont.
Niles-Bement-Pond Co., New York.
Pratt & Whitney, Hartford, Conn.

Milling Machines, Horizontal.
Baxter, Paterson & Co. Montreal.
Becker-Brainard Milling Machinery Co. Hyde Park, Mass.
John Bertram & Sons Co., Dundas, Ont.
London Mach. Tool Co., Hamilton, Ont.
Niles-Bement-Pond Co., New York.
Pratt & Whitney, Hartford, Conn.

Milling Machines, Plain.
American Tool Works Co., Cincinnati.
Becker-Brainard Milling Machine Co. Hyde Park, Mass.
John Bertram & Sons Co., Dundas, Ont.
Canada Machinery Agency, Montreal.
The Canadian Fairbanks Co., Montreal.
London Mach. Tool Co., Hamilton, Ont.
Niles-Bement-Pond Co., New York.
H. W. Petrie, Toronto.
Pratt & Whitney Co., Hartford, Conn.
Williams & Wilson, Montreal.

Milling Machines, Universal.
American Tool Works Co., Cincinnati.
Becker-Brainard Milling Machine Co., Hyde Park, Mass.
John Bertram & Sons Co., Dundas, Ont.
Canada Machinery Agency, Montreal.
The Canadian Fairbanks Co., Montreal.
London Mach. Tool Co., Hamilton, Ont
Niles-Bement-Pond Co., New York.
H. W. Petrie, Toronto.
Williams & Wilson, Montreal.

Milling Machines, Vertical.
Becker-Brainard Milling Machine Co., Hyde Park, Mass.
Brown & Sharpe, Providence, R.I.
John Bertram & Sons Co., Dundas, Ont.
Canada Machinery Agency, Montreal.
London Mach. Tool Co., Hamilton, Ont.
Niles-Bement-Pond Co., New York.

Milling Tools.
Wm. Abbott, Montreal.
Becker-Brainard Milling Machine Co., Hyde Park, Mass.
Geometric Tool Co., New Haven, Conn.
Hamilton Tool Co., Hamilton, Ont.
London Mach. Tool Co., Hamilton, Ont.
Pratt & Whitney Co., Hartford, Conn.
Standard Tool Co., Cleveland.
Standard Bearings, Ltd., Niagara Falls

Mining Machinery.
Allis-Chalmers-Bullock, Limited, Montreal
Canadian Rand Drill Co., Montreal.
Chicago Pneumatic Tool Co., Chicago.
Jeffrey Mfg. Co., Columbus, Ohio.
Laurie Engine & Machine Co., Montreal.
Rice Lewis & Son, Toronto.
John McDougall, Caledonian Iron Works Co., Montreal.
Miller Bros. & Toms, Montreal.
T. & H. Electric Co. Hamilton.

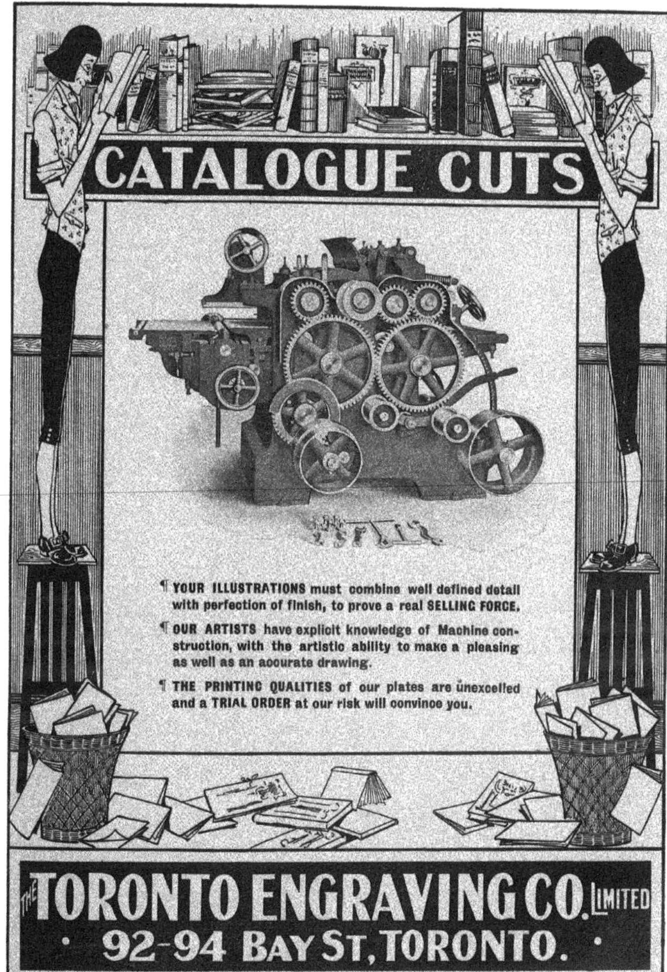
CATALOGUE CUTS
¶ YOUR ILLUSTRATIONS must combine well defined detail with perfection of finish, to prove a real SELLING FORCE.
¶ OUR ARTISTS have explicit knowledge of Machine construction, with the artistic ability to make a pleasing as well as an accurate drawing.
¶ THE PRINTING QUALITIES of our plates are unexcelled and a TRIAL ORDER at our risk will convince you.
THE TORONTO ENGRAVING CO. LIMITED
· 92-94 BAY ST, TORONTO. ·

Mixing Machines, Dough.

Greey, Wm. & J. G., Toronto
Paxson, J. W., Co., Philadelphia

Mixing Machines, Special.

Greey, Wm. & J. G., Toronto
Paxson, J. W., Co., Philadelphia

Model Tools.

Globe Machine & Stamping Co., Cleveland, Ohio.
Standard Bearings, Ltd., Niagara Falls.
Wells Pattern and Model Works, Toronto

Motors, Electric.

Allis-Chalmers-Bullock Limited, Montreal
Canadian General Electric Co., Toronto
Canadian Westinghouse Co., Hamilton.
Electrical Machinery Co., Toronto.
Consolidated Electric Co., Toronto
Gas & Electric Power Co., Toronto
Hall Engineering Works, Montreal.
Mechanics Supply Co., Quebec Que.
The Packard Electric Co., St. Catharines.
T. & H. Electric Co., Hamilton.

Motors, Air.

Canadian Rand Drill Co., Montreal.
Chicago Pneumatic Tool Co., Chicago.

Molders' Supplies.

Buffalo Foundry Supply Co., Buffalo.
Detroit Foundry Supply Co., Windsor.
Dominion Foundry Supply Co., Montreal
Hamilton Facing Mill Co., Hamilton.
Mechanics Supply Co., Quebec, Que.
Paxson J. W., Co., Philadelphia

Molders' Tools.

Buffalo Foundry Supply Co., Buffalo.
Detroit Foundry Supply Co., Windsor
Dominion Foundry Supply Co., Montreal
Hamilton Facing Mill Co., Hamilton.
Mechanics Supply Co., Quebec, Que.
Paxson, J. W., Co., Philadelphia

Molding Machines.

Ph. Bonvillain & E. Ronceray, Philadelphia
Buffalo Foundry Supply Co., Buffalo.
Dominion Foundry Supply Co., Montreal
Hamilton Facing Mill Co., Hamilton.
J. W. Paxson Co., Philadelphia, Pa.

Molding Sand.

T. W. Barnes, Hamilton.
Buffalo Foundry Supply Co., Buffalo.
Detroit Foundry Supply Co., Windsor
Dominion Foundry Supply Co., Montreal
Hamilton Facing Mill Co., Hamilton.
Paxson, J. W., Co., Philadelphia

Nut Tappers.

John Bertram & Sons Co., Dundas, Ont
A. B. Jardine & Co., Hespeler.
London Mach. Tool Co., Hamilton.
National Machinery Co., Tiffin, Ohio.

Nuts.

Canada Nut Co., Toronto

Oatmeal Mill Machinery.

Greey, Wm. & J. G., Toronto
The Goldie & McCulloch Co., Galt

Oilers, Gang.

Sight Feed Oil Pump Co., Milwaukee, Wis.

Oils, Core.

Buffalo Foundry Supply Co., Buffalo.
Dominion Foundry Supply Co., Montreal
Hamilton Facing Mill Co., Hamilton.
Paxson, J. W., Co., Philadelphia

Oil Extractors.

Darling Bros., Ltd., Montreal

Paint Mill Machinery.

Greey, Wm. & J. G., Toronto

Pans, Lathe.

Cleveland Wire Spring Co., Cleveland
Paxson, J. W., Co., Philadelphia

Pans, Steel Shop.

Cleveland Wire Spring Co., Cleveland

Parting Compound.

Doggett, Stanley, New York

Patent Solicitors.

Hanbury A. Budden, Montreal.
Fetherstonhaugh & Blackmore, Montreal
Marion & Marion, Montreal.
Ridout & Maybee, Toronto.

Patterns.

John Carr, Hamilton.
Galt Malleable Iron Co., Galt
Hamilton Pattern Works, Hamilton.
John McDougall, Caledonian Iron Works Co., Montreal.
Miller Bros. & Toms, Montreal.
Wells' Pattern and Model Works, Toronto.

Pig Iron.

Hamilton Steel & Iron Co., Hamilton.

Pipe Cutting and Threading Machines.

Butterfield & Co., Rock Island, Que.
Canada Machinery Agency, Montreal.
Curtis & Curtis Co., Bridgeport, Conn.
Frothingham & Workman, Ltd., Montreal
Hart Mfg. Co., Cleveland
A. B. Jardine & Co., Hespeler, Ont.
Loew Mfg. Co., Cleveland, Ohio
London Mach. Tool Co., Hamilton, Ont.
Niles-Bement-Pond Co., New York.
Shantz, I. E., & Co., Berlin, Ont.

Pipe, Municipal.

Canadian Pipe Co., Vancouver, B.C.
Pacific Coast Pipe Co., Vancouver, B.C.

Pipe, Waterworks.

Canadian Pipe Co., Vancouver, B.C.
Pacific Coast Pipe Co., Vancouver, B.C.

Planers, Standard.

American Tool Works, Cincinnati.
John Bertram & Sons Co., Dundas, Ont.
Canada Machinery Agency, Montreal.
The Canadian Fairbanks Co., Montreal.
Rice Lewis & Son, Toronto.
London Mach. Tool Co., Hamilton, Ont.
Niles-Bement-Pond Co., New York.
H. W. Petrie, Toronto.
Pratt & Whitney Co., Hartford, Conn.
Williams & Wilson, Montreal.

Planers, Rotary.

John Bertram & Sons Co., Dundas, Ont.
London Mach. Tool Co., Hamilton, Ont.
Niles-Bement-Pond Co., New York.

Planing Mill Fans.

Dominion Heating & Ventilating Co., Hespeler
Sheldons Limited, Galt, Ont.

Plumbago.

Buffalo Foundry Supply Co., Buffalo.
Detroit Foundry Supply Co., Windsor
Doggett, Stanley, New York
Dominion Foundry Supply Co., Montreal
Hamilton Facing Mill Co., Hamilton,
Mechanics Supply Co., Quebec, Que.
J. W. Paxson Co., Philadelphia, Pa.
United States Graphite Co., Saginaw, Mich.

Pneumatic Tools.

Allis-Chalmers-Bullock, Montreal.
Canadian Rand Drill Co., Montreal.
Chicago Pneumatic Tool Co., Chicago
Hamilton Facing Mill Co., Hamilton.
Hanna Engineering Works, Chicago.
Independent Pneumatic Tool Co, Chicago, New York
Paxson, J. W., Co., Philadelphia

Power Hack Saw Machines.

Baxter, Paterson & Co., Montreal.
Frothingham & Workman, Ltd., Montreal

Power Plants.

John McDougall Caledonian Iron Works Co., Montreal.
The Smart-Turner Mach. Co., Hamilton

Power Plant Equipments.

Darling Bros., Ltd., Montreal

Presses, Drop.

W. H. Banfield & Son, Toronto.
E. W. Bliss Co., Brooklyn, N.Y.
Canada Machinery Agency, Montreal.
Laurie Engine & Machine Co., Montreal
Miller Bros. & Toms, Montreal.
Niles-Bement-Pond Co., New York.

Presses, Hand.

E. W. Bliss Co., Brooklyn, N.Y.

Presses, Hydraulic.

John Bertram & Sons Co., Dundas, Ont.
Laurie Engine & Machine Co., Montreal.
London Mach. Tool Co., Hamilton, Ont.
John McDougall Caledonian Iron Works Co. Montreal.
Miller Bros. & Toms, Montreal.
Niles-Bement-Pond Co., New York.
Perrin, Wm. R., & Co., Toronto

Presses, Power.

E. W. Bliss Co., Brooklyn, N.Y.
Canada Machinery Agency, Montreal.
Laurie Engine & Machine Co., Montreal
London Mach Tool Co., Hamilton, Ont.
John McDougall Caledonian Iron Works Co., Montreal.
Niles-Bement-Pond Co., New York.

Presses Power Screw.

Perrin, Wm. R., & Co., Toronto

Pressure Regulators.

Darling Bros., Ltd., Montreal

Producer Plants.

Canada Foundry Co., Toronto

Pulp Mill Machinery.

Greey, Wm. & J. G., Toronto
Jeffrey Mfg Co., Columbus, Ohio.
Laurie Engine & Machine Co., Montreal.
John McDougall Caledonian Iron Works Co., Montreal.
Waterous Engine Works Co., Brantford

Pulleys.

Baxter, Paterson & Co., Montreal.
Canada Machinery Agency, Montreal.
The Canadian Fairbanks Co., Montreal.
The Goldie & McCulloch Co., Galt.
Greey, Wm. & J. G., Toronto
Laurie Engine & Machine Co., Montreal.
Link-Belt Co., Philadelphia.
John McDougall Caledonian Iron Works Co., Montreal.
Miller Bros. & Toms, Montreal.
Owen Sound Iron Works Co., Owen Sound
H. W. Petrie, Toronto.
The Smart-Turner Mach. Co., Hamilton.
Standard Bearings, Ltd., Niagara Falls.
Williams & Wilson, Montreal.
Waterous Engine Co., Brantford.

Pumps.

Laurie Engine & Machine Co., Montreal
Ontario Wind Engine & Pump Co., Toronto

Pump Governors.

Darling Bros., Ltd. Montreal

Pumps, Hydraulic.

Ph. Bonvillain & E. Ronceray, Philadelphia
Laurie Engine & Machine Co., Montreal
Perrin, Wm. R. & Co., Toronto

Pumps, Oil.

Sight Feed Oil Pump Co., Milwaukee, Wis.

Pumps, Steam.

Allis-Chalmers-Bullock, Limited, Montreal
Canada Foundry Co., Toronto.
Canada Machinery Agency, Montreal.
Darling Bros., Ltd., Montreal
The Goldie & McCulloch Co., Galt.
John McDougall Caledonian Iron Works, Montreal.
H. W. Petrie, Toronto.
The Smart-Turner Mach. Co., Hamilton.
Standard Bearings, Ltd., Niagara Falls,
Waterous Engine Co. Brantford.

Pumping Machinery.

Canada Foundry Co., Limited, Toronto
Canada Machinery Agency, Montreal.
Canadian Rand Drill Co., Montreal.
Chicago Pneumatic Tool Co., Chicago.
Darling Bros., Ltd., Montreal
Hall Engineering Works, Montreal, Que.
Laurie Engine & Machine Co., Montreal.
London Mach. Tool Co., Hamilton, Ont.
John McDougall Caledonian Iron Works Co., Montreal.
The Smart-Turner Mach. Co., Hamilton
Standard Bearings, Ltd., Niagara Falls,

Punches and Dies.

W. H. Banfield & Sons, Toronto.
Butterfield & Co., Rock Island.
Globe Machine & Stamping Co.
A. B. Jardine & Co., Hespeler, Ont.
London Mach Tool Co., Hamilton, Ont.
Miller Bros. & Toms, Montreal.
Pratt & Whitney Co., Hartford, Conn.
H. W. Petrie, Toronto.
Standard Bearings, Ltd., Niagara Falls,

Punches, Hand.

Mechanics Supply Co., Quebec, Que.

Punches, Power.

John Bertram & Sons Co., Dundas, Ont.
E. W. Bliss Co., Brooklyn, N.Y.
Canada Machinery Agency, Montreal.
Canadian Buffalo Forge Co., Montreal.
London Mach. Tool Co., Hamilton, Ont.
Niles-Bement-Pond Co., New York.

Punches, Turret.

London Mach. Tool Co., London, Ont.

Punching Machines, Horizontal.

John Bertram & Sons Co., Dundas, Ont.
London Mach. Tool Co., Hamilton, Ont.
Niles-Bement-Pond Co., New York.

Quartering Machines.

John Bertram & Sons Co., Dundas, Ont.
London Mach. Tool Co., Hamilton, Ont.

Railway Spikes and Washers.

Hamilton Steel & Iron Co., Hamilton

Rammers, Bench and Floor.

Buffalo Foundry Supply Co., Buffalo.
Detroit Foundry Supply Co., Windsor
Hamilton Facing Mill Co., Hamilton.
Paxson, J. W., Co., Philadelphia

Rapping Plates.

Detroit Foundry Supply Co., Windsor
Hamilton Facing Mill Co., Hamilton.
Paxson, J. W., Co., Philadelphia

Raw Hide Pinions.

Brown, David & Sons, Huddersfield, Eng

Reamers.

Wm. Abbott, Montreal.
Baxter Paterson & Co., Montreal.
Butterfield & Co., Rock Island.
Frothingham & Workman Ltd., Montreal
Hamilton Tool Co., Hamilton
Hanna Engineering Works, Chicago.
A. B. Jardine & Co., Hespeler, Ont.
Mechanics Supply Co. Quebec, Que.
John Millen & Son, Ltd., Montreal.
Morse Twist Drill and Machine Co., New Bedford, Mass
Pratt & Whitney Co., Hartford, Conn.
Standard Tool Co., Cleveland.

Reamers, Steel Taper.

Butterfield & Co., Rock Island.
Chicago Pneumatic Tool Co., Chicago.
A. B. Jardine & Co., Hespeler, Ont.
John Millen & Son Ltd., Montreal.
Pratt & Whitney Co., Hartford, Conn.
Standard Tool Co., Cleveland.
Whitman & Barnes Co., St. Catharines.

Rheostats.

Canadian General Electric Co., Toronto.
Canadian Westinghouse Co., Hamilton.
Hall Engineering Works, Montreal, Que.
T. & H. Electric Co., Hamilton,

Riddles.

Buffalo Foundry Supply Co., Buffalo.
Detroit Foundry Supply Co., Windsor
Hamilton Facing Mill Co., Hamilton.
Dominion Foundry Supply Co., Montreal
J. W. Paxson Co., Philadelphia, Pa.

Riveters, Pneumatic.

Hanna Engineering Works, Chicago.
Independent Pneumatic Tool Co., Chicago, Ill.

Rolls, Bending.

John Bertram & Sons Co., Dundas, Ont.
London Mach. Tool Co., Hamilton, Ont.
Niles-Bement-Pond Co., New York.

Rolls, Chilled Iron.

Greey, Wm. & J. G., Toronto

Rolls, Sand Cast.

Greey, Wm. & J. G., Toronto

Rotary Blowers.

Paxson, J. W., Co., Philadelphia

Rotary Converters.

Allis-Chalmers-Bullock, Ltd., Montreal.
Canadian Westinghouse Co., Hamilton
Paxson, J. W., Co., Philadelphia
Toronto and Hamilton Electric Co., Hamilton.

Rubbing and Sharpening Stones.

Norton Co., Worcester, Mass.

Safes.

Baxter Paterson & Co., Montreal
The Goldie & McCulloch Co., Galt,

Sand, Bench.

Buffalo Foundry Supply Co., Buffalo.
Detroit Foundry Supply Co., Windsor
Dominion Foundry Supply Co., Montreal
Hamilton Facing Mill Co., Hamilton.
Paxson, J. W., Co., Philadelphia

Sand Blast Machinery.

Canadian Rand Drill Co., Montreal.
Chicago Pneumatic Tool Co., Chicago.
Detroit Foundry Supply Co., Windsor
Dominion Foundry Supply Co., Montreal
Hamilton Facing Mill Co., Hamilton.
J. W. Paxson Co., Philadelphia, Pa.

Sand, Heavy, Grey Iron.

Buffalo Foundry Supply Co., Buffalo.
Detroit Foundry Supply Co., Windsor
Dominion Foundry Supply Co., Montreal
Hamilton Facing Mill Co., Hamilton.
Paxson, J. W., Co., Philadelphia

Sand, Malleable.

Buffalo Foundry Supply Co., Buffalo,
Detroit Foundry Supply Co., Windsor
Dominion Foundry Supply Co., Montreal
Hamilton Facing Mill Co., Hamilton.
Paxson, J. W., Co., Philadelphia

Sand, Medium Grey Iron.

Buffalo Foundry Supply Co., Buffalo.
Detroit Foundry Supply Co., Windsor
Dominion Foundry Supply Co., Montreal
Hamilton Facing Mill Co., Hamilton.
Paxson, J. W., Co., Philadelphia

Sand Sifters.

Buffalo Foundry Supply Co., Buffalo.
Detroit Foundry Supply Co., Windsor
Dominion Foundry Supply Co., Montreal
Hamilton Facing Mill Co., Hamilton.
Hanna Engineering Works, Chicago.
Paxson, J. W., Co. Philadelphia

Saw Mill Machinery.

Allis-Chalmers-Bullock, Limited, Montreal
Baxter, Paterson & Co., Montreal.
Canada Machinery Agency Montreal.
Goldie & McCulloch Co., Galt.
Greey, Wm. & J. G. Toronto
Miller Bros. & Toms, Montreal
Owen Sound Iron Works Co., Owen Sound
H. W. Petrie, Toronto.
Robb Engineering Co. Amherst,
Standard Bearings, Ltd., Niagara Falls
Waterous Engine Works, Brantford
Williams & Wilson, Montreal.

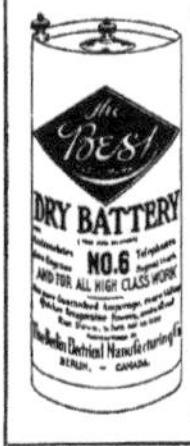

"Best" Dry Cells

For your Ignition in

Automobile
Motor Boat
Gas Engines

The kind the largest Telephone Company in Canada is using to-day, because they give the longest service.

Ask your dealer for them, or

The Berlin Electrical Mfg. Co.,
Limited
474 King Street West, TORONTO

"Economo" Emery Wheel Dresser

A Substitute for the Black Diamond

"ECONOMO" ECONOMO COMBINATION

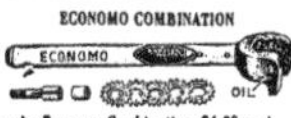

PRICE: Economo, $4.00 each; Economo Combination, $4.50 each.

Will true or shape any wheel except the very hardest and carborundum. The most efficient Roughing Dresser on the market—never gets dull from use.

Made of an abrasive nearly as hard as the Black Diamond. The Combination dresser is the roughing dresser placed in the "handle end" of the "Regular." It is made with corrugated tool steel discs and excels in wear and efficiency any dresser heretofore made. These tools are guaranteed satisfactory or money refunded. Sent on 10 day trial.

INTERNATIONAL SPECIALTY CO.
363 Holden Ave. DETROIT, MICH.

Persons addressing advertisers will kindly mention having seen their advertisement in this paper.

We should have your business.

If prompt delivery, reasonable prices and best of workmanship appeal to you, we should have your pattern orders.

SEND FOR PATTERN CLASSIFICATION CARD

THE HAMILTON PATTERN WORKS
134 Bay St. N., HAMILTON, ONT.
Phone, 2664.

WM. KENNEDY & SONS

HIGH-CLASS

Steel Gears of all kinds
Iron Gears
Grey Iron, Steel and
Brass Castings

MARINE WORK receives every attention

WM. KENNEDY & SONS, Limited
Owen Sound, Ont.

LUMEN BEARING COMPANY

Brass Founders

Buffalo, U.S.A.
Toronto, Can.

Canadian National EXHIBITION

1907—TORONTO, ONT.—1907

Saturday, Aug. 24th *to* **Monday, Sept. 9th**

$400,000	In New Buildings	$400,000
In Premiums	$45,000	In Premiums
$40,000	For Attractions	$40,000

Historical Art Loan in New Art Gallery

Extra Fine Exhibit of Pure Bred Live Stock

Five Acres of Industrial Exhibits

Grand Musical Festival
Brilliant Military Spectacle
Highest Class Special Features
Battle of Badajos (Nightly)

Largest Fire-Proof Grand Stand in America

Six Tickets for One Dollar	Reduced Rates on all Lines of Travel	Great Dog and Cat Shows

The First Week is the Best Week for Comfort

Sawing Machines, Metal.

Niles-Bement-Pond Co. New York.
Paxson, J. W., Co., Philadelphia

Saws, Hack.

Baxter Paterson & Co., Montreal.
Canada Machinery Agency, Montreal.
Detroit Foundry Supply Co., Windsor
Frothingham & Workman, Ltd., Montreal
London Mach. Tool Co., Hamilton.
Mechanics Supply Co., Quebec, Que.
Rice Lewis & Son, Toronto.
John Millen & Son, Ltd., Montreal.
L. S. Starrett Co., Athol, Mass.

Screw Cutting Tools

Hart Manufacturing Co., Cleveland, Ohio
Mechanics Supply Co., Quebec, Que.

Screw Machines, Automatic.

Canada Machinery Agency, Montreal.
Cleveland Automatic Machine Co., Cleveland, Ohio
London Mach. Tool Co., Hamilton, Ont.
National-Acme Mfg. Co., Cleveland.
Pratt & Whitney Co., Hartford, Conn.

Screw Machines, Hand.

Canada Machinery Agency Montreal.
A. B. Jardine & Co., Hespeler.
London Mach. Tool Co., Hamilton, Ont.
Mechanics Supply Co. Quebec, Que.
Pratt & Whitney Co., Hartford, Conn.
Warner & Swasey Co., Cleveland, O.

Screw Machines, Multiple Spindle.

National-Acme Mfg. Co., Cleveland

Screw Plates.

Butterfield & Co., Rock Island, Que.
Frothingham & Workman, Ltd., Montreal
Hart Manufacturing Co. Cleveland, Ohio
A. B. Jardine & Co., Hespeler
Mechanics Supply Co., Quebec, Que.

Screws, Cap and Set.

National-Acme Mfg. Co., Cleveland

Screw Slotting Machinery, Semi-Automatic.

National-Acme Mfg. Co., Cleveland

Second-Hand Machinery.

American Tool Works Co., Cincinnati.
Canada Machinery Agency, Montreal.
The Canadian Fairbanks Co., Montreal.
Goldie & McCulloch Co., Galt.
Machinery Exchange, Montreal.
Niles-Bement-Pond Co., New York.
H. W. Petrie, Toronto.
Robb Engineering Co., Amherst, N S.
Waterous Engine Co., Brantford.
Williams & Wilson, Montreal.

Shafting.

Baxter, Paterson & Co., Montreal.
Canada Machinery Agency, Montreal
The Canadian Fairbanks Co., Montreal.
Frothingham & Workman Ltd., Montreal
The Goldie & McCulloch Co., Galt, Ont
Greey, Wm., & J. G. Toronto
Kennedy Wm., & Sons, Owen Sound
Niles-Bement-Pond Co., New York.
Owen Sound Iron Works Co., Owen Sound
H. W. Petrie, Toronto.
Smart-Turner Machine Co., Hamilton.
Union Drawn Steel Co., Hamilton.
Waterous Engine Co., Brantford.

Shapers.

American Tool Works Co., Cincinnati.
John Bertram & Sons Co., Dundas, Ont
Canada Machinery Agency, Montreal.
The Canadian Fairbanks Co., Montreal.
Rice Lewis & Son, Toronto.
London Mach. Tool Co., Hamilton, Ont.
Niles-Bement-Pond Co., New York.
H. W. Petrie, Toronto.
Potter & Johnston Machine Co., Pawtucket, R.I.
Pratt & Whitney Co., Hartford, Conn.
Williams & Wilson, Montreal.

Sharpening Stones.

Carborundum Co., Niagara Falls, N.Y.

Shearing Machine, Bar.

John Bertram & Sons Co., Dundas, Ont.
A. B. Jardine & Co., Hespeler.
London Mach. Tool Co., Hamilton, Ont.
Niles-Bement-Pond Co., New York.

Shears, Power.

John Bertram & Sons Co., Dundas, Ont
Canada Machinery Agency, Montreal.
A. B. Jardine & Co., Hespeler, Ont.
Niles-Bement-Pond Co., New York.
Paxson, J. W., Co., Philadelphia

Sheet Metal Goods

Globe Machine & Stamping Co., Cleveland, Ohio.

Sheet Steel Work.

Owen Sound Iron Works Co., Owen Sound

Shingle Mill Machinery.

Owen Sound Iron Works Co., Owen Sound

Shop Trucks.

Greey, Wm. & J. G., Toronto
Owen Sound Iron Works Co., Owen Sound
Paxson, J. W., Co., Philadelphia

Shovels.

Baxter, Paterson & Co., Montreal.
Buffalo Foundry Supply Co., Buffalo.
Detroit Foundry Supply Co., Detroit.
Doggett, Stanley, New York
Frothingham & Workman, Ltd., Montreal
Dominion Foundry Supply Co., Montreal
Hamilton Facing Mill Co., Hamilton.
Mechanics Supply Co., Quebec, Que.
Paxson, J. W., Co., Philadelphia

Shovels, Steam.

Allis-Chalmers-Bullock, Montreal.

Sieves.

Buffalo Foundry Supply Co. Buffalo.
Detroit Foundry Supply Co., Windsor
Dominion Foundry Supply Co., Montreal
Hamilton Facing Mill Co., Hamilton.
Paxson, J. W., Co., Philadelphia

Silver Lead.

Buffalo Foundry Supply Co., Buffalo.
Detroit Foundry Supply Co., Windsor
Doggett, Stanley, New York
Dominion Foundry Supply Co., Montreal
Hamilton Facing Mill Co., Hamilton.
Paxson, J. W., Co., Philadelphia

Sleeves, Reducing.

Chicago Pneumatic Tool Co., Chicago.

Snap Flasks

Buffalo Foundry Supply Co., Buffalo.
Detroit Foundry Supply Co., Windsor
Dominion Foundry Supply Co., Montreal
Hamilton Facing Mill Co., Hamilton.
Paxson, J. W., Co., Philadelphia

Soap Stones & Talc Crayons & Pencils.

Doggett, Stanley, New York

Soapstone.

Buffalo Foundry Supply Co., Buffalo.
Detroit Foundry Supply Co., Windsor
Doggett, Stanley, New York
Dominion Foundry Supply Co., Montreal
Hamilton Facing Mill Co., Hamilton.
Paxson, J. W. Co., Philadelphia

Soap Stone Facings.

Doggett, Stanley, New York

Solders.

Lumen Bearing Co., Toronto

Special Machinery.

W. H. Banfield & Sons, Toronto.
Baxter, Paterson & Co., Montreal
John Bertram & Sons Co. Dundas, Ont
Globe Machine & Stamping Co., Cleveland Ohio.
Greey, Wm. & J. G., Toronto
Hanna Engineering Works, Chicago.
Laurie Engine & Machine Co., Montreal.
London Mach. Tool Co., Hamilton, Ont.
Pendrith Machinery Co., Toronto.
H. W. Petrie, Toronto.
The Smart-Turner Mach. Co., Hamilton
Standard Bearings, Ltd., Niagara Falls,
Waterous Engine Co., Brantford.
Woelfle Bros., Berlin, Ont

Special Machines and Tools.

Paxson, J. W., Philadelphia
Pratt & Whitney, Hartford, Conn.
Standard Bearings, Ltd., Niagara Falls,

Special Milled Work.

National-Acme Mfg. Co., Cleveland

Speed Changing Countershafts.

The Canadian Fairbanks Co., Montreal.

Spike Machines.

National Machinery Co., Tiffin, O.
The Smart-Turner Mach. Co., Hamilton

Spray Cans.

Detroit Foundry Supply Co., Windsor
Dominion Foundry Supply Co., Montreal
Hamilton Facing Mill Co., Hamilton.
Paxson, J. W., Co., Philadelphia

Springs, Automobile.

Cleveland Wire Spring Co., Cleveland

Springs, Coiled Wire.

Cleveland Wire Spring Co., Cleveland

Springs, Machinery.

Cleveland Wire Spring Co., Cleveland

Springs, Upholstery.

Cleveland Wire Spring Co., Cleveland

Sprue Cutters.

Detroit Foundry Supply Co., Windsor
Dominion Foundry Supply Co., Montreal
Hamilton Facing Mill Co., Hamilton.
Paxson J. W., Co., Philadelphia

Stamp Mills.

Allis-Chalmers-Bullock, Limited, Montreal
Paxson, J. W., Co., Philadelphia

Steam Separators.

Darling Bros., Ltd., Montreal.
Dominion Heating & Ventilating Co., Hespeler
Robb Engineering Co., Montreal.
Sheldon's Limited, Galt.
Smart-Turner Mach. Co., Hamilton
Waterous Engine Co., Brantford.

Steam Specialties.

Darling Bros., Ltd., Montreal
Dominion Heating & Ventilating Co., Hespeler
Sheldon's Limited, Galt.

Steam Traps.

Canada Machinery Agency, Montreal.
Darling Bros., Ltd., Montreal
Dominion Heating & Ventilating Co., Hespeler
Mechanics Supply Co., Quebec, Que.
Sheldons Limited, Galt.

Steam Valves.

Darling Bros., Ltd., Montreal

Steel Pressure Blowers

Buffalo Foundry Supply Co., Buffalo.
Dominion Foundry Supply Co., Montreal
Dominion Heating & Ventilating Co., Hespeler
Hamilton Facing Mill Co., Hamilton.
J. W. Paxon Co., Philadelphia, Pa.
Sheldon's Limited, Galt.

Steel Tubes.

Baxter, Paterson & Co., Montreal.
Mechanics Supply Co., Quebec Que.
John Millen & Son, Montreal.

Steel, High Speed.

Wm. Abbott, Montreal.
Canadian Fairbanks Co., Montreal.
Frothingham & Workman, Ltd., Montreal
Alex. Gibb, Montreal.
Jessop, Wm., & Sons, Sheffield, Eng.
B. K. Morton Co., Sheffield, Eng.
Williams & Wilson, Montreal.

Stocks and Dies

Hart Manufacturing Co., Cleveland, Ohio
Mechanics Supply Co., Quebec, Que.
Pratt & Whitney Co., Hartford, Conn.

Stone Cutting Tools, Pneumatic

Allis-Chalmers-Bullock, Ltd., Montreal
Canadian Rand Drill Co., Montreal.

Stone Surfacers

Chicago Pneumatic Tool Co., Chica

Stove Plate Facings.

Buffalo Foundry Supply Co., Buffalo
Detroit Foundry Supply Co., Windsor.
Dominion Foundry Supply Co., Montreal
Hamilton Facing Mill Co., Hamilton.
Paxson, J. W., Co., Philadelphia

Swage, Block.

A. B. Jardine & Co., Hespeler, Ont.

Switchboards.

Allis-Chalmers-Bullock, Limited, Montreal
Canadian General Electric Co., Toronto.
Canadian Westinghouse Co., Hamilton
Hall Engineering Works, Montreal, Que.
Mechanics Supply Co., Quebec, Que.
Toronto and Hamilton Electric Co., Hamilton.

Talc.

Buffalo Foundry Supply Co., Buffalo.
Detroit Foundry Supply Co., Windsor.
Doggett, Stanley New York
Dominion Foundry Supply Co., Montreal
Hamilton Facing Mill Co., Hamilton.
Paxson, J. W., Co., Philadelphia

Tanks, Oil.

Sight Feed Oil Pump Co., Milwaukee, Wis.

Tapping Machines and Attachments.

American Tool Works Co., Cincinnati.
John Bertram & Sons Co., Dundas, Ont.
Bickford Drill & Tool Co., Cincinnati.
The Geometric Tool Co., New Haven.
A. B. Jardine & Co., Hespeler.
London Mach. Tool Co., Hamilton, Ont
Murchey Machine & Tool Co., Detroit.
Niles-Bement-Pond Co., New York.
H. W. Petrie, Toronto.
Pratt & Whitney, Cincinnati, O.
L. S. Starrett Co., Athol, Mass.
Williams & Wilson, Montreal.

Tapes, Steel.

Frothingham & Workman, Ltd., Montreal
Rice Lewis & Son, Toronto.
Mechanics Supply Co., Quebec, Que.
John Millen & Son, Ltd., Montreal.
L. S. Starrett Co., Athol, Mass.

Taps.

Baxter, Paterson & Co., Montreal.
Mechanics Supply Co., Quebec, Que.

Taps, Collapsing.

The Geometric Tool Co., New Haven.

Taps and Dies.

Wm. Abbott, Montreal.
Baxter, Paterson & Co., Montreal.
Butterfield & Co., Rock Island, Que.
Frothingham & Workman Ltd., Montreal
The Geometric Tool Co., New Haven.
A. B. Jardine & Co., Hespeler, Ont.
Rice Lewis & Son, Toronto.
Mechanics Supply Co., Quebec, Que.
John Millen & Son, Ltd., Montreal.
Pratt & Whitney Co., Hartford, Conn.
Standard Tool Co., Cleveland.
L. S. Starrett Co., Athol, Mass.

Testing Laboratory.

Detroit Testing Laboratory Detroit

Testing Machines.

Detroit Foundry Supply Co., Windsor.
Dominion Foundry Supply Co., Montreal
Hamilton Facing Mill Co., Hamilton.
Paxson, J. W., Co., Philadelphia

Thread Cutting Tools.

Hart Manufacturing Co., Cleveland, Ohio
Mechanics Supply Co., Quebec Que.

Tiling, Opal Glass.

Toronto Plate Glass Importing Co., Toronto.

Time Switches, Automatic, Electric.

Berlin Electric Mfg. Co., Toronto.

Tinware Machinery

Canada Machinery Agency Montreal.

Tire Upsetters or Shrinkers

A. B. Jardine & Co., Hespeler, Ont.

Tool Cutting Machinery

Canadian Rand Drill Co., Montreal.

Tool Holders.

Armstrong Bros. Tool Co., Chicago.
Baxter, Paterson & Co., Montreal.
John Millen & Son, Ltd., Montreal.
H. W. Petrie, Toronto.
Pratt & Whitney Co., Hartford, Conn.

Tool Steel.

Wm. Abbott, Montreal.
Flockton, Tompkin & Co., Sheffield, Eng.
Frothingham & Workman, Ltd., Montreal.
Wm. Jessop, Sons & Co., Toronto.
Canadian Fairbanks Co., Montreal.
B. K. Morton & Co., Sheffield, Eng.
Williams & Wilson, Montreal.

Torches, Steel.

Baxter, Paterson & Co., Montreal.
Detroit Foundry Supply Co., Windsor.
Dominion Foundry Supply Co., Montreal
Hamilton Facing Mill Co., Hamilton.
Paxson, J. W., Co., Philadelphia

Transformers and Convertors

Allis-Chalmers-Bullock, Montreal.
Canadian General Electric Co., Toronto.
Canadian Westinghouse Co., Hamilton.
Hall Engineering Works, Montreal, Que.
T. & H. Electric Co., Hamilton.

Transmission Machinery

Allis-Chalmers-Bullock, Montreal.
The Canadian Fairbanks Co., Montreal.
Greey, Wm. & J. G., Toronto.
Laurie Engine & Machine Co., Montreal.
Link-Belt Co., Philadelphia.
Miller Bros. & Toms, Montreal.
Paxson, J. W., Co., Philadelphia
H. W. Petrie, Toronto.
The Smart-Turner Mach. Co., Hamilton.
Waterous Engine Co., Brantford.

Transmission Supplies

Baxter, Paterson & Co., Montreal.
The Canadian Fairbanks Co., Montreal.
Wm. & J. G. Greey, Toronto.
The Goldie & McCulloch Co., Galt.
Miller Bros. & Toms, Montreal.
H. W. Petrie, Toronto.

Trolleys

Canadian Rand Drill Co., Montreal.
Dominion Foundry Supply Co., Montreal
John Millen & Son, Ltd., Montreal.
Miller Bros. & Toms, Montreal.
Northern Engineering Works, Detroit

Trolley Wheels.

Lumen Bearing Co., Toronto

Trucks, Dryer and Factory

Dominion Heating & Ventilating Co., Hespeler
Greey, Wm. & J. G., Toronto.
Northern Engineering Works, Detroit
Sheldon's Limited, Galt, Ont.

Tube Expanders (Rollers)

Chicago Pneumatic Tool Co., Chicago.
A. B. Jardine & Co., Hespeler.
Mechanics Supply Co., Quebec, Que.

Turbines, Steam

Allis-Chalmers-Bullock, Limited, Montreal.
Canadian General Electric Co., Toronto.
Canadian Westinghouse Co., Hamilton.
Kerr Turbine Co., Wellsville, N.Y.

Turntables

Detroit Foundry Supply Co., Windsor.
Dominion Foundry Supply Co., Montreal
Hamilton Facing Mill Co., Hamilton.
Northern Engineering Works, Detroit
Paxson, J. W., Co., Philadelphia

Turret Machines.

American Tool Works Co., Cincinnati.
John Bertram & Sons Co., Dundas, Ont.
The Canadian Fairbanks Co., Montreal.
London Mach. Tool Co., Hamilton, Ont.
Niles-Bement-Pond Co., New York.
H. W. Petrie, Toronto.
Pratt & Whitney Co., Hartford, Conn.
Warner & Swasey Co., Cleveland, Ohio.
Williams & Wilson, Montreal.

Upsetting and Bending Machinery.

John Bertram & Sons Co., Dundas, Ont.
A. B. Jardine & Co., Hespeler.
London Mach. Tool Co., Hamilton, Ont.
National Machinery Co., Tiffin, O.
Niles-Bement-Pond Co., New York.

Valves, Blow-off.

Chicago Pneumatic Tool Co., Chicago.
Mechanics Supply Co., Quebec, Que.

Valves, Back Pressure.

Darling Bros., Ltd., Montreal
Dominion Heating & Ventilating Co., Hespeler
Mechanics Supply Co., Quebec, Que.
Sheldon's Limited, Galt.

Valve Reseating Machines.

Darling Bros. Ltd., Montreal

Ventilating Apparatus.

Canada Machinery Agency, Montreal
Darling Bros., Ltd., Montreal
Dominion Heating & Ventilating Co., Hespeler, Ont.
Sheldon's Ltd., Galt

Vises, Planer and Shaper.

American Tool Works Co., Cincinnati, O.
Frothingham & Workman, Ltd., Montreal
A. B. Jardine & Co., Hespeler, Ont.
John Millen & Son, Ltd., Montreal.
Niles-Bement-Pond Co., New York.

Washer Machines.

National Machinery Co., Tiffin, Ohio.

Water Tanks.

Ontario Wind Engine & Pump Co., Toronto

Water Wheels.

Allis-Chalmers-Bullock Co., Montreal.
Barber, Chas., & Sons, Meaford.
Canada Machinery Agency, Montreal.
The Goldie & McCulloch Co., Galt, Ont.
Greey, Wm. & J. G., Toronto.
Wm. Kennedy & Sons, Owen Sound.
John McDougall Caledonian Iron Works Co., Ltd., Montreal

Water Wheels, Turbine.

Barber, Chas., & Sons, Meaford, Ont.
Kennedy, Wm., & Sons, Owen Sound

Wheelbarrows.

Baxter, Paterson & Co., Montreal.
Buffalo Foundry Supply Co., Buffalo
Frothingham & Workman Ltd., Montreal
Detroit Foundry Supply Co., Windsor.
Dominion Foundry Supply Co., Montreal
Hamilton Facing Mill Co., Hamilton.
Paxson, J. W., Co., Philadelphia

Winches.

Canadian Pilling Co., Montreal
Frothingham & Workman, Ltd., Montreal

Wind Mills.

Ontario Wind Engine & Pump Co., Toronto

Window Wire Guards.

Expanded Metal and Fireproofing Co. Toronto.
B. Greening Wire Co., Hamilton, Ont.
Paxson, J. W., Co., Philadelphia

Wire Chains.

The B. Greening Wire Co., Hamilton.

Wire Cloth and Perforated Metals.

Expanded Metal and Fireproofing Co. Toronto.
B. Greening Wire Co., Hamilton, Ont.
Paxson, J. W., Co., Philadelphia.

Wire Guards and Railings.

Expanded Metal and Fireproofing Co. Toronto.
B. Greening Wire Co. Hamilton, Ont.

Wire Nail Machinery.

National Machinery Co., Tiffin, Ohio.

Wire Rope.

Frothingham & Workman, Ltd., Montreal
B. Greening Wire Co., Hamilton, Ont.

Wood Boring Machines. Pneumatic.

Independent Pneumatic Tool Co., Chicago, Ill.

Wood-working Machinery.

Baxter, Paterson & Co., Montreal.
Canada Machinery Agency, Montreal.
The Canadian Fairbanks Co., Montreal.
Goldie & McCulloch Co., Galt.
H. W. Petrie, Toronto.
Waterous Engine Works Co., Brantford
Williams & Wilson, Montreal.

ALPHABETICAL INDEX.

PROFESSIONAL DIRECTORY

CONSULTING ENGINEERS, PATENT ATTORNEYS, ARCHITECTS, CONTRACTORS, ETC.

T. Pringle & Son

HYDRAULIC, MILL & ELECTRICAL ENGINEERS

FACTORY & MILL CONSTRUCTION A SPECIALTY.

Coristine Bldg., St. Nicholas St., Montreal.

RODERICK J. PARKE

A.M. Can. Soc. C.E. A.M. Amer. Inst. E.E.

CONSULTING ELECTRICAL ENGINEER

INDUSTRIAL STEAM AND ELECTRIC POWER PLANTS DESIGNED. TESTS. REPORTS.

51-53 JANES BLDG., TORONTO, CAN.
Long Distance Telephones—Office and Residence.

J. A. DeCew

Chemical Engineer
Sun Life Building, MONTREAL

Industrial Plants and Processes sold and installed.
Apparatus and Materials for Water Purification.
Free Analyses to prospective purchasers.

CONSULTING ENGINEERS

should have their card in this page. It will be read by the manufacturers of Canada :: :: :: ::

CANADIAN MACHINERY
Montreal Toronto Winnipeg

JULES DE CLERCY, M. E. Gas Engineer

Expert knowledge of all classes of Engines and Producer Plants. If you are thinking of installing a Gas Engine or Producer Plant write me. If your plant is not working satisfactorily I can help you.

413 Dorchester St. - MONTREAL, QUE.

HANBURY A. BUDDEN

Advocate Patent Agent.
New York Life Building MONTREAL.
Cable Address, BREVET MONTREAL.

THE ELECTRIC BLUE PRINT CO.

Blue Prints Positive Blue Prints
Black Prints Multi-color Prints

Largest and finest plant.
Best finished work in Canada at lowest prices.

40 Hospital St., - MONTREAL

PATENTS *TRADE MARKS AND DESIGNS*

PROCURED IN ALL COUNTRIES

Special Attention given to Patent Litigation

Pamphlet sent free on application.

RIDOUT & MAYBEE *103 BAY STREET TORONTO*

PATENTS PROMPTLY SECURED

We solicit the business of Manufacturers, Engineers and others who realize the advisability of having their Patent business transacted by Experts. Preliminary advice free. Charges moderate. **Our Inventor's Adviser** sent upon request. Marion & Marion, New York Life Bldg, Montreal; and Washington, D.C., U.S.A.

PATENTS THAT PROTECT

FETHERSTONHAUGH & CO.

Patent Solicitors & Experts

Fred. B. Fetherstonhaugh, M.E., Barrister-at-law and Counsel and expert in Patent Causes. Charles W. Taylor, B.Sc., formerly Examiner in Can. Patent Office.

MONTREAL CAN. LIFE BLDG.
TORONTO HEAD OFFICE, CAN. BANK COMMERCE BLDG.

SOMERVILLE & VAN EVERY

CONSULTING AND SUPERVISING ENGINEERS

Reports and Valuations, Power Developments
Manufacturing Plants Designed and Erected
Equipments Installed

Federal Life Bldg., HAMILTON, ONT.

THE DETROIT TESTING LABORATORY

1111 Union Trust Building
DETROIT, MICH.

Experienced Foundry Chemists and Metallurgists.
Iron Mixtures for all classes of Castings our Specialty.
Reasonable charges for Analyses.

CARR & SPEER

SUCCESSORS TO

LANE & CARR

ENGINEERS

Rolling Mills,
Steel Foundries,
Grey Iron Foundries,
Malleable Iron Foundries,

Open Hearth Furnaces
Plant Designs
Metallurgical Consultation.

120 Liberty Street, New York.

Canadian Machinery and Manufacturing News

is a medium through which you can talk to 95 per cent. of the citizens of Canada who must be consulted when Machinery, Tools, Power Equipment and Industrial Supplies are to be bought. If your announcement does not appear regularly you are missing opportunities for increased business.

Let us Send you Rates and Further Particulars

FIRE BRICK
"Every Quality," "For Every Purpose."
The TORONTO POTTERY Co., Limited
FACTORIES IN OHIO
TORONTO, CANADA

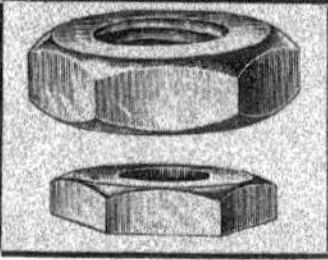

Satisfies the Most Exacting Buyers—

THE OLD RELIABLE

GREEN POSITIVE ROTARY PRESSURE BLOWER

(NEW CENTURY PATTERN)

THOUSANDS IN USE IN ALL PARTS OF THE WORLD

Write for Catalogue. **Prompt Shipments.**

Wilbraham-Green Blower Co.

2518 FRANKFORD AVENUE

1854 PHILADELPHIA, PA. 1907

Our Triple Hem

Steel Shop Barrels and Boxes

Pay for themselves

Write us for catalogue and prices.

High Grade Coiled Wire Springs for all purposes

The Cleveland Wire Spring Co.
Cleveland, Ohio.

The **"ECONOMY"**

Wire Wheel Brush

For cleaning castings and metal patterns, for removing scales from hot forgings, axes, shovels, etc., for cleaning sheet brass and copper, and almost every variety of metal work. Special sizes and shapes for special work. Acquaint us with your requirements.

BRANTFORD, CAN., May 17th, 1906

THE OSBORN MANUFACTURING CO.,
Cleveland, O.

GENTLEMEN,—We are pleased to advise you in reply to your recent letter that we have been using the Economy Steel Wire Wheel Brushes for several years and are well pleased with them.

These Wheels have given us excellent results and we firmly believe can be used to advantage in any stove foundry. Yours very truly,

THE WILLIAM BUCK STOVE CO., Limited
C. J. Parker, Sec.-Treas.

THE OSBORN MANUFACTURING CO.
CLEVELAND, O.

Don't let your moulders swear, get your patterns made by

JOHN CARR

14 Mary St., H3MILTON
Phone 2755

NORTHERN CRANES

The Newton Cupola

NORTHERN ENGINEERING WORKS, - 16 Chene St., DETROIT, MICH., U.S.A.

BOYER AND KELLER RIVETING AND SHIPPING HAMMERS AND "LITTLE GIANT" DRILLS

HAVE DEMONSTRATED THEIR SUPERIORITY OVER ALL OTHERS

Examine the vital parts—note the quality of material used in their construction and simplicity of mechanical parts, then give them a trial on your own work. Tools shipped free for demonstrative purposes to responsible parties.

DUNTLEY

AIR COOLED PORTABLE

ELECTRIC TOOLS

HAVE A RECORD OF

100 PER CENT. EARNING POWER

HAVE SAVED THEIR FIRST COST

in a few months for many purchasers.

EVERY TOOL GUARANTEED

to perform the service for which it is designed.

Write for catalogue describing DUNTLEY ELECTRIC DRILLS, HOISTS, GRINDERS, BUFFERS AND BLOWERS.

Single Motor Type Drill

CANADIAN PNEUMATIC TOOL COMPANY
TORONTO MONTREAL

BRANCH OF

CHICAGO PNEUMATIC TOOL COMPANY
CHICAGO NEW YORK

SALES AGENTS FOR CANADA:

N. J. HOLDEN & CO., Limited, MONTREAL

DOGGETT'S PERFECTION PARTING

GUARANTEED TO PERFECTLY PART UNDER ANY AND ALL CONDITIONS

For Straight and Difficult Lifts
Sticky Patterns
Slanting Copes and Deep Chasings

Permits Swabbing and Slicking.

Will not Burn off the Face of the Mould, Wash or Cut.

SUITABLE FOR BRASS, BRONZE, IRON, ALUMINUM

Manufactured by

STANLEY DOGGETT

101 Beekman Street, New York City, U. S. A.

DISTRIBUTING AGENTS

DETROIT FOUNDRY SUPPLY CO., Windsor, Ont., and Detroit, Mich.

ALSO MANUFACTURE

CHARCOAL and CHARCOAL FACINGS AND ALL GRADES OF SOAPSTONE FACINGS

If the experience gained from 20 years of Air Compressor designing and building is of advantage to you, drop us a line.

We build Air Compressors in 1000 styles and sizes, for all duties.

DO YOU WANT A CATALOG?

Canadian Rand Company,

LIMITED

MONTREAL, QUE.

P. O. BOX 383
W. H. WIGGS
TOOLS AND SUPPLIES
for
Plumbers, Gas and Steam Fitters
Machinists and Electricians
APPAREILS ETC
pour
Plombiers, Gaziers, Mecaniciens
Ferblantiers et Electriciens
WHOLESALE
AND RETAIL
MECHANICS SUPPLY CO.
EN GROS ET
EN DETAIL
Showrooms: 80-90 St. Paul St.
QUEBEC
Entrepot: 59-67 Rue St. Andre

CANADIAN PIPE COMPANY LTD.
HEAD OFFICE VANCOUVER B.C. P.O. BOX 915.

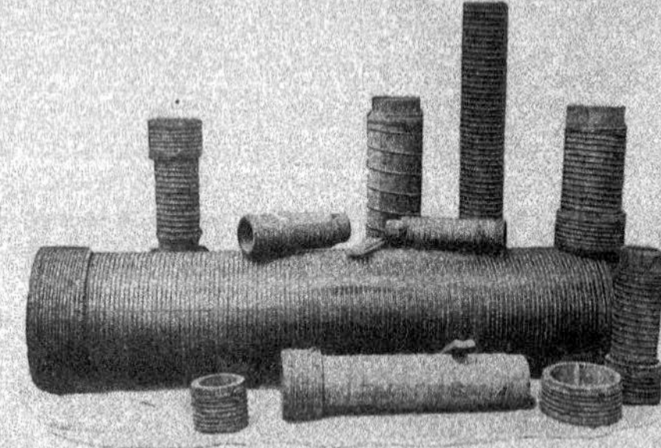

RUGGLES Perfection Flue Cleaner

Made on strictly mechanical lines of the best material. Self adjusting, light, strong and simple. Heads or ends are drop forged steel, undeniably the strongest Flue Cleaner made, though no screws or rivets are used in its construction. The Knives are so made that they are always sharp, and like a plane have the tendency to work deeper, and on account of their long shearing cut, it is impossible to tell when passing through welded flues of the worst character, as the design of the Cutter is such that it is always in direct contact with the metal of the flue.

SOLD ON ABSOLUTE GUARANTEE

Rice Lewis & Son, Limited

Sole Agents, TORONTO

EDGE'S CHAIN—Best by Test

For ***Cranes, Dredges, Derricks, Anchors, Hoists, Lumbering*** and ***Railroad*** purposes, and for all severe uses where a reliable chain is required.

Stock Sizes 3/16 to 1 1/4 Inch

We have handled this chain for about 30 years, and have found it always uniform in quality and strength.

Let us quote you some figures.

FROTHINGHAM & WORKMAN, Limited

Wholesale Hardware and Iron Merchants

Founded 1809 ***MONTREAL, CANADA***

ALLIS-CHALMERS Bullock LIMITED MONTREAL
ELECTRIC LIGHT AND POWER
ALLIS-CHALMERS Bullock LIMITED MONTREAL
Direct Current Generators of the type here shown are designed for general lighting and power service wherever machines suitable for direct connection to steam, gas, or oil engines are required. They are fully described in our new
BULLETIN 1059
ALLIS-CHALMERS-BULLOCK, Limited
Head Office and Works, Montreal; District Offices; Toronto, Traders Bank Bldg; Montreal, Sovereign Bank Bldg.; Winnipeg, 251 Notre Dame Ave.; Nelson, Josephine St.; Vancouver, 416 Seymour St.; New Glasgow, N. S., Telephone Bldg.

CANADIAN MACHINERY
AND MANUFACTURING NEWS

A monthly newspaper devoted to the manufacturing interests, covering in a practical manner the mechanical, power, foundry and allied fields. Published by The MacLean Publishing Company, Limited, Toronto, Montreal, Winnipeg, and London, Eng.

OFFICE OF PUBLICATION: 10 FRONT STREET EAST, TORONTO

Vol. III. **SEPTEMBER, 1907** No. 9

FIVE SIZES, ⅝ x 4½-inch, 1 x 10-inch, 1½ x 18-inch, 2 x 26-inch, 3 x 36-inch.
The Turret Lathe is indispensable in the shop where duplicate pieces are produced and greatly increases production at a much lower cost.

WRITE FOR NEW CATALOGUE "TURRET LATHES."

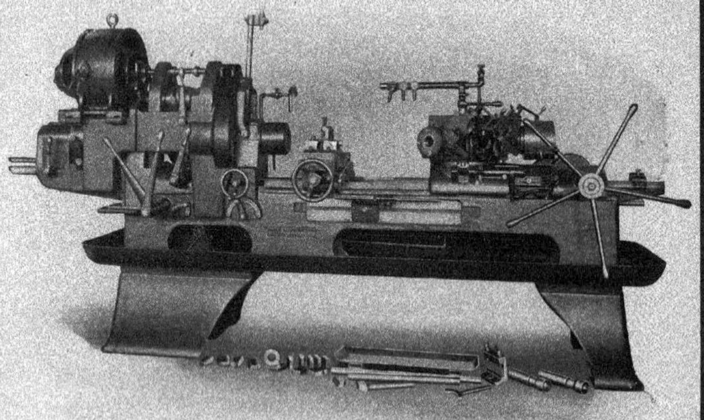

3 x 36-inch Pratt & Whitney Turret Lathe, driven by 5 h.p. motor.

The motor drives through a speed-box which, with the double friction back-gears, affords a wide range of speeds.

Cross and longitudinal power feeds are positive geared, reversible, and have automatic stops in either direction.

The cross slide permits the use of larger forming tools than can be held in a turret and makes possible the use of three turret tools at one time.

PRATT & WHITNEY COMPANY

WORKS:—HARTFORD, CONN., U.S.A.

THE CANADIAN FAIRBANKS CO., Limited, Agents for Canada

MONTREAL TORONTO WINNIPEG VANCOUVER

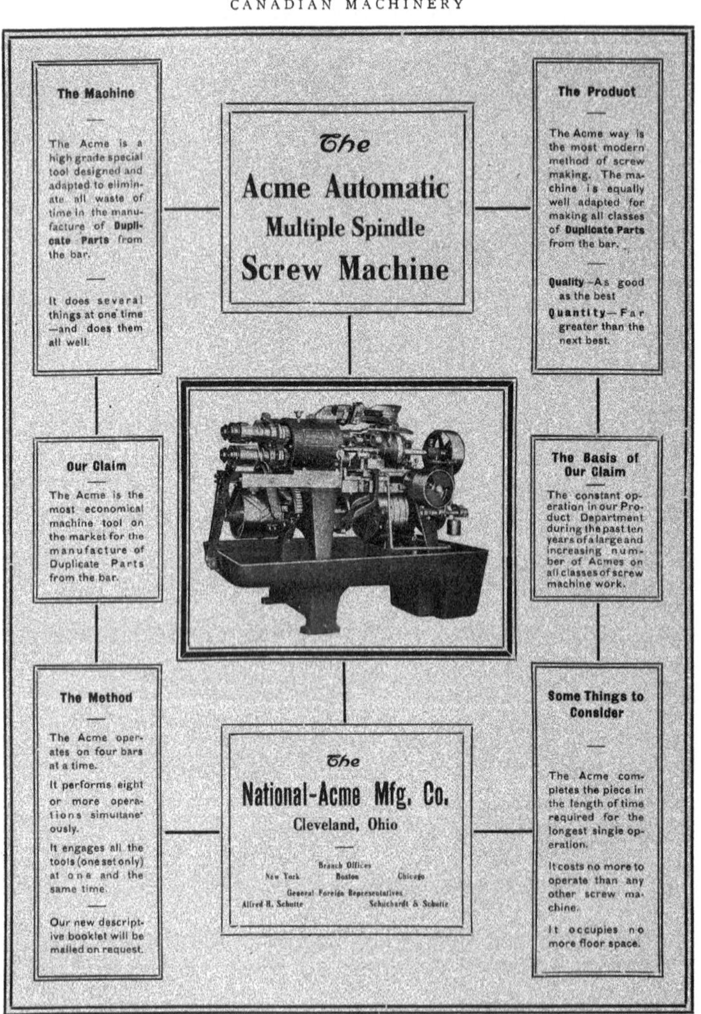

The Machine
The Acme is a high grade special tool designed and adapted to eliminate all waste of time in the manufacture of Duplicate Parts from the bar.
It does several things at one time—and does them all well.
The
Acme Automatic
Multiple Spindle
Screw Machine
The Product
The Acme way is the most modern method of screw making. The machine is equally well adapted for making all classes of Duplicate Parts from the bar.
Quality—As good as the best
Quantity—Far greater than the next best.
Our Claim
The Acme is the most economical machine tool on the market for the manufacture of Duplicate Parts from the bar.
The Basis of Our Claim
The constant operation in our Product Department during the past ten years of a large and increasing number of Acmes on all classes of screw machine work.
The Method
The Acme operates on four bars at a time.
It performs eight or more operations simultaneously.
It engages all the tools (one set only) at one and the same time.
Our new descriptive booklet will be mailed on request.
The
National-Acme Mfg. Co.
Cleveland, Ohio
Branch Offices
New York Boston Chicago
General Foreign Representatives
Alfred H. Schutte Schuchardt & Schutte
Some Things to Consider
The Acme completes the piece in the length of time required for the longest single operation.
It costs no more to operate than any other screw machine.
It occupies no more floor space.

You get quality as well as quantity from the "Cleveland"---two essentials in any machine manufacturing plant---

Let us prove the superiority of the Cleveland on your work.

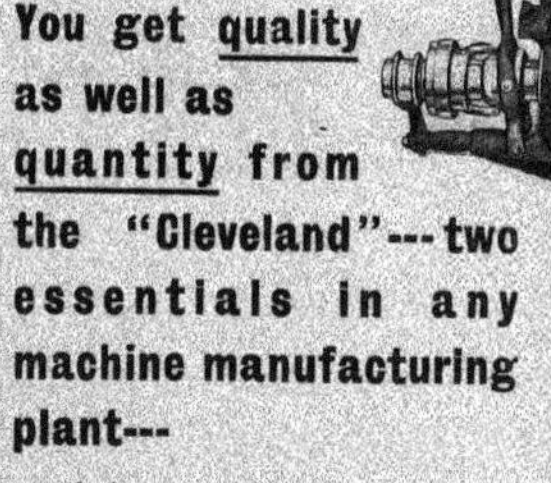

Piece No. 142.---Material, Brass Casting, made on a 2¾-inch Cleveland Automatic Turret Machine; drawing full size; time of all operations for completing this piece 4 minutes; actual cost of labor from 4 to 5 mills.

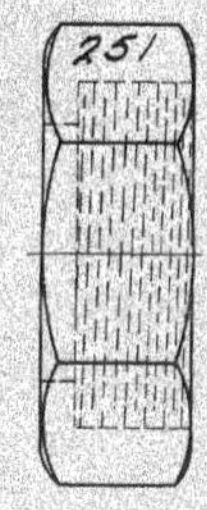

Piece No. 251.---Material, Brass Casting, using our Tilting Magazine Attachment on a 2-inch 3-Hole Cleveland Automatic Turret Machine; drawing full size; output per hour 35; this means all operations for completing this piece; actual cost of labor from 2 to 2½ mills for each piece. Send along samples or drawings. We guarantee our outputs. One man can operate from 4 to 6 machines.

AS WE HAVE NO CANADIAN REPRESENTATIVE, WE SOLICIT CORRESPONDENCE AND ORDERS DIRECT.

CLEVELAND AUTOMATIC MACHINE CO.

CLEVELAND, OHIO, U.S.A.

EASTERN REPRESENTATIVE—J. B. ANDERSON, 2450 North Thirtieth Street, Philadelphia, Pa.

WESTERN REPRESENTATIVE—H. E. NUNN, 22 Fifth Avenue, Chicago, Ill.

FOREIGN REPRESENTATIVES—CHAS. CHURCHILL & CO., London, Manchester, Birmingham, Newcastle-on-Tyne and Glasgow. MESSRS. SCHUCHARDT & SCHUTTE, Berlin, Vienna, St. Petersburg and Stockholm. ALFRED H. SCHUTTE, Cologne, Brussels, Liege, Paris, Milan and Bilbao.

Planing | Milling
105¼ Minutes | 51½ Minutes

No. 4 Plain "Cincinnati" Motor-Driven Miller.

These are grey iron castings, 10¼" wide 14" long. The operation shown takes a cut $\frac{3}{16}$" deep across the top surface and the two edges, and also cuts the ⅜x1" slot from the solid with cutters 8", 3½" and 5¾" diameter, all at one time, at a table travel of 4.2" per minute. The time on this operation, including chucking, is 15.6 minutes. The average time for milling one piece complete is 51½ minutes Former planing time, 105¼ minutes. ***The miller saves more than half the planer time.***

Are you using Cincinnati Millers? **We are Milling Specialists.**

THE CINCINNATI MILLING MACHINE COMPANY, Cincinnati, Ohio, U. S. A.

Canada Agent: H. W. PETRIE, Toronto and Montreal.

POTTER & JOHNSTON
MANUFACTURING AUTOMATICS

will machine your duplicate parts from all kinds of castings, forgings and the bar with extreme accuracy and more economically than upon any other tools in the market.

Catalogue showing interesting specimens of parts finished on these tools free upon request.

ALSO UNIVERSAL SHAPERS

POTTER & JOHNSTON MACHINE CO.

PAWTUCKET, R.I. - - U.S.A.

These gang drills not only drill, but they also ream, bore, face, tap, mill and do a lot of other operations that you are in the habit of buying expensive machinery for. Send description of your work and we will submit proposals.

B. F. BARNES COMPANY

ROCKFORD, ILL.

European Branch:

149 QUEEN VICTORIA STREET, LONDON, E.C.

LIFTING MAGNETS

After several years of quiet preparation we announce our readiness to fill orders for Lifting Magnets.

Cutler-Hammer Lifting Magnets are designed by engineers who for years have devoted themselves to problems involving electric and magnetic control. They possess **distinct advantages** not to be found in any other Magnet now on the market.

Under rigid service tests the records established by these Magnets have been so far in advance of anything previously accomplished that we are prepared to accept orders under the following:

"Three-Way" Guarantee

1.—AS TO WEIGHT: The Magnet we will build for you will be as light as is consistent with proper design and its weight will be guaranteed **not to exceed** that of any comparable Magnet.

2.—AS TO EFFICIENCY: The lifting capacity of our Magnet will be guaranteed to **exceed** that of any comparable Magnet now in use.

3.—AS TO OPERATING COST: The operating cost per ton of material moved will be guaranteed to be **less** with our Magnet than with any comparable Magnet.

Send for Illustrated Booklet. When writing, state size of Magnet desired and nature of material to be handled.

CUTLER-HAMMER CLUTCH CO.

Makers of Lifting Magnets and Magnetic Clutches

MILWAUKEE, - WISCONSIN

NEW YORK OFFICE, 136 Liberty St. CHICAGO OFFICE, Monadnock Block. PITTSBURG OFFICE, Farmers' Bank Bldg. BOSTON OFFICE, 176 Federal St.

FOR

Machinery, Mill Supplies and Hardware

J. R. BAXTER & CO.

SUCCESSORS TO

BAXTER, PATERSON & CO.

Agents for Landis Bolt Cutters and Nut Tapping Machinery; Prentice Bros. Co.'s Drilling Machines and Engine Lathes; Safety Emery Wheel Co.'s Grinding Machinery; Portable Universal Shears; Gear Hobbing Machines, etc., etc.

PROMPT ATTENTION TO ALL ENQUIRIES

MILLING CUTTERS
HIGH SPEED AND CARBON STEEL
—Standard and Special
We also manufacture DRILL PRESSES and DRILL CHUCKS
Write for Catalogue
HAMILTON TOOL CO.,
Hamilton, Ont. Limited

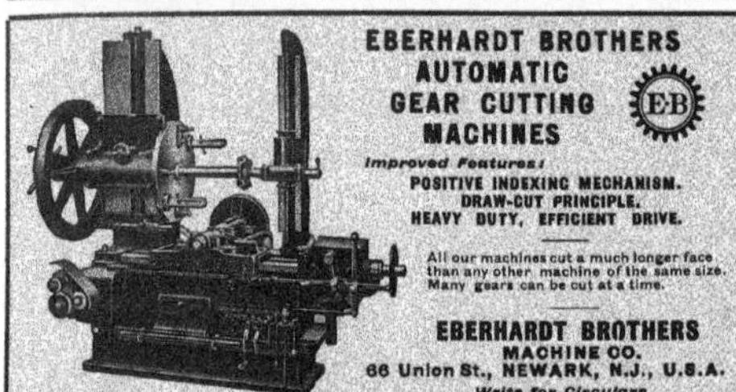
EBERHARDT BROTHERS
AUTOMATIC
GEAR CUTTING
MACHINES
E.B.
Improved Features:
POSITIVE INDEXING MECHANISM.
DRAW-CUT PRINCIPLE.
HEAVY DUTY. EFFICIENT DRIVE.
All our machines cut a much longer face than any other machine of the same size. Many gears can be cut at a time.
EBERHARDT BROTHERS
MACHINE CO.
66 Union St., NEWARK, N.J., U.S.A.
Write for Circulars

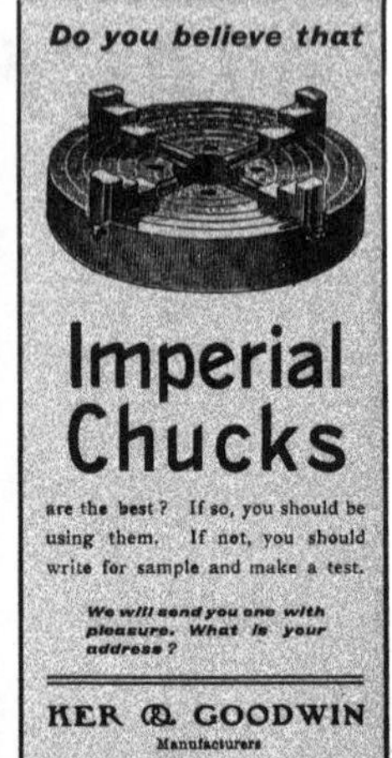
Do you believe that
Imperial
Chucks
are the best? If so, you should be using them. If not, you should write for sample and make a test.
We will send you one with pleasure. What is your address?
KER & GOODWIN
Manufacturers
BRANTFORD, CANADA

GEARS
ACCURATELY MACHINE CUT.
DAVID BROWN & SONS Ltd
HUDDERSFIELD. ENGLAND.
RAW
HIDE
PINIONS.
WORM
REDUCING
GEARS.
HIGH EFFICIENCY
& RELIABILITY
Agents
for CANADA.
W. B. McLEAN & Co
301 St James St., Montreal.

TOOL AND DIE MAKING

Model, Experimental and Special Machine Work of all Kinds.

WOELFLE BROS., - BERLIN, ONT.

Pipe Cutting and Threading Machine

For Hand or Power, Bench or Trench

FORBES PATENT DIE STOCKS will do either from ¼ inch I.D. to 15 inch O.D. *ON THE SPOT*

No expense or delay of carting your pipe to the machine. We keep regular sizes in stock and can make prompt shipment. Write for Catalog.

THE CURTIS & CURTIS CO., - *36 Garden Street, Bridgeport, Conn.*

PIPE CUTTING and THREADING MACHINE

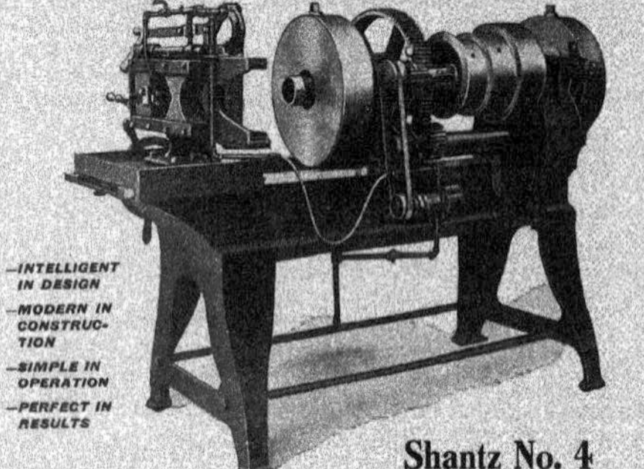

—INTELLIGENT IN DESIGN
—MODERN IN CONSTRUCTION
—SIMPLE IN OPERATION
—PERFECT IN RESULTS

Shantz No. 4

¶ The "**Borden Patent Solid Adjustable Die Head**" with which this machine is fitted is the easiest operated and most perfect mechanism on any Pipe Threading Machine. Any degree of accuracy may be obtained simply by the adjustment of an adjusting screw upon which the operating lever rests. Any variation may be obtained to suit any make of fitting, whether standard or not.

¶ How long have you been "getting along" with the old machine, losing time and money and your patience every time you have any pipe-threading to do? No necessity to do this any longer. Our machine is a time and money saver and we can prove this to your satisfaction. It will stand up to its work—no matter how hard you work it or how hard the work is. We will be glad to tell you about the many special features.

Write *TO-DAY* for Circular fully describing this machine

I. E. SHANTZ & CO., BERLIN, Ont.

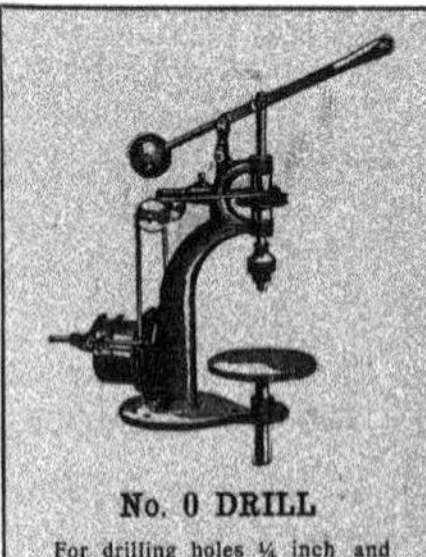

No. 0 DRILL

For drilling holes ¼ inch and smaller. Can be furnished either in bench form, as shown in cut, or mounted on pedestal. For very light drilling users will find this a most convenient tool.

The price is in its favor too.

Send for our complete catalogue.

THE

Dwight Slate Machine Co.,

HARTFORD, CONN., U.S.A.

IT PAYS TO INSTALL THE BEST

POWER PRESSES

The many exclusive features embodied in the type shown above make it so much superior to the next best that there is no comparison.

Write for prices and particulars on **Presses, Dies, Tools and Special Machinery.**

W. H. BANFIELD & SONS

120 Adelaide Street West, TORONTO.

THE UNIVERSAL SYSTEM OF MACHINE MOLDING

The Milling Machine Vise

Some suggestions for the use of the Milling Machine Vise on ordinary work and work requiring special arrangement of jaws.

By JOHN EDGAR.

In the July number under the above title, the various kinds of vises that are used on the milling machine were described; methods were also given by which the truth of the jaws might be tested for accurate work. In this article it is the purpose to give some examples of work and a few suggestions for the use of the vise, both on ordinary work and work that requires some special arrangement of the jaws. What may be said in this article applies to the flat vise in particular, but in most cases the methods described may be used in connection with the other vises, either without variation or with some slight change which will in each case become obvious.

Holding Rectangular Work.

In holding plane rectangular work in the vise the only precaution necessary is to see that it is well seated in the bottom of the stationary jaw, so that the cut may be parallel with the opposite side. When work that is in the rough is held in the vise it is advisable to place between the jaws and the work a slip of heavy paper or thin paste board which will increase the holding power of the jaws and remove the danger of the work shifting under the cut. After the work has been placed between the jaws and held by a slight pressure, it should then be rapped down onto the bottom surface of the jaw by a heavy mall or hammer by two or three sharp blows until it shows by the sound of the blow that it is well seated, then the jaws may be securely clamped.

To Hold Thin Strips.

When it is desired to hold thin strips of work between the jaws parallel of sufficient height must be used to bring the face of the work to be milled up above the top edges of the steel jaws. When taking finishing cuts on work of this character it is best to insert pieces of tissue paper under the four corners of the work to test the fact of its being seated upon the parallels or the bottom of the jaw. With the work held between the jaws firmly the hammer or mall is used to rap it down until all four pieces of tissue paper hold with the same "bit." By using this precaution it is possible to mill a piece of work to within narrow limits.

Method of Holding Round Work.

In the former article mention was made of the means made use of in holding work of other than the plain rectangular form.

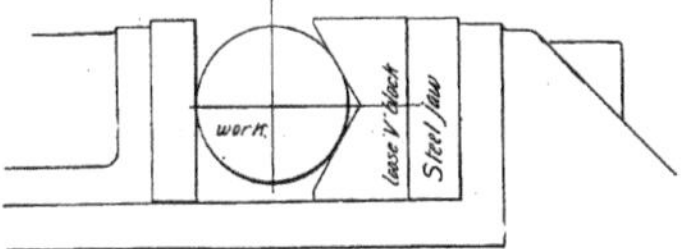

Fig. 1.—Use of "V" Block for Holding Round Work in the Vise.

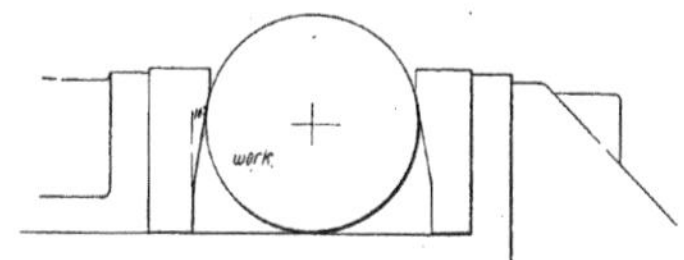

Fig. 2.—Special Jaws Fitted to Vise for Holding Round and Rectangular Work.

Fig. 1 shows the vise arranged to grip round work, such as shafting, etc. In this case a loose "V" block is dropped in between the jaws and used to hold the work as shown. This method is only suitable for small work, as the block takes up considerable of the room that coould otherwise be occupied by the work itself. However, for work that can be held in this manner it is a good method and one that is simple and inexpensive to make up. For a larger range of work the rig shown in Fig. 2 will be more satisfactory. This arrangement consists of special shaped jaws used in place of the plain steel jaws and held securely in place by the regular jaw screws. The angle of the beveled side is 10 degrees. The jaw is made with a straight portion at both top and bottom of the jaw and parallel as in the plain jaw. With jaws of this shape the work is held down against the bottom surface of the opening by the beveled faces. The work would have small chance of moving either lengthwise or by rotating about its axis.

With jaws of this shape two round pieces of work may be held with the same security as that of one piece. Work of small diameter may be rested or a

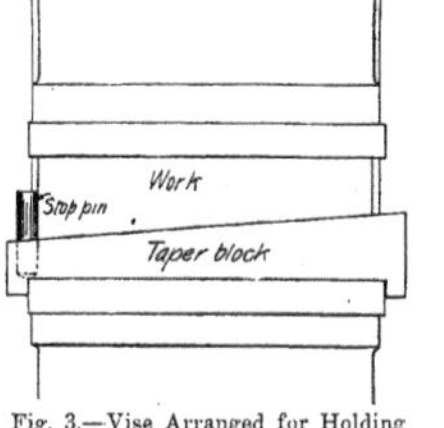

Fig. 3.—Vise Arranged for Holding Taper Work.

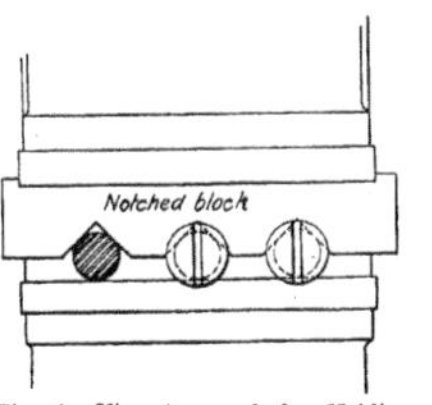

Fig. 4.—Vise Arranged for Holding Small Screws.

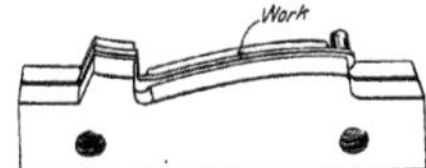

Fig. 5.—False Jaws for Sewing Machine Part.

Fig. 6.—Back False Jaw for Sewing Machine Part.

parallel strip high enough to bring it in to the range of the beveled surface. A vise fitted with these special jaws is not limited to round work, but may be used with the same facility as one fitted with the regular jaw.

It is one of the faults of the vise to tend to force the work out of the jaws, especially when the steel facings are badly worn. Now, with the design of jaw shown in Fig. 2 the pressure is exerted at the upper portion of the jaw so that the tendency to force the work out should be greatly decreased. Jaws of this shape should find a large range of usefulness and should become more common.

Vise for Holding Taper Work.

A variety of work that gives lots of trouble to the user of the vise is taper work. Fig. 3 shows the logical way out of the trouble. This is merely in using a block, one side of which is planed to the required angle and the other fitted over the back jaw to hold it in place, as the tendency would be to shoot out at one side. The stop pin is put in the position shown to hold the work from shifting endwise. This arrangement makes a very easy way to avoid trouble with such work.

To Hold Screw in Vertical Position.

Can anyone think of a more bothersome bit of work than the holding of a screw in a vertical position between the jaws of a vise in the process of slotting the head? Fig. 4 shows an arrangement which does away with all trouble from the shifting of the screw to an angular position, and also facilitates placing it with regard to the position of the slotting saw. In the figure is shown a block with three notches of 90 degrees each on one side, and which is fitted over the back jaw on the other side, holding it in position laterally. By using such a device it is possible to hold a number of screws in a predetermined position in the path of the milling cutters used in slotting the head so that time is saved by it, not being necessary to set the slide each time, and also making it possible to slot the screws in gangs, which is also an advantage.

The use to which this arrangement may be put can be extended to cover any work of this character. Take, for instance, milling the heads of square headed screws. By using heading mills or side milling cutters in gangs, milling two opposite sides of the square at one pass of the cutters, then, loosening up the vise jaw, under the screws by lining up the milled side parallel with the vise jaw by means of a parallel strip or straight edge held against the flat side of the head, clamp the screws in this position and then make another pass of the cutters. The head milled in this manner might not stand very close inspection as to being absolutely square, but would answer all practical requirements. Further uses for this arrangement could very likely be found if one cared to go deeply into the matter.

Great Variety of Work Can Be Done.

A great variety of cuts may be made upon irregular work held in the vise if suitable fixtures are made for same. Nowhere is the vise used to such advantage as in the shops of the large sewing machine manufacturies, of small arms works and typewriter factories. The many small parts made from rough forgings and from the bar, in many cases, in these factories are, to a large extent, held in vises with specially arranged jaws. As mentioned in the former article, the jaw is made to conform to the shape of the edge of the work, being milled so that it is held securely against chattering. These jaws are also used to set the work in the proper position for the operation. In Fig. 5 is shown the jaws used to hold a sewing machine part during the milling operation. It will be seen that they follow the general outline of the edge of the work, while not absolutely conforming to the shape of the piece they do support it along the edge operated upon. It would be rather unhandy, and a time-wasting operation to set the piece properly in the jaws unaided by some means or other.

The proper position that the work should take in the jaws is located by the pocket that is made in the back jaw. This pocket is so shaped that when the work is dropped into the pocket it will take the proper position with relation to the axis of the milling cutters. The pocket is clearly shown in Fig. 6, which represents the back jaw standing by itself.

The depth of the pocket is just sufficient to allow the sliding jaws to pinch the work when brought into contact with it. These false jaws are hardened so as to withstand rough usage and to prevent burrs from forming and also to prevent wear.

The vise can, with a little ingenuity, be converted into a very efficient fixture for holding irregular as well as plain work by working somewhat along the above lines different methods of holding the work suggesting themselves with each job that is tackled.

In describing the different styles and types of vises used in connection with the milling machine mention has not been made of some forms that are in the same category as the vises described, but are better classed as jigs since their use is, in many cases, limited to a singe piece.

---

*Firing With Soft Coal**

Bringing out the importance to manufacturers of having good firemen, and the necessity for firemen to study laws of combustion.

By Geo. D. Fowler.

It is too generally assumed that in firing steam boilers the fuel was burned under conditions over which the engineer or fireman had little or no control and that any man who could keep up steam was equally as good as any other man. This idea was erroneous. In a steam plant consisting of 300 h.p. a test showed that one fireman evaporated less than 8 pounds of water per pound of coal burned, while another fireman evaporated over 9 pounds, the difference between the two results being exactly 2 pounds of water per pound of coal in favor of the latter. A good deal of the waste generally attributed to the steam engine was in reality due to lack of knowledge and skill in the boiler room. It was now well known that a certain quantity of air was necessary to perfect combustion, and that too much air detracts from the economy, and injures the boiler. A glance at his fires, a knowledge of his chimney draft, a look at his dampers, and an understanding of the work his boilers were doing were quite sufficient to guide the skilled and experienced engineer without the use of the thermometer. He will have the fire kept at from 6 to 9 inches thick, a little thicker at the back end of the grate and along the furnace walls and in the corners, because the heat radiated from the side walls and the bridge causes the coal in these places to burn faster than that on the rest of the grate. The fire is kept solid and in good form by quickly sprinkling a thin, uniform layer of coal on alternate sides of the furnace at frequent intervals and by filling in such parts at may burn hollow, and through which the cool air in the ash pit will pour up freely and will chill the gases of combustion, thus lessening the efficiency of the boiler. The careful man will guard against such conditions occurring. He will have the grate surface properly covered with an incandescent bed of fuel, except near the doors, where a row about 15 to 18 inches wide, and built of fresh coal, extends across the front

* Presented before convention of Canadian Association of Stationary Engineers, Guelph, August 13, 14 and 15.

end of the furnace. The heat to which this row of fresh coal is exposed, causes it to coke, and to give off the inflammable gases that it contains, which are burned as they pass back over the incandescent bed of fuel. When brisk fuel is needed, this mass of coke is broken up and distributed over the grate. When the fire has again become incandescent, fresh coal is put to coke, and so the firing continues. The paper discouraged the practice of firing heavily only at considerable intervals, as it did that of heaping the coal up in the centre of the grate, leaving sides and corners practically bare. After putting on a fresh fire, it is a good plan to leave the furnace door open—about a half or three-quarters of an inch—until the fire has burned up a little, so as to admit an extra supply of air—that which passes up through the grate being checked for a few moments by the fresh fuel. The air thus admitted will greatly assist in consuming the gases that are given off by the fresh coal. If the door is kept wide open, the boiler will be cooled down and may be severely strained. As soon as the gases are burned off, the door should be kept tightly shut. A majority of the machines to consume or prevent smoke have turned out not to be very durable. Smoke may be prevented if only perfect combustion can be attained, and most of the smoke may be abated with careful hand-firing. If firemen would give close attention to the laws of combustion and put their knowledge to practice they could more than double the value of their services and their employers would no doubt be quite glad to pay for these.

The Art of Handling Men

Facts, Fancies and Opinions concerning the Handling of Men in Industrial Life—Gathered from various sources, including the writings of James H. Collins in Saturday Evening Post, and some happenings in Canadian Shops.

As to the relative importance of system and the personal or human element in dealing with employees in industrial life, there is as much diversity of opinion as there is diversity of human nature. System is a great thing, but so is personal magnetism. System without personality is pretty dead and is not a breeder of enthusiasm; but personal treatment without system will surely lead to confusion.

When it comes to the human element in handling employees there is a wide difference in methods.

Mrs. Ella Rawles Reader, the "woman financier," once conducted a typewriting office in New York, with sixty girls. To insure sympathy she made it a rule to kiss them all the first thing every morning.

Captain Bill Jones, who was America's greatest steel maker in his day, sometimes started the morning's work by discharging a lot of his bullies and hiring them all over again in the afternoon.

That's carrying the human element pretty far. But look at some of the indoor organizations, and it often appears as though routine ran them. When men are able to make reports, take their cue from an authority chart and put a grievance into a typewritten statement, things often run so smoothly that personality seems a secondary factor. But it seldom is. Time and again have experts worked out a mathematically accurate system for running a great organization, and set it going under their own supervision on a mathematical basis. It runs until they leave and then runs down. Everything was there but personality.

When the Human Element Counted.

In a Canadian machine shop a superintendent was discharged because he did not mix with his men. He hadn't the gift of being one of them and superintendent at the same time. He was superintendent only. He did not bring the human element into play. System did not make up for it. He did not make his men produce as much as some other men could.

Another case of where the art of handling men counted is that of a superintendent of a railway shop in Canada. Some wondered why he held the place. His knowledge of the work which he controlled was not striking, nor was his executive ability of a type to arouse comment. But when you saw him employing labor, or noted the attitude of his mechanics towards him, the problem solved itself. He knew the art of handling men. He brought personality to play and it proved a winning card.

This human element counts for an awful lot. Look among the contractors, the builders of bridges, tunnels, dams and skyscrapers, and it will be found that every man on one of their jobs is playing a game. These employers fascinate the mind. They can organize a working force among Bedouins of the desert and reconcile warring Hindu castes. They can take the thick-skulled African native, who for centuries has had no conception of time, and could not comprehend the purpose of a watch, and interest him in breaking a world record in digging, or in beating some other contractor half-way round the globe. These men make work a game. They can set 10,000 laborers straining to finish a contract a few days ahead of a rival who is often only a figment of the imagination. The job starts off with a "Hurrah!" component that seems to be the best possible cohesive for an organization of elemental men. Frequently an application of fists is necessary, or a display of revolvers. The work is usually the most dirty and disagreeable in the world and to a high degree dangerous, and seldom any too-well paid. It calls for the sudden organization of great working forces in places where men are scarce or have never learned how to work. Yet the contractor carries on his enterprises with a vim, speed and certainty that make other industrial problems appear mighty small in comparison.

Theodore Starrett is one of the pioneer skyscraper builders in this country, and credited with a vast amount of steel construction. It is a saying among steel workers that "every floor of a modern skyscraper costs a human life." Records of accidental mortality among steel workers partly bear this out. Starrett has been building skyscrapers for more than twenty years, and has had constantly under him a force varying from one thousand to fifteen thousand men. In all that period his mortality bill will not amount to a dozen lives lost, and for more than ten years he had the record of not a single fatality on one of his jobs.

Men lose their lives in steel construction chiefly through the pressure that is put on them for speed. Every steel construction job is a race. Starrett has been as speedy as the rest in putting up skyscrapers. He holds some world records. But he has been able to make work a game and a race, and still safeguard his men by good management. Steel workers do not always take kindly to safeguards. They are daredevils at heart, and will steal a ride on a girder while the boss' back is turned, straight up 200 feet in the air on a single strand of steel rope. "Beat him to it" is a maxim of this trade, and nothing is an obstacle in a race. A railroad car was derailed and thrown where a wall was to go on one of

Starrett's jobs. "Brick it in," said the foreman, and but for the arrival of a railroad wrecking crew the car would have become part of the building. A trainload of brick was sidetracked fifteen miles away and the men halted for lack of material. The railroad company put forward its stereotyped plea of being unable to handle traffic. "Go to it," said the foreman. A yard engine was taken away from its crew and that train hauled up on to the job in two hours.

The eight-hour-day is being asked for by machinists in Canada just now. Manufacturers oppose it because of the decreased output it would mean in their shops. One superintendent thought an eight-hour-day was of sufficient length, and thought that he could get as much work done in the eight as he did in the ten. Now his men are working eight hours, and the output of the shop has not suffered. The shop is run on the same system, so system is not the cause. Again, the human element—the act of handling men—comes into play. There is very seldom any labor trouble in that shop, because the men are dealt with in a way which appeals.

The Human Element in Large Organizations.

But what about this human element when the organization is too large for personal supervision! Return to Starrett and his steel buildings.

Formerly Starrett worked by the old system, one building at a time, and the job under his own eye. But now he has thirty to forty jobs going simultaneously. Yet his casualties are almost negligible.

Ask him how he does it, and, although he is a Bachelor of Arts, he will tap the nearest wood and say that it is bad luck to talk about good luck. Bad luck means the death of twenty men on a fifteen-storey building. What Starrett really has done is extend his system of individual supervision. When he hired and bossed the men himself it was a matter of judgment to pick the most agile and intelligent—it would probably not do to say the most careful in this business, though that might count, too, in a man's care for others, if not for himself. He picked good foremen—not the sort of man who cursed a steel worker from the sixth storey and then threw a bolt-key at him as is said actually to have happened in the construction of a building where nineteen men lost their lives. To-day Starrett is out of immediate touch with men. But he supervises the selection of foremen, supervises promotion, and keeps up the morale of his organization. When this is maintained, the force can be driven as fast as any other and with safety.

Work as a Game.

At the Edgar Thompson Works, belonging to Carnegie, in Braddock, Pennsylvania, an enormous broom was formerly to be seen over one of the furnaces. This broom shifted from stack to stack, and wherever placed indicated that the furnace beneath held the world's record in steel production. By and by other steel plants as far away as Chicago became interested in this little game of solitaire, and set out to break records, too. When a new one was established the figures were telegraphed, and the victorious furnace hoisted a broom. In a few years the game became more important than the candle, and brooms were abolished because they could not be shifted fast enough.

This idea was introduced into the steel industry by Captain Bill Jones, a famous character in the steel country around Pittsburg. Jones was born in Pennsylvania, of Welsh parents, and rose from private to captain in the Civil War. Beginning in Andrew Carnegie's employ at two dollars a day, he was eventually made superintendent of the Braddock plant. In less than four months he had doubled its output with the same equipment. A year later he was making steel six times as fast, and later still he doubled that record. His control of men was wonderful. He mixed profanity with quotations from Shakespeare, and would discharge his best workers right and left when in a temper. But the crack steel makers of all that district flocked to Jones.

He had three formulae for men: 1. They must be young and ambitious and able to make steel in strong but pleasant rivalry. 2. They must have an eight-hour day to keep them fit. 3. To make steel fast and well, he believed, a mixed nationality of workers was essential. Captain Jones lost his life among his men, dying of injuries received in a furnace accident and a Hungarian laborer was killed alongside him.

Consulting Employees.

Sometimes a manufacturer finds it a good scheme to find out the opinion of employees upon how things should be run. He finds that if the employee knows that his opinion has some weight with the boss, things run smoothly.

A large industrial plant had friction until the committee plan was adopted —each department head by a committee of not more than six, which could take up difficulties of its own work, and also be called into general advisory councils. There was little trouble after that, for the men who knew things had a voice, and all work went on under arrangements previously discussed, understood and agreed to, and workers in different departments got acquainted with one another—perhaps the most essential point of all. The idea of "get together" is a regular Hague tribunal of peace in industrial affairs. One organization has carried it so far that a conference must not only be held in all difficulties, but those who come to the conference are never permitted to leave in pique, anger or dissatisfaction. Such discussions seldom run along like the afternoon meeting of a sewing circle. Energy comes to confer—high-priced energy—and sometimes politics and personal rancor. but the difficulty is threshed out in the open and no matter how the decision finally falls, this rule brings everything to a close in sweetness and light, largely because it is a rule for just such occasions.

Where a System of Military Principles is Effective.

A certain contractor has adopted military principles in handling men, laying down definite lines of authority in a little handbook carried by all his superintendents and foremen. This book, or "field system," corresponds to the army manual in many ways.

With scattered parties working on contracts in different States, hiring labor under local conditions, it was found necessary to establish some standard system of procedure to insure uniform efficiency and cost. As a beginning, each superintendent was asked to write out rules for his own kind of work —not ideal ones, but rules covering methods actually in use. The first lot was crude, but it formed the basis upon which to build better. Prizes were offered for good working rules and suggestions, and, for months, the system worked lamely because there were many changes. Eventually fewer and fewer prizes were paid, because the system had reached basic principles. To-day it makes work on contracts independent of local conditions.

All employes are required to follow the field system rule for performing specified work until something better is adopted, but each has the right to suggest other methods and can get a personal hearing. Superintendends and foremen go from the home office to a contract a thousand miles away, and employ local labor. But they disregard local customs. There are, for instance, many different ways of mixing mortar. But this field system specifies that all mortar shall be at least two weeks old before it is used. When local custom says, "But this is the way we always do it," the field system backs up the foreman. It may be necessary to explain that the boss is a crank, and introduce the little red book as a compilation of his whims, but, in the end, the work goes on according to tactics. "He pays the wages, you know," is a knockdown argument.

This system contains complete lists of tools to be taken to each kind of job. It cites twenty-five different kinds of work to be performed on rainy days, when the main job is suspended. It gives forms for making out requisitions, reports, tool-lists, pay-rolls, receipts, etc. It deals minutely with steps to be taken for the care of machinery, materials and equipment.

This contractor's forces have used large quantities of dynamite. Plain rules cover the storing and handling of high explosives. Notwithstanding numerous accidents that are constantly occuring all over the country through the use of dynamite, manufacturers of explosives maintain that judicious handling will make it as safe as sawdust. And the experience of the men who work under this field system bears them out. No fatality has ever occurred on one of this contractor's blasting jobs, and only one accidental explosion, in which some excellent sprinting was done, but nobody hurt.

The book gives rules for selecting men, and is reinforced with a card system by which an office record of every competent man is kept, with data as to class of work he is best adapted to. System in selecting men has made possible a record of only four fatalities in ten years. Men seldom kill themselves, says this contractor, but their carelessness leads to the death of others. Only one man ever killed himself, so far as the records go, and he, very curiously, furnished an indorsement for the field system.

Under "Derricks" appears a rule specifying that all derrick-irons shall be procured through the home office. This employee disregarded that rule and purchased equipment from a local concern. Several days after, a derrick-iron broke, dropping a boom upon him, crushing his skull. Investigation showed a hidden blowhole in the iron. Derrick-irons sent from headquarters are made trebly thick to guard against just this danger.

The book gives formulae for mixing every variety of mortar, cement, etc.; for measuring materials and work; calculating weights; tables showing amount of work to be expected from different types of machinery. When a rule cannot be put into two or three lines of text, then diagrams and photographs are used to make the idea clear. Another notable feature is the index, which is so exhaustive that each six pages of text has a page of index, and any subject may be found instantly.

The important element to be incorporated in such a manual of authority is good sense—practically. Bureaucracy will not do—"the heavy curtain drawn between the right thing to be done and the right man to do it," as Balzac said.

The Social Feature Counts Too.

In highly efficient organizations will usually be found a vigorous social life that counts for a good deal. They not only have baseball teams, but the teams that win. They celebrate birthdays, anniversaries of long service, completion of important contracts. They begin things by laying a corner-stone and wind them up with the presentation of a loving-cup. The employer who fancies that these social functions are wanton frivolity, is commonly the man whose business pays twenty per cent. dividends to stockholders at one end and is harried by strikes on the other.

Five hundred drillers and "sand-hogs" sat down to a dinner at Sherry's a few months ago to celebrate, with engineers and capitalists, the completion of a tunnel under the North River, New York. And another striking bit of practicality growing out of such social life was the recent capture of a bond clerk who disappeared from a metropolitan trust company one Saturday morning with a large sum in securities. The police did not know him, so the president scattered his office force to depots and ferries. The culprit was captured through his wife, whom all the employees knew by reason of their association with him on an office bowling team, and the securities were recovered.

Training and Industrial Schools a Factor

Training of apprentices is a very live industrial question to-day. The Pennsylvania Railroad has invested three hundred and fifty thousand dollars in such a training school to attract boys from the high schools. The Steel Trust has another. A certain western manufacturing concern has not only an apprentice school, but other training departments that educate machinists and assemblers before they are sent to regular departments.

A large automobile factory now being built in the United States will have its own training school in a separate building, and boys entering it from the public schools will be taught trades in one-third the time needed in the factory, and with more thoroughness, while the work they turn out when at school will be paid for and used in the factory.

A manufacturing company in Canada has arranged to take boys from the high school for shop training. By this method it is hoped to sort the boy who has natural mechanical tendencies from the one who has not. The first will be encouraged, and in this way a good class of mechanics will be obtained.

Germany's industrial system is to-day regarded by many observers as the ideal one in its organization, its manual training features, its efficiency. It is continually being held up for the emulation of British and American employers. Whatever merits or demerits it may embody, Germany's system is military to the point of paternalism. She trains her worker, puts him into the factory, keeps him employed through state bureaus, insures his life, pensions him in old age or disability, boards and lodges him while out of work, letting him pay in either money or labor, assigns him to relief construction when dull times come, helps him secure a home in prosperity, treats him in sickness and loans him money on his belongings. In fact, the state goes almost to the point of coming round every night to tuck the German workman into his bed.

TORONTO EXHIBITION OPENED.

Once again the Canadian National Exhibition has been opened at Toronto, and its reputation of being better than ever each year is still maintained.

The appearance of the grounds has been greatly enhanced by the erection of the three new buildings, the horticultural building, the railways building, and the new grand stand.

The machinery exhibits, in which readers of Canadian Machinery will be chiefly interested, are more representative than they have been for some years. Last year there were only two exhibits of machine tools in the machinery hall, while two American manufacturers, the Gisholt Machine Co. and the Landis Tool Co., exhibited tools in the process building. This year there are four exhibits of machine tools in the machinery hall, the Gisholt Machine Co., the London Machine Tool Co., H. W. Petrie and the A. R. Williams Machinery Co. The London Machine Tool did not exhibit last year. Other new exhibitors in the machinery hall are: Bordon Canadian Co., Cook Motor Co., Evans Rotary Engine Co., Schumacher & Boyle, Hornby & Son, Crossley Bros.

In the way of producer gas plants, there are more firms represented this year than last, although one has dropped out and a good feature this year is that these exhibits are in the machinery hall instead of in the process building.

The rotary engine is in evidence this year, there being two engines exhibited, one of the Manson Manufacturing Co. and the other, the Evans Engine Co.

It is to be hoped that before another year comes round there will be a new machinery hall. That is the one big necessity at the exhibition now. Then instead of four exhibits of machine tools there will be a dozen, and instead of being three exhibits of wood-working machinery there will be six or eight.

CANADIAN MACHINERY
and Manufacturing News

A monthly newspaper devoted to machinery and manufacturing interests, mechanical and electrical trades, the foundry, technical progress, construction and improvement, and to all users of power developed from steam, gas, electricity, compressed air and water in Canada.

The MacLean Publishing Company, Limited

JOHN BAYNE MACLEAN - - President
W. L. EDMONDS - - - Vice-President
H. V. TYRRELL - - - Business Manager
J. C. ARMER, B.A.Sc., - Managing Editor

OFFICES:

CANADA
MONTREAL - 232 McGill Street, Phone Main 1255
TORONTO - 10 Front Street East, Phone Main 2701
WINNIPEG, 511 Union Bank Building, Phone 3726, F. R. Munro
BRITISH COLUMBIA - Vancouver, Geo. S. B. Perry

UNITED STATES
CHICAGO - 1001 Teutonic Bldg., J. Roland Kay

GREAT BRITAIN
LONDON - 88 Fleet Street, E.C., Phone Central 12960, J. Meredith McKim
MANCHESTER - 92 Market Street, H. S. Ashburner

FRANCE
PARIS - Agence Havas, 8 Place de la Bourse

SWITZERLAND
ZURICH - - - Louis Wolf, Orell Fussli & Co.

SUBSCRIPTION RATE.

Canada. United States, $1.00, Great Britain, Australia and other colonies 4s. 6d., per year; other countries, $1.50. Advertising rates on request.

Subscribers who are not receiving their paper regularly will confer a favor on us by letting us know. We should be notified at once of any change in address, giving both old and new.

Vol. III. SEPTEMBER, 1907 No. 9

CONTENTS

MACHINERY MARKET CONDITIONS.

At the beginning of the fall months, when business men are again getting down to the hard grind after the holiday season and the warm summer months (not so warm this year, though), seems a fitting time to review the machinery market conditions in Canada, and discover how business has been during the past year, and what prospects are for the coming year.

As everyone knows, Canada has just entered upon a period of great agricultural and industrial expansion. The whole world looks to Canada as a country of almost unlimited resources. People from all parts are flocking into the country. The population is increasing at a tremendous rate. The great Northwest and the vast mineral areas provide means for prosperity. There simply must be immense development in Canada during many years to come, and thus there simply must be an ever-increasing demand for all kinds of machinery for many years to come.

Naturally, in a country which is developing rapidly, there will be need of all available capital, and there may be occasional times when money will be pretty tight; but the only thing which could materially delay this prosperity would be a decided crop failure. Although crop returns this season will not be as good per acre under cultivation as in some years in the past, yet the yield will be fairly good. There is a lot of talk about the tightness of money at the present time, and there are varied opinions as to the result it will have; yet in all probability things will come back to normal conditions before the situation is felt seriously. Foreign capital has great confidence in Canadian enterprises, which is well for a country needing so much money for development purposes as Canada does.

During the past year, machinery manufacturers and machinery dealers have found business exceptionally good, so much so, that there are very few manufacturers in Canada who have not found it necessary to add to their plants in some way. Many plants have been working overtime, and all are busy.

Following are some expressions of opinion as regards the machinery market conditions by some Canadian manufacturers and American manufacturers who are directly interested in the Canadian market:

The Colonial Engineering Co., Montreal: "Business has been exceedingly good with us, considering that we opened up in Canada only on May 1st, since which time we have obtained a number of large contracts. The money market is certainly in a depressed condition. We think it is because there is more work to be done than there is actual funds to do it with. The general prosperity is exceedingly marked both in the United States and Canada. The ratio between the actual currency in circulation in the two countries and the stocks and bonds outstanding is too great. Over-capitalization in the United States is probably the direct cause of this condition."

A Montreal machinery supply firm: "With regard to deliveries we may say that generally speaking these at present are very slow both from the States and from England. With regard to monetary conditions we share the experience of most of the dealers in so far that we find it very difficult, indeed, to get in our collections."

The Packard Electric Co., St. Catharines: "We have been considerably handicapped in our work during the past year on account of slow deliveries of raw material of all kinds. Copper, which enters largely into the manufacture of our products, has been and is still very uncertain, delivery depending on the sizes and quantity ordered. We are obliged to place our orders some time in advance, and for larger quantities than would be necessary if more prompt deliveries could be depended upon. In consequence, this means the tying up of more capital than is the case when deliveries are prompt.

"Business for the past year has been exceptionally good, but some of the large propositions contemplated may be delayed or postponed on account of the stringency of the money market."

An American machine tool manufacturer: "The export business—especially European demand—has been larger than ever, while Canada is unquestionably a promising future market for American machine tools.

"With reference to the long deliveries, this, of course, has been a considerable handicap, but the buyers now realize that this is a necessary condition of the market, and are making their plans far enough ahead to accommodate themselves as to deliveries.

"The speculative feature of the money market makes all business more or less on the surface, but if the underlying bases of prosperity are there, beginning with good crops—manufacturers need not be especially concerned with regard to the manipulations on Wall Street."

Cleveland Automatic Machine Co.: "We wish to say that we have been handicapped very much on account of actually receiving more business than we were able to handle.

"We do not expect any increased business, because, as far as we can grasp the situation, times are not as good as they were six months, or a year ago. We do not see any possible chance of a great slump in business, but, nevertheless, it has settled down to a great extent.

"We are sorry to say that our business in Canada has been exceptionally poor. The reason for this we cannot explain. We have sold more or less tools in your territory during the last fifteen years, but even though business is so good the world over, Canadian manufacturers seem exceedingly slow in purchasing our automatic machines. Whether it is on account of the tariff, or some other cause, we are unable to determine. We have, from time to time, in years gone by, tried to sell our machines through agencies who handle American machinery in your country; we have also had a representative handling our tools independent of agencies, and, latterly, we are using advertising space to assist us in our business, and you would be surprised at the small number of inquiries we receive."

The American Tool Works Co., Cincinnati: "We are running our factory to its full capacity, with plenty of orders ahead at present. There is no doubt a lull in business generally throughout the United States owing to the financial situation and unless there is an improved condition of same ere long, it will have its influence on general business. We find our foreign business increasing, which will doubtless fill the gap that may be occasioned by a falling off in the States. Our Canadian business is large and promises well for the future."

There is a feature in the machine tool business in Canada at present which deserves attention. This is the efforts which are being made by Canadian machine tool builders to make their product as good as is obtainable. The complaint has been made in the past that the majority of Canadian-made tools did not come up to the standard set by American manufacturers. It was claimed by buyers that careful work was not put on them, and that they could not be depended upon for very fine and accurate machine work. Then the minor point of finish was also complained of.

This is now being all changed. The necessity of being up to date in every particular is being fully recognized, and considerable effort and money is being spent in this respect. As the market broadens this will be more and more noticeable, and after a while specialization will take place, providing opportunity for the best and most accurate work. Even now certain firms are making specialties of certain tools in anticipation of the future market condition.

AUGUST METAL MARKET.

Never before in the history of the metal trades has there been a depression so acute and of such long duration as that which has prevailed in the Canadian metal markets during the past two months. With the exception of pig iron, the business in all metal commodities is characterized by complete stagnation. The only thing which will induce consumers to buy metal is dire need of that commodity to keep their factories in operation. Now that this period of depression has become so prolonged, it has taken on a very serious aspect and a little more than the usual apprehension expressed by dealers is noticeable in metal circles.

The present stagnation is attributed to the usual annual idleness of the furnaces during the summer months and mostly to the stringent conditions existing in the money markets, due to speculation and a more than ordinary foreign investment.

Pig iron during the past month has experienced a steadily increasing demand. Opinion was expressed by a Montreal dealer recently that upwards of 10,000 tons of pig iron have been sold in local circles within the past fortnight. Canadian prices have been firm and subject to few or no fluctuations. The English market has been variable. One week prices are high; the next week they fall to a minimum. Absolutely no buying has been done in the American market, partly because of the prolonged idleness of the furnaces and partly because of the uncertainty prevalent in speculative circles.

Ingot tin continues very weak. The buying public have no confidence in the market. Prices fluctuate, and until a level of prices has been reached which is low and which will guarantee steadiness, consumers will remain coldly indifferent to buying. The price ruling throughout the past month for Lamb & Flag and Straits is $44. Minimum premiums for immediate delivery are offered, and little activity exists.

Copper is still very weak and little demand is experienced. Market conditions in Canadian circles are unsatisfactory, due to stringency of money and due to the demoralized conditions in the American market. Regarding this, the Iron Age (August 15) says: There is no exaggeration in the statement that the market is demoralized. The difference between the figures given to the public by selling agencies of the leading producers and the figures at which consumers can buy and are buying, though their wants are very meagre, is so great that it would be futile to name producers' alleged quotations. Transactions are of such limited proportions that a sale of 100 tons is nowadays considered a large quantity. Since our last report the London market for spot has dropped from £82 10s. to £76 10s., while futures fell from £80 15s. to £73 10s.

Although H. H. Rodgers' arrival in New York City did much to stimulate buying with the immediate decline in the prices, the stimulus he administered was only temporary and it will require an event of this importance to again enliven the market.

Antimony has continued weak throughout the month. Little business in it is being transacted. The price for Cookson's continues at 8c.

Machine Shop Time and Cost System

A simple and effective Time and Cost System as used in the shops of John Bertram & Sons, Limited—Premium System of arranging work found satisfactory.

The time and costs systems used in the plant of John Bertram & Sons, Ltd., Dundas, Ont., are very simple and effective, and a description of these systems will probably prove interesting and instructive to readers of Canadian Machinery.

Time Recording System.

The time system is so arranged that each man keeps his own time. He registers his time of commencing and leaving off work each day on a time register, which is placed in a convenient position in the department in which he works. Then there is the Daily Time Card, on which he registers the time of commencing and leaving off each operation during the day. These daily time cards are made out for all the operations which are carried on in the shop, such as turning, boring, milling, planing, etc., and Fig. 1 shows one for turning. This turning card is used as an example throughout.

Upon commencing an operation the mechanic fills in the date, his name, his clock number, the order number, the number of his machine, and the number of the drawing from which he is working, all shown in Fig. 1. In the Bertram shops every order which is executed, be it very large or very small, has its number, and each piece contained in the machine or device constituting that order has its number. Also every operating machine in the shops has its number. Then he marks his card with a check mark under the hour at which he commenced, say 7 a.m. Upon completing this operation, say 11.15 a.m., he places a cross under eleven and also opposite the 15-minute mark, as shown. Thus it is seen that nothing less than a quarter-hour is recorded. This shows clearly that 4¼ hours were consumed on this operation, and this is recorded in the lower right-hand corner of the card under "hours on job," as shown in Fig. 1. The operator then proceeds with his next operation, filling out a second card as before (one card is used for each operation), but puts his check mark under 11 and opposite 15 minutes, as shown in Fig. 2, to show time of commencing, and then continues as before. If the job on which he is working is not completed on that day, he marks his card with the time occupied during that day, as shown in Fig. 2, and commences work the next morning with another card, as shown in Fig. 3 finishing at 10 a.m. This there is a record that 7¾ hours' work was put on drawing number 615 for 4007, 4¾ hours, on Aug. 2, and 3 hours on Aug. 3.

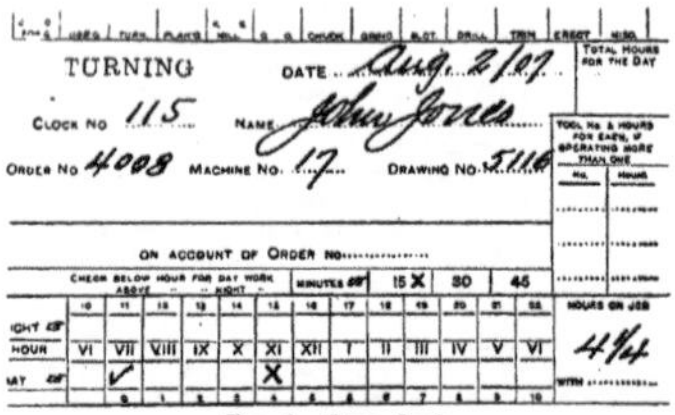

Fig. 1.—Time Card.

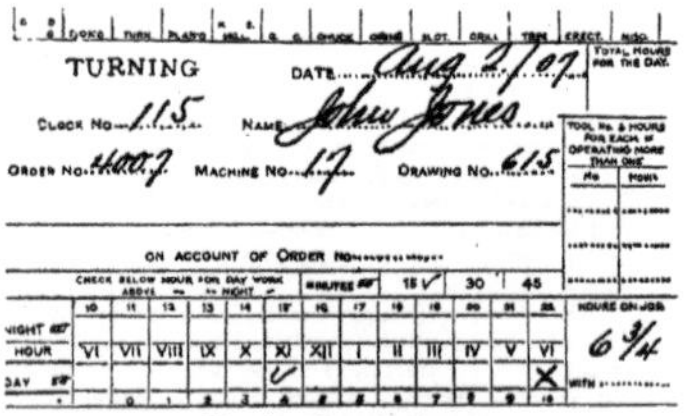

Fig. 2.—Time Card.

In regard to jobs occupying less than 15 minutes, it is desired to keep a record of these, but not to charge less than 15 minutes against any one job, and hence if two or more jobs are done in the fifteen minutes, the time of commencing and leaving off will be the same on the two or more time cards, but the 2nd and 3rd, etc., cards are marked in the right-hand lower corner, ¼ "with No. —," the number of the first order being filled in.

It will be seen on the right-hand side of the time cards that provision has been made for one man running more than one machine; and at the end of the day this column on all the cards for that day can be filled in with the numbers of the machines and the time of operation

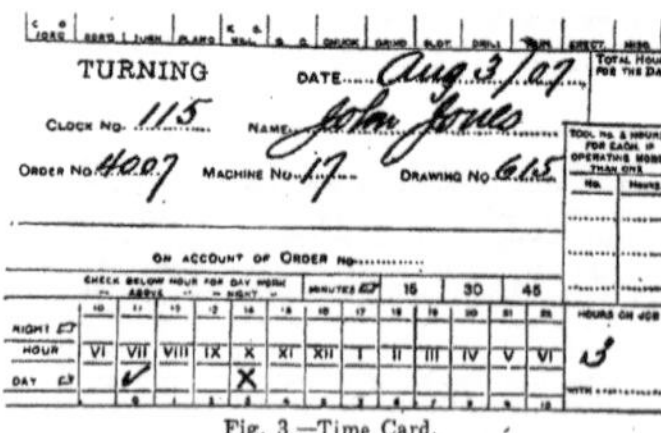

TURNING DATE Aug. 3/07

Clock No. 115 Name John Jones

Order No. 4007 Machine No. 17 Drawing No. 615

On Account of Order No.

Hours on Job 3

Fig. 3.—Time Card.

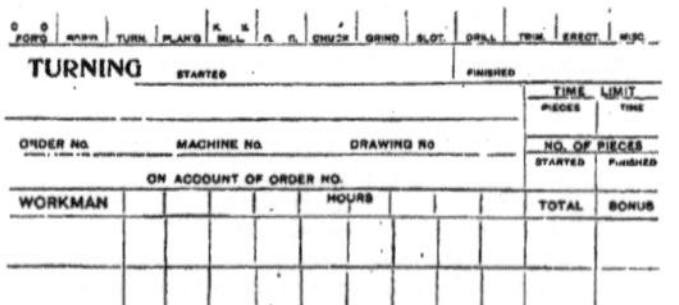

Fig. 7.—Premium System Time Card.

for each during the day. The total hours for the day is also filled in.

Similar cards to these are made out by all the mechanics in the shop, and

ORDER NO. 4008	DESCRIPTION			TIME SUMMARY									DATE OF ORDER Aug 5	CARD NO. 1
	HOURS WEEK ENDING							HOURS WEEK ENDING						
CLASS OF WORK	MON	TUES	WED.	THURS	FRI	SAT.	TOTAL TO DATE	MON	TUES	WED	THURS	FRI	SAT	TOTAL TO DATE
FORGING AND CUT OFF	2	5					5							
BORING														
TURNING	3	7		15		20	20							
PLANING	4	6		13		19	19							
MILLING & KEY SEATING														
GEAR-CUTTING														
TRIMMING														
GRINDING														
SLOTTING	1	3		5		7	7							
DRILLING				4			4							
CHUCKING					6		6							
ERECTING						12	12							
TOTAL HOURS ON ORDER							73	TOTAL HOURS ON ORDER						

Fig. 4.—Time Summary Card.

in the morning all cards for the previous day are sent to the time department, where they are checked with each man's time register. Then they are sorted into their respective operations, and listed on a time summary sheet. On this sheet is recorded the hours of each operation on each order for the day, and there is a summary of the total hours of each operation for the day as well as the total machine hours. This is carried out with the aid of a Burrough's adding machine.

From this sheet the time is transferred to a time summary card, Fig. 4. On this card is summarized all the time spent in the different operations for each order. Looking at Fig. 4, it will be seen that the order number is filled in, and a description of order given; date of order is filled in and number of card, according to the length of time required to finish the order. From the time summary sheet each day the time spent on each operation is entered on this card, and is entered additively, so that as the columns marked Monday, Tuesday, etc., are filled in they will show the total number of hours put on that operation up to that time. For instance, supposing an order, No. 4008, were started on Monday, and on that day 2 hours of forging were done; this is entered as shown. Then on Tuesday 3 hours of forging are done. Instead of entering 3 under Tuesday, 5 is entered as being the total hours of forging up to Tuesday. This is carried out until Saturday. For the other operations the same thing is done, and at the end of the week there is a complete record of hours for the week on each operation, and the total hours on the order for the week is reckoned. A similar record is kept each week until the order is completed, when the assembled weekly cards give a complete record of time spent on the order, the last card showing total time.

Description of Cost System.

The cost system naturally follows after the time system, and in this system a record of costs is kept on cards, there being a card for each order or article. Fig. 5 shows the front of this cost record card, and Fig. 6 shows the back of the same card. In Fig. 5 it will be noted that the order number and the number of the machine as sold are recorded. From the latest time summary card of this order number the total number of hours on the different operations can be filled in, and the cost computed according to the rate of wages. The total labor cost is carried forward to the back of the card, Fig. 6. For convenience of comparison the labor cost of the same machine executed in the shop at a different time and under different order numbers can be entered from previous cost record cards. In Fig. 6, the back of card, the amount and cost

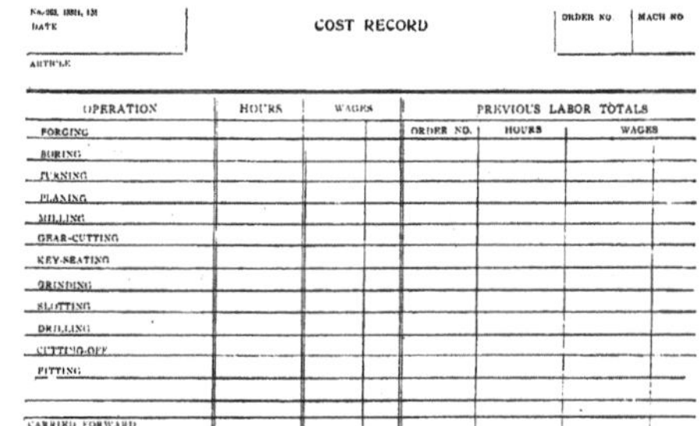

No. 262, 1001, 131
DATE

ARTICLE

COST RECORD

ORDER NO.	MACH NO.

OPERATION	HOURS	WAGES	PREVIOUS LABOR TOTALS		
			ORDER NO.	HOURS	WAGES
FORGING					
BORING					
TURNING					
PLANING					
MILLING					
GEAR-CUTTING					
KEY-SEATING					
GRINDING					
SLOTTING					
DRILLING					
CUTTING-OFF					
FITTING					
CARRIED FORWARD					

Fig. 5.—Cost Record Card.

MATERIAL			TOTAL		REMARKS
BROUGHT FORWARD					
POUNDS CAST IRON					
POUNDS MACHINE STEEL					
POUNDS TOOL STEEL					
POUNDS STEEL CASTINGS					
POUNDS BRASS					
POUNDS BABBITT					
SCREWS, NUTS, OILERS, ETC.					

SOLD TO	ADDRESS	
SERIAL NO.	DATE SHIPPED	VIA
NET PRICE	INVOICED	INVOICE NO.

LABOR COST	MATERIAL COST	TOTAL COST
WAGES......	MATERIAL......	LABOR COST......
Per Cent......	Per Cent......	MATERIAL COST......
TOTAL......	TOTAL......	TOTAL......

Fig. 6.—Back of Cost Record Card.

of materials are recorded, data for which is obtained from the stock room. Thus the complete cost of the machine is arrived at. At the bottom of the card the selling price of the machine is computed by addition of percentage of profit which the company allows themselves. A record of the disposal of the machine is also kept, and the completeness of information regarding cost which is contained in the lower three sections of the card will be noticed.

Premium Labor System.

The premium system of paying for labor is adopted in these shops, and is giving good satisfaction. A time limit for each piece of work has been set, information for the setting of this limit being obtained when the shops were being operated by day labor. The time limit set appears to be very satisfactory to the machinists and other mechanics, and never mind how much less time than the time limit is consumed in doing certain pieces the time limit remains unchanged. The system is simply this: Five hours is allowed for a certain piece; a machinist does it in 3 hours, that is in 2 hours less than the time allowed. The machinist is allowed the 5 hours plus one-half the time which he saved, i.e., one-half of 2 hours; and thus he is credited with 6 hours' work, having worked for three hours. In case the time taken to do the work exceeded 5 hours, he would be allowed only for the 5 hours. Should a man attempt to work too quickly and do careless work, the time required to remedy his careless work is deducted from his time to the amount of the bonus time, but no farther. This man and the man who exceeds the time limit are watched, however, and necessary steps taken to make them improve if such is done persistently.

The manner of recording this is shown in Fig. 7, which is a time card given to the machinist with the foregoing explanation of the system, the manner of recording on this card can easily be understood by following the headings.

The Manufacture of Double Ball Bearings

Description of the manufacture of the Chapman Double Ball Bearings as carried on in the new factory in Toronto.

An interesting line of manufacture is being carried on by the Chapman Double Ball Bearing Co. in their new plant in Sorauren Ave., Toronto. The factory is built with every provision for extension, for which there is also plenty of ground space. Of the equipment an adequate idea can be obtained from the accompanying illustration, showing the different departments. In speaking of the manufacture of the parts, this equipment will be further referred to.

View of Furnace Room.

The balls and their caging are the chief parts of the bearing, and one of the first things is to have these balls absolutely true. These balls are brought from the United States, and every ball, before it is received, is micrometered. The cagings come from the presses, and the balls are assembled into the cages

The outside housing of the bearing is an iron casting. These castings come direct from the foundry, and in a few minutes on the lathe a groove is cut inside to fit the ring, and one on the outside to receive the dust cap.

The steel ring that fits into the housing is one of the most interesting items in a double ball bearing. This is the outside track on which the balls must race. These rings are the product of a special steel composition, and all steel used for this purpose goes through a special analysis. These rings are first reamed and trued on the lathe. This is done fourteen at a time by placing on an arbor.

From the lathe the rings are taken to the furnace room, where they are hardened and tempered. This furnace room is a very interesting department, and is very well equipped. The furnaces are gas-fired and the temperature is kept constantly by means of a pyrometer. The rings are placed in the furnace by the attendant, and when they become the same temperature as the furnace, which is according to the judgment of the operator, they are removed. Each ring as it is removed is doused in a vat of salt water, where it is left about 15 seconds. The rings are then placed on top of the furnace to dry after which they are put into a bath of special tempering oil. After a thorough soaking they are removed and thoroughly scoured with benzine.

After this each ring is tested separately by the tone test. The ring is struck, and according to the tone the attendant can tell whether there are any flaws. After this test the rings are taken to the grinding machine, where the minute uneven surfaces are removed. From there the rings go to the assembling department, and are fitted into the groove in the outer casing.

The conscientious care which is taken with the ring is duplicated with the cone, which fits over the sleeve and forms the inside contact surface for the balls. The cone is machined, hardened, tempered and ground in the same way as the ring was.

The check nut, which screws in under

the cone and keeps it in place on the sleeve, is also an important part, and after it is screwed into position, it is held on by a ring of set screws.

After being assembled, the bearings are all tested. The testing is done on a hydraulic machine, by which any desired pressure may be exerted upon the bearings, and the bearings are mounted upon a shaft which can be rotated at any desired speed. Thus a complete and thorough test can be given the bearing.

We have followed the different operations in a general way from the foundry or stock room right to the tested bearings, and this can be followed with greater interest if we glance at the accompanying illustrations.

Everything turned out by the company is built to standard, and the systems used throughout the shop are very simple and comprehensive.

A View of the Machine Shop.

SCOTTISH NATIONAL EXHIBITION.

The arrangements for the Scottish National Exhibition to be held in Edinburgh next year have made good progress. The plans, subject to any alteration which may be deemed advisable, have been accepted, and work on the grounds and buildings will commence either at the end of this or the beginning of next month.

In the accepted designs the industrial hall, of 100,000 square feet and the machinery hall, of 20,000 square feet, are included in one magnificent building, and everything, both in connection with the structures and the grounds, has been planned with a view to charm the eye and at the same time to facilitate the commercial requirements.

SOUVENIR WATCH FOB.

The souvenir watch fob which the E. H. Mumford Co., Philadelphia, Pa., are sending out is very neat. This could probably be obtained from the company upon application by mentioning this paper.

NEW STEEL FOR TAPS.

Jonas & Colver, Limited., Sheffield, Eng., have lately introduced a new kind of tool steel upon the market, known as "Intra" steel. This steel is specially recommended by the makers for taps, dies, punches, chisels, etc. It is handled in Canada by Wm. Abbot, Montreal.

Showing Method of Assembling Bearings.

STANDARD PROPORTIONS OF MACHINE SCREWS.

At the Indianapolis meeting of the American Society of Mechanical Engineers, a new report was submitted on the standard proportions of machine screws, which embodies several changes over the report last handed in. Only three sizes remain the same as before, and additional sizes have been added to the list.

The included angle remains 60 degrees as before, and the flat at top and bottom of thread is still one-eighth of the pitch for the basic or standard diameter. There is a uniform increment of 0.013 inch, between all sizes from 0.06 to 0.19 and of 0.026 inch in the remaining sizes. This change has been made in the interest of simplicity and because the resulting pitch diameters are more nearly in accord with the pitch-diameters of screws in present use.

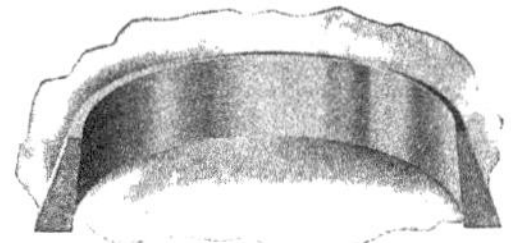

A Section of Bearing Ring Used in Ball Bearing.

The pitches are a function of the diameter as expressed by the formula

$$\text{Threads per inch} = \frac{6.5}{D+0.02},$$

with the results given approximately so as to avoid the use of fractional threads.

The diagram, which is much less complicated than the one first used shows the various sizes for both 16 and 72 threads per inch, and shows, among other things, the allowable difference in the flat surface, between the maximum tap and the minimum screw, this variation being from one-eighth to one-sixteenth.

The minimum tap conforms to the basic standard in all respects except diameter. The difference between the minimum tap and the maximum screw provides an allowance for error in pitch and for wear of tap in service.

The form of tap thread shown is recommended as being stronger and more serviceable than the so-called V-thread, but as some believe a strict adherence to the form shown might add to the cost of small taps, they have decided that taps having the correct angle and pitch diameter are permissible even with the V-thread. This will allow a large proportion of the taps now in stock to be utilized.

Tables of standard proportions have been prepared by the committee, and readers of Canadian Machinery may secure copies of these tables from the American Society of Mechanical Engineers, New York; or more compact tables have been prepared by the Corbin Screw Corporation, and probably copies of these could be secured through the society.

Proportions of Heads.

The accompanying illustrations show the four standard heads, and proportions of parts. Tables of proportions of these heads may be obtained in the same way as tables of proportions of screws.

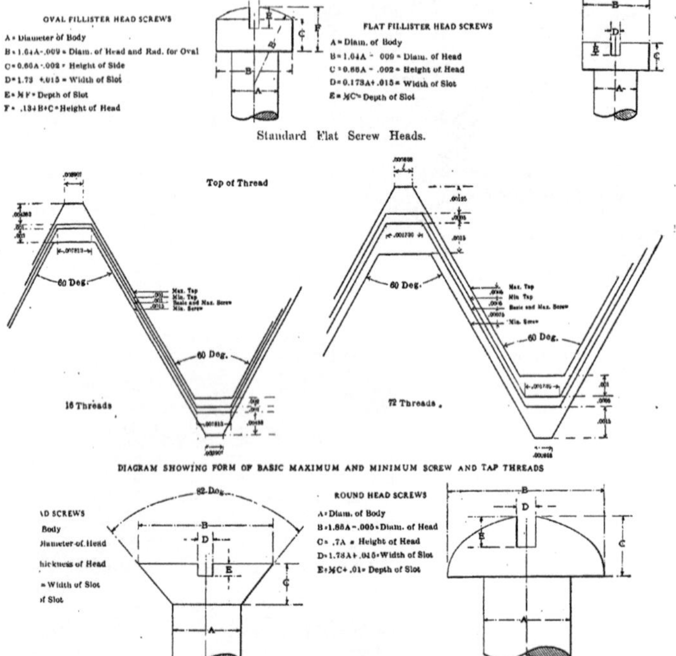

Standard Flat Screw Heads.

DIAGRAM SHOWING FORM OF BASIC MAXIMUM AND MINIMUM SCREW AND TAP THREADS

Standard Round Screw Heads.

A commission of German electrical experts and railroad officials, which was appointed to visit United States and Canada, to study interurban transportation facilities in these countries, are now on their way.

*Improvements on Compressive Riveters**

A riveting machine designed by the author in which he gets the desired toggle joint movement, and at the same time secures automatic adjustment, in which the difficulties with the hydro-pneumatic machine have been eliminated—Improvement in pneumatic hammers.

By CHESTER B. ALBREE.

In compressive riveter work there are two or three types which are quite familiar, the oldest type being the straight hydraulic machine invented by Tweddel in England, and later on, the pneumatic riveter by Allen of New York, who was perhaps the first to make it a success. Later came the hydro-pneumatic riveters. With the hydro-pneumatic riveter we have been making some experiments, and it was found to be advisable for several reasons, notably for greater economy of air, simplicity of construction and better action, to try to improve the methods that had been in use. In driving rivets the pressure required differs from punching materials in that in punching, the greatest pressure comes at the beginning of the stroke, when the punch comes down on the material. In riveting, however, especially hot riveting, the easiest work is when the die first strikes the rivet and the greatest pressure is required to finally form the head. That being the case, it can readily be perceived that a constantly increasing pressure would be the theoretically correct pressure to drive rivets. This pressure is most easily obtained by means of the toggle joint, which theoretically gives an infinite pressure with an infinitesimal movement at the end of the stroke. In practice, of course, we do not get an infinite pressure; but, as most riveters are of horse shoe or yoke type, the limit pressure is the yielding point or the bending point in the yoke. In straight toggle joint machines the general idea is shown in Fig. 1.

In practice we find that when the cylinder has made seven-eighths of its stroke the pressure line of the rivet dies rises up to about 15 times the pressure in the cylinder. By that time we would have traveled within about 1-16 inch of the final stroke of the machine, and beyond that point the probabilities are that there would be spring in the yoke. If we made the yoke strong enough not to spring at all, it would be so heavy as to be utterly unmanageable. So it is only necessary that deflection should not occur at a pressure below that necessary to drive the rivet. Hence in the toggle joint arrangement we have the best possible arrangement for driving rivets.

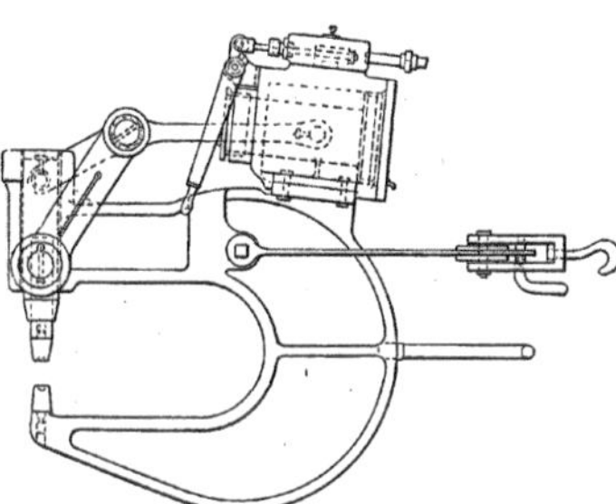

FIG. I. GENERAL ARRANGEMENT OF RIVETER

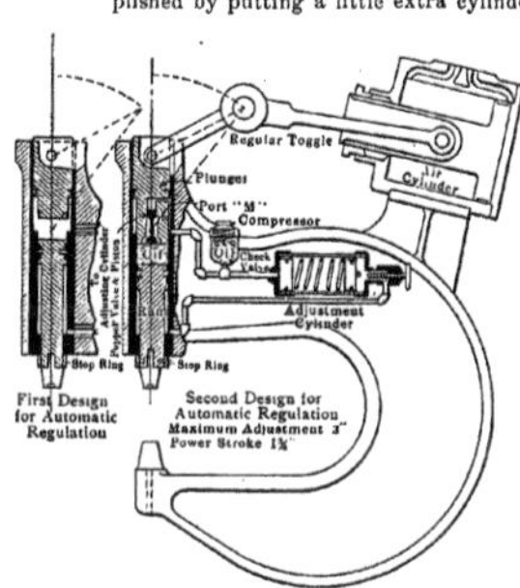

FIG 2. FIRST ATTEMPT AND PERFECTED DESIGN

But there are certain drawbacks in the practical application of the toggle pressure. The principal one is that its stroke is absolutely fixed. It never varies for a given leverage. In riveting you are liable to have 1 inch or 2 inches, or maybe only ½ inch thickness of plate, and in order to have the maximum pressure just as the die comes to the surface of the plate it is necessary to adjust the distance between the die and the point of maximum pressure by means of a screw actuated by hand. In work that does not vary it makes little difference, but in ordinary structural work, and boiler work, you have constantly to drive rivets through different thicknesses of material, and each time it will be necessary for the operator to adjust the screw. That requires a certain amount of skill, and if it is not done correctly the chances are that you will not drive the rivet sufficiently tight, if you do not close it with maximum pressure.

To overcome this difficulty of adjustment the hydro-pneumatic machine was devised, which is nothing more nor less than a hydraulic intensifier. The ram alone gives a very small but powerful motion, and it is necessary in riveting to have clearances, in order to go over angles, stiffeners, etc., so it is desirable to have a longer stroke. Of course we do not need high pressure over a longer distance than, say, 1¾ inches. The question, then, was how to get a clearance movement. This was accomplished by putting a little extra cylinder below the air cylinder. The air pressure acting on its piston forces liquid into the ram cylinder at low pressure, and by this means we get three or four inches of preliminary adjustment.

The objections to this form are: in the first place, it is difficult to pack; and second, that in order to get two inches of die motion with a maximum pressure of 50 tons, it requires a 15 to 20 inch stroke in the air cylinder, and a very high pressure throughout this stroke in the ram cylinder and plunger cylinder; sufficient for the final closing pressure, thus wasting power. The practical advantage is that it does not require

*A paper presented before the Engineers' Society of Western Pennsylvania, by Chester B. Albree, president Chester B. Albree Iron Works, Pittsburg, Pa.

skilled workmen to adjust the dies; they adjust themselves. It occurred to us that we might get the desired toggle joint effect and yet have an automatic adjustment, and what I desire to bring out is the device for the accomplishment of this, which may be of interest to you.

In Fig. 2 is shown our first attempt and also our perfected form. As far as the toggle joint action goes, it is practically the same as the first machine described. The pressure from the toggle, in the first form of machine, is transmitted from the large area of the plunger to the top of the ram and also through a pipe to the adjusting cylinder. The ram being smaller in area and free to move, advances more rapidly than the plunger and continues until the rivet die on its extension strikes the projecting rivet. As the plunger continues, the pressure in the cylinder is limited by the pressure due to the spring in the adjusting cylinder, which is only 20 pounds per square inch, insufficient to upset the rivet beneath the ram. Hence the liquid will now displace the piston in the adjusting cylinder, the ram remaining stationary. As soon, however, as the projection on the plunger enters the ram cylinder the full toggle pressure is transmitted through the incompressible liquid to the ram, forcing down the rivet, and the differential area above forces the remaining liquid in the large plunger bore into the adjusting cylinder.

During the downward motion of the ram the liquid beneath it is forced into the opposite end of the adjusting cylinder, against the spring pressure. It is obvious that the ram may move its whole adjusting stroke; or not at all, up to the time that the projection on the plunger enters the smaller area; after which the further travel of the ram is that of the plunger, until the ram meets opposition greater than the pressure of the toggle, when it will stop. This arrangement, therefore, automatically adjusts the point of maximum pressure to suit the work. On the return stroke we have the direct pressure beneath the ram, as well as the suction of the plunger, to raise the ram to its original position.

Theoretically, this design was correct, and it worked very well indeed for about two strokes. At the end of the second or third stroke our packing was gone and we have found it impossible to hold the pressures. The trouble lay in the fact that we put cup leathers at the end of the plunger, and when the cup leathers entered the chamber, the moment the pressure rose to a high point it tended to cut the leather right out. So in order to make the device practically, as well as theoretically, successful, it was necessary to devise some scheme to have the leather cups, which hold better than any other hydraulic packing, move always surrounded by the walls of the cylinders, and pass no ports whatever. To do this and yet allow the liquid to pass freely from the upper to the lower part was rather a difficult proposition. We accomplished it in this manner.

Referring to the latter form, in Fig. 2, it will be noted that the extension of the plunger, when fully up, projects into the smaller area of the ram cylinder; and the cup leathers are used to pack it. In the interior of this extension is a valve of the poppet type, but having a stem carrying on its end a small piston.

This valve is normally held open by a spring. So long as the pressure above and below this small piston is the same, the spring holds the valve open, but when the pressure below is greater than above, the piston will move up, closing the poppet valve. This occurs only when the port M leading into the space below the small piston is closed, due to its passing from the large diameter bore to the smaller ram bore. When closed, the toggle pressure acts on the liquid below the plunger extension, raising the pressure sufficiently to move the small piston and connected valve, and later exerting very high pressure on the poppet valve, shutting it very tight. The adjusting action is precisely the same as in the first type, except that the liquid flows through the plunger extension instead of around it, during the adjustment part of the stroke. With this arrangement it will be noticed that all of the plunger and ram packings are continually enclosed by cylinder walls and pass no ports or openings, so that the packing leathers are not injured. I would say that the adjusting device is patented and the poppet valve device is now being patented.

In any device of this kind there is always a certain loss of liquid, due to a film in the ram (although, theoretically, the quantity of liquid is constant), and in the course of a little while there would be a partial vacuum inside and pressure on a cavity would not give good hydraulic pressure on the rivet.

It was, therefore, necessary to provide a constant source of supply of the liquid, so arranged that when the pressure rose in the confined liquid it would not blow out, but when there was a vacuum in the system additional liquid would run in.

This loss of liquid is made up from a small storage or compensating cylinder, full of liquid, having a piston with a spring behind it, connected to a larger bore of the plunger by a pipe, having a check valve in it. Whenever there is pressure in the plunger cylinder the check valve remains closed; but when the toggle is fully back, and the piston in the adjusting cylinder is against the cylinder head, so that no pressure due to its spring is exerted on the liquid, any loss of liquid will tend to create a vacuum in the plunger cylinder, and then the check valve will open and oil flow out of the compensating cylinder, under the pressure of the spring acting in its piston, to replace that lost.

Pneumatic Hammers.

We have been working on pneumatic hammers and we have now perfected a hammer in which there are one or two novel features.

In pneumatic hammers of nearly all makes, one of the sources of trouble has been that if the workman picked up the hammer and put his finger on the trigger when there was no chisel or rivet set in it, the piston would begin to reciprocate, and not having any tool to strike at the lower end, it would strike the cylinder head; and in a matter of a minute or two it would smash the piston or cylinder. About 75 or 80 per cent. of the breakages of pneumatic hammers are due to the carelessness of the workmen in pressing on the trigger when there is no work to do. In other words, the little piston strikes the cylinder head with disastrous results. We have devised a method of obviating this trouble that is very simple.

The admission port is located near, but not at the end, of the larger cylinder bore. When no tool is placed in the end of the hammer the lower end of the large piston diameter passes and closes the admission port, thus preventing air from acting upon the differential area to lift the piston. Any compressed air below the large diameter escapes by a small leakage port to the exhaust; and this leakage port is only open when the admission port is closed. In hammers actuated by valves exterior to the piston, it seems impossible to use this device, and attempts have been made to mechanically close such valves, but they do not appear to be very successful. The same effect is obtained, but at the expense of loss of air, when the leakage port is designed to open when the piston is at the extreme end of the stroke, but does close the admission port.

Patents are now pending for these improvements.

This very simple device is very effective. One can pick up a tool with 100 pounds air pressure in the pipe, put a finger on the button, opening the throttle, and it will not start unless the tool is fully in place, whereas, with all other tools I know of, the moment the trigger is pressed the tool begins to work, and would be useless in a few minutes.

Good wages come to the men that earn them—there are no exceptions.

Machine Shop Methods and Devices

Unique Ways of Doing Things in the Machine Shop—Readers' Ideas and Opinions Concerning Shop Practice—Useful Data for the Machinist. Remuneration Will be Given for Contributions.

PROPER METHOD OF BORING GAS-ENGINE CYLINDERS.

By Gordon A. Ronan.

Many are the methods devised by modern engine practice, to produce properly bored cylinders. To do this accurately, and at the same time economically, is not always easy.

The following article has been written with a view of assisting small manufacturers of gasoline engines, who are anxious to improve their present product, to whom the initial cost of a cylinder boring machine would be almost prohibitory.

Most small manufacturers pay too little heed to the essential features of their cylinders, which leads to trouble, when the engine is completed. A cylinder when finished should be bored absolutely cylindrical and parallel, the interior should present an even smooth surface, just as it leaves the tool. Under no circumstance should grinding with emery or other abrasive materials, take place.

To insure successful results, a builder of gasoline engines should have access to a well made engine lathe, of suitable size and proportions for work undertaken, the only other requisites being, one inexpensive cylinder jig pattern as shown in the accompanying cut, from which two castings should be taken, and a cutter-bar of the largest possible diameter, suitable to the bore and length of cylinder.

This cutter-bar should be made of machine steel turned its entire length, fitted with a cutter of circular section (see A in cut), inserted through cutter-bar leaving ample room for cylinder to clear itself at end of roughing cut. Three of these cutters should be used on each diameter cylinder required.

The cutters 1, 2, 3 should each in turn be placed in hole through boring bar, centrally located and each countersunk to accommodate a taper nosed binding screw. Cutter No. 1 is made first, the outer diameter is turned to the roughing diameter required, and is there backed off on either end to form two cutting edges, which should cut opposite the centre of the cylinder bore. The No. 2 cutter is formed in similar manner, only slightly larger in diameter, forming a lighter intermediate cut. The No. 3 cutter should receive the utmost care throughout its preparation. It should be about 1-10,000 inch over size. No. 2 cutter, and should leave the cylinder bored exactly the finished size. In the cut shown herewith 3.5 inches represents the bore.

The jig pattern should be carefully planed on the bottom, and the three bases, which take the binding screws, should be perfectly square with the base. The jig should be securely bolted to the saddle of lathe, with one casting on either side of cross slide, by means of T bolts, size of lathe slots in saddle. In most lathes the jig can be slotted in such a manner as to admit a greater variation in length, thus admitting longer cylinders, and a greater range of work.

After the jig has been centred on saddle, by means of lathe centres, the cylinder is placed in position resting on the two lower binding screws, the boring-bar is next inserted through cylinder, so when cutter No. 3 is inserted only 1-10,000 of an inch has to be removed. This final cut should be taken very slowly and with a fine feed to insure a smooth, bright surface. The cutter should be carefully honed with oil-stone on the start. This leaves a surface void of ridges and the roughness so often found in so-called high-grade makes.

The cylinder, which up to this stage has not been removed from its position, has not been distorted, strained or forced out of shape in anyway, is now finished ready to be placed upon arbor, on which any later work on the exterior may be completed.

It is obvious with the jig presented herewith, many other similar classes of work such as pumps, bushings, may easily be done in like fashion, thus broadening the range of work and usefulness of this tool, making it an indespensable adjunct to every modern, up-to-date shop.

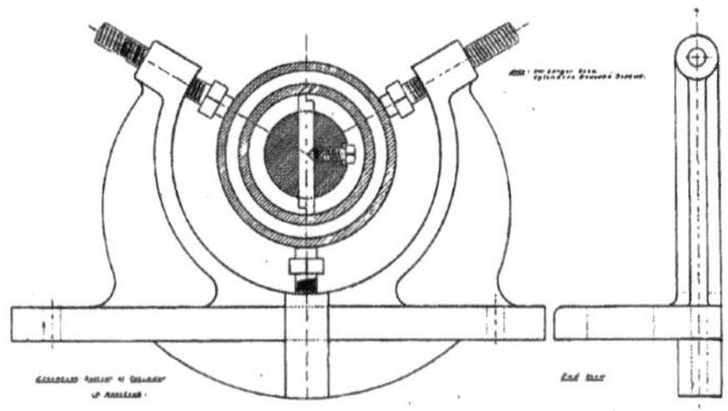

Proper Method for Boring Gas Engine Cylinders.

placed between the centres and driven from the face plate by lathe dog. By means of this revolving cutter-bar, the cylinder is quickly centred and the six binding screws ⅝-inch in diameter hold the cylinder securely in position.

The roughing cut is now taken with No. 1 cutter, which takes the heaviest feed and fastest speed the lathe will stand. Should a number of these cylinders be put through this first operation, it is advisable to lay the cylinders to one side for a day or so, to allow the molecules of the iron ample chance to regain their former position. If in haste, the second operation may be proceeded with, using cutter No. 2, with slower feed and speed. This is important, as a smoother surface will be made,

This method is not an experiment, but an actual success, having been found to operate very successfully in our experimental gasoline engine laboratory, and we recommend it to all our fellow-gasoline builders knowing that satisfactory results must follow.

ATTACHMENT AS DIVIDING HEAD.

By J. H. R., Hamilton.

I send you a sketch of a small attachment which could be placed on any lathe or similar machine to act as a dividing head.

I have had occasion many times to

lay off equal distant holes on jigs, dies, etc., and finding the large gear H on most lathes were made with an odd number of teeth, it was necessary to scribe a circle and divide it into as many parts as is desired.

The accompanying sketch shows an appliance which could be used in many ways, such as laying of holes on a turned surface, as shown, or holding reamers in place while grinding clearance, and for many different odd jobs.

A graduated circle with an indicating finger would be very handy, but it would not lock the lathe head.

A circle divided into 72 parts would be very acceptable as a division of 2, 3, 4, 6, 8, 9, and 12, could be obtained. A circle could be scribed the desired diameter and a scriber in the tool post or a surface gauge could be used to divide the circle into as many parts as is desired.

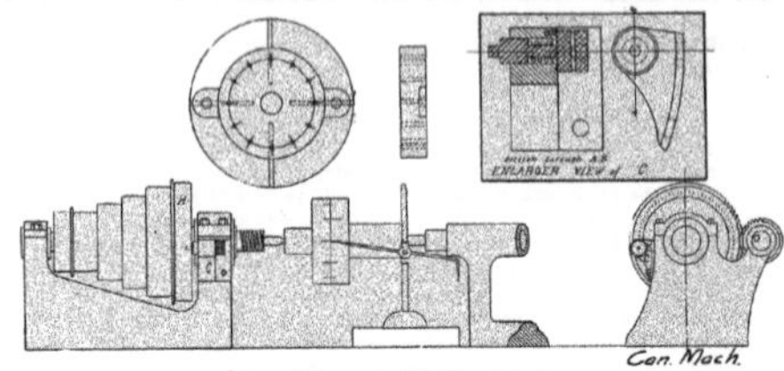

Attachment As Dividing Head.

MILLING KEY-SEATS.

By. C. S. Gingrich.

Every machine shop has some key-seating to do, and it is fast becoming the custom to consider this as distinctly milling machine work. One of the difficulties encountered in doing such work is the tendency of the machine to chatter. This is mainly due to the use of cutters poorly adapted for such work.

Milling Key Seats.

The accompanying illustration shows a key-seating job done on a No. 2 plain "Cincinnati" miller. The material is about 90 point carbon crucible steel, and the key-seat 13-16 in. wide x 3-8 in. deep is milled at a single cut with a 3½ in. diameter cutter running 44 r.p.m. with a feed of .075 in. per turn, giving a table travel of 3 1-3 in. per minute.

The second cutter shown on the arbor is for milling a part of this key-way slightly larger, so that the parts to which the key is fitted can be conveniently slipped into place.

The cutters, it will be noted, are "plain milling cutters." The teeth are spaced very wide (1⅜ in.) and although the cutters are only 13-16 in. face, the teeth are alternately nicked to break the chips. This is the form of milling cutter which has been demonstrated to cut the freest and the fastest.

It has been the practice in the past to use "side milling cutters," that is to say cutters with side teeth in addition to the peripheral teeth for such work. Experience has shown that such cutters are almost sure to chatter. A little thought will show that the side teeth are not needed, because all the cutting is done with the peripheral teeth. The chattering of side milling cutters can very often be materially reduced by reducing the angle of clearance on the side teeth. A better plan, however, is to not use such cutters at all for such work as key-seating, but use instead, cutters like those described above.

DATA ARRANGED IN HANDY FORM.

By J. H. R., Hamilton.

Herewith is shown a table of formulas and data, which is very often used by machinists and other mechanics, that I devised for ready reference, and which I have found very useful. I have taken off blue prints, which I have sold for 20 cents apiece. This table has met with such general favor that I thought it would be of interest to readers of Canadian Machinery. (Any readers desiring to secure a blue print of this table, can obtain J. H. R.'s name and address by sending us a postcard.—Editor.)

A SPECIAL RIVETER.

By Chester B. Albree.

Herewith is described a riveter built for some special work. The work to be done was the riveting of some concrete mixers, they being in the shape of two cones placed together. The problem was to reach into a very limited space and drive the rivets. It had been done by hand and something was wanted which would be a little quicker than the pneumatic hammer and something which would give tighter rivets, and be a power machine. It was necessary to have a reach of some 50 inches, and we had an extremely small opening. At the same time it was necessary to be able to adjust for different thicknesses of ma-

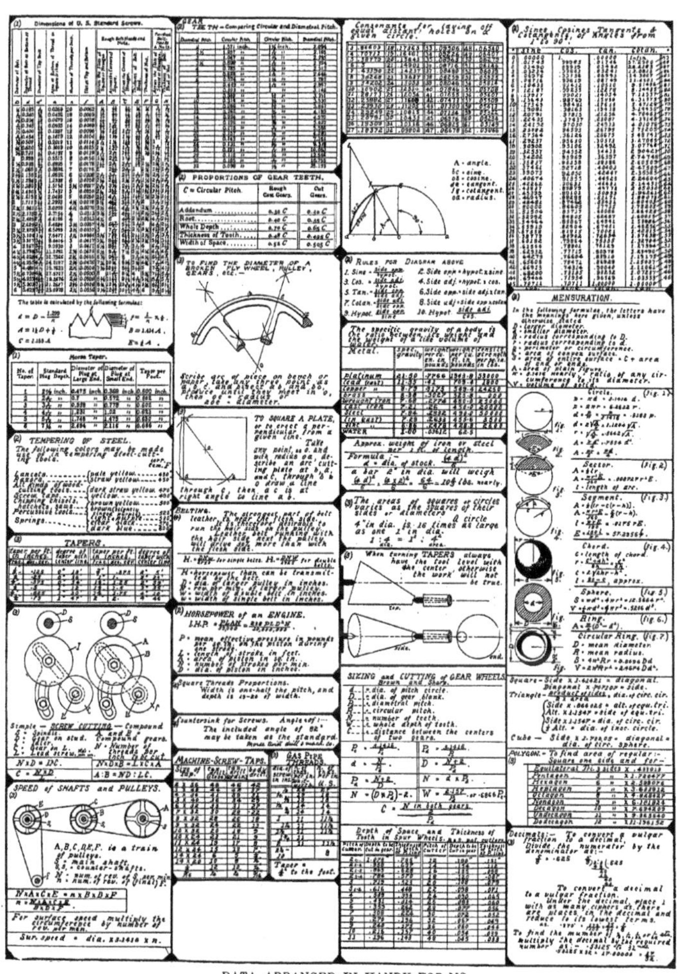

DATA ARRANGED IN HANDY FOR MS.

terial, and to give what we call an alligator motion to the jaws of the machine. The ordinary riveter could not be used for this work.

To secure adjustment of the projecting alligator jaws of the machine, we inserted a screw carrying the trunnions of the fulcrum on one end—a hand wheel with thread in the hub, serves to raise or lower the screw—thus adjusting the distance between the ends of the jaws as desired, and at the same time, not interfering with their clearance or action.

We also designed a special carriage to

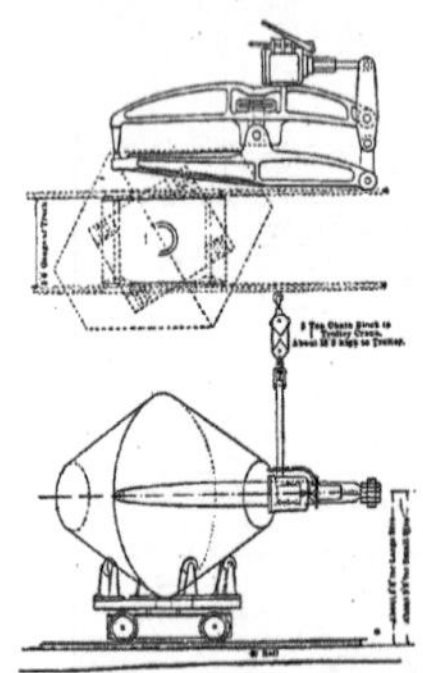

Special Riveter for Riveting Concrete Mixer.

hold the double cones, rendering it unnecessary to raise either riveter or cones, but only to revolve the cones on their axis.

I might also mention a horizontal method of riveting boilers. We made a wheeled truck carrying on its bed three sets of rollers running in the opposite direction. We suspended from a trestle a riveter large enough to do boiler work, from a bail attached through a system of sheaves and tackles to a counterweight of one-fourth the weight of the riveter, and having four times the travel. Then all that was necessary to do was to have a small chain block on the trestle to overcome the friction of the sheaves and tackle. In that way we could raise this machine with a chain block for any diameter of boiler.

We have in very successful operation a machine with a ten-foot six-inch gap, weighing about 25,000 lbs. We are now installing several other machines of this character and it seemed to be quite a feasible plan. It has several advantages over the ordinary Tower system. It takes up very much less room and the initial cost of installing the plant is very much smaller. This system requires no hoisting of the boiler, as it is simply rolled on the floor or on the rollers. Thus the power plant of the machine is limited to that necessary to actuate the toggle.

JIG FOR DRILLING BRAKE SHOE HANGERS.

There is in use in the Toronto Railway shops a special jig which permits of drilling brake shoe hangers for eight different types of trucks. The device is here illustrated. By boring brake hangers in this device, absolute uniformity is secured, insuring correct hang of the brake shoes and consequent longer wear of shoes and wheels. The undrilled hanger is first drilled on one end, centred by the eye only. It is then placed in the jig and the pin inserted in the hole, governing the length required. By means of a set-screw at the back of the device, the hanger takes the correct position without further adjustment. The hole is then drilled, the drill being guided by a steel bushing.

PNEUMATIC TOOL CLAMP FOR WHEEL LATHE.

The average wheel lathe man consumes from five to six minutes each time he changes tools, and as it is not

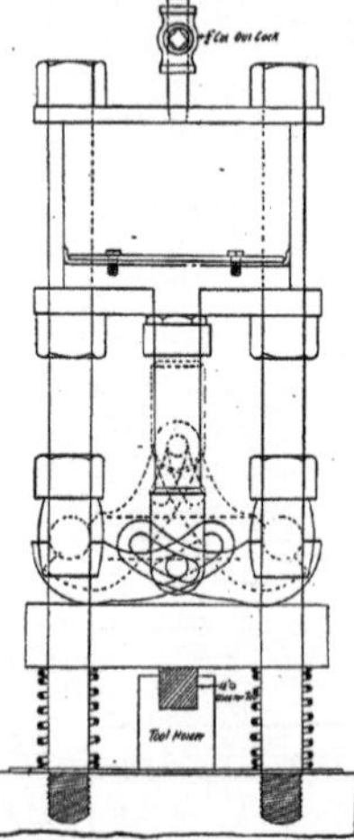

Pneumatic Tool for Wheel Lathe.

unusual to change five times for a pair of wheels, thirty minutes is used in the operation which is practically lost time. In order to reduce this time as much as

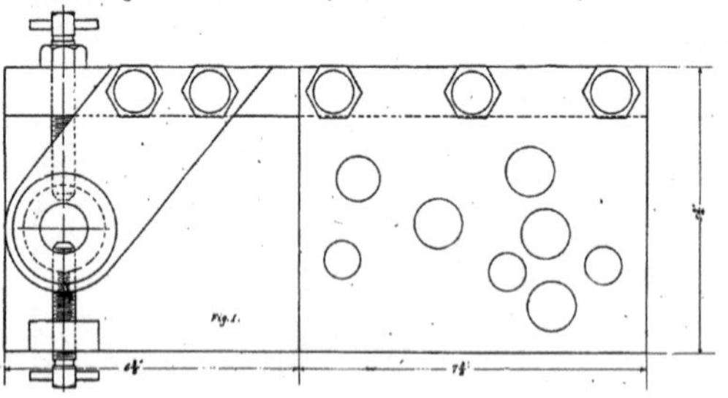

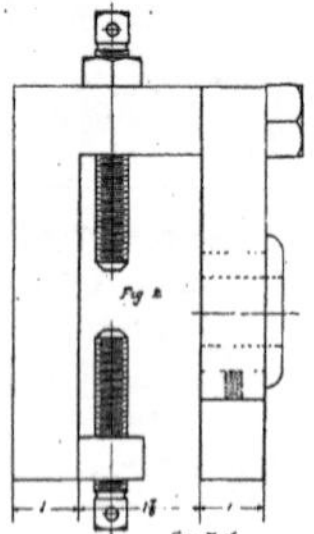

Jig for Drilling Brake Shoe Hangers.

possible, Mr. F. C. Pickard, machine foreman of the Pere Marquette shops at Grand Rapids, has devised a pneumatic appliance for raising and lowering the tool holder plate on a driving wheel lathe, to take the place of the usual four studs and nuts which are tightened with a wrench.

The device as shown by the illustration consists of a 7½ by 5½-inch air cylinder, anchored by two stud bolts to the tool column, the piston operating two slotted cams which raise or lower the tool holder plate, according to the position of the piston. When the piston is down, the plate is forced against the tool, holding it firmly in position by air pressure, and when raised to the position as shown by the dotted lines the tool is released and can be freely removed. The coiled springs under the tool plate force it upward when pressure is released. The piston moves the tool holder plate ½-inch.

For a tool rest or holder, a piece of steel 2¾ by 3 inches is used and is slotted lengthwise along the top for receiving a piece of self-hardening steel 1¼ inch square. This not only saves steel, but gives a greater bearing on the tool rest.

By using this appliance, lathe tools can be changed easily in one minute without the exertion required in the usual method, and the trouble resulting from the wrench slipping and studs breaking is entirely avoided.—Modern Machinery.

IMPROVED METHOD OF FASTENING THE LATHE CHUCK TO THE FACEPLATE.

To the left of the cut is shown an improved method of fastening a lathe chuck to the faceplate instead of fastening it as shown to the right of the cut, which is the usual way. Two important advantages are obtained by this change of method. The chuck will come nearer the bearing and a much stronger construction is possible. The method, as is plainly shown, consists of screwing the inside face of the faceplate to the chuck and allowing the hub to fit the inside of the chuck, the faceplate being finished all over and simply reversed from its usual position, which is to have the hub toward the lathe spindle and the face of the plate toward the chuck.—"Winamac," in Machinery.

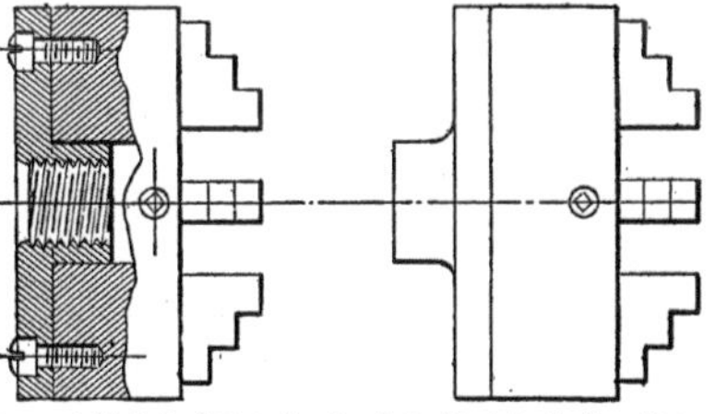

Improved Method of Fastening the Lathe Chuck in the Face Plate.

ONE USE FOR ELECTRIC DRIVE.

In a structural machine shop there is very great difficulty in arranging the machinery so that there will be constant progression of material through the shop and so that there will be no moving material over the same ground twice.

Thus in laying out a new shop there are mistakes made almost certainly, and the positions of machines have to be changed.

In view of this fact the Hamilton Bridge Works, Hamilton, have adopted a policy in connection with the extensions which they have made and are still making, which provides for just such contingencies, by driving all their machines, with individual motor, and instead of bolting the machines down to a permanent foundation, bolt them only to a wooden slab or something of like nature. Then, in case it is found advisable to change the position of a machine it can be done with the minimum amount of work.

They find that with electric drive such a foundation is quite firm enough.

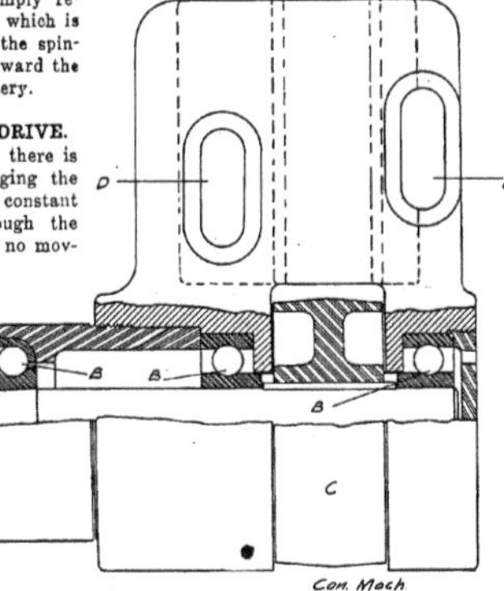

Ball Bearing Grinding Spindle.

BALL-BEARING GRINDING SPINDLE.

By J. C. Armer.

The accompanying illustration is of a grinding head devised in the shops of the Chapman Double Ball-Bearing Co., Toronto, to replace the grinding head supplied with their grinder. They were persuaded that a ball-bearing would give better satisfaction for the high speeds they desired than an ordinary bearing, and designed the head as shown. It is giving excellent satisfaction. At A is shown the grinder; at B, B. B. are shown the bearings, showing clearly the balls and their runways; C is the driving pulley; and D, D, are slots for bolting the head to the grinding machine.

REVISED DUTY ON BRICK.

The Board of Customs has declared the rates of duties on the undermentioned articles as follows:

Fire brick (9-in.x4½-in.x2½-in.) valued at $10 and upwards per thousand at place of export, free under tariff item No. 281.

Pressed brick and other brick for building purposes, including fire brick valued at less than $10 per thousand at place of export—and stove linings manufactured from fire clay, are subject to duty under tariff item No. 282.

Developments in Machinery

Metal Working ***Wood Working*** ***Power and Transmission***

THE UNIVERSAL INDEX AND SPIRAL HEAD.

This head is the result of a desire to fill the need of a spiral head that will at once answer the requirements of the heavier duty now imposed on the milling machine and still retain the fine points of accuracy expected from such a tool. In designing this head it has been the object to produce a design that would answer the requirements of the wide range of work met with in ordinary practice without sacrificing any of the desirable features of the older heads, and to add those features that good practice shows would greatly increase the usefulness of the head on general work.

The worm gear is made in two sections so that one section may be rotated on the other to take up the wear on

the teeth. This method has long been used and is the most accurate method of adjusting for wear and also in the hobbing of the teeth in the blank.

The stiffness and rigidity of the construction is shown plainly in the various views, This increase in strength has in no way impaired the ease of handling nor made the head at all clumsy or awkward to operate. The design is of an approved type, having the swivel block housed between heavy uprights in which the block swings in a vertical plane. The block is held in any position by means of clamping bolts which draw the outside plates securely against the uprights, making a secure bind, holding the head in position against the heaviest cuts.

The principal feature of the design is the large worm wheel that has been obtained. This gear has been made as large as the respective swing of each size head would allow and by the design followed it has been made much larger than that of any head so far offered.

The view showing the back plate removed gives one a good impression of the extraordinary diameter of the dividing wheel. This large diameter adds much to the life of the wheel and insures greater accuracy in the work than would be possible with the smaller di-

ameter common to other dividing heads. The large diameter and resulting coarse pitch allows of much heavier spiral cuts being taken without the danger of impairing the accuracy, or of distorting the teeth in the gear. The increase of stiffness of the head in general makes possible the taking of much heavier cuts at faster feeds and speeds, obtaining the best possible results with high speed steels.

In bringing out this new head, the idea of making the differential indexing mechanism a component part of the head, has been carried out so that the head may be used as an index or dividing head in any position along the platen with the spindle either parallel with or at right angle to the main spindle of the machine, or in any intermediate position. This has been accomplished by placing the change gears used in differential indexing on the rear side of the head as shown in the view where the gears are set in position. As seen, the gears have no connection with the table at all as is the case with the

ordinary index head fitted for differential indexing.

With the gears used in differential indexing arranged on the head as is done on this head, it is possible to swing the spindle into position for cutting bevel gears or teeth on any conical work. This at once broadens the scope of differential indexing from straight cylindrical work to that which requires the angular setting of the spindle in the vertical plane.

In order that the application of the differential indexing may be universal, it is necessary that it be made available for use on work with helical or spiral grooves, such as spiral gears. It has been done in this new head. The manner in which it is accomplished is as follows:

The principle on which the differential system of indexing works, makes it necessary to have the spindle and index plate so connected by means of change gearing, that the movement of the spindle will cause a movement of the index plate in one direction or the other as the case may be. This makes it necessary that the index plate be free to move on its axis independent of the index crank during the indexing operation. In cutting spirally the plate is geared to the lead screw by suitable change gears. The connection between lead screw and index plate must be broken when making the division in order that the index plate may be free to make the differential movement with the index crank. This breaking of the connection is accomplished by means of an adjustable clutch which is withdrawn during the indexing operation. After the division has been made, the teeth in the clutch will be found to be in such a position in relation to the corresponding spaces that it is impossible to engage same. In order to bring the teeth and spaces opposite each other, one half of the clutch is made adjustable so that it may be rotated the required amount to bring the two portions in proper position for engagement. This adjustment is accomplished by means of the knurled knobs attached to the clutch.

The connection between the index crank through the worm, worm gear, spindle and change gears of the differential indexing mechanism and the index plate when the index pin is in mesh with a hole in the plate would form a locked train, which must be released during the spiral cutting operation. This release is accomplished by means of the knurled knob back of the index plate, which operates a friction clutch.

Frequently it is desired to roll the work on its axis a small amount without shifting the dog or losing the position of the index pin or the amount of roll over may be so that should it be accomplished by rotating the crank, the pin would not come exactly over a hole as would be necessary. Should it be attempted to move both plate and crank in conjunction it would be found that the back pin of the ordinary head would not engage with a back hole. In this head the back pin is done away with and the plate is held in position when resorting to plain indexing by a frictional hold on the hub of the plate gear which is clamped or released by a suitable bolt conveniently located. By this means, work may be set regardless of the position of the plate, and the plate can then be securely held in the position it takes when the work is so set. All those who have used the ordinary head will realize the advantage to be gained in doing away with the back pin and substituting the more flexible holding device as is done in this head.

In work requiring the head to be connected up for spiral cutting, the roll over of work is made more convenient by the presence of the adjustable clutch, which, as explained above, allows the disconnection of the spiral cutting train so that the spindle and work may be revolved or rolled over without changing the position in relation to the cutter in a direction parallel with the feed motion.

This head is made by the Becker-Brainard Milling Machine Co., Hyde Park, Mass.

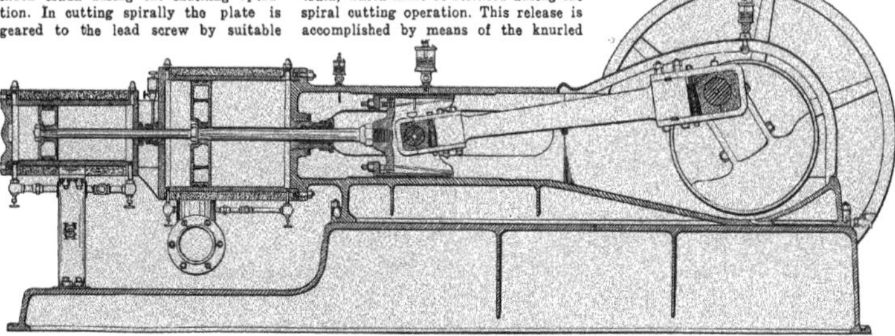

Fig. 1.—The Hardill Engine.

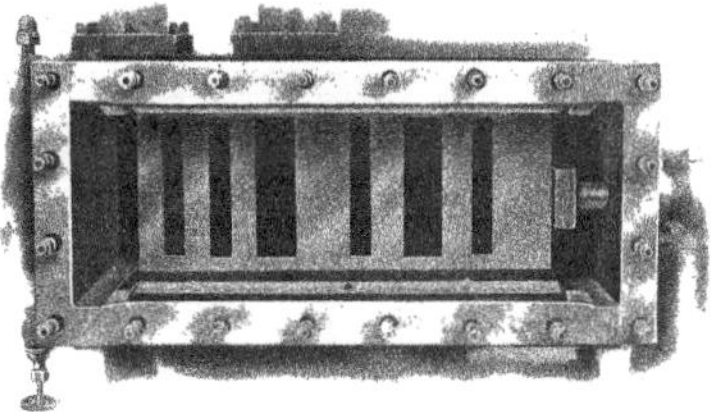

Fig. 2.—Valve Chest.

Fig. 3.—Valve and Rod.

Things don't just happen. They are the result of a universal fixed law. The good is the result of harmony and the ill is the result of discord—so get in harmony.

THE HARDILL COMPOUND ENGINE.

The engine which is herewith illustrated and described is a compound engine with some rather unique features which tend to simplicity of construction and steam economy. This engine was invented by Joseph Hardill some years ago, but since then he has been making improvements, and the one here described is the one which has just been completed. It is now being manufactured by the Berlin Foundry Co., Berlin, Ont.

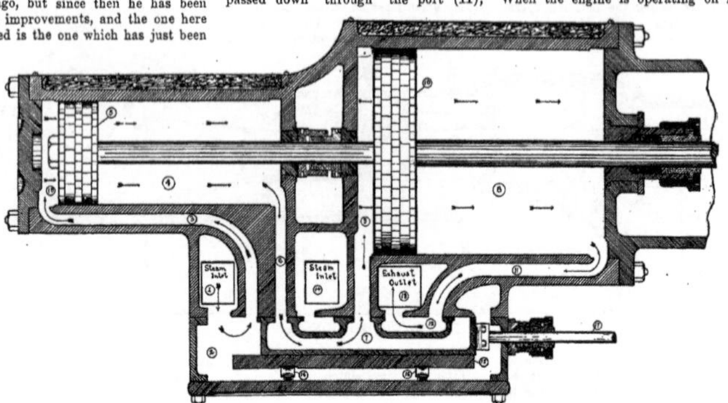

Fig. 4.—Sectional View of Cylinders, Steam Chest and Valve

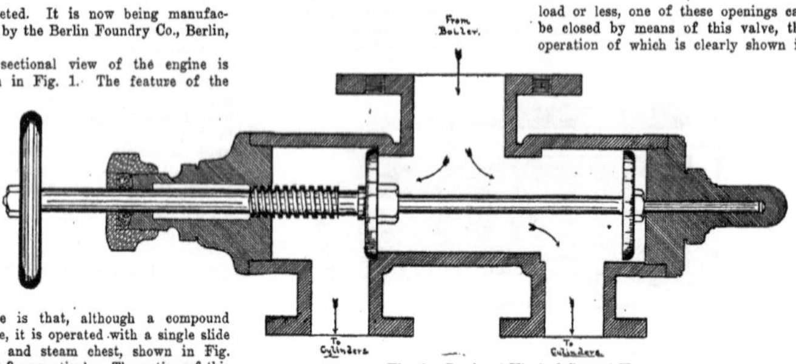

A sectional view of the engine is shown in Fig. 1. The feature of the engine is that, although a compound engine, it is operated with a single slide valve and steam chest, shown in Fig. 2 and 3 respectively. The action of this valve can be seen from Fig. 3. In this view the cylinders are just taking live steam at the head end, the steam being admitted into the steam chest at (1), filling the space in the steam chest (2), and also behind the pressure plate (15). From thence we follow the steam by aid of the arrows and pass through the port (3) into the space (18) behind the small piston (5). Here the steam does its first work, forcing the piston forward; while the partly expanded steam of the previous stroke in the small cylinder at (4) is forced down through the port (6), through the passage in the valve (7), up into the space (9), behind the large piston (10). Here the steam again does work, and the thoroughly expanded steam in the large cylinder at (8), from the previous stroke, is passed down through the port (11), through the valve at (12), and thence out through the exhaust outlet at (13), and into the atmosphere or condenser.

In this way a complete cycle of operation for one stroke, the piston moving and the valve also in proper relation so that when the piston has reached the end of the stroke, the valve has moved so as to admit live steam from a second independent inlet at (14); and then the same cycle of operations is gone through for the return stroke.

That second independent inlet brings up a second distinctive feature of the engine, the operating valve, shown in Fig. 5. From this and from the other views it will be shown that there are two live steam openings (1) and (14). When the engine is operating on half-load or less, one of these openings can be closed by means of this valve, the operation of which is clearly shown in Fig. 5, thus making the engine a single acting engine. This prevents the necessity of the abnormal throttling of steam by the governor and throttling valve, which is one of the chief sources of loss of economy in throttling engines running under light load.

Fig. 5.—Sectional View of Control Valve

Thus, in this engine, the valve mechanism has been simplified, receiver or intermediate piping between cylinders being done away with, and the abnormal throttling of steam is prevented under continuous light loads.

The general style of engine to which these new features have been attached is a self-contained centre crank engine of fairly high speed, 200 to 275 revolutions per minute.

Very low steam consumption is claimed for this engine, and favorable tests have been made on the engine by Prof. Nicholson and others.

NEW ADJUSTABLE REAMERS.

Accompanying is an illustration of the new adjustable reamer upon which the Lapointe Machine Tool Co., Hudson, Mass., have secured patents.

The blades of this reamer are made from half-round drawn steel, which can be obtained to the exact size, and does not require any machining to fit them to the body of the reamer, and permits the reamer to be made with interchangeable blades, which could not very well be done otherwise without enormous expense and accuracy in the machining of the blades. By taking the blades from stock as mentioned, it is only necessary to mill the top of the blade before fitting them into the reamer.

The grooves in the body of the reamer are first milled with a half-round cutter, after which the metal is pressed over. This, of course, is shown when the diameter of the body is left in the rough state about 1-16-inch over size, so that the same can be turned down to smooth surface. The collar is under cut so as to fit the end of the blades which are turned on a bevel. This has a tendency to retain the blades in the bottom of the groove. The groove in the body is milled on an incline tapered to the front, which is an advantage,

New Adjustable Reamer.

thereby leaving more stock at threaded part, and also admitting a larger hole in the shell.

The most interesting feature of this reamer is, as stated before, that the blades can be made in large quantities, and, having grooves in the bodies closed up on the mastered blade, makes all the grooves of uniform size to admit in blades of the same radius. By this method the blades can be replaced when worn out without refitting, and will make it necessary to return the bodies of the reamers. The bodies of the reamers are made of machinery steel and the blades of highgrade carbon steel.

A "BLISS" DOUBLE SEAMER.

The double seaming machine illustrated has been recently built by the E. W. Bliss Company, 20 Adams St., Brooklyn, N.Y., and is especially designed for use in connection with the manufacture of large articles in ironware and enamelware, where the articles to be manufactured are round, such as foot tubs, wash tubs, buckets and similar articles.

The bottom or clamping plate is operated by toggles which respond to the action of the treadle and the treadle links. By depressing the treadle the toggles are locked and form a positive clamp between the chuck and body and allow the operator to take the foot pressure off the treadle. After double seaming is done the pressure is relieved by a slight movement of a handle on the right side of the machine, after which bottom plate drops to bottom position allowing article to be taken off and another placed in readiness for double seaming.

The adjustment for different heights of bodies is made by rack and pinion on left side, while adjustment for different diameters of articles and depth of seams is made by means of screws with hand wheels attached. One hand wheel can be seen in back of arm for operating double seaming roll while the one for adjusting for different heights of seams does not show in photo of machine illustrated. These adjustments are all easy of access and quickly made.

A point of note is the smoothing attachment on left hand side. In many articles the top edge of body is wired in a die and difficulty is often experienced, especially on tapering work, in making the metal hug the wire ring closely all the way round. With this attachment it is possible, while double seaming bottom, to smooth out the wrinkles formed in the process of wiring and roll the metal close to the wire and add considerable to the appearance of the finished article. This is also of advantage in enameling and galvanizing pails.

The machine will take work up to 26 inches in diameter and 36 inches in height and weighs about 3,000 pounds.

REPORT OF B.C. MINES.

The annual report of mining operations for gold, coal, etc., in British Columbia for the year ending December 31, 1906, has been issued by the minister of mines. Besides the detailed statistical tables and data concerning mineral production during the year, the report contains a large number of half-tone illustrations of mining centres, and picturesque spots in British Columbia.

Several very interesting curve sheets of mineral production during the years since 1858 are given.

In 1858 there was no platen gold mined; in 1863 there was a yield of

Bliss Double Seamer.

$3,900,000, the high water level; last year there was in the vicinity of $900,000.

The value of the coal mined has gradually increased from $50,000 in 1860 to $4,500,000 in 1906, although there have been some big fluctuations. The production of copper has shown a very big increase during 1905 and 1906, the value jumping from $4,600,000 to $8,250,000. The year 1894 was the first in which copper was taken out.

The mining of lode gold has seen a big increase since 1894, when it was $50,000. It reached the $4,900,000 mark in 1905, but dropped to $4,650,000 in 1906. Lead has seen wild fluctuations, reaching $2,700,000 in 1900, dropping to $750,000 in 1903, and rising again to $2,650,000 in 1906. There has also been a big variation in production of silver. In 1897, the value was $3,200,000; in 1899, it was $1,700,000; in 1901, it was $2,900,000; and in 1906 it was $1,900,000.

Foundry Practice

MODERN IDEAS FOR FOUNDRYMEN

Defects in Iron Castings*

Method Pursued in Ascertaining Defects in Locomotive Castings and the Remedies by which they were Overcome.

Robert Job.

Every manufacturer and user of iron from time to time meets with casting difficulties owing to defects in the iron, and in order to reduce fractures, discards and unsatisfactory service to the minimum, it is of great importance to know definitely the cause of the defects in order to apply the proper remedy.

In the course of everyday practice for many years, the writer has been interested in making a careful study of these matters as an aid in working out the most economical methods of manufacture and in securing the greatest efficiency in the output, and it may be of interest to give some details which have been observed from time to time.

Purchase of Iron.

In many foundries to-day, as was the case in one which first came under attention fifteen years ago, the practice is to purchase a number of different grades of iron—not by analysis, merely by brand—and to mix these in the cupola, with the object of obtaining special virtue from the combined properties of the metals. The underlying reason for this plan is simply that one brand runs high in silicon while another contains an excess of manganese, and still another is high in phosphorus making the iron very fluid, and very weak. The inherent defects of this method of mixing are well known to every foundryman, for difficulty is found in keeping up shipments of each of the different brands, and even when they are in stock, it is by no means an easy matter to get a correct mixture in the cupola even under the most favorable conditions. Consequently the foundryman is troubled by irregular quality of castings, and it is difficult to maintain proper control of the output, simply because there is great difference in the composition of the iron from day to day and even in different parts of the heat, when the intention has been to make a uniform grade of castings.

Variations in Composition.

Part of this variation is due to the difference in the composition of different lots of a given brand of iron. For example, we have known of different shipments of one brand which have ranged from 0.8 per cent. of phosphorus to 1.5 per cent. of phosphorus, while another showed an equal range in its manganese contents, and still another was varied from 2 to 5 per cent. in silicon. Under such conditions it is obviously out of the question to expect uniform results in the product, for at one time the iron is strong and tough, while soon afterward it is weak or brittle, or is full of blowholes and porous, and the foreman, who under the best of conditions, is none too free from troubles, finds himself in serious difficulty.

In order to secure uniformly good results it is obviously essential at the start to have in the finished casting a composition which will give the properties needed in service, and also adapted to the conditions of casting. Many service failures in the writer's experience have been caused directly by this neglect to obtain a uniform composition suited to the service. We remember particularly instances in which locomotive cylinders and wheel centres failed within a short time, and on investigation it was found that the proportion of phosphorus averaged nearly one per cent., while the silicon was over two and one-half per cent. As a result the iron was so weak that little tenacity under impact could have been expected, although the price paid for this grade was considerably higher than the market rate for quality far better adapted to the service.

In order to put a stop to these conditions under our own practice, a careful study was made to determine both the composition and physical condition and structure which gave the best service under different classes of requirements, and to determine the means necessary to secure this quality in the output.

In many cases of failure we found that the difficulty was due wholly, or in large part, to the presence of blowholes or to porosity or sponginess of the iron, and at times to the presence of considerable proportions of oxide of iron and cinder in the iron. Any of the latter conditions, of course, were direct indications of defective foundry practice, and where they existed defective composition was also often present.

In the daily routine of our foundry, general locomotive castings were made ranging in size from locomotive cylinders and wheel centres to small castings about one-quarter of an inch thick, and most of the castings required machining in some part. In view of this wide variation in size of and conditions of service, some doubt was felt whether a single grade of pig iron would meet the requirements. At the outset, however, we drew up specifications for a strong medium iron and upon receipt of each shipment, sampled and tested each carload before acceptance, to be certain that it was of the desired quality. The same practice was also adopted with reference to our coke supply, and the proportion of ash and of sulphur were held down to reasonable amounts. At the same time methods of treatment of the iron in the ladle and in the cupola were introduced to decrease hardness and to remove oxide of iron and blowholes, and to increase the fluidity and density of the iron.

Improvement in the Quality of Castings.

As a result of these changes excellent results were obtained from the very start. The single grade of iron with the careful control of the quality, gave a degree of uniformity which had never before been possible and by means of systematic treatment the properties of the iron could be varied as far as was desirable for the different purposes.

In the machine shops the change had an immediate effect, for hard castings, "porous iron" and blowholes almost entirely disappeared. Within a few months breakages in service had fallen off to a very marked extent owing to the toughening of the iron, and at the end of the year the scrap coming in was insufficient for the needs of the foundry. In order to keep track of the service, we stamped each wheel centre as it was cast, with the date,

*Presented at the June meeting of the American Society for Testing Materials.

and after a lapse of three years not a single one had been broken in service.

In the course of our routine work we found comparatively little difficulty in maintaining the desired quality in the shipments of the pig iron and the coke, though we soon learned that the systematic test of each carload of the raw materials was essential to insure proper quality of the product, for at one time a carload of weak iron would be kept from service, and at another time iron which would have produced hard and brittle castings. Still again, coke would not be accepted which proved to be deficient in heating value or was of such quality that hard castings would have resulted.

Cost of Pig Iron Decreased.

By means of the method of purchase under specifications the cost of the pig iron was decreased, and in the foundry the proportion of discards was cut down owing to the fluidity of the iron resulting from the cleansing of the metal from oxides and slag, and to the increased strength and toughness of the product.

From what has been said it will be evident that there is a definite cause for each of the defects which commonly cause difficulty in service or in the foundry, and it will be seen that these may be very largely removed by means of systematic control.

A PRINTLESS CORE.

By H. J. McCaslin.

As the name implies, it is a form of a dry sand core, which may be used in a mold without a core print setting, that is, it is located and attached to the pattern in its correct position as shown in two views in the accompanying illustration.

The fitness of the arrangement will be readily appreciated when coreing a series of holes at the bottom of a deep flange. Not only need the core be applied in this form, but it may assume various forms to suit the requirements and conditions.

When using this form of core, the pattern is placed within the flask and rammed up to that portion of the pattern which is to receive the core. This part of the pattern being provided with the slot or opening A, as shown, the core is set in place, and the ramming up of the pattern proceeded with.

It now only remains to draw the pattern from the core in the usual way, the core remaining in its rammed up position, and forming that part of the mold for which it was intended.

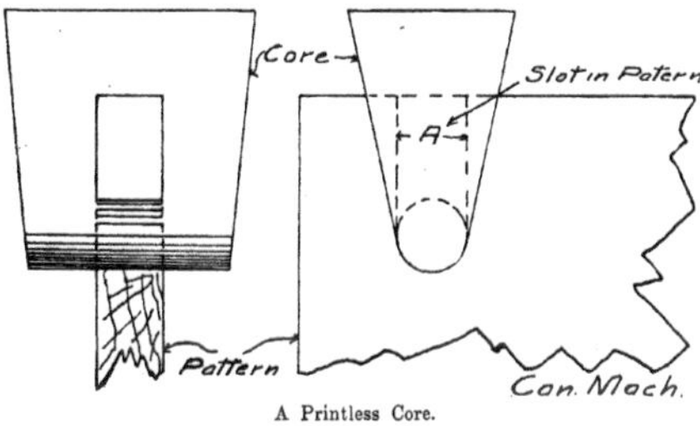

A Printless Core.

*The Steel Foundry as a Business Venture**

An interesting and instructive article, in view of the rapid strides steel castings are making for many purposes.

By W. M. Carr, New York.

In recent years the steel casting industry has made rapid strides and for many purposes cast steel is gradually superseding cast iron. The situation is recognized by all engineers as one requiring the attention of leading and progressive cast iron founders. That the steel foundry business is an established industry is incontrovertible and one constantly improving and developing. The pioneers had their failures and losses but ultimate successes are obtainable, so that prospective investors in steel foundries will not have thorny paths to travel in order to get financial returns, as in the early days of the industry. The capitalist has better assurance of dividends than at any time in the past.

Capital.

The first question is : How much capital should be invested for a given output? The answer cannot be particularized, because local conditions affect each individual problem, but the following remarks covering the subject broadly will serve as a guide to prospective investors :

Large tonnages of course yield large returns, but the net profit on a given amount of capital does not vary directly with the size of the plant. Small plants can show as great a percentage of profit as larger ones. The demand will largely control the size of the plant. To illustrate, we will consider a plant yielding 150 tons daily, or 4,000 tons monthly. Such a plant would employ about 1,800 hands, would require a large amount of ground with good shipping and receiving facilities. There should be at least five 25-ton open hearth furnaces, with four of them in regular operation. The furnaces would have to be served with electric charging machines, the molding and casting floors served by several 35-ton cranes. Hoists and light travelers would be required in cleaning, finishing and shipping departments. It should have a well equipped power house with engines and generators, pattern shop, machine shop, yard locomotives and liberal molding machine installations.

Such a plant as decribed could not well be started without $1,000,000 capital, and the cost of erection of suitable buildings with crane runways and complete equipment would amount to at least $750,000. The following estimate on the basis of ton capacity will cover plants varying from 80 tons per day with a two 25-ton furnace plant to one with five 25-ton furnaces producing 150 tons daily :

Furnaces only	$1,200
Gas producers, gas mains or oil tanks and accessories...	600
Buildings, 800 square feet per ton at $1.25......	1,000
Power, machinery, cranes, hoists, ladles, drying ovens, etc	2,300
Total cost per ton......	$5,100

Such an equipment with a good demand for casting can show 25 per cent. on the investment.

General Practice.

In general practice it is better to have two open-hearth furnaces if a modest plant is desired. Keep one in reserve in case the active one should go out of commission. The larger furnaces, 25-ton capacity, are best suited for heavy castings, although some plants run successfully on pieces averag-

* Presented at Philadelphia convention of American Foundrymen's Association.

ing 30 pounds each. A shop should be designed with a view to the character of work to be produced. If small work, pieces averaging ten pounds, a small open-hearth furnace is suitable.

Light Castings.

For very light castings, weighing ounces, the Tropenas process is in use, but it will not reach the development promised by the open-hearth process producing basic steel castings. The pronounced limitations are the heavy losses and the high price of raw material. It is admitted by the inventor of the Tropenas process that the loss in melting is 17 per cent., a loss which in comparison with any open-hearth process is wasteful. In open-hearth practice the loss in melting only averages 7 per cent., and under favorable conditions a much smaller loss can be shown. The pig iron for the Tropenas process is imported and is of a special composition and is scarcely used for any other purpose. The pig iron for the basic process is cheap, domestic and plentiful in comparison, and, further, permits the use of almost any kind of steel scrap; the latter material for a Tropenas vessel must be like the pig iron of special composition. The converter process is a comparatively new one and but very few operators can be found who are skilled. Investors who may place their money in Tropenas plants may find greater difficulties in their way than did the early users of the open-hearth processes, before commercial results can be obtained. The Tropenas process, however, is well suited for special purposes when certain kinds of steel are wanted in small quantities, but for general uses its own weaknesses are detrimental, preventing competition with open-hearth methods.

Basic or Acid Furnaces.

The question may arise as to whether an open-hearth plant should be fitted with basic or acid furnaces. The basic process is cheaper in the first place and the stock for it is more plentiful than for the acid. The chemical analyses of melting basic stock are wide and liberal. Those for the acid narrow and limited. The scrap for the basic process is almost only confined by its shape rather than its chemical composition. The scrap for the acid process is produced in the first instance by a basic process and will be represented by bloom ends, billet croppings, axle butts, plate clippings, etc., and is sold on analyses. It is stated that low phosphorous ores are getting scarce while high phosphorous ores are quite plentiful, so in regard to the material from which pig iron is made, the basic process has the greater promise. Physically basic steel in castings is just as reliable in every respect as acid steel.

Department for Gray Iron Foundries.

To the gray iron founder who feels the inroads of steel castings in his business and recognizes the demand, but is not disposed to invest a million dollars in a 150-ton plant, I would like to suggest to him the erection in his shop of an open-hearth furnace anywhere from two to ten tons in capacity. The furnace to be lined with a basic bottom and designed to burn either natural gas or fuel oil. Take as an average size one of five-tons capacity. Such would fit in ideally with existing equipment. Being basic, there would be no need of buying any fancy-priced imported stock; any pig iron with silicon below 1.25 per cent., sulphur not over 0.08 per cent., and phosphorous with 1.00 per cent. or less can be used and such analyses would cover some grades of cast iron scrap in the shape of old car wheels, malleable, etc., and, charged with 50 per cent. of steel scrap, such as springs, rails, agricultural scrap, defective steel castings, etc., would give an opportunity to produce steel castings of good quality from low grade stock. Being of a moderate tonnage there would not be any heavy powered machinery to handle the heats. The only important change would be in regard to molding sand, a special grade being necessary. The basic furnace could be used to put to profit machine shop turnings and borings by mixing them in moderate quantities with the charges.

In some cases it might be advisable to line the furnace for acid melting and if desired it could be used for turning out "air-furnace-iron" heats.

Cost of a Small Furnace.

A five-ton furnace, basic lined, erected in an active iron foundry, would cost about $6,000, and allowing $4,000 for special bottom, pouring ladles, oil storage tanks, burners, etc., the total would be $10,000 exclusive of buildings. It would be capable of producing three heats daily and would yield 12¾ tons of castings at the rate of 15 per cent. in melting loss, gates, heads, skulls and defective castings, an average yield in a well ordered plant. Producing such tonnage steadily should show good profits. The single furnace however, has the advantage of going out of commission for repairs. It should turn out 400 heats before a general overhauling, or say a run of 133 working days. Assuming that the furnace was only productive nine months of a year at 25 working days per month, there would then be 225 days of yielding, a total output of 2,868¾ tons of product. Allowing an average profit of one-half cent per pound, there would be a gross profit of $28,-687.50 annually.

But losses will arise in spite of best intention and not to make the venture look too roseate the set-backs will be considered which might be caused by breakdowns, strikes, or errors on practice. These conditions might reduce the good castings to one-half of the metal charged, so that 7½ tons were the output. With the same productive period of 225 days per year, and with a decided shrinkage, because of losses, between manufacturing costs and selling prices, so that the gross profit was only one-eighth cent per pound, the tonnage per year would be 1,687½ tons, and the returns therefrom $4,218.75. Out of this 25 per cent. can be deducted for interest on the investment, depreciation, etc., and there would remain a net profit of $3,164.07, which is equivalent to 31.6 per cent on the original $10,000.

The estimate in this case is very conservative and the low yield is an extreme. With careful practice the results can be made better than shown. There will be pitfalls and irregularities, but they will be greater in a large plane because greater units are concerned and heavier risks. The small furnace will be profitable as well as the large one, and should times be lean and the demand fall off, the small plant can bear the brunt as well as the bigger one because of the smaller amount of capital involved. There would not be in the small plant a top heavy organization to maintain or the demand for dividends to be met.

Profitable Investment.

A steel foundry in prosperous times is a good earner, but because it is a steel foundry with great possibilities, like any industry, it is not devoid of conditions of a negative character. An open-hearth plant with basic furnaces or acid furnaces does not have to pay patent royalties. The former type of furnace can take impure stock and refine it to steel. In that respect it has flexibility and a promising future. The acid process requires special composition stock of a limited nature in the market. The converter process requires skilled operators, hard to find; has heavy melting losses, requires high-priced stock, differing some from open-hearth acid stock in composition. The crucible process does not refine, but simply remelts the product of the basic process in the shape of waste material. There are other evidences in favor of the basic open-hearth process as meeting the needs of a growing industry, and the foregoing points mentioned are to be taken into consideration when looking into the question of engaging in the manufacture of steel castings.

Where Should the Blame Rest? *

The question of "rat-tailing" is exasperating in foundries doing light work—Several assertions are made here as to the cause and cure.

By R. H. McDowell.

Among the troubles occurring in the foundry to-day, there are none so exasperating or, apparently, so hard to overcome as that known as "seaming" or "rat-tailing," which occurs so frequently in those branches of the foundry industry covered by the light plate molding, such as stoves and hollow-ware.

I have seen, in numerous instances, where the work in the foundry was coming out as smooth and perfect as was possible to make it; when suddenly, without warning or any apparent cause, dollars upon dollars, worth of casting will have to be thrown into the scrap as worthless, being covered with unsightly seams or ridges which make the castings unfit for the purpose for which they were intended.

Almost invariably the foreman goes after the molder with sword drawn. The molder strenuously insists that he is making his work exactly as he made it last week and the week before. Investigation begins. First it is blamed on the facing, and I have seen barrel after barrel of number one facing condemned in stove shops and shipped back to the dealer as "no good," because of this seaming on the face of stove castings; while the facts are that the facing had no more to do with this trouble than had the book-keeper in the office.

The next step usually taken is to make considerable of the work, on which they are having the most trouble, without any facing. The following day the work comes out of the mills with a rough, half-cleaned appearance and having the same unsightly "seams" on the surface. The fact that these seams appear when no facing was used carried no weight. The facing was pronounced "no good" and must go back.

As the facing question is settled and the "seams" continue to show on the castings, investigation starts anew. The sand is too wet, then too dry, and finally discarded as "no good." New sand is obtained; still the "seaming" goes on. The work is then poured with "hot" iron, then with "dead" iron. Every day new suggestions come up, new experiments are tried, and while these experiments are being tried the trouble generally disappears as it came, suddenly and without warning.

The foreman and the molder look at each other and wonder what they did to stop the seaming. If you ask them what caused the trouble, they generally say, "Oh! Our sand was burned out and we got into a batch of bad facing; but everything is all right now, we are getting in a new facing and have renewed our sand."

The cause of the trouble must be placed somewhere, and, not knowing where to place it, this is the easiest way out of it. "The facing and sand men are big fellows anyway, and can stand it."

In analyzing this trouble I am going to make four assertions, and I want those who are having this kind of trouble to watch and see if their experience will not bear out these assertions.

1. That where this trouble occurs, and disappears, the occurrence and disappearance always take place at a time when you start in on a new carload of coke or iron.

2. That when you are having trouble with "seaming" your iron is either hard and brittle, or if it is soft, you will find, on holding your face over a ladle of molten metal, you will almost be suffocated with the sulphur arising therefrom.

3. If you break a light casting, on which you have "seams," you willl find it either has the white appearance of iron high in combined carbon or sulphur, or a dead, cinder-like appearance of iron high in graphite carbon or free sulphur, with, perhaps, an excess of silicon in either case.

4. When you have a good grade of foundry iron, which to be good must of necessity be low in carbon, sulphur, phosphorus, and silicon, and melt the same with a good grade of coke, which must be low in sulphur and phosphorus, you will never be troubled with "seams."

When certain molders are turning out first-class work, day after day, with a certain grade of facing and sand, and suddenly it becomes impossible for the same molders, with the same material to make their molds, to make a good piece of work, reason should tell us that a radical change has taken place somewhere aside from the molding department.

As the only "elsewhere" about the foundry is the cupola, turn your experimenting in this direction, and, if you are sincere in your desire to know the right cause of this trouble, you will soon learn the facts and apply a remedy.

Space permitting, I would like to consider the cause of this trouble from a technical view-point, but if you who are having trouble in this respect will study carefully the above assertions and compare them with your own actual experience, I am satisfied that the results of your future experiments will be far more satisfactory; and when the "facing" and "sand" men come around again, you will greet them with a smile instead of a scowl.

* From the Obermayer Bulletin.

"SCRUFF" IN PIG IRON.

By T. N. Burman.

I wish to say a few words in your paper in regard to pig iron, and the mixture for gray iron castings for radiators and boilers. We have had about 25 years' experience in this line of work, and we have been troubled very much in the last few years with a dirty, spongy matter in the mixture, which, under the magnfiying glass, looks like a cinder. It runs with fluid iron, and stops in heavy places, and with us at such places as cross bars of a radiator section, and in boiler work at the rod holes or near the heavy hubs.

Now, a kind, old blast furnace superintendent gave me a name for this, "scruff." Now we all wish to get rid of this "scruff," for its presence costs any company who makes radiators and boilers a large pile of money. We test our boilers and radiators by the hydraulic test to the amount of 80 to 100 pounds, and this "scruff" is liable to make them leak under these tests. Once the "scruff" gets in the iron, it is difficult to get rid of it, and the only way is to pick out the iron which is the cause of the trouble in the yard before sending it to the cupolo. I am able now to pick these pieces out without fail. I have found that analysis does not tell all about the iron; and I wish to tell you something which may help you in getting rid of this "scruff," which, in my experience, is the worst trouble-maker we have had. I hope that some other brother foundryman will take up this same subject and tell us his experiences along this line.

Having a whole plant under my charge, I had a good chance to make tests. I made tests with different grades of coke, sands, blackening, and in the molding, but still we had the trouble. I cut out the "scruff" iron, and in a few heats we had no more trouble with leaks, and I proved to my own satisfaction that the trouble was not due to coke, blackening, sand or melting conditions.

Now, the way I find the pig iron with "scruff" in it is as follows: These pigs are usually covered with a dirty, rusty, spongy mass, and these I find to be the trouble-makers. There is another fact which may be interesting, which is, that the iron which shows very gray and brittle will give trouble when mixed with coke high in sulphur.

I hope that these lines will be of use to my fellow foundrymen.

AMONG THE SOCIETIES

SOCIETY OFFICERS.

Canadian Mining Institute.

President, George R. Smith, Thetford Mines, Quebec; secretary, H. Mortimer Lamb, Victoria, B.C.; treasurer, J. Stevenson Brown, Montreal.

Toronto Branch A. I. E. E.

Chairman, R. G. Black; vice-chairman, K. L. Aitken, secretary, R. T. McKeen.

National Association of Marine Engineers of Canada.

Grand president, Alex. L. De Martigny, Varennes, P.Q.; grand secretary, Neil J. Morrison, St. John, N.B.

Canadian Electrical Association.

President, R. G. Black, Toronto; vice-presidents, R. S. Kelch, Montreal, W. R. Ryerson, Niagara Falls, Ont.; secretary-treasurer, T. S. Young, Confederation Life Building, Toronto.

Canadian Association of Stationary Engineers.

President, Ed. Grandbois, Chatham; vice-president C. Kelly, Chatham; secretary, W. A. Crocket, Mt. Hamilton P.O., Ont.; Treasurer, A. M. Wickens, Toronto.

Canadian Railway Club.

President, W. D. Robb; secretary, James Powell, chief draughtsman motive power department, G. T. R.; treasurer, S. S. Underwood, chief draughtsman car department, G. T. R.

Engineers' Club of Toronto.

President, C. B. Smith; secretary, C. M. Canniff, 100 King Street W.; treasurer, John S. Fielding, 15 Toronto Street.

Canadian Society of Civil Engineers.

President, W. McLean Walbank; secretary-treasurer, C. H. McLeod, Rooms, 877 Dorchester street, Montreal.

Society of Chemical Industry.

Chairman, Prof. W. H. Ellis; vice-chairmen, H. H. Van Der Linde, Milton, and H. Hersey and A. McGill; hon.-secy., Alfred Burton, 44 York Street, Toronto.

Ontario Land Surveyors.

President, Thos. Fawcett, Niagara Falls; vice-pres., A. J. Van Nostrand, Toronto; secy.-treas., Capt. Killaly Gamble, Toronto.

STATIONARY ENGINEERS' CONVENTION.

At Guelph, on August 13, 14 and 15, the eighteenth convention of the Canadian Association of Stationary Engineers was held. The features of the convention included the reception by the civic authorities, business meetings, band concert in the park, trip to agricultural college and manufacturing plants in the city, and the banquet at which the city was represented by the mayor and several of the councilmen.

Opening Session.

After the civic reception on the 13th, the association got down to business, a few outstanding features being:

The new lodges since last convention are: Kingston, Petrolea, Windsor, London, St. Thomas and Stratford.

The secretary's and treasurer's reports showed the association to be in a splendid position, both financially and otherwise.

The following are the standing committees:

Legislative Committee—Messrs. Marr, Toronto; Clarke, Hamilton; Bean, Waterloo; Crocket, Hamilton; Bain, Toronto; Sansbury, Owen Sound; Scrimgeour, Stratford.

Mileage and Per Diem—Messrs. Moseley, Toronto; Bradt, Windsor; Grandboro, Chatham; Lyons, London; Johnston, Hamilton.

Good of Order—Messrs. Kelly, Chatham; Tait, Toronto; Day, Petrolea, Hegg, Guelph; McGhie, Toronto; Fowler, London; Borbridge, St. Thomas.

Auditors—Messrs. Noble, St. Thomas; Heathe, Hamiton; Gordon, Chatham.

Canadian Exhibitors' Association.

While the morning session of the 14th was in progress, the supply men and manufacturers' representatives formed the Canadian Exhibitors' Association, which is to look after the making of exhibits in conjunction with the conventions of the engineers. The officers are:

President—John T. Carlind, 36 Lombard St., Toronto.

1st Vice-president—Wm. Baird, Galt.

2nd Vice-president—H. W. Cook, 112 Bay St., Toronto.

Treasurer—Archibald W. Smith, Toronto.

Secretary—E. F. Heatherington, Galt.

Assistant Secretary—H. G. Fletcher, Toronto.

Papers and Speeches.

Papers presented included: Firing With Soft Coal, by Geo. D. Fowler; Is Electricity One of the Lost Arts? by W. A. Sweet, Hamilton; The Relation Between Publisher, Advertiser and Subscriber, by Arch. W. Smith, Engineering Journal, Toronto. Papers by W. A. Crocket and A. M. Wickens.

The address by Geo. D. Fowler is published in full in another part of this issue.

W. A. Sweet's paper was a humorous account of how it might be that electricity was known of and used in the days of Moses and Aaron.

Arch. W. Smith built up his address upon this fact: "In treating the subject of the relationship between publisher, advertiser and subscriber, it is assumed that advertising is an all-round paying proposition."

At the Banquet.

At the banquet, the secretary, W. A. Crocket, said that if manufacturers understood more thoroughly the aims and objects of the organization, they would give it more hearty support, and that they would give the association more help in its attempt to elevate the standing of the stationary engineer by providing means for better technical education and by getting the government to issue certificates.

A. M. Wickens claimed that in the case where steam was used for heating purposes that a good steam plant would prove more satisfactory and economical than producer gas plants or electric power; but he gave no figures in support of his argument. He thought that the manufacturer should pay more attention to the boiler and engine room and employ capable men so that there should be no waste of fuel by poor handling of plant. He should also give his engineer and fireman every encouragement in their efforts to improve themselves in knowledge.

Closing Session.

The closing session was taken up almost altogether in discussion of ways of improving and helping the association.

Earl Heatherington, of Goldie & McCulloch, Galt, was made honorary member of the association.

The new officers are as follows:

President—Edward Granbois, Chatham.

Vice-president—C. Kelly, Chatham.

Executive Secretary—W. A. Crocket, Mt. Hamilton P.O., Ont.

Executive Treasurer—A. M. Wickens, Toronto.

Conductor—Wm. McGhie, Toronto.

Door Keeper—Jos. Hegg, Guelph, Ont.

The next annual convention of the association will be held in Windsor.

Exhibits and Supply Men.

Among the manufacturers represented at the convention were: Goldie & McCulloch, Galt, by E. H. Heatherington; Waterous Engine Works; The Philip Carey Manufacturing Co., by H. M. Cook and H. E. Rowell; Dunlop Tire and Rubber Goods Co., by H. C. Austen; the Dearborn Drug and Chemical Works, Buffalo, by Jos. N. Gregory; The U. D. Anderson Co., Cleveland, O., steam traps, by L. H. Rumage; Albert E. Hawker, Toronto, steam traps; McLeod & Henry Co., Troy, W. H. Aderhold; The Electric Boiler Compound Co., Guelph; Frank L. Patterson & Co., Toronto.

ELECTRICAL CONVENTION.

The annual convention of the Canadian Electrical Association will be held in the rooms of the Canadian Society of Civil Engineers, 413 Dorchester St. W., Montreal, on Wednesday, Thursday and Friday, September 11, 12 and 13. This convention promises to be of unusual interest.

"N.A.S.E." CONVENTION.

The next annual convention of the National Association of Stationary Engineers will be held at Niagara Falls, N.Y., September 9 to 14. The meeting will be held in the Auditorium of the National Food Company's plant and at the Cataract House, and an entire floor of the plant mentioned will be devoted to exhibits of the machines, appliances and supplies used by steam engineers.

The National Exhibitors' Association, under whose auspices the exhibit is held, have arranged what is said to be a very elaborate program for the entertainment of delegates and visitors.

PERSONAL MENTION.

On July 31, P. E. Durst, President of the Canadian Steam Boiler Equipment Co., Ltd., Toronto, passed to the great beyond. His death came very suddenly, he being at business the day before in his usual good health.

J. Cecil Nuckols, for the past three years advertising manager of the S. Obermayer Co., Cincinnati, Chicago, Pittsburg, has recently received the additional appointment of advertising manager of the Cincinnati Electrical Tool Co.

Emil Hallman has but recently returned from an extended trip out to the Pacific Coast for H. W. Petrie. The trip turned out to be most successful. Mr. Hallman spent a good deal of his time this side of the Rockies, and also a good deal in the mining districts of British Columbia. Upon his return Mr. Hallman took a trip down to Cleveland and other American cities, visiting H. W. Petrie's American agencies.

T .B. Colborne, who has been connected with the Pond Machine Tool Co., and with Wm. Sellers, Inc., and who is late chief draftsman for the Ridgeway Machine Tool Co., Ridgeway, Pa., is now designer for the London Machine Tool Co., Hamilton. This is an indication of the progressiveness of the London Machine Tool Co. Mr. Colborne will bring many up-to-date ideas to his new work, and we may expect to see them brought out in the tools of the London Machine Tool Co. This company is to be congratulated upon securing the services of a man so well acquainted with the best machine tool practice in the United States.

TO GET THE BEST RESULTS.

In order to get the best results the following four conditions are necessary:

1 Complete and exact knowledge of the best way of doing the work, proper appliances and materials. 2. An instructor competent and willing to teach the workman how to make use of this information. 3. Wages for efficient work high enough to make a competent man feel that they are worth striving for. 4. A distinct loss in wages in case a certain degree of efficiency is not maintained.

These four conditions for efficient work were first enunciated by Fred W. Taylor, president of the American Society of Mechanical Engineers, and when they are understood their truth seems almost axiomatic. They are worthy of a very careful consideration. These conditions are really the steps that must be taken to get any piece of work done efficiently:

1. Learn how to do it right, and how long it should take. 2. Teach a workman to do it in the manner and time set. 3. Award the workman greater compensation for doing it in the manner and time set than he can ordinarily earn in any other manner. 4. Make the conditions of pay such that if he fails to do the work either in the manner or time set he gets only his day's pay.

LETTERS THAT PULL.

Here is some good advice given by Sherwin Cody, the authority on letter writing:

Letter writing is a distinct art, built principally on applied psychology. A good letter makes a sharp impression at the right place and at the right time. A bad letter lessens the impression that may have been created by a first and stronger one. Two weak letters following one strong one will make no impression whatever.

This is what Mr. Cody says:

;'Write a long letter to

"A farmer,

"A woman,

"A customer who has asked a question,

"A customer who is angry and needs quieting down and will be made only more angry if you seem to slight him.

"A man who is interested but must be convinced before he will buy your goods.

"Write a short letter to

"A business man,

"An indifferent man upon whom you want to make a sharp impression,

"A person who has written you about a trivial matter for which he cares little,

"A person who only needs the slightest reminder of something he has forgotten or of something he may have overlooked."—The Business Monthly Magazine.

ABOUT CATALOGUES.

BORING MILLS.—A circular issued by the Gisholt Machine Co., Madison, Wis., describing their 30 to 36-inch boring and turning mills. There are two 6x6 illustrations showing their standard types.

CRANES.—A booklet illustrating the hand-power and electric cranes manufactured by the Northern Engineering Works, Detroit. It is very neat throughout and the different types of cranes are shown up to good advantage.

TOOLS.—Catalogue No. 18 of the L. S. Starrett Co., Athol, Mass. A complete, neatly-arranged catalogue of all kinds of tools and measuring instruments. There are 230 pages 7x5 inches, with detailed illustrations and complete price lists.

[illegible]ARDT'S SHAPERS.—A very handsomely illustrated catalogue of Gould & Eberhardt, Newark, N.J., about their shapers. This is a catalogue all interested in shapers should have, and they will be furnished upon application to the company.

MOTORS.—Circular No. 1118, issued by the Canadian Westinghouse Co., Hamilton, Ont., with descriptions and illustrations of their polyphase induction motors. It also contains a list of the circulars in force up to April, 1907, issued by this company.

PIPE CUTTING AND THREADING MACHINERY.—A catalogue gotten out by the Curtis and Curtis Co., Bridgeport, Conn., illustrating and giving price lists of the different hand and power cutting and threading machines manufactured by this firm.

FAN MOTORS.—A beautiful catalogue, issued by the Canadian Westinghouse Co., Ltd., Hamilton, Ont., describing with very clear and highly-colored illustrations, their different types of fan motors. The catalogue is of standard size, with a very pretty cover.

PULLEY BLOCKS.—A 40-page catalogue of the Steel Rope Pulley-Block Co., Sheffield, Eng., showing their different types of pulley-blocks, and emphasizing the exclusive features of each. They also show their chief cranes, hoists, and sheave blocks. It is well arranged and fully illustrated.

ARC LAMPS.—A booklet, of the Canadian Westinghouse Co., Hamilton, Ont., bringing out prominently the different points for consideration when purchasing arc lamps. It also contains a paper read by Brewer Griffin, before the Ohio Electric Light Association, which is interesting and instructive.

LIGHTNING TRANSFORMERS.—Bulletin No. 300 of the Allis-Chalmers-Bullock Co., giving a detailed description of the transformers made by them. Different sizes are shown and illustrated with elevations and cross sections. Several unassembled transformers are also shown, giving a clear idea of the parts.

GEAR CUTTING MACHINERY.—A very complete and handsome catalogue of Gould & Eberhardt, Newark, N.J., dealing with their automatic gear cutting machinery. The catalogue is 6x9 ins., and contains 60 pages. Illustrations are splendid, and information in detail. Company will send copies to anyone interested.

THE DWIGHT SLATE MACHINE CO.—A catalogue of machinery and machine tools manufactured by the Dwight Slate Machine Co., Hartford, Conn., including drills, automatic pinion cutters, marking machines, etc. It describes very fully every machine or tool, and also contains revised price lists of the different articles enumerated.

GOLDIE & McCULLOCH ENGINES.—Among the souvenir catalogues issued by Goldie & McCulloch, Galt, at convention of Canadian Association of Stationary Engineers was one of their "Ideal" high-speed engines. This catalogue is very handsomely gotten up, and the illustrating is excellent. A copy can be obtained by request to the company.

PNEUMATIC TOOLS.—An 80-page catalogue of the Independent Pneumatic Tool Co., Chicago, Ill., with cloth binding. The numerous pneumatic tools made by the company are shown in detail, and there are 30 photographs, showing the innumerable uses for them. The catalogue is arranged with the greatest taste and is indeed a credit to the company.

LIFTING MAGNETS.—A circular issued by the Cutler-Hammer Clutch Co., Milwaukee, Wis., describing their lifting magnets and claiming superior features for them. The circular is accompanied by eight photographs showing the different types of magnets in different uses. These cuts give a good idea of the many uses to which these magnets can be put.

Industrial Progress

CANADIAN MACHINERY will be pleased to receive, as confidential, industrial news of any description, the incorporation of companies, establishment or enlargement of plants, railway, mining or municipal news.

News Plants and Additions.

The town of Woodstock, N.B., suffered by fire.

The Mitchell Electric and Machinery Co., Winnipeg, is insolvent.

The Alberta Biscuit Co. will erect a factory in Edmonton, Alta.

T. Lawson will erect a blacksmith shop in Ottawa, to cost $2,000.

The St. Clair foundry, in Preston Springs, Ont., is being enlarged.

The Doty Engine Co., Goderich, Ont., are erecting a new foundry.

Fire did damage to the town of Oxbow, Sask., to the extent of $175,000.

The Hanbury Mfg. Co. will erect a factory at Brandon, to cost $4,500.

The Brandon Mfg. Co. will erect a factory in that city to cost $2,000.

Moose Jaw is in need of an addition to its power and lighting plant.

W. C. Edwards' sash and door factory, Ottawa, was destroyed by fire.

Smith, Runciman & Co., Toronto, will erect a warehouse to cost $45,000.

Williams & Wilson, Montreal, are erecting an addition to their warehouse.

The Hull Electric Co., Hull, Que., are erecting an addition to their plant.

The Canada Steel Goods Co., Hamilton, will erect a factory to cost $75,000.

The Ontario Iron and Steel Co., Welland, will begin smelting operations soon.

The stock of the Fruitland Brick & Supply Co., Fruitland, Ont., is for sale.

The assets of the Good Roads Machinery Co., Hamilton, are advertised for sale.

Fire damaged the Flemming foundry, St. John, N.B., to the extent of $1,000.

The Guelph Spring and Axle Co., Guelph, are making an addition to their works.

The Canadian branch of the Royal Mint will cost, when completed, about $500,000.

The Hawes-Yougol Mfg. Co., New York, are looking for a site in Brantford, Ont.

The Chestnut Canoe Co., Fredericton, N.B., are erecting a factory to cost $13,000.

A. I. Isaac's cigar factory at St. John, N. B., was destroyed by fire; loss, $4,000.

The H. H. Dryden Co., Sussex, N.B., will considerably enlarge their tinware plant.

The Massey-Harris Co. will erect a large factory at Brantford, Ont., to cost $20,000.

An American brass manufacturing concern are anxious to obtain a site in Sarnia, Ont.

The Western Bag Co., Winnipeg, will erect a warehouse to cost $200,000 in that city.

The Canadian Cutlery Co. want a loan of $25,000 to erect a plant at Grimsby, Ont.

The Brantford Linen Mfg. Co. are going to transfer their factory to Tillsonburg, Ont.

The planing mill owned by Geo. Wood & Son, Dundalk, Ont., was destroyed by fire.

The sash and door factory of Henry Henks, Toronto, was destroyed by fire; loss, $6,000.

A gasoline engine and cream separator manufacturing concern may locate in Chatham, Ont.

An Illinois concern may erect a factory at Edmonton for the manufacturing of furniture.

The sash and door factory of V. E. Traversy, Montreal, was destroyed by fire; loss, $20,000.

R. T. Godman, Vancouver, B.C., will erect a mill between Point Atkinson and the Narrows.

The Superior Oil Co., Sault Ste. Marie, Ont., suffered damage by fire to the extent of $1,200.

The buildings of the Manitoba Peat Works, Fort Francis, Man., will be rebuilt immediately.

The Bloomfield Packing Co. will erect a large canning factory at Hillier, Ont., costing $25,000.

The Canadian Bag Co., Winnipeg, will erect a new bag factory in that city, to cost $10,000.

A firm manufacturing heaters, stoves and furnaces is anxious to obtain a site in Sarnia, Ont.

H. Proctor and C. W. Hughes, Toronto, have formed a contracting firm in Fort William, Ont.

A large boiler works will be established at Halifax, N.S.

The Canada Woodenware Co. have been offered strong inducements to locate in Chatham, N.B.

The Cobalt Concentrators, Ltd., are asking for tenders for a concentrating mill at Cobalt.

The Dominion Car and Foundry Co., Montreal, will enlarge their works at a cost of $2,000.

The Dominion Paint Co.'s premises at Hamilton were damaged by fire to the extent of $1,200.

The storage tank of the Imperial Oil Co., at Sarnia, Ont., was destroyed by lightning recently.

The factory of the Dominion Furniture Co., St. Therese, Que., was destroyed by fire; loss, $100,000.

The International Turpentine Co. will install a plant in New Westminster, B.C., to cost $30,000.

The North American Timber Co., St. Paul, Minn., will erect six large sawmills in British Columbia.

The Canadian Shovel & Tool Co., Hamilton, Ont., will erect an addition to their plant, to cost $1,000.

The Adams River Lumber Co., Shuswap, B. C., will erect a large sawmill on Little Shuswap Lake.

The Toronto Bolt & Forging Co., Swansea, will erect a plant at their present works, to cost $30,000.

The Great Northern Supply Co., Swift Current, Sask., are putting up a new block, to cost $20,000.

Work has commenced on the erection of the steel plant of the Coughlan Company, at Vancouver, B.C.

A firm manufacturing stoves for burning soft coal wants a free site and loan of $10,000 from Sarnia, Ont.

The Kensington Furniture Co., Goderich, Ont., are making extensive additions to their furniture factory.

The pumping plant at Well 5 of Fort William's waterworks system was destroyed by fire; loss, $12,000.

A company has been formed for making cast iron culverts and will establish a plant in Port Arthur, Ont.

The Miramichi Lumber Company are negotiating for the purchase of the Clark Spool Mill, Newcastle, N.B.

The Delaware Seamless Tube Co., Reading, Pa., may build a large plant at Sarnia, Ont., to cost $200,000.

The Canadian Machine Telephone Co. will erect a building in Brantford for their automatic exchange.

The British Columbia Electric Co. suffered damage to the extent of $1,000 in the recent fire at Victoria.

The Otis-Fensom Elevator Co. will build a branch for British Columbia at Vancouver, B. C., to cost $35,000.

A new steam heating plant and ventilating system is to be installed in St. Joseph's hospital, Guelph, Ont.

The shingle mill of the Valley Shingle Co., at Padden, B.C., was destroyed by fire, entailing a loss of $7,000.

The Canadian Pin Co., a new concern, will erect a factory at Woodstock, Ont., for the manufacture of pins.

Pressed Bricks, Limited, are anxious to obtain a site in Strathcona, Alta. They want an exemption from taxes.

An electric power plant may be established at Medicine Hat to operate the C.P.R. works and the pusher engines.

The pulp and wood mills of the North River Lumber Co., Murray, N.S., were totally destroyed by fire recently.

The Ladysmith Lumber Company will erect a saw mill near Nanaimo, B.C., to cut 35,000 feet of lumber per day.

The Ottawa Steel Casting Co., Ottawa, will increase the capacity of their furnace to 8 tons per day. A new addition will be erected and a new plant installed.

The representatives of the Ross Oscillating Pump are in Calgary looking for a purchaser for the Canadian rights.

A large American firm manufacturing agricultural implements will make their western headquarters at Regina, Sask.

A brick company has been formed by some residents of Lacombe, Alta., and a large industry will be established.

The Victor Wood Works, Co., Amherst, N.S., have gone into liquidation. The Maritime Heating Co. have also failed.

The saw and shingle mills of the Miller & Seim firm at Hampden, Ont., were destroyed by fire recently; loss, $7,000.

The Jackson Wagon Works, Galt, have been purchased by D. Clement, Ayton, Ont., who will erect a sawmill there.

The plant of the Midland Engine Works is again in operation. The capacity of the molding room has been doubled.

Work has commenced on the factory of the Aluminium and Crown Stopper Co., Toronto. The factory will cost $50,000.

Hamilton purchasers have acquired the timber limits, mill, and machinery of McManus & McKelvie, New Liskeard, Ont.

J. Davidson, Millbrook, Ont., has purchased the lighting plant and has been awarded the contract for lighting the town.

The huge steam engine in the Ogilvie Flour Mills, Winnipeg, has been replaced by an electric motor of 1,200 horse-power.

Gates & Carpenter, of Salt Lake City, were in Haileybury. They propose the erection of a large smelter there, to cost $6,000,000.

The cement brick works at Radisson, Sask., have started operations and the product has been found satisfactory in every way.

The W. P. Demond Upholstering Co. wants to secure a loan of $12,000 and free water, before erecting a plant in Strathroy, Ont.

The saw mill and woodworking factory belonging to J. H. Simonson, at Spragues Mills, N.B., was destroyed by fire; loss, $10,000.

The saw mill and assets of the Manitoba Lumber Co., at Eburne, Man., have been purchased by P. D. Roe, Port Moody, Man.

The Rogers Mfg. Co., Kansas City, will erect a large plant at Strathroy, Ont., and manufacture malleable iron and journal boxes.

There is a chance of a large cut glass manufacturing concern being established at London, Ont. Mr. Wm. Gray is at the back of it.

The machinery is being installed in the new buildings of the St. Thomas Canning Co., St. Thomas, and will be running in three weeks.

The Union Lumber Co. has been incorporated in Edmonton to supply lumber to union men at a little above cost. The capital is $250,000.

R. Forbes Co., Hespeler, Ont., has secured a Dominion charter and will manufacture textile goods. The capital of the company is $1,000,000.

The Eells Lime Co., at Rockport, N.B., and the Rockport Ice Co., suffered heavily by fire recently. The damage was estimated at $75,000.

The Ryan Storm Canopy Co., Sault Ste. Marie, want to locate in Western Ontario and ask a free site and a loan of $15,000 for two years.

W. J. Campbell, Ottawa, will erect a boiler works in that city, to cost $20,000. A new bridge will be erected at Lac du Bonnett, to cost $40,000.

The W. C. Edwards Co., Ottawa, will immediately commence the erection of their mill, which was destroyed by fire. The new mill will cost $300,000.

The barns of the Southwestern Traction Co., London, Ont., were destroyed by fire, with twenty valuable motors. The total loss was about $150,000.

A large iron and steel plant will be constructed at Kootenay, B.C. Construction will commence shortly, and the initial expenditure will be $2,500,000.

The Rocky Mountain Cement Co., Toronto, will receive tenders for engines, boilers, mills, etc., to cost $90,300. E. H. Keating is the managing director.

The buildings of the new plant of Jenkins Bros., Montreal, have been completed, and the machinery is being installed. They will make valves of all kinds.

The work of building the plant of the cement works in Welland, Ont., is progressing rapidly. The buildings will be completed by the end of the year.

The first installment of machinery for the new pulp mill of the Gordon Paper Co., Dryden, Ont., is on the ground and installation will commence immediately.

The Library Bureau, of Ottawa, will not rebuild their factory on the old site, but will move to another part of the town, where a modern plant will be erected.

H. New, Hamilton, will form a company and manufacture vitrified brick and other materials. A modern plant will be erected and a high grade product will be turned out.

The Chapman Double Ball Bearing Co., Toronto, have been awarded the contract for the entire equipment of the new plant of the Standard Fitting & Valve Co., Guelph.

Negotiations are going on for the purchase of the Ottawa Electric Railway, the Ottawa Electric Light Co., and the Ottawa Gas Co. American capitalists are interested.

The International Heating and Lighting Co. have obtained a franchise from Fort Saskatchewan, and will erect a plant there. Machinery has been ordered, to cost $100,000.

The Northumberland-Durham Power Co. are considering transmitting power from Healey Falls to Deseronto and Kingston, and plans for the transmission lines are being prepared.

Fox & Co., mantel and woodwork manufacturers, will erect a factory in Windsor, Ont., to cost $40,000, if the city will give them exemption from taxation and free light and water.

The large plant of the Red Cliff Brick Co., near Medicine Hat, is in full operation. The plant cost $100,000, and will turn out 60,000 bricks a day. This number will be rapidly increased.

The Eadie-Douglas Co., Montreal, have been appointed Canadian agents for the waterproofing for concrete, plaster, wood, steel and iron manufactured by the Preservating Products Co., New York.

The North American Bent Chair Co., Owen Sound, Ont., have awarded the contract to the Dominion Heating and Ventilating Co., Hespeler, Ont., for an extensive shaving handling system.

An experiment is going on in Medicine Hat with natural gas. It is proposed to compress it in tanks and sell it for use in place of gasoline. It is claimed to be very much cheaper and easier to handle.

The Uxbridge Organ and Piano Co., Uxbridge, Ont., suffered severely by fire recently. The building in which the greater part of their machinery was kept, was destroyed by fire, entailing a loss of $25,000.

The contract for the construction of the power house of the Sydney and Glace Bay Railways at Dominion No. 4 has been let to Rhodes, Curry & Co., Amherst. The plant and machinery will cost $25,000.

The Dominion Car & Foundry Co., Montreal, will erect a large addition to their plant, to cost $2,000,000. A cast steel plant will cost $1,000,000, and a malleable iron plant and three steel car works will be erected.

The Schaake Machine Co., New Westminster, B.C., have secured the right to manufacture and sell a line of lath machines manufactured on a large scale by the Bolton Lath & Shingle Machinery Co., Minneapolis, Minn.

The Canadian Fairbanks Co. have just completed the installation of machinery, including a 6 h.p. gasoline engine, for a 20,000 grain elevator of the Canada Grain Co., Beaton, Ont. This work was in charge of T. W. Ellis.

The North Arm Lumber Co., Vancouver, have obtained a very suitable mill site and will at once erect a mill, at a cost of $150,000. The machinery has been ordered and the mill will be in running order about the beginning of the year.

The Beach Mfg. Co., manufacturing the Beach Triple Expansion Engine and Cast Iron Culverts, are anxious to locate in Port Arthur, Ont., and want a free site and twenty-five per cent. of capital subscribed. The proposed capital is $100,000.

A company to be known as the Monitor Manufacturing Co., has been formed in Fredericton, N.B. The company will manufacture the Monitor Acetylene Generator. The capital of the company is $25,000. They have secured a site and will begin operations immediately. Lighting plants of all sizes will be manufactured.

The fifteenth annual meeting of the shareholders of the Page Wire Fence Co. was held recently. The year's report showed an enormous increase, the volume of sales being the largest in the history of the company. The officers were all re-elected as follows: President, Walter Clement; vice-president, N. L. Clement; secretary-treasurer, Merton Church.

The Smart-Turner Machine Co., Hamilton, are supplying the following with standard duplex pumps: The Hamilton asylum, the Temiscaming & Northern Ontario Railway, the Exeter Canning and Preserving Co., Exeter, Ont.; the Canadian Canners, Leamington, Ont. The Lehigh Portland Cement Co., Allentown, Pa., have ordered three duplex plunger pumps.

Companies Incorporated.

The Peterborough Boiler and Radiator Co. has changed its name to Canadian Boiler and Radiator Co.

McEwen Bros., New York: capital, $25,000; have secured an Ontario license and will make and deal in steam engines and machinery fittings.

The Watford Milling Co., Watford, Ont.: capital, $40,000; to carry on a general milling business. Provisional directors: A. Dunlop, G. A. Dunlop and S. Rivers, all of Watford.

Weston Tool and Novelties, Limited: to manufacture and sell tools, cutlery, etc. Provisional directors: C. M. M. Colquhoun, H. E. Erwin, J. C. Webster, all of Toronto.

The Spanish River Navigation Co., Massey, Ont.: capital, $40,000; to build ships and vessels. Provisional directors: J. Errington, J. Sheets, J. S. Lowe, all of Massey, Ont.

Bain & Cubitt, Toronto; capital, $40,000; to manufacture paper boxes, envelopes and printers' machinery. Provisional directors: D. Bain, W. C. Cubitt, J. S. Denison, all of Toronto.

Imperial Rubber Co., Montreal; capital, $20,000; to make and deal in rubber and rubber goods. Incorporators: R. C. McMichael, D. J. Angus, R. O. McMurty, F. G. Bush, all of Montreal.

James L. Burton & Son Lumber Co., Barrie, Ont.; capital, $250,000; to cut and deal in lumber. Provisional directors: J. L. Burton, F. L. Burton, F. C. Lett, A. Alexander, W. H. Walter, all of Barrie.

O'Keefe-Sanford, Limited, Toronto; capital, $40,000; to manufacture and deal in mantels, tiles and fire place fittings. Provisional directors: W. Webb, M. Whelan, T. Main, B. F. Bennet, all of Toronto.

The Canadian Jack Co., Windsor, Ont.; capital, $25,000; to manufacture and sell a combination lifting jack and farmers' tool. Provisional directors: J. W. Yakey, M. Riddle, H. H. Calkins, all of Windsor.

Onaping Iron Ore Co., Toronto; capital, $200,000; to carry on in all its branches mining, milling and reducing of iron ores. Provisional directors: F. Denton, A. R. Cochrane, G. J. Valin, all of Toronto.

General Industries Construction Co., Toronto; capital, $100,000; to carry on a general construction business. Provisional directors: J. A. Patterson, G. F. McFarland, A. McKenzie, W. H. Templeton, all of Toronto.

Guelph Oil Clothing Co., Guelph: capital, $50,000: to manufacture and deal in oil clothing, tents, awnings, etc. Provisional directors: G. McPherson, J. T. McPherson, R. E. McPherson, G. A. McPherson, all of Guelph.

The Wiarton Steamboat Co., Wiarton, Ont.: capital, $20,000; to own and operate steamboats and carry on a wrecking business. Incorporators: S. Rutherford, T. C. Allan, J. A. Acres, W. H. Buchan, all of Wiarton.

The Dominion Wheel Co., Lindsay, Ont.: capital, $40,000; to manufacture and deal in all kinds of wheels and turned goods. Provisional directors: J. D. Flavelle, W. Mc. Flavelle, J. Carew, and T. Stewart, all of Lindsay.

The Cobbler-Sexton Co., Woodstock, Ont.: capital, $1,024,000; to carry on mining and milling reduction. Provisional directors: Wm. A. Hayward, J. C. Hovey, E. L. Greer, J. McClement, I. Draper, all of Woodstock.

The Bottle Exchange Company of Canada, Toronto: capital, $20,000; to manufacture and sell milk cans and dairy supplies. Incorporators: J. H. Lock, R. W. Dockeray, E. Grace, A. Anderson, V. E. Vansant, all of Toronto.

The Great Northern Petroleum & Asphaltum Co., Ottawa: capital, [illegible]: to prospect for and deal in oil and asphaltum. Incorporators: J. G. Gibson, H. H. Williams, M. C. Edey, W. C. Perkins, R. M. Perkins, all of Ottawa.

The Crown Oil Refining Co., Hamilton, Ont.: capital, $10,000; to manufacture and deal in oil refining apparatus and refiners' supplies. Provisional directors: E. Hull, W. Perkins, J. A. Hull, T. Barnes, G. F. Hull, all of Hamilton.

The Tecumseh and Walkerville Oil and Gas Co., Walkerville, Ont.: capital, $40,000; to carry on an oil and gas business. Provisional directors: J. Dugal, P. Dugal, H. A. Walker, H. C. Walker, R. J. Colloton, all of Walkerville.

Victor Automatic Carriers, Montreal; capital, $20,000; to manufacture and deal in electric motors and dynamos and other power machinery. Incorporators: F. Filteau, F. H. Markey, R. C. Grant, G. G. Hyde, all of Montreal.

The Benson Lumber Truck Co., Port Arthur, Ont.; capital, $20,000; to manufacture and dispose of heavy farm and lumber machinery. Incorporators: A. W. Benson, N. O. Werner, E. L. Mattson, all of Minneapolis, Minn., and G. S. Clark and F. H. Keefer, both of Port Arthur.

The Interlocking Piling and Engineering Co., Limited, Toronto; capital, $200,000; to manufacture and deal in interlocking, piling and piling of all kinds, and carry on the business of engineers. Provisional directors: H. E. Pearce, A. Gate, W. H. Smith, G. Kerwin, M. Irving, all of Toronto.

The Natural Gas Supplies Co., Montreal; capital, $18,000; to carry on a plumbing and gas fitting business and deal in oil and gas wells. Incorporators: H. G. Eadie, H. P. Douglas, W. L. Bond, E. Chamberland, M. D. Barclay, all of Montreal.

E. and T. Fairbanks & Co., Sherbrooke, Que.: capital, $150,000: to manufacture and deal in scales, weighing instruments, and machinery of all kinds. Incorporators: H. N. Turner, J. C. Clark, P. F. Hazen, C. H. Turner, all of St. Johnsbury, Vermont, U.S.A.

William Hamilton Co., Peterborough, Ont.; capital, $300,000: to deal in machinery, contractors' and builders' supplies and carry on the business of machinists and engineers. Provisional directors: W. G. Ferguson, W. S. Davidson, C. L. Hay, R. M. Glover, J. D. Clarke, all of Peterborough.

The New Liskeard Clock Co., New Liskeard, Nip.: capital, $40,000: to manufacture clocks, novelties, dies, tools and sheet metal work and to carry on the business of machinists and engineers. Provisional directors: S. D. Briden, J. Armstrong, M. McLeod, D. McKelvie, J. Redpath, all of New Liskeard.

The Lachute Graphite Mining Co., Township of Wentworth, Que.: capital, $20,000: to mine, prepare and deal in graphite, plumbago, and silver, clay bricks, marbles, and all kinds of artificial stone. Incorporators: A. Guilbault, Lachute; Z. A. Fournier, St. Andrews, Que.; J. R. Hyer, A. T. Woeltje, F. B. Kelly, all three of Waterdown, N.Y.

Building Notes.

A large armory is to be erected at Fernie, B.C.

The new court house for Vancouver will cost $400,000.

Tenders are invited for an armory for Brandon, Man.

A hotel will be erected at Kenora, Ont., to cost $225,000.

The Bank of Montreal will erect a block in Sudbury, Ont.

A new church will be erected at Lorette, Que., costing $100,000.

The new C.P.R. sample room at Vancouver will cost $28,000.

A fish hatchery will be erected at Newcastle, B.C., to cost $5,000.

A mission school will be erected at Kitamaat, B.C., to cost $25,000.

H. Robb, Winnipeg, will erect an apartment block, to cost $58,000.

J. Murphy will erect a business block in Fort William, to cost $15,000.

The new Seamen's Institute Building, at St. John, N.B., will cost $15,000.

A morgue and ambulance house will be erected in Toronto, to cost $30,000.

A police station will be erected in Toronto on Queen street, to cost $25,000.

D. Gibb & Son, Vancouver, will erect an apartment house, to cost $115,000.

The new medical building at McGill University, Montreal, will cost $500,000.

A building will be erected by the Y.M.C.A. in St. John, N.B., to cost $60,000.

Bushnell and Varty will erect an apartment block in Vancouver, to cost $20,000.

The new college building to be erected at Point Grey, B.C., will cost $100,000.

The Robert Crean Co., Toronto, will erect a warehouse in that city to cost $10,500.

A. E. Carter, Vancouver, will erect an apartment block in that city to cost $40,000.

A new wing is being erected at the Hotel Dieu hospital, Campbellton, N.B., to cost $10,000.

The contract for the building of the new court house at Victoria, B.C., has been awarded to McDonald, Wilson & Snyder, of that city. It will cost about $400,000.

The trustees of the University of Toronto will erect a residence in that city, at a cost of $150,000.

The grand stand at the Ottawa exhibition grounds was destroyed by fire, entailing a loss of $60,000.

Business Systems, Ltd., will erect a large building in Toronto for a commercial school, to cost $50,000.

The congregation of Annette street Methodist church, Toronto Junction, will build an edifice to cost $50,000.

The corner stone of the new high school at Picton, Ont., was laid recently. The building will cost $50,000.

The new edifice for the congregation of St. Andrew's Presbyterian church, Fort William, will cost $946,500.

The Edmonton Steam Laundry Co., Edmonton, Alta., are erecting an addition to their plant, to cost $2,000.

The contract for the new Bank of Commerce building, Brantford, has been let to Schultz Bros. It will cost $30,000.

The British Columbia Permanent Loan & Savings Co. will erect an office building in New Westminster, to cost $40,000.

Municipal Undertakings.

The new jail at Sydney, N.S., will cost $13,550.

Gracefield, Que., will install a waterworks system.

A court house will be erected at Saskatoon, Sask.

Massey, Ont., will spend $10,000 in waterworks.

Halifax, N.S., will install water meters, at a cost of $50,000.

Whitewater, Man., will install a municipal telephone system.

The village of Hintonburg, Ont., will raise $10,000 for waterworks.

A new pump house and equipment will be installed at Ottawa East.

A provincial jail will be erected at Moosemin, Sask., to cost $50,000.

The town of Sudbury, Ont., will [illegible] 000 in waterworks extension.

New municipal buildings will be erected at Stirling, Alta., to cost $12,000.

Nelson, B.C., is considering the installation of a municipal telephone system.

A municipal waterworks system will be installed at South Vancouver, B.C.

The ratepayers of St. Felicien, Que., will vote on a by-law for a waterworks system.

A by-law has been passed at Welland, Ont., to raise $11,940 for waterworks extension.

A town hall and public library will be erected at Hanover, Ont., at a cost of [illegible].

The ratepayers of Wellington, Ont., will vote on a by-law for $1,000 fot sidewalk construction.

The ratepayers of Amherstburg, Ont., will spend $2,500 in improving the waterworks system.

Work on the hydro-electric power survey between Brantford and Hamilton has been commenced.

A by-law will be submitted to the ratepayers of Hamilton, Ont., to raise $500,000 for waterworks pumps.

A Worthington steam pump has been installed at the Point St. Charles station of the Montreal waterworks.

The Portage la Prairie council will ask for a grant for $50,000 for the completion of the waterworks system.

The Pelham, Ont., town council have given permission to two power companies to furnish electricity to that town.

The town of Ingersoll, Ont., will purchase the waterworks for $95,000, subject to the passing of the necessary by-law.

The corporation of Berlin, Ont., have received permission to acquire the interests of the Berlin & Waterloo Street Railway.

The ratepayers of Battleford, Sask., will vote on a by-law to raise $10,000 for a bonus to the Battleford Milling and Elevator Co.

The contract for the municipal lighting system for Battleford, Sask., has been awarded to J. Stewart & Co., Winnipeg, agents for the Canadian Westinghouse Co.

The official tests on the new eight hundred kilowatt plant at Morrisburg, Ont., were made recently by Mr. Willis Chipman, C.E., and Mr. K. L. Aitken, C.E., both of Toronto.

There is a by-law before the municipality of Alliston, Ont., to grant the Lloyd & Buchanan Mfg. Co. a free site and a loan grant of $500. A factory will be erected to cost $10,000.

The council of the town of Morden, Man., have awarded the contracts for the municipal electric lighting plant. The steam plant will be installed by the Robb Engineering Co., Amherst, N.S., and the electric portion by the Allis-Chambers-Bullock, Limited.

Railroad Construction.

The Canadian Northern will erect a bridge over the Assiniboia river at Winnipeg.

The C.P.R. have given $30,000 for a railroa Y.M.C.A. to be erected at Kenora, Ont.

The permit has been issued for the new C.N.R. shops at Winnipeg, to cost $200,000.

The C.P.R. will erect a coal handling plan at Fort William, Ont., to cost $1,250,000.

Track-laying will be commenced on the Mackenzie & Mann line at Garneau Junction, Que.

The Canadian Northern have been asked to build a line from Cobalt to Parry Sound, Ont.

The G.T.R. have commenced the construction of the line from Kitamaat to Kitsalas Canyon, B.C.

The Canadian Northern line from St. Jerome to Montford, Que., will be in operation this fall.

The C.P.R. will immediately take steps to double-track their line from Brandon to Winnipeg.

The new coal chute for the C.P.R. at Lethbridge is completed and giving complete satisfaction.

The Sydney and Glace Bay Railway, Sydney, N.S., will erect their central power station at Dominion No. 4.

The G.T.R. will spend $150,000 in enlarging and improving their shops and roundhouse at St. Thomas, Ont.

Saskatoon and Goose Lake, Man., have petitioned Mackenzie & Mann for a branch line between those towns.

The C.P.R. has let the contract for the line running northwest from Moose Jaw, by way of Lacombe and Edmonton.

The C.P.R. will erect new roundhouses at Coleridge and Strathcona, Alta.; Swift Current, Sask., and Cranbrook, B.C.

The contract for the stores building for the T. & N.O. railroad at North Bay has been let to O'Boyle Bros. Construction Co., North Bay.

The Canadian Northern have placed the contract for two switching locomotives with the American Locomotive Works Co., Philadelphia.

The farmers around Goose Lake, Sask., have given the Canadian Northern 52 miles of free right of way to run a line from Saskatoon to Wigins.

The Canadian Northern Railway and the municipal government of Saskatchewan will build jointly a bridge over the Saskatchewan river at Prince Albert.

The Temiskaming and Northern Ontario Railway commissioners recently decided to purchase ten new locomotives from the Canadian Locomotive Co., of Kingston.

The plans for the new Central Station for the G.T.R. at Ottawa have been approved, an work will be commenced immediately. The cost will be a million dollars.

The new bridge which the C.P.R. proposes to build at Lethbridge will be one of the most remarkable structures of its kind in the world. For two-thirds of its length the bridge will be over 300 feet high. There will be 22 spans 100 feet long; 44 sixty-seven feet long, and one 167 feet long. The steel used in its construction will weigh 10,000 tons, and will be supplied by the Canada Bridge Company, Walkerville, Ont.

Mining News.

The Silver Queen mine, Cobalt, Ont., will erect a smelter.

A large two-stand converting plant will be erected at the Boundary mines, B.C.

A valuable deposit of silver and lead has been discovered near Loon Lake, N.S.

There is a scheme on foot for the erection of a large smelter at Cobalt. Guggenheims are interested.

Spokane capitalists have formed a company to operate the Sure copper [illegible], near Bella Coola, B.C. Machinery and supplies have been purchased and are being shipped.

A deal was completed recently by which a syndicate of Nelson men secured control of the Goodenough and Blue Bird properties at Sandon. A fund of $25,000 was subscribed, and development will begin immediately. J. A. Whittier will be in charge of the operations, and L. Pratt is treasurer.

J. B. Hall Moves.

J. B. Hall, machinist, Toronto, has moved from 110 Adelaide street W., to 115½ Simcoe street.

Canada Nut Co. Moved.

The Canada Nut Co. have moved from Toronto to Niagara Falls their plant in that place being now completed.

Molding Shop Addition.

The Gilson Mfg. Co., Guelph, Ont., are putting up a molding shop, as an addition to their plant, which has but recently been completed. The shop will be 110x67 feet, and will be equipped with modern foundry equipment for molding gas engine cylinders and other engine parts.

Gasoline Motor Cars.

The Union Pacific Railway are placing in service on their lines twelve gasoline motor cars for fast and frequent service. These cars, which are equipped each with a 200 h.p. engine, will make 60 miles an hour. They will reach high speed within six car lengths, and can be stopped within 120 feet.

Standard Fittings Plant Completed.

The plant of the Standard Fittings and Valve Co., Guelph, has now been completed, and manufacturing operations have been started. They at present are making cast iron fittings altogether, but later brass fittings and valves will be made. The management of the company say that the product of the plant will be sold to dealers only, and not direct to consumers.

Brantford Screw Co. Installing Gas Plant.

The Brantford Screw Co., Brantford, are installing a Pintsch producer gas plant as a reserve in case that natural gas gives out. At present their new plant is operated by gas engine running on natural gas. There is always the danger of this giving out, and the Economic Power, Light & Heat Supply Co., Toronto, are installing a producer gas plant as a reserve.

To Build Factory in Toronto.

A company to manufacture a new kind of incandescent electric light are thinking of establishing a plant in Toronto for the manufacture of the "Helion" light, which differs from the ordinary incandescent light in the composition of the filament. The lighting abilities of these lights will be demonstrated at the Toronto exhibition by W. W. Campbell, representative of the company, who has an office in the Stair building, Toronto.

Further Addition to Hamilton Bridge Works.

A further addition is being made to the structural shop of the Hamilton Bridge Works, Hamilton. The new shop will be 72x180 feet, and building operations have been commenced. This new plant will be equipped with 90-foot span, 400-foot runways, electric traveling crane.

The building will be steel fireproof construction. All the machinery will be driven direct by electric motor.

Gas Made From Straw.

A way in which to utilize the immense amount of straw which accumulates in the Northwest has apparently been discovered by Russell Coutts, Cleveland. Mr. Coutts is a graduate of Toronto University in chemistry, and while at Carman, Man., recently, he made a discovery which led him to investigate the possibility of producing gas from, otherwise, waste straw. This investigation was carried on at Oxford University and in New York, where he used an experimental gas plant loaned him for that purpose by the New York Gas Consolidated Co. In this plant he succeeded in obtaining 16,000 feet of gas from a ton of straw. At first, considerable trouble was had with carbon dioxide, but this was finally overcome by treatment with incandescent carbon and oil enrichment. The gas finally obtained equalled in richness the average run of coal gas.

The patents of this process are held by the International Heating & Lighting Co., of Cleveland; and this company have in operation now a $100,000 plant in Beatrice, Neb.

Chapman Double Ball Bearings

Are Specially Adapted to be Used Wherever Friction Exists

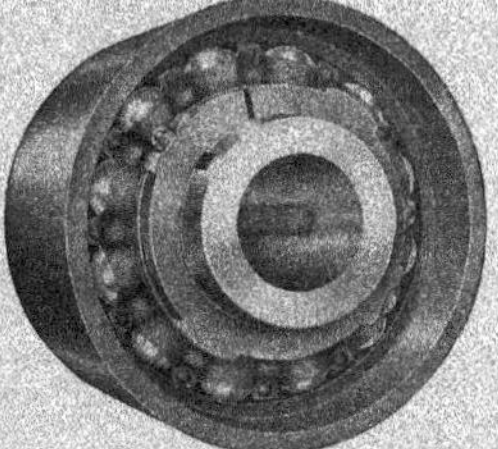

Loose Pulley Bearing

Chapman Double Ball Bearing for Shafting

Transmission—	*Vehicles—*	*Trucks—*	*Cars—*	*Machines—*	*Special—*
Clutch Couplings	Automobiles	Express	Dumping Cars	Emery Grinders	Baby Carriages
Counter Shafts	Carriages	Factory	Electric Cars	Exhaust Fans	Cotton Spindles
Jack Shafts	Express Waggons	Mill	Locomotives	Farming Implements	Lawn Mowers
Line Shafting	Road Waggons	Railroad	Mining Cars	Flour Mills	Safe Doors
Loose Pulleys	Trotting Sulkies	Turn Table	Ore Cars	Laundry Machines	Steering Gear
Mule Stands	Trucks and Drays	Warehouse	Railway Cars	Travelling Cranes	Thrust Bearings

Section of Carriage Hub Fitted With Chapman Double Ball Bearings

Car Wheel Bearing

Experience Has Proved that Twice the Load May be Moved With the Same Effort, Where Have Been Used

Chapman Double Ball Bearings

☞ *See Our Exhibits in Machinery Hall and Carriage Building, Canadian National Exhibition, Toronto*

The Chapman Double Ball Bearing Company of Canada
LIMITED

339 to 351 Sorauren Avenue — **Toronto, Canada**

At present it is proposed to build a plant at Brandon, Man., to cost [illegible]. It is also proposed to erect similar plants at Edmonton and other western points.

Gas Plant Installation Decided Upon.

The long-talked-of producer gas plant installation for the McClary Mfg. Co., London, Ont., has been quite decided upon. The installation will consist of a 200 h.p. suction producer gas plant and engine. The plant is being installed under the supervision of J. C. Royce, consulting engineer for the McClary Mfg. Co. The plant and engine are being built by the Weber Gas Engine Co., Kansas City; and the engine is of the vertical triple cylinder type, arranged to be direct connected to a Westinghouse generator.

The plant, when completed, will be tested thoroughly by the engineer in charge of installation; and the results of this test will be published later.

Steel Plant for British Columbia.

J. T. Shadforth, ironmaster, Newcastle-on-Tyne, one of the organizers of the North-Pacific Iron & Steel Corporation, is at present in Vancouver, B.C., arranging for the amalgamation of the coal and iron interests preparatory to the erection of a modern steel works. It will make steel of all grades, including steel rails and ship plate, with a big ship-building plant. The initial capital will be raised in British Columbia, and a greater amount in Manchester and London.

The company will be registered for fifteen million, of which two million will be initial expenditure on plant, with a site near Vancouver. All the raw material is now found in British Columbia. The value and extent of the iron ores of the coast is only lately being fully realized.

Electrical Equipment of Quebec Bridge.

The bridge which is being built at Quebec for the Quebec Bridge and Railway Co. is arousing considerable interest, it being the longest single span bridge in the world. All the erecting work is being done by electric motor. Alternating current is supplied from the Chaudiere Falls at 2,400 volts, to two motor-generator sets made by the Allis-Chalmers-Bullock, Ltd., Montreal. These are situated in a sub-station on the approach span, and they deliver a 550 volt direct current to the motors on the traveler and elsewhere.

This being the first time that electric power has been used on structural steel erection work of any magnitude, the outcome of the experiment has been watched with interest, and the fact that no delays or break-downs have yet been experienced speaks well for electric power for this use, and for this installation in particular.

Transformer Building Complete.

The large new transformer building for the Canadian General Electric Co., Peterboro', Ont., has been handed over by the contractors, the Dominion Engineering and Construction Co. The structure is entirely fireproof, being of concrete and steel. It is 80 feet broad, 52 feet high and 305 feet long. Along the east side of the building there is a large gallery, on which are located the machinery for winding the coils of the transformers. There is an electric crane, which travels the length of the building and has a capacity of 50 tons. There is a smaller crane of ten tons capacity. The oil for cooling the transformers is forced through pipes by an electric pump from a tank 500 feet from the building. The steel was supplied by the Canada Foundry Co., and the measurements were so accurate and the filling of the contract so faithfully done that practically no cutting was done on the ground.

Power From the Mine.

One of the greatest departures in solving the ever-present problem of obtaining cheap power was recently inaugurated in Nova Scotia. The plan is to develop power at the mouth of the coal mine and thus eliminate the necessity of transporting coal.

The Maritime Coal, Railway and Power Co., Amherst, N.S., acted upon a suggestion of Thos. A. Edison's, made recently in England, to build a power house at the mouth of their mine at Chignecto mines, N.S. Here they have installed a modern steam plant and propose to transmit the energy developed to several manufacturing centres.

The power plant consists principally of a Robb-Armstrong cross-compound, vertically-enclosed engine, developing 750 horse-power, under most economic conditions, but capable of carrying 1,000 horse-power on an overload. It was manufactured by the Robb Engineering Co., Amherst, N.S. The steam is supplied by four 200 horse-power, return tubular boilers, built also by the Robb Engineering Co. They are fed automatically by Jones underfeed stokers, and are arranged with induced draught.

Directly connected to the engine is a Canadian Westinghouse alternating current generator with a rated capacity of 500 kilowatts, delivering current at eleven thousand volts. It was manufactured by the Canadian Westinghouse Co., Hamilton, Ont.

The power will be transmitted at first to Amherst, a distance of 6½ miles, at a pressure of 11,000 volts. Here it enters a sub-station and is transformed down to 2,000 volts. At the sub-station are three Westinghouse transformers of 150 kilowatts each. There are three transformers of 40 kilowatts each, which reduce the pressure in some cases to 220 volts.

It is hoped that the cheap power generated at this station will be the means of bringing many industries to Amherst and vicinity.

A fully illustrated description will be given next month.

Addition to Armstrong Tool Plant.

Owing to the rapid increase of its business Armstrong Bros. Tool Co., of Chicago, have been compelled to make further additions to the large, modern plant which it erected about two years ago. These consist of two buildings of steel and brick constructions, one 50 feet by 105 feet, the other 90 feet by 105 feet, with brick smokestack 60 in. diameter, 115 feet high. In these buildings the company is now installing a modern power plant of 300 h.p., consisting of two 150 h.p. water tube boilers equipped with automatic stokers, direct connected engine and generator, etc., and an up-to-date drop forge department of large capacity with steam drop hammers of the latest improved design, and other high-grade equipment, including a complete die shop, with machine tools especially adapted to that important work.

The machinery is now being set in position and will be in operation about October 1st.

A New Machine Tool Company.

It is gratifying to watch the rapid development of the machinery business in Canada; and among the latest additions in this line is the Bawden Machine & Tool Co., Toronto. This company started out in a small way some months ago, but have now the beginnings of a fine machine shop. It is the idea of this company to build up a name for themselves on some standard line, and with this object in view, the company has started to manufacture a standard upright drill. The best of workmanship is being put on these tools, and a fine finish. The company expect this tool to give them their reputation.

Job work is done to fill in with.

There is plenty of room in Canada and a good future for a firm of this kind, and prosperity will surely be theirs if the business is built up along correct lines, chief of which is fine and accurate workmanship.

Canadian Cement Production.

The manufacture of cement in Canada dates from the year 1891, when operations were begun in a very small way at Marlbank and Shallow Lake, both in the Province of Ontario. The first year's output was 2,033 bbls., which is only a little more than the daily production of one of the modern plants. In 1892 the output was 20,247 bbls., and 31,924 bbls. in 1893.

The next ten years in the cement industry of Canada witnessed an increase of 50 per cent. each year over the production of the preceding year. Four plants were in operation in 1901, and now there are nineteen.

In 1906 there were fifteen plants in operation, with a total daily capacity of about 10,500 bbls., and according to the published figures of the Canadian Geological Survey, during the year 2,152,562 bbls. of Canadian cement were made, which represents an increase of 610,994 bbls., or 39.6 per cent. over the production of 1905.

Very little cement is exported from Canada. The consumption is due to the use each year of better and more improved machinery and equipment. An excellent example of a cement plant whose equipment will, when completely installed, be among the best in Canada, is that of the Western Canada Cement and Coal Company, which, some time ago, purchased through Allis-Chalmers-Bullock, Ltd., of Montreal, three Allis-Chalmers 1000 k.w. steam turbine generator units for use on the company's properties, located fifty-seven miles west of Calgary, Canada, on the Canadian Pacific Railway.

The units, which are now being erected, will generate current of 60 cycles, 3 phase, at 600 volts, and the coal to be used in their operation will be taken from the Bankhead mines of the Canadian Pacific Railway and delivered at the mills.

The adoption of electrical drive for the operation of these cement mills, whose current will be derived from a steam turbine power plant, which will rank among the foremost installed for industrial purposes, is a frank acknowledgement of the well recognized superiority of electrical drive over mechanical forms in the cement-making industry.

The current generated from the turbo-unit will be devoted entirely to power purposes; the lighting load, consisting of some forty arc and four hundred incandescent lamps, will be carried by the exciters, also of Allis-Chalmers build, in addition to their work of exciting the turbo-alternators.

The equipment, built by Allis-Chalmers Company, for the Western Canada Cement and Coal Company, is not confined, however, to the two turbines described. Allis-Chalmers' cement-making machinery, comprising a 7½ style "K" and No. 4 style "K" Gates breaker, and sixteen tube mills are also in use there.

BOOK REVIEWS.

CAMBRIA STEEL.—The eighth edition of this handbook, published by the Cambria Steel Co., Philadelphia, Pa.

This edition contains most of the matter of the prior edition, with such corrections and revisions as have been found necessary to bring it up to date. The tables of sheared and edged plates have been revised and amplified to show the larger range of sizes which the company are now prepared to furnish, particularly in edged in universal plates. Several new sections have been added, and some revisions made in the tables of rounds, flats, billets, blooms and slabs.

The table of "Properties and Principal Dimensions of T-Rails" has been revised, so that it now covers a full range of the American Society of Civil Engineers' standard sections.

ON THE ART OF CUTTING METALS.—The address made at the opening of the annual meeting of the American Society of Mechanical Engineers in New York, December, 1906, by F. W. Taylor, published in book form by the society; illustrated, with tables and diagrams; price, $3.

This address has created an immense stir in the mechanical world, and mechanical papers the world over have published more or less complete extracts from it. This is a book which every person interested in the cutting of metals should have, it being the most complete work ever published of experiments along this line.

BALANCING OF ENGINES.—An elementary text book, using principally graphic methods for the use of students, draftsmen and designers on this subject of balancing of steam, gas and other engines; by Archibald Sharp; 212 pages, 9x6 ins.; published by Longmans, Green & Co., 39 Paternoster Row, London, Eng.; price, 6s.

This book is very complete in itself, starting with simple mechanics and leading up to [illegible] higher forms of mathematics necessary for the solution of problems of this kind. The subject itself has for the most part been treated graphically. Many useful tables are contained in the back.

TESTS ON REINFORCED CONCRETE BEAMS.—Bulletin No. 148 of the University of Wisconsin contains a report of tests on reinforced concrete beams made by E. A. Moritz, C.E., instructor in mathematics; published by the University of Wisconsin, Madison, Wis.; price, 30 cents.

The report contains a general description of conditions of tests, and a detailed account of the results, including tables and curves, etc.

RAILROAD MEN'S CATECHISM.—An instruction book for enginemen, trainmen, signalmen and every person having to do with the moving of trains; by Angus Sinclair; 200 pages, 4½x6½ ins.; published by Angus Sinclair, Pub. Co., New York.

The questions used in this book are a series for the examination of engineers and firemen by a certain railroad system. Twenty questions for engineers and firemen have been added, the questions and answers being based on the standard code of the American Railway Association.

MANUFACTURERS AND MANUFACTURERS' AGENTS.

It will pay you to watch our condensed column each month. There are many money-making propositions brought to your attention here. You may find just what you are looking for.

RATES

One insertion—25c. for 20 words; 1c. a word for each additional word.

Yearly rate—$2.50 for twenty words or less, 10c a word for each additional word.

The above does not apply to notices under the head of "Machinery Wanted." These notices are inserted free for subscribers.

FOR SALE.

BELTING, RUBBER, CANVAS AND LEATHER, Hose, Packing, Blacksmith's and Mill Supplies at lowest prices. N. Smith, 138 York Street, Toronto. (2tf)

AUTOMATIC stocker, with fan and engine complete, in good running order; snap; reason for selling, compelled to burn shavings. Apply A. J. H. Eckardt, National Casket Co., 107 Niagara St., Toronto. (9)

GASOLINE engine, 12 h.p., good as new; price, $275. Krug & Crosley, Hamilton. (9)

FOR SALE—The Canadian patent rights on a valveless pneumatic hammer, proved by actual test to be superior to any hammer at present on the market; an arrangement can be made to manufacture same on royalty for Canadian market. Apply Box 39C, CANADIAN MACHINERY, Toronto. (9)

MANUFACTURERS' AGENTS WANTED.

SHEFFIELD manufacturers of tool steel, spring steel, section bars, etc., as advertised on page 16, wish to correspond with reliable Canadian firms with view to sole agencies in certain territories. Address "Newhall," care CANADIAN MACHINERY, 88 Fleet Street, London, Eng.

SITUATION VACANT.

WANTED—Young man well experienced in machinery supply business able to sell and knowing thoroughly all requirements of the trade. Salary, commission and share of profits. Apply Box 32, CANADIAN MACHINERY, Toronto.

WANTED AT ONCE—Superintendent and mechanical engineer for machine works manufacturing gasoline engines and general machine work; one familar with modern methods, capable of making estimates, drawings and producing work at minimum cost; give full particulars of qualifications, experience, age, nationality and salary. Apply to Brandon Machine Works Co., Ltd., Brandon, Man. [9]

SITUATIONS WANTED.

ADVERTISER, shortly emigrating, seeks position as correspondence clerk and accountant. Well up in all details of engineering and allied business; ledger balancing, shorthand and typewriting; excellent refrences. Lister, Ipswich Road, Ardleigh, Colchester, Eng.

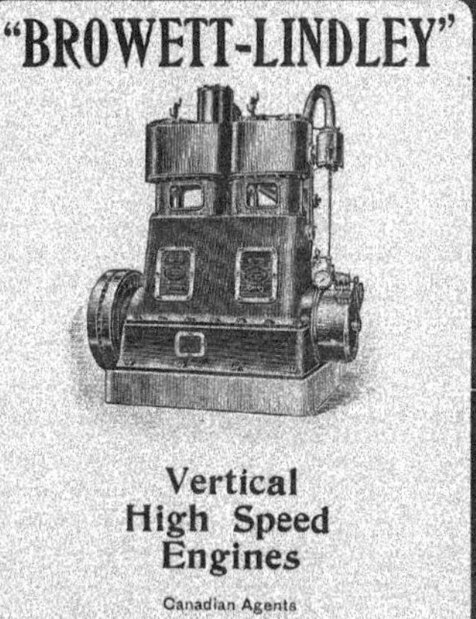

CANADIAN MACHINERY BUYERS' DIRECTORY

To Our Readers—Use this directory when seeking to buy any machinery or power equipment. You will often get information that will save you money.

To Our Advertisers—Send in your name for insertion under the heading of the lines you make or sell.

To Non-Advertisers—A nominal rate of $1 per line a year is charged non-advertisers.

Acids.
Canada Chemical Mfg. Co., London

Abrasive Grains & Powders
Carborundum Co., Niagara Falls, N.Y.

Abrasive Materials.
Baxter, J. R., & Co., Montreal.
The Canadian Fairbanks Co., Montreal.
Rice Lewis & Son, Toronto.
Norton Co., Worcester, Mass.
H. W. Petrie, Toronto.
Carborundum Co., Niagara Falls, N.Y.
Williams & Wilson, Montreal.

Air Brakes.
Canada Foundry Co., Toronto.
Canadian Westinghouse Co., Hamilton.

Air Receivers.
Allis-Chalmers-Bullock Montreal.
Canada Foundry Co., Toronto.
Canadian Rand Drill Co., Montreal.
Chicago Pneumatic Tool Co., Chicago.
John McDougall Caledonian Iron Works Co., Montreal.

Alloy, Ferro-Silicon.
Paxon, J. W., Co., Philadelphia.

Alundum Scythe Stones
Norton Company, Worcester, Mass.

Arbor Presses.
Niles-Bement-Pond Co., New York.

Augers.
Chicago Pneumatic Tool Co., Chicago.
Frothingham & Workman, Ltd., Montreal
Rice Lewis & Son, Toronto.

Automatic Machinery.
Cleveland Automatic Machine Co. Cleveland.
National-Acme Mfg. Co., Cleveland
Potter & Johnston Machine Co., Pawtucket, R. I.

Automobile Parts
Globe Machine & Stamping Co., Cleveland, Ohio.

Axle Cutters.
Butterfield & Co., Rock Island, Que.
A. B. Jardine & Co., Hespeler, Ont.

Axle Setters and Straighteners.
A. B. Jardine & Co., Hespeler, Ont.
Standard Bearings, Ltd., Niagara Falls

Babbit Metal.
Baxter, J. R., & Co., Montreal.
Canada Metal Co., Toronto.
Canada Machinery Agency, Montreal.
Frothingham & Workman, Ltd., Montreal
Greey, Wm. & J. G., Toronto.
Rice Lewis & Son, Toronto.
Lumen Bearing Co., Toronto.
Mechanics Supply Co., Quebec, Que.

Bakers' Machinery.
Greey, Wm. & J. G., Toronto.

Baling Presses.
Perrin, Wm. R., & Co., Toronto.

Barrels, Steel Shop.
Cleveland Wire Spring Co., Cleveland.

Barrels, Tumbling.
Buffalo Foundry Supply Co., Buffalo.
Detroit Foundry Supply Co., Windsor
Dominion Foundry Supply Co., Montreal
Hamilton Facing Mill Co., Hamilton.
Globe Machine & Stamping Co., Cleveland, Ohio.
John McDougall Caledonian Iron Works Co., Montreal.
Northern Engineering Works, Detroit.
J. W. Paxson Co., Philadelphia, Pa.
H. W. Petrie, Toronto.
Sly, W. W., Mfg. Co., Cleveland
The Smart-Turner Mach. Co., Hamilton

Bars, Boring.
Hall Engineering Works, Montreal.
Niles-Bement-Pond Co., New York.
Standard Bearings, Ltd., Niagara Falls

Batteries, Dry.
Berlin Electrical Mfg. Co., Toronto.
Mechanics Supply Co., Quebec, Que.

Batteries, Flashlight.
Berlin Electrical Mfg. Co., Toronto.
Mechanics Supply Co., Quebec, Que.

Batteries, Storage.
Canadian General Electric Co., Toronto
Chicago Pneumatic Tool Co., Chicago.
Rice Lewis & Son, Toronto.
John Millen & Son, Montreal.

Bearing Metals.
Lumen Bearing Co., Toronto.

Bearings, Roller.
Standard Bearings, Ltd. Niagara Falls

Bell Ringers.
Chicago Pneumatic Tool Co., Chicago.
Mechanics Supply Co., Quebec, Que.

Belting, Chain.
Baxter, J. R., & Co., Montreal.
Canada Machinery Exchange, Montreal.
Greey, Wm. & J. G., Toronto.
Jeffrey Mfg. Co., Columbia, Ohio.
Link-Belt Eng. Co., Philadelphia.
Waterous Engine Works Co., Brantford.

Belting, Cotton.
Baxter, J. R., & Co., Montreal.
Canada Machinery Agency, Montreal.
Dominion Belting Co., Hamilton.
Rice Lewis & Son, Toronto.

Belting, Leather.
Baxter, J. R., & Co., Montreal.
Canada Machinery Agency, Montreal.
The Canadian Fairbanks Co., Montreal.
Frothingham & Workman, Ltd., Montreal
Greey, Wm. & J. G., Toronto.
McLaren, J. C., Montreal.
Rice Lewis & Son, Toronto.
H. W. Petrie, Toronto.
Williams & Wilson, Montreal.

Belting, Rubber.
Baxter, J. R., & Co., Montreal.
Canada Machinery Agency, Montreal.
Frothingham & Workman, Ltd., Montreal
Greey, Wm. & J. G., Toronto.

Belting Supplies.
Baxter, J. R., & Co., Montreal.
Greey, Wm. & J. G., Toronto.
Rice Lewis & Son, Toronto.
H. W. Petrie, Toronto.

Bending Machinery.
John Bertram & Sons Co., Dundas, Ont.
Chicago Pneumatic Tool Co., Chicago.
Rice Lewis & Son, Toronto.
London Mach. Tool Co., Hamilton, Ont.
National Machinery Co., Tiffin, Ohio.
Niles-Bement-Pond Co., New York.

Benders, Tire.
A. B. Jardine & Co., Hespeler, Ont.
London Mach. Tool Co., Hamilton, Ont.

Blowers.
Buffalo Foundry Supply Co., Montreal.
Canada Machinery Agency, Montreal.
Detroit Foundry Supply Co., Windsor
Dominion Foundry Supply Co., Montreal
Hamilton Facing Mill Co., Hamilton
Kerr Turbine Co., Wellsville, N.Y.
Mechanics Supply Co., Quebec, Que
J. W. Paxson, Philadelphia, Pa.
Sheldon's Limited, Galt.
Wilbraham-Green Blower Co., Philadelphia

Blast Gauges—Cupola.
Paxson, J. W., Co. Philadelphia.
Sheldons, Limited, Galt

Blocks, Tackle.
Frothingham & Workman, Ltd., Montreal

Blocks, Wire Rope.
Frothingham & Workman, Ltd., Montreal

Blue Printing.
The Electric Blue Print Co., Montreal.

Blow-Off Tanks.
Darling Bros., Ltd., Montreal.

Boilers.
Canada Foundry Co., Limited, Toronto.
Canada Machinery Arency, Montreal.
Goldie & McCulloch Co., Galt.
John McDougall Calenonian Iron Works, Montreal.
Manitoba Iron Works, Winnipeg.
Mechanics Supply Co., Quebec, Que.
Owen Sound Iron Works Co., Owen Sound.
H. W. Petrie, Toronto.
Robb Engineering Co. Amherst, N.S.
The Smart-Turner Mach. Co., Hamilton.
Waterous Engine Works Co., Brantford.
Williams & Wilson, Montreal.

Boiler Compounds.
Canada Chemical Mfg. Co., London, Ont.
Hall Engineering Works, Montreal.
Lake Erie Boiler Compound Co., Toronto

Bolt Cutters.
John Bertram & Sons Co., Dundas, Ont.
London Mach. Tool Co., Hamilton.
Mechanics Supply Co., Quebec, Que.
National Machinery Co., Tiffin, Ohio.
Niles-Bement-Pond Co., New York.

Bolt and Nut Machinery.
John Bertram & Sons Co., Dundas, Ont.
Canada Machinery Agency, Montreal.
Rice Lewis & Son, Toronto.
London Mach. Tool Co., Hamilton.
National Machinery Co., Tiffin, Ohio.
Niles-Bement-Pond Co. New York.

Bolts and Nuts, Rivets.
Baxter, J. R., & Co., Montreal.
Mechanics Supply Co., Quebec, Que.

Boring & Drilling Machines
American Tool Works Co., Cincinnati.
B. F. Barnes Co., Rockford, Ill.
John Bertram & Sons Co., Dundas, Ont.
Canada Machinery Agency, Montreal.
A. B. Jardine & Co., Hespeler, Ont.
London Mach. Tool Co., Hamilton.
Niles-Bement-Pond Co., New York.

Boring Machine, Upright.
American Tool Works Co., Cincinnati.
John Bertram & Sons Co., Dundas, Ont.
London Mach. Tool Co., Hamilton.
Niles-Bement-Pond Co., New York.

Boring Machine, Wood.
Chicago Pneumatic Tool Co., Chicago.
London Mach. Tool Co., Hamilton.

Boring and Turning Mills.
American Tool Works Co., Cincinnati.
John Bertram & Sons Co., Dundas, Ont.
Gisholt Machine Co., Madison, Wis.
Canada Machinery Agency, Montreal.
Rice Lewis & Son, Toronto.
London Mach. Tool Co., Hamilton.
Niles-Bement-Pond Co., New York.
H. W. Petrie, Toronto.

Box Puller.
A. B. Jardine & Co., Hespeler, Ont.

Boxes, Steel Shop.
Cleveland Wire Spring Co., Cleveland.

Boxes, Tote.
Cleveland Wire Spring Co., Cleveland.

Brass Foundry Equipment.
Ph. Bonvillain & E. Ronceray, Philadelphia
Detroit Foundry Supply Co., Detroit.
Dominion Foundry Supply Co., Montreal
Paxson, J. W., Co., Philadelphia

Brass Working Machinery.
Warner & Swasey Co., Cleveland, Ohio.

Brushes, Foundry and Core.
Buffalo Foundry Supply Co., Buffalo.
Detroit Foundry Supply Co., Windsor.
Dominion Foundry Supply Co., Montreal
Hamilton Facing Mill Co., Hamilton.
Mechanics Supply Co., Quebec, Que.
Paxson, J. W., Co., Philadelphia

Brushes, Steel.
Buffalo Foundry Supply Co., Buffalo.
Dominion Foundry Supply Co., Montreal
Paxson, J. W., Co., Philadelphia

Bulldozers.
John Bertram & Sons Co., Dundas, Ont.
London Mach. Tool Co., Hamilton, Ont
National Machinery Co., Tiffin, Ohio.
Niles-Bement-Pond Co., New York.

Calipers.
Baxter, J. R., & Co., Montreal
Frothingham & Workman, Ltd., Montreal
Rice Lewis & Son, Toronto.
Mechanics Supply Co., Quebec, Que.
John Millen & Son, Ltd., Montreal, Que.
L. S. Starrett & Co., Athol, Mass.
Williams & Wilson Montreal.

Carbon.
Dominion Foundry Supply Co., Montreal

Cars, Foundry.
Buffalo Foundry Supply Co., Buffalo.
Detroit Foundry Supply Co., Windsor
Dominion Foundry Supply Co., Montreal
Hamilton Facing Mill Co., Hamilton.
Paxson, J. W., Co., Philadelphia

Castings, Aluminum.
Lumen Bearing Co., Toronto

Castings, Brass.
Chadwick Bros., Hamilton.
Greey, Wm. & J. G. Toronto.
Hall Engineering Works, Montreal.
Kennedy, Wm., & Sons, Owen Sound.
Lumen Bearing Co., Toronto
Niagara Falls Machine & Foundry Co. Niagara Falls, Ont.
Owen Sound Iron Works Co., Owen Sound.
Robb Engineering Co., Amherst, N.S.

Castings, Grey Iron.
Allis-Chalmers-Bullock Montreal.
Greey, Wm. & J. G., Toronto.
Hall Engineering Works, Montreal.
Kennedy, Wm., & Sons, Owen Sound.
Laurie Engine & Machine Co., Montreal.
Maxwell, David, & Sons, St. Marys.
John McDougall Caledonian Iron Works Co., Montreal.
Niagara Falls Machine & Foundry Co., Niagara Falls, Ont.
Owen Sound Iron Works Co., Owen Sound.
Robb Engineering Co., Amherst, N.S.
Smart-Turner Machine Co., Hamilton.

Castings, Phosphor-Bronze.
Lumen Bearing Co., Toronto

Castings, Steel.
Kennedy, Wm., & Sons, Owen Sound.

Castings, Semi-Steel.
Robb Engineering Co., Amherst, N.S.

Cement Machinery.
Allis-Chalmers-Bullock, Limited, Montreal
Greey, Wm. & J. G., Toronto.
Jeffrey Mfg. Co., Columbus, Ohio.
John McDougall Caledonian Iron Works, Co., Montreal
Owen Sound Iron Works Co., Owen Sound

Centreing Machines.
John Bertram & Sons Co., Dundas, Ont.
Jeffrey Mfg. Co., Columbia, Ohio
London Mach. Tool Co., Hamilton, Ont.
Niles-Bement-Pond Co., New York.
Pratt & Whitney Co., Hartford, Conn.
Standard Bearings, Ltd., Niagara Falls

Centres, Planer.
American Tool Works Co., Cincinnati.

Centrifugal Pumps.
John McDougall Caledonian Iron Works Co., Montreal.
Pratt & Whitney Co., Hartford, Conn.

Centrifugal Pumps—Turbine Driven
Kerr Turbine Co., Wellsville, N.Y.

Chain, Crane and Dredge.
Frothingham & Workman, Ltd., Montreal

Chaplets.
Buffalo Foundry Supply Co., Buffalo.
Detroit Foundry Supply Co., Windsor.
Dominion Foundry Supply Co., Montreal
Hamilton Facing Mill Co., Hamilton.
Paxson, J. W., Co., Philadelphia

Charcoal.
Buffalo Foundry Supply Co., Buffalo.
Detroit Foundry Supply Co., Windsor.
Doggett, Stanley, New York
Dominion Foundry Supply Co., Montreal
Hamilton Facing Mill Co., Hamilton.
Paxson, J. W., Co., Philadelphia

Charcoal Facings.
Doggett, Stanley, New York

Chemicals.
Baxter, J. R., & Co., Montreal.
Canada Chemical Co., London.

Chemists' Machinery.
Greey, Wm & J. G., Toronto.

Chemists, Industrial.
Detroit Testing Laboratory, Detroit.

Chemists, Metallurgical.
Detroit Testing Laboratory, Detroit.

Chemists, Mining.
Detroit Testing Laboratory, Detroit.

Chucks, Ring Grinding.
A. B. Jardine & Co., Hespeler, Ont.
Chicago Pneumatic Tool Co., Chicago.

Chucks, Drill and Lathe.
American Tool Works Co., Cincinnati.
Baxter, J. R., & Co., Montreal.
John Bertram & Sons Co., Dundas, Ont.
Canada Machinery Agency, Montreal.
Frothingham & Workman, Ltd., Montreal
Hamilton Tool Co., Hamilton, Ont.
Ker & Goodwin, Brantford.
A. B. Jardine & Co., Hespeler, Ont.
London Mach. Tool Co., Hamilton.
John Millen & Son, Ltd., Montreal.
Niles-Bement-Pond Co., New York.
H. W. Petrie, Toronto.
Rice Lewis & Son, Toronto.
Standard Tool Co., Cleveland.

Chucks, Planer.
American Tool Works Co., Cincinnati.
Canada Machinery Agency, Montreal.
Niles-Bement-Pond Co., New York.

Chucking Machines.
American Tool Works Co., Cincinnati.
Niles-Bement-Pond Co., New York.
H. W. Petrie, Toronto.
Warner & Swasey Co., Cleveland, Ohio

Circuit Breakers.
Allis-Chalmers-Bullock, Limited, Montreal
Canadian General Electric Co., Toronto.
Canadian Westinghouse Co., Hamilton.

Clippers, Bolt.
Frothingham & Workman, Ltd., Montreal
A. B. Jardine & Co., Hespeler, Ont.

Cloth and Wool Dryers.
Dominion Heating and Ventilating Co., Hespeler.
B. Greening Wire Co., Hamilton.
Sheldons Limited, Galt

Collectors, Pneumatic.
Sheldons Limited, Galt

Compressors, Air.
Allis-Chalmers-Bullock, Limited, Montreal
Canada Foundry Co., Limited, Toronto.
Canada Machinery Agency, Montreal.
Canadian Rand Drill Co., Montreal.
Canadian Westinghouse Co., Hamilton.
Chicago Pneumatic Tool Co., Chicago.
Darling Bros., Ltd., Montreal
Detroit Foundry Supply Co., Windsor.
Gas & Electric Power Co., Toronto.
John McDougall, Caledonian Iron Works Co., Montreal
H. W. Petrie, Toronto.
The Smart-Turner Mach. Co., Hamilton.
Hall Engineering Works, Montreal, Que.
London Mach. Tool Co., Hamilton.
Niles-Bement-Pond Co., New York.
H. W. Petrie, Toronto.
Pratt & Whitney Co., Hartford, Conn.
Williams & Wilson, Montreal.

Concentrating Plant.
Allis-Chalmers-Bullock, Montreal.
Greey, Wm. & J. G., Toronto.

Concrete Mixers.
Jeffrey Mfg. Co., Columbus, Ohio.
Link-Belt Co., Philadelphia.

Condensers.
Canada Foundry Co., Limited, Toronto.
Canada Machinery Agency, Montreal.
Hall Engineering Works Montreal.
Smart-Turner Machine Co., Hamilton.
Waterous Engine Co., Brantford.

Confectioners' Machinery.
Baxter, J. R., & Co., Montreal.
Greey, Wm & J. G., Toronto.

Consulting Engineers.
Hall Engineering Works, Montreal.
Jules De Clercy, Montreal.
Roderick J. Parke, Toronto.
Plews & Trimingham, Montreal.
T. Pringle & Son, Montreal.
Somerville & Van Every, Hamilton
Standard Bearings, Ltd., Niagara Falls

Contractors.
Expanded Metal and Fireproofing Co. Toronto.
Hall Engineering Works Montreal.
Laurie Engine & Machine Co., Montreal.
John McDougall Caledonian Iron Works Co., Montreal.
Robb Engineering Co., Amherst N S.
The Smart-Turner Mach. Co., Hamilton.

Contractors' Plant.
Allis-Chalmers-Bullock, Montreal.
John McDougall, Caledonian Iron Works Co., Montreal.
Niagara Falls Machine & Foundry Co., Niagara Falls, Ont.

Controllers and Starters Electric Motor.
Allis-Chalmers-Bullock, Montreal.
Canadian General Electric Co., Toronto.
Canadian Westinghouse Co., Hamilton.
T. & H. Electric Co., Hamilton.

Conveying Systems.
Sheldons Limited, Galt

Converters, Steel.
Northern Engineering Works, Detroit.
Paxson, J. W., Co., Philadelphia

Conveyor Machinery.
Baxter, J. R., & Co., Montreal.
Greey Wm & J. G., Toronto
Jeffrey Mfg. Co., Columbus, Ohio.
Rice Lewis & Son, Toronto
Link-Belt Co., Philadelphia.
John McDougall Caledonian Iron Works Co., Montreal.
Smart-Turner Machine Co., Hamilton.
Laurie Engine & Machine Co., Montreal.
Waterous Engine Works Co.; Brantford.
Williams & Wilson, Montreal.

Coping Machines.
John Bertram & Sons Co., Dundas, Ont.
London Mach. Tool Co., Hamilton.
Niles-Bement-Pond Co., New York.

Core Compounds.
Buffalo Foundry Supply Co., Buffalo.
Detroit Foundry Supply Co., Windsor.
Dominion Foundry Supply Co., Montreal
Hamilton Facing Mill Co., Hamilton.
Paxson, J. W., Co., Philadelphia

Core-Making Machines.
Paxson, J. W., Co., Philadelphia

Core Ovens.
Detroit Foundry Supply Co., Windsor.
Dominion Foundry Supply Co., Montreal
Hamilton Facing Mill Co., Hamilton.
Paxson, J. W., Co., Philadelphia
Sheldons Limited, Galt

Core Oven Bricks.
Buffalo Foundry Supply Co., Buffalo.
Detroit Foundry Supply Co., Windsor.
Dominion Foundry Supply Co., Montreal
Hamilton Facing Mill Co., Hamilton.
Paxson, J. W., Co., Philadelphia

Core Sand Cleaners.
Sly, W. W., Mfg. Co., Cleveland

Core Wash.
Buffalo Foundry Supply Co., Buffalo.
Detroit Foundry Supply Co., Windsor.
Dominion Foundry Supply Co., Montreal
Hamilton Facing Mill Co., Hamilton.
Paxson, Co., J. W., Philadelphia

Couplings.
Owen Sound Iron Works Co., Owen Sound

Cranes, Electric and Hand Power.
Canada Foundry Co., Limited, Toronto
Canadian Pilling Co., Montreal
Dominion Foundry Supply Co., Montreal
Hamilton Facing Mill Co., Hamilton.
Link-Belt Co., Philadelphia.
John McDougall, Caledonian Iron Works Co., Montreal.
Niles-Bement-Pond Co., New York.
Northern Engineering Works, Detroit.
Owen Sound Iron Works Co., Owen Sound
Paxson, J. W., Co., Philadelphia
Smart-Turner-Machine Co., Hamilton.

Crank Pin.
Sight Feed Oil Pump Co., Milwaukee, Wis.

Crankshafts.
The Canada Forge Co., Welland.
St. Clair Bros., Galt

Crabs.
Frothingham & Workman, Ltd., Montreal

Crank Pin Turning Machine
London Mach. Tool Co., Hamilton.
Niles-Bement-Pond Co., New York.

Cross Head Pin.
Sight Feed Oil Pump Co., Milwaukee, Wis.

Crucibles.
Buffalo Foundry Supply Co., Buffalo.
Detroit Foundry Supply Co., Windsor
Dominion Foundry Supply Co., Montreal
Hamilton Facing Mill Co., Hamilton.
J. W. Paxson Co., Philadelphia.

Crucible Caps
Dominion Foundry Supply Co., Montreal
Hamilton Facing Mill Co., Hamilton.
Paxson, J. W., Co., Philadelphia

Crushers, Rock or Ore.
Allis-Chalmers-Bullock, Montreal.
Jeffrey Mfg. Co., Columbus, Ohio.

Cup Grease.
Mechanics Supply Co., Quebec

Cupolas.
Buffalo Foundry Supply Co., Buffalo.
Detroit Foundry Supply Co., Windsor
Dominion Foundry Supply Co., Montreal
Hamilton Facing Mill Co., Hamilton.
Northern Engineering Works, Detroit
J. W. Paxson Co., Philadelphia.
Sheldons Limited, Galt.

Cupola Blast Gauges.
Dominion Foundry Supply Co., Montreal
Paxson, J. W., Co., Philadelphia
Sheldons Limited, Galt

Cupola Blocks.
Detroit Foundry Supply Co., Detroit.
Hamilton Facing Mill Co., Hamilton
Northern Engineering Works, Detroit
Ontario Lime Association, Toronto
Paxson, J. W., Co., Philadelphia
Toronto Pottery Co., Toronto

Cupola Blowers.
Buffalo Foundry Supply Co., Buffalo.
Canada Machinery Agency, Montreal.
Detroit Foundry Supply Co., Windsor
Dominion Foundry Supply Co., Montreal.
Dominion Heating and Ventilating Co., Hespeler
Hamilton Facing Mill Co., Hamilton.
Northern Engineering Works, Detroit
Paxson, J. W., Co., Philadelphia
Sheldon's Limited, Galt.

Cutters, Flue
Chicago Pneumatic Tool Co., Chicago.
J. W. Paxson Co., Philadelphia.

Cutter Grinder Attachment.
Cincinnati Milling Machine Co., Cincinnati

Cutter Grinders, Plain.
Cincinnati Milling Machine Co., Cincinnati

Cutter Grinders, Universal.
Cincinnati Milling Machine Co., Cincinnati

Cutters, Milling.
Becker, Brainard Milling Machine Co. Hyde Park, Mass.
Frothingham & Workman Ltd., Montreal
Hamilton Tool Co., Hamilton, Ont.
Pratt & Whitney Co., Hartford, Conn.
Standard Tool Co., Cleveland.

Cutting-off Machines.
John Bertram & Sons Co., Dundas, Ont.
Canada Machinery Agency, Montreal.
London Mach. Tool Co., Hamilton.
H. W. Petrie, Toronto.
Pratt & Whitney Co., Hartford, Conn.

Cutting-off Tools.
Armstrong Bros. Tool Co., Chicago.
Baxter, J. R., & Co., Montreal.
London Mach. Tool Co., Hamilton.
Mechanics Supply Co., Quebec, Que.
H. W. Petrie, Toronto.
Pratt & Whitney, Hartford, Conn.
Rice Lewis & Son, Toronto.
L. S. Starrett Co., Athol, Mass.

Damper Regulators.
Darling Bros., Ltd., Montreal

Dies
Brown, Boggs Co., Hamilton
Globe Machine & Stamping Co., Cleveland, Ohio.
Woelfle Bros., Berlin, Ont.

Die Stocks
Curtis & Curtis Co., Bridgeport, Conn.
Hart Manufacturing Co., Cleveland, Ohio
Loew Mfg. Co., Cleveland, Ohio
Mechanics Supply Co., Quebec

Dies, Opening
W. H. Banfield & Sons, Toronto
Globe Machine & Stamping Co., Cleveland, Ohio.
Pratt & Whitney Co., Hartford, Conn.
Woelfle Bros., Berlin, Ont.

Dies, Sheet Metal.
W. H. Banfield & Sons Toronto.
Bliss, E. W., Brooklyn, N.Y.
Globe Machine & Stamping Co., Cleveland, Ohio

Dies, Threading.
Frothingham & Workman, Ltd., Montreal
Hart Mfg. Co., Cleveland
Mechanics Supply Co., Quebec, Que.
John Millen & Son, Ltd., Montreal.

Draft, Mechanical.
W. H. Banfield & Sons, Toronto.
Butterfield & Co., Rock Island, Que.
A. B. Jardine & Co., Hespeler
Mechanics Supply Co., Quebec, Que.
Pratt & Whitney Co., Hartford, Conn
Sheldon's Limited, Galt.

Drawing Instruments.
Rice Lewis & Son, Toronto.
Mechanics Supply Co., Quebec, Que.

Drawing Supplies.
The Electric Blue Print Co., Montreal

Drawn Steel, Cold.
Baxter, J. R., & Co., Montreal.
Greey, Wm. & J. G., Toronto
Union Drawn Steel Co., Hamilton.

Drilling Machines, Arch Bar.
John Bertram & Sons Co., Dundas, Ont.
London Mach. Tool Co., Hamilton
Niles-Bement-Pond Co., New York.

Drilling Machines, Boiler.
American Tool Works Co., Cincinnati.
John Bertram & Sons Co., Dundas, Ont.
Bickford Drill and Tool Co., Cincinnati.
The Canadian Fairbanks Co., Montreal.
A. B. Jardine & Co., Hespeler, Ont.
London Mach. Tool Co., Hamilton, Ont.
Niles-Bement-Pond Co., New York.
H. W. Petrie, Toronto.
Williams & Wilson, Montreal

Drilling Machines Connecting Rod.
John Bertram & Sons Co., Dundas, Ont.
London Mach. Tool Co., Hamilton.
Niles-Bement-Pond Co. New York.

Drilling Machines, Locomotive Frame.
American Tool Works Co., Cincinnati.
B. F. Barnes Co., Rockford, Ill.
John Bertram & Sons Co., Dundas, Ont.
London Mach. Tool Co., Hamilton, Ont.
Niles-Bement-Pond Co., New York.

Drilling Machines, Multiple Spindle.
American Tool Works Co., Cincinnati.
B. F. Barnes Co., Rockford, Ill.
Baxter, J. R., & Co., Montreal.
John Bertram & Sons Co., Dundas, Ont.
Bickford Drill & Tool Co., Cincinnati.
Canada Machinery Agency, Montreal.
London Mach. Tool Co., Hamilton, Ont.
Niles-Bement-Pond Co., New York.
H. W. Petrie, Toronto.
Williams & Wilson, Montreal.

Drilling Machines, Pneumatic
Canada Machinery Agency, Montreal.

Drilling Machines, Portable
Baxter, J. R., & Co., Montreal.
A. B. Jardine & Co., Hespeler, Ont.
Mechanics Supply Co., Quebec, Que.
Niles-Bement-Pond Co., New York.

Drilling Machines, Radial.
American Tool Works Co., Cincinnati.
Baxter, J. R., & Co., Montreal
John Bertram & Sons Co., Dundas, Ont.
Bickford Tool & Drill Co., Cincinnati.
The Canadian Fairbanks Co., Montreal.
London Mach. Tool Co., Hamilton.
Mechanics Supply Co., Quebec, Que.
Niles-Bement-Pond Co., New York.
H. W. Petrie, Toronto.
Williams & Wilson, Montreal.

Drilling Machines, Suspension.
John Bertram & Sons Co., Dundas, Ont.
Canada Machinery Agency, Montreal.
London Mach. Tool Co., Hamilton.
Niles-Bement-Pond Co. New York

Drilling Machines, Turret.
John Bertram & Sons Co., Dundas, Ont.
London Mach. Tool Co., Hamilton.
Niles-Bement-Pond Co., New York.

Drilling Machines, Upright.
American Tool Works Co., Cincinnati.
B. F. Barnes Co., Rockford, Ill.
Baxter, J. R., & Co., Montreal.
John Bertram & Sons Co., Dundas, Ont.
Dwight Slate Machine Co., Hartford,
Hamilton Tool Co., Hamilton, Ont.
A. B. Jardine & Co., Hespeler, Ont.
Rice Lewis & Son, Toronto.
London Mach. Tool Co., Hamilton.
Mechanics Supply Co., Quebec, Que.
Niles-Bement-Pond Co., New York.
W. H. Petrie, Toronto.
Williams & Wilson Montreal.

Drills, Bench.
B. F. Barnes Co., Rockford, Ill.
Baxter, J. R. & Co., Montreal.
Hamilton Tool Co., Hamilton, Ont.
London Mach. Tool Co., Hamilton.
Mechanics Supply Co., Quebec, Que.
Pratt & Whitney Co., Hartford, Conn.

Drills, Blacksmith.
Canada Machinery Agency, Montreal.
Frothingham & Workman, Ltd., Montreal
A. B. Jardine & Co., Hespeler, Ont.
London Mach. Tool Co., Hamilton.
Mechanics Supply Co., Quebec, Que.
Standard Tool Co., Cleveland.

Drills, Centre.
Mechanics Supply Co. e, Que.
Pratt & Whitney Co., Hartford, Conn.
Standard Tool Co., Cleveland, O.
L. S. Starrett Co., Athol, Mass.

Drills, Electric
B. F. Barnes Co., Rockford, Ill.
Baxter J. R., & Co., Montreal.
Canadian Filling Co., Montreal
Chicago Pneumatic Tool Co., Chicago.
Niles-Bement-Pond Co., New York.

Drills, Gang.
American Tool Works Co., Cincinnati.
B. F. Barnes Co., Rockford, Ill.
John Bertram & Sons Co., Dundas, Ont.
Pratt & Whitney Co., Hartford, Conn.

Drills, High Speed.
Wm. Abbott, Montreal.
American Tool Works Co., Cincinnati.
B. F. Barnes Co., Rockford, Ill.
Frothingham & Workman, Ltd., Montreal
Pratt & Whitney Co., Hartford, Conn.
Standard Tool Co., Cleveland, O.

Drills, Hand.
A. B. Jardine & Co., Hespeler, Ont.

Drills, Horizontal.
B. F. Barnes Co., Rockford, Ill.
John Bertram & Sons Co., Dundas, Ont.
Canada Machinery Agency, Montreal.
London Mach. Tool Co., Hamilton.
Niles-Bement-Pond Co., New York.

Drills, Pneumatic.
Canada Machinery Agency, Montreal.
Chicago Pneumatic Tool Co., Chicago.
Independent Pneumatic Tool Co., Chicago, New York
Niles-Bement-Pond Co., New York.

Drills, Radial.
American Tool Works Co., Cincinnati.
John Bertram & Sons Co., Dundas, Ont.
Bickford Drill & Tool Co., Cincinnati
London Mach. Tool Co., Hamilton, Ont.
Niles-Bement-Pond Co., New York.

Drills, Ratchet.
Armstrong Bros. Tool Co., Chicago.
Frothingham & Workman, Ltd., Montreal
A. B. Jardine & Co., Hespeler.
Mechanics Supply Co., Quebec, Que.
Pratt & Whitney Co., Hartford, Conn.
Standard Tool Co., Cleveland.

Drills, Rock.
Allis-Chalmers-Bullock, Montreal.
Canadian Rand Drill Co., Montreal.
Chicago Pneumatic Tool Co., Chicago.
Jaffrey Mfg. Co., Columbus, Ohio.

Drills, Sensitive.
American Tool Works Co., Cincinnati.
B. F. Barnes Co., Rockford, Ill.
Canada Machinery Agency, Montreal.
Dwight Slate Machine Co., Hartford.
Niles-Bement-Pond Co., New York.

Drills, Twist.
Baxter, J. R., & Co., Montreal.
Chicago Pneumatic Tool Co., Chicago
Frothingham & Workman, Ltd., Montreal
Alex. Gibb, Montreal.
A. B. Jardine & Co., Hespeler, Ont.
Mechanics Supply Co., Quebec, Que.
John Millen & Son, Ltd., Montreal.
Morse Twist Drill and Machine Co., New Bedford, Mass.
Pratt & Whitney Co., Hartford, Conn.
Standard Tool Co., Cleveland.

Drying Apparatus of all Kinds.
Dominion Heating & Ventilating Co., Hespeler
Greey, Wm. & J. G., Toronto
Sheldons Limited, Galt

Dry Kiln Equipment.
Sheldons Limited. Galt

Dry Sand and Loam Facing.
Buffalo Foundry Supply Co., Buffalo.
Dominion Foundry Supply Co., Montreal
Hamilton Facing Mill Co., Hamilton.
Paxson, J. W., Co., Philadelphia

Dump Cars.
Canada Foundry Co., Limited, Toronto
Dominion Foundry Supply Co., Montreal
Greey, Wm. & J. G., Toronto
Hamilton Facing Mill Co., Hamilton.
John McDougall, Caledonian Iron Works Co., Montreal.
Niles-Bement-Pond Co., New York.
Standard Bearings, Ltd., Niagara Falls.
Link-Belt Eng. Co., Philadelphia.
John McDougall Caledonian Iron Works Co., Montreal.
Owen Sound Iron Works Co., Owen Sound
Paxson, J. W., Co., Philadelphia
Waterous Engine Co., Brantford.

Dust Arresters.
Sly, W. W., Mfg. Co., Cleveland

Dust Separators.
Greey, Wm. & J. G., Toronto
Dominion Heating and Ventilating Co., Hespeler.
Paxson, J. W., Co., Philadelphia
Sheldon's Limited, Galt.

Dynamos.
Allis-Chalmers-Bullock, Montreal.
Canadian General Electric Co., Toronto.
Canadian Westinghouse Co., Hamilton.
Consolidated Electric Co., Toronto
Electrical Machinery Co., Toronto.
Gas & Electric Power Co., Toronto
Hall Engineering Works, Montreal, Que.
Mechanics Supply Co., Quebec, Que.
John Millen & Son, Ltd., Montreal.
Packard Electric Co., St. Catharines.
H. W. Petrie, Toronto.
T. & H. Electric Co., Hamilton.

Dynamos—Turbine Driven.
Kerr-Turbine Co., Wellsville, N.Y.

Economizer, Fuel.
Dominion Heating & Ventilating Co., Hespeler
Standard Bearings, Ltd., Niagara Falls.

Electrical Instruments.
Canadian Westinghouse Co., Hamilton.
Mechanics Supply Co., Quebec, Que.

Electrical Steel.
Baxter, Patterson & Co., Montreal.

Electrical Supplies.
Canadian General Electric Co., Toronto.
Canadian Westinghouse Co., Hamilton.
London Mach. Tool Co., Hamilton, Ont.
Mechanics Supply Co., Quebec, Que.
John Millen & Son, Ltd., Montreal.
Packard Electric Co., St. Catharines.
T. & H. Electric Co., Hamilton.

Electrical Repairs
Canadian Westinghouse Co., Hamilton.
T. & H. Electric Co., Hamilton.

Elevator Buckets.
Greey, Wm. & J. G., Toronto
Jeffrey Mfg. Co., Columbus, Ohio.

Elevators, Foundry.
Dominion Foundry Supply Co., Montreal
Northern Engineering Works, Detroit

Elevators—For any Service.
Darling Bros., Ltd., Montreal

Emery and Emery Wheels.
Baxter, J. R., & Co., Montreal.
Dominion Foundry Supply Co., Montreal
Frothingham & Workman Ltd., Montreal
Hamilton Facing Mill Co., Hamilton.
Paxson J. W. Co. Philadelphia

Emery Wheel Dressers.
Baxter J. R. & Co., Montreal.
Canada Machinery Agency, Montreal.
Desmond-Stephan Mfg. Co., Urbana, Ohio
Dominion Foundry Supply Co., Montreal
Frothingham & Workman Ltd., Montreal
Hamilton Facing Mill Co., Hamilton.
International Specialty Co., Detroit.
Mechanics Supply Co. Quebec, Que.
John Millen & Son, Ltd., Montreal.
H. W. Petrie, Toronto.
Paxson, J. W., Co., Philadelphia
Standard Tool Co., Cleveland.

Engineers and Contractors.
Canada Foundry Co., Limited, Toronto.
Darling Bros., Ltd., Montreal
Greey, Wm. & J. G., Toronto
Hall Engineering Works, Montreal.
Laurie Engine & Machine Co., Montreal.
Link-Belt Co., Philadelphia.
John McDougall, Caledonian Iron Works Co., Montreal.
Robb Engineering Co., Amherst. N S.
The Smart-Turner Mach. Co., Hamilton.

Engineers' Supplies.
Baxter, J. R., & Co., Montreal.
Frothingham & Workman, Ltd., Montreal
Greey, Wm. & J. G., Toronto
Hall Engineering Works, Montreal.
Mechanics Supply Co., Quebec, Que.
Rice Lewis & Son, Toronto.

Engines, Gas and Gasoline.
Baxter, J. R., & Co., Montreal.
Canada Foundry Co., Toronto.
Canada Machinery Agency, Montreal.
The Canadian Fairbanks Co., Montreal.
Gas & Electric Power Co., Toronto
Gilson Mfg. Co., Guelph
The Goldie & McCulloch Co., Galt, Ont.
Rice Lewis & Son, Toronto.
Ontario Wind Engine & Pump Co. Toronto
H. W. Petrie, Toronto.
The Smart-Turner Mach. Co., Hamilton
Standard Bearings, Ltd., Niagara Falls

Engines, Steam.
Allis-Chalmers-Bullock, Montreal
Belliss & Marcom, Birmingham, Eng.
Canada Machinery Agency, Montreal.
The Goldie & McCulloch Co., Galt, Ont.
Rice Lewis & Son, Toronto.
Laurie Engine & Machine Co., Montreal.
John McDougall Caledonian Iron Works Montreal.
Miller Bros. & Toms, Montreal.
Robb Engineering Co., Amherst, N.S.
Sheldons Limited, Galt.
The Smart-Turner Mach. Co., Hamilton.
Waterous Engine Works Co., Brantford.

Exhaust Heads.
Darling Bros., Ltd., Montreal.
Dominion Heating & Ventilating Co., Hespeler
Sheldons Limited, Galt, Ont.

Expanded Metal.
Expanded Metal and Fireproofing Co. Toronto

Expanders.
A. B. Jardine & Co., Hespeler, Ont.

Fans, Electric.
Canadian General Electric Co., Toronto.
Canadian Westinghouse Co., Hamilton.
Dominion Heating & Ventilating Co. Hespeler
Mechanics Supply Co., Quebec, Que.
Sheldons Limited Galt, Ont.
The Smart-Turner Mach. Co., Hamilton.

Fans, Exhaust.
Canadian Buffalo Forge Co., Montreal.
Detroit Foundry Supply Co., Windsor.
Dominion Foundry Supply Co., Montreal
Dominion Heating & Ventilating Co., Hespeler.
Greey, Wm. & J. G., Toronto
Hamilton Facing Mill Co., Hamilton.
Paxson, J. W., Co., Philadelphia
Sheldons Limited, Galt.
B. F. Sturtevant Co., Hyde Park, Mass.

Feed Water Heaters.
Darling Bros., Montreal
Laurie Engine & Machine Co., Montreal
John McDougall, Caledonian Iron Works Co., Montreal.
The Smart-Turner Mach. Co., Hamilton

Files and Rasps.
Baxter, Paterson & Co., Montreal.
Frothingham & Workman, Ltd., Montreal
Mechanics Supply Co., Quebec, Que.
John Millen & Son, Ltd., Montreal.
Rice Lewis & Son, Toronto.
Nicholson File Co., Port Hope
H. W. Petrie, Toronto

Fillet, Pattern.
Baxter, Paterson & Co., Montreal.
Buffalo Foundry Supply Co., Buffalo.
Detroit Foundry Supply Co., Windsor.
Dominion Foundry Supply Co., Montreal
Hamilton Facing Mill Co., Hamilton.

Fire Apparatus.
Waterous Engine Works Co., Brantford.

Fire Brick and Clay.
Baxter Paterson & Co., Montreal.
Buffalo Foundry Supply Co., Buffalo.
Detroit Foundry Supply Co., Windsor.
Dominion Foundry Supply Co., Montreal
Hamilton Facing Mill Co., Hamilton.
Ontario Lime Association, Toronto
J. W. Paxson Co., Philadelphia.
Toronto Pottery Co., Toronto

Flash Lights.
Berlin Electrical Mfg. Co., Toronto.
Mechanics Supply Co., Quebec, Que.

Flour Mill Machinery.
Allis-Chalmers-Bullock, Montreal.
Greey, Wm. & J. G., Toronto
The Goldie & McCulloch Co., Galt, Ont.
John McDougall, Caledonian Iron Works Co., Montreal.

Forges.
Canada Foundry Co., Limited, Toronto.
Frothingham & Workman, Ltd., Montreal
Hamilton Facing Mill Co., Hamilton.
Mechanics Supply Co., Quebec, Que.
H. W. Petrie, Toronto.
Sheldons Limited, Galt, Ont.

Forgings, Drop.
John McDougall, Caledonian Iron Work Co., Montreal.
H. W. Petrie, Toronto.
St. Clair Bros, Galt

Forgings, Light & Heavy.
The Canada Forge Co., Welland.
Hamilton Steel & Iron Co., Hamilton

Forging Machinery.
John Bertram & Sons Co., Dundas, Ont.
London Mach. Tool Co., Hamilton, Ont.
National Machinery Co., Tiffin, Ohio
Niles-Bement-Pond Co., New York.

Founders.
Greey, Wm. & J. G., Toronto
John McDougall, Caledonian Iron Works Co., Montreal.
Niagara Falls Machine & Foundry Co., Niagara Falls, Ont.
Maxwell, David, & Sons, St. Marys
The Smart-Turner Mach. Co., Hamilton.

Foundry Coke.
Baird & West, Detroit

Foundry Equipment.
Ph. Bonvillain & E. Ronceray, Philadelphia
Detroit Foundry Supply Co., Windsor.
Dominion Foundry Supply Co., Montreal
Hamilton Facing Mill Co., Hamilton
Hanna Engineering Works, Chicago.
Northern Engineering Works, Detroit
Paxson, J. W., Co., Philadelphia

Foundry Parting.
Doggett, Stanley, New York
Dominion Foundry Supply Co., Montreal
Partomol Co., New York
Foundry Specialty Co., Cincinnati
Stanley Doggett, New York

Foundry Facings.
Buffalo Foundry Supply Co., Buffalo.
Detroit Foundry Supply Co., Windsor.
Doggett Stanley, New York
Dominion Foundry Supply Co., Montreal
Hamilton Facing Mill Co., Hamilton.
J. W. Paxson Co., Philadelphia, Pa.

Friction Clutch Pulleys, etc.
The Goldie & McCulloch Co., Galt.
Greey, Wm. & J. G., Toronto
Link-Belt Co., Philadelphia.

Furnaces.
Detroit Foundry Supply Co., Windsor.
Dominion Foundry Supply Co., Montreal
Hamilton Facing Mill Co., Hamilton.
Paxson, J. W., & Philadelphia Co.,

Galvanizing
Canada Metal Co., Toronto.
Ontario Wind Engine & Pump Co., Toronto

Gas Blowers and Exhausters.
Dominion Heating & Ventilating Co. Hespeler
Sheldons Limited, Galt.

Gas Plants, Suction and Pressure.
Baxter J. R., & Co., Montreal
Colonial Engineering Co., Montreal
Gas & Electric Power Co., Toronto
Standard Bearings, Ltd., Niagara Falls
Williams & Wilson, Montreal

Gauges, Standard.
Mechanics Supply Co., Quebec, Que.
Pratt & Whitney Co., Hartford, Conn.

Gearing.
Eberhardt Bros. Machine Co., Newark,
Greey, Wm. & J. G., Toronto

Gear Cutting Machinery.
Baxter, J. R., & Co., Montreal
Becker-Brainard Milling Mach. Co., Hyde Park, Mass.
Bickford Drill & Tool Co., Cincinnati.
Dwight Slate Machine Co., Hartford
Eberhardt Bros. Machine Co., Newark
Greey, Wm. & J. G., Toronto
Kennedy Wm., & Sons, Owen Sound
London Mach. Tool Co., Hamilton.
Niles-Bement-Pond Co., New York.
H. W. Petrie, Toronto.
Pratt & Whitney Co., Hartford. Conn.
Williams & Wilson, Montreal.

Gears, Angle.
Chicago Pneumatic Tool Co., Chicago.
Greey, Wm. & J. G., Toronto
Laurie Engine & Machine Co., Montreal.
John McDougall, Caledonian Iron Works Co., Montreal.
Waterous Engine Co., Brantford.

Gears, Cut.
Kennedy, Wm., & Sons, Owen Sound

Gears, Iron.
Greey, Wm. & J. G., Toronto
Kennedy, Wm., & Sons, Owen Sound

Gears, Mortise.
Greey, Wm. & J. G., Toronto
Kennedy, Wm., & Sons, Owen Sound

Gears, Reducing.
Brown, David & Sons, Huddersfield, Eng
Chicago Pneumatic Tool Co., Chicago.
Greey, Wm. & J. G., Toronto
John McDougall, Caledonian Iron Works Co., Montreal.

Generators, Electric.
Allis-Chalmers-Bullock, Limited, Montreal
Canadian General Electric Co., Toronto
Canadian Westinghouse Co., Hamilton.
Gas & Electric Power Co., Toronto
Ball Engineering Works, Montreal.
H. W. Petrie, Toronto.
Toronto & Hamilton Electric Co. Hamilton.

Generators, Gas.
H. W. Petrie, Toronto.

Graphite Paints.
P. D. Dods & Co., Montreal.

Graphite.
Detroit Foundry Supply Co., Windsor.
Doggett, Stanley, New York
Dominion Foundry Supply Co., Montreal
Hamilton Facing Mill Co., Hamilton.
Mechanics Supply Co., Quebec, Que.
Paxson, J. W., Co., Philadelphia

Grinders, Automatic Knife.
W. H. Banfield & Son, Toronto.

Grinders, Centre.
Niles-Bement-Pond Co., New York.
H. W. Petrie, Toronto.

Grinders, Cutter.
Becker-Brainard Milling Mach. Co., Hyde Park, Mass.
John Millen & Son, Ltd., Montreal.
Pratt & Whitney Co., Hartford, Conn.

Grinders, Tool.
Armstrong Bros. Tool Co., Chicago.
B. F. Barnes Co., Rockford, Ill.
H. W. Petrie, Toronto.
Williams & Wilson, Montreal.

Grinding Machines.
The Canadian Fairbanks, Montreal.
Rice Lewis & Son, Toronto.
Niles-Bement-Pond Co., New York.
Paxson, J. W., Co., Philadelphia
H. W. Petrie, Toronto.

Grinding & Polishing Machines
The Canadian Fairbanks Co., Montreal.
Greey, Wm. & J. G., Toronto
Independent Pneumatic Tool Co., Chicago, Ill.
John Millen & Son, Ltd., Montreal.
Niles-Bement-Pond Co., New York.
H. W. Petrie, Toronto.
Pendrith Machinery Co., Toronto.
Standard Bearings, Ltd., Niagara Falls

Grinding Wheels
Carborundum Co., Niagara Falls
Norton Company, Worcester, Mass.

Hack Saws.
Baxter, J. R., & Co., Montreal.
Canada Machinery Agency, Montreal.
The Canadian Fairbanks Co., Montreal.
Frothingham & Workman, Ltd., Montreal
Mechanics Supply Co., Quebec, Que.
John Millen & Son, Ltd., Montreal.
Niles-Bement-Pond Co., New York.
H. W. Petrie, Toronto.
Williams & Wilson, Montreal.

Hack Saw Frames.
Mechanics Supply Co., Quebec, Que.

Hammers, Drop.
London Mach. Tool Co., Hamilton, Ont.
Niles-Bement-Pond Co., New York.

Hammers, Power.
Rice Lewis & Son, Toronto.

Hammers, Steam.
John Bertram & Sons Co. Dundas, Ont.
London Mach. Tool Co., Hamilton, Ont.
Niles-Bement-Pond Co., New York.

Hand Stocks.
Borden-Canadian Co., Toronto

Hangers.
The Goldie & McCulloch Co., Galt.
Greey, Wm., & J. G., Toronto
Kennedy, Wm., & Sons, Owen Sound
Owen Sound Iron Works Co., Owen Sound
The Smart-Turner Mach. Co., Hamilton.
Waterous Engine Co., Brantford.
Standard Bearings Ltd., Niagara Falls

Hardware Specialties.
Mechanics Supply Co. Quebec, Que.

Heating Apparatus.
Darling Bros., Ltd., Montreal
Dominion Heating & Ventilating Co., Hespeler
Mechanics Supply Co., Quebec, Que.
Sheldons Limited, Galt.

Hoisting and Conveying Machinery.
Allis-Chalmers-Bullock, Limited, Montreal
Greey, Wm. & J. G., Toronto
Link-Belt Co., Philadelphia.
Niles-Bement-Pond Co., New York.
Northern Engineering Works, Detroit
The Smart-Turner Mach. Co., Hamilton.
Waterous Engine Co., Brantford.

Hoists, Chain and Rope.
Frothingham & Workman, Ltd., Montreal

Hoists, Electric.
Northern Engineering Works, Detroit

Hoists, Pneumatic.
Canadian Rand Drill Co., Montreal.
Chicago Pneumatic Tool Co. Chicago.
Dominion Foundry Supply Co., Montreal
Hamilton Facing Mill Co., Hamilton.
Northern Engineering Works, Detroit

Hoists, Portable & Stationery.
Canadian Pilling Co., Montreal

Hose.
Baxter, J. R., & Co., Montreal
Mechanics Supply Co., Quebec, Que.

Hose, Air.
Canadian Rand Drill Co., Montreal.
Canadian Westinghouse Co., Hamilton.
Chicago Pneumatic Tool Co. Chicago.
Paxson, J. W., Co., Philadelphia

Hose Couplings.
Canadian Rand Drill Co., Montreal.
Canadian Westinghouse Co., Hamilton.
Chicago Pneumatic Tool Co., Chicago.
Mechanics Supply Co., Quebec, Que.
Paxson, J. W., Co., Philadelphia

Hose, Steam.
Allis-Chalmers-Bullock, Montreal.
Canadian Rand Drill Co., Montreal.
Canadian Westinghouse Co., Hamilton.
Mechanics Supply Co., Quebec, Que.
Paxson, J. W. Co., Philadelphia

Hydraulic Accumulators.
Ph. Bonvillain & E. Ronceray, Philadelphia
Niles-Bement-Pond Co., New York.
Perrin, Wm. R., Co., Toronto.
The Smart-Turner Mach. Co., Hamilton.

Hydraulic Machinery.
Allis-Chalmers-Bullock, Montreal
Barber, Chas., & Sons, Meaford

India Oil Stones.
Norton Company, Worcester, Mass.

Indicators, Speed.
Mechanics Supply Co., Quebec, Que.
L. S. Starrett Co., Athol, Mass.

Injectors.
Canada Foundry Co., Toronto.
The Canadian Fairbanks Co., Montreal.
Desmond-Stephan Mfg. Co., Urbana, Ohio
Frothingham & Workman Ltd., Montreal
Mechanics Supply Co. Quebec, Que.
Rice Lewis & Son, Toronto.
Penberthy Injector Co., Windsor, Ont.

Iron and Steel.
Frothingham & Workman, Ltd., Montreal

Iron and Steel Bars and Bands.
Hamilton Steel & Iron Co., Hamilton

Iron Cements.
Buffalo Foundry Supply Co., Buffalo.
Detroit Foundry Supply Co., Windsor.
Dominion Foundry Supply Co., Montreal
Hamilton Facing Mill Co., Hamilton.
Paxson, J. W., Co., Philadelphia

Iron Filler.
Buffalo Foundry Supply Co., Buffalo.
Detroit Foundry Supply Co., Windsor.
Dominion Foundry Supply Co. Montreal
Hamilton Facing Mill Co., Hamilton.
Paxson, J. W., Co., Philadelphia.

Jacks.
Frothingham & Workman, Ltd., Montreal
Norton, A. O., Coaticook, Que.

Kegs, Steel Shop.
Cleveland Wire Spring Co., Cleveland

Key-Seating Machines.
B. F. Barnes Co., Rockford, Ill.
Niles-Bement-Pond Co., New York.

Lace Leather.
Baxter, J. R., & Co., Montreal.

Ladles, Foundry.
Dominion Foundry Supply Co., Montreal
Frothingham & Workman Ltd., Montreal
Northern Engineering Works, Detroit

Lamps, Arc and Incandescent.
Canadian General Electric Co., Toronto.
Canadian Westinghouse Co., Hamilton.
Gas & Electric Power Co., Toronto
The Packard Electric Co., St. Catharines.

Lathe Dogs.
Armstrong Bros., Chicago
Baxter, J. R., & Co., Montreal
Pratt & Whitney Co., Hartford, Conn.

Lathes, Engine.
American Tool Works Co., Cincinnati.
B. F. Barnes Co., Rockford, Ill.
Baxter, J. R., & Co., Montreal.
John Bertram & Sons Co., Dundas, Ont.
Canada Machinery Agency, Montreal.
The Canadian Fairbanks Co., Montreal.
London Mach. Tool Co., Hamilton, Ont.
Niles-Bement-Pond Co., New York.
H. W. Petrie, Toronto.
Pratt & Whitney Co., Hartford, Conn.

Lathes, Foot-Power.
American Tool Works Co., Cincinnati.
B. F. Barnes Co., Rockford, Ill.

Lathes, Screw Cutting.
B. F. Barnes Co., Rockford, Ill.
Baxter, J. R., & Co., Montreal.
Niles-Bement-Pond Co., New York.

Lathes, Automatic, Screw-Threading.
John Bertram & Sons Co., Dundas, Ont
London Mach. Tool Co., Hamilton, Ont
Pratt & Whitney Co., Hartford, Conn.

Lathes, Bench.
B. F. Barnes Co., Rockford, Ill.
London Mach. Tool Co., London, Ont.
Pratt & Whitney Co., Hartford, Conn.

Lathes, Turret.
American Tool Works Co., Cincinnati.
Baxter, J. R., & Co., Montreal
John Bertram & Sons Co., Dundas, Ont.
Gisholt Machine Co., Madison, Wis.
London Mach. Tool Co., Hamilton, Ont.
Niles-Bement-Pond Co., New York.
The Pratt & Whitney Co., Hartford, Conn.
Warner & Swasey Co., Cleveland O.

Leather Belting.
Baxter, J. R., & Co., Montreal.
Canada Machinery Agency, Montreal.
The Canadian Fairbanks Co., Montreal
Greey, Wm. & J. G., Toronto

Lifting Magnets.
Cutler-Hammer Clutch Co., Milwaukee

Lime Stone Flux.
Hamilton Facing Mill Co., Hamilton.
Paxson, J. W., Co., Philadelphia

Locomotives, Air.
Canadian Rand Drill Co., Montreal.

Locomotives, Electrical.
Canadian Westinghouse Co., Hamilton
Gas & Electric Power Co., Toronto
Jeffrey Mfg. Co., Columbus, Ohio.

Locomotives, Steam.
Canada Foundry Co., Toronto.
Canadian Rand Drill Co., Montreal

Locomotive Turntable Tractors.
Canadian Pilling Co., Montreal

Lubricating Graphite.
Dominion Foundry Supply Co. Montreal

Lubricating Plumbago.
Detroit Foundry Supply Co., Detroit.
Dominion Foundry Supply Co., Montreal
Hamilton Facing Mill Co., Hamilton.
Paxson, J. W., Co., Philadelphia

Lubricators, Force Feed.
Sight Feed Oil Pump Co., Milwaukee, Wis

Lumber Dry Kilns.
Dominion Heating & Ventilating Co., Hespeler.
H. W. Petrie, Toronto.
Sheldons Limited, Galt, Ont.

Machinery Dealers.
Baxter, J. R., & Co., Montreal.
Canada Machinery Agency, Montreal.
The Canadian Fairbanks Co., Montreal.
H. W. Petrie, Toronto.
The Smart-Turner Mach. Co., Hamilton.
Williams & Wilson Montreal.

Machinery Designers.
Greey, Wm. & J. G., Toronto
Standard Bearings, Ltd., Niagara Falls.

Machinists.
W. H. Banfield & Sons, Toronto.
Greey, Wm. & J. G. Toronto
Hall Engineering Works Montreal.
Link-Belt Co., Philadelphia.
John McDougall, Caledonian Iron Works Co., Montreal.
Paxson, J. W., Co., Philadelphia
Robb Engineering Co., Amherst, N.S.
The Smart-Turner Mach. Co., Hamilton.
Standard Bearings, Ltd. Niagara Falls
Waterous Engine Co., Brantford.

Machinists' Small Tools.
Armstrong Bros., Chicago.
Butterfield & Co., Rock Island, Que.
Frothingham & Workman Ltd., Montreal
Mechanics Supply Co., Quebec, Que.
Rice Lewis & Son, Montreal.
Pratt & Whitney Co., Hartford, Conn.
Standard Tool Co., Cleveland.
L. S. Starrett Co., Athol, Mass.
Williams & Wilson, Montreal.

Malleable Flask Clamps.
Buffalo Foundry Supply Co., Buffalo.
Dominion Foundry Supply Co., Montreal
Paxson, J. W., Co., Philadelphia

Malleable Iron Castings.
Galt Malleable Iron Co., Galt

Mallet, Rawhide and Wood.
Buffalo Foundry Supply Co., Buffalo.
Detroit Foundry Supply Co., Detroit.
Dominion Foundry Supply Co., Montreal
Paxson, J. W., Co., Philadelphia

Mandrels.
A. B. Jardine & Co., Hespeler, Ont.
The Pratt & Whitney Co., Hartford, Conn.
Standard Tool Co., Cleveland.

Marking Machines.
Dwight Slate Machine Co., Hartford.

Marine Work.
Kennedy, Wm., & Sons, Owen Sound

Metallic Paints.
P. D. Dods & Co., Montreal.

Meters, Electrical.
Canadian Westinghouse Co., Hamilton.
Mechanics Supply Co., Quebec, Que.

Mill Machinery.
Baxter, J. R., & Co., Montreal.
Greey Wm. & J. G., Toronto
The Goldie & McCulloch Co., Galt, Ont.
John McDougall, Caledonian Iron Works Co., Montreal.
H. W. Petrie, Toronto.
Robb Engineering Co., Amherst, N.S.
Waterous Engine Co., Brantford.
Williams & Wilson, Montreal.

Milling Attachments.
Becker-Brainard Milling Machine Co. Hyde Park, Mass.
John Bertram & Sons Co., Dundas, Ont.
Cincinnati Milling Machine Co., Cincinnati.
Niles-Bement-Pond Co., New York,
Pratt & Whitney, Hartford, Conn.

Milling Machines, Horizontal.
Baxter, J. R., & Co., Montreal.
Becker-Brainard Milling Machinery Co. Hyde Park, Mass.
John Bertram & Sons Co., Dundas, Ont.
London Mach. Tool Co., Hamilton, Ont.
Niles-Bement-Pond Co., New York.
Pratt & Whitney, Hartford, Conn.

Milling Machines, Motor Driven.
Cincinnati Milling Machine Co., Cincinnati

Milling Machines, Plain.
American Tool Works Co., Cincinnati.
Becker-Brainard Milling Machine Co. Hyde Park, Mass.
John Bertram & Sons Co., Dundas, Ont.
Canada Machinery Agency, Montreal
The Canadian Fairbanks Co., Montreal.
Cincinnati Milling Machine Co., Cincinnati
London Mach. Tool Co., Hamilton, Ont.
Niles-Bement-Pond Co., New York.
H. W. Petrie, Toronto.
Pratt & Whitney Co., Hartford, Conn.
Williams & Wilson, Montreal.

Milling Machines, Universal.
American Tool Works Co., Cincinnati.
Becker-Brainard Milling Machine Co., Hyde Park, Mass.
John Bertram & Sons Co., Dundas, Ont.
Canada Machinery Agency, Montreal.
The Canadian Fairbanks Co., Montreal.
Cincinnati Milling Machine Co., Cincinnati
London Mach. Tool Co., Hamilton, Ont.
Niles-Bement-Pond Co., New York.
H. W. Petrie, Toronto.
Williams & Wilson, Montreal.

Milling Machines, Vertical.
Becker-Brainard Milling Machine Co., Hyde Park, Mass.
Brown & Sharpe, Providence, R. I.
John Bertram & Sons Co., Dundas, Ont.
Canada Machinery Agency, Montreal.
London Mach. Tool Co., Hamilton, Ont.
Niles-Bement-Pond Co., New York.

Milling Tools.
Wm. Abbott, Montreal.
Becker-Brainard Milling Machine Co., Hyde Park, Mass.
Geometric Tool Co., New Haven, Conn.
Hamilton Tool Co., Hamilton, Ont.
London Mach. Tool Co., Hamilton, Ont.
Pratt & Whitney Co., Hartford, Conn.
Standard Tool Co., Cleveland.
Standard Bearings, Ltd., Niagara Falls

Mining Machinery.
Allis-Chalmers-Bullock, Limited, Montreal
Canadian Rand Drill Co., Montreal.
Chicago Pneumatic Tool Co., Chicago.
Jeffrey Mfg. Co., Columbus, Ohio.
Laurie Engine & Machine Co., Montreal.
Rice Lewis & Son, Toronto.
John McDougall, Caledonian Iron Works Co., Montreal.
T. & H. Electric Co. Hamilton.

Mixing Machines, Dough.

Greey, Wm. & J. G., Toronto
Paxson, J. W., Co., Philadelphia

Mixing Machines, Special.

Greey, Wm. & J. G., Toronto
Paxson, J. W., Co., Philadelphia

Model Tools.

Globe Machine & Stamping Co., Cleveland, Ohio.
Standard Bearings, Ltd., Niagara Falls.
Wells Pattern and Model Works, Toronto

Motors, Electric.

Allis-Chalmers-Bullock Limited, Montreal
Canadian General Electric Co., Toronto
Canadian Westinghouse Co., Hamilton.
Electrical Machinery Co., Toronto.
Consolidated Electric Co., Toronto
Gas & Electric Power Co., Toronto
Hall Engineering Works, Montreal.
Mechanics Supply Co., Quebec, Que.
The Packard Electric Co., St. Catharines.
T. & H. Electric Co., Hamilton.

Motors, Air.

Canadian Rand Drill Co., Montreal.
Chicago Pneumatic Tool Co., Chicago.

Molders' Supplies.

Buffalo Foundry Supply Co., Buffalo.
Detroit Foundry Supply Co., Windsor.
Dominion Foundry Supply Co., Montrea
Hamilton Facing Mill Co., Hamilton.
Mechanics Supply Co., Quebec, Que.
Paxson J. W., Co., Philadelphia

Molders' Tools.

Buffalo Foundry Supply Co., Buffalo.
Detroit Foundry Supply Co., Windsor
Dominion Foundry Supply Co., Montreal
Hamilton Facing Mill Co., Hamilton.
Mechanics Supply Co., Quebec, Que.
Paxson, J. W., Co., Philadelphia

Molding Machines.

Ph. Bonvillain & E. Ronceray, Philadelphia
Buffalo Foundry Supply Co., Buffalo.
Dominion Foundry Supply Co., Montreal
Hamiton Facing Mill Co., Hamilton.
J. W. Paxson Co., Philadelphia, Pa.

Molding Sand.

T. W. Barnes, Hamilton.
Buffalo Foundry Supply Co., Buffalo.
Detroit Foundry Supply Co., Windsor
Dominion Foundry Supply Co., Montreal
Hamilton Facing Mill Co., Hamilton.
Paxson, J. W., Co., Philadelphia

Nut Tappers.

John Bertram & Sons Co., Dundas, Ont
A. B. Jardine & Co., Hespeler.
London Mach. Tool Co., Hamilton.
National Machinery Co., Tiffin, Ohio.

Nuts.

Canada Nut Co., Toronto

Oatmeal Mill Machinery.

Greey, Wm. & J. G., Toronto
The Goldie & McCulloch Co., Galt

Oilers, Gang.

Sight Feed Oil Pump Co., Milwaukee, Wis.

Oils, Core.

Buffalo Foundry Supply Co., Buffalo.
Dominion Foundry Supply Co., Montreal
Hamilton Facing Mill Co., Hamilton.
Paxson, J. W., Co., Philadelphia

Oil Extractors.

Darling Bros., Ltd., Montreal

Paint Mill Machinery.

Greey, Wm. & J. G., Toronto

Pans, Lathe.

Cleveland Wire Spring Co., Cleveland
Paxson, J. W., Co., Philadelphia

Pans, Steel Shop.

Cleveland Wire Spring Co., Cleveland

Parting Compound.

Doggett, Stanley, New York

Patent Solicitors.

Hanbury A. Budden, Montreal.
Fetherstonhough & Blackmore, Montreal
Marion & Marion, Montreal.
Ridout & Maybee, Toronto.

Patterns.

John Carr, Hamilton.
Galt Malleable Iron Co., Galt
Hamilton Pattern Works, Hamilton.
John McDougall, Caledonian Iron Works Co., Montreal.
Wells' Pattern and Model Works, Toronto.

Pig Iron.

Hamilton Steel & Iron Co., Hamilton

Pipe Cutting and Threading Machines.

Borden-Canadian Co., Toronto
Butterfield & Co., Rock Island, Que.
Canada Machinery Agency, Montreal.
Curtis & Curtis Co., Bridgeport, Conn.
Frothingham & Workman, Ltd., Montreal
Hart Mfg. Co., Cleveland
A. B. Jardine & Co., Hespeler, Ont.
Loew Mfg. Co., Cleveland, Ohio
London Mach. Tool Co., Hamilton, Ont.
Niles-Bement-Pond Co., New York.
Shantz, I. E., & Co., Berlin, Ont.

Pipe, Municipal.

Canadian Pipe Co., Vancouver, B.C.
Pacific Coast Pipe Co., Vancouver, B.C.

Pipe, Waterworks.

Canadian Pipe Co., Vancouver, B.C.
Pacific Coast Pipe Co., Vancouver, B.C

Planers, Standard.

American Tool Works, Cincinnati.
John Bertram & Sons Co., Dundas, Ont.
Canada Machinery Agency, Montreal.
The Canadian Fairbanks Co., Montreal.
Rice Lewis & Son, Toronto.
London Mach. Tool Co., Hamilton, Ont.
Niles-Bement-Pond Co., New York.
H. W. Petrie, Toronto.
Pratt & Whitney Co., Hartford, Conn.
Williams & Wilson, Montreal.

Planers, Rotary.

John Bertram & Sons Co., Dundas, Ont.
London Mach. Tool Co., Hamilton, Ont.
Niles-Bement-Pond Co., New York.

Planing Mill Fans.

Dominion Heating & Ventilating Co., Hespeler
Sheldons Limited, Galt, Ont.

Plumbago.

Buffalo Foundry Supply Co., Buffalo.
Detroit Foundry Supply Co., Windsor
Doggett, Stanley, New York
Dominion Foundry Supply Co., Montreal
Hamilton Facing Mill Co., Hamilton,
Mechanics Supply Co., Quebec, Que.
J. W. Paxson Co., Philadelphia, Pa.

Pneumatic Tools.

Allis-Chalmers-Bullock, Montreal.
Canadian Rand Drill Co., Montreal.
Chicago Pneumatic Tool Co., Chicago
Hamilton Facing Mill Co., Hamilton.
Hanna Engineering Works, Chicago.
Independent Pneumatic Tool Co. Chicago, New York
Paxson, J. W., Co., Philadelphia

Power Hack Saw Machines.

Baxter, J. R., & Co., Montreal.
Frothingham & Workman, Ltd., Montreal

Power Plants.

John McDougall Caledonian Iron Works Co., Montreal.
The Smart-Turner Mach. Co., Hamilton

Power Plant Equipments.

Darling Bros., Ltd., Montreal

Presses, Drop.

W. H. Banfield & Son, Toronto.
E. W. Bliss Co., Brooklyn, N.Y.
Brown, Boggs Co., Hamilton
Canada Machinery Agency, Montreal.
Laurie Engine & Machine Co., Montreal
Miller Bros. & Toms, Montreal.
Niles-Bement-Pond Co., New York

Presses, Hand.

E. W. Bliss Co., Brooklyn, N.Y.
Brown, Boggs Co., Hamilton

Presses, Hydraulic.

John Bertram & Sons Co., Dundas, Ont.
Laurie Engine & Machine Co., Montreal.
London Mach. Tool Co., Hamilton, Ont.
John McDougall Caledonian Iron Works Co., Montreal.
Niles-Bement-Pond Co., New York.
Perrin, Wm. R., & Co., Toronto

Presses, Power.

E. W. Bliss Co., Brooklyn, N.Y.
Brown, Boggs Co., Hamilton
Canada Machinery Agency, Montreal.
Laurie Engine & Machine Co., Montreal
London Mach. Tool Co., Hamilton, Ont.
John McDougall Caledonian Iron Works Co., Montreal.
Niles-Bement-Pond Co., New York.

Presses Power Screw.

Brown, Boggs Co., Hamilton
Perrin, Wm. R., & Co., Toronto

Pressure Regulators.

Darling Bros., Ltd., Montreal

Producer Plants.

Canada Foundry Co., Toronto

Pulp Mill Machinery.

Greey, Wm. & J. G., Toronto
Jeffrey Mfg. Co., Columbus, Ohio.
Laurie Engine & Machine Co., Montreal.
John McDougall Caledonian Iron Works Co., Montreal.
Waterous Engine Works Co., Brantford

Pulleys.

Baxter, J. R., & Co., Montreal.
Canada Machinery Agency, Montreal.
The Canadian Fairbanks Co., Montreal.
The Goldie & McCulloch Co., Galt.
Greey, Wm. & J. G., Toronto
Laurie Engine & Machine Co., Montreal.
Link-Belt Co., Philadelphia.
John McDougall Caledonian Iron Works Co., Montreal.
Owen Sound Iron Works Co., Owen Sound
H. W. Petrie, Toronto.
The Smart-Turner Mach. Co., Hamilton.
Standard Bearings, Ltd., Niagara Falls.
Williams & Wilson, Montreal.
Waterous Engine Co., Brantford.

Pumps.

Laurie Engine & Machine Co., Montreal
Ontario Wind Engine & Pump Co., Toronto

Pump Governors.

Darling Bros., Ltd. Montreal

Pumps, Hydraulic.

Ph. Bonvillain & E. Ronceray, Philadelphia
Laurie Engine & Machine Co., Montreal
Perrin, Wm. R. & Co., Toronto

Pumps, Oil.

Sight Feed Oil Pump Co., Milwaukee, Wis.

Pumps, Steam.

Allis-Chalmers-Bullock, Limited, Montreal
Canada Foundry Co., Toronto.
Canada Machinery Agency, Montreal.
Darling Bros., Ltd., Montreal
Gas & Electric Power Co., Toronto
The Goldie & McCulloch Co., Galt.
John McDougall Caledonian Iron Works, Montreal.
H. W. Petrie, Toronto.
The Smart-Turner Mach. Co., Hamilton.
Standard Bearings, Ltd., Niagara Falls.
Waterous Engine Co. Brantford.

Pumping Machinery.

Canada Foundry Co., Limited, Toronto
Canada Machinery Agency, Montreal.
Canadian Rand Drill Co., Montreal.
Chicago Pneumatic Tool Co., Chicago.
Darling Bros., Ltd., Montreal
Gas & Electric Power Co., Toronto.
Hall Engineering Works, Montreal, Que.
Laurie Engine & Machine Co., Montreal.
London Mach. Tool Co., Hamilton, Ont.
John McDougall Caledonian Iron Works Co., Montreal.
The Smart-Turner Mach. Co., Hamilton
Standard Bearings, Ltd., Niagara Falls.

Punches and Dies.

W. H. Banfield & Sons, Toronto.
Bliss, E. W., Co., Brooklyn, N.Y.
Butterfield & Co., Rock Island.
Globe Machine & Stamping Co.
A. B. Jardine & Co., Hespeler, Ont.
London Mach. Tool Co., Hamilton, Ont.
Pratt & Whitney Co., Hartford, Conn.
H. W. Petrie, Toronto.
Standard Bearings, Ltd., Niagara Falls.

Punches, Hand.

Mechanics Supply Co., Quebec, Que.

Punches, Power.

John Bertram & Sons Co., Dundas, Ont.
E. W. Bliss Co., Brooklyn, N.Y.
Canada Machinery Agency, Montreal.
Canadian Buffalo Forge Co., Montreal.
London Mach. Tool Co., Hamilton, Ont.
Niles-Bement-Pond Co., New York.

Punches, Turret.

London Mach. Tool Co., London, Ont.

Punching Machines, Horizontal.

John Bertram & Sons Co., Dundas, Ont.
London Mach. Tool Co., Hamilton, Ont.
Niles-Bement-Pond Co., New York.

Quartering Machines.

John Bertram & Sons Co., Dundas, Ont.
London Mach. Tool Co., Hamilton, Ont.

Railway Spikes and Washers.

Hamilton Steel & Iron Co., Hamilton

Rammers, Bench and Floor.

Buffalo Foundry Supply Co., Buffalo.
Detroit Foundry Supply Co., Windsor
Hamilton Facing Mill Co., Hamilton.
Paxson, J. W., Co., Philadelphia

Rapping Plates.

Detroit Foundry Supply Co., Windsor
Hamilton Facing Mill Co., Hamilton.
Paxson, J. W., Co., Philadelphia

Raw Hide Pinions.

Brown, David & Sons, Huddersfield, Eng

Reamers.

Wm. Abbott, Montreal.
Baxter, J. R., & Co., Montreal.
Butterfield & Co., Rock Island.
Frothingham & Workman, Ltd., Montreal
Hamilton Tool Co., Hamilton
Hanna Engineering Works, Chicago.
A. B. Jardine & Co., Hespeler, Ont.
Mechanics Supply Co., Quebec, Que.
Morse Twist Drill and Machine Co., New Bedford, Mass.
Pratt & Whitney Co., Hartford, Conn.
Standard Tool Co., Cleveland.

Reamers, Steel Taper.

Butterfield & Co., Rock Island.
Chicago Pneumatic Tool Co., Chicago.
A. B. Jardine & Co., Hespeler, Ont.
John Millen & Son, Ltd., Montreal.
Pratt & Whitney Co., Hartford, Conn.
Standard Tool Co., Cleveland.

Rheostats.

Canadian General Electric Co., Toronto.
Canadian Westinghouse Co., Hamilton.
Hall Engineering Works, Montreal, Que.
T. & H. Electric Co., Hamilton,

Riddles.

Buffalo Foundry Supply Co., Buffalo.
Detroit Foundry Supply Co., Windsor
Hamilton Facing Mill Co., Hamilton.
Dominion Foundry Supply Co., Montreal
J. W. Paxson Co., Philadelphia, Pa.

Riveters, Pneumatic.

Hanna Engineering Works, Chicago.
Independent Pneumatic Tool Co., Chicago, Ill.

Rolls, Bending.

John Bertram & Sons Co., Dundas, Ont.
London Mach. Tool Co., Hamilton, Ont.
Niles-Bement-Pond Co., New York.

Rolls, Chilled Iron.

Greey, Wm. & J. G., Toronto

Rolls, Sand Cast.

Greey, Wm. & J. G., Toronto

Rotary Blowers.

Paxson, J. W., Co., Philadelphia

Rotary Converters.

Allis-Chalmers-Bullock, Ltd., Montreal.
Canadian Westinghouse Co., Hamilton.
Paxson, J. W., Co., Philadelphia
Toronto and Hamilton Electric Co., Hamilton.

Rubbing and Sharpening Stones.

Norton Co., Worcester, Mass.

Safes.

Baxter, Paterson & Co., Montreal.
The Goldie & McCulloch Co., Galt.

Sand, Bench.

Buffalo Foundry Supply Co., Buffalo.
Detroit Foundry Supply Co., Windsor
Dominion Foundry Supply Co., Montreal
Hamilton Facing Mill Co., Hamilton.
Paxson J. W., Co., Philadelphia

Sand Blast Machinery.

Canadian Rand Drill Co., Montreal.
Chicago Pneumatic Tool Co., Chicago.
Detroit Foundry Supply Co., Windsor
Dominion Foundry Supply Co., Montreal
Hamilton Facing Mill Co., Hamilton.
J. W. Paxson Co., Philadelphia, Pa.

Sand, Heavy, Grey Iron.

Buffalo Foundry Supply Co., Buffalo.
Detroit Foundry Supply Co., Windsor
Dominion Foundry Supply Co., Montreal
Hamilton Facing Mill Co., Hamilton.
Paxson, J. W., Co., Philadelphia

Sand, Malleable.

Buffalo Foundry Supply Co., Buffalo.
Detroit Foundry Supply Co., Windsor
Dominion Foundry Supply Co., Montreal
Hamilton Facing Mill Co., Hamilton.
Paxson, J. W., Co., Philadelphia

Sand, Medium Grey Iron.

Buffalo Foundry Supply Co., Buffalo.
Detroit Foundry Supply Co., Windsor
Dominion Foundry Supply Co., Montreal
Hamilton Facing Mill Co., Hamilton.
Paxson, J. W., Co., Philadelphia

Sand Sifters.

Buffalo Foundry Supply Co., Buffalo.
Detroit Foundry Supply Co., Windsor
Dominion Foundry Supply Co., Montreal
Hamilton Facing Mill Co., Hamilton.
Hanna Engineering Works, Chicago.
Paxson, J. W., Co., Philadelphia

Saw Mill Machinery.

Allis-Chalmers-Bullock, Limited, Montreal
Baxter, J. R., & Co., Montreal
Canada Machinery Agency, Montreal.
Goldie & McCulloch Co., Galt.
Greey, Wm. & J. G., Toronto
Owen Sound Iron Works Co., Owen Sound
H. W. Petrie, Toronto.
Robb Engineering Co., Amherst,
Standard Bearings, Ltd., Niagara Falls
Waterous Engine Works, Brantford
Williams & Wilson, Montreal.

Sawing Machines, Metal.

Niles-Bement-Pond Co. New York.
Paxson, J. W., Co., Philadelphia

Saws, Hack.

Baxter, J. R., & Co., Montreal.
Canada Machinery Agency, Montreal.
Detroit Foundry Supply Co., Windsor
Frothingham & Workman, Ltd., Montreal
London Mach. Tool Co., Hamilton
Mechanics Supply Co, Quebec, Que.
Rice Lewis & Son, Toronto.
John Millen & Son, Ltd., Montreal.
L. S. Starrett Co., Athol, Mass.

Screw Cutting Tools

Hart Manufacturing Co., Cleveland, Ohio
Mechanics Supply Co., Quebec, Que.

Screw Machines, Automatic.

Canada Machinery Agency, Montreal.
Cleveland Automatic Machine Co., Cleveland, Ohio
London Mach. Tool Co., Hamilton, Ont.
National-Acme Mfg. Co., Cleveland.
Pratt & Whitney Co., Hartford, Conn.

Screw Machines, Hand.

Canada Machinery Agency Montreal.
A. B. Jardine & Co., Hespeler.
London Mach. Tool Co., Hamilton, Ont.
Mechanics Supply Co. Quebec, Que.
Pratt & Whitney Co., Hartford, Conn.
Warner& Swasey Co., Cleveland, O.

Screw Machines, Multiple Spindle.

National-Acme Mfg. Co, Cleveland

Screw Plates.

Butterfield & Co., Rock Island, Que.
Frothingham & Workman, Ltd, Montreal
Hart Manufacturing Co. Cleveland, Ohio
A B. Jardine & Co., Hespeler
Mechanics Supply Co., Quebec, Que

Screws, Cap and Set.

National-Acme Mfg. Co., Cleveland

Screw Slotting Machinery, Semi-Automatic.

National-Acme Mfg. Co., Cleveland

Second-Hand Machinery.

American Tool Works Co., Cincinnati.
Canada Machinery Agency, Montreal.
The Canadian Fairbanks Co., Montreal.
Goldie & McCulloch Co., Galt.
Machinery Exchange, Montreal.
Niles-Bement-Pond Co., New York.
H. W. Petrie, Toronto.
Robb Engineering Co., Amherst, N S.
Waterous Engine Co., Brantford.
Williams & Wilson, Montreal.

Shafting.

Baxter, J. R., & Co., Montreal.
Canada Machinery Agency, Montreal
The Canadian Fairbanks Co., Montreal.
Frothingham & Workman, Ltd., Montreal
The Goldie & McCulloch Co., Galt, Ont
Greey, Wm. & J. G., Toronto
Kennedy, Wm., & Sons, Owen Sound
Niles-Bement-Pond Co., New York.
Owen Sound Iron Works Co., Owen Sound
H. W. Petrie, Toronto.
Smart-Turner Machine Co., Hamilton.
Union Drawn Steel Co., Hamilton.
Waterous Engine Co., Brantford.

Shapers.

American Tool Works Co., Cincinnati.
John Bertram & Sons Co., Dundas, Ont
Canada Machinery Agency, Montreal.
The Canadian Fairbanks Co., Montreal.
Rice Lewis & Son, Toronto.
London Mach. Tool Co., Hamilton, Ont.
Niles-Bement-Pond Co., New York.
H. W. Petrie, Toronto.
Potter & Johnston Machine Co., Pawtucket, R.I.
Pratt & Whitney Co., Hartford, Conn.
Williams & Wilson, Montreal.

Sharpening Stones.

Carborundum Co., Niagara Falls, N.Y.

Shearing Machine, Bar.

John Bertram & Sons Co., Dundas, Ont.
A. B. Jardine & Co., Hespeler.
London Mach. Tool Co., Hamilton, Ont.
Niles-Bement-Pond Co., New York.

Shears, Power.

John Bertram & Sons Co., Dundas, Ont.
Canada Machinery Agency, Montreal.
A. B. Jardine & Co., Hespeler, Ont.
Niles-Bement-Pond Co., New York.
Paxson, J. W., Co., Philadelphia

Sheet Metal Goods

Globe Machine & Stamping Co., Cleveland, Ohio.

Sheet Meal Working Tools.

Brown, Boggs Co., Hamilton

Sheet Steel Work.

Owen Sound Iron Works Co., Owen Sound

Shingle Mill Machinery.

Owen Sound Iron Works Co., Owen Sound

Shop Trucks.

Greey, Wm. & J. G., Toronto
Owen Sound Iron Works Co., Owen Sound
Paxson, J. W., Co., Philadelphia

Shovels.

Baxter, J. R., & Co., Montreal.
Buffalo Foundry Supply Co., Buffalo.
Detroit Foundry Supply Co., Detroit.
Doggett, Stanley, New York
Frothingham & Workman, Ltd., Montreal
Dominion Foundry Supply Co., Montreal
Hamilton Facing Mill Co., Hamilton.
Mechanics Supply Co., Quebec, Que.
Paxson, J. W., Co., Philadelphia

Shovels, Steam.

Allis-Chalmers-Bullock, Montreal.

Sieves.

Buffalo Foundry Supply Co., Buffalo.
Detroit Foundry Supply Co., Windsor
Dominion Foundry Supply Co., Montreal
Hamilton Facing Mill Co., Hamilton.
Paxson, J. W., Co., Philadelphia

Silver Lead.

Buffalo Foundry Supply Co., Buffalo.
Detroit Foundry Supply Co., Windsor
Doggett, Stanley, New York
Dominion Foundry Supply Co., Montreal
Hamilton Facing Mill Co., Hamilton.
Paxson, J. W., Co., Philadelphia

Sleeves, Reducing.

Chicago Pneumatic Tool Co., Chicago.

Snap Flasks

Buffalo Foundry Supply Co., Buffalo.
Detroit Foundry Supply Co., Windsor
Dominion Foundry Supply Co., Montreal
Hamilton Facing Mill Co., Hamilton.
Paxson, J. W., Co., Philadelphia

Soap Stones & Talc Crayons & Pencils.

Doggett, Stanley, New York

Soapstone.

Buffalo Foundry Supply Co., Buffalo.
Detroit Foundry Supply Co., Windsor
Doggett, Stanley, New York
Dominion Foundry Supply Co., Montreal
Hamilton Facing Mill Co., Hamilton.
Paxson, J. W., Co., Philadelphia

Soap Stone Facings.

Doggett Stanley, New York

Solders.

Lumen Bearing Co., Toronto

Special Machinery.

W. H. Banfield & Sons, Toronto.
Baxter, J. R., & Co., Montreal
John Bertram & Sons Co. Dundas, Ont
Brown, Boggs Co., Hamilton
Globe Machine & Stamping Co., Cleveland Ohio.
Greey, Wm. & J. G., Toronto
Hanna Engineering Works, Chicago.
Laurie Engine & Machine Co., Montreal.
London Mach. Tool Co., Hamilton, Ont.
H. W. Petrie, Toronto.
The Smart-Turner Mach. Co., Hamilton
Standard Bearings, Ltd., Niagara Falls.
Waterous Engine Co., Brantford.
Woelfle Bros., Berlin, Ont.

Special Machines and Tools.

Brown, Boggs Co., Hamilton
Paxson, J. W., Philadelphia
Pratt & Whitney, Hartford, Conn.
Standard Bearings, Ltd., Niagara Falls.

Special Milled Work.

National-Acme Mfg. Co., Cleveland

Speed Changing Countershafts.

The Canadian Fairbanks Co., Montreal

Spike Machines.

National Machinery Co., Tiffin, O.
The Smart-Turner Mach. Co., Hamilton

Spray Cans.

Detroit Foundry Supply Co., Windsor
Dominion Foundry Supply Co., Montreal
Hamilton Facing Mill Co., Hamilton.
Paxson, J. W., Co., Philadelphia

Springs, Automobile.

Cleveland Wire Spring Co., Cleveland

Springs, Coiled Wire.

Cleveland Wire Spring Co., Cleveland

Springs, Machinery.

Cleveland Wire Spring Co., Cleveland

Springs, Upholstery.

Cleveland Wire Spring Co., Cleveland

Sprue Cutters.

Detroit Foundry Supply Co., Windsor
Dominion Foundry Supply Co., Montreal
Hamilton Facing Mill Co., Hamilton.
Paxson, J. W., Co., Philadelphia

Stamp Mills.

Allis-Chalmers-Bullock, Limited, Montreal
Paxson, J. W., Co., Philadelphia

Steam Separators.

Darling Bros., Ltd., Montreal.
Dominion Heating & Ventilating Co., Hespeler
Robb Engineering Co., Montreal.
Sheldon's Limited, Galt.
Smart-Turner Mach. Co., Hamilton
Waterous Engine Co., Brantford.

Steam Specialties.

Darling Bros., Ltd., Montreal
Dominion Heating &, Ventilating Co., Hespeler
Sheldon's Limited, Galt.

Steam Traps.

Canada Machinery Agency, Montreal.
Darling Bros., Ltd., Montreal
Dominion Heating & Ventilating Co., Hespeler.
Mechanics Supply Co., Quebec, Que.
Sheldons Limited, Galt.

Steam Valves.

Darling Bros., Ltd., Montreal

Steel Pressure Blowers

Buffalo Foundry Supply Co., Buffalo.
Dominion Foundry Supply Co., Montreal
Domininion Heating & Ventilating Co., Hespeler
Hamilton Facing Mill Co., Hamilton.
J. W. Paxon Co., Philadalphia, Pa.
Sheldon's Limited, Galt.

Steel Tubes.

Baxter, J. R., & Co., Montreal.
Mechanics Supply Co., Quebec, Que.
John Millen & Son, Montreal.

Steel, High Speed.

Wm. Abbott, Montreal.
Canadian Fairbanks Co., Montreal.
Frothingham & Workman, Ltd., Montreal
Alex. Gibb, Montreal.
Jessop, Wm., & Sons, Sheffield, Eng.
B. K. Morton Co., Sheffield, Eng.
Williams & Wilson, Montreal.

Stocks and Dies

Hart Manufacturing Co., Cleveland, Ohio
Mechanics Supply Co., Quebec, Que.
Pratt & Whitney Co., Hartford, Conn.

Stone Cutting Tools, Pneumatic

Allis-Chalmers-Bullock, Ltd., Montreal.
Canadian Rand Drill Co., Montreal.

Stone Surfacers

Chcago Pneumatic Tool Co., Chica

Stove Plate Facings.

Buffalo Foundry Supply Co., Buffalo
Detroit Foundry Supply Co., Windsor.
Dominion Foundry Supply Co., Montreal
Hamilton Facing Mill Co., Hamilton.
Paxson, J. W., Co., Philadelphia

Swage, Block.

A. B. Jardine & Co., Hespeler, Ont.

Switchboards.

Allis-Chalmers-Bullock, Limited, Montreal
Canadian General Electric Co., Toronto.
Canadian Westinghouse Co., Hamilton
Hall Engineering Works, Montreal, Que.
Mechanics Supply Co., Quebec, Que.
Toronto and Hamilton Electric Co., Hamiton.

Talc.

Buffalo Foundry Supply Co., Buffalo.
Detroit Foundry Supply Co., Windsor.
Doggett, Stanley. New York
Dominion Foundry Supply Co., Montreal
Hamiton Facing Mill Co., Hamilton.
Paxson, J. W., Co., Philadelphia

Tanks, Oil.

Sight Feed Oil Pump Co., Milwaukee, Wis.

Tapping Machines and Attachments.

American Tool Works Co., Cincinnati.
John Bertram & Sons Co., Dundas, Ont.
Bickford Drill & Tool Co., Cincinnati.
The Geometric Tool Co., New Haven,
A. B. Jardine & Co., Hespeler.
London Mach. Tool Co., Hamilton, Ont
Murchey Machine & Tool Co., Detroit.
Niles-Bement-Pond Co., New York.
H. W. Petrie, Toronto.
Pratt & Whitney, Cincinnati, O.
L. S. Starrett Co., Athol, Mass.
Williams & Wilson, Montreal.

Tapes, Steel.

Frothingham & Workman, Ltd., Montreal
Rice Lewis & Son, Toronto.
Mechanics Supply Co., Quebec, Que.
John Millen & Son, Ltd., Montreal.
L. S. Starrett Co., Athol, Mass.

Taps.

Baxter, J. R., & Co., Montreal.
Mechanics Supply Co., Quebec, Que.

Taps, Collapsing.

The Geometric Tool Co., New Haven.

Taps and Dies.

Wm. Abbott, Montreal.
Baxter, J. R., & Co., Montreal.
Butterfield & Co., Rock Island, Que.
Frothingham & Workman, Ltd., Montreal
The Geometric Tool Co., New Haven,
A. B. Jardine & Co., Hespeler, Ont.
Rice Lewis & Son, Toronto.
Mechanics Supply Co., Quebec, Que.
John Millen & Son, Ltd., Montreal.
Pratt & Whitney Co., Hartford, Conn.
Standard Tool Co., Cleveland.
L. S. Starrett Co., Athol, Mass.

Testing Laboratory.

Detroie Testing Laboratory, Detroit

Testing Machines.

Detroit Foundry Supply Co., Windsor.
Dominion Foundry Supply Co., Montreal
Hamilton Facing Mill Co., Hamilton.
Paxson, J. W., Co., Philadelphia

Thread Cutting Tools.

Hart Manufacturing Co., Cleveland, Ohio
Mechanics Supply Co., Quebec, Que.

Tiling, Opal Glass.

Toronto Plate Glass Importing Co., Toronto.

Time Switches, Automatic, Electric.

Berlin Electric Mfg. Co., Toronto.

Tinware Machinery

Canada Machinery Agency, Montreal.

Tire Upsetters or Shrinkers

A. B. Jardine & Co., Hespeler, Ont.

Tool Cutting Machinery

Canadian Rand Drill Co., Montreal.

Tool Holders.

Armstrong Bros. Tool Co., Chicago.
Baxter, J. R., & Co., Montreal.
John Millen & Son, Ltd., Montreal.
H. W. Petrie, Toronto.
Pratt & Whitney Co., Hartford, Conn.

Tool Steel.

Wm. Abbott, Montreal.
Flockton, Tompkin & Co., Sheffield, Eng.
Frothingham & Workman, Ltd., Montreal
Wm. Jessop, Sons & Co., Toronto.
Canadian Fairbanks Co., Montreal
B. K. Morton & Co., Sheffield, Eng.
Williams & Wilson, Montral.

Torches, Steel.

Baxter, J. R., & Co., Montreal.
Detroit Foundry Supply Co., Windsor.
Dominion Foundry Supply Co., Montreal
Hamilton Facing Mill Co., Hamilton.
Paxson, J. W., Co., Philadelphia

Transformers and Convertors

Allis-Chalmers-Bullock, Montreal.
Canadian General Electric Co., Toronto.
Canadian Westinghouse Co., Hamilton.
Hall Engineering Wo'ks, Montreal, Que.
T. & H. Electric Co., Hamilton.

Transmission Machinery

Allis-Chalmers-Bullock, Montreal.
The Canadian Fairbanks Co., Montreal.
Greey, Wm. & J. G., Toronto.
Laurie Engine & Machine Co., Montreal.
Link-Belt Co., Philadelphia.
Paxson, J. W., Co., Philadelphia
H.W. Petrie, Toronto.
The Smart-Turner Mach. Co., Hamilton.
Waterous Engine Co., Brantford.

Transmission Supplies

Baxter, J. R., & Co., Montreal.
The Canadian Fairbanks Co., Montreal.
Wm. & J. G. Greey, Toronto.
The Goldie & McCulloch Co., Galt.
H. W. Petrie, Toronto.

Trolleys

Canadian Rand Drill Co., Montreal.
Dominion Foundry Supply Co., Montreal
John Millen & Son, Ltd., Montreal.
Northern Engineering Works, Detroit

Trolley Wheels.

Lumen Bearing Co., Toronto

Trucks, Dryer and Factory

Dominion Heating & Ventilating Co., Hespeler
Greey, Wm. & J. G., Toronto.
Northern Engineering Works, Detroit
Sheldon's Limited, Galt, Ont.

Tube Expanders (Rollers)

Chicago Pneumatic Tool Co., Chicago.
A. B. Jardine & Co., Hespeler.
Mechanics Supply Co., Quebec, Que.

Turbines, Steam

Allis-Chalmers-Bullock, Limited, Montreal
Canadian General Electric Co., Toronto.
Canadian Westinghouse Co., Hamilton.
Kerr Turbine Co., Wellsville, N.Y.

Turntables

Detroit Foundry Supply Co., Windsor.
Dominion Foundry Supply Co., Montreal
Hamilton Fa ing Mill Co., Hamilton.
Northern Engineering Works, Detroit
Paxson, J. W., Co., Philadelphia

Turret Machines.

American Tool Works Co., Cincinnati.
John Bertram & Sons Co., Dundas, Ont.
The Canadian Fairbanks Co., Montreal.
London Mach. Tool Co., Hamilton, Ont.
Niles-Bement-Pond Co., New York.
H. W. Petrie, Toronto.
Pratt & Whitney Co., Hartford, Conn.
Warner & Swasey Co., Cleveland, Ohio.
Williams & Wilson, Montreal.

Upsetting and Bending Machinery.

John Bertram & Sons Co., Dundas, Ont.
A. B. Jardine & Co., Hespeler.
London Mach. Tool Co., Hamilton, Ont.
National Machinery Co., Tiffin, O.
Niles-Bement-Pond Co., New York.

Valves, Blow-off.

Chicago Pneumatic Tool Co., Chicago.
Mechanics Supply Co., Quebec, Que.

Valves, Back Pressure.

Darling Bros., Ltd., Montreal
Dominion Heating & V.ntilating Co. Hespeler
Mechanics Supply Co., Quebec, Que.
Sheldon's Limited, Galt.

Valve Reseating Machines.

Darling Bros. Ltd., Montreal

Ventilating Apparatus.

Canada Machinery Agency, Montreal
Darling Bros., Ltd., Montreal
Dominion Heating & Ventilating Co, Hespeler, Ont.
Sheldon's Ltd., Galt

Vises, Planer and Shaper.

American Tool Works Co., Cincinnati, O.
Frothingham & Workman, Ltd., Montreal
A. B. Jardine & Co., Hespeler, Ont.
John Millen & Son, Ltd., Montreal.
Niles-Bement-Pond Co., New York.

Washer Machines.

National Machinery Co., Tiffin, Ohio.

Water Tanks.

Ontario Wind Engine & Pump Co., Toronto

Water Wheels.

Allis-Chalmers-Bullock Co., Montreal.
Barber, Chas., & Sons, Meaford.
Canada Machinery Agency, Montreal.
The Goldie & McCulloch Co., Galt, Ont.
Greey, Wm. & J. G., Toronto.
Wm. Kennedy & Sons, Owen Sound.
John McDougall Caledonian Iron Works Co., Ltd., Montreal

Water Wheels, Turbine.

Barber, Chas., & Sons, Meaford, Ont.
Kennedy, Wm., & Sons, Owen Sound

Wheelbarrows.

Baxter, J. W., & Co., Montreal.
Buffalo Foundry Supply Co., Buffalo
Frothingham & Workman Ltd., Montreal
Detroit Foundry Supply Co., Windsor.
Dominion Foundry Supply Co, Montreal
Hamilton Facing Mill Co., Hamilton.
Paxson, J. W., Co., Philadelphia

Winches.

Canadian Pilling Co., Montreal
Frothingham & Workman, Ltd, Montreal

Wind Mills.

Ontario Wind Engine & Pump Co, Toronto

Window Wire Guards.

Expanded Metal and Fireproofing Co Toronto.
B. Greening Wire Co., Hamilton, Ont.
Paxson, J. W., Co. Philadelphia

Wire Chains.

The B. Greening Wire Co., Hamilton.

Wire Cloth and Perforated Metals.

Expanded Metal and Fireproofing Co. Toronto.
B. Greening Wire Co., Hamilton. Ont.
Paxson, J. W., Co, Philadelphia.

Wire Guards and Railings.

Expanded Metal and Fireproofing Co. Toronto.
B. Greening Wire Co. Hamilton, Ont.

Wire Nail Machinery.

National Machinery Co., Tiffin, Ohio.

Wire Rope.

Frothingham & Workman, Ltd. Montreal
B. Greening Wire Co., Hamilton, Ont.

Wood Boring Machines. Pneumatic.

Independent Pneumatic Tool Co., Chicago, Ill.

Wood-working Machinery.

Baxter, J. R., & Co., Montreal
Canada Machinery Agency, Montreal.
The Canadian Fairbanks Co., Montreal.
Goldie & McCulloch Co., Galt.
H. W. Petrie, Toronto.
Waterous Engine Works Co., Brantford
Williams & Wilson, Montreal.

ALPHABETICAL INDEX.

PROFESSIONAL DIRECTORY
CONSULTING ENGINEERS, PATENT ATTORNEYS, ARCHITECTS, CONTRACTORS, ETC.

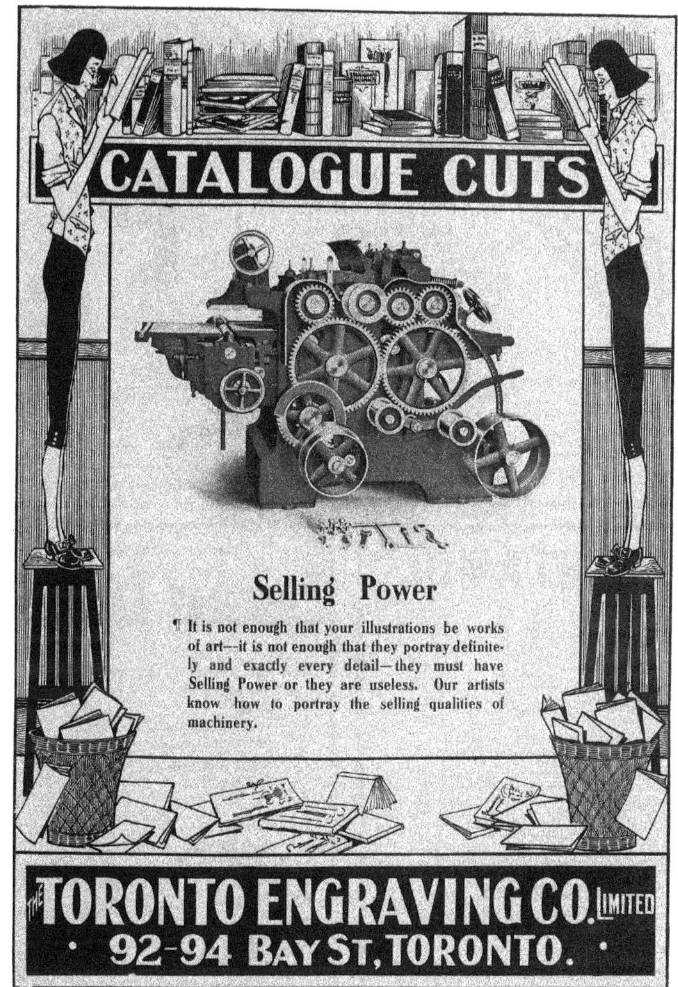
CATALOGUE CUTS
Selling Power
¶ It is not enough that your illustrations be works of art—it is not enough that they portray definitely and exactly every detail—they must have Selling Power or they are useless. Our artists know how to portray the selling qualities of machinery.
THE TORONTO ENGRAVING CO. LIMITED
• 92-94 BAY ST, TORONTO. •

MALLEABLE IRON CASTINGS

WE ALSO MAKE METAL PATTERNS

Galt Malleable Iron Co., Limited - Galt, Ontario

Partine

THE UNIVERSAL PARTING

Trade Mark Reg. U. S. Patent Off.

Preferred by Leading Foundrymen to Lycopodium

Partine does not run, cut, wash or burn off the face of the mould.

Guaranteed to produce, in every instance, a PATTERN TRUE casting, displaying the most delicate lines and features of an artistic design, accompanied by an absolute smoothness of the edges of the metal.

Its price is so low as to make its use imperative in every modern foundry.

Write for free sample, descriptive booklet and full particulars.

Sole Canadian Agents: **THE HAMILTON FACING MILL COMPANY, Limited**
Foundry Outfitters, HAMILTON, ONTARIO and MONTREAL, QUEBEC.

Cable Address: PARTINE

THE FOUNDRY SPECIALTY CO., Cincinnati, O., U.S.A.

THIS TUMBLER HAS BUT ONE LEG

—to stand on—a mighty good feature in itself. It has so many points of superiority that we won't attempt to enumerate all. Admits of perfect action on the barrel's contents without stoop-

ing. No spilling and picking up—requiring time. Occupies very little floor space. Send for "The Silent Partner,"—that famous little magazine of cleverness and our new Tumbler Book. Address

THE GLOBE MACHINE & STAMPING CO.,
899 Hamilton St., Cleveland, O.

Persons addressing advertisers will kindly mention having seen advertisement in this paper.

"PARTAMOL"

The Acknowledged Standard Parting for Parting Molds

The Money Saver and Profit Producer

"Success means willingness to put aside the old for something better."

Partamol is the something "best" in parting compounds. It turns out a perfect mold with every line true to pattern. Allows of easy and perfect "slicking" of the mold. The results from its use will astonish you. The price will please you. Sample on request.

The Dominion Foundry Supply Company

MONTREAL AND TORONTO

SELLING AGENTS FOR CANADA

The Partamol Company, 415 Broadway, New York

What do you require in

Plumbago Facings?

We can not only fill your wants but do so to your entire satisfaction.

Our range includes 50 grades from 1c. to 10c. per pound.

A pleasure to give you further information.

J. W. PAXSON CO.

PHILA., Pa., U.S.A.

MANUFACTURERS

FOUNDRY FACINGS SUPPLIES

BOYER AND KELLER RIVETING AND CHIPPING HAMMERS AND "LITTLE GIANT" DRILLS

HAVE DEMONSTRATED THEIR SUPERIORITY OVER ALL OTHERS

Examine the vital parts—note the quality of material used in their construction and simplicity of mechanical parts, then give them a trial on your own work. Tools shipped free for demonstrative purposes to responsible parties.

DUNTLEY

AIR COOLED PORTABLE

ELECTRIC TOOLS

Single Motor Type Drill

HAVE A RECORD OF

100 PER CENT. EARNING POWER

HAVE SAVED THEIR FIRST COST

in a few months for many purchasers.

EVERY TOOL GUARANTEED

to perform the service for which it is designed.

Write for catalogue describing DUNTLEY ELECTRIC DRILLS, HOISTS, GRINDERS, BUFFERS AND BLOWERS.

CANADIAN PNEUMATIC TOOL COMPANY
TORONTO MONTREAL

BRANCH OF

CHICAGO PNEUMATIC TOOL COMPANY
CHICAGO NEW YORK

SALES AGENTS FOR CANADA:

N. J. HOLDEN & CO., Limited, MONTREAL

DOGGETT'S PERFECTION PARTING

GUARANTEED TO PERFECTLY PART UNDER ANY AND ALL CONDITIONS

For Straight and Difficult Lifts
Sticky Patterns
Slanting Copes and Deep Chasings

Permits Swabbing and Slicking.

Will not Burn off the Face of the Mould, Wash or Cut.

SUITABLE FOR BRASS, BRONZE, IRON, ALUMINUM

Manufactured by

STANLEY DOGGETT

101 Beekman Street, New York City, U. S. A.

DISTRIBUTING AGENTS

DETROIT FOUNDRY SUPPLY CO., Windsor, Ont., and Detroit, Mich.

ALSO MANUFACTURE

CHARCOAL and CHARCOAL FACINGS AND ALL GRADES OF SOAPSTONE FACINGS

If the experience gained from 20 years of Air Compressor designing and building is of advantage to you, drop us a line.

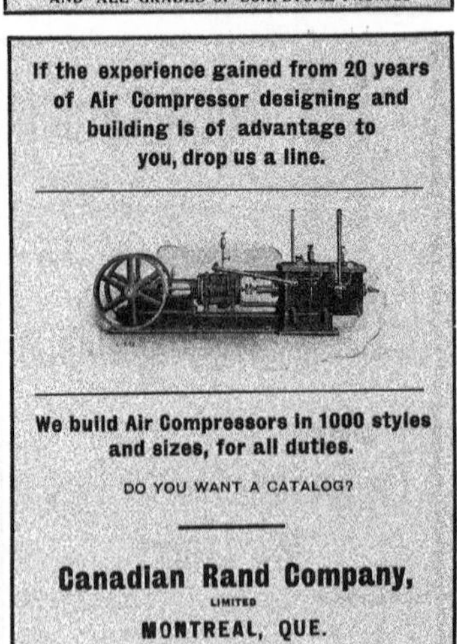

We build Air Compressors in 1000 styles and sizes, for all duties.

DO YOU WANT A CATALOG?

Canadian Rand Company,

LIMITED

MONTREAL, QUE.

Technical Books

Dies

Their Construction and Use for the Modern Working of Sheet Metals.

By Joseph V. Woodsworth

Large 8vo. ; 384 pages ; 505 illustrations ; bound in cloth ; price, **$3.00**

The New Metal Worker Pattern Book

A Treatise on Pattern-Cutting as Applied to all Branches of Sheet Metal Work.

By Geo. W. Kittredge

Cloth bound ; price postpaid, **$5.00**

Modern Machine Shop Tools

Their Construction, Operation and Manipulation, including both Hand and Machine Tools.

By W. H. Vandervoort, M.E.

Large 8vo. ; 555 pages ; 673 illustrations ; bound in cloth ; price, **$4.00**

Modern Steam Engineering

By Gardner D. Hiscox

Price, **$3.00**

This is the most complete practical book issued for the Stationery Engineer. Just issued.

Compressed Air

Its Production, Uses and Appliances.

By Gardner D. Hiscox

Large 8vo. ; 820 pages ; 545 illustrations ; bound in cloth ; price, **$5.00**

Locomotive Catechism

By Robert Grimshaw

Illustrated by 222 engravings and 12 folding plates ; price, **$2.00**

Address Technical Book Dept. :
The MacLean Publishing Co., Limited
Montreal Toronto Winnipeg

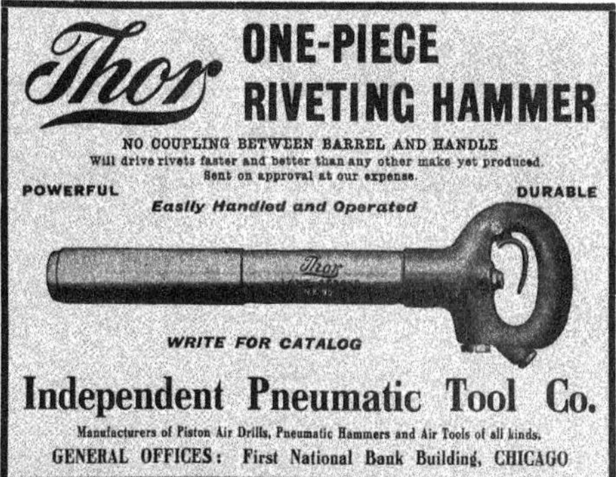

The Injector Sand Blast Apparatus

Patented in the Dominion of Canada May 9, 1905. No. 93,054

MANUFACTURED AND SOLD BY

CANADIAN RAND CO. LIMITED

Montreal, Que. Toronto, Ont. Halifax, N.S.
Kenora, Ont. Rossland, B.C. Vancouver, B.C.

PILLING PNEUMATIC DEVICES

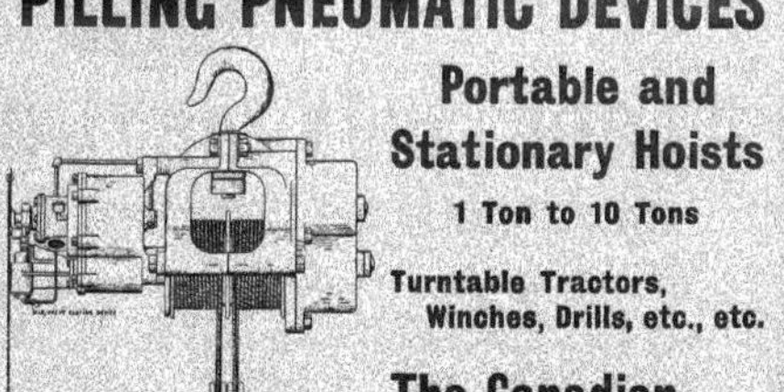

Portable and Stationary Hoists

1 Ton to 10 Tons

Turntable Tractors, Winches, Drills, etc., etc.

The Canadian Pilling Co.

180 St. James St., MONTREAL

ISLAND CITY

SUCCESSFUL

Draftsmanship

Taught personally and individually by the

CHIEF DRAFTSMAN

of a large concern, who guarantees you a first-class drafting-room knowledge, experience and a high-salaried position in a few months. Home instruction. Complete Drawing Outfit, highest quality, free.

Engineers' Equipment Co. (Inc.) Chicago, Ill., U.S.A.

VALUE, $13.85

FREE TO MY STUDENTS
Delivered at once

THE CANADIAN TURBINE WATER WHEEL

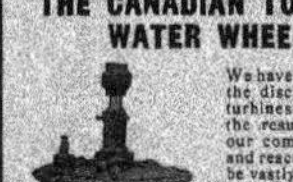

We have not expanded the discharge of our turbines, as we find the results given by our combined impact and reaction runner to be vastly superior to those obtained from turbines of the expanded discharge type. Do not sacrifice your water power. We can save you 20 to 30 per cent. of your water and at the same time give you a much more satisfactory drive.

CHAS. BARBER & SONS, Meaford, Ont.

Turbine Water Wheels, Governers and Gearing only

WM. KENNEDY & SONS

HIGH-CLASS

Steel Gears of all kinds
Iron Gears
Gray Iron, Steel and
Brass Castings

MARINE WORK receives every attention

WM. KENNEDY & SONS, Limited
Owen Sound, Ont.

NORTHERN ENGINEERING WORKS

10 Chene St.
Detroit, Mich., U.S.A.

Cranes—Traveling and Jib, Electric and Hand

Ladles — Electric and Air Hoists—Foundry Machinery

The Newton Cupola

"Economo" Emery Wheel Dresser

A Substitute for the Black Diamond

"ECONOMO" ECONOMO COMBINATION

PRICE: Economo, $4.00 each; Economo Combination, $4.50 each.

Will true or shape any wheel except the very hardest and carborundum. The most efficient Roughing Dresser on the market—never gets dull from use.

Made of an abrasive nearly as hard as the Black Diamond. The Combination dresser is the roughing dresser placed in the "handle end" of the "Regular." It is made with corrugated tool steel discs and excels in wear and efficiency any dresser heretofore made. These tools are guaranteed satisfactory or money refunded. Sent on 10 day trial.

INTERNATIONAL SPECIALTY CO.
363 Holden Ave. *DETROIT, MICH.*

American Pattern Shop Practice

By H. J. McCASLIN

JUST ISSUED. An up-to-date, practical work on pattern-making, based upon the author's own experience, treating of molding and core making as well as patternmaking. Special attention is given to core box work. The following subjects are fully dealt with in a clear, concise way:

Engine Patterns	**Gearing**
Molding and Cores	**Representative Patterns**
Sweep Work	**Hints, Suggestions and Rigs**

BOUND IN CLOTH. FULLY ILLUSTRATED

PRICE $3.00, POSTPAID.

Technical Book Dept., MACLEAN PUBLISHING CO.
10 Front St. E. - Toronto

"Best" Dry Cells

For your Ignition in

Automobile
Motor Boat
Gas Engines

The kind the largest Telephone Company in Canada is using to-day, because they give the longest service.

Ask your dealer for them, or

The Berlin Electrical Mfg. Co.,
Limited
474 King Street West, TORONTO

UNION DRAWN STEEL COMPANY, Limited

Manufacturers of

BRIGHT FINISHED STEEL

Largest in Canada. Best Equipped on the Continent.
Capacity twice present requirements of the trade.

Rounds, Flats, Squares and Hexagons

LARGE STOCK

Send for Price List

Office and Works :

HAMILTON, CANADA

WE CONTEND THAT

The Busy Man's Magazine

IS the best publication in the magazine line. But don't take our word for it. Read the unsolicited opinions of its subscribers—

"I consider The Busy Man's Magazine the best I have yet come in contact with." Henry Wilson, Toronto.

"It is the most interesting publication of its kind I have ever subscribed to." G. F. Haworth, Toronto.

"There is no better printed in the world to-day for busy men." B. T. Petch, Regina, Sask.

"Your magazine is the best I have ever had the privilege of perusing." S. W. Wynn, Yorkton.

"We would not like to miss the future copies of The Busy Man's Magazine." Ingersoll Packing Co., Ingersoll.

"I am pleased to say that each number is becoming more valuable and interesting." A. B. Musselman, Berlin.

"I am more than pleased with my investment in The Busy Man's Magazine, believing it to be the best magazine I have had the pleasure of perusing." A. B. Burrell, Amigari.

"I have been delighted with the Busy Man's Magazine. The articles are well chosen, are concise and ably written. For a busy man I know of no other magazine that will compare with it." Geo. Anderson, Toronto.

"The articles in The Busy Man's Magazine are just what we want, short and so well written that a busy man can afford time to read them and feel benefited for the short time given to each article." Wm Thoburn, Almonte.

We wish to give you further proof of the value of the ***BUSY MAN'S.*** Send us a postcard, giving your address and we will send you a sample copy.

Bear in mind every timely, interesting and instructive article appearing in any of the current magazines and periodicals appears also in the ***BUSY MAN'S***—a hundred magazines for the price of one.

WRITE US TO-DAY

The Busy Man's Magazine

TORONTO

Books for Engineers Draughtsmen Mechanics, Etc.

Post Free to any address at home or abroad, at published prices.

	NET PRICE s.	d.
Workshop Costs, for Engineers, etc., by S. and F. Pearn. 2nd Edition	12	6
How to Use Water Power (Chatley), illustrated	2	6
Sand and Clay: Their Analysis and Physical Properties	2	0
Air Compressors and Blowing Engines, by Chas. H. Innes, M.A.	4	6
The Fan; including the Theory and Practice of Centrifugal and Axial Fans, by Ch. H. Innes, M.A.	4	0
The Proportions and Movement of Slide Valves, by W. D. Wansbrough	4	6
Machines and Tools Employed in the Working of Sheet Metals, by R. B. Hodgson	4	6
Governors and Governing Mechanism, by Hall	4	0
Specification of a Modern Lancashire Boiler and its Seating (200 lbs. pressure), by "Inspector" M.E.	5	0
Continuous Current Dynamos and Motors and their Control, by W. R. Kelsey	5	0
Notes on Construction and Working of Pumps, Marks	4	6
Modern Ironfoundry Practice : Part I., Hand Moulding, Bale	5	0
Part II., Machine Moulding, Bale	3	6
Modern Gas and Oil Engines, by F. Grover. 4th Edition	5	0
The Indicator and its Diagrams, by Chas. Day. 3rd Edition	4	6
The Chemistry of Materials of Engineering, by A. H. Sexton. 2nd Edition	5	0
The Management of Small Engineering Workshops, Barker	7	6
Problems in Machine Design, by Chas. Innes	4	6
Heat and Heat Engines; a Treatise on Thermo-Dynamics. Popplewell	6	0
Centrifugal Pumps, Turbines and Water Motors.	4	6
Application of Graphic Methods to the Design of Structures. 2nd Edition	5	0
Engineering Estimates and Cost Accounts, Burton.	3	0
Graphic Methods of Engine Design, Barker. 2nd Ed.	3	6
Injectors : Theory, Construction and Working, Pullen.	3	6
Construction of Cranes and Lifting Machinery, Marks.	3	6
Marine Engineers : Their Qualifications and Duties	5	0
A.B.C. of the Differential Calculus, Wansbrough	3	0
The Theta-Phi Diagram Applied to Steam, Gas, Oil, and Air Engines	3	0
Mechanical Engineering Materials, by Marks	2	6
The Naval Engineer and Command of the Sea, Burton	2	6

The Technical Publishing Co., Limited

287 Deansgate, MANCHESTER, ENGLAND

No Frost Breaks, no Corrosion No Electrolysis

It is easily and cheaply laid. Its carrying capacity is never decreased by rust.

GALVANIZED WIRE WOUND WOODEN PIPE

PHONE 1642 **VANCOUVER, B.C.** **P.O. BOX 915**

WOOD STAVE WATER PIPE

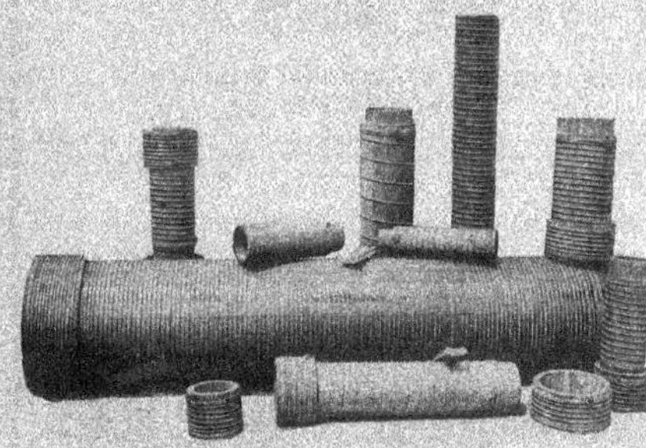

Wood Pipe Designed for Different Pressures.

Machine-banded made from choice British Columbia or Douglas Fir, absolutely clear and free from knots or defects, wound with heavy gauge galvanized steel wire. For

Water Works, Power Plants, Irrigation, Mining and all Hydraulic Propositions, where conduits or pressure pipe are required.

Reliable, durable, economical, efficient, easy to instal. Costs very much less than any other sort of water pipe, and with greater efficiency in use.

Write for Catalogue and Circulars.

Estimates also given on continuous stave pipe of large diameters.

PACIFIC COAST PIPE COMPANY

P.O. Box 563 **Limited**

1551 Granville Street (Next Bridge) - VANCOUVER, B.C.

THREAD YOUR WAY

into larger profits by doing your threading with our

Style "D" Self-Opening Screw-Cutting Die Heads

They cut any length and from smallest size up to 4½ inch bar, right or left hand threads—standard or special pitch. They save time and allow much larger output, because there is no running back after each cut. Variations in diameter of work can be made quickly and accurately. Why not increase your capacity or do the same work in less time?

ASK FOR PARTICULARS.

The Geometric Tool Co., New Haven, Conn.

Canadian Agents: WILLIAMS & WILSON, Montreal, Que. (Westville Station) U.S.A.

NORTON JACKS

ARE MADE IN CANADA AND SAVE YOU DELAY AND DUTY OVER AMERICAN JACKS

They are Ratchet Screw Jacks with Ball Bearings to reduce the friction. Working parts are protected from grit and rust. Superior to Hydraulic Jacks for all classes of service. Simple, durable, and guaranteed in every respect. Absolutely safe. Will work in any position. Standard on the leading railroads.

Made in 50 styles; 8 to 70 tons capacity

Send for complete illustrated catalogue and discount.

Sectional View of 70-Ton Bridge Jack

25-Ton Foot Lift Jack

35-Ton Plain Jack

Traversing Jack 10 to 60 Tons

Track Jack 10 and 15 Tons

Journal Box Jack 15 Tons

MADE ONLY BY

A. O. NORTON, Inc., Coaticook, Prov. Quebec

Stock carried by MUSSENS, Limited, Montreal

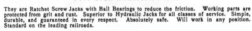

ALLIS-CHALMERS-Bullock-LIMITED MONTREAL
LIGHTING TRANSFORMERS
ALLIS-CHALMERS-Bullock-LIMITED MONTREAL
This is a view of one of our 10-K.W. Lighting Transformers. Our different styles of Transformers are fully described in Bulletin 300, just issued. Among their good points are:
1. Low core loss
2. High efficiency at all loads
3. Low temperature rise
4. Good regulation
5. Strong mechanical design
ALLIS-CHALMERS-BULLOCK, Limited
Head Office and Works, Montreal; District Offices; Toronto, Traders Bank Bldg; Montreal, Sovereign Bank Bldg.; Winnipeg, 251 Notre Dame Ave.; Nelson, Josephine St.; Vancouver, 416 Seymour St.; New Glasgow, N. S., Telephone Bldg.

YOU WILL SAVE MONEY IN BUYING IF YOU REFER TO OUR BUYERS' DIRECTORY

CANADIAN MACHINERY AND MANUFACTURING NEWS

A monthly newspaper devoted to the manufacturing interests, covering in a practical manner the mechanical, power, foundry and allied fields. Published by The MacLean Publishing Company, Limited, Toronto, Montreal, Winnipeg, and London, Eng.

OFFICE OF PUBLICATION: 10 FRONT STREET EAST, TORONTO

Vol. III. OCTOBER, 1907 No. 10

Five Sizes—$\frac{5}{8}$ x $4\frac{1}{2}$ in., 1 x 10 in., $1\frac{1}{2}$ x 18 in., 2 x 26 in., 3 x 36 in.

Some sizes for immediate delivery

The Turret Lathe is now considered indispensable in the shop where duplicate work is done. We build a variety of sizes, enabling the manufacturer to install a machine to suit his particular requirements.

2 x 26-in. Turret Lathe shown with Locomotive Equipment.

The great success of these machines is largely due to their rigidity, convenience and rapidity in operation, and facility in making ready for given work, making them economical in the production of small lots of work.

On bar work these machines will do the work of from 3 to 6 engine lathes.

Write for our new 64-page catalogue, "Turret Lathes," with complete descriptions and illustrations of all five sizes and tool equipments.

PRATT & WHITNEY COMPANY

WORKS:—HARTFORD, CONN., U.S.A.

THE CANADIAN FAIRBANKS CO., Limited, Agents for Canada

MONTREAL TORONTO WINNIPEG VANCOUVER

Radial Simplicity---

A Matter of Importance!

1.—You know pretty well what a good Radial ought to be. It should have a large number of speed changes—to adapt it to metals of all degrees of hardness—a wide range of feeds, from one of sufficient fineness for drilling small holes in steel, to one of suitable coarseness for boring large holes in cast iron ; a multiple automatic trip, a zero dial depth gauge, and such other features as would enable it to produce a maximum in output.

2.—But these features, necessary as they seem to be, are of little practical value unless they are of simple design and easily accessible.

3.—There's little gain, if any, in having a wide range of speeds or great variety of feeds, if the operator is put to the necessity (meaning delay) of shifting his position every time a lever is to be manipulated. Nor can you afford to have him waste time trying to fix a complicated machine.

4.—Your cost records are your drilling records, from which it would appear that the machine of greatest productive capacity would be one of such features as are simple in construction and of extreme convenience to the operator.

5.—Bickford Radials are rigid, durable and accurate. They embody such cost-cutting and labor-saving devices as enable them to produce a maximum in output, and their design is of the simplest. It is next to impossible for them to get out of order, and they handle so conveniently that even a boy without experience can run them and get good results.

6.—For an honest and detailed description of these drills send for our catalog. It tells all about them and contains valuable "Radial" pointers for the man who is interested. A postal with your name and address is all that's necessary.

BICKFORD IMPROVED PLAIN RADIAL.

The Bickford Drill and Tool Company

CINCINNATI, OHIO, U. S. A.

FOREIGN AGENTS Schuchardt & Schutte, Berlin, Vienna, Stockholm, St Petersburg, New York. Alfred H. Schutte, Cologne, Brussels, Liege, Paris, Milan, Bilbao, New York. Charles Churchill & Co., Ltd., London, Birmingham, Manchester, Newcastle-on-Tyne and Glasgow. H. W. Petrie, Toronto, Canada. Williams & Wilson, Montreal, Canada.

Series A-No. 2.

2-MONEY MAKERS-2
"AMERICAN'S"
New
High Power
Tools
Catalog
"C"
Now
Ready
DOUBLE BACK GEARED LATHE
Gives Wide Belt Surface—Greater Power
Extra Gear Ratio—18 to 27 Spindle Speeds
NOTICE—To receive new catalogue when issued, machine tool users please send postal with company's name and mechanical officer in charge thereof.
MULTI-SPEED PLANER
Gives the Right Cutting Speed for all Classes of Work and Maintains a Constant Speed Return.
THE AMERICAN TOOL WORKS CO.
700 to 750 CULVERT STREET - CINCINNATI, OHIO, U.S.A.
LATHES — PLANERS — SHAPERS — RADIAL DRILLS
Write Direct or to Agents: WILLIAMS & WILSON, Montreal; A. R. WILLIAMS MACHINERY CO., Toronto, Winnipeg, Vancouver
9

Machine Tools

¶ In stock or for immediate shipment—Lathes, Planers, Shapers, Drills, Emery Stands, Pumps, and Gas or Gasoline Engines in certain sizes, and good deliveries on all sizes in Metal Working Machine Tools and Power Machinery.

¶ We can now furnish Gas or Gasoline Engines in standard units up to three hundred horse-power.

¶ Let us figure with you on your factory equipment.

Address Inquiries care Machinery Dept.

RICE LEWIS & SON, LIMITED, TORONTO

Courtesy of
COWAN & CO.,
Galt, Ont.

Fine Vignetted Half Tone
Plate from
Retouched Photo.

" Good Half Tone plates are the prime essential of a successful catalogue."

Pipe Cutting and Threading Machine

For Hand or Power, Bench or Trench

FORBES PATENT DIE STOCKS will do either from ¼ inch I.D. to 15 inch O.D. **ON THE SPOT**

No expense or delay of carting your pipe to the machine. We keep regular sizes in stock and can make prompt shipment. Write for Catalog.

THE CURTIS & CURTIS CO., - *36 Garden Street, Bridgeport, Conn.*

THIS IS THE NEW SHANTZ No. 6.

BAD THREADS—LOST MONEY

That, in brief, is the tale which could be told by many a man who still sticks to his old-fashioned pipe-threading machine. It is surely the worst form of **extravagance** to keep an **obsolete, money-wasting** machine in the shop, especially when, at a reasonable price, the

SHANTZ No. 6

fitted with the **Borden Patent Solid Adjustable Die Head,** may be installed. This machine **cuts absolutely perfect threads** on 1½-inch to 6-inch pipe. It does the work **with one cut,** and does it quicker and cleaner than any other machine in America. **Absolute accuracy** can be obtained by the mere adjustment of an adjusting screw. This machine weighs 4,300 pounds, and is **so immovably rigid** that **a wavy thread is an impossibility.** And there are other points about it, which, with those above stated, place the "**Shantz No. 6**" in a class by itself. **Write for our free descriptive circular.**

I. E. SHANTZ & CO., Berlin, Ont.

DO YOU DO IT?

When you read an advertisement that interests you, why not write the advertiser and get more information? It will help you to keep posted. Try it. The advertisers in **Canadian Machinery** will appreciate it.

WRITE THEM. TRY IT.

IT PAYS TO INSTALL THE BEST

POWER PRESSES

The many exclusive features embodied in the type shown above make it so much superior to the next best that there is no comparison.

Write for prices and particulars on **Presses, Dies, Tools and Special Machinery.**

W. H. BANFIELD & SONS

120 Adelaide Street West, TORONTO.

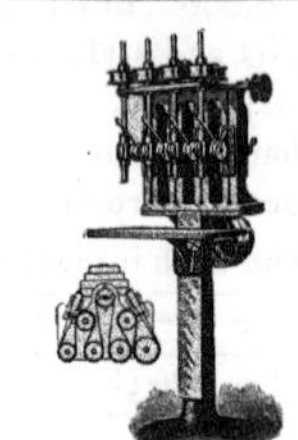

FOUR SPINDLE 13-IN. SWING DRILL

Capacity ½-in. holes and smaller. Endless belt drive. Provision for adjustment of belt tension. Interchangeable spindle pulleys. Adjustable heads. Adjustable table. Spring balanced spindles.

Also built with 2 or 3 spindles.

Send for Catalogue

THE

Dwight Slate Machine Co.,

HARTFORD, CONN., U.S.A.

A Dozen to One

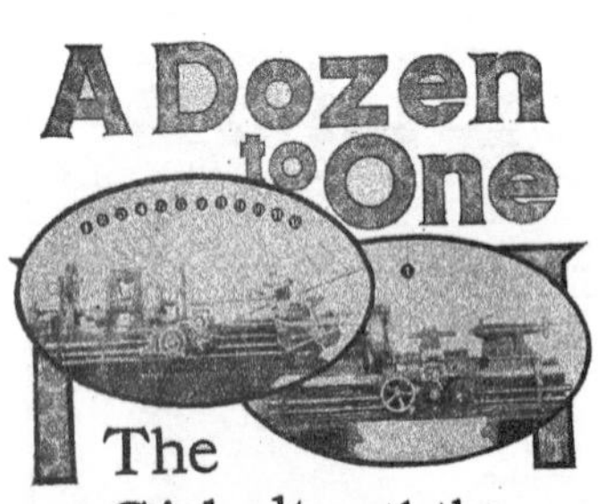

The Gisholt and the Engine Lathe

THINK what a dozen cutters on the Gisholt means to the single tool on the engine lathe. And even a dozen isn't the limit on the Gisholt. There are stations for 10 different tool holders any one of which may hold from one to six tools. Think of the metal you can remove by these methods.

Is it any wonder that the Gisholt will do work so rapidly?

Gisholt Machine Company

WORKS:
Madison, Wis. and Warren, Pa.

GENERAL OFFICES:
1309 Washington Ave., Madison, Wis.

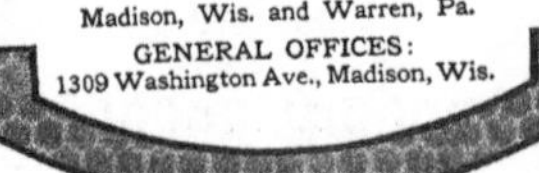

These gang drills not only drill, but they also ream, bore, face, tap, mill and do a lot of other operations that you are in the habit of buying expensive machinery for. Send description of your work and we will submit proposals.

B. F. BARNES COMPANY

ROCKFORD, ILL.

European Branch:
149 QUEEN VICTORIA STREET, LONDON, E.C.

POTTER & JOHNSTON

MANUFACTURING AUTOMATICS

will machine your duplicate parts from all kinds of castings, forgings and the bar with extreme accuracy and more economically than upon any other tools in the market.

Catalogue showing interesting specimens of parts finished on these tools free upon request.

ALSO UNIVERSAL SHAPERS

POTTER & JOHNSTON MACHINE CO.

PAWTUCKET, R.I. - - U.S.A.

1017 B.

WE MAKE IT
BABBITT METAL
Imperial Babbitt
IS BEST FOR HIGH-SPEED AND HIGH-PRESSURE WORK.
IT CAN'T BE BEAT.
Every Pound Backed by a Guarantee.
WE ARE SPECIALISTS
and can give you a bearing metal to suit any class of work.
WRITE US.
TRY US.
THE CANADA METAL CO., LIMITED
Toronto, Ont.
If you want anything in CRANES, PUMPS or CONDENSERS, drop us a line and you may have the benefit of our experience.
THE
Smart-Turner Machine Co.
Limited
HAMILTON, ONT.

J. R. BAXTER & CO.

(Successors to Baxter, Paterson & Co.)

MONTREAL

Agents for

Landis Bolt Cutters

Safety Emery Wheel Co.
(Grinding Machinery.)

Prentice Bros., Machinery

U.S. Electrical Tool Co.
(Portable Electric Tools),
Etc., Etc.

KYNOCH
Suction Gas Engines
and
Producer Plants

We have been operating one of these Engines and Plants for a year. We propose installing a larger Engine and Plant in our new warerooms now being built. The saving in expense will **Pay for the Engine in Two Years.**

WRITE FOR CATALOGUES AND PRICES

Williams & Wilson

320-326 St. James St., Montreal

The Canada Chemical Manufacturing Company, Limited

MANUFACTURERS OF

Commercial Quality **Acids and Chemicals** Chemically Pure Quality

ACIDS—Sulphuric, Muriatic, Nitric, Mixed, Acetic, Phosphoric, Hydrofluoric.

CHEMICALS—Salt Cake, Glauber's Salts, Soda Hypo, Silicate, Sulphide, Epsom Salts, Blue Vitriol, Alumina Sulphate, Lime Bisulphite, Nitrite of Iron, C.T.S. and Calcium Acid Phosphate.

Chemical Works and Head Office	Sales Office	Warehouses
LONDON	TORONTO	TORONTO and MONTREAL

Return Tubular and Locomotive Fire-Box Boilers Tanks, Stacks, Rivetted Pipe.

THE MANITOBA IRON WORKS, Limited, WINNIPEG, MAN.

KERR TURBINE BLOWERS, for gas works, foundries and forge shops. Air pressure 4 to 16 ounces. For any capacity.

TURBINE PUMPS, for water works, condenser equipment and boiler feeding. These pumps are more economical than direct acting piston pumps and require less attention, lubrication and space. No belts or gears. Speed adjustable to pressure desired. Regulation close. Start and stop quickly. For any capacity.

TURBINE GENERATORS, 2 kw. to 150 kw. direct current and alternating current 25 kw. to 150 kw., 60 cycle.

The illustration shows a 12-inch Kerr steam turbine with pressure governor coupled to a No. 7 Buffalo volume blower for forced draft in service for the Canadian Shipbuilding Co. of Toronto.

Main Office:	Built in Canada by
KERR TURBINE CO.	**WM. STEPHENSON**
Wellsville, N.Y.	34 Bay St. N., Hamilton, Ont.

"PECK" OVERLAPPING BUCKET CARRIER

Conveys Without Spill

To test every claim we make for the utility and economy of this Carrier for conveying

COAL, COKE, ASHES, STONE, CEMENT, HOT AND COLD CLINKER, ORES, ETC.

Ask us where you can see it in every-day operation.

LINK-BELT COMPANY

PHILADELPHIA, CHICAGO, INDIANAPOLIS

New York, 299 Broadway.	St. Louis, Missouri Trust Bldg.
Denver, Lindrooth, Shubart & Co.	Pittsburgh, 1501 Park Bldg.
Seattle, 440 New York Block.	New Orleans, Wilmot Machinery Co.

Machine Tools

We have a well-assorted stock of Machine Tools and Wood-working Machinery on hand for prompt delivery.

Bertram 42-in. Boring and Turning Mill

Pond Rigid Turret Lathe

WE ARE SELLING AGENTS FOR THE FOLLOWING MANUFACTURERS

Miles-Bement-Pond Co.
The John Bertram & Sons Co., Ltd.
Pratt & Whitney
R. A. Kelly Co.
J. J. McCabe
Bignall & Keeler
Wilmarth & Morman
Niagara Machine Tool Co.
Gardner Machine Co.
S. A. Woods Machine Co.
American Wood-working Machinery Co.
Baker Bros.
Merrill Mfg. Co.
Seneca Falls Mfg. Co.
Brown & Sharpe
E. W. Bliss Co.
The R. McDougall Co., Ltd.

Bertram 72-in. Motor-Driven Planer

The Canadian Fairbanks Co., Ltd.

MONTREAL TORONTO WINNIPEG VANCOUVER

Elements of Crane Construction

By C. C. MAISON

In no structures built by engineers is the question of due relation of strength to stress of greater importance than in cranes. Yet in few does more empiricism exist, in few is the accumulated experience of success and failure of greater value—a case which has notable parallels in the history of the development of the locomotive crane, and of machine tools, neither of which are much indebted to theory. There are certain crane elements in regard to the strength of which calculations are of much value, because the stresses are readily obtainable by the methods of graphic statics. These are the elements of which the cranes are built, and the strains on chains, hooks and snatch blocks. But the main side frame castings, and the plated frame castings, are not readily calculated, and, in fact, are almost invariably copied or modified from previous designs that have stood

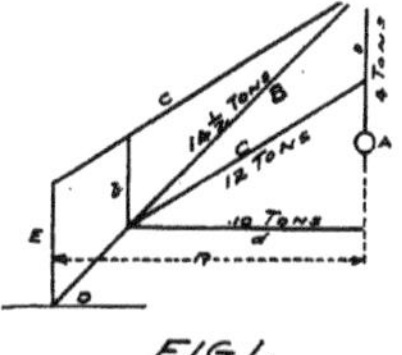

FIG. 1.
Raking Jib Type.

successful service. In the drawing room of crane shops new designs are got out without much direct calculation, because previous practice is drawn upon. The more highly the work of a firm is specialized, the more easily can modified designs be produced. There are tables kept in the office, giving loads for various sizes of chains, wire rope, and the strength of the different standard hooks used, the strength of rods of various cross sections, and a deal more of the same character, by which direct, often repeated calculations are rendered unnecessary. Then there are certain points of gears and drums which have been previously used, and these can be taken en bloc, and put on other cranes that vary in details of design. Jibs are standardized for different radii and power, and these need not be re-calculated. So are trucks, posts, ground wheels, and then, too, there are standard superstructures that can be taken bodily and put on either portable or fixed bases. The reason why calculations are so greatly modified is that stresses can be obtained for certain elements with absolute precision, and in others with a fair approximation thereto.

In the writer's experience, there is no single section of a crane which has not been overworked, whether checks, posts, jibs, chains, the rods, trucks, both cast and plated, traveler girders, tooth wheels and drums. Accidents occur because the machine is called upon to perform work which it was not designed to do, and very often the cause is the reliance placed in past actions, which a certain crane did over or nearly half a century ago. There are few machines, speaking indiscriminately, that are more ill used than these.

The checks of cranes are subject to great variations in design. In small cranes they are of cast iron; in heavy ones, steel plated. But many small cranes are cheaply made with steel plate, while for permanent way cranes this type is always employed. Plated work is to-day thirty per cent. cheaper than formerly, and is by far more reliable than castings. But castings cost still less, because the bearings are in one with the frame, while in plated work the bearings must be prepared separately and bolted or riveted on. But these are often cheaply fitted in the form of round bosses in place of the more expensive divided bearings, according to the purchaser, of course.

In the simplest types of the true cranes the elementary frame is a triangle, composed of post, jib aad tie. It is embodied both in pieced and in portable cranes, of the wharf, whip, wall, derrick and many other types, and includes cranes in which jibs are curved, or cranked to clear loads beneath. The relative disposition of the three members governs their relative strength, and these dispositions are controlled by the nature of the duties which cranes have to perform. The magnitude and direction and nature of the forces in a triangular frame are shown in Fig. 1. This is the most common type, called the raking jib arrangement. The diagram is only correct when the load to be raised on a single chain and the line of same, from the head pulley to the drum, coincides with the centre of the tie. This type is used for variable radius. The load A runs over a fixed pulley, at the top of the jib B, the radius R of which is unalterable, except by derricking. No trolley nor jinney can run along this crane to vary the radius of the load lifted; hence the reason why many of these are made to derrick. That is, the jib is hinged at D in order to permit of effecting variations in the radius of lift, accomplished both in the derrick cranes proper and in many ordinary cranes that differ from the true derricks in most details, excepting in the capacity for this particular range of movement.

The post, pillar or mast, as it may be termed, is a vital element in all triangular framed cranes, with those few exceptions in which a wall depends on fulfilling the function. The stability of a crane depends in the first place on that of the post. If this fails, it will do so either by breaking off at or near

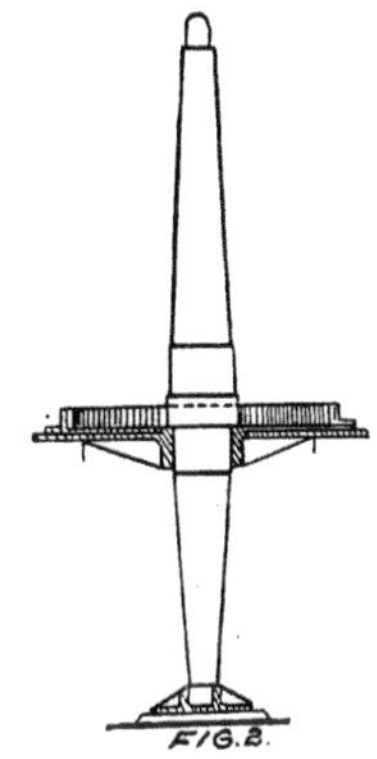

FIG. 2.
Wharf Crane.

the ground line, where the maximum bending stress occurs in cranes that are supported in a footstep only. In cranes that have top pivots in addition, the post would fail somewhere between the top and bottom pivots—a determining factor in which would be the locality where the jib happens to be stepped into the post. From this point of view the nearer the jib is brought down to the ground line, the better, though having regard to the racking of loads inwards, this is the least favorable position. Posts are made of timber, cast iron and steel, sometimes wrought iron, either solidly or built up like girder work. Fig. 2 shows a solid post built up of steel or wrought iron, as used in wharf or warehouse. Jibs, struts and ties are grouped together because they are mutually dependent. A jib, with a few exceptions, is not self-supporting, but is sustained by ties or with struts. From this point of view various forms of jibs become grouped naturally under three broad types—the cantilever, strut supported, and the tie supported. The first are not numerous, the second only moderately so; the greater number of examples come under the last-named head. Cantilever jibs are used chiefly on some forms of long-armed cranes, usually of the traveling type, with or without provision for rotation, as on the Brown cranes and related types. Having no extraneous support, they are built of girders of semi-parabolic outlines.

Crank Jib Crane.

Jib Head.

Under traveling cranes there is included all those which come under the class of overland travelers of Goliaths and gantries. The term is applied exclusively to these, notwithstanding that the portable cranes with trucks travel on rails. But the term portable, or the term locomotive, also applied to these, distinguishes them from the traveling cranes. The distinction is purely conventional, but is always understood. There are a number of designs, as those which exist in the frames of traveling cranes, and are too numerous to make mention of all of them.

Fig. 3 shows a cranked jib which is laced, and is of a composite type, in which the fitting of the rods at some distance away from the end leaves the portion between the anchorage and the foot in compression, while the length beyond the anchorage is subject to bending as a cantilever. This jib is widened at the dividing section to afford strength to the cantilever end.

In Fig. 4 a simple jib head fitting for a chain pulley is shown. Two pieces of steel plate, AA, are bolted between the H sections, and cranked to receive a distance piece, B, which also prevents the chain from jumping outside the pulley pin D. The castings are prolonged into bosses C, over which the tie rod eyes fit, and on which they are retained sideways with washers, EE. In many cases, also, are castings fitted instead of steel plates.

As over the stresses must be translated through truss rods, timber of reasonable depth alone is sufficient. The method of trussing, moreover, affects the result. A truss may be simple; see Fig. 5. Then the weight of the girders themselves, and of the crab and its load, and its position, has to be considered. The length, also, of the wheel base D affects the stresses on the end cradles C. Stresses on the wheels and their journals are calculated, as indicated in Fig. 6. Each wheel is taken as loaded with half the load of the main girders, and is wholly in compression.

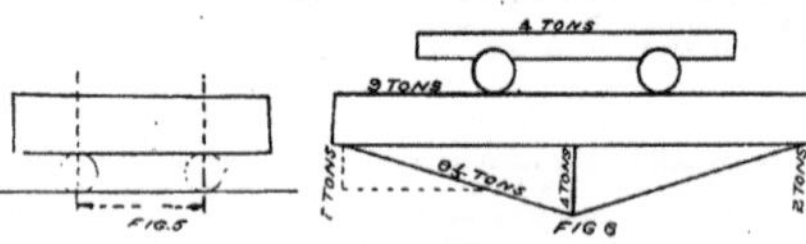

The stress on wheels, due to cross working of the frames, sometimes causes their axles to grind hard in their bearings. This may be eliminated by making the axles fast in their bearings and letting the wheels run loosely on them. The employment of iron and steel for traveler frames affords scope for a large variety in design and dimensions. They are suitable alike for travelers, the power of which ranges from half a ton to a hundred tons or more. Iron is, however, as already remarked, little used now, being mostly displaced by steel.

In use it might be supposed that over much attention has been given to the co-relation of material to stresses; it is well to point out that the question is less one of cost than of weight. It costs more to build up girders than to use rolled joists. It would generally cost more than the value of the metal saved. But that is hardly the point at issue. The dominating fact is the reduction in the dead weight of the traveling crane. By lessening this the stresses on the traveling wheels and axles are lessened, and less power is required for operating—a point of much importance with high speed travelers, which are now so common.

Girders are not quite so simple a matter as this. In designing there are two points which have to be settled—rigidity depthwise, that is, as opposed to buckling or crumpling, and rigidity sideways, to assist the lateral stresses, which are produced by the rigid longitudinal movement of the crane.

THE HUMOR OF STRIKES.

"I see th' strike has been called off" said Mr. Hennesy. "Which wan?" asked Mr. Dooley. "I can't keep thrack ix thim. Somebody is sthrikin' all th' time. Wan day th' horseshoers are out, an' another day th' teamsters. Th' Brotherhood iv Molasses Candy Pullers sthrikes, an' th' Amalgymated Union iv Pickle Sorters quits in sympathy. Th' carpinter that has been puttin' up a chicken coop f'r Hogan knocked off wurruk whin he found that Hogan was shavin' himself without a card fr'm th' Barbers' Union. Hogan fixed it with th' walkin' dillygate iv th' barbers, an' th' carpenter quit wurruk because he found that Hogan was wearin' a pair iv non-union pants. Hogan wint down town an' had his pants unionized an' come home to find that th' carpinter had sthruck because Hogan's hens was layin' eggs without th' union label. Hogan injooced th' hens to jine th' union. But wan iv thim laid an egg two days in succession, an' th' others sthruck, th' rule iv th' union bein' that no hen shall lay more eggs thin th' most reluctant hen in th' bunch."

Boring Bars and Cutters

By JOHN EDGAR

Of the many operations that the machinist performs, the operation of boring is one that requires an amount of skill seldom called into action in the performance of any of the other machine shop operations. The boring of a hole calls for accuracy in more than one direction. The hole must be straight, of a standard size, and must have a degree of finish depending on the class in which the job ranks.

Where the equipment is poor the deficiency must be made up by an excess of skill on the part of the operator. A job of boring cannot be rushed. The best of facilities are not always at hand for the proper handling of the work. Makeshifts are often necessary.

Were one to try to describe the many variations in which we find this operation performed, he would have contracted a very big job. The best that can be done is to skim a "little off the top" and present what may be called the cream of the many methods practiced.

Boring may be accomplished on a standard boring machine, which may have a horizontal spindle or it may have a vertical spindle. The class of machine which we will deal with will be those in which the operation of boring is done with a boring bar, and cutters set in this bar. The operation as performed in a lathe on the boring mill, where the work is made to revolve about the axis of the hole to be bored, does not call for a bar in the common sense of the word, but the hole is enlarged with a boring tool and is what we may call internal turning, to distinguish the two methods.

In the operation, as we are here considering it, the bar is made to revolve and is supported at both ends; one end may be supported in the spindle and the other in a tail support, or both ends may be supported in supports independent of the machine or spindle. In the latter case the bar is given its motion by means of some flexible connection with the spindle. This connection usually consists of two pairs of universal joints.

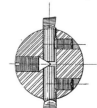

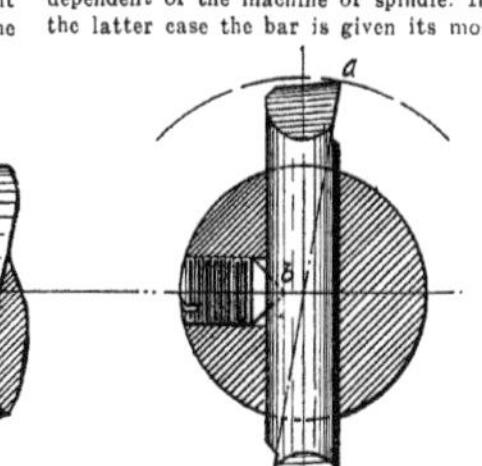

Fig. 1—Simple Round Cutters.

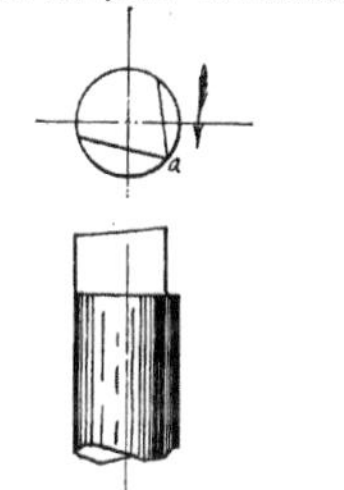

Fig. 2—Simple Boring Cutter.

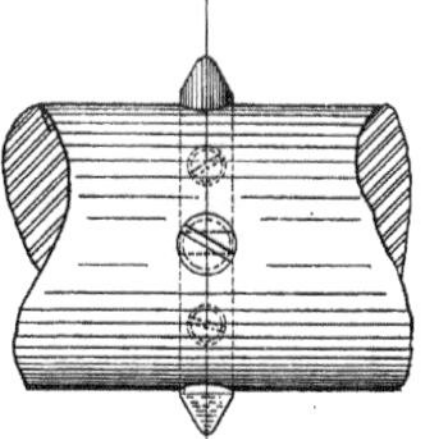

Fig. 3—Adjustible Boring Bar Cutters.

Boring bars may be classed in two divisions, viz., general utility bars and special bars.

The first class may be said to cover all bars that are made for general work, and may be also classed with other handy shop tools. We will not deal with this class to any great extent, merely referring to it in the discussion of the more important class, that of special bars.

The second class is that in which we place all bars that are made up to do one job correctly and to the best advantage, without the necessity of spoiling its effectiveness on one job in order that it may be of use on a general run of work. It may be used in connection with a boring jig, in which the bar is supported by means of suitable fixed bushings, or it may be used as a boring bar, pure and simple.

When a boring bar is to be used in conjunction with a jig, the bar generally runs in hardened steel bushings. Great care must be used in the handling and sufficient lubrication must be supplied in order that the bar does not rough up and seize in the bushing. In order to avoid this, the bar must be kept free from dirt and a generous supply of heavy vaseline used as a lubricant. Any —even the slightest—bur on the bar will be sufficient to cause trouble from this source.

To overcome the tendency of the bar to stick in the bushings when used as above described, the bushings are sometimes made to run in the fixture and are splined to the bar and made to revolve with same, the bar being free in a direction parallel with its axis. The bushings in this case are made with an external taper and are provided with chich nuts on each end so that they can be adjusted lengthwise, to take up any wear that is liable to appear. This style is not apt to be of such a degree of accuracy as the stationary bushing, because the wear on the cast iron housing would be much greater than would occur with the steel bar and the hardened bushing. However, the freedom from the tendency to seize may in some cases offset the small inaccuracy that might be present. This method of support is well worth looking into, and could be used in many cases where the other style is at present used to a disadvantage.

The simplest form of boring cutter is shown in Fig. 1. This is made of a piece of drill rod and shaped at each end as shown. The length is left a trifle longer than the size of the hole, and the cutter is then placed in the bar and the set screw spotted, and a chip is taken off the ends in the lathe to bring them central; it is then backed off with a file and hardened. The hardened cutter is then stoned to obtain the requisite cutting qualities.

Fig. 2 shows a variation in the cut-

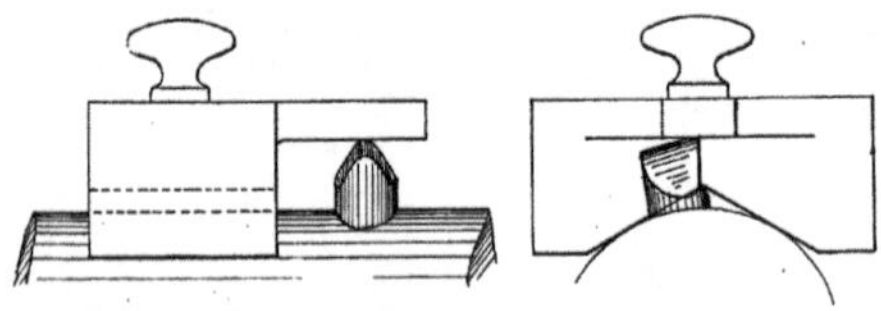

Fig. 5—A Fixture for Setting Adjustible Boring Bars to Size.

ting edge. This style is much used and is equally common with that shown in Fig. 1. These cutters are, of course, useless after dull, as they lose their size in sharpening. It is well to have a good supply of cutters on hand, where a bar with this style is to be kept at work.

Fig. 3 shows a bar fitted with adjustable cutters. The cutters in this case are single pointed affairs. The adjustment is obtained by the movement of the cone-pointed screw, which acts on the beveled surface on the inner end of the tools. The cutters are held in place by the set screws, as shown. To adjust the cutters, the set screws are loosened up a little, but still have considerable binding pressure on the cutters; the adjusting screw is then screwed inward, forcing the cutters outward an equal amount, so that each cutter still takes its share of the cut. This style will be useful where one is fitting a hole to a shaft or plug and where the holes vary in size a trifle in each separate case.

one would have to use to pass through a cored hole would be of such a size as to be very weak and springy. Fig. 4 shows a suggestion of a way in which the conditions may be altered so as to obtain a stiffer construction of the bar. The bar in this case is made with two sizes. The leading part, that is that

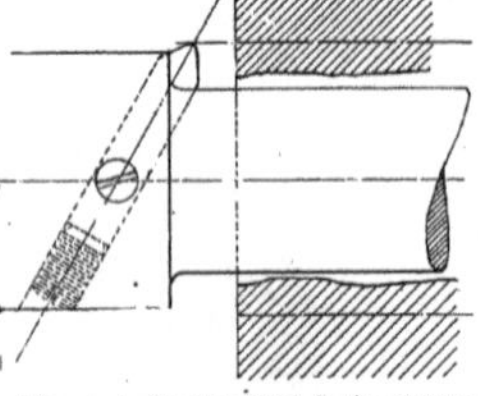

Fig. 1—A Single-pointed Boring Cutter for Small Cored Holes.

part which leads the cutter and has to pass through the rough hole, is made as small in diameter as is necessary, while the part of the bar that follows the cutter is made nearly as large as the hole when finished, leaving a small space for chips. This style bar will

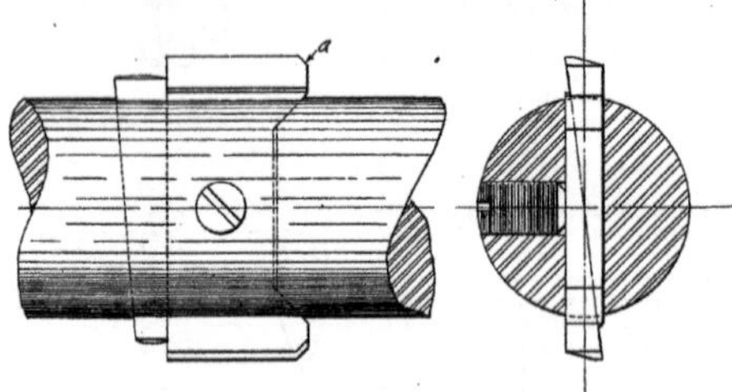

Fig. 6.—A Self-centering Boring Cutter.

cause trouble from the interference of the chips unless care is taken to keep the hole as clear of them as possible by the frequent use of the blower. As seen, the cutter is a singled-ended one, and is set at an angle through the bar, the cutting point coming through at the shoulder formed by the larger diameter. This cutter is adjustable for size, the adjustment being made by the movement of the screw set at the end. The cutter is held in position by means of a set screw as in the case with the cutters in Fig. 3.

When adjustable cutters are used and the job requires that the holes be of one size within certain limits, a gauge similar to that shown in Fig. 5 will be found to be a handy tool. This gauge, as seen, is set on the bar, and the cutter adjusted until it comes in contact with the under side of the projecting lip. The use of a piece of tissue paper or a piece of sheet steel which has been hardened, ground and polished in the setting, will insure greater accuracy. The paper or steel is used as a feeler to test the pressure of contact between the cutter and block. It is understood that settings cannot be transferred from one bar to another without the necessity of having both bars of exactly the same diameter.

When a bar is to be used to any extent, on one job, the inconvenience of the round cutters described in Figs. 1 and 2 losing their size so rapidly, is one well worth trying to overcome. This is done in the style cutter shown in Fig. 6. This bar is much more expensive to construct, but where the extent of its use is great, the extra trouble is soon paid for in the better results obtained.

The cutter used is made of flat stock, which may be of either the high speed variety, or of the ordinary carbon stock. It has been proved that the former holds its edge much longer in cast iron than does the latter. The cutter is let into the bar by the slot, which has the front end beveled at the periphen of the bar and the rear end is tapered to accommodate the wedge used to force the cutter home against the bevels. The bevels centre the cutter with the axis of the bar. The cutter is further secured by means of a set screw which passes against the side of same as shown.

The advantage of this style cutter is in the length of the cutting edge. The boring takes place at the front end at (a), and the cutter that follows smooths out the hole and acts as a reamer in bringing it to size. When the cutter becomes dull it is sharpened by grinding the face (a). The cutter is fitted to the bar and then turned to grinding size in the lathe, then backed off by hand with a file or by milling in a milling machine. It is then hardened and the cutting edge ground and stoned to a fine cutting edge. The front of the cutter is either shaped as shown, to give it a radial cutting face, or it may be left parallel with the blade, giving the cutting edge a negative rake, producing a scraping cut, which may be of advantage on some work, as, otherwise, the tendency is to chatter, due to the broad surface in contact. This style cutter produces results that cannot be found fault with.

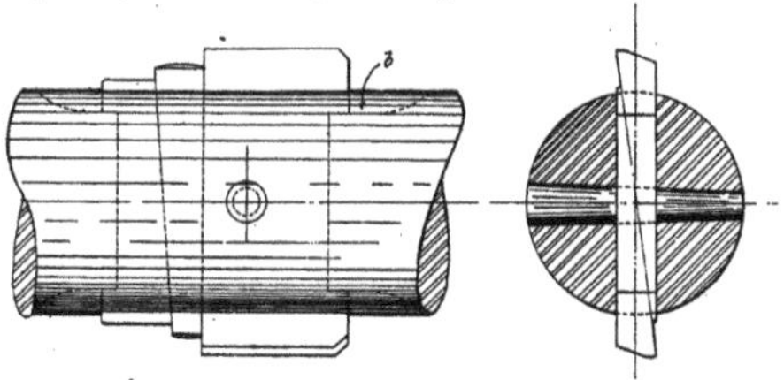

Fig. 7.—An Elaborate Design of Boring Bar with Flat Cutter.

Fig. 7 shows a variation from the style previously shown. There the slot is square with the sides of the bar. The cutter is held central by means of the shoulders formed by milling into the bar, as shown at (b). A double wedge is used, which forces the bar against the opposite end of the slot. The taper pin is used as a precaution in case the wedge should become loose. Some means might be made for holding the wedge in place, such as a set screw.

This design is much more expensive to construct than that shown in Fig. 6, but it has its admirers and is well worth investigating.

A means of holding the flat cutter is shown in Fig. 8. In this case the cutter is forced against its seat by the threaded collar. The collar is checked in position by the set screw, which presses the soft metal plug against the thread, checking any movement of the nut without injuring the thread on the bar. The centreing seat for the cutter is obtained by either filing or milling a flat on the bar at right angles with the sides of the slot. This design should not be any more expensive to carry out than that shown in Fig. 6, and has some points in its favor that do not appear in the former design.

Steel is not the only material that is used for boring tools. Sapphire is often used when boring babbitt and other soft bearing metals. The sapphire stone is set in a holder and is shaped as in Fig. 9, which is a view looking at the face of the point. The diamond is also used as a boring cutter and many are used to cut materials that are hard enough to ruin the best of steel tools. It is set in a holder in much the same manner as in the case of the sapphire point. Fig. 10 gives a suggestion of how it is held.

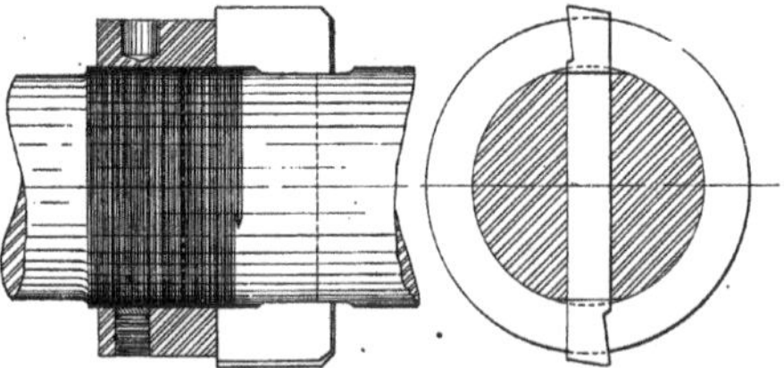

Fig. 8—A Simple and Reliable Cutter Bar.

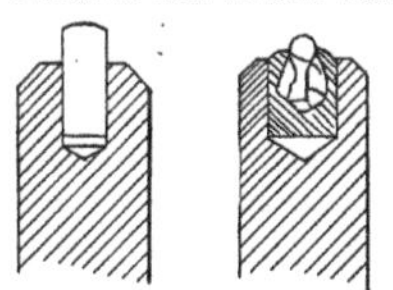

Fig. 9—A Sapphire Cutter-holder. Fig. 10—A Diamond Cutter-holder.

As mentioned in the beginning, boring is an art that may be made to attain a high degree. It calls for skill that would be but wasted on other operations. It may be said to be the king of machine shop performances. To be able to set up a difficult job and accurately bore a number of holes in same so that each one holds a proper relation with its neighbor and also with other important features of the main casting, is quite a trick. It becomes a fine art when the holes must be kept to a standard and when facilities are not of the best.

METHODS OF ANNEALING NOVO STEEL.

A great many times a machinist wants to use a piece of Novo steel for some special job, but, not having a piece that is soft enough to work, he uses carbon steel instead, generally because he does not know how to anneal high-speed steel, and imagines that an elaborate heating system must be used. I have met dozens of blacksmiths, tool-dressers and machinists, who declared that the thing was impossible, but who quickly changed front when shown how. Not long ago, a writer in one of the prominent mechanical journals, who said he had done nothing but worked with steel all his life, described his process of annealing Novo steel. He packed it in a tube or iron box in ashes, clay or charcoal, put it in the furnace, and kept it at 800 degrees F. for five hours, then at 900 degrees F. for seven hours, and then he let the fire go out. He removed his steel from the furnace twelve hours later.

This method seems rather inconvenient, however, and the writer would propose the following: If you have a piece of Novo steel that you want to soften so that you can drill, file or tap it, get a barrel of old slacked lime and a piece of pipe, threaded on both ends. Get a pipe large enough to allow the steel plenty of room. Put a cap on one end and put in about an inch of lime at the bottom, drop in the steel and pack lime around it, keeping the steel in the middle of the pipe, and fill the other cap with lime and screw on. It will be well to explain here that the whole secret of annealing this or any other steel is to keep it from coming into contact with the air and thereby chilling while in the process of cooling.

Next put the pipe and contents into the fire, and heat slowly and evenly to a white heat, then take out the pipe and bury it quickly in the middle of your barrel of lime and leave it twelve or fourteen hours. For ordinary sized pieces the time taken to heat will not be over twenty minutes, and if you have a good-sized forge, it will be considerably less. Be sure the steel is heated through before taking it from the fire. I have annealed hundreds of small pieces of Novo steel in this way, and have never yet met with a failure.—E. Viall, in Machinery, N.Y.

Patent Double Cone Roller Bearing

A new style of bearing patented by J. Dove Smith and F. E. Lauer, Niagara Falls, Ont.

By F. E. Lauer.

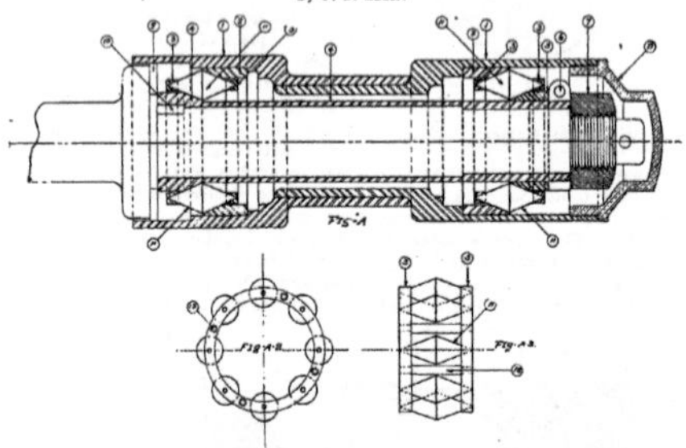

Patent Double Cone Roller Bearing.

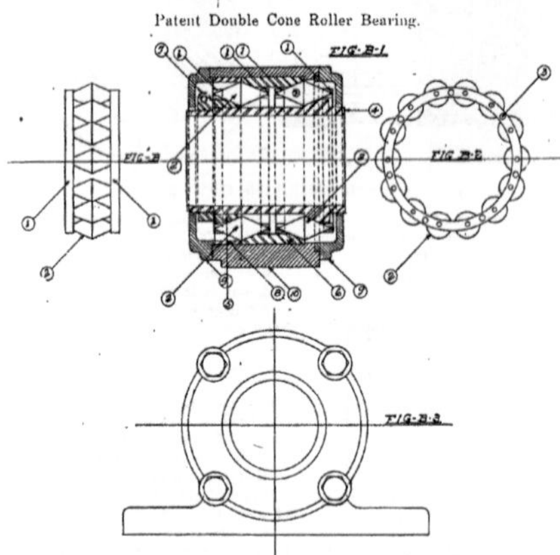

Patent Double Cone Roller Bearing.

The following is a description of this bearing as applied to a wagon axle and pillow block:

Fig. A shows the drawing of a wagon axle bearing, in which I, is the boxing which presses in the hub of the wheel, and is made of mallable iron; 2 is the outside band made of steel tubing, hardened and ground; 3 and fig. A-3, the cage rings which carry the rollers; 4 is a steel sleeve made the full length of the bearing, keyed a sliding fit to the axle end, and which carries the inside path of the bearing; 5 is the adjusting path, which is controlled by the locknut 6; 7 is a cast iron nut which allows the wheel and bearing to be taken off without disturbing the bearing, which is self-contained; 8 is a brass dust-proof cap; 9 is a fibre washer and used as a dust washer; 10 is the key which allows the sleeve to move laterally, but not around; 11 and fig. A3, are the rollers; 12. fig. a-2, and fig. a-3, show the tie rods which hold the cage rings with the rollers together.

The principle of the bearing is that the load bears on one side of each cone of the rollers, and as the opposite side of the cone of each roller does not touch anything, hence does not serge, when the wheel revolves it carries around the boxing 1-fig.-a, also the outside path 2-fig.-a, which in turn revolves the rollers which roll on the inside sleeve 4-fig.-a. The adjustment for wear is taken up by the lock nut 6, which engages the adjusting path 5. It will be readily seen that this bearing handles end thrust, or lateral motion perfectly, as the pressure is being always exerted against the paths of the inside sleeve, which are of the angle coinciding with the angle of the roller, and which demonstrates that lost motion will be taken care of mechanically itself.

Pillow Block Bearing.

Fig. B-3, and B-1, shows the plane and section view of a pillow block bearing, or truck bearing, in which fig. B-1 and fig.-B-7 are the cage rings; 2 fig.-b, and b-1 are the rollers 3 fig. b-2, are the rods, for locking cage rings and rollers together; 4 fig. b-1 is the inside sleeve and revolving path; 5 fig. b-1 is the adjusting path; 6 fig. b-1 is the outside path; 7 fig.-b-1, is the adjusting nut; 8 fig.-b-1, is the spacing piece to prevent lateral motion of outside path; 9 fig.-b-I, is the bearing caps; and 10 is the main body of the bearing.

The motion in this bearing is the reverse of the wagon, axle bearing, as 4 fig.-b-1 revolves with the shaft in a pillow block bearing, or with the axle of trucks, which, in turn, transmits the load through the rollers to the resisting and outside path 6 fig.-b-1. Fig. 1, is a plane view of the cage rings and rollers, and fig.-b-2 is an elevation view of cage rings and rollers. All working parts of these bearings are hardened and ground so as to insure proper contact and results.

*Preparing Work for the Grinding Machine**

A Treatise Showing the Practical and Economic Use of the Grinding Machine for the Finishing of Work in the Machine Shop.

By H. DARBYSHIRE.

The amount of stock to be removed by grinding is in the first place dependent upon the class of machine employed, and if this be of a light type the grinding allowance must be as small as possible and the previous machine-work must be produced by highly skilled labor.

Here we have the least economic side of grinding for, in addition to the higher cost of labor for previous work, the time taken in grinding by reason of the

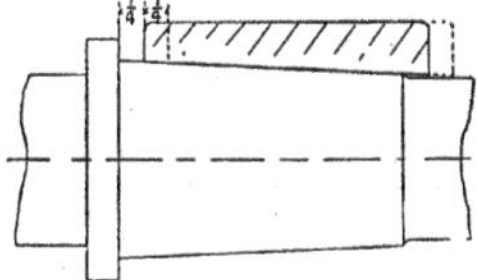

Fig. 1—Conical Spindle Neck.

narrow wheel-face and slow feed, is greater than otherwise necessary. Where, however, machines stand on a good foundation, thus preventing vibration, the actual grinding is done more quickly, and the question of cost of previous labor for the most part resolves itself into one of organization. A rapid means of grinding allows us to give the previous machineman a greater tolerance and to increase his output, or a cheaper class of labor may be employed. With the present day results obtained by the use of high speed steel, it is absurd to assert that the grinding wheel is its superior in removing stock, but the assertion may be qualified by saying that it is superior where the amount of stock to be removed is limited, taken in conjunction with the better results obtained. At the present time it is a disputed question among engineers as to whether high speed steel is suitable for finishing purposes, but even where it is successfully used the finish obtained cannot be compared for accuracy with ground work, and we may affirm that where good work is required, grinding must be resorted to. Given this, we may next consider what conditions are involved that affect the matter from the point of view of economy. Excessive grinding allowances mean a greater amount expended in wheels; fine grinding allowances entail a high cost of labor for previous work, and a risk of work not cleaning up to finished dimensions; therefore, most of the question depends on the class of labor employed in preparing the work, and the limit of error a man can be depended on to work, too; if we take .015 inch to .025 inch for a general grinding allowance where machines are powerful and rapid, we may be said to hit the medium between lowest cost of wheel and previous labor, and this latter would be very poor if it could not be trusted to work safely to this limit. Some departure from the limit given would be sometimes necessary for safety, more especially for hardened work of peculiar shapes, but the amount to leave here must be arrived at by experience.

Long cylindrical pieces are ground more cheaply and safely from black stock up to a diameter of about 2 inches, and even above that size it is much cheaper than turning and grinding if bars were straight and the amount to be removed not about 1-16 of an inch.

We may take a case to illustrate this: A black bar, 2 1-16 inches x 10 feet, would be reduced to 2.015 inches in half an hour in a modern machine, and finished outright in a total of one and a quarter hours by a man with some little experience; this same shaft, if it had been previously turned to 2.020 inches, would probably take the same or longer time, and the reason is not difficult to understand.

In the first case, the operator can attack the piece with confidence, because he knows that the heaviest cut he can take will not lead him into error, and while the main body of the stock is being removed, he is both making all necessary adjustments and obviating all the little precautions which are necessary but somewhat inexplicable when previous machinery has been thought necessary. On the other hand, where the piece has been previously turned, he is alive to the fact that this turning has induced little crooks or bends to develop, and (though hidden in the straightening necessary before grinding), they are liable to develop themselves at any moment, and so his depth of cut is so limited that it may be termed, on the whole, a kind of semi-roughing, and the caution which is necessary is an expenditure of time which had better have been utilized in substituting the grinding machine for the lathe. There should be a constant understanding between the grinding and machine departments, so that the work may be prepared in such a manner as to be adaptable to the grinders' facilities for holding or chucking it and this applies mostly to cylindrical work, and more especially when it is prepared in the turret lathe. For example, bushings that must be ground externally and have a clearance hole, should have this hole true within certain limits, and be bored to a standard size so as to accommodate the grinders' arbors.

It is a mistake to finish one portion of a piece of work in the lathe and the other in the grinding machine: errors in concentricity are the result, and the contrast between the two parts make it look like a slovenly job. This is sometimes excusable where the grinding is costly, as in the light type machine, but with a properly constructed machine it is far cheaper to grind even if the parts are finished for appearance only, and a better result is obtained.

Grinding a Taper.

In preparing taper work for grinding, errors which are the result of want of thought often creep in, and sometimes with disastrous results. Work which has been carbonized may often have so much stock left for removal (through want of thought), that all the hardened surface is ground away before finished dimensions are reached, and yet a little calculation of an elementary nature will quickly decide the right grinding allowance. Fig 1 is a conical spindle neck, and the ring gauge A must come to distance shown from collar when finished; the whole rise in taper is 1 inch in the foot and .020 inch-1.000 inch= 50, and the distance to let the ring

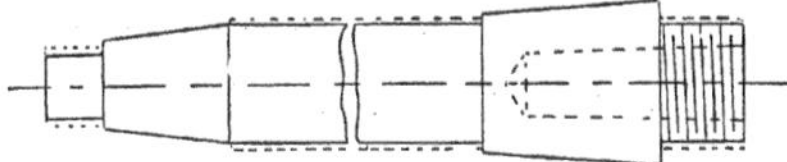

Fig. 2—Roughing Spindle for Hardening.

*Extract from H. Darbyshire's book on "Precision Grinding," published by Hill Publishing Co., New York.

gauge up in turning would be 1-50 of a foot added to the distance shown in sketch, or about ⅓-inch, roughly speaking. It is within the writer's experience that many lathe hands look on all tapers as being adaptable to the same grinding allowance, and let their ring gauge up to a certain distance, irrespective of the rise in the taper; it is a stupid error, of course, and would be prevented if shop drawings had these limits plainly marked.

The same may be said of cases in which special grinding allowances must be made for distortion in hardening or for depth of carbonizing, and, after experience has proved what is the amount to leave, the proper dimensions could be shown on the drawings. Some of these special cases are more a matter of supervision and of the laying out of work, and we may give an example.

Fig. 2 is a spindle which must be carbonized on its taper-journals only; the front end must be threaded and bored after hardening: if it were turned all over within a limit of .020 it would probably be bent so much that it would not clean up to size, nor would the front end be soft enough for the threading and borning tools; but if prior to carbonizing we leave a larger amount, as shown by dotted lines, and turn the journals only with a grinding allowance, we can first straighten on the soft portion in the middle or draw the centres, if necessary. and turn to grinding size, then rough grind all over and down to soft metal on the front end; after this we can bore, thread and finish grinding.

Care in Hardening.

The preparation of hardened work for grinding is a subject that will bear some careful consideration; nothing is more annoying than to find a piece of work over which much valuable time has been spent unfit for the purpose it was meant for after the hardening opperation. Work may be so bent or distorted as to require straightening, and then we have the ever present danger that it may resume its crooked form after still more time is expended in finishing; yet this is only too often the result of a want of some reasonable care in preparing the steel. It is, of course, impossible to foresee what is going to happen in any case when hardening is concerned, but much trouble may be averted by systematic annealing; a piece of steel as it comes from the mill or forge is always liable to contain hidden strains that can only be released by reheating. When a piece of work is made from bar steel it should be centred and straightened as true as possible. This is necessary, because the bar in its rough condition is decarbonized for an uniform distance below the surface. The straightening should not be done cold, as this only leaves opposite and antagonistic strains on two opposite sides, and these may release and cause warping in the hardening process. After the bar has been heated to a bright red, it may be straightened and then a cut taken over it to remove all the outer surface. It should then be carefully annealed, and if it comes out fairly straight, it may be machined to within the grinding allowance; if, on the contrary, it should be badly crooked, the annealing process should be repeated after it has been again straightened out. The object here in annealing is to make sure that all strains which are developed in the rolling of the bar are removed, and until this is done it is not safe to proceed with the work; and the reason we take pains to make the bar quite true in the first case, is to ensure that its surface is homogeneous.

It is necessary that all keyways be cut before grinding, and apart from the question of distortion and the damage which may be done by handling, it is another instance of the advantages accruing from grinding. We cannot finish work in the lathes where keyways are present, so they must be machined after the cylindrical finish; this springs the work slightly, and it will not run sweet without straightening, which sometimes means damaging the piece, and, in any event, entails extra care and time on the part of the operator; then the burrs must be removed from the keyways with a file, which disfigures the work and costs money for files and time. On the other hand, if we finish the parts by grinding we may do all this work previously, and the springing from keywaying is rarely more than what the amount left for grinding will cover, so that we save straightening, files and time on each piece a considerable economy, and we also get a far better finished result. What we mean by distortion in keywaying is, of course, due to the stress on two sides of a piece of work being disturbed through removing a portion of the metal on one side, the bend developing on the side opposite to the keyway.

Whilst on the subject of keyways, it may not be out of place to advise the grinder to look out for errors in any piece he may have to grind with a key inserted. If it is tight on the ends it will spring the piece over in the opposite way to that which has just been described.

Where there are shoulders with square corners they should be slightly undercut below the finished dimension or else squared out in the lathe after grinding; as it is preferable to have the grinding operation last of all to prevent danger of damage to finish, undercutting is better when possible, and for internal work where the holes are blind or have shoulders the same applies. Where there is a small radius the wheel will conform to its shape, and it may be finished before grinding; but when the radius is large it is more economical to get it out in the lathe. If we have to turn a large radius on a wheel, it imperils the diamond tool, and we have to waste a great deal of the wheel in bringing it back to its flat face. Hardened work and special cases may prove an exception to this rule, but they generally warrant the expense necessary in providing a special wheel and the time required for changing it. Where these cases do occur, the radius must first be turned good and as little material left for removal as possible, or the wheel will lose its shape.

Where large radii are to be very accurate and the material is not hardened, we may compromise by making a special fixture for finishing them with a spring tool in the grinding machine after the parallel portion has been ground, care being taken to catch the chips if large; this has the merit of saving time otherwise occupied in transporting heavy pieces back to the lathe, such as car axles, and lessens the danger of damage.

Have Standards of Grinding Allowance.

To assist the grinder as much as possible, it is desirable that some uniformity be present in the amount of grinding allowance where there are a number of duplicate pieces; it will save him much time in stopping to take dimensions and enable him to gauge his wear of wheel and make compensation for it in working.

The same uniformity in lengths will absolve him from the necessity of adjusting his stops or dogs for each piece, and so allow him to work more expeditiously, and it will lessen the danger of his running into shoulders to the damage of his wheel, the work, or himself. Both these points have some reference to an always desirable method of manufacturing, which may be much simplified by providing mechanical means which will ensure the requirements stated being easily obtained within close limits, and where possible it is well to adopt them.

In the preparation of cylindrical work for grinding or lathe work, the centre holes should be the first consideration; it is the foundation of quality in anything that pertains to accuracy. It is an instance of one of the inconsistencies in the engineering trade that the matter of centres often receives little attention, many who look for and demand good work seem to look on this important point with indifference; stabbing the end of a piece of work with a centre punch is a method frequently used and is little more primitive than the use of that abominable tool the square centre. When these methods are practiced, it is seldom that angles are kept uniform, because they depend on the individual even if gauges are provided; it costs no more to make a good centre hole than a bad one, and if proper tools are provided, it costs less; these would consist of properly made centring drills issued and controlled by the toolroom. The facing of the end of work when centred is also often neglected, though its omission is often difficult to understand; when it is not done, it is unreasonable to expect to turn out work which is round and true.

The grinding machine is not a machine that is likely to lead to the displacing of other machines, but rather to add to their utility or lead at most to a change in their design; it has superseded the engine lathe in many instances, but has made the high speed lathe and turret lathe more serviceable and desirable, the former, because the grinding machine will accept and finish work from it that has occupied it less time than was formerly the case, for let the surface be turned at any pitch or the surface be ever so irregular, it will remedy these defects more quickly and with better results than is possible by any other known means when accuracy of work is necessary. Used in conjunction with the turret lathe, it has developed the utility of that valuable machine and given an impulse to its design and manufacture which is entirely profitable to those who adopt it.

Many pieces may be quickly machined from plain black stock by using the turret lathe for making the larger reductions on the ends and leaving a safe margin for grinding, of which a few examples are given in Fig. 3. When locking threads are necessary on the ends they may be returned to the turret lathe to be screwed after grinding, and then any round corners left by the wheel may also be squared out if necessary.

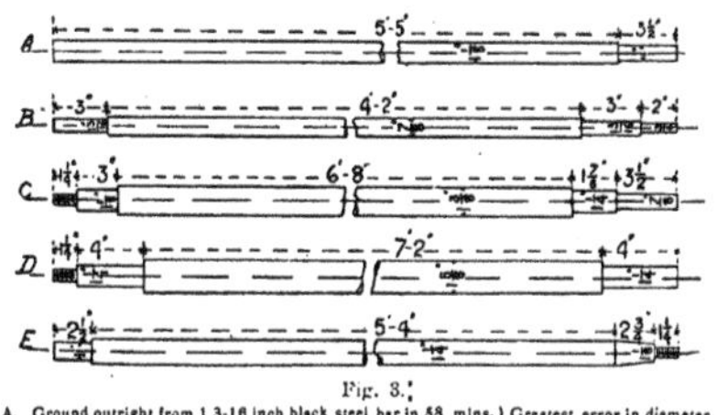

Fig. 3.

A. Ground outright from 1 3-16 inch black steel bar in 58 mins. } Greatest error in diameter
B. Ground outright from 1 5-16 inch black steel bar in 50 mins. } .0005 inch.
C. Material 1 7-16 inch black steel bar, end roughed down in turret lathe in 16 mins., afterwards ground to a limit of .0005 inch in 82 mins. Total time, 98 mins.
D. Material 1 11-16 inch black steel bar, end roughed down in turret lathe in 18 mins., afterwards ground to a limit of .001 inch in 70 mins. Total time, 88 mins.
E. Material 1 5-16 inch black steel bar, ends roughed down in turret lathe in 14 mins., afterwards ground to a limit of .001 inch in 62 minutes. Total time, 76 mins.

The grinding allowances for plane surfaces may generally be less than for cylindrical, and, in the case where there is no water supply to the machine, they must be so. For hardened work, experience alone can be the guide. Where the grinder has no magnetic holding appliances it is better to machine some small pieces in long strips and cut off to proper lengths afterwards, this, of course, where their shape will allow of it. Let all drilling, slotting, and chipping, or other operations be prior to grinding, and, where possible, contrive that the grinding shall be the final operation. With a properly constructed surface grinding machine, the milling machine may be utilized in a somewhat similar manner to that described for the turret lathe, and some work may be ground from the black if the casting or forging is good. Plain adjusting strips are an example of the latter, or light stampings or castings, where their shape will allow. All surfaces of castings or other parts that are finished bright for the sake of appearance are better done on the surface grinder than by any other means, and the result is a beautiful uniformity of finish, which is no small assistance to the selling of the finished product.

Grinding Cast Iron and the Alloys.

We cannot close this treatise without a further reference to cast iron. Grinding allowances for cast iron must never be too large, for we are compelled by the nature of the material to use a soft bonded wheel, and much waste of wheel is likely to occur if this is not watched. Fortunately it is not so liable to spring in machining as steel is, so that we can afford to work closer, nor is it so liable to tear in machining, but leaves marks with a very shallow depth. As an offset to this a close limit in preparing it for the grinding machine means a higher degree of labor employed, and the discrimination which is necessary for economic working must be more acute. There is a special advantage in grinding cast iron of a porous quality, for the act of grinding seems to close the pores and make the work more acceptable in regard to finish than any other method.

Copper, brass, and other alloys should have a fine grinding allowance; the cuttings have some value, more or less, and at present there is no method known of separating them from grinding residue that is of practical value, so they are better collected from the lathe or other machine which may be previously employed.

The grinder should receive his work ready straightened, so that he may not be tempted or compelled to use his machine for the purpose; or delay be caused through waiting for it. Let all hardened work have two grindings, at least, so as to release all hardening strains and to test the under skins before finally finishing. Improved methods of preparing work will make themselves apparent if all departments concerned will co-operate, and the very gratifying results in cheapening the cost of production by grinding methods will only be equalled by the more inviting appearance and accuracy of the collective details and the finished, built-up product.

GRANTED.

A Boston man who just returned from a hunting trip in Maine tells of an experience he had while there.

It was his misfortune to get lost one day, and, to make matters worse, just as a heavy rain set in. He started out in what he thought was the right direction and wandered on through the wet and darkness, finally stumbling on to a narrow road leading in another direction; this he took and presently came to a house. The place was all dark, but he went up and thumped loudly on the door. There was no answer, so he tried again. This time a window went up and a voice growled:

"Well, who is it?"

"A friend," the man replied.

"Well, what do you want?"

"I would like to stay here all night."

"Well, stay there, then."

And the window fell with a bang.

Ball Bearing Design*

A few of the more important results of tests on ball-bearings conducted by the well-known German engineer, Professor Stribeck.

By HENRY HESS.

Sliding bearings wear out by abrasion of the carrying surfaces. Ball bearings do not give out from wear and do not wear. They may be ground out by admitting grit, but that is as illegitimate a condition for ball bearings as it is for sliding bearings.

The only legitimate cause for the giving out of ball bearings is the stressing of their material beyond the limit of

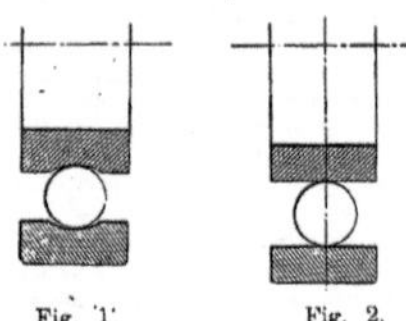

Fig. 1 Fig. 2.

proportionality. Lightly loaded bearings can be so designed as to eliminate this cause and so insure practical indestructibility. For heavily loaded bearings this condition is not realizable within practicable dimensions, but the proportions may be so chosen that the over stressing does not result in breakdown within the lifetime of any mechanism to which the ball bearing is applied.

A knowledge of the elastic qualities of the materials at the hardness under which they are used is imperative. It being the elastic behavior that is important with ball bearings as with all other egineering structures, tests of balls, such as are commonly made to determine ultimate rupture when pressed into a steel plate and using the depth of indentation of the plate, or load at which rupture occurs as a measure of ball quality, are not only of no value, but are misleading.

Equations for Load Capacity of Ball Bearings.

The quality of balls and of ball races must be determined from their behavior

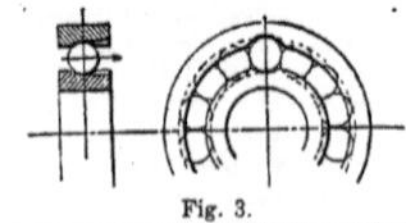

Fig. 3.

under loads in the neighborhood of the elastic limit. Balls may be subjected to loads increasing as the shape of the supporting surface more nearly becomes complemental to that of the ball, i.e., much better carrying capacity is had with a race as shown in Fig. 2 than with a flat one as shown in Fig. 1. This groove, naturally, must never have a curvature equal to that of the ball, since that would substitute sliding for rolling contact.

The frictional resistance of a ball bearing is lower, the less the number of balls. Usually bearings can be designed to have between 10 and 20 balls. The following equations have been deduced from experiments, which will be useful in designing bearings:

$$P_o = \frac{5}{Z} P_b$$

where

P_b = total load on a bearing consisting of one row of balls ;

P_o = greatest load on one ball ;

Z = number of balls.

The load carrying capacity of a ball is

$$P_o = kd^2,$$

where

d = diameter of ball,

k = constant dependent upon the material and the shape of the ball supporting surface.

From the first equation

$$P_b = P_o \frac{Z}{5}$$

and substituting for P_o from

$$P_o = kd^2$$

gives

$$P_b = kd^2 \frac{Z}{5}$$

Taking one-eighth of an inch as unity, the following table gives combination values of "k" for different shapes of races, both for the material formerly used in Germany for ball making and for the improved steel alloys used later. These constants give the value of P_b in kilograms.

Shape of Race	R=infinity	R=⅔d
k—for older materials	3 to 5	10
k—for imprv'd "	5 to 7 5	15

Here R=radius of the curvature of the ball race; and d=diameter of ball; where R is infinity the race is flat, as shown in Fig. 1. Fig. 2 shows R as 2-3d.

As an example of this, given a ball bearing with a race having a radius of two-thirds the diameter of the balls, having 15 balls each 1 inch in diameter; required the total load which can be carried by the bearing, the improved material being used in the balls.

$$P_b = kd^2 \frac{Z}{5}$$

Substituting values,

$$P_b = 15 \times 8^2 \times \frac{15}{5}$$

(notice that "d" is 8, since the unit is ⅛ in., and the diameter of the ball is 1 in.)

$$= 2880 \text{ kilograms,}$$

or

P_b = 6350.46 pounds, since 1 kilogram = 2.205 lbs.

If the centimeter is preferred as the unit for ball diameter, then the value of "k" in the above table must be multiplied by 10, or, more accurately,

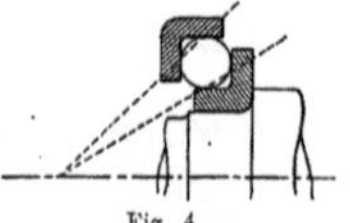

Fig. 4.

9.92 and 1 will be the unit when diameter of ball is given in centimeters, instead of ⅛ in., as when the diameter is given in inches.

Conditions Which Modify This Formula for Carrying Capacity.

Speed of rotation, in so far as it is uniform, does not affect carrying capacity. (This applies to radial bearings, but not to thrust bearings of the collar type; in these the carrying capacity decreases with increase of speed.)

But speed is rarely uniform; variations cut down the carrying capacity; sharp variations of small amplitude, particularly at high speed, have the more marked effect. Their reducing action is similar to the battering effect of sharp load variations.

Load variations reduce carrying capacity, the effect increasing with the amount of the load change and the rapidity of such change.

Accumulated experience with various classes of mechanisms is so far the only available guide for estimating the reductions in the constants k that must

* This report was translated from the German text in full by Henry Hess, and read before the American Society of Mechanical Engineers in June.

be made to take these influences into account.

The carrying capacity of a complete bearing is no greater than that of the weakest cross section that comes under the load. This applies to all those forms which have curved race sections of maximum sustaining capacity, except at a point where an opening is cut to permit the introduction of the balls; such bearings are, as to load carrying capacity, governed by the weaker cross-section at that point.

The calculated carrying capacity can be realized only if all balls sustain their share of the load. It is obvious enough that if a ball is smaller than those on either side of it, it will not carry its share of the load; should it be larger it will carry more than its share and may be overloaded. Uniformity of ball diameter is essential. The permissible variation in ball diameter will be governed by the deformation produced by a relatively small part of the total bearing load, so that the balance of the load may be distributed over the several balls. Such permissible variations of ball diameters amount to

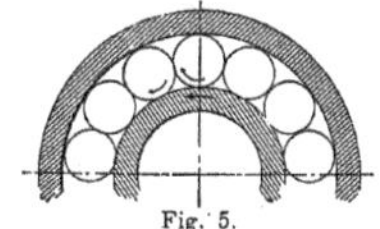

Fig. 5.

but little more than one ten-thousandth part of an inch.

High finish of both ball and ball sustaining surfaces is essential. The presence of grinding scratches will very materially cut down the cited values of the constant k. Of course this presupposes true surfaces underlying the high polish. It follows from this requirement that rust and acid must be carefully avoided, as they are destructive of finish and truth of shape.

It may not be amiss to point out that uniformity of quality of material, of hardness and of structure throughout are essential. The mischief of using balls having different values of k is not simply confined to the individual ball; if, for instance, one ball were materially harder and so deformed less than its mates, it would take more load and might, therefore overload the material of the race, which would yet be entirely suitable under a division of the bearing load among a larger number of balls.

A good ball bearing will have a coefficient of friction, independent of the speed within wide limits, and approximating 0.0015. This coefficient will rise to approximately 0.0030 under a reduction of the load to about one-tenth of the maximum.

Material Requirements.

Any material can be used that will not, under the working load, be stressed sufficiently beyond the proportional limit as to bring about its destruction before the lapse of a desired working life.

These conditions permit the use of practically all of the materials known to mechanical engineering. With very few exceptions, however, the load conditions are such as to demand steels of the highest grades and these most carefully tempered. For automobile use, no others can be considered. That puts out of the running all merely case hardened materials. In these there is always a more or less sharply defined change of structure at some distance below the surface. Continued working will cause a loosening of the hard shell from the softer core, soon followed by a breaking up and very characteristic flaking of the surface. Usually this flaking is local; its action is increasingly progressive, soon involving the entire bearing.

What has been said of case hardened materials holds true also for those carbon steels in which the hardening is not carried substantially and equally through the entire mass.

What is true and required of the ball material, is even more so for the races. With time, the ball presents its entire surface to the load, the small vibrations and changes of load being sufficient to frequently bring in a new axis of rotation. Not so the race; that is fixed and so always exposes the same surface element to the load attack.

Requirements of a Good Ball.

Truth of shape and size is the first requisite. All requirements will be met if the balls are true to shape within one ten-thousandth part of an inch, and if all the balls used in each individual bearing have a similar small error in size. It must not be inferred that for materials of lower grade, larger inaccuracies are permissible.

Surface finish to a very high degree is also essential. What is usually considered a very good finish, indeed, may be characterized as totally inadequate. The detection of grinding or polishing marks, with the eye or even with a pocket reading lens, condemns balls utterly, that is, for long life under high loads and speeds.

The elastic limit should be high; and the hardness and uniformity of hardness throughout the mass of the ball to the highest attainable degree is essential.

Correct knowledge and uniformity are more important than even these requirements of high elastic limit and hardness. It will not do to say that, though some balls of a lot may do better than others, the design may be based on the poorer ones. That would result in the better balls carrying more than their share of the load, much as and with the same bad effects described while considering truth of shape and size. Lower quality, provided it is uniform, can be allowed for. It will then merely affect dimensions.

Ball making machinery has arrived at a very considerable state of perfection; but balls within a limit of one ten-thousandth of an intended size are not yet being made without the sacrifice of other qualities. That is, however, not important beyond having some slight bearing on cost, since it is perfectly feasible to select and grade balls within the desired limit; but the hardware dealer's word for uniformity of size is not a safe guide; he is perfectly honest in throwing odd lots of ⅜-inch balls into one box and in thinking the customer who objects because they vary a

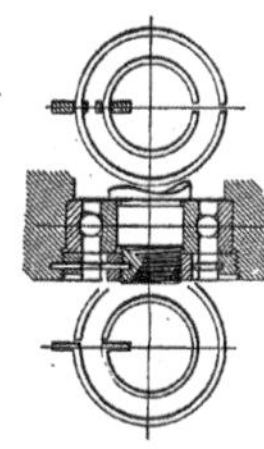

Fig. 6.

half-thousandth, or even two, a "finicky crank."

The Assembling of Balls in Bearing.

Cutting a local groove from the side into a race for the purpose of assembling the balls between the two races is general, as shown in Fig. 3, but is not good practice. If such cut is confined to one race which is then so held in the mounting as always to keep the opening at the unloaded side of the journal, this is at least defensible practice. The carrying capacity is then not decreased, as the load is carried by cross sections of maximum sustaining ability. Unfortunately, this demands the use of two differing designs; the one with the cut in the outer race, the other, with that placed in the inner race, according as the shaft or the housing rotates; the first case is the usual one, of an ordinary journal; the second is found in wheel hubs, etc.

Fig. 4 shows a typical four point contact bearing. In order to secure rolling, the contact points of balls and races should form points of a cone of rotation, whose apex lies in the centre line of the shaft, or they may form points on the surface of an imaginary cylindrical roller that is parallel to the shaft. The defect in all of these forms is their adjustable feature. This places them absolutely at the mercy of every one capable of wielding a wrench; a bearing that has been properly proportioned with reference to a certain load, will be enormously overloaded by a little extra effort applied to the wrench. Or the bearing may be adjusted with too much slack with consequent rattle and early demise. The prevalent idea that these bearings may be adjusted to compensate for wear is erroneous. Wear will form a groove on the loaded side of the race, deepest at the point of maximum load.

It is held by many designers of ball and roller bearings, that in such bearings adjacent balls or rollers are pressed one against another, with considerable force. The surfaces of the balls or rollers, roll in opposite directions, as shown in Fig. 5, and, therefore, with sliding friction. This is assumed to be a serious defect, by those who reason, that these surfaces contact under pressure. A cure often adopted, is the introduction of a smaller ball, between two larger ones. This method, however, is fallacious, as it is based on a failure to recognize an axiom in mechanics according to which a force whose direction is normal to the supporting surface, has no component in any other direction.

Correct Ball Bearing Mounting.

Ball bearings do not differ from other elements of mechanism, in that they must be used in conformity with their individual characteristics. A strict adherence to the following directions will result in a reliability far beyond that of any other form of journal:

(a) The proper size selection for the load must be made. Rated capacities are usually for steady loads and speeds. Variations from these conditions demand recognition by a suitable cutting down of the listed capacity.

(b) Bearings must be lubricated. The oft-repeated statement that ball bearings can be run without lubricant is pernicious.

(c) Bearings must be kept free of grit, moisture and acid. This prohibits the use of lubricants that contain or develop free acids. It is entirely practicable, by very simple means, to respond to b and c.

(d) The inner race must be firmly secured to the shaft. It is best to do so by a light drive fit, re-enforced by binding between a substantial shoulder and a nut.

(e) The outer race must be a slip fit in its seat.

(f) Thrust ought always to be taken up, whether in one or opposite directions, by the same bearing. That avoids all strains due to flexture of the shaft or of the housing, or due to temperature variation, and, while doing away with the considerable shop costs, inseparable from correct lengthwise dimensioning, avoids the danger of excessive end loads from forcible assembly consequent on an inaccurate lengthwise location of parts.

(g) Bearings should never be dismembered; or at least, never more than one at a time; that will avoid the danger of mixing balls from different bearings. Such balls from different bearings are apt to vary far more than is permissible for the individual bearing.

Fig. 4 will serve to show the arrangement of the ball bearings, and give an idea of a simple and useful locknut.

Particularly with machinery that is subject to vibration, it is necessary to lock firmly the nut that clamps the inner race. It is unsafe to rely on any form of a threaded lock; that also is likely to jar loose. Castellated nuts, etc., are unsightly and only permissible at the end of a shaft. The arrangement recommended consists of a split wire ring that is sprung into a groove turned circumferentially into the nut. A pin dropped into a hole drilled through the nut and partly into the shaft will be prevented from falling out by this guard ring. To permit of easily getting the pin out it is sometimes allowed to project and the ring is passed through a hole near the outer end of the pin. Not infrequently pin and ring are combined by bending the end of the ring over to act as the pin. The same arrangement is useful also when the nut is threaded on the outside to lock the outer race of a ball bearing, as here shown.

The ball bearing in its application to heavy work and serious engineering, is of recent development, for it is only in the past few years that material progress has been made. A great amount of work has already been done, testing materials, both chemically and physically, and there is still a wide field for research along this line.

Economy in Shop Suppjies

The Attention of Superintendents and Foremen is Directed to Several Sources of Loss in the Factory, and their Remedy.

By George D. MacKinnon, B.A. Sc.

The object of the writer of this article is to draw the attention of proprietors and managers of manufacturing establishments to one phase of the business which is very frequently overlooked, either because it appears unimportant or because it is difficult to get time to look into it. I refer to the use of shop supplies. such as oils, waste, belts, slushing compounds, boring and cutting compounds, emery wheels, coke, sand, coal, etc., and as it will be realized that in addition to other beneficial results, any economy in the matter of these supplies increases the net profits as well, it is something worth looking after. For instance, the economy in the use of coke in a foundry capola, when that economy is produced through a study of the processes in the cupola, or through some arrangement whereby the hot and otherwise escaping gases are utilized to assist in the melting of the iron, generally better and hotter iron is the result, the castings are more easily machined, less power is required to drive the blower, less work for the men in charge, etc., the resulting economy is not only due to the saving in coke, but is the total of all the indirect benefits. This is not only true of coke, but of practically all other shop supplies. Take the case of emery wheels. An emery wheel specially selected for a certain kind of work cuts far faster, lasts longer and requires less time for toning up—and this latter item is often a large one—when the wheels are used in special grinding machines, such as the Norton plain grinder or the Sellers' universal tool grinder.

Belt Drive.

The matter of belts is sometimes given very little consideration, it being taken for granted that belts, must break occasionally and must be fixed, and it does not seem to enter into the minds of some that one special method of joining a belt may be better than another, or that another brand or quality of belt might give better satisfaction than the one in use. Neither is the cost per year of belts for a certain machine often figured up; while in a great many cases this expenditure for new belts and for time spent and lost in fixing them up when they break, would be almost entirely saved through some other system of driving.

The writer has in mind at the present time, one mill in a machine shop for which the yearly cost in belts is upwards of $100, while the cost of the time lost, etc., in repairing these belts almost equals the first cost of the belts, or a total for one machine of from $150 to $200. How much better it might be for all concerned to do away with these troublesome belts and install another drive, the interest upon which investment would be only a fraction of the yearly cost of the belts.

Oil Consumption.

One of the large steel plants in the United States was using in connection with its machine shops something like 800 gallons of machine oil per month. The question of economy in the use of this oil had never entered the head of the superintendent, as he, like many others, believed, since he had a good, general system in force in the shops, that this detail was being looked after as it should be. However, while speaking to the agent of the oil company during one of his periodic visits, the question of the quantity of oil used was brought forcibly to his attention by the agent asking him if he had ever attempted the regulation of the supply. The superintendent replied that he did not consider that the supply was very great, considering the number and size of the machines, and the amount of work turned out, but as he was one who believed that a dollar saved was a dollar earned, he instituted an investigation, the results of which were so gratifying that other investigations followed with equal success. The investigation of the oil supply was begun on an exceptionally heavy 12 ft. plainer. All the principal bearings were fitted with sight feed oil cups, which were regulated by the machine tender, and observations were made each day for a period of two weeks, records being kept of the quantities of oil issued to planer, number of drops per oil cup per minute, and actual time machine was in use. It was found that as much as four (4) quarts of machine oil were used some days, while one and one-half quarts sufficed on others—the average being about two and nine-tenths (2 9-10) quarts per day. With judicious use it was found that one and one-half (1½) quarts was ample—the planer being run double turn—and the supply from the tool room was consequently limited to this amount.

General observations were also made of the quantity used on each machine in the shops, with the result that a considerable saving was found possible. For example, some boring mills which were running almost constantly, were using from three to four quarts per day, while others used only one and one-half quarts. One large ingot lathe. which had probably six changes of work per day, used four quarts, while another doing exactly the same work, but a newer machine, used six quarts. It can be inferred that this was not an economical use of oil, and the quantities were reduced to one and one-half, and two quarts, respectively, on which quantities the machines continued to run satisfactorily. The oil supply for each of the more important machines was regulated and table No. 1 gives comparative results. This table shows the quantities of oil used during the four months preceeding and the four following May 1st, when limits were put into effect. No data is at hand for the period previous to January, 1901, and 1903, is not included, because there had been in that year great increases in plant, while during the period given the amount of work done was practically constant from month to month. A saving of 27.22 per cent. is, of course, worth considering, amounting as it does in this case to practically $450 per year. All oil was issued from the tool rooms but there was no idea of limiting the supply to the detriment of a machine, and if the supply at any time fell short of the amount required, an additional quantity could be obtained on an order from the foreman—but, as a rule, these orders were very few.

A	JAN.	FEB.	MAR.	APR.	TOTAL	AVERAGE
1901	814	710	922	805	3251	812¾
1902	625	530	641	570	2366	591½
DIFF.	189	180	281	235	885	221¼
B	MAY	JUNE	JULY	AUG.	TOTAL	AVERAGE
1901	712	720	626	705	2763	690¾
1902	650	644	665	640	2599	649¾
DIFF	62	76	39	65	164	41
GAIN B-1901 OVER A-1901 = 488 GALS. = 15.01 %.						
" A-1902 " A-1901 = 885 " = 27.22 "						
" B-1902 " A-1901 = 652 " = 20.06 "						

Table No. 1.

Slushing Compounds.

Another supply used to quite a large extent in many shops is slushing compound. In many places this is a mixture of white lead and suet made by rendering the suet and mixing in the lead while hot, adding a quantity of machine oil to thin the mixture, if necessary. This makes a very good though very expensive compound. A much cheaper article may be made by mixing in suitable quantities, cylinder oil, asphaltun paint and Japan dryers. This may be made for about 70c per 1,000 sq. ft. slushed, against $2.60 for suet and white lead, and in the run of a year the saving is very considerable.

Emery Wheels.

As should be the case in connection with the use of almost all shop supplies, accurate records should be kept of all emery wheels, of the total time each wheel is in use, the number of pieces or tools ground—if on special work—the kind or hardness, etc., of articles ground and the grade, make and general character of the wheel, while in use. Some wheels require to be dressed much more frequently than others—some on account of being too soft, and some by reason of their glazing and refusing to cut. One instance occurs to the writer where 9-inch. dia. x ½-inch thick wheels, were being used by the dozen for nicking the cutting edges of heavy boring cutters. It occurred to mind that they must be of very poor quality, even though the men using them considered them o.k. A change was made, and instead of from 40 to 50 tools ground by each wheel, the number went up to 110 and 120. While with a wheel

¼-inch thick the number increased from 10 and 12 (much larger tools) to each wheel to 65 and 70, and this latter number with only about ¼-inch worn from the wheel.

A wheel ¾-inch thick, which lasted on an average one week, was replaced by one of another grade, which lasted five weeks. By some experimenting, the number of tools ground on a No. 1 Sellers' tool grinder was increased one-half by a change in the wheel, and as the wheels for these machines are very expensive, a considerable saving was effected. Other results might be mentioned, but these will show the advisability of looking into the matter of the emery wheel supply.

Cutting Compound.

In many places oils are used for bolt cutting, turret lathe work, boring, drilling, etc., while in almost all cases one of the many cutting compounds now on the market might be substituted, and a considerable saving thereby made.

A change was recently made in the construction of the fire places in a foundry core oven, with the result that at an outlay of less than $20, from $400 to $500 per year has been saved as a result of the decreased amount of coke necessary.

Most of the economies effected in a manufacturing establishment are due to the keeping of accurate records, and in the case of shop supplies, as in other cases, a competent manager is able to tell very quickly from a perusal of these records, whether or not the quantities used are too great, and may, consequently, take measures to regulate the consumption. Without such a record, however, he is not in nearly as good a position to judge, with the result that much is wasted, the cost of which, with the addition of the indirect benefits, would otherwise increase the net profits of the company.

Electric Power from Mouth of Coal Mine

The new power plant of the Maritime Coal & Railway Co., at Chignecto, Mines, N.S., is unique in many ways. Some years ago Thomas A. Edison, when in England, predicted that the ultimate solution of the problem of cheap power was the utilization of the waste products of the coal mine and the situation of the power house at the mouth of the mine, when the necessity of expensive transportation of coal over long distances would be eliminated.

This idea has been fully carried out by the Maritime Coal & Railway Co., and it is of great credit to Canada in general, and Nova Scotia in particular, that the first adoption of Mr. Edison's idea has taken place in the oldest Province of the fair Dominion.

In developing power, the first great saving is in the use of fuel which would otherwise go to waste. As there are no vertical shafts at the mines, the coal is hauled up a steep slope in cars, which are pulled by a cable. The cars hold fifteen hundred pounds each, and there are six cars to a train. When the cars reach the surface they continue on their journey up a slope to the top of the bank head. Here a complete system of tracks and switches places each car exactly where it is wanted. Its contents are automatically dumped into a series of screens and hoppers, where the material of commercial value is separated from the waste screenings or "culm." The good coal works its way to the railway cars waiting to receive it. The culm is dumped on an endless conveyer, which carries it to the power-house, 300 feet away. Here it is deposited in a huge bin above the boilers, which has a capacity of 85 tons. From this bin chutes, through which the coal is carried, lead to Jones' Underfeed Stokers, which automatically feed the fires. The boilers are four 200 horse-power return tubular, arranged in banks of two each. They supply steam at a pressure of 150 lbs., and the arrangement is such that any two can supply steam to the engine. A Cochrane open type feed water heater is provided, and the induced draft system, consisting of two blowers, complete with engines, supplied by the Canadian Buffalo Forge Co., has been installed. This does away with a high smoke stack, one only twenty-five feet high being used.

Water is supplied to the boilers by a pond made by the damming of a small stream. The pond is 1,000 feet away from the power house, and the water is conveyed by gravity to a pit underneath the boiler room, whence feed water and cooling water for the condenser is taken.

The power house is built of brick and is 75 by 50 feet, and is divided into the boiler and engine rooms. The engine room is 30 feet wide and runs parallel to the boiler room.

The engine is a 17-inch and 33-inch x 16-inch 750 h.p. Robb, Armstrong centre crank cross compound vertical condensing engine, of the English high-speed type, using the pressure oiling system, which carries oil to all bearings continuously at a pressure of 10 to 15 pounds. This is a new departure for Canadian engine builders, and has been successfully carried out by the Robb Engineering Co., Amherst, N.S., for the past year, a great many installations of this type of engine having been made throughout Canada by the company.

Directly connected to the engine is a 500 k.w. 11,000 volt three-phase alter-

Maritime Coal, Railway & Power Company, Chignecto, N.S.—Panoramic View.

nator, built by the Canadian Westinghouse Co., Hamilton, Ont. On the same shaft as the generator is the small direct current 125 volt exciter, for exciting the generator fields, making a very compact and substantial unit.

Power From the Mine.

Provision has been made for a future unit when the demand for power warrants its installation.

At the end of the engine room the switchboard is placed, which at present consists of a generator panel containing an 11,000 volt oil switch, with voltmeter, ammeters and wattmeters. There is also provided a synchronizing outfit to be used when the second unit is installed. The line is also fused with a set of high voltage fuse switches, and the whole equipment is protected by lightning arresters and choke coils of the standard type, manufactured by the Canadian Westinghouse Company.

The transmission line extends to Amherst, a distance of 6½ miles. Several long spans have been necessary in crossing the tidal rivers, varying from 400 to 700 feet. To cross these two poles were set at either end, guyed by steel ropes. The transmission wires are No. 4 B & S guage hard drawn copper wire supported on 15,000 volt glass insulators.

In general, the poles are 125 feet apart. Provision has been made for running a second line, as room for three more wires has been left on the cross arms.

At Amherst, a brick sub-station has been erected two storeys high. The upper portion is occupied by the three Westinghouse oil-cooled, self-contained transformers, which are used to reduce the voltage from 11,000 volts to 2,400 volts. Here also are located the fuses and lightning arresters. The secondary circuits are carried to the ground floor, where an oil switch is installed.

From the sub-station the power is distributed locally to the various consumers in Amherst and vicinity at a pressure of 2,400 volts.

To date, 400 horse power in motors has been installed, and it is expected that more will be in use in a short time. It is hoped that this will be the means of bringing many more industries to Amherst and vicinity.

Maritime Coal, Railway & Power Company—View of Generator.

Maritime Coal, Railway & Power Company—View of Boilers and Tubes for Conveying Fuel by Gravitation.

Power From the Mine.

The largest consumers of power are the Robb Engineering Co. and the Rhodes, Curry Co., Limited, who have found that Central Station power is the most satisfactory solution of the power problem, and, when all charges for the isolated plant are fully accounted for, is cheaper and infinitely more satisfactory.

Power generated under such conditions cannot be affected by shortage of coal from the mine, strikes or any other cause, as the refuse from the mine is sufficient to carry the power house over a very long-protracted strike when it would be impossible to ship coal to outside points.

So this company, in embarking on a project comparatively new to the world, and particularly new to the American continent, has, besides having ample cheap power for its own use, found an unlimited market for supplying power to distant centres. This provides a means of utilizing at a good profit what would otherwise be waste culm, and might occasion expense for removal.

The consulting engineer, in connection with the installation, was Julian C. Smith, C.E., general superintendent of the Shawinigan Water & Power Co., Montreal. The mechanical engineer was Mr. Philip A. Freeman, of Halifax, and to the wide experience of these men in power plant practice is largely due the success of this undertaking.

Machine Shop Methods and Devices

Unique Ways of Doing Things in the Machine Shop—Readers' Ideas and Opinions Concerning Shop Practice—Useful Data for the Machinist. Remuneration Will be Given for Contributions.

TAPER TURNING.

By J. H. R., Hamilton.

Here are a few rules which are very useful when turning tapers: Always have cutting edge of tool level with the centres of lathe or boring-mill table, otherwise there will be an error corresponding to the elevation or depression of the tool as shown in Fig. 1.

Tapers are either a cone or frustrum of a cone, and the only straight line that can be drawn on a cone is from.

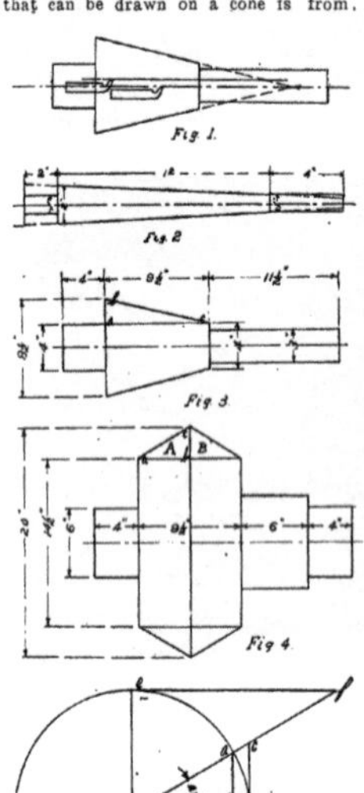

the apex to the base, therefore the cutting edge of tool must travel on this line.

If the taper is a certain amount per foot, the tail stock should be moved over one-half the taper per foot (where no taper attachment is used). Example: The taper is ¼ in. per foot; move tail stock over ⅛ in. per foot.

When taper does not run full length of work it must be calculated the same as though the taper was the whole length, as in Fig. 2.

Taper 1 in. in 12 in., therefore, in 18 in. would be 1½ in. taper.

Where taper is too quick for tail stock adjustment, the compound rest must be used. Sometimes the angle is known, at others it must be found.

In Fig. 5 the trigonometric functions are shown for the solution of triangles.

a o b is the angle.
a b is sine of angle.
o b is cosine of angle.
c d is tangent of angle.
e f is cotangent of angle.

Rule 1—Sine $= \dfrac{\text{side opposite}}{\text{hypotenuse}}$

Rule 2—Cosine $= \dfrac{\text{side adjacent}}{\text{hypotenuse}}$

Rule 3—Tangent $= \dfrac{\text{side opposite}}{\text{side adjacent}}$

In Fig. 3 angle b is required. Using Rule 3:

Tangent of angle b $= \dfrac{9.5}{2.5} = 3.80000$

Looking in table of tangents for number 3.80000, we find 3.80276 opposite angle 75° 16', which is the angle to place the compound rest.

In Fig. 4 angle c is required. Using Rule 3:

Tangent of angle c $= \dfrac{4.75}{2.75} = 1.72727$

Looking in table for number 1.72727, we find 1.72741 opposite angle 59° 56', or, say, 60°, which is the angle to place the compound rest.

As triangles A and B are similar, move the rest to the same position right or left.

Table of sines, cosines, tangents and cotangents for even degrees, can be found in "Data Arranged in Handy Form" in the September issue of Canadian Machinery.

AIR TANK AND OPEN HEATER.

By R. Manly Orr.

Some time ago my firm added to their power plant a 50 horse power gas engine. Now an engine of this size is too large to be started by spinning round the fly wheels by hand, as is the practice with smaller machines, and, consequently, one of the accessories furnished with the engine was a steel air tank constructed for a pressure of over 200 pounds per square inch, and to be used for that special purpose. After installing the engine and connecting up the tank as shown in the accompanying cut, I found that with the two valves, A and B, closed, the pressure would go down considerably during the night, and, the pressure necessary to start the engine being 80 pounds, I found myself almost unable to start the engine on the remaining air pressure on several mornings. I found it impossible to keep the small ¼-inch fittings between the tank

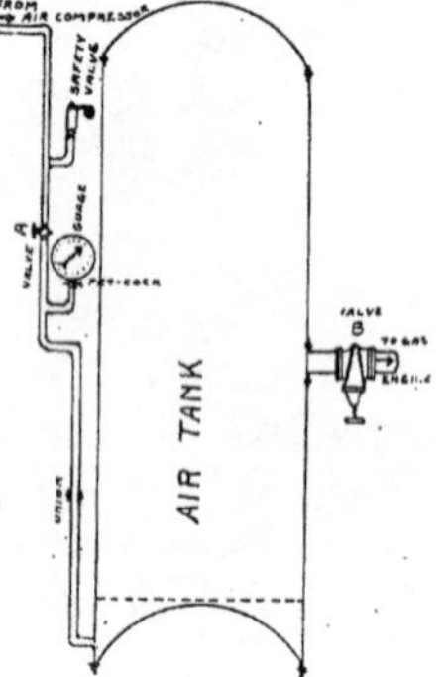

Oil Used to Prevent Lowering of Pressure.

and valve, A, from leaking, under the 200 pounds air pressure, and I experienced considerable trouble with them until I thought of the following plan.

I put about one gallon of ordinary machine oil into the bottom of the tank, filling it up to the dotted line in the cut. Of course, when the air compressor had finished pumping up the pressure, and the valve, A, was shut, the oil backed up the pipe, filling it to the valve. Where the air would escape, the oil would not leak a drop. Now I can pump my tank up and let it stand over two nights and a Sunday and the pressure gauge will not go back a single pound.

Open Heaters.

I would like also to add a point in connection with the piping up of an open heater, that experience has taught me. Many engineers are in charge of open heaters into which an overloaded engine exhausts; that is, an engine which, for more or less of the time, takes steam full stroke without cutting off. I have charge of such a heater and engine at present and well know what it means to have all the packing in the filtering chambers torn to pieces and piled up in a heap at one side, with the other side of the perforated floor of the chamber left bare. I have put weights on the packing and wired the top plates, but to no avail.

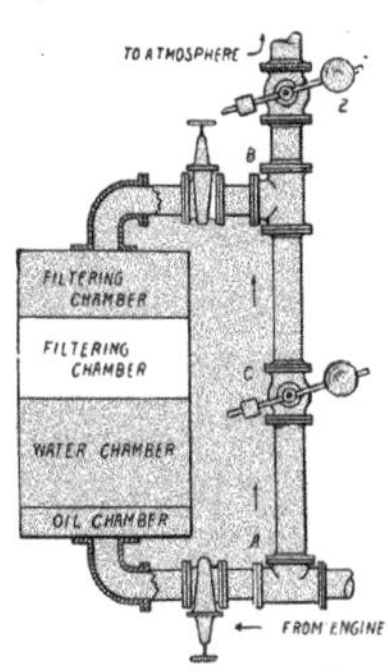

Arrangement of Piping and Valves.

When the exhaust valve opens a cylinderful of steam at boiler pressure bursts into the heater, and the result is somewhat similar to an explosion. The accompanying diagram gives an idea of my own which I believe is the best cure for this trouble; that is, except a bigger engine. The idea is to run with the valve, A, open and the valve, B, almost shut. B should be so nearly shut that a big rush of steam could not pass through the heater but will be by-passed through relief valve C. It will be easily seen that the easiest escape for the steam is through the heater, only one relief valve being on that route, thereby insuring a sufficient supply of steam for feed-water heating. The two relief valves can be weighted as the engineer sees fit. By closing A and B the heater can be cut out.

GANG-MILLING AS COMPARED WITH PLANING.

By C. S. Gingrich, M.E.

The illustration, Fig. 1, shows a machine part of very simple form and well adapted to setting many pieces at a time on the table of a planer for machining. It is the sort of a job on which a first-class planer hand would expect to make good time—in fact, these pieces have been planed repeatedly on a 36 in. x 36 in. planer, using two heads, running 55 feet per minute, finishing the two sides and the top, and at the same time, the slot, ⅝ in. wide by ⅝ in. deep, in 36 minutes each.

It was afterward decided to put this work on a miller. Fig. 2 shows how the job was set up. Two pieces were placed side by side. The machine was fitted with a gang of cutters, 8¼ in., 4½ in., and 3 in. diameter, high speed steel. The cut was approximately 3-16 in. deep all over, and in addition there was also the

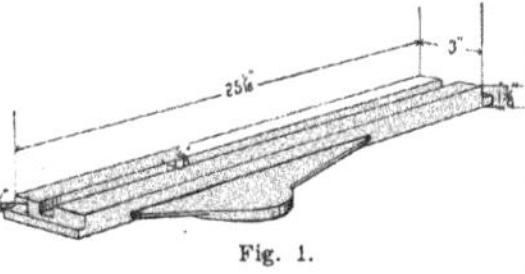

Fig. 1.

⅝ in. x ⅝ in. slot to cut from the solid. The speed of the cutters was 38 r.p.m., and the feed .1 in. per turn of cutter, which resulted in a table travel 3.8 in. per minute. The total time for finishing the two pieces, including the chucking, was 18 minutes, or 9 minutes on each piece.

The miller was driven by a direct-connected motor, so that it worked up to its full efficiency at all times and removed the metal at the rate of nine cubic inches per minute.

The slot was required to be accurate as to width, and this necessitated the use of a special adjustable planer tool, which could be kept up to size after repeated grindings, and this, together with the regular tools required, brought the maintenance of the planer tools up to about the same expense as the maintenance of the milling cutters. It will be noted that the milling cutters worked at a moderate speed, at which they can be used for a long time without being sharpened.

UTILIZING THE EXHAUST HEAT OF THE GAS ENGINE.

Very few of the users of small gas engines realize the vast amount of heat that is carried away in the exhaust. The gases in the cylinder easily reach a temperature of 1,500 degrees Fahrenheit, and fully one-third of their heat is wasted. Steel manufacturers are now utilizing blast furnace gas, which formerly went to waste. Why is it not just as practical to use the hot exhaust gases from an engine cylinder?

The man who has a small pattern or machine shop, deriving his power from, say, a 20 horse power gas engine, can economize materially on his winter's fuel bills for heating. Usually he keeps three or four large open stoves going all the time, fairly eating up the coal, and only heating the shop in spots. If he will use his exhaust, he can obtain an even temperature all over the room,

Fig. 2.

No. 4 Plain Cincinnati Miller removing 9 cubic inches of grey iron per minute.

and, at the same time, not burn any solid fuel.

If the shop is not too large and spread over too much area, it can be piped for hot water, with the circulating pipes running into a coil. This coil is placed within a metal drum, or box, properly protected with asbestos or some other heat insulator. If the engine exhaust is carried into this box or heater, it will heat up the water in the coils, thus starting through the pipes a circulation which will be constant. After the gases have become cooled in the heater, they can escape to the atmosphere through a waste pipe.

In a small pattern or woodworking shop, where the size of the engine is too small to be used for heating purposes, it can be utilized for a number of things, for example, the heating of glue. To accomplish this result, lead the exhaust into a larger pipe or water-tight box surrounded by the glue pot water. The hot gases will pass into this space, give up their heat to the water, and, finally, escape to the atmosphere through the outlet.

There is another advantage in thus using the waste heat. When the engine exhausts into a large box or receptacle, placed as near as possible to the engine, it acts as a muffler and silences the explosion without producing any back pressure in the cylinder.

The examples enumerated are but two of several uses to which the heat of this waste gas can be put by adapting apparatus to each local condition. It can even generate steam if a suitable boiler is attached. By utilizing this heat, men who previously were barred from using the gas engine on account of the necessity of having steam for their work, can install gas engines in their plants, and at the same time have steam for drying, seasoning, or any other use, without maintaining a separate boiler.—A. K. Reading, in American Machinist.

THE NECESSITY OF A PERFECT SYSTEM OF TOOL GRINDING.

By B. Pavey.

Few employers take notice of the great amount of time wasted by employes around the emery wheel; in the first place, waiting for their turn for the use of the wheel, and, secondly, the time taken in grinding tools.

Have you ever thought of the hours a three dollars a day man spends in grinding? Of course what I am speaking of is in regard to grinding tools for smaller lathes. A shop with which the writer is acquainted has a system as near perfection as it is possible to be obtained. Practically all work is turned with ½-in. square tools in holders, with a standard shape of tool. Each lathe hand is supplied with two, a right and a left, as shown in sketch. A round nose tool is supplied as well, if requested, but very few men are found to use them.

Steel is cut off in four-inch lengths by the smith in large quantities; a man

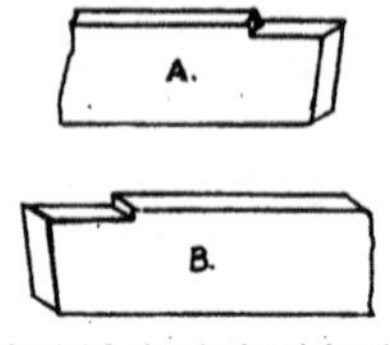

or boy grinds these in shape before they are hardened, so that after hardening they are merely touched up on the wheel with no danger of the temper deteriorating.

There are always tools in a finished condition kept in stock, so that a lathe hand, instead of grinding tools, takes his two tools to the toolroom window and is supplied with two fresh ones by a boy who looks after this part of the business.

By grinding the tool as shown in A and B it may also be used as a side finishing tool on almost all work and for roughing down there is not a better shaped tool.

The point of the tool will stand, being dead hard, and by setting the tool at an angle, a first-class finish is easily obtained.

SECURING PUNCHES IN THE PLATE.

In making multiple-punch tools, it is often necessary to locate the punches as close as possible to prevent handling the

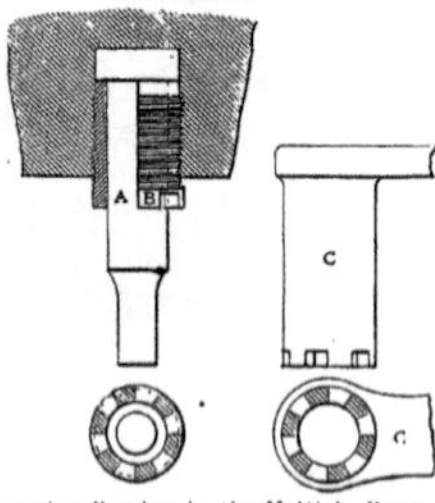

Securing Punches in the Multiple Press.

work more than once. The design shown is one I have found to be most efficient on heavy work.

The punch A is held in place by the threaded sleeve B, which is tightened up with the special wrench C.

This arrangement also permits of any punch being removed or replaced without disturbing the others.—G. Riley, in American Machinist.

HOW JOHNNY SUCCEEDED IN GETTING AN INCREASE IN WAGES.

Johnny had been waiting for nearly a week to catch the boss when he was feeling good-natured to "strike him" for more pay. One morning the boss came around feeling in the best of spirits, and Johnny promptly took advantage of his opportunity. "Let's see. how many times have I given you a raise during the past year?"

"Twice," said Johnny.

"You are now receiving $2.50 per day; don't you think that is pretty good pay for a boy that has been 'out of his time' only a year?"

"No, sir; not when I notice that you pay green men $3.50 per day, and after they spoil the job you discharge them, and turn the job over to me to make. If a new man is worth $3.50 a day to spoil work, I surely must be worth $2.75 to satisfactorily complete the job."

"Those men you speak of hired out as first-class men; they lied to me and were promptly discharged, and I fail to see wherein their inefficiency has anything to do with your claim for more pay. What you need is more experience, and then more pay will be forthcoming."

Said Johny: "My argument is, that it is not experience that counts so much as one's ability, for in this particular case these men have certainly worked at the business longer than I, but were unable to do the class of work that I am doing. For instance," continued Johnny, "suppose that the very best tool-maker in the United States should come along here and hire out for $2.50 per day claiming that he had not been at the business very long, how long would he be obliged to work here to receive $3.50 per day? How often would you raise him—every six months, as you do me?"

"If he was worth more money I should promptly give it to him."

"Well, cutting out this 'experience' part of it and talking from the standpoint of a man's worth, don't you think that I am worth $2.75 per day?"

"Well, I guess you are, Johnny. I'll start you next week at $2.75."

S. E. F., in Machinery.

Gas Engine Pistons and Piston Rings

By Gordon A. Ronan.

Nothing, perhaps, about a gas or gasoline engine is of such vital importance to its successful operation as to have well made, well designed and properly fitting piston rings.

A gas engine may be well designed, and care taken to select only first-class materials for its manufacture, and yet if these essential features, viz., properly fitting piston and piston rings, are lost sight of, a loose, faulty compression will follow, with a resultant loss of efficiency in the power development of the engine room when it is completed and in operation. This failure to realize the importance of these moving parts in a gas or gasoline engine has caused many small manufacturers no end of trouble, when their products were assembled and ready for trial.

Attention should be drawn to the fact, also, that not a few makers believe, and adhere to, the practice of making a piston trunk, just a neat working fit for the cylinder bore, having the walls of the trunk parallel their entire length. Now, that this is wrong, will at once appear obvious, when we consider the element of "heat conditions" which enters into all internal combustion engines and which is so often lost sight of. Hence, heat plays an important part, and calls for consideration in the piston trunk design. The heat caused by the numerous internal explosions of the engine, gives rise to an excessive amount of heat in the compression end of all combustion engines, causing the piston to expand unequally at this end, and to overcome this a piston should always be machined slightly tapered. If this fact is not taken care of, the piston, if machined parallel, will expand unevenly and cause a tight and improper fit, which will prove most detrimental to any engine, causing undue friction, and is apt to lead to piston seizing and scouring of the cylinder walls. Too loose fitting pistons and piston rings have, on the other hand, their evil effect, allowing, as they do, the compression for that stroke to leak past both the trunk and the piston rings, and when the stroke is completed, only a partial compression remains.

Another feature of the piston so often neglected is one arising out of poorly designed pistons having a surplus weight of metal, when only the lightest possible trunk, consistent with the force acting upon it, should be used. Especially is this true in single cylinder gasoline engines, where the moving parts are all running at high rates of speed, and where unevenness in thickness or weight tends largely to produce undue vibration in the engine. In the accompanying illustrations, 1 and 2 represent types of piston trunks and rings used largely in gasoline engines, and show clearly the design of rings, allowing ample spring when placed upon trunk and inserted in cylinder.

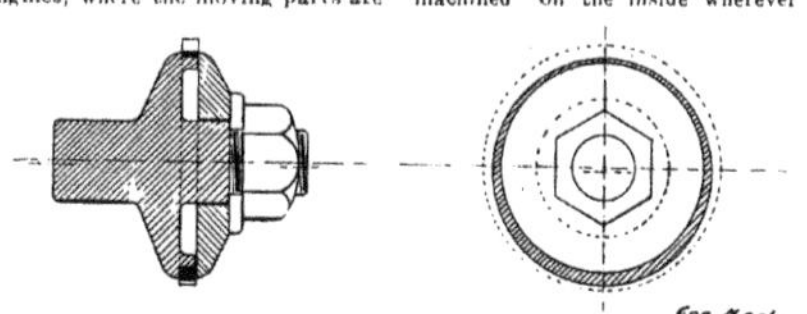

Fig. 3

Mandrel for Finishing Rings.

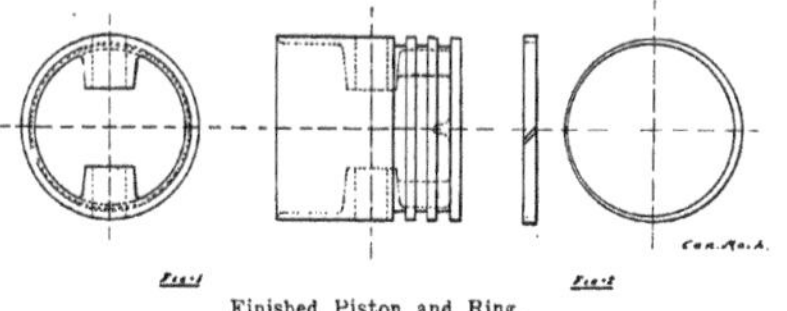

Fig. 1 Fig. 2

Finished Piston and Ring.

Mandril for Finishing Rings.

In Fig. 3 a cut is shown of a mandrel for accurately machining and leaving finished an eccentric ring suitable for all gas or gasoline engines. This mandrel is very simple in construction and operation, and, when carefully made, will be found invaluable to all gas engine makers.

The main portion of mandrel is of machine steel, turned on both ends, one end being threaded to take hexagon nut. This is done on the original centres, after which operation they are destroyed, and offset centres inserted, the required throw being given to allow for the taper piston ring. In this case shown the offset is 1-16 inch throw, the remaining parts of mandrel being a flange, a loose washer, and a hexagon nut, by means of which the eccentric ring may be held at the proper offset, while the finishing operation is going on.

Machining the Piston.

In machining the piston trunk, care should be exercised to centre it by the core in casting, to insure a uniform thickness of the walls when the trunk is completed. Each piston should be machined on the inside wherever possible, to lighten and balance it equally.

The outside of piston trunk should be turned and finished smooth. While on the chuck, the three grooves are cut for the rings, after which the trunk should be slightly tapered. In this case the cylinder bore is 3.5-inch, and an allowance of one one-hundredth part of an inch is made, which will prove ample for the piston trunk expansion.

Turning and Finishing Rings.

The piston rings should be bolted to face plate of lathe by means of two lugs cast on one end of the casting. The inside of this casting should be bored to the required diameter to fit the bottom of groove, and a roughing cut taken on the outside. One edge should be carefully squared each time before rings are parted from casting and allowance made on each for refacing the opposite side, whether turned, ground or lapped.

All rings should be ground to a uniform thickness—1¼ inch—on either a regular ring grinder, or by means of lapping them, until two perfectly square and parallel walls are secured; there the rings may be split, ready for the mandrel.

By placing the rings upon the mandrel shown in Fig. 3, and holding the ring securely in an eccentric position over the shoulder left on mandrel, the whole is locked together by means of a hexagon nut, and a roughing and smooth cut leaves a perfect eccentric ring, ready for use in engine. This method has been thoroughly worked out in practice. By adopting this method, many troubles will be avoided, and satisfaction to all obtained.

Developments in Machinery

Metal Working **Wood Working** **Power and Transmission**

A NEW RADIAL DRILL.

The mechanical means for changing speeds and feeds are the features of most interest in the new radial drill here illustrated. There are 24 changes of spindle speed and eight changes of automatic feed to each spindle speed, any of which can be instantly obtained while the constant speed driving pulley is in motion. It is claimed that the changes are made without noise or shock, and being entirely gear driven, they are positive.

The two long levers shown in front of the speed box in the general view, Fig. 1, control four changes of speed, and to prevent accidental misuse of these handles they are interlocked by the small locking lever between them, so that it is possible to move only one lever at a time from its neutral position. A sectional view is given in Fig. 2 of the gear box, showing the gears,

Fig. 3.—Top View of the Spindle Feeding Mechanism.

shafts and friction clutches. The latter are operated by the long levers referred to and cause the driving shaft to drive the intermediate shaft, which in turn drives the upper shaft in the speed box. The small lever on the right side of the speed box is used to bring the gears, which are shown in the sectional view, into mesh with the proper gears, D, E, and F, on the intermediate shaft, and three more changes of speed are obtained for each speed of the first driven shaft, making 12 changes of speed possible up to this point. This range is doubled by throwing the back gears in or out, which are located on top of the column, and are operated by the lever shown at the base of the column in Fig. 1. This lever is also used to bring the gear on the centre shaft in mesh with the gear on the elevating screw when required. When the elevating screw is reversed while lowering the arm, the gears D and G (Fig. 2) in the speed box are engaged without any intermediate gear, which causes the upper shaft in the speed box to reverse at an increased speed.

The column is in one piece, is bolted to the base and does not revolve. It is of heavy section throughout and is reinforced by four inside webs extending its entire length, making it capable of resisting the heavy strains imposed, particularly when the arm is at the top of the column and the spindle at the outer end of the arm. The arm is of hollow rectangular section, stiffened by an upper brace close to the face carrying the spindle head and a lower brace almost diagonally opposite, near the rear face. It will be appreciated that this provides admirably for resisting upward pressure of the spindle when drilling, by preventing twisting of the arm. A top cap resting on roller bearings supports the arm, and both may be swung completely around the column or instantly locked in any position by binder levers. The arm is lowered at almost three times the elevating speed by a screw having ball thrust bearings.

Fig. 1.—A new 4 ft. Radial Drill, built by the Mueller Machine Tool Co., Cincinnati, Ohio.

The spindle is of crucible steel and is ground. It is counterbalanced, has quick advance and return and provision for taking up wear. For tapping it is stopped, started and reversed by the long lever shown in front of the head, which operates two self-adjusting, noiseless friction clutches that are located on back of the head. At such times it is impossible to accidentally engage the automatic lever feed. This guards against breaking taps, as does also an adjustable gauge nut, which causes the spindle to slip when a tap reaches the bottom of a hole.

A positive automatic feed is provided for the drill spindle when high speed drills and reamers are in use, or a friction feed may be used when desirable.

The change from one to the other is easily made by simply turning a nut. An ingenious mechanism, shown in Fig. 3, accomplishes the changing of feeds. A round plate having eight circles of steel pins in located above the spindle gear, and the pins engage with a steel involute toothed pinion, feathered on a horizontal shaft, which carried a worm to transmit the feeding movement. The speed of this shaft is varied by means of a knob below the lower hand wheel. This knob is fastened to a rod which extends through the centre of the vertical feed shaft, and has teeth cut on its upper end to engage with a sliding rack and move the steel pinion into engagement with any desired circle of pins for fast or slow feed. These eight changes of feed may be quickly made while the drill is at work. The upper worm wheel has a split hub, and by means of a ring nut can be locked to the vertical feed shaft for positive feed, or slightly released if friction feed be required.

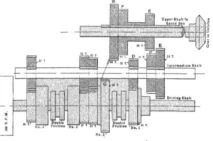

Fig. 2—Diagram of the Speed Box Gearing in the New Mueller Radial Drill.

The automatic trip is provided with a safety stop, which prevents the feeding of the spindle after it reaches the limit of its travel. A graduated bar on the counterbalance weight is set to zero when the drill enters the work, and by means of several adjustable dogs on the bar the feed may be tripped as often as desired ; these do not interfere with the spindle travel. The feed can also be tripped by a lever on the vertical feed rod.

Letters and numbers cast near the various levers on the machine indicate to the operator what levers to move and in what direction to move them, for proper speeds, which are given on a bronze plate fastened to the arm.

This machine is built by the Mueller Machine Tool Co., Cincinnati, Ohio.

AN OPEN-SIDE SHAPER.

The illustration shows a 15 in. x 60 in. open-side shaper, the smallest of which planes 15 in. in width. It is of the well known Richards type, with improvements, and for certain classes of work can be used to better advantage than either a pillar shaper, a traverse shaper, or a planer.

The machine is driven by a single screw and bronze nut, and without the intervention of any gears. The screw is 2¼ in. in diameter, and is made of .50 carbon steel. The saddle has a long and wide bearing on the column, with, however, a narrow and deep guiding surface to prevent binding, particularly when the tool is at the outer end of the rail, and is provided with a taper gib lengthwise, adjustable by screws at each end, affording uniform contact on both sides of the gib.

The cross feed is by power or by hand, is positive, and adjustable ; the down feed is by hand. The reversing mechanism is of our design is new, and a decided improvement over existing methods. We depend on the turning of the shifting rod both for reversing and motion of the saddle, and for the automatic cross feed to the head, the rod being automatically turned by cams on the saddle coming in contact with dogs adjustable on the rod, or by hand from the saddle, if desired.

There are ball bearings for the shifting rod, as well as under the elevating screw to the table. There are taper gibs to the cross rail and to the head. The head has a down feed of 6½ in., swivels, is graduated, and has a micrometer reading to .001 inches. The tables are raised and lowered by means of crank handles, or may be removed entirely and work bolted upon the aprons ; or both tables and aprons may be removed entirely and work bolted against the face of the machine. Keyseating or shafting of any diameter may be done on the machine. All flat bearings are hand scraped to surface plates, and all steel slots cut from the solid. Ample means are made for oiling.

Length of stroke, 60 inches ; width planed, 15 inches ; maximum distance head to table, 23¼ inches ; width of table, 15¾ inches ; vertical adjustment of table, 19 inches ; depth of table, 15 inches ; down movement of tool slide, 6½ inches.

This line of open-side shapers is built by the Cincinnati Shaper Co., Cincinnati, Ohio.

The Open-side Shaper, built by the Cincinnati Shaper Co., Cincinnati.

IMPROVED CONTACTOR.

The Cutler-Hammer Mfg. Co., of Milwaukee, makers of electrical controlling devices, have recently placed on the market an improved type of contactor. These contactors are used for handling main line currents where the nature of service is severe. In such cases controlling sliding contacts cannot be relied upon to handle the main line current, and it is customary to employ a controlling panel consisting of a number of contactors, this panel in turn being controlled by a master controller designed to regulate the secondary current which energizes the solenoids of the contactors.

The improved contactor here illustrated is a compact and strongly constructed piece of apparatus, and is provided with an exceptionally powerful blow-out magnet. The main line circuit is closed by the solenoid raising a pivoted arm carrying a thick copper plate to a point where contact is made with a pair of stationary, laminated copper brushes. Arcing on this contact is prevented by providing an auxiliary copper and carbon contact in the field of a powerful blow-out magnet, which instantly extinguishes the arc incident to the breaking of the circuit. This auxiliary contact closes before the main contact is made and opens after the main contact is broken, thus effectually preventing any sparking on the main contact.

A noteworthy improvement is the pivoting of the blow-out shields, permitting these to be raised (as shown in illustration) so as to expose the auxiliary carbon and copper contact. In earlier types of contactors these shields were rigidly fastened in their normal position, completely covering the copper and carbon contacts, and rendering access to these difficult.

The present construction makes renewal of either contact, or of the coiled spring (visible just above the carbon contact) a matter of a few moments only. At a recent test at one of the largest Pittsburg steel mills a 220-volt circuit was opened and closed by a contactor of this type 88,000 times, before renewal of the copper and carbon contacts became necessary, and—on a test to determine time required for repairs—the old contacts were removed and new ones inserted in less than two minutes.

—

NEW TYPE METAL SAW.

The illustration herewith shown is of a new type of metal saw that the Quincy, Manchester, Sargent Co. have recently added to their line. The machine has been designed to meet a demand for a somewhat smaller machine than those they have been manufacturing in the past, and while embodying all the strength and wearing qualities of their regular line, the machine can be placed on the market at a figure that will be within the reach of many small shops that are not warranted in purchasing a larger and more expensive machine.

The method of driving the blade is rather novel and entirely different than any other method now in use. The blade, as on the Bryant type of machine, is driven from the periphery, but instead of the sprocket drive, steel rollers are used, these being hardened and ground and journaled in removable steel bushings, which are held securely in the double driving gear. By this method of drive a much larger diameter of the blade is available for cutting than can be obtained from a blade of the same size arbor driven, where about one-third of the diameter of the blade is necessarily occupied by the driving collars. It also has an advantage of economy in repairs.

New Type of Metal Saw.

The machine has a capacity for cutting rounds up to six inches in diameter; I-beams in a vertical position up to ten inches, at any angle up to 45 degrees. The feed is of the variable friction type, adjustable with the machine in motion, is powerful and continuous in its action throughout its entire range, and greatly superior to a ratchet feed; which necessarily is intermittent in its action.

When desired the machine will be furnished arranged for direct connected motor drive.

IMPROVED CARS DUE TO AMERICAN TOOLS.

From Paris correspondent of New York Herald.

Interviewed by a Herald correspondent on the question of American-made machinery in French automobile factories, the Marquis De Dion, head of the De Dion-Bouton Company, declared that if it had not been for the fact that American houses had been in a position to supply certain types of lathes, drilling machines, gear-cutting appliances and other intricate pieces of mechanism some years back, when the automobile movement began to expand, a popular vehicle could never have been considered by the makers at that time, while all classes of automobiles would necessarily have remained at extravagant prices.

"I'm not sure, to-day," continued the Marquis, "that we are still entirely dependent upon the American supply of machine tools to enable us to continue our production on the present lines.

"Great strides have been made by European constructors of such implements, and to-day we can purchase almost everything we want in Germany and England, and to some extent in France, at prices varying very slightly from those quoted by American firms. At the present time I am interested in a French company which constructs such implements and I have been led more or less to interest myself in the affair by reason of the difficulty, frequently experienced, of getting delivery of goods we want from American firms.

Deliveries Delayed.

"These firms have been so occupied that they have been in a position to afford to neglect orders from Europe, and I have met with cases where a year or eighteen months has passed before an order has been executed. At one time, not very long ago, we were entirely dependent upon the machines supplied by American constructors.

"All French automobile makers who wished to put their automobiles upon the market at a reasonable figure were compelled to consider what American firms had to offer. These machines are still used in my factory as well as in others, and to keep abreast with things we are compelled to turn to America every year for modern appliances and new, ingenious machines which enable us to produce higher grade automobiles at less cost.

"There are certain things about some of the American tools we have purchased which are not entirely satisfactory. Castings, as a rule, are not as resisting as they should be, and I doubt, in some cases, whether their machines turn out their work with such precision and accuracy as machines made in Europe."

Walking through the great works at Puteau the Marquis pointed out different types of American machines employed.

Many American Names.

"Here," he said, pointing to a row of implements at work on boring cylinders and finishing castings, "are machines supplied by Brown & Sharpe. On the other side are automatic machines by Pratt & Whitney. We also use stamping machines by the same firm.

"Those drilling machines there are by W. F. & John Barnes Co., while these metal planing machines come from the works of the Whitcomb Manufacturing Company.

"One of the most important things in automobile construction is gear-cutting. We have machines for this purpose supplied by various firms. Brown & Sharpe, the Gleason Tool Company, and Potter & Johnson have sent us most useful machines.

"For grinding up certain parts we have bought instruments made by the Norton Company, but it is difficult to mention all the American firms to which we are indebted for the various classes of work. We have tried almost all of them in turn.

"You see, the majority of our machines are of American construction, but as we constantly are renewing it is impossible to say how long this will be the case. I think when the French company of which I spoke gets to work we shall be considerably less dependent upon American goods."

THE SCHANTZ POWER PIPE THREADING MACHINE.

A machine, just designed, and recently put on the market, which attracted a good deal of attention from the visitors to Machinery Hall at the recent Canadian National Exposition, was the Schantz Power Pipe-Threading Machine, manufactured by the S. E. Schantz Co., of Berlin. The accompanying illustration shows the special features of the machine, a description of which follows.

The No. 6 machine will thread pipe from 1¼ to 6 inches. It is arranged for nine speeds, varying from 2¼ to 40 possible rapidity. The head is fitted with two universal chucks, thus dividing the strain and providing additional clamping power. The back chuck steadies the pipe and assists the dies in their work.

The length between clutches is a noteworthy feature. The machine will accommodate a full length of 6-inch pipe without a third bearing.

Schantz Power Pipe Threading Machine

The cut-off attachment is fitted with self-centreing V-blocks, and with two cut-off tools, thus giving considerable advantage in speed. The tools operate very close to the blocks and thereby provide for steadiness.

The machine is fitted with a Borden r.p.m., and is specially designed to do the heaviest work with the greatest patent solid adjustable die head. All dies are adjustable and chasers are inserted.

The matter of lubrication is provided for by a Brown & Sharpe rotary oil pump, connected with the cut-off tools and the die with flexible steel hose.

The machine, as the photograph shows, is designed particularly for heavy work. Its great weight, 4,300 pounds, and the fact that all gearing is cut from solid metal, are factors in its reliability.

CANADIAN MACHINERY
and Manufacturing News

A monthly newspaper devoted to machinery and manufacturing interests, mechanical and electrical trades, the foundry, technical progress, construction and improvement, and to all users of power developed from steam, gas, electricity, compressed air and water in Canada.

The MacLean Publishing Company, Limited

JOHN BAYNE MACLEAN - - President
W. L. EDMONDS - - - Vice-President
H. V. TYRRELL - - - Business Manager
J. C. ARMER, B.A.Sc., - Managing Editor

OFFICES:

CANADA
MONTREAL - 232 McGill Street, Phone Main 1255
TORONTO - 10 Front Street East, Phone Main 2701
WINNIPEG, 511 Union Bank Building, Phone 3726, F. R. Munro
BRITISH COLUMBIA - Vancouver, Geo. S. B. Perry

UNITED STATES
CHICAGO - 1001 Teutonic Bldg., J. Roland Kay

GREAT BRITAIN
LONDON - 88 Fleet Street, E.C., Phone Central 12960, J. Meredith McKim
MANCHESTER - 92 Market Street, H. S. Ashburner

FRANCE
PARIS - Agence Havas, 8 Place de la Bourse

SWITZERLAND
ZURICH - - - Louis Wol. Orell Fussli & Co.

SUBSCRIPTION RATE.

Canada, United States, $1.00, Great Britain, Australia and other colonies 4s. 6d., per year; other countries, $1.50. Advertising rates on request.

Subscribers who are not receiving their paper regularly will confer a favor on us by letting us know. We should be notified at once of any change in address, giving both old and new.

Vol. III. OCTOBER, 1907 No. 10

CONTENTS

EDISON'S DREAM REALIZED.

The throwing of a lever by Lieutenant-Governor Fraser, of Nova Scotia, causing power to be flashed along the lines from the mouth of the Chignecto Mines and the wheels to revolve in the factory of the Rhodes, Curry Manufacturing Co., Amherst, N.S., nine miles distant, was a great event, not only for Amherst and vicinity, but for all coal mining districts. Owners of coal mines throughout Canada and United States can now follow this example with confidence and generate their own power and supply cheap power to all factories in the vicinity of their mines. Amherst now has facilities which are bound to make it, in a few years, a great industrial centre. A word of commendation is due to the men who so bent their energies to the scheme that they now see the realization of their expectations of five years ago. It was a dream of the great inventor, Edison, that power would, in a few years, be generated at the mouth of the coal mines and be transmitted over wires to the various industries. It remained for Canada to develop Edison's suggestion, and now, in Amherst, N.S., it is an accomplished fact. The refuse or culm from the coal mine is the fuel used and the primary object of the whole system is economy. To use a refuse product, which was previously an expense, and transform it into power to run industries miles away, is a worthy achievement. Thomas A. Edison recognizes that this is a national success and sent the following message to H. J. Logan, Chairman Board of Trade Committee, Amherst, N.S.: "Permit me to congratulate your Board of Trade and Senator Mitchell on the inauguration of the first plant on the American continent for the generation of electricity at the mouth of a coal mine and the distribution of the same to distant commercial centres. It is a bold attempt and I never thought it would be first accomplished in Nova Scotia, where my father was born over one hundred years ago."

ABOUT OURSELVES.

It is with justifiable pride that we present to our readers and advertisers our October number. From the very first when Canadian Machinery was launched into a distinctly new field, the paper has been a success. From a sixty-four-page paper it grew to eighty pages, and the continued steady growth necessitated an addition of twenty-four pages this month, making one hundred and four pages inside the covers. With every issue during the year 1907 our paper has had increased business. Discriminating advertisers who know where to get results are placing their contracts with us, so that each month's business this year shows a distinct gain of about 40 per cent. over the corresponding month last year. This shows the confidence our patrons have in the selling powers of Canadian Machinery.

We have been alive to the needs of both advertisers and readers. Evidence of this fact can be gained by comparison of the different issues and a study of the progress shown. Advertisers are the backbone and mainstay of a publication, and it requires them to make it a success. To make a good medium for advertising, one that will bring results, a large subscription list is essential. This we have realized and we are building up a paper for the readers. With the increased advertising we have added a corresponding amount of reading matter. We spare no expense to get what the mechanical public want, and give it to them, and we are continually on the lookout for good, live, reliable articles. Our editorial matter is of the best and every issue has a fund of good information and ideas for manufacturers, superintendents, foremen and mechanics. As a result, subscriptions have come in from all over Canada and our representatives in the different provinces are meeting with excellent results.

Your suggestions and criticisms will aid us materially to reach the end towards which we are aiming, to give the readers matter best fitted for their needs and thus to build up a good advertising medium, one that will be of credit not to ourselves only, but to the profession and business that Canadian Machinery is endeavoring to assist. We trust our patrons and readers will do their share in helping this along.

Quebec Bridge Disaster

The news of the collapse of the Quebec bridge on August 29th, with the loss of over seventy lives, came as a shock to the whole engineering world. Engineers from all over the globe have watched with interest the progress of this great undertaking. This structure was expected to be the finest in the world and engineers of high standing and wide experience were employed in the designing of the bridge and the carrying on of the work. The designs were by the Phoenix Bridge Co.'s designing engineer, Peter L. Salapka, and the Phoenix Bridge Company had the contract of erecting the superstructure. The design and workmanship were, so far as yet apparent, of a high order. The erection methods were carefully thought out and the work had reached the fourth panel from the south end, or within two panels from the centre of the suspended span, when the blow fell. A large traveler weighing 1,000 tons stood at the end of the cantilever span. It had been dismantled to about one-third of its weight, so that the weight here was considerably reduced, but in the immediate vicinity was a small crane weighing about 200 tons, and on the end of the suspended span was piled an enormous quantity of structural steel. The accident occurred just as a locomotive with two cars of chord sections moved out on the erected portion. The trainload was not an unusually heavy one and it was the unexpected that happened when the structure went down, carrying with it over seventy workmen.

Dimensions of the Bridge.

This bridge is intended to be the longest single span cantilever bridge in the world, the length of the centre span being 1,800 feet, and the total length between abutments is 3,220 feet. It consists of two deck truss approach spans, each 210 feet long; two anchor arms, each 500 feet; two cantilever arms, each 562½ feet, and one suspended span 675 feet long. This will be the longest single truss span ever built. For a width of 1,200 feet there is a clear headway at high tide of 150 feet, to allow free passage of vessels. The weight of the heaviest single piece of steel for the bridge is 100 tons, and the total weight is 38,500 tons. The length of the longest single section is 105 feet. The eyebars are the largest yet used, with a maximum of 56 on one pin. The diameter of the pins is from 9 to 24 inches and up to 10 feet in length. The total number of field rivets to drive is about 550,000. The bridge will have a total width of 82 feet, including the cantilever sidewalks. It is designed to carry two railroad tracks, two highways, two electric car tracks, and two foot walks on a single deck.

The main piers were of concrete, faced with granite and sunk by pneumatic caissons, 150 x 49 feet and 25 feet high. The tops of the piers measure 133 x 30 feet and they contain 35,000 cubic yards of masonry. The anchor piers are of concrete, faced with granite, and are 30 x 111 feet at the base, and 56 feet high, from the bottom to the anchorage. They are 24 x 105 feet at the coping and contain 14,400 cubic yards of masonry. The abutments are of concrete, faced with granite, and are 80 feet wide, 40 feet deep, and contain 4,000 cubic yards of masonry.

The work of constructing the superstructure was under the direction of Theodore Cooper, consulting engineer, of New York, and was progressing rapidly. On account of the severe climate, the work was carried on between April and November and the best possible use was made of the time. On account of this rushing, the work of construction, it is thought that perhaps all of the rivets were not placed in position as the work progressed, but as yet the commission has found no evidence to justify them in this belief. There was no loud report at the time of the disaster, but the bridge sank with a grinding sound, followed by the snapping of some of the tie rods. The fall of the bridge appears to have been due to a failure of the compression members, rather than in the tension members.

The Government Commission.

The Canadian Government has appointed a commission, consisting of Henry Holgate, consulting engineer, Montreal; Prof. J. G. Kerry, McGill University, and Prof. J. T. Galbraith, of Toronto University, to make a thorough investigation into the cause of the disaster. The commission is acting in harmony with the engineers of the bridge and it is of the utmost importance that these men come to some definite conclusion in order that public confidence may be restored in the scheme. Failing this, the question of restoring the bridge will be viewed with apprehension both by the public and the bridge engineer. The government realizes this and through the commission is trying to fix the cause of the disaster

The Latest Developments.

As the work of the commission goes on, new facts are brought to light. Daniel B. Haley, of Sarnia, gave his testimony that he had seen several panels bent while standing on the third panel of chord nine. He also noticed the lacing was bent in chords eight and nine. The splice of the chords, he said, had not been properly riveted, and this matter was reported to the foremen. Conflicting testimony has been given in regard to one of the bridge plates. The workmen testify that there was a crack in one of these from 18 to 20 inches long and three-quarters of an inch wide. The deflected chord led to the sending of a message to Phoenixville and Mr. Cooper's asking them to look into this matter. Owing to the operators' strike, messages were delayed. Mr. Yenser, the head foreman in charge of the work, wrote, asking for a reply by telegraph, to say whether or not he should continue work. He was called on the long distance telephone and asked further about the matter. He said as there appeared to be no immediate danger, he was proceeding with the work.

It will in all probability take some time for the commission to locate the trouble. In any case, there will necessarily be a delay in the resumption of the work of erecting the bridge. The Phoenix Bridge Company wish to complete the work, provided the Canadian Government does not object, and at least two years will be necessary to reach the stage of completion of the bridge at the time of the accident.

PLANING MILL BURNED.

The Kennedy & Davis Milling Co., of Lindsay, whose planing mills and stave factory was entirely destroyed by fire on June 2, are rebuilding. They will install all new machinery, with the exception of the boiler and one large planer, which were but slightly damaged. They are in the market for cars, for the purpose of conveying the staves into the dry kiln.

Foundry Practice

MODERN IDEAS FOR FOUNDRYMEN

Moulding Cast Steel Lead Bowls

By H. J. McCaslin.

When a lead bowl reaches such proportions as the one illustrated, the molds are usually swept up, this being accomplished by attaching to and revolving about a spindle, as shown in Fig. 2—a sweep so arranged that the striking edge contains the outer and inner outline of the bowl to be cast.

While sweeping up a mold for an object of this form is of a very simple nature, it nevertheless may embody some points of interest for the young patternmaker or molder.

Beginning the work, a full-size radial section layout (as shown to the right of the centre line; Fig. 1) is very essential, this layout being required in getting out the sweeps. The sweep may be constructed somewhat after the manner shown in Fig. 2, being secured to the spindle with collar and sleeves. Shown in greater detail, to the right of this illustration is a cross section of sweep on the line, BB, showing the metal thickness strip, A, which is detachable, being secured to the sweep proper with screws.

There is a difference of opinion existing between molders as to the most practical manner of using the strip, A. Some prefer to sweep up the outer form of the object to be cast remove strip A, and then sweep upon this surface a thickness of sand equivalent to the metal thickness. This thickness of sand is subsequently removed, following the ramming up of the cope. Others prefer to sweep up the inner surface of the object, ram up the cope and lift it off, detach strip A, and strike off the metal thickness. The latter method is the one preferred by the writer and the one that will be discussed in this article.

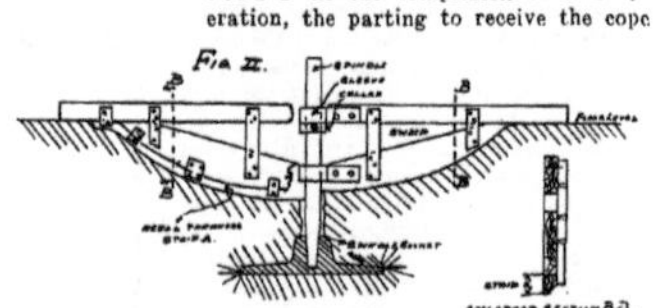

Fig. 2 illustrates a cross section of the mold at two stages of completion. At the right of the centre line is shown a cross section of mold during the operation of sweeping up the exterior form of the object. It is upon this surface the cope is rammed up and lifted off in the usual way. To the left of the centre line is shown the second operation of sweeping, at which time a thickness of sand equivalent to the metal thickness is removed. Operations are started upon the mold by digging a hole in the foundry floor, of ample depth and diameter, and the firm setting and plumbing of the socket and spindle. Next, the sweep, with strip A removed, is attached to the spindle as shown, being adjusted to the required height by the collar.

Heap sand is now firmly rammed in and struck off to conform to the sweep —as shown to the right of centre line, Form 2—as the completion of this operation, the parting to receive the cope flask is struck off, this being accomplished with the aid of a straight edge, and that portion of the parting formed by the sweep. The sweep and spindle being removed, the hole left vacant by spindle is filled with waste or other material, and a plate placed over it to prevent any sand from running or falling below. Parting sand or paper (paper preferred) is now applied over the entire surface and work upon the cope is begun. To facilitate the securing and lifting out of this body of sand forming the upper surface of the mold, a lifting iron is generally introduced near the bottom, as shown in position, Fig. 3. Extending up from the lifting iron are four bolts, which are used in securing this portion of the cope to the flask. Supported upon blocks, their depressions being filled up during the finishing of mold; the lifting iron is placed in position. Silica sand of about 1 inch in thickness is now applied over that portion of the surface which comes in contact with the metal. Heap sand, well secured with rods, is then firmly rammed in to the height of the parting line. Next the flask is placed in position and staked; the runner stick and the riser head are arranged, followed by the firm ramming up of the flask, as shown to the right, Fig. 3. Before lifting off and

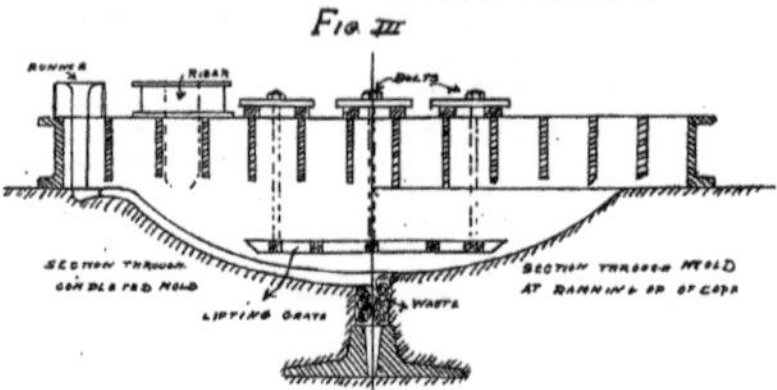

FIG. III

blocking up the cope, the two parts of this portion of the mold are securely bolted together, as previously mentioned. Operations upon the drag portion of mold are resumed by the roughing out with a shovel ample space to allow for a thickness of silica sand, and placing back in position the spindle and sweep with strip, A, attached. Silica sand is next applied, firmly rammed down and struck off to conform to the sweep, as shown to the left of Fig. 2.

With the removing of the spindle and sweep, the bedding in of the four lifting lugs, (the pattern being shown in greater detail to the right of Fig. 2) is proceeded with, this being accomplished by dividing the outer circumference into four equal parts, and the bedding in of the pattern at these points.

Finishing a mold for steel castings differs from that of iron, as nails are inserted over such surfaces of mold where scabbing must be prevented as far as possible. In this instance, the cope mold requires nailing. A silica wash is also applied over the surface of mold, and the mold is thoroughly dried by a coke fire or a gas burning arrangement. All that now remains is to blow out the mold, try the cope on to ascertain if the metal thickness is correct, this being accomplished with balls of clay set at several points in the mold. If everything is found all right, the cope is returned to place and ample weight applied to prevent raising when being poured.

An Improvement in Cupola Tuyeres

Cupola melting is one of the things in foundry practice that has not received the careful and scientific attention that it should have had in the past, and in many cases the cupola is considered fool-proof and expected to run itself. Foundry owners frequently wonder why they are having trouble with their iron, when, if they only knew it, the troubles are resulting in improper cupola work, due largely to the improper cupola design or construction.

Fig. 1.—Vertical Section of a Cupola Equipped with the Knoeppel Tuyere.

Mr. J. C. Knoeppel, of 577 E. Ferry St., Buffalo, N.Y., has had more than twenty years' experience as a foundry manager, and given a very large amount of attention to melting problems. He has devised a tuyere system for which patents are now pending in the United States and Canada

The fundamental principle of good cupola work is to introduce the required amount of oxygen with as little disturbance as possible, and in such a manner as to give an equal temperature at all points around the circumference of the cupola; in other words, to avoid cold spots. To accomplish this, the blast must be uniformly distributed, and the tuyere area must be large. If a cupola lining is allowed to take its own course and assume the form which it naturally would where the cupola is normally slagged, it will be found that the lining will build out slightly above the tuyere and form a bosh above. Recognizing these facts and also recognizing the fact that large openings were necessary to deliver the air at the lowest possible pressure so as to avoid irregularity, Mr. Knoeppel has arranged his tuyeres as shown in the accompanying photograph. The castings composing the tuyeres form a continuous belt about the cupola, and in large cupolas the upper and lower portion of the tuyeres are composed of separate castings, while in the smaller ones they are all one casting.

The line drawing, Fig. 1, represents a vertical section through the cupola. Fig. 2 represents one of the tuyeres, which is provided with a safety notch in the centre, so that when the iron rises to this point it will flow through the depression in the centre and escape through a soft metal plate, thus warning the melter before the iron flows into the tuyeres proper.

Fig. 3 shows one of the key tuyeres. In these the upper flange overhangs the other or adjacent tuyeres, while the lower portion is made with parallel sides so that it can be drawn in to the centre of the cupola, thus releasing the

Fig. 2.—Knoeppel Tuyeres with one of the Tuyere Keys Removed.

other tuyere blocks. The tuyere blocks above the spout and above the slag hole usually have their lower tier of openings stopped with brick, so as to avoid any liability of chilling the outflowing slag or metal.

One of the most important points in

Fig. 3.—Section of Cupola Showing Tuyeres and Vertical Air Pipes.

connection with this form of tuyere, however, is the fact that it may be said to have two independent wind belts which will give the air a chance to equalize in pressure both inside and outside of the shell, thus insuring equal pressure at every tuyere opening. The outside wind belt is a metal casing, as in the better style of cupola mountings, while the inside belt is formed by cutting back the vertical partitions of the tuyere blocks at the side next the cupola shell, as shown in detail in Figs. 2 and 3 and also in Fig. 1. A series of openings through the shell admit air to the back of the tuyere blocks, and this air then flows about through the inside air space. By making the tuyere area ample, the air flows at a relatively slow velocity and has opportunity to lag or rest in the tuyeres for a short space of time, thus enabling it to take

Fig. 4.—Cupola Lining with Shell Removed, Showing Vertical Pipe and Inside Wind Belt.

up some heat and to enter in the best possible condition for rapid combustion.

One important point to be observed in connection with the overhanging upper edge of the tuyere blocks is that this form effectually prevents the slag or metal in its downward course from flowing into the tuyeres and blocking them. It also prevents coke from sliding into and closing the tuyeres, thus insuring an ample opening at all times during the heat.

In the usual cupola construction, the lining is carried up vertically for a short distance above the tuyere blocks and then caused to flare out, forming the bosh. The brick lining is, of course, supported by angle irons riveted to the shell, as usual, so that the tuyeres can be removed if necessity should require any change in their arrangement.

One of the features covered in Mr. Knoeppel's patent is the provision for the introduction of air above the melting point, the object being to utilize to the fullest possible extent the heat contained in the gases. These openings are arranged by providing notches in the top of the tuyere blocks, which connect with vertical pipes next the shell on the inside of the cupola. These are provided with nozzles leading through the lining. In Fig. 3 there is shown a section of the vertical pipe taken through one of the nozzles. The nozzles are secured to the pipes with dovetail joints, so that they can be easily replaced. In operation during melting a small amount of air is admitted through these vertical pipes and is supposed to combine with the carbonic oxide gas, burning it to carbon acid. The heat thus set free will be to a considerable extent absorbed by the material as it settles toward the melting point, thus preparing the metal for rapid melting. These vertical pipes can be closed with dampers or valves which can be used to adjust the amount of air admitted above the melting point.

At the close of the heat the dampers in the vertical pipes can be opened wide and a volume of air allowed to pass up through the pipes so as to cool the lining.

In the case of very small cupolas, the amount that the tuyeres overhang on the inside would be reduced, and, possibly, in some cases, this overhanging would be entirely dispensed with and flush or vertical tuyeres used. In such cases, Mr. Knoeppel provides cast iron plates which can be placed on top of the tuyeres so as to form an overhang of from one to two inches, thus providing for a small amount of boshing.

Fig. A is an outside view of the cupola with the shell removed, showing cupola lining. It shows the vertical pipe and gives a good idea of the tuyere blocks arrangement.

In Fig. 5 we have a sectional view on a larger scale, showing the tuyeres and the inside wind chamber.

The principal advantages of this tuyere system may be summed up as follows:

First, and most important, the introduction of a very large amount of air at a low pressure, thus insuring a low melting point and uniform conditions in the cupola.

Second, by the use of the overhanging upper portion of the tuyeres, the liability to the stoppage of tuyeres during the operation of the cupola is very greatly reduced, thus enabling much longer heats to be taken from the cupola.

Third, the inside wind belt provides an opportunity for thorough equalization of the blast pressure, thus insuring equal melting at all points around the circumference of the cupola, which is a condition that cannot be fully attained when the tuyeres are not continuous, and are supplied only with an upper wind belt.

Fourth, the form of the tuyeres, to-

Fig. 5.—Section of Cupola Lining Showing Tuyere Blocks and Inside Wind Belt.

gether with the use of the safety tuyeres, reduces to a minimum the liability of stoppage caused by metal rising into the tuyeres.

Fifth, the vertical pipes with the supplementary openings into the cupola enable the operator to control the conditions above the melting point as he may see fit.

Sixth. The fact that the tuyeres project from two to sixteen inches, delivers the air much nearer the centre of the cupola than in cases where the tuyeres are flush with the lining.

NICKEL IN CAST IRON.*

By James F. Webb, Elkhart, Ind.

In reporting on the tests authorized by this association last year, I beg to submit the following results: I made twenty test bars, ten with the nickel added and ten without, all from the same ladle holding 1,400 pounds, and of good hot iron. Nickel thermit was put into the bottom of the hand ladles, the aim being to get varying proportions of nickel into the metal and then to note results of the tests.

The determinations were made by H. E. Smith, of the Lake Shore railroad, and are presented herewith:

TABLE I.

Physical Tests.

Cast Iron Test Pieces from Elkhart, Ind.

Mark.	Diam. inches.	Cross breaking weight.	De-flection.	Modulus of rupture.
1 N	1.25	3,000	0.11	46,870
1 N	1.24	2,970	0.09	47,520
2 N	1.24	2,510	0.08	40,160
2 N	1.26	2,820	0.10	43,010
3 N	1.26	3,000	0.10	45,760
3 N	1.25	2,690	0.09	42,020
4 N	1.26	3,130	0.09	47,710
4 N	1.23	3,010	0.09	49,330
5 N	1.25	2,370	0.08	37,030
5 N	1.23	2,570	0.09	42,130
1 X	1.26	2,990	0.08	45,600
I X	1.26	2,790	0.09	42,550
2 X	1.26	2,740	0.10	41,790
2 X	1.25	2,890	0.10	45,140
3 X	1.26	2,750	0.08	41,950
3 X	1.23	2,690	0.10	44,100
4 X	1.25	2,610	0.09	40,780
4 X	1.23	2,770	0.10	45,410
5 X	1.25	2,470	0.09	38,600
5 X	1.25	2,760	0.09	43,130

Tests on 1¼-inch round bars, supports 12 inches apart.

TABLE II.

Chemical Analysis.

Mark.	Man-ganese.	Phos-phorus.	Sul-phur.	Sili-con.	Nickel.
1 N	0.460	0.63	0.090	2.07	0.67
2 N	0.455	0.63	0.089	2.08	1.18
3 N	0.450	0.63	0.091	2.05	2.07
4 N	0.444	0.62	0.087	2.02	3.25
5 N	0.444	0.63	0.088	2.10	6.65
3 X	0.450	0.62	0.086	2.12	——

The results would not indicate a marked improvement in the physical strength of the castings, and would bear out the supposition that nickel in cast iron is either not distributed uniformly enough to do much good, or else will find its best use in special classes of the metal freer from the high percentages of impurities incident to the ordinary casting.

It is to be hoped that further experiments can be made to prove just what does happen when nickel is added to cast iron.

*Paper presented before Philadelphia convention of American Foundrymen's Association.

Industrial Progress

CAADIAN MACHINERY will be pleased to receive, as confidential, industrial news of any description, the incorporation of companies, establishment or enlargement of plants, railway, mining or municipal news.

New Plants and Additions.

The Canadian Bag Co. are building a factory at Winnipeg.

A bait freezer will be erected at Lingan, N.S., to cost $3,500.

A biscuit factory will be erected in Toronto to cost $10,000.

A large, modern laundry will soon be erected at Vancouver, B.C.

Peers Bros., Vancouver, will erect a sawmill at Port Moody, B.C.

The power plant for Saskatoon, Sask., will soon be in operation.

The sawmill of Mohr & Co., Killaloe, Ont., was destroyed by fire.

Alterations to G. S. Britnell's factory, Toronto, will cost $12,000.

McCoskerie's sawmill, at Hartley Bay, B.C., will double its capacity.

The Bell Telephone Co. will install a central energy system in Galt, Ont.

Fisher Bros., Toronto, will erect a galvanized iron warehouse to cost $3,000.

Laidlaw & Watson Shoe Co., Toronto, will erect a factory to cost $12,000.

Fire recently did considerable damage to the power house at Stratford, Ont.

Work is progressing on the new high-pressure plant for the Winnipeg waterworks.

The Eagle Knitting Co., Hamilton, are going to erect a new factory in that city.

The Alberta Portland Cement Co., Calgary, wish to supply power to that city.

The Empire Mfg. Co., Vancouver, will erect a foundry, in that city, to cosy $2,500.

The Cataract Power Co. have the contract for supplying power to Brantford, Ont.

Fire did damage to a mill and elevator at Russell, Man., to the extent of $25,000.

R. Bigley, stove manufacturer, Toronto, will build a new stove foundry in that city.

A two-storey biscuit factory will be built by Symons & Rae, Toronto, to cost $10,000.

The machinery is being installed in the plant of the Western Canneries, at Medicine Hat.

The Bell Telephone Co. have installed a metallic line from Tara to Owen Sound, Ont.

The Ontario Power Co.'s transformer house near Welland, Ont., was burned out recently.

The new foundry at Lunenburg, N.S., is nearing completion and will soon be in operation.

Plans are under way for a big iron and steel plant for Vancouver, B.C., to cost $15,000,000.

Fire recently did $15,000 damage to the planing mills of S. Hill & Son, Saskatoon, Sask.

The official tests of the new 800-kilowatt plant at Morrisburg, Ont., were completed recently.

The Bell foundry, St. George, Ont., has been sold to Chapman & Fleury, the price being $25,000.

Work has commenced on the erection of the steel plant of the Coughlan Co., at Vancouver, B.C.

The Brantford Linen Manulacturing Co. are going to transfer their factory to Tillsonburg, Ont.

The new plant for the Hespeler Hoisting Machinery Co., Hespeler, will soon be in operation.

The Canadian Ornamental Iron Co., Toronto, will build a one-storey brick factory, to cost $5,000.

The plant and buildings of the Intervest Peat Fuel Co., Winnipeg, were destroyed by fire recently.

The Stony Lake Navigation Co., Peterboro, Ont., will build a new boat for service on the lakes.

The new power plant at Woodstock was recently put in operation, and is giving entire satisfaction.

The Miramichi Lumber Co. are negotiating for the purchasing of the Clark Spool Mill, Newcastle, N.B.

The electric plant on the Wabi river will soon be in operation, supplying power to New Liskeard, Ont.

It is said that the power capable of [illegible] veloped at Kakabeka Falls, Ont., is 100,000 horse-power.

The capital stock of the Canadian General Electric Co., Peterboro, Ont., has been increased to $8,000,000.

A local telephone company has been formed at Vancouver, B.C., and will immediately apply for a charter.

The Northumberland & Durham Power Co. are contemplating the transmission of power to Kingston, Ont.

Fire did damage in Janesville, near Ottawa, to the extent of $9,000. The paint shop of A. Proulx suffered.

The Ladysmith Lumber Co. will erect a sawmill near Nanaimo, B.C., to cut 35,000 feet of lumber per day.

The Cataract Power Co. will run a high-potential transmission line from Hamilton to Brantford, Ont.

A telephone signal service is being established on the St. Lawrence river, between Montreal and Quebec.

Cleveland, Ohio, parties will commence to manufacture a full line of paint products at Winnipeg, in 1908.

Consignments of harvesting machinery aggregating three hundred tons, were shipped to Australia recently.

The warehouse of Shim & Son, Saskatoon, was destroyed by fire, and $15,000 worth of goods was destroyed.

The Canadian Pin Co., a new concern, will erect a factory at Woodstock, Ont., for the manufacture of pins.

E. C. Atkins & Co., Indianapolis, Ind., will erect a Canadian branch factory at Hamilton, Ont., to cost $50,000.

Fire did damage to the plant of the Partington Pulp & Paper Co., at Union Point, N.B., to the extent of $3,000.

It is expected that No. 1. factory of the new Siliker Car Works, at Halifax, N.S., will be opened in a few days.

The Canadian Independent Telephone Co., Toronto, is organized, and will erect works in Hamilton or Toronto

The Sechelt Brick & Tile Co., a new company, will erect a brick-making plant at Sechelt, to cost $50,000.

A large force of men are engaged in laying underground cables for the Bell Telephone Company, at Peterboro, Ont.

The New Liskeard Concrete Co., New Liskeard, Ont., are installing a new mixer, and a concrete block machine.

Fire recently destroyed the Maritime Mfg. Co.'s plant at Pugwash, N.S. Loss. . . . with insurance of $10,000.

A big addition will be made to the coal and furnace docks of the Canadian Northern Coal Co., at Port Arthur, Ont.

The cost of the power plant of the Stave Lake Power Co., Stave Lake, B.C., will be in the neighborhood of $2,500,000.

The Ottawa Pulp & Paper Co. has been formed, to make paper out of refuse spruce and hemlock. The capital is $150,000.

The entire plant of the Standard Brick &

Tile Co., New Glasgow, N.S., was destroyed by fire, entailing a loss of $300,000.

Hamilton purchasers have acquired the timber limits, mill and machinery of McManus & McKelvie, New Liskeard, Ont.

The British Columbia Electric Railway Co., will supply Ladner, B.C., with light. The power is generated at Lake Buntzen.

J. Davidson, Millbrook, Ont., has purchased the lighting plant and has been awarded the contract for lighting the town.

The Kennedy Carbon Light Co. will erect a plant for the manufacture of gasoline lamps and heaters at Port Arthur, Ont.

The Canadian Machine Telephone Co., Brantford, may erect a plant for the manufacture of [illegible] telephones for use in Canada.

T. Gray & Co., New Westminster, B.C., will erect a sawmill on the Pitt river. The machinery for the mill has been purchased.

Messrs. Mooney, Toronto, are considering the establishment of a large plant in that city for the manufacture of Portland cement.

A company has been formed at Minnedosa, Man., for the development of power there. A hydro-electric plant will be installed.

The Dominion Iron & Steel Co., Sydney, N.S., have received an order from an American concern for one million tons of ore.

The Railway Paint Co., Edmonton, Alta., will build a factory in that city at a cost of $100,000. The company's capital is $250,000.

Operations have commenced on the construction of the new factory for the Imperial Steel & Wire Nail Co., at Fort William, Ont.

The Record Stove & Furnace Co., capitalized at $40,000, will carry on business in Canada. Their headquarters will be at Winnipeg.

A large power plant is contemplated on the Yukon river about 50 miles below Dawson. The initial capacity will be 1,000 horse-power.

The Great West Coal Company has been formed at Port Arthur, to develop 12,000 acres of coal lands recently purchased in Alberta.

W. H. Tamm, of Minneapolis, Minn., has been appointed to have charge of the Port Arthur, Ont., branch of the Barnett-McQueen Co.

The Graham Co., Vancouver, will erect a large mill at the Queen Charlotte Islands. It will have a capacity of 200,000 feet per day.

The power situation in Moose Jaw is reaching an alarming condition. The present plant is incapable of keeping up with the demand.

The Canadian branch of the Royal mint will cost, when completed, about $500,000. The salaries will run as high as $80,000 per annum.

J. B. Pauze & Company, Montreal, have been awarded the contract for building a new jail at Montreal, the contract price bein

E. H. Keating and Wm. H. Breithaupt, Toronto, have formed an engineering partnership, with offices at Aberdeen Chambers, Toronto.

The Moreton Truck and Storage Co.'s building, Toledo, Ohio, occupied by the International Harvester Co., was destroyed by fire recently.

The Copeland-Chatterson Co., Toronto, will take over the entire business of the Elliott Fisher Billing Machine Co., for the Dominion.

R. Forbes Co., Hespeler, Ont., has secured a Dominion charter and will manufacture textile goods. The capital of the company is $1,000,000.

The Consumers' Gas Company, Toronto, will build a one-storey brick broiler house, condenser house and smokestack, at a cost of $[illegible],000.

It is said that the lumber operators on the branches of the Miramichi river, N.B., will handle 120,000,000 feet of timber this coming winter.

The Wapella Roller Mills, Wapella, Sask., owned by R. J. Lund, were destroyed by fire last week. Loss, $22,000, partly covered by insurance.

The puddling furnaces and rolling mills recently established at Winnipeg make the fourteenth iron and steel plant now in operation in Canada.

The New Brunswick Government has decided to grant to the New Brunswick Petroleum Co. a lease of 10,000 acres of land under the usual restrictions.

M. B. Perine & Co., have taken over the business of the Doon Twine & Cordage Co., Doon, Ont., and will improve the plant at a cost of about $55,000.

United States capitalists are considering the erection of a large dry dock, capable of holding the large vessels on the lakes, near St. Catharines, Ont.

The contract has been awarded to the Canadian Iron & Foundry Co., Montreal, for 21 hydrants and 140 tons of pipe, by the town of St. Mary's, N.B.

The new plant of the Canadian Brass Manufacturing Co., at Galt, is now in full operation, and will manufacture plumbing supplies and brass castings.

The Pennsylvania Railroad Co.'s Altoona shops have added to their equipment two Northern 124-ton 3-motor electric traveling cranes, 84-foot span.

The Development & Finishing Co., New York, will install a plant in St. Catharines, Ont., for the manufacture of chlorine and alkali, by the Townsend process.

A large plant for the manufacture of glass, tiles and pipes will be established at Morinville, twenty-four miles north of Edmonton, by a San Francisco syndicate.

The Hespeler Furniture Co. are building a large addition to their works, which is now nearly completed and will be in operation in the course of a few weeks.

The second brick plant for Rosthern, Sask., will be erected soon. A company has been formed and a plant with a daily capacity of 60,000 bricks will be erected.

The Canadian Machine Telephone Co. intend erecting a factory which will employ two hundred men. Brantford, Hamilton and Toronto are all after the new industry.

A large Wisconsin wagon company will erect a Canadian branch factory at Winnipeg, to supply what has been their export trade in western Canada for past three years.

The Saraguay Electric Light & Power Co., Montreal, are erecting a new 11,500-volt three-phase transmission line to Notre Dame de Grace, a distance of nine miles.

A syndicate, headed by Walsh & Arnaud, Vancouver, B.C., will erect a large brick plant in Barnaby, near New Westminster, which will have a capacity of 50,000 bricks daily.

Clare Brothers & Co., Preston, are building a new warehouse near their stove foundry. The new building will be 100 feet by 30 feet, and will be used for storage purposes.

The new blast furnace of the Atikokan Iron Co., Port Arthur, Ont., will soon be running at its full capacity. The company is considering the doubling of its plant at Port Arthur.

Plans are being made for the erection of an enormous plant in the Kootenay district for the manufacture of steel rails. It is hoped that the plant will be able to fill orders in two years.

The Canada Woodenware Company, which has been looking for a site for their factory since they were burned out at Hampton recently, has decided to locate at South Bay, near Fairville, N.B.

A St. Paul, Minn., agricultural implement firm, who have enjoyed a large export trade in western Canada for the past four years, will erect extensive factory buildings at Winnipeg, Man.

The North Atlantic Colliers Co., Port Morien, N.S., are enlarging their plant. They have ordered a Babcock & Wilcox watertube boiler. A pump and hoisting engine will be installed shortly.

O. & C. Clarke, Montreal, propose to establish a factory in that city for the manufacture of brass goods, providing the city will grant them a free site, with exemptions and other privileges.

C. J. Wright, Montreal, is proposing to erect a factory for the manufacture of brass goods in St. Catharines, Ont. A free site is wanted, with exemptions from taxation, and other privileges.

The Silica Brick and Lime Co., Parson's [illegible], is turning out 15,000 bricks per day, and is filling an order for one million bricks for the Spencer block being erected at Vancouver.

The British Columbia Electric Railway Co., Vancouver, B.C., are pushing a scheme which supplies electric power to portable sawmills in the forests. It is claimed to reduce the cost of cutting timber.

The Excelsior Factory, operated at Chilliwack B.C., by Messrs. Kipp & Sons, has been purchased by the Barbara Mattress Company, of Vancouver, and the machinery has been removed to that city.

An American syndicate is considering the erection of a smelting plant at Sherbrooke, Que.

The property and effects of the Windsor Foundry & Machine Co., Windsor, N.S., are offered for sale by tender.

A. F. Baeber & Co., Vancouver, have been awarded the contract for the electric appliances for the new C.P.R. Empress hotel in that city. It is one of the biggest contracts of its kind in western Canada.

The installation of a lighting plant at the new factory of the Gilson Company, Guelph, Ont., is being carried out by A. C. Lyons & Company, Brantford, and will be completed by the end of this week.

The rail mill of the Superior Corporation, at Sault Ste. Marie, was opened recently, after a shut-down of two weeks for repairs. One of the furnaces has been relined and the capacity of the plant increased.

Winnipeg is the location selected by New York State glove and mitten manufacturers. A factory is to be started immediately, equipped with the latest labor-saving machinery, with a capacity of 100 dozen per day.

At the organization meeting of the Wolvering Brass Co., Chatham, Ont., the following officers were elected: L. A. Cornelius, president; H. C. Cornelius, vice-president. Claud Cornelius will be manager and superintendent.

It is reported that a large English firm will compete with the Standard Oil Co. for western Canada business. The new company will build a large oil refinery at Vancouver, B.C., with a capacity of one thousand barrels a day.

A telephone company has been formed in Princeton, Ont., to be known as the Princeton-Drumbo Rural Telephone Co. A line will be run connecting Princeton, Drumbo, Eastwood and Gobles. F. J. Daniel, Princeton, is president.

Fort William, Ont., has made definite arrangements with the International Snow Plow Mfg. Co., whereby the company will receive a free site and ten years' tax exemption, in return for the establishment of steel car shops in that city.

The Sutherland Rifle Sight Co., Ltd., who have just completed a new slow-burning building at New Glasgow, N.S., have started part of their equipment and have toolmakers making tools for manufacturing the Sutherland military rifle sight.

Crowland township last week passed a by-law granting a fixed assessment of $20,000 per year to the Bemis Bag Co., of Boston, who will build a million and a half-dollar factory close to Welland, Ont., and agree to employ fifteen hundred hands.

D. Hall and Alex. Robertson have gone into the lumber business, and will build a large lumber and shingle mill at Chilliwack, B.C. They also intend putting in machinery for the equipment of an up-to-date box factory, and will manufacture fruit boxes.

A new company, the Dominion Smoke Consuming Company, is being organized; capitalization $50,000, with head offices in Toronto. The company has applied for an Ontario charter and has purchased the Canadian patents from the National Smoke-Consuming Company, of Buffalo, N.Y.

The largest fire that has happened in New Glasgow, N.S., for years occurred recently, when the magnificent building, plant and machinery of the Standard Brick & Tile Company were burned to the ground, and the great industrial establishment is now a mass of ruins. The loss is $300,000.

The Crow's Nest Coal Co. are said to have exported large quantities of coke to American smelters. As there is a great shortage at home, Canadian industries are suffering for want of coke and the Yale and Kootenay districts, B.C., are calling for enforcement of the penalty clause in the company's charter.

Recent improvements have made the Nova Scotia Steel Company's plant at Sydney, N.S., one of the finest and most modernly-equipped institutions on the American continent. The company's rod mill at Sydney has a record of production greater than that of any rod mill of the United States Steel Company.

A syndicate of wealthy men, headed by A. D. McRae, of Winnipeg, and Peter Jansen, of Jansen, Neb., have purchased a controlling interest in the Fraser River Saw Mills, Millside, B. C. Large timber holdings have been bought and the total investment made by the McRae-Jansen syndicate is said to exceed $2,500,000.

R. D. Isaacs, St. John, N.B., has submitted to the common council a request for a free site and exemption from taxation for twenty years for a proposed car works to be erected there. He says a company, with $250,000 capital, which will be increased to a million, is [illegible] scheme. The council referred the matter to a committee.

The Canadian Machine Telephone Co., Brantford, have filed plans for the laying of two miles of underground telephone cable. The cable will be laid in ducts, some eight miles of which will ultimately be constructed. The manholes will all be constructed of concrete. The plans of the company involve considerable expenditure.

The Utah Railway Paint Co., Kansas City, Mo., capitalized at over a quarter of a million of dollars, proposes to establish an immense industry of a similar nature at Edmonton, Alta. The company are said to have located in the vicinity of Edmonton immense beds of a mineral clay, called kaolin, from which paint can be manufactured. The proposed plant will cost $100,000.

The Chicago and St. Lawrence Steam Navigation Company's new steamer, the E. B. Osler, was launched at Bridgeburg, Ont. The Osler has the distinction of being the largest ever constructed in Canada. It has a capacity of 9,000 tons. It is 510 feet long and 56 feet wide, and is built throughout after the most approved modern plans. The launching was witnessed by quite a party of Toronto people.

The new works of the Parkin Elevator Co. at Hespeler, Ont., are completed, and the machinery is now being installed. The machine shop is 60 feet by 150 feet, two storeys high: the foundry, 50x60 feet, in which they have installed a cupola of eight tons capacity. The new works are most complete, and the company will be prepared to do wood-work, electrical work, ornamental iron work on all passenger and freight elevators.

The American Electric Furnace Company is establishing a plant at St. Catharines, where an 80 h.p. Colby induction furnace will be installed, and later on another furnace of the R. Jellin type, requiring 200 electric horse-power, will also be installed. This furnace will yield 1,000 pounds of steel at a single heat, and has a capacity of six heatings during twenty-four hours. A large crane of five tons capacity is being built for the plant by the Niagara Falls Machine & Foundry Company.

John Ballantyne & Co., Galt, have invented and are having patented a self-locking device for the movable side pressure and chip-breaker on molding machines. This invention has proved a great time-saver and will greatly enhance the value of their machines. At first used only on their 12-inch molder on base, a massive machine of 7,200 pounds, it proved such an unqualified success that this enterprising firm are placing it on all their molders and have taken the necessary steps to protect their invention.

Companies Incorporated.

Record Stove & Furnace Co., Winnipeg; capital, $40,000. To deal in and manufacture stoves and furnaces. Incorporators: J. Peters, A. E. Peters, D. I. Welsh, all of Moncton, N.B.

The Peele Oil & Gas Company, Toronto; capital, $10,000. To develop oil and natural gas wells. Provisional directors: F. Watts, H. A. Menet, J. L. Galloway, J. C. Colling, and W. E. Sampson, all of Toronto.

Canadian Gypsum Co., Toronto; capital, $20,000. To manufacture gypsum, gypsum plaster and kindred substances. Incorporators: J. S. Lovell, H. Chambers, Robert Gowans, S. G. Crowell, Walter Gow, all of Toronto.

The Tyrell Cooler & Filter Co., Ottawa; capital, $100,000. To manufacture coolers, filters, and similar articles. Provisional directors: P. D. Herbert, H. W. Tyrell, D. T. Smith, James Herbert and J. R. Osborne, all of Ottawa.

Union Brass Goods Co., Toronto; capital, $150,000. To carry on business as brass and iron founders. Provisional directors: Morley Punshon, Vander Voort, Francis Joseph Stanley, and William Andrew Smiley, all of Toronto.

The Ozone Sterilization Co., Haileybury, Ont.; capital, $100,000. To acquire and operate any processes and inventions for purifying liquids, ores, etc. Provisional directors: W. A. Gordon, R. O. Morrow, F. A. Day, all of Haileybury.

The Dickson Bridge Works Co., Campbellford, Ont.; capital, $40,000. To carry on the business of bridge builders and structural steel manufacturers. Provisional directors: James H. Caskey, Campbellford; William Coates Macann, F. C. Downey, Toronto.

Keystone Lorrain Mining Co., Haileybury, Ont.; capital, $1,000,000. To prospect for and deal in minerals. Incorporators: E. J. DuMee, C. J. Suplee, G. T. Armitage, George G. Thompson, all of Philadelphia, Pa., and G. M. Davis, Wilmington, Del.

The Dominion Oil Co., Hamilton; capital, $100,000. To prospect for and deal in oil, gas, salt and minerals. Incorporators: William Melville McClement, H. Harry Bicknell, D. P. Kappele, Florence Austin and Lizzie Eldon Anderson, all of Hamilton, Ont.

G. A. Rudd & Co., Toronto; capital, $100,000. To manufacture and deal in harness, saddlery, leather and kindred articles. Provisional directors: George A. Rudd, James McGregor Young, John F. Elliott, Gordon C. Rudd and Gregory S. Hodgson, all of Toronto.

Hamilton Steel & Iron Co., Hamilton, Ont.; capital, $5,000,000. To carry on a general rolling mills and smelting business. Incorporators: Robert Hobson, William Southam, John Milne, Albert E. Carpenter, Charles E. Doolittle, Geo. L. Staunton, all of Hamilton.

Standard Sanitary Manufacturing Co., of Pittsburg and Montreal; capital, $250,000. To manufacture plumbers' enamelware, brass goods and supplies. Incorporators: T. C. Collins, J. N. Collins, F. J. M. Collins, P. M. Robertson, H. J. Trihey, all of Montreal.

The Algoma Co-operative Co., Sault Ste. Marie, Ont.; capital, $40,000. To sell, manufacture and deal in general wares, hardware, paints, oils, etc. Provisional directors: W. Stringer, J. Cleland, D. Robertson, D. Dewar and D. Donald, all of Sault Ste. Marie.

The Canadian Lash Steel Process Co., Toronto; capital, $100,000. To manufacture and sell iron, steel and other metals and their alloys. Provisional directors: J. A. Macintosh, B. W. Essery, Eleanor J. Potts, J. G. Adair, and John Adair, all of Toronto.

Anthes Foundry, Limited, Toronto; capital, $100,000. To manufacture castings, iron pipe, soil pipe and fittings, and plumbers' and steamfitters' supplies. Provisional directors: Lawrence L. Anthes, William Wallbridge Vickers, and Herbert C. Sparling, all of Toronto.

Ingersoll Sergeant, of Canada, Limited, Montreal; capital, $20,000. To manufacture air compressors, rock drills, pumps and pneumatic tools. Incorporators: H. D. Lawrence, William Morris, A. F. Plant, R. F. Morris, Sherbrooke, Que.; W. E. McIver, Richmond, Que.

Monterey Plumbing Co., Toronto; capital, $50,000. To carry on a general plumbing, steamfitting, heating and lighting business, and to deal in general hardware. Incorporators: Gerard Ruel, G. F. Macdonnell, A. J. Mitchell, F. C. Annesley, Robert P. Ormsby, all of Toronto.

National Oxide Paint & Color Co., Hamilton, Ont.; capital, $50,000. To manufacture paint pigments, colors, oils, paints, varnishes, and japans. Provisional directors: George Stroud, G. F. Webb, Helena O'Sullivan, T. J. O'Sullivan, and Alfred Stroud, all of Hamilton, Ont.

The Dominion Nickel Copper Co., Toronto; capital, $10,000,000. To carry on the operations of a mining, milling, reduction, and development company. Incorporators: James Houston Spence, Ada May Duncan, Lillian M. Heal, Charles E. Freeman, Gertrude E. Jamieson, all of Toronto.

L. H. Hebert & Co., Montreal; capital, $350,000. To carry on the business of ironmongers and merchants in hardware, paints, oils and chemical compounds. Incorporators: Louis H. Hebert, Alfred Jeannotte, Eugene Poitevin, Gustave Busseau, Montreal; Joseph E. Theriault, Joliette, Que.

The New Liskeard Concrete Co., New Liskeard, Ont.; capital, $40,000. To manufacture concrete cement, artificial stone and kindred materials. Provisional directors: S. Jewell, V. E. Taplin, W. H. Carruthers, W. V. Cragg, A. McKelvie, J. E. Whyte and F. L. Smiley, all of New Liskeard, Ont.

The Jenks Dresser Company, Sarnia, Ont.; capital, $50,000. To carry on the general manufacture and sale of iron and steel work. Provisional directors: William G. Jenks, Andrew A. Dresser, Roy M. Norton, all of Port Huron, Mich.; Morton Fuller, Richmond, Mich., and Hiram Manning, Sarnia, Ont.

Ideal Foundry Co., Toronto; capital, $[illegible]. To engage in the business of general foundrymen and to manufacture cast or wrought iron, steel, brass, aluminum or composition specialties. Provisional directors: Henry Edward Pearce, William Henry Smith, Arthur Gate, Matthew Irving and William Baggs, all of Toronto.

Simplex Gas Co., Toronto; capital, $40,000. To manufacture under Colwell's patents the Simplex gas machine; to manufacture gas burners, stoves, heaters, ranges and hardware, and mechanical specialties. Provisional directors: Walter Ernest Colwell, Harry Herbert Colwell, Oakville, Ont., and Henry H. York, Toronto.

Building Notes.

An armory will be erected at Medicine Hat, Alta.

St. Joseph's church, Sydney, N.S., will be rebuilt.

H. Gleiser, Oxbow, Sask., will build a new hotel.

The new collegiate school at Ottawa will cost $225,000.

Robt. Shaw, Oxbow, Sask., will erect a new business block.

R. Cassidy, Vancouver, will erect a block, to cost $250,000.

A. Peters will erect a hotel in New Westminster, to cost $10,000.

A Masonic hall will be erected at St. Thomas, Ont., to cost $10,000.

A new Anglican church will be erected at Elmwood, Man., to cost $5,000.

An Oddfellows' hall, costing $30,000, will be erected at Strathcona, Alta.

Seymons & Roe, Toronto, will erect a brick dwelling, at a cost of $10,000.

Rogers & McKay, Vancouver, will build a residence, at a cost of $12,000.

Yuen Chong, Vancouver, B.C., will erect a $12,000 business block in that city.

An addition will be built to the Windsor hotel, Peterboro, Ont., to cost $24,000.

The Imperial Bank will erect a new building at Edmonton, Alta, to cost $90,000.

E. Hobson, Vancouver, B.C., will build an apartment house, at a cost of $12,000.

H. Dorenwend, Toronto, will erect an apartment house in that city to cost $20,000.

A new home, to be erected for the Seamen's Mission, St. John, N.B., will cost $15,000.

The Anglican parish of St. Albans, Burnby, B. C., will erect a church, to cost $2,000.

Tenders are invited by T. Gillespie, Peterboro, Ont., for the erection of a Baptist church.

The new Protestant school, which was recently opened at Delorimier, Que., cost $15,000.

P. Roach, Toronto, will build three attached two-storey brick dwellings at a cost of $10,000.

A new business block will be erected in Vancouver by W. Hepburn, at a cost of $22,000.

J. Y. Griffin Company will erect a cold storage block at Edmonton, Alta., to cost $100,000.

The Crystal Theatre Picture Show will erect a theatre in Moncton, N.B., to accommodate 2,000.

The Bank of Commerce will build a branch on Yonge street, opposite College street, Toronto.

T. Gingras, Toronto, will build four pair semi-detached houses, at an aggregate cost of $14,500.

S. J. Castleman, Vancouver, will erect a six-storey business block, to cost approximately $100,000.

C. R. S. Dinnick, Toronto, will erect six pair semi-detached brick residences, at a total cost of $30,000.

The congregation of Paisley St. Methodist church, Guelph, will erect a new edifice, at a cost of $10,000.

W. O. McTaggart, Toronto, will build three three-storey brick stores and dwellings, at a cost of $12,000.

Harry Weinberg, Toronto, will erect a large store on the corner of Yonge and Edward Sts., at a cost of $30,000.

The Nelson Theatre Co., Nelson, B.C., will erect a large theatre in that city. C. W. Busk is one of the directors.

The J. Y. Griffin Co., Edmonton, Alta., will erect a large office block and cold storage warehouse, at an estimated cost of $100,000.

L. C. Sheppard, Toronto, will build seven pairs semi-detached two and one-half-storey brick dwellings, at a total cost of $25,000.

The J. Y. Griffin Company will erect a large office block and storage house in Edmonton, Alta., to cost in the neighborhood of $100,000.

The barns of the Macdonald Agricultural College, St. Anne de Bellevue, Que., which were recently destroyed, at a loss of $35,000, will be rebuilt.

The value of Vancouver building permits for the month of August was $708,775, being an advance of 131 per cent. over the $306,925 recorded for the corresponding month last year.

Municipal Undertakings.

A new school building will be erected at New Westminster, B.C.

Public buildings will be erected at Neepawa, Man., and Selkirk, Man.

Moose Jaw, Sask., will raise $90,000 for a municipal lighting system.

The town of Gracefield, Ont., will install a municipal waterworks system.

A by-law was passed in Toronto, providing $781,121 for waterworks extension.

The foundations have been laid for the new civic buildings at Saskatoon, Sask.

Surveys are being made for the new Government telephone system for Winnipeg.

The town council of Hull, Que., has decided to erect three new fire halls, to cost $25,000.

The ratepayers of Cornwallis, Man., are petitioning for a municipal telephone system.

The by-law for [illegible],000 for electric light extension in Moose Jaw passed the first reading.

J. B. Beverage, Chatham, N.B., is considering the erection of a pulp mill in Newcastle, N.B.

The Government will build a large concrete dam near Trenton, in connection with the Trent canal.

The town of Orillia, Ont., will make a profit this year of $9,000 from its light and power plant.

The city of Revelstoke, B.C., is calling tenders for the enlargement of the city hydro-electric plant.

The council of Bracebridge, Ont., are considering the propositions of a firm manufacturing steel chairs.

The ratepayers of Prince Albert, Sask., will vote on a by-law for $29,000 for sewerage improvements.

The new light plant for Claresholm, Man., has been completed. It replaces the one destroyed by fire recently.

A by-law was passed at Islington, Ont., for the issue of debentures to the extent of $18,000 for a school house.

Tenders will be asked by the Winnipeg Board of Control for a building in connection with the waterworks, to cost $20,000.

A provincial asylum will be erected by the Government of British Columbia, at Coquitlam, B.C. It will cost $200,000.

Bids will be asked by the Winnipeg Board of Control for a $40,000 bridge in connection with the Point du Bois power plant.

The ratepayers of North Battleford, Sask., have sanctioned a by-law for $10,000 for a bonus for a new flour mill and elevator.

The town of Kenora, Ont., is contemplating the installation of an arc lighting system, at a cost of $4,000, and an electric pump for the waterworks.

The council of Ottawa, Ont., are considering the erection of a municipal lighting plant, to be put in operation when the Hydro-electric contract expires.

The ratepayers of the village of Tetreauville, Que., will vote on a by-law for issuing $15,000 of debentures and establishing a municipal waterworks system.

The ratepayers of Aylmer, Ont., have petitioned the council to raise $10,800 for additional water supply, and to furnish a site for the Canadian Condensed Milk Co.

The contract for the erection of the telephone exchange for the Manitoba Government, at Winnipeg, has been awarded to J. J. Kelley, Winnipeg, the figure being $97,172.

London, Ont., has decided to submit the Komoka water scheme to the people in two by-laws, one to provide $293,500 for extending the domestic water supply, and one for $182,000, to provide for a hydraulic power plant and reservoir.

A new court house and a new registry office will be built at Sudbury, Ont. The contract for the court house has been awarded to M. Healey, Toronto. The registry office will be built by the O'Boyle Construction Company, North Bay.

Mining News.

Part of the plant of the Irondale Smelters, Port Townsend, B.C., was destroyed by an explosion.

The Dominion Iron & Steel Co., Sydney, N. S., will work the Wabana iron ore fields, near Sydney.

A large ore crusher will be added to the plant of the Granby Co., at the Boundary mines.

The Salt Lake Smelting & Refining Co. will commence the erection of their smelter at North Cobalt, soon.

During the first month of the last half of the year the Boundary smelters, B.C., reduced 150,000 tons of copper ore.

The plant of the Algoma Steel Co., Sault Ste Marie, has shut down on account of the lack of ore and need of repairs.

Owing to the shortage of coke in the west, it is feared that the furnaces at the Trail smelter, near Rossland, will have to shut down.

It has been found possible to turn out pig iron capable of being turned into the best quality of Bessemer steel, at the Atikokan furnace, Port Arthur, Ont.

The successful experiments made by the Dominion Government with the Heroult electric smelter, at Heroult, Cal., have been corroborated. The process is found satisfactory and much cheaper than usual methods.

The Canadian Smelting and Refining Co., Sault Ste. Marie, will erect a large smelter and expect to have the plant in operation within six months. $1,000,000 is behind the enterprise.

A large and enormously rich deposit of iron ore has been discovered at Lakevale, Antigonishe, N. S. The property is said to be one of the best yet found and is owned by American capitalists.

The directors of the Golden Peak Mining Company, Larder Lake, Ont., have decided to install a mill of 20 stamps on their property. The mill will be put in as soon as the road from Boston to Larder Lake is completed. This will be in about a month.

It is stated authoritatively that a company, called the Muggley Concentrators, Limited, will put in a hundred-ton concentrating mill shortly at Cobalt. The mill will use twenty screens and crushing rolls to increase the capacity. Owing to the delay in getting in machinery it is probable that the plant will not be running before January.

Railroad Construction.

The C.P.R. will spend $100,000 in additions to their shops at Winnipeg.

A new electric line is proposed between Wallaceburg and Petrolea, Ont.

The roadbed of the G.T.P., east of Edmonton, is ready for the steel.

Work on the Goose Lake, Sask., branch of the C.N.R., was commenced recently.

Work has commenced on the construction of the new C.P.R. depot at Saskatoon, Sask.

Construction has commenced on the electric line from Eburne, B.C., to New Westminster.

Work is progressing rapidly on the construction of the Brandon-Regina line of the C.N.R.

The Michigan Central will construct a line from Charing Cross, Ont., to Chatham, Ont.

The G.T.R. will proceed without delay to construct the line from Kingston to Ottawa Ont.

The C.P.R. are installing telephones along the line between Fort William, Brandon and Winnipeg.

The B. C. Electric Railway will spend $350,000 in improvements to their New Westminster premises.

A new flour shed is about to be erected by the I.C.R. at St. John, N.B., at an expenditure of $20,000.

The grading is almost completed on the C. N.R. branch from St. Jerome, Que., to the main line.

The Canadian Northern is considering the building of an air line from Brockville to Ottawa, Ont.

The contract for the new I.C.R. station at Amherst, N.S., has been let to LeBabco, of Moncton, N.B.

There is a rumor that the proposed Hamilton-Galt electric line will be pushed through without delay.

The new Guelph to Goderich line of the C. P.R. was opened recently. Trains are now running regularly.

Construction will begin on the Kootenay Central branch of the C.P.R. from Golden, B.C., to Crow's Nest Pass.

The Ottawa Electric Railway Company propose to extend their line to the Experimental Farm, near that city.

Active operations are in full swing on the construction of the new addition to the G.T.R. shops at Stratford, Ont.

The contract for the new roundhouse for the C.N.R., at Brandon, has been let to the Sharp Construction Co., Winnipeg.

The Canadian Pacific will erect an extensive terminal station at Fort William, Ont. They will build a large dock.

The Canadian Northern will build a line from Vancouver to northern British Columbia, to connect with the main line.

The Chicago, Milwaukee and St. Paul Railway will build to the coast and run a line from Seattle to Vancouver.

The C.P.R. and Brandon, Saskatchewan and Hudson Bay Railroads will build a line connecting their yards through Brandon.

The contract for the G.T.P. bridge over the Kaministikwia river has been awarded to Wylie & Belfour and the Canadian Bridge Co.

The construction has commenced on the line of the B. C. Electric Railroad, from New Westminster to Chilliwack, to cost $2,500,000.

The contract for the G.T.P. bridge over the Kaministikwia river, Fort William, has been awarded to the Canadian Bridge Company.

The N. St. C. & T. Ry. Company's new power house at Thorold, Ont., is now under construction. It will cost $50,000, and develop 1,500 horse-power.

Contracts for the erection of a [illegible] addition to the car shops at London, have been let to Mr. John Hayman, and work will be commenced immediately.

The Ontario Municipal and Railway Board have approved the plans for the completion of the line from Pembroke to Gold Lake, on the Pembroke and Gold Lake Railroad.

The rebuilding of the railway bridge across Rice Lake, near Port Hope, Ont., and the rehabilitation of the old Cobourg and Peterboro Railway, is being seriously discussed.

The Intercolonial Railway have recently given to the Rhodes, Curry Co., Amherst, N.S., the contract for 260 flat cars, 400 box cars, 25 refrigerator cars and four conductor's vans.

The Southwestern Traction Company's line, after being tied up for three weeks, on account of the recent fire in the car barns, is now in full operation between London and St. Thomas.

The Canadian Railway Commission have issued new regulations regarding the watching of trestles and bridges, fire protection for same, watching of tracks and fire protection for cars.

The C.P.R. will build two large steamers of the Princess Victoria type for service on the Pacific. The builders are from the Fairfield Shipbuilding & Engineering Co., Govan, Scotland.

The British Columbia Electric Railway Co. Vancouver, B.C., are pushing a scheme which supplies electric power to portable sawmills in the forests. It is claimed to reduce the cost of cutting timber.

The British Columbia Electric Railway have ordered 26 cars for city service. In order to turn these out, the number of men employed at the car shops will be doubled and two shifts organized to work night and day. The cars cost $6,000 each.

The C.P.R. are fitting up a plant at Vancouver, B.C., where they intend to generate Pintsch gas, to be utilized in illuminating passenger coaches on their western lines.
completed for the C.P.R.

The C.P.R. are contemplating building a million-ton coal dock at Fort William. The entire yards of the company will be remodelled at great expense, and the plans, when carried out, will make Fort William the finest inland steamship and railway terminal in the world.

The C.P.R. will spend over $1,000,000 in constructing two tunnels through the Rockies near Field, B.C. One will be to the north of the Kicking Horse river and will be 3,400 feet long. The other will be south of the river and will form practically a complete circle through the heart of Cathedral mountain. It will be 3,800 feet long.

Eight new engines of the "consolidated" type have been received at the London, Ont., roundhouse of the Grand Trunk, and are now being operated on that division. As a result, the situation has been greatly relieved. The engines are of the heaviest type in Canada. They are able to haul 90 loaded cars and have four drive wheels. They are much heavier than the locomotives of the "900" and "800" class and will be a great addition to the freight service.

Costly Lawsuit.

In the lawsuit of the Dominion Iron & Steel Co. against the Dominion Coal Co., Sydney, fore the courts amounted to $[illegible]
N.S., it was decided that the Dominion Coal Co. committed a breach of contract in supplying a grade of coal unfit for the purposes for which it was required.

It is announced that the cost of this case before the courts... decision provides for an arbitrator to determine the damages that the Coal Company must pay, and it is said that the Coal Company will appeal. For every hour the court sat, the suit cost $1,000. The case was purely a business disagreement and had it been settled by arbitration the costs would have been much less and the results eminently more satisfactory.

New Plant for Fort William.

The Imperial Steel & Wire Company's new plant, at Fort William, is now under construction, the site being alongside the new works of the Canada Iron Foundry Company. The main building is to be 600x70 feet, while a warehouse will also be erected, 60x200 feet. The buildings will be of cement, and will be constructed in time to have machinery installed next spring. It is expected that 250 men will be employed at the works. Alderman R. S. Piper, hardware merchant, presented a handsomely decorated shovel, which was used by the wife of the general manager, Major J. A. Currie, in turning the first sod.

New Foundry and Machine Shop.

McLean & Holt are just completing a new foundry and machine shop. It will be equipped to carry on a very extensive business. The cost is not mentioned. With the installation of an up-to-date plant in this building the firm will be in a better position than ever before to turn out "Glenwood" ranges and all lines of stoves and furnaces.

A New Departure in Building.

Dr. Alexander Graham Bell has just completed, on Lookout Mountain, at his summer residence, near Baddeck, N.S., a lookout tower, built on the tetrahedral principle, said to be the first tower of its kind in the world. A tetrahedron is simply a three-sided pyramid, its three sides and base all being equilateral triangles. Those in the tower are made each of six pieces of galvanized half-inch pipe, four feet long, and four nuts, into which the pipes are screwed, and the tower is simply a giant tripod built up of these. Such a structure, Mr. Bell says, is lighter and stronger than any other, more quickly and more cheaply built, and requires no skilled labor. Patents on the principle and on the nuts used at the corners, are being taken out.

New Cars for C. P. R.

The C.P.R. will build an entirely new type of car for use on the transcontinental trains. This car will be modelled somewhat on the lines of a new type on the Southern Pacific Railway. One great feature will be windows 54 inches wide and reaching to the roof; in fact, the whole sides of the car will be practically made up of plate glass, so as to give the best possible opportunity to travelers of seeing the sights along the railway. There will, in addition, be a big observation platform, where people can sit in the open in easy chairs in fine weather.

The interior of the car will be divided into three compartments. At one end it will be furnished with easy chairs very much after the style of the present parlor cars. In the middle will be a library, with all the latest books, magazines and newspapers, and also a counter where light refreshments can be obtained; at the other end will be a very large, airy, and comfortable smoking car, of the same size and fitted up in very much the same way, observation windows and all, as the parlor end of the car. These cars will be known as "Library, Club and Observation Cars," and enough of them are to be built this winter to put one on one through train a day each way next season.

To Prevent Collisions.

Recently the Intercolonial Railway have been making exhaustive tests of the new device invented by Mr. H. W. Price, of Toronto, for the prevention of both head-on and rear-end collisions. The tests were carried on at Moncton and two trains fitted up with the instruments were started from stations some distance apart. When they came to the block of the signal system the brakes were automatically applied and the train stopped. The tests were satisfactory in every way, and the adoption of the system by the I.C.R. is quite probable.

ABOUT CATALOGUES

FURNACES.—Two bulletins on rod and bolt heating furnaces, along with prices, manufactured by Rockwell Engineering Co., New York city.

FINISHING GAS ENGINE PARTS.—Sheet No. 76 of Gisholt Machine Co., Madison, Wis., showing operations for finishing pistons and piston rings.

DROP FORGING.—A handsome booklet of E. N. Bliss Co., Brooklyn, N.Y., illustrating and treating of their forging drop-hammers, and also presses.

E. N. BLISS MEMORANDA.—A very neat little vest-pocket booklet illustrating and describing Bliss presses, and containing pages for memoranda at the back.

[illegible] VISES.—Booklet issued by Emmert Mfg. Co., Waynesboro, Pa., containing illustrations, descriptions and price lists of the Emmert patent universal vises.

WRIGHT'S TAPER-ROLLER BEARINGS.—Handsome catalogue of these bearings, made by Canadian Bearings, Ltd Hamilton. It is 6x9 inches, and is well illustrated.

POWER TRANSMISSION MACHINERY.—Catalogue No. 73 of the Link Belt Co, Philadelphia, concerning their conveying machinery. It is well illustrated, 15 pages, 6x9 ins.

[illegible] GAS POWER CO.—Catalogue of producer gas plants of the Smith Gas Power Co., Lexington, O., of whom E. S. Cooper, 12[illegible] Adelaide St. E., Toronto, is Canadian agent.

PNEUMATIC TOOLS.—A 48-page catalogue of pneumatic tools issued by the Dayton Pneumatic Tool Co., Dayton, Ohio. The different [illegible] of hammers, riveters and caulking hammers are clearly illustrated.

[illegible]CUTTERS.—A neat, 16-page pamphlet issued by the Landis Machine Co., Waynesboro, Pa., describing their bolt-cutters and threaders. The many exclusive features are fully explained and each type is clearly illustrated.

[illegible] ENGINES.—A booklet of the De La Vergne Machine Co., New York, describing their two-cycle vertical oil engines, 7½ and 15-horse-power. There are several illustrations. The engine is designed for kerosene and cheap oils.

[illegible]ING WHEELS.—Complete and handsome catalogue of the Norton Co., Worcester, Mass., containing illustrations, descriptions and price lists of all kinds of grinding wheels and machinery. Any interested in this class of machinery should not be without it.

WIRE DRAWING MACHINES.—Two catalogues issued by the German firm, W. Gerhardi, Ludenscheid, Germany. There is an 8-page booklet describing their No. 0 machine, fully describing it, and with several cuts. They also issue a complete catalogue of wire-drawing, galvanizing and rolling mill machinery. It is very tastefully arranged and profusely illustrated.

Canadian Electrical Exhibition

The Canadian Electrical Exhibition, the first of its kind in seventeen years, was held in Montreal from Sept. 2 to 14. In the brilliancy and interest of exhibits, in the elaborateness of arrangement, and in the large and distinguished attendance, it is seriously doubted whether this exhibition has been surpassed ever in America. Everything electrical from the smallest electric globe to the gigantic generator of Allis-Chalmers-Bullock, Limited, was on exhibit, and every spectator demonstrated deep interest in the various applications of electricity displayed. The Exhibition was officially opened on Monday, Sep. 2, and formally opened Thursday evening following, by Hon. Lomer Gouin, Premier of Quebec.

Amongst the specially interesting features of the show were the Dancing Skeleton (an ingenious arrangement and manipulation of electric lights), artificial lightning, Marconi wireless telegraph stations, with band concerts every evening. All the leading manufacturers of electrical apparatus and supplies in Canada were represented.

CANADIAN ELECTRICAL EXHIBITION—Exhibit of Allis-Chalmers-Bullock, Montreal.

Allis-Chalmers-Bullock, Limited.

The feature of the display made by Allis-Chalmers-Bullock, Limited, was its solid and substantial appearance, recalling the merits of motor, generators, and other machinery made by this company in Montreal. For example, there was a 225 K.W. belted alternating current generator with direct connected exciter in operation. The company had also a number of workmen at work building a 300 K.W. waterwheel type alternator to operate under the enormous pressure of 11,000 volts.

The company brought, too, a number of the winding department girls to show how induction motors are built up from the point where the wire is insulated until the motors are ready for operation, which attracted the attention of both engineers and the general public.

A practical use to which one of their induction motors was put was to drive a Worthington centrifugal pump, built by the John McDougall Caledonian Iron Works Co., Limited, showing on a small scale the motor and pump of 5,000,000 gallons capacity, at the McTavish street station, of the Montreal Water Works. The fountain in the centre of the exhibition, with its multicolored rays, was still another example of the use their constant speed induction motors may be put. The lighting transformers, which disfigure our streets and walls are, however, a necessary evil under modern conditions, and of these the company showed an interesting line, running from one to 25 K.W.

Canadian General Electric Co.

Probably one of the most interesting exhibits at the show was that of the Canadian General Electric Co. for the reason that not only did their exhibit demonstrate their ability to manufacture almost every line of electrical apparatus in use, but in this exhibit they catered to the average householder. Their electric kitchen was certainly a wonderful exhibit. Their demonstrators were brought up from Jamestown Exhibition and were demonstrating what could be done in the way of cooking and baking all kinds of delicacies for the table. Their electrically driven bench drills and breast drills were certainly marvelous, and show how rapidly machine shops could increase the efficiency of their plants.

Probably one of the most remarkable exhibits of the Canadian General Electric Co. was that of their lighting. They not only showed the G. E. M. high efficiency lamps in all candle powers, but are also showing the Nernst lamp, and in addition to this the Tungstein lamp.

They also exhibited all types of alternating current and direct current motors as used in the city of Montreal in sizes from 1-16 of 1 h.p. up to 10,000 h.p.

They also exhibited the Mercury arc rectifier panel, which converts alternating current into direct current, and which is one of the very latest systems of modern arc lighting in use to-day.

Canadian Westinghouse Co.

The name "Westinghouse" is almost synonomous with progress in the most important branches of electricity, and this is one of the many reasons why the booths of the Canadian Westinghouse Company attracted so much attention at the big show.

One of the first things that attracted the casual visitor to the Westinghouse booth was a huge 1,500 horse-power railway type generator, which the company has manufactured for the Montreal Street Railway's new big powerhouse at Hochelaga. There was also on view a 4,000 horse-power transformer.

Then, again, the Westinghouse Company showed some very splendid types of electric lamps, pre-eminent among which is the one known as the Nernst. This lamp, in addition to being placed throughout the large booths of the Westinghouse Company, was used for the general lighting of the whole Drill Hall.

In the centre of the Westinghouse booth was an immense static electric sign which was operated with a pressure of 40,000 volts.

Near the main entrance the Westinghouse Company had constructed the fore part of an electric car equipped with all the Westinghouse air brakes and other apparatus that is now so commonly used on all electric cars and railway trains. Mr. H. D. Bayne, manager of the Canadian Westinghouse Co., was also a director of the Canadian Electrical Exhibition Company, and much of the success of the first show was mainly due to his efforts and enterprise.

Belliss & Morcom.

A well-known firm of engine builders for electric service in England are Messrs. Belliss & Morcom, Limited of Birmingham. Independent statistics show that more than half of the steam-driven electric stations in Great Britain are using Belliss engines—that this firm are realizing the importance of the Dominion as an engineering field was evidenced by their exhibit at the Electrical Show. Their booth was in charge of Messrs. Laurie & Lamb, 212 Board of Trade Building, Montreal, a firm of engineers whose members have been closely identified with power station work for nearly 25 years.

The Belliss exhibit was unfortunately without the engine which they had expected to have ready to show there, but in its place they had a very suggestive display of photographs of some large installations.

The special features of the Belliss engine are well known, for instance the system of forced lubrication originated by this firm in 1890 by which it is continuously supplied to the bearings under a presure of 15 or 20 lbs. to the square inch, thus ensuring a constant film of oil between the bearing surfaces.

This system has since been fitted to upwards of 3,300 Belliss engines, aggregating 600,000 b.h.p. The results are remarkable, as the wear on brasses has been reduced to practically nil—a case in point being a 220 b.h.p. engine which after five and a half years' constant service in Croydon Electric station, showed an actual wear on main bearing of only .0025 inch.

The space occupied for horse power by the Belliss engine is very small, a 500 K.W. Belliss set taking up no more floor space than a steam turbine.

The Northern Electric & Mfg. Co. Exhibit.

The exhibit of the Northern Electric & Mfg. Co., of Montreal, at the Exhibition, was one of exceptional interest to those who were connected in any way with telephone or fire alarm affairs.

The telephone apparatus was a complete exhibit of modern switchboards from a small ten line magneto board for rural exchanges to the very latest type of central energy switchboard in use in all of the largest cities. Telephones of several types designed for use under all sorts of conditions from a rural party line of twenty-five subscribers to one for use on a large ten thousand line system was the most attractive feature of the exhibit.

This company also manufactures a very complete line of fire alarm apparatus, consisting of signal boxes of several types, suited to systems of all sizes. Registers, gongs, indicator and central office equipment. The city of Montreal is equipped with apparatus manufactured by this company.

Another interesting feature of the exhibit was the complete line of condit oil, immersed switches and circuit breakers. The Northern Electric & Mfg. Co. are the Canadian selling agents for these switches.

Another feature of this exhibit was the line of Vulcan electric soldering tools, shown in sizes and shapes especially adapted to various classes of work in the telephone business, where uniform and reliable results are absolutely necessary. The Northern Electric & Manufacturing Company has secured the Canadian general agency and manufacturing rights for these thoroughly practical and high-grade tools, the advantages of which are so fundamental that their general adoption in place of furnace-heated coppers is secured. A 24-inch soldering tool of only $\frac{1}{2}$-inch diameter, for use in congested multible switchboard repairs, is a very interesting example of the unequaled adaptability of these electrically self-heated tools.

Robb Engineering Co.

This company's booth at the Electrical Exhibition attracted a great deal of attention, it being the only steam engine exhibit at the show. They exhibited one of their Standard 12 x 12 Robb-Armstrong Automatic Side Crank Engines, direct connected to 50-k.w. generator, which is to be used as an exciter unit in connection with four 1,000 h.p. cross compound Corliss Engine direct connected to alternating generator, being installed at the Canadian Pacific Railway Co.'s Angus Shops.

Also one of their Standard Robb-Armstrong Vertical Single Crank High Speed Engines, which they manufacture in various sizes, was exhibited. The engine is of the enclosed, self oiling type, and especially arranged for direct connecting to generators, fans and centrifugal pumps, as well as for belted operation.

They also showed a working model of the Robb-Armstrong Corliss Valve Gear, which was especially interesting to engineers, it being a positive-driven motion, requiring no latches, springs, or delicate parts, such as are used in the old type Corliss Valve, while at the same time a motion is obtained precisely the same as that invented by Geo. H. Corliss. An engine fitted with this valve gear can be run from 75 to 175 revolutions per minute.

The engine parts and photographs of some of the numerous plants installed by this company are worthy of mention and show the extent of their work.

CANADIAN ELECTRICAL EXHIBITION—Exhibit of Northern Electric Mfg. Co., Montreal.

The exhibit was in charge of Alister MacLean, resident engineer, Robb Engineering Company, Montreal.

Canadian Rand Co.

The practical demonstrations of working machinery which the Canadian Rand Company are placing before the public, served to make this enterprising concern one of the central points of interest at the Electrical Show. There was shown, in constant operation, a small motor-driven air compressor, for use in situations where compressed air can be utilized as power, or as a means of cleaning out motors. The latter point was emphasized by the official in control of the exhibit. The great cause of motors burning out is the presence of dirt and dust, mixed up with the wiring of the motor. As has been the experience of a great many, this foreign substance sooner or later catches fire, and the electrician is called in for the expensive job of re-widing. By an occasional blowing out with the air compressor all this could be avoided. The work of this remarkable power formed one of the best features of the show.

J. A. Dawson & Co.

One of the most attractive and busy booths at the show was that of J. A. Dawson & Co., admirably located near the front entrance of the hall. As the picture will indicate, their booth was a mass of well displayed electrical novelties of many varieties. Flowers, bunting and prettily arranged lights and globes made a very handsome effect. Their especial lines, such as electrical railway materials, insulating joints and electric specialties were prominently displayed. The well known Macallen goods, for which they are the agents, received much attention. Among these goods were the giant strain insulators, all the various curve suspensions, and cap and cone material, porcelain guard wire insulators, splicing sleeves, trolley frogs, turnbuckles, etc. They also carry a full line of gears and pinions of the different standard types. In fact, no electrical man could pass without calling, as a competent staff of attendants were busy. Mill supplies were on exhibition, also motors and generators of various sizes and special lighting systems to suit all sized institutions and factories. A splendid array of car heating apparatus—that is, electrical apparatus—was on view and interested large numbers. Added to these was a fine display of the Crouse-Hinds Imperial Arc Headlights. Many smaller articles were noticeable, such as wrenches, the celebrated Jackson brand with interchangeable parts, etc.; also the D. & T. anchor drive and twist—in fact, a diversity too numerous to mention. A catalogue for any of these lines will be sent to all applicants, and these are worthy of immediate attention. The Dawson Co. began business in Montreal many years ago and have steadily grown in public favor. Their address is 148 McGill Street, right in the heart of the best business section, and all enquiries receive prompt and careful attention.

CANADIAN ELECTRICAL EXHIBITION—Exhibit of J. A. Dawson & Co., Montreal.

Exhibit of the Packard Electric Co.

The chief features of the Packard Electric Co.'s exhibit were the Jandus alternating current series arc-lighting system, with regulators for street lighting; Jandus line of ceiling and column gyrofans; Crocker-Wheeler dynamos and motors, with alternating and direct current and controlling devices; a full line of America switchboard type and portable electrical measuring instruments. The Packard Electric Co. have the agency for all the above lines. They showed their own new design of alternating current, induction motors, and core type transformers, the Standard Incandescent lamp in all types, also recording Wattmeters of switchboard type, and regular house type.

Canadian Fairbanks Co.

Another exhibit which attracted a great deal of attention was that of Canadian Fairbanks Co., one of the features of which was a four-ton Yale and Towne electric 220-volt, 22-foot life hoist, suitable for use in factories and warehouses. Besides this were a small Fairbanks-Morse engine running a dynamo and pump adapted to lighting and water supply, and Fairbanks-Morse dynamos.

Canadian Buffalo Forge Co.

The exhibit of the Canadian Buffalo Forge Company was a very interesting one, as it gave the visitor a chance of inspecting some of the latest devices used in ventilating purposes. The Canadian Buffalo Forge Company has already equipped a number of buildings in Montreal, including the St. Paul's Hospital, the Western Hospital, the McGill Union and the new McDonald Engineering Building.

In the Buffalo air washer and humidifier the air is brought into direct contact with water spray. After leaving the intake grating the air is led directly to the air washer and there is sprayed by closely spaced nozzles, ejecting each between one and one-half and two pounds of water per minute, in the direction of air travel. The air then encounters the eliminator plates, which remove the free moisture from the atmosphere, allowing it to retain only the amount dictated by the regulating devices to keep the humidity at the proper point. The first portion of the eliminator baffle plates presents to the water merely a water covered surface, and contact with this is what removes the solid particles which may be in suspension in the incoming air. The second portion of the plates is provided with vertical gutters which break the continuity of the water film and prevent its being blown from the farther edge of the plates and on into the fan and up into the building. The free moisture is removed by these gutters and runs into the solid matter removed from the air into a settling tank, beneath the washing chamber. Here there is provision for removing the accumulation of dirt, and then the water is pumped again to the spraying nozzles by a small brass centrifugal pump. The resistance to the air passage by the baffle plates is less than one-tenth of an inch water pressure, and hence is negligible. The design of the nozzles is unique, it may be added, because the water is atomized by centrifugal force acquired by its rotative action acquired by the direction of its passage in leaving the nozzles. The setting tank is usually built in a concrete foundation for the washing chamber, it being found that this answers well and is both inexpensive and neat.

Another office that air washers are capable of filling during the summer is that of an air cooler. This is an important application for theatres in particular, and, of course, is used so in all installations during the warm weather. The Canadian Buffalo Forge Company has made installations and guaranteed that the final difference of temperature between the air and the water shall not exceed 20 per cent. of their initial difference in temperature.

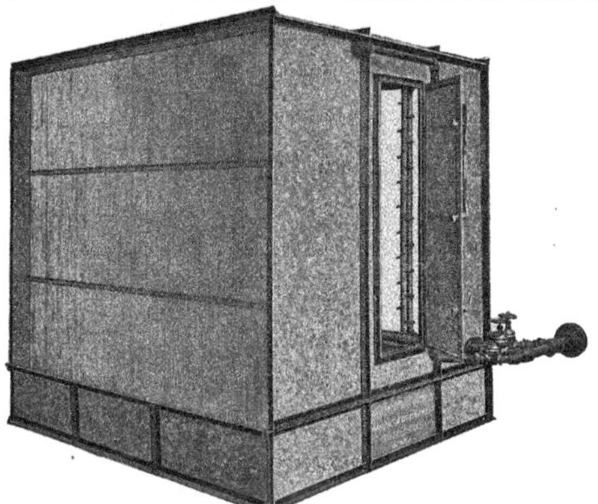

View Showing Face of Elminator, Buffalo Air Washer—Exhibited at Electrical Show, Montreal.

Amongst the other exhibitors were: Adams-Bagnel Electric Co.; Aluminum Co.; American Conduit Co.; Babcock & Wilcox; Bell Telephone Co.; Canadian Pneumatic Tool Co.; Canadian Rand Co.; Dominion Electric Mfg. Co.; Dossert & Co.; Economical Electric Lamp Co.; Fibre Conduit Co.; John Forman; The Garth Co.; Jandus Electric Co.; Laurie & Lamb; Locke Insulator Co.; Marconi Wireless Telegraph Co.; Midland Electric Co.; Montreal Light, Heat & Power Co.; Montreal Steel Works; Munderloh & Co.; Canadian Sunbeam Lamp Co.; W. J. O'Leary & Co.; R. E. T. Pringle Co.; Phillips Electrical Works; Renolds-Dull Flasher Co.; Sayer Electric Co.; Stratton Engine Co.; Thomson & Co.; Wire & Cable Co.

In conjunction with the Exhibition were held the Canadian Electrical Convention and the Canadian Street Railway Convention, delegates to the both of which paid frequent visits to the Exhibition. The Exhibition closed on Friday at 11 p.m.

ONTARIO'S RICH MICA MINE.

At Sydenham, Ontario, sixteen miles from Kingston, is located the largest mica mine in the world. The product is mostly amber mica, with some silver amber, the highest quality mined. The mine is one mile from the upper end of Sydenham Lake, and the mica is transported in bulk from the mine by barge to the railroad st Sydenham, where it is shipped to Ottawa for trimming for the market. From Ottawa it is exported to the United States and other points. This valuable mine is owned and worked by Americans, and the output is almost entirely taken by one of the largest American electric companies.

CANADIAN MANUFACTURERS' ASSOCIATION.

The thirty-sixth annual convention of the Canadian Manufacturers' Association opened on Tuesday, September 24, with a large attendance. From east and west all the well-known men in industrial circles gathered at the King Edward Hotel, Toronto, and spent the week discussing papers relative to the progress of all manufacturing lines in Canada. This is the largest convention yet held, over six hundred members o'clock, among others, that of engines having registered.

The various sections met at ten and boilers. Mr. C. H. Waterous, the president of this division, was in the chair. The chief topic of discussion was a Boiler Inspection Act. At present there is a Dominion Act governing marine boilers, and absence of such a law was keenly felt in regard to the various other stationary and portable boilers. Another topic for discussion was the terms of selling engines and boilers, guarantees, cheque deposits and terms of payment. As it is now, some of the manufacturers quote f.o.b. cars and some erected, and steps were taken to remedy this.

Mr. Leonard, of London, was elected president of the division for the ensuing year and Wm. Inglis, vice-president. In our November number will appear a full account of the conference.

PERSONAL MENTION.

B. Welbourn, contract manager of the British Insulated & Helsby Cables Co., Prescott, Eng., has been making an extended tour through Canada, looking up trade conditions. He attended the Canadian Electrical Convention in Montreal.

J. E. Millen, of John Millen & Son, Montreal, is paying a visit this week to their Toronto branch. Owing to a steadily expanding trade in Toronto the firm has found it imperative to double the capacity of that branch.

G. W. Prouty, secretary of the Macallen Co., Boston, was a visitor at the electrical exhibition in Montreal. A. J. Dawson & Co., Montreal, who had an exhibit at the exhibition, are agents of the Macallen Co.

G. Walter Green, of Peterboro, has moved into his new machine shop. The new shop is on McDonnell St., directly opposite the old one, which will be used for foundry purposes.

Machinery at Canadian National Exhibition

Some Very Fine Exhibits, particularly in Machine Tools—New Machinery Hall is Coming, but When?

As a compiled exhibition of Canada's products and resources, the Canadian National Exhibition is, without doubt, unsurpassed. At no other place nor time in Canada is there gathered together so varied a representation of the country's resources and the nation's products. And in most lines the showing is one that Canadians may be exceedingly proud of. The Exhibition, in later years, however, has only miserably represented one of the most important departments of Canadian industry, that of the growth and development of machinery. This regrettable fact can be immediately accounted for. It is not the result of lack of enterprise on the part of Canadian manufacturers. It is not on account of the lack of displayable Canadian machinery. It is, solely and directly, the result of the lack of an adequate building for exhibiting and of adequate means of handling this machinery.

Last year, it may be remembered, Canadian Machinery published, shortly after the Fair, the opinions of the heads of many of the leading machinery manufacturing firms in Canada regarding this matter. The opinions expressed were universal in confirming the statement expressed above. Several of the largest of these manufacturers, who do not now exhibit at the Exhibition, spoke of their willingness and eagerness to do so when adequate accommodation was provided.

This was the situation last year. It was hoped that a year would see some betterment of the situation, but matters seem to be in about the same position to-day. During the year a new Agricultural Building was erected, a new grand stand grew up out of the ashes of the old, and many other improvements were made. These were all needed, of course, but not more than a new Machinery Building.

Then, with the unprecedented success of the recent Exhibition, it was hoped that the long-neglected machinery department might be favored during the year to come. With this in view, a representative interviewed Dr. Orr, the Exhibition's manager, and one or two of the men who ought to know what is to be done. It was rather discouraging to find that the prevailing opinion is that the old Machinery Hall will have to do service for at least another year. Dr. Orr tells us, it is true, that an architect is working on provisional plans for a new structure, the work being carried forward on the basis of suggestions gathered from various sources, and that a new Machinery Hall is bound to come in time. But when is that to be? It may be another year; it may be two. At present there seems to be nothing definite concerning the matter. There is some excuse, of course, for the Exhibition management. The fire of last year crippled them somewhat in the matter of resources, and interfered materially with very excellent plans. This fact must be considered.

When the building does come it will, without doubt, be entirely worthy of the important department whose interests it will serve. Dr. Orr states that the plans now being prepared provide for a building several times the size of the present structure, to be placed, probably, further east than the present location. It will be designed to allow for the introduction of railway spurs into and through the building and will be fitted with electric and jib cranes and adequate means for handling the heaviest machinery. This is only what is required and has been expected.

This Year's Exhibits.

Notwithstanding all the drawbacks, inconveniences and lack of space, the exhibits of machinery at the Exhibition this year were surprisingly good, considering the circumstances, and the display, as a whole, was appreciably better than that of any previous year. This fact was particularly evident in regard to exhibits of machine tools. Several firms showed extensive lines of machines of both American and Canadian manufacture. One American firm went to the expense of bringing a large exhibit from Madison, Wis., and one enterprising Toronto firm carried on during the different days of the Exhibition a series of exhaustive tests of the tools it exhibited. Three producer gas plants were exhibited, only one of which, however, was in operation. The exhibit of motive power, in electrical and gas units, was very large and showed the recent development of this type of power. Gas power, in stationary, portable and marine engines, was particularly in evidence, some fifteen firms, showing different types of motors, having exhibits either in Machinery Hall, the Process Building, or on the grounds. The exhibits of electrical apparatus, always an interesting feature of the Fair, was larger than usual, and attracted a good deal of interest. Transmission machinery was well shown in several different exhibits and a couple of rotary engines created a good deal of interest.

The Dodge Mfg. Co.

An exhibit of various transmission devices was shown by the Dodge Mfg. Co., of Toronto, in the east end of Machinery Hall. The firm's well-known wood-split pulleys were shown in a variety of sizes, and also a new line, a split iron pulley, which combines the advantages of the split-pulley with the weight necessary for heavy work. A clutch connection was shown and also a system of rope drive transmission.

Canada Foundry Exhibit.

In their usual large space in Machinery Hall the Canada Foundry Company showed various products and machines which they manufacture. The working feature was a 500-gallon Underwriters' Pump, which supplied a miniature waterfall. Photographs were shown of large machinery and some of the large bridges built during the year, and surrounding the exhibit was a course of cast iron water pipe of various sizes.

In the Process Building the company exhibited a producer gas plant, a large Blackstone Oil Engine in operation, and two or three sizes of a simple and very powerful rotary pump.

Gisholt Machine Tool Co.

Considerable credit is due this American firm, who exhibited a very worthy representation of the lines of their manufacture.

Their exhibit included one of their well-known vertical mills, a tool grinder of special design, and a Gisholt lathe. This latter machine is worthy of special mention, being adapted for a surprisingly large variety of work on medium castings and forgings and for boring, turning and facing. It attracted a good deal of attention from machinery devotees.

Dominion Belting Co.

This well-known Hamilton company made an extensive showing of their Maple Leaf Belting. This brand, which the company claim to be the strongest and most durable driving belt on the market, is made from stitched cotton duck in all widths from 1½ to 60 inches, and in any length, without joints. It is lock-stitched with special cord, and is waterproofed by a special method. The company has recently made some extensive shipments to outside points, including Australia and South Africa.

H. W. Petrie's Exhibit.

Nothing could be of more interest to a practical machinist than machine tools in operation under actual working conditions. This is one reason why the exhibit of H. W. Petrie, in a prominent corner space at the west end of Machinery Hall, attracted more than usual attention. At specified hours during the Fair, tests were made of the various machines on exhibition. The results of several of these, with descriptions of the machines, follow herewith:

Just off the west aisle was placed an Improved Bickford Plain Radial Drill. This machine has a new drive, and is fitted with gear box. Under test, it drilled a 1½-inch hole 4½ inches deep in cast iron in 1.43 seconds, or an inch in 24 seconds.

At the rear of the exhibit stood a No. 2 Cincinnati Milling Machine. This tool turned out 45-10 pitch 28-tooth cast iron spiral gears of ⅞-inch face, in 11 minutes, the cutter running at 90 revolutions. The feed used was .063 per revolution of the spindle, and the work turned out was entirely free from chatter marks.

A 24-inch Improved Cincinnati Upright Drilling Machine, with tapping attachment, was drilling a ¾-inch hole 2 5-6 inches in cast iron in 39 seconds.

Alongside the north aisle was an 18-inch by 10-foot Lodge & Shipley 5-step Cone Standard Engine Lathe. This machine was reducing 3-inch diameter nickel steel at a periphial speed of 70 feet per minute, with a cut of ¼-inch and 3-16-inch feed.

A 15 by 48-inch Cincinnati Open Side Shaper was taking a cut of ¼-inch with ⅛-inch feed, without the slightest chatter, and this with the cutting done at the extreme end of the crossrail.

3 1¼-inch Simplex 4-Spindle Automatic Lathe, especially designed for all kinds of automatic work, was here making ¾x2¼-inch wagon head cap screws at a rate of 30 per hour.

A No. 1 Universal Cincinnati Cutter and Tool Grinder was shown, grinding or milling cutters and doing other work.

These tests were made by stop-watch, under the direction of J. L. Bishop, an expert young machinist of Cincinnati, whom the Petrie Company brought from Cincinnati for this especial purpose.

Representing the firm's supply department, an attractive display of engineers' and millmen's supplies was made. A line of Pickering governors was exhibited, as was also a Vancouver 50-inch inserted tooth saw.

The exhibit was under the direction of Mr. Hallman, whose enthusiasm in explaining the features of the machinery and whose genial welcome to visitors aided materially in the success of the exhibit.

CANADIAN NATIONAL EXHIBITION—Exhibit of H. W. Petrie, Toronto.

London Machine Tool Company.

One of the most interesting exhibits to machinery men who were interested in details, was that of the London Machine Tool Co., who had a prominent space in the middle aisle of Machinery Hall.

Perhaps the most interesting feature of the exhibit was a 22-inch All-geared Lathe, a "made in Canada" product, being manufactured at the company's works at Hamilton and, as such, is a machine tool of which they may be justly proud. In this lathe the head consists of six gears of wide face and coarse pitch, working into positive clutches, giving four changes of speed in thrust is taken on ball bearings. This method of construction maintains the position of the spindle securely against the heaviest duty. The bearings are of bronze, carefully scraped and fitted.

The screw-cutting mechanism consists of a cone of gears into which rushes an idler gear, this being operated by a handle in front of the head. By means of this and inverting tumbler gears, screw-cutting feeds from one thread to 64 threads per inch, 54 changes altogether, may be obtained.

The tail stock is strongly built, and is provided with a set-over for turning tapers. Being of a cut-away type, it allows the compound rest to be set in is built to stand highest duty and all equipment is of the highest class.

A couple of smaller engine lathes made by the Von Wyck Machine Tool Company, Cincinnati, for which the London Tool Company are the exclusive Canadian agents, was shown. One of these was a 15-inch machine, a particularly high-grade lathe, embracing all the most improved forms of design and construction. The headstock is particularly massive to provide for heavy work. The spindle is made from high-grade crucible steel with large bearings journaled in phosphor bronze. A feature of the carriage is the presence of "T" slots, allowing for numerous con-

CANADIAN NATIONAL EXHIBITION—Exhibit of London Machine Tool Co., Hamilton.

geometrical progression. With a three-speed countershaft, this gives twelve speeds. The bed is very heavy, of ample depth and width and particularly well braced. The racks, of accurately cut steel, are, excepting the long beds, in one piece.

The spindle of this lathe is a particular feature. It is manufactured from high carbon crucible steel, is first bored from solid stock, and finally put on a grinder and brought true to standard sizes. Then threads are chased on the nose, and bored to receive taper, bush and centre, which are very accurately ground and fitted by test. The end a plain parallel to the bed. Two heavy bolts secure the tailstock against movement, and the tailstock spindle is especially ground and fitted. Altogether the lathe is planned to cover the largest range of work possible with such a machine. The fact that all gears are encased is an assurance of absolute safety in operation.

Another fine machine shown was one of the company's No. 2 cold cut-off saw, which is designed for cutting off bars, angles, eyepieces and other forms of steel, with the greatest rapidity. The saw is 26 inches in diameter, with ½-inch face and runs in oil. The frame venient operations. The chasing bial permits threads to be chased without stopping the lathe or reversing the head screw, the cross feeds are graduated to one-thousandth of an inch, and the tailstock is designed so that the compound rest may swing at right angles.

The company manufacture only the finest line of machine tools at their works in Hamilton, and are selling agents in Canada for the Von Wyck Machine Tool Co., 15 in. lathes; the Kempsmith Mfg. Co., milling machines; Potter & Johnson Machine Co., automatic chucking machines and universal shapers; T. C. Dill Machine Co., Dill shapers; Murchey Machine & Tool Co., pipe threading machines and automatic taps and dies; Sibley Machine Tool Co., upright drills, and W. M. Sonnesson & Co., Sweden, shapers; with sales offices in the Traders Bank Building, Toronto.

The exhibit was in charge of Mr. A. E. Jubler, and was visited by other members of the firm from Hamilton, almost daily.

The A. R. Williams Machinery Co.

One of the most attractive exhibits in Machinery Hall was that of the A. R. Williams Machinery Company, Toronto. One reason for this is that machinery in operation always attracts attention. Another was that the exhibit was composed largely of machines and machine tools of Canadian manufacture and was thus an evidence of the growth Canada is making along these lines.

One of the largest machines shown was a 30-inch surface planer, of the Eclipse sectional roll type, manufactured by Major Harper & Son, of Whitby, who build a particularly high grade line of woodworking machinery. Samples of work done by the machine at one operation were shown, which were remarkably smooth and well finished.

A power pipe-cutting machine, built by McDougall & Co., Galt, was shown, which will cut from 1½ to 6 inch pipe either right or left hand. This machine is in use in several of the largest works in Canada.

Another Canadian machine was a Hall safety and automatic power press, which is made by J. B. Hall & Son, in Toronto.

An example of economic power was shown in the Hardill engine, which was running constantly. This engine, which is the invention of John Hardill, of Berlin, is of the tandem compressed type, and is operated by an ingenious mechanism with only one slide valve and steam chest, giving remarkable efficiency with great simplicity. The engine is handled by the A. R. Williams Company only.

Another "Made in Canada" machine shown was a bolt-threading and nut-tapping machine made by McGregor, Gourlay & Co., of Galt.

The A. R. Williams Company also represent some of the largest and best known American firms. Several fine machines of their manufacture were shown.

A power metal saw, built by Cochrane & Bly, of Rochester, one of the finest machines built to-day, cuts off a 3-inch steel shaft in 9 minutes, running at the slow speed, and in 5 minutes on the high speed. The machine has a capacity of 4 inches and turns out a beautifully-finished cut.

A 4-spindle multiple drill for the lighter classes of work, built by the Groton Machine Co., of Groton, N.Y., attracted a good deal of attention, as did also a 4-foot radial drill manufactured by the Fosdick Co., of Cincinnati.

A machine which has had large sale not only in Canada, but also in England, was a 22-inch sliding head power drill, with back gear, self feed and automatic stop. This is made by the W. F. & Jno. Barnes Co., of Rockford, Ill.

The firm's exhibit was under the direction of Mr. Cronk, who, with his assistants, was indefatigable in welcoming visitors and explaining the features of the various machines.

CANADIAN NATIONAL EXHIBITION—Exhibit of A. R. Williams Machinery Co., Toronto.

Exhibit of Rice Lewis & Son, Limited.

The Machinery Department of Rice Lewis & Son, Limited, Toronto, was very worthily represented by a large space in Machinery Hall. The department, but recently organized, aims to supply the Canadian trade with the best grade of machinery manufactured by British and American firms in lines of machine tools, power, and transmission. The feature of the exhibit, without doubt, was a beautifully finished, heavy duty engine lathe, size 24 inches x 10 feet, manufactured by Schumacher & Boye, of Cincinnati, some of the special points of which are worthy of mention. The lathe is built with a three-step cone drive, giving a 4½-inch belt width for power, and at the same time giving 18 changes of speed without the introduction of a single extra gear in either counter shaft or head. It also possesses a special gear box, very rapid in operation and of broad range. A particularly broad and heavy carriage with double plate apron, provides a maximum of stiffness. These features on the lathe may be seen by reference to the accompanying photograph.

A Steptoe Shaper was also shown, built by the John Steptoe Shaper Co., of Cincinnati, who devote their entire attention to the production of high-grade, high-duty shapers.

A compact little emery stand, manufactured by Heinbuch Bros., of Stratford, Ont., was another tool exhibited, showing how such a tool can be improved by careful design and workmanship.

Motive power was represented by the Cook Motor Co., which was shown in two sizes, 3½ and 7 h.p., both running continuously during the Exhibition. The larger engine was belted to a generator which lighted a large illuminated sign above the booth. These engines are built both for lighting and general power purposes. For lighting, they are arranged with a throttling governor for closer regulation, but for general power purposes they take a full explosion at all speeds and govern by patented governor of simple and effective design, so as not to reduce explosive power under varying speed. The 7 horse power shown was arranged for general power purposes, and, consequently, of coarser regulation, but Mr. Kellogg, who had charge of the exhibit, cut in a volt meter to show the close regulation of the engine, even under these conditions, and the absence of throw in the engine. There was no balance wheel on the dynamo to absorb fluctuations in the engine. The results under these conditions spoke well for the machine. The completeness of combustion was well demonstrated by the fact that these engines exhausted through their mufflers within the exhibit with absolutely no preceptible trace of gasoline vapor. These engines are built in vertical types only, up to 20-horse power, and are particularly well adapted for isolated lighting plants and for general power under varying constant speed. They won first award at a series of tests recently conducted at Ohio State University for economy in fuel consumption, and attracted a good deal of favorable comment at the Exhibition. The exhibit also included a display of "Conqueror" high-speed steel in fluted sections. The idea of the section is to reduce unnecessary weight and enable the mechanic to make most of his tools by grinding instead of forging. A remarkable saving in weight is effected in this manner.

CANADIAN NATIONAL EXHIBITION—Exhibit of Messrs. Rice Lewis & Son, Toronto.

The Canadian Fairbanks Co.

One of the largest and most complete exhibits in Machinery Hall was that of the Canadian Fairbanks Co., whose comprehensive display of various lines of the firm's extensive products, attracted a very great deal of attention.

The working features of the exhibit were particularly interesting. A vertical 6 h.p. Fairbanks Morse Gas Engine operating on kerosense, and ignited by magnetos, was belted to a Fairbanks Morse dynamo, which lighted 160 lamps in a large sign on the wall above the exhibit.

A 5 h.p. Horizontal Fairbanks Morse Gas Engine was driving a line shaft. This was fitted with a line of the firm's famous wood-split pulleys, and ran to the corner of the exhibit, where it was connected to another line of shafting by an interesting bit of mechanism, the Almond Right Angle Transmission. This coupling does away with troublesome bevel gearing and mule pulley stands, and is claimed to be a great economizer of power. The second line shaft was fitted with a line of steel pulleys and was connected in turn to another shaft at an angle of 30 degrees by a flexible coupling. This was certainly a novel and complete exhibit of transmission apparatus.

A 2 h.p. Frairbanks-Morse "Jack of All Trades" Gasoline Engine was in operation, as were also two of the firm's marine engines of the 2-cycle, 3-port type, a single cylinder of 4 h.p. and a double cylinder of 8 h.p. capacity.

A much-visited feature was a beautifully finished Fairbanks Personal Scale, fitted with Fan patent type registering beam, and with cast brass columns supporting beam. Visitors were weighed on this machine and their weight automatically registered on a small card.

A new cement testing apparatus, fitted with patent bearing, was shown, and a curious feature was a chair made up entirely of Fairbanks valves.

A new feature of the exhibit was a display of small tools made by Pratt & Whitney Dundas, for which the Canadian Fairbanks Co. have control of the entire Canadian business.

The Canadian Fairbanks Co. is showing surprising development. A large factory is now being erected at Sherbrooke, where a couple of hundred hands wil be employed early in the year in scale construction, and the branches of the company in the various Canadian cities are all doing largely increased business.

CANADIAN NATIONAL EXHIBITION—Exhibit of Canadian Fairbanks Co., Toronto.

Consolidated Electric Co.

The Consolidated Electric Co., of Toronto, showed a line of their King Edward motors, for both alternating and direct current, from ¼ to 25 h.p. One of their 25 h.p. multipolar generators was employed in lighting the exhibit. An interesting little machine shown was the company's new Monarch motor, which is designed especially for individual drive for printing machinery, and may be hung on the wall or ceiling.

Canadian Bearings, Limited.

A comprehensive exhibit of Wright's taper-roller bearings, which are claimed to be particularly suitable to line-shafting and for use in cranes and other bearings, where heavy weight and thrust is to be carried, was made by Canadian Bearings, Limited. They showed rollers which had been under heavy service for years with no perceptible sign of wear, and claim that the bearing will carry a thrust of 180,000 lbs.

Ontario Wind Engine & Pump Co.

The Ontario Wind Engine & Pump Co., in a large corner booth in the Implement Building, exhibited a line of Stickney gasoline engines, from 2 to 25 h.p. The features of the engine are a novel method of cooling, the exceedingly small water tank being an extension of the cylinder jacket, and improved methods of governing and ignition. They also showed the Erickson hot air pumping engine, the Canadian airmotor windmill and a line of iron pumps.

The T. & H. Electric Co.

One of the best-arranged and most complete exhibits in Machinery Hall was that of the Toronto & Hamilton Electric Co., in a corner space on the north aisle. A couple of the larger machines shown were 10 and 12½ k.w. direct current generators, a line which the company build in all sizes, from 4 k.w. up. A line of beautifully-finished induction motors was shown. These were built for various frequencies, and in 1, 2 and 3 phase type. The company make a specialty of motors designed particularly for direct connection to steam, water, gas and other prime movers, and are able to supply these in any size and for both direct and alternating current, and from 4 k.w. up. A line of beautifully-finished induction motors was shown from ½ to 25 h.p. These were of 1, 2 and 3 phase type, and built for various frequencies.

The company make a specialty of building generators for belting and direct connection to steam, water, gas and gasoline or other prime movers. One of these machines was shown running, fitted especially for belting to a gas or gasoline engine.

An interesting feature was a selection switch for disconnecting high potential lines from transformers when repairs or changes in the line are required. This was designed for 40,000 volts, 600 amps.

A very compact little outfit shown was a motor-generator set, designed for charging storage batteries. This consisted of an induction motor, single-phase type, directly connected to a direct-current generator.

The switchboard at the rear of the exhibit was very complete as to detail, and the display also included medals and diplomas awarded the company at the Exposition in previous years.

Borden-Canadian Co.

This exhibit was at the west end of Machinery Hall, where threading machines, both power and hand, were shown.

The main feature was a new Beaver die stock, the machine illustrating the new way of threading pipe. This stock threads pipe from 1 to 2 inches without changing the dies, which is a revolution in pipe threading, doing away with the annoyance and time required when changing from one size pipe to another. The ease of operation of this machine is striking. A man with one hand is able to thread a 2-inch pipe without any difficulty, whereas under the old system it really called for the strength of two men. The reason of this working easily is that the dies recede as the stock commences to advance on the pipe, thus, the further the operation is proceeded with the easier it works.

Evans Rotary Engine.

A new exhibit which attracted considerable attention was the Evans Rotary Engine. Placed on a small platform, not even bolted to the flooring, was a small cylinder about 18 inches in diameter and about 2 feet long. This, with a governor and steam pipe connection furnished 25 h.p. to drive a 10 k.w. generator, which lighted a large sign. Mr. Evans, of New York, the inventor of the engine, was present for a couple of days at the Fair. He claims that the engine will furnish the same amount of power on half the amount of steam required by a reciprocating or turbine engine. A number of these engines were sold during the Fair, and the company, which has only been established in Canada since June, received a good deal of encouragement.

CANADIAN NATIONAL EXHIBITION—Electrical Exhibit of the T. & H. Electrical Co., Hamilton.

The Jones & Moore Co.

In the middle aisle of Machinery Hall the Jones & Moore Electric Co., Toronto, had a very complete exhibit, showing lines of the machines they manufacture. The floor space was well covered with different sizes of their motors, both direct and alternating type, in the former from 1 to 80, and the latter from 1 to 60 h.p. In the rear of the exhibit two large direct current generators of the firm's standard type were running from the line of shafting above. These supplied current to several of the other exhibits, and lighted a string of vari-colored lamps around the company's exhibit.

The Strycker Pattern Works.

The earlier processes of the making of a machine are generally unknown and scarcely thought of by the general public. On this account the exhibit of the Strycker Pattern Works, on the north side of the Process Building, was a revelation to most visitors, and it was usually surrounded by a crowd who eagerly watched the operations of pattern making. In the exhibit several of the company's employes were doing the regular lines of work turned out in the factory at 87 Jarvis Street, Toronto. To facilitate this the exhibit was equipped with a lathe for wood turning, a band saw, circular saw, trimmer, and, of course, a full outfit of pattern-makers' tools. Around the booth also were finished patterns of the company's manufacture, including propellers, sprocket wheels, and even coils for chain heaters. The largest pattern exhibited was an electric motor case, this class of work being a specialty with the firm, as well as patterns for dynamos, gas engines, stoves, etc.

The company has a separate metal department, specially equipped for turning out patterns in brass, white metal and iron. This, with their wood pattern department, covers every range of demand in the pattern line. The firm employ the best class of labor to be found in this line, use only the best material, and, by insisting on high grade work, have built up an extensive local and Canadian trade during their three years of existence.

CANADIAN NATIONAL EXHIBITION—Exhibit of Strycker Pattern Works, Toronto.

The Canadian Westinghouse Co.

The Canadian Westinghouse Company had an attractive exhibit in the Process Building. Their installation included a motor-generator set, a type C, C.L. induction motor, driving a type S generator, this in turn driving a line of direct current motors. A line of induction motors were wired to a large starting panel, 3,300 volts, 200 amps, and one of the firm's modern switchboards controlled the direct current motors. In the other department of the exhibit a line of Nernst lamps for office and outside fixtures was shown, and hanging in the dome of the building, just above the exhibit, was a Nernst lamp of 30 gloires, the largest one of its kind ever made, which lighted in a notable way that section of the building.

Galt Electric Co.

A big winking sign on the west side of the Process Building made known the whereabouts of the Galt Electric Co., and incidentally made an exceedingly good advertisement. The company showed a couple of complete lighting plants of small capacity, for hotel, club and store lighting, a line in which they specialize. They manufacture these isolated plants in standard sizes, with a capacity of from 20 to 100 lamps of 16 c.p., and are finding ready sale for these outfits.

Gould, Shapley & Muir Co.'s Exhibit.

The Gould, Shapley & Muir Co., of Brantford, exhibited a line of their Ideal Gasoline Engines, from 4 to 25 h.p., also pumping and sawing outfits, concrete mixers, feed cutters and windmills.

The I. E. Schantz Co.

In a commodious space on the north side of Machinery Hall an exhibit of power pipe-threading machinery attracted, and was worthy of, a good deal of attention. Three power machines were shown, a No. 6, with capacity from 2½ to 6 inch pipe; a No. 4, which will cut from 1 to 4 inch, and a No. 2, cutting from 1 to 2 inch. All three machines were in operation.

The largest machine, No. 6, which is of new design, and numerous improvements which make it superior to anything of the kind previously built, is, by the way, described in detail, with a photograph, on another page of this issue.

In connection with this machine, Mr. Schantz, who was meeting the trade at the booth during the Fair, tells an interesting story. "Only a short time ago," says Mr. Schantz, "this machine did not exist. The design was but recently finished, and the patterns were made, castings completed, and the parts put together only a short time before the Exhibition." The facility with which this was done tells another story about the efficiency of the company's plant in Berlin. When assembled and tested, the big machine proved so surprisingly satisfactory that it was carried down bodily to the Exhibition and while there excited a good deal of favorable comment from machinery men. The machine cuts a perfect thread on a six-inch pipe at one cutting, whether the dies be new or worn. It will also cut smaller sizes of pipe down to 2½ inches. The machine has weight and power enough to do the heaviest work satisfactorily, and produces no wavy thread, even under the hardest conditions. The first thread cut, a section on a big 6-inch pipe, which was run through the machine immediately after it was assembled, was on exhibition.

The smaller machines were built after the same design and with practically the same features. The fact that several of these were sold during the two weeks of the Fair is an evidence of the impressiveness of the exhibit.

The Schantz Co. have already built up a large business in pipe-threading machinery, of which they make a specialty, and this new type of machine should bring many orders to their factory.

CANADIAN NATIONAL EXHIBITION—Exhibit of I. E. Shantz, Berlin, Ont.

Beaver Marine Engines.

A line of Beaver marine engines was exhibited by the Sherman-Cooper Co., of Toronto. An interesting feature was a section of one of their single-cylinder engines, cut down to show the fine interior finish, the excellence of the casting and the positive oiling device used in their engines. The wall at the rear of the exhibit was hung with a display of propellers and small boat fittings.

The James Morrison Brass Exhibit.

The James Morrison Brass Mfg. Co. had their usual fine exhibit of their products. This included a line of very fine lighting fixtures, engineers' supplies, including extensive lines of injectors, brass and cast iron valves, steam gauges, engine indicators, oil cups, water gauges and locomotive bells and whistles.

N. J. Holden Company.

The only exhibit of pneumatic tools at the Fair was made by the N. J. Holden Company, of Toronto and Montreal. On a table in the front of their exhibit, on the south side of Machinery Hall, was a line of these tools, including pneumatic hammers and rammers, electric drills and mining tools. One of the most interesting of these was a magnetic "Old Man," which, wired to a lighting circuit, clamped itself instantly to a steel plate or any surface to be handled. The tool was exceedingly simple, being provided with a powerful electro-magnet in the base, which did the work. Many of the tools in the exhibit were made by the Chicago Pneumatic Tool Co., of which the N. J. Holden Co. are the Canadian representatives. This company also manufacture air compressors in 100 sizes and various styles, of which many are placed in Canada. A large air-compressing plant of the company's standard type was sold, by the way, during the Fair, to Norton Bros., of Campbellford, for use in their new bridge works there. The company also make a line of rock drills and all sizes of stonecutters' tools.

The N. J. Holden Company also represent the McCord Co., of Chicago, who manufacture axle boxes for railway cars, spring dampeners, and bumping posts. Along with other lines of railway supplies they handle Priest Snow Flanges, the McKim Gasket, Fewing's Car and Engine Replacer, and the Gilmour-Brown Emergency Knuckle. The firm handle a full line of the cold metal saws made by the Quincy, Manchester Sargent Company, of New York. To complete their line of railway supplies, the N. J. Holden Company carry the goods of the Pantasote Company, and are their Canadian agents. This company manufacture Pantasote, the substitute for leather which has had such wide usage in recent years, and also an exceedingly fine line of curtain goods, car blinds, etc.

CANADIAN NATIONAL EXHIBITION—Exhibit of the N. J. Holden Co., Montreal.

Chapman Ball Bearings.

The visitors interested in frictionless transmission kept Mr. C. M. Murry and a staff of able assistants busy from morning till night showing the good points of the Chapman Double Ball Bearings and the many machines in which they may be economically used. A large sign in the centre showed where the exhibit was, and on it was a cut of the now well-known Chapman double ball bearings. An interesting test was carried on and showed conclusively the superiority of Chapman's double ball bearings over the ordinary kind. Besides a wheel which had seen a long, active service and still ran as if new, there were exhibited ball bearings for almost every industrial use. A Sheldon blower was equipped with ball bearings, and blew skyward streamers of red, white and blue, helping to make the exhibit quite an attractive one.

The Smart-Turner Company.

The Smart-Turner Company, of Hamilton, evidently believe in doing one thing and doing that well. Their exhibit was made up of lines of the various styles of pumps manufactured in their works. These included a full line of boiler feed pumps, a heating system for boiler feeds, and a large centrifugal pump in operation. This pump was employed in raising sludge, and is specially adapted for pumping sewage and heavy material.

Nicholls Bros.

In the east end of the Implement Building, Nicholls Bros. exhibited a full line of ignition outfits, including magnetos, spark coils and spark plugs. They make a specialty of the "Syntic" ignition system, and showed a system in operation. Their display of small ammeters and voltmeters for small electric plants, was also attractive.

The Automatic in Manufacturing

Comparisons Between Setting Up Automatic and Hand-Screw Machines.

By J. P. Brophy, Cleveland Automatic Machine Co.

The operations of putting in chuck end shell and bar stock, are, of course, the same in either case. Also, this is so in regard to placing and adjusting the tools in turret and on cross slide. Now comes the difference: on the hand machine there are adjustable stops to be set for each tool in turret and on cross slide; on the automatic machine we have adjustable cams to set, which should take the operator no longer to set than the stops on the hand machine. As far as the above argument goes, both machines stand about equal, but we have yet the most important advantage in favor of the automatic, which is that the operator of automatics sets one of his machines while all the others in his charge are running, and, assuming that he has six machines in his care, only one-sixth of his output is stopped temporarily; while, in the case of the hand machine operator, his entire output is stopped while setting his machine, and there are more reasons why the setting-up features of the automatic screw machine have been improved to the extent that it is simpler and easier to set than a hand machine. The numbering of regulating drum, cross slide drum and turret holes makes it possible to make out a test card for each set of tools to be used by the operator in changing from one piece of work to another. The test card enables a man with very little experience to place the cams and tools in their proper working positions, and in a shorter time than would be possible to set up a hand machine.

In turning out parts, the automatic leads all other ways. This shows up especially in the large diameter parts, such as transmission gears, ball bearings, roller bearings, pistons, thrust washers, bushings, equalizing gears, etc., either castings, forgings, or bar stock; also, for parts of all kinds for the following; typewriters, sewing machines cash registers, adding machines, agricultural machinery, railroads, tool builders, elevator, electrical, wood working and textile machinery; in fact, in manufacturing parts of all descriptions the automatic can be used to advantage. In using automatic machines on any class of work, the element that we call "human nature" in the operators, by which we mean all things which come up distracting the attention of the operator, decreasing the actual working time of the tools, and, consequently, the output is all done away with, for the obvious reason that the automatic machine is in active operation continually, losing no time from actual cutting.

Manufacturers who are wide-awake are steadily coming to the front in the use of labor-saving machinery However, in the turning out of the larger parts, the manufacturer who has not installed automatics has to depend on lathe work, or hand machines, and one of the most expensive operations in the factory to-day is lathe work. An automatic operated by a medium-priced man can do high-class, accurate work, that on a lathe would require a fine mechanic receiving the highest pay, and the automatic operator can care for from four to eight machines. The manufacturer depending on such methods is not only losing money, but is making the cost of his product so high that he is not in a position to compete with the wide-awake man who enjoys the saving he is able to make by taking advantage of automatics. He is also losing money by not having a machine that will overcome the great difficulty of men working on hand machines being absent when the parts are most needed, resulting, possibly, in the closing down of some parts of the factory, which means additional expense to the company. It is very easy for a busy manufacturer, whose time and energy is continually being expended in keeping his design and construction up to the latest ideas and improvements, to keep putting into the future the installation of automatic machines for his work, even though he is dimly conscious that he is paying out money for wages, which is unnecessary, and the reason that he has been so passive in this respect is because he has no adequate idea of the unbelievable difference in the labor cost of work produced on the automatic as against the machines spoken of above. Of course, automatic machines have been used on small work for years, but the purpose of this article is to bring to the attention of users of engine lathes, turret lathes and hand screw machines, the saving to be effected by these money-saving machines, which are especially adapted to finishing parts, large and small, up to 6 inches in diameter.

I want here to get the particular attention of shop superintendents in regard to duplicate parts. I venture to say that every superintendent who makes his own parts has been time and again exasperated almost to the bursting point by the continual habit of lathe hands getting things too long or too short, or too small, or too large. Now, we come to one of the most essential characteristics of the automatic machine. You can rest assured that the parts will be exactly alike in size and shape. This brings up another thought along this line. Some people claim that accurately finished work cannot be produced on an automatic machine, and that all the work has to be gone over on the lathe. This is not the case, as the machines are made to finish the work to size. The modern, large automatic machine is built rigid and heavy wherever great strain comes and the forming tools come up solid against a stop, giving exact size. Also, there is a sufficient volume of oil flowing on the tools while working to keep same thoroughly lubricated and the work cool, which tends towards accuracy. Nothing is left to the eye sight or feel of the operator to obtain the size. The machine takes care of that; consequently the "human element" is no longer there to be contended with, and one of your greatest difficulties eliminated. There can be no mistake on this point since the tools travel to proper distances with positive cams and adjustable stops. Accuracy must be the result after the first piece has been made to gauge. You may continue till parts are completed and the variations will be found within the required limit.

The regularity and steadiness and power of the tool feed allows of long-continued working of the tools without attention. The feed of the tools is not forced at one moment and light at the next, as is often the case of the hand machine, but is always advancing into the work at the rate it should be. In this way we obtain a great output with the minimum attention being paid to the tools. Tools on automatic machines require less attention than hand machines, because they are self-operated and do not have to contend with the different temperaments of man. The automatic works the same on "blue Monday" as on Saturday, which means a great saving on tools. Hand machine tools are very often broken by the operator conveying the tool to the part to be machined, and jamming into it, which is almost sure to result in the work being spoiled, or a broken tool. The automatic is positive in its idle movements, and can be relied upon to carry each tool direct to the part to be worked without losing any time or endangering the tools. After the tool feed has been properly adjusted on the automatic, each tool will be subject to the same strain, and will invariably take the same time to remove the stock, which will result in the tools lasting longer and will produce more accurate work than would be possible on a hand machine. Also, the great volume of oil continually pouring over the tools and the work, and through the internal oil tube of drills, counterbores and boring tools, adds greatly to the length of the time the cutting tools will stand up. If an automatic machine were to be set up to produce in one day the exact same number and kind of pieces that it would be possible to produce on a hand machine in the same time, the automatic machine would run slower, the feed of the cutting tools easier, and the tools on the automatic would stand up without re-grinding very much longer than on a hand machine. All this on account of the ceasless "hammering away" every minute of the day, which is characteristic of automatic machines.

In the race to keep up with the procession, the installation of labor-saving machinery has been neglected. In order to turn out work rapidly and smoothly, it is necessary to have parts made so that they will go together without extra labor, which has the beneficial result of less cost to assemble, and far greater output. A considerable part of the time spent in assembling, under the present conditions, is due to variation of sizes of the parts, necessitating extra labor before it is possible to put the parts together. Automatic machines will overcome this condition, and the manufacturer who is first to equip his factory with modern methods of producing the parts which he must have, is destined to have the greatest hold on the industry.

It is only within the last few years that large bars could be had without long delay. Now it is carried in stock and can be had on short notice. This was formerly one of the handicaps of the large automatic machine. It was thought that an automatic machine, to handle large bars, would be a ponderous affair, too clumsy and slow to be of much use and, also, calling for too large an outlay of money to instal them. The large automatics of to-day, however, after the past few years of hard application and experiment, are models of compactness and simplicity. They are not cumbersome, and you would not even call them large, when you consider the size of the work they produce. To come to the point in this argument, increasing the capacity of the machine does not increase the floor space very much; though, of course, it increased the weight very rapidly as the large machines are made exceedingly rigid to withstand the heavy strain. The mechanism for obtaining the automatic movements is not clumsy and awkward, but remains practically the same size as the same mechanism on smaller machines. Thus, the movements are all quick, light and easy. These large machines are no harder to operate than smaller ones, and an operator can take care of just as many as he can of the smaller sizes. Consider for a moment what this means. Take, for example, the steel sliding gear of a transmission for an automobile, which has taken a lathe hand four hours to make, representing a labor cost of about $1.10 to $1.20. Now, put this same job on an automatic machine and it would take the machine approximately one hour to make the piece. The operator is running, say, four machines, which would be a low estimate; the price paid the operator for producing one piece would be from six to seven cents, against the total of $1.10 to $1.20 paid to the lathe hand.

In a large automatic room it is only necessary to have one first-class man to look after the grinding and setting of tools, and all the other help is of the cheapest variety.

Viewed from the state of perfection of automatics a few years ago the class of work done to-day was an impossibility, but the thought of automatic builders has kept pace with the improvements in other lines, and has resulted in the present high state of development which automatics have reached.

AMONG THE SOCIETIES

CANADIAN ELECTRICAL CONVENTION.

The annual convention of the Canadian Electrical Association was held in Montreal on Sept. 11, 12 and 13. The more interesting features of the convention were the President's address on opening day, a visit of delegates as a body to the Electrical Exhibition; question box discussion, a theatre party, visits to Blue Bonnets Race Track and Dominion Park.

Opening Session.

The President, R. G. Black, opened the convention by an interesting address on the importance of producers of electric energy extending their markets in order to make any profit at present rates. During the opening day a handsome gold watch was presented by the members to Mr. Mortimer, who, it was stated, had taken an active part in the work of the society ever since it was started, sixteen years ago—for many years its secretary.

The report of the Secretary-Treasurer, T. S. Young, showed that the society was flourishing, the membership being 320, and there being a good sum to the society's credit in the bank.

On Thursday took place the election of officers for the ensuing year. The following is the result: President, R. S. Kelsch, Montreal; First Vice-President, W. N. Ryerson, Niagara Falls; Second Vice-President, R. M. Wilson, Montreal; Secretary-Treasurer, T. S. Young, Toronto. The Management Committee is composed of A. A. Dion, Ottawa; B. F. Reesor, Lindsay; C. B. Hunt, London; J. M. Robertson, Montreal; J. J. Wright, Toronto; J. G. Glasco, Hamilton; R. B. Black, Toronto; W. Williams, Sarnia; H. O. Fisk, Peterboro; J. W. Purcell, Walkerville.

Papers and Addresses.

President's address; A. B. Lambe, "Electric Heating and Cooking Devices"; M. A. Sammett, "Trials of the Operating Man"; B. T. McCormick, "Three Wire Generators"; Clarence E. Delafield, "High Tension Insulators from an Engineering and Commercial Standpoint"; A. E. Fleming, "The Value of the Nernot Lamp to the Central Station"; J. M. Robertson, "Incandescent Lamps"; John Murphy, "Frazil and Anchor Ice"; R. M. Wilson, "The Load Factor"; G. P. Cole, "Modern Lighting Transformers"; G. H. Montgomery, "The Responsibility of Electric Companies for Accidents."

The papers read by John Murphy and G. H. Montgomery were especially interesting and instructive. It was left to the Management Committee to name the next place of convening, and to the Secretary to send formal votes of thanks to all who had aided in the entertainment of the delegates and to those who had read papers at the convention. The Montreal Street Railway were very kind in placing special cars at the disposal of the delegates to attend the horse races and see the city. The session closed at 12.30 on Friday afternoon. Allis - Chalmers - Bullock, Limited, had issued and distributed among the delegates a very nicely printed programme of the convention.

CANADIAN MACHINERY BUYERS' DIRECTORY

To Our Readers—Use this directory when seeking to buy any machinery or power equipment. You will often get information that will save you money.
To Our Advertisers—Send in your name for insertion under the heading of the lines you make or sell.
To Non-Advertisers—A nominal rate of $1 per line a year is charged non-advertisers.

Acids.
Canada Chemical Mfg. Co., London

Abrasive Grains & Powders
Carborundum Co., Niagara Falls, N.Y.

Abrasive Materials.
Baxter J. R., & Co., Montreal.
The Canadian Fairbanks Co., Montreal.
Rice Lewis & Son, Toronto.
Norton Co., Worcester, Mass.
H. W. Petrie, Toronto.
Carborundum Co., Niagara Falls, N.Y.
Williams & Wilson, Montreal.

Air Brakes.
Canada Foundry Co., Toronto.
Canadian Westinghouse Co., Hamilton.

Air Receivers.
Allis-Chalmers-Bullock Montreal.
Canada Foundry Co., Toronto.
Canadian Rand Drill Co., Montreal.
Chicago Pneumatic Tool Co., Chicago.
John McDougall Caledonian Iron Works Co., Montreal.

Alloy, Ferro-Silicon.
Paxson, J. W., Co., Philadelphia.

Alundum Scythe Stones
Norton Company, Worcester, Mass.

Arbor Presses.
Niles-Bement-Pond Co., New York.

Automatic Machinery.
Cleveland Automatic Machine Co. Cleveland.
National-Acme Mfg. Co., Cleveland
Potter & Johnston Machine Co., Pawtucket, R. I.

Automobile Parts
Globe Machine & Stamping Co., Cleveland, Ohio.

Axle Cutters.
Butterfield & Co., Rock Island, Que.
A. B. Jardine & Co., Hespeler, Ont.

Axle Setters and Straighteners.
A. B. Jardine & Co., Hespeler, Ont.
Standard Bearings, Ltd, Niagara Falls

Babbit Metal.
Baxter J. R., & Co. Montreal.
Canada Metal Co., Toronto.
Canada Machinery Agency, Montreal.
Frothingham & Workman Ltd., Montreal
Greey Wm. & J. G., Toronto.
Rice Lewis & Son, Toronto.
Lumen Bearing Co., Toronto.
Mechanics Supply Co, Quebec, Que.

Bakers' Machinery.
Greey, Wm. & J. G., Toronto.

Baling Presses.
Perrin, Wm. R., & Co, Toronto.

Barrels, Steel Shop.
Cleveland Wire Spring Co., Cleveland.

Barrels, Tumbling.
Buffalo Foundry Supply Co., Buffalo.
Detroit Foundry Supply Co., Windsor
Dominion Foundry Supply Co., Montreal
Hamilton Facing Mill Co., Hamilton.
Gilmour J. New York.
Globe Machine & Stamping Co, Cleveland, Ohio.
John McDougall Caledonian Iron Works Co., Montreal.
Northern Engineering Works, Detroit.
J. W. Paxson Co., Philadelphia, Pa.
H. W. Petrie, Toronto.
Sly, W. W., Mfg. Co., Cleveland
The Smart-Turner Mach. Co., Hamilton

Bars, Boring.
Hall Engineering Works, Montreal.
Niles-Bement-Pond Co., New York.
Standard Bearings, Ltd., Niagara Falls

Batteries, Dry.
Berlin Electrical Mfg. Co., Toronto.
Mechanics Supply Co., Quebec, Que.

Batteries, Flashlight.
Berlin Electrical Mfg. Co., Toronto.
Mechanics Supply Co., Quebec Que.

Batteries, Storage.
Canadian General Electric Co. Toronto
Chicago Pneumatic Tool Co., Chicago.
Rice Lewis & Son, Toronto.
John Millen & Son, Montreal.

Bearing Metals.
Lumen Bearing Co., Toronto.

Bearings, Roller.
Standard Bearings, Ltd. Niagara Falls

Bell Ringers.
Chicago Pneumatic Tool Co., Chicago.
Mechanics Supply Co., Quebec, Que.

Belting, Chain.
Baxter, J. R., & Co. Montreal.
Canada Machinery Exchange, Montreal.
Greey, Wm. & J. G., Toronto.
Jeffrey Mfg. Co., Columbia, Ohio.
Link-Belt Eng. Co., Philadelphia.
Waterous Engine Works Co., Brantford.

Belting, Cotton.
Baxter, J. R., & Co., Montreal.
Canada Machinery Agency, Montreal.
Dominion Belting Co., Hamilton.
Rice Lewis & Son, Toronto.

Belting, Leather.
Baxter, J. R., & Co., Montreal.
Canada Machinery Agency, Montreal.
The Canadian Fairbanks Co., Montreal.
Frothingham & Workman Ltd., Montreal
Greey, Wm. & J. G., Toronto.
McLaren, J. C., Montreal.
Rice Lewis & Son, Toronto.
H. W. Petrie, Toronto.
Williams & Wilson, Montreal.

Belting, Rubber.
Baxter, J. R., & Co., Montreal.
Canada Machinery Agency, Montreal.
Frothingham & Workman, Ltd., Montreal
Greey, Wm. & J. G., Toronto.

Belting Supplies.
Baxter J. R., & Co, Montreal.
Greey Wm. & J. G., Toronto.
Rice Lewis & Son. Toronto.
H. W. Petrie, Toronto.

Bending Machinery.
John Bertram & Sons Co., Dundas, Ont.
Chicago Pneumatic Tool Co., Chicago.
Rice Lewis & Son, Toronto.
London Mach. Tool Co., Hamilton, Ont.
National Machinery Co., Tiffin, Ohio.
Niles-Bement-Pond Co., New York.

Benders, Tire.
A. B. Jardine & Co., Hespeler, Ont.
London Mach. Tool Co., Hamilton, Ont.

Blowers.
Buffalo Foundry Supply Co., Montreal.
Canada Machinery Agency, Montreal.
Detroit Foundry Supply Co, Windsor
Dominion Foundry Supply Co., Montreal
Gilmour, J., New York
Hamilton Facing Mill Co., Hamilton.
Kerr Turbine Co., Wellsville, N.Y.
Mechanics Supply Co, Quebec, Que.
J. W. Paxson, Philadelphia, Pa.
Sheldon's Limited, Galt
Wilbraham-Green Blower Co., Philadelphia

Blast Gauges—Cupola.
Paxson, J. W., Co. Philadelphia.
Sheldons, Limited, Galt

Blocks, Tackle.
Frothingham & Workman, Ltd, Montreal

Blocks, Wire Rope.
Frothingham & Workman, Ltd., Montreal

Blue Printing.
The Electric Blue Print Co., Montreal.

Blow-Off Tanks.
Darling Bros., Ltd., Montreal.

Boilers.
Canada Foundry Co., Limited, Toronto.
Canada Machinery Agency, Montreal.
Goldie & McCulloch Co., Galt.
John McDougall Calenonian Iron Works, Montreal.
Manitoba Iron Works, Winnipeg.
Mechanics Supply Co., Quebec, Que
Owen Sound Iron Works Co., Owen Sound
H. W. Petrie, Toronto.
Robb Engineering Co. Amherst, N.S.
The Smart-Turner Mach. Co., Hamilton.
Waterous Engine Works Co., Brantford.
Williams & Wilson, Montreal.

Boiler Compounds.
Canada Chemical Mfg. Co., London, Ont.
Hall Engineering Works, Montreal.
Lake Erie Boiler Compound Co., Toronto

Bolt Cutters.
John Bertram & Sons Co., Dundas, Ont.
London Mach. Tool Co., Hamilton.
Mechanics Supply Co, Quebec, Que
National Machinery Co., Tiffin, Ohio.
Niles-Bement-Pond Co., New York.

Bolt and Nut Machinery.
John Bertram & Sons Co., Dundas, Ont.
Canada Machinery Agency, Montreal.
Rice Lewis & Son, Toronto.
London Mach. Tool Co., Hamilton.
National Machinery Co., Tiffin, Ohio.
Niles-Bement-Pond Co. New York.

Bolts and Nuts, Rivets.
Baxter, J. R., & Co., Montreal.
Mechanics Supply Co., Quebec, Que.

Boring & Drilling Machines
American Tool Works Co., Cincinnati.
B. F. Barnes Co., Rockford, Ill.
John Bertram & Sons Co., Dundas, Ont.
Canada Machinery Agency Montreal.
A. B. Jardine & Co., Hespeler, Ont.
London Mach. Tool Co., Hamilton.
Niles-Bement-Pond Co., New York.

Boring Machine, Upright.
American Tool Works Co., Cincinnati.
John Bertram & Sons Co., Dundas, Ont.
London Mach Tool Co., Hamilton.
Niles-Bement-Pond Co., New York.

Boring Machine, Wood.
Chicago Pneumatic Tool Co., Chicago.
London Mach. Tool Co., Hamilton.

Boring and Turning Mills.
American Tool Works Co., Cincinnati.
John Bertram & Sons Co., Dundas, Ont.
Gisholt Machine Co., Madison, Wis.
Canada Machinery Agency, Montreal.
Rice Lewis & Son, Toronto.
London Mach. Tool Co., Hamilton.
Niles-Bement-Pond Co., New York.
H. W. Petrie, Toronto.

Box Puller.
A. B. Jardine & Co., Hespeler, Ont.

Boxes, Steel Shop.
Cleveland Wire Spring Co, Cleveland.

Boxes, Tote.
Cleveland Wire Spring Co., Cleveland

Brass Foundry Equipment.
Ph. Bonvillain & E. Ronceray, Philadelphia
Detroit Foundry Supply Co., Detroit.
Dominion Foundry Supply Co., Montreal
Paxson, J. W., Co., Philadelphia

Brass Working Machinery.
Warner & Swasey Co., Cleveland, Ohio.

Brushes, Foundry and Core.
Buffalo Foundry Supply Co., Buffalo.
Detroit Foundry Supply Co, Windsor.
Dominion Foundry Supply Co., Montreal
Hamilton Facing Mill Co., Hamilton.
Mechanics Supply Co., Quebec, Que.
Paxson, J. W., Co., Philadelphia

Brushes, Steel.
Buffalo Foundry Supply Co., Buffalo.
Dominion Foundry Supply Co. Montreal
Paxson, J. W., Co., Philadelphia

Bulldozers.
John Bertram & Sons Co., Dundas, Ont.
London Mach. Tool Co., Hamilton, Ont
National Machinery Co., Tiffin, Ohio.
Niles-Bement-Pond Co., New York.

Calipers.
Baxter, J. R., & Co., Montreal
Frothingham & Workman, Ltd., Montreal
Rice Lewis & Son, Toronto.
Mechanics Supply Co., Quebec, Que.
John Millen & Son, Ltd., Montreal, Que
L. S. Starrett & Co., Athol, Mass.
Williams & Wilson Montreal.

Carbon.
Dominion Foundry Supply Co., Montreal

Cars, Foundry.
Buffalo Foundry Supply Co., Buffalo.
Detroit Foundry Supply Co., Windsor
Dominion Foundry Supply Co., Montreal
Hamilton Facing Mill Co., Hamilton.
Paxson, J. W., Co., Philadelphia

Castings, Aluminum.
Lumen Bearing Co., Toronto

Castings, Brass.
Chadwick Bros., Hamilton.
Greey, Wm. & J. G. Toronto.
Hall Engineering Works, Montreal.
Kennedy, Wm., & Sons, Owen Sound
Lumen Bearing Co., Toronto
Niagara Falls Machine & Foundry Co. Niagara Falls, Ont.
Owen Sound Iron Works Co., Owen Sound.
Robb Engineering Co, Amherst, N.S.

Castings, Grey Iron.
Allis-Chalmers-Bullock Montreal.
Greey, Wm. & J. G, Toronto.
Hall Engineering Works, Montreal.
Kennedy, Wm, & Sons, Owen Sound.
Laurie Engine & Machine Co., Montreal.
Maxwell, David, & Sons, St. Marys.
John McDougall Caledonian Iron Works Co., Montreal.
Niagara Falls Machine & Foundry Co., Niagara Falls, Ont
Owen Sound Iron Works Co., Owen Sound.
Robb Engineering Co., Amherst, N.S.
Smart-Turner Machine Co., Hamilton.

Castings, Phosphor Bronze.
Lumen Bearing Co., Toronto

Castings, Steel.
Kennedy, Wm., & Sons, Owen Sound.

Castings, Semi-Steel.
Robb Engineering Co., Amherst, N.S.

Cement Machinery.
Allis-Chalmers-Bullock, Limited, Montreal
Greey, Wm. & J. G., Toronto
Jeffrey Mfg. Co., Columbus, Ohio
John McDougall Caledonian Iron Works, Co., Montreal
Owen Sound Iron Works Co., Owen Sound

Centreing Machines.
John Bertram & Sons Co., Dundas, Ont.
Jeffrey Mfg Co., Columbus, Ohio
London Mach. Tool Co., Hamilton, Ont.
Niles-Bement-Pond Co., New York.
Pratt & Whitney Co., Hartford, Conn.
Standard Bearings, Ltd., Niagara Falls

Centres, Planer.
American Tool Works Co., Cincinnati.

Centrifugal Pumps.
Gas & Electric Power Co., Toronto.
John McDougall Caledonian Iron Works Co, Montreal.
Pratt & Whitney Co., Hartford, Conn.

Centrifugal Pumps—Turbine Driven
Kerr Turbine Co., Wellsville, N.Y.

Chain, Crane and Dredge.
Frothingham & Workman, Ltd., Montreal

Chaplets.

Buffalo Foundry Supply Co., Buffalo.
Detroit Foundry Supply Co., Windsor.
Dominion Foundry Supply Co., Montreal
Hamilton Facing Mill Co., Hamilton.
Paxson, J. W., Co., Philadelphia

Charcoal.

Buffalo Foundry Supply Co., Buffalo.
Detroit Foundry Supply Co., Windsor.
Doggett, Stanley, New York
Dominion Foundry Supply Co., Montreal
Hamilton Facing Mill Co., Hamilton.
Paxson, J. W., Co., Philadelphia

Charcoal Facings.

Doggett, Stanley, New York

Chemicals.

Baxter, J. R., & Co., Montreal.
Canada Chemical Co., London.

Chemists' Machinery.

Greey Wm & J. G., Toronto.

Chemists, Industrial.

Detroit Testing Laboratory, Detroit.

Chemists, Metallurgical.

Detroit Testing Laboratory, Detroit.

Chemists, Mining.

Detroit Testing Laboratory, Detroit.

Chucks, Ring Grinding.

A. B. Jardine & Co., Hespeler, Ont.
Chicago Pneumatic Tool Co., Chicago.

Chucks, Drill and Lathe.

American Tool Works Co., Cincinnati.
Baxter, J. R., & Co., Montreal.
John Bertram & Sons Co., Dundas, Ont.
Canada Machinery Agency, Montreal.
Frothingham & Workman Ltd., Montreal
Hamilton Tool Co., Hamilton, Ont.
Ker & Goodwin, Brantford.
A. B. Jardine & Co., Hespeler, Ont.
London Mach. Tool Co., Hamilton.
John Millen & Son, Ltd., Montreal.
Niles-Bement-Pond Co., New York.
H. W. Petrie, Toronto.
Rice Lewis & Son, Toronto.
Standard Tool Co., Cleveland.

Chucks, Planer.

American Tool Works Co., Cincinnati.
Canada Machinery Agency, Montreal.
Niles-Bement-Pond Co., New York.

Chucking Machines.

American Tool Works Co., Cincinnati.
Niles-Bement-Pond Co., New York.
H. W. Petrie, Toronto.
Warner & Swasey Co., Cleveland, Ohio

Circuit Breakers.

Gas & Electric Power Co., Toronto.
Allis-Chalmers-Bullock, Limited, Montreal
Canadian General Electric Co., Toronto.
Canadian Westinghouse Co., Hamilton.

Clippers, Bolt.

Frothingham & Workman Ltd. Montreal
A. B. Jardine & Co., Hespeler, Ont.

Cloth and Wool Dryers.

Dominion Heating and Ventilating Co., Hespeler.
B. Greening Wire Co., Hamilton.
Sheldons Limited, Galt

Collectors, Pneumatic.

Sheldons Limited, Galt

Compressors, Air.

Allis-Chalmers-Bullock, Limited, Montreal
Canada Foundry Co., Limited, Toronto.
Canada Machinery Agency, Montreal.
Canadian Rand Drill Co., Montreal.
Canadian Westinghouse Co., Hamilton.
Chicago Pneumatic Tool Co., Chicago.
Darling Bros., Ltd., Montreal.
Detroit Foundry Supply Co. Windsor.
Gas & Electric Power Co., Toronto.
John McDougall, Caledonian Iron Works Co. Montreal
H. W. Petrie, Toronto.
The Smart-Turner Mach. Co., Hamilton.
Hall Engineering Works, Montreal, Que.
London Mach. Tool Co., Hamilton.
Niles-Bement-Pond Co., New York.
H. W. Petrie, Toronto.
Pratt & Whitney Co., Hartford, Conn.
Williams & Wilson, Montreal.

Concentrating Plant.

Allis-Chalmers-Bullock, Montreal.
Greey, Wm. & J. G., Toronto.

Concrete Mixers.

Jeffrey M g. Co., Columbus, Ohio.
Link-Belt Co., Philadelphia.

Condensers.

Canada Foundry Co., Limited, Toronto.
Canada Machinery Agency, Montreal.
Hall Engineering Works Montreal.
Smart-Turner Machine Co., Hamilton.
Waterous Engine Co., Brantford.

Confectioners' Machinery.

Baxter, J. R., & Co., Montreal.
Greey, Wm. & J. G. Toronto.

Consulting Engineers.

Hall Engineering Works, Montreal.
Jules De Clercy, Montreal.
Roderick J. Parke, Toronto.
Plews & Trimingham, Montreal.
T. Pringle & Son, Montreal.
Somerville & Van Every, Hamilton
Standard Bearings, Ltd., Niagara Falls

Contractors.

Gas & Electric Power Co., Toronto.
Expanded Metal and Fireproofing Co. Toronto.
Hall Engineering Works Montreal.
Laurie Engine & Machine Co., Montreal.
John McDougall Caledonian Iron Works Co., Montreal
Robb Engineering Co Amherst N S.
The Smart-Turner Mach. Co., Hamilton.

Contractors' Plant.

Allis-Chalmers-Bullock, Montreal.
John McDougall, Caledonian Iron Works Co., Montreal
Niagara Falls Machine & Foundry Co., Niagara Falls, Ont.

Controllers and Starters Electric Motor.

Gas & Electric Power Co., Toronto.
Allis-Chalmers-Bullock, Montreal.
Canadian General Electric Co., Toronto.
Canadian Westinghouse Co., Hamilton.
T. & H. Electric Co., Hamilton.

Converters, Steel.

Northern Engineering Works, Detroit.
Paxson, J. W., Co., Philadelphia

Conveyor Machinery.

Baxter, J. R., & Co., Montreal.
Greey Wm & J. G., Toronto
Jeffrey Mfg. Co., Columbus, Ohio.
Rice Lewis & Son, Toronto.
Link-Belt Co., Philadelphia.
John McDougall Caledonian Iron Works Co., Montreal.
Smart-Turner Machine Co., Hamilton.
Laurie Engine & Machine Co., Montreal.
Waterous Engine Works Co., Brantford.
Williams & Wilson, Montreal.

Coping Machines.

John Bertram & Sons Co., Dundas, Ont.
London Mach. Tool Co., Hamilton.
Niles-Bement-Pond Co., New York.

Core Compounds.

Buffalo Foundry Supply Co., Buffalo.
Detroit Foundry Supply Co., Windsor.
Dominion Foundry Supply Co., Montreal
Hamilton Facing Mill Co., Hamilton.
Paxson, J. W., Co., Philadelphia

Core-Making Machines.

Paxson, J. W., Co., Philadelphia

Core Ovens.

Detroit Foundry Supply Co., Windsor.
Dominion Foundry Supply Co., Montreal
Hamilton Facing Mill Co., Hamilton.
Paxson, J. W., Co., Philadelphia
Sheldons Limited, Galt

Core Oven Bricks.

Buffalo Foundry Supply Co., Buffalo.
Detroit Foundry Supply Co., Windsor.
Dominion Foundry Supply Co., Montreal
Hamilton Facing Mill Co., Hamilton.
Paxson, J. W., Co., Philadelphia

Core Sand Cleaners.

Sly, W. W., Mfg. Co, Cleveland

Core Wash.

Buffalo Foundry Supply Co., Buffalo.
Detroit Foundry Supply Co., Windsor.
Dominion Foundry Supply Co., Montreal
Hamilton Facing Mill Co., Hamilton.
Paxson, Co., J. W., Philadelphia

Couplings.

Owen Sound Iron Works Co., Owen Sound

Cranes, Electric and Hand Power.

Gas & Electric Power Co. Toronto
Canada Foundry Co., Limited Toronto
Canadian Pilling Co., Montreal
Dominion Foundry Supply Co., Montreal
Gilmour, J., New York
Hamilton Facing Mill Co., Hamilton.
Link-Belt Co., Philadelphia.
John McDougall, Caledonian Iron Works Co., Montreal.
Niles-Bement-Pond Co., New York.
Northern Engineering Works Detroit
Owen Sound Iron Works Co, Owen Sound
Paxson, J. W., Co, Philadelphia
Smart-Turner-Machine Co., Hamilton.

Crank Pin.

Sight Feed Oil Pump Co., Milwaukee, Wis

Crankshafts.

The Canada Forge Co., Welland.
St. Clair Bros., Galt

Crabs.

Frothingham & Workman, Ltd., Montreal

Crank Pin Turning Machine

London Mach. Tool Co., Hamilton.
Niles-Bement-Pond Co., New York.

Cross Head Pin.

Sight Feed Oil Pump Co., Milwaukee, Wis.

Crucibles.

Buffalo Foundry Supply Co., Buffalo.
Detroit Foundry Supply Co., Windsor
Dominion Foundry Supply Co., Montreal
Hamilton Facing Mill Co., Hamilton.
J. W. Paxson Co., Philadelphia.

Crucible Caps

Dominion Foundry Supply Co Montreal
Hamilton Facing Mill Co., Hamilton.
Paxson, J. W., Co., Philadelphia

Crushers, Rock or Ore.

Allis-Chalmers-Bullock, Montreal.
Jeffrey Mfg. Co., Columbus, Ohio.

Cup Grease.

Mechanics Supply Co., Quebec

Cupolas.

Buffalo Foundry Supply Co., Buffalo.
Detroit Foundry Supply Co., Windsor
Dominion Foundry Supply Co., Montreal
Gilmour J. New York.
Hamilton Facing Mill Co., Hamilton.
Northern Engineering Works, Detroit
J. W. Paxson Co., Philadelphia.
Sheldons Limited, Galt.

Cupola Blast Gauges.

Dominion Foundry Supply Co., Montreal
Paxson, J. W., Co., Philadelphia
Sheldons Limited, Galt

Cupola Blocks.

Detroit Foundry Supply Co., Detroit.
Hamilton Facing Mill Co., Hamilton
Gilmour, J., New York.
Northern Engineering Works Detroit
Ontario Lime Association Toronto
Paxson, J. W. Co, Philadelphia
Toronto Pottery Co., Toronto

Cupola Blowers.

Buffalo Foundry Supply Co., Buffalo.
Canada Machinery Agency, Montreal.
Detroit Foundry Supply Co., Windsor
Dominion Foundry Supply Co Montreal
Dominion Heating and Ventilating Co., Hespeler
Hamilton Facing Mill Co., Hamilton.
Northern Engineering Works, Detroit
Paxson J. W. Co., Philadelphia
Sheldons Limited, Galt.

Cutters, Flue

Chicago Pneumatic Tool Co., Chicago.
J. W. Paxson Co., Philadelphia.

Cutter Grinder Attachment.

Cincinnati Milling Machine Co., Cincinnati

Cutter Grinders, Plain.

Cincinnati Milling Machine Co., Cincinnati

Cutter Grinders, Universal.

Cincinnati Milling Machine Co., Cincinnati

Cutters, Milling.

Becker, Brainard Milling Machine Co. Hyde Park, Mass.
Frothingham & Workman Ltd. Montreal
Hamilton Tool Co. Hamilton, Ont.
Pratt & Whitney Co. Hartford, Conn.
Standard Tool Co., Cleveland.

Cutting-off Machines.

John Bertram & Sons Co., Dundas, Ont.
Canada Machinery Agency, Montreal.
London Mach. Tool Co., Hamilton.
H. W. Petrie, Toronto.
Pratt & Whitney Co., Hartford, Conn.

Cutting-off Tools.

Armstrong Bros. Tool Co., Chicago.
Baxter J. R., & Co., Montreal.
London Mach. Tool Co., Hamilton.
Mechanics Supply Co. Quebec Que.
H. W. Petrie, Toronto.
Pratt & Whitney, Hartford, Conn.
Rice Lewis & Son, Toronto.
L. S. Starrett Co., Athol, Mass.

Damper Regulators.

Darling Bros., Ltd., Montreal

Dies

Brown, Boggs Co., Hamilton
Globe Machine & Stamping Co., Cleveland, Ohio.
Woelfle Bros., Berlin, Ont.

Die Stocks

Curtis & Curtis Co., Bridgeport, Conn.
Hart Manufacturing Co., Cleveland, Ohio
Mechanics Supply Co., Quebec

Dies, Opening

W. H. Banfield & Sons, Toronto
Globe Machine & Stamping Co., Cleveland Ohio.
Pratt & Whitney Co., Hartford, Conn.
Woelfle Bros Berlin, Ont.

Dies, Sheet Metal.

W. H. Banfield & Sons Toronto.
Bliss, E. W., Brooklyn, N.Y.
Globe Machine & Stamping Co., Cleveland, Ohio

Dies, Threading.

Frothingham & Workman, Ltd., Montreal
Hart Mfg. Co., Cleveland
Mechanics Supply Co., Quebec, Que.
John Millen & Son, Ltd., Montreal.

Draft, Mechanical.

W. H. Banfield & Son, Toronto.
Butterfield & Co. Rock Island, Que.
A. B. Jardine & Co., Hespeler
Mechanics Supply Co., Quebec, Que.
Pratt & Whitney Co., Hartford, Conn.
Sheldon's Limited, Galt.

Drawing Instruments.

Rice Lewis & Son, Toronto.
Mechanics Supply Co., Quebec, Que.

Drawing Supplies.

The Electric Blue Print Co., Montreal

Drawn Steel, Cold.

Baxter J. R., & Co., Montreal.
Greey Wm. & J. G., Toronto
Union Drawn Steel Co., Hamilton.

Drilling Machines, Arch Bar.

John Bertram & Sons Co., Dundas, Ont.
London Mach. Tool Co., Hamilton
Niles-Bement-Pond Co., New York.

Drilling Machines, Boiler.

American Tool Works Co., Cincinnati.
John Bertram & Sons Co., Dundas, Ont.
Bickford Drill and Tool Co., Cincinnati.
The Canadian Fairbanks Co., Montreal.
A. B. Jardine & Co., Hespeler, Ont.
London Mach. Tool Co., Hamilton, Ont.
Niles-Bement-Pond Co., New York.
H. W. Petrie, Toronto.
Williams & Wilson, Montreal

Drilling Machines Connecting Rod.

John Bertram & Sons Co., Dundas, Ont.
London Mach. Tool Co., Hamilton.
Niles-Bement-Pond Co. New York.

Drilling Machines, Locomotive Frame.

American Tool Works Co., Cincinnati.
B. F. Barnes Co., Rockford, Ill.
John Bertram & Sons Co., Dundas, Ont.
London Mach. Tool Co., Hamilton, Ont.
Niles-Bement-Pond Co., New York.

Drilling Machines, Multiple Spindle.

American Tool Works Co., Cincinnati.
B. F. Barnes Co., Rockford, Ill.
Baxter J. R., & Co., Montreal.
John Bertram & Sons Co., Dundas, Ont.
Bickford Drill & Tool Co., Cincinnati.
Canada Machinery Agency Montreal.
London Mach. Tool Co., Hamilton, Ont.
Niles-Bement-Pond Co., New York.
H. W. Petrie, Toronto.
Williams & Wilson, Montreal.

Drilling Machines, Pneumatic

Canada Machinery Agency, Montreal.

Drilling Machines, Portable

Baxter, J. R., & Co., Montreal.
A. B. Jardine & Co., Hespeler, Ont.
Mechanics Supply Co., Quebec, Que.
Niles-Bement-Pond Co., New York.

Drilling Machines, Radial.

American Tool Works Co., Cincinnati.
Baxter J. R., & Co. Montreal
John Bertram & Sons Co., Dundas, Ont.
Bickford Tool & Drill Co., Cincinnati.
The Canadian Fairbanks Co., Montreal.
London Mach. Tool Co., Hamilton.
Mechanics Supply Co., Quebec, Que.
Niles-Bement-Pond Co., New York.
H. W. Petrie, Toronto.
Williams & Wilson, Montreal.

Drilling Machines, Suspension.

John Bertram & Sons Co., Dundas, Ont.
Canada Machinery Agency Montreal.
London Mach. Tool Co., Hamilton.
Niles-Bement-Pond Co. New York

Drilling Machines, Turret.

John Bertram & Sons Co., Dundas, Ont.
London Mach. Tool Co., Hamilton.
Niles-Bement-Pond Co., New York.

Drilling Machines, Upright.

American Tool Works Co., Cincinnati.
B. F. Barnes Co., Rockford, Ill.
Baxter, J. R., & Co., Montreal.
John Bertram & Sons Co., Dundas, Ont
Dwight Slate Machine Co., Hartford
Hamilton Tool Co. Hamilton Ont.
A. B. Jardine & Co., Hespeler, Ont.
Rice Lewis & Son, Toronto.
London Mach. Tool Co., Hamilton.
Mechanics Supply Co., Quebec, Que.
Niles-Bement-Pond Co., New York.
W. H. Petrie, Toronto.
Williams & Wilson Montreal.

Drills, Bench.

B. F. Barnes Co., Rockford, Ill.
Baxter, J. R. & Co., Montreal.
Hamilton Tool Co., Hamilton, Ont.
London Mach. Tool Co., Hamilton.
Mechanics Supply Co., Quebec Que.
Pratt & Whitney Co., Hartford, Conn.

Drills, Blacksmith.

Canada Machinery Agency, Montreal.
Frothingham & Workman, Ltd., Montreal
A. B. Jardine & Co., Hespeler, Ont.
London Mach. Tool Co., Hamilton.
Mechanics Supply Co., Quebec, Que.
Standard Tool Co., Cleveland.

Drills, Centre.

Mechanics Supply Co. ... e, Que.
Pratt & Whitney Co., Hartford, Conn.
Standard Tool Co., Cleveland, O.
L. S. Starrett Co., Athol, Mass.

Drills, Electric

Gas & Electric Power Co., Toronto.
B. F. Barnes Co., Rockford, Ill.
Baxter J. R., & Co., Montreal.
Canadian Pilling Co., Montreal
Chicago Pneumatic Tool Co., Chicago.
Niles-Bement-Pond Co., New York.

Drills, Gang.

American Tool Works Co., Cincinnati.
B. F. Barnes Co., Rockford, Ill.
John Bertram & Sons Co., Dundas, Ont.
Pratt & Whitney Co., Hartford, Conn.

Drills, High Speed.

Wm. Abbott, Montreal.
Frothingham & Workman, Ltd., Montreal
Alexander Gibb, Montreal
Pratt & Whitney Co., Hartford, Conn.
Standard Tool Co., Cleveland, O.

Drills, Hand.

A. B. Jardine & Co., Hespeler, Ont.

Drills, Horizontal.

B. F. Barnes Co., Rockford, Ill.
John Bertram & Sons Co., Dundas, Ont.
Canada Machinery Agency, Montreal.
London Mach. Tool Co., Hamilton.
Niles-Bement-Pond Co., New York.

Drills, Pneumatic.

Canada Machinery Agency, Montreal.
Chicago Pneumatic Tool Co., Chicago.
Independent Pneumatic Tool Co., Chicago, New York
Niles-Bement-Pond Co., New York.

Drills, Radial.

American Tool Works Co., Cincinnati.
John Bertram & Sons Co., Dundas, Ont.
Bickford Drill & Tool Co., Cincinnati
London Mach. Tool Co., Hamilton, Ont.
Niles-Bement-Pond Co., New York.

Drills, Ratchet.

Armstrong Bros. Tool Co., Chicago.
Frothingham & Workman, Ltd., Montreal
A. B. Jardine & Co., Hespeler.
Mechanics Supply Co., Quebec, Que.
Pratt & Whitney Co., Hartford, Conn.
Standard Tool Co., Cleveland.

Drills, Rock.

Allis-Chalmers-Bullock, Montreal.
Canadian Rand Drill Co., Montreal.
Chicago Pneumatic Tool Co., Chicago.
Jaffrey Mfg. Co., Columbus, Ohio.

Drills, Sensitive.

American Tool Works Co., Cincinnati.
B. F. Barnes Co., Rockford, Ill.
Canada Machinery Agency, Montreal.
Dwight Slate Machine Co., Hartford.
Niles-Bement-Pond Co., New York.

Drills, Twist.

Baxter, J. R., & Co., Montreal.
Chicago Pneumatic Tool Co., Chicago
Frothingham & Workman, Ltd., Montreal
Alex. Gibb, Montreal.
A. B. Jardine & Co., Hespeler, Ont.
Mechanics Supply Co., Quebec, Que.
John Millen & Son, Ltd., Montreal.
Morse Twist Drill and Machine Co., New Bedford, Mass.
Pratt & Whitney Co., Hartford, Conn.
Standard Tool Co., Cleveland.

Drying Apparatus of all Kinds.

Dominion Heating & Ventilating Co., Hespeler
Greey, Wm. & J. G., Toronto
Sheldons Limited, Galt

Dry Kiln Equipment.

Sheldons Limited, Galt

Dry Sand and Loam Facing.

Buffalo Foundry Supply Co., Buffalo.
Dominion Foundry Supply Co., Montreal
Hamilton Facing Mill Co., Hamilton.
Paxson, J. W., Co., Philadelphia

Dump Cars.

Canada Foundry Co., Limited, Toronto
Dominion Foundry Supply Co., Montreal
Greey, Wm. & J. G., Toronto
Hamilton Facing Mill Co., Hamilton.
John McDougall, Caledonian Iron Works Co., Montreal.
Niles-Bement-Pond Co., New York.
Standard Bearings, Ltd., Niagara Falls.
Link-Belt Eng. Co., Philadelphia.
John McDougall Caledonian Iron Works Co., Montreal
Owen Sound Iron Works Co., Owen Sound

Paxson, J. W., Co., Philadelphia
Waterous Engine Co., Brantford.

Dust Arresters.

Sly, W. W., Mfg. Co., Cleveland

Dust Separators.

Greey, Wm. & J. G., Toronto
Dominion Heating and Ventilating Co., Hespler.
Paxson, J. W., Co., Philadelphia
Sheldon's Limited, Galt.

Dynamos.

Allis-Chalmers-Bullock, Montreal.
Canadian General Electric Co., Toronto.
Canadian Westinghouse Co., Hamilton.
Consolidated Electric Co., Toronto
Electrical Machinery Co., Toronto.
Gas & Electric Power Co., Toronto
Hall Engineering Works, Montreal, Que.
Mechanics Supply Co., Quebec, Que.
John Millen & Son, Ltd., Montreal.
Packard Electric Co., St. Catharines.
H. W. Petrie, Toronto.
T. & H. Electric Co., Hamilton.

Dynamos—Turbine Driven.

Gas & Electric Power Co., Toronto.
Kerr-Turbine Co., Wellsville, N.Y.

Economizer, Fuel.

Dominion Heating & Ventilating Co., Hespeler
Standard Bearings, Ltd., Niagara Falls.

Electrical Instruments.

Gas & Electric Power Co., Toronto.
Canadian Westinghouse Co., Hamilton.
Mechanics Supply Co., Quebec, Que.

Electrical Steel.

Baxter, Patterson & Co., Montreal.

Electrical Supplies.

Gas & Electric Power Co., Toronto.
Canadian General Electric Co., Toronto.
Canadian Westinghouse Co., Hamilton.
London Mach. Tool Co., Hamilton, Ont.
Mechanics Supply Co., Quebec, Que
John Millen & Son, Ltd., Montreal.
Packard Electric Co., St. Catharines.
T. & H. Electric Co., Hamilton.

Electrical Repairs

Canadian Westinghouse Co., Hamilton.
T. & H. Electric Co., Hamilton.

Elevator Buckets.

Greey, Wm. & J. G., Toronto
Jeffrey Mfg. Co., Columbus, Ohio.

Emery and Emery Wheels.

Baxter, J. R., & Co., Montreal.
Dominion Foundry Supply Co., Montreal
Frothingham & Workman, Ltd., Montreal
Hamilton Facing Mill Co., Hamilton.
Paxson J. W. Co. Philadelphia

Emery Wheel Dressers.

Baxter J. R. & Co., Montreal.
Canada Machinery Agency, Montreal.
Desmond-Stephan Mfg. Co., Urbana, Ohio
Dominion Foundry Supply Co., Montreal
Frothingham & Workman Ltd., Montreal
Hamilton Facing Mill Co., Hamilton.
Mechanics Supply Co., Quebec, Que.
John Millen & Son, Ltd., Montreal.
H. W. Petrie, Toronto.
Paxson, J. W., Co., Philadelphia
Standard Tool Co., Cleveland.

Engineers and Contractors.

Gas & Electric Power Co., Toronto.
Canada Foundry Co., Limited, Toronto.
Darling Bros., Ltd., Montreal
Greey, Wm. & J. G., Toronto
Hall Engineering Works, Montreal,
Laurie Engine & Machine Co., Montreal.
Link-Belt Co., Philadelphia.
John McDougall, Caledonian Iron Works Co., Montreal.
Robb Engineering Co., Amherst, N.S.
The Smart-Turner Mach. Co., Hamilton.

Engineers' Supplies.

Baxter, J. R., & Co., Montreal.
Frothingham & Workman, Ltd., Montreal
Greey, Wm. & J. G., Toronto
Hall Engineering Works, Montreal.
Mechanics Supply Co., Quebec, Que.
Rice Lewis & Son, Toronto.

Engines, Gas and Gasoline.

Baxter, J. R., & Co., Montreal.
Canada Foundry Co., Toronto.
Canada Machinery Agency, Montreal.
The Canadian Fairbanks Co., Montreal.
Gas & Electric Power Co., Toronto
Gilson Mfg. Co., Guelph
The Goldie & McCulloch Co., Galt, Ont.
Rice Lewis & Son, Toronto
Ontario Wind Engine & Pump Co. Toronto
H. W. Petrie, Toronto.
The Smart-Turner Mach. Co., Hamilton

Engines, Steam.

Allis-Chalmers-Bullock, Montreal
Bellis & Marcom, Birmingham, Eng.
Canada Machinery Agency, Montreal.
The Goldie & McCulloch Co., Galt, Ont.
Rice Lewis & Son, Toronto.
Laurie Engine & Machine Co., Montreal.

John McDougall Caledonian Iron Works Montreal.
Miller Bros. & Toms, Montreal.
Robb Engineering Co., Amherst, N.S.
Sheldons Limited, Galt.
The Smart-Turner Mach. Co., Hamilton.
Waterous Engine Works Co., Brantford.
Gas & Electric Power Co., Toronto.

Exhaust Heads.

Darling Bros., Ltd., Montreal.
Dominion Heating & Ventilating Co., Hespeler
Sheldons Limited, Galt, Ont.

Expanded Metal.

Expanded Metal and Fireproofing Co. Toronto

Expanders.

A. B. Jardine & Co., Hespeler, Ont.

Fans, Electric.

Gas & Electric Power Co., Toronto.
Canadian General Electric Co., Toronto.
Canadian Westinghouse Co., Hamilton.
Dominion Heating & Ventilating Co. Hespeler
Mechanics Supply Co., Quebec, Que.
Sheldons Limited Galt, Ont.
The Smart-Turner Mach. Co., Hamilton.

Fans, Exhaust.

Gas & Electric Power Co., Toronto.
Canadian Buffalo Forge Co., Montreal.
Detroit Foundry Supply Co., Windsor.
Dominion Foundry Supply Co., Montreal
Dominion Heating & Ventilating Co., Hespeler.
Greey, Wm. & J. G., Toronto
Hamilton Facing Mill Co., Hamilton.
Paxson, J. W., Co., Philadelphia
Sheldons Limited, Galt.
B. F. Sturtevant Co., Hyde Park, Mass.

Feed Water Heaters.

Darling Bros., Montreal
Laurie Engine & Machine Co., Montreal
John McDougall, Caledonian Iron Works Co., Montreal.
The Smart-Turner Mach. Co., Hamilton

Files and Rasps.

J. R. Baxter, & Co., Montreal.
Frothingham & Workman, Ltd., Montreal
Mechanics Supply Co., Quebec, Que.
John Millen & Son, Ltd., Montreal.
Rice Lewis & Son, Toronto.
Nicholson File Co., Port Hope
H. W. Petrie, Toronto

Fillet, Pattern.

J. R. Baxter, & Co., Montreal.
Buffalo Foundry Supply Co., Buffalo.
Detroit Foundry Supply Co., Windsor.
Dominion Foundry Supply Co., Montreal
Hamilton Facing Mill Co., Hamilton.

Fire Apparatus.

Waterous Engine Works Co., Brantford.

Fire Brick and Clay.

J. R. Baxter & Co., Montreal.
Buffalo Foundry Supply Co., Buffalo.
Detroit Foundry Supply Co., Windsor.
Dominion Foundry Supply Co., Montreal
Gilmour J., New York.
Hamilton Facing Mill Co., Hamilton.
Ontario Lime Association, Toronto
J. W. Paxson Co., Philadelphia.
Toronto Pottery Co., Toronto

Flash Lights.

Berlin Electrical Mfg. Co., Toronto.
Mechanics Supply Co., Quebec, Que.

Flour Mill Machinery.

Allis-Chalmers-Bullock, Montreal.
Greey, Wm. & J. G., Toronto
The Goldie & McCulloch Co., Galt, Ont.
John McDougall, Caledonian Iron Works Co., Montreal.

Forges.

Canada Foundry Co., Limited, Toronto.
Frothingham & Workman Ltd., Montreal
Hamilton Facing Mill Co., Hamilton.
Mechanics Supply Co., Quebec, Que.
H. W. Petrie, Toronto.
Sheldons Limited, Galt, Ont.

Forgings, Drop.

John McDougall, Caledonian Iron Work Co., Montreal.
H. W. Petrie, Toronto.
St. Clair Bros., Galt

Forgings, Light & Heavy.

The Canada Forge Co., Welland.
Hamilton Steel & Iron Co., Hamilton

Forging Machinery.

John Bertram & Sons Co., Dundas, Ont.
London Mach. Tool Co., Hamilton, Ont.
National Machinery Co., Tiffin, Ohio
Niles-Bement-Pond Co., New York.

Founders.

Greey Wm. & J. G., Toronto
John McDougall, Caledonian Iron Works Co., Montreal.

Niagara Falls Machine & Foundry Co. Niagara Falls, Ont.
Maxwell, David, & Sons, St. Marys
The Smart-Turner Mach. Co., Hamilton.

Foundry Coke.

Baird & West, Detroit

Foundry Equipment.

Ph. Bonvillain & E. Ronceray, Philadelphia
Detroit Foundry Supply Co., Windsor.
Dominion Foundry Supply Co., Montreal
Gilmour, J., New York.
Hamilton Facing Mill Co., Hamilton
Hanna Engineering Works, Chicago.
Northern Engineering Works, Detroit
Paxson, J. W., Co., Philadelphia

Foundry Parting.

Doggett, Stanley, New York
Dominion Foundry Supply Co., Montreal
Partomol Co., New York
Foundry Specialty Co., Cincinnati
Stanley Doggett, New York

Foundry Facings.

Buffalo Foundry Supply Co., Buffalo.
Detroit Foundry Supply Co., Windsor.
Doggett Stanley, New York
Dominion Foundry Supply Co., Montreal
Hamilton Facing Mill Co., Hamilton.
J. W. Paxson Co., Philadelphia, Pa.

Friction Clutch Pulleys, etc.

The Goldie & McCulloch Co., Galt.
Greey, Wm. & J. G., Toronto
Link-Belt Co., Philadelphia.

Furnaces.

Detroit Foundry Supply Co., Windsor.
Dominion Foundry Supply Co., Montreal
Hamilton Facing Mill Co., Hamilton.
Paxson, J. W., & Philadelphia Co.,

Galvanizing

Canada Metal Co., Toronto.
Ontario Wind Engine & Pump Co., Toronto

Gas Blowers and Exhausters.

Dominion Heating & Ventilating Co. Hespeler
Sheldons Limited, Galt.

Gas Plants, Suction and Pressure.

Baxter J. R., & Co., Montreal
Colonial Engineering Co., Montreal
Gas & Electric Power Co., Toronto
Williams & Wilson, Montreal

Gauges, Standard.

Mechanics Supply Co., Quebec, Que.
Pratt & Whitney Co., Hartford, Conn.

Gearing.

Eberhardt Bros. Machine Co., Newark.
Greey, Wm. & J. G. Toronto

Gear Cutting Machinery.

Baxter, J. R., & Co., Montreal
Becker-Brainard Milling Mach. Co., Hyde Park, Mass.
Bickford Drill & Tool Co., Cincinnati.
Dwight Slate Machine Co., Hartford
Eberhardt Bros. Machine Co., Newark
Greey, Wm. & J. G., Toronto
Kennedy Wm. & Sons, Owen Sound
London Mach. Tool Co., Hamilton.
Niles-Bement-Pond Co., New York.
H. W. Petrie, Toronto.
Pratt & Whitney Co., Hartford, Conn.
Williams & Wilson, Montreal.

Gears, Angle.

Chicago Pneumatic Tool Co., Chicago.
Greey, Wm. & J. G., Toronto
Laurie Engine & Machine Co., Montreal.
John McDougall, Caledonian Iron Works Co., Montreal.
Waterous Engine Co., Brantford.

Gears, Cut.

Kennedy, Wm., & Sons, Owen Sound

Gears, Iron.

Greey, Wm. & J. G., Toronto
Kennedy, Wm., & Sons, Owen Sound

Gears, Mortise.

Greey, Wm. & J. G., Toronto
Kennedy, Wm., & Sons, Owen Sound

Gears, Reducing.

Brown, David & Sons, Huddersfield, Eng
Chicago Pneumatic Tool Co., Chicago.
Greey, Wm. & J. G., Toronto
John McDougall, Caledonian Iron Works Co., Montreal.

MANUFACTURERS AND MANUFACTURERS' AGENTS.

It will pay you to watch our condensed column each month. There are many money-making propositions brought to your attention here. You may find just what you are looking for.

RATES

One insertion—25c. for 20 words; 1c. a word for each additional word.

Yearly rate $2.50 for twenty words or less, 10c a word for each additional word.

The above does not apply to notices under the head of "Machinery Wanted." These notices are inserted free for subscribers.

BUSINESS CHANCE.

FOUNDRY FOR SALE—In a thriving Ontario town of five thousand inhabitants; is a going concern, employing from ten to fifteen hands, and does a general foundry business; no opposition within ten miles, and there are numerous orders now on hand; owing to a dissolution of the partnership the business must be sold at once. Apply Box 45, CANADIAN MACHINERY, Toronto.

FOR SALE.

BELTING, RUBBER, CANVAS AND LEATHER, Hose, Packing, Blacksmith's and Mill Supplies at lowest prices. N. Smith, 138 York Street, Toronto. (2tf)

AUTOMATIC stocker, with fan and engine complete, in good running order; snap; reason for selling, compelled to burn shavings. Apply A. J. H. Eckardt, National Casket Co., 107 Niagara St., Toronto. (9)

GASOLINE engine, 12 h.p., good as new; price, $275. Krug & Crosley, Hamilton. (9)

FOR SALE—The Canadian patent rights on a valveless pneumatic hammer, proved by actual test to be superior to any hammer at present on the market; an arrangement can be made to manufacture same on royalty for Canadian market. Apply Box 39C, CANADIAN MACHINERY, Toronto. (9)

SITUATION VACANT.

WANTED—Young man well experienced in machinery supply business able to sell and knowing thoroughly all requirements of the trade. Salary, commission and share of profits. Apply Box 32, CANADIAN MACHINERY, Toronto.

WANTED AT ONCE—Superintendent and mechanical engineer for machine works manufacturing gasoline engines and general machine work; one familar with modern methods, capable of making estimates, drawings and producing work at minimum cost; give full particulars of qualifications, experience, age, nationality and salary. Apply to Brandon Machine Works Co., Ltd., Brandon, Man. [9]

SITUATIONS WANTED.

ADVERTISER, shortly emigrating, seeks position as correspondence clerk and accountant. Well up in all details of engineering and allied business, ledger balancing, shorthand and typewriting; excellent references. Lister, Ipswich Road, Ardleigh, Colchester, Eng.

Mixing Machines, Dough.

Greey, Wm. & J. G., Toronto
Paxson, J. W., Co., Philadelphia

Mixing Machines, Special.

Greey, Wm. & J. G., Toronto
Paxson, J. W., Co., Philadelphia

Model Tools.

Globe Machine & Stamping Co., Cleveland, Ohio.
Standard Bearings, Ltd., Niagara Falls.
Wells Pattern and Model Works, Toronto

Motors, Electric.

Allis-Chalmers-Bullock Limited, Montreal
Canadian General Electric Co., Toronto
Canadian Westinghouse Co., Hamilton.
Electrical Machinery Co., Toronto.
Consolidated Electric Co., Toronto
Gas & Electric Power Co., Toronto
Hall Engineering Works, Montreal.
Mechanics Supply Co., Quebec, Que.
The Packard Electric Co., St. Catharines.
T. & H. Electric Co., Hamilton.

Motors, Air.

Canadian Rand Drill Co., Montreal.
Chicago Pneumatic Tool Co., Chicago.

Molders' Supplies.

Buffalo Foundry Supply Co., Buffalo.
Detroit Foundry Supply Co., Windsor.
Dominion Foundry Supply Co., Montreal
Hamilton Facing Mill Co., Hamilton.
Mechanics Supply Co., Quebec, Que.
Paxson, J. W., Co., Philadelphia

Molders' Tools.

Buffalo Foundry Supply Co., Buffalo.
Detroit Foundry Sup ly Co., Windsor
Dominion Foundry Supply Co., Montreal
Hamilton Facing Mill Co., Hamilton.
Mechanics Supply Co., Quebec, Que.
Paxson, J. W., Co., Philadelphia

Molding Machines.

Ph. Bonvillain & E. Ronceray, Philadelphia
Buffalo Foundry Supply Co., Buffalo.
Dominion Foundry Supply Co., Montreal
Hamilton Facing Mill Co., Hamilton.
J. W. Paxson Co., Philadelphia, Pa.

Molding Sand.

T. W. Barnes, Hamilton.
Buffalo Foundry Supply Co., Buffalo.
Detroit Foundry Supply Co., Windsor
Dominion Foundry Supply Co., Montreal
Hamilton Facing Mill Co., Hamilton.
Paxson, J. W., Co., Philadelphia

Nut Tappers.

John Bertram & Sons Co., Dundas, Ont.
A. B. Jardine & Co., Hespeler.
London Mach. Tool Co., Hamilton.
National Machinery Co., Tiffin, Ohio.

Nuts.

Canada Nut Co., Toronto.

Oatmeal Mill Machinery.

Greey, Wm. & J. G., Toronto
The Goldie & McCulloch Co., Galt

Oilers, Gang.

Sight Feed Oil Pump Co., Milwaukee, Wis.

Oils, Core.

Buffalo Foundry Supply Co., Buffalo.
Dominion Foundry Supply Co., Montreal
Hamilton Facing Mill Co., Hamilton.
Paxson, J. W., Co., Philadelphia

Oil Extractors.

Darling Bros., Ltd., Montreal

Paint Mill Machinery.

Greey, Wm. & J. G., Toronto

Pans, Lathe.

Cleveland Wire Spring Co., Cleveland
Paxson, J. W., Co., Philadelphia

Pans, Steel Shop.

Cleveland Wire Spring Co., Cleveland

Parting Compound.

Doggett, Stanley, New York

Patent Solicitors.

Hanbury A. Budden, Montreal.
Fetherstonhough & Blackmore, Montreal
Marion & Marion, Montreal.
Ridout & Maybee, Toronto.

Patterns.

John Carr, Hamilton.
Galt Malleable Iron Co., Galt
Hamilton Pattern Works, Hamilton.
John McDougall Caledonian Iron Works Co., Montreal
Wells' Pattern and Model Works, Toronto.

Pig Iron.

Hamilton Steel & Iron Co., Hamilton

Pipe Cutting and Threading Machines.

Borden-Canadian Co., Toronto
Butterfield & Co., Rock Island, Que.
Canada Machinery Agency, Montreal.
Curtis & Curtis Co., Bridgeport, Conn.
Frothingham & Workman, Ltd., Montreal
Hart Mfg. Co., Cleveland
A. B. Jardine & Co., Hespeler, Ont.
London Mach. Tool Co., Hamilton, Ont.
Niles-Bement-Pond Co., New York.
Shantz, I. E., & Co., Berlin, Ont.

Pipe, Municipal.

Canadian Pipe Co., Vancouver, B.C.
Pacific Coast Pipe Co., Vancouver, B.C.

Pipe, Waterworks.

Canadian Pipe Co., Vancouver, B.C.
Pacific Coast Pipe Co., Vancouver, B.C.

Planers, Standard.

American Tool Works, Cincinnati.
John Bertram & Sons Co., Dundas, Ont.
Canada Machinery Agency, Montreal.
The Canadian Fairbanks Co., Montreal.
Rice Lewis & Son, Toronto.
London Mach. Tool Co., Hamilton, Ont.
Niles-Bement-Pond Co., New York.
H. W. Petrie, Toronto.
Pratt & Whitney Co., Hartford, Conn.
Williams & Wilson, Montreal.

Planers, Rotary.

John Bertram & Sons Co., Dundas, Ont.
London Mach. Tool Co., Hamilton, Ont.
Niles-Bement-Pond Co., New York.

Planing Mill Fans.

Dominion Heating & Ventilating Co., Hespeler
Sheldons Limited, Galt, Ont.

Plumbago.

Buffalo Foundry Supply Co., Buffalo.
Detroit Foundry Supply Co., Windsor
Doggett Stanley, New York
Dominion Foundry Supply Co., Montreal
Hamilton Facing Mill Co., Hamilton.
Mechanics Supply Co., Quebec, Que.
J. W. Paxson Co., Philadelphia, Pa.

Pneumatic Tools.

Allis-Chalmers-Bullock, Montreal.
Canadian Rand Drill Co., Montreal.
Chicago Pneumatic Tool Co., Chicago
Hamilton Facing Mill Co., Hamilton.
Hanna Engineering Works, Chicago.
Independent Pneumatic Tool Co., Chicago, New York
Paxson, J. W., Co., Philadelphia

Power Hack Saw Machines.

Baxter, J. R., & Co., Montreal.
Frothingham & Workman, Ltd., Montreal

Power Plants.

Gas & Electric Power Co., Toronto
John McDougall Caledonian Iron Works Co., Montreal.
The Smart-Turner Mach. Co., Hamilton

Power Plant Equipments.

Gas & Electric Power Co., Toronto.
Darling Bros., Ltd., Montreal

Presses, Drop.

W. H. Banfield & Son, Toronto.
E. W. Bliss Co., Brooklyn, N.Y.
Brown, Boggs Co., Hamilton
Canada Machinery Agency, Montreal.
Laurie Engine & Machine Co., Montreal
Miller Bros. & Toms, Montreal.
Niles-Bement-Pond Co., New York

Presses, Hand.

E. W. Bliss Co., Brooklyn, N.Y.
Brown Boggs Co., Hamilton

Presses, Hydraulic.

John Bertram & Sons Co., Dundas, Ont.
Laurie Engine & Machine Co., Montreal.
London Mach. Tool Co., Hamilton, Ont.
John McDougall Caledonian Iron Works Co., Montreal.
Niles-Bement-Pond Co., New York.
Perrin, Wm. R., & Co., Toronto

Presses, Power.

E. W. Bliss Co., Brooklyn, N.Y.
Brown, Boggs Co., Hamilton
Canada Machinery Agency, Montreal.
Laurie Engine & Machine Co., Montreal
London Mach. Tool Co., Hamilton, Ont.
John McDougall Caledonian Iron Works Co., Montreal.
Niles-Bement-Pond Co., New York.

Presses Power Screw.

Brown, Boggs Co., Hamilton
Perrin, Wm. R., & Co., Toronto

Pressure Regulators.

Darling Bros., Ltd., Montreal

Producer Plants.

Canada Foundry Co., Toronto

Pulp Mill Machinery.

Greey, Wm. & J. G., Toronto
Jeffrey Mfg. Co., Columbus, Ohio.
Laurie Engine & Machine Co., Montreal.
John McDougall Caledonian Iron Works Co., Montreal.
Waterous Engine Works Co., Brantford

Pulleys.

Baxter J. R., & Co., Montreal.
Canada Machinery Agency, Montreal.
The Canadian Fairbanks Co., Montreal.
The Goldie & McCulloch Co., Galt.
Greey, Wm. & J. G., Toronto
Laurie Engine & Machine Co., Montreal.
Link-Belt Co., Philadelphia
John McDougall Caledonian Iron Works Co., Montreal.
Owen Sound Iron Works Co., Owen Sound
H. W. Petrie, Toronto.
The Smart-Turner Mach. Co., Hamilton.
Standard Bearings Ltd., Niagara Falls,
Williams & Wilson, Montreal.
Waterous Engine Co., Brantford.

Pumps.

Laurie Engine & Machine Co., Montreal
Ontario Wind Engine & Pump Co., Toronto

Pump Governors.

Darling Bros., Ltd. Montreal

Pumps, Hydraulic.

Ph. Bonvillain & E. Ronceray, Philadelphia
Laurie Engine & Machine Co., Montreal
Perrin, Wm. R. & Co., Toronto

Pumps, Oil.

Sight Feed Oil Pump Co., Milwaukee, Wis.

Pumps, Steam.

Allis-Chalmers-Bullock, Limited, Montreal
Canada Foundry Co., Toronto.
Canada Machinery Agency, Montreal.
Darling Bros., Ltd., Montreal
Gas & Electric Power Co., Toronto
The Goldie & McCulloch Co., Galt.
John McDougall Caledonian Iron Works, Montreal.
H. W. Petrie, Toronto.
The Smart-Turner Mach. Co., Hamilton.
Standard Bearings, Ltd., Niagara Falls,
Waterous Engine Co. Brantford

Pumping Machinery.

Canada Foundry Co., Limited, Toronto
Canada Machinery Agency, Montreal.
Canadian Rand Drill Co., Montreal.
Chicago Pneumatic Tool Co., Chicago.
Darling Bros., Ltd., Montreal
Gas & Electric Power Co., Toronto.
Hall Engineering Works, Montreal, Que.
Laurie Engine & Machine Co., Montreal.
London Mach. Tool Co., Hamilton, Ont.
John McDougall Caledonian Iron Works Co., Montreal.
The Smart-Turner Mach. Co., Hamilton
Standard Bearings, Ltd., Niagara Falls,

Punches and Dies.

W. H. Banfield & Sons, Toronto.
Bliss, E. W., Co., Brooklyn, N.Y.
Butterfield & Co., Rock Island.
Globe Machine & Stamping Co.
A. B. Jardine & Co., Hespeler, Ont.
London Mach. Tool Co., Hamilton, Ont.
Pratt & Whitney Co., Hartford, Conn.
H. W. Petrie, Toronto.
Standard Bearings, Ltd., Niagara Falls,

Punches, Hand.

Mechanics Supply Co., Quebec, Que.

Punches, Power.

John Bertram & Sons Co., Dundas, Ont.
E. W. Bliss Co., Brooklyn, N.Y.
Canada Machinery Agency, Montreal.
Canadian Buffalo Forge Co., Montreal.
London Mach. Tool Co., Hamilton, Ont.
Niles-Bement-Pond Co., New York.

Punches, Turret.

London Mach. Tool Co., London, Ont.

Punching Machines, Horizontal.

John Bertram & Sons Co., Dundas, Ont.
London Mach. Tool Co., Hamilton Ont.
Niles-Bement-Pond Co., New York.

Quartering Machines.

John Bertram & Sons Co., Dundas, Ont.
London Mach. Tool Co., Hamilton, Ont.

Railway Spikes and Washers.

Hamilton Steel & Iron Co., Hamilton

Rammers, Bench and Floor.

Buffalo Foundry Supply Co., Buffalo.
Detroit Foundry Supply Co., Windsor
Hamilton Facing Mill Co., Hamilton.
Paxson, J. W. Co., Philadelphia

Rapping Plates.

Detroit Foundry Supply Co., Windsor
Hamilton Facing Mill Co., Hamilton.
Paxson, J. W., Co., Philadelphia

Raw Hide Pinions.

Brown, David & Sons, Huddersfield, Eng

Reamers.

Wm. Abbott, Montreal.
Baxter J. R. & Co. Montreal.
Butterfield & Co., Rock Island.
Frothingham & Workman Ltd., Montreal
Hamilton Tool Co., Hamilton
Hanna Engineering Works Chicago.
A. B. Jardine & Co., Hespeler, Ont.
Mechanics Supply Co., Quebec, Que.
Morse Twist Drill and Machine Co., New Bedford, Mass.
Pratt & Whitney Co., Hartford, Conn.
Standard Tool Co., Cleveland.

Reamers, Steel Taper.

Butterfield & Co., Rock Island.
Chicago Pneumatic Tool Co., Chicago.
A. B. Jardine & Co., Hespeler, Ont.
John Millen & Son Ltd., Montreal.
Pratt & Whitney Co., Hartford, Conn.
Standard Tool Co., Cleveland.

Rheostats.

Gas & Electric Power Co., Toronto.
Canadian General Electric Co., Toronto.
Canadian Westinghouse Co., Hamilton.
Hall Engineering Works Montreal, Que.
T. & H. Electric Co., Hamilton,

Riddles.

Buffalo Foundry Supply Co., Buffalo.
Detroit Foundry Supply Co., Windsor
Hamilton Facing Mill Co., Hamilton.
Dominion Foundry Supply Co., Montreal
J. W. Paxson Co., Philadelphia, Pa.

Riveters, Pneumatic.

Hanna Engineering Works, Chicago.
Independent Pneumatic Tool Co., Chicago, Ill.

Rolls, Bending.

John Bertram & Sons Co., Dundas, Ont.
London Mach. Tool Co., Hamilton, Ont.
Niles-Bement-Pond Co., New York.

Rolls, Chilled Iron.

Greey, Wm. & J. G., Toronto

Rolls, Sand Cast.

Greey, Wm. & J. G., Toronto

Rotary Converters.

Allis-Chalmers-Bullock, Ltd., Montreal.
Canadian Westinghouse Co., Hamilton.
Paxson, J. W., Co. Philadelphia
Toronto and Hamilton Electric Co., Hamilton.

Rubbing and Sharpening Stones.

Norton Co., Worcester, Mass.

Safes.

Baxter Paterson & Co., Montreal.
The Goldie & McCulloch Co., Galt,

Sand Blast Machinery.

Canadian Rand Drill Co., Montreal.
Chicago Pneumatic Tool Co., Chicago.
Detroit Foundry Supply Co., Windsor
Dominion Foundry Supply Co., Montreal
Hamilton Facing Mill Co., Hamilton.
J. W. Paxson Co., Philadelphia, Pa.

Sand, Heavy, Grey Iron.

Buffalo Foundry Supply Co., Buffalo.
Detroit Foundry Supply Co., Windsor
Dominion Foundry Supply Co., Montreal
Hamilton Facing Mill Co., Hamilton.
Paxson, J. W., Co., Philadelphia

Sand, Malleable.

Buffalo Foundry Supply Co., Buffalo.
Detroit Foundry Supply Co., Windsor
Dominion Foundry Supply Co., Montreal
Hamilton Facing Mill Co., Hamilton.
Paxson, J. W., Co., Philadelphia

Sand, Medium Grey Iron.

Buffalo Foundry Supply Co., Buffalo.
Detroit Foundry Supply Co., Windsor
Dominion Foundry Supply Co., Montreal
Hamilton Facing Mill Co., Hamilton.
Paxson, J. W., Co., Philadelphia

Sand Sifters.

Buffalo Foundry Supply Co., Buffalo.
Detroit Foundry Supply Co., Windsor
Dominion Foundry Supply Co., Montreal
Hamilton Facing Mill Co., Hamilton.
Hanna Engineering Works, Chicago.
Paxson, J. W., Co., Philadelphia

Saw Mill Machinery.

Allis-Chalmers-Bullock, Limited, Montreal
Baxter, J. R., & Co., Montreal
Canada Machinery Agency, Montreal.
Goldie & McCulloch Co., Galt.
Greey, Wm. & J. G., Toronto
Owen Sound Iron Works Co., Owen Sound
H. W. Petrie, Toronto.
Robb Engineering Co. Amherst,
Standard Bearings, Ltd., Niagara Falls
Waterous Engine Works, Brantford
Williams & Wilson, Montreal.

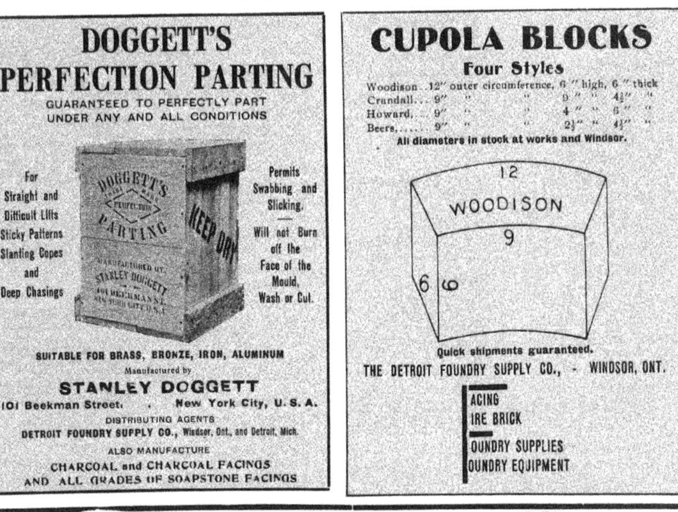

GILMOUR'S

20th Century Foundry Equipment

Cupolas, Tumblers, Trucks, Cranes, Ladles and Piqua Blowers

Gilmour's Double Spur Geared Tumbler is the longest lived and most practical mill ever put on the market. Let us tell you more about it.

Fire Brick and Cupola Blocks a Specialty. We claim that our Fire Brick is superior to any other on the market. Write for samples and prices.

J. GILMOUR, Bennett Building, NEW YORK

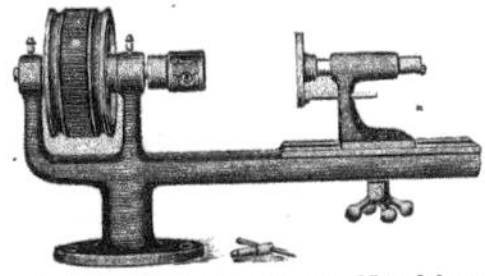

Champion Tapping Machine

Especially adapted for high speed tapping in light work. Taps holes either through or to depth. Is automatic in operation, very rapid and reduces the [illegible] of taps to the minimum. Capacity up to [illegible]-in. holes.

MANUFACTURED BY THE

Blair Tool and Machine Works

24 West St. NEW YORK CITY

SUBSCRIPTION FORM

To the Publishers of

Canadian Machinery and Manufacturing News

10 Front St. East Toronto

Gentlemen:—

Please send one copy of Canadian Machinery and Manufacturing News, commencing with next issue, for twelve months, for which I enclose $1.00, to

Name ....

Address ....

Write name and address plainly

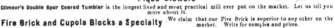

Sawing Machines, Metal.

Niles-Bement-Pond Co. New York.
Paxson, J. W., Co., Philadelphia

Saws, Hack.

Baxter, J. R., & Co., Montreal.
Canada Machinery Agency, Montreal.
Detroit Foundry Supply Co., Windsor
Frothingham & Workman Ltd., Montreal
London Mach. Tool Co., Hamilton.
Mechanics Supply Co., Quebec, Que.
Rice Lewis & Son, Toronto.
John Millen & Son, Ltd., Montreal.
L. S. Starrett Co., Athol, Mass.

Screw Cutting Tools

Hart Manufacturing Co., Cleveland, Ohio
Mechanics Supply Co., Quebec, Que.

Screw Machines, Automatic.

Canada Machinery Agency, Montreal,
Cleveland Automatic Machine Co., Cleveland, Ohio
London Mach. Tool Co., Hamilton, Ont.
National-Acme Mfg. Co., Cleveland.
Pratt & Whitney Co., Hartford, Conn.

Screw Machines, Hand.

Canada Machinery Agency Montreal.
A. B. Jardine & Co., Hespeler.
London Mach. Tool Co., Hamilton, Ont.
Mechanics Supply Co. Quebec, Que.
Pratt & Whitney Co., Hartford, Conn.
Warner& Swasey Co., Cleveland, O.

Screw Machines, Multiple Spindle.

National-Acme Mfg. Co., Cleveland

Screw Plates.

Butterfield & Co., Rock Island, Que.
Frothingham & Workman, Ltd., Montreal
Hart Manufacturing Co. Cleveland, Ohio
A. B. Jardine & Co., Hespeler
Mechanics Supply Co., Quebec, Que.

Screws, Cap and Set.

National-Acme Mfg. Co., Cleveland

Screw Slotting Machinery, Semi-Automatic.

National-Acme Mfg. Co., Cleveland

Second-Hand Machinery.

American Tool Works Co., Cincinnati.
Canada Machinery Agency, Montreal.
The Canadian Fairbanks Co., Montreal.
Goldie & McCulloch Co., Galt.
Machinery Exchange, Montreal.
Niles-Bement-Pond Co., New York.
H. W. Petrie, Toronto.
Robb Engineering Co., Amherst, N S.
Waterous Engine Co., Brantford.
Williams & Wilson, Montreal.

Shafting.

Baxter, J. R., & Co., Montreal.
Canada Machinery Agency, Montreal.
The Canadian Fairbanks Co., Montreal.
Frothingham & Workman Ltd., Montreal
The Goldie & McCulloch Co., Galt, Ont
Greey Wm. & J. G. Toronto
Kennedy Wm., & Sons, Owen Sound
Niles-Bement-Pond Co., New York.
Owen Sound Iron Works Co., Owen Sound
H. W. Petrie, Toronto.
Smart-Turner Machine Co., Hamilton.
Union Drawn Steel Co., Hamilton.
Waterous Engine Co., Brantford.

Shapers.

American Tool Works Co., Cincinnati.
John Bertram & Sons Co., Dundas, Ont
Canada Machinery Agency, Montreal.
The Canadian Fairbanks Co., Montreal.
Rice Lewis & Son, Toronto.
London Mach. Tool Co., Hamilton, Ont.
Niles-Bement-Pond Co., New York.
H. W. Petrie, Toronto.
Potter & Johnston Machine Co., Pawtucket, R.I.
Pratt & Whitney Co., Hartford, Conn.
Williams & Wilson, Montreal.

Sharpening Stones.

Carborundum Co., Niagara Falls, N.Y.

Shearing Machine, Bar.

John Bertram & Sons Co., Dundas, Ont.
A. B. Jardine & Co., Hespeler.
London Mach. Tool Co., Hamilton, Ont.
Niles-Bement-Pond Co., New York.

Shears, Power.

John Bertram & Sons Co., Dundas, Ont.
Canada Machinery Agency, Montreal,
A. B. Jardine & Co., Hespeler, Ont.
Niles-Bement-Pond Co., NewYork.
Paxson, J. W., Co., Philadelphia

Sheet Metal Goods

Globe Machine & Stamping Co., Cleveland, Ohio.

Sheet Meal Working Tools.

Brown, Boggs Co., Hamilton

Sheet Steel Work.

Owen Sound Iron Works Co., Owen Sound

Shingle Mill Machinery.

Owen Sound Iron Works Co., Owen Sound

Shop Trucks.

Greey, Wm. & J. G., Toronto
Owen Sound Iron Works Co., Owen Sound
Paxson, J. W, Co., Philadelphia

Shovels.

Baxter, J. R., & Co., Montreal.
Buffalo Foundry Supply Co., Buffalo.
Detroit Foundry Supply Co., Detroit.
Doggett, Stanley, New York
Frothingham & Workman, Ltd., Montreal
Dominion Foundry Supply Co., Montreal
Hamilton Facing Mill Co., Hamilton.
Mechanics Supply Co., Quebec, Que.
Paxson, J. W., Co., Philadelphia

Sieves.

Buffalo Foundry Supply Co., Buffalo.
Detroit Foundry Supply Co., Windsor
Dominion Foundry Supply Co., Montreal
Hamilton Facing Mill Co., Hamilton.
Paxson, J. W., Co., Philadelphia

Silver Lead.

Buffalo Foundry Supply Co., Buffalo.
Detroit Foundry Supply Co., Windsor
Doggett, Stanley, New York
Dominion Foundry Supply Co., Montreal
Hamilton Facing Mill Co., Hamilton.
Paxson, J. W., Co., Philadelphia

Snap Flasks

Buffalo Foundry Supply Co., Buffalo.
Detroit Foundry Supply Co., Windsor
Dominion Foundry Supply Co., Montreal
Hamilton Facing Mill Co., Hamilton.
Paxson, J. W., Co., Philadelphia

Soap Stones & Talc Crayons & Pencils.

Doggett, Stanley, New York

Soapstone.

Buffalo Foundry Supply Co., Buffalo.
Detroit Foundry Supply Co., Windsor
Doggett, Stanley, New York
Dominion Foundry Supply Co., Montreal
Hamilton Facing Mill Co., Hamilton.
Paxson, J. W., Co., Philadelphia

Soap Stone Facings.

Doggett Stanley, New York

Solders.

Lumen Bearing Co., Toronto

Special Machinery.

W. H. Banfield & Sons, Toronto.
Baxter, J. R. & Co., Montreal
John Bertram & Sons Co. Dundas, Ont
Brown, Boggs Co., Hamilton
Globe Machine & Stamping Co., Cleveland Ohio.
Greey, Wm. & J. G., Toronto
Hanna Engineering Works, Chicago.
Laurie Engine & Machine Co., Montreal.
London Mach. Tool Co., Hamilton, Ont.
H. W. Petrie, Toronto.
The Smart-Turner Mach. Co., Hamilton
Standard Bearings, Ltd., Niagara Falls.
Waterous Engine Co., Brantford.

Special Machines and Tools.

Brown, Boggs Co., Hamilton
Paxson, J. W., Philadelphia
Pratt & Whitney, Hartford, Conn.
Standard Bearings, Ltd., Niagara Falls,

Special Milled Work.

National-Acme Mfg. Co., Cleveland

Speed Changing Countershafts.

The Canadian Fairbanks Co., Montreal

Spike Machines.

National Machinery Co., Tiffin, O.
The Smart-Turner Mach. Co., Hamilton

Spray Cans.

Detroit Foundry Supply Co., Windsor
Dominion Foundry Supply Co., Montreal
Hamilton Facing Mill Co., Hamilton.
Paxson, J. W., Co., Philadelphia

Springs, Automobile.

Cleveland Wire Spring Co., Cleveland

Springs, Coiled Wire.

Cleveland Wire Spring Co., Cleveland

Springs, Machinery.

Cleveland Wire Spring Co., Cleveland

Springs, Upholstery.

Cleveland Wire Spring Co., Cleveland

Sprue Cutters.

Detroit Foundry Supply Co., Windsor
Dominion Foundry Supply Co., Montreal
Hamilton Facing Mill Co., Hamilton.
Paxson, J. W., Co., Philadelphia

Stamp Mills.

Allis-Chalmers-Bullock, Limited, Montreal
Paxson, J. W., Co., Philadelphia

Steam Separators.

Darling Bros., Ltd., Montreal.
Dominion Heating & Ventilating Co., Hespeler
Robb Engineering Co., Montreal.
Sheldon's Limited, Galt.
Smart-Turner Mach. Co., Hamilton
Waterous Engine Co., Brantford.

Steam Specialties.

Darling Bros., Ltd., Montreal
Dominion Heating & Ventilating Co., Hespeler
Sheldon's Limited, Galt.

Steam Traps.

Canada Machinery Agency, Montreal.
Darling Bros., Ltd., Montreal
Dominion Heating & Ventilating Co., Hespeler.
Mechanics Supply Co., Quebec, Que.
Sheldons Limited, Galt.

Steam Valves.

Darling Bros., Ltd., Montreal

Steel Pressure Blowers

Buffalo Foundry Supply Co., Buffalo.
Dominion Foundry Supply Co., Montreal
Dominion Heating & Ventilating Co., Hespeler
Hamilton Facing Mill Co., Hamilton.
J. W. Paxon Co., Philadelphia, Pa.
Sheldon's Limited, Galt.

Steel Tubes.

Baxter, J. R., & Co., Montreal.
Mechanics Supply Co., Quebec, Que.
John Millen & Son, Montreal.

Steel, High Speed.

Wm. Abbott, Montreal.
Canadian Fairbanks Co., Montreal.
Frothingham & Workman, Ltd., Montreal
Alex. Gibb, Montreal.
Jessop, Wm., & Sons, Sheffield, Eng.
B. K. Morton Co., Sheffield, Eng.
Williams & Wilson, Montreal.

Stocks and Dies

Hart Manufacturing Co., Cleveland, Ohio
Mechanics Supply Co., Quebec, Que.
Pratt & Whitney Co., Hartford, Conn.

Stone Cutting Tools, Pneumatic

Allis-Chalmers-Bullock, Ltd., Montreal.
Canadian Rand Drill Co., Montreal.

Stone Surfacers

Chicago Pneumatic Tool Co., Chicago.

Stove Plate Facings.

Buffalo Foundry Supply Co., Buffalo
Detroit Foundry Supply Co., Windsor.
Dominion Foundry Supply Co., Montreal
Hamilton Facing Mill Co., Hamilton.
Paxson, J. W., Co., Philadelphia

Swage, Block.

A. B. Jardine & Co., Hespeler, Ont.

Switchboards.

Gas & Electric Power Co., Toronto.
Allis-Chalmers-Bullock, Limited, Montreal
Canadian General Electric Co., Toronto.
Canadian Westinghouse Co., Hamilton
Hall Engineering Works, Montreal, Que.
Mechanics Supply Co., Quebec, Que.
Toronto and Hamilton Electric Co., Hamilton.

Talc.

Buffalo Foundry Supply Co., Buffalo.
Detroit Foundry Supply Co., Windsor.
Doggett, Stanley, New York
Dominion Foundry Supply Co., Montreal
Hamiton Facing Mill Co., Hamilton.
Paxson, J. W., Co., Philadelphia

Tanks, Oil.

Sight Feed Oil Pump Co., Milwaukee, Wis.

Tapping Machines and Attachments.

American Tool Works Co., Cincinnati.
John Bertram & Sons Co., Dundas, Ont.
Bickford Drill & Tool Co., Cincinnati.
Blair Tool & Machine Works, New York.
The Geometric Tool Co., New Haven,
A. B. Jardine & Co., Hespeler.
London Mach. Tool Co., Hamilton, Ont
Murchey Machine & Tool Co., Detroit.
Niles-Bement-Pond Co., New York.
H. W. Petrie, Toronto.
Pratt & Whitney, Cincinnati, O.
L. S. Starrett Co., Athol, Mass.
Williams & Wilson, Montreal.

Tapes, Steel.

Frothingham & Workman, Ltd., Montreal
Rice Lewis & Son, Toronto.
Mechanics Supply Co., Quebec, Que.
John Millen & Son, Ltd., Montreal.
L. S. Starrett Co., Athol, Mass.

Taps.

Baxter, J. R., & Co., Montreal.
Mechanics Supply Co., Quebec, Que.

Taps, Collapsing.

The Geometric Tool Co., New Haven.

Taps and Dies.

Wm. Abbott, Montreal.
Baxter, J. R., & Co., Montreal.
Butterfield & Co., Rock Island, Que.
Frothingham & Workman, Ltd., Montreal
The Geometric Tool Co., New Haven.
A. B. Jardine & Co., Hespeler, Ont.
Rice Lewis & Son, Toronto.
Mechanics Supply Co., Quebec, Que.
John Millen & Son, Ltd., Montreal.
Pratt & Whitney Co., Hartford, Conn.
Standard Tool Co., Cleveland.
L. S. Starrett Co., Athol, Mass.

Testing Laboratory.

Detroit Testing Laboratory, Detroit

Testing Machines.

Detroit Foundry Supply Co., Windsor.
Dominion Foundry Supply Co., Montreal
Hamilton Facing Mill Co., Hamilton.
Paxson, J. W., Co., Philadelphia

Thread Cutting Tools.

Hart Manufacturing Co., Cleveland, Ohio
Mechanics Supply Co., Quebec Que.

Tiling, Opal Glass.

Toronto Plate Glass Importing Co., Toronto.

Time Switches, Automatic, Electric.

Berlin Electric Mfg. Co., Toronto.

Tinware Machinery

Canada Machinery Agency, Montreal.

Tire Upsetters or Shrinkers

A. B. Jardine & Co., Hespeler, Ont.

Tool Cutting Machinery

Canadian Rand Drill Co., Montreal.

PROFESSIONAL DIRECTORY
CONSULTING ENGINEERS, PATENT ATTORNEYS, ARCHITECTS, CONTRACTORS, ETC.

Tool Holders.

Armstrong Bros. Tool Co., Chicago.
Baxter, J. R., & Co., Montreal.
John Millen & Son, Ltd., Montreal.
H. W. Petrie, Toronto.
Pratt & Whitney Co., Hartford, Conn.

Tool Steel.

Wm. Abbott, Montreal.
Flockton, Tompkin & Co., Sheffield, Eng.
Frothingham & Workman Ltd., Montreal
Alex. Gibb, Montreal
Wm. Jessop, Sons & Co., Toronto.
Canadian Fairbanks Co., Montreal
B. K. Morton & Co., Sheffield, Eng.
Williams & Wilson, Montreal.

Torches, Steel.

Baxter, J. R., & Co., Montreal.
Detroit Foundry Supply Co., Windsor.
Dominion Foundry Supply Co., Montreal
Hamilton Facing Mill Co., Hamilton.
Paxson, J. W., Co., Philadelphia

Transformers and Convertors

Gas & Electric Power Co., Toronto.
Allis-Chalmers-Bullock, Montreal.
Canadian General Electric Co., Toronto.
Canadian Westinghouse Co., Hamilton.
Hal Engineering Works, Montreal, Que.
T. & H. Electric Co., Hamilton.

Transmission Machinery

Gas & Electric Power Co., Toronto.
Allis-Chalmers-Bullock, Montreal.
The Canadian Fairbanks Co., Montreal.
Greey, Wm. & J. G., Toronto
Laurie Engine & Machine Co., Montreal.
Link-Belt Co., Philadelphia.
Paxson, J. W., Co., Philadelphia
H. W. Petrie, Toronto.
The Smart-Turner Mach. Co., Hamilton.
Waterous Engine Co., Brantford.

Transmission Supplies

Baxter, J. R., & Co., Montreal.
The Canadian Fairbanks Co., Montreal.
Wm. & J. G. Greey, Toronto.
The Goldie & McCulloch Co., Galt.
H. W. Petrie, Toronto.

Trolleys

Canadian Rand Drill Co., Montreal.
Dominion Foundry Supply Co., Montreal
John Millen & Son, Ltd., Montreal.
Northern Engineering Works, Detroit

Trolley Wheels.

Lumen Bearing Co., Toronto

Trucks, Dryer and Factory

Dominion Heating & Ventilating Co., Hespeler
Greey, Wm. & J. G., Toronto.
Northern Engineering Works, Detroit
Sheldon's Limited, Galt, Ont.

Tube Expanders (Rollers)

Chicago Pneumatic Tool Co., Chicago.
A. B. Jardine & Co., Hespeler.
Mechanics Supply Co., Quebec, Que.

Turbines, Steam

Gas & Electric Power Co., Toronto.
Allis-Chalmers-Bullock Limited, Montreal.
Canadian General Electric Co., Toronto.
Canadian Westinghouse Co., Hamilton.
Kerr Turbine Co., Wellsville, N.Y.

Turntables

Detroit Foundry Supply Co., Windsor.
Dominion Foundry Supply Co., Montreal
Hamilton Facing Mill Co., Hamilton.
Northern Engineering Works, Detroit
Paxson, J. W., Co., Philadelphia

Turret Machines.

American Tool Works Co., Cincinnati.
John Bertram & Sons Co., Dundas, Ont.
The Canadian Fairbanks Co., Montreal.
Gisholt Machine Co., Madison, Wis.
London Mach. Tool Co., Hamilton, Ont.
Niles-Bement-Pond Co., New York.
H. W. Petrie, Toronto.
Pratt & Whitney Co., Hartford, Conn.
Warner & Swasey Co., Cleveland, Ohio.
Williams & Wilson, Montreal.

Upsetting and Bending Machinery.

John Bertram & Sons Co., Dundas, Ont.
A. B. Jardine & Co., Hespeler.
London Mach. Tool Co., Hamilton, Ont.
National Machinery Co., Tiffin, O.
Niles-Bement-Pond Co., New York.

Valves, Blow-off.

Chicago Pneumatic Tool Co., Chicago.
Mechanics Supply Co., Quebec, Que.

Valves, Back Pressure.

Darling Bros., Ltd., Montreal
Dominion Heating & Ventilating Co. Hespeler
Mechanics Supply Co., Quebec, Que.
Sheldon's Limited, Galt.

Valve Reseating Machines.

Darling Bros. Ltd., Montreal

Ventilating Apparatus.

Canada Machinery Agency, Montreal
Darling Bros. Ltd., Montreal
Dominion Heating & Ventilating Co., Hespeler, Ont.
Sheldon's Ltd., Galt

Vises, Planer and Shaper.

American Tool Works Co., Cincinnati, O.
Frothingham & Workman Ltd., Montreal
A. B. Jardine & Co., Hespeler, Ont.
John Millen & Son, Ltd., Montreal.
Niles-Bement-Pond Co., New York.

Washer Machines.

National Machinery Co., Tiffin, Ohio.

Water Tanks.

Ontario Wind Engine & Pump Co., Toronto

Water Wheels.

Allis-Chalmers-Bullock Co., Montreal.
Barber, Chas., & Sons, Meaford.
Canada Machinery Agency, Montreal.
The Goldie & McCulloch Co., Galt, Ont.
Greey, Wm. & J. G., Toronto.
Wm. Kennedy & Sons, Owen Sound.
John McDougall Caledonian Iron Works Co., Ltd., Montreal

Water Wheels, Turbine.

Barber, Chas., & Sons, Meaford, Ont.
Kennedy, Wm., & Sons, Owen Sound

Wheelbarrows.

Baxter, J. W. & Co., Montreal.
Buffalo Foundry Supply Co., Buffalo
Frothingham & Workman Ltd., Montreal
Detroit Foundry Supply Co., Windsor.
Dominion Foundry Supply Co., Montreal
Hamilton Facing Mill Co., Hamilton.
Paxson, J. W., Co., Philadelphia

Winches.

Canadian Pilling Co., Montreal
Frothingham & Workman, Ltd., Montreal

Wind Mills.

Ontario Wind Engine & Pump Co., Toronto

Window Wire Guards.

Expanded Metal and Fireproofing Co. Toronto.
B. Greening Wire Co., Hamilton, Ont.
Paxson, J. W., Co. Philadelphia

Wire Chains.

The B. Greening Wire Co., Hamilton.

Wire Cloth and Perforated Metals.

Expanded Metal and Fireproofing Co. Toronto.
B. Greening Wire Co., Hamilton, Ont.
Paxson, J. W., Co., Philadelphia.

Wire Guards and Railings.

Expanded Metal and Fireproofing Co. Toronto.
B. Greening Wire Co. Hamilton, Ont.

Wire Nail Machinery.

National Machinery Co., Tiffin, Ohio.

Wire Rope.

Frothingham & Workman Ltd., Montreal
B. Greening Wire Co., Hamilton, Ont.

Wood Boring Machines. Pneumatic.

Independent Pneumatic Tool Co., Chicago, Ill.

Wood-working Machinery.

Baxter, J. R., & Co., Montreal
Canada Machinery Agency, Montreal.
The Canadian Fairbanks Co., Montreal.
Goldie & McCulloch Co., Galt.
H. W. Petrie, Toronto.
Waterous Engine Works Co., Brantford
Williams & Wilson, Montreal.

ALPHABETICAL INDEX.

PATTERNS
WELLS' PATTERN AND MODEL WORKS
Tel. Main 3581. (HARRY WELLS, Proprietor.)
Patterns and Models made in wood and metal for Engines, Pumps, Furnaces, Agricultural, Electrical and Architectural Works and Machines of every description.
35 Richmond St. E., Toronto

JOHN J. GARTSHORE
83 Front St. W., Toronto
RAILS and SUPPIIES, New and Second-hand
For RAILWAYS, TRAMWAYS, Etc.
Old material bought and sold.

OPAL GLASS TILING
FOR WALLS OF
MACHINERY AND POWER HOUSES
Most approved material.
TORONTO PLATE GLASS IMPORTING CO'Y
PLATE AND WINDOW GLASS
135 to 143 Victoria St., - Toronto

CHADWICK BROS.
BRASS FOUNDERS
METAL SPINNERS
STAMPERS, PLATERS
and all kinds of Special Metal Workers
LET US QUOTE PRICES ON YOUR WORK.
CHADWICK BROS.
Factory: HAMILTON, OAK AVE.

Are you interested in any of the lines that are advertised?
A Post Card will bring you price list and full information.
Don't forget to mention Canadian Machinery.

FOUNDRY FACINGS and SUPPLIES

Anything required in your Foundry ?
WE HAVE IT

THE HAMILTON FACING MILL CO., Limited
FOUNDRY OUTFITTERS
Head Office and Works, **HAMILTON, Ontario.**
Eastern Office and Warehouse, **MONTREAL, Quebec.**

FIRE BRICK

"Every Quality." "For Every Purpose."
The TORONTO POTTERY CO., Limited
FACTORIES IN OHIO TORONTO, CANADA

The Strongest Nut in the World

This NUT is destined to be used EXCLUSIVELY *the world over* by all manufacturers of Machinery and Engines on account of its immense superiority in strength over the pressed iron nut formerly in general use. Write at once for Catalog and Price List to

CANADA NUT CO. - Niagara Falls, Ont.
Sole Manfrs. of Turned Steel Hexigon Nuts.

Delays are Dangerous

They often read serious financial loss. You cannot afford to place your orders for
PIPE, CORE, STOVE PLATE and MACHINERY MOLDING SAND and LAKE ROOFING GRAVEL
with any old concern. Sometimes you want your sand orders filled mighty quick. Send for prices and samples. Quick deliveries a specialty.

T. W. BARNES, HAMILTON, ONT.

MALLEABLE IRON CASTINGS

WE ALSO MAKE METAL PATTERNS

Galt Malleable Iron Co., Limited - Galt, Ontario

Partine

THE UNIVERSAL PARTING

Trade Mark Reg. U. S. Patent Off.

Preferred by Leading Foundrymen to Lycopodium

Partine does not run, cut, wash or burn off the face of the mould.

Guaranteed to produce, in every instance, a PATTERN TRUE casting, displaying the most delicate lines and features of an artistic design, accompanied by an absolute smoothness of the edges of the metal.

Its price is so low as to make its use imperative in every modern foundry.

Write for free sample, descriptive booklet and full particulars.

Sole Canadian Agents: THE HAMILTON FACING MILL COMPANY, Limited
Foundry Outfitters, HAMILTON, ONTARIO and MONTREAL, QUEBEC.

Cable Address: PARTINE

THE FOUNDRY SPECIALTY CO., Cincinnati, O., U.S.A.

IF WE ASKED $10.00 A YEAR SOME WOULD GLADLY PAY IT.

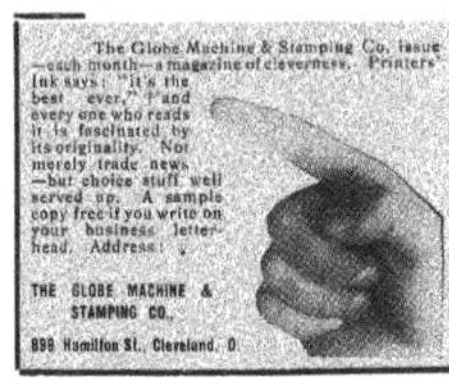

Persons addressing advertisers will kindly mention having seen advertisement in this paper.

Perfect Satisfaction

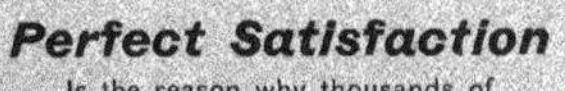

Is the reason why thousands of

"THE OLD RELIABLE"

GREEN ROTARY POSITIVE PRESSURE BLOWER

are in use to-day in every part of the world. Have **you** got one?

Write for our Catalogue.

Wilbraham-Green Blower Co.

2518 FRANKFORD AVENUE

PHILADELPHIA, PA.

What do you require in

Plumbago Facings?

We can not only fill your wants but do so to your entire satisfaction.

Our range includes 50 grades from 1c. to 10c. per pound.

A pleasure to give you further information.

J. W. PAXSON CO.

PHILA., Pa., U.S.A.

MANUFACTURERS

FOUNDRY FACINGS SUPPLIES

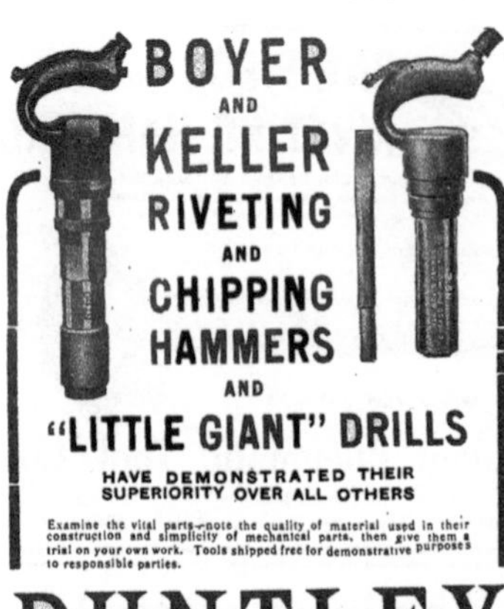

AIR COOLED PORTABLE

ELECTRIC TOOLS

HAVE A RECORD OF

100 PER CENT. EARNING POWER

HAVE SAVED THEIR FIRST COST

in a few months for many purchasers.

Single Motor Type Drill

EVERY TOOL GUARANTEED

to perform the service for which it is designed.

Write for catalogue describing DUNTLEY ELECTRIC DRILLS, HOISTS, GRINDERS, BUFFERS AND BLOWERS.

CANADIAN PNEUMATIC
TORONTO **TOOL COMPANY** MONTREAL

BRANCH OF

CHICAGO PNEUMATIC
CHICAGO **TOOL COMPANY** NEW YORK

SALES AGENTS FOR CANADA:

N. J. HOLDEN & CO., Limited, MONTREAL

If the experience gained from 20 years of Air Compressor designing and building is of advantage to you, drop us a line.

We build Air Compressors in 1000 styles and sizes, for all duties.

DO YOU WANT A CATALOG?

Canadian Rand Company,
LIMITED

MONTREAL, QUE.

DO YOU KNOW that there is no other magazine in the world that covers the fields of engineering and technology to the same extent or in such a useful literary way as TECHNICAL LITERATURE, and that no other appeals equally to specialists and the general reader in so many branches of the profession? It has become a most valuable companion to such professional men as

Architects	**Engineering Students**	**Hydraulic Engineers**	**Mechanics**	**Patent Attorneys**
Chemists	**Factory Managers**	**Iron and Steel Makers**	**Metallurgists**	**Power Consumers**
Civil Engineers	**Foundrymen**	**Irrigation Engineers**	**Mining Engineers**	**Railroad Engineers**
Contractors	**Gas Engineers**	**Machine Tool Makers**	**Municipal Engineers**	**Sanitary Engineers**
Draftsmen	**Heating, Lighting and Ventilating Engineers**	**Marine Engineers**	**Oil, Paint and Varnish Makers**	**Steam Engineers**
Electrical Engineers		**Mechanical Engineers**		**Water Works Engineers**

AND OTHERS INTERESTED IN TECHNICAL WORK OR STUDY

Such men may be considered as forming the backbone of the citizenship of the country. They are the men who do things, and who control the expenditure of millions of dollars, involving the purchase of all kinds of material and machinery. If you have anything to offer them for business or pleasure, advertise it in TECHNICAL LITERATURE. *Let us show you the scope, vitality and the power of our circulation.*

ONE ADVERTISER WRITES: **"Your magazine is a good one and sure of success. It is going to be looked over by a good class of readers, and our advertisement will not be lost in a mass of ill-digested matter."**

ADVERTISING RATES SENT ON APPLICATION

Subscription price, $2.00 a year to United States or Mexico; $2.25 to Canada; foreign countries, $2.50. Single copies, 20 cents.

TECHNICAL LITERATURE, 905 St. Paul Building, NEW YORK

"The New Way."

Let our Beaver do your Work

You Don't have to Change Dies.
Cuts 1, 1¼, 1½ and 2 inches, perfect threads, all with one set of chasers.
Other sizes of Beavers will be born soon—¼ to ¾—2½ to 4—4½ to 8—8 to 12.

Write for our special 10-day trial offer.

You cannot afford to take time to thread pipe by hand in any other way.

Manufactured by

Borden-Canadian Company

166 Adelaide Street West, Toronto

OTIS ELEVATORS

FOR ALL DUTIES

Electric, Hydraulic, Belt, Steam and Hand Power

MANUFACTURED BY

OTIS-FENSOM ELEVATOR COMPANY
LIMITED

Head Office: TORONTO, ONT.

Works: HAMILTON, ONT.

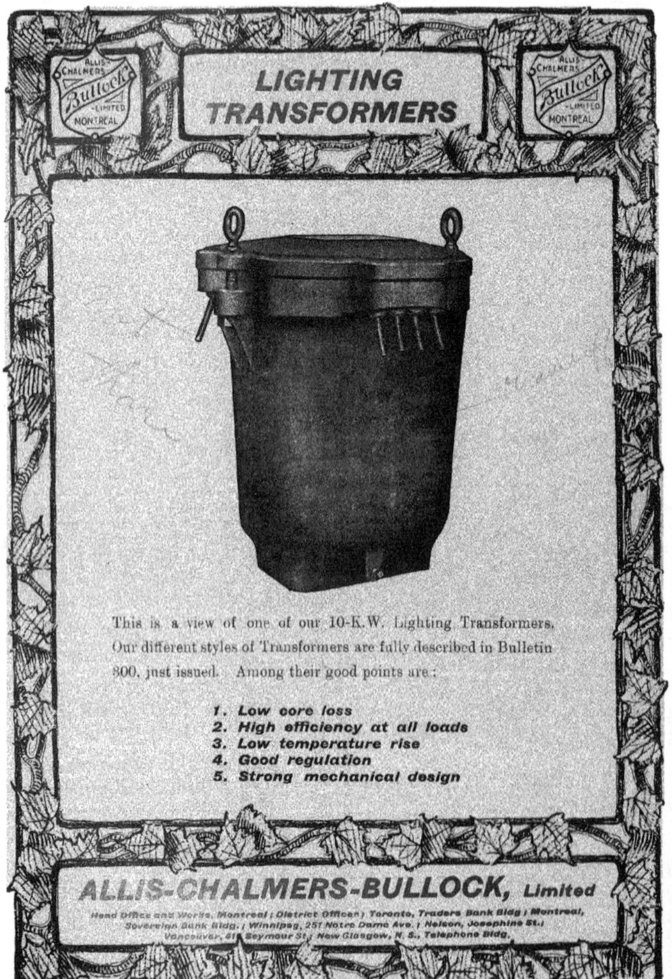
ALLIS-CHALMERS Bullock LIMITED MONTREAL
LIGHTING TRANSFORMERS
ALLIS-CHALMERS Bullock LIMITED MONTREAL
This is a view of one of our 10-K.W. Lighting Transformers. Our different styles of Transformers are fully described in Bulletin 800, just issued. Among their good points are:
1. Low core loss
2. High efficiency at all loads
3. Low temperature rise
4. Good regulation
5. Strong mechanical design
ALLIS-CHALMERS-BULLOCK, Limited
Head Office and Works, Montreal; District Offices; Toronto, Traders Bank Bldg; Montreal, Sovereign Bank Bldg.; Winnipeg, 251 Notre Dame Ave.; Nelson, Josephine St.; Vancouver, 41 Seymour St.; New Glasgow, N. S., Telephone Bldg.

CANADIAN MACHINERY AND MANUFACTURING NEWS

A monthly newspaper devoted to the manufacturing interests, covering in a practical manner the mechanical, power, foun dry and allied fields. Published by The MacLean Publishing Company, Limited, Toronto, Montreal, Winnipeg, and London, Eng.

OFFICE OF PUBLICATION: 10 FRONT STREET EAST, TORONTO

Vol. III. NOVEMBER, 1907 No. 11

SMALL P&W TOOLS

ADJUSTABLE REAMERS

(Patented December 22, 1903)

Size Accurately Adjusted Without Regrinding.

Will go to the bottom of a blind hole, as there is no obstruction at front end.

VARIETY OF SIZES

Blades of either Carbon or High Power Steel

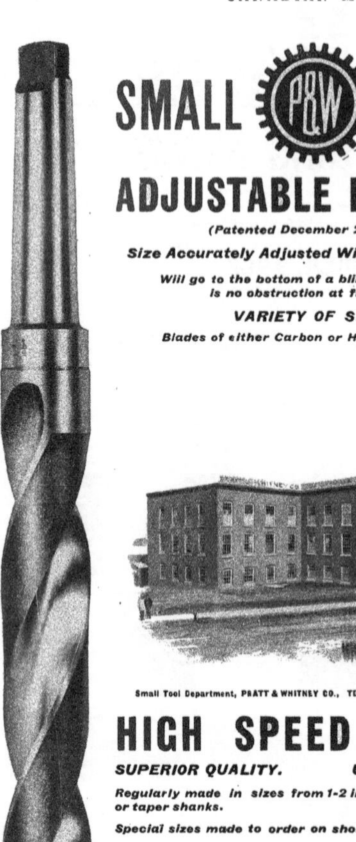

Small Tool Department, PRATT & WHITNEY CO., TD., OF CANADA, Dundas, Ontario

HIGH SPEED DRILLS

SUPERIOR QUALITY. UNIFORM TEMPER.

Regularly made in sizes from 1-2 in. to 3 in., with straight or taper shanks.

Special sizes made to order on short notice.

These tools are shown in our New Small Tool Catalog.

PRATT & WHITNEY CO., LTD. OF CANADA

DUNDAS - CANADA

Agents: THE CANADIAN FAIRBANKS CO., Ltd., Montreal, Toronto, Winnipeg and Vancouver

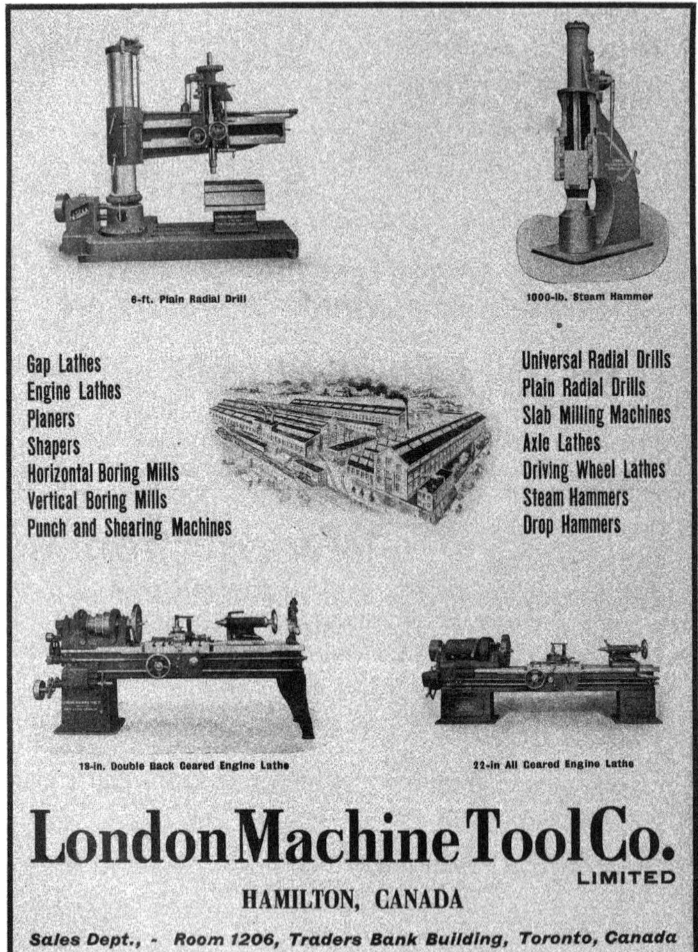
6-ft. Plain Radial Drill
1000-lb. Steam Hammer
Gap Lathes
Engine Lathes
Planers
Shapers
Horizontal Boring Mills
Vertical Boring Mills
Punch and Shearing Machines
Universal Radial Drills
Plain Radial Drills
Slab Milling Machines
Axle Lathes
Driving Wheel Lathes
Steam Hammers
Drop Hammers
18-in. Double Back Geared Engine Lathe
22-in All Geared Engine Lathe
London Machine Tool Co.
LIMITED
HAMILTON, CANADA
Sales Dept., - Room 1206, Traders Bank Building, Toronto, Canada

Warner & Swasey Turret Lathes—

For Brass and Iron Work—complete equipments

Universal Turret Lathe

for the manufacture of Plumbers' and Steamfitters' fittings.

Every modern facility provided for rapid and economical production.

One type of machine here illustrated — adapted for general work.

Styles and sizes for every requirement.

1907 General Catalogue—recently published—now ready for distribution.

THE WARNER & SWASEY CO., Cleveland, Ohio, U.S.A.

Canadian Agents: A. R. Williams Machinery Co., Toronto and Williams & Wilson, Montreal.

Do You Know This Man?

If not, you should make his acquaintance. He has done more for the improvement of the cleaning room than any other man in the world. Write him, and tell him your troubles in cleaning your castings, and doing away with your dust, and he will show you some of the most improved devices ever put on the market. He will save you Money. Write for catalog and prices.

The W. W. Sly Mfg. Co., Cleveland, O.

NOTICE—To receive new catalogue when issued, machine tool users please send postal with company's name and mechanical officer in charge thereof.

THE AMERICAN TOOL WORKS CO

700-750 CULVERT STREET, CINCINNATI, U. S. A.

Write Direct or to Agents: WILLIAMS & WILSON, Montreal; A. R. WILLIAMS MACHINERY CO., Toronto, Winnipeg, Vancouver. 5

THOUSANDS OF MANUFACTURERS IN THIS COUNTRY AND ABROAD ARE USING

BECKER-BRAINARD MILLING MACHINES and MILLING CUTTERS

Because they have found them to be Practical, Reliable, Economical and Durable. Years of practical test have proved the efficiency of our product. "Show us your work, and we will show you our method."

Becker-Brainard Milling Machine Co.

Hyde Park, Mass., U.S.A.

AGENTS:—McDowell, Stocker & Co., Chicago. Chas. G. Smith Co., Pittsburg. J. L. Osgood, Buffalo. A. B. Bowman, St. Louis. A. R. Williams Machinery Co., Toronto, Montreal, Winnipeg and Vancouver, Canada. Ludw. Loewe & Co., Berlin. Bevan & Edwards Propty, Ltd., Melbourne. Selig Sonnenthal & Co., London. Schuchardt & Schutte, Berlin, Stockholm, Vienna, St. Petersburg. A. H. Schutte, Cologne, Liege, Milan, Barcelona, Brussels, Paris, Bilbao. 366

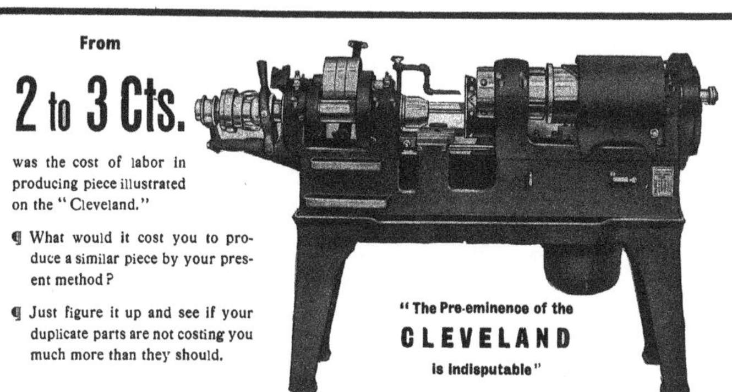

Why not let us **show you** what the "Cleveland" will do on your work? Mail us drawings and we'll tell you.

Piece No. 166.—Material, machine steel bar, made on a 2-inch Cleveland Automatic Plain Machine, 15-inch feed; drawing full size; time of all operations for completing this piece, 18 minutes; actual cost of labor, from 2 to 3 cents.

AS WE HAVE NO CANADIAN REPRESENTATIVE, WE SOLICIT CORRESPONDENCE AND ORDERS DIRECT.

CLEVELAND AUTOMATIC MACHINE CO.

CLEVELAND, OHIO, U.S.A.

EASTERN REPRESENTATIVE— J. B. ANDERSON, 2450 North Thirtieth Street, Philadelphia, Pa. ***WESTERN REPRESENTATIVE***— H. E. NUNN, 22 Fifth Avenue, Chicago, Ill. ***FOREIGN REPRESENTATIVES***—CHAS. CHURCHILL & CO., London, Manchester, Birmingham, Newcastle-on-Tyne and Glasgow. MESSRS. SCHUCHARDT & SCHUTTE, Berlin, Vienna, St. Petersburg and Stockholm. ALFRED H. SCHUTTE, Cologne, Brussels, Liege, Paris, Milan and Bilbao.

CARBORUNDUM

GRINDING WHEELS

MADE AT NIAGARA

USED THE WORLD OVER

There are something like two hundred thousand different sizes and shapes and grades of Carborundum Grinding Wheels—

There are little bits of wheels for the use of jewelers and dentists and there are wheels 8 feet high for grinding car wheels and other big work of that sort—

There are wheels for working marble and granite that do more work and better work than steel tools ever did, and that have practically replaced steel tools in many establishments—

And then there are wheels to do every other sort of grinding work that anybody ever thought of—Carborundum

Does the work faster and better—
Accomplishes more work in a day—
And lasts more days before wearing out than any other abrasive material on earth.

We are constantly making special wheels for special purposes—

We have experts here whose sole duty it is to solve grinding problems —

If you will write us about your grinding requirements we will be glad to tell you just what Carborundum can do for you.

The Carborundum Company
Niagara Falls, N.Y.

Pipe Cutting and Threading Machine

For Hand or Power, Bench or Trench

FORBES PATENT DIE STOCKS will do either from ¼ inch I.D. to 15 inch O.D. **ON THE SPOT**

No expense or delay of carting your pipe to the machine. We keep regular sizes in stock and can make prompt shipment. Write for Catalog.

THE CURTIS & CURTIS CO., - *36 Garden Street, Bridgeport, Conn.*

THIS IS THE NEW SHANTZ No. 6.

BAD THREADS—LOST MONEY

That, in brief, is the tale which could be told by many a man who still sticks to his old-fashioned pipe-threading machine. It is surely the worst form of **extravagance** to keep an **obsolete, money-wasting** machine in the shop, especially when, at a reasonable price, the

SHANTZ No. 6

fitted with the **Borden Patent Solid Adjustable Die Head,** may be installed. This machine **cuts absolutely perfect threads** on 1½-inch to 6-inch pipe. It does the work **with one cut,** and does it quicker and cleaner than any other machine in America. **Absolute accuracy** can be obtained by the mere adjustment of an adjusting screw. This machine weighs 4,300 pounds, and is **so immovably rigid** that **a wavy thread is an impossibility.** And there are other points about it, which, with those above stated, place the "**Shantz No. 6**" in a class by itself. **Write for our free descriptive circular.**

I. E. SHANTZ & CO., Berlin, Ont.

DO YOU DO IT?

When you read an advertisement that interests you, why not write the advertiser and get more information? It will help you to keep posted. Try it. The advertisers in **Canadian Machinery** will appreciate it.

WRITE THEM. **TRY IT.**

IT PAYS TO INSTALL THE BEST

POWER PRESSES

The many exclusive features embodied in the type shown above make it so much superior to the next best that there is no comparison.

Write for prices and particulars on **Presses, Dies, Tools and Special Machinery.**

W. H. BANFIELD & SONS

120 Adelaide Street West, TORONTO.

No. 1 Auto. Gear Cutter

Rapidity and accuracy are the strong points of our Automatic Gear Cutters. Style No. 1, as shown in cut, is Full Automatic and will handle work on all kinds of material ranging from pinion rods 1-8-in. diam. 50 pitch, to 5-in. diam. 16 pitch. Will cut 3 1-2-in. face.

Send for Catalogue

THE

Dwight Slate Machine Co.,

HARTFORD, CONN., U.S.A.

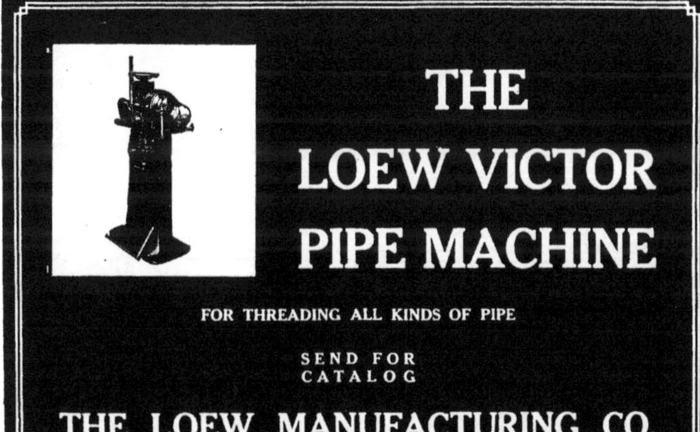

THE
LOEW VICTOR
PIPE MACHINE
FOR THREADING ALL KINDS OF PIPE
SEND FOR
CATALOG
THE LOEW MANUFACTURING CO.
CLEVELAND : OHIO

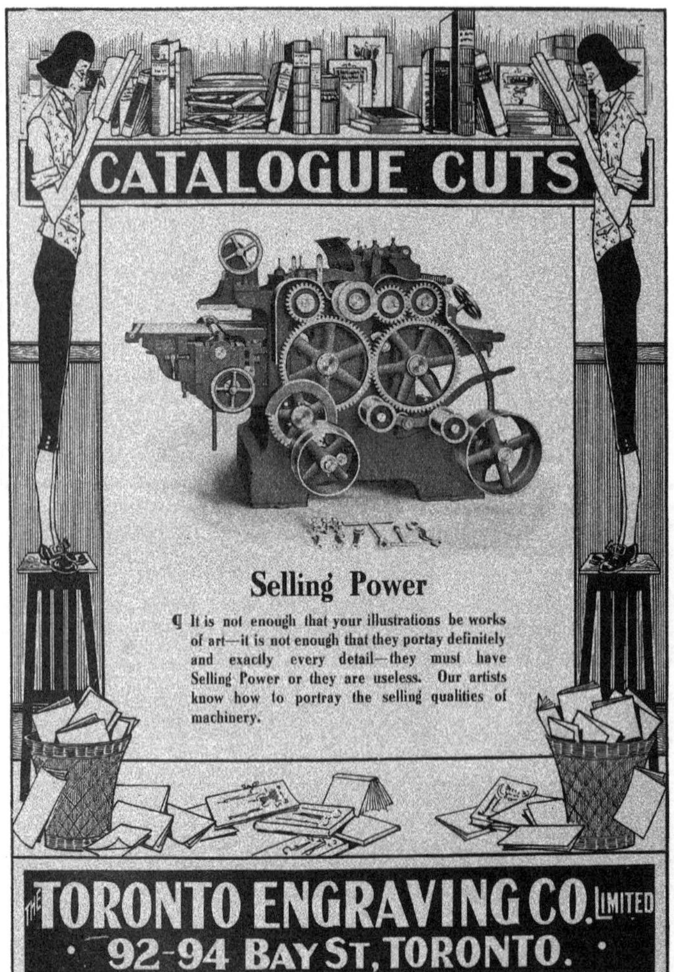

CATALOGUE CUTS
Selling Power
¶ It is not enough that your illustrations be works of art—it is not enough that they portay definitely and exactly every detail—they must have Selling Power or they are useless. Our artists know how to portray the selling qualities of machinery.
THE TORONTO ENGRAVING CO. LIMITED
· 92-94 BAY ST, TORONTO. ·

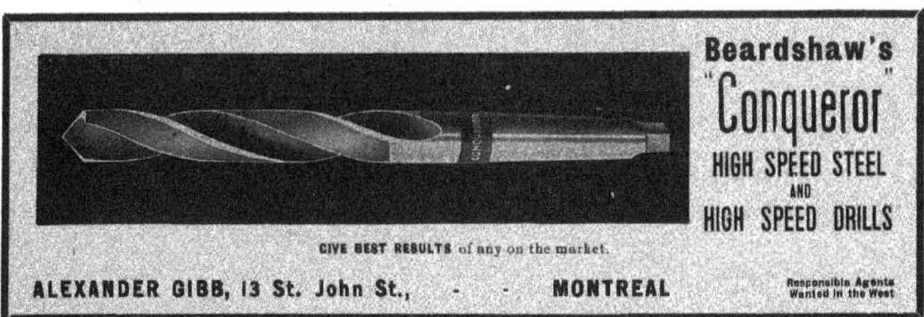

Beardshaw's
"Conqueror"
HIGH SPEED STEEL
AND
HIGH SPEED DRILLS
GIVE BEST RESULTS of any on the market.
ALEXANDER GIBB, 13 St. John St., - - MONTREAL
Responsible Agents Wanted in the West

WELL KNOWN BRANDS MADE IN CANADA BY
AMERICAN
ARCADE
KEARNEY and FOOT
McCLELLAN
Nicholson Co.
GLOBE
EAGLE
GREAT WESTERN
J. B. SMITH
Dominion Works, Port Hope, Ont.

UNION DRAWN STEEL COMPANY, Limited
Manufacturers of
BRIGHT FINISHED STEEL
Largest in Canada. Best Equipped on the Continent.
Capacity twice present requirements of the trade.
Rounds, Flats, Squares and Hexagons
LARGE STOCK
Send for Price List
Office and Works:
HAMILTON, CANADA

"PECK" OVERLAPPING BUCKET CARRIER

Conveys Without Spill

To test every claim we make for the utility and economy of this Carrier for conveying

COAL, COKE, ASHES, STONE, CEMENT, HOT AND COLD CLINKER, ORES, ETC.

Ask us where you can see it in every-day operation.

LINK-BELT COMPANY

PHILADELPHIA, CHICAGO, INDIANAPOLIS

New York: 299 Broadway. Pittsburgh: 1501 Park Bldg.
Seattle: 410 New York Block. St. Louis: Missouri Trust Bldg.
Denver: Lindrooth, Shubart & Co. New Orleans: Wilmot Machinery Co.

12" KERR STEAM TURBINE 25 b.h.b., at 3000 to 3200 r.p.m., coupled to No. 00 RAYMOND PULVERIZER AND AIR SEPARATOR for handling lime, cement, mica, rags, scraps, corn, cottonseed hulls, dry colors and the like.

KERR TURBINE BLOWERS, for gas works, foundries and forge shops. Air pressure 4 to 16 ounces. For any capacity.

TURBINE PUMPS, for water works, condenser equipment and boiler feeding. These pumps are more economical than direct acting piston pumps and require less attention, lubrication and space. No belts or gears. Speed adjustable to pressure desired. Regulation close. Start and stop quickly. For any capacity.

TURBINE GENERATORS, 2 kw. to 150 kw. direct current and alternating current 25 k.w. to 150 kw. 60 cycle.

Built in Canada by WM. STEPHENSON, 34 Bay St. N., Hamilton, Ont.
Main Office: KERR TURBINE CO., Wellsville, N.Y.

STEEL TANKS

— Manufacturers of —

STEEL TANKS, Round, Oblong, Square
STEEL SHOP BOXES and BARRELS
LATHE BOXES

Manufacturers and Contractors write us for prices.

The Steel Trough & Machine Co., Ltd.
20 James St., Tweed, Ont.

OWEN SOUND IRON WORKS, LIMITED
OWEN SOUND, ONT.

Cement Mill Machinery, Boiler and Stee Tank Work of all kinds, Grey Iron and Brass Castings

Tell them you saw it in Canadian Machinery.

The advertiser will appreciate it and so will we.

An intelligent Engineer uses nothing but the **GENUINE PENBERTHY INJECTOR.** He has had experience with just-as-good kind before.

Flame Arc Lamps

FOR

Indoor and Outdoor Service

Cheapest form of Light.

Standard and Ornamental Designs

Prices and trade discount on application.

Gas and Electric Power Co.

Stair Bldg., Bay Street, Toronto

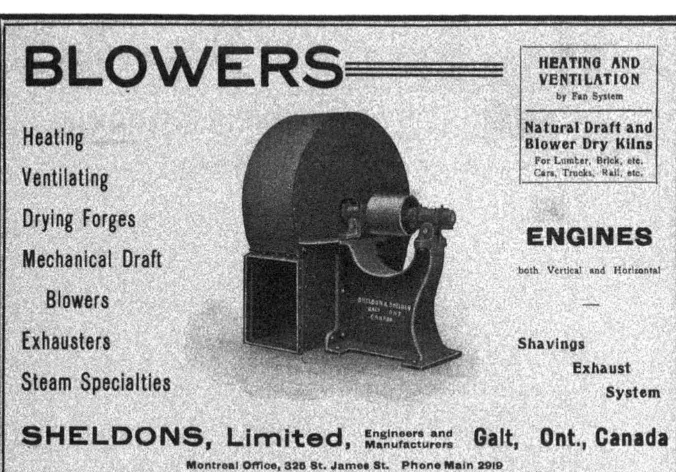
BLOWERS
HEATING AND VENTILATION
by Fan System
Natural Draft and Blower Dry Kilns
For Lumber, Brick, etc. Cars, Trucks, Rail, etc.
Heating
Ventilating
Drying Forges
Mechanical Draft Blowers
Exhausters
Steam Specialties
ENGINES
both Vertical and Horizontal
Shavings Exhaust System
SHELDONS, Limited, Engineers and Manufacturers Galt, Ont., Canada
Montreal Office, 325 St. James St. Phone Main 2919

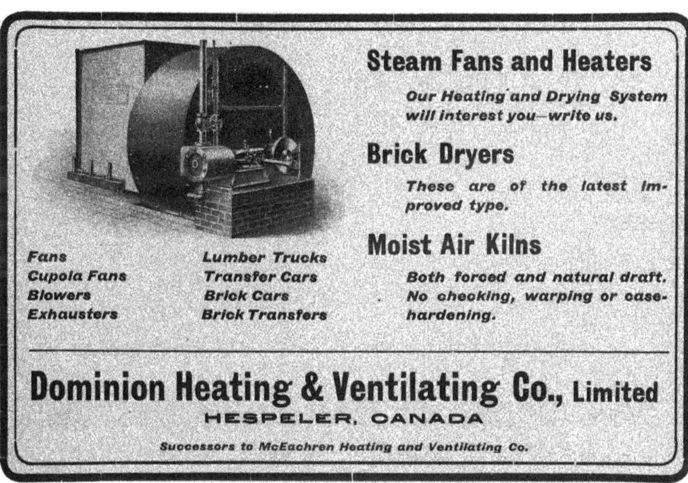
Steam Fans and Heaters
Our Heating and Drying System will interest you—write us.
Brick Dryers
These are of the latest improved type.
Moist Air Kilns
Both forced and natural draft. No checking, warping or case-hardening.
Fans
Cupola Fans
Blowers
Exhausters
Lumber Trucks
Transfer Cars
Brick Cars
Brick Transfers
Dominion Heating & Ventilating Co., Limited
HESPELER, CANADA
Successors to McEachren Heating and Ventilating Co.

Modern Canadian Manufacturing Plants.

ARTICLE XXIII.—John Bertram & Sons Co., Limited, Dundas, Ont.

The demand for machine tools in Canada has very greatly increased in the last few years and Canadian manufacturers of this line are doing their utmost to keep up with the great expansion of business. To be better able to cater to the trade John Bertram & Sons Co. entered into an agreement with the Niles-Bement-Pond Company in 1905 whereby the experience of the oldest and largest machine tool builders in the United States was obtained for Canadian users of this machinery. This arrangement gave the John Bertram & Sons Company a largely increased trade and additional floor space for handling it was necessitated. Under the direction of L. A. Somerville, supervising engineer, of Hamilton, new and large additions have just been completed which enable the company to better meet the requirements of the trade. It shows the confidence that the industrial men of Canada are placing in its future and the result in this case is a modern up-to-date plant.

The plant is favorably located at Dundas, Ont., a short trolley ride from Hamilton and connected to the Toronto, Hamilton and Buffalo by means of fie switches which we show in the general plan in Fig. 1. It is seen from this illustration that the company has ample store room and manufacturing space. The site covers altogether about fifteen acres and the plant is one of the most notable engaged in the manufacture of machine tools in Canada and United States in the variety and ranges of sizes of its products.

Fig. 2—Bird's-eye View of the Plant.

History of the Company's Growth.

The business was started in 1861 as McKecknie & Bertram in a small building located on the site of the present office. In 1865 these men assumed the name of the Canada Tool Works and branched out into special lines, erecting a large building 24x40 feet to carry on the work. This name, The Canada Tool Works, has still been retained, though the company is now more popularly known throughout the Dominion as The John Bertram & Sons, Company, Limited. A few years later this building was burned and was immediately replaced by a rough-cast building 60x40 feet. In 1868 a two-storey brick building was added and then extended in the front. Later a molding shop was added in the rear and finally the triangle was closed by the erection of a two-storey brick structure, leaving in the centre an area of 80x100 feet. Messrs. McKecknie & Bertram worked together unitl 1886 when the partnership was dissolved and from that time on it has been known as John Bertram & Sons. Building operations were continued in 1899, when this company erected a steel structure, converting the hollow area enclosed by the original buildings into an erecting shop, equipped with a 20-ton traveling crane and a railroad track communicating with all the other departments. This structure is one storey with a clear height of 28 feet overhead and is roofed with heavy glass and corrugated iron. The progress of the works is shown in the number of men employed. Six years ago 150 were given employment and there are now 425 enrolled.

The directors of the company are Henry and Alex. Bertram, sons of the founder, John Bertram, who died April 4, 1906.

The Machine and Erecting Shop.

To the person uninitiated in the manufacture of large machinery it would seem that the manufacture of these immense planers, boring mills, turning lathes, etc., is a large problem, and so it is, but a look into the Bertram machine shop shows that every facility available is used for the handling and machining of the work. One striking feature of this department is that nearly all the manufacturing machinery is the output of their own shop. In Fig. 1 is shown a general plan with the relations of the buildings to each other and in Fig. 2 we show a bird's eye view of the plant, giving the reader an idea of the extensive works. Fig. 3 is a plan of the machine shop and its gallery, showing the arrangement of the tools. The tools are grouped following a scheme of keeping close together those employed in the construction of each type of product. Erecting shop No. 1 is devoted to the erection of boring mills up to 48 inches and planers; large boring mills, drill presses, wheel lathes, etc., are erected in No. 2 and large and special machines

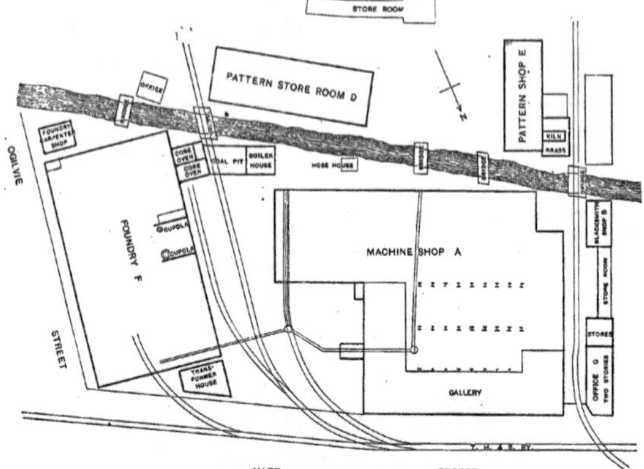

Fig. 1—General Layout of John Bertram & Sons Co., Limited.

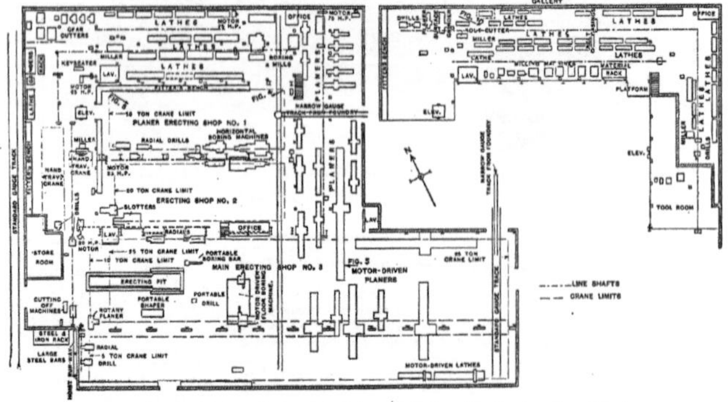

Fig. 3—Plan of Machine and Erecting Shops and Gallery Showing Location of Tools.

in the main erecting shop No. 3. This main erecting shop has just been completed and extends about 84 feet beyond the older buildings and is much higher, making it particularly convenient for the erection of large machines. The new buildings are brick and steel on concrete foundations.

The shops are roomy and well lighted. The present floor space is 67,000 square feet, including 12,000 square feet in the gallery. Illustrations 4, 5, 6 and 7 give a good conception of the machine department. Fig. 7 is taken from the position marked Fig. 10 in our plan No. 3. Fig. 4 shows best the erecting shops Nos. 1 and 2. No. 3 may be partly seen in the left background. A better view of the shop is given in Fig. 5, which shows a large 10-foot planer in progress of construction. All standard planers are erected in the section of the shop shown in Fig. 6. In general, the large work is done on the ground floor and the machining of the small parts is carried on as far as possible in the gallery, a view of which is given in Fig. 7. By a study of Fig. 3 one may get a good idea of the layout of the shop. The arrows indicate the direction the camera was placed on taking the photographs. On the left of the west entrance is a store-room for special sizes of steel and cap screws. To the left in Fig. 5 may be seen the office of the shop engineer and the assistant-superintendent. From this lookout these officers have a complete view of the machine and erecting departments.

As far as possible the machinery is arranged along the walls of the building, leaving the central space free for assembling purposes. The machinery requiring crane services is arranged as close to the centre but does not encroach on the erecting space. The planers are arranged as shown in Fig. 3. Small shapers and fast planers are placed together and the planers get larger towards the south part of the building. One planer is a 60-inch, is 60 feet long and planes 36-foot stock. Larger planers, 96-inch and 106-inch, are also in service. With all the tools the shop is not overcrowded, although it contains over 170 separate pieces of equipment exclusive of benches. In all, there are probably about 60 lathes, including engine, hand, turret, gap, boring, chuck and pulley lathes; 22 planers and a rotary planer, 7-gear cutting machines, 12 milling machines and a complete equipment of radial and upright drills, grind-stones and tool and drill grinders shapers, horizontal boring machines and vertical boring mills, universal and surface grinders, slotters, key seating machines, bolt cutters, power hack saws, cutting-off machines, etc. Some of the machines are very large and include a lathe with 28-foot bed

Fig. 4—View from Northeast Gallery, Showing Erecting Shops Nos. 1, 2 and 3.

swinging 72-inches, short lathes of 12 feet swinging 36 inches, and 26 feet swing 28 inches. The largest planer is a 72 x 94½ inch x 36 foot machine. One larger than this for their own use is now in course of erection. Another large planer 36 feet long, is 60 x 48 inches between the housings and there are several 36 x 36 and one 72 x 72 x 20 feet. All the machinery except that in the gallery is on concrete foundations.

Some of the machinery is very interesting. The machinery for cutting the mitre gears are unique and are automatic the boring tool is kept cool and the chips carried out. A spiral gear machine is next with an equipment of rope pulleys so arranged that either spiral or straight gears may be cut. One man attends these two lathes. Polishing is all done on machines for the purpose, so that there is a uniformity in all small parts. All the gears from 18 to 24 inches are finished upstairs while up to 60 inches are finished on the ground floor. All the machinery for the cutting of gears is automatic-dividing and one man takes care of several machines. Small parts, such as lathe sadders, heads required were sufficient to warrant it direct connected motors are employed.

Handling the Work.

The floor of the foundry is one foot above the level of the machine shop and a narrow gauge track equipped with turntables permits the work to be moved from the foundry to any part of the machine department. In Fig. 1 we show the narrow gauge track and the two ways in which castings are brought into the machine shop. The heavy castings are taken to the main erecting shop and the smaller to Nos. 1 and 2, where they are

Fig. 5—View from East End of Main Erecting Shop.

shapers. These are situated in the northwest corner of the building. A series of cams the bevel of the teeth to be cut and curves corresponding to these teeth give shape to the teeth. The tool travels absolutely towards the centre giving a correctly shaped tooth that cannot be obtained by rotary cutters. All the gears are divided automatically and the large gears are finished on the ground floor.

In the gallery each turret is equipped with special features for special classes of work. Here the lathe spindles are bored and by an arrangement of a stuffing box, oil pump and open centre drill, and aprons are assembled on the west end of the gallery.

Throughout the shops are several interesting portable tools, including a portable boring bar, a special portable shaper and a portable electric drill. The machines are group and individual driven suitable for the purposes for which they are intended. In the plan of the machine shop the lines of shafting are indicated and the location of the driving motors with their sizes are given. All the tools in the main erecting shop are driven by individual motors and in other cases where the location or the power distributed among the machines according to the operations to be performed on them. The castings are passed along the shop from one end to the other on the different machines and pass out at the west side a completed article, so that the work is one of progression, little handling being necessary. Electric cranes facilitate the handling of the heavy machinery. In erecting shop No. 1 thre is a 10-ton crane, in No. 2 a 15-ton crane, in the main assembling room, No. 3, one 25-ton crane, with a 5-ton auxiliary hoist and one 10-ton crane, and in the bay adjoining the latter shop, a 5-ton crane. West of erecting shop, No.

2, are two hand cranes for use there. A system of compressed air is used for hoists and chipping hammers. Babbit for bearings is melted by an electric furnace situated in the bay adjoining erecting shop No. 2.

Tool Room.

This department is situated in the east side of the gallery as indicated in Fig. 3. A check system is used in connection with this department and when a man requires a tool or gauge he hands in a check with his number and becomes responsible for the tool. The tool room is responsible for the standard sizes of all tools. All shafts and bearings are turned to fixed gauges and then ground and lapped. All the reamers, taps, dies, drills, milling cutters are kept to standard and jigs are made in the tool room. Standard lead screws for the lathes are kept and if the lathes are found to vary from this then the lead screw is replaced. All the machinery, in like manner, is accurate and before a machine leaves the works it is thoroughly tested. Drawings of each part are made with every piece numbered and dimensions given. An elaborate cost system has been established. Men charge their time against the several jobs each night. Small pieces which are done in quantity, are charged to stock and when used for a machine are charged against it. As soon as the work is finished the cost keeper can tell absolutely the cost of the machine or the detailed cost.

The location of the plant has a great advantage, as it comes within the zone of electric current distribution from the various Niagara water-power developments. The power used by Bertrams' is generated at De Cew Falls, near St. Catharines and is purchased from the Dominion Power and Traction Company, of Hamilton. Alternating current is transmitted at 60,000 volts to Hamilton, and from there to a point near Dundas at 10,000 volts, where it is transformed to 2,400 volts at which voltage it is received at the plant and finally stepped down to 220 volts. Alternating current at this pressure is used directly for lighting and the constant speed motors and is converted to 220-volt direct current for variable speed motors on individually driven tools and on all cranes. All of the substation apparatus is contained in the transformer house an entirely fireproof structure and includes four 250-KW static transformers which reduce the pressure to 220 volts, a 175-KW, motor-generator, which takes 220-volt alternating current and delivers 220 volts direct current, a 175-H.P. induction motor driving a Canadian Rand air compressor supplying air at 100 pounds pressure for use in various parts of the plant, and a 1,000-gal. triplex fire pump, which, when occasion requires, may be instantly connected by a clutch to the same motor, the air compressor being disconnected. The pump supplies, three 6-inch mains extending about the works for high pressure fire protection. The motor-generator set was furnished by the Canadian General Electric Co., and the motor and transformer by the Canadian Westinghouse Co. A large 4-panel switch-board with General Electric equipment, controls the distribution to different parts of the plant. The boiler house is 25 by 30 feet and furnishes heat to all the building, two large boilers being run during the winter. Pumps return the condensation to the boilers. A convenient arrangement for reducing the handling of coal has been effected by placing the firing floor of the boilers below the charging floor. Coal is received in cars on a spur track between the foundry and machine shop and dumped directly into a pit beside the boiler-house. The latter holds a considerable supply and delivers

Fig. 6—View in Planer Section from Northwest Corner of the Gallery.

it through grates directly upon the firing floor, by gravity.

The photographs will give the reader an idea of the natural lighting. Both natural and electric lighting are exceptional. All buildings are well-lighted in the daytime. The artificial illumination for dark days, early dusk or night work when necessary is such as to render working practically as easy as by bright sunlight and is afforded by Nernst lamps, no separate incandescent lights are used at the individual machines. This is the first installation of Nernst lamps on a large scale in Canada and the experience with them has been very satisfactory. The lights are so arranged that light is distributed uniformly.

The Foundry.

The equipment is modern in every respect, and everything is arranged to be most favorable to economy of time and labor. The cupolas are arranged along the west side and are a Collion 66 inches in diameter of 14 tons per hour capacity and a Whitney Foundry & Equipment Company's cupola 40 inches inside diameter of 7 tons per hour capacity. There is a full equipment of core ovens, a No. 6 Root blower driven by a 40 h.p. Westinghouse induction motor, Hamilton Facing Mill Company's sand sifters and powder mixers,and an elevating and conveying system supplied by the Waterous Engine Works Co., Brantford, Ont.

One of the most interesting features is the means provided for holding down large floor moulds. This consists of a system of grillage embedded 7½ feet below the floor surface, and binding hooks are inserted to hold down copes of large moulds made in the floor. Air rammers are provided, cranes and natural gas to harden the skin of the moulds, which is necessary in large moulds. In our May issue, 1906, we gave a detailed description of the foundry with its equipment.

Pattern Shop and Vault.

The pattern shop consists of the usual equipment of a drying kiln and a full line of saws, planers and surfacers, lathes, bandsaws and jointers. The stock patterns are well made from seasoned lumber, and when not in use are stored in the pattern vault. Here the patterns with core boxes are piled away in bins and shelves, according to number. A look through this building gives one an idea of the amount of money invested and the years of labor to place on the market Bertram's list of machines and tools.

Other Buildings.

Attached to the pattern shop is a brass foundry. The blacksmith shop is 25x45 feet equipped with forges, annealing and tempering furnaces, and an 800-pound steam hammer. The office is a two-storey building, accommodating the business offices on the ground floor and drafting, blue printing, and photographic departments on the second floor.

This is the important work in connection with any establishment. The output of the John Bertram & Sons Company is varied, but the workmanship is of the best and the greatest care is exercised in the machining of all parts. The company make a specialty of railroad machinery and have recently shipped an immense driving wheel lathe with individual drive to the Canadian Locomotive Company, Kingston; a 100-inch swing lathe of the same type weighing 115,000 pounds to the Intercolonial Works, Moncton; two coach wheel lathes, 42-inch swing for the C.P.R. at Winnipeg and one of the same type for their Angus shops. In course of construction are four large multiple punches and

Fig. 7—View of Machine Shop Gallery.

shears for the Dominion Bridge Co., Montreal. These are 8 and 12 punches up to 15-16-inch diameter with individual electric drive and the shears are for angles 8x8 inches by 1 inch, cutting from 60 to 90 degree angles. Another large machine in course of erection is a large planer 10x10 feet and 20 feet long. It has four cutting tools and these are raised and lowered by four Canadian Westinghouse motors. The planer is individually driven and is for the Canadian Westinghouse Company.

Although making a specialty of bridge builders' and locomotive machinery, the company include in its list of standard tools manufactured, shaping, slotting, boring and drilling machines, vertical boring and turning mills, cutting-off machines, bolt cutting and nut-tapping machinery, punching and shearing machines, plate planing machines, plate bending rolls, steam hammers and horizontal bending and forming machines. Though the range of product is so extensive the excellence of the individual tool has been kept up and a name for quality has been established on the market.

EVANS ROTARY ENGINE.

Considerable experimenting has been done in the line of rotary engines, and Mr. Evans has now placed on the market an engine for commercial use, built along that line. A complete equipment of this type, with a 20-horse power engine, takes up very little space. The boiler is 22 inch x 22 inch x 19 inch. The engine is 18 inch x 18 inch, and underneath the engine shaft is an Evans rotary pump occupying a space 6 inch x 6 inch. Engine and pump weigh only 550 pounds. One sight-feed lubricator supplies the engine with oil, and a full way valves controls the direction of rotation.

The cylinder is in three compartments, of 120 degrees, and there are three admissions in one revolution. The cylinder heads fit into the cylinder, and are slotted to support the dividing walls. These dividing walls shown in Fig. 2 keep steam in one chamber from mingling with the steam in another. The steam valve fits on to the shaft by means of a taper key so that it can be placed in one position only, thus the valve makes the same number of revolutions per minute as the shaft and piston. The steam valve consists of two parts. The disc and a sliding annular plate containing the same number of openings as the disc upon which it is carried. Similar valves set in opposite directions are at either end. The three compartments each have a separate steam entry and exhaust port. These ports are similar, each being on alternate sides. The dividing walls are supported on both sides for the whole length by the grooves in the cylinder heads, and also by the whole width of the pockets, the cylinder walls having a motion up and down. A rocking shoe is provided at the inner end of each slide. To these shoes, which work in annular grooves in the piston, the walls are attached. The shaft and cylinder are concentric.

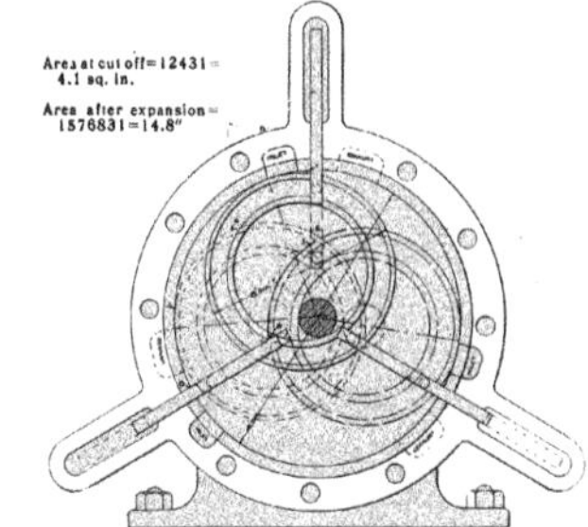

Fig. 2—Positions of Cylinder at the Points of Cut-off and Exhaust.

Fig. 1—25-h.p. Directly Connected to Electric Generator.

The piston is in three parts, two annular rings one-eighth thick interlock with serrated edges over an inner core, bored to receive the crank pin. They are turned down over the sides of the piston so as to carry the grooves for the shoes of the slides and are fastened to the piston by flat annular rings of copper, the outer circumference of the rings being anchored to the inside of the moveable shells and the inner circumference clamped to the piston by clamping rings and screws. The motion of the piston is hypocycloidal.

The action of the engine is interesting and is fully shown in Fig. 2. Steam is taken in for about 100 degrees, and then cut-off occurs. When this position is reached in one, admission begins in chamber next it, and exhaust in the other one in reverse cyclic order. From this it is seen that the M. E. P. is always constant for a given steam pressure. For this reason no fly-wheel is required and the whole equipment occupies very little space. Fig. 1 shows a 25 horse power Evans rotary engine direct connected to an electric generator.

An adjustable cut-off is a feature of the engine. Weights are tested for different speeds, and these weights regulate the speed by giving an early cut-off, this governor acting centrifugally inside the cylinder head.

This engine is being placed on the market by the Evans Rotary Engine Company, Manning Arcade, King street west, Toronto, Ontario.

CANADIAN MACHINERY
and Manufacturing News

A monthly newspaper devoted to machinery and manufacturing interests, mechanical and electrical trades, the foundry, technical progress, construction and improvement, and to all users of power developed from steam, gas, electricity, compressed air and water in Canada.

The MacLean Publishing Company, Limited

JOHN BAYNE MACLEAN - - President
W. L. EDMONDS - - - Vice-President
H. V. TYRRELL - - - Business Manager
J. C. ARMER, B.A.Sc., - Managing Editor

OFFICES:

CANADA
MONTREAL - 232 McGill Street, Phone Main 1255
TORONTO - 10 Front Street East, Phone Main 2701
WINNIPEG, 511 Union Bank Building, Phone 3726, F. R. Munro
BRITISH COLUMBIA - Vancouver, Geo. S. B. Perry

UNITED STATES
CHICAGO - 1001 Teutonic Bldg. J. Roland Kay

GREAT BRITAIN
LONDON - 88 Fleet Street, E.C. Phone Central 12960, J. Meredith McKim
MANCHESTER - 92 Market Street, H. S. Ashburner

FRANCE
PARIS - Agence Havas, 8 Place de la Bourse

SWITZERLAND
ZURICH - - - Louis Wol, Orell Fussli & Co

SUBSCRIPTION RATE.

Canada, United States, $1.00, Great Britain, Australia and other colonies 4s. 6d., per year; other countries, $1.50. Advertising rates on request.

Subscribers who are not receiving their paper regularly will confer a favor on us by letting us know. We should be notified at once of any change in address, giving both old and new.

Vol. III. NOVEMBER, 1907 No. 11

CONTENTS

PUBLICITY.

The total membership of the Canadian Manufacturers' Association is now 2,189. Only eight years ago this organization had a membership of 132. There must be some good reason for the rapid growth, not only of this Association, but of all the engineering societies and industrial organizations, as well. Making a study of the progress of the different manufacturing countries, the reason appears to be that with the increased membership of these organizations has come increased progress, more business and a great advance in machine shops and engineering practice hitherto unthought of. The conference of the Electrical Association in Montreal has suggested ideas to hundreds of brains, the C.M.A. convention in Toronto has stimulated manufacturers to do greater things in the future, and with a united interest to work for the welfare and progress of Canada and Canadian industries. But we believe there is another potent factor working for the improvement of our machinery and an increase in trade. We believe that the trade journals have done much to develop industries. With the aid of employers and engineers they are distributing with a free hand the experiments, results, data and tests that are being made. To this is due much of the development and growth. Take a country where there is no publicity, where trade secrets are carefully guarded and you see a country where little progress is being made. At the C.M.A., Mr. Emery, of New York, pointed out that Germany, in the last quarter of a century, had made wonderful progress. This is partly because of her industrial educational system, but it is in no small measure due to the publicity given in the last few years to working methods. It may assist one particular manufacturer to keep a trade secret for a time, but it will retard development and trade, and if a nation is to forge ahead it must be by mechanical men, superintendents, foremen and employes gathering together in the different organizations and exchanging their ideas. Manufacturers have realized this and are no longer keeping the information to themselves, but by giving and receiving are helping others and benefiting themselves. Trade papers are the greatest aid in this connection. They reach men who cannot attend these conventions and good ideas that would otherwise be lost are brought to the notice of the other manufacturers, superintendents, foremen and employes. Canadian Machinery devotes a large amount of space each month to machine shop practice, and we believe we are assisting in the work of improving the design and construction of machinery, making workmen more efficient by the new methods and devices which appear each month and stimulating that spirit of industry and progress that should be instilled into every manufacturer, every superintendent or foreman, every engineer and employe, if we are to make Canada a great industrial nation. If any reader has a good idea we trust he will assist by passing it along.

PUBLIC PRESS VERSUS TECHNICAL JOURNALS.

At the regular meeting of the Engineers' Club, of Toronto, on Thursday evening, October 10th, the question arose whether public press or technical journals should be resorted to for the discussion of engineering problems. As a great majority of people do not read the technical journals, questions of general interest and those involving public monies should be discussed in the public press. The public press, however, is not a place for engineering controversy. Engineers should look to the technical press for information, and if the technical press is to be most efficient and be of greatest service to the engineering profession, engineers should use its columns for the dissemination of information. The technical press is for the engineering profession, and to be of greatest value, the engineer and the mechanic and the superintendent must take an interest in it. If he has had a interesting experience, if he has accomplished an unusual piece of work, then a description of the work in the technical journals will be of great assistance to the engineering profession. The engineers outside the cities have not the same facilities for the exchange of ideas, they have not the privileges of a society, and they must depend on the technical journals if they are to keep in touch with the rapid strides being made in the engineering world. One of the objects of a society is to facilitate the acquirement and interchange of professional knowledge, and to promote and stimulate investigation into the several branches connected with the profession. In this the object of the society and technical press is mutual, and the columns of Canadian Machinery are open to the mechanic or engineer who has anything of interest to pass along to his brothers in the profession.

QUEBEC BRIDGE.

The Government Commission is making a thorough investigation into the cause of the disaster and have examined Mr. E. A. Hoare, the Quebec Bridge Company's engineer, and Mr. Collingwood Schrieber, consulting engineer of the railway department, that department having approved of the plans. The following statement has been made by the Commission as a result of their investigation so far:

"As far as we have been able to find out, there has been the greatest care taken all along the line to insure the safety and permanency of the structure. The specifications and plans were prepared by the best engineers on the continent, and wonderful care and accuracy have been shown throughout. We have found absolutely no trace of dishonesty or graft in connection with the construction of the bridge. It seems to be really a case of the best engineering brains on the continent, and the best engineering methods being on trial. In so far as the Government's connection with the enterprise is concerned, every thing seems to be quite regular, and what should have been done has been done."

They have since examined Mr. Cooper, the consulting engineer of the Quebec bridge, in New York, and visited Phoenixville. They will again visit the bridge before submitting their final report on the cause of the disaster. The Montreal Shipping Federation and Board of Trade are agitating that the bridge be rebuilt at a height of two hundred feet, instead of one hundred and fifty feet, which would allow the large steamers with high masts to travel up the St. Lawrence to Montreal.

Mr. Theodore Cooper, before the Commission in New York, declared:

"It was not the best bridge, but the best bridge that could be built with the money. The amount to build it with was limited to a certain extent, and the bridge had to be planned to meet this amount."

Furthermore, Mr. Cooper to a certain extent, emphatically blames both companies; the Quebec Bridge Company, and the Phoenix Company, for the collapse, inasmuch as he is of the opinion that neither had proper officials at the work, men with sufficient technical knowledge to superintend a construction of such magnitude.

"If prompt action had been taken to protect chord 9. west, from further deflection, when the bend was discovered, the Quebec bridge would not have gone down. This would have been only about three hours' work, and the expense would only have been about one hundred dollars in timber and bolts."

If the Commission of enquiry find this to be true, there will be little difficulty in locating the cause of the disaster.

TORONTO UNIVERSITY'S NEW PRESIDENT.

Dr. Robert A. Falconer, of Halifax, was installed as President of Toronto University on September 26, in Convocation Hall, before a large audience of men representing the business, professional, educational and religious men of the country. Dr. Falconer is held in high esteem by those who know him and his opening address showed the firm but sympathetic, inspiring character of the new head of the University. On Wednesday, October 9, he addressed the University of Toronto Engineering Society and referred to the work of the mechanic and the engineer, pointing out that accuracy and a thorough knowledge are essential; in bridge building or any work a man must show a character and this can be done by having an accurate knowledge of details. The selection and installation of such a strong man as Dr. Falconer is of great importance, not only to Ontario as president of the Provincial University, but to the large field where the influence of the university is felt.

TORONTO POWER QUESTION.

The people of Toronto are agitated over the question of Niagara power. At present the Toronto Electric Light Co. is supplying the power for the city, but the Hydro-Electric Commission proposes to supply it at the city limits at $17.75 per horse-power year. Mr. Wright, of the Toronto Electric Light Co., declares that their charge is equivalent to a flat rate of $8.75 per horse-power-per year. The public are anxious for cheap power, and care not whether it comes from a private or a public enterprise. The rate of the Electric Company appears low when operating expenses, depreciation, etc., are all considered. However, the question is worth considering carefully, as it is of great importance to the users of power.

CAUTION AND CONFIDENCE.

The Canadian banking institutions, by their foresight and caution, have averted the possibility of a repetition of New York's experiences in the past week. The retrenchment may have caused a little inconvenience in some quarters, but it has curbed unhealthy speculation and exerted a steadying influence, prudently restraining unsafe expansion in times of prosperity and seeing that retrenchment does not precipitate depression. Canadian trade is still sound, the flurry in the stock market will soon pass over and we will then better appreciate the policy of our financial corporations in their cautious course, which now sustains confidence and courage in Canadian markets.

INDUSTRIAL GROWTH.

The recent census as to the progress of Canadian industries during the period of 1900-1905 gives some valuable information. Capital and output have greatly increased. In the year previous to 1900 the value of manufactured products was 481 million dollars and the merchandise imports 181 million dollars. In 1905 there was an increase of 234 million dollars in manufactures, making 715 million dollars for that year. The imports in 1905 amounted to 256 million dollars. The growth of trade in a period of five years has been enormous and a great many large increases are noted. The manufacture of electrical machines and accessories and other power machines has trebled. The smelting industry has quadrupled and the stone-cutting industry has very greatly increased its output. Canada's industries are growing rapidly each year, and, though they have not displaced the imports to a very great extent, show large returns for the investment.

Canadian Manufacturers' Association

The 36th Annual Conference in King Edward Hotel, Toronto, a Great Success.

The annual convention of the Canadian Manufacturers' was opened Thursday morning, Sept. 24, and was largely attended by the captains of industry who direct Canadian manufacturing enterprises. The session opened with the meeting of the various sections, the larger assemblage afterwards dealing with all matters of general interest. The following officers were elected for the ensuing year:

President—Hon. J. D. Rolland, of the Rolland Paper Co., Montreal.

Vice-President—John Hendry, of the British Columbia Mills, Timber & Trading Co., Vancouver.

Treasurer—George Booth, of the Booth Copper Co., Toronto.

Technical Education Committee—S. Morley Wickett, of Wickett & Craig, tanners, Toronto.

At the meeting of the engine and boiler section, Mr. Leonard, of London, was elected president, and William Inglis, vice-president. At the meeting of this division, Mr. C. H. Waterous occupied the chair, and the chief topic for discussion was an inspection act to govern stationary boilers. The need of a Dominion law was keenly felt, and a long discussion took place on how to obtain such a law and the terms of it. At present a different law exists in each Province, and it makes it difficult for a company in one Province to supply the trade in another Province, and comply with the regulations of the Province in which he is doing business. Steps were taken to overcome this difficulty, and a committee was appointed to confer with experts, find out what was best to do under the circumstances, and report at a special meeting of this section. A special committee will then be elected to confer with the Federal authorities, and, if possible, bring about this new and much needed Dominion Act. At present a Dominion Marine Act governs marine boilers, but this in no way regulates the various other stationary and portable boilers. In our March issue of this year we published an editorial and letters from a number of well-known manufacturers on this topic and all are of the one opinion that a Dominion law is an urgent necessity.

Another matter that the engine and boiler branch took up was the terms of selling engines and boilers, guarantees, check deposits and terms of payment. As it is at present, manufacturers of engines and boilers are quoted F.O.B. cars and others erected. It was desirable that some agreement be arrived at in this matter so that builders of this line of goods would be on equal footing.

Presentation to Ex-President Ballantyne.

An interesting feature was the presentation by Mr. Cockshutt on behalf of the association of an illuminated address to C. C. Ballantyne, of Montreal, ex-president of the association, in recognition of his valuable services.

Another feature of the convention was an address by Archibald Blue, chief officer of the Dominion census, illustrative of the growth of Canada in the twentieth century. Capital and products showed large increases in the five years for every Province of the Dominion, except Prince Edward Island, Ontario and Quebec showing the largest development.

HON. J. D. ROLLAND
President of the Canadian Manufacturers' Association.

The President's Address.

President H. Cockshutt of Brantford, in his annual address, made a strong plea for higher protection, as well as for a more vigorous forestry policy, providing for an export duty on pulpwood, so that this industry would be developed in the country to which the raw material belongs. After touching upon the fishing and mining industries, he pointed out that manufacturing showed progress and activity, the output of the factories in the last five years having increased from $481,053,871 to $717,118,092.

Reception by Toronto Branch in the Evening.

The reception tendered the visiting manufacturers and their wives by the Toronto colleagues at the King Edward was brilliant and pleasant. A delightful programme was given, contributed to by Mrs. Freyseng, and Messrs. Arthur Blight and Howard Massey Frederick, and promenading was enjoyed to sweet strains of the orchestra.

WEDNESDAY, SEPTEMBER 25TH.

Prof. Fernow, Professor of Forestry at Toronto University, spoke to the association in the morning on "Reforestration," and said that a forester's work did not begin when the country was deforested. It should begin when the first tree was cut, and so render the laborious and expensive system of artificial reforesting unnecessary. For the Government to take control of the timber holdings would be a proper method of treatment of our timber resources.

The ladies, who had come with their husbands to the city, spent the afternoon at the Lampton Golf Links. The Tariff Committee gave their report at the afternoon session, and the keynote was adequate protection for Canadian industries.

In the evening the delegates were the guests of the Toronto branch at the Royal Alexandra Theatre.

THURSDAY, SEPTEMBER 26TH.

Industrial Education.

The needs of industrial education in Canada were ably expressed before the Canadian Manufacturers' Association at this morning's session by Mr. James A. Emery, of New York.

Mr. Emery pointed out how much more necessary efficient workmanship is to the manufacturer every day.

"Remember this," he said, "that both labor and capital are dependent for profit on increased production by superior methods. President Roosevelt says, 'The man behind the gun is the man who makes the way,' and the man behind the machine is the man who makes the industry. Gentlemen, there never was more need in the land for industrial efficiency."

Mr. Emery then illustrated how Germany had risen in thirty years from the bottom of the scale to the second place among manufacturing nations by industrial education. "Canada should not be any more dependent on other nations for men, than for material," he said.

Then followed a paper by Mr. L. G. Read, of Montreal, on "Power as a Fixed Charge," which we reprint in full on another page.

A garden party was given at the Government House in the afternoon, and in the evening the ladies attended a theatre party at Shea's.

ANNUAL BANQUET.

Nothing could have indicated in a more suggestive way the progress and prosperity of the country than the scene in the beautiful King Edward dining-room Thursday evening, where 350 guests had gathered to hear the foremost statesmen in the Canadian public life. Union Jacks were everywhere and national sentiment permeated the entire gathering.

In proposing the royal toast, Hon. J. D. Rolland, of Montreal, the newly-elected president, thanked the association for the honor they had done him, and referred to the national character of the organization.

Lieutenant-Governor Mortimer Clark replied to the toast, and spoke along the lines of Technical Education, a subject of such vital importance to our manufacturers that we give it in full.

Lieut.-Governor's Speech.

The Lieut.-Governor, who had a great reception, said he had read with a very great deal of interest the many subjects which the convention had taken up, and he congratulated the members of the association upon the success which had attended the convention. He felt very much pleasure to find that a paper was read to the convention on the subject of technical education. (Cheers.) "You all know," went on his Honor, "what a very great deal of success amongst the German manufacturers has been the result of technical education to the young in that country. You know also that the result has been to give in some branches of manufactures a very great pre-eminence and a consequent loss of trade to other countries. In Great Britain and here, too, we have suffered from lack of technical education." (Cheers).

He hoped that manufacturers would have their sons avail themselves of the scientific education provided in our great universities, such as Toronto and McGill. He hoped that the Manufacturers' Association would throw the weight of their great influence in seeing that proper technical instruction was afforded to our youth. Our agricultural colleges have largely increased the value of the products of the farm. Technical and scientific schools would do the same for manufacturers. His Honor recommended greater interest in this subject, and said that Canada owed a great deal to the manufacturers. "You have very largely changed the whole complexion of this country. A visit to our Toronto Exhibition would abundantly show this. In past years it was but an exhibit of fine cattle, horses and dairy produce, while to-day it is also a wonderful display of fine machinery and manufactures of great excellence. He remarked that in the past we have too long regarded Canada as a purely agricultural country. We have been slow to admit that agriculture alone will not build up a great country." (Cheers.) His Honor quoted the case of Venice, which for hundreds of years, although built only on sand and piles in the midst of the sea, acquired enormous wealth and commanded the Mediterranean by its navies, and, proceeding, said that agriculture was not alone necessary to build up a great nation. Agriculture was not the source of the prosperity of Great Britain; but it was manufacture which was the source of building up the head of our great Empire. However, at the back of manufacture, we are fortunate in having a magnificent agricultural country to provide an ever-increasing market. For years and years the daily papers toward harvest time were filled with accounts of the condition of the crops. They heard of rust and mildew and the smut and they all seemed to feel that the whole country was going down. Now, however, they hardly heard of anything of the kind; and the manufacturers by building up industries, had done much to do away with this. It would be improper for him to refer to the great question of tariff, "But," added the Lieut-Governor, "I have a very great deal of sympathy with you in your desire to give pre-eminence to the manufacturing industries of this country. (Cheers.) How that may be brought about it is not for me to say." His Honor made a suitable reference to the visit of the British journalists; and, concluding, said it was the duty of the people of Canada to place before the people of other nations the advantages which Canada offers, remarking that there was plenty of British money ready to come to Canada, provided the capitalists were assured of the honesty of the schemes submitted to them and the reliability of our legislation.

Mr. George Tate Blackstock, K.C., proposed the toast to "Canada and the British Empire," and spoke of the relation between Canada and the Mother Country, and referred to our progress and our share in the defence of the Empire.

The Premier Warmly Welcomed.

Sir Wilfrid Laurier responded to the toast and was received with a tumultous applause. He spoke on many topics most prominent before us, militarism, French treaty and the Asiatic problem. In speaking of diplomatic relations, Sir Wilfrid said that our diplomatic relations carried on by the Mother Country had not been as successful as we would wish them, though the diplomacy had been in the hands of capable men. "We have suffered on the Atlantic, we have suffered on the Pacific, we have suffered on the lakes," he remarked. "Well, we have come to a decision, whether in our relations with foreign countries, it would not be better to manage them ourselves rather than entrust them to the ministers in Great Britain. This long-looked-for reform has at last come to be a living reality." (Cheers.)

Sir Wilfrid spoke of the "All-Red" line, and said he had the fullest confidence in the project. In the undertaking it was necessary to have the assistance of Great Britain, and it was not Canada's place to dictate to the Mother Country.

He did not know that there would be an immediate success for the "All Red" line. It was surrounded with difficulties, but they had no terror for him at all. He had had them all his life, he had them in connection with this scheme and would still have them, but they would be overcome, and this project must, and would, succeed. (Great applause.)

Major George W. Stephens, of Montreal, replied to the transportation toast.

An interesting souvenir given to the members of the C.M.A. at the convention was a paper-weight, consisting of a large ball revolving on a pedestal on ball bearings. With it was a card bearing the following, "May your life have as little friction as this bearing, is the wish of the Chapman Double Ball Bearing Company of Canada, Limited, Toronto."

FRIDAY, SEPTEMBER 27TH.

The business finished, a day was reserved for recreation, and the visitors enjoyed an excursion to Hamilton and Niagara Falls. A visit was paid to the shops of the International Harvester Co., and the Canadian Westinghouse Co., after which the entire party enjoyed a visit around the city. After lunch the party proceeded to Niagara Falls, where the members visited the huge power plants now in operation there. As a close to the successful convention, the Niagara Falls Board of Trade tendered the guests a supper at the Clifton House.

Ice Breaking Car Ferry Steamer

The Machinery Equipment of a Modern Steel Ice-Breaker Being Built at the Polson Iron Works, Limited, Toronto.

On the second of April, 1908, the Canadian Pacific Car and Passenger Transfer Company, Limited, of Prescott, Ontario, will have added to their equipment an up-to-date ice-breaking car ferry steamer now in course of construction in the Polson Iron Works, Toronto. The equipment is in itself a complete factory with its power machinery, electric engine and generator, steering gear, fire equipment, capstans, compasses, engineers' firemens' tools and all the useful fittings for the successful operation of a first-class vessel of this type, complete in every respect.

Principal Dimensions.

Length over all, 280 feet; breadth molded, 40 feet; depth molded, centre, 22 feet; and draft full load, 12 feet. The frames are of channels, 7 inches by 3½ inches, 21 pounds. On every floor plate is an angle bar 3½ inches by 3 inches, 9 pounds. The floor plates are the full depth of the centre keelson, 12 pounds. The floor plates under the boilers are 16 pounds. The centre continuous keelson plates are 36 inches wide and of 18 to 16-pound plate. It is secured to the floor by double vertical angles, 3 inches by 3 inches by 7 pounds. At the bottom of the keelson are two continuous angle bars, 5 inches by 3½ inches by 13 pounds riveted to the keelson and flat steel plate. Two continuous angles, 3½ inches by 3½ inches by 8 pounds are riveted to top of keelson and an 18-pound rider plate will rivet to top of these. Four side intercostal keelsons of 16-pound plate are secured to the floors. Stringers of 15-pound plate are riveted to the steel plating between each frame. The deck beams are channels 10 inches by 3½ inches by 28 pounds.

The ice belt is laid so that lower edge is 40 inches below the load water line, and the upper edge is 20 inches above load water-line and is 30-pound plate. All holes are punched ⅛-inch small and reamed to size. The steamer will have ten bulkheads, which will conform with the requirements of the American Bureau of shipping. All will be water-tight with water-tight manholes and doors. On the deck will be arranged four coaling scuttles with covers and iron gratings so arranged that a hopper-car can dump directly into the coal bunkers. There are two coal bunkers, one at each end of the boiler space. Each bunker has two doors at the fire room.

The Electric Lighting Plant.

This consists of one direct connected engine and generator, 110 volts of 15 kilowatts capacity. The engine and generator are secured on a common bedplate, in the engine room. The bedplate rests in a metal pan with edges turned up two inches, to catch the oil and water and the pan is secured to a wooden base to deaden vibration. The electric generator, switchboard and accessories are furnished by the Canadian Westinghouse Company, of Hamilton. The switchboard consists of three panels of slate 1¼ inches thick and on it are independent switches for each current, one Weston voltmeter, one ammeter, one shunt rheostat to regulate voltage from 10 volts below normal to 10 volts above, one two-light ground detector, one voltmeter switch, one ammeter switch, automatic circuit breaker, two 110-volt pilot lamps, one magneto bell, and one 15 to 150-volt portable voltmeter. There are 166 16-candle-power lamps and two

Fig. 1—Low Pressure Cylinder for 1,000 H.P. Engine.

searchlights of 35 amperes capacity. The lamps are designed for the 110-volt system. The wiring throughout the ship is the 2-wire watertight conduit system.

Propelling Engines.

There are two sets of vertical inverted compound jet condensing engines, each driving an independent propeller, one at the forward end and one at the after end. The cylinders are each 24 inches and 48 inches by 30 inches stroke. In Fig. 1 we show one of the low pressure 48-inch cylinders as it was being machined on the planer. This cylinder weighs 8,560 pounds and one pair complete with cylinder covers weigh 20,000 pounds. Each set is capable of developing 1,000 indicated horse-power at about 125 revolutions per minute with a steam pressure of 130 pounds at the boilers. These engines are made especially strong to withstand the severe shocks to which they will be subject. Each engine has an independent jet condenser and air pump, each having two independent suction and discharge. The cylinders are made of hard, close-grained cast-iron. Both cylinders are fitted with relief valves. The cylinder walls are 1¼ inches thick in the barrel, 1⅝ inches thick over the counter-bore at the bottom, and 1½ inches thick over counter-bore at the top. The cylinders of each engine are supported by four cast-iron columns and the bedplate is one casting. Fig. 2 shows the eight columns for the two compound engines. These columns are twelve feet high. The crank shaft bearings in the bedplate are 13 and 15 inches long. The bearing sleeve is of strong bronze lined with the best babbit metal. The cast iron caps are held down by two steel bolts 2¾ inches in diameter with locknuts. The piston is of the box type. All cast-iron is at least 22,000 pounds tensile strength.

Forge Steel Parts.

All forgings are of mild steel 60,000 pounds tensile strength and at least 30 per cent. elongation in 2 inches. The piston rods are 4¾ inches in diameter in the body with 3½ inches diameter of the thread at top and bottom. They are fitted to the cross heads and pistons with cones, collars and steel nuts, case hardened. The cross heads have each two gudgeons 5 inches in diameter by 5½ inches long and the square or block part of the cross head is about 9 inches by 9 inches. The guide slippers are cast-iron and the guide surfaces are babbited.

The connecting rods are 6 feet 3 inches from centre to centre. The body of the rod is tapered from 4 inches at the top to 5½ inches at the bottom. The bearing at the upper end of the rod is brass, the crank pin bearing is of steel lined with babbit metal. The bolts at the crank pin end are 2¾ inches and those at the upper end are 1¾ inches in diameter.

The valves of both cylinders are operated by Stevenson's double link motion. The valve sterns are forge steel, the bearings at the lower end are 3½ inches in diameter by 3 inches long; where the valve stem passes through the guides it is a square section, the sides measuring 2½ inches, where the stem passes through the stuffing box they are 2½ inches in diameter. The links are 4 inches by 1¼ inches with a link travel of 20 inches. The eccentric rods work on pins at the end of the links. The pin for reversing rods was forced in and riveted over. The valve stem guides are cast iron and each have an adjustable brass lined bearing.

Reversing Gear.

The reversing shaft is wrought steel 4½ inches in diameter. The reversing engine has a cast-iron cylinder 10 inches in diameter by 16 inches stroke. The piston is of cast-iron with spring rings. The cylinder is attached to the low pressure front column and can be seen on the frame in Fig. 2 Control is by means of a flat slide valve actuated by a floating link motion.

The crank webs are cast-steel 7½ inches thick. The crank pins are 10½ inches in diameter by 12 inches long. The thrush shaft is 10½ inches in diameter; has 8 thrust collars 18 inches in diameter, 1⅞ inches thick, forged on. The intermediate shafting is 10 inches and the propeller shaft 11 inches in diameter. The thrust bearing consists of a cast-iron pedestal which is securely bolted to the main engine foundation plate. The thrust is taken up by seven cast-steel horseshoe collars lined with babbit metal and mounted on two Tobin bronze rods 2½ inches in diameter. There is a bearing at each end of the thrust block.

The propellers are cast-steel 9 feet 6 inches in diameter and 10 feet pitch. They are fitted to the shaft with a taper and held in place by a key and nut.

The Boilers.

There are four single-ended return tubular cylindrical boilers built for a working pressure of 130 pounds per square inch. The boilers are operated with Howden's system of forced draft. The air heaters are arranged for this system. The tubes are 2¾ inches in diameter by 4 feet 6 inches long between the tube plates. The ratio of heating surface in heater tubes to grate surface is 18¼ to 1. The tube plates are ¼-inch thick. A Buffalo forge steel plate full housing pulley fan is used with a blast wheel 85 inches in diameter and 42¾ inches wide at centre, inlet 55 inches in diameter and outlet 44½ inches square. The engine for use with this fan is a Buffalo Forge Co. class B of single vertical, double-acting cylinder type direct connected to fan. The cylinder is 7 inches in diameter by 7 inches stroke, crankpin diameter 2¾ inches, length of crankpin 2½ inches, diameter of shaft 3 inches, length of main bearing 6¾ inches, initial steam pressure 125 pounds. This equipment is of sufficient capacity to burn 5,040 pounds of coal per hour and maintain ½-inch air pressure in the ashpits of the boilers.

Duplex single acting air pumps are direct connected to the jet condensers and air pump and condenser are of the Washington type, having cylinders 12 and 28 by 18 inches. Ballast pumps are of the Morris Machine Company's centrifugal direct connected type, with 6-inch suction and 6-inch discharge, the engine is 6 by 6 inches automatic governing type. The arrangement is such that any compartment can be flooded or drained independent of the others. There are two double-acting, duplex brass lined feed pumps of the Worthington type, with cylinders 9 by 5¼ by 10 inches. These pumps are arranged to draw from hotwell or sea and discharge to boilers. The equipment also includes a general service double-acting, duplex brass lined pump of Worthington type with cylinders 7½ by 5 by 6 inches. There are two injectors connected for feeding the boilers. A complete fire equipment insures safety in case of fire.

The engine room is fitted with pipe vise, and bench vise and a complete workman's outfit of hammers, files, wrenches, taps, drills, dies, etc., and the steamer when it reaches Prescott will be complete in every detail. It was designed by M. C. Furstenau, of Philadelphia, but all the working drawings were made at the Polson Works where large contracts are easily and competently handled. The present steamer now under course of construction is for the use of the C.P.R. and New York Central and will ply between Ogdensburg and Prescott, making a closer link between the Canadian and United States lines.

Fig. 2—Frames for the 1,000 H.P. Engines.

There is a good deal of difference between publicity and advertising—you can shoot a man and get publicity.

NEW PUBLICATIONS.

Journal of Foundry Practice.

With October comes the first issue of a new journal on the art of founding, "Castings," published monthly by the Gardner Printing Company, Cleveland, Ohio, and edited by H. M. Lane and R. I. Clegg. It treats a variety of subjects restricted to one trade, foundry, and is a general foundry publication.

Construction Magazine.

The first issue of Construction has been received from Toronto, a monthly magazine, published by H. Gagnier, Limited, and edited by Ivan S. Macdonald. It wil be devoted to the building and engineering interests.

NEW TOOL ROOM LATHE.

Among the machines exhibited at the Toronto National Exhibition by the A. R. Williams Company, Limited, Toronto, was a 16-inch tool room lathe, manufactured by the American Tool Works Company, Cincinnati, Ohio. This lathe created considerable favorable comment and was purchased at the Exhibition by a prominent Canadian manufacturer. It is equipped with relieving and improved taper attachment, and a draw in collect mechanism and pan. Owing to the fact that this lathe arrived after our October issue was in the press, a description of the "American" tool lathe was not included in the machinery exhibit.

Subaqueous Tunnel

Railroad Tunnel Joining United States and Canada

The Michigan Central Railroad, by means of a tunnel, is overcoming the gap in its line where the Detroit River forms the national boundary between United States and Canada. This is the last serious break to be overcome in the whole roadbed of the New York Central lines. The river separating Detroit and Windsor is about three miles wide, and powerful ferry boats make the connection between the two towns.

The bed of the river is soft blue clay and makes the building of a railroad tunnel at this point a very serious undertaking. The vessel traffic up and down the Detroit River is very heavy, and when reckoned in tonnage is about equal to the famous Suez Canal. In considering the project, W. J. Wilgus, vice-president of the New York Central, suggested that a ditch or channel be dredged across the bed of the river, fill it with concrete and then drive the tunnel through it. This would practically be equivalent to making rock and then boring it, and at once an easier method suggested itself to his mind. The form of tunnel decided upon was that of twin tubes, and the method of construction devised by Mr. Wilgus may roughly be described as the "cut and cover" method, as applied to tunnels under water. The method thus evolved consists of dredging a channel or ditch across the bed of the river between the two shores, laying sections of tube in the ditch, connecting them together when in place, filling around and over them with concrete, and then pumping out the water in each of the continuous tubes.

From the illustrations it is seen that the tunnel is composed of twin tubes of steel. The part beneath the water is 2,622 feet in length. The minimum depth over the roof of the tubes is at no place less than 41 feet, which is slightly more than the depth required by navigation laws. On either shore the tubes connect with tunnels by open cuts and through these cars will be drawn by electric locomotives. The work altogether will comprise an electric zone on each side and under the river about 23,000 feet long. A double track enters the cut at the Detroit end, 1,540 feet long, then through a land tunnel 2,129 feet. The subaqueous tube itself comes next, 2,622 feet, and on the Canadian side there is a land tunnel 3,192 feet long and an open cut 3,300 feet in length.

Side View of Tunnel Sections.
(Photo by Detroit News-Tribune.)

Front View of Tunnel Tubes Connecting Diaphragms.
(Photo by Detroit News-Tribune.)

Our half-tone illustrations show sections of the tubes as they stand in the contractor's yard at St. Clair, Mich. Each section consists of a circular ring of ⅜-inch steel plate, with a central fin or diaphragm all around it. The tube sections are 250 feet long, and are each 23 feet 4 inches in diameter, and have a concrete lining 20 inches thick which gives a clear diameter of 20 feet. Each tube contains one track and the roof of the tunnel is 18 feet above the rails. Running along the sides of the tunnel are concrete platforms, 5 feet 3 inches above the rail and 3 feet 10⅜ inches wide on top. In these concrete platforms are contained conduits for signal, lighting and electric power cables, telephone and telegraph. The platforms provide a walk way for passengers in case of necessity, and room for the workmen.

The line cut shows the method of carrying on the work. The dredge, equipped with powerful clam-shell buckets for excavating the trench in the bed of the river, is seen in the distance nearing the Canadian shore. The trench is 48 feet wide at the bottom and about 32 feet deep, with a slope of ½ to 1, so that the ditch is between 50 and 60 feet wide at the top. Behind the dredges are the pile drivers. Temporary piles are driven into the trench, and the tubes are located on these prior to placing them in the concrete. The piles are driven on either side of the trench in two rows, and are long enough to come up to the surface of the water. These piles guide the tubes so that they

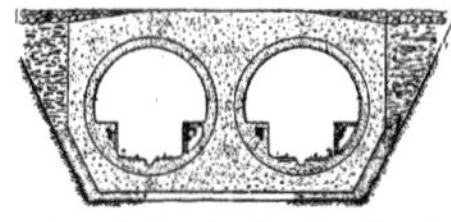

Section of Tunnel, Showing Trench, Filling Concrete, and Tubes.

settle in the required direction. The piles are then removed and used for the further carrying on of the work.

Placing the Tubes.

After the pile drivers come the scows, with derricks, air machinery, hoisting apparatus, drivers' equipments, etc., for placing the tubes. A section of the tunnel, about 250 feet long containing two tubes and 23 diaphragms, is blocked up at each of the twin tubes with bulkheads of wood and made ready to float on the river. This method makes the tubes air tight. Bolted to the diaphragm is three-inch planking, so that when made ready the section is like a huge wooden vessel with wooden sides, having steel ribs across, holding the tubes. Four temporary floating cylinders, each about 10 feet in diameter, and 60 feet long, are chained on the top of the tubes. These cylinders are filled with compressed air, and with the enclosed air in the main tubes, and the outside planking, the whole section is made buoyant enough to float. The section, thus prepared, is lowered into position between the guide piles in the river and the air allowed to escape from the tubes. Final adjustments are made by the compressed air cylinders. When in position in the trench, the section is joined to the one already down. The tubes shoulder in heavy rubber gaskets at the joints, in each side of which are partially cylindrical chambers. Into these chambers the best grade of cement grout is introduced. The joints are finally locked with heavy pins fitting into corresponding sockets in the adjoining section, and securely bolted. The forward end of each of the tubes has a sleeve 17 inches long, which is fitted over the end of the section already sunk.

Filling in the Concrete.

A concrete mixer, with a set of tremies, distributes the concrete under and around the tubes. These tremies are long steel tubes from 8 to 12 inches in diameter, with a hopper at the upper end. The concrete is filled into the hopper and the concrete is distributed along the tubes evenly by moving the lower end along the bottom. In this manner the cement is deposited in the bed of the river, with very little loss. The scow for this work is about 135 feet long by 35 feet wide, with three tremies on each side. Concrete made of one part cement, three of sand and six of broken stone is poured in between the 3-inch planking and the steel tubes. This concrete fills around the tubes completely and covers them to a depth of about 4 feet 6 inches. The space between the slope sides of the trench and the planking is filled in with sand, gravel and a covering of stone. When the tubes are finally in position, the wooden bulkhead at the end of the section is removed and the water pumped out. The danger of the undertaking has been lessened by a great deal of the work being done at the surface. On account of locating the tunnel on the river bed, the grade is 2 per cent. at the Detroit end, and 1.5 at the Windsor

Detroit—Windsor Tunnel, Looking Towards Canada.
(Courtesy of Railway and Locomotive Engineering.)

end, much less than if the tunnel had been built below the bed of the river. The tunnel will cost $10,000,000, and it is expected that it will be completed by June, 1909.

DEATH OF MR. W. H. WIGGIN.

In the death of William H. Wiggin, superintendent of the International Harvester Company of Canada, Hamilton, Ontario, the company has lost a man of exceptional executive ability and an able mechanical expert.

Mr. Willin was born in Dracut, Mass., and served his apprenticeship in his native State. He there gained a reputation as an expert mechanic, and his motto, "It Can Be Done," is worthy of being followed by every young man starting out in the battle of life. Mr. Wiggin was ill for several months, and though able professional men from Canada and the United States were called in consultation, he passed away on Oct. 2. His last remains were conveyed to Worcester, Mass., the scene of his early boyhood, where interment took place on Friday, Oct. 4.

INTERNATIONAL JAPANESE EXHIBITION.

Construction work has already been started for the International Japanese exhibition, which is to be held in Tokio next year. The total cost of the exhibition has been estimated at $10,000,000; one-half of this amount has been set aside by the Government of Japan, while private citizens have subscribed the remainder. The exhibition is to open April 1st, and will last until October 15th. Some of the principal exhibits will be of a scientific character.

Power as a Fixed Charge*

The Load Factor is Pointed out to be of Utmost Importance in Measuring Power Used and in Calculating True Costs

By L. G. READ

To give full consideration to the great question of power—the backbone of all industry, the one most vital requisite to material progress—would require time far beyond that which is at my disposal on this occasion. Nor can we view the broader and more interesting subject to the evolution of power—the advent of steam—with its quickening of the world's pulse and its development of its tremendous resources within almost a single generation. We are, therefore, compelled to deal briefly—to deliver, as it were, a little address on a big subject.

We must deal directly with the facts as we see them today. On the one hand are the manufacturers, with their almost countless thousands of employes, engaged in the most highly legitimate pursuit conceivable—the conversion of raw material into finished product for the infinite needs of mankind. On the other, the coal supply. Or let it be called by its proper name, the Coal Trust, which, like a gigantic octopus, extends a thirsty tentacle to every manufacturing plant on this continent, and whose fingers are slowly tightening their clutch upon your earning powers. The cost of power as a fixed charge in manufacturing is, therefore, a subject which now knocks at your door and demands admittance.

The cost of power is a subject which has gone so long without expert treatment by the average manufacturer that he has come to look upon his coal bill as an item of expense with which little or nothing can be done except pay it. I take it for granted that most of you are generating your own power in your own premises. That some of you are purchasing power from hydroelectric sources; and that all of you are interested in the question: "Can I produce my own power in my own premises at a lesser cost than it can be purchased for from the outside?" The manufacturer is in intimate touch with all his other departments. He knows exactly the outlay for advertising, and he employs the best of talent in this department. He knows exactly his costs of selling, and in this department he employs the highest ability within his reach. In fact, at the end of the year he knows the exact percentage which his advertising, his selling and his other departments constitute in the total outlay for the year. But as for power—beyond his coal bill and what he pays his engineer he does not know and, in my opinion, does not make a proper effort to know the exact cost.

A Concrete Example.

Since nearly every manufacturing plant which produces its own power is equipped with a steam plant, let us take as an illustration a moderate sized ordinary steam plant and let us assume that it operates on a basis of 10 hours per day for 300 days in the year, and that the average actual power required throughout each day of 10 hours is 200 h.p., and let us take coal at $3.50 per ton. It is safe to say that the coal bill for one year in such a plant will be not less than $4,200. That you will pay your engineer $1,000, a fireman $720, and after the incidentals have been added—for oil, packing, waste, etc.—the total will amount to $6,200. Then, interest and depreciation will easily bring this amount up to at least $7,000, or, say, $35.00 per h.p. for the average of 200 h.p.

It is probably safe to say that not one manufacturing plant in fifty is making its own power for so low a cost as $35. Now, let us see what this $7,000 per year as a fixed charge for power really stands for. Seven thousand represents the net profits on, perhaps, from $75,000 to $100,000 of finished product. It represents a fixed charge against your business—equivalent to $140,000 of 5 per cent. bonds, and it is needless to say that you would give much serious thought to the question of the issuance of such an amount of bonds. It is needless to say that you would analyze all the vital elements in your business before you decided to incur such a burden upon your earnings. And yet, you pay $7,000 a year for power, without having given anything like a corresponding consideration to the question as to whether or not $7,000 per year for an average of 200 h.p. is the right price.

Look only at the coal bill—of $4,200. Perhaps most of you do not know that of this $4,200 burned under your boilers, $3,800 are lost—absolutely non-productive; that out of every dollar's worth of coal burned, over 90 cents goes up the chimney and out of the exhaust, without doing any work of any kind.

If you contracted to pay a man a salary of $4,200 per year and should discover that the actual service he gave you was equal to only one hour out of each day, would you consider that good business? Or, suppose you owed $25,000 borrowed money and suppose you had to pay 17 per cent. interest on this money, would you consider that good finance? And yet, if yours is the average plant, I assure you that these are exactly the things which are happening in your coal bill.

It then becomes obvious that to get the cost of your power down to a point where the outlay in that department is consistent with the economies demanded in the science of modern manufacturing, you must either secure a reduction in the price of coal or you must generate your power on less coal than you are now consuming. The first alternative is out of the question, at least, so far as we can see. To materially reduce the amount of coal required to produce your present horse power is easily obtainable, if you will give this department the expert treatment which it deserves.

Horse Power Hour.

Let us now understand what one h.p. is. You hear on every hand "So much per annum per horse power," and "Cost per horse power hour." What is a horse power hour? If all the power in this little piece of coal could be utilized, if all its latent energy could be converted into actual work, it would lift 1,980,000 lbs. one foot high in one hour, or nearly 1,000 tons of dead weight off the ground for one hour. It weighs but 2¼ ounces, and yet could lift a weight twelve million times greater than itself. In other words, it contains the net equivalent of one horse power hour. But, when you undertake to convert this amount of latent energy into the actual driving of your machinery, I will not say that you must burn, but that you do burn, a piece this size (5 lbs.). It seems like a vast distance from this piece to that piece. But we cannot convert all of this latent energy into actual work, because nature demands a premium on what she gives us.

On the other hand, however, she does not ask us to pay any such price as this (5 lbs.) for one b.h.p. hour—this 17 per cent. interest. And since under modern methods we know that a brake horse power hour of actual work can be produced for a piece this size (2 lbs.), it certainly becomes apparent that in the average plant there is being burned, unnecessarily, and without return, the dif-

* Paper read before the Canadian Manufacturers' Association at Toronto, September 26th, 1907, by L. G. Read, Consulting Engineer, Montreal.

ference between 2 and 5 lbs., or 60 per cent. more coal than is required by refined engineering of to-day.

If 60 per cent. of this $4,200 coal bill can be saved, or, say, $2,500 per year, you may then consider that $2,500 per year saved is equivalent to your having retired an obligation of $41,500. Would you not consider a department worthy of expert treatment which unnecessarily places upon your business the equivalent of a $41,500 obligation at 6 per cent.?

Load Factor is Important.

If you will treat your power as a department and if you will give to that department the attention it deserves, you will easily confirm my claim that modern practice and actual results will place this great economy within your reach.

It would be impracticable, considering the limits of time on this occasion, and considering the great variety of local conditions, for me to undertake to detail the actual ways and means by which you may attain these ends; but, at least, I may say that you must approach this subject of power cost as a fixed charge in your manufacturing with a serious mind. You must ascertain what your load factor is. You must know what actual horse power it takes to drive your plant and when you know this you have got the key which will open the door to an important department in your business—a department which you have hitherto neglected.

To ascertain your load factor a diagram covering your daily run must be made by an expert. This diagram must show the h.p. output from your engines at every hour during the day's run and, preferably, for a number of days in succession, so that you may arrive at your load factor—that is to say, the maximum load and at what hour, the minimum load and at what time, and from these two you will obtain the load factor—the average actual h.p. you require. Then open a power account. Charge into that account that part of your capital invested in engines, boilers, pumps, dynamos, power house, chimneys, foundations, piping, shafting, belting and every part of your equipment which either generates or transmits power—together with all their accessories. This item will show you how much of your capital you have invested in your power department and upon it you will know what to charge for interest and depreciation. Charge into this account your fuel, engineers and firemen's salaries, oil, water, repairs, upkeep and all the incidentals which in the average manufacturing plant are lost in other accounts.

Some of you may say, "Yes, but all these items come out at the end of the year in some other account. So, what is the difference?" The difference is simply this: that in order to manage your power department in a way that will insure your getting your power at the right price, you must know all about that department—just as you know and insist upon knowing all about all the other departments in your business, and you cannot know all about your power department unless you follow it in this manner.

With the load factor known, and with all these items charged into this account, it then becomes merely a matter of dividing the total number of dollars footed up by the average h.p. taken from your power diagram—for one year—and the quotient will be your cost per annum per h.p.

It may not be for me to set down an arbitrary price per annum per h.p., as representing the price at which you would be justified in generating your own power instead of purchasing it from outside sources, but I will assume the responsibility of saying that unless you can purchase from outside sources—after charging interest and depreciation and everything which is properly chargeable to such motor equipment and appliances as may be necessary for you to utilize outside power—you must not pay more than $25 per annum per h.p. and only upon your average load and not upon your maximum load.

The Maximum Cost.

I will, however, place a limit upon your load factor; that is to say, your average h.p. requirement should be at least 60 per cent. of your maximum load. In other words, with a maximum of 100 h.p. and a minimum of 20 h.p.—or a load factor of 60 per cent in your plant, you would contract to pay $1,500 per year and no more. For, obviously, if it can be shown that you can produce your power on this basis, at this price, why should you pay a higher price than this for it from some outside source?

Whatever you may pay to outside sources in excess of this price, represents just that much dead loss in your business, and when you consider that a reduction in your fixed charges of even $2 per day is equivalent to your going to your bank and paying off $10,000 of outstanding 6 per cent. paper, you will realize that the earning power of money, like the latent heat in coal, is fraught with great possibilities, when given the proper treatment.

Now, let us take the point of view of the power company that sells its power as a public utility. Some think that the development of water powers will solve this great problem of power cost. But, if you will look closely, you will find that the limits of these possibilities are soon reached. In the first place, water powers are not always located where we want them, and where we need them. Let us take a hypothetical case. Suppose we develop a water power with a maximum of 100,000 h.p. Our total investment will amount to at least $65 per h.p. initial cost, or, say, $6,500,000.

We contract to deliver power to thousands of users. Our whole equipment is designed and installed, with all our transmission lines, our transforming stations; with all the local equipment required at points of destination; our management and our organization are based upon $6,500,000 actual investment. We will say that our peak load; that is, the top notch of our output each 24 hours, reaches the 100,000 h.p. mark. It is safe to say that the load plot—that is, a diagram showing the horse power output at each hour during the 24—will average not more than 40,000 h.p. In other words, the total amount of power we sell equals a load factor of 40 per cent.

Now, 40,000 multiplied by $65 per h.p. (initial cost) equals only $2,600,000, and you will see at once that the price which must be charged per annum per h.p. must be an amount sufficiently high so that an average output of 40,000 h.p. (or $2,600,000 of our investment) will earn a return sufficient to carry the entire $6,500,000. And this is the condition which usually prevails when a water power is developed and expanded into a public utility.

On the other hand, if a water power is developed for local use, by an individual concern; obviating the necessity of long transmission losses and the low average h.p. output, you will readily see that, under these conditions, water power serves its best and most practicable purpose. The ten best water power developments in the world—including the cheap water powers of Canada—show an average cost of $10 to $12 per annum per h.p. This is cost—not the price at which the power is sold. Suppose, after allowances for transmission, transformer losses, fixed charges and a fair profit, it were delivered to the consumer for $20 per h.p. on the usual flat rate basis.

You contract to pay for a certain amount of power whether you use it or not. (Say 100 h.p.) At the end of the year you will find that since your own load factor will probably not average over 60 per cent. (for you, as well as the big power company, have your peak load and your minimum load, and, consequently, your average horse power requirement), what you have actually paid for what you have actually gotten equals $33.33, instead of $20. Because

—$20 multiplied by 100 h.p. equals $2,-000, and $2,000 divided by your actual average load of 60 h.p. equals $33.33 per h.p.—or 65 per cent. more per annum per h.p. than the rate named in your contract. In other words, you pay the power company (60 h.p. × $20) $1,200 for power you do get, and $800 for power you do not get.

Some Comparisons.

The gas company—a public utility—charges you for the exact amount of gas you use and no more. The water company—a public utility—does the same. Now, when a power company elects to expand itself into a public utility, on what theory has it the right to demand this premium of $800, this 65 per cent. for something it does not deliver? Is it because "the power is there, if you want it"? Then, so is the gas "there if you want it"! So is the water "there if you want it"! Is it because of the power company's low per cent. of load factor? Then why have not the gas company and the water company an equal right to make the consumer pay the difference between their average output and their maximum capacity?

The answer is that the sale of power as a public utility is a new enterprise. The consumer has overlooked the importance of this department. Hence the power companies make hay while the sun shines. But there are exceptions; where the average load factor in a plant is equal to and more than the minimum amount of power contracted for; but these exceptions are few, and you will find that a public power company avoids such contracts.

The Key to the Power Question.

Now, just one final word. The key to this whole question is the load factor—the average actual h.p. required throughout the day's run. Do not confound this with your maximum h.p. requirement. When you talk of "so much per h.p. per annum," I insist that it be based upon the average h.p. and not the maximum rating. And I insist that with a load factor of 60 per cent. it should not cost you—whether you make it yourself or purchase it from the outside—more than $25.

Make up your minds now that you will make a department of your power; that you will open a power account; that you will study this question, and if you do these things, the small expense involved will come back multiplied tenfold, as profit.

I am indebted to the officers of this association for the opportunity of even presenting so briefly the principal claims which I think the question of power has to the consideration of manufacturers.

Since, however, justice cannot be done to so broad and so important a subject within such time limit, I shall be glad to send to any member of this association, more explicit details by which he may be guided in making a department of his power and the giving to that department the attention which, in this day of refined engineering, is easily within his reach.

Welding by Thermit

The Stern-Post Shoe of the S.S. Corunna Successfully welded.

An interesting experiment and one that marks the progress of Canada, was the steel welding process that was tried for the first time in Canada on the

Fig. 1—A view of the fracture of the S.S. Corunna's stern-post shoe before welding.

steamer Corunna. This vessel plys between Montreal and the Upper Lakes, and had her rudder post and shoe broken recently in an accident in the

Fig. 2—The stern-post after the welding process by Thermit was concluded.

Cote St. Paul Locks. It was successfully mended without removing the broken parts, and the work was watched with great interest by a large number of shipping men, as the success attending the demonstration was of great importance to marine repairing in Canada. A comparison of the two methods of repairing will show the saving in the repairs to the Corunna. Under ordinary circumstances, the owners would first have to undergo an expense of $1,000 in having the steamer towed to Cleveland before commencing the work of repairing. Both time and expense were thus saved, and, in future, work of this character instead of going to United States, as formerly, will be done in Canada. The method is marvelously simple, as is shown in the accompanying illustrations. Fig. 1 is a view of the broken stern-post shoe of the s.s. Corunna before welding.

The parts to be welded were thoroughly cleaned, a wax pattern was made over the broken parts, and then a mould was built around the shoe, a hole being left in the bottom of the mould to allow the wax to run out, leaving the space for the molten steel. A powerful gasoline torch was inserted, and brought the broken parts to a red heat. As the wax melted it flowed out the hole in the bottom of the mould, and the opening was then closed up. In the meantime, a sheet iron shell crucible was swung over the mould and the charge of thermit powder placed in it. A teaspoonful of ignition powder was placed on top of the thermit in the crucible, and ignited with a match. In thirty seconds the crucible was full of molten steel at a temperature of 5,500 degrees Fahrenheit. A hole was then tapped in the bottom of the crucible, allowing the steel to flow into the mould. The superheated thermit steel fused the broken parts together, making the shoe as strong as before the break. Fig. 2 shows the stern-post shoe after the welding was concluded. Mr. William Abbott, agent for the Goldschmidt Thermit Company, New York, undertook and successfully completed the repair of the broken stern-post shoe.

You cannot learn more than you now know without venturing something that you have not tried.

To underpay or overwork men is to deprive them of facility, time and repose for self-improvement. We create ignorance, we put a brick in their hands to throw at our heads.

Bearings and Their Lubrication

Overcoming Friction in Bearings and Their Design.

By JOHN EDGAR

The bearing is one of the machine elements that we meet on every side. They are absent only in few instances. It is hard to conceive of a machine that has no bearings. In the strict technical sense a bearing is any area that takes a pressure or carries a load. In this article we shall restrict the term to those cases where the bearing carries a rotating journal or is a rotating member itself, and is supported by a stationary journal. A bearing in machine design is the internal surface of the cylindrical housing for the journal. The journal being the shaft or other cylindrical object that rotates in the bearing.

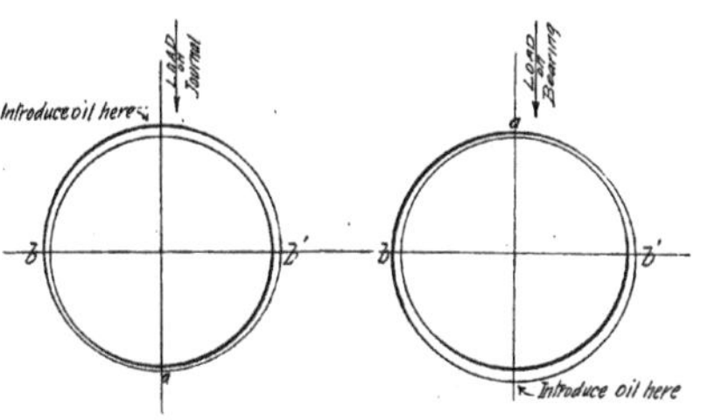

Fig. 1—Greatest Pressure at Lowest Point.

Fig. 2—Greatest Pressure at top.

The friction caused by the movement of one surface upon another is the cause of most of the trouble in the operation of machinery, that is due to the bearings. The object is to reduce this friction as much as possible, and in properly designed bearings this friction is reduced to a minimum, the load per square inch of the projected area being such that experience has shown can be used without causing the bearings to heat and seize the shaft. Lubrication is necessary, and, in most cases, is accomplished by introducing between the surface of the journal and that of the bearing oil, grease or other lubricant. The pressure per square inch that a bearing will stand depends on the lubricant and its method of introduction. The object of the lubricant is to separate the surfaces as much as possible, and thus reduce the friction to a minimum. Were it possible to produce surfaces that were absolutely smooth the movement of one such surface upon another would not cause friction, therefore trouble would not be experienced and a lubricant would not be necessary. It is one of the offices of the lubricant to fill up the hollows in the imperfect surfaces and thus produce a good substitute for the perfectly smooth surface. It is obvious that any pressure sufficient in extent to squeeze the oil or grease out from between the surfaces must be avoided. For this reason heavier oils and the greases are used where the heavy pressures have to be carried at moderate speeds. Should the heavier lubricants be used in fast running light bearings the friction of the lubricant within itself would be excessive, so that for this class of bearing a lighter weight oil is necessary.

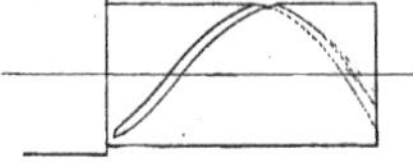

Fig. 3—Spiral Oil Groove.

Bearings Design.

The proportion of the bearing with respect to diameter and length is generally made as follows: For hangers, supporting line shafting, when the bearing is self aligning, the length is made about four times the diameter of the shaft. When, however, the bearing is not self-aligning, as in the case where the bracket and bearing are cast in one piece, or the latter bolted to the bracket the diameter should be only about one-half as long, being about two diameters in length. The reason for so shortening the bearing in the latter case is to reduce the pressure at the end of the bearing, due to the deflection of the shaft which causes no harm in a box that is self-aligning, because the latter adjusts itself to the general deflection and takes a position tangent to the elastic curve; in the former case, however, bending of the shaft causes it to bear harder, and with more pressure at the edge near the load, because the centre line of the box cannot adjust itself to the general direction of that of the shaft, therefore the length must be shorter in order that the pressure may be better distributed over the whole length.

Other bearings, for uses outside of the supporting of line shafting, are so designed that the pressure per square inch of the projected area does not exceed certain prescribed amounts. These pressures range from 15 lbs. to 600 lbs. The former pressure is used in the design of line shafting bearings of cast iron, and is rather a small value, while the latter is found in examples of steam engine bearings, where the journal runs slowly. Sixty pounds pressure is a good figure to use when space and conditions

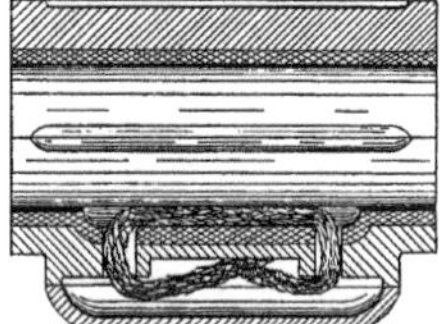

Fig. 4—Wick Oiling Bearing.

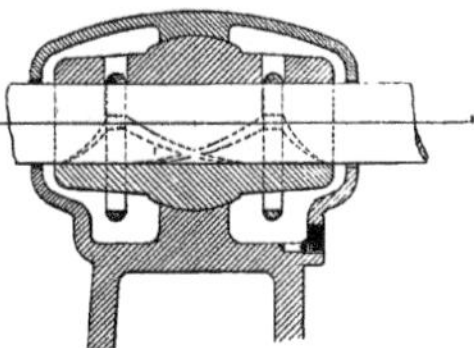

Fig. 5—Dynamo Ring Bearing.

will permit, and may be run up to one hundred pounds without any serious results.

Applying the Lubricant.

The pressure that a bearing will stand depends much on the manner in which the lubricant is applied. A bearing properly proportioned and all right as to

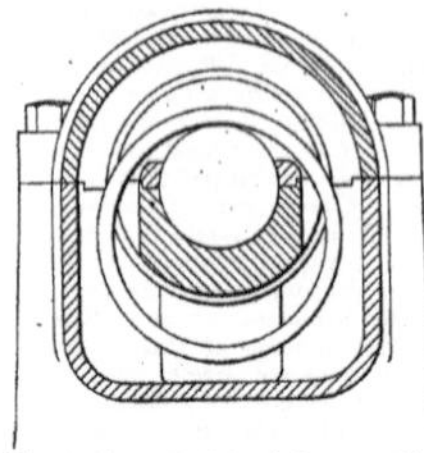

Fig. 6—Cross Section of Dynamo Ring Bearing.

design in other respects, may act badly from lack of consideration on the point of lubrication. The application of the lubricant is then as great in importance as the general design of the bearing. The same bearing when inadequately lubricated has been known to seize at a pressure of 300 lbs. per square inch, when altered and oil applied by an oily pad the pressure permissible before seizure was raised to over 500 lbs. The introduction of the oil by means of small oil holes or cups is a very unsatisfactory way of lubricating a bearing and should be used only in cases where the bearing is subjected to very light pressures. Oil should never be fed into the bearing at the point where the load is heaviest, but at the point where the pressure is at a minimum. When an attempt is made to introduce

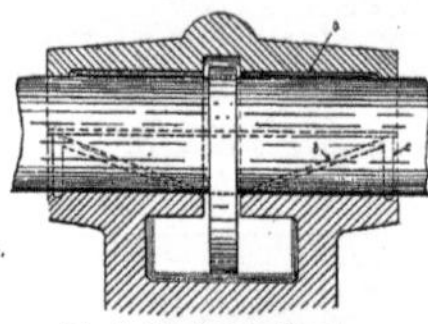

Fig. 7—Single Ring Bearing.

the oil at this point, the bearing is only oiled at times when the load reaches a minimum, or zero, or when the bearing is stationary, because the pressure of the oil between the surfaces at that point is greater than that in the reservoir, so that it is impossible for the latter to enter the bearing under those conditions. It has also the bad quality of allowing a vent for the escape of the oil that is already between the surfaces, so that the result is worse than bad. The above point is shown by Figs. 1 and 2—Fig. 1 represents a bearing in which the greatest pressure occurs at (a) the lowest point this pressure gradually decreases until at the horizontal diameter (b, b') the pressure becomes zero, and is zero at all other points in the circumference. The lubricant in this case should be introduced at the top as indicated, where it has a chance to enter and is carried to the effective point by the motion of the shaft. Fig. 2 shows a similar bearing, but in this case the bearing carries the load and the greatest pressure then comes at the top and the oil should be introduced at the bottom.

It is the general practice to groove the bearings carrying the groove along the bearing nearly the full length. This groove is invariably cut at the point (a) and acts as a vent for the oil to escape, by thus defeating its purpose.

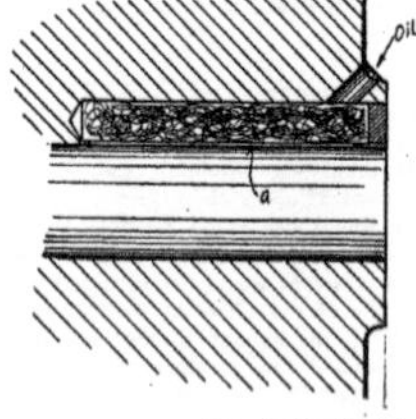

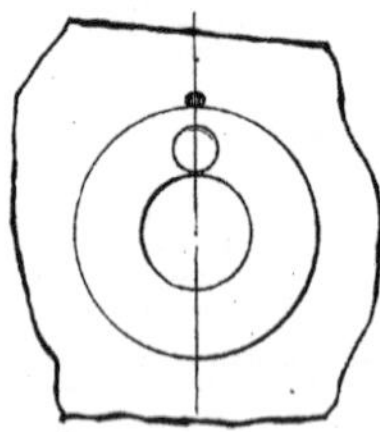

Fig. 8—Special Bearing with Reservoir.

The proper place for the groove is at the point of introduction of the lubricant and at the points (b and b') where they act as reservoirs and are then very serviceable. It is sometimes the practice to groove the journal. The groove shown in Fig. 3 in this case being cut spirally around the shaft forming one continuous groove. For slow running journals that have to carry a considerable load it is a mistake to groove them in this manner. The groove in this case acting as a vent and also carrying off the oil instead of holding it in its proper place. For fast running shafts, it is good to groove them spirally, because the pressure on such bearings is generally light and the groove in that case keeps a fresh supply of oil on the whole surface.

Where the load is intermittent, as in the case of presses and such machinery, and where it is first in one direction and then in the other, as on the crank pin of a steam engine, the lubrication is more perfect and the siphon feed will be more effective.

Special Oiling Devices.

Fig. 4 shows a bearing fitted with a wick oiling device. In this case the oil reservoir is in the cap and the wick is threaded through the holes in this cap and in the groove provided for it. As long as there is any oil in the reservoir the wick will, through the aid of capillary attraction, feed it to the shaft which it wipes. It will be noted that this bearing is lined with a bearing metal.

The high speeds and continuous running necessary in the cases of the dynamo and motor, have been responsible for an entirely new form of bearing, that, the ring oiling type. In this type the bearing is lubricated by oil carried onto the shaft by a ring which rolls on the shaft and through a reservoir of oil from which it carries oil and deposits it on the shaft at the point where it comes in contact with same. Fig. 5 shows a longitudinal section through the pedestal of a dynamo fitted with such a bearing. It will be noted that this bearing is of the self-aligning type. The shaft being oiled by two rings, the oil is distributed in all directions by the grooves. Provision is made for drawing off the oil spent in vitality, and also means for replenishing the supply. Fig. 6 shows a cross section through the same bearing.

Fig. 7 shows a single ring, ring-oiling bearing, made solid with the pedestal. The reservoir is cored out under the centre of the bearing, a slot opening into it in which the ring runs. In this type bearing where the supply of the lubricant is so steady, and plentiful precaution must be taken to hinder the oil from flying about as it runs out at the ends of the bearing. The design shown in Figs. 5 and 6 takes care of this by having the bearing encased in the covering shown. But in that of Fig. 7 no such casing is provided. To stop the oil from flying the grooves (c) are cut in the lower half of the bearing at the ends. These grooves catch the oil and the groove b leads it to the reservoir. The groove a distributes the oil along the length of the bearing. The ring oiler is the highest type of modern self-oiling bearing, and is being applied to many varieties of uses, notable among which is that of countershafts and line shafts.

Lubrication under pressure is also being developed to a certain extent, especially in engine and kindred work. Here the oil is forced into the bearing by the aid of a pump. The oil being used over and over again, and a fresh supply is thus continually fed into the bearing. Sometimes a bearing that gives trouble can be made to work satisfactorily by doing as suggested in Fig. 8. Here a hole of considerable size is drilled into the metal surrounding the bearing as close to same as can be drilled without breaking through. A slot (a) is then cut connecting the hole with the bearing. An oil hole is then drilled into the hole from the outside. The large hole is filled with felt or waste and the end plugged. This forms a good reservoir for oil and gives a constant feed to the bearing. Vertical bearings that are bored out from the solid frame can be oiled by this method the hole running along the bearing in the vertical position and the feeding being done from the top.

Machine Shop Cost System

A Simple Cost System That Is Efficient.

Machine shop superintendents and cost keepers are constantly on the lookout

ORDER SHEET **BAWDEN MACHINE & TOOL CO., LIMITED**

Binder No.

Date Billed

Order No.

Requisition No.

Order From..

....

Remarks: ....

Description of Order

TIME				TIME				MATERIAL		
Date	Hours	Rate	Amount	Date	Hours	Rate	Amount	Date	From	Amount

Fig. 1.

for a simple and efficient cost system. For the benefit of those readers of Canadian Machinery we give below a description of the system employed by the Bawden Machine Tool Co., Limited, Toronto. F. M. Bawden has realized that simplicity and efficiency are the chief features of a cost system and with that idea, designed his time and cost cards. We have reproduced those used in his cost system.

When an order comes to the works for a certain number of engines or engine parts, card No. 1 is filled out. To illustrate, we will take a concrete example: "Order from," Gas Engine Co., Toronto; "Order No.," 285; and under "Description of order" is written, Gas governors—6 wanted. Drawings G—18-24. A copy of this order is then sent to the foreman, who issues cards similar to Fig. No. 2, which would be filled in as follows:

Job sheet for Gas Engine Co.
Employe's No.—35.
Job No.—285.
Drawing No.—G-20.
Name of part.—Ferrule.

Under operation is written boring or whatever the operation may be and the number of pieces, and opposite this the workman fills in the number of hours he is on the work each day. Before starting to work on the castings they are weighed and under material is written the number of pounds of material used. Suppose a man was turning ferrules and it took him four hours on Monday, nine on Tuesday, and five on Wednesday, he would have a total of eighteen hours, or, dividing by six, it takes three hours to turn each ferrule. In the same way may be had the weight of each ferrule. A workman gets only one of these cards at a time. He must hand in one card before starting another job, and a card is given out for each operation done on each part, thus there is a card for drilling, boring, turning, etc., for each part going to make up the machine. If a duplicate order comes in, then, by looking up former cards, the foreman knows at once how long a job should take, and if more time is being spent on the work than in the first case, he can at once look into the matter and find out the cause. Then, too, if he thinks he can improve on the time, on some of the work, he can experiment and find a means for simplifying the work and increasing the working capacity of his mechanic.

When the cards are returned to the office the items on card No. 1, date, hours, rate and amount, are filled in from card No. 2, and the amount of material, with its cost, is filled in so that the superintendent can tell at a glance the exact cost of each part or the total cost of each order going through the shop. Very little clerical work is required in this system, and it is found to give all the desired information. No. 1

BAWDEN MACHINE TOOL CO.
LIMITED

Job Sheet for..
Jig Number ....
Employe's No. ....
Job No. ....
Drawing No. ....
Name of Part..

Operation	Mon. Hrs	Tues.	Wed.	Thur.	Frid.	Sat.

Materials ....

Fig. 2.

cards are filed according to the order number, and No. 2 cards are filed away alphabetically, thus the cards we have described come under G. They are there arranged according to drawing numbers, so it is an easy matter to use them for reference. It is, therefore, obvious, that while simple, this system is complete and all the details are under the direct notice of foreman, superintendent and cost-clerk.

Machine Shop Methods and Devices

Unique Ways of Doing Things in the Machine Shop—Readers' Ideas and Opinions Concerning Shop Practice—Useful Data for the Machinist. Remuneration Will be Given for Contributions.

TURRET FOR ENGINE LATHE.

By J. H. R., Hamilton.

The turret attachment shown in the illustration can be placed on any engine lathe and will greatly add to its output of nuts and screws or any other small fittings made from a bar. It can also be used for small castings requiring the operations of turning, boring, drilling, or tapping.

A is the main spindle (in the form of a crank shaft) fitted in the tail-stock, with (a) as the crank or centre piece for turret B.

B is the turret which revolves on (a) and contains as many holes as are desired.

C, C, C, are cams which revolve the turret by contact on roller R.

D is a bracket bolted to tail-stock carrying trip T, and roller R.

E is a lock pin place in flange of A and held in position by spring S.

F is a lever to release pin E.

G is a movable piece held in position by spring S, and revolving on pin P.

B is a shell to keep turret B in position and fastened to B with screws, I, I.

J, J, are screws to lock tools, etc., in turret.

R is a roller bolted to D.

T is a trip to release pin E.

Arranging tools in turret to suit work and chucking work by gauging with stop M, turret is taken back by hand wheel N (or for light work, lever O); lever F comes in contact with trip T, releasing pin E, cam strikes roller R, revolving turret to next position, and then advances for next operation, roller R passing next cam by piece G swinging on pin P; operation is completed and continued as before. Suitable stops could be arranged so as to insure duplicate work.

Piece B could be cored out as shown to avoid weight.

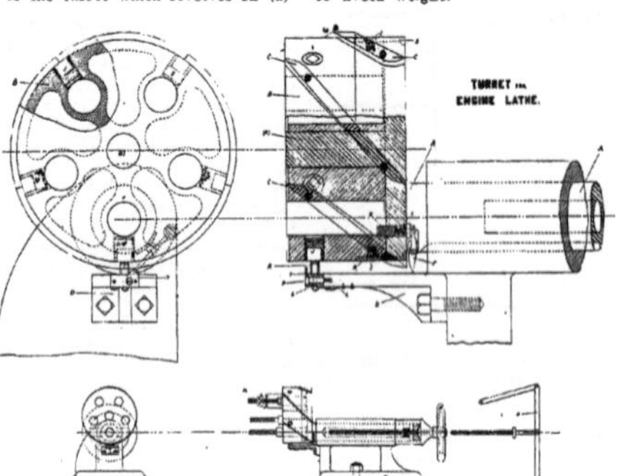

Turret Equipment for Engine Lathe.

HOLDING DEVICE FOR TRIANGULAR RULES.

By a Draughtsman.

The accompanying sketch shows a block I have used with a triangular rule and found a great convenience. Each time I laid down the rule I would lose the particular scale I was using, whether 1-8, 1-16, 1-32, 1-64, etc. By taking a small block of wood about two inches long and one inch high and cutting a V-shaped hole to conform with the shape of the rule, as shown in the sketch, I could keep the scale always before me and use a pair of dividers without any waste time.

LATHE DEVICE FOR THREADING PIPES.

By A. E. D.

Considerable difficulty is often encountered when a pipe is to be threaded

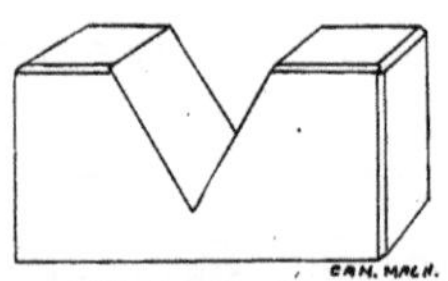

Holder for Triangular Ruler.

or cut in a lathe. One end is put in a chuck and a board or wooden plug is fastened in the end to be threaded and a rest is used to keep the pipe running true. Anyone who has used this arrangement knows how very unsatisfactory it is and how long it takes to set up a pipe. In our illustration we show a device which we think will save time and obviate the trouble. This has been used in several shops with success. It is a self-centre made of cast iron and the size depends on the size of the pipe to be threaded, but one centre will accommodate a great number of sizes of pipe. It is a cast iron cone with a spindle, turned to take the place of the centre in the tail-stock, and the outside face of the cone is machined. In small sizes the centre is cast solid, but for the larger size pipes, where a large centre is necessary, it can be lightened by casting it with a hollow centre, as shown by the dotted lines. If it is

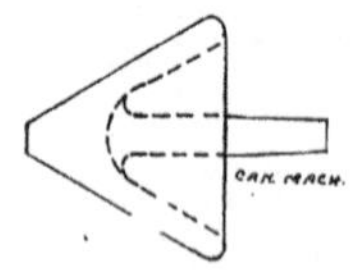

Lathe Centre for Threading Pipe.

thought that more strength in the centre is required, it can be cast with webbs. By means of such a device a pipe can be easily and quickly centred for cutting or threading a pipe in a lathe.

THREAD GAUGE.

By James C. Moore.

A simple, but very handy tool, is a thread gauge with bevel attached for setting the cutting tool to the required angle, clearing the rake of the thread while cutting coarse-pitch. The rake of

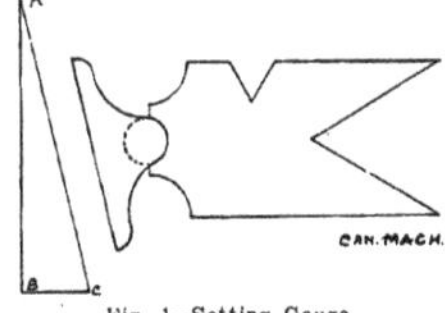

Fig. 1—Setting Gauge.

the thread is obtained by a piece of paper or tin placed around the circumference to be threaded and cut the exact length of the circumference and as wide as the pitch to be cut. This is shown in Fig. 1: A, B, is the line of circumference B, C, the pitch line, and

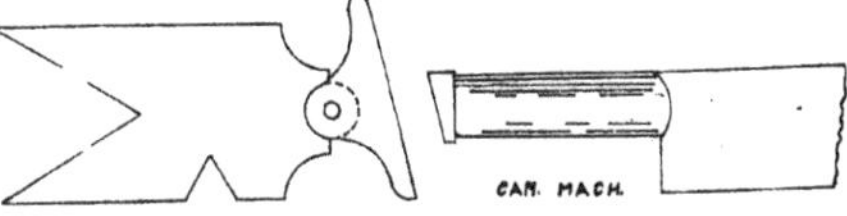

Fig. 2—Setting Tool for Nut.

the line A, C, is the rake of the thread. The bevel piece of the gauge is set to this line and the tool is placed on tool-holder to clear. Round steel will be found most convenient, as it can be placed at any level to suit rake. If flat or square steel is used the tool must be ground to suit. This tool is particularly valuable in cutting a nut. The accompanying illustrations show the use of the thread gauge. Fig. 1 shows the rake of the thread and the method of setting the gauge. Fig. 2 shows the manner of setting the tool to thread a nut.

GRINDING FRAME FOR LATHE.

Herewith is a grinder frame designed to be placed on the carriage of a lathe, the work being held in a chuck, face plate or by other means. It does heavy grinding with good satisfaction. It is simple in construction, very strong and rigid. It has adjustable bearings, and is absolutely dust proof.

The frame can be secured to the lathe carriage in any position or, as shown, with open end of box frame up so as to allow the belting from above. The box frame, Figs. 1-2, is a plain box finished all over outside and recessed as shown to receive bearing yoke. The yoke is of cast iron bored to receive bronze bushings, the outside turned and threaded to receive brass adjusting nuts. The pulley is also of cast iron. A fibre washer is put between pulley ends and bearing yokes. A dust proof collar, A, is secured with headless screw. A full size end of shaft is given in Fig. 3, showing taper end, also wheel bushing and clamping collars. We have in use other grinders, but this serves the purpose the best.—Machinist, in Modern Machinery.

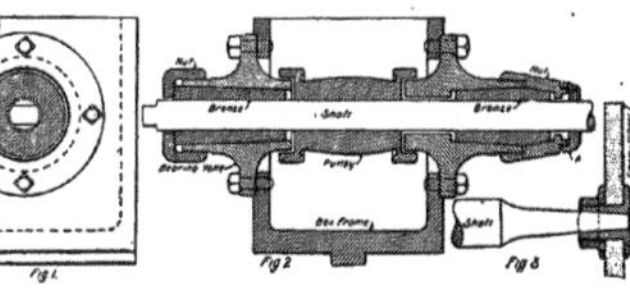

Grinding Frame for Lathe.

ERRATA.

In the article, Taper Turning, by J. H. R., Hamilton, on page 42 of our October issue, the example in the third paragraph should read: "The taper is three-quarter inches per foot, move tail stock over ⅜ inches per foot.

A QUICK AND ACCURATE BORING MACHINE.

F. A. Bawden.

The illustration shows a plan of a boring and facing machine used by the Bawden Machine & Tool Company, Limited, Toronto, for the machining of their drill presses. This boring machine is bolted to a planer and the drill press rests on two V-blocks, which are securely bolted to the planer bed to which the

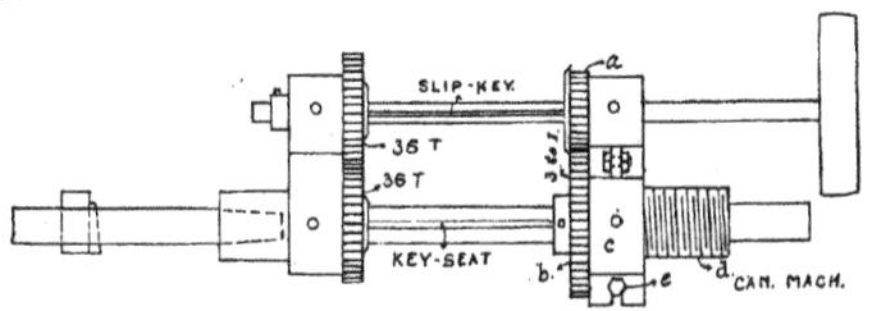

A quick and accurate boring machine.

drill press is also bolted. The tool is in a shaft, as shown in sketch, and the spindle works in two babbited bearings fastened to the planer. Vibration is thus eliminated and the whole arrangement is fool proof. All the operation of boring, facing and babbiting are done on the planer. The drill press is thus handled only once, and the facing of the holes at right angles for the gears are made absolutely squared with the bore. Gear a is for throwing the drill in and out of gear. When facing, this gear, which runs on a slip key, is pushed forward and the facing shaft is turned by hand by means of a special wrench used on the gear b. The gear b and feeding sleeve d are in one piece and work in a keyseat in the shaft, c is a bearing threaded in the top only so that when the cutters feed up to the limit, by loosening bolt e the bearing c is lifted upon a hinge and the sleeve pushed back. The collar in front of the gear b is fastened firmly to the shaft. The power is obtained from a countershaft above the planer, and a quarter-turn belt transmits the power to the driving pulley.

Mr. W. H. Munro, of Peterboro, has joined the staff of McDougall & McRae, consulting engineers, Ottawa.

Mr. J. W. Smith, C.E., of the firm of Galt & Smith, Toronto, and Mr. R. S. Lea, C.E., of Montreal, have been appointed to investigate and report on a sewerage system for Regina.

Developments in Machinery

Metal Working ***Wood Working*** ***Power and Transmission***

A NEW LIFTING MAGNET.

After much experimenting and many severe tests, there has been placed on the market a lifting magnet that is meeting with success. Most of the features in the design are original, but the fundamental principle is that of any ordinary magnet, a steel core and coils of wire to carry the exciting current.

Wherever pig iron, metal plates, tubes, rails, beams, scrap or heavy castings of iron or steel are handled, lifting magnets can be advantageously employed. There is no waste time adjusting block and tackle to the object to be lifted. For lifting scrap and pig iron the lifting magnet is of special service. The magnet is lowered onto the piles of scrap or pig iron and without any further work of handling the iron, it is lifted onto or from the cars, as the case may be. In foundries and rolling mills, magnets are useful for lifting and transporting metal too hot to be touched with the hands.

Fig. 1—Lifting Pig Iron.

Lifting Capacity of Magnets.

The lifting capacity depends to a great extent on the nature of the material to be handled. A 50-inch magnet, under favorable circumstances, will lift as much as 20,000 pounds, but under more adverse circumstances the lifting capacity might drop to 1,000 pounds, or even less. In one test, a 52-inch magnet readily picked up a steel skull weighing 5,500 pounds, and this in spite of the fact that the surface presented to the magnet was very uneven and extremely dirty, being partially covered with slag.

Fig. 2—10-inch Magnet Weighing 75 lbs. Lifting 800 lbs.

Fig. 1 was taken during a test at the Carnegie Steel Co.'s Donora works, when a 52-inch magnet lifted from the ground 32 sand-cast manganese iron pigs, averaging 65 pounds each in weight, making a total lift of 2,080 pounds.

Weight and Construction.

A 10-inch magnet weighs about 75 pounds, and Fig. 2 is a view of one of these small magnets lifting a Bliss Electric Car Lighting Generator weighing 800 pounds, or over ten times its own weight. A 50-inch magnet weighs about 5,000 pounds. The construction consists of a hollow steel casting in which the magnetizing coil, on the design and construction of which the efficiency of the magnet depends, is placed. The coil is most carefully built up of alternate layers of copper and asbestos, and is insulated from the cast steel frame by thick sheets of mica. There is no danger of the magnet being damaged if, by accident, it should be left in the circuit over night. Under test, these coils have been heated to 470 degrees F., without injury. It will be noticed by reference to Figs. 1 and 3, that the frame of the large magnets is corrugated. This construction provides a greater surface for heat radiation and at the same time forms niches to protect from injury the heads of the through bolts which fasten the removable pole piece to the magnet frame. To aid the heat radiation, the magnet frame has a central aperture through which the air circulates freely. This opening reduces the weight of the magnet and prevents the hot air from forming in the concave under the surface of the magnet. In lifting plates with plain surfaces this concavity becomes objectionable owing to the fact that an air gap intervenes between the inner pole and the object to be lifted. This is overcome by inserting in the central aperture an auxiliary pole piece so proportioned as to extend downward to the level of the outer pole, thus insuring intimate contact of both poles when an object is being lifted. Another feature worthy of note is the through bolts which are used to secure the pole

Fig. 3—Magnet Handling Steel Stampings.

shoe to the magnet frame. These are used to overcome the difficulty sometimes experienced with cap screws that have rusted in, for if a through rod becomes rusted, it is an easy matter to chisel off a nut and knock out the through bolt. It is thus a comparatively simple matter to replace a pole shoe, should it wear out. The removable pole shoe is clearly illustrated in Fig. 3, in which the magnet is shown handling steel stampings. In this case the magnet was not lowered into the stampings, but was brought to within about three feet of them, when the perforated steel plates rose to the magnet as shown in the illustration.

Magnets such as we have described are manufactured by Cutler-Hammer Clutch Company, Milwaukee, Wis.

NEW POWER HACK SAW.

A new power hacksaw has just been placed on the market, called Kwiv-Kut on account of its quick cutting ability. An adjustment on the saw arm allows the saw-blade to be used from end to end. A feature of the machine is the quick-return motion after cutting. The arm connects the saw by means of pin joints at either end. A cross-head runs on ways in the lever, and as the return motion begins the cross-head is at the lowest end of the ways. The leverage is thus very short and a quick-return is the result. The return stroke is three times as fast as the forward stroke.

It is constructed on strong mechanical lines and the machine automatically adjusts its stroke to all sizes of cuts. It is absolutely automatic, being equipped with an attachment which throws a clutch and stops machine as soon as the cut is made. The saw frame is mounted on babbited bearings. The frame may be thrown back, leaving the vise absolutely clear. Hand-power may be used if desired. A split finger holds the blade in perfect alignment.

Improved Power Hack Saw.

The construction is simple. A friction clutch is used for applying power, a feature used in the saw we here illustrate. The vise is equipped with right and left-handed screws, and holds work firmly. It is manufactured in three sizes and will accommodate cuts up to 8x8 inches. A No. 2, with capacity 5x5 inches, uses a floor space 36x17 inches, a blade 14 inches long, a pully 14x2½ inches, runs at 50 revolutions per minute, and weighs 180 pounds.

This power hacksaw is manufactured by E. C. Atkins & Co., Indianapolis, Indiana.

A New Universal Radial Drill.

NEW UNIVERSAL RADIAL.

The chief characteristic claimed for this machine by its makers lies in the design of the arm which offers great resistance to the combined stresses of twisting and bending. This feature. combined with the exceptional facilities which the open form of arm offers for the introduction of a driving mechanism commensurate with the strength. power and durability obtainable in the balance of the parts, marks a most noteworthy advance in universal radial drill construction, containing, for the first time in its history, a degree of efficiency equal to that of a plain machine.

The sleeve is mounted on a stationary stump which extends up to and has a bearing at the top of the machine. This is equivalent to a double column, and affords that stiffness which is so essential to true work. The arm may be rotated through a complete circle on its girdle and the head through a complete circle on its saddle, which permits drilling at all angles radiating from the centre of a sphere. The back gears, which are located back of the saddle, and may be engaged or disengaged from the front of the machine while running, furnish three changes of speed, each of which exerts at the spindle more than two and one-half times the pulling power of the next faster one. The spindle has fifteen changes of speed with the cone drive and twenty-four with the gear drive, and is provided with both hand and power feed, quick advance and return, safety stop, automatic trip, dial depth gauge and hand lever reverse. An engraved plate attached to the speed box shows how to obtain the proper speeds for different diameters of drills.

The depth gauge answers a double purpose: besides enabling the operator to read all depths from zero, which does away with the usual delays concomitant to scaling or calipering, it supplies a convenient means for setting the automatic trip, the graduations showing exactly where each dog should be located in order to disengage the feed at the desired points. The automatic trip operates at as many different points as there are depths to be drilled at one setting of the work; in addition, it leaves the spindle free, after any intermediate tripping, to be advanced, or raised and advanced, or traversed its full length, without disturbing the setting of the dogs; it also throws out the feed when the spindle reaches its limit of movement. The feeding mechanism furnishes eight rates of feed, ranging in geometrical progression from .007 inch to .064 inch per revolution of spindle, each of which is instantly available, eliminating all loss of time incident to shifting a belt, or to operating under a feed of unnecessary fineness. An engraved plate attached to the head shows the operator how to obtain each of the feeds. The tapping mechanism is located on the head, and permits the backing out of taps at any speed with which the machine is provided, regardless of the speed used in driving them in. It is fitted with a friction clutch operated by a lever, the handle of which extends around under the arm within convenient reach of the operator. The driving mechanism is incased in a box made fast to the base of the machine, and consists essentially of a pulley, a cone of seven gears, a ratchet, a ratchet gear and operating lever, the mere shifting of which from one notch to another furnishes any one of eight speeds. This box, taken in connection with the back gears on the head, gives the operator a choice of twenty-four carefully selected speeds, each of which is instantly available. The machine is furnished as either a half or full universal, each of which styles is made in three sizes, 4, 5 and 6 feet.

This new line of radial drills is manufactured by the Bickford Drill & Tool Company, Cincinnati, Ohio, U.S.A.

A BORING, DRILLING AND MILLING MACHINE.

The accompanying engraving illustrates a new machine, designed for boring, drilling, milling or tapping at any angle.

The machine is equipped with a constant speed drive, which can either be driven by a single belt or geared direct to a constant speed motor. A five-step cone pulley can be substituted for this constant speed drive.

The machine consists of a 42 inch x 127 inch bed plate tongued and bolted to the main bed. The wide upper surface of this bed carries the column with its saddle, which is gibbed down to a sliding fit on a wide scraped bearing. The longitudinal travel of the column on this bed is 62 inch. On this column the spindle head is mounted and is also gibbed to a sliding fit on the columns wide scraped surface. The vertical travel of the spindle head is 54 inches. The spindle head is fully counterbalanced and can be adjusted by the pilot wheel or can be raised and lowered or fed by power. Eight feeds in all directions are provided for this. The eight feeds referred to can be used as follows: To feed the column horizontally on the bed, the spindle head vertically on the column and the spindle horizontally. All these movements can be in-

New Fosdick Boring, Drilling and Milling Machine.

stantly reversed. The pilot wheel shown can be used to run the column forward and back, raise or lower the spindle head or run the spindle forward and back rapidly. For heavy milling, the column can be fed backward or forward on the bed. This throws the strain of the cut upon a screw of suitable diameter and lead with its ball thrust bearings. This feed can be controlled by the lever shown on semi-circular guide on the front of the machine.

The drive consists of a speed box with positive gear changes which are easily and noiselessly changed to their different speeds. This drive has six changes of speed, and with two changes on the head allows of twelve changes of spindle speed from 4 to 260 R.P.M. A friction device is placed between the back gear and the speed changing device, which can be used for starting, stopping or reversing the entire machine. The new friction is of the toggle friction type and has proven itself to be very powerful.

The spindle sleeve bearings are of very large diameter and length, and are fit into adjustable taper bronze bushings, which allow of easy adjustment for wear. The spindle sleeve is fitted with a threaded collar of large diameter, which carries a device used to securely tighten the spindle bar centrally in any position desired. The spindle bar is 4 inches in diameter, is fitted with a No. 6 Morse taper, and has 30 inches travel by hand or power.

The bed plate, which carries the universal table and outboard bearing, has a rack inserted in which the pinions (operated by ratchet wrench on the outboard bearing and universal table) mesh to operate these respective parts to and from the spindle. The outboard bearing's horizontal movement is operated by a rack and pinion, while the vertical movement is controlled by a crank and screw arrangement. Both horizontal and vertical movement of the column, spindle head, outboard bearing and also horizontal movement of the spindle are provided with scales for quick settings.

The universal table has a 40 inch x 46 inch platen, which can be tilted and swiveled to any angle. It has a hole bored through its centre to allow of passing the boring bar through to the outboard bearing. The tilting motion, as well as the swiveling motion, is imparted to the platen through a worm and worm gear in conjunction with a circular rack and pinion. The table is provided with tighteners to rigidly clamp all parts to any angle. The weight of this machine is 21,000 lbs. This machine is designed as style D, No. 2. Horizontal boring, drilling and milling machine by the Fosdick Machine Tool Co., of Cincinnati, Ohio.

BEAVER DIE STOCK.

The die stock here illustrated differs from the ordinary die stock in the manner of setting the tool for different sizes of pipe. By changing the lever on the

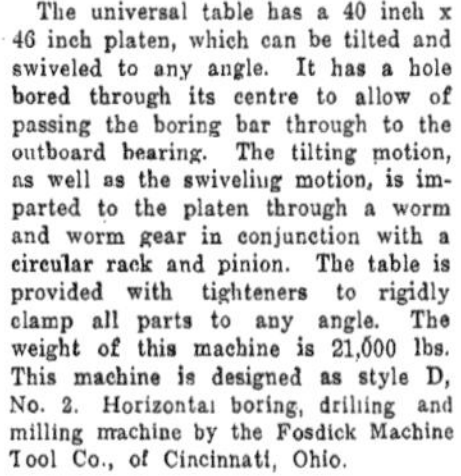

Assembled View of Beaver Die Stock.

dial plate on a No. 2 Beaver stock, the chasers are set to thread four different sizes of pipe, 1 in., 1¼ inch, 1½ inch and 2 inch. A cam with four slots open up the four segments to cut the different sizes. These four segments have each three full threads, and thus the area of the cutting surface is small. The cam and dial plate are held rigidly together by means of a clamp with thumbscrew. By a simple mechanical principle they automatically recede while threading to form the tapered thread, so that they cut less depth with each turn. Small lugs on the dial plate travel in slots in the side post. These slots are 16 degrees from the perpendicular, so that the segments recede from the centre, giving a pipe-taper of ¾ inches to the foot. The first tooth of each chaser takes a roughing cut, each succeeding tooth cuts a little deeper and the last tooth makes a finishing cut forming an especially clean and perfect thread. A ring locks the side post and rear nut together. Bushings are inserted in the rear nut for different sizes of pipe. A top-locking nut holds the different parts together. Two thumbscrews and one drop-handle screw fasten the rear nut to the pipe. Chips and grit cannot work in because the tension screw collar prevents and takes up all looseness and wear. Should the pipe split the dies can be released by shifting the lever. The Beaver die stock, in six sizes, ranging from ⅛-inch to 12-inch, are manufactured by the Borden-Canadian Company, 66 Richmond street, east, Toronto.

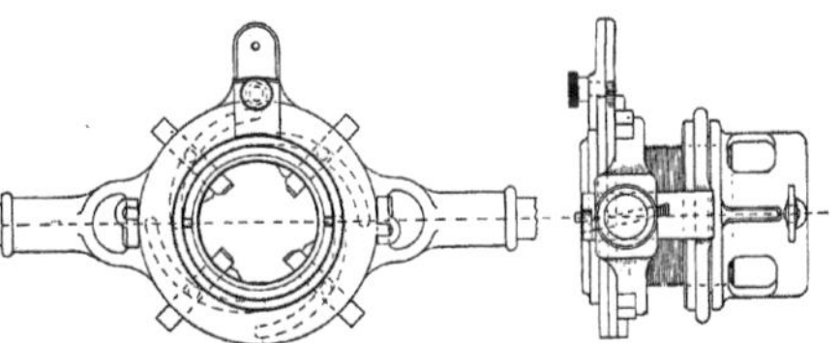

Die Stock Showing Cams and Adjustable Chasers.

THE DOMINION IRON & STEEL COMPANY'S OUTLOOK.

The annual meeting of the Dominion Iron & Steel Company was held in Montreal on Oct. 9. President Plummer, in his address, referring to earnings, prices and outlook, pointed out that great progress had been made in the surplus earnings. In 1905 it was $71,000, in 1906 $652,000, and it is expected that if figures keep growing through the year, the surplus earnings will show something like $2,000,000. Business is prosperous at present, and there is enough on the books and in sight to absorb all the output to the end of next season.

In regard to bounties, since the company began its operations in a finished plant in 1905, the costs have steadily decreased and have just about offset the decrease in the bounties. The president also referred to the property at Wabana, which will furnish an inexhaustible supply of good quality ore at a moderate cost.

DECIMAL EQUIVALENT AND DRILL SPEED INDICATOR.

A very handy device for machinists and foremen is here illustrated. Both sides are shown; on one side are the decimal equivalents of fractions from 1-64 inch to 1 inch inclusive, by 64ths; and the drill speed of carbon and high speed steel on the other, together with information pertaining to the feed of drills.

It is substantially made of three heavy discs of celluloid, pivotted at the centres so that they can be rotated with respect to each other. Fig. 1 gives the decimal equivalents of all the 64th parts of an inch. To determine the equivalent of any of these fractions, the smaller or upper disc is revolved until the notch in it exposes the decimal corresponding to the required fraction which is shown on the outer circumference of the centre disc.

On the reverse side, as shown by Fig. 2, the outer circle of figures on the centre disc gives all the standard sizes of drills from 1 inch to 3 inch, varying by 16ths. When the smaller disc is revolved until the notch is adjacent to

Fig. 1—Decimal Equivalents.

the given size of the drill, the proper drilling speed in revolutions per minute is at once indicated for both carbon and high speed drills. This speed is adapted in drilling wrought iron, machinery steel, or soft tool steel.

Recommendations as to the feed per revolution for the various sizes of drills are given on the smaller disc as well as information to determine by observation when the maximum rate of drilling is being accomplished. Its convenient size makes it easy to be carried in the pocket and act as a ready reference. As it cannot be injured by ordinary usage it should, therefore, last a considerable time and prove a valuable souvenir from this progressive concern. The Cleveland Twist Drill Co., Cleveland, Ohio, are distributing these gratuitously and will send them to machinists or anyone interested on application.

Fig. 2—Drill Speed and Feed Indicator.

ABOUT CATALOGUES

By mentioning Canadian Machinery to show that you are in the trade, a copy of any of these catalogues will be sent by the firm whose address is given.

PUBLICITY ENGINEERING.—A booklet from Walter B. Snow, 170 Summer Street, Boston, Mass., on productive publicity and publicity engineering.

FINISHING GEAR BLANKS.—Sheet from Gisholt Machine Co., Madison, Wis., describing minutely, with illustrations, the operations for finishing forged steel gear blanks.

ALGOR BOILER SYRUP.—Folder from the Liverpool Borax Co., Ltd., Liverpool, explaining the use, with tests, of Algor Syrup, for stationary, locomotive and marine boilers and heating plant equipments.

OIL ENGINES.—Catalogue showing the system of working and specifications of the Blackstone Oil Engines. The Canada Foundry Company, Limited, Toronto, are the sole agents in Canada for these engines.

RIGHT ANGLE TRANSMISSION.—A folder showing the construction of the Right Angle Transmission Coupling, manufactured by the T. R. Almond Mfg. Co., 83 and 85 Washington Street, Brooklyn, N.Y.

ARC LAMPS.—A folder from the Western Electric Co., Hawthorne, Ill., giving data for their multiple alternating and multiple series direct current enclosed arc lamps. The folder illustrates a few of their types of arc lamps.

BEARING ALLOYS.—A circular issued by the Lumen Bearing Co., Buffalo, N.Y., and Toronto. It deals very fully with the various bearing alloys manufactured by this firm, and is well illustrated throughout.

STEAM SPECIALTIES.—A catalogue from Darling Bros., Limited, 2 Toronto Street, Toronto, Ontario, illustrating and describing their feed-water heaters, low and high pressure regulating valves, steam traps, pumps, oil separators and exhaust heads.

SHAPER AND AUTOMATIC TWO-SPINDLE MILLER.—A folder from the Burkes Machinery Co., of Cleveland, Ohio, describing and illustrating their Malterner hand shaper, drill presses and milling machines, with automatic rise and fall to the table.

PNEUMATIC AND HYDRAULIC RIVETERS.—A folder of the Hanna Engineering Works, 820 Elston Ave., Chicago, Ill., illustrates their full line of various styles and sizes of riveters suitable for bridges, towers, boilers, and the many works where riveters may be used.

MECHANICAL DRAFT.—Sheldon's, Limited, Galt, Ontario, Catalogue No. 22 illustrate their various blowers and engines, mill exhausters, down draft equipments for blacksmith shops and electric fans. Tables of capacities are given and a treatise on mechanical draft.

ELECTRIC MOTORS.—The B. F. Sturtevant Company, Hyde Park, Mass., have issued two bulletins, No. 144, describing eight-pole electric motors, and No. 147, describing their bi-polar and four-pole types. They describe the closed and semi-enclosed motors, and give sizes and details of each class.

OIL ENGINE.—A folder describing the new "De La Vergne" vertical oil engine, run without carburetor, spark plugs or electric ignitor. It is manufactured for both stationary and marine trade. This folder may be had by application to the De La Vergne Machine Co., foot of East 138th Street, New York City.

ARC LAMPS.—Circular No. 1084, issued by the Canadian Westinghouse Company, Limited, Hamilton, Ontario, illustrates clearly the Westinghouse Series Alternating Arc Light System, and also a pamphlet dealing with the Westinghouse Protective Apparatus for alternating and direct current circuits.

TOOLS FOR CUTTING SCREW THREADS.—J. M. Carpenter Tap & Die Co., Pawtucket, R.I. A 160-page catalogue and price list No. 17 of tools for cutting screw threads, taps, dies, screw plates, die stocks and tap wrenches. Machine nut taps, screw taps, taper taps, hobs, dies, die sets, stocks, are all described with illustrations and prices.

BRIDGE AND ROOF PAINT.—Catalogues from the Canada Paint Company, Limited, of Montreal, Toronto, and Winnipeg, describing their various paints and varnishes. A booklet on bridge and roof paint describes with colors their well-known bridge, girder and roof paints. These are intensely strong combinations of the finest weather-resisting pigments known for painting steel and iron girders, bridges and metallic shingles. Samples of their bridge paint will be furnished free on application to architects, engineers and large corporations. Their diamond graphite paints and fillers are used for finishing iron and steel castings, engines, lathes, boilers, electrical apparatus, and all kinds of machinery.

Foundry Practice

MODERN IDEAS FOR FOUNDRYMEN

*Selling for the Foundry**

By Elliott A. Kebler.

While much has been written about the importance of keeping the costs of the product of the foundry, the absolute necessity of knowing the approximate cost of castings before bidding on the work has not been sufficiently dwelt upon. It is the custom of founders to glance at a pile of blue prints, and if they appear similar to work which has been done before, the price is made the same as was obtained for the other work, and too often, when it reaches the foundry, unexpected difficulties arise which add greatly to the cost.

While it is often difficult to have a careful estimate made in advance by reason of lack of time or insufficient data, if a foundryman absolutely refuses to quote until he has made an estimate, he will find in nearly every instance his customer will allow him sufficient time to make his estimate, and his profits at the end of the year will be greatly increased.

Get Correct Estimates.

To sell satisfactorily a foundryman must know that his estimates are correct, for this will give him a confidence in the price at which he is offering his castings that cannot be obtained in any other way, in fact, an estimate which is not afterward verified by the actual cost is of but little value. As an instance of this, a foundry which based its quotations on an estimate sheet, compiled by its superintendent in consultation with his foreman, adopted one on which the detailed actual cost was also entered after the work was completed, so that at a glance the correctness of each item in the estimate could be seen. When this new form was first used the estimates were found to have been very unreliable, but after the differences between the costs and estimates had been pointed out, the superintendent and his assistants soon learned to make correct estimates.

A serious mistake is made by some foundrymen who base their quotations on the prices named by others. A mill whose castings, while apparently similar to that of other plants, but were really much more expensive to make, secured its repairs for the first year at less than cost, and continued doing so by placing its orders for succeeding years with different foundries at approximately the same price by showing the bidders the figures which they had paid in preceding years. In fact unless a foundryman knows by actual experience, or can figure the costs of the castings required, it is very dangerous to take an indefinite order unless a large margin of profit is allowed. These yearly repair contracts are also unsatisfactory, for if prices of castings advance, the mill may stock up repair parts, while on a falling market, the repairs are often inconsiderable. As the profits vary on different classes of castings, a foundry salesman should keep constantly before him the desirability of securing orders which will yield the most profit, as his value to the company largely depends on this. One foundry makes a careful record of this so that a salesman who, on a smaller tonnage, obtains more returns is appreciated more than a man who fills up the shop with orders which show but little profit.

Securing Orders.

Too often foundrymen are not sufficiently alert to secure orders for work which they had not previously made, or for which castings might be used. A foundry salesman overhearing the president of a company telling his purchasing agent that some steel work which they were using was oxidizing rapidly, suggested the substitution of cast iron, and thus secured large orders from this plant and others for a class of work which heretofore had never been made in a foundry.

In fact, a decided change has taken place in the art of selling material, and to be a good foundry salesman a man must not only be a good talker, but must be thoroughly conversant with the class of castings he is selling, and often must know how they should be used to get the best results. As the buyer often relies on the representations of the seller, it is absolutely essential that the latter be honest in his statements if he wishes to hold his trade, in fact, a first order often hinges on this very consideration. Recently a buyer was in the market for some material, and clearly showed the bidders that he wished the material to have a certain peculiar characteristic. He placed his sample order not with those who flatly stated that their material would be as specified, for the buyer knew that the requirements were not fully understood, but with the salesman who frankly told him that he could not promise to fill the bill, but would sell him, subject to a sample which could be tested.

The successful salesman should also know what can be used successfully under varying conditions. He must also keep in touch with the requirements of his customers, for by so doing he may be able to convince them that it is to their advantage to purchase material which under ordinary conditions would be made in their own works.

* Read at the September meeting of the Pittsburg Foundrymen's Association.

MOLDER'S TOOL SUPPLIES.

By an Old Molder.

It is necessary to furnish molders with the best in the way of tools and other supplies. A molder loses much time hunting for things which should be within easy reach.

The above sentence has a variety of meanings or interpretations. The one which I wish to call attention to is the practice in different foundries of first putting new molders to work, as well as the way they have to work afterwards. What I should term well-regulated foundry practice would be to start in a new molder and show him his floor and pattern, or patterns, with some instructions as regards facing, gating, risers, etc. Then furnish him with shovel, rammer, strike-off, riddles, bellows, water sponge-pot, parting-sand pot or box, brushes, facing-bags, in fact everything necessary to do the work with, and hints thrown in in regard to the amount of work expected after a few days' experience. I forgot to mention gaggers, nails and clamps, which, in most cases have been carried off the floor. It is certainly annoying, as well as poor foundry practice, to have a new man hunting for clamps and gaggers in any foundry, most particularly a new one.

In the sense in which I am endeavoring to write in regard to Foundry Practice, the furnishing of the foregoing articles to a new, as well as the old, molder, would be called good foundry practice, and will, I presume, be disputed by none. I am pleased to say it is the practice carried out by many foundries, but there are many where it is entirely neglected, to the detriment of the proprietors as well as annoyance and decreased output by the molder, whose time is frittered away looking for tools which should be at his hand. I have in view several large foundries which are well equipped with cupolas, electric cranes, good sand, facings and almost everything necessary to doing good work, but are sadly deficient in this one respect.

It must be the fault of someone—certainly not the proprietors, who have too much at stake. I don't see anyone to saddle it on unless it be the foreman or assistant, who have been chosen to fill the position often on account of their ability to successfully turn out large castings, and who have never done side floor work, and were accustomed to a helper handing them the tools wanted, regardless of how he got them. True, most molders, or at least quite a number, become accustomed to the way of working, and in candor I am compelled to say, like it. It's no fault of theirs and it's as easy to be looking 'round for something you can't find as it it to be ramming or shoveling sand.

I worked in a large steel works which has since become extinct, in my opinion due to this kind of foundry practice as much as other causes and probably more. I think they had over a half-million dollars invested in buildings, furnaces, electric cranes, etc.; worked a large number of molders and ran day and night gangs. When you went in in the morning the assistant foreman gave you a pattern. You had to hunt up a flask, temper up the sand (if there was any to temper). Then began a hunt for water-buckets, of which there were only a few, and they were to be found in all kinds of improbable places, and so you went hunting for everything the whole day long. You picked up a rammer or shovel or riddle and laid it down. The next time you wanted it someone was using it. So it was with brushes, bellows. Everything belonged to everybody. While the foreman stood looking on or rushing around seeing nothing and caring nothing for the waste of time, which is great where a molder is working on side floor on smaller work.

Some folks might think this a fancy sketch of foundry practice, especially some of the proprietors, who never look into such small matters. I called a foreman's attention to it once and he told me he had to keep down foundry expenses to hold his position.

F. B. Polson, president of Polson Iron Works, Toronto, died suddenly on the afternoon of October 28th.

OCTOBER METAL MARKETS.

Pig iron throughout the past month and, in fact, throughout the past two or three months, has maintained a splendid activity. Not even stringent financial conditions have been sufficiently serious to affect the demand for it. Canadian consumers have come into the field and have bought very liberally considering the existing circumstances. The steadiness of the British market since the middle of the year is a matter of surprise, especially when it became plainly apparent that the end of demand from the United States was imminent. Quotations on Scotch and English iron have been very firm, owing solely to the strong demand. Statistics published by a prominent Glasgow firm show that on September 19 of this year only 166,951 tons of Scotch iron were available, as against 582,272 tons on September 20, 1906.

Antimony has gradually become a little firmer. Very little business has been transacted. The market does not yet perfectly satisfy the consumers and, although American prices have slightly advanced, dealers show a willingness to make concessions to stimulate heavy buying.

Copper has been affected more seriously by recent developments in the money market than any other metal, perhaps, as it is more subjected to speculative fluctuations. Throughout the past month quotations have been very weak and transactions have been limited to small volume. The majority of the American producers have adopted the curtailment of production policy with the result that the output has been reduced on the basis of 25,000,000 pounds per month. Although prices have been actually cut in two in the past six months, dealers console themselves by thinking that the prices cannot very easily be further reduced.

Ingot tin has been subject to wide fluctuations in the past five weeks, the general tendency being downward. In the first week of October there was a decline of £8, in the second week £10, and in the third week £10. Local prices have been very weak and a limited volume of business has been transacted.

FOUNDRY ECONOMY.

We are told of a recent case where a bright young engineer approached the manager of a certain foundry and offered to secure work for them on a percentage basis, at the same time to look after their buying and such other deals where he could assist them. The contract was closed and the engineer promptly got busy and filled the foundry with orders. He also installed a system of controlling the mixtures by analysis and insisted that every shipment of pig iron and coke should be analyzed. By taking advantage of the pig iron market he succeeded in saving the company a large sum of money on material.

The old management, however, did not see this, but looked mainly at the figures representing 5 per cent. of the gross selling price of the castings, which would go to the engineer each month. For several months they simply complained and then began to look around for some method to get rid of this expensive engineer. The contract was finally canceled and the foundry left to carry on its work as it saw fit. The plant was full of work and everything went smoothly for a short time, but among other things they just cut out that costly chemical analysis, then they forgot that they lacked the services of the engineer who used to interpret the analysis and apply the results for them.

Some of the castings were expected to fill difficult specifications and it was not long before complaints began pouring in at a goodly rate. Then work was withdrawn and withdrawn so that it could never be regained. In place of operating on high-class work in which there was a good margin, the foundry had to drop back into its old rut of standard, cheap jobbing castings. In the meanwhile the engineer found other foundry connections and started to build up a good business once more.

All of which goes to show that you get out of a foundry something akin to what is put into it.—Castings.

CUPOLA PRACTICE.

By T. U. Burman.

Having erected and had the care of thirteen different cupolas for melting soft grey iron in different parts of the country, I have found that the cupola furnace is about the same as a pot with air forced into it, only some require a high set of tuyeres and some a low set, some require a low pressure of blast and some high. I have learned that all other parts of the practice being equal, I can get more good results out of the coke and iron of the kind made these days, though both coke and iron have more impurities in them than they did 25 years ago. All foundrymen are having some trouble with the coke and iron, so if I can tell any brother foundryman anything to aid him, I want to do so. In our plant we make light castings to stand a water pressure of 80 pounds, and some to stand 180 pounds. We use a southern coke, fairly good hard coke, but all the cars are not the same, some come in with coke softer and lighter with more sulphur, and some come in a bright heavy coke with less sulphur, all from the same company and region. Having been among the coke ovens and men who work at them for two years, I know how the unevenness in coke comes. The good or best coke has been burned more hours. I had a car come into the yard near Pittsburg, and I said to a laborer I had, who one time worked coke ovens: "That looks good, it must be 72-hour coke." "No, sir," he said, "I drew many an oven at 36 hours burning." I find that the heavy and bright with very little black coal on the large pieces proves the burning of it. Now I hope no good brother will let happen what did to me. We had plenty of coke, and our iron was full of scruff and sulphur. My melter had a heavy heat of 26 tons in a 42-inch cupola all hot iron. Our man, Jack, the coke man, got hold of all large pieces, and, as it happened, the big pieces were very much topped with coal not burned well. It seemed almost impossible that heat to keep hot iron and the loss ran away up. We discovered this on the next heat, and had fine iron and have had ever since. Our coke man does not get all the coke from the same pile now. We have him get his loads from three piles, so he cannot get all black-heads. We use a flux and the flux brings out with the slag a yellow, bright, spongy mass, almost all sulphur. Before we used the flux, we did not get that, and sulphur went with our iron into our radiator and they were spongy. I find the larger coke in the first seven charges in the cupola will gain speed, and the small coke in the last nineteen charges would melt as fast and at the very last faster. The large coke about the tuyeres and slag hole is a great help to get slag to run free, and easy to manage. All these things have been gained by years of hard work, and now anyone can get it by reading. I want to answer a question I saw in the Foundry some weeks ago. Someone wanted to know why a cupola boiled slag over into the tuyeres. When a cupola is properly arranged with tuyeres and slag hole, and the slag allowed to flow out it will not go into the tuyeres. If one day or many days the cupola is not run dry and the bottom is dropped with some pig and coke in it, as it should be, and on some other heat the blast is kept on a long time after the iron is all melted out, the slag and sulphur on the walls of the cupola will melt and run down like molasses and go into all the places it can get in. So be careful and not melt out your brick work as it costs money.

COST OF A SMALL STEEL FURNACE.

The increasing inroads which cast-steel is making in the business of the manufacture of light gray iron castings have caused some founders to install a small steel furnace running 5 or 6 tons per day, and they have found, in the main, that it was a profitable investment.

A 5-ton furnace, basic lined, erected in an active iron foundry, would cost about $6,000, and allowing $4,000 for special bottom, pouring ladles, oil storage tanks, burners, etc., the total would be $10,000, exclusive of buildings. It would be capable of producing three heats daily and would yield 12¾ tons of castings at the rate of 15 per cent. in melting loss, gates heads, skulls and defective castings, an average yield in a well ordered plant. Producing such tonnage steadily should show good profits. The single furnace, however, has the disadvantage of going out of commission for repairs. It should turn out 400 heats before a general overhauling, or say a run of 133 working days. Assuming that the furnace was only productive nine months of a year at 25 working days per month, there would then be 225 days yielding a total output of 2,868¾ tons of product. Allowing an average profit of one-half cent per pound, there would be a gross profit of $28,687.50 annually.

But losses will arise in spite of best intention and not to make the venture look too roseate the setbacks will be considered which might be caused by breakdowns, strikes, or errors on practice. These conditions might reduce the good castings to one-half of the metal charged, so that 7½ tons were the output. With the same productive period of 225 days per year, and with a decided shrinkage, because of losses, between manufacturing costs and selling prices so that the gross profit was only one-eighth cent per pound, the tonnage per year would be 1,687½ tons, and the returns therefrom $4,218.75. Out of this, 25 per cent. can be deducted for interest on the investment, depreciation, etc., there would remain a net profit of $3,164.07, which is equivalent to 31.6 per cent. on the original $10,000.—Popular Mechanics.

BOOK REVIEWS.

A COURSE OF STRUCTURAL DRAFTING.—Compiled and arranged by W. D. Browning, M.E., Collingwood, Ohio, describing material used in structural design and drafting room practice, relating to it, with fifteen plates for students; price $1 net. Published by the Industrial Magazine, Collingwood, Ohio.

This is the first treatise of its kind and it is fully illustrated throughout to make all points clear. Tables are inserted giving sizes of standard I-beams, angles, channels, etc., rivets are taken up in detail and signs relating to same; also the design of bolts, eyebars and pins, turnbuckles, anchors; the strength of different materials, beams, girders, columns, trusses, drawing-room practice and plate lettering.

As it is a good thing for every mechanic to be able to understand drawings, and sometimes to make a drawing, this work is of considerable value to one wishing to make a study of drafting.

5,000 FACTS ABOUT CANADA.—A 70-page booklet, compiled by Frank Yeigh, Toronto, and published by the Canadian Facts Publishing Co., of that city. It has chapters dealing with every subject of national interest and industrial and commercial progress. Some of the following chapters will be of special interest to our readers:—Mining, Milling, Manufacturies, Railways, Trade and Tariff, Water Powers. Official figures are given where interesting, thus making it a valuable book to industrial men.

OPEN-HEARTH STEEL CASTINGS.—An exposition of the methods involved in the manufacture of open-hearth steel castings by the basic and acid processes, by W. M. Carr; 118 pages, 5x7½ inches; illustrated, cloth; published by the Penton Publishing Co., Cleveland, Ohio.

This book is compiled from a series of articles by the author appearing in the Foundry and the Iron Trade Review. The author being an authority on this subject, the information is very reliable.

HOW TO USE WATER POWER.—By Herbert Chatley, B.Sc.; fully illustrated; 87 pages, Technical Pub. Co., Manchester, Eng., and by D. Van Nostrand Co., New York; price, 2s 6d.

This is a short and clear account of the methods and principles of hydraulic engineering, produced in a form that can be easily grasped by those with limited knowledge of mechanics and hydraulics. The illustrations are clear and to the point.

THE 20TH CENTURY TOOLSMITH AND STEELWORKER.—A complete treatment of the manufacture of steel into tools, written in the interests of blacksmiths, toolmakers, etc.; by H. Holford; 240 pages, 5x7½ inches; fully illustrated; cloth; published by Frederick J. Drake & Co., Chicago.

This book is written by a man who has had practical experience in many shops, and is written in a simple style. The illustrations are simple and easily understood.

Industrial Progress

CANADIAN MACHINERY will be pleased to receive, as confidential, industrial news of any description, the incorporation of companies, establishment or enlargement of plants, railway, mining or municipal news.

New Plants and Additions.

A heavy flow of natural gas has been struck near Calgary, Alta.

The Canada Boiler & Radiator Co., Hastings, Ont., has assigned.

McDonald, Man., is to have a new court house, costing $40,000.

The Livingstone flax mill at Brussels, Ont., was destroyed by fire.

The Chatham Gas Co., Chatham, Ont., have completed their plant.

Windsor Hotel, Winnipeg, is to be enlarged, at a cost of about $25,000.

The J. Y. Griffin Co. are planning to erect a $100,000 warehouse and office.

A new fire hall, costing $25,000, is being erected at St. Boniface, Man.

The Brandon Brewing Co.'s new warehouse, at Brandon, Man., will cost $25,000.

The Standard Valve & Fitting Co., Guelph, Ont., will soon enlarge their works.

The Brunswick-Balke-Collender Co. will erect a factory in Toronto, to cost $30,000.

A Detroit firm, manufacturing malleable iron, wishes to locate in Parry Sound, Ont.

An electric plant will be installed at Torbrook Mines, N.S., at a cost of $40,000.

The planing mill of Wm. Drader, Chatham, Ont., was destroyed by fire; loss, $50,000.

A large sawmill, with a daily capacity of 45,000 feet, will be erected at Nelson, B.C.

The Lake Huron Lumber Co., Ottawa, have assigned, with liabilities amounting to $45,000.

The Delaware Seamless Tube Co., Auburn, Pa., will erect a plant at Sarnia, Ont., to cost $200,000.

The bridges and locks on the Welland canal will be operated by electric motors next season.

The Canadian Portland Cement Co., Welland, Ont., are installing four rotary kilns, 150 feet long.

Plans have been prepared for a $6,000 machine shop for Matthew McNaughton, Calgary, Alta.

Valuable deposits of radium have been found in the Simplon tunnel, between Switzerland and Italy.

The Lloyd, Thompson Wire Co., Toronto, have gone into liquidation, with liabilities of about $33,000.

The newly organized Dominion Tool Company, Peterboro, have purchased a site for their new factory.

David Spencer, Limited, Vancouver, is erecting an eight-storey warehouse, at a cost of $120,000.

The Prince Albert Lumber Co., Prince Albert, Sask., suffered loss by fire to the extent of $10,0000.

A system of marine telephones has been installed between seven stations on the St. Lawrence.

The works of the Standard Drain and Pipe Co., New Glasgow, N.S., will be rebuilt immediately.

Geo. White & Sons, London, Ont., are considering the removal of their plant to St. Catharines.

An American syndicate has proposed to establish a large cement works on the St. John river, N.B.

The affairs of the Truro Foundry & Machine Co., Truro, N.S., will be wound up and the plant sold.

The British Columbia Bedding & Upholstering Co., Vancouver, B.C., suffered heavy damage by fire recently.

The capital stock of the McLaughlin Carriage Co., Oshawa, has been increased from $400,000 to $1,500,000.

The new factory being erected at Hull, Que., by J. W. Woods is nearing completion. It will cost $100,000.

Zuelsdorf Bros., Berlin, Ont., will erect a factory in that city for the manufacture of high-grade furniture.

C.N.R. have awarded the contract for a new roundhouse at Brandon, Man., to the Sharp Construction Co.

J. E. Doak, Saskatoon, Alta., will erect a factory for the manufacture of all kinds of house furnishings.

The Orangeville Woolen Mills, Orangeville, Ontario, were destroyed by fire recently, entailing a loss of $12,000.

A new invention makes it possible to reverse steam turbines. This will be appreciated in shipbuilding circles.

The Peterboro Lock Co. will, it is expected, erect a factory addition to their present premises, Peterboro, Ont.

J. E. Doak, Doaktown, N.B., will erect a factory in Saskatoon, Sask., for the manufacture of house furnishings.

A company manufacturing furniture will locate in New Hamburg, Ont. A loan has been granted by the town.

A firm manufacturing engines is considering the erection of a plant at Leamington, Ont., to cost about $38,000.

A firm manufacturing engines wishes to locate in London, Ont. They want $25,000 or $30,000 as an inducement.

A company formed chiefly of Chinamen, and called the Sydney Brick & Tile Co., will erect works at Sydney, B.C.

The new factory of the King Radiator Co., Toronto, is nearing completion. The whole plant will cost $250,000.

Clark Brothers have opened up a machine shop at 164 Duke street, for the manufacture of their friction clutches.

The new post office building at Edmonton, Alta., will be fireproof throughout. It will be in use early in the spring.

Plans have been completed for a corrugated iron warehouse for the Paris Plow Co., Paris, Ont., to cost about $9,000.

Canadian Condensed Milk Company, Limited, Hamilton, have a large plant in course of erection at Aylmer, Ontario.

It has been announced that MacKenzie & Mann will double the capacity of their blast furnaces at Port Arthur, Ont.

George Brown, of San Francisco, Cal., will at once commence the erection of a large lumber mill at Prince Rupert, B.C.

The grist and planing mills of T. A. Vaughan, St. Martin's, N.B., were destroyed by fire recently, entailing a loss of $5,000.

The International Harvester Co., of Hamilton, are contemplating the erection of a distributing warehouse at Port Arthur, Ont.

P. Venables, R. G. Venables and another local party are preparing to build a large cold storage plant in New Westminster, B.C.

Shipments of machinery have been arriving lately for the Sarnia Brass Works. The company will manufacture plumbers' supplies.

An American company, manufacturing farm implements, garden swings, ladders, etc., is looking for a site for a Canadian factory.

The Fawcett Manufacturing Company's foundry, at Sackville, N.B., was damaged to the extent of $3,000, by fire. There is no insurance.

F. W. Birks & Son, makers of Paroid roofing and waterproofs, are looking for available sites for manufacturing goods for the Canadian market.

Walker & Sons, Walkerville, Ont., will erect an addition to their planing mill. Additions will be made to the boiler and engine equipment.

A ferro-nickel smelter is now under construction at Sault Ste. Marie. White metal for cutlery will be manufactured by the ferro-nickel process.

The Interwest Peat Fuel Company's plant, Lac du Bonnet, Man., was destroyed by fire last week. Loss, $40,000, partly covered by insurance.

Defective electric wiring is supposed to have been the cause of a fire that destroyed the building and plant of Parker's laundry works, Peterboro.

The power plant of the Maine & New Brunswick Electrical Power Company was opened recently. The plant is situated at Aroostook, Falls, N.B.

The Grand Bay lumber mill, St. John, N.B., was destroyed by fire on Monday. The loss of $15,000 was covered by insurance to the amount of $11,000.

The Automatic Telephone Co. is considering the erection of a factory at Guelph, Brantford or Welland. The factory at Toronto is not large enough.

The Sherwin-Williams Paint Co., Montreal, took out a permit for the erection of a new three-storey warehouse on St. Patrick street, to cost $20,000.

Messrs. Shurly & Dietrich, Galt, Ont., will erect a sheet metal building 163 feet long, equipped with every modern device for the tempering of saws.

The new plant of the Berlin Manufacturing Company, of Hamilton, is nearing completion. They will manufacture a full line of wood-working machinery.

The Standard Contracting Co., Toronto, have bought a building in that city and will install a plant to cost $30,000 for the manufacture of machinery and tools.

J. J. Murray & Co., Cayuga, Ont., will erect a gas pumping station and reservoir for the Dominion Gas Co., near Canfield, Ont. The contract price is $35,000.

The Tacoma Construction Co. are asking New Westminster, B.C., for a free site on Lytton Square, in order to erect a hotel and business block, at a cost of $200,000.

The Industrial Development Company, of Canada, manufacturers of tar, turpentine, alcohol, etc., have been offered ten years' exemption from taxation by Hull, Que., council.

The Evans Rotary Engine Co. Toronto, has the contract to install one of their 40 horse-power rotary engines for the lighting and heating of the Rossin House, Toronto.

M. L. Aubert, Montreal, a member of a firm of French capitalists, is negotiating with Hull, Que., for the erection of a large biscuit and confectionery factory in that city.

The Colonial Engineering Co., Montreal, have the contract to install a 75 horse-power Hornsby-Stockport gas engine in the plant of the Queen City Printing Ink Co., Toronto.

The Windsor Foundry, Windsor, N.S., which was put up for auction recently, was purchased by Geo. Mounce, of Avondale, for $18,000. The foundry will commence work immediately.

Mr. Chapman, head of the Ontario Wind, Engine & Pump Co., Toronto, and Mr. Fleury, of J. Fleury & Sons, Aurora, Ont., have taken over the Bell Foundry, St. George, Ont.

The American Electrical Furnace Co. will erect a Canadian factory at Niagara Falls, Ont. They will manufacture iron and steel castings, the melting being done by electricity.

The Porto Rico Co., Nelson, B.C., will erect a lumber mill, with a daily capacity of 45,000 feet. The limits possessed by the company are conservatively estimated at one hundred million feet.

The Southern Cross Mining & Smelting Co., Victoria, B.C., propose to have erected before next spring a smelter to cost at least $500,000. H. Cecil, of that city, has the matter in hand.

Mayor Lyle, of Smith's Falls, has replaced his saw-mill, which was destroyed by fire a few months ago, with a large, new fire-proof mill. New, up-to-date machinery will be installed.

The Albert Sawmills, Barachois, Que., were recently totally destroyed by fire, together with the wharf and all the buildings connected with the mills. Five million feet of lumber also disappeared.

An English company is said to be behind the locomotive works to be erected at Lachine, Que. The works will cost $3,000,000. The company wants a bonus of $50,000 and exemption from taxation.

A. C. Flumerfelt and H. N. Galer have acquired a large area of rich coal deposits near Lethbridge, Alta. A company has been formed to develop the deposits and a large plant will be installed at once.

The Fleming Aerial Ladder Co. have submitted a proposition to the town of Barrie, Ont., for the erection of a large plant there to cost $26,000. They want a loan of $20,000, and a fixed assessment of $5,000.

I. S. G. VanWart, president Albert Biscuit Co., of Calgary, and the manager, E. B. Johnson, are considering the establishment of a branch factory in Edmonton, Alta., at an estimated cost of $150,000.

A company with a capital of $50,000 proposes to erect a factory for the manufacture of cement bricks at New Westminster, B.C., providing suitable deposits of sand and clay can be found in the vicinity.

The Canadian Marble & Granite Works have recently been opened in Edmonton, Alta. The machinery for cutting and polishing the marble has been installed. Stone of the best quality is quarried near Nelson, B.C.

The Berlin Machine Works, of Beloit, Wis., has installed a Newton patent cupola, having a capacity of sixteen tons per hour. A cupola of the same make will be installed in their Canadian plant at Hamilton.

The Fairchild Co., Winnipeg, will sell all interests to the John Deere Plow Co., Moline, Ill., under a Dominion charter, with a capitalization of $1,000,000. It is to be known as the John Deere Plow Co. of Canada.

Large shipments of machinery have recently been arriving for the plant of the Central Foundry Co., Port Hope, Ont. Over $100,000 has been spent already and the plant will be ready for work in a short time.

Architect Chas. Mills, of Hamilton, Ont., has awarded the contract for the erection of a large addition to the Eagle Spinning Co.'s works there, to the Provincial Construction Co., of Toronto, at a cost of $50,000.

The strike at the Springhill, N.S., Coal Mines has now been on for two months, and the Government of Nova Scotia will on Wednesday take charge of the work heretofore done by members of the Mechanics' Lodge of the Collieries.

The Thames River Oil and Gas Company wants a franchise from the town of Aylmer, Ont., for power to enable them to pipe the town for purposes of supplying natural gas for power, heating and manufacturing purposes.

The new exchange building for the automatic telephone system now being installed in Edmonton is almost completed, the machinery being installed. The Mortimer automatic system is being used. It will be ready by November.

B. Ledoux & Co., Osborne street, Montreal, have been granted a building permit for the erection of an addition to their carriage factory. The structure will be [illegible] storeys with a frontage of 75 feet, and about 90 feet deep.

The Industrial Financial Co., Toronto, propose establishing a piling and engineering works at Kincardine, Ont. The company wants a loan of $50,000 and will manufacture the Hunter Patent Interlocking Pile, having the rights for Canada.

Traces of graphite in Haliburton county, Ontario, are being investigated by Eugene McSweeney, president of the United States Graphite Co., Saginaw, Mich. It is said the deposits are rich along the line of the I. B. & O. Railway.

The Shortells, Limited, Wood Alcohol and Charcoal Company are considering the erection of a plant at Parry Sound, Ont. The plant will cost $50,000, and 500 hands will be employed at first. They want a loan of $30,000, which the town has endorsed.

Three industries are looking for a location in Berlin. They will manufacture collapsible go-carts, concrete machinery and washing machines. The outlay will be $100,000, and 300 hands would be employed. Certain conditions will have to be fulfilled.

The organization of a company for the manufacture of railway cars in the Canadian west is being effected in Toronto. Winnipeg is being considered as a location for the factory. The company will have a capital of $2,000,000, and will employ 1,000 hands.

The Georgian Bay Power Co. are developing power at Eugenia Falls, near Flesherton, Ont. They propose building a large dam about 40 rods above the falls and piping the river through a large hill to the Beaver valley. About 3,000 h.p. will be developed at first.

For the first eight months of this year the total immigration to Canada was 216,865, an increase of 50,058 as compared with the first eight months of 1906. The total for the eight months is more than the total immigration during the whole six years, 1896 to 1902.

Canadian Rand Co., Limited, of Montreal, have opened up a show-room at 11 St. Nicholas street, where they display a complete line of air-compressors, rock drills and Imperial pneumatic tools. A stock of repair parts will also be carried for the convenience of local customers.

The Dominion Heat & Vent Co., of Hespeler, have closed contracts for the heating and ventilating of the Parkin elevator works, of Hespeler. They have also supplied a Toronto firm with one 160-inch double discharge steam fan, and 8,000-foot heater, and also one 120-inch fan and 4,800-foot heater.

The Canadian Rubber Company, of Montreal, has been appointed sole agents in Canada for "Rainbow" red sheet packing and "Eclipse" gaskets, manufactured by the Peerless Rubber Manufacturing Co., New York. Orders for these goods may be sent direct to any of the company's sales branches.

The International Heating and Lighting Co., Edmonton, Alta., will erect a plant for the manufacture of producer gas from straw. The plant will cost $100,000, and mains will be laid to pipe the gas to all parts of the city. Over thirty miles of these mains will be laid, entailing an expenditure of $200,000.

The Canadian Steel Specialty Co., of Gravenhurst, Ont., will erect a two-storey factory in that town, together with a boiler house and dry kiln, at a cost of $10,000. The company will manufacture steel furniture, electric fixtures, and novelties. The town has granted a free site and exemption from taxation.

The Edmonton Produce Co., Edmonton, Alta., proposes to erect a cold storage building and install therein a refrigerator plant, the whole to cost not less than $50,000, if the by-law voted on by the taxpayers of that place passes. If arrangements are satisfactory, building operations will be commenced not later than January, 1908.

The Hanover Portland Cement Company is enlarging its plant, the following contracts having been placed: Canada Foundry Company, for two 250 horse-power boilers, condensers and pumps; Kilmer & Pullen, Toronto, for one 400 kilowatt generator; the Robb Engineering Co., Amherst, N.S., for one 500 horse-power engine.

The new plant of the Standard Chain Company, of Canada, now being erected at Sarnia, is progressing very rapidly. The dock has been entirely completed and the main building will shortly be under roof. The blast system is being furnished by Sheldons, Limited, of Galt, Ont. It is anticipated that the plant will be in operation the latter part of October.

The new machinery at the Canadian Locomotive Works, Kingston, Ont., is nearly all installed and the buildings are fast nearing completion. The new power house is practically completed and ready to have the steam turned on. The new switch boards being installed by the Canadian Westinghouse Company, is in place, and is one of the best ever installed in that city.

The Dominion Bridge Company, of Lachine, of which James Ross is president, will increase its capital from $1,000,000 to $1,500,000. This course has been found necessary by the great expansion of business and the large additions to the company's several plants. New shops are being built in Toronto and Winnipeg, and very extensive additions to the plant at Lachine are likewise being undertaken.

It is announced from Owen Sound that the Northern Navigation Company are inviting tenders for a new passenger and freight vessel 259 feet in length, to take the place of the wrecked Monarch. Plans and specifications have been prepared by Hugh Calderwood, marine architect. It is intended that the steamer will be ready to go into commission on the opening of navigation next spring.

London Fence, Ltd., Portage la Prairie, have decided to rebuild their factory, which was destroyed by fire early in the year. For some time there was uncertainty as to the future plans of the company, owing to the dispute with the insurance companies as to the amount of the fire losses. However, the fire losses have been adjusted, and at a meeting of the shareholders, held last week, it was unanimously decided to rebuild.

The Dominion Wrought Iron Wheel Works, Orillia, is enlarging its factory premises by the addition of a 40x50 brick building with metal roof to be used as a foundry, and a 40x20 metal-clad storehouse. The new buildings are west of the present factory, and close to the railway switch, which will be convenient for handling supplies as well as shipping the finished products. The Wheel Works is one of Orillia's rapidly expanding industries.

It has been discovered that the new C.P.R. Assiniboia, which has been cut in two at the graving dock at Levis, Que., for the purpose of taking her through the canals to do service on the lakes, would not float. Her ballast will, therefore, have to be shifted, and this operation was begun this morning. As soon as this vessel leaves port her sister ship, the Keewatin, will undergo the same operation of being severed in two, and also leave to do service on the lakes.

The Peel Oil and Gas Co. has let contracts for the sinking of oil wells on the property of the company at Cooksville, 16 miles west of Toronto. This organization has over 1,000 acres of land under option, and it is expected that deposits of oil or natural gas will be encountered. Some years ago natural gas was obtained in a shaft sunk a few miles west of Mimico, but, owing to a lack of facilities for handling the product, the well was plugged. The above company expects that if the venture proves successful it will be able to supply natural gas at a cheap rate to Toronto.

The Smart-Turner Machine Company, Limited, 191 Barton street, east, Hamilton, Ont., report a brisk demand for their steam pumps. Among the firms who have been supplied are: E. Leonard & Sons, London: Johnston Bros., Jarvis; Canada Paint Co., Toronto; Fred. Armstrong, Toronto; McColl Bros. & Co., Toronto; Hiram Walker & Sons, Walkerville; the John Whitfield Co., Toronto; the Biggs Fruit and Produce Co., Limited, Burlington; J. W. Holmes, Selkirk; H. B. Kinster & Co., Woodslee, and E. Van Allan & Co., Hamilton. These pumps include the Smart-Turner duplex, automatic feed and side suction centrifugal pumps, with accessories.

The National Light & Manufacturing Co. of London, Ont., have applied for incorporation, with a capital of $50,000. The company have acquired the patent rights for Canada of the Cody inverted gas lights, and inverted arc lights, and the Keller gas governing devices, and have also become owners of the Cody oil gas patents, including the only known method of generating a non-carbonizing gas from common coal oil. The latter, on account of its absolute safety, freedom from odor, and lower cost, is calculated to entirely replace gasoline and wick-using lamps for both lighting and cooking, and the firm will handle a full line of stoves and arc lights constructed on this principle. The premises, leased for six years, cover some 10,000 feet of floor space, and a staff of about 25 men will be employed as soon as the work of equipment is completed. The firm is already shipping out gas lights and will be accepting orders for the oil-gas lights and stoves in a few days.

Companies Incorporated.

The Joliette Light, Heat and Power Company, Limited, has been incorporated.

Monarch Electric Co., Montreal: capital, $20,000; to manufacture electrical machinery. Incorporators, J. R. Lewis, J. W. Schlieffers and H. S. Poole, all of Montreal.

The Nanton Coal Fields, Limited, Ottawa; capital, $100,000. To prospect, reduce ores, etc. Provisional directors: A. B. West, A. N. McLean and J. A. Rawley, of Ottawa.

Shurly & Derrett, Toronto: capital, $75,000; to manufacture and deal in twine, hemp, flax, etc. Provisional directors, C. J. Shurly, T. F. Shurly, Galt, Ont.; G. D. McAllister, Toronto.

Cobalt Silver Fountain Mines, Toronto; capital, $500,000; to carry on mining, milling and reducing ores. Provisional directors: P. S. Hairston, J. H. Stephens and R. McKay, Toronto.

The Chatham Carriage Co., Chatham, Ont.; capital, $100,000; to manufacture all kinds of vehicles. Provisional directors, Ira Teeter, Arthur Cooke, Frank E. Fisher, all of Chatham, Ont.

Wilbur Iron Ore Company, Toronto; capital, $500,000; to prospect for and deal in ores, metals and minerals. Incorporators, C. L. Dunbar, E. A. Dunbar, and H. C. Scholfield, all of Guelph, Ont.

Ahern Safe Co., Montreal; capital, $75,000; to manufacture and deal in locks, keys, safes, vaults and hardware pertaining to same. Incorporators, R. M. Ahern, A. Ahern and A. D. Ahern, all of Montreal.

The Lovering Lumber Co., Toronto, have been incorporated, with a capital of $100,000, to manufacture lumber, timber, etc. The provisional directors include, J. F. Hollis, T. H. Wilson and W. J. Lovering, Toronto.

The P. Hymmen Company, Berlin, Ont.; capital, $60,000; to take over the hardware and plumbing business of Peter Hymmen, Berlin. Provisional directors, P. Hymmen, H. Hymmen and H. S. Hymmen, all of Berlin.

The Producers' Natural Gas Co., Hamilton, Ont., have been incorporated with a capital of $100,000, to manufacture oil, gas, ores, etc. The provisional directors include, W. Southam, J. Milne and F. A. Magee, Hamilton.

Wm. G. Hartranft Cement Co., Montreal; capital, $25,000; to manufacture and deal in Portland cement, gypsum, etc. Incorporators, G. W. MacDougall, L. Macfarlane, C. A. Pope, and A. Swindlehurst, all of Montreal.

The Right-Process Co., Toronto, have been incorporated with a capacity of $40,000, to manufacture washing compounds, etc. The provisional directors include A. H. Brother, W. McLean and W. C. Ormsby, Barrie, Ont.

The Canadian Thermos Bottle Co., Montreal, have been incorporated, with a capital of $300,000, to manufacture bottles, glass, metals, etc. The charter members include, L. Macfarlane, J. B. Schwabacher and C. A. Pope, Montreal.

Queen City Foundry Co., Toronto; capital, $40,000; to manufacture and sell boilers, hardware, ranges, locks, etc. Provisional directors, E. Gillespie, W. C. Burt, J. W. Clark, J. Brooks and R. J. Smythe, all of Toronto.

The Spears Mining Corporation, Toronto, have been incorporated, with a capital of $50,000, to carry on a mining, milling and reduction business. The provisional directors include, D. A. Rose, F. W. Rose and R. S. Gilpin, Toronto.

Dunelin, Limited, Hamilton, Ont., have been incorporated, with a capital of $100,000, to carry on a navigation and transportation business. The provisional directors include, R. O. MacKay, J. P. Steedman and G. Somerville, Hamilton.

The Toronto Yarn Spinning Co., Toronto, have been incorporated, with a capital of $100,000, to manufacture woolen, worsted, cotton yarns, etc. The provisional directors include, J. A. Murray, J. P. Murray and F. B. Hayes, Toronto.

Dominion Crown Cork Co., Toronto; capital, $15,000; to manufacture and deal in crown corks, wire, tin and other metals, metal stamping and cork. Provisional directors, W. F. Hayes, Gordon Russel, Violet Waldock, all of Toronto.

The Canadian Smelting & Refining Co., Toronto, have been incorporated, with a capital of $2,500,000, to carry on a smelting and refining business. The provisional directors include, W. M. Wallace, A. G. Robertson and J. D. Pringle, Toronto.

The Chatham Carriage Co., Chatham, Ont., have been incorporated, with a capital of $100,000, to manufacture carriages, trucks, wagons, sleighs, automobiles, etc. The provisional directors include, I. Teeter, A. Cooke and F. E. Fisher, Chatham.

The Standard Contracting Co., Toronto, have been incorporated, with a capital of $40,000, to carry on a foundry and machine shop and contracting business. The provisional directors include, C. A. Campbell, J. W. DuLaney and S. Campbell, Toronto.

The Dominion Tool Co., Peterboro, Ont., have been incorporated, with a capital of $100,000, to manufacture augurs, bits, gimlets, bolts, screws, rivets, etc. The provisional directors include, W. R. G. Higgins, P. J. Creedon and F. C. Cubitt, Peterboro, Ont.

The Haileybury Brick & Tile Co., Haileybury, Ont., have been incorporated, with a capital of $50,000, to manufacture brick, tile, clay, etc. The provisional directors include, A. J. Murphy, B. C. Beach, Haileybury, Ont., and D. McArthur, Kenmore, Ont.

The Standard Metal Manufacturing Co., Montreal; capital, $18,000; to take over as a going concern the Standard Metal Manufacturing Co. Incorporators, F. G. Robinson, L. Seymour, H. W. Cooper, H. S. Vipond and F. T. Enright, all of Montreal.

The Gilmour Mining Co., Belleville, Ont., have been incorporated, with a capital of $300,000, to carry on a mining, milling and reduction business. The provisional directors include, F. Landenberger, W. Carnew, Belleville, Ont., and L. P. Smith, Syracuse, N.Y.

The Plantagenet Woolen Mills Co., Plantagenet, Ont., have been incorporated, with a capital of $20,000, to manufacture woolen and textile goods, etc. The provisional directors include, T. A. VanBridger, A. A. Fraser and D. M. Viau, Plantagenet, Ont.

The Canadian Smelting & Refining Co., Toronto; capital, $2,500,000; to prospect for and deal in ores, metals and minerals. Provisional directors, J. D. Pringle, D'Arcy Grierson, Victoria Morrison, Alexander G. Robertson, W. Maxime Wallace, all of Toronto.

Haileybury Brick & Tile Co., Haileybury, Ont.; capital, $50,000; to manufacture and sell building material of all kinds. Provisional directors, A. J. Murray, Benson Clothier Beach, Haileybury, Ont.; Chas. F. McArthur, E. E. Wilson, Duncan McArthur, all of Kenmore, Ont.

Canadian Northern Systems Terminals, Toronto, have been incorporated, with a capital of $2,000,000, to construct oil tanks, pipe lines, steamboats, piers, docks, warehouses, factories, etc. The provisional directors include, A. J. Mitchell, F. C. Annesley and L. W. Mitchell, Toronto.

The Canadian Northern System Terminals, Toronto; capital, $2,000,000; to build and operate terminals, stock yards, oil tanks, elevators, smelters and ore furnaces. Incorporators, Gerard Ruel, A. J. Mitchell, J. B. Robertson, R. P. Ormsby, F. C. Annesley, L. W. Mitchell, all of Toronto.

The Canadian Smelting and Refining Company, Limited, Toronto; capital, $2,500,000; to prospect for, own and develop mining lands and refine ores. Provisional directors: John Davidson Pringle, D'Arcy D. Grierson, Victoria Morrison, Alexander Godfrey Robertson, and William Maxime Wallace, all of Toronto.

The Dominion Tool Company, Limited, Peterboro; capital, $100,000; to carry on the manufacture of augurs, bits and other tools and appliances for boring and drilling, bolts, rivets, screws, nails, and machinery for the manufacture of the same. Provisional directors: James Davidson, Francis Randolph, Joseph MacPherson, Patrick Joseph Creedon, Frederick Cromwell Cubitt and Francis Patrick McNulty, all of Peterboro.

Montreal Engineering Company, Limited: capital, $100,000; to carry on the business of electrical, mechanical and civil engineers and contractors to install machinery, equip plants, erect bridges, dams, etc., build canals, telephone and telegraph lines and general contracting. Incorporators: F. C. Clarke, A. J. Nesbitt, C. C. Giles, J. W. Killam and H. A. Porter, all of Montreal.

The Calkins Tile & Mosaic Co., Montreal: capital, $20,000; to manufacture and trade in any and every kind of marble, tile, terrasse and mosaic, concrete, granite, sandstones, limestones, clays, slates, plaster, terra cotta, lumber and all classes of structural iron and steel and building supplies. Incorporators, W. J. Henderson, Allan L. Smith, J. W. Hannah, A. C. Calder, John W. Graham, all of Montreal.

The Dominion Tool Company, Peterboto: capital, $100,000; to carry on the manufacture and sale of augurs, bits, gimlets, and all other tools or appliances used for boring or drilling purposes; also to manufacture and sell bolts, rivets, screws, and the necessary machinery for the manufacture of the same. Provisional directors: James Davidson, F. R. J. MacPherson, P. J. Creedon, F. C. Cubitt, F. P. McNulty, all of Peterboro, Ont.

Municipal Undertakings.

Paris, Ont., will install a duplicate lighting plant.

Ottawa will spend $28,000 in sewage construction.

Mimico, Ont., will borrow $7,000 for school purposes.

Clinton, Ont., passed a by-law for $53,000 for waterworks.

Listowel, Ont., will install its own electric lighting plant.

A new hospital will be erected by the town of Welland, Ont.

Olds, Alta., has passed a by-law to raise $12,000 for fire protection.

The new municipal power plant of Saskatoon, Sask., was put in commission recently.

The ratepayers of Estevan, Sask., will vote on a by-law to raise $92,000 for waterworks.

The civic asphalt plant for Ottawa will cost $14,800. A Buffalo firm's offer has been accepted.

The ratepayers of Kenora, Ont., will vote on a by-law to raise $75,000 for water power development.

A number of capitalists have applied to St. John, N.B., for the establishment of a car works in that city.

Public Works Department, Ontario Government, will erect a court house and registry office at Sudbury, Ont.

Neepawa, Man.—The Public Works Department, Ottawa, will erect and complete a public building at Neepawa, Man.

The Hamilton Electric Light & Power Company has made an announcement of the long-promised reduction in rates.

N. J. Ker, city engineer, Ottawa, is calling for bids for supplying and erecting an asphalt and bituminous paving plant.

Three money by-laws were passed in Vancouver: For new sewers, $300,000; for new roads, $100,000; for schools, $45,000.

The waterworks at Oshawa, Ont., recently added to their equipment a pump manufactured by the Smart-Turner Machine Co., Hamilton.

The town of Gravenhurst, Ont., is considering the granting of certain concessions to Hess & Co., who will erect a factory to cost $10,000.

The Kingston city council is to ask the Hydro-Electric Power Commission to quote prices for 1,000 to 2,000 horse-power of electric energy, to be supplied at the city power station ready for distribution.

Two electric pumps, costing $50,000 and having a capacity of 10,000,000 gallons per day, will be installed in the city waterworks plant at Hamilton, Ont.

Brandon, Man.—Messrs. Dumas & Lochrane, of Ottawa, Ont., are the successful contractors for the $50,000 armory to be erected here. The proposed structure will be of brick and stone.

Winnipeg, Man.—The Winnipeg council has viewed favorably the establishing of public swimming baths. A by-law will be considered at a future date which, if sanctioned by the people, will authorize the city council to issue debentures for $30,000, for the purpose of providing public baths.

The new municipal lighting plant for the town of Gravenhurst, Ont., was opened recently at South Falls, Ont. The plant cost $45,000, and consists of a 450 k.w. generator, driven by a water turbine of 750 horse-power capacity. Power can be supplied to Gravenhurst, at a cost of $7 per horse-power. This should be a strong inducement for industries seeking desirable locations.

The Ottawa electric commission will be ready to take over the lighting of the streets on December 20th, the date fixed in the notice given by the city to the Ottawa Electric Company, but it will be necessary to commence the work of changing the various circuits several weeks before that, as the civic plant will not operate the direct current system now in use by the electric company and the alternating current lamps will have to be installed circuit by circuit. It is expected that the installation will be fully complete by December 20.

Mining News.

Large shipments of iron are being made from the Atikokan district.

The Dominion Copper Co. will enlarge its plant at Boundary Falls, B.C.

The smelter of the Montreal Reduction and Refining Co., North Bay, Ont., will be in operation in a year's time.

The Pacific Coal Company, of Banff, Alberta, has let a contract for the erection of 240 coke ovens at the colliery at Hosmer.

The Del Nord Co. have located their head office at Sarnia, Ont. They are considering the erection of a large ore reducer in Quebec.

A. A. Cole, mining engineer, reports the discovery of promising quartz five miles from Temagami station, on the T. & N.O. Railway.

Charters have been issued to the Monessen Cobalt Mining Company, Limited, and the Oxford Prospecting & Mining Company, Limited, of Cobalt.

The Granby smelters are operating at fullest capacity at both mines and smelter, the entire battery of eight blast furnaces being in operation.

The Cape Breton Coal, Iron and Railway Company's colliery at Broughton, C.B., is being abandoned and a new boring is being made at a point four miles distant.

The Crescent Mines, Limited, Boundary, B.C., have installed an Allis-Chalmers-Bullock, Limited, electric compressor. Power is being furnished by the Greenwood City Waterworks Company.

Sault Ste. Marie is to have another large smelter, with a capacity of 125 tons per day. The capital of the new company, which is being promoted by O. Stalmann, Salt Lake City, is $1,000,000.

Another smelter is in prospect for Vancouver Island. The enterprise, it is said, will be undertaken by the Southern Cross Mining and Smelting Company, and £100,000 has already been subscribed for the work.

A Prospectors' and Mineral Interests Exchange has been organized in New Liskeard, with J. R. Lawless and V. E. Taplin, of that place, as president and secretary, respectively. The object of the exchange is to eliminate middlemen's profits and brokers' options.

Canada's Yukon coal fields cannot be bought outright, but are leased for twenty-one years, and no concession of more than four miles square is granted on any terms. Only ten tons of coal per acre of the concession can be mined in any one year, and coal at the pit's mouth must be sold to actual settlers at $1.75 per ton.

A number of American capitalists and mineral experts are proposing to erect electric furnaces at Sydney, C.B., such as are now being used in France and Germany, to develop the iron ore mountain at St. George's, Newfoundland. The mountain rises to a height of 200 feet, and the assay of the ore, which is of a black magnetic variety, has shown as high as 60 to 65 per cent. metallic iron.

The Wilbur iron ore mine, north of Kingston, Ont., will be opened again to supply a Sault Ste. Marie firm for the manufacture of Bessemer steel rails. This ore is a high-grade Bessemer iron ore, low in phosphorus, and was one of the richest iron producing mines in the country till the Lake Superior mines were opened up. The better shipping facilities near Lake Superior made the mines there more accessable.

Building Notes.

A new armory will be erected at Strathroy, Ont.

James Worts, Toronto, will build a residence to cost $18,000.

The Toronto Rowing Club will erect a club house to cost $7,000

Building operations in the city of Hull, Que, for 1907, cost $275,000.

The Bank of New Brunswick will open a branch in Halifax, N.S.

Building operations at Barrie, Ont., for 1907, will approximate $150,000,

A Presbyterian church will be erected at Stettler, Sask., to cost $2,500.

New armories will shortly be erected in Calgary, Alta., at a cost of $40,000.

A. J. Strathy, Toronto, will build a four-storey store at a cost of $12,000.

A new police court building will be erected at Port Arthur, Ont., to cost $11,225.

Improvements to the Y.W.C.A. building, Simcoe street, Toronto, will cost $12,000.

Building permits were issued in Toronto from Oct. 7 to 15, to the value of $175,000.

The offices of the Traders Bank, at Guelph, will be enlarged, at a cost of $15,000.

The Y.W.C.A., Toronto, will erect an addition to their premises on Simcoe street, to cost $12,000.

The Bank of Montreal, Moncton, N.B., are calling for tenders for the construction of a branch there.

Montreal's new post office will cost in the neighborhood of $500,000. It will be completed in May, 1909.

The Royal Northwest Mounted Police barracks at Lethbridge, Alta., will be improved at a cost of $10,000.

The Canadian Bank of Commerce are erecting a new building at Kamsack, Sask., at a cost of about $10,000.

W. & D. Dineen, Toronto, will erect a factory at the rear of their premises, for the cutting and making of furs.

Architects have prepared plans for a $20,000 printing office for the Compagnie Action Sociale Catholique, Quebec, Que.

The corner-stone of the new school at Westmount, Que., was laid recently. The building will cost about $100,000.

The new building for the Imperial Bank at Edmonton will cost $90,000. It will be completed in the near future.

The congregation of St. Joseph's church, Sydney, N.S., will erect a large edifice to replace the one recently destroyed by fire.

D. Spencer, Limited, Vancouver, B.C., have taken out a permit for the erection of a new store and arcade, at a cost of $240,000.

Building permits to the value of $385,700 were issued in Stratford, Ont., during September. The new G.T.R. shops, alone, will cost $360,000.

The Bank of Hamilton has applied for a permit to erect a three-storey bank building at the corner of Ossington avenue and College street, Toronto, to cost $25,000.

The Masons of Alpha Lodge, Toronto, have voted to join with other lodges in the erection of a new Masonic hall, to be built in the northwestern part of that city, at a cost of $60,000.

The number and value of building permits in Toronto for September has decreased considerably in comparison with September, 1906, although the aggregate for the year, so far, is well above that for the same period last year. The total in September was $550,000, but the $200,000 permit for the Alexandra theatre had been added to this, and thus the total is $750,000. This is $152,803 below the figure for September, 1906, which was $902,803. The permits issued so far this year amount to $12,204,000, as compared with $9,566,000 for the same period of 1906.

Railroad Construction.

Valuable coal beds have been found by the G.T.P. in British Columbia.

The C.N.R. are preparing to build a new station at New Westminster, B.C.

The C.P.R. has commenced the re-timbering of the trestle of the Interprovincial bridge.

The Esquimalt and Nanaimo Railroad is spending $100,000 in improvements on their road.

The Michigan Central Railway Co. is to construct a line from Charing Cross, Ont., to Chatham, Ont.

The Grand Trunk Pacific has 140 miles of track ready for Government inspection west of Portage la Prairie.

The C.N.R. intend increasing their coal docks at Port Arthur, Ont. The capacity of the docks when completed, will be 600,000 tons.

Work on the proposed railway from Port Simpson, B.C., to Port Churchill, on the Hudson Bay, will be commenced next summer.

P. & J. Powers & Co., Ottawa, have completed six large locomotive boilers for the C.N.R. These are the first locomotive boilers made at Ottawa.

The Canadian Pacific has decided to make Grand Forks, B.C., a divisional point and will erect roundhouses, repair shops and yards, at an expense of $100,000.

Welland Pipe Works.

The Page-Hersey Company have started operations on their new plant at Welland, Ontario. This new mill will manufacture pipe in sizes from four to eight inches. The work of construction is being carried on by the Berlin Construction Co. The buildings adjoin the works of the Ontario Iron & Steel Co., and the products of these works will be used in the manufacture of pipe.

The building for the lap weld furnaces will be 200 feet long, with wings at each end 380 feet long and 50 feet wide. The wings will contain the storehouse, thread-testing plant and overhead crane runway. Between the two wings will be a separate building for machine shop, 33x60 feet, and socket shop, 33x180 feet. Additional to this will be the gas producer house, 35x66 feet.

A Steady Growth.

Quebec city, in spite of the set-back owing to the Quebec bridge disaster, is showing fairly good progress. A strong evidence of this growth is a walk through the various departments of the extensive establishment of the Mechanics' Supply Co., in St. Paul street, Quebec. The steady progress made by this firm is evident in the large stocks carried and in the additions to their building. The electrical department contains many exclusive lines, and their showrooms throughout are the essence of taste; depicting many uncommon lines.

New Locomotive Works for Lachine.

A by-law to give a bonus of $50,000, exemption from taxes for twenty years, and reduction in the water tax to an English syndicate to establish a locomotive works in the town of Lachine was voted on, on October 21, and carried by 288 to 21. For this cash bonus, the syndicate will build a plant costing $2,250,000 and employ 500 men at the end of the first year, and 1,500 at the end of the third year.

Big Power Plant on the St. Lawrence.

Another immense power plant was discussed before the International Waterways Commission when a proposition for the development of electrical power on the St. Lawrence river, near Cornwall, at a cost of $20,000,000, was put forward by the St. Lawrence River Co. of Canada, and the Long Sault Development Co., of United States. The St. Lawrence Co. at present supplies Cornwall with light power, but its capacity of 1,250 horse-power is exhausted and the new plan provides for 50,000 horse-power. The Canadian concern will invest $5,000,000, and the American $15,000,000 in the scheme.

Large Smelter for Toronto.

D. D. Mann, of the firm of Mackenzie & Mann, has sent a definite offer to the city of Toronto to build a 1,400-ton smelter in Toronto, in return for a grant in fee simple of some waste lands known as Ashbridge's Bay. This marsh or bog extends to 1,377 acres, and Mr. Mann has made an offer to use part of this marsh to establish works, which will eventually employ 15,000 men. The Moose Mountain Mining Company is behind the scheme. The capital of this company is $2,500,000 paid up. At Moose Mountain, in Hutton township, there is an unlimited supply of raw material of good quality. The intention of the company is to make a beginning in smelting and to put up a thoroughly equipped modern plant for the smelting and making of pig iron, and to add to it gradually so that in the end a very large iron and steel industry would be established. Eventually about 1,000 acres would be required for the entire plant, as the smelter works will be followed by a steel plant, rolling mills, steel car buildings. The allied industries would take up the manufacture of steel billets, steel bars, angle iron, rolled billets for shipbuilding, and steel and iron of all kinds for various industries, while the manufacture of steel rails is definitely in view. As soon as the smelter works are in operation, the company will have 700 tons of slag per day to be used in filling-in, but the filling for the smelter works would have to be taken in by rail and water. This is an opportunity to make Toronto the Pittsburg of Canada. The work of construction will begin at once should the city accept the offer.

PERSONAL MENTION.

Mr. J. W. Prophie, formerly master mechanic for the Atlantic Steel Co., Atlanta, Ga., has accepted a similar position with the Ontario Iron & Steel Co., Welland, Ontario.

Mr. Walter B. Snow, for many years connected with the publicity department of the B. F. Sturtevant Company, has resigned and opened offices at 170 Summer St., Boston Mass., as a publicity engineer. He will do all kinds of work in the way of publicity for manufacturers of machinery and allied products.

AMERICAN FOUNDRYMEN'S ASSOCIATION.

The annual convention of the American Foundrymen's Association will meet in Toronto about the first week in June, 1908. Over fifteen hundred attended the convention in Philadelphia this year and a profitable time was spent. Full information in regard to the Canadian conference will be given in a succeeding issue.

ENGINEERS' CLUB OF TORONTO.

At the meeting of the Engineers' Club on Oct. 10, the subject for discussion was, "Principles of Professional Conduct for Guidance of Engineers," introduced by Mr. A. B. Barry, who said that professional ethics simply consists in the exercise of that honorable feeling and courteous manner which gentlemen extend to, and expect from one another and there is no code, written or unwritten law, which can justify or absolve an engineer from doing that which his own sense of honor and integrity tells him is unprofessional. The engineer is, or should be, from the nature of his calling, a strong-minded man, one who seeks to control the forces of nature for the benefit of man. An engineer should obtain all the information he could, but should not tell trade secrets while under salary. A discussion arose as to the proper place for controversy, and while some thought the public press was the best place, yet a general feeling prevailed that the proper place is in the technical press and trade journals.

At a meeting of the Toronto branch of the American Institute of Electrical Engineers, held at the Engineers' Club, Toronto, on Friday, October 11th, the following were elected: Chairman, K. L. Aitken; vice-chairman, W. A. Bucke; secretary, L. W. Pratt; executive committee, H. W. Price, Edw. Richards and W. G. Chace.

MANUFACTURERS AND MANUFACTURERS' AGENTS.

It will pay you to watch our condensed column each month. There are many money-making propositions brought to your attention here. You may find just what you are looking for.

RATES

One insertion—25c. for 20 words; 1c. a word for each additional word.

Yearly rate—$2.50 for twenty words or less, 10c a word for each additional word.

The above does not apply to notices under the head of "Machinery Wanted." These notices are inserted free for subscribers.

BUSINESS CHANCE.

FOUNDRY FOR SALE—In a thriving Ontario town of five thousand inhabitants; is a going concern, employing from ten to fifteen hands, and does a general foundry business; no opposition within ten miles, and there are numerous orders now on hand; owing to a dissolution of the partnership the business must be sold at once. Apply Box 45, CANADIAN MACHINERY, Toronto.

FOR SALE.

FOR SALE—The Canadian patent rights on a valveless pneumatic hammer, proved by actual test to be superior to any hammer at present on the market; an arrangement can be made to manufacture same on royalty for Canadian market. Apply Box 39C, CANADIAN MACHINERY, Toronto. (9)

FOR SALE.

BELTING, RUBBER, CANVAS AND LEATHER, Hose, Packing, Blacksmith's and Mill Supplies at lowest prices. N. Smith, 138 York Street, Toronto. (2tf)

MARINE gasoline engine castings, with blue print and full instructions, etc.; 2½, 4, 6 h.p.; also complete finished outfits at $65 up; catalogue, Krug & Crosby, Hamilton. [10tf]

OWING to the installation of a new water-power plant, the town of Gravenhurst has for sale the following electrical machinery:—One 100 k.w. 2,400 volt, 66 cycle, 2 phase, 1,000 R.P.M. belted type S.K.C. alternator, complete with rheostat, phase regulators and field rheostat.

ONE 2½ K.W., 125, volt 1900 R.KM.. shunt wound, C.G.E., exciter; complete with pulley, base and rheostat.

ONE blue Vermont marble switchboard panel, 48 x 60 x 1¼, complete with pilot lamps, ammeters, voltmeter, switches, fuses, rheostat mountings, etc., for the control of the generator and the exciter.

ONE tandem compound, belted type Wheelock engine, 13" diameter, L.P. 22" x 30" stroke; this engine complete with Inglis independent condenser, feed water heater, and one Smart-Turner duplex boiler feed pump, 5¼ x 3½ x 6 in., all necessary oil cups, lubricators, line shaft pulleys, etc.

ONE 14" double-ply leather belt; one 12" ditto, and one 18".

ONE belt drum Goldie & M'Culloch condenser.

FOR further paeticulars apply to W. H. Cross, Secretary Electric Light Commission.

SITUATION VACANT.

WANTED AT ONCE—Superintendent and mechanical engineer for machine works manufacturing gasoline engines and general machine work; one familar with modern methods, capable of making estimates, drawings and producing work at minimum cost; give full particulars of qualifications, experience, age, nationality and salary. Apply to Brandon Machine Works Co., Ltd., Brandon, Man. [9]

SITUATIONS WANTED.

ADVERTISER, shortly emigrating, seeks position as correspondence clerk and accountant. Well up in all details of engineering and allied business; ledger balancing, shorthand and typewriting; excellent references. Lister, Ipswich Road, Ardleigh, Colchester, Eng.

WANTED—Position as purchasing agent or correspondent for manufacturing firm by a young man at present employed in a similar capacity by one of the largest manufacturing concerns in Ontario; has had long experience in these lines, thoroughly understands all branches of office work and can furnish best of references. Address Box 43, CANADIAN MACHINERY, Toronto. [10]

MECHANICAL engineer (25) seeks position; experienced in construction of modern copper, zinc, smelting works, machinery furnaces, etc.; general engineering, gas, steam engines, suction and pressure gas plants; structural steel, etc.; reliable, energetic; good references. Box 27, CANADIAN MACHINERY, 88 Fleet St., London, Eng. [10]

AGENTS WANTED.

A Firm of Chemical Engineers

are requiring first-class agents in the large cities of Canada. Apply to W. J. Fraser & Co., Ltd., Chemical Engineers, 98 Commercial Road East, London, England. [11]

"MORSE"

stamped on tools is the guarantee to look for.

Arbors, Chucks, Counterbores, Countersinks, Cutters, Dies, Drills, Gauges, Machines, Mandrels, Mills, Reamers, Screw Plates, Sleeves, Taps, Taper Pins, Wrenches.

Morse Twist Drill & Machine Co.

NEW BEDFORD, MASS., U.S.A.

LARGEST STOCK IN CANADA

TOOLS OF ALL KINDS

Send for our Illustrated Sheets, Catalogues and Prices

MECHANICS SUPPLY CO., 80-90 St. Paul Street, QUEBEC

NEW BOILERS IN STOCK

For 100 Pounds

Two 42"x12', 35 h.p.
Two 48"x12', 45 h.p.
Two 48"x14', 50 h.p.
Two Locomotive, 25 h.p.
One Locomotive, 40 h.p.

For 125 Pounds

Five 60"x16', 90 h.p.
Two 66"x14', 100 h.p.
Three 66"x16', 110 h.p.
Five 72"x16', 130 h.p.
Five 72"x18', 150 h.p.

Made in Canada by

CANADA FOUNDRY COMPANY, Limited

Head Office and Works: TORONTO, ONT.

District Offices: Montreal Halifax Ottawa Winnipeg Vancouver Rossland

CANADIAN MACHINERY BUYERS' DIRECTORY

To Our Readers—Use this directory when seeking to buy any machinery or power equipment. You will often get information that will save you money.
To Our Advertisers—Send in your name for insertion under the heading of the lines you make or sell.
To Non-Advertisers—A nominal rate of $1 per line a year is charged non-advertisers.

Acids.
Canada Chemical Mfg. Co., London.

Abrasive Grains & Powders
Carborundum Co., Niagara Falls, N.Y.

Abrasive Materials.
Baxter, J. R., & Co., Montreal.
The Canadian Fairbanks Co., Montreal.
Rice Lewis & Son, Toronto.
Norton Co., Worcester, Mass.
H. W. Petrie, Toronto.
Carborundum Co., Niagara Falls, N.Y.
Williams & Wilson, Montreal.

Air Brakes.
Canada Foundry Co., Toronto.
Canadian Westinghouse Co., Hamilton.

Air Receivers.
Allis-Chalmers-Bullock Montreal.
Canada Foundry Co., Toronto.
Canadian Rand Drill Co., Montreal.
Chicago Pneumatic Tool Co., Chicago.
John McDougall Caledonian Iron Works Co., Montreal.

Alundum Scythe Stones
Norton Company, Worcester, Mass.

Arbor Presses.
Niles-Bement-Pond Co., New York.

Automatic Machinery.
Cleveland Automatic Machine Co. Cleveland.
National-Acme Mfg. Co., Cleveland
Potter & Johnston Machine Co., Pawtucket, R. I.

Automobile Parts
Globe Machine & Stamping Co., Cleveland, Ohio.

Axle Cutters.
Butterfield & Co., Rock Island, Que.
A. B. Jardine & Co., Hespeler, Ont.

Axle Setters and Straighteners.
A. B. Jardine & Co., Hespeler, Ont.
Standard Bearings, Ltd., Niagara Falls

Babbit Metal.
Baxter, J. R., & Co., Montreal.
Canada Metal Co., Toronto.
Canada Machinery Agency, Montreal.
Frothingham & Workman Ltd., Montreal
Greey, Wm. & J. G., Toronto.
Rice Lewis & Son, Toronto.
Lumen Bearing Co., Toronto.
Mechanics Supply Co., Quebec, Que.

Bakers' Machinery.
Greey, Wm. & J. G., Toronto.

Baling Presses.
Perrin, Wm. R., & Co., Toronto.

Barrels, Steel Shop.
Cleveland Wire Spring Co., Cleveland.
Steel Trough & Machine Co., Tweed, Ont.

Barrels, Tumbling.
Buffalo Foundry Supply Co., Buffalo.
Detroit Foundry Supply Co., Windsor
Dominion Foundry Supply Co., Montreal
Hamilton Facing Mill Co., Hamilton.
Gilmour, J., New York.
Globe Machine & Stamping Co., Cleveland, Ohio.
John McDougall Caledonian Iron Works Co., Montreal.
Northern Engineering Works, Detroit.
J. W. Paxson Co., Philadelphia, Pa.
H. W. Petrie, Toronto.
Sly, W. W., Mfg. Co., Cleveland
The Smart-Turner Mach. Co., Hamilton

Bars, Boring.
Hall Engineering Works, Montreal.
Niles-Bement-Pond Co., New York.
Standard Bearings, Ltd., Niagara Falls

Batteries, Dry.
Berlin Electrical Mfg. Co., Toronto.
Mechanics Supply Co., Quebec, Que.

Batteries, Flashlight.
Berlin Electrical Mfg. Co., Toronto.
Mechanics Supply Co., Quebec Que

Batteries, Storage.
Canadian General Electric Co. Toronto
Chicago Pneumatic Tool Co., Chicago.
Rice Lewis & Son, Toronto.
John Millen & Son, Montreal.

Bearing Metals.
Lumen Bearing Co., Toronto.

Bearings, Roller.
Standard Bearings, Ltd. Niagara Falls

Bell Ringers.
Chicago Pneumatic Tool Co., Chicago.
Mechanics Supply Co., Quebec, Que.

Belting, Chain.
Baxter, J. R., & Co. Montreal.
Canada Machinery Exchange, Montreal
Greey, Wm. & J. G., Toronto.
Jeffrey Mfg. Co., Columbia, Ohio.
Link-Belt Eng. Co., Philadelphia.
Waterous Engine Works Co., Brantford.

Belting, Cotton.
Baxter, J. R., & Co., Montreal.
Canada Machinery Agency, Montreal.
Dominion Belting Co., Hamilton.
Rice Lewis & Son, Toronto.

Belting, Leather.
Baxter, J. R., & Co., Montreal.
Canada Machinery Agency, Montreal.
The Canadian Fairbanks Co., Montreal.
Frothingham & Workman Ltd., Montreal
Greey, Wm. & J. G., Toronto.
McLaren, J. C., Montreal.
Rice Lewis & Son, Toronto.
H. W. Petrie, Toronto.
Williams & Wilson, Montreal.

Belting, Rubber.
Baxter, J. R., & Co., Montreal.
Canada Machinery Agency, Montreal.
Frothingham & Workman, Ltd., Montreal
Greey, Wm. & J. G., Toronto.

Belting Supplies.
Baxter, J. R., & Co., Montreal.
Greey, Wm. & J. G., Toronto.
Rice Lewis & Son, Toronto.
H. W. Petrie, Toronto.

Bending Machinery.
John Bertram & Sons Co., Dundas, Ont.
Chicago Pneumatic Tool Co., Chicago.
Rice Lewis & Son, Toronto.
London Mach. Tool Co., Hamilton, Ont.
National Machinery Co., Tiffin, Ohio.
Niles-Bement-Pond Co., New York.

Benders, Tire.
A. B. Jardine & Co., Hespeler, Ont.
London Mach. Tool Co., Hamilton, Ont.

Blowers.
Buffalo Foundry Supply Co., Montreal.
Canada Machinery Agency, Montreal.
Detroit Foundry Supply Co., Windsor
Dominion Foundry Supply Co., Montreal
Gilmour, J., New York.
Hamilton Facing Mill Co., Hamilton.
Kerr Turbine Co., Wellsville, N.Y.
Mechanics Supply Co., Quebec, Que.
J. W. Paxson, Philadelphia, Pa.
Sheldon's Limited, Galt.

Blast Gauges—Cupola.
Paxson, J. W., Co. Philadelphia.
Sheldons, Limited, Galt

Blocks, Tackle.
Frothingham & Workman Ltd., Montreal

Blocks, Wire Rope.
Frothingham & Workman, Ltd., Montreal

Blow-Off Tanks.
Darling Bros., Ltd., Montreal.

Boilers.
Canada Foundry Co., Limited, Toronto.
Canada Machinery Agency, Montreal.
Goldie & McCulloch Co., Galt.
John McDougall Caledonian Iron Works, Montreal.
Manitoba Iron Works, Winnipeg.
Mechanics Supply Co., Quebec, Que.
Owen Sound Iron Works Co., Owen Sound.
H. W. Petrie, Toronto.
Robb Engineering Co., Amherst, N.S.
The Smart-Turner Mach. Co., Hamilton.
Waterous Engine Works Co., Brantford.
Williams & Wilson, Montreal.

Boiler Compounds.
Canada Chemical Mfg. Co., London, Ont.
Hall Engineering Works, Montreal.

Bolt Cutters.
John Bertram & Sons Co., Dundas, Ont.
London Mach. Tool Co., Hamilton.
Mechanics Supply Co., Quebec, Que.
National Machinery Co., Tiffin, Ohio.
Niles-Bement-Pond Co., New York.

Bolt and Nut Machinery.
John Bertram & Sons Co., Dundas, Ont.
Canada Machinery Agency, Montreal.
Rice Lewis & Son, Toronto.
London Mach. Tool Co., Hamilton.
National Machinery Co., Tiffin, Ohio.
Niles-Bement-Pond Co. New York.

Bolts and Nuts, Rivets.
Baxter, J. R., & Co., Montreal.
Mechanics Supply Co., Quebec, Que.

Boring & Drilling Machines
American Tool Works Co., Cincinnati.
John Bertram & Sons Co., Dundas, Ont.
Canada Machinery Agency, Montreal.
A. B. Jardine & Co., Hespeler, Ont.
London Mach. Tool Co., Hamilton.
Niles-Bement-Pond Co., New York.

Boring Machine, Upright.
American Tool Works Co., Cincinnati.
John Bertram & Sons Co., Dundas, Ont.
London Mach. Tool Co., Hamilton.
Niles-Bement-Pond Co., New York.

Boring Machine, Wood.
Chicago Pneumatic Tool Co., Chicago.
London Mach. Tool Co., Hamilton.

Boring and Turning Mills.
American Tool Works Co., Cincinnati.
John Bertram & Sons Co., Dundas, Ont.
Gisholt Machine Co., Madison, Wis.
Canada Machinery Agency, Montreal.
Rice Lewis & Son, Toronto.
London Mach. Tool Co., Hamilton.
Niles-Bement-Pond Co., New York.
H. W. Petrie, Toronto.

Box Puller.
A. B. Jardine & Co., Hespeler, Ont.

Boxes, Lathe.
Steel Trough & Machine Co., Tweed, Ont.

Boxes, Steel Shop.
Cleveland Wire Spring Co., Cleveland.
Steel Trough & Machine Co., Tweed, Ont.

Boxes, Tote.
Cleveland Wire Spring Co., Cleveland.

Brass Foundry Equipment.
Ph. Bonvillain & E. Ronceray, Philadelphia
Detroit Foundry Supply Co., Detroit.
Dominion Foundry Supply Co., Montreal
Paxson, J. W., Co., Philadelphia

Brass Working Machinery.
Warner & Swasey Co., Cleveland, Ohio.

Brushes, Foundry and Core.
Buffalo Foundry Supply Co., Buffalo.
Detroit Foundry Supply Co., Windsor.
Dominion Foundry Supply Co., Montreal
Hamilton Facing Mill Co., Hamilton.
Mechanics Supply Co., Quebec, Que.
Paxson, J. W., Co., Philadelphia

Brushes, Steel.
Buffalo Foundry Supply Co., Buffalo.
Dominion Foundry Supply Co., Montreal
Paxson, J. W., Co., Philadelphia

Bulldozers.
John Bertram & Sons Co., Dundas, Ont.
London Mach. Tool Co., Hamilton, Ont
National Machinery Co., Tiffin, Ohio.
Niles-Bement-Pond Co., New York.

Calipers.
Baxter, J. R., & Co., Montreal
Frothingham & Workman, Ltd., Montreal
Rice Lewis & Son, Toronto.
Mechanics Supply Co., Quebec, Que.
John Millen & Son, Ltd., Montreal, Que.
L. S. Starrett & Co., Athol, Mass.
Williams & Wilson Montreal.

Carbon.
Dominion Foundry Supply Co., Montreal

Cars, Foundry.
Buffalo Foundry Supply Co., Buffalo.
Detroit Foundry Supply Co., Windsor
Dominion Foundry Supply Co., Montreal
Hamilton Facing Mill Co., Hamilton,
Paxson, J. W., Co., Philadelphia

Castings, Aluminum.
Lumen Bearing Co., Toronto

Castings, Brass.
Chadwick Bros., Hamilton.
Greey, Wm. & J. G., Toronto.
Hall Engineering Works, Montreal.
Kennedy, Wm., & Sons, Owen Sound.
Lumen Bearing Co., Toronto
Niagara Falls Machine & Foundry Co. Niagara Falls, Ont.
Owen Sound Iron Works Co., Owen Sound.
Robb Engineering Co., Amherst, N.S.

Castings, Grey Iron.
Allis-Chalmers-Bullock Montreal.
Greey, Wm. & J. G., Toronto.
Hall Engineering Works, Montreal.
Kennedy, Wm., & Sons, Owen Sound.
Laurie Engine & Machine Co., Montreal.
Maxwell, David, & Sons, St. Marys.
John McDougall Caledonian Iron Works Co., Montreal.
Niagara Falls Machine & Foundry Co., Niagara Falls, Ont.
Owen Sound Iron Works Co., Owen Sound.
Robb Engineering Co., Amherst, N.S.
Smart-Turner Machine Co., Hamilton.

Castings, Phosphor Bronze.
Lumen Bearing Co., Toronto

Castings, Steel.
Kennedy, Wm., & Sons, Owen Sound.

Castings, Semi-Steel.
Robb Engineering Co., Amherst, N.S.

Cement Machinery.
Allis-Chalmers-Bullock, Limited, Montreal
Greey, Wm. & J. G., Toronto.
Jeffrey Mfg. Co., Columbus, Ohio.
John McDougall Caledonian Iron Works, Co., Montreal
Owen Sound Iron Works Co., Owen Sound.

Centreing Machines.
John Bertram & Sons Co., Dundas, Ont.
Jeffrey Mfg Co., Columbia, Ohio
London Mach. Tool Co., Hamilton, Ont.
Niles-Bement-Pond Co., New York.
Pratt & Whitney Co., Hartford, Conn.
Standard Bearings, Ltd., Niagara Falls

Centres, Planer.
American Tool Works Co., Cincinnati.

Centrifugal Pumps.
Gas & Electric Power Co., Toronto.
John McDougall Caledonian Iron Works Co., Montreal.
Pratt & Whitney Co., Hartford, Conn.

Centrifugal Pumps—Turbine Driven.
Kerr Turbine Co., Wellsville, N.Y.

Chain, Crane and Dredge.
Frothingham & Workman, Ltd., Montreal

Drills, Blacksmith.
Canada Machinery Agency, Montreal.
Frothingham & Workman, Ltd., Montreal
A. B. Jardine & Co., Hespeler, Ont.
London Mach. Tool Co., Hamilton.
Mechanics Supply Co., Quebec, Que.
Standard Tool Co., Cleveland.

Drills, Centre.
Mechanics Supply Co. Quebec, Que.
Pratt & Whitney Co., Hartford, Conn.
Standard Tool Co., Cleveland, O.
L. S. Starrett Co., Athol, Mass.

Drills, Electric
Gas & Electric Powe Co., Toronto.
Baxter, J. R., & Co., Montreal.
Canadian Pilling Co., Montreal
Chicago Pneumatic Tool Co., Chicago.
Niles-Bement-Pond Co., New York.

Drills, Gang.
American Tool Works Co., Cincinnati.
John Bertram & Sons Co., Dundas, Ont.
Pratt & Whitney Co., Hartford, Conn.

Drills, High Speed.
Wm. Abbott, Montreal.
Frothingham & Workman, Ltd., Montreal
Alexander Gibb, Montreal.
Pratt & Whitney Co., Hartford, Conn.
Standard Tool Co., Cleveland, O.

Drills, Hand.
A. B. Jardine & Co., Hespeler, Ont.

Drills, Horizontal.
John Bertram & Sons Co., Dundas, Ont.
Canada Machinery Agency, Montreal.
London Mach. Tool Co., Hamilton.
Niles-Bement-Pond Co., New York.

Drills, Pneumatic.
Canada Machinery Agency, Montreal.
Chicago Pneumatic Tool Co., Chicago.
Independent Pneumatic Tool Co., Chicago, New York
Niles-Bement-Pond Co., New York.

Drills, Radial.
American Tool Works Co., Cincinnati.
John Bertram & Sons Co., Dundas, Ont.
Bickford Drill & Tool Co., Cincinnati
London Mach. Tool Co., Hamilton, Ont.
Niles-Bement-Pond Co., New York.

Drills, Ratchet.
Armstrong Bros. Tool Co., Chicago.
Frothingham & Workman, Ltd., Montreal
A. B. Jardine & Co., Hespeler,
Mechanics Supply Co., Quebec, Que.
Pratt & Whitney Co., Hartford, Conn.
Standard Tool Co., Cleveland.

Drills, Rock.
Allis-Chalmers-Bullock, Montreal.
Canadian Rand Drill Co., Montreal.
Chicago Pneumatic Tool Co., Chicago.
Jaffrey Mfg. Co., Columbus, Ohio.

Drills, Sensitive.
American Tool Works Co., Cincinnati.
Canada Machinery Agency, Montreal.
Dwight Slate Machine Co., Hartford.
Niles-Bement-Pond Co., New York.

Drills, Twist.
Baxter, J. R., & Co., Montreal.
Chicago Pneumatic Tool Co., Chicago
Frothingham & Workman, Ltd., Montreal
Alex. Gibb, Montreal.
A. B. Jardine & Co., Hespeler, Ont.
Mechanics Supply Co., Quebec, Que.
John Millen & Son, Ltd., Montreal.
Morse Twist Drill and Machine Co., New Bedford, Mass.
Pratt & Whitney Co., Hartford, Conn.
Standard Tool Co., Cleveland.

Drying Apparatus of all Kinds.
Dominion Heating & Ventilating Co., Hespeler
Greey, Wm. & J. G., Toronto
Sheldons Limited, Galt

Dry Kiln Equipment.
Sheldons Limited, Galt

Dry Sand and Loam Facing.
Buffalo Foundry Supply Co., Buffalo.
Dominion Foundry Supply Co., Montreal
Hamilton Facing Mill Co., Hamilton.
Paxson, J. W., Co., Philadelphia

Dump Cars.
Canada Foundry Co., Limited, Toronto
Dominion Foundry Supply Co., Montreal
Greey, Wm. & J. G., Toronto
Hamilton Facing Mill Co., Hamilton.
John McDougall, Caledonian Iron Works Co., Montreal.
Niles-Bement-Pond Co., New York.
Standard Bearings, Ltd., Niagara Falls.
Link-Belt Eng. Co., Philadelphia.
John McDougall Caledonian Iron Works Co., Montreal.
Owen Sound Iron Works Co., Owen Sound

Paxson, J. W., Co., Philadelphia
Waterous Engine Co., Brantford.

Dust Arresters.
Sly, W. W., Mfg. Co., Cleveland

Dust Separators.
Greey, Wm. & J. G., Toronto
Dominion Heating and Ventilating Co., Hespeler.
Paxson, J. W., Co., Philadelphia
Sheldon's Limited, Galt.

Dynamos.
Allis-Chalmers-Bullock, Montreal.
Canadian General Electric Co., Toronto.
Canadian Westinghouse Co., Hamilton.
Consolidated Electric Co., Toronto
Electrical Machinery Co., Toronto.
Gas & Electric Power Co., Toronto
Hall Engineering Works, Montreal, Que.
Mechanics Supply Co., Quebec, Que.
John Millen & Son, Ltd., Montreal.
Packard Electric Co., St. Catharines.
H. W. Petrie, Toronto.
T. & H. Electric Co., Hamilton.

Dynamos—Turbine Driven.
Gas & Electric Power Co., Toronto.
Kerr-Turbine Co., Wellsville, N.Y.

Economizer, Fuel.
Dominion Heating & Ventilating Co., Hespeler
Standard Bearings, Ltd., Niagara Falls.

Electrical Instruments.
Gas & Electric Power Co., Toronto.
Canadian Westinghouse Co., Hamilton.
Mechanics Supply Co., Quebec, Que.

Electrical Steel.
Baxter, Patterson & Co., Montreal.

Electrical Supplies.
Gas & Electric Power Co., Toronto
Canadian General Electric Co., Toronto.
Canadian Westinghouse Co., Hamilton.
London Mach. Tool Co., Hamilton, Ont.
Mechanics Supply Co., Quebec, Que.
John Millen & Son, Ltd., Montreal.
Packard Electric Co., St. Catharines.
T. & H. Electric Co., Hamilton.

Electrical Repairs
Canadian Westinghouse Co., Hamilton.
T. & H. Electric Co., Hamilton.

Elevator Buckets.
Greey, Wm. & J. G., Toronto
Jeffrey Mfg. Co., Columbus, Ohio.

Emery and Emery Wheels.
Baxter, J. R., & Co., Montreal.
Dominion Foundry Supply Co., Montreal
Frothingham & Workman, Ltd., Montreal
Hamilton Facing Mill Co., Hamilton.
Paxson J. W. Co. Philadelphia

Emery Wheel Dressers.
Baxter, J. R. & Co., Montreal.
Canada Machinery Agency, Montreal.
Dominion Foundry Supply Co., Montreal
Frothingham & Workman Ltd., Montreal
Hamilton Facing Mill Co., Hamilton.
Mechanics Supply Co. Quebec, Que.
John Millen & Son, Ltd., Montreal.
H. W. Petrie, Toronto.
Paxson, J. W., Co., Philadelphia
Standard Tool Co., Cleveland.

Engineers and Contractors.
Gas & Electric Power Co., Toronto.
Canada Foundry Co., Limited, Toronto.
Darling Bros., Ltd., Montreal
Greey, Wm. & J. G., Toronto
Hall Engineering Works, Montreal.
Laurie Engine & Machine Co., Montreal.
Link-Belt Co., Philadelphia.
John McDougall, Caledonian Iron Works Co., Montreal.
Robb Engineering Co., Amherst, N.S.
The Smart-Turner Mach. Co., Hamilton.

Engineers' Supplies.
Baxter, J. R., & Co., Montreal.
Frothingham & Workman, Ltd., Montreal
Greey, Wm. & J. G., Toronto
Hall Engineering Works, Montreal.
Mechanics Supply Co., Quebec, Que.
Rice Lewis & Son, Toronto.

Engines, Gas and Gasoline.
Baxter, J. R., & Co., Montreal.
Canada Foundry Co., Toronto.
Canada Machinery Agency, Montreal.
The Canadian Fairbanks Co., Montreal.
Gas & Electric Power Co., Toronto
Gilson Mfg. Co., Guelph
The Goldie & McCulloch Co., Galt, Ont.
Rice Lewis & Son, Toronto.
Ontario Wind Engine & Pump Co. Toronto
H. W. Petrie, Toronto.
The Smart-Turner Mach. Co., Hamilton

Engines, Steam.
Allis-Chalmers-Bullock, Montreal
Belliss & Marcom, Birmingham, Eng.
Canada Machinery Agency, Montreal.
The Goldie & McCulloch Co., Galt, Ont.
Rice Lewis & Son, Toronto.
Laurie Engine & Machine Co., Montreal.

John McDougall Caledonian Iron Works Montreal.
Robb Engineering Co., Amherst, N.S.
Sheldons Limited, Galt.
The Smart-Turner Mach. Co., Hamilton.
Waterous Engine Works Co., Brantford.
Gas & Electric Power Co., Toronto.

Exhaust Heads.
Darling Bros., Ltd., Montreal.
Dominion Heating & Ventilating Co., Hespeler
Sheldons Limited, Galt, Ont.

Expanded Metal.
Expanded Metal and Fireproofing Co. Toronto

Expanders.
A. B. Jardine & Co., Hespeler, Ont.

Fans, Electric.
Gas & Electric Power Co., Toronto.
Canadian General Electric Co., Toronto.
Canadian Westinghouse Co., Hamilton.
Dominion Heating & Ventilating Co. Hespeler
Mechanics Supply Co., Quebec, Que.
Sheldons Limited Galt, Ont.
The Smart-Turner Mach. Co., Hamilton.

Fans, Exhaust.
Gas & Electric Power Co., Toronto.
Canadian Buffalo Forge Co., Montreal.
Detroit Foundry Supply Co., Windsor.
Dominion Foundry Supply Co., Montreal
Dominion Heating & Ventilating Co., Hespeler.
Greey, Wm. & J. G., Toronto
Hamilton Facing Mill Co., Hamilton.
Paxson, J. W., Co., Philadelphia
Sheldons Limited, Galt.

Feed Water Heaters.
Darling Bros., Montreal
Laurie Engine & Machine Co., Montreal
John McDougall, Caledonian Iron Works Co., Montreal.
The Smart-Turner Mach. Co., Hamilton

Files and Rasps.
J. R. Baxter, & Co., Montreal.
Frothingham & Workman, Ltd., Montreal
Mechanics Supply Co., Quebec, Que.
John Millen & Son, Ltd., Montreal.
Rice Lewis & Son, Toronto.
Nicholson File Co., Port Hope
H. W. Petrie, Toronto

Fillet, Pattern.
J. R. Baxter, & Co., Montreal.
Buffalo Foundry Supply Co., Buffalo.
Detroit Foundry Supply Co., Windsor.
Dominion Foundry Supply Co., Montreal
Hamilton Facing Mill Co., Hamilton.

Fire Apparatus.
Waterous Engine Works Co., Brantford.

Fire Brick and Clay.
J. R. Baxter & Co., Montreal.
Buffalo Foundry Supply Co., Buffalo.
Detroit Foundry Supply Co., Windsor.
Dominion Foundry Supply Co., Montreal
Gilmour, J., New York.
Hamilton Facing Mill Co., Hamilton.
Ontario Lime Association, Toronto
J. W. Paxson Co., Philadelphia.
Toronto Pottery Co., Toronto

Flash Lights.
Berlin Electrical Mfg. Co., Toronto.
Mechanics Supply Co., Quebec, Que.

Flour Mill Machinery.
Allis-Chalmers-Bullock, Montreal.
Greey, Wm. & J. G., Toronto
The Goldie & McCulloch Co., Galt, Ont.
John McDougall, Caledonian Iron Works Co., Montreal.

Forges.
Canada Foundry Co., Limited, Toronto.
Frothingham & Workman, Ltd., Montreal
Hamilton Facing Mill Co., Hamilton.
Mechanics Supply Co., Quebec, Que.
H. W. Petrie, Toronto.
Sheldons Limited, Galt, Ont.

Forgings, Drop.
John McDougall, Caledonian Iron Works Co., Montreal.
H. W. Petrie, Toronto.
St. Clair Bros., Galt

Forgings, Light & Heavy.
Hamilton Steel & Iron Co., Hamilton

Forging Machinery.
John Bertram & Sons Co., Dundas, Ont.
London Mach. Tool Co., Hamilton, Ont
National Machinery Co., Tiffin, Ohio
Niles-Bement-Pond Co., New York.

Founders.
Greey, Wm. & J. G., Toronto
John McDougall, Caledonian Iron Works Co., Montreal.

Niagara Falls Machine & Foundry Co., Niagara Falls, Ont.
Maxwell, David. & Sons, St. Marys
The Smart-Turner Mach. Co., Hamilton.

Foundry Coke.
Baird & West, Detroit

Foundry Equipment.
Ph. Bonvillain & E. Ronceray, Philadelphia
Detroit Foundry Supply Co., Windsor.
Dominion Foundry Supply Co., Montreal
Gilmour, J., New York.
Hamilton Facing Mill Co., Hamilton
Hanna Engineering Works, Chicago.
Northern Engineering Works, Detroit
Paxson, J. W., Co., Philadelphia

Foundry Parting.
Doggett, Stanley, New York
Dominion Foundry Supply Co., Montreal
Partomol Co., New York
Foundry Specialty Co., Cincinnati
Stanley Doggett, New York

Foundry Facings.
Buffalo Foundry Supply Co., Buffalo.
Detroit Foundry Supply Co., Windsor.
Doggett Stanley, New York
Dominion Foundry Supply Co., Montreal
Hamilton Facing Mill Co., Hamilton.
J. W. Paxson Co., Philadelphia, Pa.

Friction Clutch Pulleys, etc.
The Goldie & McCulloch Co., Galt.
Greey, Wm. & J. G., Toronto
Link-Belt Co., Philadelphia.

Furnaces.
Detroit Foundry Supply Co., Windsor.
Dominion Foundry Supply Co., Montreal
Hamilton FacinMill Co., Hamilton.
Paxson, J. W., & Philadelphia Co.,

Gas Blowers and Exhausters.
Dominion Heating & Ventilating Co. Hespeler
Sheldons Limited, Galt.

Gas Plants, Suction and Pressure.
Baxter J. R., & Co., Montreal
Colonial Engineering Co., Montreal
Gas & Electric Power Co., Toronto
Williams & Wilson, Montreal

Gauges, Standard.
Mechanics Supply Co., Quebec, Que.
Pratt & Whitney Co., Hartford, Conn.

Gearing.
Eberhardt Bros. Machine Co., Newark.
Greey, Wm. & J. G., Toronto

Gear Cutting Machinery.
Baxter, J. R., & Co., Montreal
Becker-Brainard Milling Mach. Co., Hyde Park, Mass.
Bickford Drill & Tool Co., Cincinnati.
Dwight Slate Machine Co., Hartford
Eberhardt Bros. Machine Co., Newark
Greey, Wm. & J. G., Toronto
Kennedy, Wm., & Sons, Owen Sound
London Mach. Tool Co., Hamilton.
Niles-Bement-Pond Co., New York.
H. W. Petrie, Toronto.
Pratt & Whitney Co., Hartford. Conn.
Williams & Wilson, Montreal.

Gears, Angle.
Chicago Pneumatic Tool Co., Chicago.
Greey, Wm. & J. G., Toronto
Laurie Engine & Machine Co., Montreal.
John McDougall, Caledonian Iron Works Co., Montreal.
Waterous Engine Co., Brantford.

Gears, Cut.
Kennedy, Wm., & Sons, Owen Sound

Gears, Iron.
Greey, Wm. & J. G., Toronto
Kennedy, Wm., & Sons, Owen Sound

Gears, Mortise.
Greey, Wm. & J. G., Toronto
Kennedy, Wm., & Sons, Owen Sound

Gears, Reducing.
Brown, David & Sons, Huddersfield, Eng
Greey, Wm. & J. G., Toronto
John McDougall, Caledonian Iron Works Co., Montreal.

Generators, Electric.

Allis-Chalmers-Bullock, Limited, Montreal
Canadian General Electric Co., Toronto
Canadian Westinghouse Co., Hamilton.
Gas & Electric Power Co., Toronto
Hall Engineering Works, Montreal.
H. W. Petrie, Toronto.
Toronto & Hamilton Electric Co. Hamilton.

Graphite Paints.

P. D. Dods & Co., Montreal.

Graphite.

Detroit Foundry Supply Co., Windsor.
Doggett, Stanley, New York
Dominion Foundry Supply Co., Montreal
Hamilton Facing Mill Co., Hamilton.
Mechanics Supply Co., Quebec, Que.
Paxson, J. W., Co., Philadelphia

Grinders, Automatic Knife.

W. H. Banfield & Son, Toronto.

Grinders, Centre.

Niles-Bement-Pond Co., New York.
H. W. Petrie, Toronto.

Grinders, Cutter.

Becker-Brainard Milling Mach. Co., Hyde Park, Mass.
John Millen & Son, Ltd., Montreal.
Pratt & Whitney Co., Hartford, Conn.

Grinders, Tool.

Armstrong Bros. Tool Co., Chicago.
Gisholt Machine Co Ma.ison, Wis.
H. W. Petrie, T.ronto.
Williams & Wilson, Montreal.

Grinding Machines.

The Canadian Fairbanks, Montreal.
Rice Lewis & Son, Toronto.
Niles-Bement-Pond Co., New York.
Paxson, J. W., Co., Philadelphia
H. W. Petrie, Toronto.

Grinding & Polishing Machines

The Canadian Fairbanks Co., Montreal.
Greey, Wm. & J. G., Toronto
Independent Pneumatic Tool Co., Chicago, Ill.
John Millen & Son, Ltd., Montreal.
Niles-Bement-Pond Co., New York.
H. W. Petrie, Toronto.
Standard Bearings, Ltd., Niagara Falls

Grinding Wheels

Carborundum Co., Niagara Falls
Norton Company, Worcester, Mass.

Hack Saws.

Baxter, J. R., & Co., Montreal.
Canada Machinery Agency, Montreal.
The Canadian Fairbanks Co., Montreal.
Frothingham & Workman, Ltd., Montreal
Mechanics Supply Co., Quebec, Que.
John Millen & Son, Ltd., Montreal.
Niles-Bement-Pond Co., New York.
H. W. Petrie, Toronto.
Williams & Wilson, Montreal.

Hack Saw Frames.

Mechanics Supply Co., Quebec, Que.

Hammers, Drop.

London Mach. Tool Co., Hamilton, Ont.
Niles-Bement-Pond Co., New York.

Hammers, Steam.

John Bertram & Sons Co. Dundas, Ont.
London Mach. Tool Co., Hamilton, Ont.
Niles-Bement-Pond Co., New York.

Hand Stocks.

Borden-Canadian Co., Toronto

Hangers.

The Goldie & McCulloch Co., Galt.
Greey, Wm. & J. G., Toronto
Kennedy, Wm., & Sons, Owen Sound
Owen Sound Iron Works Co., Owen Sound
The Smart-Turner Mach. Co., Hamilton.
Waterous Engine Co., Brantford.
Standard Bearings, Ltd., Niagara Falls

Hardware Specialties.

Mechanics Supply Co., Quebec, Que.

Heating Apparatus.

Darling Bros., Ltd., Montreal
Dominion Heating & Ventilating Co., Hespeler
Mechanics Supply Co., Quebec, Que.
Sheldons Limited, Galt.

Hoisting and Conveying Machinery.

Allis-Chalmers-Bullock, Limited, Montrea
Greey, Wm & J. G., Toronto
Link-Belt Co., Philadelphia.
Niles-Bement-Pond Co., New York.
Northern Engineering Works, Detroit
The Smart-Turner Mach. Co., Hamilton.
Waterous Engine Co., Brantford.

Hoists, Electric.

Northern Engineering Works, Detroit

Hoists, Pneumatic.

Canadian Rand Drill Co., Montreal.
Dominion Foundry Supply Co., Montreal
Hamilton Facing Mill Co., Hamilton.
Northern Engineering Works, Detroit

Hoists, Portable & Stationery.

Canadian Piling Co., Montreal

Hose.

Baxter, J. R., & Co., Montreal
Mechanics Supply Co., Quebec, Que.

Hose, Air.

Canadian Rand Drill Co., Montreal.
Canadian Westinghouse Co., Hamilton.
Independent Pneumatic Tool Co., Chicago
Paxson, J. W., Co., Philadelphia

Hose Couplings.

Canadian Rand Drill Co., Montreal.
Canadian Westinghouse Co., Hamilton.
Mechanics Supply Co., Quebec, Que.
Paxson, J. W., Co., Philadelphia

Hose, Steam.

Allis-Chalmers-Bullock, Montreal.
Canadian Rand Drill Co., Montreal.
Canadian Westinghouse Co., Hamilton.
Mechanics Supply Co., Quebec, Que.
Paxson, J. W. Co., Philadelphio

Hydraulic Accumulators.

Ph. Bonvillain & E. Ronceray, Philadelphia
Niles-Bement-Pond Co., New York.
Perrin, Wm. R., Co., Toronto.
The Smart-Turner Mach. Co., Hamilton.

Hydraulic Machinery.

Gas & Electric Power Co., Toronto.
Allis-Chalmers-Bullock, Montreal.
Barber, Chas., & Sons, Meaford

India Oil Stones.

Norton Company, Worcester, Mass.

Indicators, Speed.

Mechanics Supply Co., Quebec, Que.
L. S. Starrett Co., Athol, Mass.

Injectors.

Canada Foundry Co., Toronto.
The Canadian Fairbanks Co., Montreal.
Desmond-Stephan Mfg. Co., Urbana, Ohio
Frothingham & Workman Ltd., Montreal
Mechanics Supply Co. Quebec, Que.
Rice Lewis & Son, Toronto.
Penberthy Injector Co., Windsor, Ont.

Iron and Steel.

Frothingham & Workman, Ltd., Montreal

Iron and Steel Bars and Bands.

Hamilton Steel & Iron Co., Hamilton

Iron Cements.

Buffalo Foundry Supply Co., Buffalo.
Detroit Foundry Supply Co., Windsor.
Dominion Foundry Supply Co. Montreal
Hamilton Facing Mill Co., Hamilton.
Paxson, J. W, Co., Philadelphia

Iron Filler.

Buffalo Foundry Supply Co., Buffalo.
Detroit Foundry Supply Co., Windsor.
Dominion Foundry Supply Co., Montreal
Hamilton Facing Mill Co., Hamilton.
Paxson, J. W., Co, Philadelphia.

Jacks.

Frothingham & Workman, Ltd., Montreal
Norton, A. O., Coaticook, Que.

Kegs, Steel Shop.

Cleveland Wire Spring Co., Cleveland

Lace Leather.

Baxter, J. R., & Co., Montreal.

Ladles, Foundry.

Dominion Foundry Supply Co., Montreal
Frothingham & Workman Ltd, Montreal
Northern Engineering Works, Detroit

Lamps, Arc and Incandescent.

Canadian General Electric Co., Toronto.
Canadian Westinghouse Co., Hamilton.
Gas & Electric Power Co., Toronto
The Packard Electric Co., St. Catharines.

Lathe Dogs.

Armstrong Bros., Chicago
Baxter, J. R., & Co., Montreal.
Pratt & Whitney Co., Hartford, Conn.

Lathes, Engine.

American Tool Work Co., Cincinnati.
Baxter, J. R., & Co., Montreal.
John Bertram & Sons Co., Dundas, Ont.
Canada Machinery Agency, Montreal.
The Canadian Fairbanks Co., Montreal.
London Mach. Tool Co., Hamilton, Ont.
Niles-Bement-Pond Co., New York.
H. W. Petrie, Toronto.
Pratt & Whitney Co., Hartford, Conn.

Lathes, Foot-Power.

American Tool Works Co., Cincinnati.

Lathes, Screw Cutting.

Baxter, J. R., & Co., Montreal.
Niles-Bement-Pond Co., New York.

Lathes, Automatic, Screw-Threading.

John Bertram & Sons Co., Dundas, Ont
London Mach. Tool Co., Hamilton, Ont
Pratt & Whitney Co., Hartford, Conn.

Lathes, Bench.

London Mach. Tool Co., London, Ont.
Pratt & Whitney Co., Hartford, Conn.

Lathes, Turret.

American Tool Works Co., Cincinnati.
Baxter, J. R., & Co., Montreal
John Bertram & Sons Co., Dundas, Ont.
Gisholt Machine Co., Madison, Wis.
London Mach. Tool Co., Hamilton, Ont.
Niles-Bement-Pond Co., New York.
The Pratt & Whitney Co., Hartford, Conn.
Warner & Swasey Co., Cleveland O.

Leather Belting.

Baxter, J. R., & Co., Montreal.
Canada Machinery Agency, Montreal.
The Canadian Fairbanks Co., Montreal
Greey, Wm. & J. G., Toronto

Locomotives, Air.

Canadian Rand Drill Co., Montreal.

Locomotives, Electrical.

Canadian Westinghouse Co., Hamilton
Gas & Electric Power Co., Toronto
Jeffrey Mfg. Co., Columbus, Ohio.

Locomotives, Steam.

Canada Foundry Co., Toronto.
Canadian Rand Drill Co., Montreal.

Locomotive Turntable Tractors.

Canadian Pilling Co., Montreal

Lubricating Plumbago.

Detroit Foundry Supply Co., Detroit.
Dominion Foundry Supply Co., Montreal
Hamilton Facing Mill Co., Hamilton.
Paxson, J. W, Co., Philadelphia

Lubricators, Force Feed.

Sight Feed Oil Pump Co., Milwaukee, Wis

Lumber Dry Kilns.

Dominion Heating & Ventilating Co., Hespeler.
H. W. Petrie, Toronto.
Sheldons Limited, Galt, Ont.

Machinery Dealers.

Baxter, J. R., & Co. Montreal.
Canada Machinery Agency, Montreal.
The Canadian Fairbanks Co., Montreal.
H. W. Petrie, Toronto.
The Smart-Turner Mach. Co., Hamilton.
Williams & Wilson Montreal.

Machinery Designers.

Greey, Wm. & J. G., Toronto
Standard Bearings, Ltd., Niagara Falls.

Machinists.

W. H. Banfield & Sons, Toronto.
Greey, Wm. & J. G. Toronto
Hall Engineering Works, Montreal.
Link-Belt Co., Philadelphia.
John McDougall, Caledonian Iron Works Co., Montreal.
Paxson, J. W., Co, Philadelphia
Robb Engineering Co., Amherst, N.S.
The Smart-Turner Mach. Co., Hamilton.
Standard Bearings, Ltd, Niagara Falls
Waterous Engine Co., Brantford.

Machinists' Small Tools.

Armstrong Bros., Chicago.
Butterfield & Co., Rock Island, Que.
Frothingham & Workman Ltd., Montrea
Mechanics Supply Co., Quebec, Que.
Rice Lewis & Son, Montreal.
Pratt & Whitney Co., Hartford, Conn.
Standard Tool Co., Cleveland.
L. S. Starrett Co., Athol, Mass.
Williams & Wilson, Montreal.

Malleable Flask Clamps.

Buffalo Foundry Supply Co., Buffalo.
Dominion Foundry Supply Co., Montreal
Paxson, J. W., Co., Philadelphia

Malleable Iron Castings.

Galt Malleable Iron Co., Galt

Mallet, Rawhide and Wood.

Buffalo Foundry Supply Co., Buffalo.
Detroit Foundry Supply Co., Detroit.
Dominion Foundry Supply Co., Montreal
Paxson, J. W., Co, Philadelphia

Mandrels.

A. B. Jardine & Co., Hespeler, Ont.
The Pratt & Whitney Co., Hartford, Conn.
Standard Tool Co., Cleveland.

Marking Machines.

Dwight Slate Machine Co, Hartford,

Metallic Paints.

P. D. Dods & Co., Montreal.

Meters, Electrical.

Gas & Electric Power Co., Toronto.
Canadian Westinghouse Co., Hamilton.
Mechanics Supply Co., Quebec, Que.

Mill Machinery.

Baxter, J. R., & Co., Montreal.
Greey Wm. & J. G., Toronto
The Goldie & McCulloch Co., Galt, Ont.
John McDougall, Caledonian Iron Works Co., Montreal.
H. W. Petrie, Toronto.
Robb Engineering Co., Amherst, N.S.
Waterous Engine Co., Brantford.
Williams & Wilson, Montreal.

Milling Attachments.

Becker-Brainard Milling Machine Co. Hyde Park, Mass.
John Bertram & Sons Co., Dundas, Ont.
Cincinnati Milling Machine Co., Cincinnati.
Niles-Bement-Pond Co., New York,
Pratt & Whitney, Hartford, Conn.

Milling Machines, Horizontal.

Baxter, J. R., & Co., Montreal.
Becker-Brainard Milling Machinery Co. Hyde Park, Mass.
John Bertram & Sons Co., Dundas, Ont.
London Mach. Tool Co., Hamilton, Ont.
Niles-Bement-Pond Co., New York.
Pratt & Whitney, Hartford, Conn.

Milling Machines, Motor Driven.

Cincinnati Milling Machine Co., Cincinnati

Milling Machines, Plain.

American Tool Works Co., Cincinnati.
Becker-Brainard Milling Machine Co. Hyde Park, Mass.
John Bertram & Sons Co., Dundas, Ont.
Canada Machinery Agency, Montreal.
The Canadian Fairbanks Co., Montreal.
Cincinnati Milling Machine Co., Cincinnati
London Mach. Tool Co., Hamilton, Ont.
Niles-Bement-Pond Co., New York.
H. W. Petrie, Toronto.
Pratt & Whitney Co., Hartford, Conn.
Williams & Wilson, Montreal.

Milling Machines, Universal.

American Tool Works Co., Cincinnati.
Becker-Brainard Milling Machine Co., Hyde Park, Mass.
John Bertram & Sons Co., Dundas, Ont.
Canada Machinery Agency, Montreal.
The Canadian Fairbanks Co., Montreal.
Cincinnati Milling Machine Co., Cincinnati
London Mach. Tool Co., Hamilton, Ont
Niles-Bement-Pond Co., New York.
H. W. Petrie, Toronto.
Williams & Wilson, Montreal.

Milling Machines, Vertical.

Becker-Brainard Milling Machine Co., Hyde Park, Mass.
Brown & Sharpe, Providence, R.I.
John Bertram & Sons Co., Dundas, Ont.
Canada Machinery Agency, Montreal.
London Mach. Tool Co., Hamilton, Ont.
Niles-Bement-Pond Co., New York.

Milling Tools.

Wm. Abbott, Montreal.
Becker-Brainard Milling Machine Co., Hyde Park, Mass.
Geometric Tool Co., New Haven, Conn.
Hamilton Tool Co., Hamilton, Ont.
London Mach. Tool Co., Hamilton, Ont.
Pratt & Whitney Co., Hartford, Conn.
Standard Tool Co., Cleveland.
Standard Bearings, Ltd., Niagara Falls

Mining Machinery.

Gas & Electric Power Co., Toronto.
Allis-Chalmers-Bullock, Limited, Montreal
Canadian Rand Drill Co., Montreal.
Jeffrey Mfg. Co., Columbus, Ohio.
Laurie Engine & Machine Co., Montreal.
Rice Lewis & Son, Toronto.
John McDougall, Caledonian Iron Works Co., Montreal.
T. & H. Electric Co. Hamilton.

PROFESSIONAL DIRECTORY

CONSULTING ENGINEERS, PATENT ATTORNEYS, ARCHITECTS, CONTRACTORS, ETC.

T. Pringle & Son

HYDRAULIC, MILL & ELECTRICAL ENGINEERS

FACTORY & MILL CONSTRUCTION A SPECIALTY.

Coristine Bldg., St. Nicholas St., Montreal.

RODERICK J. PARKE

A.M. Can. Soc. C.E. A.M. Amer. Inst. E.E.

CONSULTING ELECTRICAL ENGINEER

INDUSTRIAL STEAM AND ELECTRIC POWER PLANTS DESIGNED. TESTS. REPORTS.

51-53 JANES BLDG., TORONTO, CAN.
Long Distance Telephones—Office and Residence.

J. A. DeCew

Chemical Engineer

Sun Life Building, MONTREAL

Industrial Plants and Processes sold and installed.
Apparatus and Materials for Water Purification.
Free Analyses to prospective purchasers.

JULES DE CLERCY, M. E. Gas Engineer

Expert knowledge of all classes of Engines and Producer Plants. If you are thinking of installing a Gas Engine or Producer Plant write me. If your plant is not working satisfactorily I can help you.

413 Dorchester St. - MONTREAL, QUE.

HANBURY A. BUDDEN

Advocate Patent Agent.

New York Life Building MONTREAL.

Cable Address, BREVET MONTREAL.

PATENTS TRADE MARKS AND DESIGNS

PROCURED IN ALL COUNTRIES

Special Attention given to Patent Litigation

Pamphlet sent free on application.

RIDOUT & MAYBEE 103 BAY STREET TORONTO

PATENTS PROMPTLY SECURED

We solicit the business of Manufacturers, Engineers and others who realize the advisability of having their Patent business transacted by Experts. Preliminary advice free. Charges moderate. **Our Inventor's Adviser** sent upon request. Marion & Marion, New York Life Bldg, Montreal; and Washington, D.C., U.S.A.

PATENTS THAT PROTECT

FETHERSTONHAUGH & CO.

Patent Solicitors & Experts

Fred. B. Fetherstonhaugh, M.E., Barrister-at-law and Counsel and expert in Patent Causes. Charles W. Taylor, B.Sc., formerly Examiner in Can. Patent Office.

MONTREAL CAN. LIFE BLDG.

TORONTO HEAD OFFICE, CAN. BANK COMMERCE BLDG.

T. A. SOMERVILLE

CONSULTING AND SUPERVISING ENGINEER

Reports and Valuations, Factories Rearranged, Manufacturing Plants Designed and Erected Equipments Installed

Federal Life Bldg., HAMILTON, ONT.

THE DETROIT TESTING LABORATORY

1111 Union Trust Building

DETROIT, MICH.

Experienced Foundry Chemists and Metallurgists.
Iron Mixtures for all classes of Castings our Specialty.
Reasonable charges for Analyses.

THE ELECTRIC BLUE PRINT CO.

Blue Prints Positive Blue Prints
Black Prints Multi-color Prints

Largest and finest plant.
Best finished work in Canada at lowest prices.

40 Hospital St., - MONTREAL

CONSULTING ENGINEERS

should have their card in this page. It will be read by the manufacturers of Canada

CANADIAN MACHINERY

Montreal Toronto Winnipeg

THE CANADIAN TURBINE WATER WHEEL

We have not expanded the discharge of our turbines, as we find the results given by our combined impact and reaction runner to be vastly superior to those obtained from turbines of the expanded discharge type. Do not sacrifice your water power. We can save you 20 to 30 per cent. of your water and at the same time give you a much more satisfactory drive.

CHAS. BARBER & SONS, Meaford, Ont.

Turbine Water Wheels, Governors and Gearing only

WM. KENNEDY & SONS

HIGH-CLASS

Steel Gears of all kinds
Iron Gears
Gray Iron, Steel and
Brass Castings

MARINE WORK receives every attention

WM. KENNEDY & SONS, Limited
Owen Sound, Ont.

A Labor Saving Combination

Couch and forming rolls with special attachments for economical manufacture of pipes, complete machines made to meet your requirements (patent 94746), estimates furnished upon application.

Stuart B. Playfair

Traders Bank Building

TORONTO, - CANADA

Mixing Machines, Dougn.

Greey, Wm. & J. G., Toronto
Paxson, J. W., Co., Philadelphia

Mixing Machines, Special.

Greey, Wm. & J. G., Toronto
Paxson, J. W., Co., Philadelphia

Model Tools.

Globe Machine & Stamping Co., Cleveland, Ohio.
Standard Bearings, Ltd., Niagara Falls.
Wells Pattern and Model Works, Toronto

Motors, Electric.

Allis-Chalmers-Bullock Limited, Montreal
Canadian General Electric Co., Toronto
Canadian Westinghouse Co., Hamilton.
Electrical Machinery Co., Toronto.
Consolidated Electric Co., Toronto
Gas & Electric Power Co., Toronto
Hall Engineering Works, Montreal.
Mechanics Supply Co., Quebec, Que.
The Packard Electric Co., St. Catharines.
T. & H. Electric Co., Hamilton.

Motors, Air.

Canadian Rand Drill Co., Montreal.

Molders' Supplies.

Buffalo Foundry Supply Co., Buffalo.
Detroit Foundry Supply Co., Windsor.
Dominion Foundry Supply Co., Montrea
Hamilton Facing Mill Co., Hamilton
Mechanics Supply Co., Quebec, Que.
Paxson J. W., Co., Philadelphia

Molders' Tools.

Buffalo Foundry Supply Co., Buffalo.
Detroit Foundry Supply Co., Windsor
Dominion Foundry Supply Co., Montreal
Hamilton Facing Mill Co., Hamilton.
Mechanics Supply Co., Quebec, Que.
Paxson, J. W., Co., Philadelphia

Molding Machines.

Ph. Bonvillain & E. Ronceray Philadelphia
Buffalo Foundry Supply Co., Buffalo.
Dominion Foundry Supply Co., Montreal
Hamiton Facing Mill Co., Hamilton.
J. W. Paxson Co., Philadelphia, Pa.

Molding Sand.

T. W. Barnes, Hamilton.
Buffalo Foundry Supply Co., Buffalo.
Detroit Foundry Supply Co., Windsor
Dominion Foundry Supply Co., Montreal
Hamilton Facing Mill Co., Hamilton.
Paxson, J. W., Co., Philadelphia

Nut Tappers.

John Bertram & Sons Co., Dundas, Ont
A. B. Jardine & Co., Hespeler.
London Mach. Tool Co., Hamilton.
National Machinery Co., Tiffin, Ohio.

Nuts.

Canada Nut Co., Toronto.

Oatmeal Mill Machinery.

Greey, Wm. & J. G., Toronto
The Goldie & McCulloch Co., Galt

Oilers, Gang.

Sight Feed Oil Pump Co., Milwaukee, Wis.

Oils, Core.

Buffalo Foundry Supply Co., Buffalo.
Dominion Foundry Supply Co., Montreal
Hamilton Facing Mill Co., Hamilton.
Paxson, J. W., Co., Philadelphia

Oil Extractors.

Darling Bros., Ltd., Montreal

Paint Mill Machinery.

Greey, Wm. & J. G., Toronto

Pans, Lathe.

Cleveland Wire Spring Co., Cleveland
Paxson, J. W., Co., Philadelphia

Pans, Steel Shop.

Cleveland Wire Spring Co., Cleveland

Parting Compound.

Doggett, Stanley, New York

Patent Solicitors.

Hanbury A. Budden, Montreal.
Fetherstonhough & Blackmore, Montreal
Marion & Marion, Montreal.
Ridout & Maybee, Toronto.

Patterns.

John Carr, Hamilton.
Galt Malleable Iron Co., Galt
Hamilton Pattern Works, Hamilton.
John McDougall, Caledonian Iron Works Co., Montreal.
Wells' Pattern and Model Works, Toronto.

Pig Iron.

Hamilton Steel & Iron Co., Hamilton

Pipe Cutting and Threading Machines.

Bordon-Canadian Co., Toronto
Butterfield & Co., Rock Island, Que.
Canada Machinery Agency, Montreal.
Curtis & Curtis Co., Bridgeport, Conn.
Frothingham & Workman, Ltd., Montreal
Hart Mfg. Co., Cleveland
A. B. Jardine & Co., Hespeler, Ont.
London Mach. Tool Co., Hamilton, Ont.
Niles-Bement-Pond Co., New York.
Shantz, I. E., & Co., Berlin, Ont.

Pipe, Municipal.

Canadian Pipe Co., Vancouver, B.C.
Pacific Coast Pipe Co., Vancouver, B.C.

Pipe, Waterworks.

Canadian Pipe Co., Vancouver, B.C.
Pacific Coast Pipe Co., Vancouver, B.C.

Planers, Standard.

American Tool Works, Cincinnati.
John Bertram & Sons Co., Dundas, Ont.
Canada Machinery Agency, Montreal.
The Canadian Fairbanks Co., Montreal.
Rice Lewis & Son, Toronto.
London Mach. Tool Co., Hamilton Ont.
Niles-Bement-Pond Co., New York.
H. W. Petrie, Toronto.
Pratt & Whitney Co., Hartford, Conn.
Williams & Wilson, Montreal.

Planers, Rotary.

John Bertram & Sons Co., Dundas, Ont.
London Mach. Tool Co., Hamilton, Ont.
Niles-Bement-Pond Co., New York.

Planing Mill Fans.

Dominion Heating & Ventilating Co., Hespeler
Sheldons Limited, Galt, Ont.

Plumbago.

Buffalo Foundry Supply Co., Buffalo.
Detroit Foundry Supply Co., Windsor
Doggett, Stanley, New York
Dominion Foundry Supply Co., Montreal
Hamilton Facing Mill Co., Hamilton,
Mechanics Supply Co., Quebec, Que.
J. W. Paxson Co., Philadelphia, Pa.

Pneumatic Tools.

Allis-Chalmers-Bullock, Montreal.
Canadian Rand Drill Co., Montreal.
Hamilton Facing Mill Co., Hamilton.
Hanna Engineering Works, Chicago.
Independent Pneumatic Tool Co. Chicago, New York
Paxson, J. W., Co., Philadelphia

Power Hack Saw Machines.

Baxter, J. R., & Co., Montreal.
Frothingham & Workman, Ltd., Montreal

Power Plants.

Gas & Electric Power Co., Toronto
John McDougall Caledonian Iron Works Co., Montreal.
The Smart-Turner Mach. Co., Hamilton

Power Plant Equipments.

Gas & Electric Power Co., Toronto.
Darling Bros., Ltd., Montreal

Presses, Drop.

W. H. Banfield & Son, Toronto.
E. W. Bliss Co., Brooklyn, N.Y.
Brown, Boggs Co., Hamilton
Canada Machinery Agency, Montreal.
Laurie Engine & Machine Co., Montreal
Niles-Bement-Pond Co., New York.

Presses, Hand.

E. W. Bliss Co., Brooklyn, N.Y.
Brown, Boggs Co., Hamilton

Presses, Hydraulic.

John Bertram & Sons Co., Dundas, Ont.
Laurie Engine & Machine Co., Montreal.
London Mach. Tool Co., Hamilton, Ont.
John McDougall Caledonian Iron Works Co., Montreal.
Niles-Bement-Pond Co., New York.
Perrin, Wm. R., & Co., Toronto

Presses, Power.

E. W. Bliss Co., Brooklyn, N.Y.
Brown, Boggs Co., Hamilton
Canada Machinery Agency, Montreal.
Laurie Engine & Machine Co., Montreal
London Mach. Tool Co., Hamilton, Ont.
John McDougall Caledonian Iron Works Co., Montreal.
Niles-Bement-Pond Co., New York.

Presses Power Screw.

Brown, Boggs Co., Hamilton
Perrin, Wm. R., & Co., Toronto

Pressure Regulators.

Darling Bros., Ltd., Montreal

Producer Plants.

Canada Foundry Co., Toronto

Pulp Mill Machinery.

Greey, Wm. & J. G., Toronto
Jeffrey Mfg. Co., Columbus, Ohio.
Laurie Engine & Machine Co., Montreal.
John McDougall Caledonian Iron Works Co., Montreal.
Waterous Engine Works Co., Brantford

Pulleys.

Baxter, J. R., & Co., Montreal.
Canada Machinery Agency, Montreal.
The Canadian Fairbanks Co., Montreal.
The Goldie & McCulloch Co., Galt.
Greey, Wm. & J. G., Toronto
Laurie Engine & Machine Co., Montreal.
Link-Belt Co., Philadelphia.
John McDougall Caledonian Iron Works Co., Montreal.
Owen Sound Iron Works Co., Owen Sound
H. W. Petrie, Toronto.
The Smart-Turner Mach. Co., Hamilton.
Standard Bearings Ltd., Niagara Falls.
Williams & Wilson, Montreal.
Waterous Engine Co., Brantford.

Pumps.

Laurie Engine & Machine Co., Montreal
Ontario Wind Engine & Pump Co., Toronto

Pump Governors.

Darling Bros., Ltd. Montreal

Pumps, Hydraulic.

Ph. Bonvillain & E. Ronceray, Philadelphia
Laurie Engine & Machine Co., Montreal
Perrin, Wm. R. & Co., Toronto

Pumps, Oil.

Sight Feed Oil Pump Co., Milwaukee, Wis.

Pumps, Steam.

Allis-Chalmers-Bullock, Limited, Montreal
Canada Foundry Co., Toronto.
Canada Machinery Agency, Montreal.
Darling Bros., Ltd., Montreal
Gas & Electric Power Co., Toronto
The Goldie & McCulloch Co., Galt.
John McDougall Caledonian Iron Works, Montreal.
H. W. Petrie, Toronto.
The Smart-Turner Mach. Co., Hamilton.
Standard Bearings, Ltd., Niagara Falls,
Waterous Engine Co. Brantford.

Pumping Machinery.

Canada Foundry Co., Limited, Toronto
Canada Machinery Agency, Montreal.
Canadian Rand Drill Co., Montreal.
Darling Bros., Ltd., Montreal
Gas & Electric Power Co., Toronto.
Hall Engineering Works, Montreal, Que.
Laurie Engine & Machine Co., Montreal.
London Mach. Tool Co., Hamilton, Ont.
John McDougall Caledonian Iron Works Co., Montreal.
The Smart-Turner Mach. Co., Hamilton
Standard Bearings, Ltd., Niagara Falls,

Punches and Dies.

W. H. Banfield & Sons, Toronto.
Bliss, E. W., Co., Brooklyn, N.Y.
Butterfield & Co., Rock Island.
Globe Machine & Stamping Co.
A. B. Jardine & Co., Hespeler, Ont.
London Mach. Tool Co., Hamilton, Ont.
Pratt & Whitney Co., Hartford, Conn.
H. W. Petrie, Toronto.
Standard Bearings, Ltd., Niagara Falls,

Punches, Hand.

Mechanics Supply Co., Quebec, Que.

Punches, Power.

John Bertram & Sons Co., Dundas, Ont.
E. W. Bliss Co., Brooklyn, N.Y.
Canada Machinery Agency, Montreal.
London Mach. Tool Co., Hamilton, Ont.
Niles-Bement-Pond Co., New York.

Punches, Turret.

London Mach. Tool Co., London, Ont.

Punching Machines, Horizontal.

John Bertram & Sons Co., Dundas, Ont.
London Mach. Tool Co., Hamilton, Ont.
Niles-Bement-Pond Co., New York.

Quartering Machines.

John Bertram & Sons Co., Dundas, Ont.
London Mach. Tool Co., Hamilton, Ont.

Railway Spikes and Washers.

Hamilton Steel & Iron Co., Hamilton

Rammers, Bench and Floor.

Buffalo Foundry Supply Co., Buffalo.
Detroit Foundry Supply Co., Windsor
Hamilton Facing Mill Co., Hamilton.
Paxson, J. W., Co., Philadelphia

Rapping Plates.

Detroit Foundry Supply Co., Windsor
Hamilton Facing Mill Co., Hamilton.
Paxson, J. W., Co., Philadelphia

Raw Hide Pinions.

Brown, David & Sons, Huddersfield, Eng

Reamers.

Wm. Abbott, Montreal.
Baxter, J. R., & Co., Montreal.
Butterfield & Co., Rock Island.
Frothingham & Workman, Ltd., Montreal
Hamilton Tool Co., Hamilton
Hanna Engineering Works, Chicago.
A. B. Jardine & Co., Hespeler, Ont.
Mechanics Supply Co., Quebec, Que.
Morse Twist Drill and Machine Co., New Bedford, Mass
Pratt & Whitney Co., Hartford, Conn.
Standard Tool Co., Cleveland.

Reamers, Steel Taper.

Butterfield & Co., Rock Island.
A. B. Jardine & Co., Hespeler, Ont.
John Millen & Son Ltd., Montreal.
Pratt & Whitney Co., Hartford, Conn.
Standard Tool Co., Cleveland.

Rheostats.

Gas & Electric Power Co., Toronto.
Canadian General Electric Co., Toronto.
Canadian Westinghouse Co., Hamilton.
Hall Engineering Works, Montreal, Que.
T. & H. Electric Co., Hamilton,

Riddles.

Buffalo Foundry Supply Co., Buffalo.
Detroit Foundry Supply Co., Windsor
Hamilton Facing Mill Co., Hamilton.
Dominion Foundry Supply Co., Montreal
J. W. Paxson Co., Philadelphia, Pa.

Riveters, Pneumatic.

Hanna Engineering Works, Chicago.
Independent Pneumatic Tool Co., Chicago, Ill.

Rolls, Bending.

John Bertram & Sons Co., Dundas, Ont.
London Mach. Tool Co., Hamilton, Ont.
Niles-Bement-Pond Co., New York.

Rolls, Chilled Iron.

Greey, Wm. & J. G., Toronto

Rolls, Sand Cast.

Greey, Wm. & J. G., Toronto

Rotary Converters.

Allis-Chalmers-Bullock, Ltd., Montreal.
Canadian Westinghouse Co., Hamilton
Paxson, J. W., Co., Philadelphia
Toronto and Hamilton Electric Co., Hamilton.

Rubbing and Sharpening Stones.

Norton Co., Worcester, Mass.

Safes.

Baxter, Paterson & Co., Montreal
The Goldie & McCulloch Co., Galt,

Sand Blast Machinery.

Canadian Rand Drill Co., Montreal.
Detroit Foundry Supply Co., Windsor
Dominion Foundry Supply Co., Montreal
Hamilton Facing Mill Co., Hamilton.
J. W. Paxson Co., Philadelphia, Pa.

Sand, Heavy, Grey Iron.

Buffalo Foundry Supply Co., Buffalo.
Detroit Foundry Supply Co., Windsor
Dominion Foundry Supply Co., Montreal
Hamilton Facing Mill Co., Hamilton.
Paxson, J. W., Co., Philadelphia

Sand, Malleable.

Buffalo Foundry Supply Co., Buffalo.
Detroit Foundry Supply Co., Windsor
Dominion Foundry Supply Co., Montreal
Hamilton Facing Mill Co., Hamilton.
Paxson, J. W., Co., Philadelphia

Sand, Medium Grey Iron.

Buffalo Foundry Supply Co., Buffalo.
Detroit Foundry Supply Co., Windsor
Dominion Foundry Supply Co., Montreal
Hamilton Facing Mill Co., Hamilton.
Paxson, J. W., Co., Philadelphia

Sand Sifters.

Buffalo Foundry Supply Co., Buffalo.
Detroit Foundry Supply Co., Windsor
Dominion Foundry Supply Co., Montreal
Hamilton Facing Mill Co., Hamilton.
Hanna Engineering Works, Chicago.
Paxson, J. W., Co., Philadelphia

Saw Mill Machinery.

Allis-Chalmers-Bullock, Limited, Montreal
Baxter, J. R., & Co., Montreal.
Canada Machinery Agency, Montreal.
Goldie & McCulloch Co., Galt.
Greey, Wm. & J. G., Toronto
Owen Sound Iron Works Co., Owen Sound
H. W. Petrie, Toronto.
Robb Engineering Co., Amherst.
Standard Bearings, Ltd., Niagara Fall
Waterous Engine Works, Brantford
Williams & Wilson, Montreal.

PRACTICAL LITERATURE

For Electricians, Scientists and Engineers

All orders payable in advance.

Locomotive Breakdowns

Emergencies and their Remedies

by Geo. L. Fowler, M.E.

Price $1.50, Postpaid.
250 Pages, 96 Illustrations.

This book is an up-to-date catechism treating on all classes of accidents which may happen to locomotive engines and giving the remedy to be applied. Accidents to the Valves, Cylinders, Valve Motion, Running Gear, Compound Locomotives, Boiler Troubles are treated separately and fully. It tells what should be done in any emergency.

Liquid Air

and the

Liquefaction of Gases

by T. O'Conor Sloane, A.M., E.M., Ph.D.

Price $2.00, Postpaid.
365 Pages, Fully Illustrated.

This book contains the full theory, history, biography and bibliography of the subject from the earliest times, as well as a description of all the wonderful experiments performed with it, showing how liquid air is carried hundreds of miles and handled in open buckets. It also tells what may be expected from it in the near future. Although a work of scientific interest and authority, it is intended for the general reader.

Engine Runner's

Catechism

by Robert Grimshaw, M.E.

Price $2.50, Postpaid.
354 pages, Profusely Illustrated.

This work gives complete descriptions of the general characteristics of various well-known makes of engines, followed by details of shipping, making foundations, erecting and starting, and the specific instructions as to the adjustment of special makes. Every engineer will find in this volume more really practical and useful information than is to be found anywhere else within a like compass.

Steam Engine

Catechism

by Robert Grimshaw, M.E.

Price $2.00, Postpaid.
413 Pages, Fully Illustrated.

This book is a series of nearly 1000 direct, practical questions and answers; in fact it contains every essential question on the Steam Engine with its answer. Leading engineers have recommended it as a valuable educator for the beginner as well as a reference book for the engineer. It is intended for

EVERYONE INTERESTED IN STEAM.

How to Become a Successful Electrician

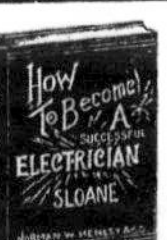

by Prof. T. O'Conor Sloane, A.M., E.M., Ph. D.

Price $1.00, Postpaid
200 Pages, Bound in Cloth.

Every young man whose ambition it is to become a successful electrician should read this book. It is the clearest and most practical book on the subject ever published. Its 19 Chapters include Mathematics, Physics, Chemistry, Electricity at Home, Drawing, Electrical Factory Work, Mechanical and Steam Engineering, the Manufacturing and Constructing Engineer, Inventing, Original Investigation, etc.

Gas, Gasoline and Oil Engines

by G. D. Hiscox, M.E.,

Price $2.50, Postpaid.
450 Pages, 351 Illustrations.

This book is a thorough treatise on the construction, installation, operation and maintenance of gas, gasoline, kerosene and crude petroleum engines. It is the new rewritten enlarged and revised 15th edition of this work giving general information for everyone interested in the popular motive power, and its adaptation to the increasing demand for a cheap and easily managed motor requiring no licensed engineer.

The MacLean Publishing Co.

Technical Book Department

10 Front St. East, - - TORONTO, CANADA

Sawing Machines, Metal.

Niles-Bement-Pond Co. New York.
Paxson, J. W., Co., Philadelphia

Saws, Hack.

Baxter, J. R., & Co., Montreal.
Canada Machinery Agency, Montreal.
Detroit Foundry Supply Co., Windsor
Frothingham & Workman, Ltd., Montreal
London Mach. Tool Co., Hamilton.
Mechanics Supply Co., Quebec, Que.
Rice Lewis & Son, Toronto.
John Millen & Son, Ltd., Montreal.
L. S. Starrett Co., Athol, Mass.

Screw Cutting Tools.

Hart Manufacturing Co., Cleveland, Ohio
Mechanics Supply Co., Quebec, Que.

Screw Machines, Automatic.

Canada Machinery Agency, Montreal.
Cleveland Automatic Machine Co., Cleveland, Ohio
London Mach. Tool Co., Hamilton, Ont.
National-Acme Mfg. Co., Cleveland.
Pratt & Whitney Co., Hartford, Conn.

Screw Machines, Hand.

Canada Machinery Agency, Montreal.
A. B. Jardine & Co., Hespeler.
London Mach. Tool Co., Hamilton, Ont.
Mechanics Supply Co. Quebec, Que.
Pratt & Whitney Co., Hartford, Conn
Warner & Swasey Co., Cleveland, O.

Screw Machines, Multiple Spindle.

National-Acme Mfg. Co., Cleveland

Screw Plates.

Butterfield & Co., Rock Island, Que.
Frothingham & Workman, Ltd., Montreal
Hart Manufacturing Co. Cleveland, Ohio
A. B. Jardine & Co., Hespeler
Mechanics Supply Co., Quebec, Que.

Screws, Cap and Set.

National-Acme Mfg. Co., Cleveland

Screw Slotting Machinery, Semi-Automatic.

National-Acme Mfg. Co., Cleveland

Second-Hand Machinery.

American Tool Works Co., Cincinnati.
Canada Machinery Agency, Montreal.
The Canadian Fairbanks Co., Montreal.
Goldie & McCulloch Co., Galt.
Machinery Exchange, Montreal.
Niles-Bement-Pond Co., New York.
H. W. Petrie, Toronto.
Robb Engineering Co., Amherst, N S.
Waterous Engine Co., Brantford.
Williams & Wilson, Montreal.

Shafting.

Baxter, J. R., & Co., Montreal.
Canada Machinery Agency, Montreal
The Canadian Fairbanks Co., Montreal.
Frothingham & Workman, Ltd., Montreal
The Goldie & McCulloch Co., Galt, Ont
Greey, Wm., & J. G. Toronto
Kennedy, Wm., & Sons, Owen Sound
Niles-Bement-Pond Co., New York.
Owen Sound Iron Works Co., Owen Sound
H. W. Petrie, Toronto.
Smart-Turner Machine Co., Hamilton.
Union Drawn Steel Co., Hamilton.
Waterous Engine Co., Brantford.

Shapers.

American Tool Works Co., Cincinnati.
John Bertram & Sons Co., Dundas, Ont
Canada Machinery Agency, Montreal.
The Canadian Fairbanks Co., Montreal.
Rice Lewis & Son, Toronto.
London Mach. Tool Co., Hamilton, Ont.
Niles-Bement-Pond Co., New York.
H. W. Petrie, Toronto.
Potter & Johnston Machine Co., Pawtucket, R.I.
Pratt & Whitney Co., Hartford, Conn.
Williams & Wilson, Montreal.

Sharpening Stones.

Carborundum Co., Niagara Falls, N.Y.

Shearing Machine, Bar.

John Bertram & Sons Co., Dundas, Ont.
A. B. Jardine & Co., Hespeler.
London Mach. Tool Co., Hamilton, Ont.
Niles-Bement-Pond Co., New York.

Shears, Power.

John Bertram & Sons Co., Dundas, Ont.
Canada Machinery Agency, Montreal.
A. B. Jardine & Co., Hespeler, Ont.
Niles-Bement-Pond Co., New York.
Paxson, J. W., Co., Philadelphia

Sheet Metal Goods

Globe Machine & Stamping Co., Cleveland, Ohio.

Sheet Meal Working Tools.

Brown, Boggs Co., Hamilton

Sheet Steel Work.

Owen Sound Iron Works Co., Owen Sound

Shingle Mill Machinery.

Owen Sound Iron Works Co., Owen Sound

Shop Trucks.

Greey, Wm. & J. G., Toronto
Owen Sound Iron Works Co., Owen Sound
Paxson, J. W., Co., Philadelphia

Shovels.

Baxter, J. R., & Co., Montreal.
Buffalo Foundry Supply Co., Buffalo.
Detroit Foundry Supply Co., Detroit.
Doggett, Stanley, New York
Frothingham & Workman, Ltd., Montreal
Dominion Foundry Supply Co., Montreal
Hamilton Facing Mill Co., Hamilton.
Mechanics Supply Co., Quebec, Que.
Paxson, J. W., Co., Philadelphia

Sieves.

Buffalo Foundry Supply Co., Buffalo.
Detroit Foundry Supply Co., Windsor
Dominion Foundry Supply Co., Montreal
Hamilton Facing Mill Co., Hamilton.
Paxson, J. W., Co., Philadelphia

Silver Lead.

Buffalo Foundry Supply Co., Buffalo.
Detroit Foundry Supply Co., Windsor
Doggett, Stanley, New York
Dominion Foundry Supply Co., Montreal
Hamilton Facing Mill Co., Hamilton.
Paxson, J. W., Co., Philadelphia

Snap Flasks

Buffalo Foundry Supply Co., Buffalo.
Detroit Foundry Supply Co., Windsor
Dominion Foundry Supply Co., Montreal
Hamilton Facing Mill Co., Hamilton.
Paxson, J. W., Co., Philadelphia

Soap Stones & Talc Crayons & Pencils.

Doggett, Stanley, New York

Soapstone.

Buffalo Foundry Supp'y Co., Buffalo.
Detroit Foundry Supply Co., Windsor
Doggett, Stanley, New York
Dominion Foundry Supply Co., Montreal
Hamilton Facing Mill Co., Hamilton.
Paxson, J. W. Co., Philadelphia

Soap Stone Facings.

Doggett Stanley, New York

Solders.

Lumen Bearing Co., Toronto

Special Machinery.

W. H. Banfield & Sons, Toronto.
Baxter, J. R. & Co., Montreal
John Bertram & Sons Co. Dundas, On
Brown, Boggs Co. Hamilton
Globe Machine & Stamping Co., Cleveland Ohio.
Greey, Wm. & J. G., Toronto
Hanna Engineering Works, Chicago.
Laurie Engine & Machine Co., Montreal.
London Mach. Tool Co., Hamilton, Ont.
H. W. Petrie, Toronto.
The Smart-Turner Mach. Co., Hamilton
Standard Bearings, Ltd., Niagara Falls.
Waterous Engine Co., Brantford.

Special Machines and Tools.

Brown, Boggs Co., Hamilton
Paxson, J. W., Philadelphia
Pratt & Whitney, Hartford, Conn.
Standard Bearings, Ltd., Niagara Falls

Special Milled Work.

National-Acme Mfg. Co., Cleveland

Columbian Core Wash

for Cores, Dry Sand and Loam Work

leads all others. Is better and cheaper than Plumbago.

Order a trial barrel and be convinced of its efficiency.

BUFFALO FOUNDRY SUPPLY CO.
BUFFALO, N.Y.

Mfrs. of finest Foundry Facings and Foundry Supplies of all kinds.

Our Triple Hem

Steel Shop Barrels and Boxes

Pay for themselves

Write us for catalogue and prices.

High Grade Coiled Wire Springs for all purposes

The Cleveland Wire Spring Co.
Cleveland, Ohio.

GILSON MFG. CO., Ltd., 110 York St., GUELPH, ONT.

We should have your business.

If prompt delivery, reasonable prices and best of workmanship appeal to you, we should have your pattern orders.

SEND FOR PATTERN CLASSIFICATION CARD

THE HAMILTON PATTERN WORKS
134 Bay St. N. HAMILTON, ONT.
Phone, 2664.

Partine — THE UNIVERSAL PARTING — Trade Mark Reg. U. S. Patent Off.

Preferred by Leading Foundrymen to Lycopodium

Partine does not run, cut, wash or burn off the face of the mould.

Guaranteed to produce, in every instance, a PATTERN TRUE casting, displaying the most delicate lines and features of an artistic design, accompanied by an absolute smoothness of the edges of the metal.

Its price is so low as to make its use imperative in every modern foundry.

Write for free sample, descriptive booklet and full particulars.

Sole Canadian Agents: **THE HAMILTON FACING MILL COMPANY, Limited**
Foundry Outfitters, HAMILTON, ONTARIO and MONTREAL, QUEBEC.

Cable Address: PARTINE — **THE FOUNDRY SPECIALTY CO.,** Cincinnati, O., U.S.A.

GILMOUR'S

20th Century Foundry Equipment

Cupolas, Tumblers, Trucks, Cranes, Ladles and Piqua Blowers

Gilmour's Double Spur Geared Tumbler is the longest lived and most practical mill ever put on the market. Let us tell you more about it.

Fire Brick and Cupola Blocks a Specialty We claim that our Fire Brick is superior to any other on the market. Write for samples and prices.

J. GILMOUR, Bennett Building, NEW YORK

Speed Changing Countershafts.

The Canadian Fairbanks Co., Montreal.

Spike Machines.

National Machinery Co., Tiffin, O.
The Smart-Turner Mach. Co., Hamilton

Spray Cans.

Detroit Foundry Supply Co., Windsor
Dominion Foundry Supply Co., Montreal
Hamilton Facing Mill Co., Hamilton.
Paxson, J. W., Co., Philadelphia

Springs, Automobile.

Cleveland Wire Spring Co., Cleveland

Springs, Coiled Wire.

Cleveland Wire Spring Co., Cleveland

Springs, Machinery.

Cleveland Wire Spring Co., Cleveland

Springs, Upholstery.

Cleveland Wire Spring Co., Cleveland

Sprue Cutters.

Detroit Foundry Supply Co., Windsor
Dominion Foundry Supply Co., Montreal
Hamilton Facing Mill Co., Hamilton.
Paxson, J. W., Co., Philadelphia

Stamp Mills.

Allis-Chalmers-Bullock, Limited, Montreal
Paxson. J. W., Co., Philadelphia

Steam Separators.

Darling Bros., Ltd., Montreal.
Dominion Heating & Ventilating Co., Hespeler
Robb Engineering Co., Montreal.
Sheldon's Limited, Galt.
Smart-Turner Mach. Co., Hamilton
Waterous Engine Co., Brantford.

Steam Specialties.

Darling Bros., Ltd., Montreal
Dominion Heating & Ventilating Co., Hespeler
Sheldon's Limited, Galt.

Steam Traps.

Canada Machinery Agency, Montreal.
Darling Bros., Ltd., Montreal
Dominion Heating & Ventilating Co., Hespeler.
Mechanics Supply Co., Quebec, Que.
Sheldons Limited, Galt.

Steam Valves.

Darling Bros., Ltd., Montreal

Steel Pressure Blowers

Buffalo Foundry Supply Co., Buffalo.
Dominion Foundry Supply Co., Montreal
Domininion Heating & Ventilating Co., Hespeler
Hamilton Facing Mill Co., Hamilton.
J. W. Paxon Co., Philadalphia, Pa.
Sheldon's Limited, Galt.

Steel Tubes.

Baxter, J. R., & Co., Montreal.
Mechanics Supply Co., Quebec, Que.
John Millen & Son, Montreal.

Steel, High Speed.

Wm. Abbott, Montreal.
Canadian Fairbanks Co., Montreal.
Frothingham & Workman, Ltd., Montreal
Alex. Gibb, Montreal.
Jessop, Wm., & Sons, Sheffield, Eng.
B. K. Morton Co., Sheffield, Eng.
Williams & Wilson, Montreal.

Stocks and Dies

Hart Manufacturing Co., Cleveland, Ohio
Mechanics Supply Co., Quebec, Que.
Pratt & Whitney Co., Hartford, Conn.

Stone Cutting Tools, Pneumatic

Allis-Chalmers-Bullock, Ltd., Montreal.
Canadian Rand Drill Co., Montreal.

Stove Plate Facings.

Buffalo Foundry Supply Co., Buffalo
Detroit Foundry Supply Co., Windsor.
Dominion Foundry Supply Co., Montreal
Hamilton Facing Mill Co., Hamilton.
Paxson, J. W., Co., Philadelphia

Swage, Block.

A. B. Jardine & Co., Hespeler, Ont.

Switchboards.

Gas & Electric Power Co., Toronto.
Allis-Chalmers-Bullock, Limited, Montreal
Canadian General Electric Co., Toronto.
Canadian Westinghouse Co., Hamilton.
Hall Engineering Works, Montreal, Que.
Mechanics Supply Co., Quebec, Que.
Toronto and Hamilton Electric Co., Hamiton.

Talc.

Buffalo Foundry Supply Co., Buffalo.
Detroit Foundry Supply Co., Windsor.
Doggett, Stanley, New York
Dominion Foundry Supply Co., Montreal
Hamiton Facing Mill Co., Hamilton.
Paxson, J. W., Co., Philadelphia

Tanks, Oil.

Sight Feed Oil Pump Co., Milwaukee, Wis.

Tanks, Steel.

Steel Trough & Machine Co., Tweed, Ont.

Tapping Machines and Attachments.

American Tool Works Co., Cincinnati.
John Bertram & Sons Co., Dundas, Ont.
Bickford Drill & Tool Co., Cincinnati.
Blair Tool & Machine Works, New York.
The Geometric Tool Co., New Haven.
A. B. Jardine & Co., Hespeler.
London Mach. Tool Co., Hamilton, Ont
Murchey Machine & Tool Co., Detroit.
Niles-Bement-Pond Co., New York.
H. W. Petrie, Toronto.
Pratt & Whitney, Cincinnati, O.
L. S. Starrett Co., Athol, Mass.
Williams & Wilson, Montreal.

Tapes, Steel.

Frothingham & Workman, Ltd., Montreal
Rice Lewis & Son, Toronto.
Mechanics Supply Co., Quebec, Que.
John Millen & Son, Ltd., Montreal.
L. S. Starrett Co., Athol, Mass.

Taps.

Baxter, J. R., & Co., Montreal.
Mechanics Supply Co., Quebec, Que.

Taps, Collapsing.

The Geometric Tool Co., New Haven.

Taps and Dies.

Wm. Abbott, Montreal.
Baxter, J. R., & Co., Montreal.
Butterfield & Co., Rock Island, Que.
Frothingham & Workman, Ltd., Montreal
The Geometric Tool Co., New Haven.
A. B. Jardine & Co., Hespeler, Ont.
Rice Lewis & Son, Toronto.
Mechanics Supply Co., Quebec, Que.
John Millen & Son, Ltd., Montreal.
Pratt & Whitney Co., Hartford, Conn.
Standard Tool Co., Cleveland.
L. S. Starrett Co., Athol, Mass.

Testing Laboratory.

Detroit Testing Laboratory, Detroit

Testing Machines.

Detroit Foundry Supply Co., Windsor.
Dominion Foundry Supply Co., Montreal
Hamilton Facing Mill Co., Hamilton.
Paxson, J. W., Co., Philadelphia

Thread Cutting Tools.

Hart Manufacturing Co., Cleveland, Ohio
Mechanics Supply Co., Quebec Que.

Tiling, Opal Glass.

Toronto Plate Glass Importing Co., Toronto.

Time Switches, Automatic, Electric.

Berlin Electric Mfg. Co., Toronto.

Tinware Machinery

Canada Machinery Agency, Montreal.

Tire Upsetters or Shrinkers

A. B. Jardine & Co., Hespeler, Ont.

Tool Cutting Machinery

Canadian Rand Drill Co., Montreal.

Tool Holders.

Armstrong Bros. Tool Co., Chicago.
Baxter, J. R., & Co., Montreal.
John Millen & Son, Ltd., Montreal.
H. W. Petrie, Toronto.
Pratt & Whitney Co., Hartford, Conn.

Tool Steel.

Wm. Abbott, Montreal.
Flockton, Tompkin & Co., Sheffield, Eng.
Frothingham & Workman Ltd., Montreal
Alexander Gibb, Montreal.
Wm. Jessop, Sons & Co., Toronto.
Canadian Fairbanks Co., Montreal
B. K. Morton & Co., Sheffield, Eng.
Williams & Wilson, Montral.

Torches, Steel.

Baxter, J. R., & Co., Montreal.
Detroit Foundry Supply Co., Windsor.
Dominion Foundry Supply Co., Montreal
Hamilton Facing Mill Co., Hamilton.
Paxson, J. W., Co., Philadelphia

Transformers and Convertors

Gas & Electric Power Co., Toronto.
Allis-Chalmers-Bullock, Montreal.
Canadian General Electric Co., Toronto.
Canadian Westinghouse Co., Hamilton.
Hall Engineering Works, Montreal, Que.
T. & H. Electric Co., Hamilton.

Transmission Machinery

Gas & Electric Power Co., Toronto.
Allis-Chalmers-Bullock, Montreal.
The Canadian Fairbanks Co., Montreal.
Greey, Wm. & J. G., Toronto.
Laurie Engine & Machine Co., Montreal.
Link-Belt Co., Philadelphia.
Paxson, J. W., Co., Philadelphia
H. W. Petrie, Toronto.
The Smart-Turner Mach. Co., Hamilton.
Waterous Engine Co., Brantford.

Transmission Supplies

Baxter, J. R., & Co., Montreal.
The Canadian Fairbanks Co., Montreal.
Wm. & J. G. Greey, Toronto.
The Goldie & McCulloch Co., Galt.
H. W. Petrie, Toronto.

Trolleys

Canadian Rand Drill Co., Montreal.
Dominion Foundry Supply Co., Montreal
John Millen & Son, Ltd., Montreal.
Northern Engineering Works, Detroit

Trolley Wheels.

Lumen Bearing Co., Toronto

Trucks, Dryer and Factory

Dominion Heating & Ventilating Co., Hespeler
Greey, Wm. & J. G., Toronto.
Northern Engineering Works, Detroit
Sheldon's Limited, Galt, Ont.

Tube Expanders (Rollers)

A. B. Jardine & Co., Hespeler.
Mechanics Supply Co., Quebec, Que.

Turbines, Steam

Gas & Electric Power Co., Toronto.
Allis-Chalmers-Bullock, Limited, Montreal
Canadian General Electric Co., Toronto.
Canadian Westinghouse Co., Hamilton.
Kerr Turbine Co., Wellsville, N.Y.

Turntables

Detroit Foundry Supply Co., Windsor.
Dominion Foundry Supply Co., Montreal
Hamilton Facing Mill Co., Hamilton.
Northern Engineering Works, Detroit
Paxson, J. W., Co., Philadelphia

Turret Machines.

American Tool Works Co., Cincinnati.
John Bertram & Sons Co., Dundas, Ont.
The Canadian Fairbanks Co., Montreal.
Gisholt Machine Co., Madison, Wis.
London Mach. Tool Co., Hamilton, Ont.
Niles-Bement-Pond Co., New York.
H. W. Petrie, Toronto.
Pratt & Whitney Co., Hartford, Conn.
Warner & Swasey Co., Cleveland, Ohio.
Williams & Wilson, Montreal.

Upsetting and Bending Machinery.

John Bertram & Sons Co., Dundas, Ont.
A. B. Jardine & Co., Hespeler.
London Mach. Tool Co., Hamilton, Ont.
National Machinery Co., Tiffin, O.
Niles-Bement-Pond Co., New York.

Valves, Blow-off.

Mechanics Supp'y Co., Quebec, Que.

Valves, Back Pressure.

Darling Bros., Ltd., Montreal
Dominion Heating & Ventilating Co. Hespeler
Mechanics Supply Co., Quebec, Que.
Sheldon's Limited, Galt.

Valve Reseating Machines.

Darling Bros., Ltd., Montreal

Ventilating Apparatus.

Canada Machinery Agency, Montreal
Darling Bros., Ltd., Montreal
Dominion Heating & Ventilating Co., Hespeler, Ont.
Sheldon's Ltd., Galt

Vises, Planer and Shaper.

American Tool Works Co., Cincinnati, O.
Frothingham & Workman, Ltd., Montreal
A. B. Jardine & Co., Hespeler, Ont.
John Millen & Son, Ltd., Montreal.
Niles-Bement-Pond Co., New York.

Washer Machines.

National Machinery Co., Tiffin, Ohio.

Water Tanks.

Ontario Wind Engine & Pump Co., Toronto

Water Wheels.

Allis-Chalmers-Bullock Co., Montreal.
Barber, Chas., & Sons, Meaford.
Canada Machinery Agency, Montreal.
The Goldie & McCulloch Co., Galt, Ont.
Greey, Wm. & J. G., Toronto.
Wm. Kennedy & Sons, Owen Sound.
John McDougall Caledonian Iron Works Co., Ltd., Montreal

Water Wheels, Turbine.

Barber, Chas., & Sons, Meaford, Ont.
Kennedy, Wm., & Sons, Owen Sound

Wheelbarrows.

Baxter, J. W., & Co., Montreal.
Buffalo Foundry Supply Co., Buffalo
Frothingham & Workman Ltd., Montrea
Detroit Foundry Supply Co., Windsor.
Dominion Foundry Supply Co., Montrea
Hamilton Facing Mill Co., Hamilton.
Paxson, J. W., Co., Philadelphia

Winches.

Canadian Pilling Co., Montreal
Frothingham & Workman, Ltd., Montreal

Wind Mills.

Ontario Wind Engine & Pump Co., Toronto

Window Wire Guards.

Expanded Metal and Fireproofing Co Toronto.
B. Greening Wire Co., Hamilton, Ont.
Paxson, J. W., Co. Philadelphia

Wire Chains.

The B. Greening Wire Co., Hamilton.

Wire Cloth and Perforated Metals.

Expanded Metal and Fireproofing Co. Toronto.
B. Greening Wire Co., Hamilton. Ont.
Paxson, J. W., Co., Philadelphia.

Wire Guards and Railings.

Expanded Metal and Fireproofing Co. Toronto.
B. Greening Wire Co. Hamilton, Ont.

Wire Nail Machinery.

National Machinery Co., Tiffin, Ohio.

Wire Rope.

Frothingham & Workman, Ltd. Montreal
B. Greening Wire Co., Hamilton, Ont.

Wood Boring Machines. Pneumatic.

Independent Pneumatic Tool Co., Chicago, Ill.

Wood-working Machinery.

Baxter, J. R., & Co., Montreal.
Canada Machinery Agency, Montreal.
The Canadian Fairbanks Co., Montreal.
Goldie & McCulloch Co., Galt.
H. W. Petrie, Toronto.
Waterous Engine Works Co., Brantford.
Williams & Wilson, Montreal.

The Acknowledged Standard Parting for Parting Molds

The Money Saver and Profit Producer

"Success means willingness to put aside the old for something better."

Partamol is the something "best" in parting compounds. It turns out a perfect mold with every line true to pattern. Allows of easy and perfect "slicking" of the mold. The results from its use will astonish you. The price will please you. Sample on request.

The Dominion Foundry Supply Company

MONTREAL AND TORONTO

SELLING AGENTS FOR CANADA

The Partamol Company, 415 Broadway, New York

FOR RIDDLING FOUNDRY SAND

THERE IS NOTHING CAN COMPARE WITH

HANNA SHAKERS

OPERATED BY

AIR ELECTRICITY STEAM

DON'T SCREEN YOUR SAND BY HAND

HANNA ENGINEERING WORKS

820 ELSTON AVE., CHICAGO, ILL.

THE UNIVERSAL SYSTEM OF MACHINE MOLDING

CAN BE SEEN IN OPERATION

AT PH. BONVILLAIN & E. RONCERAY'S, 1315 RACE STREET, PHILADELPHIA, PA.

AS GOOD FOR SMALL AS FOR VERY LARGE WORK.
AS GOOD FOR SIMPLE AS FOR INTRICATE WORK.

THE UNIVERSAL SYSTEM OF MACHINE MOLDING

ALPHABETICAL INDEX

SCOTCH AND AMERICAN

FULL STOCK CARRIED OF ALL SHAPES
PROMPT SHIPMENTS

ONTARIO LIME ASSOCIATION

BUILDERS' SUPPLIES — TORONTO

A Cupola unequaled for excellent results—14 sizes

NORTHERN ENGINEERING WORKS
16 Chene St., Detroit, Mich., U.S.A.

SOLVAY FOUNDRY COKE

HIGH CARBON
LOW SULPHUR

Can Make Prompt Shipment

BAIRD & WEST

Sole Selling Agents. — DETROIT, MICH.

THE HAMILTON STEEL & IRON CO.

Hamilton, - Canada LIMITED

PIG IRON

FOUNDRY—BASIC—MALLEABLE

Specialty of Iron for Machine Castings requiring finishing.

FORGINGS

ALL KINDS AND WEIGHTS

In ROUGH or ROUGH TURNED

DELIVERIES MADE PROMPTLY.

OPEN HEARTH BAR STEEL — HIGH GRADE BAR IRON

Technical Books

Dies

Their Construction and Use for the Modern Working of Sheet Metals.

By Joseph V. Woodsworth

Large 8vo. ; 384 pages ; 50 ; illustrations bound in cloth ; price, **$3.00**

The New Metal Worker Pattern Book

A Treatise on Pattern-Cutting as Applied to all Branches of Sheet Metal Work.

By Geo. W. Kittredge

Cloth bound ; price postpaid, **$5.00**

Modern Machine Shop Tools

Their Construction, Operation and Manipulation, including both Hand and Machine Tools.

By W. H. Vandervoort, M.E.

Large 8vo. ; 555 pages ; 673 illustrations ; bound in cloth ; price, **$4.00**

Modern Steam Engineering

By Gardner D. Hiscox

Price, **$3.00**

This is the most complete practical book issued for the Stationery Engineer. Just issued.

Compressed Air

Its Production, Uses and Appliances.

By Gardner D. Hiscox

Large 8vo. ; 820 pages ; 545 illustrations ; bound in cloth ; price, **$5.00**

Locomotive Catechism

By Robert Grimshaw

Illustrated by 222 engravings and 12 folding plates ; price, **$2.00**

Address Technical Book Dept. :

The MacLean Publishing Co., Limited

Montreal Toronto Winnipeg

Pilling Air Devices
Portable and Stationary Hoists
3/4 Ton to 10 Tons
Turntable Tractors, Winches, Cranes, etc., etc.
The Canadian Pilling Co.
180 St. James St.,
MONTREAL

If the experience gained from 20 years of Air Compressor designing and building is of advantage to you, drop us a line.

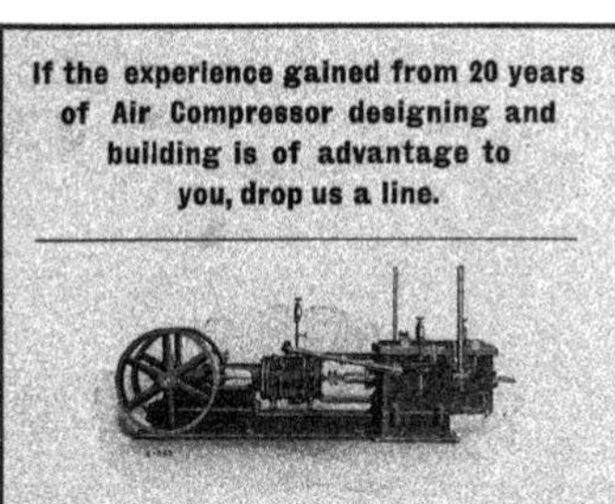

We build Air Compressors in 1000 styles and sizes, for all duties.

DO YOU WANT A CATALOG?

Canadian Rand Company,
LIMITED
MONTREAL, QUE.

JEFFREY

SWING-HAMMER
PULVERIZERS

With New Automatic Feed, Worm Gear and Screw Lowering Device.

For Full Description Ask for Catalogue No. 31.

THE JEFFREY MFG. COMPANY
COLUMBUS, OHIO, U.S.A.
Canadian Office and Works: MONTREAL

Rugged Design and Construction.

High Efficiency and Power Factor.

Satisfactory Operation under most exacting conditions.

5 HP. Three-Phase Motor

The Packard Electric Co. *Lim.*
Works: St. Catharines
Montreal Office: 127-129 Bell Telephone Building
Winnipeg Office: Somerset Block

American Pattern Shop Practice

By H. J. McCASLIN

JUST ISSUED. An up-to-date, practical work on pattern-making, based upon the author's own experience, treating of molding and core making as well as patternmaking. Special attention is given to core box work. The following subjects are fully dealt with in a clear, concise way:

Engine Patterns	**Gearing**
Molding and Cores	**Representative Pattens**
Sweep Work	**Hints, Suggestions and Rigs**

BOUND IN CLOTH. FULLY ILLUSTRATED

PRICE $3.00, POSTPAID.

Technical Book Dept., MACLEAN PUBLISHING CO. 10 Front St. E. - Toronto

"Best" Dry Cells

For your Ignition in

Automobile
Motor Boat
Gas Engines

The kind the largest Telephone Company in Canada is using to-day, because they give the longest service.

Ask your dealer for them, or

The Berlin Electrical Mfg. Co., Limited
474 King Street West, TORONTO

No Frost Breaks, no Corrosion No Electrolysis

It is easily and cheaply laid. Its carrying capacity is never decreased by rust.

GALVANIZED WIRE WOUND WOODEN PIPE

CANADIAN PIPE COMPANY, Limited

PHONE 1642 ***VANCOUVER, B.C.*** ***P.O. BOX 915***

WOOD STAVE WATER PIPE

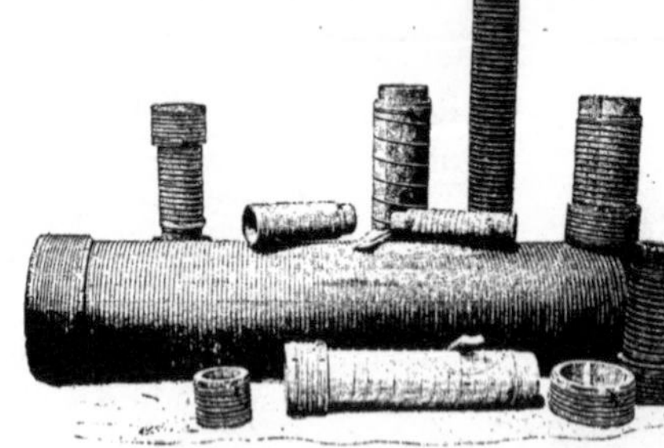

Wood Pipe Designed for Different Pressures.

Machine-banded made from choice British Columbia or Douglas Fir, absolutely clear and free from knots or defects, wound with heavy gauge galvanized steel wire. For

Water Works, Power Plants, Irrigation, Mining and all Hydraulic Propositions, where conduits or pressure pipe are required.

Reliable, durable, economical, efficient, easy to instal. Costs very much less than any other sort of water pipe, and with greater efficiency in use.

Write for Catalogue and Circulars.

Estimates also given on continuous stave pipe of large diameters.

PACIFIC COAST PIPE COMPANY Limited

P.O. Box 563

1551 Granville Street (Next Bridge) - VANCOUVER, B.C.

ISLAND CITY ANTI-RUST PAINTS

To Architects and Structural Steel Manufacturers

Our Anti-Rust Paints are made specially to prevent the effects of Electrolysis on iron and steel. They are guaranteed to stand moisture, heat, frost and contraction, and will keep the metal covered to its standard strength - being made expressly for Bridges and Structural Work. Prices and Samples for testing can be had by applying to

P. D. DODS & CO.,

162-164 McGILL STREET, - - MONTREAL

NORTON JACKS

ARE MADE IN CANADA AND SAVE YOU DELAY AND DUTY OVER AMERICAN JACKS

They are Ratchet Screw Jacks with Ball Bearings to reduce the friction. Working parts are protected from grit and rust. Superior to Hydraulic Jacks for all classes of service. Simple, durable, and guaranteed in every respect. Absolutely safe. Will work in any position. Standard on the leading railroads.

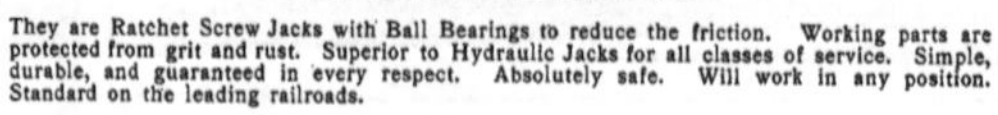

Made in 50 styles; 8 to 70 tons capacity

Send for complete illustrated catalogue and discount.

Sectional View of 70-Ton Bridge Jack | 25-Ton Foot Lift Jack | 35-Ton Plain Jack | Traversing Jack 16 to 60 Tons | Track Jack 10 and 15 Tons | Journal Box Jack 15 Tons

MADE ONLY BY

A. O. NORTON, Inc., Coaticook, Prov. Quebec

Stock carried by MUSSENS, Limited, Montreal

HART'S ADJUSTABLE DIE STOCKS

For Pipe "DUPLEX" For Bolts

Their instant adjustability, requiring only the hand and no wrench, makes tight joints always obtainable. Perfect threads are made with a single cut. After cutting a thread the dies may be opened instantly and the stock removed, avoiding the running back over the threads, an operation which frequently strips the threads just cut. The cutting edge of the dies is on an individual tool by itself, and can be sharpened by simply grinding the face, as you do a chisel. These stocks are fitted with self-centering guides. We can supply them with a cutting off device if so desired.

No. 3½
Range ½ to 2-in., for Pipe.

Ask for our prices

Ratchet Pattern. No. 8
Range 1 to 2 in. Pipe, one or two handles may be used.

FROTHINGHAM & WORKMAN, Limited

MONTREAL, CANADA

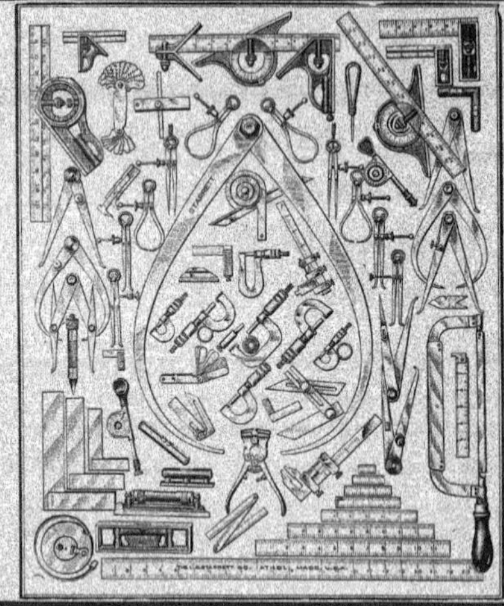

TOOLS! TOOLS! TOOLS!

Tools of the Finest Quality. The most delicate Mechanism. The Best Finish.

Tools that cannot be surpassed, that cannot be equalled. Tools for all kinds of Mechanical Work.

Rice Lewis & Son
LIMITED

TORONTO

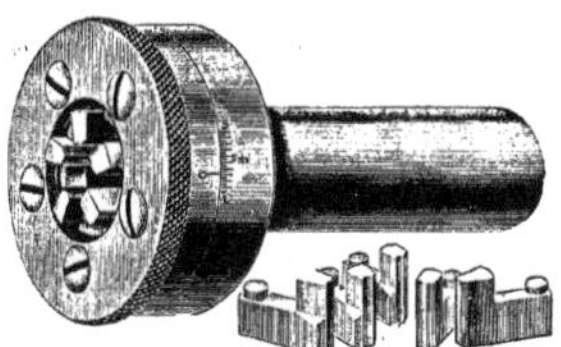

GEOMETRIC ADJUSTABLE HOLLOW MILLING TOOLS

If you wish to secure economy in Brass Finishing, it will pay you to investigate the labor saving qualities of our ADJUSTABLE HOLLOW MILLING TOOLS. No other device will so rapidly produce accurate uniform diameters. They are made in various sizes, and can be used on Automatic Screw Machines, Hand Turret Lathes, or rotated in any way desired. The index permits rapid setting to any diameter within its capacity and equally quick change from one diameter to another. They will reduce work of any length, will mill close to shoulder, and as the sharpening of the blades is done only at the outer edges, there is very little time lost in regrinding. If you are finishing brass work of any kind you will be interested in a full description of this tool.

The Geometric Tool Co., New Haven, Conn.

Canadian Agents: WILLIAMS & WILSON, Montreal, Que. U.S.A.

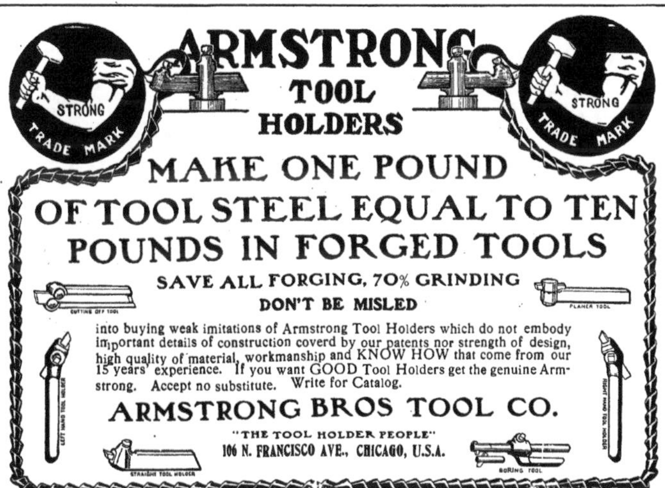

The Place to Secure THE BEST

CASTINGS

GREY IRON
BRASS BRONZE
ALUMINUM
ROUGH AND MACHINED

Hoisting Engines
Power Hoists

ELECTRIC, STEAM, BELT AND HAND POWER

Derricks and Contractors' Machinery

FOR FURTHER PARTICULARS ADDRESS

NIAGARA FALLS MACHINE AND FOUNDRY COMPANY, LIMITED
NIAGARA FALLS, Ont.

329 for the price of 1

Last month three hundred and twenty-nine magazines and periodicals reached the Editorial Department of the BUSY MAN'S MAGAZINE. Each of these was carefully ransacked by a competent editorial staff, and the most INTERESTING, LIVE, UP-TO-DATE articles selected and reproduced in the

BUSY MAN'S MAGAZINE

Think of the value of this single number of the BUSY MAN'S to you—containing, as it does, the cream of the world's periodical press.

You will not have to turn over a dozen pages to get one that interests you. Every article is interesting, selected, as it was, from a leading magazine because it was the best in that publication.

Yet this many-in-one magazine is only $2 per year.

We have a special offer which will interest you.

Write Us To-day *A Post Card Will Do*

THE BUSY MAN'S MAGAZINE
Toronto

BOOKS
for
Engineers
Draughtsmen, Mechanics, etc.

Post free by registered post to any address in the world, at published prices.

	NET PRICE s.	d.
Workshop Costs, for Engineers, etc., by S. and F. Pearn. 2nd Edition	12	6
How to Use Water Power (Chatley), illustrated	2	6
Modern Steam Traps (English and American) by Gordon Stewart	3	0
Sand and Clay: Their Analysis and Physical Properties	2	0
Air Compressors and Blowing Engines, by Chas. H. Innes, M.A.	4	6
The Fan; including the Theory and Practice of Centrifugal and Axial Fans, by Ch. H. Innes, M.A.	4	0
The Proportions and Movement of Slide Valves, by W. D. Wansbrough	4	6
Machines and Tools Employed in the Working of Sheet Metals, by R. B. Hodgson	4	6
Governors and Governing Mechanism, by Hall	4	0
Specification of a Modern Lancashire Boiler and its Seating (200 lbs. pressure), by "Inspector" M.E.	5	0
Continuous Current Dynamos and Motors and their Control, by W. R. Kelsey	5	0
Notes on Construction and Working of Pumps, Marks	4	6
Modern Ironfoundry Practice: Part I., Hand Moulding, Bale	5	0
Part II., Machine Moulding, Bale	3	6
Modern Gas and Oil Engines, by F. Grover. 4th Edition	5	0
The Indicator and its Diagrams, by Chas. Day. 3rd Edition	4	6
The Chemistry of Materials of Engineering, by A. H. Sexton. 2nd Edition	5	0
The Management of Small Engineering Workshops, Barker	7	6
Problems in Machine Design, by Chas. Innes	4	6
Heat and Heat Engines; a Treatise on Thermo-Dynamics. Popplewell	6	0
Centrifugal Pumps, Turbines and Water Motors.	4	6
Application of Graphic Methods to the Design of Structures. 2nd Edition	5	0
Engineering Estimates and Cost Accounts, Burton.	3	0
Graphic Methods of Engine Design, Barker. 2nd Ed.	3	6
Injectors: Theory, Construction and Working, Pullen.	3	6
Construction of Cranes and Lifting Machinery, Marks.	3	6
Marine Engineers: Their Qualifications and Duties	5	0
A.B.C. of the Differential Calculus, Wansbrough	3	0
The Theta-Phi Diagram Applied to Steam, Gas, Oil, and Air Engines	3	0
Mechanical Engineering Materials, by Marks	2	6
The Naval Engineer and Command of the Sea, Burton	2	6

Send post card for complete Catalogue, post free.

The Technical Publishing Co., Limited

30 Cross St., MANCHESTER, ENGLAND

DO YOU KNOW that there is no other magazine in the world that covers the fields of engineering and technology to the same extent or in such a useful literary way as TECHNICAL LITERATURE, and that no other appeals equally to specialists and the general reader in so many branches of the profession? It has become a most valuable companion to such professional men as

Architects	Engineering Students	Hydraulic Engineers	Mechanics	Patent Attorneys
Chemists	Factory Managers	Iron and Steel Makers	Metallurgists	Power Consumers
Civil Engineers	Foundrymen	Irrigation Engineers	Mining Engineers	Railroad Engineers
Contractors	Gas Engineers	Machine Tool Makers	Municipal Engineers	Sanitary Engineers
Draftsmen	Heating, Lighting and Ventilating Engineers	Marine Engineers	Oil, Paint and Varnish Makers	Steam Engineers
Electrical Engineers		Mechanical Engineers		Water Works Engineers

AND OTHERS INTERESTED IN TECHNICAL WORK OR STUDY

Such men may be considered as forming the backbone of the citizenship of the country. They are the men who do things, and who control the expenditure of millions of dollars, involving the purchase of all kinds of material and machinery. If you have anything to offer them for business or pleasure, advertise it in TECHNICAL LITERATURE. *Let us show you the scope, vitality and the power of our circulation.*

ONE ADVERTISER WRITES: "Your magazine is a good one and sure of success. It is going to be looked over by a good class of readers, and our advertisement will not be lost in a mass of ill-digested matter."

ADVERTISING RATES SENT ON APPLICATION

Subscription price, $2.00 a year to United States or Mexico; $2.25 to Canada; foreign countries, $2.50. Single copies, 20 cents.

TECHNICAL LITERATURE, 905 St. Paul Building NEW YORK

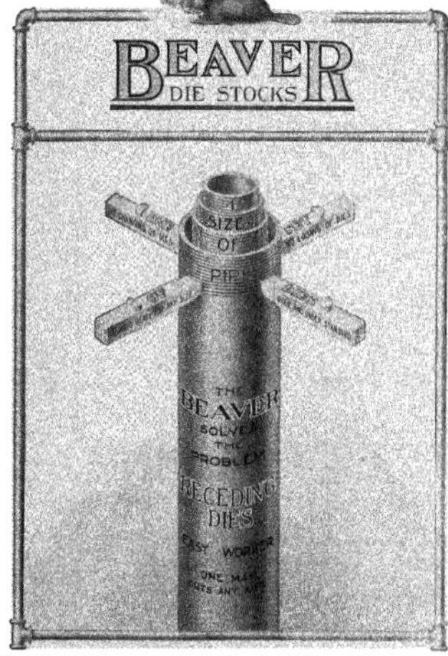

Manufactured by ***BORDEN-CANADIAN CO.***
66 Richmond St. East, Toronto

OTIS ELEVATORS
FOR ALL DUTIES
Electric, Hydraulic, Belt,
Steam and Hand Power
MANUFACTURED BY
OTIS-FENSOM ELEVATOR COMPANY
LIMITED
Head Office: TORONTO, ONT.
Works: HAMILTON, ONT.

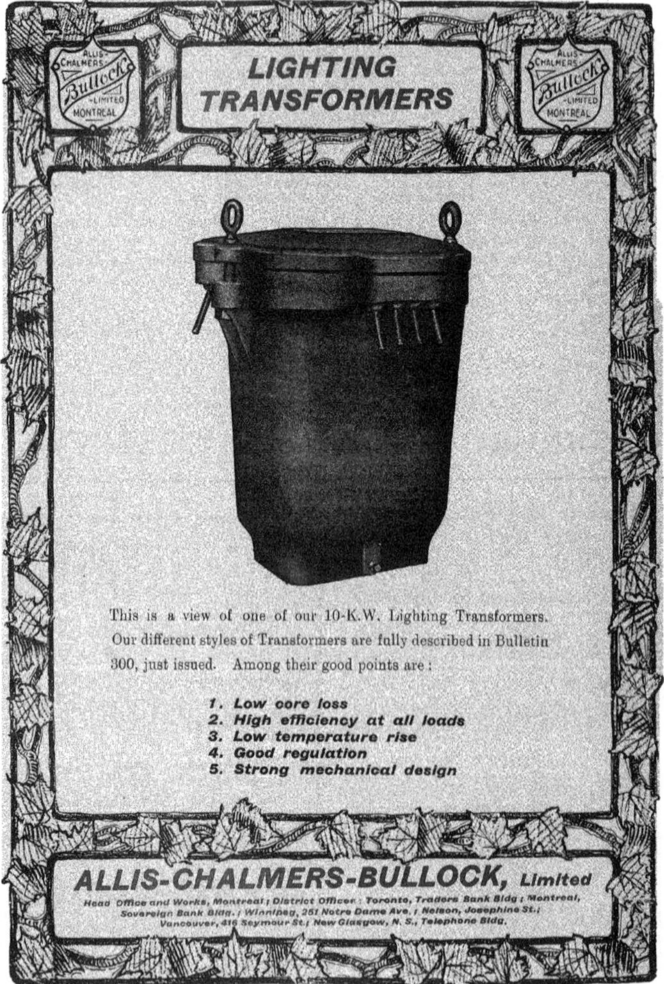
ALLIS-CHALMERS-Bullock-LIMITED MONTREAL
LIGHTING TRANSFORMERS
ALLIS-CHALMERS-Bullock-LIMITED MONTREAL
This is a view of one of our 10-K.W. Lighting Transformers. Our different styles of Transformers are fully described in Bulletin 300, just issued. Among their good points are:
1. Low core loss
2. High efficiency at all loads
3. Low temperature rise
4. Good regulation
5. Strong mechanical design
ALLIS-CHALMERS-BULLOCK, Limited
Head Office and Works, Montreal; District Offices: Toronto, Traders Bank Bldg; Montreal, Sovereign Bank Bldg.; Winnipeg, 251 Notre Dame Ave.; Nelson, Josephine St.; Vancouver, 416 Seymour St.; New Glasgow, N. S., Telephone Bldg.

CANADIAN MACHINERY AND MANUFACTURING NEWS

A monthly newspaper devoted to the manufacturing interests, covering in a practical manner the mechanical, power, foundry and allied fields. Published by The MacLean Publishing Company, Limited, Toronto, Montreal, Winnipeg, and London, Eng.

OFFICE OF PUBLICATION : 10 FRONT STREET EAST, TORONTO

Vol. III. DECEMBER, 1907 No. 12

New Type 18" Double Back Geared

ENGINE LATHE

If you are looking for the most complete and up-to-date **Lathe** on the market, investigate our new line of **Double Back Geared Engine Lathes.** For accuracy and handiness they are unexcelled. They are made extra heavy in all parts. Spindles are of extra large diameter and made of Crucible Steel. Cone is of large diameter and wide face. Quick Change Feed Mechanism gives four instantaneous changes of feed by simply throwing over lever in front of machine. These changes can be obtained without stopping machine

We can make prompt shipment of 18" Lathes in 8 and 10-ft. beds

LONDON MACHINE TOOL CO., Limited

HAMILTON, CANADA

Sales Office: Traders' Bank Building, TORONTO, Canada

Warner & Swasey Turret Lathes—

For Brass and Iron Work—complete equipments

for the manufacture of Plumbers and Steamfitters' fittings.

Every modern facility provided for rapid and economical production.

One type of machine here illustrated — adapted for general work.

Styles and sizes for every requirement.

Universal Turret Lathe

1907 General Catalogue—recently published—now ready for distribution.

THE WARNER & SWASEY CO., Cleveland, Ohio, U.S.A.

Canadian Agents: A. R. Williams Machinery Co., Toronto and Williams & Wilson, Montreal.

¶ The illustration shown is one of our latest. The success of the W. W. Sly Mfg Co., in manufacturing cleaning room equipments has attracted the attention of people without the ability to create or design goods that are required for that purpose—(we consider this a compliment). If you compare their catalogue with ours, you will see that they have even copied our catalogue. If you compare their Mills with ours, you will see that it is the Mill that we built a number of years ago, or a similar mill to our present Mill, with all the vital improvements left out. We are constantly imroving our Mills. Buying them means that you are from one to five years in advance of our competitors.

Let us figure on your Cleaning Room Requirements.

The W. W. Sly Mfg. Co., Cleveland, O.

"American" Double-Back-Geared Lathe
WITH
3-STEP CONE
REPRESENTS THE
HIGHEST DEVELOPMENT
IN AN
"All-Round Lathe"
3-STEP CONE
gives an Extra wide Belt and Great Power.
DOUBLE-BACK-GEARS
afford an Extra Gear Ratio with Increase of Spindle Speeds.
Built in 14, 16, 18, 20, 24, 27, 30 and 36 inch sizes
WRITE NOW for Catalog "C"
INVESTIGATE
THE AMERICAN TOOL WORKS CO.
700-750 CULVERT STREET, CINCINNATI, U. S. A.
12
LATHES PLANERS SHAPERS RADIAL DRILLS
Write direct or to Agents: WILLIAMS & WILSON, Montreal; A. R. WILLIAMS MACHINERY CO., Toronto, Winnipeg, Vancouver.

THOUSANDS OF MANUFACTURERS
IN THIS COUNTRY AND ABROAD ARE USING
BECKER-BRAINARD MILLING
MACHINES and MILLING CUTTERS
Because they have found them to be Practical, Reliable, Economical and Durable. Years of practical test have proved the efficiency of our product. "Show us your work, and we will show you our method."
Becker-Brainard Milling Machine Co.
Hyde Park, Mass., U.S.A.
AGENTS:—McDowell, Stocker & Co., Chicago. Chas. G. Smith Co., Pittsburg. J. L. Osgood, Buffalo. A. B. Bowman, St. Louis. A. R. Williams Machinery Co., Toronto, Montreal, Winnipeg and Vancouver, Canada. Ludw. Loewe & Co., Berlin. Bevan & Edwards Propty, Ltd., Melbourne. Selig Sonnenthal & Co., London. Schuchardt & Schutte, Berlin, Stockholm, Vienna, St. Petersburg. A. H. Schutte, Cologne, Liege, Milan Barcelona, Brussels, Paris, Bilbao.
366

REPETITION WORK

A great many shops are proving to their own satisfaction, that it no longer pays to finish small machine parts on the old-fashioned Lincoln type of miller. We can tell you of some large shops that have replaced their Lincoln pattern miller equipment with Cincinnati No. 2 Plains, and are getting out more work and better work, because these machines are more powerful and more convenient to operate. Note for example, this grey iron casting $1\frac{1}{8}$ in. wide. Twenty-four of these are held in a string jig and finished at one cut with a gang of cutters $4\frac{1}{2}$ in. and 3 in. diameter as shown. The total width of cut is $3\frac{3}{4}$ in., depth of cut $\frac{1}{8}$ in. —table travel $1\frac{7}{8}$ in. per minute. The finished pieces are accurate within .001, ready for assembling without additional handwork.

This machine finishes 24 of these pieces in 15 minutes. Let us figure on millers for your work. WE ARE MILLING SPECIALISTS.

The Cincinnati Milling Machine Company

CINCINNATI, OHIO, U.S.A.

CANADA AGENT: H. W. PETRIE, Toronto and Montreal.

"DON'T BORE—USE A TUBE."

SHELBY SEAMLESS STEEL TUBES

USED FOR:—

ELEVATOR CAGES	BUSHINGS
CREAM SEPARATORS	HAND RAILINGS
DISPLAY FRAMES	TROLLEY POLES
HOSE MANDRILS	LOCOMOTIVE BELLS
ROLLER BOOK SHELVES	AGRICULTURAL MACHINERY
HOLLOW SHAFTING	FLAG STAFFS
HANDLES	POLES AND MASTS
UMBRELLA RODS	PISTON RODS

and a thousand other things.

Don't waste time boring from the solid—use SHELBY SEAMLESS STEEL TUBING

WRITE FOR CATALOGUE AND PRICES.

CANADIAN DISTRIBUTORS:

JOHN MILLEN & SON, Limited

MONTREAL TORONTO VANCOUVER

IT PAYS TO INSTALL THE BEST

POWER PRESSES

The many exclusive features embodied in the type shown above make it so much superior to the next best that there is no comparison.

Write for prices and particulars on **Presses, Dies, Tools and Special Machinery.**

W. H. BANFIELD & SONS

120 Adelaide Street West, TORONTO.

No. 3 Marking Machine

This machine is designed to mark by means of steel dies letters, numbers, trade-marks, monograms, etc., on either flat or round metal surfaces, such as axes, pistol barrels, files, table cutlery, twist drills, taps, dies, reamers and a great variety of work.

Our principle of bringing a slight line contact of die with work insures light wear on dies and makes it possible to mark hollow tubes, handle, chuck and valve work which would not stand the blow of a punch or drop press.

Correspondence and sample of work solicited. Machine sent on approval to responsible parties.

Send for Catalogue

THE

Dwight Slate Machine Co.,

HARTFORD, CONN., U.S.A.

Opening Die Head

To Cut ¼" to 1".

Applied to our Bolt Threaders or sold separately.

SAVES TIME. SAVES DIES.

A. B. JARDINE & CO.,

HESPELER, ONT.

NORTON COMPANY

MANUFACTURERS OF

NORTON GRINDING WHEELS

MADE OF

ALUNDUM

Works at WORCESTER, MASS., U.S.A., and NIAGARA FALLS, N. Y.

NORTON GRINDING WHEELS are made of ALUNDUM and are fast cutting and long wearing, an economy which works both ways. Your grinding can be done in the least time—a saving in labor, in the workman's time, and an increased output—and a Norton Grinding Wheel of the proper grain and grade will wear longer than an ordinary wheel.

Write us your requirements and we will tell you the grade of wheel best suited to your work.

NORTON COMPANY, WORCESTER, MASSACHUSETTS

CANADIAN AGENTS:

THE CANADIAN FAIRBANKS CO., LIMITED

Montreal Toronto Winnipeg Vancouver

F. H. ANDREWS & SON

Quebec, P.Q. 60

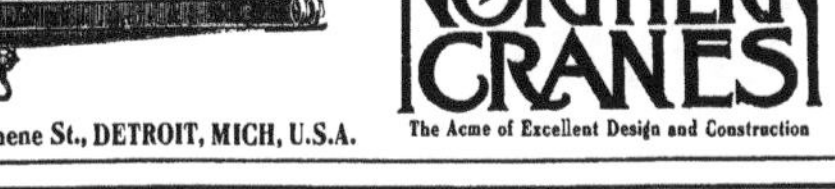

NORTHERN ENGINEERING WORKS, 16 Chene St., DETROIT, MICH, U.S.A.

Pipe Cutting and Threading Machine

For Hand or Power, Bench or Trench

FORBES PATENT DIE STOCKS will do either from ⅛ inch I.D. to 15 inch O.D. ***ON THE SPOT***

No expense or delay of carting your pipe to the machine. We keep regular sizes in stock and can make prompt shipment. Write for Catalog.

THE CURTIS & CURTIS CO., ***36 Garden Street, Bridgeport, Conn.***

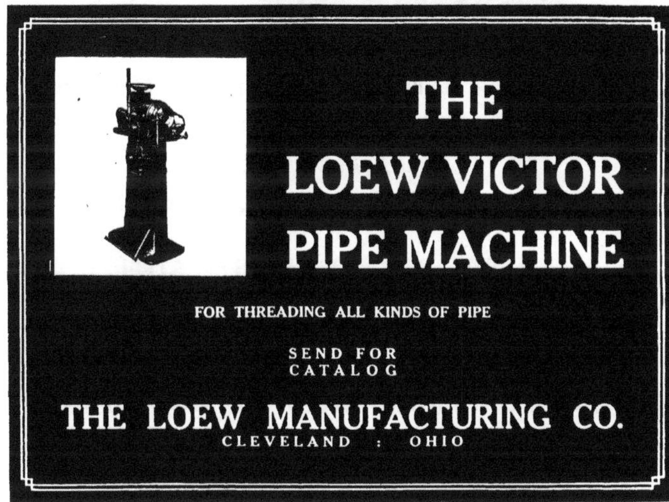

HOW DO YOU HEAT YOUR HIGH SPEED STEELS ?

No doubt in a coal forge, and suffer the same losses and have much the same troubles as other manufacturers and shop owners who do likewise.

Why not do all your heating of high speed steels absolutely correct? Why not cut out the losses?

A Stewart High Speed Steel Furnace will do it.

Burns any kind of gas or crude oil.

Send for complete descriptive circular and catalogue showing over 60 styles of furnaces. Something for every metal worker.

Chicago Flexible Shaft Co.

144 La Salle Ave., Chicago

"RAYO" FILES

Made by

B. K. MORTON & CO.
Sheffield, Eng.

Canadian Representative: D W. CLARK, P.O Box 52, Toronto

"INTRA" STEEL

Made by

Messrs. Jonas & Colver, Ltd., SHEFFIELD, ENG.

Specially recommended for

Taps, Dies, Punches, Chisels, Screw-Cutting Tools, Etc.

Combines toughness and durability with a cutting power superior to highest grades of carbon steel, at less price.

IN STOCK

WILLIAM ABBOTT
334 St. James St., MONTREAL

JESSOP'S BEST TOOL STEEL

"ARK" High-Speed Steel

THE FAVORITE BRANDS WITH USERS OF GOOD STEEL.
A LARGE ASSORTMENT OF SIZES IN STOCK.
JESSOP'S HIGH-GRADE FILES AND RASPS.

80 Bay, St., Toronto, Ontario,
Chas. L. Bailey, Agent.

Jas. Robertson Co., Ltd.
Montreal, Quebec

Reid-Newfoundland Company
St John's, Newfoundland.

Jas. Robertson Co., Ltd.,
St. John, New Brunswick.

WM. JESSOP & SONS, Limited, Manufactory, SHEFFIELD, ENGLAND.

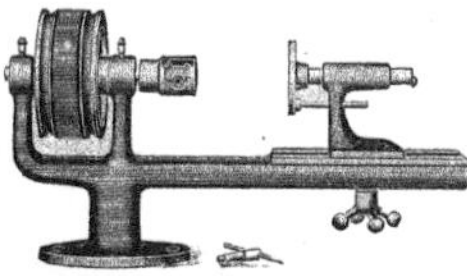

Champion Tapping Machine

Especially adapted for high speed tapping in light work.

Taps holes either through or to depth. Is automatic in operation, very rapid and reduces the breakage of taps to the minimum. Capacity up to ½-in. holes.

MANUFACTURED BY THE

Blair Tool and Machine Works
West St. NEW YORK CITY

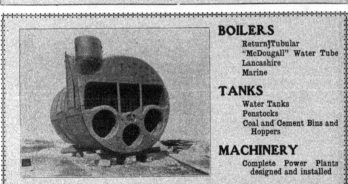

Telegraphic Address:
"HALLTYNE"

WORKS
Phone Main 256

THE HALL ENGINEERING WORKS

14 COTE ST., MONTREAL

ENGINEERS, BOILERMAKERS, BRASS & IRON FOUNDERS, COPPER SMITHS & BLACKSMITHS, SHIP REPAIRS A SPECIALTY

Agents for J. & E. Hall's Refrigerating Machinery.
Atlas Preservative Co. & United Asbestos Co.
of LONDON, ENG.

THOS. HALL, M. N. E. C. I. E. S. — Phone West 1737

Late Supt. Engineer Messrs. Elder Dempster & Co. & Can. Pac. Rly. Co., London

T. O. SINCLAIR, M. I. M. E. — Phone West 917

The Heaviest Line of
Automatic Valves and Traps
For High Pressure Steam and Water Service Ever Presented to Users.

No Chattering or Hammering in the

ANDERSON

Everlasting Cushioned Non-Return
"ANGLE" or "GLOBE"
VALVES

"The Life and Property Insurance Valves"

Protect Your Power Station With the Valves that are Absolutely Guaranteed

GOLDEN-ANDERSON
VALVE SPECIALTY CO.
1025 Fulton Building, Pittsburg, Pa.

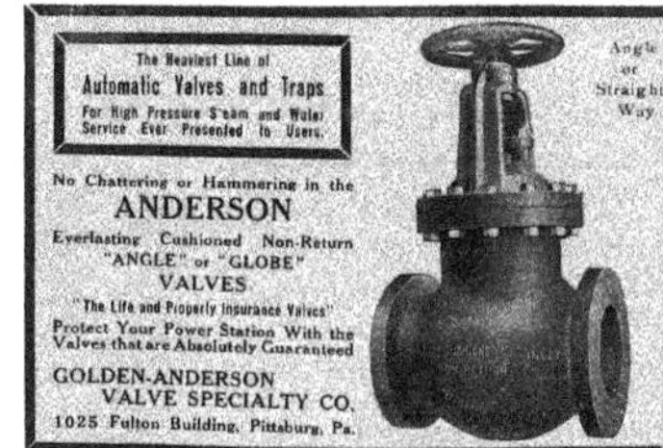

Angle or Straight Way

CHEAP POWER

by securing one of our

"STICKNEY" GASOLINE ENGINES

Sizes, 1 h.p. to 25 h.p.

Has less parts than any other engine.

Made to Stand Hard Work

We guarantee power stated in catalogue.
Write us for Illustrated Catalogue.

ONTARIO WIND ENGINE and PUMP CO., Limited
Winnipeg — TORONTO — Montreal

JUST WHAT YOU ARE LOOKING FOR!

INJECTORS, CONSTRUCTION, CARE AND MANAGEMENT

Useful information on the principles of their construction, use and practical operation as steam boiler feeders and other purposes; intended for the use of engineers, firemen and steam users. By FREDERICK KEPPY, M. E.

84 pp., 6x9 in. 45 engravings, 16 tables. Price, 25c., postpaid

We supply any book by mail, postage prepaid, on receipt of the publication price. Send for catalog.

The American Industrial Publishing Co.
Publishers, Booksellers and Importers
Box B, Bridgeport, Conn., U.S.A.

PULP MILL MACHINERY

SUCCESS WOOD PULP GRINDER

WE MAKE

- "SUCCESS" SCREENS
- GRINDERS
- WET MACHINES
- BARKERS
- WOOD SPLITTERS
- CONVEYORS
- SPECIAL CUTTING-OFF RIGS

Our Special Catalogue will give full information about these Machines.

Waterous Engine Works Co., Limited

Brantford, - Canada

A Year's Record of 18 Coaling Stations:

Time to coal per locomotive, 3 min.
Average cost per ton, 4 cts.

Locomotive Coal and Cinder Stations

designed and built according to our specifications "typify efficiency." Such is railroad testimony.

LINK-BELT COMPANY

PHILADELPHIA, CHICAGO, INDIANAPOLIS

New York: 299 Broadway.
Seattle: 440 New York Block.
Denver: Lindrooth, Shubart & Co.
Pittsburgh: 1501 Park Bldg.
St. Louis: Missouri Trust Bldg.
New Orleans: Wilmot Machinery Co.

The new Metallic Packing. Fits all rods and stuffing boxes, which can be packed from the same can, avoids carrying a stock of various sizes. No Friction—no Binding. No scoring of rods. Stands "HIGH PRESSURE" and SPEED. 600 degrees Fahrenheit won't hurt it. Costs only half what other metallic packings cost.

"Ambest" is put up in 2, 5, 10, 25 and 50 pound cans.

The original and only genuine. Beware of worthless imitations.

Manufactured only by
Canfield Mfg. Co., Ltd., Philadelphia, Pa.

OTTAWA FIREPROOF SUPPLY CO.
OTTAWA, CANADA
Sole Agents for Canada

This cut shows one of the six Robb-Armstrong Corliss Engines in the Plant of J. R. Booth, Ottawa.

Robb Engineering Co., Limited, Amherst, N.S.

DISTRICT OFFICES
320 Ossington Avenue, TORONTO; Wm. McKay, Manager.
Bell Telephone Building, MONTREAL; Watson Jack, Manager.
355 Carlton Street, WINNIPEG; J. F. Porter, Manager

Accurate Drilling and Reaming on a Manufacturing Basis

The Development and Rise of Jigs in the Machine Shop—A Milling Machine Knee as an Example.

BY JOHN EDGAR

The progress that has been made in the production of small parts for various kinds of machinery has placed accuracy and low cost of production on a level. Up to a few years ago they did not in but few instances travel hand in hand, and in those few scattered cases the methods were considered trade secrets.

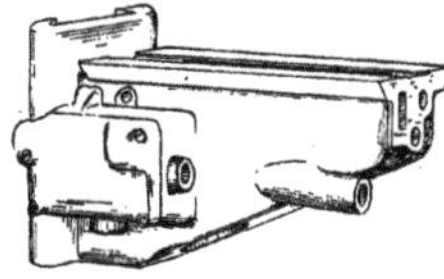
Fig. 1—The Knee.

The large run that the bicycle had and the competition that grew up in that industry are responsible for many of the modern forms of manufacturing practices and were in the main the real beginning of the new era in the manufacturing world in relation to the production of high class work at a low cost.

How has all this been accomplished? In the broad sense, by employing manufacturing methods in the production of the machinery, that is, by producing as many parts of a peculiar type as possible at a time. It has been found that by following that course and by the introduction of special tools and fixtures, the work was not only produced much quicker and cheaper, but much more accurate than by the old method of producing each part separately and fitting same in position. Similar parts, though supposed to be duplicates, were more often not duplicates when fitted in this manner. In the new system this is taken into consideration and each part is a duplicate of any one of a hundred similar parts, and anyone may be replaced by another due to the narrow limits that they are allowed to differ in passing the inspector. The new method is then based on the interchangeability of the details which go to make up a machine.

Where, on the introduction of this system we heard grave murmurs against it, we now find it to be one of the greatest recommendations that a manufacturer can have: "That he manufactures on the interchangeable plan." In fact, it is such a recommendation that we find that those who were loud in words against the system on its introduction and who have in no way changed their methods other than by a few minor jigs, if they may be called such, supporting the new system.

What a vast difference this system has made in all branches of manufacture. It is touching every nook and corner, and the firm that tries nowadays to live outside of its border finds itself drifting further and further backward until it must either change its methods or drop out of sight completely.

Progress in the New System.

While this system had its inception in the small arms plant it, as mentioned above, was really nourished into robust existence by the bicycle industry. In the small arms plant and also in the typewriter factory, it was used in the production of very small parts that go to make up these machines, and its introduction made it possible to produce good working machines at a reasonable price, but in the bicycle business it was applied to much larger parts and to parts that required much greater accuracy in their make up. This opened the eyes of manufacturers in other lines, and so we have to-day the modern system so much in vogue that one wonders how it was that it was not instituted earlier. But such is progress.

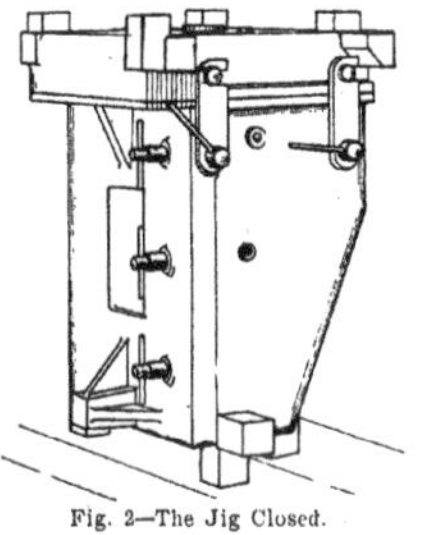
Fig. 2—The Jig Closed.

The machine tool manufacturer, strange to say, was among the last to take up this system, and when we think that in his line where everything is made for accuracy, we may ask why he should be among the last to employ such an advance in methods. However, he has at last laid aside his badge of conservatism and come forth with some wonderful changes in his large line of manufacture, both in his own methods and in the methods he offers for sale in the machines he produces. His first advance was in the use of drill jigs and fixtures. With these he was able to produce parts in which the different holes were located properly by means of fixed bushing, thus automatically locating them in relation to a proper surface and with each other. From this we find him using them (jigs and fixtures), on his milling machine, in conjunction with gang cutters, thus producing duplicate parts in this relation. The work being located from some important bearing and the cutters being set producing surfaces in proper relation to that bearing. He then goes still further and uses similar methods for holding his heavier work on the planer. Here he was in a different position with regard to the actual machining operation, because he must work with a single pointed tool capable of producing none but plain surfaces. In order to overcome this handicap he uses a dummy gauge by which to set his planer tool, thus, had he to plane a lathe bed-ways. He first had a dummy made of a profile exactly similar to that desired on the work in hand and which clamped to the plate of the machine, the cutting edge of the tool was set to this, ensuring, within narrow limits, duplication in each series of parts planed under the guidance of any gauge.

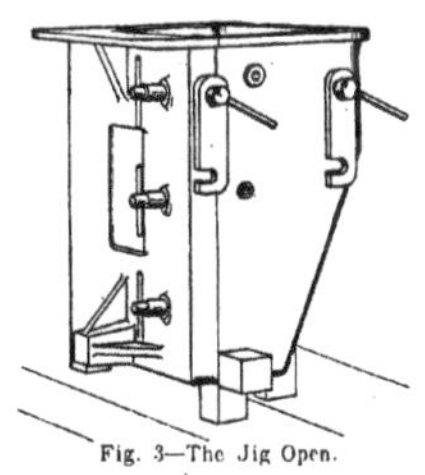
Fig. 3—The Jig Open.

The system has so increased in application that it is now used in the manufacture of the heaviest classes of machinery.

The Milling Machine Knee Shows the Advancement.

So much dependence is placed upon holes and their location in the design and construction of machinery that the branch of the trade which has to do with producing them has received more attention than any other single branch. For that reason we find it far in advance of all other departments in the quality of work produced under the new system of things. As an example, of what is being done in the drilling and reaming of holes accurate as to size and alignment by means of jigs and special tools, the case of the milling machine knee, shown in Fig. 1, will be described and the methods used and the results obtained pointed out.

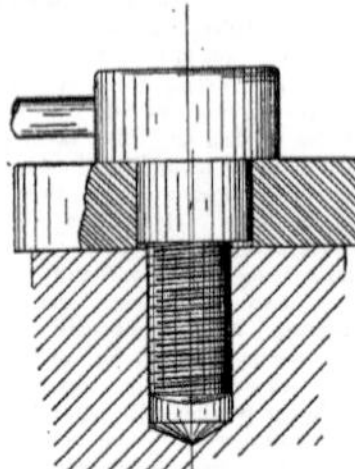

Fig. 4—The Eccentric Clamp.

This casting is rather heavy for easy handling and on account of its size it is liable to lead one into the temptation of trying to do the drilling by using a number of detached jigs that might be made to care for one or two sets of holes irrespective of the proper relation with holes in other locations. When possible the most economical method to pursue, in almost any case of drilling, where a number of holes have to be drilled, is to use a box jig, the drill bushings being fixed in the sides of the box. This style jig construction gives the greatest rigidity and reduces the liability of having the holes thrown out of line due to the casting changing in seasoning or from rough usage. Figs. 2 and 3 show the jig used in our example, the former figure showing the box closed and ready for use, while the latter shows the jig with cover removed and realy to receive the work.

Operations on the Knee.

Before being placed in the jig the knee is planed all over and the cover for the box on the side is drilled on. As there is a bearing in this cap which carries a shaft and gear which latter meshes with a train of gears on the shaft in the box, the bearing in the cap must bear a fixed relation to the hole in the box. Were the cap or cover drill otherwise the joining surfaces of both cap and box would require an extra amount of caution which is eliminated. This is one place where the principle of interchangability is not carried into practice, and such being the case the cover and knee must go in pairs. The two holes shown through the front end of the knee are for the cross screw and centre shaft, and line with holes in the rear end. These holes bear a certain relation to the carriage that carries the table and which slides on the top face of the knee. Since the motion to the lead screw is obtained from the centre shaft through a vertical shaft connecting both screw and centre shaft, we have a hole in the carriage which must line with the holes supporting the centre shaft. For that reason the top face and the angles at the sides are chosen as locating points. As in practice, we allow certain variations, it is obvious that we must allow some in the width of the top face of the knee. As the centre shaft must be parallel with the angles on the side of the knee, as well as with the top in order that the carriage may have adjustment along the top of the knee without cramping the centre shaft, we must locate sideways from one of the angles. This angle is called the "work" angle, because all measurements are taken from it. It is necessarily the fact that this angle is not the one on the side with the adjusting gib used in the carriage. This corresponding angle in the carriage is used to locate all holes in that part, so that when the carriage is assembled in place on the knee, the running parts go into place without trouble. In order to bring this within narrow limits the top of the knee and the angles are plated, that is, hand-scraped to special surface plates so that the desired results may be obtained. Should the scraping be left until after and the knee slid into the jig as from the planer, the variation due to scraping would cause considerable trouble. When

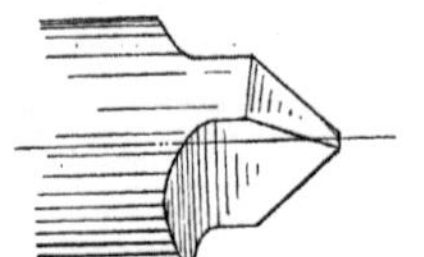

Fig. 5—The Spotting Tool.

we try to consider this done in both the case of the knee and the carriage, the error due to the after scraping would be doubled. Therefore we scrape the carriage or top bearing on the knee before placing in the jig.

Construction of Jig.

Fig. 2 shows that the jig is built up of three castings, the cover, which is made detachable being doweled and clamped to the jig proper by dowels in the upper face of the latter and by the clamps shown at the side, there being four of these clamps.

The jig proper is composed of two castings, one of which has the dovetailed slide on its inner surface for receiving the knee which is slid front end down into the box when the latter is in the position shown in Fig. 3. The slide is provided with a clamping gib which serves the double purpose of clamping the knee and forcing it over into the work angle. This gib is clamped by means of the three handles shown in the front of the jig. When the knee has been placed into the jig and clamped, the cover is put on and the clamps are brought into use. These clamps have the peculiar property of drawing the

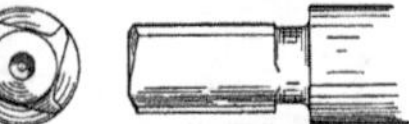

Fig. 6—The Reamer.

hood down onto its seat and of clamping or binding themselves against loosening. This is accomplished by means of an eccentric bodied screw. Fig. 4 shows a section through one of the clamps, showing the screw and from which its working is plainly seen. In clamping the cover on, the turning of the handle causes the eccentric shoulder to draw the clamp toward its centre and the thread is at the same time drawing the clamp against the face of the boss, binding it in position.

The Use of Bushings in Jigs.

It is quite common practice to make for each hole to be drilled a double set of bushings, one set to guide the drill and the other to guide the reamer. This practice is not only expensive, but is liable to inaccuracy, due to wear of the bushing from constant removal. In incompetent hands trouble is liable to be caused by chips and grit becoming lodged between the bushings and their seats, throwing them out of line.

The above mentioned method is overcome by using but one set of bushings which are driven into their seats and are thus fixed in position, all wear being on the inside of them. The holes in these bushings are made equal in size to that of the hole or up to the size of the reamer.

The drill shown in Fig. 5 is used to start the hole and is called a spotting drill, the point is ground to the same angle as the twist drills that are used to drill the hole. The body of this spotting drill is made to fit the bushing, so that the hole is started right and if care is used in having the twist drill carefully ground and in good condition, no difficulty will be had in obtaining a straight hole. Where the hole is large and is drilled from the solid two or more sizes of drills are used. The drilled hole is then reamed out to size by the reamer shown in Fig. 6. This reamer is made with four flutes or lips, and is made to cut on the principle that a rose reamer acts. The lips are backed off as in the ordinary reamer, but all the cutting is done on the ends so that by grinding the ends, the reamer is kept sharp and in good condition. The trouble with the rose reamer is that the corners of the teeth become worn and the body of the reamer tears the hole leaving a rough hole much over size. The particular virtue of the reamer shown is that while its cutting is similar to that of the rose reamer—which is considered to be the best when in good working condition. when the corners of this reamer become dull the sharp lips then do the cutting, producing a nice, smooth hole, even under conditions due to carelessness. The formation of the flutes also allows plenty of chip space.

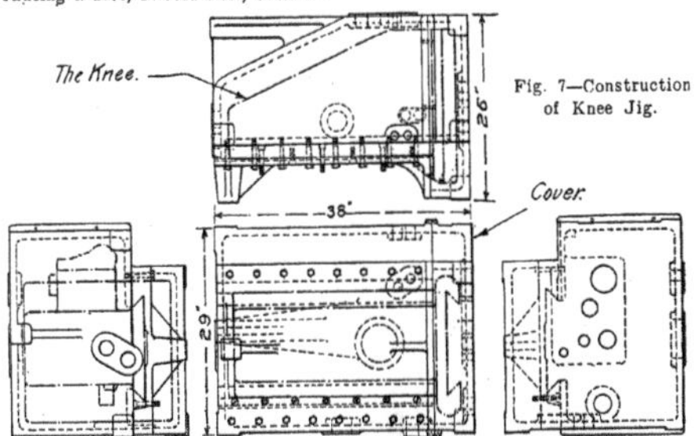

Fig. 7—Construction of Knee Jig.

The drilling in this case is all done under a radial drill and the jig placed in the several positions necessary to bring the bushings on the upper side.

The elevating shaft horn is not done in the jig in this particular case, but is performed at another setting. The hole is produced from the solid stock. A large hole is first drilled through as nearly in the proper position as can be assumed, the knee which is placed on an inclined plane is then readjusted so that the machine spindle is directly above the hole for the elevating screw. This hole is too long to allow a boring bar being used and the cut taken clear through because the bar would strike the inner wall of the knee. In order to bore this hole a telescoping bar was made.

The holes drilled in this jig a line within a half thousandth in the length of the knee and the time to drill all the holes was reduced to something like one-quarter of the time used to set one up on the boring machine, layout and bore the holes. The results in the latter case being anything but satisfactory as gears had to be turned into place, and extra work was placed on the assemblers due to poor alignments.

PRODUCER GAS POWER

Mr. L. G. Read, Consulting Engineer for the Colonial Engineering Co., Montreal, speaks of Cheap Power

The number of producer gas plants has very greatly increased in Canada. The city of Chatham is now erecting a municipal plant to be equipped entirely with gas engines and with an absolute guarantee that its street lighting plant shall cost only $25 per annum per lamp. Two years ago there were practically no gas engines in Canada. To-day there are probably twenty-five successful plants in operation and perhaps a hundred contracted for. I am putting in nine plants, now, myself in the province of Quebec alone.

The Lackawanna Steel Company, of Buffalo, has recently installed 40,000,-horse-power of gas engines, and this too within a stone's throw of Niagara Falls. United States Steel Co. is spending three millions of dollars substituting gas engines for steam engines. A single English gas engine firm built and shipped last year alone over eleven hundred gas engines. A gas engine with the highest priced fuel will produce a kilowatt hour (one and a half-horse-power) for a cost of only one cent.

The Peak Load and the Meter.

If the power commission will sell Niagara power at 2-10 of 1% per horse-power per hour through a meter, instead of by the peak load as they propose to do, said Mr. Read addressing the Stratford Board of Trade, it is the cheapest power that can be manufactured. This is equivalent to $17.50 per horse-power for 8,750 hours of power or 365 days of 24-hour power. With the present proposed arrangement, however, what you will really pay for and what you will really get is this:

Your load factor—that is your actual average horse-power consumption (public or domestic lighting or power) will likely not average over fifty per cent. of the peak (or say 2 500 horse-power). In this case you are paying just exactly twice $17.50 or $35 per annum per horse-power for the actual power you consume. If your average horse-power consumption is sixty per cent., then (based on $17.50 peak rate) you are actually paying $29; if seventy per cent., you are actually paying $25, and so on.

But, he said, electricity is a general utility and should be placed in the same class as other public utilities. The proposition of the commission was that "you simply pay according to the highest peak load in any twenty minutes during the month." This was the most dangerous feature in the whole subject and the one least understood by the public.

Producer Plant.

To the Toronto Council Mr. Read said he was prepared to hand over the construction of a producer gas plant to any one of four English engineering firms who would guarantee to produce electricity at 1c per kilowatt hour. He was also prepared to construct a steam plant which would produce power for Stratford at 2 cents per kilowatt hour. One cent per kilowatt hour for a producer gas plant meant $30 per kilowatt for 300 days of 10 hours each. This meant $20 per horse-power, the other $8 or $10 for interest and expense.

Long Service of Producer Plants.

Producer gas holds the record for continuous service. A producer gas engine will run as steadily as a steam engine. It is sufficiently safe to warrant its use. Of the engines failing in England last year there was 1 in 8 steam, and one in 12 gas. The reason so many gas engines failed in Canada was that every large machine shop in Canada was trying to make them whether it understood how to build them or not.

The Head of a Great Industry

Charles B. Frost, President of the Frost & Wood Company, Limited, Smith's Falls, Canada.

Many heads of large industries are unknown to the thousands with whom they do business, and to very few is it the privilege to see their face and receive the hearty handshake. Mr. Frost is respected and esteemed by all who know him and he has been called in business circles the premier of the captains of Canadian industry. He was born in Smith's Falls on August 26, 1840, and with his brother, the Hon. F. T. Frost, for one cannot be mentioned without the other, broad and deep foundations have been laid for a great industrial enterprise which finds its expression in the high chimneys and great buildings erected throughout the Dominion.

The business of which Mr. Frost is at present the honored head, was founded by his father, Ebenezer Frost, away back in 1839, the business at that time consisting of the manufacture of a few ploughs and stoves. Those were not the days when the proprietor of a shop sat in his glass-enclosed office and dictated letters into his phonograph and exercised a general supervision of the plant, etc. They were the days when the proprietor was foreman, moulder, office boy and everything else, all combined. Perseverance was an innate quality of the early pioneers and perseverance and hard work have developed that small and rudely equipped shop into the huge modern manufacturing establishment, which, while it is not the largest in size, is second to none in equipment and facilities for turning out the very highest grade of farm machinery.

The Firm "Frost & Wood."

About the year 1846, Ebenezer Frost formed a partnership with Alexander Wood, under the name of "Frost & Wood," and shortly after this the firm began branching out and adding new lines of manufactured goods. At that time there were no railroads to aid in the transportation of raw materials and finished products, so everything had to be teamed from "the front," or, speaking more definitely, from Brockville, or else brought up by boat. Rolling mills and blast furnaces were also unknown quantities in Canada, and all steel had to be imported from England.

In 1863 Ebenezer Frost died, and his two sons, Chas. B. and Francis T. Frost, with the late Alexander Wood as a partner, carried on the business. This partnership continued until 1885, when the Messrs. Frost took over the business, but still maintained the name of "Frost & Wood." Canada was growing very rapidly and the ever increasing demand for farm machinery necessitated a larger selling organization, so an office was opened in Montreal. That organization has grown from one branch office to a dozen, and covers every province in the Dominion, as well as many of the foreign countries.

Reorganization.

In 1899 the Messrs. Frost decided to form a joint stock company and with increased capital to push harder than ever the sale of Frost & Wood machines. There was a market for the goods which greatly exceeded the supply, and since 1899 the plant of the Frost & Wood Company, Limited, has been growing at a remarkable rate. Each year either new buildings were put up or substantial additions were made to the old. Warehouses were also erected in the different cities, in which branch offices are located, in which to store the finished product and have it ready for immediate shipment when called for.

CHARLES B. FROST.
President of the Frost & Wood Company, Limited.

Plant Lost by Fire.

The year 1906 was truly a disastrous one for this company, because on Feb. 8 of that year fire destroyed practically their entire manufacturing plant. It was a hard crack, but there is an old saying which states "that a person never knows what he can do until he has to do it." It certainly must have looked to these men like a hopeless task to try and replace their plant and still keep on the business. However, that is what they had to do and that is what they did. The smoke had not cleared away from the ruins when they had a gang of several hundred men cleaning away the debris and making preparations for a plant just twice the size of the one that had been destroyed. It took a lot of nerve and a lot of hard work, but the energetic officials of the company, backed by a progressive directorate and a loyal staff, not only put up a plant inside of one short year, but also supplied the demands of their customers and filled all orders for goods, with the single exception of binder orders. Temporary shops were improvised in old warehouses and mills in different parts of the town, and while manufacturing costs for 1906 were run up to an abnormal figure, still the organization of the company was held intact and ready to do business on a larger scale in the future.

This was a time when Mr. Frost's experience of almost fifty years stood his firm in good stead, because he knew every defect in the old plant and was able to avoid a repetition of such in the new buildings. To a stranger walking through the present plant, it appears that everything is just as nearly perfect as it can be. The different shops are light, airy and conveniently situated. They are now as nearly fireproof as automatic sprinkler systems, underwriters' fire pumps and a complete system of town hydrants can make them.

While Mr. Chas. B. Frost and his brother, the Honorable F. T. Frost, are both men well up in years, and have laid aside the responsibility of the minor details of the business, still every day finds them in their offices and just as interested in everything that is going on, as years ago, when their own shoulders had to carry the entire load.

W. S. Chase, sales manager of the National Acme Manufacturing Co., of Cleveland, Ohio, and W. C. Long, their Canadian traveler, were in Toronto during the last fortnight and paid Canadian Machinery a visit.

At The Noon Hour

Cheap Labor.

While an Irishman was gazing in the window of a Washington bookstore the following sign caught his eye: "Dicken's works all this week for only $4.00." "The divvle he does!" exclaimed Pat in disgust. "The dirty scab."

* * *

Considerate of Father.

"Johnnie," said a mother threateningly to a naughty son, "I am going to tell your father to whip you when he comes home to-night."

"Please don't do that, mother," said the lad penitently. "Dad's always so tired when he comes home!"—Philadelphia Inquirer.

* * *

Rich.

Blobbs—Why do they call Pittsburg the smoky city?

Slobbs—Because there are so many millionaires there who seem to have money to burn.—Philadelphia Record.

* * *

Needed, But Where?

"Did you see the discouraged-looking young fellow who was in a while ago trying to get a job?"

"Yes."

"That was J. Filber Browne, who read an essay at his college commencement on 'The World Needs Us.'"

* * *

A Drop in Copper.

"You look worried. What's the matter?"

"Oh, we're all upset at our house. There's been another drop in copper."

"But I thought you never speculated?"

"I don't. Our cook let the coffee pot fall on her foot, and she has had to go to the hospital."—Chicago Record-Herald.

* * *

Making a Fortune.

"With $100,000," said the man of expansive ideas, "I could make a fortune in Wall Street."

"Yes," rejoined the piker, "but whose fortune would you make?"—Washington Star.

* * *

Hope for Them.

A rich young man who was trying to learn to work had fallen in love with the daughter of his employer, but he found that his path was by no means clear of obstacles.

"You tell me your father objects to your marrying me," he said, in a crestfallen tone. "Is it because I am in his employ? I can leave it and go back to a life of idleness if he prefers."

"Oh, no, that isn't what he wants," said the object of his choice. "He says I may marry you just as soon as you're valuable enough to have your salary raised."—Youth's Companion.

* * *

A Good Time.

"Well, Pat," said the sympathetic employer, "did you go to your friend's funeral?"

"Did I go, is it sor?" returned Pat, enthusiastically. "Sure an' I did, sor, an' had the time o' me life sor. That wake wor a drame!"

* * *

Must Be Twins.

"See here, Pat," said his employer, "didn't you tell me that when you was out west the Indians scalped you. and now you have your hat off I see you have an extraordinary quantity of hair? You certainly told me so, didn't you, Pat?"

"Oi did, sor," answered Pat, "but Oi bear in moind now that it was me bruder, Mike. It's thot much we be aloike that Oi think Oi'm Moike an' Moike be me."—New York Times.

TORONTO'S MANUFACTURING DISTRICT.

A visit to the district north of the Dundas bridge, in the northwest of Toronto, presents to the eye of a visitor a scene of great industry and a walk around this section immediately adjacent to the C.P.R. and G.T.R. tracks presents a scene of great activity, which is bound to make Toronto a hub of industry. Here new factories are going up on almost every street. Some are nearing completion and others are either in course of erection or preparations are being made for construction, and more factories are promised for next spring. Cheap lands and room for expansion has attracted manufacturers to this district, until it now is a very bee-hive of industry. A brief survey of this district will prove interesting. It is situated along the G.T.R. and C.P.R. tracks and nearly every factory in the district has at least one spur line. The buildings are for the most part, large, well-built brick or reinforced concrete, the workshops are bright, spacious and airy and all are busy.

Beginning at the south we find the new plant of the Dominion Bridge Co., Ltd., a large building, 407x70 feet, now ready for construction work, the first of which will be for the Robert Simpson Co. Starting northward, we visit the new plant of the Chapman Double Ball Bearing Co., now complete in every particular; Robt. Watson Co., manufacturer of cough drops; the Toronto Laundry Machine Co., Ltd., makers of laundry machinery and supplies.

Old and New Plants.

Continuing northward we come to the Standard Meter Co., the plant of the Toronto Furnace & Crematory Co., Ltd., makers of furnaces, and the Johnston Oil Engine Co., Ltd., situated close to the Furnace & Crematory Co., R. Laidlaw, makers of frames, sashes, doors, moldings, etc.; the Sharples Separator Company, manufacturers of tubular cream separators; the Cyclone Woven Wire Fence Co., the Canada Malleable & Steel Range Co., Ltd., and the Bowman Gas Range Mfg. Co. Crossing the track eastward, we see the large plant of the Canada Foundry, general founders. Adjacent to the Canada Foundry is the Wilkinson plow foundry, and in this vicinity the Dominion Radiator Company has a large building in course of construction to keep pace with their increasing business. Near here, on Sterling avenue, the Fleer Co., of Philadelphia, will erect two large gum factories. South of this is Hancock Bros., manufacturers of sashes, doors, etc. The large plant of the Canadian Fairbanks-Morse Mfg. Co., Ltd., occupies several buildings, in which they manufacture gasoline, gas and oil engines, steam and power pumps. Following along is the Griffin Curled Hair Co., and on St. Helen's avenue is a fine block of new buildings. These include the plant of the Somerville Limited, manufacturers of plumbers' and steamfitters' brass supplies, and the King Radiator Co., which ran their first heat in the foundry in November 20th. Another row of factories consists of the new factory of T. A. Lytles for pickle manufacturing, Cowan's cocoa and chocolate works, Matthew's Bros., makers of moldings, frames, etc., and across the street is the General Brass Works, Ltd., a large new plant, 180x75 feet, where the manufacture of plumbers' and steamfitters' brass goods will be carried on.

Still in the distance, and outside these narrow limits we have outlined, arise numerous smokestacks which betoken more activity, and the large buildings spoke to us of numerous men busily employed, of machinery being built, strong and durable; of ideas being perfected and a name for Canada being built up on the good workmanship of the busy artisans. A few good sites are still open and Commissioner Thompson is in correspondence with a few large firms looking for Canadian sites, from which they will be better able to look after their Canadian customers.

CANADIAN MACHINERY
and Manufacturing News

A monthly newspaper devoted to machinery and manufacturing interests, mechanical and electrical trades, the foundry, technical progress, construction and improvement, and to all users of power developed from steam, gas, electricity, compressed air and water in Canada.

The MacLean Publishing Company, Limited

JOHN BAYNE MACLEAN - - President
W. L. EDMONDS - - - Vice-President
H. V. TYRRELL - - - Business Manager
J. C. ARMER, B.A.Sc., - Managing Editor

OFFICES:

CANADA.
MONTREAL - 232 McGill Street, Phone Main 1255
TORONTO - 10 Front Street East, Phone Main 2701
WINNIPEG, 511 Union Bank Building, Phone 3726, F. R. Munro
BRITISH COLUMBIA - Vancouver, Geo. S. B. Perry

UNITED STATES
CHICAGO - 1001 Teutonic Bldg. J. Roland Kay

GREAT BRITAIN
LONDON - 88 Fleet Street, E.C. Phone Central 12960, J. Meredith McKim

FRANCE
PARIS - Agénce Havas, 8 Place de la Bourse

SWITZERLAND
ZURICH - - - Louis Wol. Orell Fussli & Cot

SUBSCRIPTION RATE.

Canada, United States, $1.00, Great Britain, Australia and other colonies 4s. 6d., per year; other countries, $1.50. Advertising rates on request.

Subscribers who are not receiving their paper regularly will confer a favor on us by letting us know. We should be notified at once of any change in address, giving both old and new.

Vol. III. DECEMBER, 1907 No. 12

CONTENTS

TECHNICAL NORMAL COLLEGE.

In providing for the establishment of technical schools at Quebec and Montreal, the Quebec Government has taken a practical step which commands general approval and which will bear good fruit before long. But there is an important point in this connection which should not be overlooked, and which is the desirability, sooner or later, of having an institution in the nature of a technical normal school or college for the training of teachers in technical work in order to meet the demand for such teachers.

Technical instruction is being introduced in the schools of several of the more progressive Ontario towns; it would be more generally introduced but for the lack of qualified instructors. Before long it will be necessary to have an institution for the training of teachers in technical work in order to meet the demand for such teachers.

It is proposed to establish a technical normal college in Hamilton and it is to be hoped that a provincial institution will be established. All Ontario teachers who desire to qualify as technical instructors will have to take the course prescribed, and the college will increase in importance and the scope of its work as the years go on and technical instruction is popularized.

TORONTO'S POWER SITUATION.

The ratepayers of Toronto will be asked on January 1st to vote $2,500,000 for a plant for the distribution of electrical power from Niagara Falls by a Government-built transmission line. Cheap power is a vital necessity for the industrial growth of Toronto and it is Commissioner Thompson's greatest problem to overcome when the question of power is mentioned by prospective new industries for that city.

According to the present plan the city can construct a distributing system which will provide for: Fifteen hundred street lights; power delivered to pumping station; lighting of public buildings; underground conductors in central area, three-quarters of a mile square, and main streets leading therefrom; supply of power to factories along railroad lines and all principal factory areas; supply of light and incidental power in suitable manner in all parts of the city where there is a reasonable demand; facilities for extension, namely—spare ducts in conduits, spare arms on pole lines, and spare floor space in sub-stations, the estimate being based on the receipts of say 12,000 horsepower, and these spaces being sufficient for an increase to say 20,000 horsepower.

CANADIAN FINANCIAL STABILITY.

That the financial stringency in the United States has not reached Canada to an appreciable extent is no doubt due to our system of banking which is entirely different from that in the country to the south of us. In the United States the revenues are locked up in the treasury vaults, while in Canada the revenues are deposited in the bank, just enough being reserved to cover saving bank deposits. As a result of our superior banking system, the currency is always ready for immediate use, the Canadian financial institutions are still issuing currency for the carrying on of legitimate industrial business. The Canadian banking system has shown its efficiency in withstanding the strain of the past two months, and United States financiers are calling for a new banking law modelled after the system in Canada.

At the present time Canadian banks have over $63,000,000 on "call" loans on Wall Street, and this cannot be secured except in the form of non-negotiable and easily counterfeited clearing house certificates. Canadian bankers have all along contended that it was a source of strength to the Canadian banking system to have from fifty to one hundred million in cash on loan in New York, subject to an immediate call, the effect of which would not be felt by Canadian industries.

The bankers were supposed to know their business and their expressed views were agreed to by many. Others, however, contended that the money should not go out of the country to be used in gambling on Wall Street. Many lessons can be learned from the present financial flurry, one of them undoubtedly being the necessity of restricting the territory in which our banks can legally loan the money placed in their hands by Canadian depositors. Some peculiar incidents have happened in the United States, due to their financial stringency. The people of Detroit are depositing their money in Canadian banks and the Detroit Railway Company is paying its employes with Canadian currency. In another city the street railway paid out one payday, over $30,000 in nickles. In many places wages are being paid by cheque or clearing-house certificates. Some of the banks are also using clearing-house certificates in "cashing" cheques. With the help of the secretary of the treasury, this situation is being overcome and many banks report that they are ready to pay in currency as soon as New York sets the example.

QUEBEC BRIDGE DISASTER.

As the work of the Government Commission progresses, new developments arise and the cause of the disaster now appears to be due to modifications in the original design of the bridge made by the consulting engineer, Mr. Theodore Cooper. The testimony of the officers of the Phoenix Bridge Company was that Mr. Cooper changed the length of the original span from 1,600 to 1,800 feet, without making adequate changes in the support to compensate for the increased length.

"The fall of the bridge is to be laid directly to the change in the unit stresses as made by Mr. Cooper," said David Reaves, president of the Phoenix Bridge Co, in his sworn testimony, and he continued: "He made modifications in the unit stresses to be employed upon the various members, which very much increased them beyond any precedent, and by so doing placed the whole deisgn in a field outside the benefit of experience. Such high stresses had never before been used."

If this be true, the fault was in the design of the bridge of the bridge itself, not due to a flaw or defect in the material. When the Commission again interview Mr. Cooper, we anticipate a definite statement from them showing the cause of the disaster.

DOMINION BOILER ACT.

Pursuing the policy adopted by the engine and boiler section of the Canadian Manufacturers' Association on September 24th, at their meeting in Toronto, a second meeting of this section was held in Toronto on November 13th, and further steps were taken towards a uniform boiler construction act for all the Provinces of the Dominion, that would be satisfactory to the manufacturers and at the same time protect the consumer.

Canadian Machinery has noted the lack of uniformity in Provincial boiler regulations, and has been a warm advocate of a Dominion law to overcome the difficulties that must be met by both manufacturer and purchaser on account of the Provinces having their own acts. We are glad to see that the C.M.A. has taken up this matter and we hope to see in the near future a Dominion act that will meet the needs of all boiler manufacturers and purchasers.

The meeting was called at the instigation of the president of the Engine and Boiler Section of the C.M.A., Mr. F. E. Leonard, of London, and active steps are being taken to bring about the much-needed legislation. Since the meeting in September, the members have been collecting information regarding boiler construction and legislation in different countries, and a committee of experts, representing the largest boiler makers in Canada, will meet and frame a measure that will meet the requirements and interests of all localities. A united effort will then be made to secure a Dominion act to cover all stationary and portable boilers.

SAFE ELECTRIC WIRING.

The importance of entrusting electrical work to none but men who have a thorough knowledge of the business, and the dangers of allowing electric wiring to be done by incompetents is being impressed in Winnipeg. The City Electrician, in a report, says:

"While the legitimate contractors have lived up to the requirements as to safe wiring in a fairly satisfactory manner, this department is considerably handicapped by the fact that at present anyone may obtain a permit to install electric appliances, whether possessing any qualifications or not.

"Cities elsewhere are having the same difficulty in dealing with the matter, in consequence of which efforts are being made to regulate the matter by licensing master electricians and by the institution of a board of examiners."

Canadian Machinery has reviewed the question of electric wiring in past issues and pointed out the results of incompetent work. It seems to us that such important work as wiring, which has been the cause of so many fires, attended with loss of life, ought to be regulated by examinations of qualifications of the electricians. The insurance companies recognize this fact and very stringent regulations are enforced in this work. The city councils should unite with them to secure satisfactory workmanship.

MAKE GOOD USE OF OPPORTUNITIES.

A young man starts on his apprenticeship looking forward in anticipation to the opportunities that will open up when his long term of apprenticeship is over. Yet it is not the opportunity that makes a man of him, it is the use he makes of it and he may be ever so faithful and industrious, but if he has not the intelligence to make good use of his opportunities he is not fit for promotion.

Every young man cannot have a college course, but every young apprentice can give one night a week for study if he wishes to make a success in life. Recreation is essential, but to spend night after night in amusement is neglecting a great opportunity open to every young man. He will sometime be thankful for the time spent in study. There is some satisfaction in "knowing how," and knowledge is not burdensome. The Government may establish technical and industrial schools in every city and town, but it rests with the young man to apply himself. Thus preparing himself by a few hours' study each week, he will be better able to make use of his opportunities when they present themselves to him. If an apprentice is promoted and called upon to make a drawing or design a machine, would it not be a great satisfaction to know how to do it. Theoretical, combined with practical knowledge is a great benefit to a man. To be able to make the drawing, to be able to design the machine, is to make use of his opportunities and the knowledge that he can do the work he is called upon to do adds greatly to a man's self-respect. To be able to make use of the opportunities is success. Success is not gauged by wealth, but by the honesty, faithfulness and intelligence which command the respect of others.

The Art of Making Machinists' Tools

Practical and Scientific Methods of Manufacturing Steel into Tools

By H. Holford

Connected with the machinist's trade is a wide range of fine and complicated tools, among them being lathe and planer tools, milling cutters, taps, dies, reamers, etc., and the toolsmith in a large shop who can forge, harden and temper these tools satisfactorily will have a great many friends among the machinists, but should the toolsmith be otherwise, there is nothing that tries the temper and patience so much as to have a number of machinists continually around the fire and each one making a complaint that the tool would not do this, and it would not do that, etc.

When making lathe and planer tools, there is no definite rule to go by "as to the shape," with the exception of a few standard or ordinary tools, as shown in Figures 1 to 4, as the machinist has so many jobs of a different nature that he requires a tool of a special shape to suit the work, which he explains to the toolsmith or gives him a drawing of it, and sometimes a pattern of it made from wood. Steel of 1 per cent. or 100-point carbon is best for making

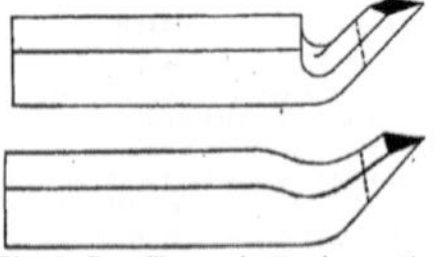

Fig. 1—Two Ways of Forging a Diamond-point Lathe Tool.

lathe and planer tools, as these tools do their work by steady pressure. But nevertheless they must be properly forged, hardened and tempered in order to stand the great strain that is continually bearing against them, and also to hold a good cutting edge in order to save time, and do work of a very exact and skilful nature, especially in the finishing stage.

Lathe tools of a flat surface, such as bent cutting-off tools and side tools, must be forged to shape and the offset put in before the last hammering is done. When making cutting-off tools and other tools of a similar nature, always give plenty of clearance, as illustrated in Fig. 2, which shows the cutting edge a little wider than (b), also wider than (c) as shown in front view of the same figure, and bear in mind when forging the tool to leave it a little wider at the cutting edge than the exact width, as the hammering on the flat surface after the offset is put in, will bring the cutting edge to the right width. To more fully explain, let us suppose the cutting edge of the tool is to be ¼ of an inch in width, but in order to refine and pack the steel, we will have the cutting edge 5-16 of an inch wide before undertaking to hammer on the flat surface. If the hammering does not flatten the steel to exactly the right size and so leave the cutting edge a trifle wider, so much the better, as the tool will have a little stock for grinding.

All lathe and planer tools for ordinary work should be quenched or hardened about one inch back and according to the cutting edge, as illustrated by dotted lines in Figures 1 to 4, and tempered to a dark straw or copper color. But should the tools have to cut very hard cast iron or other very hard material, harden and draw no temper. Should the tool still fail to cut the metal, use equal parts of powdered cyanide of potassium and prussiate of potash. To use this compound heat the tool as hot as if it was to be hardened, then place the heated cutting edge of the tools into the powder, reheat again to a proper hardening heat and plunge into the hardening bath and cool off entirely.

When hardening large tools, such as square corrugating tools, and which have very fine teeth in the cutting end, make sure when heating that the centre of the cutting face is as hot as the outside or corners. And never heat carelessly or too fast, so that the corners will be at a white heat, while the centre is barely red. But heat slowly and very evenly, until the whole face of the cutting end is just hot enough to harden, then plunge into the hardening bath and cool off entirely. Draw no temper.

Heat slowly and evenly until the whole face of the cutting end is just hot enough to harden, then plunge in the hardening bath and cool off. Ordinary corrugating tools require no temper drawn.

Air-Hardening Steels for Lathe and Plainer Tools.

Cast steel for lathe and planer tools is to a certain extent done away with in large shops by the use of air or self-hardening steels. A certain amount of this steel is manufactured in sizes so that no forging of tools is required, and by the use of a patent tool-holder (made especially for certain sizes of the steel) is superior to a hand forged tool in some respects. But the toolsmith has an important work to do connected with air-hardening steel, as he is called upon to cut it off in lengths and harden it, besides. As there is a limit to the sizes that are manufactured for immediate use, the steel must be forged to the shape of the tool, while the forging and hardening of air-hardening steel is somewhat of a different nature to cast tool steel. Air-hardening steel is not, as a rule, used for anything but for roughing lathe and planer tools, it being too hard to make into any tool which is to do its work by the use of a hammer and also to make into any expensive tool such as milling cutters, as on account of its extreme hardness it cannot be machined or worked satisfactorily. There are several makes and brands of air-hardening steel, but some of the leading makes may be mentioned, as Sanderson's, Jessop's, Novo, Mushet

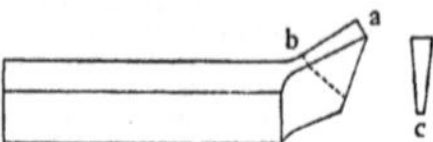

Fig. 2—The Left-hand Bent Cutting-off Tool.

and Blue Chip. When forging these brands into tools, do not heat the steel above a bright yellow, especially Mushet or Sanderson's, but the steel must be heated evenly clear through the bar, and unless the tool is to be of a fine nature do the forging under the steam hammer and sledge. The forging should be done as quickly as possible, while the heat is in the steel, and never attempt to bend or crook the steel when at a low heat, as it will be apt to crack or break in two, so be careful to have the steel at least at a deep red heat when it is to be bent.

When any new brand comes into the shop, be careful to look at the directions on the bar, as some brands are hardened a little different from others. For example, the cutting edge of a tool made from Blue Chips is hardened by being heated to a white heat until it commenced to melt, when small bubbles or blisters will form on the steel, then

* Extract from the 20th Century Toolsmith and Steelworker, by H. Holford, copyrighted and published by permission of Frederick J. Drake & Co., 350 Wabash avenue, Chicago.

it is placed in a blast of cold air. To harden Mushet steel, heat to a deep yellow and cool off in a blast of cold air; Sanderson's and Jessop's after the same process. Novo steel is hardened by heating it to a white heat, then cooled off in a blast of cold air, or may be quenched in oil.

When hardening steel with a blast of cold air, have the pipe or nozzle (which

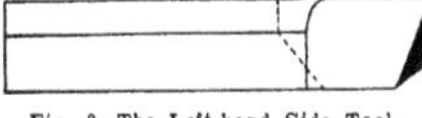

Fig. 3—The Left-hand Side Tool.

conveys the air) as close to the fire as possible, and when cooling off the tool, have same arranged to hold the point or cutting edge directly in front of the blast. If this is not done the tool is apt to become turned to one side by the force of the blast. Also bear in mind if the steel is heated to a melting heat, as Blue Chip, be careful not to put the blast on too strong at first or it will blow the point off the tool. Instead, put on the blast light and gradual until the steel begins to cool a little. Then turn on the blast to its full capacity and keep on the tool until it is perfectly cold.

A great deal is to be learnt in working air-hardening steel in order to get the best results. But for the beginner who has had no experience, follow up the directions given by the manufacturer. Should the directions fail to give good results, do some experimenting. For illustration, Novo brand of steel will give good results if heated to a white heat, then plunged into oil or boiling water, while Blue Chip will sometimes give better results if (after being heated to a melting point and cooled off in a blast of cold air) it is heated to a very low black red heat and allowed to cool of its own accord. Yet there are other brands of eteel that are hardened by heating to a light yellow heat, then placed in a cool place and allowed to cool slowly of its own accord.

Never allow Mushet steel to come in contact with water or it will crack, although no visible heat can be seen in the steel. Figure 5 illustrates a heavy roughing tool, made from air-hardening steel, for turning locomotive tires, car axles, etc. Dotted line at (a) indicates how far back from the cutting edge, as (b), the steel is to be heated when hardening. All roughing tools are made after the same shape. Bear in mind when forging air-hardening steel, never try to forge it below a low cherry red, and if at any time it is necessary to cut a bar into lengths, do not try to cut it cold, as it must be heated.

How to Anneal Air-Hardening Steel.

Sometimes, although seldom, the toolsmith is called upon to anneal air-hardening steel, so that it may be turned and planed. This class of steel is very difficult to anneal on account of its extreme hardness. It must not be packed directly in slacked lime or ashes, as cast tool steel, as it will cool off too quickly. To anneal air-hardening steel, use an air-tight, heavy iron box, place the steel inside the box, then heat altogether to a deep red heat, then pack the box deep into slacked lime or ashes. If an iron box cannot be obtained, use a heavy iron pipe or band large enough to accommodate the steel, without having the steel project out through the end. Also have two heavy flat iron plates large enough to cover the end of the pipe. Place the steel in the pipe and heat the pipe, likewise the plates, as formerly mentioned, then pack the pipe into the lime in an upright position, having one of the plates directly under the pipe, the other on top, which will keep the steel from coming in contact with the lime. By following this method the steel will keep hot a very long time, and thus give good results. To anneal a piece of 1-inch square air-hardening steel will require from twelve to fifteen hours' time.

Milling Cutters.

These tools are very valuable to the machinist's trade, and the making of them requires great skill, valuable time and good steel, which makes them very expensive tools. If they are not properly forged and hardened there is a great loss.

When forging a blank for a milling cutter, (steel about 90 point carbon is best) be sure and leave the blank a little larger every way than the exact size of the milling cutter when finished,

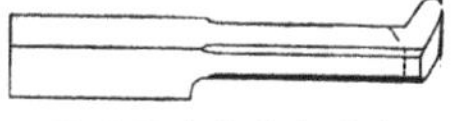

Fig. 4—The Inside Boring Tool.

as a machinist always prefers a little extra stock, which enables him to machine the tool with greater ease and less caution than if forged to the exact size of the tool. When forging a milling cutter, heat the steel evenly to a good yellow, but do not heat too fast or the outside or corners will be at a white heat while the inside is barely red, and by unevenly heating will cause strains which will produce a tendency to crack when hardening the milling cutter.

If the milling cutter is large, forge it to shape under the steam hammer, being careful to forge it evenly from all sides alike, and so work the steel clear through, but as it becomes finished, reduce the heat. Tools such as milling cutters, which are either round or almost of an equal size, cannot be refined, as the steel will remain at its natural state, consequently it is not desirable to finish at quite so low a heat as those of a flat surface. But if the milling cutter is very thin and flat then the

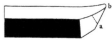

Fig. 5—Heavy Roughing Tool.

steel may be refined by using a flatter on the flat surface while at a low heat.

Annealing Milling Cutter.

After forging the milling cutter, it must be annealed, which operation must not be neglected, as the whole blank must be heated again slow and evenly to a blood red, (never exceeding a cherry red) then packed deep into slacked lime, allowing it to remain there until cold. Have the annealing box large enough so that it will contain plenty of lime to keep the heated steel away from the air. If there have been any slight strains left in the steel by forging, the annealing will take them out and make the steel soft so that it can be worked with ease by the machinist.

The hardening of a milling cutter after it has been machined to shape, is the process that must not be overlooked, as in the hardening the toolsmith must either succeed or spoil the tool. The milling cutter must be hardened properly without cracking, so that when put into use it will do a great amount of cutting by holding a good cutting edge, and so it is necessary that the toolsmith use great care when heating for hardening. If there is no heating furnace in the shop, and the milling cutter is to be heated for hardening in the coal or open fire, too much care cannot be used.

When heating to harden in the open fire, have the coal well charred and the fire plenty large enough. The top of the fire should be perfectly flat and the whole surface a perfectly and very evenly heated mass, place the cutter (if after the shape as (b), Figure 6) flatway on the top or surface of the fire, now heat very slowly and evenly and turning it over occasionally until it is heated to a very even, low cherry red, or just enough to harden, then plunge it into the hardening bath edgeways from a vertical position, allowing it to remain deep in the bath until quite cold, then dry off and polish bright.

Tempering Milling Cutter.

To temper the milling cutter, take two round bars of iron about two feet long and just large enough to go through the hole in the centre of the milling cutter, put one end of each iron into the fire and heat to a white heat to about two inches back from the end or according to the thickness of the tool. Now take one of the irons (leaving the other in the fire) and place the heated end directly in the centre of the milling cutter, holding there until the temper in the teeth has drawn to a dark straw color. If the iron cools to a low heat before the temper is drawn to the exact color, use the other iron which was left in the fire to finish drawing the temper, then cool off. Should the milling cutter have very heavy teeth, or if it is to cut very hard metal, it will not be necessary to draw any temper.

If the milling cutter be to the shape and size, say 4 inches long and 2¼ inches in diameter (or larger sizes), as a, Fig. 6, a good way to heat for hardening in the open fire is to have a heavy iron pipe about 6 inches long and plenty large enough for the cutter to go inside. A pipe about 1 inch wider than the diameter of the milling cutter will be about right. Place the pipe in the fire and building the coal on top of it, but do not build the fire over either end of the pipe. Instead leave the ends of the pipe open and do not allow the coal to get inside the pipe. By this method an opening is left clear through the fire Heat the pipe to a bright cherry red the whole length, insert the milling cutter and heat it to the necessary heat to harden, then plunge into the hardening bath from a vertical position. But when heating to harden a milling cutter which has very fine teeth after this method, be very careful not to bruise the fine cutting edges of the teeth against the pipe. It is better to have something arranged to hold the milling cutter up from the pipe. An iron bar placed through the centre of the cutter with a bearing under each end, will keep the cutter from coming in contact with the pipe.

The Use of Asbestos and Clay, When Hardening Milling Cutters and Other Tools.

Very often a milling cutter is made with a thread through the centre of the tool which must be kept soft, while the outside or teeth are hardened, and the way this process is accomplished is by the use of asbestos, which is packed well into the inside or thread, but make sure that the outside ends of the thread are well padded over without allowing the asbestos to come in contact with the cutting edges of the teeth of the milling cutter. The asbestos is kept in place while hardening by the use of fine pliable wire, wrapped around the tool. After the hardening has been done and the asbestos taken from the inside the

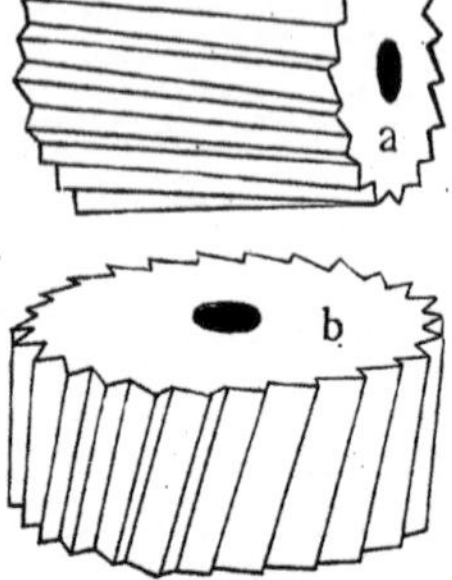

Fig. 6—Plain or Ordinary Milling Cutters.

thread will be quite soft. The reason the thread has remained soft while the teeth are hardened is because the water could not come in contact with the thread when being quenched, on account of the presence of the asbestos.

I have saved a great many delicate and expensive milling cutters from cracking when hardening, by the use of asbestos. Take, for example an angle end milling cutter. That is made with a thin or delicate part extended from the main body of the tool. Now, although the tool may be very evenly heated and properly hardened, it is still very liable to crack, and in some cases the thin or extended part will crack off in a solid ring. To stop a milling cutter of this kind from cracking, fill the hollow in the end with asbestos, being careful not to cover any of the cutting edges of the teeth and hold the asbestos in place by the use of fine

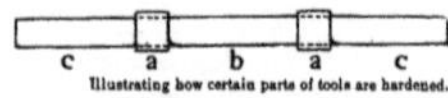

Illustrating how certain parts of tools are hardened.

Fig. 7—Hardening Certain Parts.

wire, while hardening. The reason the tool will not crack is, when quenching to harden only the teeth side comes in contact with the water which hardens, while the other side is kept soft as the asbestos keeps the water from coming in contact with the hot steel.

Another example: Take a piece of steel 3 inches long and 1 inch thick, now an inch on each end is to be hardened, while the remaining inch in the centre is to be kept soft, and to accomplish this process, wrap the centre well with asbestos, keeping it in place by winding some fine wire around it, or instead of using asbestos, wrap the steel around with clay, keeping it in place by the use of a thin piece of sheet metal wound around it, then heat to harden and the results will be as formerly explained.

Hardening Hollow Tools.

When hardening milling cutters as a, Fig. 6, spring threading dies or any similar tool, always quench them from a vertical or upright position, which will allow the steam and water to come up through the tool and cause the steel to be hardened more evenly. Should the tool be quenched from a horizontal position it will be impossible for the steam to escape, and which will keep the water from coming in contact with the hot steel. Thus, when the water is held back by the steam there is a tendency for soft spots in the tool.

As a rule, steel workers never pay any attention to the steam when hardening, which is a great mistake, as many tools are partially, if not altogether, spoiled (more especially if the tools are of a delicate nature) by the great amount of steam which rises as soon as the hot steel come in contact with the water. Delicate or fine tools of a hollow nature will sometimes warp or even crack, caused by steam and improper methods of quenching.

When hardening a spring threading die, it is not necessary to harden the whole tool, but just far enough back from the thread to allow the temper to be drawn with safety. To draw the temper, after being hardened as just mentioned, hold the end or thread part of the tool above the fire and draw the temper very slowly and evenly (by keeping the tool turned around) to a dark straw color.

The Hardening and Tempering of Hob Taps, Stay Bolt Taps and Similar Tools.

In forging, annealing and hardening of long, slender tools, such as hob taps, stay bolt taps, etc., too much care cannot be exercised, although, as a rule, these tools do not have to be forged, as the steel is generally obtained from the manufacturer the right size to allow it to be machined into the tool. However, the steel should be well and evenly annealed, should it come direct from the manufacturer or should it be forged by the toolsmith.

In annealing as well as hardening long, slender tools, they must be carefully handled when the tool is heated the whole length of itself or it will warp easily, also pack the tool very carefully when annealing so that it will have an equal bearing.

When hardening get the tool to a very even heat, enough to harden the whole length of the thread, and when quenching dip deep in the centre of the hardening bath from a perfectly upright position, allowing it to remain in the bath until perfectly cold. Bear in mind that when quenching a long slender tool, any variation from a perfectly upright position will have a tendency to warp the tool. When hardening do not harden the shank, as all that is required to be hardened is the cutting teeth or thread.

To draw the temper, polish bright the grooves in the thread from end to end, have a couple of heavy wrought iron pipes or bands. Heat both in the fire to almost a white heat, then remove one from the fire and put it in a convenient place, place the tap in the heated pipe and draw the tap back and forth to insure a very even temper from end to end. If one pipe is not sufficient to draw the temper, replace with the hot pipe that is in the fire, then cool off the tool. For all kinds of taps draw the temper to a dark straw color. Be sure when drawing the temper not to use too small a pipe, or the extreme fine points of the thread will draw too quickly. For all ordinary taps use a pipe about five inches long and 3 inches inside diameter, while the thickness of the pipe should not be less than ½ an inch or it will cool off too quickly.

Sometimes when hardening a long slender tool only a certain part of it is to be hardened. For example, supposing we have a long slender tool 18 inches long and 1 inch thick. Now 6 inches in the centre is to be hardened, while 6 inches at each end is to be kept soft. In a case of this kind take an iron pipe 7 inches long and 2 inches inside diameter, build the pipe into the fire a little above the surface of the forge, and heat the pipe evenly all around and from end to end. Now place the tool through the pipe, having the part which is to to be hardened directly in the heated pipe, while the ends which are to be left unhardened will project from each end of the pipe, which will prevent them from becoming hot enough to harden. To keep the tool from warping or bending (while being heated) place something under each end close to the pipe to form a bearing and also to keep the tool in the centre of the pipe. If the tool is to be hardened to a very exact length wrap with asbestos at the ends of the part which is to be hardened, as the asbestos will prevent the steel from becoming hardened while being quenched. To illustrate this more clearly, a a, Fig. 7, represents the asbestos, b the part of the tool which is to be hardened, and c, c the unhardened ends.

Heating Furnaces.

In large shops and factories where tools are made in great quantities, furnaces are used to heat steel, but principally for hardening purposes, and in a great many respects a furnace is superior to an open fire, as steel can be heated in a furnace very evenly and with less danger of it becoming overheated than if heated in an open fire. Tools of a long slender shape. such as stay bolt taps, and tools of a very wide and flat surface such as milling cutters, are best heated in a furnace. Different kinds of fuel are used for heating furnaces.

The principal ones used are gas and oil. Gas, however, is preferred, as the furnace can be very readily regulated in order to heat the steel to any degree of temperature. There are many different makes of furnaces and different methods of operating them.

Heated Lead for Hardening Purposes.

The lead bath is extensively used for heating steel when hardening, and has many advantages that a furnace does not possess, as in the lead bath certain parts of tools may be heated in order to harden with ease, and the temper drawn in many ways which could not be accomplished with a furnace.

When heating steel in lead, be sure to use a chemically pure lead, containing as little sulphur as possible. Sometimes when heating in lead there is danger of it sticking to the tool when hardening, but to overcome this difficulty make use of the following compound: Take a pound of powdered cyanide and dissolve it in a gallon of boiling water afterwards allowing it to cool. Now dip the articles to be hardened in the liquid, remove them and allow to dry before placing them in the lead. The liquid when allowed to dry on the tools will form a moisture on the tools when in the heated lead and prevent the lead from sticking. If lead is allowed to stick to the tool while hardening it will cause soft spots where the lead remains. When heating tools with fine projections, have a fine brush and clean off the tool should any lead happen to stick to it.

To obtain the best results when heating in lead keep the lead stirred up, as it will always naturally be the hottest at the bottom.

A lead bath is preferred to an ordinary heating furnace, as steel heated in lead will not raise a scale; also if the lead is heated to the proper degree it will be impossible to overheat the steel, consequently the steel will be very evenly heated. Should there be a great many small tools to harden at once, place them in a heavy wire basket or sieve and lower into the lead; after being heated they may be all quenched together.

AN IMMENSE WHEEL.

A pair of the largest wheels ever built in Canada have just been completed in the works of the Laurie Engine and Machine Co., Limited, Montreal. These are for the Canada Tin Plate and Sheet Steel Co., of Morrisburg, and are immense pieces of work. Some idea of the magnitude of these large wheels can be obtained from a description of them. They are 30 feet pitch diameter, 30 feet 3 inches outside diameter, 48½ inches face and weigh, including shaft. 180,000 pounds.

Each wheel is grooved for twenty-two 1¾ inch ropes. The rim is made up of twelve sections supported by twelve arms and two hub plates. The securing of these pieces together necessitated the use of nearly 200 bolts. In order to secure the accurate fitting of parts, jigs and gauges were made and all parts were accurately machined and fitted to these before being assembled.

The hubs were pressed and keyed to the shaft, after which the shaft was centred in a large lathe. On account of the weight, supports were used at either end to keep the wheel on true centre in the lathe. The inside of each hub plate was then forced to gauge for the arms so that when the arms were inserted they were exactly perpendicular to the shaft centre. The arms were then placed in position each one being accurately spaced by means of special gauges. The rim was put on last and when all the parts were assembled the bolt holes were reamed and fitted with turned steel bolts, a rigid construction being the result. The finishing operation was the turning of the grooves for the ropes. This was done without removing the wheel from the assembled position in the lathe so that a finished wheel, concentric with the shaft and running perfectly true was secured.

H. L. Hibbard, electrical expert to the Bureau of Construction and Repair, U.S. Navy Department, has resigned that position, to enter the employ of the Cutler-Hammer Mfg. Co., of Milwaukee, makers of electric controlling devices.

A New Departure in Lubrication

Mr. Edward G. Atcheson, Niagara Falls, N.Y., in a Paper Before New York Electrical Society, October 23, 1907, Describes his Latest Achievement

Mr. Acheson's address was a revelation. He told of persistency, of determination, of the careful workings of an inventive mind in search of the secrets of nature. He opened his remarks with a reference to his discovery in 1901 of the effects of tannin on clay, his effort at that time being to make common clay plastic and of greater tenure of strength. His words carried his audience back to the time when the Egyptians used straw in making bricks, telling that he conceived the use of the stubble was not for any benefits derived from the fibres, but from the extract. The use of this extract of straw or tannin, he found, increased the strength of the clay very much, some times as high as 300 per cent., but he also observed that clay treated with tannin remained suspended in water indefinitely. For this there is no known explanation. It is simply an "effect," and, as it was discovered by Mr. Acheson, it will probably be called the "Acheson effect."

Perhaps its great importance did not at first appear, but along in 1906 Mr. Acheson discovered a method for making a fine unctuous graphite of the refuse matter, the culm piles of the anthracite coal regions. For years he had been in search of a suitable material from which to make a graphite applicable to lubrication. Every part of Europe and America was canvassed, but right at home, in the great culm piles of Pennsylvania, he found it, and because of this discovery the world is to benefit, the unctuous graphite made at Niagara being almost chemically pure, or over 90 per cent. in purity.

Discovery of the "Effect."

After Mr. Acheson had made this unctuous graphite the thought occurred to him that tannin might have the same effect on it that he had found it to have on clay. He tried it with successful results, the graphite being reduced to an extremely fine condition, believed to be molecular. It remained suspended indefinitely in water and will pass through the finest filter paper. Mr. Acheson related how he had found the "effect" was also produced by the extract of sumac, oak bark, spruce bark, and even tea leaves, and that not only was the effect produced on graphite and clay, but apparently on all non-metallic amorphous bodies generally, having been secured with silica, alumina, lampblack and even his new product, siloxicon. He considered the wide distribution of these re-agents, admitting it is difficult to appreciate or realize the important part that this "effect" may have in nature's economy.

Mr. Acheson became impressively interesting when he told that it was this "effect" that prepared the clay for the potter's use, and that he considered the effect on amphorous bodies to be one of deflocculation. "As found in nature," he said, "amorphous bodies are in what is termed a flocculated state; namely, the molecules of the body are grouped together in masses of appreciable size, and to deflocculate them is to separate them into the molecular condition," when they assume the colloidal state. He admitted that in assuming this state the deflocculated body apparently sets at naught the laws of gravitation, for in this "collodial" state deflocculated graphite, weighing approximately two and one-quarter times more than the water in which it is suspended, will remain indefinitely in that suspension without any noticeable disposition to settle. Waters of rivers, lakes and deep wells have so many impurities that they are unsuitable for the production of this state, but he had found the addition of a little ammonia corrected the ordinary water.

Aquadag is the name Mr. Acheson has given to the new lubricant made by suspending graphite in the water, while Oildag is the name he has selected for the lubricant made by suspending graphite in oil. This deflocculated graphite in water has been found to make an excellent lubricant for light work, the graphite preventing rust of any of the parts, while for heavier work the Oildag lubricant has been found ideal, he said. He went on to tell how both Professor C. H. Benjamin, dean of the engineering schools of Purdue University, and himself had made exhaustive tests, each one of which demonstrated the wonderful efficiency of these new lubricants.

A Practical Test.

Possibly the most satisfying test of all was that made at the works of the General Electric Company at Schenectady, N.Y., under the supervision of W. L. R. Emmett. These tests were not made to include the co-efficient of friction, but of the temperature and the surface speed of the shafts in the bearing. The shaft measured seven and a half inches in diameter, resting in a bearing twenty-one inches in length. The test covered both forced lubrication and oil ring lubrication. Not a great deal of advantage was shown in the case of the forced feed lubrication, the presence of the graphite holding down the temperatures but a very little. In the test on the oil ring feed, however, very pronounced advantages were shown in favor of the graphite.

It is astonishing to think that the graphite content as used in these tests was 0.35 of one per cent. of the weight of the oil, but the tests showed that with the same pressure and temperature " a shaft can be run from 30 to 100 per cent. faster with the graphite in the oil than with the plain oil."

The supported statements of the address will bring happiness to automobilists, for Mr. Acheson stated he had run a heavy car lubricated with Oildag over 4,000 miles without the necessity of cleaning the spark plugs, and, what is still more remarkable, without the necessity of grinding the valves, while materially reducing the consumption of oil. The results would indicate that the use of Oildag in the gas engine will eliminate the pitting of the valve seats.

The surfaces produced on the valve seats are remarkable, being much finer than is possible of attainment by any mechanical finishing, the graphite being incorporated in the body of the metal. The compression is so perfect and remains so that it usually starts on the compression after standing over night.

The address closed with the intimation that the use of this new and modern lubricant will greatly reduce the consumption of lubricating oils, and the audience realized another leg is about to be knocked from under the Standard Oil Company, the major part of whose business has been the sale of lubricanting oils, but the tests made give assurances that the oil consumption will be reduced over 20 per cent., which will be a tremendous volume considering the immense quantity of lubricating oils used annually.

"I will lubricate the bearings of the world," is the slogan adopted by Mr. Acheson.

Practical Questions and Answers

Ques.—I have a planer on which I wish to do some milling. What speed should the planer be run, also, what speed should I run the milling cutters?

Ans.—The speed of the cutter depends on the work to be done and on the shape and width of the cutters. No standard rule has yet been arrived at for the varied conditions that arise, the different amounts of stock to be removed and the rigidity of the planer. It has been found good practice to give the cutter a lineal velocity of 40 to 50 feet per minute for cast iron and a feed of two to three hundredths of an inch to one revolution of the cutter; a velocity of 30 feet per minute for wrought iron or soft steel and a feed of from one to two hundredths to each revolution of the cutter. These speeds, however, may be increased or decreased to suit the particular class of work to be machined.

Ques.—I wish to case-harden machinery steel. Will you tell me the process followed to case-harden them to a depth of say 1-64 or 1-32 of an inch?

Ans.—Cyanide of potassium is sometimes used, but as it is very poisonous great care should be exercised in handling it. The work to be case-hardened is put in the case-hardening box and heated in a furnace. When the work becomes a cherry red sprinkle with cyanide and after fifteen minutes sprinkle again and dump the steel into cold water. A better method and one that will give more satisfactory results in hardening the steel is to pack the steel in raw bone and wood charcoal in parts one to five and heat in the furnace for about three to twelve hours after it becomes a bright red. Let the box cool and then reheat the steel, without the bone and charcoal, to a cherry red and dump in cold water. This second heating will increase the depth of the hardening. Mere hardening of the thin outer surface will do very little if any good.

Ques.—Why is oil used in turning and drilling wrought iron and steel and is not used when machining cast iron?

Ans.—The use of oil and compounds in machine shops is the result of experience which shows how different materials can be worked to the best advantage. These prevent the overheating of the tools and so facilitate the cutting. The best machine shops practice is to use water when machining wrought iron or malleable iron and nothing on steel or cast iron. Oil is sometimes used for turning and drilling wrought iron or steel, but the above is the general practice now adopted.

Ques.—What is the meaning of lead and lap in a slide valve engine?

Ans.—The amount of opening measured in a fraction of an inch, which allows steam to enter at the instant the crank is on dead centre is called the "lead." This amount varies from almost nothing up to ⅛ of an inch or even more. Slow speed stationary engines may have from 1-64 to 1-16 of an inch lead. When the valve is in this central position, the distance the valve overlaps the steam edge of the port is called the outside lap or steam lap. The distance the inner or exhaust edge of the valve laps over the exhaust edge of the port is called the inside or exhaust lap. The two terms inside and outside lap are often confusing, especially in valves where the steam is admitted by the inside and exhausted by the outside edges. The steam lap is much larger than the exhaust lap, which is sometimes zero and is often made negative. In this latter case the distance between the edge of the valve and the edge of the port is sometimes called inside or exhaust clearance.

Ques.—How do you figure the gears on a lathe for cutting threads?

Ans.—In a simple geared lathe where the stud and the spindle run at the same speed, select any gear for the screw, multiplying the number of teeth by the number of threads per inch in the lead screw and divide the result by the number of threads you wish to cut. This will give you the number of teeth in the gear for the stud. If this number is a fraction or you do not possess a gear with that number of teeth, try a different gear on the screw. In a compound lathe first select at random all the driving gears, multiply the number of their teeth together and this product by the number of threads you wish to cut. Second select at random all the driven gears except one, multiply the numbers of their teeth together and this product by the number of threads per inch in the lead screw. Third, divide the first result by the second and you will get the number of teeth in the remaining driven gear. Sometimes the gears on the compounding stud are fast together, but there is usually a definite relation between the gears. The driver has usually half the number of teeth as the driven gear and this can be allowed for by multiplying the number of threads in the lead screw and using this amount in the calculation. When the stud on which the first is placed revolves at half the revolutions of the lead spindle allowance is made by doubling the number of threads on the lead screw. If both these conditions are present, multiply the number of threads on the lead screw by four. For a fractional thread, or the pitch of the lead screw is a fractional one, reduce them to a common denominator and use the numerators of the fractions as the pitch of the screw to be cut and the lead screw respectively. If we wanted to cut a thread 25-64 inch pitch and the lead screw is four threads to the inch, that is, the pitch is 1-4 or 16-64. The screws will be in proportion of 25 to 16 and the gears are figured out using 25 per inch as the number of threads on the lead screw and 16 as the number of threads per inch to be cut.

THE CITY OF WELLAND, ONT.

We are in receipt of a neat little book published by the Board of Trade, Welland, showing the growth of the town since the development of Niagara and Cataract Power. Welland is twelve miles from four of the largest electrical power plants in the world. Three are at Niagara Falls—the Canadian Niagara Power Co., whose development consists of 110,000 horse power; the Ontario Power Co., 220,000 h.p.; the Electrical Development Co., 125,000 horse power, and the Cataract Power Co., of De Cew Falls

Welland is a railroad centre and has therefore great facilities for transportation. The Michigan Central, Toronto, Hamilton & Buffalo Railway, the Welland division of the G.T.R. the Wabash, Pere Marquette, C.P.R. and Toronto, Niagara and St. Catharines Electric Railway. Welland is in the centre of the gas belt and smelting companies and foundries are using natural gas from their own mills.

The book gives a list of the present industries, and among others are the Plymouth Cordage Co., Robertson Machinery Co., makers of hoisting engines, the Canada Forge Co., makers of forgings, the Canadian Billings & Spencer, Limited, makers of drop forgings, the Ontario Iron & Steel Co., makers of steel castings, bar steel and rails, the Electro-Metal Makers of Steel, and Page-Hersey Iron Tube and Pipe Company.

J. R. Nelson, who is in charge of the Toronto and Smith's Falls section of the C.P.R., has been appointed superintendent of Muskoka, Owen Sound and Elora divisions. W. K. Thompson, of White River division, succeeds Mr. Nelson.

Gas Engine Flywheels, Their Function and Application

A Treatise on the Importance of Flywheels of Correct Design.

BY GORDAN A. RONAN.

Since the advent of the internal combustion engine, it has been found that a good flywheel is an indispensable feature, in that it provides a means of carrying the piston, crank-shaft, and all other moving parts over the "Dead Centre," and also gives the momentum ne-

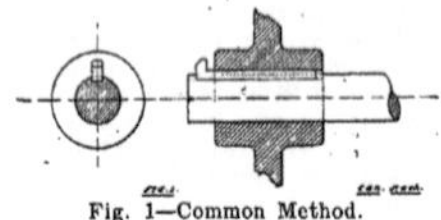

Fig. 1—Common Method.

cessary to carry on the cycle of operations, during the idle intervals, when the engine is preparing for its next impulse.

In addition, the flywheel, when correctly designed, has an element which tends to steady the engine in its operations, and regulates the smooth, even, running of the engine. The flywheel is still further instrumental in absorbing the shocks or impulses which occur at regular intervals in each cycle, and thus eliminates much of the vibration and jar, which would otherwise result, owing to improper flywheel construction.

Flywheel Construction.

A flywheel to be correctly designed should be of the largest possible diameter, proportionate to the engine capacity. This is essential as it gives a great leverage on the crank shaft, and hence more power and effective energy to the engine, which enables the piston to be more easily carried over the point of compression. The flywheel rim should receive the utmost attention, care to have the maximum amount of metal distributed throughout the rim. This should

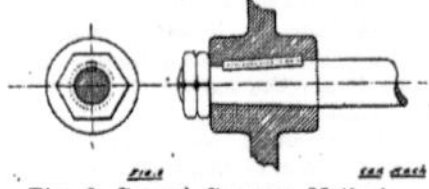

Fig. 2—Second Common Method.

have a wide face and ample side thickness, in combination with the lightest possible web and hub, consistent with safe and efficient construction.

All first-class engine builders have realized the fact, that it does not pay to save metal in their flywheels, but put in every ounce of leverage metal they can and still keep within the above limits of safety and construction.

Some few makers have, however, skimped their flywheels, rendering them much less effective and causing both trouble and annoyance to users when cranking their engines. Their flywheels are too small in diameter, too light in the rim or the weight is placed around the hub or in the web, where it is lost. The result is that these flywheels will not carry the piston over the compression more than a couple of times at most when starting the engine.

Methods—Correct and Incorrect of Application.

The engine buying public is to-day confronted with the almost universal "defective" methods of applying flywheels.

In the following, Figs. 1 and 2, are descriptive of the two most common methods of applying flywheels, both of which are more or less defective, in that each tends in time to wear sidewise, as shown exaggerated in Fig. 4 at A. It will at once appear obvious that a parallel shaft shown in Fig. 1 and fitted with a parallel key to a key-seat, will in the course of time, even with the best of care and atention, tend to assume the above mentioned shape. This is directly due to the fact that the terrific strain thrown upon the crank shaft and bearings, by the desire on the part of the flywheel to go in one direction compelled by the momentum, while the impulse of the piston, when advancing the spark tends to act in the opposite direction.

More especially is this true of marine engines, where the almost universal practice, in small h.p. units is to have them reversible. The key-way must, therefore, tend to wear sideways, and finally wears the key-seat from its original form, marring both the shaft and flywheel oftimes past repair.

Fig. 4—Result of Wear Using Methods 1 and 2.

A striking example of this has come before the writer's notice during the past week. B, in Fig. 4, shows in exaggerated form what actually occurred to a flywheel keyed as in Fig. 1, when it was suddenly wrenched, by an obstruction taking place in the crank case, the momentum of the flywheel twisted the key, key-seat and shaft one-quarter of its circumference out of the original straight line of key-seat.

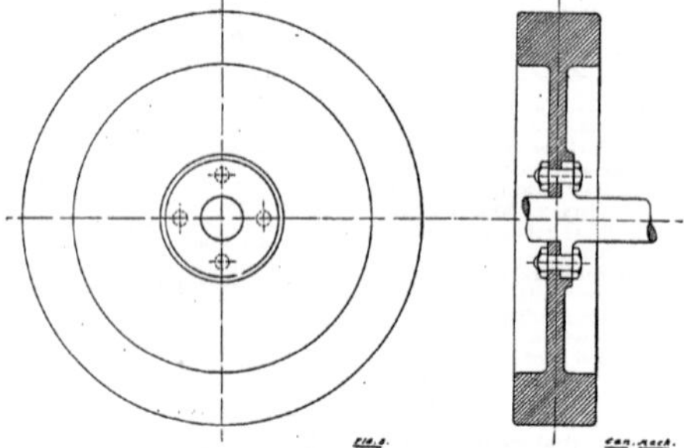

Fig. 3—Ideal Method.

The principle of using a taper while much better than the above, is far from being ideal. Some makers anxious to do the public a great favor, have adopted this method in the hope of making it a talking point in their engines. However, this method, too, is subject to loosening of the flywheel by the nuts and key becoming constantly jarred by the engine vibration and the terrific strains on the crank shaft while being reversed. Actual tests have shown the taper to give this annoying trouble also.

A few successful manufacturers in the engine industry who have led the onward march of progress, have given us the ideal method of application as shown in Fig. 3, in which the flanged crank shaft is used, and inserted into an accommodating recess in flywheel. The four or six holes are first accurately bored to template. and are afterwards carefully reamed. The bolts are machined a neat fit, and the flywheel is then placed in position and locked tight.

The large flange inserted as it is, eliminates entirely the danger of the other two plans, and is more accessible, and easily detached by the amateur, without the marring effect of the chisel, hammer or mall in removing the flywheel when required.

Rope Transmission in the Factory

The Use and Life of Rope for Transmission and Hoisting

Rope drive is coming into favor in some factories with the advent of machinery equipped with change gears. In years past it has not been practicable to use rope drive for lathes, drills or shapers where variable speed is required and when it was obtained by cone pulleys. In cotton mills manila rope has been replacing spur gearing though this latter form is the most exact method of transmitting velocity ratios and gives the highest efficiency when properly made. The rope drive has an advantage of increasing the power delivered to the cutting tool by high initial belt speed. It is an ideal means of supplying power to constant speed shafts. The sheaves occupy less space as they are narrower than those used for belts and this will allow bearings to be placed more closely together, eliminating spring in the shaft. There will be less loss from slipping and the quietness of rope drive is a great contrast to other methods.

Ropes may be arranged in two ways. In the first way there are as many separate ropes as grooves on the pulley In the second method a single rope is wound progressively around the two pulleys, as in Fig. 1, being delivered from the last pulley on one side to the first groove on the other side of the pulley by means of a guide pulley. In the first method if there is one rope used an idler can be inserted to take up the tension, but if more than one rope is required, it is difficult to get the same tension on each rope and one rope transmits more power than another. In the second method the tension is practically equal since the rope is continuous. A device used to keep a tension on the continuous rope drive is an arrangement of three guide pulleys. One is called a "take-up" and its shaft works in blocks in guides. A weight is attached to it and as it is free to move vertically a tension is kept on the rope. This arrangement, shown in Fig. 2, is substituted for "a" in Fig. 1 and one pulley guides the rope from the system to the take-up and the other back to the driving system. Ropes usually used for this drive are 1½ to 2 inches. The net section of the rope is about 9-10 of a circumscribed circle for a 3 strand rope, and in practice a stress of from 140 to 240 pounds per square inch may be allowed on the driving side of the rope. The breaking strain of good cotton ropes is about 4 tons. The speed varies from 3,000 to 5,000 feet per minute. In practice a sheave with a groove of about 45 degrees is used, which gives about three times the friction of the rope running on a flat surface. A rope is made endless by splicing and as every machinist can lace a belt and few can

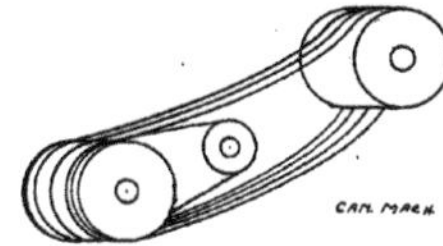

Fig. 1—Method of Keeping Tension on an Endless Drive System.

splice a rope, this is one objection, but the ropes very seldom have to be spliced. One expert rope splicer could attend to more work than a large industrial establishment using rope drive would furnish him.

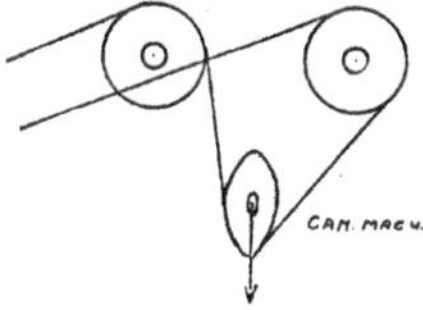

Fig. 2—Loaded Pulley for Tension on Ropes.

The Life of a Rope.

The pulley should never be less than thirty times the diameter of the rope. If smaller sheaves than this are used the power transmitted will be less in proportion. Well cared for, a rope will last fifteen years, though the average life of a rope is about ten to twelve years. With an increase in the diameter of the sheaves there is a material increase in the life of a rope. Lubricating the rope with plumbago mixed with tallow will prevent internal chafing and wear.

Horse-power Transmitted by Good Cotton Ropes.

Diam. of Rope in Inches	Velocity in feet per second: 600	1,500	3,000	4,000	5,000
	Horsepower transmitted				
½	.8	2.1	3.7	4.3	4.4
¾	1.9	4.7	8.3	9.8	10.0
1	3.4	8.3	14.9	17.5	18.0
1½	7.7	18.7	33.4	39.2	49.3
2	13.5	32.9	58.7	68.9	70.7

Manila Rope for Hoisting.

In the manufacture of rope the fibres are spun into a yarn which is twisted right hand. From twenty to eighty of these, depending on the size of the rope, are twisted in the opposite direction into a strand. Three or four of these are twisted right hand into a rope. This twisting the parts in opposite directions keeps the rope in its proper form.

When a rope is first used in hoisting there is a tendency for it to untwist and become longer. The first day or two the rope will untwist a certain amount, but after that the length of the rope should remain substantially the same. The four strand rope is most commonly used, both for transmission and hoisting. Hoisting ropes are never spliced, as it is difficult to make a splice that will not pull out.

Practical experience for many years has settled the most economical sizes of rope that should be used. The amount of coal for instance that can be hoisted with a rope varies greatly. With care a rope will hoist from 12,000 to 20,000 tons of coal, but with only ordinary care, only from 6,000 to 9,000 tons. A 1½ inch rope, running on 14 inch sheaves with a strain of 600 pounds per square inch, is found to be economical. Hoisting rope is ordered by the circumference; transmission rope by the diameter.

CAR FERRY READY.

The new car ferry, Ontario No. 1, built by the Canadian Shipbuilding Company for the Ontario Car Ferry Company, which is to be operated in the interests of the Grand Trunk Railway Company and the Buffalo, Rochester & Pittsburg Railway Company between Cobourg and Charlotte during the year round, left for Cobourg last week to go into commission on the new route.

The big boat will carry coal going north and iron ore and lumber going south, the trip being about sixty miles across the lake one way. The Ontario No. 1 is valued at $375,000.

ACCIDENT ON TORONTO ISLAND.

Cable Snaps, Probably as a result of Crystalizing.

An unfortunate accident, attended with a loss of four lives, occurred on Toronto Island the evening of Nov. 19. A shaft about one hundred feet deep has been sunk on the island to carry a waterworks tunnel under Toronto Bay. A cable snapped when seven men were within fifteen feet of the top, dashing them to the bottom

One end of the false cable was attached to the end of the boom which hung over the shaft. Then the steel cable went through a pulley on the bucket, leading again to the boom and then to the drums of the hoisting engine. It was the portion of the steel cable between the pulley and the end of the boom which gave way. The rigging is that which is commonly employed in all such shaft work, and is considered sufficiently safe.

The broken cable is supposed to have a working tensile strain of twenty-one tons. The cables were examined hourly and if one strand was broken it was immediately replaced by a new cable, one being always kept on hand.

Causes of Cables Breaking.

The manner in which the cable snapped would lead one to believe that the steel in the cable crystallized. The writer has seen within the past few months a practically new cable draw apart, precipitating five men a distance of about thirty feet and injuring all five. Sometimes the cause of cables breaking is due to running over small sheaves, causing internal strains. These break the inner strands and this can only be recognized by the creaking sounds of the cable. The parting of the cable when this is the cause is by the drawing apart of the strands.

It is a well known fact that if a bar of iron be tapped on the end the molecules of iron will arrange themselves in such manner that the bar becomes a magnet. Something of the same takes place in the iron posts supporting a building. From the break, straight across the strands we should judge that the molecules in the steel had in some such manner arranged themselves. In other words, the steel had crystallized. The crystallizing may be caused by the constant bending of the cable, causing the molecules to arrange themselves in a certain position, the carbon changing its physical state. When a rearrangement takes place an unusual strain will fracture the steel. This is what probably happened in this case. The carbon has crystallized out forming a coarser structure with a lower tensile strength and with the strain coming upon it the cable snapped at this weak spot thus formed.

DEATH OF MR. F. B. POLSON.

The President of the Polson Iron Works, Limited, Toronto, died suddenly on October 28th at his residence, 6 Beaumont road. Mr. Polson occupied a prominent position in commercial circles as president and general manager of the well known ship-building and engineering firm, the Polson Iron Works. He was born at Port Hope, Ontario, on February 10, 1858, so that he lacked but a few months of being 50 years of age. He was a son of William Polson, who was mechanical superintendent of the Grand Junction Railway at Belleville before founding the present business in conjunction with his son.

THE LATE MR. F. B. POLSON.

The Polson Iron Works, Limited, was established in 1883, under the name of William Polson & Co., the members of the firm being William and his son Franklin Bates Polson.

The firm started work on the Esplanade at the foot of Jarvis St., where a general machinery trade and yacht building business was carried on. The business grew so rapidly, however, that in 1885 a joint stock company was formed. A new site was obtained at the foot of Sherbourne St. and suitable buildings erected thereon for carrying on the business of engineers, boiler makers and ship builders.

Mr. F. B. Polson received his education in the Cobourg Public School and Collegiate Institute. He was a mechanical engineer of thirty years' standing, held marine engineers' and masters' papers, and was a member of the Institute of Naval Architects of London, England; Engineers' Club, New York; Canadian Society of Civil Engineers; and the Toronto Board of Trade. He took a very active interest in technical education, being for some time chairman of the Toronto Technical School Board. He was a member of the National Trades Association and the Employers' Association of Toronto. Mr. Polson served a lengthy apprenticeship with railway and manufacturing companies, and gained an experience in ten years that fitted him for the position of mechanical superintendent of the Victoria Railway at Lindsay. This position he resigned to found with his father the shipbuilding business. They undertook and completed important contracts, such as the magnificent steel steamer Manitoba for the C.P.R., the Petrel, the J. Israël Tarte, and many others. The industry has grown to be one of the largest of its class in Canada.

Mr. Polson was laid to rest in the family plot, St. James Cemetery, October 30.

The business will continue under the presidency of Mr. J. B. Miller, and the management of Mr. J. J. Main.

PERSONALS.

M. S. Blaiklock has been appointed engineer of maintenance of way for the G.T.R., with office in Montreal.

H. E. Whittenberger has been appointed superintendent of the eastern division of the G.T.R., with office in Montreal.

Cecil B. Smith, of Toronto, has been engaged by the city council of Port Arthur to make a preliminary survey of Dog Lake, which will give a head of 340 feet. It is, therefore, of much larger power than Kakabeka Falls, which gives a head of 185 feet.

Richard D. Hurley, manager of the Pittsburg office of the Independent Pneumatic Tool Co., of Chicago, died suddenly in Chicago on Nov. 5th. He was a young man, but had been with the company over ten years and made a wide circle of friends.

Harvey Graham, assistant general manager and director of the Nova Scotia Steel Company, died at his home in New Glasgow, Thursday morning, Oct. 31st. He was connected with the large coal, iron and steel industries for many years.

E. Hallman has just returned from a business trip to northern and western Ontario for H. W. Petrie, Limited, Toronto, and reports business in machinery lines active in Fort William, Port Arthur, Sudbury, North Bay and Cobalt.

MACHINE SHOP METHODS AND DEVICES

Unique Ways of Doing Things in the Machine Shop—Readers' Ideas and Opinions Concerning Shop Practice—Useful Data for the Machinist—Remuneration will be Given for Contributions.

REPAIRING A CRACKED WATER JACKET.

By E. S. Cooper.

Just at this season of the year, when you are apt to think it is not going to freeze, is the time that nearly all splitting of water jackets on gasolene engines occur. In the middle of winter it is generally taken for granted that there is danger from freezing all the time the water is left on when the engine is not running and it becomes a habit to drain off the jacket water when shutting down the engine for the night or for any great length of time. In spite of all personal precautions on the part of the owner, or careful and oft-repeated instructions to employes, there is apt to come a time when the jacket water is left in the cylinder on a cold night and the inevitable happens.

The cracks in the jacket take all kinds of shapes and occur on all sides of it. Sometimes they run into the face of the cylinder against which the cylinder head is bolted and this allows the casting often to spring so badly that the job of repairing it is almost impossible.

Repairing Crack in a Gas Engine Water Jacket.

The usual method of repairing a cracked water jacket is to bolt a band around it and draw the edge together tight enough to prevent a leak. This accomplishes its purpose, but makes an unsightly job and one that detracts from the value of the engine if it is ever wished to sell it. If the job is done so that the repair does not stand up above the body of the casting it can be painted over and never show.

One method we used in order to do this was to cut a dovetail groove all along the crack, leaving the crack in the centre of it. A piece of copper wire was then taken, annealed by bringing it to a red heat in the blacksmith's fire, and plunging it into cold water, hammered to a square so as to just fit the groove. annealed again and driven into place. It was left so that when hard on the bottom of the groove the top of the copper stands slightly above the casting, and it was then peened until the copper metal flowed into and filled the dovetail of the groove.

The top can then be dressed off and painted over, and, if the job has been properly done, will hold as well as ever. We know of some jobs done this way that have stood for 3 or 4 years and seem to be in as good shape now as when first done.

One of our repair men hit on a plan of filling the groove that was much simpler than the one described above and which proved quite a practical success. After the groove was cut. instead of all the fitting of the copper. it was simply filled with smooth-on iron cement and allowed to stand over night. In the morning it was dressed off, painted. and as far as we know, these jobs have held and stayed right.

When you see some of the large patches put on some gasolene engine cylinders just try and figure out the amount of time it took to drill and tap all the holes in the casting and to fit and drill the patch. Not only is it an unsightly job but the groove can be cut and filled with the iron cement in one-quarter of the time and is much neater when done.

The above sketch shows very plainly how the groove was made in relation to the crack in the wall of the water jacket. A 3-16 inch or ¼ inch capechisel is used to cut the groove to the proper depth and a side cut chisel easily undercuts the dovetail. It is not necessary to smooth the groove out at all, and the fact of not having to do any fitting on it, when using the iron cement, makes it possible to do a quick job with the simplest of tools.

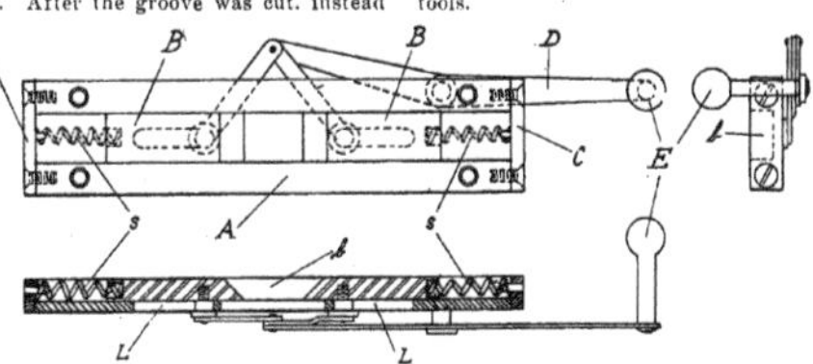

Fig. 1—For Tail-stock.

ATTACHMENT FOR DRILLING.

By R. Prosser.

The accompanying drawings show a fixture which could be applied to an engine lathe, where there is any great amount of drilling done with the tail-stock.

In Fig. 1 is shown part of the fixture to be fastened to the tail-stock by screws, A A, Fig. 4. A is a plain block shaped out at B, to receive the sliding blocks B B.

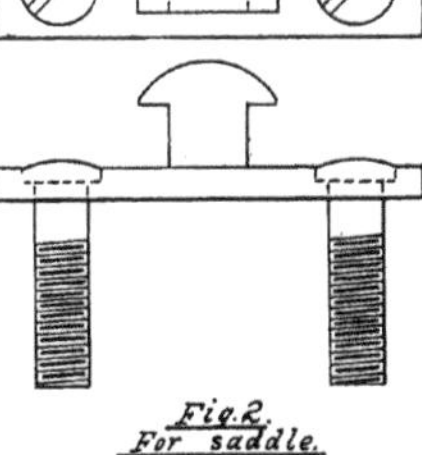

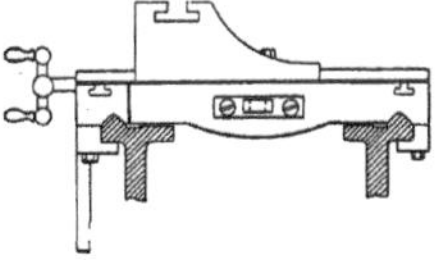

Fig. 3—Attachment on Saddle.

C C are strips fastened on the ends of A to back up springs S S which are kept in place by recess and dowel-pin as shown. D is a lever (moved by handle E) to act on toggle joint to move blocks B B back and forward. Slots L L act as a guide to keep blocks in position. Fig. 2 shows part to be fastened to saddle as shown in Fig. 3. If possible, Fig. 1 could be placed on the inside of tail-stock to avoid removing it when not in use.

When drilling is to be done bolt Fig. 2 to saddle, start drill in the usual way by centreing, then take saddle back and snap saddle and tail-stock together, thus locking them and proceed to drill with saddle feed. To unlock saddle from tail-stock, raise handle E and remove saddle. Gib G could be used to prevent tail-stock getting out of line, as the king bolt in tail-stock must be a

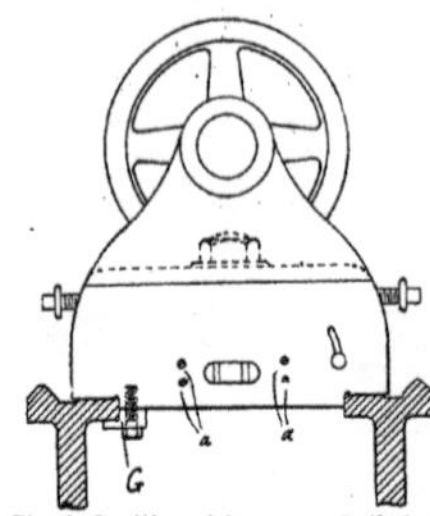

Fig. 4—Position of Screws on Tail-stock

little slack when drilling in this manner. The appliance must be modified to suit different makes of lathes.

AN INSERTED-TOOTH FACE-MILLING CUTTER.

The accompanying illustration shows a 14-inch face mill devoid of all frills and ruffles, but which has been in constant use for the past six months and has given entire satisfaction. The system of holding the cutters is as simple

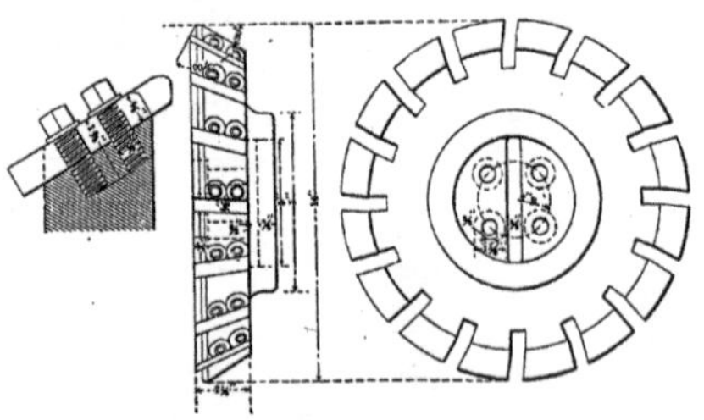

An Inserted-tooth Face-milling Cutter.

as it is possible to get and have them hold, and the fact that we have broken several ½-inch square cutters and never had one slip back is evidence enough that the cutters cannot move. The slots are milled a good snug fit for ½-inch square tool steel and the holes for the collar screws are counterbored about 1-32 inch below the top of the cutter so the screw will tighten up on the cutter before it strikes its seat in the body. The cutters are ground to a gauge and set in the body to another gauge so that when one set gets dull another can be put in with the assurance that each cutter will do its share of the work. The cutters are ground like a lathe roughing tool, leaving a round nose and doing away with all sharp corners, which, I believe, is the cause of most face mills getting dull so soon: the sharp corner burns off and leaves the cutter without any cutting edge.

With a cutter ground in the shape here shown, it is possible to remove ½ inch of cast steel at a cut with a feed of 2 inches per minute, and if some good drill compound be used to flood the cutter, it is remarkable how long a set of cutters will last without sharpening.—H. M. L. in American Machinist.

USEFUL CONSTANTS.

By J. H. R., Hamilton.

Some times, after a man has spent considerable time on a job and cut away a lot of stock from the same, the foreman comes along and asks him for the weight of the material, or when he comes to fill out his time slip, he either has to tell the foreman that he forgot to weigh the stock or (if no table of weights are at hand), make a guess at it.

Now, here are a few simple constants which only vary a fraction of a pound in a hundred:

Weight in lbs. per 1 ft. of length:

For Square Iron—Square one side and multiply by 3.333.

For Round Iron—Square the diamter and multiply by 2.6175.

For Machine Steel—Add about 2 per cent.

The areas of squares or circles vary as the squares of their sides or diameters.

A square 3 inches on a side is 9 times as large as one 1 inch on a side.

A circles 5 inches in diameter is 25 times as large as one 1 inch in diameter.

Therefore, the weights of material vary in the same proportions.

HOME-MADE CALIPERS.

By a Subscriber.

I had much difficulty measuring the thickness of iron when a flange was cast

A Handy Pair of Calipers.

along the edge, and made a pair of calipers, as shown in sketch, to overcome the trouble. On the top of one arm of the calipers I filed out a circular plate and on the other an arrow, fastening the two arms together with a small piece of paper brass between to make them work smoothly. To get the thickness marks on the plate, I closed the calipers and marked with a line opposite the arrow; then I took a scale, opened the calipers to each one-eighth and put a permanent mark opposite the arrow each time. If anyone wishes smaller divisions, the plate can be marked in sixteenths. By direct reading I could then tell at once the thickness of the metal. The sketch shows the calipers in use, showing the thickness of the iron to be one-eighth of an inch.

HOW SHOULD GLOBE VALVES BE PLACED?

All globe or angle valves for steam or gas pressures should be placed so that the pressure comes on the under side of the disk. The main reason is that when the pressure is below the disk, the valve will almost invariably seat better, as, when the steam enters above the disk and the valve is nearly seated, the rushing steam has a tendency to blow a little harder through one side, owing to the whirling motion it attains, and this causes the disk to seat improperly by one side seating a little ahead of the other.

Some engineers argue that the pressure helps to press the disk down on the valve seat when the pressure is above the disk, but if the valve leaks when turned down as tightly as it can be turned by the handle, it is time to regrind the valve, or put in a new disk, as the case may be.

I have had several experiences with the disk becoming detached from the stem, and in one case, it was in the water-supply pipe between the feed-pump and the boiler, or, rather, between the check-valve and the boiler. It immediately blew the packing out of the feed-pump, and it being the only source of supply to the boiler, we had to shut down the plant for nearly a whole day, so the valve could be repaired.

Had the valve been placed the other way the disk would have acted as a check-valve and we could have kept running until a convenient time for repairing.—M. Johnson, in Power.

PARALLEL BARS FOR PLANER.

By J. H. R., Hamilton.

The accompanying sketch shows an adjustable parallel strip which has been used in the shop for quite a long while. As it is a very useful tool for supporting planed pieces on the planer or shaper, and does away with a lot of packing up with washers and pieces of tin, etc., being very easily and quickly adjusted to any desired size within the limit of the adjusting screw.

As is shown in the sketch, the tool consists of 2 pieces of cast iron, a and a, planed and dovetailed at an angle with the faces about ½ inch or 1 inch taper to the foot, making a fairly tight fit, thus avoiding a gib. The adjustment is made by turning screw S, which works in nut N, fastened to top block by two screws. Collar C is placed in a recess in A, and plate P fastened on the end to support screw S and avoid end play.

The two in the shop have been in use for a number of years and are as good as new. Such a strip with careful handling would last a long time.

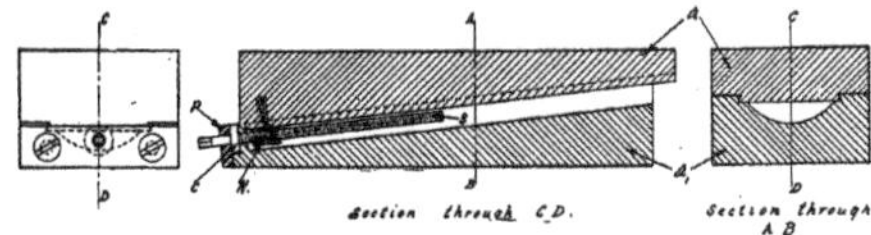

Parallel Bar for Planer.

POWER REQUIRED FOR MILLING MACHINES.

The following experiments by S. Streiff, in "Werkstatts-Technik," for finding the power required for driving milling cutters have been carried out on a special milling machine. The results are given in the accompanying table. Only the results of such experiments where the power required was determined definitely are taken into consideration. A number of experiments, where either the time of cutting was insufficient to get a good average, or where some other disturbing factor entered in the experiment, have been left out of consideration. The milling cutters used were made from high-speed steel. It will be seen from the table that a proportionally higher amount of power is required for light milling than for heavy, a result that is rather natural, as the running of the machine itself, irrespective of the work being performed, would require practically the same amount of power in either case. It will also be seen from the table that the depth of the cut does not increase the power required in the same proportion as does the width of the cut, and that work with a heavy feed and a deep but comparatively narrow cut requires far less power for the amount of metal removed than does a slow feed, and a cut of moderate depth, but wide. In general, it seems as if a slow feed is particularly uneconomical in the use of the milling machine. The figures in the table have been transformed into the English system of measurements, to facilitate comparisons.

Number of Experiment.	Number of Revolutions of Cutter per Minute.	Feed. Per Minute, Inches.	Feed. Per Revolution, Inches.	Cutting Speed of Cutter per Minute, Feet.	Depth of Cut, Inches.	Width of Cut, Inches.	Power Required, H.P.	Metal Removed per Hour, Pounds.	Power Required per Pound Hour, H.P.
1	24	2·48	0·10	37	0·25	23·8	25	245	0·102
2	24	3·60	0·15	37	0·26	10·2	17	150	0·113
3	24	4·35	0·18	37	0·14	9·8	17	97	0·175
4	24	3·60	0·15	37	0·49	9·8	27	490	0·055
5	19	4·33	0·23	29·5	0·28	9·3	17	331	0·051
6	23	4·17	0·18	36	0·25	20·5	27	386	0·070
7	23	4·17	0·18	36	0·28	9·8	20	183	0·109
8	40	1·89	0·05	64	0·24	10·2	17	74	0·230
9	40	3·94	0·10	64	0·37	13·8	21	331	0·063
10	40	5·79	0·14	64	0·16	16·5	17	193	0·138

Table of Milling Machine Experiments, Showing Relation Between Feed, Depth of Cut, and Power Required.

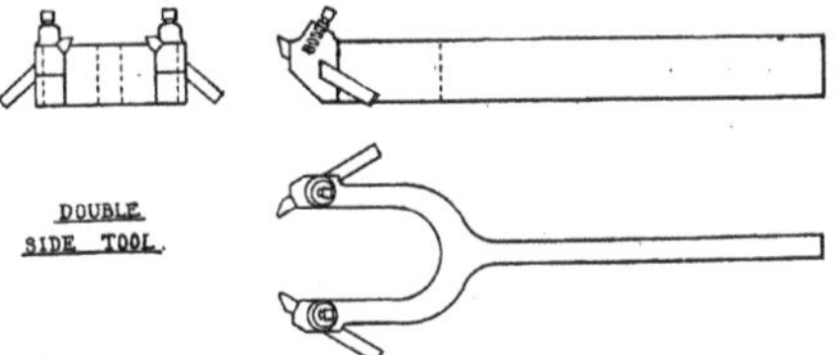

THE DOUBLE SIDE-TOOL.

By R. Prosser.

The double side-tool is a most useful tool, and is very easily made. In doing a job that has to be faced off on both sides, use the double side-tool. It does both sides at once and takes no more time than to do one side. The tool will do different sized jobs by loosening the set screws and moving tools to desired size.

TO BORE A BLIND HOLE.

By A. E. D.

When boring a small cylinder or cutting a blind hole where there is no clearance and the thread ends against an internal shoulder, it is a hard matter to watch the tool itself. I have found it to advantage to run the tool in against the shoulder and chalk a mark on the lathe saddle and another on the bed, and then put a line with a lead pencil or a scriber, if one does not mind marking up the machine. When the two lines on saddle and bed coincide in feeding, the cutting tool has reached the end of its travel.

A HANDY JACK.

By Machinist.

A device that I have used in the machine shop for leveling work when drilling and planing is a small lifting jack that we cast in our own foundry. We made them in several sizes and the illustration will show the construction. The top is hexagonal so that a wrench may be used to prevent the jack from turning when the bolt is turned to raise or lower the work. An ordinary hexagonal-headed bolt is used with the head rounded on the top. Work can be easily and quickly leveled up by means of this device. For one job on the planer where we had a number of pieces

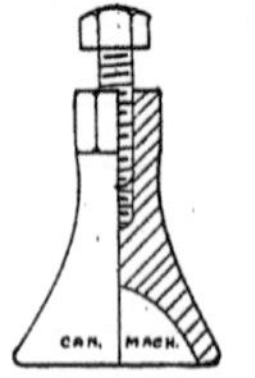

A Handy Jack.

to plane we bored holes in the base of the jack and bolted it in position on the planer.

SCREWING FLANGES ON PIPE.

It is unnecessary to tie the handle of the chain tongs to a post. The practical way to arrange the pipe and tongs would be to place the tongs as shown in Fig. 1, so that they will hold the pipe from turning, and then allow the handle of the tongs to rest on the floor,

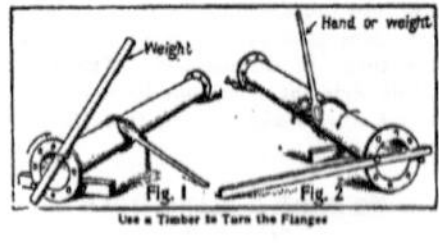

Use a Timber to Turn the Flange

using the timber to turn the flange, says a correspondent in the Metal Worker.

Another way would be to allow the end of the plank to rest on the floor, and the workmen could use the tongs to turn the pipe into the flange, the flange being held stationary by the timber, as in Fig. 2.

SQUARE ARRANGED FOR RULING.

By a Draftsman.

When inking in drawings and tracings I was often troubled with the ink flowing along the edge of the square. I successfully overcame this difficulty by glueing on the square three small pieces

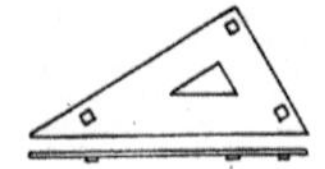

A Handy Square Attachment.

of cardboard as shown in the illustration. The cardboard pieces are about one-sixteenth of an inch from the edge of the square. I am never troubled with the ink flowing along the square now. In addition with this arrangement I can ink in lines crossing each other without fear of blotting.

WESTERN MACHINERY MARKET.

Winnipeg, Nov. 22.—The year just closing has not been the most prosperous on record for the machinery trade of western Canada, but has been quite as good as could have been expected under the particular conditions obtaining this year.

The machine tool trade in the west is good and selling agents generally report good sales. Business in this line has kept up, especially on account of the railroad shops, which are kept busy.

Production in the lumber mills has been curtailed owing to the lessened demand for building material, due to various causes, which are well understood and which, therefore, need not be dealt with here. Dull times in the lumbering industry has meant a decrease in the orders for wood-working machinery from the mills at the coast and in the north country. At the present time the coast lumbering industry is not in the best shape and the machinery interests are unable to secure their usual large orders from that quarter. It is recognized that the depression in this industry is merely temporary and conditions are bound to improve very soon, but it is none the less a fact that at the present time the machinery dealers find their business with the lumber and allied industries far from satisfactory.

Sales of mining machinery have been affected by the condition of the copper mining industry in southern British Columbia. The closing of the Granby smelter and the reduced output of the mines at Rossland and Grand Forks, due to the low market price of the metal have had a corresponding effect on the demand for mining machinery. In this case also it is believed that present conditions are temporary and relief may soon be expected, but there is depression at present, and the machinery trade feels the effects. On the other hand the coal mining industry is in fairly satisfactory condition.

Elevators and flour mills provide a good market in the west for machinery. This year sales of this class of machinery have been large, but owing to the unsatisfactory crop outlook they have scarcely been as large as in previous years.

Machinery trade in the west is, therefore, not in the most satisfactory condition at present. It must not, however, be thought that the outlook is discouraging. An examination of the causes responsible for the slight depression now felt shows clearly that these causes are temporary in their nature. Once removed, their effect will disappear and business will once again be on a normal basis. It has been an exceptional year, a year in which a combination of unfortunate circumstances has checked the natural expansion of all classes of business and the machinery trade has suffered with the rest. But the country is in sound condition and the prospects for 1908 are more encouraging.

BOOK REVIEWS.

GAS AND OIL ENGINES AND GAS PRODUCERS—By L. S. Marks and S. S. Wyer. 143 pages. $6\frac{1}{4}$x$9\frac{1}{4}$ inches. Illustrated. Published by the American School of Correspondence, Chicago, Ill. Price, $1.00.

Part I, by Prof. Marks, reviews the developments of the gas, gasoline and oil engines and the Thermodynamics of the Otto cycle and its later development, the Diesel cycle. Part II, by Prof. Wyer, treats gaseous fuels and producer gas and its manufacture. To those who wish to avail themselves of the developments along these lines we recommend this book.

ELEMENTS OF MECHANICS.—Published by the Macmillan Company, of Canada. Limited, 27 Richmond street west, Toronto, Ontario, 280 pages, cloth covers, 6x9 inches, price net $1.50. The book is written by W. S. Franklin and Barry Macnutt, and illustrated throughout, over 180 cuts being used. The work covers the measurement of length, angle, mass and time, physical arithmetic, simple statics, translatory motion, friction, work and energy, rotatory motion, elasticity of materials, hydrostatics, hydraulics, wave and oscillatory motion. This last topic is treated in minutest detail, and is one of the most valuable chapters in the book. The work is prepared to be of service for both colleges and technical schools. Practical problems with explanations are given on locomotive curves, engines, friction, the gyroscope and on all the various topics treated. The work is comprehensive, and explains in a thorough manner, the fundamental principles of mechanics, with their applications to actual apparatus in daily use and to engineering problems. The endeavor of the authors has been to make a dry subject interesting, and in this they have succeeded, giving both a mathematical and descriptive treatise of the laws governing the field of mechanics.

DEVELOPMENTS IN MACHINERY

Metal Working Wood Working Power and Transmission

LARGEST PLANER BUILT IN CANADA.

Some months ago there was dismantled for shipment in the works of John Bertram & Sons, Limited, Dundas, Ont., a planer of such immense proportions that one was naturally led to make enquiries concerning it. This planer is now doing service in the plant of the Canadian Westinghouse Co., Hamilton, and is unique among Canadian manufacturers for its size and detail construction. Not only is it the largest planer built in Canada, but it is also the largest in use in Canada.

The planer is 10 ft. 2 in. between housings and under cross-rail, and is equipped with a 20-ft. table. The construction and equipment is modern in every respect. It is motor driven with a 50 h.-p. variable speed D.C. motor. The cross-rail is raised and lowered by means of separate motor. There is also a motor provided for quick movement of side head, and also one for movement of cross-rail tool head.

The main drive is through gearing, and the planer is controlled by means of a pneumatic clutch. This pneumatic clutch is something new in Canadian shops, and is only used in the States on large planers. The gearing and rack are of cast steel, and are thoroughly encased to ensure against damage.

The table rests on one "V" and one flat surface, a common style of construction in large planers. The V and flat surface are provided with rollers for lubricating.

When working under full load this planer is capable of removing, in cast iron, 4 cuts, 1½-in. deep. The planer can be reversed to a line, showing the exactness of the pneumatic clutch.

The making of this planer occupied no little time and no little space in the Bertram shops, and was completely erected and operated for a month before it was dismantled and shipped.

The Largest Planer Built in Canada—John Bertram & Sons Co., Dundas, Int.

Additional ideas of the size of this machine can be obtained from the weight, 200,000 pounds.

THE BATEMAN HIGH SPEED PLANER.

In the construction of all kinds of engines, machinery, and mechanical appliances there are many flat surfaces to be machined, and manufacturers have a choice of several types of machines for doing this work. These machines can be divided into two leading classes, i.e., reciprocating machines, such as planers, slotters and shapers and rotary machines, comprising all the different kinds of milling machines.

The introduction of high-speed steels has completely upset old standards, and made possible greater and better production from planers equally with other tools. A modern planing machine must now be capable of taking full advantage of the best brands of tool steel, and cutting speeds must be aimed at which are at least equal to the cutting speed of a lathe using tools made from the same brands of high-speed steel. The return speed should also be rapid, so that very little time be wasted over the idle stroke, and the time lost at reversal should be reduced to a minimum.

The makers of the Bateman High-speed Planers put upon the market an exceptionally rigid and massively designed tool, and by means of two patent attachments have succeeded in attaining cutting speeds as high as high-speed steels will allow in conjunction with quick return speeds. This is accomplished without the consumption of an excessive amount of power.

The table of this machine varies from the usual construction, in that the rack, instead of being rigidly fixed to the table is attached through springs. The baffer action of these springs enables the table to be stopped and reversed without any jar, and at the same time so lessens the shock upon the gearing that very high speeds are attained without risk of failure in the gearing.

The reversing at high speeds of a heavy moving mass like a loaded planer table calls for increased energy, but the makers ingeniously meet this difficulty by fitting two heavy flywheels at either

The Bateman Planer—Leeds, Eng.

side of the machine, instead of ordinary loose pulleys. These flywheels at each reversal are coupled to the fixed pulley by the belt, which overlaps both pulleys, and by the contract of cone clutch faces, between the fast pulleys and the fly wheels. The energy stored in the fly-wheels is utilized effectively in stopping and reversing the table, thus equalizing the power required for driving.

The general excellence of the design and build fits this planer for very heavy work, while the high cutting and return speeds enormously increase its earning capacity. The importance of the absence of chatter is generally recognized, and the makers, with this object in view, devote special attention to the bedding down of the v-slides and other parts concerned.

This illustration is one of three 84-in. x 72-in. x 16 feet motor-driven machines working at a cutting speed of 45 feet per minute, with a return speed of 110 feet per minute.

This planer is manufactured by The Bateman Machine Tool Co., Leeds, England.

VICTOR PIPE MACHINE.

The Loew Victor Pipe Machines, manufactured in both hand and power types, have many features of interest. The die head is designed on the principle of the universal chuck, and an automatic device releases the dies after the thread is cut and removes burrs from the thread. Power is stransmitted in such

Victor Pipe Machine.

a manner that insures a maximum amount of power and renders the machine noiseless in operation.

All the Loew Victor Pipe Machines are now being equipped with cut-off attachment, which extends down through the solid portion of the die head in a manner similar to that of the dies. When the cut-off is in use the pipe is centred by an adjustable self-locking guide, similar to the self-locking feature of the die head, which obviates the necessity of using bushings.

One of the new styles, just placed on the market, is the No. 0 power machine, for cutting and threading 1/4 to 2-inch pipe, shown in the accompanying cut. This machine has been designed principally for pipe machine users who really desire a power driven machine that will take small sizes and yet do not care to go to the expense of a large lathe-bed outfit. The No. 0 machine is moderate in price, in fact, so inexpensive that everyone having use for a machine of this type can afford to own one.

Additional new types will be placed on the market to complete their line in pipe threading machinery. A catalogue of these machines may be had by applying to the manufacturers, the Leow Manufacturing Co., Cleveland, Ohio.

NEW TYPE UNIVERSAL MILLING MACHINE.

The accompanying cuts represent a milling machine with some new and commendable features. The distinctly new feature of the machine is the absence of a counter shaft, the machine being directly connected from the line shaft to a pulley shown on the machine. On this account the machine can be brought into operation very quickly and the expense and inconvenience of overhead pulleys is done away with. The driving pulley may be replaced by direct motor drive if such is desired.

New Type Universal Milling Machine.

Three styles are being placed on the market, manufacturer's, plain and universal. In the manufacturer's milling machine there is no up and down feed to the knee, the plain has all automatic feeds and the universal as shown in our illustration is the same as the plain with a table that swivels, making it a universal. The three styles are made from the same pattern, differing in the attachments.

The speed change gears are enclosed in an oil tight reservoir in the bottom of the machine frame and from this the machine oil is pumped to all gearing and bearings, keeping them flooded with oil. This pump is a smple spur-gear and pumps the oil through channels in the framework to parts requiring lubrication.

The three types have automatic cross vertical and table feed, and are equipped with stops at the end of the stroke for all feeds and an adjustable stop trips the feed at any point desired. No two feeds can be used at the same time and the stops cannot be removed from the machine so that the operator cannot omit them.

The Universal Milling machine is installed in the Allis Chalmers, of Milwaukee, and Warney & Swasey, Cleveland, and is used largely in tool room work for milling tool steel. A pump is supplied with each machine as the proper lubrication of tools largely increases the output.

The feed is constant speed, independent of the spindle speed. The feed can be manipulated in combination with the spindle speeds to a fractional degree. Vertical milling attachments can be furnished with the universal, if desired, thus combining all the features of a horizontal and vertical milling machine.

The milling machine may be used to cut large gears. The index rotary table is made strong and rigid for the cutting of these large gears. The indexing plate interchanges with the dividing centres. Accurate divisions can be made to within one minute of arc. The rotary table has power feed for circular milling.

These milling machines are manufactured by Kearney & Trecker, and are now being placed on the Canadian market by A. R. Williams Machinery Co., Montreal, Toronto and Winnipeg.

SAFETY YIELDABLE SLIP GEAR.

The Yieldable Slip Gear is a safety device for use in machines of many types and is intended to replace friction clutches and shearing pins. It is used in any mechanism where it is necessary to have some point yield to prevent other parts from breaking.

Fig. 1 shows the disassembled slip gear and gives an idea of the construction. The cast iron body in the centre is keyed to the shaft and has four holes in which are seated springs contained within hollow bushings, which they press outward. The construction is better shown in the line drawing, Fig. 3. These hollow bushings have seats at the outer end in which rest balls which are pressed against the face of the gear mounted on the shaft. The twelve seats for the balls are shown in the gear in Fig. 1. The shaft drives the gear, or the gear the shaft unless the driving power becomes great enough to force the balls from the seats in the web of the gear against the pressure of the springs which hold the balls in the seats. The power which may be transmitted without displacing the balls is regulated by a nut which bears against the plungers on which the springs are seated and which increases or decreases the compression.

The safety gear is applicable to any feed or speed device and is especially for use on mechanism liable to shock or sudden over-load, electrically driven machines and automatic machinery. It has an advantage over the crank pin and is always ready in case of emergency to counteract any undue strain. It can be easily applied to any machine without altering its design and as it is adjustable it can be changed to suit any new conditions. Fig. 2 shows an assembled view of the yieldable gear.

Patent rights have been granted in United States and foreign patents are pending. The safety yieldable gear is manufactured by the Yieldable Gear Co., Springfield, Ohio.

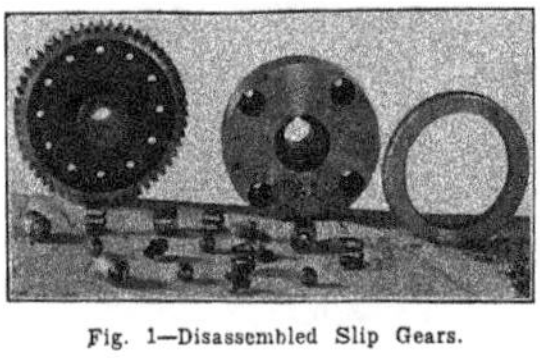

Fig. 1—Disassembled Slip Gears.

Fig. 2—Safety Slip Gear.

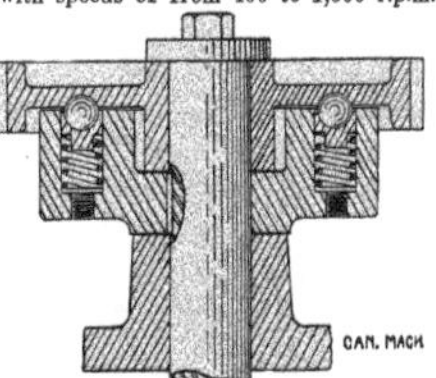

Fig. 3—Section of Yieldable Gear.

The Gisholt Motor-driven Boring Mill.

MOTOR-DRIVEN BORING AND TURNING MILL.

The boring and turning mill here shown with motor drive, is one of the latest placed on the market. The motor is a 5 h.-p., variable speed, 4 to 1 motor with speeds of from 400 to 1,600 r.p.m. On this form of drive a single friction pulley controlled by a lever, as shown, is substituted for the three step cone pulley used on the belt-driven machine. Through the friction device provided on the motor-driven machine, the table is under the absolute control of the operator and may be started, stopped or moved any fractional part of a revolution. Owing to the position of the motor as mounted no additional floor space is required, the machne being self-contained.

tion without the use of a scale. By means of the feed tripping device any feed may be positively or automatically stopped at any predetermined point; the dials for controlling feed trip may be plainly seen at the end of the cross rail. This device will also trip any feed

Fig. 1—Front of the Automatic Gear Cutter.

Fig. 2—Rear View of the Automatic Gear Cutter.

This machine is provided with 8 feeds, gear-driven, and is equipped with a friction device reducing to a minimum the possibility of stripping gears through careless handling. Machine also has micrometer index dials reading to

Fig. 3—Cutting a Worm Gear.

.001 inch on all feed screws, by means of which the tool in the turret may be moved .001 inch or more in any direc-

positively at either end of feed traverse, whether dials are set or not. The machine may be fitted with either plain table or chuck and has an extreme swing of 36½ inches.

This boring and turning mill is manufactured by the Gisholt Machine Co., Madison, Wis.

GRANT-LEES AUTOMATIC GEAR CUTTER.

This new gear-cutter is the first successful machine built on the principle of cutting gears with a hob, hat has been placed on the market. It is different from the European machines for generating gear teeth and is the result of the combined effort of John J. Grant and Earnest J. Lees.

The machine is designed to automatically cut gears up to 20 inches in dia-

Fig. 4—Cutting an Ordinary Spiral Gear.

meter. It will cut spur and spiral gears up to 5 inches pitch and worm gears, any pitch up to ¾-inch. It will cut spur.

worm, spiral and helical gears with only a slight adjustment of machine for each gear. It is specially used in cutting cream separator, gas engine gears and helical gears. The use of a hob necessitates a continuous cut. Only one hob is necessary for each pitch for spur gears. By means of a special hob both spiral and spur gears can be cut with one hob. It is only necessary to machine the face of the blank to 1-32-inch of exact size as the bottom of the tooth of the hob will cut the blank to the exact diameter automatically determined by a micrometer stop.

Cutting the Gears.

As shown in Fig. 5, the hob is set at an angle in cutting spur gears. The blank revolving at the proper ratio, automatically feeds vertically through the teeth of the hob and the gear is cut in one continuous operation. When the gears are cut the quick return automatically brings them to their original position and a gong summons the operator to replace the finished gears with blanks. The arbor will accommodate a gang of gears or gear with six inches face. The cone gives cutting speeds of 29, 52 and 84 feet per minute. The vertical speeds are determined by material and pitch required. For spiral gears the hob is set at the angle desired as shown in Fig. 4 and the blanks are rotated to give the correct increase or decrease per revolution to generate the correct angle. No indexing is required and the process is one continuous cut. The hob used is a patented diametral pitch hob, the same used to cut spur gears. To cut a worm gear the hob is set horizontally as shown in Fig. 3. The blanks revolve automatically and feed horizontally into the hob, until the proper centre distances are obtained. No nicking of the blank is required.

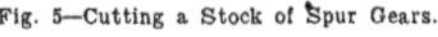

Fig. 5—Cutting a Stock of Spur Gears.

Fig. 6—Cutting Plain Gear Showing S'eady Rest.

Figs. 1 and 2 show the front and rear views, respectively, giving an idea of the general construction and the manner in which the driving motion is conveyed to the hob, giving a powerful drive and still allowing it to be swiveled at any angle. Fig. 6 shows the milling of a plain gear. In this illustration is shown an ingenuous rest used for large gears. By this mechanism the rest takes the thrust immediately under the cut and moves simultaneously with the main spindle.

The machine weighs 2,600 pounds and occupies a floor space of 4 feet by 4 feet 6 inches. The complete equipment includes a full set of change gears, countershaft and wrenches. As an extra an oil pump is attached. The automatic gear cutter is manufactured by the Grant-Lees Machine Company, Cleveland, Ohio.

THE GAS ENGINE AS A SOURCE OF POWER.

At a meeting of the International Association of Stationary Engineers, held in the Temple Building, Toronto, Mr. W. S. Haggas, of the Canada Gas. Power & Launches Company, Limited, illustrated and explained the modern types of gas engines and suction gas producers.

In taking up the cost of producing electrical power by means of large gas engines operating on suction or blast furnace, gas as compared with steam power, he stated that in order to produce, say, 100 electrical horse-power per hour the Carliss engine, belted to the generator, would require at least 400 pounds of coal per hour. The compound condensing engine on the steam turbine would be doing well if the coal consumption did not exceed 300 pounds per hour. It was, therefore, very gratifying to state that with the modern gas engine and suction gas producer official records had been obtained showing the consumption of coal to be 95 pounds per hour for 100 electrical horse-power, measured at the switchboard, or, allowing 15 per cent. loss from engine to switchboard, a full rate of 783 pounds of dry coal per B. H.P., or, roughly speaking, about ¾ pound per B.H.P. per hour. So that, figuring all possible items of cost, such as labor, oil, water, interest, depreciation, etc., power could be produced at from $14 to $15 per horsepower per year. Figure this out on the flat rate basis to the ordinary consumer, and it should not exceed $5 or $6 per horsepower per year, as it is a well known fact that the average amount of power actually used does not exceed one-third that of the possible amount that may be required occasionally.

Speaking of the flat rate, Mr. Haggas stated that he knew of fully 100 gas engines in Toronto operating on city gas, and could say postively that the average cost for all of these did not exceed $12 per horsepower per year, figuring on the flat rate basis.

Figure this out at a flat rate basis and you have about $8.25 per horsepower per year. As to the satisfactory working of large gas engines, there was abundance of proof. Take, for example the Lackawanna Iron & Steel Company, of Buffalo, N.Y., operating ten two-cylinder gas engines of 1,000 horse-power each for electrical service, 16 engines of the same power for blowing their blast furnaces, also engines for other purposes, making a total of some 42,000 horse-power. This installation constitutes one of the most important power plants in the world.

Catalogues of the Trade

By mentioning Canadian Machinery to show that you are in the trade, a copy of any of these catalogues will be sent by the firm whose addresses is given.

IMPERIAL CHUCKS—Catalogue and price list of chucks for lathes, boring mills, etc., manufactured by Ker & Goodwin, Brantford, Ont.

STEEL SUPPLIES—Catalogue No. 27, from the Cleveland Wire Spring Co., Cleveland, Ohio, well illustrated, describing their steel shop boxes, barrels, waste cans, supply racks and other steel equipments for factories.

ELECTRIC TROLLEY HOIST—Pamphlet from the Northern Engineering Works, Detroit, Mich., showing photographs of a few types of their cranes with descriptions of them.

POWER PUMPS AND PAPER MILL MACHINERY—Catalogue from the Sandusky Foundry and Machine Co., Sandusky, Ohio., well illustrated, giving descriptions and full specifications of their line of pumps.

ALUNDUM—A ten-page treatise on alundum for grinding—Norton Company, Worcester, Mass., U.S.A.

METAL WORKING MACHINERY—Catalogue describing the "American" lathes, planers, shapers, radial drills, etc. The A. R. Williams Machinery Co., Limited, Toronto, agents.

NAVY BULLETIN, NO. 22—A book of information designed for the information of manufacturers of electric motors and motor-driven machinery, describing panels, controllers, repair parts, etc. The Cutter-Hammer Mfg. Co., Milwaukee, Wis., U.S.A.

CROCKER-WHEELER BULLETIN—Issued by the Crocker-Wheeler Company, Ampere, N.J. A full set of up-to-date bulletins, bound in covers describing with illustrations, their generator and motor sets, switch-board equipments and electrical accessories.

ADAMANTINE THREADING TOOLS—Catalogue from the American Tap and Die Co., Greenfield, Mass., illustrating and describing screw, plates, taps and dies, wrenches and pipe cutters. These are made in standard and special sizes and their catalogue covers fully this line of work.

ENGINES.—Catalogue from Belliss & Morcom, Birmingham, England, illustrating and describing their patent self-lubricating, quick revolution engines for electric lighting, power, traction and mill driving. Sectional views are shown, results of tests are given, and the action of the governor is explained with diagrams and cuts. There is also a treatise with diagrams on the effect of superheat on steam and heat consumption, and several views of their engines now in use, showing them direct-connected to generators, and equipped for belt and rope drive. Messrs. Laurie & Lamb, consulting and contracting engineers, 212 Board of Trade Building, Montreal, are the sole agents for these engines in Canada. In three years this firm have installed these engines in ten plants in Canada, totalling 12,925 horse power.

AIR COMPRESSORS AND DRILLS.—The Chicago Pneumatic Tool Co., Fisher Building, or 95 Liberty Street, Chicago, are mailing two new catalogues, each 100 pages, Nos. 23 and 24. No. 23 is devoted exclusively to Franklin Air Compressors and is well illustrated. No. 24 deals with the company's widely known line of pneumatic tools and appliances, including Boyer and Keller hammers, Little Giant drills, sand reamers and hoists. Both books are printed in colors, conveniently indexed and strongly bound, making good volumes for references.

GRINDING AND POLISHING MACHINERY.—Builders' Iron Foundry, Providence, R.I., U.S.A. A catalogue describing their emery grinders, with sizes and materials used in construction, illustrating and explaining special features; emery wheel dressers for truing solid emery wheels running at high speed; patent countershafts with self-contained belt shifters; ring oiling polishing machines and leather and canvas polishing wheels.

HYDRAULIC MACHINERY.—An extensive catalogue issued by the Waterbury Farrel Foundry and Machine Co., Waterbury, Conn., describing and illustrating their makes of hydraulic machinery, used in various industrial operations, such as the manufacture of tubing of all kinds, drawn rods and shafting, cups for tanks, boilers, etc., embossed and stamped coins, punching thick blanks from sheet metal, etc. It is of standard size and very neat throughout.

ELECTRICAL EQUIPMENT.—Bulletins from Canadian Westinghouse Company, Limited, Hamilton, Ontario. No. 1092 describes the Westinghouse Multiple Alternating Arc Lamp; 1102, direct-current multiple arc lamp; 1128, small power motors for alternating and direct-current circuits; 1143, regulating and reversing controllers for direct-current motors; 1144, mill motors for direct-current service. Full descriptions and illustrations are given of all details. Accompanying these was a book of forty pages, containing an article on the Westinghouse electrical machinery for the operation of Mexican mines, by Charles V. Allen.

UNIVERSAL PROPELLER.—Folder describing the universal propeller, manufactured by R. G. & E. Holland, 13 Oakwood lane, Barnton, Northwich, Chesire, England. This propeller is manufactured both as fixed and reversible, and its construction makes it suitable for all classes of vessels from the small motor boat to the large ocean liner.

STEAM TURBINE BLOWERS.—Bulletin No. 3, from the Kerr Turbine Company, Wellsville, N.Y., describing and illustrating in detail steam turbine blowers, forge and cupola blowers, centrifugal pumps, etc. The Kerr turbine is built on the lines of the Pelton water wheel, which has been so successful. Pressure tables are given and also tables of dimensions and shipping weights.

STEAM AND WATER SPECIALTIES—Catalogue No. 9 from the Golden-Anderson Valve Specialty Co., Pittsburg, Pa., containing descriptions and illustrations of their line of globe and angle valves, traps, locomotive valves, standpipes and automatic float valves.

INDIA OIL STONES—Pamphlet from Norton Emery Wheel Co., Worcester, Mass., describing oil stones for sharpening ordinary tools, such as chisels, planes, etc.

LO—SWING LATHE—A neat volume on good paper, well bound and illustrated, with detailed description of and full instructions for running the lo-swing lathe manufactured by Fitchburg Machine Works, Fitchburg, Mass.

BOOK REVIEWS.

TOOL-MAKING—By E. R. Markham, 200 pages. Illustrated. Published by the American School of Correspondence, Chicago, Ill., U.S.A. Price, $1.50.

This is an exhaustive treatise on how to make tools and should appeal to every tool maker and mechanic. It treats of all tools, dies, milling cutters, gauges, dull jigs, etc., and is a volume of practical information for practical men.

PATTERN-MAKING—By James Ritchey. 150 pages. Illustrated. Published by the American School of Correspondence, Chicago. Ill. Price, $1.00.

For metal workers and pattern-makers this work, by a practical man, is both interesting and instructive. It is a valuable compilation of data on patterns, tools and methods of molding.

FOUNDRY PRACTICE AND EQUIPMENT

Practical Articles for Canadian Foundrymen and News of Associations. Contributions Invited.

VENTILATING A BRASS FOUNDRY*

By Walter B. Snow

In much of the early work done for the welfare of the employe there was a strange confusion of motives. Even though the project was primarily humanitarian in its spirit, the advertising department got in its work in proclaiming this spirit to all the world. But out of this confusion has now grown a definite recognition of the purely economic advantages of surrounding the workman with healthful conditions. While some other industries are more directly harmful to the health than is the brass industry, there is, nevertheless, ample opportunity within its field to greatly improve the conditions. Although the heat and fumes are primarily uncomfortable and only secondarily injurious, the greatest harm is done by the dust which is inhaled. This dust is usually of mineral or metallic origin resulting from the grinding, polishing, tumbling, and sand blasting processes, and also from the shaking out of the castings.

Injurious Industries.

It is commonly recognized that life is shortened by working in a dust-laden atmosphere, but the extent to which some industries are injurious is startling. In the cutlery and tool industry, which is declared to be one of the most dangerous of trades in this class, the average age of the operatives at death is exceedingly low and in establishments conducted without proper hygienic precautions sound men are rare after a few years' work. The prevailing cause of death is consumption, which usually overtakes a susceptible worker so early that his period of usefulness does not extend much beyond five or six years, except where the health is properly safeguarded.

The testimony of physicians is, that of those employed in this industry nearly all who reach the age of 40 die of consumption, excepting those who succumb to some acute disease. As proof of this statement, it is instructive to note that in Northampton, Mass., an important seat of the cutlery industry, the death rate from tuberculosis for the entire male adult population was 2.9 per thousand, while that for the cutlers of that town was four times as great; namely, 11.8 per thousand. The trouble lies not so much in any directly poisonous results from inhaling the dust as in its power to bring about constant irritation, which produces such a condition of the mucous surfaces that they more readily admit of invasion by disease germs. Fortunately brass is less irritating than steel, and consequently the results in the brass industry are not as disastrous as they are among the cutlers. But the dust of corundum and emery is peculiarly irritating, and the brass workers' surroundings are therefore susceptible of marked improvement.

Grinding Discs Equipped with Special Hoods Designed to give Freedom to Workmen.

Laws to Safeguard Health.

The unhygienic conditions existing in the various industries have received the attention of State Boards of Health, whose official investigations are bringing about the passage and enforcement of more stringent laws looking to the safeguarding of the health of the employes in all industrial establishments.

In a word, advance has been made from a matter of individual interest to one of almost national importance. The statute books of the leading States already contain laws, usually somewhat vague in their expression, which require cleanliness, light, warmth, ventilation, and the introduction of specific devices for removing dust, fumes, and the like. While the first impulse of the manufacturer may be to resent the enactment of further laws, yet his compliance with them is not without eventual advantage. Not only will a better class of men prefer to work for him if improved conditions are provided, but there will be far less interference with work because of sickness, more energy in the work which is done, and less loss by death of the potential value possessed by the man who has become thoroughly skilled in a given line of work. Continued sickness and death naturally mean constant replacement of individuals, with the loss of knowledge and skill gained by those who have gone. As a result there is far less stability of employment in an unhealthy industry.

Ventilating Appliances.

Experience has shown, and the reports of investigations confirm the fact, that mechanical means are absolutely necessary to maintain a rapid air change or to insure proper removal of dust. In fact, the fan blower figures everywhere as the only device adapted to secure these results. It is manifest that the action must be positive, and of sufficient intensity to create ample movement of air. Where there is but little dust or the requirements of ventilation are slight, a fan applied for mere renewal of air throughout the entire extent of a room will meet the requirements. When warranted by the size of the plant the fan may form part of a blower heating system, by means of which warm air from a centralized heater is delivered under pressure through pipes to all parts of the building. In overheated rooms and particularly for summer ventilation the disc or propeller type of fan meets the requirements if placed in wall or ceiling.

Wherever dust or fumes are formed locally, as in connection with grinding

*A paper read by Walter B. Snow before the American Foundrymen's Association at Philadelphia.

and polishing wheels, tumbling barrels, or furnaces, the exhaust should be direct from hoods which enclose the objectionable source as completely as possible. In a word, prevention is better than cure. The objection which is often shown by workmen to hoods and similar contrivances—even to the extent of actual destruction—is largely due to their improper construction. In fact, the cause for condemnation or criticism of many exhausting systems lies in the method of application of the fan, and not in the fan itself. The success of the fan not only depends upon its speed and its proper proportioning to the work, but also upon the system of piping and hoods which would give the greatest efficiency. It seems so simple to employ a local tinsmith to rig up an exhausting system that it is not strange that unsatisfactory conditions result.

It is far better policy to secure the best advice, which will always be freely given by blower manufacturers, and then have the thing done right. It must not be overlooked that the installation of an economical exhausting arrangement requires definite engineering ability, and should only be undertaken by one who makes a specialty of this particular class of work.

Cost of Installation.

Because of the lower first cost the user is always strongly tempted to buy the smallest apparatus that can be made to do the work. But first cost is only one of the factors in the total cost. Large slow running fans with ample pipe areas are conducive to small power expenditures. It is easy to save enough in power in six months to pay the additional cost of a more efficient outfit or system. Thereafter its economy is all clear gain.

Even though the fan be of ample size when first installed, it may, as a result of speeding up to meet added requirements, frequently demand from 50 to 100 per cent. more power than would be necessary to do the work with a proper outfit. It is none too generally understood that the power required to drive a fan increases as the cube of the speed. In other words, that doubling the speed calls for an eight-fold increase in power, while 27 times the power is required at three times the speed; an increase of only 25 per cent. in speed calls for nearly double the power, and yet such an increase is common enough. How long would it take to pay for a new outfit from the money thus squandered in excessive power?

The designs of hoods for grinding, polishing, or buffing wheels are many and varied. Each must be arranged to suit the particular class of work for which the wheel is used. In some cases it is even necessary to have several different types in the same room. This is true where the pieces of such shape and size that it is impossible to get very close to the wheel, the result being that at one time the operator uses the wheel at a point near the top, and again at a point directly underneath. Under these conditions especial care must be taken to provide the most effective type of hood and maintain the maximum blast. In heavy work of this type the air suction pipe should be five inches in diameter for wheels up to and including 16 inches in diameter by three-inch face. In ordinary grinding and buffing rooms the suction pipes should be four inches in diameter for wheels 2½ inches or less in width, and from 10 to 18 inches in diameter. Wheels ranging from 19 inches to 28 inches should have 5-inch or 6-inch pipes according to class of work for which they are used.

All hoods should be so designed that the velocity through the openings should not be less than 5,000 feet per minute, which is usually sufficient to create the draft necessary to carry away the particles. The best general type of hood is of patented form, provided with a receptacle below to trap out all heavy particles, as well as the threads from the buffing wheels, while allowing the finer dust to pass through the pipe. The result is that the metallic particles are left in clean condition ready for resmelting, and the wear on the pipes and the fan is greatly reduced. This arrangement also prevents the annoyance caused by the dust from the rag wheel adhering to the fan wheel and throwing it out of balance. The trapping-out feature furthermore permits of the ready recovery from the bottom of the hood of any small piece of work or other material, which with other types of hoods might get into the main trunk line or up into the fan.

All properly designed systems should have clean-cut caps so as to provide free access to the interior of the piping for the removal of anything that might possibly tend to clog it. The main suction pipe should be proportionately increased in size as each connection is made to it.

Fan Installation.

To secure the most economical results a fan should be chosen which has an area of inlet about twice the combined area of the inlet pipes. This proportion will give the maximum velocity through the branch pipes and hoods. The fan should then be operated at about 1½-ounce speed, under which condition it would consume about ½-horse-power for each 4-inch opening.

The most work is done, and consequently the most power is required by a fan when it is discharging with free inlet and outlet. The more extended the system of piping, the smaller the area of inlet or outlet; and the greater the friction, the less will be the volume delivered by the fan; and consequently the less will be the power required to drive it. It is therefore manifest that the fact that the fan is consuming but little power is not always evidence of its successful operation, for it may be doing little effective work.

The dust which is collected by the fan should be discharged into a centrifugal dust collector. Here the dust is separated from the air by centrifugal force; the air escapes from the top practically free from dust, while the dust itself drops out of the bottom through a pipe. It should be periodically removed. The dust from wheels grinding iron and steel should not be mixed with that from rag wheels, for in some cases fire will result. Separate fans and systems should be used.

Exhausting From Tumbling Barrels.

The same general principles hold in connection with systems exhausting from tumbling barrels. If a maximum effect of the fan is desired on tumbling barrels equipped with hollow trunnions, the area of fan inlet should be about double that of the sum of the openings in the trunnions. The sizes of pipes and the speeds of fans to be applied in connection with housed rattlers must depend largely upon the conditions, but a 6-inch pipe connection will usually serve for each tumbling barrel, if the same is tightly enclosed. A fan running at about one ounce speed will give sufficient draft.

No general rules can be given for the application of the fan system to sand-blast rooms or apparatus. The arrangement must depend entirely upon the local conditions, but the provisions must be generous if successful results are to be obtained.

With installations such as are here described it is possible to maintain a relatively healthy atmosphere, which is bound to insure better work.

PATTERN MAKING FOR THE FOUNDRY*

Showing the Several Operations in the Construction of a Pattern

BY JAMES RITCHIE

A good example of the manner in which patterns are built and glued up is shown in the construction of the pat-

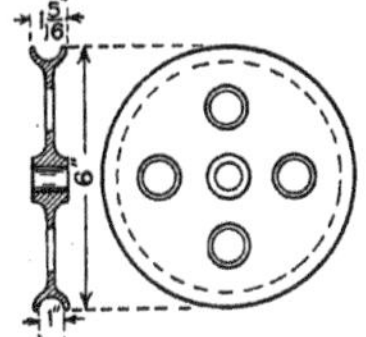

Fig. 1—Sheave Pulley.

tern for the 6-inch sheave pulley shown in Fig. 1. The groove is a semi-circle 1-inch wide, and the rim containing the groove is connected with the hub by a solid web ¼-inch in thickness, and having four or six holes, each 1-inch in diameter, this web taking the place of arms. If there is to be no finish on the sheave, as is usual, the only allowance to be made on the pattern, which must be parted, will be for shrinkage and for draft. A cross-section through the finished pattern for this casting is shown in Fig. 2.

In all large patterns of this kind, the web is first glued up in sectors, six, eight or more in number, according to the size of the sheave (see Fig. 3). The sectors are fitted by hand or on the trimmer, the ends are glue-sized, and when the sizing is dry the joints are carefully scraped smooth, and the whole glued together. After drying four or five hours, it is sawed to a circle of ½-inch greater diameter than the finished pattern, and the block for the hub is glued over the centre. Six segments to form the outer rim are glued around the outer edge, care being taken to break joints as shown in Fig. 3. If the groove is to be large, the six segments of like thickness are glued over the first, breaking joints not only with the first set, but also with sectors of the web. In other words, in all glued-up rims, no two joints should be directly over each other. All joints must be so broken and so distributed as to give the greatest possible strength to the rim. In the present case, our pattern is so small that it is only necessary to use a thin board, ¼-inch in thickness, for each half of the web. After sawing to 6½ inches in diameter, ½-inch for turning, a block ½-inch in thickness is glued on the centre of each to form the hub; and six segments, 1¼ inches wide and ½-inch in thickness, are glued around on the outer surface of each to form the rim and groove, as shown in Fig. 4.

Care must be taken to place the segments so that the grain of the web

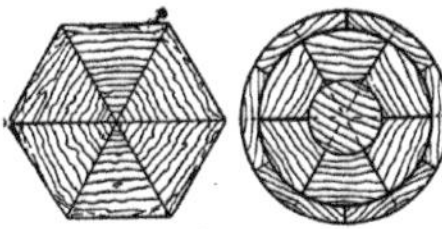

Fig. 3—The Sectors.

will be crossed by two of the segments as shown in the drawings. On the second half of the pattern, a thin circular block ⅛-inch in thickness, is glued on the inside opposite to the hub block, to form the projection (⅛-inch) which will keep the two halves of the pattern in alignment, as shown in the cross-sectional drawing in Fig. 2. Having glued up the stock as described, and as shown in Fig. 4, the outside must be planed to a level surface, or so that the six segments forming the rim and the centre hub block will be in the same plane.

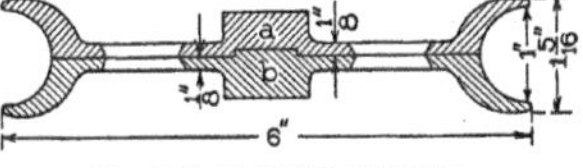

Fig. 2—Cross Section of Pattern.

Turning the Pattern.

The half pattern is now screwed on the screw chuck of the lathe as illustrated in Fig. 5, and the inside, or the parting face C, is turned perfectly straight and true. The edge is turned down to 6 inches in diameter, and the quartered circle shown by the dotted lines is carefully shaped. A template, made as shown at D, will assist greatly at this stage of the work. A recess is turned at the centre, and in the face

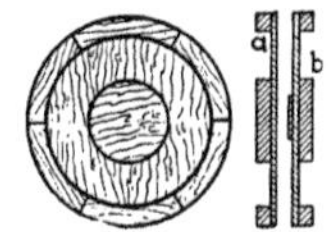

Fig. 4—Making Rim.

of A, Fig. 2, 1½ inches in diameter and ⅛-inch deep, to receive the corresponding projection of the half pattern B which is to keep the two halves in alignment. The half pattern A, is now removed from the screw chuck, and the second half, B, is screwed on and turned in the same manner except that the central projection is carefully turned to fit in the recess in A. Before removing B from the chuck, test by trying the second half A, and change B until a perfect fit is obtained between the two halves, not only in the central recess and projection, but also in the two curves which form the semi-circular groove of the rim. A cross-section of the pattern at this stage of construction is shown in Fig. 6.

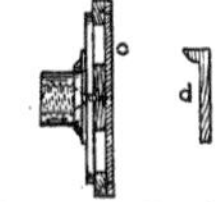

Fig. 5—Fastening Pattern for Turning.

A disc or chuck of wood 5½ inches in diameter is now screwed to the iron face-plate, or the screw chuck, and turned off true on the face with a projection ⅛-inch high, which will fit into the recess in the middle of the parting

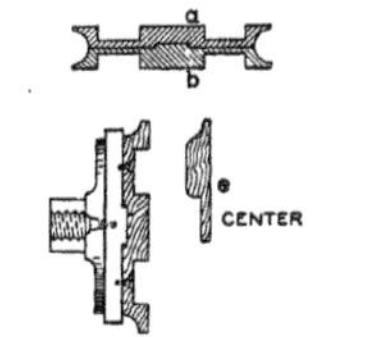

Fig. 6—Section and Pattern Fastened on Chuck.

face of A. This projection will centre the half pattern A on the face-plate, and it can be held in position by two or four short wood screws driven through the web into the wooden chuck, as shown in Fig. 6. Care must be taken to place the screws in such a position that the screw holes will be cut or bored out when making the four or six openings 1-inch in diameter in the finished web of the pulley. The screws must be small and slender and the heads well countersunk out of reach of the turning tools. The face of the half pattern is now turned to the required shape, the tem-

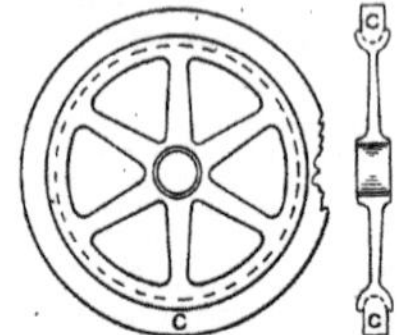

Fig. 7—Periphery Core Print.

plate shown at E in Fig. 6 being used for the purpose. Having finished with fine sandpaper, remove the half pattern and turn off the projection on the centre of the wooden chuck; turn a recess instead to receive the projection on b, and proceed with this second half as with the first. If the wood has been well seasoned, and the work carefully done a perfect 6-inch sheave pulley will be obtained, such as shown in Fig. 2.

Completing the Operations.

The pattern for a sheave pulley has been explained because it embraces so many profitable points and conditions, not only in gluing and building up, but especially in chucking and turning, all of which must be done with great care and accuracy. The 1 inch holes in the web are bored out with a 1 inch center bit, which, when well sharpened, will not split or splinter the thin webs of the two halves of the pattern, if care is taken

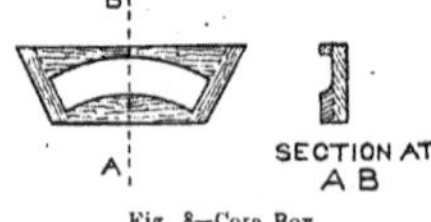

Fig. 8—Core Box.

to reverse the bore from the opposite side when the points of the center bit come through. The holes should be given a slight draft as shown in Fig. 2, with a small half-round cabinet file. When very large sheave pulleys, having arms, are to be made, such as are common for power transmission by rope or cable, the patterns are not halved but are made in one piece, and the groove is cored around the rim. Such a pattern is illustrated in Fig. 7, with a wide core print c c extending entirely around the periphery of the pattern.

A segmental core box is made for one sixth or one eighth the circumference of the wheel, as shown in Fig. 8, and here again only half of the core box for a full core is needed. When coring the rim as above, the core print must be made wide, at least two to three times the depth of the groove, so that the core may rest firmly and remain in position without tilting while the metal is being poured into the mould.

NEW PROCESS FOR MALLEABLE IRON AND STEEL.

Two Australians have evolved a process for directly converting iron ore into malleable iron or steel by a continuous system. It is claimed that the new discovery will effect a saving of 25 per cent. in the manufacture. The ore is simply concentrated by ordinary methods, or if it is magnetic it is separated electrically until the pure oxide of iron is obtained. The oxide of iron is passed through a revolving cylinder heated by waste gases from subsequent operations, and brought in that cylinder to a dull red heat. It drops from the cylinder to a second similar cylinder, and in the latter it is brought into contact with the deoxidizing gas, which is forced through and brought into contact with the heated ore. The heated ore is thus converted into a pure iron. Accompanied by and protected by the deoxidizing gas, it is passed into a third chamber or melting hearth, where it falls into a bath of molten iron, and is converted directly into steel or balled up as malleable iron.

ENTERPRISE CLUB.

The Enterprise Club, that is the name of an organization recently formed by the men of the Enterprise Foundry, of Sackville, N.B. The foundry company has always taken much interest in its employes and the club has been formed largely the result of that interest. The club rooms, three in number, are in the Enterprise office building. One is fitted up as a reading room, with current magazines and papers; another is a smoking room, while the third will be the lecture room. In the last, lectures will be given every fortnight.

DOMINION STEEL.

As yet the Steel people say there is no great falling off in orders. The two leading Canadian railways have still ample funds and steel rails are in demand.

UTILITY OF FAN BLOWERS.

With the possible exception of the steam pump, writes Walter B. Snow, it is doubtful if any other mechanical device has so varied a field of direct application as the fan blower; these two machines really being in the same class, the one handling liquids at pressures measured in pounds, while the other creates movement of air and gases under pressure, differences measured in ounces.

Fan blower applications may be broadly divided into two classes. First, those in which the object is to actually transport the air or gas, and second, those in which the air serves as a medium for the transportation of other materials. Ventilation and the supply of air for combustion are properly included under the first classification. The conveyance of light material and the transmission of heat or moisture in the air come within the second classification.

Ventilation whether secured by natural or artificial means results only from the creation of a pressure difference between two points such as to cause a flow of the liquid from the higher and the lower pressure area. In the case of natural ventilating systems maximum efficiency is secured only through the employment of flues and ducts of large area, through which the air flows at low velocity. At best, the mechanical efficiency of the heater flue is very low. With a fan, however, much greater pressure difference can easily be created and smaller pipes with higher air velocities employed. With the fan, the heat expenditure necessary to produce a given air movement is surprisingly less than with natural draft in a flue or chimney.

According to the degree of activity of the occupant the velocity of the air admitted to the room may be high or low, but it is easily regulated where a fan is employed. In the sparsely-occupied building the per capita space is large, the air volume therefor required is relatively small, and the rate of flow may be high. In the crowded hall of audience the minimum of floor space is provided for each occupant and the maximum of air supply is required. Here the utmost refinement is necessary in the tempering and distribution of the air.

In the ordinary building, ventilation is very properly a process of substitution of fresh air for foul air, but the fan does not absolutely control the conditions, for the walls are usually porous and there are leaks around the windows and doors. In the mine, however, with its practically air-tight chambers and its vertical shaft, there is no appreciable inward or outward leakage. All air which passes through the mine likewise passes through the fan, very positive control being had over its distribution.

The De La Vergne Oil Engine

An oil engine specially designed for the use of ordinary kerosene or fuel oil known as the De la Vergne Oil Engine, has been placed on the market and a number of them have been operating satisfactorily for many months. In Fig. 1, we show a fifteen brake horse power De la Vergne Oil Engine, direct connected to a ten kilowatt generator. Fig. 2 is a twin cylinder engine of the De la Vergne type, with one-half in section. Air port A and exhaust port E are in the middle of the cylinder, indicating at once that the engine operates on the two-stroke cycle and is single acting; hence in a twin cylinder engine, the crankshaft receives in each revolution two impulses, the result being a uniform turning moment.

Fig. 1—A Twin Cylinder 15-h.p. De La Vergne Direct Connected to a 10-k.w. Electric Generator.

Fig. 2—A Twin Cylinder with one-half in Section.

There are no batteries sparking dynamo, carbureter, or mixing valve. The cylinder head V (Fig. 2) is shaped as a bulb, kept hot enough by successive explosions to ignite the combustible mixture. This bulb or vaporizer is made of gun iron, heated up at the start by a kerosene blow lamp.

Pre-ignitions cannot occur for on the up-stroke, there is nothing but pure air in the cylinder which has entered under slight pressure from the enclosed crank case through port A. Just as the piston is reversing its notion, oil is sprayed into vaporizer V by nozzle N. The heat of the walls of the vaporizer and the hot air resulting from the high compression, at once vaporize the oil and burn it rapidly, thus giving the descending piston an impulse. Just before the end of the down stroke, exhaust port E is uncovered by the piston to let the burnt gases escape to the atmosphere. No exhaust valve is required.

A small plunger pump forces the oil into the vaporizer suddenly, at the right time and in proper quantity to suit the load on the engine. This pump is under the control of an ingenious and simple throttling governor, fastened to and revolving with the flywheel. This arrangement is shown in Fig. 3. A frame F is fastened with two studs, concentrically, to the inside of the flywheel. To this frame is hinged the cam ring R, which, on the flywheel side, has a projection B, and back of that cam projection C, raising the roller A once each revolution of the flywheel, thus moving the oil plunger P. The stroke imparted to the pump plunger is gauged by the lever L, pivoted to frame F. L has a wedge W, which separates the cam ring from frame F. When the engine is started, the governor does not come into action until the normal speed is approched. Under light load, the speed will increase, when by the action of centrifugal force, the counterweight K will overcome the tension of spring S, thus withdrawing wedge W. Spring E always keeps roller A in contact with cam ring R, and when the wedge is withdrawn, buffer G cannot actuate the pump plunger and no oil will be injected. Knob K' at the head of the throttle bar T is used at the start for pumping oil by hand. Lockout N serves to limit the stroke of the oil pump.

The two concentric slots in frame F allow adjustment when it is desired to greatly change the normal speed of the engine. For smaller speed variations the spring S and weight K can be shifted in the holes of lever L. The close regulation obtained with this governor and the high rotative speed make this engine admirably suited to electric work.

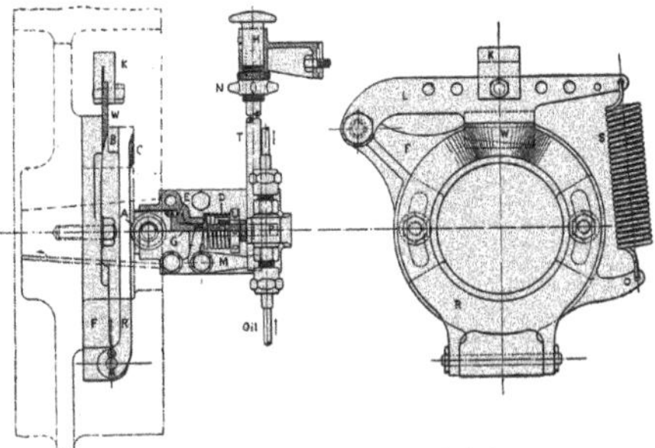

Fig. 3—Details of the Governor and Fuel Pump.

The steel oil spray nozzle is so arranged that it very rarely requires cleaning out, deep grooves being cut helically into the spray pin. The pressure under which the oil is forced in keeps these grooves clean. A steel ball check valve prevents the explosion from firing back into the oil pipe. The entire nozzle is inserted in a section of the vaporizer, which is cooled by circulating water. A small rotary pump driven by spur gear on the shaft circulates the water first through the crank case and around the three bearings, then through the two cylinders and around the spray nozzle.

A 15 h.p. complete weighs 1,180 lbs., and a $7\frac{1}{2}$ h.p. weighs 750 lbs., with cylinder 7 inch x $7\frac{1}{2}$ inch running at 450 R.P.M. This engine is manufactured by the De la Vergne Machine Co., New York City.

PERSONALS.

R. A. Brown, Calgary, has been appointed city electrician of Nelson, B.C

H. C. Philpott, general sales agent of the Dominion Car and Foundry Co., of Montreal, died suddenly of heart failure on Nov. 21st.

G. G. S. Lindsay has been elected president of the Crow's Nest Pass Coal Company, and James D. Hurd succeeds him as general manager.

Geo. A. Gallinger has been appointed manager of the Pittsburg, Pa., office of the Independent Pneumatic Tool Co., at 1210 Farmers' Bank building, Pittsburg.

Alan Owen Leach, of Toronto, has just finished an appraisal of the Great Northern Railway, and is at present on a tour of inspection of that road with the general manager, Louis J. Hill.

J. C. Boyd, superintendent of the Canadian Soo canal. is being transferred to the Williamsburg canal, with headquarters at Morrisburg. J. L. Ross, C.E., Morrisburg, will take Mr. Boyd's place at the Soo.

Albert Elsstrom, manager of the Gothenberg plant of the General Electric Company of Sweden, is in Canada on business. Kilmer & Pullen, Toronto, are the representatives of this company in Canada.

AWARDED GOLD MEDAL.

J. W. Cumming, manufacturer of modern patented coal and plaster drills, high grade miners' tools, mine cars and hitchings, New Glasgow, N.S., has been awarded the gold medal at the Halifax Provincial Exposition, 1906 and 1907.

BOOK REVIEW.

THE BLACKSMITH'S GUIDE—By J. Sallows, published by The Technical Press, Brattleboro, Vt., U.S.A. 150 pages. Price, $1.50. It is a book of six chapters treating in a very practical way with illustrations, machine forging, tool forging hardening and tempering, high-speed steel, casehardening and coloring, brazing and general blacksmithing. On every page is valuable information for foremen and blacksmiths.

SOCIETY OFFICERS.

Canadian Mining Institute.

President, George R. Smith, Thetford Mines, Quebec; secretary, H. Mortimer Lamb, Victoria, B.C.; treasurer, J. Stevenson Brown, Montreal.

Toronto Branch A. I. E. E.

Chairman, R. G. Black; vice-chairman, K. L. Aitken, secretary, R. T. McKeen.

National Association of Marine Engineers of Canada.

Grand president, Alex. L. De Martigny, Varennes, P.Q.; grand secretary, Neil J. Morrison, St. John, N.B.

Canadian Electrical Association.

President, R. S. Kelsch, Montreal; 1st Vice-Pres., W. N. Ryerson, Niagara Falls; 2nd Vice-Pres., R. M. Wilson, Montreal; Sec.-Treas., T. S. Young, Toronto.

Canadian Association of Stationary Engineers.

President, W. A. Sweet, Hamilton, Ont.; Vice-Pres. Jos. Ironsides, Hamilton, Ont.; Secretary, W. Outhwaite, Toronto, Ont; Treasurer, J. M. Dixon, Toronto Ont.

Canadian Railway Club.

President, W. D. Robb; secretary, James Powell, chief draughtsman motive power department, G. T. R., treasurer, S. S. Underwood, chief draughtsman car department, G. T. R.

Engineers' Club of Toronto.

President, C. B. Smith; secretary, C. M. Canniff, 100 King Street W.; treasurer, John S. Fielding, 15 Toronto Street.

Canadian Society of Civil Engineers.

President, W. McLean Walbank; secretary-treasurer, C. H. McLeod. Rooms, 877 Dorchester street, Montreal.

Society of Chemical Industry.

Chairman, Prof. W. H. Ellis; vice-chairmen, H. H. Van Der Linde, Milton, and H. Hersey and A. McGill; hon.-secy., Alfred Burton, 44 York Street, Toronto.

Ontario Land Surveyors.

President, Thos. Fawcett, Niagara Falls; vice-pres., A. J. Van Nostrand, Toronto; secy.-treas., Capt. Killaly Gamble, Toronto.

The Canadian Clay Products Manufacturers, in convention in Ottawa last month, passed a resolution asking the Dominion and Provincial Governments to make appropriations for the technical training of students in the development of non-metallic minerals, which would include the manufacture of different kinds of brick, tile, pottery, porcelain, glass, cements, plaster of Paris, etc. The development was hampered by lack of investigation into the possibilities of these minerals, and lack of the technical knowledge necessary to the progressive development of the clays, shales and marls. These resources were more important than any other mineral owing to their intimate association with Canadian domestic and social needs. A technical school devoted to these subjects, said the resolution, would encourage individual enterprises in the utilization of these resources and do away with the necessity of importing foreign goods and foreign skilled labor.

UNIVERSITY ENGINEERING SOCIETY.

J. H. Hall, representative of the Railway Union at Ottawa, spoke to the members of Toronto University Engineering Society on "Railway Accidents," in which he defended the conductors. He recommended the students to go into railroad work, where there are great opportunities for young men in Canada. He said the block system should prevail on the railways, and there is no reason why it should not be on the double-track systems.

SELLING MAGAZINE ANNIVERSARY.

The first number of volume four of Selling Magazine marks another milestone of this successful journal. This special November issue contains a fund of information for sales and publicity men. The special feature of the magazine is a directory of technical, trade and class publications, giving size of page, rate per page and circulation of trade journals which may be employed in advertising machinery, tools, industrial equipment and supplies.

TO KEEP TOOLS FROM RUSTING.

Take two ounces of tallow and one ounce of resin; melt together and strain, while hot, to remove the specks which are in the resin. Apply a slight coat on the tools with a brush and it will keep off the rust for any length of time.—Practical Carpenter.

CHEAP POWER.

At a cost of $45,000 the town of Gravenhurst has installed a power plant at South Falls, a picturesque spot eight miles distant from the town. Situated on the south bank of the Muskoka river, the power building is a substantial brick structure with concrete foundations and floor. The electrical equipment was built by Allis-Chalmers-Bullock, Limited, of Montreal, and includes a 450 k. w., 60-cycle. 3-phase, 6,600 volt water wheel type, alternating current generator, running at 600 r.p.m., with a 15 k. w. direct current generator as an exciter; two 250 k.w. transformers, and the generating station switchboard, instruments and other auxiliary apparatus. The generator is directly connected with and driven by a horizontal water wheel, made of bronze, with a capacity of 750 horse-power. The plant is so constructed that the capacity can be doubled at a cost of about $20,000, and it has been estimated that the flow of the Muskoka at South Falls is such that during low water it is possible to generate 4,000 horse-power. The water wheel was installed by the Jenkes Machine Co., Sherbrooke, P.Q. At the banquet celebrating the opening of the new power plant, T. T. Simpson, consulting engineer, of Ottawa, who had charge of the work, stated that the cost to the town of developing the power worked out at $7 per horse-power per annum.

INDUSTRIAL PROGRESS

CANADIAN MACHINERY AND MANUFACTURING NEWS will be pleased to receive from any authoritative source industrial news of any sort, the formation or incorporation of companies, establishment or enlargement of mills, factories or foundries, railway or mining news, etc. All such correspondence will be treated as confidential when desired.

New Plants and Additions.

Tolton Bros. will erect a box factory at Guelph.

Mossy River, Man., will erect a bridge near Winnipegosis.

T. E. Essery, Toronto, will erect a brick factory to cost $20,000.

The Paris Plow Co., Paris, Ont., will erect a warehouse to cost $9,000.

Mathew & McNaughton, Calgary, will erect a machine shop to cost $6,000.

The Raven Lake Portland Cement Company has gone into liquidation.

R. Beach, Winchester, Ont., will establish a mattress factory in that town.

A brick plant will be erected at Claysmore Siding, Alta., to cost $25,000.

Messrs. McKeough & Trotter, Chatham, Ont., are erecting a machine shop.

The Western Fire Clay Products will erect a plant near Yellow Grass, Sask.

Rider & Kitchener's sawmill at Lindsay, Ont., was destroyed by fire recently.

The Peterboro Lock Co., Peterboro, will erect an addition to present premises.

The Winnipeg Power Committee have decided to erect a bridge to cost $40,000.

The Ridgetown, Ont., carriage works suffered loss by fire to the extent of $800.

The waterworks and lighting plant for Carman, Man., were completed recently.

The repairs necessary on the electric light plant, Wingham, Ont., will cost $10,000.

The Shuttleworth Chemical Company, Toronto, will erect a building to cost $25,000.

The Aluminum Crown Stopper Company, Toronto, will erect a factory to cost $30,000.

Shurly & Dietrich, Galt, will erect a building and plant for the tempering of steel saws.

The main building of Kreutziger's planing mill, Waterloo, Ont., was destroyed by fire.

The Eagle Spinning Co., Hamilton, will erect an addition to its works at a cost of $50,000.

A large lumber mill will be erected at Prince Rupert, B.C., by G. Brown, of San Francisco.

The Edmonton Produce Co., Edmonton, Alta., will erect a cold storage plant to cost $50,000.

The foundry in connection with the Stevenson Boiler Works, Petrolea, was destroyed by fire.

Dickson Bros., Campbellford, Ont., will build the Riverdale park footbridge, Toronto, for $3,421.

The Albert Biscuit Company, Calgary, are considering the erection of a factory in Edmonton.

Fire in the plant of the Diamond Glass Works, Montreal, did damage estimated at $6,000.

The large tannery of E. Julien at Limoilou, Que., was destroyed by fire, the loss being $15,000.

The Atikokan Blast Furnace recently shipped 2,000 tons of pig iron to Sault Ste. Marie, Ont.

Rapid progress is being made with the plant of the Electric Smelting Company at Welland, Ont.

The Grand Trunk Railway has laid off 250 hands from its shops at Point St. Charles, Que.

The American-Abell Engine Company, Toronto, will establish a warehouse in Calgary, Alta.

The sawmill of J. Stormont, West Lorne, Ont., was destroyed by fire, with a loss of $3,000.

Ex-Ald. Armstrong, London, Ont., announces that he will erect a new brass works in East London.

The Century Telephone Company will transfer its Canadian branch from Toronto to Bridgeburg, Ont.

The Rideau Foundry & Malleable Castings Company, Smith's Falls, Ont., recently opened their new plant.

A Montreal syndicate has secured possession of the St. Lawrence Power Company's big plant at Cornwall.

The Wabasso Cotton Company, Three Rivers, Que., is going on rapidly with the construction of its new plant.

The plant of Blakeney & Company, Hull, Que., was completely destroyed by fire, entailing a loss of $8,000.

The planing mills of R. Leeder & Sons, Toronto, were destroyed by fire, causing a loss of more than $20,000.

The Union Foundry and Machine Works, West St. John, N.B., are making extensive improvements to their plant.

The stave and heading mill, owned by J. Greenless and M. Kennedy, at Forest, Ont., was destroyed by fire, loss $3,000.

The Cleveland Cobalt Company have under consideration the question of establishing an electric plant at Cobalt, Ont.

The factory of the Seaman Kent Co., Meaford, Ont., manufacturers of hardwood flooring, was totally destroyed by fire.

New telephone buildings are to be erected in Charlottetown, P.E.I., and a power plant and a complete equipment installed.

The Intercolonial freight sheds were destroyed by fire, at Campbellton, N.B., Nov. 1st, with a total loss of about $30,000.

The Dominion Power and Transmission Company will erect a new freight shed and large car shops at Hamilton in the spring.

The Granby smelter, at Greenwood, B.C., with the entire blast of eight furnaces, is closed on account of the low price of copper.

Frank H. Fleer Company, Toronto, will erect two reinforced concrete gum factories on Sterling road, in that city, at a cost of $65,000.

The new power house for the Montreal Street Railway at the corner of Notre Dame street east and Raymond street, is nearing completion.

The Victoria Machinery Depot, Victoria, B.C., has been awarded the contract for repairing the C.P.R. steamer Tartar, the price being $10,000.

The Kaufman Rubber Company, recently formed in Berlin, elected the following officers recently: President, J. Kaufman; secretary, A. R. Kauffman.

The Robb Engineering Company has been awarded the contract for placing three new boilers in the wheel house of the Montreal waterworks.

The warehouses of the Singer Sewing Machine Company and the Pioneer Fruit Company, Brandon, Man., were damaged by fire to the extent of $10,000.

The Western Fuel Company, of Nanaimo, B.C., have recently purchased a 220 kw. d.c. C. G. E. generator, direct connected to a Robb-Armstrong engine.

G. Walter Green has just added to his plant in Peterboro, a warehouse 40x75 feet, and a new machine shop 35x120 feet, which is being equipped with up-to-date machinery.

The electric furnaces at the plant of the Electro-Metals, Limited, at Welland, Ont., were recently put into operation, and the results are said to have been very satisfactory.

Ernest Scott has started a machine shop at 91 Bleury street, Montreal. He has installed considerable machinery and is making a specialty of die-making and experimental work.

Fire did damage to the premises of the Toronto Electric Light Company to the extent of $5,000. The blacksmith shop was destroyed, along with the stables and a storehouse.

The Britannia Copper Company, at Howe Sound, have recently installed two 30 h.p. 850 r.p.m. Allis-Chalmers-Bullock induction motors for the operation of their conveyor systems.

A seventy-five horse-power Hornsby-Stockport gas engine has been installed in the premises of the Queen City Printing Ink Company, Toronto, by the Colonial Engineering Co.

The Assessment Commissioner, Toronto, has recommended that the city lease the Don Foundry Company, 120 feet of land on Ashbridge's Bay, at $150 per year, and taxes for 21 years.

It has been announced by the Canadian Pacific Railway that a substantial reduction will take place in freight rates between eastern Canada and points west of Fort William, Ont.

Of 51 different machines required by the Cobalt Concentrators for the Nipissing Mill, 50 were supplied by Canadian dealers. The plant when completed will treat 100 tons of ore per day.

Plans are being prepared for the new half-million technical school to be erected on Bloor street, Toronto. It will be complete in every detail and will be planned to allow extensive enlargements.

The new machine shop for the Canadian Rand Company, at Sherbrooke, Que., is completed and the machinery is partly installed. Rock drills, air compressors and mining machinery are being manufactured.

Two 75 kilowatt belted type alternators with two generator panels, two lighting panels and a 25-light series a.c. arc light outfit, all of C. G. E. make, were recently purchased for the Hosimer mines, Hosimer, B.C.

The contract for the new pattern storage vault, 80x50 feet, for the Paris Plow Co., Limited, Paris, Ont., has been awarded to John Carnie, contractor, Paris. It will be of fire-proof construction throughout.

E. Warren and A. G. Penman, of Toronto, representing the Canadian Smelting & Refining Company, Toronto, were at Sault Ste Marie, Ont., arranging for the construction of a mammoth smelting works at that place.

The Waterman Company has purchased property at St. Lambert, Que., and a large plant for the company's sole Canadian manufacture of pens will be erected. The factory will be of brick and stone, and five storeys high.

The Hinton Electric Company, of Vancouver, B.C., are installing a 500-light electric plant at the Canadian Sulphite Pulp Company's mill at Swanson Bay, B.C. They are using a C.G.E. 25 kw. direct current belted type generator.

The electrical equipment of the C.P.R. shops, Vancouver, has recently been augmented by the purchase from the C. G. E. Company of three 75 kw. transformers, and thirteen induction motors, ranging from 75 h.p. down to 5 h.p.

The Fort William Car Company purposes to build a car works at Fort William. The promoters have guaranteed a plant to employ about 800 hands at the outset, the number to be increased to 1,500 in the course of five years.

To meet with their increased business, the Shawinigan Lake Lumber Company, of Victoria, B.C., have recently installed a new circular saw and edger; these machines being driven by a 30 h.p. 900 r.p.m. Allis-Chalmers-Bullock induction motor.

The Fraser River Sawmills, Millside, B.C., will install a complete new mechanical equipment in the engine-room and to increase the capacity of the plant to a quarter of a million feet of rough lumber per day. The changes to be carried out will aggregate in value $100,000.

A factory for the manufacture of Mitchellite, the new explosive, will be erected near Victoria, B.C. This new invention has given highly satisfactory results during recent trials. Its high explosive power, combined with its safety and absence of poisonous gases, recommends it to builders and contractors.

The Columbia River Lumber Company, of Golden, B.C., have duplicated their present power plant. The new outfit consists of a 75 kw. 3 phase 60 cycle 2,300 volt generator, 4 kw. exciter and two panel switchboard, all of Allis-Chalmers-Bullock manufacture; also a 14-inch by 14-inch Robb-Armstrong horizontal engine.

Messrs. Smith, Kerry & Chace, of Toronto, have been engaged by the city council of Port Arthur, Ont., to make a survey of Silver Falls, Dog Lake, and prepare plans and estimates for a power plant. The ultimate capacity of the works will be 30,000 horse-power, although the present development will probably be confined to about 6,000 h.p.

The C.P.R. Company has erected a cold storage plant at Liverpool, covering an area of 60,000 square feet. The power for working the compressors is obtained from two Crossley gas engines of 45 h.p. each. These engines are supplied with fuel by one of Crossley's patent suction gas plants. Crossley's Canadian agency is Laurie & Lamb, 212 Board of Trade building, Montreal.

The Aetna Machine Co., 214 St. James St., Montreal, have made arrangements with C. Richard & Co., Montreal, for the building of Eclipse band saws, of which the Aetna Co. hold the patents for Canada. The feature of this machine is the improved treadle, which eliminates dead centres and saves lost motion. It will appeal to all wood-working shops where power is not used.

The British Columbia Electric Railway Company, of Vancouver, recently received a 200-light series alternating current arc lighting outfit for the city of Vancouver. This makes a total of 850 arc lights in that city. The outfit was purchased from the Canadian General Electric Company. They have also ordered, from the same company, twenty-six "C. G. E., 67" four-motor equipments.

On November 15th the railway and manufacturers' committee decided to bring in a by-law at the next meeting of the council of Guelph, Ont., to provide $52,000 for the distribution of Niagara power throughout the city. The net annual gain to the city is figured at $10,000, on the basis of a 1,500 horse-power contract. Of this amount 750 horse-power will be used by the public utilities of the city.

The Goldie & McCulloch Co. are building, and will soon have installed a 200 h.p. producer gas plant at their north works, Galt. The prospects are that they will largely go into the manufacturing of this system of power production—an important addition to Galt industries. Another Galt firm is also negotiating with the builders for a 175 h.p. gas producer plant, which the builders guarantee will not exceed $8 h.p. per annum.

The National-Acme Manufacturing Co., of Cleveland, Ohio, one of the largest companies in the United States, manufacturing the Acme automatic multiple spindle screw machines, as well as set screws, cap screws, and special milled products, etc., have practically decided to establish a branch of their works in Canada, which will enable them to better supply the growing demand for their product in this country. The location of the plant has not been finally decided on.

A model non-lifting railway spike has recently been patented by John McNeil, of Reserve, C.B. According to the patentee, the new rail-holding device will be a boon to the railway companies. The spike is so constructed that after it has been driven into the sleeper a small spring is wedged out under the bottom of the rail, thereby doing away with all possibility of the nail drawing away from its original fastening. Most of the big railway wrecks are accounted for on investigation to have been caused through rail-spreading, owing to the spikes becoming loose. Mr. McNeil claims that where his invention is used, the chance of accident through such cause would be greatly lessened. It is understood that Sir Wm. Van Horne is interested in Mr. McNeil's invention, and will give the new spike a tryout on the C.P.R. The article has been patented in Canada and the United States.

Companies Incorporated.

Glenn Stove and Furnace Company, Toronto, capital, $50,000; to manufacture and deal in stoves and plumbers' fittings. Directors, W. G. Glenn, J. McQuaker, E. G. Morris, all of Toronto.

McDougall and Cuzner, Ottawa; capital, $30,000; to conduct a retail hardware business. Provisional directors, W. F. Cuzner, J. P. Brown, W. Robertson and J. Bishop, all of Ottawa.

The Automatic Grain Shocker Machine Company, Hamilton; capital, $100,000; to manufacture farm machinery. Provisional directors, C. T. Grantham, A. E. Osler, A. Zimmerman, all of Hamilton.

B. Bell & Son Company, Toronto; capital, $200,000; to manufacture and deal in machinery, implements and supplies. Provisional directors, S. H. Chapman, H. H. Hurd, C. Morris, all of Toronto.

Toronto Brass Mills, Toronto; capital, $500,000; to manufacture and deal in brass and other castings. Provisional directors, A. E. J. Blackman, A. Munro, J. E. Fennell, A. J. Algate, all of Toronto.

Toronto Iron Works, Toronto; capital, $40,000; to carry on the work and business of a foundry, machine shop and boat-building establishment. Provisional directors, W. A. Manlon, H. B. Malone and J. H. Malone, all of Toronto.

Standard Automobile Company, Toronto; capital, $40,000; to deal in automobiles and carry on the business of mechanical engineers. Provisional directors, A. R. Bickerstaff, T. A. Silverthorn, M. G. Carroll, Nora Corcoran, all of Toronto.

Soss Invisible Hinge Company, Toronto; capital, $40,000; to manufacture and deal in hardware specialties and obtain the rights for the Soss invisible hinge. Provisional directors, J. Soss, New York; S. King and F. Watts, both of Toronto.

North Star Oil & Gas Company, Chatham, Ont.; capital, $36,000; to carry on a general mining and development business. Provisional directors, J. H. Teall, E. C. Jackson, and J. W. Cuthbertson, all of Tillsonburg, and F. T. Merrill, Chatham.

Bice Regulator Company, London, Ont.; capital, $70,000; to manufacture and deal in boiler feeds, regulators, valves and all kinds of boiler supplies. Provisional directors, G. Russell, Ethel M. Lindsay, both of Toronto, and A. W. Bice and M. Owen, both of London.

Municipal Undertakings.

A waterworks system will be installed at Port Moody, B.C.

Brandon, Man., is installing a lighting plant for its pumping station.

A waterworks plant will be installed at Elmira, Ont., to cost $25,000.

The council of Kelowna, B.C., has passed a by-law to raise ..,000 for waterworks.

Montreal will try to bring natural gas from St. Maurice county to supply the city with lights.

The road commission of Montreal has asked for the council to vote $40,000 for a sewer on St. Denis street.

Several citizens of Humboldt, Sask., are organizing a company to supply the town with a telephone system.

A by-law was carried in Chesley, Ont., for $38,000 for waterworks. The supply will come from artesian wells.

The Minneapolis Street & Town Lighting Company, Minneapolis, Minn., is seeking a franchise to supply Carnduff, Sask., with electric light.

The shareholders of the Sherbrooke Light, Heat and Power Company, Sherbrooke, Que., are in favor of accepting the city's offer to buy the plant.

Lewis P. Nott, of Montreal, has been awarded the contract for the construction of section 1 of the Trent Valley canal by the Department of Railways and Canals.

The city council, London, has decided to submit a by-law at the January elections, providing for the expenditure of $235,000 for transmission of Niagara power to that city.

The by-law for the new waterworks improvements at Welland has been prepared. For this purpose $15,800 will be required. A report has also been received on a new sewerage system.

The Canadian General Electric Co. has been awarded the contract for $24,672, by the city of Ottawa, for electrical supplies necessary for the taking over of the system of street lighting.

The township of Stamford, near the city of Niagara Falls, have decided to light the highways by electricity, and have entered into a contract with one of the lighteng companies to that end.

The city council of Quebec, Que., received a communication from the St. Laurent Saw & Steel Works, Sorel, Que., stating that the company would establish a works in that city if certain privileges were granted.

Thamesville is voting this week on a proposition to bonus the Thamesville Box & Handle Co. to the extent of $7,500 with a fixed assessment of $2,000 for 10 years. The factory is to cost not less than $12,000, and to employ 25 hands the year round. Jas. Galloway and D. J. Davies are behind the enterprise.

The city of Victoria has recently purchased through the Vancouver office of Allis-Chalmers-Bullock a complete rock crushing plant, including a No. 5-D Gates breaker, a No. 5 single head 25-foot elevator, a 30 h.p. induction motor, and a complete set of screens. This outfit will be used in the extensive street reconstruction which is about to be begun.

The Department of Marine and Fisheries, Ottawa, will shortly call for tenders for the construction of a large ice-breaking steamer to be used in keeping the channel between Prince Edward Island and the mainland open in the winter time. The new steamer will cost about $600,000, and will be one of the largest and most powerful ice breakers in the world.

The Ontario Government has awarded to Purdy, Mansell & Co., Toronto, the contracts for the heating, plumbing, and ventilating of the new Normal Schools at Hamilton, Stratford, Peterboro, and North Bay. The contracts for the electric wiring in the same buildings go to Fred. Armstrong & Company, Toronto. The aggregate amount of all the contracts is in the neighborhood of $63,000.

Railroad Construction.

The C.P.R. is installing its own water-pumping plant at Brandon, Man. The offer of the city was declined.

The C.P.R. will make Grand Forks, B.C., a divisional point, and will spend $100,000 in erecting roundhouses, shops and yards.

Work on the proposed railway from Port Simpson, B.C., to Fort Churchill, on the Hudson Bay, will be commenced next summer.

Now that the Dominion Copper Co.'s smelter is out of commission for the time being, the B.C. Copper Co. is able to get additional supplies of coke from Coleman, Alberta, much improving the situation at that smelter.

Boundary Copper Mines, in British Columbia, have shipped more than five and a half million tons of ore, running over a million for this year alone. This is equivalent to more than 100,000,000 pounds of refined copper.

The contract for the construction of the Copper River and Northwestern Railway, in Alaska, has been awarded to M. J. Heney, of Seattle. The cost is estimated at about $25,000,000. E. C. Hawkins, of Victoria, will be chief engineer.

About fifty miles of rails have now been laid on the National Transcontinental on the Quebec section, and grading is well advanced for a long distance on the various divisions now under contract. It is thought that by next fall about four hundred miles of road will be graded, and the track will be laid on a large portion of it. Tenders for the greater portion of the remaining thousand miles between Winnipeg and Moncton will be called for next spring. The contractors on the sections now under construction are reported to be making satisfactory progress.

Everything is now booming at the collieries and at the blast furnaces of the Nova Scotia Steel & Coal Company. During the month of October, 57,071 tons of coal were raised, an increase of 8,415 over September. In the steel department, great activity prevails, and the output of every branch of the plant shows an increase over the previous month. The output of pig iron was 5,755 tons, an increase of 300 tons over September; the steel output reached 6,986 tons, an increase of 1,476 tons over September; while 8,772 tons of coke were produced, being 1,100 tons more than during the previous month. This company will also open up another colliery.

Speaking of the general progress of the construction of the lines of the company in the west, Wm. Whyte, second vice-president of the C.P.R., stated that during the past season there had been but one difficulty and this had been the lack of railway ties. Through traffic to Edmonton would be possible over the new line early next season. The Pheasant Hills branch, which will run through Edmonton in connection with the Yorkton line, is being operated to Nokomis. The steel has been laid to Lanigan, the point of junction with the Yorkton branch, and steel has also been laid for a distance of twenty miles beyond Lanigan. Work was being pushed in the more distant west with the greatest energy and it was hoped that the company would be able to operate the line into Saskatoon by the end of the present month. The grading of the line would also be completed into Lanigan from the east on the extension of the Yorkton line this fall.

Building Notes.

A large rink is being erected at Orillia.

A post office will be erected in Kincardine, to cost $17,000.

E. J. Evans, Toronto, will erect a hotel to cost $18,000.

The value of buildings erected in Regina, Sask., during 1907, is $1,103,435.

The new building for the Montreal Sailors' Institute will cost $60,000.

Loo Gee Wing, Vancouver, B.C., will erect a business block to cost $80,000.

A new Oddfellows' temple will be erected at Niagara Falls, Ont., to cost $25,000.

The I.O.O.F. lodge, of Hamilton, will improve their hall at a cost of $10,000.

Charlotte St. Methodist church, Peterboro, will erect an edifice to cost $30,000.

The chartered congregation of the Temple of Solomon, Montreal, to cost $52,000.

The West End Baptist church, Halifax, N.S., was destroyed by fire, the loss being $10,000.

The Canadian Birkbeck Investment and Savings Co., Toronto, will erect an office building to cost $120,000.

Building permits in Victoria, B.C., for October aggregated $81,875. This is an increase of $20,000 over September.

W. J. Cavanagh, Vancouver, B.C., has plans prepared for a large hotel to be erected in that city at a cost of $100,000.

The International Harvester Co., Hamilton, are contemplating the erection of a distributing warehouse at Port Arthur, Ont.

The contract for the new armory for Medicine Hat, Alta., has been awarded to A. Burns, of that city. The cost is $160,000.

A transportation building will be erected at the Exhibition Grounds, Toronto, to cost $95,000. It will have a floor space of 40,000 square feet.

The contract for a Church of England cathedral in Halifax, N.S., has been awarded to S. M. Brookfield, of that city. The buiiing will cost $125,000.

The directors of the Montreal Dominion Park Company have decided to start at once the rebuilding of that portion of their amusement buildings recently destroyed by fire.

Tenders are invited by the chairman of the Farm Committee, Shawbridge, Que., for an industrial school to be erected on the farm of the Boys' Home, Montreal.

The building total in Winnipeg, Man., for October is $132,450. There were ninety-five permits issued, covering ninety-six buildings. In October, 1906, there were 306 permits issued, covering 355 buildings, at a cost of $1,091,950.

Building permits in Edmonton, Alta., for October, total $78,560, a large falling off from previous months. This has been the case in nearly all the larger cities in the west, as well as in the east, and is an indication of a slight curtailment of buildings, as the results of the financial stringency. This curtailment, however, is far less here than in rival cities of Edmonton.

Trade Notes.

The Evans Rotary Engine Company, Limited, Toronto, is supplying Jonas Byer, Stouffville, with a 75 h.p. rotary engine for his electric lighting plant.

The Chapman Double Ball Bearing Company has secured the contract to equip the works of the Wabasso Cotton Co., Three Rivers, P.Q., with their double ball bearings.

The new terminal station at Hamilton is now occupied by thirteen electrical companies. It was erected by the Dominion Power and Transmission Company, under the supervision of Charles Mills, Hamilton, at a cost of $250,000.

The Colonial Engineering Co., of Montreal, has offered to install producer gas plants in London and Toronto, with a guarantee that the city's light shall not cost over $30 per lamp, including all operating costs, maintenance, interest and depreciation.

The Ferracute Machine Co., of Bridgeton, N. J., manufacturers of presses and dies for sheet metal work, made shipments of their presses last week to India, Austria and England. They intend giving more attention to their Canadian trade and their advertisement appears in this issue.

The Smart-Turner Machine Co., of Hamilton, Ont., report supplying a large number of companies with their feed and vacuum pumps. These include the Brantford Roofing Co., Brantford, Ont.; Steamwinder Gold & Coal Mining Co., Fairview, B.C.; Beamsville Preserving Co., Beamsville; Canadian Asbestos Co., Montreal; Sherlock Manning Organ Co., London, Ont.; Somerville Ltd., Toronto; Victoria Industrial School, Mimico; Helena Costume Co., London, Ont.; two to the G.T.R. and six to the I.C.R., Moncton.

Expanding Their Business.

In this issue is the announcement of the H. W. Petrie, Limited. This expansion is due to the increase in trade, and in order to extend their rapidly growing business more capital has been interested, and the company is in a better position than ever to look after the needs of its customers.

New Factory for Stratford.

A Canadian branch of the Allen P. Boyer Company, of Goshen, is being opened in Stratford, Ontario. A two-storey brick building, 100 by 68 feet, is being erected on a ten-lot site on King street, and the company will employ 25 to 40 hands at the outset. They will manufacture various kinds of swings, ladders, hay slings, carriers and forks. It is intended to put $40,000 into the company here.

Tremendous Power to be Harnessed.

A company has been formed in Vancouver, the British Columbia Power and Electric Company, who have obtained a site on the Chekamus river, forty-eight miles from Vancouver, where the river has a drop of 520 feet in 600 feet. It is estimated that there is 250,000 horse-power, and it is proposed to make provision for a plant that will supply 100,000 horse-power in Vancouver.

Car Shop for Hamilton.

The Dominion Power and Transmission Company, of Hamilton, will begin work in the spring on an immense new freight shed, car shops and sheds for the cars used by the various electrical companies using the new terminal station. The scheme includes a car manufacturing industry, which will turn out the cars required for the different lines. The site will be on Sanford avenue, and plans are now being prepared.

New Factory at Galt Starts Work.

The Car & Coach Company, of Galt, have started business and are in need of more men. The engine room, smoke stack and coal pit are completed and are permanent. As soon as weather conditions will permit in the spring a main work shop, car-erecting shop, transfer table, an office and two additional buildings will be erected. At present a temporary main building, 160x76 feet, will be used to carry out the orders now in the hands of the company. This will be replaced in the spring by a reinforced concrete structure, 120x65 feet, and three storeys high.

New Plant for Montreal.

The Metal Shingle & Siding Co., Montreal, manufacturers of herringbone lath and truss beam fabric for concrete reinforcement have installed in their new plant some up-to-date machinery. They intend manufacturing, in addition to their special line dies, tools and gauges. Among other machinery they have installed a Brown & Sharpe No. 3 surface grinder with magnetic chuck, and a No. 3 Universal grinder and milling machine, fitted with differential driving head. They have also a modern hardening and tempering plant for either carbon or high speed steels.

Hamilton Steel & Iron Co.'s Plant in Full Swing.

With the starting of the new furnace at Hamilton the capacity of this company has been doubled. The new furnace was installed at a cost of over half a million dollars, and has a capacity of 300 tons of foundry iron, or 400 tons of basic iron, used in the manufacture of steel. The location of this second furnace in Hamilton means a large addition to the number of men employed. The furnace is equipped with many labor saving devices for handling ore and pig iron and the company will be able to turn out much more steel than in the past and will thus be in a better position to supply the demand.

Electric Power Plant at Middle Falls.

On Oct. 25th the ratepayers of Campbellford, by a vote of 296 to 6, carried a by-law to build a $60,000 power plant at Middle Falls. The contract for the dam has been let to Brown & Alymer, and Bogue & Buchanan, of Peterboro, will erect the power house.

The power house is being erected as a municipal enterprise to supply power users in the town, and also to generate electricity for street lighting. The plant will be a modern one in every respect, and will be placed in operation at as early a date as possible.

Electrical Development at Healey Falls.

Construction operations on the Northumberland-Durham Power Co. at Healey Falls and rapids is advancing under the supervision of Managing Director Culverwell. The company is now constructing two different power developments, one being a dam at the foot of the lower rapids to flood the water back on the top and several chains above the big falls, this being done to hold the Ontario Government lease. This will be completed if the Dominion Government do not settle with the company as to their portion of the upper rapids which intervenes the Ontario Government lease portion and portions of the upper rapids owned by the company.

The other where work is going on also, which the company hopes they will be allowed to finish, is a four thousand-foot head-race to extend from above the bridge along the high table and on the southwest side, to a point nearly opposite the foot of the rapids, and from that point the water will be carried down the hill to the power house by two twelve-foot pipe lines. This head-race would give a total head or fall of seventy-three feet.

This power, with reservoir above, would give fourteen thousand horse-power during working hours of day in dry season, which the company wish to supply to the whole of south midland Ontario.

New York capitalists visited Healey Falls recently, with Managing Director Culverwell, and their engineer drove over the projected electric railway to connect the C.P.R. at Havelock with Campbellford, Warkworth and Cobourg, and Port Hope, with branches—one of which will connect Hastings with [illegible], making a junction at Warkworth. Mr. Culverwell and his associates will apply for this charter at the session of the local Legislature.

Brass Mills to Re-open.

President Menzies, of the newly-formed Toronto Brass Rolling Mills, with mills at New Toronto, states that the mills will be re-opened as soon as arrangements are made for a new manager for the works. The plant has been closed down for a couple of years, the old company, the Canada Brass Rolling Mills, having been wound up by foreclosure in August. The new tariff is favorable to the operation of the mills, the duty on brass sheets, rods and tubing having been increased in November, 1906, since which time users of brass have been forced to import and pay the duty, no supply being available from a Canadian mill.

Proposal to Enlarge Doty Engine Works.

The Doty Engine Works Company, of Goderich, Ont., has made a proposal to the council of that city that in return for $35,000, repayable in yearly instalments, with interest at 5 per cent. per annum, also exemption from taxes for a term of ten years, including school taxes, they will erect extensions to their plant and employ 100 hands continually. The new plant will be devoted to the manufacture of boilers and steel furnaces for steamers, and structural materials, and will be located on a site convenient to the railways. The company will accept the town debentures and finance them.

New Machine Shop.

The American Auto-Engine Co. are a new firm who have recently opened up a machine shop in Montreal. They have erected a one-storey building at the corner of St. Louis and Bonsecours streets. This is 42x60 feet, and will all be used as a work shop, with the exception of a space 18x20 feet, which has been separated off to make two rooms, an office and pattern room. This firm have also procured a building at the rear, which they will convert into an automobile garage. Considerable up-to-date machinery has already been installed and more will be put in later on. They are manufacturing marine and stationary gas and gasoline engines, and are also making a specialty of fine repair work. At present the power used is steam, but they have under construction a large type of engine of their own particular make.

Berlin Machine Works, Limited, Hamilton.

This company has almost completed its new plant in Hamilton and by January 1st will give employment to 150 men. The company will manufacture a complete line of wood-working machinery, including planers, surfacers, matchers, sanders, saws, band-saws and molders.

The company purchased twenty acres of land at the intersection of the G.T.R. main line to Buffalo and the T. H. & B. railway. The plant, including machine shop, foundry and power house, was erected under the supervision of Mr. G. W. Robinson, secretary and resident manager. Motor drive will be employed, lighting will be by Cooper-Hewitt and Nernst lamps, while for babbiting and tempering furnaces, natural gas will be used.

The power house, 90x50 feet, is of brick and steel construction, and contains heating plant, Westinghouse motor generators, Cataract power transformers and Curtis air compressor. A chimney 116 feet high, is being erected by the Weber Steel Concrete Chimney Co.

The foundry, 200x72 feet, has a capacity of 30 tons per day. A reinforced concrete gallery, back of the cupolas, affords storing space for sand and coke. The foundry equipment consists of Whiting and Newton cupolas and Root blowers, the Sly system of tumbling barrels and dust collector and a Powling & Harnischfeger 15-ton electric traveling crane. There are also twenty indepentent jib cranes for use of the molders.

The machine shop is also of brick and steel, 300x200 feet. The machinery will be driven by what is known as group drive. A gallery, 200 x48 feet, will be used for a tool room. The machine shop will be equipped with over seventy machine tools run by Westinghouse motors from 5 to 75 h.p. capacity. Two Pawling & Harnischfeger 10-ton electric traveling cranes have been installed. These buildings will accommodate over 700 men. Plans have been prepared for a number of other buildings, which will be erected in the spring. These include a pattern shop, pattern storage vault and a chemical laboratory.

More Power for Montreal.

The big project of the Montreal and Provincial Light, Heat and Power Companies, at the Coteau de Lac rapids, thirty miles from Montreal, is nearing completion and power will be used from this source in Montreal this autumn. A million and a half dollars have been spent on this plant, which is expected to develop 15,000 horse-power. The plant is complete in every detail and includes a canal, half a mile long, an immense reinforced concrete dam, penstocks and draft tubes, as well as a large power and wheel house.

Canada's Peat Beds.

James Murray, of Woodstock, has discovered a valuable peat bog on his farm, at West Zorra, and will immediately take steps towards its development. An expert has been consulted and on examination the bog was found to be an inexhaustable bed of peat. A plant will be put in operation and peat briquettes manufactured for general use.

It has been estimated that Canada has 37,000 square miles of peat lands. At one time a Government commission was appointed to examine and report on the principal peat industries in Europe, with a view to turning the extensive peat bogs to good account. So far the reports have not been made public and this resource has not been utilized. The wide-spread coal fields of America have been supplying us with fuel, and this can be had with less work than digging, drying and pressing the peat. Until the coal supply runs short we can hope for little advance unless some cheaper method is devised for curing the peat. In Europe peat is converted into paper and wood alcohol is extracted from it, so that progress is being made along this line. A company, manufacturing peat at Orlando, Florida, lets the peat dry by natural processes, and then breaks it into lumps for sale. Should the Canadian peat-beds be developed and the output sold at a moderate cost, it will, no doubt, find a ready market.

C.P.R. Land Holdings.

The Canadian Pacific Railway occupies a unique position in the railroad world, in the possession of valuable assets, altogether distinct from the properties and equipment necessary for the conduct of its business. At the close of the last fiscal year there remained unsold out of the original grant of lands by the Government, 8,905,823 acres of agricultural land, besides 3,419,673 acres in British Columbia, and 2,500,000 acres, also in British Columbia, which the company is to receive through the Columbia & Western Railway. The Canadian Pacific, then, after all its profit-taking from land sales extending over 20 years, still has possession of the enormous amount of 14,825,496 acres of land, free from bonded debt, which is rapidly appreciating in value.

CANADIAN MACHINERY BUYERS' DIRECTORY

To Our Readers—Use this directory when seeking to buy any machinery or power equipment. You will often get information that will save you money.
To Our Advertisers—Send in your name for insertion under the heading of the lines you make or sell.
To Non-Advertisers—A nominal rate of $1 per line a year is charged non-advertisers.

Acids.
Canada Chemical Mfg. Co., London.

Abrasive Grains & Powders
Carborundum Co., Niagara Falls, N.Y.

Abrasive Materials.
The Canadian Fairbanks Co., Montreal.
Rice Lewis & Son, Toronto.
Norton Co., Worcester, Mass.
H. W. Petrie, Toronto.
Carborundum Co., Niagara Falls, N.Y.
Williams & Wilson, Montreal.

Air Brakes.
Canada Foundry Co., Toronto.
Canadian Westinghouse Co., Hamilton.

Air Receivers.
Allis-Chalmers-Bullock Montreal.
Canada Foundry Co., Toronto.
Canadian Rand Drill Co., Montreal.
Chicago Pneumatic Tool Co., Chicago.
John McDougall Caledonian Iron Works Co., Montreal.

Alundum Scythe Stones
Norton Company, Worcester, Mass.

Arbor Presses.
Niles-Bement-Pond Co., New York.

Automatic Machinery.
Cleveland Automatic Machine Co. Cleveland.
National-Acme Mfg. Co., Cleveland
Potter & Johnston Machine Co., Pawtucket, R. I.

Automobile Parts
Globe Machine & Stamping Co., Cleveland, Ohio.

Axle Cutters.
Butterfield & Co., Rock Island, Que.
A. B. Jardine & Co., Hespeler, Ont.

Axle Setters and Straighteners.
A. B. Jardine & Co., Hespeler, Ont.
Standard Bearings, Ltd, Niagara Falls

Babbit Metal.
Canada Metal Co., Toronto.
Canada Machinery Agency, Montreal.
Frothingham & Workman Ltd., Montreal
Greey Wm. & J. G., Toronto.
Rice Lewis & Son, Toronto.
Lumen Bearing Co., Toronto.

Bakers' Machinery.
Greey, Wm. & J. G., Toronto.

Baling Presses.
Perrin, Wm. R., & Co., Toronto.

Barrels, Steel Shop.
Cleveland Wire Spring Co., Cleveland.
Steel Trough & Machine Co., Tweed, Ont.

Barrels, Tumbling.
Buffalo Foundry Supply Co., Buffalo.
Detroit Foundry Supply Co., Windsor
Dominion Foundry Supply Co., Montreal
Hamilton Facing Mill Co., Hamilton.
Gilmour, J. New York.
Globe Machine & Stamping Co, Cleveland, Ohio.
John McDougall Caledonian Iron Works Co., Montreal.
Northern Engineering Works, Detroit.
J. W. Paxson Co., Philadelphia, Pa.
H. W. Petrie, Toronto.
Sly, W. W., Mfg. Co., Cleveland
The Smart-Turner Mach. Co., Hamilton

Bars, Boring.
Hall Engineering Works, Montreal.
Niles-Bement-Pond Co., New York.
Standard Bearings Ltd., Niagara Falls

Bars, Grate.
Wilson, J. C., & Co., Glenora, Ont.

Batteries, Storage.
Canadian General Electric Co. Toronto
Rice Lewis & Son, Toronto.
John Millen & Son, Montreal.

Bearing Metals.
Lumen Bearing Co., Toronto.

Bearings, Roller.
Standard Bearings, Ltd. Niagara Falls

Bearings, Self-Oiling.
Wilson, J. C., & Co., Glenora, Ont.

Belting, Chain.
Canada Machinery Exchange, Montreal.
Greey, Wm. & J. G., Toronto.
Jeffrey Mfg. Co., Columbia, Ohio.
Link-Belt Eng. Co., Philadelphia.
Waterous Engine Works Co., Brantford.

Belting, Cotton.
Canada Machinery Agency, Montreal.
Dominion Belting Co., Hamilton.
Rice Lewis & Son, Toronto.

Belting, Leather.
Canada Machinery Agency, Montreal.
The Canadian Fairbanks Co., Montreal.
Frothingham & Workman Ltd., Montreal
Greey, Wm. & J. G., Toronto.
McLaren, J. C., Montreal.
Rice Lewis & Son, Toronto.
H. W. Petrie, Toronto.
Williams & Wilson, Montreal.

Belting, Rubber.
Canada Machinery Agency, Montreal.
Frothingham & Workman, Ltd., Montreal
Greey, Wm. & J. G., Toronto.

Belting Supplies.
Greey, Wm. & J. G., Toronto.
Rice Lewis & Son, Toronto.
H. W. Petrie, Toronto.

Bending Machinery.
John Bertram & Sons Co., Dundas, Ont.
Chicago Pneumatic Tool Co., Chicago.
Rice Lewis & Son, Toronto.
London Mach. Tool Co., Hamilton, Ont.
National Machinery Co., Tiffin, Ohio.
Niles-Bement-Pond Co., New York.

Benders, Tire.
A. B. Jardine & Co., Hespeler, Ont.
London Mach. Tool Co., Hamilton, Ont.

Blowers.
Buffalo Foundry Supply Co., Montreal.
Canada Machinery Agency, Montreal.
Detroit Foundry Supply Co., Windsor
Dominion Foundry Supply Co., Montreal
Gilmour, J., New York.
Hamilton Facing Mill Co., Hamilton.
Kerr Turbine Co., Wellsville, N.Y.
J. W. Paxson, Philadelphia, Pa.
Sheldon's Limited, Galt.

Blast Gauges—Cupola.
Paxson, J. W., Co. Philadelphia.
Sheldons, Limited, Galt

Blocks, Tackle.
Frothingham & Workman, Ltd, Montreal

Blocks, Wire Rope.
Frothingham & Workman, Ltd., Montreal

Blow-Off Tanks.
Darling Bros., Ltd., Montreal.

Boilers.
Canada Foundry Co., Limited, Toronto.
Canada Machinery Agency, Montreal.
Goldie & McCulloch Co., Galt.
John McDougall Calenonian Iron Works, Montreal.
Manitoba Iron Works, Winnipeg.
Owen Sound Iron Works Co., Owen Sound
H. W. Petrie, Toronto.
Robb Engineering Co., Amherst, N.S.
The Smart-Turner Mach. Co., Hamilton.
Waterous Engine Works Co., Brantford.
Williams & Wilson, Montreal.

Boiler Compounds.
Canada Chemical Mfg. Co., London, Ont.
Hall Engineering Works, Montreal.

Bolt Cutters.
John Bertram & Sons Co., Dundas, Ont.
London Mach. Tool Co., Hamilton.
National Machinery Co., Tiffin, Ohio.
Niles-Bement-Pond Co., New York.

Bolt and Nut Machinery.
John Bertram & Sons Co., Dundas, Ont.
Canada Machinery Agency, Montreal.
Rice Lewis & Son, Toronto.
London Mach. Tool Co., Hamilton.
National Machinery Co., Tiffin, Ohio.
Niles-Bement-Pond Co. New York.

Bolts and Nuts, Rivets.
Mechanics Supply Co., Quebec, Que.

Boring & Drilling Machines
American Tool Works Co., Cincinnati.
John Bertram & Sons Co., Dundas, Ont.
Canada Machinery Agency, Montreal.
A. B. Jardine & Co., Hespeler, Ont.
London Mach. Tool Co., Hamilton.
Niles-Bement-Pond Co., New York.

Boring Machine, Upright.
American Tool Works Co., Cincinnati.
John Bertram & Sons Co., Dundas, Ont.
London Mach Tool Co., Hamilton.
Niles-Bement-Pond Co., New York.

Boring Machine, Wood.
Chicago Pneumatic Tool Co., Chicago.
Independent Pneumatic Tool Co., Chicago, Ill.
London Mach. Tool Co., Hamilton.

Boring and Turning Mills.
American Tool Works Co., Cincinnati.
John Bertram & Sons Co., Dundas, Ont.
Canada Machinery Agency, Montreal.
Gisholt Machine Co., Madison, Wis.
Rice Lewis & Son, Toronto.
London Mach. Tool Co., Hamilton.
Niles-Bement-Pond Co., New York.
H. W. Petrie, Toronto.

Box Puller.
A. B. Jardine & Co., Hespeler, Ont.

Boxes, Lathe.
Steel Trough & Machine Co., Tweed, Ont.

Boxes, Steel Shop.
Cleveland Wire Spring Co, Cleveland.
Steel Trough & Machine Co., Tweed, Ont.

Boxes, Tote.
Cleveland Wire Spring Co., Cleveland.

Brass Foundry Equipment.
Ph. Bonvillain & E. Ronceray, Philadelphia
Detroit Foundry Supply Co., Detroit.
Dominion Foundry Supply Co., Montreal
Paxson, J. W., Co., Philadelphia

Brass Working Machinery.
Warner & Swasey Co., Cleveland, Ohio.

Brushes, Foundry and Core.
Buffalo Foundry Supply Co., Buffalo.
Detroit Foundry Supply Co, Windsor.
Dominion Foundry Supply Co., Montreal
Hamilton Facing Mill Co., Hamilton.
Paxson, J. W., Co., Philadelphia

Brushes, Steel.
Buffalo Foundry Supply Co., Buffalo.
Dominion Foundry Supply Co, Montreal
Paxson, J. W., Co., Philadelphia

Bulldozers.
John Bertram & Sons Co., Dundas, Ont.
London Mach. Tool Co., Hamilton, Ont
National Machinery Co., Tiffin. Ohio.
Niles-Bement-Pond Co., New York.

Calipers.
Frothingham & Workman Ltd., Montreal
Rice Lewis & Son, Toronto.
John Millen & Son, Ltd, Montreal, Que.
L. S. Starrett & Co., Athol, Mass.
Williams & Wilson Montreal.

Canners' Machinery.
Wilson, J. C., & Co., Glenora, Ont.

Carbon.
Dominion Foundry Supply Co., Montreal

Carborundum Paper and Cloth.
Carborundum Co., Niagara Falls, N.Y.

Cars, Foundry.
Buffalo Foundry Supply Co., Buffalo.
Detroit Foundry Supply Co., Windsor
Dominion Foundry Supply Co., Montrea
Hamilton Facing Mill Co., Hamilton.
Paxson, J. W., Co., Philadelphia

Castings, Aluminum.
Lumen Bearing Co., Toronto

Castings, Brass.
Chadwick Bros., Hamilton.
Greey, Wm. & J. G. Toronto.
Hall Engineering Works, Montreal.
Kennedy, Wm., & Sons, Owen Sound.
Lumen Bearing Co., Toronto
Niagara Falls Machine & Foundry Co. Niagara Falls, Ont.
Owen Sound Iron Works Co., Owen Sound.
Robb Engineering Co, Amherst, N.S.
Wilson, J. C., & Co., Glenora, Ont.

Castings, Grey Iron.
Allis-Chalmers-Bullock Montreal.
Crowe's Iron Works Guelph, Ont.
Greey, Wm. & J. G., Toronto.
Hall Engineering Works, Montreal.
Kennedy, Wm., & Sons, Owen Sound.
Laurie Engine & Machine Co., Montreal.
Maxwell, David, & Sons, St. Marys.
John McDougall Caledonian Iron Works Co., Montreal.
Niagara Falls Machine & Foundry Co., Niagara Falls, Ont.
Owen Sound Iron Works Co., Owen Sound.
Robb Engineering Co., Amherst, N.S.
Smart-Turner Machine Co., Hamilton.
Wilson, J. C., & Co., Glenora, Out.

Castings, Phosphor Bronze.
Lumen Bearing Co., Toronto

Castings, Steel.
Kennedy, Wm., & Sons, Owen Sound.

Castings, Semi-Steel.
Robb Engineering Co., Amherst, N.S.

Cement Machinery.
Allis-Chalmers-Bullock, Limited, Montreal
Greey, Wm. & J. G., Toronto.
Jeffrey Mfg. Co., Columbus, Ohio
John McDougall Cal donian Iron Works, Co., Montrea
Owen Sound Iron Works Co., Owen Sound.

Centreing Machines.
John Bertram & Sons Co., Dundas, Ont.
Jeffrey Mfg Co., Columbia, Ohio
London Mach. Tool Co., Hamilton, Ont.
Niles-Bement-Pond Co., New York.
Pratt & Whitney Co., Hartford, Conn.
Standard Bearings, Ltd., Niagara Falls

Centres, Planer.
American Tool Works Co., Cincinnati.

Centrifugal Pumps.
Gas & Electric Power Co., Toronto.
John McDougall Caledonian Iron Works Co. Montreal.
Pratt & Whitney Co., Hartford, Conn.

Centrifugal Pumps—Turbine Driven.
Kerr Turbine Co., Wellsville, N.Y.

Chain, Crane and Dredge.
Frothingham & Workman, Ltd., Montrea

Drills, Blacksmith.

Canada Machinery Agency, Montreal.
Frothingham & Workman, Ltd., Montreal
A. B. Jardine & Co., Hespeler, Ont.
London Mach. Tool Co., Hamilton.
Standard Tool Co., Cleveland.

Drills, Centre.

Pratt & Whitney Co., Hartford, Conn.
Standard Tool Co., Cleveland, O.
L. S. Starrett Co., Athol, Mass.

Drills, Coal and Plaster.

Cumming, J. W., New Glasgow, N.S.

Drills, Electric

Canadian Pilling Co., Montreal
Chicago Pneumatic Tool Co., Chicago.
Gas & Electric Powe Co., Toronto.
Niles-Bement-Pond Co., New York.

Drills, Gang.

American Tool Works Co., Cincinnati.
John Bertram & Sons Co., Dundas, Ont.
Pratt & Whitney Co., Hartford, Conn.

Drills, High Speed.

Wm. Abbott, Montreal.
Frothingham & Workman, Ltd., Montreal
Alexander Gibb, Montreal.
Pratt & Whitney Co., Hartford, Conn.
Standard Tool Co., Cleveland, O.

Drills, Hand.

A. B. Jardine & Co., Hespeler, Ont.

Drills, Horizontal.

John Bertram & Sons Co., Dundas, Ont.
Canada Machinery Agency, Montreal.
London Mach. Tool Co., Hamilton.
Niles-Bement-Pond Co., New York.

Drills, Pneumatic.

Canada Machinery Agency, Montreal.
Chicago Pneumatic Tool Co., Chicago.
Independent Pneumatic Tool Co., Chicago, New York
Niles-Bement-Pond Co., New York.

Drills, Radial.

American Tool Works Co., Cincinnati.
John Bertram & Sons Co., Dundas, Ont.
Bickford Drill & Tool Co., Cincinnati
London Mach. Tool Co., Hamilton, Ont.
Niles-Bement-Pond Co., New York.

Drills, Ratchet.

Armstrong Bros. Tool Co., Chicago.
Frothingham & Workman, Ltd., Montreal
A. B. Jardine & Co., Hespeler,
Pratt & Whitney Co., Hartford, Conn.
Standard Tool Co., Cleveland.

Drills, Rock.

Allis-Chalmers-Bullock, Montreal.
Canadian Rand Drill Co., Montreal.
Chicago Pneumatic Tool Co., Chicago.
Jaffrey Mfg. Co., Columbus, Ohio.

Drills, Sensitive.

American Tool Works Co., Cincinnati.
Canada Machinery Agency, Montreal.
Dwight Slate Machine Co., Hartford.
Niles-Bement-Pond Co., New York.

Drills, Twist.

Chicago Pneumatic Tool Co., Chicago
Frothingham & Workman, Ltd., Montreal
Alex. Gibb, Montreal.
A. B. Jardine & Co., Hespeler, Ont.
John Millen & Son, Ltd., Montreal.
Morse Twist Drill and Machine Co., New Bedford, Mass.
Pratt & Whitney Co., Hartford, Conn.
Standard Tool Co., Cleveland.

Drying Apparatus of all Kinds.

Dominion Heating & Ventilating Co., Hespeler
Greey, Wm. & J. G., Toronto
Sheldons Limited, Galt

Dry Kiln Equipment.

Sheldons Limited, Galt

Dry Sand and Loam Facing.

Buffalo Foundry Supply Co., Buffalo.
Dominion Foundry Supply Co., Montreal
Hamilton Facing Mill Co., Hamilton.
Paxson, J. W., Co., Philadelphia

Dump Cars.

Canada Foundry Co., Limited, Toronto
Dominion Foundry Supply Co., Montreal
Greey, Wm. & J. G., Toronto
Hamilton Facing Mill Co., Hamilton.
John McDougall, Caledonian Iron Works Co., Montreal.
Niles-Bement-Pond Co., New York.
Standard Bearings, Ltd., Niagara Falls.
Link-Belt Eng. Co., Philadelphia.
John McDougall Caledonian Iron Works Co., Montreal
Owen Sound Iron Works Co., Owen Sound
Paxson, J. W., Co., Philadelphia
Waterous Engine Co., Brantford.

Dust Arresters.

Sly, W. W., Mfg. Co., Cleveland

Dust Separators.

Dominion Heating and Ventilating Co., Hespler.
Greey, Wm. & J. G., Toronto
Paxson, J. W., Co., Philadelphia
Sheldon's Limited, Galt.

Dynamos.

Allis-Chalmers-Bullock, Montreal.
Canadian General Electric Co., Toronto.
Canadian Westinghouse Co., Hamilton.
Consolidated Electric Co., Toronto
Electrical Machinery Co., Toronto.
Gas & Electric Power Co., Toronto
Hall Engineering Works, Montreal, Que.
John Millen & Son, Ltd., Montreal.
Packard Electric Co., St. Catharines.
H. W. Petrie, Toronto.
T. & H. Electric Co., Hamilton.

Dynamos—Turbine Driven.

Gas & Electric Power Co., Toronto.
Kerr-Turbine Co., Wellsville, N.Y.

Economizer, Fuel.

Dominion Heating & Ventilating Co., Hespeler
Standard Bearings, Ltd., Niagara Falls.

Electrical Instruments.

Canadian Westinghouse Co., Hamilton.
Gas & Electric Power Co., Toronto.

Electrical Supplies.

Canadian General Electric Co., Toronto.
Canadian Westinghouse Co., Hamilton.
Gas & Electric Power Co., Toronto
London Mach. Tool Co., Hamilton, Ont.
John Millen & Son, Ltd., Montreal.
Packard Electric Co., St. Catharines.
T. & H. Electric Co., Hamilton.

Electrical Repairs

Canadian Westinghouse Co., Hamilton.
T. & H. Electric Co., Hamilton.

Elevator Buckets.

Greey, Wm. & J. G., Toronto
Jeffrey Mfg. Co., Columbus, Ohio.

Emery and Emery Wheels.

Dominion Foundry Supply Co., Montreal
Frothingham & Workman Ltd., Montreal
Hamilton Facing Mill Co., Hamilton.
Paxson J. W. Co. Philadelphia

Emery Wheel Dressers.

Canada Machinery Agency, Montreal.
Dominion Foundry Supply Co., Montreal
Frothingham & Workman Ltd., Montreal
Hamilton Facing Mill Co., Hamilton.
John Millen & Son, Ltd., Montreal.
H. W. Petrie, Toronto.
Paxson, J. W., Co., Philadelphia
Standard Tool Co., Cleveland.

Engineers and Contractors.

Canada Foundry Co., Limited, Toronto.
Darling Bros., Ltd., Montreal
Gas & Electric Power Co., Toronto
Greey, Wm. & J. G., Toronto
Hall Engineering Works, Montreal
Laurie Engine & Machine Co., Montreal.
Link-Belt Co., Philadelphia.
John McDougall, Caledonian Iron Works Co., Montreal.
Robb Engineering Co., Amherst, N.S.
The Smart-Turner Mach. Co., Hamilton.

Engineers' Supplies.

Frothingham & Workman, Ltd., Montreal
Greey, Wm. & J. G., Toronto
Hall Engineering Works, Montreal.
Rice Lewis & Son, Toronto.

Engines, Gas and Gasoline.

Canada Foundry Co., Toronto.
Canada Machinery Agency, Montreal.
The Canadian Fairbanks Co., Montreal.
Gas & Electric Power Co., Toronto
Gilson Mfg. Co., Guelph
The Goldie & McCulloch Co., Galt, Ont.
Rice Lewis & Son, Toronto
Ontario Wind Engine & Pump Co. Toronto
H. W. Petrie, Toronto.
The Smart-Turner Mach. Co., Hamilton

Engines, Steam.

Allis-Chalmers-Bullock, Montreal
Belliss & Marcom, Birmingham, Eng.
Canada Machinery Agency, Montreal.
The Goldie & McCulloch Co., Galt, Ont.
Rice Lewis & Son, Toronto.
Laurie Engine & Machine Co., Montreal.
Gas & Electric Power Co., Toronto
John McDougall Caledonian Iron Works Montreal.
Robb Engineering Co., Amherst, N.S.
Sheldons Limited, Galt.
The Smart-Turner Mach. Co., Hamilton.
Waterous Engine Works Co., Brantford.

Exhaust Heads.

Darling Bros., Ltd., Montreal.
Dominion Heating & Ventilating Co., Hespeler
Sheldons Limited, Galt, Ont.

Expanded Metal.

Expanded Metal and Fireproofing Co. Toronto

Expanders.

A. B. Jardine & Co., Hespeler, Ont.

Fans, Electric.

Canadian General Electric Co., Toronto.
Canadian Westinghouse Co., Hamilton.
Dominion Heating & Ventilating Co. Hespeler
Gas & Electric Power Co., Toronto.
Mechanics Supply Co., Quebec, Que.
Sheldons Limited Galt, Ont.
The Smart-Turner Mach. Co., Hamilton.

Fans, Exhaust.

Canadian Buffalo Forge Co., Montreal.
Detroit Foundry Supply Co., Windsor.
Dominion Foundry Supply Co., Montreal
Dominion Heating & Ventilating Co., Hespeler.
Gas & Electric Power Co., Toronto.
Greey, Wm. & J. G., Toronto
Hamilton Facing Mill Co., Hamilton.
Paxson, J. W., Co., Philadelphia
Sheldons Limited, Galt.

Feed Water Heaters.

Darling Bros., Montreal
Laurie Engine & Machine Co., Montreal
John McDougall, Caledonian Iron Works Co., Montreal.
The Smart-Turner Mach. Co., Hamilton

Files and Rasps.

Frothingham & Workman, Ltd., Montreal
John Millen & Son, Ltd., Montreal.
Rice Lewis & Son, Toronto.
Nicholson File Co., Port Hope
H. W. Petrie, Toronto

Fillet, Pattern.

Buffalo Foundry Supply Co., Buffalo.
Detroit Foundry Supply Co., Windsor.
Dominion Foundry Supply Co., Montreal
Hamilton Facing Mill Co., Hamilton.

Fire Apparatus.

Waterous Engine Works Co., Brantford.

Fire Brick and Clay.

Buffalo Foundry Supply Co., Buffalo.
Detroit Foundry Supply Co., Windsor.
Dominion Foundry Supply Co., Montreal
Gilmour J. New York.
Hamilton Facing Mill Co., Hamilton.
Ontario Lime Association Toronto
J. W. Paxson Co. Philadelphia.
Toronto Pottery Co., Toronto

Flash Lights.

Berlin Electrical Mfg. Co., Toronto.
Mechanics Supply Co., Quebec, Que.

Flour Mill Machinery.

Allis-Chalmers-Bullock, Montreal.
Greey, Wm. & J. G., Toronto
The Goldie & McCulloch Co., Galt, Ont.
John McDougal, Caledonian Iron Works Co., Montreal.

Forges.

Canada Foundry Co., Limited, Toronto.
Frothingham & Workman Ltd., Montreal
Hamilton Facing Mill Co., Hamilton.
Independent Pneumatic Tool Co., Chicago, Ill.
H. W. Petrie, Toronto.
Sheldons Limited, Galt, Ont.

Forgings, Drop.

John McDougall, Caledonian Iron Works Co., Montreal.
H. W. Petrie, Toronto.
St. Clair Bros., Galt
Wilson J. C., & Co, Glenora, Ont.

Forgings, Light & Heavy.

Hamilton Steel & Iron Co., Hamilton

Forging Machinery.

John Bertram & Sons Co., Dundas, Ont.
London Mach. Tool Co., Hamilton, Ont
National Machinery Co., Tiffin, Ohio
Niles-Bement-Pond Co., New York.

Founders.

Greey, Wm. & J. G., Toronto
John McDougall, Caledonian Iron Works Co., Montreal
Niagara Falls Machine & Foundry Co., Niagara Falls, Ont.
Maxwell, David & Sons, St. Marys
The Smart-Turner Mach. Co., Hamilton.
Wilson J. C., & Co., Glenora, Ont.

Foundry Coke.

Baird & West, Detroit

Foundry Equipment.

Ph. Bonvillain & E. Ronceray, Philadelphia
Detroit Foundry Supply Co., Windsor.
Dominion Foundry Supply Co., Montreal
Gilmour, J., New York.
Hamilton Facing Mill Co., Hamilton
Hanna Engineering Works, Chicago.
Northern Engineering Works, Detroit
Paxson, J. W., Co., Philadelphia

Foundry Parting.

Doggett, Stanley, New York
Dominion Foundry Supply Co., Montreal
Partomol Co., New York
Foundry Specialty Co., Cincinnati
Stanley Doggett, New York

Foundry Facings.

Buffalo Foundry Supply Co., Buffalo
Detroit Foundry Supply Co., Windsor.
Doggett Stanley, New York
Dominion Foundry Supply Co., Montreal
Hamilton Facing Mill Co., Hamilton.
J. W. Paxson Co., Philadelphia, Pa.

Friction Clutch Pulleys, etc.

The Goldie & McCulloch Co., Galt.
Greey, Wm. & J. G., Toronto
Link-Belt Co., Philadelphia.

Furnaces.

Detroit Foundry Supply Co., Windsor.
Dominion Foundry Supply Co., Montreal
Hamilton Facing Mill Co., Hamilton.
Paxson, J. W., & Philadelphia Co.

Gas Blowers and Exhausters.

Dominion Heating & Ventilating Co. Hespeler
Sheldons Limited, Galt.

Gas Furnaces.

Chicago Flexible Shaft Co., Chicago

Gas Plants, Suction and Pressure.

Colonial Engineering Co., Montreal
Gas & Electric Power Co., Toronto
Williams & Wilson, Montreal

Gauges, Standard.

Pratt & Whitney Co., Hartford, Conn.

Gearing.

Eberhardt Bros. Machine Co., Newark.
Greey, Wm. & J. G. Toronto
Wilson, J. C., & Co., Glenora, Ont

Gear Cutting Machinery.

Becker-Brainard Milling Mach. Co., Hyde Park, Mass.
Bickford Drill & Tool Co., Cincinnati.
Dwight Slate Machine Co., Hartford
Eberhardt Bros. Machine Co., Newark
Greey, Wm. & J. G., Toronto
Kennedy Wm., & Sons, Owen Sound
London Mach. Tool Co., Hamilton.
Niles-Bement-Pond Co., New York.
H. W. Petrie, Toronto.
Pratt & Whitney Co., Hartford, Conn.
Williams & Wilson, Montreal.
Wilson, J. C., & Co., Glenora, Ont.

Gears, Angle.

Chicago Pneumatic Tool Co., Chicago.
Greey, Wm. & J. G., Toronto
Laurie Engine & Machine Co., Montreal.
John McDougall, Caledonian Iron Works Co., Montreal.
Waterous Engine Co., Brantford.
Wilson, J. C., & Co., Glenora, Ont.

Gears, Cut.

Kennedy, Wm., & Sons, Owen Sound
Wilson, J. C., & Co., Glenora, Ont.

Gears, Iron.

Greey, Wm. & J. G., Toronto
Kennedy, Wm., & Sons, Owen Sound
Wilson, J. C., & Co., Glenora, Ont.

Gears, Mortise.

Greey, Wm. & J. G., Toronto
Kennedy, Wm. & Sons, Owen Sound
Wilson, J. C., & Co., Glenora, Ont.

Gears, Reducing.

Brown, David & Sons, Huddersfield, Eng
Greey, Wm. & J. G., Toronto
John McDougall, Caledonian Iron Works Co., Montreal.
Wilson, J. C., & Co., Glenora, Ont.

Gears, Worm.

Wilson, J. C., & Co., Glenora Ont.

Gears and Pinions.

Wilson, J. C., & Co., Glenora, Ont.

MANUFACTURERS AND MANUFACTURERS' AGENTS.

It will pay you to watch our condensed column each month. There are many money-making propositions brought to your attention here. You may find just what you are looking for.

RATES

One insertion—25c. for 20 words; 1c. a word for each additional word.

Yearly rate - $2.50 for twenty words or less, 10c a word for each additional word.

The above does not apply to notices under the head of "Machinery Wanted." These notices are inserted free for subscribers.

FOR SALE.

BELTING, RUBBER, CANVAS AND LEATHER, Hose Packing, Blacksmith's and Mill Supplies at lowest price. N. Smith, 138 York Street, Toronto. (2tf)

MARINE gasoline engine castings, with blue print and full instructions, etc.; 2½, 4, 6 h.p.; also complete finished outfits at $65 up; catalogue, Krug & Crosby, Hamilton. [10tf]

FOR SALE—Complete outfit of modern hardwood flooring machinery; Bellot and Hoyt matchers, Fay-Egan double planer, Cowan broken roll surfacer, Sherman end matcher and boring machines; also dry kiln outfit, with steel cars; latest style of tools and practically new; no scrap iron. J. S. Finlay, Owen Sound.

FOR SALE—All machinery, tools, motors and equipment contained in machine shop, pattern shop, molding shop and boiler shop of William Hamilton Company, Limited, manufacturers of saw mill machinery, water wheels, mining machinery and contractors' supplies, Peterborough; will be sold in parcels or en bloc. Apply to R. R. Hall, Peterborough, Ont.

HOISTING ENGINES—We have for immediate shipment three 6½ x 8 and three 7 x 10 double cylinder, double drum steam hoists; also six single drum horse power hoisting machines. The Manson Mfg. Co., Limited, Thorold, Ont.

FOR SALE—The Canadian patent rights on a valveless pneumatic hammer, proved by actual test to be superior to any hammer at present on the market; an arrangement can be made to manufacture same on royalty for Canadian market. Apply Box 39C, CANADIAN MACHINERY, Toronto. (12)

SITUATIONS WANTED.

MOLDER—Machinery, first-class man, 15 years' experience in best shops in Detroit and Chicago. R. A. Nash, Sarnia, Ont.

WANTED—Position as purchasing agent or correspondent for manufacturing firm by a young man at present employed in a similar capacity by one of the largest manufacturing concerns in Ontario; has had long experience in these lines, thoroughly understands all branches of office work and can furnish best of references. Address Box 43, CANADIAN MACHINERY, Toronto. [12]

MECHANICAL engineer (25) seeks position; experienced in construction of modern copper, zinc, smelting works, machinery furnaces, etc.; general engineering, gas, steam engines, suction and pressure gas plants; structural steel, etc.; reliable, energetic; good references. Box 27, CANADIAN MACHINERY, 88 Fleet St., London, Eng. [12]

LARGER PREMISES WANTED.

THE VALLEY CITY SEATING CO., LIMITED, is desirous of enlarging their premises, or would consider proposition to purchase furniture or woodworking plant if modern and located in good town. Address all communications to J. D. Pennington, Dundas, Ont.

AGENTS WANTED.

A Firm of Chemical Engineers are requiring first-class agents in the large cities of Canada. Apply to W. J. Fraser & Co., Ltd., Chemical Engineers, 98 Commercial Road East, London, England. [11]

Persons addressing advertisers will kindly mention having seen advertisement in this paper.

ANNIVERSARY NUMBER.

Thirty years have elapsed since the inception of the American Machinist in 1877 and the occasion is celebrated by issuing a 532 page anniversary number. This number is their special quintennial edition and is marked by a survey of the progress of machinery development during the past five years and the amount of advertising carried in this number of the journal, 433 pages being occupied by machine tool advertisers.

To the publisher, Mr. Hill, and his splendid organization, Canadian Machinery extends its hearty congratulations and hopes that the prosperity which attends them, and which they richly deserve, may long continue. It is an issue that is worth having because it contains so much real good matter of interest to every manufacturer. Unfortunately, the asinine policy of some of our Post Office officials shuts out a lot of this class of most valuable matter—information that means dollars and cents to Canadian manufacturers and to Canadian people.

Generators, Electric.
Allis-Chalmers-Bullock, Limited, Montreal
Canadian General Electric Co., Toronto
Canadian Westinghouse Co., Hamilton.
Gas & Electric Power Co., Toronto
Hall Engineering Works, Montreal.
H. W. Petrie, Toronto.
Toronto & Hamilton Electric Co. Hamilton.

Governors, Water Wheel.
Wilson, J. C., & Co., Glenora, Ont.

Graphite Paints.
P. D. Dods & Co., Montreal.

Graphite.
Detroit Foundry Supply Co., Windsor.
Doggett, Stanley, New York
Dominion Foundry Supply Co., Montreal
Hamilton Facing Mill Co., Hamilton.
Mechanics Supply Co., Que ec, Que.
Paxson, J. W., Co, Philadelphia

Grinders, Automatic Knife.
W. H. Banfield & Son, Toronto.

Grinders, Centre.
Niles-Bement-Pond Co., New York.
H. W. Petrie, Toronto.

Grinders, Cutter.
Becker-Brainard Milling Mach. Co., Hyde Park, Mass.
John Millen & Son, Ltd., Montreal.
Pratt & Whitney Co., Hartford, Conn.

Grinders, Tool.
Armstrong Bros. Tool Co., Chicago.
Gisholt Machine Co Madison, Wis.
H. W. Petrie, Toronto.
Williams & Wilson, Montreal.

Grinding Machines.
The Canadian Fairbanks, Montreal.
Independent Pneumatic Tool Co., Chicago, Ill.
Rice Lewis & Son, Toronto.
Niles-Bement-Pond Co., New York.
Paxson, J. W., Co., Philadelphia
H. W. Petrie, Toronto.

Grinding & Polishing Machines
The Canadian Fairbanks Co., Montreal.
Greey, Wm. & J. G., Toronto
Independent Pneumatic Tool Co., Chicago, Ill.
John Millen & Son, Ltd., Montreal.
Niles-Bement-Pond Co., New York.
H. W. Petrie, Toronto.
Standard Bearings, Ltd., Niagara Falls

Grinding Wheels
Carborundum Co., Niagara Falls
Norton Company, Worcester, Mass.

Hack Saws.
Canada Machinery Agency, Montreal.
The Canadian Fairbanks Co., Montreal.
Frothingham & Workman, Ltd. Montreal
John Millen & Son, Ltd., Montreal.
Niles-Bement-Pond Co., New York.
H. W. Petrie, Toronto.
Williams & Wilson, Montreal.

Hammers, Drop.
London Mach. Tool Co., Hamilton, Ont.
Niles-Bement-Pond Co., New York.

Hammers, Steam.
John Bertram & Sons Co. Dundas, Ont.
London Mach. Tool Co., Hamilton, Ont.
Niles-Bement-Pond Co., New York.

Hand Stocks.
Borden-Canadian Co., Toronto

Hangers.
The Goldie & McCulloch Co., Galt.
Greey, Wm. & J. G., Toronto
Kennedy, Wm., & Sons, Owen Sound
Owen Sound Iron Works Co., Owen Sound
The Smart-Turner Mach. Co., Hamilton.
Standard Bearings Ltd., Niagara Falls
Waterous Engine Co., Brantford.
Wilson, J. C., & Co., Glenora, Ont.

Hardware Specialties.
Mechanics Supply Co., Quebec, Que.

Heating Apparatus.
Darling Bros., Ltd., Montreal
Dominion Heating & Ventilating Co., Hespeler
Sheldons Limited, Galt.

Hoisting and Conveying Machinery.
Allis-Chalmers-Bullock, Limited, Montrea
Greey, Wm. & J. G., Toronto
Link-Belt Co., Philadelphia.
Niles-Bement-Pond Co., New York.
Northern Engineering Works, Detroit
The Smart-Turner Mach. Co., Hamilton.
Waterous Engine Co., Brantford.
Wilson, J. C., & Co., Glenora, Ont.

Hoists, Electric.
Northern Engineering Works, Detroit

Hoists, Pneumatic.
Canadian Rand Drill Co., Montreal.
Dominion Foundry Supply Co., Montreal
Hamilton Facing Mill Co., Hamilton.
Northern Engineering Works, Detroit

Hoists, Portable & Stationary.
Canadian Pilling Co., Montreal

Hose, Air.
Canadian Rand Drill Co., Montreal.
Canadian Westinghouse Co., Hamilton.
Independent Pneumatic Tool Co., Chicago
Paxson, J. W., Co., Philadelphia

Hose Couplings.
Canadian Rand Drill Co., Montreal.
Canadian Westinghouse Co., Hamilton.
Mechanics Supply Co., Quebec, Que.
Paxson, J. W., Co., Philadelphia

Hose, Steam.
Allis-Chalmers-Bullock, Montreal.
Canadian Rand Drill Co., Montreal.
Canadian Westinghouse Co., Hamilton.
Independent Pneumatic Tool Co., Chicago, Ill.
Paxson, J. W. Co, Philadelphia

Hydraulic Accumulators.
Ph. Bonvillain & E. Ronceray, Philadelphia
Niles-Bement-Pond Co, New York.
Perrin, Wm. R., Co., Toronto.
The Smart-Turner Mach. Co., Hamilton.

Hydraulic Machinery.
Allis-Chalmers-Bullock, Montreal.
Barber, Chas., & Sons, Meaford
Gas & Electric Power Co., Toronto.
Wilson, J. C., & Co., Glenora, Ont.

India Oil Stones.
Norton Company, Worcester, Mass.

Indicators, Speed.
L. S. Starrett Co., Athol, Mass.

Injectors.
Canada Foundry Co., Toronto.
The Canadian Fairbanks Co., Montreal.
Desmond-Stephan Mfg. Co., Urbana, Ohio
Frothingham & Workman Ltd., Montreal
Mechanics Supply Co. Quebec, Que.
Rice Lewis & Son, Toronto.
Penberthy Injector Co., Windsor, Ont.

Iron and Steel.
Frothingham & Workman, Ltd., Montreal

Iron and Steel Bars and Bands.
Hamilton Steel & Iron Co., Hamilton

Iron Cements.
Buffalo Foundry Supply Co., Buffalo.
Detroit Foundry Supply Co., Windsor.
Dominion Foundry Supply Co. Montreal
Hamilton Facing Mill Co., Hamilton.
Paxson, J. W., Co., Philadelphia

Iron Filler.
Buffalo Foundry Supply Co., Buffalo.
Detroit Foundry Supply Co., Windsor.
Dominion Foundry Supply Co. Montreal
Hamilton Facing Mill Co., Hamilton.
Paxson, J. W., Co., Philadelphia.

Jacks.
Frothingham & Workman, Ltd., Montreal
Norton, A. O., Coaticook, Que.

Kegs, Steel Shop.
Cleveland Wire Spring Co., Cleveland

Ladles, Foundry.
Dominion Foundry Supply Co., Montreal
Frothingham & Workman Ltd., Montreal
Northern Engineering Works, Detroit

Lamps, Arc and Incandescent.
Canadian General Electric Co., Toronto.
Canadian Westinghouse Co., Hamilton.
Gas & Electric Power Co., Toronto
The Packard Electric Co., St. Catharines.

Lathe Dogs.
Armstrong Bros., Chicago
Pratt & Whitney Co., Hartford, Conn.

Lathes, Engine.
American Tool Work Co., Cincinnati.
John Bertram & Sons Co., Dundas, Ont.
Canada Machinery Agency, Montreal.
The Canadian Fairbanks Co., Montreal.
London Mach. Tool Co., Hamilton, Ont.
Niles-Bement-Pond Co., New York.
H. W. Petrie, Toronto.
Pratt & Whitney Co., Hartford, Conn.

Lathes, Foot-Power.
American Tool Works Co., Cincinnati.

Lathes, Screw Cutting.
Niles-Bement-Pond Co., New York.

Lathes, Automatic, Screw-Threading.
John Bertram & Sons Co., Dundas, Ont
London Mach. Tool Co., Hamilton, Ont
Pratt & Whitney Co., Hartford, Conn.

Lathes, Bench.
London Mach. Tool Co., London, Ont.
Pratt & Whitney Co., Hartford, Conn.

Lathes, Turret.
American Tool Works Co., Cincinnati.
John Bertram & Sons Co., Dundas, Ont.
Gisholt Machine Co., Madison Wis.
London Mach. Tool Co., Hamilton, Ont.
Niles-Bement-Pond Co., New York.
The Pratt & Whitney Co., Hartford, Conn
Warner & Swasey Co., Cleveland, O.

Leather Belting.
Canada Machinery Agency, Montreal.
The Canadian Fairbanks Co., Montreal
Greey, Wm. & J. G., Toronto

Locomotives, Air.
Canadian Rand Drill Co., Montreal.

Locomotives, Electrical.
Canadian Westinghouse Co., Hamilton
Gas & Electric Power Co., Toronto
Jeffrey Mfg. Co., Columbus, Ohio.

Locomotives, Steam.
Canada Foundry Co., Toronto.
Canadian Rand Drill Co., Montreal.

Locomotive Turntable Tractors.
Canadian Pilling Co., Montreal

Lubricating Plumbago.
Detroit Foundry Supply Co., Detroit.
Dominion Foundry Supply Co., Montreal
Hamilton Facing Mill Co., Hamilton.
Paxson, J. W, Co., Philadelphia

Lubricators, Force Feed.
Sight Feed Oil Pump Co., Milwaukee, Wis

Lumber Dry Kilns.
Dominion Heating & Ventilating Co., Hespeler.
H. W. Petrie, Toronto.
Sheldons Limited, Galt, Ont.

Machinery Dealers.
Canada Machinery Agency, Montreal.
The Canadian Fairbanks Co., Montreal.
H. W. Petrie, Toronto.
The Smart-Turner Mach. Co., Hamilton.
Williams & Wilson Montreal.

Machinery Designers.
Greey, Wm. & J. G., Toronto
Standard Bearings, Ltd., Niagara Falls.
Wilson, J. C., & Co., Glenora, Ont.

Machinists.
W. H. Banfield & Sons, Toronto.
Greey, Wm. & J. G. Toronto
Hall Engineering Works Montreal.
Link-Belt Co., Philadelphia.
John McDougall, Caledonian Iron Works Co., Montreal.
Paxson, J. W., Co, Philadelphia
Robb Engineering Co., Amherst, N.S.
The Smart-Turner Mach. Co., Hamilton.
Standard Bearings, Ltd. Niagara Falls
Waterous Engine Co., Brantford.
Wilson, J. C., & Co., Glenora, Ont.

Machinists' Small Tools.
Armstrong Bros., Chicago.
Butterfield & Co., Rock Island, Que.
Frothingham & Workman, Ltd., Montrea
Rice Lewis & Son, Montreal.
Pratt & Whitney Co., Hartford, Conn.
Standard Tool Co., Cleveland.
L. S. Starrett Co., Athol, Mass.
Williams & Wilson, Montreal.

Malleable Flask Clamps.
Buffalo Foundry Supply Co, Buffalo.
Dominion Foundry Supply Co, Montreal
Paxson, J. W, Co., Philadelphia

Malleable Iron Castings.
Galt Malleable Iron Co., Galt

Mallet, Rawhide and Wood.
Buffalo Foundry Supply Co., Buffalo.
Detroit Foundry Supply Co., Detroit.
Dominion Foundry Supply Co., Montreal
Paxson, J. W., Co, Philadelphia

Mandrels.
A. B. Jardine & Co., Hespeler, Ont.
The Pratt & Whitney Co., Hartford, Conn.
Standard Tool Co., Cleveland.

Maple Cogs, Blank Face.
Wilson, J. C., & Co., Glenora, Ont.

Maple Cogs, Machine Dressed.
Wilson, J. C., & Co., Glenora, Ont.

Marking Machines.
Dwight Slate Machine Co, Hartford.

Metallic Paints.
P. D. Dods & Co., Montreal.

Meters, Electrical.
Canadian Westinghouse Co., Hamilton.
Gas & Electric Power Co., Toronto.

Mill Machinery.
Greey Wm. & J. G., Toronto
The Goldie & McCulloch Co., Galt, Ont.
John McDougall Caledonian Iron Works Co., Montreal.
H. W. Petrie, Toronto.
Robb Engineering Co., Amherst, N.S.
Waterous Engine Co., Brantford.
Williams & Wilson, Montreal.
Wilson, J. C. & Co. Glenora, Ont.

Milling Attachments.
Becker-Brainard Milling Machine Co. Hyde Park, Mass.
John Bertram & Sons Co., Dundas, Ont.
Cincinnati Milling Machine Co., Cincinnati.
Niles-Bement-Pond Co., New York,
Pratt & Whitney, Hartford, Conn.

Milling Machines, Horizontal.
Becker-Brainard Milling Machinery Co. Hyde Park, Mass.
John Bertram & Sons Co., Dundas, Ont.
London Mach. Tool Co., Hamilton, Ont.
Niles-Bement-Pond Co., New York.
Pratt & Whitney, Hartford, Conn.

Milling Machines, Motor Driven.
Cincinnati Milling Machine Co., Cincinnati

Milling Machines, Plain.
American Tool Works Co., Cincinnati.
Becker-Brainard Milling Machine Co. Hyde Park, Mass.
John Bertram & Sons Co., Dundas, Ont.
Canada Machinery Agency, Montreal.
The Canadian Fairbanks Co., Montreal.
Cincinnati Milling Machine Co., Cincinnati
London Mach. Tool Co., Hamilton, Ont.
Niles-Bement-Pond Co., New York.
H. W. Petrie, Toronto.
Pratt & Whitney Co., Hartford, Conn.
Williams & Wilson, Montreal.

Milling Machines, Universal.
American Tool Works Co., Cincinnati.
Becker-Brainard Milling Machine Co., Hyde Park, Mass.
John Bertram & Sons Co., Dundas, Ont.
Canada Machinery Agency, Montreal.
The Canadian Fairbanks Co., Montreal.
Cincinnati Milling Machine Co., Cincinnati
London Mach. Tool Co., Hamilton, Ont
Niles-Bement-Pond Co., New York.
H. W. Petrie, Toronto.
Williams & Wilson, Montreal.

Milling Machines, Vertical.
Becker-Brainard Milling Machine Co., Hyde Park, Mass.
Brown & Sharpe, Providence, R.I.
John Bertram & Sons Co., Dundas, Ont.
Canada Machinery Agency, Montreal.
London Mach. Tool Co., Hamilton, Ont.
Niles-Bement-Pond Co., New York.

Milling Tools.
Wm. Abbott, Montreal.
Becker-Brainard Milling Machine Co., Hyde Park, Mass.
Geometric Tool Co., New Haven, Conn.
Hamilton Tool Co., Hamilton, Ont.
London Mach. Tool Co., Hamilton, Ont.
Pratt & Whitney Co., Hartford, Conn.
Standard Tool Co., Cleveland.
Standard Bearings, Ltd., Niagara Falls

Mine Cars and Hitchings.
Cumming, J. W., New Glasgow, N.S.

Miners' Copper Needles and Stemmers.
Cumming, J. W., New Glasgow, N.S.

Mining Machinery.
Allis-Chalmers-Bullock, Limited, Montreal
Canadian Rand Drill Co., Montreal.
Gas & Electric Power Co., Toronto.
Jeffrey Mfg. Co., Columbus, Ohio.
Laurie Engine & Machine Co., Montreal.
Rice Lewis & Son, Toronto.
John McDougall, Caledonian Iron Works Co., Montreal.
T. & H. Electric Co. Hamilton.

T. Pringle & Son

HYDRAULIC, MILL & ELECTRICAL ENGINEERS

FACTORY & MILL CONSTRUCTION A SPECIALTY.

Coristine Bldg., St. Nicholas St., Montreal.

RODERICK J. PARKE

A.M. Can. Soc. C.E. A.M. Amer. Inst. E.E.

CONSULTING ELECTRICAL ENGINEER

INDUSTRIAL STEAM AND ELECTRIC POWER PLANTS DESIGNED. TESTS. REPORTS.

51-53 JANES BLDG., TORONTO, CAN.
Long Distance Telephones—Office and Residence.

J. A. DeCew

Chemical Engineer

Sun Life Building, MONTREAL

Industrial Plants and Processes sold and installed. Apparatus and Materials for Water Purification. Free Analyses to prospective purchasers.

JULES DE CLERCY, M. E. Gas Engineer

Expert knowledge of all classes of Engines and Producer Plants. If you are thinking of installing a Gas Engine or Producer Plant write me. If your plant is not working satisfactorily I can help you.

413 Dorchester St. - MONTREAL, QUE.

HANBURY A. BUDDEN

Advocate Patent Agent.
New York Life Building MONTREAL.
Cable Address, BREVET MONTREAL.

PATENTS *TRADE MARKS AND DESIGNS*

PROCURED IN ALL COUNTRIES

Special Attention given to Patent Litigation

Pamphlet sent free on application.

RIDOUT & MAYBEE *103 BAY STREET TORONTO*

PATENTS PROMPTLY SECURED

We solicit the business of Manufacturers, Engineers and others who realize the advisability of having their Patent business transacted by Experts. Preliminary advice free. Charges moderate. **Our Inventor's Adviser** sent upon request. Marion & Marion, New York Life Bldg, Montreal; and Washington, D.C., U.S.A.

PATENTS THAT PROTECT

FETHERSTONHAUGH & CO.

Patent Solicitors & Experts

Fred. B. Fetherstonhaugh, M.E., Barrister-at-law and Counsel and expert in Patent Causes. Charles W. Taylor, B.Sc., formerly Examiner in Can. Patent Office.

MONTREAL CAN. LIFE BLDG.
TORONTO HEAD OFFICE, CAN. BANK COMMERCE BLDG.

T. A. SOMERVILLE

CONSULTING AND SUPERVISING ENGINEER

Reports and Valuations, Factories Rearranged, Manufacturing Plants Designed and Erected Equipments Installed

Federal Life Bldg., HAMILTON, ONT.

THE DETROIT TESTING LABORATORY

1111 Union Trust Building
DETROIT, MICH.

Experienced Foundry Chemists and Metallurgists. Iron Mixtures for all classes of Castings our Specialty. Reasonable charges for Analyses.

CONSULTING ENGINEERS

should have their card in this page. It will be read by the manufacturers of Canada.

Canadian Machinery

Montreal Toronto Winnipeg

THE CANADIAN TURBINE WATER WHEEL

We have not expanded the discharge of our turbines, as we find the results given by our combined impact and reaction runner to be vastly superior to those obtained from turbines of the expanded discharge type. Do not sacrifice your water power. We can save you 20 to 30 per cent. of your water and at the same time give you a much more satisfactory drive.

CHAS. BARBER & SONS, Meaford, Ont.

Turbine Water Wheels, Governors and Gearing only

A Labor Saving Combination

Couch and forming rolls with special attachments for economical manufacture of pipes, complete machines made to meet your requirements (patent 94746), estimates furnished upon application.

Stuart B. Playfair

Traders Bank Building

TORONTO, - CANADA

Improved "LITTLE GIANT" TURBINE WATER WHEEL

Complete in self-contained case. No expensive flume required. 18 sizes carried in stock, prompt shipment insured. Highest efficiency at full or partial gate opening.

Construction, acme of simplicity.

Unequalled durability.

Write for descriptive catalog.

J. C. Wilson & Co.

GLENORA, ONT.

Builders of Power Transmission Machinery

Machine-cut Wood and Iron Tooth Gearing our specialty.

Mixing Machines, Dough.

Greey, Wm. & J. G., Toronto
Paxson, J. W., Co., Philadelphia

Mixing Machines, Special.

Greey, Wm. & J. G., Toronto
Paxson, J. W., Co., Philadelphia

Model Tools.

Globe Machine & Stamping Co., Cleveland, Ohio.
Standard Bearings, Ltd., Niagara Falls.
Wells Pattern and Model Works, Toronto

Motors, Electric.

Allis-Chalmers-Bullock Limited, Montreal
Canadian General Electric Co., Toronto
Canadian Westinghouse Co., Hamilton.
Consolidated Electric Co., Toronto
Electrical Machinery Co., Toronto.
Gas & Electric Power Co., Toronto
Hall Engineering Works, Montreal.
The Packard Electric Co., St. Catharines.
T. & H. Electric Co., Hamilton.

Motors, Air.

Canadian Rand Drill Co., Montreal.

Molders' Supplies.

Buffalo Foundry Supply Co., Buffalo.
Detroit Foundry Supply Co., Windsor
Dominion Foundry Supply Co., Montreal
Hamilton Facing Mill Co., Hamilton
Paxson J. W., Co., Philadelphia

Molders' Tools.

Buffalo Foundry Supply Co., Buffalo.
Detroit Foundry Supply Co., Windsor
Dominion Foundry Supply Co., Montreal
Hamilton Facing Mill Co., Hamilton.
Paxson, J. W., Co., Philadelphia

Molding Machines.

P. Bonvillain & E. Ronceray, Philadelphia
Buffalo Foundry Supply Co., Buffalo.
Dominion Foundry Supply Co., Montreal
Hamilton Facing Mill Co., Hamilton.
J. W. Paxson Co., Philadelphia, Pa.

Molding Sand.

T. W. Barnes, Hamilton.
Buffalo Foundry Supply Co., Buffalo.
Detroit Foundry Supply Co., Windsor
Dominion Foundry Supply Co., Montreal
Hamilton Facing Mill Co., Hamilton.
Paxson, J. W., Co., Philadelphia

Nut Tappers.

John Bertram & Sons Co., Dundas, Ont
A. B. Jardine & Co., Hespeler.
London Mach. Tool Co., Hamilton.
National Machinery Co., Tiffin, Ohio.

Nuts.

Canada Nut Co., Toronto

Oatmeal Mill Machinery.

Greey, Wm. & J. G., Toronto
The Goldie & McCulloch Co., Galt

Oilers, Gang.

Sight Feed Oil Pump Co., Milwaukee, Wis.

Oils, Core.

Buffalo Foundry Supply Co., Buffalo.
Dominion Foundry Supply Co., Montreal
Hamilton Facing Mill Co., Hamilton.
Paxson, J. W., Co., Philadelphia

Oil Extractors.

Darling Bros., Ltd., Montreal

Oil Stones.

Carborundum Co., Niagara Falls N.Y.

Packing Metallic.

Canfield Mfg. Co., Philadelphia

Paint Mill Machinery.

Greey, Wm. & J. G., Toronto

Pans, Lathe.

Cleveland Wire Spring Co., Cleveland
Paxson, J. W., Co., Philadelphia

Pans, Steel Shop.

Cleveland Wire Spring Co., Cleveland

Parting Compound.

Doggett, Stanley, New York
Dominion Foundry Supply Co., Montreal
Partamol Co., New York

Patent Solicitors.

Hanbury A. Budden, Montreal.
Fetherstonhaugh & Blackmore, Montreal
Marion & Marion, Montreal.
Ridout & Maybee, Toronto.

Patterns.

John Carr, Hamilton.
Galt Malleable Iron Co., Galt
Hamilton Pattern Works, Hamilton.
John McDougall, Caledonian Iron Works Co., Montreal.
Wells' Pattern and Model Works, Toronto.

Piano Plates.

Crowe's Iron Works, Guelph, Ont.

Pig Iron.

Hamilton Steel & Iron Co., Hamilton

Pipe Cutting and Threading Machines.

Borden-Canadian Co., Toronto
Butterfield & Co., Rock Island, Que.
Canada Machinery Agency, Montreal.
Curtis & Curtis Co., Bridgeport, Conn.
Frothingham & Workman, Ltd., Montreal
Hart Mfg. Co., Cleveland
A. B. Jardine & Co., Hespeler, Ont.
London Mach. Tool Co., Hamilton, Ont.
Niles-Bement-Pond Co., New York.
Shantz, I. E., & Co., Berlin, Ont.

Pipe, Municipal.

Canadian Pipe Co., Vancouver, B.C.
Pacific Coast Pipe Co., Vancouver, B.C.

Pipe, Waterworks.

Canadian Pipe Co., Vancouver, B.C.

Planers, Standard.

American Tool Works, Cincinnati.
Bateman's Machine Tool Co., Leeds, Eng.
John Bertram & Sons Co., Dundas, Ont.
Canada Machinery Agency, Montreal.
The Canadian Fairbanks Co., Montreal.
Rice Lewis & Son, Toronto.
London Mach. Tool Co., Hamilton Ont.
Niles-Bement-Pond Co., New York.
H. W. Petrie, Toronto.
Pratt & Whitney Co., Hartford, Conn.
Williams & Wilson, Montreal.

Planers, Rotary.

John Bertram & Sons Co., Dundas, Ont.
London Mach. Tool Co., Hamilton, Ont.
Niles-Bement-Pond Co., New York.

Planing Mill Fans.

Dominion Heating & Ventilating Co., Hespeler
Sheldons Limited, Galt, Ont.

Plumbago.

Buffalo Foundry Supply Co., Buffalo.
Detroit Foundry Supply Co., Windsor
Doggett, Stanley, New York
Dominion Foundry Supply Co., Montreal
Hamilton Facing Mill Co., Hamilton,
J. W. Paxson Co., Philadelphia, Pa.

Pneumatic Tools.

Allis-Chalmers-Bullock, Montreal.
Canadian Rand Drill Co., Montreal.
Hamilton Facing Mill Co., Hamilton.
Hanna Engineering Works, Chicago.
Independent Pneumatic Tool Co. Chicago, New York
Paxson, J. W., Co., Philadelphia

Power Hack Saw Machines.

Frothingham & Workman, Ltd., Montreal

Power Plants.

Gas & Electric Power Co., Toronto
John McDougall Caledonian Iron Works Co., Montreal.
The Smart-Turner Mach. Co., Hamilton

Power Plant Equipments.

Darling Bros., Ltd., Montreal
Gas & Electric Power Co., Toronto

Presses, Drop.

W. H. Banfield & Son, Toronto.
E. W. Bliss Co., Brooklyn, N.Y.
Brown, Boggs Co., Hamilton
Canada Machinery Agency, Montreal.
Ferracute Machine Co., Bridgeton, N.J.
Laurie Engine & Machine Co., Montreal
Niles-Bement-Pond Co., New York

Presses, Hand.

E. W. Bliss Co., Brooklyn, N.Y.
Brown Boggs Co., Hamilton
Ferracute Machine Co., Bridgeton, N.J.

Presses, Hydraulic.

John Bertram & Sons Co., Dundas, Ont.
Laurie Engine & Machine Co., Montreal.
London Mach. Tool Co., Hamilton, Ont.
John McDougall Caledonian Iron Works Co., Montreal.
Niles-Bement-Pond Co., New York.
Perrin, Wm. R., & Co., Toronto

Presses, Power.

E. W. Bliss Co., Brooklyn, N.Y.
Brown, Boggs Co., Hamilton
Canada Machinery Agency, Montreal.
Ferracute Machine Co., Bridgeton, N.J.
Laurie Engine & Machine Co., Montreal
London Mach. Tool Co., Hamilton, Ont.
John McDougall Caledonian Iron Works Co., Montreal.
Niles-Bement-Pond Co., New York.

Presses Power Screw.

Brown, Boggs Co., Hamilton
Ferracute Machine Co., Bridgeton, N.J.
Perrin, Wm. R., & Co., Toronto

Pressure Regulators.

Darling Bros., Ltd., Montreal

Producer Plants.

Canada Foundry Co., Toronto

Pulp Mill Machinery.

Greey, Wm. & J. G., Toronto
Jeffrey Mfg. Co., Columbus, Ohio.

Laurie Engine & Machine Co., Montreal.
John McDougall Caledonian Iron Works Co., Montreal.
Waterous Engine Works Co., Brantford

Pulleys.

Canada Machinery Agency, Montreal.
The Canadian Fairbanks Co., Montreal.
The Goldie & McCulloch Co., Galt.
Greey, Wm. & J. G., Toronto
Laurie Engine & Machine Co., Montreal.
Link-Belt Co., Philadelphia.
John McDougall Caledonian Iron Works Co., Montreal.
Owen Sound Iron Works Co., Owen Sound
H. W. Petrie, Toronto.
The Smart-Turner Mach. Co., Hamilton.
Standard Bearings, Ltd., Niagara Falls,
Waterous Engine Co., Brantford.
Williams & Wilson, Montreal.
Wilson, J. C., & Co., Glenora, Ont.

Pumps.

Laurie Engine & Machine Co., Montreal
Ontario Wind Engine & Pump Co., Toronto

Pump Governors.

Darling Bros., Ltd. Montreal

Pumps, Hydraulic.

Ph. Bonvillain & E. Ronceray, Philadelphia
Laurie Engine & Machine Co., Montreal
Perrin, Wm. R. & Co., Toronto

Pumps, Oil.

Sight Feed Oil Pump Co., Milwaukee, Wis.

Pumps, Steam.

Allis-Chalmers-Bullock, Limited, Montreal
Canada Foundry Co., Toronto.
Canada Machinery Agency, Montreal.
Darling Bros., Ltd., Montreal
Gas & Electric Power Co., Toronto
The Goldie & McCulloch Co., Galt.
John McDougall Caledonian Iron Works, Montreal.
H. W. Petrie, Toronto.
The Smart-Turner Mach. Co., Hamilton.
Standard Bearings, Ltd., Niagara Falls,
Waterous Engine Co. Brantford.

Pumping Machinery.

Canada Foundry Co., Limited, Toronto
Canada Machinery Agency, Montreal.
Canadian Rand Drill Co., Montreal.
Darling Bros., Ltd., Montreal
Gas & Electric Power Co., Toronto.
Hall Engineering Works, Montreal, Que.
Laurie Engine & Machine Co., Montreal.
London Mach. Tool Co., Hamilton, Ont.
John McDougall Caledonian Iron Works Co., Montreal.
The Smart-Turner Mach. Co., Hamilton
Standard Bearings, Ltd., Niagara Falls,

Punches and Dies.

W. H. Banfield & Sons, Toronto.
Bliss, E. W., Co., Brooklyn, N.Y.
Butterfield & Co., Rock Island.
Ferracute Machine Co., Bridgeton, N.J.
Globe Machine & Stamping Co.
A. B. Jardine & Co., Hespeler, Ont.
London Mach. Tool Co., Hamilton, Ont.
Pratt & Whitney Co., Hartford, Conn.
H. W. Petrie, Toronto.
Standard Bearings, Ltd., Niagara Falls,

Punches, Hand.

Mechanics Supply Co., Quebec, Que.

Punches, Power.

John Bertram & Sons Co., Dundas, Ont.
E. W. Bliss Co., Brooklyn, N.Y.
Canada Machinery Agency, Montreal.
Ferracute Machine Co., Bridgeton, N.J.
London Mach. Tool Co., Hamilton, Ont.
Niles-Bement-Pond Co., New York.

Punches, Turret.

London Mach. Tool Co., London, Ont

Punching Machines, Horizontal.

John Bertram & Sons Co., Dundas, Ont.
London Mach. Tool Co., Hamilton, Ont.
Niles-Bement-Pond Co., New York.

Quartering Machines.

John Bertram & Sons Co., Dundas, Ont.
London Mach. Tool Co., Hamilton, Ont.

Railway Spikes and Washers.

Hamilton Steel & Iron Co., Hamilton

Rammers, Bench and Floor.

Buffalo Foundry Supply Co., Buffalo.
Detroit Foundry Supply Co., Windsor
Hamilton Facing Mill Co., Hamilton.
Paxson, J. W., Co., Philadelphia

Rapping Plates.

Detroit Foundry Supply Co., Windsor
Hamilton Facing Mill Co., Hamilton.
Paxson, J. W., Co., Philadelphia

Raw Hide Pinions.

Brown, David & Sons, Huddersfield, Eng.

Reamers.

Wm. Abbott, Montreal.
Butterfield & Co., Rock Island.
Frothingham & Workman Ltd., Montreal
Hamilton Tool Co., Hamilton
Hanna Engineering Works, Chicago
A. B. Jardine & Co., Hespeler, Ont.
Morse Twist Drill and Machine Co., New Bedford, Mass.
Pratt & Whitney Co., Hartford, Conn.
Standard Tool Co., Cleveland.

Reamers, Steel Taper.

Butterfield & Co., Rock Island.
A. B. Jardine & Co., Hespeler, Ont.
John Millen & Son Ltd., Montreal.
Pratt & Whitney Co., Hartford, Conn.
Standard Tool Co., Cleveland.

Rheostats.

Canadian General Electric Co., Toronto.
Canadian Westinghouse Co., Hamilton.
Gas & Electric Power Co., Toronto.
Hall Engineering Works, Montreal, Que.
T. & H. Electric Co., Hamilton,

Riddles.

Buffalo Foundry Supply Co., Buffalo.
Detroit Foundry Supply Co., Windsor
Hamilton Facing Mill Co., Hamilton.
Dominion Foundry Supply Co., Montreal
J. W. Paxson Co., Philadelphia, Pa.

Riveters, Pneumatic.

Hanna Engineering Works, Chicago.
Independent Pneumatic Tool Co., Chicago, Ill.

Rolls, Bending.

John Bertram & Sons Co., Dundas, Ont.
London Mach. Tool Co., Hamilton, Ont.
Niles-Bement-Pond Co., New York.

Rolls, Chilled Iron.

Greey, Wm. & J. G., Toronto

Rolls, Sand Cast.

Greey, Wm. & J. G., Toronto

Rotary Converters.

Allis-Chalmers-Bullock, Ltd., Montreal.
Canadian Westinghouse Co., Hamilton
Paxson, J. W., Co. Philadelphia
Toronto and Hamilton Electric Co., Hamilton.

Rubbing and Sharpening Stones.

Norton Co., Worcester, Mass.

Safes.

The Goldie & McCulloch Co., Galt,

Sand Blast Machinery.

Canadian Rand Drill Co., Montreal.
Detroit Foundry Supply Co., Windsor
Dominion Foundry Supply Co., Montreal
Hamilton Facing Mill Co., Hamilton.
J. W. Paxson Co., Philadelphia, Pa.

Sand, Heavy, Grey Iron.

Buffalo Foundry Supply Co., Buffalo.
Detroit Foundry Supply Co., Windsor
Dominion Foundry Supply Co., Montreal
Hamilton Facing Mill Co., Hamilton.
Paxson, J. W., Co., Philadelphia

Sand, Malleable.

Buffalo Foundry Supply Co., Buffalo.
Detroit Foundry Supply Co., Windsor
Dominion Foundry Supply Co., Montreal
Hamilton Facing Mill Co., Hamilton.
Paxson, J. W., Co., Philadelphia

Sand, Medium Grey Iron.

Buffalo Foundry Supply Co., Buffalo.
Detroit Foundry Supply Co., Windsor
Dominion Foundry Supply Co., Montreal
Hamilton Facing Mill Co., Hamilton.
Paxson, J. W., Co., Philadelphia

Sand Sifters.

Buffalo Foundry Supply Co., Buffalo.
Detroit Foundry Supply Co., Windsor
Dominion Foundry Supply Co., Montreal
Hamilton Facing Mill Co., Hamilton.
Hanna Engineering Works, Chicago.
Paxson, J. W., Co., Philadelphia

Saw Mill Machinery.

Allis-Chalmers-Bullock, Limited, Montreal
Canada Machinery Agency, Montreal.
Goldie & McCulloch Co., Galt.
Greey, Wm. & J. G., Toronto
Owen Sound Iron Works Co., Owen Sound
H. W. Petrie, Toronto.
Robb Engineering Co., Amherst, Ont.
Standard Bearings, Ltd., Niagara Falls
Waterous Engine Works, Brantford
Williams & Wilson, Montreal.
Wilson, J. C., & Co., Glenora, Ont.

Sawing Machines, Metal.

Niles-Bement-Pond Co. New York.
Paxson, J. W., Co., Philadelphia

Saws, Hack.

Canada Machinery Agency, Montreal.
Detroit Foundry Supply Co., Windsor
Frothingham & Workman, Ltd., Montreal
London Mach. Tool Co., Hamilton.
Rice Lewis & Son, Toronto.
John Millen & Son, Ltd., Montreal.
L. S. Starrett Co., Athol, Mass.

Screw Cutting Tools

Hart Manufacturing Co., Cleveland, Ohio

Screw Machines, Automatic.

Canada Machinery Agency, Montreal.
Cleveland Automatic Machine Co., Cleveland, Ohio
London Mach. Tool Co.; Hamilton, Ont.
National-Acme Mfg. Co., Cleveland.
Pratt & Whitney Co., Hartford, Conn.

Screw Machines, Hand.

Canada Machinery Agency, Montreal.
A. B. Jardine & Co., Hespeler.
London Mach. Tool Co., Hamilton, Ont.
Pratt & Whitney Co., Hartford, Conn.
Warner & Swasey Co., Cleveland, O.

Screw Machines, Multiple Spindle.

National-Acme Mfg. Co., Cleveland

Screw Plates.

Butterfield & Co., Rock Island, Que.
Frothingham & Workman, Ltd., Montreal
Hart Manufacturing Co. Cleveland, Ohio
A B. Jardine & Co., Hespeler

Screws, Cap and Set.

National-Acme Mfg. Co., Cleveland

Screw Slotting Machinery, Semi-Automatic.

National-Acme M'g. Co., Cleveland

Second-Hand Machinery.

American Tool Works Co., Cincinnati.
Canada Machinery Agency, Montreal.
The Canadian Fairbanks Co., Montreal.
Goldie & McCulloch Co., Galt.
Machinery Exchange, Montreal.
Niles-Bement-Pond Co., New York.
H. W. Petrie, Toronto.
Robb Engineering Co., Amherst, N S.
Waterous Engine Co., Brantford.
Williams & Wilson, Montreal.

Shafting.

Canada Machinery Agency, Montreal.
The Canadian Fairbanks Co., Montreal.
Frothingham & Workman, Ltd., Montreal
The Goldie & McCulloch Co., Galt, Ont
Greey, Wm., & J. G. Toronto
Kennedy, Wm., & Sons, Owen Sound
Niles-Bement-Pond Co., New York.
Owen Sound Iron Works Co., Owen Sound
H. W. Petrie, Toronto.
Smart-Turner Machine Co., Hamilton.
Union Drawn Steel Co., Hamilton.
Waterous Engine Co., Brantford.
Wilson, J. C., & Co., Glenora, Ont.

Shapers.

American Tool Works Co., Cincinnati.
John Bertram & Sons Co., Dundas, Ont
Canada Machinery Agency, Montreal.
The Canadian Fairbanks Co., Montreal.
Rice Lewis & Son, Toronto.
London Mach. Tool Co., Hamilton, Ont.
Niles-Bement-Pond Co., New York.
H. W. Petrie, Toronto.
Potter & Johnston Machine Co., Pawtucket, R.I.
Pratt & Whitney Co., Hartford, Conn.
Williams & Wilson, Montreal.

Sharpening Stones.

Carborundum Co., Niagara Falls, N.Y.

Shearing Machine, Bar.

John Bertram & Sons Co., Dundas, Ont.
A. B. Jardine & Co., Hespeler.
London Mach. Tool Co., Hamilton, Ont.
Niles-Bement-Pond Co., New York.

Shears, Power.

John Bertram & Sons Co., Dundas, Ont.
Canada Machinery Agency, Montreal.
A. B. Jardine & Co., Hespeler, Ont.
Niles-Bement-Pond Co., New York.
Paxson, J. W., Co., Philadelphia

Sheet Metal.

Ferracute Machine Co., Bridgeton, N.J.

Sheet Metal Goods

Globe Machine & Stamping Co., Cleveland, Ohio.

Sheet Metal Working Tools.

Brown, Boggs Co., Hamilton
Ferracute Machine Co., Bridgeton, N.J.

Sheet Steel Work.

Owen Sound Iron Works Co., Owen Sound

Shingle Mill Machinery.

Owen Sound Iron Works Co., Owen Sound

Shop Trucks.

Greey, Wm. & J. G., Toronto
Owen Sound Iron Works Co., Owen Sound
Paxson, J. W., Co., Philadelphia

Shovels.

Buffalo Foundry Supply Co., Buffalo.
Detroit Foundry Supply Co., Detroit.
Doggett, Stanley, New York
Frothingham & Workman, Ltd., Montreal
Dominion Foundry Supply Co., Montreal
Hamilton Facing Mill Co., Hamilton.
Paxson, J. W., Co., Philadelphia

Sieves.

Buffalo Foundry Supply Co. Buffalo.
Detroit Foundry Supply Co., Windsor
Dominion Foundry Supply Co., Montreal
Hamilton Facing Mill Co., Hamilton.
Paxson, J. W., Co.. Philadelphia

Silver Lead.

Buffalo Foundry Supply Co., Buffalo.
Detroit Foundry Supply Co., Windsor
Doggett, Stanley, New York
Dominion Foundry Supply Co., Montreal
Hamilton Facing Mill Co., Hamilton.
Paxson, J. W., Co., Philadelphia

Snap Flasks

Buffalo Foundry Supply Co., Buffalo.
Detroit Foundry Supply Co., Windsor
Dominion Foundry Supply Co., Montreal
Hamilton Facing Mill Co., Hamilton.
Paxson, J. W., Co., Philadelphia

Soap Stones & Talc Crayons & Pencils.

Doggett, Stanley, New York

Soapstone.

Buffalo Foundry Supply Co., Buffalo.
Detroit Foundry Supply Co., Windsor
Doggett, Stanley, New York
Dominion Foundry Supply Co., Montreal
Hamilton Facing Mill Co., Hamilton.
Paxson, J. W. Co., Philadelphia

Soap Stone Facings.

Doggett, Stanley, New York

Solders.

Lumen Bearing Co., Toronto

Special Machinery.

W. H. Banfield & Sons, Toronto.
John Bertram & Sons Co. Dundas, Ont.
Brown, Boggs Co., Hamilton
Globe Machine & Stamping Co., Cleveland, Ohio.
Greey, Wm. & J. G., Toronto
Hanna Engineering Works, Chicago.
Laurie Engine & Machine Co., Montreal.
London Mach. Tool Co., Hamilton, Ont.
H. W. Petrie, Toronto.
The Smart-Turner Mach. Co., Hamilton
Standard Bearings, Ltd., Niagara Falls.
Waterous Engine Co., Brantford.
Wilson, J. C., & Co., Glenora, Ont.

Special Machines and Tools.

Brown, Boggs Co., Hamilton
Paxson, J. W., Philadelphia
Pratt & Whitney, Hartford, Conn.
Standard Bearings, Ltd., Niagara Falls,

Special Milled Work.

National-Acme M'g. Co., Cleveland

THE UNIVERSAL SYSTEM OF MACHINE MOLDING

Speed Changing Countershafts.

The Canadian Fairbanks Co., Montreal

Spike Machines.

National Machinery Co., Tiffin, O.
The Smart-Turner Mach. Co., Hamilton

Spray Cans.

Detroit Foundry Supply Co., Windsor
Dominion Foundry Supply Co., Montreal
Hamilton Facing Mill Co., Hamilton.
Paxson, J. W., Co., Philadelphia

Springs, Automobile.

Cleveland Wire Spring Co., Cleveland

Springs, Coiled Wire.

Cleveland Wire Spring Co., Cleveland

Springs, Machinery.

Cleveland Wire Spring Co., Cleveland

Springs, Upholstery.

Cleveland Wire Spring Co., Cleveland

Sprocket Wheels.

Wilson, J. C., & Co., Glenora, Ont.

Sprue Cutters.

Detroit Foundry Supply Co., Windsor
Dominion Foundry Supply Co., Montreal
Hamilton Facing Mill Co., Hamilton.
Paxson, J. W., Co., Philadelphia

Spur Wheels.

Wilson, J. C., & Co., Glenora, Ont.

Stamp Mills.

Allis-Chalmers-Bullock, Limited, Montrea
Paxson, J. W., Co, Philadelphia

Steam Separators.

Darling Bros., Ltd., Montreal.
Dominion Heating & Ventilating Co., Hespeler.
Robb Engineering Co., Montreal.
Sheldon's Limited, Galt.
Smart-Turner Mach. Co., Hamilton
Waterous Engine Co., Brantford.

Steam Specialties.

Darling Bros., Ltd., Montreal
Dominion Heating & Ventilating Co., Hespeler
Sheldon's Limited, Galt.

Steam Traps.

Canada Machinery Agency, Montreal.
Darling Bros., Ltd., Montreal
Dominion Heating & Ventilating Co., Hespeler.
Sheldons-Limited, Galt.

Steam Valves.

Darling Bros., Ltd., Montreal

Steel Pressure Blowers

Buffalo Foundry Supply Co., Buffalo.
Dominion Foundry Supply Co., Montreal
Domininion Heating & Ventilating Co., Hespeler
Hamilton Facing Mill Co., Hamilton.
J. W. Paxson Co., Philadalphia, Pa.
Sheldon's Limited, Galt.

Steel Tubes.

John Millen & Son, Montreal.

Steel, High Speed.

Wm. Abbott, Montreal.
Canadian Fairbanks Co., Montreal.
Frothingham & Workman, Ltd., Montreal
Alex. Gibb, Montreal.
Jessop, Wm., & Sons, Sheffield, Eng.
B. K. Morton Co., Sheffield, Eng.
Williams & Wilson, Montreal.

Stocks and Dies

Hart Manufacturing Co., Cleveland, Ohio
Pratt & Whitney Co., Hartford, Conn.

Stone Cutting Tools, Pneumatic

Allis-Chalmers-Bullock, Ltd., Montreal.
Canadian Rand Drill Co., Montreal.

Stove Plate Facings.

Buffalo Foundry Supply Co., Buffalo
Detroit Foundry Supply Co., Windsor.
Dominion Foundry Supply Co., Montreal
Hamilton Facing Mill Co., Hamilton.
Paxson, J. W., Co., Philadelphia

Swage, Block.

A. B. Jardine & Co., Hespeler, Ont.

Switchboards.

Allis-Chalmers-Bullock, Limited, Montreal
Canadian General Electric Co., Toronto
Canadian Westinghouse Co., Hamilton
Gas & Electric Power Co., Toronto
Hall Engineering Works, Montreal, Que.
Toronto and Hamilton Electric Co., Hamiton.

Talc.

Buffalo Foundry Supply Co., Buffalo.
Detroit Foundry Supply Co., Windsor.
Doggett, Stanley, New York
Dominion Foundry Supply Co., Montreal
Hamiton Facing Mill Co., Hamilton.
Paxson, J. W., Co., Philadelphia

Tanks, Oil.

Sight Feed Oil Pump Co., Milwaukee, Wis.

Tanks, Steel.

Steel Trough & Machine Co., Tweed, Ont.

Tapping Machines and Attachments.

American Tool Works Co., Cincinnati.
John Bertram & Sons Co., Dundas, Ont.
Bickford Drill & Tool Co., Cincinnati.
Blair Tool & Machine Works, New York.
The Geometric Tool Co., New Haven.
A. B. Jardine & Co., Hespeler.
London Mach. Tool Co., Hamilton, Ont
Murchey Machine & Tool Co., Detroit.
Niles-Bement-Pond Co., New York.
H. W. Petrie, Toronto.
Pratt & Whitney, Cincinnati, O.
L. S. Starrett Co., Athol, Mass.
Williams & Wilson, Montreal.

Tapes, Steel.

Frothingham & Workman, Ltd., Montreal
Rice Lewis & Son, Toronto.
Mechanics Supply Co., Quebec, Que.
John Millen & Son, Ltd., Montreal.
L. S. Starrett Co., Athol, Mass.

Taps, Collapsing.

The Geometric Tool Co., New Haven.

Taps and Dies.

Wm. Abbott, Montreal.
Butterfield & Co., Rock Island, Que.
Frothingham & Workman Ltd., Montreal
The Geometric Tool Co., New Haven.
A. B. Jardine & Co., Hespeler, Ont.
Rice Lewis & Son, Toronto.
John Millen & Son, Ltd., Montreal.
Pratt & Whitney Co., Hartford, Conn.
Standard Tool Co., Cleveland.
L. S. Starrett Co., Athol, Mass.

Testing Laboratory.

Detroit Testing Laboratory, Detroit

Testing Machines.

Detroit Foundry Supply Co., Windsor.
Dominion Foundry Supply Co., Montreal
Hamilton Facing Mill Co., Hamilton.
Paxson, J. W., Co, Philadelphia

Thread Cutting Tools.

Hart Manufacturing Co, Cleveland, Ohio

Tiling, Opal Glass.

Toronto Plate Glass Importing Co., Toronto.

Tinware Machinery

Canada Machinery Agency Montreal.
Ferracute Machine Co., Bridgeton, N.J

Tire Upsetters or Shrinkers

A. B. Jardine & Co., Hespeler, Ont.

Tool Cutting Machinery

Canadian Rand Drill Co., Montreal.

Tool Holders.

Armstrong Bros. Tool Co., Chicago.
John Millen & Son, Ltd., Montreal.
H. W. Petrie, Toronto.
Pratt & Whitney Co., Hartford, Conn.

Tool Steel.

Wm. Abbott, Montreal.
Flockton, Tompkin & Co., Sheffield, Eng.
Frothingham & Workman Ltd., Montreal
Alexander Gibb, Montreal.
Wm. Jessop, Sons & Co., Toronto.
Canadian Fairbanks Co., Montreal
B. K. Morton & Co., Sheffield, Eng.
Williams & Wilson, Montral.

Torches, Steel.

Detroit Foundry Supply Co., Windsor.
Dominion Foundry Supply Co., Montreal
Hamilton Facing Mill Co., Hamilton.
Paxson, J. W., Co., Philadelphia

Transformers and Convertors

Allis-Chalmers-Bullock, Montreal.
Canadian General Electric Co., Toronto.
Canadian Westinghouse Co., Hamilton.
Gas & Electric Power Co., Toronto.
Hall Engineering Works, Montreal, Que.
T. & H. Electric Co., Hamilton.

Transmission Machinery

Allis-Chalmers-Bullock, Montreal.
The Canadian Fairbanks Co., Montreal.
Gas & Electric Power Co., Toronto.
Greey, Wm. & J. G., Toronto.
Laurie Engine & Machine Co., Montreal.
Link-Belt Co., Philadelphia.
Paxson, J. W., Co., Philadelphia
H. W. Petrie, Toronto.
The Smart-Turner Mach. Co., Hamilton.
Waterous Engine Co., Brantford.
Wilson, J. C., & Co., Glenora, Ont.

Transmission Supplies

Baxter, J. R., & Co., Montreal.
The Canadian Fairbanks Co., Montreal.
Wm. & J. G. Greey, Toronto.
The Goldie & McCulloch Co., Galt.
H. W. Petrie, Toronto.

Trolleys

Canadian Rand Drill Co., Montreal.
Dominion Foundry Supply Co., Montreal
John Millen & Son, Ltd., Montreal.
Northern Engineering Works Detroit

Trolley Wheels.

Lumen Bearing Co., Toronto

Trucks, Dryer and Factory

Dominion Heating & Ventilating Co., Hespeler
Greey, Wm. & J. G., Toronto.
Northern Engineering Works, Detroit
Sheldon's Limited, Galt, Ont.

Tube Expanders (Rollers)

A. B. Jardine & Co, Hespeler.

Turbines, Steam

Allis-Chalmers-Bullock, Limited, Montreal
Canadian General Electric Co., Toronto.
Canadian Westinghouse Co., Hamilton.
Gas & Electric Power Co., Toronto.
Kerr Turbine Co., Wellsville, N.Y.
Wilson, J. C. & Co., Glenora, Ont.

Turntables

Detroit Foundry Supply Co., Windsor.
Dominion Foundry Supply Co., Montreal
Hamilton Facing Mill Co., Hamilton.
Northern Engineering Works, Detroit
Paxson, J. W., Co., Philadelphia

Turret Machines.

American Tool Works Co., Cincinnati.
John Bertram & Sons Co., Dundas, Ont.
The Canadian Fairbanks Co., Montreal.
Gisholt Machine Co., Madison, Wis.
London Mach. Tool Co., Hamilton, Ont.
Niles-Bement-Pond Co., New York.
H. W. Petrie, Toronto.
Pratt & Whitney Co., Hartford, Conn.
Warner & Swasey Co., Cleveland, Ohio.
Williams & Wilson, Montreal.

Upsetting and Bending Machinery.

John Bertram & Sons Co., Dundas, Ont.
A. B. Jardine & Co., Hespeler.
London Mach. Tool Co., Hamilton, Ont.
National Machinery Co., Tiffin, O.
Niles-Bement-Pond Co., New York.

Valves.

Golden-Anderson Valve Specialty Co., Pittsburg, Pa.

Valves, Back Pressure.

Darling Bros., Ltd., Montreal
Dominion Heating & Ventilating Co. Hespeler
Mechanics Supply Co., Quebec, Que.
Sheldon's Limited, Galt.

Valve Reseating Machines.

Darling Bros., Ltd., Montreal

Ventilating Apparatus.

Canada Machinery Agency, Montreal
Darling Bros., Ltd., Montreal
Dominion Heating & Ventilating Co., Hespeler, Ont.
Sheldon's Ltd., Galt

Vises, Planer and Shaper.

American Tool Works Co., Cincinnati, O.
Frothingham & Workman, Ltd., Montreal
A. B. Jardine & Co., Hespeler, Ont.
John Millen & Son, Ltd., Montreal.
Niles-Bement-Pond Co., New York.

Washer Machines.

National Machinery Co., Tiffin, Ohio.

Water Tanks.

Ontario Wind Engine & Pump Co., Toronto

Water Wheels.

Allis-Chalmers-Bullock Co., Montreal.
Barber, Chas., & Sons, Meaford.
Canada Machinery Agency, Montreal.
The Goldie & McCulloch Co., Galt, Ont.
Greey, Wm. & J. G., Toronto.
Wm. Kennedy & Sons, Owen Sound.
John McDougall Caledonian Iron Works Co., Ltd., Montreal
Wilson, J. C., & Co., Glenora, Ont.

Water Wheels, Turbine.

Barber, Chas., & Sons, Meaford, Ont.
Kennedy, Wm., & Sons, Owen Sound

Wheelbarrows.

Buffalo Foundry Supply Co., Buffalo
Frothingham & Workman Ltd., Montrea
Detroit Foundry Supply Co., Windsor.
Dominion Foundry Supply Co., Montrea
Hamilton Facing Mill Co., Hamilton.
Paxson, J. W., Co., Philadelphia

Winches.

Canadian Pilling Co., Montreal
Frothingham & Workman, Ltd, Montreal

Wind Mills.

Ontario Wind Engine & Pump Co, Toronto

Window Wire Guards.

Expanded Metal and Fireproofing Co Toronto.
B. Greening Wire Co., Hamilton, Ont.
Paxson, J. W., Co. Philadelphia

Wire Chains.

The B. Greening Wire Co., Hamilton.

Wire Cloth and Perforated Metals.

Expanded Metal and Fireproofing Co. Toronto.
B. Greening Wire Co., Hamilton. Ont.
Paxson, J. W., Co, Philadelphia.

Wire Guards and Railings.

Expanded Metal and Fireproofing Co. Toronto.
B. Greening Wire Co. Hamilton, Ont.

Wire Nail Machinery.

National Machinery Co., Tiffin, Ohio.

Wire Rope.

Frothingham & Workman, Ltd., Montreal
B. Greening Wire Co., Hamilton, Ont.

Wood Boring Machines. Pneumatic.

Independent Pneumatic Tool Co., Chicago, Ill.

Wood-working Machinery.

Canada Machinery Agency, Montreal.
The Canadian Fairbanks Co., Montreal.
Goldie & McCulloch Co., Galt.
H. W. Petrie, Toronto.
Waterous Engine Works Co., Brantford
Williams & Wilson, Montreal.

ALPHABETICAL INDEX

Light and Heavy Malleable Iron Castings

PROMPT DELIVERY

Galt Malleable Iron Co., Limited - Galt, Ontario

ELECTRICAL MACHINERY CO.
OF TORONTO
ELECTRICAL ENGINEERS
We manufacture—design and repair motors and dynamos and all electrical apparatus.
Office and Works
113-115 Simcoe St., Toronto
Type B—Direct Current 5 to 75 Horse Power

EXPANDED METAL

IS HANDIEST CHEAPEST STRONGEST FOR

Floors, Roofs, Partitions, and all Varieties of Concrete Slab Work

CATALOGUES—SAMPLES FREE

Expanded Metal & Fireproofing Co. Limited

100 King Street West, TORONTO

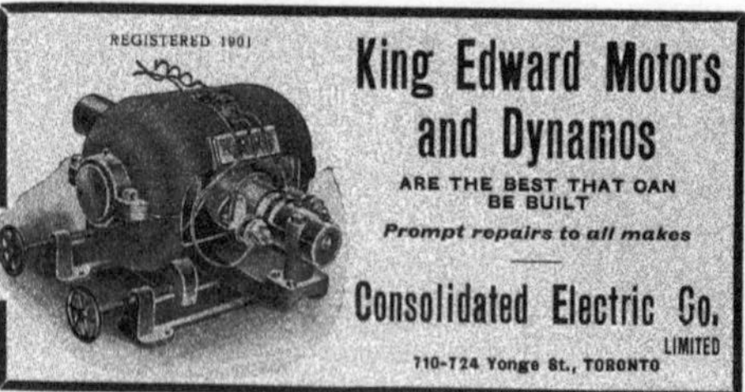

REGISTERED 1901
King Edward Motors and Dynamos
ARE THE BEST THAT CAN BE BUILT
Prompt repairs to all makes
Consolidated Electric Co.
LIMITED
710-724 Yonge St., TORONTO

DIRECT CURRENT MOTORS AND DYNAMOS

FOR ALL PURPOSES

TORONTO & HAMILTON ELECTRIC CO.

99-103 McNab Street, HAMILTON, ONT.

EUGENE F. PHILLIPS ELECTRICAL WORKS, LTD.

Bare and Insulated Electric Wires

TORONTO STORE,
67 ADELAIDE STREET
EAST.

GENERAL OFFICES
AND FACTORY,
MONTREAL, Canada.

PRACTICAL LITERATURE

For Electricians, Scientists and Engineers

All orders payable in advance.

Locomotive Breakdowns

Emergencies and their Remedies

by Geo. L. Fowler, M.E.

Price $1.50, Postpaid.
250 Pages, 96 Illustrations.

This book is an up-to-date catechism treating on all classes of accidents which may happen to locomotive engines and giving the remedy to be applied. Accidents to the Valves, Cylinders, Valve Motion, Running Gear, Compound Locomotives, Boiler Troubles are treated separately and fully. It tells what should be done in any emergency.

Liquid Air

and the

Liquefaction of Gases

by T. O'Conor Sloane, A.M., E.M., Ph.D.

Price $2.00, Postpaid.
365 Pages, Fully Illustrated.

This book contains the full theory, history, biography and bibliography of the subject from the earliest times, as well as a description of all the wonderful experiments performed with it, showing how liquid air is carried hundreds of miles and handled in open buckets. It also tells what may be expected from it in the near future. Although a work of scientific interest and authority, it is intended for the general reader.

Engine Runner's

Catechism

by Robert Grimshaw, M.E.

Price $2.50, Postpaid.
354 pages, Profusely Illustrated.

This work gives complete descriptions of the general characteristics of various well-known makes of engines, followed by details of shipping, making foundations, erecting and starting, and the specific instructions as to the adjustment of special makes. Every engineer will find in this volume more really practical and useful information than is to be found anywhere else within a like compass.

Steam Engine

Catechism

by Robert Grimshaw, M.E.

Price $2.00, Postpaid.
413 Pages, Fully Illustrated.

This book is a series of nearly 1000 direct, practical questions and answers; in fact it contains every essential question on the Steam Engine with its answer. Leading engineers have recommended it as a valuable educator for the beginner as well as a reference book for the engineer. It is intended for

EVERYONE INTERESTED IN STEAM.

How to Become a Successful Electrician

by Prof. T. O'Conor Sloane, A.M., E.M., Ph. D.

Price $1.00, Postpaid
200 Pages, Bound in Cloth.

Every young man whose ambition it is to become a successful electrican should read this book. It is the clearest and most practical book on the subject ever published. Its 19 Chapters include Mathematics, Physics, Chemistry, Electricity at Home, Drawing, Electrical Factory Work, Mechanical and Steam Engineering, the Manufacturing and Constructing Engineer, Inventing, Original Investigation, etc.

Gas, Gasoline and Oil Engines

by G. D. Hiscox, M.E.,

Price $2.50, Postpaid.
450 Pages, 351 Illustrations.

This book is a thorough treatise on the construction, installation, operation and maintenance of gas, gasoline, kerosene and crude petroleum engines. It is the new rewritten enlarged and revised 15th edition of this work giving general information for everyone interested in the popular motive power, and its adaptation to the increasing demand for a cheap and easily managed motor requiring no licensed engineer.

The MacLean Publishing Co.

Technical Book Department

10 Front St. East, - - TORONTO, CANADA

NEW AND UP-TO-DATE

BOLT AND NUT MACHINERY

INCLUDING

Bolt Cutters, Nut Tappers, Bolt Headers, Upsetting and Forging Machines, Wire Nail and Spike Machines and Bulldozers.

Send for Catalogue M.

NATIONAL MACHINERY CO., **Tiffin, Ohio, U.S.A.**

Canadian Agents: H. W. PETRIE, Toronto, Ont. WILLIAMS & WILSON, Montreal, Que.

MODERN EXAMINATIONS OF STEAM ENGINEERS

BY W. H. WAKEMAN

The Fifty-three Chapters of excellent reading in this book fully answer THREE HUNDRED Practical Questions about Engines, Boilers, Safety Valves, Belting, Shafting, Steam Heating, Boiler Bracing, etc. It is an excellent book of Reference, especially written for Engineers and Firemen, and is UNIVERSALLY ACKNOWLEDGED as the BEST and MOST COMPLETE work on the subject.

THIRD EDITION NOW READY.

272 Pages. Fully Illustrated. Price $2.00, Postpaid.

MACLEAN PUBLISHING COMPANY

Technical Book Dept. 10 FRONT ST. EAST, TORONTO

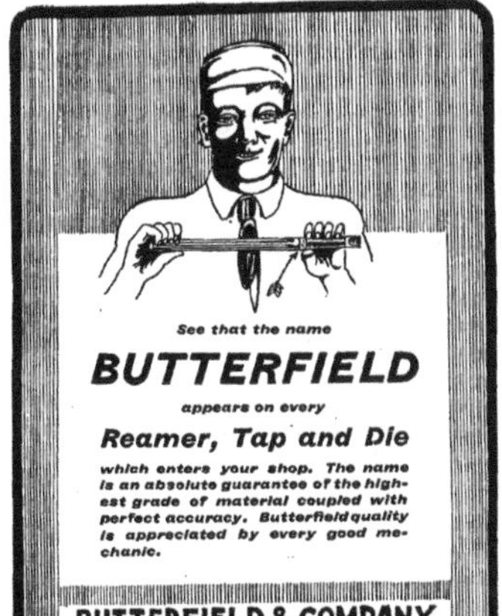

THE STANDARD TOOL CO'S Twist Drills

If you use double grooved Shank Drills, try the Standard. We manufacture them in both the short and long lengths.

Highest Quality.

General Office and Factory: **CLEVELAND, OHIO**

Eastern Salesroom: **94 Reade St., New York**

THE L. S. STARRETT CO., ATHOL, MASS., U. S. A.

Style "D" Self-Opening Screw Cutting Die Heads

This style of Die Head is arranged for use on the turrets of hand and automatic screw machines —cuts threads of any diameter, pitch, form and length, on steel, iron, brass or other metal.

As its name implies, it is self-opening; the dies opening automatically at the end of the cut, clearing the thread and saving the time required in backing off solid dies, which leaves a good, smooth thread and does away with the stripping and tearing, which is so common in the use of solid dies.

With the aid of extra sets of dies these die heads will cover a very wide range of diameters, an extra set being required for each different diameter cut.

Upon receipt of samples or blue prints, we are always glad to submit quotation and time of delivery on a suitable equipment of tools to cover the requirements.

Shall we send you a copy of our new complete catalogue?

The Geometric Tool Co., NEW HAVEN, CONN., U.S.A.

Canadian Agents, Williams & Wilson, Montreal, Que.

Lightning Source UK Ltd.
Milton Keynes UK
UKHW012034121218
333853UK00008B/558/P

9 780265 832677